Handbook of

SOLVENTS

Second Edition

Handbook of SOLVENTS Second Edition

Compiled by

Michael and Irene Ash

Synapse Information Resources

Published by
Synapse Information Resources, Inc.
1247 Taft Ave.
Endicott, NY 13760

Library of Congress Control Number: 2003092567

ISBN Number **1-890595-63-2**

Printed in the U.S.A.

Contents

Preface

Handbook of Solvents, Second Edition provides comprehensive information on approximately 3000 trade name and generic solvent chemicals. Solvents are essential components in formulating a wide range of products because of their ability to clean, dissolve, suspend, and extract other chemicals or materials without changing their composition. They are used extensively in the paint and coating industry as well as in cleaning and degreasing, printing inks, pharmaceutical and chemical processing, cosmetic and personal care products, adhesives, fuels, electronics, agrochemicals, food and beverages, and many other industries

When choosing or blending a solvent, factors such as performance, health and safety, cost and the environment all must be weighed. This reference contains comprehensive coverage of the many solvents that are used in industry throughout the world and will serve as an invaluable ingredient selection tool for formulators, industrial and environmental chemists, manufacturers, product managers, safety regulators, purchasing agents, and chemical distributors as well as any professional associated with the processing and formulation industries.

This Second Edition has been greatly expanded and contains more than 50% new products as well as expanded entries for solvents contained in the First Edition.

This book is divided into the following sections:

Part I — *Trade Name Reference* provides an alphabetical listing of more than 1500 trade name chemicals that function as solvents in all aspects of industry. Each entry contains a comprehensive profile of the trade name chemical including: manufacturer's name; chemical/material description; detailed functions and applications; physical properties, such as form, molecular weight, density, solubility, boiling point, cloud point, flash point, pH, freezing point, activity; toxicology, storage, precautions, etc.

Part II — *Discontinued/Unverified Trade Name Reference* provides an alphabetical listing of more than 500 trade name products that have been discontinued by a manufacturer, associated with a manufacturer that is no longer in business, or has not been verified by the manufacturer as an active product for a number of years. This section is important for anyone who is trying to identify the composition of a solvent product for storage or handling concerns, or trying to find a substitute for this product in a specific application.

Part III — *Chemical Component Cross-Reference* covers more than 1000 chemicals and materials that are contained in the trade name products profiled in Part I or generic chemicals that are not linked to trade names but are used as solvents or in formulating solvents. Each monograph entry contains information including the following: CAS, EINECS/ELINCS, and UN/DOT reference numbers, classification, definition, chemical synonyms, empirical and molecular formulas, properties, toxicology, precautions, storage, uses, use level, regulatory details, manufacturers and distributors. Thousands of chemical/material synonyms are cross-referenced back to the main entry. The chemical monographs also contain a listing of the trade name products that are either equivalent to this chemical or contain the chemical as one of its ingredients.

Part IV — *Functional/Application Index* is a powerful tool for locating the trade names and chemicals based on their function and/or industrial application area. By searching for key functional words such as solvent, diluent, intermediate, carrier in specific application areas such as cosmetics, pesticides, dry cleaning, metal cleaning, etc., the user is directed to the trade names and chemicals that have that specific functional/application attribute.

Part V — *Manufacturers Directory* contains detailed contact information for the more than 1000 worldwide manufacturers and their branches of the solvent chemicals contained in this handbook including telephone, telefax, toll-free numbers, e-mail and internet addresses, and complete mailing addresses.

Appendixes
CAS Number Index contains CAS number entries followed by

a listing of their trade name product and chemical equivalents in alphabetical order. The chemical name is in boldfaced type.

EINECS/ELINCS Number Index contains EINECS/ELINCS number entries followed by a listing of their trade name product and generic chemical equivalents in alphabetical order. The chemical name is in boldfaced type.

FDA Regulatory Number Index contains the trade names and chemicals that have been approved for use as solvents in foods, food coatings, and food processing.

KB Value Index contains trade name hydrocarbon solvents that have KB values for solvency ranging from 20 (poor) to more than 1000 (excellent), as provided by the manufacturer.

Glossary contains definitions of terminology associated with solvents and their areas of application.

We are confident that those involved in any aspect of the chemical industry and need ready access to detailed information on solvents that are used throughout the world will find this book an important addition to their reference library. We are also pleased to provide this reference as an electronic product.

This reference is the culmination of many years of research, investigation of product sources acquired through personal contacts and correspondences with major chemical manufacturers worldwide, as well as toxicological databases, chemical reference books, trade magazines and journals. We would especially like to express our gratitude to Roberta Dakan for her contribution to standardizing the entry format and managing the trade name and chemical database that represents the basis of this and future chemical references. Her untiring efforts have been instrumental in the production of this source book.

M. & I. Ash

NOTE:

Abbreviations

ABS	acrylonitrile-butadiene-styrene
abs.	absolute
absorp.	absorption
ACC	Abel closed cup
ACGIH	American Conference of Governmental Industrial Hygienists
ACN	acrylonitrile
act.	active
ADI	acceptable daily intake (FAO/WHO)
ADR	adverse drug reactions
adsorp.	adsorption
AEL	acceptable exposure limit
agric.	agricultural
agrichem.	agrichemical(s)
agrochem.	agrochemical
a.i.	active ingredient
AIHA	American Industrial Hygiene Association
alc.	alcohol
Am., Amer.	American
amts.	amounts
anhyd.	anhydrous
ANSI	American National Standards Institute
APHA	American Public Health Association
API	Active Pharmceutical Ingredients
applic(s).	application(s)
aq.	aqueous
ASA	acrylic-styrene-acrylonitrile
ASBC	Am. Society of Brewing Chemists
ASTM	American Society for Testing and Materials
ATH	alumina trihydrate
atm	atmosphere
at.wt.	atomic weight
autoignit.	autoignition
aux.	auxiliary
avail.	available
avg.	average
a.w.	atomic weight
BATF	Bureau of Alcohol, Tobacco, and Firearms (U.S.)
BDG	butyl diglycol
BfArM	Bundesinstitut für Arzneimittel und Medizinprodukte (Federal Institute for Drugs and Medical Devices, Germany)
BGA	German Federal Health Office (obsolete)
BgVV	Bundesinsititut für Gesundheitlichen Verbraucherschutz und Veterinärmedizin (Federal Institute for Consumer Health Protection and Veterinary Medicine) (Germany)
BHA	butylated hydroxyanisole
BHT	butylated hydroxytoluene
biochem.	biochemical
biodeg.	biodegradable
bldg.	building
blk.	black
BMC	bulk molding compound
BOD	biochemical oxygen demand
BP	British Pharmacopeia
b.p.	boiling point
BR	butadiene rubbers, polybutadienes
B&R	Ball & Ring
br., brn.	brown
brnsh.	brownish
BS	British Standards

B/S	butadiene/styrene
BSI	British Standards Institute
BSS	British Standard Sieve
Btu	British thermal unit
B.U.	Brabender units (viscosity)
BVC	British Veterinary Codex
BVO	brominated vegetable oil
byprod.	byproduct(s)
C	degrees Centigrade
CAA	Clean Air Act
CAB	cellulose acetate butyrate
calcd.	calculated
cap.	capillary
CAP	Color Additive Petition (U.S.)
CAS	Chemical Abstracts Service
CC	closed cup
cc	cubic centimeter(s)
CCFAC	Codex Committee on Food Additives and Contaminants
CCl_4	carbon tetrachloride
CD	completely denatured
CDA	completely denatured alcohol
CE	Council of Europe
CEL	corporate exposure limit
CERCLA	Comprehensive Environmental Response, Compensation, & Liability Act (U.S.)
CFC	chlorofluorocarbon
CFN	Council on Food & Nutrition (Am. Medical Assoc.)
CFR	Code of Federal Regulations (U.S.)
cGMP	current good manufacturing practice (U.S. FDA)
ch.	Chapter
char.	characteristic, characterized
chel.	chelation
chem(s).	chemical(s)
CI	Color Index
CIIR	chlorobutyl rubber
CIR	Cosmetic Ingredient Review
cks	centistoke(s)
cl	clear
CL	ceiling concentration
cm	centimeter(s)
cm^3	cubic centimeter(s)
CMC	carboxymethylcellulose
CMC	critical Micelle concentration
c.m.p.	capillary melting point
CNS	central nervous system
CO	carbon monoxide
COC	Cleveland Open Cup
COD	chemical oxygen demand
coeff.	coefficient
COF	coefficient of friction
compat.	compatible
compd(s).	compound(s)
compr.	compression
conc(s).	concentrated, concentration
conduct.	Conductive, conductivity
const.	constant
contg.	containing
cosolv.	cosolvent
CP	Canadian Pharmacopeia
cp	centipoise(s)
CPE	chlorinated polyethylene

cps	centipoise(s)
CPVC	chlorinated polyvinyl chloride
CR	chloroprene rubber, polychloroprene
cryst.	crystalline, crystallization
cs	centistoke(s)
cSt	centistoke(s)
CTFA	Cosmetic, Toiletry and Fragrance Association
ctks	centistoke(s)
CWA	Clean Water Act
cwt	hundred weight
DAB	Deutsche. Arzneibuch
DAC	Deutscher Arneimittel Codex (German Pharmacopoeia Codex)
dc	direct current
D&C	Drugs & Cosmetics
DE	dextrose equivalent
DEA	diethanolamide, diethanolamine
dec.	decomposes
decomp.	decomposition
DEG	diethylene glycol
deliq.	deliquescent
dens.	density
deriv(s).	derivative(s)
descrip.	description
dg	decigram(s)
DI	deionized
diam.	diameter
dielec.	dielectric
dil.	dilute
DIN	Deutsches Institut fuer Normung (German Standards Organization)
disp.	dispersible, dispersion
dissip.	dissipation
dist.	distilled
distort.	distortion
distrib.	distributor
dk.	dark
DMF	dimethyl formamide
DMF	Drug Master Files
DMSO	dimethyl sulfoxide
D.O.	dissolved oxygen
DOC	dissolved organic carbon
DOM	dissolved organic matter
DOP	dioctyl phthalate
DOT	Department of Transportation (U.S.)
DPG	diphenyl guanidine, dipropylene glycol
DSB	dry solids basis
DSI	Canadian Provisional Domestic Substance list
DTUL	deflection temperature under load
DVB	divinylbenzene
DW	distilled water, deionized water
DWV	drainage, waste and vent
eb, EB	electron beam
EC	European Community
EC50	environmental concentration, 50%
EDTA	ethylenediamine tetraacetic acid
EE	epoxy equivalent
EEC	European Economic Community
EED	environmental endocrine disrupter
EEW	epoxide equivalent weight
e.g.	for example
EINECS	European Inventory of Existing Commercial Chemical Substances
elec.	electrical
ELINCS	European List of Notified Chemical Substances
elong.	elongation
EMI	electromagnetic interference
EMS	electromagnetic shielding
ENB	5-ethylidene-2-norbornene
EO	ethylene oxide
EP	European Pharmacopoeia
EP	extreme pressure
EPA	Environmental Protection Agency (U.S.)
EPDM	ethylene-propylene-diene rubber, ethylene-propylene terpolymer
EPR	ethylene-propylene rubber
EPS	expandable polystyrene
equip.	equipment
equiv.	equivalent
ESCR	environmental stress crack resistance
ESD	electrostatic discharge
ESO	epoxidized soybean oil
ESP	electrostatic protection
esp.	especially
EU	European Union
Eur.Ph.	European Pharmacopeia
EVA	ethylene vinyl acetate
exc.	excellent
F	degrees Fahrenheit
FA	fatty acid
FAC	Food Advisory Committee
FAO	Food and Agriculture Organization (United Nations)
FAP	Food Additive Petition (U.S.)
FCC	Food Chemicals Codex
FDA	Food and Drug Administration (U.S.)
FD&C	Foods, Drugs, and Cosmetics
FDD	Food and Drug Directorate
FEMA	Flavor and Extract Manufacturers' Association (U.S.)
FEP	fluorinated ethylene propylene
FFA	free fatty acid
FFDCA	Federal Food, Drug, and Cosmetic Act
FG	food grade
FIFRA	Federal Insecticide, Fungicide, and Rodenticide Act (U.S. EPA)
FKM	fluoroelastomer
fl	fluid
flamm.	flammable, flammability
flex.	flexural
FNB	Food and Nutrition Board
f.p.	freezing point
FPC	fish-protein concentrate
FR	flame retardant; Federal Register
FR-ABS	flame retardant ABS
FRP	fiberglass-reinforced plastics
F-T	Fischer-Tropsch
ft	foot, feet
f.w.	formula weight
G	giga
g	gram(s)
gal	gallon(s)
g/d	gram/dyne
GFRP	glass fiber-reinforced plastic
G-H	Gardner-Holdt
GI	gastrointestinal
glac.	glacial
GLP	good laboratory practice
GLY	glycine
GMP	good manufacturing practice; guanosine monophosphate
gpd	gallons per day
gpm	gallons per minute
gpt	gallons per ton
gr.	gravity
gran.	granules, granular
GRAS	generally regarded as safe
grn.	green
GRP	glass-reinforced plastics, glass-reinforced polyester
GVS	Gardner varnish scale (color)
h	hour(s)
HAF	high abrasion furnace carbon black
HALS	hindered amine light stabilizer
HAP's	hazardous air pollutants
HB	horizontal burning
HC	hydrocarbon
HCFC	hydrochlorofluorocarbon
HCl	hydrochloride, hydrochloric acid
HDPE	high-density polyethylene
HDT	heat distortion (deflection) temp.
HFC	hydrofluorocarbon
HFCS	high fructose corn syrup
Hg	mercury
HIPS	high-impact polystyrene
HLB	hydrophilic lipophilic balance
HMIS	Hazardous Material Identification System
HPLC	high performance liquid chromatography
hr	hour(s)
HTST	high temperature short-time pasteurization
HVP	hydrolyzed vegetable protein
hyd.	hydroxyl
hydrog.	hydrogenated
Hz	hertz

IARC	International Agency for Research on Cancer (United Nations)
i.b.p.	initial boiling point
IDLH	immediately dangerous to life and health
I&I	industrial and institutional
IIR	isobutylene-isoprene rubber
IM	intramuscular
immisc.	immiscible
IMP	inosine monophosphate
in.	inch(es)
Inc.	Incorporated
inc.	increases, increased
INCI	International Nomenclature Cosmetic Ingredient
incl.	including
incompat.	incompatible
incorp.	Incorporated, incorporation
indent.	indentation
ing.	ingestion
ingred(s).	ingredient(s)
inh.	inhalation
inj.	injection
inorg.	inorganic
INS	International Numbering System for Food Additives
insol.	insoluble
Int'l.	International
IOFI	International Organization of the Flavor Industry
IP	intraperitoneal
IPA	isopropyl alcohol
IPM	isopropyl myristate
IPP	isopropyl palmitate
IR	isoprene rubber (synthetic), polyisoprene
irreg.	irregular
IU	International Unit
IV	intravenous
J	joule
JCIC	Japanese Cosmetic Ingredients Codex
JCID	Japanese Cosmetic Ingredients Dictionary
JECFA	Joint Expert Committee on Food Additives
JP	Japanese Pharmacopoeia
JSCI	Japanese Standard of Cosmetic Ingredients
JSFA	Japan Standards for Food Additives
k	kilo
KB	Kauri-Butanol
kg	kilogram(s)
KTPP	potassium tripolyphosphate
KU	Krebs units
l	liter(s)
lb	pound(s)
LC50	lethal concentration 50%
LCLo	lethal concentration low
LD0	lethal dose 0%
LD50	lethal dose 50%
LDLo	lowest published lethal dose
LDPE	low-density polyethylene
LED	light-emitting diode
LEL	lower explosive level
LFL	lower flammability limit
lg.	large
liq.	liquid
LLDPE	linear low-density polyethylene
LOEL	lowest observed effect level
lt.	light
Ltd.	Limited
LVP	low vapor pressure
M	mega
M	mole
m	milli
m	meter(s)
m-	meta
manuf.	manufacturer
max.	maximum
mbar	millibar
MCF	methyl chloroform
MCT	medium chain triglycerides
MD	machine direction, mold direction
MEA	monoethanolamine, monoethanolamide
mech.	mechanial
med.	medium
MEK	methyl ethyl ketone
MEL	maximum exposure limit
mfg.	manufacture
mg	milligram(s)
MIBK	methyl isobutyl ketone
microcryst.	microcrystalline
microgran.	microgranules, microgranular
MID	Meat Inspection Division (USDA)
MIL	Military Specifications
mil	$^{1}/_{1000}$th inch
min	minute(s)
min.	mineral
min.	minimum
MIPA	monoisopropanolamine, monoisopropanolamide
misc.	miscible, miscellaneous
MITI	Japanese Inventory of Chemical Substances (list)
mixt(s).	mixture(s)
ml	milliliter(s)
MLD	minimum lethal dose
mm	millimeter(s)
MMW-HDPE	medium molecular weight high density polyethylene
mN	millinewton(s)
mo, mos	month(s)
mod.	moderately
mod.	modulus
monocl.	monoclinic
m.p.	melting point
mPa·s	millipascal-second(s)
mppcf	million particles per cubic foot
MRL	maximum residual limits
MSDS	Material Safety Data Sheet
MSG	monosodium glutamate
MT	medium thermal
mus	mouse
MVTR	moisture vapor transmission rate
m.w.	molecular weight
N	normal
nat.	natural
NB	nonbreaking
N/B	nitrile-butadiene
NBR	nitrile rubber, nitrile-butadiene rubber
NC	nitrocellulose
NCI	National Cancer Institute
need.	needles
neg.	negative
neut.	neutral, neutralized
NF	National Formulary
NFPA	National Fire Protection Association
ng	nanogram
NIOSH	National Institute for Occupational Safety and Health (U.S.)
nm	nanometer
no.	number
N_2O	nitrous oxide
NOEL	no observed effect level
NOI	not otherwise indexed (U.S. DOT)
nonalc.	nonalcoholic
nonaq.	nonaqueous
nonbiodeg.	nonbiodegradable
nonflamm.	nonflammable
nonyel.	nonyellowing
N.O.S.	not otherwise specified (transport regulations)
NO_x	nitrogen oxides
NPRI	National Pollutant Release Inventory (Canada)
NR	natural rubber, isoprene rubber (natural)
NSF	National Sanitation Foundation, National Standards Foundation
NTP	National Toxicology Program (U.S.)
NV	nonvolatiles
o-	ortho
OBPA	oxybisphenoxarsine
OC	open cup
ODC	ozone-depleting compound
ODP	ozone-depletion potential
OEL	occupational exposure limit
OEM	original equipment manufacturer
OMS	odorless mineral spirits
org.	organic
OSHA	Occupational Safety and Health Administration (U.S.)
OTC	over-the-counter
o/w	oil-in-water
oz	ounce
p-	para

Pa	Pascal
PBT	polybutylene terephthalate
pbw	parts by weight
PC	polycarbonate
PCB	polychlorinated biphenyl
pcf	pounds per cubic foot
PCP	Pest Control Product Act, 1972 (Canada)
PCTFE	polychlorotrifluoroethylene
PE	polyethylene
PEEK	polyetheretherketone
PEG	polyethylene glycol
PEK	polyetherketone
PEL	permissible exposure level
perc	perchloroethylene
percut.	percutaneous
PES	polyether sulfone
PET	polyethylene terephthalate
petrol.	petroleum
PFPE	perfluoropolyether
PG	propylene glycol
pH	hydrogen-ion concentration
Ph.	Pharmacopoeia
pharm.	pharmaceutical
Ph.Eur.	European Pharmacopoeia
phr	parts per hundred of rubber or resin
PIB	polyisobutylene
PIN	product identification number
PIR	polyisocyanurate
Pk	peak concentration
pkg.	packaging
PM, P-M	Pensky-Martens
PMCC	Pensky-Martens closed cup
PMMA	polymethyl methacrylate
PMOC	Pensky-Martens open cup
PNS	peripheral nervous system
PO	propylene oxide
POC	particulate organic carbon
POE	polyoxyethylene, polyoxyethylated
polyunsat.	polyunsaturated
PoM	prescription-only medicine
POM	polyoxymethylene
POP	polyoxypropylene, polyoxypropylated
POP's	persistant organic pollutants
powd.	powder
PP	polypropylene
ppb	parts per billion
PPE	polyphenylene ether
PPG	polypropylene glycol
pph	parts per hundred (percent)
ppm	parts per million
PPO	polyphenylene oxide
PPS	polyphenylene sulfide
ppt	parts per trillion
pract.	practically
prep(s).	preparation(s)
prod.	product(s), production
props.	properties
PS	polystyrene
ps	poise
psi	pounds per square inch
psia	pounds per square inch absolute
psig	pounds per square inch gauge
pt.	point
Pt-Co	platinum-cobalt
PTFE	polytetrafluoroethylene
PTMEG	polytetramethylene ether glycol
PU	polyurethane
PUF	polyurethane foam
PUR	polyurethane
PVA	polyvinyl alcohol
PVAc	polyvinyl acetate
PVAL	polyvinyl alcohol
PVB	polyvinyl butyral
PVC	polyvinyl chloride
PVC-P	plasticized polyvinyl chloride
PVC-U	unplasticized polyvinyl chloride
PVDC, PVdC	polyvinylidene chloride
PVDF	polyvinylidene fluoride
PVM	polyvinyl methyl ether
PVM/MA	polyvinyl methyl ether/maleic anhydride
PVP	polyvinylpyrrolidone
qt	quart
quat.	quaternary
R&B	Ring & Ball
rbt	rabbit
RCRA	Resource Conservation and Recovery Act (U.S. EPA 40CFR §261)
R&D	research and development
RDA	recommended daily allowances
rdsh.	reddish
rec.	recommended
redsh.	reddish
ref.	refractive
reg.	registry
regs.	regulations
REL	recommended exposure limit
rep.	represents
resist.	resistance, resistant, resistivity
resp.	respectively
RFI	radio frequency interference
r.h.	relative humidity
rhomb.	rhombic
RIM	reaction injection molded/molding
RO	reverse osmosis
rpm	revolutions per minute
RQ	reportable quantity
R.T.	room temperature
RTECS	Registry of Toxic Effects of Chemical Substances (U.S.)
RTM	resin transfer molding
RTV	room temperature vulcanizing
s	second(s)
s-	secondary
SADT	self accelerating decomposition temp.
SAN	styrene-acrylonitrile
sapon.	saponification
SARA	Superfund Amendments & Reauthorization Act (U.S.)
sat.	saturated
S/B	styrene/butadiene
SBR	styrene/butadiene rubber
SBS	styrene-butadiene-styrene
SD	specially denatured
SDA	specially denatured alcohol
SE	self-emulsifying
SEBS	styrene-ethylene/butylene-styrene
sec.	secondary
semicryst.	semicrystalline
semiliq.	semiliquid
semisyn.	semisynthetic
sl.	slight, slightly
sm.	small
SMA	styrene maleic anhydride
SMC	sheet molding compound
SMG	succinylated monoglycerides
SNAP	Significant New Alternative Policy (U.S. EPA)
soften.	softening
sol.	soluble, solubility
solid.	solidification
sol'n.	solution
solv(s).	solvent(s)
sp.	specific
spec.	specification, specialty
SPF	sun protection factor
spp.	non-specified species
SS	stainless steel
SSU	Saybolt Universal Seconds
std.	standard
STEL	short term exposure limit
Stod.	Stoddard solvent
STP	standard temperature and pressure
str.	strength
subcut.	subcutaneous
subl.	sublimes
surf.	surface
SUS	Saybolt Universal Seconds
susp.	suspension
syn.	synthetic
t	tertiary
TAPPI	Technical Association of the Pulp & Paper Industry
TBHQ	tert-butyl hydroquinone
TCC	Tag closed cup

TCLo	toxic concentration low
TDI	toluene diisocyanate
TDLo	toxic dose low
TDS	total dissolved solids
TEA	triethanolamine, triethanolamide
tech.	technical
temp.	temperature
tens.	tensile, tension
tert	tertiary
THF	tetrahydrofuran
ThOD	theoretical oxygen demand
thru	through
TIPA	triisopropanolamine
TKPP	tetrapotassium pyrophosphate
TLV	Threshold Limit Value
TLV-CL	Threshold Limit Value/ceiling limit
TLV-STEL	Threshold Limit Value/short therm exposure limit
TLV-TWA	Threshold Limit Value/time weighted average
TMC	thick molding compound
TOC	Tag open cup, total organic carbon
TPE	thermoplastic elastomer
TPU	thermoplastic polyurethane
TRI	Toxic Release Inventory
TSCA	Toxic Substances Control Act
tsp	teaspoon
TSS	total suspended solids
TWA	time weighted average
TWC	time weighted concentration
typ.	typical
UEL	upper explosive limits
UF	urea formaldehyde
UFL	upper flammability limit
UHF	ultra high frequency
UHMW	ultra high molecular weight
UHMWPE	ultra high molecular weight polyethylene
UHT	ultra high temperature
UL	Underwriter's Laboratory
UN No.	United Nations Substance Identification Number (for transport purposes)
unsat.	unsaturated
UPVC	unplasticized polyvinyl chloride
USDA	U.S. Department of Agriculture
USFA	United States Food Additives
USP	United States Pharmacopeia
uv, UV	ultraviolet

V	volt
VA	vinyl acetate
VAE	vinyl acetate ethylene
VC	vinyl chloride
VdC, VDC	vinylidene chloride
veg.	vegetable
visc.	viscous, viscosity
VM&P	Varnish Makers and Painters
VOC	volatile organic compounds
vol.	volume
vol.%	percent by volume
v/v	volume by volume
WEEL	Workplace Environmental Exposure Level (U.S.)
WFC	World Food Council
wh.	white
WHMIS	Workplace Hazardous Materials Information System (Canada)
WHO	World Health Organization (United Nations)
wks	weeks
w/o	water-in-oil
wt.	weight
w/v	weight by volume
w/w	weight by weight
XLPE	crosslinked polyethylene
X-PE	crosslinked polyethylene
yel.	yellow
ylsh.	yellowish
yr	year
#	number
%	percent
±	plus or minus
<	less than
>	greater than
≤	less than or equal to
≥	greater than or equal to
@	at
α	alpha
β	beta
δ, Δ	delta
ε	epsilon
γ	gamma
ω	omega
μ	micron, micrometer
μg	microgram
≈	approximately equal to

Part I
Trade Name Reference

A

Abzol® EG [Albemarle]
Uses: Cleaning solvent for electronics, PCs and ionics removal
Features: Low ozone-depleting; non-CFC
Regulatory: DOT nonregulated
Properties: APHA 15 max. colorless clear liq.; sp.gr. 1.267; b.p. 70 C; flash pt. (TCC, TOC, COC) none; 200 ppm max. water content
Toxicology: Avoid prolonged or repeated skin contact
Precaution: Material forms flamm. mixt. with air over narrow conc. range (3-8 vol.%); do not use near open flame or high intensity source of heat
Storage: Store in well-ventilated, cool, dry area

Abzol® JEG [Albemarle]
Uses: Cleaning solvent for electronics, PC boards, and ionics removal
Features: Low ozone-depleting
Regulatory: DOT nonregulated
Properties: APHA 15 clear, colorless liq.; sol. 0.25 g/100 ml in water (20 C); sp.gr. 1.280; b.p. 70 C; flash pt. (TCC, TOC, COC) none; 0.001% max. NV; 250 ppm water
Toxicology: LD50 (oral, rat) 4000 mg/kg, (inh., rat, 30 min.) 253 g/m^3
Environmental: LC50 (96 h, flathead minnow) 67,300 mg/l
Precaution: Material forms flamm. mixt. with air over narrow conc. range (3-8 vol.%); do not use near open flame or high intensity source of heat
Storage: Store in well-ventilated area; prevent spills or leaks from entering groundwater

Abzol® LS [Albemarle]
Uses: Cleaning solvent for textiles, blends, and formulations
Features: Low ozone-depleting; non-CFC
Regulatory: DOT nonregulated
Properties: APHA 15 max. colorless clear liq.; sp.gr. 1.35; b.p. 71 C; flash pt. (TCC, TOC, COC) none; 100 ppm max. water content
Toxicology: Avoid prolonged or repeated skin contact
Precaution: Material forms flamm. mixt. with air over narrow conc. range (3-8 vol.%); do not use near open flame or high intensity source of heat
Storage: Store in well-ventilated, cool, dry area

Abzol® VG [Albemarle]
Uses: Cleaning solvent for precision and general purpose cleaning in vapor degreaser or ultrasonic bath
Features: Low ozone-depleting; non-CFC
Regulatory: DOT nonregulated
Properties: APHA 15 max. colorless clear liq.; sp.gr. 1.328; b.p. 71 C; flash pt. (TCC, TOC, COC) none; 100 ppm max. water content
Toxicology: LD50 (oral, rat) 4000 mg/kg, (inh., rat, 30 min.) 253 g/m^3; avoid prolonged or repeated skin contact; CNS depressant at high doses; toxic to liver and lung
Precaution: Material forms flamm. mixt. with air over narrow conc. range (3-8 vol.%); do not use near open flame or high intensity source of heat
Storage: Store in well-ventilated, cool, dry area

Acetol® 1706 [Cognis/Care Chems.; Cognis Canada]
Chem. Descrip.: Cetyl acetate, acetylated lanolin alcohol
Uses: Water repellent, emollient, penetrant, lubricant, and cosolvent in hair care prods., creams, lotions, bath preps., suntan preps., and baby prods.
Features: Strongly hydrophobic
Properties: Lt. yel. liq.; sol. in IPM, min., castor, and olive oil; water-insol.

Acetulan® [Amerchol]
Chem. Descrip.: Cetyl acetate, acetylated lanolin alcohol
Uses: Binder for pressed powds.; emollient, plasticizer, cosolvent, NV and sebum solvent for personal care prods.; lubricant for clay, talc, and starch; stabilizer for lanolin; solubilizer in aerosols; penetrant; spreading agent
Properties: Pale yel. thin oily liq.; odorless; sol. in ethanol, min., castor and veg. oil, IPM, IPP, IPA, silicone, butyl stearate, sulfonated castor oil, ethyl acetate, and org. solv.; insol. in water; sp.gr. 0.850-0.880; visc. 10 cps; acid no. 1.0 max.; sapon. no. 180-200; hyd. 8.0 max.; pH neutral.

Actrel® 1111L [ExxonMobil]
Chem. Descrip.: Petroleum distillates, hydrotreated light
CAS 64742-47-8; EINECS/ELINCS 265-149-8
Uses: Solvent cleaner for oils and greases, OEM and maintenance applics., oil-sol. metalworking (cutting) oils, quenching oils, electro-discharge machining oils, rust preventative oils, drawing/stamping oils, rolling/forming oils, silicone oils, water-sol. metalworking fluids, honing or lapping oils, buffing compds., carbonized oils and residues, lubricating greases, waxes, dye penetrants
Features: No ODP, no HAPs
Regulatory: SARA nonreportable
Properties: Very low odor; dens. 0.81 kg/l; flash pt. (CC) 114 C; surf. tens. 27 dynes/cm
Toxicology: OEL 300 ppm; very low acute, oral, and inh. toxicity

Actrel® 1138L [ExxonMobil]
Chem. Descrip.: Aliphatic hydrocarbon with oxygenated hydrocarbons
Uses: Solvent cleaner for oils and greases, OEM and maintenance applics., oil-sol. metalworking (cutting) oils, quenching oils, electro-discharge machining oils, rust preventative oils, drawing/stamping oils, rolling/forming oils, silicone oils, water-sol. metalworking fluids, honing or lapping oils, buffing compds., carbonized oils and residues, lubricating greases, waxes, dye penetrants
Features: No ODP, no HAPs
Regulatory: SARA nonreportable
Properties: Moderate odor; dens. 0.77 kg/l; flash pt. (CC) 42 C; surf. tens. 24 dynes/cm
Toxicology: OEL 300 ppm; very low acute, oral, and inh. toxicity

Actrel® 1160L [ExxonMobil; Exxon Canada]
Chem. Descrip.: Petroleum distillates, hydrotreated light
CAS 64742-47-8; EINECS/ELINCS 265-149-8
Uses: Solvent cleaner for oils and greases, OEM and maintenance applics., oil-sol. metalworking (cutting) oils, quenching oils, electro-discharge machining oils, rust preventative oils, drawing/stamping oils, rolling/forming oils, silicone oils, water-sol. metalworking fluids, honing or lapping oils, buffing compds., carbonized oils and residues, lubricating greases, waxes, dye penetrants; diluent; chem. feedstock; fuel
Features: No ODP, no HAPs
Regulatory: SARA nonreportable; Canada DSL listed; NPRI nonreportable
Properties: Colorless clear liq.; sl. odor; sol. $< 0.01\%$ in water; m.w. 155; dens. 0.79 kg/l; visc. 1.64 cSt; vapor pressure 0.07 kPa (20 C); f.p. -50 C; b.p. 188-209 C; flash pt. (TCC) 64 C; autoignition temp. $\approx$ 227 C; surf. tens. 25 dynes/cm; 100% act.
Toxicology: REL TWA 184 ppm; LD50 (oral, rat) > 5 g/kg, (skin, rabbit) >

3 g/kg; very low acute, oral, and inh. toxicity; inh. of high vapor concs. may cause eye/respiratory tract irritation, headaches, dizziness, anesthesia, drowsiness, unconsciousness, other CNS effects, death; sl. eye irritant; frequent/prolonged skin contact may cause irritation, dermatitis; if aspirated, may cause mild to severe pulmonary injury, death; chronic exposure may cause stomach/liver/kidney damage
Environmental: Prevent entry into sewers, watercourses, low areas
Precaution: Combustible liq.; flamm. limits 1.4-9.3 vol.%; volatile; gives off invisible vapors; liq. or vapor may settle in low areas; distant ignition/explosion possible; electrostatic accumulation hazard; use proper grounding procedures; incompat. with strong oxidizing agents
Hazardous decomp. prods.: None
Storage: Store in cool, well-ventilated area away from incompat. materials, heat, open flame, ignition sources, direct sunlight; keep container closed

Actrel® 3338L [ExxonMobil; Exxon Canada]
Chem. Descrip.: Naphtha, hydrotreated heavy
CAS 64742-48-9; EINECS/ELINCS 265-150-3
UN 1268
Uses: Solvent cleaner for oils and greases, OEM and maintenance applics., oil-sol. metalworking (cutting) oils, quenching oils, electro-discharge machining oils, rust preventative oils, drawing/stamping oils, rolling/forming oils, silicone oils, water-sol. metalworking fluids, honing or lapping oils, buffing compds., carbonized oils and residues, lubricating greases, waxes, dye penetrants; diluent; chem. feedstock; fuel
Features: No ODP, no HAPs
Regulatory: SARA nonreportable; Canada DSL listed; NPRI nonreportable
Properties: Colorless clear liq.; faint petrol. odor; sol. < 0.1% in water (20 C); m.w. 149; dens. 0.75 kg/l; visc. 1.07 cSt; vapor pressure 0.23 kPa (20 C); f.p. < -60 C; b.p. 154-177 C; flash pt. (TCC) > 38 C; autoignition temp. ≈ 293 C; surf. tens. 24 dynes/cm; 100% act.
Toxicology: REL TWA 196 ppm; very low acute, oral, and inh. toxicity; inh. of high vapor concs. may cause eye/respiratory tract irritation, headaches, dizziness, anesthesia, drowsiness, unconsciousness, other CNS effects, death; sl. eye irritant; frequent/prolonged skin contact may cause irritation, dermatitis; if aspirated, may cause mild to severe pulmonary injury, death
Environmental: Prevent entry into sewers, watercourses, low areas
Precaution: Combustible liq.; flamm. limits 1.2-9.6 vol.%; volatile; gives off invisible vapors; liq. or vapor may settle in low areas; distant ignition/explosion possible; electrostatic accumulation hazard; use proper grounding procedures; incompat. with strong oxidizing agents
Hazardous decomp. prods.: None
Storage: Store in cool, well-ventilated area away from incompat. materials, heat, open flame, ignition sources, direct sunlight; keep container closed

Actrel® 3349L [ExxonMobil]
Chem. Descrip.: Aliphatic hydrocarbon with oxygenated hydrocarbons
Uses: Solvent cleaner for oils and greases, OEM and maintenance applics., oil-sol. metalworking (cutting) oils, quenching oils, electro-discharge machining oils, rust preventative oils, drawing/stamping oils, rolling/forming oils, silicone oils, water-sol. metalworking fluids, honing or lapping oils, buffing compds., carbonized oils and residues, lubricating greases, waxes, dye penetrants
Features: No ODP, no HAPs
Regulatory: SARA nonreportable
Properties: Very low odor; dens. 0.76 kg/L; flash pt. (CC) 53 C; surf. tens. 25 dynes/cm
Toxicology: OEL 300 ppm; very low acute, oral, and inh. toxicity

Actrel® 3360L [ExxonMobil]
Chem. Descrip.: Naphtha, hydrotreated heavy
CAS 64742-48-9; EINECS/ELINCS 265-150-3
Uses: Solvent cleaner for oils and greases, OEM and maintenance applics., oil-sol. metalworking (cutting) oils, quenching oils, electro-discharge machining oils, rust preventative oils, drawing/stamping oils, rolling/forming oils, silicone oils, water-sol. metalworking fluids, honing or lapping oils, buffing compds., carbonized oils and residues, lubricating greases, waxes, dye penetrants
Features: No ODP, no HAPs
Regulatory: SARA nonreportable
Properties: Colorless clear liq.; essentially odorless; sol. < 0.1% in water; dens. 0.77 kg/l; vapor pressure 2 mm Hg (70 F); f.p. < 0 C; b.p. > 185 C; flash pt. (TCC) 61 C; surf. tens. 26 dynes/cm
Toxicology: OEL 300 ppm; very low acute, oral, and inh. toxicity; inh. of high vapor concs. may cause eye/respiratory tract irritation, headaches, dizziness, anesthesia, drowsiness, unconsciousness, other CNS effects, death; sl. eye irritant; frequent/prolonged skin contact may cause irritation, dermatitis; if aspirated, may cause mild to severe pulmonary injury, death
Precaution: Combustible liq.; LEL 1.3%; UEL 8.8%; volatile; gives off invisible vapors; liq. or vapor may settle in low areas; incompat. with strong oxidizing agents
Hazardous decomp. prods.: None

Actrel® 4493L [ExxonMobil]
Chem. Descrip.: Aliphatic hydrocarbon with oxygenated hydrocarbons
Uses: Solvent cleaner for oils and greases, OEM and maintenance applics., oil-sol. metalworking (cutting) oils, quenching oils, electro-discharge machining oils, rust preventative oils, drawing/stamping oils, rolling/forming oils, silicone oils, water-sol. metalworking fluids, honing or lapping oils, buffing compds., carbonized oils and residues, lubricating greases, waxes, dye penetrants
Features: No ODP, no HAPs
Regulatory: SARA nonreportable
Properties: Essentially odorless; dens. 0.76 kg/l; flash pt. (CC) 93 C; surf. tens. 26 dynes/cm
Toxicology: OEL 300 ppm; very low acute, oral, and inh. toxicity

Aerothene MM [Dow; Dow UK]
Chem. Descrip.: Methylene chloride, specially inhibited (> 99%), propylene oxide (> 0.1%)
CAS 75-09-2; EINECS/ELINCS 200-838-9
UN 1593
Uses: Solvent for aerosol pkg.
Features: Highly stabilized; provides special protection for aerosol pkg. both in liq. and vapor space of the container; longer shelf life
Properties: Colorless liq.; ethereal odor; sol. 2% in water; sp.gr. 1.32; vapor pressure 340-355 mm Hg (20 C); f.p. -96.7 C; b.p. 40 C; flash pt. (TCC) none; autoignition temp. 640 C
Toxicology: UK MEL TWA (8 h) 100 ppm, STEL (15 min) 300 ppm; LD50 (oral, rat) 1500-2500 mg/kg; low toxicity by ing.; if aspirated, may be rapidly absorbed thru lungs and injure other organs; prolonged/repeated contact may cause skin irritation/burns; may cause mod. eye irritation; vapors can readily accumulate in confined areas causing unconsciousness, death; may cause narcotic or anesthetic effects; overexposure may cause carboxyhemoglobinemia; may cause cancer
Environmental: Biodegrad. may occur slowly; LC50 > 100 mg/l in most sensitive species; not classified as dangerous to aquatic organisms; prevent contamination of soil, surf. water, ground water
Precaution: Flamm. limits 14-22 vol.%; vapors can form flamm. mixts. with air @ ambient temps.; vapors are heavier than air, will collect in low areas; incompat. with amines, aluminum, magnesium, potassium, sodium; hydrolysis producing sm. amts. of HCl possible with gross water contamination
Hazardous decomp. prods.: Hydrogen chloride and sm. amts. of chlorine and phosgene if exposed to heat of fire or on thermal decomp.
Storage: Store in cool, dry, well-ventilated area away from heat, ignition sources

AG 101 [Amity UK]
Chem. Descrip.: n-Propyl bromide (> 95%)
CAS 106-94-5; EINECS/ELINCS 203-445-0
Uses: Solvent, degreasing solvent, solubilizer for adhesives, sealants, inks, paints, textile grease spot removal
Features: High performance; fast evaporating; CFC-free; replacement for 1,1,1-trichloroethane; safer nonflamm. alternative to methylene chloride
Regulatory: EINECS, MITI listed
Properties: Colorless liq.; sweet odor; sol. 2.4 g/l in water; sp.gr. 1.35; visc. 0.42 cps; vapor pressure 110.8 mm Hg (20 C); m.p. -110 C; b.p. 69 C; decomp. pt. 204.4 C; flash pt. (TCC) none; KB value 126
Toxicology: LD50 (oral, rat) 4260 mg/kg; LC50 (inh., rat, 0.5 h) 253,000 mg/m^3; PEL/TLV 100 ppm; harmful by inh.; vapors may irritate eyes, nose, respiratory tract; mild irritant to skin, eyes; may cause CNS effects, irregular pulse, systemic effectsTSCA listed
Environmental: Environmentally acceptable; very low global warming potential; low ODP; sinks in water; low bioaccumulation potential; not

readily biodeg.; large vols. may penetrate soil and contaminate groundwater
Precaution: LEL 4.0%, UEL 7.8%; vapors are heavier than air and may travel considerable distances; incompat. with strong oxidizing agents, and calcium, finely powd. aluminum, alkali metals
Hazardous decomp. prods.: Hydrogen halides, CO; generates toxic smoke in fire
Storage: Store in cool, well-ventilated area in tightly closed original containers; do not store in open sunshine; keep away from heat

Agnique® 2237 [Cognis/Chems. Group]
Chem. Descrip.: Methyl soyate
CAS 67784-80-9
Uses: Agric. solvent and adjuvant used in tank mixes
Features: Exc. low temp. props.

Agrisol PX401 [Akzo Nobel Surf. Chem. AB]
Chem. Descrip.: Aromatic alkoxylates
Uses: Cosolvent, penetrant, flow promoter for agrochem. toxicants
Properties: Clear liq., faint odor; insol. in water; sp.gr. 1.098; visc. 27 cs; pour pt. -15 C; flash pt. (COC) > 100 C; pH 7.0 (1% aq.); nonionic; 100% act.
Toxicology: Damaging to eyes

Agrisol PX413 [Akzo Nobel Surf. Chem. AB]
Chem. Descrip.: Aromatic alkoxylate
Uses: Cosolvent, penetrant, flow promoter for agrochem. toxicants
Properties: Clear liq., faint odor; sol. in water; sp.gr. 1.121; visc. 64 cs; pour pt. -6 C; flash pt. (COC) > 150 C; pH 6.4 (1% aq.); nonionic; 100% act.

Agrisynth® BLO [ISP]
Chem. Descrip.: γ-Butyrolactone
CAS 96-48-0; EINECS/ELINCS 202-509-5
Uses: Solvent for PAN, PS, fluorinated hydrocarbons, cellulose triacetate, shellac, used in paint removers, petrol. processing, hectograph process, specialty inks; intermediate for aliphatic and cyclic compds.; reaction and diluent solvent for pesticides; dyeing of acetate; wetting agent for cellulose acetate films, fibers; solvent welding of plastic films in adhesive applics.

Agsolex® 1 [ISP]
Chem. Descrip.: N-Methyl-2-pyrrolidone
CAS 872-50-4; EINECS/ELINCS 212-828-1
Uses: Solvent for agric. act. ingreds. to formulate emulsifiable concs., microemulsions, ultralow volume sprays; synthesis medium; cleaning agent for chemical reactors; translocation enhancer in adjuvant systems
Features: Nonvolatile; noncorrosive; good stability; low fire hazard; chemically inert
Regulatory: EPA 40CFR §180.1001(d) exempt
Properties: Clear liq.; completely misc. with water, most org. solvs. incl. alcohols, ketones, aromatic and chlorinated hydrocarbons; insol. with high b.p. aliphatic hydrocarbons; clear liq.; dens. 1.03 g/cm^3; visc. 1.65 mPa·s; f.p. -24.4 C; b.p. 202 C; flash pt. (OC) 95 C
Toxicology: Nontoxic
Environmental: Biodeg.

Agsolex® 8 [ISP]
Chem. Descrip.: N-Octyl-2-pyrrolidone
CAS 2687-94-7
Uses: Solvent for agric.

Agsolex® 12 [ISP]
Chem. Descrip.: N-Dodecyl-2-pyrrolidone
CAS 2687-96-9; EINECS/ELINCS 403-730-1
Uses: Solvent for agric.

Agsolex® BLO [ISP]
Chem. Descrip.: γ-Butyrolactone
CAS 96-48-0; EINECS/ELINCS 202-509-5
Uses: Solvent for agric. act. ingreds., to formulate tech. concs., emulsifiable concs., microemulsions; solvent for synthesis of pesticides; stabilizer for fertilizers; cleaning agent for chemical reactors
Features: Thermally stable; low fire hazard; noncorrosive
Regulatory: 40 CFR §180.1001(d) exempt
Properties: Clear liq.; completely misc. with water, methanol, acetone, benzene, CCl_4, ethyl ether, monochlorobenzene; insol. in aliphatic hydrocarbons; dens. 1.12 g/cm^3; visc. 1.7 mPa·s; f.p. -44 C; b.p. 204 C; flash pt (OC) 98 C
Environmental: Biodeg.

Akomed R [Jarchem Ind.]
Chem. Descrip.: Caprylic/capric triglyceride
Uses: Penetrant, spreading agent, dispersant for personal care, cosmetics, pharmaceuticals; solvent for sunscreens, vitamins, actives
Features: Exc. oxidation stability; low cloud pt.
Regulatory: EP, BP, DAB, USP/NF compliance
Properties: Low color and odor

Albrite® TBP [Huntsman Surf. Sciences]
Chem. Descrip.: Tributyl phosphate
CAS 126-73-8; EINECS/ELINCS 204-800-2
Uses: Plasticizer for CR, CPE, Hypalon elastomers, PVC, PVAc, cellulosics, PS, ABS, ink systems; lubricant additive; antioxidant for cellulosics, nylon, polyester, PP, PVC; antifoam; ink solvent; paints/coatings additive
Features: Nonstaining
Properties: Clear; low odor

Albrite® Tributyl Phosphate [Huntsman Surf. Sciences]
Chem. Descrip.: Tri (n-butyl) phosphate
CAS 126-73-8; EINECS/ELINCS 204-800-2
Uses: Solvent, plasticizer for polymers; antifoam for polymer systems, low temp., low flamm. hydraulic fluids; metal complexing agent in solv. extraction; plasticizer for cellulosic, nitrocellulose, vinyl resins; defoamer in adhesives, textile sizes, vat dyeing; pigment wetting agent, rheology modifier, coalescing aid in coatings
Properties: Colorless clear liq., mild odor; sl. sol. in water; sp.gr. 0.98; visc. 3.4 cps; vapor pressure 13.7 mm Hg (20 C); f.p. -80 C; b.p. 289 C (dec.); acid no. < 0.1; flash pt. 160 C; surf. tens. 28 dynes/cm (20 C); 11.7% P
Toxicology: LD50 1.8 ml/kg, mod. toxic; avoid prolonged/repeated eye and skin contact

Aldo® MCT KFG [Lonza]
Chem. Descrip.: Caprylic/capric triglyceride
CAS 73398-61-5; EINECS/ELINCS 265-724-3
Uses: Dietary supplement; calorie source; used to produce energy; emollient in personal care prods.; carrier or suspension medium for oil-sol. antibiotics and drugs; lubricant for gel capsules, confectionery coatings, antisticking agent; carrier for spray coatings; solvent, diluent, and fixing agent for flavors
Regulatory: Kosher
Properties: Clear liq., neutral odor and taste; acid no. 0.10 max.; iodine no. 1.0 max.; sapon. no. 335-360; hyd. no. 10 max.; 0.1% max. moisture

Alfol® 4 [Sasol N. Am.]
Chem. Descrip.: n-Butyl alcohol
CAS 71-36-3; EINECS/ELINCS 200-751-6
UN 1120
Uses: Intermediate for amines, plasticizers, fatty acid esters, detergents; process solvent
Regulatory: SARA §302 nonreportable, §31/312 acute health/fire hazard, §313 reportable; Calif. Prop. 65 reportable [ethanol (< 0.1%)]; Canada DSL, EINECS, Japan MITI, Australia AICS listed
Properties: Colorless liq.; char. sweet pungent penetrating odor; sol. 7.8 g/100 g in water; dens. 6.76; sp.gr. 0.805 (27 C); visc. 4 cSt (20 C); vapor pressure 4.85 mm Hg; m.p. -89 C; i.b.p. 117.7 C; flash pt. (PM) 38 C; autoignition temp. 343 C; 99-99.8% act.
Toxicology: ACGIH TLV CL 50 ppm; LD50 (oral, rat) 2.4-4.4 g/kg, (skin, rabbit) 2.0-5.3 g/kg; LC50 (inh., rat, 4 h) 7.4 mg/l; strong contact irritant; causes eye/skin/respiratory tract irritation; may cause corneal inflamm., dermatitis; vapors may cause nose/throat irritation, dizziness, headache; ing. may cause CNS depression; TSCA listed
Environmental: Biodegrades to 99% in an inoculated medium after 1 day; expected to biodegrade in soils; LC50 (fathead minnow, 96 h) 1740 mg/l; prevent contamination of lakes, streams, ponds, groundwater, soil
Precaution: Flamm. liq. and vapor; flamm. limits in air 1.4-11.2 vol.%; vapors may flash back; container may explode in heat of fire; prevent electrostatic discharge; avoid high temps.; can react with strong oxidizers, inorg. acids, halogens
Hazardous decomp. prods.: None expected
HMIS: Health 2, Flammability 3, Reactivity 0
Storage: Store @ ambient pressure; inert gas blanket and breathing system rec. to maintain color stability; use dry inert gas with -40 F min. dew pt.; prevent moisture contamination

Alfol® 14 [Sasol N. Am.]
Chem. Descrip.: Myristyl alcohol (96-98.4%), 1-dodecanol (0.9-2%), 1-hexadecanol (0.7-2%)
CAS 112-72-1; EINECS/ELINCS 204-000-3
Uses: Detergent intermediate; process solvent
Regulatory: DOT nonregulated; SARA §302 nonreportable, §311/312 acute health hazard, §313 nonreportable; Calif. Prop. 65 nonreportable; EINECS, Canada DSL, Japan MITI, Australia AICS listed
Properties: Wh. cryst. solid; sweet pungent odor; colorless liq. when melted; nil sol. in water; m.w. 214; sp.gr. 0.815 (22 C); visc. 15 cSt (38 C); vapor pressure 0.0015 mm Hg (38 C); m.p. 38 C; b.p. 297-301 C; iodine no. 0.03; sapon. no. 0.07; hyd. no. 258; flash pt. (PM) 143 C; autoignition temp. 254 C; 96-98.4% act.
Toxicology: LD50 (oral, rat) > 20 g/kg, (skin, rabbit) ≈ 8 g/kg; mod. contact irritant; may cause eye irritation, corneal inflamm.; repeated/prolonged skin contact may cause redness, irritation, dermatitis; in high concs., inh. of vapor may cause eye/nose/throat irritation, headache, dizziness; ing. may cause discomfort, nausea, vomiting; TSCA listed
Environmental: Biodeg. (70% after 2 days in inoculated medium); non-toxic to fathead minnows at its sat. conc. in water; prevent contamination of lakes, streams, ponds, groundwater, soil
Precaution: May burn but will not ignite readily; flamm. limits in air ≈ 0.5-3.5 vol.%; spillages may be slippery; prevent electrostatic discharge; avoid high temps.; can react with strong oxidizers, inorg. acids, halogens
Hazardous decomp. prods.: None expected
HMIS: Health 1, Flammability 1, Reactivity 1
Storage: Store @ ambient pressure; hot water system rec. for temp. control; avoid temps. > 43 C for long periods; inert gas blanket and breathing system rec.; use dry inert gas with -40 F min. dew pt.; prevent moisture contamination

Alfol® 1412 [Sasol N. Am.]
Chem. Descrip.: 1-Dodecanol (34-42%), 1-tetradecanol (55-63%)
Uses: Lubricant in polymer processing, metal rolling and forming; emollient, lubricant for cosmetics; defoamers; intermediate for surfactants for cosmetic, household, and industrial use; intermediate for esters for additives for plastics, rubber, metalworking, lubricants; intermediate for quaternaries for germicides and fabric softeners; process solvent
Regulatory: DOT nonregulated; SARA §302 nonreportable, §311/312 acute health hazard, §313 nonreportable; Calif. Prop. 65 nonreportable; EINECS, Canada DSL, Japan MITI, Australia AICS listed
Properties: Colorless liq.; sweet pungent odor; nil sol. in water; m.w. 205; sp.gr. 0.838 (22 C); visc. 14.4 cSt (38 C); vapor pressure < 0.1 mm Hg (21 C); m.p. 22-24 C; b.p. 274-307 C; iodine no. 0.1; sapon. no. < 1; hyd. no. 274; flash pt. (PM) 132 C; autoignition temp. 216 C; 99.7% act.
Toxicology: Mod. contact irritant; may cause eye irritation, corneal inflamm.; repeated/prolonged skin contact may cause redness, irritation, dermatitis; in high concs., inh. of vapor may cause eye/nose/upper respiratory tract irritation, headache, dizziness; ing. may cause discomfort, nausea, vomiting; TSCA listed
Environmental: Biodeg.; prevent contamination of lakes, streams, ponds, groundwater, soil
Precaution: Flamm. limits in air ≈ 0.6-3.4 vol.%; spillages may be slippery; prevent electrostatic discharge; avoid high temps.; can react with strong oxidizers, inorg. acids, halogens
Hazardous decomp. prods.: None expected
HMIS: Health 2, Flammability 1, Reactivity 1
Storage: Store @ ambient pressure; hot water system rec. for temp. control; avoid temps. > 43 C for long periods; inert gas blanket and breathing system rec.; use dry inert gas with -40 F min. dew pt.; prevent moisture contamination

Alkasurf® OD-410 [Rhodia Canada]
Chem. Descrip.: Phenol ethoxylate (1-1.5 EO)
Uses: Solvent for vinyl, phenolic, polyester, alkyd, cellulose acetate, nitrocellulose, and other resins, in hot caustic paint strippers, solv.-based cleaners and inks
Properties: Nonionic

Alkylate 215 [Huntsman]
Chem. Descrip.: C8-16 alkylbenzene
CAS 68648-87-3
Uses: Intermediate for mfg. of anionic surfactants for hand dishwash, hard surf. cleaning, industrial cleaning; solvent; lubricant oil additive intermediate; sec. plasticizer; intermediate for enhanced oil recovery
Features: Low vapor pressure
Regulatory: EINECS, DSL, MITI listed
Properties: Water-wh. liq.; m.w. 235; sp.gr. 0.855-0.870; dens. 7.2 lb/gal; anionic; 100% conc.
Toxicology: Eye and skin irritantTSCA listed
Environmental: Biodeg.
HMIS: Health 1, Flammability 1, Reactivity 0

Alkylate 229 [Huntsman]
Chem. Descrip.: C8-16 alkylbenzene
CAS 68648-87-3
Uses: Intermediate for mfg. of anionic surfactants for heavy-duty, spray-dried laundry detergents; solvent; lubricant oil additive intermediate; intermediate for enhanced oil recovery
Regulatory: EINECS, DSL, MITI listed
Properties: Liq.; m.w. 254; sp.gr. 0.862; 100% conc.
Toxicology: TSCA listed
Environmental: Biodeg.
HMIS: Health 1, Flammability 1, Reactivity 0

Alkylate L210-L [Huntsman]
Chem. Descrip.: Linear alkylbenzene coprod.
Uses: Surfactant intermediate; sec. plasticizer; solvent for chem. processes and dyes
Properties: Sp.gr. 0.890 (15.5/15.5 C); visc. 3 cSt (38 C); i.b.p. 247 C; flash pt. 118 C

Altene® [Atofina SA]
Chem. Descrip.: Trichlorethylene, stabilized
CAS 79-01-6; EINECS/ELINCS 201-167-4
Uses: Solvent

AM 230 Solvent [Americhem Sales]
Chem. Descrip.: Petroleum distillates, hydrotreated light (60-90%), alkanes, C14-16 (10-40%)
Uses: Solvent
Regulatory: DOT nonregulated; SARA §311/312/313 nonreportable; Calif. Prop. 65 nonreportable
Properties: Lt. straw to water-wh. clear oily liq.; mild hydrocarbon odor; sol. < 1% in water; sp.gr. 0.815 (60 F); vapor pressure < 0.1 mm Hg (68 F); m.p. 0 F; b.p. 470-531 F; flash pt. (PMOC) 206-226 F; autoignition temp. 459 F
Toxicology: LD50 (oral, rat) > 5 g/kg, (skin, rabbit) 2-4 g/kg; may cause minimal eye/mod. skin irritation/discomfort/defatting; ing. may cause GI tract irritation/discomfort; inh. of high vapor concs. may cause headaches, stupor, throat/eye irritation, kidney effects; extreme aspiration into lungs may cause pneumonia, death; TSCA listed
Environmental: Inherently biodeg.; prevent runoff or release to ground; prevent spread to waterways
Precaution: Flamm. limits 0.5-7.0%; electrostatic accumulation hazard; bond and ground equip. during transfer; avoid high temps.; incompat. with strong oxidizing agents
Hazardous decomp. prods.: Combustion: CO_x, noxious fumes
HMIS: Health 1, Flammability 1, Reactivity 0
Storage: Store unopened containers in cool, dry, ventilated area away from oxidizing agents, heat, sparks, flame; store @ 0 F min.; keep containers closed when not in use

AM 321 Fluid [Americhem Sales]
Chem. Descrip.: Petroleum distillates, hydrotreated light
CAS 64742-47-8; EINECS/ELINCS 265-149-8
UN 1268
Uses: Solvent
Regulatory: SARA §311/312 fire hazard, §313 nonreportable; Calif. Prop. 65 nonreportable
Properties: Colorless clear liq.; mild hydrocarbon odor; negligible sol. in water; sp.gr. 0.811 (60 F); vapor pressure 0.29 mm Hg (20 C); m.p. < -90 F; b.p. 369-498 F; flash pt. (PMCC) 160 F; autoignition temp. 434 F; 100% act.
Toxicology: OSHA PEL TWA 100 ppm; LD50 (oral, rat) > 39.9 g/kg, (skin, rabbit) 2-4 g/kg; LC50 (inh., 1 h) > 3.62 mg/l; may cause minor eye, mod. skin irritation/discomfort/defatting/redness/dermatitis/toxic systemic effects; ing. may cause GI tract irritation/discomfort; inh. may cause dizziness, impaired coordination, headaches, loss of consciousness, severe respiratory tract irritation, toxic systemic effects; TSCA listed
Environmental: Inherently biodeg.

Precaution: Combustible; flamm. limits 0.6-7.0%; may accumulate static electricity; bond and ground equip.; incompat. with strong oxidizing agents; avoid high temps.
Hazardous decomp. prods.: Combustion: CO_x, noxious fumes
HMIS: Health 1, Flammability 2, Reactivity 0
Storage: Store unopened containers @ ambient temps. in cool, dry, ventilated area away from heat, sparks, flames; keep containers closed when not in use

AM OMS 75 Solvent [Americhem Sales]
Chem. Descrip.: Heavy alkylate naphtha (75%), Stoddard solvent (25%)
UN 1268
Uses: Solvent
Regulatory: SARA §311/312 fire/acute health hazard, §313 nonreportable; Calif. Prop. 65 reportable (benzene, toluene)
Properties: Colorless clear liq.; odorless; negligible sol. in water; sp.gr. 0.76 (15.5/15.5 C); dens. 6.34 lb/gal; vapor pressure 1.2 mm Hg; b.p. 346-406 F; flash pt. (TCC) 125 F; autoignition temp. 680 F; 100% volatile
Toxicology: May cause mod. eye burning/tearing/redness/swelling, mod. to severe skin irritation/redness/burning/drying/cracking/dermatitis; ing. may cause significant digestive tract irritation; inh. of vapors may cause mild lung irritation; overexposure may cause dizziness, headache, excitation, CNS depression; repeated/prolonged exposure may cause permanent brain and nervous system damage; may be harmful or fatal by inh. abuse; TSCA listed
Environmental: Prevent entry into sewers, waterways
Precaution: Flamm.; flamm. limits 1.0-6.0%; vapor accumulations may flash/explode if ignited; avoid formation of excessive mists/fumes; keep fumes away from open flames, ignition sources; can accumulate static charge; vapors can be ignited by static discharge; open containers slowly to relieve any pressure; bond and ground all equip.; incompat. with strong oxidizing agents
Hazardous decomp. prods.: Combustion: CO_x
HMIS: Health 1, Flammability 2, Reactivity 0
Storage: Store unopened containers in cool, dry, ventilated area away from heat, sparks, flames; keep containers closed when not in use

Ameroxol® OE-2 [Amerchol; Amerchol Europe]
Chem. Descrip.: Oleth-2
CAS 9004-98-2
Uses: Solubilizer, emulsifier, dispersant, stabilizer, cosolvent for creams/lotions, shampoos, detergents, fluid and gelled transparent emulsions, fragrance prods., aerosols, pharmaceuticals
Features: Lipophilic
Properties: Pale straw-colored clear liq.; bland odor; sol. in min. oil, isopropyl esters, anhyd. ethanol; HLB 5.0; acid no. 0.2 max.; sapon. no. 2 max.; pH 4.5-7.0; nonionic; 100% conc.

Amgard® DMMP [Rhodia/Phosphorus Perf. Derivs.]
Chem. Descrip.: Dimethyl methylphosphonate
CAS 756-79-6; EINECS/ELINCS 212-052-3
Uses: Flame retardant for rigid PU foams, unsat. polyester resins, and water-borne latexes; solvent
Properties: Liq.; 25% P

Amgard® TBPO^4 [Rhodia/Phosphorus Perf. Derivs.]
Chem. Descrip.: Tributyl phosphate
CAS 126-73-8; EINECS/ELINCS 204-800-2
Uses: Flame retardant, plasticizer in plastics and coatings; solvent extractant; flame retardant in hydraulic fluids
Properties: 11.6% P

Amscour MS-100 [Am. Emulsions]
Uses: Solvent scouring agent for the toughest dirt, oil, and grease problems
Properties: Nonionic

Amsol 10 [Americhem Sales]
Chem. Descrip.: Petroleum distillates, hydrotreated light, low-boiling
CAS 68410-97-9; EINECS/ELINCS 270-093-2
UN 1268
Uses: Solvent
Regulatory: SARA §311/312 fire/acute health/chronic health hazard, §313 reportable [benzene (0.35%), toluene (2.20%)]; may contain cyclohexane (3.53%), xylene (3.55%), hexane (10.50%)
Properties: Colorless clear liq.; mild hydrocarbon odor; sp.gr. 0.72; visc. < 100 SUS (100 F); vapor pressure 200 mm Hg (20 C); b.p. > 100 F; flash pt. (TCC) -30 F; autoignition temp. > 650 F; pH neutral; 100% act.
Toxicology: May cause mod. eye burning/tearing/redness/swelling, mod. skin irritation/discomfort, defatting, redness, dermatitis; ing. may cause GI tract irritation/discomfort; inh. may cause dizziness, impaired coordination, headaches, loss of consciousness, severe respiratory tract irritation, toxic systemic effects; prolonged exposure may cause brain dysfunction; ing. may cause nausea, vomiting, diarrhea; aspiration into lungs can cause possibly fatal chem. pneumonitis; carcinogen
Environmental: Prevent entry into sewers, waterways
Precaution: Flamm.; flamm. limits 1-7%; vapor accumulations may flash/explode if ignited; avoid formation of excessive mists/fumes; keep fumes away from open flames, ignition sources; can accumulate static charge; bond and ground equip.; incompat. with strong oxidizing agents
Hazardous decomp. prods.: Combustion: CO_x
HMIS: Health 1, Flammability 4, Reactivity 0
Storage: Store unopened containers in cool, dry, ventilated area away from excessive heat, sparks, open flames; keep containers closed when not in use

Amsol 100 [Americhem Sales]
Chem. Descrip.: Petroleum distillates, light hydrocracked
CAS 64741-77-1; EINECS/ELINCS 265-078-2
Uses: Solvent
Regulatory: DOT nonregulated; SARA §311/312 acute health hazard, §313 nonreportable
Properties: Lt. yel. clear liq.; mild hydrocarbon odor; negligible sol. in water; sp.gr. 0.86 (60 F); visc. 4.2 cSt; vapor pressure 0.08 mm Hg (20 C); b.p. > 400 F; flash pt. (COC) 250 F; 100% act.
Toxicology: OSHA PEL 5 ppm (oil mist); may cause minimal eye, sl. skin irritation on direct contact; aspiration may cause chem. pneumonitis, pulmonary edema, hemorrhage and may be fatal; excessive exposure to mists/vapors by inh. may cause eye/nose/throat/lung irritation; TSCA listed
Precaution: Mists/sprays may ignite at temps. below flash pt.; keep fumes away from ignition sources, open flames; avoid formation of excessive mists/fumes; incompat. with strong oxidizing agents
Hazardous decomp. prods.: Combustion: CO_x
HMIS: Health 0, Flammability 1, Reactivity 0
Storage: Store in cool, dry, ventilated area away from excessive heat, sparks, open flames; keep containers closed when not in use

Amsol 230-K Solvent [Americhem Sales]
Chem. Descrip.: Hydrotreated light petroleum distillates (> 90%), severely hydrotreated kerosene (< 10%)
Uses: Solvent
Regulatory: DOT nonregulated; SARA §311/312 nonhazardous, §313 nonreportable
Properties: Colorless clear liq.; mild hydrocarbon odor; flash pt. 200-210 F
Toxicology: May cause mod. eye burns/tearing/redness/swelling, sl. skin irritation; ing. may cause GI tract irritation/discomfort; inh. may cause mod. respiratory discomfort, headachesTSCA listed
Environmental: Prevent entry into sewers, waterways
Precaution: Flamm. limits 1.0-6.0%; incompat. with strong oxidizing agents; vapor accumulations may flash/explode if ignited; keep fumes away from ignition sources, open flames
Hazardous decomp. prods.: None
NFPA: Health 0, Flammability 1, Reactivity 0
Storage: Store unopened containers in cool, dry, ventilated area away from heat, sparks, flames, oxidizing agents

Amsol 400 [Americhem Sales]
Chem. Descrip.: Petroleum distillates, hydrotreated middle
CAS 64742-46-7; EINECS/ELINCS 265-148-2
Uses: Solvent
Regulatory: DOT nonregulated; SARA §311/312 nonhazardous, §313 nonreportable
Properties: Colorless clear liq.; mild hydrocarbon odor; negligible sol. in water; sp.gr. 0.83 (60 F); visc. < 100 SUS (100 F); vapor pressure < 0.1 mm Hg (70 F); b.p. > 490 F; flash pt. (COC) > 250 F; 100% act.
Toxicology: ACGIH TLV TWA 5 ppm, STEL 10 ppm; LD50 (oral, rat) > 5000 mg/kg, (skin, rabbit) > 2000 mg/kg; may cause mod. eye/skin irritation/discomfort, skin defatting; ing. may cause GI tract irritation/discomfort; inh. may cause mod. respiratory discomfort, headaches; chronic exposure may cause dermatitis, pneumonitis, pulmonary edema; TSCA

listed
Environmental: Prevent entry into sewers, waterways
Precaution: Mists/sprays may ignite at temps. below flash pt.; avoid formation of excessive mists/fumes; keep fumes away from open flames, ignition sources; incompat. with strong oxidizing agents
Hazardous decomp. prods.: Combustion: CO_x
HMIS: Health 0, Flammability 1, Reactivity 0
Storage: Store in cool, dry, ventilated area away from excessive heat, sparks, open flames; keep containers closed when not in use

Amsol 500 [Americhem Sales]
Chem. Descrip.: Petroleum distillates, hydrotreated middle
CAS 64742-46-7; EINECS/ELINCS 265-148-2
Uses: Solvent
Regulatory: DOT nonregulated; SARA §311/312 nonhazardous, §313 nonreportable
Properties: Nearly colorless to colorless clear liq.; mild hydrocarbon odor; insol. in water; sp.gr. 0.84 (60 F); visc. 4.3-4.6 cps (40 C); vapor pressure < 0.1 mm Hg (70 F); b.p. 560 C; pour pt. ≈ 20 C; flash pt. (COC) 285 F; 100% act.
Toxicology: ACGIH TLV TWA 5 ppm, STEL 10 ppm; LD50 (oral, rat) > 5000 mg/kg, (skin, rabbit) > 2000 mg/kg; may cause mod. eye/skin irritation/discomfort, skin defatting; ing. may cause GI tract irritation/discomfort; inh. may cause mod. respiratory discomfort, headaches; TSCA listed
Environmental: Prevent entry into sewers, waterways
Precaution: Mists/sprays may ignite at temps. below flash pt.; avoid formation of excessive mists/fumes; keep fumes away from open flames, ignition sources; incompat. with strong oxidizing agents
Hazardous decomp. prods.: Combustion: CO_x
HMIS: Health 0, Flammability 1, Reactivity 0
Storage: Store unopened containers in cool, dry, ventilated area away from excessive heat, sparks, open flames; keep containers closed when not in use

Amsol 600 [Americhem Sales]
Chem. Descrip.: Petroleum distillates, hydrotreated middle
CAS 64742-46-7; EINECS/ELINCS 265-148-2
Uses: Solvent
Regulatory: DOT nonregulated; SARA §311/312 nonhazardous, §313 nonreportable
Properties: Nearly colorless to colorless clear liq.; mild hydrocarbon odor; insol. in water; sp.gr. 0.85 (60 F); visc. 4.5-6.0 cSt (40 C); b.p. 493-563 F; pour pt. ≤ -9 C; flash pt. > 285 F; autoignition temp. 455 F; 100% act.
Toxicology: ACGIH TLV TWA 5 ppm, STEL 10 ppm; may cause minor eye/skin irritation; ing. may cause GI tract irritation/discomfort; inh. may cause mod. respiratory discomfort, headaches; aspiration into lungs may cause chem. pneumonitis; TSCA listed
Environmental: Prevent entry into sewers, waterways
Precaution: Containers may explode in heat of fire; avoid formation of excessive mists/fumes; incompat. with strong oxidizing agents
Hazardous decomp. prods.: Combustion: CO_x
HMIS: Health 0, Flammability 0, Reactivity 0
Storage: Store unopened containers in cool, dry, ventilated area away from excessive heat, sparks, open flames; keep containers closed when not in use

Amsol 650 [Americhem Sales]
Chem. Descrip.: Petroleum distillates, hydrotreated light
CAS 64742-47-8; EINECS/ELINCS 265-149-8
Uses: Solvent
Regulatory: DOT nonregulated; SARA §311/312 nonhazardous, §313 nonreportable
Properties: Colorless clear liq.; mild hydrocarbon odor; negligible sol. in water; sp.gr. 0.83-0.84; b.p. 490-680 F; flash pt. > 250 F; 100% act.
Toxicology: May cause minimal eye, sl. skin irritation on direct contact; aspiration may cause chem. pneumonitis, pulmonary edema, hemorrhage and may be fatalTSCA listed
Environmental: Prevent runoff or release to ground
Precaution: Incompat. with strong oxidizing agents
Hazardous decomp. prods.: Combustion: CO_x, noxious fumes
HMIS: Health 0, Flammability 0, Reactivity 0
Storage: Store unopened containers in cool, dry, ventilated area away from excessive heat, sparks, open flames, oxidizing agents; keep containers closed when not in use

Amsol 3800 [Americhem Sales]
Chem. Descrip.: Petroleum distillates, hydrotreated light
CAS 64742-47-8; EINECS/ELINCS 265-149-8
Uses: Solvent
Regulatory: DOT nonregulated; SARA §311/312 nonhazardous, §313 nonreportable
Properties: Lt. straw to water-wh. clear liq.; mild hydrocarbon odor; negligible sol. in water; m.w. 190; sp.gr. 0.88 (60 F); visc. 38.4 SUS (100 F); vapor pressure < 0.04 mm Hg (20 C); b.p. > 380 F; flash pt. > 205 F; autoignition temp. > 650 F; pH neutral; 100% act.
Toxicology: May cause mod. eye burning/tearing/redness/swelling, mod. skin irritation/discomfort, defatting; ing. may cause GI tract irritation/discomfort; inh. may cause mod. respiratory discomfort, headachesTSCA listed
Environmental: Prevent entry into sewers, waterways
Precaution: Mists/sprays may ignite at temps. below flash pt.; avoid formation of excessive mists/fumes; keep fumes away from open flames, ignition sources; incompat. with strong oxidizing agents
Hazardous decomp. prods.: Combustion: CO_x
HMIS: Health 0, Flammability 1, Reactivity 0
Storage: Store unopened containers in cool, dry, ventilated area away from excessive heat, sparks, open flames; keep containers closed when not in use

Antiblaze® DMMP [Rhodia/Phosphorus Perf. Derivs.]
Chem. Descrip.: Dimethyl methylphosphonate
CAS 756-79-6; EINECS/ELINCS 212-052-3
Uses: Flame retardant for rigid urethane foams, unsat. polyester resins, and water-borne latexes; solvent
Properties: Liq.; 25% P

Aqualose W15 [Croda/Westbrook]
Chem. Descrip.: Laneth-15
CAS 61791-20-6
Uses: O/w emulsifier, solubilizer, wetting agent, dispersant; solubilizer for fragrances and other oil-sol. ingreds. in aq. (clear liq. shampoos) or aq.-alcoholic systems; solvent for bromo acid dyes; foam stabilizer for detergent systems
Features: Produces soft anhyd. gels
Properties: Golden yel. soft waxy solid; faint char. odor; sol. in water, anhyd. and 40% ethanol; sl. sol. in min. oil, IPM; drop pt. 41-49 C; HLB 12.7; acid no. 6 max.; sapon. no. 12 max.; cloud pt. 70-88 C (aq.); flash pt. > 100 C; pH 3.5-7.0 (5% aq.); nonionic
Toxicology: LD50 (oral, rat) > 16 g/kg; very low acute oral toxicity; nonirritating to skin and eyes
Environmental: Substantially biodeg.
Precaution: Nonflamm. but will burn if strongly heated
Storage: 2 yrs shelf life under good storage conditions; avoid prolonged heating above 80 C; contains 150 ppm BHT

Arconate® PC-1000 [Lyondell]
Chem. Descrip.: Propylene carbonate
CAS 108-32-7; EINECS/ELINCS 203-572-1
Uses: Solvent for broad range of applics.; reactive diluent for wood binders, urethane foams and coatings, foundry sand binders, in textile and syn. fiber industry, natural gas treating; lubricant in cosmetics
Features: High boiling pt., low toxicity
Properties: Amber liq.; mild ether-like odor; m.w. 102.09; sp.gr. 1.203-1.210 (20/20 C); dens. 10 lb/gal (20 C); visc. 2.4 cps; vapor pressure 0.03 mm Hg (20 C); f.p. -50 C; i.b.p. 170 C; decomp. pt. 320 F; flash pt. (Seta) 226 C; ref. index 1.419; pH 4.5-6.5 (10% aq.); 99% min. assay

Arconate® PC-HP [Lyondell]
Chem. Descrip.: Propylene carbonate
CAS 108-32-7; EINECS/ELINCS 203-572-1
Uses: Reactive diluent for wood binders, urethane foams and coatings, foundry sand binders, in textile and syn. fiber industry, natural gas treating; lubricant in cosmetics; polar additive for montmorillonite or bentonite clay gellants; solvent for mascara, lipstick, deodorant, nail polish remover; suitable for certain food-contact applics.
Features: High purity; high boiling pt., low toxicity, broad range of applics.
Properties: Crystal clear, mild odor; sp.gr. 1.203-1.210 (20/20 C); dens. 10.1 lb/gal (20 C); visc. 4 cs; vapor pressure 0.02 mm Hg (20 C); f.p. -50 C; b.p. 242 C; flash pt. (TOC) 132 C; ref. index 1.419; pH 6.5-7.5 (10% aq.); 99.6% min. assay
Toxicology: Low acute systemic toxicity; may cause mod. skin irritation in

undiluted form

Arcopure® MTBE [Lyondell]

Chem. Descrip.: Methyl t-butyl ether

CAS 1634-04-4; EINECS/ELINCS 216-653-1

Uses: Extraction solvent, reaction medium in pharmaceuticals, for polymerizations, Grignard reactions; octane enhancer for gasoline

Features: Highly resist. to peroxide formation

Regulatory: SARA §313

Properties: Clear liq., terpene-like odor; sl. sol. in water; misc. with most org. solvs.; m.w. 88.15; sp.gr. 0.744; dens. 6.2 lb/gal (20 C); visc. 0.350 cps (20 C); vapor pressure 204 mm Hg (20 C); f.p. -108 C; b.p. 55 C (760 mm Hg); flash pt. -30 C; ref. index 1.366; surf. tension 19.2 dynes/cm; > 99.9% purity

HMIS: Health 1, Flammability 3, Reactivity 0

Arcosolv® DPM [Lyondell; Lyondell Europe]

Chem. Descrip.: Dipropylene glycol monomethyl ether

CAS 34590-94-8; EINECS/ELINCS 252-104-2

Uses: Solvent for coatings, cleaners, inks, agric. prods., cosmetics, chemical intermediate applics., epoxy laminates, adhesives, floor polishes, fuel additives, oil field, mining, and electronic chemicals; wetting agent; solubilizer

Features: Surf. tens. reduction and slow evaporation in cleaning formulations; HAPs-free

Regulatory: SARA §313 and CERCLA nonreportable

Properties: APHA 15 max. liq.; mild, pleasant char. odor; sol. 100% in water (20 C); misc. with a number of org. solvs.; m.w. 148.2; sp.gr. 0.950-0.953; dens. 7.91 lb/gal; visc. 3.6 cstk; vapor pressure 0.17 mm Hg (20 C); f.p. -80 C; b.p. 188.3 (760 mm); flash pt. (Seta) 167 F; ref. index 1.422; surf. tens. 28.2 dynes/cm; sp. heat 0.54 cal/g/°C

Toxicology: OSHA TWA 100 ppm (inh.); low acute toxicity; sl. eye irritant; may be absorbed thru skinTSCA listed

Precaution: Combustible; combustible residue remains after emptying containers

HMIS: Health 1, Flammability 2, Reactivity 0

Storage: Store in tightly closed, properly vented containers away from heat, sparks, open flame, or strong oxidizing agents; rec. storage in carbon steel vessels

Arcosolv® DPMA [Lyondell]

Chem. Descrip.: Dipropylene glycol methyl ether acetate

CAS 88917-22-0

Uses: Solvent for coatings, cleaners, inks, epoxy laminates, agric., adhesives, floor polishes, fuel additives, oil field, mining, electronics; coalescent in waterborne emulsion systems

Features: Slow evaporating; nonhydroxylic

Properties: APHA 15 max. color; sol. 12.3% in water (20 C); m.w. 190.2; sp.gr. 0.972; dens. 8.14 lb/gal; visc. 2.2 cstk; vapor pressure < 0.10 mm Hg (20 C); b.p. 205 C (760 mm Hg); ref. index 1.4142; flash pt. (Seta) 186 F; surf. tens. 28.3 dynes/cm

HMIS: Health 2, Flammability 2, Reactivity 0

Arcosolv® DPNB [Lyondell; Lyondell Europe]

Chem. Descrip.: Dipropylene glycol n-butyl ether

CAS 29911-28-2; EINECS/ELINCS 249-951-5

Uses: Solvent for coating resins, cleaners, cosmetics, agric., electronics, inks, textiles, and adhesives; wetting agent; solubilizer; coalescing performance in coatings based on acrylic, styrene-acrylic, PVAc; film-former

Features: Slow evaporating; good solvency; low volatility; HAPs-free

Regulatory: SARA §313 and CERCLA nonreportable

Properties: APHA 20 max. color; mild odor; sol. 5% in water (20 C); sp.gr. 0.908-0.915; dens. 7.6 lb/gal; visc. 4.8 cstk; vapor pressure 0.03 mm Hg (77 F); b.p. 215 C (760 mm); flash pt. (Seta) 212 F; ref. index 1.42; surf. tens. 29 dynes/cm

Toxicology: Low toxicity by acute oral or dermal exposure; sl. irritating to eyes; only mildly irritating to skin on repeated exposuresTSCA listed

HMIS: Health 1, Flammability 1, Reactivity 0

Storage: Store in tightly closed, properly vented containers away from heat, sparks, open flame, or strong oxidizing agents; rec. storage in carbon steel vessels

Arcosolv® DPNP [Lyondell; Lyondell Europe]

Chem. Descrip.: Dipropylene glycol n-propyl ether

CAS 29911-27-1; EINECS/ELINCS 249-949-4

Uses: Solvent for I&I and consumer cleaners

Features: Reduced causticity; performance features; HAPs-free

Regulatory: SARA §313 nonreportable

Properties: Sol. 18% in water (20 C); m.w. 176.3; sp.gr. 0.922 (20 C); dens. 7.7 lb/gal; visc. 4.4 cstk; vapor pressure 0.05 mm Hg (77 F); b.p. 212 C; flash pt.(Seta) 190 F; ref. index 1.422; surf. tens. 25.8 dynes/cm

HMIS: Health 1, Flammability 2, Reactivity 0

Arcosolv® DPTB [Lyondell]

Chem. Descrip.: Dipropylene glycol t-butyl ether

CAS 132739-31-2

Uses: Solvent for coating resins

Features: Slow evaporating; HAPs-free

Regulatory: SARA nonreportable

Properties: APHA 20 max. color; mild odor; sol. 12% in water (20 C); sp.gr. 0.890-0.190; dens. 7.6 lb/gal; visc. 4.9 cstk; vapor pressure 0.08 mm Hg (77 F); b.p. 200 C (760 mm); flash pt. (Seta) 190 F; ref. index 1.42; surf. tens. 26 dynes/cm

Precaution: Ether cleavage can occur in acidic formulations; stable in neutral and alkaline conditions

Arcosolv® PE [Lyondell Europe]

Chem. Descrip.: Propylene glycol monoethyl ether

CAS 52125-53-8

Uses: Solvent for coatings, cleaners, cosmetics, agric., electronics, inks, textiles, and adhesives; wetting agent; solubilizer; surf. tens. reducer in cleaning formulations

Features: Fast evaporating; HAPs-free

Regulatory: SARA §313 nonreportable

Properties: Colorless liq.; mild pleasant ether-like odor; sol. 100% in water; misc. with a number of solvs.; m.w. 104.1; sp.gr. 0.890-0.905; dens. 7.5 lb/gal; visc. 2.08 cstk; vapor pressure 4.40 mm Hg (20 C); b.p. 148 C (760 mm); flash pt. (Seta) 92 F; ref. index 1.405; surf. tens. 29.7 dynes/cm

Toxicology: Low toxicity on acute oral, dermal, or inh. exposure; can cause severe but reversible eye irritation; mildly irritating to skinTSCA listed

HMIS: Health 1, Flammability 2, Reactivity 0

Storage: Store in tightly closed, properly vented containers away from heat, sparks, open flame, or strong oxidizing agents; rec. storage in carbon steel vessels

Arcosolv® PEA [Lyondell]

Chem. Descrip.: Propylene glycol ethyl ether acetate

CAS 98516-30-4

Uses: Solvent for coatings, cleaners, agric., cosmetics, electronics, inks, textiles, adhesives

Features: Highly misc.; broad solvency

Properties: APHA 15 max. color; mild odor; sol. 10% in water (20 C); m.w. 146; sp.gr. 0.942-0.948; dens. 7.5 lb/gal; visc. 1.3 cstk; vapor pressure 1.5 mm Hg (68 F); b.p. 158 C (760 mm); flash pt. (Seta) 129 F; ref. index 1.40; surf. tens. 26.3 dynes/cm

Toxicology: Low toxicity

HMIS: Health 2, Flammability 2, Reactivity 0

Arcosolv® PM [Lyondell; Lyondell Europe]

Chem. Descrip.: Propylene glycol methyl ether

CAS 107-98-2; EINECS/ELINCS 203-539-1

Uses: Solvent for coatings, cleaners, inks, agric. prods., cosmetics, chemical intermediate applics., epoxy laminates, adhesives, floor polishes, fuel additives, oil field, mining, and electronic chemicals; wetting agent; solubilizer; surf. tens. reducer in cleaning formulations

Features: Fast evaporating; HAPs-free

Regulatory: SARA §313 and CERCLA nonreportable

Properties: Colorless liq.; mild, pleasant char. odor; sol. 100% in water @ 20 C; misc. with a number of org. solvs.; m.w. 90.1; sp.gr. 0.918-0.921; dens. 7.65 lb/gal; visc. 1.8 cstk; vapor pressure 8.1 mm Hg (20 C); f.p. -95 C; b.p. 120.1 C (760 mm); ref. index 1.404; flash pt. (Seta) 89 F; surf. tens. 26.5 dynes/cm; sp. heat 0.57 cal/g/°C

Toxicology: Low acute toxicity; only mild adverse effects on skin and eye contactTSCA listed

Precaution: Flamm.

HMIS: Health 1, Flammability 3, Reactivity 0

Storage: Store in tightly closed, properly vented containers away from heat, sparks, open flame, or strong oxidizing agents; rec. storage in carbon steel vessels

Arcosolv® PMA [Lyondell; Lyondell Europe]

Chem. Descrip.: Propylene glycol methyl ether acetate

CAS 108-65-6; EINECS/ELINCS 203-603-9

Uses: Solvent for coating resins (acrylics, NC, urethanes), lacquers, aq. paints, cleaners, electronics (photoresists, semiconductor processing, cleaning/degreasing circuit boards, removing solder flux), agric., cosmetics, inks, textiles, adhesives
Features: Slow-evaporating; broad solvency; HAPs-free
Regulatory: SARA nonreportable
Properties: APHA 10 max. liq.; mild, ester-like odor; sol. 18.5% in water (20 C); m.w. 132.2; sp.gr. 0.963-0.966; dens. 8.02 lb/gal; visc. 1.1 cstk; vapor pressure 3.7 mm Hg (20 C); b.p. 146 C; flash pt. (Seta) 114 F; ref. index 1.400; surf. tens. 27.4 dynes/cm
Environmental: Biodeg.
Precaution: Combustible
HMIS: Health 2, Flammability 2, Reactivity 0

Arcosolv® PNB [Lyondell; Lyondell Europe]
Chem. Descrip.: Propylene glycol n-butyl ether
CAS 5131-66-8; EINECS/ELINCS 225-878-4
Uses: Solvent, coupling agent for coating resins, cleaners, cosmetics, agric., electronics, inks, textiles, and adhesives; wetting agent; solubilizer; surf. tens. reducer in water-based cleaners
Features: Good solvency; HAPs-free
Regulatory: SARA §313 and CERCLA nonreportable
Properties: Colorless color; low ether-like odor; sol. 6.4% in water (20 C); sp.gr. 0.874-0.881; dens. 7.3 lb/gal; visc. 3.5 cstk; vapor pressure 0.92 mm Hg (77 F); b.p. 160 C (760 mm); flash pt. (Seta) 138 F; ref. index 1.42; surf. tens. 26 dynes/cm
Toxicology: Low toxicity by acute oral, dermal, or inh. exposure; mod. irritating to eyes; may cause skin irritation; repeated exposures to high levels may cause liver injuryTSCA listed
Precaution: Combustible
HMIS: Health 2, Flammability 2, Reactivity 0
Storage: Store in tightly closed, properly vented containers away from heat, sparks, open flame, or strong oxidizing agents; rec. storage in carbon steel vessels

Arcosolv® PNP [Lyondell; Lyondell Europe]
Chem. Descrip.: Propylene glycol n-propyl ether
CAS 1569-01-3; EINECS/ELINCS 250-069-8
Uses: Coalescing agent, solvent for coatings, cleaners, cosmetics, agric., electronics, inks, textiles, and adhesives; wetting agent; solubilizer
Features: Low volatility; HAPs-free
Regulatory: SARA §313 nonreportable
Properties: APHA 10 max. color; low odor; sol. 100% in water; m.w. 118.2; sp.gr. 0.880-0.890; dens. 7.4 lb/gal; visc. 2.3 cstk; vapor pressure 2.5 mm Hg (77 F); b.p. 140 C (760 mm); flash pt. (Seta) 119 F; ref. index 1.410; surf. tens. 27 dynes/cm
Toxicology: Sl. to mod. toxic by ing. or dermal exposure; mod. irritating to eyes; can cause skin irritation, esp. on prolonged/repeated exposure; inh. of high vapor concs. causes sensory irritation and sedationTSCA listed
Precaution: Flamm.
HMIS: Health 2, Flammability 2, Reactivity 0
Storage: Store in tightly closed, properly vented containers away from heat, sparks, open flame, or strong oxidizing agents; rec. storage in carbon steel vessels

Arcosolv® PTB [Lyondell]
Chem. Descrip.: Propylene glycol mono-t-butyl ether
CAS 57018-52-7
Uses: Solvent, cosolvent, coupling agent for coatings, cleaners, electronics, inks, lt.-duty and hard-surf. cleaners, water-reducible polyester and alkyd resins, agric., cosmetics, textiles, adhesives; wetting agent; solubilizer; chemical intermediate in the synthesis of monomeric and polymeric prods.
Features: Offers a blend of hydrophobicity and hydrophilicity; HAPs-free
Regulatory: SARA §313 nonreportable
Properties: Colorless liq.; char. eucalyptus-like odor; sol. 17.5% in water (20 C); misc. with many org. solvs.; m.w. 132.2; sp.gr. 0.870-0.874; visc. 3.8 cstk; vapor pressure 1.9 mm Hg (20 C); f.p. -56 C; b.p. 151 C (760 mm Hg); flash pt. (Seta) 113 F; ref. index 1.4116; surf. tens. 24.2 dynes/cm
Toxicology: LD50 (oral) 3771 mg/kg, (dermal) > 2.0 g/kg; sl. toxic orally; mildly irritating to skin, severe-but-reversible eye irritant in pure formTSCA listed
Precaution: Flamm.; ether cleavage can occur in acidic formulations; stable in neutral and alkaline conditions
HMIS: Health 2, Flammability 2, Reactivity 0
Storage: Store in tightly closed, properly vented containers away from heat, sparks, open flame, or strong oxidizing agents; rec. storage in carbon steel vessels

Arcosolv® TPM [Lyondell; Lyondell Europe]
Chem. Descrip.: PPG-3 methyl ether
CAS 25498-49-1; EINECS/ELINCS 243-734-9
Uses: Solvent for coatings, cleaners, inks, agric. prods., cosmetics, chemical intermediate applics., epoxy laminates, adhesives, floor polishes, fuel additives, oil field, mining, and electronic chemicals; wetting agent; solubilizer; surf. tens. reducer in cleaning formulations
Features: Slow evaporating; HAPs-free
Regulatory: SARA §313 and CERCLA nonreportable
Properties: APHA 15 max. liq.; mild, pleasant char. odor; sol. 100% in water (20 C); misc. with a number of org. solvs.; m.w. 206.3; sp.gr. 0.962-0.965; dens. 8.03 lb/gal; visc. 5.8 cstk; vapor pressure 0.03 mm Hg (20 C); f.p. -79 C; b.p. 242.4 C (760 mm); ref. index 1.430; flash pt. (Seta) 232 F; surf. tens. 29.0 dynes/cm; sp. heat 0.51 cal/g/°C
Toxicology: Low acute toxicity; sl. eye irritant; prolonged/repeated skin contact with excessive amts. may cause kidney damageTSCA listed
HMIS: Health 1, Flammability 1, Reactivity 0
Storage: Store in tightly closed, properly vented containers away from heat, sparks, open flame, or strong oxidizing agents; rec. storage in carbon steel vessels

Arcosolv® TPNB [Lyondell Europe]
Chem. Descrip.: Tripropylene glycol n-butyl ether
CAS 55934-93-5; EINECS/ELINCS 259-910-3
Uses: Solvent for coatings (aerosol, automotive, industrial maintenance, metal containers, plastic substrates, architectural, coil, machinery, equip., metal furniture and fixtures, wood furniture)
Properties: Sol. 3% in water (20 C); sp.gr. 0.934; vapor pressure < 0.01 mm Hg (20 C); b.p. 276 C; flash pt. (Seta) 276 F; surf. tens. 29.9 dynes/cm
Toxicology: Low toxicity
Storage: Avoid contact with air when storing for extended periods; do not allow contact with copper or copper alloys, aluminum or aluminum alloys, or galvanized steel

Argobase 125T [Croda/Westbrook]
Chem. Descrip.: Lanolin alcohol, mineral oil, and octyldodecanol with 150 ppm BHT (antioxidant)
Uses: Liq. absorption base, enhancing appearance and elegance of cosmetics and pharmaceuticals, skin and hair care prods.; stabilizer for w/o or o/w emulsions, liq. make-up, pigmented systems; plasticizer in hair sprays; solvent in aerosols
Properties: Almost colorless to pale yel. clear oily liq., faint char. odor; sol. in min. oil, IPA, IPM; partly sol. in ethanol; insol. in water; sp.gr. 0.84-0.87; HLB 8.0; acid no. 1 max.; iodine no. 12 max.; sapon. no. 3 max.; hyd. no. 11-18; flash pt. > 100 C; ref. index 1.46-1.476
Toxicology: Extremely low toxicity; LD50 (oral, rat) > max. pract. dose of 16 g/kg; nonirritating to eyes and skin
Environmental: Nonhazardous
Precaution: Nonflamm., but will burn if strongly heated
Storage: 2 yrs shelf life when stored in well-closed containers under good conditions; avoid prolonged heating above 80 C; if cloudiness occurs, gently warm and stir

Arlamol® E [Uniqema; Uniqema Am.]
Chem. Descrip.: PPG-15 stearyl ether
CAS 25231-21-4
Uses: Emollient, solvent for personal care
Properties: Colorless oily liq.; sol. in alcohol, cottonseed and min. oils, isopropyl esters; insol. in water; sp.gr. 0.95; visc. 80 cps; pour pt. typ. < 0 C; nonionic; 100% act.

Arlasolve® DMI [Uniqema; Uniqema Am.]
Chem. Descrip.: Dimethyl isosorbide
CAS 5306-85-4; EINECS/ELINCS 226-159-8
Uses: Surfactant, emollient, solubilizer for personal care; solvent for cosmetics and pharmaceuticals
Properties: Colorless liq.; sol. in water, alcohol, cottonseed oil, propylene glycol; visc. 6 cps; nonionic

Arlatone® T [Uniqema; Uniqema Am.]
Chem. Descrip.: PEG-40 sorbitan peroleate

Uses: W/o emulsifier, solvent, solubilizer, antistat, lubricant, spreading agent for bath oils, household prods., textiles, pharmaceuticals
Properties: Yel. liq.; sol. in veg., min. oils, IPM, IPP; water disp.; sp.gr. 1; visc. 175 cps; HLB 9.5; flash pt. > 300F; nonionic; 100% act.

Arlypon® DV [Cognis]
Chem. Descrip.: Triglycerides of med. chain fatty acids
Uses: Oil component for cutting oils, release emulsions; solvent for fats; consistency agent; release agent

Arlypon LSE [Cognis]
Chem. Descrip.: Linoleic acid ethyl ester
CAS 544-35-4; EINECS/ELINCS 208-868-4
Uses: Oil component for cutting oils, release emulsions; solvent for fats; consistency agent; release agent

Arlypon OSO [Cognis]
Chem. Descrip.: Oleyl erucate
CAS 3687-45-4; EINECS/ELINCS 222-980-4
Uses: Oil component for cutting oils, release emulsions; solvent for fats; consistency agent; release agent

Arlypon SO [Cognis]
Chem. Descrip.: Glycerol trioleate
CAS 122-32-7; EINECS/ELINCS 204-534-7
Uses: Oil component for cutting oils, release emulsions; solvent for fats; consistency agent; release agent

Arlypon SV [Cognis]
Chem. Descrip.: Triglycerides of med. chain fatty acids
Uses: Oil component for cutting oils, release emulsions; solvent for fats; consistency agent; release agent

Aromasol® 12 [Huntsman Petrochems. UK]
Chem. Descrip.: Hydrocarbon byprod.
Uses: Solvent for creosote mfg.
Features: General aromatic solvent

Aromasol® H [Huntsman Petrochems. UK]
Chem. Descrip.: Light aromatic naphtha
CAS 64742-95-6; EINECS/ELINCS 265-199-0
Uses: Solvent for agric., paints
Properties: Clear

Aromatic 100 [ExxonMobil]
Chem. Descrip.: Naphtha, light aromatic hydrocarbon fluid
CAS 64742-95-6; EINECS/ELINCS 265-199-0
Uses: Solvent, diluent for paints/coatings, printing inks, agrochems.; cosolvent, coalescing agent for acrylic emulsions; azeotrope; solvent for additives, pesticides, industrial cleaning, mastics, sealants, process fluids, acrylic polymerization
Features: Very low HAPs; slow evaporation rates for good flow and leveling; high elec. resist.; good solvency
Regulatory: FDA 40CFR §180.1001(c)(d)(e)
Properties: Bright clear appearance; m.w. 120; sp.gr. 0.868-0.878 (15.6/15.6 C); dens. 7.27 lb/gal (60 F); visc. 0.8 cP; vapor pressure 4.4 mm Hg (20 C); i.b.p. 154 C min.; flash pt. 42 C min.; ref. index 1.4993 (20 C); surf. tens. 30 dynes/cm; KB value 94; 99.8% aromatic
Toxicology: TLV/OEL 50 ppm

Aromatic 100 Anti-Static [ExxonMobil]
Chem. Descrip.: Hydrocarbon fluid with antistatic additive (15 ml/1000 gal)
Uses: Solvent for additives, paints/coatings, pesticides, industrial cleaning, mastics, sealants, process fluids
Features: ≤ 3.4% HAPs
Properties: Sp.gr. 0.868-0.878 (15.6/15.6 C); i.b.p. 154 C min.; flash pt. > 42 C; KB value 90 min.; 98% min. aromatics

0-Aromatic 142 Solvent [Americhem Sales]
Chem. Descrip.: Stoddard solvent
CAS 8052-41-3; EINECS/ELINCS 232-489-3
UN 1268
Uses: Solvent
Regulatory: SARA §311/312 fire/acute health hazard, §313 nonreportable; Calif. Prop. 65 nonreportable
Properties: Colorless clear liq.; mild hydrocarbon odor; negligible sol. in water; sp.gr. 0.79 (15.5/15.5 C); dens. 6.60 lb/gal (60 F); vapor pressure 2 mm Hg (20 C); b.p. 370-405 F; flash pt. 146 F; autoignition temp. 440 F; 100% volatile
Toxicology: ACGIH TLV TWA 100 ppm; may cause mod. eye/skin irritation/discomfort, skin defatting; inh. may cause GI tract irritation/discomfort; inh. may cause mod. respiratory discomfort, headaches; chronic overexposure may cause permanent brain/nervous system damage; may be harmful or fatal by inh. abuse; TSCA listed
Environmental: Prevent entry into sewers, waterways
Precaution: Combustible; flamm. limits 1.0-7.0%; vapor accumulations may flash/explode if ignited; avoid formation of excessive mists/fumes; keep fumes away from open flames, ignition sources; can accumulate static charge; vapors can be ignited by static discharge; open containers slowly to relieve any pressure; bond and ground all equip.; incompat. with strong oxidizing agents
Hazardous decomp. prods.: Combustion: CO_x
HMIS: Health 1, Flammability 2, Reactivity 0
Storage: Store unopened containers in cool, dry, ventilated area away from heat, sparks, flames; keep containers closed when not in use

Aromatic 150 [ExxonMobil]
Chem. Descrip.: Heavy aromatic naphtha hydrocarbon fluid
CAS 64742-94-5; EINECS/ELINCS 265-198-5
Uses: Solvent, diluent for paints/coatings, printing inks, agrochems.; cosolvent, coalescing agent for acrylic emulsions; azeotrope; solvent for additives, pesticides, industrial cleaning, mastics, sealants, process fluids, acrylic polymerization
Features: Very low HAPs; slow evaporation rates for good flow and leveling; high elec. resist.; good solvency
Regulatory: FDA 40CFR §180.1001(b)
Properties: Dens. 7.45 lb/gal (60 F); visc. 1.2 cP; vapor pressure 0.5 mm Hg (20 C); b.p. 184-205 C; flash pt. 151 F; surf. tens. 31.4 dynes/cm; KB value 97; 99.8% aromatic
Toxicology: TLV/OEL 100 ppm

Aromatic 200 [ExxonMobil; Exxon Canada]
Chem. Descrip.: Heavy aromatic naphtha hydrocarbon fluid, BHT (25 ppm)
CAS 64742-94-5; EINECS/ELINCS 265-198-5
Uses: Solvent, diluent for paints/coatings, agrochems.; cosolvent, coalescing agent for acrylic emulsions; solvent for additives, pesticides, industrial cleaning, mastics, sealants, process fluids, acrylic polymerization; azeotrope; active tail solvent for most polyester coil coatings; antisagging agent, antiblistering agent during high temp. curing; chem. feedstock; fuel
Features: Slow evaporation rates for good flow and leveling; high elec. resist.; good solvency; complies with criteria for LVP solvent
Regulatory: FDA 40CFR §180.1001(b); Canada DSL listed; NPRI reportable [naphthalene (13%)]
Properties: Pale yel. clear liq.; aromatic hydrocarbon odor; sol. < 0.1% in water; m.w. 161; sp.gr. 1.0 (15.6 C); dens. 8.33 lb/gal (60 F); visc. 2.6 cP; vapor pressure < 0.1 mm Hg (20 C); f.p. -19 C; b.p. 231-277 C; flash pt. (PMCC) 95 C; autoignition temp. 491 C; surf. tens. 38 dynes/cm; KB value 101; 99.8% aromatic
Toxicology: REL 15 ppm; inh. of high vapor concs. irritating to eyes, respiratory tract, may cause headaches, dizziness, anesthesia, drowsiness, unconsciousness, other CNS effects, death; sl. eye irritant; frequent/prolonged skin contact may cause irritation, dermatitis; if aspirated, may cause mild to severe pulmonary injury, death; chronic exposure may cause liver/thyroid/urinary bladder damage; possible carcinogen
Environmental: Prevent entry into sewers, watercourses, low areas
Precaution: Flamm. limits 0.7-5.3 vol.%; liq. may burn at flash pt.; electrostatic accumulation hazard; use proper grounding procedure; incompat. with nitric acid, sulfuric acid, strong oxidizing agents
Hazardous decomp. prods.: None
Storage: Store in cool, well-ventilated area away from incompat. materials, direct sunlight; keep container closed

0-Aromatic Mineral Spirits [Americhem Sales]
Chem. Descrip.: Stoddard solvent
CAS 8052-41-3; EINECS/ELINCS 232-489-3
UN 1268
Uses: Solvent
Regulatory: SARA §311/312 fire/acute health/chronic health hazard
Properties: Colorless clear liq.; mild hydrocarbon odor; nil sol. in water; sp.gr. 0.78 (20 C); vapor pressure 2.6 mm Hg (20 C); b.p. ≈ 320 F; flash pt. (TCC) ≈ 110 F; 100% act.
Toxicology: ACGIH TLV TWA 100 ppm; may cause mod. eye burning/tearing/redness/swelling, mod. skin irritation/discomfort, defatting, redness, dermatitis; ing. may cause GI tract irritation/discomfort; inh. may

cause mod. respiratory discomfort, headaches; serious aspiration hazard possible if swallowed; higher concs. of vapors may cause CNS depression, headaches, dizziness, nausea
Environmental: Prevent entry into sewers, waterways; prevent runoff or release to ground
Precaution: Combustible; flamm. limits 1.0-7.0%; vapor accumulations may flash/explode if ignited; avoid formation of excessive mists/fumes; keep fumes away from open flames, ignition sources; incompat. with strong oxidizing agents
Hazardous decomp. prods.: Combustion: CO_x
HMIS: Health 2, Flammability 2, Reactivity 0
Storage: Store unopened containers in cool, dry, ventilated area away from heat, sparks, flames, oxidizing agents; keep containers closed when not in use

Arosolve 570-HF [Arol Chem. Prods.]
Chem. Descrip.: Aromatic solvs./detergent blend
Uses: Solvent scouring agent, detergent for textiles, syn. and nat. fibers; roller cleaner; tar remover
Features: Mod. to high foam; non-red label
Properties: Water-wh. to pale yel. thin liq., solv.-type odor; disp. in water; sp.gr. 0.88 g/cc; dens. 7.39 lb/gal; flash pt. (COC) 55 C; pH 8.5 ± 0.5; anionic/nonionic; 100% act.

Arosolve 570 Special [Arol Chem. Prods.]
Chem. Descrip.: Aromatic solvs./detergent blend
Uses: Solvent scouring agent for textiles; roller cleaner; tar remover
Properties: Water-wh. to pale yel. thin liq., solv.-type odor; disp. in water; dens. 7.39 lb/gal; flash pt. (COC) 117 F; pH 8.5 ± 0.5; anionic/nonionic; 100% act.
Precaution: Combustible

Arosolve B-950 [Arol Chem. Prods.]
Chem. Descrip.: Chlorinated solvs./emulsifiers blend
Uses: Solvent scouring agent for removal of grease, graphite and oil stains on polyester, nylon and other syn. and natural fibers; degreaser; tar remover
Features: Stable to most acids, alkalis, metal salts
Properties: Lt. straw liq.; pH 7.5 (0.5%); anionic/nonionic; 100% act.
Use Level: 2-6% ow

Arosolve CON [Arol Chem. Prods.]
Chem. Descrip.: Chlorinated aromatic solvs./alcohols/emulsifiers blend
Uses: Solvent scouring agent for natural and syn. fibers for batch processing in becks or jigs
Features: Removes oil, sizes, and waxes for subsequent dyeing
Properties: Yel. to amber clear liq., chlorinated aromatic-type odor; sol. in water to wh. emulsion; sp.gr. 1.028 g/cc; dens. 8.58 lb/gal; pH 6.6-7.8; anionic/nonionic
Use Level: 1-3% ow

Arosolve CZU [Arol Chem. Prods.]
Uses: Emulsifier, solvent blend for dark colors
Features: Low-foaming; yields excellent depth of shade

Arosolve DBM [Arol Chem. Prods.]
Chem. Descrip.: Aromatic hydrocarbon/emulsifiers blend
Uses: Carrier auxiliary producing level dyeings on syn. and natural fibers
Properties: Pale yel. to amber thin liq.; sol. in water forming stable blue-wh. emulsion; sp.gr. 0.94 g/cc; dens. 7.84 lb/gal; flash pt. (COC) 78 C; pH 8.0 ± 0.5 (5%); anionic/nonionic
Use Level: 6-15%
Toxicology: Low toxicity

Arosolve DMI-F [Arol Chem. Prods.]
Chem. Descrip.: Solvs./detergents blend
Uses: Solvent scouring agent, machine cleaner for textile industry
Features: High foaming
Properties: Water-wh. to pale straw clear thin liq., mild solv.-type odor; disp. in warm water; pH 8-9; 100% act.

Arosolve FX-6 [Arol Chem. Prods.]
Chem. Descrip.: Surfactant/solv. blend
Uses: One-bath desizing agent, solvent scouring agent, and dyeing auxiliary for syn., natural and blended fabrics; antiredeposition aid for soil and sizings
Features: Does not interfere with subsequent dyeing operations
Properties: Yel. clear liq., mild aromatic solv.-type odor; sp.gr. 0.935 g/cc; dens. 7.8 lb/gal; flash pt. (COC) 106 F; pH 7.0 ± 0.2
Use Level: 2-4% ow

Arosolve H99B [Arol Chem. Prods.]
Chem. Descrip.: Blend of chlorinated solvent and aromatic solvent
Uses: Solvent dye carrier

Arosolve IWS [Arol Chem. Prods.]
Uses: Solvent scouring agent for removal of soils, oils, and sizes from cellulosics and syn. fabrics and their blends; wetting agent
Features: Good high caustic stability
Properties: Water-wh. sl. visc. liq., sl. aromatic odor; sol. in water; sp.gr. 0.925; dens. 7.72 lb/gal; pH 7.2 ± 0.3; nonionic/anionic

Arosolve KJ [Arol Chem. Prods.]
Uses: Solvent carrier for syn. fabrics and blends; emulsifier, solvent for frame drying
Properties: Nonionic

Arosolve LOC [Arol Chem. Prods.]
Uses: Carrier for atmospheric equipment, poor ventilation conditions
Properties: Low odor
Toxicology: Nontoxic

Arosolve MN [Arol Chem. Prods.]
Uses: Polyester dye carrier for jets, high pressure dyeing equipment
Features: Low-foaming; excellent depth of shade yield on most colors

Arosolve MN-LF [Arol Chem. Prods.]
Chem. Descrip.: Methylnaphthalene
CAS 1321-94-4
Uses: Polyester dye carrier for use in jets and other high-pressure dyeing equip.
Features: Low foaming; produces level and bright shades on polyester
Properties: Pale yel. to amber liq., mild aromatic odor; sp.gr. 1.002 g/cc; dens. 8.36 lb/gal; flash pt. (PMCC) 190 F; pH 8.0-8.3; 82% volatile.

Arosolve MRC [Arol Chem. Prods.]
Chem. Descrip.: Solvs./detergents blend
Uses: Solvent cleaner for textile rollers
Features: Low foaming
Properties: Pale yel. clear thin liq., mild solv. and ammonia odor; disp. in water; 95+% act.

Arosolve MRC-HF [Arol Chem. Prods.]
Chem. Descrip.: Aromatic and oxygenated solvs./detergents blend
Uses: Solvent cleaner for textile rollers and machines
Features: Does not exhibit the flamm. associated with competitive prods.
Properties: Pale yel. clear thin liq., mild solv. odor; disp. in water; nonionic/anionic; 95+% act.

Arosolve ONF [Arol Chem. Prods.]
Chem. Descrip.: Blend of nonpetroleum distillate-based solvs. and low-foaming detergents
Uses: Roller cleaner, solubilizer for pigment, disperse, and other padded dyestuffs
Features: Will not cause rubber swelling normally associated with petroleum-based prods.

Arosolve RCB [Arol Chem. Prods.]
Chem. Descrip.: Low-foaming emulsifiers/aromatic petrol. distillate/biphenyl blend
Uses: Solvent carrier for syn. fabrics and blends
Features: Low foaming for high shear jets
Properties: Lt. yel. to pale amber liq., mild naphthenic odor; sol. in warm water to stable blue-wh. emulsion; dens. 8.15 lb/gal; flash pt. (PMCC) 130 F; pH 6-7; anionic/nonionic
Precaution: Combustible

Arosolve SH-180 LF [Arol Chem. Prods.]
Chem. Descrip.: Solvs./low foaming detergent blend
Uses: Solvent scouring agent, equip. cleaner for rubber rollers; tar remover; dye dispersant
Features: Stable to inorg. builders, alkalis, and aqua ammonia
Properties: Water-wh. to pale yel. thin liq., mild odor; misc. with water; sp.gr. 0.93; flash pt. (COC) 204 F; pH 7.7 ± 0.3; 95+% act.

Arosolve TBP [Arol Chem. Prods.]
Chem. Descrip.: Perchloroethylene-based
Uses: Solvent scouring agent for synthetics and blends; removal aid for oils, greases, contaminants
Features: Rec. when heavy soiling and oil content is required, and sufficient ventilation is available

Arosolve TMB [Arol Chem. Prods.]
Chem. Descrip.: Chlorinated/aromatic hydrocarbons/aromatic ester blend
Uses: Polyester dye carrier for pressure jet procedures

Properties: Lt. amber clear thin liq., mild chlorinated odor; disp. in warm water forming stable low foaming emulsions; pH 7-8

Arosolve TUV [Arol Chem. Prods.]
Uses: Solvent carrier for syn. fabrics and blends; emulsifier; leveling agent; solvent
Features: Low-foaming; produces high color yields
Properties: Anionic/nonionic

Arosolve VNO-LF [Arol Chem. Prods.]
Uses: Polyester dye carrier imparting levelness and depth of shade to syn. fabrics dyed with disperse dyes
Features: Low foaming
Properties: Pale yel. to lt. amber thin liq., mild aromatic odor; sp.gr. 1.05 g/cc; dens. 8.76 lb/gal; flash pt. 115 F; pH 6.6
Precaution: Combustible

Arosolve XNF-1 [Arol Chem. Prods.]
Chem. Descrip.: Solvs./low foaming detergent blend
Uses: Pressure jet solvent scouring agent for difficult grease, graphite and oil stains on polyester, nylon and other syn. and natural fibers; general degreaser and tar remover
Features: Low foaming; stable to most acids and alkalis
Properties: Clear liq.; disp. in warm water; sp.gr. 0.91; dens. 7.6 lb/gal; pH 7.5-8.0; nonionic; 95+% act.
Use Level: 2-6% ow

Axarel® 52 [Petroferm]
Chem. Descrip.: Petroleum distillates, hydrotreated light (70-90%), alkyloxy polyethylene oxyethanol (CAS 84133-50-6, 4.5-9.5%), and disobutyl glutarate/diisobutyl adipate/diisobutyl succinate (4-9%)
Uses: Solvent precision cleaner for cleaning and removal of oils, greases, and high melting waxes from metal, glass, and plastic substrates
Features: For immersion cleaning; HAPs-free; zero ODP; low GWP
Regulatory: SARA nonreportable; DOT nonregulated; SNAP approved
Properties: Colorless to lt. yel. liq.; mild hydrocarbon odor; sol. < 0.1% in water; sp.gr. 0.84; visc. 2.9 cps; vapor pressure < 0.1 mm Hg (20 C); b.p. 221-295 C; flash pt. (PMCC) 96 C; VOC 837 g/l
Toxicology: May irritate skin, eyes, upper respiratory tract; ing. may cause mouth/throat/GI tract irritation, mild nausea, headache, diarrhea, thirst, weakness, and abdominal discomfortTSCA listed
Precaution: Combustible; incompat. with strong bases, strong acids, strong oxidizing agents; avoid temps. > 221 C; keep away from heat, sparks, and open flame
NFPA: Health 1, Flammability 1, Reactivity 0
Storage: Unlimited shelf life in orig. sealed container in cool, ventilated, fire-resist. building; avoid overheating or freezing

Axarel® 56 [Petroferm]
Chem. Descrip.: Petroleum distillates, hydrotreated light (70-90%), alkyloxy polyethylene oxyethanol (CAS 84133-50-6, 4.5-9.5%), and diisobutyl glutarate/diisobutyl adipate/diisobutyl succinate (4-9%)
Uses: Solvent precision cleaner for removing oils, greases, and high melting waxes from metal, glass, and plastic
Features: For immersion cleaning; contains no rust inhibitors; HAPs-free; zero ODP; low GWP
Regulatory: SARA nonreportable; DOT nonregulated; SNAP approved
Properties: Colorless to lt. yel. liq.; mild hydrocarbon odor; sol. < 0.1% in water; sp.gr. 0.84 ± 0.01; visc. 2.9 cps; vapor pressure 0.1 mm Hg (20 C); b.p. 221-295 C; flash pt. (PMCC) < 93 C; VOC 837 g/l
Toxicology: May irritate skin, eyes, upper respiratory tract; ing. may cause mouth/throat/GI tract irritation, mild nausea, headache, diarrhea, thirst, weaknessTSCA listed
Precaution: Keep away from heat, sparks, open flame; incompat. with strong bases, acids, oxidizing agents
NFPA: Health 1, Flammability 1, Reactivity 0
Storage: Unlimited shelf life in orig. sealed container in cool, ventilated, fire-resist. building; avoid overheating or freezing

Axarel® 2200 [Petroferm]
Chem. Descrip.: C9-12 isoparaffin (30-60%) and 1-propoxy-2-propanol (30-60%)
UN 1993
Uses: Solvent precision cleaner for metal cleaning, general purpose cleaning and removal of oils, lubricating fluids, metal cuttings, and fines prior to painting, bonding, plating, or coating
Features: For immersion, ultrasonic, and spray-under-immersion systems, and recirculating-type parts washers; HAPs-free; zero ODP; essentially zero GWP
Regulatory: SARA §313 nonreportable; SNAP approved
Properties: Colorless liq.; mild alcohol odor; sol. 7% in water; sp.gr. 0.82; visc. 2.2 cps; vapor pressure 3.54 mm Hg (68 F); b.p. > 152 C; flash pt. (PMCC) 44 C; KB > 150
Toxicology: May irritate skin, eyes, upper respiratory tract; ing. may cause severe irritation of GI tract, nausea, vomiting, abdominal cramps, diarrheaTSCA listed
Precaution: Flamm.; incompat. with strong bases, strong acids, strong oxidizing agents; avoid temps. > 152 C
NFPA: Health 1, Flammability 2, Reactivity 0
Storage: Unlimited shelf life in orig. sealed container in cool, ventilated, fire-resist. building; avoid overheating or freezing

Axarel® 4100 [Petroferm]
Chem. Descrip.: Isoparaffinic hydrocarbon: C9-12 isoalkanes (88-93%) and 1-propoxy-2-propanol (7-12%)
UN 1993
Uses: Solvent precision cleaner for general purpose cleaning and removal of oils, lubricating fluids, metal cuttings, and fines prior to painting, bonding, plating, or coating
Features: For immersion, ultrasonic, and spray-under-immersion systems, and recirculating-type parts washers; HAPs-free; zero ODP; essentially zero GWP
Regulatory: SARA nonreportable; SNAP approved
Properties: Colorless liq.; mild hydrocarbon odor; sol. < 10% in water; sp.gr. 0.76; visc. 1.0 cps; vapor pressure 2 mm Hg (20 C); b.p. > 152 C; flash pt. (PMCC) 47 C; KB 35; VOC 790 g/l
Toxicology: May irritate skin, eyes, upper respiratory tract; ing. may cause severe GI irritation, nausea, vomiting, abdominal cramps, diarrheaTSCA listed
Precaution: Flamm.; incompat. with strong bases, strong acids, strong oxidizing agents; avoid temps. > 152 C
NFPA: Health 1, Flammability 2, Reactivity 0
Storage: Unlimited shelf life in orig. sealed container in cool, ventilated, fire-resist. building; avoid overheating or freezing

Axarel® 6100 [Petroferm]
Chem. Descrip.: Petroleum distillates, hydrotreated light (85-95%) and dimethyl glutarate (5-15%)
Uses: Solvent precision cleaner for general purpose cleaning and removal of oils, lubricating fluids, metal cuttings, and fines prior to painting, bonding, plating, or coating
Features: For immersion, ultrasonic, and spray-under-immersion systems, and recirculating-type parts washers; HAPs-free; zero ODP; low GWP; not a hazardous waste (EPA 40 CFR §261)
Regulatory: SARA nonreportable; DOT nonregulated; SNAP approved
Properties: Colorless to lt. yel. liq.; mild hydrocarbon odor; sol. < 0.1% in water; sp.gr. 0.82; visc. 1.6 cps; vapor pressure 1 mm Hg (20 C); b.p. 182-295 C; flash pt. (PMCC) 61 C; VOC 820 g/l
Toxicology: May irritate skin, eyes, upper respiratory tract; ing. may cause mouth/throat/GI tract irritation, nausea, headache, diarrhea, thirst, and weaknessTSCA listed
Environmental: Biodeg.
Precaution: Incompat. with strong bases, strong acids, strong oxidizing agents; avoid temps. > 182 C; keep away from heat, sparks, and open flame
NFPA: Health 1, Flammability 2, Reactivity 0
Storage: Unlimited shelf life in orig. sealed container in cool, ventilated, fire-resist. building; avoid overheating or freezing

Axarel® 9100 [Petroferm]
Chem. Descrip.: Petroleum distillates, hydrotreated light (88-93%) and diisobutyl glutarate/diisobutyl adipate/diisobutyl succinate (7-12%)
Uses: Solvent precision cleaner for general metal cleaning and degreasing
Features: Replaces handwipe or immersion isopropanol, toluene, xylene, acetone, and min. spirits; HAPs-free; zero ODP; low GWP
Regulatory: SARA §313 nonreportable; DOT nonregulated; SNAP approved
Properties: Colorless to lt. yel. liq.; mild hydrocarbon odor; sol. < 0.1% in water; sp.gr. 0.83; visc. 2.2 cps; vapor pressure 0.1 mm Hg (20 C); b.p. 221-295 C; flash pt. (PMCC) 96 C; VOC 830 g/l
Toxicology: May irritate skin, eyes, upper respiratory tract; low toxicity by ing.; ing. may cause mild nausea and abdominal discomfortTSCA listed
Environmental: Biodeg.

Precaution: Incompat. with strong bases, strong acids, strong oxidizing agents; avoid temps. > 221 C; keep away from heat, sparks, and flames
NFPA: Health 1, Flammability 1, Reactivity 0
Storage: Unlimited shelf life in orig. sealed container in cool, ventilated, fire-resist. building; avoid overheating or freezing

Axol® E 61 [Goldschmidt AG]
Chem. Descrip.: Acetylated hydrogenated lard glyceride
CAS 8029-91-2; EINECS/ELINCS 297-461-5
Uses: Food emulsifier, lubricant, solvent, plasticizer and coating material for foodstuffs (meat prods., cheese, dried fruit, nuts), and cosmetics; can be combined with hydrophilic emulsifiers to increase whipping props. in shortenings/topping
Regulatory: EU E472a, FAO/WHO compliance
Properties: Ivory waxy slabs, faint char. odor; misc. with fats and oils; insol. in water; m.p. 34-38 C; HLB 2-3; iodine no. 3 max.; sapon. no. 310-340; hyd. no. 75-95; flash pt. > 100 C; nonionic; 100% conc.
Toxicology: Nonhazardous

Axol® E 66 [Goldschmidt AG]
Chem. Descrip.: Acetic acid ester of mono/diglycerides
CAS 68990-55-6
Uses: Lubricant, solubilizer, solvent, plasticizer for foodstuffs, cosmetics; solubilizer for spice oils, essential oils, antioxidants; plasticizer for fats
Regulatory: EU E472a compliance
Properties: Pale yel. liq., faint char. odor; misc. with fats and oils; insol. in water; sp.gr. 0.900; HLB 2-3; iodine no. ≈ 30; sapon. no. 360-390; hyd. no. 20 max.; cloud pt. 10-15 C; flash pt. > 100 C; nonionic; 100% conc.
Toxicology: Nonhazardous

B

Balmix® [Atofina SA]
Chem. Descrip.: 1,1,1-Trichlorethane, stabilized
CAS 71-55-6; EINECS/ELINCS 200-756-3
Uses: Solvent

Baltane® [Atofina SA]
Chem. Descrip.: 1,1,1-Trichlorethane, stabilized
CAS 71-55-6; EINECS/ELINCS 200-756-3
Uses: Solvent

Bayol 35 [Exxon Canada]
Chem. Descrip.: Petroleum distillates, hydrotreated light, BHT (10 ppm)
CAS 64742-47-8; EINECS/ELINCS 265-149-8
Uses: Solvent; diluent; chem. feedstock; fuel
Regulatory: Canada DSL listed; NPRI nonreportable
Properties: Colorless clear liq.; faint petrol. odor; sol. < 0.01% in water; m.w. 189; sp.gr. 0792 (15.6 C); vapor pressure < 0.15 kPa (20 C); f.p. < -57 C; b.p. 224-254 C; flash pt. (PMCC) > 81 C; autoignition temp. ≈ 250 C; > 99.9% act.
Toxicology: REL TWA 152 ppm; LD50 (oral, rat) > 5 g/kg, (skin, rabbit) > 3 g/kg; inh. of high vapor concs. may cause eye/respiratory tract irritation, headaches, dizziness, anesthesia, drowsiness, unconsciousness, other CNS effects, death; sl. eye irritant; frequent/prolonged skin contact may cause irritation, dermatitis; if aspirated, may cause mild to severe pulmonary injury, death
Environmental: Prevent entry into sewers, watercourses, low areas
Precaution: Combustible liq.; flamm. limits 0.5-4.9 vol.%; volatile; gives off invisible fumes; liq. or vapor may settle in low areas; distant ignition/explosion possible; electrostatic accumulation hazard; use proper grounding procedures; incompat. with strong oxidizing agents
Hazardous decomp. prods.: None
Storage: Store in cool, well-ventilated area away from incompat. materials, heat, open flame, ignition sources, direct sunlight; keep container closed

BDGE [INEOS Oxide]
Chem. Descrip.: Butoxydiglycol
CAS 112-34-5; EINECS/ELINCS 203-961-6
Uses: Solvent for resins (NC, acrylic, epoxy, polyamide, alkyd), oils, waxes, fats, dyestuffs, insecticides, herbicides, grease/grime, surf. coatings, lacquers, enamels, waterborne coatings, phenolic varnishes, epoxy coatings, non-grain-raising wood stains; gloss aid, flow aid in lacquers; dye solvent/stabilizer/penetration aid, color enhancer for textiles, leather, printing; mutual solvent, coupling agent in sol. oils, metal/glass cleaners, varnish removers; diluent in hydraulic brake fluids
Features: Very stable; low f.p.; relatively low vapor pressure
Properties: Pt-Co 15 max. color; mild odor; misc. with wide range of solvs., water; m.w. 162.23; dens. 0.950-0.953 g/ml (20 C); visc. 6.5 mPa·s (20 C); vapor pressure 0.02 hPa (20 C); i.b.p. 225 C min. (101.3 kPa); pour pt. -68.1 C; flash pt. (CC) 101 C; autoignition temp. 228 C; ref. index 1.4316 (20 C); 99% min. purity
Precaution: Flamm. limits 0.85-24.6 vol.% (20 C); not a significant flamm. hazard @ normal storage temps.; very reactive with zinc (hydrogen may be produced); contact with zinc, tin, copper/alloys can cause discoloration

Benzoflex® 2-45 [Velsicol]
Chem. Descrip.: Diethylene glycol dibenzoate
CAS 120-55-8; EINECS/ELINCS 204-407-6
Uses: Plasticizer producing adhesives with quick grab and set times; solvator for PVC; also in latex coatings; food pkg. adhesives, paper/paperboard
Features: Compatible with PVAc homopolymer and copolymer emulsions, PVC
Regulatory: FDA 21CFR §175.105, 176.170, 176.180
Properties: APHA 100 max., mild ester odor; sol. in aliphatic and aromatic hydrocarbons; sol. < 0.01% in water; m.w. 314.3; sp.gr. 1.178; dens. 9.8 lb/gal; f.p. 16 and 28 C (two cryst. forms); b.p. 240 C; flash pt. (COC) 199 C; ref. index 1.5424; dielec. const. 7.16 (10,000 Hz); 98.0% min. assay
Toxicology: LD50 (oral, rat) 2562 mg/kg, (dermal, rabbit) > 200 mg/l; not an eye or primary skin irritant; repeated exposure may cause defatting of skin and mild irritation to eyes and nose

Benzoflex® 284 [Velsicol]
Chem. Descrip.: Propylene glycol dibenzoate
CAS 19224-26-1
Uses: Solvating plasticizer for PVC applic., stain-resist. flooring, resilient vinyl flooring top coats, latex caulk formulations, PVAc emulsion adhesives, and castable PU; coalescing agent in latex paints
Properties: APHA 100 max., mild ester odor; sol. in aliphatic and aromatic hydrocarbons; sol. 0.01% in water; m.w. 284.3; sp.gr. 1.146; dens. 9.55 lb/gal; vapor pressure 1.9 x 10^7 mm Hg; f.p. -3 C; b.p. 233 C (5 mm); pour pt. -16 C; flash pt. (COC) 199 C min.; ref. index 1.544; 98.0% min. assay
Toxicology: LD50 (oral, rat) 3593 mg/kg, (skin, rabbit) > 2000 mg/kg, (inh., rat, 4 h) > 200 mg/l; not an eye or primary skin irritant; repeated exposure may cause defatting of skin and mild irritation to eyes and nose

Bernel® Ester DCM [Bernel]
Chem. Descrip.: Dicaprylyl maleate
CAS 97298-48-1; EINECS/ELINCS 205-524-5
Uses: Emollient , oxybenzone solubilizer, cleansing agent, wax solvent in cosmetics; gloss aid in hair conditioners
Features: Noncomedogenic
Regulatory: USA, EEC, Japan permitted
Properties: Colorless liq.; sol. in oil
Use Level: 2-10%

Berol® 02 [Akzo Nobel Surf. Chem.; Akzo Nobel Surf. Chem. AB]
Chem. Descrip.: Nonylphenol ethoxylate
CAS 68412-54-4
Uses: Emulsifier; solvent cleaner; emulsion polymerization
Properties: Hazen < 200 clear liq.; sol. in ethanol, xylene, trichloroethylene, lt. fuel oil; disp. in water, paraffin oil, wh. spirit; dens. 1040 kg/m^3; visc. 350 mPa·s; HLB 10.9; cloud pt. 62-66 C (5 g in 25 ml of 25% butyl diglycol); flash pt. > 100 C; pour pt. < 0 C; pH 6-7 (1% aq.); surf. tens. 29 mN/m (0.1%); Draves wetting 32 s (0.1%); Ross-Miles foam 20 mm (initial, 0.05%, 50 C); nonionic; 100% conc.

BGE [INEOS Oxide]
Chem. Descrip.: Butoxyethanol
CAS 111-76-2; EINECS/ELINCS 203-905-0
Uses: Solvent for resins (NC, acrylic, epoxy, polyamide, alkyd), oils, waxes, fats, dyestuffs, insecticides, herbicides, grease/grime, surf. coatings, lacquers, enamels, waterborne coatings, phenolic varnishes, epoxy coatings, non-grain-raising wood stains; gloss aid, flow aid in lacquers; dye solvent/stabilizer/penetration aid, color enhancer for textiles, leather,

printing; mutual solvent, coupling agent in sol. oils, metal/glass cleaners, varnish removers; diluent in hydraulic brake fluids
Features: Very stable; low f.p.; relatively low vapor pressure
Properties: Pt-Co 10 max. color; mild odor; misc. with wide range of solvs., water; m.w. 118.18; dens. 0.899-0.902 g/ml (20 C); visc. 6.4 mPa·s (20 C); vapor pressure 1.3 hPa (20 C); i.b.p. 168 C min. (101.3 kPa); pour pt. -70 C; flash pt. (CC) 60 C; autoignition temp. 245 C; ref. index 1.4193 (20 C); 99% min. purity
Precaution: Flamm. limits 1.1-12.7 vol.% (20 C); not a significant flamm. hazard @ normal storage temps.; very reactive with zinc (hydrogen may be produced); contact with zinc, tin, copper/alloys can cause discoloration

BIO-SOFT® N-411 [Stepan]
Chem. Descrip.: Isopropylamine linear alkylbenzene sulfonate
Uses: Surfactant, emulsifier, solvent degreaser for emulsion degreasers, laundry prespotters, dry cleaning solvs., paint strippers, waterless hand cleaners; emulsifier for d-limonene; dry cleaning detergent
Properties: Liq.; oil-sol.; anionic; 46% conc.

BLO® [ISP]
Chem. Descrip.: γ-Butyrolactone
CAS 96-48-0; EINECS/ELINCS 202-509-5
Uses: Solvent for PAN, PS, fluorinated hydrocarbons, cellulose triacetate, shellac; used in paint removers, petrol. processing, hectograph process, specialty inks; reaction and diluent solvent for pesticides; intermediate for aliphatic and cyclic compds., prod. of Vitamins B1 and E, alkaloids, analgesics, anesthetics; used in dyeing of acetate; wetting agent for cellulose acetate films, fibers, solvent welding of plastic films in adhesive applics.
Properties: APHA 40 clear liq.; sol. in acetone, benzene, CCl_4, ethyl ether, methanol, chlorobenzene, water; m.w. 86; sp.gr. 1.124; visc. 1.7 cp; f.p. -44 C; b.p. 204 C; flash pt. (TCC) 93 C; fire pt. 99 C; ref. index 1.435; pH 4.5 (10% aq.); sp. heat 0.40 cal/g/°C; dielec. const. 39 (20 C)
Toxicology: LD50 (oral, rat) 1.5 cc/kg; severe eye irritant; nonprimary skin irritant, nonsensitizer; noncarcinogenic

Brazilian Gum Turpentine-NT [PDM]
Chem. Descrip.: Turpentine gum
CAS 9005-90-7
UN UN 1299
Uses: Solvent cleaner for artist supplies, paint removal, and pharmaceuticals
Regulatory: DOT regulated
Properties: Water wh. liq.; turpentine odor; negligible sol. in water; sp.gr. 0.860-0.875; vapor pressure 4 mm Hg (20 C); b.p. 155 C; flash pt. (TCC) 35 C; ref. index 1.465-1.478 (20 C); 99% volatile
Toxicology: May produce primary skin irritation, contact dermatitis, or chem. burns; causes eye irritation and damage, possible corneal burns; overexposure to vapor causes headaches, dizziness, vertigo, chest pain, bronchitis, pulmonary edema, cyanosis
Precaution: Flamm.; avoid strong oxidizing agents
Hazardous Ingredients: α-pinene (80-90%), β-pinene (3-10%), other terpenes (3-10%)
Hazardous decomp. prods.: Thermal degradation causes CO, CO_2
NFPA: Health 2, Flammability 3, Reactivity 0
Storage: Store in cool, well-ventilated place in closed containers away from heat or sources of ignition

Burcolev DLS [Burlington Chem.]
Uses: Dyeing aux., leveling agent, diffusion accelerator, migrating agent for disperse dyes on polyester, esp. in reworking damaged or off-shade goods; emulsifier, solvent to reduce trimer buildup on machine interiors and to provide prescouring and one-bath scour-dye props.
Features: Prevents unlevel dyeings caused by different strike rates and preferential absorp. of dyes to various parts of the fiber; eliminates need for carrier; solvent-free; low foaming; suitable for pkg. and jet dyeing
Properties: Lt. amber clear liq.; no objectionable odor; sp.gr. 0.90; dens. 7.49 lb/gal; pH 8.0 (1%); nonionic
Environmental: More environmentally friendly

Butoxyl® [Celanese; Celanese GmbH]
Chem. Descrip.: 3-Methoxy-1-butyl acetate
CAS 4435-53-4; EINECS/ELINCS 224-644-9
UN NA 1993
Uses: Solvent, thinner, antiblush agent, gloss aid for paints, spray paints; solvent, visc. reducer in high-solids paints; solvent, flow aid, gloss aid in stoving enamels, typically alkyds; solvent in isocyanate- and epoxy-contg. systems
Features: Low volatility
Regulatory: SARA §311 acute health/fire hazard; EINECS, Canada, Australia, China, Korea, Philippines, Japan listed
Properties: Colorless clear liq.; faint char. ester odor; sol. 30 g/l in water (20 C); misc. with common org. solvs.; sp.gr. 0.954-0.956 (20 C); dens. 7.94 lb/gal; visc. 1.43 mPa·s (20 C); vapor pressure 1.125 mm Hg (20 C); f.p. -80 C; b.p. 169-173 C; acid no. 0.05 max.; flash pt. (CC) 62 C; autoignition temp. 410 C; ref. index 1.408-1.410 (20 C); water absorp. 4% (20 C); sp. heat 1.93 kJ/kg°K; dielec. const. 8.0 (20 C); elec. conduct. ≈ 0.8 x 10^{-8} S·cm^{-1}; 98.5% min. act.
Toxicology: LD50 (oral, rat) 4210 mg/kg; causes eye irritation/burning sensation/pain/watering/change of vision; harmful by inh.; inh. may cause respiratory tract irritation, nasal discharge, hoarseness, coughing, chest pain, breathing difficulty, nausea, headache, dizziness; essentially nontoxic by ing.; may cause nausea, vomiting, loss of appetite, GI irritation, diarrhea; overexposure may cause CNS depression, respiratory tract/ digestive tract irritation; TSCA listed
Environmental: > 70% biodeg.; EC0 (daphnia) 10-100 ppm; avoid run-off into storm sewers, ditches leading to waterways
Precaution: Combustible liq. and vapor; flamm. limits in air 0.8-4.7 vol.%; may generate a static charge; prevent buildup of electrostatic charges during transfer; ground/bond equip.; incompat. with sulfuric and other strong inorg. acids, aluminum, lead, oxidizing agents such as peroxides, nitric acid, perchloric acid, chromium trioxide
Hazardous decomp. prods.: Thermal decomp. and combustion: CO_x
HMIS: Health 1, Flammability 2, Reactivity 0
Storage: Virtually unlimited shelf life if stored in cool., dry place away from air, water, direct sunlight; keep containers tightly closed when not in use; open containers slowly to allow excess pressure to vent

Butyl Carbitol® [Dow]
Chem. Descrip.: Butoxydiglycol
CAS 112-34-5; EINECS/ELINCS 203-961-6
Uses: Solvent, coalescing agent for water-disp., water-sol., and high-solids coatings
Properties: Sol. in water (20 C); m.w. 162.23; sp.gr. 0.954 (20/20 C); dens. 7.96 lb/gal (20 C); visc. 4.2 cP (20 C); vapor pressure 0.01 mm Hg (20 C); b.p. 230.6 C; flash pt. (CC) 214 F

Butyl Carbitol® Acetate [Dow]
Chem. Descrip.: Diethylene glycol butyl ether acetate
CAS 124-17-4; EINECS/ELINCS 204-685-9
Uses: Solvent for high-solids coatings, NC emulsions; coalescing aid for water-disp. coatings
Properties: Colorless transparent liq.; mild ethereal fruity odor; sol. 6.5% in water (20 C); m.w. 204.27; sp.gr. 0.980 (20/20 C); dens. 8.16 lb/gal (20 C); visc. 3.6 cP (20 C); vapor pressure 0.04 mm Hg (20 C); b.p. 246.7 C; flash pt. (CC) 221 F

Butyl Cellosolve® [Dow]
Chem. Descrip.: Butoxyethanol
CAS 111-76-2; EINECS/ELINCS 203-905-0
UN NA 1993
Uses: Solvent for high-solids coatings (NC, butyrate, ethyl cellulose); solvent, lacquer retardant for NR, SBR, NC coatings; coalescing agent for water-disp. and water-sol. coatings
Regulatory: DOT nonregulated; SARA §302 nonreportable, §311/312 acute health/chronic health/fire hazard, §313 reportable; Calif. Prop. 65 nonreportable; Canada DSL listed
Properties: Colorless liq.; ether odor; sol. in water (20 C); m.w. 118.18; sp.gr. 0.902 (20/20 C); dens. 7.51 lb/gal (20 C); visc. 3.8 cP (20 C); vapor pressure 0.4 mm Hg (20 C); f.p. -65 C; b.p. 171.2 C; flash pt. (PMCC) 68 C; autoignition temp. 244 C; VOC 899 g/l; ≥ 99% act.
Toxicology: ACGIH TWA (8 h) 20 ppm; LD50 (oral, rat) 470-3000 mg/kg, (skin, rabbit) 220 mg/kg; LC50 (inh., rat) 700 ppmmay cause severe eye irritation, mod. corneal injury; harmful if swallowed; may cause respiratory tract irritation, headache; may cause blood, kidney, liver effects on inh. or ing.; repeated skin contact may cause irritation, burns; massive ing. may cause metabolic acidosis, hemolysis, CNS/kidney effects; nonsensitizing; TSCA listed
Environmental: Readily biodeg.; BOD5 5.2%; LC50 (daphnia magna) 835 mg/l, (fathead minnow) 1900mg/l; pract. nontoxic to aquatic organisms; prevent entry into soil, ditches, sewers, waterways, ground water

Precaution: Combustible liq. and vapor; flamm. limits in air 1.1-10.6 vol.%; container may rupture from gas generation in a fire; can oxidize at elevated temps.; incompat. with strong acids, strong bases, strong oxidizers
Hazardous decomp. prods.: Combustion: CO, CO_2; decomp.: aldehydes, ketones, org. acids
NFPA: Health 2, Flammability 2, Reactivity 0
Storage: Keep away from heat, sparks, flame

Butyl Cellosolve® Acetate [Dow]
Chem. Descrip.: Butoxyethanol acetate
CAS 112-07-2; EINECS/ELINCS 203-933-3
Uses: Solvent for high-solids coatings (NC, butyrate, ethyl cellulose acrylic); coalescing agent for water-disp. coatings
Properties: Sol. 1.5% in water (20 C); m.w. 160.21; sp.gr. 0.942 (20/20 C); dens. 7.84 lb/gal (20 C); visc. 1.8 cP (20 C); vapor pressure 0.9 mm Hg (20 C); b.p. 192.3 C; flash pt. (CC) 165 F

Butyl Diglyme [Ferro/Grant]
Chem. Descrip.: Diethylene glycol dibutyl ether
CAS 112-73-2; EINECS/ELINCS 204-001-9
Uses: Solvent used in electrochemistry, polymer and boron chemistry, physical processes such as gas absorption, extraction, stabilization, industrial prods. such as fuels, lubricants, textiles, pharmaceuticals, pesticides; solvent, drug carrier, stabilizer in pharmaceuticals
Features: Tends to solvate cations
Properties: Colorless clear liq.; mild nonresidual odor; sol. 0.3% in water; misc. with ethanol, acetone, benzene, diethyl ether, octane; m.w. 218.34; sp.gr. 0.8814; dens. 7.36 lb/gal; visc. 2.4 cP; f.p. -60.2 C; b.p. 256 C; flash pt. (CC) 118 C; ref. index 1.4235; pH neutral; surf. tens. 27 dynes/cm (20 C); sp. heat 0.495 cal/g/°C; 98.5% min. purity
Environmental: LD50 3900 mg/kg; low to mod. acute toxicity; chronic exposure may cause reproductive effects
Precaution: Slowly biodeg.

Butyl Di-Icinol [Huntsman]
Chem. Descrip.: 2-(2-Butoxyethoxy)-ethanol
CAS 112-34-5; EINECS/ELINCS 203-961-6
Uses: Solvent for protective coatings, inks, cleaning prods., agric. chems.; aids wetting, penetration, and soil removal; coupling solvent

Butyl Dioxitol® [Shell]
Chem. Descrip.: Butoxydiglycol
CAS 112-34-5; EINECS/ELINCS 203-961-6
Uses: Solvent for nitrocellulose, phenolics, Epon resins, alkyds, and acrylics, oils, dyes, lacquers, inks, enamels; coupling solvent in preparation of specialized cleaning sol'ns. and cutting oils; coalescing agent in latex paints; component of brake fluids in automotive industry
Features: Slow evaporating; improved flow-out and gloss
Properties: Colorless liq.; mild odor; water-misc.; sp.gr. 0.949-0.952; 0.10% max. water

Butyl Icinol [Huntsman]
Chem. Descrip.: 2-Butoxyethanol
CAS 111-76-2; EINECS/ELINCS 203-905-0
Uses: Solvent for lacquers, enamels, water-borne systems, inks, industrial cleaners, waterless hand cleaners
Properties: Sp.gr. 0.901-0.904; visc. 6.4 cps; b.p. 166 C; flash pt. (PMCC) 68 C; ref. index 1.419

Butyl Oxitol® [Shell]
Chem. Descrip.: Butoxyethanol
CAS 111-76-2; EINECS/ELINCS 203-905-0
Uses: Solvent for nitrocellulose, alkyds, phenolics, acrylics, Epon resins, nat. resins, dyes, waxes, oils, and org. materials; solvent, lacquer retardant for NR, SR, NC coatings; gloss aid, leveling agent in surf. coating formulations such as lacquers and enamels; coupling agent in cleaners and cutting oils
Properties: Colorless liq.; mild odor; water-misc.; sp.gr. 0.898-0.901; 0.20% max. water

C

Cachalot® AR-20 [M. Michel]
Chem. Descrip.: Arachidyl alcohol
CAS 629-96-9; EINECS/ELINCS 211-119-4
Uses: Adhesive tackifier; antifoam; antistat; bactericide; corrosion inhibitor; cosmetic expedient; detergent; emulsifier; evaporation control agent; flotation agent; extraction agent; fungicide; ink solvent; lubricant; coatings; plasticizer; paint flatting agent; pharmaceutical vehicle; polymerization modifier; quaternary ammonium compds.; soap emollient; textile processing aux.; viscosity index improver; water repellent; soil repellent; waxes

Cachalot® BE-22 [M. Michel]
Chem. Descrip.: Behenyl alcohol
CAS 661-19-8; EINECS/ELINCS 211-546-6
Uses: Adhesive tackifier; antifoam; antistat; bactericide; corrosion inhibitor; cosmetic expedient; detergent; emulsifier; evaporation control agent; flotation agent; extraction agent; fungicide; ink solvent; lubricant; coatings; plasticizer; paint flatting agent; pharmaceutical vehicle; polymerization modifier; quaternary ammonium compds.; soap emollient; textile processing aux.; viscosity index improver; water repellent; soil repellent; waxes

Cachalot® C-50 [M. Michel]
Chem. Descrip.: Cetyl alcohol NF/USP/BP/EP
CAS 36653-82-4; EINECS/ELINCS 253-149-0
Uses: Emollient in cosmetics, pharmaceuticals; adhesive tackifier; antifoam; antistat; bactericide; corrosion inhibitor; cosmetic expedient; detergent; emulsifier; evaporation control agent; flotation agent; extraction agent; fungicide; ink solvent; lubricant; coatings; plasticizer; paint flatting agent; pharmaceutical vehicle; polymerization modifier; quaternary ammonium compds.; soap emollient; textile processing aux.; viscosity index improver; water repellent; soil repellent; waxes
Properties: Hazen 20 color; sol. in acetone, alcohol, aromatic hydrocarbons, carbon disulfide, chloroform, glycol and diglycol ethers; m.w. 241-255; sp.gr. 0.820 (50 C); visc. 7 cps (70 C); m.p. 47-49 C; b.p. 310-330 C; acid no. 0.1 max.; iodine no. 1 max.; sapon. no. 0.5 max.; hyd. no. 225-235; flash pt. 165 C; ref. index 1.4320-1.4322 (70 C)

Cachalot® L-90 [M. Michel]
Chem. Descrip.: Lauryl alcohol NF/USP/BP/EP
CAS 112-53-8; EINECS/ELINCS 203-982-0
Uses: Adhesive tackifier; antifoam; antistat; bactericide; corrosion inhibitor; cosmetic expedient; detergent; emulsifier; evaporation control agent; flotation agent; extraction agent; fungicide; ink solvent; lubricant; coatings; plasticizer; paint flatting agent; pharmaceutical vehicle; polymerization modifier; quaternary ammonium compds.; soap emollient; textile processing aux.; viscosity index improver; water repellent; soil repellent; waxes

Cachalot® M-43 [M. Michel]
Chem. Descrip.: Myristyl alcohol NF/USP/BP/EP
CAS 112-72-1; EINECS/ELINCS 204-000-3
Uses: Adhesive tackifier; antifoam; antistat; bactericide; corrosion inhibitor; cosmetic emollient; detergent; emulsifier; evaporation control agent; flotation agent; extraction agent; fungicide; ink solvent; lubricant; coatings; plasticizer; paint flatting agent; pharmaceutical vehicle; polymerization modifier; quaternary ammonium compds.; soap emollient; textile processing aux.; viscosity index improver; water repellent; soil repellent; waxes
Properties: Hazen 20 color; sol. in acetone, alcohol, aromatic hydrocarbons, carbon disulfide, chloroform, glycol and diglycol ethers; m.w. 212-220; sp.gr. 0.825 (40 C); visc. 9 cps (50 C); b.p. 280-295 C; solid. pt. 36-38 C; acid no. 0.3 max.; iodine no. 0.3 max.; sapon. no. 0.5 max.; hyd. no. 255-265; flash pt. 150 C; ref. index 1.4334-1.4336 (50 C)

Cachalot® S-56 [M. Michel]
Chem. Descrip.: Stearyl alcohol NF/USP/BP/EP
CAS 112-92-5; EINECS/ELINCS 204-017-6
Uses: Emollient in cosmetics, pharmaceuticals; adhesive tackifier; antifoam; antistat; bactericide; corrosion inhibitor; cosmetic emollient; detergent; emulsifier; evaporation control agent; flotation agent; extraction agent; fungicide; ink solvent; lubricant; coatings; plasticizer; paint flatting agent; pharmaceutical vehicle; polymerization modifier; quaternary ammonium compds.; soap emollient; textile processing aux.; viscosity index improver; water repellent; soil repellent; waxes
Properties: Hazen 20 color; sol. in acetone, alcohol, aromatic hydrocarbons, carbon disulfide, chloroform, glycol and diglycol ethers; m.w. 267-281; sp.gr. 0.815 (60 C); visc. 9 cps (70 C); m.p. 56-58 C; b.p. 330-350 C; acid no. 0.3 max.; iodine no. 1 max.; sapon no. 1 max.; hyd. no. 200-210; flash pt. 190 C; ref. index 1.4347-1.4349 (70 C)

Calprint 35 [Calumet Lubricants]
Chem. Descrip.: Aliphatic solvent
Uses: Solvent
Properties: Saybolt +30 color; sp.gr. 0.7936 (60 F); dens. 6.607 lb/gal; kinematic visc. 2.25 cSt (100 F); i.b.p. 452 F; aniline pt. 183 F; flash pt. (PMCC) 235 F; KB value 25.5; < 0.01 vol.% aromatics; < 1 ppm benzene

Calprint 38 [Calumet Lubricants]
Chem. Descrip.: Aliphatic solvent
Uses: Solvent
Properties: Saybolt +30 color; sp.gr. 0.809 (60 F); dens. 6.736 lb/gal; kinematic visc. 3.84 cSt (100 F); i.b.p. 500 F; aniline pt. 197 F; flash pt. (COC) 285 F; KB value 23; < 0.01 vol.% aromatics; < 1 ppm benzene

Calumet 142 Flash Naphtha Rule 66 [Calumet Lubricants]
Chem. Descrip.: Naphtha
EINECS/ELINCS 232-443-2
Uses: Solvent
Properties: Saybolt +30 color; sp.gr. 0.7766 (60 F); dens. 6.47 lb/gal; kinematic visc. 1.3 cSt (100 F); i.b.p. 365 F; aniline pt. 162 F; flash pt. (TCC) 146 F; KB value 29; 6 vol.% aromatics; < 1 ppm benzene

Calumet 180/210 Naphtha [Calumet Lubricants]
Chem. Descrip.: Naphtha
EINECS/ELINCS 232-443-2
Uses: Solvent
Properties: Saybolt +30 color; sp.gr. 0.71 (60 F); dens. 5.964 lb/gal; i.b.p. 180 F; aniline pt. 137 F; flash pt. (TCC) < 0 F; KB value 36; < 0.01 vol.% aromatics; < 1 ppm benzene

Calumet 420/460 [Calumet Lubricants]
Chem. Descrip.: Aliphatic solvent
Uses: Solvent
Properties: Saybolt +30 color; sp.gr. 0.7844 (60 F); dens. 6.53 lb/gal; kinematic visc. 1.72 cSt (100 F); i.b.p. 421 F; aniline pt. 178 F; flash pt. (PMCC) 189 F; KB value 26.5; < 0.01 vol.% aromatics; < 1 ppm benzene

Calumet Hexane [Calumet Lubricants]

Chem. Descrip.: Hexane
CAS 110-54-3; EINECS/ELINCS 203-777-6
Uses: Solvent
Properties: Saybolt +30 color; sp.gr. 0.672 (60 F); dens. 5.652 lb/gal; i.b.p. 146 F; aniline pt. 146 F; flash pt. (TCC) < 0 F; KB value 31; < 0.01 vol.% aromatics; < 1 ppm benzene

Calumet Isohexane [Calumet Lubricants]
Chem. Descrip.: Isohexane
CAS 107-83-5; EINECS/ELINCS 203-523-4
Uses: Solvent
Properties: Saybolt +30 color; sp.gr. 0.6594 (60 F); dens. 5.489 lb/gal; i.b.p. 120 F; flash pt. (TCC) < 0 F; KB value 28; < 0.01 vol.% aromatics; < 1 ppm benzene

Calumet Mineral Spirits < 1% [Calumet Lubricants]
Chem. Descrip.: Mineral spirits
CAS 64475-85-0; EINECS/ELINCS 232-453-7
Uses: Solvent
Properties: Saybolt +30 color; sp.gr. 0.7563 (60 F); dens. 6.296 lb/gal; i.b.p. 305 F; aniline pt. 162 F; flash pt. (TCC) 105 F; KB value 30; < 0.01 vol.% aromatics; < 1 ppm benzene

Calumet VM&P Naphtha < 1% [Calumet Lubricants]
Chem. Descrip.: VM&P naphtha
CAS 8032-32-4; EINECS/ELINCS 232-453-7
Uses: Solvent
Properties: Saybolt +30 color; sp.gr. 0.7354 (60 F); dens. 6.122 lb/gal; i.b.p. 252 F; aniline pt. 151 F; flash pt. (TCC) 64 F; KB value 31.5; < 0.01 vol.% aromatics; < 1 ppm benzene

Capmul® MCM [ABITEC]
Chem. Descrip.: Glyceryl caprylate/caprate
Uses: Solvent, dispersant, emulsifier, solubilizer, vehicle, carrier, penetrant, bacteriostat for pharmaceuticals (aerosols, clinical nutrition, coating, delivery/absorp. enhancement, dermatologicals, microemulsions, nutritional/sports supplements, suppositories); dispersant, emulsifier, solubilizer, carrier for cosmetics (creams/lotions, lip, foundation, sun care preps.); solubilizer, solvent, carrier, visc. modifier in flavors, fragrances
Properties: Lovibond 4R max. liq. to semisolid; sol. in oil and alcohol; HLB 5.5-6.0; acid no. 2.5 max.; iodine no. 2 max.; nonionic; 55% min. alpha mono

Capmul® MCM-90 [ABITEC]
Chem. Descrip.: Glyceryl caprylate/caprate
Uses: W/o emulsifier, solvent for cosmetics, pharmaceuticals
Properties: Liq.; acid no. 2.5 max.; iodine no. 2 max.; nonionic; 80+% alpha mono

Capmul® MCMC8 [ABITEC]
Chem. Descrip.: Glyceryl caprylate
CAS 26402-26-6; EINECS/ELINCS 247-668-1
Uses: Dispersant, emulsifier, solubilizer, vehicle, carrier, penetrant, bacteriostat for pharmaceuticals (delivery/absorp. enhancement, dermatologicals, microemulsions, suppositories); dispersant, emulsifier, solubilizer, carrier for cosmetics (creams/lotions, lip, foundation, sun care preps.); solubilizer, solvent, carrier, visc. modifier in flavors, fragrances
Properties: Lovibond 5R max. liq. to semisolid; acid no. 2.5 max.; iodine no. 2 max.; nonionic; 55% min. alpha mono

Capmul® MCMC10 [ABITEC]
Chem. Descrip.: Glyceryl monocaprate
CAS 26402-22-2; EINECS/ELINCS 247-667-6
Uses: Dispersant, solubilizer, emulsifier, vehicle, carrier, penetrant, bacteriostat in pharmaceuticals (coating, microemulsions, soft gelatin capsules, suppositories); moisturizer, solubilizer, visc. modifier for cosmetics (lipsticks, creams/lotions, ointments, makeup removers, eye pencil, bath/shower preps.); solubilizer, solvent, carrier, visc. modifier in flavors, fragrances
Properties: Solid; sol. in propylene glycol, dipropylene glycol, IPM, octyl stearate, octyl palmitate, SDA alcohol, ethoxydiglycol, phenyl trimethicone; nonionic
Use Level: 10% (lipsticks)

Capmul® POE-O [ABITEC]
Chem. Descrip.: Polysorbate 80
CAS 9005-65-6
Uses: Food emulsifier for frozen desserts; solubilizer for oils into water systems, for flavors, fragrances, vitamins, pharmaceuticals; solvent
Properties: Gardner 6 max. liq.; water-sol.; HLB 15.0; acid no. 2 max.; sapon. no. 45-55; hyd. no. 65-80; nonionic; 100% conc.

Captex® 200 [ABITEC]
Chem. Descrip.: Propylene glycol dicaprylate/dicaprate
CAS 68583-51-7; EINECS/ELINCS 271-516-3
Uses: Vehicle, carrier, solubilizer, coupling agent, bioavailability enhancer, solvent for flavors, fragrance oil, sol. colorants, vitamins in pharmaceuticals, medicinals, cosmetics, foods; emollient, moisturizer, solubilizer, visc. modifier for cosmetics (creams/lotions, lip, mascara, makeup, foundation, sun care, hair, bath/shower preps.)
Regulatory: Conforms to hydrogenated veg. oil NF
Properties: APHA 100 liq.; sol. in alcohol, oils, hydrocarbons, ketones; visc. 7-13 mPa·s; acid no. 0.1 max.; iodine no. 1.0 max.; sapon. no. 315-335; cloud pt. < -20 C; ref. index 1.4393

Captex® 300 [ABITEC]
Chem. Descrip.: Caprylic/capric triglyceride
CAS 65381-09-1; EINECS/ELINCS 265-724-3
Uses: Solvent for colors and perfumes; emollient, moisturizer, solubilizer, carrier, visc. modifier, lubricant for cosmetics (creams/lotions, lip, mascara, sun care, hair, bath/shower preps.); plasticizer; emollient, lubricant, moisturizer, diluent, solubilizer, vehicle, carrier, visc. modifier for pharmaceuticals, aerosols, clinical nutrition, coating, delivery/absorp. enhancement, dermatologicals, infant formulas, soft gelatin capsules, suspensions
Features: Med. chain triglyceride
Properties: Gardner 2 max. liq., bland odor and flavor; sol. in alcohol, oils, hydrocarbons, ketones; visc. 24-30 mPa·s; acid no. 0.1 max.; iodine no. 0.5; sapon. no. 335-350; cloud pt. < -5 C; ref. index 1.4481

Captex® 350 [ABITEC]
Chem. Descrip.: Caprylic/capric/lauric triglyceride; conforms to hydrogenated veg. oil NF
CAS 68991-68-4
Uses: Diluent, emollient, moisturizer, solvent, vehicle, carrier, fixing agent, extender for pharmaceuticals (clinical nutrition, coating, delivery/absorption enhancement, dermatologicals, infant formulas, nutritional/sports supplements); emollient, moisturizer, carrier for cosmetics (creams/lotions, lip, mascara, sun care, bath/shower preps.)
Features: Med. chain triglyceride
Properties: Gardner 2 max. liq., bland odor and flavor; visc. 36-42 mPa·s; acid no. 0.1 max.; iodine no. 1.52 max.; sapon. no. 290-310; cloud pt. 0 C max.; ref. index 1.4582

Captex® 355EP [ABITEC]
Chem. Descrip.: Propylene glycol diethylhexanoate
CAS 56519-71-2; EINECS/ELINCS 301-185-3
Uses: Emollient, solvent, fixing agent, extender, carrier, solubilizer in pharmaceuticals

Captex® 500 [ABITEC]
Chem. Descrip.: Glyceryl triacetate
CAS 102-76-1; EINECS/ELINCS 203-051-9
Uses: Solvent, fixing agent for flavors and fragrances in foods and pharmaceuticals
Regulatory: FDA 21CFR §184.1901, GRAS; FCC, USP approved
Properties: Sol. in water; misc. with alcohol and oils

Captex® 500P [ABITEC]
Chem. Descrip.: Triacetin
CAS 102-76-1; EINECS/ELINCS 203-051-9
Uses: Emollient, solvent, fixing agent, solubilizer, extender, carrier in pharmaceuticals

Captex® 810D [ABITEC]
Chem. Descrip.: Caprylic/capric/linoleic triglyceride; conforms to hydrogenated veg. oil NF
Uses: Diluent, emollient, moisturizer, solubilizer, solvent, vehicle, carrier, fixing agent, extender in pharmaceutical, nutritional, and cosmetic applics. incl. lipsticks, lip glosses, makeup bases, bath oils, aftershave lotions, aerosols; carrier for flavors and fragrances
Properties: Gardner 2 max. liq.; acid no. 0.1 max.; iodine no. 85; sapon. no. 235-253

Captex® 8227 [ABITEC]
Chem. Descrip.: Triundecanoin
CAS 13552-80-2; EINECS/ELINCS 236-935-8
Uses: Emollient, lubricant, moisturizer, visc. modifier, solvent, fixing agent, extender in pharmaceuticals, cosmetics, nutritional applics. incl. lipsticks, lip glosses, makeup bases, bath oils, aftershave lotions, aerosols; car-

rier for flavors and fragrances
Properties: APHA 150 solid; acid no. 0.5 max.; iodine no. 5 max.; sapon. no. 270-290; cloud pt. 21 C

Carbitol® [Dow]
Chem. Descrip.: Ethoxydiglycol
CAS 111-90-0; EINECS/ELINCS 203-919-7
Uses: Solvent for high-solids coatings
Features: High b.p.; low vapor pressure; promotes uniform disps.
Properties: M.w. 134.18; dens. 8.25 lb/gal (20 C); visc. 4.5 cP (20 C); vapor pressure < 1 mm Hg (20 C); b.p. 202.7 C; flash pt. (CC) 182 F

Carbitol® PM-600 [Dow]
Uses: Solvent, coalescing agent for water-disp. and water-sol. coatings
Properties: Sol. in water; sp.gr. 1.027 (20/20 C); dens. 8.55 lb/gal (20 C); b.p. 195.0 C; flash pt. (CC) 204 F

Carbitol® Acetate [Dow]
Chem. Descrip.: Ethoxydiglycol acetate
CAS 112-15-2; EINECS/ELINCS 203-940-1
Uses: Solvent

Carbitol® Low Gravity [Dow]
Chem. Descrip.: Ethoxydiglycol
CAS 111-90-0; EINECS/ELINCS 203-919-7
Uses: Solvent for waterborne coatings
Properties: Sol. in water (20 C); m.w. 134.18; sp.gr. 0.990 (20/20 C); dens. 8.25 lb/gal (20 C); visc. 4.4 cP (20 C); vapor pressure 0.08 mm Hg (20 C); b.p. 202.7 C; flash pt. (CC) 182 F

Carbopol® 940 [Noveon]
Chem. Descrip.: Carbomer 940 NF
CAS 9003-01-4
Uses: Emulsifier, thickener, stabilizer, suspending agent for cosmetics, topical pharmaceuticals, for die-casting and forging lubricants, thixotropic paints; solvent thickening with or without neutralizing; food pkg. adhesives, coatings, paper/paperboard, polymers; defoamer in food-contact coatings
Regulatory: FDA 21CFR §175.105, 175.300, 175.320, 176.170, 176.180, 176.200, 177.1210, 177.2260; 40CFR §180.1001; NF, BP, JSPI compliance; EPA hazardous waste; SARA §312 chronic health hazard, §313 reportable (benzene)
Properties: Wh. fluffy powd.; sl. acetic odor; sol. in water, polar solvs., many nonpolar solvs. blends; sp.gr. 1.41; bulk dens. 208 kg/m^3; visc. 40,000-60,000 cps (0.5%); pH 2.5-3.0 (1% aq. disp.); anionic; 100% conc., < 2% moisture
Toxicology: LD50 (oral, rat) > 2500 mg/kg, (skin, rabbit) > 3000 mg/kg; dust may cause eye irritation, pain; dust inh. may cause coughing, mucous prod., shortness of breath; prolonged/repeated exposure may cause contact dermatitisTSCA listed
Environmental: LC50 (daphnia magna, 96 h) 168-280 mg/l; low aquatic toxicity; BOD=0; nonbiodeg.; removed with the biomass in typ. wastewater treatment
Precaution: Take precautions against dust explosion; heat may be generated in contact with strong basic materials, e.g., ammonia, NaOH, KOH, or strongly basic amines
Hazardous Ingredients: Benzene (< 0.3%)
Hazardous decomp. prods.: CO, CO_2, hydrocarbons, irritating vapors
NFPA: Health 2, Flammability 1, Reactivity 0
Storage: 2 yr min. shelf life stored in sealed containers, protected from moisture and extreme temps.; hygroscopic

Carbopol® 980 [Noveon]
Chem. Descrip.: Carbomer 940 NF
CAS 9003-01-4
Uses: Solvent, thickener for cosmetics, topical pharmaceuticals, sparkling clear water or hydroalcoholic topical gels, aq. or solv. systems; food pkg. adhesives, coatings, paper/paperboard, polymers; defoamer in food-contact coatings
Features: For use with or without neutralizing
Regulatory: FDA 21CFR §175.105, 175.300, 175.320, 176.170, 176.180, 176.200, 177.1210, 177.2260; 40CFR §180.1001; NF, DAB, JSPI compliance; SARA §312 chronic health hazard
Properties: Wh. fluffy powd.; sl. acetic odor; sol. in water; sp.gr. 1.41; bulk dens. 208 kg/m^3; visc. 13,000-30,000 cps (0.2%), 40,000-60,000 cps (0.5%); pH 2.5-3.0 (1% aq. disp.); < 2% moisture
Toxicology: LD50 (oral, rat) > 2500 mg/kg, (skin, rabbit) > 3000 mg/kg; sl. irritating to skin; minimal eye irritant; dust inh. may cause coughing, mucous prod., shortness of breath; prolonged/repeated exposure may cause contact dermatitisTSCA listed
Environmental: LC50 (daphnia magna, 96 h) 168-280 mg/l; nonbiodeg.; removed with the biomass in typ. wastewater treatment
Precaution: Take precautions against dust explosion; heat may be generated in contact with strong basic materials, e.g., ammonia, NaOH, KOH, or strongly basic amines
Hazardous decomp. prods.: CO, CO_2, hydrocarbons, irritating vapors
NFPA: Health 2, Flammability 1, Reactivity 0
Storage: 2 yr min. shelf life stored in sealed containers, protected from moisture and extreme temps.; hygroscopic

Carbowax® E200 [Dow]
Chem. Descrip.: PEG-4
CAS 25322-68-3; EINECS/ELINCS 203-989-9
Uses: Carrier, modifier, solvent, lubricant for personal care prods. (makeup, bath preps., toothpaste), foam control agents, syn. lubricants, rubber and plastics, metalworking fluids, chem. intermediates; controls cure rates in clay-contg. rubber formulations; improves flexibility, imparts antistatic props. to thermoplastics; heat transfer fluid in plastic thermoforming; mold release
Properties: Liq.; m.w. 200; sp.gr. 1.124; dens. 9.35 lb/gal; visc. 40 cSt; f.p. supercools; flash pt. (PMCC) 340 F; ref. index 1.459; sp. heat 0.524 cal/g/°C

Carbowax® E300 [Dow]
Chem. Descrip.: PEG-6
CAS 25322-68-3; EINECS/ELINCS 203-989-9
Uses: Thickener, carrier, modifier, solvent, lubricant for personal care prods. (makeup, bath preps., toothpaste), foam control agents, syn. lubricants, rubber and plastics, metalworking fluids, chem. intermediates; controls cure rates in clay-contg. rubber formulations; improves flexibility, imparts antistatic props. to thermoplastics; heat transfer fluid in plastic thermoforming; mold release
Properties: Liq.; m.w. 300; sp.gr. 1.125; dens. 9.36 lb/gal; visc. 69 cSt; f.p. -10 C; flash pt. (PMCC) > 400 F; ref. index 1.463; sp. heat 0.508 cal/g/°C

Carbowax® E300 NF [Dow]
Chem. Descrip.: PEG-6 NF, FCC
CAS 25322-68-3; EINECS/ELINCS 220-045-1
Uses: Pharmaceuticals (carrier for ointments for antiseptics and other medicaments, plasticizer in tablet coatings, base for suppositories, carrier, solvent, suspending agent in liq. preps., vehicle in gelatin capsules); thickener; defoamer in fermentation and food processing
Properties: Clear visc. liq.; sol. in water; m.w. 300; sp.gr. 1.125; dens. 9.36 lb/gal; f.p. -10 C; visc. 69 cSt; flash pt. (PMCC) > 400 F; ref. index 1.463; sp. heat 0.508 cal/g/°C
Toxicology: Extremely low acute oral toxicity; little or no irritation to eyes and skin; not recommended for use on abraded skin; may generate irritating vapors on decomp.

Carbowax® E400 [Dow]
Chem. Descrip.: PEG-8
CAS 25322-68-3; EINECS/ELINCS 225-856-4
Uses: Thickener, carrier, modifier, solvent, lubricant for personal care prods. (makeup, bath preps., toothpaste), foam control agents, syn. lubricants, rubber and plastics, metalworking fluids, chem. intermediates; controls cure rates in clay-contg. rubber formulations; improves flexibility, imparts antistatic props. to thermoplastics; heat transfer fluid in plastic thermoforming; mold release
Properties: Liq.; m.w. 400; sp.gr. 1.125; dens. 9.36 lb/gal; f.p. 6 C; visc. 90 cSt; flash pt. (PMCC) > 450 F; ref. index 1.465; sp. heat 0.498 cal/g/°C

Carbowax® E400 NF [Dow]
Chem. Descrip.: PEG-8 NF, FCC
CAS 25322-68-3; EINECS/ELINCS 225-856-4
Uses: Pharmaceuticals (carrier for ointments for antiseptics and other medicaments, plasticizer in tablet coatings, base for suppositories, carrier, solvent, suspending agent in liq. preps., vehicle in gelatin capsules); thickener; defoamer in fermentation and food processing
Properties: Clear visc. liq.; m.w. 400; sp.gr. 1.125; dens. 9.36 lb/gal; f.p. 6 C; visc. 90 cSt; flash pt. (PMCC) > 450 F; ref. index 1.465; sp. heat 0.498 cal/g/°C
Toxicology: Extremely low acute oral toxicity; little or no irritation to eyes and skin; not recommended for use on abraded skin; may generate irri-

tating vapors on decomp.

Carbowax® E600 [Dow]
Chem. Descrip.: PEG-12
CAS 25322-68-3; EINECS/ELINCS 229-859-1
Uses: Thickener, carrier, modifier, solvent, lubricant for personal care prods. (makeup, bath preps., toothpaste), foam control agents, syn. lubricants, rubber and plastics, metalworking fluids, chem. intermediates; controls cure rates in clay-contg. rubber formulations; improves flexibility, imparts antistatic props. to thermoplastics; heat transfer fluid in plastic thermoforming; mold release
Properties: Liq.; sol. in water, ethanol, cyclomethicone, sunscreens, lactic acid; m.w. 600; sp.gr. 1.126; dens. 9.37 lb/gal; f.p. 22 C; visc. 131 cSt; flash pt. (PMCC) > 450 F; ref. index 1.466; sp. heat 0.490 cal/g/°C

Carbowax® E600 NF [Dow]
Chem. Descrip.: PEG-12 NF, FCC
CAS 25322-68-3; EINECS/ELINCS 229-859-1
Uses: Pharmaceuticals (carrier for ointments for antiseptics and other medicaments, plasticizer in tablet coatings, base for suppositories, carrier, solvent, suspending agent in liq. preps., vehicle in gelatin capsules); thickener; defoamer in fermentation and food processing
Properties: Clear visc. liq.; sol. in water, ethanol, cyclomethicone, sunscreens, lactic acid; m.w. 600; sp.gr. 1.126; dens. 9.37 lb/gal; f.p. 22 C; visc. 131 cSt; flash pt. (PMCC) > 450 F; ref. index 1.466; sp. heat 0.490 cal/g/°C
Toxicology: Extremely low acute oral toxicity; little or no irritation to eyes and skin; not recommended for use on abraded skin; may generate irritating vapors on decomp.

Carbowax® E900 [Dow]
Chem. Descrip.: PEG
CAS 25322-68-3; EINECS/ELINCS 203-989-9
Uses: Thickener, carrier, modifier, solvent, lubricant for personal care prods. (makeup, bath preps., toothpaste), foam control agents, syn. lubricants, rubber and plastics, metalworking fluids, chem. intermediates; controls cure rates in clay-contg. rubber formulations; improves flexibility, imparts antistatic props. to thermoplastics; heat transfer fluid in plastic thermoforming; mold release
Properties: Wh. waxy solid; m.w. 900; sp.gr. 1.204; f.p. 34 C; visc. 100 cSt (100 F); flash pt. (PMCC) > 450 F

Carbowax® E900 NF [Dow]
Chem. Descrip.: PEG NF, FCC
CAS 25322-68-3
Uses: Pharmaceuticals (carrier for ointments for antiseptics and other medicaments, plasticizer in tablet coatings, base for suppositories, carrier, solvent, suspending agent in liq. preps., vehicle in gelatin capsules); thickener; defoamer in fermentation and food processing
Properties: Wh. waxy solid; m.w. 900; sp.gr. 1.204; f.p. 34 C; visc. 100 cSt (100 F); flash pt. (PMCC) > 450 F
Toxicology: Extremely low acute oral toxicity; little or no irritation to eyes and skin; not recommended for use on abraded skin; may generate irritating vapors on decomp.

Carbowax® E1000 [Dow]
Chem. Descrip.: PEG-20
CAS 25322-68-3; EINECS/ELINCS 203-989-9
Uses: Thickener, carrier, modifier, solvent, lubricant for personal care prods. (makeup, bath preps., toothpaste), foam control agents, syn. lubricants, rubber and plastics, metalworking fluids, chem. intermediates; controls cure rates in clay-contg. rubber formulations; improves flexibility, imparts antistatic props. to thermoplastics; heat transfer fluid in plastic thermoforming; mold release
Properties: Wh. waxy solid; m.w. 1000; sp.gr. 1.214; f.p. 37 C; visc. 18 cSt (210 F); flash pt. (PMCC) > 450 F

Carbowax® E1000 NF [Dow]
Chem. Descrip.: PEG-20 NF, FCC
CAS 25322-68-3; EINECS/ELINCS 203-989-9
Uses: Pharmaceuticals (carrier for ointments for antiseptics and other medicaments, plasticizer in tablet coatings, base for suppositories, carrier, solvent, suspending agent in liq. preps., vehicle in gelatin capsules); thickener; defoamer in fermentation and food processing
Properties: Wh. waxy solid; m.w. 1000; sp.gr. 1.214; f.p. 37 C; visc. 18 cSt (210 F); flash pt. (PMCC) > 450 F
Toxicology: Extremely low acute oral toxicity; little or no irritation to eyes and skin; not recommended for use on abraded skin; may generate irritating vapors on decomp.

Carbowax® E1450 [Dow]
Chem. Descrip.: PEG-32
CAS 25322-68-3; EINECS/ELINCS 203-989-9
Uses: Thickener, carrier, modifier, solvent, lubricant for personal care prods. (makeup, bath preps., toothpaste), foam control agents, syn. lubricants, rubber and plastics, metalworking fluids, chem. intermediates; controls cure rates in clay-contg. rubber formulations; improves flexibility, imparts antistatic props. to thermoplastics; heat transfer fluid in plastic thermoforming; mold release
Properties: Wh. waxy solid; m.w. 1450; sp.gr. 1.214; f.p. 44 C; visc. 29 cSt (210 F); flash pt. (PMCC) > 450 F

Carbowax® E1450 NF [Dow]
Chem. Descrip.: PEG-32 NF, FCC
CAS 25322-68-3; EINECS/ELINCS 203-989-9
Uses: Pharmaceuticals (carrier for ointments for antiseptics and other medicaments, plasticizer in tablet coatings, base for suppositories, carrier, solvent, suspending agent in liq. preps., vehicle in gelatin capsules); thickener; defoamer in fermentation and food processing
Properties: Wh. waxy solid; m.w. 1450; sp.gr. 1.214; f.p. 44 C; visc. 29 cSt (210 F); flash pt. (PMCC) > 450 F
Toxicology: Extremely low acute oral toxicity; little or no irritation to eyes and skin; not recommended for use on abraded skin; may generate irritating vapors on decomp.

Carbowax® E3350 [Dow]
Chem. Descrip.: PEG-75
CAS 25322-68-3; EINECS/ELINCS 203-989-9
Uses: Thickener, carrier, modifier, solvent, lubricant for personal care prods. (makeup, bath preps., toothpaste), foam control agents, syn. lubricants, rubber and plastics, metalworking fluids, chem. intermediates; controls cure rates in clay-contg. rubber formulations; improves flexibility, imparts antistatic props. to thermoplastics; heat transfer fluid in plastic thermoforming; mold release
Properties: Wh. waxy solid, pract. odorless; sol. > 100 g/100 g in water; m.w. 3350; sp.gr. 1.224; visc. 93 cSt (210 F); f.p. 54 C; b.p. dec.; flash pt. (PMCC) > 232 C; pH 4.5-7.5 (5% aq.)

Carbowax® E3350 NF [Dow]
Chem. Descrip.: PEG-75 NF, FCC
CAS 25322-68-3; EINECS/ELINCS 203-989-9
Uses: Pharmaceuticals (carrier for ointments for antiseptics and other medicaments, plasticizer in tablet coatings, base for suppositories, carrier, solvent, suspending agent in liq. preps., vehicle in gelatin capsules); thickener; defoamer in fermentation and food processing
Properties: Wh. waxy solid, pract. odorless; sol. > 100 g/100 g in water; m.w. 3350; sp.gr. 1.224; visc. 93 cSt (210 F); f.p. 54 C; b.p. dec.; flash pt. (PMCC) > 232 C; pH 4.5-7.5 (5% aq.)
Toxicology: Single dose oral toxicity believed to be very low; may cause sl. transient eye irritation; avoid prolonged/repeated contact with abraded skin
Precaution: Dusts may present explosive hazard; incompat. with oxidizers, conc. min. acids

Carbowax® E4500 NF [Dow]
Chem. Descrip.: PEG-100 NF, FCC
CAS 25322-68-3; EINECS/ELINCS 203-989-9
Uses: Pharmaceuticals (carrier for ointments for antiseptics and other medicaments, plasticizer in tablet coatings, base for suppositories, carrier, solvent, suspending agent in liq. preps., vehicle in gelatin capsules); thickener; defoamer in fermentation and food processing
Properties: Wh. waxy solid, sl. polyether odor; colorless visc. liq. above 136 F; sol. > 100 g/100 g in water; m.w. 4500; sp.gr. 1.224; visc. 180 cSt (210 F); f.p. 58 C; b.p. dec.; flash pt. (PMCC) > 232 C; pH 4.5-7.5 (5% aq.)
Toxicology: LD50 (oral, rat) > 50,000 mg/kg; extremely low oral toxicity; may cause sl. transient eye irritation; avoid prolonged/repeated exposure on abraded skin; contact with heated material may cause thermal burns
Precaution: Supports combustion; do not breathe smoke; incompat. with oxidizers and strong acids

Carbowax® E8000 [Dow]
Chem. Descrip.: PEG-150
CAS 25322-68-3; EINECS/ELINCS 203-989-9
Uses: Thickener, carrier, modifier, solvent, lubricant for personal care prods.

(makeup, bath preps., toothpaste), foam control agents, syn. lubricants, rubber and plastics, metalworking fluids, chem. intermediates; controls cure rates in clay-contg. rubber formulations; improves flexibility, imparts antistatic props. to thermoplastics; heat transfer fluid in plastic thermoforming; mold release
Properties: Wh. waxy solid, clear liq. above 65 C, pract. odorless; sol. > 100 g/100 g in water; m.w. 8000; sp.gr. 1.224; f.p. 60 C; visc. 800 cSt (210 F); flash pt. (PMCC) > 260 C; pH 4.5-7.5 (5% aq.)

Carbowax® E8000 NF [Dow]
Chem. Descrip.: PEG-150 NF, FCC
CAS 25322-68-3; EINECS/ELINCS 203-989-9
Uses: Pharmaceuticals (carrier for ointments for antiseptics and other medicaments, plasticizer in tablet coatings, base for suppositories, carrier, solvent, suspending agent in liq. preps., vehicle in gelatin capsules); thickener; defoamer in fermentation and food processing
Properties: Wh. waxy solid, clear liq. above 65 C, pract. odorless; sol. > 100 g/100 g in water; m.w. 8000; sp.gr. 1.224; f.p. 60 C; visc. 800 cSt (210 F); flash pt. (PMCC) > 260 C; pH 4.5-7.5 (5% aq.); > 99% act.
Toxicology: LD50 (oral, rat) > 50 g/kg (low oral toxicity); may cause sl. transient temporary eye irritation; avoid prolonged/repeated exposure on abraded skin
Precaution: Dusts may present explosive hazard; incompat. with oxidizers, conc. min. acids

Carbowax® PEG 300 [Dow]
Chem. Descrip.: PEG-6
CAS 25322-68-3; EINECS/ELINCS 220-045-1
Uses: Coupling agent, solvent, vehicle, humectant, lubricant, binder, base in adhesives, agric., ceramics, cosmetics/toiletries, electroplating/electropolishing, food processing, household prods., lubricants, metal fabrication, paints, paper, pharmaceuticals, printing, rubber and elastomers, textiles, wood processing; chem. intermediate; food pkg. adhesives and food-contact articles
Regulatory: FDA 21CFR §175.105, 178.3750, 178.3910; EPA regulated
Properties: Water-wh. visc. liq.; sol. in water, alcohols, glycerin, glycols; m.w. 285-315; sp.gr. 1.1250; dens. 9.38 lb/gal; visc. 5.8 cSt (99 C); f.p. -15 to -8 C; hyd. no. 356-394; flash pt. (PMCC) > 180 C; ref. index 1.463; pH 4.5-7.5 (5% aq.); surf. tens. 44.5 dynes/cm; nonionic

Carbowax® Sentry® PEG 300 [Dow]
Chem. Descrip.: PEG-6 FCC, USP/NF
CAS 25322-68-3; EINECS/ELINCS 220-045-1
Uses: Coupling agent, solvent, vehicle, humectant, lubricant, binder, base, bodying agent, dispersant, conditioner for cosmetics/toiletries (antiperspirants, creams/lotions, hand cleaners, toothpaste, makeup, shampoo, shave cream, sunscreens), food processing (citrus coatings, defoamer, pkg. component), pharmaceuticals (coating, binder, plasticizer, and lubricant in tablets); glycerin replacement
Regulatory: FDA 21CFR §172.210, 172.820, 173.310, 173.340
Properties: Water-wh. visc. liq.; sol. in water, alcohols, glycerin, glycols; m.w. 285-315; sp.gr. 1.1250; dens. 9.38 lb/gal; visc. 5.8 cSt (99 C); f.p. -15 to -8 C; hyd. no. 356-394; flash pt. (PMCC) > 180 C; ref. index 1.463; pH 4.5-7.5 (5% aq.); surf. tens. 44.5 dynes/cm; nonionic

Carbowax® Sentry® PEG 400 [Dow]
Chem. Descrip.: PEG-8 FCC, USP/NF
CAS 25322-68-3; EINECS/ELINCS 225-856-4
Uses: Coupling agent, solvent, vehicle, humectant, lubricant, binder, base for cosmetics/toiletries (bath oil, cologne, creams/lotions, toothpaste, hair dressing, lipsticks, makeup/removers, nail polish remover, shampoo, shave cream, soap, sunscreens), food processing (citrus coatings, bodying agent, dispersant, defoamer, pkg. component), pharmaceuticals (coating, binder, plasticizer, and lubricant in tablets)
Regulatory: FDA 21CFR §172.210, 172.820, 173.310, 173.340
Properties: Water-wh. visc. liq.; sol. in water, methanol, ethanol, acetone, trichloroethylene, Cellosolve®, Carbitol®, dibutyl phthalate, toluene; m.w. 380-420; sp.gr. 1.1254; dens. 9.39 lb/gal; visc. 7.3 cSt (99 C); f.p. 4-8 C; hyd. no. 267-295; flash pt. (PMCC) > 180 C; ref. index 1.465; pH 4.5-7.5 (5% aq.); surf. tens. 44.5 dynes/cm; nonionic

Carbowax® Sentry® PEG 540 Blend [Dow]
Chem. Descrip.: PEG-6 and PEG-32 FCC, USP/NF
Uses: Coupling agent, solvent, vehicle, humectant, lubricant, binder, base for cosmetics/toiletries (hair dressings, hand cleaners, ointments, antiperspirant, creams/lotions, shaving preps., sunscreens); food processing (citrus coatings, bodying agent, dispersant, defoamer, pkg. component), pharmaceuticals (coating, binder, plasticizer, and lubricant in tablets)
Regulatory: FDA 21CFR §172.210, 172.820, 173.310, 173.340
Properties: Wh. soft waxy solid; sol. in methylene chloride, 73% in water, 50% in trichloroethylene, 48% in methanol; m.w. 500-600; sp.gr. 1.0930; dens. 9.17 lb/gal (55 C); visc. 15.1 cSt (99 C); m.p. 38-41 C; hyd. no. 187-224; flash pt. (PMCC) > 180 C; pH 4.5-7.5 (5% aq.)

Carbowax® Sentry® PEG 600 [Dow]
Chem. Descrip.: PEG-12 FCC, USP/NF
CAS 25322-68-3; EINECS/ELINCS 229-859-1
Uses: Coupling agent, solvent, vehicle, humectant, lubricant, binder, base for cosmetics/toiletries (antiperspirants, colognes, creams/lotions, dentifrices, lipsticks, makeup, shampoo, shave creams, sunscreens); food processing (citrus coatings, bodying agent, dispersant, defoamer, pkg. component), pharmaceuticals (coating, binder, plasticizer, and lubricant in tablets)
Regulatory: FDA 21CFR §172.210, 172.820, 173.310, 173.340
Properties: Water-wh. visc. liq.; sol. in water, alcohols, glycols; m.w. 570-630; sp.gr. 1.1257; dens. 9.40 lb/gal; visc. 10.8 cSt (99 C); f.p. 20-25 C; hyd. no. 178-197; flash pt. (PMCC) > 180 C; ref. index 1.46; pH 4.5-7.5 (5% aq.); surf. tens. 44.5 dynes/cm; nonionic

Carbowax® Sentry® PEG 900 [Dow]
Chem. Descrip.: PEG-20 FCC, USP/NF
CAS 25322-68-3; EINECS/ELINCS 203-989-9
Uses: Coupling agent, solvent, vehicle, humectant, lubricant, binder, base for cosmetics/toiletries (antiperspirants, creams/lotions, shave creams); food processing (citrus coatings, bodying agent, dispersant, defoamer, pkg. component), pharmaceuticals (coating, binder, plasticizer, and lubricant in tablets)
Regulatory: FDA 21CFR §172.210, 172.820, 173.310, 173.340
Properties: Wh. soft waxy solid; sol. 86% in water; m.w. 855-945; sp.gr. 1.0927 (60 C); dens. 9.16 lb/gal (55 C); visc. 15.3 cSt (99 C); m.p. 32-36 C; hyd. no. 119-131; flash pt. (PMCC) > 180 C; pH 4.5-7.5 (5% aq.); nonionic

Carbowax® Sentry® PEG 1000 [Dow]
Chem. Descrip.: PEG-20 FCC, USP/NF
CAS 25322-68-3; EINECS/ELINCS 203-989-9
Uses: Coupling agent, solvent, vehicle, humectant, lubricant, binder, base for cosmetics/toiletries (antiperspirants, creams/lotions, ointments, shave creams); food processing (citrus coatings, bodying agent, dispersant, defoamer, pkg. component), pharmaceuticals (coating, binder, plasticizer, and lubricant in tablets)
Regulatory: FDA 21CFR §172.210, 172.820, 173.310, 173.340
Properties: Wh. soft waxy solid; sol. 80% in water; m.w. 950-1050; sp.gr. 1.0926 (60 C); dens. 9.16 lb/gal (55 C); visc. 17.2 cSt (99 C); m.p. 37-40 C; hyd. no. 107-118; flash pt. (PMCC) > 180 C; pH 4.5-7.5 (5% aq.); nonionic

Carbowax® Sentry® PEG 1450 [Dow]
Chem. Descrip.: PEG-32 FCC, USP/NF
CAS 25322-68-3; EINECS/ELINCS 203-989-9
Uses: Coupling agent, solvent, vehicle, humectant, lubricant, binder, base for cosmetics/toiletries (antiperspirants, creams/lotions, ointments, shave creams); food processing (citrus coatings, bodying agent, dispersant, defoamer, pkg. component), pharmaceuticals (coating, binder, plasticizer, and lubricant in tablets)
Regulatory: FDA 21CFR §172.210, 172.820, 173.310, 173.340
Properties: Wh. soft waxy solid or flake; sol. 72% in water; m.w. 1300-1600; sp.gr. 1.0919 (60 C); dens. 9.17 lb/gal (55 C); bulk dens. 30 lb/ft^3 (flake); visc. 26.5 cSt (99 C); m.p. 43-46 C; hyd. no. 70-86; flash pt. (PMCC) > 180 C; pH 4.5-7.5 (5% aq.); nonionic

Carbowax® Sentry® PEG 3350 [Dow]
Chem. Descrip.: PEG-75 FCC, USP/NF
CAS 25322-68-3; EINECS/ELINCS 203-989-9
Uses: Coupling agent, solvent, vehicle, humectant, lubricant, binder, base for cosmetics/toiletries (antiperspirants, bath powds., creams/lotions, toothpaste, hair dressing, makeup, nail polish remover, soap); food processing (citrus coatings, bodying agent, dispersant, defoamer, pkg. component), pharmaceuticals (coating, binder, plasticizer, and lubricant in tablets)
Regulatory: FDA 21CFR §172.210, 172.820, 173.310, 173.340
Properties: Wh. hard waxy flake or powd.; sol. 67% in water; m.w. 3000-3700; sp.gr. 1.0926 (60 C); dens. 8.94 lb/gal (80 C); bulk dens. 30 lb/ft^3

(flake), 40 lb/ft³ (powd.); visc. 90.8 cSt (99 C); m.p. 54-58 C; hyd. no. 30-37; flash pt. (PMCC) > 180 C; pH 4.5-7.5 (5% aq.); nonionic

Carbowax® Sentry® PEG 4600 [Dow]

Chem. Descrip.: PEG-100 FCC, USP/NF

CAS 25322-68-3; EINECS/ELINCS 203-989-9

Uses: Coupling agent, solvent, vehicle, humectant, lubricant, binder, base for cosmetics/toiletries (antiperspirants, creams/lotions); food processing (citrus coatings, bodying agent, dispersant, defoamer, pkg. component), pharmaceuticals (coating, binder, plasticizer, and lubricant in tablets)

Regulatory: FDA 21CFR §172.210, 172.820, 173.310, 173.340

Properties: Wh. hard waxy flake or powd.; sol. 65% in water; m.w. 4400-4800; sp.gr. 1.0926 (60 C); dens. 8.95 lb/gal (80 C); bulk dens. 30 lb/ft³ (flake), 40 lb/ft³ (powd.); visc. 184 cSt (99 C); m.p. 57-61 C; hyd. no. 23-26; flash pt. PMCC) > 180 C; pH 4.5-7.5 (5% aq.); nonionic

Carbowax® Sentry® PEG 8000 [Dow]

Chem. Descrip.: PEG-150 FCC, USP/NF

CAS 25322-68-3; EINECS/ELINCS 203-989-9

Uses: Coupling agent, solvent, vehicle, humectant, lubricant, binder, base for cosmetics/toiletries (antiperspirants, creams/lotions, hair dressing, nail polish remover); food processing (citrus coatings, bodying agent, dispersant, defoamer, pkg. component), pharmaceuticals (coating, binder, plasticizer, and lubricant in tablets)

Regulatory: FDA 21CFR §172.210, 172.820, 173.310, 173.340

Properties: Wh. hard waxy flake or powd.; sol. 63% in water; m.w. 7000-9000; sp.gr. 1.0845 (60 C); dens. 8.96 lb/gal (80 C); bulk dens. 30 lb/ft³ (flake), 40 lb/ft³ (powd.); visc. 822 cSt (99 C); m.p. 60-63 C; hyd. no. 13-16; flash pt. (PMCC) > 180 C; pH 4.5-7.5 (5% aq.); nonionic

Cardolite® NC-510 [Cardolite]

Chem. Descrip.: 3-(n-Pentadecyl) phenol

CAS 501-24-6

Uses: Raw material for surfactants, antioxidants; coupling agent for pigments and dyes; cosolvent for insecticides, germicides; resin modifier for phenolic-aldehyde, polyester, PC polymers; in photographic industry

Properties: Sl. pink to wh. waxy solid; sol. in org. solvs. incl. aliphatic hydrocarbons; m.w. 304; sp.gr. > 1.0; m.p. 49-51 C (760 mm); b.p. 190-195 C (1 mm); ref. index 1.4750

CDA 19-3 [Equistar]

Chem. Descrip.: Completely denatured ethanol

CAS 64-17-5; EINECS/ELINCS 200-578-6

Uses: Solvent

Cekasol BOD [Cekal Spec.]

Uses: Solvent scouring agent for prep. of all substrates and blends without causing foam, and for removing oils, waxes, greases, and silicone fluids from machinery

Environmental: Environmentally friendly

Cellosolve® [Dow]

Chem. Descrip.: Ethoxyethanol

CAS 110-80-5; EINECS/ELINCS 203-804-1

Uses: Solvent for coatings (NC, ethyl acetate, butyrate, acrylic, water-sol.)

Properties: Sol. in water (20 C); m.w. 90.12; sp.gr. 0.931 (20/20 C); dens. 7.74 lb/gal (20 C); visc. 2.3 cP (20 C); vapor pressure 4.1 mm Hg (20 C); b.p. 135.6 C; flash pt. (CC) 108 F

Cellosolve® Acetate [Dow]

Chem. Descrip.: Ethoxyethanol acetate

CAS 111-15-9; EINECS/ELINCS 203-839-2

Uses: Solvent for coatings (NC, cellulose acetate, ethyl cellulose, butyrate, acrylic, urethane)

Properties: Sol. 22.9% in water (20 C); m.w. 132.16; sp.gr. 0.975 (20/20 C); dens. 8.10 lb/gal (20 C); visc. 1.1 cP (20 C); vapor pressure 2.0 mm Hg (20 C); b.p. 156.3 C; flash pt. (CC) 126 F

Celtone [Celanese; Celanese GmbH]

Chem. Descrip.: Acetone (50%), methanol (50%), 2,2-dimethoxypropane (8%), water (1%)

UN 1993

Uses: Solvent for coatings, cosmetics, pharmaceuticals

Regulatory: SARA §3131/312 acute health/chronic health/fire hazard, §313 reportable (methanol); EINECS, Australia, China, Canada, Korea, Philippines, Japan listed

Properties: Colorless clear mobile liq.; mild acetone odor; completely sol. in water, common org. solvs.; sp.gr. 0.78 (20 C); vapor pressure 140 mm Hg (20 C); f.p. -97 C; b.p. 60 C; flash pt. (CC) 0 C

Toxicology: Prolonged/repeated skin contact can cause death, blindness; may be fatal or cause blindness if swallowed; eye irritant; may cause respiratory tract irritation; harmful if inhaled; may cause CNS depression, nausea, dizziness, headache, stupor, unconsciousness, nasal discharge, coughing, chest pain, breathing difficulty, GI irritation, diarrhea; possible birth defect hazard; overexposure may cause kidney/liver damage; TSCA listed

Environmental: Avoid run-off into storm sewers and ditches leading to waterways

Precaution: Flamm. liq. and vapor; flamm. limits in air 2-36 vol.%; vapor is heavier than air, may flash back; may generate static charge; ground/bond during transfer; closed containers > 49 C may develop excessive vapor pressure; incompat. with caustic soda, lime, other strong alkalis, hydrochloric, sulfuric, and other strong inorg. acids, amines, oxidizing agents, peroxides, nitric acid, perchloric acid, chromic acid, chromium trioxide, copper/alloys, pyridines

Hazardous decomp. prods.: Thermal decomp. or combustion: CO_x

HMIS: Health 2, Flammability 3, Reactivity 0

Storage: Store in cool, dry, well-ventilated area under dry air or dry nitrogen @ 15-38 C, atm. pressure, away from direct sunlight, incompat. materials; keep containers tightly closed when not in use; open containers slowly to allow excess pressure to vent

Ceraphyl® 28 [ISP Van Dyk; S. Black]

Chem. Descrip.: Cetyl lactate

CAS 35274-05-6; EINECS/ELINCS 252-478-7

Uses: Lubricant, emollient for cosmetics and topical pharmaceuticals; binder for pressed powds.; imparts sheen and silkiness to skin and hair; reduces tack in deodorant sticks; solvent for dyes in lipstick formulas

Properties: Wh. solid, faint char. odor; sol. @ 5% in min. oil, peanut oil, IPM, oleyl alcohol, and 95% ethanol; partly sol. in water, propylene glycol; sp.gr. 0.893-0.905; HLB 13-15; acid no. 2 max.; sapon. no. 174-189

Toxicology: LD50 (oral, rat) > 20 ml/kg; nonirritating to eyes and skin

Ceraphyl® 140 [ISP Van Dyk; S. Black]

Chem. Descrip.: Decyl oleate

CAS 3687-46-5; EINECS/ELINCS 222-981-6

Uses: Emollient, cosolvent, slip agent, lubricant for cosmetic/pharmaceutical creams, lotions, moisturizers, skin care prods., bath oils, liq. makeup; binder for pressed powds.; pigment dispersant

Properties: Wh. to straw-colored liq., char. mild odor; sol. @ 5% in peanut oil, 95% ethanol, IPM, oleyl alcohol; insol. in water; sp.gr. 0.858-0.864; acid no. 5 max.; iodine no. 55-65; sapon. no. 130-145; ref. index 1.4540-1.4560; nonionic; 100% act.

Toxicology: LD50 (oral, rat) > 40 ml/kg; sl. eye irritant; very sl. skin irritant; nonsensitizing

Ceraphyl® 140-A [ISP Van Dyk; S. Black]

Chem. Descrip.: Isodecyl oleate

CAS 59231-34-4; EINECS/ELINCS 261-673-6

Uses: Emollient, cosolvent, solubilizer for cosmetic systems, topical pharmaceuticals, liq. makeup; binder for pressed powd.; wetting agent for iron oxides; cleansing agent for emulsions

Features: Lighter, drier skin feel than Ceraphyl 140

Properties: Wh. to straw-colored liq., char. mild odor; sol. @ 5% in peanut oil, 95% ethanol, IPM, oleyl alcohol; insol. in water; sp.gr. 0.858-0.864; acid no. 5 max.; iodine no. 50-65; sapon. no. 130-145; ref. index 1.4540-1.4560; nonionic; 100% act.

Toxicology: LD50 (oral, rat) > 40 ml/kg; nonirritating to eyes; mild skin irritant; nonsensitizing

Ceraphyl® ICA [ISP Van Dyk; S. Black]

Chem. Descrip.: Isocetyl alcohol

CAS 36311-34-9; EINECS/ELINCS 252-964-9

Uses: Emollient for creams/lotions, topical pharmaceuticals; pigment dispersant, binder for pressed powds.; hair and skin conditioner; detackifier in deodorant sticks; solvent for lipstick dyes; carrier and extender for flavors and fragrances

Features: Nongreasy; ideal for oil-free cosmetics and high-pH applics.

Properties: Colorless clear liq., low odor; sol. @ 5% in peanut and min. oils, 95% ethanol, IPM, oleyl alcohol, castor oil, cyclomethicone; insol. in water; m.w. 242; sp.gr. 0.830-0.840; HLB 12-14; acid no. 5 max.; iodine no. 10 max.; sapon. no. 10 max.; hyd. no. 195-230

Toxicology: LD50 (oral, rat) > 5 g/kg; mild eye and skin irritant; noncomedogenic

Cerette V [Undesa; S. Black]
Chem. Descrip.: Decyl oleate
CAS 3687-46-5; EINECS/ELINCS 222-981-6
Uses: Emollient in bath oils, creams/lotions, sun creams, shaving prods., insect repellents, antiperspirants, deodorants; oily component for cleansing creams, nail polish removers; plasticizer for nail polishes; pigment wetting agent; softener in creams, hair care; solvent for fatty cosmetic ingreds.; vehicle for lipsticks; solubilizer for perfume oils, vitamin oils, hormones; solubilizer, vehicle for medicines; excipient for injectable substances; lubricant for PVC
Features: Changes skin feel, rub out props. and greasiness; veg. oil substitute; vaseline substitute

Cetiol® B [Cognis KGaA/Care Chems.]
Chem. Descrip.: Dibutyl adipate
CAS 105-99-7; EINECS/ELINCS 203-350-4
Uses: Emollient, oily component for skin oils, massage oils, sun protection oils, pharmaceuticals, bathing preps., day creams, liq. emulsions; penetrant on skin; solvent for lipoid-sol. substances; superfatting agent and plasticizer for hair sprays
Properties: Colorless oily liq., faint intrinsic odor; sp.gr. 0.958-0.962; visc. 5-7 mPa·s; solid. pt. < -30 C; acid no. 0.5 max.; iodine no. 1 max.; sapon. no. 420-440; hyd. no. 1 max.; cloud pt. < -25 C; flash pt. 150 C min.; ref. index 1.434-1.437; 100% act.
Storage: 1 yr storage if in sealed original containers at temps. below 40 C, protected from moisture

Chemax E-200 MO [Chemax]
Chem. Descrip.: PEG-5 oleate
CAS 9004-96-0
Uses: Emulsifier for min. and fatty oils; solvent; degreaser, dispersant, detergent, lubricant for metal, textiles, cosmetics, plastisol formulations
Properties: Liq.; oil-sol.; HLB 8.3; sapon. no. 118; nonionic; 100% conc.

Chemax E-1000 MO [Chemax]
Chem. Descrip.: PEG-20 oleate
CAS 9004-96-0
Uses: Emulsifier for min. and fatty oils; solvent; degreaser; dispersant; detergent; lubricant
Properties: Solid; nonionic; 100% conc.

Chemsil 994 [Chemsil Silicones]
Chem. Descrip.: Cycloctasiloxane
Uses: Solvent for personal care
Features: Nongreasy
Properties: Visc. 2.3 cps; flash pt. 54 C; ref. index 1.394

Chemsil 995 [Chemsil Silicones]
Chem. Descrip.: Cyclopentasiloxane
Uses: Solvent for personal care
Features: Nongreasy
Properties: Visc. 4 cps; flash pt. 77 C; ref. index 1.396

Chemsil 1585 [Chemsil Silicones]
Chem. Descrip.: Cyclomethicone
Uses: Solvent for personal care
Features: Nongreasy
Properties: Visc. 3 cps; flash pt. 52 C; ref. index 1.395

Chemsil 6535 [Chemsil Silicones]
Chem. Descrip.: Cyclomethicone
Uses: Solvent for personal care
Features: Nongreasy
Properties: Visc. 6 cps; flash pt. 76 C; ref. index 1.398

Chemsil 8515 [Chemsil Silicones]
Chem. Descrip.: Cyclomethicone
Uses: Solvent for personal care
Features: Nongreasy
Properties: Visc. 3.5 cps; flash pt. 76 C; ref. index 1.398

Chemsil DM 0.65 [Chemsil Silicones]
Chem. Descrip.: Hexamethyldisiloxane
CAS 107-46-0; EINECS/ELINCS 203-492-7
Uses: Solvent for personal care
Features: Nongreasy
Properties: Visc. 0.65 cps; flash pt. -1 C; ref. index 1.375

CHP® [ISP]
Chem. Descrip.: N-Cyclohexyl pyrrolidone
CAS 6837-24-7; EINECS/ELINCS 229-919-7
Uses: Solvent for plating engineering plastics, esp. EMS applics.; dye penetrant for aramid fibers; photo resists; gas separation; elec. insulating coatings; chem. synthesis; absorp. of sulfur dioxide and nitric acid; aromatic extraction; high temp. polyimide processing aid
Features: Superior resist. to hydrolysis
Properties: Liq.; water-sol.; m.w. 167; sp.gr. 1.026; visc. 11.5 cps; f.p. 12 C; b.p. 284 C; flash pt. (CC) 293 F; ref. index 1.4950; 98% min. purity
Toxicology: LD50 (oral, rat) 0.37 g/kg; primary skin irritant; may cause permanent eye damage
Environmental: Biodeg.

Cithrol 4DL [Croda Chem. Ltd; Croda Oleochems.]
Chem. Descrip.: PEG-8 dilaurate
CAS 9005-02-1
Uses: Dispersant, w/o emulsifier, wetting agent, cosolvent in cosmetic and industrial applics.
Properties: Liq.; HLB 10.2; nonionic; 97% conc.

Cithrol 4DO [Croda Chem. Ltd; Croda Oleochems.]
Chem. Descrip.: PEG-8 dioleate
CAS 9005-07-6
Uses: Dispersant, w/o emulsifier, wetting agent, cosolvent in cosmetic and industrial applics.
Properties: Amber liq.; HLB 8.3; nonionic; 97% conc.

Citroflex® 2 [Morflex]
Chem. Descrip.: Triethyl citrate
CAS 77-93-0; EINECS/ELINCS 201-070-7
Uses: Plasticizer for cellulose acetate, cellulose acetate butyrate, cellulose nitrate, chlorinated rubber, ethyl cellulose, PVAc, PVB, PVC, PVP, PVdC, PET, poly (acrylate/methacrylate), dammar, ester gums, lacquer formulations; fragrance carrier in perfumes, deodorants, shampoos, detergents; solvent in household, I&I cleaners; film-former, plasticizer, solvent for inks, adhesives, coatings; food-contact paper/paperboard
Features: Enhances grease resist.; lightfast
Regulatory: FDA 21CFR §175.300, 175.320, 175.380, 175.390, 176.170, 177.1210, 181.27, 182.1911, 184.1911; kosher
Properties: Clear liq., essentially odorless; sol. in toluene, 6.5 g/100 ml in water; insol. in heptane; m.w. 276.3; sp.gr. 1.135-1.139; visc. 35 cps; pour pt. -45 C; flash pt. (COC) 155 C; ref. index 1.440-1.442; 99% assay
Toxicology: LD50 (oral, rat) 7 cc/kg, (IP, mice) 1750 mg/kg
Environmental: Environmentally friendly
Precaution: Incompat. with oxidizing materials

Citroflex® 4 [Morflex]
Chem. Descrip.: Tri-n-butyl citrate
CAS 77-94-1; EINECS/ELINCS 201-071-2
Uses: Plasticizer for cellulose nitrate, ethyl cellulose, chlorinated rubber, cellulosic coatings, PVAc, PVB, PVC, PVP, PVdC, poly(acrylate/methacrylate); defoamer in proteinaceous sol'ns.; improved lt. stability in cellulose acetate; solvent for household, I&I cleaners; solvent, film-former, plasticizer in inks, adhesives, coatings; food pkg. adhesives
Regulatory: FDA 21CFR §175.105; kosher
Properties: Clear liq., essentially odorless; sol. in toluene, heptane, < 0.1 g/100 ml water; m.w. 360; sp.gr. 1.037-1.045; visc. 32 cps; vapor pressure 1 mm Hg (170 C); pour pt. -62 C; flash pt. (COC) 185 C; ref. index 1.443-1.445; 99% assay
Toxicology: LD50 (oral, rat) > 30 ml/kg, (IP, mice) 2900 mg/kg; avoid breathing vapors or mist
Environmental: Environmentally friendly
Precaution: Incompat. with oxidizing materials
Hazardous decomp. prods.: Combustion produces CO, CO_2
Storage: Store in well ventilated area in tightly closed containers; protect from moisture

Citroflex® A-2 [Morflex]
Chem. Descrip.: Acetyltriethyl citrate
CAS 77-89-4; EINECS/ELINCS 201-066-5
Uses: Plasticizer for more polar resins such as cellulosics, polyacrylates, PVAc, polyvinyl butyral, PVC, PVP, PVdC, chlorinated rubber; lightfastness for lacquer formulations; solvent in household, I&I cleaners; solvent, film-former, plasticizer for inks, adhesives, coatings, food pkg. adhesives/coatings/paper/paperboard/polymers
Regulatory: FDA 21CFR §175.105, 175.300, 175.320, 175.380, 176.170, 177.1210, 178.3910; kosher
Properties: Clear liq., essentially odorless; sol. in toluene, 0.72 g/100 ml water; m.w. 318.3; sp.gr. 1.135-1.139; visc. 54 cps; pour pt. -43 C; flash pt. (COC) 188 C; ref. index 1.432-1.441; 99% assay

Toxicology: LD50 (oral, rat) 7 cc/kg, (IP, mice) 1150 mg/kg; sl. but transient eye irritant; nonirritating to skin
Environmental: Environmentally friendly

Citroflex® A-4 [Morflex]
Chem. Descrip.: Acetyl tri-n-butyl citrate
CAS 77-90-7; EINECS/ELINCS 201-067-0
Uses: Plasticizer for PVC, PVdC, esp. food films, medical articles, cellulose nitrate, chlorinated rubber, ethyl cellulose, PVAc, PVB, PVP, PU, poly(acrylate/methacrylate); sol'n. coatings on paperboard and foil; vinyl toys; solvent in household, I&I cleaners; solvent, film-former, and plasticizer for inks, adhesives, coatings; food pkg. adhesives, coatings, paper, polymers
Regulatory: FDA 21CFR §172.515, 175.105, 175.300, 175.320, 175.380, 175.390, 176.170, 177.1210, 178.3910, 181.27; kosher
Properties: Clear liq., essentially odorless; sol. in toluene, heptane, < 0.1 g/100 ml water; m.w. 402.5; sp.gr. 1.045-1.055; visc. 33 cps; pour pt. -59 C; flash pt. (COC) 104 C; ref. index 1.441-1.4425; 99% min. assay
Toxicology: LD50 (oral, rat) 30 cc/kg, (IP, mice) 4000 mg/kg; sl. but transient eye irritant; nonirritating to skin
Environmental: Environmentally friendly

Citrofol® AI [Jungbunzlauer; Jungbunzlauer Int'l. AG]
Chem. Descrip.: Triethyl citrate FCC, USP/NF, EP
CAS 77-93-0; EINECS/ELINCS 201-070-7
Uses: Plasticizer for cellulose derivs., explosives, film coated tablets (pharmaceutical); plasticizer, humectant for cigarette filters; solv. for nat. resins; solv., diluent, fixative in perfumery; carrier for fragrances; mfg. of lacquers; deodorant act.; cosmetics; antitackifier in low volatile hair sprays; food additive and solv. (carrier for flavors, whipping aid in dried egg white); plasticizer for plastic materials in contact with foodstuffs; mfg. of specialty inks for food pkg.
Features: Stable to light
Regulatory: FDA 21CFR §184.1911 GRAS; USP approved; E 1505; DOT nonregulated; CERCLA nonhazardous; SARA nonhazardous
Properties: Colorless/translucent sl. oily liq., free of suspended matter; sl. odor; sol. 6.5 g/100 ml in water; m.w. 276.3; sp.gr. 1.137; visc. 35.2 mPa·s; vapor pressure 1.33 hPa (107 C); b.p. 294 C; sapon. no. 609.20; pour pt. -45.5 C; flash pt. 150 C; ref. index 1.440; 99% min. assay
Toxicology: LD50 (oral, rat) > 3200 mg/kg, (skin, rabbit) > 5000 mg/kg; LC50 (inh., rat, 6 h) 1300 mg/kg; may cause eye and skin irritationTSCA listed
Environmental: Readily biodeg.
Precaution: Incompat. with strong bases and oxidizing agents
Hazardous decomp. prods.: Thermal decomp. may lead to release of irritating gases and vapors
HMIS: Health 1, Flammability 1, Reactivity 0
Storage: Unlimited storage when stored @ R.T. in tightly closed original container in cool, dry place

Citrofol® AII [Jungbunzlauer]
Chem. Descrip.: Acetyltriethyl citrate USP/NF
CAS 77-89-4; EINECS/ELINCS 201-066-5
Uses: Plasticizer, carrier, solvent, fixing agent in cosmetics (aerosol hair sprays, nail polish, perfume), pharmaceuticals (film coated tablets), industrial applics. (can coatings, inks, transparency films); plasticizer for polymers; hardness reducer and antitackifier for resins; solv. and plasticizer for plastic materials in contact with foodstuffs
Features: High b.p.; low volatility
Properties: Clear liq.; odorless; sol. 0.7 g/100 ml in water; dens. 1.037; visc. 55 mPa·s; b.p. 132 C (1 mm); pour pt. -43.5 C; flash pt. 188 C; ref. index 1.438
Toxicology: LD50 (oral, rat) 5900 mg/kg; nontoxic; may cause skin and eye irritation
Environmental: Readily biodeg.
Storage: Store @ R.T. in tightly closed original container in dry, cool place

Citrofol® BI [Jungbunzlauer]
Chem. Descrip.: Tributyl citrate USP/NF
CAS 77-94-1; EINECS/ELINCS 201-071-2
Uses: Plasticizer for cosmetics (aerosol hair sprays), pharmaceuticals (film coated tablets), industrial applics. (adhesives, coatings, floor coverings, inks, transparency films); hardness reducer; solvent, plasticizer for plastic materials in contact with foodstuffs
Features: High b.p.
Properties: Liq.; insol. in water; dens. 1.140; visc. 32 mPa·s; b.p. 164 C (1 mm); pour pt. -63.5 C; flash pt. 185 C; ref. index 1.441
Toxicology: LD50 (oral, rat) 26,300 mg/kg; toxicologically harmless; may cause skin and eye irritation
Environmental: Readily biodeg.
Storage: Store @ R.T. in tightly closed original container in dry, cool place

Citrofol® BII [Jungbunzlauer; Jungbunzlauer Int'l. AG]
Chem. Descrip.: Acetyltributyl citrate USP/NF
CAS 77-90-7; EINECS/ELINCS 201-067-0
Uses: Plasticizer for medical, pharmaceuticals (dental forming masses, prostheses, film coating tablets), foods, cosmetics (aerosol hair sprays, nail polish), industrial applics. (foil coatings, inks, safety glass, etc.), cellulose nitrate films, PVC and PVC-PVA copolymers; flexibilizer for resins at low temps.; UV resist. aid, heat stabilizer for resins; fixing agent; hardness reducer; solvent, plasticizer for plastic materials in contact with foodstuffs; plasticizer for inks for food pkg.
Features: May be used with vinyl latexes; high b.p.
Regulatory: DOT nonregulated; CERCLA nonhazardous; SARA nonhazardous
Properties: Colorless/translucent sl. oily liq., free of suspended matter; sl. odor; insol. in water; m.w. 402.5; sp.gr. 1.045; visc. 33 mPa·s; vapor pressure 1.33 hPa (173 C); b.p. 327 C; sapon. no. 550-558; pour pt. -59 C; flash pt. 204 C; ref. index 1.445; 99% min. assay
Toxicology: LD50 (oral, rat) 26,300 mg/kg; may cause eye and skin irritation with susceptible personsTSCA listed
Environmental: Readily biodeg.
Precaution: Incompat. with strong bases and oxidizing agents
Hazardous decomp. prods.: Thermal decomp. can lead to release of irritating gases and vapors
HMIS: Health 1, Flammability 1, Reactivity 0
Storage: Unlimited storage when stored @ R.T. in tightly closed original container in cool, dry place

Citrus Stripper Oil [PDM]
Chem. Descrip.: d-Limonene
Chem. Analysis: D-Llimonene (95%), terpene hydrocarbons (3%), and oxygenated terpenes (2%)
CAS 5989-27-5; EINECS/ELINCS 227-813-5
UN 2319 (DOT)
Uses: Essential oil; solvent
Regulatory: GRAS; DOT regulated
Properties: Water-wh. to lt. tint clear liq.; char. orange odor; negligible sol. in water; sp.gr. 0.838-0.843; b.p. 176 C; flash pt. 50 C; ref. index 1.471-1.474 (20 C); KB 64; VOC 95% (solv.); 0.37-1.50% aldehyde
Toxicology: LD50 (oral, rat) > 5000 mg/kg, (skin, rabbit) > 5000 mg/kg; causes eye, skin, and mucous membrane irritation
Precaution: Combustible; avoid strong oxidizing agents and exposure to heat, sparks, and open flame
Hazardous Ingredients: D-limonene (95%), terpene hydrocarbons (3%), and oxygenated terpenes (2%)
Hazardous decomp. prods.: CO, CO_2
NFPA: Health 1, Flammability 2, Reactivity 0
Storage: Store in closed containers

Citrus Terpenes [PDM]
Chem. Descrip.: d-Limonene (70-8%), γ-terpinene (9-13%), terpinolene (4-6%), myrcene (1-2%)
UN 2319
Uses: Essential oil; solvent
Regulatory: GRAS; DOT regulated
Properties: Water-wh. clear liq.; mild lemon/lime odor; negligible sol. in water; sp.gr. 0.844; vapor pressure 1.0 mm Hg (14 C); b.p. 176 F; flash pt. (TCC) 48.3 C; ref. index 1.4690-1.4730 (20 C); VOC 95% (as solv.)
Toxicology: LD50 (oral, rat) > 5000 mg/kg, (Skin, rabbit) > 5000 mg/kg; irritates eyes, skin, and mucous membranes
Precaution: Combustible; avoid strong oxidizing agents
Hazardous Ingredients: D-limonene (70-78%), gamma terpinene (9-13%), terpene hydrocarbons (5-10%), oxygenated terpenes (0-2%)
Hazardous decomp. prods.: Thermal degradation causes CO, CO_2
NFPA: Health 1, Flammability 2, Reactivity 0
Storage: Store in closed containers away from heat or sources of ignition

Clean Wiz-9 [Axel Plastics Research Labs]
Chem. Descrip.: Proprietary solv. blend
Uses: Solvent system for cleaning all types of molds (removes resins, waxes, grease, rubber, etc.)

Properties: Colorless liq.; dens. 8 lb/gal; flash pt. 200 F

Coasol™ [Haltermann Prods. UK]

Chem. Descrip.: Diisobutyl glutarate (55-65%), diisobutyl succinate (15-25%), diisobutyl adipate (12-23%)

Uses: Coalescing agent, solvent in paints/coatings, resin cleaners, water-based coatings, hand cleaners, polishes, floor waxes, ink cleaners, textile chems., solv. degreasers, carpet shampoos

Features: Alternative for high VOC solvs.

Properties: Hazen 15 max. clear liq.; very low odor; misc. with most common solvs.; m.w. 244; sp.gr. 0.958-0.960; visc. 5.3 mPa·s; vapor pressure 0.003 max. mm Hg (20 C); f.p. - 60 C; acid no. 1.0 max.; flash pt. 131 C; ref. index 1.427-1.428; surf. tens. 27.2 dynes/cm (24 C); 99% min. diester; 0.2% max. water

Environmental: Readily biodeg.

Coasol B® [Aceto]

Chem. Descrip.: Diisobutyl esters of dicarboxylic blend

Uses: Solvent for coatings, inks, adhesives, plastics, textiles

Coasol C [Aceto]

Chem. Descrip.: Mixt.

Uses: Solvent for coatings, inks, adhesives

Comperlan® COD [Cognis KGaA/Care Chems.; Cognis Iberica]

Chem. Descrip.: Cocamide DEA and glycerin

Uses: Foaming agent, visc. builder, emulsifier, conditioner, solvent for shampoos and other cosmetic and pharmaceutical applics.

Properties: Yel. clear liq.; solid. pt. -5 to 5 C; pH 9.5-10.5 (2%); 85% conc.

Conap® S-1 [Cytec Olean]

Uses: Solvent, thinner for dilution of Conap® AD-1146, AD-1147 primer series for brush, dip, and spray applics., and for clean up

Features: Fairly short retention time

Properties: Water-wh. liq.; sp.gr. 0.825-0.845

Precaution: Flamm. liq.

Storage: 18 mos shelf life when stored in original unopened containers

Conap® S-13 [Cytec Olean]

Uses: Solvent, thinner for dilution of Conap® CE-1170 conformal coating for dip and spray applics., and for clean up

Features: Fairly short retention time

Properties: Water-wh. liq.; sp.gr. 0.85-0.88; flash pt. (CC) 80 F

Toxicology: Mod. toxic by ing. and inh.; sl. skin irritant

Precaution: Flamm. liq.

Storage: 1 yr shelf life when stored in original unopened containers

Conap® S-22 [Cytec Olean]

Uses: Solvent, thinner for dilution of Conathane® urethane conformal coatings for dip and spray applics.

Regulatory: Calif. EPA Rule 66 compliance

Properties: Water-wh. liq.; sp.gr. 0.90-0.93; vapor pressure < 10 mm Hg; flash pt. (CC) 65 F

Precaution: Flamm. liq.

Storage: 2 yr shelf life when stored in original unopened containers @ R.T.

Condea Vista 47 [Sasol N. Am.]

Chem. Descrip.: Hydrotreated mixts. of mainly isoparaffins and naphthenes with very low levels of aromatics

Uses: Solvent for printing inks, esp. suited as an ink oil in heat-set printing formulations; in food-contact applics.; foods (defoamer, coating on egg shells, float on fermentation fluids, froth flotation cleaning of vegetables)

Regulatory: FDA 21CFR §172.884, 178.3650

Properties: Saybolt Universal +20 color liq.; m.w. 197; sp.gr. 0.812 (60/60 F); dens. 6.78 lb/gal (60 F); visc. 3.8 cSt (70 F); vapor pressure 0.10 mm Hg (100 F); f.p. 0 F; i.b.p. 464 F; pour pt. 0 F; flash pt. (PM) 228 F; KB value 27

Storage: Store in carbon steel tanks

Condea Vista LPA [Sasol N. Am.]

Chem. Descrip.: Hydrotreated mixts. of mainly isoparaffins and naphthenes with very low levels of aromatics

Uses: Chemical process solvent; carrier in pesticide formulations, waterless hand cleaners; solvent for dry-cleaning, household/industrial cleaners, metal rolling oils, paints, coatings, thinners, paper and mining chems., water treatment chems., food; as charcoal lighter fluids, degreasers, freeze pt. depressants, lamp oils, lubricating oils, printing ink oils; food pkg. adhesives, paper, cellophane, films, rubber, textiles; defoamer in food-contact coatings, paper/paperboard

Features: High purity, med. flash pt.; broad boiling range

Regulatory: FDA 21CFR §172.884, 173.340, 175.105, 176.180, 176.200, 176.210, 177.1200, 177.1400, 177.2260, 177.2600, 177.2800, 178.3120, 178.3570, 178.3620(b), 178.3650, 178.3910(b)(2)

Properties: Clear liq., mild odor; m.w. 167; sp.gr. 0.809 (60/60 F); dens. 6.75 lb/gal (60 F); visc. 2.2 cSt (70 F); vapor pressure 1 mm Hg (100 F); f.p. -90 F; pour pt. -95 F; i.b.p. 362 F; flash pt. (TCC) 148 F; KB value 32

Precaution: DOT: combustible liq.

Condea Vista LPA-142 [Sasol N. Am.]

Chem. Descrip.: Hydrotreated mixts. of mainly isoparaffins and naphthenes with very low levels of aromatics

Uses: Solvent for industrial cleaners, pesticides, coatings, household prods., chemical process solvs., ink oils, water treatment, paper, and mining chems., and food applics.; diluent in polymer processing; in charcoal lighter fluid; food pkg. adhesives, paper, cellophane, films, rubber, textiles; defoamer in food-contact coatings, paper/paperboard

Regulatory: FDA 21CFR §172.884, 173.340, 175.105, 176.180, 176.200, 176.210, 177.1200, 177.1400, 177.2260, 177.2600, 177.2800, 178.3120, 178.3570, 178.3620(b), 178.3650, 178.3910(b)(2)

Properties: Clear liq., mild odor; m.w. 152; sp.gr. 0.809 (60/60 F); dens. 6.75 lb/gal (60 F); visc. 1.8 cSt (70 F); vapor pressure 1.1 mm Hg (100 F); f.p. < -103 F; i.b.p. 368 F; pour pt. < -30 F; flash pt. (TCC) 146 F; KB value 35

Precaution: DOT: combustible liq.

Condea Vista LPA-170 [Sasol N. Am.]

Chem. Descrip.: Hydrotreated mixts. of mainly isoparaffins and naphthenes with very low levels of aromatics

Uses: Solvent for mfg. of polymer emulsions, in mining solvent extraction processes, metal lubricants, specialty coatings, pesticides, household prods., chemical process solvs., ink oils, water treatment, paper, and mining chems.; food pkg. adhesives, paper, cellophane, films, rubber, textiles; defoamer in food-contact coatings, paper/paperboard

Regulatory: FDA 21CFR §172.884, 173.340, 175.105, 176.180, 176.200, 176.210, 177.1200, 177.1400, 177.2260, 177.2600, 177.2800, 178.3120, 178.3570, 178.3620(b), 178.3650, 178.3910(b)(2)

Properties: Clear liq., mild odor; m.w. 171; sp.gr. 0.811 (60/60 F); dens. 6.77 lb/gal (60 F); visc. 2.4 cSt (70 F); vapor pressure 0.37 mm Hg (100 F); f.p. -81 F; i.b.p. 413 F; pour pt. < -30 F; flash pt. (TCC) 178 F; KB value 32

Precaution: DOT: combustible liq.

Condea Vista LPA-210 [Sasol N. Am.]

Chem. Descrip.: Hydrotreated mixts. of mainly isoparaffins and naphthenes with very low levels of aromatics

Uses: Solvent for chem. processing, printing ink oils, metal lubricants, coatings, household prods., water treatment, paper, and mining chems.; pesticide carrier; food pkg. applics.

Features: Rec. for applics. where odor is critical

Regulatory: FDA 21CFR §178.3620(c), 178.3910(a)(3)

Properties: Clear liq., mild odor; m.w. 194; sp.gr. 0.823 (60/60 F); dens. 6.87 lb/gal (60 F); visc. 3.8 cSt (70 F); vapor pressure 0.10 mm Hg (100 F); f.p. -43 F; i.b.p. 465 F; pour pt. < -30 F; flash pt. (PM) 226 F; KB value 29

Conosol 215 [Penreco]

Chem. Descrip.: Hydrotreated light distillate (primarily C12-15 isoparaffinic and cycloparaffinic hydrocarbons)

CAS 64742-47-8; EINECS/ELINCS 265-149-8

Uses: Solvent for chem. processing, degreasers, household/industrial cleaners, insecticides, metal rolling oils, printing ink oils, water treatment chems.

Features: High purity; higher solv. str. than competitive aliphatic solvs.; controlled flash pt. and boiling range; low reactivity; low olefin content

Regulatory: FDA approvals for direct/indirect food additives; meets Calif. LVP VOC exemption; DOT nonregulated; SARA §311/312/313 nonreportable; Canada DSL, EINECS, Australia AICS, China, Japan ENCS, Korea KCL, Philippines PICCS listed

Properties: Colorless clear liq.; negligible odor; insol. in water; sp.gr. 0.8-0.81; b.p. 221.1-282.2 C; flash pt. (PMCC) 101.7-107.2 C); 100% act.; < 0.5% aromatics; < 1 ppm benzene

Toxicology: ACGIH TLV/TWA 5 mg/m^3, STEL 10 mg/m^3 (oil mist); prolonged/repeated contact may cause irritation to eyes, nose, throat, lungs, skin; extreme overexposure or aspiration may cause lung damage, death; overexposure may cause weakness, headache, nausea, confusion, blurred vision, drowsiness, other nervous system effects, unconscious-

ness, convulsions; TSCA listed
Environmental: Environmentally friendly; VOC exempt; do not allow spills to enter sewers or watercourses
Precaution: May react with strong oxidizing agents; spillages may be slippery
Hazardous decomp. prods.: Combustion: dense smoke, CO, CO_2, other oxides
HMIS: Health 1, Flammability 1, Reactivity 0
Storage: Store in cool, well-ventilated area in closed containers away from heat, sparks, open flame, oxidizing materials

Conosol 260 [Penreco]
Chem. Descrip.: Petroleum distillates, hydrotreated middle (primarily C13-20 isoparaffinic and cycloparaffinic hydrocarbons)
CAS 64742-46-7; EINECS/ELINCS 265-148-2
Uses: Solvent for chem. processing, degreasers, household/industrial cleaners, insecticides, metal rolling oils, printing ink oils, water treatment chems., automotive prods., mineral seal oils
Features: High purity; higher solvent str. than competitive aliphatic solvs.; controlled flash pt. and boiling range; low reactivity; low olefin content
Regulatory: FDA approvals for direct/indirect food additives; meets Calif. LVP VOC exemption; DOT nonregulated; SARA §311/312/313 nonreportable; Canada DSL, EINECS, Australia AICS, China, Japan ENCS, Korea KCL, Philippines PICCS listed
Properties: Clear liq.; hydrocarbon odor; insol. in water; sp.gr. 0.83-0.84; b.p. 260-329.4 C; flash pt. (PMCC) > 121.1 C; 100% act.; < 0.5% aromatics
Toxicology: ACGIH TLV/TWA 5 mg/m^3, STEL 10 mg/m^3 (oil mist); prolonged/repeated contact may cause irritation to eyes, nose, throat, lungs, skin; extreme overexposure or aspiration may cause lung damage, death; overexposure may cause weakness, headache, nausea, confusion, blurred vision, drowsiness, other nervous system effects, unconsciousness, convulsions; TSCA listed
Environmental: Environmentally friendly; VOC exempt; do not allow spills to enter sewers or watercourses
Precaution: Combustible; may react with strong oxidizing agents; spillages may be slippery
Hazardous decomp. prods.: Combustion: dense smoke, CO, CO_2, other oxides
HMIS: Health 1, Flammability 1, Reactivity 0
Storage: Store in cool, well-ventilated area in closed containers away from heat, sparks, open flame, oxidizing materials

Conosol 340 [Penreco]
Chem. Descrip.: Petroleum distillates, hydrotreated middle (primarily C17-22 isoparaffinic and cycloparaffinic hydrocarbons)
CAS 64742-46-7; EINECS/ELINCS 265-148-2
Uses: Solvent for chem. processing, degreasers, household/industrial cleaners, insecticides, metal rolling oils, printing ink oils, water treatment chems., automotive prods.
Features: High purity; higher solv. str. than competitive aliphatic solvs.; controlled flash pt. and boiling range; low reactivity; low olefin content
Regulatory: FDA approvals for direct/indirect food additives; meets Calif. LVP VOC exemption; DOT nonregulated; SARA §311/312/313 nonreportable; Canada DSL, EINECS, Australia AICS, China, Japan ENCS, Korea KCL, Philippines PICCS listed
Properties: Clear liq.; hydrocarbon odor; insol. in water; sp.gr. 0.83-0.84; b.p. 310-360 C; flash pt. (PMCC) > 165.6 C; 100% act.; < 0.5% aromatics; < 1 ppm benzene
Toxicology: ACGIH TLV/TWA 5 mg/m^3, STEL 10 mg/m^3 (oil mist); prolonged/repeated contact may cause irritation to eyes, nose, throat, lungs, skin; extreme overexposure or aspiration may cause lung damage, death; overexposure may cause weakness, headache, nausea, confusion, blurred vision, drowsiness, other nervous system effects, unconsciousness, convulsions
Environmental: Environmentally friendly; VOC exempt; do not allow spills to enter sewers or watercourses
Precaution: Combustible; may react with strong oxidizing agents; spillages may be slippery
Hazardous decomp. prods.: Combustion: dense smoke, CO, CO_2, other oxides
HMIS: Health 1, Flammability 1, Reactivity 0
Storage: Store in cool, well-ventilated area in closed containers away from heat, sparks, open flame, oxidizing materials

Conosol C-145 [Penreco]
Chem. Descrip.: Hydrotreated light distillate (primarily C10-13 cycloparaffinic and isoparaffinic hydrocarbons)
CAS 64742-47-8; EINECS/ELINCS 265-149-8
UN 1268
Uses: Solvent for charcoal lighter fluids, chem. processing, degreasers, household/industrial cleaners, insecticides, lamp oils, automotive prods., water treatment chems., waterless hand cleaners
Features: High purity; higher solv. str. than competitive 140+ flash aliphatic solvs.; controlled flash pt. and boiling range; low reactivity
Regulatory: FDA approvals for direct/indirect food additives; SARA §311/312 acute/chronic health hazard, fire hazard, §313 nonreportable; Canada DSL, EINECS, Australia AICS, China, Japan ENCS, Korea KCL, Philippines PICCS listed
Properties: Water-wh. liq.; mild odor; insol. in water; sp.gr. 0.8-0.82 (16 C); vapor pressure < 0.5 mm Hg (68 F); b.p. 187.8-243.3 C; flash pt. (TCC) > 62.8 C; 100% act.; < 0.5% aromatics; < 1 ppm benzene
Toxicology: OSHA PEL/TWA 500 ppm; may be harmful or fatal by ing., inh. in very high concs.; inh. of high vapor/mist concs. can cause respiratory tract/eye irritation, CNS effects (drowsiness, dizziness, nausea, headache, unconsciousness); prolonged/repeated contact can cause irritation to eyes, nose, throat, lungs, skin; extreme overexposure or aspiration may cause lung damage, death; TSCA listed
Environmental: Environmentally friendly; do not allow spills to enter sewers or watercourses
Precaution: Combustible; flamm. limits in air 1-6%; may react with strong oxidizing agents; spillages may be slippery
Hazardous decomp. prods.: Combustion: dense smoke, CO, CO_2, other oxides
HMIS: Health 1, Flammability 2, Reactivity 0
Storage: Store in cool, well-ventilated area in closed containers away from heat, sparks, open flame, oxidizing materials

Conosol C-170 [Penreco]
Chem. Descrip.: Hydrotreated light distillate (primarily C10-15 cycloparaffinic and isoparaffinic hydrocarbons)
CAS 64742-47-8; EINECS/ELINCS 265-149-8
UN 1268
Uses: Solvent for chem. processing, degreasers, household/industrial cleaners, insecticides, metal rolling oils, water treatment chems.
Features: High purity; controlled flash pt.; low reactivity; higher solvency than comparative aliphatics
Regulatory: FDA approvals for direct/indirect food additives; SARA §311/312 acute/chronic health hazard, fire hazard, §313 nonreportable; Canada DSL, EINECS, Australia AICS, China, Japan ENCS, Korea KCL, Philippines PICCS listed
Properties: Colorless liq.; mild odor; insol. in water; sp.gr. 0.8-0.82 (16 C); b.p. 187.8-287.8 C; flash pt. (PMCC) 76.7 C; < 0.5% aromatics; < 1 ppm benzene
Toxicology: OSHA PEL/TWA 500 ppm; may be harmful or fatal by ing., inh. in very high concs.; inh. of high vapor/mist concs. can cause respiratory tract/eye irritation, CNS effects (drowsiness, dizziness, nausea, headache, unconsciousness); prolonged/repeated contact can cause irritation to eyes, nose, throat, lungs, skin; extreme overexposure or aspiration may cause lung damage, death; TSCA listed
Environmental: Environmentally friendly; do not allow spills to enter sewers or watercourses
Precaution: Combustible; flamm. limits in air 1-6%; may react with strong oxidizing agents; spillages may be slippery
Hazardous decomp. prods.: Combustion: dense smoke, CO, CO_2, other oxides
HMIS: Health 1, Flammability 2, Reactivity 0
Storage: Store in cool, well-ventilated area in closed containers away from heat, sparks, open flame, oxidizing materials

Conosol C-200 [Penreco]
Chem. Descrip.: Hydrotreated light distillate (primarily C12-16 cycloparaffinic and isoparaffinic hydrocarbons)
CAS 64742-47-8; EINECS/ELINCS 265-149-8
Uses: Solvent for chem. processing, degreasers, household/industrial cleaners, insecticides, metal rolling oils, printing ink oils, water treatment chems., automotive prods.
Features: High purity; controlled flash pt. and boiling range; higher solvency than comparable aliphatics; low reactivity

Regulatory: FDA approvals for direct/indirect food additives; DOT nonregulated; SARA §311/312/313 nonreportable; Canada DSL, EINECS, Australia AICS, China, Japan ENCS, Korea KCL, Philippines PICCS listed
Properties: Colorless clear liq.; negligible odor; insol. in water; sp.gr. 0.82-0.83; vapor pressure < 0.1 mm Hg (68 F); b.p. 221.1-287.8 C; pour pt. < -40 F; flash pt. (PMCC) > 93.3 C; < 0.5% aromatics; < 1 ppm benzene
Toxicology: ACGIH TLV/TWA 5 mg/m³, STEL 10 mg/m³ (oil mist); prolonged/repeated exposure may cause irritation to eyes, nose, throat, lungs, skin; extreme overexposure or aspiration may cause lung damage, death; overexposure may cause weakness, headache, nausea, confusion, blurred vision, drowsiness, other nervous system effects, unconsciousness, convulsions; TSCA listed
Environmental: Do not allow spills to enter sewers or watercourses
Precaution: Flamm. limits in air 1-6%; may react with strong oxidizing agents; spillages may be slippery
Hazardous decomp. prods.: Combustion: dense smoke, CO, CO_2, other oxides
HMIS: Health 1, Flammability 1, Reactivity 0
Storage: Store in cool, well-ventilated area in closed containers away from heat, sparks, open flame, oxidizing materials

Consoscour P [Consos]
Chem. Descrip.: Aromatic solv. mixture with emulsifier and surfactants
Uses: Cleaner for textile printing and dyeing equip.; solvent scour
Features: Solvent-based
Properties: Liq.; anionic; 100% conc.

Consoscour TKB [Consos]
Chem. Descrip.: Solvent/emulsifier blend
Uses: Solvent scouring agent, penetrant for nylon, spandex, and other syn. fiber
Features: Low foam, low temp.; removes heavy deposits of grease, oil, dirt, and graphite
Properties: Liq.; nonionic
Environmental: Biodeg.

Corial® [BASF AG]
Uses: Solvent for leather finishes
Features: Org.

Corn Oil, Refined [Natural Oils Int'l.; Tri-K Ind.]
Chem. Descrip.: Corn oil, refined
CAS 8001-30-7; EINECS/ELINCS 232-281-2
Uses: Antistat, emollient, solvent in cosmetics
Properties: Amber clear oil, bland odor and taste; insol. in water; sp.gr. 0.914-0.921; iodine no. 102-130; sapon. no. 187-193; flash pt. (OC) 580 F; 0.05% max. moisture
Toxicology: Nonhazardous

Corona Lanolin [Croda Oleochems.]
Chem. Descrip.: Anhyd. lanolin
CAS 8006-54-0; EINECS/ELINCS 232-348-6
Uses: Conditioner, emollient, moisturizer, cosolvent, plasticizer, w/o emulsifier, superfatting agent, wetting agent, dispersant in cosmetics, pharmaceuticals; pharmaceutical base
Properties: Paste; oil-sol.; nonionic; 100% conc.

Coronet Lanolin [Croda Oleochems.]
Chem. Descrip.: Cosmetic grade lanolin
CAS 8006-54-0; EINECS/ELINCS 232-348-6
Uses: Conditioner, emollient, moisturizer, cosolvent, plasticizer, w/o emulsifier, superfatting agent, wetting agent, dispersant for cosmetics, pharmaceuticals; pharmaceutical base
Features: Minimum odor, color, batch variation in this grade; slightly higher m.p. than USP and BP standards
Properties: Paste; nonionic; 100% conc.

Cosmol 525 [Nisshin Oillio]
Chem. Descrip.: Neopentyl glycol dioctanoate
CAS 28510-23-8; EINECS/ELINCS 249-060-1
Uses: Emollient, solvent, pigment dispersant in cosmetics

Cremophor® EL [BASF; BASF AG]
Chem. Descrip.: PEG-35 castor oil
CAS 61791-12-6
Uses: Emulsifier, solubilizer, solvent for essential oils, steroids and fat-sol. vitamins in pharmaceuticals (oral sol'ns., topicals, parenterals), cosmetics, veterinary medicine; emulsifier for topical creams and parenterally administered cyclosporine; dispersant, bioavailability enhancer for vitamins in feed and veterinary medicines
Features: Stable towards electrolytes (except mercury II chloride)
Regulatory: EP, USP/NF compliance
Properties: Pale yel. oily liq.; faint char. odor; sol. in water, ethanol, propanol, ethyl acetate, chloroform, CCl_4, benzene, toluene, and xylene; sp.gr. 1.05-1.06; visc. 700-850 cps; vapor pressure 6. 2mm Hg (20 C); HLB 12-14; acid no. < 2; iodine no. 25-35; sapon no. 65-70; hyd. no. 65-78; flash pt. 250 C; ref. index 1.471; pH 6-8 (10% aq.); nonionic; 100% act.; ≤ 2.8% water
Toxicology: LD50 (IV, mouse) 6500 mg/kg; may be harmful by ing., inh., or skin absorption; may cause eye/skin/respiratory tract irritation; may cause sensitization by inh. and skin contact; may cause anaphylactic reactions
Precaution: Avoid very acidic or basic substances
Storage: 2 yrs shelf life if stored in unopened original containers @ R.T.; prolonged storage is not rec. unless the containers are completely full

Cremophor® ELP [BASF; BASF AG]
Chem. Descrip.: PEG-35 castor oil
CAS 61791-12-6
Uses: Solubilizer, solvent, emulsifier for pharmaceutical injection sol'ns.
Features: Extra pure grade
Properties: Nonionic

Crill 1 [Croda Food Services Ltd; Croda Oleochems.]
Chem. Descrip.: Sorbitan laurate
CAS 1338-39-2; EINECS/ELINCS 215-663-3
Uses: Emulsifier, pigment dispersant, cosolvent, wetting agent, antifoam, visc. reducer, mold release, antiblock agent, corrosion inhibitor, lubricant, antistat for cosmetics, foods, food pkg., insecticides, herbicides, leather treatment, metalworking fluids, oil slick dispersing, paints, inks, pharmaceuticals, plastics, polishes, textiles, paper, adhesives; antifoam for syrups, calf starters; food pkg. adhesives, coatings, paper; emulsifier in mfg. of food-contact articles
Regulatory: FDA 21CFR §175.105, 175.300, 175.320, 175.380, 175.390, 176.170, 176.210, 178.3120, 178.3400; 40CFR §180.1001(c)(e) exempt; EEC, UK clearance; DOT nonregulated/nonhazardous; SARA §313 nonreportable
Properties: Pale yel. clear visc. liq.; sol. in ethanol, oleyl alcohol, min. oil; disp. in water; sp.gr. 1.00; visc. 39-49 poise; HLB 8.6; acid no. 4-7; sapon. no. 160-175; hyd. no. 330-358; nonionic; 98% conc.
Toxicology: LD50 (acute oral) > 20 g/kg; essentially nonirritating to eyes; may be mildly irritating to skin; no significant adverse health effects are known; may aggravate pre-existing dermatological conditions
Precaution: Incompat. with strong oxidizing agents
Hazardous decomp. prods.: Burning will produce CO_x
Storage: Store under cool, dry conditions

Crill 2 [Croda Oleochems.]
Chem. Descrip.: Sorbitan palmitate
CAS 26266-57-9; EINECS/ELINCS 247-568-8
Uses: Emulsifier; pigment dispersant; cosolvent; wetting agent; antifoam; visc. reducer; mold release; antiblock agent; corrosion inhibitor; lubricant; antistat; used for cosmetics, food and food pkg., polishes and cleaners, adhesives; antifoam for syrups, calf starters; food pkg. adhesives, coatings; emulsifier in mfg. of food-contact articles; defoamer in food-contact paper/paperboard
Regulatory: FDA 21CFR §175.105, 175.320, 176.210, 178.3120, 178.3400; 40CFR §180.1001(c)(e) exempt; EEC compliance; FDA nonregulated/nonhazardous; SARA §313 nonreportable
Properties: Pale tan hard waxy solid; partially sol. in propylene glycol, ethyl and oleyl alcohols, olive oil, oleic acid; disp. in water; vapor pressure negligible; m.p. 46 C; HLB 6.7; sapon. no. 140-150; nonionic; 98% conc.
Toxicology: Acute oral toxicity > 15.9 g/kg; essentially nonirritating to eyes (30% sol'n.), and to skin (50% sol'n.); no adverse health effects are known; may aggravate pre-existing dermatological conditions
Precaution: Incompat. with strong oxidizing agents
Hazardous decomp. prods.: Burning will produce CO_x
Storage: Store under cool, dry conditions

Crillet 6 [Croda Inc; Croda Oleochems.]
Chem. Descrip.: PEG-20 sorbitan isostearate
CAS 66794-58-9
Uses: O/w emulsifier, solvent, and wetting agent for topical pharmaceuticals; o/w emulsifier, solubilizer for fragrances and perfumes; surfactant

for creams, lotions, ointments, pesticides, polishes, cleaners, textiles; pigment dispersant
Features: Used with Crills; improved resistance to oxidation
Properties: Yel. clear liq. to soft solid; sol. in water, ethanol, IPA, oleyl alcohol, oleic acid, xylene, trichlorethylene; partly sol. in IPM; insol. in propylene glycol, min. oil, olive oil; HLB 14.9; acid no. 2 max.; sapon. no. 40-50; hyd. no. 65-85; surf. tens. 38.6 dynes/cm (0.1%); nonionic; 97% conc.; 3% max. water
Use Level: 0.5-5%

Crodamol AB [Croda Oleochems.]
Chem. Descrip.: C12-15 alkyl benzoate
CAS 68411-27-8; EINECS/ELINCS 270-112-4
Uses: Emollient, vehicle for antiperspirants, skin care prods., makeup, sunscreens; solvent and perfume fixative; reduces greasy/oily texture and promotes spreading
Features: Fine dry emollience
Properties: Colorless to pale straw liq.; sol. in castor oil, corn oil, min. oil, oleyl alcohol, ethanol; sp.gr. 0.92; visc. 13.7 cst

Crodamol CAP [Croda Inc; Croda Oleochems.]
Chem. Descrip.: Cetearyl octanoate and isopropyl myristate
Uses: Emollient, spreading agent, water repellent, reducing stickiness and occlusivity, and imparting smooth after-feel to skin care and makeup preps.; solvent, emollient, wetting agent for pharmaceuticals (burn creams, acne creams and lotions, antibiotic ointments)
Features: Simulates props. of preen gland oil; nonoily feel
Properties: Colorless clear liq., low odor; sol. in castor oil, corn oil, min. oil, oleyl alcohol, ethanol, IPA; insol. in water, propylene glycol; sp.gr. 0.848-0.853; visc. 11.9 cst; HLB 8.22; cloud pt. 0 C
Use Level: 5-10%
Toxicology: LD50 (oral, rat) 13.6 g/kg; moderate skin irritant; nonirritating to eyes

Crodamol DA [Croda Oleochems.]
Chem. Descrip.: Diisopropyl adipate
CAS 6938-94-9; EINECS/ELINCS 230-072-0
Uses: Solvent for aromatic oils and perfumes; superfatting agent for aq. alcoholic lotions, aftershave lotions, colognes, skin fresheners; solvent for lipstick dyestuffs; emollient in skin and bath preps; plasticizer in nail varnishes, hair sprays
Properties: Water-wh. liq., almost odorless; sol. in castor oil, corn oil, min. oil, oleyl alcohol, ethanol; sp.gr. 0.950-0.962; visc. 3.9 cst; f.p. -1 C; acid no. 0.5 max.; sapon. no. 480-500; ref. index 1.422-1.424

Crodamol EO [Croda Inc]
Chem. Descrip.: Ethyl oleate NF
CAS 111-62-6; EINECS/ELINCS 203-889-5
Uses: Vehicle, solvent, visc. reducer for drug delivery systems, for parenteral and topical pharmaceuticals
Properties: Colorless liq.; sol. in min. oil, IPA; insol. in water, propylene glycol; HLB 10.75
Use Level: 2-50%

Crodamol GTCC [Croda Inc; Croda Food Services Ltd; Croda Oleochems.]
Chem. Descrip.: Glyceryl tricaprylate/caprate
CAS 65381-09-1; EINECS/ELINCS 265-724-3
Uses: Emollient for skin care; solvent, vehicle, diluent for active ingreds. in cosmetics and pharmaceuticals; nutritional source of lipids; bakery lubricant, release agent, glazing agent for sugar confectionery, flour confectionery; solvent carrier for flavors and fragrances; prevents excessive dehydration in dried fruit; hypoallergenic baby food formulations
Features: Exc. dermatological props.; imparts fine, soft emollience; exc. oxidative stability; nongreasy; substitute for min. hydrocarbon oils
Properties: Colorless to pale straw clear liq.; sol. in castor oil, corn oil, min. oil, oleyl alcohol, ethanol; sp.gr. 0.950; visc. 30 cst; acid no. 0.2 max.; iodine no. 1 max.; sapon. no. 325-345; nonionic; 100% conc.
Use Level: 2-15%
Storage: Store under cool, dry conditions

Crodamol GTCC-PN [Croda Inc]
Chem. Descrip.: Caprylic/capric triglyceride
CAS 65381-09-1; EINECS/ELINCS 265-724-3
Uses: Nutritional source of lipids; solvent, vehicle, stabilizer, emollient, spreading agent in oral, topical, and parenteral pharmaceuticals
Features: Nongreasy; low visc.; exc. oxidative stability
Properties: Colorless liq.; sol. in min. oil, IPA; insol. in water, propylene glycol; HLB 10.81
Use Level: 2-15%

Crodamol IPM [Croda Inc; Croda Oleochems.]
Chem. Descrip.: Isopropyl myristate, perfumery grade
CAS 110-27-0; EINECS/ELINCS 203-751-4
Uses: Spreading agent, emollient, cosolvent, vehicle for cosmetic raw materials
Properties: Water wh. liq.; odorless.; misc. with veg. and min. oils; sp.gr. 0.847-0.853; visc. 5.3 cst; acid no. 2 max.; ref. index 1.432-1.434; 92% act.

Crodamol IPP [Croda Oleochems.]
Chem. Descrip.: Isopropyl palmitate
CAS 142-91-6; EINECS/ELINCS 205-571-1
Uses: Spreading agent, emollient, cosolvent, vehicle for cosmetics
Properties: Colorless liq.; sol. in castor oil, corn oil, min. oil, oleyl alcohol, ethanol; sp.gr. 0.850-0.855; visc. 7.7 cSt

Crodamol PC-PN [Croda Inc]
Chem. Descrip.: Propylene glycol dicaprylate/caprate DAB
CAS 68583-51-7; EINECS/ELINCS 271-516-3
Uses: Solvent, vehicle for foods and oral, topical, and parenteral pharmaceuticals
Features: Low visc.
Regulatory: FDA 21CFR §172.856
Properties: Colorless liq.; sol. in min. oil, IPA; insol. in water, propylene glycol; HLB 10.93
Use Level: 2-40%

Crodamol PMP [Croda Inc; Croda Oleochems.]
Chem. Descrip.: PPG-2 myristyl ether propionate
CAS 111497-87-1
Uses: Emollient, lubricant, solvent, emulsion stabilizer for bath oils, creams, moisturizers, emulsions, topical pharmaceuticals (burn creams, acne creams and lotions, antibiotic ointments); spreading agent, antitackifier in ointments and creams; coupling agent; solvent for sunscreen actives
Features: Nonoily; dry, lt. greaseless feel; can partially replace min. oil to reduce greasy feel; enhances freeze/thaw stability
Properties: Gardner 1 max. colorless clear liq., very mild char. odor; sol. (1%) in min. oil, IPM, oleyl alcohol, ethanol/water, lanolin, cetyl alcohol; sp.gr. 0.870-0.880; visc. 8.9 cSt; HLB 8.03; cloud pt. -5 C; acid no. 0.5 max.; iodine no. 1.0 max.; sapon. no. 140-155
Use Level: 5-20%
Toxicology: LD50 (oral, rat) > 5 g/kg; minimal skin irritant, nonirritating to eyes; esp. low in comedogenic potential

Crodamol W [Croda Inc; Croda Oleochems.]
Chem. Descrip.: Stearyl heptanoate
CAS 66009-41-4; EINECS/ELINCS 266-065-4
Uses: Emollient, water repellent for cosmetics and toiletries, esp. stick formulations, and topical pharmaceuticals (burn creams, acne creams and lotions, antibiotic ointments); lubricant, solvent for household/industrial applics.
Features: Nongreasy; melts rapidly on applic. to skin; syn. preen gland wax
Properties: Wh. waxy solid; mild odor; sol. in castor oil, corn oil, min. oil, oleyl alcohol, ethanol, IPA; sp.gr. 0.850-0.855; m.p. 23-27 C
Use Level: 2-20%
Toxicology: LD50 (oral, rat) > 16 g/kg; mild skin and eye irritant

Cromollient DP3A [Croda Inc]
Chem. Descrip.: Di-PPG-3 myristyl ether adipate
Uses: Skin emollient in sunscreens, makeup, sticks, creams/lotions; skin care vehicle; sunscreen solvent; pigment dispersant/wetting agent
Features: Elegant feel; nongreasy
Regulatory: SARA §313 nonreportable; DOT nonregulated
Properties: Colorless oily liq.; odorless; sol. in min. oil, almond oil, cyclomethicone; insol. in water, 75/25 alcohol/water, glycerin, and PG; acid no. 1; sapon. no. 98-126; pH 6-7
Precaution: Incompat. with strong oxidizing agents
Hazardous Ingredients: None
Hazardous decomp. prods.: CO_2
Storage: Store under cool, dry conditions

Cropure® Almond [Croda Inc]
Chem. Descrip.: Sweet almond (Prunus amygdalus dulcis) oil and BHT (preservative)
CAS 8007-69-0
Uses: Emollient, spreading agent, lubricant, moisturizer, skin protectant,

solvent for creams and lotions, and bath oils
Properties: APHA 60 max. clear liq.; essentially odorless; iodine no. 95-105; sapon. no. 190-200; hyd. no. 5.0 max.

Cropure® Babassu [Croda Oleochems.]
Chem. Descrip.: Babassu (Orbignya oleifera) oil, BHT (preservative)
CAS 91078-92-1; EINECS/ELINCS 293-376-2
Uses: Emollient, gloss aid for sun preps., lipsticks, skin creams and lotions; sunscreen solvent
Features: Nongreasy
Properties: APHA 50 max. color, clear oily liq./wh. waxy solid; essentially odorless; acid no. 1 max.; iodine no. 10-20; sapon. no. 230-265; hyd. no. 5 max.

Cropure® Menhaden [Croda Inc]
Chem. Descrip.: Menhaden oil and BHT (preservative)
CAS 8002-50-4; EINECS/ELINCS 232-311-4
Uses: Emollient, spreading agent, lubricant, moisturizer, skin protectant, solvent for skin care
Properties: Gardner 2.0 max. clear liq.; faint char. odor; acid no. 1.0 max.; iodine no. 165-185; sapon. no. 180-200; hyd. no. 5.0 max.

Cropure® Olive [Croda Inc]
Chem. Descrip.: Olive (Olea Europaea) oil and BHT (preservative)
CAS 8001-25-0; EINECS/ELINCS 232-277-0
Uses: Emollient, spreading agent, lubricant, moisturizer, skin protectant, solvent for hair care, skin care
Features: Highly resist. to oxidation
Properties: APHA 40 max. clear liq.; essentially odorless; acid no. 1.0 max.; iodine no. 75-95; sapon. no. 180-200; hyd. no. 5.0 max.

Cropure® Sesame [Croda Oleochems.]
Chem. Descrip.: Sesame (Sesamom indicum) oil, BHT (preservative)
CAS 8008-74-0; EINECS/ELINCS 232-370-6
Uses: Emollient, lubricant, solvent for skin care
Properties: APHA 50 max. color, clear oily liq.; essentially odorless; iodine no. 103-116; sapon. no. 188-195; hyd. no. 5 max.

Cropure® Soybean [Croda Inc]
Chem. Descrip.: Soybean (Glycine soja) oil and BHT (preservative)
CAS 8001-22-7; EINECS/ELINCS 232-274-4
Uses: Emollient, lubricant, solvent for skin care
Properties: APHA 40 max. clear liq.; essentially odorless; acid no. 1.0 max.; iodine no. 120-141; sapon. no. 180-200; hyd. no. 5.0 max.

Crossential® EL90 [Croda Inc]
Chem. Descrip.: Ethyl linoleate
CAS 544-35-4; EINECS/ELINCS 208-868-4
Uses: Solvent for topical pharmaceuticals
Features: Props. similar to ethyloleate
Properties: Yel. liq.; sol. in min. oil, IPA; insol. in water, propylene glycol; HLB 15.95; 90% as ethyl ester
Use Level: 0-100%

Crystal® O [CasChem; Costec]
Chem. Descrip.: Refined castor (Ricinus communis) oil USP
CAS 8001-79-4; EINECS/ELINCS 232-293-8
Uses: Emulsifier, emollient, pigment dispersant/wetter/cosolvent, solvent, plasticizer, lubricant, film-former for cosmetics, makeup, antiperspirant sticks, medicine, pharmaceuticals, soaps
Features: Mild; nondrying; stable
Regulatory: FDA approval
Properties: Gardner 1- color; sp.gr. 0.959; visc. 7.5 stokes; pour pt. -23 C; acid no. 2; iodine no. 86; sapon. no. 180; hyd. no. 164; 0.20% moisture

Crystal® Crown [CasChem; Costec; C.P. Hall]
Chem. Descrip.: Refined castor (Ricinus communis) oil USP (99.98%), with BHA (0.02%)
CAS 8001-79-4; EINECS/ELINCS 232-293-8
Uses: Emulsifier, carrier, emollient, pigment dispersant/wetting agent/cosolvent, lubricant, plasticizer, solvent, surfactant, film-former for cosmetics, toiletries, lipstick, makeup, antiperspirant sticks, medicine, pharmaceuticals, soaps
Regulatory: FDA approval; DOT nonregulated; SARA §302/311/312/313 nonreportable; Calif. Prop. 65 reportable [BHA (0.02%)]; Canada DSL listed
Properties: Pale yel. visc. liq.; char. odor; insol. in water; sp.gr. 0.959; visc. 7.5 stokes; f.p. 15 F; i.b.p. 313 C; pour pt. -23 C; acid no. 2; iodine no. 86; sapon. no. 180; hyd. no. 164; flash pt. (PMCC) 282 C; autoignition temp. 840 F; 0.20% moisture
Toxicology: LD50 (oral, rat) > 5000 mg/kg; prolonged/repeated contact may cause skin irritation in some individuals; may cause sl. eye irritation; not expected to be toxic by ing.; single inh. exposure to vapors or mist not likely to be hazardousTSCA listed
Environmental: Minimize entry into sewers, drainage systems
Precaution: May burn, but does not readily ignite; avoid high temps.; no sensitivity to explosion expected by mech. impact or static discharge; can react vigorously with oxidizing materials
Hazardous decomp. prods.: Incomplete combustion: CO, CO_2, dense smoke; combustion may generate toxic gases, vapors, fumes
HMIS: Health 1, Flammability 1, Reactivity 0
Storage: Store in cool, dry place away from strong oxidizers and acids

Cyanex® 272 [Cytec Ind.]
Chem. Descrip.: Bis (2,4,4,-trimethylpentyl) phosphinic acid
CAS 83411-71-6; EINECS/ELINCS 280-445-7
Uses: Solvent extraction reagent for separating cobalt and nickel from either sulfate or chloride media; selective extraction solvent for many other metals depending on sol'n. pH
Features: Rejects calcium and magnesium
Properties: Colorless to lt. amber liq.; sol. 16 mg/l in water @ pH 2.6, 38 mg/l in water @ pH 3.7; sp.gr. 0.92; visc. 42 cP; b.p. > 300 C; pour pt. -32 C; flash pt. (CC) > 108 C; sp. heat 0.48 cal/g/°C; thermal conduct. 2.7×10^{-4} cal/cm/s/°C

Cyanex® 301 [Cytec Ind.]
Chem. Descrip.: Bis (2,4,4,-trimethylpentyl) dithiophosphinic acid
CAS 107667-02-7
Uses: Solvent extraction reagent for extracting many metals at low pH (< 2), esp. for cobalt and nickel and selective recovery of zinc from effluents
Features: High degree of selectivity for heavy metals vs. alkaline and alkali earths
Properties: Grn. mobile liq.; sol. 7 mg/l in wwater; sp.gr. 0.95 (24 C); visc. 78 cp (24 C); b.p. dec. @ 220 C; flashpt. (CC) 146 C

Cyanex® 302 [Cytec Ind.]
Chem. Descrip.: Bis (2,4,4,-trimethylpentyl) monothiophosphinic acid
CAS 132767-86-3
Uses: Solvent extraction reagent for extracting many metals at low pH (< 2)
Features: High degree of selectivity for heavy metals vs. alkaline and alkali earths; more hydrolytically stable than Cyanex™ 301
Properties: Lt. yel. mobile liq.; sol. 3 mg/l in water; sp.gr. 0.93 (24 C); visc. 195 cp (24 C); b.p. dec. @ 337 C; flash pt. (CC) > 96 C; 85% act.

Cyanex® 471X [Cytec Ind.]
Chem. Descrip.: Triisobutylphosphine sulfide
CAS 3982-87-4; EINECS/ELINCS 223-620-5
Uses: Solvent extraction reagent for selective recovery of silver, separation of palladium from platinum, extraction of mercury and gold
Properties: Off-wh. cryst. solid; sol. 43 mg/l in dist. water (24 C); sp.gr. 0.91 (22 C); m.p. 58-59 C; decomp. temp. > 249 C; flash pt. (CC) 152 C; 97% act.

Cyanex® 921 [Cytec Ind.]
Chem. Descrip.: Trioctylphosphine oxide
CAS 78-50-2; EINECS/ELINCS 201-121-3
Uses: Solvent extraction reagent used with D2EHPA to recover uranium from wet process phosphoric acid, and for many other metals and some organics, e.g., acetic acid and phenol
Features: High extraction coeff.
Properties: Off-wh. waxy solid; sol. < 3 mg/l in water; sp.gr. 0.88; visc. 15 cp (55 C); m.p. 47-52 C; 93% act.

Cyanex® 923 [Cytec Ind.]
Chem. Descrip.: Trioctylphosphine oxide, dioctylmonohexylphosphine oxide, dihexylmonooctylphosphine oxide, trihexylphosphine oxide
Uses: Solvent extraction reagent for recovery of carboxylic acids, phenol, and ethanol from effluent streams, and removal of arsenic from copper electrolytes
Properties: Colorless mobile liq.; misc. with all commonly used hydrocarbon diluents; sol. < 10 mg/l in water; sp.gr. 0.88; visc. 40 cp (25 C), 13.7 cp (50 C); f.p. 0 to -5 C; flash pt. 182 C; autoignition temp. 281 C

D

DA-405B [ONDEO Nalco]
Chem. Descrip.: Solvent/detergent blend (glycol ether, water)
Uses: Solvent, detergent for spray washers
Features: Low foam; exc. rinsing; suitable @ 85-150 F; enhances performance of iron phosphates and alkaline cleaners
Regulatory: DOT nonregulated; SARA §302/313 nonreportable, §311/312 acute health hazard
Properties: Clear liq.; mild odor; completely sol. in water; sp.gr. 0.98; b.p. 102.8 C; flash pt. 224 F; pH neutral
Use Level: 5-15% (on conc. of phosphate or cleaner)
Toxicology: Mod. irritating to eyes; may cause skin irritation on prolonged contact; not expected to be a sensitizerTSCA listed
Environmental: Biodeg.
Hazardous decomp. prods.: May evolve CO_x under fire conditions
NFPA: Health 2, Flammability 1, Reactivity 0
Storage: Store in tightly closed containers; keep containers closed when not in use

DA-425 [ONDEO Nalco]
Chem. Descrip.: 2-Ethylhexyl sodium sulfate (5-10%), trideceth (5-10%), sodium xylene sulfonate (1-5%), limonene (1-5%)
UN NA 1993
Uses: Solvent, detergent, emulsifier for hard-to-remove soils in alkaline systems
Features: Free rinsing; mod. foaming; does not contain any petrol. or chlorinated solvs.; hard water stability up to 500 ppm
Regulatory: SARA §302/313 nonreportable, §311/312 acute health and fire hazard
Properties: Off-wh. to lt. yel. clear liq.; lemon citrus odor; disp. in water; sp.gr. 1.04; dens. 8.66 lb/gal; m.p. -8.8 C; b.p. 101.7 C; flash pt. (TCC) 63.8 C; pH 7.5-8.0; 5.5% VOC
Use Level: 1-2%
Toxicology: May cause mild skin and eye irritation; ing. not likely route of exposure, but may cause irritation to GI tract, nausea, vomiting; repeated/prolonged inh. may irritate respiratory tract; not expected to be a sensitizerTSCA listed
Environmental: Biodeg.
Precaution: Combustible; empty containers may contain residue; incompat. with strong oxidizers (contact may generate heat, fires, explosions, toxic vapors), strong acids; ground connections to avoid elec. charges
Hazardous decomp. prods.: May evolve CO_x, SO_x under fire conditions
NFPA: Health 2, Flammability 2, Reactivity 0
Storage: Store in tightly closed containers; keep containers closed when not in use; keep away from heat, ignition sources, static discharges

Dalpad A [Dow]
Chem. Descrip.: Ethylene glycol phenyl ether (90%), diethylene glycol phenyl ether (10%)
Uses: Solvent, emulsifier, coupling agent providing improved surf. wetting, soil penetration in household, commercial and industrial cleaning prods.; coalescent in adhesives, water-based architectural and industrial coatings; plasticizer for cellulosics; solvent for printing inks; carrier solvent for textile dyes; intermediate for soap/perfume fixatives; antimicrobial in cosmetics and fragrances; paint removers; ingred. in metal corrosion protection slurries
Regulatory: DOT nonregulated; SARA §311/312 acute/chronic health hazard, §313 reportable
Properties: Colorless liq.; pleasant odor; sol. 2.3 ml/100 ml in water; m.w. 138.2; sp.gr. 1.104; dens. 9.2 lb/gal; vapor pressure 0.007 mm Hg; f.p. 12 C; b.p. 245.6 C; flash pt. (PMCC) 121 C; autoignition temp. 495 C; surf. tens. 42 dynes/cm; VOC 1104 g/l
Toxicology: LD50 (oral, rat) 1400-4000 mg/kg, (skin, rabbit) > 2000 mg/kg; may cause mod. eye irritation, corneal injury; prolonged/repeated exposure may cause hemolysis; may cause more severe response if confined to skin; may cause liver, kidney, thyroid, and blood effects; TSCA listed
Environmental: Readily biodeg.; low bioconcentration potential; LC50 > 100 mg/l in most sensitive species; pract. nontoxic to aquatic organisms; prevent contamination of soil, surf. water, ground water
Precaution: Lower flamm. limit 0.9 vol.%; incompat. with oxidizing materials
Hazardous decomp. prods.: Combustion: toxic fumes, dense smoke
NFPA: Health 1, Flammability 1, Reactivity 0
Storage: Keep containers tightly closed when not in use

Dantogard® Plus Liq. [Lonza]
Chem. Descrip.: DMDM hydantoin, hydroxymethyl-5,5-dimethyl hydantoin, 5,5-dimethylhydantoin, iodopropynyl butylcarbamate (2.6%), 1,3-butylene glycol (4.5%), water (23%)
Uses: Preservative for household and industrial applics.; biocide, fungicide, solvent, preservative, lubricant for pharmaceuticals; pesticide
Features: Stable under most normal conditions, @ pH 4-9
Regulatory: EPA reg. no. 6836-271; DOT nonregulated; SARA §302/304 nonreportable, §311/312 acute health hazard, §313 reportable (iodopropynyl butylcarbamate); Calif. Prop. 65 reportable (formaldehyde)
Properties: Colorless clear liq.; sl. formaldehyde odor; sol. in water; readily disp. in surfactant systems; sp.gr. 1.19; flash pt. > 200 F; pH 7
Toxicology: LD50 (oral, rat) 3141 mg/kg, (skin, rabbit) > 2000 mg/kg; LC50 (inh., rat, 4 h) > 2.09 mg/l; expected to be mod. eye and mild skin irritant on direct contact; inh. may irritate mucous membranes; may cause allergic skin response in certain individuals; formaldehyde can cause sensitization reactions and is possible human carcinogen by inh.; TSCA listed
Environmental: Keep spills out of sewers and open bodies of water
Precaution: Avoid temps. above 90 C (to avoid decomp. and release of formaldehyde); incompat. with strong acids and alkalis; spillages may be slippery
Hazardous Ingredients: Contains up to 0.09% free formaldehyde in sol'n.
Hazardous decomp. prods.: Toxic combustion prods.; heating will release formaldehyde; thermal decomp. may produce toxic vapors/fumes of iodine, other org. materials, CO_x, NO_x
Storage: Store @ or near R.T.; keep containers tightly closed until used; do not store below 60 F to avoid formation of crystals

DBE [DuPont Nylon]
Chem. Descrip.: Dimethyl glutarate (60%), dimethyl adipate (15%), dimethyl succinate (24%)
Uses: Solvent for coatings, cleaners, inks, textile lubricants, urethane prod.; plasticizer for flexible thermoset polyester; polymer intermediate for polyester polyols for urethanes, wet-str. paper resins, polyester resins; specialty chemical intermediate
Regulatory: DOT nonregulated; SARA nonreportable
Properties: Clear colorless liq., mild sweet odor; sol. in alcohols, ketones, ethers, most hydrocarbons; sl. sol. in water and higher paraffinic hydro-

carbons; m.w. 159; sp.gr. 1.092 (20 C); dens. 9.09 lb/gal; visc. 2.4 cSt; vapor pressure 0.2 mm Hg (20 C); f.p. -20 C; b.p. 196-225 C; acid no. 0.3 max.; flash pt. (TCC) 100 C; surf. tens. 35.6 dynes/cm; 99% min. esters

Toxicology: LD50 (oral, rat) 8191 mg/kg, (dermal, rabbit) > 2250 mg/kg; mild to severe skin irritant, moderate eye irritant; may cause nose and throat irritation and blurry visionTSCA listed

Environmental: Aquatic toxicity: LC50 (fathead minnows, 96 h) 18-24 mg/l (mod. toxic), (daphnia magna, 48 h) 112-150 mg/l

Precaution: Incompat. with strong oxidants, acids, or alkalis; dec. with heat

HMIS: Health 1, Flammability 1, Reactivity 0

Storage: Store in well ventilated place in tightly closed container

DBE-2 [DuPont Nylon]

Chem. Descrip.: Dimethyl glutarate (76%), dimethyl adipate (24%), dimethyl succinate (0.3%)

Uses: Solvent for coatings, cleaners, inks, textile lubricants, urethane prod.; plasticizer for flexible thermoset polyester; polymer intermediate for polyester polyols for urethanes, wet-str. paper resins, polyester resins; specialty chemical intermediate

Properties: Clear colorless liq., mild odor; m.w. 163; sol. in alcohols, ketones, ethers, most hydrocarbons; sl. sol. in water and higher paraffinic hydrocarbons; sp.gr. 1.081; dens. 9.00 lb/gal; visc. 2.5 cSt; f.p. -13 C; b.p. 210-225 C; acid no. 1 max.; flash pt. (TCC) 104 C; 99% min. esters

Toxicology: Low health risk; may cause eye irritation

Environmental: Low environmental risk

DBE-3 [DuPont Nylon]

Chem. Descrip.: Dimethyl adipate (89%), dimethyl glutarate (10%), dimethyl succinate (0.2%)

Uses: Solvent for coatings, cleaners, inks, textile lubricants, urethane prod.; plasticizer for flexible thermoset polyester; polymer intermediate for polyester polyols for urethanes, wet-str. paper resins, polyester resins; specialty chemical intermediate

Properties: Clear colorless liq., mild odor; m.w. 173; sol. in alcohols, ketones, ethers, most hydrocarbons; sl. sol. in water and higher paraffinic hydrocarbons; sp.gr. 1.068; dens. 8.89 lb/gal; visc. 2.5 cSt; f.p. 8 C; b.p. 215-225 C; acid no. 1 max.; flash pt. (TCC) 102 C; 99% min. esters

Toxicology: Low health risk; may cause eye irritation

Environmental: Low environmental risk

DBE-4 [DuPont Nylon]

Chem. Descrip.: Dimethyl succinate

CAS 106-65-0; EINECS/ELINCS 203-419-9

Uses: Solvent for coatings, cleaners, inks, textile lubricants, urethane prod.; plasticizer for flexible thermoset polyester; polymer intermediate for polyester polyols for urethanes, wet-str. paper resins, polyester resins; specialty chemical intermediate

Properties: Clear colorless liq., mild odor; m.w. 146; sol. in alcohols, ketones, ethers, most hydrocarbons; sl. sol. in water and higher paraffinic hydrocarbons; sp.gr. 1.121; dens. 9.33 lb/gal; visc. 2.5 cSt; f.p. 19 C; b.p. 196 C; flash pt. (TCC) 94 C; 98.5% min. esters

Toxicology: Low health risk; may cause eye irritation

Environmental: Low environmental risk

DBE-5 [DuPont Nylon]

Chem. Descrip.: Dimethyl glutarate

CAS 1119-40-0; EINECS/ELINCS 214-277-2

Uses: Solvent for coatings, cleaners, inks, textile lubricants, urethane prod.; plasticizer for flexible thermoset polyester; polymer intermediate for polyester polyols for urethanes, wet-str. paper resins, polyester resins; specialty chemical intermediate

Properties: Clear colorless liq., mild odor; m.w. 160; sol. in alcohols, ketones, ethers, most hydrocarbons; sl. sol. in water and higher paraffinic hydrocarbons; sp.gr. 1.091; dens. 9.08 lb/gal; visc. 2.5 cSt; f.p. -37 C; b.p. 210-215 C; acid no. 1 max.; flash pt. (TCC) 107 C; 99% min. esters

Toxicology: Low health risk; may cause eye irritation

Environmental: Low environmental risk

DBE-6 [DuPont Nylon]

Chem. Descrip.: Dimethyl adipate

CAS 627-93-0; EINECS/ELINCS 211-020-6

Uses: Solvent for coatings, cleaners, inks, textile lubricants, urethane prod.; plasticizer for flexible thermoset polyester; polymer intermediate for polyester polyols for urethanes, wet-str. paper resins, polyester resins; specialty chemical intermediate

Properties: Clear colorless liq., mild odor; m.w. 174; sol. in alcohols, ketones, ethers, most hydrocarbons; sl. sol. in water and higher paraffinic hydrocarbons; sp.gr. 1.064; dens. 8.86 lb/gal; visc. 2.5 cSt; f.p. 10 C; b.p. 227-230 C; acid no. 1 max.; flash pt. (TCC) 113 C; 99% min. esters

Toxicology: Low health risk; may cause eye irritation

Environmental: Low environmental risk

DBE-9 [DuPont Nylon]

Chem. Descrip.: Dimethyl glutarate ((66%), dimethyl succinate (33%), dimethyl adipate (0.2%)

Uses: Solvent for coatings, cleaners, inks, textile lubricants, urethane prod.; plasticizer; polymer intermediate for polyester polyols for urethanes, wet-str. paper resins, polyester resins; specialty chemical intermediate

Properties: Clear colorless liq., mild odor; m.w. 156; sol. in alcohols, ketones, ethers, most hydrocarbons; sl. sol. in water and higher paraffinic hydrocarbons; sp.gr. 1.099; dens. 9.15 lb/gal; visc. 2.4 cSt; f.p. -10 C; b.p. 196-215 C; flash pt. (TCC) 94 C; 99% min. esters

Toxicology: Low health risk; may cause eye irritation

Environmental: Low environmental risk

DBE-IB [DuPont Nylon]

Chem. Descrip.: Diisobutyl adipate (15%), diisobutyl glutarate (58%), diisobutyl succinate (27%) mixt.

Uses: Solvent for waterborne and solvent coatings, as coalescing aid; solvent for printing ink, cleaning; textile lubricant

Regulatory: DOT nonregulated; SARA §311/312 acute hazard

Properties: Colorless liq.; sweet odor; misc. with most org. solvs.; sol. in water < 0.1% wt.; m.w. 242 (avg.); sp.gr. 0.955 g/ml; dens. 7.97 lb/gal; visc. 18.8 cSt; vapor pressure 0.01 mm Hg (20 C); f.p. ≈ -55 C; b.p. 275-295 C; decomp. pt. 350 C; flash pt. (TCC) 133 C; ref. index 1.427-1.428; surf. tens. 27.2 dynes/cm

Toxicology: LD50 (oral, rat) 16,426 mg/kg; LC50 (inh., rat, 4 h) > 31.9 mg/l; may cause eye/skin irritation; skin contact may cause discomfort or rash; eye contact may cause discomfort, tearing; not a skin sensitizer in animalsTSCA listed

Environmental: LC50 (fathead minnow, 96 h) 18-24 mg/l—mod. toxic; LC50 (daphnia magna, 48 h) 112-150 mg/l; readily biodeg.; prevent material from entering sewers, waterways, or low areas

Precaution: Incompat. or can react with strong oxidizers, acids, or alkalis

HMIS: Health 1, Flammability 1, Reactivity 0

Storage: Not classified as flamm.; store in well-ventilated area; keep container tightly closed; stable at normal storage conditions of temp. and humidity

DCP [Raschig; Raschig AG]

Chem. Descrip.: 1,3-Dichloro-2-propanol

CAS 96-23-1; EINECS/ELINCS 202-491-9

Uses: Solvent; reactive chloro intermediate for the modification of amines, alcohols and carboxylic acids; C3-building block

Decalin [DuPont]

Chem. Descrip.: Decahydronaphthalene

CAS 91-17-8; EINECS/ELINCS 202-046-9

Uses: Solvent, stabilizer for shoe creams and floor waxes; solvent in paints, lacquers, oils, resins, rubber, and asphalt

Properties: Colorless to pale yel. liq.; m.w. 138.2; f.p. -45 C; b.p. 186 C; negligible sol. in water; sp.gr. 0.876; dens. 7.39 lb/gal; flash pt. 57 C (TCC); 97.0% min. conc.

Decap [Dynaloy]

Chem. Descrip.: DMSO-based

Uses: Solvent for electronic depotting, deflashing, and decapsulation of epoxy castings, transfer moldings, and for general adhesive and polymer removal

Features: Used hot (150 C); nonselective

Degreez [Alzo]

Chem. Descrip.: Methyl oleate, methyl stearate, methyl palmitate, methyl laurate, methyl myristate

Uses: Degreasing agent for removing oil stains from metal, textiles, etc.

Properties: Yel. liq., mild typ. almost fruity odor; disp. in water; sp.gr. 0.87; b.p. > 400 F (760 mm); f.p. 18 F; cloud pt. 27 F; flash pt. (COC) 259 F

Toxicology: Nontoxic, nonirritating; LD50 (rat, acute oral) > 20 g/kg; mild transient eye irritation

Environmental: Biodeg.

Dekalin® [Cognis KGaA/Coatings]

Chem. Descrip.: Decahydronaphthalene

CAS 91-17-8; EINECS/ELINCS 202-046-9

Uses: Solvent; flow aid
Features: High boiling; strong solv. power
Use Level: 5-10% on solv.

Delios® [Grünau GmbH]
Chem. Descrip.: Medium chain triglycerides
Uses: Nutrient, solvent for rapid absorp. and metabolization of fat for dietetic foods; absorbent, solvent for foods (baby food, health foods, diet foods, sports food); solvent for aromas, vitamins, colorants; base for pan release compds.; glazing agent, antisticking agent for food, esp. confectionery; carrier for pharmaceutical lipophilic active ingreds.
Features: Exc. oxidation stability

Delios® S [Grünau GmbH]
Chem. Descrip.: Caprylic/capric triglyceride
CAS 65381-09-1; EINECS/ELINCS 265-724-3
Uses: Solvent for flavors and oil-sol. food additives for dietetic prods., sport foods; glazing and release agent for dried fruit and confectionery; lubricant, release agent for baked goods; antidusting agent; visc. reducer
Features: Also avail. as kosher grade (Delios SU); very good oxidation stability
Properties: Colorless to sl. ylsh. liq., neutral odor, sl. coconut taste; misc. with ethanol; visc. 27 mPa·s; cloud pt. 0 C max.; acid no. 0.1 max.; iodine no. 0.5 max.; sapon. no. 335-350; hyd. no. 5 max.; 70% caprylic acid, 30% capric acid
Storage: Store in tightly closed containers @ 0-30 C; avoid contact with water; 18 mos storage life in original pkg.

DePEG 200 [DeForest Enterprises]
Chem. Descrip.: PEG-4
CAS 25322-68-3; EINECS/ELINCS 203-989-9
Uses: Surfactant; intermediate for PEG fatty acid esters; humectant in certain printing inks; plasticizer for starch pastes; solvent for personal care prods.; binder, lubricant for compressed tablets; plasticizer in PE sheet; lubricant for rubber molds; humectant, solvent for ink pad dyes; water softener and additive in paper processing; solubilizer; coupling agent; food pkg. adhesives and food-contact articles
Regulatory: FDA 21CFR §172.820, 173.310, 173.340, 175.105, 178.3750; DOT nonregulated
Properties: Water-wh. clear liq.; sol. in water; insol. in most solvs. and oils; m.w. 190-210; sp.gr. 1.13 g/ml; f.p. < -5 C; flash pt. (COC) 385 F; pH 5.0-8.0 (5% in DW); nonionic; 100% conc.
Toxicology: Nontoxic; LD50 (oral, mouse) 29,500 mg/kg
Precaution: Incompat. with strong oxidizers
Hazardous Ingredients: None
Hazardous decomp. prods.: Thermal decomp. prods.: CO_x
HMIS: Health 0, Flammability 1, Reactivity 0

DePEG 300 [DeForest Enterprises]
Chem. Descrip.: PEG-6
CAS 25322-68-3; EINECS/ELINCS 220-045-1
Uses: Surfactant; intermediate for PEG fatty acid esters; humectant in certain printing inks; plasticizer for starch pastes; solvent for personal care prods.; binder, lubricant for compressed tablets; plasticizer in PE sheet; lubricant for rubber molds; humectant, solvent for ink pad dyes; water softener and additive in paper processing; solubilizer; coupling agent; food pkg. adhesives and food-contact articles
Regulatory: FDA 21CFR §172.820, 173.310, 173.340, 175.105, 178.3750; DOT nonregulated
Properties: Water-wh. clear liq.; sol. in water; insol. in most solvs. and oils; m.w. 285-315; sp.gr. 1.13 g/ml; f.p. < 0 C; flash pt. (COC) 385 F; pH 5.0-8.0 (5% in DW); nonionic; 100% conc.
Toxicology: Nontoxic; LD50 (oral, mouse) 29,500 mg/kg
Precaution: Incompat. with strong oxidizers
Hazardous decomp. prods.: Thermal decomp. prods.: CO_x
HMIS: Health 0, Flammability 1, Reactivity 0

DePEG 400 [DeForest Enterprises]
Chem. Descrip.: PEG-8
CAS 25322-68-3; EINECS/ELINCS 225-856-4
Uses: Surfactant; intermediate for PEG fatty acid esters; humectant in certain printing inks; plasticizer for starch pastes; solvent for personal care prods.; binder, lubricant for compressed tablets; plasticizer in PE sheet; lubricant for rubber molds; humectant, solvent for ink pad dyes; water softener and additive in paper processing; coupling agent in creams/lotions; food pkg. adhesives and food-contact articles
Regulatory: FDA 21CFR §172.820, 173.310, 173.340, 175.105, 178.3750, 181.30; DOT nonregulated
Properties: Water wh. clear liq.; sol. in water; insol. in most solvs. and oils; m.w. 380-420; sp.gr. 1.13 g/ml; f.p. < 4-8 C; flash pt. (PMCC) 350 F; pH 5.0-8.0 (5% in DW); nonionic; 100% conc.
Toxicology: Nontoxic; LD50 (oral, mouse) 29,500 mg/kg
Precaution: Incompat. with strong oxidizers
Hazardous decomp. prods.: Thermal decomp. prods.: CO_x
HMIS: Health 0, Flammability 1, Reactivity 0

DePEG 600 [DeForest Enterprises]
Chem. Descrip.: PEG-12
CAS 25322-68-3; EINECS/ELINCS 229-859-1
Uses: Surfactant; intermediate for PEG fatty acid esters; humectant in certain printing inks; plasticizer for starch pastes; solvent for personal care prods.; binder, lubricant for compressed tablets; plasticizer in PE sheet; food pkg. adhesives; lubricant for rubber molds; humectant, solvent for ink pad dyes; water softener and additive in paper processing; coupling agent in creams/lotions; color stabilizer for fuel oils; leveling agent in electropolishing formulas
Regulatory: FDA 21CFR §172.820, 173.310, 173.340, 175.105, 178.3750, 178.3910; DOT nonregulated
Properties: Water wh. clear liq.; sol. in water; insol. in most solvs. and oils; m.w. 570-630; sp.gr. 1.13 g/ml; f.p. < 20-25 C; flash pt. (COC) 475 F; pH 5.0-8.0 (5% in DW); nonionic; 100% conc.
Toxicology: Nontoxic; LD50 (oral, mouse) 29,500 mg/kg
Precaution: Incompat. with strong oxidizers
Hazardous decomp. prods.: Thermal decomp. prods.: CO_x
HMIS: Health 0, Flammability 1, Reactivity 0
Storage: Will become hazy just below 25 C and will solidify at lower temps.

Dermol ICSA [Alzo]
Chem. Descrip.: Isocetyl salicylate
CAS 138208-68-1
Uses: Emollient; solvent, plasticizer for benzophenone in sunscreen formulations, waxes, resins, polymers
Properties: Pale yel. clear liq., sl. typ. odor; sp.gr. 1.02; acid no. 3; sapon. no. 145-160; flash pt. (PMCC) > 200 F; ref. index 1.4865 ± 0.005; nonionic; 100% conc.

DF-2000™ [ExxonMobil]
Chem. Descrip.: Syn. high purity hydrocarbon solv.
Uses: Dry cleaning fluid; replacement for Stoddard solvent
Features: Safe; effective; virtually HAPs-free; good cleaning power, low residue, and low corrosion; rapid drying; high flash pt.; mild on sensitive fabrics, trims, sequins, etc.
Properties: Water-wh. clear appearance; essentially odorless; low water-sol.; sp.gr. 0.77 (15.6/15.6 C); vapor pressure 0.8 mm Hg (68 F); flash pt. 147 F; KB value 27
Toxicology: OEL 300 ppm

Diadol 18G-C [Mitsubishi Chem.]
Chem. Descrip.: Isostearyl alcohol
CAS 27458-93-1; EINECS/ELINCS 248-470-8
Uses: Surfactant; lubricant; solvent

Diglycolamine® Agent (DGA®) [Huntsman]
Chem. Descrip.: 2-(2-Aminoethoxy) ethanol
CAS 929-06-6; EINECS/ELINCS 213-195-4
Uses: Solvent for removal of CO_2 or H_2S from gases, for recovery of aromatics from refinery streams; surfactant intermediate for prep. of foam stabilizers, wetting agents, emulsifiers, and condensation polymers
Properties: Clear colorless slightly visc. liq.; mild amine odor; misc. with water, alcohols, aromatic hydrocarbons, ethyl ether; m.w. 105.14; sp.gr. 1.0572; dens. 8.7 lb/gal; b.p. 221 C; f.p. -12.5 C; flash pt. (PMCC) 255 F; ref. index 1.4598; 98% min. act.

Dipentene-C [PDM]
Chem. Descrip.: Terpinolene (39%), dl-limonene (32%), α-pinene (6.3%), p-cymene (4.6%), terpene alcohols (2%), α-terpineol (1.3%)
UN 2052
Uses: Solvent, wetting agent, dispersant for rubber processing and reclaiming, paints, varnishes, cleaners, waxes, polishes, and in oil drilling operations
Regulatory: DOT regulated
Properties: APHA < 50 liq.; mild pine odor; insol. in water; sp.gr. 0.850-0.875; vapor pressure < 2 mm Hg (20 C); b.p. 177.2 C; flash pt. (TCC) 46 C; ref. index 1.468-1.477 (20 C); KB 100

Toxicology: Causes eye, skin, and mucous membrane irritation
Precaution: Combustible; avoid strong oxidizing agents and exposure to heat, sparks, and open flame
Hazardous Ingredients: P-menthadienes (89%), terpene hydrocarbons (6%), α-pinene (5%)
Hazardous decomp. prods.: Burning produces CO, CO_2
NFPA: Health 1, Flammability 2, Reactivity 0
Storage: Store in closed containers

Dipentene Extra [Millennium/F&F]
Chem. Descrip.: Terpene hydrocarbons (terpinolene)
CAS 68956-56-9; EINECS/ELINCS 209-578-0
UN 2052; UN 2319
Uses: Chem. intermediate, solvent, fragrance in personal care prods.
Features: Rec. where good odor and color are desirable
Regulatory: DOT regulated
Properties: APHA 50 max. clear liq.; m.w. 154.25; sp.gr. 0.863-0.873; dens. 7.24 lb/gal; b.p. 170 C; flash pt. (TCC) 43.3 C; ref. index 1.472 (20 C); KB 113
Toxicology: TSCA listed

Dipentene No. 122 [Hercules]
Chem. Descrip.: Dipentene
CAS 138-86-3; EINECS/ELINCS 205-341-0
Uses: Solvent

DMSO [Atofina; Atofina SA]
Chem. Descrip.: Dimethyl sulfoxide
CAS 67-68-5; EINECS/ELINCS 200-664-3
Uses: Solvent used as reaction medium in mfg. of pesticides, pharmaceuticals, dyes, inks; solvent in polymers, electronics, refining; chem. intermediate for pesticides
Features: Aprotic

Dow B100-4800 [Dow]
Chem. Descrip.: Polybutylene glycol
Uses: Solvent providing low odor, low irritation, low toxicity to personal care prods., cosmetics
Properties: Liq.; m.w. 4800; dens. 8.10 lb/gal; visc. 731.4 cSt (40 C); pour pt. -21 C; flash pt. (PMCC) > 350 F; ref. index 1.4551

Dow EP530 [Dow]
Chem. Descrip.: Poloxamer-181
CAS 53637-25-5
Uses: Solvent providing low odor, low irritation, low toxicity to personal care prods., cosmetics; defoamer in scald baths for poultry defeathering and hog dehairing machines; defoamer for food-grade papers and coatings and sanitizing sol'ns. for food processing equip.
Regulatory: 21CFR §172.808, 176.200, 176.210, 178.1010
Properties: Liq.; m.w. 2000; sp.gr. 1.017; dens. 8.46 lb/gal; visc. 321 cSt; pour pt. -32 C; flash pt. (PMCC) > 420 F; ref. index 1.452; surf. tens. 42.6 dynes/cm (100 ppm); nonionic

Dowanol® DMM [Dow]
Chem. Descrip.: Dipropylene glycol dimethyl ether
CAS 111109-77-4; EINECS/ELINCS 404-640-5
Uses: Solvent, coupling agent for cosmetics, toiletries

Dowanol® DPM [Dow; Dow Germany]
Chem. Descrip.: Dipropylene glycol methyl ether
CAS 34590-94-8; EINECS/ELINCS 252-104-2
Uses: Solvent, coupling agent providing improved surf. wetting, soil penetration in household, commercial and industrial cleaning prods., grease/paint removers, metal cleaners; stabilizer for agric. herbicides; solvent, coupling agent for water- and solv.-based coatings, printing inks, vat dyeing fabrics; solvent, coupling agent, emollient in cosmetics, toiletries
Features: HAPs-free
Regulatory: SARA §311/312 acute health/fire hazard, §313 nonreportable
Properties: Colorless clear liq.; sl. ether odor; sol. in water; m.w. 148.2; sp.gr. 0.951; dens. 7.91 lb/gal; vapor pressure 0.41 mm Hg; f.p. -117 F; b.p. 184 C; flash pt. (TCC) 79.4 C; surf. tens. 28.8 dynes/cm; VOC 951 g/l; 99% act.
Toxicology: ACGIH TLV 100 ppm; LD50 (oral, rat) 5.4 ml/kg, (skin, rabbit) > 20 ml/kg; may cause sl. transient eye irritation; prolonged skin contact with very large amts. may cause drowsiness; extremely low toxicity by ing.; excessive inh. exposure may cause upper respiratory tract irritation, anesthetic or narcotic effects; may cause minor liver or kidney effects; TSCA listed
Environmental: Readily biodeg.; low bioconcentration potential; LC50 (fathead minnow) > 10,000 mg/l; prevent contamination of soil, surf. water, ground water
Precaution: Combustible; LFL 1.1 vol.% (100 C); UFL 14 vol.% (150 C); flamm. vapors can be released at elevated temps.; avoid static discharge; incompat. with oxidizing materials
Hazardous decomp. prods.: Combustion: CO, CO_2
NFPA: Health 0, Flammability 2, Reactivity 0

Dowanol® DPMA [Dow]
Chem. Descrip.: PPG-2 methyl ether acetate
CAS 88917-22-0
Uses: Solvent, coupling agent providing improved surf. wetting, soil penetration in household, commercial and industrial cleaning prods.; solvent for solv.-based coatings and printing inks; coalescing agent for architectural water-borne coatings
Features: HAPs-free
Properties: Colorless liq.; sweet ether odor; sol. 19.4 ml/100 ml water; m.w. 190.2; sp.gr. 0.976; dens. 8.12 lb/gal; vapor pressure < 1.0 mm Hg (20 C); f.p. < -67 F; b.p. 209.3 C; flash pt. (TCC) 86 C; surf. tens. 28.6 dynes/cm; VOC 976 g/l; 99% act.
Toxicology: LD50 (oral, rat) > 5000 mg/kg, (skin, rabbit) > 5000 mg/kg; may cause sl. transient eye irritation; prolonged skin contact with very lg. amts. may cause drowsiness; low toxicity by ing.TSCA listed
Environmental: LC50 (fathead minnow) 151 mg/l
Precaution: Combustible; flamm. limits 1.2-5.35 vol.% (150 C); flamm. vapors can be released @ elevated temps.; vapor explosion hazard; ground equip.; minimize ignition sources, static buildup, heat, spark, flame; avoid static discharge; incompat. with oxidizing materials
Hazardous Ingredients: High biodeg.; BOD28 1.30 p/p; low bioconcentration potential; pract. nontoxic to aquatic organisms; prevent entry into sewers
Hazardous decomp. prods.: Combustion: CO, CO_2
NFPA: Health 0, Flammability 2, Reactivity 0
Storage: Keep containers tightly closed when not in use

Dowanol® DPnB [Dow; Dow Germany]
Chem. Descrip.: Dipropylene glycol n-butyl ether
CAS 29911-28-2; EINECS/ELINCS 249-951-5
Uses: Solvent, coupling agent providing improved surf. wetting, soil penetration, surfactant props. in household, commercial and industrial cleaning prods., grease/paint removers, metal cleaners, hard surf. cleaners; coalescent in water-borne latex coatings; solvent for solv.-based coatings; chem. intermediate for prod. of epoxides, acid ester derivs., solvents, and plasticizers; solvent, coupling agent in cosmetics, toiletries
Features: HAPs-free
Regulatory: DOT nonregulated; SARA §311/312/313 nonreportable
Properties: Colorless clear liq.; sl. ethereal odor; sol. 5 g/100 ml water; m.w. 190.3; sp.gr. 0.9; dens. 7.56 lb/gal; visc. 4.9 cps; vapor pressure 0.06 mm Hg (20 C); f.p. -75 C; b.p. 230 C; flash pt. (Seta) 100 C; autoignition temp. 194 C; ref. index 1.425; surf. tens. 28.8 dynes/cm; sp. heat 0.59 cal/g/C; VOC 906 g/l; > 98% act.
Toxicology: LD50 (oral, rat) 4400 mg/kg, (skin, rat) > 2000 mg/kg; may cause very sl. transient corneal injury; prolonged/repeated contact may cause sl. skin irritation; low toxicity by ing.; may cause liver effects, nasal irritation; nonhazardous by EC criteria; TSCA listed
Environmental: Readily biodeg.; low bioconcentration potential; LC50 (guppy) 841 mg/l; pract. nontoxic to aquatic organisms; prevent contamination of soil, surf. water, ground water; nonhazardous by EC criteria
Precaution: Flamm. limits in air 0.6% (145 C) to 20.4% (180 C); can oxidize @ elevated temps.; incompat. with oxidizing materials, strong acids or bases
Hazardous decomp. prods.: Aldehydes, ketones, org. acids
NFPA: Health 0, Flammability 1, Reactivity 0
Storage: Keep containers tightly closed when not in use

Dowanol® DPnP [Dow]
Chem. Descrip.: Dipropylene glycol n-propyl ether
CAS 29911-27-1; EINECS/ELINCS 249-949-4
Uses: Coalescing agent in water-borne latex coatings; active solvent for solv.-borne coatings; solvent, coupling agent in cosmetics, toiletries
Regulatory: SARA §311/312 fire hazard, §313 nonreportable
Properties: Colorless clear liq.; sl. ether odor; mod. water sol.; m.w. 176.2; sp.gr. 0.922; dens. 7.70 lb/gal; visc. 4.3 cSt; vapor pressure 0.12 mm Hg; b.p. 213 C; flash pt. (Seta CC) 88 C; surf. tens. 27.6 dynes/cm; VOC

921 g/l; 99% act.
Toxicology: LD50 (oral, rat) 2000-2500 mg/kg, (skin, rabbit) > 2000 mg/kg; may cause mod. eye irritation, sl. corneal injury; prolonged/repeated contact may cause sl. skin irritation; low toxicity by ing.TSCA listed
Environmental: Readily biodeg.; low bioconcentration potential; LC50 (daphnia magna) > 100 mg/l; pract. nontoxic to aquatic organisms; prevent entry into sewers
Precaution: Combustible; flamm. vapors can be released @ elevated temps.; vapor explosion hazard; ground equip.; minimize ignition sources, static buildup, heat, spark, flame; avoid static discharge; incompat. with oxidizing materials
Hazardous decomp. prods.: Combustion: toxic fumes, CO, CO_2
Storage: Keep containers tightly closed when not in use

Dowanol® EPh [Dow]
Chem. Descrip.: Ethylene glycol phenyl ether (90%), diethylene glycol phenyl ether (10%)
Uses: Solvent, emulsifier, coupling agent providing improved surf. wetting, soil penetration in household, commercial and industrial cleaning prods.; coalescent in adhesives, water-based architectural and industrial coatings; plasticizer for cellulosics; solvent for printing inks; carrier solvent for textile dyes; intermediate for soap/perfume fixatives; antimicrobial in cosmetics and fragrances; solvent, coupling agent in cosmetics; paint removers; ingred. in metal corrosion protection slurries
Regulatory: SARA §311/312 acute/chronic health hazard, §313 reportable [glycol ethers (100%)]
Properties: Colorless liq.; pleasant rose-like odor; sol. 2.3 ml/100 ml in water; m.w. 138.2; sp.gr. 1.104 (20/4 C); dens. 9.2 lb/gal; vapor pressure 0.007 mm Hg; f.p. 51 F; b.p. 245.6 C; flash pt. (PMCC) 121 C; autoignition temp. 495 C; surf. tens. 42 dynes/cm; VOC 1104 g/l
Toxicology: LD50 (oral, rat) 1400-4000 mg/kg, (skin, rabbit) > 2000 mg/kg; may cause mod. eye irritation and corneal injury; repeated skin exposure may result in absorp. in harmful amts. and cause hemolysis; low toxicity by ing.; may cause liver, kidney, thyroid, and blood effects; TSCA listed
Environmental: Readily biodeg.; low bioconcentration potential; LC50 (fathead minnow) 366 mg/l; pract. nontoxic to fish and aquatic invertebrate
Precaution: Ufl 0.9 vol.%; can dec. at elevated temps.; incompat. with oxidizing materials
Hazardous decomp. prods.: Combustion: toxic fumes
NFPA: Health 1, Flammability 1, Reactivity 0
Storage: Keep containers tightly closed when not in use

Dowanol® PGDA [Dow]
Chem. Descrip.: Propylene glycol diacetate
CAS 623-84-7; EINECS/ELINCS 210-817-6
Uses: Solvent
Properties: Sol. 7.4% in water; m.w. 160; sp.gr. 1.056; dens. 8.77 lb/gal; visc. 2.6 cP; vapor pressure 0.23 mm Hg (20 C); b.p. 161 C; flash pt. (TCC) 86 C; surf. tens. 32.5 dynes/cm

Dowanol® PM [Dow; Dow Germany]
Chem. Descrip.: Propylene glycol methyl ether
CAS 107-98-2; EINECS/ELINCS 203-539-1
Uses: Solvent, coupling agent providing improved surf. wetting, soil penetration in household, commercial and industrial cleaning prods., rust removers, hard surf. cleaners; carrier solvent for writing pen inks; solvent for agric. pesticides; deactivator and emollient for livestock pesticides; solvent, coupling agent for solv.-based coatings, printing inks, cosmetics, toiletries
Features: HAPs-free; fast evaporating; hydrophilic
Properties: Sol. in water; m.w. 90.1; sp.gr. 0.919; dens. 7.65 lb/gal; vapor pressure 12.5 mm Hg; f.p. -142 F; b.p. 120.1 C; flash pt. (Seta) 90 F; surf. tens. 27.7 dynes/cm
Toxicology: ACGIH TLV 100 ppm; may cause CNS depression, liver effects
Environmental: LC50 (fathead minnow) 20,800 mg/l

Dowanol® PMA [Dow]
Chem. Descrip.: Propylene glycol methyl ether acetate (99%), 2-methoxy-1-acetoxypropane (< 0.5%)
CAS 108-65-6; EINECS/ELINCS 203-603-9
Uses: Solvent, coupling agent providing improved surf. wetting, soil penetration in household, commercial and industrial cleaning prods.; solvent for solv.-based coatings and printing inks
Features: HAPs-free
Regulatory: SARA §311/312 acute health/chronic health/fire hazare, §313 nonreportable
Properties: Colorless liq.; sweet ether odor; sol. 19.8 g/100 g water; m.w. 132.2; sp.gr. 0.966; dens. 8.03 lb/gal; vapor pressure 3.7 mm Hg (20 C); f.p. < -67 F; b.p. 145.8 C; flash pt. (Seta CC) 46 C; surf. tens. 26.4 dynes/cm; VOC 966 g/l; 99% in. act.
Toxicology: AIHA WEEL TWA 100 ppm; LD50 (oral, female rat) 8532 mg/kg, (skin, rabbit) > 5000 mg/kg; may cause sl. eye irritation, sl. corneal injury; repeated skin contact may cause irritation, local redness; prolonged skin contact with very lg. amts. may cause dizziness, drowsiness; low toxicity by ing.; may cause respiratory tract irritation, liver or kidney effects; TSCA listed
Environmental: Readily biodeg.; low bioconcentration potential; LC50 (fathead minnow) 161 mg/l; pract. nontoxic to aquatic organisms
Precaution: Combustible; LFL 1.5 vol.% (200 C); UFL 7.0 vol.% (200 C); vapor explosion hazard; flamm. vapors can be released @ elevated temps.; gases may accumulate in low areas; minimize ignition sources, static buildup, heat, sparks, flames; ground and bond containers and equip.; incompat. with oxidizing materials
Hazardous decomp. prods.: Combustion: CO, CO_2
NFPA: Health 1, Flammability 2, Reactivity 0
Storage: Keep containers tightly closed when not in use

Dowanol® PnB [Dow; Dow Germany]
Chem. Descrip.: Propylene glycol n-butyl ether
CAS 5131-66-8; EINECS/ELINCS 225-878-4
Uses: Solvent, coupling agent providing improved surf. wetting, soil penetration, surfactant props. in household, commercial and industrial cleaning prods., grease/paint removers, metal cleaners, hard surf. cleaners; solvent, coupling agent, coalescent in water-reducible and water-borne coatings; active solvent in solv.-based coatings; chem. intermediate for prod. of epoxides, acid ester derivs., solvents, and plasticizers; solvent, coupling agent in cosmetics
Features: HAPs-free
Properties: Low odor; sol. 6.4 g/100 g water; m.w. 132.2; sp.gr. 0.9; dens. 7.31 lb/gal; visc. 3.1 cps; vapor pressure 0.6 mm Hg (20 C); f.p. < -85 C; b.p. 170 C; flash pt. (TCC) 138 F; ref. index 1.416; surf. tens. 26.3 dynes/cm; sp. heat 0.63 cal/g/C
Toxicology: May cause skin irritation, liver and kidney effects
Environmental: LC50 (guppy) 560-1000 mg/l
Precaution: Flamm. limits in air 1.1 (80 C) to 8.4 (145 C)

Dowanol® PnP [Dow]
Chem. Descrip.: Propylene glycol n-propyl ether
CAS 1569-01-3; EINECS/ELINCS 250-069-8
Uses: Coalescing agent in water-borne latex coatings; coupling agent, cosolvent in water-reducible aerosol paints; coupling agent, solvent in household and industrial cleaners, grease/paint removers, metal cleaners, hard surface cleaners; solvent, coupling agent in cosmetics, toiletries
Regulatory: SARA §311/312 acute health/chronic health/fire hazard, §313 nonreportable
Properties: Colorless clear liq.; sl. ether odor; completely water sol.; m.w. 118.2; sp.gr. 0.885; dens. 7.38 lb/gal; visc. 2.2 cSt; vapor pressure 1.5 mm Hg (20 C); b.p. 149.8 C; flash pt. (Seta) 48 C; surf. tens. 25.9 dynes/cm; VOC 885 g/l; 99% act.
Toxicology: LD50 (oral, rat) 2000-4350 mg/kg, (skin, rabbit) 4-5 ml/kg; may cause mod. eye irritation and corneal injury; prolonged skin contact may cause sl. irritation, repeated exposure may cause irritation, burns; excessive skin exposure may cause CNS effects; low toxicity by ing.; excessive inh. exposure may cause upper respiratory tract irritation, anesthetic or narcotic effects; may cause liver, kidney, CNS effects; TSCA listed
Environmental: High biodeg.; BOD5 0.20 p/p; low bioconcentration potential; LC50 (fathead minnow) < 1000 mg/l; pract. nontoxic to aquatic organisms; prevent entry into sewers
Precaution: Combustible; LFL 1.1%; vapor explosion hazard; vapors are heavier than air, vapors may accumulate in low areas; vapors may flash back; flamm. vapors can be released @ elevated temps.; eliminate ignition sources; avoid static discharge; incompat. with oxidizing materials
Hazardous decomp. prods.: Combustion: toxic fumes, CO, CO_2
Storage: Keep containers tightly closed when not in use

Dowanol® PPh [Dow]

Chem. Descrip.: Phenoxyisopropanol (93% min.), dipropylene glycol phenyl ether (7% max.)
Uses: Latex coalescing agent in aq. architectural and industrial coatings; carrier solvent for textile dyes; solvent for inks, textile printing pastes; paint removers; coalescent for latex adhesives; solvent, coupling agent in cosmetics, toiletries
Features: HAPs-free
Regulatory: SARA §311/312 §acute health hazard, 313 nonreportable; EINECS, Korea ECL, Japan MITI, Canada DSL listed
Properties: Colorless to lt. yel. clear liq.; odorless to sl. odor; sol. 1.1 ml/100 ml in water; m.w. 152.2; sp.gr. 1.063; dens. 8.8 lb/gal; vapor pressure < 0.01 mm Hg (20 C); f.p. 55 F; b.p. 242.7 C; flash pt. (TCC) 120 C; autoignition temp. 490 C; surf. tens. 38.1 dynes/cm; VOC 1063 g/l
Toxicology: LD50 (oral, male rat) 2830 mg/kg, (skin, rabbit) > 2000 mg/kg; may cause severe eye irritation, sl. corneal injury; prolonged/repeated skin contact may cause sl. irritation; low toxicity by ing.; inh. of heated material or mist may be hazardousTSCA listed
Environmental: Readily biodeg.; BOD5 0.08 p/p; low bioconcentration potential; LC50 (fathead minnow) 280 mg/l; pract. nontoxic to aquatic organisms; prevent entry into sewers, watercourses
Precaution: LFL 0.8 vol.%; eliminate ignition sources; ground and bond containers and equip.; can oxidize @ elevated temps.; incompat. with oxidizing materials
Hazardous decomp. prods.: Combustion: toxic fumes
NFPA: Health 2, Flammability 1, Reactivity 0
Storage: Keep containers tightly closed when not in use

Dowanol® TPM [Dow]
Chem. Descrip.: PPG-3 methyl ether
CAS 25498-49-1; EINECS/ELINCS 243-734-9
Uses: Solvent, coupling agent providing improved surf. wetting, soil penetration in household, commercial and industrial cleaning prods., rust, paint, and varnish removers, penetrating oils, stamp pad and pen inks; solvent, coupling agent in cosmetics, toiletries; solvent in high-solids, solv.-based coatings
Features: HAPs-free
Regulatory: SARA §311/312 acute health hazard
Properties: Colorless liq.; sl. odor; sol. in water; m.w. 206.3; sp.gr. 0.965; dens. 8.03 lb/gal; vapor pressure 0.02 mm Hg; f.p. -110 F; b.p. 242.4 C; flash pt. (Seta) 111 C; autoignition temp. 277 C; surf. tens. 30 dynes/cm; VOC 965 g/l; 99% act.
Toxicology: LD50 (oral, rat) 3200 mg/kg, (skin, rabbit) > 20 ml/kg; no significant immediate hazards; may cause sl. transient eye irritation; mild skin irritant; prolonged/repeated skin exposure to very lg. amts. may cause narcosis; low toxicity by ing.; excessive exposure may cause anesthetic or narcotic effects; TSCA listed
Environmental: Readily biodeg.; low bioconcentration potential; LC50 (fathead minnow) 11,600 mg/l; pract. nontoxic to aquatic organisms; prevent contamination of soil, surf. water, ground water
Precaution: LFL 1.5 vol.% (151 C); UFL 10.9 vol.% (151 C); avoid contact with vapors; incompat. with oxidizing materials; avoid contact with metals such as aluminum
NFPA: Health 0, Flammability 1, Reactivity 0

Dowanol® TPnB [Dow]
Chem. Descrip.: Tripropylene glycol n-butyl ether (95%), PPG butyl ether (5%)
Uses: Solvent for heavy-duty cleaners; possible additive for floor wax strippers; plasticizer in selected resin systems; coupling agent and grease penetrant in oven cleaners; solvent for inks in ball-point and felt tip pens, stamp pads; solvent, coupling agent in cosmetics, toiletries; textile printing pastes; suitable for indoor decorative paints
Features: HAPs-free
Regulatory: SARA §311/312/313 nonreportable
Properties: Colorless clear low visc. liq.; mild odor; sol. 3.0 g/100 g in water; m.w. 248.4; sp.gr. 0.932; dens. 7.77 lb/gal; visc. 7.3 cks; vapor pressure < 0.01 mm Hg (20 C); b.p. 276 C; f.p. < -75 C; flash pt. (Seta) 126 C; autoignition temp. 202 C; ref. index 1.433; surf. tens. 29.9 dynes/cm; VOC 932 g/l
Toxicology: LD50 (oral, male rat) 3100 mg/kg, (skin, rabbit) > 2000 mg/kg; may cause mod. eye irritation, sl. corneal injury, lacrimation; prolonged/repeated skin exposure may cause sl. irritation; low toxicity by ing.; swallowing larger amts. may cause injury, ataxia, tremors, convulsions; TSCA listed
Environmental: Readily biodeg; low bioconcentration potential; LC50 (guppy) 564 mg/l; pract. nontoxic to aquatic organisms; prevent contamination of soil, surf. water, ground water
Precaution: Can dec. @ elevated temps.; incompat. with oxidizing materials, strong acids, strong bases
Hazardous decomp. prods.: Aldehydes, ketones, org. acids; combustion: CO, CO_2
Storage: Keep containers tightly closed when not in use

Dowanol® TPnP [Dow]
Chem. Descrip.: Tripropylene glycol n-propyl ether
CAS 96077-04-2
Uses: Solvent, coupling agent for water-based coatings and cleaners
Features: Slow evaporating; compat. with wide range of resins
Regulatory: SARA §311/312/313 nonreportable
Properties: Clear liq.; sl. ether odor; sol. 17% in water; m.w. 234.3; sp.gr. 0.938; dens. 7.8 lb/gal; visc. 6.3 mPa·s; vapor pressure 0.00486 mbar (20 C); f.p. -112 C; b.p. 260.9 C; flash pt. (Seta) 128.7 C; surf. tens. 29.8 dynes/cm; 90-100% act.
Toxicology: LD50 (oral, male rat) 2675 mg/kg; may cause mod. eye irritation, sl. corneal injury; prolonged/repeated contact may cause sl. skin irritation; low toxicity by ing.; swallowing larger amts. may cause injuryTSCA listed
Environmental: Readily biodeg.; low bioconcentration potential; LC50 (daphnia, 48 h) 2400 mg/l; EC50 (green algae) 1400 mg/l; prevent contamination of soil, surf. water, ground water
Precaution: Can dec. @ elevated temps.; incompat. with strong acids, strong oxidizers
Hazardous decomp. prods.: Aldehydes, ketones, org. acids; combustion: toxic fumes, CO, CO_2
Storage: Keep containers tightly closed when not in use

Dow Corning® L910 [Dow Corning]
Chem. Descrip.: PPG-14 butyl ether
CAS 9003-13-8
Uses: Solvent providing low odor, low irritation, low toxicity to personal care prods., cosmetics
Properties: Liq.; m.w. 910; sp.gr. 0.9833; dens. 8.23 lb/gal; visc. 83 cSt; pour pt. -43 C; flash pt. (PMCC) 345 F; ref. index 1.444

Dow Corning® L1150 [Dow Corning]
Chem. Descrip.: PPG-18 butyl ether
CAS 9003-13-8
Uses: Solvent providing low odor, low irritation, low toxicity to personal care prods., cosmetics
Properties: Liq.; m.w. 1150; sp.gr. 0.9888; dens. 8.28 lb/gal; visc. 115 cSt; pour pt. -40 C; flash pt. (PMCC) > 400 F; ref. index 1.446

Dow Corning® MPEG350 [Dow Corning]
Chem. Descrip.: PEG-6 methyl ether
CAS 9004-74-4
Uses: Solvent providing low odor, low irritation, low toxicity to personal care prods., cosmetics
Properties: Liq.; m.w. 350; sp.gr. 1.097; dens. 9.14 lb/gal; visc. 27 cSt; f.p. 0 C; flash pt. (PMCC) > 350 F; ref. index 1.455

Dow Corning® MPEG550 [Dow Corning]
Chem. Descrip.: PEG-10 methyl ether
CAS 9004-74-4
Uses: Solvent providing low odor, low irritation, low toxicity to personal care prods., cosmetics
Properties: Liq.; m.w. 550; sp.gr. 1.101; dens. 9.17 lb/gal; visc. 56 cSt; f.p. 20 C; flash pt. (PMCC) > 450 F; ref. index 1.461

Dow Corning® MPEG750 [Dow Corning]
Chem. Descrip.: PEG-16 methyl ether
CAS 9004-74-4
Uses: Solvent providing low odor, low irritation, low toxicity to personal care prods., cosmetics
Properties: Solid; m.w. 750; sp.gr. 1.102 (50 C); dens. 9.04 lb/gal (50 C); visc. 53 cSt (100 F); f.p. 30 C; flash pt. (PMCC) > 450 F; ref. index 1.463

Dow Glycerine 96% USP [Dow]
Chem. Descrip.: Glycerin FCC, USP
CAS 56-81-5; EINECS/ELINCS 200-289-5
Uses: Humectant, solvent in foods, personal care prods., pharmaceuticals
Regulatory: FDA GRAS; E422 compliant; avail. in kosher grade
Properties: Colorless visc. liq.; odorless; sweet taste; misc. with ethanol,

IPA, water; immiscible with benzene, chloroform, ethyl ether, heptane; sp.gr. 1.25170; dens. 1.24770g/cm^3; visc. 624 cps (20 C); f.p. 8 C; b.p. 175 C; flash pt. (PMCC) > 198.8 C; ref. index 1.46752 (20 C); 96% act., 4% water

Dow Glycerine 99.5% USP/EP [Dow Europe]
Chem. Descrip.: Glycerin FCC, USP, EP
CAS 56-81-5; EINECS/ELINCS 200-289-5
Uses: Humectant, solvent in foods, personal care prods., pharmaceuticals
Regulatory: FDA GRAS; E422 compliant; avail. in kosher grade
Properties: Colorless visc. liq.; odorless; sweet taste; misc. with ethanol, IPA, water; immiscible with benzene, chloroform, ethyl ether, heptane; m.w. 92.09; sp.gr. 1.26201; dens. 1.25802 g/cm^3; visc. 1410 cps (20 C); vapor pressure 0.0025 mm Hg (50 C); f.p. 17 C; b.p. 290 C; flash pt. (PMCC) 195.5 C; ref. index 1.47399 (20 C); surf. tens. 63 dynes/cm (20 C); 99.5% act.

Dowper [Dow]
Chem. Descrip.: Perchloroethylene, specially inhibited
CAS 127-18-4; EINECS/ELINCS 204-825-9
Uses: Solvent for dry cleaning
Features: High purity; nonflamm.

DPG lo [Lyondell]
Chem. Descrip.: Dipropylene glycol
CAS 110-98-5; EINECS/ELINCS 203-821-4
Uses: Solvent for fragrances, deodorant
Regulatory: SARA §313 nonreportable
Properties: Colorless liq.; low odor; 100% sol. in water; m.w. 134.2; sp.gr. 1.020-1.025 (20/20 C); dens. 8.53 lb/gal (20 C); visc. 107 cps; vapor pressure < 0.1 mm Hg (20 C); i.b.p. 228 C min.; HLB 9.3; flash pt. (Seta) 264 F; ref. index 1.441; surf. tens. 35 dynes/cm; sp. heat 0.582 Cal/g/C; 99.5% assay
HMIS: Health 0, Flammability 1, Reactivity 0

Drakeol® 7 [Penreco]
Chem. Descrip.: Lt. mineral oil NF
CAS 8042-47-5; EINECS/ELINCS 232-384-2
Uses: Carrier, base in ointments, lotions, baby oils, sun tan lotions, makeup, topical pharmaceutical ointments; solvent, emollient in creams, waterless hand cleaners; protective coating for foods, eggs, fruits, and vegetables; divider oil, mold release lubricant in foods, food pkg., meat packing; pigment dispersant, lubricant for plastics; lubricant for gelatin capsules
Regulatory: FDA 21CFR §172.878, 178.3620, 573.680; DOT nonregulated; OSHA nonhazardous; SARA §311/312/313 nonreportable; Australia, Canada DSL, China, EC, Japan, Korea listed
Properties: Water-wh. transparent liq., odorless; sol. in hydrocarbons; negligible sol. in water; sp.gr. 0.828-0.843; dens. 6.94-7.08 lb/gal; visc. 10.8-13.6 cSt (40 C); pour pt. -9 C; flash pt. 177 C; ref. index 1.4632
Toxicology: STEL 10 mg/m^3 (as oil mist); relatively nontoxic by ingestion unless aspiration occurs (which may cause pneumonitis and may be fatal); laxative props. may result in abdominal cramps, diarrhea; minimally irritating to eyes on direct contactTSCA listed
Precaution: May react with strong oxidizing agents; spillages may be slippery
Hazardous decomp. prods.: Combustion may produce dense smoke, CO, CO_2, other oxides
HMIS: Health 0, Flammability 1, Reactivity 0
Storage: Store in closed containers away from heat, sparks, open flame, oxidizers

Drakesol® 165 [Penreco]
Chem. Descrip.: Petroleum distillates, hydrotreated light
CAS 64742-47-8; EINECS/ELINCS 265-149-8
Uses: Solvent for carbonless copy paper for office forms, industrial cleaning solvs., waterless hand cleaners, fragrance bases, automotive and furniture polishes, insecticide bases, lamp oils; defoamer in paper and coatings industries; coating on egg shells; defoamer in foods; froth flotation cleaning of vegetables; electrostatic discharge machine fluids; magnetic particle inspection fluids; food-contact applics.
Regulatory: FDA 21CFR §172.884, 178.3650, 573.740
Properties: Sp.gr. 0.786 (60/60 F); dens. 6.56 lb/gal (60 F); visc. 1.68 cSt (100 F); i.b.p. 191 C min.; pour pt. -40 C; flash pt. 74 C; KB value 29; < 1% aromatics
Precaution: DOT: combustible liq.

Drakesol® 165AT [Penreco]
Chem. Descrip.: Petroleum distillates
CAS 64742-14-9; EINECS/ELINCS 232-298-5
Uses: Solvent for carbonless copy paper for office forms, industrial cleaning solvs., waterless hand cleaners, fragrance bases, automotive and furniture polishes, insecticide bases, lamp oils, food processing; defoamer in paper and coatings industries; froth flotation cleaning of vegetables; electrostatic discharge machine fluids; magnetic particle inspection fluids; food-contact applics.
Features: High purity hydrocarbon
Regulatory: FDA 21CFR §172.884, 178.3650, 573.740
Properties: Sp.gr. 0.786 (60/60 F); dens. 6.56 lb/gal (60 F); visc. 1.66 cSt (100 F); i.b.p. 191 C min.; pour pt. -40 C; flash pt. 74 C; KB value 28; < 1% aromatics
Precaution: DOT: combustible liq.

Drakesol® 205 [Penreco]
Chem. Descrip.: Petroleum distillates, hydrotreated middle
CAS 64742-46-7; EINECS/ELINCS 265-148-2
Uses: Solvent for carbonless copy paper for office forms, industrial cleaning solvs., waterless hand cleaners, fragrance bases, automotive and furniture polishes, insecticide bases, lamp oils, food processing; defoamer in paper and coatings industries; froth flotation cleaning of vegetables; electrostatic discharge machine fluids, magnetic particle inspection fluids; food-contact applics.
Features: High purity
Regulatory: FDA 21CFR §172.884, 178.3650, 573.740
Properties: Sp.gr. 0.795 (60/60 F); dens. 6.64 lb/gal (60 F); visc. 2.18 cSt (100 F); vapor pressure < 0.1 mm Hg; i.b.p. 221 C min.; pour pt. -26 C; flash pt. 99 C; KB value 26.5; < 1% aromatics

Drakesol® 220 [Penreco]
Chem. Descrip.: Mineral oil
CAS 8042-47-5; EINECS/ELINCS 232-384-2
Uses: Solvent for carbonless copy paper for office forms, industrial cleaning solvs., waterless hand cleaners, fragrance bases, automotive and furniture polishes, insecticide bases, lamp oils; defoamer in paper and coatings industries; coating on egg shells; defoamer in foods; froth flotation cleaning of vegetables; electrostatic discharge machine fluids; magnetic particle inspection fluids; food-contact applics.
Regulatory: FDA 21CFR §172.884, 178.3650, 573.740
Properties: Sp.gr. 0.804 (60/60 F); dens. 6.69 lb/gal (60 F); visc. 2.28 cSt (100 F); vapor pressure < 0.1 mm Hg; i.b.p. 235 C min.; pour pt. -18 C; flash pt. 102 C; KB value 24.5; < 1% aromatics

Drakesol® 260 [Penreco]
Chem. Descrip.: Mineral oil
CAS 8042-47-5; EINECS/ELINCS 232-384-2
Uses: Solvent for carbonless copy paper for office forms, industrial cleaning solvs., waterless hand cleaners, fragrance bases, automotive and furniture polishes, insecticide bases, lamp oils; defoamer in paper and coatings industries; coating on egg shells; defoamer in foods; froth flotation cleaning of vegetables, electrostatic discharge machine fluids, magnetic particle inspection fluids; food-contact applics.
Regulatory: FDA 21CFR §172.884, 178.3620(b), 178.3650, 573.740
Properties: Sp.gr. 0.812 (60/60 F); dens. 6.76 lb/gal (60 F); visc. 4.3 cSt (100 F); vapor pressure < 0.1 mm Hg; i.b.p. 266 C min.; pour pt. -4 C; flash pt. 127 C; KB value 22.6; < 1% aromatics

Drakesol® 260 AT [Penreco]
Chem. Descrip.: White mineral oil, tech.
CAS 8042-47-5; EINECS/ELINCS 232-384-2
Uses: Solvent for carbonless copy paper for office forms, industrial cleaning solvs., waterless hand cleaners, fragrance bases, automotive and furniture polishes, insecticide bases, lamp oils; electrostatic discharge machine fluids; magnetic particle inspection fluids; defoamer in paper and coatings industries; froth flotation cleaning of vegetables; food-contact applics.
Regulatory: FDA 21CFR §172.884, 178.3620(b), 178.3650, 573.740
Properties: M.w. 230; sp.gr. 0.8146 (15.6 C); dens. 6.78 lb/gal (15.6 C); visc. 3.90 cSt (100 F); vapor pressure 0.01 mm Hg; i.b.p. 268 C; pour pt. -6.67 C; cloud pt. -2.8 C; flash pt. (CC) 127 C; autoignition temp. 220 C; ref. index 1.4498 (20 C); KB value 22; sp. heat 0.492; < 1% aromatics

Drakesol® 305 [Penreco]
Chem. Descrip.: White mineral oil, tech.
CAS 8042-47-5; EINECS/ELINCS 232-384-2

Uses: Solvent for carbonless copy paper for office forms, industrial cleaning solvs., waterless hand cleaners, fragrance bases, automotive and furniture polishes, insecticide bases, lamp oils; electrostatic discharge machine fluids; magnetic particle inspection fluids; defoamer in paper and coatings industries; froth flotation cleaning of vegetables; food-contact applics.
Regulatory: FDA 21CFR §172.884, 178.3620(b), 178.3650, 573.740; DOT nonregulated; SARA §311/312/313 nonreportable; Canada DSL, EINECS, Australia AICS, China, Japan ENCS, Korea KCL, Philippines PICCS listed
Properties: Clear liq.; sl. hydrocarbon odor; sol. in hydrocarbons; insol. in water; m.w. 266; sp.gr. 0.8299 (15.6 C); dens. 6.91 lb/gal (15.6 C); visc. 6.27 cSt (100 F); vapor pressure < 0.01 mm Hg; i.b.p. 304 C; pour pt. 4.4 C; cloud pt. 10 C; flash pt. (PMCC) 151.7 C; autoignition temp. 220 C; ref. index 1.4570 (20 C); KB value 20; sp. heat 0.493; 100% act.
Toxicology: ACGIH TLV/TWA 5 mg/m^3, STEL 10 mg/m^3 (oil mist); minimal eye/minimal to sl. skin irritant; not expected to be inh. hazard @ ambient conditions; aspiration may cause chem. pneumonitis, pulmonary edema, hemorrhage, death; ing. may cause GI discomfort, vomiting, diarrhea; laxative props.; TSCA listed
Environmental: Do not allow spills to enter sewers or watercourses
Precaution: May react with strong oxidizing agents; spillages may be slippery
Hazardous decomp. prods.: Combustion: dense smoke, CO, CO_2, other oxides
HMIS: Health 0, Flammability 1, Reactivity 0
Storage: Store in closed containers away from heat, sparks, open flame, oxidizing materials

Dymel® A [DuPont; DuPont Canada]
Chem. Descrip.: Dimethyl ether
CAS 115-10-6; EINECS/ELINCS 204-065-8
Uses: Aerosol propellant; solvent
Features: Low flamm.; chemically stable; hydrolytically stable over wide pH range in aq. sol'ns.
Properties: M.w. 46.1; high sol. in water; misc. with most conventional solvs.; sp.gr. 0.66; vapor pressure 63 psig (70 F); b.p. -12.7 F; KB value 60
Toxicology: REL 1000 ppm; low acute and chronic toxicity; weak anesthetic at high inhaled levels; may cause weak cardiac sensitization
Environmental: Environmentally safe
Precaution: Flamm. limits in air 3.4-18.0 vol.%

Dynacerin® 660 [Sasol Germany; Sasol N. Am.]
Chem. Descrip.: Oleyl erucate
CAS 17673-56-2; EINECS/ELINCS 241-654-9
Uses: Emollient, lubricant, dispersant, solubilizer for cosmetics, pharmaceuticals; carrier, vehicle, solvent for topical ointments, creams, lotions
Features: Oxidation-stable; environmentally compat.; jojoba oil substitute
Properties: Ylsh. liq. wax; sol. in alcohol, min. oil, acetone; visc. 15-18 mPa·s (40 C); acid no. 1 max.; iodine no. 95 max.; sapon. no. 95-110; cloud pt. 15 C

Dynaflush [Dynaloy]
Uses: Solvent for flushing and cleaning residues from urethane processing equip. (mixing and metering equip., nozzles, spray guns, RIM molds)
Features: Nonflamm., noncorrosive, non-ozone depleting; usable at R.T.; can be used hot (95 C) for faster results
Properties: Clear liq.; sp.gr. 1.06; b.p. 202 C; flash pt. (CC) 191 F; pH 4-6
Toxicology: Noncarcinogenic

Dynasolve 100 [Dynaloy]
Chem. Descrip.: Dimethylformamide-based solvent
Uses: Solvent for removal of adhesives, polymers, and fiber optics coatings, for electronic depotting, decapsulation on cured acrylics, epoxies, and urethanes
Features: Nonselective

Dynasolve 165 [Dynaloy]
Chem. Descrip.: Methylene chloride (80-90%), acetic acid (1-10%), formic acid (5-15%), dodecylbenzene sulfonic acid (0.5-5%)
UN 1760
Uses: Decapsulation solvent for dissolving cast epoxy resins, for general adhesive and polymer removal, paint stripping, conformal coating removal, depotting, deflashing
Features: For R.T. use
Regulatory: SARA §313 reportable (methylene chloride)
Properties: Clear liq.; strong acetic acid-chlorinated solv. odor; sol. 2 g/100 g in water; sp.gr. 1.288; vapor pressure 355 mm Hg (20 C); b.p. 104 F; VOC 211 g/l; 100 vol.% volatile
Toxicology: May cause skin/eye/mouth burns, respiratory tract irritation, dizziness, headache, nausea, drowsiness, narcosis, unconsciousness, arrhythmia, lung damage, wt. loss, blindness; chronic: dermatitis, digestive and CNS effects, kidney/liver damage
Precaution: Corrosive liq.; forms flamm. vapor-air mixts.; conc. vapors can be ignited by high intensity heat source; flamm. limits in air 14-22% by vol.; incompat. with acids, alkalis, oxidizers, amines, calcium hypochlorite, hydrogen peroxide, aluminum, etc.
Hazardous decomp. prods.: Hydrogen chloride, CO, CO_2, phosgene, formic acid vapors
HMIS: Health 3, Flammability 1, Reactivity 0
Storage: Store in cool, dry, ventilated area; vapors will collect in low areas; keep container sealed

Dynasolve 180 [Dynaloy]
Chem. Descrip.: N-Methyl pyrrolidone base
Uses: Solvent for removal of adhesives and inks from rollers, general polymer removal, paint stripping, cured epoxy removal, electronic depotting, deflashing, decapsulation, and deinking applics., surf. prep. and photoresist stripping
Features: Use hot (105-107 C); not selective

Dynasolve 185 [Dynaloy]
Chem. Descrip.: N-Methyl pyrrolidone (78-98%) and 2-phenoxy ethanol (5-15%)
UN NA 1993
Uses: Solvent for removal of adhesives and inks from rollers, general polymer removal, paint stripping, removal of fiber optic coatings, cured epoxy removal, electronic depotting, deflashing, decapsulation, and deinking applics., surf. prep. and photoresist stripping
Features: Use hot (105 C); not selective; modified version of Dynasolve 180
Regulatory: SARA §313 nonreportable
Properties: Ylsh. transparent liq.; mild amine-like odor; completely sol. in water; sp.gr. 1.038; vapor pressure 0.29 mm Hg (20 C); b.p. 202 C; flash pt. (CC) 199 F; pH 9.3; 100% volatile
Toxicology: May cause irritation, inflamm., or burns to eyes, irritation to mouth, nose, throat, mucous membranes, coughing, difficulty breathing, skin irritation, nausea, vomiting, convulsions; high concs. may cause mental confusion; dermatitis (chronic)
Precaution: Combustible liq.; flamm. limits in air 1.3-9.5%; mod. fire hazard exposed to heat or flame; incompat. with strong oxidizing or reducing agents
Hazardous decomp. prods.: Heated to decomp., emits CO and NO_x fumes
HMIS: Health 2, Flammability 2, Reactivity 0
Storage: Keep container closed

Dynasolve 190 [Dynaloy]
Chem. Descrip.: Methanol base
Uses: Solvent for electronic depotting, conformal coating removal, cable assembly repair, elec. connector repair, cured urethane and silicone removal
Features: For R.T. use only; stronger than Uresolve Blue but less selective
Precaution: Flamm.

Dynasolve 200 [Dynaloy]
Chem. Descrip.: Methylene chloride-based
Uses: Solvent for polysulfide removal from aircraft fuel tanks, in adhesives, polymers, inks
Features: Selective; effective at R.T.

Dynasolve 210 [Dynaloy]
Chem. Descrip.: Methylene chloride (75-95%), dodecylbenzenesulfonic acid (5-25%)
UN 1760
Uses: Solvent which dissolves RTV silicones and silicone conformal coatings effectively at R.T.; solvent for depotting, decapsulation, removal of RTVs from CRT tubes, removal of buffer coatings from optical fibers, paint stripping
Features: May attack epoxies and urethanes
Regulatory: SARA §313 reportable (methylene chloride)
Properties: Amber liq.; chlorinated solv. odor; sol. 2.0 g/100g water; sp.gr.

1.272; pH 2.14; 100% volatile
Toxicology: Acute overexposure: burns to skin or eyes, respiratory tract irritation, dizziness, headache, vertigo, nausea, drowsiness, narcosis, unconsciousness, arrhythmia, carboxyhemoglobinemia; chronic overexposure: skin irritation, CNS effects
Precaution: Corrosive liq.; incompat. with water, alkalis, oxidizers, amines, aluminum
Hazardous Ingredients: Methylene chloride (75-95%), dodecylbenzenesulfonic acid (5-25%)
Hazardous decomp. prods.: Hydrogen chloride, phosgene, chlorine
HMIS: Health 2, Flammability 1, Reactivity 0
Storage: Store in cool, dry, ventilated area; vapors will collect in low areas; keep container sealed

Dynasolve 220 [Dynaloy]
Chem. Descrip.: Hydrotreated heavy naphtha (60-80%), dodecylbenzene sulfonic acid (20-40%)
UN 1760
Uses: Solvent which dissolves RTV silicones and silicone conformal coatings effectively at R.T.; solvent for removal of mold releases, parting agents, adhesives, polymers; solvent for electronic depotting, decapsulation, removal of RTVs from CRT tubes
Features: May attack epoxies and urethanes
Regulatory: SARA §313 nonreportable
Properties: Amber liq.; faint petroleum hydrocarbon odor; sol. < 0.1% in water; sp.gr. 0.828; flash pt. 110 F; pH 1.70; 100% volatile
Toxicology: Acute overexposure: burns to skin or eyes, respiratory tract irritation, headaches, dizziness, anesthesia, CNS effects; chronic overexposure: skin irritation, dermatitis
Precaution: Corrosive liq.; incompat. with water, alkalis, oxidizers, liq. chlorine, conc. O_2, sodium hypochlorite, calcium hypochlorite
Hazardous Ingredients: Hydrotreated heavy naphtha, petroleum (60-80%, CAS 64742-48-9), dodecylbenzene sulfonic acid (20-40%)
Hazardous decomp. prods.: Fumes, smoke, CO, aldehydes
HMIS: Health 2, Flammability 2, Reactivity 0
Storage: Store in cool, dry, ventilated area; keep containers closed; use proper grounding for all containers

Dynasolve 225 [Dynaloy]
Chem. Descrip.: Hydrotreated heavy naphtha (60-80%), dodecylbenzene sulfonic acid (20-40%)
UN 2924
Uses: Solvent which dissolves RTV silicones and silicone conformal coatings effectively at R.T.; solvent for removal of mold releases, parting agents, adhesives, polymers; solvent for electronic depotting, decapsulation, removal of RTVs from CRT tubes
Features: May attack epoxies and urethanes
Regulatory: SARA §313 nonreportable
Properties: Amber liq.; faint petroleum hydrocarbon odor; sol. < 0.1% in water; sp.gr. 0.806; flash pt. 52 F; pH 1.60; 100% volatile
Toxicology: Acute overexposure: burns to skin or eyes, respiratory tract irritation, headaches, dizziness, anesthesia, CNS effects; chronic overexposure: skin irritation, dermatitis
Precaution: Corrosive, flamm. liq.; incompat. with water, alkalis, oxidizers, liq. chlorine, conc. O_2, sodium hypochlorite, calcium hypochlorite
Hazardous decomp. prods.: Fumes, smoke, CO, aldehydes
HMIS: Health 2, Flammability 3, Reactivity 0
Storage: Store in cool, dry, ventilated area; keep containers closed; use proper grounding for all containers

Dynasolve 230 [Dynaloy]
Chem. Descrip.: Hydrotreated heavy naphtha (60-80%), dodecylbenzene sulfonic acid (20-40%)
UN 2924
Uses: Solvent which dissolves RTV silicones and silicone conformal coatings effectively at R.T.; solvent for removal of mold releases, parting agents, adhesives, polymers; solvent for electronic depotting, decapsulation, removal of RTVs from CRT tubes
Features: May attack epoxies and urethanes
Regulatory: SARA §313 nonreportable
Properties: Amber liq.; faint petroleum hydrocarbon odor; sol. < 0.1% in water; sp.gr. 0.707; flash pt. 16 F; pH 1.30; 100% volatile
Toxicology: Acute overexposure: burns to skin or eyes, respiratory tract irritation, headaches, dizziness, anesthesia, CNS effects; chronic overexposure: skin irritation, dermatitis
Precaution: Corrosive, flamm. liq.; incompat. with water, alkalis, oxidizers, liq. chlorine, conc. O_2, sodium hypochlorite, calcium hypochlorite; becomes more corrosive when mixed with water
Hazardous Ingredients: Hydrotreated heavy naphtha, petroleum (60-80%, CAS 74742-48-9), dodecylbenzenesulfonic acid (20-40%)
Hazardous decomp. prods.: Fumes, smoke, CO, aldehydes
HMIS: Health 2, Flammability 3, Reactivity 0
Storage: Store in cool, dry, ventilated area; keep containers closed

Dynasolve 699 [Dynaloy]
Chem. Descrip.: N-Methyl pyrrolidone (58-78%), γ-butyrolactone (20-40%)
UN NA 1993
Uses: Solvent for use at R.T. for thin acrylic coatings, and hot (120-150 C) for acrylic adhesives and epoxy powd. coatings; solvent for removal of conformal coatings, fiber optics coatings, adhesives, and polymers, and for paint stripping, electronic depotting applics.
Regulatory: SARA §313 nonreportable
Properties: Water-like liq.; mild pine oil odor; misc. with water; sp.gr. 1.054; flash pt. (CC) 199 F
Toxicology: Acute overexposure: irritation, inflamm., or burns to eyes; conjunctivitis, opaqueness of cornea, possibly vision loss; headache, mental confusion, nausea, narcosis in high concs.; chronic overexposure: severe dermatitis
Precaution: Combustible liq.; mod. fire hazard exposed to heat or flame; incompat. with strong oxidizing or reducing agents
Hazardous decomp. prods.: Heated to decomp., emits CO and NO_x fumes
HMIS: Health 3, Flammability 2, Reactivity 0
Storage: Keep container closed

Dynasolve 711 [Dynaloy]
Chem. Descrip.: Propylene glycol ether base
Uses: Solvent for electronic depotting, decapsulation, deinking, surf. prep. and photo resist stripping applics., suitable for cured epoxies, urethanes, and silicones
Features: Similar to Uresolve Plus SG; can be used in contact with aluminum

Dynasolve 750 [Dynaloy]
Chem. Descrip.: Propylene glycol ether base
Uses: Solvent for electronic depotting, decapsulation, deinking, and surf. prep. applics., suitable for cured epoxies, urethanes, and silicones
Features: Stronger than Dynasolve 710, but less selective; similar in str. to Uresolve Plus 500; slow at R.T., more effective at 65-110 C

Dynasolve CU-5 [Dynaloy]
Chem. Descrip.: N-Methyl-2-pyrrolidone (55-70%), butyrolactone (25-40%), ethyl 3-ethoxypropionate (5-20%)
Uses: Solvent for cleaning urethane residues and crystallized isocyanates from various types of polyurethane processing equip.
Features: Nonchlorinated, non-ozone depleting; replaces methylene chloride, acetone, MEK, 1,1,1-trichloroethane; can be used for extended periods, tolerates high resin loading; recyclable via vacuum distillation
Regulatory: DOT nonregulated; SARA §313 nonreportable
Properties: Clear liq.; sweet odor; completely sol. in water; sp.gr. 1.060; vapor pressure < 1 mm Hg (20 C); b.p. 202 C; flash pt. (CC) 191 F; pH 4-6; nonflamm.; 0 vol.% volatile
Toxicology: Harmful if inhaled or swallowed; acute overexposure: skin irritation or dermatitis, eye irritation or inflamm., pallor, nausea, incoordination, stupor; chronic overexposure: dermatitis; noncarcinogenic
Precaution: Incompat. with strong oxidizing or reducing agents
Hazardous decomp. prods.: CO, CO_2, NO_x, smoke
HMIS: Health 3, Flammability 2, Reactivity 0
Storage: Keep container closed; keep away from heat and flame

Dynasolve CU-6 [Dynaloy]
Chem. Descrip.: N-Methyl-2-pyrrolidone (45-60%), butyrolactone (15-30%), PPG-3 methyl ether (5-20%), ethyl 3-ethoxypropionate (3-20%)
Uses: Solvent for cleaning urethane residues and crystallized isocyanates from various types of polyurethane processing equip.
Features: Nongelling; nonchlorinated, non-ozone depleting; replaces methylene chloride, acetone, MEK, 1,1,1-trichlororethane; can be used for extended periods, tolerates high resin loading; recyclable via vacuum distillation
Regulatory: DOT nonregulated; SARA §313 nonreportable
Properties: Clear liq.; sweet odor; completely sol. in water; sp.gr. 1.055; vapor pressure < 1 mm Hg (20 C); b.p. 202 C; flash pt. (CC) 210 F; pH 4-6; nonflamm.

Toxicology: Harmful if inhaled or swallowed; may cause skin irritation or dermatitis, eye irritation or inflamm., pallor, nausea, incoordination, stupor; noncarcinogenic
Precaution: Incompat. with strong oxidizing or reducing agents
Hazardous decomp. prods.: CO, CO_2, NO_x, smoke
HMIS: Health 3, Flammability 1, Reactivity 0
Storage: Keep container closed; keep away from heat and flame

Dynasolve M-10 [Dynaloy]
Chem. Descrip.: Hydrocarbon base
Uses: Solvent for silicone oils and parting agents at R.T., and for general adhesive and polymer removal
Features: For quicker removal, heat to 65 C

Dynasolve M-35 [Dynaloy]
Uses: Solvent for general cleaning applics., esp. for cleaning of uncured polymers such as epoxies, urethanes, and silicones, and for dissolving cured cyanoacrylate instant glues
Features: Low-toxicity, low volatility

Dynasolve MP-500 [Dynaloy]
Chem. Descrip.: Ethylene glycol monomethyl ether base
Uses: Solvent for cleaning residues from mixing, metering equip., nozzles, RIM molds, for removal of mold releases, adhesives and inks from rollers, for general polymer removal, suitable for cured epoxies, silicones, and urethanes
Features: Stronger than Dynasolve MP-AL GR, but less selective; slow at R.T.; more effective at 65-110 C

Dynasolve MP-AL [Dynaloy]
Chem. Descrip.: Ethylene glycol monomethyl ether base
Uses: Solvent for dissolving cured urethane (polyester, modified polyester urethane, polyether urethane), varnishes, RTV rubbers
Features: Aluminum grade
Properties: Blue-green liq.; sp.gr. 0.999; b.p. 256 F; flash pt. (CC) 103 F

Dynasolve XD 16-3 [Dynaloy]
Chem. Descrip.: Aq. sol'n.
Uses: Solvent for removal of silicone oils and uncured silicone polymers from molds, molded parts, and processing equip., esp. effective for cleaning mold release residue from plastic molded parts
Properties: Lt. amber liq.; sp.gr. 1.010; b.p. 212 F; flash pt. none; pH 6-8
Toxicology: Avoid contact with eyes and skin, breathing of vapors

Dynasolve XD 22-1 [Dynaloy]
Chem. Descrip.: Proprietary
UN NA 1993
Uses: Solvent for cleaning and degreasing applics.
Features: Nonchlorinated, non-ozone depleting; low vapor pressure; minimizes VOC emissions; noncorrosive to metals; leaves no residue; replaces methylene chloride, acetone, MEK, min. spirits, kerosene, 1,1,1-trichloroethane; recyclable via vacuum distillation
Regulatory: SARA §313 nonreportable
Properties: Clear liq.; no odor; $< 0.1\%$ sol. in water; sp.gr. 0.780; vapor pressure 3 mm Hg (100 F); b.p. 360-419 F; flash pt. (CC) 143 F; VOC 780 g/l; nonflamm.; 100 vol.% volatile
Toxicology: Harmful if inhaled or swallowed; acute overexposure: irritation to eyes and respiratory tract, headaches, dizziness, anesthesia, drowsiness, unconsciousness, CNS effects, skin irritation, dermatitis; noncarcinogenicTSCA listed
Precaution: Combustible liq.; incompat. with strong oxidizers
Hazardous decomp. prods.: None
HMIS: Health 1, Flammability 2, Reactivity 0
Storage: Keep containers closed; keep away from heat and flame

Dytek® EP Diamine [DuPont Nylon; DuPont Canada]
Chem. Descrip.: 1,3-Pentanediamine
CAS 589-37-7
UN 2734
Uses: Intermediate; epoxy curative; solvent for gas treatment; surfactant (asphalt emulsifiers, textiles); corrosion/scale inhibitor; additive for fuel, oil, plastic, petrol. and chem. processing; extender, catalyst for PU; polyamide monomer for coatings, adhesives, sealants, elastomers
Regulatory: CERCLA reportable
Properties: Colorless liq., sl. odor; misc. with water; m.w. 102; dens. 0.855 g/ml; visc. 1.89 cp; vapor pressure 1 mm Hg (20 C); m.p. -121 C; b.p. 164 C; flash pt. (CC) 59 C; ref. index 1.4512 (20/20 C); pH 12.3 (10%); surf. tens. 32.2 dynes/cm; 99% act.
Toxicology: Causes eye burns; damage may be permanent and may incl. blindness; causes skin sensitization and burns; may cause nose, throat, and lung irritationTSCA listed
Environmental: Biodeg.
Precaution: Corrosive, flamm.; incompat. with strong oxidants
Hazardous decomp. prods.: Thermal decomp. produces toxic fumes of NO_x
HMIS: Health 3, Flammability 2, Reactivity 1
Storage: Store in well ventilated area in tightly closed containers; keep away from heat, spark, flames

E

Eastman® C-7 Ketone [Eastman]
Chem. Descrip.: Ketone mixt.
Uses: Solvent
Properties: Pt-Co 25 max. color; sp.gr. 0.817; dens. 6.82 lb/gal; f.p. -57 C; b.p. 140-146 C; flash pt. (TCC) 33 C

Eastman® C-11 Ketone [Eastman]
Chem. Descrip.: Ketone mixt.
CAS 71808-49-6
Uses: Retarder, solvent for coatings, lacquers, screen inks
Features: Very slow-evaporating
Properties: Sol. 0.2% in water (20 C); sp.gr. 0.84 (20/20 C); dens. 7.02 lb/gal (20 C); visc. 2.12 cP (neat); f.p. 11 F; b.p. 200-240 C; flash pt. (Seta) 84 C; ref. index 1.4355 (20 C); surf. tens. 27.5 dynes/cm (24 C)

Eastman® CDA-19 [Eastman]
Chem. Descrip.: Completely denatured alcohol: ethanol denatured with MIBK (4%), heptane (1%)
Uses: Solvent for protective coatings, printing inks, paint and varnish removers, antifreeze sol'n.; as fuel for light, heat, or power prod.
Properties: Pt-Co 10 max. color; sp.gr. 0.792-0.796 (anhyd.), 0.808-0.820 (95%); dens. 6.62 lb/gal (anhyd., 20 C), 6.78 lb/gal (95%, 20 C); b.p. 76-82 C; flash pt. (TCC) 7 C (anhyd.), 11 C (95%); avail. 95% act. and anhyd.

Eastman® DB [Eastman]
Chem. Descrip.: Diethylene glycol monobutyl ether
CAS 112-34-5; EINECS/ELINCS 203-961-6
Uses: Solvent, gloss aid in baking enamels to promote increased flow-out; coalescing aid for architectural and industrial latex paints; organosol dispersant; diluent in hydraulic brake fluids; solvent for fragrances, household and general industrial cleaners, detergents, soaps, screen printing inks, textile dyeing, cutting oils, waterborne coatings, cosmetics, fragrances; deactivator in pesticides
Regulatory: SARA and CERCLA reportable
Properties: Colorless clear liq., sl. odor; completely sol. in water; m.w. 162.23; sp.gr. 0.955 (20/20 C); dens. 7.94 lb/gal; visc. 4.74 cP; vapor pressure 0.02 mm Hg (20 C); f.p. -76 C; i.b.p. 230 C; flash pt. (COC) 111 C; fire pt. 117 C; ref. index 1.4316 (20 C); 99.2% min. purity
Toxicology: LD50 (oral, rat) 7292 mg/kg, (dermal, rabbit) 2764 mg/kg; causes eye irritation; expected to be a low ingestion hazardTSCA listed
Environmental: Low biochem. oxygen demand; little potential to cause oxygen depletion in aq. systems; low potential to affect aquatic organisms; BOD-5 0.25 g oxygen/g; COD 2.08 g oxygen/g
Precaution: Can react with oxidizing materials; peroxide former; flamm. limits in air 0.85 vol.% (135 C) to 24.6 vol.% (199 C)
Hazardous decomp. prods.: Hazardous combustion prods.: CO, CO_2
HMIS: Health 2, Flammability 1, Reactivity 0
Storage: Store in closed container away from heat and light; after opening, purge container with nitrogen before reclosing; minimize exposure to air; do not allow to evaporate to near dryness

Eastman® DB Acetate [Eastman]
Chem. Descrip.: Diethylene glycol monobutyl ether acetate
CAS 124-17-4; EINECS/ELINCS 204-685-9
Uses: Solvent for cosmetics, printing inks, high-bake enamels, esp. screen inks and coil coatings; selective solvent in separation of alcohols and ketone; coalescing aid in low-pH latex paints
Features: Very slow evaporation rate; limited water sol.
Regulatory: SARA and CERCLA reportable
Properties: Colorless clear liq., pleasant odor; mod. sol. in water; m.w. 204.27; sp.gr. 0.98 (20/20 C); dens. 8.16 lb/gal; visc. 3.20 cP; vapor pressure 0.04 mm Hg (20 C); f.p. -32 C; i.b.p. 235 C; flash pt. (TCC) 105 C; fire pt. 128 C; ref. index 1.43 (20 C)
Toxicology: LD50 (oral, rat) 7.1-12.2 ml/kg, (dermal, rabbit) 5.5-14.48 ml/kg; prolonged/repeated skin contact may cause drying, cracking, or irritation; expected to be low ingestion hazardTSCA listed
Precaution: Can react with strong oxidizing materials, strong bases, strong acids; peroxide former; flamm. limits in air 0.8 vol.% (135 C) to 10.7 vol.% (203 C)
Hazardous decomp. prods.: Hazardous combustion prods.: CO, CO_2
HMIS: Health 1, Flammability 1, Reactivity 0
Storage: Store in closed container away from heat and light; after opening, purge container with nitrogen before reclosing; minimize exposure to air; do not allow to evaporate to near dryness

Eastman® DBP [Eastman]
Chem. Descrip.: Dibutyl phthalate
CAS 84-74-2; EINECS/ELINCS 201-557-4
Uses: Plasticizer, solvent for nitrocellulose lacquers, architectural latex coatings, adhesives; plasticizer for rubbers, vinyl prods., CAB, ethyl cellulose, PVAc, and syn. resins; solvent for oil-sol. dyes, insecticides, peroxides, org. compds.; antifoam, fiber lubricant in textile mfg.; plasticizer in cosmetics, nail care; solvent, fixative for fragrances
Regulatory: SARA §311/312 chronic health hazard, §313 reportable; Canada DSL, EINECS, Australia AICS, Japan MITI, Korea ECL listed
Properties: Colorless liq., odorless; sol. 11.2 mg/l in water; m.w. 278.35; sp.gr. 1.047 (20/20 C); dens. 8.72 lb/gal (20 C); vapor pressure 0.0000142 mm Hg (20 C); f.p. -35 C; b.p. 340 C; flash pt. (COC) 191 C; 100% act.
Toxicology: ACGIH TLV/TWA 5 mg/m^3; OSHA PEL 5 mg/m^3; LD50 (oral, rat) 20-25 g/kg, (dermal, rabbit) > 20 ml/kg; low hazard by inh., skin and eye contact; expected to be a low ingestion hazard; high oral doses cause adverse reproductive effects in lab animals; TSCA listed
Environmental: Mod. BOD; may cause oxygen depletion in aq. systems; high potential to affect some aquatic organisms; low potential to affect germination/early growth of some plants; expected to biodegrade; COD 1.71 g O_2/g; BOD-5 0.34 g O_2/g
Precaution: Reacts with strong oxidizing agents; LEL 0.47 vol.% (236 C)
Hazardous decomp. prods.: Combustion: CO, CO_2
HMIS: Health 1, Flammability 1, Reactivity 0
Storage: Store in closed containers

Eastman® DB Stabilized [Eastman]
Chem. Descrip.: Diethylene glycol monobutyl ether
CAS 112-34-5; EINECS/ELINCS 203-961-6
Uses: Solvent for cosmetics, personal care prods.
Properties: Pt-Co color 10; m.w. 162.23; sol. in water (20 C); sp. gr. 0.955 (20 C); dens. 7.94 lb/gal; vapor pressure 0.02 mm Hg; f.p. -76 C; flash pt. (COC) 111 C; autoignition temp. 204 C; fire pt. 117 C; ref. index 1.4316 (20 C)

Eastman® DE [Eastman]
Chem. Descrip.: Diethylene glycol monoethyl ether
CAS 111-90-0; EINECS/ELINCS 203-919-7
Uses: Coupling agent, solvent for CAB, nitrocellulose, epoxy resins, polyamide resins, and other coating materials; coalescing aid for latex paints and some PVA emulsions; in non-grain-raising wood stains; stamp-pad

and printing inks; gloss aid in floor polishes based on modified acrylic polymers; component in heavy-duty brake fluids, textile dyeing applics.; deactivator in pesticides; diluent in brake fluids; solvent for fragrances, cosmetics, hair care, skin care
Features: Very polar, slow-evaporating
Regulatory: SARA and CERCLA reportable
Properties: Colorless clear liq., mild odor; completely sol. in water; m.w. 134.17; sp.gr. 0.990 (20/20 C); dens. 8.25 lb/gal; visc. 4.5 cP; vapor pressure 0.12 mm Hg (20 C); f.p. -90 C; i.b.p. 198 C; flash pt. (TCC) 91 C; fire pt. 96 C; ref. index 1.4260 (20 C); 100% act.
Toxicology: LD50 (oral, rat) 10,502 mg/kg, (dermal, rabbit) 9143 mg/kg; low hazardTSCA listed
Environmental: May cause oxygen depletion in aq. systems; low potential to affect aquatic organisms, sec. waste treatment microbial metabolism; high potential to biodegrade; BOD-5 0.14 g oxygen/ml; COD 1.9 g O_2/g
Precaution: Combustible liq. and vapor; incompat. with strong oxidizing materials; peroxide former; flamm. limits in air 1.20 vol.% (135 C) to 23.5 vol.% (182 C)
Hazardous decomp. prods.: Hazardous combustion prods.: CO, CO_2
HMIS: Health 1, Flammability 2, Reactivity 0
Storage: Store in closed container away from heat and light; minimize exposure to air; do not allow to evaporate to near dryness

Eastman® DE Acetate [Eastman]
Chem. Descrip.: Diethylene glycol monoethyl ether acetate
CAS 112-15-2; EINECS/ELINCS 203-940-1
Uses: Coalescing agent in latex paints; solvent in coatings, automotive paints, plastic coatings, latex paints, textile and silk-screen printing inks, slow-drying lacquers; solvent, dispersant, diluent in cosmetics, personal care
Features: High-boiling
Regulatory: DOT nonregulated; SARA §313 nonreportable; Canda DSL, EINECS, Australia AICS, Japan MITI, Korea ECL listed
Properties: Colorless clear liq., sl. odor; completely sol. in water; m.w. 176.21; sp.gr. 1.012 (20/20 C); dens. 8.42 lb/gal; visc. 4.5 cP; vapor pressure 0.04 mm Hg (20 C); f.p. -25 C; i.b.p. 214 C; flash pt. (COC) 107 C; fire pt. 110 C; ref. index 1.42 (20 C); 100% act.
Toxicology: LD50 (oral, rat) 11 g/kg, (dermal, rabbit) 15 g/kg; low hazard; very sl. eye/skin irritantTSCA listed
Environmental: Mod. BOD; may cause oxygen depletion in aq. systems; high potential to biodegrade; BOD-5 0.42 g oxygen/g; prevent runoff from entering drains, sewers, streams
Precaution: Incompat. with strong oxidizing materials; peroxide former; flamm. limits in air 0.98 vol.% (135 C) to 19.4 vol.% (185 C)
Hazardous decomp. prods.: Combustion prods.: CO, CO_2; forms peroxides of unknown stability
HMIS: Health 1, Flammability 1, Reactivity 1
Storage: Store in closed container away from heat and light; after opening, purge container with nitrogen before reclosing; minimize exposure to air; do not allow to evaporate to near dryness

Eastman® DE-HG [Eastman]
Chem. Descrip.: Diethylene glycol monoethyl ether (70%), ethylene glycol (30%)
Uses: Solvent for coatings, plastic coatings, automotive paints, stripper formulations; coalescing agent
Regulatory: SARA §311/312 acute health hazard, §313 reportable (diethylene glycol monoethyl ether, ethylene glycol); Canada DSL, EINECS, Australia AICS, Japan MITI listed
Properties: Colorless liq.; mild odor; completely sol. in water; sp.gr. 1.027 (20/20 C); dens. 8.56 lb/gal; vapor pressure 0.05 mm Hg (20 C); f.p. -70 C; b.p. 190-205 C; flash pt. (TCC) 96 C; fire pt. 104 C; autoignition temp. 311 C; ref. index 1.4281 (20 C)
Toxicology: Harmful or fatal if swallowed; can cause kidney damage; at elevated temps., vapor may cause eye/respiratory tract irritation; may cause skin sensitization; may cause adverse reproductive effects, e.g., birth defects, miscarriages, infertility; TSCA listed
Environmental: Prevent runoff from entering drains, sewers, streams
Precaution: Potential peroxide former; LEL 56.3 vol.% @ 121 C; can react with strong oxidizing agents
Hazardous decomp. prods.: Combustion: CO, CO_2; may form peroxides of unknown stability
HMIS: Health 1, Flammability 1, Reactivity 0
Storage: Store in tightly closed containers away from heat, light, moisture

Eastman® DEP [Eastman]
Chem. Descrip.: Diethyl phthalate
CAS 84-66-2; EINECS/ELINCS 201-550-6
Uses: Solvent; chem. raw material and flavor extractant for flavors and fragrances; plasticizer; fragrance fixing agent; nail care
Regulatory: SARA §311/312 acute health hazard, §313 nonreportable; Canada DSL, EINECS, Australia AICS, Japan MITI, Korea ECL listed
Properties: Colorless clear liq.; odorless; sol. 0.12 g/l in water; sp.gr. 1.12 (20/20 C); dens. 9.32 lb/gal (20 C); visc. 9.5 mPa·s; vapor pressure 0.0005 mbar (20 C); f.p. < -50 C; b.p. 298 C; flash pt. (COC) 161 C; autoignition temp. 457 C; 100% act.
Toxicology: LD50 (oral, rat) 9000 mg/kg, (skin, guinea pig) > 20 ml/kg; LC50 (inh., rat, 6 h) > 4.64 mg/l; ACGIH TLV/TWA 5 mg/m^3; low hazard; sl. eye/skin irritantTSCA listed
Environmental: BOD5 2000 mg/g; COD 1660-2100 mg/g; LC50 (rainbow trout, 96 h) 12 mg/l, NOEC 3.8 mg/l, (fathead minnow, 96 h) 17 mg/l, NOEC 4 mg/l; EC50 (daphnia, 48 h) 86 mg/l; prevent runoff from entering drains, sewers, streams
Precaution: Reacts with strong oxidizing agents
Hazardous decomp. prods.: Combustion: CO, CO_2
HMIS: Health 1, Flammability 1, Reactivity 0
Storage: Store in closed containers

Eastman® DEP, Fragrance Grade [Eastman]
Chem. Descrip.: Diethyl phthalate
CAS 84-66-2; EINECS/ELINCS 201-550-6
Uses: Plasticizer; fixing agent, solvent for fragrances, cosmetics, nail care; wetting agent in grinding pigments; pigment-disp. medium in cellulose acetate sol'ns. and plastics; solvent for natural resins and polymers
Features: Seldom used in PVC prods. due to relatively high volatility
Regulatory: SARA §313 nonreportable; Canada DSL, EINECS, Australia AICS, Japan MITI, Korea ECL listed
Properties: Colorless liq.; odorless; sol. 0.12 g/l in water; m.w. 222; sp.gr. 1.120; dens. 9.32 lb/gal (20 C); visc.9.5 cP; vapor pressure 0.0005 mbar (20 C); f.p. < -50 C; b.p. 298 C (760 mm); ref. index 1.4990; flash pt. (COC) 161 C; fire pt. (COC) 171 C; autoignition temp. 457 C; dissip. factor 0.10 x 10^{-2} (1 MHz); dielec. const. 6.7 (1 MHz); vol. resist. 1.45 x 10^9 ohm cm; 100% act.
Toxicology: LD50 (oral, rat) 9000 mg/kg, (skin, guinea pig) > 20 ml/kg; LC50 (inh., rat, 6 h) > 4.64 mg/l; ACGIH TLV/TWA 5 mg/m^3; low hazard; sl. eye/skin irritantTSCA listed
Environmental: BOD5 2000 mg/g; COD 1660-2100 mg/g; LC50 (rainbow trout, 96 h) 12 mg/l, NOEC 3.8 mg/l, (fathead minnow, 96 h) 17 mg/l, NOEC 4 mg/l; EC50 (daphnia, 48 h) 86 mg/l; prevent runoff from entering drains, sewers, streams
Precaution: Reacts with strong oxidizing agents
Hazardous decomp. prods.: Combustion: CO, CO_2
HMIS: Health 1, Flammability 1, Reactivity 0
Storage: Store in closed containers

Eastman® DM [Eastman]
Chem. Descrip.: Diethylene glycol methyl ether
CAS 111-77-3; EINECS/ELINCS 203-906-6
Uses: Solvent for coatings incl. nongrain-raising wood stains, printing inks, dye pastes for textiles; coalescing aid for PVAc latex paints; solvent in stamp pad and stencil inks; diluent for hydraulic brake fluids; deicer in jet fuel; deactivator in pesticides; coalescent for floor polishes; solvent in cosmetics, personal care
Regulatory: SARA §311/312 chronic health/fire hazard, §313 reportable; Canada DSL, EINECS, Australia AICS, Japan MITI, Korea ECL listed
Properties: Colorless liq.; mild odor; sol. in water; m.w. 120.15; sp.gr. 1.023 (20/20 C); dens. 8.51 lb/gal; visc. 3.90 cP (neat); vapor pressure 0.2 torr (20 C); f.p. -121 F; b.p. 191-198 C; flash pt. (TCC) 86 C; ref. index 1.4268 (20 C); surf. tens. 34.8 dynes/cm; 100% act.
Toxicology: LD50 (oral, rat) 7128 mg/kg, (skin, rabbit) 9404 mg/kg; TLV 100 ppm; harmful by skin absorp.; sl. eye/skin irritant; may cause adverse reproductive effects, birth defects, miscarriages, infertilityTSCA listed
Environmental: BOD5 20 mg/g; BOD20 80 mg/g; LC50 (goldfish, 24 h) > 5000 mg/l, (bluegill sunfish, 96 h) 7500 mg/l; prevent runoff from entering drains, sewers, streams
Precaution: Combustible liq. and vapor; flamm. limits in air 1.38 vol.% (135 C) to 22.7 vol.% (168 C); peroxide former; reacts with strong oxidizing agents

Hazardous decomp. prods.: Combustion: CO, CO_2; forms peroxides of unknown stability
HMIS: Health 2, Flammability 2, Reactivity 1
Storage: Store in closed containers away from heat and light; after opening, purge container with nitrogen before reclosing

Eastman® DP [Eastman]
Chem. Descrip.: Diethylene glycol monopropyl ether
CAS 6881-94-3; EINECS/ELINCS 229-985-7
Uses: Coupling agent, solvent for sol'n. and water-dilutable coatings; active solvent for NC, acrylic copolymers, epoxy resins, chlorinated rubber, alkyd resins; coalescing agent in architectural latex coatings, floor polishes, liq. cleaners; diluent in hydraulic brake fluids; household and general industrial cleaners
Features: Very slow-evaporating
Regulatory: SARA and CERCLA reportable
Properties: Colorless clear liq., mild odor; completely sol. in water; m.w. 148.2; sp.gr. 0.963 (20/20 C); dens. 8.04 lb/gal; visc. 4.1 cP; vapor pressure 0.05 mm Hg (20 C); f.p. < -68 C; i.b.p. 210 C; flash pt. (TCC) 93 C; fire pt. 103 C; ref. index 1.429 (20 C); 99% min. purity
Toxicology: LD50 (oral, rat) 6661 mg/kg, (dermal, guinea pig) 5048 mg/kg; causes eye irritation; low hazard by inh., ingestion or skin contactTSCA listed
Environmental: May cause oxygen depletion in aq. systems; low potential to affect aquatic organisms or germination/early growth of some plants; mod. potential to biodegrade; COD 2 g O_2/g; BOD5 0.14 g O_2/g
Precaution: Incompat. with strong oxidizing materials; potential peroxide former; lower flamm. limit in air 0.85 vol.% (122 C)
Hazardous decomp. prods.: Hazardous combustion prods.: CO, CO_2
HMIS: Health 2, Flammability 1, Reactivity 0
Storage: Store in closed container away from heat and light; minimize exposure to air

Eastman® DPG, Fragrance Grade [Eastman]
Chem. Descrip.: Dipropylene glycol
CAS 110-98-5; EINECS/ELINCS 203-821-4
Uses: Solubilizer, humectant for perfumes; solvent, fixative for fragrances
Properties: Pt-Co 10 max. color; sp.gr. 1.02 (20 C); vapor pressure 0.005 mm Hg (20 C); b.p. 232 C; flash pt. (COC) 118 C; 99% min. assay

Eastman® EAA [Eastman]
Chem. Descrip.: Ethyl acetoacetate
CAS 141-97-9; EINECS/ELINCS 205-516-1
Uses: Solvent for fragrances, cosmetics, personal care, pharmaceuticals, coatings, pesticides, pigments; copromoter for unsat. polyesters; formaldehyde scavenger; chemical intermediate
Regulatory: SARA §311/312 acute health/fire hazard, §313 nonreportable; Canada DSL, EINECS, Australia AICS, Japan MITI, Korea ECL listed
Properties: Colorless liq.; ester odor; sol. in alcohol, ether, 12% in water (20 C); m.w. 130.14; sp. gr. 1.03 (20 C); dens. 8.54 lb/gal; vapor pressure 1.3 mbar (14 C); f.p. -43 C; b.p.170 C; flash pt. (Seta CC) 68 C; 100% act.
Toxicology: LD50 (oral, rat) > 6400 mg/kg, (skin, guinea pig) > 20 ml/kg; LC50 (inh., rat, 6 h) > 1129 ppm; causes mod. eye irritation; sl. skin irritant; nonsensitizing to skinTSCA listed
Environmental: BOD5 0.60 g/g; COD 1.71 g/g; LC50 (fathead minnow, 96 h) > 100 mg/l, (golden orfe, 48 h) 275-515 mg/l; EC50 (daphnia, 24 h) 800 mg/l; prevent runoff from entering drains, sewers, streams
Precaution: Combustible liq. and vapor; reacts with strong oxidizing agents
Hazardous decomp. prods.: Combustion: CO, CO_2
HMIS: Health 2, Flammability 2, Reactivity 0
Storage: Store in closed container

Eastman® EB [Eastman]
Chem. Descrip.: Ethylene glycol monobutyl ether
CAS 111-76-2; EINECS/ELINCS 203-905-0
Uses: Solvent for alkyd, phenolic, maleic, epoxy, and NC resins; solvent, gloss aid and flow-out improver, orange peel reducer in water-dilutable coatings, waterborne latex emulsions or disps., pesticides, inks, resin mfg.; solvent in household and general industrial cleaners, fuel additives, cosmetics, hair care; coupling agent for oil-sol. components with water; coalescing aid; retarder solvent for lacquers
Regulatory: SARA and CERCLA reportable
Properties: Colorless clear liq., sl. odor; completely sol. in water; m.w. 118.7; sp.gr. 0.902 (20/20 C); dens. 7.51 lb/gal; visc. 6.4 cP; vapor pressure 0.6 mm Hg (20 C); f.p. -75 C; i.b.p. 169 C; flash pt. (TCC) 62 C; fire pt. 70 C; ref. index 1.4193 (20 C); 99.4% min. purity
Toxicology: LD50 (oral, rat) 1746 mg/kg, (dermal, rabbit) 435 mg/kg; ACGIH TLV 25 ppm TWA, skin; harmful if inhaled, absorbed thru skin, or swallowed; may cause blood disorders based on animal data; causes skin and eye irritationTSCA listed
Environmental: High BOD; potential to cause oxygen depletion in aq. systems; low potential to affect aquatic organisms; high potential to biodegrade; COD 2.18 g O_2/g
Precaution: Combustible liq. and vapor; can react violently with strong oxidizing materials; peroxide former; flamm. limits in air 1.1 vol.% (93 C) to 12.7 vol.% (135 C)
Hazardous decomp. prods.: Combustion: CO, CO_2
HMIS: Health 2, Flammability 2, Reactivity 0
Storage: Store in closed containers away from heat and light; minimize exposure to air; after opening, purge container with nitrogen before reclosing; do not allow to evaporate to near dryness

Eastman® EB Acetate [Eastman]
Chem. Descrip.: Ethylene glycol monobutyl ether acetate
CAS 112-07-2; EINECS/ELINCS 203-933-3
Uses: Retarder, solvent in automotive paints and coil coatings; flow aid, leveling agent in high-bake and air-dry coatings incl. epoxies; solvent in moisture-sensitive paints, e.g. PU coatings, screen and stamp pad inks, cosmetics; coalescing aid for latex paints
Features: High-boiling; slow evaporating
Regulatory: SARA and CERCLA reportable
Properties: Colorless liq.; mild odor; sol. 1.1% in water; m.w. 160.21; sp.gr. 0.941 (20/20 C); dens. 7.84 lb/gal; visc. 1.8 cP; vapor pressure 0.29 mm Hg (20 C); f.p. -64 C; i.b.p. 186 C; flash pt. (TCC) 71 C; fire pt. 82 C; ref. index 1.4142 (20 C); 99% min. assay
Toxicology: LD50 (oral, rat) 7.46 ml/kg, (dermal, rabbit) 1.5 g/kg; may cause blood disorders, kidney damage; harmful if inhaled or absorbed thru skin; expected to be low ingestion hazardTSCA listed
Precaution: Combustible liq. and vapor; can react with strong oxidizing agents; peroxide former; flamm. limits in air 0.88 vol.% (93 C) to 8.54 vol.% (135 C)
Hazardous decomp. prods.: Combustion: CO, CO_2
HMIS: Health 2, Flammability 2, Reactivity 0
Storage: Store in closed containers away from heat and light; minimize exposure to air; after opening, purge container with nitrogen before reclosing

Eastman® EEH [Eastman]
Chem. Descrip.: Ethylene glycol 2-ethylhexyl ether (85%), diethylene glycol 2-ethylhexyl ether (15%)
Uses: Retarder, solvent, coalescing agent in coatings incl. architectural and industrial emulsions/disps.; solvent for polyesters, alkyds, polyamides, and other coatings, cathodic electrodeposition primers
Features: Slow-evaporating; low surf. tens., high elec. resist.
Regulatory: SARA reportable
Properties: Colorless clear liq., ether odor; sol. 0.2% in water (20 C); sp.gr. 0.892; dens. 7.42 lb/gal; vapor pressure 0.08 mm Hg (20 C); f.p. < -45 C; i.b.p. 224 C; flash pt. (Seta CC) 98 C; ref. index 1.4361 (20 C)
Toxicology: LD50 (oral, male rat) 4674 mg/kg, (dermal, guinea pig) > 20 ml/kg; causes skin and eye irritation; ingestion may cause GI tract irritationTSCA listed
Environmental: Low BOD; little potential to cause oxygen depletion in aq. systems; mod. potential to affect some aquatic organisms and germination/early plant growth of some plants; some potential to biodegrade; COD 1.00-1.03 g O_2/g
Precaution: Can react with strong oxidizing materials; potential peroxide former; flamm. limits in air 0.61 vol.% (121 C) to 8.02 vol.% (174 C)
Hazardous decomp. prods.: Combustion: CO, CO_2
HMIS: Health 2, Flammability 1, Reactivity 0
Storage: Store in closed containers away from heat and light; minimize exposure to air

Eastman® EEP [Eastman]
Chem. Descrip.: Ethyl 3-ethoxypropionate (> 99.5%), formaldehyde (< 0.02%), BHT as inhibitor (< 0.02%)
CAS 763-69-9
Uses: Solvent for coatings (aerospace, automotive OEM and refinish, metal, coil, maintenance, marine, photoresist); solvent for thinning syn. coating resins, esp. moisture-sensitive resins for PU coatings, formulating air-dry and baking finishes; retarder solvent for high-bake enamels;

polymerization solvent for high-solids acrylic resins used in photoresist inks; process solvent
Features: Let-down solvent; provides easier handling, storage, shipping, resin stability
Regulatory: SARA §311/312 fire hazard, §313 nonreportable; Canada DSL, EINECS, Australia AICS, Japan MITI, Korea ECL listed
Properties: Colorless liq.; pungent ester odor; sol. 2.9% in water (20 C); m.w. 146.19; sp.gr. 0.95 (20/20 C); dens.7.91 lb/gal (20 C); visc. 1.2 mPa·s; vapor pressure 1.5 mm Hg; f.p. < -50 C; b.p. 165-172 C; flash pt. (Seta CC) 58 C; autoignition temp. 377 C; ref. index 1.4074 (20 C); surf. tens. 27.0 dynes/cm (23 C); elec. resist. 20 megohms; 99.5% min. assay
Toxicology: LD50 (oral, male rat) > 5000 mg/kg, (skin, guinea pig) > 20 ml/kg; LC50 (inh., rat, 6 h) > 1000 ppm; high vapor concs. may cause drowsiness; low toxicity; sl. eye/skin irritantTSCA listed
Environmental: BOD5 370 mg/g; BOD20 560 mg/g; COD 1920 mg/g; LC50 (fathead minnow, 96 h) 50 mg/l, (daphnia magna, 48 h) > 480 mg/l; prevent runoff from entering drains, sewers, streams
Precaution: Combustible liq. and vapor; forms peroxides of unknown stability if material becomes uninhibited; keep inhibited; reacts with strong oxidizing agents
Hazardous decomp. prods.: Combustion: CO, CO_2
HMIS: Health 1, Flammability 2, Reactivity 0
Storage: Store in closed containers away from heat, flame, light; after opening, purge container with nitrogen before reclosing

Eastman® EP [Eastman]
Chem. Descrip.: Ethylene glycol monopropyl ether
CAS 2807-30-9; EINECS/ELINCS 220-548-6
Uses: Coupling solvent for resin/water systems incl. waterborne coatings; visc. control agent for waterborne resins; retarder solvent for coatings based on NC, acrylic, epoxy, polyamide, and alkyd resins; solvent for coatings, plastic coatings, enamels, lacquers, household and industrial cleaners
Features: Slow-evaporating
Regulatory: SARA §311/312 acute health/chronic health/fire hazard, §313 reportable; CERCLA reportable; Canada DSL, EINECS, Australia AICS, Japan MITI, Korea ECL listed
Properties: Colorless clear liq.; mild odor; completely misc. with water over wide temp. range; m.w. 104.15; sp.gr. 0.913 (20/20 C); dens. 7.59 lb/gal; visc. 2.42 cP; vapor pressure 1.3 mm Hg (20 C); f.p. < -90 C; i.b.p. 149.5 C; flash pt. (TCC) 49 C; fire pt.56 C; ref. index 1.4136 (20 C); 100% act.
Toxicology: LD50 (oral, rat) 3089 mg/kg, (dermal, rabbit) 1337 mg/kg; LC50 (inh., rat, 6 h) > 2132 ppm; causes eye irritation; harmful if inhaled, absorbed thru skin, or swallowed; may cause blood disordersTSCA listed
Environmental: BOD5 1.3 g/g; potential to cause oxygen depletion in aq. systems; low potential to affect aquatic organisms or germination/early growth of some plants; high potential to biodegrade; COD 2.04 g O_2/g; prevent runoff to drains, sewers, streams
Precaution: Combustible liq. and vapor; reacts violently with strong oxidizing agents; peroxide former; flamm. limits in air 1.26 vol.% (69 C) to 15.8 vol.% (127 C)
Hazardous decomp. prods.: Combustion: CO, CO_2
HMIS: Health 2, Flammability 2, Reactivity 0
Storage: Store in closed containers away from heat and light; minimize exposure to air; after opening, purge container with nitrogen before reclosing

Eastman® IBIB [Eastman]
Chem. Descrip.: Isobutyl isobutyrate
CAS 97-85-8; EINECS/ELINCS 202-612-5
Uses: Solvent for nat. resins incl. rosin, acaroids (except shellac), and syn. resins incl. PS, PVC copolymers, urea, melamines, and maleic; solvent for coatings, automotive coatings, plastic coatings
Features: Swells both nat. and syn. rubbers, i.e., polyisobutylene
Regulatory: SARA §311/312 acute health/fire hazard, §313 nonreportable; Canada DSL, EINECS, Australia AICS, Japan MITI, Korea ECL listed
Properties: Colorless clear liq.; sweet odor; sol. < 0.1% in water (20 C); m.w. 142.22; sp.gr. 0.855 (20/20 C); dens. 7.13 lb/gal; visc. 0.9 mPa·s; vapor pressure 3.2 mm Hg (20 C); f.p. -80 C; b.p. 144-151 C; flash pt. (TCC) 40 C; fire pt. 46 C; autoignition temp. 432 C; ref. index 1.4 (20 C); elec. resist. > 20 megohms; 100% act.
Toxicology: LD50 (oral, rat) 6400-12,800 mg/kg, (skin, guinea pig) 8520mg/kg; high vapor concs. may cause drowsiness, eye/respiratory tract irritation; prolonged/repeated skin contact may cause drying, cracking, irritation; TSCA listed
Environmental: LC50 (fathead minnow, 96 h) 12.54 mg/l; EC50 (daphnia, 48 h) 55.8 mg/l; prevent runoff from entering drains, sewers, streams
Precaution: Combustible liq. and vapor; vapors may cause flash fire, ignite explosively, or flash back; peroxide former; reacts with strong oxidizing agents
Hazardous decomp. prods.: Combustion: CO, CO_2
HMIS: Health 1, Flammability 2, Reactivity 0
Storage: Store in tightly closed containers in well-ventilated area away from heat, flame, light; after opening, purge with nitrogen before reclosing

Eastman® MAA [Eastman]
Chem. Descrip.: Methyl acetoacetate
CAS 105-45-3; EINECS/ELINCS 203-299-8
Uses: Solvent for fragrances, cosmetics, personal care, pharmaceuticals, pesticides; copromoter for unsat. polyesters; pharmaceutical intermediate
Regulatory: SARA §311/312 acute health/chronic health/fire hazard, §313 nonreportable; Canada DSL, EINECS, Australia AICS, Japan MITI, Korea ECL listed
Properties: Colorless liq.; pleasant odor; sol. (@ 20 C) in water (44.5%), alcohol, ether, ethyl acetate; m.w. 116.12; sp. gr. 1.078 (20 C); dens. 8.95 lb/gal (20 C); visc. 1.74 mPa·s (20 C); vapor pressure 1.32 mbar (20 C); f.p. -55 C; b.p. 170 C; flash pt. (COC) 74 C; pH 4.1; 100% act.
Toxicology: LD50 (oral, rat) 3200 mg/kg, (skin, rabbit) > 10 ml/kg; LC50 (inh., rat, 6 h) > 6.5 mg/l; harmful by ing.; may cause blindness if swallowed; causes eye irritation; high vapor concs. may cause eye/respiratory tract irritation; sl. skin irritant; mod. to strong eye irritant; TSCA listed
Environmental: BOD5 0.31 g/g; BOD20 0.77 g/g; COD 1.45 g/g; LC50 (fathead minnow, 96 h) > 111.4 mg/l, NOEC 111.4 mg/l; EC50 (daphnia, 48 h) > 236.9 mg/l, NOEC 236.9 mg/l; prevent runoff from entering drains, sewers, streams
Precaution: Combustible liq. and vapor; reacts violently with strong oxidizing agents
Hazardous decomp. prods.: Combustion: CO, CO_2
HMIS: Health 2, Flammability 2, Reactivity 0
Storage: Store in closed containers away from heat and flame

Eastman® MIBK [Eastman]
Chem. Descrip.: Methyl isobutyl ketone
CAS 108-10-1; EINECS/ELINCS 203-550-1
Uses: Solvent for cosmetics, personal care prods., coatings, plastic coatings, printing inks, fungicides, germicides
Regulatory: SARA §311/312 acute health/fire hazard, §313 reportable; Canada DSL, EINECS, Australia AICS, Japan MITI, Korea ECL listed
Properties: Colorless liq.; ketone odor; m.w. 100.16; sol. in water (20 C) 2% wt; sp. gr. 0.802 (20 C); dens. 6.67 lb/gal; vapor pressure 15 mm Hg; f.p. -84 C; flash pt. (TCC) 16 C; autoignition temp. 449 C; ref. index 1.3958 (20 C); surf. tens. 23.6 dynes/cm (20 C); 100% act.
Toxicology: LD50 (oral, rat) 1600-3200 mg/kg, (skin, rabbit) > 10 ml/kg; LC50 (inh., rat, 4 h) 2000-4000 ppm; ACGIH TLV/TWA 50 ppm; high vapor concs. may cause drowsiness, eye/respiratory tract irritation; sl. to mod. eye/skin irritant; prolonged/repeated skin contact may cause drying, cracking, irritation; TSCA listed
Environmental: BOD5 1940-2060 mg/g; COD 2160-2460 mg/g; LC50 (golden orfe, 48 h) 672-744 mg/l, (daphnia, 24 h) 4280 mg/l; prevent runoff from entering drains, sewers, streams
Precaution: Flamm. liq. and vapor; vapors may cause flash fire, ignite explosively, or flash back; forms explosive peroxides which may be shock-sensitive; reacts with bases, halogenated materials, acids, amines; reacts violently with strong oxidizing agents, strong reducing agents
Hazardous decomp. prods.: Combustion: CO, CO_2
HMIS: Health 1, Flammability 3, Reactivity 1
Storage: Store in tightly closed containers in well-ventilated area, away from heat and light; after opening, purge container with nitrogen before reclosing

Eastman® PM [Eastman]
Chem. Descrip.: Propylene glycol methyl ether
CAS 107-98-2; EINECS/ELINCS 203-539-1
Uses: Solvent for coatings, plastic coatings, automotive paints, printing inks, cosmetics, personal care, nail care

Regulatory: SARA §311/312 acute health/fire hazard, §313 nonreportable; Canada DSL, EINECS, Australia AICS, Japan MITI listed
Properties: Colorless liq.; ether odor; sol. in water; m.w. 90.12; sp.gr. 0.923 (20/20 C); dens. 7.69 lb/gal; visc. 1.90 cP (neat); vapor pressure 8 mm Hg (20 C); f.p. -139 F; b.p. 121 C; flash pt. (TCC) 33 C; ref. index 1.4036 (20 C); surf. tens. 28.3 dynes/cm; 97-100% act.
Toxicology: LD50 (oral, rat) 6040 mg/kg, (skin, rabbit) 13,000-14,000 mg/kg; LC50 (inh., rat, 4 h) > 10,000 ppm; ACGIH TLV/TWA 100 ppm; harmful by ing.; may enter lungs if swallowed or vomited; high vapor concs. may cause drowsiness, eye/respiratory tract irritation; sl. eye/skin irritant; TSCA listed
Environmental: BOD20 1140 mg/g; COD 1840 mg/g; LC50 (fathead minnow, 96 h) 20,800 mg/l, NOEC 11,300 mg/l; EC50 (daphnia, 48 h) 23,300 mg/l, NOEC 1412 mg/l; prevent runoff from entering drains, sewers, streams
Precaution: Flamm. liq. and vapor; vapors may cause flash fire, ignite explosively, or flash back; residual vapors may explode on ignition; peroxide former; reacts with strong oxidizing agents
Hazardous decomp. prods.: Combustion: CO,CO_2
HMIS: Health 1, Flammability 3, Reactivity 1
Storage: Store in closed containers away from heat, sparks, flame, light; after opening, purge container with nitrogen before reclosing

Eastman® PM Acetate [Eastman]
Chem. Descrip.: Propylene glycol monomethyl ether acetate
CAS 108-65-6; EINECS/ELINCS 203-603-9
Uses: Retarder, solvent for gravure inks, lacquers, thinners, baking enamels; solvent for screen inks, cosmetics, personal care, food-contact applics., coating polymers (CAB, NC, epoxy, acrylic copolymers, phenoxy resins), cleaning fluids, coatings, automotive paints, plastic coatings; process solvent for resin mfg., urethanes
Features: Slow evaporating
Regulatory: SARA §311/312 acute health/fire hazard, §313 nonreportable; Canada DSL, EINECS, Australia AICS, Japan MITI, Korea ECL listed
Properties: Colorless liq.; mild sweet odor; sol. 20% in water (20 C); m.w. 132.2; sp.gr. 0.969 (20/20 C); dens. 8.06 lb/gal; visc. 1.1 cP; vapor pressure 2.8 mm Hg (20 C); f.p. < -67 C; b.p. 140-150 C; flash pt. (Seta CC) 45 C; ref. index 1.4 (20 C); surf. tens. 26.4 dynes/cm (20 C); 100% act.
Toxicology: LD50 (oral, male rat) > 10,000 mg/kg, (skin, rabbit) > 5000 mg/kg; WEEL TWA 100 ppm; high vapor concs. may cause eye/respiratory tract irritation; sl. eye/skin irritantTSCA listed
Environmental: LC50 (fathead minnow, 96 h) 161 mg/l, (daphnia, 48 h) 408 mg/l; prevent runoff from entering drains, sewers, streams
Precaution: Combustible liq. and vapor; peroxide former; reacts with strong oxidizing agents
Hazardous decomp. prods.: Combustion: CO, CO_2
HMIS: Health 1, Flammability 2, Reactivity 0
Storage: Store in closed containers away from heat, flame, light; after opening, purge container with nitrogen before reclosing

Eastman® TEP [Eastman]
Chem. Descrip.: Triethyl phosphate
CAS 78-40-0; EINECS/ELINCS 201-114-5
Uses: Intermediate for insecticides/pesticides; in floor polishes; plasticizer for cellulose acetate; solvent for textiles; dyeing assistant; in sizes; flame retardant synergist in unsat. polyesters; in lubricants, hydraulic fluids; catalyst; visc. suppressant for general polymer applics.
Properties: Colorless clear liq., mild pleasant cider-like odor; misc. with water, ethanol, ethyl acetate, ethyl ether, acetone, chloroform, benzene, xylene, castor oil; partly sol. in min. oil, kerosene, CS_2; sp.gr. 1.067-1.072 (20/20 C); dens. 8.91 lb/gal; visc. 1.76 cP (20 C); vapor pressure 1 mm Hg (40 C); m.p. -56.4 C; b.p. 209 C; flash pt. (PMCC) 99 C; fire pt. (COC) 141 C; ref. index 1.4019
Toxicology: LD50 (oral, rat) 1311 mg/kg (sl. toxic), (dermal, guinea pig) > 20 ml/kg; causes eye irritation; sl. skin irritant; emits highly toxic fumes of phosphorus oxides when heated to decomp.
Precaution: Incompat. with oxidizing materials

Eastman® Acetone, High Purity Sales Grade [Eastman]
Chem. Descrip.: Acetone
CAS 67-64-1; EINECS/ELINCS 200-662-2
Uses: Solvent for cosmetics, personal care prods., nail care, coatings, automotive coatings, plastic coatings
Regulatory: SARA §311/312 acute health/fire hazard, §313 nonreportable; Canada DSL, EINECS, Australia AICS, Japan MITI, Korea ECL listed
Properties: Colorless liq.; fruity odor; sol. in water (20 C); m.w. 58.08; sp. gr. 0.792 (20 C); visc. (liq.) 0.3 mPa·s; dens. 6.6 lb/gal (20 C); vapor pressure 185 mm Hg (20 C); f.p. -95 C; b.p. 55.5-57.1 C (760 mm Hg); flash pt. (TCC) -20 C; autoignition temp. 538 C; ref. index 1.3591 (20 C); surf. tens. 22.3 dynes/cm (20 C); 100% act.
Toxicology: LD50 (oral, rat) 6715-8453 mg/kg, (skin, rabbit) > 15,800 mg/kg; LC50 (inh., rat, 4 h) > 16,000 ppm; ACGIH TLV/TWA 500 ppm; causes eye irritation; high vapor concs. may cause drowsiness, eye/respiratory tract irritation; prolonged/repeated skin contact may cause drying, cracking, irritation; not expected to cause nervous system damage; TSCA listed
Environmental: BOD5 310-1850 mg/g; BOD20 1780 mg/g; COD 1120-2070 mg/g; LC50 (golden orfe, 48 h) 7505-11,300 mg/l, (bluegill sunfish, 96 h) 8300 mg/l, (daphnia, 24 h) > 10,000 mg/l; prevent runoff from entering drains, sewers, or streams
Precaution: Extremely flamm. liq. and vapor; vapor may cause flash fire or ignite explosively; vapors may flash back; residual vapors may explode on ignition; reacts violently with strong oxidizing agents, strong acids
Hazardous decomp. prods.: CO, OC_2
HMIS: Health 2, Flammability 3, Reactivity 0
Storage: Store in tightly closed container in well-ventilated area, away from heat, sparks, flames

Eastman® n-Butanol [Eastman]
Chem. Descrip.: n-Butanol
CAS 71-36-3; EINECS/ELINCS 200-751-6
Uses: Solvent for cosmetics, personal care prods., fragrances, acrylates, coatings, automotive coatings, plastic coatings, nail care, plasticizers
Regulatory: SARA §311/312 acute health/fire hazard, §313 reportable; Canada DSL, EINECS, Australia AICS, Japan MITI listed
Properties: Colorless liq.; alcohol odor; sol. in water (20 C) at 7.9% wt.; m.w. 74.12; sp. gr. 0.81 (20 C); dens. 6.75 lb/gal (20 C); visc. (liq.) 3 mPa·s; vapor pressure 5.5 mm Hg (20 C); f.p. 36 C; b.p. 116-119 C (760 mm Hg); flash pt. (TCC) 36 C; autoignition temp. 355 C; ref. index 1.4 (20 C); surf. tens. 24.6 dynes/cm (20 C); 100% act.
Toxicology: LD50 (oral, rat) 2500 mg/kg, (skin, rabbit) 5300 mg/kg; LC50 (inh., rat, 4 h) > 8000 ppm; ACGIH TLV/TWA 20 ppm, STEL 50 ppm; causes eye irritation; high vapor concs. may cause drowsiness, eye/respiratory tract irritation; prolonged/repeated skin contact may cause drying, cracking, irritation; TSCA listed
Environmental: BOD5 1710 mg/g; BOD20 1890 mg/kg; COD 2460 mg/g; LC50 (goldfish, 24 h) 1855 mg/l, (golden orfe, 48 h) 1770 mg/l, (daphnia, 24 h) 1855 mg/l
Precaution: Flamm. liq. and vapor; vapors may cause flash fire or ignite explosively; vapors may flash back; residual vapors may explode on ignition; reacts violently with strong oxidizing agents
Hazardous decomp. prods.: Combustion: CO, CO_2
HMIS: Health 2, Flammability 3, Reactivity 0
Storage: Store in closed container in well-ventilated area, away from heat, sparks, flames

Eastman® n-Butyl Acetate, Urethane Grade [Eastman]
Chem. Descrip.: n-Butyl acetate
CAS 123-86-4; EINECS/ELINCS 204-658-1
Uses: Solvent for fragrances, cosmetics, personal care prods.
Properties: Pt-Co color 10 max.; sol. in water (0.7%, 20 C); m.w. 116.16; sp. gr. 0.88 (20 C); visc. (liq.) 0.7 mPa·s (20 C); dens. 7.35 lb/gal (20 C); vapor pressure 10 mm Hg (20 C); f.p. -74 C; b.p. 124-129 C; flash pt. (TCC) 27 C; autoignition pt. 407 C; ref. index 1.39 (20 C); surf. tens. 25.1 dynes/cm (20 C)

Eastman® Diethyl Phthalate [Eastman]
Chem. Descrip.: Diethyl phthalate
CAS 84-66-2; EINECS/ELINCS 201-550-6
Uses: Fixing agent in perfumes, fragrances, colognes, toilet waters; denaturant in specially denatured alcohols 39-B and 39-C; solvent/carrier; plasticizer
Properties: Colorless clear liq.; odorless; m.w. 222; sp.gr. 1.118-1.122 (20 C); vapor pressure 0.0004 mm Hg (20 C); ref. index 1.498-1.501; 99.5% min. assay; 0.18% max. water

Eastman® Ethyl Acetate, Food Grade [Eastman]
Chem. Descrip.: Ethyl acetate
CAS 141-78-6; EINECS/ELINCS 205-500-4
Uses: Solvent for flavors, fragrances, foods, cosmetics, personal care

prods., nail care
Regulatory: SARA §311/312 acute health/fire hazard, §313 nonreportable; Canada DSL, EINECS, Australia AICS, Japan MITI listed
Properties: Colorless liq.; sweet ester odor; sol. in water (20 C) at 8% wt.; m.w. 88.11; sp. gr. 0.894-0.898 (25 C); dens. 7.51 lb/gal; vapor pressure 86 mm Hg (20 C); f.p. -84 C; b.p. 76-77.5 C (760 mm Hg); flash pt. (TCC) -4 C; ref. index 1.37 (20 C); 100% act.
Toxicology: LD50 (oral, rat) 5600 mg/kg, (skin, rabbit) > 20 ml/kg; LC50 (inh., rat, 6 h) 16,000 ppm; ACGIH TLV/TWA 400 ppm; high vapor concs. may cause drowsiness, eye/respiratory tract irritation; prolonged/repeated skin contact may cause drying, cracking, irritation; TSCA listed
Environmental: BOD5 1240 mg/g; BOD20 1240-1430 mg/g; COD 1540 mg/g; LC50 (golden orfe, 48 h) 270 mg/l, (daphnia, 24 h) 3090 mg/l; prevent runoff from entering drains, sewers, streams
Precaution: Flamm. liq. and vapor; vapors may cause flash fire, ignite explosive, or flash back; residual vapors may explode on ignition; reacts with strong oxidizing agents, strong acids, strong bases
Hazardous decomp. prods.: Combustion: CO, CO_2
HMIS: Health 1, Flammability 3, Reactivity 0
Storage: Store in tightly closed containers away from heat, sparks, flame

Eastman® Glacial Acetic Acid [Eastman; C.P. Hall]
Chem. Descrip.: Glacial acetic acid
CAS 64-19-7; EINECS/ELINCS 200-580-7
Uses: Solvent, pH control agent for water treatment
Regulatory: DOT regulated; SARA §311/312 acute health/fire hazard, §313 nonreportable; Canada DSL, EINECS, Australia AICS, Japan listed
Properties: Colorless liq.; pungent vinegar odor; completely sol. in water; m.w. 60.06; sp.gr. 1.05 (20 C); visc. 1.20 mPa·s (20 C); vapor pressure 11.4 mm Hg (20 C); m.p. 17 C; b.p. 118 C; flash pt. (TCC) 39 C; autoignition temp. 516 C; pH 2.4 (60 g/l water); 100% act.
Toxicology: LD50 (oral, rat) 3310-3530 mg/kg, (skin, rabbit) 1060 mg/kg; LC50 (inh., mouse, 1 h) 5620 ppm; ACGIH TLV/TWA 10 ppm, STEL 15 ppm; causes severe skin/eye burns; harmful if swallowed; vapor irritating to eyes, respiratory tract; ing. may cause burns of GI tract; TSCA listed
Environmental: May cause adverse environmental effects; high BOD; potential to cause oxygen depletion in aq. systems; low potential to affect aq. organisms, sec. waste treatment microbial metabolism; prevent runoff from entering drains, sewers, streams
Precaution: Combustible liq. and vapor; LEL 4 vol.% (59 C); UEL 19.3 vol.% (93 C); can react violently with strong oxidizing agents; can react with metals, strong bases, amines
Hazardous decomp. prods.: Combustion: CO, CO_2
HMIS: Health 3, Flammability 2, Reactivity 0
Storage: Store in closed containers away from heat, flame, toxic substances

Eastman® Glacial Acetic Acid Food Grade, Kosher [Eastman]
Chem. Descrip.: Glacial acetic acid FCC
CAS 64-19-7; EINECS/ELINCS 200-580-7
Uses: Acidulant, flavor enhancer, carrier, formulating aid, curing agent, preservative in food processing (baked goods, fats, oils, milk prods., frozen dairy prods., processed fruit, condiments, relishes, soft candy, sweet sauce, gelatin, puddings, snack foods, soft beverages); solvent
Regulatory: FDA 21CFR §184.1005, GRAS; kosher; DOT regulated; SARA §311/312 fire hazard, immediate (acute) health hazard; Canada DSL listed
Properties: Colorless liq.; pungent vinegar odor; completely sol. in water; m.w. 60.06; sp.gr. 1.051 (20 C); visc. 1.22 cps (20 C); vapor pressure 11.4 mm Hg (20 C); f.p. 16.4 C; b.p. 118 C; flash pt. (TCC) 42 C; ref. index 1.3698; pH 2.4 (60 g/l water); 99.5-100.5% assay
Toxicology: LD50 (oral, rat) 3310-3530 mg/kg, (skin, rabbit) 1060 mg/kg; causes severe eye/skin burns; vapor is irritating by inh., eye contact; harmful if swallowed; ing. may cause burns of GI tractTSCA listed
Environmental: COD 1.03 g oxygen/g; BOD-5 0.34-0.88 g oxygen/g; LC50 (daphnia, 96 h) > 100 mg/l; biodeg.; strongly acidic aq. sol'n.; may cause adverse environmental effects; high BOD; potential to cause oxygen depletion in aq. systems
Precaution: Combustible liq. and vapor; can react violently with strong oxidizing agents; can react with metals, strong bases, amines
Storage: Keep container closed; keep away from heat and flame

Eastman® Isobutyl Acetate [Eastman]
Chem. Descrip.: Isobutyl acetate
CAS 110-19-0; EINECS/ELINCS 203-745-1
Uses: Solvent for fragrances, cosmetics, personal care prods., nail care, coatings, plastic coatings
Regulatory: SARA §311/312 acute health/fire hazard, §313 nonreportable; Canada DSL, EINECS, Australia AICS, Japan MITI listed
Properties: Colorless liq.; ester sweet odor; m.w. 116.2; sol. in water (20 C) at 0.7% wt.; sp. gr. 0.87 (20 C); dens. 7.25 lb/gal; vapor pressure 12.5 mm Hg; f.p. -99 C; b.p. 112-119 C (760 mm Hg); flash pt. (TCC) 20 C; fire pt. 31 C; ref. index 1.37 (20 C); surf. tens. 23.7 dynes/cm (20 C); 100% act.
Toxicology: LD50 (oral, rat) > 4350 mg/kg, (skin, rabbit) > 17,4000 mg/kg; LC50 (inh., rat, 4 h) 3500-3800 ppm; ACGIH TLV/TWA 150 ppm; high vapor concs. may cause drowsiness, eye/respiratory tract irritation; sl. eye/skin irritant; prolonged/repeated skin contact may cause drying, cracking, irritation; TSCA listed
Environmental: Prevent runoff from entering drains, sewers, streams
Precaution: Flamm. liq. and vapor; vapors may cause flash fire, ignite explosively, or flash back; residual vapors may explode on ignition; reacts violently with strong oxidizing agents
Hazardous Ingredients: Combustion: CO, CO_2
HMIS: Health 1, Flammability 3, Reactivity 0
Storage: Store in tightly closed containers in well-ventilated area, away from heat, sparks, flame

Eastman® Isopropanol [Eastman]
Chem. Descrip.: Isopropyl alcohol
CAS 67-63-0; EINECS/ELINCS 200-661-7
Uses: Solvent for fragrances, cosmetics, personal care, pharmaceuticals, coatings, automotive coatings, plastic coatings; gasoline additive
Regulatory: SARA §311/312 acute health/fire hazard, §313 nonreportable; Canada DSL, EINECS, Australia AICS, Japan MITI, Korea ECL listed
Properties: Colorless liq.; alcohol odor; sol. in water (20 C); m.w. 60.1; sp. gr. 0.79 (20 C); dens. 6.54 lb/gal (20 C); visc. 2.4 cp (20 C); vapor pressure 32.8 mm Hg (20 C); f.p. -88 C; b.p. 80.8-83.8 C (760 mm Hg); flash pt. (TCC) 13 C; fire pt. 16 C; autoignition temp. 360 C; surf. tens. 21.3 dynes/cm (20 C); 100% act.
Toxicology: LD50 (oral, rat) 5800 mg/kg, (skin, rabbit) 16.4 ml/kg; LC50 (inh., rat, 8 h) 12,000 ppm; ACGIH TLV/TWA 400 ppm; mod. eye irritant; high vapor concs. may cause drowsiness, eye/respiratory tract irritation; sl. skin irritant; prolonged/repeated skin contact may cause drying, cracking, irritation; TSCA listed
Environmental: BOD5 1190-1720 mg/g; BOD20 1680 mg/g; COD 2230 mg/g; LC50 (fathead minnow, 96 h) > 1000 μl/l, (golden orfe, 48 h) 8970-9280 mg/l, (daphnia, 96 h) > 1000 μl/l; prevent runoff from entering drains, sewers, streams
Precaution: Flamm. liq. and vapor; flamm. limits 2.31-12.75 vol.% in air (93 C); vapors may cause flash fire, ignite explosively, or flash back; residual vapors may explode on ignition; forms explosive peroxides which may be shock-sensitive; material may accumulate a static charge which could act as an ignition source; reacts violently with strong oxidizing agents
Hazardous decomp. prods.: Combustion: CO, CO_2
HMIS: Health 2, Flammability 3, Reactivity 1
Storage: Store in tightly closed container in well-ventilated area away from heat, sparks, flame, light; after opening, purge container with nitrogen before reclosing

Eastman® Isopropyl Acetate [Eastman]
Chem. Descrip.: Isopropyl acetate
CAS 108-21-4; EINECS/ELINCS 203-561-1
Uses: Solvent for fragrances, cosmetics, personal care prods., nail care, cleaning fluids, coatings, automotive coatings, plastic coatings, printing inks
Regulatory: SARA §311/312 acute heath/fire hazard, §313 nonreportable; Canada DSL, EINECS, Australia AICS, Japan MITI, Korea ECL listed
Properties: Colorless liq.; sweet ester odor; sol. in water (20 C) at 2.9% wt.; m.w. 102.13; sp. gr. 0.873 (20 C); dens. 7.26 lb/gal; vapor pressure 47.5 mm Hg; f.p. -73 C; b.p. 85-90 C (760 mm Hg); flash pt. (TCC) 2 C; autoignition temp. 479 C; ref. index 1.38 (20 C); 100% act.
Toxicology: LD50 (oral, rat) 3000 mg/kg, (skin, rabbit) > 20 ml/kg; LC50 (inh., rat, 4 h) 16,000-32,000 ppm; ACGIH TLV/TWA 250 ppm; high vapor concs. may cause drowsiness, eye/respiratory tract irritation; sl. eye/skin irritant; prolonged/repeated skin contact may cause drying, cracking, irritation; expected to below ing. hazard; TSCA listed

Environmental: ThOD 2040 mg/g; LC50 (golden orfe, 48 h) 265-360 mg/l, (daphnia, 24 h) 1260 mg/l; prevent runoff from entering drains, sewers, streams
Precaution: Flamm. liq. and vapor; vapors may cause flash fire, ignite explosively, or flash back; residual vapors may explode on ignition; reacts violently with strong oxidizing agents
Hazardous Ingredients: Combustion: CO, CO_2
HMIS: Health 1, Flammability 3, Reactivity 0
Storage: Store in tightly closed containers in well-ventilated area, away from heat, sparks, flame

Eastman® Methyl Acetate [Eastman]
Chem. Descrip.: Methyl acetate (> 96%), methanol (< 2.5%), water (< 1.5%)
Uses: Solvent for cosmetics, personal care prods., nail care, coatings, plastic coatings, printing inks
Regulatory: SARA §311/312 acute health/chronic health/fire hazard, §313 reportable (methanol); Canada DSL, EINECS, Australia AICS, Japan MITI, Korea ECL listed
Properties: Colorless liq.; pleasant odor; m.w. 74.09; sol. in water (20 C) at 22% wt.; sp. gr. 0.93 (20 C); dens. 7.8 lb/gal; vapor pressure 178.3 mm Hg; f.p. -98 C; b.p. 55.8-58.2 C (760 mm Hg); flash pt. (SCC) -15 C; autoignition temp. 501 C; ref. index 1.36 (20 C); surf. tens. 25.8 dynes/cm (20 C)
Toxicology: LD50 (oral, rat) > 5000 mg/kg, (skin, rabbit) > 5000 mg/kg; LC50 (inh., rat, 4 h) > 16,000 ppm; ACGIH TLV/TWA 200 ppm (methyl acetate); harmful by inh., skin absorp.; may be fatal or cause blindness on ing.; causes eye/skin irritation; high vapor concs. may cause drowsiness, eye/respiratory tract irritation; TSCA listed
Environmental: BOD5 393 mg/g; ThOD 1510 mg/g; LC50 (fathead minnow, 96 h) 320-399 mg/l; prevent runoff from entering drains, sewers, streams
Precaution: Extremely flamm. liq. and vapor; vapors may cause flash fire, ignite explosively, or flash back; residual vapors may explode on ignition; reacts with strong oxidizing agents
Hazardous decomp. prods.: Combustion: CO, CO_2
HMIS: Health 2, Flammability 3, Reactivity 0
Storage: Store in tightly closed containers in well-ventilated area, away from heat, sparks, flame

Eastman® Methyl Acetate, Cosmetic Grade [Eastman]
Chem. Descrip.: Methyl acetate (> 98%), methanol (< 1%), water (< 1%)
Uses: Solvent for cosmetics, personal care, hair care, nail care
Regulatory: SARA §311/312 acute health/chronic health/fire hazard, §313 nonreportable; Canada DSL, EINECS, Australia AICS, Japan MITI listed
Properties: Colorless liq.; pleasant odor; sol. 22% in water (20 C); completely sol. in ethanol; m.w. 74.09; sp.gr. 0.934 (20/20 C); vapor pressure 178 mm Hg (20 C); b.p. 55.8-58.2 C; flash pt. (TCC) -13 C; autoignition temp. 501 C; ref. index 1.36 (20 C); VOC 1% max.; 98% min. assay
Toxicology: LD50 (oral, rat) > 5000 mg/kg, (skin, rabbit) > 5000 mg/kg; LC50 (inh., rat, 4 h) > 16,000 ppm; ACGIH TLV/TWA 200 ppm (methyl acetate); harmful by inh., skin absorp.; may be fatal or cause blindness on ing.; causes eye/skin irritation; high vapor concs. may cause drowsiness, eye/respiratory tract irritation; TSCA listed
Environmental: BOD5 393 mg/g; LC50 (fathead minnow, 96 h) > 320 mg/l; prevent runoff from entering drains, sewers, streams
Precaution: Extremely flamm. liq. and vapor; LEL 3.1%; UEL 16%; vapors may cause flash fire, ignite explosively, or flash back; residual vapors may explode on ignition; reacts with strong oxidizing agents
Hazardous decomp. prods.: Combustion: CO, CO_2
HMIS: Health 2, Flammability 3, Reactivity 0
Storage: Store in tightly closed containers in well-ventilated area, away from heat, sparks, flame

Eastman® Methyl Acetate, High Purity [Eastman]
Chem. Descrip.: Methyl acetate
CAS 79-20-9; EINECS/ELINCS 201-185-2
Uses: Solvent for coating and ink resins incl. cellulose acetate butyrate, NC, vinyl chloride acetate, melamine, isocyanate, acrylic, epoxy, polyamide, vinyl butyral, phenolic, alkyd, polyester, and urea, coatings, moisture-sensitive paints, e.g., PU coatings, plastic coatings, printing inks, fragrances, cosmetics, personal care
Features: Fast evaporating; exempt from regulation as VOC
Regulatory: SARA §311/312 acute health/chronic health/fire hazard, §313 nonreportable; Canada DSL, EINECS, Australia AICS, Japan MITI listed
Properties: Colorless liq.; pleasant odor; sol. 22.7% in water (20 C); m.w. 74.09; sp.gr. 0.933 (20/20 C); visc. 0.38 mPa·s; vapor pressure 179.5 mm Hg (20 C); f.p. -98 C; b.p. 55.8-58.2 C; flash pt. (Seta CC) -15 C; autoignition temp. 501 C; ref. index 1.3589 (20 C); surf. tens. 25.2 dynes/cm (20 C); 100% act.
Toxicology: LD50 (oral, rat) > 5000 mg/kg, (skin, rabbit) > 5000 mg/kg; LC50 (inh., rat, 4 h) > 16,000 ppm; ACGIH TLV/TWA 200 ppm; harmful by inh., skin absorp.; may be fatal or cause blindness on ing.; causes eye/skin irritation; high vapor concs. may cause drowsiness, eye/respiratory tract irritation; TSCA listed
Environmental: BOD5 393 mg/g; ThOD 1510 mg/g; LC50 (fathead minnow, 96 h) 320-399 mg/l; prevent runoff from entering drains, sewers, streams
Precaution: Extremely flamm. liq. and vapor; LEL 3.1%, UEL 16%; vapors may cause flash fire, ignite explosively, or flash back; residual vapors may explode on ignition; reacts with strong oxidizing agents
Hazardous decomp. prods.: Combustion: CO, CO_2
HMIS: Health 2, Flammability 3, Reactivity 0
Storage: Store in tightly closed containers in well-ventilated area, away from heat, sparks, flame

Eastman® n-Propanol [Eastman]
Chem. Descrip.: n-Propanol
CAS 71-23-8; EINECS/ELINCS 200-746-9
Uses: Solvent for cosmetics, personal care prods., nail care, coatings, automotive coatings, plastic coatings, printing inks; chemical intermediate
Regulatory: SARA §311/312 acute health/fire hazard, §313 nonreportable; Canada DSL, EINECS, Australia AICCS, Japan MITI, Korea ECL listed
Properties: Colorless liq.; alcohol odor; sol. in water (20 C); m.w. 60.1; sp. gr. 0.8 (20 C); dens. 6.71 lb/gal; vapor pressure 14.5 mm Hg; f.p. -127 C; flash pt. (TCC) 23 C; autoignition temp. 413 C; ref. index 1.39 (20 C); surf. tens. 23.8 dynes/cm (20 C); 100% act.
Toxicology: LD50 (oral, rat) 1.87-6.5 g/kg, (skin, rabbit) 4.0 g/kg; LC50 (inh., rat, 1 h) > 20,000 ppm; ACGIH TLV/TWA 200 ppm; cause eye irritation; high vapor concs. may cause drowsiness and eye/respiratory tract irritation; prolonged/repeated skin contact may cause drying, cracking, irritation; sl. skin/mod. eye irritant; TSCA listed
Environmental: BOD5 1.43-1.6 g/g; BOD20 < 2.0 g/g; COD 1.4 g/g; LC50 (fathead minnow, 96 h) > 1000 mg/l, (golden orfe, 48 h) 4320-4560 mg/l; prevent runoff from entering drains, sewers, streams
Precaution: Flamm. liq. and vapor; vapors may cause flash fire, ignite explosively, or flash back; residual vapors may explode on ignition; reacts violently with strong oxidizing agents
Hazardous decomp. prods.: Combustion: CO, CO_2
HMIS: Health 2, Flammability 3, Reactivity 0
Storage: Store in tightly closed container in well-ventilated area away from heat, sparks, flame

Eastman® n-Propyl Acetate [Eastman]
Chem. Descrip.: n-Propyl acetate
CAS 109-60-4; EINECS/ELINCS 203-686-1
Uses: Solvent for cosmetics, personal care, fragrances, nail care, coatings, automotive coatings, plastic coatings, printing inks
Regulatory: SARA §311/312 acute health/fire hazard, §313 nonreportable; Canada DSL, EINECS, Australia AICCS, Japan MITI, Korea ECL listed
Properties: Colorless liq.; sweet ester odor; sol. 2.3% in water (20 C); m.w. 102.14; sp. gr. 0.889 (20 C); dens. 7.39 lb/gal; vapor pressure 23 mm Hg (20 C); f.p. -92 C; b.p. 99-103 C (760 mm Hg); flash pt. (TCC) 13 C; autoignition temp. 457 C; fire pt. 21 C; ref. index 1.38 (20 C); surf. tens. 24.3 dynes/cm (20 C); 100% act.
Toxicology: LD50 (oral, rat) 9370 mg/kg, (skin, rabbit) > 20 ml/kg; LC50 (inh., rat, 4 h) 8000 ppm; ACGIH TLV/TWA 200 ppm; high vapor concs. may cause drowsiness and eye/respiratory tract irritation; prolonged/repeated skin contact may cause drying, cracking, irritation; TSCA listed
Environmental: ThOD 2040 mg/g; LC50 (golden orfe, 48 h) 97-194 mg/l, (daphnia, 96 h) 511 mg/l; prevent runoff from entering drains, sewers, streams
Precaution: Flamm. liq. and vapor; vapors may cause flash fire, ignite explosively, or flash back; residual vapors may explode on ignition; reacts violently with strong oxidizing agents
Hazardous decomp. prods.: Combustion: CO, CO_2
HMIS: Health 1, Flammability 3, Reactivity 0

Storage: Store in tightly closed containers in well-ventilated area, away from heat, sparks, flame

Eastman® Triacetin, Food Grade [Eastman]
Chem. Descrip.: Triacetin
CAS 102-76-1; EINECS/ELINCS 203-051-9
Uses: Food additive; humectant; solvent; flavorings; food pkg.; food grade plasticizer; colored cosmetics; skin care; animal feeds; pet foods
Regulatory: FDA 21CFR §184.1901, GRAS; avail. in kosher grade; DOT nonregulated; SARA §313 nonreportable; Canada DSL listed
Properties: Colorless liq.; sl. odor; mod. sol. in water; sp.gr. 1.16 (24 C); visc. 17.4 cps; vapor pressure 1.33 mbar (100 C); f.p. 3 C; b.p. 258 C; decomp. pt. 265 C; flash pt. (COC) 153.3 C; ref. index 1.429-1.431; 99.5% assay
Toxicology: LD50 (oral, rat) 6400-12,800 mg/kg, (skin, guinea pig) > 20 ml/kg; sl. skin/eye irritant; nonsensitizing to skinTSCA listed
Environmental: LC50 (fathead minnow, 96 h) > 165 mg/l; EC50 (daphnia, 48 h) > 811 mg/l
Precaution: Reacts with strong oxidizing agents
Hazardous decomp. prods.: Combustion: CO, CO_2
HMIS: Health 0, Flammability 1, Reactivity 0
Storage: Keep container closed

Eastman® Triethyl Phosphate [Eastman]
Chem. Descrip.: Triethyl phosphate
CAS 78-40-0; EINECS/ELINCS 201-114-5
Uses: Chemical intermediate; solvent; catalyst; flame-resist. unsat. polyester resins; hydraulic brakes
Regulatory: DOT nonregulated; SARA §311/312 acute health hazard, §313 nonreportable; Canada DSL, EINECS, Australia AICS, Japan MITI, Korea ECL listed
Properties: Colorless clear liq.; mild sweet cider-like odor; m.w. 182.16; sp.gr. 1.067-1.072 (20/20 C); visc. 1.76 cP (20 C); vapor pressure 1.33 mbar (40 C); m.p. -56.4 C; b.p. 209 C; flash pt. (PMCC) 99 C; fire pt. 141 C; autoignition temp. 227 C; ref. index 1.4037; 100% act.
Toxicology: LD50 (oral, rat) 1311 mg/kg, (skin, guinea pig) > 21,400 mg/kg; LC50 (inh., rat, 6 h) > 2.05 mg/l; mod. eye irritant; high vapor concs. may cause drowsiness; sl. skin irritant; nonsensitizing to skinTSCA listed
Environmental: COD 1.55 mg/g; LC50 (fathead minnow, 96 h) > 1070 mg/l; EC50 (daphnia, 96 h) 353.1 mg/l; prevent runoff from entering drains, sewers, streams
Precaution: Reacts with strong oxidizing agents
Hazardous decomp. prods.: Combustion: CO, CO_2, PO_x
HMIS: Health 2, Flammability 1, Reactivity 0

Eastone FSE [Eastman]
Chem. Descrip.: Glycol ether
Uses: Solvent for lithography
Properties: Pt-Co 10 max. color; sp.gr. 0.913 (20.20 C); dens. 7.60 lb/gal; f.p. -90 C; b.p. 150-159 C; flash pt. (TCC) 49 C; 99% min. assay

Eccobrite Cleaner AK [Eastern Color & Chem.]
Uses: Solvent cleaner, scouring agent for textiles, pigment printing cleaning, leather processing/finishing
Features: Nonhazardous; noncombustible; nonstaining; removes greases, waxes, oils, and dirt; removes old dyes and colors from leather skins
Properties: Lt. amber clear liq.; mild odor; sol. < 50 g/l in water; sp.gr. 1.09 (15.5 C); visc. 2.4-2.5 cps; f.p. -25 C; flash pt. (PMCC) 235 C; 100% act.
Use Level: 4-5 gal/100 lb (leather)
Environmental: Highly biodeg.; environmentally friendly

Eccoclean RPW [Eastern Color & Chem.]
Uses: Solvent cleaner providing rapid penetration and solvency for difficult cleaning problems
Features: Self-emulsifying
Properties: Combustible

Ecco Cleaner 930 Conc. [Eastern Color & Chem.]
Chem. Descrip.: Organic ester sol'n.
Uses: Solvent cleaner, detergent, and scouring agent for nat. and syn. fabrics and fibers, esp. for removal of spinning oils from woolen fabrics
Features: Contains ester-type solvs. most useful in general scouring needs for removing oils, greases, and soils
Regulatory: DOT nonregulated; SARA nonreportable
Properties: Lt. amber clear liq.; mild odor; sol. in water; sp.gr. 1.04; dens. 8.2 lb/gal; b.p. 212 F; flash pt. none; pH 8.5 (2%); anionic; 60% act.
Use Level: 0.25-0.75% (wool scour), full str. (cleaner)
Toxicology: Nontoxic; will cause irritation and possible tearing on eye contact with undiluted form; may cause mild skin irritation, defatting of tissues, possible corneal damage; may be harmful if swallowed; may cause nausea, vomiting, diarrheaTSCA listed
Environmental: Biodeg.
Precaution: Incompat. with acids; spills may be slippery
Hazardous decomp. prods.: Thermal decomp. yields CO, CO_2, and other noxious vapors
Storage: Store indoors; keep from freezing

Ecco Cleaner AK [Eastern Color & Chem.]
Uses: Solvent cleaner, spot cleaner for textiles, machinery and equip., print screens, leather; wax stripper; degreaser for elec. components and motors; solvent for adhesives, inks, coatings
Features: Rule 442-exempt; replaces common solv. cleaners, chlorinated solvs., aromatics, glycol ethers; stable in air @ normal room conditions; dries completely without leaving residue
Properties: Lt. amber clear liq.; mild odor; m.w. 160; sp.gr. 1.09 (15.5 C); visc. 2.4-2.5 cps; vapor pressure 0.1 max. mm Hg (20 C); f.p. -25 C; i.b.p. 200 C; acid no. 0.5 max.; flash pt. (PMCC) 235 C; ref. index 1.423-1.425 (20 C); 100% act.; 0.1% max. water
Environmental: Biodeg.; environmentally friendly

Ecco Cleaner HR-LO [Eastern Color & Chem.]
Chem. Descrip.: Organic ester/terpene alcohol sol'n.
Uses: Solvent cleaner for removing oils, greases, adhesives, inks and coatings from metal and nonmetallic surfaces, textile equip.
Regulatory: SARA nonreportable
Properties: Lt. amber clear liq.; mild odor; forms stable emulsions in water; sp.gr. 0.99; dens. 8.3 lb/gal; b.p. 212 F; flash pt. (COC) 167 F; pH 6.2-7.2; nonionic; 100% act.
Toxicology: Generally nontoxic; will cause irritation and possible tearing on eye contact with undiluted form; may cause mild skin irritation, defatting of tissues, possible corneal damage; may be harmful if swallowed; may cause nausea, vomiting, diarrheaTSCA listed
Environmental: Biodeg.; environmentally friendly
Precaution: Incompat. with acids; spills may be slippery
Hazardous decomp. prods.: Thermal decomp. yields CO, CO_2, and other noxious vapors
Storage: Store indoors; keep from freezing

Eccoflex 090 [Eastern Color & Chem.]
Chem. Descrip.: Mixed methyl esters
Uses: Solvent cleaner, spot cleaner for textiles, machinery and equip., print screens, leather; wax stripper; degreaser for elec. components and motors; solvent for adhesives, inks, and coatings
Features: Environmentally friendly; replaces common solvent cleaners, chlorinated solvs., aromatic, glycol ethers
Properties: Lt. amber clear liq., mild odor; m.w. 160; sp.gr. 1.09 (15.5 C); visc. 2.4-2.5 cps; vapor pressure 0.1 max. mm Hg (20 C); f.p. -25 C; i.b.p. 200 C; acid no. 0.5 max.; flash pt. (PMCC) 235 C; ref. index 1.423-1.425; 100% act.
Environmental: Highly biodeg.; environmentally friendly; BOD 0.9 g/g (23 days)

Ecco Strip LS-W [Eastern Color & Chem.]
Chem. Descrip.: Emulsifiable ester sol'n. contg. methyl pyrrolidone (< 30%) and ethylene glycol butyl ether (< 10%)
Uses: Detergent, stripping agent, solvent cleaner, spot cleaner for textiles, machinery and equip., print screens, leather; wax stripper; degreaser for elec. components and motors; solvent for adhesives, inks and coatings
Features: Syn.; based on org. esters that can be used in place of common solvent cleaners, chlorinated solvs., aromatic and glycol esters; dries completely without leaving residue under controlled conditions
Regulatory: Exempt from Rule 442; SARA reportable
Properties: Lt. amber clear liq.; mild odor; water-reducible; forms stable emulsions in water; sp.gr. 1.05 (15.5 C); visc. 2.0 cps; b.p. 212 F; acid no. 0.5 max.; flash pt. (COC) 198 F; VOC 76.5%; 100% act.
Toxicology: Generally nontoxic; eye contact will cause irritation and possible tearing in undiluted state; may cause mild skin irritation, defatting of tissues, possible corneal damage; may cause nausea, vomiting, diarrheaTSCA listed
Environmental: Environmentally friendly; biodeg.
Precaution: Incompat. with acids
Hazardous decomp. prods.: Thermal decomp. yields CO, CO_2, and other noxious vapors

Storage: Store indoors; keep from freezing

Ecco Tipping Solution #3 [Eastern Color & Chem.]
Chem. Descrip.: Solvent blend
Uses: Solvent providing exc. shoelace tipping with cellulose acetate
Features: Improved economy

Ecosoft™ PB Solvent [Dow]
Chem. Descrip.: Ethylene glycol butyl ether mixt.
Uses: Solvent for solv.-based coatings, paint/floor strippers, well drilling water-based muds, mining industry metal extraction, tint bases, specialty inks; coalescing agent for industrial water-based coatings; coupling agent, solvent in household/industrial cleaners, rust removers, hard surf. cleaners, disinfectants; octane/cetane booster in fuels; deicer, moisture control agent for diesel fuels; chem. intermediate
Features: High b.p.; low volatility; powerful solvency for oils
Properties: Completely sol. in water; misc. with alcohol, ketones, esters, most common org. solvs. incl. aliphatic and aromatic hydrocarbons; sp.gr. 1.001 (20/20 C); dens. 8.33 lb/gal (20 C); visc. 10.4 cP; vapor pressure < 0.01 mm Hg (20 C); f.p. -33 C; b.p. 311.1 C; flash pt. (PMCC) 149 C; surf. tens. 39.9 dynes/cm
Environmental: Readily biodeg.

Ecosoft™ PE Solvent [Dow]
Chem. Descrip.: Ethylene glycol ethyl ether mixt.
Uses: Solvent for solv.-based coatings, paint/floor strippers, well drilling water-based muds, mining industry metal extraction, tint bases, specialty inks; coalescing agent for industrial water-based coatings; coupling agent, solvent in household/industrial cleaners, rust removers, hard surf. cleaners, disinfectants; octane/cetane booster in fuels; deicer, moisture control agent for diesel fuels; chem. intermediate
Features: High b.p.; low volatility; powerful solvency for oils
Properties: Completely sol. in water; misc. with alcohol, ketones, esters, most common org. solvs. incl. aliphatic and aromatic hydrocarbons; sp.gr. 1.041 (20/20 C); dens. 8.66 lb/gal (20 C); visc. 13.2 cP; vapor pressure < 0.01 mm Hg (20 C); f.p. -15 C; b.p. 310 C; flash pt. (PMCC) 143 C; surf. tens. 43 dynes/cm
Environmental: Readily biodeg.

Ecosoft™ PH Solvent [Dow]
Chem. Descrip.: Ethylene glycol hexyl ether mixt.
Uses: Solvent for solv.-based coatings, paint/floor strippers, well drilling water-based muds, mining industry metal extraction, tint bases, specialty inks; coalescing agent for industrial water-based coatings; coupling agent, solvent in household/industrial cleaners, rust removers, hard surf. cleaners, disinfectants; octane/cetane booster in fuels; deicer, moisture control agent for diesel fuels; chem. intermediate
Features: High b.p.; low volatility; powerful solvency for oils
Properties: Completely sol. in water; misc. with alcohol, ketones, esters, most common org. solvs. incl. aliphatic and aromatic hydrocarbons; sp.gr. 0.986 (20/20 C); dens. 8.20 lb/gal (20 C); visc. 15.3 cP; vapor pressure < 0.01 mm Hg (20 C); f.p. -33 C; b.p. 289.9 C; flash pt. (PMCC) 157 C; surf. tens. 39.8 dynes/cm
Environmental: Readily biodeg.

Ecosoft™ Solvent IK [Dow]
Chem. Descrip.: Isobutyl heptyl ketone
CAS 123-18-2; EINECS/ELINCS 204-607-3
Uses: Solvent
Properties: Sol. 0.002% in water; m.w. 184.32; sp.gr. 0.820 (20/20 C); dens. 6.83 lb/gal; visc. 1.4 cP; vapor pressure 0.03 mm Hg (20 C); b.p. 217.2 C; flash pt. 170 F; surf. tens. 26.8 dynes/cm (20 C)

EcoSolv™ Dry Cleaning Fluid [ChevronPhillips]
Chem. Descrip.: C10-13 isoparaffins, preservative BHT (10 ppm)
CAS 68551-17-7; EINECS/ELINCS 271-366-9
UN 3295
Uses: Dry cleaning fluid
Regulatory: SARA C313 nonreportable; Canada DSL listed
Properties: Colorless liq.; odorless; negligible sol. in water; sp.gr. 0.758-0768 (60.60 F); visc. 1.55 cSt (38 C); vapor pressure 1.5 mm Hg (38 C); b.p. 181-209 C; flash pt. (PMCC) 62.2 C; 100% act.
Toxicology: LD50 (oral, rat) > 34.6 g/kg, (skin, rabbit) 15.4 g/kg; LC50 (inh., rat, 6 h, 14 day post exposure) > 1277 ppm; REL (8 h TWA) 300 ppm as total petrol. hydrocarbon; may be sl. irritating to eyes, skin; inh. may cause general weakness, ataxia, rapid respiration, nausea, tremors; ing. may irritate stomach, intestines; if swallowed, may be aspirated into lungs resulting in inflamm., possible fluid accumulation; may cause kidney damage; TSCA listed
Environmental: Keep out of water sources and sewers
Precaution: Combustible liq.; LEL 1.1% estimated; UEL 6.0% estimated; highly flamm. vapors may accumulate in low areas; may flash back; incompat. with oxygen and strong oxidizing agents
Hazardous decomp. prods.: CO_x formed when burned
NFPA: Health 0, Flammability 2, Reactivity 0
Storage: Store in closed containers in well-ventilated area, away from heat, sparks, flame; bond and ground during transfer

Edenor® ME C6-10 [Cognis/Chems. Group]
Chem. Descrip.: Methyl caprylate/caprate
Chem. Analysis: Methyl caprylate (55%), methyl caprate (40%), methyl caproate (4%), methyl laurate (1%)
CAS 67762-39-4
Uses: Agric. solvent; surfactants; lubricants; oil components
Properties: Acid no. 0.5 max.; iodine no. 1 max.; sapon. no. 320-340; 0.1% max. moisture

Edenor® ME C18-90 [Cognis/Chems. Group]
Chem. Descrip.: Methyl stearate
CAS 112-61-8; EINECS/ELINCS 203-990-4
Uses: Detergent intermediate; solvent for pesticides, herbicides; lubricants; oil components
Regulatory: EPA exempt
Properties: Gardner 1 color; m.p. 36 C; acid no. 0.5 max.; iodine no. 1.0 max.; sapon. no. 188-192; flash pt. 307 F; 95% C18 ester

EDGE [INEOS Oxide]
Chem. Descrip.: Ethoxydiglycol
CAS 111-90-0; EINECS/ELINCS 203-919-7
Uses: Solvent for resins (NC, acrylic, epoxy, polyamide, alkyd), oils, waxes, fats, dyestuffs, insecticides, herbicides, grease/grime, surf. coatings, lacquers, enamels, waterborne coatings, phenolic varnishes, epoxy coatings, non-grain-raising wood stains; gloss aid, flow aid in lacquers; dye solvent/stabilizer/penetration aid, color enhancer for textiles, leather, printing; mutual solvent, coupling agent in sol. oils, metal/glass cleaners, varnish removers; diluent in hydraulic brake fluids
Features: Very stable; low f.p.; relatively low vapor pressure
Properties: Pt-Co 10 max. color; mild odor; misc. with wide range of solvs., water; m.w. 134.18; dens. 0.987-0.991 g/ml (20 C); visc. 4.5 mPa·s (20 C); vapor pressure 0.13 hPa (20 C); i.b.p. 197.5 C min. (101.3 kPa); pour pt. -78 C; flash pt. (CC) 83C; autoignition temp. 204 C; ref. index 1.4273 (20 C); 99% min. purity
Precaution: Flamm. limits 1.2-11.6 vol.% (20 C); not a significant flamm. hazard @ normal storage temps.; very reactive with zinc (hydrogen may be produced); contact with zinc, tin, copper/alloys can cause discoloration

EGE [INEOS Oxide]
Chem. Descrip.: Ethoxyethanol
CAS 110-80-5; EINECS/ELINCS 203-804-1
Uses: Solvent for resins (NC, acrylic, epoxy, polyamide, alkyd), oils, waxes, fats, dyestuffs, insecticides, herbicides, grease/grime, surf. coatings, lacquers, enamels, waterborne coatings, phenolic varnishes, epoxy coatings, non-grain-raising wood stains; gloss aid, flow aid in lacquers; dye solvent/stabilizer/penetration aid, color enhancer for textiles, leather, printing; mutual solvent, coupling agent in sol. oils, metal/glass cleaners, varnish removers; diluent in hydraulic brake fluids
Features: Very stable; low f.p.; relatively low vapor pressure
Properties: Pt-Co 10 max. color; mild odor; misc. with wide range of solvs., water; m.w. 90.12; dens. 0.927-0.930 g/ml (20 C); visc. 2.1 mPa·s (20 C); vapor pressure 5 hPa (20 C); i.b.p. 134 C min. (101.3 kPa); pour pt. -90 C; flash pt. (CC) 42 C; autoignition temp. 235 C; ref. index 1.4076 (20 C); 99.5% min. purity
Precaution: Flamm. limits 1.8-15.7 vol.% (20 C); not a significant flamm. hazard @ normal storage temps.; very reactive with zinc (hydrogen may be produced); contact with zinc, tin, copper/alloys can cause discoloration

EmCon™ CO [Fanning]
Chem. Descrip.: Castor oil USP
CAS 8001-79-4; EINECS/ELINCS 232-293-8
Uses: Skin conditioner, solvent for lipsticks, eye shadow, blushes, makeup, nail polish, pharmaceuticals
Features: Occlusive
Properties: Gardner 1 oil; sp.gr. 0.957-0.961; acid no. 2 max.; iodine no.

83-88; sapon. no. 176-182; hyd. no. 160-168

EmCon™ Olive [Fanning]
Chem. Descrip.: Olive oil USP
CAS 8001-25-0; EINECS/ELINCS 232-277-0
Uses: Hair conditioner, skin conditioner, solvent for shampoos, tonics, cleansing prods., face/body/hand lotions, suntan gels, creams, lotions, pharmaceutical ointments
Features: Occlusive
Properties: Pale straw clear liq., bland odor; sp.gr. 0.910-0.915; acid no. 3.5 max.; iodine no. 79-88; sapon. no. 190-195

EmCon™ Rice Bran [Fanning]
Chem. Descrip.: Rice bran oil
CAS 68553-81-1; EINECS/ELINCS 271-397-8
Uses: Skin conditioner, solvent for face powds., cleansing prods., moisturizing creams and lotions, pharmaceutical ointments
Features: Occlusive
Properties: Sp.gr. 0.971-0.973; acid no. 0.10 max.; iodine no. 102-110; sapon. no. 185-195

EmCon™ SES [Fanning]
Chem. Descrip.: Sesame (Sesamum indicum) oil
CAS 8008-74-0; EINECS/ELINCS 232-370-6
Uses: Hair conditioner, skin conditioner, solvent in cosmetics
Features: Occlusive
Properties: Gardner 8 max. liq.; sp.gr. 0.910-0.925; iodine no. 100-116

EmCon™ SUN [Fanning]
Chem. Descrip.: Sunflower (Helianthus annuus) seed oil
CAS 8001-21-6; EINECS/ELINCS 232-273-9
Uses: Hair conditioner, skin conditioner, solvent in cosmetics
Features: Occlusive
Properties: Gardner 3 max. liq.; iodine no. 130-140; smoke pt. 400 F

EmCon™ W [Fanning]
Chem. Descrip.: Wheat germ oil USP
CAS 8006-95-9
Uses: External lubricant, hair conditioner, skin conditioner, solvent for hair conditioners, shampoos, lipsticks, cleansers, moisturizing creams and lotions, skin care prods., pharmaceuticals; lubricant for topical ointments
Features: Occlusive
Properties: Gardner 10 max. oil; acid no. 0.5 max.; iodine no. 120-135; sapon. no. 180-195

Emerest® 2452 [Cognis; Cognis Canada]
Chem. Descrip.: Polyglyceryl-3 diisostearate
CAS 66082-42-6; EINECS/ELINCS 291-548-1
Uses: Emulsifier, solubilizer, dye/pigment wetting agent, emollient, thickener, solvent for creams, lotions, lip prods.
Properties: Amber visc. liq.; sol. @ 5% in min. oil, triolein, IPA, IPM; disp. in glycerin; dens. 8.2 lb/gal; visc. 990 cSt (100 F); HLB 6.7; acid no. 20 max.; iodine no. 0.5 max.; sapon. no. 165-175; pour pt. 4 C; flash pt. 455 F; nonionic; 100% conc.

Emery® 912 [Cognis/Chems. Group]
Chem. Descrip.: Glycerin CP/USP
CAS 56-81-5; EINECS/ELINCS 200-289-5
Uses: Skin softener, visc. modifier, flavor enhancer, moisturizer, solvent, humectant, thickener, solubilizer in cosmetics, drug vehicles, foods, glass, ceramics, agric., and adhesives
Regulatory: EPA-exempt; SARA nonreportable
Properties: Water-wh. visc. liq.; odorless; sol. in water; sp.gr. 1.2517; m.p. 18 C; b.p. 171 C; pour pt. 18 C; flash pt. (CC) 199 C; 96.0% min. glycerol
Toxicology: TLV:TWA 10 mg/m^3 (mist); LD50 (oral, rat) 12.6 g/kg; mild eye irritant, moderate skin irritant; inhalation of mist may cause respiratory irritationTSCA listed
Storage: Store in cool, dry, well-ventilated area; avoid excessive heat, open flames; incompat. with strong acids and oxidizers

Emery® 916 [Cognis/Chems. Group]
Chem. Descrip.: Glycerin CP/USP
CAS 56-81-5; EINECS/ELINCS 200-289-5
Uses: Skin softener, visc. modifier, flavor enhancer, moisturizer, solvent, humectant, thickener, solubilizer in cosmetics, drug vehicles, foods, glass, ceramics, agric., and adhesives
Regulatory: EPA exempt
Properties: Water-wh. visc. liq.; odorless; sol. in water; sp.gr. 1.2607; m.p. 18 C; b.p. 171 C; pour pt. 18 C; flash pt. (CC) 199 C; 99.5% min. glycerol
Toxicology: TLV:TWA 10 mg/m^3 (mist); LD50 (oral, rat) 12.6 g/kg; mild eye irritant, moderate skin irritant; inhalation of mist may cause respiratory irritation
Storage: Store in cool, dry area; avoid excessive heat, open flames; incompat. with strong acids and oxidizers

Emery® 917 [Cognis/Chems. Group]
Chem. Descrip.: Glycerin CP/USP kosher
CAS 56-81-5; EINECS/ELINCS 200-289-5
Uses: Skin softener, visc. modifier, flavor enhancer, moisturizer, solvent, humectant, thickener, solubilizer in cosmetics, drug vehicles, foods, glass, ceramics, agric., and adhesives
Properties: Water-wh. liq.; sol. in water; sp.gr. 1.2607; m.p. 18 C; b.p. 171 C; flash pt. (CC) 199 C; 99.7% min. glycerol
Toxicology: TLV:TWA 10 mg/m^3 (mist); LD50 (oral, rat) 12.6 g/kg; mild eye irritant, moderate skin irritant; inhalation of mist may cause respiratory irritation
Storage: Store in cool, dry area; avoid excessive heat, open flames; incompat. with strong acids and oxidizers

Emery® 2214 [Cognis/Chems. Group]
Chem. Descrip.: Methyl myristate
CAS 124-10-7; EINECS/ELINCS 204-680-1
Uses: Detergent intermediate; solvent for pesticides, herbicides; lubricants; oil components
Regulatory: EPA exempt
Properties: Gardner 1 color; dens. 7.2 lb/gal; visc. 3.4 cSt (100 C); m.p. 17 C; acid no. 1.0 max.; iodine no. 0.6 max.; sapon. no. 230-234; flash pt. 300 F; 95% C14 ester

Emery® 2216 [Cognis/Chems. Group]
Chem. Descrip.: Methyl palmitate
CAS 112-39-0; EINECS/ELINCS 203-966-3
Uses: Detergent intermediate; solvent for pesticides, herbicides
Regulatory: EPA exempt
Properties: Gardner 1 color; m.p. 27 C; acid no. 0.2 max.; iodine no. 0.2 max.; sapon. no. 206-210; flash pt. 330 F; 95% C14 ester

Emery® 2230 [Cognis/Chems. Group]
Chem. Descrip.: Methyl sunflowerate
Uses: Solvent, adjuvant for pesticides, herbicides
Features: Exc. low temp. props.
Regulatory: EPA exempt
Properties: Gardner 1 color; dens. 7.3 lb/gal; visc. 4.6 cSt (100 C); m.p. -2 C; flash pt. 345 F

Emery® 2232 [Cognis]
Chem. Descrip.: Methyl canolate
CAS 73891-99-3
Uses: Solvent for pesticides, herbicides
Regulatory: EPA exempt
Properties: Gardner 10 color; dens. 7.3 lb/gal; visc. 5.0 cSt (100 C); m.p. < -10 C; flash pt. 345 F

Emery® 2253 [Cognis/Chems. Group]
Chem. Descrip.: Methyl coconate
CAS 61788-59-8; EINECS/ELINCS 262-988-1
Uses: Detergent intermediate; solvent for pesticides, herbicides
Regulatory: EPA exempt
Properties: Gardner 1 color; dens. 7.5 lb/gal; visc. 2.8 cSt (100 C); m.p. 4 C; acid no. 0.5 max.; iodine no. 7-11; sapon. no. 250-260; flash pt. 250 F

Emery® 2254 [Cognis/Chems. Group]
Chem. Descrip.: Stripped methyl coconate
CAS 61788-59-8; EINECS/ELINCS 262-988-1
Uses: Detergent intermediate; solvent for pesticides, herbicides; lubricants; oil components
Regulatory: EPA exempt
Properties: Gardner 1 color; acid no. 1.0 max.; iodine no. 5-10; sapon. no. 237-247; flash pt. 295 F

Emery® 2255 [Cognis/Chems. Group]
Chem. Descrip.: Methyl palm kernelate
Uses: Detergent intermediate; agric. solvent; lubricants; oil components
Properties: Acid no. 1.0 max.; iodine no. 14-20; sapon. no. 230-240

Emery® 2270 [Cognis/Chems. Group]
Chem. Descrip.: Methyl laurate
CAS 111-82-0; EINECS/ELINCS 203-911-3

Uses: Detergent intermediate; solvent for pesticides, herbicides; lubricants; oil components
Regulatory: EPA exempt
Properties: Gardner 1 color; dens. 7.2 lb/gal; visc. 2.8 cSt (100 C); m.p. -1 C; acid no. 0.5 max.; iodine no. 0.5 max.; sapon. no. 251-255; flash pt. 270 F; 70% C12, 28% C14.

Emery® 2290 [Cognis/Chems. Group]
Chem. Descrip.: Methyl laurate
CAS 111-82-0; EINECS/ELINCS 203-911-3
Uses: Detergent intermediate; solvent for pesticides, herbicides
Regulatory: EPA exempt
Properties: Gardner 1 color; dens. 7.2 lb/gal; visc. 2.5 cSt (100 C); m.p. 2 C; acid no. 0.5 max.; iodine no. 0.5 max.; sapon. no. 258-262; flash pt. 265 F; 90% C12 ester

Emery® 2296 [Cognis/Chems. Group]
Chem. Descrip.: Methyl laurate
CAS 111-82-0; EINECS/ELINCS 203-911-3
Uses: Detergent intermediate; solvent for pesticides, herbicides; lubricants; oil components
Regulatory: EPA exempt
Properties: Gardner 1 color; dens. 7.2 lb/gal; visc. 2.3 cSt (100 C); m.p. 5 C; acid no. 0.5 max.; iodine no. 0.5 max.; sapon. no. 260-264; flash pt. 305 F; 96% C12 ester

Emery® 5794 [Cognis]
Chem. Descrip.: PEG-3 triethanolamine
CAS 36936-60-4
Uses: Solvent neutralizer for acid dyes
Properties: Dk. brn. liq.; m.w. 281; hyd. no. 590-630; amine no. 196-204; flash pt. (COC) 315 F

Emery® 6705 [Cognis]
Chem. Descrip.: Phenoxyethanol
CAS 122-99-6; EINECS/ELINCS 204-589-7
Uses: Solvent for NC, cellulose acetate, ethylcellulose, and vinyl, alkyl, and ester type resins in printing inks and special formulations for removing industrial finishes and coatings
Properties: Gardner 2 liq.; sol. 5% in Stod., xylene; water-insol.; dens. 9.1 lb/gal; visc. 10 cSt (100 F); pour pt. 13 C; flash pt. 250 F

Emulsifier Q [Consos]
Chem. Descrip.: Solvent/emulsifier blend
Uses: Solvent scouring agent for heavily oiled and waxed textiles
Properties: Liq.; anionic
Environmental: Biodeg.

Envirolite® Solvent [Shell]
Chem. Descrip.: Solvent naphtha (petroleum), medium aliphatic
CAS 64742-88-7; EINECS/ELINCS 265-191-7
UN 1268
Uses: Solvent for use as charcoal lighter fluid
Features: Complies with air emissions test of US EPA
Regulatory: SARA §311/312 acute health/chronic health/fire hazard
Properties: Light-colored liq.; hydrocarbon odor; negligible sol. in water; m.w. 140; sp.gr. 0.77 (60 F); vapor pressure < 2 mm Hg (68 F); b.p. 320-351 F; flash pt. 40 C; autoignition temp. 240 C; VOC 6.4 lb/gal (60 F)
Toxicology: LD50 (skin, rabbit) 5 ml/kg; if aspirated, can cause severe lung damage, chem. pneumonitis; may be fatal if swallowed; causes skin irritation; may cause CNS depression (dizziness, lightheadedness, headache, nausea, incoordination, unconsciousness, death); vapors may be irritating to respiratory system; may cause eye irritation/temporary discomfort, skin irritation/burning sensation/redness/swelling/defatting; TSCA listed
Environmental: Prevent entry into waterways, sewer
Precaution: Flamm. liq.; vapors are heavier than air; vapors may flash back; flamm. in air 1-7 vol.%; prevent vapor accumulation; static electricity may accumulate and create a fire hazard; bond and ground equip. and transfer containers; avoid contact with strong oxidizing agents; prevent entry into basements, confined areas
Hazardous decomp. prods.: Complex mixt. of airborne solids, liqs., and gases incl. CO, CO_2, other org. compds.
HMIS: Health 2, Flammability 2, Reactivity 0
Storage: Keep containers closed when not in use; keep away from heat, sparks, flames

Epiol M-0 [NOF]
Chem. Descrip.: α-Monochlorohydrine
Uses: Intermediate for pharmaceuticals; solvent for agric. chems.; accelerator for hardening epoxy resins

Escaid® 90 [Exxon Canada]
Chem. Descrip.: Petroleum distillates, hydrotreated light
CAS 64742-47-8; EINECS/ELINCS 265-149-8
Uses: Solvent; diluent; chem. feedstock; fuel
Regulatory: Canada DSL listed; NPRI nonreportable
Properties: Colorless clear liq.; sol. < 0.01% in water; m.w. 155; dens. 0.79 kg/l (15 C); visc. 1.64 cSt; vapor pressure 0.07 kPa (20 C); f.p. -50 C; b.p. 188-209 C; flash pt. (TCC) 64 C; autoignition temp. ≈ 227 C; 100% act.
Toxicology: REL TWA 184 ppm; LD50 (oral, rat) > 5 g/kg, (skin, rabbit) > 3 g/kg; inh. of high vapor concs. may cause eye/respiratory tract irritation, headaches, dizziness, anesthesia, drowsiness, unconsciousness, other CNS effects, death; sl. eye irritant; frequent/prolonged skin contact may cause irritation, dermatitis; if aspirated, may cause mild to severe pulmonary injury, death; chronic exposure may cause stomach/liver/kidney damage
Environmental: Prevent entry into sewers, watercourses, low areas
Precaution: Combustible liq.; flamm. limits 1.4-9.3 vol.%; volatile; gives off invisible vapors; liq. or vapor may settle in low areas; distant ignition/explosion possible; electrostatic accumulation hazard; use proper grounding procedures; incompat. with strong oxidizing agents
Hazardous decomp. prods.: None
Storage: Store in cool, well-ventilated area away from incompat. materials, heat, open flame, ignition sources, direct sunlight; keep container closed

Escaid® 95 [Exxon Canada]
Chem. Descrip.: Petroleum distillates, hydrotreated light
CAS 64742-47-8; EINECS/ELINCS 265-149-8
Uses: Solvent; diluent; chem. feedstock; fuel
Regulatory: Canada DSL listed; NPRI nonreportable
Properties: Colorless clear liq.; sol. < 0.01% in water; m.w. 162; sp.gr. 0.793 (15.6 C); visc. 1.63 cSt; vapor pressure 0.07 kPa (20 C); f.p. -46 C; b.p. 187-208 C; flash pt. (TCC) 63 C; autoignition temp. > 253 C; 100% act.
Toxicology: REL TWA 184 ppm; LD50 (oral, rat) > 5 g/kg, (skin, rabbit) > 3 g/kg; inh. of high vapor concs. may cause eye/respiratory tract irritation, headaches, dizziness, anesthesia, drowsiness, unconsciousness, other CNS effects, death; sl. eye irritant; frequent/prolonged skin contact may cause irritation, dermatitis; if aspirated, may cause mild to severe pulmonary injury, death; chronic exposure may cause stomach/liver/kidney damage
Environmental: Prevent entry into sewers, watercourses, low areas
Precaution: Combustible liq.; flamm. limits 0.7-5.3 vol.%; volatile; gives off invisible vapors; liq. or vapor may settle in low areas; distant ignition/explosion possible; electrostatic accumulation hazard; use proper grounding procedures; incompat. with strong oxidizing agents
Hazardous decomp. prods.: None
Storage: Store in cool, well-ventilated area away from incompat. materials, heat, open flame, ignition sources, direct sunlight; keep container closed

Escaid® 110 [ExxonMobil]
Chem. Descrip.: Petroleum distillates, hydrotreated light
CAS 64742-47-8; EINECS/ELINCS 265-149-8
UN 1268
Uses: Solvent
Regulatory: SARA §311/312 fire hazard, §313 nonreportable
Properties: Colorless clear liq.; sol. < 0.01% in water; m.w. 177; sp.gr. 0.80 (15.6 C); dens. 6.65 lb/gal; vapor pressure 0.2 mm Hg (20 C); f.p. -22 F; b.p. 408-442 F; flash pt. (PMCC) 78.9 C; autoignition temp. ≈ 421 F
Toxicology: Inh. of high vapor concs. may cause eye/respiratory tract irritation, headaches, dizziness, anesthesia, drowsiness, unconsciousness, other CNS effects, death; sl. eye irritant; frequent/prolonged skin contact may cause irritation, dermatitis; if aspirated, may cause mild to severe pulmonary injury, death; chronic exposure may cause liver/kidney damage; TSCA listed
Environmental: Prevent entry into sewers, watercourses, low areas
Precaution: Combustible liq.; flamm. limits 1.3-8.1%; volatile; gives off invisible vapors; liq. or vapor may settle in low areas; distant ignition/explosion possible; electrostatic accumulation hazard; use proper ground-

ing procedures; incompat. with strong oxidizing agents
Hazardous decomp. prods.: None
HMIS: Health 1, Flammability 2, Reactivity 0
Storage: Store @ ambient temps., atmospheric pressure in cool, well-ventilated area away from incompat. materials, heat, open flame, ignition sources, direct sunlight; keep container closed

Estalan 430 [Lanaetex Prods.]
Chem. Descrip.: Diethylene glycol dioctanoate/diisononanoate
Uses: Emollient, penetrant, spreading agent, perfume solvent in antiperspirants, deodorants, preshave lotions, and other liqs.
Properties: Clear liq.; typ. clean odor; acid no. 5 max.; sapon. no. 265-285; ref. index 1.430-1.440

Estalan DIA [Lanaetex Prods.]
Chem. Descrip.: Diisopropyl adipate
CAS 6938-94-9; EINECS/ELINCS 230-072-0
Uses: Emollient, lubricant, penetrant, solvent in cosmetic creams and lotions, shave lotions, suntan prods., lipsticks, hand lotions, grooming prods., and bath oils
Features: Compat. in hydroalcoholic systems; rec. for high oil prods. contg. lanolin, min. oil, castor oil, IPM, and petrolatum
Properties: Colorless clear liq.; faint typ. odor; sol. in min. oil, ketones, esters, aromatic, aliphatic, and chlorinated hydrocarbons; insol. in water; sp.gr. 0.960; acid no. 1 max.; iodine no. nil; sapon. no. 480-500

Estalan-DIPS [Lanaetex Prods.]
Chem. Descrip.: Diisopropyl sebacate
CAS 7491-02-3; EINECS/ELINCS 231-306-4
Uses: Emollient, lubricant, solvent, coupling agent for special face preparations such as pre- and after-shave lotions; emollient for use in bath oils, lip prods., men's grooming aids, and hand lotions
Features: Nonoily
Properties: Clear colorless liq.; faint characteristic odor; sol. in alcohols, esters, ketones, higher glycols, silicones, min. and nat. oils; aromatic, aliphatic and chlorinated hydrocarbons; insol. in water, ethylene glycol, propylene glycol, glycerol; sp.gr. 0.932; acid no. 1.0 max.; iodine no. nil; sapon no. 380-400; ref. index 1.4320

Estalan DO [Lanaetex Prods.]
Chem. Descrip.: Decyl oleate
CAS 3687-46-5; EINECS/ELINCS 222-981-6
Uses: Cosmetic ingred. enhancing smooth rub-out chars. in creams and lotions; pigment dispersant/wetting agent; consistency agent for liq. makeup and pressed powds.; solvent in makeup remover prods.
Features: Imparts creamy consistency
Properties: Yel. liq.; mild typ. odor; sol. in peanut oil, min. oil, 95% ethanol, IPM; insol. in water, glycerin, propylene glycol; acid no. 2 max.; iodine no. 45-60; sapon. no. 130-145; ref. index 1.454

Estalan-IDO [Lanaetex Prods.]
Chem. Descrip.: Isodecyl oleate
CAS 59231-34-4; EINECS/ELINCS 261-673-6
Uses: Pigment dispersant for liq. makeup, pressed powds., creams/lotions; solvent for makeup remover prods.
Properties: Pale yel. liq.; typ. odor; sol. in peanut oil, 95% ethyl alcohol, isopropyl myristate, min. oil; insol. in water, glycerin, and propylene glycol; f.p. < 0 C; acid no. 2.0 max.; iodine no. 45-60; sapon. no. 120-140; ref. index 1.455 ± 0.009; nonionic

Estalan-ISI [Lanaetex Prods.]
Chem. Descrip.: Isostearyl isostearate
CAS 41669-30-1; EINECS/ELINCS 255-485-3
Uses: Emollient in creams/lotions; cosolvent, solubilizer in perfumes
Properties: Pale yel. clear oily liq.; bland odor; sol. in min. oil, isopropyl myristate, castor oil, ketones, isopropanol, and aldehydes; insol. in water, propylene glycol, and ethanol; sp.gr. 0.863-0.873; f.p. -5 C; acid no. 1.0 max.; iodine no. 13.0 max.; sapon. no. 95-115; hyd. no. 4.0 max.; ref. index 1.461-1.463; nonionic

Estalan M-5 [Lanaetex Prods.]
Chem. Descrip.: Caprylic/capric triglyceride
CAS 65381-09-1; EINECS/ELINCS 265-724-3
Uses: Carrier cosolvent/coupling agent, emollient in creams/lotions
Regulatory: FDA GRAS
Properties: Gardner 5 max.; bland odor; acid no. 1.0 max.; iodine no. 2.0 max.; sapon. no. 338-358; hyd. no. 10 max.; 0.2% max. moisture

Estalan-PGDP [Lanaetex Prods.]
Chem. Descrip.: Propylene glycol dipelargonate
CAS 41395-83-9; EINECS/ELINCS 255-350-9
Uses: Emollient, cosolvent for perfumes, bath oils, creams/lotions, makeup remover prods.; binder, emollient in pressed powds.
Properties: Water-wh. clear liq.; essentially odorless; f.p. -20 C; acid no. 3.0 max.; iodine no. 1.0 max.; sapon. no. 295-320

Estasan™ GT 8-40 3578 [Uniqema; Uniqema Am.]
Chem. Descrip.: Caprylic/capric triglyceride (40% C8 + 60% C10)
CAS 65381-09-1; EINECS/ELINCS 265-724-3
Uses: Solvent, carrier for drugs, capsules, flavors, fragrances, food ingreds.; lubricant, mold release for foods; fat source in dietetic prods., enteral and parenteral feeding formulas; spreading agent, penetrant, solvent for cosmetics, toiletries, creams, lotions; lubricant for soft gelatin capsules
Regulatory: BP, DAB compliance; Japan approval
Properties: Lovibond 5Y/0.5R oil, odorless, bland taste; visc. 29 mPa·s; acid no. 0.1 max.; iodine no. 2 max.; sapon. no. 300-315; hyd. no. 5 max.; cloud pt. 10 C
Environmental: Readily biodeg.
Storage: Store @ R.T. in dry atmosphere

Estasan™ GT 8-60 3575 [Uniqema; Uniqema Am.]
Chem. Descrip.: Caprylic/capric triglyceride (60% C8 + 40% C10)
CAS 65381-09-1; EINECS/ELINCS 265-724-3
Uses: Solvent, carrier for drugs, capsules, flavors, fragrances, food ingreds.; lubricant for mfg. of soft gelatin capsules; nutrient, fat source in dietetic prods., adult and infant nutrition, enteral and parenteral feeding formulas; lubricant, mold release for food applics.; emollient, spreading agent, penetrant, solvent for topical pharmaceuticals, cosmetics, toiletries, creams, lotions; lubricant for soft gelatin capsules
Features: Wide compat. with many drugs; min. oil replacement
Regulatory: Ph.Eur., BP, DAB compliance; Japan approval
Properties: Lovibond 5Y/0.5R oil, odorless, bland taste; visc. 23 mPa·s; acid no. 0.1 max.; iodine no. 0.5 max.; sapon. no. 325-345; hyd. no. 5 max.; cloud pt. -8 C
Environmental: Readily biodeg.
Storage: Store @ R.T. in dry atmosphere

Estasan™ GT 8-60 3580 [Uniqema; Uniqema Am.]
Chem. Descrip.: Caprylic/capric triglyceride (60% C8 + 40% C10)
CAS 65381-09-1; EINECS/ELINCS 265-724-3
Uses: Solvent, carrier for drugs, capsules, flavors, fragrances, food ingreds.; lubricant, mold release for food applics.; fat source in dietetic prods., enteral and parenteral feeding formulas; spreading agent, penetrant, solvent for cosmetics and toiletries, creams, lotions; emollient, solvent for topical pharmaceuticals; lubricant for soft gelatin capsules
Regulatory: BP, DAB compliance; Japan approval; kosher
Properties: Lovibond 5Y/0.5R oil, odorless, bland taste; visc. 23 mPa·s; acid no. 0.1 max.; iodine no. 0.5 max.; sapon. no. 325-345; hyd. no. 5 max.; cloud pt. -8 C
Environmental: Readily biodeg.
Storage: Store @ R.T. in dry atmosphere

Estasan™ GT 8-65 3577 [Uniqema; Uniqema Am.]
Chem. Descrip.: Caprylic/capric triglyceride (65% C8 + 35% C10)
CAS 65381-09-1; EINECS/ELINCS 265-724-3
Uses: Solvent, carrier for drugs, capsules, flavors, fragrances, food ingreds.; lubricant for mfg. of soft gelatin capsules; nutrient, fat source in dietetic prods., adult and infant nutrition, enteral and parenteral feeding formulas; lubricant, mold release for foods; emollient, spreading agent, penetrant, solvent for topical pharmaceuticals, cosmetics, toiletries, creams, lotions; lubricant for soft gelatin capsules
Features: Wide compat. with many drugs; min. oil replacement
Regulatory: Ph.Eur., BP, DAB compliance; Japan approval
Properties: Lovibond 5Y/0.5R oil, odorless, bland taste; visc. 23 mPa·s; acid no. 0.1 max.; iodine no. 0.5 max.; sapon. no. 325-360; hyd. no. 5 max.; cloud pt. -5 C
Environmental: Readily biodeg.
Storage: Store @ R.T. in dry atmosphere

Estasan™ GT 8-65 3581 [Uniqema; Uniqema Am.]
Chem. Descrip.: Caprylic/capric triglyceride (65% C8 + 35% C10)
CAS 65381-09-1; EINECS/ELINCS 265-724-3
Uses: Solvent, carrier for drugs, capsules, flavors, fragrances, food ingreds.; lubricant, mold release for foods; fat source in dietetic prods., enteral and parenteral feeding formulas; spreading agent, penetrant, solvent for cosmetics and toiletries, creams, lotions; lubricant for soft

gelatin capsules
Regulatory: BP, DAB compliance; Japan approval; kosher
Properties: Lovibond 5Y/0.5R oil, odorless, bland taste; visc. 23 mPa·s; acid no. 0.1 max.; iodine no. 0.5 max.; sapon. no. 335-360; hyd. no. 5 max.; cloud pt. -5 C
Environmental: Readily biodeg.
Storage: Store @ R.T. in dry atmosphere

Estasan™ GT 8-70 3579 [Uniqema; Uniqema Am.]
Chem. Descrip.: Caprylic/capric triglyceride (70% C8 + 25-30% C10)
CAS 65381-09-1; EINECS/ELINCS 265-724-3
Uses: Solvent, carrier for drugs, capsules, flavors, fragrances, food ingreds.; lubricant for mfg. of soft gelatin capsules; nutrient, fat source in dietetic prods., adult and infant nutrition, enteral and parenteral feeding formulas; lubricant, mold release for foods; emollient, spreading agent, penetrant, solvent for topical pharmaceuticals, cosmetics, toiletries, creams, lotions; lubricant for soft gelatin capsules
Features: Wide compat. with many drugs; min. oil replacement
Regulatory: Ph.Eur., BP, DAB compliance; Japan approval
Properties: Lovibond 5Y/0.5R oil, odorless, bland taste; visc. 23 mPa·s; acid no. 0.1 max.; iodine no. 0.5 max.; sapon. no. 335-360; hyd. no. 5 max.; cloud pt. -12 C
Environmental: Readily biodeg.
Storage: Store @ R.T. in dry atmosphere

Estasan™ GT 8-99 3596 [Uniqema]
Chem. Descrip.: MCT multifunctional oil
Uses: Solvent, suspension medium for various drugs; fat source in formulation of dietetic prods., enteral and parenteral feeding formulas
Features: Very stable towards oxidation
Properties: APHA 70 oil; odorless; visc. 22 mPa·s; m.p. 8 C; acid no. 0.1 max.; iodine no. 0.5 max.
Environmental: Biodeg.

Estasan™ GT 10-99 3599 [Uniqema]
Chem. Descrip.: MCT multifunctional oil
Uses: Solvent, suspension medium for various drugs; fat source in formulation of dietetic prods., enteral and parenteral feeding formulas
Features: Very stable towards oxidation
Properties: APHA 70 oil; odorless; visc. 17 mPa·s (40 C); m.p. 31 C; acid no. 0.1 max.; iodine no. 0.5 max.
Environmental: Biodeg.

Estasol™ [Haltermann Prods. UK]
Chem. Descrip.: Dimethyl glutarate (55-65%), dimethyl succinate (15-25%), dimethyl adipate (12-23%)
Uses: Solvent for inks, coatings, cleaning chems., paint strippers, foundry core binders, hand cleaners, textile lubricants, pesticides, insecticides, sealants and waxes
Features: Alternative to high VOC solvs.; esp. effective replacement for acetone and methylene chloride
Properties: Mild odor; misc. with most solvs.; m.w. 160 max.; sp.gr. 1.085-1.095; visc. 2.4-2.5 cSt; vapor pressure 0.1 mm Hg (20 C); f.p. - 25 C; acid no. 0.5; flash pt. 108 C; ref. index 1.423-1.425; 99% min. diester; 0.2% water
Environmental: Biodeg.

Estasol™ MD10 [Haltermann Prods. UK]
Chem. Descrip.: Methyl decanoate (95%)
CAS 110-42-9; EINECS/ELINCS 203-766-6
Uses: Solvent for screen inks, coil coatings, paint strippers, hand cleaners, graffiti removers; intermediate for detergents, lubricants, emulsifiers, wetting agents, plasticizers; replacement for wh. spirit, isoparaffins, glycols, glycol ethers, ketones, aromatic compds.
Properties: Water-wh. to ylsh. liq.; readily sol. in ethanol and ether; sp.gr. 0.87; vapor pressure < 1 mm Hg (20 C); m.p. - 18 C; b.p. - 177 C; acid no. 0.5; flash pt. 93 C; 0.15% max. water
Environmental: Environmentally friendly

Estemol N-01 [Nisshin Oillio]
Chem. Descrip.: Neopentyl glycol dicaprate
CAS 27841-06-1; EINECS/ELINCS 248-688-3
Uses: Emollient, solvent in cosmetics

Estol IPM [Uniqema]
Chem. Descrip.: Isopropyl myristate
CAS 110-27-0; EINECS/ELINCS 203-751-4
Uses: Emollient, consistency agent, solvent for topical pharmaceuticals

Estol IPM 1509 (BIO-IPM) [Uniqema]
Chem. Descrip.: Isopropyl myristate, veg. derived
CAS 110-27-0; EINECS/ELINCS 203-751-4
Uses: Penetrant, emollient, solvent, spreading agent for cosmetics, topical medicinals, skin creams, sun care preps., aerosols, bath oils, preshave lotions; gloss aid, conditioner in hair sprays
Features: Reduces heaviness and oiliness of prods. containing triglycerides, liq. waxes, and min. oils
Properties: Colorless low-visc. liq., odorless; sp.gr. 0.850-0.855; acid no. 0.1 max.; iodine no. 0.5 max.; sapon. no. 206-211; cloud pt. 0 C max.; 98% min. act.

Ethylan® HB1 [Akzo Nobel Surf. Chem. AB]
Chem. Descrip.: Short chain ethoxylate
Uses: Coalescing agent for surface coatings; solvent for acrylic and vinyl acetate copolymers; plasticizer; preservative effective against a variety of microorganisms
Properties: Lt. amber clear liq., mild odor; sol. hazy in water; sp.gr. 1.098; visc. 27 cs; pour pt. -15 C; flash pt. (COC) > 100 C; pH 8.8 (1% aq.); nonionic; 100% act.
Toxicology: LD50 (oral, rat) 2-4 ml/kg; sl. toxic; sl. skin irritant, severe eye irritant

Ethylan® HB4 [Akzo Nobel Surf. Chem. AB]
Chem. Descrip.: Aromatic ethoxylate
Uses: Penetrant, hydrotrope, cosolvent, solubilizer, coupling agent, detergent, emulsifier, and wetting agent for iodophors, rinse aids, detergent sterilizers, glass cleaners
Properties: Water-wh. clear liq., faint odor; sol. in water; sp.gr. 1.121; visc. 64 cs; pour pt. < 0 C; cloud pt. > 100 C (1% aq.); flash pt. (COC) > 180 C; pH 6.4 (1% aq.); nonionic; 100% conc.

Ethyl Diglyme [Ferro/Grant]
Chem. Descrip.: Diethylene glycol diethyl ether
CAS 112-36-7; EINECS/ELINCS 203-963-7
Uses: Solvent for electrochemistry, polymer and boron chemistry, physical processes such as gas absorption, extraction, stabilization, industrial prods. such as fuels, lubricants, textiles, pharmaceuticals, pesticides
Features: Tends to solvate cations
Properties: Colorless; mild, nonresidual odor; water-sol.; misc. with ethanol, acetone, benzene, diethyl ether, octane; m.w. 162.23; sp.gr. 0.9082; dens. 7.56 lb/gal; visc. 1.4 cP; f.p. -44.3 C; b.p. 189 C; flash pt. (CC) 90 C; ref. index 1.4115; pH neutral; surf. tens. 27.2 dynes/cm; 98.0% min. purity
Toxicology: LD50 5000 mg/kg; low to mod. acute toxicity; chronic exposure may cause reproductive effects
Environmental: Slowly biodeg.

Ethyl Glyme [Ferro/Grant]
Chem. Descrip.: Ethylene glycol diethyl ether
CAS 629-14-1; EINECS/ELINCS 211-076-1
Uses: Solvent used in electrochemistry, polymer and boron chemistry, physical processes such as gas absorption, extraction, stabilization, industrial prods. such as fuels, lubricants
Features: Tends to solvate cations
Properties: Colorless; mild ethereal, nonresidual odor; water-sol.; misc. with ethanol, acetone, benzene, diethyl ether, octane; m.w. 118.18; sp.gr. 0.8417; dens. 7.0 lb/gal; visc. 0.7 cP; f.p. -74 C; b.p. 121 C; flash pt. (CC) 27 C; ref. index 1.3922; pH neutral; 97.0% min. purity
Toxicology: LD50 4400 mg/kg; low to mod. acute toxicity; chronic exposure may cause reproductive effects
Environmental: Slowly biodeg.

Ethyl Icinol [Huntsman]
Chem. Descrip.: 2-Ethoxyethanol
CAS 110-80-5; EINECS/ELINCS 203-804-1
Uses: Solvent for protective coatings, inks, agric. chems.; aids wetting, penetration, and soil removal; coupling solvent

Ethyl Proxitol [Shell Netherlands]
Chem. Descrip.: Propylene glycol ethyl ether
CAS 1569-02-4
UN 1987
Uses: Solvent, flow control aid, leveling agent, coalescent for surf. coatings, printing inks, felt pens and stamp pad inks, leather finishes, wood stains, furniture polishes; extractant in pharmaceuticals, cosmetics, adhesives, printing inks, agrochems., household/industrial cleaners
Features: Does not contain detectable quantities of polycyclic aromatic

hydrocarbons, heavy metals, or chlorinated compds.
Regulatory: EINECS, Canada DSL, Australia AICS, China IECSC, Japan MITI, Philippines PICCS, Korea TCCL listed
Properties: Colorless clear liq.; mild ethereal odor; completely misc. with water, many org. solvs.; m.w. 104; dens. 910 kg/m^3 (20 C); dynamic visc. 2.1 mPa·s; vapor pressure 1.0 kPa (20 C); b.p. 129-136 C; pour pt. -90 C; flash pt. (Abel) 42 C; autoignition temp. 255 C; ref. index 1.404; surf. tens. 25.6 mN/m; sp. heat 2.7 kJ/kg/°C; dielec. const. 10.4
Toxicology: Expected LD50 (oral) > 2000 mg/kg, (skin) > 2000 mg/kg; LC50 (inh.) > 5 mg/l; narcotic at high vapor concs.; causes CNS depression; expected to be sl. eye and skin irritant
Environmental: Readily biodeg.; does not bioaccumulate significantly; nonhazardous under EC criteria; LC50 (fish) > 1000 mg/l; pract. nontoxic to fish, algae, bacteria; prevent contamination of soil and water; prevent entry into drains, ditches, rivers
Precaution: Flamm.; LEL 1.3 vol.%; UEL 12 vol.%; may form flamm./explosive vapor-air mixt.; vapor is heavier than air, spreads along ground; distant ignition is possible; reacts with strong oxidizing agents; oxidizes on contact with air; avoid exposure to air
Hazardous decomp. prods.: Combustion: CO
Storage: Hygroscopic; 12 mos shelf life under proper storage; store in tightly closed containers @ ambient temps. in well-ventilated area, away from direct sunlight, other heat/ignition sources

Etocas 35 NF [Croda Inc]
Chem. Descrip.: PEG-35 castor oil NF
CAS 61791-12-6
Uses: Surfactant, emulsifier, solvent, wetting agent, emollient, lubricant for oral, topical, and parenteral pharmaceuticals
Features: Tends to yield clear sol'ns.
Properties: Pale yel. liq.; sol. in water, IPA; disp. in propylene glycol; insol. in min. oil; HLB 12.5; nonionic
Use Level: 0.5-5%

Exceparl BB [Kao Corp.]
Chem. Descrip.: Behenyl behenate
CAS 17671-27-1; EINECS/ELINCS 241-646-5
Uses: Raw material for spin finishes, surfactants; oiling agent for textiles; processing aid for rubber; lubricant, plasticizer for plastics; additive in paints, inks; base for cosmetics; solvent for pharmaceuticals
Properties: Wh. solid; insol. in ethanol; very sl. sol. in n-hexane, xylene; m.p. 73 C; acid no. 1 max.; iodine no. 2 max.; sapon. no. 84-90; hyd. no. 7 max.

Exceparl BS [Kao Corp.]
Chem. Descrip.: Butyl stearate
CAS 123-95-5; EINECS/ELINCS 204-666-5
Uses: Raw material for spin finishes, surfactants; oiling agent for textiles; processing aid for rubber; lubricant, plasticizer for plastics; additive in paints, inks; base for cosmetics; solvent for pharmaceuticals
Properties: APHA color 25 max. solid; sol. in n-hexane, xylene, ethanol; m.p. 21-24 C; visc. 7.7 cp (30 C); acid no. 1 max.; iodine no. 2 max.; sapon. no. 167-178

Exceparl EH-P [Kao Corp.]
Chem. Descrip.: 2-Ethyl hexyl palmitate
CAS 29806-73-3; EINECS/ELINCS 249-862-1
Uses: Raw material for spin finishes, surfactants; oiling agent for textiles; processing aid for rubber; lubricant, plasticizer for plastics; additive in paints, inks; base for cosmetics; solvent for pharmaceuticals
Properties: APHA color 100 max. liq.; sol. in n-hexane, xylene (10%); very sl. sol. in ethanol (10%); visc. 11 cst (30 C); acid no. 1 max.; sapon. no. 147-157

Exceparl EH-S [Kao Corp.]
Chem. Descrip.: 2-Ethyl hexyl stearate
CAS 22047-49-0; EINECS/ELINCS 244-754-0
Uses: Raw material for spin finishes, surfactants; oiling agent for textiles; processing aid for rubber; lubricant, plasticizer for plastics; additive in paints, inks; base for cosmetics; solvent for pharmaceuticals
Properties: APHA color 100 max. liq.; sol. in n-hexane, xylene (10%); very sl. sol. in ethanol (10%); visc. 13 cst (30 C); acid no. 1 max.; sapon. no. 140-150; hyd. no. 3 max.

Exceparl IB-OL [Kao Corp.]
Chem. Descrip.: Isobutyl oleate
CAS 84988-79-4; EINECS/ELINCS 284-868-8
Uses: Raw material for spin finishes, surfactants; oiling agent for textiles; processing aid for rubber; lubricant, plasticizer for plastics; additive in paints, inks; base for cosmetics; solvent for pharmaceuticals
Properties: G5 max. color liq.; sol. in n-hexane, xylene, ethanol; f.p. 10 C max.; visc. 9.0 cp (20 C); acid no. 1 max.; iodine no. 78-88; sapon. no. 163-165 (ester v.); hyd. no. 5 max.

Exceparl L-OL [Kao Corp.]
Chem. Descrip.: Lauryl oleate
CAS 36078-10-1; EINECS/ELINCS 252-862-4
Uses: Raw material for spin finishes, surfactants; oiling agent for textiles; processing aid for rubber; lubricant, plasticizer for plastics; additive in paints, inks; base for cosmetics; solvent for pharmaceuticals
Properties: APHA color 250 max. liq.; sol. in n-hexane, xylene; very sl. sol. in ethanol; visc. 17 cst (30 C); acid no. 5 max.; iodine no. 49-50; sapon. no. 120-130; hyd. no. 10 max.

Exceparl MC [Kao Corp.]
Chem. Descrip.: Methyl coconate
CAS 61788-59-8; EINECS/ELINCS 262-988-1
Uses: Raw material for spin finishes, surfactants; oiling agent for textiles; processing aid for rubber; lubricant, plasticizer for plastics; additive in paints, inks; base for cosmetics; solvent for pharmaceuticals
Properties: APHA color 120 max. liq.; sol. in ethanol, xylene (10%); clearly sol. to hazy in n-hexane (10%); visc. 2.9 cp; m.p. -5 C max.; acid no. 0.1 max.; iodine no. 10 max.; sapon. no. 244-260

Exceparl ML-85 [Kao Corp.]
Chem. Descrip.: Methyl laurate
CAS 111-82-0; EINECS/ELINCS 203-911-3
Uses: Raw material for spin finishes, surfactants; oiling agent for textiles; processing aid for rubber; lubricant, plasticizer for plastics; additive in paints, inks; base for cosmetics; solvent for pharmaceuticals
Properties: APHA color 20 max. liq.; sol. in ethanol, xylene, n-hexane (10%); visc. 2.6 cp; m.p. -5 C max.; acid no. 0.1 max.; iodine no. 0.5 max.; sapon. no. 258-268

Exceparl M-OL [Kao Corp.]
Chem. Descrip.: Methyl oleate
CAS 112-62-9; EINECS/ELINCS 203-992-5
Uses: Raw material for spin finishes, surfactants; oiling agent for textiles; processing aid for rubber; lubricant, plasticizer for plastics; additive in paints, inks; base for cosmetics; solvent for pharmaceuticals
Properties: APHA color 100 max. liq.; sol. in ethanol, xylene, n-hexane (10%); visc. 5.5 cp; m.p. -5 C max.; acid no. 1 max.; iodine no. 80-90; sapon. no. 189-199

Exceparl MS [Kao Corp.]
Chem. Descrip.: Methyl stearate
CAS 112-61-8; EINECS/ELINCS 203-990-4
Uses: Raw material for spin finishes, surfactants; oiling agent for textiles; processing aid for rubber; lubricant, plasticizer for plastics; additive in paints, inks; base for cosmetics; solvent for pharmaceuticals
Properties: APHA color 100 max. solid; sol. in ethanol, xylene, n-hexane (10%); visc. 3.8 cp (45 C); m.p. 31-35 C; acid no. 1 max.; iodine no. 0.7 max.; sapon. no. 189-195

Exceparl MT [Kao Corp.]
Chem. Descrip.: Methyl tallowate
CAS 61788-61-2; EINECS/ELINCS 262-989-7
Uses: Raw material for spin finishes, surfactants; oiling agent for textiles; processing aid for rubber; lubricant, plasticizer for plastics; additive in paints, inks; base for cosmetics; solvent for pharmaceuticals
Properties: APHA color 150 max. liq.; sol. in ethanol, xylene, n-hexane (10%); iodine no. 45-55; sapon. no. 190-200

Exceparl MY-M [Kao Corp.]
Chem. Descrip.: Myristyl myristate
CAS 3234-85-3; EINECS/ELINCS 221-787-9
Uses: Raw material for spin finishes, surfactants; oiling agent for textiles; processing aid for rubber; lubricant, plasticizer for plastics; additive in paints, inks; base for cosmetics; solvent for pharmaceuticals
Properties: APHA color 50 max. solid; sl. sol. in ethanol; m.p. 36-46 C; acid no. 1 max.; iodine no. 1 max.; sapon. no. 115-135; hyd. no. 6 max.

Exceparl OD-B [Kao Corp.]
Chem. Descrip.: Octyldodecyl behenate
Uses: Raw material for spin finishes, surfactants; oiling agent for textiles; processing aid for rubber; lubricant, plasticizer for plastics; additive in paints, inks; base for cosmetics; solvent for pharmaceuticals
Properties: Lt. yel. solid; sol. in n-hexane, xylene; m.p. 20 C; acid no. 0.4;

sapon. no. 90.7; hyd. no. 3.2

Exceparl OL-OL [Kao Corp.]
Chem. Descrip.: Oleyl oleate
CAS 3687-45-4; EINECS/ELINCS 222-980-4
Uses: Raw material for spin finishes, surfactants; oiling agent for textiles; processing aid for rubber; lubricant, plasticizer for plastics; additive in paints, inks; base for cosmetics; solvent for pharmaceuticals
Properties: GP max. color liq.; sol. in n-hexane, xylene; very sl. sol. in ethanol; visc. 24 cst (30 C); acid no. 1 max.; iodine no. 70-79; sapon. no. 101-111; hyd. no. 10 max.

Exceparl O-OL [Kao Corp.]
Chem. Descrip.: Octyl oleate
Uses: Raw material for spin finishes, surfactants; oiling agent for textiles; processing aid for rubber; lubricant, plasticizer for plastics; additive in paints, inks; base for cosmetics; solvent for pharmaceuticals
Properties: G10 max. color liq.; sol. in ethanol, n-hexane, xylene; acid no. 5-10; iodine no. 61-71; sapon. no. 140-150

Exceparl PE-MO [Kao Corp.]
Chem. Descrip.: Pentaerythritol monooleate
Uses: Raw material for spin finishes, oiling agents, lubricants; solvent for agrochemicals
Properties: Ylsh. brn. liq.; sol. in ethanol, n-hexane, xylene; acid no. 2 max.

Exceparl SS [Kao Corp.]
Chem. Descrip.: Stearyl stearate
CAS 2778-96-3; EINECS/ELINCS 220-476-5
Uses: Raw material for spin finishes, surfactants; oiling agent for textiles; processing aid for rubber; lubricant, plasticizer for plastics; additive in paints, inks; base for cosmetics; solvent for pharmaceuticals
Properties: APHA color 100 max. solid; sl. sol. in n-hexane, xylene; very sl. sol. in ethanol; m.p. 56-66 C; acid no. 1 max.; iodine no. 1 max.; sapon. no. 97-107 (ester v.); hyd. no. 7 max.

Exceparl TD-S [Kao Corp.]
Chem. Descrip.: Isotridecyl stearate
CAS 31565-37-4; EINECS/ELINCS 250-703-3
Uses: Raw material for spin finishes, surfactants; oiling agent for textiles; processing aid for rubber; lubricant, plasticizer for plastics; additive in paints, inks; base for cosmetics; solvent for pharmaceuticals
Properties: APHA color 100 max. clear liq.; sol. in n-hexane, xylene; very sl. sol. in ethanol; visc. 24 cst (30 C); acid no. 1 max.; sapon. no. 115-125

EXK-385-MSO [Magie Bros. Oil]
Chem. Descrip.: Sweetened middle distillate
CAS 64741-86-2; EINECS/ELINCS 265-088-7
Uses: Solvent for heat-set inks, sheet-fed inks
Regulatory: DOT nonregulated; SARA §311/312/313 nonreportable; Canada DSL, EINECS, Australia AICS, China, Japan ENCS, Korea KCL, Philippines PICCS listed
Properties: Lt. yel. liq.; distillate odor; sol. in hydrocarbons; insol. in water; m.w. 243; sp.gr. 0.8203 (15.6 C); dens. 6.83 lb/gal (15.6 C); visc. 41.5 cSt (100 F); vapor pressure 0.06 mm Hg (70 F); i.b.p. 252 C; pour pt. -3.9 C; cloud pt.1.7 C; flash pt. (PMCC) 115.6 C; autoignition temp. 218 C; ref. index 1.4526 (20 C); KB value 22; 100% act.
Toxicology: ACGIH TLV/TWA 5 mg/m^3 (oil mist); minimal eye/sl. skin irritant; prolonged/repeated contact may cause contact dermatitis; not expected to be inh. hazard @ ambient conditions; acute/chronic overexposure may cause respiratory tract irritation, drowsiness, dullness, numbness, headache, dizziness, weakness, nausea; aspiration may cause chem. pneumonitis, pulmonary edema, hemorrhage, death; ing. may cause GI discomfort, burning in throat, unconsciousness, convulsions; TSCA listed
Environmental: Do not allow spill to enter sewers or watercourses
Precaution: May react with strong oxidizing agents; spillages may be slippery
Hazardous decomp. prods.: Combustion: dense smoke, CO, CO_2, other oxides
HMIS: Health 1, Flammability 1, Reactivity 0
Storage: Store in closed containers away from heat, sparks, open flame, oxidizing materials

Exxate® 600 [ExxonMobil; Exxon Canada]
Chem. Descrip.: Oxo-hexyl acetate
CAS 88230-35-7
Uses: Solvent for high-solids, maintenance, marine, and industrial waterborne coatings, acrylic resin polymerization, metalworking fluids, urethane foam, industrial cleaners/degreasers, paint strippers, pigment dispersions, pesticide emulsifiable concs., textile scouring aids, NC or screen printing inks, penetrating/lubricating oils, vinyl plastisols; wetting agent; lubricant; visc. modifier; latex coalescer
Features: Good solvency, low surf. tens. and interfacial tens., no HAPs
Regulatory: SARA nonreportable
Properties: Pt-Co 10 color, sweet ester odor; sol. 0.02% in water; misc. with common petrol. solvs.; m.w. 144; sp.gr. 0.874 (20/20 C); dens. 7.27 lb/gal (20 C); visc. 1.04 cp; vapor pressure 1.4 mm Hg (20 C); f.p. < -60 C; i.b.p. 164 C; pour pt. < -70 F; cloud pt. < -70 F; flash pt. (TCC) 134 F; ref. index 1.4095 (20 C); surf. tens. 25 dynes/cm; 99% purity
Toxicology: LD50 (oral, rat) ≥ 5 g/kg, (dermal, rabbit) > 3 g/kg; low toxicity by ing. or skin contact; frequent/prolonged skin contact may cause irritation and dermatitis; sl. to mod. eye irritant; high concs. of vapor are irritating to mucous membranes
Environmental: Biodeg.
Precaution: DOT: combustible liq.
Storage: Keep containers closed; keep away from heat, sparks, and flames

Exxate® 700 [ExxonMobil; Exxon Canada]
Chem. Descrip.: Oxo-heptyl acetate
CAS 90438-79-2
Uses: Solvent for high-solids, maintenance, marine, and industrial waterborne coatings, acrylic resin polymerization, metalworking fluids, urethane foam, industrial cleaners/degreasers, paint strippers, pigment dispersions, pesticide emulsifiable concs., textile scouring aids, NC or screen printing inks, penetrating/lubricating oils, vinyl plastisols; wetting agent; lubricant; visc. modifier; latex coalescer
Features: Good solvency, low surf. tens. and interfacial tens.; no HAPs
Regulatory: SARA nonreportable
Properties: Pt-Co 10 color, sweet ester odor; sol. 0.01% in water; misc. with common petrol. solvs.; m.w. 158; sp.gr. 0.874 (20/20 C); dens. 7.27 lb/gal (20 C); visc. 1.24 cp; vapor pressure 0.8 mm Hg (20 C); f.p. < -60 C; i.b.p. 176 C; pour pt. < -70 F; cloud pt. < -70 F; flash pt. (TCC) 150 F; ref. index 1.4158 (20 C); surf. tens. 26 dynes/cm
Toxicology: LD50 (oral, rat) ≥ 5 g/kg, (dermal, rabbit) > 3 g/kg; low toxicity by ing. or skin contact; frequent/prolonged skin contact may cause irritation and dermatitis; sl. to mod. eye irritant; high concs. of vapor are irritating to mucous membranes
Environmental: Biodeg.
Precaution: DOT: combustible liq.
Storage: Keep containers closed; keep away from heat, sparks, and flames

Exxate® 800 [ExxonMobil; Exxon Canada]
Chem. Descrip.: C8 alkyl acetate
CAS 108419-32-5
Uses: Solvent for high-solids, maintenance, marine, and industrial waterborne coatings, acrylic resin polymerization, metalworking fluids, urethane foam, industrial cleaners/degreasers, paint strippers, pigment dispersions, pesticide emulsifiable concs., textile scouring aids, NC or screen printing inks, penetrating/lubricating oils, vinyl plastisols; wetting agent; lubricant; visc. modifier; latex coalescer
Features: Good solvency, low surf. tens. and interfacial tens., no HAPs
Regulatory: SARA nonreportable
Properties: Pt-Co 10 color, mild ester odor; sol. 0.02% in water; misc. with common petrol. solvs.; m.w. 172; sp.gr. 0.875 (20/20 C); dens. 7.28 lb/gal (20 C); visc. 1.74 cp; vapor pressure 0.75 mm Hg (20 C); f.p. < -60 C; i.b.p. 186 C; pour pt. < -70 F; cloud pt. < -70 F; flash pt. (TCC) 171 F; ref. index 1.420 (20 C); surf. tens. 26 dynes/cm
Toxicology: LD50 (oral, rat) ≥ 5 g/kg, (dermal, rabbit) > 3 g/kg; low toxicity by ing. or skin contact; frequent/prolonged skin contact may cause irritation and dermatitis; sl. to mod. eye irritant; high concs. of vapor are irritating to mucous membranes
Environmental: Biodeg.
Precaution: DOT: combustible liq.
Storage: Keep containers closed; keep away from heat, sparks, and flames

Exxate® 900 [ExxonMobil; Exxon Canada]
Chem. Descrip.: C9 alkyl acetate
CAS 108419-33-6
Uses: Solvent for high-solids, maintenance, marine, and industrial waterborne coatings, acrylic resin polymerization, metalworking fluids, urethane foam, industrial cleaners/degreasers, paint strippers, pigment dis-

persions, pesticide emulsifiable concs., textile scouring aids, NC or screen printing inks, penetrating/lubricating oils, vinyl plastisols; wetting agent; lubricant; visc. modifier; latex coalescer

Features: Good solvency, low surf. tens. and interfacial tens., no HAPs

Regulatory: SARA nonreportable

Properties: Pt-Co 10 color, mild ester odor; sol. 0.02% in water; misc. with common petrol. solvs.; m.w. 186 ; sp.gr. 0.873 (20/20 C); dens. 7.26 lb/gal (20 C); visc. 2.22 cp; vapor pressure 0.24 mm Hg (20 C); f.p. < -60 C; i.b.p. 205 C; pour pt. < -70 F; cloud pt. < -70 F; flash pt. (TCC) 194 F; ref. index 1.425 (20 C); surf. tens. 27 dynes/cm

Toxicology: LD50 (oral, rat) ≥ 5 g/kg, (dermal, rabbit) > 3 g/kg; low toxicity by ing. or skin contact; frequent/prolonged skin contact may cause irritation and dermatitis; sl. to mod. eye irritant; high concs. of vapor are irritating to mucous membranes

Environmental: Biodeg.

Precaution: DOT: combustible liq.

Storage: Keep containers closed; keep away from heat, sparks, and flames

Exxate® 1000 [ExxonMobil; Exxon Canada]

Chem. Descrip.: C10 alkyl acetate

CAS 108419-34-7; EINECS/ELINCS 270-817-7

Uses: Solvent for high-solids, maintenance, marine, and industrial waterborne coatings, acrylic resin polymerization, metalworking fluids, urethane foam, industrial cleaners/degreasers, paint strippers, pigment dispersions, pesticide emulsifiable concs., textile scouring aids, screen printing inks, penetrating/lubricating oils, vinyl plastisols; wetting agent; lubricant; visc. modifier; latex coalescer; diluent; chem. feedstock; fuel

Features: Low vapor pressure; low surf. tens. and interfacial tens.; no HAPs; meets LVP exemption (Calif.)

Regulatory: SARA nonreportable; Canada DSL listed

Properties: Pt-Co 10 liq.; mild ester odor; nil sol. in water; misc. with common petrol. solvs.; m.w. 200; sp.gr. 0.873 (20/20 C); dens. 7.26 lb/gal (20 C); visc. 2.60 cp; vapor pressure 0.1 mm Hg (20 C); f.p. < -60 C; b.p. 220-250 C; pour pt. < -70 F; cloud pt. < -70 F; flash pt. (TCC) 100 C; autoignition temp. 300 C; ref. index 1.429 (20 C); surf. tens. 27.5 dynes/cm; KB value > 100

Toxicology: OEL TWA 50 ppm (vapor), 10 mg/m^3 (aerosol); LD50 (oral, rat) ≥ 5 g/kg, (dermal, rabbit) > 3 g/kg; low toxicity by ing. or skin contact; frequent/prolonged skin contact may cause irritation and dermatitis; sl. to mod. eye irritant; high concs. of vapor are irritating to mucous membranes

Environmental: Biodeg.; prevent entry into sewers, watercourses, low areas

Precaution: Flamm. limits 1-8 vol.%; liqs. may burn upon heating to flash pt.; electrostatic accumulation hazard; use proper grounding procedures; incompat. with caustics, strong oxidizing agents

Hazardous decomp. prods.: None

Storage: Store in cool, well-ventilated area away from incompat. materials, direct sunlight; keep containers closed; keep away from heat, sparks, and flames

Exxate® 1200 [ExxonMobil]

Chem. Descrip.: Oxo-dodecyl acetate

CAS 108419-35-8

Uses: Solvent for architectural waterborne coatings, industrial coatings, traffic marking, maintenance coatings, household cleaners/degreasers, automotive (cleaners/degreasers, lubricants/penetrants), insecticides, herbicides

Features: Meets LVP exemption (Calif.)

Properties: Pt-Co 15 max. color; sp.gr. 0.871-0.877 (20/20 C); dens. 7.27 lb/gal; i.b.p. 220 C; flash pt. (TCC) 245 F; 99% min. purity

Exxate® 1300 [ExxonMobil; Exxon Canada]

Chem. Descrip.: C13 alkyl acetate

CAS 108419-35-8

Uses: Solvent for high-solids, maintenance, marine, and industrial waterborne coatings, acrylic resin polymerization, metalworking fluids, urethane foam, industrial cleaners and degreasers, paint strippers, pigment dispersions, pesticide emulsifiable concs., textile scouring aids, screen printing inks, penetrating/lubricating oils, vinyl plastisols; wetting agent; lubricant; visc. modifier; latex coalescer

Features: Low vapor pressure; low surf. tens. and interfacial tens.; no HAPs

Regulatory: SARA nonreportable

Properties: Pt-Co 10 color, nearly odorless; nil sol. in water; misc. with common petrol. solvs.; m.w. 242; sp.gr. 0.879 (20/20 C); dens. 7.28 lb/gal (20 C); visc. 4.63 cp; vapor pressure 0.03 mm Hg (20 C); f.p. < -60 C; i.b.p. 240 C; pour pt. < -70 F; cloud pt. < -70 F; flash pt. (TCC) 261 F; ref. index 1.438 (20 C); surf. tens. 28 dynes/cm; KB value 96

Toxicology: LD50 (oral, rat) ≥ 5 g/kg, (dermal, rabbit) > 3 g/kg; low toxicity by ing. or skin contact; frequent/prolonged skin contact may cause irritation and dermatitis; sl. to mod. eye irritant; high concs. of vapor are irritating to mucous membranes

Environmental: Biodeg.

Storage: Keep containers closed; keep away from heat, sparks, and flames

ExxonMobil IPA [ExxonMobil Europe]

Chem. Descrip.: Isopropyl alcohol

CAS 67-63-0; EINECS/ELINCS 200-661-7

Uses: Solvent; intermediate

Properties: Pt-Co 5 color; sol. in water; sp.gr. 0.786 (20/20 C); visc. 2.08 mPa·s; i.b.p. 82 C; flash pt. 12 C; ref. index 1.377 (20 C); surf. tens. 20.6 mN/m; > 99% act.

ExxonMobil MEK [ExxonMobil; ExxonMobil Europe]

Chem. Descrip.: Methyl ethyl ketone

CAS 78-93-3; EINECS/ELINCS 201-159-0

Uses: Solvent for automotive refinishes, industrial coatings, maintenance coatings, traffic marking, aerosol wood furniture coatings

Features: Exc. solvency; fast visc. reduction; cost effective; potential HAPs delisting

Properties: Pt-Co 10 max. liq.; sp.gr. 0.801-0.803; i.b.p. 78.5 C min.; 99.5% min. purity; 0.002% max. NV

Precaution: DOT flamm. liq.

ExxonMobil T 500-100 [Exxon Canada]

Chem. Descrip.: Light aromatic solvent naphtha (60-100%), xylenes (10-30%), ethyl benzene (1-5%)

UN 1993

Uses: Solvent; thinner; inert ingred. in agric. chems.

Regulatory: Canada DSL listed; NPRI reportable [1,2,4-trimethylbenzene (26.5%), cumene (1%), ethyl benzene (2.4%), xylenes (16.3%)]

Properties: Colorless clear liq.; aromatic hydrocarbon odor; sol. 0.02% in water; m.w. 120; sp.gr. 0.87 (15.6 C); visc. 0.92 cSt; vapor pressure 0.4 kPa (20 C); f.p. -60 C; b.p. 153-170 C; flash pt. (TCC) 44 C; autoignition temp. ≈ 471 C

Toxicology: REL TWA 20 ppm; LC50 (rat) 5000 ppm; inh. of high vapor concs. may cause eye/respiratory tract irritation, headaches, dizziness, anesthesia, drowsiness, unconsciousness, other CNS effects, death; eye irritant; frequent/prolonged skin contact may cause irritation, dermatitis; if aspirated, may cause mild to severe pulmonary injury, death; chronic exposure may cause effects in embryo/fetus; contains ethyl benzene, possible human carcinogen

Environmental: Prevent entry into sewers, watercourses, low areas

Precaution: Flamm. liq.; flamm. limits 1.9-12.6 vol.%; volatile; gives off invisible vapors; liq. or vapor may settle in low areas; distant ignition/explosion possible; electrostatic accumulation hazard; use proper grounding procedures; incompat. with nitric acid, sulfuric acid, strong oxidizing agents

Hazardous decomp. prods.: None

Storage: Store in cool, well-ventilated area away from incompat. materials, heat, open flame, ignition sources, direct sunlight; keep container closed

Exxpar 451 [Exxon Canada]

Chem. Descrip.: Paraffins, normal C5-20

CAS 64771-72-8

Uses: Solvent; diluent; chem. feedstock; fuel

Regulatory: Canada DSL listed; NPRI nonreportable

Properties: Colorless clear liq.; mild hydrocarbon odor; insol. in water; sp.gr. 0.77 (15.6 C); visc. 3.24 cSt; vapor pressure 0 kPa (4 C), 0.007 kPa (38 C); f.p. 9 C; b.p. 204-316 C; flash pt. (PMCC) 118 C; autoignition temp. > 200 C

Toxicology: REL TWA 140 ppm; inh. of high vapor concs. may cause eye/respiratory tract irritation, headaches, dizziness, anesthesia, other CNS effects; sl. eye irritant; frequent/prolonged skin contact may cause irritation, dermatitis; if aspirated, may cause severe health effects (bronchopneumonia, pulmonary injury)

Environmental: Prevent entry into sewers, watercourses, low areas

Precaution: Estimated flamm. limits 0.6-7 vol.%; use proper grounding procedures; incompat. with strong oxidizing agents

Hazardous decomp. prods.: None
Storage: Store in cool, well-ventilated area away from incompat. materials, heat, open flame, ignition sources, direct sunlight; keep container closed; protect from freezing

Exx-Print® 176 D [ExxonMobil]
Chem. Descrip.: Petroleum distillates, hydrotreated light
CAS 64742-47-8; EINECS/ELINCS 265-149-8
UN 1268
Uses: Solvent for printing inks
Regulatory: SARA §311/312 fire hazard, §313 nonreportable
Properties: Water-wh. clear liq.; mild min. spirits odor; low odor; negligible sol. in water; sp.gr. 0.80 (15.6 C); dens. 6.65 lb/gal (60 F); vapor pressure 0.2 mm Hg (20 C); b.p. 207-234 C; pour pt. < -18 C; flash pt. (PMCC) 71 C; autoignition temp. ≈ 216 C; surf. tens. 27.6 dynes/cm; 100% act.; 53% paraffin; 47% cycloparaffin; < 0.2% aromatic
Toxicology: REL (8 h) 300 ppm; LD50 (oral, rat) > 5 g/kg, (skin, rabbit) > 3.16 g/kg; inh. of high vapor concs. may cause eye/respiratory tract irritation, headaches, dizziness, anesthesia, drowsiness, unconsciousness, other CNS effects, death; may cause eye irritation; prolonged/repeated skin contact may cause irritation, dermatitis; if aspirated, may cause mild to severe pulmonary injury, death; TSCA listed
Environmental: Prevent entry into sewers, watercourses
Precaution: Combustible liq.; flamm. limits in air 1.3-8.1 vol.%; volatile; gives off invisible vapors; liq. or vapor may settle in low areas; distant ignition/explosion possible; static accumulation hazard; ground containers, equip.; incompat. with strong oxidants (liq. chlorine, conc. oxygen, sodium or calcium hypochlorite)
Hazardous decomp. prods.: Combustion: fumes, smoke, CO, SO_x, aldehydes, other decomp. prods.
HMIS: Health 1, Flammability 2, Reactivity 0
Storage: Store away from heat, sparks, flame, strong oxidants; keep containers closed when not in use

Exx-Print® 274 A [ExxonMobil]
Chem. Descrip.: Hydrocarbon fluid
Uses: Solvent for printing inks
Properties: Sp.gr. 0.822 (15.6 C); dens. 6.85 lb/gal (15.6 C); visc. 2.8 cps; flash pt. (TCC) 113 C

Exx-Print® 281 A [ExxonMobil]
Chem. Descrip.: Hydrocarbon fluid
Uses: Solvent for printing inks
Properties: Sp.gr. 0.81 (15.6 C); dens. 6.73 lb/gal (15.6 C); visc. 2.8 cps; flash pt. (TCC) 117 C

Exx-Print® 283 D [ExxonMobil; Exxon Canada]
Chem. Descrip.: Petroleum distillates, hydrotreated light
CAS 64742-47-8; EINECS/ELINCS 265-149-8
Uses: Solvent for printing inks; diluent; chem. feedstock; fuel
Features: Very good tack stability
Regulatory: Canada DSL listed; NPRI nonreportable
Properties: Colorless clear liq.; sol. < 0.01% in water; sp.gr. 0.81 (15.6 C); dens. 6.77 lb/gal (60 F); visc. 3.46 cSt; vapor pressure < 0.1 (68 F); f.p. -16 C; b.p. 249-269 C; flash pt. (PMCC) 105 C; autoignition temp. ≈ 211 C; surf. tens. 28.6 dynes/cm; < 0.5% aromatic
Toxicology: REL TWA 143 ppm; inh. of high vapor concs. may cause eye/respiratory tract irritation, headaches, dizziness, anesthesia, drowsiness, unconsciousness, other CNS effects, death; sl. eye irritant; frequent/prolonged skin contact may cause irritation, dermatitis; if aspirated, may cause mild to severe pulmonary injury, death
Environmental: Prevent entry into sewers, watercourses, low areas
Precaution: Flamm. limits 1.6-10.2 vol.%; electrostatic accumulation hazard; use proper grounding procedures; incompat. with strong oxidizing agents
Hazardous decomp. prods.: None
Storage: Store in cool, well-ventilated area away from incompat. materials, heat, open flame, ignition source, direct sunlight; keep container closed

Exx-Print® 287 D [ExxonMobil]
Chem. Descrip.: Paraffins (petroleum), normal C5-20
CAS 64771-72-8
Uses: Solvent for printing inks
Regulatory: DOT nonregulated; SARA §301/304 nonreportable, §311/312 nonhazardous, §313 nonreportable
Properties: Water-wh. clear liq.; faint petrol. hydrocarbon odor; negligible sol. in water; m.w. 189; sp.gr. 0.76 (15.6 C); dens. 6.35 lb/gal (60 F); vapor pressure 0.2 mm Hg (20 C); b.p. 222-243 C; pour pt. -4 C; flash pt. (PMCC) > 93.3 C; autoignition temp. ≈ 210 C; pH essentially neutral; surf. tens. 26.7 dynes/cm; 100% paraffin; < 0.01% aromatic
Toxicology: LD50 (oral, rat) > 5 g/kg, (skin, rabbit) > 3.16 g/kg; inh. of high vapor concs. may cause eye/respiratory tract irritation, headaches, dizziness, anesthesia, drowsiness, unconsciousness, other CNS effects, death; may cause eye irritation; prolonged/repeated skin contact may cause irritation, dermatitis; if aspirated, may cause mild to severe pulmonary injury, death; TSCA listed
Environmental: Prevent entry into sewers, watercourses
Precaution: Flamm. limits 1.4-8.9%; volatile; gives off invisible vapors; liq. or vapor may settle in low areas; distant ignition/explosion possible; static accumulation hazard; ground containers, equip.; incompat. with strong oxidants (liq. chlorine, conc. oxygen, sodium or calcium hypochlorite)
Hazardous decomp. prods.: Combustion: fumes, smoke, CO, SO_x, aldehydes, other decomp. prods.
NFPA: Health 1, Flammability 1, Reactivity 0
Storage: Store away from heat, sparks, flame, strong oxidants; keep containers closed when not in use

Exx-Print® 392 D [ExxonMobil]
Uses: Solvent for printing inks
Features: Low smoke tendency; limited solvency
Properties: Low odor; sp.gr. 0.77 (15.6 C); dens. 6.44 lb/gal (60 F); b.p. 255-279 C; flash pt. (TCC) 248 F; surf. tens. 28.9 dynes/cm; 100% paraffin; 0.01% aromatic

Exx-Print® 493 D [ExxonMobil]
Uses: Solvent for printing inks
Features: Exc. tack stability
Properties: Odorless; sp.gr. 0.82 (15.6 C); dens. 6.82 lb/gal (60 F); b.p. 273-312 C; flash pt. (TCC) 265 F; surf. tens. 26.9 dynes/cm; 66% paraffin; 34% cycloparaffin; < 0.5% aromatic

Exx-Print® 588 D [ExxonMobil; Exxon Canada]
Chem. Descrip.: Petroleum distillates, hydrotreated middle
CAS 64742-46-7; EINECS/ELINCS 265-148-2
Uses: Solvent for printing inks; diluent; chem. feedstock; fuel
Features: Good tack stability
Regulatory: Canada DSL listed; NPRI nonreportable
Properties: Colorless clear liq.; mild min. spirits odor; sol. < 0.01% in water; sp.gr. 0.81-0.83 (15.6 C); dens. 6.99 lb/gal (60 F); visc. 6.89 cSt; vapor pressure 0 kPa (20 C); f.p. -43 C; b.p. 275-313 C; flash pt. (PMCC) 129 C; autoignition temp. ≈ 241 C; surf. tens. 26.9 dynes/cm; < 0.5% aromatic
Toxicology: REL 5 mg/m^3 or lower (mist); inh. of high vapor concs. may cause eye/respiratory tract irritation, headaches, dizziness, anesthesia, drowsiness, unconsciousness, other CNS effects, death; sl. eye irritant; frequent/prolonged skin contact may cause irritation, dermatitis; if aspirated, may cause mild to severe pulmonary injury, death
Environmental: Prevent entry into sewers, watercourses, low areas
Precaution: Flamm. limits 0.4-4.3 vol.%; electrostatic accumulation hazard; use proper grounding procedures; incompat. with strong oxidizing agents
Hazardous decomp. prods.: None
Storage: Store in cool, well-ventilated area away from incompat. materials, heat, open flame, ignition source, direct sunlight; keep container closed

Exxsol® D 30 [ExxonMobil Europe]
Chem. Descrip.: Dearomatized hydrocarbon fluid
Uses: Solvent, thinner, cleaner for industrial coatings, solv.-based decorative coatings, maintenance finishes
Features: Low aromatics content; low surf. tens. for better substrate wetting
Properties: Low odor; dens. 0.764 kg/dm^3 (15 C); visc. 0.78 mPa·s; i.b.p. 142 C; flash pt. 29 C; ref. index 1.422 (20 C); surf. tens. 23.7 mN/m; KB value 33

Exxsol® D 40 [ExxonMobil; ExxonMobil Europe]
Chem. Descrip.: Petroleum distillates, hydrotreated light
CAS 64742-47-8; EINECS/ELINCS 265-149-8
UN 1268
Uses: Solvent for solv.-based architectural coatings and maintenance finishes, printing inks, cleaning and degreasing formulations, agric. pesticides, herbicides; diluent; food-contact paper/paperboard, rubber ar-

ticles
Features: Cost-effective; no SARA reportables; < 100 ppm HAPs; improved solvency
Regulatory: FDA 21CFR §176.180, 176.210, 177.2600, 178.3570, 178.3620(b); SARA §311/312 acute health/chronic health/fire hazard, §313 nonreportable
Properties: Water-wh. clear liq.; min. spirit odor; negligible sol. in water; sp.gr. 0.778 (15/4 C); visc. < 1.0 cSt (40 C); vapor pressure < 5.0 mm Hg (20 C); b.p. > 149 C; pour pt. -18 C; aniline pt. 60-72 C; flash pt. > 38 C; surf. tens. 26 dynes/cm; KB value 34; 100% act.; 42% paraffin, 57% cycloparaffin, 0.4% aromatics
Toxicology: REL TWA (8 h) 1200 mg/m^3; pract. nontoxic by ing., skin contact, inh.; pract. nonirritating to eyes; inh. of high vapor concs. may cause eye/respiratory tract irritation, anesthesia; may cause nose/throat/lung irritation, dizziness, nausea, loss of consciousness; prolonged/repeated skin contact may cause irritation, dermatitis; if aspirated, may cause serious or fatal lung damage
Environmental: Expected to be biodeg.; may be harmful to aquatic organisms, toxic to algae, daphnia; prevent entry into sewers, watercourses, low areas
Precaution: Combustible liq. and vapor; flamm. limits 1.2-9.6 vol.%; liq. can release vapors that readily form flamm. mixts. at or above flash pt.; static accumulation hazard; use proper grounding procedures; incompat. with halogens, strong acids, alkalis, oxidizers
Hazardous decomp. prods.: Combustion: fumes, smoke, CO, SO_x, aldehydes, other decomp. prods.
NFPA: Health 1, Flammability 2, Reactivity 0
Storage: Store in cool, well-ventilated area away from incompat. materials, heat, open flame, ignition sources

Exxsol® D 40 Anti-Static Fluid [ExxonMobil; Exxon Canada]
Chem. Descrip.: Petroleum distillates, hydrotreated light with antistatic additive (30 ml/1000 gal)
CAS 64742-47-8; EINECS/ELINCS 265-149-8
UN 1268
Uses: Solvent; diluent; chem. feedstock; fuel
Regulatory: Canada DSL listed; NPRI nonreportable
Properties: Colorless clear liq.; sol. < 0.01% in water; m.w. 141; sp.gr. 0.770-0.800 (15.6/15.6 C); visc. 1.28 cSt; vapor pressure 0.289 kPa (20 C); f.p. < -60 C; b.p. 159-194 C; aniline pt. 60-72 C; flash pt. (TCC) 42 C; autoignition temp. ≈ 232 C; 100% act.; 1.0% aromatics
Toxicology: REL TWA 197 ppm; LD50 (oral, rat) > 5 g/kg, (skin, rabbit) > 3 g/kg; inh. of high vapor concs. may cause eye/respiratory tract irritation, headaches, dizziness, anesthesia, drowsiness, unconsciousness, other CNS effects, death; sl. eye irritant; frequent/prolonged skin contact may cause irritation, dermatitis; if aspirated, may cause mild to severe pulmonary injury, death
Environmental: Prevent entry into sewers, watercourses, low areas
Precaution: Combustible liq.; flamm. limits 1.4-8.9 vol.%; volatile; gives off invisible vapors; liq. or vapor may settle in low areas; distant ignition/explosion possible; electrostatic accumulation hazard; use proper grounding procedures; incompat. with strong oxidizing agents
Hazardous decomp. prods.: None
Storage: Store in cool, well-ventilated area away from incompat. materials, heat, open flame, ignition source, direct sunlight; keep container closed

Exxsol® D 40 Naphtha [ExxonMobil]
Chem. Descrip.: Naphtha
EINECS/ELINCS 232-443-2
Uses: Solvent
Properties: Bright clear appearance; sp.gr. 0.770-0.800 (15.6/15.6 C); i.b.p. 152 C min.; flash pt. > 40 C; aniline pt. 60-72 C; 0.85% max. aromatics

Exxsol® D 60 [ExxonMobil; Exxon Canada; ExxonMobil Europe]
Chem. Descrip.: Petroleum distillates, hydrotreated light
CAS 64742-47-8; EINECS/ELINCS 265-149-8
Uses: Solvent for solv.-based architectural coatings and maintenance finishes, printing inks, agrochems., cleaning and degreasing formulations; diluent; food-contact paper, rubber articles; defoamer in food-contact paper/paperboard
Features: Cost-effective; < 100 ppm HAPs; improved solvency
Regulatory: FDA 21CFR §176.180, 176.210, 177.2600, 178.3570, 178.3620(b), 40CFR §180.1001(c)(e); SARA nonreportable; Canada DSL listed; NPRI nonreportable
Properties: Colorless clear liq.; mild pleasant odor; sol. < 0.01% in water; m.w. 155; sp.gr. 0.79 (15.6 C); dens. 6.59 lb/gal (60 F); visc. 1.2 cP; vapor pressure 1.1 mm Hg (20 C); f.p. -50 C; b.p. 187-210 C; aniline pt. 65-75 C; flash pt. (TCC) 64 C; autoignition temp. ≈ 227 C; surf. tens. 27.3 dynes/cm; KB value 32; 100% act.; 49% paraffin, 51% cycloparaffin, 0.4% aromatics
Toxicology: REL TWA 184 ppm; LD50 (oral, rat) > 5 g/kg, (skin, rabbit) > 3 g/kg; inh. of high vapor concs. may cause eye/respiratory tract irritation, headaches, dizziness, anesthesia, drowsiness, unconsciousness, other CNS effects, death; sl. eye irritant; frequent/prolonged skin contact may cause irritation, dermatitis; if aspirated, may cause mild to severe pulmonary injury, death; chronic exposure may cause stomach/liver/kidney damage
Environmental: Prevent entry into sewers, watercourses, low areas
Precaution: Combustible liq.; flamm. limits 1.4-9.3 vol.%; volatile; gives off invisible vapors; liq. or vapor may settle in low areas; distant ignition/explosion possible; electrostatic accumulation hazard; use proper grounding procedures; incompat. with strong oxidizing agents
Hazardous decomp. prods.: None
Storage: Store in cool, well-ventilated area away from incompat. materials, heat, open flame, ignition source, direct sunlight; keep container closed

Exxsol® D 60 Anti-Static Fluid [ExxonMobil; Exxon Canada]
Chem. Descrip.: Hydrotreated light petroleum distillates with antistatic additive (30 ml/1000 gal)
CAS 64742-47-8; EINECS/ELINCS 265-149-8
Uses: Solvent; diluent; chem. feedstock; fuel
Regulatory: Canada DSL listed; NPRI nonreportable
Properties: Colorless clear liq.; sol. < 0.1% in water (20 C); m.w. 162; sp.gr. 0.768-0.820 (15.6/15.6 C); visc. 1.63 cSt; vapor pressure 0.4 kPa (38 C); f.p. < 0 C; b.p. 182-220 C; aniline pt. 65-75 C; flash pt. (TCC) 62 C min.; autoignition temp. > 200 C; 100% act.; 1.0% max. aromatics
Toxicology: REL TWA 184 ppm; LD50 (oral, rat) > 5 g/kg, (skin, rabbit) > 3 g/kg; inh. of high vapor concs. may cause eye/respiratory tract irritation, headaches, dizziness, anesthesia, drowsiness, unconsciousness, other CNS effects, death; sl. eye irritant; frequent/prolonged skin contact may cause irritation, dermatitis; if aspirated, may cause mild to severe pulmonary injury, death; chronic exposure may cause stomach/liver/kidney damage
Environmental: Prevent entry into sewers, watercourses, low areas
Precaution: Combustible liq.; flamm. limits 1.4-9.3 vol.%; volatile; gives off invisible vapors; liq. or vapor may settle in low areas; distant ignition/explosion possible; electrostatic accumulation hazard; use proper grounding procedures; incompat. with strong oxidizing agents
Hazardous decomp. prods.: None
Storage: Store in cool, well-ventilated area away from incompat. materials, heat, open flame, ignition source, direct sunlight; keep container closed

Exxsol® D 60 Naphtha [ExxonMobil]
Chem. Descrip.: Naphtha
EINECS/ELINCS 232-443-2
Uses: Solvent
Properties: Bright clear appearance; sp.gr. 0.768-0.820 (15.6/15.6 C); i.b.p. 182 C min.; flash pt. > 61.5 C; 0.85% max. aromatics

Exxsol® D 60 Naphtha Anti-Static [ExxonMobil; Exxon Canada]
Chem. Descrip.: Petroleum distillates, hydrotreated light with antistatic additive (16 ml/1000 gal)
CAS 64742-47-8; EINECS/ELINCS 265-149-8
Uses: Solvent; diluent; chem. feedstock; fuel
Regulatory: Canada DSL listed; NPRI nonreportable
Properties: Colorless bright clear liq.; sol. < 0.1% in water (20 C); m.w. 162; sp.gr. 0.768-0.820 (15.6/15.6 C); visc. 1.63 cSt; vapor pressure 0.4 kPa (38 C); f.p. < 0 C; b.p. 182-220 C; flash pt. (TCC) > 62 C; autoignition temp. > 200 C; 100% act.; 0.85% max. aromatics
Toxicology: REL TWA 184 ppm; LD50 (oral, rat) > 5 g/kg, (skin, rabbit) > 3 g/kg; inh. of high vapor concs. may cause eye/respiratory tract irritation, headaches, dizziness, anesthesia, drowsiness, unconsciousness, other CNS effects, death; sl. eye irritant; frequent/prolonged skin contact may cause irritation, dermatitis; if aspirated, may cause mild to severe pulmonary injury, death; chronic exposure may cause stomach/liver/kidney damage

Environmental: Prevent entry into sewers, watercourses, low areas
Precaution: Combustible liq.; flamm. limits 1.4-9.3 vol.%; volatile; gives off invisible vapors; liq. or vapor may settle in low areas; distant ignition/explosion possible; electrostatic accumulation hazard; use proper grounding procedures; incompat. with strong oxidizing agents
Hazardous decomp. prods.: None
Storage: Store in cool, well-ventilated area away from incompat. materials, heat, open flame, ignition sources, direct sunlight; keep container closed

Exxsol® D 70 [Exxon Canada]
Chem. Descrip.: Petroleum distillates, hydrotreated light
CAS 64742-47-8; EINECS/ELINCS 265-149-8
Uses: Solvent; diluent; chem. feedstock; fuel
Regulatory: Canada DSL listed; NPRI nonreportable
Properties: Colorless clear liq.; sol. < 0.01% in water; m.w. 162; sp.gr. 0.793 (15.6 C); visc. 1.63 cSt; vapor pressure 0.07 kPa (20 C); f.p. -46 C; b.p. 187-208 C; flash pt. (TCC) 64 C; autoignition temp. > 253 C; 100% act.
Toxicology: REL TWA 184 ppm; LD50 (oral, rat) > 5 g/kg, (skin, rabbit) > 3 g/kg; inh. of high vapor concs. may cause eye/respiratory tract irritation, headaches, dizziness, anesthesia, drowsiness, unconsciousness, other CNS effects, death; sl. eye irritant; frequent/prolonged skin contact may cause irritation, dermatitis; if aspirated, may cause mild to severe pulmonary injury, death; chronic exposure may cause stomach/liver/kidney damage
Environmental: Prevent entry into sewers, watercourses, low areas
Precaution: Combustible liq.; flamm. limits 0.7-5.3 vol.%; volatile; gives off invisible vapors; liq. or vapor may settle in low areas; distant ignition/explosion possible; electrostatic accumulation hazard; use proper grounding procedures; incompat. with strong oxidizing agents
Hazardous decomp. prods.: None
Storage: Store in cool, well-ventilated area away from incompat. materials, heat, open flame, ignition sources, direct sunlight; keep container closed

Exxsol® D 75/100 [Exxon Canada]
Chem. Descrip.: Naphtha, hydrotreated light
CAS 64742-49-0; EINECS/ELINCS 203-777-6
UN 1268
Uses: Solvent; diluent; chem. feedstock; fuel
Regulatory: Canada DSL listed; NPRI reportable [cyclohexane (12%), n-hexane (10.6%)]
Properties: Colorless clear liq.; sol. < 0.01% in water; m.w. 95; dens. 0.72 kg/l (15 C); visc. 0.63 cSt; vapor pressure 11.62 kPa (20 C); f.p. < -63 C; b.p. 78-99 C; flash pt. (TCC) < -18 C; autoignition temp. ≈ 257 C; 100% act.
Toxicology: REL TWA 206 ppm; inh. of high vapor concs. may cause eye/respiratory tract irritation, headaches, dizziness, anesthesia, drowsiness, unconsciousness, other CNS effects, death; sl. eye irritant; frequent/prolonged skin contact may cause irritation, dermatitis; if aspirated, may cause mild to severe pulmonary injury, death; contains n-hexane (may cause peripheral nervous system damage)
Environmental: Prevent entry into sewers, watercourses, low areas
Precaution: Flamm. liq.; flamm. limits 1.8-11.6 vol.%; volatile; gives off invisible vapors; liq. or vapor may settle in low areas; distant ignition/explosion possible; electrostatic accumulation hazard; use proper grounding procedures; incompat. with strong oxidizing agents
Hazardous decomp. prods.: None
Storage: Store in cool, well-ventilated area away from incompat. materials, heat, open flame, ignition sources, direct sunlight; keep container closed

Exxsol® D 75/100 Naphtha [ExxonMobil]
Chem. Descrip.: Naphtha, hydrotreated light
CAS 64742-49-0; EINECS/ELINCS 203-777-6
Uses: Solvent for architectural coatings, OEM and auto refinish coatings, industrial metal and wood coatings, maintenance and marine coatings
Properties: Sp.gr. 0.72 (15.6/15.6 C); dens. 5.99 lb/gal (60 F); vapor pressure 75 mm Hg (20 C); b.p. 76-96 C; flash pt. (TCC) < -18 C; surf. tens. 20.4 dynes/cm; KB value 36; 92% paraffin, 8% cycloparaffin, < 0.01% aromatics
Toxicology: OEL 300 ppm

Exxsol® D 80 [ExxonMobil; Exxon Canada; ExxonMobil Europe]
Chem. Descrip.: Petroleum distillates, hydrotreated light
CAS 64742-47-8; EINECS/ELINCS 265-149-8
Uses: Solvent for coatings, cleaners, engine degreasers, agric. pesticides, herbicides, car dewaxing; diluent; chem. feedstock; fuel; food pkg. paper, rubber articles; defoamer in food-contact paper/paperboard
Features: Cost-effective; 50% compliance with criteria for LVP solvent; < 100 ppm HAPs; improved solvency; good solvency for greases
Regulatory: FDA 21CFR §176.180, 176.210, 177.2600, 178.3570, 178.3620(b), 40CFR §180.1001(c)(e); SARA nonreportable
Properties: Liq.; mild pleasant odor; sol. < 0.01% in water; m.w. 179; sp.gr. 0.79 (15.6/15.6 C); dens. 6.65 lb/gal (60 F); visc. 1.7 cP; vapor pressure 0.2 mm Hg (20 C); f.p. -39 C; b.p. 207-234 C; flash pt. (PMCC) 77 C; autoignition temp. ≈ 251 C; surf. tens. 27.6 dynes/cm; KB value 28; 100% act.; 53% paraffin, 47% cycloparaffin
Toxicology: REL TWA 165 ppm; LD50 (oral, rat) > 5 g/kg, (skin, rabbit) > 3 g/kg; inh. of high vapor concs. may cause eye/respiratory tract irritation, headaches, dizziness, anesthesia, drowsiness, unconsciousness, other CNS effects, death; sl. eye irritant; frequent/prolonged skin contact may cause irritation, dermatitis; if aspirated, may cause mild to severe pulmonary injury, death; chronic exposure may cause liver/kidney damage
Environmental: Prevent entry into sewers, watercourses, low areas
Precaution: Combustible liq.; flamm. limits 0.6-5 vol.%; volatile; gives off invisible vapors; liq. or vapor may settle in low areas; distant ignition/explosion possible; electrostatic accumulation hazard; use proper grounding procedures; incompat. with strong oxidizing agents
Hazardous decomp. prods.: None
Storage: Store in cool, well-ventilated area away from incompat. materials, heat, open flame, ignition sources, direct sunlight; keep container closed

Exxsol® D 90/120 [Exxon Canada]
Chem. Descrip.: Petroleum distillates, hydrotreated light
CAS 64742-47-8; EINECS/ELINCS 265-149-8
UN 1268
Uses: Solvent; diluent; chem. feedstock; fuel
Regulatory: Canada DSL listed; NPRI reportable [cyclohexane (1.3%)]
Properties: Colorless clear liq.; mild petrol. odor; sol. < 0.01% in water; m.w. 103; sp.gr. 0.735 (15.6 C); vapor pressure 5.7 kPa (20 C); f.p. less than ambient; b.p. 95-119 C; flash pt. (TCC) < -4 C; 100% act.
Toxicology: REL TWA 332 ppm; LD50 (oral, rat) > 5 g/kg, (skin, rabbit) > 3 g/kg; inh. of high vapor concs. may cause eye/respiratory tract irritation, headaches, dizziness, anesthesia, drowsiness, unconsciousness, other CNS effects, death; sl. eye irritant; frequent/prolonged skin contact may cause irritation, dermatitis; if aspirated, may cause mild to severe pulmonary injury, death
Environmental: Prevent entry into sewers, watercourses, low areas
Precaution: Flamm. liq.; electrostatic accumulation hazard; use proper grounding procedures; incompat. with halogens, molten sulfur, strong oxidizing agents
Hazardous decomp. prods.: Combustion: toxic gases, fumes, smoke, CO
Storage: Store in cool, well-ventilated area away from incompat. materials, heat, open flame, ignition sources, direct sunlight; keep container closed

Exxsol® D 90/120 Anti-Static Solvent [Exxon Canada]
Chem. Descrip.: Petroleum distillates, hydrotreated light
CAS 64742-47-8; EINECS/ELINCS 265-149-8
UN 1268
Uses: Solvent; diluent; chem. feedstock; fuel
Regulatory: Canada DSL listed; NPRI reportable [cyclohexane (1.3%)]
Properties: Colorless clear liq.; mild petrol. odor; sol. < 0.01% in water; m.w. 103; sp.gr. 0.735 (15.6 C); vapor pressure 5.7 kPa (20 C); f.p. less than ambient; b.p. 95-119 C; 100% act.
Toxicology: REL TWA 332 ppm; LD50 (oral, rat) > 5 g/kg, (skin, rabbit) > 3 g/kg; inh. of high vapor concs. may cause eye/respiratory tract irritation, headaches, dizziness, anesthesia, drowsiness, unconsciousness, other CNS effects, death; sl. eye irritant; frequent/prolonged skin contact may cause irritation, dermatitis; if aspirated, may cause mild to severe pulmonary injury, death
Precaution: Flamm. liq.

Exxsol® D 95 [ExxonMobil; Exxon Canada]
Chem. Descrip.: Petroleum distillates, hydrotreated light, dearomatized
CAS 64742-47-8; EINECS/ELINCS 265-149-8
Uses: Solvent for household prods., general-purpose cleaners/degreasers,

floor polishes, adhesives, spot removers, automotive prods. (cleaners/degreasers, lubricants, penetrants, waxes, polishes, sealants), insecticides, herbicides; diluent; chem. feedstock; fuel
Features: Meets LVP VOC exemption (Calif.)
Regulatory: Canada DSL listed; NPRI nonreportable
Properties: Saybolt 30 min. liq.; sol. < 0.01% in water; m.w. 177; sp.gr. 0.79-0.81 (15.6/15.6 C); visc. 2.13 cSt; vapor pressure < 0.133 kPa (20 C); f.p. < -25 C; b.p. 221-271 C; flash pt. (PMCC) > 94 C; autoignition temp. ≈ 216 C
Toxicology: REL TWA 166 ppm; inh. of high vapor concs. may cause eye/respiratory tract irritation, headaches, dizziness, anesthesia, drowsiness, unconsciousness, other CNS effects, death; sl. eye irritant; frequent/prolonged skin contact may cause irritation, dermatitis; if aspirated, may cause mild to severe pulmonary injury, death
Environmental: Prevent entry into sewers, watercourses, low areas
Precaution: Flamm. limits 1.3-8.1 vol.%; volatile; gives off invisible vapors; liq. or vapor may settle in low areas; distant ignition/explosion possible; electrostatic accumulation hazard; use proper grounding procedures; incompat. with strong oxidizing agents
Hazardous decomp. prods.: None
Storage: Store in cool, well-ventilated area away from incompat. materials, heat, open flame, ignition sources, direct sunlight; keep container closed

Exxsol® D 100 [ExxonMobil Europe]
Chem. Descrip.: Dearomatized hydrocarbon fluid
Uses: Solvent for agric., pesticides, herbicides
Properties: Dens. 0.822 kg/dm^3 (15 C); visc. 3.23 mPa·s; i.b.p. 240 C; flash pt. 103 C; ref. index 1.451 (20 C); surf. tens. 27.3 mN/m; KB value 27; aniline pt. 77 C; 0.4% aromatics

Exxsol® D 100 S [ExxonMobil Europe]
Chem. Descrip.: Dearomatized hydrocarbon fluid
Uses: Solvent for agric., pesticides, herbicides
Properties: Dens. 0.820 kg/dm^3 (15 C); visc. 3.22 mPa·s; i.b.p. 239 C; flash pt. 103 C; ref. index 1.450 (20 C); surf. tens. 27.3 mN/m; KB value 27; aniline pt. 78 C; 0.05% aromatics

Exxsol® D 110 [ExxonMobil; Exxon Canada; ExxonMobil Europe]
Chem. Descrip.: Petroleum distillates, hydrotreated light
CAS 64742-47-8; EINECS/ELINCS 265-149-8
Uses: Solvent for coatings, carburetor cleaners, brake cleaners, degreasers, insecticides; diluent; chem. feedstock; fuel; food pkg. paper, rubber articles; defoamer in food-contact paper/paperboard
Features: Cost-effective; complies with criteria for LVP solvent; < 100 ppm HAPs; improved solvency
Regulatory: FDA 21CFR §176.180, 176.210, 177.2600, 178.3570, 178.3620(b), 40CFR §180.1001(c)(e); SARA nonreportable; Canada DSL listed; NPRI nonreportable
Properties: Colorless clear liq.; mild pleasant odor; sol. < 0.01%; m.w. 215; sp.gr. 0.81 (15.6/15.6 C); dens. 6.77 lb/gal (60 F); visc. 2.8 cP; vapor pressure < 0.1 mm Hg (20 C); f.p. -48 C; b.p. 251-269 C; flash pt. (PMCC) 105 C; autoignition temp. ≈ 243 C; surf. tens. 28.6 dynes/cm; KB value 26; 59% paraffin, 41% cycloparaffin
Toxicology: REL TWA 143 ppm; inh. of high vapor concs. may cause eye/respiratory tract irritation, headaches, dizziness, anesthesia, drowsiness, unconsciousness, other CNS effects, death; sl. eye irritant; frequent/prolonged skin contact may cause irritation, dermatitis; if aspirated, may cause mild to severe pulmonary injury, death; contains ethylbenzene, possible carcinogen
Environmental: Prevent entry into sewers, watercourses, low areas
Precaution: Flamm. limits 0.5-4.6 vol.%; electrostatic accumulation hazard; use proper grounding procedures; incompat. with strong oxidizing agents
Hazardous decomp. prods.: None
Storage: Store in cool, well-ventilated area away from incompat. materials, heat, open flame, ignition sources, direct sunlight; keep container closed

Exxsol® D 115/145 [Exxon Canada]
Chem. Descrip.: Naphtha, hydrotreated light
CAS 64742-49-0; EINECS/ELINCS 265-151-9
UN 1268
Uses: Solvent; diluent; chem. feedstock; fuel
Regulatory: Canada DSL listed; NPRI nonreportable
Properties: Colorless clear liq.; mild petrol. odor; sol. < 0.01% in water; m.w. 114; dens. 0.75 kg/l (15 C); visc. 0.83 cSt; vapor pressure 2.3 kPa (20 C); f.p. < -62 C; b.p. 116-144 C; flash pt. (TCC) 7 C; autoignition temp. ≈ 250 C; 100% act.
Toxicology: REL TWA 300 ppm; inh. of high vapor concs. may cause eye/respiratory tract irritation, headaches, dizziness, anesthesia, drowsiness, unconsciousness, other CNS effects, death; sl. eye irritant; frequent/prolonged skin contact may cause irritation, dermatitis; if aspirated, may cause mild to severe pulmonary injury, death
Environmental: Prevent entry into sewers, watercourses, low areas
Precaution: Flamm. liq.; flamm. limits 1.3-10.2 vol.%; volatile; gives off invisible vapors; liq. or vapor may settle in low areas; distant ignition/explosion possible; electrostatic accumulation hazard; use proper grounding procedures; incompat. with strong oxidizing agents
Hazardous decomp. prods.: Combustion: toxic gases, fumes, smoke, CO
Storage: Store in cool, well-ventilated area away from incompat. materials, heat, open flame, ignition sources, direct sunlight; keep container closed

Exxsol® D 115/145 Anti-Static Solvent [Exxon Canada]
Chem. Descrip.: Naphtha, hydrotreated light with antistatic agent (16 ml/1000 Imp. gal)
CAS 64742-49-0; EINECS/ELINCS 265-151-9
UN 1268
Uses: Solvent; diluent; chem. feedstock; fuel
Regulatory: Canada DSL listed; NPRI nonreportable
Properties: Colorless clear liq.; mild petrol. odor; sol. < 0.01% in water; m.w. 114; dens. 0.75 kg/l (15 C); visc. 0.83 cSt; vapor pressure 2.3 kPa (20 C); f.p. < -62 C; b.p. 116-144 C; flash pt. (TCC) 7 C; autoignition temp. ≈ 250 C; 100% act.
Toxicology: REL TWA 300 ppm; inh. of high vapor concs. may cause eye/respiratory tract irritation, headaches, dizziness, anesthesia, drowsiness, unconsciousness, other CNS effects, death; sl. eye irritant; frequent/prolonged skin contact may cause irritation, dermatitis; if aspirated, may cause mild to severe pulmonary injury, death
Environmental: Prevent entry into sewers, watercourses, low areas
Precaution: Flamm. liq.; flamm. limits 1.3-10.2 vol.%; volatile; gives off invisible vapors; liq. or vapor may settle in low areas; distant ignition/explosion possible; electrostatic accumulation hazard; use proper grounding procedures; incompat. with strong oxidizing agents
Hazardous decomp. prods.: Combustion: toxic gases, fumes, smoke, CO
Storage: Store in cool, well-ventilated area away from incompat. materials, heat, open flame, ignition sources, direct sunlight; keep container closed

Exxsol® D 115/145 Naphtha [ExxonMobil]
Chem. Descrip.: Naphtha, hydrotreated light
CAS 64742-49-0; EINECS/ELINCS 265-151-9
Uses: Solvent for architectural coatings, OEM and auto refinish coatings, industrial metal and wood coatings, maintenance and marine coatings
Properties: Sp.gr. 0.75 (15.6/15.6 C); dens. 6.22 lb/gal (60 F); vapor pressure 13 mm Hg (20 C); b.p. 116-144 C; flash pt. (TCC) 10 C; surf. tens. 22.1 dynes/cm; KB value 36; 59% paraffi, 41% cycloparaffin, 0.01% aromatics
Toxicology: OEL 300 ppm

Exxsol® D 120 [ExxonMobil Europe]
Chem. Descrip.: Dearomatized hydrocarbon fluid
Uses: Solvent for agric., pesticides, herbicides
Properties: Dens. 0.834 kg/dm^3 (15 C); visc. 4.83 mm^2/s; i.b.p. 260 C; flash pt. 119 C; ref. index 1.457 (20 C); surf. tens. 28.5 mN/m; KB value 24; aniline pt. 81 C; 0.6% aromatics

Exxsol® D 130 [ExxonMobil; Exxon Canada]
Chem. Descrip.: Petroleum distillates, hydrotreated middle
CAS 64742-46-7; EINECS/ELINCS 265-148-2
Uses: Solvent for agric., pesticides, architectural coatings, OEM and auto refinish coatings, industrial metal and wood coatings, maintenance and marine coatings, cleaning and degreasing formulations; diluent; chem. feedstock; fuel; food pkg. paper, rubber articles; defoamer in food-contact paper/paperboard
Regulatory: FDA 21CFR §176.180, 176.210, 177.2600, 178.3570, 178.3620(b), 40CFR §180.1001(c)(e); DOT nonregulated; SARA §311/312 acute/chronic hazard, §313 nonreportable; Canada DSL listed; NPRI nonreportable
Properties: Colorless clear liq.; mild min. spirits odor; sol. < 0.01% in water; m.w. 200; sp.gr. 0.84 (15.6/15.6 C); dens. 6.97 lb/gal (60 F); visc.

6.89 cSt; vapor pressure < 0.1 mm Hg (20 C); pour pt. -43 C; b.p. 276-316 C; flash pt. (PMCC) > 129 C; autoignition temp. ≈ 241 C; surf. tens. 29.7 dynes/cm; KB value 24; 100% act.
Toxicology: REL 5 mg/m³ or lower (mist); inh. of high vapor concs. may cause eye/respiratory tract irritation, headaches, dizziness, anesthesia, drowsiness, unconsciousness, other CNS effects, death; sl. eye irritant; frequent/prolonged skin contact may cause irritation, dermatitis; if aspirated, may cause mild to severe pulmonary injury, death
Environmental: Prevent entry into sewers, watercourses, low areas, contact with soil
Precaution: Flamm. limits 0.4-4.3 vol.%; low flamm./explosion hazard; electrostatic accumulation hazard; use proper grounding procedures; incompat. with strong oxidizing agents
Hazardous decomp. prods.: CO
NFPA: Health 0, Flammability 1, Reactivity 0
Storage: Store in cool, well-ventilated area away from incompat. materials, heat, open flame, ignition sources, direct sunlight; keep container closed

Exxsol® D 140 [ExxonMobil Europe]
Chem. Descrip.: Dearomatized hydrocarbon fluid
Uses: Solvent for agric., pesticides, herbicides
Properties: Dens. 0.837 kg/dm³ (15 C); visc. 7 mPa·s; i.b.p. 285 C; flash pt. 138 C; ref. index 1.459 (20 C); surf. tens. 28.9 mN/m; KB value 22; aniline pt. 87 C; 1.2% aromatics

Exxsol® D 155/170 [ExxonMobil Europe]
Chem. Descrip.: Dearomatized hydrocarbon fluid
Uses: Solvent
Properties: Dens. 0.765 kg/dm³ (15 C); visc. 0.89 mPa·s; i.b.p. 159 C; flash pt. 45 C; ref. index 1.424 (20 C); surf. tens. 24.2 mN/m; KB value 31; aniline pt. 67 C; 0.01% aromatics

Exxsol® D 180/200 SP [ExxonMobil Europe]
Chem. Descrip.: Dearomatized hydrocarbon fluid
Uses: Solvent for solv.-based decorative coatings, maintenance finishes; diluent, cleaner for decorative paints
Features: Low aromatics content; low surf. tens. for better substrate wetting
Properties: Low odor; dens. 0.793 kg/dm³ (15 C); visc. 1.26 mPa·s; i.b.p. 186 C; flash pt. 64 C; ref. index 1.437 (20 C); surf. tens. 25.6 mN/m; KB value 31; 0.01% aromatic content

Exxsol® D 220/230 [ExxonMobil Europe]
Chem. Descrip.: Dearomatized hydrocarbon fluid
Uses: Solvent
Properties: Dens. 0.810 kg/dm³ (15 C); visc. 2.47 mPa·s; i.b.p. 222 C; flash pt. 90 C; ref. index 1.446 (20 C); surf. tens. 26.9 mN/m; KB value 27; aniline pt. 74 C; 0.1% aromatics

Exxsol® D 3135 [ExxonMobil; Exxon Canada]
Chem. Descrip.: Petroleum distillates, hydrotreated light
CAS 64742-47-8; EINECS/ELINCS 265-149-8
UN 1268
Uses: Solvent for architectural coatings, OEM and auto refinish coatings, industrial metal and wood coatings, maintenance and marine coatings, cleaning and degreasing formulations; diluent; chem. feedstock; fuel; food pkg. paper, rubber articles; defoamer in food-contact paper/paperboard
Regulatory: FDA 21CFR §176.180, 176.210, 177.2600, 178.3570, 178.3620(b); Canada DSL listed; NPRI nonreportable
Properties: Colorless clear liq.; mild petrol. odor; sol. < 0.01% in water; m.w. 136; sp.gr. 0.77 (15.6/15.6 C); dens. 6.43 lb/gal (60 F); visc. 1.17 cSt; vapor pressure 1.1 mm Hg (20 C); f.p. < -61 C; b.p. 157-177 C; flash pt. (TCC) 41 C; autoignition temp. ≈ 216 C; surf. tens. 23.7 dynes/cm; KB value 34; 100% act.; 52% paraffin, 48% cycloparaffin, 0.02% aromatics
Toxicology: REL 213 ppm; DL50 (oral, rat) > 5 g/kg, (skin, rabbit) > 3 g/kg; inh. of high vapor concs. may cause eye/respiratory tract irritation, headaches, dizziness, anesthesia, drowsiness, unconsciousness, other CNS effects, death; sl. eye irritant; frequent/prolonged skin contact may cause irritation, dermatitis; if aspirated, may cause mild to severe pulmonary injury, death
Environmental: Prevent entry into sewers, watercourses, low areas
Precaution: Combustible liq.; flamm. limits 1.4-8.1 vol.%; volatile; gives off invisible vapors; liq. or vapor may settle in low areas; distant ignition/explosion possible; electrostatic accumulation hazard; use proper grounding procedures; incompat. with strong oxidizing agents
Hazardous decomp. prods.: Combustion: toxic gases, fumes, smoke, CO
Storage: Store in cool, well-ventilated area away from incompat. materials, heat, open flame, ignition sources, direct sunlight; keep container closed

Exxsol® D 3135 Anti-Static Solvent [Exxon Canada]
Chem. Descrip.: Petroleum distillates, hydrotreated light
CAS 64742-47-8; EINECS/ELINCS 265-149-8
UN 1268
Uses: Solvent; diluent; chem. feedstock; fuel
Regulatory: Canada DSL listed; NPRI nonreportable
Properties: Colorless clear liq.; mild petrol. odor; sol. < 0.01% in water; sp.gr. 0.774 (15.6 C); visc. 2.13 cSt; vapor pressure 0.3 kPa (20 C); f.p. -54 C; b.p. 158-177 C; flash pt. (TCC) 41 C; autoignition temp. ≈ 216 C; 100% act.
Toxicology: REL TWA 213 ppm; LD50 (oral, rat) > 5 g/kg, (skin, rabbit) > 3 g/kg; inh. of high vapor concs. may cause eye/respiratory tract irritation, headaches, dizziness, anesthesia, drowsiness, unconsciousness, other CNS effects, death; sl. eye irritant; frequent/prolonged skin contact may cause irritation, dermatitis; if aspirated, may cause mild to severe pulmonary injury, death
Environmental: Prevent entry into sewers, watercourses, low areas
Precaution: Combustible liq.; flamm. limits 1.4-8.1 vol.%; liq. or vapor may settle in low areas; distant ignition/explosion possible; electrostatic accumulation hazard; use proper grounding procedures; incompat. with strong oxidizing agents
Hazardous decomp. prods.: Combustion: toxic gases, fumes, smoke, CO
Storage: Store in cool, well-ventilated area away from incompat. materials, heat, open flame, ignition sources, direct sunlight; keep container closed

Exxsol® D 3135 Naphtha [ExxonMobil]
Chem. Descrip.: Naphtha
EINECS/ELINCS 232-443-2
Uses: Solvent
Properties: Sp.gr. 0.771-0.776 (15.6/15.6 C); i.b.p. 152 C min.; flash pt. 38 C min.; 0.3% max. aromatics

Exxsol® D 3135 Naphtha Anti-Static [ExxonMobil; Exxon Canada]
Chem. Descrip.: Petroleum distillates, hydrotreated light with antistatic additive (10 ml/1000 gal)
CAS 64742-47-8; EINECS/ELINCS 265-149-8
UN 1268
Uses: Solvent
Regulatory: Canada DSL listed; NPRI nonreportable
Properties: Colorless bright clear liq.; mild petrol. odor; sol. < 0.01% in water; sp.gr. 0.771-0.776 (15.6/15.6 C); visc. 2.13 cSt; vapor pressure 0.3 kPa (20 C); f.p. -54 C; b.p. 158-177 C; flash pt. (TCC) 41 C; autoignition temp. ≈ 216 C; 100% act.; 0.3% max. aromatics
Toxicology: REL TWA 213 ppm; LD50 (oral, rat) > 5 g/kg, (skin, rabbit) > 3 g/kg; inh. of high vapor concs. may cause eye/respiratory tract irritation, headaches, dizziness, anesthesia, drowsiness, unconsciousness, other CNS effects, death; sl. eye irritant; frequent/prolonged skin contact may cause irritation, dermatitis; if aspirated, may cause mild to severe pulmonary injury, death
Environmental: Prevent entry into sewers, watercourses, low areas
Precaution: Combustible liq.; flamm. limits 1.4-8.1 vol.%; volatile; gives off invisible vapors; liq. or vapor may settle in low areas; distant ignition/explosion possible; electrostatic accumulation hazard; use proper grounding procedures; incompat. with strong oxidizing agents
Hazardous decomp. prods.: Combustion: toxic gases, fumes, smoke, CO
Storage: Store in cool, well-ventilated area away from incompat. materials, heat, open flame, ignition sources, direct sunlight; keep container closed

Exxsol® DSP 60/95 S [ExxonMobil Europe]
Chem. Descrip.: Dearomatized hydrocarbon fluid
Uses: Solvent
Properties: Dens. 0.700 kg/dm³ (15 C); visc. 0.41 mPa·s; i.b.p. 65 C; flash pt. < 0 C; ref. index 1.391 (20 C); surf. tens. 20.2 mN/m; KB value 32; aniline pt. 61 C; 0.0001% aromatics

Exxsol® DSP 65/100 [ExxonMobil Europe]
Chem. Descrip.: Dearomatized hydrocarbon fluid
Uses: Solvent
Properties: Dens. 0.690 kg/dm³ (15 C); visc. 0.42 mPa·s; i.b.p. 69 C; flash

pt. < 0 C; ref. index 1.386 (20 C); surf. tens. 19.1 mN/m; KB value 31; aniline pt. 62 C; 0.0001% aromatics

Exxsol® DSP 80/110 [ExxonMobil Europe]
Chem. Descrip.: Dearomatized hydrocarbon fluid
Uses: Solvent, thinner, cleaner for industrial coatings
Features: Low aromatics content
Properties: Dens. 0.723 kg/dm^3 (15 C); visc. 0.48 mPa·s; i.b.p. 87 C; flash pt. < 0 C; ref. index 1.401 (20 C); surf. tens. 20.8 mN/m; KB value 35

Exxsol® DSP 90/120 Naphtha [Exxon Canada; ExxonMobil Europe]
Chem. Descrip.: Petroleum distillates, hydrotreated light
CAS 64742-47-8; EINECS/ELINCS 265-149-8
UN 1268
Uses: Solvent; diluent; chem. feedstock; fuel
Regulatory: Canada DSL listed; NPRI reportable [cyclohexane (1.3%)]
Properties: Colorless bright clear liq.; mild petrol. odor; sol. < 0.01% in water; m.w. ≈ 103; sp.gr. 0.735 (15.5 C); vapor pressure 5.7 kPa (20 C); f.p. < ambient; b.p. 95-119 C; flash pt. (TCC) < -4 C; 100% act.; 0.2% max. aromatics
Toxicology: REL TWA 332 ppm; LD50 (oral, rat) > 5 g/kg, (skin, rabbit) > 3 g/kg; inh. of high vapor concs. may cause eye/respiratory tract irritation, headaches, dizziness, anesthesia, drowsiness, unconsciousness, other CNS effects, death; sl. eye irritant; frequent/prolonged skin contact may cause irritation, dermatitis; if aspirated, may cause mild to severe pulmonary injury, death
Environmental: Prevent entry into sewers, watercourses, low areas
Precaution: Flamm. liq.; may release vapors that form flamm. mixts. at or above flash pt.; incompat. with halogens, molten sulfur, strong oxidizing agents
Hazardous decomp. prods.: Combustion: toxic gases, fumes, smoke, CO
Storage: Store in cool, well-ventilated area away from incompat. materials, heat, open flame, ignition sources, direct sunlight; keep container closed

Exxsol® DSP 100/120 [ExxonMobil Europe]
Chem. Descrip.: Dearomatized hydrocarbon fluid
Uses: Solvent, thinner, cleaner for industrial coatings
Features: Low aromatics content
Properties: Dens. 0.737 kg/dm^3 (15 C); visc. 0.55 mPa·s; i.b.p. 103 C; flash pt. < 0 C; ref. index 1.408 (20 C); surf. tens. 21.7 mN/m; KB value 36; aniline pt. 57 C; 0.00004% aromatics

Exxsol® DSP 100/140 [ExxonMobil Europe]
Chem. Descrip.: Dearomatized hydrocarbon fluid
Uses: Solvent, thinner, cleaner for industrial coatings
Features: Low aromatics content
Properties: Dens. 0.741 kg/dm^3 (15 C); visc. 0.56 mPa·s; i.b.p. 106 C; flash pt. 0 C; ref. index 1.411 (20 C); surf. tens. 21.8 mN/m; KB value 34

Exxsol® DSP 100/160 [ExxonMobil Europe]
Chem. Descrip.: Dearomatized hydrocarbon fluid
Uses: Solvent, thinner, cleaner for industrial coatings
Features: Low aromatics content
Properties: Dens. 0.745 kg/dm^3 (15 C); visc. 0.60 mPa·s; i.b.p. 107 C; flash pt. 1 C; ref. index 1.413 (20 C); surf. tens. 22.0 mN/m; KB value 34

Exxsol® DSP 115/145 Naphtha [ExxonMobil Europe]
Chem. Descrip.: Naphtha
EINECS/ELINCS 232-443-2
Uses: Solvent
Properties: Bright clear appearance; i.b.p. 115 C min.; flash pt. > 7 C; 1.0% max. aromatics

Exxsol® DSP 145/160 [ExxonMobil Europe]
Chem. Descrip.: Dearomatized hydrocarbon fluid
Uses: Solvent, thinner, cleaner for industrial coatings
Features: Low aromatics content
Properties: Dens. 0.766 kg/dm^3 (15 C); visc. 0.75 mPa·s; i.b.p. 147 C; flash pt. 32 C; ref. index 1.423 (20 C); surf. tens. 23.8 mN/m; KB value 33

Exxsol® DSP 90/120 Naphtha Anti-Static [ExxonMobil]
Chem. Descrip.: Naphtha with antistatic additive (13 ml/1000 gal)
EINECS/ELINCS 232-443-2
Uses: Solvent
Properties: I.b.p. 90 C min.; 0.2% max. aromatics

Exxsol® HP-95 [ExxonMobil; Exxon Canada]
Chem. Descrip.: Cyclopentane (60-100%)
CAS 287-92-3; EINECS/ELINCS 206-016-6
UN 1146
Uses: Solvent; blowing agent for PU foam
Regulatory: Canada DSL listed; NPRI nonreportable
Properties: Colorless clear liq.; naphthenic hydrocarbon odor; sol. < 0.1% in water (20 C); m.w. 73; sp.gr. 0.733-0.756 (15.6/15.6 C); visc. 0.59 cSt; vapor pressure 67.99 kPa (38 C); f.p. < -50 C; b.p. 49 C; flash pt. (TCC) -37 C; autoignition temp. ≈ > 399 C; 95% min. act.
Toxicology: REL TWA 482 ppm; inh. of high vapor concs. may cause eye/respiratory tract irritation, headaches, dizziness, anesthesia, drowsiness, unconsciousness, other CNS effects, death; eye irritant; frequent/prolonged skin contact may cause irritation, dermatitis; if aspirated, may cause severe health effects (bronchopneumonia, pulmonary edema)
Environmental: Prevent entry into sewers, watercourses, low areas
Precaution: Extremely flamm. liq.; flamm. limits 1.5-8.7 vol.%; readily ignites at normal temps.; electrostatic accumulation hazard; use proper grounding procedures; incompat. with strong oxidizing agents
Hazardous decomp. prods.: None
Storage: Store in cool, well-ventilated area away from incompat. materials, heat, open flame, ignition sources, direct sunlight; keep container closed

Exxsol® S-K Premium Solvent [Exxon Canada]
Chem. Descrip.: Petroleum distillates, hydrotreated light
CAS 64742-47-8; EINECS/ELINCS 265-149-8
Uses: Solvent; diluent; chem. feedstock; fuel
Regulatory: Canada DSL listed; NPRI nonreportable
Properties: Colorless clear liq.; sol. < 0.1% in water (20 C); m.w. 162; sp.gr. 0.79 (15 C); visc. 1.6 cSt; f.p. -43 C; b.p. 193-211 C; flash pt. (TCC) 66 C; autoignition temp. ≈ 227 C; 100% act.
Toxicology: REL TWA 184 ppm; LD50 (oral, rat) > 5 g/kg, (skin, rabbit) > 3 g/kg; inh. of high vapor concs. may cause eye/respiratory tract irritation, headaches, dizziness, anesthesia, drowsiness, unconsciousness, other CNS effects, death; sl. eye irritant; frequent/prolonged skin contact may cause irritation, dermatitis; if aspirated, may cause mild to severe pulmonary injury, death
Environmental: Prevent entry into sewers, watercourses, low areas
Precaution: Flamm. limits 1.4-9.3 vol.%; volatile; gives off invisible vapors; liq. or vapor may settle in low areas; distant ignition/explosion possible; electrostatic accumulation hazard; use proper grounding procedures; incompat. with strong oxidizing agents
Hazardous decomp. prods.: None
Storage: Store in cool, well-ventilated area away from incompat. materials, heat, open flame, ignition sources, direct sunlight; keep container closed

Exxsol® Cyclopentane S [ExxonMobil Europe]
Chem. Descrip.: Cyclopentane
CAS 287-92-3; EINECS/ELINCS 206-016-6
Uses: Solvent
Properties: Sp.gr. 0.747 (15 C); visc. 0.42 mPa·s; i.b.p. 49 C; ref. index 1.402 (20 C); surf. tens. 22.2 mN/m; KB value 54; aniline pt. 21 C; 0.0001% max. aromatics

Exxsol® Heptane [ExxonMobil; Exxon Canada; ExxonMobil Europe]
Chem. Descrip.: Naphtha, hydrotreated light
CAS 64742-49-0; EINECS/ELINCS 265-151-9
UN 1206
Uses: Solvent for printing inks; solvent, thinner, cleaner for industrial coatings; diluent; chem. feedstock; fuel
Features: Low aromatics content
Regulatory: SARA §311/312 chronic health/fire hazard, §313 nonreportable; Canada DSL listed; NPRI nonreportable
Properties: Water-wh. clear liq.; bland petrol. odor; negligible sol. in water; m.w. 99; sp.gr. 0.70 (15.6/15.6 C); dens. 5.79 lb/gal (60 F); visc. 0.5 cP; vapor pressure 76 mm Hg (20 C); pour pt. -18 C; b.p. 94-98 C; flash pt. (TCC) -8 C; autoignition temp. ≈ 257 C; surf. tens. 21.2 dynes/cm; KB value 29; 100% act.; 96% paraffin, 4% cycloparaffin, < 0.01% aromatic
Toxicology: REL (8 h) 1600 mg/m^3; LD50 (oral, rat) > 5 g/kg, (skin, rabbit) > 3.16 g/kg; inh. of high vapor concs. may cause eye/respiratory tract irritation, headaches, dizziness, anesthesia, drowsiness, unconsciousness, other CNS effects, death; may cause eye irritation; prolonged/repeated skin contact may cause irritation, dermatitis; if aspirated, may cause mild to severe pulmonary injury, death; TSCA listed
Environmental: Prevent entry into sewers, watercourses

Precaution: Extremely flamm. liq.; flamm. limits in air 1.9-11.6 vol.%; volatile; gives off invisible vapors; liq. or vapor may settle in low areas; distant ignition/explosion possible; static accumulation hazard; use proper grounding procedures; incompat. with strong oxidants (liq. chlorine, conc. oxygen, sodium or calcium hypochlorite)
Hazardous decomp. prods.: Combustion: fumes, smoke, CO, SO_x, aldehydes, other decomp. prods.
HMIS: Health 1, Flammability 3, Reactivity 0
Storage: Store away from heat, sparks, flame, strong oxidants; keep containers closed when not in use

Exxsol® Heptane Anti-Static [ExxonMobil]
Chem. Descrip.: Heptane with antistatic additive (30 ml/1000 gal)
CAS 142-82-5; EINECS/ELINCS 205-563-8
Uses: Solvent
Properties: Sp.gr. 0.69-0.72 (15.6/15.6 C); i.b.p. 92 C min.; aniline pt. 60-71 C; 0.1% max. aromatics

Exxsol® Hexane [ExxonMobil; ExxonMobil Europe]
Chem. Descrip.: Naphtha, hydrotreated light
CAS 64742-49-0; EINECS/ELINCS 265-151-9
UN 1208
Uses: Solvent
Regulatory: SARA §311/312 chronic health/acute health/fire hazard, §313 reportable [n-hexane (55%), cyclohexane (3%)]
Properties: Colorless liq.; odorless; negligible sol. in water; m.w. 86; sp.gr. 0.67 (15.6/15.6 C); dens. 5.59 lb/gal (60 F); visc. 0.4 cSt (40 C); vapor pressure 125 mm Hg (20 C); f.p. -95 C; b.p. 66-69 C; flash pt. (TCC) < -18 C; autoignition temp. 266 C; surf. tens. 19.4 dynes/cm; KB value 29; 92% paraffin, 8% cycloparaffin, < 0.01% aromatic
Toxicology: Pract. nontoxic by ing., inh., skin contact; may cause respiratory irritation, dizziness, nausea, loss of consciousness; may cause narcotic effects; overexposure may cause CNS damage; prolonged/repeated skin contact may cause irritation, dermatitis; if swallowed, can cause serious or fatal lung damage; risk of impaired fertility; contains n-hexane (may cause peripheral nervous system/kidney effects, CNS damage); TSCA listed
Environmental: Prevent entry into sewers, drains, contact with soil
Precaution: Extremely flamm. liq. and vapor; flamm. limits in air 1.8-11.6 vol.%; may cause flash fire; vapor accumulation can flash/explode; prevent buildup of static electricity; ground/bond equip./containers; incompat. with strong oxidizers, calcium hypochlorite, act. chlorine
Hazardous decomp. prods.: CO, CO_2
NFPA: Health 1, Flammability 3, Reactivity 0
Storage: Store in cool, well-ventilated area in bonded/grounded containers away from ignition sources

Exxsol® Hexane Anti-Static [ExxonMobil; Exxon Canada]
Chem. Descrip.: Naphtha, hydrotreated light with antistatic additive (16-30 ml/1000 gal)
CAS 64742-49-0; EINECS/ELINCS 265-151-9
UN 1208
Uses: Solvent; diluent; chem. feedstock; fuel
Regulatory: Canada DSL listed; NPRI reportable [n-hexane (55%)]
Properties: Colorless clear liq.; sol. < 0.01% in water; m.w. 86; sp.gr. 0.664-0.684 (15.6/15.6 C); visc. 0.47 cSt; vapor pressure 19.14 kPa (20 C); f.p. < -60 C; b.p. 57-84 C; aniline pt. 57-68.9 C; flash pt. (TCC) < -18 C; autoignition temp. ≈ 266 C; 100 vol.% act.; 0.01% max. aromatics
Toxicology: REL TWA 85 ppm; inh. of high vapor concs. may cause eye/respiratory tract irritation, headaches, dizziness, anesthesia, drowsiness, unconsciousness, other CNS effects, death; may cause peripheral nervous system damage; sl. eye irritant; frequent/prolonged skin contact may cause irritation, dermatitis; if aspirated, may cause mild to severe pulmonary injury, death; contains n-hexane
Environmental: Prevent entry into sewers, watercourses, low areas
Precaution: Extremely flamm. liq.; flamm. limits 1.8-11.6 vol.%; readily ignites @ normal temps.; volatile; gives off invisible vapors; may release vapors that form flamm. mixts. at or above flash pt.; liq. or vapor may settle in low areas; distant ignition/explosion possible; electrostatic accumulation hazard; use proper grounding procedure; incompat. with strong oxidizing agents
Hazardous decomp. prods.: Combustion: toxic gases, fumes, smoke, CO
Storage: Store in cool, well-ventilated area away from incompat. materials, heat, open flame, ignition source, direct sunlight; keep container closed

Exxsol® Hexane Naphtha [ExxonMobil]
Chem. Descrip.: Naphtha
EINECS/ELINCS 232-443-2
Uses: Solvent
Properties: Bright clear appearance; i.b.p. 52.8 C min.; 45-65 vol.% hexane; 0.01% max. aromatics

Exxsol® Hexane Naphtha Anti-Static [ExxonMobil]
Chem. Descrip.: Naphtha, hydrotreated light with antistatic additive (13 ml/1000 gal)
CAS 64742-49-0; EINECS/ELINCS 265-151-9
UN 1208
Uses: Solvent; diluent; chem. feedstock; fuel
Regulatory: Canada DSL listed; NPRI reportable [n-hexane (55%)]
Properties: Colorless bright clear liq.; sol. < 0.01% in water; m.w. 86; dens. 0.67 kg/l (15 C); visc. 0.47 cSt; vapor pressure 19.14 kPa (20 C); f.p. < -60 C; b.p. 57-84 C; flash pt. (TCC) < -18 C; autoignition temp. ≈ 266 C; 100% act.; 0.01% max. aromatics
Toxicology: REL TWA 85 ppm; inh. of high vapor concs. may cause eye/respiratory tract irritation, headaches, dizziness, anesthesia, drowsiness, unconsciousness, other CNS effects, death; may cause peripheral nervous system damage; sl. eye irritant; frequent/prolonged skin contact may cause irritation, dermatitis; if aspirated, may cause mild to severe pulmonary injury, death; contains n-hexane
Environmental: Prevent entry into sewers, watercourses, low areas
Precaution: Extremely flamm. liq.; flamm. limits 1.8-11.6 vol.%; readily ignites @ normal temps.; volatile; gives off invisible vapors; may release vapors that form flamm. mixts. at or above flash pt.; liq. or vapor may settle in low areas; distant ignition/explosion possible; electrostatic accumulation hazard; use proper grounding procedures; incompat. with strong oxidizing agents
Hazardous decomp. prods.: Combustion: toxic gases, fumes, smoke, CO
Storage: Store in cool, well-ventilated area away from incompat. materials, heat, open flame, ignition source, direct sunlight; keep container closed

Exxsol® Isoheptane [ExxonMobil]
Chem. Descrip.: Isoheptane, dearomatized
CAS 591-76-4
Uses: Solvent
Properties: Sp.gr. 0.680-0.710 (15.6/15.6 C); i.b.p. 79 C min.

Exxsol® Isohexane [ExxonMobil Europe]
Chem. Descrip.: Isohexane, dearomatized
CAS 107-83-5; EINECS/ELINCS 203-523-4
Uses: Solvent
Properties: Dens. 0.670 kg/dm³ (15 C); visc. 0.33 mPa·s; i.b.p. 56 C; flash pt. < 0 C; ref. index 1.377 (20 C); surf. tens. 18.2 mN/m; KB value 30

Exxsol® Isohexane Naphtha [Exxon Canada]
Chem. Descrip.: Naphtha, hydrotreated light
CAS 64742-49-0; EINECS/ELINCS 265-151-9
UN 1208
Uses: Solvent; diluent; chem. feedstock; fuel
Regulatory: Canada DSL listed
Properties: Colorless clear liq.; sol. < 0.01% in water; m.w. 81; dens. 0.66 kg/l (15 C); visc. 0.45 cSt; vapor pressure 23.5 kPa (20 C); f.p. < -60 C; b.p. 45-72 C; flash pt. (TCC) < -18 C; autoignition temp. ≈ 254 C; 100 vol.%
Toxicology: REL TWA 353 ppm; inh. of high vapor concs. may cause eye/respiratory tract irritation, headaches, dizziness, anesthesia, drowsiness, unconsciousness, other CNS effects, death; may cause peripheral nervous system damage; sl. eye irritant; frequent/prolonged skin contact may cause irritation, dermatitis; if aspirated, may cause mild to severe pulmonary injury, death; contains n-hexane
Environmental: Prevent entry into sewers, watercourses, low areas
Precaution: Extremely flamm. liq.; flamm. limits 1.7-9 vol.%; readily ignites @ normal temps.; volatile; gives off invisible vapors; may release vapors that form flamm. mixts. at or above flash pt.; liq. or vapor may settle in low areas; distant ignition/explosion possible; electrostatic accumulation hazard; use proper grounding procedures; incompat. with strong oxidizing agents
Hazardous decomp. prods.: Combustion: toxic gases, fumes, smoke, CO
Storage: Store in cool, well-ventilated area away from incompat. materials, heat, open flame, ignition source, direct sunlight; keep container closed

Exxsol® Isopentane [ExxonMobil]
Chem. Descrip.: Dearomatized isopentane (60-100%), pentane (0.5-1.5%)
UN 1265
Uses: Solvent; diluent; chem. feedstock; fuel
Regulatory: Canada DSL listed; NPRI nonreportable
Properties: Liq.; sol. < 0.01% in water; m.w. 72; sp.gr. 0.62 (15.6/15.6 C); dens. 5.21 lb/gal (60 F); visc. 0.32 cSt; vapor pressure 620 mm Hg (20 C); f.p. < -60 C; b.p. 28 C; flash pt. (TCC) < -18 C; autoignition temp. > 399 C; surf. tens. 14.5 dynes/cm; 100% paraffin, < 0.01% aromatics
Toxicology: REL TWA 600 ppm; inh. may cause liver/kidney/brain disorders/damage; sl. eye irritant; frequent/prolonged skin contact may cause irritation, dermatitis; if aspirated, may cause severe health effects (bronchopneumonia, pulmonary edema)
Environmental: Prevent entry into sewers, watercourses, low areas
Precaution: Flamm. liq.; flamm. limits 1.4-8.3 vol.%; electrostatic accumulation hazard; use proper grounding procedures; incompat. with strong oxidizing agents
Hazardous decomp. prods.: None
Storage: Store in cool, well-ventilated area away from incompat. materials, heat, open flame, ignition source, direct sunlight; keep container closed

Exxsol® Isopentane S [ExxonMobil Europe]
Chem. Descrip.: Pentane
CAS 109-66-0; EINECS/ELINCS 203-692-4
Uses: Solvent
Properties: Dens. 0.625 kg/dm^3 ((15 C; visc. 0.24 mPa·s; i.b.p. 26 C; flash pt. < 0 C; ref. index 1.354 (20 C); surf. tens. 15.5 mN/m; KB value 26; 0.0001% aromatics

Exxsol® Methylpentane Naphtha [ExxonMobil; Exxon Canada]
Chem. Descrip.: Hexane, dearomatized
CAS 110-54-3; EINECS/ELINCS 203-777-6
UN 1208
Uses: Solvent; diluent; chem. feedstock; fuel
Regulatory: Canada DSL listed
Properties: Colorless clear liq.; sol. < 0.01% in water; m.w. 81; sp.gr. 0.66 (15.6/15.6 C); dens. 5.5 lb/gal (60 F); visc. 0.45 cSt; vapor pressure 200 mm Hg; f.p. < -60 C; b.p. 45-72 C; flash pt. (TCC) < -18 C; surf. tens. 17.0 dynes/cm; KB value 28; 99% paraffin, 1% cycloparaffin, < 0.01% aromatics
Toxicology: REL TWA 353 ppm; inh. of high vapor concs. may cause eye/respiratory tract irritation, headaches, dizziness, anesthesia, drowsiness, unconsciousness, other CNS effects, death; may cause peripheral nervous system damage; sl. eye irritant; frequent/prolonged skin contact may cause irritation, dermatitis; if aspirated, may cause mild to severe pulmonary injury, death; contains n-hexane
Environmental: Prevent entry into sewers, watercourses, low areas
Precaution: Extremely flamm. liq.; flamm. limits 1.7-9 vol.%; readily ignites @ normal temps.; volatile; gives off invisible vapors; may release vapors that form flamm. mixts. at or above flash pt.; liq. or vapor may settle in low areas; distant ignition/explosion possible; electrostatic accumulation hazard; use proper grounding procedures; incompat. with strong oxidizing agents
Hazardous decomp. prods.: Combustion: toxic gases, fumes, smoke, CO
Storage: Store in cool, well-ventilated area away from incompat. materials, heat, open flame, ignition source, direct sunlight; keep container closed

Exxsol® Pentane 80 [ExxonMobil Europe]
Chem. Descrip.: Pentane
CAS 109-66-0; EINECS/ELINCS 203-692-4
Uses: Solvent
Properties: Dens. 0.631 kg/dm^3 (15 C); visc. 0.23 mPa·s; i.b.p. 33 C; flash pt. < 0 C; ref. index 1.357 (20 C); surf. tens. 16.1 mN/m; KB value 27; 0.0001% aromatics

EZA® [Albemarle]
Chem. Descrip.: Zeolite A, a hydrated sodium aluminosilicate
CAS 1344-00-9; EINECS/ELINCS 215-684-8
Uses: Cryst. molecular sieve which selectively adsorbs molecules and exchange ions; detergent builder; solvent; anticaking agent for detergents and desiccants; gas separation; wh. pigment in paper, paint, and plastics; also in carpet cleaners, dentifrices, flame retardants
Properties: Wh. cryst. powd.; 3 μ mean particle size; pH 10-12 (1% slurry); 100% act., 18.5% moisture

F

FA® [Penn Spec. Chems.]
Chem. Descrip.: Furfuryl alcohol
CAS 98-00-0; EINECS/ELINCS 202-626-1
Uses: Used in the prod. of foundry sand binders and corrosion-resistant resins; intermediate for esterification and etherification; impregnating sol'n. and carbon binder; wood adhesive component; solvent and temporary plasticizer for phenolic resins in the mfg. of cold-molded abrasive wheels; visc. reducer, cure promoter, and carrier in amine-cured epoxy resins
Properties: Pale yel. liq.; sol. in water and many common org. solvs.; m.w. 98.10; sp.gr. 1.14 (20 C); dens. 9.45 lb/gal (20 C); visc. 5 cps; f.p. -14.6 C; b.p. 170 C; cloud pt. 10 C max.; flash pt. (TCC) 77 C; ref. index 1.4869; 98% min. assay

FAME [Cargill Ind. Oils & Lubes]
Chem. Descrip.: Fatty acid methyl ester
Uses: Cleaning solvent in metal cleaning and precision cleaning; cosolvent with other prods. such as d-limonene
Properties: Sp.gr. 0.88 ± 0.02; dens. 7.3 lb/gal (23 C); vapor pressure < 0.01 mm Hg (100 C); b.p. > 204 C; iodine no. 95-110; sapon. no. 201.5; cloud pt. 15 C; flash pt. > 150 C; VOC 59 g/l; 0.01-0.1% moisture

Fancol™ ALA [Fanning]
Chem. Descrip.: Cetyl acetate, acetylated lanolin alcohol
Uses: Conditioner, emollient, pigment binder, spreading agent, oil coupling agent, solvent for cosmetics, toiletries, skin care prods., hand and body lotions, suntan and bath oils, pharmaceutical ointments; gloss aid and plasticizer for hairsprays; lubricant for topical pharmaceutical ointments
Properties: Clear lt. oily liq.; acid no. 1 max.; iodine no. 12 max.; sapon. no. 200 max.; hyd. no. 8 max.

Fancol™ CB [Fanning]
Chem. Descrip.: Cocoa butter obtained from roasted seeds of *Theobroma cacao*
CAS 8002-31-1; EINECS/ELINCS 310-127-6
Uses: Skin conditioner, solvent, skin protectant for OTC drug prods., makeup, moisturizing creams and lotions, suntan preps.
Features: Occlusive
Properties: Ylsh. wh. solid; iodine no. 35-40; sapon. no. 192-197; ref. index 1.4560-1.4580

Fancol™ CB Extra [Fanning]
Chem. Descrip.: Cocoa butter USP obtained from roasted seeds of *Theobroma cacao*
CAS 8002-31-1; EINECS/ELINCS 310-127-6
Uses: Skin conditioner, solvent, skin protectant for OTC drug prods., makeup, moisturizing creams and lotions, suntan preps.
Features: Occlusive
Properties: Ylsh. wh. solid; iodine no. 35-43; sapon. no. 188-195; ref. index 1.4560-1.4580

Fancol™ Karite Butter [Fanning]
Chem. Descrip.: Shea butter
CAS 977026-99-5; EINECS/ELINCS 293-515-7
Uses: Ointment base, skin anti-irritant, skin conditioner, UV absorber, solvent, gloss aid, superfatting agent for suntan preps., body lotions, winter sports prods., wrinkle creams, soaps, shave foams, shampoos, balsams
Features: Occlusive; melts at body temp.; good spreadability
Properties: Pale off-wh. to ylsh. pasty solid; odorless; insol. in water; m.p. 32-45 C; b.p. > 500 F; acid no. 0.5 max.; iodine no. 60-70; sapon. no. 178-190; hyd. no. 5; flash pt.> 200 F
Toxicology: Fine oil mists may be hazardous (nuisance particulates)
Precaution: Oil-soaked rags may spontaneously heat and ignite when exposed to air; incompat. with strong oxidizing agents
Hazardous decomp. prods.: None

Fancorsil A [Fanning]
Chem. Descrip.: Cyclomethicone and dimethicone
Uses: Hair conditioner, skin conditioner, emollient, solvent, lubricant, gloss aid in hair conditioners, makeup, moisturizing creams and lotions, topical pharmaceutical ointments
Properties: Clear visc. liq.; sp.gr. 0.960; visc. 5000 cps min.; 100% solids

Fanoliv™ Oil [Fanning]
Chem. Descrip.: Olive oil
CAS 8001-25-0; EINECS/ELINCS 232-277-0
Uses: Hair conditioner, skin conditioner, solvent in shampoos, tonics, hair dressings, conditioners, cleansers, face/body/hand lotions, suntan gels/creams/lotions
Features: Occlusive
Properties: Pale straw clear oily liq.; bland odor; negligible sol. in water; sp.gr. 0.910-0.915; iodine no. 79-88; sapon. no. 190-195; flash pt. (CC) > 300 C
Precaution: Incomplete burning can produce CO, CO_2, and other harmful prods.
Storage: Store in cool, dry area away from excessive heat or moisture

Felosan FDO NEW [CHT R. Beitlich GmbH]
Chem. Descrip.: Emulsified terpene hydrocarbon
Uses: Grease solvent, wet stain removing agent for dissolving oils, fats, and other preps.
Features: Low foaming
Properties: Liq.; nonionic
Environmental: Biodeg.

Filmex® 3C-1 190 Proof [Equistar]
Chem. Descrip.: Ethyl alcohol (83.3%), IPA (9.0%), MIBK (1.0%), water (6.8%)
UN 1170
Uses: Solvent
Regulatory: SARA §311/312 acute health/chronic health/fire hazard, §313 reportable (MIBK, IPA); Calif. Prop. 65 reportable (ethanol, benzene); TSCA export notification required (MIBK)
Properties: Colorless liq.; invisible vapor; sweet alcohol-like odor; sol. in water; sp.gr. 0.78-0.85, 1.59 (vapor); vapor pressure 44.6 mm Hg; f.p. -144 C; b.p. 78.4 C; flash pt. (TCC) 16 C; autoignition temp. 363 C
Toxicology: May be irritating to eyes, upper respiratory tract; short-term exposure above 1000 ppm by inh. may cause CNS effects, headache, eye/nose/throat irritation; exposure > 1 h may cause dizziness, drowsiness, loss of appetite, inability to conc., GI effects, nausea, vomiting; prolonged/repeated skin contact may cause dermatitis; ing. may cause drunkenness, CNS depression, nausea, vomiting, diarrhea, liver damage, death; TSCA listed
Environmental: On land, apt to volatilize, biodegrade, or leach to ground water; in water, photolysis, oxidation, biodegradation expected to occur; not expected to bioconcentrate in aquatic organisms; LC50 (rainbow trout, 96 h) 10,400 ppm; prevent entry into sewers

Precaution: Flamm. liq.; flamm. limits 3.3-19 vol.%; material can burn with little or no visible flame; vapors are heavier than air, may flash back; prevent entry into basements, confined areas; ground and bond containers; avoid static discharges; avoid strong oxidizers, excessive heat, sparks, open flame; contact with acetyl chloride or other oxidizing agents may result in violent reaction
Hazardous decomp. prods.: Combustion: CO
HMIS: Health 1, Flammability 3, Reactivity 0
Storage: 1 yr shelf life; store under ambient conditions out of direct sunlight

Filmex® A-1, 190 Proof [Equistar]
Chem. Descrip.: Ethyl alcohol (79.6%), methyl alcohol (4.2%), MIBK (0.9%), IPA (8.8%), water (6.5%)
UN NA 1987
Uses: Solvent, thinner for general-purpose industrial applics.
Regulatory: FDA 27CFR §20.112; SARA §311/312 acute health/chronic health/fire hazard, §313 reportable (IPA, methyl alcohol, MIBK); Calif. Prop. 65 reportable
Properties: Colorless liq.; invisible vapor; mild sweet alcohol-like nonresidual odor; sol. in conc. sulfuric acid, water; misc. in alcohol, chlorine, benzene; sp.gr. 0.812-0.8132 (60/60 F); dens. 6.762-6.772 lb/gal (60 F); vapor pressure 44.6 mm Hg (20 C); f.p. -144 C; b.p. 78.4 C; flash pt. (TCC) 16 C; autoignition temp. 363 C
Toxicology: May be irritating to eyes, upper respiratory tract; may affect CNS (headache, dizziness, drowsiness, loss of appetite, inability to concentrate); may cause GI effects (nausea, vomiting); may cause skin defatting, irritation, dryness, cracking; chronic exposure may cause memory loss, mental retardation, liver damage, dementia, somnolence, cardiac myopathy, hepatotoxicity, GI bleeding, pancreatitis; TSCA listed
Environmental: LC50 (rainbow trout, 96 h), 10,4000 ppm, (fathead minnow, 96 h) 15.3 g/l; will neither adsorb to soil nor bioconcentrate in aquatic organisms; prevent entry into sewers, confined areas
Precaution: Flamm. liq.; flamm. limits 3.3-19 vol.%; material can burn with little or no visible flame; vapors are heavier than air, may flash back; avoid strong oxidizers, excessive heat, sparks, open flame; violent reaction with acetyl chloride, other oxidizing agents
Hazardous decomp. prods.: Combustion: CO

Filmex® A-2 190 Proof [Equistar]
Chem. Descrip.: Ethyl alcohol (79.6%), methyl alcohol (13.0%), MIBK (0.9%), water (6.5%)
UN 1987
Uses: Solvent for flexographic printing for cleaning equip., formulating and thinning inks, producing rotogravure etchings; solvent in chemical/ pharmaceutical processing; solvent, adhesion promoter, dye penetrant, fiber softener in textiles; in chemical specialties prod., such as latex coagulants and lacquers
Regulatory: FDA 27CFR §20.112; SARA §311/312 acute health/chronic health/fire hazard, §313 reportable (MIBK, methanol); Calif. Prop. 65 reportable (ethanol, benzene); TSCA export notification required (MIBK)
Properties: Colorless liq.; invisible vapor; mild sweet alcohol-like nonresidual odor; sol. in water; sp.gr. 0.8060-0.8072 (77/77 F), 1.59 (vapor); dens. 6.768-6.778 lb/gal (60 F); vapor pressure 44.6 mm Hg; f.p. -144 C; b.p. 78.4 C; flash pt. (TCC) 16 C; autoignition temp. 363 C
Toxicology: May be irritating to eyes, upper respiratory tract; short-term exposure above 1000 ppm by inh. may cause CNS effects, headache, eye/nose/throat irritation; exposure > 1 h may cause dizziness, drowsiness, loss of appetite, inability to conc., GI effects, nausea, vomiting; prolonged/repeated skin contact may cause dermatitis; ing. may cause drunkenness, CNS depression, nausea, vomiting, diarrhea, liver damage, death; TSCA listed
Environmental: On land, apt to volatilize, biodegrade, or leach to ground water; in water, photolysis, oxidation, biodegradation expected to occur; not expected to bioconcentrate in aquatic organisms; LC50 (rainbow trout, 96 h) 10,400 ppm; prevent entry into sewers
Precaution: Flamm. liq.; flamm. limits 3.3-19 vol.%; material can burn with little or no visible flame; vapors are heavier than air, may flash back; prevent entry into basements, confined areas; ground and bond containers; avoid static discharges; avoid strong oxidizers, excessive heat, sparks, open flame; contact with acetyl chloride or other oxidizing agents may result in violent reaction
Hazardous decomp. prods.: Combustion: CO
HMIS: Health 1, Flammability 3, Reactivity 0

Filmex® A-2 Anhydrous [Equistar]
Chem. Descrip.: Ethyl alcohol (85.8%), methyl alcohol (13.3%), MIBK (1.0%)
UN 1987
Uses: Solvent for flexographic printing for cleaning equip., formulating and thinning inks, producing rotogravure etchings; solvent in chemical/ pharmaceutical processing; solvent, adhesion promoter, dye penetrant, fiber softener in textiles; in chemical specialties prod., such as latex coagulants and lacquers
Regulatory: SARA §311/312 acute health/chronic health/fire hazard, §313 reportable (MIBK, methanol); Calif. Prop. 65 reportable (ethanol, benzene); TSCA export notification required (MIBK)
Properties: Colorless liq.; invisible vapor; sweet alcohol-like odor; sol. in water; sp.gr. 0.7936-0.7946 (60/60 F), 1.59 (vapor); dens. 6.768-6.778 lb/gal (60 F); vapor pressure 44.6 mm Hg; f.p. -144 C; b.p. 78.4 C; flash pt. (TCC) 13 C; autoignition temp. 363 C
Toxicology: May be irritating to eyes, upper respiratory tract; short-term exposure above 1000 ppm by inh. may cause CNS effects, headache, eye/nose/throat irritation; exposure > 1 h may cause dizziness, drowsiness, loss of appetite, inability to conc., GI effects, nausea, vomiting; prolonged/repeated skin contact may cause dermatitis; ing. may cause drunkenness, CNS depression, nausea, vomiting, diarrhea, liver damage, death; TSCA listed
Environmental: On land, apt to volatilize, biodegrade, or leach to ground water; in water, photolysis, oxidation, biodegradation expected to occur; not expected to bioconcentrate in aquatic organisms; LC50 (rainbow trout, 96 h) 10,400 ppm; prevent entry into sewers
Precaution: Flamm. liq.; flamm. limits 3.3-19 vol.%; material can burn with little or no visible flame; vapors are heavier than air, may flash back; prevent entry into basements, confined areas; ground and bond containers; avoid static discharges; avoid strong oxidizers, excessive heat, sparks, open flame; contact with acetyl chloride or other oxidizing agents may result in violent reaction
Hazardous decomp. prods.: Combustion: CO
HMIS: Health 1, Flammability 3, Reactivity 0
Storage: 1 yr shelf life; store under ambient conditions out of direct sunlight

Filmex® B 190 Proof [Equistar]
Chem. Descrip.: Ethyl alcohol (79.6%), methyl alcohol (8.6%), isopropyl alcohol (4.4%), MIBK (0.9%), water (6.5%)
UN 1987
Uses: Solvent for flexographic printing for cleaning equip., formulating and thinning inks, producing rotogravure etchings; solvent in chemical/ pharmaceutical processing; solvent, adhesion promoter, dye penetrant, fiber softener in textiles; in chemical specialties prod., such as latex coagulants and lacquers
Regulatory: FDA 27CFR §20.112; SARA §311/312 acute health/chronic health/fire hazard, §313 reportable (MIBK, methanol, IPA); Calif. Prop. 65 reportable (ethanol, benzene); TSCA export notification required (MIBK)
Properties: Colorless liq.; invisible vapor; mild sweet alcohol-like nonresidual odor; sol. in water; sp.gr. 0.8058-0.8070 (77/77 F), 1.59 (vapor); dens. 6.766-6.776 lb/gal (60 F); vapor pressure 44.6 mm Hg; f.p. -144 C; b.p. 78.4 C; flash pt. (TCC) 16 C; autoignition temp. 363 C
Toxicology: May be irritating to eyes, upper respiratory tract; short-term exposure above 1000 ppm by inh. may cause CNS effects, headache, eye/nose/throat irritation; exposure > 1 h may cause dizziness, drowsiness, loss of appetite, inability to conc., GI effects, nausea, vomiting; prolonged/repeated skin contact may cause dermatitis; ing. may cause drunkenness, CNS depression, nausea, vomiting, diarrhea, liver damage, death; TSCA listed
Environmental: On land, apt to volatilize, biodegrade, or leach to ground water; in water, photolysis, oxidation, biodegradation expected to occur; not expected to bioconcentrate in aquatic organisms; LC50 (rainbow trout, 96 h) 10,400 ppm; prevent entry into sewers
Precaution: Flamm. liq.; flamm. limits 3.3-19 vol.%; material can burn with little or no visible flame; vapors are heavier than air, may flash back; prevent entry into basements, confined areas; ground and bond containers; avoid static discharges; avoid strong oxidizers, excessive heat, sparks, open flame; contact with acetyl chloride or other oxidizing agents may result in violent reaction
Hazardous decomp. prods.: Combustion: CO

HMIS: Health 1, Flammability 3, Reactivity 0

Filmex® B Anhydrous [Equistar]

Chem. Descrip.: Ethyl alcohol (85.8%), methyl alcohol (8.8%), isopropyl alcohol (4.5%), MIBK (1.0%)

UN 1987

Uses: Solvent for flexographic printing for cleaning equip., formulating and thinning inks, producing rotogravure etchings; solvent in chemical/pharmaceutical processing; solvent, adhesion promoter, dye penetrant, fiber softener in textiles; in chemical specialties prod., such as latex coagulants and lacquers

Regulatory: FDA 27CFR §20.112; SARA §311/312 acute health/chronic health/fire hazard, §313 reportable (MIBK, methanol, IPA); Calif. Prop. 65 reportable (ethanol, benzene); TSCA export notification required (MIBK)

Properties: Colorless liq.; invisible vapor; mild sweet alcohol-like nonresidual odor; sol. in water; sp.gr. 0.7869-0.7879 (77/77 F), 1.59 (vapor); dens. 6.608-6.616 lb/gal (60 F); vapor pressure 44.6 mm Hg; f.p. -144 C; b.p. 78.4 C; flash pt. (TCC) 13 C; autoignition temp. 363 C

Toxicology: May be irritating to eyes, upper respiratory tract; short-term exposure above 1000 ppm by inh. may cause CNS effects, headache, eye/nose/throat irritation; exposure > 1 h may cause dizziness, drowsiness, loss of appetite, inability to conc., GI effects, nausea, vomiting; prolonged/repeated skin contact may cause dermatitis; ing. may cause drunkenness, CNS depression, nausea, vomiting, diarrhea, liver damage, death; TSCA listed

Environmental: On land, apt to volatilize, biodegrade, or leach to ground water; in water, photolysis, oxidation, biodegradation expected to occur; not expected to bioconcentrate in aquatic organisms; LC50 (rainbow trout, 96 h) 10,400 ppm; prevent entry into sewers

Precaution: Flamm. liq.; flamm. limits 3.3-19 vol.%; material can burn with little or no visible flame; vapors are heavier than air, may flash back; prevent entry into basements, confined areas; ground and bond containers; avoid static discharges; avoid strong oxidizers, excessive heat, sparks, open flame; contact with acetyl chloride or other oxidizing agents may result in violent reaction

Hazardous decomp. prods.: Combustion: CO

HMIS: Health 1, Flammability 3, Reactivity 0

Filmex® C 190 Proof [Equistar]

Chem. Descrip.: Ethyl alcohol (83.36%), methyl alcohol (4.4%), ethyl acetate (4.47%), MIBK (0.93%), water (6.84%)

UN 1170

Uses: Solvent for flexographic printing for cleaning equip., formulating and thinning inks, producing rotogravure etchings; solvent in chemical/pharmaceutical processing; solvent, adhesion promoter, dye penetrant, fiber softener in textiles; in chemical specialties prod., such as latex coagulants and lacquers

Regulatory: SARA §311/312 acute health/chronic health/fire hazard, §313 reportable (MIBK, methanol); Calif. Prop. 65 reportable (ethanol, benzene); TSCA export notification required (MIBK, ethyl acetate)

Properties: Colorless liq.; invisible vapor; sweet alcohol-like odor; sol. in water; sp.gr. 0.8175-0.8187 (60/60 F), 1.59 (vapor); vapor pressure 44.6 mm Hg; f.p. -144 C; b.p. 78.4 C; flash pt. (TCC) 13 C; autoignition temp. 363 C

Toxicology: May be irritating to eyes, upper respiratory tract; short-term exposure above 1000 ppm by inh. may cause CNS effects, headache, eye/nose/throat irritation; exposure > 1 h may cause dizziness, drowsiness, loss of appetite, inability to conc., GI effects, nausea, vomiting; prolonged/repeated skin contact may cause dermatitis; ing. may cause drunkenness, CNS depression, nausea, vomiting, diarrhea, liver damage, death; TSCA listed

Environmental: On land, apt to volatilize, biodegrade, or leach to ground water; in water, photolysis, oxidation, biodegradation expected to occur; not expected to bioconcentrate in aquatic organisms; LC50 (rainbow trout, 96 h) 10,400 ppm; prevent entry into sewers

Precaution: Flamm. liq.; flamm. limits 3.3-19 vol.%; material can burn with little or no visible flame; vapors are heavier than air, may flash back; prevent entry into basements, confined areas; ground and bond containers; avoid static discharges; avoid strong oxidizers, excessive heat, sparks, open flame; contact with acetyl chloride or other oxidizing agents may result in violent reaction

Hazardous decomp. prods.: Combustion: CO

HMIS: Health 1, Flammability 3, Reactivity 0

Storage: 1 yr shelf life; store under ambient conditions out of direct sunlight

Filmex® C Anhydrous [Equistar]

Chem. Descrip.: Ethyl alcohol (90.0%), methyl alcohol (4.5%), ethyl acetate (4.6%), MIBK (1.0%)

UN 1170

Uses: Solvent for flexographic printing for cleaning equip., formulating and thinning inks, producing rotogravure etchings; solvent in chemical/pharmaceutical processing; solvent, adhesion promoter, dye penetrant, fiber softener in textiles; in chemical specialties prod., such as latex coagulants and lacquers

Regulatory: FDA 27CFR §20.112; SARA §311/312 acute health/chronic health/fire hazard, §313 reportable (MIBK, methanol); Calif. Prop. 65 reportable (ethanol, benzene); TSCA export notification required (MIBK, ethyl acetate)

Properties: Colorless liq.; invisible vapor; mild sweet alcohol-like nonresidual odor; sol. in water; sp.gr. 0.7914-0.7923 (77/77 F), 1.59 (vapor); dens. 6.645-6.653 lb/gal (60 F); vapor pressure 44.6 mm Hg; f.p. -144 C; b.p. 78.4 C; flash pt. (TCC) 13 C; autoignition temp. 363 C

Toxicology: May be irritating to eyes, upper respiratory tract; short-term exposure above 1000 ppm by inh. may cause CNS effects, headache, eye/nose/throat irritation; exposure > 1 h may cause dizziness, drowsiness, loss of appetite, inability to conc., GI effects, nausea, vomiting; prolonged/repeated skin contact may cause dermatitis; ing. may cause drunkenness, CNS depression, nausea, vomiting, diarrhea, liver damage, death; TSCA listed

Environmental: On land, apt to volatilize, biodegrade, or leach to ground water; in water, photolysis, oxidation, biodegradation expected to occur; not expected to bioconcentrate in aquatic organisms; LC50 (rainbow trout, 96 h) 10,400 ppm; prevent entry into sewers

Precaution: Flamm. liq.; flamm. limits 3.3-19 vol.%; material can burn with little or no visible flame; vapors are heavier than air, may flash back; prevent entry into basements, confined areas; ground and bond containers; avoid static discharges; avoid strong oxidizers, excessive heat, sparks, open flame; contact with acetyl chloride or other oxidizing agents may result in violent reaction

Hazardous decomp. prods.: Combustion: CO

HMIS: Health 1, Flammability 3, Reactivity 0

Filmex® D-1 [Equistar]

Chem. Descrip.: Ethyl alcohol (100 gal), methyl alcohol (5 gal), MIBK (1.05 gal), isopropyl alcohol (15.75 lb)

Uses: Solvent for flexographic printing for cleaning equip., formulating and thinning inks, producing rotogravure etchings; solvent in chemical/pharmaceutical processing; solvent, adhesion promoter, dye penetrant, fiber softener in textiles; in chemical specialties prod., such as latex coagulants and lacquers

Filmex® D-2 190 Proof [Equistar]

Chem. Descrip.: Ethyl alcohol (76.2%), methyl alcohol (16.7%), MIBK (0.9%), water (6.3%)

UN 1170

Uses: Solvent for flexographic printing for cleaning equip., formulating and thinning inks, producing rotogravure etchings; solvent in chemical/pharmaceutical processing; solvent, adhesion promoter, dye penetrant, fiber softener in textiles; in chemical specialties prod., such as latex coagulants and lacquers

Regulatory: SARA §311/312 acute health/chronic health/fire hazard, §313 reportable (MIBK, methanol); Calif. Prop. 65 reportable (ethanol, benzene); TSCA export notification required (MIBK)

Properties: Colorless liq.; invisible vapor; mild sweet alcohol-like nonresidual odor; sol. in water; sp.gr. 0.7871-0.7882 (77/77 F), 1.59 (vapor); dens. 6.609-6.618 lb/gal (60 F); vapor pressure 44.6 mm Hg; f.p. -144 C; b.p. 78.4 C; flash pt. (TCC) 16 C; autoignition temp. 363 C

Toxicology: May be irritating to eyes, upper respiratory tract; short-term exposure above 1000 ppm by inh. may cause CNS effects, headache, eye/nose/throat irritation; exposure > 1 h may cause dizziness, drowsiness, loss of appetite, inability to conc., GI effects, nausea, vomiting; prolonged/repeated skin contact may cause dermatitis; ing. may cause drunkenness, CNS depression, nausea, vomiting, diarrhea, liver damage, death; TSCA listed

Environmental: On land, apt to volatilize, biodegrade, or leach to ground water; in water, photolysis, oxidation, biodegradation expected to occur; not expected to bioconcentrate in aquatic organisms; LC50 (rainbow

trout, 96 h) 10,400 ppm; prevent entry into sewers
Precaution: Flamm. liq.; flamm. limits 3.3-19 vol.%; material can burn with little or no visible flame; vapors are heavier than air, may flash back; prevent entry into basements, confined areas; ground and bond containers; avoid static discharges; avoid strong oxidizers, excessive heat, sparks, open flame; contact with acetyl chloride or other oxidizing agents may result in violent reaction
Hazardous decomp. prods.: Combustion: CO
HMIS: Health 1, Flammability 3, Reactivity 0

Filmex® D-2 Anhydrous [Equistar]
Chem. Descrip.: Ethyl alcohol (82.1%), methyl alcohol (17.1%), MIBK (0.9%)
UN 1987
Uses: Solvent for flexographic printing for cleaning equip., formulating and thinning inks, producing rotogravure etchings; solvent in chemical/pharmaceutical processing; solvent, adhesion promoter, dye penetrant, fiber softener in textiles; in chemical specialties prod., such as latex coagulants and lacquers
Regulatory: FDA 27CFR §20.112; SARA §311/312 acute health/chronic health/fire hazard, §313 reportable (MIBK, methanol); Calif. Prop. 65 reportable (ethanol, benzene); TSCA export notification required (MIBK)
Properties: Colorless liq.; invisible vapor; mild sweet alcohol-like nonresidual odor; sol. in water; sp.gr. 0.7871-0.7882 (77/77 F), 1.59 (vapor); dens. 6.609-6.618 lb/gal (60 F); vapor pressure 44.6 mm Hg; f.p. -144 C; b.p. 78.4 C; flash pt. (TCC) 13 C; autoignition temp. 363 C
Toxicology: May be irritating to eyes, upper respiratory tract; short-term exposure above 1000 ppm by inh. may cause CNS effects, headache, eye/nose/throat irritation; exposure > 1 h may cause dizziness, drowsiness, loss of appetite, inability to conc., GI effects, nausea, vomiting; prolonged/repeated skin contact may cause dermatitis; ing. may cause drunkenness, CNS depression, nausea, vomiting, diarrhea, liver damage, death; TSCA listed
Environmental: On land, apt to volatilize, biodegrade, or leach to ground water; in water, photolysis, oxidation, biodegradation expected to occur; not expected to bioconcentrate in aquatic organisms; LC50 (rainbow trout, 96 h) 10,400 ppm; prevent entry into sewers
Precaution: Flamm. liq.; flamm. limits 3.3-19 vol.%; material can burn with little or no visible flame; vapors are heavier than air, may flash back; prevent entry into basements, confined areas; ground and bond containers; avoid static discharges; avoid strong oxidizers, excessive heat, sparks, open flame; contact with acetyl chloride or other oxidizing agents may result in violent reaction
Hazardous decomp. prods.: Combustion: CO
HMIS: Health 1, Flammability 3, Reactivity 0

Finsolv® 137 [Finetex]
Chem. Descrip.: Behenyl benzoate
CAS 103403-38-9
Uses: Structurizing agent, heat stabilizer, solvent in cosmetic stick formulations
Features: High m.p.; compat. with lipophilic materials; aids reduction of sweating and syneresis problems
Properties: M.p. 52-55 C

Finsolv® BOHS-111 [Finetex]
Chem. Descrip.: Octyl hydroxystearate benzoate
CAS 199277-69-5
Uses: Emollient, pigment wetting agent/dispersant in skin care, dry skin and therapeutic prods., bath care; solvent in makeup removers, cleansers; dispersant for org. and inorg. sunscreen actives in sun care; pigment wetting agent, gloss aid, moisturizer, color enhancer in lipsticks, lip glosses

Finsolv® PG-22 [Finetex]
Chem. Descrip.: Dipropylene glycol dibenzoate
Uses: Skin conditioner, humectant, emollient in skin care; humectant, moisturizer, gloss aid in hair care; solvent, solubilizer in clear cosmetic prods., antiperspirants, deodorants
Features: Leaves skin soft, smooth, supple; reduces flaking

Finsolv® TN [Finetex]
Chem. Descrip.: C12-15 alkyl benzoate
CAS 68411-27-8; EINECS/ELINCS 270-112-4
Uses: Emollient, solubilizer, conditioner, sunscreen vehicle, deoiler for cosmetic creams and oils; binder in pressed powds.; solvent for UV filters; detackifier for min. oils
Features: Good spreading performance; noncomedogenic
Properties: Liq.; 100% act.

Finsolv® TN-O [Finetex]
Chem. Descrip.: C12-15 alkyl benzoate
CAS 68411-27-8; EINECS/ELINCS 270-112-4
Uses: Emollient, solvent in decorative cosmetics, fragrance-free prods., sunscreens, after-sun prods., hand and body lotions, bath prods., after-bath prods., shampoos, conditioners, hair fixatives, hair sprays, fine fragrances
Features: Good spreading performance
Properties: Liq.; odorless
Toxicology: Extremely mild to skin and eyes

Finsolv® TPP [Finetex]
Chem. Descrip.: C12-15 alkyl benzoate, dipropylene glycol dibenzoate, PPG-15 stearyl ether benzoate
Uses: Solvent, solubilizer for sunscreen actives (benzophenone-3, avobenzone), sun care; emollient, moisturizer, skin conditioner in skin care; reduces wh. residue on skin in antiperspirants, deodorants
Features: Exc. tactile and performance props.; imparts light, dry, nonoily skinfeel

FoamFlush® Urethane Remover [ISP]
Chem. Descrip.: N-Methyl-2-pyrrolidone, butyrolactone, other ingreds.
Uses: Solvent cleaner for urethane removal from cured and uncured polymers in the foam industry, as well as coatings and adhesives used in foam cutting and fabrication
Features: Safer alternative to methylene chloride, ketones, alcohols
Properties: Water-thin clear liq., sweet odor; misc. with water and most org. solvs.; sp.gr. 1.06; visc. 2.09 cps; vapor pressure 0.29 mm Hg; f.p. -20 C; b.p. 202 C; flash pt. (Seta CC) 88 C; pH 4-6 (10% aq.)
Toxicology: TLV 100 ppm; low oral and inhalation toxicity; mod. dermal toxicity; mod. eye irritant; contains org. solvs.; avoid breathing vapors or mist
Precaution: Keep away from excessive heat and open flames

Foamid AME-70 [Alzo]
Chem. Descrip.: Acetamide MEA
CAS 142-26-7; EINECS/ELINCS 205-530-8
Uses: Wetting agent for cosmetics and health prods.; solvent for org. and inorg. materials
Features: Low visc.; broad solvent power; high b.p.
Properties: Gardner 2 clear liq., mild organoleptic odor; sol. in most alcohols, glycols, diols, polyols, glycol ethers, ketones, and water; sp.gr. 1.100; dens. 9.2 lb/gal; acid no. 15 max.; alkali no. 15 max.; flash pt. (OC) > 180 C; ref. index 1.4400; pH 6.0-8.5 (50% aq.); nonionic; 70% min. act. in water
Toxicology: Low toxicity

Foamid AME-75 [Alzo]
Chem. Descrip.: Acetamide MEA
CAS 142-26-7; EINECS/ELINCS 205-530-8
Uses: Wetting agent for cosmetics and health prods.; solvent for org. and inorg. materials
Features: Low visc.; broad solvent power; high b.p.
Properties: Gardner 2 clear liq., mild organoleptic odor; sol. in most alcohols, glycols, diols, polyols, glycol ethers, ketones, and water; sp.gr. 1.100; dens. 9.2 lb/gal; acid no. 15 max.; alkali no. 15 max.; flash pt. (OC) > 180 C; ref. index 1.4410; pH 6.0-8.5 (50% aq.); nonionic; 75% min. act. in water
Toxicology: Low toxicityTSCA listed
Hazardous decomp. prods.: Burning after dehydration may produce ammonia, CO, and/or CO_2
HMIS: Health 0, Flammability 0, Reactivity 0

Foamid AME-100 [Alzo]
Chem. Descrip.: Acetamide MEA
CAS 142-26-7; EINECS/ELINCS 205-530-8
Uses: Wetting agent for cosmetics and health prods.; solvent for org. and inorg. materials
Features: Low visc.; broad solvent power; high b.p.
Properties: Gardner 2 clear liq., mild organoleptic odor; sol. in most alcohols, glycols, diols, polyols, glycol ethers, ketones, and water; sp.gr. 1.120; dens. 9.3 lb/gal; acid no. 15 max.; alkali no. 15 max.; flash pt. (OC) > 180 C; ref. index 1.4700; pH 6.0-8.5 (50% aq.); nonionic; 95% min. act.
Toxicology: Low toxicity

Foamid LM2E [Alzo]
Chem. Descrip.: Lactamide DGA
CAS 123792-29-0
Uses: Humectant, emollient, o/w emulsifier, cosolvent for hair/skin creams/lotions; gloss aid, conditioner, moisturizer, and antistat for hair preps. esp. for dull, overtreated hair; solvent, freeze pt. depressant, coupling agent, wetting agent in 'oil-free' pigmented prods.
Features: Oil-free; imparts light, smooth, nongreasy feel to the skin
Properties: Straw to lt. yel. clear liq., mild organoleptic odor; sol. in most low m.w. alcohols, diols, triols, polyols, glycol ethers, and water; sp.gr. 1.150; dens. 9.58 lb/gal; acid no. 30 max.; alkali no. 35 max.; flash pt. (OC) > 180 C; ref. index 1.4350; nonionic; 80% min. act. in water
Storage: 1 yr min. shelf life in close containers

Fomblin® Perfluorosolv PFS-1 [Solvay Solexis]
Chem. Descrip.: Perfluoropolyether
CAS 60164-51-4
Uses: Solvent, diluent in electronic and semiconductor cleaning applics. (pump cleaning, degreasing, dewatering), vapor blanketing, reducing flash pt., eliminating trace residue
Features: Zero ODP; replacement for CFC-113 and other ozone-depleting solvs.; low m.w.; safe; inert; contains no CFCs, chlorine, bromine; compat. with metals, rubbers, elastomers, plastics; relatively high b.p.; low evaporation rate
Properties: Colorless; odorless; misc. with perfluoropolyether oils and greases, perfluorocarbon fluids, chlorotrifluoroethylene oils (above 45 C); dens. 1.69 g/cm^3; visc. 0.75 cSt; vapor pressure 100 torr; b.p. 90 C; flash pt. none; surf. tens. 14 dynes/cm
Toxicology: Nontoxic

Forane 141b DGX [Atofina SA]
Chem. Descrip.: 1,1-Dichloro-1-fluoroethane, stabilizers (300 ppm)
CAS 1717-00-6
Uses: Solvent for removal of flux, cleaning, degreasing metals, drying
Features: Nonflamm.; substitute for Forane® 113
Properties: Colorless liq.; sl. ethereal odor; sol. 0.5% in water; m.w. 117; sp.gr. 1.236; vapor pressure 0.81 bar; b.p. 32 C (< 1.013 bar); flash pt. none; autoignition temp. 532 C; ref. index 1.37; surf. tens. 18.4 mN.m^{-1}; ≥ 99.5% purity
Toxicology: WEEL TWA (8 h) 500 ppm; low toxicity; low acute and subchronic inh. toxicity; may cause drowsiness, danger of anesthesia if inhaled in high concs. or in confined spaces; may cause benign tumors with long-term exposure to high concs.; not a developmental toxicant; nonmutagenic
Precaution: Limited flamm. when mixed with air @ concs. between 5.6-17.7% of vapor to air

Forane 141b MGX [Atofina SA]
Chem. Descrip.: 1,1-Dichloro-1-fluoroethane (95.6%), methanol (4%), inhibitor (0.4%)
Uses: Solvent for removal of flux, cleaning
Features: Nonflamm.
Properties: Colorless liq.; sl. ethereal odor; sp.gr. 1.224; vapor pressure 0.845 bar; b.p. 29.4 C (< 1.013 bar); flash pt. none; autoignition temp. 532 C; ref. index 1.37; surf. tens. 18.4 mN.m^{-1}; KB value 77; sp.heat 1.137 kJ/kg/°C; thermal conduct. 0.101 W/m°C
Toxicology: WEEL TWA (8 h) 500 ppm; low toxicity; low acute and subchronic inh. toxicity; may cause drowsiness, danger of anesthesia if inhaled in high concs. or in confined spaces
Precaution: Limited flamm. when mixed with air @ concs. between 5.9-20.1% of vapor to air

Formcel® [Celanese; Celanese GmbH]
Chem. Descrip.: Sol'n. of formaldehyde (53%) in butanol (40%)/water (7%)
UN 1198
Uses: Raw material, intermediate for phenolic, urea, furfural alcohol, resorcinol, or melamine formaldehyde resins (some of the butanol reacts chemically with the polymers, the rest of the butanol becomes a solvent for them), used in industrial coatings, wood prods., textiles, foundry resins
Regulatory: SARA §311 acute health/chronic health/fire hazard, §313 reportable (butanol, formaldehyde); EINECS, Australia, China, Canada, Korea, Philippines, Japan listed
Properties: Colorless clear liq.; strong pungent char. formaldehyde odor; misc. with water, ethanol, ethyl ether; sp.gr. 0.963 (40 C); vapor pressure 43 mm Hg (40 C); f.p. < 65 C; b.p. 104.5 C; flash pt. (Seta CC) 57 C; pH 6.23
Toxicology: Causes skin/digestive tract burns, respiratory tract/eye irritation; harmful by inh., ing., skin absorp.; may cause allergic respiratory and skin reaction; may cause CNS depression, headache, stupor, unconsciousness, nausea, dizziness, nasal discharge, hoarseness, coughing, chest pain, breathing difficulty, pulmonary edema, GI irritation, diarrhea, loss of appetite; suspect cancer hazard; TSCA listed
Environmental: Prevent run-off into storm sewers and ditches leading to waterways
Precaution: Flamm. liq. and vapor; may generate static charge; ground/bond equip. during transfer; avoid heat, sparks, flames, ignition sources; incompat. with caustic soda, lime, other strong alkalis, sodium, potassium, other alkali metals, hydrochloric, sulfuric, other strong inorg. acids, nitrogen oxides, amines, oxidizing agents, peroxides, nitric acid, perchloric acid, chromium trioxide, phenols, urea
Hazardous decomp. prods.: Thermal decomp. or combustion: CO_x
HMIS: Health 3, Flammability 2, Reactivity 0
Storage: Store away from incompat. materials, direct sunlight; keep containers tightly closed when not in use; open containers slowly to allow excess pressure to vent

G

GAFsolver™ TMP [ISP]
Chem. Descrip.: N-Methyl-2-pyrrolidone
CAS 872-50-4; EINECS/ELINCS 212-828-1
Uses: Solvent for solv.-based cleaners, degreasers, and stripping agents, adhesives, PVC cements, PVC pipe and hose mfg., removal of carbon deposits (automotive and jet engine parts), aq. household and industrial cleaners, equip. (resin removal) cleaning, general maintenance solv.-based cleaners; base solvent for specialty cleaners
Properties: APHA < 200 water-thin liq.; completely misc. with water and most org. solvs. incl. alcohols, ethers, ketones, glycol ethers, aromatic and chlorinated hydrocarbons, and veg. oils; sp.gr. 1.027; dens. 8.55 lb/gal; visc. 165 cps; f.p. -24.4 C; b.p. 202 C; flash pt. (TCC) 93 C; fire pt. (COC) 99 C; 99% min. purity
Toxicology: Relatively low toxicity; severe eye irritant
Environmental: Biodeg.
Precaution: Combustible
Storage: Hygroscopic; avoid extended exposure to moist air

Galaster EL [Galactic]
Chem. Descrip.: Ethyl lactate
CAS 97-64-3; EINECS/ELINCS 202-598-0
Uses: Solvent for paints, lacquers, varnishes, photoresist compositions; cleaning agent for removal of silicone oils and greases, machining coolants, flux residues, tapping oils, lithium grease, layout inks, fingerprints, in precision cleaning, mfg. of electronic circuits/chips/boards; plasticizer for cellulose ethers; mfg. of thermoplastic adhesives, nontoxic pesticides, inks; softener for rubber; component in liq. crystal displays
Features: High b.p.; CFC substitute
Regulatory: FDA approved
Environmental: Biodeg.

GAM [Raschig AG]
Chem. Descrip.: Glyceryl-α-monochlorohydrin
CAS 96-24-2; EINECS/ELINCS 202-492-4
UN 2689
Uses: Reactive chem. intermediate for the modification of amines, alcohols, and carboxylic acids; C3-building block; solvent; synthesis of pharmaceutical and cosmetic prods.; reactive solvent for cellulose derivatives
Regulatory: MITI listed
Properties: Colorless clear liq.; sol. in water and polar solvs., sl. sol. in nonpolar solvs.; sp.gr. 1.32; pH 4 (50% aq.); 98% assay
Toxicology: Toxic; irritantTSCA listed
Storage: Keep tightly closed in cool, dry place

GBL [Lyondell]
Chem. Descrip.: γ-Butyrolactone
CAS 96-48-0; EINECS/ELINCS 202-509-5
Uses: Spinning and coagulating solvent for textiles; solvent in nail polish removers, detergents, sunscreens, household cleaners, paint removers, agric., polymers, petrol. industry, mfg. of electronics, dyes; chem. process solvent; solvent in synthesis of vitamins; derivs. as relaxants, antispasmodics, antibacterial agents, surgical sutures, and injectable sedative, hypnotic, and anesthetic sol'ns.
Features: No ODP; HAPs-free, recyclable; replacement for regulated solvs.; exc. chem. and thermal stability
Properties: Completely misc. with water; sp.gr. 1.13 (20 C); vapor pressure < 1.5 mm Hg (20 C); m.p. 43 C; b.p. 204 C; flash pt. 98 C; nonflamm.
HMIS: Health 1, Flammability 1, Reactivity 0

Genesolv® 2000 [Honeywell Spec. Chems.]
Chem. Descrip.: 1,1-Dichloro-1-fluoroethane
CAS 1717-00-6
UN NA3082
Uses: Solvent
Features: Replacement for CFC-113 solvs.; compat. with acetal, nylon, PVC, HDPE, PTFE, epoxy, PP, CPVC, polyetherimide, polyester, PPS, PVdF, PEEK, PET, vinyl esters; incompat. with acrylic, PS, PC, ABS, cellulose acetate; nonflamm.
Regulatory: DOT nonregulated; SARA §311 immediate/delayed hazard, §313 reportable; EINECS, Japan, Korea, Australia listed
Properties: Colorless volatile liq.; ethereal faint sweetish odor; sol. 1800 ppm in water; m.w. 116.95; sp.gr. 1.24 (21.1 C); visc. 0.430 cP; vapor pressure 11.47 psia; b.p. 32.1 C; flash pt. none; pH neutral; surf. tens. 19.3 dynes/cm; KB value 56; dielec. const. 8.07; 99.5% assay
Toxicology: LC50 (inh., rat, 4 h) 62,000 ppm; WEEL (TWA 8 h) 500 ppm; low overall toxicity; overexposure may cause dizziness, loss of concentration, and at higher levels, CNS depression, cardiac arrhythmia; vapors displace air and can cause asphyxiation in confined spaces; may cause skin irritation, defatting on prolonged/repeated contact; may cause eye irritation, conjunctivitis; TSCA listed
Environmental: Ozone depleting; minimal biodeg.; mod. toxic to fish; not toxic up to 44 mg/l in algae; prevent contamination of ground- and surf. waters
Precaution: Flamm. limits 7.6-17.7 vol.%; solv. vapors in the flamm. range, esp. in confined or poorly ventilated spaces, can be ignited with a flame or strong ignition sources; contact with some finely divided reactive metals may cause explosive/exothermic reactions; incompat. with strong acids and alkalis, reactive metals, strong oxidizing agents; may react with some desiccants
Hazardous decomp. prods.: At high temps. (> 250 C), decomp. prods. may incl. hydrochloric acid, hydrofluoric acid, carbonyl halides such as phosgene
HMIS: Health 1, Flammability 1, Reactivity 0
Storage: Store in tightly closed original containers in cool, dry, well-ventilated area away from sunlight, heat, sparks, open flame; store below 120 F; keep closed when not in use

Genesolv® 2004 [Honeywell Spec. Chems.]
Chem. Descrip.: 1,1-Dichloro-1-fluoroethane (95.8%), methanol (3.9%), nitromethane (0.3%)
UN NA3082
Uses: Solvent
Features: Replacement for CFC-113/alcohol blends; compat. with acetal, nylon, PVC, HDPE, PTFE, epoxy, PP, CPVC, polyester, PPS, PVDF, PEEK, PET, vinyl esters; nonflamm.
Regulatory: DOT nonregulated; SARA §311 immediate/delayed hazard, §313 reportable (dichlorofluoroethane, methanol)
Properties: Colorless volatile liq.; ethereal faint sweetish odor; sol. 0.17% in water; sp.gr. 1.20 (21.1 C); dens. 10.1 lb/gal; visc. 0.450 cP; vapor pressure 10.2 psia; b.p. 29.4 C; flash pt. none; pH neutral; surf. tens. 18.5 dynes/cm; KB value 76
Toxicology: Low overall toxicity; overexposure may cause dizziness, loss of concentration, and at higher levels, CNS depression, cardiac arrhythmia; vapors displace air and can cause asphyxiation in confined spaces; may cause skin irritation, defatting on prolonged/repeated contact; may

cause eye irritation, conjunctivitis; TSCA listed
Environmental: Ozone depleting; minimal biodeg.; mod. toxic to fish; not toxic up to 44 mg/l in algae; prevent contamination of ground- and surf. waters
Precaution: Flamm. limits 6.0-20.3 vol.%; solv. vapors in the flamm. range, esp. in confined or poorly ventilated spaces, can be ignited with a flame or strong ignition sources; contact with some finely divided reactive metals may cause explosive/exothermic reactions; incompat. with strong acids and alkalis, reactive metals, strong oxidizing agents; may react with some desiccants
Hazardous decomp. prods.: At high temps. (> 250 C), decomp. prods. may incl. hydrochloric acid, hydrofluoric acid, carbonyl halides such as phosgene
HMIS: Health 1, Flammability 1, Reactivity 0
Storage: Store in tightly closed containers in cool, dry, well-ventilated area away from sunlight, heat, sparks, open flame; store below 120 F; keep closed when not in use

Genesolv® D [Honeywell Spec. Chems.]
Chem. Descrip.: Trichlorotrifluoroethane
CAS 76-13-1; EINECS/ELINCS 200-936-1
UN NA3082
Uses: Solvent for removal/suspension of oils/particulate matter in vapor degreasing, ultrasonic cleaning, cold cleaning, cleaning of metal parts, electronic components, integrated circuits, gyroscopes, guidance systems, aerospace/missile hardware; solvent for mfg. automatic temp. control valves, aerosol electronic contact cleaner and silicone-dispersed coating materials, biomedical parts, S/B latex finishes; isolates viruses in biomedical; as dielectric; solvent cleaner for polyester fibers
Features: Selective solvency; compat. with most metals, plastics, elastomers; less toxic than chlorinated solvs.; nonflamm.; nonexplosive; chemically stable; economical; recoverable; avail. in electronic grade
Regulatory: DOT nonregulated below 5000 lb; SARA §311 immediate/delayed hazard, §313 reportable; subject to EPA Clean Air Act Regs.; EINECS, Japan MITI, Korea MOE, Canada DSL, Australia AICS listed
Properties: Colorless volatile liq.; ethereal faint sweetish odor; sol. 0.028% in water (21.1 C); m.w. 187.4; sp.gr. 1.47 (21.1 C); dens. 31.1 lb/gal (21.1 C, liq.); f.p. -35 C; b.p. 47.6 C; KB value 32; sp. heat 0.21 Btu/lb° F (liq.); dielec. str. 37 kV (0.1 in.); dielec. const. 2.41 (60 Hz); vol. resist. 10^{16} ohm-cm; 100% act.
Toxicology: LC50 (inh., rat, 2 h) 110,000 ppm; ACGIH TLV/TWA (8 h) 1000 ppm, STEL 1250 ppm; overexposure may cause dizziness, loss of concentration, and at higher levels, CNS depression, cardiac arrhythmia; vapors displace air and can cause asphyxiation in confined spaces; may cause skin irritation, defatting on prolonged/repeated contact; may cause eye irritation, conjunctivitis; TSCA listed
Environmental: LC50 (flathead minnow, 96 h) 1250 ppm; ozone depleting; minimal biodeg.; prevent contamination of ground- and surf. waters; do not vent to the atmosphere
Precaution: Vapors in the flamm. range, esp. in confined or poorly ventilated spaces, can be ignited with a flame or strong ignition sources; contact with some finely divided reactive metals may cause explosive/exothermic reactions; incompat. with strong acids and alkalis, reactive metals (e.g., powd. aluminum, sodium, potassium, calcium, magnesium, zinc, barium, lithium), strong oxidizing agents
Hazardous decomp. prods.: At high temps. (> 250 C), decomp. prods. may incl. hydrochloric acid, hydrofluoric acid, carbonyl halides such as phosgene
HMIS: Health 1, Flammability 0, Reactivity 0
Storage: Store in closed containers in cool, dry, well-ventilated area away from direct flame, heat, sparks

Genesolv® SF [Honeywell Spec. Chems.]
Chem. Descrip.: 1,1,1,3,3-Pentafluoropropane
CAS 460-73-1; EINECS/ELINCS 419-170-6
Uses: Solvent for industrial aerosols, specialty cleaning, flushing, deposition
Features: Replacement for HCFC-141b and other halogenated solvs.; low cost; nonflamm.; high thermal and hydrolytic stability
Regulatory: DOT nonregulated; SARA §311 immediate/pressure hazard, §313 nonreportable; EINECS, Japan MITI listed
Properties: Colorless liq.; ethereal, faint sweetish odor; misc. with HFC-134a, HFC-152a, DME, hydrocarbon aerosol propellants, many common org. solvs. incl. other fluorocarbons, chlorine solvs. (methylene chloride), hydrocarbons (pentanes, hexanes), lower alcohols; sol. 7200 ppm in water; m.w. 134.0; sp.gr. 1.32 (20 C); vapor pressure 4 psig (70 F); f.p. < -256 F; b.p. 15 C; flash pt. none; pH neutral; 100% act.
Toxicology: LD50 (skin, rabbit) > 2000 mg/kg; LC50 (inh., rat, 4 h) > 200,000 ppm; WEEL (TWA 8 h) 300 ppm; low toxicity; overexposure may cause dizziness, loss of concentration, and at higher levels, CNS depression, cardiac arrhythmia; vapors displace air and can cause asphyxiation in confined spaces; nonirritating to skin; may cause eye irritation; nonmutagenic; nonteratogenic; TSCA listed
Environmental: EC50 (daphnia magna, 48 h) > 97.9 mg/l, (rainbow trout, 96 h) > 81.8 mg/l; nonozone-depleting; relatively low global warming potential; non-VOC
Precaution: Nonflamm.; however, will become combustible when mixed with air under pressure and exposed to strong ignition sources; contact with some finely divided reactive metals may cause explosive/exothermic reactions; incompat. with strong acids and alkalis, reactive metals (e.g., powd. aluminum, sodium, potassium, calcium, magnesium, zinc, barium, lithium), strong oxidizing agents
Hazardous decomp. prods.: Halogens, halogen acids; at high temps. (> 250 C), decomp. prods. may incl. hydrofluoric acid, carbonyl halides
HMIS: Health 2, Flammability 0, Reactivity 1
Storage: Store in approved container in cool, well-ventilated area; avoid excessive heat, direct sunlight, open flames

Genesolv® S-TZ [Honeywell Spec. Chems.]
Chem. Descrip.: 1,1,1,3,3-Pentafluoropropane (87%), trans-1,2-dichloroethylene (13%)
UN 3082
Uses: Solvent for industrial aerosols, specialty cleaning, flushing, deposition
Features: Azeotropic; enhanced solvency; replacement for HCFC-141b and other halogenated solvs.; nonflamm.; good sol. for fluorosilicones, perfluoropolyether lubricant, min. oils, many refrigerant oils, other lubricants
Regulatory: DOT nonregulated; SARA §311 immediate/pressure hazard, §313 nonreportable; EINECS listed
Properties: Colorless liq.; ethereal faint sweetish odor; misc. with HFC-134a, HFC-152a, DME, hydrocarbon aerosol propellants, common solvs. incl. other fluorocarbons, chlorinated solvs., hydrocarbons, alcohols, ketones; sol. ≈ 7.18 g/l in water; sp.gr. 1.34 (21 C); vapor pressure 4.2 psig (70 F); f.p. < -256 F; b.p. 15 C; flash pt. none; pH neutral
Toxicology: WEEL (TWA 8 h) 300 ppm; low toxicity; overexposure may cause dizziness, loss of concentration, and at higher levels, CNS depression, cardiac arrhythmia; vapors displace air and can cause asphyxiation in confined spaces; mildly irritating to skin; may cause mod. eye irritation; nonmutagenic; nonteratogenic; TSCA listed
Environmental: Nonozone-depleting; relatively low global warming potential
Precaution: Nonflamm.; however, will become combustible when mixed with air under pressure and exposed to strong ignition sources; contact with some finely divided reactive metals may cause explosive/exothermic reactions; incompat. with strong acids and alkalis, reactive metals (e.g., powd. aluminum, sodium, potassium, calcium, magnesium, zinc, barium, lithium), strong oxidizing agents
Hazardous decomp. prods.: Halogens, halogen acids; at high temps. (> 250 C), decomp. prods. may incl. hydrofluoric acid, carbonyl halides
HMIS: Health 2, Flammability 0, Reactivity 1
Storage: Store in approved container in cool, well-ventilated area; avoid excessive heat, direct sunlight, open flames

Glicopol 2-G [Undesa; S. Black]
Chem. Descrip.: Glyceryl dilaurate
CAS 27638-00-2; EINECS/ELINCS 248-586-9
Uses: Dispersant, solvent for act. ingreds.; vehicle for injectable substances; emulsifier for textiles; emulsifier, opacifier, thickener, dispersant for emulsions; greasing agent in shampoos, foam baths; food lubricant; perfume diluent; intermediate in textiles and tanning industry
Properties: Nonionic

Glidsol® 66-2 [Millennium/F&F]
Chem. Descrip.: Sat. terpene hydrocarbon
Uses: Solvent
Properties: Flash pt. 123 F

Glidsol® 140 [Millennium/F&F]
Chem. Descrip.: Terpene alcohol

Uses: Solvent for ore flotation
Properties: Flash pt. 199 F

Glidsol® 175 [Millennium/F&F]
Chem. Descrip.: Terpene alcohol
Uses: Solvent for cleaning formulations
Properties: Flash pt. 183 F

Glidsol® 180 [Millennium/F&F]
Uses: Solvent
Features: High flash pt. limonene replacement
Properties: Flash pt. 142 F

Glidsol® 5140 [Millennium/F&F]
Uses: Limonene extender
Properties: Low odor; flash pt. 108 F

Glucam® E-10 [Amerchol]
Chem. Descrip.: Methyl gluceth-10
CAS 68239-42-9
Uses: Humectant, moisturizer, gloss aid, conditioner for personal care prods., emulsions, toilet articles, shampoos, shaving creams, shower gels, eye gels, facial cleansers; emollient in aq. and hydroalcoholic prods.; foam modifier in detergents and shampoos; humectant, solvent, and solubilizer for topical pharmaceuticals; film plasticizer; freezing pt. depressant
Properties: Pale yel. med. visc. syrup; pract. odorless; sol. in water, alcohol, hydroalcoholic systems; acid no. 1.5 max.; iodine no. 1 max.; sapon. no. 1.5 max.; hyd. no. 350-370; nonionic; 100% conc.
Toxicology: LD50 (acute oral) > 5 ml/kg; nonirritating to eyes and skin

Glucam® E-20 [Amerchol]
Chem. Descrip.: Methyl gluceth-20
CAS 68239-43-0
Uses: Humectant, moisturizer for personal care prods., sunscreens, soap bars, dry skin creams, shampoos, liq. soaps; emollient in aq. and hydroalcoholic prods.; foam modifier in detergent and shampoo systems; humectant, solvent, and solubilizer for topical pharmaceuticals; freezing pt. depressant
Properties: Pale yel. thin syrup; pract. odorless; sol. in water, alcohol, hydroalcoholic systems; acid no. 1.0 max.; iodine no. 1 max.; sapon. no. 1.0 max.; hyd. no. 205-225; nonionic; 100% conc.
Toxicology: LD50 (acute oral) > 5 ml/kg; nonirritating to eyes and skin

Glucam® P-10 [Amerchol]
Chem. Descrip.: PPG-10 methyl glucose ether
CAS 61849-72-7
Uses: Humectant, moisturizer, emollient for personal care prods., eye makeup removers, hair conditioners, styling prods., makeup, moisturizing creams; emollient in aq. and hydroalcoholic prods.; foam modifier in detergents and shampoos; solvent and solubilizer for topical pharmaceuticals; freezing pt. depressant
Properties: Pale yel. heavy visc. syrups; pract. odorless; sol. in water, alcohol, and hydroalcoholic systems, castor oil, IPM, IPP; visc. 8500 cps; acid no. 1.0 max.; iodine no. 1 max.; sapon. no. 1.0 max.; hyd. no. 285-305; nonionic; 100% conc.
Toxicology: LD50 (acute oral) > 5 ml/kg; nonirritating to skin; mild transient irritation to eyes

Glucam® P-20 [Amerchol]
Chem. Descrip.: PPG-20 methyl glucose ether
CAS 61849-72-7
Uses: Humectant, moisturizer, emollient for personal care prods., face cleansing toners, nail conditioners, fragrances, aftershaves; emollient in aq. and hydroalcoholic prods.; foam modifier in detergents, shampoos; solvent and solubilizer for topical pharmaceuticals; freezing pt. depressant
Properties: Pale yel. med. visc. syrup; pract. odorless; sol. in water, alcohol, and hydroalcoholic systems, castor oil, IPM, IPP; visc. 1700 cps; acid no. 1.0 max.; iodine no. 1 max.; sapon. no. 1.0 max.; hyd. no. 160-180; nonionic; 100% conc.
Toxicology: LD50 (acute oral) > 3 ml/kg; nonirritating to skin; mild transient irritation to eyes

Glyezin® A [BASF AG]
Uses: Dye solvent/fixing agent for textile finishing

H

HallBrite® BHB [C.P. Hall]
Chem. Descrip.: Butyloctyl salicylate
CAS 190085-41-7; EINECS/ELINCS 431-090-3
Uses: Emollient, moisturizer, solvent, carrier for hair and skin care, sunscreens, moisturizers, creams/lotions, hair conditioners, lipstick, and lip balms; gloss aid, pigment wetting agent for personal care; solubilizer, stabilizer for sunscreen actives
Regulatory: DOT nonhazardous; SARA §311/312/313 nonreportable; Calif. Prop. 65 nonreportable
Properties: Pale yel. to colorless clear liq.; practically odorless; sol. in isopropanol, min. oil, isopropyl myristate, di- and cyclomethicone, capric/caprylic triglyceride, octyl methoxycinnamate; insol. in glycerin and water; m.w. 306; sp.gr. 0.9660; f.p. < -20 C; acid no. < 1; sapon. no. 181.1; cloud pt. < -20 C; ref. index 1.4909; pH 5.8
Use Level: 2-10%
Toxicology: LD50 (oral) > 5 g/kg, (skin) > 2000 mg/kg; may be harmful if swallowed; may irritate mouth, throat, stomach; inh. may cause dizziness; minimal eye, mod. skin irritant; nonsensitizing
Environmental: NOEC (daphnia magna) ≥ 0.38 μg/l; not found to be acutely toxic @ limit of aq. sol.; may cause long-term adverse effects in aquatic environment; avoid release to environment; keep spills out of sewers and watercourses
Precaution: Incompat. with oxidizing materials
HMIS: Health 1, Flammability 1, Reactivity 0
Storage: Store in tightly closed containers in dry, well-ventilated area away from oxidizing materials

HallBrite® OS [C.P. Hall]
Chem. Descrip.: Ethylhexyl salicylate
CAS 118-60-5; EINECS/ELINCS 204-263-4
Uses: Gloss aid, solvent, moisturizer, UV absorber for personal care, sunscreens, hair sprays, lip balms, lotions
Regulatory: DOT nonhazardous; SARA §311/312/313 nonreportable; Calif. Prop. 65 nonreportable; Canada DSL, Australia AICS, EINECS listed
Properties: Clear liq.; mild odor; insol. in water; sp.gr. 1.014; b.p. 400 F; flash pt. (COC) 310 F
Toxicology: LD50 (oral) > 5 g/kg; may be harmful if swallowed; may irritate mouth, throat, stomach; inh. may cause dizziness; minimal eye, sl. skin irritant; nonsensitizingTSCA listed
Environmental: No ozone depleting compds.; keep spills out of sewers and watercourses
Precaution: Incompat. with strong oxidizing agents
HMIS: Health 1, Flammability 0, Reactivity 0
Storage: Store in tightly closed containers in dry, well-ventilated area, away from oxidizing materials

Hallcomid® M-8-10 [C.P. Hall]
Chem. Descrip.: Dimethyl octanamide, dimethyl decanamide
Uses: Mutual solvent for polar and nonpolar ingreds. in cosmetics, cleaners, corrosion inhibitor; polar solvent for nat. and syn. resins and elastomers; modifier for polymer coatings
Regulatory: DOT nonhazardous; SARA §313 nonreportable; Calif. Prop. 65 nonreportable; Australia AICS, EINECS listed
Properties: Colorless to pale yel. liq.; sol. 0.17% in water; sol. in most common organic solvs., hydrocarbons, plasticizers, fatty acids, veg. oils; m.w. 171-199; sp.gr. 0.88; visc. 10.5 cps; f.p. -21 C; b.p. 115-170 C (3 mm Hg); flash pt. 245 F; nonionic; 95% conc.
Toxicology: Severely irritating to eyes, skin; corrosive; may produce corneal opacity, iritis, conjunctival irritationTSCA listed
Environmental: LC50 (daphnia magna, 48 h) 7.7 mg/l, (fathead minnow, 96 h) 19.0 mg/l; no ozone depleting compds.; may cause detrimental effect on some vegetation; keep spills out of sewers, watercourses
Precaution: Incompat. with strong oxidizing agents
Hazardous decomp. prods.: Combustion: CO_x
Storage: Store in tightly closed containers in dry, well-ventilated area

HallStar™ BST [C.P. Hall]
Chem. Descrip.: Butyl stearate
CAS 123-95-5; EINECS/ELINCS 204-666-5
Uses: Emollient, solvent, pigment wetting agent/dispersant, lubricant, slip agent, color suspending agent, dye solvent, water-repelling plasticizer, gloss aid, spreading agent for deodorants, creams/lotions, preshave preps., makeup, makeup removers; plasticizer, gloss aid in hair sprays, nail varnishes; wetting agent, solubilizer for dyestuffs in lipsticks
Features: Nongreasy and nontacky
Properties: Colorless clear liq.; sl. odor; sol. in ethanol, min. oil, silicone oils, esters, fats and waxes, ethyl acetate, ketones and ethers; sl. sol. in water; insol. in glycerin; m.w. 340.57; sp.gr. 0.855; m.p. 19 C; acid no. 1.0 max.; sapon. no. 167; iodine no. 2.4; ref. index 1.441; 100% act.
Toxicology: LD50 (oral) > 64 ml/kg; low toxicity; nonirritating to eyes, skin

HallStar™ DIPA [C.P. Hall]
Chem. Descrip.: Diisopropyl adipate
CAS 6938-94-9; EINECS/ELINCS 230-072-0
Uses: Emollient, solvent, coupling agent, pigment wetting agent/dispersant, lubricant, plasticizer for perfumes, colognes, toilet waters, aftershave, bath oils, cleansers, tonics, fresheners, makeup, sunscreens, skin creams/lotions; solvent, coupling agent for fragrances; cosolvent for lipophilic components; plasticizer for hair spray resins
Features: Nonoily; modifies tactile and flow props.
Properties: Colorless clear liq.; faint, char. odor; sol. 50% in ethanol, 5% in min. oil, isopropanol, isopropyl myristate, and propylene glycol; insol. in water; m.w. 230.3; sp.gr. 0.9600; HLB 9; acid no. 1.0 max.; sapon. no. 475-495; ref. index 1.4230
Use Level: 1-25%
Toxicology: LD50 (oral) > 15 g/kg; nontoxic; nonirritating to eyes, skin; nonsensitizing

HallStar™ DIPS [C.P. Hall]
Chem. Descrip.: Diisopropyl sebacate
CAS 7491-02-3; EINECS/ELINCS 231-306-4
Uses: Emollient, solvent, hydroalcoholic coupling agent, lubricant, plasticizer, pigment wetting agent/dispersant for clear sticks/gels, perfumes, colognes, toilet waters, aftershaves, cleansers, tonics, fresheners, makeup, sunscreen, skin creams/lotions; spreading agent for oil blends
Features: Nonoily; nonocclusive; mildly drying
Properties: Colorless clear liq.; mild, char. odor; sol. in alcohols, higher glycols, ketones, esters, aromatic/aliphatic/chlorinated hydrocarbons; insol. in water; m.w. 286.4; sp.gr. 0.9360; acid no. 0.5; sapon. no. 390; ref. index 1.4312

HallStar™ DOA [C.P. Hall]
Chem. Descrip.: Dioctyl adipate
CAS 103-23-1; EINECS/ELINCS 203-090-1
Uses: Emollient, moisturizer, solvent, coupling agent, plasticizer, pigment wetting agent/dispersant for bath prods., sunscreen, colognes, toilet

waters, lipstick, makeup, nail polishes and removers, deodorants; solvent, hydroalcoholic coupling agent for fragrances; cosolvent for lipophilic components
Features: Imparts shine; modifies tactile and flow props.
Properties: Colorless clear liq.; faint odor; sol. in ethanol, isopropanol, min. oil, castor oil, and corn oil; insol. in glycerin and water; m.w. 370.6; sp.gr. 0.9240; acid no. < 1.0; sapon. no. 303; ref. index 1.4550
Use Level: 1-25%
Toxicology: LD50 (oral) > 6 g/kg; nonirritating to skin and eyes; nonsensitizing

HallStar™ DOM [C.P. Hall]
Chem. Descrip.: Di-2-ethylhexyl maleate
CAS 142-16-5; EINECS/ELINCS 205-524-5
Uses: Emollient, solvent, cleanser, gloss aid for hair sprays, sunscreens, creams/lotions, skin cleansers, hair conditioners; solvent for waxes
Features: Noncomedogenic; easily absorbed; lt., nonoily feel
Regulatory: FDA, EEC, and Japan compliant
Properties: APHA 15 clear liq.; practically odorless; sol. in cyclomethicone, capric/caprylic triglyceride, IPA, min. oil, IPM, dimethicone, and octyl methoxycinnamate; insol. in glycerin and water; m.w. 340.5; sp.gr. 0.9405; acid no. 0.05; ref. index 1.4527; 0.05% moisture; 99% min. purity

HallStar™ DOS [C.P. Hall]
Chem. Descrip.: Dioctyl sebacate
CAS 122-62-3; EINECS/ELINCS 204-558-8
Uses: Emollient, solvent, lubricant, plasticizer, pigment wetting agent/dispersant for cleansers, tonic, fresheners, makeup, sunscreens, skin creams/lotions; spreading agent for oil blends
Features: Noncomedogenic; nonoily; silky feel
Properties: Colorless to pale yel. clear liq.; mild, char. odor; sol. in alcohol (95%), safflower oil (95%), soybean oil (95%), lanolin (95%), min. oil (10%), IPM (10%), dimethicone (10%), castor oil (101%); insol. in glycerin, PG, and water; m.w. 426; sp.gr. 0.912; acid no. 0.3 max.; sapon. no. 262; ref. index 1.4490
Toxicology: LD50 (oral) > 5 g/kg; nonirritating to skin and eyes; noncomedogenic

HallStar™ IPM [C.P. Hall]
Chem. Descrip.: Isopropyl myristate
CAS 110-27-0; EINECS/ELINCS 203-751-4
Uses: Emollient, spreading agent, penetrant for personal care prods. incl. bath oils, perfumes, creams/lotions, lipsticks, hair preps., shaving lotions, aerosol toiletries, pharmaceutical ointments; solvent, carrier for cosmetic additives; solubilizer for lanolin
Properties: Colorless clear liq.; pract. odorless; sol. in ethanol, IPA, cyclomethicone, most org. solvs. and oils; insol. in water, glycerin, propylene glycol; sp.gr. 0.850; HLB 11-12; acid no. 1.0; sapon. no. 210; ref. index 1.4320

HallTress™ DIBA Special [C.P. Hall]
Chem. Descrip.: Diisobutyl adipate
CAS 141-04-8; EINECS/ELINCS 205-450-3
Uses: Emollient, skin conditioner, solvent, penetrant, superfatting agent, hydroalcoholic coupler, pigment wetter/dispersant in hair sprays, nail polish and enamels, colognes, toilet waters, bath prods., skin lotions; plasticizer for hair fixatives and nail enamels
Features: Nonoily; nonocclusive; mildly drying
Regulatory: DOT nonhazardous; SARA §311/312/313 nonreportable; Calif. Prop. 65 nonreportable; Canada DSL, Australia AICS, EINECS, Japan ENCS listed
Properties: Colorless to pale yel. clear liq.; mild, char. odor; sol. in ethanol, IPA, IPM, cyclomethicone, capric/caprylic triglyceride; insol. in glycerin, water; m.w. 262; sp.gr. 0.9511; HLB 9; acid no. 0.15; sapon. no. 435; flash pt. (COC) 400 F; ref. index 1.4302
Toxicology: Repeated/prolonged contact may cause skin irritation; may cause sl. eye irritation; may be harmful if swallowed; may irritate mouth, throat, stomach; may cause dizziness by inh.TSCA listed
Environmental: No ozone-depleting compds.; keep spills out of sewers and watercourses
Storage: Store in tightly closed containers in dry, well-ventilated area

Halso® 99 [Occidental]
Chem. Descrip.: Monochlorotoluene
CAS 25168-05-2
Uses: Solvent, extender, diluent, substitute for other org. solvents for dye carrier, fuel oil additive, sludge solvent, in paint thinners, metal parts cleaners, adhesives
Properties: Dens. 9 lb/gal; visc. 0.747 cps (100 C); f.p. -32 C; b.p. 159-162 C; flash pt. (TCC) 123 F; surf. tens. 33 dynes/cm (20 C); KB value 110

HAN® 906 Solvent [Exxon Canada]
Chem. Descrip.: Naphtha, heavy aromatic, BHT (25 ppm)
CAS 64742-94-5; EINECS/ELINCS 265-198-5
Uses: Solvent; diluent; chem. feedstock; fuel
Regulatory: Canada DSL listed; NPRI reportable [naphthalene (12.3%), 1,2,4-trimethylbenzene (1.7%)]
Properties: Colorless clear liq.; mild aromatic odor; sol. < 0.01% in water; m.w. 142; sp.gr. 0.9 (15.6 C); visc. 1.15 cSt; vapor pressure 0.067 kPa (20 C); f.p. -18 C; b.p. 184-205 C; flash pt. (TCC) 63 C; autoignition temp. 454 C; 100% act.
Toxicology: REL TWA 17 ppm; LD50 (oral, rat) 3 g/kg, (skin, rabbit) > 3 g/kg; inh. of high vapor concs. may cause eye/respiratory tract irritation, headaches, dizziness, anesthesia, drowsiness, unconsciousness, other CNS effects, death; sl. eye irritant; frequent/prolonged skin contact may cause irritation, dermatitis; if aspirated, may cause mild to severe pulmonary injury, death; contains naphthalene, possible human carcinogen
Environmental: Prevent entry into sewers, watercourses, low areas
Precaution: Combustible liq.; flamm. limits 0.8-5.7 vol.%; volatile; gives off invisible vapors; liq. or vapor may settle in low areas; distant ignition/explosion possible; electrostatic accumulation hazard; use proper grounding procedures; incompat. with nitric acid, sulfuric acid, strong oxidizing agents
Hazardous decomp. prods.: None
Storage: Store in cool, well-ventilated area away from incompat. materials, heat, open flame, ignition sources, direct sunlight; keep container closed

Hartosolve OL [Hart Prods.]
Chem. Descrip.: Emulsified solvents
Uses: Solvent scouring agent, cleaner
Properties: Liq.; anionic; 100% conc.

HC-250 [Tosoh]
Chem. Descrip.: Paraffin hydrocarbon
Uses: Solvent for steel and nonferrous metals, machinery, precision equip., elec. machinery and electronics industries, transportation equip.
Features: Nonaq.; exc. solvency; substitute for 1,1,1-trichloroethane and CFC-113
Properties: Sp.gr. 0.729; visc. 0.0008 Pa·s (30 C); b.p. 172 C; flash pt. 53 C; surf. tens. 0.024 N/m (20 C)
Environmental: Environmentally friendly

HC-370 [Tosoh]
Chem. Descrip.: Paraffin hydrocarbon
Uses: Solvent for steel and nonferrous metals, machinery, precision equip., elec. machinery and electronics industries, transportation equip.
Features: Nonaq.; exc. solvency; substitute for 1,1,1-trichloroethane and CFC-113
Properties: Sp.gr. 0.740; visc. 0.0011 Pa·s (30 C); b.p. 196 C; flash pt. 70 C; surf. tens. 0.025 N/m (20 C)
Environmental: Environmentally friendly

HC-DCF Low Flash™ Dry Cleaning Solvent [ChevronPhillips]
Chem. Descrip.: C9-11 isoalkanes, preservative BHT (10 ppm)
CAS 68551-16-6; EINECS/ELINCS 271-365-3
UN 3295
Uses: Dry cleaning solvent
Regulatory: SARA §313 nonreportable
Properties: Colorless liq.; mild odor; negligible sol. in water; sp.gr. 0.746 (60/60 F); visc. 1.12 cSt (38 C); vapor pressure 0.1 psia; b.p. 160-167 C; flash pt. (TCC) 41 C; 100% act.
Toxicology: REL (8 h TWA) 100 mg/m^3 as total hydrocarbon; may be sl. irritating to eyes, skin; inh. may cause headache, dizziness, nausea, unconsciousness; ing. may irritate stomach, intestines; if swallowed, may be aspirated into lungs, resulting in inflamm., possible fluid accumulation; may cause kidney damage; TSCA listed
Environmental: Keep out of water sources and sewers
Precaution: Flamm. liq.; incompat. with oxygen and strong oxidizing agents; highly flamm. vapors may accumulate in low areas; may flash back
Hazardous decomp. prods.: CO_x formed when burned
NFPA: Health 1, Flammability 2, Reactivity 0

Storage: Store in closed containers in well-ventilated area, away from heat, sparks, flame; bond and ground during transfer

HC-FX50 [Tosoh]
Chem. Descrip.: Paraffin hydrocarbon
Uses: Solvent for steel and nonferrous metals, machinery, precision equip., elec. machinery and electronics industries, transportation equip.
Features: Nonaq.; exc. solvency; substitute for 1,1,1-trichloroethane and CFC-113
Properties: Sp.gr. 0.760; visc. 0.0011 Pa·s (30 C); b.p. 170 C; flash pt. 50 C; surf. tens. 0.024 N/m (20 C)
Environmental: Environmentally friendly

HC-FX70 [Tosoh]
Chem. Descrip.: Paraffin hydrocarbon
Uses: Solvent for steel and nonferrous metals, machinery, precision equip., elec. machinery and electronics industries, transportation equip.
Features: Nonaq.; exc. solvency; substitute for 1,1,1-trichloroethane and CFC-113
Properties: Sp.gr. 0.784; visc. 0.0015 Pa·s (30 C); b.p. 194 C; flash pt. 76 C; surf. tens. 0.026 N/m (20 C)
Environmental: Environmentally friendly

Hedislide RS [Consos]
Chem. Descrip.: Blend of silicone lubricant solvs. and rust inhibitors
Uses: Lubricant, solvent, rust inhibitor for cleaning of machine parts
Features: Removes lint, oil, grease, wax, and size from metal surfaces, leaving clean, lubricated surf. that will resist rust and other corrosion

HEP® [ISP]
Chem. Descrip.: N-Hydroxyethyl pyrrolidone
CAS 3445-11-2; EINECS/ELINCS 222-359-4
Uses: Solvent for microelectronics, photoresists, to clean silk screens, prep. of polymer membranes for gas scrubbing, anticlogging jet printing ink, aq. veterinary tetracycline antibiotic; reaction solvent for prep. of retinoid esters for skin treatment; mfg. of haloalkyl lactams; alkaline paint stripper; solvent plant growth regulator; intermediate in lube oil additives
Features: Offers reactive hydroxyl group
Properties: Clear liq.; water-sol.; m.w. 129; sp.gr. 1.139; visc. 53 cps; f.p. 20 C; b.p. 295 C; flash pt. (CC) 320 F; ref. index 1.4951; surf. tens. 49 dynes/cm; 98% min. purity
Toxicology: LD50 (oral, rat) > 14.43 g/kg; low acute toxicity; minimally irritating to eyes; nonirritating to skin
Environmental: Biodeg.

Herco® Pine Oil [Hercules]
Chem. Descrip.: Pine oil
CAS 8002-09-3
Uses: Solvent for household/industrial cleaners; disinfectant; antifoam; textile specialties
Features: Various grades
Properties: Disp. in water

Hetester PMA [Global-Seven; Bernel]
Chem. Descrip.: Propylene glycol myristyl ether acetate
CAS 93385-02-5
Uses: Emollient, solvent, plasticizer, SPF booster, oxybenzone solubilizer in cosmetics, anhyd. oil systems, emulsions
Features: Acetylated lanolin alcohol replacement
Regulatory: FDA, EU, and Japanese compliances
Properties: Colorless clear liq.; sol. @ 5% in most cosmetic oils; insol. in water, glycols, 70% ethanol; acid no. 2.0 max.; sapon. no. 140-160; hyd. no. 10.0 max.; pH 6.0-7.0 (5% in 50/50 IPA/water); cloud pt. ≈ 0 C
Use Level: 2-10%

Hetoxide MPC [Global-Seven]
Chem. Descrip.: m,p-Cresol hydrophobe
Uses: Solvent for lacquers, coatings
Properties: Gardner 5 max. liq.; sol. in IPA, min. oil; disp. in water; hyd. no. 275-295

Hexalin® [Cognis KGaA/Coatings]
Chem. Descrip.: Cyclohexanol
CAS 108-93-0; EINECS/ELINCS 203-630-6
Uses: Solvent, flow aid in coatings
Features: High boiling; strong solvent power
Use Level: 2-20% on solv.

Hexyl Carbitol® [Dow]
Chem. Descrip.: Diethylene glycol n-hexyl ether
CAS 112-59-4; EINECS/ELINCS 203-988-3
Uses: Solvent for waterborne coatings
Properties: Sol. 1.7% in water (20 C); m.w. 190.28; sp.gr. 0.935 (20/20 C); dens. 7.78 lb/gal (20 C); visc. 8.5 cP (20 C); vapor pressure < 0.01 mm Hg (20 C); b.p. 259.1 C; flash pt. (CC) 271 F

Hexyl Cellosolve® [Dow]
Chem. Descrip.: Ethylene glycol n-hexyl ether
CAS 112-25-4; EINECS/ELINCS 203-951-1
UN 2922
Uses: Solvent, coalescing agent for water-disp. coatings
Regulatory: SARA §311/312 acute health/chronic health hazard, §313 reportable; EINECS, Canada DSL listed
Properties: Colorless transparent liq.; pungent odor; sol. 0.99% in water (20 C); m.w. 146.23; sp.gr. 0.889 (20/20 C); dens. 7.40 lb/gal (20 C); visc. 5.2 cP (20 C); vapor pressure 0.05 mm Hg (20 C); f.p. -45 C; b.p. 208.1 C; flash pt. (PMCC) 99 C; pH 6.3; VOC 87 g/l; 98-100% act.
Toxicology: REL TWA (8 h) 20 ppm; LD50 (oral, rat) 1.67 ml/kg, (skin, rabbit) 0.81 ml/kg; causes eye/skin burns; harmful, corrosive if swallowed; may cause mouth/throat/esophagus/stomach irritation, pain, discomfort, nausea, vomiting, diarrhea, thirst, dizziness, drowsiness, weakness, unconsciousness; harmful if absorbed thru skin; aspiration may cause lung damage; TSCA listed
Environmental: BOD5 72%; COD 1.89 mg/mg; LC50 (daphnia, 48 h) 305 mg/l, (fathead minnow, 96 h) 139.5 mg/l
Precaution: Flamm. limits in air 1.4-9.0 vol.%; may produce floating fire hazard; incompat. with strong alkalis, high temps. in presence of strong bases, acids, strong oxidizing agents
Hazardous decomp. prods.: Combustion: CO, CO_2
HMIS: Health 3, Flammability 1, Reactivity 0
Storage: Keep container closed

HFE-71DA [3M/Perf. Materials]
Chem. Descrip.: Methyl nonafluorobutyl ether (52.7%), trans-1,2-dichloroethylene (44.6%), ethanol (2.7%) azeotrope formulation
Uses: Solvent cleaner, rinsing agent, drying agent for cleaning of oils, greases, waxes, handling oils, and solder flux residue
Features: Zero ozone-depleting potential
Regulatory: US EPA SNAP approved
Properties: Clear colorless liq.; sp.gr. 1.33; visc. 0.45 cps; vapor pressure 381 mm Hg; f.p. -29 C; b.p. 40 C; surf. tens. 16.4 g/ml; KB 33; flash pt. none
Storage: Highly resist. to thermal breakdown and hydrolysis

HFE-71DE [3M/Perf. Materials]
Chem. Descrip.: Methyl nonafluorobutyl ether (50%) and trans-1,2-dichloroethylene (50%) azeotrope formulation
Uses: Solvent cleaner, degreasing agent, rinsing agent, drying agent for specialty solv. applics., and for certain coating and lubricant deposition applics.
Features: Replaces ozone depleting materials
Regulatory: EPA approved
Properties: Clear, colorless fluid; sp.gr. 1.37; visc. 0.45 cps; vapor pressure 473 mm Hg; f.p. -24 C; b.p. 41 C; flash pt. none; surf. tens. 16.6 dynes/cm; methyl nonafluorobutyl ethers (50%), trans-1,2-dichloroethylene (50%)
Storage: Highly resist. to thermal breakdown and hydrolysis in storage

HFE-7100 [3M/Perf. Materials]
Chem. Descrip.: 1-Methoxy-nonafluorobutane (99% min.)
CAS 163702-07-6
Uses: Solvent cleaner; rinsing agent; lubricant carrier; spot-free water drying agent; specialty solvs. dispersion medium; heat transfer medium; spray contact cleaner
Features: Replaces ozone depleting materials (CFC, HCFC, HFC, and PFC)
Regulatory: EPA approved
Properties: Clear, colorless fluid; low odor; m.w. 250; sol. < 20 ppm in water; sp.gr. 1.52; visc. 0.61 cps; vapor pressure 210 mm Hg; f.p. -135 C; b.p. 60 C; flash pt. none; surf. tens. 13.6 dynes/cm; 2.0 ppm max. NVR
Toxicology: LD50 (oral) > 5 g/kg; pract. nontoxic orally; pract. nonirritating to skin, eyes; low overall toxicity

HFE-7200 [3M/Perf. Materials]
Chem. Descrip.: Ethoxy-nonafluorobutane (99% min.)
CAS 163702-05-4
Uses: Solvent cleaner, rinsing agent for vapor degreasing, lt. duty clean-

ing; lubricant carrier for fluorocarbons, hydrocarbons, silicones; cold cleaner (movie film, wipe solv.); specialty solvent; dispersion medium; reaction medium; spray contact cleaner
Features: Replacement for CFC, HCFC, HFC, and PFC
Properties: Clear colorless liq.; low odor; m.w. 264; sol. < 20 ppm in water; sp.gr. 1.43; visc. 0.61 cps; vapor pressure 109 mm Hg; f.p. -138 C; b.p. 78 C; surf. tens. 13.6 g/ml; flash pt. none; NV 1 ppm max.
Toxicology: Low toxicity; practically nonirritating to eyes, min. irritating to skin; practically nontoxic thru inh.

Hipochem MS-BW [Boehme Filatex]
Chem. Descrip.: Emulsified aromatic and aliphatic solvs.
Uses: Detergent, solvent scouring agent for removal of oils, grease, waxes, and fatty materials in textile operations
Features: Low foaming
Properties: Colorless clear liq., mild aromatic odor; pH 6-8 (1%); nonionic; 100% conc.

Hiposcour® MS-BW [High Point]
Uses: Solvent scouring agent for textile prep.
Features: Low foam

Hi-Sol® 10 [Ashland]
Chem. Descrip.: Naphtha, light aromatic
CAS 64742-95-6; EINECS/ELINCS 265-199-0
UN 1255
Uses: Solvent for rubber processing
Properties: Water-wh. liq.; sp.gr. 0.884-0.864 (60 F); dens. 7.18 lb/gal (15.6 C); i.b.p. 152 C; flash pt. (TCC) 40.6 C; KB value 87 min.

Hi-Sol® 15 [Ashland]
Chem. Descrip.: Naphtha
EINECS/ELINCS 232-443-2
UN 1256
Uses: Solvent for rubber processing
Properties: Water-wh. liq.; sp.gr. 0.880-0.910 (60 F); dens. 7.43 lb/gal (15.6 C); i.b.p. 176.7 C; flash pt. (TCC) 61.1 C; KB value 85 min.

Hi-Sol® 70 [Ashland]
Chem. Descrip.: Naphtha
CAS 8030-30-6; EINECS/ELINCS 232-443-2
Uses: Solvent

Hisolve DM [Toho Chem. Ind.]
Chem. Descrip.: Diethylene glycol monomethyl ether
CAS 111-77-3; EINECS/ELINCS 203-906-6
Uses: Solvent for lacquer thinners, inks, dyes, brake fluids, metal surf. finishing agents; cleaner for kitchen, oven, bath
Features: High b.p.; good solubility
Properties: Liq.; m.w. 120

Hisolve MC [Toho Chem. Ind.]
Chem. Descrip.: Ethylene glycol monomethyl ether
CAS 109-86-4; EINECS/ELINCS 203-713-7
Uses: Solvent for paints, inks, adhesives, plastics and plasticizers, cleaners, paint removers, nail polish removers
Features: Good solubility for resin, castor oil
Properties: Liq.; m.w. 76

Hisolve MDM [Toho Chem. Ind.]
Chem. Descrip.: Diethylene glycol dimethyl ether
CAS 111-96-6; EINECS/ELINCS 203-924-4
Uses: Reaction solvent in reduction, isomerization, substitution, alkylation condensation, polymerization, and metal including reaction; separation and extraction solvent; acid gas absorbent
Properties: Liq.; m.w. 134

Hisolve PM [Toho Chem. Ind.]
Chem. Descrip.: Polyethylene glycol monomethyl ether
CAS 9004-74-4
Uses: Solvent for brake fluids
Properties: Liq.

Hisolve TM [Toho Chem. Ind.]
Chem. Descrip.: Triethylene glycol monomethyl ether
CAS 112-35-6
Uses: Solvent for brake fluids, wet process duplicator liqs.
Properties: Liq.

Hi-Tri [Dow]
Chem. Descrip.: Trichloroethylene
CAS 79-01-6; EINECS/ELINCS 201-167-4
Uses: Solvent for chem. processing and blending
Features: High purity; min. amt. of stabilizer
Regulatory: SARA §311/312 acute health/chronic health hazard, 313 reportable; Calif. Prop. 65 reportable
Properties: Colorless liq.; irritating odor; sol. 0.1 g/100 g in water; sp.gr. 1.46; vapor pressure 60 mm Hg (20 C); b.p. 87 C; flash pt. (TCC) none; autoignition temp. 420 C; 99.9% act.
Toxicology: ACGIH TLV TWA 50 ppm, STEL 100 ppm; harmful by inh., possibly fatal; may cause unconsciousness, upper respiratory tract irritation, irregular heartbeat; may cause eye pain, sl. eye irritation; irritant on prolonged/repeated skin contact; may be absorbed thru skin; may cause numbness in fingers immersed in liq.; low toxicity by ing.; swallowing larger amts. may cause serious injury, death; if aspirated, may cause injury to other body systems; may cause tumors; TSCA listed
Environmental: Biodeg.; low bioconcentration potential; LC50 (daphnia magna) 2.2-100 mg/l, (fathead minnow) 41-67 mg/l; mod. toxic to aquatic organisms; prevent contamination of soil, surf. water, ground water
Precaution: LFL 8.0%; UFL 10.5%; possible fire/explosion hazard; vapors are heavier than air, may accumulate in low areas; containers may rupture in fire; avoid open flames, high temp. sources, high energy sources; avoid contact with amines, strong bases, strong oxidizers, contact with metals such as Al powd., Mg powd., K, Na, Zn powd.; reaction with strong bases may form dichloroacetylene
Hazardous decomp. prods.: Hydrogen chloride, trace amts. of chlorine, phosgene; combustion: toxic fumes, hydrogen chloride, trace amts. of phosgene, chlorine, CO
NFPA: Health 2, Flammability 1, Reactivity 0
Storage: Store in dry place; keep containers tightly closed when not in use

Hodag® S-35 FGK [Lambent Tech.]
Chem. Descrip.: Mono- and diglycerides of fatty acids with propylene glycol
Uses: Emulsifier, dough strengthener, flavoring agent, formulation aid, lubricant, release agent, solvent, vehicle, stabilizer, thickener, surfactant, surface-finishing agent, texturizer for foods incl. shortenings, whipped toppings, coffee whiteners, frozen desserts, baked goods; antifoam, defoamer in food-contact paper
Regulatory: FDA 21CFR §170.3, 176.210, 184.1(b)(1), 184.666, 184.1505, GRAS; FCC compliant; kosher; SARA §311/312/313 nonreportable
Properties: Yel. clear liq.; bland odor; sol. in alcohols, veg. and min. oils; disp. in water; sp.gr. 0.97; dens. 8.3 lb/gal; visc. 140 cps; vapor pressure < 1 mm Hg; b.p. > 200 C; HLB 2.8; acid no. 5 max.; iodine no. 83; sapon. no. 135; pour pt. -17 C; clear pt. 24 C; flash pt. (COC) 110 C; nonionic; 100% conc.; 1% max. moisture
Toxicology: May cause eye irritationTSCA listed
Precaution: Incompat. with strong oxidizing agents; keep away from excessive heat, ignition sources; spillages may be slippery
Hazardous decomp. prods.: Combustion: CO, CO_2, thick smoke
NFPA: Health 1, Flammability 1, Reactivity 0
Storage: 1 yr shelf life when stored in closed original containers below 32 C; store and use in well-ventilated areas

Hydrol H660 [Huntsman]
Chem. Descrip.: Polyoxyalkylene glycol ether
Uses: Lubricant; defoamer; solvent
Features: Syn.; high visc. index
Properties: Liq.; water-sol.; visc. 660 SUS

Hydrol H660YA [Huntsman]
Chem. Descrip.: Polyoxyalkylene glycol ether
Uses: Lubricant; defoamer; solvent
Features: Syn.; high visc. index; inhibited version of Icinol H660

Hydrol H5100 [Huntsman]
Chem. Descrip.: Polyoxyalkylene glycol ether
Uses: Lubricant; defoamer; solvent
Features: Syn.; high visc. index
Properties: Liq.; water-sol.; visc. 5100 SUS

Hydrol L385 [Huntsman]
Chem. Descrip.: Polyoxyalkylene glycol ether
Uses: Lubricant; defoamer; solvent
Features: Syn.; high visc. index
Properties: Liq.; water-insol.; visc. 385 SUS

Hydroseal G2 [TotalFina-Elf]
Chem. Descrip.: Petroleum distillates, hydrotreated middle (C14-19)
CAS 64742-46-7; EINECS/ELINCS 265-148-2

Uses: Solvent for lubricants
Features: Desulfurized (< 2 ppm S)
Properties: Saybolt 25 color; dens. 815 kg/m^3 (15 C); visc. 6 mm^2/s (20 C); vapor pressure < 0.01 mbar (20 C); i.b.p. 270 C; pour pt. 20 C; aniline pt. 88 C; flash pt. (PM) 120 C; ref. index 1.4490 (20 C); surf. tens. 28.90 dynes/cm (20 C); 7% aromatics; 1 ppm benzene

Hydroseal G3H [TotalFina-Elf]
Chem. Descrip.: Petroleum distillates, acid-treated middle (C14-19)
CAS 64742-13-8; EINECS/ELINCS 265-113-1
Uses: Solvent for lubricants, textiles (coating of fibers)
Features: Desulfurized (< 2 ppm S); low aromatics; exc. thermal stability; UV resist.; high flash pt.; high visc.
Properties: Saybolt 30 color; low odor; dens. 812 kg/m^3 (15 C); visc. 7 mm^2/s (20 C); vapor pressure < 0.01 mbar (20 C); i.b.p. 277 C; pour pt. -20 C; aniline pt. 93 C; flash pt. (PM) 135 C; ref. index 1.4502 (20 C); surf. tens. 28.70 dynes/cm (20 C); 0.01% aromatics; 0 ppm benzene
Environmental: Biodeg.

Hydroseal G232H [TotalFina-Elf]
Chem. Descrip.: Petroleum distillates, acid-treated middle (C13-15)
CAS 64742-13-8; EINECS/ELINCS 265-113-1
Uses: Solvent for lubricants, material lubrication, degreasing, electro-discharge machining
Features: Desulfurized (< 2 ppm S); low aromatics; exc. thermal stability; UV resist.; high flash pt.; high visc.
Properties: Saybolt 30 color; low odor; dens. 812 kg/m^3 (15 C); visc. 3.7 mm^2/s (20 C); vapor pressure < 0.01 mbar (20 C); i.b.p. 235 C; pour pt. 20 C; aniline pt. 80 C; flash pt. (PM) 102 C; ref. index 1.4470 (20 C); surf. tens. 28.20 dynes/cm (20 C); 0.01% aromatics; 0 ppm benzene
Environmental: Biodeg.

Hydroseal G240H [TotalFina-Elf]
Chem. Descrip.: Petroleum distillates, acid-treated middle (C13-15)
CAS 64742-13-8; EINECS/ELINCS 265-113-1
Uses: Solvent for lubricants
Features: Desulfurized (< 2 ppm S); low aromatics; exc. thermal stability; UV resist.; high flash pt.; high visc.
Properties: Saybolt 30 color; low odor; dens. 810 kg/m^3 (15 C); visc. 4.5 mm^2/s (20 C); vapor pressure < 0.01 mbar (20 C); i.b.p. 248 C; pour pt. -40 C; aniline pt. 84 C; flash pt. (PM) 112 C; ref. index 1.4467 (20 C); surf. tens. 28.20 dynes/cm (20 C); 0.01% aromatics; 0 ppm benzene
Environmental: Biodeg.

Hydroseal G250H [TotalFina-Elf]
Chem. Descrip.: Petroleum distillates, acid-treated middle (C14-19)
CAS 64742-13-8; EINECS/ELINCS 265-113-1
Uses: Solvent for electric-discharge machining
Features: Desulfurized (< 2 ppm S); low aromatics; exc. thermal stability; UV resist.; high flash pt.; high visc.
Properties: Saybolt 30 color; low odor; dens. 815 kg/m^3 (15 C); visc. 5.75 mm^2/s (20 C); vapor pressure < 0.01 mbar (20 C); i.b.p. 255 C; pour pt. 27 C; aniline pt. 91 C; flash pt. (PM) 122 C; ref. index 1.4500 (20 C); surf. tens. 28.70 dynes/cm (20 C); 0.01% aromatics; 0 ppm benzene
Environmental: Biodeg.

Hydroseal G400H [TotalFina-Elf]
Chem. Descrip.: Petroleum distillates, acid-treated middle (C16-21)
CAS 64742-13-8; EINECS/ELINCS 265-113-1
Uses: Solvent for lubricants, textiles, metal processing
Features: Desulfurized (< 2 ppm S); low aromatics; exc. thermal stability; UV resist.; high flash pt.; high visc.
Properties: Saybolt 30 color; low odor; dens. 8915 kg/m^3 (15 C); visc. 11 mm^2/s (20 C); vapor pressure < 0.01 mbar (20 C); i.b.p. 305 C; pour pt. 0 C; aniline pt. 101 C; flash pt. (PM) 160 C; ref. index 1.4510 (20 C); surf. tens. 29.30 dynes/cm (20 C); 0.01% aromatics; 0 ppm benzene
Environmental: Biodeg.

Hydrosol 75/95 N [TotalFina-Elf]
Chem. Descrip.: Naphtha, hydrotreated light (C6-7)
CAS 64742-49-0; EINECS/ELINCS 265-151-9
Uses: Solvent for adhesives, degreasing
Properties: Saybolt 30 color; dens. 760 kg/m^3 (15 C); visc. 0.85 mm^2/s (20 C); vapor pressure 90 mbar (20 C); i.b.p. 82 C; aniline pt. 39 C; flash pt. (Abel) < 0 C; ref. index 1.4185 (20 C); surf. tens. 25.00 dynes/cm (20 C); 0.01% aromatics; < 1 ppm benzene; 2% n-hexane

Hydrosol 100/130 N [TotalFina-Elf]
Chem. Descrip.: Naphtha, hydrotreated light (C7-9)
CAS 64742-49-0; EINECS/ELINCS 265-151-9
Uses: Solvent for paints, varnishes, rubber processing, degreasing
Properties: Saybolt 30 color; dens. 765 kg/m^3 (15 C); visc. 0.9 mm^2/s (20 C); vapor pressure 30 mbar (20 C); i.b.p. 100 C; aniline pt. 47 C; flash pt. (Abel) < 0 C; ref. index 1.4210 (20 C); surf. tens. 25.50 dynes/cm (20 C); 0.01% aromatics; < 1 ppm benzene; 0.01% n-hexane

Hydrosol Essence 60/95 [TotalFina-Elf]
Chem. Descrip.: Naphtha, hydrotreated light (C6-7)
CAS 64742-49-0; EINECS/ELINCS 265-151-9
Uses: Solvent for glues, adhesives, household prods., industrial cleaning, spot removal
Properties: Saybolt 30 color; dens. 680 /m^3 (15 C); visc. 0.5 mm^2/s (20 C); vapor pressure 140 mbar (20 C); i.b.p. 62 C; aniline pt. 69 C; flash pt. (Abel) < 0 C; ref. index 1.385 (20 C); surf. tens. 19.50 dynes/cm (20 C); 0.001% aromatics; 3 ppm benzene; 1.5% n-hexane

Hydrosol Essence 80/110 [TotalFina-Elf]
Chem. Descrip.: Naphtha, hydrotreated light (C6-8)
CAS 64742-49-0; EINECS/ELINCS 265-151-9
Uses: Solvent for glues, adhesives, rubber processing, printing
Properties: Saybolt 30 color; dens. 695 kg/m^3 (15 C); visc. 0.7 mm^2/s (20 C); vapor pressure 90 mbar (20 C); i.b.p. 83 C; aniline pt. 68 C; flash pt. (Abel) < 0 C; ref. index 1.391 (20 C); surf. tens. 20.50 dynes/cm (20 C); 0.001% aromatics; < 1 ppm benzene; 1% n-hexane

Hydrosol Essence 100/140 [TotalFina-Elf]
Chem. Descrip.: Naphtha, hydrotreated light (C7-9)
CAS 64742-49-0; EINECS/ELINCS 265-151-9
Uses: Solvent for rubber processing
Properties: Saybolt 30 color; dens. 730 kg/m^3 (15 C); visc. 0.7 mm^2/s (20 C); vapor pressure 30 mbar (20 C); i.b.p. 103 C; aniline pt. 60 C; flash pt. (Abel) < 0 C; ref. index 1.405 (20 C); surf. tens. 23.00 dynes/cm (20 C); 0.001% aromatics; < 1 ppm benzene; 0.01% n-hexane

Hydrosol Essence A [TotalFina-Elf]
Chem. Descrip.: Naphtha, hydrotreated light (C5-7)
CAS 64742-49-0; EINECS/ELINCS 265-151-9
Uses: Solvent for household prods., industrial cleaning, stain removal
Properties: Saybolt 30 color; dens. 660 /m^3 (15 C); visc. 0.5 mm^2/s (20 C); vapor pressure 150 mbar (20 C); i.b.p. 45 C; aniline pt. 67 C; flash pt. (Abel) < 0 C; ref. index 1.374 (20 C); surf. tens. 18.10 dynes/cm (20 C); 0.001% aromatics; 3 ppm benzene; 2% n-hexane

Hydrosol Essence C [TotalFina-Elf]
Chem. Descrip.: Naphtha, hydrotreated light (C6-7)
CAS 64742-49-0; EINECS/ELINCS 265-151-9
Uses: Solvent for glues, adhesives, household cleaners, rubber mfg.; fuel for catalytic burners
Properties: Saybolt 30 color; dens. 700 /m^3 (15 C); visc. 0.6 mm^2/s (20 C); vapor pressure 110 mbar (20 C); i.b.p. 73 C; aniline pt. 58 C; flash pt. (Abel) < 0 C; ref. index 1.392 (20 C); surf. tens. 21.10 dynes/cm (20 C); 0.001% aromatics; 3 ppm benzene; 25% n-hexane

Hydrosol Essence E<5 [TotalFina-Elf]
Chem. Descrip.: Naphtha, hydrotreated light (C7-8)
CAS 64742-49-0; EINECS/ELINCS 265-151-9
Uses: Solvent for paints, varnishes, rubber processing, household prods., industrial cleaning
Properties: Saybolt 30 color; dens. 725 kg/m^3 (15 C); visc. 0.7 mm^2/s (20 C); vapor pressure 30 mbar (20 C); i.b.p. 103 C; aniline pt. 59 C; flash pt. (Abel) < 0 C; ref. index 1.404 (20 C); surf. tens. 22.60 dynes/cm (20 C); 0.001% aromatics; < 1 ppm benzene; 0.01% n-hexane

Hydrosol Essence F<5 [TotalFina-Elf]
Chem. Descrip.: Naphtha, hydrotreated light (C7-9)
CAS 64742-49-0; EINECS/ELINCS 265-151-9
Uses: Solvent for paints, varnishes, rubber processing, household prods., industrial cleaning, printing, degreasing, dry cleaning
Properties: Saybolt 30 color; dens. 735 kg/m^3 (15 C); visc. 0.8 mm^2/s (20 C); vapor pressure 20 mbar (20 C); i.b.p. 105 C; aniline pt. 60 C; flash pt. (Abel) < 0 C; ref. index 1.409 (20 C); surf. tens. 23.50 dynes/cm (20 C); 0.001% aromatics; < 1 ppm benzene; 0.01% n-hexane

Hydrosol Heptane [TotalFina-Elf]
Chem. Descrip.: Heptane
CAS 142-82-5; EINECS/ELINCS 205-563-8
Uses: Solvent for glues, adhesives, rubber processing, household prods., industrial cleaning, tires
Properties: Saybolt 30 color; dens. 695 /m^3 (15 C); visc. 0.6 mm^2/s (20 C);

vapor pressure 70 mbar (20 C); i.b.p. 90 C; aniline pt. 68 C; flash pt. (Abel) < 0 C; ref. index 1.390 (20 C); surf. tens. 20.50 dynes/cm (20 C); 0.001% aromatics; 3 ppm benzene; 0.5% n-hexane

Hydrosol Hexane [TotalFina-Elf]
Chem. Descrip.: Hexane, branched and linear
CAS 92112-69-1; EINECS/ELINCS 295-570-2
Uses: Solvent for glues, adhesives, oil seed extraction, extraction of essential oils for perfumes, process fluids
Properties: Saybolt 30 color; dens. 670 /m^3 (15 C); visc. 0.5 mm^2/s (20 C); vapor pressure 200 mbar (20 C); i.b.p. 64 C; aniline pt. 67 C; flash pt. (Abel) < 0 C; ref. index 1.377 (20 C); surf. tens. 19.30 dynes/cm (20 C); 0.001% aromatics; 3 ppm benzene; 38% n-hexane

Hydrosol Isohexane [TotalFina-Elf]
Chem. Descrip.: Naphtha, hydrotreated light (C6)
CAS 64742-49-0; EINECS/ELINCS 265-151-9
Uses: Solvent for glues, adhesives, fat extraction, aerosols, process fluids
Properties: Saybolt 30 color; dens. 660 /m^3 (15 C); visc. 0.5 mm^2/s (20 C); vapor pressure 250 mbar (20 C); i.b.p. 50 C; aniline pt. 70 C; flash pt. (Abel) < 0 C; ref. index 1.373 (20 C); surf. tens. 18.40 dynes/cm (20 C); 0.001% aromatics; 3 ppm benzene; 2.5% n-hexane

Hydrosol Pentane [TotalFina-Elf]
Chem. Descrip.: n-Pentane
CAS 109-66-0; EINECS/ELINCS 203-692-4
Uses: Solvent for chemicals, expandable PS, aerosols
Properties: Saybolt 30 color; dens. 630 kg/m^3 (15 C); visc. 0.4 mm^2/s (20 C); vapor pressure 650 mbar (20 C); i.b.p. 34 C; flash pt. (Abel) < 0 C; ref. index 1.358 (20 C); surf. tens. 17.10 dynes/cm (20 C); 0.001% aromatics; 3 ppm benzene; 0.01% n-hexane

Hydrothene™ [Amity UK]
Uses: Precision cleaning solvent for circuit board cleaning in ultrasonic and batch operations, removal of solder flux and light oils from electronic components
Features: Alternative to CFC, HFC, HCFC; high performance; mod. dry rate; nonflamm.; exc. solvency; cost effective; compat. with all plastics incl. PC and PS; patent pending
Properties: Colorless clear liq.; sl. ether odor; completely water-sol.; sp.gr. 0.96-0.98; vapor pressure 0.30 mm Hg; b.p. 150 C
Toxicology: LD50 (oral, rat) 2000-2500 mg/kg, (skin, rabbit) > 2000 mg/kg; low toxicity by ing.; may cause mod. eye irritation, sl. corneal injury; prolonged/repeated skin contact may cause sl. irritation; not classified as carcinogenTSCA listed
Environmental: Environmentally safe; easily recycled; zero ozone depletion potential; HAPs-free; pract. nontoxic to aquatic organisms (LC50 > 100 mg/l in most sensitive species)
Precaution: Avoid high temps.; incompat. with oxidizers
Hazardous decomp. prods.: Releases toxic fumes in fire situations; combustion may produce CO, CO_2
Storage: Keep containers tightly closed when not in use; store in carbon steel, SS, Teflon

I

Igepal® OD-410 [Rhodia HPCII; Rhodia HPCII France]
Chem. Descrip.: Phenoxyethanol, phenoxyglycol, and phenoxydiglycol
Uses: Solvent for vinyl, phenolic, polyester, alkyd, NC, and cellulose acetate resins; ingred. of metal cleaners, paint strippers, and cleaning compds. where solvs. and solv. boosters are required; vehicle for inks
Properties: Liq.; nonionic; 100% act.

Imwitor® 408 [Sasol N. Am.]
Chem. Descrip.: Propylene glycol caprylate
CAS 31565-12-5
Uses: Coemulsifier, solubilizer, absorp. promoter for cosmetics, topical and rectal pharmaceuticals; solvent for lipophilic drugs
Properties: Sl. yel. liq.; acid no. 2 max.; iodine no. 1 max.; sapon. no. 250-275; 50% monoglycerides
Environmental: Environmentally compat.

Imwitor® 412 [Sasol N. Am.]
Chem. Descrip.: Propylene glycol laurate
CAS 142-55-2; EINECS/ELINCS 205-542-3
Uses: Coemulsifier, solubilizer, absorp. promoter for cosmetics, topical and rectal pharmaceuticals; solvent for lipophilic drugs
Properties: Sl. yel. solid; acid no. 2 max.; iodine no. 1 max.; sapon. no. 220-235; nonionic; 50% monoglycerides
Environmental: Environmentally compat.

Imwitor® 742 [Sasol Germany; Sasol N. Am.]
Chem. Descrip.: Caprylic/capric glycerides
CAS 85409-09-2; EINECS/ELINCS 287-075-5
Uses: Plasticizer for hard fats; solvent for lipophilic ingreds., active substances; cosmetic emollient; coemulsifier, solubilizer, carrier for lipophilic pharmaceuticals (oral, topical, rectal); compressing aid, lubricant in tablets; compressing aid in prod. of compressed sweets; dispersant; surfactant; absorp. promoter; bacteriostat; filler for hard and soft gelatin capsules; release agent and bioavailability enhancer for drugs
Regulatory: German pharmacopoeia compliance
Properties: Wh-ylsh. cryst. solid, sl. coconut odor, fatty bitter taste; sol. in ethanol, acetone, ether, heptane, hexane; misc. with oils and fats; m.p. 25 C; HLB 3-4; acid no. 2 max.; iodine no. 1 max.; sapon. no. 250-280; nonionic; 45% min. monoglycerides
Toxicology: LD50 (oral, rat) > 5 g/kg; moderate skin irritant (undiluted)
Environmental: Environmentally compat.

Incrocas 30 [Croda Inc]
Chem. Descrip.: PEG-30 castor oil
CAS 61791-12-6
Uses: Emulsifier, solubilizer, emollient, superfatting agent, lubricant for personal care prods., detergents, metalworking fluids, insecticides, herbicides, household prods.; lubricant, antistat, softener, dye leveling agent for textiles; lime soap dispersant in alkaline scouring systems; surfactant, solvent, wetting agent, emollient, emulsifier, lubricant for topical pharmaceuticals
Features: Tends to yield clear sol'ns.
Properties: Pale yel. liq.; sol. in water, ethanol, oleyl alcohol, naphtha, MEK, oleic acid, trichloroethylene; HLB 11.7; cloud pt. 35-40 C (1% in 10% brine); sapon. no. 72-82; pH 6.0-7.5 (3% aq.); surf. tens. 41.5 dynes/cm (0.1% DW); nonionic; 100% act.
Use Level: 0.5-5%

Incrocas 40 [Croda Inc]
Chem. Descrip.: PEG-40 castor oil
CAS 61791-12-6
Uses: Emulsifier, solubilizer, emollient, superfatting agent, lubricant for personal care prods., detergents, metalworking fluids, insecticides, herbicides, household prods.; lubricant, antistat, softener, dye leveling agent for textiles; surfactant, solvent, wetting agent, emollient, emulsifier, lubricant for topical and parenteral pharmaceuticals
Properties: Pale yel. liq.; sol. @ 10% water, ethanol, oleyl alcohol, naphtha, MEK, oleic acid, trichloroethylene; HLB 13.0; cloud pt. 50 C (1% in 10% brine); sapon. no. 60-65; pH 6.0-7.5 (3% aq.); surf. tens. 40.90 dynes/cm (0.1% DW); nonionic; 100% act.
Use Level: 0.5-5%

Incromectant AMEA-70 [Croda Inc; Croda Oleochems.]
Chem. Descrip.: Acetamide MEA
CAS 142-26-7; EINECS/ELINCS 205-530-8
Uses: Solvent, humectant, moisturizer for cosmetics; clarifying agent for shampoos; conditioner for creme rinses
Properties: Clear liq.; sp.gr. 1.07-1.17; dens. 9.3 lb/gal; pH 6.0-8.5; nonionic; 70% min. act.
Use Level: 0.5-15%

Intravon® SOL-W [Yorkshire Am.]
Chem. Descrip.: Aromatic solv./surfactant blend
Uses: Fast wetting agent/penetrant, scouring detergent for syn. and natural fibers and their blends, in one-bath dyeing/scouring processes; oil solubilizer
Features: Solv.-based; alkali stable
Properties: Off-wh. to yel. gel; self-emulsifiable in water; flash pt. (PMCC) 52 C; pH 6.0 ± 0.5; anionic/nonionic
Toxicology: Defatting to skin on prolonged contact
Environmental: Biodeg.

Isosol RS Solvent [Exxon Canada]
Chem. Descrip.: Naphtha, hydrotreated light
CAS 64742-49-0; EINECS/ELINCS 265-151-9
UN 1268
Uses: Solvent; diluent; chem. feedstock; fuel
Regulatory: Canada DSL listed; NPRI reportable [toluene (10%), n-hexane (19.7%)]
Properties: Colorless clear liq.; mild petrol. odor; negligible sol. in water; m.w. 89; sp.gr. 0.693 (15.5 C); visc. 0.49 cSt; vapor pressure 18.45 kPa (20 C); f.p. -54 C; b.p. 66-103 C; flash pt. (TCC) -18 C; autoignition temp. ≈ 254 C; 100% act.
Toxicology: REL TWA 138 ppm; inh. of high vapor concs. may cause eye/respiratory tract irritation, headaches, dizziness, anesthesia, drowsiness, unconsciousness, other CNS effects, death; sl. eye irritant; frequent/prolonged skin contact may cause irritation, dermatitis; if aspirated, may cause mild to severe pulmonary injury, death; contains hexane (may cause peripheral nervous system damage on chronic exposure)
Environmental: Prevent entry into sewers, watercourses, low areas
Precaution: Extremely flamm. liq.; flamm. limits 1-7 vol.%; readily ignites @ normal temps.; volatile; gives off invisible vapors; liq. or vapor may settle in low areas; distant ignition/explosion possible; electrostatic accumulation hazard; use proper grounding procedures; incompat. with strong oxidizing agents
Hazardous decomp. prods.: Combustion: toxic gases, fumes, smoke, CO
Storage: Store in cool, well-ventilated area away from incompat. materials, heat, open flame, ignition sources, direct sunlight; keep container

closed

i-PGE [INEOS Oxide]
Chem. Descrip.: 2-Isopropoxyethanol
CAS 109-59-1; EINECS/ELINCS 203-685-6
Uses: Solvent for resins (NC, acrylic, epoxy, polyamide, alkyd), oils, waxes, fats, dyestuffs, insecticides, herbicides, grease/grime, surf. coatings, lacquers, enamels, waterborne coatings, phenolic varnishes, epoxy coatings, non-grain-raising wood stains; gloss aid, flow aid in lacquers; dye solvent/stabilizer/penetration aid, color enhancer for textiles, leather, printing; mutual solvent, coupling agent in sol. oils, metal/glass cleaners, varnish removers; diluent in hydraulic brake fluids
Features: Very stable; low f.p.; relatively low vapor pressure
Properties: Pt-Co 5 max. color; mild odor; misc. with wide range of solvs., water; m.w. 104.15; dens. 0.902-0.905 g/ml (20 C); visc. 2.5 mPa·s (20 C); vapor pressure 3.5 hPa (20 C); i.b.p. 139.5 C min. (101.3 kPa); pour pt. -90 C; flash pt. (CC) 49 C; autoignition temp. 240 C; ref. index 1.4082 (20 C); 99% min. purity
Precaution: Flamm. limits 1.6-13.0 vol.% (20 C); not a significant flamm. hazard @ normal storage temps.; very reactive with zinc (hydrogen may be produced); contact with zinc, tin, copper/alloys can cause discoloration

Isane IP 130 [TotalFina-Elf]
Chem. Descrip.: C7-10 isoalkane
CAS 90622-56-3; EINECS/ELINCS 292-458-5
Uses: Solvent for inks, metalworking, cutting fluids, cosmetics, hand soaps, sprays, photocopier prods., household cleaning prods., low odor alkyd paint resins; diluent for paints; odorless fuel for petrol. burners
Features: Desulfurized (< 2 ppm S); low surf. tens.; high interfacial tens.; low elec. conduct.; very low pour pt.
Regulatory: FDA 21CFR §176.210, 172.882, 172.884, 178.3530, 178.3650
Properties: Saybolt 30 color; dens. 728 kg/m^3; visc. 0.87 mm^2/s (20 C); vapor pressure 11 mbar (20 C); i.b.p. 134 C; pour pt. -60 C; aniline pt. 75 C; flash pt. (PM) 21 C; ref. index 1.4091 (20 C); surf. tens. 23.20 dynes/cm (20 C); 0.0001% aromatics; < 1% benzene

Isane IP 155 [TotalFina-Elf]
Chem. Descrip.: C9-12 isoalkane (C9-11)
CAS 90622-57-4; EINECS/ELINCS 292-459-0
Uses: Solvent for degreasing, inks, metalworking, cutting fluids, sprays, photocopier prods., household cleaning prods., low odor alkyd paint resins; diluent for paints; odorless fuel for petrol. burners
Features: Desulfurized (< 2 ppm S); low surf. tens.; high interfacial tens.; low elec. conduct.; very low pour pt.
Regulatory: FDA 21CFR §176.210, 172.882, 172.884, 178.3530, 178.3650
Properties: Saybolt 30 color; dens. 745 kg/m^3; visc. 1.20 mm^2/s (20 C); vapor pressure 3 mbar (20 C); i.b.p. 160 C; pour pt. -60 C; aniline pt. 78 C; flash pt. (PM) 45 C; ref. index 1.4160 (20 C); surf. tens. 23.90 dynes/cm (20 C); 0.0001% aromatics; < 1% benzene

Isane IP 165 [TotalFina-Elf]
Chem. Descrip.: C9-12 isoalkane (C10-12)
CAS 90622-57-4; EINECS/ELINCS 292-459-0
Uses: Solvent for degreasing, inks, metalworking, cutting fluids, sprays, photocopier prods., household cleaning prods., low odor alkyd paint resins, wood preservatives; diluent for paints; odorless fuel for petrol. burners
Features: Desulfurized (< 2 ppm S); low surf. tens.; high interfacial tens.; low elec. conduct.; very low pour pt.
Regulatory: FDA 21CFR §176.210, 172.882, 172.884, 178.3530, 178.3650
Properties: Saybolt 30 color; dens. 755 kg/m^3; visc. 1.4 mm^2/s (20 C); vapor pressure 2 mbar (20 C); i.b.p. 166 C; pour pt. -60 C; aniline pt. 80 C; flash pt. (PM) 48 C; ref. index 1.4230 (20 C); surf. tens. 24.50 dynes/cm (20 C); 0.0001% aromatics; < 1% benzene

Isane IP 175 [TotalFina-Elf]
Chem. Descrip.: C9-12 isoalkane (C10-12)
CAS 90622-57-4; EINECS/ELINCS 292-459-0
Uses: Solvent for degreasing, inks, metalworking, cutting fluids, cosmetics, hand soaps, sprays, photocopier prods., household cleaning prods., low odor alkyd paint resins, wood preservatives, lubricants; diluent for paints; odorless fuel for petrol. burners
Features: Desulfurized (< 2 ppm S); low surf. tens.; high interfacial tens.; low elec. conduct.; very low pour pt.
Regulatory: FDA 21CFR §176.210, 172.882, 172.884, 178.3530, 178.3650
Properties: Saybolt 30 color; dens. 762 kg/m^3; visc. 1.67 mm^2/s (20 C); vapor pressure 1 mbar (20 C); i.b.p. 180 C; pour pt. -60 C; aniline pt. 81 C; flash pt. (PM) 62 C; ref. index 1.4260 (20 C); surf. tens. 25.50 dynes/cm (20 C); 0.0001% aromatics; < 1% benzene

Isane IP 185 [TotalFina-Elf]
Chem. Descrip.: C9-12 isoalkane (C11-14)
CAS 90622-57-4; EINECS/ELINCS 292-459-0
Uses: Solvent for degreasing, inks, metalworking, cutting fluids, sprays, photocopier prods., household cleaning prods., low odor alkyd paint resins, wood preservatives, lubricants; diluent for paints; odorless fuel for petrol. burners
Features: Desulfurized (< 2 ppm S); low surf. tens.; high interfacial tens.; low elec. conduct.; very low pour pt.
Regulatory: FDA 21CFR §176.210, 172.882, 172.884, 178.3530, 178.3650
Properties: Saybolt 30 color; dens. 764 kg/m^3; visc. 1.8 mm^2/s (20 C); vapor pressure < 1 mbar (20 C); i.b.p. 186 C; pour pt. -60 C; aniline pt. 81 C; flash pt. (PM) 65 C; ref. index 1.4270 (20 C); surf. tens. 25.50 dynes/cm (20 C); 0.0001% aromatics; < 1% benzene

Isodragol 2/050300 [Dragoco]
Chem. Descrip.: Triisononanoin
CAS 56554-53-1; EINECS/ELINCS 260-257-1
Uses: Skinfeel enhancer in skin care, decorative cosmetics; pigment wetting agent; solvent for waxes; lubricant, color enhancer, softener, stabilizer in lipsticks
Features: Nongreasy

Isoform [Dow]
Chem. Descrip.: Perchloroethylene
CAS 127-18-4; EINECS/ELINCS 204-825-9
Uses: Solvent for chem. processing, esp. to reactivate catalyst where the catalyst is not sensitive to oxygen- and nitrogen-contg. compds.
Features: Reforming grade
Regulatory: SARA §311/312 acute/chronic health hazard, §313 reportable; Calif. Prop. 65 reportable
Properties: Colorless liq.; irritating odor; sol. 0.015 g/100 g in water; sp.gr. 1.619; vapor pressure 13 mm Hg (20 C); b.p. 121.1 C; flash pt. (TCC) none; 99.9% act.
Toxicology: ACGIH TLV TWA 25 ppm, STEL 100 ppm; harmful by inh., possibly fatal; may cause dizziness, nasal irritation, nausea, irregular heartbeat, unconsciousness; may cause eye pain, sl. transient irritation; skin irritant; may cause skin burns; extremely low toxicity by ing.; swallowing larger amts. may cause injury; if aspirated, may cause injury to other body systems; may cause CNS/anesthetic/narcotic/liver/kidney effects; possible carcinogen; TSCA listed
Environmental: ThOD 0.19 p/p; low bioconcentration potential; LC50 (daphnia magna) 3.2-123 mg/l, (rainbow trout) 4.8-5.8 mg/l; mod. toxic to aquatic organisms; prevent contamination of soil, surf. water, ground water
Precaution: Does not burn; container may rupture in fire; vapors are heavier than air, may accumulate in low areas; denser than water; avoid direct sunlight, UV sources, open flames, welding arcs, high temp. sources; avoid contact with amines, strong bases, strong oxidizers, metals (Al powd., Mg powd., K, Na, Zn powd.)
Hazardous decomp. prods.: Hydrogen chloride, trace amts. of chlorine, phosgene; combustion: toxic fumes, hydrogen chloride, trace amts. of phosgene, chlorine
NFPA: Health 2, Flammability 0, Reactivity 0
Storage: Store in dry place; keep containers tightly closed when not in use

Isohexane LNH [Shell Netherlands]
Chem. Descrip.: Hexane
CAS 64742-49-0; EINECS/ELINCS 265-151-9
UN 1208
Uses: Solvent for adhesives, rubber; extraction solvent
Features: Narrow boiling range; very low aromatics and olefins content; does not contain detectable quantities of polycyclic aromatics, heavy metals, or chlorinated compds.
Regulatory: EINECS, Canada DSL, Australia AICS, China IECSC, Japan MITI, Philippines PICCS, Korea TCCL listed
Properties: Colorless liq.; paraffinic odor; negligible sol. in water; m.w. 86; dens. 665 kg/m^3 (15 C); kinematic visc. 0.44 mm^2/s; vapor pressure 24,000 Pa (20 C); i.b.p. 57 C; b.p. 57-63 C; pour pt. -150 C; flash pt. -33 C; autoignition temp. 405 C; KB value 28; surf. tens. 17.2 mN/m (20 C)
Toxicology: Expected to be of low toxicity by ing., inh., skin contact; aspi-

ration into lungs may cause chem. pneumonitis which can be fatal; narcotic at high vapor concs.; causes CNS depression, headache, dizziness, nausea, skin dryness; skin irritant; prolonged/repeated contact may cause dermatitis; TSCA listed
Environmental: Inherently biodeg.; potential to bioaccumulate; toxic to aquatic organisms; may cause long-term adverse effects in aquatic environment; prevent contamination of soil, water; prevent entry into drains, ditches, rivers
Precaution: Highly flamm.; LEL 1 vol.%; UEL 7.4 vol.%; vapor readily forms explosive mixt. in air; electrostatic charges may be generated; prevent static discharge; ground equip.
Hazardous decomp. prods.: Combustion: CO
Storage: 12 mos shelf life under proper storage; store @ ambient temps. in tightly closed container in well-ventilated area, away from direct sunlight, other heat/ignition sources

Isopar® C [ExxonMobil; Exxon Canada; ExxonMobil Europe]
Chem. Descrip.: Light alkylation naphtha
CAS 64741-66-8; EINECS/ELINCS 265-068-8
UN 1268
Uses: Solvent, thinner, cleaner for solv.-based architectural, automotive OEM and refinish coatings, odorless alkyd and acrylic paints, plastics, food pkg.; diluent; chem. feedstock; fuel
Features: Minimizes VOC; virtually HAPs-free; extremely low surf. tens. to improve flow and wetting
Regulatory: FDA 21CFR §172.882, 178.3530; 40CFR §180.1001(c)(e); SARA §311/312 fire hazard, §313 nonreportable; Canada DSL listed; NPRI nonreportable
Properties: Colorless clear liq.; mild hydrocarbon odor; negligible sol. in water; m.w. 114; sp.gr. 0.70 (15.6 C); dens. 5.83 lb/gal (60 F); visc. 0.5 cP; vapor pressure 63 mm Hg (20 C); b.p. 98-104 C; pour pt. -57 C; flash pt. > -8 C; autoignition temp. 442 C; surf. tens. 20.3 dynes/cm; KB value 27; 100% act.
Toxicology: OEL 400 ppm; high vapor concs. may cause eye/respiratory irritation, headaches, dizziness, anesthesia, drowsiness, unconsciousness, other CNS effects, death; sl. eye irritant; prolonged/repeated skin contact may cause irritation, dermatitis; if swallowed, may enter lungs and cause lung damage
Environmental: Minimize effects on groundwater; prevent entry into sewers, watercourses, low areas
Precaution: Extremely flamm. liq. and vapor; flamm. limits in air 0.9-6.3 vol.%; liq. can release vapors below ambient temps., which readily form flamm. mixts.; volatile; gives off invisible vapors; vapors may settle; distant ignition/explosion possible; avoid heat, sparks, flame; static accumulation hazard; ground and bond containers; incompat. with strong oxidizers
Hazardous decomp. prods.: Combustion: CO, CO_2, water; does not dec. @ ambient temps.
NFPA: Health 1, Flammability 3, Reactivity 0
Storage: Store in cool, well-ventilated area in bonded/grounded containers away from ignition sources, direct sunlight; keep container closed

Isopar® E Fluid [ExxonMobil; Exxon Canada; ExxonMobil Europe]
Chem. Descrip.: Light alkylation naphtha
CAS 64741-66-8; EINECS/ELINCS 265-068-8
UN 1268
Uses: Solvent for coatings, odorless alkyd and acrylic paints; thinner; cleaner; diluent; chem. feedstock; fuel; food pkg.
Features: Minimizes VOC; virtually HAPs-free; extremely low surf. tens. to improve flow and wetting
Regulatory: FDA 21CFR §172.882, 178.3530, 573.220; 40CFR §180.1001(c)(e); Canada DSL listed; NPRI nonreportable
Properties: Colorless clear liq.; odorless; sol. < 0.01% in water; m.w. 123; sp.gr. 0.72 (15.6 C); dens. 6.02 lb/gal (60 F); visc. 0.6 cP; vapor pressure 36 mm Hg (20 C); f.p. -63 C; b.p. 118-137 C; flash pt. 45 F; surf. tens. 22.1 dynes/cm; KB value 29; 100% act.; 99% paraffin, 1% cycloparaffin
Toxicology: REL TWA 281 ppm; low toxicity; inh. of high vapor concs. may cause eye/respiratory tract irritation, headaches, dizziness, anesthesia, drowsiness, unconsciousness, other CNS effects, death; sl. eye irritant; frequent/prolonged skin contact may cause irritation, dermatitis; if aspirated, may cause mild to severe pulmonary injury, death
Environmental: Prevent entry into sewers, watercourses, low areas
Precaution: Flamm. liq.; flamm. limits 0.9-6.2 vol.%; may release vapors that form flamm. mixts. at or above flash pt.; volatile; gives off invisible vapors; liq. or vapor may settle in low areas; distant ignition/explosion possible; electrostatic accumulation hazard; use proper grounding procedures; incompat. with strong oxidizing agents
Hazardous decomp. prods.: None
Storage: Store in cool, well-ventilated area away from incompat. materials, heat, open flame, ignition sources, direct sunlight; keep container closed

Isopar® G Fluid [ExxonMobil; Exxon Canada; ExxonMobil Europe]
Chem. Descrip.: Naphtha, hydrotreated heavy
CAS 64742-48-9; EINECS/ELINCS 265-150-3
UN 1268
Uses: Solvent, diluent, carrier for personal care sprays; solvent for coatings, odorless alkyd/acrylic paints, thinners, cleaners, degreasers; flow aid; wetting agent; chem. feedstock; fuel; food-contact paper, rubber articles; defoamer in food-contact paper/paperboard
Features: Minimizes VOC; virtually HAPs-free; extremely low surf. tens.
Regulatory: FDA 21CFR §172.882, 172.884, 176.180, 176.210, 177.2600, 178.3530, 178.3570, 178.3620(b), 178.3650; 40CFR §180.1001(c)(e); SARA nonreportable; Canada DSL listed; NPRI nonreportable
Properties: Liq.; faint petrol. odor; sol. < 0.01% in water; m.w. 145; sp.gr. 0.75 (15.6 C); dens. 6.23 lb/gal (60 F); visc. 1.0 cP; vapor pressure 2.3 mm Hg (20 C); f.p. -57 C; b.p. 160-176 C; flash pt. (TCC) > 40 C; autoignition temp. 365 C; surf. tens. 23.8 dynes/cm; KB value 28; 100% act.; paraffin (95%), cycloparaffin (5%)
Toxicology: REL TWA 196 ppm; inh. of high vapor concs. may cause eye/respiratory tract irritation, headaches, dizziness, anesthesia, drowsiness, unconsciousness, other CNS effects, death; sl. eye irritant; frequent/prolonged skin contact may cause irritation, dermatitis; if aspirated, may cause mild to severe pulmonary injury, death
Environmental: Prevent entry into sewers, watercourses, low areas
Precaution: Combustible liq.; flamm. limits 0.7-5.6 vol.%; volatile; gives off invisible vapors; liq. or vapor may settle in low areas; distant ignition/explosion possible; electrostatic accumulation hazard; use proper grounding procedures; incompat. with strong oxidizing agents
Hazardous decomp. prods.: None
Storage: Store in cool, well-ventilated area away from incompat. materials, heat, open flame, ignition sources, direct sunlight; keep container closed

Isopar® H Fluid [ExxonMobil; Exxon Canada; ExxonMobil Europe]
Chem. Descrip.: Naphtha, hydrotreated heavy
CAS 64742-48-9; EINECS/ELINCS 265-150-3
UN 1268
Uses: Solvent, diluent, carrier for personal care sprays; solvent for coatings, odorless alkyd/acrylic paints , thinners, cleaners, degreasers, agric. pesticides/herbicides; solvent, flow aid, wetting agent for coatings; chem. feedstock; fuel; food-contact paper, rubber articles; defoamer in food-contact paper/paperboard
Features: Extremely low surf. tens. to improve flow and wetting; minimizes VOC; virtually HAPs-free
Regulatory: FDA 21CFR §172.882, 172.884, 176.180, 176.210, 177.2600, 178.3530, 178.3570, 178.3620(b), 178.3650; 40CFR §180.1001(c)(e); SARA nonreportable; Canada DSL listed; NPRI nonreportable
Properties: Liq.; faint petrol. odor; sol. < 0.01% in water; m.w. 152; sp.gr. 0.759 (15.6 C); dens. 6.32 lb/gal (60 F); visc. 1.3 cP; vapor pressure 1.0 mm Hg (20 C); f.p. < -63 C; b.p. 178-188 C; flash pt. (TCC) 49 C; autoignition temp. 365 C; surf. tens. 24.1 dynes/cm; KB value 27; 100% act.; paraffin (95%), cycloparaffin (5%)
Toxicology: REL TWA 177 ppm; inh. of high vapor concs. may cause eye/respiratory tract irritation, headaches, dizziness, anesthesia, drowsiness, unconsciousness, other CNS effects, death; sl. eye irritant; frequent/prolonged skin contact may cause irritation, dermatitis; if aspirated, may cause mild to severe pulmonary injury, death
Environmental: Prevent entry into sewers, watercourses, low areas
Precaution: Combustible liq.; flamm. limits 0.7-5.4 vol.%; volatile; gives off invisible vapors; liq. or vapor may settle in low areas; distant ignition/explosion possible; electrostatic accumulation hazard; use proper grounding procedures; incompat. with strong oxidizing agents
Hazardous decomp. prods.: None
Storage: Store in cool, well-ventilated area away from incompat. materials, heat, open flame, ignition sources, direct sunlight; keep container closed

Isopar® I Fluid [ExxonMobil Europe]
Chem. Descrip.: Isoparaffin
Uses: Solvent for odorless/low odor solv.-based decorative coatings, plastics; diluent, cleaner for decorative paints; defoamer
Features: Low surf. tens. for better flow, leveling, and adhesion; reduced VOCs
Properties: Very low odor; dens. 0.768 kg/dm^3 (15 C); visc. 1.23 mPa·s; i.b.p. 188 C; flash pt. 65 C; ref. index 1.428 (20 C); surf. tens. 24.1 mN/m; KB value 28; 0.002% aromatic content

Isopar® J Fluid [ExxonMobil Europe]
Chem. Descrip.: Isoparaffin
Uses: Solvent for odorless/low odor solv.-based decorative coatings, plastics; diluent, cleaner for decorative paints; defoamer
Features: Low surf. tens. for better flow, leveling, and adhesion; reduced VOCs
Properties: Very low odor; dens. 0.766 kg/dm^3 (15 C); visc. 1.23 mPa·s; i.b.p. 185 C; flash pt. 61 C; ref. index 1.427 (20 C); surf. tens. 24.1 mN/m; KB value 28; 0.002% aromatic content

Isopar® K Fluid [ExxonMobil; Exxon Canada; ExxonMobil Europe]
Chem. Descrip.: Naphtha, hydrotreated heavy
CAS 64742-48-9; EINECS/ELINCS 265-150-3
UN 1268
Uses: Solvent, diluent, carrier for personal care sprays; solvent for coatings, odorless alkyd/acrylic paints, thinners, cleaners, degreasers; flow aid; wetting agent; chem. feedstock; fuel; food-contact paper, rubber articles; defoamer in food-contact paper/paperboard
Features: Minimizes VOC; virtually HAPs-free; extremely low surf. tens.
Regulatory: FDA 21CFR §172.882, 172.884, 176.180, 176.210, 177.2600, 178.3530, 178.3570, 178.3620(b), 178.3650; 40CFR §180.1001(c)(e); SARA nonreportable; Canada DSL listed; NPRI nonreportable
Properties: Liq.; faint petrol. odor; sol. < 0.01% in water; m.w. 154; sp.gr. 0.76 (15.6 C); dens. 6.34 lb/gal (60 F); visc. 1.4 cP; vapor pressure 0.9 mm Hg (20 C); f.p. < -60 C; b.p. 177-197 C; flash pt. (TCC) > 54 C; autoignition temp. ≈ 349 C; surf. tens. 24.2 dynes/cm; KB value 27; > 99.9% act.; paraffin (94%), cycloparaffin (6%)
Toxicology: REL TWA 175 ppm; inh. of high vapor concs. may cause eye/respiratory tract irritation, headaches, dizziness, anesthesia, drowsiness, unconsciousness, other CNS effects, death; sl. eye irritant; frequent/prolonged skin contact may cause irritation, dermatitis; if aspirated, may cause mild to severe pulmonary injury, death
Environmental: Prevent entry into sewers, watercourses, low areas
Precaution: Combustible liq.; flamm. limits 1.2-9.3 vol.%; volatile; gives off invisible vapors; liq. or vapor may settle in low areas; distant ignition/explosion possible; electrostatic accumulation hazard; use proper grounding procedures; incompat. with strong oxidizing agents
Hazardous decomp. prods.: None
Storage: Store in cool, well-ventilated area away from incompat. materials, heat, open flame, ignition sources, direct sunlight; keep container closed

Isopar® K Naphtha [Exxon Canada]
Chem. Descrip.: Naphtha, hydrotreated heavy
CAS 64742-48-9; EINECS/ELINCS 265-150-3
UN 1268
Uses: Solvent; diluent; chem. feedstock; fuel
Regulatory: Canada DSL listed; NPRI nonreportable
Properties: Colorless clear liq.; faint petrol. odor; sol. < 0.01% in water; m.w. 154; sp.gr. 0.76 (15.5 C); visc. 1.84 cSt; vapor pressure 0.1 kPa (20 C); f.p. < -60 C; b.p. 177-17 C; flash pt. (TCC) > 54 C; autoignition temp. ≈ 349 C; > 99.9 vol.%
Toxicology: REL TWA 175 ppm; inh. of high vapor concs. may cause eye/respiratory tract irritation, headaches, dizziness, anesthesia, drowsiness, unconsciousness, other CNS effects, death; sl. eye irritant; frequent/prolonged skin contact may cause irritation, dermatitis; if aspirated, may cause mild to severe pulmonary injury, death
Environmental: Prevent entry into sewers, watercourses, low areas
Precaution: Combustible liq.; flamm. limits 1.2-9.3 vol.%; volatile; gives off invisible vapors; liq. or vapor may settle in low areas; distant ignition/explosion possible; electrostatic accumulation hazard; use proper grounding procedures; incompat. with strong oxidizing agents
Hazardous decomp. prods.: None
Storage: Store in cool, well-ventilated area away from incompat. materials, heat, open flame, ignition sources, direct sunlight; keep container closed

Isopar® L Fluid [ExxonMobil; ExxonMobil Europe]
Chem. Descrip.: Naphtha, hydrotreated heavy
CAS 64742-48-9; EINECS/ELINCS 265-150-3
Uses: Solvent for skin care, coatings, odorless alkyd/acrylic paints, agrochems., thinners, cleaners, and degreasers; chem. feedstock; fuel; food-contact paper, rubber articles; defoamer in food-contact paper/paperboard
Features: Minimizes VOC; virtually HAPs-free; extremely low surf. tens. to improve flow, wetting
Regulatory: FDA 21CFR §172.882, 172.884, 176.180, 176.210, 177.2600, 178.3530, 178.3570, 178.3620(b), 178.3650, 573.680; 40CFR §180.1001(c)(e); SARA nonreportable; Canada DSL listed; NPRI nonreportable
Properties: Colorless clear liq.; faint petrol. odor; sol. < 0.01% in water; m.w. 163; sp.gr. 0.768 (15.6 C); dens. 6.40 lb/gal (60 F); visc. 1.6 cP; vapor pressure 0.8 mm Hg (20 C); f.p. -57 C; b.p. 189-207 C; flash pt. (CC) 62 C; autoignition temp. > 260 C; surf. tens. 25.1 dynes/cm; KB value 27; 100% act.; 89% paraffin, 11% cycloparaffin
Toxicology: REL TWA 171 ppm; inh. of high vapor concs. may cause eye/respiratory tract irritation, headaches, dizziness, anesthesia, drowsiness, unconsciousness, other CNS effects, death; sl. eye irritant; frequent/prolonged skin contact may cause irritation, dermatitis; if aspirated, may cause mild to severe pulmonary injury, death
Environmental: Prevent entry into sewers, watercourses, low areas
Precaution: Combustible liq.; flamm. limits 1.3-8.8 vol.%; volatile; gives off invisible vapors; liq. or vapor may settle in low areas; distant ignition/explosion possible; electrostatic accumulation hazard; use proper grounding procedures; incompat. with strong oxidizing agents
Hazardous decomp. prods.: None
Storage: Store in cool, well-ventilated area away from incompat. materials, heat, open flame, ignition sources, direct sunlight; keep container closed

Isopar® M Fluid [ExxonMobil; Exxon Canada; ExxonMobil Europe]
Chem. Descrip.: Petroleum distillates, hydrotreated light
CAS 64742-47-8; EINECS/ELINCS 265-149-8
Uses: Solvent for hair spray, coatings, odorless alkyd/acrylic paints, antiperspirants, laundry prewash, furniture/floor polish, waterless hand cleaner, agrochems, thinners, cleaners, degreasers; carrier for insecticide actives; dispersant; flow aid; wetting agent; food-contact paper, rubber articles; diluent; chem. feedstock; fuel; defoamer in food-contact paper/paperboard
Features: Complies with criteria for LVP solvent; minimizes VOC; virtually HAPs-free; extremely low surf. tens.
Regulatory: FDA 21CFR §172.882, 172.884, 176.180, 176.210, 177.2600, 178.3530, 178.3570, 178.3620(b), 573.680; 40CFR §180.1001(c)(e); SARA §311/312 fire hazard, §313 nonreportable; Canada DSL listed; NPRI nonreportable
Properties: Water-wh. clear liq.; essentially odorless; m.w. 191; sp.gr. 0.79 (15.6 C); dens. 6.57 lb/gal (60 F); visc. 2.7 cP; vapor pressure < 0.1 mm Hg (20 C); b.p. 223-254 C; flash pt. (PMCC) 80.5 C; autoignition temp. ≈ 338 C; surf. tens. 26.4 dynes/cm; KB value 27; 100% act.; 80% paraffin, 21% cycloparaffin
Toxicology: REL (8 h) 300 ppm; LD50 (oral, rat) > 5 g/kg, (skin, rabbit) > 3.16 g/kg; low acute toxicity; inh. of high vapor concs. may cause eye/respiratory tract irritation, headaches, dizziness, anesthesia, drowsiness, unconsciousness, other CNS effects, death; may cause eye irritation; frequent/prolonged skin contact may cause irritation, dermatitis; if aspirated, may cause mild to severe pulmonary injury, death; TSCA listed
Environmental: Prevent entry into sewers, watercourses, low areas
Precaution: Combustible liq.; flamm. limits in air 1.3-8.8 vol.%; volatile; gives off invisible vapors; liq. or vapor may settle in low areas; distant ignition/explosion possible; electrostatic accumulation hazard; use proper grounding procedures; incompat. with strong oxidizing agents
Hazardous decomp. prods.: Combustion: fumes, smoke, CO, SO_x, aldehydes, other decomp. prods.
HMIS: Health 1, Flammability 2, Reactivity 0

Isopar® P [ExxonMobil Europe]
Chem. Descrip.: Isoparaffin
Uses: Solvent for agric., pesticides, herbicides
Properties: Liq.; dens. 0.804 kg/dm^3 (15 C); visc. 4.82 mm^2/s; i.b.p. 244 C; flash pt. 108 C; ref. index 1.445 (20 C); surf. tens. 26.2 mN/m; KB value

26; aniline pt. 90 C; 0.05% max. aromatics

Isopar® V Fluid [ExxonMobil; Exxon Canada]

Chem. Descrip.: Petroleum distillates, hydrotreated middle, BHT (10 ppm)
CAS 64742-46-7; EINECS/ELINCS 265-148-2

Uses: Solvent for household applics. (polishes, cleaning compds., degreasers, waterless hand cleaners, stain removers, air fresheners, insecticides), inks, and elastomer-based adhesives; diluent; chem. feedstock; fuel; food-contact paper and applics.; defoamer in food-contact paper/paperboard; coating on egg shells; froth flotation cleaning of vegetables

Features: Complies with criteria for LVP solvent

Regulatory: FDA 21CFR §172.884, 176.180, 176.210, 178.3620(b), 178.3650, 573.220, 573.680; 40CFR §180.1001(c)(e); SARA nonreportable; Canada DSL listed; NPRI nonreportable

Properties: Colorless clear liq.; faint petrol. odor; misc. with other hydrocarbons; sol. < 0.01% in water; m.w. 197; sp.gr. 0.82 (15.6 C); dens. 6.82 lb/gal (60 F); visc 7.5 cps; vapor pressure 0.3 mm Hg (38 C); f.p. -53 C; b.p. 263-329 C; flash pt. (PMCC) 118 C; autoignition temp. 335 C; surf. tens. 26.9 dynes/cm; KB value 25

Toxicology: REL 5 mg/m^3 or lower (mist); inh. of high vapor concs. may cause eye/respiratory tract irritation, headaches, dizziness, anesthesia, drowsiness, unconsciousness, other CNS effects, death; sl. eye irritant; frequent/prolonged skin contact may cause irritation, dermatitis; if aspirated, may cause mild to severe pulmonary injury, death

Precaution: Flamm. limit 0.5-4.4 vol.%; volatile; vapors can be ignited causing explosions or flash fires; electrostatic accumulation hazard; use proper grounding procedures; incompat. with strong oxidizing agents (liq. chlorine, conc. oxygen, sodium or calcium hypochlorite)

Hazardous decomp. prods.: None

Storage: Store in cool, well-ventilated area away from incompat. materials, heat, open flame, ignition sources, direct sunlight; keep container closed

Isopropyl Oxitol [Shell Netherlands]

Chem. Descrip.: Ethylene glycol isopropyl ether
CAS 109-59-1; EINECS/ELINCS 203-685-6
UN 3271

Uses: Solvent for nat. and syn. resins such as colophony, dammar, alkyd, phenolic and epoxy resins, NC, ethyl cellulose, surf. coatings, printing inks, household/industrial cleaners, cosmetics, agric.; coalescing agent, flow aid for water-based paints

Regulatory: DIN 55999, BS 509 compliance; EINECS, Australia AICS, China IECSC, Japan MITI, Philippines PICCS, Korea TCCL listed

Properties: Colorless clear liq.; mild ethereal odor; completely misc. with water, aromatic and aliphatic hydrocarbons, many other org. solvs.; m.w. 104; dens. 902-905 kg/m^3 (20 C); dynamic visc. 2.55 mPa·s; vapor pressure 0.38 kPa (20 C); f.p. -60 C; b.p. 139.5-144.5 C; flash pt. (Abel) 46 C; autoignition temp. 240 C; ref. index 1.410; surf. tens. 37.2 mN/m; sp. heat 2.22 kJ/kg/°C; elec. conduct. 7 x 10^6 pS/m; dielec. const. 9.9

Toxicology: ACGIH TLV/TWA (8 h) 25 ppm; LD50 (oral) ≤ 2000 mg/kg, (skin) ≤ 2000 mg/kg; LC50 (inh.) > 5 mg/l; skin/respiratory tract/severe eye irritant; harmful by inh., ing., skin contact; ing. may cause digestive tract irritation; causes CNS depression; may cause hemolysis; TSCA listed

Environmental: Readily biodeg.; does not bioaccumulate significantly; not classified as dangerous under EC criteria; pract. nontoxic to fish, invertebrates; prevent contamination of soil, water; prevent entry into drains, ditches, rivers

Precaution: Flamm.; LEL 1.6 vol.%; UEL 13.0 vol.%; may form flamm./explosive vapor-air mixts.; vapor is heavier than air, spreads along ground; distant ignition is possible; ground equip.

Hazardous decomp. prods.: Combustion: CO

Storage: Hygroscopic; 12 mos shelf life under proper storage; store @ ambient temps. in tightly closed containers in well-ventilated area away from direct sunlight, other heat/ignition sources

Isostriol IC [Undesa; S. Black]

Chem. Descrip.: Isocetyl isostearate
CAS 52006-45-8; EINECS/ELINCS 257-598-3

Uses: Emollient in bath oils, creams/lotions, sun creams, shaving prods., insect repellents, antiperspirants, deodorants; oily component for cleansing creams, nail polish removers; plasticizer for nail polishes; pigment wetting agent; softener in creams, hair care; solvent for fatty cosmetic ingreds.; vehicle for lipsticks; solubilizer for perfume oils, vitamin oils, hormones; solubilizer, vehicle for medicines; excipient for injectable substances; lubricant for PVC

Features: Changes skin feel, rub out props. and greasiness; veg. oil substitute; vaseline substitute

J

Jarcol™ I-16 [Jarchem Ind.]
Chem. Descrip.: 2-Hexyl-decanol (97% min.)
CAS 2425-77-6; EINECS/ELINCS 219-370-1
Uses: Extender, solvent for cosmetics, pharmaceuticals, detergents, lubricants, emulsifiers, solvs., rubber, plastics, polymers, metal processing, paper coating, and textiles; solubilizer in water-misc. cutting fluids; lubricant; coupling agent; emollient for clear formulations
Properties: APHA 20 max. liq.; m.w. ≈ 242; sp.gr. 0.836; visc. 38 mPa·s; m.p. -21 to -15 C; b.p. 193-197 C; acid no. 0.1 max.; iodine no. 1.0 max.; sapon. no. 1.0 max.; hyd. no. 225-235; flash pt. 156 C; ref. index 1.450; 0.1% max. water

Jarcol™ I-18EX [Jarchem Ind.]
Chem. Descrip.: Isostearyl alcohol
CAS 27458-93-1; EINECS/ELINCS 248-470-8
Uses: Solvent
Regulatory: DOT nonregulated; SARA §311/312/313 nonreportable
Properties: Clear liq.; negligible sol. in water; sp.gr. 0.85; visc. 49 cps; acid no. 0.1 max.; iodine no. 1.0 max.; sapon. no. 1.0 max.; hyd. no. 200-212; pour pt. 0 C; flash pt. (OC) 310 F; fire pt. 325 F; 100% act.; 0.2% max. moisture
Toxicology: LD50 (oral, rat) > 5 g/kg for similar prod.; expected to cause minor to mod. eye irritation; may cause minor to mod. skin irritation on prolonged/repeated contactTSCA listed
Environmental: Avoid uncontrolled releases to environment
Precaution: Incompat. with oxidizing agents, strong acids
Hazardous decomp. prods.: CO, CO_2
HMIS: Health 1, Flammability 1, Reactivity 0

Jarcol™ I-24 [Jarchem Ind.]
Chem. Descrip.: 2-Decyl-tetradecanol (97% min.)
CAS 58670-89-6; EINECS/ELINCS 216-385-0
Uses: Solvent, solubilizer for cosmetics, pharmaceuticals, detergents, lubricants, emulsifiers, solvs., rubber, plastics, polymers, metal processing, paper coating, and textiles
Properties: APHA 20 max. liq.; m.w. 354; sp.gr. 0.842; visc. 86 mPa·s; m.p. 17-20 C; b.p. 271-275 C; acid no. 0.1 max.; iodine no. 1.0 max.; sapon. no. 1.0 max.; hyd. no. 154-190; flash pt. 230 C; ref. index 1.457; 0.1% max. water

Jarysol® [Atofina SA]
Uses: Solvent for microencapsulation

Jayflex® 215 [ExxonMobil; Exxon Canada]
Chem. Descrip.: C5-20 normal paraffins
CAS 64771-72-8
Uses: Sec. plasticizer; solvent; diluent; chem. feedstock; fuel; food pkg. adhesives, cellophane, rubber articles
Regulatory: FDA 21CFR §175.105, 177.1200, 177.2600; Canada DSL listed; NPRI nonreportable
Properties: Wh. clear liq.; odorless; sol. < 0.01% in water; sp.gr. 0.762 (20/20 C); dens. 6.43 lb/gal (20 C); visc. 3 cSt; vapor pressure 5 mm Hg (20 C); f.p. 7 C; b.p. 271 C (5 mm); pour pt. -48 C; flash pt. (PMCC) > 121 C; autoignition temp. ≈ 210 C
Toxicology: REL TWA 140 ppm; inh. of high vapor concs. may cause eye/respiratory tract irritation, headaches, dizziness, anesthesia, drowsiness, unconsciousness, other CNS effects, death; sl. eye irritant; frequent/prolonged skin contact may cause irritation, dermatitis; if aspirated, may cause mild to severe pulmonary injury, death
Environmental: Prevent entry into sewers, watercourses, low areas
Precaution: Flamm. limits 1.3-8.7 vol.%; low flamm./explosion hazard; volatile; gives off invisible vapors; liq. or vapor may settle in low areas; distant ignition/explosion possible; electrostatic accumulation hazard; use proper grounding procedures; incompat with strong oxidizing agents
Hazardous decomp. prods.: Combustion: toxic gases, fumes, smoke, CO
Storage: Store in cool, well-ventilated areas away from incompat. materials, heat, open flame, ignition sources, direct sunlight; keep container closed

Jeffamine® D-230 [Huntsman]
Chem. Descrip.: Polyoxypropylenediamine
Uses: Epoxy curing agent in surface coatings, heat-cured sol'n. coatings, laminates, potting, casting, reinforced composites, adhesives, polymer concrete, syntactic foam, reactive solvent
Properties: Colorless to sl. yel sl. hazy liq.; sol. in most hydrocarbons and oxygenated solvs., misc. in water; m.w. 230; sp.gr. 0.9480; dens. 7.9 lb/gal; visc. 9 cps; vapor pressure 1 mm Hg (101 C); flash pt. (PMCC) 121 C; ref. index 1.4466; pH 11.7 (5% aq.); distort. temp 60-75 C (264 psi); 0.1% water
Use Level: 32 phr
Toxicology: LD50 (oral, rat) 1.66 g/kg, (dermal, rabbit) 0.76 g/kg; can cause burns to skin and eyes, toxic effects if absorbed thru skin or swallowed
Storage: Nitrogen blanket suggested for all storage in case of accidental high temp.; pronounced discoloration is likely to occur above 140 F

Jeffox® PPG-400 [Huntsman]
Chem. Descrip.: PPG 400
CAS 25322-69-4
Uses: Intermediate yielding esters useful as lubricants, defoamers in rubber and pharmaceuticals, solvs. and humectant modifiers for inks, plasticizers, and functional fluids
Properties: Water-wh. visc. liq., faint ether-like odor; m.w. 400; dens. 8.40 lb/gal; visc. 150-175 SUS (100 F); flash pt. (PMCC) 320 F; pour pt. -35 F; pH 5-7

Jeffox® PPG-2000 [Huntsman]
Chem. Descrip.: PPG-2000
CAS 25322-69-4
Uses: Intermediate yielding esters useful as lubricants, defoamers in rubber and pharmaceuticals, solvs. and humectant modifiers for inks, plasticizers, and functional fluids
Properties: Water-wh. visc. liq., faint ether-like odor; m.w. 2000; dens. 8.37 lb/gal; flash pt. (PMCC) 370 F; pour pt. -25 F; pH 5-7

Jeffox® WL-440 [Huntsman]
Chem. Descrip.: EO/PO deriv.
Uses: Functional fluid; lubricant, plasticizer, solvent, coupling agent, frothing agent, heat transfer fluid, intermediate, defoamer in solder reflow applics., boiler defoaming, ore flotation, inks, dyes
Properties: Clear liq.; m.w. 488; water-sol.; sp.gr. 1.1515; dens. 9.55 lb/gal; visc. 100 cSt (100 F); flash pt. 580 F (COC); pour pt. -30 F; ref. index 1.4721.

Jeffox® WL-660 [Huntsman]
Chem. Descrip.: EO/PO deriv.
Uses: Functional fluid; lubricant, plasticizer, solvent, coupling agent, frothing agent, heat transfer fluid, intermediate, defoamer in solder reflow applics., boiler defoaming, ore flotation, inks, dyes

Properties: Clear liq.; m.w. 1800; water-sol.; sp.gr. 1.0502; dens. 8.75 lb/gal; visc. 143 cSt (100 F); cloud pt. 60 C (1% aq.); flash pt. 475 F (COC); pour pt. -45 F; ref. index 1.4586.

Jeffox® WL-1400 [Huntsman]
Chem. Descrip.: EO/PO deriv.
Uses: Functional fluid; lubricant, plasticizer, solvent, coupling agent, frothing agent, heat transfer fluid, intermediate, defoamer in solder reflow applics., boiler defoaming, ore flotation, inks, dyes
Properties: Clear liq.; m.w. 2500; water-sol.; sp.gr. 1.095; dens. 9.15 lb/gal; visc. 306 cSt (100 F); cloud pt. > 99 C (1% aq.); flash pt. 475 F (COC); pour pt. 50 F; ref. index 1.4669.

Jeffox® WL-5000 [Huntsman]
Chem. Descrip.: EO/PO deriv.
Uses: Functional fluid; lubricant, plasticizer, solvent, coupling agent, frothing agent, heat transfer fluid, intermediate, defoamer in solder reflow applics., boiler defoaming, ore flotation, inks, dyes
Properties: Clear liq.; m.w. 4365; water-sol.; sp.gr. 1.0557; dens. 8.80 lb/gal; visc. 1080 cSt (100 F); cloud pt. 53 C (1% aq.); flash pt. 460 F (COC); pour pt. -10 F; ref. index 1.4600.

Jeffsol® 1,2-Butylene Carbonate [Huntsman]
Chem. Descrip.: 1,2-Butylene carbonate
CAS 4437-85-8; EINECS/ELINCS 224-651-7
Uses: Solvent for org. and inorg. materials in surf. cleaners, degreasers, dyes, fibers, plastics, batteries, aromatic hydrocarbons, nat. gas; gellant for clays; binder for foundry sand molds
Regulatory: DOT nonregulated; OSHA nonhazardous; SARA §302/304/311/313 nonreportable
Properties: Colorless clear mobile liq.; nearly odorless to pleasant odor; sol. 0.1-1% in water; m.w. 116.12; sp.gr. 1.141 (20/4 C); visc. 2.77 cSt; vapor pressure 0.0285 mm Hg; f.p. -45 C; b.p. 102 C (5 mm Hg), 251 C (760 mm Hg); flash pt. (PMCC) 275 F; pH 6-8 (1% aq.); VOC 18%; 99% assay; 0.1% max. water
Toxicology: LD50 (oral, rat) > 5 g/kg; aspiration hazard by ing.; can enter lungs and cause damage; may cause minimal eye irritation; prolonged skin contact may cause defatting/irritation; inh. may cause nose/throat irritation, headache, nausea, drowsinessTSCA listed
Precaution: Contamination may result in dangerous CO_2 pressure buildup, rupture of container
Hazardous decomp. prods.: Toxic levels of CO, CO_2, irritating aldehydes and ketones may be formed on burning; heating in air may produce irritating aldehydes, acids, ketones
HMIS: Health 0, Flammability 1, Reactivity 0
Storage: Avoid exposure to high temps., water contamination

Jeffsol® EC [Huntsman]
Chem. Descrip.: Ethylene carbonate
CAS 96-49-1; EINECS/ELINCS 202-510-0
Uses: Solvent, reactant, plasticizer for fibers, textiles, dyeing, plastics and resins, aromatic hydrocarbon extraction, electrolytes, photosensitive compositions for preparing lithographic plates; fuel for jet engines; hydraulic brake fluids
Properties: Colorless solid, pract. odorless; sol. in benzene, butyl acetate, chloroform, ethanol, ethyl acetate, methanol, nonylphenol, toluene, water; m.w. 88.06; sp.gr. 1.3218 (39/4 C); dens. 10.87 lb/gal (40 C); visc. 1.9 cs (40 C); m.p. 36.4 C; b.p. 238 C (760 mm); flash pt. 320 F; ref. index 1.4158 (50 C); dielec. const. 95.3; 99.5% min. act.
Toxicology: LD50 (oral, rat) > 5 g/kg, (dermal, rat) > 3 g/kg; pract. nontoxic; minimally irritating to skin
Precaution: May decompose with evolution of CO_2 in presence of an acid, base, metal oxide, or salt; protect from sources of ignition or static elec. discharge

Jeffsol® EC-25 [Huntsman]
Chem. Descrip.: Blend of Jeffsol® Ethylene Carbonate (25%) and Jeffsol® Propylene Carbonate (75%)
Uses: Solvent for org. and inorg. materials in surf. cleaners, degreasers, dyes, fibers, plastics, batteries, aromatic hydrocarbons, nat. gas; gellant for clays; binder for foundry sand molds
Properties: Pt-Co 50 max. clear mobile liq.; sp.gr. 1.24 (20/20 C); dens. 10.32 lb/gal (20 C); visc. 2.52 cps; vapor pressure < 0.02 mm Hg (20 C); f.p. < -51.1 C; b.p. 243-244 C; flash pt. (PMCC) 133 C; 0.1% max. water
Storage: Hygroscopic

Jeffsol® EC-50 [Huntsman]
Chem. Descrip.: Blend of Jeffsol® Ethylene Carbonate (50%) and Jeffsol® Propylene Carbonate (50%)
Uses: Solvent for org. and inorg. materials in surf. cleaners, degreasers, dyes, fibers, plastics, batteries, aromatic hydrocarbons, nat. gas; gellant for clays; binder for foundry sand molds
Properties: Pt-Co 50 max. clear mobile liq.; sp.gr. 1.27 (20/20 C); dens. 10.51 lb/gal (20 C); visc. 2.52 cps; vapor pressure < 0.03 mm Hg (20 C); f.p. -10 C; b.p. 244-245 C; flash pt. (PMCC) 154 C; 0.1% max. water
Storage: Hygroscopic

Jeffsol® EC-75 [Huntsman]
Chem. Descrip.: Blend of Jeffsol® Ethylene Carbonate (75%) and Jeffsol® Propylene Carbonate (25%)
Uses: Solvent for org. and inorg. materials in surf. cleaners, degreasers, dyes, fibers, plastics, batteries, aromatic hydrocarbons, nat. gas; gellant for clays; binder for foundry sand molds
Properties: Pt-Co 50 max. clear mobile liq.; sp.gr. 1.30 (20/20 C); dens. 10.78 lb/gal (20 C); visc. 2.55 cps; vapor pressure < 0.03 mm Hg (20 C); f.p. 17 C; b.p. 246-247 C; flash pt. (PMCC) 285-290 C; 0.1% max. water
Storage: Hygroscopic

Jeffsol® PC [Huntsman]
Chem. Descrip.: Propylene carbonate
CAS 108-32-7; EINECS/ELINCS 203-572-1
Uses: Solvent, reactant, plasticizer for fibers, textiles, dyeing, plastics and resins, gas treating, aromatic hydrocarbon extraction, metal extraction, surf. coatings (paints, varnishes, adhesives, plastics, epoxies, mastics), foundry sand binder, lubricants, electrolytes, cosmetics (polar additive for montmorillonite or bentonite clay gellants); fire extinguishing compds.; antifoam for antifreeze; hydraulic brake fluids; as plating medium
Properties: Clear mobile liq., pract. odorless; sol. in acetone, benzene, chloroform, glycols, ethanol, methanol, MEK, toluene, xylene; sol. 25 g/100 g water; m.w. 102.09; sp.gr. 1.2057 (20/4 C); dens. 10.05 lb/gal (20 C); visc. 1.9 cs (20 C); f.p. -49 C; b.p. 242 C (760 mm); pour pt. -100 F (supercools easily); flash pt. 275 F; fire pt. 280 F; ref. index 1.4210; dielec. const. 65; 99.8% min. act.
Toxicology: LD50 (dermal, rabbit) > 20 g/kg; pract. nontoxic; nonirritating to skin; ingestion may cause renal damage and death
Precaution: May decompose with evolution of CO_2 in presence of an acid, base, metal oxide, or salt; protect from sources of ignition or static elec. discharge

Jeffstrip™ FST [Huntsman]
Chem. Descrip.: Activated alkylene carbonate sol'n.
Uses: Solvent for formulation of 'safer' paint strippers and cleaners
Properties: Clear mobile liq.; dens. 1.1695 g/ml; visc. 1.712 cs; vapor pressure 20 ± 2.2 psig (93 C); m.p. -28.4 C; flash pt. (TCC) 92-93 C
Use Level: 40-50% (paint strippers)
Precaution: LEL 5.3% (160-170 C); UEL 29.2% (160-170 C); should not be formulated with acetone, MEK, diethyl ketone, or similar ketones

Jefftreat® M Solvent [Huntsman]
Chem. Descrip.: Methyldiethanolamine
CAS 105-59-9; EINECS/ELINCS 203-312-7
Uses: Solvent for removal of CO_2 and H_2S from gas streams in tail gas treating, hydrogen sulfide enrichment, natural gas plants
Features: Absorbs @ lower temps. and releases the acid gases @ higher temps.; low vapor pressure
Regulatory: DOT nonhazardous
Properties: Water-wh. to pale amber clear liq.; ammoniacal odor; completely sol. in water; sp.gr. 1.0431 (20/20 C); dens. 8.69 lb/gal (20 C); visc. 36.8 cSt (100 F); vapor pressure < 0.01 mm Hg (20 C); f.p. -21 C; i.b.p. 242 C min.; flash pt. (PMCC) 240 F; 99% min. act.
Use Level: 40-50%
Toxicology: LD50 (oral, rat) 4.78 g/kg, (skin, rabbit) 6.24 g/kg; sl. toxic by single oral dose; pract. nontoxic by single dermal applic.; mod. irritating to eyes; sl. irritating to skin
Precaution: Reacts with copper to form complex salts; swells elastomers such as Buna N, neoprene, nitrile
Storage: Hygroscopic

Jefftreat® M 45% [Huntsman]
Chem. Descrip.: Methyldiethanolamine-based
Uses: Solvent for removal of CO_2 from gas processing applics., e.g., hydrogen prod., ammonia mfg., cryogenic gas processing, natural gas upgrading, LNG prod.
Features: Exc. chem. stability; longer prod. life under severe operating

conditions; higher reaction rates; rec. where H_2S is not present
Properties: Dens. 1.035 g/ml (38 C); visc. 4.3 cP (38 C); f.p. -21.6 C; 45% act.
Use Level: 40-50%
Toxicology: Mod. irritating to eyes; sl. irritating to skin

Jefftreat® M-504 [Huntsman]
Chem. Descrip.: Methyldiethanolamine (35-49.99%), proprietary components
UN 3267
Uses: Solvent for gas treating
Regulatory: SARA §311 acute hazard, §313 nonreportable; Calif. Prop. 65 nonreportable; EINECS, Canada DSL, Australia AICS, Japan MITI listed or exempt
Properties: Lt. yel. liq.; ammonia-like odor; sol. > 10% in water; sp.gr. 1.0469; visc. 24.67 cSt (37.8 C); f.p. -41 C; flash pt. (PMCC) > 132.2 C; pH 11.3 (1% aq.)
Toxicology: LD50 (oral, rat) 2-5 g/kg, (skin, rabbit) > 2 g/kg; causes respiratory tract irritation, possible damage; corrosive, causes eye/skin burns; harmful or fatal if swallowed; aspiration hazard if swallowed, can damage lungs; may cause severe eye damage, blindness; inh. of vapors may cause nasal discharge, coughing, nose/throat discomfort; ing. may cause mouth/throat/stomach burning, abdominal/chest pain, nausea, vomiting, diarrhea, thirst, weakness, collapse; TSCA listed
Environmental: Prevent entry into sewers, waterways
Precaution: Reacts violently with acids
Hazardous decomp. prods.: Toxic levels of ammonia, combustion prods. of nitrogen, CO, CO_2, irritating aldehydes and ketones may be formed on burning in limited air supply
HMIS: Health 3, Flammability 1, Reactivity 0
Storage: Minimize exposure to high temps.; avoid water contamination

Jefftreat® M-505 [Huntsman]
Chem. Descrip.: Proprietary
UN 3267
Uses: Solvent for gas treating
Regulatory: SARA §311 acute hazard, §313 nonreportable; Calif. Prop. 65 nonreportable; EINECS, Canada DSL, Australia AICS, Japan MITI listed or exempt
Properties: Wh., lt. yel. or amber liq.; amine odor; sol. > 10% in water; sp.gr. 1.04-1.05; visc. 25-35 cSt (37.8 C); flash pt. (PMCC) 135 C; pH 11.2 (1% aq.)
Toxicology: LD50 (oral, rat) 2-5 g/kg, (skin, rabbit) > 2 g/kg; causes respiratory tract irritation, possible damage; corrosive, causes eye/skin burns; harmful or fatal if swallowed; aspiration hazard if swallowed, can damage lungs; may cause severe eye damage, blindness; inh. of vapors may cause nasal discharge, coughing, nose/throat discomfort; ing. may cause mouth/throat/stomach burning, abdominal/chest pain, nausea, vomiting, diarrhea, thirst, weakness, collapse; TSCA listed
Environmental: Prevent entry into sewers, waterways
Precaution: Reacts violently with acids
Hazardous decomp. prods.: Toxic levels of ammonia, combustion prods. of nitrogen, CO, CO_2, irritating aldehydes and ketones may be formed on burning in limited air supply
HMIS: Health 3, Flammability 1, Reactivity 0
Storage: Minimize exposure to high temps.; avoid water contamination

Jefftreat® M-507 [Huntsman]
Chem. Descrip.: Methyldiethanolamine (65-79.99%), proprietary components
UN 3267
Uses: Solvent for gas treating
Regulatory: SARA §311 acute hazard, §313 nonreportable; Calif. Prop. 65 nonreportable; EINECS, Canada DSL, Australia AICS, Japan MITI listed or exempt
Properties: Lt. yel. liq.; ammonia-like odor; sol. > 10% in water; sp.gr. 1.04; visc. 32 cSt (37.8 C); f.p. -38 C; flash pt. (PMCC) > 132.2 C; pH 11 (1% aq.)
Toxicology: LD50 (oral, rat) 2-5 g/kg, (skin, rabbit) > 2 g/kg; causes respiratory tract irritation, possible damage; corrosive, causes eye/skin burns; harmful or fatal if swallowed; aspiration hazard if swallowed, can damage lungsTSCA listed
Environmental: Prevent entry into sewers, waterways
Precaution: Reacts violently with acids
Hazardous decomp. prods.: Toxic levels of ammonia, combustion prods. of nitrogen, CO, CO_2, irritating aldehydes and ketones may be formed on burning in limited air supply
HMIS: Health 3, Flammability 1, Reactivity 0
Storage: Minimize exposure to high temps.; avoid water contamination

Jefftreat® M-510 45% [Huntsman]
Chem. Descrip.: Methyldiethanolamine-based
Uses: Solvent for removal of CO_2 from gas processing applics., e.g., hydrogen prod., ammonia mfg., cryogenic gas processing, natural gas upgrading, LNG prod.
Features: Exc. chem. stability; longer prod. life under severe operating conditions; higher reaction rates; rec. where H_2S is not present
Properties: Dens. 1.031 g/ml (38 C); visc. 3.9 cP (38 C); 45% act.
Use Level: 40-50%
Toxicology: Mod. irritating to eyes; sl. irritating to skin

Jefftreat® M-510 100% [Huntsman]
Chem. Descrip.: Methyldiethanolamine (65-79.99%), proprietary components
UN 3267
Uses: Solvent for removal of CO_2 from gas processing applics., e.g., hydrogen prod., ammonia mfg., cryogenic gas processing, natural gas upgrading, LNG prod.
Features: Exc. chem. stability; longer prod. life under severe operating conditions; higher reaction rates; rec. where H_2S is not present
Regulatory: SARA §311 acute hazard, §313 nonreportable; Calif. Prop. 65 nonreportable; EINECS, Canada DSL, Australia AICS, Japan MITI listed or exempt
Properties: Lt. yel. liq.; ammonia-like odor; sol. > 10% in water; sp.gr. 1.0423; visc. 32.26 cSt (37.8 C); f.p. -38.2 C; flash pt. (PMCC) > 132.2 C; pH 11.1 (1% aq.); 100% act.
Use Level: 40-50%
Toxicology: Causes respiratory tract irritation, possible damage; corrosive, causes eye/skin burns; harmful or fatal if swallowed; aspiration hazard if swallowed, can cause lung damage; may cause severe eye damage, blindness; excessive inh. of vapors may cause nasal discharge, coughing, nose/throat discomfort; ing. may cause mouth/throat/stomach burning, abdominal/chest pain, nausea, vomiting, diarrhea, thirst, weakness, collapse; TSCA listed
Environmental: Prevent entry into sewers, waterways
Precaution: Reacts violently with acids
Hazardous decomp. prods.: Toxic levels of ammonia, combustion prods. of nitrogen, CO, CO_2, irritating aldehydes and ketones may be formed on burning in limited air supply
HMIS: Health 3, Flammability 1, Reactivity 0
Storage: Minimize exposure to high temps.; avoid water contamination

Jefftreat® MS-100 [Huntsman]
Chem. Descrip.: Methyldiethanolamine
CAS 105-59-9; EINECS/ELINCS 203-312-7
Uses: Solvent for removal of H_2S and bulk removal of CO_2 for high pressure nat. gas, refinery gas, or Claus off-gas treating

Jefftreat® MS-200 Series [Huntsman]
Chem. Descrip.: Methyldiethanolamine-based
Uses: Solvent for removal of CO_2 and H_2S from gas and gas-liq. streams for gas treating (natural gases and gas liqs., refinery gases and gas liqs., coal and waste gasification gases, petrochem. gases and gas liqs.)
Features: Low heat of reaction; high acid gas loading; minimal corrosivity; negligible degradation
Regulatory: DOT nonhazardous
Toxicology: Mod. irritating to eyes; sl. irritating to skin; DOT noncorrosive

Jefftreat® MS-205 [Huntsman]
Chem. Descrip.: Diethanolamine (50-64.99%), methyldiethanolamine (35-49.99%)
UN 3082
Uses: Solvent for gas treating
Regulatory: SARA §311 acute/chronic hazard, §313 reportable (diethanolamine); Calif. Prop. 65 nonreportable; EINECS, Canada DSL, Australia AICS, Japan MITI listed or exempt
Properties: Pale liq.; amine-like odor; sol. > 10% in water; sp.gr. 1.06; visc. > 37 cSt (37.7 C); vapor pressure < 0.01 mm Hg (20 C); b.p. > 246.1 C; flash pt. > 115.5 C; pH 11.5
Toxicology: LD50 (oral, rat) 0.50-2.00 g/kg, (skin, rabbit) > 2.0 g/kg; causes eye irritation; may cause skin irritation; excessive inh. of vapors may cause nose/throat irritation, headache, nausea, drowsiness; mod. toxic

by ing.; may cause abdominal discomfort, nausea, vomiting, diarrhea; may cause blood effects, liver/kidney damage; may form suspected cancer causing nitrosamines; TSCA listed

Environmental: Environmentally hazardous; prevent entry into sewers, waterways

Precaution: Incompat. with strong acids, oxidizers, acid anhydrides, halides; may be corrosive to copper, zinc, galvanized iron; do not add or formulate with nitrites

Hazardous decomp. prods.: Toxic levels of ammonia, combustion prods. of nitrogen, CO, CO_2, irritating aldehydes and ketones may be formed on burning in limited air supply

HMIS: Health 3, Flammability 1, Reactivity 0

Storage: Minimize exposure to high temps.; avoid water contamination

Jefftreat® MS-205 LFG [Huntsman]

Chem. Descrip.: Diethanolamine (35-64.99%), methyldiethanolamine (20-49.99%), water (10-19.99%)

UN 3082

Uses: Solvent for gas treating

Regulatory: SARA §311 acute/chronic hazard, §313 reportable (diethanolamine); Calif. Prop. 65 nonreportable; EINECS, Canada DSL, Australia AICS, Japan MITI listed or exempt

Properties: Pale liq.; amine-like odor; sol. > 10% in water; sp.gr. 1.079; visc. 169 cSt; b.p. > 246.1 C; flash pt. (PMCC) > 137.8 C; pH 11-12

Toxicology: LD50 (oral, rat) 2-5 g/kg, (skin, rabbit) 2 g/kg; may cause eye irritation, blood effects, liver/kidney damage; may cause possibly corneal injury; may cause sl. skin irritation; excessive inh. of vapors may cause nose/throat irritation, headache, nausea, drowsiness ; ing. may cause abdominal discomfort, nausea, diarrhea; do not add nitrites—may form suspected cancer-causing nitrosamines; TSCA listed

Environmental: Environmentally hazardous; prevent entry into sewers, waterways

Precaution: Incompat. with strong acids, oxidizers, acid anhydrides, halides; may be corrosive to copper, zinc, galvanized iron; do not add or formulate with nitrites

Hazardous decomp. prods.: Toxic levels of ammonia, combustion prods. of nitrogen, CO, CO_2, irritating aldehydes and ketones may be formed on burning in limited air supply

HMIS: Health 2, Flammability 1, Reactivity 0

Storage: Minimize exposure to high temps.; avoid water contamination

Jefftreat® MS-300 [Huntsman]

Chem. Descrip.: Methyldiethanolamine

CAS 105-59-9; EINECS/ELINCS 203-312-7

Uses: Solvent for tail-gas-treating units or other units requiring very low residual H_2S and additional CO_2 slippage

K

Kemester® 205 [Crompton]
Chem. Descrip.: Methyl oleate
CAS 112-62-9; EINECS/ELINCS 203-992-5
Uses: Intermediate for mfg. of superamides, in metalworking lubricants; solvent; opacifier; visc. control agent
Properties: Gardner 1 liq.; acid no. 4.0 max.; iodine no. 80-90; sapon. no. 194-203; 100% conc.

Kemester® 213 [Crompton]
Chem. Descrip.: Methyl oleate/linoleate
Uses: Intermediate for mfg. of superamides, in metalworking lubricants; solvent
Properties: Gardner 3 max.; acid no. 10.0 max.; iodine no. 130-145; sapon. no. 194-204

Kemester® 226 [Crompton]
Chem. Descrip.: Methyl soyate
CAS 67784-80-9
Uses: Intermediate for mfg. of superamides, in metalworking lubricants; solvent; opacifier; visc. control agent
Properties: Gardner 4 max. liq.; acid no. 7.0 max.; iodine no. 125-135; sapon. no. 195-200; 100% conc.

Kemester® 2050 [Crompton]
Chem. Descrip.: Methyl eicosenate
CAS 1120-28-1; EINECS/ELINCS 214-304-8
Uses: Intermediate for mfg. of alkanolamides, in metalworking lubricants; solvent; foam depressant and nutrient in fermentation
Properties: Gardner 14 max. color; acid no. 25 max.; iodine no. 90-110; sapon. no. 180-190

Kemester® EGDS [Crompton]
Chem. Descrip.: Glycol distearate
CAS 627-83-8; EINECS/ELINCS 211-014-3
Uses: Intermediate for mfg. of superamides, in metalworking lubricants; solvent; industrial lubricant; opacifier, pearlescent, thickener for cosmetics and pharmaceuticals; internal/external lubricant for PVC
Regulatory: FDA accepted
Properties: Gardner 2 max. solid; m.p. 60-63 C; acid no. 6 max.; iodine no. 1 max.; sapon. no. 190-200; nonionic; 100% conc.

Kemstrene® 96.0% USP [Crompton]
Chem. Descrip.: Refined glycerin USP
CAS 56-81-5; EINECS/ELINCS 200-289-5
Uses: Humectant, solvent in cosmetics, pharmaceuticals, liq. soaps, confections, inks, and lubricants; intermediate used in polyester and PU formulations
Properties: Sp.gr. 1.25165 min; 96.0% purity

Kemstrene® 99.7% USP [Crompton]
Chem. Descrip.: Refined glycerin USP
CAS 56-81-5; EINECS/ELINCS 200-289-5
Uses: Humectant, solvent in cosmetics, pharmaceuticals, liq. soaps, confections, inks, and lubricants; intermediate used in polyester and PU formulations; plasticizer, softener for NR,SR; mold release agent; lubricant; inc. tack, improves processing
Properties: Hazen 12 max. color; sp.gr. 1.26124; 99.7% purity

Kemstrene® Reagent Grade [Crompton]
Chem. Descrip.: Refined glycerin
CAS 56-81-5; EINECS/ELINCS 200-289-5
Uses: Humectant, solvent in cosmetics, liq. soaps, confections, inks, and lubricants; intermediate in polyester and PU formulations
Properties: Sp.gr. 1.2587.

Kenflex® N1 [Kenrich Petrochems.]
Chem. Descrip.: Aromatic hydrocarbon polymeric plasticizer (50% alkylnaphthalene formalin, CAS 52613-22-6 and 50% petrol. naphthenic distillates, CAS 64742-53-6)
Uses: Plasticizer, softener, processing aid, flow control agent, mold release agent for chlorosulfonated PE, polysulfide, neoprene, butyl rubber and polar thermoplastics; solvent; coupling agent for poly blends
Features: Discoloring; provides superior processing, improved extrusion, exc. electricals, high temp. stability, permanence, nonextractability, nonmigrance, nonvolatility
Properties: Dk. brn. liq.; sp.gr. 1.01; dens. 8.41 lb/gal; visc. 12,000 cps; i.b.p. 176 C; flash pt. (COC) 190 ± 20 C; VOC none
Use Level: 5-20 phr (Hypalon)
Toxicology: Potential skin, eye, and respiratory tract irritant; may be harmful if swallowed; may contain trace amts. of formaldehyde, a suspected carcinogen; may emit toxic fumes on thermal decomp.
Precaution: Incompat. with oxidizers, acids; avoid excessive heat, ignition sources; thermal dec. may emit toxic fumes
Hazardous Ingredients: 50% alkylnaphthalene formalin, 50% petrol. naphthenic distillate, trace formaldehyde
Hazardous decomp. prods.: CO, SO_x, aldehydes
Storage: Store in well-ventilated area in tightly closed containers

Kenplast® A-450 [Kenrich Petrochems.]
Chem. Descrip.: Petroleum distillates, catalytic reformer fractionator residue, low boiling (60-100%), methylnaphthalenes (15-40%), naphthalene (3-7%)
UN 1993; 3082
Uses: Plasticizer, diluent, solvent for urethanes, epoxies, PVC, cellulosics
Properties: Yel. clear liq.; acetic odor; negligible sol. in water; sp.gr. 0.98; vapor pressure 0.15 mm Hg (20 C); b.p. 210-288 C; flash pt. > 180 F
Toxicology: LD50 (oral, rat) > 2622 mg/kg, (dermal, rabbit) > 2000 mg/kg; LC50 (inh., 1 h, rat) > 1.67 mg/l (no death); can be harmful if high concs. are inhaled; harmful or fatal if liq. is aspirated into lungs; causes mild skin irritation
Precaution: Combustible
Hazardous decomp. prods.: Incomplete combustion can produce CO, CO_2, or other harmful prods.
Storage: Store away from heat, ignition sources, open flame

Kenplast® ES-2HP [Kenrich Petrochems.]
Chem. Descrip.: Cumylphenyl acetate (95%), p-cumyl phenol (5%)
Uses: Plasticizer for urethanes; reactive diluent for epoxy, reducing odor and irritation; visc. reducer for highly filled systems; ideal for polyamide-cured epoxy floorings and coal tar pipe coatings; comonomer and impact modifier for phenolics; improves adhesion of vinyl plastisols coating metals and polar plastics; solvent
Features: High flash pt.
Properties: Lt. yel. clear liq.; sp.gr. 1.08; dens. 9.0 lb/gal; visc. 50-200 cps; i.b.p. 326 C; dist. range 326-343 C; flash pt. (COC) 370 ± 30 F; pour pt. -13 C; acid no. < 1; sapon. no. 154; hyd. no. 13; pH 7 ± 1.0; VOC none
Toxicology: LD50 (rat) 6 g/kg; mild skin irritant; potential eye and respiratory tract irritant; may be harmful if swallowed

Precaution: Incompat. with oxidizers, acids; may emit toxic fumes on thermal decomp.

Kenplast® ESB [Kenrich Petrochems.]
Chem. Descrip.: Cumylphenyl benzoate
Uses: Primary plasticizer for PVC, PVC/nitrile; process aid for semirigid PVC extrudates; solvates conductive polyester and acrylic inks to improve impact
Properties: Gardner > 18 liq.; sp.gr. 1.13 dens. 9.4 lb/gal; visc. 15,000 cps; dist. range 250-427 C; pour pt. 16 C; acid no. < 1; sapon. no. 124; flash pt. (COC) 171 C; 70% act. in α-methyl styrene oligomer

Kenplast® ESN [Kenrich Petrochems.]
Chem. Descrip.: Cumylphenyl neodecanoate
Uses: Primary plasticizer for PVC, PVC/nitrile; process aid for semirigid PVC extrusions; solvates and impact modifies conductive polyester and acrylic inks
Properties: Gardner > 18 liq.; sp.gr. 1.05; dens. 8.75 lb/gal; visc. 5800 cps; pour pt. 1.66 C; acid no. < 1; sapon. no. 130; flash pt. (COC) 196 C; 85% act. in a-methyl styrene oligomer.

Kenplast® G [Kenrich Petrochems.]
Chem. Descrip.: Mixed alkylated phenanthrenes: petroleum distillates, heavy thermal cracked (93-95%), molten sulfur (1-5%), 1,2-benzanthracene (< 3%)
Uses: Primary plasticizer for dark PVC, nitrile, urethane compds., millable gum urethanes; flexibilizer, diluent for epoxy resins; solvent
Features: Nonreactive; discoloring
Properties: Dk. straw-colored liq.; sp.gr. 1.08 (60 F); visc. 25 cps (100 F); i.b.p. 420 F; flash pt. (COC) 196 C
Toxicology: LD50 (oral, rat) 2.5 g/kg; exposure may cause eye irritation, mild skin irritation, nose, throat, and respiratory tract irritation, CNS depression, death; ing. of lg. amts. may be harmful; nonmutagenic; 1,2-benzanthracene is carcinogenic
Precaution: Incompat. with strong oxidizing agents
Hazardous decomp. prods.: CO_2, CO, various hydrocarbons, sulfur compds.
Storage: Will crystallize during storage below 50 F; warm before use to eliminate crystals; do not store above 70 F

Kerdane D65 [TotalFina-Elf]
Chem. Descrip.: Petroleum distillates, hydrotreated light (C10-13)
CAS 64742-47-8; EINECS/ELINCS 265-149-8
Uses: Solvent for use in oil burners
Features: Meets requirements of NF D35300
Properties: Saybolt 30 color; dens. 815 kg/m^3 (15 C); visc. 2 mm^2/s (20 C); vapor pressure 1 mbar (20 C); i.b.p. 190 C; aniline pt. 77 C; flash pt. (PM) 64 C; ref. index 1.440 (20 C); surf. tens. 27.20 dynes/cm (20 C); KB value 29; 0.003% aromatics; 1 ppm benzene

Kerdane D75 [TotalFina-Elf]
Chem. Descrip.: Petroleum distillates, hydrotreated light (C10-15)
CAS 64742-47-8; EINECS/ELINCS 265-149-8
Uses: Solvent for use in oil burners
Features: Meets requirements of NF D35300
Properties: Saybolt 30 color; dens. 815 kg/m^3 (15 C); visc. 2.4 mm^2/s (20 C); vapor pressure 0 mbar (20 C); i.b.p. 190 C; aniline pt. 77 C; flash pt. (PM) > 62 C; ref. index 1.443 (20 C); surf. tens. 27.70 dynes/cm (20 C); KB value 30; 1% aromatics; 1 ppm benzene

Kerosol [Shell]
Chem. Descrip.: Naphtha (petroleum), hydrodesulfurized heavy
CAS 64742-81-0; EINECS/ELINCS 265-185-4
Uses: Solvent
Features: Slow evaporating; < 1% HAPs
Properties: Water-wh.; m.w. 160; vapor pressure 0.07 kPa (20 C); i.b.p. 188 C; pour pt. -39 C; cloud pt. -34 C; flash pt. (TCC) > 40 C; autoignition temp. 287 C; KB value 35

Ketrul 212 [TotalFina-Elf]
Chem. Descrip.: Kerosene, hydrodesulfurized (C12-14)
CAS 64742-81-0; EINECS/ELINCS 265-184-9
Uses: Solvent, diluent for paints, varnishes, lubricants, wood preservatives, resins, cleaning agents, degreasing, dewaxing
Features: High solvency
Properties: Saybolt 28 color; dens. 805 kg/m^3 (15 C); visc. 2.6 mm^2/s (20 C); vapor pressure 0 mbar (20 C); i.b.p. 185 C; aniline pt. 64 C; flash pt. (PM) 90 C; ref. index 1.450 (20 C); surf. tens. 27.80 dynes/cm (20 C); KB value 31; 20 vol.% aromatics; 1 ppm benzene

Ketrul D70 [TotalFina-Elf]
Chem. Descrip.: Petroleum distillates, hydrotreated light (C10-13)
CAS 64742-47-8; EINECS/ELINCS 265-149-8
Uses: Solvent for paints, varnishes, lubricants, wood preservatives
Features: Low aromatic
Properties: Saybolt 30 color; dens. 810 kg/m^3 (15 C); visc. 1.8 mm^2/s (20 C); vapor pressure 0.7 mbar (20 C); i.b.p. 185 C; aniline pt. 72 C; flash pt. (PM) 65 C; ref. index 1.434 (20 C); surf. tens. 26.90 dynes/cm (20 C); KB value 29; 0.01% aromatics; 1 ppm benzene

Ketrul D80 [TotalFina-Elf]
Chem. Descrip.: Petroleum distillates, hydrotreated light (C11-14)
CAS 64742-47-8; EINECS/ELINCS 265-149-8
Uses: Solvent for paints, varnishes, lubricants, wood preservatives, degreasing, dewaxing
Features: Low aromatic
Properties: Saybolt 30 color; dens. 815 kg/m^3 (15 C); visc. 2.2 mm^2/s (20 C); vapor pressure 0.3 mbar (20 C); i.b.p. 198 C; aniline pt. 72 C; flash pt. (PM) 75 C; ref. index 1.448 (20 C); surf. tens. 27.50 dynes/cm (20 C); KB value 27; 0.01% aromatics; 1 ppm benzene

Ketrul D85 [TotalFina-Elf]
Chem. Descrip.: Petroleum distillates, hydrotreated light (C12-13)
CAS 64742-47-8; EINECS/ELINCS 265-149-8
Uses: Solvent for paints, varnishes, lubricants
Features: Low aromatic
Properties: Saybolt 30 color; dens. 810 kg/m^3 (15 C); visc. 2.4 mm^2/s (20 C); vapor pressure 0 mbar (20 C); i.b.p. 206 C; aniline pt. 73 C; flash pt. (PM) 81 C; ref. index 1.445 (20 C); surf. tens. 27.50 dynes/cm (20 C); KB value 29; 0.01% aromatics; 1 ppm benzene

Ketrul D211 [TotalFina-Elf]
Chem. Descrip.: Petroleum distillates, hydrotreated light (C12-14)
CAS 64742-47-8; EINECS/ELINCS 265-149-8
Uses: Solvent for paints, varnishes, wood preservatives, degreasing, dewaxing
Features: Low aromatic
Properties: Saybolt 30 color; dens. 815 kg/m^3 (15 C); visc. 2.3 mm^2/s (20 C); vapor pressure 0 mbar (20 C); i.b.p. 210 C; aniline pt. 78 C; flash pt. (PM) 90 C; ref. index 1.438 (20 C); surf. tens. 27.40 dynes/cm (20 C); KB value 27; 0.01% aromatics; 1 ppm benzene

Ketrul D220 [TotalFina-Elf]
Chem. Descrip.: Petroleum distillates, hydrotreated light (C12-14)
CAS 64742-47-8; EINECS/ELINCS 265-149-8
Uses: Solvent for paints, varnishes, wood preservatives, degreasing, dewaxing
Features: Low aromatic
Properties: Saybolt 30 color; dens. 805 kg/m^3 (15 C); visc. 2.5 mm^2/s (20 C); vapor pressure 0 mbar (20 C); i.b.p. 220 C; aniline pt. 76 C; flash pt. (PM) 90 C; ref. index 1.439 (20 C); surf. tens. 27.50 dynes/cm (20 C); KB value 28; 0.03% aromatics; 1 ppm benzene

Ketrul HT [TotalFina-Elf]
Chem. Descrip.: Kerosene, hydrodesulfurized (C10-13)
CAS 64742-81-0; EINECS/ELINCS 265-184-9
Uses: Solvent, diluent for paints, varnishes, lubricants, wood preservatives, resins, cleaning agents, degreasing, dewaxing
Properties: Saybolt 30 color; dens. 800 kg/m^3 (15 C); visc. 1.8 mm^2/s (20 C); vapor pressure 1 mbar (20 C); i.b.p. 86 C; aniline pt. 64 C; flash pt. (PM) 70 C; ref. index 1.450 (20 C); surf. tens. 27.40 dynes/cm (20 C); KB value 32; 20 vol.% aromatics; 1 ppm benzene

Kristalloel 21 [Shell Netherlands]
Chem. Descrip.: Naphtha (petroleum), hydrodesulfurized heavy, contains xylene (6%), ethylbenzene (1.5%), 1,2,4-trimethylbenzene (2%)
UN 1300
Uses: Solvent
Features: Fast evaporating
Regulatory: EINECS, Canada DSL, Australia AICS, China IECSC, Japan MITI, Philippines PICCS, Korea TCCL listed
Properties: Water-wh. liq.; paraffinic odor; sol. 40 mg/l in water (20 C); m.w. 131; dens. 0.775 kg/l (15 C); kinematic visc. 0.98 mm^2/s (20 C); vapor pressure 0.8 kPa (20 C); m.p. < -15 C; b.p. 132-175 C; pour pt. < -30 C; flash pt. 23 C; autoignition temp. 324 C; surf. tens. 25.5 mN/m (20 C); KB value 39; elec. conduct. 7 pS/m (20 C); dielec. const. 2.1 (20 C)
Toxicology: TWA TLV (8 h) 100 ppm (UK); LD50 (oral) > 2000 mg/kg, (skin) > 2000 mg/kg; LC50 (inh.) > 5 mg/l; expected to be of low toxicity

by ing., inh., skin contact; aspiration into lungs may cause chem. pneumonitis which can be fatal; may cause headache, dizziness, nausea, narcosis, skin irritation; prolonged/repeated contact may cause defatting of skin, dermatitis; TSCA listed

Environmental: Expected to be readily biodeg.; potential to bioaccumulate; toxic to aquatic organisms; may cause long-term adverse effects in aquatic environment; prevent contamination of soil, water; prevent entry into drains, ditches, rivers

Precaution: Flamm.; LEL 0.7 vol.%; UEL 6.5 vol.%; may form flamm./explosive vapor-air mixt.; vapor is heavier than air, spreads along the ground; distant ignition is possible; electrostatic charges may be generated; prevent static discharge; ground equip.; risk of explosion

Hazardous decomp. prods.: Incomplete combustion: CO

Storage: Store @ ambient temps. in tightly closed containers in well-ventilated area, away from direct sunlight, other heat/ignition sources

Kristalloel 30 [Shell Netherlands]

Chem. Descrip.: Petroleum naphtha, hydrodesulfurized heavy, xylene (1%), mesitylene (1%), 1,2,4-trimethylbenzene (3%)

UN 1300

Uses: Solvent

Features: Faster evaporating

Regulatory: EINECS, Canada DSL, Australia AICS, China IECSC, Japan MITI, Korea TCCL, Philippines PICCS listed

Properties: Water-wh. liq.; paraffinic odor; sol. ≈ 20 mg/l in water (20 C); m.w. 140; dens. 788 kg/m^3 (15 C); kinematic visc. 1.23 mm^2/s (20 C); vapor pressure 300 Pa (20 C); f.p. < -15 C; i.b.p. 158 C; b.p. 153-198 C; pour pt. < -30 C; flash pt. 41 C; autoignition temp. 235 C; KB value 36

Toxicology: OEL TWA (8 h) 350 mg/m^3; expected to be of low toxicity by ing., inh., skin contact; aspiration into lungs may cause chem. pneumonitis which can be fatal; may cause headache, dizziness, nausea, narcosis, skin irritation; prolonged/repeated contact may cause defatting of skin, dermatitis

Environmental: Expected to be readily biodeg.; potential to bioaccumulate; toxic to aquatic organisms; may cause long-term adverse effects in aquatic environment; prevent contamination of soil, water; prevent entry into drains, ditches, rivers

Precaution: Flamm.; LEL 0.7 vol.%; UEL 6.5 vol.% (100 C); vapor is heavier than air, spreads along the ground; distant ignition is possible; electrostatic charges may be generated; prevent static discharge; ground equip.

Hazardous decomp. prods.: Combustion: CO

Storage: Store @ ambient temps. in tightly closed containers in well-ventilated area, away from direct sunlight, other heat/ignition sources

Kristalloel 60 [Shell Netherlands]

Chem. Descrip.: Petroleum naphtha, hydrodesulfurized heavy

CAS 64742-82-1; EINECS/ELINCS 265-185-4

UN 3082

Uses: Solvent

Features: Slower evaporating

Regulatory: EINECS, Canada DSL, Australia AICS, China IECSC, Japan MITI, Korea TCCL, Philippines PICCS listed

Properties: Water-wh. liq.; mild odor; sol. ≈ 42 mg/l in water (20 C); m.w. 158; dens. 799 kg/m^3 (15 C); kinematic visc. 1.64 mm^2/s (20 C); vapor pressure ≈ 100 Pa (20 C); f.p. < -15 C; i.b.p. 182 C; b.p. 182-212 C; pour pt. < -30 C; flash pt. 64 C; KB value 33

Toxicology: Expected to be of low toxicity by ing., inh., skin contact; aspiration into lungs may cause chem. pneumonitis which can be fatal; may cause sl. skin irritation; prolonged/repeated contact may cause defatting of skin, dermatitis; TSCA listed

Environmental: Expected to be readily biodeg.; potential to bioaccumulate; environmentally hazardous; LC50 (fish) ≤ 100 mg/l; sl. toxic to fish, algae; prevent contamination of soil, water; prevent entry into drains, ditches, rivers

Precaution: Combustible; LEL 0.7 vol.%; UEL 6.5 vol.% (100 C); may form flamm./explosive vapor-air mixt. above 60 C; electrostatic charges may be generated; prevent static discharge

Hazardous decomp. prods.: Combustion: CO

Storage: Store @ ambient temps.

Krystal Kleer Castor Oil [Süd-Chemie Inc/Rheologicals]

Chem. Descrip.: Castor oil USP contg. an antioxidant

CAS 8001-79-4; EINECS/ELINCS 232-293-8

Uses: Gloss aid, emollient, dye solvent in personal care prods., lipsticks, pharmaceuticals, and urethane applics.

Features: High quality oil

Properties: Gardner 1 color; sp.gr. 0.959; dens. 7.98 lb/gal; visc. 7.5 stokes; acid no. 1; iodine no. 86; sapon. no. 180; hyd. no. 164

Kwik Dri [Ashland]

Chem. Descrip.: Aliphatic naphtha

EINECS/ELINCS 232-443-2

Uses: Solvent for rubber, latexes

Properties: Water-wh. liq.; sp.gr. 0.7883-0.7711; b.p. 310-340 F; flash pt. (TCC) 105 F; KB no. 35

L

Labrafac® Hydrophile [Gattefosse; Gattefosse SA]
Chem. Descrip.: Caprylic/capric triglyceride PEG-4 complex
Uses: Bioavailability enhancer, emulsifier, solvent, solubilizer for pharmaceutical orals, topicals, vaginals, rectals, and nasals, cosmetics (creams, lotions, shampoos, bath oils), veterinary prods.
Features: Hydrophilic; rancidless
Properties: Gardner < 5 oily liq., faint odor; sol. in veg. oil, alcohol; partly disp. in water; HLB 4-5; acid no. < 2; iodine no. < 3; sapon. no. 265-285; nonionic
Toxicology: LD0 (oral, rat) > 20 ml/kg; sl. skin and eye irritant

Labrafac® Lipophile [Gattefosse; Gattefosse SA]
Chem. Descrip.: Caprylic/capric triglyceride
CAS 65381-09-1; EINECS/ELINCS 265-724-3
Uses: Excipient for pharmaceutical gelatin capsules; oily vehicle for oral, topical, rectal, and parenteral uses; solvent, superfatting agent for cosmetic, pharmaceutical, or veterinary emulsions and microemulsions, aerosols, capsules, gels, creams/lotions, ointments, shampoos, massage and sun oils
Features: Nonrancidifying; temp. stable
Regulatory: DAB/BP/EP compliance
Properties: Gardner < 1 oily liq., odorless to faint odor; HLB 1; acid no. < 0.2; iodine no. < 2; sapon. no. 320-340
Toxicology: LD50 (oral, rat) > 10 ml/kg; sl. eye irritant, nonirritating to skin

Labrafil® Isostearique [Gattefosse; Gattefosse SA]
Chem. Descrip.: Triisostearin PEG-6 esters
Uses: Solvent, amphiphilic agent for improving drug delivery; excipient for aerosols, nasal preps., creams, massage oils, cosmetics, veterinary preps.
Features: Hydrophilic; nonrancidifying
Properties: Gardner < 5 oily liq., faint odor; HLB 3-4; acid no. < 2; iodine no. < 15; sapon. no. 150-170; nonionic
Toxicology: LD50 (oral, rat) > 20 ml/kg; nonirritating to skin and eyes

Labrasol® [Gattefosse; Gattefosse SA]
Chem. Descrip.: PEG-8 caprylic/capric glycerides
CAS 85536-07-8
Uses: Solvent, excipient, solubilizer, bioavailability enhancer for pharmaceuticals (oral liq., capsules), cosmetics, veterinary preps.; surfactant for microemulsions; wetting agent, penetration enhancer, absorp. enhancer for topical, rectal, and oral pharmaceuticals; superfatting agent for bath preps., shampoos
Features: Hydrophilic; amphiphilic
Regulatory: EP compliance; DMF no. 5734
Properties: Gardner < 5 oily liq., faint odor; sol. in water, veg. oil, propylene glycol; very sol. in ethanol, chloroform, methylene chloride; sp.gr. 1.060-1.070; visc. 80-110 mPa·s; HLB 14.0; acid no. < 1; iodine no. < 2; sapon. no. 85-105; ref. index 1.450-1.470; nonionic; 100% conc.
Toxicology: LD50 (oral, rat) 22 g/kg; sl. ocular irritant at 0.1 ml; nonirritating to skin
Storage: Preserve in org. container; prevent exposure to air, light, heat, and moisture

Lacolene™ [Ashland]
Chem. Descrip.: Aliphatic naphtha
EINECS/ELINCS 232-443-2
Uses: Solvent for rubber, latexes
Properties: Water-wh. liq.; sp.gr. 0.725-0.7467; b.p. 200-226 F; flash pt. (TCC) 18 F; KB no. 35

Lactol Spirits 2300 [CITGO]
Chem. Descrip.: Aliphatic hydrocarbon primarily C6-8
Uses: Solvent
Features: Low-boiling; meets definition of non-photochemically reactive as defined by 66/3 Rule requirements; sp.gr. 0.725-0.740 (60/60 F); dens. 6.09 lb/gal (15.6 C); i.b.p. 200 F min.; flash pt. (TCC) < 20 F; ref. index 1.413; KB value 39.5
Properties: Colorless clear liq.; lt. aliphatic odor; misc. with most common org. solvs.
Precaution: Flamm. liq.

Lanesta L [Croda/Westbrook]
Chem. Descrip.: Isopropyl lanolate
CAS 63393-93-1; EINECS/ELINCS 264-119-1
Uses: Emollient, lubricant, gloss aid, water repellent for baby oils, cleansers, foam baths/gels, sunscreens; carrier for actives; solvent for oil-sol. dyes for makeup
Features: Nonsticky
Properties: Pale yel. low-visc. liq.; faint char. odor; sol. in acetone, ethanol, IPA, min. oil, IPM; insol. in water; sp.gr. 0.850-0.870; acid no. 7 max.; sapon. no. 160-190; pour pt. 10 C; flash pt. > 100 C
Toxicology: LD50 (oral, rat) > max. pract. dose of 16 g/kg; nontoxic; nonirritating to skin and eyes
Environmental: Nonhazardous
Precaution: Nonflamm. but will burn if strongly heated
Storage: 2 yrs shelf life when stored under normal conditions; may develop some sediment on prolonged storage; redissolve by warming; contains 150 ppm BHT

Leksol™ [Amity UK]
Chem. Descrip.: n-Propyl bromide (< 95%)
CAS 106-94-5; EINECS/ELINCS 203-445-0
UN 1993
Uses: Precision cleaning solvent for cleaning printed circuit boards, elec. components, motors, and other temp. sensitive items; solvent for coatings, adhesives, aerosols, and cold cleaning
Features: Nonflamm., non-CFC; drop-in replacement for 1,1,1 trichloroethane and other chlorinated solvs., incl. TCE, methylene chloride, and also HCFCs
Properties: Colorless clear liq.; sweet odor; negligible sol. in water; sp.gr. 1.32; vapor pressure 110.4 mm Hg (20 C); f.p. -166 C; b.p. 69.5 C; flash pt. (TCC) none; KB value 126; surf. tens. 25.9 dynes/cm; sp.heat 0.25
Toxicology: May irritate eyes, skin, respiratory tract; inh. may cause headache, dizziness, nausea, vomiting, narcosis in confined or poorly ventilated areas; prolonged skin contact may cause dermatitis; ing. may irritate digestive tract; noncarcinogenicTSCA listed
Precaution: Avoid temps. > 204.4 C, strong acids and oxidizers
Hazardous Ingredients: Halogenated hydrocarbon mixt.
Hazardous decomp. prods.: CO, CO_2, and hydrogen bromide
HMIS: Health 1, Flammability 0, Reactivity 0
Storage: Store tightly closed; protect from extreme temps.

Leksol™ AL [Amity UK]
Chem. Descrip.: Propyl bromide (< 95%)
CAS 106-94-5; EINECS/ELINCS 203-445-0
Uses: Precision cleaning solvent for vapor degreasing, ultrasonic cleaning, cold cleaning, PCB's, electronic components, elec. equip., metal

components, optical lenses, instruments, aerospace parts and equip.
Features: Removes ionic contamination, nonresin fluxes, oil, grease, resins, inks, adhesives, silicone oils; fast drying; no residue; CFC-free; reduced emissions
Properties: Colorless liq.; low sweet odor; sol. 2.4 g/l in water; sp.gr. 1.31; visc. 0.42 cps; vapor pressure 110.8 mm Hg (20 C); m.p. -110 C; b.p. 69.5 C; flash pt. none; KB value 120; surf. tens. 25.9 dynes/cm
Toxicology: LD50 (oral, rat) 4260 mg/kg; LC50 (rat, inh., 0.5 h) 253,000 mg/m^3; PEL TLV 100 ppm; low toxicity; harmful by inh.; vapors may cause eye/nose/respiratory tract irritation; sl. irritating to eyes/skin; may cause irregular pulse, CNS/systemic effectsTSCA listed
Environmental: Prevent spills from entering drains or water courses; low bioaccumulation potential; not readily biodeg., but low atmospheric lifetime
Precaution: LEL 4.0%, UEL 7.8%; incompat. with strong oxidizing agents; compat. with most metals except calcium, finely divided Al, alkali metals
Hazardous decomp. prods.: Hydrogen halides, CO
Storage: Store in cool, well-ventilated space away from heat, open sunlight; keep in original tightly closed container

Leosolb 703B [Lion]
Uses: Detergent, solvent for paints and inks; urethane visc. reducer
Properties: Liq.; visc. 9.2 cP

LexFeel® 7 [Inolex]
Chem. Descrip.: Neopentyl glycol diheptanoate
CAS 68855-18-5; EINECS/ELINCS 272-469-1
Uses: Emollient, solvent, coupling agent, lubricant for skin care creams/lotions; emollient for antiperspirants
Features: Drier skin feel; less oily after-feel; low visc.; exc. oxidative stability; does not contain animal-derived raw materials
Regulatory: DOT nonregulated; SARA §302/311/312/313 nonreportable; Calif. Prop. 65 nonreportable; Canada DSL, EINECS, Korea ECL, Australia AICS, Philippines PICCS listed
Properties: Yel.-amber clear liq.; odorless; insol. in water; sp.gr. 0.943; dens. 7.86 lb/gal; kinematic visc. 5.5 cSt (40 C); m.p. < 25 C; b.p. > 200 C; acid no. 0.05; sapon. no. 340; flash pt. (COC) 180 C; 0.1% moisture
Toxicology: May cause minimal eye irritation; nonirritating to skin; relatively low single-dose oral toxicity; ing. of sm. amts. not likely to cause injury; ing. of larger amts. may cause injuryTSCA listed
Precaution: Containers can build up pressure if exposed to heat (fire) and may rupture
Hazardous decomp. prods.: No unusual decomp. prods. known
HMIS: Health 1, Flammability 1
Storage: Keep container closed when not in use

Lexol® 3975 [Inolex; S. Black]
Chem. Descrip.: Isopropyl palmitate, isopropyl myristate, isopropyl stearate
Uses: Emollient, solvent for cosmetics and topical pharmaceuticals
Features: Replacement for IPM or IPP; oxidative resist.
Properties: Colorless liq., odorless; sol. in acetone, benzene, CCl_4, castor oil, chloroform, ethanol, heptane, IPA; insol. in water; sp.gr. 0.850-0.858; visc. 7 cps; f.p. 11 C; acid no. 1 max.; iodine no. 1 max.; sapon. no. 184-194; flash pt. (PMCC) 169 C; ref. index 1.4367; 100% conc.

Lexol® GT-855 [Inolex]
Chem. Descrip.: Caprylic/capric triglyceride
CAS 65381-09-1; EINECS/ELINCS 265-724-3
Uses: Emollient, moisturizer for creams/lotions, bath oils, lipstick, makeup; solvent for perfume and flavor ingreds.; vehicle for medicinals, antibiotics, vitamins; solubilizer
Features: Nonoily skin feel; oxidative stability
Properties: APHA 100 max. clear liq., odorless; tasteless; alcohol-sol.; sp.gr. 0.943; visc. 27 cps; f.p. -19 C; acid no. 0.1 max.; iodine no. 1 max.; sapon. no. 325-355; flash pt. (PMCC) 224 C; ref. index 1.4479; 100% conc.

Lexol® GT-865 [Inolex; S. Black]
Chem. Descrip.: Caprylic/capric triglyceride
CAS 65381-09-1; EINECS/ELINCS 265-724-3
Uses: Emollient, moisturizer, emulsion stabilizer, vehicle for cosmetic creams/lotions, bath oils, preshave lotions, lipsticks, makeup; solvent for perfume; carrier for flavors; vehicle for medicinals, antibiotics, vitamins, org. sunscreens
Properties: APHA 100 max. clear liq., odorless, tasteless; sol. in alcohol; sp.gr. 0.947; visc. 25 cps; f.p. -19 C; acid no. 0.1 max.; iodine no. 1 max.; sapon. no. 335-355; flash pt. (PMCC) 233 C; ref. index 1.4471; 100% conc.

Lexol® IPL [Inolex]
Chem. Descrip.: Isopropyl laurate
CAS 10233-13-3; EINECS/ELINCS 233-560-1
Uses: Emollient, solvent, coupling agent, lubricant for skin care creams and lotions; lubricant for antiperspirant systems
Features: Dry skin feel; less oily after-feel; exc. hydrolytic stability
Regulatory: DOT nonregulated; SARA §302/311/312/313 nonreportable; Calif. Prop. 65 nonreportable; Canada DSL, EINECS, Korea ECL, Australia AICS, Philippines PICCS listed
Properties: Colorless clear liq.; mild, char. odor; sol. in alcohol, ether, acetone; insol. in water; sp.gr. 0.85; dens. 7.09-7.13 lb/gal; visc. 4 cSt; f.p. -8 C; m.p. < 25 c; b.p. < 180 C (20 mm Hg); acid no. 0.2; sapon. no. 230; flash pt. (COC) 140 C; ref. index 1.43; pH 6 (5% aq.); VOC 100%; 100% act.; 0.1% moisture content
Toxicology: May cause mild eye/skin irritation; low toxicityTSCA listed
Precaution: Containers can build up pressure if exposed to fire; vapors may form ignitable mixt. with air
Hazardous decomp. prods.: None known
HMIS: Health 1, Flammability 1
Storage: Store in original container below 40 C

Lexol® IPM [Inolex; S. Black]
Chem. Descrip.: Isopropyl myristate NF
CAS 110-27-0; EINECS/ELINCS 203-751-4
Uses: Emollient, solvent, penetrant, cloud pt. depressant in personal care prods.
Properties: Colorless clear liq.; odorless; sol. in acetone, benzene, CCl_4, castor oil, chloroform, ethanol, heptane, IPA; insol. in water; m.w. 270; sp.gr. 0.847-0.854; dens. 7.1 lb/gal; visc. 4.8 cp; f.p. 3 C; b.p. 170 C; acid no. 1 max.; iodine no. 1 max.; sapon. no. 202-212; flash pt. 305 F; ref. index 1.433

Lexol® IPM-NF [Inolex; S. Black]
Chem. Descrip.: Isopropyl myristate NF
CAS 110-27-0; EINECS/ELINCS 203-751-4
Uses: Emollient, solvent, penetrant for cosmetics (bath oils, elec. preshaves), pharmaceuticals
Features: Low visc., low f.p., outstanding spreading, good sol., high oxidative stability
Properties: Colorless clear liq., odorless; sol. in acetone, benzene, CCl_4, castor oil, chloroform, ethanol, heptane, IPA; insol. in water, glycerol, propylene glycol; sp.gr. 0.852; visc. 0.5 cps; f.p. 2 C; acid no. 0.1; iodine no. 0.3; sapon. no. 208; flash pt. (COC) 155 C; ref. index 1.4327; 100% conc.

Lexol® PG-800 [Inolex]
Chem. Descrip.: Propylene glycol dioctanoate
CAS 56519-71-2; EINECS/ELINCS 301-185-3
Uses: Emollient, moisturizer for cosmetic/pharmaceutical creams/lotions, topicals, lipsticks, glossers, makeup bases, bath oils, aftershaves; solvent, carrier, vehicle for flavors, fragrance, vitamins, antibiotics, medicinals
Features: Nonoily feel, oxidation stability
Properties: Colorless clear liq., char. odor; sol. in alcohol, min. oil, acetone; sp.gr. 0.918; visc. 10 cps; f.p. -34 C; acid no. 1 max.; iodine no. 1 max.; sapon. no. 320-340; flash pt. (COC) 272 C; ref. index 1.4350; 100% conc.

Lexol® PG-865 [Inolex; S. Black]
Chem. Descrip.: Propylene glycol dicaprylate/dicaprate
CAS 68583-51-7; EINECS/ELINCS 271-516-3
Uses: Emollient, moisturizer, lubricant for creams, lotions, makeup, bath oils, pre-electric shave lotions, aerosol systems; solubilizer, solvent, coupling agent, vehicle for flavors, fragrances, pigmented cosmetics, vitamins, antibiotics, medicinals
Features: Nonoily skin deposition
Properties: APHA 100 max. clear liq., odorless; sol. in alcohol, min. and veg. oil, acetone; sp.gr. 0.922; visc. 10 cps; f.p. -38 C; acid no. 0.1 max.; iodine no. 1 max.; sapon. no. 315-335; flash pt. (PMCC) 195 C; ref. index 1.4391; 100% conc.
Storage: Store in tightly closed original containers @ 13-38 C to avoid deterioration or contamination; store bulk liq. under dry inert atmosphere in SS or aluminum vessels

Linpar® 15-V Normal Paraffin [Sasol N. Am.]
Chem. Descrip.: C14-16 paraffins

CAS 90622-46-1; EINECS/ELINCS 292-448-0
Uses: Solvent
Regulatory: DOT nonhazardous; SARA §302/31/312/313 nonreportable; Calif. Prop. 65 nonreportable; EINECS, Canada DSL, Australia AICS, Japan MITI listed
Properties: Colorless oily liq.; hydrocarbon odor; negligible sol. in water; sp.gr. 0.768 (60/60 F); visc. 2.1 cSt (40 C); vapor pressure 0.09 mm Hg (20 C); m.p. 4 C; b.p. 244-251 C; flash pt. (PM) 114 C; autoignition temp. 210 C; 100% act.
Toxicology: ACGIH TWA (8 h) 5 mg/m^3 (mists); LD50 (oral, rat) > 5 g/kg, (skin, rabbit) > 2 g/kg; LC50 (inh, rat, 4 h) 5.8-8.7 mg/l; may cause eye/skin irritation; high vapor concs. may cause headache, stupor, dizziness, throat/eye irritation; prolonged breathing of mist or vomiting following ing. may cause serious lung injury, possibly death; TSCA listed
Precaution: Flamm. limits in air 0.5-7.0 vol.%; electrostatic accumulation hazard; ground equip.; avoid high temps.; incompat. with strong oxidizers
Hazardous decomp. prods.: Combustion: CO_x
HMIS: Health 1, Flammability 1, Reactivity 0
Storage: Store @ ambient temps. and pressure

Linpar® 1416-V Normal Paraffin [Sasol N. Am.]
Chem. Descrip.: C14-16 paraffins
CAS 90622-46-1; EINECS/ELINCS 292-448-0
Uses: Chemical intermediate for chlorination (flame retardants, plasticizers, EP additives), oxidation to alcohols, nitration to nitro-alkanes and amines, sulfoxidation or sulfochlorination for surfactant or plasticizer mfg.; raw material for microbiological oxidation or fermentation process; solvent; lamp fuel; direct and indirect food contact applics.
Features: High purity; LVP-VOC
Regulatory: FDA approvals for direct and indirect food contact; DOT nonhazardous; SARA §302/31/312/313 nonreportable; Calif. Prop. 65 nonreportable; EINECS, Canada DSL, Australia AICS, Japan MITI listed as its components (tetradecane, pentadecane, hexadecane)
Properties: Water-wh. oily liq.; mild hydrocarbon odor; negligible sol. in water; m.w. 205; sp.gr. 0.773 (16/16 C); dens. 6.45 lb/gal (60 F); visc. 2.5-2.7 cSt (40 C); vapor pressure 0.05 mm Hg (20 C); f.p. 8 C; b.p. 254-283 C; flash pt. (PM) 121 C; autoignition temp. 216 C; KB value 20; VOC 6.45 lb/gal; 100% act.
Toxicology: ACGIH TWA (8 h) 5 mg/m^3 (mists); LD50 (oral, rat) > 5 g/kg, (skin, rabbit) > 2 g/kg; LC50 (inh, rat, 4 h) > 5.8 mg/l; may cause eye/skin irritation; high vapor concs. may cause headache, stupor, dizziness, throat/eye irritation; prolonged breathing of mist or vomiting following ing. may cause serious lung injury, possibly death; TSCA listed
Precaution: Flamm. limits in air 0.5-7.0 vol.%; electrostatic accumulation hazard; ground equip.; avoid high temps.; incompat. with strong oxidizers
Hazardous decomp. prods.: Combustion: CO_2, CO, other org. compds.
HMIS: Health 1, Flammability 1, Reactivity 0
Storage: Store @ ambient temps. and pressure

Lipocol P-15 [Lipo]
Chem. Descrip.: PPG-15 stearyl ether
CAS 25231-21-4
Uses: Surfactant, emulsifier, dispersant, solubilizer, visc. control agent, emollient, solvent, lubricant for cosmetics, pharmaceuticals
Properties: Clear liq.; mild char. odor; sl. sol. in water; acid no. 1.5 max.; hyd. no. 58-70; flash pt. 300 F; 0.5% max. moisture
Toxicology: Not considered a toxic substanceTSCA listed
Environmental: Biodeg.
Precaution: Container may be hazardous when emptied
Hazardous decomp. prods.: Thermal decomp. may produce CO, CO_2
Storage: Keep containers closed until use; do not store in heat or direct sunlight

Lipolan R [Lipo]
Chem. Descrip.: Lanolin oil
CAS 70321-63-0; EINECS/ELINCS 274-559-6
Uses: Emollient, emulsifier, solubilizer, stabilizer, conditioner, and moisturizer for cosmetics, pharmaceuticals, skin care prods., makeup, lipstick, shampoos, rinses, soap/bath prods., shaving preps., sun prods., ointments, acne preps., veterinary prods.; spreading agent, cosolvent, plasticizer, lubricant for personal care prods.; dispersant for pigments
Properties: Yel. to amber liq.; mild char. odor; sol. in min. and veg. oils, isopropyl esters, and anhyd. IPA; insol. in water; acid no. 2 max.; iodine no. 18-36; sapon. no. 85-110; cloud pt. 18 C max.; nonionic

Liponate GC-K [Lipo]
Chem. Descrip.: Caprylic/capric triglyceride
CAS 65381-09-1; EINECS/ELINCS 265-724-3
Uses: Emollient, thickener, visc. control agent, pigment dispersant for cosmetic creams/lotions, bath preps.; clouding agent, diluent for vitamins, solvent for oil-disp. materials, useful in flavor systems and dietary formulations
Features: Adjusts rub-in and afterfeel in cosmetics
Regulatory: FDA 21CFR §170.30; kosher grade
Properties: Colorless to sl. yel. oily liq., bland odor and taste; sol. in anhyd. alcohol, min. and veg. oils; insol. in water; sp.gr. 0.935-0.955; acid no. 0.1 max.; iodine no. 1 max.; sapon. no. 335-360; nonionic

Liponate PC [Lipo]
Chem. Descrip.: Propylene glycol dicaprylate/dicaprate
CAS 68583-51-7; EINECS/ELINCS 271-516-3
Uses: Emollient, thickener, visc. control agent, pigment dispersant for cosmetic creams/lotions, bath preps.; clouding agent, diluent for vitamins, solvent for oil-disp. materials, useful in flavor systems and dietary formulations
Features: Adjusts rub-in and afterfeel
Regulatory: FDA 21CFR §172.860, 172.856; kosher grade avail.
Properties: Colorless clear liq., bland odor and taste; sol. in min. and veg. oils, ethanol; water-insol.; acid no. 0.1 max.; iodine no. 1 max.; sapon. no. 315-335; nonionic

Lipovol SES [Lipo]
Chem. Descrip.: Sesame (Sesamum indicum) oil
CAS 8008-74-0; EINECS/ELINCS 232-370-6
Uses: Emollient, solvent, vehicle in cosmetics, toiletries, pharmaceuticals, foods; gloss aid, spreading agent for pigmented sticks and pot prods.
Features: Offers lt., nontacky feel
Properties: Yel. clear liq.; bland char. odor; sol. in isopropyl esters and min. oil; insol. in water; sp.gr. 0.916-0.921; acid no. 0.2 max.; iodine no. 103-116; sapon. no. 188-195

Liquid Absorption Base T [Croda Inc]
Chem. Descrip.: Mineral oil, lanolin alcohol
Uses: Surfactant; primary oil phase ingred. in o/w emulsions; solvent for oil-sol. dyes; solubilizer for oil-sol. actives in pharmaceuticals
Features: Mild
Properties: Clear yel. liq.
Use Level: 2-10%

Liquid Medilan™ [Croda/Westbrook]
Chem. Descrip.: Lanolin oil
EINECS/ELINCS 274-559-6
Uses: Solvent for actives in pharmaceutical preps.; visc. reducer for petrolatum bases
Properties: Lt. amber liq.; sol. in min. oil; insol. in water, IPA, propylene glycol; HLB 9.64
Use Level: 2-10%
Toxicology: LD50 (oral, rat) > 20 g/kg; mild skin/eye irritant

Liquid Medilan™ Ultra [Croda/Westbrook]
Chem. Descrip.: Lanolin oil
EINECS/ELINCS 274-559-6
Uses: Solvent for actives in pharmaceutical preps.; visc. reducer for petrolatum bases
Properties: Clear liq.; odorless; sol. in min. oil; insol. in water, IPA, propylene glycol
Use Level: 2-10%
Toxicology: LD50 (oral, rat) > 20 g/kg; mild skin/eye irritant

Lonzest® 143-S [Lonza]
Chem. Descrip.: Myristyl propionate
CAS 6221-95-0; EINECS/ELINCS 226-300-9
Uses: Emollient, penetrant, spreading agent in hair and skin cosmetics; perfume solvent in bath oils, antiperspirants, preshave lotions, sunburn lotions; solvent/emollient in lipsticks, waterless hand cleaners, makeup, facial cleaners; humectant
Properties: APHA 10 clear liq., odorless; sol. in acetone, CCl_4, castor, lanolin, peanut, silicone, and cottonseed oils, ethanol, ethyl acetate, heptane, IPA, toluene; sp.gr. 0.852; visc. 5 cps; f.p. -5 C; acid no. 0.5 max.; sapon. no. 195-210; ref. index 1.433; pH 6-8 (10% aq. disp.); nonionic; 100% act.

Toxicology: LD50 (oral) > 39 g/kg; not a primary skin or eye irritant; non-comedogenic

Lowenol Solvent 402 [Lowenstein Dyes & Cosmetics]
Uses: Solvent, visc. builder for semipermanent liq. hair color systems
Properties: Nonionic

Lumulse® GMO [Lambent Tech.]
Chem. Descrip.: Glyceryl oleate
CAS 25496-72-4; EINECS/ELINCS 203-827-7
Uses: Emulsifier, opacifier, stabilizer for foods, pharmaceuticals, and cosmetics; stabilizer in pet foods, margarine; dough conditioner for tacos and tortillas; lubricant for textiles; anti-icing fuel additive; solvent; mold release agent; clarifier and antiblocking agent in plastics (food pkg.); in food pkg. adhesives, coatings, paper/paperboard, textiles; defoamer in food-contact paper/paperboard
Features: Improves overrun in frozen desserts
Regulatory: FDA 21CFR §166, 175.105, 175.320, 176.210, 177.2800, 181.27; also avail. in kosher grade
Properties: Yel. opaque liq., bland odor; insol. in water; sol. in ethanol, propylene glycol, org. solvs., veg. and min. oils; sp.gr. 0.95; dens. 8.0 lb/gal; visc. 90 cSt (40 C); b.p. > 200 C (760 mm); HLB 3.4; acid no. 2 max.; iodine no. 96; sapon. no. 160-170; pour pt. 0 C; flash pt. 215 C; nonionic; 100% conc.
Use Level: 0.1-0.5%
Toxicology: May cause eye irritation
Precaution: Avoid strong oxidizing agents
Storage: Store in well-ventilated areas @ 50-120 F; keep away from oxidizing agents, excessive heat, ignition sources

Lumulse® GMT [Lambent Tech.]
Chem. Descrip.: Glycerol monotallate
CAS 61789-12-6
Uses: Emulsifier, pigment dispersant, lubricant for industrial applics.; rust preventive in oils; anti-icing fuel additive; textile lubricant for syn. fibers and spin finishes; solvent; mold release agent; w/o emulsifier for cutting fluids and drawing compds.
Properties: Amber liq.; sol. in ethyl alcohol, propylene glycol, vegetable and min. oils, and org. solvs.; dens. 7.97 lb/gal; visc. 130 cps; HLB 3.4; acid no. 5 max.; iodine no. 73; sapon. no. 165; pour pt. -18 C; flash pt. (COC) > 250 C; nonionic
Storage: Avoid storage at temps. > 90 F

Lumulse® GMT-40 [Lambent Tech.]
Chem. Descrip.: Glycerol monotallate
CAS 61789-12-6
Uses: Emulsifier, pigment dispersant, lubricant for industrial applics.; rust preventive in oils; anti-icing fuel additive; textile lubricant for syn. fibers and spin finishes; solvent; mold release agent; w/o emulsifier for cutting fluids and drawing compds.
Properties: Amber liq.; sol. in ethyl alcohol, propylene glycol, vegetable and min. oils, and org. solvs.; dens. 7.97 lb/gal; visc. 11 cSt (210 F); HLB 3.4; acid no. 6 max.; iodine no. 73; sapon. no. 160; pour pt. -18 C; flash pt. (COC) > 250 C; nonionic
Storage: Avoid storage at temps. > 90 F

Lusolvan® FBH [BASF AG]
Chem. Descrip.: Diisobutyl ester of a mixt. of dicarboxylic acids
Uses: Coalescing agent, solvent for architectural coatings, interior paints, wood paints and stains
Features: High boiling

Lutrol® E 300 [BASF AG]
Chem. Descrip.: PEG-6 USP/NF, Ph.Eur.
CAS 25322-68-3; EINECS/ELINCS 220-045-1
Uses: Emulsifier, emollient, lubricant, and solvent for liq. preps.; solvent, solubilizer for act. substances and excipient in pharmaceuticals (liq. and semisolid preps.); carrier for dissolved or suspended drugs in soft gelatin capsules; base for ointments, suppositories; solvent for removal of phenol, cresol, aniline from surf. of skin burnt by any of these toxic compds.
Properties: Colorless liq.; pract. odorless and tasteless; sol. in water, ethanol, acetone, glycols, chloroform; insol. in ether, paraffin, oils, fats; m.w. 285-315; dens. 1.13 g/cm^3; visc. 80-105 mPa·s (20 C); hyd. no. 340-394; pH 4.5-7.5 (5% aq.); nonionic; 100% conc.; ≤ 1% water
Storage: 24 mos shelf life when stored dry in closed containers

Lutrol® E 400 [BASF AG]
Chem. Descrip.: PEG-8 USP/NF, Ph.Eur.
CAS 25322-68-3; EINECS/ELINCS 225-856-4
Uses: Emulsifier, emollient, lubricant, and solvent for liq. preps.; solvent, solubilizer for act. substances and excipient in pharmaceuticals (liq. and semisolid preps.); carrier for dissolved or suspended drugs in soft gelatin capsules; base for ointments, suppositories; solvent for removal of phenol, cresol, aniline from surf. of skin burnt by any of these toxic compds.
Properties: Colorless liq.; pract. odorless and tasteless; sol. in water, ethanol, acetone, glycols, chloroform; insol. in ether, paraffin, oils, fats; m.w. 380-420; dens. 1.13 g/cm^3; visc. 105-130 mPa·s (20 C); hyd. no. 264-300; pH 4.5-7.5 (5% aq.); nonionic; 100% conc.
Storage: 24 mos shelf life when stored dry in closed containers

Lutrol® E 600 [BASF AG]
Chem. Descrip.: PEG-12
CAS 25322-68-3; EINECS/ELINCS 229-859-1
Uses: Solvent, solubilizer for act. substances, excipient in pharmaceuticals (liq. and semisolid preps.); carrier for dissolved or suspended drugs in soft gelatin capsules; base for ointments, suppositories; solvent for removal of phenol, cresol, aniline from surf. of skin burnt by any of these toxic compds.
Properties: Liq. or paste; sol. in water, ethanol, acetone, glycols, chloroform; insol. in ether, paraffin, oils, fats; m.w. 570-630; visc. 9.9-11.3 cSt; hyd. no. 178-197; pH 4.5-7.5 (5% aq.)
Storage: 24 mos shelf life when stored dry in closed containers

Lutrol® E 1500 [BASF AG]
Chem. Descrip.: PEG-32 USP/NF, FCC, EP
CAS 25322-68-3; EINECS/ELINCS 203-989-9
Uses: Solvent, emulsifier, carrier, solubilizer, absorp. enhancer, bioavailability enhancer for drugs that are insol. or sparingly sol. in water, e.g., steroids, fat-sol. vitamins; binder in mfg. of sugar-coated and uncoated tablets; controls release in sustained-release formulations; used with buffers in laxative formulations; antisticking agent , gloss aid for sugar-coated tablets; film-former, plasticizer in tablet coatings; base for ointments and suppositories
Properties: Colorless microbeads; 100% < 1000 µm, 90% < 500 µm; almost odorless and tasteless; sol. in water, ethanol, acetone, glycols, chloroform; insol. in ether, paraffin, oils, fats; m.w. 1350-1650; bulk dens. 0.65 g/cm^3; visc. 34-50 mPa·s (20 C, 50% aq.); f.p. 42-48 C; hyd. no. 70-80; pH 4.5-7.5 (5% aq.); nonionic; 100% conc.; ≤ 1% water
Storage: 24 mos shelf life when stored dry in closed containers

Lutrol® E 3350 [BASF AG]
Chem. Descrip.: PEG-75 USP/NF, FCC, EP
CAS 25322-68-3; EINECS/ELINCS 203-989-9
Uses: Solvent, emulsifier, carrier, solubilizer, absorp. enhancer, bioavailability enhancer for drugs that are insol. or sparingly sol. in water, e.g., steroids, fat-sol. vitamins; binder in mfg. of sugar-coated and uncoated tablets; controls release in sustained-release formulations; used with buffers in laxative formulations; antisticking agent , gloss aid for sugar-coated tablets; film-former, plasticizer in tablet coatings; base for ointments and suppositories
Properties: Colorless microbeads; 99% min. < 710 µm, 85 ± 15% < 250 µm; almost odorless and tasteless; sol. in water, ethanol, acetone, glycols, chloroform; insol. in ether, paraffin, oils, fats; m.w. 3015-3685; bulk dens. 0.60 g/cm^3; visc. 83-120 mPa·s (20 C, 50% aq.); f.p. 53-57 C; hyd. no. 30-38; pH 4.5-7.5 (5% aq.)
Storage: 24 mos shelf life when stored dry in closed containers

Lutrol® E 4000 [BASF AG]
Chem. Descrip.: PEG-90 USP/NF, FCC, EP
CAS 25322-68-3; EINECS/ELINCS 203-989-9
Uses: Solvent, emulsifier, carrier, solubilizer, absorp. enhancer, bioavailability enhancer for drugs that are insol. or sparingly sol. in water, e.g., steroids, fat-sol. vitamins; binder in mfg. of sugar-coated and uncoated tablets; controls release in sustained-release formulations; used with buffers in laxative formulations; antisticking agent , gloss aid for sugar-coated tablets; film-former, plasticizer in tablet coatings; base for ointments and suppositories
Properties: Colorless microbeads; 95% < 1000 µm, 50% < 500 µm; almost odorless and tasteless; sol. in water, ethanol, acetone, glycols, chloroform; insol. in ether, paraffin, oils, fats; m.w. 3600-4400; bulk dens. 0.60 g/cm^3; visc. 110-170 mPa·s (20 C, 50% aq.); f.p. 53-59 C; hyd. no. 25-32; pH 4.5-7.5 (5% aq.); nonionic; 100% conc.; ≤ 1% water
Storage: 24 mos shelf life when stored dry in closed containers

Lutrol® E 6000 [BASF AG]
Chem. Descrip.: PEG-125 USP/NF, FCC, EP
CAS 25322-68-3; EINECS/ELINCS 203-989-9
Uses: Solvent, emulsifier, carrier, solubilizer, absorp. enhancer, bioavailability enhancer for drugs that are insol. or sparingly sol. in water, e.g., steroids, fat-sol. vitamins; binder in mfg. of sugar-coated and uncoated tablets; controls release in sustained-release formulations; used with buffers in laxative formulations; antisticking agent , gloss aid for sugar-coated tablets; film-former, plasticizer in tablet coatings; base for ointments and suppositories
Properties: Colorless microbeads; 90% < 1000 μm, 50% < 500 μm; almost odorless and tasteless; sol. in water, ethanol, acetone, glycols, chloroform; insol. in ether, paraffin, oils, fats; m.w. 5400-6600; bulk dens. 0.60 g/cm^3; visc. 200-270 mPa·s (20 C, 50% aq.); f.p. 55-61 C; hyd. no. 16-22; pH 4.5-7.5 (5% aq.); nonionic; 100% conc.; ≤ 1% water
Storage: 24 mos shelf life when stored dry in closed containers

Luvitol® BD 10 P [BASF]
Chem. Descrip.: PPG-10 butanediol
Uses: Emollient in skin care, creams/lotions, sunscreen oils, shave prods., antiperspirant/deodorant gels, sticks or roll-ons; solvent for sunscreen actives
Features: Nongreasy
Properties: APHA color ≲ 40 max. clear liq.; sol. in water, SD alcohol 40B, butylene glycol, isopropyl myristate, benzyl laurate, nat. oils; disp. in cyclomethicone, min. oil, and jojoba oil; insol. in glycerin, dimethicone; m.w. 670; visc. 85-105 cst; acid no. 0-0.6; hyd. no. 175-225
Storage: 1 yr shelf life when stored in original, unopened containers; store in a cool, dry place

Lyondell Tetrahydrofuran (THF) [Penn Spec. Chems.; Lyondell]
Chem. Descrip.: Tetrahydrofuran
CAS 109-99-9; EINECS/ELINCS 203-726-8
Uses: Industrial solvent; prod. of heterocyclic and open-chain compds.; chemical intermediate
Features: High solvency for broad range of materials. low boiling pt.
Properties: Colorless liq.; misc. with water; b.p. 66 C

M

M-1421-7B, 12B Nylon Tipping Solutions [Eastern Color & Chem.]
Uses: Solvent system tipping prods. provide exc. nylon tips to nylon cord with proper heat/mold procedures

Macol® 57 [BASF/Perf. Chems.]
Chem. Descrip.: PPG-10 butanediol
Uses: Emollient, nonwhitening aid, spreading agent for colognes, aftershaves, skin fresheners, antiperspirants, nail enamel removers; gloss aid for hair; solvent for sunscreen actives; crystal growth inhibitor
Features: Nongreasy; high ref. index
Properties: Water-wh. clear liq., mild char. odor; sol. in water, SD alcohol 40B, butylene glycol, IPM, benzyl laurate, almond oil; disp. in cyclomethicone, min. and jojoba oils; insol. in glycerin; sp.gr. 1.005; visc. < 100 cps; m.p. < -25 C; b.p. > 150 C; cloud pt. 42 C; flash pt. (CC) > 95 C; ref. index 1.450; pH 6.0
Toxicology: LD50 (oral) < 5 g/kg; nontoxic inhalation hazard; safe for external use

Macol® 85 [BASF/Perf. Chems.]
Chem. Descrip.: m,p-Cresol propoxylate
CAS 9064-13-5
Uses: Wetting agent, dispersant, antistat, defoamer, gellant, solubilizer, lubricant base for cosmetic, medical, paper, pharmaceutical and textile industries; coalescing solvent for plasticizing acrylic emulsion polymers, latex caulks, epoxies, alkyds, polyesters, phenolics; increases flexibility and toughness, improves film formation
Regulatory: SARA nonreportable
Properties: Clear liq., bland odor; sol. in aromatic solvs.; disp. in water; m.w. 4600; sp.gr. 1.00; visc. 100 cps; vapor pressure < 1 mm Hg; b.p. > 300 F; HLB 16.0; acid no. 1 max.; hyd. no. 120-138; pour pt. < 10 C; cloud pt. 85 C (1% aq.); flash pt. (PMCC) > 200 F; ref. index 1.475; surf. tens. 42.5 dynes/cm (0.1%); Ross-Miles foam > 600 mm (0.1%); nonionic; 100% conc.
Toxicology: Contact with skin or eyes may cause temporary irritationTSCA listed
Precaution: Incompat. with strong oxidizing materials
Hazardous Ingredients: None
Hazardous decomp. prods.: Thermal decomp. may produce CO_x
NFPA: Health 1, Flammability 1, Reactivity 0
Storage: Store in well-ventilated area below 120 F

Magie 470 Oil [Magie Bros. Oil]
Chem. Descrip.: Sweetened middle distillate
CAS 64741-86-2; EINECS/ELINCS 265-088-7
Uses: Solvent for heat-set inks
Features: Narrow boiling range; consistent sol. chars.
Regulatory: DOT nonregulated; SARA §311/312/313 nonreportable; Canada DSL, EINECS, Australia AICS, China, Japan ENCS, Korea KCL, Philippines PICCS listed
Properties: Lt. straw-colored liq.; hydrocarbon odor; sol. in hydrocarbons; insol. in water; m.w. 207; sp.gr. 0.8086 (15.6 C); dens. 6.73 lb/gal (15.6 C); visc. 2.48 cSt (100 F); vapor pressure 0.02 mm Hg (70 F); i.b.p. 241 C; pour pt. -23 C; cloud pt. -6.7 C; flash pt. (PMCC) 98.9 C; autoignition temp. 217.2 C; ref. index 1.4497 (20 C); KB value 26.5; sp.heat 0.48; 100% act.
Toxicology: ACGIH TLV/TWA 5 mg/m^3 (oil mist); minimal eye/sl. skin irritant; prolonged/repeated contact may cause contact dermatitis; not expected to be inh. hazard @ ambient conditions; acute/chronic overexposure may cause respiratory tract irritation, drowsiness, dullness, numbness, headache, dizziness, weakness, nausea; aspiration may cause chem. pneumonitis, pulmonary edema, hemorrhage, death; ing. may cause GI discomfort, burning in throat, unconsciousness, convulsions; TSCA listed
Environmental: Do not allow spills to enter sewers or watercourses
Precaution: Flamm. limits in air 1.1-6%; may react with strong oxidizing agents
Hazardous decomp. prods.: Combustion: dense smoke, CO, CO_2, other oxides
HMIS: Health 1, Flammability 1, Reactivity 0
Storage: Store in closed containers away from heat, sparks, open flame, oxidizing materials

Magie 500 Oil [Magie Bros. Oil]
Chem. Descrip.: Sweetened middle distillate
CAS 64741-86-2; EINECS/ELINCS 265-088-7
Uses: Solvent for heat-set inks
Features: Narrow boiling range; consistent sol. chars.
Regulatory: DOT nonregulated; SARA §311/312/313 nonreportable; Canada DSL, EINECS, Australia AICS, China, Japan ENCS, Korea KCL, Philippines PICCS listed
Properties: Clear liq.; hydrocarbon odor; sol. in hydrocarbons; insol. in water; m.w. 228; sp.gr. 0.8203 (15.6 C); dens. 6.83 lb/gal (15.6 C); visc. 3.79 cSt (100 F); vapor pressure < 0.01 mm Hg (70 F); i.b.p. 271 C; pour pt. -6.7 C; cloud pt. -5.6 C; flash pt. (PMCC) 121.1 C; autoignition temp. 218 C; ref. index 1.4555 (20 C); KB value 24; sp.heat 0.470; 100% act.
Toxicology: ACGIH TLV/TWA 5 mg/m^3 (oil mist); minimal eye/sl. skin irritant; prolonged/repeated contact may cause contact dermatitis; not expected to be inh. hazard @ ambient conditions; acute/chronic overexposure may cause respiratory tract irritation, drowsiness, dullness, numbness, headache, dizziness, weakness, nausea; aspiration may cause chem. pneumonitis, pulmonary edema, hemorrhage, death; ing. may cause GI discomfort, burning in throat, unconsciousness, convulsions; TSCA listed
Environmental: Do not allow spills to enter sewers or watercourses
Precaution: Flamm. limits in air 1.1-6%; may react with strong oxidizing agents
Hazardous decomp. prods.: Combustion: dense smoke, CO, CO_2, other oxides
HMIS: Health 1, Flammability 1, Reactivity 0
Storage: Store in closed containers away from heat, sparks, open flame, oxidizing materials

Magie 4700 Oil [Magie Bros. Oil]
Chem. Descrip.: Sweetened middle distillate (70-80%) and white mineral oil, tech. (20-30%)
Uses: Solvent for heat-set inks
Features: Narrow boiling range; consistent sol. chars.
Regulatory: DOT nonregulated; SARA §311/312/313 nonreportable; Canada DSL, EINECS, Australia AICS, China, Japan ENCS, Korea KCL, Philippines PICCS listed
Properties: Clear liq.; hydrocarbon odor; sol. in hydrocarbons; insol. in water; m.w. 206; sp.gr. 08072 (15.6 C); dens. 6.72 lb/gal (15.6 C); visc. 2.51 cSt (100 F); vapor pressure 0.02 mm Hg (70 F); i.b.p. 240 C; pour

pt. -20.6 C; cloud pt. -16.7 C; flash pt. (PMCC) 98.9 C; autoignition temp. 217 C; ref. index 1.4480 (20 C); KB value 26; sp.heat 0.484

Toxicology: ACGIH TLV/TWA 5 mg/m³ (oil mist); minimal eye/sl. skin irritant; prolonged/repeated contact may cause contact dermatitis; not expected to be inh. hazard @ ambient conditions; acute/chronic overexposure may cause respiratory tract irritation, drowsiness, dullness, numbness, headache, dizziness, weakness, nausea; aspiration may cause chem. pneumonitis, pulmonary edema, hemorrhage, death; ing. may cause GI discomfort, burning in throat, unconsciousness, convulsions; TSCA listed

Environmental: Do not allow spills to enter sewers or watercourses

Precaution: Flamm. limits in air 1.1-6%; may react with strong oxidizing agents

Hazardous decomp. prods.: Combustion: dense smoke, CO, CO_2, other oxides

HMIS: Health 1, Flammability 1, Reactivity 0

Storage: Store in closed containers away from heat, sparks, open flame, oxidizing materials

Magie 5200-HTX [Magie Bros. Oil]

Chem. Descrip.: Petroleum distillates, hydrotreated middle

CAS 64742-46-7; EINECS/ELINCS 265-148-2

Uses: Solvent for sheet-fed inks

Features: Unique carbon distribution; narrow boiling range; consistent solvency power

Regulatory: DOT nonregulated; SARA §311/312/313 nonreportable; Canada DSL, EINECS, Australia AICS, China, Japan ENCS, Korea KCL, Philippines PICCS listed

Properties: Clear liq.; mild petrol. odor; sol. in hydrocarbons; insol. in water; m.w. 224; sp.gr. 0.8478 (15.6 C); dens. 7.06 lb/gal (15.6 C); visc. 4.18 cSt (100 F); vapor pressure < 0.1 mm Hg (70 F); i.b.p. 264 C; pour pt. -21 C; flash pt. (PMCC) 107 C; ref. index 1.4640 (20 C); KB value 28; 100% act.

Toxicology: ACGIH TLV/TWA 5 mg/m³ (oil mist); minimal eye/skin irritant; prolonged/repeated contact may cause contact dermatitis; not expected to be inh. hazard @ ambient conditions; acute/chronic overexposure may cause respiratory tract irritation, drowsiness, dullness, numbness, headache, dizziness, weakness, nausea; aspiration may cause chem. pneumonitis, pulmonary edema, hemorrhage, death; ing. may cause GI discomfort, burning in throat, unconsciousness, convulsions; TSCA listed

Environmental: Do not allow spills to enter sewers or watercourses

Precaution: May react with strong oxidizing agents

Hazardous decomp. prods.: Combustion: dense smoke, CO, CO_2, other oxides

HMIS: Health 1, Flammability 1, Reactivity 0

Storage: Store in cool, well-ventilated area in closed containers away from heat, sparks, open flame, oxidizing materials

Magie 5200 Oil [Magie Bros. Oil]

Chem. Descrip.: Sweetened middle distillate (70-80%) and white mineral oil, tech. (20-30%)

Uses: Solvent for heat-set inks

Features: Narrow boiling range; consistent sol. chars.

Regulatory: DOT nonregulated; SARA §311/312/313 nonreportable; Canada DSL, EINECS, Australia AICS, China, Japan ENCS, Korea KCL, Philippines PICCS listed

Properties: Clear liq.; hydrocarbon odor; sol. in hydrocarbons; insol. in water; m.w. 236; sp.gr. 0.8184 (15.6 C); dens. 6.81 lb/gal (15.6 C); visc. 3.86 cSt (100 F); vapor pressure < 0.01 mm Hg (70 F); i.b.p. 270 C; pour pt. -6.7 C; cloud pt. -2.8 C; flash pt. (PMCC) 121.1 C; autoignition temp. 218 C; ref. index 1.4540 (20 C); KB value 23.5; sp.heat 0.477

Toxicology: ACGIH TLV/TWA 5 mg/m³ (oil mist); minimal eye/sl. skin irritant; prolonged/repeated contact may cause contact dermatitis; not expected to be inh. hazard @ ambient conditions; acute/chronic overexposure may cause respiratory tract irritation, drowsiness, dullness, numbness, headache, dizziness, weakness, nausea; aspiration may cause chem. pneumonitis, pulmonary edema, hemorrhage, death; ing. may cause GI discomfort, burning in throat, unconsciousness, convulsions; TSCA listed

Environmental: Do not allow spills to enter sewers or watercourses

Precaution: Flamm. limits in air 1.1-6%; may react with strong oxidizing agents

Hazardous decomp. prods.: Combustion: dense smoke, CO, CO_2, other oxides

HMIS: Health 1, Flammability 1, Reactivity 0

Storage: Store in closed containers away from heat, sparks, open flame, oxidizing materials

MagieSol 44 [Magie Bros. Oil]

Chem. Descrip.: Petroleum distillates, hydrotreated light

CAS 64742-47-8; EINECS/ELINCS 265-149-8

Uses: Solvent for heat-set inks

Features: Narrow boiling range; consistent sol. chars.

Regulatory: DOT nonregulated; SARA §311/312/313 nonreportable; Canada DSL, EINECS, Australia AICS, China, Japan ENCS, Korea KCL, Philippines PICCS listed

Properties: Clear liq.; odorless; sol. in hydrocarbons; insol. in water; m.w. 199; sp.gr. 0.7952 (15.6 C); dens. 6.64 lb/gal (15.6 C); visc. 2.01 cSt (100 F); vapor pressure 0.005 mm Hg (70 F); i.b.p. 224 C; pour pt. -28.9 C; cloud pt. -23.3 C; flash pt. (PMCC) 93.3 C; ref. index 1.4401 (20 C); KB value 26.5; sp.heat 0.485; 100% act.

Toxicology: ACGIH TLV/TWA 5 mg/m³ (oil mist); minimal eye/skin irritant; prolonged/repeated contact may cause contact dermatitis; not expected to be inh. hazard @ ambient conditions; acute/chronic overexposure may cause respiratory tract irritation, drowsiness, dullness, numbness, headache, dizziness, weakness, nausea; aspiration may cause chem. pneumonitis, pulmonary edema, hemorrhage, death; ing. may cause GI discomfort, burning in throat, unconsciousness, convulsions; TSCA listed

Environmental: Do not allow spills to enter sewers or watercourses

Precaution: Flamm. limits in air 1.1-6%; may react with strong oxidizing agents; spillages may be slippery

Hazardous decomp. prods.: Combustion: dense smoke, CO, CO_2, other oxides

HMIS: Health 1, Flammability 1, Reactivity 0

Storage: Store in cool, well-ventilated area in closed containers away from heat, sparks, open flame, oxidizing materials

MagieSol 47 [Magie Bros. Oil]

Chem. Descrip.: Mineral oil, white, tech. grade

CAS 8042-47-5

Uses: Solvent for heat-set inks

Features: Narrow boiling range; consistent sol. chars.

Regulatory: DOT nonregulated; SARA §311/312/313 nonreportable; Canada DSL, EINECS, Australia AICS, China, Japan ENCS, Korea KCL, Philippines PICCS listed

Properties: Clear liq.; sl. hydrocarbon odor; sol. in hydrocarbons; insol. in water; m.w. 203; sp.gr. 0.8035 (15.6 C); dens. 6.69 lb/gal (15.6 C); visc. 2.51 cSt (100 F); vapor pressure 0.02 mm Hg (70 F); i.b.p. 239 C; pour pt. -20.6 C; cloud pt. -16.7 C; flash pt. (PMCC) 101.7 C; autoignition temp. 220 C; ref. index 1.4442 (20 C); KB value 25; sp.heat 0.480; 100% act.

Toxicology: ACGIH TLV/TWA 5 mg/m³, STEL 10 mg/m³ (oil mist); min. eye irritant; min. to sl. skin irritant; not expected to be inh. hazard @ ambient conditions; aspiration may cause chem. pneumonitis, pulmonary edema, hemorrhage, death; ing. may cause GI discomfort, vomiting, diarrhea; laxative props.; TSCA listed

Environmental: Do not allow spills to enter sewers or watercourses

Precaution: May react with strong oxidizing agents; spillages may be slippery

Hazardous decomp. prods.: Combustion: dense smoke, CO, CO_2, other oxides

HMIS: Health 0, Flammability 1, Reactivity 0

Storage: Store in closed containers away from heat, sparks, open flame, oxidizing materials

MagieSol 47LX [Magie Bros. Oil]

Chem. Descrip.: Petroleum distillates, hydrotreated light

CAS 64742-47-8; EINECS/ELINCS 265-149-8

Uses: Solvent for heat-set inks

Features: Narrow boiling range; consistent sol. chars.

Regulatory: FDA 21CFR §178.3620(b); DOT nonregulated; SARA §311/312/313 nonreportable; Canada DSL, EINECS, Australia AICS, China, Japan ENCS, Korea KCL, Philippines PICCS listed

Properties: Clear liq.; mild petrol. odor; sol. in hydrocarbons; insol. in water; m.w. 190; sp.gr. 0.8193 (15.6 C); dens. 6.82 lb/gal (15.6 C); visc. 2.45 cSt (100 F); vapor pressure 0.02 mm Hg (70 F); i.b.p. 234 C; pour pt. -40 C; cloud pt. -23.3 C; flash pt. (PMCC) 98.9 C; ref. index 1.4521 (20 C); KB value 29; 100% act.

Toxicology: ACGIH TLV/TWA 5 mg/m³ (oil mist); minimal eye/skin irritant;

prolonged/repeated contact may cause contact dermatitis; not expected to be inh. hazard @ ambient conditions; acute/chronic overexposure may cause respiratory tract irritation, drowsiness, dullness, numbness, headache, dizziness, weakness, nausea; aspiration may cause chem. pneumonitis, pulmonary edema, hemorrhage, death; ing. may cause GI discomfort, burning in throat, unconsciousness, convulsions; TSCA listed
Environmental: Do not allow spills to enter sewers or watercourses
Precaution: Flamm. limits in air 1-6%; may react with strong oxidizing agents; spillages may be slippery
Hazardous decomp. prods.: Combustion: dense smoke, CO, CO_2, other oxides
HMIS: Health 1, Flammability 1, Reactivity 0
Storage: Store in cool, well-ventilated area in closed containers away from heat, sparks, open flame, oxidizing materials

MagieSol 52 [Magie Bros. Oil]
Chem. Descrip.: Mineral oil, white, tech. grade
CAS 8042-47-5
Uses: Solvent for heat-set inks
Features: Narrow boiling range; consistent sol. chars.
Regulatory: DOT nonregulated; SARA §311/312/313 nonreportable; Canada DSL, EINECS, Australia AICS, China, Japan ENCS, Korea KCL, Philippines PICCS listed
Properties: Clear liq.; sl. hydrocarbon odor; sol. in hydrocarbons; insol. in water; m.w. 230; sp.gr. 0.8146 (15.6 C); dens. 6.78 lb/gal (15.6 C); visc. 3.90 cSt (100 F); vapor pressure 0.01 mm Hg (70 F); i.b.p. 268 C; pour pt. -6.67 C; cloud pt. -2.8 C; flash pt. (PMCC) 129.4 C; autoignition temp. 220 C; ref. index 1.4498 (20 C); KB value 22; sp.heat 0.492; 100% act.
Toxicology: ACGIH TLV/TWA 5 mg/m^3, STEL 10 mg/m^3 (oil mist); min. eye irritant; min. to sl. skin irritant; not expected to be inh. hazard @ ambient conditions; aspiration may cause chem. pneumonitis, pulmonary edema, hemorrhage, death; ing. may cause GI discomfort, vomiting, diarrhea; laxative props.; TSCA listed
Environmental: Do not allow spills to enter sewers or watercourses
Precaution: May react with strong oxidizing agents; spillages may be slippery
Hazardous decomp. prods.: Combustion: dense smoke, CO, CO_2, other oxides
HMIS: Health 0, Flammability 1, Reactivity 0
Storage: Store in closed containers away from heat, sparks, open flame, oxidizing materials

MagieSol 55LX [Magie Bros. Oil]
Chem. Descrip.: Petroleum distillates, hydrotreated middle
CAS 64742-46-7; EINECS/ELINCS 265-148-2
Uses: Solvent for sheet-fed inks
Features: Unique carbon distribution; narrow boiling range; consistent solvency power
Regulatory: DOT nonregulated; SARA §311/312/313 nonreportable; Canada DSL, EINECS, Australia AICS, China, Japan ENCS, Korea KCL, Philippines PICCS listed
Properties: Clear liq.; faint hydrocarbon odor; sol. in hydrocarbons; insol. in water; sp.gr. 0.8324 (15.6 C); dens. 6.94 lb/gal (15.6 C); visc. 4.3 cSt (100 F); i.b.p. 271 C; pour pt. -51 C; flash pt. (PMCC) 121 C; ref. index 1.4571 (20 C); KB value 23.8; 100% act.
Toxicology: ACGIH TLV/TWA 5 mg/m^3 (oil mist); minimal eye/skin irritant; prolonged/repeated contact may cause contact dermatitis; not expected to be inh. hazard @ ambient conditions; acute/chronic overexposure may cause respiratory tract irritation, drowsiness, dullness, numbness, headache, dizziness, weakness, nausea; aspiration may cause chem. pneumonitis, pulmonary edema, hemorrhage, death; ing. may cause GI discomfort, burning in throat, unconsciousness, convulsions; TSCA listed
Environmental: Do not allow spills to enter sewers or watercourses
Precaution: May react with strong oxidizing agents; spillages may be slippery
Hazardous decomp. prods.: Combustion: dense smoke, CO, CO_2, other oxides
HMIS: Health 1, Flammability 1, Reactivity 0
Storage: Store in cool, well-ventilated area in closed containers away from heat, sparks, open flame, oxidizing materials

MagieSol 60 [Magie Bros. Oil]
Chem. Descrip.: Mineral oil, white, tech. grade
CAS 8042-47-5
Uses: Solvent for sheet-fed inks
Features: Unique carbon distribution; narrow boiling range; consistent solvency power
Regulatory: DOT nonregulated; SARA §311/312/313 nonreportable; Canada DSL, EINECS, Australia AICS, China, Japan ENCS, Korea KCL, Philippines PICCS listed
Properties: Clear liq.; sl. hydrocarbon odor; sol. in hydrocarbons; insol. in water; m.w. 266; sp.gr. 0.8299 (15.6 C); dens. 6.91 lb/gal (15.6 C); visc. 6.27 cSt (100 F); vapor pressure < 0.01 mm Hg (70 F); i.b.p. 304 C; pour pt. 4.4 C; cloud pt. 10 C; flash pt. (PMCC) 151.7 C; ref. index 1.4570 (20 C); KB value 20; sp.heat 0.493; 100% act.
Toxicology: ACGIH TLV/TWA 5 mg/m^3, STEL 10 mg/m^3 (oil mist); min. eye irritant; min. to sl. skin irritant; not expected to be inh. hazard @ ambient conditions; aspiration may cause chem. pneumonitis, pulmonary edema, hemorrhage, death; ing. may cause GI discomfort, vomiting, diarrhea; laxative props.; TSCA listed
Environmental: Do not allow spills to enter sewers or watercourses
Precaution: May react with strong oxidizing agents; spillages may be slippery
Hazardous decomp. prods.: Combustion: dense smoke, CO, CO_2, other oxides
HMIS: Health 0, Flammability 1, Reactivity 0
Storage: Store in closed containers away from heat, sparks, open flame, oxidizing materials

MagieSol 65LX [Magie Bros. Oil]
Chem. Descrip.: Petroleum distillates, hydrotreated middle
CAS 64742-46-7; EINECS/ELINCS 265-148-2
Uses: Solvent for sheet-fed inks
Features: Unique carbon distribution; narrow boiling range; consistent solvency power
Regulatory: DOT nonregulated; SARA §311/312/313 nonreportable; Canada DSL, EINECS, Australia AICS, China, Japan ENCS, Korea KCL, Philippines PICCS listed
Properties: Clear liq.; hydrocarbon odor; sol. in hydrocarbons; insol. in water; sp.gr. 0.8654 (15.6 C); dens. 7.22 lb/gal (15.6 C); visc. 10.2 cSt (40 C); i.b.p. 327 C; pour pt. -45 C; flash pt. (PMCC) 171 C; ref. index 1.4714 (20 C); KB value 22.4; 100% act.
Toxicology: ACGIH TLV/TWA 5 mg/m^3 (oil mist); minimal eye/skin irritant; prolonged/repeated contact may cause contact dermatitis; not expected to be inh. hazard @ ambient conditions; acute/chronic overexposure may cause respiratory tract irritation, drowsiness, dullness, numbness, headache, dizziness, weakness, nausea; aspiration may cause chem. pneumonitis, pulmonary edema, hemorrhage, death; ing. may cause GI discomfort, burning in throat, unconsciousness, convulsions; TSCA listed
Environmental: Do not allow spills to enter sewers or watercourses
Precaution: May react with strong oxidizing agents; spillages may be slippery
Hazardous decomp. prods.: Combustion: dense smoke, CO, CO_2, other oxides
HMIS: Health 1, Flammability 1, Reactivity 0
Storage: Store in cool, well-ventilated area in closed containers away from heat, sparks, open flame, oxidizing materials

Marlasol EX-1 [Lenmar]
Uses: Solvent scouring agent for removal of oils, waxes, and unfixed dyes from goods

Marthene [Amity UK]
Chem. Descrip.: Blend of nonaromatic hydrocarbons
Uses: Solvent cleaner for electronic components, engineering, fabrication (flux and adhesive remover), printing (excess ink remover), coatings (excess resin remover), aerospace (manual cold degasser)
Features: Evaporates to leave no residue; very high flash pt., useful as nonflamm. solv. in most applics.
Properties: Colorless liq.; sweet odor; sp.gr. 1.06 (20 C); visc. 2.4 cSt; vapor pressure 0.02 mm Hg (20 C); b.p. 196-225 C; flash pt. (CC) 101 C
Toxicology: LD50 (oral, rat) 8191 mg/kg, (skin, rabbit) 2259 mg/kg; irritating to eyes, skin, mucous membranes; inh. may cause respiratory irritation; may cause blurred vision
Environmental: LC50 (fish, 96 h) 18 mg/kg; LC (bluegill sunfish, 96 h) 30.9 mg/l; EC50 (daphnia, 48 h) 112 mg/l; readily biodeg.; BOD5 44; BOD10 55; do not allow runoff to sewer, waterway, or ground
Precaution: Pressure may build up; vapors may form explosive mixts. with air; avoid sparks, flames, smoking; flamm. limits 0.9-8%; dec. on heating; incompat. with strong oxidizing agents, acids, bases

Hazardous decomp. prods.: CO
Storage: Store in cool, dry, ventilated area in closed original containers; keep away from heat, sparks, open flame, oxidizing materials

Mazol® 300 K [BASF/Perf. Chems.]
Chem. Descrip.: Glyceryl oleate
CAS 111-03-5; EINECS/ELINCS 203-827-7
Uses: Dispersant for oil or solv. systems; emulsifier, antioxidant, solvent in foods; antifoam for sugar and protein processing; coemulsifier with T-Maz 60 or 80
Regulatory: FDA 21CFR §184.1323, 184.1666, GRAS; kosher
Properties: Amber liq.; sol. in min. oil, toluene, IPA, ethanol, veg. oil; disp. in min. spirits, propylene glycol; insol. in water; m.p. 20 C; HLB 3.8; acid no. 2 max.; iodine no. 80; sapon. no. 145-155; flash pt. (PMCC) > 350 F; nonionic; 100% conc., 40% alpha mono

MDGE [INEOS Oxide]
Chem. Descrip.: Methoxydiglycol
CAS 111-77-3; EINECS/ELINCS 203-906-6
Uses: Solvent for resins (NC, acrylic, epoxy, polyamide, alkyd), oils, waxes, fats, dyestuffs, insecticides, herbicides, grease/grime, surf. coatings, lacquers, enamels, waterborne coatings, phenolic varnishes, epoxy coatings, non-grain-raising wood stains; gloss aid, flow aid in lacquers; dye solvent/stabilizer/penetration aid, color enhancer for textiles, leather, printing; mutual solvent, coupling agent in sol. oils, metal/glass cleaners, varnish removers; diluent in hydraulic brake fluids
Features: Very stable; low f.p.; relatively low vapor pressure
Properties: Pt-Co 10 max. color; mild odor; misc. with wide range of solvs., water; m.w. 120.15; dens. 1.019-1.023 g/ml (20 C); visc. 3.9 mPa·s (20 C); vapor pressure 0.16 hPa (20 C); i.b.p. 191 C min. (101.3 kPa); pour pt. -85 C; flash pt. (CC) 87 C; autoignition temp. 240 C; ref. index 1.4263 (20 C); 99% min. purity
Precaution: Flamm. limits 1.5-18.0 vol.% (20 C); not a significant flamm. hazard @ normal storage temps.; very reactive with zinc (hydrogen may be produced); contact with zinc, tin, copper/alloys can cause discoloration

Mecloran® [Atofina SA]
Chem. Descrip.: 1,1,1-Trichloroethane, stabilized
CAS 71-55-6; EINECS/ELINCS 200-756-3
Uses: Solvent

Melio® S-4830 [Clariant]
Uses: Solvent for aliphatic PU lacquers

Methyl Carbitol® [Dow]
Chem. Descrip.: Methoxydiglycol
CAS 111-77-3; EINECS/ELINCS 203-906-6
Uses: Solvent, coalescing agent for water-disp. and water-sol. coatings
Properties: Sol. in water (20 C); m.w. 120.16; sp.gr. 1.021 (20/20 C); dens. 8.51 lb/gal (20 C); visc. 3.9 cP (20 C); vapor pressure 0.1 mm Hg (20 C); b.p. 194.0 C; flash pt. (CC) 188 F

Methyl Cellosolve® [Dow]
Chem. Descrip.: Methoxyethanol
CAS 109-86-4; EINECS/ELINCS 203-713-7
Uses: Solvent for coatings (NC, cellulose acetate, ethyl cellulose, butyrate, acrylic)
Properties: Sol. in water (20 C); m.w. 76.10; sp.gr. 0.966 (20/20 C); dens. 8.04 lb/gal (20 C); b.p. 124.5 C; flash pt. (CC) 103 F

Methyl Cellosolve® Acetate [Dow]
Chem. Descrip.: Methoxyethanol acetate
CAS 110-49-6; EINECS/ELINCS 203-772-9
Uses: Solvent for coatings (NC, cellulose acetate, ethyl cellulose, butyrate, acrylic)
Properties: Sol. in water (20 C); m.w. 118.14; sp.gr. 1.006 (20/20 C); dens. 8.37 lb/gal (20 C); b.p. 145.0 C; flash pt. (CC) 121 F

Methyl Diproxitol [Shell Netherlands]
Chem. Descrip.: PPG-2 methyl ether
CAS 34590-94-8; EINECS/ELINCS 252-104-2
Uses: Solvent for NC, dyestuffs
Features: High boiling
Regulatory: EINECS, Canada DSL, Australia AICS, China IECSC, Japan MITI, Philippines PICCS, Korea TCCI listed
Properties: Colorless clear liq.; mild ethereal odor; misc. with water and many org. solvs.; m.w. 148; dens. 0.952-0.960 g/ml (20 C); kinematic visc. 3.71 mPa·s (20 C); vapor pressure 52 Pa (20 C); f.p. -83 C; i.b.p. 184 C; b.p. 184-190 C; pour pt. -83 C; flash pt. 79 C; autoignition temp. 205 C; surf. tens. 29.4 mN/m (20 C); elec. conduct. 1 μS/m (20 C); 98.5% min. purity
Toxicology: ACGIH TLV/TWA (8 h) 100 ppm, STEL (15 min) 150 ppm; LD50 (oral) > 2000 mg/kg, (skin) > 2000 mg/kg; LC50 (inh.) > 5 mg/l; contact may cause skin dryness, eye irritation; inh. may cause mucous membrane irritation; causes CNS depression; not a human skin sensitizer; TSCA listed
Environmental: Expected to be readily biodeg.; does not bioaccumulate significantly; not classified as dangerous under EC criteria; expected to be pract. nontoxic to fish, algae, bacteria; prevent contamination of soil, water; prevent entry into drains
Precaution: Combustible; LEL 1.3 vol.%; UEL 8.7 vol.%; may form flamm./explosive vapor-air mixts. above 79 C; ground equip.
Hazardous decomp. prods.: Combustion: CO
Storage: Hygroscopic; store @ 40 C max.

Methylhexalin® [Cognis KGaA/Coatings]
Chem. Descrip.: Methylcyclohexanol
CAS 25639-42-3; EINECS/ELINCS 247-152-6
Uses: Solvent, flow aid in coatings
Features: High boiling; strong solvent power
Use Level: 2-20% on solv.

Methylnaphthalene Fraction [Koppers Ind.]
Chem. Descrip.: Alkylnaphthalene (54-59%), acenaphthene (5-10%), quinoline (3-7%), dibenzofuran (3-7%), biphenyl (3-7%), naphthalene (1-5%), isoquinoline (1-5%), fluorene (1-5%), phenanthrene, anthracene
Uses: Solvent
Regulatory: SARA §313 and DOT reportable
Properties: Bluish brn. oil; coal tar or mothball odor; negligible sol. in water; sp.gr. 1.01 (15.5 C); vapor pressure < 1 mm Hg (20 C); b.p. 232-271 C; flash pt. (CC) 93 C; pH alkaline
Toxicology: Harmful if inh. or absorbed thru skin; may be fatal if swallowed; may irritate eyes
Precaution: Flamm.; incompat. with strong oxidizing agents
Hazardous Ingredients: Alkylnaphthalene, acenaphthene, quinoline, dibenzofuran, biphenyl, naphthalene, isoquinoline, fluorene, phenanthrene, anthracene
Hazardous decomp. prods.: Hazardous decomp. produces oxides of carbon
HMIS: Health 1, Flammability 2, Reactivity 1
Storage: Keep in closed container in cool, well-ventilated area

Methyl Proxitol [Shell]
Chem. Descrip.: Methoxyisopropanol
CAS 107-98-2; EINECS/ELINCS 203-539-1
UN 3092
Uses: Solvent
Regulatory: SARA §311/312 acute health/fire hazard
Properties: Colorless liq.; sharp sweet ether-like odor; completely sol. in water; m.w. 90; dens. 0.920-0.923 g/ml (20 C); vapor pressure 8.9 mm Hg (68 F); m.p. -140 F; i.b.p. 117 C min.; b.p. 243-257 F; pour pt. -96 C; flash pt. 30 C; VOC 7.7 lb/gal; 100% act.
Toxicology: ACGIH TLV/TWA 100 ppm, STEL 150 ppm; if aspirated, can cause severe lung damage, chem. pneumonitis; may be fatal if swallowed; eye irritant (burning sensation, redness, swelling, blurred vision); may be harmful by ing.; vapors may be irritating to respiratory system; may cause CNS depression (dizziness, lightheadedness, headache, nausea, incoordination, unconsciousness, death); sl. skin irritation, defatting; may cause kidney effects; TSCA listed
Environmental: Prevent entry into waterways, sewer
Precaution: Flamm.; LEL 1.9 vol.%; UEL 13.1 vol.%; vapors are heavier than air; vapors may flash back; prevent vapor accumulation; possible peroxide former; potential for runaway reaction at elevated temps. in presence of strong bases/salts; incompat. with strong oxidizing agents; static electricity may accumulate and create fire hazard; bond and ground equip./containers; avoid contact with Al; prevent entry into basements, confined areas
Hazardous decomp. prods.: Complex mixt. of airborne solids, liqs., and gases, incl. CO, CO_2, other org. compds.
HMIS: Health 1, Flammability 3, Reactivity 0
Storage: Hygroscopic; keep closed when not in use; keep away from heat, sparks, flame

Methyl Proxitol Acetate [Shell; Shell Netherlands]
Chem. Descrip.: Propylene glycol methyl ether acetate

CAS 108-65-6; EINECS/ELINCS 203-603-9
UN 3272
Uses: Solvent
Regulatory: SARA §311/312 acute health/chronic health/fire hazard; EINECS, China IECSC, Philippines PICCS, Korea TCCL listed
Properties: Colorless clear liq.; mild ethereal odor; sol. 23% in water; m.w. 132; dens. 0.965-0.969 kg/l (20 C); dynamic visc. 1.31 mPa·s; vapor pressure 420 Pa (20 C); f.p. -65 C; i.b.p. 143 C min.; b.p. 143-149 C; pour pt. -65 C; flash pt. (Abel) 45 C; autoignition temp. 315 C; VOC 8.04 lb/gal; elec. conduct. 0.2 µD/m (20 C); 100% act.
Toxicology: TLV/TWA (8 h) 50 ppm, STEL (15 min) 100 ppm; LD50 (oral) > 2000 mg/kg, (skin) > 2000 mg/kg; LC50 (inh.) > 5 mg/l; may be harmful by skin absorp.; eye irritant; inh. may cause nasal irritation, mucous membrane damage; may cause skin dryness; may cause CNS depression (dizziness, lightheadedness, headache, nausea, incoordination, unconsciousness); if aspirated, can cause severe lung damage, chem. pneumonitis; may be fatal by ing.; target organ: liver; TSCA listed
Environmental: Inherently biodeg.; does not bioaccumulate significantly; not classified as dangerous under EC criteria; pract. nontoxic to fish, algae, bacteria; prevent contamination of soil, water; prevent entry into drains, ditches, rivers
Precaution: Flamm.; LEL 1.5 vol.%; UEL 7.0 vol.%; may form flamm./explosive vapor-air mixt.; vapor is heavier than air, spreads along ground; distant ignition possible; prevent vapor accumulation; risk of explosion; ground equip.; reacts with strong oxidizing agents; oxidizes on contact with air; avoid exposure to air
Hazardous decomp. prods.: Complex mixt. of airborne solids, liqs., and gases incl. CO, CO_2, other org. compds.
HMIS: Health 1, Flammability 2, Reactivity 0
Storage: Hygroscopic; store @ ambient temps. away from direct sunlight, other heat/ignition sources; keep container closed when not in use

MGE [INEOS Oxide]
Chem. Descrip.: Methoxyethanol
CAS 109-86-4; EINECS/ELINCS 203-713-7
Uses: Solvent for resins (NC, acrylic, epoxy, polyamide, alkyd), oils, waxes, fats, dyestuffs, insecticides, herbicides, grease/grime, surf. coatings, lacquers, enamels, waterborne coatings, phenolic varnishes, epoxy coatings, non-grain-raising wood stains; gloss aid, flow aid in lacquers; dye solvent/stabilizer/penetration aid, color enhancer for textiles, leather, printing; mutual solvent, coupling agent in sol. oils, metal/glass cleaners, varnish removers; diluent in hydraulic brake fluids
Features: Very stable; low f.p.; relatively low vapor pressure
Properties: Pt-Co 15 max. color; mild odor; misc. with wide range of solvs., water; m.w. 76.10; dens. 0.961-0.965 g/ml (20 C); visc. 1.7 mPa·s (20 C); vapor pressure 8 hPa (20 C); i.b.p. 123.5 C min. (101.3 kPa); pour pt. -85.1 C; flash pt. (CC) 39 C; autoignition temp. 285 C; ref. index 1.4021 (20 C); 99.5% min. purity
Precaution: Flamm. limits 1.8-20 vol.% (20 C); not a significant flamm. hazard @ normal storage temps.; very reactive with zinc (hydrogen may be produced); contact with zinc, tin, copper/alloys can cause discoloration

Miglyol® 810 [Sasol Germany; Sasol N. Am.]
Chem. Descrip.: Caprylic/capric triglyceride
CAS 65381-09-1; EINECS/ELINCS 265-724-3
Uses: Emollient, dispersant, lubricant, suspending agent, solubilizer for cosmetics, pharmaceuticals (oral, topical, rectal, parenteral); carrier, vehicle, solvent for inj. prods., topical ointments, creams, lotions, suppositories; dietetic prods.; spreading agent, penetrant on skin
Properties: Colorless liq., neutral odor; sol. in diethyl ether, petrol. ether, chloroform, IPA, toluene, alcohol, min. oil, acetone; sp.gr. 0.94-0.95; visc. 25-35 mPa·s; acid no. 0.1 max.; iodine no. 0.5 max.; sapon. no. 335-355; pH neutral; cloud pt. 0 C; nonionic
Environmental: Environmentally compat.

Miglyol® 812 [Sasol N. Am.]
Chem. Descrip.: Caprylic/capric triglyceride
CAS 65381-09-1; EINECS/ELINCS 265-724-3
Uses: Dispersant, lubricant, anticaking agent, carrier, solvent, solubilizer, suspending agent for cosmetics, pharmaceuticals (oral, external, suppositories), dietetic prods.; solvent for lipophilic drugs; carrier, filler in capsules; spreading agent, penetrant on skin
Features: Stable to oxidation
Properties: Colorless liq.; sol. in alcohol, min. oil, acetone; visc. 25-35 mPa·s; acid no. 0.1 max.; iodine no. 0.5 max.; sapon. no. 325-345; cloud pt. 10 C; nonionic; 100% conc.
Environmental: Environmentally compat.

Miglyol® 812 N [Sasol Germany; Sasol N. Am.]
Chem. Descrip.: Glyceryl tricaprylate/tricaprate
CAS 65381-09-1; EINECS/ELINCS 265-724-3
Uses: Solvent, solubilizer in pharmaceuticals (oral, topical, rectal, parenteral), creams/lotions, oil preps.; spreading agent, penetrant on skin; special oil for the prod. of dietary foods containing MCT; fat component for medical nutrition
Features: Oxidation-stable; no visible greasy coating of skin surf.
Properties: Thin oily liq.; dens. 0.95 kg/dm^3; visc. 25-35 mPa·s; turbidity pt. 10 C; acid no. 0.2 max.; iodine no. 1 max.; sapon. no. 325-345

Miglyol® 840 [Sasol Germany; Sasol N. Am.]
Chem. Descrip.: Propylene glycol dicaprylate/dicaprate
CAS 68583-51-7; EINECS/ELINCS 271-516-3
Uses: Emollient, dispersant, lubricant, suspending agent, solubilizer for cosmetics, pharmaceuticals; carrier, vehicle, solvent for pharmaceuticals (orals, parenterals, rectals, inj. prods., topical ointments, creams, lotions, oils, suppositories), dietetic prods.; flavor fixing agent; spreading agent, penetrant on skin
Features: Oxidation-stable; no visible greasy coating of skin surf.
Properties: Colorless liq.; sol. in alcohol, min. oil, acetone; visc. 8-14 mPa·s; acid no. 0.1 max.; iodine no. 0.5 max.; sapon. no. 320-340; cloud pt. -30 C
Environmental: Environmentally compat.

Miglyol® 8810 [Sasol Germany; Sasol N. Am.]
Chem. Descrip.: Butylene glycol dicaprylate/dicaprate
Uses: Emollient, dispersant, lubricant, suspending agent, solubilizer for cosmetics, pharmaceuticals; carrier, vehicle, solvent for inj. prods., topical ointments, creams, lotions, esp. baby skin care; penetrant and spreading agent on skin
Features: Hypoallergenic; environmentally compat.
Properties: Colorless liq., neutral odor; sol. in alcohol, diethyl ether, n-hexane, fatty oils; insol. in water; visc. 8-13 mPa·s; acid no. 0.1 max.; iodine no. 1.0 max.; sapon. no. 295-315; hyd. no. 5 max.

Milopol G-7 [Undesa; S. Black]
Chem. Descrip.: PEG-7 glyceryl cocoate
CAS 68201-46-7
Uses: Dispersant, solvent for act. ingreds.; vehicle for injectable substances; emulsifier for textiles; emulsifier, opacifier, thickener, dispersant for emulsions; greasing agent in shampoos, foam baths; food lubricant; perfume diluent; intermediate in textiles and tanning industry

Mineral Spirits 066/3 [CITGO]
Chem. Descrip.: Stoddard solvent
CAS 8052-41-3; EINECS/ELINCS 232-489-3
UN 1268
Uses: Solvent
Features: Med. boiling; < 1% aromatics; hydrotreated to reduce aromatics and olefins; meets definition of non-photochemically reactive as defined by 66/3 Rule requirements
Regulatory: FDA 21CFR §172.882, 172,884, 175.105, 178.3910; SARA §302/304 nonreportable, §311/312 acute health/chronic health/fire hazard, §313 nonreportable; Calif. Prop. 65 nonreportable; may contain detectable amts. 1,3,5-trimethylbenzene
Properties: Colorless transparent liq.; lt. paraffinic hydrocarbon odor; sl. sol. in cold water (< 0.05%); sp.gr. 0.770-0.810 (60/60 F); dens. 6.48 lb/gal (15.6 C); visc. 1.36 cst (20 C); vapor pressure 2.6 mm Hg (20 C); f.p. -65 to -25 C; b.p. 160-200 C; flash pt. (TCC) 42-45 C; autoignition temp. 230 C; ref. index 1.4262 (20 C); KB value 33; VOC 777 g/l; 100% act.
Toxicology: ACGIH TLV TWA 100 ppm; LD50 (oral, rat) > 34,600 mg/kg; mist/vapor may irritate eyes, mucous membranes, respiratory tract; liq. contact may cause minimal to mod. eye irritation, mod. to severe skin irritation/inflamm.; may be harmful by inh., skin absorp.; overexposure may cause CNS depression, other target organ effects; may be harmful or fatal by ing.; aspiration into lungs can cause pulmonary edema, chem. pneumonia; TSCA listed
Environmental: May cause oxygen depletion in waterways; potentially toxic to water ecosystems; may be hazardous to aquatic life; prevent entry into sewers, waterways; not a DOT marine pollutant
Precaution: Combustible liq.; flamm. limits 0.6-6.0%; vapor may cause flash fire; vapors are heavier than air, may flash back; may create vapor/

air explosion hazard in confined spaces; spills may create a slipping hazard; floats on water; container can rupture in fire; prevent entry into basements, confined areas; bond and ground equip.; incompat. with strong acids, alkalis, oxidizers (liq. chlorine, other halogens, hydrogen peroxide, oxygen)
Hazardous decomp. prods.: Combustion: smoke, CO, CO_2, possibly other harmful gases/vapors
HMIS: Health 1, Flammability 2, Reactivity 0
Storage: Store in tightly closed containers in cool, dry, well-ventilated place away from heat, sparks, ignition sources, strong oxidizing conditions

Mineral Spirits 075 [CITGO]
Chem. Descrip.: Stoddard solvent
CAS 8052-41-3; EINECS/ELINCS 232-489-3
Uses: Solvent
Features: Med. boiling; < 8% aromatics; meets definition of non-photochemically reactive as defined by 66/3 Rule requirements
Properties: Sp.gr. 0.7802 (60/60 F); dens. 6.54 lb/gal (15.6 C); vapor pressure 0.66 torr (isoten, 68 F); i.b.p. 300 F min.; flash pt. (TCC) > 105 F; KB value 36

Mineral Spirits 150 [Americhem Sales]
Chem. Descrip.: Stoddard solvent
CAS 8052-41-3; EINECS/ELINCS 232-489-3
UN 1268
Uses: Solvent
Regulatory: SARA §311/312 fire/acute health hazardous, §313 nonreportable; Calif. Prop. 65 nonreportable
Properties: Colorless clear liq.; mild hydrocarbon odor; negligible sol. in water; sp.gr. 0.80 (60 F); dens. 6.62 lb/gal (60 F); vapor pressure 2 mm Hg; b.p. 370-405 F; flash pt. (TCC) 152 F; autoignition temp. 440 F; 100% act.
Toxicology: ACGIH TLV TWA 100 ppm; may cause mod. eye/skin irritation/discomfort, skin defatting; ing. may cause GI tract irritation/discomfort; inh. may cause mod. respiratory discomfort, headaches; chronic overexposure may cause permanent brain/nervous system damage; may be harmful or fatal by inh. abuse; TSCA listed
Environmental: Prevent entry into sewers, waterways
Precaution: Combustible; vapor accumulations may flash/explode if ignited; avoid formation of excessive mists/fumes; keep fumes away from open flames, ignition sources; can accumulate static charge; vapors can be ignited by static discharge; incompat. with strong oxidizing agents
Hazardous decomp. prods.: Combustion: CO_x
HMIS: Health 1, Flammability 2, Reactivity 0
Storage: Store unopened containers in cool, dry, ventilated area away from excessive heat, sparks, flames; keep containers closed when not in use

Mineral Spirits 150 66/3 [CITGO]
Chem. Descrip.: Stoddard solvent
CAS 8052-41-3; EINECS/ELINCS 232-489-3
UN 1268; UN NA 199
Uses: Solvent
Features: Hydrotreated to reduce aromatics and olefins; meets Rule 66/3 requirements
Regulatory: FDA 21CFR §175.105, 178.3620, 178.3910; SARA §302/304 nonreportable, §311/312 acute health/chronic health/fire hazard, §313 nonreportable; Calif. Prop. 65 nonreportable; may contain detectable amts. 1,3,5-trimethylbenzene
Properties: Colorless transparent liq.; lt. paraffinic hydrocarbon odor; negligible sol. in cold water (< 0.01%); sp.gr. 0.7954 (60/60 F); dens. 6.62 lb/gal (15.6 C); visc. 1.74 cst (20 C); vapor pressure 2.9 mm Hg (20 C); f.p. -65 to -25 C; b.p. 188-210 C; flash pt. (TCC) 67-68 C; autoignition temp. 229 C; KB value 31; VOC 795 g/l; 100% act.
Toxicology: ACGIH TLV TWA 100 ppm; LD50 (oral, rat) > 34,600 mg/kg; mist/vapor may irritate eyes, mucous membranes, respiratory tract; liq. contact may cause mild to mod. eye irritation, mod. to severe skin irritation/inflamm.; may be harmful by inh., skin absorp.; overexposure may cause CNS depression, other target organ effects; may be harmful or fatal by ing.; aspiration into lungs can cause pulmonary edema, chem. pneumonia; TSCA listed
Environmental: May cause oxygen depletion in waterways; potentially toxic to water ecosystems; may be hazardous to aquatic life; may contribute to smog; prevent entry into sewers, waterways; not a DOT marine pollutant
Precaution: Combustible liq.; flamm. limits 0.8-6.0%; vapor may cause flash fire; vapors are heavier than air, may flash back; may create vapor/air explosion hazard in confined spaces; spills may create a slipping hazard; floats on water; container can rupture in fire; prevent entry into basements, confined areas; bond and ground equip.; incompat. with strong acids, alkalis, oxidizers (liq. chlorine, other halogens, hydrogen peroxide, oxygen)
Hazardous decomp. prods.: Combustion: smoke, CO, CO_2, possibly other harmful gases/vapors
HMIS: Health 1, Flammability 2, Reactivity 0
Storage: Store in tightly closed containers in cool, dry, well-ventilated place away from heat, sparks, ignition sources, strong oxidizing conditions

MN [Taiwan Surf.]
Chem. Descrip.: Methyl naphthalene
CAS 90-12-0; EINECS/ELINCS 201-966-8
Uses: Raw material for mfg. of textile dye carrier; solvent for pesticides

Mod Acid [Rütgers Org.]
Chem. Descrip.: Toluene sulfonic acid/xylene sulfonic acid
Uses: Hydrotrope; solvent; intermediate; crosslinking catalyst for phenolic adhesives, melamine-formaldehyde and acrylic polymers; curing catalyst in polymerization of furan for foundry molds or cores; vulcanization of rubber; prep. of acetals, esters
Properties: Cryst. range 0-15 C; 92% min. assay
Toxicology: Corrosive; avoid contact with skin and eyes

Monestriol B [Undesa; S. Black]
Chem. Descrip.: Butyl stearate
CAS 123-95-5; EINECS/ELINCS 204-666-5
Uses: Emollient in bath oils, creams/lotions, sun creams, shaving prods., insect repellents, antiperspirants, deodorants; oily component for cleansing creams, nail polish removers; plasticizer for nail polishes; pigment wetting agent; softener in creams, hair care; solvent for fatty cosmetic ingreds.; vehicle for lipsticks; solubilizer for perfume oils, vitamin oils, hormones; solubilizer, vehicle for medicines; excipient for injectable substances; lubricant for PVC
Features: Changes skin feel, rub out props. and greasiness; veg. oil substitute; vaseline substitute

Monestriol C [Undesa; S. Black]
Chem. Descrip.: Cetyl stearate
CAS 1190-63-2; EINECS/ELINCS 214-724-1
Uses: Emollient in bath oils, creams/lotions, sun creams, shaving prods., insect repellents, antiperspirants, deodorants; oily component for cleansing creams, nail polish removers; plasticizer for nail polishes; pigment wetting agent; softener in creams, hair care; solvent for fatty cosmetic ingreds.; vehicle for lipsticks; solubilizer for perfume oils, vitamin oils, hormones; solubilizer, vehicle for medicines; excipient for injectable substances; lubricant for PVC
Features: Changes skin feel, rub out props. and greasiness; veg. oil substitute; vaseline substitute

Monestriol DM [Undesa; S. Black]
Chem. Descrip.: Acetylated glycerine monostearate
Uses: Dispersant, solvent for act. ingreds.; vehicle for injectable substances; emulsifier for textiles; emulsifier, opacifier, thickener, dispersant for emulsions; greasing agent in shampoos, foam baths; food lubricant; perfume diluent; intermediate in textiles and tanning industry
Properties: Nonionic

Monestriol GCT [Undesa; S. Black]
Chem. Descrip.: Polyglyceryl monostearate
Uses: Dispersant, solvent for act. ingreds.; vehicle for injectable substances; emulsifier for textiles; emulsifier, opacifier, thickener, dispersant for emulsions; greasing agent in shampoos, foam baths; food lubricant; perfume diluent; intermediate in textiles and tanning industry
Properties: Nonionic

Monestriol GE [Undesa; S. Black]
Chem. Descrip.: Glyceryl stearate
Uses: Dispersant, solvent for act. ingreds.; vehicle for injectable substances; emulsifier for textiles; emulsifier, opacifier, thickener, dispersant for emulsions; greasing agent in shampoos, foam baths; food lubricant; perfume diluent; intermediate in textiles and tanning industry
Properties: Nonionic

Monestriol GP [Undesa; S. Black]

Chem. Descrip.: Glyceryl stearate
CAS 123-94-4
Uses: Dispersant, solvent for act. ingreds.; vehicle for injectable substances; emulsifier for textiles; emulsifier, opacifier, thickener, dispersant for emulsions; greasing agent in shampoos, foam baths; food lubricant; perfume diluent; intermediate in textiles and tanning industry
Properties: Nonionic

Monestriol GP-35 [Undesa; S. Black]
Chem. Descrip.: Glyceryl stearate
CAS 123-94-4
Uses: Dispersant, solvent for act. ingreds.; vehicle for injectable substances; emulsifier for textiles; emulsifier, opacifier, thickener, dispersant for emulsions; greasing agent in shampoos, foam baths; food lubricant; perfume diluent; intermediate in textiles and tanning industry
Properties: Nonionic

Monestriol GP-35-3% [Undesa; S. Black]
Chem. Descrip.: Glyceryl stearate SE
CAS 11099-07-3
Uses: Dispersant, solvent for act. ingreds.; vehicle for injectable substances; emulsifier for textiles; emulsifier, opacifier, thickener, dispersant for emulsions; greasing agent in shampoos, foam baths; food lubricant; perfume diluent; intermediate in textiles and tanning industry
Properties: Anionic

Monestriol GP-35-AE [Undesa; S. Black]
Chem. Descrip.: Glyceryl stearate SE
CAS 11099-07-3
Uses: Dispersant, solvent for act. ingreds.; vehicle for injectable substances; emulsifier for textiles; emulsifier, opacifier, thickener, dispersant for emulsions; greasing agent in shampoos, foam baths; food lubricant; perfume diluent; intermediate in textiles and tanning industry
Properties: Anionic

Monestriol GP-40 [Undesa; S. Black]
Chem. Descrip.: Glyceryl stearate
CAS 123-94-4
Uses: Dispersant, solvent for act. ingreds.; vehicle for injectable substances; emulsifier for textiles; emulsifier, opacifier, thickener, dispersant for emulsions; greasing agent in shampoos, foam baths; food lubricant; perfume diluent; intermediate in textiles and tanning industry
Properties: Nonionic

Monestriol GP-40-CF [Undesa; S. Black]
Chem. Descrip.: Glyceryl stearate
Uses: Dispersant, solvent for act. ingreds.; vehicle for injectable substances; emulsifier for textiles; emulsifier, opacifier, thickener, dispersant for emulsions; greasing agent in shampoos, foam baths; food lubricant; perfume diluent; intermediate in textiles and tanning industry
Properties: Nonionic

Monestriol GP-40-DC [Undesa; S. Black]
Chem. Descrip.: Glyceryl stearate
CAS 123-94-4
Uses: Dispersant, solvent for act. ingreds.; vehicle for injectable substances; emulsifier for textiles; emulsifier, opacifier, thickener, dispersant for emulsions; greasing agent in shampoos, foam baths; food lubricant; perfume diluent; intermediate in textiles and tanning industry
Properties: Nonionic

Monestriol GP-45 [Undesa; S. Black]
Chem. Descrip.: Polyglyceryl monostearate
Uses: Dispersant, solvent for act. ingreds.; vehicle for injectable substances; emulsifier for textiles; emulsifier, opacifier, thickener, dispersant for emulsions; greasing agent in shampoos, foam baths; food lubricant; perfume diluent; intermediate in textiles and tanning industry
Properties: Nonionic

Monestriol GP-45-DC [Undesa; S. Black]
Chem. Descrip.: Glyceryl stearate
CAS 123-94-4
Uses: Dispersant, solvent for act. ingreds.; vehicle for injectable substances; emulsifier for textiles; emulsifier, opacifier, thickener, dispersant for emulsions; greasing agent in shampoos, foam baths; food lubricant; perfume diluent; intermediate in textiles and tanning industry
Properties: Nonionic

Monestriol GP-60 [Undesa; S. Black]
Chem. Descrip.: Glyceryl stearate
CAS 123-94-4
Uses: Dispersant, solvent for act. ingreds.; vehicle for injectable substances; emulsifier for textiles; emulsifier, opacifier, thickener, dispersant for emulsions; greasing agent in shampoos, foam baths; food lubricant; perfume diluent; intermediate in textiles and tanning industry
Properties: Nonionic

Monestriol GP-AS [Undesa; S. Black]
Chem. Descrip.: Glyceryl stearate
CAS 123-94-4
Uses: Dispersant, solvent for act. ingreds.; vehicle for injectable substances; emulsifier for textiles; emulsifier, opacifier, thickener, dispersant for emulsions; greasing agent in shampoos, foam baths; food lubricant; perfume diluent; intermediate in textiles and tanning industry
Properties: Nonionic

Monestriol IC [Undesa; S. Black]
Chem. Descrip.: Isocetyl stearate
CAS 25339-09-7; EINECS/ELINCS 246-868-6
Uses: Emollient in bath oils, creams/lotions, sun creams, shaving prods., insect repellents, antiperspirants, deodorants; oily component for cleansing creams, nail polish removers; plasticizer for nail polishes; pigment wetting agent; softener in creams, hair care; solvent for fatty cosmetic ingreds.; vehicle for lipsticks; solubilizer for perfume oils, vitamin oils, hormones; solubilizer, vehicle for medicines; excipient for injectable substances; lubricant for PVC
Features: Changes skin feel, rub out props. and greasiness; veg. oil substitute; vaseline substitute

Monestriol IS-C [Undesa; S. Black]
Chem. Descrip.: Isopropyl stearate
CAS 112-10-7; EINECS/ELINCS 203-934-9
Uses: Emollient in bath oils, creams/lotions, sun creams, shaving prods., insect repellents, antiperspirants, deodorants; oily component for cleansing creams, nail polish removers; plasticizer for nail polishes; pigment wetting agent; softener in creams, hair care; solvent for fatty cosmetic ingreds.; vehicle for lipsticks; solubilizer for perfume oils, vitamin oils, hormones; solubilizer, vehicle for medicines; excipient for injectable substances; lubricant for PVC
Features: Changes skin feel, rub out props. and greasiness; veg. oil substitute; vaseline substitute

Monoglyme [Ferro/Grant]
Chem. Descrip.: Ethylene glycol dimethyl ether
CAS 110-71-4; EINECS/ELINCS 203-794-9
Uses: Solvent used in electrochemistry, polymer and boron chemistry, physical processes such as gas absorp., extraction, stabilization, industrial prods. such as fuels, lubricants, textiles, pesticides; solvent vehicle for pharmaceuticals; reaction solvent for pharmaceutical mfg.
Features: Tends to solvate cations
Properties: Colorless; ethereal, nonresidual odor; water-sol.; misc. with ethanol, acetone, benzene, diethyl ether, octane; m.w. 90.12; sp.gr. 0.8683 (20/20 C); dens. 7.24 lb/gal; visc. 1.1 cP (20 C); vapor pressure 54 mm Hg (20 C); f.p. -69 C; b.p. 85.2 C; flash pt. (CC) -6 C; ref. index 1.3792; pH neutral; surf. tens. 22.9 dynes/cm (20 C); sp. heat 0.438 cal/g/°C; 99.6% min. purity
Toxicology: LD50 5370 mg/kg; low to mod. acute toxicity; avoid exposure to vapors, skin contact; TLV/8h TWA: 5 ppm; STEL 25 ppm; LD50 5100 mg/kg); chronic exposure may cause reproductive effects
Environmental: Slowly biodeg.

Morcryl® 535 [Rohm & Haas]
Chem. Descrip.: Water-based resin sol'n. of Morez 500
Uses: Solvent, extender, gloss aid, rewetting agent, pigment dispersant for inks, coatings, overprints, food pkg. adhesives, paper
Features: Improves holdout; solv.-free; no VOC
Regulatory: FDA 21CFR §175.105, 175.300, 175.320, 176.170, 176.180 compliant
Properties: Sl. yel. clear liq.; m.w. 3000-5000; dens. 8.8 lb/gal; visc. 2000 cps; acid no. 230; pH 8.5; 0% VOC; 34% solids

Morflex® DEIP [Morflex]
Chem. Descrip.: Diethyl isophthalate
CAS 636-53-3
Uses: Chemical intermediate; solvent
Properties: Liq.; sol. in most org. solvs.; insol. in water; sp.gr. 1.123 (20/20 C); f.p. 12 C; b.p. 16 C (10 mm Hg); flash pt. (COC) 325 F; ref. index 1.504-.1506; 99% min. assay; 0.05% max. moisture

MPDiol® Glycol [Lyondell]

Chem. Descrip.: 2-Methyl-1,3-propanediol
CAS 2163-42-0; EINECS/ELINCS 412-350-5
Uses: Intermediate for mfg. of urethanes, unsat. polyesters, gel coats, sat. polyester and alkyd coatings, polymeric plasticizers; food-contact coatings; solvent, compatibilizer for personal care, adjuvants, fabric softener carriers, nonionic surfactants, pigment dispersions; solubilizer and emulsifier in personal care and air freshener applics.; stabilizer; humectant; neutralizer; emollient
Features: Produces unsat. polyester resins with light color, exc. styrene misc., exc. balance of tens. str., elong., flexibility
Regulatory: FDA 21CFR §175.300; SARA and CERCLA nonreportable; listed on TSCA, Korean, and Japanese inventories
Properties: Colorless clear low-visc. liq.; sol. in water; m.w. 90.1; sp.gr. 1.015; dens. < 8.47 lb/gal; visc. 174 cps; vapor pressure 4.3 mm Hg (100 C); m.p. -91 C; b.p. 213 C; hyd. no. 1200; flash pt. (OC) > 230 F; 98% assay
Toxicology: LD50 (oral, rat) > 5000 mg/kg, (dermal, rabbit) > 2000 mg/kg; may cause sl. eye irritation; not a skin sensitizerTSCA listed
Environmental: Inherently biodeg.; low aquatic toxicity
Precaution: Store away from strong min. acids, dehydrating agents, strong oxidizing agents
HMIS: Health 1, Flammability 1, Reactivity 0
Storage: Avoid excessive exposure to moisture; keep away from heat, sparks, and flame

M-Pyrol® [ISP]
Chem. Descrip.: N-Methyl-2-pyrrolidone
CAS 872-50-4; EINECS/ELINCS 212-828-1
Uses: Solvent, stabilizer for vinyls, urethanes, acrylics, difficultly sol. resins and waxes; spinning solvent for syn. fibers; in commercial petrochemical processing; for agric. chemicals; in removal of strippable coatings; for lube oil processing; as nonaq. syn. medium for polyol reactions; residue cleaning solvent for PVC and urethane prod. equipment; in adhesives
Features: Replaces toxic solvs.; reduces VOC emissions
Properties: Liq., mild amine-like odor; misc. with water; m.w. 99.1; sp.gr. 1.027; dens. 1.03 g/cc (20 C); visc. 1.7 cp (20 C); vapor pressure 1.0 torr (40 C); f.p. -24.4 C; b.p. 202 C; flash pt. (OC) 93 C; ref. index 1.4700; surf. tens. 41 dynes/cm; KB > 300; 99.8% min. purity
Toxicology: Low toxicity
Environmental: Rapidly biodeg.; low toxicity to aquatic organisms
Precaution: Combustible in bulk containers; when ignited, will sustain a fire; flamm. limits in air 1.3-9.5 vol.%
Storage: Stable but hygroscopic

Myritol® 318 [Cognis/Care Chems.]
Chem. Descrip.: Caprylic/capric triglyceride
CAS 73398-61-5; EINECS/ELINCS 265-724-3
Uses: Spreading agent, emollient, solvent, solubilizer for pharmaceutical and cosmetic skin care preps. in emulsion form (bath oils, lotions); oil binder for pressed powd. cosmetics
Features: Exc. oxidative stability; compressibility; oily component with solv. capacity
Regulatory: JCIC, DAB compliance
Properties: Sl. ylsh. clear low-visc. oily liq., odorless; m.w. 500; sp.gr. 0.945-0.949; visc. 27-33 mPa·s (20 C); acid no. 0.1 max.; iodine no. 0.5 max.; sapon. no. 335-350; hyd. no. 5 max.; cloud pt. < -5 C; ref. index 1.448-1.450 (20 C); nonionic; 100% conc.
Storage: 1 yr min. storage life in original containers at temps. below 30 C in a dry environment

Myritol® PC [Cognis/Care Chems.]
Chem. Descrip.: Propylene glycol dicaprylate/dicaprate
CAS 68583-51-7; EINECS/ELINCS 271-516-3
Uses: Emollient, spreading agent, cosolvent for cosmetics, pharmaceuticals, skin care preps., bath oils, aerosols, and antiperspirants; solvent for active ingreds.; vehicle for pigmented cosmetics
Features: Stable
Regulatory: DAB, JCIC compliance
Properties: Colorless clear low visc. oil, faint char. odor; m.w. 340; sp.gr. 0.91-0.93 (20 C); visc. 9-12 mPa·s (20 C); acid no. 0.1 max.; iodine no. 0.5 max.; sapon. no. 325-335; hyd. no. 5 max.; cloud pt. < -20 C; ref. index 1.440-1.442 (20 C)
Storage: 1 yr min. storage stability in sealed original containers at temps. below 30 C in a dry environment

Myvacet® 9-45K [Quest Int'l.]
Chem. Descrip.: Acetylated hydrogenated soybean oil glyceride
Uses: Emulsifier, lubricant, solvent for icings and shortenings
Properties: Liq.; sp.gr. 0.94 (80 C); m.p. 4-12 C; HLB 3.8-4.0; acid no. 3 max.; iodine no. 43-53; sapon. no. 370-382; hyd. no. 0-15; nonionic; 96% min. acetylation

N

Nadone® [Honeywell Perf. Polymers]
Chem. Descrip.: Cyclohexanone
CAS 108-94-1; EINECS/ELINCS 203-631-1
Uses: Solvent for coating resins, printing inks, adhesives, fats, oils, waxes, agric., textile applics.
Properties: Sol. in acetone, ethanol, benzene, ethylene glycol

Naetex-L [Lanaetex Prods.]
Chem. Descrip.: PEG-2 lactamide
Uses: Coupling agent, lubricant, emollient, humectant for personal care prods.; enhances wet combing, imparts softness and antistatic props. to hair; solvent in cosmetic creams and lotions; hygroscopic nature prevents dry skin

Naetex-LAM [Lanaetex Prods.]
Chem. Descrip.: Lactic acid MEA
CAS 5422-34-4; EINECS/ELINCS 226-546-1
Uses: Coupling agent in cosmetics, esp. shampoos; lubricant, emollient in creams/lotions; water scavenger, humectant; possible antistat in hair care; solvent in creams/lotions; skin protectant (helps to prevent skin dryness)
Properties: Yel. to straw-colored clear liq.; typ. odor; completely sol. in water; acid no. 45 max.; alkali no. 45 max.

1520 Naphtha [ExxonMobil]
Chem. Descrip.: Naphtha
EINECS/ELINCS 232-443-2
Uses: Solvent
Properties: Bright clear liq.; sp.gr. 0.67 (15.6/15.6 C); dens. 5.61 lb/gal (60 F); b.p. 64-94 C; flash pt. (TCC) < -18 C; surf. tens. 18.2 dynes/cm; KB value 28; 98% paraffin, 2% cycloparaffin, < 0.5% aromatics
Toxicology: OEL 100 ppm

2024 Naphtha [ExxonMobil]
Chem. Descrip.: C isoparaffin hydrocarbon solvent
CAS 64742-48-9
Uses: Solvent
Properties: Sp.gr. 0.74 (15.6/15.6 C); dens. 6.15 lb/gal (60 F); b.p. 97-117 C; flash pt. (TCC) 3 C; surf. tens. 21.3 dynes/cm; KB value 39; 52% paraffin, 42% cycloparaffin, 6% aromatics
Toxicology: OEL 400 ppm

2024 Naphtha Anti-Static [ExxonMobil]
Chem. Descrip.: Hydrocarbon fluid with antistatic additive (13 ml/1000 gal)
Uses: Solvent
Properties: Bright clear liq.; i.b.p. 90 C min.; KB value 37-42

Naphthalene Depleted Aromatic 150 Fluid [ExxonMobil; Exxon Canada]
Chem. Descrip.: Heavy aromatic solvent naphtha, BHT (16 ppm)
CAS 64742-94-5; EINECS/ELINCS 265-198-5
Uses: Solvent for additives, paints/coatings, pesticides, industrial cleaning, mastics, sealants, process fluids; diluent; chem. feedstock; fuel
Regulatory: Canada DSL listed; NPRI reportable [naphthalene (0.5%)]
Properties: Colorless clear liq.; mild aromatic odor; m.w. 140; sol. < 0.01% in water; sp.gr. 0.89 (15.5 C); visc. 1.24 cSt; vapor pressure 0.09 kPa (20 C); f.p. -30 C b.p. 183-202 C; flash pt. (TCC) > 63 C; autoignition temp. ≈ 443 C; 100% act.; 95% min. aromatics
Toxicology: REL TWA 17 ppm; LD50 (oral, rat) 3 g/kg, (skin, rabbit) > 3 g/kg; inh. of high vapor concs. may cause eye/respiratory tract irritation, headaches, dizziness, anesthesia, drowsiness, unconsciousness, other CNS effects, death; sl. eye irritant; frequent/prolonged skin contact may cause irritation, dermatitis; if aspirated, may cause mild to severe pulmonary injury, death; contains naphthalene, possible human carcinogen
Environmental: Prevent entry into sewers, watercourses, low areas
Precaution: Combustible liq.; flamm. limits 1.8-11.7 vol.%; volatile; gives off invisible vapors; liq. or vapor may settle in low areas; distant ignition/explosion possible; electrostatic accumulation hazard; use proper grounding procedures; incompat. with nitric acid, sulfuric acid, strong oxidizing agents
Hazardous decomp. prods.: Combustion: toxic gases, fumes, smoke, CO
Storage: Store in cool, well-ventilated area away from incompat. materials, heat, open flame, ignition sources, direct sunlight; keep container closed

Naphthalene Depleted Aromatic 200 Fluid [ExxonMobil; Exxon Canada]
Chem. Descrip.: Heavy aromatic solvent naphtha, BHT (17 ppm)
CAS 64742-94-5; EINECS/ELINCS 265-198-5
Uses: Solvent for additives, paints/coatings, pesticides, industrial cleaning, mastics, sealants, process fluids; diluent; chem. feedstock; fuel
Features: High solvency; controlled evaporation chars.
Regulatory: Canada DSL listed; NPRI nonreportable
Properties: Pale yel. bright clear liq.; aromatic hydrocarbon odor; m.w. ≈ 120; sol. < 0.01% in water; sp.gr. 01 (15.6 C); visc. 2.7 cSt; vapor pressure < 0.1 kPa (20 C); f.p. -10 C; b.p. 235-278 C; flash pt. (PMCC) 106 C; autoignition temp. ≈ 484 C; > 97 min. vol.% aromatics; 0.9% max. naphthalene
Toxicology: Inh. of high vapor concs. may cause eye/respiratory tract irritation, headaches, dizziness, anesthesia, drowsiness, unconsciousness, other CNS effects, death; sl. eye irritant; frequent/prolonged skin contact may cause irritation, dermatitis; if aspirated, may cause mild to severe pulmonary injury, death
Environmental: Prevent entry into sewers, watercourses, low areas
Precaution: Flamm. limits 1.8-11.8 vol.%; electrostatic accumulation hazard; use proper grounding procedures; incompat. with nitric acid, sulfuric acid, strong oxidizing agents
Hazardous decomp. prods.: Combustion: toxic gases, fumes, smoke, CO
Storage: Store in cool, well-ventilated area away from incompat. materials, heat, open flame, ignition sources, direct sunlight; keep container closed

Naphthol Spirits 66/3 [CITGO]
Chem. Descrip.: Stoddard solvent
CAS 8052-41-3; EINECS/ELINCS 232-489-3
UN 1268
Uses: Solvent
Features: Med. boiling; hydrotreated to reduce aromatics and olefins; meets definition of non-photochemically reactive as defined by 66/3 Rule requirements
Regulatory: FDA 21CFR §175.105, 178.3620, 178.3910; SARA §302/304 nonreportable, §311/312 acute health/chronic health/fire hazard, §313 nonreportable; Calif. Prop. 65 nonreportable; may contain detectable amts. 1,3,5-trimethylbenzene
Properties: Colorless transparent liq.; lt. paraffinic hydrocarbon odor; sl. sol. in cold water (< 0.05%); sp.gr. 0.776 (60/60 F); dens. 6.46 lb/gal (15.6 C); visc. 1.19 cst (20 C); vapor pressure 2.9 mm Hg (20 C); f.p. -65 to -25 C; b.p. 160-185 C; flash pt. (TCC) 41-45 C; autoignition temp. 230

C; ref. index 1.4261 (20 C); KB value 33.6; VOC 775 g/l; 100% act.
Toxicology: ACGIH TLV TWA 100 ppm; LD50 (oral, rat) > 34,600 mg/kg; mist/vapor may irritate eyes, mucous membranes, respiratory tract; liq. contact may cause minimal to mod. eye irritation, mod. to severe skin irritation/inflamm.; may be harmful by inh., skin absorp.; overexposure may cause CNS depression, other target organ effects; may be harmful or fatal by ing.; aspiration into lungs can cause pulmonary edema, chem. pneumonia; TSCA listed
Environmental: May cause oxygen depletion in waterways; potentially toxic to water ecosystems; may be hazardous to aquatic life; may contribute to smog; prevent entry into sewers, waterways; not a DOT marine pollutant
Precaution: Combustible liq.; flamm. limits 0.8-6.0%; vapor may cause flash fire; vapors are heavier than air, may flash back; may create vapor/air explosion hazard in confined spaces; spills may create a slipping hazard; floats on water; container can rupture in fire; prevent entry into basements, confined areas; bond and ground equip.; incompat. with strong acids, alkalis, oxidizers (liq. chlorine, other halogens, hydrogen peroxide, oxygen)
Hazardous decomp. prods.: Combustion: smoke, CO, CO_2, possibly other harmful gases/vapors
HMIS: Health 1, Flammability 2, Reactivity 0
Storage: Store in tightly closed containers in cool, dry, well-ventilated place away from heat, sparks, ignition sources, strong oxidizing conditions

Nappar™ 6 [ExxonMobil Europe]
Chem. Descrip.: Naphthenic hydrocarbon fluid
Uses: Solvent for paints/coatings
Properties: Liq.; dens. 0.764 kg/dm³ (15 C); visc. 0.70 mPa·s; i.b.p. 78 C; flash pt. < 0 C; ref. index 1.418 (20 C); surf. tens. 23.2 mN/m; KB value 48; aniline pt. 36 C; 0.0001% max. aromatics

Nappar™ 10 [ExxonMobil Europe]
Chem. Descrip.: Naphthenic hydrocarbon fluid
Uses: Solvent
Properties: Liq.; dens. 0.815 kg/dm³ (15 C); visc. 1.12 mPa·s; i.b.p. 165 C; flash pt. 43 C; ref. index 1.445 (20 C); surf. tens. 26.6 mN/m; KB value 41; aniline pt. 54 C; 0.2% max. aromatics

Naturechem® GTR [CasChem; Costec]
Chem. Descrip.: Glyceryl triacetyl ricinoleate
CAS 101-34-8; EINECS/ELINCS 202-935-1
Uses: Emollient, pigment wetting agent/dispersant/cosolvent, lubricant, moisturizer, emulsifier, visc. builder for creams/lotions, lipstick, lip balm, sunscreen, soaps; softener for waxes and resins
Features: Mild
Properties: Gardner 2+ color; m.p. -40 C; acid no. < 1; iodine no. 76; hyd. no. 5; pour pt. - 40 C

Naturechem® MAR [CasChem; Costec]
Chem. Descrip.: Methyl acetyl ricinoleate
CAS 140-03-4; EINECS/ELINCS 205-392-9
Uses: Emollient, cosolvent for personal care prods., creams/lotions, lipstick, lip balm, sunscreens, soaps; solubilizer for benzophenone-3
Features: Light emolliency; reduces greasiness of emollients such as min. oil; superior freeze/thaw props.
Properties: Gardner 1 color; m.p. -25 C; acid no. < 1; iodine no. 77; hyd. no. < 5

Naxol® [Honeywell Perf. Polymers]
Chem. Descrip.: Cyclohexanol
CAS 108-93-0; EINECS/ELINCS 203-630-6
Uses: Solvent for fats, oils, waxes, resins, soaps, dry cleaning agents, detergents; coupling agent
Properties: Misc. with ether, ethanol, aromatic hydrocarbons.

NE™ [ANGUS]
Chem. Descrip.: Nitroethane
CAS 79-24-3; EINECS/ELINCS 201-188-9
Uses: Intermediate, stabilizer for halogenated solvs.; as fuel; as explosive; solvent for coatings or industrial processes; raw material for synthesis of act. pharmaceutical compds., for α-methyldopa, a hypertensive drug, and for phenylpropanolamine used in bronchial decongestants and wt. control agents
Properties: Liq.; sol. 4.6% in water (20 C); m.w. 75.1; dens. 1.051 g/ml (20 C), 8.75 lb/gal; visc. 0.677 cp (20 C); vapor pressure 20.93 mm Hg; f.p. -89.52 C; b.p. 114 C; flash pt. (TCC) 87 F; ref. index 1.39193 (20 C); pH 6.0 (0.01M aq.); surf. tens. 32.66 dynes/cm (20 C); dielec. const. 28.06 (30 C); 98% min. purity
Toxicology: LD50 (oral, rat) 1625 ± 193 mg/kg; sl. toxic by ing.; nonirritating to eyes and skin; prolonged/repeated skin exposure may cause defatting; inh.: PEL/TWC 100 ppm, mild irritant, headache, nausea, vomiting; avoid prolonged breathing of vapors
Environmental: Biodeg.; classified as solv. exempt under Rule 66
Precaution: Flamm.; keep away from heat, sparks, flame
Storage: Keep container closed

NE™ 101 [ANGUS]
Chem. Descrip.: Nitroethane
CAS 79-24-3; EINECS/ELINCS 201-188-9
Uses: Intermediate; solvent for coatings, adhesives, inks
Features: Low resistivity and med. evaporation rate; improves solvent release without blocking; not HAPs listed
Properties: Colorless liq., char. odor; sol. 4.5 vol.% in water (20 C); m.w. 75.068; sp.gr. 1.052 (20/20 C); visc. 0.677 cp (20 C); vapor pressure 15.6 mm Hg (20 C); f.p. -89.52 C; b.p. 114 C; flash pt. (TCC) 87 F; ref. index 1.39193 (20 C); pH 6.0 (0.01m aq.); surf. tens. 32.66 dynes/cm (20 C); dielec. const. 28.06 (30 C); 98% min. act.
Toxicology: LD50 (oral, rat) 1624 mg/kg; TLV/PEL 100 ppm; sl. toxic by ing.; harmful if swallowed; vapors may cause headache, nausea, vomiting, narcosis; chronic overexposure may cause liver/kidney damage; defatting skin irritant; nonirritating to eyes
Precaution: Flamm.; salts of nitroparaffins with inorg. bases or amines are thermally unstable; violent decomp. can occur; lower explosive limit 4%; incompat. with lead, copper and alloys, strong alkalis; produces CO_x, NO_x
Storage: Keep container closed; keep away from heat, sparks, flame

NEOBEE® 18 [Stepan]
Chem. Descrip.: Hybrid safflower oil, refined, deodorized
CAS 8001-23-8; EINECS/ELINCS 232-276-5
Uses: Oil for margarine, mayonnaise, salad dressings; emollient base for cosmetics and pharmaceuticals; solubilizer; solvent; lubricant; nutritional fluid
Features: Highly stable
Regulatory: FDA 21CFR §170.30, GRAS; avail. in kosher grade
Properties: Sl. yel. liq., sl. fatty odor, bland taste; sol. in alcohol, min. oil, acetone; sp.gr. 0.915; visc. 70 cps; iodine no. 90; sapon. no. 190; surf. tens. 31.6 dynes/cm.
Toxicology: Nonirritating
Storage: Avoid prolonged storage above 90 F

NEOBEE® M-5 [Stepan]
Chem. Descrip.: Caprylic/capric triglyceride
CAS 73398-61-5; EINECS/ELINCS 265-724-3
Uses: Diluent, vehicle, carrier for flavors, vitamins, antibiotics, nutritional fluids, medicinals, essential oils, colors; solubilizer, cosolvent for fragrances, flavors, medicinals, food colors; clouding agent for beverages; source of nutritional MCT; fat source in nutritional prods.; emollient for cosmetics (bath oils, creams, lotions, lipsticks, makeup base, shaving preps.), pharmaceuticals; moisture barrier; release agent; polishing agent
Features: Min. oil alternative
Regulatory: FDA 21CFR §170.30, GRAS; kosher; Europe, Japan, Canada, Australia listed
Properties: Water-wh. oily liq.; bland odor and taste; sol. in alcohol, min. oil, acetone; sp.gr. 0.945; visc. 14.9 cSt (100 F); f.p. -5 C; acid no. 0.10 max.; iodine no. 0.006; sapon. no. 345; hyd. no. 1.4; flash pt. (MCC) 260 C; ref. index 1.447; surf. tens. 28.8 mN/m; 0% RVOC; 200 ppm moisture
Toxicology: LD50 (rat) > 30 ml/kg; may cause very mild transient eye but no skin irritation if applied undiluted
Environmental: Biodeg.
Storage: Store in sealed containers below 32 C; avoid prolonged storage above 32 C; avoid overheating

NEOBEE® M-20 [Stepan]
Chem. Descrip.: Propylene glycol dicaprylate/dicaprate
CAS 68583-51-7; EINECS/ELINCS 271-516-3
Uses: Emollient for creams, lotions, and ointments; solvent, solubilizer, vehicle for antibiotics, medicinals; diluent for essential oils, injectables; lubricant; deposition agent on skin; diluent for essential oils in foods
Features: Low visc.
Regulatory: FDA 21CFR§170.30, 172.856; avail. in kosher grade
Properties: Pale yel. oily liq., bland odor; sol. in alcohol containing up to

20% water; iodine no. 1 max.; sapon. no. 326; hyd. no. 0.6; 0.04% moisture
Storage: Avoid prolonged storage above 90 F

Neosolv™ 2 [Shell]
Chem. Descrip.: C_{12} unsat. aliphatic hydrocarbon
Uses: Solvent for adhesives, air fresheners, automotive brake, carburetor, and choke cleaners, bathroom and tile cleaners, dusting aids, engine degreasers, fabric protection sprays, floor polishes/waxes, furniture polishes/waxes, hair sprays/mousses/styling gels, insecticides, insect repellents, laundry aids, oven cleaners, shaving aids, waterless hand cleaners
Features: Environmentally friendly; meets VOC restrictions
Properties: Pt-Co 5 max. color; low odor; m.w. 168; sp.gr. 0.764 (60/60 F); dens. 6.36 lb/gal (60 F); visc. 1.42 cP (60 F); vapor pressure 0.0984 mm Hg (20 F); f.p. -33 F; b.p. 415-419 F; flash pt. (Seta) 168 F; KB value 27; VOC 0 lb/gal
Environmental: Readily biodeg.
Precaution: Reacts readily with air or oxygen to form trace peroxides and other oxygenates
Storage: Store under nitrogen blanketing

Neosolv™ 4 [Shell]
Chem. Descrip.: Unsat. aliphatic hydrocarbon
Uses: Solvent for adhesives, air fresheners, automotive brake, carburetor, and choke cleaners, bathroom and tile cleaners, dusting aids, engine degreasers, fabric protection sprays, floor polishes/waxes, furniture polishes/waxes, hair sprays/mousses/styling gels, insecticides, insect repellents, laundry aids, oven cleaners, shaving aids, waterless hand cleaners
Features: Environmentally friendly; meets VOC restrictions
Properties: Pt-Co 10 max. color; low odor; sp.gr. 0.770-0.780 (60/60 F); dens. 6.46 lb/gal (60 F); visc. 1.42 cP (60 F); vapor pressure 0.0128 mm Hg (20 F); f.p. 6 F; b.p. 467-475 F; flash pt. (PMCC) 210 F min.; KB value 24-26; VOC 0 lb/gal
Environmental: Readily biodeg.
Precaution: Reacts readily with air or oxygen to form trace peroxides and other oxygenates
Storage: Store under nitrogen blanketing

Neosolv™ 6 [Shell]
Chem. Descrip.: Unsat. aliphatic hydrocarbon
Uses: Solvent for adhesives, air fresheners, automotive brake, carburetor, and choke cleaners, bathroom and tile cleaners, dusting aids, engine degreasers, fabric protection sprays, floor polishes/waxes, furniture polishes/waxes, hair sprays/mousses/styling gels, insecticides, insect repellents, laundry aids, oven cleaners, shaving aids, waterless hand cleaners
Features: Environmentally friendly; meets VOC restrictions
Properties: Pt-Co 10 max. color; low odor; sp.gr. 0.780-0.790 (60/60 F); dens. 6.54 lb/gal (60 F); visc. 3.10-3.30 cP (60 F); vapor pressure 0.00129 mm Hg (20 F); f.p. 41 F; b.p. 514-536 F; flash pt. (PMCC) 240 F min.; KB value 22-24; VOC 0 lb/gal
Environmental: Readily biodeg.
Precaution: Reacts readily with air or oxygen to form trace peroxides and other oxygenates
Storage: To keep prod. fluid in cold weather, store at least 5 F above f.p., but < 100 F; store under nitrogen blanketing

Neosolv™ 8 [Shell]
Chem. Descrip.: Unsat. aliphatic hydrocarbon
Uses: Solvent for adhesives, air fresheners, automotive brake, carburetor, and choke cleaners, bathroom and tile cleaners, dusting aids, engine degreasers, fabric protection sprays, floor polishes/waxes, furniture polishes/waxes, hair sprays/mousses/styling gels, insecticides, insect repellents, laundry aids, oven cleaners, shaving aids, waterless hand cleaners
Features: Environmentally friendly; meets VOC restrictions
Properties: Pt-Co 10 max. color; low odor; sp.gr. 0.787-0.797 (60/60 F); dens. 6.59 lb/gal (60 F); visc. 4.51-4.61 cP (60 F); vapor pressure < 0.001 mm Hg (20 F); f.p. 37 F; b.p. 517-602 F; flash pt. (PMCC) 250 F min.; KB value 20-22; VOC 0 lb/gal
Environmental: Readily biodeg.
Precaution: Reacts readily with air or oxygen to form trace peroxides and other oxygenates
Storage: To keep prod. fluid in cold weather, store at least 5 F above f.p., but < 100 F; store under nitrogen blanketing

Neosolv™ 24 [Shell]
Chem. Descrip.: Unsat. aliphatic hydrocarbon
Uses: Solvent for adhesives, air fresheners, automotive brake, carburetor, and choke cleaners, bathroom and tile cleaners, dusting aids, engine degreasers, fabric protection sprays, floor polishes/waxes, furniture polishes/waxes, hair sprays/mousses/styling gels, insecticides, insect repellents, laundry aids, oven cleaners, shaving aids, waterless hand cleaners
Features: Environmentally friendly; meets VOC restrictions
Properties: Pt-Co 10 max. color; low odor; sp.gr. 0.763-0.773 (60/60 F); dens. 6.40 lb/gal (60 F); visc. 1.70-1.90 cP (60 F); vapor pressure 0.0589 mm Hg (20 F); f.p. -9 F; b.p. 430-473 F; flash pt. (PMCC) 190 F min.; KB value 26-28; VOC 0 lb/gal
Environmental: Readily biodeg.
Precaution: Reacts readily with air or oxygen to form trace peroxides and other oxygenates
Storage: Store under nitrogen blanketing

Neosolv™ 150 [Shell]
Chem. Descrip.: Unsat. aliphatic hydrocarbon
Uses: Solvent for adhesives, air fresheners, automotive brake, carburetor, and choke cleaners, bathroom and tile cleaners, dusting aids, engine degreasers, fabric protection sprays, floor polishes/waxes, furniture polishes/waxes, hair sprays/mousses/styling gels, insecticides, insect repellents, laundry aids, oven cleaners, shaving aids, waterless hand cleaners
Features: Environmentally friendly; meets VOC restrictions
Properties: Pt-Co 10 max. color; low odor; sp.gr. 0.763-0.773 (60/60 F); visc. 1.66-1.91 cP (60 F); vapor pressure 0.097-0.161 mm Hg (20 F); f.p. 9 F max.; b.p. 389-471 F; flash pt. (PMCC) 150 F min.; KB value 21-23; VOC 2.5 lb/gal max.
Environmental: Readily biodeg.
Storage: Store below 100 F; nitrogen blanketing is rec. to prevent oxidation to peroxides and carbonyls

Neosolvent NSG [Toho Chem. Ind.]
Chem. Descrip.: Polypropylene polyethylene glycol monomethyl ether
Uses: Solvent for brake fluids
Properties: Liq.

NEP® [ISP]
Chem. Descrip.: N-Ethyl pyrrolidone
CAS 2687-91-4
Uses: Reaction solvent for chem. reactions down to -70 C; antibiotic prod.; elec. capacitors; mfg. of polyamide fibers and film; jet printing ink; photo resist stripper; polymerization (adipamide and hexamethylene diamine); synthesis of ethylene glycol; working fluid in advanced heat transformers
Properties: Clear liq.; water-sol.; m.w. 113; sp.gr. 0.993; visc. 3.5 cps; f.p. -77 C; b.p. 212 C; flash pt. (CC) 199 F; ref. index 1.4664; surf. tens. 36 dynes/cm; 98% min. purity
Toxicology: LD50 (oral, rat) > 14.43 g/kg; low acute toxicity; minimally irritating to eyes; nonirritating to skin
Environmental: Biodeg.

Neu-Tri E [Dow]
Chem. Descrip.: Trichloroethylene
CAS 79-01-6; EINECS/ELINCS 201-167-4
Uses: Solvent for metal cleaning/degreasing, vapor degreasing, chemical processing
Features: Highly stabilized; stabilizer system prevents buildup of acid in the degreaser and protects against metal corrosion and reaction in the solv.; effective for long-term use; non-ozone depleting
Properties: Sol. 0.1% in water; m.w. 131.4; dens. 1.465 g/cm^3 (20 C); visc. 0.58 mPa·s (20 C); vapor pressure 72.3 hPa (20 C); f.p. -87 C; b.p. 87 C; flash pt. (PMCC) none; autoignition temp. 420 C; surf. tens. 32 mN/m (20 C); KB value 128; sp.heat 0.93 kJ/kg/K; dielec. str. 30,000 V; dielec. const. 3.39 (24 C, 100 MHz); resist. 6.7 x 10^9 ohm-cm
Toxicology: ACGIH TLV 50 ppm; may be harmful to human health if used improperly
Environmental: May be harmful if used improperly; very low impact on smog formation, global warming, acid rain
Precaution: Explosive limits in air 8.0-44.8 vol.%

Nikkol BWA-5 [Nikko Chems. Co. Ltd]
Chem. Descrip.: PEG-5 lanolin alcohol

CAS 61791-20-6
Uses: Cosolvent, emulsifier, solubilizer, and softener for shampoos
Features: Provides hydrophilic props. to oil phase
Regulatory: JSCI listed
Properties: Ylsh. brn. paste; HLB 12.5; nonionic; 100% conc.
Toxicology: TSCA listed

Nikkol BWA-10 [Nikko Chems. Co. Ltd]
Chem. Descrip.: PEG-10 lanolin alcohol
CAS 61791-20-6
Uses: Cosolvent, emollient for creams, milk lotions
Features: Provides hydrophilic props. to oil phase
Regulatory: JSCI listed
Properties: Ylsh. brn. solid; HLB 15.5; nonionic
Toxicology: TSCA listed

Nikkol SO-30 [Nikko Chems. Co. Ltd]
Chem. Descrip.: Sorbitan trioleate
CAS 26266-58-0; EINECS/ELINCS 247-569-3
Uses: W/o emulsifier for cosmetics; solvent; oil phase regulator; antirust
Regulatory: JSCI listed
Properties: Yel. liq.; HLB 1.7; nonionic; 100% conc.
Toxicology: TSCA listed

Nikkol SO-30R [Nikko Chems. Co. Ltd]
Chem. Descrip.: Sorbitan trioleate
CAS 26266-58-0; EINECS/ELINCS 247-569-3
Uses: W/o emulsifier for cosmetics; antirust; solvent; oil phase regulator
Features: Refined grade (low color/odor)
Regulatory: JSCI listed
Properties: Pale yel. liq.; HLB 4.0; nonionic
Toxicology: TSCA listed

Nikkol Sefsol 218 [Nikko Chems. Co. Ltd]
Chem. Descrip.: Propylene glycol monocaprylate
CAS 31565-12-5
Uses: Solvent, solubilizer for sl. sol. ingreds. in cosmetics
Features: Good compat. with oily materials
Properties: Colorless liq.

Nikkol Sefsol 228 [Nikko Chems. Co. Ltd]
Chem. Descrip.: Propylene glycol dicaprylate
CAS 7384-98-7; EINECS/ELINCS 230-962-9
Uses: Solvent, solubilizer for sl. sol. ingreds. in cosmetics
Features: Good compat. with oily materials
Regulatory: JSCI listed
Properties: Colorless liq.
Toxicology: TSCA listed

NiPar S-10™ [ANGUS]
Chem. Descrip.: 1-Nitropropane
CAS 108-03-2; EINECS/ELINCS 203-544-9
Uses: Intermediate, solvent for coatings; pharmaceutical intermediate
Properties: Liq.; sol. 1.5% in water (20 C); m.w. 89.1; dens. 1.0 g/ml (20 C), 8.35 lb/gal; visc. 0.844 cp (20 C); vapor pressure 10.23 mm Hg; f.p. -104 C; b.p. 131 C; flash pt. (TCC) 96 F; ref. index 1.40160 (20 C); pH 6.0 (0.01M aq.); surf. tens. 30.64 dynes/cm (20 C); dielec. const. 23.24 (30 C); 98.5% min. purity
Toxicology: Harmful if inhaled; overexposure can cause liver damage and death; OSHA PEL 25 ppm; avoid breathing vapors
Precaution: Flamm.; keep away from heat, sparks, flame
Storage: Keep container closed

NM™ [ANGUS]
Chem. Descrip.: Nitromethane
CAS 75-52-5; EINECS/ELINCS 200-876-6
Uses: Intermediate, stabilizer for halogenated solvs.; as fuel; as explosive; solvent for coatings or industrial processes; cosmetic and personal care raw material; raw material for synthesis of act. pharmaceutical compds., e.g. for 1,1-bis(methylthio)-2-nitroethane used to mfg. ranitidine, an anti-ulcer drug, or serinol, a raw material for iopamidol, an injectable radio-opaque x-ray contrast medium
Properties: Sol. 10.5% in water (20 C); m.w. 61.0; sp.gr. 1.124-1.135; dens. 1.138 g/ml (20 C), 9.4 lb/gal; visc. 0.647 cp (20 C); vapor pressure 36.66 mm Hg; f.p. -28.55 C; b.p. 101 C; flash pt. (TCC) 96 F; ref. index 1.38188 (20 C); pH 6.4 (0.01M aq.); surf. tens. 37.48 dynes/cm (20 C); dielec. const. 35.87 (30 C); 98% min. purity
Toxicology: LD50 (oral, rat) 1210 ± 322 mg/kg; nonirritating to skin and eyes; prolonged/repeated skin contact may cause defatting; inh.: PEL/TWC 100 ppm; avoid prolonged breathing of vapors
Environmental: Biodeg.; classified as solv. exempt under Rule 66
Precaution: Flamm. liq. and vapor; keep away from heat, sparks, flame; forms shock-sensitive mixts. with amines; compds. formed with strong alkalis are explosive when dry; avoid mixing with amines or strong alkalis
Storage: Store in original container; do not clean drum with lye or caustic

NMP [Dynaloy]
Chem. Descrip.: N-Methyl pyrrolidone-based
Uses: Solvent for removal of adhesives and inks from rollers, general polymer removal, paint stripping, cured epoxy removal, electronic depotting, deflashing, decapsulation, and deinking applics., surf. prep. and photoresist stripping
Features: Use hot (105-175 C); not selective; available filtered thru 0.2 μ filter

NMP [Lyondell]
Chem. Descrip.: N-Methyl pyrrolidone
CAS 872-50-4; EINECS/ELINCS 212-828-1
Uses: Solvent, cosolvent in coatings, stripping/cleaning of paints/varnishes, industrial cleaning, mold cleaning, petrochem. processing; solvent for adhesives, inks, dyes, soaps, metal complexes, agric.; solvent for PVAc, PVDF, PS, vinyl copolymers, nylon and aromatic polyamides and polyimides, polyesters, acrylics, PC, cellulose derivs., syn. elastomers, waxes
Features: Low flamm., noncorrosive, nonchlorine; low evaporation rate
Regulatory: SARA reportable
Properties: Colorless mobile liq., mild amine-like odor; completely sol. in water; m.w. 99.13; sp.gr. 1.028; dens. 8.57 lb/gal (20 C); vapor pressure 0.2 mm Hg (20 C); f.p. -25 C; b.p. 202-205 C; flash pt. 93 C; ref. index 1.4680-1.4695; surf. tens. 45.8 dynes/cm; VOC 99.7%; 99.9+% assay
Toxicology: LD50 (oral, rat) 3.9 g/kg; relatively nontoxic; low inhalation hazard; mod. eye irritant; not a skin absorption hazard; may cause delayed skin irritation and blistering; sl. heath hazard by ing. of large quantities
Environmental: Environmentally friendly; readily biodeg.; EC50 (rainbow trout, 96 h) ≥ 500 mg/l, (algae, 72 h) ≥ 500 mg/l
Precaution: DOT: combustible; avoid contact with strong acids or alkalis; incompat. with reducing agents, strong oxidizing agents
Hazardous Ingredients: N-Methyl pyrrolidone
Hazardous decomp. prods.: Incomplete combustion may produce CO and other toxic gases
HMIS: Health 2, Flammability 1, Reactivity 0
Storage: Store in dry area, out of direct sunlight; protect from weather; hygroscopic; air and moisture exclusion will prolong shelf life

NMP-EL [Lyondell]
Chem. Descrip.: N-Methyl-2-pyrrolidone
CAS 872-50-4; EINECS/ELINCS 212-828-1
Uses: Photoresist stripper, solvent in electronics applics. incl. semiaq. defluxing, degreasing and coatings
Features: HAPs-free
Properties: Completely water misc.; sp.gr. 1.028; vapor pressure < 0.3 mm Hg (20 C); m.p. -25 C; b.p. 202 C; flash pt. 93 C
Toxicology: Causes irritation to skin and eyes
Storage: Air and moisture exclusion will prolong storage life; store in clean, mild steel or alloy drums

Norfox® B-54 [Norman, Fox]
Chem. Descrip.: Butyl stearate
CAS 123-95-5; EINECS/ELINCS 204-666-5
Uses: Solvent, spreading agent, softener in cosmetics

Norfox® MCS [Norman, Fox]
Chem. Descrip.: Methyl cottonseed
CAS 61788-60-1
Uses: Solvent, cosolvent, penetrant for industrial, food processing, and agric. applics., in textile processing aids, printing inks
Properties: Acid no. 2 max.; iodine no. 108; sapon. no. 188-198

Norfox® MTL [Norman, Fox]
Chem. Descrip.: Methyl tallowate
CAS 61788-61-2; EINECS/ELINCS 262-989-7
Uses: Solvent, surfactant, lubricant, antiwear agent in metalworking compds.; process aid, release agent for concrete form oils and metal rolling oils
Features: No VOCs

Properties: NDT 20Y/2.0R (Lovibond) color; sp.gr. 0.849 (40 C); iodine no. 55-70; sapon. no. 188-200; cloud pt. 14 C; flash pt. > 300 F; 1% max. free fatty acid

Norpar® 5 S [ExxonMobil Europe]
Chem. Descrip.: Normal paraffin hydrocarbon fluid
Uses: Solvent
Properties: Liq.; dens. 0.632 kg/dm^3 (15 C); visc. 0.23 mPa·s; i.b.p. 35 C; flash pt. < 0 C; ref. index 1.358 (20 C); surf. tens. 16.3 mN/m; KB value 27; 0.0001% max. aromatics

Norpar® 12 Fluid [ExxonMobil; Exxon Canada]
Chem. Descrip.: C5-20 normal paraffins hydrocarbon solvent
CAS 64771-72-8
Uses: Solvent for aluminum rolling oils; diluent solvent in carbonless copy paper and in spark erosion machining; as clean fuel in lamp oil and barbecue lighter fluids; solvent in agric. pesticides/herbicides, cleaners, degreasers; chem. feedstock; fuel; food-contact paper/paperboard, rubber articles; defoamer in food-contact paper/paperboard
Regulatory: FDA 21CFR §176.180, 176.210, 177.2600, 178.3570, 178.3620(b); 40CFR §180.1001(c)(e); Canada DSL listed; NPRI nonreportable
Properties: Colorless clear liq.; little/no odor; sol. < 0.01% in water; m.w. 161; sp.gr. 0.749 (60/60 F); dens. 6.24 lb/gal (60 F); visc. 1.26 cps; vapor pressure 0.09 kPa (20 C); f.p. -21 C; b.p. 187-221 C; flash pt. (PMCC) > 66 C; autoignition temp. ≈ 231 C; surf. tens. 26.9 dynes/cm; KB value 23; 100% act.; < 0.01% aromatics
Toxicology: REL TWA 180 ppm; LD50 (oral, rat) > 5 g/kg, (skin rabbit) > 3 g/kg; inh. of high vapor concs. may cause eye/respiratory tract irritation, headaches, dizziness, anesthesia, drowsiness, unconsciousness, other CNS effects, death; sl. eye irritant; frequent/prolonged skin contact may cause irritation, dermatitis; if aspirated, may cause mild to severe pulmonary injury, death
Environmental: Prevent entry into sewers, watercourses, low areas
Precaution: Combustible; flamm. limits 0.7-5.3 vol.%; volatile; gives off invisible vapors; liq. or vapor may settle in low areas; distant ignition/explosion possible; electrostatic accumulation hazard; use proper grounding procedures; incompat. with strong oxidizing agents
Hazardous decomp. prods.: Combustion: toxic gases, fumes, smoke, CO
Storage: Store in cool, well-ventilated area away from incompat. materials, heat, open flame, other ignition sources, direct sunlight; keep container closed

Norpar® 13 Fluid [ExxonMobil; Exxon Canada]
Chem. Descrip.: C5-20 normal paraffins hydrocarbon solvent
CAS 64771-72-8
Uses: Solvent for aluminum rolling oils; diluent solvent in carbonless copy paper and in spark erosion machining; as clean fuel in lamp oil and barbecue lighter fluids; solvent in agric. pesticides/herbicides, furniture polishes, cleaners, degreasers; chem. feedstock; fuel; food pkg. rubber articles; defoamer in food-contact paper/paperboard
Features: Meets LVP exemption (Calif.)
Regulatory: FDA 21CFR §176.210, 177.2600, 178.3570, 178.3620(b); 40CFR §180.1001(c)(e); Canada DSL listed; NPRI nonreportable
Properties: Liq.; little/no odor; sol. < 0.01% in water; m.w. 189; sp.gr. 0.764 (60/60 F); dens. 6.35 lb/gal (60 F); visc. 1.84 cps; vapor pressure 0.01 kPa (20 C); f.p. 0 C; b.p. 225-243 C; flash pt. (PMCC) 97 C; autoignition temp. 229 C; surf. tens. 26.7 dynes/cm; KB value 22; < 0.01% aromatics
Toxicology: REL TWA 158 ppm; inh. of high vapor concs. may cause eye/respiratory tract irritation, headaches, dizziness, anesthesia, drowsiness, unconsciousness, other CNS effects, death; negligible inh. hazard below 38 C; sl. eye irritant; frequent/prolonged skin contact may cause irritation, dermatitis; if aspirated, may cause mild to severe pulmonary injury, death
Environmental: Prevent entry into sewers, watercourses, low areas
Precaution: Flamm. limits 0.6-4.9 vol.%; low flamm. hazard; volatile; gives off invisible vapors; liq. or vapor may settle in low areas; distant ignition/explosion possible; electrostatic accumulation hazard; use proper grounding procedures; incompat. with strong oxidizing agents
Hazardous decomp. prods.: None
Storage: Store in cool, well-ventilated area away from incompat. materials, heat, open flame, other ignition sources, direct sunlight; keep container closed

Norpar® 14 Fluid [ExxonMobil]
Chem. Descrip.: Normal paraffin hydrocarbon fluid
Uses: Solvent for agric., pesticides, herbicides, household prods., cleaners/degreasers, automotive lubricants/penetrants
Features: Meets LVP exemption (Calif.)
Properties: Liq.; sp.gr. 0.762-0.772 (15.6/15.6 C); i.b.p. 230 C min.; flash pt. > 93.5 C; 97% min. normal paraffin content; 0.03% max. aromatics

Norpar® 15 Fluid [ExxonMobil; Exxon Canada]
Chem. Descrip.: C5-20 normal paraffins hydrocarbon solvent
CAS 64771-72-8
Uses: Solvent for aluminum rolling oils; diluent solvent in carbonless copy paper and in spark erosion machining; as clean fuel in lamp oil and barbecue lighter fluids; solvent in pesticides, floor polishes/cleaners, liq. candles, cleaners/degreasers; solvent for waxes; chem. feedstock; fuel; food-contact paper, rubber articles; defoamer in food-contact paper/paperboard
Features: Meets LVP exemption (Calif.)
Regulatory: FDA 21CFR §176.180, 176.210, 177.2600, 178.3570, 178.3620(b); 40CFR §180.1001(c)(e); Canada DSL listed; NPRI nonreportable
Properties: Wh. clear liq.; odorless; sol. < 0.01% in water; m.w. 212; sp.gr. 0.773 (60/60 F); dens. 6.44 lb/gal (60 F); visc. 2.53 cps; vapor pressure < 0 kPa (20 C); f.p. 10 C; b.p. 240-292 C; flash pt. (PMCC) > 116 C; autoignition temp. ≈ 233 C; surf. tens. 28.9 dynes/cm; KB value 28; 0.01% aromatics
Toxicology: REL TWA 140 ppm; low acute toxicity; inh. of high vapor concs. may cause eye/respiratory tract irritation, headaches, dizziness, anesthesia, drowsiness, unconsciousness, other CNS effects, death; sl. eye irritant; frequent/prolonged skin contact may cause irritation, dermatitis; if aspirated, may cause mild to severe pulmonary injury, death
Environmental: Prevent entry into sewers, watercourses, low areas
Precaution: Flamm. limits 0.5-4.6 vol.%; low flamm. hazard; volatile; gives off invisible vapors; liq. or vapor may settle in low areas; distant ignition/explosion possible; electrostatic accumulation hazard; use proper grounding procedures; incompat. with strong oxidizing agents
Hazardous decomp. prods.: Combustion: toxic gases, fumes, smoke, CO
Storage: Store in cool, well-ventilated area away from incompat. materials, heat, open flame, other ignition sources, direct sunlight; keep container closed

Novec™ HFE-7100 [3M/Spec. Materials]
Chem. Descrip.: Methyl nonafluoroisobutyl ether (20-80%), methyl nonafluorobutyl ether (20-80%)
Uses: Carrier solvent for fingerprint formulations; cleaning solvent for vapor degreasing, wet cleaning, electronic equip. cleaning; spray contact cleaner; surf. wetting agent, flow aid, solvent for PFPE lubricants; heat transfer fluid in various industries, semiconductor industry (automated test equip., wafer processing equip.); lubricant carrier; water drying agent; dielec. test media
Features: Replaces CFCs, HFCs, HCFCs, flamm. solvents; zero ozone depleting; low global warming potential; non-VOC; SNAP compliant; nonflamm.; low visc.; low surf. tens.; chem. and thermal stability; resist. to hydrolysis
Regulatory: SARA §311/312 nonhazardous; ELINCS, Canada, Japan, Korea, Australia, PICCS compliance
Properties: Colorless clear liq.; sl. ethereal odor; sol. < 12 ppm in water; m.w. 250; sp.gr. 1.5; visc. 0.6 cps (23 C); vapor pressure 202 mm Hg; f.p. -135 C; b.p. 61 C; flash pt. none; autoignition temp. 405 C; surf. tens. 13.6 dynes/cm; 100% volatile
Toxicology: Low toxicity; pract. nonirritating to eyes; pract. nontoxic by inh.; minimally irritating to skin; if thermal decomp. occurs, inh. may cause respiratory effects (coughing, sneezing, shortness of breath, chest tightness, nasal discharge, wheezing); not a skin sensitizer; TSCA listed
Environmental: Only partially biodeg.; potential to bioconcentrate likely to disappear rapidly from aerobic environments; LC50 (fathead minnow, 96 h) > 7.9 mg/l; very low toxicity to aquatic organisms; prevent direct release to environment
Precaution: Flamm. in air: none; will not burn; incompat. with strong bases
Hazardous decomp. prods.: Hydrogen fluoride, perfluoroisobutylene, perfluorinated acid fluorides
HMIS: Health 0, Flammability 1, Reactivity 0
Storage: Store in tightly closed container in well-ventilated area, away from heat, strong bases

Novec™ HFE-7100DL [3M/Spec. Materials]

Chem. Descrip.: Methyl nonafluoroisobutyl ether (20-80%), methyl nonafluorobutyl ether (20-80%)
Uses: Cleaning solvent for vapor degreasing, wet cleaning, electronic equip. cleaning; solvent for disk lube deposition applics., esp. precleaning of disks prior to lubrication deposition; surf. wetting agent, flow aid, solvent for PFPE lubricants; diluent for fluorochemical acrylate polymer coatings for electronics and medical applics.
Features: Higher purity version of HFE-7100; tightly controlled for ions, metals, NV residue; replaces CFCs, HFCs, HCFCs, chlorinated solvs.; non-ozone depleting; non-VOC; SNAP compliant; low surf. tens.; nonflamm.; resist. to hydrolysis, thermal breakdown
Regulatory: SARA §311/312 nonhazardous; ELINCS, Canada, Japan, Korea, Australia, PICCS compliance
Properties: Colorless clear liq.; sl. ethereal odor; sol. < 12 ppm in water; m.w. 250; sp.gr. 1.52; visc. 0.58 cps; vapor pressure 202 mm Hg; f.p. -135 C; b.p. 61 C; flash pt. none; autoignition temp. 405 C; surf. tens. 13.6 dynes/cm; sp. heat 0.28 cal/g °C
Toxicology: Pract. nonirritating to eyes; minimally irritating to skin; pract. nontoxic by ing.; if thermal decomp. occurs, inh. may cause respiratory effects (coughing, sneezing, shortness of breath, chest tightness, nasal discharge, wheezing); not a skin sensitizer; TSCA listed
Environmental: Only partially biodeg.; potential to bioconcentrate likely to disappear rapidly from aerobic environments; LC50 (fathead minnow, 96 h) > 7.9 mg/l; very low toxicity to aquatic organisms; prevent direct release to environment
Precaution: Flamm. in air: none; will not burn; incompat. with strong bases
Hazardous decomp. prods.: Hydrogen fluoride, perfluoroisobutylene, perfluorinated acid fluorides
HMIS: Health 0, Flammability 1, Reactivity 0
Storage: 1 yr shelf life when stored @ R.T. in original unopened containers; store in tightly closed container in well-ventilated area, away from heat, strong bases

Novec™ HFE-7200 [3M/Spec. Materials]
Chem. Descrip.: Ethyl nonafluoroisobutyl ether (20-80%), ethyl nonafluorobutyl ether (20-80%)
Uses: Cleaning solvent for vapor degreasing, wet cleaning, cold cleaning; surf. wetting agent, flow aid, solvent for deposition applics.; heat transfer fluid for various industries; lubricant carrier; dispersion medium; reaction medium; extraction solvent; spray contact cleaner
Features: Non-ozone depleting; low global warming potential; non-VOC; SNAP compliant; replaces CFCs, HCFCs, HFCs, PFCs, chlorinated solvs.; low visc.; low surf. tens.; higher b.p. reduces evaporative losses; chem. and thermally stable; nonflamm.
Regulatory: SARA §311/312 nonhazardous; ELINCS, Canada, Japan, Philippines listed
Properties: Colorless clear liq.; faint odor; sol. < 20 ppm in water; m.w. 264; sp.gr. 1.43; visc. 0.61 cps; vapor pressure 109 mm Hg; f.p. -138 C; b.p. 76 C; flash pt. none; autoignition temp. 375 C; surf. tens. 13.6 dynes/cm; sp. heat 0.29 cal/g °C
Toxicology: Low toxicity; TWA (8 h) 200 ppm; LC50 (inh., rat, 4 h) > 92,000 ppm; pract. nontoxic by ing.; minimally irritating to eyes; nonirritating to skin; not a skin sensitizer; may cause target organ effects; inh. may cause cardiac sensitization, heart arrhythmia, faintness, chest pain, death; if thermal decomp. occurs, inh. may cause upper respiratory tract irritation, coughing, sneezing, nasal discharge, headache, nose/throat pain; nonmutagenic; TSCA listed
Environmental: Only partially biodeg.; BOD28 nil; potential to bioconcentrate likely to disappear rapidly from aerobic environments; LC50 (fathead minnow, 96 h) > 2.75 mg/l; low aquatic toxicity; prevent direct release to environment
Precaution: Flamm. limits in air 2.4-12.4%; will not burn; incompat. with strong bases
Hazardous decomp. prods.: Hydrogen fluoride, perfluoroisobutylene, perfluorinated acid fluorides
HMIS: Health 0, Flammability 1, Reactivity 0
Storage: Store in tightly closed container in well-ventilated area, away from heat, strong bases

Novec™ HFE-7200DL [3M/Spec. Materials]
Chem. Descrip.: Ethyl nonafluoroisobutyl ether (20-80%), ethyl nonafluorobutyl ether (20-80%)
Uses: Cleaning solvent for vapor degreasing, wet cleaning, electronics industry cleaning; solvent for disk lube deposition applics., medical industry applics.; surf. wetting agent, flow aid for PFPE lubricants
Features: High purity version of HFE-7200; tightly controlled for ions, metals, NV residue; non-ozone depleting; non-VOC; SNAP compliant; low surf. tens.; lower evaporative losses
Regulatory: SARA §311/312 nonhazardous; ELINCS, Canada, Jpana, Philippines listed
Properties: Colorless clear liq.; faint odor; sol. < 20 ppm in water; m.w. 264; sp.gr. 1.42; visc. 0.58 cps; vapor pressure 118 mm Hg; f.p. -138 C; b.p. 76 C; flash pt. none; autoignition temp. 375 C; surf. tens. 13.6 dynes/cm; sp. heat 0.29 cal/g °C
Toxicology: Low toxicity; TWA (8 h) 200 ppm; LC50 (inh., rat, 4 h) > 92,000 ppm; pract. nontoxic by ing.; minimally irritating to eyes; nonirritating to skin; not a skin sensitizer; may cause target organ effects; inh. may cause cardiac sensitization, heart arrhythmia, faintness, chest pain, death; if thermal decomp. occurs, inh. may cause upper respiratory tract irritation, coughing, sneezing, nasal discharge, headache, nose/throat pain; nonmutagenic; TSCA listed
Environmental: Only partially biodeg.; BOD28 nil; potential to bioconcentrate likely to disappear rapidly from aerobic environments; LC50 (fathead minnow, 96 h) > 2.75 mg/l; very low aquatic toxicity; prevent direct release to environment
Precaution: Flamm. limits in air 2.4-12.4%; will not burn; incompat. with strong bases
Hazardous decomp. prods.: Hydrogen fluoride, perfluoroisobutylene, perfluorinated acid fluorides
HMIS: Health 0, Flammability 1, Reactivity 0
Storage: 2 yr shelf life when stored @ R.T. in original unopened containers; store in tightly closed container in well-ventilated area, away from heat, strong bases

Novol NF [Croda Inc; Croda Oleochems.]
Chem. Descrip.: Super refined oleyl alcohol NF
CAS 143-28-2; EINECS/ELINCS 205-597-3
Uses: Emollient, emulsion stabilizer, solubilizer, superfatting agent, penetrant, solvent, pigment suspending agent in cosmetics, personal care prods., pharmaceuticals, lipsticks, sunscreens, antiperspirants, bath oils; solvent and penetrant for topical and transdermal pharmaceuticals
Properties: Gardner 1 max. liq.; mild odor; sol. in min. oil, IPA, propylene glycol; misc. with fat, oil and wax mixts.; sp.gr. 0.845-0.855 (15 C); visc. 24-32 cps; acid no. 0.1 max.; sapon. no. 0.3 max.; hyd. no. 195-210
Use Level: 2-30%
Toxicology: Mild skin irritant, nonirritating to eyes

N-Terpinal RS-25 [WSI Chem.]
Uses: Solvent for equip. cleaning, part washers, floor cleaning, tank reconditioning, label adhesive removal, and spray gun clean-up
Features: Environmentally safe; replaces 1,1,1-trichloroethane, MEK, toluene, xylene, etc.; noncorrosive; compat. with most plastic prods.; low VOC; slow evaporating
Regulatory: DOT nonregulated; SARA §313 reportable (N-methyl-2-pyrrolidone); WHMIS noncontrolled
Properties: Clear liq.; very faint odor; sol. 100% in water (20 C); sp.gr. 1.0444 (21 C); visc. 5.5 cps; vapor pressure < 0.0733 mm Hg; b.p. 98 C; flash pt. (CC) 96 C; pH 6.5 (26 C); nonflamm.; VOC 0.22 lb/gal; 92% volatiles (105 C)
Toxicology: Low acute and chronic toxicity; not a primary irritant or sensitizer to humans; inh. may cause mucous membrane irritation; skin contact may cause skin irritation, dermatitis; eye contact may cause stinging, burning, damage (chronic exposure); ing. may cause digestive system irritation
Environmental: Fully biodeg.
Precaution: Incompat. with strong oxidizing agents
Hazardous Ingredients: Methyl-N-2-pyrolidone
Hazardous decomp. prods.: Combustion: CO, NO_x
Storage: Store away from oxidizing agents

N-Terpinal RS-50 [WSI Chem.]
Uses: Solvent for paint stripping, ink cleaning, urethane removal, resin cleaning, epoxy and alkyd removal, o/w based systems, varnish and adhesives, and degreasing
Features: Environmentally safe; replaces 1,1,1-trichloroethane, MEK, toluene, xylene, etc.; noncorrosive; low VOC; slow evaporating
Regulatory: DOT nonregulated; SARA §313 reportable (N-methyl-2-pyrrolidone); WHMIS noncontrolled
Properties: Clear liq.; very faint odor; sol. 100% in water (20 C); sp.gr.

1.0641 (21 C); visc. 8.2 cps; vapor pressure < 0.0733 mm Hg; b.p. 99 C; flash pt. (CC) 96 C; pH 5.5 (26 C); VOC 0.38 lb/gal; 88% volatiles (105 C)

Toxicology: Low acute and chronic toxicity; not a primary irritant or sensitizer to humans; inh. may cause mucous membrane irritation; skin contact may cause skin irritation, dermatitis; eye contact may cause stinging, burning, damage (chronic exposure); ing. may cause digestive system irritation

Environmental: Fully biodeg.

Precaution: Incompat. with strong oxidizing agents

Hazardous Ingredients: Methyl-N-2-pyrolidone

Hazardous decomp. prods.: Combustion: CO, NO_x

Storage: Store away from oxidizing agents

N-Terpinal RS-75 [WSI Chem.]

Chem. Descrip.: Conc. formula acidified with citric acid

Uses: Solvent for removal of powd. coat and e-coat systems, and for manual cleaning of floors, prod. mixing vats, various parts and equip.

Features: Unstable; requires heat or mixing prior to sue

Regulatory: SARA §313 reportable (N-methyl-2-pyrrolidone); WHMIS noncontrolled

Properties: Lit. amber clear liq.; very faint odor; sol. 40% in water (20 C); sp.gr. 1.0821 (21 C); visc. 6.8 cps; vapor pressure < 0.0733 mm Hg; b.p. 98 C; flash pt. 96 C; pH 3.5 (22 C); VOC 0.38 lb/gal; 92% volatiles (105 C)

Toxicology: Inh. may cause mucous membrane irritation; skin contact may cause skin irritation, dermatitis; eye contact may cause stinging, burning, damage (chronic exposure); ing. may cause digestive system irritation

Environmental: Biodeg.

Precaution: Incompat. with strong oxidizing agents

Hazardous Ingredients: Methyl-N-2-pyrolidone

Hazardous decomp. prods.: Combustion: CO, NO_x

Storage: Store away from oxidizing agents

N-Terpinal RS-98 [WSI Chem.]

Chem. Descrip.: Nonchlorinated, non-petrol.-based solvent

Uses: Solvent for industrial tank washing, agitated dip tanks, floor and pressure washers, parts cleaning units, and ultrasonic systems

Features: Environmentally safe; replaces 1,1,1-trichloroethane, MEK, toluene, xylene, etc.; nonflamm., no reportable VOC's; noncorrosive

Regulatory: DOT nonregulated; SARA §313 reportable (N-methyl-2-pyrrolidone); WHMIS noncontrolled

Properties: Lt. amber clear liq.; very faint odor; sol. 40% in water (20 C); sp.gr. 1.047; visc. 11.1 cps; vapor pressure < 0.733 mm Hg; b.p. 198 C; flash pt. (TCC) 93 C; pH 8.4; VOC 0.87 lb/gal; 2% surfactant, chelating agents; 92% volatiles (105 C)

Use Level: Dilution rate 75-90% with water

Toxicology: Low acute and chronic toxicity; not a primary irritant or sensitizer to humans; inh. may cause mucous membrane irritation; skin contact may cause skin irritation, dermatitis; eye contact may cause stinging, burning, damage (chronic exposure); ing. may cause digestive system irritation

Environmental: Fully biodeg.

Precaution: Incompat. with strong oxidizing agents

Hazardous Ingredients: Methyl-N-2-pyrolidone

Hazardous decomp. prods.: Combustion: CO, NO_x

Storage: 1 yr shelf life; store away from oxidizing agents

N-Terpinal RS-GEL [WSI Chem.]

Uses: Solvent for equip. cleaning, mold release, printing press cleaning, graffiti removal, elec. part cleaning, and paint stripping

Features: Environmentally safe; replaces 1,1,1-trichloroethane, MEK, toluene, xylene, etc.; noncorrosive; low VOC

Regulatory: DOT nonregulated; SARA §313 reportable (N-methyl-2-pyrrolidone); WHMIS noncontrolled

Properties: Wh. clear liq.; faint odor; sol. 100% in water (20 C); sp.gr. 1.94 (21 C); vapor pressure < 0.0733 mm Hg; b.p. 98 C; flash pt. (CC) 96 C; pH 7.8 (26 C); VOC 7.42 lb/gal; 92% volatiles (105 C)

Toxicology: Low acute and chronic toxicity; not a primary irritant or sensitizer to humans; inh. may cause mucous membrane irritation; skin contact may cause skin irritation, dermatitis; eye contact may cause stinging, burning, damage (chronic exposure); ing. may cause digestive system irritation

Environmental: Fully biodeg.

Precaution: Incompat. with strong oxidizing agents

Hazardous Ingredients: Methyl-N-2-pyrolidone

Hazardous decomp. prods.: Combustion: CO, NO_x

Storage: Store away from oxidizing agents

O

#15 Oil [CasChem]
Chem. Descrip.: Polymerized castor oil
Uses: Pigment wetting agent/dispersant; plasticizer for resins, gums, polymers; lubricant; penetrant; coupling solvent; adhesion promoter; for cellulose lacquers, inks, adhesives, industrial lubricants, polishes, caulks, leather dressing, hydraulic fluids; rubber compding.
Regulatory: FDA approval
Properties: Gardner 11 color; sp.gr. 1.013; visc. 250 stokes; pour pt. 35 F; acid no. 14; iodine no. 64; sapon. no. 220; hyd. no. 137

#30 Oil [CasChem]
Chem. Descrip.: Polymerized castor oil
Uses: Pigment wetting agent/dispersant; plasticizer for resins, gums, polymers; lubricant, penetrant, coupling solvent, adhesion promoter for cellulose lacquers, inks, adhesives, industrial lubricants, polishes, caulks, leather dressing, hydraulic fluids, rubber compding.
Regulatory: FDA approval
Properties: Gardner 13 visc. liq.; sp.gr. 1.019; visc. 500 stokes; pour pt. 45 F; acid no. 13; iodine no. 63; sapon. no. 220; hyd. no. 136

#40 Oil [CasChem]
Chem. Descrip.: Polymerized castor oil
Uses: Pigment wetting agent/dispersant; plasticizer for resins, gums, polymers; lubricant, penetrant, coupling solvent, adhesion promoter for cellulose lacquers, inks, adhesives, industrial lubricants, polishes, caulks, leather dressing, hydraulic fluids, rubber compding.
Regulatory: FDA approval
Properties: Gardner 14 color; sp.gr. 1.020; visc. 800 stokes; pour pt. 50 F; acid no. 13; iodine no. 60; sapon. no. 225; hyd. no. 135

Oleocal® C-102 [Lambent Tech.]
Chem. Descrip.: Rapeseed oil, low erucic acid, refined, bleached, deodorized
CAS 120962-03-0; EINECS/ELINCS 232-299-0
Uses: Lubricant base oil for environmentally friendly lubricants, hydraulic fluids, chain and bar lubricants, marine lubricants, greases, gear oils, esp. for environmentally sensitive forestry and mining applics.; solvent, carrier for paints/coatings; oil for foods (baking, blending, frying, salad dressings)
Features: Exceptional shelf life
Regulatory: FDA 21CFR §176.210, 177.2800, 184.1555; DOT nonreportable; SARA §311/312/313 nonreportable; Calif. Prop. 65 nonreportable; Canada DSL, Korea ECL listed
Properties: Lt. amber clear liq.; faint fatty odor; bland flavor; insol. in water; sp.gr. 0.92; dens. 7.6 lb/gal; visc. 67 cSt; vapor pressure < 1 mm Hg; b.p. > 200 C; iodine no. 118; pour pt. - 32 C; flash pt. (PMCC) > 200 C; ref. index 1.46; 100% act.; 0.15% moisture
Toxicology: Negligible effect by inh. unless heated to produce vapors; vapors may irritate mucous membranes and cause dizziness, nausea; may cause eye irritation; thermal burns possible on skin contactTSCA listed
Precaution: Spillages may be slippery; incompat. with strong oxidizing agents
Hazardous decomp. prods.: Combustion: CO, CO_2, thick smoke
NFPA: Health 1, Flammability 1, Reactivity 0
Storage: Store in closed containers @ 10-49 C in well-ventilated area, away from oxidizing agents, excessive heat, ignition sources, sunlight

Oleocal® C-102 A [Lambent Tech.]
Chem. Descrip.: Rapeseed oil, low erucic acid, from food-grade canola oil, refined, bleached, deodorized with additive to inc. oxidative stability
CAS 120962-03-0; EINECS/ELINCS 232-299-0
Uses: Lubricant base oil for environmentally friendly lubricants, hydraulic fluids, chain and bar lubricants, marine lubricants, greases, gear oils, esp. for environmentally sensitive forestry and mining applics.; solvent, carrier for paints/coatings; oil for foods (baking, blending, frying, salad dressings)
Features: Increased oxidative stability; exc. thermal stability; good compat. with other base fluids and additives; high flash pt.; low volatility; exc. shelf life
Regulatory: DOT nonregulated; SARA §311/312/313 nonreportable; Calif. Prop. 65 nonreportable; Canada DSL, Korea ECL listed
Properties: Lt. amber liq.; faint fatty odor; bland flavor; sol. in min. oil; disp. in propylene glycol, hexylene glycol; insol. in ethanol, water; sp.gr. 0.92; dens. 7.6 lb/gal; visc. 67 cSt; vapor pressure < 1 mm Hg; b.p. > 200 C; iodine no. 118; pour pt. -32 C; smoke pt. 250 C; flash pt. (COC) 321 C; ref. index 1.46; 99.9% act.; 0.15% max. moisture
Toxicology: Pract. nontoxic; negligible effect by inh. unless heated to produce vapors; vapors may irritate mucous membranes and cause dizziness, nausea; may cause eye irritation; thermal burns possible on skin contactTSCA listed
Environmental: Biodeg.
Precaution: Combustible liq.; spillages may be slippery; incompat. with strong oxidizing agents
Hazardous decomp. prods.: Combustion: CO, CO_2, thick smoke
NFPA: Health 1, Flammability 1, Reactivity 0
Storage: 9 mos shelf life when stored in closed, factory-sealed containers @ 10-49 C, in well-ventilated area, away from oxidizing agents, excessive heat, ignition sources, sunlight

Oleocal® C-105 [Lambent Tech.]
Chem. Descrip.: Winterized hydrogenated canola oil
CAS 68334-28-1
Uses: Lubricant base oil for environmentally friendly lubricants, hydraulic fluids, chain and bar lubricants, marine lubricants, greases, gear oils, esp. for environmentally sensitive forestry and mining applics.; solvent, carrier for paints/coatings; dielec. fluid in oil-filled capacitors; incidental food contact applics.
Features: Exc. thermal and oxidative stability; remains clear @ refrigerated temps.; good compat. with other base fluids and additives; low pour pt.; high flash pt.; produced from renewable resources
Regulatory: FDA 21CFR §184.1555; kosher; DOT nonregulated; SARA §311/312/313 nonreportable; Calif. Prop. 65 nonreportable; Canada DSL, EINECS, Australia AICS, Korea ECL, Philippines PICCS listed
Properties: Lt. amber liq.; faint fatty odor; insol. in water; sp.gr. 0.92; dens. 7.6 lb/gal; visc. 43 cSt (40 C); vapor pressure < 1 mm Hg; b.p. > 200 C; iodine no. 88; sapon. no. 190; pour pt. -18 C; smoke pt. 250 C; flash pt. (COC) 338 C; ref. index 1.47; dielec. const. 3.15 (21.4 C); dielec. str. 75 kV; vol. resist. 24 x 10^{12} ohm-cm; 0.15% max. moisture
Toxicology: Pract. nontoxic; negligible effect by inh. unless heated to produce vapors; vapors may irritate mucous membranes and cause dizziness, nausea; may cause eye irritation; thermal burns possible on skin contactTSCA listed
Environmental: Biodeg.
Precaution: Combustible liq.; spillages may be slippery; incompat. with

strong oxidizing agents
Hazardous decomp. prods.: Combustion: CO, CO_2, thick smoke
NFPA: Health 1, Flammability 1, Reactivity 0
Storage: 6 mos shelf life when stored in closed, factory-sealed containers @ 10-49 C, in well-ventilated area, away from oxidizing agents, excessive heat, ignition sources, sunlight

Oleocal® IVO-114 [Lambent Tech.]
Chem. Descrip.: Soybean oil, refined, bleached, deodorized
CAS 8001-22-7; EINECS/ELINCS 232-274-4
Uses: Lubricant base oil for environmentally friendly lubricants, hydraulic fluids, chain and bar lubricants, marine lubricants, greases, gear oils; drying oil for inks, paints, varnishes; solvent, carrier, pigment dispersant in paints/coatings
Features: Exc. thermal stability; good compat. with other base fluids and additives; high flash pt.; low volatility; produced from renewable resources; min. oil replacement
Regulatory: DOT nonregulated; SARA §311/312/313 nonreportable; Calif. Prop. 65 nonreportable; Canada DSL, EINECS, Australia AICS, Korea ECL, Philippines PICCS, Switzerland SWISS listed
Properties: Lt. amber liq.; faint fatty odor; insol. in water; sp.gr. 0.92; dens. 7.6 lb/gal; visc. 55.5 cSt; vapor pressure < 1 mm Hg; b.p. > 200 C; pour pt. -7 C; smoke pt. 235 C; flash pt. (COC) 337 C; ref. index 1.46; 100% act.
Toxicology: Pract. nontoxic; negligible effect by inh. unless heated to produce vapors; vapors may irritate mucous membranes and cause dizziness, nausea; may cause eye irritation; thermal burns possible on skin contactTSCA listed
Environmental: Biodeg.
Precaution: Spillages may be slippery; incompat. with strong oxidizing agents
Hazardous decomp. prods.: Combustion: CO, CO_2, thick smoke
NFPA: Health 1, Flammability 1, Reactivity 0
Storage: 6 mos shelf life when stored in closed, factory-sealed containers @ 10-49 C, in well-ventilated area, away from oxidizing agents, excessive heat, ignition sources, sunlight

Oleocal® ME-70 [Lambent Tech.]
Chem. Descrip.: Methyl esters of animal fat, primarily methyl oleate
Uses: Lubricant, coupling agent for metalworking; intermediate for EP additives, surfactants, oleochem. derivs.; wetting agent in agric. adjuvants; base fluid in environmentally friendly solvs. and cleaners; carrier, solvent for pesticides, inks, leather treatment; food coatings; food pkg. adhesives
Regulatory: FDA 21 CFR §172.225, 175.105
Properties: Yel. liq.; mild odor; sp.gr. 0.88; dens. 7.3 lb/gal; visc. 5 cSt (40 F); acid no. 1 max.; iodine no. 60-68; sapon. no. 190-195; pour pt. 4 C; cloud pt. 11 C; flash pt. (COC) 188 C; 95% act.; 0.1% max. moisture
Toxicology: Nontoxic

Oleocal® ME-112 [Lambent Tech.]
Chem. Descrip.: Methyl canolate
CAS 112-62-9; EINECS/ELINCS 203-992-5
Uses: Lubricant, coupling agent for metalworking; intermediate for EP additives, surfactants, oleochem. derivs.; wetting agent in agric. adjuvants; base fluid in environmentally friendly solvs. and cleaners; carrier, solvent for pesticides, inks, leather treatment, plasticizing waxes, antifoams; food coatings; food pkg. adhesives
Regulatory: FDA 21 CFR §172.225, 175.105
Properties: Yel. liq.; mild odor; sp.gr. 0.88; dens. 7.3 lb/gal; visc. 5 cSt (40 F); acid no. 1 max.; iodine no. 110-118; sapon. no. 182-193; pour pt. 4 C; cloud pt. 11 C; flash pt. (COC) 188 C; nonionic; 95% act.; 0.1% max. moisture
Environmental: Biodeg.

Oleocal® ME-130 [Lambent Tech.]
Chem. Descrip.: Methyl soyate
CAS 67784-80-9
Uses: Lubricant, coupling agent for metalworking; intermediate for EP additives, surfactants, oleochem. derivs.; wetting agent in agric. adjuvants; base in environmentally friendly solvs. and cleaners; carrier, solvent for inks, leather treatment compds.; in antifoams, plasticizing waxes, diesel/biodiesel fuel blends; food coatings; food pkg. adhesives
Regulatory: FDA 21CFR §172.225, 175.105
Properties: Yel. liq.; mild odor; sp.gr. 0.88; dens. 7.3 lb/gal; visc. 4.6 cSt (40 F); acid no. 1 max.; iodine no. 128-136; sapon. no. 189-194; pour pt. -15 C; cloud pt. -5 C; flash pt. (COC) 190 C; 95% act.; 0.1% max. moisture
Toxicology: Pract. nontoxic
Environmental: Biodeg.

Olepal ISO [Gattefosse SA]
Chem. Descrip.: PEG-6 isostearate
CAS 56002-14-3
Uses: Solvent, emulsifier for cosmetic/pharmaceutical emulsions and microemulsions
Properties: Liq.; HLB 12-13; acid no. < 6; iodine no. < 15; sapon. no. 95-125; nonionic

Olive Oil, Refined [Natural Oils Int'l.; Tri-K Ind.]
Chem. Descrip.: Olive oil
CAS 8001-25-0; EINECS/ELINCS 232-277-0
Uses: Lubricant, emollient, cosolvent, conditioner in cosmetics
Properties: Pale straw clear oil, bland odor and taste; immiscible in water; sp.gr. 0.910-0.915; iodine no. 79-88; sapon. no. 190-195; flash pt. 640 F; trace moisture
Toxicology: Nonhazardous; edible

Optim™ Glycerine 99.7 [Dow]
Chem. Descrip.: Glycerin FCC, USP
CAS 56-81-5; EINECS/ELINCS 200-289-5
Uses: Humectant, solvent in foods, personal care prods., pharmaceuticals
Regulatory: FDA GRAS; E422 compliant; kosher grade; SARA §311/312 nonhazardous
Properties: Colorless visc. liq.; odorless; sweet taste; misc. with ethanol, IPA, water; immiscible with benzene, chloroform, ethyl ether, heptane; m.w. 92.09; sp.gr. 1.26201; dens. 1.25802 g/cm^3; visc. 1410 cps (20 C); vapor pressure 0.0025 mm Hg (50 C); f.p. 17 C; b.p. 290 C; flash pt. (PMCC) 195.5 C; autoignition temp. 370 C; ref. index 1.47399 (20 C); surf. tens. 63 dynes/cm (20 C); dielec. const. 41.14 (2 x 10^6 Hz, 20 C); 99.7% act.
Toxicology: LD50 (oral, rat) 17,000-27,200 mg/kg, (skin, rabbit) > 10,000 mg/kg; may cause sl. transient eye irritation; may be absorbed thru skin in potentially harmful amts. when applied in lg. quantities to severe burns; extremely low toxicity by ing.; swallowing larger amts. may cause injury, CNS effects, inc. blood sugar levels; inh. of heated material or mist may cause upper respiratory irritation; may cause kidney/liver/GI effects in very lg. doses
Environmental: High biodeg.; BOD5 0.54 p/p; low bioconcentration potential; LC50 (fish) > 100 mg/l; pract. nontoxic to fish; prevent entry into sewers, storm drains, surf. water, soil
Precaution: Incompat. with strong oxidizing agents (e.g., sodium hypochlorite, hypochlorous acid)
Hazardous decomp. prods.: Acrolein; combustion: aldehydes, CO, CO_2
Storage: Store @ 64-130 F

Orange Terpenes [PDM]
Chem. Descrip.: d-Limonene, food grade
Chem. Analysis: D-Limonene (94.5-97.5%), myrcene (1.4-2.7%)
CAS 68647-72-3; EINECS/ELINCS 227-813-5
UN 2319
Uses: Essential oil, solvent for foods, industrial applics.
Regulatory: GRAS; DOT regulated
Properties: Water-wh. to lt. tint clear liq.; mild orange odor; negligible sol. in water; sp.gr. 0.844; vapor pressure 1.0 mm Hg (14 C); b.p. 176 F; flash pt. (TCC) 50 C; ref. index 1.472 (20 C); VOC 95% (as solv.)
Toxicology: Irritates eyes, skin, and mucous membranes
Precaution: Combustible; avoid strong oxidizing agents
Hazardous Ingredients: D-limonene (95%), terpene hydrocarbons (3%), oxygenated terpenes (2%)
Hazardous decomp. prods.: Thermal degradation causes CO, CO_2
NFPA: Health 1, Flammability 2, Reactivity 0
Storage: Store in closed containers away from heat or sources of ignition

Orfom® SX-1 [ChevronPhillips]
Chem. Descrip.: Naphtha, heavy aliphatic C11-16
CAS 64742-96-7; EINECS/ELINCS 265-200-4
UN 1268
Uses: Solvent extraction diluent
Regulatory: SARA §313 nonreportable; Canada DSL listed
Properties: Pale yel.-straw liq.; mild odor; negligible sol. in water; sp.gr. 0.80-0.85 (16/16 C); b.p. 191-305 C; flash pt. (PMCC) 79-81 C; 100%

active

Toxicology: REL (8 h TWA) 400 ppm as total hydrocarbon vapor; may be sl. irritating to eyes; may cause severe skin irritation, esp. on repeated contact; inh. may cause headache, dizziness; ing. may cause sl. irritation to intestines; if swallowed, may be aspirated into lungs, resulting in inflamm., possible fluid accumulation; long-term exposure to high oil mist concs. may cause nondebilitating lung changes; TSCA listed

Environmental: Keep out of water sources and sewers

Precaution: Combustible liq.; incompat. with oxygen, strong oxidizing agents

Hazardous decomp. prods.: CO_x and various hydrocarbons formed when burned

NFPA: Health 1, Flammability 2, Reactivity 0

Storage: Store in closed container in well-ventilated area, away from ignition sources; bond and ground during liq. transfer

Orfom® SX 7 [ChevronPhillips]

Chem. Descrip.: Petroleum distillates, hydrotreated light, C9-16

CAS 64742-47-8; EINECS/ELINCS 265-149-8

UN 1268

Uses: Solvent extraction diluent for solvent extraction of metals, esp. copper; carrier for solvent extraction processes

Features: High performance; exc. phase disengagement props.; rapid extraction kinetics; high flash pt.; kerosene replacement; high Cu/Fe selectivity

Regulatory: SARA §313 nonreportable; Canada DSL listed

Properties: Colorless liq.; mild char. odor; negligible sol. in water; sp.gr. 0.81-0.82 (16/16 C); dens. 0.82 kg/l (15.6 C); vapor pressure 0.2 psia; f.p. -40 C; b.p. 187-274 C; flash pt. (PMCC) 71-74 C; 100% act.

Toxicology: REL (8 h TWA) 100 ppm; may be sl. irritating to eyes; may cause severe skin irritation, esp. on repeated contact; inh. may cause headache, dizziness; ing. may cause sl. irritation to intestines; if swallowed, may be aspirated into lungs, resulting in inflamm., possible fluid accumulation; long-term exposure to high oil mist concs. may cause nondebilitating lung changes; TSCA listed

Environmental: Keep out of water sources and sewers

Precaution: Combustible liq.; LEL 0.7%; UEL 5.0%; incompat. with oxygen, strong oxidizing agents

Hazardous decomp. prods.: CO_x and various hydrocarbons formed when burned

NFPA: Health 1, Flammability 2, Reactivity 0

Storage: Store in closed container in well-ventilated area, away from heat, sparks, flames; bond and ground during liq. transfer

Orfom® SX 10 [ChevronPhillips]

Chem. Descrip.: C12-14 isoparaffin

CAS 68551-19-9; EINECS/ELINCS 271-369-5

UN 3295

Uses: Solvent extraction diluent for solvent extraction of metals; carrier for solvent extract processes

Features: High flash pt.; rapid phase disengagement props.; rapid kinetics; kerosene replacement

Regulatory: SARA §313 nonreportable

Properties: Colorless liq.; mild odor; negligible sol. in water; sp.gr. 0.778 (15.6/15.6 C); visc. 38 SUS; vapor pressure 0.012 psia (0.6 mm Hg); pour pt. < -46 C; flash pt. (TCC) 88.9 C; 100% act.

Toxicology: REL (8 h TWA) 400 ppm; may be sl. irritating to eyes, skin; inh. may cause headache, dizziness, nausea, unconsciousness; ing. may irritate stomach, intestines; if swallowed, may be aspirated into lungs, resulting in inflamm., possible fluid accumulation; may cause kidney damage; TSCA listed

Environmental: Keep out of water sources and sewers

Precaution: Combustible liq.; incompat. with oxygen, strong oxidizing agents

Hazardous decomp. prods.: CO_x formed when burned

Storage: Store in closed container in well-ventilated area, away from heat, sparks, flames; bond and ground during transfer

Orfom® SX 11 [ChevronPhillips]

Chem. Descrip.: C13-16 isoparaffin

CAS 68551-20-2; EINECS/ELINCS 271-370-0

Uses: Solvent extraction diluent for solvent extraction of metals; carrier for solvent extract processes

Features: High flash pt.; exc. phase disengagement props.; rapid kinetics; kerosene replacement; esp. effective in high-temp. circuits

Regulatory: SARA §313 nonreportable; Canada DSL listed

Properties: Colorless liq.; mild odor; negligible sol. in water; sp.gr. 0.809 (15.6/15.6 C); visc. 39.2 SUS; vapor pressure 0.004 psia (0.2 mm Hg); b.p. 232-288 C; flash pt. (COC) 106 C; 100% act.

Toxicology: REL (8 h TWA) 400 ppm; may be sl. irritating to eyes, skin; inh. may cause headache, dizziness, nausea, unconsciousness; ing. may irritate stomach, intestines; if swallowed, may be aspirated into lungs, resulting in inflamm., possible fluid accumulation; may cause kidney damage; TSCA listed

Environmental: Keep out of water sources and sewers

Precaution: Incompat. with oxygen, strong oxidizing agents

Hazardous decomp. prods.: CO_x formed when burned

NFPA: Health 2, Flammability 1, Reactivity 0

Storage: Store in closed container in well-ventilated area

Orfom® SX 12 [ChevronPhillips]

Chem. Descrip.: Petroleum distillates, hydrotreated light, C9-16

CAS 64742-47-8; EINECS/ELINCS 265-149-8

UN 1268

Uses: Solvent extraction diluent for solvent extraction of metals; carrier for solvent extract processes

Features: High flash pt.; exc. phase disengagement props.; rapid kinetics; kerosene replacement; high Cu/Fe selectivity

Regulatory: SARA §313 nonreportable; Canada DSL listed

Properties: Colorless liq.; mild char. odor; negligible sol. in water; sp.gr. 0.81-0.82 (15.6/15.6 C); visc. 1.0-1.9 cSt (40 C); vapor pressure 0.2 psia; f.p. -40 C; b.p. 187-274 C; flash pt. (PMCC) > 66 C; 100% act.

Toxicology: REL (8 h TWA) 100 ppm; may be sl. irritating to eyes; may cause severe skin irritation, esp. on repeated contact; inh. may cause headache, dizziness; ing. may cause sl. irritation to intestines; if swallowed, may be aspirated into lungs, resulting in inflamm., possible fluid accumulation; long-term exposure to high oil mist concs. may cause nondebilitating lung changes; TSCA listed

Environmental: Keep out of water sources and sewers

Precaution: Combustible liq.; LEL 0.7%; UEL 5.0%; incompat. with oxygen, strong oxidizing agents

Hazardous decomp. prods.: CO_x and various hydrocarbons formed when burned

NFPA: Health 1, Flammability 2, Reactivity 0

Storage: Store in closed container in well-ventilated area, away from heat, sparks, flames; bond and ground during liq. transfer

Orfom® SX 18 [ChevronPhillips]

Chem. Descrip.: C9-11 isoparaffin

CAS 68551-16-6; EINECS/ELINCS 271-365-3

UN 1268

Uses: Solvent extraction diluent

Regulatory: SARA §313 nonreportable; Canada DSL listed

Properties: Colorless liq.; mild odor; negligible sol. in water; sp.gr. 0.746 (15.6/15.6 C); visc. 1.12 cSt (38 C); vapor pressure 0.1 psia; b.p. 160-167 C; flash pt. (TCC) 41 C; 100% act.

Toxicology: LD50 (oral, rat) > 34.6 g/kg, (skin, rabbit) 15.4 g/kg; LC50 (inh., rat, 4 h, 14 day post exposure) > 3684 ppm; REL 400 ppm; may be sl. irritating to eyes, skin; inh. may cause headache, dizziness, nausea, unconsciousness; ing. may irritate stomach, intestines; if swallowed, may be aspirated into lungs, resulting in inflamm., possible fluid accumulation; may cause kidney damage

Environmental: Keep out of water sources and sewers

Precaution: Flamm. liq.; incompat. with oxygen, strong oxidizing agents; highly flamm. vapors may accumulate in low areas; may flash back

Hazardous decomp. prods.: CO_x formed when burned

NFPA: Health 0, Flammability 2, Reactivity 0

Storage: Store in tightly closed container in well-ventilated area, away from heat, sparks, flames; bond and ground during transfer

Oxsol® 100 [Makhteshim Agan Ind.]

Chem. Descrip.: p-Chlorobenzotrifluoride

CAS 98-56-6; EINECS/ELINCS 202-681-1

UN 2234

Uses: Solvent for coatings, inks, industrial solvent cleaning, aerosols, adhesives, electronic applics.

Features: Non-VOC, non-ozone depleting; 1,1,1-trichloroethane alternative; VOC-exempt status

Regulatory: SARA nonreportable

Properties: Water-wh. clear liq., char. chlorinated aromatic odor; sp.gr.

1.33-1.35; dens. 11.2 lb/gal; visc. 0.79 cps; vapor pressure 8 mm Hg; f.p. -36 C; b.p. 139 C; flash pt. (TCC) 109 F; surf. tens. 25 dynes/cm; KB value 64

Toxicology: CEL 25 ppm; LD50 (oral, rat) > 6.8 g/kg, (rabbit, dermal) > 2.7 g/kg; may cause mild to moderate eye, skin, and respiratory tract irritation; may cause CNS depression; not highly toxic if swallowed; prolonged ing. may cause liver/kidney damage

Precaution: Combustible; incompat. with oxidizing agents

Hazardous Ingredients: O-chlorobenzotrifluoride (0.5-3%), 3,4-dichlorobenzotrifluoride (0.05-1%)

Hazardous decomp. prods.: Decomp. by burning may yield toxic hydrogen chloride and hydrogen fluoride gases

HMIS: Health 1, Flammability 2, Reactivity 1

Storage: Store containers in cool, dry, ventilated, fire-resist. area; keep away from heat, sparks, open flame; ground equip. to avoid static charges; vent indoor tanks to outdoors

Oxsol® 1000 [Makhteshim Agan Ind.]

Chem. Descrip.: 3,4-Dichlorobenzotrifluoride

CAS 328-84-7

Uses: Solvent for industrial solvent cleaning, aerosols, adhesives, coatings, and inks

Features: Non-ozone depleting; 1,1,1-trichloroethane alternative; not hazardous waste

Properties: Odor; dens. 12.3 lb/gal; visc. 1.52 cps; vapor pressure 2 mm Hg; f.p. -12 C; b.p. 174 C; flash pt. (TCC) 170 F; surf. tens. 29 dynes/cm; KB value 69

Toxicology: CEL 5 ppm; may cause mild to severe eye irritation, sl. to severe skin irritation, respiratory tract irritation, CNS depression

Precaution: Combustible

Oxsol® 2000 [Makhteshim Agan Ind.]

Chem. Descrip.: Benzotrifluoride

CAS 98-08-8

Uses: Reaction solvent for precision and elec. cleaning, aerosols, wipe cleaning

Features: Replacement for hexane and toluene, or blended with HFC and HCFCs; HAPs-free; not an ozone depleter

Regulatory: SARA nonreportable

Properties: Water-wh. clear liq.; aromatic odor; sp.gr. 1.19; dens. 9.88 lb/gal; visc. 0.56 cps; vapor pressure 30 mm Hg; f.p. -29 C; b.p. 102 C; flash pt. (TCC) 54 F; surf. tens. 23 dynes/cm; KB value 49

Toxicology: CEL 100 ppm; LD50 (oral, rat) 15 g/kg; may cause mild to mod. eye, skin, and respiratory tract irritation; may cause CNS depression

Precaution: Flamm.

Oxypure® 50% [FMC; C.P. Hall]

Chem. Descrip.: Hydrogen peroxide

CAS 7722-84-1; EINECS/ELINCS 231-765-0

UN 2014

Uses: Disinfectant, oxidizer for drinking water treatment, removal of iron or hydrogen sulfide in ground water, enhancing taste/odor removal efficiency of ozone, destroying ozone residuals in potable water; solvent

Regulatory: SARA §302 extremely hazardous substance, §311 acute health/fire hazard, §313 nonreportable; NSF certified for drinking water treatment

Properties: Colorless clear liq.; odorless; sol. 100% in water; sp.gr. 1.19 (20/4 C); dens. 10.0 lb/gal (20 C); vapor pressure 18.3 mm Hg (30 C); f.p. -52 C; b.p. 114 C; flash pt. none; pH 1-3 (as is), 5-6 (1%); 23.5% act. oxygen

Toxicology: ACGIH TWA 1 ppm; LD50 (oral, rat) > 225 mg/kg; corrosive to eyes, skin, nose, throat, lungs; may cause irreversible tissue damage to eyes, blindness; severe irritant/corrosive to GI tract; inh. of mist or vapors may be severely irritating to nose, throat, lungs; TSCA listed

Environmental: Dec. into water and oxygen in aquatic environment; LC50 (channel catfish, 96 h) 37.4 mg/l, (fathead minnow, 96 h) 16.5 mg/l; EC50 (daphnia magna, 24 h) 7.7 mg/l

Precaution: Noncombustible; strong oxidizer; contact with combustibles may cause fire; dec. yielding oxygen that supports combustion of org. matters, can cause overpressure if confined; avoid excessive heat and contamination; incompat. with reducing agents, wood, paper, other combustibles, iron and other heavy metals, copper alloys, caustic; avoid dirt, cyanides, organics

Hazardous decomp. prods.: Oxygen

HMIS: Health 2, Flammability 0, Reactivity 1

Storage: Store in cool areas away from direct sunlight, combustibles

P

Pale® 4 [CasChem]
Chem. Descrip.: Polymerized castor oil
Uses: Pigment wetting agent/dispersant; plasticizer for resins, gums, polymers; lubricant, penetrant, coupling solvent, adhesion promoter for cellulose lacquers, inks, adhesives, industrial lubricants, polishes, caulks, leather dressing, hydraulic fluids, rubber compding., gasket cement
Regulatory: FDA approval
Properties: Gardner 4 color; sp.gr. 0.998; visc. 32 stokes; pour pt. 5 F; acid no. 16; iodine no. 70; sapon. no. 212; hyd. no. 158

Pale® 16 [CasChem]
Chem. Descrip.: Polymerized castor oil
Uses: Pigment wetting agent/dispersant; plasticizer for resins, gums, polymers; lubricant; penetrant; coupling solvent; adhesion promoter; for cellulose lacquers, inks, adhesives, industrial lubricants, polishes, caulks, leather dressing, hydraulic fluids; rubber compding., gasket cement
Regulatory: FDA approval
Properties: Gardner 9 color; sp.gr. 1.025; visc. 250 stokes; pour pt. 25 F; acid no. 24; iodine no. 56; sapon. no. 237; hyd. no. 136

Pale® 170 [CasChem]
Chem. Descrip.: Polymerized castor oil
Uses: Pigment wetting agent/dispersant; plasticizer for resins, gums, polymers; lubricant; penetrant; coupling solvent; adhesion promoter; for cellulose lacquers, inks, adhesives, industrial lubricants, polishes, caulks, leather dressing, hydraulic fluids; rubber compding., gasket cement
Regulatory: FDA approval
Properties: Gardner 2 color; sp.gr. 0.970; visc. 11 stokes; pour pt. -5 F; acid no. 4; iodine no. 80; sapon. no. 184; hyd. no. 160

Pale® 1000 [CasChem]
Chem. Descrip.: Polymerized castor oil
Uses: Pigment wetting agent/dispersant; plasticizer for resins, gums, polymers; lubricant; penetrant; coupling solvent; adhesion promoter; for cellulose lacquers, inks, adhesives, industrial lubricants, polishes, caulks, leather dressing, hydraulic fluids; rubber compding., gasket cement
Regulatory: FDA approval
Properties: Gardner 9 color; sp.gr. 1.018; visc. 120 stokes; pour pt. 25 F; acid no. 20; iodine no. 59; sapon. no. 230; hyd. no. 139

Palmil C [Undesa; S. Black]
Chem. Descrip.: Cetyl palmitate
CAS 540-10-3; EINECS/ELINCS 208-736-6
Uses: Emollient in bath oils, creams/lotions, sun creams, shaving prods., insect repellents, antiperspirants, deodorants; oily component for cleansing creams, nail polish removers; plasticizer for nail polishes; pigment wetting agent; softener in creams, hair care; solvent for fatty cosmetic ingreds.; vehicle for lipsticks; solubilizer for perfume oils, vitamin oils, hormones; solubilizer, vehicle for medicines; excipient for injectable substances; lubricant for PVC
Features: Changes skin feel, rub out props. and greasiness; veg. oil substitute; vaseline substitute

Palmil IS [Undesa; S. Black]
Chem. Descrip.: Isopropyl palmitate
CAS 142-91-6; EINECS/ELINCS 205-571-1
Uses: Emollient in bath oils, creams/lotions, sun creams, shaving prods., insect repellents, antiperspirants, deodorants; oily component for cleansing creams, nail polish removers; plasticizer for nail polishes; pigment wetting agent; softener in creams, hair care; solvent for fatty cosmetic ingreds.; vehicle for lipsticks; solubilizer for perfume oils, vitamin oils, hormones; solubilizer, vehicle for medicines; excipient for injectable substances; lubricant for PVC
Features: Changes skin feel, rub out props. and greasiness; veg. oil substitute; vaseline substitute

Palmil O [Undesa; S. Black]
Chem. Descrip.: Octyl palmitate
CAS 29806-73-3; EINECS/ELINCS 249-862-1
Uses: Emollient in bath oils, creams/lotions, sun creams, shaving prods., insect repellents, antiperspirants, deodorants; oily component for cleansing creams, nail polish removers; plasticizer for nail polishes; pigment wetting agent; softener in creams, hair care; solvent for fatty cosmetic ingreds.; vehicle for lipsticks; solubilizer for perfume oils, vitamin oils, hormones; solubilizer, vehicle for medicines; excipient for injectable substances; lubricant for PVC
Features: Changes skin feel, rub out props. and greasiness; veg. oil substitute; vaseline substitute

Parabase [Shell Chems. Canada]
Chem. Descrip.: Petroleum distillates, alkylate
CAS 64741-73-7; EINECS/ELINCS 265-074-0
Uses: Solvent for agric. sprays, waterless hand cleaners, disinfectants, surf. coatings
Features: Slow evaporating; good color chars.; high unsulfonated mineral residue
Properties: Saybolt 18 color; low odor; m.w. 198; i.b.p. 199 C min.; flash pt. (COC) 86 C

Partsprep™ Degreaser [ISP]
Chem. Descrip.: N-Methyl-2-pyrrolidone and surfactants
Uses: Solvent cleaner for removal of grease, cutting oils, buffing compds., and waxes in metal, photographic, aerospace industries, and precision mechanics
Features: Safer alternative to CFCs, trichloroethylene, semiaq. cleaners
Properties: Clear liq.; water-sol.; sp.gr. 1.02-1.04; visc. 2.49 cps; m.p. -24.4 C; b.p. 202 C; flash pt. 89.4 C; ref. index 1.46-1.48
Toxicology: TLV 100 ppm; low oral and inhalation toxicity; mod. dermal toxicity; mod. eye irritant

Paxsolv CS [Pax Group]
Chem. Descrip.: Dibasic ester
Uses: Solvent for VOC-free formulations
Features: High b.p.; exc. solvency; replacement for conventional toxic and hazardous chlorinated hydrocarbons
Properties: Liq.; 98% act.
Environmental: Biodeg.

Paxsolv PC [Pax Group]
Chem. Descrip.: Dibasic ester
Uses: Solvent, solubilizer, leveling agent, plasticizer, lubricant in paper/pulp
Features: Very high purity; high b.p. permits VOC-free formulations; replacement for toxic and hazardous chlorinated hydrocarbons
Properties: Liq.; 98% act.
Environmental: Biodeg.

PCHP [Lyondell]
Chem. Descrip.: Propylene carbonate, high purity
CAS 108-32-7; EINECS/ELINCS 203-572-1
Uses: Solvent for mascara, lipstick, deodorant, nail polish remover

Regulatory: SARA §313 nonreportable
Properties: Sol. 21% in water (20 C); m.w. 102.1; sp.gr. 1.20 (20/20 C); dens. 10 lb/gal (20 C); visc. 2.44 cps; vapor pressure 0.03 mm Hg (20 C); b.p. 242 C; HLB 13.7; flash pt. (Seta) 226 F; ref. index 1.419; surf. tens. 40.9 dynes/cm; sp. heat 0.40 Cal/g °C
HMIS: Health 1, Flammability 1, Reactivity 0

PCPL Solvent [Shell Chems. Canada]
Chem. Descrip.: Petroleum distillates, hydrotreated middle
CAS 64742-46-7; EINECS/ELINCS 265-148-2
Uses: Solvent
Features: High flash; slow evaporating
Properties: Pt-Co 2 max. color; m.w. 198; dens. 850 kg/m^3 (15 C); vapor pressure < 0.3 mm Hg (20 C); i.b.p. 200 C; flash pt. (TCC) 66 C

PCP Solvent [Shell]
Chem. Descrip.: Petroleum distillates, hydrotreated middle
CAS 64742-46-7; EINECS/ELINCS 265-148-2
Uses: Solvent, diluent, carrier for pentachlorophenol wood preservatives
Features: Slow evaporating; high proportion of aromatics; low visc.
Properties: Pt-Co 2.5 max. color; m.w. 198; dens. 850 kg/m^{23} (15 C); vapor pressure < 0.3 mm Hg (20 C); i.b.p. 205 C; flash pt. (TCC) 86 C; autoignition temp. > 450 C

PD-23 [Crompton/Witco Refined Prods.]
Chem. Descrip.: Deodorized kerosene
CAS 8008-20-6; EINECS/ELINCS 232-366-4
Uses: Solvent, defoamer, dust control agent, quench oil, diluent for fiber lubricants, metal foil and sheet rolling oils, in froth flotation of vegetables, mfg. of paper/paperboard, in waterless hand cleaners, liq. candles, charcoal lighter fluid, polishes, cleaners, cosmetic/pharmaceutical creams/lotions, water treatment, food-contact applics.; pesticide diluent; coating on egg shells
Features: Refined aliphatic solvent; non-VOC
Regulatory: FDA 21CFR §172.884, 178.3650
Properties: Colorless clear liq., nearly odorless; sp.gr. 0.800 (60/60 C); visc. 2.6 cSt (40 C); vapor pressure < 0.01 mm Hg (20 C); b.p. 452 F; pour pt. 0 F; flash pt. (COC) 230 F; ref. index 1.442

PD-25 [Crompton/Witco Refined Prods.]
Chem. Descrip.: Petroleum distillates
Uses: Solvent, defoamer, dust control agent, quench oil, diluent for fiber lubricants, metal foil and sheet rolling oils, in froth flotation of vegetables, mfg. of paper/paperboard, in waterless hand cleaners, liq. candles, charcoal lighter fluid, polishes, cleaners, pharmaceutical creams/lotions, water treatment, food-contact applics.; pesticide diluent; coating on egg shells
Features: Refined aliphatic solvent; non-VOC
Regulatory: FDA 21CFR §172.884, 178.3650
Properties: Colorless clear liq., nearly odorless; sp.gr. 0.810 (60/60 C); visc. 3.9 cSt (40 C); vapor pressure < 0.01 mm Hg (20 C); b.p. 510 F; pour pt. 20 F; flash pt. (COC) 260 F; ref. index 1.446

PD-28 [Crompton/Witco Refined Prods.]
Chem. Descrip.: Petroleum distillates
Uses: Solvent, defoamer, dust control agent, quench oil, diluent for fiber lubricants, metal foil and sheet rolling oils, in froth flotation of vegetables, mfg. of paper/paperboard, in waterless hand cleaners, liq. candles, charcoal lighter fluid, polishes, cleaners, pharmaceutical creams and lotions, water treatment, food-contact applics.; pesticide diluent; protective coating on foods
Features: Refined aliphatic solvent; non-VOC
Regulatory: FDA 21CFR §172.878, 178.3620a
Properties: Colorless liq., nearly odorless; sp.gr. 0.820 (60/60 C); visc. 4.6 cSt (40 C); vapor pressure < 0.01 mm Hg (20 C); b.p. 526 F; pour pt. 30 F; flash pt. (COC) 285 F; ref. index 1.450

Pecosil® DCU [Phoenix]
Chem. Descrip.: Dimethicone copolyol undecylenate
CAS 159100-33-1
Uses: Emollient in creams/lotions, cosmetics, sunscreens, perms, lip prods., aerosol systems, fragrances, bath prods., nail conditioning treatments, pharmaceutical topicals; coupling agent, cosolvent for water and oil systems
Properties: Off-wh. to yel. clear to hazy liq.; sol. in water, castor oil; misc. with ethanol; disp. in cyclomethicone, min. oil, propylene glycol, IPM; sp.gr. 1.01 g/ml; acid no. 15 max.; flash pt. (TCC) > 200 F; pH 5-7 (10%); 100% act.
Toxicology: Nonirritating to eyes; nonprimary skin irritant; nontoxic orally; noncomedogenic; skin exposure can cause irritation; ingestion may cause nausea, vomiting, abdominal pain
Precaution: Incompat. with strong oxidizers and reducers
Hazardous Ingredients: None
Hazardous decomp. prods.: Thermal decomp. prods.: $(CO)_x$, $(SiO)_x$
HMIS: Health 1, Flammability 1, Reactivity 0

PEG 200 [Huntsman]
Chem. Descrip.: PEG 200
CAS 25322-68-3; EINECS/ELINCS 203-989-9
Uses: Binder for glazing; intermediate for PEG esters, methacrylate resins, PU foams; plasticizer, solvent for cork; toiletries; metalworking lubricants; paints/resins; paper/film; printing inks; textile emulsifier
Properties: APHA 25 color; sol. in water and most polar org. solvs.; m.w. 190-210; sp.gr. 1.127; visc. 4.2 cst (99 C); pour pt. -50 C; flash pt. (OC) > 170 C; ref. index 1.459; pH 4.5-7.5; surf. tens. 44.6 dynes/cm; nonionic; 99% act.
Environmental: Biodeg.

PEG 300 [Huntsman]
Chem. Descrip.: PEG 300 USP
CAS 25322-68-3; EINECS/ELINCS 220-045-1
Uses: Solubilizer, carrier, plasticizer, solvent for cosmetics, pesticides, metalworking lubricants, paints, resins, paper, film, printing inks, liq. pharmaceutical preps., ointment bases; rubber release; textile aux.; visc. modifier for brake fluids; intermediate for PEG esters, PU foams; binder for capsules/pills
Properties: APHA 25 color; sol. in water and most polar org. solvs.; m.w. 285-315; sp.gr. 1.128; visc. 5.8 cst (99 C); pour pt. -12 C; flash pt. (OC) > 190 C; ref. index 1.463; pH 4.5-7.5; surf. tens. 44.6 dynes/cm; nonionic; 99% act.
Environmental: Biodeg.

PEG 400 [Huntsman]
Chem. Descrip.: PEG 400 USP
CAS 25322-68-3; EINECS/ELINCS 225-856-4
Uses: Emulsifier, antistat for textiles; rubber; inks; cosmetics; paper; film; pesticide solubilizer/carrier; intermediate for PEG esters, PU foams; plasticizer, solvent for cork; metalworking lubricants; paints/resins; pharmaceutical liq. preps., ointment bases, suppository bases; binder for capsules/pills; preservative for pathological specimens
Properties: APHA 25 color; sol. in water and most polar org. solvs.; m.w. 380-420; sp.gr. 1.130; visc. 7.3 cst (99 C); pour pt. 6 C; flash pt. (OC) > 215 C; ref. index 1.465; pH 4.5-7.5; surf. tens. 44.6 dynes/cm; nonionic; 99% act.
Environmental: Biodeg.

Pelemol® G7A [Phoenix]
Chem. Descrip.: Glycereth-7 triacetate
CAS 57569-76-3
Uses: Humectant, emollient, coupling agent, solubilizer in facial cleansers, creams/lotions, splash-on prods., shaving prods., hair prods., pharmaceutical topicals; solvent for sunscreen ingreds.
Properties: Sl. yel. clear visc. liq., bland odor; sol. in water, alcohol, hydroalcoholic sol'ns.; misc. with propylene glycol; insol. in cyclomethicone, min. oil, IPM; sp.gr. ≈ 1.16; acid no. 7 max.; sapon. no. 305-330; hyd. no. 20 max.; flash pt. (PMCC) > 150 C; 100% act.
Toxicology: Nonirritating to eyes; nonprimary skin irritant; LD50 (oral) > 5 g/kg, nontoxic orally; noncomedogenic; skin exposure can cause irritation; ingestion may cause nausea, vomiting, abdominal pain
Precaution: Incompat. with oxidizing agents
Hazardous Ingredients: None
Hazardous decomp. prods.: May form oxides of carbon
HMIS: Health 1, Flammability 0, Reactivity 0
Storage: Store in cool, dry place

Penreco 150-B Solvent [Penreco]
Chem. Descrip.: Petroleum raffinates, sorption process (primarily C9-15 cycloparaffinic, isoparaffinic, and aromatic hydrocarbons, derived from highly refined kerosene)
CAS 64741-85-1; EINECS/ELINCS 265-087-1
UN NA1993
Uses: Solvent for chem. processing, degreasers, household/industrial cleaners, insecticides
Features: Broad boiling range; controlled flash pt.; higher solvency than comparable aliphatics; low reactivity
Regulatory: SARA §311/312 acute/chronic health hazard, fire hazard, §313

nonreportable; Canada DSL, EINECS, Australia AICS, China, Japan ENCS, Korea KCL, Philippines PICCS listed
Properties: Colorless clear liq.; petrol. odor; sol. in hydrocarbons; insol. in water; sp.gr. 0.81-0.82; vapor pressure < 2 mm Hg (68 F); b.p. 165.6-287.8 C; flash pt. (TCC) 65.6 C; 100% act.
Toxicology: OSHA PEL/TWA 500 ppm (8 h); may be harmful or fatal by ing., inh. in very high concs.; inh. of high vapor/mist concs. may cause respiratory tract/eye irritation, CNS effects (drowsiness, dizziness, nausea, headache, unconsciousness); prolonged/repeated contact can cause irritation to eyes, nose, throat, lungs, skin; extreme overexposure or aspiration can cause lung damage, death; TSCA listed
Environmental: Do not allow spills to enter sewers or watercourses
Precaution: Combustible liq.; flamm. limits in air 0.5-6.0%; may react with strong oxidizing agents; spillages may be slippery
Hazardous decomp. prods.: Combustion: dense smoke, CO, CO_2, other oxides
HMIS: Health 1, Flammability 2, Reactivity 0
Storage: Store in cool, well-ventilated area in closed containers away from heat, sparks, open flame, oxidizing materials

Penreco 170 Exempt Solvent [Penreco]
Chem. Descrip.: Hydrotreated light petroleum distillates (primarily C10-15 cycloparaffinic and isoparaffinic hydrocarbons)
CAS 64742-47-8; EINECS/ELINCS 265-149-8
UN 1268
Uses: Solvent for chem. processing, degreasers, household/industrial cleaners, insecticides, metal rolling oils, mining extraction
Features: High purity; controlled flash pt. and boiling range; higher solvency than comparable aliphatics; low reactivity
Regulatory: FDA approvals for direct/indirect food additives; SARA §311/312 acute/chronic health hazard, fire hazard, §313 nonreportable; Canada DSL, EINECS, Australia AICS, China, Japan ENCS, Korea KCL, Philippines PICCS listed
Properties: Colorless clear liq.; negligible odor; sol. in hydrocarbons; insol. in water; sp.gr. 0.80-0.82; vapor pressure < 0.5 mm Hg (68 F); b.p. 187.8-287.8 C; flash pt. (PMCC) 76.7 C; 100% act.; < 8% aromatics; < 1 ppm benzene
Toxicology: OSHA PEL/TWA 500 ppm (8 h); may be harmful or fatal by ing., inh. in very high concs.; inh. of high vapor/mist concs. can cause respiratory tract/eye irritation, CNS effects (drowsiness, dizziness, nausea, headache, unconsciousness); prolonged/repeated contact can cause irritation to eyes, nose, throat, lungs, skin; extreme overexposure or aspiration can cause lung damage, death; TSCA listed
Environmental: Environmentally friendly; do not allow spills to enter sewers or watercourses
Precaution: Combustible; flamm. limits in air 1-6%; may react with strong oxidizing agents; spillages may be slippery
Hazardous decomp. prods.: Combustion: dense smoke, CO, CO_2, other oxides
HMIS: Health 1, Flammability 2, Reactivity 0

Penreco Snow [Penreco]
Chem. Descrip.: Wh. petrolatum USP
CAS 8027-32-5; EINECS/ELINCS 232-373-2
Uses: Emollient, base, solvent, carrier for cosmetics/pharmaceuticals (creams/lotions, petrol. jellies, ointments, dental adhesives), foods (animal feed supplement, fruit/veg. coatings); sanitary lubricant in food prod. machinery; rust-preventive coating in food processing equipment; carrier in adhesive tapes and compds.; lubricant for textile, paper, PVC; shoe polishes; food-contact applics.
Regulatory: FDA 21CFR 172.880, 178.3700, 573.720; OSHA nonhazardous; DOT nonhazardous
Properties: Wh. opaque semisolid, odorless; sol. in hydrocarbons; sp.gr. 0.86 (60/60 F); visc. 64-75 SUS (210 F); vapor pressure < 1 mm Hg (70 F); m.p. 122-135 F; b.p. 650 F; congeal pt. 123 F; solid. pt. 121 F; flash pt. (COC) 400 F
Toxicology: Pract. nontoxic by ingestion, but laxative props. may cause abdominal cramps, diarrhea; minimally irritating to eyes and skin on direct contact; STEL 10 mg/m^3 (as oil mist)TSCA listed
Precaution: May react with strong oxidizing agents
Hazardous decomp. prods.: Combustion: dense smoke, CO, CO_2, other oxides
Storage: Store in closed containers away from heat, sparks, open flame, oxidizers

Pentane 60/40 [Shell]
Chem. Descrip.: n-Pentane (60%), isopentane (40%)
Uses: Solvent; blowing agent in mfg. and processing of expandable PS
Features: Does not contain detectable quantities of polycyclic aromatics, heavy metals, or chlorinated compds.
Properties: Saybolt 30 color; dens. 0.627 kg/l (15 C); i.b.p. 32 C
Precaution: Highly flamm.; vapor can readily form explosive mixt. in air

Pentane 80/20 [Shell]
Chem. Descrip.: n-Pentane (76% min.) blend
UN 1265
Uses: Solvent; blowing agent in mfg. and processing of expandable PS; propellant in aerosol sprays
Features: Very volatile; does not contain detectable quantities of polycyclic aromatics, heavy metals, or chlorinated compds.
Properties: M.w. 72; dens. 0.63 kg/l (15 C); visc. 0.34 mm^2/s; vapor pressure 75 kPa (20 C); i.b.p. 26 C; pour pt. < 100 C; flash pt. < -50 C; autoignition temp. 410 C; surf. tens. 15.8 mN/m (20 C); KB value 27; elec. conduct. 1.1 pS/m (20 C); dielec. const. 1.8 (20 C)
Precaution: Highly flamm.; LEL 1.4 vol.%; UEL 7.8 vol.%; vapor can readily form explosive mixt. in air
Storage: 12 mos min. shelf life under proper storage

Pentanol 45 [Celanese; Celanese GmbH]
Chem. Descrip.: 1-Pentanol (45%), 2-methyl-1-butanol (55%)
UN 1105
Uses: Raw material for plasticizers (e.g., phthalates), pesticides, flotation collectors for sulfide ore (e.g., amyl xanthates), paint stabilizers, corrosion inhibitors for gear lubricants and crankcase oil; latent cosolvent for NC; extraction solvent in separation of bituminous components from coal tar, separation of uranium, neptunium, platinum, and zinc salts, and separation of penicillin; flow aid in insecticides; corrosion inhibitor
Regulatory: SARA §311 acute health/fire hazard
Properties: Colorless clear liq.; char. fusel oil odor; sol. 26 g/l in water; misc. with common org. solvs.; sp.gr. 0.815 (20 C); visc. ≈ 4.3 mPa·s (20 C); vapor pressure 2.25 mm Hg (20 C); b.p. 135 C; acid no. 0.10 max.; pour pt. < -60 C; flash pt. (CC) 43 C; autoignition temp. 365 C; surf. tens. 25.6 mN/m (20 C); dielec. const. 15.3 (20 C)
Toxicology: LD50 (oral, rat) 2.20-4.01 g/kg, (skin, rabbit) > 2000 mg/kg; LC50 (inh., rat, 4 h) 2-20 mg/l; harmful by inh.; causes respiratory tract/eye irritation; mod. to severe skin irritant; may be harmful absorbed thru skin; may cause CNS depression, headache, stupor, unconsciousness, nausea, dizziness, chest pain, breathing difficulty; severe mucous membrane irritant; narcotic effects if inh. in fairly high concs.; may be harmful by ing.; TSCA listed
Environmental: Readily biodeg.; BOD5 0.59-0.87; low potential to bioaccumulate; EC50 (daphnia magna, 24 h) 423-580 ppm, (E. coli) 961-1058 ppm; low acute toxicity to aquatic species; avoid run-off to storm sewers, ditches leading to waterways
Precaution: Flamm. liq. and vapor; flamm. limits in air 1.2-12.4 vol.% (1013 hPa); floats on water; may generate static charge; ground/bond equip. during transfer; avoid heat, sparks, flame, ignition sources; incompat. with sulfuric and other strong inorg. acids, aluminum, lead, oxidizing agents, peroxides, nitric acid, perchloric acid, chromium trioxide
Hazardous decomp. prods.: Thermal decomp. or combustion: CO_x
HMIS: Health 2, Flammability 2, Reactivity 0
Storage: Store away from moisture, direct sunlight, incompat. materials; keep tightly closed when not in use; open containers slowly to allow excess pressure to vent

Permethyl® 97A [Presperse]
Chem. Descrip.: Isooctane
CAS 540-84-1; EINECS/ELINCS 203-892-1
Uses: Solvent, film-former for cosmetics and personal care prods.
Features: Quick drying

Permethyl® 99A [Presperse]
Chem. Descrip.: Isododecane
CAS 31807-55-3; EINECS/ELINCS 250-816-8
Uses: Cosolubilizer for nonhydrocarbon materials; solvent for mascara, eyeliner, antiperspirant; solvent for debris on skin
Features: Clean feel, very little residual; exc. where residual film is not desirable
Regulatory: JSCI/JCID approved
Properties: Colorless clear liq.; sol. in cyclomethicone, hydrocarbons, isoparaffins, min. spirits; m.w. 170; sp.gr. 0.747; visc. 1 cps (20 C); ref.

index 1.424
Toxicology: Acute oral > 5 g/kg; pract. nonirritating to eyes; mild skin irritant; noncomedogenic

Perstabil® [Solvay SA]
Uses: Solvent
Properties: Colorless

Pertene® [Atofina SA]
Chem. Descrip.: Perchloroethylene, stabilized
CAS 127-18-4; EINECS/ELINCS 204-825-9
Uses: Solvent

Petrinex® ASB [Huntsman Petrochems. UK]
Chem. Descrip.: Heavy aromatic naphtha
CAS 64742-94-5; EINECS/ELINCS 265-198-5
Uses: Solvent for oil extraction, refinery cleaning chems., creosote mfg., other solvent applics.
Features: General purpose; relatively high b.p.; good pouring chars.
Properties: Dk. brn. color

PF-5050 [3M/Perf. Materials]
Chem. Descrip.: Perfluoropentane
CAS 678-26-2
Uses: Surf. wetting agent; dispersant for lubricants, mold release agents, protective coatings; reaction medium for polymerizations, purification, separation processes; solvent for removal of halogenated lubricants, oils, greases; carrier for halogenated material; reduces aggressiveness of other solvs.
Features: Performance fluid offering low surf. tension, high chem./thermal stability, no VOC for improved coating flow and performance; zero ozone depletion potential
Properties: Liq.; m.w. 288; dens. 1.63 g/ml; visc. 0.45 cp; vapor pressure 11.80 psia; b.p. 30 C; flash pt. none; surf. tens. 9.5 dynes/cm

PF-5060 [3M/Perf. Materials]
Chem. Descrip.: Perfluorohexane
CAS 355-42-0; EINECS/ELINCS 206-585-0
Uses: Surf. wetting agent; dielec. coolant for GE Vapor Tran™ transformers; dispersant for lubricants, mold release agents, protective coatings; reaction medium for polymerizations, purification, separation processes; solvent for removal of halogenated lubricants, oils, greases; carrier for halogenated material; reduces aggressiveness of other solvs.
Features: Performance fluid; offers low surf. tension, high chem./thermal stability, no VOC for improved coating flow and performance; replacement for CFC-113; nonflamm.; highly resist. to thermal breakdown/hydrolysis
Properties: Colorless clear liq.; m.w. 338; dens. 1.68 g/ml; visc. 0.67 cp; vapor pressure 232 mm Hg; b.p. 56 C (1 atm); pour pt. -90 C; flash pt. none; surf. tens. 12.0 dynes/cm; dielec. str. 38 kV (0.1 in. gap); vol. resist. 1 x 10^{15} ohm-cm
Toxicology: Nonirritating to eyes and skin; pract. nontoxic orally; very low acute and sub-chronic in. toxicity; nonmutagenic; not a cardiac sensitizer
Environmental: Zero ozone depletion potential; non-VOC; high global warming potential and long atmospheric lifetime

PF-5070 [3M/Perf. Materials]
Chem. Descrip.: Perfluoroheptane
CAS 335-57-9; EINECS/ELINCS 206-392-1
Uses: Surf. wetting agent; dispersant for lubricants, mold release agents, protective coatings; reaction medium for polymerizations, purification, separation processes; solvent for removal of halogenated lubricants, oils, greases; carrier for halogenated material; reduces aggressiveness of other solvs.
Features: Performance fluid offering low surf. tension, high chem./thermal stability, no VOC for improved coating flow and performance; zero ozone depletion potential
Properties: Liq.; m.w. 388; dens. 1.73 g/ml; visc. 0.95 cp; vapor pressure 1.53 psia; b.p. 80 C; flash pt. none; surf. tens. 13.0 dynes/cm

Pharmasolve™ [ISP]
Chem. Descrip.: N-Methyl-2-pyrrolidone, GMP grade
CAS 872-50-4; EINECS/ELINCS 212-828-1
Uses: Solvent, solubilizer for pharmaceuticals; chem. reaction medium; reaction solvent in many syntheses incl. vitamin E precursor; solubilizer for topicals for applic. to human or animal hair, skin; bioavailability enhancer for topicals; skin penetration enhancer; extraction solvent; bioadhesive; spray bandages
Features: Nonaq.; increases water sol. of chloramphenicol, procaine, other therapeutic compds.; increases sol'n. stability of drugs in aq. sol'n.
Properties: Clear liq., amine-like odor; misc. with water, most org. solvs. incl. alcohols, ketones, aromatic and chlorinated hydrocarbons; sp.gr. 1.027 (25/4 C); visc. 1.65 cp; vapor pressure 0.29 mm Hg (20 C); m.p. -24.4 C; b.p. 202 C; flash pt.(OC) 95 C; ref. index 1.465-1.470; pH sl. alkaline; 99.85% min. purity
Toxicology: CEL 100 ppm; low toxicity; LD50 (oral, rat) 4200 mg/kg, (dermal, rabbit) 8000 mg/kg; causes eye irritation; not a primary skin irritant; nonsensitizing; prolonged or repeated exposure causes severe skin irritation
Environmental: Biodeg.
Precaution: Combustible; flamm. limits 1.3-9.5%; noncorrosive; incompat. with strong oxidizing or reducing agents; heated to decomp., emits CO and NO_x fumes
HMIS: Health 2, Flammability 2, Reactivity 0
Storage: Keep container and vapors away from heat, sparks, flame; keep container closed

Phosflex® 4 [Akzo Nobel; Harwick Std. Distrib.]
Chem. Descrip.: Tributyl phosphate
CAS 126-73-8; EINECS/ELINCS 204-800-2
Uses: Flame retardant, plasticizer for nitrocellulose, cellulose acetate, chlorinated rubber, and vinyls; antifoam for paper coatings, water-based adhesives, inks, textile sizes, detergent sol'ns.; high boiling solvent for lithographic inks, latex paints; also suitable for epoxy, PS, PVAc
Properties: Colorless liq.; sp.gr. 0.980 (20/20 C); visc. 6 mPa·s; 11.7% P

Phytol™ AVO [Custom Ingreds.]
Chem. Descrip.: Avocado oil
CAS 8024-32-6; EINECS/ELINCS 232-428-0
Uses: Emollient, solvent for personal care
Features: Nongreasy
Regulatory: EU, Japan approved
Properties: Yel. clear liq.; bland taste; odorless; sp.gr. 0.908-0.920; acid no. 3.0 max.; iodine no. 65-95; sapon. no. 177-198; ref. index 1.460-1.479

Phytol™ CAN [Custom Ingreds.]
Chem. Descrip.: Refined, bleached, deodorized canola oil, TBHQ
Uses: Emollient, solvent for hair care, skin care
Features: Nongreasy; occlusive
Regulatory: EU, Japan approved
Properties: Gardner 3 max. color; bland odor; 59% monostaurates, 33% polyunsaturates, 8% saturates; 0.05% free fatty acid

Pilot® SXS-40 [Pilot]
Chem. Descrip.: Sodium xylene sulfonate
CAS 1300-72-7; EINECS/ELINCS 215-090-9
Uses: Coupling agent, hydrotrope, solubilizer, solvent in org. polymers and dyestuffs, petrol. industry, pulping, animal glues; solubilizer, hydrotrope for textiles; dispersant and stabilizer for lowering cloud pt. and visc. in liq. detergents
Properties: Lt. yel. clear liq.; dens. 9.9 lb/gal; pH 7.5; anionic; 42% total solids

Pine Oil 150 [Millennium/F&F]
Chem. Descrip.: α-Terpineol, terpene alcohols, and terpene hydrocarbons
UN NA 1993
Uses: Used in disinfectants, cleaners and solvs.
Regulatory: DOT regulated
Properties: Clear liq.; sp.gr. 0.930; dens. 7.76 lb/gal; b.p. 195 C; flash pt. (TCC) 75 C; ref. index 1.479-1.489
Toxicology: LD50 (oral, rat) 2.76 g/kg; eye and skin irritantTSCA listed
Precaution: Combustible

Pluracol® E200 [BASF/Perf. Chems.]
Chem. Descrip.: PEG-4
CAS 25322-68-3; EINECS/ELINCS 203-989-9
Uses: Intermediate for mfg. of nonionic surfactants; binder, base, coating, stabilizer, solvent, vehicle, extender, coupling agent for cosmetics, toiletries; lubricant for metal applics., rubber industry; wood treatment; textile conditioner, antistat, sizing agent, softener
Properties: Colorless clear liq.; m.w. 200; sol. in water and org. solvs. except aliphatic hydrocarbons; dens. 9.4 lb/gal; sp.gr. 1.12; visc. 4.36 cs (210 F); flash pt. 360 F; surf. tens. 57.2 dynes/cm (1%); pH 6.5 (5% aq.)

Pluracol® E300 [BASF/Perf. Chems.]
Chem. Descrip.: PEG-6
CAS 25322-68-3; EINECS/ELINCS 220-045-1
Uses: Intermediate for mfg. of nonionic surfactants; binder, base, coating, stabilizer, solvent, vehicle, extender, coupling agent for pharmaceuticals, cosmetics, toiletries; lubricant for metal applics., rubber industry; wood treatment; textile conditioner, antistat, sizing agent, softener; dispersant in food tablets and preps.; plasticizer
Properties: Colorless clear liq.; m.w. 300; sol. in water and org. solvs. except aliphatic hydrocarbons; dens. 9.4 lb/gal; sp.gr. 1.12; visc. 5.9 cs (99 C); flash pt. (COC) 210 C; pour pt. -13 C; surf. tens. 62.9 dynes/cm (1%); pH 5.7 (5% aq.); nonionic

Pluracol® E400 [BASF/Perf. Chems.]
Chem. Descrip.: PEG-8
CAS 25322-68-3; EINECS/ELINCS 225-856-4
Uses: Intermediate for mfg. of nonionic surfactants; binder, base, coating, stabilizer, solvent, vehicle, extender, coupling agent for pharmaceuticals, cosmetics, toiletries; lubricant for metal applics., rubber industry; wood treatment; textile conditioner, antistat, sizing agent, softener; dispersant in food tablets and preps.; plasticizer
Properties: Colorless clear liq.; sol. in water and org. solvs. except aliphatic hydrocarbons; dens. 9.4 lb/gal; sp.gr. 1.12; visc. 7.39 cs (210 F); flash pt. 460 F; surf. tens. 66.6 dynes/cm (1%); pH 6.2 (5% aq.); nonionic; 100% act.

Pluracol® E400 NF [BASF/Perf. Chems.]
Chem. Descrip.: PEG-8
CAS 25322-68-3; EINECS/ELINCS 225-856-4
Uses: Base, coupling agent, thickener, lubricant, mold release agent, defoamer, softener, conditioner, antistat, sizing agent, dispersant for cosmetic and oral care preps., metal polishing and cleaning formulations, rubber prods., paper and wood prods., textile processing, ink formulations; excipient, carrier, solvent in pharmaceuticals; chemical intermediate
Properties: Liq.; m.w. 400; visc. 7.4 cs (99 C); pour pt. 5 C; flash pt. (COC) 182 C; nonionic

Pluracol® E600 [BASF/Perf. Chems.]
Chem. Descrip.: PEG-12
CAS 25322-68-3; EINECS/ELINCS 229-859-1
Uses: Intermediate for mfg. of nonionic surfactants; binder, base, coating, stabilizer, solvent, vehicle, extender, coupling agent for pharmaceuticals, cosmetics, toiletries; lubricant for metal applics., rubber industry; wood treatment; textile conditioner, antistat, sizing agent, softener; dispersant in food tablets and preps.; plasticizer
Properties: Colorless clear liq.; sol. in water and org. solvs. except aliphatic hydrocarbons; dens. 9.4 lb/gal; sp.gr. 1.12; visc. 10.83 cs (210 F); flash pt. 480 F; surf. tens. 65.2 dynes/cm (1%); pH 5.3 (5% aq.); nonionic; 100% act.

Pluracol® E600 NF [BASF/Perf. Chems.]
Chem. Descrip.: PEG-12
CAS 25322-68-3; EINECS/ELINCS 229-859-1
Uses: Base, coupling agent, thickener, lubricant, mold release agent, defoamer, softener, conditioner, antistat, sizing agent, dispersant for cosmetic and oral care preps., metal polishing and cleaning formulations, rubber prods., paper and wood prods., textile processing, ink formulations; excipient, carrier, solvent for pharmaceuticals; chemical intermediate
Properties: Liq.; m.w. 600; visc. 10.8 cs (99 C); pour pt. 20 C; flash pt. (COC) 249 C; nonionic

Pluracol® E1450 NF [BASF/Perf. Chems.]
Chem. Descrip.: PEG-32
CAS 25322-68-3; EINECS/ELINCS 203-989-9
Uses: Chemical intermediate, base, coupling agent, thickener, lubricant, mold release agent, defoamer, softener, conditioner, antistat, sizing agent, dispersant, excipient, carrier, solvent for pharmaceuticals, cosmetics, oral care preps., in metal polishing and cleaning formulations, rubber prods., paper and wood prods., textile processing, inks
Properties: Solid; m.w. 600; visc. 28.5 cs (99 C); m.p. 45 C; flash pt. (COC) 255 C

Pluracol® E4000 [BASF/Perf. Chems.]
Chem. Descrip.: PEG-75
CAS 25322-68-3; EINECS/ELINCS 203-989-9
Uses: Intermediate for mfg. of nonionic surfactants; binder, base, coating, stabilizer, solvent, vehicle, extender, coupling agent for pharmaceuticals, cosmetics, toiletries; lubricant for metal applics., rubber industry; wood treatment; textile conditioner, antistat, sizing agent, softener; dispersant in food tablets and preps.; plasticizer
Properties: Wh. waxy solid; sol. in water and org. solvs. except aliphatic hydrocarbons; dens. 10.0 lb/gal; sp.gr. 1.20; m.p. 59.5 C; flash pt. > 490 F; surf. tens. 61.9 dynes/cm (1%); pH 6.7 (5% aq.); nonionic; 100% act.

Pluracol® E6000 [BASF/Perf. Chems.]
Chem. Descrip.: PEG-150
CAS 25322-68-3; EINECS/ELINCS 203-989-9
Uses: Intermediate for mfg. of nonionic surfactants; binder, base, coating, stabilizer, solvent, vehicle, extender, coupling agent for pharmaceuticals, cosmetics, toiletries; lubricant for metal applics., rubber industry; wood treatment; textile conditioner, antistat, sizing agent, softener
Properties: Wh. waxy solid; sol. in water and org. solvs. except aliphatic hydrocarbons; dens. 10.0 lb/gal; sp.gr. 1.21; m.p. 61.0 C; flash pt. > 490 F; surf. tens. 62.1 dynes/cm (1%); pH 6.7 (5% aq.)

Pluracol® SRF-2 [BASF/Perf. Chems.]
Chem. Descrip.: Polyalkylene glycol deriv.
Uses: Lubricant, wetting agent, detergent in metalworking fluids (cutting, drawing, stamping, rolling, etc.); hydraulic fluid for boundary lubrication and low-temp. applics.; lubricant in motor vehicle brake fluids; lubricant in high temp., compressor gear, rubber, textile fiber, mill, and calendar lubricants; chemical intermediate; defoamer, plasticizer, solvent for inks and dyes
Features: Stable; resist. to bacterial and chem. deterioration; noncorrosive
Properties: Liq.; water-sol.; sp.gr. 1.140 (20/20 C); dens. 9.51 lb/gal (20 C); visc. 460 SUS (38 C); pour pt. -37 C; flash pt. (COC) 293 C

Pluracol® W3520N [BASF/Perf. Chems.]
Chem. Descrip.: PPG-28-buteth-35
CAS 9038-95-3
Uses: Lubricant, wetting agent, detergent in metalworking fluids (cutting, drawing, stamping, rolling, etc.); hydraulic fluid for boundary lubrication and low-temp. applics.; lubricant in motor vehicle brake fluids; lubricant in high temp., compressor gear, rubber, textile fiber, mill, and calendar lubricants; chemical intermediate; defoamer, plasticizer, solvent for inks and dyes
Features: Stable; resist. to bacterial and chem. deterioration; noncorrosive
Properties: APHA 40 max. visc. liq.; sol. in water, alcohols, ketones, esters, benzene, toluene, glycol ethers, chlorinated solvs.; sp.gr. 1.059 (20/20 C); dens. 8.83 lb/gal (20 C); visc. 3520 SUS (38 C); pour pt. -29 C; flash pt. (COC) 226 C; pH 6.0-7.5 (10% aq.); nonionic

Pluracol® WI 285 [BASF/Perf. Chems.]
Chem. Descrip.: Polyalkylene glycol deriv.
Uses: Lubricant, wetting agent, detergent in metalworking fluids (cutting, drawing, stamping, rolling, etc.); hydraulic fluid for boundary lubrication and low-temp. applics.; lubricant in motor vehicle brake fluids; lubricant in high temp., compressor gear, rubber, textile fiber, mill, and calendar lubricants; chemical intermediate; defoamer, plasticizer, solvent for inks and dyes
Features: Stable; resist. to bacterial and chem. deterioration; noncorrosive
Properties: Liq.; water-insol.; sp.gr. 0.991 (20/20 C); dens. 8.25 lb/gal (20 C); visc. 285 SUS (38 C); visc. index 183; pour pt. -48 C; flash pt. (COC) 218 C

Pluracol® WI 625 [BASF/Perf. Chems.]
Chem. Descrip.: Polyalkylene glycol deriv.
Uses: Lubricant, wetting agent, detergent in metalworking fluids (cutting, drawing, stamping, rolling, etc.); hydraulic fluid for boundary lubrication and low-temp. applics.; lubricant in motor vehicle brake fluids; lubricant in high temp., compressor gear, rubber, textile fiber, mill, and calendar lubricants; chemical intermediate; defoamer, plasticizer, solvent for inks and dyes
Features: Stable; resist. to bacterial and chem. deterioration; noncorrosive
Properties: Liq.; water-insol.; sp.gr. 0.996 (20/20 C); dens. 8.29 lb/gal (20 C); visc. 625 SUS (38 C); visc. index 186; pour pt. -39 C; flash pt. (COC) 227 C

Pluracol® WI 1715 [BASF/Perf. Chems.]
Chem. Descrip.: Polyalkylene glycol deriv.

Uses: Lubricant, wetting agent, detergent in metalworking fluids (cutting, drawing, stamping, rolling, etc.); hydraulic fluid for boundary lubrication and low-temp. applics.; lubricant in motor vehicle brake fluids; lubricant in high temp., compressor gear, rubber, textile fiber, mill, and calendar lubricants; chemical intermediate; defoamer, plasticizer, solvent for inks and dyes
Features: Stable; resist. to bacterial and chem. deterioration; noncorrosive
Properties: Liq.; water-insol.; sp.gr. 1.003 (20/20 C); dens. 8.35 lb/gal (20 C); visc. 1715 SUS (38 C); visc. index 218; pour pt. -23 C; flash pt. (COC) 221 C

Pluracol® WS 170 [BASF/Perf. Chems.]
Chem. Descrip.: Polyalkylene glycol deriv.
Uses: Lubricant, wetting agent, detergent in metalworking fluids (cutting, drawing, stamping, rolling, etc.); hydraulic fluid for boundary lubrication and low-temp. applics.; lubricant in motor vehicle brake fluids; lubricant in high temp., compressor gear, rubber, textile fiber, mill, and calendar lubricants; chemical intermediate; defoamer, plasticizer, solvent for inks and dyes
Features: Stable; resist. to bacterial and chem. deterioration; noncorrosive
Properties: Liq.; water-sol.; sp.gr. 1.035 (20/20 C); dens. 8.62 lb/gal (20 C); visc. 170 SUS (38 C); visc. index 218; pour pt. -61 C; flash pt. (COC) 210 C

Pluracol® WS 260 [BASF/Perf. Chems.]
Chem. Descrip.: Polyalkylene glycol deriv.
Uses: Lubricant, wetting agent, detergent in metalworking fluids (cutting, drawing, stamping, rolling, etc.); hydraulic fluid for boundary lubrication and low-temp. applics.; lubricant in motor vehicle brake fluids; lubricant in high temp., compressor gear, rubber, textile fiber, mill, and calendar lubricants; chemical intermediate; defoamer, plasticizer, solvent for inks and dyes
Features: Stable; resist. to bacterial and chem. deterioration; noncorrosive
Properties: Liq.; water-sol.; sp.gr. 1.038 (20/20 C); dens. 8.64 lb/gal (20 C); visc. 260 SUS (38 C); visc. index 205; pour pt. -62 C; flash pt. (COC) 227 C

Pluracol® WS 660 [BASF/Perf. Chems.]
Chem. Descrip.: Polyalkylene glycol deriv.
Uses: Lubricant, wetting agent, detergent in metalworking fluids (cutting, drawing, stamping, rolling, etc.); hydraulic fluid for boundary lubrication and low-temp. applics.; lubricant in motor vehicle brake fluids; lubricant in high temp., compressor gear, rubber, textile fiber, mill, and calendar lubricants; chemical intermediate; defoamer, plasticizer, solvent for inks and dyes
Features: Stable; resist. to bacterial and chem. deterioration; noncorrosive
Properties: Liq.; water-sol.; sp.gr. 1.052 (20/20 C); dens. 8.75 lb/gal (20 C); visc. 660 SUS (38 C); visc. index 228; pour pt. -42 C; flash pt. (COC) 227 C

Pluracol® WS 2000 [BASF/Perf. Chems.]
Chem. Descrip.: Polyalkylene glycol deriv.
Uses: Lubricant, wetting agent, detergent in metalworking fluids (cutting, drawing, stamping, rolling, etc.); hydraulic fluid for boundary lubrication and low-temp. applics.; lubricant in motor vehicle brake fluids; lubricant in high temp., compressor gear, rubber, textile fiber, mill, and calendar lubricants; chemical intermediate; defoamer, plasticizer, solvent for inks and dyes
Features: Stable; resist. to bacterial and chem. deterioration; noncorrosive
Properties: Liq.; water-sol.; sp.gr. 1.058 (20/20 C); dens. 8.80 lb/gal (20 C); visc. 2000 SUS (38 C); visc. index 259; pour pt. -33 C; flash pt. (COC) 243 C

Pluracol® WS 3520 [BASF/Perf. Chems.]
Chem. Descrip.: Polyalkylene glycol deriv.
Uses: Lubricant, wetting agent, detergent in metalworking fluids (cutting, drawing, stamping, rolling, etc.); hydraulic fluid for boundary lubrication and low-temp. applics.; lubricant in motor vehicle brake fluids; lubricant in high temp., compressor gear, rubber, textile fiber, mill, and calendar lubricants; chemical intermediate; defoamer, plasticizer, solvent for inks and dyes
Features: Stable; resist. to bacterial and chem. deterioration; noncorrosive
Properties: Liq.; water-sol.; sp.gr. 1.061 (20/20 C); dens. 8.83 lb/gal (20 C); visc. 3520 SUS (38 C); visc. index 269; pour pt. -28 C; flash pt. (COC) 227 C

Pluracol® WS 5100 [BASF/Perf. Chems.]
Chem. Descrip.: Polyalkylene glycol deriv.
Uses: Lubricant, wetting agent, detergent in metalworking fluids (cutting, drawing, stamping, rolling, etc.); hydraulic fluid for boundary lubrication and low-temp. applics.; lubricant in motor vehicle brake fluids; lubricant in high temp., compressor gear, rubber, textile fiber, mill, and calendar lubricants; chemical intermediate; defoamer, plasticizer, solvent for inks and dyes
Features: Stable; resist. to bacterial and chem. deterioration; noncorrosive
Properties: Liq.; water-sol.; sp.gr. 1.063 (20/20 C); dens. 8.85 lb/gal (20 C); visc. 5100 SUS (38 C); visc. index 281; pour pt. -31 C; flash pt. (COC) 227 C

Pluracol® WT 1400 [BASF/Perf. Chems.]
Chem. Descrip.: Polyalkylene glycol
CAS 9003-11-6
Uses: Lubricant, wetting agent, detergent in metalworking fluids (cutting, drawing, stamping, rolling, etc.); hydraulic fluid for boundary lubrication and low-temp. applics.; lubricant in motor vehicle brake fluids; lubricant in high temp., compressor gear, rubber, textile fiber, mill, and calendar lubricants; chemical intermediate; defoamer, plasticizer, solvent for inks and dyes
Features: Stable; resist. to bacterial and chem. deterioration; noncorrosive
Properties: Clear liq.; mild glycol odor; completely sol. in water; m.w. 2500; sp.gr. 1.093 (20/20 C); dens. 9.14 lb/gal (20 C); visc. 1400 SUS (38 C); visc. index 238; pour pt. -6 C; f.p. 22 F; decomp. pt. > 475 F; flash pt. (COC) 204 C; pH 5.5-7.5 (10% aq.); nonionic; 100% act.; < 1% volatile
Toxicology: May cause temporary eye irritation; prolonged/repeated skin contact may cause sl. temporary irritation; ing. may cause diarrhea, weakness; inh. may cause respiratory irritationTSCA listed
Precaution: May contain trace amts. of EO, PO, and 1,4-dioxane
HMIS: Health 1, Flammability 1, Reactivity 0

Pluracol® WT 9150 [BASF/Perf. Chems.]
Chem. Descrip.: Polyalkylene glycol deriv.
Uses: Lubricant, wetting agent, detergent in metalworking fluids (cutting, drawing, stamping, rolling, etc.); hydraulic fluid for boundary lubrication and low-temp. applics.; lubricant in motor vehicle brake fluids; lubricant in high temp., compressor gear, rubber, textile fiber, mill, and calendar lubricants; chemical intermediate; defoamer, plasticizer, solvent for inks and dyes
Features: Stable; resist. to bacterial and chem. deterioration; noncorrosive
Properties: Liq.; water-thickening; sp.gr. 1.089 (20/20 C); dens. 9.15 lb/gal (20 C); visc. 9150 SUS (38 C); visc. index 283; pour pt. -2 C; flash pt. (COC) 207 C

Pluracol® WT 90,000 [BASF/Perf. Chems.]
Chem. Descrip.: Polyalkylene glycol deriv.
Uses: Lubricant, wetting agent, detergent in metalworking fluids (cutting, drawing, stamping, rolling, etc.); hydraulic fluid for boundary lubrication and low-temp. applics.; lubricant in motor vehicle brake fluids; lubricant in high temp., compressor gear, rubber, textile fiber, mill, and calendar lubricants; chemical intermediate; defoamer, plasticizer, solvent for inks and dyes
Features: Stable; resist. to bacterial and chem. deterioration; noncorrosive
Properties: Liq.; water-thickening; sp.gr. 1.095 (20/20 C); dens. 9.11 lb/gal (20 C); visc. 90,000 SUS (38 C); visc. index 432; pour pt. 3 C; flash pt. (COC) 252 C

Pogol® 400 [Huntsman]
Chem. Descrip.: PEG 400
CAS 25322-68-3; EINECS/ELINCS 225-856-4
Uses: Solubilizer; solvent; mold release agent and lubricant for natural and syn. prods.; dye carrier; antistat, softener, humectant, and lubricant for textiles, metalworking
Properties: Pt-Co < 50 color; clear liq.; m.w. 380-420; dens. 1.1254 g/ml (20 C); visc. 42 cSt (37.8 C); f.p. 4-8 C; flash pt. (PMCC) 200 C; pH 5.5-7.0 (5% aq.); nonionic; 100% act.

Polarin® Range [Grünau GmbH]
Chem. Descrip.: Glycerol and derivs.
Uses: Solvent, carrier, humectant for foods, pharmaceuticals (aerosols, capsules, syrups)

Polychol 5 [Croda Inc; Croda Chem. Ltd; Croda Oleochems.]
Chem. Descrip.: Laneth-5
CAS 61791-20-6
Uses: O/w emulsifier, dispersant, gellant, emollient, solubilizer for personal care prods., topical pharmaceuticals; bromo dye solvent; coemulsifier in prod. of w/o emulsions and microemulsions for topical pharmaceuticals; cosolvent for nonmiscible oils
Features: Reduces greasy feeling of lanolin
Properties: Soft yel. wax; sol. in IPA, disp. in water, propylene glycol, min. oil; m.p. 35-48 C; HLB 7.3; acid no. 5 max.; hyd. no. 120-135; pH 4.5-6.0 (10% aq.); nonionic; 100% conc.
Use Level: 1-10%
Toxicology: LD50 (oral, rat) 7.7 g/kg; mild skin irritant, nonirritating to eyes

Polychol 15 [Croda Inc; Croda Oleochems.]
Chem. Descrip.: Laneth-15
CAS 61791-20-6
Uses: O/w emulsifier, dispersant, solubilizer, emollient, thickener, and gellant for cosmetics and topical pharmaceuticals; cosolvent for nonmiscible oils
Features: Mild; reduces greasy feeling of lanolin
Properties: Yel. soft wax; sol. in water, IPA, propylene glycol; disp. in min. oil; m.p. 34-42 C; HLB 12.7; acid no. 5 max.; hyd. no. 82-92; pH 3.5-5.5 (10% aq.); nonionic; 100% conc.
Use Level: 1-10%
Toxicology: Nonirritating to skin and eyes

Polyglycol DME 200 [Clariant]
Uses: Solvent for flue gas cleaning systems, delacquering
Properties: M.p. -37 C; b.p. > 270 C

Polyglycol DME 250 [Clariant]
Uses: Solvent for pharmaceuticals
Properties: M.p. -23 C; b.p. > 270 C

Polyglycol DME 500 [Clariant]
Uses: Solvent for electroplating
Features: High boiling
Properties: M.p. 13 C; b.p. > 300 C

Polyglycol DME 1000 [Clariant]
Uses: Solvent for electroplating
Features: High boiling
Properties: M.p. 36 C; b.p. > 300 C

Polyglycol DME 2000 [Clariant]
Uses: Solvent for electroplating
Features: High boiling
Properties: M.p. 50 C; b.p. > 300 C

Polyglycol P15-200 [Dow]
Chem. Descrip.: PPG-24 glycereth-24
CAS 9082-00-2
Uses: Solvent providing low odor, low irritation, low toxicity to personal care prods., cosmetics; defoamer for aq. systems at elevated temps.
Properties: Liq.; m.w. 2600; sp.gr. 1.063; dens. 8.81 lb/gal; visc. 420 cSt; pour pt. -40 C; flash pt. (PMCC) > 450 F; ref. index 1.460; sp. heat 0.470 cal/g/°C; nonionic

Polyglycol P112-2 [Dow]
Chem. Descrip.: PPG-66 glycereth-12
CAS 9082-00-2
Uses: Solvent providing low odor, low irritation, low toxicity to personal care prods., cosmetics; defoamer for cutting and drilling fluids, fermentation, chem. mfg., acid gas treatment, pulp and paper, and textiles
Properties: Liq.; m.w. 4900; sp.gr. 1.028; dens. 8.56 lb/gal; visc. 1000 cSt; pour pt. -18 C; flash pt. (PMCC) 455 F; ref. index 1.455; surf. tens. 33.7 dynes/cm (100 ppm); sp. heat 0.430 cal/g/°C; nonionic

Polyglycol P425 [Dow]
Chem. Descrip.: PPG-9
CAS 25322-69-4
Uses: Solvent providing low odor, low irritation, low toxicity to personal care prods., cosmetics
Properties: Liq.; m.w. 425; sp.gr. 1.007; dens. 8.39 lb/gal; visc. 70 cSt; pour pt. -45 C; flash pt. (PMCC) 330 F; ref. index 1.447; sp. heat 0.477 cal/g/°C

Polyglycol P1000TB [Dow]
Chem. Descrip.: PPG-17
CAS 25322-69-4
Uses: Solvent providing low odor, low irritation, low toxicity to personal care prods., cosmetics; defoamer for food-grade coatings and paper prods.
Regulatory: 21CFR §176.200, 176.210
Properties: Liq.; m.w. 1000; sp.gr. 1.005; dens. 8.38 lb/gal; visc. 162 cSt; pour pt. -25 C; flash pt. (PMCC) > 360 F; ref. index 1.448; surf. tens. 40.7 dynes/cm (100 ppm); sp. heat 0.461 cal/g/°C; nonionic

Polyglycol P1200 [Dow]
Chem. Descrip.: PPG-20
CAS 25322-69-4
Uses: Solvent providing low odor, low irritation, low toxicity to personal care prods., cosmetics; boiler water additive for prep. of steam that will contact food; defoamer for beet sugar and yeast processing; defoamer for food-grade coatings and paper prods.
Regulatory: 21CFR §173.310, 173.340(a)(3), 176.200, 176.210
Properties: Liq.; m.w. 1200; sp.gr. 1.007; dens. 8.38 lb/gal; visc. 175 cSt; pour pt. -40 C; flash pt. (PMCC) 345 F; ref. index 1.448; surf. tens. 43.3 dynes/cm (100 ppm); sp. heat 0.459 cal/g/°C; nonionic

Polyglycol P2000 [Dow]
Chem. Descrip.: PPG-26
CAS 25322-69-4
Uses: Solvent providing low odor, low irritation, low toxicity to personal care prods., cosmetics; boiler water additive for prep. of steam that will contact food; defoamer for beet sugar and yeast processing; defoamer for food-grade coatings and paper prods.
Regulatory: 21CFR §173.310, 173.340(a)(3), 176.200, 176.210
Properties: Liq.; sol. in ethanol, IPM, cyclomethicone, sunscreens, lactic acid; partly sol. in water, lt. min. oil; m.w. 2000; sp.gr. 1.002; dens. 8.34 lb/gal; visc. 300 cSt; pour pt. -30 C; flash pt. (PMCC) 390 F; ref. index 1.449; surf. tens. 39.6 dynes/cm (100 ppm); sp. heat 0.452 cal/g/°C; nonionic

Polyglycol P4000 [Dow]
Chem. Descrip.: PPG-30
CAS 25322-69-4
Uses: Solvent providing low odor, low irritation, low toxicity to personal care prods., cosmetics; boiler water additive for prep. of steam that will contact food; defoamer for food-grade paper coatings
Regulatory: 21CFR §173.310, 176.200
Properties: Liq.; m.w. 4000; sp.gr. 1.005; dens. 8.36 lb/gal; visc. 800 cSt; pour pt. -26 C; flash pt. (PMCC) 365 F; ref. index 1.450; surf. tens. 39.6 dynes/cm (100 ppm); nonionic

Polyglycol PT250 [Dow]
Chem. Descrip.: PPG glyceryl ether
CAS 25791-96-2
Uses: Solvent providing low odor, low irritation, low toxicity to personal care prods., cosmetics
Properties: Liq.; m.w. 250; sp.gr. 1.091; dens. 9.07 lb/gal; visc. 850 cSt; pour pt. -18 C; flash pt. (PMCC) 390 F; ref. index 1.459

Polyglycol PT700 [Dow]
Chem. Descrip.: PPG-10 glyceryl ether
CAS 25791-96-2
Uses: Solvent providing low odor, low irritation, low toxicity to personal care prods., cosmetics
Properties: Liq.; m.w. 700; sp.gr. 1.033; dens. 8.59 lb/gal; visc. 236 cSt; pour pt. -32 C; flash pt. (PMCC) 500 F; ref. index 1.453; sp. heat 0.442 cal/g/°C

Polyglycol PT3000 [Dow]
Chem. Descrip.: PPG-55 glyceryl ether
CAS 25791-96-2
Uses: Solvent providing low odor, low irritation, low toxicity to personal care prods., cosmetics
Properties: Liq.; m.w. 3000; sp.gr. 1.010; dens. 8.40 lb/gal; visc. 480 cSt; pour pt. -27 C; flash pt. (PMCC) 430 F; ref. index 1.451; sp. heat 0.445 cal/g/°C

Polyglyme [Ferro/Grant]
Chem. Descrip.: Poly(ethylene glycol) dimethyl ether
CAS 24991-55-7
Uses: Solvent for removal of CO, CO_2, SO_2, H_2S and other sour-gas components from ammonia gas streams, for absorp. of fluorocarbons and

other halogenated liqs./gases; stack gas scrubbing; brake fluid component; solubilizer for plastic, textile, and paper

Properties: Sl. yel.; very mild nonresidual odor; water-sol.; misc. with ethanol, acetone, benzene, diethyl ether, octane; m.w. 275; sp.gr. 1.04 (20/20 C); dens. 8.6 lb/gal; visc. 12 cP (20 C); vapor pressure < 0.01 mm Hg (20 C); flash pt. (CC) > 130 C

Environmental: Slowly biodeg.

Polypax IPM [Pax Group]

Chem. Descrip.: Isopropyl myristate

CAS 110-27-0; EINECS/ELINCS 203-751-4

Uses: Solvent, emollient, lubricant, conditioner, antistat, plasticizer in cosmetics

Properties: Liq.; 99% act.

Polypax IPP [Pax Group]

Chem. Descrip.: Isopropyl palmitate

CAS 142-91-6; EINECS/ELINCS 205-571-1

Uses: Solvent, emollient, lubricant, conditioner, antistat, plasticizer in cosmetics

Properties: Liq.; 99% act.

Poly-Solv® DM [Arch Chems./Specialty Polyols]

Chem. Descrip.: Diethylene glycol monomethyl ether

CAS 111-77-3; EINECS/ELINCS 203-906-6

Uses: Solvent for brake fluids, hard-surf. cleaners, leather dyeing, paints, coatings, printing inks, textile vat dyeing and printing, adhesives, antifreeze, floor waxes/polishes, insect repellents; solubilizer for dyes; plasticizer

Properties: APHA 15 max. liq.; mild, char. odor; m.w. 120.15; f.p. -85 C; b.p. 194 C (760 mm); sol. in water; misc. with many org. solvs.; sp.gr. 1.021; dens. 8.51 lb/gal; visc. 3.9 cP; ref. index 1.4263; flash pt. (TCC) 87 C; sp. heat 2.149 joules/ g-°C

Poly-Solv® DPM [Arch Chems./Specialty Polyols]

Chem. Descrip.: Dipropylene glycol monomethyl ether

CAS 34590-94-8; EINECS/ELINCS 252-104-2

Uses: Solvent for brake fluids, hard-surf. cleaners, leather dyeing, paints, coatings, printing inks, textile vat dyeing and printing, adhesives, antifreeze, floor waxes/polishes, insect repellents; solubilizer for dyes; plasticizer

Properties: APHA 15 max. liq.; mild, char. odor; m.w. 148.20; f.p. -83 C; b.p. 187 C (760 mm); misc. with water and many org. solvs.; sp.gr. 0.954; dens. 7.94 lb/gal; visc. 3.9 cP; ref. index 1.419; flash pt. (TCC) 169 C

Poly-Solv® EM [Arch Chems./Specialty Polyols]

Chem. Descrip.: Ethylene glycol monomethyl ether

CAS 109-86-4; EINECS/ELINCS 203-713-7

Uses: Solvent for brake fluids, hard-surf. cleaners, leather dyeing, paints, coatings, printing inks, textile vat dyeing and printing, adhesives, antifreeze, floor waxes/polishes, insect repellents; solubilizer for dyes; plasticizer

Properties: APHA 10 max. liq.; mild char. odor; m.w. 76.09; f.p. -85 C; b.p. 124 C (760 mm); sol. in water; misc. with many org. solvs.; sp.gr. 0.966; dens. 8.05 lb/gal; visc. 1.7 cP; ref. index 1.4021; flash pt. (TCC) 41 C; sp. heat 2.233 joules/g/°C

Poly-Solv® MPM [Arch Chems./Specialty Polyols]

Chem. Descrip.: Monopropylene glycol monomethyl ether

CAS 107-98-2; EINECS/ELINCS 203-539-1

Uses: Solvent for brake fluids, hard-surf. cleaners, leather dyeing, paints, coatings, printing inks, textile vat dyeing and printing, adhesives, antifreeze, floor waxes/polishes, insect repellents; solubilizer for dyes; plasticizer

Properties: APHA 10 max. liq.; mild char. odor; m.w. 90.12; f.p. -96 C; b.p. 121 C (760 mm); misc. with water and many org. solvs.; sp.gr. 0.923; dens. 7.68 lb/gal; visc. 1.9 cP; ref. index 1.4036; flash pt. (TCC) 36 C

Poly-Solv® TM [Arch Chems./Specialty Polyols]

Chem. Descrip.: Triethylene glycol monomethyl ether

CAS 112-35-6

Uses: Solvent for brake fluids, hard-surf. cleaners, leather dyeing, paints, coatings, printing inks, textile vat dyeing and printing, adhesives, antifreeze, floor waxes/polishes, insect repellents; solubilizer for dyes; plasticizer

Properties: APHA 50 max. liq.; mild, char. odor; m.w. 164.20; f.p. -55 C; b.p. 249 C (760 mm); misc. with water and many org. solvs.; sp.gr. 1.048; dens. 8.74 lb/gal; visc. 7.5 cP; ref. index 1.4381; flash pt. (COC) 118 C

Poly-Solv® TPM [Arch Chems./Specialty Polyols]

Chem. Descrip.: PPG-3 methyl ether

CAS 25498-49-1; EINECS/ELINCS 243-734-9

Uses: Solvent for brake fluids, hard-surf. cleaners, leather dyeing, paints, coatings, printing inks, textile vat dyeing and printing, adhesives, antifreeze, floor waxes/polishes, insect repellents; solubilizer for dyes; plasticizer

Properties: APHA 15 max. liq.; mild char. odor; m.w. 206.28; f.p. -78 C; b.p. 242 C (760 mm); misc. with water and many org. solvs.; sp.gr. 0.969; dens. 8.06 lb/gal; visc. 6.1 cP; ref. index 1.428; flash pt. (COC) 121 C

Polysolvan® O [Celanese]

Chem. Descrip.: Butyl glycolate

CAS 7397-62-8; EINECS/ELINCS 230-991-7

UN NA 1993

Uses: Paint additive, flow aid for nitrocellulose lacquers, brush-applied paints, alkyd resin and oil paints; high-boiling solvent in stoving finishes; mfg. of printing inks

Regulatory: SARA §311 acute health/fire hazard; EINECS, Canada, Australia, China, Korea, Philippines, Japan listed

Properties: Colorless liq.; very faint ester-like odor; misc. with common org. solvs.; sol. 20 g/l in water (20 C); dens. 1.015-1.023 g/cc (20 C); visc. 4.4 mPa·s (20 C); vapor pressure 1.33 mbar (20 C); f.p. ≈ -26 C; b.p. 82-97 C; flash pt. (PM) ≈ 75 C; ref. index 1.423-1.426 (20 C); pH neutral; sp. heat ≈ 2.1 kJ/kg °K (20 C); dielec. const. 13 (20 C); 97% min. act.

Toxicology: LD50 (oral, rat) 4595 mg/kg; causes respiratory tract/eye irritation; irritant to mucous membranes; risk of serious eye damage; may cause skin irritation; harmful by inh. (may cause CNS depression, nausea, dizziness, headache, stupor, unconsciousness, nasal discharge, hoarseness, coughing, chest pain, breathing difficulty); essentially nontoxic by ing., may cause nausea, vomiting, loss of appetite, GI irritation, diarrhea; TSCA listed

Environmental: Readily biodeg.; low potential for bioaccumulation; EC50 (daphnia magna, 24 h) 280 ppm, (P. putida, 18 h) 2320 ppm; sl. toxic to aquatic species; prevent run-off to storm sewers and ditches leading to waterways

Precaution: LFL in air 1.3 vol.%; may generate static charge; ground/bond equip. during transfer; avoid heat, sparks, open flame, ignition sources; incompat. with sulfuric and other strong inorg. acids, aluminum, lead, oxidizing agents, peroxides, nitric acid, perchloric acid, chromium trioxide

Hazardous decomp. prods.: Thermal decomp. or combustion: CO_x

NFPA: Health 0, Flammability 0, Reactivity 2

Storage: Store away from incompat. materials, direct sunlight; keep containers tightly closed when not in use; open containers slowly to allow excess pressure to vent

PPG 400 [Huntsman]

Chem. Descrip.: PPG 400

Uses: Intermediate for mfg. of surfactants, ethers, esters; metalworking lubricants; solvent for paints/varnishes, printing inks, veg. oils; textile lubricants

Properties: APHA 100 color; sol. in polar org. solvs.; sp.gr. 1.009; visc. 82 cps; hyd. no. 220-250; pour pt. -26 C; flash pt. (OC) 204 C; ref. index 1.445; pH 5-8 (1% aq. disp.); surf. tens. 31.1 dynes/cm; 99% act.

Environmental: Biodeg.

PPG 1650 [Huntsman]

Chem. Descrip.: PPG 1650

Uses: Intermediate for mfg. of surfactants, ethers, esters; antifoam for ceramics, rubber; lubricant, softener for leather; metalworking lubricants; demulsifier for petrol. industry; plasticizer for plastics; solvent for cosmetics

Properties: Sol. in polar org. solvs.; sp.gr. 1.006; visc. 330 cps; hyd. no. 67-71; pour pt. -32 C; flash pt. (OC) 230 C; ref. index 1.449; pH 5-8 (1% aq. disp.); surf. tens. 31.8 dynes/cm; 99% act.

Environmental: Biodeg.

PPG 2250 [Huntsman]

Chem. Descrip.: PPG 2250

Uses: Intermediate for mfg. of surfactants, ethers, esters; hydraulic brake fluids; cosmetics; dyeing; lubricant, softener for leather; metalworking lubricant; demulsifier for petrol.; latex coagulant; rubber release agent; solvent for veg. oils

Properties: APHA 50 color; sol. in polar org. solvs.; sp.gr. 1.005; visc. 482 cps; hyd. no. 47-54; pour pt. -29 C; flash pt. (OC) 240 C; ref. index

1.449; pH 4.0-7.5 (10% in aq. methanol); surf. tens. 32.1 dynes/cm; 99% act.
Environmental: Biodeg.

PPG Perchlor [PPG Ind.]
Chem. Descrip.: Perchlorethylene
CAS 127-18-4; EINECS/ELINCS 204-825-9
Uses: Solvent for dry cleaning (unique stabilizer components are completely removed from garments, leave no residual odor, are recoverable with solvent after distilling); degreasing and general solvent for vapor degreasing, cold cleaning, or dry-cleaning applics.
Features: Also avail. as heavy duty grade which is more heavily stabilized and intended for use in more rigorous vapor degreasing applics.
Properties: Water-wh. clear liq.; m.w. 165.85; f.p. -22.3 C; b.p. 121.1 C; sol. 0.015 g sol./100 g water; sp.gr. 1.623-1.628; dens. 13.57 lb/gal; visc. 0.88 cps; ref. index 1.5053.

PPG Trichlor [PPG Ind.]
Chem. Descrip.: Trichloroethylene
CAS 79-01-6; EINECS/ELINCS 201-167-4
Uses: Solvent for general heavy-duty vapor degreasing and cold cleaning applics., LOX flushing, vapor degreasing, and cleaning electronic components and synthesizing chemicals
Features: Avail. in standard grade, degreasing and general solvent grade, stabilized
Properties: Water-wh. clear liq.; sweet odor; m.w. 131.40; f.p. -86.4 C; b.p. 86.9 C; sol. 0.11 g sol./100 g water; sp.gr. 1.460-1.470; dens. 12.2 lb/gal; visc. 0.58 cps; ref. index 1.4782; pH 6.7-7.5.

Pricerine™ 9091 [Uniqema]
Chem. Descrip.: Vegetable grade glycerin
CAS 56-81-5; EINECS/ELINCS 200-289-5
Uses: Solvent, humectant, bodying agent for cosmetics, pharmaceuticals, foods, polyurethane
Properties: APHA 10 color; odorless; sp.gr. 1.262; 99.5% min. act.

Pricerine™ 9099 [Uniqema]
Chem. Descrip.: Glycerin
CAS 56-81-5; EINECS/ELINCS 200-289-5
Uses: Solvent, humectant, bodying agent for cosmetics, pharmaceuticals, foods, polyurethane
Properties: APHA 10 color; odorless; sp.gr. 1.263; 99.7% min. act.

Procetyl 10 [Croda Inc; Croda Chem. Ltd; Croda Oleochems.]
Chem. Descrip.: PPG-10 cetyl ether
CAS 9035-85-2
Uses: Emollient, coupling agent, cosolvent, plasticizer, superfatting agent, wetting agent, spreading agent, penetrant, lubricant in cosmetics and personal care prods., alcoholic and aq. alcoholic compositions
Properties: APHA 150 max. clear liq.; faint, char. sweet odor; sol. in min. oil, acetone, IPM, lanolin oil, alcohol; acid no. 1 max.; hyd. no. 80-100; pH 6.0-7.5 (3% disp.); nonionic; 100% act.
Use Level: 5-30%
Toxicology: Nontoxic

Procetyl AWS [Croda Inc; Croda Oleochems.]
Chem. Descrip.: PPG-5-ceteth-20
CAS 9087-53-0
Uses: Emulsifier, plasticizer, coupling agent, humectant, dispersant, emollient, fragrance solubilizer in aq. and aq. alcoholic systems, personal care prods., antiperspirants, bath oils; solvent, emollient, and humectant in topical pharmaceuticals
Properties: Water-wh. turbid oily liq.; char. sweet odor; sol. in water, alcohol, IPM, PEG-200, glycerin, propylene glycol, lanolin oil; acid no. 1 max.; hyd. no. 35-50; HLB 16.0; pH 6.0-7.5 (30% aq.); nonionic; 100% act.
Use Level: 0.5-5%

Prochinor® SOL [Ceca SA]
Uses: Corrosion inhibitor, demulsifier, pour pt. depressant, paraffin inhibitor, paraffin solvent, bactericide, scale inhibitor, oxygen scavenger, antifoam for oil and gas prod.; coagulant, flocculant for water treatment; fuel additive, process additive in refining

Proglyde DMM [Dow]
Chem. Descrip.: Dipropylene glycol dimethyl ether
CAS 111109-77-4; EINECS/ELINCS 404-640-5
Uses: Solvent, coupling agent, coalescing agent, water remover for coatings (PU coatings, latexes, solv.-based), agric., resins, cleaning prods., electronics cleaners, paint removers; diluent for thixotropic systems; in extraction processes; formulation stabilizer; aprotic reaction medium for polymerization reactions
Regulatory: SARA §311/312 fire hazard, §313 nonreportable; Canada DSL, EINECS, Japan listed
Properties: Colorless liq.; sl. solvent odor; sol. 35% in water; m.w. 162; sp.gr. 0.904 (20/20 C); visc. 1.1 cps; vapor pressure 0.82 mm Hg; f.p. < -71 C; b.p. 175 C; flash pt. (PMCC) 65 C; autoignition temp. 165 C; ref. index 1.407; surf. tens. 26.8 dynes/cm; sp. heat 0.43 cal/gl; VOC 898 g/l; > 99% act.
Toxicology: LD50 (oral, rat) 3300 mg/kg, (skin, rat) > 2000 mg/kg; low acute toxicity by oral, dermal, or inh. routes; may cause mod. eye irritation; may rarely cause allergic skin response; swallowing larger amts. may cause injury; may cause sedation at very high dose levels; may cause liver/kidney/adrenal effects; nonmutagenic
Environmental: Biodeg.; low bioconcentration potential; LC50 (daphnia magna) > 1000 mg/l; pract. nontoxic to aquatic organisms; prevent entry into sewers, storm drains, surf. waters, soil
Precaution: Combustible liq. and vapor; LFL 0.85 vol.%; vapors are heavier than air, may accumulate in low areas, may flash back; vapors can form flamm. mixts. if sl. heated; will produce flamm. vapors above flash pt.; spontaneous combustion can occur with prolonged exposure to high surf. area materials; incompat. with oxidizing materials; keep away from heat, spark, flames
Hazardous decomp. prods.: Combustion: CO_2, CO
Storage: Keep containers closed when not in use

Promyr [Amerchol]
Chem. Descrip.: Isopropyl myristate
CAS 110-27-0; EINECS/ELINCS 203-751-4
Uses: Emollient, solvent for cosmetics, toiletries, makeup, topical pharmaceuticals
Features: Nongreasy rub-in
Properties: Water-wh. low visc. liq.; odorless; oil-sol.; acid no. 1.0 max.; sapon. no. 202-211.

Promyristyl PM-3 [Croda Inc; Croda Chem. Ltd; Croda Oleochems.]
Chem. Descrip.: PPG-3 myristyl ether
CAS 63793-60-2
Uses: Solubilizer, coupling agent, emollient for clear analgesic, deodorant, and fragrance sticks, bath oils, insect repellents, sunscreen preps., facial cleansers, makeup, hair and skin care prods.; cosolvent for many fatty compds.; modifier; extender; vehicle; valve lubricant and emollient in conventional sprays and lotions
Features: Low-visc.
Properties: Gardner 1 max. clear liq.; sol. in ethanol, min. oil, IPM, veg. oil, oleyl alcohol, castor oil, lanolin oil; essentially sol. in volatile silicone fluid; insol. in water, glycerin; acid no. 1 max.; hyd. no. 145-160; nonionic
Use Level: 5-15%

Propal NF [Amerchol]
Chem. Descrip.: Isopropyl palmitate
CAS 142-91-6; EINECS/ELINCS 205-571-1
Uses: Emollient, solvent for cosmetics, toiletries, makeup
Features: Nongreasy rub in
Properties: Water-wh. liq.; odorless; acid no. 1.0 max.; sapon. no. 182-191.

Propsolv I-1 190 Proof [Equistar]
Chem. Descrip.: Ethyl alcohol (84.3%), ethyl acetate (4.5%), methanol (3.6%), light aliphatic solvent naphtha (0.8%), denatonium benzoate (0.001%), water (6.9%)
UN 1170
Uses: Solvent for shellac, chemical specialties, latex coagulants
Regulatory: SARA §311/312 acute health/chronic health/fire hazard, §313 reportable (methanol); Calif. Prop. 65 reportable (ethanol, benzene); TSCA export notification required (ethyl acetate)
Properties: Colorless liq.; invisible vapor; sweet alcohol-like odor; sol. 64 ml/100 ml in water; sp.gr. 0.8168-0.8179 (60/60 F), 1.59 (vapor); vapor pressure 44.6 mm Hg; f.p. -144 C; b.p. 78.4 C; flash pt. (TCC) 16 C; autoignition temp. 363 C; ref. index 1.3606
Toxicology: LD50 (oral, rat) 7060 mg/kg; may be irritating to eyes, upper respiratory tract; short-term exposure above 1000 ppm by inh. may cause CNS effects, headache, eye/nose/throat irritation; exposure > 1 h may cause dizziness, drowsiness, loss of appetite, inability to conc., GI effects, nausea, vomiting; prolonged/repeated skin contact may cause dermatitis; ing. may cause drunkenness, CNS depression, nausea, vom-

iting, diarrhea, liver damage, death; TSCA listed
Environmental: On land, apt to volatilize, biodegrade, or leach to ground water; in water, photolysis, oxidation, biodegradation expected to occur; not expected to bioconcentrate in aquatic organisms; LC50 (rainbow trout, 96 h) 10,400 ppm; prevent entry into sewers
Precaution: Flamm. liq.; flamm. limits 3.3-19 vol.%; material can burn with little or no visible flame; vapors are heavier than air, may flash back; prevent entry into basements, confined areas; ground and bond containers; avoid static discharges; avoid strong oxidizers, excessive heat, sparks, open flame; contact with acetyl chloride or other oxidizing agents may result in violent reaction
Hazardous decomp. prods.: Combustion: CO
HMIS: Health 1, Flammability 3, Reactivity 0
Storage: 1 yr shelf life; store under ambient conditions out of direct sunlight

Propsolv I-1 Anhydrous [Equistar]
Chem. Descrip.: Ethyl alcohol (90.9%), ethyl acetate (4.6%), methyl alcohol (3.7%), light aliphatic solvent naphtha (0.8%), denatonium benzoate (trace)
UN 1170
Uses: Solvent for shellac, chemical specialties, latex coagulants
Regulatory: FDA 27CFR §20.113; SARA §311/312 acute health/chronic health/fire hazard, §313 reportable (methanol); Calif. Prop. 65 reportable (ethanol, benzene); TSCA export notification required (ethyl acetate)
Properties: Colorless liq.; invisible vapor; mild sweet alcohol-like nonresidual odor; sol. in water; sp.gr. 0.7902-0.7911 (77/77 F), 1.59 (vapor); dens. 6.635-6.643 lb/gal (60 F); vapor pressure 44.6 mm Hg; f.p. -144 C; b.p. 78.4 C; flash pt. (TCC) 16 C; autoignition temp. 363 C
Toxicology: May be irritating to eyes, upper respiratory tract; short-term exposure above 1000 ppm by inh. may cause CNS effects, headache, eye/nose/throat irritation; exposure > 1 h may cause dizziness, drowsiness, loss of appetite, inability to conc., GI effects, nausea, vomiting; prolonged/repeated skin contact may cause dermatitis; ing. may cause drunkenness, CNS depression, nausea, vomiting, diarrhea, liver damage, death; TSCA listed
Environmental: On land, apt to volatilize, biodegrade, or leach to ground water; in water, photolysis, oxidation, biodegradation expected to occur; not expected to bioconcentrate in aquatic organisms; LC50 (rainbow trout, 96 h) 10,400 ppm; prevent entry into sewers
Precaution: Flamm. liq.; flamm. limits 3.3-19 vol.%; material can burn with little or no visible flame; vapors are heavier than air, may flash back; prevent entry into basements, confined areas; ground and bond containers; avoid static discharges; avoid strong oxidizers, excessive heat, sparks, open flame; contact with acetyl chloride or other oxidizing agents may result in violent reaction
Hazardous decomp. prods.: Combustion: CO
HMIS: Health 1, Flammability 3, Reactivity 0

Propsolv III-1 190 Proof [Equistar]
Chem. Descrip.: Ethyl alcohol (86.5%), methyl alcohol (3.7%), MIBK (1.0%), ethyl acetate (1.0%), light aliphatic solvent naphtha (0.8%), denatonium benzoate (0.001% max.), water (7.1%)
UN 1170
Uses: Solvent for shellac, chemical specialties, latex coagulants
Regulatory: SARA §311/312 acute health/chronic health/fire hazard, §313 reportable (methanol, MIBK); Calif. Prop. 65 reportable (ethanol)
Properties: Colorless liq.; invisible vapor; sweet alcohol-like odor; sol. in water; sp.gr. 0.8136-0.8148 (60/60 F), 1.59 (vapor); vapor pressure 44.6 mm Hg; f.p. -144 C; b.p. 78.4 C; flash pt. (TCC) 16 C; autoignition temp. 363 C; ref. index 1.3618
Toxicology: May be irritating to eyes, upper respiratory tract; short-term exposure above 1000 ppm by inh. may cause CNS effects, headache, eye/nose/throat irritation; exposure > 1 h may cause dizziness, drowsiness, loss of appetite, inability to conc., GI effects, nausea, vomiting; prolonged/repeated skin contact may cause dermatitis; ing. may cause drunkenness, CNS depression, nausea, vomiting, diarrhea, liver damage, death; TSCA listed
Environmental: On land, apt to volatilize, biodegrade, or leach to ground water; in water, photolysis, oxidation, biodegradation expected to occur; not expected to bioconcentrate in aquatic organisms; LC50 (rainbow trout, 96 h) 10,400 ppm; prevent entry into sewers
Precaution: Flamm. liq.; flamm. limits 3.3-19 vol.%; material can burn with little or no visible flame; vapors are heavier than air, may flash back; prevent entry into basements, confined areas; ground and bond containers; avoid static discharges; avoid strong oxidizers, excessive heat, sparks, open flame; contact with acetyl chloride or other oxidizing agents may result in violent reaction
Hazardous decomp. prods.: Combustion: CO
HMIS: Health 1, Flammability 3, Reactivity 0
Storage: 1 yr shelf life; store under ambient conditions out of direct sunlight

Propsolv III-1 Anhydrous [Equistar]
Chem. Descrip.: Ethyl alcohol (93.5%), methyl alcohol (3.8%), MIBK (1.0%), ethyl acetate (1.0%), light aliphatic solvent naphtha (0.9%), denatonium benzoate (0.0012%)
UN 1170
Uses: Solvent for shellac, chemical specialties, latex coagulants
Regulatory: SARA §311/312 acute health/chronic health/fire hazard, §313 reportable (methanol, MIBK); Calif. Prop. 65 reportable (ethanol, benzene); TSCA export notification required (ethyl acetate, MIBK)
Properties: Colorless liq.; invisible vapor; sweet alcohol-like odor; sol. in water; sp.gr. 0.7932-0.7942 (60/60 F), 1.59 (vapor); vapor pressure 44.6 mm Hg; f.p. -144 C; b.p. 78.4 C; flash pt. (TCC) 16 C; autoignition temp. 363 C; ref. index 1.3590
Toxicology: May be irritating to eyes, upper respiratory tract; short-term exposure above 1000 ppm by inh. may cause CNS effects, headache, eye/nose/throat irritation; exposure > 1 h may cause dizziness, drowsiness, loss of appetite, inability to conc., GI effects, nausea, vomiting; prolonged/repeated skin contact may cause dermatitis; ing. may cause drunkenness, CNS depression, nausea, vomiting, diarrhea, liver damage, death; TSCA listed
Environmental: On land, apt to volatilize, biodegrade, or leach to ground water; in water, photolysis, oxidation, biodegradation expected to occur; not expected to bioconcentrate in aquatic organisms; LC50 (rainbow trout, 96 h) 10,400 ppm; prevent entry into sewers
Precaution: Flamm. liq.; flamm. limits 3.3-19 vol.%; material can burn with little or no visible flame; vapors are heavier than air, may flash back; prevent entry into basements, confined areas; ground and bond containers; avoid static discharges; avoid strong oxidizers, excessive heat, sparks, open flame; contact with acetyl chloride or other oxidizing agents may result in violent reaction
Hazardous decomp. prods.: Combustion: CO
HMIS: Health 1, Flammability 3, Reactivity 0
Storage: 1 yr shelf life; store under ambient conditions out of direct sunlight

Propsolv III-2 190 Proof [Equistar]
Chem. Descrip.: Ethyl alcohol (85.7%), methyl alcohol (3.6%), MIBK (1.9%), ethyl acetate (1.0%), light aliphatic solvent naphtha (0.8%), water (7.0%)
UN 1170
Uses: Solvent for shellac, chemical specialties, latex coagulants
Regulatory: FDA 27CFR §20.113; SARA §311/312 acute health/chronic health/fire hazard, §313 reportable (methanol, MIBK); Calif. Prop. 65 reportable (ethanol)
Properties: Colorless liq.; invisible vapor; mild sweet alcohol-like nonresidual odor; sol. in water; sp.gr. 0.80770-0.8082 (77/77 F), 1.59 (vapor); dens. 6.776-6.786 lb/gal (60 F); vapor pressure 44.6 mm Hg; f.p. -144 C; b.p. 78.4 C; flash pt. (TCC) 16 C; autoignition temp. 363 C
Toxicology: May be irritating to eyes, upper respiratory tract; short-term exposure above 1000 ppm by inh. may cause CNS effects, headache, eye/nose/throat irritation; exposure > 1 h may cause dizziness, drowsiness, loss of appetite, inability to conc., GI effects, nausea, vomiting; prolonged/repeated skin contact may cause dermatitis; ing. may cause drunkenness, CNS depression, nausea, vomiting, diarrhea, liver damage, death; TSCA listed
Environmental: On land, apt to volatilize, biodegrade, or leach to ground water; in water, photolysis, oxidation, biodegradation expected to occur; not expected to bioconcentrate in aquatic organisms; LC50 (rainbow trout, 96 h) 10,400 ppm; prevent entry into sewers
Precaution: Flamm. liq.; flamm. limits 3.3-19 vol.%; material can burn with little or no visible flame; vapors are heavier than air, may flash back; prevent entry into basements, confined areas; ground and bond containers; avoid static discharges; avoid strong oxidizers, excessive heat, sparks, open flame; contact with acetyl chloride or other oxidizing agents

may result in violent reaction
Hazardous decomp. prods.: Combustion: CO
HMIS: Health 1, Flammability 3, Reactivity 0

Propsolv III-2 Anhydrous [Equistar]
Chem. Descrip.: Ethyl alcohol (92.6%), methyl alcohol (3.7%), MIBK (1.9%), ethyl acetate (1.0%), light aliphatic solvent naphtha (0.9%)
UN 1170
Uses: Solvent for shellac, chemical specialties, latex coagulants
Regulatory: FDA 27CFR §20.113; SARA §311/312 acute health/chronic health/fire hazard, §313 reportable (methanol, MIBK); Calif. Prop. 65 reportable (ethanol, benzene); TSCA export notification required (ethyl acetate, MIBK)
Properties: Colorless liq.; invisible vapor; mild sweet alcohol-like nonresidual odor; sol. in water; sp.gr. 0.7872-0.7882 (77/77 F), 1.59 (vapor); dens. 6.611-6.619 lb/gal (60 F); vapor pressure 44.6 mm Hg; f.p. -144 C; b.p. 78.4 C; flash pt. (TCC) 16 C; autoignition temp. 363 C
Toxicology: May be irritating to eyes, upper respiratory tract; short-term exposure above 1000 ppm by inh. may cause CNS effects, headache, eye/nose/throat irritation; exposure > 1 h may cause dizziness, drowsiness, loss of appetite, inability to conc., GI effects, nausea, vomiting; prolonged/repeated skin contact may cause dermatitis; ing. may cause drunkenness, CNS depression, nausea, vomiting, diarrhea, liver damage, death; TSCA listed
Environmental: On land, apt to volatilize, biodegrade, or leach to ground water; in water, photolysis, oxidation, biodegradation expected to occur; not expected to bioconcentrate in aquatic organisms; LC50 (rainbow trout, 96 h) 10,400 ppm; prevent entry into sewers
Precaution: Flamm. liq.; flamm. limits 3.3-19 vol.%; material can burn with little or no visible flame; vapors are heavier than air, may flash back; prevent entry into basements, confined areas; ground and bond containers; avoid static discharges; avoid strong oxidizers, excessive heat, sparks, open flame; contact with acetyl chloride or other oxidizing agents may result in violent reaction
Hazardous decomp. prods.: Combustion: CO
HMIS: Health 1, Flammability 3, Reactivity 0

Propulsol® [Solvay SA]
Chem. Descrip.: Methylene chloride
CAS 75-09-2; EINECS/ELINCS 200-838-9
Uses: Solvent in treatment of metals, pharmaceuticals

Propyl Cellosolve® [Dow]
Chem. Descrip.: Ethylene glycol propyl ether
CAS 2807-30-9; EINECS/ELINCS 220-548-6
UN 1993
Uses: Solvent for high-solids coatings (NC, vinyls, ethyl cellulose, butyrate, acrylic, water-sol.)
Regulatory: DOT nonregulated; SARA §311/312 acute health/chronic health/fire hazard, §313 reportable; Calif. Prop. 65 reportable; EINECS, Canada DSL listed
Properties: Colorless transparent liq.; mild ethereal odor; sol. in water (20 C); m.w. 104.15; sp.gr. 0.912 (20/20 C); dens. 7.59 lb/gal (20 C); visc. 3.8 cP (20 C); vapor pressure 1.6 mm Hg (20 C); f.p. -90 C; b.p. 151.5 C; flash pt. (PMCC) 57 C; VOC 909 g/l; 100% act.
Toxicology: REL TWA (8 h) 20 ppm; LD50 (oral, rat) 4.89 g/kg; harmful by inh., skin absorp.; causes skin/severe eye irritation; may cause red blood cell damage; inh. may cause respiratory tract irritation, nasal discomfort/discharge, chest pain, coughing, hemolysis, sec. kidney injury; may cause dizziness, drowsiness, hemolysis, sec. kidney injury by skin absorp.; mod. toxic by ing.; may cause incoordination, headache, nausea, vomiting, diarrhea, abdominal discomfort; TSCA listed
Environmental: BOD5 13%; COD 1.94 mg/mg; LC50 (fathead minnow, 96 h) > 5000 mg/l
Precaution: Flamm.; flamm. limits in air 1.3-15.8 vol.%; may produce floating fire hazard; incompat. with strong alkalis, high temps. in presence of strong bases, acids, strong oxidizing agents
Hazardous decomp. prods.: Combustion: CO, CO_2
HMIS: Health 2, Flammability 2, Reactivity 0
Storage: Keep away from heat, sparks, flame

Propylene Glycol USP, FCC [Eastman]
Chem. Descrip.: 1,2-Propanediol
CAS 57-55-6; EINECS/ELINCS 200-338-0
Uses: Solvent, humectant, preservative for foods, pet foods, snacks, pharmaceuticals, and cosmetics; plasticizer in cork seals or crowns for food pkg. and bottle cap liners; solvent for high speed steam-set inks; lubricant for machinery used in pkg. foodstuffs, pharmaceuticals, and beverages; antifreeze in refrigeration units
Regulatory: FDA 21CFR §184.1666; USDA approved; Kosher
Properties: Colorless clear visc. liq.; practically odorless; sl. char. taste; completely sol. in water, acetone, ethyl ether, and methanol; m.w. 76.10; sp.gr. 1.0381; dens. 8.64 lb/gal; visc. 60.5 cP (20 C); vapor pressure < 0.1 mm Hg (20 C); f.p. -60 C; b.p. 187 C; flash pt. (TCC) 101 C; ref. index 1.4326 (20 C); 99.5% purity
Toxicology: Low toxicity
Storage: Hygroscopic; overheating for prolonged periods may result in degradation of odor and taste

Propylene Glycol USP/FCC Ultra Grade [Eastman]
Chem. Descrip.: Propylene glycol USP/FCC
CAS 57-55-6; EINECS/ELINCS 200-338-0
Uses: Solvent, humectant, preservative for foods; plasticizer for cork seals for food pkg., in cellulose coatings for sausage and cheese; solvent for inks for food pkg.; lubricant for food pkg. equip.; antifreeze; solvent for org. chems. for medicine, water-sol. vitamins, dyes, perfumes; softener, spreading agent, emollient, humectant, preservative for cosmetics and pharmaceuticals; vehicle for drugs, ointments; dispensing aid in laundry detergents
Regulatory: FDA 21CFR §184.1666, GRAS; EPA exempt from pesticide tolerances; USDA accepted; kosher
Properties: Colorless clear visc. liq., essentially odorless, sl. char. taste; misc. with water, acetone, chloroform; sp.gr. 1.0351-1.0364; i.b.p. 186 C min.; 99.5% min. purity
Storage: Hygroscopic; storage in lined steel tank recommended; becomes quite visc. at low temps.

Prostearyl 15 [Croda Chem. Ltd; Croda Oleochems.]
Chem. Descrip.: PPG-15 stearyl ether
CAS 25231-21-4
Uses: Emollient, lubricant for cosmetics, bath oils, sunscreens, hair prods., aerosol antiperspirants, hand and body lotions; coupling agent for fragrances; cosolvent for perfumes and essential oils; superfatting agent in foam baths and hair/body shampoos
Properties: APHA 100 max. clear liq.; sol. in acetone, castor oil, ethanol, cetyl alcohol, ethyl acetate, IPM, liq. lanolin, oleic acid, oleyl alcohol, olive oil, volatile silicone oil, hot wh. petrol. jelly; insol. in water, propylene glycol; acid no. 1 max.; hyd. no. 60-70; nonionic; 0.2% max. water
Use Level: 5-15%
Toxicology: Low acute oral toxicity; essentially nonirritating to skin; contact of neat material with eyes may cause mild transient irritation

Protachem™ 100 [Protameen; S. Black]
Chem. Descrip.: Polyglyceryl-4 oleate
CAS 9007-48-1
Uses: Emollient, lubricant, solvent, aux. emulsifier, stabilizer, emulsion thickener, opacifier in personal care prods.
Features: Imparts dry, silky skinfeel; reduces greasy effect of oils and petrolatum
Properties: Nonionic

Protachem™ AWS-100 [Protameen]
Chem. Descrip.: PPG-5-ceteth-20
CAS 9087-53-0
Uses: Emollient, lubricant, solvent for personal care prods.
Features: Imparts dry, silky skinfeel; reduces greasy effect of oils and petrolatum
Properties: Nonionic

Protachem™ DGS [Protameen; S. Black]
Chem. Descrip.: PEG-2 stearate
CAS 106-11-6; EINECS/ELINCS 203-363-5
Uses: Emollient, lubricant, solvent, opacifier, pearlescent in personal care prods.
Features: Imparts dry, silky skinfeel; reduces greasy effect of oils and petrolatum
Properties: Nonionic

Protachem™ EGDS [Protameen; S. Black]
Chem. Descrip.: Glycol distearate
CAS 627-83-8; EINECS/ELINCS 211-014-3
Uses: Emollient, lubricant, solvent, opacifier, pearlescent in personal care prods.
Features: Imparts dry, silky skinfeel; reduces greasy effect of oils and

petrolatum
Properties: Nonionic

Protachem™ EGMS [Protameen; S. Black]
Chem. Descrip.: Glycol stearate
CAS 111-60-4; EINECS/ELINCS 203-886-9
Uses: Emollient, lubricant, solvent, opacifier, pearlescent in personal care prods.
Features: Imparts dry, silky skinfeel; reduces greasy effect of oils and petrolatum
Properties: Nonionic

Protachem™ G-5509 [Protameen; S. Black]
Chem. Descrip.: Polyglyceryl-9 stearate
Uses: Emollient, lubricant, solvent, aux. emulsifier, stabilizer, emulsion thickener, opacifier in personal care prods.
Features: Imparts dry, silky skinfeel; reduces greasy effect of oils and petrolatum
Properties: Nonionic

Protachem™ G-5566 [Protameen; S. Black]
Chem. Descrip.: Polyglyceryl-3 stearate
CAS 37349-34-1; EINECS/ELINCS 248-403-2
Uses: Emollient, lubricant, solvent, aux. emulsifier, stabilizer, emulsion thickener, opacifier in personal care prods.
Features: Imparts dry, silky skinfeel; reduces greasy effect of oils and petrolatum
Properties: Nonionic

Protachem™ GC-7 [Protameen]
Chem. Descrip.: PEG-7 glyceryl cocoate
CAS 68201-46-7
Uses: Emollient, lubricant, solvent, aux. emulsifier, stabilizer, emulsion thickener, opacifier in personal care prods.
Features: Imparts dry, silky skinfeel; reduces greasy effect of oils and petrolatum
Properties: Nonionic

Protachem™ GC-30 [Protameen]
Chem. Descrip.: PEG-30 glyceryl cocoate
CAS 68201-46-7
Uses: Emollient, lubricant, solvent, aux. emulsifier, stabilizer, emulsion thickener, opacifier in personal care prods.
Features: Imparts dry, silky skinfeel; reduces greasy effect of oils and petrolatum
Properties: Nonionic

Protachem™ GDL [Protameen]
Chem. Descrip.: Glyceryl dilaurate
CAS 27638-00-2; EINECS/ELINCS 248-586-9
Uses: Emollient, lubricant, solvent, aux. emulsifier, stabilizer, emulsion thickener, opacifier in personal care prods.
Features: Imparts dry, silky skinfeel; reduces greasy effect of oils and petrolatum
Properties: Nonionic

Protachem™ GMS-165 [Protameen; S. Black]
Chem. Descrip.: Glyceryl stearate, PEG-100 stearate
Uses: Emollient, lubricant, solvent, aux. emulsifier, stabilizer, emulsion thickener, opacifier in personal care prods.
Features: Imparts dry, silky skinfeel; reduces greasy effect of oils and petrolatum; acid-stable

Protachem™ GMS-450 [Protameen; S. Black]
Chem. Descrip.: Glyceryl stearate
CAS 123-94-4
Uses: Emollient, lubricant, solvent, aux. emulsifier, stabilizer, emulsion thickener, opacifier in personal care prods.
Features: Imparts dry, silky skinfeel; reduces greasy effect of oils and petrolatum
Properties: Nonionic

Protachem™ GMS-AS [Protameen]
Chem. Descrip.: Glyceryl stearate and sodium lauryl sulfate
Uses: Emollient, lubricant, solvent, aux. emulsifier, stabilizer, emulsion thickener, opacifier in personal care prods.
Features: Imparts dry, silky skinfeel; reduces greasy effect of oils and petrolatum

Protachem™ GMS-D [Protameen]
Chem. Descrip.: Glyceryl stearate SE
CAS 31566-31-1
Uses: Emollient, lubricant, solvent, aux. emulsifier, stabilizer, emulsion thickener, opacifier in personal care prods.
Features: Imparts dry, silky skinfeel; reduces greasy effect of oils and petrolatum
Properties: Anionic

Protachem™ GMS-T [Protameen]
Chem. Descrip.: Glyceryl stearate SE
CAS 31566-31-1
Uses: Emollient, lubricant, solvent, aux. emulsifier, stabilizer, emulsion thickener, opacifier in personal care prods.
Features: Imparts dry, silky skinfeel; reduces greasy effect of oils and petrolatum
Properties: Anionic

Protachem™ HMS [Protameen; S. Black]
Chem. Descrip.: Glyceryl stearate
CAS 123-94-4
Uses: Emollient, lubricant, solvent, aux. emulsifier, stabilizer, emulsion thickener, opacifier in personal care prods.
Features: Imparts dry, silky skinfeel; reduces greasy effect of oils and petrolatum
Properties: Nonionic

Protachem™ MLD [Protameen; S. Black]
Chem. Descrip.: Glyceryl laurate
CAS 142-18-7; EINECS/ELINCS 205-526-6
Uses: Emollient, lubricant, solvent, aux. emulsifier, stabilizer, emulsion thickener, opacifier in personal care prods.
Features: Imparts dry, silky skinfeel; reduces greasy effect of oils and petrolatum
Properties: Nonionic

Protachem™ PGMS [Protameen; S. Black]
Chem. Descrip.: Propylene glycol monostearate
CAS 1323-39-3; EINECS/ELINCS 215-354-3
Uses: Emollient, lubricant, solvent for personal care prods.
Features: Imparts dry, silky skinfeel; reduces greasy effect of oils and petrolatum
Properties: Nonionic

Protachem™ SAS [Protameen]
Chem. Descrip.: Glycol stearate and stearamide AMP
Uses: Emollient, lubricant, solvent for personal care prods.
Features: Imparts dry, silky skinfeel; reduces greasy effect of oils and petrolatum

Protene® SC [Atofina SA]
Chem. Descrip.: Hydrofluorocarbon-based
Uses: Degreasing solvent for cold degreasing operations, high speed, manual precision, and hot zone cleaning
Features: Non-ozone depleting; no flash pt.; fast drying; can be used as an aerosol

Prothene [Amity UK]
Chem. Descrip.: Blend of isoparaffinic and nonaromatic hydrocarbons, 1-propoxy-2-propanol (10-20%)
UN 1192; UN 1993
Uses: Solvent cleaner for electronic components, engineering, fabrication (flux and adhesive remover), printing (excess ink remover), coatings (excess resin remover)
Features: Able to remove both oily and water-sol. contaminants; evaporates to leave no residue
Properties: Colorless liq.; mild odor; misc. with most org. solvs.; sol. 20% in water @ 20 C; visc. 2.5 cSt; vapor pressure 2.2 mbar (20 C), 172 mbar (100 C); b.p. > 150 C; flash pt. (OC) 48 C
Toxicology: LD50 (oral, mouse) 2500 mg/kg; LD50 (inh., rat, 8 h) 5400 mg/m^3; may cause irritation in susceptible persons; may degrease skin, cause eye irritation
Environmental: Readily biodeg.; BOC 28 = 1.43 mg O_2/mg; COD = 1.66 mg O_2/mg; bioaccumulation is unlikely; EC50 (daphnia, 48 h) 683 mg/l; LC50 (fish, 48 h) 320 mg/l; EC50 (algae) 2200 mg/l; do not allow runoff to sewer, waterway, or ground
Precaution: Flamm.; pressure may build up; vapors may form explosive mixts. with air; flamm. limits 1-10%; hydrolyzes in presence of water, acids, bases; incompat. with strong oxidizing agents, acids, bases
Hazardous decomp. prods.: CO_x; thermal decomp. can produce irritating gases and vapors
Storage: Store in tightly closed original container; keep away from heat,

sparks, open flame; ground and bond all equip.

Punctilious® CDA 19-3 190 Proof [Equistar]

Chem. Descrip.: Completely denatured alcohol 19-3: ethanol (88.2%), MIBK (3.8%), light aliphatic solvent naphtha (0.8%), water (7.2%)

UN 1170

Uses: Solvent

Regulatory: SARA §311/312 acute health/chronic health/fire hazard, §313 reportable (MIBK); Calif. Prop. 65 reportable (ethanol, benzene); TSCA export notification required (MIBK)

Properties: Colorless liq.; invisible vapor; sweet alcohol-like odor; sol. in water; sp.gr. 0.8059-0.8089 (77/77 F), 1.59 (vapor); vapor pressure 44.6 mm Hg; f.p. -144 C; b.p. 78.4 C; flash pt. (TCC) 16 C; autoignition temp. 363 C; ref. index 1.3642

Toxicology: May be irritating to eyes, upper respiratory tract; short-term exposure above 1000 ppm by inh. may cause CNS effects, headache, eye/nose/throat irritation; exposure > 1 h may cause dizziness, drowsiness, loss of appetite, inability to conc., GI effects, nausea, vomiting; prolonged/repeated skin contact may cause dermatitis; ing. may cause drunkenness, CNS depression, nausea, vomiting, diarrhea, liver damage, death

Environmental: On land, apt to volatilize, biodegrade, or leach to ground water; in water, photolysis, oxidation, biodegradation expected to occur; not expected to bioconcentrate in aquatic organisms; LC50 (rainbow trout, 96 h) 10,400 ppm; prevent entry into sewers

Precaution: Flamm. liq.; flamm. limits 3.3-19 vol.%; material can burn with little or no visible flame; vapors are heavier than air, may flash back; prevent entry into basements, confined areas; ground and bond containers; avoid static discharges; avoid strong oxidizers, excessive heat, sparks, open flame; contact with acetyl chloride or other oxidizing agents may result in violent reaction

Hazardous decomp. prods.: Combustion: CO

HMIS: Health 1, Flammability 3, Reactivity 0

Storage: 1 yr shelf life; store under ambient conditions out of direct sunlight

Punctilious® CDA 19-3 Anhydrous A [Equistar]

Chem. Descrip.: Completely denatured alcohol 19-3: ethanol (95.3%), MIBK (3.9%), light aliphatic solvent naphtha (0.8%)

UN 1170

Uses: Solvent

Regulatory: SARA §311/312 acute health/chronic health/fire hazard, §313 reportable (MIBK); Calif. Prop. 65 reportable (ethanol, benzene); TSCA export notification required (MIBK)

Properties: Colorless liq.; invisible vapor; sweet alcohol-like odor; sol. in water; sp.gr. 0.7851-0.7881 (77/77 F), 1.59 (vapor); dens. 6.603-6.609 lb/gal (60 F); vapor pressure 44.6 mm Hg; f.p. -144 C; b.p. 78.4 C; flash pt. (TCC) 13 C; autoignition temp. 363 C; ref. index 1.3642

Toxicology: May be irritating to eyes, upper respiratory tract; short-term exposure above 1000 ppm by inh. may cause CNS effects, headache, eye/nose/throat irritation; exposure > 1 h may cause dizziness, drowsiness, loss of appetite, inability to conc., GI effects, nausea, vomiting; prolonged/repeated skin contact may cause dermatitis; ing. may cause drunkenness, CNS depression, nausea, vomiting, diarrhea, liver damage, death

Environmental: On land, apt to volatilize, biodegrade, or leach to ground water; in water, photolysis, oxidation, biodegradation expected to occur; not expected to bioconcentrate in aquatic organisms; LC50 (rainbow trout, 96 h) 10,400 ppm; prevent entry into sewers

Precaution: Flamm. liq.; flamm. limits 3.3-19 vol.%; material can burn with little or no visible flame; vapors are heavier than air, may flash back; prevent entry into basements, confined areas; ground and bond containers; avoid static discharges; avoid strong oxidizers, excessive heat, sparks, open flame; contact with acetyl chloride or other oxidizing agents may result in violent reaction

Hazardous decomp. prods.: Combustion: CO

HMIS: Health 1, Flammability 3, Reactivity 0

Storage: 1 yr shelf life; store under ambient conditions out of direct sunlight

Punctilious® Ethyl Alcohol 190 Proof Pure USP [Equistar]

Chem. Descrip.: Ethyl alcohol (92.5%), water (7.5%)

CAS 64-17-5; EINECS/ELINCS 200-578-6

UN 1170

Uses: Solvent for hair spray, perfume, surfactants, detergents, pharmaceuticals, agric., cleaning sol'ns., disinfectants, explosives, plastics, polishes, shellac, stains, textiles, adhesives; extraction medium

Regulatory: SARA §311/312 acute health/chronic health/fire hazard, §313 nonreportable; Calif. Prop. 65 reportable (ethanol, benzene)

Properties: Colorless liq.; invisible vapor; sweet alcohol-like odor; sol. in methanol, water; misc. in ether, chloroform; sp.gr. 0.8090-0.8094 (77/77 F), 1.59 (vapor); vapor pressure 44.6 mm Hg; f.p. -144 C; b.p. 78.4 C; flash pt. (TCC) 16 C; autoignition temp. 363 C; ref. index 1.3619

Toxicology: LD50 (oral, rat) 7060 mg/kg; may be irritating to eyes, upper respiratory tract; short-term exposure above 1000 ppm by inh. may cause CNS effects, headache, eye/nose/throat irritation; exposure > 1 h may cause dizziness, drowsiness, loss of appetite, inability to conc., GI effects, nausea, vomiting; prolonged/repeated skin contact may cause dermatitis; ing. may cause drunkenness, CNS depression, nausea, vomiting, diarrhea, liver damage, death; TSCA listed

Environmental: On land, apt to volatilize, biodegrade, or leach to ground water; in water, photolysis, oxidation, biodegradation expected to occur; not expected to bioconcentrate in aquatic organisms; LC50 (rainbow trout, 96 h) 10,400 ppm; prevent entry into sewers

Precaution: Flamm. liq.; flamm. limits 3.3-19 vol.%; material can burn with little or no visible flame; vapors are heavier than air, may flash back; prevent entry into basements, confined areas; ground and bond containers; avoid static discharges; avoid strong oxidizers, excessive heat, sparks, open flame; contact with acetyl chloride or other oxidizing agents may result in violent reaction

Hazardous decomp. prods.: Combustion: CO

HMIS: Health 1, Flammability 3, Reactivity 0

Storage: 1 yr shelf life; store under ambient conditions out of direct sunlight

Punctilious® Ethyl Alcohol 200 Proof Pure USP [Equistar]

Chem. Descrip.: Ethyl alcohol (100%)

CAS 64-17-5; EINECS/ELINCS 200-578-6

UN 1170

Uses: Solvent for hair spray, perfume, surfactants, detergents, pharmaceuticals, agric., cleaning sol'ns., disinfectants, explosives, plastics, polishes, shellac, stains, textiles, adhesives; extraction medium

Regulatory: SARA §311/312 acute health/chronic health/fire hazard, §313 nonreportable; Calif. Prop. 65 reportable (ethanol, benzene)

Properties: Colorless liq.; invisible vapor; sweet alcohol-like odor; sol. in methanol, water; misc. in ether, chloroform; sp.gr. 0.7876 max. (77/77 F), 1.59 (vapor); vapor pressure 44.6 mm Hg; f.p. -144 C; b.p. 78.4 C; flash pt. (TCC) 13 C; autoignition temp. 363 C; ref. index 1.3596

Toxicology: LD50 (oral, rat) 7060 mg/kg; may be irritating to eyes, upper respiratory tract; short-term exposure above 1000 ppm by inh. may cause CNS effects, headache, eye/nose/throat irritation; exposure > 1 h may cause dizziness, drowsiness, loss of appetite, inability to conc., GI effects, nausea, vomiting; prolonged/repeated skin contact may cause dermatitis; ing. may cause drunkenness, CNS depression, nausea, vomiting, diarrhea, liver damage, death; TSCA listed

Environmental: On land, apt to volatilize, biodegrade, or leach to ground water; in water, photolysis, oxidation, biodegradation expected to occur; not expected to bioconcentrate in aquatic organisms; LC50 (rainbow trout, 96 h) 10,400 ppm; prevent entry into sewers

Precaution: Flamm. liq.; flamm. limits 3.3-19 vol.%; material can burn with little or no visible flame; vapors are heavier than air, may flash back; prevent entry into basements, confined areas; ground and bond containers; avoid static discharges; avoid strong oxidizers, excessive heat, sparks, open flame; contact with acetyl chloride or other oxidizing agents may result in violent reaction

Hazardous decomp. prods.: Combustion: CO

HMIS: Health 1, Flammability 3, Reactivity 0

Storage: 1 yr shelf life; store under ambient conditions out of direct sunlight

Punctilious® SDA 1-1 190 Proof [Equistar]

Chem. Descrip.: Specially denatured alcohol 1-1: ethanol (89.0%), methanol (3.8%), denatonium benzoate (0.0011%), water (7.3%)

UN 1170

Uses: Solvent, thinner for external toiletries, cosmetics, pharmaceuticals, detergents

Regulatory: SARA §311/312 acute health/chronic health/fire hazard, §313 reportable (methanol); Calif. Prop. 65 reportable (ethanol, benzene)

Properties: Colorless liq.; invisible vapor; sweet alcohol-like odor; sol. in water; sp.gr. 0.8142 (60 F), 1.59 (vapor); vapor pressure 44.6 mm Hg; f.p. -144 C; b.p. 78.4 C; flash pt. (TCC) 56 F; autoignition temp. 363 C; ref. index 1.3611

Toxicology: May be irritating to eyes, upper respiratory tract; short-term exposure above 1000 ppm by inh. may cause CNS effects, headache, eye/nose/throat irritation; exposure > 1 h may cause dizziness, drowsiness, loss of appetite, inability to conc., GI effects, nausea, vomiting; prolonged/repeated skin contact may cause dermatitis; ing. may cause drunkenness, CNS depression, nausea, vomiting, diarrhea, liver damage, death; TSCA listed

Environmental: On land, apt to volatilize, biodegrade, or leach to ground water; in water, photolysis, oxidation, biodegradation expected to occur; not expected to bioconcentrate in aquatic organisms; LC50 (rainbow trout, 96 h) 10,400 ppm; prevent entry into sewers

Precaution: Flamm. liq.; flamm. limits 3.3-19 vol.%; material can burn with little or no visible flame; vapors are heavier than air, may flash back; prevent entry into basements, confined areas; ground and bond containers; avoid static discharges; avoid strong oxidizers, excessive heat, sparks, open flame; contact with acetyl chloride or other oxidizing agents may result in violent reaction

Hazardous decomp. prods.: Combustion: CO

HMIS: Health 1, Flammability 3, Reactivity 0

Storage: 1 yr shelf life; store under ambient conditions out of direct sunlight away from ignition sources

Punctilious® SDA 1-1 Anhydrous [Equistar]

Chem. Descrip.: Specially denatured alcohol 1-1: ethanol (96.1%), methanol (3.9%), denatonium benzoate (0.0012%)

UN 1170

Uses: Solvent, thinner for external toiletries, cosmetics, pharmaceuticals, detergents

Regulatory: SARA §311/312 acute health/chronic health/fire hazard, §313 reportable (methanol); Calif. Prop. 65 reportable (ethanol, benzene)

Properties: Colorless liq.; invisible vapor; sweet alcohol-like odor; sol. in methanol, water; misc. in ether, chloroform; sp.gr. 0.7868-0.7879 (77/77 F), 1.59 (vapor); vapor pressure 44.6 mm Hg; f.p. -144 C; b.p. 78.4 C; flash pt. (TCC) 13 C; autoignition temp. 363 C; ref. index 1.3584

Toxicology: May be irritating to eyes, upper respiratory tract; short-term exposure above 1000 ppm by inh. may cause CNS effects, headache, eye/nose/throat irritation; exposure > 1 h may cause dizziness, drowsiness, loss of appetite, inability to conc., GI effects, nausea, vomiting; prolonged/repeated skin contact may cause dermatitis; ing. may cause drunkenness, CNS depression, nausea, vomiting, diarrhea, liver damage, death; TSCA listed

Environmental: On land, apt to volatilize, biodegrade, or leach to ground water; in water, photolysis, oxidation, biodegradation expected to occur; not expected to bioconcentrate in aquatic organisms; LC50 (rainbow trout, 96 h) 10,400 ppm; prevent entry into sewers

Precaution: Flamm. liq.; flamm. limits 3.3-19 vol.%; material can burn with little or no visible flame; vapors are heavier than air, may flash back; prevent entry into basements, confined areas; ground and bond containers; avoid static discharges; avoid strong oxidizers, excessive heat, sparks, open flame; contact with acetyl chloride or other oxidizing agents may result in violent reaction

Hazardous decomp. prods.: Combustion: CO

HMIS: Health 1, Flammability 3, Reactivity 0

Storage: 1 yr shelf life; store under ambient conditions out of direct sunlight away from ignition sources

Punctilious® SDA 1-2 190 Proof [Equistar]

Chem. Descrip.: Specially denatured alcohol 1-2: ethyl alcohol 190 proof (95.34%), methanol (3.72%), MIBK (0.94%)

Uses: Solvent for hair spray, perfume, surfactants, detergents, pharmaceuticals, agric., cleaning sol'ns., disinfectants, explosives, plastics, polishes, shellac, stains, textiles, adhesives

Properties: Pt-Co 10 max. color; typ. odor; sol. in water; sp.gr. 0.8076-0.8880 (77/77 F); flash pt. (TCC) 58 F; ref. index 1.36122

Storage: 1 yr shelf life; store under ambient conditions out of direct sunlight

Punctilious® SDA 1-2 [Equistar]

Chem. Descrip.: Specially denatured alcohol 1-2

CAS 64-17-5; EINECS/ELINCS 200-578-6

Uses: Solvent for hair sprays, perfumes, surfactants, detergents, pharmaceuticals, agric., cleaning sol'ns., disinfectants, explosives, plastics, polishes, shellac, stains, textiles, adhesives

Punctilious® SDA 2B-2 190 Proof [Equistar]

Chem. Descrip.: Specially denatured alcohol 2B: ethanol 190 proof (92.03%), light aliphatic solvent naphtha (0.42%), water (7.55%)

UN 1170

Uses: Solvent for hair spray, perfume, surfactants, detergents, pharmaceuticals, agric., cleaning sol'ns., disinfectants, explosives, plastics, polishes, shellac, stains, textiles, adhesives

Regulatory: SARA §311/312 acute health/chronic health/fire hazard, §313 nonreportable; Calif. Prop. 65 reportable (ethanol, benzene)

Properties: Colorless liq.; invisible vapor; sweet alcohol-like odor; sol. in water; sp.gr. 0.8078-0.8090 (77/77 F), 1.59 (vapor); vapor pressure 44.6 mm Hg; f.p. -144 C; b.p. 78.4 C; flash pt. (TCC) 16 C; autoignition temp. 363 C; ref. index 1.3618

Toxicology: May be irritating to eyes, upper respiratory tract; short-term exposure above 1000 ppm by inh. may cause CNS effects, headache, eye/nose/throat irritation; exposure > 1 h may cause dizziness, drowsiness, loss of appetite, inability to conc., GI effects, nausea, vomiting; prolonged/repeated skin contact may cause dermatitis; ing. may cause drunkenness, CNS depression, nausea, vomiting, diarrhea, liver damage, death; TSCA listed

Environmental: On land, apt to volatilize, biodegrade, or leach to ground water; in water, photolysis, oxidation, biodegradation expected to occur; not expected to bioconcentrate in aquatic organisms; LC50 (rainbow trout, 96 h) 10,400 ppm; prevent entry into sewers

Precaution: Flamm. liq.; flamm. limits 3.3-19 vol.%; material can burn with little or no visible flame; vapors are heavier than air, may flash back; prevent entry into basements, confined areas; ground and bond containers; avoid static discharges; avoid strong oxidizers, excessive heat, sparks, open flame; contact with acetyl chloride or other oxidizing agents may result in violent reaction

Hazardous decomp. prods.: Combustion: CO

HMIS: Health 1, Flammability 3, Reactivity 0

Storage: 1 yr shelf life; store under ambient conditions out of direct sunlight

Punctilious® SDA 2B-2 Anhydrous [Equistar]

Chem. Descrip.: Specially denatured alcohol 2B: ethanol (99.6%), naphtha, light aliphatic (0.4%)

UN 1170

Uses: Solvent for hair spray, perfume, surfactants, detergents, pharmaceuticals, agric., cleaning sol'ns., disinfectants, explosives, plastics, polishes, shellac, stains, textiles, adhesives

Regulatory: SARA §311/312 acute health/chronic health/fire hazard, §313 nonreportable; Calif. Prop. 65 reportable (ethanol, benzene)

Properties: Colorless liq.; invisible vapor; sweet alcohol-like odor; sol. in water; sp.gr. 0.7862-0.7871 (77/77 F), 1.59 (vapor); vapor pressure 44.6 mm Hg; f.p. -144 C; b.p. 78.4 C; flash pt. (TCC) 13 C; autoignition temp. 363 C; ref. index 1.3596

Toxicology: LD50 (oral, rat) 7060 mg/kg; may be irritating to eyes, upper respiratory tract; short-term exposure above 1000 ppm by inh. may cause CNS effects, headache, eye/nose/throat irritation; exposure > 1 h may cause dizziness, drowsiness, loss of appetite, inability to conc., GI effects, nausea, vomiting; prolonged/repeated skin contact may cause dermatitis; ing. may cause drunkenness, CNS depression, nausea, vomiting, diarrhea, liver damage, death; TSCA listed

Environmental: On land, apt to volatilize, biodegrade, or leach to ground water; in water, photolysis, oxidation, biodegradation expected to occur; not expected to bioconcentrate in aquatic organisms; LC50 (rainbow trout, 96 h) 10,400 ppm; prevent entry into sewers

Precaution: Flamm. liq.; flamm. limits 3.3-19 vol.%; material can burn with little or no visible flame; vapors are heavier than air, may flash back; prevent entry into basements, confined areas; ground and bond containers; avoid static discharges; avoid strong oxidizers, excessive heat, sparks, open flame; contact with acetyl chloride or other oxidizing agents may result in violent reaction

Hazardous decomp. prods.: Combustion: CO

HMIS: Health 1, Flammability 3, Reactivity 0

Storage: 1 yr shelf life; store under ambient conditions out of direct sunlight

Punctilious® SDA 2B-3 190 Proof [Equistar]

Chem. Descrip.: Specially denatured alcohol 2B: ethanol 190 proof (91.9%), toluene (0.5%), water (7.5%)

UN 1170
Uses: Solvent for hair spray, perfume, surfactants, detergents, pharmaceuticals, agric., cleaning sol'ns., disinfectants, explosives, plastics, polishes, shellac, stains, textiles, adhesives
Regulatory: SARA §311/312 acute health/chronic health/fire hazard, §313 reportable (toluene); Calif. Prop. 65 reportable (toluene, ethanol, benzene)
Properties: Colorless liq.; invisible vapor; sweet alcohol-like odor; sol. in water; sp.gr. 0.7862-0.7871 (77/77 F), 1.59 (vapor); vapor pressure 44.6 mm Hg; f.p. -144 C; b.p. 78.4 C; flash pt. (TCC) 16 C; autoignition temp. 363 C; ref. index 1.3596
Toxicology: May be irritating to eyes, upper respiratory tract; short-term exposure above 1000 ppm by inh. may cause CNS effects, headache, eye/nose/throat irritation; exposure > 1 h may cause dizziness, drowsiness, loss of appetite, inability to conc., GI effects, nausea, vomiting; prolonged/repeated skin contact may cause dermatitis; ing. may cause drunkenness, CNS depression, nausea, vomiting, diarrhea, liver damage, death; TSCA listed
Environmental: On land, apt to volatilize, biodegrade, or leach to ground water; in water, photolysis, oxidation, biodegradation expected to occur; not expected to bioconcentrate in aquatic organisms; LC50 (rainbow trout, 96 h) 10,400 ppm; prevent entry into sewers
Precaution: Flamm. liq.; flamm. limits 3.3-19 vol.%; material can burn with little or no visible flame; vapors are heavier than air, may flash back; prevent entry into basements, confined areas; ground and bond containers; avoid static discharges; avoid strong oxidizers, excessive heat, sparks, open flame; contact with acetyl chloride or other oxidizing agents may result in violent reaction
Hazardous decomp. prods.: Combustion: CO
HMIS: Health 1, Flammability 3, Reactivity 0
Storage: 1 yr shelf life; store under ambient conditions out of direct sunlight

Punctilious® SDA 2B-4 190 Proof [Equistar]
Chem. Descrip.: Specially denatured alcohol 2B: ethanol, light aliphatic solvent naphtha, water (7.55%)
Uses: Solvent for hair spray, perfume, surfactants, detergents, pharmaceuticals, agric., cleaning sol'ns., disinfectants, explosives, plastics, polishes, shellac, stains, textiles, adhesives
Properties: Pt-Co 10 max. color; typ. odor; sol. in water; sp.gr. 0.8078-0.8090 (77/77 F); flash pt. (TCC) 44 F; ref. index 1.36179
Storage: 1 yr shelf life; store under ambient conditions out of direct sunlight

Punctilious® SDA 2B-4 Anhydrous [Equistar]
Chem. Descrip.: Specially denatured alcohol 2B: ethanol (99.6%), light aliphatic solvent naphtha (0.4%)
UN 1170
Uses: Solvent for hair spray, perfume, surfactants, detergents, pharmaceuticals, agric., cleaning sol'ns., disinfectants, explosives, plastics, polishes, shellac, stains, textiles, adhesives
Regulatory: SARA §311/312 acute health/chronic health/fire hazard, §313 nonreportable; Calif. Prop. 65 reportable (ethanol, benzene)
Properties: Colorless liq.; invisible vapor; sweet alcohol-like odor; sol. in water; sp.gr. 1.59 (vapor); vapor pressure 44.6 mm Hg; f.p. -144 C; b.p. 78.4 C; flash pt. (TCC) 13 C; autoignition temp. 363 C
Toxicology: LD50 (oral, rat) 7060 mg/kg; may be irritating to eyes, upper respiratory tract; short-term exposure above 1000 ppm by inh. may cause CNS effects, headache, eye/nose/throat irritation; exposure > 1 h may cause dizziness, drowsiness, loss of appetite, inability to conc., GI effects, nausea, vomiting; prolonged/repeated skin contact may cause dermatitis; ing. may cause drunkenness, CNS depression, nausea, vomiting, diarrhea, liver damage, death; TSCA listed
Environmental: On land, apt to volatilize, biodegrade, or leach to ground water; in water, photolysis, oxidation, biodegradation expected to occur; not expected to bioconcentrate in aquatic organisms; LC50 (rainbow trout, 96 h) 10,400 ppm; prevent entry into sewers
Precaution: Flamm. liq.; flamm. limits 3.3-19 vol.%; material can burn with little or no visible flame; vapors are heavier than air, may flash back; prevent entry into basements, confined areas; ground and bond containers; avoid static discharges; avoid strong oxidizers, excessive heat, sparks, open flame; contact with acetyl chloride or other oxidizing agents may result in violent reaction
Hazardous decomp. prods.: Combustion: CO
HMIS: Health 1, Flammability 3, Reactivity 0
Storage: 1 yr shelf life; store under ambient conditions out of direct sunlight

Punctilious® SDA 3A 190 Proof [Equistar]
Chem. Descrip.: Specially denatured alcohol 3A: ethanol (88.1%), methanol (4.7%), water (7.2%)
UN 1170
Uses: Solvent for hair spray, perfume, surfactants, detergents, pharmaceuticals, agric., cleaning sol'ns., disinfectants, explosives, plastics, polishes, shellac, stains, textiles, adhesives
Regulatory: SARA §311/312 acute health/chronic health/fire hazard, §313 reportable (methanol)
Properties: Colorless liq.; invisible vapor; sweet alcohol-like odor; sol. in water; sp.gr. 0.8078-0.8090 (77/77 F), 1.59 (vapor); vapor pressure 44.6 mm Hg; f.p. -144 C; b.p. 78.4 C; flash pt. (TCC) 16 C; autoignition temp. 363 C; ref. index 1.36078
Toxicology: May be irritating to eyes, upper respiratory tract; short-term exposure above 1000 ppm by inh. may cause CNS effects, headache, eye/nose/throat irritation; exposure > 1 h may cause dizziness, drowsiness, loss of appetite, inability to conc., GI effects, nausea, vomiting; prolonged/repeated skin contact may cause dermatitis; ing. may cause drunkenness, CNS depression, nausea, vomiting, diarrhea, liver damage, death; TSCA listed
Environmental: On land, apt to volatilize, biodegrade, or leach to ground water; in water, photolysis, oxidation, biodegradation expected to occur; not expected to bioconcentrate in aquatic organisms; LC50 (rainbow trout, 96 h) 10,400 ppm; prevent entry into sewers
Precaution: Flamm. liq.; flamm. limits 3.3-19 vol.%; material can burn with little or no visible flame; vapors are heavier than air, may flash back; prevent entry into basements, confined areas; ground and bond containers; avoid static discharges; avoid strong oxidizers, excessive heat, sparks, open flame; contact with acetyl chloride or other oxidizing agents may result in violent reaction
Hazardous decomp. prods.: Combustion: CO
HMIS: Health 1, Flammability 3, Reactivity 0
Storage: 1 yr shelf life; store under ambient conditions out of direct sunlight

Punctilious® SDA 3A Anhydrous [Equistar]
Chem. Descrip.: Specially denatured alcohol 3A: ethanol (95.2%), methanol (4.8%)
UN 1170
Uses: Solvent for hair spray, perfume, surfactants, detergents, pharmaceuticals, agric., cleaning sol'ns., disinfectants, explosives, plastics, polishes, shellac, stains, textiles, adhesives
Regulatory: SARA §311/312 acute health/chronic health/fire hazard, §313 reportable (methanol); Calif. Prop. 65 reportable (ethanol, benzene)
Properties: Colorless liq.; invisible vapor; sweet alcohol-like odor; sol. in water; sp.gr. 0.7869-0.7879 (77/77 F), 1.59 (vapor); vapor pressure 44.6 mm Hg; f.p. -144 C; b.p. 78.4 C; flash pt. (TCC) 16 C; autoignition temp. 363 C; ref. index 1.35939
Toxicology: May be irritating to eyes, upper respiratory tract; short-term exposure above 1000 ppm by inh. may cause CNS effects, headache, eye/nose/throat irritation; exposure > 1 h may cause dizziness, drowsiness, loss of appetite, inability to conc., GI effects, nausea, vomiting; prolonged/repeated skin contact may cause dermatitis; ing. may cause drunkenness, CNS depression, nausea, vomiting, diarrhea, liver damage, death; TSCA listed
Environmental: On land, apt to volatilize, biodegrade, or leach to ground water; in water, photolysis, oxidation, biodegradation expected to occur; not expected to bioconcentrate in aquatic organisms; LC50 (rainbow trout, 96 h) 10,400 ppm; prevent entry into sewers
Precaution: Flamm. liq.; flamm. limits 3.3-19 vol.%; material can burn with little or no visible flame; vapors are heavier than air, may flash back; prevent entry into basements, confined areas; ground and bond containers; avoid static discharges; avoid strong oxidizers, excessive heat, sparks, open flame; contact with acetyl chloride or other oxidizing agents may result in violent reaction
Hazardous decomp. prods.: Combustion: CO
HMIS: Health 1, Flammability 3, Reactivity 0
Storage: 1 yr shelf life; store under ambient conditions out of direct sunlight

Punctilious® SDA 3C 190 Proof [Equistar]

Chem. Descrip.: Specially denatured alcohol 3C: ethanol (88.2%), isopropyl alcohol (4.6%), water (7.2%)
UN 1170
Uses: Solvent for hair spray, perfume, surfactants, detergents, pharmaceuticals, agric., cleaning sol'ns., disinfectants, explosives, plastics, polishes, shellac, stains, textiles, adhesives
Regulatory: SARA §311/312 acute health/chronic health/fire hazard, §313 reportable (IPA); Calif. Prop. 65 reportable (ethanol, benzene)
Properties: Colorless liq.; invisible vapor; sweet alcohol-like odor; sol. in water; sp.gr. 0.8074-0.8086 (77/77 F), 1.59 (vapor); vapor pressure 44.6 mm Hg; f.p. -144 C; b.p. 78.4 C; flash pt. (TCC) 16 C; autoignition temp. 363 C; ref. index 1.36289
Toxicology: May be irritating to eyes, upper respiratory tract; short-term exposure above 1000 ppm by inh. may cause CNS effects, headache, eye/nose/throat irritation; exposure > 1 h may cause dizziness, drowsiness, loss of appetite, inability to conc., GI effects, nausea, vomiting; prolonged/repeated skin contact may cause dermatitis; ing. may cause drunkenness, CNS depression, nausea, vomiting, diarrhea, liver damage, death; TSCA listed
Environmental: On land, apt to volatilize, biodegrade, or leach to ground water; in water, photolysis, oxidation, biodegradation expected to occur; not expected to bioconcentrate in aquatic organisms; LC50 (rainbow trout, 96 h) 10,400 ppm; prevent entry into sewers
Precaution: Flamm. liq.; flamm. limits 3.3-19 vol.%; material can burn with little or no visible flame; vapors are heavier than air, may flash back; prevent entry into basements, confined areas; ground and bond containers; avoid static discharges
Hazardous decomp. prods.: Combustion: CO
HMIS: Health 1, Flammability 3, Reactivity 0
Storage: 1 yr shelf life; store under ambient conditions out of direct sunlight

Punctilious® SDA 12A-3 190 Proof [Equistar]
Chem. Descrip.: Specially denatured alcohol 12A: ethanol, toluene, water
Uses: Solvent for hair spray, perfume, surfactants, detergents, pharmaceuticals, agric., cleaning sol'ns., disinfectants, explosives, plastics, polishes, shellac, stains, textiles, adhesives
Properties: Pt-Co 10 max. color; typ. odor; sol. in water; sp.gr. 0.8111-0.8123 (77/77 F); flash pt. (TCC) 52 F; ref. index 1.36787
Storage: 1 yr shelf life; store under ambient conditions out of direct sunlight

Punctilious® SDA 12A-3 Anhydrous [Equistar]
Chem. Descrip.: Specially denatured alcohol 12A: ethanol (94.8%), toluene (5.2%)
UN 1170
Uses: Solvent for hair spray, perfume, surfactants, detergents, pharmaceuticals, agric., cleaning sol'ns., disinfectants, explosives, plastics, polishes, shellac, stains, textiles, adhesives
Regulatory: SARA §311/312 acute health/chronic health/fire hazard, §313 reportable (toluene); Calif. Prop. 65 reportable (toluene, ethanol, benzene)
Properties: Colorless liq.; invisible vapor; sweet alcohol-like odor; sol. in water; sp.gr. 0.7907-0.7921 (77/77 F), 1.59 (vapor); vapor pressure 44.6 mm Hg; f.p. -144 C; b.p. 78.4 C; flash pt. (TCC) 13 C; autoignition temp. 363 C; ref. index 1.3660
Toxicology: May be irritating to eyes, upper respiratory tract; short-term exposure above 1000 ppm by inh. may cause CNS effects, headache, eye/nose/throat irritation; exposure > 1 h may cause dizziness, drowsiness, loss of appetite, inability to conc., GI effects, nausea, vomiting; prolonged/repeated skin contact may cause dermatitis; ing. may cause drunkenness, CNS depression, nausea, vomiting, diarrhea, liver damage, death; TSCA listed
Environmental: On land, apt to volatilize, biodegrade, or leach to ground water; in water, photolysis, oxidation, biodegradation expected to occur; not expected to bioconcentrate in aquatic organisms; LC50 (rainbow trout, 96 h) 10,400 ppm; prevent entry into sewers
Precaution: Flamm. liq.; flamm. limits 3.3-19 vol.%; material can burn with little or no visible flame; vapors are heavier than air, may flash back; prevent entry into basements, confined areas; ground and bond containers; avoid static discharges
Hazardous decomp. prods.: Combustion: CO
HMIS: Health 1, Flammability 3, Reactivity 0
Storage: 1 yr shelf life; store under ambient conditions out of direct sunlight

Punctilious® SDA 13A 190 Proof [Equistar]
Chem. Descrip.: Specially denatured alcohol 13A: ethanol (85.1%), ethyl ether (7.9%), water (7.0%)
UN 1170
Uses: Solvent for hair spray, perfume, surfactants, detergents, pharmaceuticals, agric., cleaning sol'ns., disinfectants, explosives, plastics, polishes, shellac, stains, textiles, adhesives
Regulatory: SARA §311/312 acute health/chronic health/fire hazard, §313 nonreportable; Calif. Prop. 65 reportable (ethanol, benzene)
Properties: Colorless liq.; invisible vapor; sweet alcohol-like odor; sol. in water; sp.gr. 0.8022-0.8034 (77/77 F), 1.59 (vapor); vapor pressure 44.6 mm Hg; f.p. -144 C; b.p. 78.4 C; flash pt. (TCC) 16 C; autoignition temp. 363 C; ref. index 1.36198
Toxicology: May be irritating to eyes, upper respiratory tract; short-term exposure above 1000 ppm by inh. may cause CNS effects, headache, eye/nose/throat irritation; exposure > 1 h may cause dizziness, drowsiness, loss of appetite, inability to conc., GI effects, nausea, vomiting; prolonged/repeated skin contact may cause dermatitis; ing. may cause drunkenness, CNS depression, nausea, vomiting, diarrhea, liver damage, death; TSCA listed
Environmental: On land, apt to volatilize, biodegrade, or leach to ground water; in water, photolysis, oxidation, biodegradation expected to occur; not expected to bioconcentrate in aquatic organisms; LC50 (rainbow trout, 96 h) 10,400 ppm; prevent entry into sewers
Precaution: Flamm. liq.; flamm. limits 3.3-19 vol.%; material can burn with little or no visible flame; vapors are heavier than air, may flash back; prevent entry into basements, confined areas; ground and bond containers; avoid static discharges
Hazardous decomp. prods.: Combustion: CO
HMIS: Health 1, Flammability 3, Reactivity 0
Storage: 1 yr shelf life; store under ambient conditions out of direct sunlight

Punctilious® SDA 19 190 Proof [Equistar]
Chem. Descrip.: Specially denatured alcohol 19: ethanol (50%), ethyl ether (45.7%), water (4.3%)
UN 1170
Uses: Solvent for hair spray, perfume, surfactants, detergents, pharmaceuticals, agric., cleaning sol'ns., disinfectants, explosives, plastics, polishes, shellac, stains, textiles, adhesives
Regulatory: SARA §311/312 acute health/chronic health/fire hazard, §313 nonreportable; Calif. Prop. 65 reportable (ethanol, benzene)
Properties: Colorless liq.; invisible vapor; sweet alcohol-like odor; sol. in water; sp.gr. 0.7678-0.7698 (77/77 F), 1.59 (vapor); vapor pressure 44.6 mm Hg; f.p. -144 C; b.p. 78.4 C; flash pt. (TCC) 16 C; autoignition temp. 363 C; ref. index 1.35394
Toxicology: May be irritating to eyes, upper respiratory tract; short-term exposure above 1000 ppm by inh. may cause CNS effects, headache, eye/nose/throat irritation; exposure > 1 h may cause dizziness, drowsiness, loss of appetite, inability to conc., GI effects, nausea, vomiting; prolonged/repeated skin contact may cause dermatitis; ing. may cause drunkenness, CNS depression, nausea, vomiting, diarrhea, liver damage, death; TSCA listed
Environmental: On land, apt to volatilize, biodegrade, or leach to ground water; in water, photolysis, oxidation, biodegradation expected to occur; not expected to bioconcentrate in aquatic organisms; LC50 (rainbow trout, 96 h) 10,400 ppm; prevent entry into sewers
Precaution: Flamm. liq.; flamm. limits 3.3-19 vol.%; material can burn with little or no visible flame; vapors are heavier than air, may flash back; prevent entry into basements, confined areas; ground and bond containers; avoid static discharges
Hazardous decomp. prods.: Combustion: CO
HMIS: Health 1, Flammability 3, Reactivity 0
Storage: 1 yr shelf life; store under ambient conditions out of direct sunlight

Punctilious® SDA 20 190 Proof [Equistar]
Chem. Descrip.: Specially denatured alcohol 20: ethanol, chloroform, water
Uses: Solvent for hair spray, perfume, surfactants, detergents, pharmaceuticals, agric., cleaning sol'ns., disinfectants, explosives, plastics, polishes, shellac, stains, textiles, adhesives
Properties: Pt-Co 10 max. color; typ. odor; sol. in water; sp.gr. 0.8399-

0.8429 (77/77 F); flash pt. (TCC) 56 F; ref. index 1.36593
Storage: 1 yr shelf life; store under ambient conditions out of direct sunlight

Punctilious® SDA 20 Anhydrous [Equistar]
Chem. Descrip.: Specially denatured alcohol 20: ethanol (91.5%), chloroform (8.5%)
UN 1170
Uses: Solvent for hair spray, perfume, surfactants, detergents, pharmaceuticals, agric., cleaning sol'ns., disinfectants, explosives, plastics, polishes, shellac, stains, textiles, adhesives
Regulatory: SARA §311/312 acute health/chronic health/fire hazard, §313 nonreportable; Calif. Prop. 65 reportable (ethanol, chloroform)
Properties: Colorless liq.; invisible vapor; sweet alcohol-like odor; sol. in water; sp.gr. 0.8199-0.8216 (77/77 F), 1.59 (vapor); vapor pressure 44.6 mm Hg; f.p. -144 C; b.p. 78.4 C; flash pt. (TCC) 13 C; autoignition temp. 363 C; ref. index 1.36329
Toxicology: May be irritating to eyes, upper respiratory tract; short-term exposure above 1000 ppm by inh. may cause CNS effects, headache, eye/nose/throat irritation; exposure > 1 h may cause dizziness, drowsiness, loss of appetite, inability to conc., GI effects, nausea, vomiting; prolonged/repeated skin contact may cause dermatitis; ing. may cause drunkenness, CNS depression, nausea, vomiting, diarrhea, liver damage, death; TSCA listed
Environmental: On land, apt to volatilize, biodegrade, or leach to ground water; in water, photolysis, oxidation, biodegradation expected to occur; not expected to bioconcentrate in aquatic organisms; LC50 (rainbow trout, 96 h) 10,400 ppm; prevent entry into sewers
Precaution: Flamm. liq.; flamm. limits 3.3-19 vol.%; material can burn with little or no visible flame; vapors are heavier than air, may flash back; prevent entry into basements, confined areas; ground and bond containers; avoid static discharges; avoid strong oxidizers, excessive heat, sparks, open flame; contact with acetyl chloride or other oxidizing agents may result in violent reaction
Hazardous decomp. prods.: Combustion: CO
HMIS: Health 1, Flammability 3, Reactivity 0
Storage: 1 yr shelf life; store under ambient conditions out of direct sunlight

Punctilious® SDA 23A 190 Proof [Equistar]
Chem. Descrip.: Specially denatured alcohol 23A: ethanol (85.7%), acetone (7.2%%), water (7.0%)
UN 1170
Uses: Solvent for hair spray, perfume, surfactants, detergents, pharmaceuticals, agric., cleaning sol'ns., disinfectants, explosives, plastics, polishes, shellac, stains, textiles, adhesives
Regulatory: SARA §311/312 acute health/chronic health/fire hazard, §313 nonreportable; Calif. Prop. 65 reportable (ethanol, benzene); TSCA export notification required (acetone)
Properties: Colorless liq.; invisible vapor; sweet alcohol-like odor; sol. in water; sp.gr. 0.8078-0.8090 (77/77 F), 1.59 (vapor); vapor pressure 44.6 mm Hg; f.p. -144 C; b.p. 78.4 C; flash pt. (TCC) 16 C; autoignition temp. 363 C; ref. index 1.36204
Toxicology: May be irritating to eyes, upper respiratory tract; short-term exposure above 1000 ppm by inh. may cause CNS effects, headache, eye/nose/throat irritation; exposure > 1 h may cause dizziness, drowsiness, loss of appetite, inability to conc., GI effects, nausea, vomiting; prolonged/repeated skin contact may cause dermatitis; ing. may cause drunkenness, CNS depression, nausea, vomiting, diarrhea, liver damage, death
Environmental: On land, apt to volatilize, biodegrade, or leach to ground water; in water, photolysis, oxidation, biodegradation expected to occur; not expected to bioconcentrate in aquatic organisms; LC50 (rainbow trout, 96 h) 10,400 ppm; prevent entry into sewers
Precaution: Flamm. liq.; flamm. limits 3.3-19 vol.%; material can burn with little or no visible flame; vapors are heavier than air, may flash back; prevent entry into basements, confined areas; ground and bond containers; avoid static discharges; avoid strong oxidizers, excessive heat, sparks, open flame; contact with acetyl chloride or other oxidizing agents may result in violent reaction
Hazardous decomp. prods.: Combustion: CO
HMIS: Health 1, Flammability 3, Reactivity 0
Storage: 1 yr shelf life; store under ambient conditions out of direct sunlight

Punctilious® SDA 23A Anhydrous [Equistar]
Chem. Descrip.: Specially denatured alcohol 23A: ethanol (92.6%), acetone (7.4%)
UN 1170
Uses: Solvent for hair spray, perfume, surfactants, detergents, pharmaceuticals, agric., cleaning sol'ns., disinfectants, explosives, plastics, polishes, shellac, stains, textiles, adhesives
Regulatory: SARA §311/312 acute health/chronic health/fire hazard, §313 nonreportable; Calif. Prop. 65 reportable (ethanol, benzene); TSCA export notification required (acetone)
Properties: Colorless liq.; invisible vapor; sweet alcohol-like odor; sol. in water; sp.gr. 0.7874-0.7884 (77/77 F), 1.59 (vapor); vapor pressure 44.6 mm Hg; f.p. -144 C; b.p. 78.4 C; flash pt. (TCC) 13 C; autoignition temp. 363 C; ref. index 1.35958
Toxicology: May be irritating to eyes, upper respiratory tract; short-term exposure above 1000 ppm by inh. may cause CNS effects, headache, eye/nose/throat irritation; exposure > 1 h may cause dizziness, drowsiness, loss of appetite, inability to conc., GI effects, nausea, vomiting; prolonged/repeated skin contact may cause dermatitis; ing. may cause drunkenness, CNS depression, nausea, vomiting, diarrhea, liver damage, death; TSCA listed
Environmental: On land, apt to volatilize, biodegrade, or leach to ground water; in water, photolysis, oxidation, biodegradation expected to occur; not expected to bioconcentrate in aquatic organisms; LC50 (rainbow trout, 96 h) 10,400 ppm; prevent entry into sewers
Precaution: Flamm. liq.; flamm. limits 3.3-19 vol.%; material can burn with little or no visible flame; vapors are heavier than air, may flash back; prevent entry into basements, confined areas; ground and bond containers; avoid static discharges; avoid strong oxidizers, excessive heat, sparks, open flame; contact with acetyl chloride or other oxidizing agents may result in violent reaction
Hazardous decomp. prods.: Combustion: CO
HMIS: Health 1, Flammability 3, Reactivity 0
Storage: 1 yr shelf life; store under ambient conditions out of direct sunlight

Punctilious® SDA 23H 190 Proof [Equistar]
Chem. Descrip.: Specially denatured alcohol 23H: ethanol (84.6%), acetone (7.1%), MIBK (1.4%), water (6.9%)
UN 1170
Uses: Solvent for hair spray, perfume, surfactants, detergents, pharmaceuticals, agric., cleaning sol'ns., disinfectants, explosives, plastics, polishes, shellac, stains, textiles, adhesives
Regulatory: SARA §311/312 acute health/chronic health/fire hazard, §313 reportable (MIBK); Calif. Prop. 65 reportable (ethanol, benzene); TSCA export notification required (acetone, MIBK)
Properties: Colorless liq.; invisible vapor; sweet alcohol-like odor; sol. in water; sp.gr. 0.8074-0.8086 (77/77 F), 1.59 (vapor); vapor pressure 44.6 mm Hg; f.p. -144 C; b.p. 78.4 C; flash pt. (TCC) 16 C; autoignition temp. 363 C; ref. index 1.36229
Toxicology: May be irritating to eyes, upper respiratory tract; short-term exposure above 1000 ppm by inh. may cause CNS effects, headache, eye/nose/throat irritation; exposure > 1 h may cause dizziness, drowsiness, loss of appetite, inability to conc., GI effects, nausea, vomiting; prolonged/repeated skin contact may cause dermatitis; ing. may cause drunkenness, CNS depression, nausea, vomiting, diarrhea, liver damage, death; TSCA listed
Environmental: On land, apt to volatilize, biodegrade, or leach to ground water; in water, photolysis, oxidation, biodegradation expected to occur; not expected to bioconcentrate in aquatic organisms; LC50 (rainbow trout, 96 h) 10,400 ppm; prevent entry into sewers
Precaution: Flamm. liq.; flamm. limits 3.3-19 vol.%; material can burn with little or no visible flame; vapors are heavier than air, may flash back; prevent entry into basements, confined areas; ground and bond containers; avoid static discharges; avoid strong oxidizers, excessive heat, sparks, open flame; contact with acetyl chloride or other oxidizing agents may result in violent reaction
Hazardous decomp. prods.: Combustion: CO
HMIS: Health 1, Flammability 3, Reactivity 0
Storage: 1 yr shelf life; store under ambient conditions out of direct sunlight

Punctilious® SDA 25A-1 190 Proof [Equistar]
Chem. Descrip.: Specially denatured alcohol 25A: ethanol (86.1%), io-

dine (2.7%), potassium iodide (2.1%), water (9.1%)
UN 1170
Uses: Solvent for hair spray, perfume, surfactants, detergents, pharmaceuticals, agric., cleaning sol'ns., disinfectants, explosives, plastics, polishes, shellac, stains, textiles, adhesives
Regulatory: SARA §311/312 acute health/chronic health/fire hazard, §313 nonreportable; Calif. Prop. 65 reportable (ethanol, benzene)
Properties: Deep red-brn. liq.; invisible vapor; sweet alcohol-like odor; sol. in water; sp.gr. 0.8471-0.8500 (77/77 F), 1.59 (vapor); vapor pressure 44.6 mm Hg; f.p. -144 C; b.p. 78.4 C; flash pt. (TCC) 16 C; autoignition temp. 363 C
Toxicology: May be irritating to eyes, upper respiratory tract; short-term exposure above 1000 ppm by inh. may cause CNS effects, headache, eye/nose/throat irritation; exposure > 1 h may cause dizziness, drowsiness, loss of appetite, inability to conc., GI effects, nausea, vomiting; prolonged/repeated skin contact may cause dermatitis; ing. may cause drunkenness, CNS depression, nausea, vomiting, diarrhea, liver damage, death; TSCA listed
Environmental: On land, apt to volatilize, biodegrade, or leach to ground water; in water, photolysis, oxidation, biodegradation expected to occur; not expected to bioconcentrate in aquatic organisms; LC50 (rainbow trout, 96 h) 10,400 ppm; prevent entry into sewers
Precaution: Flamm. liq.; flamm. limits 3.3-19 vol.%; material can burn with little or no visible flame; vapors are heavier than air, may flash back; prevent entry into basements, confined areas; ground and bond containers; avoid static discharges; avoid strong oxidizers, excessive heat, sparks, open flame; contact with acetyl chloride or other oxidizing agents may result in violent reaction
Hazardous decomp. prods.: Combustion: CO
HMIS: Health 1, Flammability 3, Reactivity 0
Storage: 1 yr shelf life; store under ambient conditions out of direct sunlight

Punctilious® SDA 25A-2 190 Proof [Equistar]
Chem. Descrip.: Specially denatured alcohol 25A: ethanol (86.1%), iodine (2.7%), sodium iodide (2.1%), water (9.1%)
UN 1170
Uses: Solvent for hair spray, perfume, surfactants, detergents, pharmaceuticals, agric., cleaning sol'ns., disinfectants, explosives, plastics, polishes, shellac, stains, textiles, adhesives
Regulatory: SARA §311/312 acute health/chronic health/fire hazard, §313 nonreportable; Calif. Prop. 65 reportable (ethanol, benzene)
Properties: Deep red-brn. liq.; invisible vapor; sweet alcohol-like odor; sol. in water; sp.gr. 0.8476-0.8506 (77/77 F), 1.59 (vapor); vapor pressure 44.6 mm Hg; f.p. -144 C; b.p. 78.4 C; flash pt. (TCC) 16 C; autoignition temp. 363 C
Toxicology: May be irritating to eyes, upper respiratory tract; short-term exposure above 1000 ppm by inh. may cause CNS effects, headache, eye/nose/throat irritation; exposure > 1 h may cause dizziness, drowsiness, loss of appetite, inability to conc., GI effects, nausea, vomiting; prolonged/repeated skin contact may cause dermatitis; ing. may cause drunkenness, CNS depression, nausea, vomiting, diarrhea, liver damage, death; TSCA listed
Environmental: On land, apt to volatilize, biodegrade, or leach to ground water; in water, photolysis, oxidation, biodegradation expected to occur; not expected to bioconcentrate in aquatic organisms; LC50 (rainbow trout, 96 h) 10,400 ppm; prevent entry into sewers
Precaution: Flamm. liq.; flamm. limits 3.3-19 vol.%; material can burn with little or no visible flame; vapors are heavier than air, may flash back; prevent entry into basements, confined areas; ground and bond containers; avoid static discharges; avoid strong oxidizers, excessive heat, sparks, open flame; contact with acetyl chloride or other oxidizing agents may result in violent reaction
Hazardous decomp. prods.: Combustion: CO
HMIS: Health 1, Flammability 3, Reactivity 0
Storage: 1 yr shelf life; store under ambient conditions out of direct sunlight

Punctilious® SDA 28A Anhydrous [Equistar]
Chem. Descrip.: Specially denatured alcohol 28A: ethanol (99.13%), gasoline (0.87%)
Uses: Solvent
Properties: Pt-Co 10 max. color; typ. odor; sol. 85 ml/100 ml in water; sp.gr. 0.7857-0.7867 (77/77 F); flash pt. (TCC) 45 F; ref. index 1.35990
Storage: 1 yr shelf life; store under ambient conditions out of direct sunlight

Punctilious® SDA 29-2 190 Proof [Equistar]
Chem. Descrip.: Specially denatured alcohol 29-2: ethanol, ethyl acetoacetate, water
Uses: Solvent for hair spray, perfume, surfactants, detergents, pharmaceuticals, agric., cleaning sol'ns., disinfectants, explosives, plastics, polishes, shellac, stains, textiles, adhesives
Properties: Pt-Co 10 max. color; typ. odor; sol. in water; sp.gr. 0.8108-0.8119 (77/77 F); flash pt. (TCC) 58 F; ref. index 1.3622
Storage: 1 yr shelf life; store under ambient conditions out of direct sunlight

Punctilious® SDA 29-2 Anhydrous [Equistar]
Chem. Descrip.: Specially denatured alcohol 29-2: ethanol (98.7%), ethyl acetoacetate (1.3%)
UN 1170
Uses: Solvent for hair spray, perfume, surfactants, detergents, pharmaceuticals, agric., cleaning sol'ns., disinfectants, explosives, plastics, polishes, shellac, stains, textiles, adhesives
Regulatory: SARA §311/312 acute health/chronic health/fire hazard, §313 nonreportable; Calif. Prop. 65 reportable (ethanol, benzene)
Properties: Colorless liq.; invisible vapor; sweet alcohol-like odor; sol. in water; sp.gr. 0.7892-0.7902 (77/77 F), 1.59 (vapor); vapor pressure 44.6 mm Hg; f.p. -144 C; b.p. 78.4 C; flash pt. (TCC) 13 C; autoignition temp. 363 C; ref. index 1.3602
Toxicology: May be irritating to eyes, upper respiratory tract; short-term exposure above 1000 ppm by inh. may cause CNS effects, headache, eye/nose/throat irritation; exposure > 1 h may cause dizziness, drowsiness, loss of appetite, inability to conc., GI effects, nausea, vomiting; prolonged/repeated skin contact may cause dermatitis; ing. may cause drunkenness, CNS depression, nausea, vomiting, diarrhea, liver damage, death; TSCA listed
Environmental: On land, apt to volatilize, biodegrade, or leach to ground water; in water, photolysis, oxidation, biodegradation expected to occur; not expected to bioconcentrate in aquatic organisms; LC50 (rainbow trout, 96 h) 10,400 ppm; prevent entry into sewers
Precaution: Flamm. liq.; flamm. limits 3.3-19 vol.%; material can burn with little or no visible flame; vapors are heavier than air, may flash back; prevent entry into basements, confined areas; ground and bond containers; avoid static discharges; avoid strong oxidizers, excessive heat, sparks, open flame; contact with acetyl chloride or other oxidizing agents may result in violent reaction
Hazardous decomp. prods.: Combustion: CO
HMIS: Health 1, Flammability 3, Reactivity 0
Storage: 1 yr shelf life; store under ambient conditions out of direct sunlight

Punctilious® SDA 29-3 190 Proof [Equistar]
Chem. Descrip.: Specially denatured alcohol 29-3: ethanol (91.4%), ethyl acetate (1.1%), water (7.5%)
UN 1170
Uses: Solvent for hair spray, perfume, surfactants, detergents, pharmaceuticals, agric., cleaning sol'ns., disinfectants, explosives, plastics, polishes, shellac, stains, textiles, adhesives
Regulatory: SARA §311/312 acute health/chronic health/fire hazard, §313 nonreportable; Calif. Prop. 65 reportable (ethanol, benzene); TSCA export notification required (ethyl acetate)
Properties: Colorless liq.; invisible vapor; sweet alcohol-like odor; sol. in water; sp.gr. 0.8094-0.8105 (77/77 F), 1.59 (vapor); vapor pressure 44.6 mm Hg; f.p. -144 C; b.p. 78.4 C; flash pt. (TCC) 16 C; autoignition temp. 363 C; ref. index 1.3619
Toxicology: May be irritating to eyes, upper respiratory tract; short-term exposure above 1000 ppm by inh. may cause CNS effects, headache, eye/nose/throat irritation; exposure > 1 h may cause dizziness, drowsiness, loss of appetite, inability to conc., GI effects, nausea, vomiting; prolonged/repeated skin contact may cause dermatitis; ing. may cause drunkenness, CNS depression, nausea, vomiting, diarrhea, liver damage, death; TSCA listed
Environmental: On land, apt to volatilize, biodegrade, or leach to ground water; in water, photolysis, oxidation, biodegradation expected to occur; not expected to bioconcentrate in aquatic organisms; LC50 (rainbow trout, 96 h) 10,400 ppm; prevent entry into sewers
Precaution: Flamm. liq.; flamm. limits 3.3-19 vol.%; material can burn with

little or no visible flame; vapors are heavier than air, may flash back; prevent entry into basements, confined areas; ground and bond containers; avoid static discharges; avoid strong oxidizers, excessive heat, sparks, open flame; contact with acetyl chloride or other oxidizing agents may result in violent reaction
Hazardous decomp. prods.: Combustion: CO
HMIS: Health 1, Flammability 3, Reactivity 0
Storage: 1 yr shelf life; store under ambient conditions out of direct sunlight

Punctilious® SDA 29-13 190 Proof [Equistar]
Chem. Descrip.: Specially denatured alcohol 29-13: ethanol, aniline, water (7.48%)
Uses: Solvent
Properties: Pt-Co 10 max. color; typ. odor; sol. in water; sp.gr. 0.8109-0.8121 (77/77 F); flash pt. (TCC) 56 F; ref. index 1.3644
Storage: 1 yr shelf life; store under ambient conditions out of direct sunlight

Punctilious® SDA 29-13 Anhydrous [Equistar]
Chem. Descrip.: Specially denatured alcohol 29-13: ethanol (98.7%), aniline (1.3%)
UN 1170
Uses: Solvent
Regulatory: SARA §311/312 acute health/chronic health/fire hazard, §313 reportable (aniline); Calif. Prop. 65 reportable (ethanol, benzene, aniline)
Properties: Colorless liq.; invisible vapor; sweet alcohol-like odor; sol. in water; sp.gr. 0.8094-0.8105 (77/77 F), 1.59 (vapor); vapor pressure 44.6 mm Hg; f.p. -144 C; b.p. 78.4 C; flash pt. (TCC) 13 C; autoignition temp. 363 C; ref. index 1.3619
Toxicology: May be irritating to eyes, upper respiratory tract; short-term exposure above 1000 ppm by inh. may cause CNS effects, headache, eye/nose/throat irritation; exposure > 1 h may cause dizziness, drowsiness, loss of appetite, inability to conc., GI effects, nausea, vomiting; prolonged/repeated skin contact may cause dermatitis; ing. may cause drunkenness, CNS depression, nausea, vomiting, diarrhea, liver damage, death; TSCA listed
Environmental: On land, apt to volatilize, biodegrade, or leach to ground water; in water, photolysis, oxidation, biodegradation expected to occur; not expected to bioconcentrate in aquatic organisms; LC50 (rainbow trout, 96 h) 10,400 ppm; prevent entry into sewers
Precaution: Flamm. liq.; flamm. limits 3.3-19 vol.%; material can burn with little or no visible flame; vapors are heavier than air, may flash back; prevent entry into basements, confined areas; ground and bond containers; avoid static discharges; avoid strong oxidizers, excessive heat, sparks, open flame; contact with acetyl chloride or other oxidizing agents may result in violent reaction
Hazardous decomp. prods.: Combustion: CO
HMIS: Health 1, Flammability 3, Reactivity 0
Storage: 1 yr shelf life; store under ambient conditions out of direct sunlight

Punctilious® SDA 30 190 Proof [Equistar]
Chem. Descrip.: Specially denatured alcohol 30: ethanol (84.2%), methanol (8.9%), water (6.9%)
UN NA1987
Uses: Solvent for hair spray, perfume, surfactants, detergents, pharmaceuticals, agric., cleaning sol'ns., disinfectants, explosives, plastics, polishes, shellac, stains, textiles, adhesives
Regulatory: SARA §311/312 acute health/chronic health/fire hazard, §313 reportable (methanol); Calif. Prop. 65 reportable (ethanol, benzene)
Properties: Colorless liq.; invisible vapor; sweet alcohol-like odor; sol. in water; sp.gr. 0.8066-0.8080 (77/77 F), 1.59 (vapor); vapor pressure 44.6 mm Hg; f.p. -144 C; b.p. 78.4 C; flash pt. (TCC) 16 C; autoignition temp. 363 C; ref. index 1.3586
Toxicology: May be irritating to eyes, upper respiratory tract; short-term exposure above 1000 ppm by inh. may cause CNS effects, headache, eye/nose/throat irritation; exposure > 1 h may cause dizziness, drowsiness, loss of appetite, inability to conc., GI effects, nausea, vomiting; prolonged/repeated skin contact may cause dermatitis; ing. may cause drunkenness, CNS depression, nausea, vomiting, diarrhea, liver damage, death; TSCA listed
Environmental: On land, apt to volatilize, biodegrade, or leach to ground water; in water, photolysis, oxidation, biodegradation expected to occur; not expected to bioconcentrate in aquatic organisms; LC50 (rainbow trout, 96 h) 10,400 ppm; prevent entry into sewers
Precaution: Flamm. liq.; flamm. limits 3.3-19 vol.%; material can burn with little or no visible flame; vapors are heavier than air, may flash back; prevent entry into basements, confined areas; ground and bond containers; avoid static discharges; avoid strong oxidizers, excessive heat, sparks, open flame; contact with acetyl chloride or other oxidizing agents may result in violent reaction
Hazardous decomp. prods.: Combustion: CO
HMIS: Health 1, Flammability 3, Reactivity 0
Storage: 1 yr shelf life; store under ambient conditions out of direct sunlight

Punctilious® SDA 30 Anhydrous [Equistar]
Chem. Descrip.: Specially denatured alcohol 30: ethanol (90.9%), methanol (9.1%)
UN 1170
Uses: Solvent for hair spray, perfume, surfactants, detergents, pharmaceuticals, agric., cleaning sol'ns., disinfectants, explosives, plastics, polishes, shellac, stains, textiles, adhesives
Regulatory: SARA §311/312 acute health/chronic health/fire hazard, §313 reportable (methanol); Calif. Prop. 65 reportable (ethanol, benzene)
Properties: Colorless liq.; invisible vapor; sweet alcohol-like odor; sol. in water; sp.gr. 0.7868-0.7879 (77/77 F), 1.59 (vapor); vapor pressure 44.6 mm Hg; f.p. -144 C; b.p. 78.4 C; flash pt. (TCC) 13 C; autoignition temp. 363 C; ref. index 1.3566
Toxicology: May be irritating to eyes, upper respiratory tract; short-term exposure above 1000 ppm by inh. may cause CNS effects, headache, eye/nose/throat irritation; exposure > 1 h may cause dizziness, drowsiness, loss of appetite, inability to conc., GI effects, nausea, vomiting; prolonged/repeated skin contact may cause dermatitis; ing. may cause drunkenness, CNS depression, nausea, vomiting, diarrhea, liver damage, death; TSCA listed
Environmental: On land, apt to volatilize, biodegrade, or leach to ground water; in water, photolysis, oxidation, biodegradation expected to occur; not expected to bioconcentrate in aquatic organisms; LC50 (rainbow trout, 96 h) 10,400 ppm; prevent entry into sewers
Precaution: Flamm. liq.; flamm. limits 3.3-19 vol.%; material can burn with little or no visible flame; vapors are heavier than air, may flash back; prevent entry into basements, confined areas; ground and bond containers; avoid static discharges; avoid strong oxidizers, excessive heat, sparks, open flame; contact with acetyl chloride or other oxidizing agents may result in violent reaction
Hazardous decomp. prods.: Combustion: CO
HMIS: Health 1, Flammability 3, Reactivity 0
Storage: 1 yr shelf life; store under ambient conditions out of direct sunlight

Punctilious® SDA 32 190 Proof [Equistar]
Chem. Descrip.: Specially denatured alcohol 32: ethanol (88.6%), ethyl ether (4.1%%), water (7.3%%)
UN 1170
Uses: Solvent for hair spray, perfume, surfactants, detergents, pharmaceuticals, agric., cleaning sol'ns., disinfectants, explosives, plastics, polishes, shellac, stains, textiles, adhesives
Regulatory: SARA §311/312 acute health/chronic health/fire hazard, §313 nonreportable; Calif. Prop. 65 reportable (ethanol, benzene)
Properties: Colorless liq.; invisible vapor; sweet alcohol-like odor; sol. in water; sp.gr. 0.8056-0.8068 (77/77 F), 1.59 (vapor); vapor pressure 44.6 mm Hg; f.p. -144 C; b.p. 78.4 C; flash pt. (TCC) 16 C; autoignition temp. 363 C; ref. index 1.3621
Toxicology: May be irritating to eyes, upper respiratory tract; short-term exposure above 1000 ppm by inh. may cause CNS effects, headache, eye/nose/throat irritation; exposure > 1 h may cause dizziness, drowsiness, loss of appetite, inability to conc., GI effects, nausea, vomiting; prolonged/repeated skin contact may cause dermatitis; ing. may cause drunkenness, CNS depression, nausea, vomiting, diarrhea, liver damage, death; TSCA listed
Environmental: On land, apt to volatilize, biodegrade, or leach to ground water; in water, photolysis, oxidation, biodegradation expected to occur; not expected to bioconcentrate in aquatic organisms; LC50 (rainbow trout, 96 h) 10,400 ppm; prevent entry into sewers
Precaution: Flamm. liq.; flamm. limits 3.3-19 vol.%; material can burn with little or no visible flame; vapors are heavier than air, may flash back; prevent entry into basements, confined areas; ground and bond con-

tainers; avoid static discharges; avoid strong oxidizers, excessive heat, sparks, open flame; contact with acetyl chloride or other oxidizing agents may result in violent reaction
Hazardous decomp. prods.: Combustion: CO
HMIS: Health 1, Flammability 3, Reactivity 0
Storage: 1 yr shelf life; store under ambient conditions out of direct sunlight

Punctilious® SDA 32 Anhydrous [Equistar]
Chem. Descrip.: Specially denatured alcohol 32: ethanol (95.7%), ethyl ether (4.2%), water (0.03%)
UN 1170
Uses: Solvent for hair spray, perfume, surfactants, detergents, pharmaceuticals, agric., cleaning sol'ns., disinfectants, explosives, plastics, polishes, shellac, stains, textiles, adhesives
Regulatory: SARA §311/312 acute health/chronic health/fire hazard, §313 nonreportable; Calif. Prop. 65 reportable (ethanol, benzene)
Properties: Colorless liq.; invisible vapor; sweet alcohol-like odor; sol. in water; sp.gr. 0.7845-0.7855 (77/77 F), 1.59 (vapor); vapor pressure 44.6 mm Hg; f.p. -144 C; b.p. 78.4 C; flash pt. (TCC) 13 C; autoignition temp. 363 C; ref. index 1.3591
Toxicology: May be irritating to eyes, upper respiratory tract; short-term exposure above 1000 ppm by inh. may cause CNS effects, headache, eye/nose/throat irritation; exposure > 1 h may cause dizziness, drowsiness, loss of appetite, inability to conc., GI effects, nausea, vomiting; prolonged/repeated skin contact may cause dermatitis; ing. may cause drunkenness, CNS depression, nausea, vomiting, diarrhea, liver damage, death; TSCA listed
Environmental: On land, apt to volatilize, biodegrade, or leach to ground water; in water, photolysis, oxidation, biodegradation expected to occur; not expected to bioconcentrate in aquatic organisms; LC50 (rainbow trout, 96 h) 10,400 ppm; prevent entry into sewers
Precaution: Flamm. liq.; flamm. limits 3.3-19 vol.%; material can burn with little or no visible flame; vapors are heavier than air, may flash back; prevent entry into basements, confined areas; ground and bond containers; avoid static discharges; avoid strong oxidizers, excessive heat, sparks, open flame; contact with acetyl chloride or other oxidizing agents may result in violent reaction
Hazardous decomp. prods.: Combustion: CO
HMIS: Health 1, Flammability 3, Reactivity 0
Storage: 1 yr shelf life; store under ambient conditions out of direct sunlight

Punctilious® SDA 35A-1 190 Proof [Equistar]
Chem. Descrip.: Specially denatured alcohol 35A: ethanol (88.3%), ethyl acetate (4.5%%), water (7.2%)
UN 1170
Uses: Solvent
Regulatory: SARA §311/312 acute health/chronic health/fire hazard, §313 nonreportable; Calif. Prop. 65 reportable (ethanol, benzene); TSCA export notification required (ethyl acetate)
Properties: Colorless liq.; invisible vapor; sweet alcohol-like odor; sol. in water; sp.gr. 0.8119-0.8130 (77/77 F), 1.59 (vapor); vapor pressure 44.6 mm Hg; f.p. -144 C; b.p. 78.4 C; flash pt. (TCC) 16 C; autoignition temp. 363 C; ref. index 1.3627
Toxicology: May be irritating to eyes, upper respiratory tract; short-term exposure above 1000 ppm by inh. may cause CNS effects, headache, eye/nose/throat irritation; exposure > 1 h may cause dizziness, drowsiness, loss of appetite, inability to conc., GI effects, nausea, vomiting; prolonged/repeated skin contact may cause dermatitis; ing. may cause drunkenness, CNS depression, nausea, vomiting, diarrhea, liver damage, death; TSCA listed
Environmental: On land, apt to volatilize, biodegrade, or leach to ground water; in water, photolysis, oxidation, biodegradation expected to occur; not expected to bioconcentrate in aquatic organisms; LC50 (rainbow trout, 96 h) 10,400 ppm; prevent entry into sewers
Precaution: Flamm. liq.; flamm. limits 3.3-19 vol.%; material can burn with little or no visible flame; vapors are heavier than air, may flash back; prevent entry into basements, confined areas; ground and bond containers; avoid static discharges; avoid strong oxidizers, excessive heat, sparks, open flame; contact with acetyl chloride or other oxidizing agents may result in violent reaction
Hazardous decomp. prods.: Combustion: CO
HMIS: Health 1, Flammability 3, Reactivity 0
Storage: 1 yr shelf life; store under ambient conditions out of direct sunlight

Punctilious® SDA 35A-1 Anhydrous [Equistar]
Chem. Descrip.: Specially denatured alcohol 35A: ethanol (95.4%), ethyl acetate (4.6%)
UN 1170
Uses: Solvent
Regulatory: SARA §311/312 acute health/chronic health/fire hazard, §313 nonreportable; Calif. Prop. 65 reportable (ethanol, benzene); TSCA export notification required (ethyl acetate)
Properties: Colorless liq.; invisible vapor; sweet alcohol-like odor; sol. in water; sp.gr. 0.7909-0.7923 (77/77 F), 1.59 (vapor); vapor pressure 44.6 mm Hg; f.p. -144 C; b.p. 78.4 C; flash pt. (TCC) 13 C; autoignition temp. 363 C; ref. index 1.3603
Toxicology: May be irritating to eyes, upper respiratory tract; short-term exposure above 1000 ppm by inh. may cause CNS effects, headache, eye/nose/throat irritation; exposure > 1 h may cause dizziness, drowsiness, loss of appetite, inability to conc., GI effects, nausea, vomiting; prolonged/repeated skin contact may cause dermatitis; ing. may cause drunkenness, CNS depression, nausea, vomiting, diarrhea, liver damage, death; TSCA listed
Environmental: On land, apt to volatilize, biodegrade, or leach to ground water; in water, photolysis, oxidation, biodegradation expected to occur; not expected to bioconcentrate in aquatic organisms; LC50 (rainbow trout, 96 h) 10,400 ppm; prevent entry into sewers
Precaution: Flamm. liq.; flamm. limits 3.3-19 vol.%; material can burn with little or no visible flame; vapors are heavier than air, may flash back; prevent entry into basements, confined areas; ground and bond containers; avoid static discharges; avoid strong oxidizers, excessive heat, sparks, open flame; contact with acetyl chloride or other oxidizing agents may result in violent reaction
Hazardous decomp. prods.: Combustion: CO
HMIS: Health 1, Flammability 3, Reactivity 0
Storage: 1 yr shelf life; store under ambient conditions out of direct sunlight

Punctilious® SDA 36-2 190 Proof [Equistar]
Chem. Descrip.: Specially denatured alcohol 36-2: ethanol (89.5%), ammonium hydroxide (1.0%), water (9.6%)
UN 1170
Uses: Solvent
Regulatory: SARA §311/312 acute health/chronic health/fire hazard, §313 nonreportable; Calif. Prop. 65 reportable (ethanol, benzene)
Properties: Colorless liq.; invisible vapor; sweet alcohol-like odor; sol. in water; sp.gr. 0.8119-0.8130 (77/77 F), 1.59 (vapor); vapor pressure 44.6 mm Hg; f.p. -144 C; b.p. 78.4 C; flash pt. (TCC) 16 C; autoignition temp. 363 C; ref. index 1.3627
Toxicology: May be irritating to eyes, upper respiratory tract; short-term exposure above 1000 ppm by inh. may cause CNS effects, headache, eye/nose/throat irritation; exposure > 1 h may cause dizziness, drowsiness, loss of appetite, inability to conc., GI effects, nausea, vomiting; prolonged/repeated skin contact may cause dermatitis; ing. may cause drunkenness, CNS depression, nausea, vomiting, diarrhea, liver damage, death; TSCA listed
Environmental: On land, apt to volatilize, biodegrade, or leach to ground water; in water, photolysis, oxidation, biodegradation expected to occur; not expected to bioconcentrate in aquatic organisms; LC50 (rainbow trout, 96 h) 10,400 ppm; prevent entry into sewers
Precaution: Flamm. liq.; flamm. limits 3.3-19 vol.%; material can burn with little or no visible flame; vapors are heavier than air, may flash back; prevent entry into basements, confined areas; ground and bond containers; avoid static discharges; avoid strong oxidizers, excessive heat, sparks, open flame; contact with acetyl chloride or other oxidizing agents may result in violent reaction
Hazardous decomp. prods.: Combustion: CO
HMIS: Health 1, Flammability 3, Reactivity 0
Storage: 1 yr shelf life; store under ambient conditions out of direct sunlight

Punctilious® SDA 37-1 190 Proof [Equistar]
Chem. Descrip.: Specially denatured alcohol 37-1: ethanol (91.6%), thymol (0.3%), eucalyptol (0.4%), l-menthol (0.2%), water (7.5%)
UN 1170
Uses: Solvent

Regulatory: SARA §311/312 acute health/chronic health/fire hazard, §313 nonreportable; Calif. Prop. 65 reportable (ethanol, benzene)
Properties: Colorless liq.; invisible vapor; sweet alcohol-like odor; sol. in water; sp.gr. 0.78-0.85, 1.59 (vapor); vapor pressure 44.6 mm Hg; f.p. -144 C; b.p. 78.4 C; flash pt. (TCC) 16 C; autoignition temp. 363 C
Toxicology: May be irritating to eyes, upper respiratory tract; short-term exposure above 1000 ppm by inh. may cause CNS effects, headache, eye/nose/throat irritation; exposure > 1 h may cause dizziness, drowsiness, loss of appetite, inability to conc., GI effects, nausea, vomiting; prolonged/repeated skin contact may cause dermatitis; ing. may cause drunkenness, CNS depression, nausea, vomiting, diarrhea, liver damage, death; TSCA listed
Environmental: On land, apt to volatilize, biodegrade, or leach to ground water; in water, photolysis, oxidation, biodegradation expected to occur; not expected to bioconcentrate in aquatic organisms; LC50 (rainbow trout, 96 h) 10,400 ppm; prevent entry into sewers
Precaution: Flamm. liq.; flamm. limits 3.3-19 vol.%; material can burn with little or no visible flame; vapors are heavier than air, may flash back; prevent entry into basements, confined areas; ground and bond containers; avoid static discharges; avoid strong oxidizers, excessive heat, sparks, open flame; contact with acetyl chloride or other oxidizing agents may result in violent reaction
Hazardous decomp. prods.: Combustion: CO
HMIS: Health 1, Flammability 3, Reactivity 0
Storage: 1 yr shelf life; store under ambient conditions out of direct sunlight

Punctilious® SDA 38B 190 Proof Prod. Code 13601027 [Equistar]
Chem. Descrip.: Specially denatured alcohol 38B: ethanol (91.1%), methyl 2-hydroxybenzoatel (1.5%), water (7.5%)
UN 1170
Uses: Solvent
Regulatory: SARA §311/312 acute health/chronic health/fire hazard, §313 nonreportable; Calif. Prop. 65 reportable (ethanol, benzene)
Properties: Colorless liq.; invisible vapor; sweet alcohol-like odor; sol. in water; sp.gr. 0.8125-0.8136 (77/77 F), 1.59 (vapor); vapor pressure 44.6 mm Hg; f.p. -144 C; b.p. 78.4 C; flash pt. (TCC) 16 C; autoignition temp. 363 C; ref. index 1.3641
Toxicology: May be irritating to eyes, upper respiratory tract; short-term exposure above 1000 ppm by inh. may cause CNS effects, headache, eye/nose/throat irritation; exposure > 1 h may cause dizziness, drowsiness, loss of appetite, inability to conc., GI effects, nausea, vomiting; prolonged/repeated skin contact may cause dermatitis; ing. may cause drunkenness, CNS depression, nausea, vomiting, diarrhea, liver damage, death; TSCA listed
Environmental: On land, apt to volatilize, biodegrade, or leach to ground water; in water, photolysis, oxidation, biodegradation expected to occur; not expected to bioconcentrate in aquatic organisms; LC50 (rainbow trout, 96 h) 10,400 ppm; prevent entry into sewers
Precaution: Flamm. liq.; flamm. limits 3.3-19 vol.%; material can burn with little or no visible flame; vapors are heavier than air, may flash back; prevent entry into basements, confined areas; ground and bond containers; avoid static discharges; avoid strong oxidizers, excessive heat, sparks, open flame; contact with acetyl chloride or other oxidizing agents may result in violent reaction
Hazardous decomp. prods.: Combustion: CO
HMIS: Health 1, Flammability 3, Reactivity 0
Storage: 1 yr shelf life; store under ambient conditions out of direct sunlight

Punctilious® SDA 38B 190 Proof Prod. Code 13602007 [Equistar]
Chem. Descrip.: Specially denatured alcohol 38B: ethanol (90.1%), spearmint oil, terpeneless (1.1%), peppermint oil (0.4%), water (7.5%)
UN 1170
Uses: Solvent
Regulatory: SARA §311/312 acute health/chronic health/fire hazard, §313 nonreportable; Calif. Prop. 65 reportable (ethanol, benzene)
Properties: Colorless liq.; invisible vapor; sweet alcohol-like odor; sol. in water; sp.gr. 0.8100-0.8111 (77/77 F), 1.59 (vapor); vapor pressure 44.6 mm Hg; f.p. -144 C; b.p. 78.4 C; flash pt. (TCC) 16 C; autoignition temp. 363 C; ref. index 1.3626
Toxicology: May be irritating to eyes, upper respiratory tract; short-term exposure above 1000 ppm by inh. may cause CNS effects, headache, eye/nose/throat irritation; exposure > 1 h may cause dizziness, drowsiness, loss of appetite, inability to conc., GI effects, nausea, vomiting; prolonged/repeated skin contact may cause dermatitis; ing. may cause drunkenness, CNS depression, nausea, vomiting, diarrhea, liver damage, death; TSCA listed
Environmental: On land, apt to volatilize, biodegrade, or leach to ground water; in water, photolysis, oxidation, biodegradation expected to occur; not expected to bioconcentrate in aquatic organisms; LC50 (rainbow trout, 96 h) 10,400 ppm; prevent entry into sewers
Precaution: Flamm. liq.; flamm. limits 3.3-19 vol.%; material can burn with little or no visible flame; vapors are heavier than air, may flash back; prevent entry into basements, confined areas; ground and bond containers; avoid static discharges; avoid strong oxidizers, excessive heat, sparks, open flame; contact with acetyl chloride or other oxidizing agents may result in violent reaction
Hazardous decomp. prods.: Combustion: CO
HMIS: Health 1, Flammability 3, Reactivity 0
Storage: 1 yr shelf life; store under ambient conditions out of direct sunlight

Punctilious® SDA 38B 190 Proof Prod. Code 13602027 [Equistar]
Chem. Descrip.: Specially denatured alcohol 38B: ethanol (91.1%), l-menthol (0.7%), clove oil (0.7%), water (7.5%)
UN 1170
Uses: Solvent
Regulatory: SARA §311/312 acute health/chronic health/fire hazard, §313 nonreportable; Calif. Prop. 65 reportable (ethanol, benzene)
Properties: Colorless liq.; invisible vapor; sweet alcohol-like odor; sol. in water; sp.gr. 0.8105-0.8117 (77/77 F), 1.59 (vapor); vapor pressure 44.6 mm Hg; f.p. -144 C; b.p. 78.4 C; flash pt. (TCC) 16 C; autoignition temp. 363 C
Toxicology: May be irritating to eyes, upper respiratory tract; short-term exposure above 1000 ppm by inh. may cause CNS effects, headache, eye/nose/throat irritation; exposure > 1 h may cause dizziness, drowsiness, loss of appetite, inability to conc., GI effects, nausea, vomiting; prolonged/repeated skin contact may cause dermatitis; ing. may cause drunkenness, CNS depression, nausea, vomiting, diarrhea, liver damage, death; TSCA listed
Environmental: On land, apt to volatilize, biodegrade, or leach to ground water; in water, photolysis, oxidation, biodegradation expected to occur; not expected to bioconcentrate in aquatic organisms; LC50 (rainbow trout, 96 h) 10,400 ppm; prevent entry into sewers
Precaution: Flamm. liq.; flamm. limits 3.3-19 vol.%; material can burn with little or no visible flame; vapors are heavier than air, may flash back; prevent entry into basements, confined areas; ground and bond containers; avoid static discharges; avoid strong oxidizers, excessive heat, sparks, open flame; contact with acetyl chloride or other oxidizing agents may result in violent reaction
Hazardous decomp. prods.: Combustion: CO
HMIS: Health 1, Flammability 3, Reactivity 0
Storage: 1 yr shelf life; store under ambient conditions out of direct sunlight

Punctilious® SDA 38B 190 Proof Prod. Code 13602133 [Equistar]
Chem. Descrip.: Specially denatured alcohol 38B: ethanol (91.1%), 2-methoxy (propenyl) phenol (1.0%), eucalyptol (0.4%), water (7.5%)
UN 1170
Uses: Solvent
Regulatory: SARA §311/312 acute health/chronic health/fire hazard, §313 nonreportable; Calif. Prop. 65 reportable (ethanol, benzene)
Properties: Colorless liq.; invisible vapor; sweet alcohol-like odor; sol. in water; sp.gr. 0.8111-0.8126 (77/77 F), 1.59 (vapor); vapor pressure 44.6 mm Hg; f.p. -144 C; b.p. 78.4 C; flash pt. (TCC) 16 C; autoignition temp. 363 C
Toxicology: May be irritating to eyes, upper respiratory tract; short-term exposure above 1000 ppm by inh. may cause CNS effects, headache, eye/nose/throat irritation; exposure > 1 h may cause dizziness, drowsiness, loss of appetite, inability to conc., GI effects, nausea, vomiting; prolonged/repeated skin contact may cause dermatitis; ing. may cause drunkenness, CNS depression, nausea, vomiting, diarrhea, liver damage, death; TSCA listed
Environmental: On land, apt to volatilize, biodegrade, or leach to ground water; in water, photolysis, oxidation, biodegradation expected to occur; not expected to bioconcentrate in aquatic organisms; LC50 (rainbow trout, 96 h) 10,400 ppm; prevent entry into sewers

Precaution: Flamm. liq.; flamm. limits 3.3-19 vol.%; material can burn with little or no visible flame; vapors are heavier than air, may flash back; prevent entry into basements, confined areas; ground and bond containers; avoid static discharges; avoid strong oxidizers, excessive heat, sparks, open flame; contact with acetyl chloride or other oxidizing agents may result in violent reaction
Hazardous decomp. prods.: Combustion: CO
HMIS: Health 1, Flammability 3, Reactivity 0
Storage: 1 yr shelf life; store under ambient conditions out of direct sunlight

Punctilious® SDA 38B 190 Proof Prod. Code 13603998 [Equistar]
Chem. Descrip.: Specially denatured alcohol 38B: ethanol (91.1%), syn. camphor (0.3%), D-menthol (0.4%), phenol (0.8%), water (7.5%)
UN 1170
Uses: Solvent
Regulatory: SARA §311/312 acute health/chronic health/fire hazard, §313 reportable (phenol); Calif. Prop. 65 reportable (ethanol, benzene); TSCA export notification required (phenol)
Properties: Colorless liq.; invisible vapor; sweet alcohol-like odor; sol. in water; sp.gr. 0.8109-0.8121 (77/77 F), 1.59 (vapor); vapor pressure 44.6 mm Hg; f.p. -144 C; b.p. 78.4 C; flash pt. (TCC) 16 C; autoignition temp. 363 C; ref. index 1.3633
Toxicology: May be irritating to eyes, upper respiratory tract; short-term exposure above 1000 ppm by inh. may cause CNS effects, headache, eye/nose/throat irritation; exposure > 1 h may cause dizziness, drowsiness, loss of appetite, inability to conc., GI effects, nausea, vomiting; prolonged/repeated skin contact may cause dermatitis; ing. may cause drunkenness, CNS depression, nausea, vomiting, diarrhea, liver damage, death; TSCA listed
Environmental: On land, apt to volatilize, biodegrade, or leach to ground water; in water, photolysis, oxidation, biodegradation expected to occur; not expected to bioconcentrate in aquatic organisms; LC50 (rainbow trout, 96 h) 10,400 ppm; prevent entry into sewers
Precaution: Flamm. liq.; flamm. limits 3.3-19 vol.%; material can burn with little or no visible flame; vapors are heavier than air, may flash back; prevent entry into basements, confined areas; ground and bond containers; avoid static discharges; avoid strong oxidizers, excessive heat, sparks, open flame; contact with acetyl chloride or other oxidizing agents may result in violent reaction
Hazardous decomp. prods.: Combustion: CO
HMIS: Health 1, Flammability 3, Reactivity 0
Storage: 1 yr shelf life; store under ambient conditions out of direct sunlight

Punctilious® SDA 38B 200 Proof Prod. Code 14601023 [Equistar]
Chem. Descrip.: Specially denatured alcohol 38B: ethanol (98.5%), syn. camphor (1.5%)
UN 1170
Uses: Solvent
Regulatory: SARA §311/312 acute health/chronic health/fire hazard, §313 nonreportable; Calif. Prop. 65 reportable (ethanol, benzene)
Properties: Colorless liq.; invisible vapor; sweet alcohol-like odor; sol. in water; sp.gr. 0.7892-0.7902 (77/77 F), 1.59 (vapor); vapor pressure 44.6 mm Hg; f.p. -144 C; b.p. 78.4 C; flash pt. (TCC) 13 C; autoignition temp. 363 C; ref. index 1.3601
Toxicology: May be irritating to eyes, upper respiratory tract; short-term exposure above 1000 ppm by inh. may cause CNS effects, headache, eye/nose/throat irritation; exposure > 1 h may cause dizziness, drowsiness, loss of appetite, inability to conc., GI effects, nausea, vomiting; prolonged/repeated skin contact may cause dermatitis; ing. may cause drunkenness, CNS depression, nausea, vomiting, diarrhea, liver damage, death; TSCA listed
Environmental: On land, apt to volatilize, biodegrade, or leach to ground water; in water, photolysis, oxidation, biodegradation expected to occur; not expected to bioconcentrate in aquatic organisms; LC50 (rainbow trout, 96 h) 10,400 ppm; prevent entry into sewers
Precaution: Flamm. liq.; flamm. limits 3.3-19 vol.%; material can burn with little or no visible flame; vapors are heavier than air, may flash back; prevent entry into basements, confined areas; ground and bond containers; avoid static discharges; avoid strong oxidizers, excessive heat, sparks, open flame; contact with acetyl chloride or other oxidizing agents may result in violent reaction
Hazardous decomp. prods.: Combustion: CO
HMIS: Health 1, Flammability 3, Reactivity 0
Storage: 1 yr shelf life; store under ambient conditions out of direct sunlight

Punctilious® SDA 38B 200 Proof Prod. Code 14601040 [Equistar]
Chem. Descrip.: Specially denatured alcohol 38B: ethanol (98.5%), D-menthol (1.5%)
UN 1170
Uses: Solvent
Regulatory: SARA §311/312 acute health/chronic health/fire hazard, §313 nonreportable; Calif. Prop. 65 reportable (ethanol, benzene)
Properties: Colorless liq.; invisible vapor; sweet alcohol-like odor; sol. in water; sp.gr. 0.7882-0.7892 (77/77 F), 1.59 (vapor); vapor pressure 44.6 mm Hg; f.p. -144 C; b.p. 78.4 C; flash pt. (TCC) 13 C; autoignition temp. 363 C; ref. index 1.3605
Toxicology: May be irritating to eyes, upper respiratory tract; short-term exposure above 1000 ppm by inh. may cause CNS effects, headache, eye/nose/throat irritation; exposure > 1 h may cause dizziness, drowsiness, loss of appetite, inability to conc., GI effects, nausea, vomiting; prolonged/repeated skin contact may cause dermatitis; ing. may cause drunkenness, CNS depression, nausea, vomiting, diarrhea, liver damage, death; TSCA listed
Environmental: On land, apt to volatilize, biodegrade, or leach to ground water; in water, photolysis, oxidation, biodegradation expected to occur; not expected to bioconcentrate in aquatic organisms; LC50 (rainbow trout, 96 h) 10,400 ppm; prevent entry into sewers
Precaution: Flamm. liq.; flamm. limits 3.3-19 vol.%; material can burn with little or no visible flame; vapors are heavier than air, may flash back; prevent entry into basements, confined areas; ground and bond containers; avoid static discharges; avoid strong oxidizers, excessive heat, sparks, open flame; contact with acetyl chloride or other oxidizing agents may result in violent reaction
Hazardous decomp. prods.: Combustion: CO
HMIS: Health 1, Flammability 3, Reactivity 0
Storage: 1 yr shelf life; store under ambient conditions out of direct sunlight

Punctilious® SDA 38B Anhydrous Prod. Code 14680000 [Equistar]
Chem. Descrip.: Specially denatured alcohol 38B: ethanol (96.5%), diethyl phthalate (3.4%), t-butanol (0.1%)
UN 1170
Uses: Solvent
Regulatory: SARA §311/312 acute health/chronic health/fire hazard, §313 reportable (t-butanol); Calif. Prop. 65 reportable (ethanol, benzene)
Properties: Colorless liq.; invisible vapor; sweet alcohol-like odor; sol. in water; sp.gr. 0.7963-0.7972 (77/77 F), 1.59 (vapor); vapor pressure 44.6 mm Hg; f.p. -144 C; b.p. 78.4 C; flash pt. (TCC) 13 C; autoignition temp. 363 C; ref. index 1.3641
Toxicology: May be irritating to eyes, upper respiratory tract; short-term exposure above 1000 ppm by inh. may cause CNS effects, headache, eye/nose/throat irritation; exposure > 1 h may cause dizziness, drowsiness, loss of appetite, inability to conc., GI effects, nausea, vomiting; prolonged/repeated skin contact may cause dermatitis; ing. may cause drunkenness, CNS depression, nausea, vomiting, diarrhea, liver damage, death; TSCA listed
Environmental: On land, apt to volatilize, biodegrade, or leach to ground water; in water, photolysis, oxidation, biodegradation expected to occur; not expected to bioconcentrate in aquatic organisms; LC50 (rainbow trout, 96 h) 10,400 ppm; prevent entry into sewers
Precaution: Flamm. liq.; flamm. limits 3.3-19 vol.%; material can burn with little or no visible flame; vapors are heavier than air, may flash back; prevent entry into basements, confined areas; ground and bond containers; avoid static discharges; avoid strong oxidizers, excessive heat, sparks, open flame; contact with acetyl chloride or other oxidizing agents may result in violent reaction
Hazardous decomp. prods.: Combustion: CO
HMIS: Health 1, Flammability 3, Reactivity 0
Storage: 1 yr shelf life; store under ambient conditions out of direct sunlight

Punctilious® SDA 38F 190 Proof [Equistar]
Chem. Descrip.: Specially denatured alcohol 38F: ethanol (91.1%), boric acid (1.0%), peppermint oil (0.2%), methyl 2-hydroxybenzoate (0.2%), water (7.4%)
UN 1170

Uses: Solvent
Regulatory: SARA §311/312 acute health/chronic health/fire hazard, §313 nonreportable; Calif. Prop. 65 reportable (ethanol, benzene)
Properties: Colorless liq.; invisible vapor; sweet alcohol-like odor; sol. in water; sp.gr. 0.8132-0.8143 (77/77 F), 1.59 (vapor); vapor pressure 44.6 mm Hg; f.p. -144 C; b.p. 78.4 C; flash pt. (TCC) 16 C; autoignition temp. 363 C; ref. index 1.3630
Toxicology: May be irritating to eyes, upper respiratory tract; short-term exposure above 1000 ppm by inh. may cause CNS effects, headache, eye/nose/throat irritation; exposure > 1 h may cause dizziness, drowsiness, loss of appetite, inability to conc., GI effects, nausea, vomiting; prolonged/repeated skin contact may cause dermatitis; ing. may cause drunkenness, CNS depression, nausea, vomiting, diarrhea, liver damage, death; TSCA listed
Environmental: On land, apt to volatilize, biodegrade, or leach to ground water; in water, photolysis, oxidation, biodegradation expected to occur; not expected to bioconcentrate in aquatic organisms; LC50 (rainbow trout, 96 h) 10,400 ppm; prevent entry into sewers
Precaution: Flamm. liq.; flamm. limits 3.3-19 vol.%; material can burn with little or no visible flame; vapors are heavier than air, may flash back; prevent entry into basements, confined areas; ground and bond containers; avoid static discharges; avoid strong oxidizers, excessive heat, sparks, open flame; contact with acetyl chloride or other oxidizing agents may result in violent reaction
Hazardous decomp. prods.: Combustion: CO
HMIS: Health 1, Flammability 3, Reactivity 0
Storage: 1 yr shelf life; store under ambient conditions out of direct sunlight

Punctilious® SDA 38F-2 190 Proof Prod. Code 13643020 [Equistar]
Chem. Descrip.: Specially denatured alcohol 38F-2: ethanol (91.1%), boric acid (1.0%), peppermint oil (0.2%), methyl 2-hydroxybenzoate (0.2%), water (7.4%)
UN 1170
Uses: Solvent
Regulatory: SARA §311/312 acute health/chronic health/fire hazard, §313 nonreportable; Calif. Prop. 65 reportable (ethanol, benzene)
Properties: Colorless liq.; invisible vapor; sweet alcohol-like odor; sol. in water; sp.gr. 0.8134-0.8149 (77/77 F), 1.59 (vapor); vapor pressure 44.6 mm Hg; f.p. -144 C; b.p. 78.4 C; flash pt. (TCC) 16 C; autoignition temp. 363 C
Toxicology: May be irritating to eyes, upper respiratory tract; short-term exposure above 1000 ppm by inh. may cause CNS effects, headache, eye/nose/throat irritation; exposure > 1 h may cause dizziness, drowsiness, loss of appetite, inability to conc., GI effects, nausea, vomiting; prolonged/repeated skin contact may cause dermatitis; ing. may cause drunkenness, CNS depression, nausea, vomiting, diarrhea, liver damage, death; TSCA listed
Environmental: On land, apt to volatilize, biodegrade, or leach to ground water; in water, photolysis, oxidation, biodegradation expected to occur; not expected to bioconcentrate in aquatic organisms; LC50 (rainbow trout, 96 h) 10,400 ppm; prevent entry into sewers
Precaution: Flamm. liq.; flamm. limits 3.3-19 vol.%; material can burn with little or no visible flame; vapors are heavier than air, may flash back; prevent entry into basements, confined areas; ground and bond containers; avoid static discharges; avoid strong oxidizers, excessive heat, sparks, open flame; contact with acetyl chloride or other oxidizing agents may result in violent reaction
Hazardous decomp. prods.: Combustion: CO
HMIS: Health 1, Flammability 3, Reactivity 0
Storage: 1 yr shelf life; store under ambient conditions out of direct sunlight

Punctilious® SDA 38F-2 190 Proof Prod. Code 13645006 [Equistar]
Chem. Descrip.: Specially denatured alcohol 38F-2: ethanol (91.1%), boric acid (1.0%), D-menthol (0.3%), peppermint oil (0.06%), clove oil (0.06%), cinnamic aldehyde (0.05%), water (7.4%)
UN 1170
Uses: Solvent
Regulatory: SARA §311/312 acute health/chronic health/fire hazard, §313 nonreportable; Calif. Prop. 65 reportable (ethanol, benzene)
Properties: Colorless liq.; invisible vapor; sweet alcohol-like odor; sol. in water; sp.gr. 0.8130-0.8145 (77/77 F), 1.59 (vapor); vapor pressure 44.6 mm Hg; f.p. -144 C; b.p. 78.4 C; flash pt. (TCC) 16 C; autoignition temp. 363 C; ref. index 1.3633
Toxicology: May be irritating to eyes, upper respiratory tract; short-term exposure above 1000 ppm by inh. may cause CNS effects, headache, eye/nose/throat irritation; exposure > 1 h may cause dizziness, drowsiness, loss of appetite, inability to conc., GI effects, nausea, vomiting; prolonged/repeated skin contact may cause dermatitis; ing. may cause drunkenness, CNS depression, nausea, vomiting, diarrhea, liver damage, death; TSCA listed
Environmental: On land, apt to volatilize, biodegrade, or leach to ground water; in water, photolysis, oxidation, biodegradation expected to occur; not expected to bioconcentrate in aquatic organisms; LC50 (rainbow trout, 96 h) 10,400 ppm; prevent entry into sewers
Precaution: Flamm. liq.; flamm. limits 3.3-19 vol.%; material can burn with little or no visible flame; vapors are heavier than air, may flash back; prevent entry into basements, confined areas; ground and bond containers; avoid static discharges; avoid strong oxidizers, excessive heat, sparks, open flame; contact with acetyl chloride or other oxidizing agents may result in violent reaction
Hazardous decomp. prods.: Combustion: CO
HMIS: Health 1, Flammability 3, Reactivity 0
Storage: 1 yr shelf life; store under ambient conditions out of direct sunlight

Punctilious® SDA 38F-6 190 Proof [Equistar]
Chem. Descrip.: Specially denatured alcohol 38F-6: ethanol (91.1%), POE (20) sorbitan oleate (1.0%), peppermint oil (0.3%), l-menthol (0.1%), water (7.5%)
UN 1170
Uses: Solvent
Regulatory: SARA §311/312 acute health/chronic health/fire hazard, §313 nonreportable; Calif. Prop. 65 reportable (ethanol, benzene)
Properties: Colorless liq.; invisible vapor; sweet alcohol-like odor; sol. in water; sp.gr. 0.8111-0.8123 (77/77 F), 1.59 (vapor); vapor pressure 44.6 mm Hg; f.p. -144 C; b.p. 78.4 C; flash pt. (TCC) 16 C; autoignition temp. 363 C; ref. index 1.3625
Toxicology: May be irritating to eyes, upper respiratory tract; short-term exposure above 1000 ppm by inh. may cause CNS effects, headache, eye/nose/throat irritation; exposure > 1 h may cause dizziness, drowsiness, loss of appetite, inability to conc., GI effects, nausea, vomiting; prolonged/repeated skin contact may cause dermatitis; ing. may cause drunkenness, CNS depression, nausea, vomiting, diarrhea, liver damage, death; TSCA listed
Environmental: On land, apt to volatilize, biodegrade, or leach to ground water; in water, photolysis, oxidation, biodegradation expected to occur; not expected to bioconcentrate in aquatic organisms; LC50 (rainbow trout, 96 h) 10,400 ppm; prevent entry into sewers
Precaution: Flamm. liq.; flamm. limits 3.3-19 vol.%; material can burn with little or no visible flame; vapors are heavier than air, may flash back; prevent entry into basements, confined areas; ground and bond containers; avoid static discharges; avoid strong oxidizers, excessive heat, sparks, open flame; contact with acetyl chloride or other oxidizing agents may result in violent reaction
Hazardous decomp. prods.: Combustion: CO
HMIS: Health 1, Flammability 3, Reactivity 0
Storage: 1 yr shelf life; store under ambient conditions out of direct sunlight

Punctilious® SDA 39B 190 Proof [Equistar]
Chem. Descrip.: Specially denatured alcohol 39B: ethanol (89.2%), diethyl phthalate (3.3%), t-butanol (0.1%), water (7.3%)
UN 1170
Uses: Solvent
Regulatory: SARA §311/312 acute health/chronic health/fire hazard, §313 reportable (t-butanol); Calif. Prop. 65 reportable (ethanol, benzene)
Properties: Colorless liq.; invisible vapor; sweet alcohol-like odor; sol. in water; sp.gr. 0.8162-0.8172 (77/77 F), 1.59 (vapor); vapor pressure 44.6 mm Hg; f.p. -144 C; b.p. 78.4 C; flash pt. (TCC) 16 C; autoignition temp. 363 C; ref. index 1.36536
Toxicology: May be irritating to eyes, upper respiratory tract; short-term exposure above 1000 ppm by inh. may cause CNS effects, headache, eye/nose/throat irritation; exposure > 1 h may cause dizziness, drowsiness, loss of appetite, inability to conc., GI effects, nausea, vomiting; prolonged/repeated skin contact may cause dermatitis; ing. may cause drunkenness, CNS depression, nausea, vomiting, diarrhea, liver dam-

age, death; TSCA listed
Environmental: On land, apt to volatilize, biodegrade, or leach to ground water; in water, photolysis, oxidation, biodegradation expected to occur; not expected to bioconcentrate in aquatic organisms; LC50 (rainbow trout, 96 h) 10,400 ppm; prevent entry into sewers
Precaution: Flamm. liq.; flamm. limits 3.3-19 vol.%; material can burn with little or no visible flame; vapors are heavier than air, may flash back; prevent entry into basements, confined areas; ground and bond containers; avoid static discharges; avoid strong oxidizers, excessive heat, sparks, open flame; contact with acetyl chloride or other oxidizing agents may result in violent reaction
Hazardous decomp. prods.: Combustion: CO
HMIS: Health 1, Flammability 3, Reactivity 0
Storage: 1 yr shelf life; store under ambient conditions out of direct sunlight

Punctilious® SDA 39C Anhydrous [Equistar]
Chem. Descrip.: Specially denatured alcohol 39C: ethanol (98.6%), diethyl phthalate (1.4%)
UN 1170
Uses: Solvent
Regulatory: SARA §311/312 acute health/chronic health/fire hazard, §313 nonreportable; Calif. Prop. 65 reportable (ethanol, benzene)
Properties: Colorless liq.; invisible vapor; sweet alcohol-like odor; sol. in water; sp.gr. 0.7899-0.7914 (77/77 F), 1.59 (vapor); vapor pressure 44.6 mm Hg; f.p. -144 C; b.p. 78.4 C; flash pt. (TCC) 13 C; autoignition temp. 363 C; ref. index 1.3611
Toxicology: May be irritating to eyes, upper respiratory tract; short-term exposure above 1000 ppm by inh. may cause CNS effects, headache, eye/nose/throat irritation; exposure > 1 h may cause dizziness, drowsiness, loss of appetite, inability to conc., GI effects, nausea, vomiting; prolonged/repeated skin contact may cause dermatitis; ing. may cause drunkenness, CNS depression, nausea, vomiting, diarrhea, liver damage, death; TSCA listed
Environmental: On land, apt to volatilize, biodegrade, or leach to ground water; in water, photolysis, oxidation, biodegradation expected to occur; not expected to bioconcentrate in aquatic organisms; LC50 (rainbow trout, 96 h) 10,400 ppm; prevent entry into sewers
Precaution: Flamm. liq.; flamm. limits 3.3-19 vol.%; material can burn with little or no visible flame; vapors are heavier than air, may flash back; prevent entry into basements, confined areas; ground and bond containers; avoid static discharges; avoid strong oxidizers, excessive heat, sparks, open flame; contact with acetyl chloride or other oxidizing agents may result in violent reaction
Hazardous decomp. prods.: Combustion: CO
HMIS: Health 1, Flammability 3, Reactivity 0
Storage: 1 yr shelf life; store under ambient conditions out of direct sunlight

Punctilious® SDA 40-1 190 Proof [Equistar]
Chem. Descrip.: Specially denatured alcohol 40-1: ethanol (92.3%), t-butanol (0.1%), brucine alkaloid (0.01%), water (7.6%)
UN 1170
Uses: Solvent
Regulatory: SARA §311/312 acute health/chronic health/fire hazard, §313 reportable (brucine alkaloid, t-butanol); Calif. Prop. 65 reportable (ethanol, benzene)
Properties: Colorless liq.; invisible vapor; sweet alcohol-like odor; sol. in water; sp.gr. 0.8088-0.8098 (77/77 F), 1.59 (vapor); vapor pressure 44.6 mm Hg; f.p. -144 C; b.p. 78.4 C; flash pt. (TCC) 16 C; autoignition temp. 363 C; ref. index 1.3619
Toxicology: May be irritating to eyes, upper respiratory tract; short-term exposure above 1000 ppm by inh. may cause CNS effects, headache, eye/nose/throat irritation; exposure > 1 h may cause dizziness, drowsiness, loss of appetite, inability to conc., GI effects, nausea, vomiting; prolonged/repeated skin contact may cause dermatitis; ing. may cause drunkenness, CNS depression, nausea, vomiting, diarrhea, liver damage, death; TSCA listed
Environmental: On land, apt to volatilize, biodegrade, or leach to ground water; in water, photolysis, oxidation, biodegradation expected to occur; not expected to bioconcentrate in aquatic organisms; LC50 (rainbow trout, 96 h) 10,400 ppm; prevent entry into sewers
Precaution: Flamm. liq.; flamm. limits 3.3-19 vol.%; material can burn with little or no visible flame; vapors are heavier than air, may flash back; prevent entry into basements, confined areas; ground and bond containers; avoid static discharges; avoid strong oxidizers, excessive heat, sparks, open flame; contact with acetyl chloride or other oxidizing agents may result in violent reaction
Hazardous decomp. prods.: Combustion: CO
HMIS: Health 1, Flammability 3, Reactivity 0
Storage: 1 yr shelf life; store under ambient conditions out of direct sunlight

Punctilious® SDA 40-1 Anhydrous [Equistar]
Chem. Descrip.: Specially denatured alcohol 40-1: ethanol (99.9%), t-butanol (0.1%), brucine alkaloid (0.01%)
UN 1170
Uses: Solvent
Regulatory: SARA §311/312 acute health/chronic health/fire hazard, §313 reportable (brucine alkaloid, t-butanol); Calif. Prop. 65 reportable (ethanol, benzene)
Properties: Colorless liq.; invisible vapor; sweet alcohol-like odor; sol. in water; sp.gr. 0.7868-0.7879 (77/77 F), 1.59 (vapor); vapor pressure 44.6 mm Hg; f.p. -144 C; b.p. 78.4 C; flash pt. (TCC) 13 C; autoignition temp. 363 C; ref. index 1.3607
Toxicology: LD50 (oral, rat) 7060 mg/kg; may be irritating to eyes, upper respiratory tract; short-term exposure above 1000 ppm by inh. may cause CNS effects, headache, eye/nose/throat irritation; exposure > 1 h may cause dizziness, drowsiness, loss of appetite, inability to conc., GI effects, nausea, vomiting; prolonged/repeated skin contact may cause dermatitis; ing. may cause drunkenness, CNS depression, nausea, vomiting, diarrhea, liver damage, death; TSCA listed
Environmental: On land, apt to volatilize, biodegrade, or leach to ground water; in water, photolysis, oxidation, biodegradation expected to occur; not expected to bioconcentrate in aquatic organisms; LC50 (rainbow trout, 96 h) 10,400 ppm; prevent entry into sewers
Precaution: Flamm. liq.; flamm. limits 3.3-19 vol.%; material can burn with little or no visible flame; vapors are heavier than air, may flash back; prevent entry into basements, confined areas; ground and bond containers; avoid static discharges; avoid strong oxidizers, excessive heat, sparks, open flame; contact with acetyl chloride or other oxidizing agents may result in violent reaction
Hazardous decomp. prods.: Combustion: CO
HMIS: Health 1, Flammability 3, Reactivity 0
Storage: 1 yr shelf life; store under ambient conditions out of direct sunlight

Punctilious® SDA 40-2 Anhydrous [Equistar]
Chem. Descrip.: Specially denatured alcohol 40-2: ethanol (99.9%), t-butanol (0.1%), brucine sulfate (2:1) heptahydrate (0.01%)
UN 1170
Uses: Solvent
Regulatory: SARA §311/312 acute health/chronic health/fire hazard, §313 reportable (t-butanol); Calif. Prop. 65 reportable (ethanol, benzene)
Properties: Colorless liq.; invisible vapor; sweet alcohol-like odor; sol. in water; sp.gr. 0.7868-0.7879 (77/77 F), 1.59 (vapor); vapor pressure 44.6 mm Hg; f.p. -144 C; b.p. 78.4 C; flash pt. (TCC) 13 C; autoignition temp. 363 C; ref. index 1.3593
Toxicology: LD50 (oral, rat) 7060 mg/kg; may be irritating to eyes, upper respiratory tract; short-term exposure above 1000 ppm by inh. may cause CNS effects, headache, eye/nose/throat irritation; exposure > 1 h may cause dizziness, drowsiness, loss of appetite, inability to conc., GI effects, nausea, vomiting; prolonged/repeated skin contact may cause dermatitis; ing. may cause drunkenness, CNS depression, nausea, vomiting, diarrhea, liver damage, death; TSCA listed
Environmental: On land, apt to volatilize, biodegrade, or leach to ground water; in water, photolysis, oxidation, biodegradation expected to occur; not expected to bioconcentrate in aquatic organisms; LC50 (rainbow trout, 96 h) 10,400 ppm; prevent entry into sewers
Precaution: Flamm. liq.; flamm. limits 3.3-19 vol.%; material can burn with little or no visible flame; vapors are heavier than air, may flash back; prevent entry into basements, confined areas; ground and bond containers; avoid static discharges; avoid strong oxidizers, excessive heat, sparks, open flame; contact with acetyl chloride or other oxidizing agents may result in violent reaction
Hazardous decomp. prods.: Combustion: CO
HMIS: Health 1, Flammability 3, Reactivity 0
Storage: 1 yr shelf life; store under ambient conditions out of direct sun-

light

Punctilious® SDA 40A 190 Proof [Equistar]
Chem. Descrip.: Specially denatured alcohol 40A: ethanol (92.2%), sucrose octaacetate (0.2%), t-butanol (0.1%), water (7.6%)
UN 1170
Uses: Solvent
Regulatory: SARA §311/312 acute health/chronic health/fire hazard, §313 reportable (t-butanol); Calif. Prop. 65 reportable (ethanol, benzene)
Properties: Colorless liq.; invisible vapor; sweet alcohol-like odor; sol. in water; sp.gr. 0.8192-0.8104 (77/77 F), 1.59 (vapor); vapor pressure 44.6 mm Hg; f.p. -144 C; b.p. 78.4 C; flash pt. (TCC) 16 C; autoignition temp. 363 C; ref. index 1.3624
Toxicology: May be irritating to eyes, upper respiratory tract; short-term exposure above 1000 ppm by inh. may cause CNS effects, headache, eye/nose/throat irritation; exposure > 1 h may cause dizziness, drowsiness, loss of appetite, inability to conc., GI effects, nausea, vomiting; prolonged/repeated skin contact may cause dermatitis; ing. may cause drunkenness, CNS depression, nausea, vomiting, diarrhea, liver damage, death; TSCA listed
Environmental: On land, apt to volatilize, biodegrade, or leach to ground water; in water, photolysis, oxidation, biodegradation expected to occur; not expected to bioconcentrate in aquatic organisms; LC50 (rainbow trout, 96 h) 10,400 ppm; prevent entry into sewers
Precaution: Flamm. liq.; flamm. limits 3.3-19 vol.%; material can burn with little or no visible flame; vapors are heavier than air, may flash back; prevent entry into basements, confined areas; ground and bond containers; avoid static discharges; avoid strong oxidizers, excessive heat, sparks, open flame; contact with acetyl chloride or other oxidizing agents may result in violent reaction
Hazardous decomp. prods.: Combustion: CO
HMIS: Health 1, Flammability 3, Reactivity 0
Storage: 1 yr shelf life; store under ambient conditions out of direct sunlight

Punctilious® SDA 40A Anhydrous [Equistar]
Chem. Descrip.: Specially denatured alcohol 40A: ethanol (99.7%), sucrose octaacetate (0.2%), t-butanol (0.1%)
UN 1170
Uses: Solvent
Regulatory: SARA §311/312 acute health/chronic health/fire hazard, §313 reportable (t-butanol); Calif. Prop. 65 reportable (ethanol, benzene)
Properties: Colorless liq.; invisible vapor; sweet alcohol-like odor; sol. in water; sp.gr. 0.7874-0.7884 (77/77 F), 1.59 (vapor); vapor pressure 44.6 mm Hg; f.p. -144 C; b.p. 78.4 C; flash pt. (TCC) 13 C; autoignition temp. 363 C; ref. index 1.3602
Toxicology: LD50 (oral, rat) 7060 mg/kg; may be irritating to eyes, upper respiratory tract; short-term exposure above 1000 ppm by inh. may cause CNS effects, headache, eye/nose/throat irritation; exposure > 1 h may cause dizziness, drowsiness, loss of appetite, inability to conc., GI effects, nausea, vomiting; prolonged/repeated skin contact may cause dermatitis; ing. may cause drunkenness, CNS depression, nausea, vomiting, diarrhea, liver damage, death; TSCA listed
Environmental: On land, apt to volatilize, biodegrade, or leach to ground water; in water, photolysis, oxidation, biodegradation expected to occur; not expected to bioconcentrate in aquatic organisms; LC50 (rainbow trout, 96 h) 10,400 ppm; prevent entry into sewers
Precaution: Flamm. liq.; flamm. limits 3.3-19 vol.%; material can burn with little or no visible flame; vapors are heavier than air, may flash back; prevent entry into basements, confined areas; ground and bond containers; avoid static discharges; avoid strong oxidizers, excessive heat, sparks, open flame; contact with acetyl chloride or other oxidizing agents may result in violent reaction
Hazardous decomp. prods.: Combustion: CO
HMIS: Health 1, Flammability 3, Reactivity 0
Storage: 1 yr shelf life; store under ambient conditions out of direct sunlight

Punctilious® SDA 40B Anhydrous [Equistar]
Chem. Descrip.: Specially denatured alcohol 40B: ethanol (99.9%), t-butanol (0.1%), denatonium benzoate (0.0006%)
UN 1170
Uses: Solvent
Regulatory: SARA §311/312 acute health/chronic health/fire hazard, §313 reportable (t-butanol); Calif. Prop. 65 reportable (ethanol, benzene)
Properties: Colorless liq.; invisible vapor; sweet alcohol-like odor; sol. in water; sp.gr. 0.7868-0.7879 (77/77 F), 1.59 (vapor); vapor pressure 44.6 mm Hg; f.p. -144 C; b.p. 78.4 C; flash pt. (TCC) 13 C; autoignition temp. 363 C; ref. index 1.3598
Toxicology: LD50 (oral, rat) 7060 mg/kg; may be irritating to eyes, upper respiratory tract; short-term exposure above 1000 ppm by inh. may cause CNS effects, headache, eye/nose/throat irritation; exposure > 1 h may cause dizziness, drowsiness, loss of appetite, inability to conc., GI effects, nausea, vomiting; prolonged/repeated skin contact may cause dermatitis; ing. may cause drunkenness, CNS depression, nausea, vomiting, diarrhea, liver damage, death; TSCA listed
Environmental: On land, apt to volatilize, biodegrade, or leach to ground water; in water, photolysis, oxidation, biodegradation expected to occur; not expected to bioconcentrate in aquatic organisms; LC50 (rainbow trout, 96 h) 10,400 ppm; prevent entry into sewers
Precaution: Flamm. liq.; flamm. limits 3.3-19 vol.%; material can burn with little or no visible flame; vapors are heavier than air, may flash back; prevent entry into basements, confined areas; ground and bond containers; avoid static discharges; avoid strong oxidizers, excessive heat, sparks, open flame; contact with acetyl chloride or other oxidizing agents may result in violent reaction
Hazardous decomp. prods.: Combustion: CO
HMIS: Health 1, Flammability 3, Reactivity 0
Storage: 1 yr shelf life; store under ambient conditions out of direct sunlight

Punctilious® SDA 40B Modified Anhydrous [Equistar]
Chem. Descrip.: Specially denatured alcohol 40B: ethanol (97.9%), propyl alcohol (2.0%), t-butanol (0.1%), denatonium benzoate (0.0012%)
UN 1170
Uses: Solvent
Regulatory: SARA §311/312 acute health/chronic health/fire hazard, §313 reportable (t-butanol); Calif. Prop. 65 reportable (ethanol, benzene)
Properties: Colorless liq.; invisible vapor; sweet alcohol-like odor; sol. in water; sp.gr. 0.7871-0.7882 (77/77 F), 1.59 (vapor); vapor pressure 44.6 mm Hg; f.p. -144 C; b.p. 78.4 C; flash pt. (TCC) 13 C; autoignition temp. 363 C
Toxicology: May be irritating to eyes, upper respiratory tract; short-term exposure above 1000 ppm by inh. may cause CNS effects, headache, eye/nose/throat irritation; exposure > 1 h may cause dizziness, drowsiness, loss of appetite, inability to conc., GI effects, nausea, vomiting; prolonged/repeated skin contact may cause dermatitis; ing. may cause drunkenness, CNS depression, nausea, vomiting, diarrhea, liver damage, death; TSCA listed
Environmental: On land, apt to volatilize, biodegrade, or leach to ground water; in water, photolysis, oxidation, biodegradation expected to occur; not expected to bioconcentrate in aquatic organisms; LC50 (rainbow trout, 96 h) 10,400 ppm; prevent entry into sewers
Precaution: Flamm. liq.; flamm. limits 3.3-19 vol.%; material can burn with little or no visible flame; vapors are heavier than air, may flash back; prevent entry into basements, confined areas; ground and bond containers; avoid static discharges; avoid strong oxidizers, excessive heat, sparks, open flame; contact with acetyl chloride or other oxidizing agents may result in violent reaction
Hazardous decomp. prods.: Combustion: CO
HMIS: Health 1, Flammability 3, Reactivity 0
Storage: 1 yr shelf life; store under ambient conditions out of direct sunlight

Punctilious® SDA 40C 190 Proof [Equistar]
Chem. Descrip.: Specially denatured alcohol 40B: ethanol (97.1%), t-butanol (2.9%)
UN 1170
Uses: Solvent
Regulatory: SARA §311/312 acute health/chronic health/fire hazard, §313 reportable (t-butanol); Calif. Prop. 65 reportable (ethanol, benzene)
Properties: Colorless liq.; invisible vapor; sweet alcohol-like odor; sol. in water; sp.gr. 0.8082-0.8094 (77/77 F), 1.59 (vapor); vapor pressure 44.6 mm Hg; f.p. -144 C; b.p. 78.4 C; flash pt. (TCC) 16 C; autoignition temp. 363 C; ref. index 1.3630
Toxicology: May be irritating to eyes, upper respiratory tract; short-term exposure above 1000 ppm by inh. may cause CNS effects, headache, eye/nose/throat irritation; exposure > 1 h may cause dizziness, drowsiness, loss of appetite, inability to conc., GI effects, nausea, vomiting;

prolonged/repeated skin contact may cause dermatitis; ing. may cause drunkenness, CNS depression, nausea, vomiting, diarrhea, liver damage, death; TSCA listed

Environmental: On land, apt to volatilize, biodegrade, or leach to ground water; in water, photolysis, oxidation, biodegradation expected to occur; not expected to bioconcentrate in aquatic organisms; LC50 (rainbow trout, 96 h) 10,400 ppm; prevent entry into sewers

Precaution: Flamm. liq.; flamm. limits 3.3-19 vol.%; material can burn with little or no visible flame; vapors are heavier than air, may flash back; prevent entry into basements, confined areas; ground and bond containers; avoid static discharges; avoid strong oxidizers, excessive heat, sparks, open flame; contact with acetyl chloride or other oxidizing agents may result in violent reaction

Hazardous decomp. prods.: Combustion: CO

HMIS: Health 1, Flammability 3, Reactivity 0

Storage: 1 yr shelf life; store under ambient conditions out of direct sunlight

Punctilious® SDA Canada DAG-2A Anhydrous [Equistar]

Chem. Descrip.: Specially denatured alcohol: ethanol (91.1%), methanol (8.1%), MEK (0.5%), ethyl acetate (0.3%)

UN 1170

Uses: Solvent

Regulatory: SARA §311/312 acute health/chronic health/fire hazard, §313 reportable (methanol, MEK); Calif. Prop. 65 reportable (ethanol, benzene); TSCA export notification required (ethyl acetate)

Properties: Colorless liq.; invisible vapor; sweet alcohol-like odor; sol. in water; sp.gr. 0.7874-0.7884 (77/77 F), 1.59 (vapor); vapor pressure 44.6 mm Hg; f.p. -144 C; b.p. 78.4 C; flash pt. (TCC) 16 C; autoignition temp. 363 C; ref. index 1.3602

Toxicology: May be irritating to eyes, upper respiratory tract; short-term exposure above 1000 ppm by inh. may cause CNS effects, headache, eye/nose/throat irritation; exposure > 1 h may cause dizziness, drowsiness, loss of appetite, inability to conc., GI effects, nausea, vomiting; prolonged/repeated skin contact may cause dermatitis; ing. may cause drunkenness, CNS depression, nausea, vomiting, diarrhea, liver damage, death; TSCA listed

Environmental: On land, apt to volatilize, biodegrade, or leach to ground water; in water, photolysis, oxidation, biodegradation expected to occur; not expected to bioconcentrate in aquatic organisms; LC50 (rainbow trout, 96 h) 10,400 ppm; prevent entry into sewers

Precaution: Flamm. liq.; flamm. limits 3.3-19 vol.%; material can burn with little or no visible flame; vapors are heavier than air, may flash back; prevent entry into basements, confined areas; ground and bond containers; avoid static discharges; avoid strong oxidizers, excessive heat, sparks, open flame; contact with acetyl chloride or other oxidizing agents may result in violent reaction

Hazardous decomp. prods.: Combustion: CO

HMIS: Health 1, Flammability 3, Reactivity 0

Storage: 1 yr shelf life; store under ambient conditions out of direct sunlight

Punctilious® SDA Canada DAG-2A2 Anhydrous [Equistar]

Chem. Descrip.: Specially denatured alcohol: ethanol (84.9%), methanol (14.3%), MEK (0.5%), ethyl acetate (0.3%)

UN 1170

Uses: Solvent

Regulatory: SARA §311/312 acute health/chronic health/fire hazard, §313 reportable (methanol, MEK); Calif. Prop. 65 reportable (ethanol, benzene); TSCA export notification required (ethyl acetate)

Properties: Colorless liq.; invisible vapor; sweet alcohol-like odor; sol. in water; sp.gr. 0.7871-0.7882 (77/77 F), 1.59 (vapor); vapor pressure 44.6 mm Hg; f.p. -144 C; b.p. 78.4 C; flash pt. (TCC) 13 C; autoignition temp. 363 C

Toxicology: May be irritating to eyes, upper respiratory tract; short-term exposure above 1000 ppm by inh. may cause CNS effects, headache, eye/nose/throat irritation; exposure > 1 h may cause dizziness, drowsiness, loss of appetite, inability to conc., GI effects, nausea, vomiting; prolonged/repeated skin contact may cause dermatitis; ing. may cause drunkenness, CNS depression, nausea, vomiting, diarrhea, liver damage, death

Environmental: On land, apt to volatilize, biodegrade, or leach to ground water; in water, photolysis, oxidation, biodegradation expected to occur; not expected to bioconcentrate in aquatic organisms; LC50 (rainbow trout, 96 h) 10,400 ppm; prevent entry into sewers

Precaution: Flamm. liq.; flamm. limits 3.3-19 vol.%; material can burn with little or no visible flame; vapors are heavier than air, may flash back; prevent entry into basements, confined areas; ground and bond containers; avoid static discharges; avoid strong oxidizers, excessive heat, sparks, open flame; contact with acetyl chloride or other oxidizing agents may result in violent reaction

Hazardous decomp. prods.: Combustion: CO

HMIS: Health 1, Flammability 3, Reactivity 0

Storage: 1 yr shelf life; store under ambient conditions out of direct sunlight

Purasolv® BL [PURAC Am.]

Chem. Descrip.: n-Butyl-(S)-lactate

CAS 34451-19-9; EINECS/ELINCS 252-036-3

Uses: Solvent for electronics, paints/coatings esp. waterborne and high-solids systems, printing inks, cleaning agents, mold release agents; coalescing agent for aq. formulations; pharmaceutical intermediate; food additive; flavor

Features: Non-ozone depleting

Regulatory: FDA approved GRAS flavor

Properties: Colorless to lt. yel. clear liq.; mild char. odor; partly misc. with water and most org. solvs.; m.w. 146; sp.gr. 0.975-0.985 (20 C); visc. 3.9 mPa·s; vapor pressure 0.3 mbar (20 C); b.p. 189 C; flash pt. (CC) 79 C; autoignition temp. 380 C; ref. index 1.418-1.422; 97% min. assay

Toxicology: TLV 25 mg/m^3 max.; LD50 (oral, rat) > 5000 mg/kg; LC50 (inh., rat) > 5140 mg/m^3; irritating to eyes, skin; risk of serious eye damage; may degrease skin; may cause coughing, headache; high concs. of vapors may cause respiratory tract irritation, narcotic effects

Environmental: BOD5 0.76 mg O_2/mg; COD 1.97 mg O_2/mg; readily biodeg.; unlikely to bioaccumulate; EC50 (daphnia, 48 h) 423 mg/l, (algae) 0.93 g/l; LC50 (fish, 48 h) 75 mg/l

Precaution: LEL 1% (100 C); UEL 7.9% (150 C); can be oxidized; hydrolyzes in presence of water, acids, bases; avoid temps. above 79 C

Hazardous decomp. prods.: Combustion: irritating fumes

Storage: Keep container tightly closed

Purasolv® EHL [PURAC Am.]

Chem. Descrip.: 2-Ethylhexyl-S-lactate

CAS 186817-80-1; EINECS/ELINCS 228-503-2

Uses: Solvent

Features: Low volatility and high flash pt.

Properties: Colorless clear liq.; mild char. odor; sol. 0.03 g/100 ml water; sp.gr. 0.93-0.95 kg/m^3 (20 C); dens. 7.8 lb/gal; visc. 7.7 mPa·s; vapor pressure 0.02 mbar (20 C); m.p. -23 C; b.p. 246 C; decomp. pt. > 246 C; flash pt. (CC) 113 C; ref. index 1.43-1.44; surf. tens. 29.8 mN/m; 95% min. assay

Toxicology: LD50 (oral, rat) > 2000 mg/kg; irritating to eyes, skin; may degrease skin

Environmental: Inherently biodeg.; LC50 (fish, 96 h) 32 mg/l; sl. ecotoxicity

Precaution: Incompat. with oxidizing agents; can be oxidized; hydrolyzes in presence of water, acids, bases; avoid temps. above 113 C

Hazardous decomp. prods.: CO_x; combustion: irritating fumes

Storage: Keep container tightly closed

Purasolv® EL [PURAC Am.]

Chem. Descrip.: Ethyl-(S)- lactate

CAS 687-47-8; EINECS/ELINCS 211-694-1

UN 1190

Uses: Solvent for electronics, paints/coatings (solv.-based, high-solids systems, stripping formulations, as coalescing agent), printed circuit boards, metal industry; food additive; flavor

Features: Non-ozone depleting; low volatility

Regulatory: FDA approved GRAS flavor; JECFA approved food additive

Properties: Colorless clear liq.; mild char. odor; completely misc. with water, most org. solvs.; m.w. 118; sp.gr. 1.03 (20 C); visc. 2.8 cP (20 C); vapor pressure 2.7 mbar (20 C); m.p. -3 C; b.p. 154 C; flash pt. (CC) 61 C; autoignition temp. 400 C; surf. tens. 30.6 dynes/cm; KB value > 1000

Toxicology: LD50 (oral, mouse) 2500 mg/kg; LC50 (inh., rat, 8 h) > 5400 mg/m^3; eye irritant; risk of serious eye damage; irritating to respiratory system, mucous membranes; skin contact may degrease skin, cause erythema

Environmental: BOD28 1.43 mg O_2/mg; COD 1.66 mg O_2/mg; readily biodeg.; unlikely to bioaccumulate; EC50 (daphnia, 48 h) 683 mg/l, (algae) 2200 mg/l; LC50 (fish, 48 h) 320 mg/l

Precaution: Hydrolyzes in presence of water, acids, bases; avoid temps. above 61 C
Hazardous decomp. prods.: CO_x; thermal decomp. can lead to release of irritating gases and vapors
Storage: Keep container tightly closed

Purasolv® ELECT [PURAC Am.]
Chem. Descrip.: Ethyl-(S)- lactate
CAS 687-47-8; EINECS/ELINCS 211-694-1
UN 1190
Uses: Precision cleaning solvent for wide variety of contaminants; solvent for electronics, paints/coatings esp. waterborne and high-solids systems, printing inks, cleaning agents, mold release agents; food additive
Features: Photoresist grade; non-ozone depleting
Regulatory: FDA GRAS; FAO/WHO, JECFA approved
Properties: Colorless clear liq., char. mild pleasant odor; misc. with water and most org. solvs.; m.w. 118; sp.gr. 1.030-1.038; m.p. -3 C; b.p. 154 C; flash pt. (CC) 52 C; ref. index 1.410-1.420; autoignition temp. 400 C; 99% min. assay
Toxicology: LD50 (oral, mouse) 2500 mg/kg; LC50 (inh., rat, 8 h) > 5400 mg/m^3; eye irritant; risk of serious eye damage; irritating to respiratory system, mucous membranes; skin contact may degrease skin, cause erythema; inh. of high vapor concs. can cause CNS depression, narcosis; immediately degrades into lactic acid and ethanol
Environmental: BOD28 1.43 mg O_2/mg; COD 1.66 mg O_2/mg; readily biodeg.; unlikely to bioaccumulate; EC50 (daphnia, 48 h) 683 mg/l, (algae) 2200 mg/l; LC50 (fish, 48 h) 320 mg/l
Precaution: Hydrolyzes in presence of water, acids, bases; avoid temps. above 61 C
Hazardous decomp. prods.: CO_x; thermal decomp. can lead to release of irritating gases and vapors
Storage: Keep container tightly closed

Purasolv® ELS [PURAC Am.]
Chem. Descrip.: Ethyl-(S)- lactate
CAS 687-47-8; EINECS/ELINCS 211-694-1
UN 1190
Uses: Precision cleaning solvent for electronics, paints/coatings esp. waterborne and high-solids systems, printing inks, cleaning agents, mold release agents; food additive
Features: Non-ozone depleting
Regulatory: FDA GRAS; FAO/WHO, JECFA approved
Properties: Colorless clear liq., char. mild pleasant odor; misc. with water and most org. solvs.; m.w. 118; sp.gr. 1.030-1.038; m.p. -3 C; b.p. 154 C; flash pt. (CC) 52 C; ref. index 1.410-1.420; autoignition temp. 400 C; 98% min. assay
Toxicology: LD50 (oral, mouse) 2500 mg/kg; LC50 (inh., rat, 8 h) > 5400 mg/m^3; eye irritant; risk of serious eye damage; irritating to respiratory system, mucous membranes; skin contact may degrease skin, cause erythema; inh. of high vapor concs. can cause CNS depression, narcosis; immediately degrades into lactic acid and ethanol
Environmental: BOD28 1.43 mg O_2/mg; COD 1.66 mg O_2/mg; readily biodeg.; unlikely to bioaccumulate; EC50 (daphnia, 48 h) 683 mg/l, (algae) 2200 mg/l; LC50 (fish, 48 h) 320 mg/l
Precaution: Hydrolyzes in presence of water, acids, bases; avoid temps. above 61 C
Hazardous decomp. prods.: CO_x; thermal decomp. can lead to release of irritating gases and vapors
Storage: Keep container tightly closed

Purasolv® IPL [PURAC Am.]
Chem. Descrip.: Isopropyl-S-(-)-lactate
CAS 63697-00-7; EINECS/ELINCS 264-417-1
UN 3272
Uses: Solvent for electronics, paints/coatings (solv.-based, high-solids systems, stripping formulations, as coalescing agent), printed circuit boards, metal industry
Features: Non-ozone depleting
Properties: Colorless clear liq., char. odor; completely misc. with water and most org. solvs.; m.w. 132; sp.gr. 0.984-0.987 (20 C); visc. 3.2 cP (20 C); vapor pressure 1.0 mbar (20 C); m.p. 3 C; b.p. 160 C; decomp. pt. > 153 C; autoignition temp. > 380 C; flash pt. (CC) 60 C; ref. index 1.407-1.409; surf. tens. 28 dynes/cm; KB value > 1000; 97% min. assay
Toxicology: LD50 (oral, rat) > 2000 mg/kg; eye irritant; risk of serious eye damage; may degrease skin; inh. of vapors is irritating to respiratory system, may cause coughing
Environmental: BOD5 0.10 mg O_2/mg; BOD28 1.14 mg O_2/mg; ThOD 1.81 mg O_2/mg; biodeg.; bioaccumulation unlikely
Precaution: Flamm.; can be oxidized; hydrolyzes in presence of water, acids, bases; avoid temps. above 60 C
Hazardous decomp. prods.: CO_x; combustion: irritating fumes
Storage: Keep container tightly closed

Purasolv® ML [PURAC Am.]
Chem. Descrip.: Methyl-(S)-lactate
CAS 27871-49-4; EINECS/ELINCS 248-704-9
UN 3272
Uses: Solvent for electronics, paints/coatings (solv.-based, high-solids systems, stripping formulations, as coalescing agent), printed circuit boards, metal industry
Features: Non-ozone depleting
Properties: Colorless clear liq.; mild char. odor; completely misc. with water and most org. solvs.; m.w. 104; sp.gr. 1.085-1.098 (20 C); visc. 3.1 mPa·s (20 C); vapor pressure 3.4 mbar (20 C); m.p. -44 C; b.p. 145 C; decomp. pt. > 144 C; flash pt. (CC) 55 C; ref. index 1.410-1.415; surf. tens. 34.2 dynes/cm; KB value 62; 97% min. assay
Toxicology: LD50 (oral, rat) > 2000 mg/kg; LC50 (inh., rat, 8 h) > 5030 mg/m^3; severe eye irritant; risk of serious eye damage; may degrease skin; inh. of vapors is irritating to respiratory system, may cause throat pain and cough, and produce narcotic effects
Environmental: Biodeg.; bioaccumulation unlikely
Precaution: Flamm.; can be oxidized; hydrolyzes in presence of water, acids, bases; avoid temps. above 55 C
Hazardous decomp. prods.: CO_x; combustion: irritating fumes
Storage: Keep container tightly closed

2-Pyrol® [ISP]
Chem. Descrip.: 2-Pyrrolidone
CAS 616-45-5; EINECS/ELINCS 204-648-7
Uses: Monomer for polypyrrolidone; solvent for polymers, insecticides, polyhydroxylic alcohols, sugars, iodine, specialty inks; used in petrol. processing and acrylonitrile mfg.; solvent, solubilizer, intermediate for pharmaceuticals; plasticizer/coalescing agent for acrylic latexes, acrylic-styrene copolymers in emulsion-type floor coatings; intermediate for N-methylol derivs., alkaloids related to peganine, amino-butyric acid derivs.; adhesives
Properties: Liq.; solidifies below 25 C; misc. with water, ethanol, ethyl ether, chloroform, benzene, ethyl acetate, CS_2; insol. in aliphatic hydrocarbons; sp.gr. 0.11 (25/4 C); visc. 13.3 cp; m.p. 25 C; b.p. 245 C; flash pt. (OC) 130 C
Precaution: Noncorrosive; low fire hazard

Q

Qemifeltwash 8020A [Qemi Int'l.]
Chem. Descrip.: Solvent/surfactant blend
Uses: Solvent, surfactant for continuous cleaning of wet felts in paper machine press sections, e.g., in newsprint or mech. furnish sheets where pitch control is very important
Features: Acidic version; applied diluted at 0.1-1.0% conc.
Properties: Pale yel. clear liq.; sol. in water; sp.gr. 1.03; pH 2.4 (5%)
Use Level: 3-7 lb/ton paper
Storage: Store at 40-90 F; keep containers closed; store away from heat or ignition sources

Qemifeltwash 8020N [Qemi Int'l.]
Chem. Descrip.: Solvent/surfactant blend
Uses: Solvent, surfactant for continuous cleaning of wet felts in paper machine press sections, esp. on brn. paper, linerboard
Features: Neutral version; applied diluted at 0.1-1.0% conc.
Properties: Colorless clear liq.; org. solv. odor; disp. in water; sp.gr. 1.0; f.p. < 0 C; pH 6.5 (5%)
Use Level: 3-7 lb/ton paper
Precaution: May contain trace amts. of ethylene oxide, which may accumulate in the headspace of shipping/storage containers
Storage: Store at 40-90 F; keep containers closed; store away from heat or ignition sources

QO® Furfural [Penn Spec. Chems.]
Chem. Descrip.: 2-Furaldehyde
CAS 98-01-1; EINECS/ELINCS 202-627-7
Uses: Chemical intermediate for mfg. of derivatives (furan and THF); solvent for separating sat. from unsat. compds. in petrol. lubricating oil, gas oil, and diesel fuel; extractive distillation of C4 and C5 hydrocarbons for the mfg. of syn. rubber; decolorizing agent for wood rosin; solvent and processing aid for anthracene; ingred. in resins; reactive solvent and wetting agent in abrasive wheels and brake linings
Properties: Colorless when freshly distilled; industrial furfural is lt. yel. to brn. liq.; pungent almond-like odor; m.w. 96.08; dens. 1.1545; visc. 1.49 cps; b.p. 161.7 C (760 mm); f.p. -36.5 C; TCC flash pt. 61.7 C; surf. tens. 40.7 dynes/cm (29.9 C); ref. index 1.5235

R

Renex® PEG 300 [Uniqema; Uniqema Am.]
Chem. Descrip.: PEG-6 USP/NF
CAS 25322-68-3; EINECS/ELINCS 220-045-1
Uses: Surfactant, solvent for personal care, moisturizing skin care preps., suntan gels, deodorants, topical pharmaceuticals, industrial, agrochem. applics., insect repellents; suspending agent, consistency agent, visc. control agent, plasticizer for controlled-release solid drug forms, laxatives
Regulatory: USP/NF and EP compliance
Properties: Colorless liq.; sol. in water; m.w. 300; dens. 1.128 g/cc (20 C); f.p. -15 C; acid no. 0.1; hyd. no. 374; pH 5.8 (5% aq.); nonionic; 100% conc.

Renex® PEG 400 [Uniqema; Uniqema Am.]
Chem. Descrip.: PEG-8 USP/NF
CAS 25322-68-3; EINECS/ELINCS 225-856-4
Uses: Surfactant, solvent for personal care, skin care, hair care, makeup, deodorants, shave prods., toothpaste, topical pharmaceuticals, suppository bases, industrial, agrochem. applics.; suspending agent, consistency agent, visc. control agent, plasticizer for controlled-release solid drug forms, laxatives
Regulatory: USP/NF and EP compliance
Properties: Colorless liq.; sol. in water; m.w. 400; dens. 1.130 g/cc (20 C); f.p. 5 C; acid no. 0.1; hyd. no. 280; pH 5.8 (5% aq.); nonionic; 100% conc.

Renex® PEG 600 [Uniqema; Uniqema Am.]
Chem. Descrip.: PEG-12 USP/NF
CAS 25322-68-3; EINECS/ELINCS 229-859-1
Uses: Surfactant, solvent, wetting agent for personal care, hair straighteners, deodorants, aftershaves, toothpaste, topical pharmaceuticals, suppository bases, industrial, agrochem. applics., insect repellents; suspending agent, consistency agent, visc. control agent, plasticizer for controlled-release solid drug forms, laxatives
Regulatory: USP/NF and EP compliance
Properties: Wh. wax; sol. in water; m.w. 600; dens. 1.127 g/cc (20 C); f.p. 20 C; acid no. 0.1; hyd. no. 187; pH 5.8 (5% aq.); nonionic; 100% conc.

Reomol® TBP [Great Lakes]
Chem. Descrip.: Tributyl phosphate
CAS 126-73-8; EINECS/ELINCS 204-800-2
Uses: Defoamer for drilling muds, cements, fracturing fluids, plaster compositions, paper coatings, pulp bleaching, mold coating, glass luminescent coatings, aq. emulsion paints, adhesives, mercerizing liquors/dye baths; antifreeze sol'ns., fermentation, detergents; primary plasticizer and solvent; pigment dispersant
Properties: Liq.; sol. in min. oil, gasoline, most org. liqs.; sol. 0.1% in water; insol. in glycerin, glycols; sp.gr. 0.97-0.983 (20/20 C); dens. 8.14 lb/gal (20 C); visc. 3.7 cp (20 C); vapor pressure 7.3 mm Hg (150 C); f.p. < -80 C; b.p. 137-145 C (4 mm Hg); pour pt. < 80 C; flash pt. 166 C; fire pt. 182 C; surf. tens. 29 dynes/cm (20 C); 7.4% P, < 0.1% moisture
Toxicology: Severe skin irritant
Precaution: Low degree of fire hazard

Reomol® TOP [Great Lakes]
Chem. Descrip.: Trioctyl phosphate
CAS 78-42-2; EINECS/ELINCS 201-116-6
Uses: Plasticizer for PVC, rubber, paints, coatings; solvent
Features: Low temp. plasticizer; highly efficient solvent

Rewopal® MPG 10 [Goldschmidt-Rewo GmbH]
Chem. Descrip.: Phenoxyethanol
CAS 122-99-6; EINECS/ELINCS 204-589-7
Uses: Solvent; solubilizer; preservative
Properties: Liq.; nonionic; 100% conc.

Rewopal® MPG 40 [Goldschmidt-Rewo GmbH]
Chem. Descrip.: Phenol polyglycol ether
CAS 122-99-6; EINECS/ELINCS 204-589-7
Uses: Solvent
Properties: Liq.; nonionic; 100% conc.

Rhodacal® 330 [Rhodia HPCII]
Chem. Descrip.: Branched isopropylamine dodecylbenzene sulfonate
CAS 26264-05-1; EINECS/ELINCS 247-556-2
Uses: Emulsifier, wetting agent, grease/pigment dispersant, lubricant, solubilizer, solvent, penetrant, high foaming base for shampoos and cleaners, metalworking fluids, agric. pesticides, emulsion polymerization, dry cleaning, latex paints, metal cleaning, acid aluminum cleaners; oil slick emulsifier
Properties: Amber liq.; oil-sol.; sp.gr. 1.03; dens. 8.497 lb/gal; visc. 6500 cps; HLB 11.7; flash pt. > 200 C; pH 5-6.8 (5%); anionic; 90% act.

Rhodasurf® PEG 600 [Rhodia HPCII]
Chem. Descrip.: PEG-14
CAS 25322-68-3
Uses: Plasticizer; solvent; lubricant, binder for pharmaceuticals; color stabilizer for fuel oils; intermediate for processing surfactants
Properties: Wh. liq.; m.w. 570-630; water sol.; sp.gr. 1.13 (30/15.5 C); f.p. 20-25 C; pH 5-8; 100% conc.

Rhodiasolv RPDE [Rhodia HPCII]
Chem. Descrip.: Methyl glutarate (59-67%), methyl succinate (20-28%), methyl adipate (9-17%)
Uses: Solvent for cleaning and stripping, paints (coil coatings, can coatings, polyester and epoxy sealing coats, acrylic lacquers, enamel finishing coats, oven-dried industrial paints, PU packing paints, automobile paints, industrial paints), liq. soaps, pesticides, insecticides, textile lubricants, corrosion inhibitors, adhesives, resins for foundry cores, elec. varnishes, printing inks; intermediate for synthesis; plasticizer for PVC, PS
Features: Substitute solvent for ethylglycol acetate, cresylic acid, methylene chloride, cyclohexanone, acetone, diacetone alcohol, glycol ether, isophorone
Properties: APHA ≤ 10 color; misc. with most org. solvs. (alcohols, ethers, ketones, hydrocarbon solvs.); sol. 5.7% in water (20 C); m.w. 158.6; dens. 1.09 g/cm^3 (20 C); dynamic visc. 2.9 mPa·s (20 C); vapor pressure 6 Pa (20 C); f.p. -20 C; distillation pt. 200-235 C (1013 hPa); acid no. ≤ 0.3; flash pt. (CC) 100 C; ref. index 1.425 (20 C); ≥ 99.5% ester content
Toxicology: LD50 (oral) > 200 mg/kg, (dermal) > 200 mg/kg; nonirritating to eyes/skin; exc. toxicological props.
Environmental: Environmentally friendly; exc. ecotoxicological props.; complete and rapid biodegradability

Rilanit® EHK [Cognis]
Chem. Descrip.: Ethylhexyl C8-C14 fatty acid ester
Uses: Solvent substitute for cleaners and hand cleaning pastes
Properties: Liq.; pour pt. < -32 C

Ritacane ID [R.I.T.A.]

Chem. Descrip.: Isododecane
Uses: Emollient, skin conditioner, solvent in skin care, mascara, eyeliner, antiperspirants, sun care
Features: Leaves silky afterfeel on skin; replacement for min. oil and low-visc. dimethicones and cyclomethicones
Properties: Colorless liq.; misc. with min. oil, IPM, IPA; sl. misc. with castor oil; insol. in water; m.w. 170; dens. 0.74-0.75 g/cm^3 (20 C); visc. 3-5 mPa·s; acid no. < 0.5; iodine no. < 1; ref. index 1.418-1.425 (20 C); < 50 ppm water

Ritacane IH [R.I.T.A.]
Chem. Descrip.: Isohexadecane
CAS 4390-04-9; EINECS/ELINCS 224-506-8
Uses: Emollient, skin conditioner, solvent, cleansing agent in face and eye makeup
Features: Leaves silky afterfeel on skin; replacement for min. oil and low-visc. dimethicones and cyclomethicones
Properties: Colorless liq.; misc. with min. oil, IPM, IPA; sl. misc. with castor oil; insol. in water; m.w. 220; dens. 0.78-0.79 g/cm^3 (20 C); acid no. < 0.5; iodine no. < 1; ref. index 1.439-1.445 (20 C)(20 C); < 50 ppm water

Ritawax AEO [R.I.T.A.]
Chem. Descrip.: Polysorbate 80, acetylated lanolin alcohol, cetyl acetate
Uses: Emollient, lubricant, moisturizer, penetrant, solubilizer, dispersant, plasticizer, cosolvent, emulsifier for personal care prods., pharmaceuticals; hair and skin softener and conditioner; vehicle for medicaments; degreaser and detackifier for emulsions
Properties: Lemon yel. to straw-colored clear oily liq., char. bland odor; sol. in water, IPA, castor oil, min. oil, veg. oil; acid no. 2 max.; sapon. no. 65-80; hyd. no. 60-70; nonionic; 100% conc.

Ritawax ALA [R.I.T.A.]
Chem. Descrip.: Cetyl acetate, acetylated lanolin alcohol
Uses: Emollient, lubricant, moisturizer, penetrant, plasticizer, cosolvent, solubilizer, emulsifier for personal care prods., pharmaceuticals; hair and skin softener and conditioner; vehicle for medicaments; degreaser and detackifier for emulsions
Properties: Lemon yel. to straw-colored oily liq., char. bland odor; sol. in min. oil, castor oil, veg. oil, alcohol; acid no. 1 max.; iodine no. 10 max.; sapon. no. 180-200; hyd. no. 8 max.; nonionic; 100% conc.

RotoSolv [CITGO]
Chem. Descrip.: VM&P naphtha
CAS 8032-32-4; EINECS/ELINCS 232-453-7
UN 1268
Uses: Solvent for printing industry
Features: Low boiling; narrow distillation range; high solvency; Rule 66 compliance
Regulatory: SARA §302/304 nonreportable, §311/312 acute health/chronic health/fire hazard, §313 reportable [toluene (0.005-4%), xylenes (0.02-2%)]; Calif. Prop. 65 reportable (benzene, toluene)
Properties: Colorless clear liq.; lt. hydrocarbon odor; misc. with most common org. solvs.; negligible sol. in cold water; sp.gr. 0.734 (60/60 F); dens. 6.09 lb/gal (15.6 C); visc. 0.766 cst (20 C); vapor pressure 17 mm Hg (20 C); f.p. -95 to -48 C; b.p. 116-139 C; flash pt. (TCC) 7-8 C; autoignition temp. 232 C; ref. index 1.4952 (20 C); KB value 33; VOC 731 g/l; 100% act.
Toxicology: Mist/vapor may irritate eyes, mucous membranes, respiratory tract; liq. contact may cause minimal to mild eye irritation, mild to mod. skin irritation/inflamm.; may be harmful by inh., skin absorp.; overexposure may cause CNS depression, other target organ effects; may be harmful or fatal if ingested; aspiration into lungs can cause pulmonary edema, chem. pneumonia; prolonged/repeated inh. may cause heart arrhythmias; possible reproductive hazard; sensitizer; TSCA listed
Environmental: May cause oxygen depletion in waterways; potentially toxic to water ecosystems; may be hazardous to aquatic life; prevent entry into sewers, waterways; not a DOT marine pollutant
Precaution: Highly flamm.; flamm. limits 0.9-6.5%; vapors are heavier than air, may flash back; may create vapor/air explosion hazard in confined spaces; spills may create slipping hazard; floats on water; container can rupture in fire; prevent entry into basements, confined areas; bond and ground equip.; incompat. with strong acids, alkalis, oxidizers (liq. chlorine, hydrogen peroxide, oxygen); if inhibited, may cause rusting of copper/alloys
Hazardous decomp. prods.: Combustion: smoke, CO, CO_2, possibly other harmful gases/vapors
HMIS: Health 2, Flammability 3, Reactivity 0
Storage: Store in tightly closed containers in cool, dry, well-ventilated place away from heat, sparks, ignition sources, strong oxidizing conditions

RS Naphtha [ExxonMobil]
Chem. Descrip.: Hydrocarbon solvent
Uses: Solvent
Properties: Sp.gr. 0.69 (15.6/15.6 C); dens. 5.75 lb/gal (60 F); b.p. 66-102 C; flash pt. (TCC) < -18 C; surf. tens. 19.1 dynes/cm; KB value 34; 89% paraffin, 2% cycloparaffin, 9% aromatics
Toxicology: OEL 100 ppm

Rubber Solvent [Shell]
Chem. Descrip.: Solvent naphtha (petroleum), light aliphatic, toluene, n-hexane, cyclohexane
UN 1993
Uses: Solvent for printing inks, coatings, quick-dry polishes
Features: Fast evaporating; low aromatic
Regulatory: SARA §311/312 acute health/chronic health/fire hazard, §313 reportable [cyclohexane (0-11%), toluene (9%), n-hexane (19-26%)]; Calif. Prop. 65 reportable [benzene 95 ppm), toluene (9%)]
Properties: Colorless liq.; hydrocarbon odor; negligible sol. in water; m.w. 88; sp.gr. 0.7 (60 F); vapor pressure 125 mm Hg (68 F); b.p. 149-248 F; flash pt. < -18 C; VOC 5.8 lb/gal (60 F)
Toxicology: If aspirated, can cause severe lung damage, chem. pneumonitis; may be fatal if swallowed; eye irritant; may cause CNS depression (dizziness, lightheadedness, headache, nausea, incoordination, unconsciousness, death); chronic exposure may cause hearing loss; chronic abuse may cause irregular heart rhythms, cardiac arrest; may cause sl. skin irritation, defatting; target organs: nerves, heart, kidney, liver, auditory, reproductive systems; TSCA listed
Environmental: Prevent entry into waterways, sewer
Precaution: Extremely flamm.; flamm. in air 1-8 vol.%; vapors are heavier than air; vapors may flash back; prevent vapor accumulation; static electricity may accumulate and create fire hazard; bond and ground equip.; avoid contact with strong oxidizing agents; prevent entry into basements, confined areas
Hazardous decomp. prods.: Complex mixt. of airborne solids, liqs., and gases incl. CO, CO_2, other org. compds.
HMIS: Health 1, Flammability 3, Reactivity 0
Storage: Keep containers closed when not in use; keep away from heat, sparks, flame

Rubber Solvent 332 [Shell]
Chem. Descrip.: Solvent naphtha (petroleum), light aliphatic, n-hexane, cyclohexane
UN 1268
Uses: Solvent for printing inks, coatings, adhesives, camper fuel
Features: Fast evaporating; very low in aromatics content
Regulatory: SARA §311/312 acute health/chronic health/fire hazard, §313 reportable [n-hexane (10-13%), cyclohexane (0.42-9%)]
Properties: Colorless liq.; hydrocarbon odor; negligible sol. in water; m.w. 95; sp.gr. 0.69; vapor pressure 89 mm Hg (68 F); b.p. 164-271 F; flash pt. < -18 C; VOC 5.8 lb/gal (60 F); 100% act.
Toxicology: If aspirated, can cause severe lung damage, chem. pneumonitis; may be fatal if swallowed; may cause CNS depression (dizziness, lightheadedness, headache, nausea, incoordination, unconsciousness, death); chronic exposure may cause hearing loss; chronic abuse may cause irregular heart rhythms, cardiac arrest; may cause eye/sl. skin irritation/defatting; target organs: nerves, heart, auditory, reproductive systems; TSCA listed
Environmental: Prevent entry into waterways, sewer
Precaution: Extremely flamm.; flamm. in air 1-8 vol.%; vapors are heavier than air; vapors may flash back; prevent vapor accumulation; static electricity may accumulate and create fire hazard; bond and ground equip.; avoid contact with strong oxidizing agents; prevent entry into basements, confined areas
Hazardous decomp. prods.: Complex mixt. of airborne solids, liqs., and gases incl. CO, CO_2, other org. compds.
HMIS: Health 1, Flammability 3, Reactivity 0
Storage: Keep containers closed when not in use; keep away from heat, sparks, flame

S

S-327 190 Proof [Equistar]
Chem. Descrip.: Ethyl alcohol, methyl alcohol, propyl acetate, water
UN 1987
Uses: Solvent
Regulatory: FDA 27CFR §211.200
Properties: Pt-Co 10 max. color; sol. in water; sp.gr. 0.8058-0.8070 (77/77 F); dens. 6.766-6.776 lb/gal (60 F)

S-327 Anhydrous [Equistar]
Chem. Descrip.: Ethyl alcohol (80.8%), methyl alcohol (18.08%), propyl acetate (1.12%)
UN 1987
Uses: Solvent
Regulatory: FDA 27CFR §20.112; SARA §311/312 acute health/chronic health/fire hazard, §313 reportable (methanol); Calif. Prop. 65 reportable (ethanol, benzene)
Properties: Colorless liq.; invisible vapor; mild sweet alcohol-like nonresidual odor; sol. in water; sp.gr. 0.7882-0.7892 (77/77 F), 1.59 (vapor); dens. 6.618-6.627 lb/gal (60 F); vapor pressure 44.6 mm Hg; f.p. -144 C; b.p. 78.4 C; flash pt. (TCC) 13 C; autoignition temp. 363 C
Toxicology: May be irritating to eyes, upper respiratory tract; short-term exposure above 1000 ppm by inh. may cause CNS effects, headache, eye/nose/throat irritation; exposure > 1 h may cause dizziness, drowsiness, loss of appetite, inability to conc., GI effects, nausea, vomiting; prolonged/repeated skin contact may cause dermatitis; ing. may cause drunkenness, CNS depression, nausea, vomiting, diarrhea, liver damage, death; TSCA listed
Environmental: On land, apt to volatilize, biodegrade, or leach to ground water; in water, photolysis, oxidation, biodegradation expected to occur; not expected to bioconcentrate in aquatic organisms; LC50 (rainbow trout, 96 h) 10,400 ppm; prevent entry into sewers
Precaution: Flamm. liq.; flamm. limits 3.3-19 vol.%; material can burn with little or no visible flame; vapors are heavier than air, may flash back; prevent entry into basements, confined areas; ground and bond containers; avoid static discharges; avoid strong oxidizers, excessive heat, sparks, open flame; contact with acetyl chloride or other oxidizing agents may result in violent reaction
Hazardous decomp. prods.: Combustion: CO
HMIS: Health 1, Flammability 3, Reactivity 0

S-333 Anhydrous [Equistar]
Chem. Descrip.: Ethyl alcohol (89.0%), methyl alcohol (10.0%), propyl acetate (1.0%)
UN 1987
Uses: Solvent
Regulatory: FDA 27CFR §211.200; SARA §311/312 acute health/chronic health/fire hazard, §313 reportable (methanol); Calif. Prop. 65 reportable (ethanol, benzene)
Properties: Colorless liq.; invisible vapor; mild sweet alcohol-like nonresidual odor; sol. in water; sp.gr. 0.7882-0.7892 (77/77 F), 1.59 (vapor); dens. 6.618-6.627 lb/gal (60 F); vapor pressure 44.6 mm Hg; f.p. -144 C; b.p. 78.4 C; flash pt. (TCC) 13 C; autoignition temp. 363 C
Toxicology: May be irritating to eyes, upper respiratory tract; short-term exposure above 1000 ppm by inh. may cause CNS effects, headache, eye/nose/throat irritation; exposure > 1 h may cause dizziness, drowsiness, loss of appetite, inability to conc., GI effects, nausea, vomiting; prolonged/repeated skin contact may cause dermatitis; ing. may cause drunkenness, CNS depression, nausea, vomiting, diarrhea, liver damage, death; TSCA listed
Environmental: On land, apt to volatilize, biodegrade, or leach to ground water; in water, photolysis, oxidation, biodegradation expected to occur; not expected to bioconcentrate in aquatic organisms; LC50 (rainbow trout, 96 h) 10,400 ppm; prevent entry into sewers
Precaution: Extremely flamm. liq.; flamm. limits 3.3-19 vol.%; material can burn with little or no visible flame; vapors are heavier than air, may flash back; prevent entry into basements, confined areas; ground and bond containers; avoid static discharges; avoid strong oxidizers, excessive heat, sparks, open flame; contact with acetyl chloride or other oxidizing agents may result in violent reaction
Hazardous decomp. prods.: Combustion: CO
HMIS: Health 1, Flammability 3, Reactivity 0
Storage: Store in tightly closed containers in well-ventilated area away from oxidizers, other combustible materials

S-336 Anhydrous [Equistar]
Chem. Descrip.: Ethyl alcohol (85.9%), IPA (13.2%), MIBK (1.0%)
UN NA1987
Uses: Solvent
Regulatory: FDA 27CFR §20.112; SARA §311/312 acute health/chronic health/fire hazard, §313 reportable (MIBK, IPA); Calif. Prop. 65 reportable (ethanol, benzene); TSCA export notification required (MIBK)
Properties: Colorless liq.; invisible vapor; mild sweet alcohol-like nonresidual odor; sol. in water; sp.gr. 0.7865-0.7876 (77/77 F), 1.59 (vapor); dens. 6.605-6.614 lb/gal (60 F); vapor pressure 44.6 mm Hg; f.p. -144 C; b.p. 78.4 C; flash pt. (TCC) 13 C; autoignition temp. 363 C
Toxicology: May be irritating to eyes, upper respiratory tract; short-term exposure above 1000 ppm by inh. may cause CNS effects, headache, eye/nose/throat irritation; exposure > 1 h may cause dizziness, drowsiness, loss of appetite, inability to conc., GI effects, nausea, vomiting; prolonged/repeated skin contact may cause dermatitis; ing. may cause drunkenness, CNS depression, nausea, vomiting, diarrhea, liver damage, death; TSCA listed
Environmental: On land, apt to volatilize, biodegrade, or leach to ground water; in water, photolysis, oxidation, biodegradation expected to occur; not expected to bioconcentrate in aquatic organisms; LC50 (rainbow trout, 96 h) 10,400 ppm; prevent entry into sewers
Precaution: Flamm. liq.; flamm. limits 3.3-19 vol.%; material can burn with little or no visible flame; vapors are heavier than air, may flash back; prevent entry into basements, confined areas; ground and bond containers; avoid static discharges; avoid strong oxidizers, excessive heat, sparks, open flame; contact with acetyl chloride or other oxidizing agents may result in violent reaction
Hazardous decomp. prods.: Combustion: CO
HMIS: Health 1, Flammability 3, Reactivity 0

S-357 Anhydrous [Equistar]
Chem. Descrip.: Ethyl alcohol (84.71%), IPA (14.16%), n-propyl acetate (1.13%)
UN NA1987
Uses: Solvent
Regulatory: FDA 27CFR §21.200; SARA §311/312 acute health/chronic health/fire hazard, §313 reportable (IPA); Calif. Prop. 65 reportable (ethanol, benzene)
Properties: Colorless liq.; invisible vapor; mild sweet alcohol-like

nonresidual odor; sol. in water; sp.gr. 0.7868-0.7879 (77/77 F), 1.59 (vapor); dens. 6.608-6.616 lb/gal (60 F); vapor pressure 44.6 mm Hg; f.p. -144 C; b.p. 78.4 C; flash pt. (TCC) 13 C; autoignition temp. 363 C
Toxicology: May be irritating to eyes, upper respiratory tract; short-term exposure above 1000 ppm by inh. may cause CNS effects, headache, eye/nose/throat irritation; exposure > 1 h may cause dizziness, drowsiness, loss of appetite, inability to conc., GI effects, nausea, vomiting; prolonged/repeated skin contact may cause dermatitis; ing. may cause drunkenness, CNS depression, nausea, vomiting, diarrhea, liver damage, death; TSCA listed
Environmental: On land, apt to volatilize, biodegrade, or leach to ground water; in water, photolysis, oxidation, biodegradation expected to occur; not expected to bioconcentrate in aquatic organisms; LC50 (rainbow trout, 96 h) 10,400 ppm; prevent entry into sewers
Precaution: Flamm. liq.; flamm. limits 3.3-19 vol.%; material can burn with little or no visible flame; vapors are heavier than air, may flash back; prevent entry into basements, confined areas; ground and bond containers; avoid static discharges; avoid strong oxidizers, excessive heat, sparks, open flame; contact with acetyl chloride or other oxidizing agents may result in violent reaction
Hazardous decomp. prods.: Combustion: CO
HMIS: Health 1, Flammability 3, Reactivity 0

S-358 Anhydrous [Equistar]
Chem. Descrip.: Ethyl alcohol (85.8%), IPA (13.2%), n-propyl acetate (1.0%)
UN 1170
Uses: Solvent
Regulatory: SARA §311/312 acute health/chronic health/fire hazard, §313 reportable (IPA); Calif. Prop. 65 reportable (ethanol, benzene)
Properties: Colorless liq.; invisible vapor; sweet alcohol-like odor; sol. in water; sp.gr. 0.7899-0.7914 (77/77 F), 1.59 (vapor); vapor pressure 44.6 mm Hg; f.p. -144 C; b.p. 78.4 C; flash pt. (TCC) 13 C; autoignition temp. 363 C
Toxicology: May be irritating to eyes, upper respiratory tract; short-term exposure above 1000 ppm by inh. may cause CNS effects, headache, eye/nose/throat irritation; exposure > 1 h may cause dizziness, drowsiness, loss of appetite, inability to conc., GI effects, nausea, vomiting; prolonged/repeated skin contact may cause dermatitis; ing. may cause drunkenness, CNS depression, nausea, vomiting, diarrhea, liver damage, death; TSCA listed
Environmental: On land, apt to volatilize, biodegrade, or leach to ground water; in water, photolysis, oxidation, biodegradation expected to occur; not expected to bioconcentrate in aquatic organisms; LC50 (rainbow trout, 96 h) 10,400 ppm; prevent entry into sewers
Precaution: Flamm. liq.; flamm. limits 3.3-19 vol.%; material can burn with little or no visible flame; vapors are heavier than air, may flash back; prevent entry into basements, confined areas; ground and bond containers; avoid static discharges; avoid strong oxidizers, excessive heat, sparks, open flame; contact with acetyl chloride or other oxidizing agents may result in violent reaction
Hazardous decomp. prods.: Combustion: CO
HMIS: Health 1, Flammability 3, Reactivity 0
Storage: 1 yr shelf life; store under ambient conditions out of direct sunlight

S-359 Anhydrous [Equistar]
Chem. Descrip.: Ethyl alcohol (90.2%), n-propyl acetate (5.4%), IPA (4.5%)
UN 1170
Uses: Solvent
Regulatory: SARA §311/312 acute health/chronic health/fire hazard, §313 reportable (IPA); Calif. Prop. 65 reportable (ethanol, benzene)
Properties: Colorless liq.; invisible vapor; sweet alcohol-like odor; sol. in water; sp.gr. 0.7914-0.7923 (77/77 F), 1.59 (vapor); vapor pressure 44.6 mm Hg; f.p. -144 C; b.p. 78.4 C; flash pt. (TCC) 13 C; autoignition temp. 363 C
Toxicology: May be irritating to eyes, upper respiratory tract; short-term exposure above 1000 ppm by inh. may cause CNS effects, headache, eye/nose/throat irritation; exposure > 1 h may cause dizziness, drowsiness, loss of appetite, inability to conc., GI effects, nausea, vomiting; prolonged/repeated skin contact may cause dermatitis; ing. may cause drunkenness, CNS depression, nausea, vomiting, diarrhea, liver damage, death; TSCA listed
Environmental: On land, apt to volatilize, biodegrade, or leach to ground water; in water, photolysis, oxidation, biodegradation expected to occur; not expected to bioconcentrate in aquatic organisms; LC50 (rainbow trout, 96 h) 10,400 ppm; prevent entry into sewers
Precaution: Flamm. liq.; flamm. limits 3.3-19 vol.%; material can burn with little or no visible flame; vapors are heavier than air, may flash back; prevent entry into basements, confined areas; ground and bond containers; avoid static discharges
Hazardous decomp. prods.: Combustion: CO
HMIS: Health 1, Flammability 3, Reactivity 0
Storage: 1 yr shelf life; store under ambient conditions out of direct sunlight

Safety-Solv 140 Plus [Process Chems. LLC]
Chem. Descrip.: Terpene hydrocarbons and proprietary emulsifiers
Uses: Solvent to extract asphalt cement from bituminous mixes; degreaser; asphalt cleaner
Features: Nonchlorinated; nonhazardous; rinses easily with water; replaces trichloroethylene, trichloroethane
Properties: Clear liq.; pine oil-like odor; dens. 7.2 lb/gal; visc. 15 cps; flash pt. (PMCC) > 140 F
Environmental: Biodeg.

Salacos® 99 [Nisshin Oillio]
Chem. Descrip.: Isononyl isononanoate
CAS 42131-25-9; EINECS/ELINCS 261-665-2
Uses: Solvent, antistat, emollient in cosmetics

Santosol® DMA [Solutia]
Chem. Descrip.: Dimethyl adipate
CAS 627-93-0; EINECS/ELINCS 211-020-6
Uses: Solvent for coil, container, and automotive coatings, printing inks, acrylic lacquers, enamels, surf. primers for epoxies and polyesters, paint strippers, foundry core binders, resin cleanup, intermediates, industrial cleaners, wire enamel resins
Features: Offers VOC compliance, noncorrosive
Regulatory: DOT nonregulated; SARA nonreportable
Properties: APHA 50 max.; sweet odor; sol. in alcohols, esters, glycol ethers, halogenated solvs., ketones, hydrocarbons; sol. 2.1% in water; m.w. 174; sp.gr. 1.057; dens. 8.82 lb/gal; visc. 4.33 cps; f.p. 9.4 C; b.p. 216-230 C; acid no. 0.1; flash pt. 124 C; surf. tens. 35.1 dynes/cm; ≥ 99.0% act.; 0.1% max. water content
Toxicology: LD50 (oral, rat) > 5000 mg/kg, (skin, rabbit) > 5000 mg/kg; practically nontoxic; may cause eye and respiratory tract irritation; may cause blurred visionTSCA listed
Environmental: Biodeg.
Precaution: Avoid strong oxidizing agents, acids, and alkalis
Storage: May be stored/used with mild steel containers and equip.; no special precautions

Santosol® DME [Solutia]
Chem. Descrip.: Dimethyl glutarate (61-65%), dimethyl succinate (17-21%), dimethyl adipate (14-22%)
Uses: Solvent for coil, container, and automotive coatings, printing inks, acrylic lacquers, enamels, surf. primers for epoxies and polyesters, paint strippers, foundry core binders, resin cleanup, intermediates, industrial cleaners, wire enamel resins
Features: Offers VOC compliance, noncorrosive
Regulatory: DOT nonregulated; SARA 302, 313 nonreportable
Properties: APHA 50 max.; sweet odor; sol. in alcohols, esters, glycol ethers, halogenated solvs., ketones, hydrocarbons; sol. 5.5% in water; m.w. 160; sp.gr. 1.086; dens. 9.06 lb/gal; visc. 3.88 cps; f.p. -40.2 C; b.p. 200-220 C; acid no. 0.1; flash pt. (PMCC) 102 C; surf. tens. 35.4 dynes/cm; 0.1% max. water content
Toxicology: LD50 (oral, rat) 500-5000 mg/kg; sl. toxic; LD50 (skin, rabbit) > 5000 mg/kg; practically nontoxic; may cause eye and respiratory tract irritation; may cause blurred visionTSCA listed
Environmental: Biodeg.; EC50 (daphnia magna, 48 h) 137 mg/l; practically nontoxic
Precaution: Avoid strong oxidizers, acids, and alkalis
Hazardous decomp. prods.: CO and CO_2
Storage: May be stored/used with mild steel containers and equip., no special precautions

Santosol® DME-1 [Solutia]
Chem. Descrip.: Dimethyl glutarate (59-73%), dimethyl succinate (17-25%), dimethyl adipate (10-14%)
Uses: Solvent; formulating agent; chem. intermediate

Features: VOC compliance
Regulatory: DOT nonregulated; SARA 302, 313 nonreportable
Properties: APHA 50 colorless clear liq.; sweet odor; sol. 5.5% in water; m.w. 159; sp.gr. 1.091; dens. 9.10 lb/gal; visc. 3.88 cps; f.p. -40 C; b.p. 195-216 C; acid no. 0.3; flash pt. (PMCC) 100 C; surf. tens. 35.3 dynes/cm; 0.1% max. water
Toxicology: LD50 (oral, rat) 500-5000 mg/kg; sl. toxic; LD50 (skin, rabbit) > 5000 mg/kg; practically nontoxic; may cause blurred vision, eye and respiratory tract irritationTSCA listed
Precaution: Avoid strong oxidizers, acids, and alkalis
Hazardous decomp. prods.: CO and CO_2
Storage: Keep container closed

Santosol® DME-2 [Solutia]
Chem. Descrip.: Dimethyl glutarate (72-76%), dimethyl adipate (23-27%), dimethyl succinate (≤ 1.0%)
Uses: Solvent for coil, container, and automotive coatings, printing inks, acrylic lacquers, enamels, surf. primers for epoxies and polyesters, paint strippers, foundry core binders, resin cleanup, intermediates, industrial cleaners, wire enamel resins
Features: Offers VOC compliance, noncorrosive
Regulatory: DOT nonregulated; SARA 302, 313 nonreportable
Properties: APHA 50 max.; sweet odor; sol. in alcohols, esters, glycol ethers, halogenated solvs., ketones, hydrocarbons; sol. 4.3% in water; m.w. 163; sp.gr. 1.076; dens. 8.98 lb/gal; visc. 3.31 cps; f.p. -42.4 C; b.p. 203-220 C; acid no. 0.1; flash pt. (PMCC) 108 C; surf. tens. 35 dynes/cm; 0.1% max. water content
Toxicology: May cause eye and respiratory tract irritation; may cause blurred visionTSCA listed
Environmental: Biodeg.
Precaution: Avoid strong oxidizers, acids, and alkalis
Hazardous decomp. prods.: CO, CO_2
Storage: May be stored/used with mild steel containers and equip., no special precautions

Santosol® DME-3 [Solutia]
Chem. Descrip.: Dimethyl adipate (87-92%), dimethyl glutarate (8-12%)
Uses: Solvent for coil, container, and automotive coatings, printing inks, acrylic lacquers, enamels, surf. primers for epoxies and polyesters, paint strippers, foundry core binders, resin cleanup, intermediates, industrial cleaners, wire enamel resins
Features: Offers VOC compliance, noncorrosive
Regulatory: DOT nonregulated; SARA 302, 313 nonreportable
Properties: APHA 50 max.; sweet odor; sol. in alcohols, esters, glycol ethers, halogenated solvs., ketones, hydrocarbons; sol. 2.6% in water; m.w. 172; sp.gr. 1.058; dens. 8.83 lb/gal; visc. 3.71 cps; f.p. 2.8 C; b.p. 211-229 C; acid no. 0.1; flash pt. (PMCC) 113 C; surf. tens. 32.5 dynes/cm; 0.1% max. water content
Toxicology: May cause eye and respiratory tract irritation; may cause blurred visionTSCA listed
Environmental: Biodeg.
Precaution: Avoid strong oxidizers, acids, and alkalis
Hazardous decomp. prods.: CO, CO_2
Storage: May be stored/used with mild steel containers and equip., no special precautions

Santosol® DMG [Solutia]
Chem. Descrip.: Dimethyl glutarate
CAS 1119-40-0; EINECS/ELINCS 214-277-2
Uses: Solvent for coil, container, and automotive coatings, printing inks, acrylic lacquers, enamels, surf. primers for epoxies and polyesters, paint strippers, foundry core binders, resin cleanup, intermediates, industrial cleaners, wire enamel resins
Features: Offers VOC compliance, noncorrosive
Regulatory: DOT nonregulated; SARA 302, 313 nonreportable
Properties: APHA 50 max.; sweet odor; sol. in alcohols, esters, glycol ethers, halogenated solvs., ketones, hydrocarbons; sol. 5.1% in water; m.w. 160; sp.gr. 1.082; dens. 9.03 lb/gal; visc. 3.66 cps; f.p. -37.5 C; b.p. 203-214 C; acid no. 0.1; flash pt. (PMCC) 107 C; surf. tens. 35.6 dynes/cm; ≥ 99.0% act.; 0.1% max. water content
Toxicology: LD50 (oral, rat) > 5000 mg/kg, (skin, rabbit) > 5000 mg/kg; practically nontoxic; may cause blurred vision, eye and respiratory tract irritationTSCA listed
Environmental: Biodeg.
Precaution: Avoid strong oxidizers, acids, and alkalis
Hazardous decomp. prods.: CO, CO_2
Storage: May be stored/used with mild steel containers and equip., no special precautions

Santosol® DMS [Solutia]
Chem. Descrip.: Dimethyl succinate
CAS 106-65-0; EINECS/ELINCS 203-419-9
Uses: Solvent for coil, container, and automotive coatings, printing inks, acrylic lacquers, enamels, surf. primers for epoxies and polyesters, paint strippers, foundry core binders, resin cleanup, intermediates, industrial cleaners, wire enamel resins
Features: Offers VOC compliance, noncorrosive
Regulatory: DOT nonregulated; SARA 302, 313 nonreportable
Properties: APHA 50 max.; sweet odor; sol. in alcohols, esters, glycol ethers, halogenated solvs., ketones, hydrocarbons; sol. 10.3% in water; m.w. 146; sp.gr. 1.112; dens. 9.28 lb/gal; visc. 3.91 cps; f.p. 16.8 C; b.p. 192-201 C; acid no. 0.1; flash pt. (PMCC) 94 C; surf. tens. 34.6 dynes/cm; ≥ 99.0% act.; 0.1% max. water content
Toxicology: LD50 (oral, rat) 500-5000 mg/kg); sl. toxic; LD50 (skin, rabbit) > 5000 mg/kg; practically nontoxic; may cause blurred vision, eye and respiratory tract irritationTSCA listed
Environmental: Biodeg.
Precaution: Avoid strong oxidizers, acids, and alkalis
Hazardous decomp. prods.: CO, CO_2
Storage: May be stored/used with mild steel containers and equip., no special precautions

Sasol-47 [Sasol N. Am.]
Chem. Descrip.: Petroleum distillates, hydrotreated light (75-83%), C14-16 alkane (17-25%)
Uses: Solvent for printing inks, esp. heat-set printing inks
Regulatory: DOT nonregulated; SARA §302 nonreportable, §311/312 nonhazardous, §313 nonreportable; Calif. Prop. 65 nonreportable; EINECS, Canada DSL, Australia AICS listed
Properties: Lt. straw to water-wh. oily liq.; mild hydrocarbon odor; sol. < 1 ppm in water; m.w. 196; sp.gr. 0.812 (60/60 F); dens. 6.78 lb/gal (60 F); visc. 3.9 cSt (20 C); vapor pressure < 0.1 mm Hg (20 C); f.p. -18 C; b.p. 240-277 C; pour pt. -18 C; flash pt. (PM) 97-108 C; autoignition temp. 237 C; KB value 27
Toxicology: Rec. PEL TWA (8 h) 100 ppm; may cause sl. eye irritation; repeated/prolonged contact may cause skin redness, irritation, dermatitis; high vapor concs. may cause headaches, stupor, throat/eye irritation, kidney effects; aspiration into lungs may cause pneumonia or death; low toxicity by ing.; causes irritation of stomach and intestines, nausea, vomiting; TSCA listed
Environmental: Inherently biodeg.
Precaution: Flamm. limits in air 0.5-7.0 vol.%; potential for accumulation of static electricity when transferring prod.—bond and ground; avoid high temps.; may react with strong oxidizers
Hazardous decomp. prods.: Combustion: CO_x
HMIS: Health 1, Flammability 1, Reactivity 0
Storage: Store above 0 F; avoid unnecessary heating

Sasol C1316 [Sasol N. Am.]
Chem. Descrip.: Tetradecane (63%), pentadecane (23%), tridecane (5%), hexadecane (5%)
Uses: Solvent for agric., oil field drilling, lubricants; lamp fuel
Features: High purity; extremely low sulfur and nitrogen content
Regulatory: DOT nonhazardous; SARA §302 nonreportable, §311/312 acute health hazard, §313 nonreportable; Calif. Prop. 65 nonreportable; EINECS, Canada DSL, Japan MITI, Australia AICS listed
Properties: Lt. yel. to water-wh. clear low-visc. oily liq.; hydrocarbon odor; negligible sol. in water; m.w. 203; sp.gr. 0.772 (60/60 F); visc. 2.1 cSt (40 C); vapor pressure < 1 mm Hg (20 C); m.p. 3 C; i.b.p. 243 C; b.p. 243-275 C; flash pt. (PMCC) 113 C; autoignition temp. 210 C; KB value 20; VOC 100%
Toxicology: ACGIH TWA (8 h) 5 mg/m^3; LD50 (oral, rat) > 5 g/kg, (skin, rabbit) > 2 g/kg; LC50 (inh., rat, 4 h) > 5.6 mg/l; causes skin irritation; prolonged/repeated contact may cause dermatitis, reddening of skin; may cause eye irritation on contact or exposure to conc. vapors; inh. of high conc. of vapors may cause headache, stupor; prolonged breathing of mist or vomiting following ing. may cause serious lung injury, possibly death; TSCA listed
Precaution: Combustible liq.; flamm. limits in air 0.5-7.0 vol.%; electrostatic accumulation hazard; ground equip.; avoid high temps.; incompat.

with strong oxidizers
Hazardous decomp. prods.: Combustion: CO_x, other org. compds.
HMIS: Health 1, Flammability 1, Reactivity 0
Storage: Store @ ambient temp. and pressure

Sasol LPA Solvent [Sasol N. Am.]
Chem. Descrip.: Petroleum distillates, hydrotreated light
CAS 64742-47-8; EINECS/ELINCS 265-149-8
UN 1268
Uses: Solvent for household/industrial cleaners, metal rolling oils, paints/coatings, thinners, stains, paper and mining chems., pesticides, water treatment chems., waterless hand cleaners, food-contact applics. (adhesives, paper, coatings, rubber, textiles, etc.); chemical process solvent; carrier in pesticides, waterless hand cleaners; degreaser; f.p. depressant; lamp oil; lubricating oil; printing ink oil
Features: High purity; med. flash pt.; broad boiling range; virtually free of sulfur and nitrogen compds.
Regulatory: FDA 21CFR §172.884, 173.340, 175.105, 176.180, 176.200, 176.210, 177.1200, 177.1400, 177.2260, 177.2600, 177.2800, 178.3120, 178.3570, 178.3620(b), 178.3650, 178.3910(b)(2); SARA §302 nonreportable, §311/312 fire hazard, §313 nonreportable; Calif. Prop. 65 nonreportable; Canada DSL, EINECS, Japan MITI, Australia AICS listed
Properties: Water-wh. oily liq.; hydrocarbon odor; sol. 51 ppm in water (20 C); m.w. 166; sp.gr. 0.811 (16/16 C); dens. 6.77 lb/gal (60 F); visc. 1.9 cSt (20 C); vapor pressure 0.3 mm Hg (20 C); f.p. < -68 C; b.p. 187-259 C; pour pt. < -90 F; flash pt. (PMCC) 71 C; autoignition temp. 233 C; KB value 32; 100% act.
Toxicology: Rec. PEL TWA (8 h) 100 ppm; LD50 (oral, rat) > 39.9 g/kg, (skin, rabbit) 2-4 g/kg; may cause skin/sl. eye irritation; high vapor concs. may cause headaches, stupor, throat/eye irritation, kidney effects; aspiration can cause pneumonia or death; prolonged/repeated skin contact may cause redness, dermatitis; low toxicity by ing.; causes irritation of stomach and intestines, nausea, vomiting; TSCA listed
Environmental: Low biodeg.; can cause toxic effects to aquatic organisms; prevent runoff to sewers, waterways, soil
Precaution: Combustible liq. and vapor; flamm. limits in air 0.6-7.0 vol.%; vapors may flash back; container may explode in heat of fire; potential for accumulation of static electricity when transferring prod.—bond and ground; avoid high temps.; may react with strong oxidizers
Hazardous decomp. prods.: None expected
HMIS: Health 1, Flammability 2, Reactivity 0
Storage: Store @ ambient temps. and pressure

Sasol LPA-142 Solvent [Sasol N. Am.]
Chem. Descrip.: Petroleum distillates, hydrotreated light
CAS 64742-47-8; EINECS/ELINCS 265-149-8
UN 1268
Uses: Solvent for household/industrial cleaners, metal rolling oils, paints/coatings, thinners, stains, paper and mining chems., pesticides, water treatment chems., waterless hand cleaners, food-contact applics. (adhesives, paper, coatings, rubber, textiles, etc.); chemical process solvent; degreaser; f.p. depressant; lamp oil; lubricating oil; printing ink oil; diluent in polymer processing
Features: High purity; highest solvency of 140°F flash pt. aliphatic solvents; < 1% aromatic content
Regulatory: FDA 21CFR §172.884, 173.340, 175.105, 176.180, 176.200, 176.210, 177.1200, 177.1400, 177.2260, 177.2600, 177.2800, 178.3120, 178.3570, 178.3620(b), 178.3650, 178.3910(b)(2); SARA §302 nonreportable, §311/312 fire hazard, §313 nonreportable; Calif. Prop. 65 nonreportable; Canada DSL, EINECS, Japan MITI, Australia AICS listed
Properties: Water-wh. oily liq.; hydrocarbon odor; sol. 52 ppm in water (20 C); m.w. 158; sp.gr. 0.802 (16/16 C); dens. 6.69 lb/gal (60 F); visc. 2.0 cSt (20 C); vapor pressure 0.36 mm Hg (20 C); f.p. < -70 C; b.p. 193-211 C; pour pt. < -70 C; flash pt. (TCC) 64 C; autoignition temp. 233 C; KB value 34; 100% act.
Toxicology: Rec. PEL TWA (8 h) 100 ppm; LD50 (oral, rat) > 5 g/kg; LC50 (inh., rat) > 4.3 mg/l; may cause skin/sl. eye irritation; high vapor concs. may cause headaches, stupor, throat/eye irritation, kidney effects; aspiration can cause pneumonia or death; prolonged/repeated skin contact may cause redness, dermatitis; low toxicity by ing.; causes irritation of stomach and intestines, nausea, vomiting; TSCA listed
Environmental: Low biodeg.; can cause toxic effects to aquatic organisms; prevent runoff to sewers, waterways, soil
Precaution: Combustible liq. and vapor; flamm. limits in air 0.7-7.0 vol.%; vapors may flash back; container may explode in heat of fire; potential for accumulation of static electricity when transferring prod.—bond and ground; avoid high temps.; may react with strong oxidizers
Hazardous decomp. prods.: None expected
HMIS: Health 1, Flammability 2, Reactivity 0
Storage: Store @ ambient temps. and pressure

Sasol LPA-150 Solvent [Sasol N. Am.]
Chem. Descrip.: Petroleum distillates, hydrotreated light
CAS 64742-47-8; EINECS/ELINCS 265-149-8
UN 1268
Uses: Solvent for household/industrial cleaners, metal rolling oils, paints/coatings, thinners, stains, paper and mining chems., pesticides, water treatment chems., waterless hand cleaners, food-contact applics. (adhesives, paper, coatings, rubber, textiles, etc.); chemical process solvent; degreaser; f.p. depressant; lamp oil; lubricating oil; printing ink oil
Features: High purity; very low levels of aromatics; high solvency; higher flash pt.
Regulatory: FDA 21CFR §172.884, 173.340, 175.105, 176.180, 176.200, 176.210, 177.1200, 177.1400, 177.2260, 177.2600, 177.2800, 178.3120, 178.3570, 178.3620(b), 178.3650, 178.3910(b)(2); SARA §302 nonreportable, §311/312 fire hazard, §313 nonreportable; Calif. Prop. 65 nonreportable; Canada DSL, EINECS, Japan MITI, Australia AICS listed
Properties: Water-wh. oily liq.; mild hydrocarbon odor; negligible sol. in water; m.w. 164; sp.gr. 0.811 (16 C); dens. 6.74 lb/gal (60 F); visc. 2.3 cSt (20 C); vapor pressure 0.75 torr (38 C); m.p. -63 C; b.p. 192-230 C; pour pt. < -94 F; flash pt. (PMCC) 66 C; KB value 32; 100% act.
Toxicology: Rec. PEL TWA (8 h) 100 ppm; LD50 (oral, rat) > 5 g/kg, (skin, rabbit) 2-4 g/kg; may cause eye/skin irritation; high vapor concs. may cause headaches, stupor, throat/eye irritation, kidney effects; aspiration can cause pneumonia or death; prolonged/repeated skin contact may cause redness, dermatitis; low toxicity by ing.; causes irritation of stomach and intestines, nausea, vomiting; TSCA listed
Environmental: Inherently biodeg.
Precaution: Combustible liq. and vapor; flamm. limits in air 0.6-4.7 vol.%; vapors may flash back; container may explode in heat of fire; potential for accumulation of static electricity when transferring prod.—bond and ground; avoid high temps.; may react with strong oxidizers
Hazardous decomp. prods.: None expected
HMIS: Health 1, Flammability 2, Reactivity 1
Storage: Store @ ambient temps. and pressure

Sasol LPA-170 Solvent [Sasol N. Am.]
Chem. Descrip.: Petroleum distillates, hydrotreated light
CAS 64742-47-8; EINECS/ELINCS 265-149-8
UN 1268
Uses: Solvent for household/industrial cleaners, metal rolling oils, paints/coatings, thinners, stains, paper and mining chems., pesticides, water treatment chems., waterless hand cleaners, food-contact applics. (adhesives, paper, coatings, rubber, textiles, etc.); solvent for chemical processing, esp. prod. of polymer emulsions, mining extraction processes, metal lubricants, specialty coatings; degreaser; f.p. depressant; lamp oil; lubricating oil; printing ink oil
Features: Higher flash pt.; narrow boiling range
Regulatory: FDA 21CFR §172.884, 173.340, 175.105, 176.180, 176.200, 176.210, 177.1200, 177.1400, 177.2260, 177.2600, 177.2800, 178.3120, 178.3570, 178.3620(b), 178.3650, 178.3910(b)(2); SARA §302 nonreportable, §311/312 fire hazard, §313 nonreportable; Calif. Prop. 65 nonreportable; Canada DSL, EINECS, Japan MITI, Australia AICS listed
Properties: Water-wh. oily liq.; mild hydrocarbon odor; sol. 61 ppm in water (20 C); m.w. 172; sp.gr. 0.813 (16 C); dens. 6.78 lb/gal (60 F); visc. 2.6 cSt (20 C); vapor pressure 0.05 mm Hg (20 C); f.p. < -70 C; b.p. 213-23 C; pour pt. < -70 C; flash pt. (TCC) 81 C; autoignition temp. 228 C; KB value 32; 100% act.
Toxicology: Rec. PEL TWA (8 h) 100 ppm; LD50 (oral, rat) > 5 g/kg; LC50 (inh.) > 6.8 mg/l; may cause skin/sl. eye irritation; high vapor concs. may cause headaches, stupor, throat/eye irritation, kidney effects; aspiration can cause pneumonia or death; prolonged/repeated skin contact may cause redness, dermatitis; low toxicity by ing.; causes irritation of stomach and intestines, nausea, vomiting; TSCA listed

Environmental: Low biodeg.; can cause toxic effects to aquatic organisms; prevent runoff to sewers, waterways, soil
Precaution: Combustible liq. and vapor; flamm. limits in air 0.6-7.0 vol.%; vapors may flash back; container may explode in heat of fire; potential for accumulation of static electricity when transferring prod.—bond and ground; avoid high temps.; may react with strong oxidizers
Hazardous decomp. prods.: None expected
HMIS: Health 1, Flammability 2, Reactivity 0
Storage: Store @ ambient temps. and pressure

Sasol LPA-210 Solvent [Sasol N. Am.]
Chem. Descrip.: Petroleum distillates, hydrotreated light
CAS 64742-47-8; EINECS/ELINCS 265-149-8
Uses: Solvent for household/industrial cleaners, metal rolling oils, paints/coatings, thinners, stains, paper and mining chems., pesticides, water treatment chems., waterless hand cleaners, food-contact applics. (adhesives, paper, coatings, rubber, textiles, etc.); chemical process solvent; degreaser; f.p. depressant; lamp oil; metal lubricant; printing ink oil; pesticide carrier
Features: High purity; higher flash pt. for added safety
Regulatory: FDA 21CFR §172.884, 173.340, 175.105, 176.180, 176.200, 176.210, 177.1200, 177.1400, 177.2260, 177.2600, 177.2800, 178.3120, 178.3570, 178.3620(b)(c), 178.3650, 178.3910(a)(3),(b)(2); DOT nonregulated; SARA §302 nonreportable, §311/312 nonhazardous, §313 nonreportable; Calif. Prop. 65 nonreportable; Canada DSL, EINECS, Japan MITI, Australia AICS listed
Properties: Water-wh. oily liq.; mild hydrocarbon odor; sol. 1.8 ppm in water (38 C); m.w. 192; sp.gr. 0.824 (16 C); dens. 6.87 lb/gal (60 F); visc. 4.0 cSt (20 C); vapor pressure 0.03 mm Hg (20 C); f.p. < -50.5 C; b.p. 242-276 C; pour pt. -72 F; flash pt. (PM) 108 C; autoignition temp. 218 C; KB value 29; 100% act.
Toxicology: Rec. PEL TWA (8 h) 100 ppm; LC50 (inh., rat, 4 h) 0.57 mg/l; may cause skin/sl. eye irritation; high vapor concs. may cause headaches, stupor, throat/eye irritation, kidney effects; aspiration can cause pneumonia or death; prolonged/repeated skin contact may cause redness, dermatitis; low toxicity by ing.; causes irritation of stomach and intestines, nausea, vomiting; TSCA listed
Environmental: Inherently biodeg.
Precaution: Flamm. limits in air 0.5-7.0 vol.%; potential for accumulation of static electricity when transferring prod.—bond and ground; avoid high temps.; may react with strong oxidizers
Hazardous decomp. prods.: Combustion: CO_x
HMIS: Health 1, Flammability 1, Reactivity 0
Storage: Store @ ambient temps. and pressure

Sasol MR Solvent [Sasol N. Am.]
Chem. Descrip.: Petroleum raffinates, sorption process
CAS 64741-85-1; EINECS/ELINCS 265-087-1
UN 1268
Uses: Solvent for agric., vehicle coatings, solvent extract processes (copper ore, etc.), lubricants, freeze pt. depressants; base oil for drilling fluids
Features: Highly refined; low visc.; 15% aromatics content; extremely low sulfur and nitrogen; higher solv. str. and lower f.p. than competitive solvs. with equiv. b.p.; 35% VOC exemption
Regulatory: SARA §311/312 fire hazard, §302/313 nonreportable; Calif. Prop. 65 nonreportable; Canada DSL, EINECS, Japan MITI, Australia AICS listed
Properties: Water-wh. oily liq.; mild hydrocarbon odor; sol. 50 ppm in water (20 C); m.w. 170; sp.gr. 0.817 (60/60 F); dens. 6.82 lb/gal (60 F); visc. 2.1 cSt (21 C); vapor pressure 1.2 mm Hg (100 F); f.p. < -90 F; i.b.p. 370 F; pour pt. < -95 F; flash pt. (PMCC) 64-88 C; autoignition temp. 231 C; KB value 33; 100% act.
Toxicology: Rec. PEL TWA (8 h) 100 ppm; LD50 (oral, rat) > 5 g/kg, (skin, rabbit) 3-4 g/kg; may cause eye/skin irritation; high vapor concs. may cause headaches, stupor, throat/eye irritation, kidney effects; aspiration can cause pneumonia or death; prolonged/repeated skin contact may cause redness, dermatitis; low toxicity by ing.; causes irritation of stomach and intestines, nausea, vomiting; TSCA listed
Environmental: Inherently biodeg.
Precaution: Combustible liq. and vapor; flamm. limits in air 0.6-7.0 vol.%; vapors may flash back; container may explode in heat of fire; potential for accumulation of static electricity when transferring prod.—bond and ground; avoid high temps.; may react with strong oxidizers
HMIS: Health 1, Flammability 2, Reactivity 0
Storage: Store @ ambient temps. and pressure

SBP 40/65 LNH [Shell Netherlands]
Chem. Descrip.: Naphtha (petroleum), hydrotreated light , n-hexane (< 3%)
UN 3295
Uses: Solvent for adhesives, rubber; extraction solvent for medicinal purposes (skin cleaning)
Features: High volatility; very low aromatics and olefins; contains < 2% n-hexane; does not contain detectable quantities of polycyclic aromatics, heavy metals, or chlorinated compds.
Regulatory: EINECS, Canada DSL, Australia AICS, China IECSC, Japan MITI, Philippines PICCS, Korea TCCL listed
Properties: Colorless liq.; paraffinic odor; insol. in water; m.w. 82; dens. 658 kg/m^3 (15 C); kinematic visc. 0.41 mm^2/s; vapor pressure 33,000 Pa (20 C); i.b.p. 44 C; b.p. 44-62 C; pour pt. -150 C; flash pt. -43 C; autoignition temp. 392 C; surf. tens. 16.8 mN/m (20 C); KB value 28
Toxicology: Expected to be of low toxicity by ing., inh., skin contact; aspiration into lungs may cause chem. pneumonitis which can be fatal; narcotic at high vapor concs.; causes CNS depression, headache, dizziness, nausea
Environmental: Inherently biodeg.; potential to bioaccumulate; toxic to aquatic organisms; may cause long-term adverse effects in aquatic environment; prevent contamination of soil, water; prevent entry into drains, ditches, rivers
Precaution: Highly flamm.; LEL 1.1 vol.%; UEL 7.5 vol.%; readily forms explosive vapor-air mixt.; vapor is heavier than air, spreads along ground; distant ignition is possible; electrostatic charges may be generated; prevent static discharge
Hazardous decomp. prods.: Combustion: CO
Storage: 12 mos shelf life @ ambient temps. in tightly closed containers in well-ventilated area, away from direct sunlight, other heat/ignition sources

SBP 60/95 LNH [Shell Netherlands]
Chem. Descrip.: Naphtha (petroleum), hydrotreated light , n-hexane (< 3%)
UN 3295
Uses: Solvent for rubber, adhesives, printing inks; cleaning solvent in printing industry
Features: Very low aromatics and olefins; does not contain detectable quantities of polycyclic aromatics, heavy metals, or chlorinated compds.
Regulatory: EINECS, Canada DSL, Australia AICS, China IECSC, Japan MITI, Philippines PICCS, Korea TCCL listed
Properties: Colorless liq.; paraffinic odor; insol. in water; m.w. 92; dens. 693 kg/m^3 (15 C); kinematic visc. 0.45 mm^2/s; vapor pressure 19,000 Pa (20 C); i.b.p. 67.5 C; b.p. 60-95 C; pour pt. < -100 C; flash pt. -25 C; autoignition temp. 413 C; KB value 34
Toxicology: Expected to be of low toxicity by ing., inh., skin contact; aspiration into lungs may cause chem. pneumonitis which can be fatal; narcotic at high vapor concs.; causes CNS depression, headache, dizziness, nausea; may cause skin irritation/dryness/dermatitis; risk of impaired fertility; TSCA listed
Environmental: Expected to be inherently biodeg.; potential to bioaccumulate; toxic to aquatic organisms; may cause long-term adverse effects in aquatic environment; prevent contamination of soil, water; prevent entry into drains, ditches, rivers
Precaution: Highly flamm.; LEL 1.0 vol.%; UEL 7.3 vol.%; readily forms explosive vapor-air mixt.; vapor is heavier than air, spreads along ground; distant ignition is possible; electrostatic charges may be generated; prevent static discharge; ground equip.; risk of explosion
Hazardous decomp. prods.: Combustion: CO
Storage: 12 mos shelf life @ ambient temps. in tightly closed containers in well-ventilated area, away from direct sunlight, other heat/ignition sources

SBP 80/95 LNH [Shell Netherlands]
Chem. Descrip.: Naphtha (petroleum), hydrotreated light , n-hexane (< 3%)
UN 3295
Uses: Solvent for rubber, chem. processing, paper, adhesives, sealants, surf. coatings, printing inks, household cleaners
Features: Very low aromatics and olefins; does not contain detectable quantities of polycyclic aromatics, heavy metals, or chlorinated compds.

Regulatory: EINECS, Canada DSL, Australia AICS, China IECSC, Japan MITI, Philippines PICCS, Korea TCCL listed
Properties: Colorless liq.; paraffinic odor; immisc. with water; m.w. 96; dens. 715 kg/m³ (15 C); kinematic visc. 0.7 mm²/s; vapor pressure 8500 Pa (20 C); i.b.p. 86 C; b.p. 86-93 C; pour pt. -60 C; flash pt. -18 C; KB value 37
Toxicology: Expected to be of low toxicity by ing., inh., skin contact; aspiration into lungs may cause chem. pneumonitis which can be fatal; narcotic at high vapor concs.; causes CNS depression, headache, dizziness, nausea; may cause skin irritation/dryness/dermatitis; risk of impaired fertility; TSCA listed
Environmental: Expected to be inherently biodeg.; potential to bioaccumulate; toxic to aquatic organisms; may cause long-term adverse effects in aquatic environment; prevent contamination of soil, water; prevent entry into drains, ditches, rivers
Precaution: Highly flamm.; LEL 1.0 vol.%; UEL 7.3 vol.%; readily forms explosive vapor-air mixt.; vapor is heavier than air, spreads along ground; distant ignition is possible; electrostatic charges may be generated; prevent static discharge; ground equip.; risk of explosion
Hazardous decomp. prods.: Combustion: CO
Storage: 12 mos shelf life @ ambient temps. in tightly closed containers in well-ventilated area, away from direct sunlight, other heat/ignition sources

SBP 80/100 [ExxonMobil Singapore]
Chem. Descrip.: Dearomatized aliphatic hydrocarbon
Uses: Solvent
Properties: Dens. 0.708 kg/dm³ (15 C); visc. 0.59 mm²/s; i.b.p. 78 C; flash pt. < 0 C; ref. index 1.396 (20 C); surf. tens. 20.2 mN/m; KB value 34; aniline pt. 60 C; < 0.01% aromatics

SBP 80/110 LNH [Shell Netherlands]
Chem. Descrip.: Naphtha (petroleum), hydrotreated light , n-hexane (< 5%)
UN 3295
Uses: Solvent for rubber, chem. processing, paper, adhesives, sealants, surf. coatings, printing inks, household cleaners
Features: Very low aromatics and olefins; does not contain detectable quantities of polycyclic aromatics, heavy metals, or chlorinated compds.
Regulatory: EINECS, Canada DSL, Australia AICS, China IECSC, Japan MITI, Philippines PICCS, Korea TCCL listed
Properties: Colorless liq.; paraffinic odor; immisc. with water; m.w. 99; dens. 714 kg/m³ (15 C); kinematic visc. 0.61 mm²/s; vapor pressure 8500 Pa (20 C); i.b.p. 88 C; b.p. 88-105 C; pour pt. < -30 C; flash pt. -12 C; autoignition temp. 367 C; surf. tens. 21.2 mN/m (20 C); KB value 33
Toxicology: LD50 (oral) > 2000 mg/kg, (skin) > 2000 mg/kg; LC50 (inh.) > 5 mg/l; harmful by inh.; risk of serious health damage on prolonged exposure; aspiration into lungs may cause chem. pneumonitis which can be fatal; narcotic at high vapor concs.; causes CNS depression, headache, dizziness, nausea, respiratory tract irritation, skin irritation/dryness/dermatitis; risk of impaired fertility; TSCA listed
Environmental: Inherently biodeg.; potential to bioaccumulate; toxic to aquatic organisms; may cause long-term adverse effects in aquatic environment; prevent contamination of soil, water; prevent entry into drains, ditches, rivers
Precaution: Highly flamm.; LEL 1.0 vol.%; UEL 7.0 vol.%; may form explosive vapor-air mixt.; vapor is heavier than air, spreads along ground; distant ignition is possible; electrostatic charges may be generated; prevent static discharge; ground equip.; risk of explosion
Hazardous decomp. prods.: Combustion: CO
Storage: 12 mos shelf life @ ambient temps. in tightly closed containers in well-ventilated area, away from direct sunlight, other heat/ignition sources

SBP 100/140 [Shell Netherlands]
Chem. Descrip.: Naphtha (petroleum), hydrotreated light
CAS 64742-49-0; EINECS/ELINCS 265-151-9
UN 3295
Uses: Solvent for rubber, chem. processing, paper, adhesives, surf. coatings, printing inks, household cleaners, pesticides, solvent extraction; degreasing solvent
Features: Very low aromatics and olefins; does not contain detectable quantities of polycyclic aromatics, heavy metals, or chlorinated compds.
Regulatory: EINECS, Canada DSL, Australia AICS, China IECSC, Japan MITI, Philippines PICCS, Korea TCCL listed
Properties: Colorless liq.; paraffinic odor; insol. in water; m.w. 112; dens. 728 kg/m³² (15 C); kinematic visc. 0.76 mm²/s; vapor pressure 3500 Pa (20 C); i.b.p. 107 C; b.p. 107-137 C; pour pt. < -30 C; flash pt. 1 C; autoignition temp. 310 C; surf. tens. 21.9 mN/m (20 C); KB value 33
Toxicology: LD50 (oral) > 2000 mg/kg, (skin) > 2000 mg/kg; LC50 (inh.) > 5 mg/l; aspiration into lungs may cause chem. pneumonitis which can be fatal; narcotic at high vapor concs.; causes CNS depression, headache, dizziness, nausea, respiratory tract irritation, skin irritation/dryness/dermatitis; TSCA listed
Environmental: Inherently biodeg.; potential to bioaccumulate; toxic to aquatic organisms; may cause long-term adverse effects in aquatic environment; prevent contamination of soil, water; prevent entry into drains, ditches, rivers
Precaution: Highly flamm.; LEL 0.9 vol.%; UEL 6.8 vol.%; extreme risk of vapor ignition @ normal handling temps.; readily forms explosive vapor-air mixt.; vapor is heavier than air, spreads along ground; distant ignition is possible; electrostatic charges may be generated; prevent static discharge; ground equip.; risk of explosion
Hazardous decomp. prods.: Combustion: CO
Storage: 12 mos shelf life @ ambient temps. in tightly closed containers in well-ventilated area, away from direct sunlight, other heat/ignition sources

SBP 140/165 [Shell Netherlands]
Chem. Descrip.: Naphtha (petroleum), hydrotreated light
CAS 64742-49-0; EINECS/ELINCS 265-151-9
UN 3295
Uses: Solvent for rubber, chem. processing, paper, household prods., surf. coatings, printing inks, adhesives, pesticides, waxes, polishes, degreasing applics.; extraction solvent
Features: Very low aromatics and olefins; does not contain detectable quantities of polycyclic aromatics, heavy metals, or chlorinated compds.
Regulatory: EINECS, Canada DSL, Australia AICS, China IECSC, Japan MITI, Philippines PICCS, Korea TCCL listed
Properties: Colorless liq.; paraffinic odor; immisc. with water; m.w. 130; dens. 0.750 kg/l (15 C); kinematic visc. 0.91 mm²/s; vapor pressure 1.3 kPa (20 C); i.b.p. 143 C; b.p. 143-160 C; flash pt. (Abel) 25 C; autoignition temp. 287 C; ref. index 1.415; KB value 31; surf. tens. 22.2 mN/m; sp.heat 2.0 kJ/kg/°C; elec. conduct. 0.065 pS/m
Toxicology: Expected to be of low toxicity by ing., inh., skin contact; aspiration into lungs may cause chem. pneumonitis which can be fatal; narcotic at high vapor concs.; causes CNS depression, headache, dizziness, nausea; may cause skin irritation/dryness/dermatitis; TSCA listed
Environmental: Inherently biodeg.; potential to bioaccumulate; not classified as dangerous under EC criteria; sl. toxic to fish; pract. nontoxic to algae; prevent contamination of soil, water; prevent entry into drains, ditches, rivers
Precaution: Highly flamm.; LEL 0.8 vol.%; UEL 6.0 vol.%; may form explosive vapor-air mixt.; vapor is heavier than air, spreads along ground; distant ignition is possible; electrostatic charges may be generated; prevent static discharge; ground equip.; risk of explosion
Hazardous decomp. prods.: Combustion: CO
Storage: 12 mos shelf life @ ambient temps. in tightly closed containers in well-ventilated area, away from direct sunlight, other heat/ignition sources

Schercemol 1818 [Scher; Chesham Chems. Ltd]
Chem. Descrip.: Isostearyl isostearate
CAS 41669-30-1; EINECS/ELINCS 255-485-3
Uses: Substantive emollient, slip agent, lubricant, gloss aid in cosmetics; cosolvent, solubilizer in perfumes
Features: Imparts luxurious softness to skin
Properties: Yel. clear liq., slight typ. odor; sol. in org. solvs., insol. in water; m.w. 536; sp.gr. 0.865-0.875; f.p. -5 C; acid no. 2.0 max.; iodine no. 13 max.; sapon. no. 95-110; cloud pt. 17 C; flash pt. (OC) > 180 C; ref. index 1.4610

Schercemol CO [Scher; Chesham Chems. Ltd]
Chem. Descrip.: Cetyl octanoate
CAS 59130-69-7; EINECS/ELINCS 261-619-1
Uses: Solvent for make-up removers
Properties: Clear liq.; f.p. 10 C; acid no. 3 max.; iodine no. nil; sapon. no. 140-155

Schercemol DIA [Scher; Chesham Chems. Ltd]
Chem. Descrip.: Diisopropyl adipate

CAS 6938-94-9; EINECS/ELINCS 230-072-0
Uses: Penetrant, emollient, moisturizer, lubricant, solvent in hydroalcoholic cosmetics incl. bath oils, lipsticks, men's grooming aids, hand lotions; lubricant, plasticizer, solubilizer for metalworking, petrol. and plastic industries
Features: Nonoily; mild drying effects
Properties: Water-wh. clear liq., faint ester odor; sol. in alcohols, higher glycols, ketones, ester, aromatic, chlorinated, and aliphatic hydrocarbons; insol. in water; m.w. 230; sp.gr. 0.960 ± 0.01; dens. 8.0 lb/gal; vapor pressure < 0.05 mm Hg (20 C); f.p. -1 C; b.p. 232 C; acid no. 2 max.; iodine no. nil; sapon. no. 480-500; flash pt. (COC) > 132 C; ref. index 1.423 ± 0.001
Toxicology: Eye and intestinal tract irritant; prolonged contact may cause skin irritation; inh. may cause irritation to nasal passagesTSCA listed
Precaution: Spills are slippery
Hazardous decomp. prods.: CO, CO_2
Storage: Avoid overheating or freezing

Schercemol DIS [Scher; Chesham Chems. Ltd]
Chem. Descrip.: Diisopropyl sebacate
CAS 7491-02-3; EINECS/ELINCS 231-306-4
Uses: Emollient, lubricant, solubilizer in hydroalcoholic personal care prods.; solvent; coupling agent; spreading agent
Features: Nonoily; mild drying effects; fast spreading action
Properties: Colorless clear liq., bland odor; sol. in alcohols, higher glycols, ketones, esters, aromatic, aliphatic, and chlorinated hydrocarbons, min. and natural oils; insol. in water; m.w. 286; sp.gr. 0.932 ± 0.01; dens. 7.8 lb/gal; f.p. 0 C; acid no. 1.0 max.; iodine no. nil; sapon. no. 380-400; flash pt. > 170 C; ref. index 1.4320; 99% min. ester conc.

Schercemol NGDC [Scher; Chesham Chems. Ltd]
Chem. Descrip.: Neopentyl glycol dicaprate
CAS 27841-06-1; EINECS/ELINCS 248-688-3
Uses: Solvent for make-up removers, topical pharmaceuticals
Properties: Clear liq.; f.p. 2 C; acid no. 3 max.; iodine no. nil; sapon. no. 255-270

Schercemol NGDO [Scher; Chesham Chems. Ltd]
Chem. Descrip.: Neopentyl glycol dioctanoate
CAS 28510-23-8; EINECS/ELINCS 249-060-1
Uses: Emollient, solvent for makeup removers, topical pharmaceuticals
Features: Low freeze pt.
Properties: Clear liq.; f.p. < -12 C; acid no. 3 max.; iodine no. nil; sapon. no. 290-310

Schercemol OLO [Scher; Chesham Chems. Ltd]
Chem. Descrip.: Oleyl oleate
CAS 3687-45-4; EINECS/ELINCS 222-980-4
Uses: Emollient, softener, gloss aid, lubricant for cosmetics, hair care, skin care, topical pharmaceuticals; cosolvent, solubilizer in perfumes; lubricant for metalworking and wire drawing
Features: Nonoily
Properties: Amber clear liq., mild oleic odor; sol. in org. solvs., insol. in water; m.w. 332; sp.gr. 0.860 ± 0.01; dens. 7.2 lb/gal; acid no. 2 max.; iodine no. 95 max.; sapon. no. 95-110; cloud pt. 13 C; flash pt. (OC) > 180 C; ref. index 1.4630 ± 0.001

Schercemol PGDP [Scher; Chesham Chems. Ltd]
Chem. Descrip.: Propylene glycol dipelargonate
CAS 41395-83-9; EINECS/ELINCS 255-350-9
Uses: Emollient; cosolvent for perfumed bath oils, creams/lotions
Features: Low f.p.
Properties: Straw-colored clear liq., slight typ. odor; sol. in org. solvs., insol. in water; m.w. 360; sp.gr. 0.917; dens. 7.6 lb/gal; f.p. -25 C; acid no. 5 max.; iodine no. nil; sapon. no. 300-320; flash pt. (OC) > 170 C; ref. index 1.440; nonionic

Schercemol PGML [Scher; Chesham Chems. Ltd]
Chem. Descrip.: Propylene glycol laurate
CAS 27194-74-7; EINECS/ELINCS 205-542-3
Uses: Emollient, solvent, emulsion stabilizer for cosmetics, topical pharmaceuticals; solubilizer, coupling agent for perfumes, colors, flavors, sunscreen compds.; solvent for sprayed org. pesticides; plasticizer and stabilizer in vinyl copolymers made from PVAc and PVC; defoamer in PVAc emulsions
Features: Stable base; produces sprayable oils which spread well, have good adherence, and are not readily removed by rainfall
Properties: Yel. clear liq., mild odor; sol. in most org. solvs. such as alcohols, ketones, esters, glycol ethers, veg. oil, min. oil, aliphatic, aromatic, chlorinated hydrocarbons; disp. in glycols, triols, polyols; m.w. 258 (theoret.); sp.gr. 0.905 ± 0.01; dens. 7.45 lb/gal; f.p. 10 C; acid no. 5 max.; iodine no. 1 max.; sapon no. 225-240; flash pt. (OC) 160 C min.; pH 7.0 (10% disp.); nonionic

Scour NFP [Bayer/Industrial Chems./Textile]
Uses: Solvent scouring agent for textile roller cleaner, screen clear applics.
Features: High flash pt.

SDA-1 [Eastman]
Chem. Descrip.: SD alcohol 1
Uses: Solvent, astringent, diluent, visc reducer in cosmetics, personal care

SDA-2B [Eastman]
Chem. Descrip.: SD alcohol 2B
Uses: Solvent for toluene, rubber

SDA-3A [Eastman]
Chem. Descrip.: SD alcohol 3-A
CAS 977021-59-2
Uses: Solvent, astringent, diluent, visc reducer in cosmetics, personal care

SDA-3C [Eastman]
Chem. Descrip.: SD alcohol 3-C
Uses: Solvent, astringent, diluent, visc reducer in cosmetics, personal care

SDA-12A [Eastman]
Chem. Descrip.: SD alcohol 12-A
Uses: Solvent for mfg. of other chemicals, cosmetics, toiletries, and pharmaceuticals

SDA-23A [Eastman]
Chem. Descrip.: SD alcohol 23-A
CAS 977021-64-9
Uses: Solvent, astringent, diluent, visc reducer in cosmetics, personal care

SDA-23H [Eastman]
Chem. Descrip.: SD alcohol 23-H
Uses: Solvent for mfg. of other chemicals, cosmetics, toiletries, and pharmaceuticals

SDA-29-3 [Eastman]
Chem. Descrip.: SD alcohol 29-3
Uses: Solvent

SDA-29-4 [Eastman]
Chem. Descrip.: SD alcohol 29-4
Uses: Solvent

SDA-29-5 [Eastman]
Chem. Descrip.: SD alcohol 29-5
Uses: Solvent

SDA-29-8 [Eastman]
Chem. Descrip.: SD alcohol 29-8
Uses: Solvent for mfg. of other chemicals, cosmetics, toiletries, and pharmaceuticals

SDA-29-9 [Eastman]
Chem. Descrip.: SD alcohol 29-9
Uses: Solvent

SDA-29-11 [Eastman]
Chem. Descrip.: SD alcohol 29-11
Uses: Solvent

SDA-30 [Eastman]
Chem. Descrip.: SD alcohol 30
Uses: Solvent, astringent, diluent, visc reducer in cosmetics, personal care

SDA-32 [Eastman]
Chem. Descrip.: SD alcohol 32
Uses: Solvent

SDA-35A [Eastman]
Chem. Descrip.: SD alcohol 35-A
Uses: Solvent

SDA-36 [Eastman]
Chem. Descrip.: SD alcohol 36
Uses: Solvent

SDA-39C [Eastman]
Chem. Descrip.: SD alcohol 39-C
Uses: Solvent, astringent, diluent, visc reducer in cosmetics, personal

care

SDA-40 [Eastman]
Chem. Descrip.: SD alcohol 40
Uses: Solvent, astringent, diluent, visc reducer in cosmetics, personal care

SDA-40A [Eastman]
Chem. Descrip.: SD alcohol 40-A
Uses: Solvent, astringent, diluent, visc reducer in cosmetics, personal care

SDA-40B [Eastman]
Chem. Descrip.: SD alcohol 40-B
Uses: Solvent, astringent, diluent, visc reducer in cosmetics, personal care

SDA-40C [Eastman]
Chem. Descrip.: SD alcohol 40-C
Uses: Solvent, astringent, diluent, visc reducer in cosmetics, personal care

SECOSTER® A [Stepan]
Chem. Descrip.: Fatty acid ethoxylate
Uses: Dispersant, solvent, oil emulsifier for household/industrial cleaners, oil field prod.
Properties: Yel. to brn. liq.; nonionic; 100% act.

Selexol® [Dow]
Uses: Solvent for gas treating (removes CO_2, hydrogen sulfide, and other impurities from natural and refinery gases)

Sesame Oil USP/NF 16 [Natural Oils Int'l.; Tri-K Ind.]
Chem. Descrip.: Sesame oil
CAS 8008-74-0; EINECS/ELINCS 232-370-6
Uses: Emollient in cosmetics; solvent, vehicle in pharmaceuticals
Properties: Pale straw clear oil, bland odor and taste; insol. in water; sp.gr. 0.916-0.921; iodine no. 103-116; sapon. no. 188-195; flash pt. (OC) 640 F
Toxicology: Nonhazardous
Storage: Preserve in tight, light-resist. containers; avoid exposure to excessive heat

Shell OMS [Shell]
Chem. Descrip.: Odorless mineral spirits
CAS 64475-85-0; EINECS/ELINCS 232-453-7
Uses: Solvent for dry cleaning, coatings, automotive, and chemical specialties
Properties: Sp.gr. 0.747 (60 F); dens. 6.22 lb/gal (60 F); flash pt. (CC) 104 F

Shell Mineral Spirits 7 EC [Shell]
Chem. Descrip.: Mineral spirits
CAS 64475-85-0; EINECS/ELINCS 232-453-7
Uses: Solvent for dry cleaning, coatings, automotive, and chemical specialties
Features: Hydrocarbon
Properties: Sp.gr. 0.793 (60 F); dens. 6.60 lb/gal (60 F); flash pt. (CC) 103 F

Shell Mineral Spirits 135 [Shell]
Chem. Descrip.: Mineral spirits
CAS 64475-85-0; EINECS/ELINCS 232-453-7
Uses: Solvent for dry cleaning, coatings, adhesives, agric., automotive, and chemical specialties, cleaning and polishing compds., pharmaceuticals, inks, textile coatings, asphalt, concrete curing
Features: Hydrocarbon solvent
Properties: Sp.gr. 0.783 (60 F); dens. 6.52 lb/gal (60 F); flash pt. (CC) 110 F

Shellsol® 3EC [Shell]
Chem. Descrip.: Solvent naphtha (petroleum), medium aliphatic (97%), solvent naphtha (petroleum), light aromatic (3%)
UN 1268
Uses: Solvent
Regulatory: SARA §311/312 acute health/chronic health/fire hazad, §313 reportable [1,2,4-trimethylbenzene (1%)]; Calif. Prop. 65 reportable [naphthalene (0-0.003%)]
Properties: Light colored liq.; hydrocarbon odor; negligible sol. in water; m.w. 147; sp.gr. 0.769-0.790 (60/60 F); vapor pressure 1 mm Hg (68 F); i.b.p. 300 F min.; b.p. 327-406 F; flash pt. (TCC) 100 F; VOC 6.54 lb/gal (60 F)
Toxicology: If aspirated, can cause severe lung damage, chem. pneumonitis; may be harmful or fatal if swallowed; causes skin irritation/defatting, respiratory tract irritation; may cause CNS depression, dizziness, lightheadedness, headache, nausea, incoordination, unconsciousness, death, eye irritation/temporary discomfort, sl. anemia, liver/kidney effects; TSCA listed
Environmental: Prevent entry into waterways, sewers
Precaution: Combustible; flamm. in air 1-7 vol.%; vapors are heavier than air; vapors may flash back; prevent vapor accumulation; static electricity may accumulate and create a fire hazard; bond and ground equip.; avoid contact with strong oxidizing agents; prevent entry into basements, confined areas
Hazardous decomp. prods.: Complex mixt. of airborne solids, liqs. and gases, incl. CO, CO_2, other org. compds.
HMIS: Health 1, Flammability 2, Reactivity 0
Storage: Keep closed when not in use; keep away from heat, sparks, flame

Shellsol® 7EC [Shell]
Chem. Descrip.: Solvent naphtha (petroleum), medium aliphatic (93%), solvent naphtha (petroleum), light aromatic (7%)
UN 1268
Uses: Solvent
Regulatory: SARA §311/312 acute health/chronic health/fire hazad, §313 reportable [1,2,4-trimethylbenzene (2-3%)]; Calif. Prop. 65 reportable [naphthalene (0.002-0.007%)]
Properties: Light colored liq.; hydrocarbon odor; negligible sol. in water; m.w. 142; sp.gr. 0.779; vapor pressure 1.2 mm Hg (68 F); i.b.p. 300 F min.; b.p. 325-400 F; flash pt. 43.89 C; VOC 6.5 lb/gal (60 F)
Toxicology: If aspirated, can cause severe lung damage, chem. pneumonitis; may be harmful or fatal if swallowed; causes skin irritation/defatting, respiratory tract irritation; may cause CNS depression, dizziness, lightheadedness, headache, nausea, incoordination, unconsciousness, death, eye irritation/temporary discomfort; chronic exposure to high concs. may cause hearing loss; target organs: auditory system; TSCA listed
Environmental: Prevent entry into waterways, sewers
Precaution: Combustible; flamm. in air 1-6 vol.%; vapors are heavier than air; vapors may flash back; prevent vapor accumulation; static electricity may accumulate and create a fire hazard; bond and ground equip.; avoid contact with strong oxidizing agents; prevent entry into basements, confined areas
Hazardous decomp. prods.: Complex mixt. of airborne solids, liqs. and gases, incl. CO, CO_2, other org. compds.
HMIS: Health 1, Flammability 2, Reactivity 0
Storage: Keep closed when not in use; keep away from heat, sparks, flame

Shellsol® 9 [Shell]
Chem. Descrip.: Solvent naphtha (petroleum), medium aliphatic
CAS 64742-88-7; EINECS/ELINCS 265-191-7
Uses: Solvent
Features: Med. evaporating; limited aromatic content
Properties: Saybolt +30 color; sp.gr. 0.763 (15.6/15.6 C); dens. 6.36 lb/gal (15.6 C); i.b.p. 300-315 F; flash pt. (TCC) > 100 F; ref. index 1.426

Shellsol® 15 [Shell]
Chem. Descrip.: Solvent naphtha (petroleum), medium aliphatic (85%), solvent naphtha (petroleum), light aromatic (15%)
UN 1268
Uses: Solvent
Features: 15 vol.% aromatics; photochemically reactive
Regulatory: SARA §311/312 acute health/chronic health/fire hazad, §313 reportable [1,2,4-trimethylbenzene (4-6%)]; Calif. Prop. 65 reportable [naphthalene (0.004-0.015%)]
Properties: Clear liq.; hydrocarbon odor; negligible sol. in water; m.w. 139; sp.gr. 0.786; vapor pressure 1.1 mm Hg (68 F); i.b.p. 300 F min.; b.p. 320-357 F; flash pt. (TCC) 105 F; VOC 6 lb/gal (60 F)
Toxicology: If aspirated, can cause severe lung damage, chem. pneumonitis; may be harmful or fatal if swallowed; causes skin irritation/defatting, respiratory tract irritation; may cause CNS depression, dizziness, lightheadedness, headache, nausea, incoordination, unconsciousness, death, eye irritation/temporary discomfort; chronic abuse may cause irregular heart rhythms, cardiac arrest; target organs: heart; TSCA listed
Environmental: Prevent entry into waterways, sewers
Precaution: Combustible; flamm. in air 0.7-6 vol.%; vapors are heavier

than air; vapors may flash back; prevent vapor accumulation; static electricity may accumulate and create a fire hazard; bond and ground equip.; avoid contact with strong oxidizing agents; prevent entry into basements, confined areas
Hazardous decomp. prods.: Complex mixt. of airborne solids, liqs. and gases, incl. CO, CO_2, other org. compds.
HMIS: Health 1, Flammability 2, Reactivity 0
Storage: Keep closed when not in use; keep away from heat, sparks, flame

Shellsol® 16 [Shell]
Chem. Descrip.: Naphtha, medium aliphatic
CAS 64742-88-7; EINECS/ELINCS 265-191-7
Uses: Solvent
Features: Med. evaporating; typ. aromatic content; < 5% HAPs
Properties: Water-wh.; m.w. 143; i.b.p. 150 C min.; flash pt. (TCC) 42 C; KB value 35-40

Shellsol® 426 [Shell]
Chem. Descrip.: C5-8 hydrocarbon solvent
Uses: Solvent for rubber, adhesives, sealants, surf. coatings, household cleaners, printing inks
Features: Low aromatics; 12% avg. n-hexane content
Properties: Immisc. with water; dens. 0.685 kg/l (15 C); i.b.p. 45 C; flash pt. < 0 C; conduct. 450 pS/m (23 C)
Toxicology: Prolonged/repeated skin contact can cause dermatitis, defatting
Precaution: Highly flamm.; can form explosive mixts. with air; electrostatic charges may be generated
Storage: Keep away from heat, sparks, flames

Shellsol® 664 [Shell]
Chem. Descrip.: Hydrocarbon solvent
Uses: Solvent for surf. coatings
Features: Med. solvency; med. boiling; fast evaporating
Properties: Immisc. with water; dens. 0.775 kg/l (15 C); i.b.p. 45 C; flash pt. (Abel) < 0 C; conduct. 450 pS/m (23 C)
Toxicology: Irritating to eyes, skin; may be absorbed thru skin; prolonged/repeated skin contact can cause dermatitis, defatting
Precaution: Highly flamm.; can form explosive mixts. with air; electrostatic charges may be generated
Storage: Keep away from heat, sparks, flames

Shellsol® 1495 [Shell Netherlands]
Chem. Descrip.: Naphtha (petroleum), hydrodesulfurized heavy, contains ethylbenzene (0.1%)
CAS 64742-82-1; EINECS/ELINCS 265-185-4
UN 3295
Uses: Solvent
Regulatory: EINECS, Canada DSL, Australia AICS, China IECSC, Japan MITI, Philippines PICCS, Korea TCCL listed
Properties: Colorless liq.; paraffinic odor; insol. in water; dens. 780 kg/m^3 (15 C); vapor pressure 420 Pa (15 C); b.p. 152-255 C; flash pt. 41 C; autoignition temp. 250 C
Toxicology: OEL TWA (8 h) 200 mg/m^3; LD50 (oral) > 2000 mg/kg, (skin) > 2000 mg/kg; LC50 (inh.) > 5 mg/l; expected to be of low toxicity by ing., inh., skin contact; aspiration into lungs may cause chem. pneumonitis which can be fatal; may cause headache, dizziness, nausea, narcosis; prolonged/repeated skin contact can cause dermatitis, defatting; TSCA listed
Environmental: Readily biodeg.; potential to bioaccumulate; toxic to aquatic organisms; may cause long-term adverse effects in aquatic environment; prevent contamination of soil and water; prevent entry into drains, ditches, rivers
Precaution: Flamm.; explosion/flamm. limits in air 0.47-7.5 vol.%; may form flamm./explosive vapor-air mixts.; vapor is heavier than air, spreads along ground; distant ignition possible; electrostatic charges may be generated; prevent static discharge; ground equip.; risk of explosion; floats on water
Hazardous decomp. prods.: Incomplete combustion: CO
Storage: Store in tightly closed container @ ambient temps. in well-ventilated area away from direct sunlight, heat/ignition sources

Shellsol® 1626 [Shell Netherlands]
Chem. Descrip.: Naphtha (petroleum), hydrodesulfurized heavy, contains ethylbenzene (0.1%)
CAS 64742-82-1; EINECS/ELINCS 265-185-4
UN 1268
Uses: Solvent for mfg. of specialty prods.
Features: Narrow boiling range; med. evaporating
Regulatory: EINECS, Canada DSL, Australia AICS, China IECSC, Japan MITI, Philippines PICCS, Korea TCCL listed
Properties: Colorless liq.; paraffinic odor; insol. in water; dens. 0.770-0.800 kg/l (15 C); vapor pressure 200 Pa (15 C); b.p. 150-250 C; flash pt. 41 C; autoignition temp. 250 C
Toxicology: OEL TWA (8 h) 350 mg/m^3; LD50 (oral) > 2000 mg/kg, (skin) > 2000 mg/kg; LC50 (inh.) > 5 mg/l; expected to be of low toxicity by ing., inh., skin contact; aspiration into lungs may cause chem. pneumonitis which can be fatal; may cause headache, dizziness, nausea, narcosis; prolonged/repeated skin contact can cause dermatitis, defatting; TSCA listed
Environmental: Readily biodeg.; potential to bioaccumulate; toxic to aquatic organisms; may cause long-term adverse effects in aquatic environment; prevent contamination of soil and water; prevent entry into drains, ditches, rivers
Precaution: Flamm.; explosion/flamm. limits in air 0.47-3.6 vol.%; may form flamm./explosive vapor-air mixts.; vapor is heavier than air, spreads along ground; distant ignition possible; electrostatic charges may be generated; prevent static discharge; ground equip.; risk of explosion
Hazardous decomp. prods.: Incomplete combustion: CO
Storage: Store in tightly closed container @ ambient temps. in well-ventilated area away from direct sunlight, heat/ignition sources

Shellsol® 2046 [Shell Netherlands]
Chem. Descrip.: Naphtha (petroleum), hydrodesulfurized heavy
CAS 64742-82-1; EINECS/ELINCS 265-185-4
Uses: Solvent for mining industry, cleaning applics.
Features: Wide boiling range; slow evaporating
Regulatory: EINECS, Canada DSL, Australia AICS, China IECSC, Japan MITI, Philippines PICCS, Korea TCCL listed
Properties: Colorless liq.; mild odor; insol. in water; dens. 0.806 kg/l (15 C); vapor pressure ≈ 20 Pa (15 C); b.p. 198-240 C; flash pt. (PMCC) 79 C; autoignition temp. 229 C; conduct. 10 pS/m (23 C)
Toxicology: OEL TWA (8 h) 200 mg/m^3; LD50 (oral) > 2000 mg/kg, (skin) > 2000 mg/kg; LC50 (inh.) > 5 mg/l; expected to be of low toxicity by ing., inh., skin contact; aspiration into lungs may cause chem. pneumonitis which can be fatal; prolonged/repeated skin contact can cause dermatitis, defatting; TSCA listed
Environmental: Expected to be readily biodeg.; potential to bioaccumulate; sl. toxic to fish, algae; hazard to aquatic life; prevent contamination of soil and water; prevent entry into drains, ditches, rivers
Precaution: Combustible; explosion/flamm. limits in air 0.7-6 vol.%; may form flamm./explosive vapor-air mixts. above 60 C; electrostatic charges may be generated; prevent static discharge
Hazardous decomp. prods.: Incomplete combustion: CO
Storage: Store @ ambient temps.; keep away from heat, sparks, flames

Shellsol® A100 Solvent [Shell]
Chem. Descrip.: Naphtha, light aromatic
CAS 64742-95-6; EINECS/ELINCS 265-199-0
UN 1268
Uses: Solvent for paints, adhesives, printing inks; inert ingred. in pesticides
Features: Good solvency; low HAPs; very low sulfur content; meets ASTM D3734 (high-flash aromatic naphthas type I)
Regulatory: FDA 40CFR §180.1001(c) (as inert in pesticides applied to growing crops); SARA §311/312 acute health/chronic health/fire hazard, §313 reportable [xylene (1-3%), 1,2,4-trimethylbenzene (31-40%), cumene (< 2%)]
Properties: Light-colored clear liq.; aromatic hydrocarbon odor; negligible sol. in water; m.w. 123; sp.gr. 0.865-0.882 (60/60 F); dens. 7.302 lb/gal (15 C); visc. 1.2 cP; vapor pressure 2.0 mm Hg (20 C); i.b.p. 149 C; b.p. 320-349 F; flash pt. (TCC) 39 C; autoignition temp. 462 C; KB value 87; VOC 7.3 lb/gal (60 F)
Toxicology: LD50 (oral, rat) 4.7 g/kg; if aspirated, can cause severe lung damage, chem. pneumonitis; may be harmful/fatal by ing.; may cause CNS depression (dizziness, light-headedness, headache, nausea, incoordination, unconsciousness, death); irritating to respiratory system; chronic inh. may cause irregular heart rhythms, cardiac arrest; may cause eye irritation, sl. skin irritation; prolonged contact may cause defatting; target organs: heart, auditory system; TSCA listed

Environmental: Prevent entry into waterways, sewers
Precaution: Flamm. liq.; flamm. in air 1-7 vol.%; vapors are heavier than air; vapors may flash back; prevent vapor accumulation; static electricity may accumulate and create a fire hazard; bond and ground equip. and transfer containers; avoid contact with strong oxidizing agents; prevent entry into basements, confined areas
Hazardous Ingredients: Naphthalene (0.03-0.1%)
Hazardous decomp. prods.: Complex mixt. of airborne solids, liqs. and gases, incl. CO, CO_2, other org. compds.
HMIS: Health 1, Flammability 2, Reactivity 0
Storage: Keep closed when not in use; keep away from heat, sparks, flame

Shellsol® A150ND [Shell Netherlands]
Chem. Descrip.: Heavy aromatic naphtha, 1,2,4-trimethylbenzene (12%), 1,3,5-trimethylbenzene (2%)
UN 3082
Uses: Solvent for rubber, textiles, chem. processing, household prods., surf. coatings, printing inks, cleaners, pesticides, wood preservatives
Features: Low in naphthalene content; considered nonhazardous; does not contain detectable quantities of polycyclic aromatics, heavy metals, or chlorinated compds.
Regulatory: Canada DSL, Australia AICS, EINECS, China IECSC, Philippines PICCS, Korea TCCL listed
Properties: Colorless liq.; aromatic odor; immisc. in water; m.w. 126; dens. 884 kg/m^3 (15 C); kinematic visc. 1.1 mm^2/s; vapor pressure 150 Pa (20 C); i.b.p. 183 C; b.p. 183-197 C; pour pt. -16 C; flash pt. 63 C; autoignition temp. 499 C; surf. tens. 30 mN/m (20 C); KB value 95
Toxicology: Expected to be of low toxicity; aspiration into lungs may cause chem. pneumonitis which can be fatal; narcotic at high vapor concs.; may cause eye/skin irritation; prolonged/repeated contact may cause defatting of skin, dermatitis; inh. may cause mucous membrane irritation; TSCA listed
Environmental: Readily biodeg.; environmentally hazardous; potential to bioaccumulate; toxic to aquatic organisms; may cause long-term adverse effects in aquatic environment; prevent contamination of soil and water; prevent entry into drains, ditches, rivers
Precaution: Combustible; LEL 0.6 vol.%; UEL 6.0 vol.%; may form flamm./explosive vapor/air mixts. above 57 C; electrostatic charges may be generated; prevent static discharge
Hazardous decomp. prods.: Combustion: CO
Storage: 12 mos min. shelf life under proper storage @ ambient temps. away from heat, sparks, flames

Shellsol® A150 Solvent [Shell]
Chem. Descrip.: Solvent naphtha (petroleum) heavy aromatic, 1,2,4-trimethylbenzene (5-11%), naphthalene (5-8%)
UN 1268
Uses: Solvent for agric. sprays; inert ingred. in pesticides
Features: Good solvency; low HAPs; very low sulfur content; high boiling; slow evaporating; high flash pt.; meets ASTM D3734 (high-flash aromatic naphtha type II)
Regulatory: FDA 40CFR §180.1001(c) (as inert in pesticides applied to growing crops); SARA §311/312 acute health/fire hazard, §313 reportable [1,2,4-trimethylbenzene (5-11%), naphthalene (5-8%)]
Properties: Colorless clear liq.; aromatic hydrocarbon odor; negligible sol. in water; m.w. 136; sp.gr. 0.880-0.910 (60/60 F); dens. 7.485 lb/gal (15 C); visc. 3.1 cP; vapor pressure 0.53 mm Hg (20 C); i.b.p. 177 C; b.p. 359-419 F; flash pt. (TCC) 63 C; autoignition temp. 480 C; KB value 85; VOC 7.5 lb/gal (60 F)
Toxicology: LD50 (oral, rat) 10 ml/kg; if aspirated, can cause severe lung damage, chem. pneumonitis; may be fatal if swallowed; may cause CNS depression (dizziness, lightheadedness, headache, nausea, incoordination, unconsciousness, death); may cause eye/sl. skin irritation; prolonged contact can cause defatting
Environmental: Prevent entry into waterways, sewer
Precaution: Combustible; flamm. in air 0.6-6 vol.%; prevent vapor accumulation; avoid contact with strong oxidizing agents; prevent entry into basements, confined areas
Hazardous decomp. prods.: Complex mixt. of airborne solids, liqs. and gases, incl. CO, CO_2, other org. compds.
HMIS: Health 1, Flammability 2, Reactivity 0
Storage: Keep closed when not in use; keep away from heat, sparks, flame

Shellsol® A200 Solvent [Shell]
Chem. Descrip.: Solvent naphtha (petroleum) heavy aromatic, contains naphthalene (0.6%)
CAS 64742-94-5; EINECS/ELINCS 265-198-5
Uses: Solvent for agric. sprays; inert ingred. in pesticides
Features: Good solvency; low HAPs; very low sulfur content; high boiling; slow evaporating; high flash pt.
Regulatory: FDA 40CFR §180.1001(c) (as inert in pesticides applied to growing crops); SARA §311/312 acute health hazard; Calif. Prop. 65 reportable [naphthalene (0.6%), toluene (0.02%)]
Properties: Colorless liq.; aromatic hydrocarbon odor; negligible sol. in water; m.w. 150; sp.gr. 0.99-1.01 (60/60 F); dens. 8.373 lb/gal (15 C); vapor pressure < 0.04 mm Hg (20 C); i.b.p. 233 C; flash pt. (PMCC) 95 C; autoignition temp. > 450 C; KB value 97; VOC 1003 g/l; 100% act.
Toxicology: LD50 (oral, rat) 10 ml/kg, (skin, rat) 4 ml/kg; if aspirated, can cause severe lung damage, chem. pneumonitis; may be fatal if swallowed; inh. of vapors/mists may cause mild burning sensation (nose/throat/lungs); inh. of high concs. may cause CNS depression, dizziness, lightheadedness, headache, nausea, loss of coordination, unconsciousness, death; may cause temporary eye irritation, sl. skin irritation; naphthalene may cause tumors; TSCA listed
Environmental: Prevent entry into waterways, sewers
Precaution: Combustible class IIIB; prevent entry into basements, confined areas
HMIS: Health 1, Flammability 1, Reactivity 0
Storage: Keep closed when not in use

Shellsol® B6 [Shell]
Chem. Descrip.: Hydrocarbon solvent
Uses: Solvent
Features: Fast evaporating; low level of impurities such as sulfur, olefins, and benzene
Properties: M.w. 87; i.b.p. 60 C min.; flash pt. < -18 C; KB value 32-35

Shellsol® BF [Shell Netherlands]
Chem. Descrip.: Naphtha, hydrotreated light, n-hexane (< 5%)
UN 3295
Uses: Solvent for mfg. of tires, rubber sol'ns., general-purpose rubber cements, adhesives for medicinal plasters, pressure-sensitive tapes, rubberproofing, textile cleaning, industrial cleaning and degreasing; solvent, diluent in printing inks; diluent in paints, lacquers, varnishes
Features: Highly refined; very minimal aromatics; virtually free of benzene; chemically stable; noncorrosive; fast constant evaporating
Regulatory: EINECS, Canada DSL, Australia AICS, China IECSC, Japan MITI, Philippines PICCS, Korea TCCL listed
Properties: Water-wh. liq.; paraffinic odor; dens. 0.735 kg/l (15 C); i.b.p. 85 C; b.p. 88-105 C; flash pt. (Abel) < 0 C
Toxicology: LD50 (oral) > 2000 mg/kg, (skin) > 2000 mg/kg; LC50 (inh.) > 5 mg/l; low toxicity; aspiration into lungs may cause chem. pneumonitis which can be fatal; narcotic at high vapor concs.; causes CNS depression, headache, dizziness, nausea, respiratory tract irritation, skin irritation/dryness/dermatitis; risk of impaired fertility; TSCA listed
Environmental: Inherently biodeg.; potential to bioaccumulate; toxic to aquatic organisms; may cause long-term adverse effects in aquatic environment; prevent contamination of soil and water; prevent entry into drains, ditches, rivers
Precaution: Highly flamm.; flamm. limits in air 1-8 vol.%; may form flamm./explosive vapor-air mixt.; vapor is heavier than air, spreads along ground; distant ignition possible; electrostatic charges may be generated; prevent static discharge; ground equip.; risk of explosion
Hazardous decomp. prods.: Combustion: CO
Storage: Store in tightly closed container @ ambient temps. in well-ventilated area away from direct sunlight, heat/ignition sources

Shellsol® B HT [Shell]
Chem. Descrip.: Solvent naphtha (petroleum), light aliphatic, contains n-hexane (22-30%), cyclohexane (0-13%)
UN 1268
Uses: Solvent for agric. fat/oil extraction, automotive undercoating, pharmaceuticals, adhesives, inks, paints, paper coating, lighter fluids, resins, plastics, NR, SR (except NBR)
Features: Fast evaporating; hydrotreating removes unwanted compds., converts unsat. hydrocarbons, results in relatively low odor; lower hexane content than typ. commercial hexanes, thus reducing worker exposure

Regulatory: SARA §311/312 acute health/chronic health/fire hazard, §313 reportable (cyclohexane, n-hexane)
Properties: Colorless liq.; hydrocarbon odor; negligible sol. in water; sp.gr. 0.674 (60/60 F); vapor pressure 140 mm Hg (68 F); b.p. 147-176 F; flash pt. (TCC) < -17.78 C; autoignition temp. 534 C; KB value 29; VOC 5.53 lb/gal; 100% act.
Toxicology: LD50 (oral, rat) 17 g/kg, (skin, rabbit) 3.4 g/kg; LC50 (inh., rat, 4 h) 73,680 ppm; if aspirated, can cause severe lung damage, chem. pneumonitis; may be fatal if swallowed; inh. of high vapor concs. may cause CNS depression, dizziness, headache, nausea, loss of coordination, unconsciousness, death; chronic abuse may cause irregular heart rhythms, cardiac arrest; may cause temporary eye irritation, sl. skin irritation; target organs: nerves, heart, reproductive system; TSCA listed
Environmental: Prevent entry into waterways, sewers
Precaution: Extremely flamm.; flamm. in air 1-7 vol.%; vapors are heavier than air, may flash back; prevent vapor accumulation; floats on water; static electricity may accumulate and create a fire hazard; bond and ground equip. and transfer containers; avoid contact with strong oxidizing agents; prevent entry into basements, confined areas
Hazardous decomp. prods.: Complex mixt. of airborne solids, liqs. and gases, incl. CO, CO_2, other org. compds.
HMIS: Health 1, Flammability 3, Reactivity 0
Storage: Keep containers closed when not in use; keep away from heat, sparks, flame, ignition sources

Shellsol® B HT60 [Shell]
Chem. Descrip.: Solvent naphtha (petroleum), light aliphatic, n-hexane (13-18%), cyclohexane (0-10%)
UN 1268
Uses: Solvent
Features: Fast evaporating; low level of impurities e.g., sulfur, olefins, benzene, total aromatics
Regulatory: SARA §311/312 acute health/chronic health/fire hazard, §313 reportable [hexane (13-18%), cyclohexane (0-10%)]
Properties: Colorless liq.; hydrocarbon odor; negligible sol. in water; m.w. 92; sp.gr. 0.686; vapor pressure 91 mm Hg (68 F); b.p. 154-203 F; flash pt. < -18 C; VOC 5.7 lb/gal (60 F)
Toxicology: If aspirated, can cause severe lung damage, chem. pneumonitis; may be fatal if swallowed; may cause CNS depression (dizziness, lightheadedness, headache, nausea, incoordination, unconsciousness); may cause eye irritation, sl. skin irritation/defatting; chronic exposure may cause hearing loss; chronic abuse may cause irregular heart rhythms, cardiac arrest; target organs: nerves, heart, auditory system, reproductive system; TSCA listed
Environmental: Prevent entry into waterways, sewer
Precaution: Extremely flamm.; flamm. in air 1-6 vol.%; vapors are heavier than air; vapors may flash back; prevent vapor accumulation; static electricity may accumulate and create a fire hazard; bond and ground equip. and transfer containers; avoid contact with strong oxidizing agents; prevent entry into basements, confined areas
Hazardous decomp. prods.: Complex mixt. of airborne solids, liqs. and gases, incl. CO, CO_2, other org. compds.
HMIS: Health 1, Flammability 3, Reactivity 0
Storage: Keep closed when not in use; keep away from heat, sparks, flame

Shellsol® D25 [Shell Netherlands]
Chem. Descrip.: Naphtha (petroleum), hydrotreated light
CAS 64742-49-0; EINECS/ELINCS 265-151-9
UN 3295
Uses: Solvent for rubber, chem. processing, paper, household prods., surf. coatings, printing inks, adhesives, pesticides, waxes, polishes, degreasing applics.; extraction solvent
Features: Very low aromatics and olefins; does not contain detectable quantities of polycyclic aromatics, heavy metals, or chlorinated compds.
Regulatory: EINECS, Canada DSL, Australia AICS, China IECSC, Japan MITI, Philippines PICCS, Korea TCCL listed
Properties: Colorless liq.; paraffinic odor; immisc. with water; m.w. 130; dens. 0.750 kg/l (15 C); kinematic visc. 0.91 mm^2/s; vapor pressure 1.3 kPa (20 C); i.b.p. 143 C; b.p. 143-160 C; flash pt. (Abel) 25 C; autoignition temp. 287 C; ref. index 1.415; KB value 31; surf. tens. 22.2 mN/m; sp.heat 2.0 kJ/kg/°C; elec. conduct. 0.065 pS/m
Toxicology: Expected to be of low toxicity by ing., inh., skin contact; aspiration into lungs may cause chem. pneumonitis which can be fatal; narcotic at high vapor concs.; causes CNS depression, headache, dizziness, nausea; may cause skin irritation/dryness/dermatitis; TSCA listed
Environmental: Inherently biodeg.; potential to bioaccumulate; not classified as dangerous under EC criteria; sl. toxic to fish; pract. nontoxic to algae; prevent contamination of soil, water; prevent entry into drains, ditches, rivers
Precaution: Highly flamm.; LEL 0.8 vol.%; UEL 6.0 vol.%; may form explosive vapor-air mixt.; vapor is heavier than air, spreads along ground; distant ignition is possible; electrostatic charges may be generated; prevent static discharge; ground equip.; risk of explosion
Hazardous decomp. prods.: Combustion: CO
Storage: 12 mos min. shelf life @ ambient temps. in tightly closed containers in well-ventilated area, away from direct sunlight, other heat/ignition sources

Shellsol® D38 [Shell]
Chem. Descrip.: Solvent naphtha (petroleum), medium aliphatic
CAS 64742-88-7; EINECS/ELINCS 265-191-7
UN 1268
Uses: Solvent for paints/coatings, dry cleaning, commercial and industrial cleaners, insecticides, lighter fluids, resins, plastics, wood preservatives, polishes, petrol. gas absorp.; catalytic/reaction medium
Features: Narrow boiling range, slow-evaporating; hydrotreating removes unwanted compds. (e.g., sulfur, nitrogen, oxygen), converts unsat. hydrocarbons (olefins, acetylenes, aromatics), results in relatively low odor; no more than trace amts. of HAPs
Regulatory: SARA §311/312 acute health/chronic health/fire hazard; Calif. Prop. 65 nonreportable
Properties: Lt. colored liq.; low hydrocarbon odor; negligible sol. in water; sp.gr. 0.77 (60 F); vapor pressure 1.3 mm Hg (68 F); b.p. 318-356 F; flash pt. 41.11 C; autoignition temp. 243.89 C; KB value 32; VOC 6.37 lb/gal; 100% act.
Toxicology: LD50 (oral, rat) 25 ml/kg, (skin, rabbit) ≈[5 ml/kg; if aspirated, can cause severe lung damage, chem. pneumonitis; may be fatal if swallowed; causes skin irritation; may cause CNS depression, dizziness, lightheadedness, nausea, loss of coordination, unconsciousness, death; vapors may be irritating to respiratory system; may cause temporary eye irritation; TSCA listed
Environmental: Prevent entry into sewers, waterways
Precaution: Flamm. liq.; flamm. in air 1-7 vol.%; vapors are heavier than air, may flash back; prevent vapor accumulation; floats on water; static electricity may accumulate and create a fire hazard; bond and ground equip. and transfer containers; avoid contact with strong oxidizing agents; prevent entry into basements, confined areas
Hazardous decomp. prods.: Complex mixt. of airborne solids, liqs. and gases, incl. CO, CO_2, other org. compds.
HMIS: Health 1, Flammability 2, Reactivity 0
Storage: Keep closed when not in use; keep away from heat, sparks, flame

Shellsol® D40 [Shell; Shell Netherlands]
Chem. Descrip.: Naphtha (petroleum), hydrotreated heavy
CAS 64742-48-9; EINECS/ELINCS 265-150-3
UN 3295
Uses: Solvent for paints/coatings, dry cleaning, commercial and industrial cleaners, insecticides, lighter fluids, resins, plastics, adhesives, polishes, pharmaceuticals, inks, textile coatings, asphalt, concrete
Features: Hydrotreating removes unwanted compds. (sulfur, nitrogen, oxygen), converts unsat. hydrocarbons (olefins, acetylenes, aromatics), results in relatively low odor; no more than trace amts. of HAPs; relatively high solvency; slow-evaporating
Regulatory: EINECS, Canada DSL, Australia AICS, China IECSC, Japan MITI, Philippines PICCS, Korea TCCL listed
Properties: Colorless liq.; paraffinic sweet odor; insol. in water; m.w. 143-145; dens. 765-769 kg/m^3 (15 C); kinematic visc. 1.14 mm^2/s; vapor pressure 170-300 Pa (20 C); b.p. 160-197 C; flash pt. 40-45 C; autoignition temp. 270 C; surf. tens. 24.5 mN/m (20 C); KB value 32; VOC 6.36 lb/gal
Toxicology: LD50 (oral) > 2000 mg/kg, (skin) > 2000 mg/kg; LC50 (inh.) > 5 mg/l; if aspirated, may cause chem. pneumonitis which can be fatal; prolonged/repeated skin contact may cause defatting, dermatitis; may cause headache, dizziness, nausea, narcosis; TSCA listed
Environmental: Readily biodeg.; potential to bioaccumulate; not classified as dangerous under EC criteria; LC50 (fish) > 1000 mg/l, pract.

nontoxic; prevent contamination of soil and water; prevent entry into drains, ditches, rivers
Precaution: Flamm.; flamm. limits in air 0.7-6 vol.%; may form flamm./explosive vapor-air mixts.; electrostatic charges may be generated; prevent static discharge; ground equip.; floats on water; vapor is heavier than air, spreads along ground; distant ignition possible; risk of explosion
Hazardous decomp. prods.: Incomplete combustion: CO
Storage: Store in tightly closed container @ ambient temps. in well-ventilated area away from direct sunlight, heat/ignition sources

Shellsol® D43 [Shell]
Chem. Descrip.: Solvent naphtha (petroleum), medium aliphatic
CAS 64742-88-7; EINECS/ELINCS 265-191-7
UN 1268
Uses: Solvent for paints/coatings, dry cleaning, commercial and industrial cleaners, insecticides, lighter fluids, resins, plastics, adhesives, polishes, pharmaceuticals, inks, textile coating, asphalt, concrete
Features: Slow-evaporating; hydrotreated; hydrotreating removes unwanted compds. (e.g., sulfur, nitrogen, oxygen), converts unsat. hydrocarbons (olefins, acetylenes, aromatics); no more than trace amts. of HAPs; relatively high solvency
Regulatory: FDA 21CFR §172.884, 178.3650; SARA §311/312 acute health/chronic health, fire hazard
Properties: Lt. colored liq.; low hydrocarbon odor; negligible sol. in water; m.w. 146; sp.gr. 0.77; vapor pressure 1.1 mm Hg (68 F); i.b.p. 300 F min.; b.p. 325-401 F; flash pt. (TCC) 111 F; KB value 32; VOC 6.40 lb/gal (60 F); 100% act.
Toxicology: LD50 (skin, rabbit) 5 ml/kg; if aspirated, can cause severe lung damage, chem. pneumonitis; may be fatal if swallowed; causes skin irritation; may cause CNS depression (dizziness, lightheadedness, headache, nausea, incoordination, unconsciousness, death); vapors may be irritating to respiratory system; may cause eye irritation/temporary discomfort, skin irritation, burning sensation, redness, swelling, defatting; TSCA listed
Environmental: Prevent entry into waterways, sewer
Precaution: Flamm. liq.; flamm. in air 1-7 vol.%; vapors are heavier than air; vapors may flash back; prevent vapor accumulation; static electricity may accumulate and create a fire hazard; bond and ground equip. and transfer containers; avoid contact with strong oxidizing agents; prevent entry into basements, confined areas
Hazardous decomp. prods.: Complex mixt. of airborne solids, liqs. and gases, incl. CO, CO_2, other org. compds.
HMIS: Health 1, Flammability 2, Reactivity 0
Storage: Keep closed when not in use; keep away from heat, sparks, flame

Shellsol® D60 [Shell; Shell Netherlands]
Chem. Descrip.: Solvent naphtha (petroleum), medium aliphatic
CAS 64742-88-7; EINECS/ELINCS 265-191-7
UN 1268
Uses: Solvent for paints/coatings, dry cleaning, commercial and industrial cleaners, insecticides, lighter fluids, resins, plastics, polishes, textile coatings, minerals/mining
Features: Slow-evaporating; hydrotreated; hydrotreating removes unwanted compds. (e.g., sulfur, nitrogen, oxygen), converts unsat. hydrocarbons (olefins, acetylenes, aromatics), results in relatively low odor; no more than trace amts. of HAPs
Regulatory: SARA §311/312 acute health/chronic health, fire hazard
Properties: Lt. colored liq.; low hydrocarbon odor; negligible sol. in water; m.w. 160; sp.gr. 0.768-0.82 (60/60 F); vapor pressure 0.075 mm Hg (60 F); i.b.p. 350 F min.; b.p. 367-394 F; flash pt. (TCC) 63 C; autoignition temp. 451 F; KB value 30; VOC 6.5 lb/gal (60 F); 100% act.
Toxicology: LD50 (skin, rabbit) 5 ml/kg; if aspirated, can cause severe lung damage, chem. pneumonitis; may be fatal if swallowed; causes skin irritation; may cause CNS depression (dizziness, lightheadedness, headache, nausea, incoordination, unconsciousness, death); vapors may be irritating to respiratory system; may cause eye irritation/temporary discomfort, skin irritation, burning sensation, redness, swelling, defatting; TSCA listed
Environmental: Prevent entry into waterways, sewer
Precaution: Combustible; flamm. in air 1-7 vol.%; prevent vapor accumulation; avoid contact with strong oxidizing agents; prevent entry into basements, confined areas
Hazardous decomp. prods.: Complex mixt. of airborne solids, liqs. and gases, incl. CO, CO_2, other org. compds.
HMIS: Health 1, Flammability 2, Reactivity 0
Storage: Keep closed when not in use; keep away from heat, sparks, flame

Shellsol® D60 SC [Shell Netherlands]
Chem. Descrip.: C isoparaffin, (predominantly C10-12 paraffins and naphthenics)
CAS 64742-48-9; EINECS/ELINCS 265-150-3
Uses: Solvent for architectural coatings, wood preservatives, metal cleaning, car dewaxing, photocopier fluids, textile printing, household applics. (lamp oils, barbecue lighter fluid, waxes, polishes), adhesives, dewatering fluids, cosmetics, plastisols, metalworking fluids
Features: Very low aromatics; negligible reactive impurities; does not contain detectable quantities of polycyclic aromatics, heavy metals, or chlorinated compds.
Regulatory: FDA 21CFR §172.884, 178.3650
Properties: Low sweet odor; m.w. 160; dens. 0.776 kg/l (15 C); visc. 1.57 mm^2/s; vapor pressure 0.15 kPa (20 C); i.b.p. 186 C; pour pt. -55 C; flash pt. 65 C; autoignition temp. 230 C; surf. tens. 26 mN/m (20 C); KB value 32; elec. conduct. < 0.81 pS/m (20 C); dielec. const. 2.1 (20 C)
Precaution: LEL 0.63 vol.% (100 C); UEL 5.7 vol.% (100 C)
Storage: 12 mos min. shelf life under proper storage

Shellsol® D70 [Shell Netherlands]
Chem. Descrip.: Petroleum distillates, hydrotreated light
CAS 64742-47-8; EINECS/ELINCS 265-149-8
Uses: Solvent for coatings, architectural coatings, wood preservatives, metal degreasing, photocopier fluids, carbonless copy papers, food processing, household applics. (lamp oil, barbecue lighter fluid, waxes, polishes), cosmetics, metalworking fluids, crop protection formulations, drilling muds, dry cleaning, solvent extraction of metals
Features: Very low aromatics; negligible reactive impurities; does not contain detectable quantities of polycyclic aromatics, heavy metals, or chlorinated compds.
Regulatory: FDA 21CFR §172.884, 178.3650; EINECS, Canada DSL, Australia AICS, China IECSC, Japan MITI, Philippines PICCS, Korea TCCL listed
Properties: Colorless liq.; low paraffinic sweet odor; insol. in water; m.w. 174; dens. 0.788 kg/l (15 C); visc. 1.97 mm^2/s; vapor pressure 0.06 kPa (20 C); b.p. 193-249 C; pour pt. -50 C; flash pt. 73 C; autoignition temp. 236 C; surf. tens. 26 mN/m (20 C); KB value 30; elec. conduct. < 0.085 pS/m (20 C); dielec. const. 2.15 (20 C)
Toxicology: LD50 (oral) > 2000 mg/kg, (skin) > 2000 mg/kg; LC50 (inh.) > 5 mg/l; low toxicity by ing., skin contact; expected to be of low toxicity by inh.; if aspirated into lungs, may cause chem. pneumonitis which can be fatal; prolonged/repeated skin contact may cause defatting, dermatitis; TSCA listed
Environmental: Readily biodeg.; potential to bioaccumulate; LC50 (fish) > 1000 mg/l; expected to be nontoxic @ limit of water sol.; not classified as dangerous under EC criteria; prevent contamination of soil, water; prevent entry into drains, ditches, rivers
Precaution: Combustible; explosive limits 0.6-5.5 vol.% (100 C); may form flamm./explosive vapor-air mixts above 68 C; electrostatic charges may be generated; prevent static discharge
Hazardous decomp. prods.: Incomplete combustion: CO
Storage: 12 mos min. shelf life under proper storage @ ambient temps. away from heat, sparks, flame

Shellsol® D80 [Shell]
Chem. Descrip.: Petroleum distillates, hydrotreated light
CAS 64742-47-8; EINECS/ELINCS 265-149-8
Uses: Solvent for paints, wood treatment prods., printing inks, adhesives, metal cleaning/degreasing, dry cleaning, aerosols, household specialty prods., lamp oil, lighter fluids, hand cleaners, metal rolling oils, drilling
Features: Low visc.; low aromatics content
Properties: Colorless; mild odor; m.w. 175; i.b.p. 200 C min.; pour pt. -45 C; flash pt. (PMCC) 80 C

Shellsol® D90 [Shell Netherlands]
Chem. Descrip.: Petroleum distillates, hydrotreated light
CAS 64742-47-8; EINECS/ELINCS 265-149-8
Uses: Solvent for film and paper coatings, printing inks; rolling oil in prod. of aluminum foil; lamp oil
Features: Very low aromatics; negligible reactive impurities; does not con-

tain detectable quantities of polycyclic aromatics, heavy metals, or chlorinated compds.
Regulatory: FDA 21CFR §172.884, 178.3650
Properties: Low sweet odor; m.w. 190; dens. 0.796 kg/l (15 C); visc. 2.7 mm^2/s; vapor pressure 0.012 kPa; i.b.p. 225 C; pour pt. -30 C; flash pt. 95 C; autoignition temp. 231 C; surf. tens. 26 mN/m (20 C); KB value 28; elec. conduct. 2.1 pS/m (20 C)
Precaution: LEL 0.6 vol.%; UEL 5.5 vol.%
Storage: 12 mos min. shelf life under proper storage

Shellsol® D100 [Shell Netherlands]
Chem. Descrip.: Petroleum distillates, hydrotreated light
CAS 64742-47-8; EINECS/ELINCS 265-149-8
Uses: Solvent for printing inks, vehicle underbody coatings; bitumen modifier in road construction; rolling oil in aluminum foil prod.; lamp oil
Features: Very low aromatics; negligible reactive impurities; does not contain detectable quantities of polycyclic aromatics, heavy metals, or chlorinated compds.
Regulatory: FDA 21CFR §172.884, 178.3650
Properties: Low sweet odor; m.w. 206; dens. 0.797 kg/l (15 C); visc. 3.2 mm^2/s; vapor pressure 0.003 kPa (20 C); i.b.p. 238 C; pour pt. -26 C; flash pt. 105 C; autoignition temp. 232 C; surf. tens. 28 mN/m (20 C); KB value 26; dielec. const. 2.1 (20 C)
Precaution: LEL 0.5 vol.% (100 C); UEL 5.5 vol.% (150 C)
Storage: 12 mos min. shelf life under proper storage

Shellsol® D100S [Shell Netherlands]
Chem. Descrip.: Petroleum distillates, hydrotreated light
CAS 64742-47-8; EINECS/ELINCS 265-149-8
Uses: Solvent for printing inks, vehicle underbody coatings, food and related applics.; modifier for bitumen in road construction; rolling oil in aluminum foil prod.; lamp oil
Features: Slow evaporating; high flash; low levels of impurities such as sulfur, olefins, aromatics; high stability; non-VOC; does not contain detectable quantities of polycyclic aromatics, heavy metals, or chlorinated compds.; considered nonhazardous
Regulatory: FDA 21CFR §172.884, 178.3650; BGVV compliant
Properties: Low odor; m.w. 206; dens. 0.797 kg/l (15 C); visc. 3.2 mm^2/s; vapor pressure 0.003 kPa (20 C); i.b.p. 238 C; pour pt. -26 C; flash pt. 105 C; autoignition temp. 232 C; KB value 26; surf. tens. 28 mN/m (20 C); dielec. const. 2.1 (20 C)
Precaution: LEL 0.5 vol.% (100 C); UEL 5.5 vol.% (150 C)
Storage: 12 mos min. shelf life under proper storage

Shellsol® D120 [Shell]
Chem. Descrip.: Petroleum distillates, hydrotreated light paraffinic
CAS 64742-55-8; EINECS/ELINCS 265-158-7
Uses: Solvent for mfg. of carbonless copy paper; as textile auxiliary additive; as degreaser/dewaxer
Features: Very low aromatics; negligible reactive impurities; does not contain detectable quantities of polycyclic aromatics, heavy metals, or chlorinated compds.
Properties: Low sweet odor; m.w. 216; dens. 0.825 kg/l (15 C); visc. 5.9 mPa·s; i.b.p. 247 C; pour pt. < -9 C; flash pt. 125 C; autoignition temp. 226 C; surf. tens. 40 mN/m (20 C); KB value 22; elec. conduct. ≈ 1.0 pS/m (20 C); dielec. const. 2.1 (20 C)
Precaution: LEL 0.5 vol.%; UEL 5.5 vol.% (150 C)
Storage: 12 mos min. shelf life under proper storage

Shellsol® D-LAWS [Shell S. Africa]
Chem. Descrip.: Narrow cut aliphatic hydrocarbon solvent
Uses: Solvent for surf. coatings, wood preservatives, barbecue lighter fluids, insecticides, waxes, polishes
Features: Improved flash pt. level; fast drying chars.; low level of impurities such as sulfur, olefins, benzene, total aromatics; high stability
Properties: Low odor; i.b.p. 156 C; flash pt. 37 C; KB value 30
Precaution: Highly flamm.; vapor can readily form explosive mixt. in air
Storage: 12 mos min. shelf life under proper storage

Shellsol® H [Shell Netherlands]
Chem. Descrip.: Naphtha (petroleum), hydrodesulfurized heavy
CAS 64742-82-1; EINECS/ELINCS 265-185-4
UN 3082
Uses: Solvent for paints, rubber industry, surf. coatings, household prods., pesticides, wood treatment formulations, printing inks for paper and textiles, adhesives, textile dry cleaning, metal degreasing
Features: Low aromatic wh. spirit; higher flash pt.; does not contain detectable quantities of polycyclic aromatics, heavy metals, or chlorinated compds.
Regulatory: EINECS, Canada DSL, Australia AICS, China IECSC, Japan MITI, Philippines PICCS, Korea TCCL listed
Properties: Colorless liq.; aromatic odor; insol. in water; m.w. 158; dens. 0.795 kg/l (15 C); visc. 1.44 mm^2/s; vapor pressure 0.15 kPa (20 C); b.p. 187-210 C; pour pt. < -30 C; flash pt. (Abel) 64 C; autoignition temp. 281 C; surf. tens. 26.0 mN/m (20 C); KB value 33; elec. conduct. 0.40 pS/m (20 C); dielec. const. 2.1 (20 C)
Toxicology: TLV TWA (8 h) 100 ppm (UK); LD50 (oral) > 2000 mg/kg, (skin) > 2000 mg/kg; LC50 (inh.) > 5 mg/l; expected to be of low toxicity by ing., inh., skin contact; aspiration into lungs may cause chem. pneumonitis which can be fatal; prolonged/repeated contact may cause defatting of skin, dermatitis; TSCA listed
Environmental: Readily biodeg.; potential to bioaccumulate; sl. toxic to fish, algae; prevent contamination of soil and water; prevent entry into drains, ditches, rivers
Precaution: Combustible; LEL 0.7 vol.%; UEL 6.5 vol.% (100 C); may form flamm./explosive vapor-air mixts. above 59 C; floats on water; electrostatic charges may be generated; prevent static discharge
Hazardous decomp. prods.: Incomplete combustion: CO
Storage: 12 mos min. shelf life under proper storage @ ambient temps.; keep away from heat, sparks, flame

Shellsol® Heptane [Shell]
Chem. Descrip.: Heptane (C_7)
CAS 64742-89-8; EINECS/ELINCS 205-563-8
Uses: Solvent for agric. fat/oil extraction, automotive undercoating, polishes, adhesives, inks, paints, lighter fluids, resins, plastics, rubber, petrol. gas absorption, textile coatings, minerals; starting component for rubber solvs. and lacquer diluents; solvent in protective coatings on citrus fruits
Features: Hydrotreated; hydrotreating removes unwanted compds. (e.g., sulfur, nitrogen, oxygen), converts unsat. hydrocarbons (olefins, acetylenes, aromatics), results in relatively low odor; exc. for recycling; no more than trace amts. of HAPs
Regulatory: FDA 21CFR §172.250; SARA nonreportable
Properties: Saybolt 30 min color; sp.gr. 0.695 (60/60 F); vapor pressure 41 mm Hg (68 F); i.b.p. 189 F min.; distillation range 194-208 F; flash pt. 7 F; KB value 32 min.; VOC 5.71 lb/gal; 91.2 vol.% paraffins, 8.8 vol.% cycloparaffins

Shellsol® OMS [Shell]
Chem. Descrip.: Naphtha (petroleum), heavy alkylate (predominantly C9-12 hydrocarbons)
CAS 64741-65-7; EINECS/ELINCS 265-067-2
UN 1268
Uses: Solvent for surf. coatings, aerosol sprays, specialty formulations, reaction media
Features: Highly stable
Regulatory: SARA §311/312 acute health/chronic health/fire hazard
Properties: Light-colored clear liq.; odorless to hydrocarbon odor; negligible sol. in water; m.w. 166; sp.gr. 0.74-0.775 (60/60 F); vapor pressure 0.53 mm Hg (68 F); i.b.p. 340 F; b.p. 347-401 F; flash pt. (TCC) 120 F; autoignition temp. 347.8 C; KB 29; VOC 6.3 lb/gal (60 F); 100% act.
Toxicology: If aspirated, can cause severe lung damage, chem. pneumonitis; may be fatal if swallowed; may cause CNS depression (dizziness, lightheadedness, headache, nausea, incoordination, unconsciousness, death); may cause eye irritation/temporary discomfort, skin irritation, burning sensation, defatting; TSCA listed
Environmental: Prevent entry into waterways, sewers
Precaution: Flamm.; flamm. in air 1-7 vol.%; vapors are heavier than air; vapors may flash back; prevent vapor accumulation; static electricity may accumulate and create a fire hazard; bond and ground equip. and transfer containers; avoid contact with strong oxidizing agents; prevent entry into basements, confined areas
Hazardous decomp. prods.: Complex mixt. of airborne solids, liqs. and gases, incl. CO, CO_2, other org. compds.
HMIS: Health 1, Flammability 2, Reactivity 0
Storage: Keep closed when not in use; keep away from heat, sparks, flame

Shellsol® T [Shell Netherlands]
Chem. Descrip.: Naphtha (petroleum), heavy alkylate
CAS 64741-65-7; EINECS/ELINCS 265-067-2
UN 3295

Uses: Solvent for odor-free paints, printing inks, dry cleaning, wallpaper mfg., photocopying, household sprays, fragrant polishes, hand cleaners; carrier for catalysts in polymerization
Features: Low odor
Regulatory: EINECS, Canada DSL, Australia AICS, China IECSC, Japan MITI, Philippines PICCS, Korea TCCL listed
Properties: Colorless liq.; paraffinic odor; insol. in water; m.w. 172; dens. 0.761 kg/l (15 C); kinematic visc. 1.85 mm²/s; vapor pressure 100 Pa (20 C); i.b.p. 187 C; b.p. 187-213 C; pour pt. < -30 C; flash pt. (PMCC) 60 C; autoignition temp. 430 C; surf. tens. 23.5 mN/m (20 C)
Toxicology: LD50 (oral) > 2000 mg/kg, (skin) > 2000 mg/kg; LC50 (inh.) > 5 mg/l; low toxicity; aspiration into lungs may cause chem. pneumonitis which can be fatal; prolonged/repeated contact may cause defatting of skin, dermatitis; TSCA listed
Environmental: Potential to bioaccumulate; may cause long-term adverse effects in aquatic environment; prevent contamination of soil and water; prevent entry into drains, ditches, rivers; expected to be nontoxic to fish, algae, bacteria
Precaution: Combustible; flamm. limits in air 0.6-6 vol.%; may form flamm./explosive vapor/air mixts. above 55 C; electrostatic charges may be generated; prevent static discharge
Hazardous decomp. prods.: Combustion: CO
Storage: Store @ ambient temps. away from heat, sparks, flames

Shellsol® TC [Shell]
Chem. Descrip.: Naphtha (petroleum), heavy alkylate
CAS 64741-65-7; EINECS/ELINCS 265-067-2
UN 1268
Uses: Solvent for aerosols, printing inks, catalyst carriers, adhesives, surf. coatings
Features: Quick drying; extremely low olefin and aromatic content; low reactivity
Regulatory: SARA §311/312 acute health/fire hazard; Calif. Prop. 65 nonreportable
Properties: Lt. colored liq.; negligible sol. in water; m.w. 114; sp.gr. 0.7 (60 F); vapor pressure 30 mm Hg (68 F); i.b.p. 98 C; flash pt. (TCC) -3 C; autoignition temp. 417.22 C; KB value 30; VOC 5.89 lb/gal (60 F); 100% act.
Toxicology: LD50 (oral, rat) 25 ml/kg, (skin, rabbit) 5 ml/kg; if aspirated, can cause severe lung damage, chem. pneumonitis; may be fatal if swallowed; inh. of high vapor concs. may cause CNS depression, dizziness, lightheadedness, headache, nausea, loss of coordination, unconsciousness, death; may cause temporary eye irritation, sl. skin irritation; TSCA listed
Environmental: Prevent entry into waterways, sewers
Precaution: Flamm.; flamm. in air 0.8-7.1 vol.%; floats on water; static electricity may accumulate and create a fire hazard; bond and ground equip. and transfer containers; avoid contact with strong oxidizing agents; prevent entry into basements, confined areas
Hazardous decomp. prods.: Complex mixt. of airborne solids, liqs. and gases, incl. CO, CO_2, other org. compds.
HMIS: Health 1, Flammability 3, Reactivity 0
Storage: Keep containers closed when not in use; keep away from heat, sparks, flame, ignition sources

Shellsol® TD [Shell Netherlands]
Chem. Descrip.: Naphtha (petroleum), heavy alkylate
CAS 64741-65-7; EINECS/ELINCS 265-067-2
UN 3295
Uses: Solvent for household prods. (cleaners, lighter fuels, waxes, polishes), surf. coatings, printing inks, industrial and textile cleaning, photocopier toner dispersants, cosmetics, pesticides, wallpaper mfg.
Features: Does not contain detectable quantities of polycyclic aromatics, heavy metals, or chlorinated compds.
Regulatory: FDA 21CFR §172.882, 172.884, 178.3530, 178.3620, 178.3650; German DAB 10; EINECS, Canada DSL, Australia AICS, China IECSC, Japan MITI, Philippines PICCS, Korea TCCL listed
Properties: Colorless liq.; low paraffinic odor; insol. in water; m.w. 160; dens. 0.751 kg/l (15 C); kinematic visc. 1.6 mm²/s; vapor pressure 0.16 kPa (20 C); i.b.p. 172 C; b.p. 172-185 C; pour pt. < -30 C; flash pt. (Abel) 44 C; autoignition temp. 450 C; ref. index 1.420; KB value 26; surf. tens. 23 mN/m; sp. heat 2.0 kJ/kg/°C; dielec. const. 2.0
Toxicology: LD50 (oral) > 2000 mg/kg, (skin) > 2000 mg/kg; LC50 (inh.) > 5 mg/l; low toxicity; aspiration into lungs may cause chem. pneumonitis which can be fatal; narcotic at high vapor concs.; may cause eye/skin irritation; prolonged/repeated contact may cause defatting of skin, dermatitis; inh. may cause mucous membrane irritation; TSCA listed
Environmental: LC50 (fish) > 1000 mg/l; pract. nontoxic; potential to bioaccumulate; may cause long-term adverse effects in aquatic environment; prevent contamination of soil and water; prevent entry into drains, ditches, rivers
Precaution: Flamm.; LEL 0.6 vol.%; UEL 6.0 vol.%; may form flamm./explosive vapor/air mixts. above 57 C; electrostatic charges may be generated; prevent static discharge; ground equip.
Hazardous decomp. prods.: Combustion: CO
Storage: 12 mos min. shelf life if stored in tightly closed container @ ambient temps. in well-ventilated area away from direct sunlight, heat/ignition sources

Shellsol® TK [Shell Netherlands]
Chem. Descrip.: Naphtha (petroleum), heavy alkylate
CAS 64741-65-7; EINECS/ELINCS 265-067-2
UN 3295
Uses: Solvent for aerosols, specialty formulations
Features: Narrow boiling range; med. evaporating
Regulatory: EINECS, Canada DSL, Australia AICS, China IECSC, Japan MITI, Philippines PICCS, Korea TCCL listed
Properties: Colorless liq.; paraffinic odor; insol. in water; dens. 0.769 kg/l (15 C); kinematic visc. 1.68 mm²/s (20 C); i.b.p. 181 C; b.p. 185-199 C; flash pt. 57 C; autoignition temp. 417 C; surf. tens. 22.7 mN/m; conductivity 10 pS/m (23 C)
Toxicology: Expected to be of low toxicity by ing., inh., skin contact; aspiration into lungs may cause chem. pneumonitis which can be fatal; prolonged/repeated contact may cause defatting of skin, dermatitis; TSCA listed
Environmental: Potential to bioaccumulate; may cause long-term adverse effects in aquatic environment; expected to be nontoxic to fish, algae, bacteria
Precaution: Flamm.; flamm. limits in air 0.6-6 vol.%; may form flamm./explosive vapor/air mixts. above 54 C; electrostatic charges may ge generated; prevent static discharge
Hazardous decomp. prods.: Combustion: CO
Storage: Store @ ambient temps.; keep away from heat, naked flame, sparks

Shellsol® TM [Shell Netherlands]
Chem. Descrip.: Petroleum distillates, alkylate
CAS 64741-73-7; EINECS/ELINCS 265-074-0
Uses: Solvent for household prods. (cleaners, lighter fuels, waxes, polishes), surf. coatings, printing inks, industrial and textile cleaning, photocopier toner dispersants, cosmetics, pesticides, wallpaper mfg.
Features: Does not contain detectable quantities of polycyclic aromatics, heavy metals, or chlorinated compds.; considered nonhazardous
Regulatory: EINECS, Canada DSL, Australia AICS, China IECSC, Philippines PICCS, Korea TCCL listed
Properties: Colorless liq.; low paraffinic odor; insol. in water; dens. 0.779 kg/l (15 C); visc. 3.30 mm²/s; i.b.p. 215 C; b.p. 215-260 C; flash pt. (PMCC) 86 C; autoignition temp. 375 C; ref. index 1.442 (20 C); surf. tens. 24.6 mN/m
Toxicology: Expected to be of low toxicity by ing., inh., skin contact; aspiration into lungs may cause chem. pneumonitis which can be fatal; prolonged/repeated contact may cause defatting of skin, dermatitis; TSCA listed
Environmental: Potential to bioaccumulate; may cause long-term adverse effects in aquatic environment; expected to be nontoxic to fish, algae, bacteria
Precaution: Combustible; LEL 0.6 vol.%; UEL 6.0 vol.%; may form flamm./explosive vapor/air mixts. above 81 C; electrostatic charges may ge generated; prevent static discharge
Hazardous decomp. prods.: Combustion: CO
Storage: 12 mos min. shelf life under proper storage @ ambient temps.; keep away from heat, sparks, flame

Shellsol® W HT [Shell]
Chem. Descrip.: Solvent naphtha (petroleum), light aliphatic
CAS 64742-89-8; EINECS/ELINCS 265-192-2
UN 1268
Uses: Solvent for agric., fat/oil extraction, automotive undercoating, polishes, adhesives, inks, paints, paper coatings, petrol. gas absorption,

lighter fluids, resins, plastics, rubber; starting component for rubber solvs., lacquer diluents; solvent in protective coatings on citrus fruits

Features: High solvency; hydrotreating removes unwanted compds. (e.g., sulfur, nitrogen, oxygen), converts unsat. hydrocarbons (olefins, acetylenes, aromatics), results in relatively low odor; no more than trace amts. of HAPs; high cycloparaffin content

Regulatory: FDA 21CFR §172.250; SARA §311/131 acute health/chronic health/fire hazard; Calif. Prop. 65 reportable [toluene (0.2%)]

Properties: Lt. colored liq.; low hydrocarbon odor; negligible sol. in water; m.w. 101; sp.gr. 0.734 (60/60 F); vapor pressure 41 mm Hg (68 F); i.b.p. 205 F min.; b.p. 210-234 F; flash pt. -7 C; KB value 32 min.; VOC 6.1 lb/gal (60 F)

Toxicology: If aspirated, can cause severe lung damage, chem. pneumonitis; may be fatal if swallowed; may cause CNS depression (dizziness, lightheadedness, headache, nausea, incoordination, unconsciousness, death); may cause eye irritation/temporary discomfort, skin irritation, burning sensation, defatting; chronic exposure may cause hearing loss; chronic abuse may cause irregular heart rhythms, cardiac arrest; target organs: heart, auditory system; TSCA listed

Environmental: Prevent entry into waterways, sewers

Precaution: Extremely flamm.; flamm. in air 0.8-8 vol.%; vapors are heavier than air; vapors may flash back; prevent vapor accumulation; ground equip.; avoid contact with strong oxidizing agents; prevent entry into basements, confined areas

Hazardous decomp. prods.: Complex mixt. of airborne solids, liqs. and gases, incl. CO, CO_2, other org. compds.

HMIS: Health 1, Flammability 3, Reactivity 0

Storage: Keep closed when not in use; keep away from heat, sparks, flame

Shellsol® X7B [Shell S. Africa]

Chem. Descrip.: Mixt. of toluene, xylene, and aliphatic hydrocarbon solvent

Uses: Solvent, diluent in lacquer thinners, enamels, liq. printing inks, adhesives, dip coatings

Features: Meets requirements of a med. aromatic solv.; good solvent power; fast evaporating

Properties: M.w. 96; vapor pressure 5 kPa (20 C); i.b.p. 110 C; pour pt. < -40 C; flash pt. 8.1 C; KB value 71; sp. heat 1.7 kJ/kg/°C; elec. conduct. 6 pS/m (20 C); dielec. const. 2.4 (20 C)

Precaution: Highly flamm.; LEL 1.2 vol.%; UEL 7.4 vol.%; vapors can readily form explosive mixt. in air

Storage: 12 mos min. shelf life under proper storage

Shell Toluene Solvent [Shell]

Chem. Descrip.: Toluene

CAS 108-88-3; EINECS/ELINCS 203-625-9

Uses: Solvent for dry cleaning, surf. coatings, automotive, chemical specialties, paint removers, chem. synthesis; starting component for rubber solvs. and lacquer diluents

Features: Med.-evaporating; used primarily in blends

Properties: Sp.gr. 0.871 (60 F); dens. 7.25 lb/gal (60 F); flash pt. (CC) 41 F; KB value 105; VOC 7.16 lb/gal

Shell Xylene [Shell]

Chem. Descrip.: Xylene

CAS 1330-20-7; EINECS/ELINCS 215-535-7

Uses: Solvent for dry cleaning, coatings, automotive, and chemical specialties; starting component for rubber solvents and lacquer diluents

Features: Med.-evaporating aromatic; used primarily in blends

Properties: Sp.gr. 0.871 (60 F); dens. 7.25 lb/gal (60 F); flash pt. (CC) 81 F; KB value 95; VOC 7.17 lb/gal

Ship Shape® Resin Cleaner [ISP]

Chem. Descrip.: N-Methyl-2-pyrrolidone, butyrolactone, deceth-6, fragrance

Uses: Solvent cleaner for removal of adhesives, paints, varnishes, and resins (polyester, acrylic, urethane, vinyl ester, epoxy) in fiberglass fabrication, furniture refinishing, jewelry, paint and resin mfg., rubber, aircraft, and automotive industries

Features: Safer alternative to acetone, methylene chloride, MEK, alkalines; low VOC

Properties: Lt. yel. clear; water-sol.; dens. 1.07 g/ml; visc. 2.02 cps; m.p. -20 C; b.p. 202-204 C; flash pt. 91.7 C

Toxicology: TLV 100 ppm; low oral and inhalation toxicity; mod. dermal toxicity; mod. eye irritant

Skykleen® 1000 [Solutia]

Chem. Descrip.: Dimethyl glutarate (55-75%); dimethyl succinate (15-35%); dimethyl adipate (5-25%)

Uses: Aviation solvent for sealant and adhesive removal, surf. preparation for painting/bonding, general surf. cleaning and degreasing

Features: Environmentally friendly alternative to chlorinated or ozone-depleting solvs.

Regulatory: DOT nonregulated; SARA nonreportable

Properties: Colorless liq.; sweet odor; sol. 5.5% in water (20 C); sp.gr. 1.086 (20 C); dens. 9.06 lb/gal (20 C); vapor pressure 0.06 mm Hg (20 C); f.p. - 40 C; b.p. 196-226 C; flash pt. (TCC) 103 C

Toxicology: May cause mod. eye and sl. skin irritation; sl. toxic if inh. or ing.TSCA listed

Environmental: Readily biodeg.; EC50 (daphnia magna, 48 h) 136 mg/l; practically nontoxic; LC50 (fathead minnow, 96 h) 18-24 mg/l; sl. toxic

Precaution: May hydrolyze with strong acids and bases; decomposes with heat; vapor forms explosive mixt. in air

Skykleen® 2000 [Solutia]

Chem. Descrip.: Dimethyl ester mixt. (91%) and functional additives (9%)

Uses: Aviation solvent for sealant and adhesive removal, surf. preparation for painting/bonding, general surf. cleaning and degreasing

Features: Environmentally friendly alternative to chlorinated or ozone-depleting solvs.

Regulatory: DOT nonregulated; SARA nonreportable

Properties: Clear paste; sweet odor; vapor pressure 0.06 mm Hg (20 C); flash pt. (TCC) 102 C

Toxicology: May cause eye and respiratory tract irritation; nonirritating to skin; inh. cause coughing, chest tightness, runny nose; ing. sl. toxicTSCA listed

Precaution: Avoid strong oxidants, acids, and alkalis

Hazardous decomp. prods.: Burning produces CO and CO_2

Storage: Keep container closed

S&L Naphtha [Sunnyside]

Chem. Descrip.: Light aliphatic naphtha, n-hexane, other hexane isomers, n-heptane, other heptane isomers, pentane, toluene, cyclohexane, other C5-10 paraffins and cycloparaffins

UN 1268

Uses: Solvent

Regulatory: SARA §311/312 acute health/chronic health/fire hazard, §313 reportable [n-hexane (12-18%), toluene (0-6%), cyclohexane (1-3%)]; Calif. Prop. 65 reportable [toluene (trace), benzene (trace)]

Properties: Liq.; sol. < 0.1% in water; dens. 5.83 lb/gal; b.p. 114-285 F; flash pt. (TCC) 0 F; autoignition temp. 520 F; 100% volatile

Toxicology: May cause eye irritation/stinging/watering/redness/swelling, skin irritation; prolonged/repeated contact may cause skin drying, burns, severe damage; inh. of vapors may cause nose/throat/respiratory tract irritation, mild CNS depression, dizziness, headache, convulsions, loss of consciousness; overexposure may cause peripheral and central nerve damage; if aspirated, may cause pneumonitis; exposure to high concs. may cause cardiac arrhythmias

Precaution: Extremely flamm. liq.; LEL 1 vol.%; vapors may cause flash fire or ignite explosively; vapors may flash back; incompat. with strong oxidizing agents (liq. chlorine, conc. oxygen); ground containers to prevent static accumulation and discharge; empty containers may contain residue, may explode

Hazardous decomp. prods.: CO, CO_2

Storage: Keep containers closed when not in use; keep away from heat, sparks, open flame

Soltene® [Solvay SA]

Uses: Solvent

Features: Highly stabilized

Properties: Colorless

Soltrol® 10 [ChevronPhillips]

Chem. Descrip.: Aliphatic hydrocarbon mixt.: heptane isomers (CAS 31394-54-4, 28%), octanes (CAS 26635-64-3, 10%), 2,2,4-trimethylpentane (62%)

UN 3295

Uses: Solvent

Regulatory: SARA §313 nonreportable

Properties: Colorless liq.; mild odor; negligible sol. in water; sp.gr. 0.697-0.710 (60/60 F); dens. 5.827 lb/gal (60 F); visc. 0.57 cSt (38 C); i.b.p. 200 F min.; flash pt. (TCC) < 13 F; KB value 28

Toxicology: May produce sl. eye, skin irritation; inh. may cause sl. mucous membrane irritation, rapid breathing, fatigue, headache, lightheadedness, dizziness, other CNS effects; extreme exposures may cause respiratory arrest, convulsions, unconsciousness; ing. may cause diarrhea, sl. CNS depression, difficulty breathing, fatigue; if swallowed, may be aspirated into lungs resulting in inflamm. and possible fluid accumulation; may cause kidney damage; TSCA listed
Precaution: Flamm. liq.; incompat. with oxygen and strong oxidizing agents; highly flamm. vapors may accumulate in low areas
Hazardous decomp. prods.: CO_x and various hydrocarbons formed when burned
NFPA: Health 0, Flammability 3, Reactivity 0
Storage: Store in tightly closed container in well-ventilated area, away from heat, sparks, flame; bond and ground during transfer

Soltrol® 100 [ChevronPhillips]
Chem. Descrip.: C9-11 isoalkanes
CAS 68551-16-6; EINECS/ELINCS 271-365-3
UN 3295
Uses: Solvent
Regulatory: SARA §313 nonreportable
Properties: Colorless clear liq.; mild odor; negligible sol. in water; sp.gr. 0.746 (60/60 F); dens. 6.21 lb/gal (60 F); visc. 1.12 cSt (388 C); i.b.p. 300 F min.; flash pt. (TCC) 41 C; 100% act.
Toxicology: LD50 (oral, rat) > 34.6 g/kg, (skin, rabbit) 15.4 g/kg; LC50 (inh., rat, 4 h) > 3684 ppm; REL 400 ppm; may cause sl. eye/skin irritation; inh. may cause headache, dizziness, nausea, unconsciousness; ing. may irritate stomach, intestines; may be aspirated into lungs resulting in inflamm., possible fluid accumulation; may cause kidney damage; TSCA listed
Environmental: Keep out of water sources and sewers
Precaution: Flamm. liq.; incompat. with oxygen and strong oxidizing agents; highly flamm. vapors may accumulate in low areas; may flash back
Hazardous decomp. prods.: CO_x formed when burned
NFPA: Health 1, Flammability 2, Reactivity 0
Storage: Store in tightly closed container in well-ventilated area, away from heat, sparks, flame; bond and ground during transfer

Soltrol® 130 [ChevronPhillips]
Chem. Descrip.: C10-13 isoalkanes
CAS 68551-17-7; EINECS/ELINCS 271-366-9
UN 3295
Uses: Solvent
Regulatory: SARA §313 nonreportable; Canada DSL listed
Properties: Colorless clear liq.; mild odor; negligible sol. in water; sp.gr. 0.746-0.768 (60/60 F); dens. 6.34 lb/gal (60 F); visc. 1.55 cSt (38 C); vapor pressure 1.5 mm Hg (38 C); i.b.p. 355 F min.; flash pt. (PMCC) 62.2 C; KB value 26; 100% act.
Toxicology: LD50 (oral, rat) > 34.6 g/kg, (skin, rabbit) 15.4 g/kg; LC50 (inh, rat, 6 h, 14 day post exposure) > 1277 ppm; may be sl. irritating to eyes, skin; inh. may cause headache, dizziness, nausea, unconsciousness; ing. may irritate stomach, intestines; may be aspirated into lungs resulting in inflamm., possible fluid accumulation; may cause kidney damage; TSCA listed
Precaution: Combustible liq.; incompat. with oxygen and strong oxidizing agents; highly flamm. vapors may accumulate in low areas; may flash back
Hazardous decomp. prods.: CO_x formed when burned
NFPA: Health 0, Flammability 2, Reactivity 0
Storage: Store in tightly closed container in well-ventilated area, away from heat, sparks, flame; bond and ground during transfer

Soltrol® 170 [ChevronPhillips]
Chem. Descrip.: C12-14 isoalkanes
CAS 68551-19-9; EINECS/ELINCS 271-369-5
UN 3295
Uses: Solvent
Regulatory: SARA §313 nonreportable
Properties: Colorless liq.; mild odor; negligible sol. in water; sp.gr. 0.78 (60/60 F); dens. 6.53 lb/gal (60 F); visc. 2.45 cSt (38 C); vapor pressure 0.012 psia; i.b.p. 420 F min.; flash pt. (TCC) > 89 C; KB value 24.6; 100% act.
Toxicology: May be sl. irritating to eyes, skin; inh. may cause headache, dizziness, nausea, unconsciousness; ing. may irritate stomach, intestines; may be aspirated into lungs resulting in inflamm., possible fluid accumulation; may cause kidney damage; TSCA listed
Precaution: Combustible liq.; incompat. with oxygen and strong oxidizing agents; highly flamm. vapors may accumulate in low areas; may flash back
Hazardous decomp. prods.: CO_x formed when burned
NFPA: Health 0, Flammability 1, Reactivity 0
Storage: Store in tightly closed container in well-ventilated area, away from heat, sparks, flame; bond and ground during transfer

Soltrol® 220 [ChevronPhillips]
Chem. Descrip.: C13-17 isoalkanes
CAS 68551-20-2; EINECS/ELINCS 271-370-0
Uses: Solvent
Regulatory: SARA §313 nonreportable; Canada DSL listed
Properties: Colorless liq.; mild odor; negligible sol. in water; sp.gr. 0.803 (60/60 F); visc. 4.5 cSt (38 C); vapor pressure 0.01 psia (38 C); i.b.p. 425 F min.; flash pt. (PMCC) > 212 F; 100% act.
Toxicology: REL 400 ppm; may be sl. irritating to eyes, skin; inh. may cause headache, dizziness, nausea, unconsciousness; ing. may irritate stomach, intestines; may be aspirated into lungs resulting in inflamm., possible fluid accumulation; may cause kidney damage; TSCA listed
Environmental: Keep out of water sources and sewers
Precaution: Incompat. with oxygen and strong oxidizing agents; highly flamm. vapors may accumulate in low areas; may flash back
Hazardous decomp. prods.: CO_x formed when burned
NFPA: Health 0, Flammability 1, Reactivity 0
Storage: Store in tightly closed container in well-ventilated area, away from heat, sparks, flame; bond and ground during transfer

Soltrol® 300 [ChevronPhillips]
Chem. Descrip.: C13-17 isoalkanes
CAS 68551-20-2
Uses: Solvent
Regulatory: SARA §313 nonreportable
Properties: Colorless clear liq.; mild odor; negligible sol. in water; sp.gr. 0.818 (60/60 F); visc. 5 cSt (38 C); vapor pressure 0.01 psia (38 C); i.b.p. 515 F min.; flash pt. (PMCC) > 110 C; KB value 23; 100% act.
Toxicology: May be sl. irritating to eyes, skin; inh. may cause headache, dizziness, nausea, unconsciousness; ing. may irritate stomach, intestines; may be aspirated into lungs resulting in inflamm., possible fluid accumulation; may cause kidney damage; TSCA listed
Precaution: Incompat. with oxygen and strong oxidizing agents; highly flamm. vapors may accumulate in low areas; may flash back
Hazardous decomp. prods.: CO_x formed when burned
NFPA: Health 0, Flammability 1, Reactivity 0
Storage: Store in tightly closed container in well-ventilated area, away from heat, sparks, flame; bond and ground during transfer

Solulan® 16 [Amerchol; Amerchol Europe]
Chem. Descrip.: Laneth-16, ceteth-16, oleth-16 and steareth-16
Uses: Emulsifier, wetting agent, dispersant, lubricant, solvent, conditioner, plasticizer, emollient in personal care prods., dermatology prods., topical pharmaceuticals; foam stabilizer for detergent systems
Properties: Lt. tan waxy solid; sol. @ 5% in water, 30% ethanol; HLB 15; cloud pt. 64-70 C; acid no. 3 max.; sapon. no. 8 max.; pH 4.5-7.5 (10% aq.); nonionic; 100% conc.

Solulan® 75 [Amerchol; Amerchol Europe]
Chem. Descrip.: PEG-75 lanolin
CAS 61790-81-6
Uses: Emulsifier, wetting agent, dispersant, lubricant, solvent, conditioner, plasticizer, emollient in personal care prods., dermatology prods., topical pharmaceuticals; foam stabilizer for detergent systems
Properties: Lt. yel.-amber waxy solid; faint pleasant odor; sol. @ 5% in water, 30% ethanol; HLB 15; cloud pt. 80-87 C; acid no. 3 max.; sapon. no. 10-20; pH 4.5-7.5 (10% aq.); nonionic; 100% conc.

Solulan® C-24 [Amerchol; Amerchol Europe]
Chem. Descrip.: Choleth-24 and ceteth-24
Uses: Emulsifier, wetting agent, dispersant, lubricant, solvent, conditioner, plasticizer, emollient used in personal care and dermatology prods.; foam stabilizer for detergent systems
Properties: Off-wh. to pale yel. waxy solid; faint pleasant odor; sol. @ 5% in water; HLB 14; cloud pt. 88-95 C; acid no. 1; sapon. no. 3 max.; nonionic; 100% conc.

Solulan® L-575 [Amerchol; Amerchol Europe]
Chem. Descrip.: PEG-75 lanolin

CAS 61790-81-6
Uses: Emulsifier, wetting agent, dispersant, lubricant, solvent, conditioner, plasticizer, emollient in personal care prods., dermatology prods., topical pharmaceuticals; foam stabilizer for detergent systems
Properties: Lt. yel.-amber visc. liq.; faint pleasant odor; sol. @ 5% in water, 30% ethanol; HLB 15; cloud pt. 80-87 C; acid no. 1 max.; sapon. no. 10 max.; nonionic; 50% conc.

Soluphor® P [BASF; BASF AG]
Chem. Descrip.: Pyrrolidone-2
CAS 616-45-5; EINECS/ELINCS 204-648-7
Uses: Solvent, antimicrobial for pharmaceutical injectables, parenterals, orals, veterinary medicine; absorp. accelerator of active substances in topicals
Features: Effective against gram-positive and gram-negative bacteria and molds
Properties: Colorless or sl. colored liq., solidifies @ R.T.; char. odor; sol. in water, many org. solvs. incl. ethanol, IPA, aromatic hydrocarbons; visc. 1-14 cSt; solid. pt. 25-26 C; ref. index 1.480-1.490; pH 8.2-10.8 (10% aq.); ≤ 0.5% moisture
Toxicology: LD50 (oral, rat) > 5 g/kg, (IV, rabbit) ≈ 1 ml/kg, (subcut., mouse) ≈ 4 g/kg, (IP, mouse) ≈ 3.8 g/kg; nonirritating to skin, mucous membranes; nonmutagenic
Storage: 12 mos shelf life when stored in unopened original containers @ R.T.

Solutol® HS 15 [BASF AG]
Chem. Descrip.: PEG-660 hydroxystearate
Uses: Solubilizer, solvent for vitamins A, D, E, and K and other lipophilic actives in pharmaceuticals, aq. parenteral preps., human and veterinary injection sol'ns.
Features: High chem. stability
Properties: Ylsh. wh. paste @ R.T., liq. @ 30 C; sol. in water, ethanol, 2-propanol; visc. ≈ 12 mPa·s (30% aq.); solid. pt. 25-30 C; acid no. ≤ 1; sapon. no. 53-63; hyd. no. 90-110; pH 6-7 (1'0% aq.); nonionic; ≤ 0.5% water
Toxicology: LD5 (oral, rat) ≈ 20 g/kg, (IP, mouse) ≥ 8.5 mg/kg, (IV, mouse) ≥ 3.16 g/kg
Storage: Prolonged exposure to heat may cause physical separation into liq. and solid phases, reversible by homogenization

Solv G [CITGO]
Chem. Descrip.: Heavy aromatic solvent naphtha (petroleum)
CAS 64742-94-5; EINECS/ELINCS 265-198-5
UN NA 1993
Uses: Solvent
Regulatory: SARA §302/304 nonreportable, §311/312 acute health/chronic health/fire hazard, §313 reportable [1,2,4-trimethylbenzene (0.1-1.5%), naphthalene (4-7%)]; Calif. Prop. 65 nonreportable
Properties: Colorless transparent liq.; mod. to strong sweet pleasant aromatic hydrocarbon odor; sol. in methanol, diethyl ether; negligible sol. in cold water (< 0.05%); sp.gr. 0.90; visc. < 1.5 cSt (40 C); vapor pressure 0.6 mm Hg (15.56 C); f.p. -57 to -18 C; b.p. 169-218 C; flash pt. (TCC) 63-67 C; autoignition temp. 246 C; KB value 94; VOC 900 g/l; 100% act.
Toxicology: Mist/vapor may irritate eyes, mucous membranes, respiratory tract; liq. contact may cause minimal to mod. eye irritation, mod. to severe skin irritation/inflamm.; may be harmful by inh., skin absorp.; overexposure may cause CNS depression, other target organ effects; may be harmful or fatal by ing.; aspiration into lungs can cause pulmonary edema, chem. pneumonia; possible reproductive hazard; TSCA listed
Environmental: May cause oxygen depletion in waterways; potentially toxic to water ecosystems; may be hazardous to aquatic life; may contribute to smog; prevent entry into sewers, waterways
Precaution: Combustible liq.; flamm. limits 0.7-6.0%; vapor may cause flash fire; vapors are heavier than air, may flash back; may create vapor/air explosion hazard in confined spaces; spills may create a slipping hazard; floats on water; container can rupture in fire; prevent entry into basements, confined areas; bond and ground equip.; incompat. with strong acids, alkalis, oxidizers (liq. chlorine, hydrogen peroxide, oxygen)
Hazardous decomp. prods.: Combustion: smoke, CO, CO_2, possibly other harmful gases/vapors
HMIS: Health 2, Flammability 2, Reactivity 0
Storage: Store in tightly closed containers in cool, dry, well-ventilated place away from heat, sparks, ignition sources, strong oxidizing conditions

Solvaclene® [Solvay SA]
Chem. Descrip.: Methylene chloride
CAS 75-09-2; EINECS/ELINCS 200-838-9
Uses: Solvent in treatment of metals, pharmaceuticals

Solvant 474 D [Grau Aromatics]
Chem. Descrip.: Alylarylpolyglycol ether
Uses: Solvent for perfume oils
Properties: Nonionic

Solvarex 9 [TotalFina-Elf]
Chem. Descrip.: Naphtha, light aromatic
CAS 64742-95-6; EINECS/ELINCS 265-199-0
Uses: Solvent for paints, varnishes, printing, wood preservatives, textiles, pesticides, industrial cleaning, rare earth extraction; diluent
Properties: Saybolt 30 color; dens. 872 kg/m^3; visc. 0.95 mm^2/s (20 C); vapor pressure 3 mbar (20 C); i.b.p. 175 C; aniline pt. 14 C; flash pt. (Abel) 43 C; ref. index 1.500 (20 C); surf. tens. 30.4 dynes/cm (20 C); 99.5% aromatics; 0.01% benzene

Solvarex 10 [TotalFina-Elf]
Chem. Descrip.: Naphtha, heavy aromatic
CAS 64742-94-5; EINECS/ELINCS 265-198-5
Uses: Solvent for paints, varnishes, printing, degreasing, pesticides, mold release, chemicals, rare earth extraction; diluent
Properties: Saybolt 28 color; dens. 890 kg/m^3; visc. 1.3 mm^2/s (20 C); vapor pressure 0.5 mbar (20 C); i.b.p. 215 C; aniline pt. 16 C; flash pt. (Abel) 68 C; ref. index 1.512 (20 C); surf. tens. 31.7 dynes/cm (20 C); 99% aromatics; 0.0002% benzene

Solvarex 90/160 [TotalFina-Elf]
Chem. Descrip.: Naphtha, light aromatic
CAS 64742-95-6; EINECS/ELINCS 265-199-0
Uses: Solvent for paints, varnishes, degreasing, chemicals, industrial cleaning; diluent
Properties: Saybolt 30 color; dens. 872 kg/m^3; visc. 0.85 mm^2/s (20 C); vapor pressure 5 mbar (20 C); i.b.p. 168 C; aniline pt. 13 C; flash pt. (Abel) 33 C; ref. index 1.498 (20 C); surf. tens. 30.2 dynes/cm (20 C); 99.5% aromatics; 0.005% benzene

Solvarex 90/180 [TotalFina-Elf]
Chem. Descrip.: Naphtha, heavy aromatic
CAS 64742-94-5; EINECS/ELINCS 265-198-5
Uses: Solvent for paints, varnishes, industrial cleaning; diluent
Properties: Saybolt 29 color; dens. 877 kg/m^3; visc. 1 mm^2/s (20 C); vapor pressure 2 mbar (20 C); i.b.p. 195 C; aniline pt. 15 C; flash pt. (Abel) 43 C; ref. index 1.500 (20 C); surf. tens. 30.7 dynes/cm (20 C); 99% aromatics; 0.01% benzene

Solvaro AFD [TotalFina-Elf]
Chem. Descrip.: Naphtha, heavy aromatic
CAS 64742-94-5; EINECS/ELINCS 265-198-5
Uses: Solvent for paints, varnishes, printing, degreasing, pesticides, mold release, chemicals
Properties: Saybolt 30 color; dens. 866 kg/m^3; visc. 1.6 mm^2/s (20 C); vapor pressure 0.1 mbar (20 C); i.b.p. 225 C; aniline pt. 26 C; flash pt. (Abel) 73 C; ref. index 1.491 (20 C); surf. tens. 29.5 dynes/cm (20 C); 99.5% aromatics; 0.0005% benzene

Solvenol® 2 [Hercules; Hercules Ltd]
Chem. Descrip.: Mixed terpenes
Uses: Solvent for inks, coatings; constituent of industrial cleaners, deodorants; processing aid for NR, SBR, reclaim rubber
Features: Nonstaining
Properties: Colorless to pale yel. liq.; lemon-like odor; sp.gr. 0.854 (15.6/15.6 C); dens. 7.15 lb/gal; visc. 1 cps; b.p. 173-182 C; flash pt. (TCC) 115 C; KB value 80

Solvenol® B1 [Hercules]
Chem. Descrip.: Monocyclic terpenes and hydrocarbon mixt.
Uses: Solvent; processing aid for NR, SBR, reclaim rubber
Features: Nonstaining
Properties: Colorless to pale yel. liq.; lemon-like odor; sp.gr. 0.854 (15.6/15.6 C); dens. 7.15 lb/gal; visc. 1 cps; b.p. 176-184 C; flash pt. (TCC) 120 F; KB value 70

Solvenon® DPM [BASF AG]
Chem. Descrip.: Mixt. of isomeric dipropylene glycol monomethyl ethers
Uses: Solvent for domestic cleaners, acid rust remover, prod. of printing inks; film-binding assistant for aq. polymer dispersions in paints, var-

nishes

Solvenon® DPnB [BASF AG]
Chem. Descrip.: Dipropylene glycol butyl ether
CAS 29911-28-2; EINECS/ELINCS 249-951-5
Uses: Solvent for solv.-based coatings, household/industrial cleaners, metal cleaners, paint stripping
Properties: Pt-Co 20 max. color; m.w. 190.3; dens. 0.95-0.925 g/cm^3 (20 C); b.p. 225-235 C (1013 mbar); acid no. < 0.1; ref. index 1.426-1.428 (20 C); 98% min. act.

Solvenon® PC [BASF AG]
Chem. Descrip.: Propylene carbonate
CAS 108-32-7; EINECS/ELINCS 203-572-1
Uses: Solvent for pigments and dyes, in screen printing dyes; extracting agent; washing liq. for natural and syn. gases; intermediate for org. syntheses

Solvenon® PM [BASF AG]
Chem. Descrip.: 1-Methoxypropanol-2
CAS 107-98-2; EINECS/ELINCS 203-539-1
Uses: Solvent for cleaning agents, printing inks, surface coatings

Solvenon® PP [BASF AG]
Chem. Descrip.: 1-Phenoxypropanol-2 and 2-phenoxypropanol-1
Uses: Solvent for surf. coatings, e.g., electrodipping varnishes by cathodic deposition; component of cleaning agents

Solvenon® PnB [BASF AG]
Chem. Descrip.: Propylene glycol butyl ether
CAS 5131-66-8; EINECS/ELINCS 225-878-4
Uses: Solvent for solv.-based coatings, household/industrial cleaners, metal cleaners, paint stripping
Properties: Pt-Co 15 max. color; m.w. 132; dens. 0.87-0.89 g/cm^3 (20 C); b.p. 165-175 C (1013 mbar); acid no. < 0.1; ref. index 1.417-1.420 (20 C); 99% min. act.

Solvenon® TPnB [BASF AG]
Chem. Descrip.: Tripropylene glycol butyl ether
CAS 55934-93-5; EINECS/ELINCS 259-910-3
Uses: Solvent, coalescing agent for water-borne architectural and industrial paints/coatings
Properties: Pt-Co 50 max. color; m.w. 248.4; dens. 0.93-0.95 g/cm^3 (20 C); b.p. 269-332 C (1013 mbar); acid no. <0.5; ref. index 1.433-1.436(20 C)

142 Solvent 66/3 [CITGO]
Chem. Descrip.: Stoddard solvent
CAS 8052-41-3; EINECS/ELINCS 265-198-5
UN 1268; UN NA1993
Uses: Solvent
Features: Med. boiling; hydrotreated to reduce aromatics and olefins; meets definition of non-photochemically reactive as defined by 66/3 Rule requirements
Regulatory: FDA 21CFR §175.105, 178.3620, 178.3910; SARA §302/304 nonreportable, §311/312 acute health/chronic health/fire hazard, §313 nonreportable; Calif. Prop. 65 nonreportable; may contain detectable amts. 1,3,5-trimethylbenzene
Properties: Colorless transparent liq.; lt. paraffinic hydrocarbon odor; negligible sol. in cold water (< 0.01%); sp.gr. 0.768-0.820 (60/60 F); dens. 6.61 lb/gal (15.6 C); visc. 1.72 cSt (20 C); vapor pressure 2.9 mm Hg (20 C); f.p. -85 to -25 C; b.p. 188-207 C; flash pt. (TCC) 61-63 C; autoignition temp. 235 C; ref. index 1.4319 (20 C); KB value 31; VOC 793 g/l; 100% act.
Toxicology: ACGIH TLV TWA 100 ppm; LD50 (oral, rat) > 34,600 mg/kg; mist/vapor may irritate eyes, mucous membranes, respiratory tract; liq. contact may cause minimal to mod. eye irritation, mod. to severe skin irritation/inflamm.; may be harmful by inh., skin absorp.; overexposure may cause CNS depression, other target organ effects; may be harmful or fatal by ing.; aspiration into lungs can cause pulmonary edema, chem. pneumonia; TSCA listed
Environmental: May cause oxygen depletion in waterways; potentially toxic to water ecosystems; may be hazardous to aquatic life; may contribute to smog; prevent entry into sewers, waterways; not a DOT marine pollutant
Precaution: Combustible liq.; flamm. limits 0.8-6.0%; vapor may cause flash fire; vapors are heavier than air, may flash back; may create vapor/air explosion hazard in confined spaces; spills may create a slipping hazard; floats on water; container can rupture in fire; prevent entry into basements, confined areas; bond and ground equip.; incompat. with strong acids, alkalis, oxidizers (liq. chlorine, other halogens, hydrogen peroxide, oxygen)
Hazardous decomp. prods.: Combustion: smoke, CO, CO_2, possibly other harmful gases/vapors
HMIS: Health 1, Flammability 2, Reactivity 0
Storage: Store in tightly closed containers in cool, dry, well-ventilated place away from heat, sparks, ignition sources, strong oxidizing conditions

Solvent 170 [CITGO]
Chem. Descrip.: Petroleum distillates, hydrotreated light
CAS 64742-47-8; EINECS/ELINCS 265-149-8
UN 1268; UN NA1993
Uses: Solvent for uranium extraction, formulations requiring high flash pts.; lamp oil
Features: Low aromatics
Regulatory: SARA §302/304 nonreportable, §311/312 acute health/chronic health/fire hazard, §313 nonreportable; Calif. Prop. 65 nonreportable
Properties: Water-wh. transparent liq.; low paraffinic hydrocarbon odor; negligible sol. in cold water; sp.gr. 0.80-0.82; dens. 6.76 lb/gal (15.6 C); visc. 2.23 cst; vapor pressure 0.066 torr (20 C); f.p. -9 to -4 C; b.p. 205-238 C; pour pt. -80 F; flash pt. (TCC) 76-80 C; autoignition temp. 221 C; ref. index 1.445; KB value 30; VOC 808 g/l; 100% act.
Toxicology: OSHA TWA 400 ppm; mist or vapor may irritate eyes, mucous membranes, respiratory tract; liq. contact may cause minimal to mod. eye, mild to severe skin irritation/inflamm.; may be harmful if inhaled or absorbed thru skin; overexposure may cause CNS depression, other target organ effects; may be harmful or fatal if ingested; aspiration into lungs can cause pulmonary edema, chem. pneumonia; target organs: lungs, CNS, brain, blood, liver, kidneys; TSCA listed
Environmental: May cause oxygen depletion in waterways; potentially toxic to water ecosystems; may be hazardous to aquatic life; prevent entry into sewers, waterways; not a DOT marine pollutant
Precaution: Combustible liq.; flamm. limits 0.6-6.0%; vapors are heavier than air, may flash back; may create vapor/air explosion hazard in confined spaces; floats on water; spills may create slipping hazard; container can rupture in fire; static electricity charge can accumulate; prevent static discharge; ground equip. and containers; prevent entry into basements, confined areas; incompat. with strong acids, alkalis, oxidizers
Hazardous decomp. prods.: Combustion: smoke, CO, CO_2, possibly other harmful gases/vapors
HMIS: Health 1, Flammability 2, Reactivity 0
Storage: Store in tightly closed containers in cool, dry, well-ventilated place away from heat, sparks, ignition sources, oxidizable materials

Solvent 2205 [CITGO]
Chem. Descrip.: Heptane isomers (51-69%), n-heptane (25-40%), methylcyclohexane (1-2%), toluene (≈ 2%), octanes (3-5%)
UN 1206; UN 1268
Uses: Solvent
Regulatory: SARA §302/304 nonreportable, §311/312 acute health/chronic health/fire hazard, §313 reportable [toluene (2%)]; Calif. Prop. 65 reportable (benzene, toluene); may contain detectable amts. of cyclohexane
Properties: Colorless transparent liq.; lt. paraffinic hydrocarbon odor; sl. sol. in cold water (≈ 0.1%); sp.gr. 0.70; visc. < 1 cSt (40 C); vapor pressure ≈ 44 mm Hg (20 C); f.p. -135 to -90 C; b.p. 90-109 C; flash pt. (TCC) -9 to -7 C; autoignition temp. ≈ 246 C; KB value 35-40; VOC 702 g/l
Toxicology: Mist/vapor may irritate eyes, mucous membranes, respiratory tract; liq. contact may cause minimal to mild eye, mild to mod. skin irritation/inflamm.; may be harmful by inh., skin absorp.; overexposure may cause CNS depression, other target organ effects; maybe harmful or fatal by ing.; aspiration into lungs can cause pulmonary edema, chem. pneumonia; prolonged/repeated inh. may cause heart arrhythmias; possible reproductive hazard; TSCA listed
Environmental: May cause oxygen depletion in waterways; potentially toxic to water ecosystems; may be hazardous to aquatic life; may contribute to smog; prevent entry into sewers, waterways; not a DOT marine pollutant
Precaution: Extremely flamm.; flamm. limits 1.0-7.1%; vapor may cause flash fire or explosion; vapors are heavier than air, may flash back; may

create vapor/air explosion hazard in confined spaces; spills may create slipping hazard; floats on water; container can rupture in fire; prevent entry into basements, confined areas; bond and ground equip.; incompat. with strong acids, alkalis, oxidizers (liq. chlorine, hydrogen peroxide, oxygen); heptane will dissolve some plastics

Hazardous decomp. prods.: Combustion: smoke, CO, CO_2, possibly other harmful gases/vapors

HMIS: Health 2, Flammability 3, Reactivity 0

Storage: Store in tightly closed containers in cool, dry, well-ventilated place away from heat, sparks, ignition sources, strong oxidizing conditions

Solvent 2215 [CITGO]

Chem. Descrip.: Stoddard solvent

CAS 8052-41-3; EINECS/ELINCS 232-489-3

UN 1268

Uses: Solvent

Regulatory: SARA §302/304 nonreportable, §311/312 acute health/chronic health/fire hazard, §313 nonreportable; Calif. Prop. 65 nonreportable; may contain detectable amts. of C9 aromatic hydrocarbons

Properties: Colorless transparent liq.; lt. paraffinic hydrocarbon odor; sl. sol. in cold water (< 0.03%); sp.gr. 0.78-0.79; visc. < 20 cSt (40 C); vapor pressure 2.8 mm Hg (20 C); f.p. -85 to -25 C; b.p. 160-213 C; flash pt. (TCC) 53-55 C; autoignition temp. ≈ 232 C; KB value 32; VOC 787 g/l; 100% act.

Toxicology: ACGIH TLV TWA 100 ppm; LD50 (oral, rat) > 34,600 mg/kg; mist/vapor may irritate eyes, mucous membranes, respiratory tract; liq. contact may cause minimal to mod. eye, mod. to severe skin irritation/inflamm.; may be harmful by inh., skin absorp.; overexposure may cause CNS depression, other target organ effects; maybe harmful or fatal by ing.; aspiration into lungs can cause pulmonary edema, chem. pneumonia; TSCA listed

Environmental: May cause oxygen depletion in waterways; potentially toxic to water ecosystems; may be hazardous to aquatic life; prevent entry into sewers, waterways; not a DOT marine pollutant

Precaution: Combustible; flamm. limits 0.7-6.0%; vapor may cause flash fire; vapors are heavier than air, may flash back; may create vapor/air explosion hazard in confined space; spills may create slipping hazard; floats on water; container can rupture in fire; prevent entry into basements, confined areas; bond and ground equip.; incompat. with strong acids, alkalis, oxidizers (liq. chlorine, hydrogen peroxide, oxygen)

Hazardous decomp. prods.: Combustion: smoke, CO, CO_2, possibly other harmful gases/vapors

HMIS: Health 2, Flammability 2, Reactivity 0

Storage: Store in tightly closed containers in cool, dry, well-ventilated place away from heat, sparks, ignition sources, strong oxidizing conditions

Solvent 2216 [CITGO]

Chem. Descrip.: Hydrotreated light petroleum distillates (30-40%), light aromatic solvent naphtha (30-35%), xylenes, mixed isomers (26-33%), ethylbenzene (3-9%), nonanes (6-12%), n-propylbenzene (0.5-4%), trimethylbenzene (9-21%), 1,2,4-trimethylbenzene (3-15%), ethyl toluene (3-9%)

UN 1268

Uses: Solvent

Regulatory: SARA §302/304 nonreportable, §311/312 acute health/chronic health/fire hazard, §313 reportable (xylenes, ethylbenzene, 1,2,4-trimethylbenzene); Calif. Prop. 65 reportable (benzene, toluene); may contain detectable amts. of 1,3,5-trimethylbenzene and other C9 aromatic hydrocarbons

Properties: Colorless transparent liq.; aromatic hydrocarbon odor; sl. sol. in cold water (< 0.08%); sp.gr. 0.84; visc. 10-20 cSt (40 C); vapor pressure 4.1 mm Hg (20 C); f.p. -95 to 13 C; b.p. 138-213 C; flash pt. (TCC) 31-33 C; autoignition temp. ≈ 230 C; KB value 71-72

Toxicology: Mist/vapor may irritate eyes, mucous membranes, respiratory tract; liq. contact may cause minimal to mod. eye, mod. to severe skin irritation/inflamm.; may be harmful or fatal by inh., ing., skin absorp.; sensitizer; overexposure may cause CNS depression, other target organ effects, hearing loss; aspiration into lungs can cause pulmonary edema, chem. pneumonia; prolonged/repeated inh. may cause heart arrhythmias; possible cancer/reproductive hazard; TSCA listed

Environmental: May cause oxygen depletion in waterways; potentially toxic to water ecosystems; may be hazardous to aquatic life; may contribute to smog; prevent entry into sewers, waterways; DOT marine pollutant

Precaution: Flamm.; flamm. limits 0.8-7.1%; vapor may cause flash fire; vapors are heavier than air, may flash back; may create vapor/air explosion hazard in confined space; spills may create slipping hazard; container can rupture in fire; prevent entry into basements, confined areas; bond and ground equip.; incompat. with strong acids, alkalis, oxidizers (liq. chlorine, other halogens, hydrogen peroxide, oxygen)

Hazardous decomp. prods.: Combustion: smoke, CO, CO_2, possibly other harmful gases/vapors

HMIS: Health 2, Flammability 3, Reactivity 0

Storage: Store in tightly closed containers in cool, dry, well-ventilated place away from heat, sparks, ignition sources, strong oxidizing conditions

Solvent 2217 [CITGO]

Chem. Descrip.: Heavy hydrotreated naphtha (petroleum)

CAS 64742-48-9; EINECS/ELINCS 265-150-3

UN 1268

Uses: Solvent

Regulatory: SARA §302/304 nonreportable, §311/312 acute health/chronic health/fire hazard, §313 nonreportable; Calif. Prop. 65 reportable (benzene, toluene); may contain detectable amts. of C9 aromatic hydrocarbons

Properties: Colorless transparent liq.; lt. hydrocarbon odor; sl. sol. in cold water (< 0.1%); sp.gr. 0.76; visc. < 10 cSt (40 C); vapor pressure 4.7 mm Hg (20 C); f.p. -65 to -25 C; b.p. 125-185 C; flash pt. (TCC) 24-26 C; autoignition temp. ≈ 230 C; KB value 34; VOC 755 g/l; 100% act.

Toxicology: Mist/vapor may irritate eyes, mucous membranes, respiratory tract; liq. contact may cause minimal to mod. eye, mod. to severe skin irritation/inflamm.; may be harmful by inh., skin absorp.; overexposure may cause CNS depression, other target organ effects; aspiration into lungs can cause pulmonary edema, chem. pneumonia; TSCA listed

Environmental: May cause oxygen depletion in waterways; potentially toxic to water ecosystems; may be hazardous to aquatic life; may contribute to smog; prevent entry into sewers, waterways; not a DOT marine pollutant

Precaution: Flamm.; flamm. limits 0.9-6.5%; vapor may cause flash fire; vapors are heavier than air, may flash back; may create vapor/air explosion hazard in confined space; spills may create slipping hazard; floats on water; container can rupture in fire; prevent entry into basements, confined areas; bond and ground equip.; incompat. with strong acids, alkalis, oxidizers (liq. chlorine, hydrogen peroxide, oxygen)

Hazardous decomp. prods.: Combustion: smoke, CO, CO_2, possibly other harmful gases/vapors

HMIS: Health 1, Flammability 3, Reactivity 0

Storage: Store in tightly closed containers in cool, dry, well-ventilated place away from heat, sparks, ignition sources, strong oxidizing conditions

Solvent 2218 [CITGO]

Chem. Descrip.: VM&P naphtha

CAS 8032-32-4; EINECS/ELINCS 232-453-7

UN 1268

Uses: Solvent

Regulatory: SARA §302/304 nonreportable, §311/312 acute health/chronic health/fire hazard, §313 reportable [xylenes (18-24%), ethylbenzene (2-7%)]; Calif. Prop. 65 reportable (toluene); may contain detectable amts. of 1,3,5-trimethylbenzene

Properties: Colorless transparent liq.; lt. hydrocarbon odor; sl. sol. in cold water (< 0.05%); sp.gr. 0.78-0.79; visc. < 5 cSt (40 C); vapor pressure 5.6 mm Hg (20 C); f.p. -57 to -25 C; b.p. 125-151 C; flash pt. (TCC) 23-25 C; autoignition temp. ≈ 232 C; KB value 40-45; VOC 788 g/l; 100% act.

Toxicology: Mist/vapor may irritate eyes, mucous membranes, respiratory tract; liq. contact may cause mild to mod. eye, mod. to severe skin irritation/inflamm.; may be harmful by inh., skin absorp.; may be harmful or fatal by ing.; sensitizer; overexposure may cause CNS depression, other target organ effects, hearing loss; aspiration into lungs can cause pulmonary edema, chem. pneumonia; prolonged/repeated inh. may cause heart arrhythmias; possible cancer/reproductive hazard; TSCA listed

Environmental: May cause oxygen depletion in waterways; potentially toxic to water ecosystems; may be hazardous to aquatic life; may contribute to smog; prevent entry into sewers, waterways; not a DOT marine pollutant

Precaution: Flamm.; flamm. limits 0.9-6.5%; vapor may cause flash fire; vapors are heavier than air, may flash back; may create vapor/air explosion hazard in confined space; spills may create slipping hazard; floats on water; container can rupture in fire; prevent entry into basements, confined areas; bond and ground equip.; incompat. with strong acids, alkalis, oxidizers (liq. chlorine, other halogens, hydrogen peroxide, oxygen)

Hazardous decomp. prods.: Combustion: smoke, CO, CO_2, possibly other harmful gases/vapors

HMIS: Health 2, Flammability 3, Reactivity 0

Storage: Store in tightly closed containers in cool, dry, well-ventilated place away from heat, sparks, ignition sources, strong oxidizing conditions

Solvent 2222 [CITGO]

Chem. Descrip.: Stoddard solvent

CAS 8052-41-3; EINECS/ELINCS 232-489-3

UN 1268

Uses: Solvent

Regulatory: SARA §302/304 nonreportable, §311/312 acute health/chronic health/fire hazard, §313 reportable [1,2,4-trimethylbenzene (0.5-1.5%); Calif. Prop. 65 reportable (benzene, toluene); may contain detectable amts. of 1,3,5-trimethylbenzene

Properties: Colorless transparent liq.; lt. hydrocarbon odor; sl. sol. in cold water (< 0.05%); sp.gr. 0.78-0.79; visc. < 15 cSt (40 C); vapor pressure 2.1 mm Hg (20 C); f.p. -65 to -25 C; b.p. 149-204 C; flash pt. (TCC) 42-43 C; autoignition temp. ≈ 230 C; KB value 36; VOC 779 g/l; 100% act.

Toxicology: ACGIH TLV/TWA 100 ppm; LD50 (oral, rat) > 34,600 mg/kg; mist/vapor may irritate eyes, mucous membranes, respiratory tract; liq. contact may cause minimal to mod. eye, mod. to severe skin irritation/inflamm.; may be harmful by inh., skin absorp.; may be harmful or fatal by ing.; overexposure may cause CNS depression, other target organ effects; aspiration into lungs can cause pulmonary edema, chem. pneumonia; possible reproductive hazard; TSCA listed

Environmental: May cause oxygen depletion in waterways; potentially toxic to water ecosystems; may be hazardous to aquatic life; may contribute to smog; prevent entry into sewers, waterways; not a DOT marine pollutant

Precaution: Flamm.; flamm. limits 0.5-6.0%; vapor may cause flash fire; vapors are heavier than air, may flash back; may create vapor/air explosion hazard in confined space; spills may create slipping hazard; floats on water; container can rupture in fire; prevent entry into basements, confined areas; bond and ground equip.; incompat. with strong acids, alkalis, oxidizers (liq. chlorine, other halogens, hydrogen peroxide, oxygen)

Hazardous decomp. prods.: Combustion: smoke, CO, CO_2, possibly other harmful gases/vapors

HMIS: Health 2, Flammability 2, Reactivity 0

Storage: Store in tightly closed containers in cool, dry, well-ventilated place away from heat, sparks, ignition sources, strong oxidizing conditions

Solvent 2231 [CITGO]

Chem. Descrip.: Stoddard solvent

CAS 8052-41-3; EINECS/ELINCS 232-489-3

UN 1268

Uses: Solvent

Regulatory: SARA §302/304 nonreportable, §311/312 acute health/chronic health/fire hazard, §313 reportable [1,2,4-trimethylbenzene (1-7%); Calif. Prop. 65 reportable (benzene, toluene); may contain detectable amts. of 1,3,5-trimethylbenzene

Properties: Colorless transparent liq.; lt. hydrocarbon odor; sl. sol. in cold water (< 0.05%); sp.gr. 0.78-0.79; visc. < 20 cSt (40 C); vapor pressure 2.8 mm Hg (20 C); f.p. -53 to 13 C; b.p. 152-177 C; flash pt. (TCC) 41-44 C; autoignition temp. ≈ 230 C; KB value 38-39; VOC 790 g/l; 100% act.

Toxicology: ACGIH TLV/TWA 100 ppm; LD50 (oral, rat) > 34,600 mg/kg; mist/vapor may irritate eyes, mucous membranes, respiratory tract; liq. contact may cause minimal to mod. eye, mod. to severe skin irritation/inflamm.; may be harmful by inh., skin absorp.; overexposure may cause CNS depression, other target organ effects; aspiration into lungs can cause pulmonary edema, chem. pneumonia; possible reproductive hazard; TSCA listed

Environmental: May cause oxygen depletion in waterways; potentially toxic to water ecosystems; may be hazardous to aquatic life; may contribute to smog; prevent entry into sewers, waterways; delisted as DOT marine pollutants

Precaution: Combustible; flamm. limits 0.7-6.0%; vapor may cause flash fire; vapors are heavier than air, may flash back; may create vapor/air explosion hazard in confined space; spills may create slipping hazard; floats on water; container can rupture in fire; prevent entry into basements, confined areas; bond and ground equip.; incompat. with strong acids, alkalis, oxidizers (liq. chlorine, hydrogen peroxide, oxygen)

Hazardous decomp. prods.: Combustion: smoke, CO, CO_2, possibly other harmful gases/vapors

HMIS: Health 2, Flammability 2, Reactivity 0

Storage: Store in tightly closed containers in cool, dry, well-ventilated place away from heat, sparks, ignition sources, strong oxidizing conditions

Solvent 2238 [CITGO]

Chem. Descrip.: Solvent-refined light naphtha (90-95%), toluene (7-9%), heptanes (mixt.)

UN 1268

Uses: Solvent

Regulatory: SARA §302/304 nonreportable, §311/312 acute health/chronic health/fire hazard, §313 reportable [n-hexane (27-37%), cyclohexane (0.0092-2.76%, toluene (7-9%)]; Calif. Prop. 65 reportable (benzene, toluene); may contain detectable amts. of n-hexane, cyclohexane

Properties: Colorless transparent liq.; lt. hydrocarbon odor; sl. sol. in cold water (≈ 0.1%); sp.gr. 0.70; visc. 1-2 cSt (40 C); vapor pressure 107 mm Hg (20 C); f.p. -95 to -90 C; b.p. 60-111 C; flash pt. (TCC) -21 to -20 C; autoignition temp. ≈ 232 C; KB value 35; VOC 703 g/l

Toxicology: Mist/vapor may irritate eyes, mucous membranes, respiratory tract; liq. contact may cause minimal to mild eye, mild to mod. skin irritation/inflamm.; may be harmful by inh., skin absorp.; may be harmful or fatal by ing.; sensitizer; overexposure may cause CNS depression, other target organ effects, peripheral nerve system damage; aspiration into lungs can cause pulmonary edema, chem. pneumonia; prolonged/repeated inh. may cause heart arrhythmias; possible reproductive hazard; TSCA listed

Environmental: May cause oxygen depletion in waterways; potentially toxic to water ecosystems; may be hazardous to aquatic life; may contribute to smog; prevent entry into sewers, waterways; not a DOT marine pollutant

Precaution: Extremely flamm.; flamm. limits 1.0-7.1%; vapor may cause flash fire or explosion; vapors are heavier than air, may flash back; may create vapor/air explosion hazard in confined space; spills may create slipping hazard; floats on water; container can rupture in fire; prevent entry into basements, confined areas; bond and ground equip.; incompat. with strong acids, alkalis, oxidizers (liq. chlorine, other halogens, hydrogen peroxide, oxygen); heptane will dissolve some plastics

Hazardous decomp. prods.: Combustion: smoke, CO, CO_2, possibly other harmful gases/vapors

HMIS: Health 2, Flammability 3, Reactivity 0

Storage: Store in tightly closed containers in cool, dry, well-ventilated place away from heat, sparks, ignition sources, strong oxidizing conditions

Solvent 2241 [CITGO]

Chem. Descrip.: Heavy hydrotreated naphtha (45-50%), heptanes mixt. (48-54%), methylcyclohexane (0.9-1.2%)

UN 1268

Uses: Solvent

Regulatory: SARA §302/304 nonreportable, §311/312 acute health/chronic health/fire hazard, §313 nonreportable; Calif. Prop. 65 reportable (toluene); may contain detectable amts. of cyclohexane

Properties: Colorless transparent liq.; lt. paraffinic hydrocarbon odor; sl. sol. in cold water (≈ 0.1%); sp.gr. 0.72-0.73; visc. 1-4 cSt (40 C); vapor pressure 23.5 mm Hg (20 C); f.p. -135 to -25 C; b.p. 90-151 C; flash pt. (TCC) -4 to -1 C; autoignition temp. ≈ 240 C; KB value 33; VOC 730 g/l

Toxicology: Mist/vapor may irritate eyes, mucous membranes, respiratory tract; liq. contact may cause minimal to mild eye, mod. to severe skin irritation/inflamm.; may be harmful by inh., skin absorp.; sensitizer; overexposure may cause CNS depression, other target organ effects; aspiration into lungs can cause pulmonary edema, chem. pneumonia; prolonged/repeated inh. may cause heart arrhythmias; TSCA listed

Environmental: May cause oxygen depletion in waterways; potentially toxic to water ecosystems; may be hazardous to aquatic life; prevent entry into sewers, waterways; not a DOT marine pollutant
Precaution: Flamm.; flamm. limits 0.9-6.7%; vapor may cause flash fire; vapors are heavier than air, may flash back; may create vapor/air explosion hazard in confined space; spills may create slipping hazard; floats on water; container can rupture in fire; prevent entry into basements, confined areas; bond and ground equip.; incompat. with strong acids, alkalis, oxidizers (liq. chlorine, other halogens, hydrogen peroxide, oxygen); heptane will dissolve some plastics
Hazardous decomp. prods.: Combustion: smoke, CO, CO_2, possibly other harmful gases/vapors
HMIS: Health 1, Flammability 3, Reactivity 0
Storage: Store in tightly closed containers in cool, dry, well-ventilated place away from heat, sparks, ignition sources, strong oxidizing conditions

Solvent 2245 [CITGO]
Chem. Descrip.: Heavy hydrotreated naphtha (39-46%), heptanes mixt. (54-57%), methylcyclohexane (1.0-1.3%)
UN 1268
Uses: Solvent
Regulatory: SARA §302/304 nonreportable, §311/312 acute health/chronic health/fire hazard, §313 nonreportable; Calif. Prop. 65 reportable (toluene); may contain detectable amts. of cyclohexane
Properties: Colorless transparent liq.; lt. paraffinic hydrocarbon odor; sl. sol. in cold water (≈ 0.1%); sp.gr. 0.72; visc. 1-4 cSt (40 C); vapor pressure 21 mm Hg (20 C); f.p. -135 to -25 C; b.p. 90-151 C; flash pt. (TCC) -6 to -4 C; autoignition temp. ≈ 240 C; KB value 32; VOC 721 g/l
Toxicology: Mist/vapor may irritate eyes, mucous membranes, respiratory tract; liq. contact may cause minimal to mild eye, mod. to severe skin irritation/inflamm.; may be harmful by inh., skin absorp.; may be harmful or fatal by ing.; sensitizer; overexposure may cause CNS depression, other target organ effects; aspiration into lungs can cause pulmonary edema, chem. pneumonia; prolonged/repeated inh. may cause heart arrhythmias; TSCA listed
Environmental: May cause oxygen depletion in waterways; potentially toxic to water ecosystems; may be hazardous to aquatic life; prevent entry into sewers, waterways; not a DOT marine pollutant
Precaution: Flamm.; flamm. limits 0.9-6.7%; vapor may cause flash fire; vapors are heavier than air, may flash back; may create vapor/air explosion hazard in confined space; spills may create slipping hazard; floats on water; container can rupture in fire; prevent entry into basements, confined areas; bond and ground equip.; incompat. with strong acids, alkalis, oxidizers (liq. chlorine, other halogens, hydrogen peroxide, oxygen); heptane will dissolve some plastics
Hazardous decomp. prods.: Combustion: smoke, CO, CO_2, possibly other harmful gases/vapors
HMIS: Health 1, Flammability 3, Reactivity 0
Storage: Store in tightly closed containers in cool, dry, well-ventilated place away from heat, sparks, ignition sources, strong oxidizing conditions

Solvent 2249 [CITGO]
Chem. Descrip.: Solvent-refined light naphtha (85-90%), light aliphatic naphtha (8-13), toluene (≈ 2%)
UN 1268
Uses: Solvent
Regulatory: SARA §302/304 nonreportable, §311/312 acute health/chronic health/fire hazard, §313 reportable [hexane (26-36%), cyclohexane (0.0098-2.65%), toluene (2-2.005%)]; Calif. Prop. 65 reportable (benzene, toluene); may contain detectable amts. of methylcyclopentane, n-hexane, cyclohexane
Properties: Colorless transparent liq.; lt. hydrocarbon odor; sl. sol. in cold water (≈ 0.09%); sp.gr. 0.685-0.69; visc. 1-2 cSt (40 C); vapor pressure 106 mm Hg (20 C); f.p. -108 to -90 C; b.p. 67-97 C; flash pt. (TCC) -18 to -15 C; autoignition temp. ≈ 232 C; KB value 30.7; VOC 683 g/l
Toxicology: Mist/vapor may irritate eyes, mucous membranes, respiratory tract; liq. contact may cause minimal to mod. eye, mild to severe skin irritation/inflamm.; may be harmful by inh., skin absorp.; may be harmful or fatal by ing.; sensitizer; overexposure may cause CNS depression, hearing loss, peripheral nervous system damage; aspiration into lungs can cause pulmonary edema, chem. pneumonia; prolonged/repeated inh. may cause heart arrhythmias; possible reproductive hazard; TSCA listed
Environmental: May cause oxygen depletion in waterways; potentially toxic to water ecosystems; may be hazardous to aquatic life; may contribute to smog; prevent entry into sewers, waterways; not a DOT marine pollutant
Precaution: Extremely flamm.; flamm. limits 1.0-7.0%; vapor may cause flash fire or explosion; vapors are heavier than air, may flash back; may create vapor/air explosion hazard in confined space; spills may create slipping hazard; floats on water; container can rupture in fire; prevent entry into basements, confined areas; bond and ground equip.; incompat. with strong acids, alkalis, oxidizers (liq. chlorine, other halogens, hydrogen peroxide, oxygen); heptane will dissolve some plastics
Hazardous decomp. prods.: Combustion: smoke, CO, CO_2, possibly other harmful gases/vapors
HMIS: Health 2, Flammability 3, Reactivity 0
Storage: Store in tightly closed containers in cool, dry, well-ventilated place away from heat, sparks, ignition sources, strong oxidizing conditions

Solvent 2254 [CITGO]
Chem. Descrip.: Solvent-refined light naphtha (70-80%), toluene (25-26%), n-hexane (22-30%)
UN 1268
Uses: Solvent
Regulatory: SARA §302/304 nonreportable, §311/312 acute health/chronic health/fire hazard, §313 reportable [hexane (22-30%), cyclohexane (0.0075-2.25%), toluene (24-26%)]; Calif. Prop. 65 reportable (benzene, toluene); may contain detectable amts. of n-hexane, cyclohexane
Properties: Colorless transparent liq.; lt. hydrocarbon odor; sl. sol. in cold water (≈ 0.1%); sp.gr. 0.727; visc. 1-2 cSt (40 C); vapor pressure 92 mm Hg (20 C); f.p. -95 to -90 C; b.p. 60-111 C; flash pt. (TCC) -19 to -17 C; autoignition temp. ≈ 232 C; KB value 45; VOC 726 g/l
Toxicology: Mist/vapor may irritate eyes, mucous membranes, respiratory tract; liq. contact may cause minimal to mod. eye, mild to severe skin irritation/inflamm.; may be harmful by inh., skin absorp.; may be harmful or fatal by ing.; sensitizer; overexposure may cause CNS depression, peripheral nervous system damage; aspiration into lungs can cause pulmonary edema, chem. pneumonia; prolonged/repeated inh. may cause heart arrhythmias; possible reproductive hazard; TSCA listed
Environmental: May cause oxygen depletion in waterways; potentially toxic to water ecosystems; may be hazardous to aquatic life; may contribute to smog; prevent entry into sewers, waterways; not a DOT marine pollutant
Precaution: Extremely flamm.; flamm. limits 1.0-7.1%; vapor may cause flash fire or explosion; vapors are heavier than air, may flash back; may create vapor/air explosion hazard in confined space; spills may create slipping hazard; floats on water; container can rupture in fire; prevent entry into basements, confined areas; bond and ground equip.; incompat. with strong acids, alkalis, oxidizers (liq. chlorine, other halogens, hydrogen peroxide, oxygen); heptane will dissolve some plastics
Hazardous decomp. prods.: Combustion: smoke, CO, CO_2, possibly other harmful gases/vapors
HMIS: Health 2, Flammability 3, Reactivity 0
Storage: Store in tightly closed containers in cool, dry, well-ventilated place away from heat, sparks, ignition sources, strong oxidizing conditions

Solvent 2255 [CITGO]
Chem. Descrip.: Solvent-refined light naphtha (70-80%), n-hexane (22-31%)
UN 1268
Uses: Solvent
Regulatory: SARA §302/304 nonreportable, §311/312 acute health/chronic health/fire hazard, §313 reportable [hexane (22-31%), cyclohexane (0.01-2.306); Calif. Prop. 65 reportable (benzene, toluene)
Properties: Colorless transparent liq.; lt. hydrocarbon odor; sl. sol. in cold water (≈ 0.1%); sp.gr. 0.68; visc. 1-2 cSt (40 C); vapor pressure 98 mm Hg (20 C); f.p. -108 to -90 C; b.p. 60-99 C; flash pt. (TCC) -22 to -21 C; autoignition temp. ≈ 232 C; KB value 30; VOC 703 g/l
Toxicology: Mist/vapor may irritate eyes, mucous membranes, respiratory tract; liq. contact may cause minimal to mild eye, mild to severe skin irritation/inflamm.; may be harmful by inh., skin absorp.; may be harmful or fatal by ing.; sensitizer; overexposure may cause CNS depression, peripheral nervous system damage; aspiration into lungs can cause

pulmonary edema, chem. pneumonia; prolonged/repeated inh. may cause heart arrhythmias; possible reproductive hazard; TSCA listed

Environmental: May cause oxygen depletion in waterways; potentially toxic to water ecosystems; may be hazardous to aquatic life; may contribute to smog; prevent entry into sewers, waterways; not a DOT marine pollutant

Precaution: Extremely flamm.; flamm. limits 1.0-7.0%; vapor may cause flash fire or explosion; vapors are heavier than air, may flash back; may create vapor/air explosion hazard in confined space; spills may create slipping hazard; floats on water; container can rupture in fire; prevent entry into basements, confined areas; bond and ground equip.; incompat. with strong acids, alkalis, oxidizers (liq. chlorine, other halogens, hydrogen peroxide, oxygen); heptane will dissolve some plastics

Hazardous decomp. prods.: Combustion: smoke, CO, CO_2, possibly other harmful gases/vapors

HMIS: Health 2, Flammability 3, Reactivity 0

Storage: Store in tightly closed containers in cool, dry, well-ventilated place away from heat, sparks, ignition sources, strong oxidizing conditions

Solvent 2256 [CITGO]

Chem. Descrip.: Stoddard solvent

CAS 8052-41-3; EINECS/ELINCS 232-489-3

UN 1268; UN NA 1993

Uses: Solvent

Regulatory: SARA §302/304 nonreportable, §311/312 acute health/chronic health/fire hazard, §313 nonreportable; Calif. Prop. 65 nonreportable; CERCLA reportable [naphthalene (0-0.0675%)]; may contain detectable amts. of 1,3,5-trimethylbenzene

Properties: Colorless transparent liq.; lt. paraffinic hydrocarbon odor; negligible sol. in cold water (< 0.01%); sp.gr. 0.802-0.808; visc. 1.6-1.7 cSt (40 C); vapor pressure 0.5-0.8 mm Hg (20 C); f.p. -65 to -4 C; b.p. 201-222 C; flash pt. (TCC) 73-75 C; autoignition temp. ≈ 230 C; KB value 31.3; VOC 805 g/l

Toxicology: ACGIH TLV TWA 100 ppm; LD50 (oral, rat) > 34,600 mg/kg; mist/vapor may irritate eyes, mucous membranes, respiratory tract; liq. contact may cause minimal to mod. eye, mild to severe skin irritation/inflamm.; may be harmful by inh., skin absorp.; may be harmful or fatal by ing.; overexposure may cause CNS depression, other target organ effects; aspiration into lungs can cause pulmonary edema, chem. pneumonia; TSCA listed

Environmental: May cause oxygen depletion in waterways; potentially toxic to water ecosystems; may be hazardous to aquatic life; may contribute to smog; prevent entry into sewers, waterways; not a DOT marine pollutant

Precaution: Combustible; flamm. limits 0.7-6.0%; vapor may cause flash fire; vapors are heavier than air, may flash back; may create vapor/air explosion hazard in confined space; spills may create slipping hazard; floats on water; container can rupture in fire; prevent entry into basements, confined areas; bond and ground equip.; incompat. with strong acids, alkalis, oxidizers (liq. chlorine, other halogens, hydrogen peroxide, oxygen)

Hazardous decomp. prods.: Combustion: smoke, CO, CO_2, possibly other harmful gases/vapors

HMIS: Health 1, Flammability 2, Reactivity 0

Storage: Store in tightly closed containers in cool, dry, well-ventilated place away from heat, sparks, ignition sources, strong oxidizing conditions

Solvent 2257 [CITGO]

Chem. Descrip.: Stoddard solvent

CAS 8052-41-3; EINECS/ELINCS 232-489-3

UN 1268; UN NA 1993

Uses: Solvent

Regulatory: SARA §302/304 nonreportable, §311/312 acute health/chronic health/fire hazard, §313 nonreportable; Calif. Prop. 65 nonreportable; CERCLA reportable [naphthalene (0-0.045%)]; may contain detectable amts. of 1,3,5-trimethylbenzene

Properties: Colorless transparent liq.; lt. paraffinic hydrocarbon odor; negligible sol. in cold water (< 0.01%); sp.gr. 0.80-0.805; visc. 1.5-1.6 cSt (40 C); vapor pressure 0.5-0.9 mm Hg (20 C); f.p. -65 to -4 C; b.p. 194-218 C; flash pt. (TCC) 71-74 C; autoignition temp. ≈ 230 C; KB value 31.5; VOC 800 g/l

Toxicology: ACGIH TLV TWA 100 ppm; LD50 (oral, rat) > 34,600 mg/kgmist/vapor may irritate eyes, mucous membranes, respiratory tract; liq. contact may cause minimal to mod. eye, mild to severe skin irritation/inflamm.; may be harmful by inh., skin absorp.; may be harmful or fatal by ing.; overexposure may cause CNS depression, other target organ effects; aspiration into lungs can cause pulmonary edema, chem. pneumonia; TSCA listed

Environmental: May cause oxygen depletion in waterways; potentially toxic to water ecosystems; may be hazardous to aquatic life; may contribute to smog; prevent entry into sewers, waterways; not a DOT marine pollutant

Precaution: Combustible; flamm. limits 0.7-6.0%; vapor may cause flash fire; vapors are heavier than air, may flash back; may create vapor/air explosion hazard in confined space; spills may create slipping hazard; floats on water; container can rupture in fire; prevent entry into basements, confined areas; bond and ground equip.; incompat. with strong acids, alkalis, oxidizers (liq. chlorine, other halogens, hydrogen peroxide, oxygen)

Hazardous decomp. prods.: Combustion: smoke, CO, CO_2, possibly other harmful gases/vapors

HMIS: Health 1, Flammability 2, Reactivity 0

Storage: Store in tightly closed containers in cool, dry, well-ventilated place away from heat, sparks, ignition sources, strong oxidizing conditions

Solvent 2268 [CITGO]

Chem. Descrip.: VM&P naphtha

CAS 8032-32-4; EINECS/ELINCS 232-453-7

UN 1268

Uses: Solvent

Regulatory: SARA §302/304 nonreportable, §311/312 acute health/chronic health/fire hazard, §313 reportable [xylenes (37-46%), ethylbenzene (5-13%)]; Calif. Prop. 65 reportable (benzene, toluene); may contain detectable amts. of 1,3,5-trimethylbenzene

Properties: Colorless transparent liq.; lt. hydrocarbon odor; sl. sol. in cold water (< 0.05%); sp.gr. 0.81-0.82; visc. < 5 cSt (40 C); vapor pressure 5.9 mm Hg (20 C); f.p. -57 to -25 C; b.p. 125-151 C; flash pt. (TCC) 25-27 C; autoignition temp. ≈ 232 C; KB value 65-70; VOC 811 g/l; 100% act.

Toxicology: Mist/vapor may irritate eyes, mucous membranes, respiratory tract; liq. contact may cause mild to mod. eye, mod. to severe skin irritation/inflamm.; may be harmful by inh., skin absorp.; may be harmful or fatal by ing.; sensitizer; overexposure may cause CNS depression, other target organ effects, hearing loss; aspiration into lungs can cause pulmonary edema, chem. pneumonia; prolonged/repeated inh. may cause heart arrhythmias; possible cancer/reproductive hazard; TSCA listed

Environmental: May cause oxygen depletion in waterways; potentially toxic to water ecosystems; may be hazardous to aquatic life; may contribute to smog; prevent entry into sewers, waterways; not a DOT marine pollutant

Precaution: Flamm.; flamm. limits 1.0-6.0%; vapor may cause flash fire; vapors are heavier than air, may flash back; may create vapor/air explosion hazard in confined space; spills may create slipping hazard; floats on water; container can rupture in fire; prevent entry into basements, confined areas; bond and ground equip.; incompat. with strong acids, alkalis, oxidizers (liq. chlorine, other halogens, hydrogen peroxide, oxygen)

Hazardous decomp. prods.: Combustion: smoke, CO, CO_2, possibly other harmful gases/vapors

HMIS: Health 2, Flammability 3, Reactivity 0

Storage: Store in tightly closed containers in cool, dry, well-ventilated place away from heat, sparks, ignition sources, strong oxidizing conditions

Solvent 2284 [CITGO]

Chem. Descrip.: Stoddard solvent

CAS 8052-41-3; EINECS/ELINCS 232-489-3

UN 1268

Uses: Solvent

Regulatory: SARA §302/304 nonreportable, §311/312 acute health/chronic health/fire hazard, §313 reportable [1,2,4-trimethylbenzene (0.5-3%)]; Calif. Prop. 65 reportable (benzene, toluene); may contain detectable amts. of cumene, other C9 aromatic hydrocarbons

Properties: Colorless transparent liq.; lt. hydrocarbon odor; sl. sol. in cold water (< 0.05%); sp.gr. 0.78; visc. < 20 cSt (40 C); vapor pressure 2.0 mm Hg (20 C); f.p. -65 to -25 C; b.p. 155-213 C; flash pt. (TCC) 42-45 C; autoignition temp. ≈ 230 C; KB value 35; VOC 777 g/l; 100% act.

Toxicology: ACGIH TLV TWA 100 ppm; LD50 (oral, rat) > 34,600 mg/kg; mist/vapor may irritate eyes, mucous membranes, respiratory tract; liq. contact may cause minimal to mod. eye, mod. to severe skin irritation/inflamm.; may be harmful by inh., skin absorp.; may be harmful or fatal by ing.; overexposure may cause CNS depression, other target organ effects; aspiration into lungs can cause pulmonary edema, chem. pneumonia; possible reproductive hazard; TSCA listed

Environmental: May cause oxygen depletion in waterways; potentially toxic to water ecosystems; may be hazardous to aquatic life; may contribute to smog; prevent entry into sewers, waterways; not a DOT marine pollutant

Precaution: Combustible; flamm. limits 0.6-6.0%; vapor may cause flash fire; vapors are heavier than air, may flash back; may create vapor/air explosion hazard in confined space; spills may create slipping hazard; floats on water; container can rupture in fire; prevent entry into basements, confined areas; bond and ground equip.; incompat. with strong acids, alkalis, oxidizers (liq. chlorine, hydrogen peroxide, oxygen)

Hazardous decomp. prods.: Combustion: smoke, CO, CO_2, possibly other harmful gases/vapors

HMIS: Health 2, Flammability 2, Reactivity 0

Storage: Store in tightly closed containers in cool, dry, well-ventilated place away from heat, sparks, ignition sources, strong oxidizing conditions

Solvent 2292 [CITGO]

Chem. Descrip.: Stoddard solvent

CAS 8052-41-3; EINECS/ELINCS 232-489-3

UN 1268

Uses: Solvent

Regulatory: SARA §302/304 nonreportable, §311/312 acute health/chronic health/fire hazard, §313 reportable [1,2,4-trimethylbenzene (1-6%)]; Calif. Prop. 65 reportable (benzene, toluene); may contain detectable amts. of cumene, other C9 aromatic hydrocarbons

Properties: Colorless transparent liq.; lt. hydrocarbon odor; sl. sol. in cold water (< 0.05%); sp.gr. 0.79-0.80; visc. < 20 cSt (40 C); vapor pressure 2.1 mm Hg (20 C); f.p. -65 to -25 C; b.p. 155-213 C; flash pt. (TCC) 42-45 C; autoignition temp. ≈ 230 C; KB value 35-45; VOC 784 g/l; 100% act.

Toxicology: ACGIH TLV TWA 100 ppm; LD50 (oral, rat) > 34,600 mg/kg; mist/vapor may irritate eyes, mucous membranes, respiratory tract; liq. contact may cause minimal to mod. eye, mod. to severe skin irritation/inflamm.; may be harmful by inh., skin absorp.; may be harmful or fatal by ing.; overexposure may cause CNS depression, other target organ effects; aspiration into lungs can cause pulmonary edema, chem. pneumonia; possible reproductive hazard; TSCA listed

Environmental: May cause oxygen depletion in waterways; potentially toxic to water ecosystems; may be hazardous to aquatic life; may contribute to smog; prevent entry into sewers, waterways; may be a DOT marine pollutant

Precaution: Combustible; flamm. limits 0.8-6.0%; vapor may cause flash fire; vapors are heavier than air, may flash back; may create vapor/air explosion hazard in confined space; spills may create slipping hazard; floats on water; container can rupture in fire; prevent entry into basements, confined areas; bond and ground equip.; incompat. with strong acids, alkalis, oxidizers (liq. chlorine, hydrogen peroxide, oxygen)

Hazardous decomp. prods.: Combustion: smoke, CO, CO_2, possibly other harmful gases/vapors

HMIS: Health 2, Flammability 2, Reactivity 0

Storage: Store in tightly closed containers in cool, dry, well-ventilated place away from heat, sparks, ignition sources, strong oxidizing conditions

Solvent 2298 [CITGO]

Chem. Descrip.: Light aliphatic solvent naphtha (95-99%), toluene (≈ 2%)

UN 1268

Uses: Solvent

Regulatory: SARA §302/304 nonreportable, §311/312 acute health/chronic health/fire hazard, §313 reportable [toluene (2-2.05%)]; Calif. Prop. 65 reportable (benzene, toluene); may contain detectable amts. of methylcyclopentane, n-hexane, cyclohexane

Properties: Colorless transparent liq.; lt. paraffinic hydrocarbon odor; negligible sol. in cold water (< 0.01%); sp.gr. 0.70-0.705; visc. < 1 cSt (40 C); vapor pressure 44.5 mm Hg (20 C); f.p. -135 to -90 C; b.p. 90-111 C; flash pt. (TCC) -9 to -7 C; autoignition temp. ≈ 246 C; KB value 30-31; VOC 704 g/l

Toxicology: Mist/vapor may irritate eyes, mucous membranes, respiratory tract; liq. contact may cause minimal to mod. eye, mild to severe skin irritation/inflamm.; may be harmful by inh., skin absorp.; may be harmful or fatal by ing.; sensitizer; overexposure may cause CNS depression, other target organ effects; aspiration into lungs can cause pulmonary edema, chem. pneumonia; prolonged/repeated inh. may cause heart arrhythmias; possible reproductive hazard; TSCA listed

Environmental: May cause oxygen depletion in waterways; potentially toxic to water ecosystems; may be hazardous to aquatic life; may contribute to smog; prevent entry into sewers, waterways; not a DOT marine pollutant

Precaution: Extremely flamm.; flamm. limits 1.0-7.1%; vapor may cause flash fire or explosion; vapors are heavier than air, may flash back; may create vapor/air explosion hazard in confined space; spills may create slipping hazard; floats on water; container can rupture in fire; prevent entry into basements, confined areas; bond and ground equip.; incompat. with strong acids, alkalis, oxidizers (liq. chlorine, hydrogen peroxide, oxygen); heptane will dissolve some plastics

Hazardous decomp. prods.: Combustion: smoke, CO, CO_2, possibly other harmful gases/vapors

HMIS: Health 2, Flammability 3, Reactivity 0

Storage: Store in tightly closed containers in cool, dry, well-ventilated place away from heat, sparks, ignition sources, strong oxidizing conditions

Solvent 2299 [CITGO]

Chem. Descrip.: Light aliphatic solvent naphtha (90-95%), toluene (≈ 5%)

UN 1268

Uses: Solvent

Regulatory: SARA §302/304 nonreportable, §311/312 acute health/chronic health/fire hazard, §313 reportable [toluene (5-5.05%)]; Calif. Prop. 65 reportable (benzene, toluene); may contain detectable amts. of methylcyclopentane, n-hexane, cyclohexane

Properties: Colorless transparent liq.; lt. paraffinic hydrocarbon odor; negligible sol. in cold water (< 0.01%); sp.gr. 0.705-0.71; visc. < 1 cSt (40 C); vapor pressure 44 mm Hg (20 C); f.p. -135 to -90 C; b.p. 90-111 C; flash pt. (TCC) -9 to -7 C; autoignition temp. ≈ 246 C; KB value 32.1; VOC 709 g/l

Toxicology: Mist/vapor may irritate eyes, mucous membranes, respiratory tract; liq. contact may cause minimal to mod. eye, mild to severe skin irritation/inflamm.; may be harmful by inh., skin absorp.; may be harmful or fatal by ing.; sensitizer; overexposure may cause CNS depression, other target organ effects; aspiration into lungs can cause pulmonary edema, chem. pneumonia; prolonged/repeated inh. may cause heart arrhythmias; possible reproductive hazard; TSCA listed

Environmental: May cause oxygen depletion in waterways; potentially toxic to water ecosystems; may be hazardous to aquatic life; may contribute to smog; prevent entry into sewers, waterways; not a DOT marine pollutant

Precaution: Extremely flamm.; flamm. limits 1.0-7.1%; vapor may cause flash fire or explosion; vapors are heavier than air, may flash back; may create vapor/air explosion hazard in confined space; spills may create slipping hazard; floats on water; container can rupture in fire; prevent entry into basements, confined areas; bond and ground equip.; incompat. with strong acids, alkalis, oxidizers (liq. chlorine, hydrogen peroxide, oxygen); heptane will dissolve some plastics

Hazardous decomp. prods.: Combustion: smoke, CO, CO_2, possibly other harmful gases/vapors

HMIS: Health 2, Flammability 3, Reactivity 0

Storage: Store in tightly closed containers in cool, dry, well-ventilated place away from heat, sparks, ignition sources, strong oxidizing conditions

Solvent 2300 [CITGO]

Chem. Descrip.: Light aliphatic solvent naphtha (75-85%), toluene (≈ 19%)

UN 1268

Uses: Solvent

Regulatory: SARA §302/304 nonreportable, §311/312 acute health/chronic health/fire hazard, §313 reportable [toluene (19-19.05%)]; Calif. Prop. 65 reportable (benzene, toluene); may contain detectable amts. of methylcyclopentane, n-hexane, cyclohexane

Properties: Colorless transparent liq.; lt. paraffinic hydrocarbon odor; negligible sol. in cold water (< 0.01%); sp.gr. 0.73-0.735; visc. < 1 cSt (40 C); vapor pressure 41 mm Hg (20 C); f.p. -135 to -90 C; b.p. 90-111 C; flash pt. (TCC) -7 to -6 C; autoignition temp. ≈ 246 C; KB value 39.3; VOC 733 g/l

Toxicology: Mist/vapor may irritate eyes, mucous membranes, respiratory tract; liq. contact may cause minimal to severe eye, mild to severe skin irritation/inflamm.; may be harmful by inh., skin absorp.; may be harmful or fatal by ing.; sensitizer; overexposure may cause CNS depression, other target organ effects, hearing loss; aspiration into lungs can cause pulmonary edema, chem. pneumonia; prolonged/repeated inh. may cause heart arrhythmias; possible reproductive hazard; TSCA listed

Environmental: May cause oxygen depletion in waterways; potentially toxic to water ecosystems; may be hazardous to aquatic life; may contribute to smog; prevent entry into sewers, waterways; not a DOT marine pollutant

Precaution: Extremely flamm.; flamm. limits 1.0-7.1%; vapor may cause flash fire or explosion; vapors are heavier than air, may flash back; may create vapor/air explosion hazard in confined space; spills may create slipping hazard; floats on water; container can rupture in fire; prevent entry into basements, confined areas; bond and ground equip.; incompat. with strong acids, alkalis, oxidizers (liq. chlorine, hydrogen peroxide, oxygen); heptane will dissolve some plastics

Hazardous decomp. prods.: Combustion: smoke, CO, CO_2, possibly other harmful gases/vapors

HMIS: Health 2, Flammability 3, Reactivity 0

Storage: Store in tightly closed containers in cool, dry, well-ventilated place away from heat, sparks, ignition sources, strong oxidizing conditions

Solvent 2302 [CITGO]

Chem. Descrip.: Stoddard solvent

CAS 8052-41-3; EINECS/ELINCS 232-489-3

UN 1268

Uses: Solvent

Regulatory: SARA §302/304 nonreportable, §311/312 acute health/chronic health/fire hazard, §313 reportable [xylenes (3.0-3.6%), ethylbenzene (0.4-1.0%)]; Calif. Prop. 65 reportable (toluene); may contain detectable amts. of 1,3,5-trimethylbenzene

Properties: Colorless transparent liq.; lt. hydrocarbon odor; sl. sol. in cold water (< 0.05%); sp.gr. 0.78; visc. < 20 cSt (40 C); vapor pressure 2.7 mm Hg (20 C); f.p. -85 to -25 C; b.p. 149-200 C; flash pt. (TCC) 40-42 C; autoignition temp. ≈ 230 C; KB value 36; VOC 780 g/l; 100% act.

Toxicology: Mist/vapor may irritate eyes, mucous membranes, respiratory tract; liq. contact may cause mild to mod. eye, mod. to severe skin irritation/inflamm.; may be harmful by inh., skin absorp.; may be harmful or fatal by ing.; sensitizer; overexposure may cause CNS depression, other target organ effects; aspiration into lungs can cause pulmonary edema, chem. pneumonia; prolonged/repeated inh. may cause heart arrhythmias; possible cancer/reproductive hazard; TSCA listed

Environmental: May cause oxygen depletion in waterways; potentially toxic to water ecosystems; may be hazardous to aquatic life; may contribute to smog; prevent entry into sewers, waterways; not a DOT marine pollutant

Precaution: Combustible; flamm. limits 0.7-6.0%; vapor may cause flash fire; vapors are heavier than air, may flash back; may create vapor/air explosion hazard in confined space; spills may create slipping hazard; floats on water; container can rupture in fire; prevent entry into basements, confined areas; bond and ground equip.; incompat. with strong acids, alkalis, oxidizers (liq. chlorine, other halogens, hydrogen peroxide, oxygen)

Hazardous decomp. prods.: Combustion: smoke, CO, CO_2, possibly other harmful gases/vapors

HMIS: Health 2, Flammability 2, Reactivity 0

Storage: Store in tightly closed containers in cool, dry, well-ventilated place away from heat, sparks, ignition sources, strong oxidizing conditions

Solvent 2304 [CITGO]

Chem. Descrip.: Heptane isomers (28-51%), n-heptane (31-40%), methylcyclohexane (2-7%), toluene (≈ 11%), octanes (4-14%)

UN 1268

Uses: Solvent

Regulatory: SARA §302/304 nonreportable, §311/312 acute health/chronic health/fire hazard, §313 reportable [toluene (11-11.05%)]; Calif. Prop. 65 reportable (benzene, toluene); may contain detectable amts. of methylcyclopentane, n-hexane, cyclohexane

Properties: Colorless transparent liq.; lt. paraffinic hydrocarbon odor; negligible sol. in cold water (< 0.01%); sp.gr. 0.71-0.72; visc. < 1 cSt (40 C); vapor pressure 42 mm Hg (20 C); f.p. -135 to -90 C; b.p. 90-111 C; flash pt. (TCC) -8 to -6 C; autoignition temp. ≈ 246 C; KB value 34.5; VOC 719 g/l

Toxicology: Mist/vapor may irritate eyes, mucous membranes, respiratory tract; liq. contact may cause minimal to mod. eye, mild to severe skin irritation/inflamm.; may be harmful by inh., skin absorp.; may be harmful or fatal by ing.; sensitizer; overexposure may cause CNS depression, other target organ effects, hearing loss; aspiration into lungs can cause pulmonary edema, chem. pneumonia; prolonged/repeated inh. may cause heart arrhythmias; possible reproductive hazard; TSCA listed

Environmental: May cause oxygen depletion in waterways; potentially toxic to water ecosystems; may be hazardous to aquatic life; may contribute to smog; prevent entry into sewers, waterways; not a DOT marine pollutant

Precaution: Extremely flamm.; flamm. limits 1.0-7.1%; vapor may cause flash fire or explosion; vapors are heavier than air, may flash back; may create vapor/air explosion hazard in confined space; spills may create slipping hazard; floats on water; container can rupture in fire; prevent entry into basements, confined areas; bond and ground equip.; incompat. with strong acids, alkalis, oxidizers (liq. chlorine, other halogens, hydrogen peroxide, oxygen); heptane will dissolve some plastics

Hazardous decomp. prods.: Combustion: smoke, CO, CO_2, possibly other harmful gases/vapors

HMIS: Health 2, Flammability 3, Reactivity 0

Storage: Store in tightly closed containers in cool, dry, well-ventilated place away from heat, sparks, ignition sources, strong oxidizing conditions

Solvent 2307 [CITGO]

Chem. Descrip.: Solvent-refined light naphtha (75-85%), light aliphatic solvent naphtha (11-21%), toluene (≈ 4%)

UN 1268

Uses: Solvent

Regulatory: SARA §302/304 nonreportable, §311/312 acute health/chronic health/fire hazard, §313 reportable [hexane (24-32.016%, cyclohexane (0.0096-2.416%, toluene (4-4.008%)]; Calif. Prop. 65 reportable (benzene, toluene); may contain detectable amts. of methylcyclopentane, n-hexane, cyclohexane

Properties: Colorless transparent liq.; lt. hydrocarbon odor; sl. sol. in cold water (≈ 0.09%); sp.gr. 0.69; visc. 1-2 cSt (40 C); vapor pressure 100 mm Hg (20 C); f.p. -108 to -90 C; b.p. 60-125 C; flash pt. (TCC) -18 to -15 C; autoignition temp. ≈ 232 C; KB value 31.8; VOC 709 g/l

Toxicology: Mist/vapor may irritate eyes, mucous membranes, respiratory tract; liq. contact may cause minimal to mod. eye, mild to severe skin irritation/inflamm.; may be harmful by inh., skin absorp.; may be harmful or fatal by ing.; sensitizer; overexposure may cause CNS depression, hearing loss, peripheral nervous system damage; aspiration into lungs can cause pulmonary edema, chem. pneumonia; prolonged/repeated inh. may cause heart arrhythmias; possible reproductive hazard; TSCA listed

Environmental: May cause oxygen depletion in waterways; potentially toxic to water ecosystems; may be hazardous to aquatic life; may contribute to smog; prevent entry into sewers, waterways; not a DOT marine pollutant

Precaution: Extremely flamm.; flamm. limits 1.0-7.0%; vapor may cause flash fire or explosion; vapors are heavier than air, may flash back; may create vapor/air explosion hazard in confined space; spills may create slipping hazard; floats on water; container can rupture in fire; prevent entry into basements, confined areas; bond and ground equip.; incompat. with strong acids, alkalis, oxidizers (liq. chlorine, other halogens, hydrogen peroxide, oxygen); heptane will dissolve some plastics

Hazardous decomp. prods.: Combustion: smoke, CO, CO_2, possibly other harmful gases/vapors

HMIS: Health 2, Flammability 3, Reactivity 0

Storage: Store in tightly closed containers in cool, dry, well-ventilated place away from heat, sparks, ignition sources, strong oxidizing conditions

Solvent 2310 [CITGO]

Chem. Descrip.: Heavy hydrotreated naphtha (35-45%), methylcyclohexane (0.9-1.2%), heptanes mixt. (48-52%), octanes (18-23%), xylenes (3-5%), ethylbenzene (0.5-1.5%)

UN 1268

Uses: Solvent

Regulatory: SARA §302/304 nonreportable, §311/312 acute health/chronic health/fire hazard, §313 reportable [xylenes (3-5%), ethylbenzene (0.5-1.5%]; Calif. Prop. 65 reportable (benzene, toluene); may contain detectable amts. of 1,3,5-trimethylbenzene, cyclohexane

Properties: Colorless transparent liq.; lt. hydrocarbon odor; sl. sol. in cold water (≈ 0.1%); sp.gr. 0.73; visc. 2-4 cSt (40 C); vapor pressure 27 mm Hg (20 C); f.p. -135 to -25 C; b.p. 90-151 C; flash pt. (TCC) -5 to -4 C; autoignition temp. ≈ 232 C; KB value 35; VOC 731 g/l

Toxicology: Mist/vapor may irritate eyes, mucous membranes, respiratory tract; liq. contact may cause mild to mod. eye, mod. to severe skin irritation/inflamm.; may be harmful by inh., skin absorp.; may be harmful or fatal by ing.; sensitizer; overexposure may cause CNS depression, other target organ effects; aspiration into lungs can cause pulmonary edema, chem. pneumonia; prolonged/repeated inh. may cause heart arrhythmias; possible cancer/reproductive hazard; TSCA listed

Environmental: May cause oxygen depletion in waterways; potentially toxic to water ecosystems; may be hazardous to aquatic life; may contribute to smog; prevent entry into sewers, waterways; not a DOT marine pollutant

Precaution: Flamm.; flamm. limits 0.9-7.0%; vapor may cause flash fire; vapors are heavier than air, may flash back; may create vapor/air explosion hazard in confined space; spills may create slipping hazard; floats on water; container can rupture in fire; prevent entry into basements, confined areas; bond and ground equip.; incompat. with strong acids, alkalis, oxidizers (liq. chlorine, other halogens, hydrogen peroxide, oxygen); heptane will dissolve some plastics

Hazardous decomp. prods.: Combustion: smoke, CO, CO_2, possibly other harmful gases/vapors

HMIS: Health 2, Flammability 3, Reactivity 0

Storage: Store in tightly closed containers in cool, dry, well-ventilated place away from heat, sparks, ignition sources, strong oxidizing conditions

Solvent 2313 [CITGO]

Chem. Descrip.: Light aliphatic solvent naptha (80-90%), toluene (≈ 15%)

UN 1268

Uses: Solvent

Regulatory: SARA §302/304 nonreportable, §311/312 acute health/chronic health/fire hazard, §313 reportable [toluene (15-15.05%]; Calif. Prop. 65 reportable (benzene, toluene); may contain detectable amts. of methylcyclopentane, n-hexane, cyclohexane

Properties: Colorless transparent liq.; lt. paraffinic hydrocarbon odor; negligible sol. in cold water (< 0.01%); sp.gr. 0.72-0.73; visc. < 1 cSt (40 C); vapor pressure 42 mm Hg (20 C); f.p. -135 to -90 C; b.p. 90-111 C; flash pt. (TCC) -8 to -7 C; autoignition temp. ≈ 246 C; KB value 36.9; VOC 726 g/l

Toxicology: Mist/vapor may irritate eyes, mucous membranes, respiratory tract; liq. contact may cause minimal to severe eye, mild to severe skin irritation/inflamm.; may be harmful by inh., skin absorp.; may be harmful or fatal by ing.; sensitizer; overexposure may cause CNS depression, other target organ effects, hearing loss; aspiration into lungs can cause pulmonary edema, chem. pneumonia; prolonged/repeated inh. may cause heart arrhythmias; possible reproductive hazard; TSCA listed

Environmental: May cause oxygen depletion in waterways; potentially toxic to water ecosystems; may be hazardous to aquatic life; may contribute to smog; prevent entry into sewers, waterways; not a DOT marine pollutant

Precaution: Extremely flamm.; flamm. limits 1.0-7.1%; vapor may cause flash fire or explosion; vapors are heavier than air, may flash back; may create vapor/air explosion hazard in confined space; spills may create slipping hazard; floats on water; container can rupture in fire; prevent entry into basements, confined areas; bond and ground equip.; incompat. with strong acids, alkalis, oxidizers (liq. chlorine, hydrogen peroxide, oxygen); heptane will dissolve some plastics

Hazardous decomp. prods.: Combustion: smoke, CO, CO_2, possibly other harmful gases/vapors

HMIS: Health 2, Flammability 3, Reactivity 0

Storage: Store in tightly closed containers in cool, dry, well-ventilated place away from heat, sparks, ignition sources, strong oxidizing conditions

Solvent 2314 [CITGO]

Chem. Descrip.: Light aliphatic solvent naptha (85-95%), toluene (≈ 10%)

UN 1268

Uses: Solvent

Regulatory: SARA §302/304 nonreportable, §311/312 acute health/chronic health/fire hazard, §313 reportable [toluene (10-10.05%)]; Calif. Prop. 65 reportable (benzene, toluene); may contain detectable amts. of methylcyclopentane, n-hexane, cyclohexane

Properties: Colorless transparent liq.; lt. paraffinic hydrocarbon odor; negligible sol. in cold water (< 0.01%); sp.gr. 0.71-0.72; visc. < 1 cSt (40 C); vapor pressure 43 mm Hg (20 C); f.p. -135 to -90 C; b.p. 90-111 C; flash pt. (TCC) -8 to -7 C; autoignition temp. ≈ 246 C; KB value 33.7; VOC 717 g/l

Toxicology: Mist/vapor may irritate eyes, mucous membranes, respiratory tract; liq. contact may cause minimal to mod. eye, mild to severe skin irritation/inflamm.; may be harmful by inh., skin absorp.; may be harmful or fatal by ing.; sensitizer; overexposure may cause CNS depression, other target organ effects, hearing loss; aspiration into lungs can cause pulmonary edema, chem. pneumonia; prolonged/repeated inh. may cause heart arrhythmias; possible reproductive hazard; TSCA listed

Environmental: May cause oxygen depletion in waterways; potentially toxic to water ecosystems; may be hazardous to aquatic life; may contribute to smog; prevent entry into sewers, waterways; not a DOT marine pollutant

Precaution: Extremely flamm.; flamm. limits 1.0-7.1%; vapor may cause flash fire or explosion; vapors are heavier than air, may flash back; may create vapor/air explosion hazard in confined space; spills may create slipping hazard; floats on water; container can rupture in fire; prevent entry into basements, confined areas; bond and ground equip.; incompat. with strong acids, alkalis, oxidizers (liq. chlorine, hydrogen peroxide, oxygen); heptane will dissolve some plastics

Hazardous decomp. prods.: Combustion: smoke, CO, CO_2, possibly other harmful gases/vapors

HMIS: Health 2, Flammability 3, Reactivity 0

Storage: Store in tightly closed containers in cool, dry, well-ventilated place away from heat, sparks, ignition sources, strong oxidizing conditions

Solvent 2316 [CITGO]

Chem. Descrip.: Light aliphatic solvent naptha (70-80%), toluene (≈ 25%)

UN 1268

Uses: Solvent

Regulatory: SARA §302/304 nonreportable, §311/312 acute health/chronic health/fire hazard, §313 reportable [toluene (25-25.05%)]; Calif. Prop. 65 reportable (benzene, toluene); may contain detectable amts. of methylcyclopentane, n-hexane, cyclohexane

Properties: Colorless transparent liq.; lt. paraffinic hydrocarbon odor; negligible sol. in cold water (< 0.01%); sp.gr. 0.74; visc. < 1 cSt (40 C); vapor pressure 40 mm Hg (20 C); f.p. -135 to -90 C; b.p. 90-111 C; flash pt. (TCC) -6 to -4 C; autoignition temp. ≈ 246 C; KB value 43.3; VOC 743 g/l

Toxicology: Mist/vapor may irritate eyes, mucous membranes, respiratory tract; liq. contact may cause minimal to severe eye, mild to severe skin irritation/inflamm.; may be harmful by inh., skin absorp.; may be harmful or fatal by ing.; sensitizer; overexposure may cause CNS depression, other target organ effects, hearing loss; aspiration into lungs can cause pulmonary edema, chem. pneumonia; prolonged/repeated inh. may cause heart arrhythmias; possible reproductive hazard; TSCA listed

Environmental: May cause oxygen depletion in waterways; potentially toxic to water ecosystems; may be hazardous to aquatic life; may contribute to smog; prevent entry into sewers, waterways; not a DOT marine pollutant

Precaution: Flamm.; flamm. limits 1.0-7.1%; vapor may cause flash fire; vapors are heavier than air, may flash back; may create vapor/air explosion hazard in confined space; spills may create slipping hazard; floats on water; container can rupture in fire; prevent entry into basements, confined areas; bond and ground equip.; incompat. with strong acids, alkalis, oxidizers (liq. chlorine, hydrogen peroxide, oxygen); heptane will dissolve some plastics

Hazardous decomp. prods.: Combustion: smoke, CO, CO_2, possibly other harmful gases/vapors

HMIS: Health 2, Flammability 3, Reactivity 0

Storage: Store in tightly closed containers in cool, dry, well-ventilated place away from heat, sparks, ignition sources, strong oxidizing conditions

Solvent 2318 [CITGO]

Chem. Descrip.: Light aliphatic solvent naptha (65-75%), toluene (≈ 30%)
UN 1268

Uses: Solvent

Regulatory: SARA §302/304 nonreportable, §311/312 acute health/chronic health/fire hazard, §313 reportable [toluene (30-30.05%)]; Calif. Prop. 65 reportable (benzene, toluene); may contain detectable amts. of methylcyclopentane, n-hexane, cyclohexane

Properties: Colorless transparent liq.; lt. paraffinic hydrocarbon odor; negligible sol. in cold water (< 0.01%); sp.gr. 0.75; visc. < 1 cSt (40 C); vapor pressure 38.6 mm Hg (20 C); f.p. -135 to -90 C; b.p. 90-111 C; flash pt. (TCC) -5 to -3 C; autoignition temp. ≈ 246 C; KB value 46.7; VOC 752 g/l

Toxicology: Mist/vapor may irritate eyes, mucous membranes, respiratory tract; liq. contact may cause minimal to severe eye, mild to severe skin irritation/inflamm.; may be harmful by inh., skin absorp.; may be harmful or fatal by ing.; sensitizer; overexposure may cause CNS depression, other target organ effects, hearing loss; aspiration into lungs can cause pulmonary edema, chem. pneumonia; prolonged/repeated inh. may cause heart arrhythmias; possible reproductive hazard; TSCA listed

Environmental: May cause oxygen depletion in waterways; potentially toxic to water ecosystems; may be hazardous to aquatic life; may contribute to smog; prevent entry into sewers, waterways; not a DOT marine pollutant

Precaution: Flamm.; flamm. limits 1.0-7.1%; vapor may cause flash fire; vapors are heavier than air, may flash back; may create vapor/air explosion hazard in confined space; spills may create slipping hazard; floats on water; container can rupture in fire; prevent entry into basements, confined areas; bond and ground equip.; incompat. with strong acids, alkalis, oxidizers (liq. chlorine, hydrogen peroxide, oxygen); heptane will dissolve some plastics

Hazardous decomp. prods.: Combustion: smoke, CO, CO_2, possibly other harmful gases/vapors

HMIS: Health 2, Flammability 3, Reactivity 0

Storage: Store in tightly closed containers in cool, dry, well-ventilated place away from heat, sparks, ignition sources, strong oxidizing conditions

Solvent 2321 [CITGO]

Chem. Descrip.: Light aliphatic solvent naptha (60-70%), toluene (≈ 35%)
UN 1268

Uses: Solvent

Regulatory: SARA §302/304 nonreportable, §311/312 acute health/chronic health/fire hazard, §313 reportable [toluene (35-35.05%)]; Calif. Prop. 65 reportable (benzene, toluene); may contain detectable amts. of methylcyclopentane, n-hexane, cyclohexane

Properties: Colorless transparent liq.; lt. paraffinic hydrocarbon odor; negligible sol. in cold water (< 0.01%); sp.gr. 0.755-0.76; visc. < 1 cSt (40 C); vapor pressure 37.6 mm Hg (20 C); f.p. -135 to -90 C; b.p. 90-111 C; flash pt. (TCC) -4 to -2 C; autoignition temp. ≈ 246 C; KB value 49.7; VOC 760 g/l

Toxicology: Mist/vapor may irritate eyes, mucous membranes, respiratory tract; liq. contact may cause minimal to severe eye, mild to severe skin irritation/inflamm.; may be harmful by inh., skin absorp.; may be harmful or fatal by ing.; sensitizer; overexposure may cause CNS depression, other target organ effects, hearing loss; aspiration into lungs can cause pulmonary edema, chem. pneumonia; prolonged/repeated inh. may cause heart arrhythmias; possible reproductive hazard; TSCA listed

Environmental: May cause oxygen depletion in waterways; potentially toxic to water ecosystems; may be hazardous to aquatic life; may contribute to smog; prevent entry into sewers, waterways; not a DOT marine pollutant

Precaution: Flamm.; flamm. limits 1.0-7.1%; vapor may cause flash fire; vapors are heavier than air, may flash back; may create vapor/air explosion hazard in confined space; spills may create slipping hazard; floats on water; container can rupture in fire; prevent entry into basements, confined areas; bond and ground equip.; incompat. with strong acids, alkalis, oxidizers (liq. chlorine, hydrogen peroxide, oxygen); heptane will dissolve some plastics

Hazardous decomp. prods.: Combustion: smoke, CO, CO_2, possibly other harmful gases/vapors

HMIS: Health 2, Flammability 3, Reactivity 0

Storage: Store in tightly closed containers in cool, dry, well-ventilated place away from heat, sparks, ignition sources, strong oxidizing conditions

Solvent 2322 [CITGO]

Chem. Descrip.: Light aliphatic solvent naptha (55-65%), solvent-refined light naphtha (20-30%)), toluene (≈ 15%)
UN 1268

Uses: Solvent

Regulatory: SARA §302/304 nonreportable, §311/312 acute health/chronic health/fire hazard, §313 reportable [hexane (7.5-10.0%, toluene (15-15.03%)]; Calif. Prop. 65 reportable (benzene, toluene); may contain detectable amts. of methylcyclopentane, n-hexane, cyclohexane

Properties: Colorless transparent liq.; lt. paraffinic hydrocarbon odor; sl. sol. in cold water (≈ 0.09%); sp.gr. 0.72; visc. 1-2 cSt (40 C); vapor pressure 59.3 mm Hg (20 C); f.p. -108 to -90 C; b.p. 60-125 C; flash pt. (TCC) -18 to -15 C; autoignition temp. ≈ 232 C; KB value 36.5-37; VOC 720 g/l

Toxicology: Mist/vapor may irritate eyes, mucous membranes, respiratory tract; liq. contact may cause minimal to mod. eye, mild to severe skin irritation/inflamm.; may be harmful by inh., skin absorp.; may be harmful or fatal by ing.; sensitizer; overexposure may cause CNS depression, hearing loss, peripheral nervous system damage; aspiration into lungs can cause pulmonary edema, chem. pneumonia; prolonged/repeated inh. may cause heart arrhythmias; possible reproductive hazard; TSCA listed

Environmental: May cause oxygen depletion in waterways; potentially toxic to water ecosystems; may be hazardous to aquatic life; may contribute to smog; prevent entry into sewers, waterways; not a DOT marine pollutant

Precaution: Extremely flamm.; flamm. limits 1.0-7.0%; vapor may cause flash fire or explosion; vapors are heavier than air, may flash back; may create vapor/air explosion hazard in confined space; spills may create slipping hazard; floats on water; container can rupture in fire; prevent entry into basements, confined areas; bond and ground equip.; incompat. with strong acids, alkalis, oxidizers (liq. chlorine, other halogens, hydrogen peroxide, oxygen); heptane will dissolve some plastics

Hazardous decomp. prods.: Combustion: smoke, CO, CO_2, possibly other harmful gases/vapors

HMIS: Health 2, Flammability 3, Reactivity 0

Storage: Store in tightly closed containers in cool, dry, well-ventilated place away from heat, sparks, ignition sources, strong oxidizing conditions

Solvent 2324 [CITGO]

Chem. Descrip.: Heavy hydrotreated naphtha
CAS 64742-48-9; EINECS/ELINCS 265-150-3
UN 1268

Uses: Solvent

Regulatory: SARA §302/304 nonreportable, §311/312 acute health/chronic health/fire hazard, §313 nonreportable; Calif. Prop. 65 reportable (benzene, toluene); may contain detectable amts. of C9 aromatic hydrocarbons

Properties: Colorless transparent liq.; lt. paraffinic hydrocarbon odor; sl. sol. in cold water (< 0.05%); sp.gr. 0.77; visc. < 10 cSt (40 C); vapor pressure 3.5 mm Hg (20 C); f.p. -65 to -25 C; b.p. 125-185 C; flash pt. (TCC) 33-35 C; autoignition temp. ≈ 232 C; KB value 34; VOC 767 g/l; 100% act.

Toxicology: Mist/vapor may irritate eyes, mucous membranes, respiratory tract; liq. contact may cause minimal to mod. eye, mod. to severe skin irritation/inflamm.; may be harmful by inh., skin absorp.; may be harmful or fatal by ing.; overexposure may cause CNS depression, other target organ effects; aspiration into lungs can cause pulmonary edema, chem. pneumonia; TSCA listed

Environmental: May cause oxygen depletion in waterways; potentially toxic to water ecosystems; may be hazardous to aquatic life; may contribute to smog; prevent entry into sewers, waterways; not a DOT marine pollutant

Precaution: Flamm.; flamm. limits 0.9-6.2%; vapor may cause flash fire; vapors are heavier than air, may flash back; may create vapor/air explosion hazard in confined space; spills may create slipping hazard; floats on water; container can rupture in fire; prevent entry into basements, confined areas; bond and ground equip.; incompat. with strong acids, alkalis, oxidizers (liq. chlorine, hydrogen peroxide, oxygen)

Hazardous decomp. prods.: Combustion: smoke, CO, CO_2, possibly other harmful gases/vapors

HMIS: Health 1, Flammability 3, Reactivity 0

Storage: Store in tightly closed containers in cool, dry, well-ventilated place away from heat, sparks, ignition sources, strong oxidizing conditions

Solvent 2326 [CITGO]

Chem. Descrip.: Light aliphatic solvent naphtha (55-65%), toluene (≈ 40%)

UN 1268

Uses: Solvent

Regulatory: SARA §302/304 nonreportable, §311/312 acute health/chronic health/fire hazard, §313 reportable [toluene (40-40.05%)]; Calif. Prop. 65 reportable (benzene, toluene); may contain detectable amts. of methylcyclopentane, n-hexane, cyclohexane

Properties: Colorless transparent liq.; lt. paraffinic hydrocarbon odor; negligible sol. in cold water (< 0.01%); sp.gr. 0.76-0.77; visc. < 1 cSt (40 C); vapor pressure 36.5 mm Hg (20 C); f.p. -135 to -90 C; b.p. 90-111 C; flash pt. (TCC) -3 to -1 C; autoignition temp. ≈ 246 C; KB value 52.7; VOC 770 g/l

Toxicology: Mist/vapor may irritate eyes, mucous membranes, respiratory tract; liq. contact may cause minimal to severe eye, mild to severe skin irritation/inflamm.; may be harmful by inh., skin absorp.; may be harmful or fatal by ing.; sensitizer; overexposure may cause CNS depression, other target organ effects, hearing loss; aspiration into lungs can cause pulmonary edema, chem. pneumonia; prolonged/repeated inh. may cause heart arrhythmias; possible reproductive hazard; TSCA listed

Environmental: May cause oxygen depletion in waterways; potentially toxic to water ecosystems; may be hazardous to aquatic life; may contribute to smog; prevent entry into sewers, waterways; not a DOT marine pollutant

Precaution: Flamm.; flamm. limits 1.0-7.1%; vapor may cause flash fire; vapors are heavier than air, may flash back; may create vapor/air explosion hazard in confined space; spills may create slipping hazard; floats on water; container can rupture in fire; prevent entry into basements, confined areas; bond and ground equip.; incompat. with strong acids, alkalis, oxidizers (liq. chlorine, hydrogen peroxide, oxygen); heptane will dissolve some plastics

Hazardous decomp. prods.: Combustion: smoke, CO, CO_2, possibly other harmful gases/vapors

HMIS: Health 2, Flammability 3, Reactivity 0

Storage: Store in tightly closed containers in cool, dry, well-ventilated place away from heat, sparks, ignition sources, strong oxidizing conditions

Solvent 2329 [CITGO]

Chem. Descrip.: Solvent-refined light naphtha (70-80%), light aliphatic solvent naphtha (20-30%)

UN 1268

Uses: Solvent

Regulatory: SARA §302/304 nonreportable, §311/312 acute health/chronic health/fire hazard, §313 reportable [n-hexane (22-30%), cyclohexane (0.01-2.25%)]; Calif. Prop. 65 reportable (benzene, toluene); may contain detectable amts. of methylcyclopentane, n-hexane, cyclohexane

Properties: Colorless transparent liq.; lt. hydrocarbon odor; sl. sol. in cold water (≈ 0.09%); sp.gr. 0.683; visc. 1-2 cSt (40 C); vapor pressure 97 mm Hg (20 C); f.p. -108 to -57 C; b.p. 60-101 C; flash pt. (TCC) -19 to -17C; autoignition temp. ≈ 232 C; KB value 29.6; VOC 683 g/l

Toxicology: Mist/vapor may irritate eyes, mucous membranes, respiratory tract; liq. contact may cause minimal to mild eye, mild to mod. skin irritation/inflamm.; may be harmful by inh., skin absorp.; may be harmful or fatal by ing.; sensitizer; overexposure may cause CNS depression, other target organ effects, peripheral nervous system damage; aspiration into lungs can cause pulmonary edema, chem. pneumonia; prolonged/repeated inh. may cause heart arrhythmias; possible reproductive hazard; TSCA listed

Environmental: May cause oxygen depletion in waterways; potentially toxic to water ecosystems; may be hazardous to aquatic life; may contribute to smog; prevent entry into sewers, waterways; not a DOT marine pollutant

Precaution: Extremely flamm.; flamm. limits 1.0-7.0%; vapor may cause flash fire or explosion; vapors are heavier than air, may flash back; may create vapor/air explosion hazard in confined space; spills may create slipping hazard; floats on water; container can rupture in fire; prevent entry into basements, confined areas; bond and ground equip.; incompat. with strong acids, alkalis, oxidizers (liq. chlorine, other halogens, hydrogen peroxide, oxygen); heptane will dissolve some plastics

Hazardous decomp. prods.: Combustion: smoke, CO, CO_2, possibly other harmful gases/vapors

HMIS: Health 2, Flammability 3, Reactivity 0

Storage: Store in tightly closed containers in cool, dry, well-ventilated place away from heat, sparks, ignition sources, strong oxidizing conditions

Solvent 2331 [CITGO]

Chem. Descrip.: Light aliphatic solvent naphtha (65-75%), solvent-refined light naphtha (15-15%), toluene (≈ 10%)

UN 1268

Uses: Solvent

Regulatory: SARA §302/304 nonreportable, §311/312 acute health/chronic health/fire hazard, §313 reportable [n-hexane (6-8.07%), toluene (10-10.035%)]; Calif. Prop. 65 reportable (benzene, toluene); may contain detectable amts. of methylcyclopentane, n-hexane, cyclohexane

Properties: Colorless transparent liq.; lt. hydrocarbon odor; sl. sol. in cold water (≈ 0.09%); sp.gr. 0.71-0.715; visc. 1-2 cSt (40 C); vapor pressure 56.9 mm Hg (20 C); f.p. -108 to -90 C; b.p. 82-105 C; flash pt. (TCC) -17 to -15C; autoignition temp. ≈ 232 C; KB value 34.2; VOC 713 g/l

Toxicology: Mist/vapor may irritate eyes, mucous membranes, respiratory tract; liq. contact may cause minimal to mod. eye, mild to severe skin irritation/inflamm.; may be harmful by inh., skin absorp.; may be harmful or fatal by ing.; sensitizer; overexposure may cause CNS depression, hearing loss, peripheral nervous system damage; aspiration into lungs can cause pulmonary edema, chem. pneumonia; prolonged/repeated inh. may cause heart arrhythmias; possible reproductive hazard; TSCA listed

Environmental: May cause oxygen depletion in waterways; potentially toxic to water ecosystems; may be hazardous to aquatic life; may contribute to smog; prevent entry into sewers, waterways; not a DOT marine pollutant

Precaution: Extremely flamm.; flamm. limits 1.0-7.0%; vapor may cause flash fire or explosion; vapors are heavier than air, may flash back; may create vapor/air explosion hazard in confined space; spills may create slipping hazard; floats on water; container can rupture in fire; prevent entry into basements, confined areas; bond and ground equip.; incompat. with strong acids, alkalis, oxidizers (liq. chlorine, other halogens, hydrogen peroxide, oxygen); heptane will dissolve some plastics

Hazardous decomp. prods.: Combustion: smoke, CO, CO_2, possibly other harmful gases/vapors

HMIS: Health 2, Flammability 3, Reactivity 0

Storage: Store in tightly closed containers in cool, dry, well-ventilated place away from heat, sparks, ignition sources, strong oxidizing conditions

Solvent 2332 [CITGO]

Chem. Descrip.: Heptanes mixt. (83-89%), solvent-refined light naphtha (5-15%), n-hexane (2-4%)

UN 1268

Uses: Solvent

Regulatory: SARA §302/304 nonreportable, §311/312 acute health/chronic health/fire hazard, §313 reportable [n-hexane (2.4-3.6%)]; Calif. Prop. 65 reportable (benzene, toluene); may contain detectable amts. of n-hexane, cyclohexane

Properties: Colorless transparent liq.; lt. hydrocarbon odor; sl. sol. in cold water (≈ 0.1%); sp.gr. 0.69-0.70; visc. 1-2 cSt (40 C); vapor pressure 51 mm Hg (20 C); f.p. -108 to -90 C; b.p. 60-99 C; flash pt. (TCC) -14 to -13 C; autoignition temp. ≈ 232 C; KB value 30; VOC 700 g/l

Toxicology: Mist/vapor may irritate eyes, mucous membranes, respiratory tract; liq. contact may cause minimal to mild eye, mild to mod. skin irritation/inflamm.; may be harmful by inh., skin absorp.; may be harmful or fatal by ing.; sensitizer; overexposure may cause CNS depression, peripheral nervous system damage; aspiration into lungs can cause pulmonary edema, chem. pneumonia; prolonged/repeated inh. may cause heart arrhythmias; possible reproductive hazard; TSCA listed

Environmental: May cause oxygen depletion in waterways; potentially toxic to water ecosystems; may be hazardous to aquatic life; may contribute to smog; prevent entry into sewers, waterways; not a DOT marine pollutant

Precaution: Extremely flamm.; flamm. limits 1.0-7.0%; vapor may cause flash fire or explosion; vapors are heavier than air, may flash back; may create vapor/air explosion hazard in confined space; spills may create slipping hazard; floats on water; container can rupture in fire; prevent entry into basements, confined areas; bond and ground equip.; incompat. with strong acids, alkalis, oxidizers (liq. chlorine, other halogens, hydrogen peroxide, oxygen); heptane will dissolve some plastics

Hazardous decomp. prods.: Combustion: smoke, CO, CO_2, possibly other harmful gases/vapors

HMIS: Health 2, Flammability 3, Reactivity 0

Storage: Store in tightly closed containers in cool, dry, well-ventilated place away from heat, sparks, ignition sources, strong oxidizing conditions

Solvent 2335 [CITGO]

Chem. Descrip.: Solvent-refined light naphtha (85-95%), light aliphatic solvent naphtha (5-15%)

UN 1268

Uses: Solvent

Regulatory: SARA §302/304 nonreportable, §311/312 acute health/chronic health/fire hazard, §313 reportable [n-hexane (27-36%), cyclohexane (0.01-2.71%)]; Calif. Prop. 65 reportable (benzene, toluene); may contain detectable amts. of methylcyclopentane, n-hexane, cyclohexane

Properties: Colorless transparent liq.; lt. hydrocarbon odor; sl. sol. in cold water (≈ 0.09%); sp.gr. 0.68; visc. 1-2 cSt (40 C); vapor pressure 108 mm Hg (20 C); f.p. -108 to -57 C; b.p. 60-125 C; flash pt. (TCC) -20 to -17 C; autoignition temp. ≈ 232 C; KB value 29.7; VOC 679 g/l

Toxicology: Mist/vapor may irritate eyes, mucous membranes, respiratory tract; liq. contact may cause minimal to mild eye, mild to mod. skin irritation/inflamm.; may be harmful by inh., skin absorp.; may be harmful or fatal by ing.; sensitizer; overexposure may cause CNS depression, peripheral nervous system damage; aspiration into lungs can cause pulmonary edema, chem. pneumonia; prolonged/repeated inh. may cause heart arrhythmias; possible reproductive hazard; TSCA listed

Environmental: May cause oxygen depletion in waterways; potentially toxic to water ecosystems; may be hazardous to aquatic life; may contribute to smog; prevent entry into sewers, waterways; not a DOT marine pollutant

Precaution: Extremely flamm.; flamm. limits 1.0-7.0%; vapor may cause flash fire or explosion; vapors are heavier than air, may flash back; may create vapor/air explosion hazard in confined space; spills may create slipping hazard; floats on water; container can rupture in fire; prevent entry into basements, confined areas; bond and ground equip.; incompat. with strong acids, alkalis, oxidizers (liq. chlorine, other halogens, hydrogen peroxide, oxygen); heptane will dissolve some plastics

Hazardous decomp. prods.: Combustion: smoke, CO, CO_2, possibly other harmful gases/vapors

HMIS: Health 2, Flammability 3, Reactivity 0

Storage: Store in tightly closed containers in cool, dry, well-ventilated place away from heat, sparks, ignition sources, strong oxidizing conditions

Solvent 2336 [CITGO]

Chem. Descrip.: Solvent-refined light naphtha (60-75%), light aliphatic solvent naphtha (15-25%), heavy hydrotreated naphtha (5-10%), toluene (≈ 5%)

UN 1268

Uses: Solvent

Regulatory: SARA §302/304 nonreportable, §311/312 acute health/chronic health/fire hazard, §313 reportable [n-hexane (20.402-27.22%), cyclohexane (0.0088-2.06%), toluene (5-5.0125%)]; Calif. Prop. 65 reportable (benzene, toluene); may contain detectable amts. of methylcyclopentane, n-hexane, cyclohexane

Properties: Colorless transparent liq.; lt. hydrocarbon odor; sl. sol. in cold water (≈ 0.08%); sp.gr. 0.70; visc. 1-2 cSt (40 C); vapor pressure 89 mm Hg (20 C); f.p. -108 to -25 C; b.p. 63-140 C; flash pt. (TCC) -15 to -12 C; autoignition temp. ≈ 232 C; KB value 32.6; VOC 697 g/l

Toxicology: Mist/vapor may irritate eyes, mucous membranes, respiratory tract; liq. contact may cause minimal to mod. eye, mild to severe skin irritation/inflamm.; may be harmful by inh., skin absorp.; may be harmful or fatal by ing.; sensitizer; overexposure may cause CNS depression, hearing loss, peripheral nervous system damage; aspiration into lungs can cause pulmonary edema, chem. pneumonia; prolonged/repeated inh. may cause heart arrhythmias; possible reproductive hazard; TSCA listed

Environmental: May cause oxygen depletion in waterways; potentially toxic to water ecosystems; may be hazardous to aquatic life; may contribute to smog; prevent entry into sewers, waterways; not a DOT marine pollutant

Precaution: Extremely flamm.; flamm. limits 1.0-7.0%; vapor may cause flash fire or explosion; vapors are heavier than air, may flash back; may create vapor/air explosion hazard in confined space; spills may create slipping hazard; floats on water; container can rupture in fire; prevent entry into basements, confined areas; bond and ground equip.; incompat. with strong acids, alkalis, oxidizers (liq. chlorine, other halogens, hydrogen peroxide, oxygen); heptane will dissolve some plastics

Hazardous decomp. prods.: Combustion: smoke, CO, CO_2, possibly other harmful gases/vapors

HMIS: Health 2, Flammability 3, Reactivity 0

Storage: Store in tightly closed containers in cool, dry, well-ventilated place away from heat, sparks, ignition sources, strong oxidizing conditions

Solvent 2337 [CITGO]

Chem. Descrip.: Light aliphatic solvent naphtha (55-65%), heavy hydrotreated naphtha (20-30%), toluene (≈ 15%)

UN 1268

Uses: Solvent

Regulatory: SARA §302/304 nonreportable, §311/312 acute health/chronic health/fire hazard, §313 reportable [toluene (15-15.033%)]; Calif. Prop. 65 reportable (benzene, toluene); may contain detectable amts. of methylcyclopentane, n-hexane, cyclohexane

Properties: Colorless transparent liq.; lt. hydrocarbon odor; sl. sol. in cold water (≈ 0.05%); sp.gr. 0.735-0.74; visc. 0.5-1.5 cSt (40 C); vapor pressure 31.9 mm Hg (20 C); f.p. -108 to -25 C; b.p. 95-142 C; flash pt. (TCC) -6 to -4 C; autoignition temp. ≈ 246 C; KB value 37.7; VOC 740 g/l

Toxicology: Mist/vapor may irritate eyes, mucous membranes, respiratory tract; liq. contact may cause minimal to severe eye, mild to severe skin irritation/inflamm.; may be harmful by inh., skin absorp.; may be harmful or fatal by ing.; sensitizer; overexposure may cause CNS depression, other target organ effects, hearing loss; aspiration into lungs can cause pulmonary edema, chem. pneumonia; prolonged/repeated inh. may cause heart arrhythmias; possible reproductive hazard; TSCA listed

Environmental: May cause oxygen depletion in waterways; potentially toxic to water ecosystems; may be hazardous to aquatic life; may contribute to smog; prevent entry into sewers, waterways; not a DOT marine pollutant

Precaution: Flamm.; flamm. limits 1.0-7.1%; vapor may cause flash fire; vapors are heavier than air, may flash back; may create vapor/air explosion hazard in confined space; spills may create slipping hazard; floats on water; container can rupture in fire; prevent entry into basements, confined areas; bond and ground equip.; incompat. with strong acids, alkalis, oxidizers (liq. chlorine, other halogens, hydrogen peroxide, oxygen); heptane will dissolve some plastics

Hazardous decomp. prods.: Combustion: smoke, CO, CO_2, possibly other harmful gases/vapors

HMIS: Health 2, Flammability 3, Reactivity 0

Storage: Store in tightly closed containers in cool, dry, well-ventilated place away from heat, sparks, ignition sources, strong oxidizing conditions

Solvent 2338 [CITGO]

Chem. Descrip.: Solvent-refined light naphtha (65-76%), light aliphatic solvent naphtha (20-30%), toluene (≈ 5%)

UN 1268

Uses: Solvent; camping fuel

Regulatory: SARA §302/304 nonreportable, §311/312 acute health/chronic health/fire hazard, §313 reportable [n-hexane (21-28.025%), cyclohexane (0.0095-2.125%, toluene (5-5.0125%)]; Calif. Prop. 65 reportable (benzene, toluene); may contain detectable amts. of methylcyclopentane, n-hexane, cyclohexane

Properties: Colorless transparent liq.; lt. hydrocarbon odor; sl. sol. in cold water (≈ 0.09%); sp.gr. 0.69-0.70; visc. 1-2 cSt (40 C); vapor pressure 100 mm Hg (20 C); f.p. -108 to -90 C; b.p. 60-125 C; flash pt. (TCC) -18 to -15 C; autoignition temp. ≈ 232 C; KB value 31.8; VOC 710 g/l

Toxicology: Mist/vapor may irritate eyes, mucous membranes, respiratory tract; liq. contact may cause minimal to mod. eye, mild to severe skin irritation/inflamm.; may be harmful by inh., skin absorp.; may be harmful or fatal by ing.; sensitizer; overexposure may cause CNS depression, hearing loss, peripheral nervous system damage; aspiration into lungs can cause pulmonary edema, chem. pneumonia; prolonged/ repeated inh. may cause heart arrhythmias; possible reproductive hazard; TSCA listed

Environmental: May cause oxygen depletion in waterways; potentially toxic to water ecosystems; may be hazardous to aquatic life; may contribute to smog; prevent entry into sewers, waterways; not a DOT marine pollutant

Precaution: Extremely flamm.; flamm. limits 1.0-7.0%; vapor may cause flash fire or explosion; vapors are heavier than air, may flash back; may create vapor/air explosion hazard in confined space; spills may create slipping hazard; floats on water; container can rupture in fire; prevent entry into basements, confined areas; bond and ground equip.; incompat. with strong acids, alkalis, oxidizers (liq. chlorine, other halogens, hydrogen peroxide, oxygen); heptane will dissolve some plastics

Hazardous decomp. prods.: Combustion: smoke, CO, CO_2, possibly other harmful gases/vapors

HMIS: Health 2, Flammability 3, Reactivity 0

Storage: Store in tightly closed containers in cool, dry, well-ventilated place away from heat, sparks, ignition sources, strong oxidizing conditions

Solvent 2339 [CITGO]

Chem. Descrip.: Heavy hydrotreated naphtha (95-97%), xylenes mixed isomers (3-5%), ethylbenzene (0.25-1.25%)

UN 1268

Uses: Solvent

Regulatory: SARA §302/304 nonreportable, §311/312 acute health/chronic health/fire hazard, §313 reportable [xylenes (3.75-4.5%), ethylbenzene (0.25-1.25%)]; Calif. Prop. 65 reportable (benzene, toluene); may contain detectable amts. of 1,3,5-trimethylbenzene

Properties: Colorless transparent liq.; lt. hydrocarbon odor; sl. sol. in cold water (≈ 0.05%); sp.gr. 0.77-0.78; visc. < 10 cSt (40 C); vapor pressure 3.65 mm Hg (20 C); f.p. -57 to -25 C; b.p. 140-175 C; flash pt. (TCC) 33-34 C; autoignition temp. ≈ 232 C; KB value 36-37; VOC 775 g/l

Toxicology: Mist/vapor may irritate eyes, mucous membranes, respiratory tract; liq. contact may cause mild to mod. eye, mod. to severe skin irritation/inflamm.; may be harmful by inh., skin absorp.; may be harmful or fatal by ing.; sensitizer; overexposure may cause CNS depression, other target organ effects; aspiration into lungs can cause pulmonary edema, chem. pneumonia; prolonged/repeated inh. may cause heart arrhythmias; possible cancer/reproductive hazard; TSCA listed

Environmental: May cause oxygen depletion in waterways; potentially toxic to water ecosystems; may be hazardous to aquatic life; may contribute to smog; prevent entry into sewers, waterways; not a DOT marine pollutant

Precaution: Flamm.; flamm. limits 0.9-6.3%; vapor may cause flash fire; vapors are heavier than air, may flash back; may create vapor/air explosion hazard in confined space; spills may create slipping hazard; floats on water; container can rupture in fire; prevent entry into basements, confined areas; bond and ground equip.; incompat. with strong acids, alkalis, oxidizers (liq. chlorine, other halogens, hydrogen peroxide, oxygen)

Hazardous decomp. prods.: Combustion: smoke, CO, CO_2, possibly other harmful gases/vapors

HMIS: Health 2, Flammability 3, Reactivity 0

Storage: Store in tightly closed containers in cool, dry, well-ventilated place away from heat, sparks, ignition sources, strong oxidizing conditions

Solvent 2340 [CITGO]

Chem. Descrip.: Light aliphatic solvent naphtha (70-80%), n-heptane (38-46%), methylcyclohexane (2-7%), octanes (4-13%)

UN 1268

Uses: Solvent

Regulatory: SARA §302/304 nonreportable, §311/312 acute health/chronic health/fire hazard, §313 nonreportable; Calif. Prop. 65 reportable (benzene, toluene); may contain detectable amts. of methylcyclopentane, n-hexane, cyclohexane

Properties: Colorless transparent liq.; lt. hydrocarbon odor; negligible sol. in cold water (< 0.01%); sp.gr. 0.70; visc. < 1 cSt (40 C); vapor pressure 45 mm Hg (20 C); f.p. -135 to -90 C; b.p. 90-110 C; flash pt. (TCC) -9 to -7 C; autoignition temp. ≈ 246 C; KB value 29; VOC 699 g/l

Toxicology: Mist/vapor may irritate eyes, mucous membranes, respiratory tract; liq. contact may cause minimal to mild eye, mild to mod. skin irritation/inflamm.; may be harmful by inh., skin absorp.; may be harmful or fatal by ing.; sensitizer; overexposure may cause CNS depression, other target organ effects; aspiration into lungs can cause pulmonary edema, chem. pneumonia; prolonged/repeated inh. may cause heart arrhythmias; TSCA listed

Environmental: May cause oxygen depletion in waterways; potentially toxic to water ecosystems; may be hazardous to aquatic life; may contribute to smog; prevent entry into sewers, waterways; not a DOT marine pollutant

Precaution: Extremely flamm.; flamm. limits 1.0-6.7%; vapor may cause flash fire or explosion; vapors are heavier than air, may flash back; may create vapor/air explosion hazard in confined space; spills may create slipping hazard; floats on water; container can rupture in fire; prevent entry into basements, confined areas; bond and ground equip.; incompat. with strong acids, alkalis, oxidizers (liq. chlorine, hydrogen peroxide, oxygen); heptane will dissolve some plastics

Hazardous decomp. prods.: Combustion: smoke, CO, CO_2, possibly other harmful gases/vapors

HMIS: Health 1, Flammability 3, Reactivity 0

Storage: Store in tightly closed containers in cool, dry, well-ventilated place away from heat, sparks, ignition sources, strong oxidizing conditions

Solvent 2341 [CITGO]

Chem. Descrip.: Heavy hydrodesulfurized naphtha (40-50%), heavy hydrotreated naphtha (40-50%), light aliphatic solvent naphtha (5-10%), heptanes (6-8%), toluene (≈ 2%)

UN 1268

Uses: Solvent

Regulatory: SARA §302/304 nonreportable, §311/312 acute health/chronic health/fire hazard, §313 reportable [cyclohexane (0.001-0.3%), toluene (2%), 1,2,4-trimethylbenzene (1.08-3.24%); Calif. Prop. 65 reportable (benzene, toluene); may contain detectable amts. of n-hexane, cyclohexane

Properties: Colorless transparent liq.; lt. hydrocarbon odor; sl. sol. in cold water (≈ 0.02%); sp.gr. 0.785-0.79; visc. 2-3 cSt (40 C); vapor pressure 3.8 mm Hg (20 C); f.p. -91 to -25 C; b.p. 90-215 C; flash pt. (TCC) 27-30 C; autoignition temp. ≈ 232 C; KB value 38; VOC 780 g/l

Toxicology: Mist/vapor may irritate eyes, mucous membranes, respiratory tract; liq. contact may cause mild to mod. eye, mod. to severe skin irritation/inflamm.; may be harmful by inh., skin absorp.; may be harmful or fatal by ing.; sensitizer; overexposure may cause CNS depression, other target organ effects; aspiration into lungs can cause pulmonary edema, chem. pneumonia; prolonged/repeated inh. may cause heart arrhythmias; possible reproductive hazard; TSCA listed

Environmental: May cause oxygen depletion in waterways; potentially toxic to water ecosystems; may be hazardous to aquatic life; may contribute to smog; prevent entry into sewers, waterways; may be a DOT marine pollutant

Precaution: Flamm.; flamm. limits 0.6-7.0%; vapor may cause flash fire; vapors are heavier than air, may flash back; may create vapor/air explosion hazard in confined space; spills may create slipping hazard; floats on water; container can rupture in fire; prevent entry into basements, confined areas; bond and ground equip.; incompat. with strong acids, alkalis, oxidizers (liq. chlorine, other halogens, hydrogen peroxide, oxygen); heptane will dissolve some plastics

Hazardous decomp. prods.: Combustion: smoke, CO, CO_2, possibly other harmful gases/vapors

HMIS: Health 2, Flammability 3, Reactivity 0

Storage: Store in tightly closed containers in cool, dry, well-ventilated place away from heat, sparks, ignition sources, strong oxidizing conditions

Solvent 2350 [CITGO]
Chem. Descrip.: Stoddard solvent
CAS 8052-41-3; EINECS/ELINCS 232-489-3
UN 1268
Uses: Solvent
Regulatory: SARA §302/304 nonreportable, §311/312 acute health/chronic health/fire hazard, §313 nonreportable; Calif. Prop. 65 reportable (benzene, toluene); may contain detectable amts. of cumene, other C9 aromatic hydrocarbons
Properties: Colorless transparent liq.; lt. hydrocarbon odor; sl. sol. in cold water (< 0.05%); sp.gr. 0.78; visc. < 20 cSt (40 C); vapor pressure 2.5 mm Hg (20 C); f.p. -65 to -25 C; b.p. 149-204 C; flash pt. (TCC) 41-43 C; autoignition temp. ≈ 230 C; KB value 35; VOC 780 g/l; 100% act.
Toxicology: ACGIH TLV TWA 100 ppm; LD50 (oral, rat) > 34,600 mg/kg; mist/vapor may irritate eyes, mucous membranes, respiratory tract; liq. contact may cause minimal to mod. eye, mod. to severe skin irritation/inflamm.; may be harmful by inh., skin absorp.; may be harmful or fatal by ing.; overexposure may cause CNS depression, other target organ effects; aspiration into lungs can cause pulmonary edema, chem. pneumonia; possible reproductive hazard; TSCA listed
Environmental: May cause oxygen depletion in waterways; potentially toxic to water ecosystems; may be hazardous to aquatic life; may contribute to smog; prevent entry into sewers, waterways; not a DOT marine pollutant
Precaution: Combustible; flamm. limits 0.6-6.0%; vapor may cause flash fire; vapors are heavier than air, may flash back; may create vapor/air explosion hazard in confined space; spills may create slipping hazard; floats on water; container can rupture in fire; prevent entry into basements, confined areas; bond and ground equip.; incompat. with strong acids, alkalis, oxidizers (liq. chlorine, hydrogen peroxide, oxygen)
Hazardous decomp. prods.: Combustion: smoke, CO, CO_2, possibly other harmful gases/vapors
HMIS: Health 2, Flammability 2, Reactivity 0
Storage: Store in tightly closed containers in cool, dry, well-ventilated place away from heat, sparks, ignition sources, strong oxidizing conditions

Solvent 2351 [CITGO]
Chem. Descrip.: Stoddard solvent
CAS 8052-41-3; EINECS/ELINCS 232-489-3
UN 1268
Uses: Solvent
Regulatory: SARA §302/304 nonreportable, §311/312 acute health/chronic health/fire hazard, §313 reportable [1,2,4,-trimethylbenzene (0.36-1.1%)]; Calif. Prop. 65 reportable (benzene, toluene); may contain detectable amts. of cumene, other C9 aromatic hydrocarbons
Properties: Colorless transparent liq.; lt. hydrocarbon odor; sl. sol. in cold water (< 0.05%); sp.gr. 0.78; visc. < 20 cSt (40 C); vapor pressure 2.5 mm Hg (20 C); f.p. -65 to -25 C; b.p. 149-204 C; flash pt. (TCC) 41-43 C; autoignition temp. ≈ 230 C; KB value 34; VOC 780 g/l; 100% act.
Toxicology: ACGIH TLV TWA 100 ppm; LD50 (oral, rat) > 34,600 mg/kg; mist/vapor may irritate eyes, mucous membranes, respiratory tract; liq. contact may cause minimal to mod. eye, mod. to severe skin irritation/inflamm.; may be harmful by inh., skin absorp.; may be harmful or fatal by ing.; overexposure may cause CNS depression, other target organ effects; aspiration into lungs can cause pulmonary edema, chem. pneumonia; possible reproductive hazard; TSCA listed
Environmental: May cause oxygen depletion in waterways; potentially toxic to water ecosystems; may be hazardous to aquatic life; may contribute to smog; prevent entry into sewers, waterways; not a DOT marine pollutant
Precaution: Combustible; flamm. limits 0.6-6.0%; vapor may cause flash fire; vapors are heavier than air, may flash back; may create vapor/air explosion hazard in confined space; spills may create slipping hazard; floats on water; container can rupture in fire; prevent entry into basements, confined areas; bond and ground equip.; incompat. with strong acids, alkalis, oxidizers (liq. chlorine, hydrogen peroxide, oxygen)
Hazardous decomp. prods.: Combustion: smoke, CO, CO_2, possibly other harmful gases/vapors
HMIS: Health 2, Flammability 2, Reactivity 0
Storage: Store in tightly closed containers in cool, dry, well-ventilated place away from heat, sparks, ignition sources, strong oxidizing conditions

Solvent 2352 [CITGO]
Chem. Descrip.: Stoddard solvent
CAS 8052-41-3; EINECS/ELINCS 232-489-3
UN 1268
Uses: Solvent
Regulatory: SARA §302/304 nonreportable, §311/312 acute health/chronic health/fire hazard, §313 reportable [1,2,4,-trimethylbenzene (0.8-2.6%)]; Calif. Prop. 65 reportable (benzene, toluene); may contain detectable amts. of n-propylbenzene, cumene, other C9 aromatic hydrocarbons
Properties: Colorless transparent liq.; lt. hydrocarbon odor; sl. sol. in cold water (< 0.05%); sp.gr. 0.78-0.79; visc. < 20 cSt (40 C); vapor pressure 2.5 mm Hg (20 C); f.p. -65 to -25 C; b.p. 149-204 C; flash pt. (TCC) 41-43 C; autoignition temp. ≈ 230 C; KB value 35; VOC 784 g/l; 100% act.
Toxicology: ACGIH TLV TWA 100 ppm; LD50 (oral, rat) > 34,600 mg/kg; mist/vapor may irritate eyes, mucous membranes, respiratory tract; liq. contact may cause minimal to mod. eye, mod. to severe skin irritation/inflamm.; may be harmful by inh., skin absorp.; may be harmful or fatal by ing.; overexposure may cause CNS depression, other target organ effects; aspiration into lungs can cause pulmonary edema, chem. pneumonia; possible reproductive hazard; TSCA listed
Environmental: May cause oxygen depletion in waterways; potentially toxic to water ecosystems; may be hazardous to aquatic life; may contribute to smog; prevent entry into sewers, waterways; not a DOT marine pollutant
Precaution: Combustible; flamm. limits 0.6-6.0%; vapor may cause flash fire; vapors are heavier than air, may flash back; may create vapor/air explosion hazard in confined space; spills may create slipping hazard; floats on water; container can rupture in fire; prevent entry into basements, confined areas; bond and ground equip.; incompat. with strong acids, alkalis, oxidizers (liq. chlorine, hydrogen peroxide, oxygen)
Hazardous decomp. prods.: Combustion: smoke, CO, CO_2, possibly other harmful gases/vapors
HMIS: Health 2, Flammability 2, Reactivity 0
Storage: Store in tightly closed containers in cool, dry, well-ventilated place away from heat, sparks, ignition sources, strong oxidizing conditions

Solvent 2359 [CITGO]
Chem. Descrip.: Stoddard solvent
CAS 8052-41-3; EINECS/ELINCS 232-489-3
UN 1268
Uses: Solvent
Regulatory: SARA §302/304 nonreportable, §311/312 acute health/chronic health/fire hazard, §313 reportable [1,2,4,-trimethylbenzene (0.5-2%)]; Calif. Prop. 65 reportable (benzene, toluene); may contain detectable amts. of n-propylbenzene, cumene, other C9 aromatic hydrocarbons
Properties: Colorless transparent liq.; lt. hydrocarbon odor; sl. sol. in cold water (< 0.05%); sp.gr. 0.78; visc. < 20 cSt (40 C); vapor pressure 2.0 mm Hg (20 C); f.p. -65 to -25 C; b.p. 149-213 C; flash pt. (TCC) 42-45 C; autoignition temp. ≈ 230 C; KB value 35; VOC 776 g/l; 100% act.
Toxicology: ACGIH TLV TWA 100 ppm; LD50 (oral, rat) > 34,600 mg/kg; mist/vapor may irritate eyes, mucous membranes, respiratory tract; liq. contact may cause minimal to mod. eye, mod. to severe skin irritation/inflamm.; may be harmful by inh., skin absorp.; may be harmful or fatal by ing.; overexposure may cause CNS depression, other target organ effects; aspiration into lungs can cause pulmonary edema, chem. pneumonia; possible reproductive hazard; TSCA listed
Environmental: May cause oxygen depletion in waterways; potentially toxic to water ecosystems; may be hazardous to aquatic life; may contribute to smog; prevent entry into sewers, waterways; not a DOT marine pollutant
Precaution: Combustible; flamm. limits 0.6-6.0%; vapor may cause flash fire; vapors are heavier than air, may flash back; may create vapor/air explosion hazard in confined space; spills may create slipping hazard; floats on water; container can rupture in fire; prevent entry into basements, confined areas; bond and ground equip.; incompat. with strong acids, alkalis, oxidizers (liq. chlorine, hydrogen peroxide, oxygen)
Hazardous decomp. prods.: Combustion: smoke, CO, CO_2, possibly other harmful gases/vapors
HMIS: Health 2, Flammability 2, Reactivity 0

Storage: Store in tightly closed containers in cool, dry, well-ventilated place away from heat, sparks, ignition sources, strong oxidizing conditions

Solvent 2361 [CITGO]

Chem. Descrip.: Stoddard solvent
CAS 8052-41-3; EINECS/ELINCS 232-489-3
UN 1268
Uses: Solvent
Regulatory: SARA §302/304 nonreportable, §311/312 acute health/chronic health/fire hazard, §313 nonreportable; Calif. Prop. 65 nonreportable; may contain detectable amts. of n-propylbenzene, cumene, other C9 aromatic hydrocarbons
Properties: Colorless transparent liq.; lt. hydrocarbon odor; sl. sol. in cold water (< 0.05%); sp.gr. 0.78; visc. < 20 cSt (40 C); vapor pressure 2.7 mm Hg (20 C); f.p. -65 to -25 C; b.p. 149-204 C; flash pt. (TCC) 42-44 C; autoignition temp. ≈ 230 C; KB value 35; VOC 782 g/l; 100% act.
Toxicology: ACGIH TLV TWA 100 ppm; LD50 (oral, rat) > 34,600 mg/kg; mist/vapor may irritate eyes, mucous membranes, respiratory tract; liq. contact may cause minimal to mod. eye, mod. to severe skin irritation/inflamm.; may be harmful by inh., skin absorp.; may be harmful or fatal by ing.; overexposure may cause CNS depression, other target organ effects; aspiration into lungs can cause pulmonary edema, chem. pneumonia; possible reproductive hazard; TSCA listed
Environmental: May cause oxygen depletion in waterways; potentially toxic to water ecosystems; may be hazardous to aquatic life; may contribute to smog; prevent entry into sewers, waterways; not a DOT marine pollutant
Precaution: Combustible; flamm. limits 0.6-6.0%; vapor may cause flash fire; vapors are heavier than air, may flash back; may create vapor/air explosion hazard in confined space; spills may create slipping hazard; floats on water; container can rupture in fire; prevent entry into basements, confined areas; bond and ground equip.; incompat. with strong acids, alkalis, oxidizers (liq. chlorine, hydrogen peroxide, oxygen)
Hazardous decomp. prods.: Combustion: smoke, CO, CO_2, possibly other harmful gases/vapors
HMIS: Health 2, Flammability 2, Reactivity 0
Storage: Store in tightly closed containers in cool, dry, well-ventilated place away from heat, sparks, ignition sources, strong oxidizing conditions

Solvent 2363 [CITGO]

Chem. Descrip.: Stoddard solvent
CAS 8052-41-3; EINECS/ELINCS 232-489-3
UN 1268
Uses: Solvent
Regulatory: SARA §302/304 nonreportable, §311/312 acute health/chronic health/fire hazard, §313 reportable [1,2,4-trimethylbenzene (1-7%)]; Calif. Prop. 65 reportable (benzene, toluene); may contain detectable amts. of n-propylbenzene, cumene, other C9 aromatic hydrocarbons
Properties: Colorless transparent liq.; lt. hydrocarbon odor; sl. sol. in cold water (< 0.05%); sp.gr. 0.79; visc. < 20 cSt (40 C); vapor pressure 2.1 mm Hg (20 C); f.p. -65 to -25 C; b.p. 149-213 C; flash pt. (TCC) 40-43 C; autoignition temp. ≈ 230 C; KB value 41; VOC 786 g/l; 100% act.
Toxicology: ACGIH TLV TWA 100 ppm; LD50 (oral, rat) > 34,600 mg/kg; mist/vapor may irritate eyes, mucous membranes, respiratory tract; liq. contact may cause minimal to mod. eye, mod. to severe skin irritation/inflamm.; may be harmful by inh., skin absorp.; may be harmful or fatal by ing.; overexposure may cause CNS depression, other target organ effects; aspiration into lungs can cause pulmonary edema, chem. pneumonia; possible reproductive hazard; TSCA listed
Environmental: May cause oxygen depletion in waterways; potentially toxic to water ecosystems; may be hazardous to aquatic life; may contribute to smog; prevent entry into sewers, waterways; may be a DOT marine pollutant
Precaution: Combustible; flamm. limits 0.8-6.0%; vapor may cause flash fire; vapors are heavier than air, may flash back; may create vapor/air explosion hazard in confined space; spills may create slipping hazard; floats on water; container can rupture in fire; prevent entry into basements, confined areas; bond and ground equip.; incompat. with strong acids, alkalis, oxidizers (liq. chlorine, hydrogen peroxide, oxygen)
Hazardous decomp. prods.: Combustion: smoke, CO, CO_2, possibly other harmful gases/vapors
HMIS: Health 2, Flammability 2, Reactivity 0
Storage: Store in tightly closed containers in cool, dry, well-ventilated place away from heat, sparks, ignition sources, strong oxidizing conditions

Solvent 2364 [CITGO]

Chem. Descrip.: Stoddard solvent
CAS 8052-41-3; EINECS/ELINCS 232-489-3
UN 1268
Uses: Solvent
Regulatory: SARA §302/304 nonreportable, §311/312 acute health/chronic health/fire hazard, §313 reportable [1,2,4-trimethylbenzene (0.6-1.8%)]; Calif. Prop. 65 reportable (benzene, toluene); may contain detectable amts. of cumene, other C9 aromatic hydrocarbons
Properties: Colorless transparent liq.; lt. hydrocarbon odor; sl. sol. in cold water (< 0.05%); sp.gr. 0.79; visc. < 20 cSt (40 C); vapor pressure 4.0 mm Hg (20 C); f.p. -65 to -25 C; b.p. 149-204 C; flash pt. (TCC) 41-43 C; autoignition temp. ≈ 230 C; KB value 34; VOC 782 g/l; 100% act.
Toxicology: ACGIH TLV TWA 100 ppm; LD50 (oral, rat) > 34,600 mg/kg; mist/vapor may irritate eyes, mucous membranes, respiratory tract; liq. contact may cause minimal to mod. eye, mod. to severe skin irritation/inflamm.; may be harmful by inh., skin absorp.; may be harmful or fatal by ing.; overexposure may cause CNS depression, other target organ effects; aspiration into lungs can cause pulmonary edema, chem. pneumonia; possible reproductive hazard; TSCA listed
Environmental: May cause oxygen depletion in waterways; potentially toxic to water ecosystems; may be hazardous to aquatic life; may contribute to smog; prevent entry into sewers, waterways; not a DOT marine pollutant
Precaution: Combustible; flamm. limits 0.6-6.0%; vapor may cause flash fire; vapors are heavier than air, may flash back; may create vapor/air explosion hazard in confined space; spills may create slipping hazard; floats on water; container can rupture in fire; prevent entry into basements, confined areas; bond and ground equip.; incompat. with strong acids, alkalis, oxidizers (liq. chlorine, hydrogen peroxide, oxygen)
Hazardous decomp. prods.: Combustion: smoke, CO, CO_2, possibly other harmful gases/vapors
HMIS: Health 2, Flammability 2, Reactivity 0
Storage: Store in tightly closed containers in cool, dry, well-ventilated place away from heat, sparks, ignition sources, strong oxidizing conditions

Solvent 2376 [CITGO]

Chem. Descrip.: Light aliphatic solvent naphtha (70-80%), heavy hydrotreated naphtha (20-30%)
UN 1268
Uses: Solvent
Regulatory: SARA §302/304 nonreportable, §311/312 acute health/chronic health/fire hazard, §313 nonreportable]; Calif. Prop. 65 reportable (benzene, toluene); may contain detectable amts. of methylcyclopentane, n-hexane, cyclohexane
Properties: Colorless transparent liq.; lt. hydrocarbon odor; sl. sol. in cold water (≈ 0.05%); sp.gr. 0.71-0.72; visc. 0.5-1.5 cSt (40 C); vapor pressure 35 mm Hg (20 C); f.p. -135 to -67 C; b.p. 95-142 C; flash pt. (TCC) -8 to -5 C; autoignition temp. 232-246 C; KB value 31.2; VOC 715 g/l
Toxicology: Mist/vapor may irritate eyes, mucous membranes, respiratory tract; liq. contact may cause minimal to mild eye, mild to severe skin irritation/inflamm.; may be harmful by inh., skin absorp.; may be harmful or fatal by ing.; overexposure may cause CNS depression, other target organ effects; aspiration into lungs can cause pulmonary edema, chem. pneumonia; prolonged/repeated inh. may cause heart arrhythmias; TSCA listed
Environmental: May cause oxygen depletion in waterways; potentially toxic to water ecosystems; may be hazardous to aquatic life; may contribute to smog; prevent entry into sewers, waterways; not a DOT marine pollutant
Precaution: Extremely flamm.; flamm. limits 0.9-7.1%; vapor may cause flash fire or explosion; vapors are heavier than air, may flash back; may create vapor/air explosion hazard in confined space; spills may create slipping hazard; floats on water; container can rupture in fire; prevent entry into basements, confined areas; bond and ground equip.; incompat. with strong acids, alkalis, oxidizers (liq. chlorine, other halogens, hydrogen peroxide, oxygen); heptane will dissolve some plastics
Hazardous decomp. prods.: Combustion: smoke, CO, CO_2, possibly other harmful gases/vapors

HMIS: Health 1, Flammability 3, Reactivity 0
Storage: Store in tightly closed containers in cool, dry, well-ventilated place away from heat, sparks, ignition sources, strong oxidizing conditions

Solvent 2379 [CITGO]
Chem. Descrip.: Stoddard solvent
CAS 8052-41-3; EINECS/ELINCS 232-489-3
UN 1268
Uses: Solvent
Regulatory: SARA §302/304 nonreportable, §311/312 acute health/chronic health/fire hazard, §313 reportable [1,2,4-trimethylbenzene (1-4%)]; Calif. Prop. 65 reportable (benzene, toluene); may contain detectable amts. of cumene, other C9 aromatic hydrocarbons
Properties: Colorless transparent liq.; lt. hydrocarbon odor; sl. sol. in cold water (< 0.05%); sp.gr. 0.78-0.79; visc. < 20 cSt (40 C); vapor pressure 2.1 mm Hg (20 C); f.p. -65 to -25 C; b.p. 149-213 C; flash pt. (TCC) 42-45 C; autoignition temp. ≈ 230 C; KB value 35-40; VOC 779 g/l; 100% act.
Toxicology: ACGIH TLV TWA 100 ppm; LD50 (oral, rat) > 34,600 mg/kg; mist/vapor may irritate eyes, mucous membranes, respiratory tract; liq. contact may cause minimal to mod. eye, mod. to severe skin irritation/inflamm.; may be harmful by inh., skin absorp.; may be harmful or fatal by ing.; overexposure may cause CNS depression, other target organ effects; aspiration into lungs can cause pulmonary edema, chem. pneumonia; possible reproductive hazard; TSCA listed
Environmental: May cause oxygen depletion in waterways; potentially toxic to water ecosystems; may be hazardous to aquatic life; may contribute to smog; prevent entry into sewers, waterways; not a DOT marine pollutant
Precaution: Combustible; flamm. limits 0.7-6.0%; vapor may cause flash fire; vapors are heavier than air, may flash back; may create vapor/air explosion hazard in confined space; spills may create slipping hazard; floats on water; container can rupture in fire; prevent entry into basements, confined areas; bond and ground equip.; incompat. with strong acids, alkalis, oxidizers (liq. chlorine, hydrogen peroxide, oxygen)
Hazardous decomp. prods.: Combustion: smoke, CO, CO_2, possibly other harmful gases/vapors
HMIS: Health 2, Flammability 2, Reactivity 0
Storage: Store in tightly closed containers in cool, dry, well-ventilated place away from heat, sparks, ignition sources, strong oxidizing conditions

Solvent 2380 [CITGO]
Chem. Descrip.: Stoddard solvent
CAS 8052-41-3; EINECS/ELINCS 232-489-3
UN 1268; UN NA1993
Uses: Solvent
Regulatory: SARA §302/304 nonreportable, §311/312 acute health/chronic health/fire hazard, §313 reportable [1,2,4-trimethylbenzene (1.6-7.2%)]; Calif. Prop. 65 reportable (benzene, toluene, naphthalene)
Properties: Colorless transparent liq.; lt. hydrocarbon odor; sl. sol. in cold water (< 0.05%); sp.gr. 0.78-0.79; visc. < 20 cSt (40 C); vapor pressure 2.1 mm Hg (20 C); f.p. -65 to -25 C; b.p. 149-213 C; flash pt. (TCC) 41-43 C; autoignition temp. 230-300 C; KB value 35-40; VOC 779 g/l; 100% act.
Toxicology: ACGIH TLV TWA 100 ppm; mist/vapor may irritate eyes, mucous membranes, respiratory tract; liq. contact may cause minimal to mild eye, mod. to severe skin irritation/inflamm.; may be harmful by inh., skin absorp.; may be harmful or fatal by ing.; overexposure may cause CNS depression, other target organ effects; aspiration into lungs can cause pulmonary edema, chem. pneumonia; possible reproductive hazard; TSCA listed
Environmental: May cause oxygen depletion in waterways; potentially toxic to water ecosystems; may be hazardous to aquatic life; may contribute to smog; prevent entry into sewers, waterways; delisted as DOT marine pollutants
Precaution: Combustible; flamm. limits 0.9-7.0%; vapor may cause flash fire; vapors are heavier than air, may flash back; may create vapor/air explosion hazard in confined space; spills may create slipping hazard; floats on water; container can rupture in fire; prevent entry into basements, confined areas; bond and ground equip.; incompat. with strong acids, alkalis, oxidizers (liq. chlorine, other halogens, hydrogen peroxide, oxygen)
Hazardous decomp. prods.: Combustion: smoke, CO, CO_2, possibly other harmful gases/vapors
HMIS: Health 2, Flammability 2, Reactivity 0
Storage: Store in tightly closed containers in cool, dry, well-ventilated place away from heat, sparks, ignition sources, strong oxidizing conditions

Solvent 2383 [CITGO]
Chem. Descrip.: Stoddard solvent
CAS 8052-41-3; EINECS/ELINCS 232-489-3
UN 1268
Uses: Solvent
Regulatory: SARA §302/304 nonreportable, §311/312 acute health/chronic health/fire hazard, §313 reportable [xylenes (3-3.6%), ethylbenzene (0.4-1.0%)]; Calif. Prop. 65 reportable (benzene, toluene); may contain detectable amts. of 1,3,5-trimethylbenzene
Properties: Colorless transparent liq.; lt. hydrocarbon odor; sl. sol. in cold water (< 0.05%); sp.gr. 0.78; visc. < 20 cSt (40 C); vapor pressure 2.8 mm Hg (20 C); f.p. -85 to -25 C; b.p. 149-213 C; flash pt. (TCC) 39-42 C; autoignition temp. ≈ 230 C; KB value 32; VOC 780 g/l; 100% act.
Toxicology: ACGIH TLV TWA 100 ppm; mist/vapor may irritate eyes, mucous membranes, respiratory tract; liq. contact may cause mild to mod. eye, mod. to severe skin irritation/inflamm.; may be harmful by inh., skin absorp.; may be harmful or fatal by ing.; sensitizer; overexposure may cause CNS depression, other target organ effects; aspiration into lungs can cause pulmonary edema, chem. pneumonia; prolonged/repeated inh. may cause heart arrhythmias; possible cancer/reproductive hazard; TSCA listed
Environmental: May cause oxygen depletion in waterways; potentially toxic to water ecosystems; may be hazardous to aquatic life; may contribute to smog; prevent entry into sewers, waterways; not a DOT marine pollutant
Precaution: Combustible; flamm. limits 0.7-6.0%; vapor may cause flash fire; vapors are heavier than air, may flash back; may create vapor/air explosion hazard in confined space; spills may create slipping hazard; floats on water; container can rupture in fire; prevent entry into basements, confined areas; bond and ground equip.; incompat. with strong acids, alkalis, oxidizers (liq. chlorine, other halogens, hydrogen peroxide, oxygen)
Hazardous decomp. prods.: Combustion: smoke, CO, CO_2, possibly other harmful gases/vapors
HMIS: Health 2, Flammability 2, Reactivity 0
Storage: Store in tightly closed containers in cool, dry, well-ventilated place away from heat, sparks, ignition sources, strong oxidizing conditions

Solvent 2388 [CITGO]
Chem. Descrip.: Stoddard solvent
CAS 8052-41-3; EINECS/ELINCS 232-489-3
UN 1268
Uses: Solvent
Regulatory: SARA §302/304 nonreportable, §311/312 acute health/chronic health/fire hazard, §313 reportable [xylenes (4-7%), ethylbenzene (0.6-1.7%)]; Calif. Prop. 65 reportable (toluene); may contain detectable amts. of 1,3,5-trimethylbenzene
Properties: Colorless transparent liq.; lt. hydrocarbon odor; sl. sol. in cold water (< 0.05%); sp.gr. 0.78; visc. < 20 cSt (40 C); vapor pressure 2.8 mm Hg (20 C); f.p. -65 to -25 C; b.p. 149-185 C; flash pt. (TCC) 42-45 C; autoignition temp. ≈ 230 C; KB value 35-40; VOC 780 g/l; 100% act.
Toxicology: ACGIH TLV TWA 100 ppm; LD50 (oral, rat) > 34,600 mg/kg; mist/vapor may irritate eyes, mucous membranes, respiratory tract; liq. contact may cause minimal to mod. eye, mod. to severe skin irritation/inflamm.; may be harmful by inh., skin absorp.; may be harmful or fatal by ing.; overexposure may cause CNS depression, other target organ effects; aspiration into lungs can cause pulmonary edema, chem. pneumonia; possible reproductive hazard; TSCA listed
Environmental: May cause oxygen depletion in waterways; potentially toxic to water ecosystems; may be hazardous to aquatic life; may contribute to smog; prevent entry into sewers, waterways; not a DOT marine pollutant
Precaution: Combustible; flamm. limits 0.7-6.0%; vapor may cause flash fire; vapors are heavier than air, may flash back; may create vapor/air explosion hazard in confined space; spills may create slipping hazard; floats on water; container can rupture in fire; prevent entry into base-

ments, confined areas; bond and ground equip.; incompat. with strong acids, alkalis, oxidizers (liq. chlorine, other halogens, hydrogen peroxide, oxygen)

Hazardous decomp. prods.: Combustion: smoke, CO, CO_2, possibly other harmful gases/vapors

HMIS: Health 2, Flammability 2, Reactivity 0

Storage: Store in tightly closed containers in cool, dry, well-ventilated place away from heat, sparks, ignition sources, strong oxidizing conditions

Solvent 2391 [CITGO]

Chem. Descrip.: VM&P naphtha

CAS 8032-32-4; EINECS/ELINCS 232-453-7

UN 1268

Uses: Solvent

Regulatory: SARA §302/304 nonreportable, §311/312 acute health/chronic health/fire hazard, §313 reportable [xylenes (4-7%), ethylbenzene (0.6-1.7%)]; Calif. Prop. 65 reportable (toluene); may contain detectable amts. of 1,3,5-trimethylbenzene

Properties: Colorless transparent liq.; lt. hydrocarbon odor; sl. sol. in cold water (< 0.05%); sp.gr. 0.76; visc. < 5 cSt (40 C); vapor pressure 5.0 mm Hg (20 C); f.p. -57 to -25 C; b.p. 125-151 C; flash pt. (TCC) 18-21 C; autoignition temp. ≈ 232 C; KB value 38-40; VOC 757 g/l; 100% act.

Toxicology: Mist/vapor may irritate eyes, mucous membranes, respiratory tract; liq. contact may cause mild to mod. eye, mod. to severe skin irritation/inflamm.; may be harmful by inh., skin absorp.; may be harmful or fatal by ing.; sensitizer; overexposure may cause CNS depression, other target organ effects; aspiration into lungs can cause pulmonary edema, chem. pneumonia; prolonged/repeated inh. may cause heart arrhythmias; possible cancer/reproductive hazard; TSCA listed

Environmental: May cause oxygen depletion in waterways; potentially toxic to water ecosystems; may be hazardous to aquatic life; may contribute to smog; prevent entry into sewers, waterways; not a DOT marine pollutant

Precaution: Flamm.; flamm. limits 0.9-6.5%; vapor may cause flash fire; vapors are heavier than air, may flash back; may create vapor/air explosion hazard in confined space; spills may create slipping hazard; floats on water; container can rupture in fire; prevent entry into basements, confined areas; bond and ground equip.; incompat. with strong acids, alkalis, oxidizers (liq. chlorine, hydrogen peroxide, oxygen)

Hazardous decomp. prods.: Combustion: smoke, CO, CO_2, possibly other harmful gases/vapors

HMIS: Health 2, Flammability 3, Reactivity 0

Storage: Store in tightly closed containers in cool, dry, well-ventilated place away from heat, sparks, ignition sources, strong oxidizing conditions

Solvent 2409 [CITGO]

Chem. Descrip.: VM&P naphtha

CAS 8032-32-4; EINECS/ELINCS 232-453-7

UN 1268

Uses: Solvent

Regulatory: SARA §302/304 nonreportable, §311/312 acute health/chronic health/fire hazard, §313 reportable [toluene (7-9%), xylenes (0.02-1.6%)]; Calif. Prop. 65 reportable (benzene, toluene); may contain detectable amts. of cumene

Properties: Colorless transparent liq.; lt. hydrocarbon odor; sl. sol. in cold water (0.1%); sp.gr. 0.75-0.76; visc. < 5 cSt (40 C); vapor pressure 12 mm Hg (20 C); f.p. -95 to -25 C; b.p. 110-151 C; flash pt. (TCC) 12-14 C; autoignition temp. ≈ 232 C; KB value 35-40; VOC 752 g/l; 100% act.

Toxicology: Mist/vapor may irritate eyes, mucous membranes, respiratory tract; liq. contact may cause minimal to mild eye, mod. to severe skin irritation/inflamm.; may be harmful by inh., skin absorp.; may be harmful or fatal by ing.; sensitizer; overexposure may cause CNS depression, other target organ effects; aspiration into lungs can cause pulmonary edema, chem. pneumonia; prolonged/repeated inh. may cause heart arrhythmias; possible reproductive hazard; TSCA listed

Environmental: May cause oxygen depletion in waterways; potentially toxic to water ecosystems; may be hazardous to aquatic life; may contribute to smog; prevent entry into sewers, waterways; not a DOT marine pollutant

Precaution: Extremely flamm.; flamm. limits 0.9-7.1%; vapor may cause flash fire or explosion; vapors are heavier than air, may flash back; may create vapor/air explosion hazard in confined space; spills may create slipping hazard; floats on water; container can rupture in fire; prevent entry into basements, confined areas; bond and ground equip.; incompat. with strong acids, alkalis, oxidizers (liq. chlorine, other halogens, hydrogen peroxide, oxygen)

Hazardous decomp. prods.: Combustion: smoke, CO, CO_2, possibly other harmful gases/vapors

HMIS: Health 2, Flammability 3, Reactivity 0

Storage: Store in tightly closed containers in cool, dry, well-ventilated place away from heat, sparks, ignition sources, strong oxidizing conditions

Solvent 2417 [CITGO]

Chem. Descrip.: Stoddard solvent

CAS 8052-41-3; EINECS/ELINCS 232-489-3

UN 1268; UN NA1993

Uses: Solvent

Regulatory: SARA §302/304 nonreportable, §311/312 acute health/chronic health/fire hazard, §313 nonreportable; Calif. Prop. 65 nonreportable; may contain detectable amts. of 1,3,5-trimethylbenzene

Properties: Colorless transparent liq.; lt. paraffinic hydrocarbon odor; negligible sol. in cold water (< 0.01%); sp.gr. 0.80; visc. < 20 cSt (40 C); vapor pressure 1.9 mm Hg (20 C); f.p. -65 to -4 C; b.p. 188-238 C; flash pt. (TCC) 65-68 C; autoignition temp. ≈ 230 C; KB value 31; VOC 800 g/l; 100% act.

Toxicology: ACGIH TLV TWA 100 ppm; LD50 (oral, rat) > 34,600 mg/kg; mist/vapor may irritate eyes, mucous membranes, respiratory tract; liq. contact may cause minimal to mod. eye, mild to severe skin irritation/inflamm.; may be harmful by inh., skin absorp.; may be harmful or fatal by ing.; overexposure may cause CNS depression, other target organ effects; aspiration into lungs can cause pulmonary edema, chem. pneumonia; TSCA listed

Environmental: May cause oxygen depletion in waterways; potentially toxic to water ecosystems; may be hazardous to aquatic life; may contribute to smog; prevent entry into sewers, waterways; not a DOT marine pollutant

Precaution: Combustible; flamm. limits 0.7-6.0%; vapor may cause flash fire; vapors are heavier than air, may flash back; may create vapor/air explosion hazard in confined space; spills may create slipping hazard; floats on water; container can rupture in fire; prevent entry into basements, confined areas; bond and ground equip.; incompat. with strong acids, alkalis, oxidizers (liq. chlorine, other halogens, hydrogen peroxide, oxygen)

Hazardous decomp. prods.: Combustion: smoke, CO, CO_2, possibly other harmful gases/vapors

HMIS: Health 1, Flammability 2, Reactivity 0

Storage: Store in tightly closed containers in cool, dry, well-ventilated place away from heat, sparks, ignition sources, strong oxidizing conditions

Solvent 2423 [CITGO]

Chem. Descrip.: VM&P naphtha

CAS 8032-32-4; EINECS/ELINCS 232-453-7

UN 1268

Uses: Solvent

Regulatory: SARA §302/304 nonreportable, §311/312 acute health/chronic health/fire hazard, §313 reportable [toluene (0-2%), xylenes (0.2-2%)]; Calif. Prop. 65 reportable (benzene, toluene); may contain detectable amts. of cumene

Properties: Colorless transparent liq.; lt. hydrocarbon odor; negligible sol. in cold water (< 0.01%); sp.gr. 0.74-0.75; visc. < 5 cSt (40 C); vapor pressure 11 mm Hg (20 C); f.p. -95 to -25 C; b.p. 110-151 C; flash pt. (TCC) 13-14 C; autoignition temp. ≈ 232 C; KB value 35; VOC 740 g/l; 100% act.

Toxicology: Mist/vapor may irritate eyes, mucous membranes, respiratory tract; liq. contact may cause minimal to mild eye, mod. to severe skin irritation/inflamm.; may be harmful by inh., skin absorp.; may be harmful or fatal by ing.; sensitizer; overexposure may cause CNS depression, other target organ effects; aspiration into lungs can cause pulmonary edema, chem. pneumonia; prolonged/repeated inh. may cause heart arrhythmias; possible reproductive hazard; TSCA listed

Environmental: May cause oxygen depletion in waterways; potentially toxic to water ecosystems; may be hazardous to aquatic life; may contribute to smog; prevent entry into sewers, waterways; not a DOT marine pollutant

Precaution: Flamm.; flamm. limits 0.9-7.0%; vapor may cause flash fire; vapors are heavier than air, may flash back; may create vapor/air explosion hazard in confined space; spills may create slipping hazard; floats on water; container can rupture in fire; prevent entry into basements, confined areas; bond and ground equip.; incompat. with strong acids, alkalis, oxidizers (liq. chlorine, other halogens, hydrogen peroxide, oxygen)
Hazardous decomp. prods.: Combustion: smoke, CO, CO_2, possibly other harmful gases/vapors
HMIS: Health 2, Flammability 3, Reactivity 0
Storage: Store in tightly closed containers in cool, dry, well-ventilated place away from heat, sparks, ignition sources, strong oxidizing conditions

Solvent 2430 [CITGO]
Chem. Descrip.: Hexane isomers (20-25%), n-hexane (22-28%), heptanes (45-50%), methylcyclohexane (0.8-1.1%), octanes (1-2.5%)
UN 1268
Uses: Solvent
Regulatory: SARA §302/304 nonreportable, §311/312 acute health/chronic health/fire hazard, §313 reportable [n-hexane (22-28%)]; Calif. Prop. 65 reportable (benzene, toluene); may contain detectable amts. of n-hexane, cyclohexane
Properties: Colorless transparent liq.; lt. paraffinic hydrocarbon odor; sl. sol. in cold water (≈ 0.05%); sp.gr. 0.686; visc. 1-2 cSt (40 C); vapor pressure 95 mm Hg (20 C); f.p. -108 to -90 C; b.p. 60-109 C; flash pt. (TCC) -22 to -20 C; autoignition temp. ≈ 232 C; KB value 30; VOC 687 g/l
Toxicology: Mist/vapor may irritate eyes, mucous membranes, respiratory tract; liq. contact may cause minimal to mild eye, mild to mod. skin irritation/inflamm.; may be harmful by inh., skin absorp.; may be harmful or fatal by ing.; sensitizer; overexposure may cause CNS depression, other target organ effects, peripheral nervous system damage; aspiration into lungs can cause pulmonary edema, chem. pneumonia; prolonged/repeated inh. may cause heart arrhythmias; possible reproductive hazard; TSCA listed
Environmental: May cause oxygen depletion in waterways; potentially toxic to water ecosystems; may be hazardous to aquatic life; may contribute to smog; prevent entry into sewers, waterways; not a DOT marine pollutant
Precaution: Extremely flamm.; flamm. limits 1.0-7.5%; vapor may cause flash fire or explosion; vapors are heavier than air, may flash back; may create vapor/air explosion hazard in confined space; spills may create slipping hazard; floats on water; container can rupture in fire; prevent entry into basements, confined areas; bond and ground equip.; incompat. with strong acids, alkalis, oxidizers (liq. chlorine, other halogens, hydrogen peroxide, oxygen); heptane will dissolve some plastics
Hazardous decomp. prods.: Combustion: smoke, CO, CO_2, possibly other harmful gases/vapors
HMIS: Health 2, Flammability 3, Reactivity 0
Storage: Store in tightly closed containers in cool, dry, well-ventilated place away from heat, sparks, ignition sources, strong oxidizing conditions

Solvent 2439 [CITGO]
Chem. Descrip.: Xylene (49-61%), ethylbenzene (6-17%), n-propylbenzene (0.3-3.5%), cumene (0-1%), trimethylbenzene (10-22%), 1,2,4-trimethylbenzene (3-16%), ethyl toluene (3-9%), C10 alkylbenzene (1-5%)
UN 1268
Uses: Solvent
Regulatory: SARA §302/304 nonreportable, §311/312 acute health/chronic health/fire hazard, §313 reportable [xylenes (49.5-60.42%, ethylbenzene (6.6-16.704%, 1,2,4-trimethylbenzene (3.4-15.3%)]; Calif. Prop. 65 reportable (benzene, toluene, naphthalene)
Properties: Colorless transparent liq.; mod. sweet, pleasant, aromatic hydrocarbon odor; sol. in methanol, diethyl ether; sl. sol. in cold water (< 0.05%); sp.gr. 0.87; visc. 3-8 cSt (40 C); vapor pressure 2.7-6.6 mm Hg (20 C); f.p. -95 to 13 C; b.p. 136-180 C; flash pt. (TCC) 27-32 C; autoignition temp. 343-432 C; KB value ≈ 95.5; VOC 873 g/l
Toxicology: Mist/vapor may irritate eyes, mucous membranes, respiratory tract; liq. contact may cause minimal to mild eye, mild to mod. skin irritation/inflamm.; may be harmful by inh., skin absorp.; may be harmful or fatal by ing.; overexposure may cause CNS depression, other target organ effects; aspiration into lungs can cause pulmonary edema, chem. pneumonia; prolonged/repeated inh. may cause heart arrhythmias; possible reproductive hazard; tumorigen; TSCA listed
Environmental: May cause oxygen depletion in waterways; potentially toxic to water ecosystems; may be hazardous to aquatic life; may contribute to smog; prevent entry into sewers, waterways; delisted as DOT marine pollutants
Precaution: Flamm.; flamm. limits 0.9-7.1%; vapor may cause flash fire; vapors are heavier than air, may flash back; may create vapor/air explosion hazard in confined space; spills may create slipping hazard; floats on water; container can rupture in fire; prevent entry into basements, confined areas; bond and ground equip.; incompat. with strong acids, alkalis, oxidizers (liq. chlorine, other halogens, hydrogen peroxide, oxygen)
Hazardous decomp. prods.: Combustion: smoke, CO, CO_2, possibly other harmful gases/vapors
HMIS: Health 2, Flammability 3, Reactivity 0
Storage: Store in tightly closed containers in cool, dry, well-ventilated place away from heat, sparks, ignition sources, strong oxidizing conditions

Solvent 2442 [CITGO]
Chem. Descrip.: Solvent-refined light naphtha (75-85%), toluene (19-21%), n-hexane (24-32%), cyclohexane (0-2.5%)
UN 1268
Uses: Solvent
Regulatory: SARA §302/304 nonreportable, §311/312 acute health/chronic health/fire hazard, §313 reportable [n-hexane (24-32%), cyclohexane (0.008-2.4%), toluene (19-21%)]; Calif. Prop. 65 reportable (benzene, toluene); may contain detectable amts. of n-hexane, cyclohexane
Properties: Colorless transparent liq.; lt. hydrocarbon odor; sl. sol. in cold water (≈ 0.1%); sp.gr. 0.72; visc. 1-2 cSt (40 C); vapor pressure 96 mm Hg (20 C); f.p. -95 to -90 C; b.p. 60-111 C; flash pt. (TCC) -21 to -20 C; autoignition temp. ≈ 232 C; KB value 44; VOC 725 g/l
Toxicology: Mist/vapor may irritate eyes, mucous membranes, respiratory tract; liq. contact may cause minimal to mod. eye, mild to severe skin irritation/inflamm.; may be harmful by inh., skin absorp.; may be harmful or fatal by ing.; sensitizer; overexposure may cause CNS depression, other target organ effects, peripheral nervous system damage; aspiration into lungs can cause pulmonary edema, chem. pneumonia; prolonged/repeated inh. may cause heart arrhythmias; possible reproductive hazard; TSCA listed
Environmental: May cause oxygen depletion in waterways; potentially toxic to water ecosystems; may be hazardous to aquatic life; may contribute to smog; prevent entry into sewers, waterways; not a DOT marine pollutant
Precaution: Extremely flamm.; flamm. limits 1.0-7.1%; vapor may cause flash fire or explosion; vapors are heavier than air, may flash back; may create vapor/air explosion hazard in confined space; spills may create slipping hazard; floats on water; container can rupture in fire; prevent entry into basements, confined areas; bond and ground equip.; incompat. with strong acids, alkalis, oxidizers (liq. chlorine, other halogens, hydrogen peroxide, oxygen); heptane will dissolve some plastics
Hazardous decomp. prods.: Combustion: smoke, CO, CO_2, possibly other harmful gases/vapors
HMIS: Health 2, Flammability 3, Reactivity 0
Storage: Store in tightly closed containers in cool, dry, well-ventilated place away from heat, sparks, ignition sources, strong oxidizing conditions

Solvent 2461 [CITGO]
Chem. Descrip.: Stoddard solvent (85-95%), VM&P naphtha (5-15%)
UN 1268
Uses: Solvent
Regulatory: SARA §302/304 nonreportable, §311/312 acute health/chronic health/fire hazard, §313 reportable [1,2,4,-trimethylbenzene (1-7%)]; Calif. Prop. 65 reportable (benzene, toluene); may contain detectable amts. of cumene, other C9 aromatic hydrocarbons
Properties: Colorless transparent liq.; lt. hydrocarbon odor; sl. sol. in cold water (< 0.05%); sp.gr. 0.79; visc. < 20 cSt (40 C); vapor pressure 2.8-3.0 mm Hg (20 C); f.p. -65 to 13 C; b.p. 125-213 C; flash pt. (TCC) 36-37 C; autoignition temp. ≈ 230 C; KB value 40; VOC 788 g/l
Toxicology: Mist/vapor may irritate eyes, mucous membranes, respiratory tract; liq. contact may cause minimal to mod. eye, mod. to severe skin irritation/inflamm.; may be harmful by inh., skin absorp.; may be

harmful or fatal by ing.; overexposure may cause CNS depression, other target organ effects; aspiration into lungs can cause pulmonary edema, chem. pneumonia; possible reproductive hazard; TSCA listed
Environmental: May cause oxygen depletion in waterways; potentially toxic to water ecosystems; may be hazardous to aquatic life; may contribute to smog; prevent entry into sewers, waterways; may be a DOT marine pollutant
Precaution: Flamm.; flamm. limits 0.8-6.7%; vapor may cause flash fire; vapors are heavier than air, may flash back; may create vapor/air explosion hazard in confined space; spills may create slipping hazard; floats on water; container can rupture in fire; prevent entry into basements, confined areas; bond and ground equip.; incompat. with strong acids, alkalis, oxidizers (liq. chlorine, hydrogen peroxide, oxygen)
Hazardous decomp. prods.: Combustion: smoke, CO, CO_2, possibly other harmful gases/vapors
HMIS: Health 2, Flammability 3, Reactivity 0
Storage: Store in tightly closed containers in cool, dry, well-ventilated place away from heat, sparks, ignition sources, strong oxidizing conditions

Solvent 2472 [CITGO]
Chem. Descrip.: VM&P naphtha
CAS 8032-32-4; EINECS/ELINCS 232-453-7
UN 1268
Uses: Solvent
Regulatory: SARA §302/304 nonreportable, §311/312 acute health/chronic health/fire hazard, §313 reportable [xylenes (2-4%)]; Calif. Prop. 65 reportable (toluene); may contain detectable amts. of 1,3,5-trimethylbenzene
Properties: Colorless transparent liq.; lt. hydrocarbon odor; sl. sol. in cold water (< 0.05%); sp.gr. 0.76; visc. < 5 cSt (40 C); vapor pressure 5.3 mm Hg (20 C); f.p. -57 to -25 C; b.p. 125-151 C; flash pt. (TCC) 23-24 C; autoignition temp. ≈ 232 C; KB value 36-40; VOC 750 g/l; 100% act.
Toxicology: Mist/vapor may irritate eyes, mucous membranes, respiratory tract; liq. contact may cause mild to mod. eye, mod. to severe skin irritation/inflamm.; may be harmful by inh., skin absorp.; may be harmful or fatal by ing.; sensitizer; overexposure may cause CNS depression, other target organ effects; aspiration into lungs can cause pulmonary edema, chem. pneumonia; prolonged/repeated inh. may cause heart arrhythmias; possible cancer/reproductive hazard; TSCA listed
Environmental: May cause oxygen depletion in waterways; potentially toxic to water ecosystems; may be hazardous to aquatic life; may contribute to smog; prevent entry into sewers, waterways; not a DOT marine pollutant
Precaution: Flamm.; flamm. limits 0.9-6.5%; vapor may cause flash fire; vapors are heavier than air, may flash back; may create vapor/air explosion hazard in confined space; spills may create slipping hazard; floats on water; container can rupture in fire; prevent entry into basements, confined areas; bond and ground equip.; incompat. with strong acids, alkalis, oxidizers (liq. chlorine, other halogens, hydrogen peroxide, oxygen)
Hazardous decomp. prods.: Combustion: smoke, CO, CO_2, possibly other harmful gases/vapors
HMIS: Health 2, Flammability 3, Reactivity 0
Storage: Store in tightly closed containers in cool, dry, well-ventilated place away from heat, sparks, ignition sources, strong oxidizing conditions

Solvent 2495 [CITGO]
Chem. Descrip.: Toluene (50-51%), xylene mixed isomers (37-45%), ethylbenzene (5-12%)
UN 1268
Uses: Solvent
Regulatory: SARA §302/304 nonreportable, §311/312 acute health/chronic health/fire hazard, §313 reportable [toluene (50-50.9%), xylenes (37.5-45%), ethylbenzene (5-12.5%)]; Calif. Prop. 65 reportable (benzene, toluene)
Properties: Colorless transparent liq.; char. sweet, pleasant, aromatic hydrocarbon odor; sl. sol. in cold water (0.09%); sp.gr. 0.871; visc. < 4 cSt (40 C); vapor pressure 15.2 mm Hg (20 C); f.p. -95 to -25 C; b.p. 110-145 C; flash pt. (TCC) 4-7 C; autoignition temp. ≈ 432 C; KB value 101; VOC 872 g/l; 100% act.
Toxicology: Mist/vapor may irritate eyes, mucous membranes, respiratory tract; liq. contact may cause mild to severe eye, mod. to severe skin irritation/inflamm.; may be harmful by inh., skin absorp.; may be harmful or fatal by ing.; sensitizer; overexposure may cause CNS depression, other target organ effects, hearing loss; aspiration into lungs can cause pulmonary edema, chem. pneumonia; prolonged/repeated inh. may cause heart arrhythmias; possible cancer/reproductive hazard; TSCA listed
Environmental: May cause oxygen depletion in waterways; potentially toxic to water ecosystems; may be hazardous to aquatic life; may contribute to smog; prevent entry into sewers, waterways; not a DOT marine pollutant
Precaution: Flamm.; flamm. limits 1.0-7.1%; vapor may cause flash fire; vapors are heavier than air, may flash back; may create vapor/air explosion hazard in confined space; spills may create slipping hazard; floats on water; container can rupture in fire; prevent entry into basements, confined areas; bond and ground equip.; incompat. with strong acids, alkalis, oxidizers (liq. chlorine, other halogens, hydrogen peroxide, oxygen)
Hazardous decomp. prods.: Combustion: smoke, CO, CO_2, possibly other harmful gases/vapors
HMIS: Health 2, Flammability 3, Reactivity 0
Storage: Store in tightly closed containers in cool, dry, well-ventilated place away from heat, sparks, ignition sources, strong oxidizing conditions

Solvent 30 823 [Sunnyside]
Chem. Descrip.: Medium aliphatic naphtha, light aromatic naphtha, 1,2,4-trimethylbenzene (26.18%), xylene (2.45%), cumene (1.23%), ethyl benzene (0.41%)
Uses: Solvent
Regulatory: DOT nonregulated; SARA §311/312 acute health/chronic health/fire hazard, §313 reportable [1,2,4-trimethylbenzene (26.18%), xylene (2.45%), cumene (1.23%), ethylbenzene (0.41%)]; Calif. Prop. 65 reportable [benzene (trace), toluene]
Properties: Liq.; negligible sol. in water; dens. 7.15 lb/gal; b.p. 310-395 F; flash pt. (TCC) 102 F; 100% volatile
Toxicology: TLV 50 ppm; minimal eye/sl. skin irritant; high vapor concs. may irritate eyes; prolonged/repeated skin contact may cause irritation, dermatitis; inh. of vapors may cause nose/throat/respiratory tract irritation, headache, dizziness, CNS depression; if aspirated, may cause pneumonitis; contains ethyl benzene, possible human carcinogen
Environmental: Prevent entry into sewers, ditches, bodies of water
Precaution: Combustible liq.; LEL 1.0% (estimated); incompat. with strong oxidizers (liq. chlorine, conc. oxygen); ground containers to prevent static accumulation and discharge; empty containers may contain residue, may explode
Hazardous decomp. prods.: CO_2, CO
HMIS: Health 2, Flammability 2, Reactivity 0
Storage: Keep containers closed when not in use; keep away from heat, sparks, open flame

Solvent B-100 [Eastern Color & Chem.]
Uses: Solvent exhibiting exc. drying props. and quick solubilization of acetate films normally used in shoelace tipping
Features: Universally used; does not cause cloudiness or whiteness in finished tip

Solvent B-200 (B-2) [Eastern Color & Chem.]
Uses: Solvent for shoelace tipping
Features: Modified to alter drying props. without altering other desirable characteristics

Solvesso™ 100 [Exxon Canada; ExxonMobil Europe]
Chem. Descrip.: Solvent naphtha, light aromatic, BHT (25 ppm)
CAS 64742-95-6; EINECS/ELINCS 265-199-0
Uses: Solvent for automotive OEM and refinish coatings, metal coatings, electrostatic applics., additives, paints/coatings, pesticides, industrial cleaning, mastics, sealants, process fluids
Features: Cost effective; high solvency power for wide range of resins/polymers; high elec. resistivity
Regulatory: Canada DSL listed; NPRI reportable [cumene (1.1%), 1,2,4-trimethylbenzene (29.4%), xylenes (2.3%), o-xylene (1.2%)]
Properties: Colorless clear liq.; aromatic hydrocarbon odor; sol. 0.02% in water; m.w. 127; dens. 0.877 kg/dm^3 (15 C); visc. 0.89 mPa·s; vapor pressure 0.59 kPa (20 C); f.p. -60 C; b.p. 159-170 C; flash pt. (TCC) > 42 C; autoignition temp. 471 C; KB value 90; ref. index 1.502 (20 C); surf. tens. 29.7 mN/m; 100% act.; > 99% aromatic content

Toxicology: REL TWA 19 ppm; inh. of high vapor concs. may cause eye/respiratory tract irritation, headaches, dizziness, anesthesia, drowsiness, unconsciousness, other CNS effects, death; sl. eye irritant; frequent/prolonged skin contact may cause irritation, dermatitis; if aspirated, may cause mild to severe pulmonary injury, death
Environmental: Prevent entry into sewers, watercourses, low areas
Precaution: Combustible liq.; flamm. limits 1.9-12.6 vol.%; volatile; gives off invisible vapors; liq. or vapor may settle in low areas; distant ignition/explosion possible; electrostatic accumulation hazard; use proper grounding procedures; incompat. with nitric acid, sulfuric acid, strong oxidizing agents
Hazardous decomp. prods.: Combustion: toxic gases, fumes, smoke, CO
Storage: Store in cool, well-ventilated area away from incompat. materials, heat, open flame, other ignition sources, direct sunlight; keep container closed

Solvesso™ 100 Anti-Static Solvent [Exxon Canada]
Chem. Descrip.: Naphtha, light aromatic, antistatic additive (16 ml/1000 gal), BHT (25 ppm)
CAS 64742-95-6; EINECS/ELINCS 265-199-0
UN 1268
Uses: Solvent; diluent; chem. feedstock; fuel
Regulatory: Canada DSL listed; NPRI reportable [cumene (1.1%), 1,2,4-trimethylbenzene (29.4%), xylenes (2.3%), o-xylene (1.2%)]
Properties: Colorless clear liq.; aromatic hydrocarbon odor; sol. 0.02% in water; m.w. 127; sp.gr. 0.87 (15.6 C); visc. 0.92 cSt; vapor pressure 0.59 kPa (20 C); f.p. -60 C; b.p. 159-170 C; flash pt. (TCC) > 42 C; autoignition temp. 471 C; 100% act.
Toxicology: REL TWA 19 ppm; inh. of high vapor concs. may cause eye/respiratory tract irritation, headaches, dizziness, anesthesia, drowsiness, unconsciousness, other CNS effects, death; sl. eye irritant; frequent/prolonged skin contact may cause irritation, dermatitis; if aspirated, may cause mild to severe pulmonary injury, death
Environmental: Prevent entry into sewers, watercourses, low areas
Precaution: Combustible liq.; flamm. limits 1.9-12.6 vol.%; volatile; gives off invisible vapors; liq. or vapor may settle in low areas; distant ignition/explosion possible; electrostatic accumulation hazard; use proper grounding procedures; incompat. with nitric acid, sulfuric acid, strong oxidizing agents
Hazardous decomp. prods.: Combustion: toxic gases, fumes, smoke, CO
Storage: Store in cool, well-ventilated area away from incompat. materials, heat, open flame, ignition sources, direct sunlight; keep container closed

Solvesso™ 150 [ExxonMobil Europe]
Chem. Descrip.: Aromatic hydrocarbon fluid
Uses: Solvent for metal coatings, electrostatic applics., additives, paints/coatings, pesticides, industrial cleaning, mastics, sealants, process fluids
Features: Cost effective; high solvency power for wide range of resins/polymers; high elec. resistivity
Properties: Dens. 0.899 kg/dm³ (15 C); visc. 1.10 mPa·s; i.b.p. 182 C; flash pt. 65 C; ref. index 1.516 (20 C); surf. tens. 30.4 mN/m; KB value 86; > 99% aromatic content

Solvesso™ 150 Anti-Static Solvent [Exxon Canada]
Chem. Descrip.: Naphtha, heavy aromatic, antistatic additive (16-36 ml/1000 gal), BHT (25 ppm)
CAS 64742-94-5; EINECS/ELINCS 265-198-5
Uses: Solvent; diluent; chem. feedstock; fuel
Regulatory: Canada DSL listed; NPRI reportable [naphthalene (12.3%), 1,2,4-trimethylbenzene (1.7%)]
Properties: Colorless clear liq.; mild aromatic odor; sol. < 0.01% in water; m.w. 142; sp.gr. 0.9 (15.5 C); visc. 1.33 cSt; vapor pressure 0.4 kPa (38 C); f.p. -43 C; b.p. 184-205 C; flash pt. (TCC) 63 C; autoignition temp. ≈ 443 C; 100% act.
Toxicology: REL TWA 17 ppm; LD50 (oral, rat) 3 g/kg, (skin, rabbit) > 3 g/kg; inh. of high vapor concs. may cause eye/respiratory tract irritation, headaches, dizziness, anesthesia, drowsiness, unconsciousness, other CNS effects, death; sl. eye irritant; frequent/prolonged skin contact may cause irritation, dermatitis; if aspirated, may cause mild to severe pulmonary injury, death; contains naphthalene, a possible human carcinogen
Environmental: Prevent entry into sewers, watercourses, low areas
Precaution: Combustible liq.; flamm. limits 1.8-11.7 vol.%; volatile; gives off invisible vapors; liq. or vapor may settle in low areas; distant ignition/explosion possible; electrostatic accumulation hazard; use proper grounding procedures; incompat. with nitric acid, sulfuric acid, strong oxidizing agents
Hazardous decomp. prods.: None
Storage: Store in cool, well-ventilated area away from incompat. materials, heat, open flame, ignition sources, direct sunlight; keep container closed

Solvesso™ 150 ND [Exxon Canada; ExxonMobil Europe]
Chem. Descrip.: Heavy aromatic solvent naphtha, BHT (16 ppm)
CAS 64742-94-5; EINECS/ELINCS 265-198-5
Uses: Solvent for metal coatings, electrostatic applics., additives, pesticides, industrial cleaning, mastics, sealants, process fluids; diluent; chem. feedstock; fuel
Features: Naphthalene-depleted version; cost effective; high solvency power for wide range of resins/polymers; high elec. resistivity
Regulatory: Canada DSL listed; NPRI reportable [naphthalene (0.5%)]
Properties: Colorless clear liq.; mild aromatic odor; sol. < 0.01% in water; m.w. 140; dens. 0.886 kg/dm³ (15 C); visc. 1.02 mPa·s; vapor pressure 0.09 kPa (20 C); f.p.-30 C; b.p. 183-202 C; flash pt. (TCC) > 63 C; autoignition temp. ≈ 443 C; ref. index 1.506 (20 C); surf. tens. 30.8 mN/m; KB value 86; 100% act.; > 99% aromatic content
Toxicology: REL TWA 17 ppm; LD50 (oral, rat) 3 g/kg, (skin, rabbit) > 3 g/kg; inh. of high vapor concs. may cause eye/respiratory tract irritation, headaches, dizziness, anesthesia, drowsiness, unconsciousness, other CNS effects, death; sl. eye irritant; frequent/prolonged skin contact may cause irritation, dermatitis; if aspirated, may cause mild to severe pulmonary injury, death; contains naphthalene, a possible human carcinogen
Environmental: Prevent entry into sewers, watercourses, low areas
Precaution: Combustible liq.; flamm. limits 1.8-11.7 vol.%; volatile; gives off invisible vapors; liq. or vapor may settle in low areas; distant ignition/explosion possible; electrostatic accumulation hazard; use proper grounding procedures; incompat. with nitric acid, sulfuric acid, strong oxidizing agents
Hazardous decomp. prods.: Combustion: toxic gases, fumes, smoke, CO
Storage: Store in cool, well-ventilated area away from incompat. materials, heat, open flame, other ignition sources, direct sunlight; keep container closed

Solvesso™ 200 [Exxon Canada; ExxonMobil Europe]
Chem. Descrip.: Heavy aromatic solvent naphtha, BHT (25 ppm)
CAS 64742-94-5; EINECS/ELINCS 265-198-5
Uses: Solvent for metal coatings, electrostatic applics., additives, paints/coatings, pesticides, industrial cleaning, mastics, sealants, process fluids; diluent; chem. feedstock; fuel
Features: Cost effective; high solvency power for wide range of resins/polymers; high elec. resistivity
Regulatory: Canada DSL listed; NPRI reportable [naphthalene (14%)]
Properties: Colorless clear liq.; aromatic hydrocarbon odor; sol. < 0.01% in water; m.w. 154; dens. 0.994 kg/dm³ (15 C); visc. 2.92 mPa·s; vapor pressure 0.01 kPa (20 C); f.p. -8 C; b.p. 232-275 C; flash pt. (PMCC) > 99 C; autoignition temp. ≈ 484 C; ref. index 1.591 (20 C); surf. tens. 36.1 mN/m; KB value 88; > 99% aromatic content
Toxicology: REL TWA 15 ppm; inh. of high vapor concs. may cause eye/respiratory tract irritation, headaches, dizziness, anesthesia, drowsiness, unconsciousness, other CNS effects, death; sl. eye irritant; frequent/prolonged skin contact may cause irritation, dermatitis; if aspirated, may cause mild to severe pulmonary injury, death; chronic exposure may cause stomach/liver/thyroid/urinary bladder damage; contains naphthalene, possible carcinogen
Environmental: Prevent entry into sewers, watercourses, low areas
Precaution: Flamm. limits 1.8-11.8 vol.%; low flamm. hazard; electrostatic accumulation hazard; use proper grounding procedures; incompat. with nitric acid, sulfuric acid, strong oxidizing agents
Hazardous decomp. prods.: None
Storage: Store in cool, well-ventilated area away from incompat. materials, heat, open flame, other ignition sources, direct sunlight; keep container closed

Solvesso™ 200 ND [Exxon Canada; ExxonMobil Europe]
Chem. Descrip.: Heavy aromatic solvent naphtha, BHT (17 ppm)
CAS 64742-94-5; EINECS/ELINCS 265-198-5
Uses: Solvent for additives, paints/coatings, pesticides, industrial clean-

ing, mastics, sealants, process fluids; diluent; chem. feedstock; fuel
Features: Naphthalene-depleted version; high solvency; controlled evaporation chars.
Regulatory: Canada DSL listed; NPRI nonreportable
Properties: Pale yel. liq.; aromatic hydrocarbon odor; sol. < 0.01% in water; m.w. ≈ 120; dens. 1.000 kg/dm³ (15 C); visc. 3.26 mm²/s; vapor pressure < 0.1 kPa (20 C); f.p. -10 C; b.p. 235-278 C; flash pt. 108 C; autoignition temp. ≈ 484 C; ref. index 1.598 (20 C); surf. tens. 36.1 mN/m; KB value 90; aniline pt. 13 C; > 99% aromatic content
Toxicology: REL TWA 15 ppm; inh. of high vapor concs. may cause eye/respiratory tract irritation, headaches, dizziness, anesthesia, drowsiness, unconsciousness, other CNS effects, death; sl. eye irritant; frequent/prolonged skin contact may cause irritation, dermatitis; if aspirated, may cause mild to severe pulmonary injury, death
Environmental: Prevent entry into sewers, watercourses, low areas
Precaution: Flamm. limits 1.8-11.8 vol.%; electrostatic accumulation hazard; use proper grounding procedures; incompat. with nitric acid, sulfuric acid, strong oxidizing agents
Hazardous decomp. prods.: Combustion: toxic gases
Storage: Store in cool, well-ventilated area away from incompat. materials, heat, open flame, ignition sources, direct sunlight; keep container closed

Solvesso™ 200 S [ExxonMobil Europe]
Chem. Descrip.: Aromatic hydrocarbon fluid
Uses: Solvent for additives, paints/coatings, pesticides, industrial cleaning, mastics, sealants, process fluids
Features: Controlled evaporation chars.
Properties: Dens. 0.995 kg/dm³ (15 C); visc. 2.70 mm²/s; i.b.p. 231 C; flash pt. 101 C; ref. index 1.593 (20 C); surf. tens. 36.1 mN/m; KB value 87; aniline pt. 14 C; > 99% aromatic content; < 10 ppm sulfur

Solvolene E [Hart Prods.]
Chem. Descrip.: Aromatic solvent
Uses: Solvent

Sovatex C Series [Thor UK]
Chem. Descrip.: Dodecylbenzene sulfonate blends
Uses: Detergent, solvent scouring agent, wetting agent, emulsifier for textiles, laundry, dry cleaning, leather, and chemical industries
Properties: Liq., paste, solid; water-disp.; anionic

Soygold® 1000 [Ag Environmental Prods.]
Chem. Descrip.: Methyl soyate
CAS 67784-80-9
Uses: Solvent for petrol. and animal fat degreasing, precision cleaning, asphalt industry (release agent), concrete release, mold release, metal cutting (lubricant), adhesive removal, silk screen, printing, paints, glass cutting, automotive prod.; substitute for d-limonene, naphtha, min. spirits, etc.; defoamer in food-contact paper coatings; in resin-bonded filters for food contact; in surf.-lubricants for mfg. of food-contact metallic articles
Features: Environmentally responsible; contains no petrol. distillates; low VOCs; ultra low evaporative emissions; high flash pt.; EPA listed as approved replacement (SNAP); water rinsable
Regulatory: FDA 21CFR §172.225, 174.5, 176.200, 177.2260, 178.3910; SARA §313 nonreportable; DOT regulated (Class 65)
Properties: Clear to lt. yel. liq.; lt. musty odor; negligible sol. in water; sp.gr. 0.870; vapor pressure < 5 mm Hg (72 F); m.p. -1 C; b.p. 320 C; flash pt. (PMCC) 218 C; KB 58; VOC < 14%
Toxicology: Nontoxic; LD50 (oral, rat) 17.4 g/kg; nonirritating to eyes and skinTSCA listed
Environmental: Readily biodeg.; LC50 (bluegill, 96 h) > 1000 mg/l
Precaution: Avoid strong oxidizing agents
Hazardous Ingredients: None
Hazardous decomp. prods.: CO_2, CO
HMIS: Health 0, Flammability 1, Reactivity 0

Soygold® 2000 [Ag Environmental Prods.]
Chem. Descrip.: Methyl soyate with 1-3% nonoxynol surfactant (CAS 9016-45-9)
Uses: Solvent for petrol. and animal fat degreasing, precision cleaning, asphalt industry (release agent), concrete release, mold release, metal cutting (lubricant), adhesive removal, silk screen, printing, paints, glass cutting, automotive prods.; substitute for d-limonene, naphtha, min. spirits, etc.; defoamer in food-contact paper coatings; in resin-bonded filters for food contact; in surf.-lubricants for mfg. of food-contact metallic articles
Features: Contains no petrol. distillates; low VOCs; ultra low evaporative emissions; high flash pt.; EPA listed as approved replacement (SNAP); water rinsable
Regulatory: FDA 21CFR §172.225, 174.5, 176.200, 177.2260, 178.3910; SARA §313 nonreportable; DOT regulated (Class 65)
Properties: Clear to lt. yel. liq.; lt. musty odor; negligible sol. in water; sp.gr. 0.87; vapor pressure < 5 mm Hg (72 F); m.p. -1 C; b.p. > 204 C; flash pt. (PMCC) 218 C; KB 58; VOC < 14%
Toxicology: Nontoxic; LD50 (oral, rat) 17.4 g/kg; nonirritating to eyes and skinTSCA listed
Environmental: Readily biodeg.; LC50 (bluegill, 96 h) > 1000 mg/l
Precaution: Avoid strong oxidizing agents
Hazardous Ingredients: Alkyl C_{16}-$C_{18\text{-}m}$ethyl esters (97-99%); surfactant (1-3%)
Hazardous decomp. prods.: CO_2, CO
HMIS: Health 0, Flammability 1, Reactivity 0

Special Lactolite [CITGO]
Chem. Descrip.: Light aliphatic solvent naphtha (petroleum)
CAS 64742-89-8; EINECS/ELINCS 265-192-2
UN 1268
Uses: Solvent
Features: Low boiling; hydrotreated to reduce aromatics and olefins to very low level
Regulatory: SARA §302/304 nonreportable, §311/312 acute health/chronic health/fire hazard, §313 nonreportable; Calif. Prop. 65 reportable (benzene, toluene); may contain detectable amts. of n-hexane, cyclohexane, 2,2,4-trimethylpentane, methylcyclopentane
Properties: Colorless clear liq.; lt. paraffinic hydrocarbon odor; misc. with most org. solvs.; negligible sol. in cold water; sp.gr. 0.69-0.72 (60/60 F); dens. 5.825 lb/gal (15.6 C); visc. < 1 cSt (40 C); vapor pressure 2.10 psi; f.p. -135 to -90 C; b.p. 90-110 C; flash pt. (TCC) -9 to -7 C; autoignition temp. 246 C; ref. index 1.3915; KB value 29.5; VOC 700 g/l; 100% act.
Toxicology: Mist/vapor may irritate eyes, mucous membranes, respiratory tract; liq. contact may cause minimal to mild eye irritation, mild to mod. skin irritation/inflamm.; may be harmful by inh., skin absorp.; overexposure may cause CNS depression, other target organ effects; may be harmful or fatal if ingested; aspiration into lungs can cause pulmonary edema, chem. pneumonia; prolonged/repeated inh. may cause heart arrhythmias; sensitizer; TSCA listed
Environmental: May cause oxygen depletion in waterways; potentially toxic to water ecosystems; may be hazardous to aquatic life; prevent entry into sewers, waterways; not a DOT marine pollutant
Precaution: Extremely flamm. liq.; flamm. limits 1.0-6.7%; vapor may cause flash fire or explosion; vapors are heavier than air, may flash back; may create vapor/air explosion hazard in confined spaces; spills may create slipping hazard; floats on water; container can rupture in fire; prevent entry into basements, confined areas; bond and ground equip.; incompat. with strong acids, alkalis, oxidizers (liq. chlorine, hydrogen peroxide, oxygen)
Hazardous decomp. prods.: Combustion: smoke, CO, CO_2, possibly other harmful gases/vapors
HMIS: Health 1, Flammability 3, Reactivity 0
Storage: Store in tightly closed containers in cool, dry, well-ventilated place away from heat, sparks, ignition sources, strong oxidizing conditions

Special Naphtholite 66/3 [CITGO]
Chem. Descrip.: VM&P naphtha
CAS 8032-32-4; EINECS/ELINCS 232-453-7
UN 1268
Uses: Solvent
Features: Low boiling; hydrotreated to reduce aromatics and olefins; meets definition of nonphotochemically reactive per 66/3 Rule requirements
Regulatory: FDA 21CFR §172.882, 175.105, 178.3530; SARA §302/304 nonreportable, §311/312 acute health/chronic health/fire hazard, §313 nonreportable; Calif. Prop. 65 reportable (benzene, toluene)
Properties: Colorless transparent liq.; lt. hydrocarbon odor; negligible sol. in cold water (< 0.01%); sp.gr. 0.750-0.770 (60/60 F); dens. 6.32 lb/gal; visc. 0.786 cSt (20 C); vapor pressure 5.2 mm Hg (20 C); f.p. -57 to -25 C; b.p. 125-151 C; flash pt. (TCC)18-20 C; autoignition temp. ≈ 232 C; ref. index 1.4198 (20 C); KB value 36; VOC 758 g/l
Toxicology: ACGIH TLV TWA 300 ppm; LD50 (oral, rat) > 5000 mg/kg; mist/vapor may irritate mucous membranes, respiratory tract; liq. con-

tact may cause minimal eye, mod. to severe skin irritation/inflamm.; may be harmful by inh., skin absorp.; may be harmful or fatal by ing. ; overexposure may cause CNS depression, other target organ effects; aspiration into lungs can cause pulmonary edema, chem. pneumonia; TSCA listed

Environmental: May cause oxygen depletion in waterways; potentially toxic to water ecosystems; may be hazardous to aquatic life; may contribute to smog; prevent entry into sewers, waterways; not a DOT marine pollutant

Precaution: Flamm.; flamm. limits 0.9-6.7%; vapor may cause flash fire; vapors are heavier than air, may flash back; may create vapor/air explosion hazard in confined space; spills may create slipping hazard; floats on water; container can rupture in fire; prevent entry into basements, confined areas; bond and ground equip.; incompat. with strong acids, alkalis, oxidizers (liq. chlorine, other halogens, hydrogen peroxide, oxygen)

Hazardous decomp. prods.: Combustion: smoke, CO, CO_2, possibly other harmful gases/vapors

HMIS: Health 1, Flammability 3, Reactivity 0

Storage: Store in tightly closed containers in cool, dry, well-ventilated place away from heat, sparks, ignition sources, strong oxidizing conditions

Speziol C 10/2 [Cognis]

Chem. Descrip.: 1,10-Decanediol

CAS 112-47-0; EINECS/ELINCS 203-975-2

Uses: Personal care prods.; flavors; fragrances; polyesters; acrylates; epoxy diluents; plastics; coatings; inks; photographic materials; textiles; leather; lubricants; polyurethanes; agric.; solvents; modifiers in alkyd resins; pharmaceuticals; fine chems.; plasticizers

Properties: Colorless pellets; mild odor; solid. pt. 68-73 C; acid no. 0-1.0; iodine no. 0-0.7; sapon. no. 0-3.0; hyd. no. 525-645; 98-100% act.

Storage: 2 yrs shelf life when stored in sealed original bags below solid. pt.; may become pasty during longer storage

Speziol C 18/2 [Cognis]

Chem. Descrip.: 12-Hydroxystearyl alcohol

CAS 2726-73-0; EINECS/ELINCS 220-342-6

Uses: Personal care prods.; flavors; fragrances; polyesters; acrylates; epoxy diluents; plastics; coatings; inks; photographic materials; textiles; leather; lubricants; polyurethanes; agric.; solvents; modifiers in alkyd resins; pharmaceuticals; fine chems.; plasticizers

Properties: Colorless pellets; mild odor; solid. pt. 61-65 C; acid no. 0-1; iodine no. 0-0.7; sapon. no. 0-2; hyd. no. 345-360; 75-85% act.

Storage: 2 yrs shelf life when stored in sealed original bags below solid. pt.; may become pasty during longer storage

Spirdane BT [TotalFina-Elf]

Chem. Descrip.: C9-11 isoparaffin

CAS 64742-48-9; EINECS/ELINCS 265-150-3

Uses: Solvent, diluent for paints, varnishes, household prods., industrial cleaning, lubricants, wood preservatives, degreasing, textiles

Features: Good solvency, relatively low aromatic content

Properties: Saybolt 30 color; dens. 770 kg/m^3; visc. 1.3 mm^2/s (20 C); vapor pressure 3 mbar (20 C); i.b.p. 155 C; aniline pt. 68 C; flash pt. (Abel) 42 C; ref. index 1.430 (20 C); surf. tens. 25.50 dynes/cm (20 C); KB value 32; 3.5% aromatics; 1 ppm benzene

Spirdane D30 [TotalFina-Elf]

Chem. Descrip.: C8-9 isoparaffin

CAS 64742-48-9; EINECS/ELINCS 265-150-3

Uses: Solvent for paints, varnishes, lubricants, degreasing

Features: Low aromatic

Properties: Saybolt 30 color; dens. 765 kg/m^3; visc. 1.1 mm^2/s (20 C); vapor pressure 6 mbar (20 C); i.b.p. 145 C; aniline pt. 67 C; flash pt. (Abel) 37 C; ref. index 1.422 (20 C); surf. tens. 25.00 dynes/cm (20 C); KB value 33; 0.01% aromatics; 1 ppm benzene

Spirdane D30L [TotalFina-Elf]

Chem. Descrip.: C8-9 isoparaffin

CAS 64742-48-9; EINECS/ELINCS 265-150-3

Uses: Solvent for degreasing

Features: Low aromatic; fast evaporation rate

Properties: Saybolt 30 color; dens. 767 kg/m^3; visc. 1.1 mm^2/s (20 C); vapor pressure 6 mbar (20 C); i.b.p. 147 C; aniline pt. 68 C; flash pt. (Abel) 37 C; ref. index 1.410 (20 C); surf. tens. 24.80 dynes/cm (20 C); KB value 33; 0.01% aromatics; 1 ppm benzene

Spirdane D40 [TotalFina-Elf]

Chem. Descrip.: C9-11 isoparaffin

CAS 64742-48-9; EINECS/ELINCS 265-150-3

Uses: Solvent for paints, varnishes, household prods., industrial cleaning, lubricants, wood preservatives, degreasing, textiles

Features: Low aromatic

Properties: Saybolt 30 color; dens. 770 kg/m^3; visc. 1.3 mm^2/s (20 C); vapor pressure 3 mbar (20 C); i.b.p. 155 C; aniline pt. 68 C; flash pt. (Abel) 42 C; ref. index 1.430 (20 C); surf. tens. 25.50 dynes/cm (20 C); KB value 31; 0.01% aromatics; 1 ppm benzene

Spirdane D40L [TotalFina-Elf]

Chem. Descrip.: C9-10 isoparaffin

CAS 64742-48-9; EINECS/ELINCS 265-150-3

Uses: Solvent for paints, varnishes, degreasing

Features: Low aromatic; fast evaporation rate

Properties: Saybolt 30 color; dens. 770 kg/m^3; visc. 1.2 mm^2/s (20 C); vapor pressure 3 mbar (20 C); i.b.p. 158 C; aniline pt. 67 C; flash pt. (Abel) 44 C; ref. index 1.415 (20 C); surf. tens. 25.30 dynes/cm (20 C); KB value 31; 0.01% aromatics; 1 ppm benzene

Spirdane D60 [TotalFina-Elf]

Chem. Descrip.: C9-12 isoparaffin

CAS 64742-48-9; EINECS/ELINCS 265-150-3

Uses: Solvent for paints, varnishes, household prods., industrial cleaning, lubricants, wood preservatives, degreasing, textiles

Features: Low aromatic

Properties: Saybolt 30 color; dens. 785 kg/m^3; visc. 1.7 mm^2/s (20 C); vapor pressure 1 mbar (20 C); i.b.p. 180 C; aniline pt. 73 C; flash pt. (PM) 63 C; ref. index 1.435 (20 C); surf. tens. 26.30 dynes/cm (20 C); KB value 31; 0.01% aromatics; 1 ppm benzene

Spirdane D60L [TotalFina-Elf]

Chem. Descrip.: C10-11 isoparaffin

CAS 64742-48-9; EINECS/ELINCS 265-150-3

Uses: Solvent for paints, varnishes, lubricants, degreasing

Features: Low aromatic; fast evaporation rate

Properties: Saybolt 30 color; dens. 772 kg/m^3; visc. 1.6 mm^2/s (20 C); vapor pressure 1 mbar (20 C); i.b.p. 180 C; aniline pt. 72 C; flash pt. (PM) 59 C; ref. index 1.430 (20 C); surf. tens. 26.30 dynes/cm (20 C); KB value 31; 0.01% aromatics; 1 ppm benzene

Spirdane HT [TotalFina-Elf]

Chem. Descrip.: Naphtha, hydrodesulfurized heavy

CAS 64742-82-1; EINECS/ELINCS 265-185-4

Uses: Solvent, diluent for paints, varnishes, household prods., industrial cleaning, lubricants, wood preservatives, degreasing, textiles

Features: Exc. solvency

Properties: Saybolt 30 color; dens. 785 kg/m^3; visc. 1.2 mm^2/s (20 C); vapor pressure 3 mbar (20 C); i.b.p. 155 C; aniline pt. 56 C; flash pt. (Abel) 42 C; ref. index 1.435 (20 C); surf. tens. 26.00 dynes/cm (20 C); KB value 37; 19 vol.% aromatics; 40 ppm benzene

Spirdane K1 [TotalFina-Elf]

Chem. Descrip.: C9-12 isoparaffin

CAS 64742-48-9; EINECS/ELINCS 265-150-3

Uses: Solvent, diluent for paints, varnishes, wood preservatives, degreasing

Features: Good solvency, relatively low aromatic content, high flash pt.

Properties: Saybolt 30 color; dens. 785 kg/m^3; visc. 1.7 mm^2/s (20 C); vapor pressure 1 mbar (20 C); i.b.p. 180 C; aniline pt. 70 C; flash pt. (PM) 63 C; ref. index 1.435 (20 C); surf. tens. 26.30 dynes/cm (20 C); KB value 32; 3% aromatics; 1 ppm benzene

Spirdane K2 [TotalFina-Elf]

Chem. Descrip.: C9-12 isoparaffin

CAS 64742-48-9; EINECS/ELINCS 265-150-3

Uses: Solvent, diluent for paints, varnishes, lubricants, wood preservatives, textiles

Features: Exc. solvency, high flash pt.

Properties: Saybolt 30 color; dens. 795 kg/m^3; visc. 1.7 mm^2/s (20 C); vapor pressure 1 mbar (20 C); i.b.p. 180 C; aniline pt. 62 C; flash pt. (PM) 62 C; ref. index 1.445 (20 C); surf. tens. 26.90 dynes/cm (20 C); KB value 34; 20 vol.% aromatics; 3 ppm benzene

Spirdane L1 [TotalFina-Elf]

Chem. Descrip.: C8-9 isoparaffin

CAS 64742-48-9; EINECS/ELINCS 265-150-3

Uses: Solvent for paints, varnishes, household prods., industrial clean-

ing, degreasing, chemicals
Features: Low aromatic
Properties: Saybolt 30 color; dens. 750 kg/m^3; visc. 0.9 mm^2/s (20 C); vapor pressure 10 mbar (20 C); i.b.p. 135 C; aniline pt. 66 C; flash pt. (Abel) 24 C; ref. index 1.416 (20 C); surf. tens. 24.50 dynes/cm (20 C); KB value 32; 0.01% aromatics; 1 ppm benzene

SR 256 [Sartomer]
Chem. Descrip.: 2-(2-Ethoxyethoxy)-ethyl acrylate with MEHQ inhibitor (1000 ppm)
CAS 7328-17-8; EINECS/ELINCS 230-811-7
Uses: Monomer, solvent for radiation-curable inks, coatings
Features: Monofunctional; can be polymerized by the use of heat, catalysts, and/or irradiation; esp. sensitive to UV lt.
Properties: APHA 70 clear liq.; mild odor; m.w. 188; sp.gr. 1.013; dens. 9.4 lb/gal; flash pt. 216 F (PMCC); ref. index 1.4355
Storage: Store away from direct sunlight, oxidizing agents and materials which may generate free radicals; storage temps. should not be > 90 F

Stearate 1500 [Gattefosse SA]
Chem. Descrip.: PEG (1500) palmitostearate
Uses: Solvent, emulsifier for cosmetic/pharmaceutical emulsions
Properties: Drop pt. 46-50 C; HLB 16; acid no. < 6; iodine no. < 3; sapon. no. 30-45
Toxicology: Nonirritating to skin and eyes

STEPAN® IPM [Stepan]
Chem. Descrip.: Isopropyl myristate
CAS 110-27-0; EINECS/ELINCS 203-751-4
Uses: Lubricant in textile spin finish, coning oils, carding, dye bath; emollient, solubilizer, vehicle, solvent for makeup, shaving preps., bath oils, hair preps., highly pigmented prods.
Features: Replacement for min. oil
Properties: APHA 20 max. clear liq., pract. odorless; insol. in water; sol. in IPA, min. oil, peanut oil; f.p. -3 C; m.p. -3 C; b.p. 162.5 C; acid no. 1 max.; iodine no. 1 max.; sapon. no. 204-212; flash pt. (COC) 305 F; ref. index 1.433; 100% act.
Environmental: Biodeg.

STEPAN® IPM-NF [Stepan]
Chem. Descrip.: Isopropyl myristate NF
CAS 110-27-0; EINECS/ELINCS 203-751-4
Uses: Solvent, emollient, feel enhancer, hair conditioner, solubilizer, blending agent for bath oils, creams/lotions, eye shadows, dusting and talcum powds., lipsticks, makeup prods., cleansers, and personal care
Properties: Clear liq.; sol. in IPA, min. oil, peanut oil; insol. in water; sp. gr. 0.8534 g/ml; dens. 0.85 g/ml; visc. 4.5 cps (25 C); m.p. -3 C; b.p. 163 C; acid no. 0.24; iodine no. 0.19; sapon. no. 208; pour pt. 6 C; cloud pt. 4 C; flash pt. (COC) 152 C; ref. index 1.433 (25 C); pH 4.5 min. (10%); 99.9% act.
Toxicology: LD50 (oral, mouse) 49,700 mg/kg, (skin, rabbit) 5 g/kg; may be irritating to eyes, respiratory system; prolonged or excessive skin contact may cause mild skin irritationTSCA listed
Environmental: Biodegradable
Precaution: May react with strong oxidizing agents, strong alkalis
Hazardous decomp. prods.: May evolve IPA if hydrolyzed or saponified
HMIS: Health 1, Flammability 1, Reactivity 0
Storage: Store in sealed containers in a cool, dry place; prolonged storage at temps. >32 C should be avoided; avoid overheating or freezing

STEPAN® IPP [Stepan]
Chem. Descrip.: Isopropyl palmitate
CAS 142-91-6; EINECS/ELINCS 205-571-1
Uses: Lubricant in textile spin finish, coning oils, carding, dye bath; emollient, solubilizer, vehicle, solvent for makeup, shaving preps., bath oils, hair preps., highly pigmented prods.
Features: Replacement for min. oil
Properties: APHA 20 max. clear liq., pract. odorless; insol. in water; sol. in IPA, min. oil, peanut oil; sp.gr. 0.849-0.855; f.p. 13 C; b.p. 172 C (4 mm); acid no. 1 max.; iodine no. 1 max.; sapon. no. 185-191; flash pt. (COC) 323 F; ref. index 1.437; 100% act.
Environmental: Biodeg.

STEPAN® IPP-NF [Stepan]
Chem. Descrip.: Isopropyl palmitate
CAS 142-91-6; EINECS/ELINCS 205-571-1
Uses: Emollient, solvent for min. oil/silicone/lanolin in personal care
Features: Dry, soft, nonoily; economical
Properties: Liq.

STEPANATE® AXS [Stepan]
Chem. Descrip.: Ammonium xylene sulfonate
CAS 26447-10-9; EINECS/ELINCS 247-710-9
Uses: Hydrotrope, solubilizer, coupling agent in detergents, heavy-duty cleaners, wax strippers, dishwashing detergents; solvent; fluidizer; freeze-thaw stabilizer; additive for high electrolyte or brine systems
Properties: Colorless liq.; water-sol.; dens. 9.05 lb/gal; pH 6.5-7.2 (1% aq.); anionic; 40% act.

Stetlasolv SES-CONC. [Stev-Tex Labs]
Chem. Descrip.: Self-emulsifiable solvent
Uses: Solvent scouring agent for textiles for removal of dirt, silicone, stubborn solids, waxes, etc., suitable in jet machines; emulsifier for removal of silicone oils and lubricants, and removal of silicone residue from machinery; antiredeposition agent in scour/dye operations or in repairing spotted goods
Features: Mod. foaming; compat. with most anionic, cationic, and nonionic prods. used in textile processing
Properties: Lt. amber liq.; disp. in warm water; pH 5.5-6.5 (2%); pseudo nonionic
Use Level: 0.5 g/l (prescouring); 3 g/l (machine cleaning)
Environmental: Biodeg.
Storage: Protect from freezing; freezing will occur below 40 F; if freezing occurs, allow to thaw completely @ R.T. and mix well before using

Stetlasolv VJS [Stev-Tex Labs]
Chem. Descrip.: Aromatic and paraffinic hydrocarbons with emulsifier
Uses: Solvent, emulsifier for removal of nat. and syn. waxes and oils from fabrics; improves penetration of direct and reactive dyes into heavyweight cotton fabrics
Features: Very low foaming; stable to 212 F; no negative effect on sol. or strike rate of direct, reactive, or disperse dyes; compat. with conventional and high temp. stable amylase enzymes
Properties: Pale yel. to water-wh. clear liq.; sp.gr. 0.89; pH 7.0 ± 0.5; wetting 4.2 s (75 F)

Strimix® 841 [Atofina SA]
Chem. Descrip.: Dimethylsulfoxide formulation
Uses: Solvent for stripping of marine antifouling paints
Features: Does not affect gel-coat primer which acts as waterproofing sealant on polyester hulls; low volatile content; easy to apply; fast action
Properties: Creamy consistency

Sulfa-Hitech® 0382 [Atofina]
Chem. Descrip.: DMDS-based
Uses: Solvent to dissolve sulfur in the prod. of sour gas wells, in sour-gas pipelines, and in refinery and chemical plant flowlines
Properties: Sp.gr. 1.060; b.p. 107-111 C

Sulfolane W [ChevronPhillips]
Chem. Descrip.: Tetramethylene sulfone
CAS 126-33-0; EINECS/ELINCS 204-783-1
Uses: Extraction solvent for extraction of benzene, toluene, other aromatic hydrocarbons from oil refinery streams; solvent in Sulfinol process to remove acid gases; solvent for separation of low boiling alcohols, min. oils, tars; plasticizer; polymerization solvent; dielec. in elec. equipment; solvent in surf. coatings; component in hydraulic fluids
Features: Highly polar compd. with outstanding solvent properties and chemical and thermal stability
Properties: Sp.gr. 1.264; f.p. 42-50 F; b.p. 284.8 C (5%, 760 mm Hg); flash pt. 330 F; 2.6-3.0% water

Sunoco 100 [Sunoco]
Uses: Solvent for catalysts, wood preservatives, additives, inks, industrial cleaners, herbicides, insecticides
Features: Highly specialized; slow evaporating; high boiling range; rec. where VOC limits are an issue

Sunoco 150 [Sunoco]
Uses: Solvent for catalysts, wood preservatives, additives, inks, industrial cleaners, herbicides, insecticides
Features: Highly specialized; slow evaporating; high boiling range; rec. where VOC limits are an issue

Super Hi-Flash Naphtha [CITGO]
Chem. Descrip.: Light aromatic solvent naphtha
CAS 64742-95-6; EINECS/ELINCS 265-199-0
UN 1268; UN NA1993
Uses: Solvent

Regulatory: SARA §302/304 nonreportable, §311/312 acute health/chronic health/fire hazard, §313 reportable [xylenes (0-3%), ethylbenzene (0-0.6%), cumene (0-3%), 1,2,4-trimethylbenzene (10-45%)]; Calif. Prop. 65 reportable (benzene, toluene, naphthalene)
Properties: Colorless transparent liq.; mod. sweet, pleasant, aromatic hydrocarbon odor; sol. in methanol, diethyl ether; sl. sol. in cold water (< 0.05%); sp.gr. 0.87; visc. < 20 cSt (40 C); vapor pressure 2.7 mm Hg (20 C); f.p. -25 to 13 C; b.p. 136-180 C; flash pt. (TCC) 42-45 C; autoignition temp. ≈ 343 C; KB value 91; VOC 873 g/l
Toxicology: Mist/vapor may irritate eyes, mucous membranes, respiratory tract; liq. contact may cause minimal to mild eye, mild to mod. skin irritation/inflamm.; may be harmful by inh., skin absorp.; may be harmful or fatal by ing.; overexposure may cause CNS depression, other target organ effects; aspiration into lungs can cause pulmonary edema, chem. pneumonia; prolonged/repeated inh. may cause heart arrhythmias; possible reproductive hazard, tumorigen; TSCA listed
Environmental: May cause oxygen depletion in waterways; potentially toxic to water ecosystems; may be hazardous to aquatic life; may contribute to smog; prevent entry into sewers, waterways; delisted as DOT marine pollutants
Precaution: Combustible; flamm. limits 0.9-7.0%; vapor may cause flash fire; vapors are heavier than air, may flash back; may create vapor/air explosion hazard in confined space; spills may create slipping hazard; floats on water; container can rupture in fire; prevent entry into basements, confined areas; bond and ground equip.; incompat. with strong acids, alkalis, oxidizers (liq. chlorine, hydrogen peroxide, oxygen)
Hazardous decomp. prods.: Combustion: smoke, CO, CO_2, possibly other harmful gases/vapors
HMIS: Health 2, Flammability 2, Reactivity 0
Storage: Store in tightly closed containers in cool, dry, well-ventilated place away from heat, sparks, ignition sources, strong oxidizing conditions

Super Lacolene [Ashland]
Chem. Descrip.: Aliphatic
Uses: Solvent
Properties: API gr. 56.2 (60 F); i.b.p. 195 F; flash pt. (TCC) 22 F; KB value 44

Super Refined® Almond NF [Croda Inc; Croda Oleochems.]
Chem. Descrip.: Sweet almond oil NF
CAS 8007-69-0
Uses: Emollient, spreading agent for creams, lotions, bath oils; solvent, vehicle for pharmaceuticals (parenteral formulations for intramuscular administration, nutritional supplements, IV emulsions, liniments, ointments)
Features: Provides elegant skin feel
Properties: APHA 40 oil, odorless, tasteless; sol. in min. oil, IPA
Toxicology: LD50 (oral, rat) > 5 g/kg; nonirritating to skin

Super Refined® Corn NF [Croda Inc]
Chem. Descrip.: Corn oil NF
CAS 8001-30-7; EINECS/ELINCS 232-281-2
Uses: Solvent, vehicle for pharmaceuticals (orals, topicals, parenteral formulations for intramuscular administration, nutritional supplements, IV emulsions, liniments, ointments)
Features: Oleaginous
Properties: APHA 40 oil, odorless, tasteless; sol. in min. oil, IPA; insol. in water, propylene glycol; HLB 7.89
Toxicology: LD50 (oral, rat) > 30 g/kg; minimal skin irritant; nonirritating to eyes

Super Refined® Cottonseed NF [Croda Inc]
Chem. Descrip.: Cottonseed oil NF/USP
CAS 8001-29-4; EINECS/ELINCS 232-280-7
Uses: Solvent, vehicle, emollient for pharmaceuticals (orals, topicals, parenteral formulations for intramuscular administration, nutritional supplements, IV emulsions, liniments, ointments)
Features: Oleaginous
Properties: APHA 50 oil, odorless, tasteless; sol. in min. oil, IPA; insol. in water, propylene glycol; HLB 7.85
Toxicology: LD50 (oral, rat) > 30 g/kg; minimal skin irritant; nonirritating to eyes

Super Refined® Crodamol IPM [Croda Inc]
Chem. Descrip.: Isopropyl myristate NF
CAS 110-27-0; EINECS/ELINCS 203-751-4
Uses: Solvent for topical pharmaceuticals
Features: Rapid absorp. into skin; highly purified
Properties: Colorless liq.; sol. in min. oil, IPA; insol. in water, propylene glycol
Use Level: 2-10%

Super Refined® Crodamol SCO [Croda Inc]
Chem. Descrip.: Stearyl octanoate and cetyl octanoate
Uses: Emollient, spreading agent, solvent, water repellent, film-former for topical pharmaceuticals
Features: Highly purified
Properties: Colorless liq.; sol. in min. oil, IPA; insol. in water, propylene glycol
Use Level: 5-10%

Super Refined® Olive NF [Croda Inc; Croda Oleochems.]
Chem. Descrip.: Olive (Olea europaea) oil NF
CAS 8001-25-0; EINECS/ELINCS 232-277-0
Uses: Emollient for cosmetics; lubricant for hair care prods.; solvent, vehicle, emollient, lubricant for pharmaceuticals (orals, topicals, parenteral formulations for intramuscular administration, nutritional supplements, IV emulsions, liniments, ointments)
Features: Oleaginous
Properties: APHA 30 oil, odorless, tasteless; sol. in min. oil, IPA; insol. in water, propylene glycol; HLB 7.66
Toxicology: LD50 (oral, rat) > 30 g/kg; minimal skin irritant; nonirritating to eyes

Super Refined® Peanut NF [Croda Inc]
Chem. Descrip.: Peanut oil NF/USP
CAS 8002-03-7; EINECS/ELINCS 232-296-4
Uses: Emollient in skin care; solvent, vehicle for pharmaceuticals (orals, topicals, parenteral formulations for intramuscular administration, nutritional supplements, IV emulsions, liniments, ointments)
Features: Oleaginous
Properties: APHA 30 oil, odorless, tasteless; sol. in min. oil, IPA; insol. in water, propylene glycol; HLB 7.78
Toxicology: LD50 (oral, rat) > 30 g/kg; minimal skin/eye irritant

Super Refined® Safflower USP [Croda Inc; Croda Oleochems.]
Chem. Descrip.: Safflower oil USP
CAS 8001-23-8; EINECS/ELINCS 232-276-5
Uses: Solvent, vehicle, emollient for pharmaceuticals (orals, topicals, parenteral formulations for intramuscular administration, nutritional supplements, IV emulsions, liniments, ointments)
Features: Oleaginous
Properties: APHA 30 mobile oil, odorless, tasteless; sol. in min. oil, IPA; insol. in water, propylene glycol; HLB 7.72
Toxicology: LD50 (oral, rat) > 5 g/kg; minimal skin/eye irritant

Super Refined® Sesame NF [Croda Inc; Croda Oleochems.]
Chem. Descrip.: Sesame oil NF
CAS 8008-74-0; EINECS/ELINCS 232-370-6
Uses: Emollient in skin care; solvent, vehicle, emollient, lubricant for pharmaceuticals (orals, topicals, parenteral formulations for intramuscular administration, nutritional supplements, IV emulsions, liniments, ointments)
Regulatory: NF, BP compliance
Properties: APHA 30 clear oily liq., odorless, tasteless; sol. in min. oil, IPA; insol. in water, propylene glycol; sp.gr. 0.916-0.921; HLB 7.83; iodine no. 103-116; sapon. no. 188-195; hyd. no. 5 max.; ref. index 1.472-1.476; 0.05% max. moisture
Toxicology: LD50 (oral, rat) > 5 g/kg; mild skin irritant, minimal eye irritant

Super Refined® Soybean USP [Croda Inc; Croda Oleochems.]
Chem. Descrip.: Soybean oil USP
CAS 8001-22-7; EINECS/ELINCS 232-274-4
Uses: Emollient in skin care; solvent, vehicle, nutritional source, lubricant for pharmaceuticals (orals, topicals, parenteral formulations for intramuscular administration, nutritional supplements, IV emulsions, liniments, ointments)
Features: Oleaginous
Properties: APHA 60 max. clear oily liq., odorless, tasteless; sol. in min. oil, IPA; insol. in water, propylene glycol; sp.gr. 0.916-0.922; HLB 7.66; iodine no. 120-141; sapon. no. 180-200; hyd. no. 5 max.; ref. index 1.465-1.475
Toxicology: LD50 (oral, rat) > 5 g/kg; mild skin irritant, minimal eye irritant

Supersol CSS [Bayer/Industrial Chems./Textile]
Uses: Solvent scouring agent for removal of wax and oil in textiles
Features: Caustic stable

Properties: Anionic/nonionic

Sure Sol®-150ND [Koch Chem. Spec.]
Chem. Descrip.: Diethylbenzene
CAS 25340-17-4
Uses: Solvent for textile, coatings, and resin industries
Regulatory: SARA nonreportable
Properties: Sp.gr. 0.8796; flash pt. 145 F; > 98% aromatics

Sure Sol®-202 [Koch Chem. Spec.]
Chem. Descrip.: Mixed methylnaphthalenes
Uses: Solvent for textile, agric., coatings, and resin industries
Properties: Sp.gr. 1.0001; i.b.p. 458 F; pour pt. -5 F; flash pt. 220 F; KB value 98; > 98% aromatics

Sure Sol®-205 [Koch Chem. Spec.]
Chem. Descrip.: Mixed methylnaphthalenes
Uses: Solvent for textile, agric., coatings, and resin industries
Properties: Sp.gr. 1.0042; flash pt. 315 F; > 98% aromatics

Surfadone® LP-100 [ISP]
Chem. Descrip.: Caprylyl pyrrolidone
CAS 2687-94-7
Uses: Surfactant for personal care; solvent for commercial cleaning, textile processing, lithographic printing prods., water-borne coatings, inks; leveling agent; dispersant; emulsification aid; visc. builder; pigment dispersant
Features: VOC replacement
Properties: Pale yel. low visc., clear to sl. hazy liq.; sol. in polar and nonpolar solvs. incl. ethanol, acetone, xylene, heptane, paraffin oil, Stod., perchloroethylene; sparingly sol. in water; m.w. 192; sp.gr. 0.92; vapor pressure < 0.001 mm Hg; b.p. 100 C (0.3 mm Hg); solid. pt. -25 C; HLB 6; flash pt. (TCC) 119 C; surf. tens. 28 dynes/cm; Draves wetting 3.5 s (0.1%); Ross-Miles foam 25 mm (initial, 0.1%); 97% min. act.
Toxicology: LD50 (oral) 2.05 g/kg; not a primary skin or eye irritant; noncomedogenic
Environmental: Biodeg.

Surfadone® LP-300 [ISP]
Chem. Descrip.: Lauryl pyrrolidone
CAS 2687-96-9; EINECS/ELINCS 403-730-1
Uses: Surfactant, bodying agent, gloss aid, foam booster/stabilizer, visc. builder, clarifier, combability improver for shampoos; leveling agent; emulsifier; pigment dispersant; solvent for commercial cleaning, textile processing, lithographic printing prods., water-borne coatings, inks, adhesives; absorption enhancer in pharmaceuticals
Features: VOC replacement
Properties: Pale yel. low visc., clear to sl. hazy liq.; sol. in polar and nonpolar solvs. incl. ethanol, acetone, xylene, heptane, paraffin oil, Stod., perchloroethylene; sparingly sol. in water; m.w. 253; sp.gr. 0.90; b.p. 145 C (0.2 mm Hg); solid. pt. 10 C; HLB 3; flash pt. (TCC) 116 C; surf. tens. 26 dynes/cm; Draves wetting > 300 s (0.1%); Ross-Miles foam 13 mm (initial, 0.1%); nonionic; 97% min. act.
Use Level: 1-2%
Toxicology: LD50 (oral) > 5 g/kg; not a primary skin or eye irritant; noncomedogenic
Environmental: Biodeg.

Surphos 515 [Christianson SA de CV]
Chem. Descrip.: POE linear fatty alcohol phosphate (99.5%), phosphoric acid (< 0.5%)
Uses: Emulsifier for agric. toxicants, incl. chlorinated insecticides and herbicidal esters; primary emulsifier in emulsion polymerization of acrylic monomers, vinyl acetates, and vinyl acrylics; textile antistat and lubricant; cosmetics additive; principal detergent in industrial alkaline cleaners for bottles, laundry, dairy, aircraft, metal, steam, or electrolytic cleaning; paint and stripping agent; solvent cleaner
Properties: Yel. clear liq.; disp. in water; sp.gr. 1.05-1.06; acid no. 100-120 (pH 5.5); flash pt. (PMCC) > 200 F; pH 1.5-2.5 (10% aq.); anionic; 2.0% moisture
Toxicology: Skin irritant; causes eye irritation that may lead to loss of sight; avoid breathing mist
Precaution: Avoid strong oxidizing or reducing agents
Hazardous Ingredients: Polyoxyethylene linear fatty alcohol phosphate, phosphoric acid

Surphos 620 [Christianson SA de CV]
Chem. Descrip.: POE alkylphenyl ether phosphate (99.5% min.), phosphoric acid (< 0.5%)
CAS 51811-79-1
Uses: Emulsifier for agric. toxicants, incl. chlorinated insecticides and herbicidal esters; emulsifier for methylene chloride, trichloroethylene; paint and wax stripping agent; solvent cleaner; corrosion inhibitor; detergent in industrial alkaline cleaners for bottles, laundry, dairy, aircraft, metal, steam, or electrolytic cleaning; lt. and heavy-duty household detergent, wall, and floor cleaner conc.; textile assistant and dedusting agent
Properties: Yel. visc. liq.; disp. in water; sp.gr. 1.00-1.20; visc. ≈ 4500 cps; acid no. 85-105 (pH 5.5); flash pt. (OC) 440 F; pH 1.0-2.5 (5% sol'n.); surf. tens. ≈ 30 dynes/cm; Ross-Miles foam 158/146 mm (49 C); anionic; 1.0% max. moisture
Toxicology: May cause eye irritation, conjunctivitis, opaqueness of cornea; may lead to loss of sight; inh. can irritate mucous membranes, nose, throat; causes inflammation, burns, blisters on skin; ing. can burn, blister mouth, throat, digestive tract
Precaution: Incompat. with strong oxidizing or reducing agents
Hazardous Ingredients: Polyoxyethylene alkylphenyl ether phosphate, phosphoric acid

Surphos 630 [Christianson SA de CV]
Chem. Descrip.: Phosphate ester
Uses: Emulsifier for agric. toxicants, incl. chlorinated insecticides and herbicidal esters; wetting agent for metal and engineering industry for pickling and plating baths; cutting and drilling oil; paint and wax stripping agent; solvent cleaner; detergent in industrial alkaline cleaners for bottles, laundry, dairy, aircraft, metal, steam, or electrolytic cleaning; lt. and heavy-duty household detergent, wall, and floor cleaner conc.; textile assistant
Properties: Lt. yel. visc. liq.; completely sol. in water; sp.gr. 1.08-1.091; visc. ≈ 4500 cps; acid no. 49-59 (pH 5.5); surf. tens. ≈ 30 dynes/cm; Ross-Miles foam 158/146 mm (49 C); anionic; 1.0% max. moisture

Surphos 645 [Christianson SA de CV]
Chem. Descrip.: POE alkylphenyl ether phosphate (99.5% min.), phosphoric acid (< 0.5%)
CAS 51811-79-1
Uses: Emulsifier for agric. toxicants, incl. chlorinated insecticides and herbicidal esters; emulsifier for methylene chloride, trichloroethylene; paint and wax stripping agent; solvent cleaner; corrosion inhibitor; detergent in industrial alkaline cleaners for bottles, laundry, dairy, aircraft, metal, steam, or electrolytic cleaning, lt.- and heavy-duty household detergents, wall and floor cleaner concs.; textile assistant and dedusting agent
Properties: Clear visc. liq.; disp. in water; sp.gr. 1.095-1.115; visc. ≈ 4500 cps; acid no. 62-70 (pH 5.5); b.p. > 235 C; pour pt. < 0 C; flash pt. (OC) 440 F; pH 1.5-2.5 (10% sol'n.); surf. tens. ≈ 35 dynes/cm; Ross-Miles foam 168/158 mm (49 C); anionic; 0.5% max. moisture
Toxicology: May cause eye irritation, conjunctivitis, opaqueness of cornea; may lead to loss of sight; inh. can irritate mucous membranes, nose, throat; causes inflammation, burns, blisters on skin; ing. can burn, blister mouth, throat, digestive tract
Precaution: Incompat. with strong oxidizing or reducing agents
Hazardous Ingredients: Polyoxyethylene alkylphenyl ether phosphate, phosphoric acid
Hazardous decomp. prods.: None

Sylfat® 9012 [Arizona]
Chem. Descrip.: Ester of selected tall oil fatty acids
Uses: Cosolvent, visc. reducer, detackifier in soybean oil ink systems, low-VOC lithographic inks, web-offset heat-set inks, high-solids inks
Features: Good solvency power; low volatility; low visc.; low m.w.; broad compat.; aliphatic ester functionality; allows fast heat-setting of web-offset heat-set inks
Properties: Gardner 8 max. color; sp.gr. 0.91; dens. 7.56 lb/gal; visc. 7 cps; acid no. 4 max.; iodine no.(Wijs) 110; sapon. no. 170; flash pt. (CC) > 300 F; > 98% NV

Syn Fac® TEA-97 [Milliken]
Chem. Descrip.: Ethoxylated amine
Uses: Solvent, dispersant for acidic materials; solubilizer for acid dyes, pigments, and optical brighteners in dye industry; dispersant for soils in metal cleaning
Properties: Gardner < 10 liq.; sp.gr. 1.10; visc. 400 cps; acid no. 22; hyd. no. 610; cloud pt. > 100 C (1%); pH 8.4 (10% sol'n.); nonionic; 100% conc.
Toxicology: Strongly alkaline; avoid contact with skin and eyes

T

Tavoxene® [Solvay SA]
Chem. Descrip.: Trichloroethylene
CAS 79-01-6; EINECS/ELINCS 201-167-4
Uses: Solvent for hot degreasing equip. incl. ultrasonic equip. for metal cleaning and to remove soils, greases, oils, waxes, resins, soldering fluxes; as heat transfer fluid
Features: Highly stabilized; volatile; high stability in severe conditions of applic. incl. in presence of aluminum and other light metals and their alloys; high quality; nonflamm.; multipurpose; non-ozone depleting
Properties: Colorless liq.; m.w. 131.4; sp.gr. 1.450-1.460 (20/4 C); vapor pressure 0.079 bar (20 C); b.p. 87 C; KB value 130; sp.heat 0.93 kJ/kg °K; 99.5% min. purity
Toxicology: Toxic; irritating to eyes, skin; may cause dermatitis; narcotic effect by inh. of vapor; suspected carcinogen
Environmental: Avoid emission and release to the environment
Precaution: Incompat. with strong acids, bases, and oxidizing agents; vapor is heavier than air; prevent vapor from coming into contact with high temp. source

TBP [FMC]
Chem. Descrip.: Tributyl phosphate
CAS 126-73-8; EINECS/ELINCS 204-800-2
Uses: Antifoam in paper coatings, aq. adhesives, inks, casein sol'ns., textile sizes, and detergent sol'ns.; plasticizer, solvent for NC, cellulose acetate, chlorinated rubber, and vinyls; solvent for natural gums and syn. resins; flame retardant for cellulosics, epoxies, PS, PVAc, PVC; component in extraction of rare earth metal salts
Properties: Pt-Co 50 max. liq.; char. odor; m.w. 266; f.p. < -80 C; b.p. 137-145 C; sol. in org. liqs., min. oil, and gasoline; sp.gr. 0.978 ± 0.003; dens. 975 kg/m^3; visc. 3.7 cp; pour pt. < -80 C; ref. index 1.423 ± 0.001; flash pt. 166 C; fire pt. 182 C

TCD Ketone A [Celanese; Celanese GmbH]
Chem. Descrip.: Tricyclo [5.2.1.0^2,6] decane-8-one
CAS 13380-94-4; EINECS/ELINCS 236-458-5
Uses: Raw material for fragrances; solvent in inks
Properties: Colorless liq.; char. ketone odor; sol. 3 g/l in water; sp.gr. 1.05; dens.8.74 lb/gal; f.p. ≈ -70 C; b.p. ≈ 240 C; flash pt. (CC) 102 C

T-Det® N-4 [Harcros]
Chem. Descrip.: Nonoxynol-4
CAS 9016-45-9; EINECS/ELINCS 230-770-5
Uses: Detergent; wetting agent; agric. toxicant formulations; intermediate for household detergents; dry cleaning soap additive; sludge dispersant additive in fuel oils; gasoline additive; solvent cleaner for metal processing; color enhancer, stabilizer, emulsifier for latex paints; food pkg. adhesives; emulsifier in mfg. of food-contact articles
Regulatory: FDA 21CFR §175.105, 178.3400; 40CFR §180.1001c,d,e
Properties: Pale yel. oily liq.; mild aromatic odor; sol. in butyl Cellosolve, CCl_4, min. and corn oil, ethanol, kerosene, min. spirits, perchloroethylene, Stod., toluene; dens. 8.5 lb/gal; sp.gr. 1.02; visc. 52 SUS (210 F); HLB 8.9; flash pt. > 400 F (OC); pour pt. < O F; pH 7; nonionic; 99.5% min. act.

T-Det® N-6 [Harcros]
Chem. Descrip.: Nonoxynol-6
CAS 9016-45-9
Uses: Emulsifier for silicones, household waxes and polishes; wetting agent; detergent, dispersant for petrol. and fuel oils; agric. toxicant formulations; dry cleaning soap additive; leather and metal processing; solvent cleaner; deinking agent for paper; food pkg. adhesives; emulsifier in mfg. of food-contact articles
Regulatory: FDA 21CFR §175.105, 178.3400; 40CFR §180.1001c,d,e
Properties: Pale yel. liq.; mild aromatic odor; sol. in butyl Cellosolve, CCl_4, corn oil, diesel fuel, ethanol, kerosene, Stod., xylene; dens. 8.7 lb/gal; sp.gr. 1.04; visc. 58 SUS (210 F); HLB 10.9; flash pt. > 400 F (OC); pour pt. < 0 F; pH 7; nonionic; 99.5% min. act.

Tebol® [Lyondell]
Chem. Descrip.: Tert. butyl alcohol
CAS 75-65-0; EINECS/ELINCS 200-889-7
Uses: Solvent, cosolvent, compatibilizer, coupling agent, processing aid for pharmaceuticals, personal care prods., aq. coatings and adhesives, agric. formulations, polymer processing, cleaners/disinfectants; chlorinated hydrocarbon stabilizer
Properties: Sol. 100% in water; m.w. 74.12; sp.gr. 0.80 (20/20 C); dens. 6.7 lb/gal (20 C); vapor pressure 30 mm Hg (20 C); b.p. 175 F; flash pt. 52 F

Tebol® 99 [Lyondell]
Chem. Descrip.: High purity tert. butyl alcohol
CAS 75-65-0; EINECS/ELINCS 200-889-7
Uses: Solvent, cosolvent, compatibilizer, coupling agent, processing aid for pharmaceuticals, personal care prods., aq. coatings and adhesives, agric. formulations, polymer processing, cleaners/disinfectants; chlorinated hydrocarbon stabilizer
Properties: Pt-Co ≤ 10 color; sol. in water; misc. with most org. solvs.; sp.gr. 0.78; dens. 6.5 lb/gal; visc. 3.3 cps (30 C); i.b.p. 81.5 C (760 mm Hg); f.p. 24.5 C; flash pt. (CC) 11 C; 99.3% act.

Tecsol® 1 [Eastman]
Chem. Descrip.: SD alcohol 1 (100 parts) with ethyl acetate (5) and heptane (1)
Uses: Solvent for shellac, lacquers, varnishes; latent solvent in NC lacquers to reduce sol'n. visc.
Properties: Completely sol. in water; m.w. 46.07; sp.gr. 0.789-0.820 (20/20C); dens. 6.576.83 lb/gal; visc. 1.201.5 cP (20 C); f.p. -173 F; b.p. 74-82; flash pt. (COC) 49-72 F; ref. index 1.3614 (20 C); surf. tens. 22.4 dynes/cm (20 C)

Tecsol® 3 [Eastman]
Chem. Descrip.: SD alcohol 1 (100 parts) with MIBK (1), ethyl acetate (1), and heptane (1)
Uses: Solvent for shellac, lacquers, varnishes; latent solvent in NC lacquers to reduce sol'n. visc.
Properties: Completely sol. in water; m.w. 46.07; sp.gr. 0.789-0.820 (20/20C); dens. 6.576.83 lb/gal; visc. 1.201.5 cP (20 C); f.p. -173 F; b.p. 74-82; flash pt. (COC) 49-72 F; ref. index 1.3614 (20 C); surf. tens. 22.4 dynes/cm (20 C)

Tecsol® A, 95% [Eastman]
Chem. Descrip.: Ethyl alcohol (80%), IPA (9%), methanol (4%), water (6%), MIBK (1%)
Uses: Solvent for coatings, plastic coatings, cellulosic coatings, printing inks
Regulatory: SARA §311/312 acute health/fire hazard, §313 reportable (methanol, MIBK); EINECS, Australia AICS listed
Properties: Colorless clear liq.; alcohol odor; sol. in water; sp.gr. 0.81-0.815 (15.6 C/15.6 C); vapor pressure 53 mbar (19 C); b.p. 78 C; flash

pt. (TCC) 13 C (ethanol); autoignition temp. > 404 C
Toxicology: Causes eye irritation; harmful by inh., ing.; high vapor concs. may cause drowsiness, eye/respiratory tract irritation; prolonged/repeated skin contact may cause drying, cracking, irritation; may cause skin sensitization; TSCA listed
Environmental: Prevent runoff from entering drains, sewers, streams
Precaution: Flamm. liq. and vapor; vapors may cause flash fire, ignite explosively, or flash back; reacts violently with strong oxidizing agents
Hazardous decomp. prods.: Combustion: CO, CO_2
HMIS: Health 2, Flammability 3, Reactivity 0
Storage: Store in tightly closed container in well-ventilated area away from heat, sparks, flame

Tecsol® A Anhydrous [Eastman]
Chem. Descrip.: Ethanol (86%), IPA (9%), methanol (4%), MIBK (1%)
Uses: Solvent for manufacturing processes, printing inks, plastic coatings; latent solvent in cellulosic coatings
Regulatory: SARA §311/312 acute health/chronic health/fire hazard, §313 reportable (methanol, MIBK); Canada DSL, EINECS, Australia AICS, Japan MITI listed
Properties: Colorless clear liq.; alcohol odor; completely sol. in water; m.w. 46.07; sp.gr. 0.789-0.820 (20/20C); dens. 6.576.83 lb/gal; visc. 1.201.5 cP (20 C); vapor pressure 53 mbar (19 C); f.p. -173 F; b.p. 74-82; flash pt. (COC) 49-72 F; ref. index 1.3614 (20 C); surf. tens. 22.4 dynes/cm (20 C)
Toxicology: Causes eye irritation; harmful by inh., ing.; high vapor concs. may cause drowsiness, eye/respiratory tract irritation; prolonged/repeated skin contact may cause drying, cracking, irritation; may cause skin sensitization; TSCA listed
Environmental: Prevent runoff from entering drains, sewers, streams
Precaution: Flamm. liq. and vapor; vapors may cause flash fire, ignite explosively, or flash back; residual vapors may explode on ignition; reacts violently with strong oxidizing agents
Hazardous decomp. prods.: Combustion: CO, CO_2
HMIS: Health 2, Flammability 3, Reactivity 0
Storage: Store in tightly closed container in well-ventilated area away from heat, sparks, flame

Tecsol® A-2, 95% [Eastman]
Chem. Descrip.: Ethyl alcohol (80%), methyl alcohol (13%), MIBK (1%), water (6%)
Uses: Solvent for manufacturing processes, coatings, plastic coatings, printing inks; latent solvent in cellulosic coatings
Regulatory: SARA §311/312 acute health/chronic health/fire hazard, §313 reportable (methanol, MIBK); EINECS, Australia AICS listed
Properties: Colorless clear liq.; alcohol odor; sol. in water; sp.gr. 0.81-0.815 (15.6/15.6 C); vapor pressure 52.8 mbar (19 C); b.p. 78 C; flash pt. (TCC) 13 C; autoignition temp. > 404 C
Toxicology: May be fatal or cause blindness if swallowed; vapor harmful; cannot be made nonpoisonous; harmful if absorbed thru skin; causes eye irritation; high vapor concs. may cause eye/respiratory tract irritation; may cause skin sensitization; TSCA listed
Environmental: Prevent runoff from entering drains, sewers, streams
Precaution: Flamm. liq. and vapor; vapors may flash back; residual vapors may explode on ignition; reacts with strong oxidizing agents
Hazardous decomp. prods.: Combustion: CO, CO_2
HMIS: Health 2, Flammability 3, Reactivity 0
Storage: Store in tightly closed container in well-ventilated area away from heat, sparks, flame

Tecsol® A-2 Anhydrous [Eastman]
Chem. Descrip.: Ethyl alcohol (86%), methanol (13%), MIBK (1%)
Uses: Solvent for coatings, plastic coatings, cellulosic coatings, printing inks
Regulatory: SARA §311/312 acute health/chronic health/fire hazard, §313 reportable (methanol, MIBK); Canada DSL, EINECS, Australia AICS, Japan MITI listed
Properties: Colorless clear liq.; alcohol odor; sp.gr. 0.791-0.796 (15.6 C/15.6 C); vapor pressure 52.8 mbar (19 C); b.p. 77 C; flash pt. (TCC) 13 C; autoignition temp. 410 C
Toxicology: May be fatal or cause blindness if swallowed; vapor harmful; cannot be made nonpoisonous; harmful if absorbed thru skin; causes eye irritation; high vapor concs. may cause eye/respiratory tract irritation; may cause skin sensitization; TSCA listed
Environmental: Prevent runoff from entering drains, sewers, streams
Precaution: Flamm. liq. and vapor; vapors may flash back; residual vapors may explode on ignition; reacts with strong oxidizing agents
Hazardous decomp. prods.: Combustion: CO, CO_2
HMIS: Health 2, Flammability 3, Reactivity 0
Storage: Store in tightly closed container in well-ventilated area away from heat, sparks, flame

Tecsol® B [Eastman]
Chem. Descrip.: Ethyl alcohol (86%), methyl alcohol (9%), isopropyl alcohol (4%), MIBK (1%)
Uses: Solvent for manufacturing processes, coatings, plastic coatings, printing inks; latent solvent in cellulosic coatings
Regulatory: SARA §311/312 acute health/chronic health/fire hazard, §313 reportable (methanol, MIBK); Canada DSL, EINECS, Australia AICS, Japan MITI, Korea ECL listed
Properties: Colorless clear liq.; alcohol odor; completely sol. in water; m.w. 46.07; sp.gr. 0.791-0.796 (15.6 C/15.6 C); dens. 6.576.83 lb/gal; visc. 1.201.5 cP (20 C); vapor pressure 53 mbar (19 C); f.p. -173 F; b.p. 74-82; flash pt. (COC) 49-72 F; ref. index 1.3614 (20 C); surf. tens. 22.4 dynes/cm (20 C)
Toxicology: May be fatal or cause blindness if swallowed; vapor harmful; cannot be made nonpoisonous; harmful if absorbed thru skin; causes eye irritation; high vapor concs. may cause drowsiness and eye/respiratory tract irritation; may cause skin sensitization; TSCA listed
Environmental: Prevent runoff from entering drains, sewers, streams
Precaution: Flamm. liq. and vapor; vapors may cause flash fire, ignite explosively, or flash back; residual vapors may explode on ignition; reacts with strong oxidizing agents
Hazardous decomp. prods.: Combustion: CO, CO_2
HMIS: Health 2, Flammability 3, Reactivity 0
Storage: Store in tightly closed container in well-ventilated area away from heat, sparks, flame

Tecsol® C 95% [Eastman]
Chem. Descrip.: Ethyl alcohol (83%), water (7%), ethyl acetate (5%), methyl alcohol (4%), MIBK (1%)
Uses: Solvent for manufacturing processes, printing inks, coatings, plastic coatings; latent solvent in cellulosic coatings; cosolvent; surf. tens. reducer, evaporation rate control aid for waterborne, industrial, and trade sales coatings
Regulatory: SARA §311/312 acute health/chronic health/fire hazard, §313 reportable (methanol, MIBK); Canada DSL, EINECS, Australia AICS, Japan MITI listed
Properties: Colorless clear liq.; alcohol odor; sol. in water; sp.gr. 0.816-0.821 (15.6 C/15.6 C); vapor pressure 86.5 mbar (20 C); b.p. 78 C; flash pt. (TCC) 12 C
Toxicology: Harmful by inh. or ing.; causes eye irritation; high vapor concs. may cause eye/respiratory tract irritation; prolonged/repeated skin contact may cause drying, cracking, irritation; may cause skin sensitization; TSCA listed
Environmental: Prevent runoff from entering drains, sewers, streams
Precaution: Flamm. liq. and vapor; vapors may cause flash fire, ignite explosively, or flash back; residual vapors may explode on ignition; reacts with strong oxidizing agents
Hazardous decomp. prods.: Combustion: CO, CO_2
HMIS: Health 2, Flammability 3, Reactivity 0
Storage: Store in tightly closed container in well-ventilated area away from heat, sparks, flame

Tecsol® C Anhydrous [Eastman]
Chem. Descrip.: Ethyl alcohol (89%), ethyl acetate (5%), methyl alcohol (5%), MIBK (1%)
Uses: Solvent for coatings, plastic coatings, cellulosic coatings, printing inks
Regulatory: SARA §311/312 acute health/chronic health/fire hazard, §313 reportable (methanol, MIBK); Canada DSL, EINECS, Australia AICS, Japan MITI listed
Properties: Colorless clear liq.; alcohol odor; completely sol. in water; m.w. 46.01; sp.gr. 0.789 (20/20 C); visc. 1.5 mPa·s (20 C); vapor pressure 77.1 mbar (20 C); f.p. -114 C; b.p. 74-82 C; flash pt. (TCC) 10 C; autoignition temp. 419 C; ref. index 1.3594; surf. tens. 22.5 dynes/cm (20 C); elec. resist. < 0.1 megohms
Toxicology: May be fatal or cause blindness if swallowed; vapor harmful; cannot be made nonpoisonous; harmful if absorbed thru skin; causes eye irritation; high vapor concs. may cause drowsiness and eye/respira-

tory tract irritation; may cause skin sensitization; TSCA listed
Environmental: Prevent runoff from entering drains, sewers, streams
Precaution: Flamm. liq. and vapor; vapors may flash back; residual vapors may explode on ignition; reacts with strong oxidizing agents
Hazardous decomp. prods.: Combustion: CO, CO_2
HMIS: Health 2, Flammability 3, Reactivity 0
Storage: Store in tightly closed container in well-ventilated area away from heat, sparks, flame

Tecsol® D, 95% [Eastman]
Chem. Descrip.: Ethyl alcohol (76%), isopropyl alcohol (13%), methyl alcohol (4%), water (6%), MIBK (1%)
Uses: Solvent for manufacturing processes, printing inks, coatings, plastic coatings; latent solvent in cellulosic coatings
Regulatory: SARA §311/312 acute health/chronic health/fire hazard, §313 reportable (methanol, MIBK); EINECS, Australia AICS, Japan MITI listed
Properties: Colorless clear liq.; alcohol odor; completely sol. in water; sp.gr. 0.81-0.815 (15.6/15.6 C); vapor pressure 53 mbar (19 C); b.p. 78 C; flash pt. (TCC) 13 C; autoignition temp. > 404 C
Toxicology: Causes eye irritation; harmful by inh. or ing.; high vapor concs. may cause drowsiness and eye/respiratory tract irritation; prolonged/repeated skin contact may cause drying, cracking, irritation; may cause skin sensitization; TSCA listed
Environmental: Prevent runoff from entering drains, sewers, streams
Precaution: Flamm. liq. and vapor; vapors may cause flash fire, ignite explosively, or flash back; residual vapors may explode on ignition; reacts violently with strong oxidizing agents
Hazardous decomp. prods.: Combustion: CO, CO_2
HMIS: Health 2, Flammability 3, Reactivity 0
Storage: Store in tightly closed container in well-ventilated area away from heat, sparks, flame

Tecsol® D-1, 95% [Eastman]
Chem. Descrip.: Ethyl alcohol (76.3%), IPA (16.6%), water (6.3%), MIBK (0.8%)
Uses: Solvent for coatings, plastic coatings, cellulosic coatings, printing inks
Regulatory: SARA §311/312 acute health/chronic health/fire hazard, §313 nonreportable; Canada DSL, EINECS, Australia AICS listed
Properties: Colorless liq.; alcohol odor; sol. in water; vapor pressure 40 mm Hg (19 C); b.p. 78 C; flash pt. (TCC) 13 C; autoignition temp. 404 C
Toxicology: Causes eye irritation; high vapor concs. may cause drowsiness and eye/respiratory tract irritation; prolonged/repeated skin contact may cause drying, cracking, irritation; may cause skin sensitization; TSCA listed
Environmental: Prevent runoff from entering drains, sewers, streams
Precaution: Flamm. liq. and vapor; vapors may cause flash fire, ignite explosively, or flash back; residual vapors may explode on ignition; reacts with strong oxidizing agents
Hazardous decomp. prods.: Combustion: CO, CO_2
HMIS: Health 2, Flammability 3, Reactivity 0
Storage: Store in tightly closed container in well-ventilated area away from heat, sparks, flame

Tecsol® D-2, 95% [Eastman]
Chem. Descrip.: Ethyl alcohol (77%), methyl alcohol (16%), water (6%), MIBK (1%)
Uses: Solvent for manufacturing processes, coatings, plastic coatings, printing inks; latent solvent in cellulosic coatings
Regulatory: SARA §311/312 acute health/chronic health/fire hazard, §313 reportable (methanol, MIBK); EINECS, Australia AICS listed
Properties: Colorless clear liq.; alcohol odor; sol. in water; sp.gr. 0.809-0.815 (15.6 C/15.6 C); vapor pressure 53 mbar (19 C); b.p. 78 C; flash pt. (TCC) 13 C (ethanol); autoignition temp. > 404 C
Toxicology: May be fatal or cause blindness if swallowed; vapor harmful; cannot be made nonpoisonous; harmful if absorbed thru skin; causes eye irritation; high vapor concs. may cause eye/respiratory tract irritation; may cause skin sensitization; TSCA listed
Environmental: Prevent runoff from entering drains, sewers, streams
Precaution: Flamm. liq. and vapor; vapors may flash back; residual vapors may explode on ignition; reacts with strong oxidizing agents
Hazardous decomp. prods.: Combustion: CO, CO_2
HMIS: Health 2, Flammability 3, Reactivity 0
Storage: Store in tightly closed container in well-ventilated area away from heat, sparks, flame

Tecsol® D-2 Anydrous [Eastman]
Chem. Descrip.: Ethyl alcohol (82%), methanol (17%), MIBK (1%)
Uses: Solvent for coatings, plastic coatings, cellulosic coatings, printing inks
Regulatory: SARA §311/312 acute health/chronic health/fire hazard, §313 reportable (methanol, MIBK); Canada DSL, EINECS, Australia AICS, Japan MITI listed
Properties: Colorless clear liq.; alcohol odor; sol. in water; sp.gr. 0.791-0.796 (15.6/15.6 C); vapor pressure 53 mbar (19 C); b.p. 78 C; flash pt. (TCC) 13 C (ethanol); autoignition temp. 404 C
Toxicology: May be fatal or cause blindness if swallowed; vapor harmful; cannot be made nonpoisonous; harmful if absorbed thru skin; causes eye irritation; high vapor concs. may cause drowsiness and eye/respiratory tract irritation; may cause skin sensitization; TSCA listed
Environmental: Prevent runoff from entering drains, sewers, streams
Precaution: Flamm. liq. and vapor; vapors may flash back; residual vapors may explode on ignition; reacts violently with strong oxidizing agents
Hazardous decomp. prods.: Combustion: CO, CO_2
HMIS: Health 2, Flammability 3, Reactivity 0
Storage: Store in tightly closed container in well-ventilated area away from heat, sparks, flame

Tecsol® H [Eastman]
Chem. Descrip.: SD alcohol 3-A (100 parts) with isopropyl alcohol (100)
Uses: Solvent for manufacturing processes, prod. of printing inks; latent solvent in cellulosic coatings
Properties: Completely sol. in water; m.w. 46.07; sp.gr. 0.789-0.820 (20/20C); dens. 6.576.83 lb/gal; visc. 1.201.5 cP (20 C); f.p. -173 F; b.p. 74-82; flash pt. (COC) 49-72 F; ref. index 1.3614 (20 C); surf. tens. 22.4 dynes/cm (20 C)

Tegosoft® CI [Goldschmidt; Goldschmidt AG]
Chem. Descrip.: Cetearyl isononanoate
CAS 84878-33-1; EINECS/ELINCS 284-424-3
Uses: Emollient, skin softener, and moisture retention aid in hair/skin care prods., creams/lotions, hair conditioners/glosses, sunscreen preps., soaps, topical pharmaceuticals; solubilizer for act. ingreds.; solvent for lipophilic additives, e.g., UV filters
Features: Oxidation-stable
Regulatory: JCIC registered
Properties: Liq.; nonionic

Tegosoft® CT [Goldschmidt; Goldschmidt AG]
Chem. Descrip.: Caprylic/capric triglyceride
CAS 65381-09-1; EINECS/ELINCS 265-724-3
Uses: Emollient, skin softener, and moisture retention aid in cosmetics, topical pharmaceuticals, hair/skin care, creams/lotions, hair conditioners/glosses, sunscreen preps., soaps; solvent for lipophilic additives and UV filters
Features: Pleasant skinfeel; vegetable-based
Regulatory: JSCI, Ph.Eur. registered
Properties: Liq.; nonionic

Tegosoft® DO [Goldschmidt; Goldschmidt AG]
Chem. Descrip.: Decyl oleate
CAS 3687-46-5; EINECS/ELINCS 222-981-6
Uses: Emollient, refatting agent, skin softener, and moisture retention aid in hair care, skin care, dry skin treatment, creams/lotions, hair conditioners/glosses, sunscreen preps., soaps, topical pharmaceuticals; solvent for lipophilic act. ingreds.
Features: Lipophilic
Regulatory: JSCI registered
Properties: Liq.; nonionic

Tegosoft® M [Goldschmidt; Goldschmidt AG]
Chem. Descrip.: Isopropyl myristate
CAS 110-27-0; EINECS/ELINCS 203-751-4
Uses: Emollient, spreading agent, skin softener, and moisture retention aid in cosmetics, topical pharmaceuticals, hair/skin care, creams/lotions, hair conditioners/glosses, sunscreen preps., soaps; solvent for lipophilic additives, act. ingreds., oil-sol. UV filters and pigments
Regulatory: JSCI, Ph.Eur. registered
Properties: Liq.; nonionic

Tegosoft® P [Goldschmidt; Goldschmidt AG]
Chem. Descrip.: Isopropyl palmitate
CAS 142-91-6; EINECS/ELINCS 205-571-1
Uses: Emollient, spreading agent, refatting agent, skin softener, and mois-

ture retention aid in hair/skin care prods., creams/lotions, hair conditioners/glosses, sunscreen preps., soaps, topical pharmaceuticals; solubilizer for act. ingreds.; solvent for lipophilic additives
Regulatory: JSCI registered
Properties: Liq.; nonionic

Tegosoft® TN [Goldschmidt; Goldschmidt AG]
Chem. Descrip.: C12-15 alkyl benzoate
CAS 68411-27-8; EINECS/ELINCS 270-112-4
Uses: Emollient for sun care, skin care; solvent for lipophilic act. ingreds. and UV filters
Features: Leaves pleasant, nonoily skinfeel
Properties: Liq.

Tetraglyme [Ferro/Grant]
Chem. Descrip.: Tetraethylene glycol dimethyl ether
CAS 24991-55-7
Uses: Solvent for electrochemistry, polymer and boron chemistry, processes such as gas absorp., extraction, stabilization, industrial prods. such as fuels, lubricants, textiles, pesticides; solvent for pharmaceuticals; drug carrier, stabilizer
Features: Tends to solvate cations
Properties: Colorless clear; mild, nonresidual odor; water-sol.; misc. with ethanol, acetone, benzene, diethyl ether, octane; m.w. 222.28; sp.gr. 1.0132; dens. 8.45 lb/gal; visc. 4.1 cP; f.p. -29.7 C; b.p. 275 C; flash pt. (CC) 141 C; ref. index 1.4330 (20 C); pH neutral; surf. tens. 33.8 dynes/cm (20 C); 98.0% min. purity
Toxicology: LD50 5100 mg/kg; low to mod. acute toxicity; chronic exposure may cause reproductive effects
Environmental: Slowly biodeg.

Tetralin® [Cognis KGaA/Coatings]
Chem. Descrip.: Tetrahydronaphthalene
CAS 119-64-2; EINECS/ELINCS 204-340-2
Uses: Solvent, flow aid in coatings
Features: High boiling; strong solv. power
Use Level: 5-10% on solv.

Tetralin [DuPont]
Chem. Descrip.: 1,2,3,4-Tetrahydronaphthalene
CAS 119-64-2; EINECS/ELINCS 204-340-2
Uses: Solvent for oils, resins, waxes, rubber, and asphalt
Properties: Colorless to pale yel. clear liq.; m.w. 132.2; f.p. -31 C; b.p. 207 C (760 mm Hg); negligible sol. in water; sp.gr. 0.970; dens. 8.1 lb/gal; flash pt. 82 C (TCC); 97.0% min. tetrahydronaphthalene

Texanol® Ester-Alcohol [Eastman]
Chem. Descrip.: 2,2,4-Trimethyl-1,3-pentanediol monoisobutyrate
CAS 25265-77-4; EINECS/ELINCS 246-771-9
Uses: Solvent, coalescing agent for latex finishes and water-base inks, PVAc latexes, PVAc-acrylic latexes, and acrylic, EVA, and B/S latexes; solvent in electrodeposition coatings, lacquer coatings, high-bake enamels, cosmetics; defoamer in waterborne systems and drilling muds for oil industry; chemical intermediate for prod. of esters. used as stain-resistance plasticizers, lubricant basestocks, solvents, syn. detergents, and herbicides
Features: Slow-evaporating
Regulatory: SARA nonreportable
Properties: Colorless clear liq.; mild char. odor; water-insol.; m.w. 216.3; sp.gr. 0.950; dens. 0.95 kg/l; visc. 13.5 cP (20 C); vapor pressure 0.013 mm Hg; f.p. -50 C; b.p. 244 C min. (initial, 760 mm); pour pt. -57 C; ref. index 1.4423; flash pt. 120 C (COC); fire pt. 132 C; > 99 vol.% volatiles
Toxicology: LD50 (oral, rat) 6.86 ml/kg, (dermal, rabbit) > 15 ml/kg; may cause transient eye irritation; expected to be low ingestion hazardTSCA listed
Environmental: Mod. potential to affect some aquatic organisms, algal growth; low potential to affect sec. waste treatment microbial metabolism, germination or early growth of some plants; expected to biodegrade
Precaution: Incompat. with oxidizing materials
Hazardous decomp. prods.: Hazardous combustion prods.: CO, CO_2
HMIS: Health 1, Flammability 1, Reactivity 0
Storage: Store in closed containers

Texaprint® SDCE [Cognis]
Chem. Descrip.: Dicapryl ether
CAS 629-82-3; EINECS/ELINCS 211-112-6
Uses: Solvent, diluent for offset printing inks; cleaner component for printing equip.
Features: Min. oil substitute
Regulatory: EINECS, MITI, DSL, AICS listed
Properties: Clear low-visc. liq.; almost odor-free; sol. < 2 mg/l in water; dens. 0.800-0.815 g/cm³; visc. 2-5 mPa·s; acid no. 0.05 max.; iodine no. < 1; sapon. no. 5 max.; hyd. no. 0.5 max.; cloud pt. -5 C max.; ref. index 1.431-1.435 (20 C); > 96% act.
Toxicology: TSCA listed
Environmental: Biodegradable
Storage: 12 mos shelf life when stored @ R.T. in sealed containers; storage at temps. < -8 C will lead to solidification; prod. will not lose efficiency after warming back up to R.T.

Texaprint® SDM [Cognis]
Chem. Descrip.: Fatty acid methyl ester
Uses: Solvent, diluent for offset printing inks; cleaner component for printing equip.
Features: Min. oil substitute
Regulatory: EINECS, DSL, PICCS, AICS, ECL, MITI, ECS listed
Properties: Dens. 0.86-0.88 g/cm³; visc. 4-5 mPa·s (20 C); acid no. 0.5 max.; iodine no. 0.5 max.; sapon. no. 250-260; ref. index 1.432-1.435
Toxicology: TSCA listed
Environmental: Biodegradable
Storage: 24 mos shelf life when stored @ R.T. in sealed containers; do not expose to heat or frost; prod. will not lose efficiency after warming back up to room temp.

Texaprint® SJEH [Cognis]
Chem. Descrip.: Fatty acid 2-ethylhexylester
Uses: Solvent, diluent for offset printing inks; cleaner component for printing equip.
Features: Min. oil substitute
Regulatory: EINECS, DSL, AICS, ECL, MITI, ECS listed
Properties: Dens. 0.855-0.875 g/cm³ (30 C); visc. 9-10 mPa·s (20 C); b.p. 369-389 C; acid no. < 1; iodine no. 85-95; sapon. no. 140-150; cloud pt. -10 C; flash pt. ≈ 170 C; ref. index 1.432-1.435
Toxicology: TSCA listed
Environmental: Biodegradable
Storage: 24 mos shelf life when stored at room temp. in sealed containers; do not expose to heat or frost; storage at temp. below - 13 C will lead to solidification; prod. will not lose efficiency after warming back up to room temp.

Texaprint® SKEH [Cognis]
Chem. Descrip.: Fatty acid 2-ethylhexylester
Uses: Solvent, diluent for offset printing inks; cleaner component for printing equip.
Features: Min. oil substitute
Regulatory: EINECS, DSL, PICCS, AICS, ECL, ECS listed
Properties: Dens. 0.84-0.88 g/cm³ (20 C); visc. 6.5-8.0 mPa·s (20 C); b.p. 337-357 C; acid no. 0-0.3; sapon. no. 174-184; cloud pt. ≈ -20 C; flash pt. > 170 C; ref. index 1.439-1.444
Toxicology: TSCA listed
Environmental: Biodegradable
Storage: 24 mos shelf life when stored at room temp. in sealed containers; do not expose to heat or frost; storage at temp. below - 30 C will lead to solidification; prod. will not lose efficiency after warming back up to room temp.

Texaprint® SLIB EXP [Cognis]
Chem. Descrip.: Fatty acid methyl ester
Uses: Solvent, diluent for offset printing inks; cleaner component for printing equip.
Features: Min. oil substitute
Regulatory: EINECS, DSL, AICS, ENCS listed
Properties: Dens. 0.852-0.854 g/cm³ (20 C); visc. 5-7 mPa·s; b.p. 300-315 C; acid no. < 1; iodine no. < 1; sapon. no. 167-177; hyd. no. < 1; ref. index 1.4353-1.4357
Toxicology: TSCA listed
Environmental: Biodegradable
Storage: 24 mos shelf life when stored at room temp. in sealed containers; do not expose to heat or frost; storage at temp. below - 19 C will lead to solidification; prod. will not lose efficiency after warming back up to room temp.

Texaprint® SLIP [Cognis]
Chem. Descrip.: Fatty acid isopropyl ester

Uses: Solvent, diluent for offset printing inks; cleaner component for printing equip.
Features: Min. oil substitute
Regulatory: EINECS, DSL, PICCS, AICS, ECL, MITI, ECS listed
Properties: Dens. 0.850-0.870 g/cm^3 (20 C); visc. 3-5 mPa·s (20 C); b.p. 280-290 C; acid no. < 1; iodine no. < 1; sapon. no. 225-235; hyd. no. < 5; ref. index 1.4272-1.4312
Toxicology: TSCA listed
Environmental: Biodegradable
Storage: 12 mos shelf life when stored at room temp. in sealed containers; do not expose to heat or frost; storage at temps. below - 8 C will lead to solidification; prod. will not lose efficiency after warming back up to room temp.

Texaprint® SRM [Cognis]
Chem. Descrip.: Fatty acid methyl ester
Uses: Solvent, diluent for offset printing inks; cleaner component for printing equip.
Features: Min. oil substitute
Regulatory: EINECS, DSL, PICCS, AICS, ECL, MITI, ECS listed
Properties: Dens. 0.860-0.880 g/cm^3 (30 C); visc. 5-7 mPa·s; b.p. 335-365 C; acid no. < 1; iodine no. 100-115; sapon. no. 187-197; cloud pt. ≈ -8 C; flash pt. ≈ 155 C; ref. index 1.457-1.458
Toxicology: TSCA listed
Environmental: Biodegradable
Storage: 24 mos shelf life when stored at room temp. in sealed containers; do not expose to heat or frost; storage at temp. below - 19 C will lead to solidification; prod. will not lose efficiency after warming back up to room temp.

Texaprint® SSEH [Cognis]
Chem. Descrip.: Fatty acid 2-ethylhexylester
Uses: Solvent, diluent for offset printing inks; cleaner component for printing equip.
Features: Min. oil substitute
Regulatory: EINECS, DSL, PICCS, AICS, ECL, MITI, ECS listed
Properties: Dens. 0.845-0.865 g/cm^3 (30 C); visc. 11-12 mPa·s; b.p. 357-407 C; acid no. < 0.3; iodine no. < 2; sapon. no. 140-150; hyd. no. < 3; cold cloud pt. 10-15 C; flash pt. > 210 C; ref. index 1.447-1.449
Toxicology: TSCA listed
Environmental: Biodegradable
Storage: 24 mos shelf life when stored at room temp. in sealed containers; do not expose to heat or frost; storage at temp. below - 5 C will lead to solidification; prod. will not lose efficiency after warming back up to room temp.

Textile Spirits [CITGO]
Chem. Descrip.: Solvent-refined light naphtha
CAS 64741-84-0; EINECS/ELINCS 265-086-6
UN 1268
Uses: Solvent
Regulatory: SARA §302/304 nonreportable, §311/312 acute health/chronic health/fire hazard, §313 reportable [hexane (30-40%), cyclohexane (0-3%)]; Calif. Prop. 65 reportable (benzene, toluene); may contain detectable amts. of n-hexane, cyclohexane
Properties: Colorless transparent liq.; lt. hydrocarbon odor; misc. with most common org. solvs.; sl. sol. in cold water (≈ 0.1%); sp.gr. 0.677 (60/60 F); dens. 5.63 lb/gal (15.6 C); visc. 0.528 cSt (20 C); vapor pressure 115 mm Hg (20 C); f.p. -154 to -90 C; b.p. 60-98 C; flash pt. (TCC) -21 to -20 C; autoignition temp. ≈ 232 C; ref. index 1.3796 (20 C); KB value 29.5; VOC 677 g/l; 100% act.
Toxicology: ACGIH TLV TWA 400 ppm; mist/vapor may irritate eyes, mucous membranes, respiratory tract; liq. contact may cause minimal to mild eye, mild to mod. skin irritation/inflamm.; may be harmful by inh., skin absorp.; may be harmful or fatal by ing.; overexposure may cause CNS depression, other target organ effects, peripheral nervous system damage; aspiration into lungs can cause pulmonary edema, chem. pneumonia; prolonged/repeated inh. may cause heart arrhythmias; possible reproductive hazard; TSCA listed
Environmental: May cause oxygen depletion in waterways; potentially toxic to water ecosystems; may be hazardous to aquatic life; may contribute to smog; prevent entry into sewers, waterways; not a DOT marine pollutant
Precaution: Extremely flamm.; flamm. limits 1.0-7.0%; vapor may cause flash fire or explosion; vapors are heavier than air, may flash back; may create vapor/air explosion hazard in confined space; spills may create slipping hazard; floats on water; container can rupture in fire; prevent entry into basements, confined areas; bond and ground equip.; incompat. with strong acids, alkalis, oxidizers (liq. chlorine, other halogens, hydrogen peroxide, oxygen); heptane will dissolve some plastics
Hazardous decomp. prods.: Combustion: smoke, CO, CO_2, possibly other harmful gases/vapors
HMIS: Health 2, Flammability 3, Reactivity 0
Storage: Store in tightly closed containers in cool, dry, well-ventilated place away from heat, sparks, ignition sources, strong oxidizing conditions

THFA® [Penn Spec. Chems.]
Chem. Descrip.: Tetrahydrofurfuryl alcohol
CAS 97-99-4; EINECS/ELINCS 202-625-6
Uses: Solvent, carrier for pesticides; FDA approved for use in paper processing; chemical intermediate; also in industrial and consumer cleaners, leather and textile dyeing, epoxies, coatings, inks, paints, and adhesives; plasticizer and vinyl stabilizer carrier
Features: High boiling
Regulatory: FDA compliance
Properties: Colorless liq.; sol. in water; m.w. 102.13; sp.gr. 1.05 (20 C); dens. 8.79 lb/gal (20 C); visc. 6.24 cps (20 C); f.p. < -80 C; b.p. 178 C; flash pt. (TCC) 74 C; ref. index 1.4520 (20 C)
Environmental: Biodeg.

Thomil 14 [Undesa; S. Black]
Chem. Descrip.: Myristyl myristate
CAS 3234-85-3; EINECS/ELINCS 221-787-9
Uses: Emollient in bath oils, creams/lotions, sun creams, shaving prods., insect repellents, antiperspirants, deodorants; oily component for cleansing creams, nail polish removers; plasticizer for nail polishes; pigment wetting agent; softener in creams, hair care; solvent for fatty cosmetic ingreds.; vehicle for lipsticks; solubilizer for perfume oils, vitamin oils, hormones; solubilizer, vehicle for medicines; excipient for injectable substances; lubricant for PVC
Features: Changes skin feel, rub out props. and greasiness; veg. oil substitute; vaseline substitute

Thomil G [Undesa; S. Black]
Chem. Descrip.: Glyceryl myristate
CAS 589-68-4; EINECS/ELINCS 248-329-0
Uses: Dispersant, solvent for act. ingreds.; vehicle for injectable substances; emulsifier for textiles; emulsifier, opacifier, thickener, dispersant for emulsions; greasing agent in shampoos, foam baths; food lubricant; perfume diluent; intermediate in textiles and tanning industry
Properties: Nonionic

Thomil IS [Undesa; S. Black]
Chem. Descrip.: Isopropyl myristate
CAS 110-27-0; EINECS/ELINCS 203-751-4
Uses: Emollient in bath oils, creams/lotions, sun creams, shaving prods., insect repellents, antiperspirants, deodorants; oily component for cleansing creams, nail polish removers; plasticizer for nail polishes; pigment wetting agent; softener in creams, hair care; solvent for fatty cosmetic ingreds.; vehicle for lipsticks; solubilizer for perfume oils, vitamin oils, hormones; solubilizer, vehicle for medicines; excipient for injectable substances; lubricant for PVC
Features: Changes skin feel, rub out props. and greasiness; veg. oil substitute; vaseline substitute

Thomil IS-F [Undesa; S. Black]
Chem. Descrip.: Isopropyl myristate
CAS 110-27-0; EINECS/ELINCS 203-751-4
Uses: Emollient in bath oils, creams/lotions, sun creams, shaving prods., insect repellents, antiperspirants, deodorants; oily component for cleansing creams, nail polish removers; plasticizer for nail polishes; pigment wetting agent; softener in creams, hair care; solvent for fatty cosmetic ingreds.; vehicle for lipsticks; solubilizer for perfume oils, vitamin oils, hormones; solubilizer, vehicle for medicines; excipient for injectable substances; lubricant for PVC
Features: Changes skin feel, rub out props. and greasiness; veg. oil substitute; vaseline substitute

Tilol B [Undesa; S. Black]
Chem. Descrip.: Butyl oleate
CAS 142-77-8; EINECS/ELINCS 205-559-6
Uses: Emollient in bath oils, creams/lotions, sun creams, shaving prods.,

insect repellents, antiperspirants, deodorants; oily component for cleansing creams, nail polish removers; plasticizer for nail polishes; pigment wetting agent; softener in creams, hair care; solvent for fatty cosmetic ingreds.; vehicle for lipsticks; solubilizer for perfume oils, vitamin oils, hormones; solubilizer, vehicle for medicines; excipient for injectable substances; lubricant for PVC
Features: Changes skin feel, rub out props. and greasiness; veg. oil substitute; vaseline substitute

Tilol ET [Undesa; S. Black]
Chem. Descrip.: Ethyl oleate
CAS 111-62-6; EINECS/ELINCS 203-889-5
Uses: Emollient in bath oils, creams/lotions, sun creams, shaving prods., insect repellents, antiperspirants, deodorants; oily component for cleansing creams, nail polish removers; plasticizer for nail polishes; pigment wetting agent; softener in creams, hair care; solvent for fatty cosmetic ingreds.; vehicle for lipsticks; solubilizer for perfume oils, vitamin oils, hormones; solubilizer, vehicle for medicines; excipient for injectable substances; lubricant for PVC
Features: Veg. oil substitute; vaseline substitute
Properties: Low visc. oil
Toxicology: Low toxicity

Tilol GCT-R [Undesa; S. Black]
Chem. Descrip.: Polyglyceryl polyricinoleate
Uses: Dispersant, solvent for act. ingreds.; vehicle for injectable substances; emulsifier for textiles; emulsifier, opacifier, thickener, dispersant for emulsions; greasing agent in shampoos, foam baths; food lubricant; perfume diluent; intermediate in textiles and tanning industry
Properties: Nonionic

Tilol GP-O [Undesa; S. Black]
Chem. Descrip.: Glyceryl oleate
CAS 111-03-5; EINECS/ELINCS 203-827-7
Uses: Dispersant, solvent for act. ingreds.; vehicle for injectable substances; emulsifier for textiles; emulsifier, opacifier, thickener, dispersant for emulsions; greasing agent in shampoos, foam baths; food lubricant; perfume diluent; intermediate in textiles, tanning industry
Properties: Nonionic

Tilol IS [Undesa; S. Black]
Chem. Descrip.: Isopropyl oleate
CAS 112-11-8; EINECS/ELINCS 203-935-4
Uses: Emollient in bath oils, creams/lotions, sun creams, shaving prods., insect repellents, antiperspirants, deodorants; oily component for cleansing creams, nail polish removers; plasticizer for nail polishes; pigment wetting agent; softener in creams, hair care; solvent for fatty cosmetic ingreds.; vehicle for lipsticks; solubilizer for perfume oils, vitamin oils, hormones; solubilizer, vehicle for medicines; excipient for injectable substances; lubricant for PVC
Features: Changes skin feel, rub out props. and greasiness; veg. oil substitute; vaseline substitute

T-Mulz® 426 [Harcros]
Chem. Descrip.: Phosphate ester, free acid
Uses: Dry cleaning detergent; industrial formulations; solvent degreasers; textile lubricant; agric. applics.
Regulatory: FDA compliance
Properties: Amber visc. liq.; sol. in ethanol, ethylene, xylene; disp. in min. oil, kerosene; sp.gr. 1.00; dens. 8.3 lb/gal; acid no. 97 (pH 5.5); pour pt. 25 F; flash pt. (PMCC) > 350 F; pH 2-3 (1% aq.); surf. tens. 34.8 dynes/cm (0.05% aq.); anionic; 99.5% min. act.

Tolu-Sol 20 [Shell]
Chem. Descrip.: Hydrocarbon solvent
Uses: Solvent for printing inks, rubber; diluent for lacquers
Features: Narrow cut; med. evaporating; increased cycloparaffin content; low level of impurities e.g., sulfur, olefins, benzene
Properties: Saybolt 30 min. color; m.w. 101; i.b.p. 90.5 C min.; flash pt. -7 C; KB value 37

Tolu-Sol 30 [Shell]
Chem. Descrip.: Solvent naphtha (petroleum), light aliphatic (85%), toluene (15%), cyclohexane (0.08-5 vol.%)
UN 1993
Uses: Solvent for printing inks, rubber; diluent for lacquers
Features: Narrow cut; short, well regulated evaporation time
Regulatory: SARA §311/312 acute health/chronic health/fire hazard, §313 reportable [toluene (15%), cyclohexane (0.08-5 vol.%)]; Calif. Prop. 65 reportable [benzene (6 ppm), toluene (15%)]
Properties: Colorless liq.; hydrocarbon odor; negligible sol. in water; m.w. 100; sp.gr. 0.753 (60 F); vapor pressure > 100 mm Hg (100 F); i.b.p. 88 C min.; b.p. 190-235 F; flash pt. -7 C; autoignition temp. > 260 C; KB value 44-48; VOC 6.3 lb/gal (60 F)
Toxicology: If aspirated, can cause severe lung damage, chem. pneumonitis; may be fatal if swallowed; causes severe eye irritation; may cause CNS depression (dizziness, lightheadedness, headache, nausea, incoordination, unconsciousness, death); chronic exposure may cause hearing loss; chronic abuse may cause irregular heart rhythms, cardiac arrest; may cause sl. skin irritation, defatting; target organs: heart, kidney, liver, auditory system
Environmental: Prevent entry into waterways, sewers
Precaution: Extremely flamm.; flamm. in air 1-8 vol.%; vapors are heavier than air; vapors may flash back; prevent vapor accumulation; ground equip.; avoid contact with strong oxidizing agents; prevent entry into basements, confined areas
Hazardous decomp. prods.: Complex mixt. of airborne solids, liqs., and gases, incl. CO, CO_2, other org. compds.
HMIS: Health 2, Flammability 3, Reactivity 0
Storage: Keep containers closed when not in use; keep away from heat, sparks, flame

Toxfree™ [Amity UK]
Chem. Descrip.: C9-15 isoalkanes
UN 3295
Uses: Solvent cleaner for precision cleaning, industrial degreasing, elec./electronic cleaning, spot cleaner on fabrics, printing ink remover, engineering maintenance degreaser, cleaner/degreaser prior to using adhesives and prior to painting
Features: Nonflamm.; increased contact time; safe on metals and plastics; CFC-free; improved health, safety, environmental profiles; British aerospace/military approvals
Properties: Colorless clear liq.; odorless to very sl. odor; immisc. with water; dens. 0.75 g/cm^3; dielec. str. 45 kV; 100% act.
Toxicology: Harmful; ing. of spray or mist may cause lung damage, severe respiratory problems; may cause eye irritation; prolonged skin contact may cause dryness, irritation; inh. of vapor may cause dizziness, headaches, drowsiness
Environmental: Inherently biodeg.; removed easily in wastewater treatment; nontoxic to aquatic organisms; do not allow into drains, watercourses, sewers
Storage: Keep container closed; replace closure after use; do not store near heat sources, naked flames; use proper grounding procedures when pouring or pumping

Transcutol® CG [Gattefosse]
Chem. Descrip.: Ethoxydiglycol
CAS 111-90-0; EINECS/ELINCS 203-919-7
Uses: Cosurfactant, solvent for active ingreds. in cosmetic anhyd. microemulsions

Transcutol® HP [Gattefosse; Gattefosse SA]
Chem. Descrip.: Diethylene glycol monoethyl ether, purified, USP/NF
CAS 111-90-0; EINECS/ELINCS 203-919-7
Uses: Solvent, solubilizer for very poorly sol. drug actives in liq. and capsules oral pharmaceuticals
Regulatory: USP/NF, EP compliance; US Drug master file 5718
Properties: Colorless limpid liq.; faint odor; sol. in oil and water; > 99.9% purity

Transcutol® P [Gattefosse; Gattefosse SA]
Chem. Descrip.: Diethylene glycol monoethyl ether, purified
CAS 111-90-0; EINECS/ELINCS 203-919-7
Uses: Solvent, solubilizer for very poorly sol. drug actives in liqs. and hard or soft gelatin capsules; penetrant for topical and transdermal emulsions, sol'ns. , and gels
Regulatory: USP/NF, EP compliance; DMF no. 5718
Properties: Colorless limpid liq.; faint odor; sol. in water and oil; > 99.7% purity

Trax K-40 [NOF]
Chem. Descrip.: POE alkyl ether sulfate
Uses: Detergent; wetting agent; penetrant for bleach bath; scouring agent for wool; aux. solvent for dyestuffs; base material for shampoos; emulsifier for petrol. and tar industries
Properties: Lt. yel. transparent liq.; anionic; 30% conc.

Triacetina [Undesa; S. Black]
Chem. Descrip.: Glyceryl triacetate
CAS 102-76-1; EINECS/ELINCS 203-051-9
Uses: Dispersant, solvent for act. ingreds.; vehicle for injectable substances; emulsifier for textiles; emulsifier, opacifier, thickener, dispersant for emulsions; greasing agent in shampoos, foam baths; food lubricant; perfume diluent; intermediate in textiles and tanning industry
Properties: Nonionic

Triglyme [Ferro/Grant]
Chem. Descrip.: Triethylene glycol dimethyl ether
CAS 112-49-2
Uses: Solvent for electrochemistry, polymer/boron chemistry, gas absorp., extraction, stabilization, industrial prods. such as fuels, lubricants, textiles, pesticides; solvent for pharmaceuticals
Features: Tends to solvate cations
Properties: Colorless clear; mild, nonresidual odor; water-sol.; misc. with ethanol, acetone, benzene, diethyl ether, octane; m.w. 178.22; sp.gr. 0.9862; dens. 8.23 lb/gal; visc. 3.8 cP; f.p. -45 C; b.p. 216 C; flash pt. (CC) 111 C; ref. index 1.4224; pH neutral; surf. tens. 29.4 dynes/cm (20 C); 98.0% min. purity
Toxicology: LD50 5000 mg/kg; low to mod. acute toxicity; chronic exposure may cause reproductive effects
Environmental: Slowly biodeg.

Triol 91 [Undesa; S. Black]
Chem. Descrip.: Tricaprin/tricaprylin
Uses: Dispersant, solvent for act. ingreds.; vehicle for pharmaceutical injections, oral suspensions, and suppositories; emulsifier for textiles; emulsifier, opacifier, thickener, dispersant for emulsions; greasing agent in shampoos, foam baths; food lubricant; perfume diluent; intermediate in textiles and tanning industry

Trioleina [Undesa; S. Black]
Chem. Descrip.: Glyceryl trioleate
CAS 122-32-7; EINECS/ELINCS 204-534-7
Uses: Dispersant, solvent for act. ingreds.; vehicle for injectable substances; emulsifier for textiles; emulsifier, opacifier, thickener, dispersant for emulsions; greasing agent in shampoos, foam baths; food lubricant; perfume diluent; intermediate in textiles and tanning industry
Properties: Nonionic

Tristabil® [Solvay SA]
Chem. Descrip.: Trichloroethylene
CAS 79-01-6; EINECS/ELINCS 201-167-4
Uses: Solvent to dissolve or extract greases, oils, resins, etc., esp. for metal and glass cleaning performed in degreasing machines
Features: Stabilized grade; high quality; multipurpose; non-ozone depleting; nonflamm.; volatile; easy separation of the solvent from the soil by distillation; effective recovery of the fugitive vapor
Properties: Colorless liq.; m.w. 131.4; sp.gr. 1.461-1.465; vapor pressure 0.079 bar (20 C); b.p. 87 C (1 bar); pH 8 min. (aq. extract); KB value 130; sp.heat 0.93 kJ/kg °K (20 C); 99.5% min. purity
Toxicology: Toxic; irritating to eyes, skin; may cause dermatitis; narcotic effect by inh. of vapor; suspected carcinogen
Environmental: Avoid emission and release to the environment
Precaution: Incompat. with strong acids, bases, and oxidizing agents; vapor is heavier than air; prevent vapor from coming into contact with high temp. source

Trivent DIA [Trivent; S. Black]
Chem. Descrip.: Diisopropyl adipate
CAS 6938-94-9; EINECS/ELINCS 230-072-0
Uses: Lubricant, emollient, solvent in antiperspirants, hand creams, moisturizing creams/lotions, skin care prods., aftershave lotions and afterbath splashes; lubricant and detackifier in antiperspirants; drying agent in heavy oily systems
Features: Dry emolliency; hydro/alcoholic compat.
Regulatory: Approved in US, EC, Japan, Mexico, Australia, etc.
Properties: Colorless clear liq., faint typ. odor; sol. in min. oils, lower alcohols, ketones, esters, chlorinated hydrocarbons; insol. in water, glycerin; m.w. 230; sp.gr. 0.960; f.p. -2 C; acid no. 1 max.; iodine no. nil; sapon. no. 480-500; flash pt. (COC) > 170 C; 100% act.
Use Level: 2-10%
HMIS: Health 1, Flammability 1, Reactivity 0
Storage: Store under cool, dry conditions

Trivent NP-13 [Trivent]
Chem. Descrip.: Tridecyl neopentanoate
CAS 106436-39-9
Uses: Emollient for nonoily creams/lotions; SPF booster for sunscreens; binder/pigment dispersant; plasticizer for hair setting polymers; cosolvent for oils and waxes in anhyd. sticks; wetting agent for Carbopol
Features: Very dry emolliency
Properties: Colorless to pale yel. liq., sl. mild odor; sol. @ 5% in 95% ethanol, min. oil, cyclomethicone, IPM, dimethicone, isocetyl alcohol, hexylene glycol; insol. in water, glycerin, propylene glycol; m.w. 284; sp.gr. 0.850-0.860; acid no. 2 max.; sapon. no. 185-205; ref. index 1.434-1.437
Toxicology: LD50 (acute oral) > 5 g/kg; may cause eye and skin irritation; noncomedogenic

Trivent OP [Trivent; S. Black]
Chem. Descrip.: Octyl palmitate
CAS 29806-73-3; EINECS/ELINCS 249-862-1
Uses: Emollient in liq. makeup, bath oils, after-sun prods., hand creams/lotions, night creams, moisturizing lotions, skin care prods.; binder for pressed powds.; solvent for sunscreens
Regulatory: Approved in US, EC, Japan, Argentina, Australia, Canada, Mexico, etc.
Properties: Colorless clear liq.; typ. bland odor; sol. in peanut oil, min. oil, 95% ethanol, IPM, oleyl alcohol; insol. in water, glycerin, propylene glycol; m.w. 368; sp.gr. 0.854-0.858; f.p. 0 C; acid no. 3 max.; iodine no. 1 max.; flash pt. (COC) 395 F; ref. index 1.445; 100% act.
HMIS: Health 1, Flammability 1, Reactivity 0
Storage: Store in cool, dry location

U

Ucar® Acetone [Dow]
Chem. Descrip.: Acetone
CAS 67-64-1; EINECS/ELINCS 200-662-2
Uses: Solvent for higher solids coatings
Properties: M.w. 58.08; dens. 6.59 lb/gal (20 C); visc. 0.3 cps (20 C); vapor pressure 185 mm Hg (20 C); b.p. 56.3 C; flash pt. (CC) 0 F

Ucar® Butanol [Dow]
Chem. Descrip.: Butanol
CAS 71-36-3; EINECS/ELINCS 200-751-6
Uses: Solvent for higher solids coatings
Properties: M.w. 74.12; dens. 6.75 lb/gal (20 C); visc. 3.0 cps (20 C); vapor pressure 4 mm Hg (20 C); b.p. 117.7 C; flash pt. (CC) 95 F

Ucar® n-Butyl Propionate [Dow]
Chem. Descrip.: n-Butyl propionate
CAS 590-01-2; EINECS/ELINCS 209-669-5
Uses: Solvent for high-solids and waterborne coatings
Features: Provides strong solvency, high elec. resistivity, good odor chars.
Properties: Sol. 0.2% in water (20 C); m.w. 130.19; sp.gr. 0.875 (20/20 C); dens. 7.30 lb/gal (20 C); visc. 0.9 cP (20 C); vapor pressure 3.4 mm Hg (20 C); b.p. 144.7 C; flash pt. (CC) 100 F; surf. tens. 25.3 dynes/cm (20 C); elec. resist. > 100 megohm

Ucar® Diacetone Alcohol [Dow]
Chem. Descrip.: Diacetone alcohol
CAS 123-42-2; EINECS/ELINCS 204-626-7
Uses: Solvent for higher solids coatings
Properties: M.w. 116.16; dens. 7.82 lb/gal (20 C); visc. 3.2 cps (20 C); vapor pressure < 1 mm Hg (20 C); b.p. 169.2 C; flash pt. (CC) 117 F

Ucar® Diisobutyl Ketone [Dow]
Chem. Descrip.: Diisobutyl ketone
CAS 108-83-8; EINECS/ELINCS 203-620-1
Uses: Solvent for higher solids coatings
Properties: M.w. 142.24; dens. 6.72 lb/gal (20 C); visc. 1.0 cps (20 C); vapor pressure 1 mm Hg (20 C); b.p. 169.4 C; flash pt. (CC) 118 F

Ucar® Ester EEP [Dow]
Chem. Descrip.: Ethyl 3-ethoxypropionate
CAS 763-69-9
Uses: Solvent for waterborne and high-solids coatings (NC, cellulose acetate, ethyl cellulose, butyrate, acrylic, urethane)
Properties: Sol. 1.6% in water (20 C); m.w. 146.19; sp.gr. 0.950 (20/20 C); dens. 7.91 lb/gal (20 C); visc. 1.3 cP (20 C); vapor pressure 0.8 mm Hg (20 C); b.p. 169.7 C; flash pt. (CC) 136 F

Ucar® Ethanol [Dow]
Chem. Descrip.: Ethanol
CAS 64-17-5; EINECS/ELINCS 200-578-6
Uses: Solvent for higher solids coatings
Properties: M.w. 46.07; dens. 6.76 lb/gal (20 C); visc. 1.2 cps (20 C); vapor pressure 45 mm Hg (20 C); b.p. 78.3 C; flash pt. (CC) 58 F

Ucar® 2-Ethylhexanol [Dow]
Chem. Descrip.: 2-Ethylhexanol
CAS 104-76-7; EINECS/ELINCS 203-234-3
Uses: Solvent for higher solids coatings
Properties: M.w. 130.23; dens. 6.94 lb/gal (20 C); visc. 10.3 cps (20 C); vapor pressure < 1 mm Hg (20 C); b.p. 184.6 C; flash pt. (CC) 162 F

Ucar® Filmer IBT [Dow]
Chem. Descrip.: 2,2,4-Trimethyl-1,3-pentanediol monoisobutyrate
CAS 25265-77-4; EINECS/ELINCS 246-771-9
Uses: Solvent, coalescing agent for water-disp. and water-sol. coatings
Properties: Sol. < 0.1% in water (20 C); m.w. 216.30; sp.gr. 0.950 (20/20 C); dens. 7.91 lb/gal (20 C); visc. 17.8 cP (20 C); vapor pressure < 0.01 mm Hg (20 C); b.p. 244.0 C; flash pt. (CC) 248 F

Ucar® Isobutanol [Dow]
Chem. Descrip.: Isobutanol
CAS 78-83-1; EINECS/ELINCS 201-148-0
Uses: Solvent for higher solids coatings
Properties: M.w. 74.12; dens. 6.68 lb/gal (20 C); visc. 4.0 cps (20 C); vapor pressure 7 mm Hg (20 C); b.p. 107.9 C; flash pt. (CC) 82 F

Ucar® Isobutyl Acetate [Dow]
Chem. Descrip.: Isobutyl acetate
CAS 110-19-0; EINECS/ELINCS 203-745-1
Uses: Solvent for higher solids coatings
Properties: M.w. 116.16; dens. 7.24 lb/gal (20 C); visc. 0.7 cps (20 C); vapor pressure 4 mm Hg (20 C); b.p. 118 C; flash pt. (CC) 62 F

Ucar® Isophorone [Dow]
Chem. Descrip.: Isophorone
CAS 78-59-1; EINECS/ELINCS 201-126-0
Uses: Solvent for higher solids coatings
Properties: M.w. 138.21; dens. 7.67 lb/gal (20 C); visc. 2.6 cps (20 C); vapor pressure < 1 mm Hg (20 C); b.p. 215.3 C; flash pt. (CC) 179 F

Ucar® Isopropanol [Dow]
Chem. Descrip.: Isopropyl alcohol
CAS 67-63-0; EINECS/ELINCS 200-661-7
Uses: Solvent for higher solids coatings
Properties: M.w. 60.10; dens. 6.55 lb/gal (20 C); visc. 2.4 cps (20 C); vapor pressure 33 mm Hg (20 C); b.p. 82.3 C; flash pt. (CC) 53 F

Ucar® Isopropyl Acetate [Dow]
Chem. Descrip.: Isopropyl acetate
CAS 108-21-4; EINECS/ELINCS 203-561-1
Uses: Solvent for higher solids coatings
Properties: M.w. 102.13; dens. 7.26 lb/gal (20 C); visc. 0.5 cps (20 C); vapor pressure 47 mm Hg (20 C); b.p. 88.9 C; flash pt. (CC) 42 F

Ucar® Methyl Isobutyl Ketone [Dow]
Chem. Descrip.: Methyl isobutyl ketone
CAS 108-10-1; EINECS/ELINCS 203-550-1
Uses: Solvent for higher solids coatings
Properties: M.w. 100.16; dens. 6.67 lb/gal (20 C); visc. 0.6 cps (20 C); vapor pressure 15 mm Hg (20 C); b.p. 116.1 C; flash pt. (CC) 61 F

Ucar® 1-Pentanol [Dow]
Chem. Descrip.: 1-Pentanol
CAS 71-41-0; EINECS/ELINCS 200-752-1
Uses: Solvent for higher solids coatings
Properties: M.w. 88.15; dens. 6.79 lb/gal (20 C); visc. 3.9 cps (20 C); vapor pressure 2 mm Hg (20 C); b.p. 137.8 C; flash pt. (CC) 119 F

Ucar® n-Pentyl Propionate [Dow]
Chem. Descrip.: n-Pentyl propionate
CAS 624-54-4; EINECS/ELINCS 210-852-7
Uses: Solvent for high-solids and waterborne coatings
Features: Provides strong solvency, high elec. resistivity, good odor chars.
Properties: Sol. < 0.02% in water (20 C); m.w. 144.12; sp.gr. 0.874 (20/20 C); dens. 7.28 lb/gal (20 C); visc. 1.0 cP (20 C); vapor pressure 1.5 mm Hg (20 C); b.p. 164.9 C; flash pt. (CC) 135 F

Ucar® Primary Amyl Acetate [Dow]
Chem. Descrip.: Primary amyl acetate
CAS 628-63-7; EINECS/ELINCS 211-047-3
Uses: Solvent for higher solids coatings
Properties: M.w. 130.19; dens. 7.29 lb/gal (20 C); visc. 0.9 cps (20 C); vapor pressure 2 mm Hg (20 C); b.p. 146 C; flash pt. (CC) 101 F

Ucar® 1-Propanol [Dow]
Chem. Descrip.: 1-Propanol
CAS 71-23-8; EINECS/ELINCS 200-746-9
Uses: Solvent for higher solids coatings
Properties: M.w. 60.10; dens. 6.71 lb/gal (20 C); visc. 2.2 cps (20 C); vapor pressure 15 mm Hg (20 C); b.p. 97.2 C; flash pt. (CC) 74 F

Ucar® n-Propyl Acetate [Dow]
Chem. Descrip.: n-Propyl acetate
CAS 109-60-4; EINECS/ELINCS 203-686-1
Uses: Solvent for higher solids coatings
Properties: M.w. 102.13; dens. 7.39 lb/gal (20 C); visc. 0.6 cps (20 C); vapor pressure 26 mm Hg (20 C); b.p. 101.5 C; flash pt. (CC) 58 F

Ucar® n-Propyl Propionate [Dow]
Chem. Descrip.: n-Propyl propionate
CAS 106-36-5; EINECS/ELINCS 203-389-7
Uses: Solvent, visc. reducer for coatings, high-solids coatings, electrostatically sprayed coatings, automotive refinishes, OEM coatings, appliance coatings, printing inks; solvent for polymerization
Features: Fast evaporating; linear structure; improves solv. diffusion from coating films; non-HAPs; replacement for xylene
Properties: Sol. 0.5% in water (20 C); m.w. 116.16; sp.gr. 0.881 (20/20 C); dens. 7.34 lb/gal (20 C); visc. 0.7 cP (20 C); vapor pressure 10.7 mm Hg (20 C); b.p. 122.4 C; flash pt. (CC) 75 F; surf. tens. 24.7 dynes/cm (20 C); elec. resist. > 1000 M-ohm
Toxicology: Harmful by inh.; avoid skin contact
Precaution: Flamm.

Ucon® 50-HB-55 [Dow]
Chem. Descrip.: PPG-2-buteth-3
CAS 9038-95-3
Uses: Emollient for cosmetics; lubricant for compressor, rubber, greases, textiles; intermediate for prep. of resins, plasticizers, modifiers, and surfactants; solvent for inks and dyes; demulsifier; solvent, softener, plasticizer for textile and paper coatings, NC coatings; mold release agents and dye solvs. for vinyl resins; antifoam in boiler water, latex processing, washing sol'ns., paints, adhesives, salt water evaporators
Properties: Sol. in water, alcohol; m.w. 270 (avg.); sp.gr. 0.971 (20/20 C); dens. 8.08 lb/gal (20 C); visc. 55 SUS (38 C); pour pt. -65 C; flash pt. (TOC) 90 C; ref. index 1.4402 (20 C); nonionic
Toxicology: LD50 (oral, rat) 8530 mg/kg; mildly toxic by ing.; skin and severe eye irritantTSCA listed

Ucon® 50-HB-100 [Dow]
Chem. Descrip.: PPG-3-buteth-5
CAS 9038-95-3
Uses: Emollient for cosmetics; lubricant for compressor, rubber, greases, textiles; intermediate for prep. of resins, plasticizers, modifiers, and surfactants; solvent for inks and dyes; demulsifier; solvent, softener, plasticizer for textile and paper coatings, NC coatings; mold release agents and dye solvs. for vinyl resins; antifoam in boiler water, latex processing, washing sol'ns., paints, adhesives, salt water evaporators
Properties: M.w. 520 (avg.); sp.gr. 1.023 (20/20 C); dens. 8.51 lb/gal (20 C); visc. 100 SUS (38 C); pour pt. -51 C; flash pt. (PMCC) 141 C; ref. index 1.4491 (20 C); nonionic
Toxicology: LD50 (oral, rat) 9170 mg/kg, (skin, rabbit) 14 g/kg; mildly toxic by ing. and skin contact; skin irritantTSCA listed

Ucon® 50-HB-260 [Dow]
Chem. Descrip.: PPG-7-buteth-10
CAS 9038-95-3
Uses: Emollient for cosmetics; lubricant for compressor, rubber, greases, textiles; intermediate for prep. of resins, plasticizers, modifiers, and surfactants; solvent for inks and dyes; demulsifier; food pkg. adhesives; defoamer in food-contact paper; solvent, softener, plasticizer for textile and paper coatings, NC coatings; mold release agents and dye solvs. for vinyl resins; antifoam in boiler water, latex processing, washing sol'ns., paints, adhesives, salt water evaporators
Regulatory: FDA 21CFR §175.105, 176.210(d)(3), 178.3120
Properties: M.w. 970 (avg.); sp.gr. 1.035 (20/20 C); dens. 8.62 lb/gal (20 C); visc. 260 SUS (38 C); pour pt. -40 C; flash pt. (PMCC) 194 C; ref. index 1.4542 (20 C); nonionic
Toxicology: LD50 (oral, rat) 4000 mg/kg; mod. toxic by ing.; skin irritantTSCA listed

Ucon® 50-HB-400 [Dow]
Chem. Descrip.: PPG-9-buteth-12
CAS 9038-95-3
Uses: Emollient for cosmetics; lubricant for compressor, rubber, greases, textiles; intermediate for prep. of resins, plasticizers, modifiers, and surfactants; solvent for inks and dyes; demulsifier; food pkg. adhesives; defoamer in food-contact paper; solvent, softener, plasticizer for textile and paper coatings, NC coatings; mold release agents and dye solvs. for vinyl resins; antifoam in boiler water, latex processing, washing sol'ns., paints, adhesives, salt water evaporators
Regulatory: FDA 21CFR §175.105, 176.210(d)(3), 178.3120, 178.3910
Properties: Clear liq., char. odor; m.w. 1230 (avg.); sp.gr. 1.041 (20/20 C); dens. 8.67 lb/gal (20 C); visc. 400 SUS (38 C); pour pt.-38 C; flash pt. (PMCC) 204 C; ref. index. 1.4576 (20 C); pH 5.5-7.5 (10% aq.); nonionic
Toxicology: LD50 (oral, rat) 5370 mg/kg; mildly toxic by ing.; skin irritantTSCA listed

Ucon® 50-HB-660 [Dow]
Chem. Descrip.: PPG-12-buteth-16
CAS 9038-95-3
Uses: Emollient for cosmetics; lubricant for compressor, rubber, greases, textiles; intermediate for prep. of resins, plasticizers, modifiers, and surfactants; solvent for inks and dyes; demulsifier; food processing; food pkg. adhesives; solvent, softener, plasticizer for textile and paper coatings, NC coatings; mold release agents and dye solvs. for vinyl resins; antifoam in boiler water, latex processing, washing sol'ns., paints, adhesives, salt water evaporators, food contact
Regulatory: FDA 21CFR §173.310, 175.105, 176.210(d)(3), 178.3120, 178.3570, 178.3910
Properties: Clear liq., low char. odor; m.w. 1590 (avg.); sp.gr. 1.051 (20/20 C); dens. 8.75 lb/gal (20 C); visc. 660 SUS (38 C); pour pt. -34 C; flash pt. (PMCC) 202 C; ref. index 1.4576 (20 C); pH 5.5-7.5 (10% aq.); nonionic
Toxicology: LD50 (oral, rat) 18 g/kg; mildly toxic by ing.; skin irritantTSCA listed

Ucon® 50-HB-2000 [Dow]
Chem. Descrip.: PPG-20-buteth-30
CAS 9038-95-3
Uses: Emollient for cosmetics; lubricant for compressor, rubber, greases, textiles; intermediate for prep. of resins, plasticizers, modifiers, and surfactants; solvent for inks and dyes; demulsifier; food processing; food pkg. adhesives; solvent, softener, plasticizer for textile and paper coatings, NC coatings; mold release agents and dye solvs. for vinyl resins; antifoam in boiler water, latex processing, washing sol'ns., paints, adhesives, salt water evaporators, food contact
Regulatory: FDA 21CFR §173.310, 175.105, 176.210(d)(3), 178.3120, 178.3570, 178.3910
Properties: M.w. 2660 (avg.); sp.gr. 1.062 (20/20 C); dens. 8.84 lb/gal (20 C); visc. 2000 SUS (38 C); pour pt. -32 C; flash pt. (PMCC) 172 C; ref. index 1.4590 (20 C); nonionic
Toxicology: LD50 (oral, rat) 21 g/kg; mildly toxic by ing.; skin irritantTSCA listed

Ucon® 50-HB-3520 [Dow]
Chem. Descrip.: PPG-28-buteth-35
CAS 9038-95-3
Uses: Emollient for cosmetics; lubricant for compressor, rubber, greases, textiles; intermediate for prep. of resins, plasticizers, modifiers, and surfactants; solvent for inks and dyes; demulsifier; food processing; food pkg. adhesives; solvent, softener, plasticizer for textile and paper coatings, NC coatings; mold release agents and dye solvs. for vinyl resins; antifoam in boiler water, latex processing, washing sol'ns., paints, adhesives, salt water evaporators, food contact
Regulatory: FDA 21CFR §173.310, 175.105, 176.210(d)(3), 178.3120, 178.3570, 178.3910
Properties: Clear liq., char. odor; m.w. 3380 (avg.); sp.gr. 1.062 (20/20 C); dens. 8.81 lb/gal (20 C); visc. 3520 SUS (38 C); pour pt. -29 C; flash pt. (PMCC) 172 C; ref. index 1.4598 (20 C); pH 4.5-7.5 (10% aq.); nonionic
Toxicology: LD50 (oral, rat) 38 g/kg; mildly toxic by ing.; skin irritantTSCA

listed

Ucon® 50-HB-5100 [Dow]

Chem. Descrip.: PPG-33-buteth-45

CAS 9038-95-3

Uses: Emollient for cosmetics; lubricant for compressor, rubber, greases, textiles; intermediate for prep. of resins, plasticizers, modifiers, and surfactants; solvent for inks and dyes; demulsifier; food processing; food pkg. adhesives, cellophane; solvent, softener, plasticizer for textile and paper coatings, NC coatings; mold release agents and dye solvs. for vinyl resins; antifoam in boiler water, latex processing, washing sol'ns., paints, adhesives, salt water evaporators, food contact

Regulatory: FDA 21CFR §173.310, 173.340, 175.105, 176.200, 176.210(d)(3), 177.1200, 178.3120, 178.3570, 178.3910

Properties: Clear visc. liq., low char. odor; m.w. 3930 (avg.); sp.gr. 1.063 (20/20 C); dens. 8.85 lb/gal (20 C); visc. 5100 SUS (38 C); pour pt. -29 C; flash pt. (PMCC) 180 C; ref. index 1.4596 (20 C); pH 5.5-7.5 (10% aq.); nonionic

Toxicology: LD50 (oral, rat) 49 g/kg; mildly toxic by ing.; skin irritantTSCA listed

Ucon® 75-H-450 [Dow]

Chem. Descrip.: PEG/PPG-17/6 copolymer

CAS 9003-11-6

Uses: Lubricant for antiperspirants; emollient; solvent; bodying agent; lubricant for textile spin finishes; intermediate for prep. of resins, plasticizers, modifiers, and surfactants; demulsifier; solvent for inks and dyes; solvent, softener, and plasticizer for textile and paper coatings, NC coatings; mold release agents and dye solvents for vinyl resins; food pkg. adhesives, animal glues; defoamer in food-contact paper/paperboard

Regulatory: FDA 21CFR §175.105, 176.210(d)(3), 178.3120

Properties: Clear to cloudy liq., mild char. odor; m.w. 980 (avg.); sp.gr. 1.097 (20/20 C); dens. 9.13 lb/gal (20 C); visc. 450 SUS (38 C); pour pt. -15 C; flash pt. (PMCC) 188 C; ref. index 1.4655 (20 C); pH 5.5-7.5 (10% aq.)

Ucon® Fluid AP [Amerchol]

Chem. Descrip.: PPG-14 butyl ether

CAS 9003-13-8

Uses: Lubricant for antiperspirants; emollient; solvent; bodying agent

Properties: Clear liq., char. odor; visc. 85-105 cps; acid no. 0.5 max.; flash pt. (COC) 400 F min.

Ucon® LB-65 [Dow]

Chem. Descrip.: PPG-5 butyl ether

CAS 9003-13-8

Uses: Lubricant for two-stroke cycle engines, rubber, greases, textiles; reactive intermediate for resins, plasticizers, modifiers, and surfactant mfg.; antifoam in boiler water, latex processing, washing sol'ns., paints, adhesives, salt water evaporators; solvent for inks and dyes; solvent, softener, and plasticizer for textile and paper coatings, NC coatings; mold release agents and dye solvents for vinyl resins; demulsifier; cosmetic emollient

Properties: M.w. 340 (avg.); sp.gr. 0.960 (20/20 C); dens. 7.99 lb/gal (20 C); visc. 65 SUS (38 C); pour pt. -57 C; flash pt. (PMCC) 119 C; ref. index 1.3281 (20 C); nonionic

Ucon® LB-135 [Dow]

Chem. Descrip.: PPG-9 butyl ether

CAS 9003-13-8

Uses: Lubricant for two-stroke cycle engines, rubber, greases, textiles; reactive intermediate for resins, plasticizers, modifiers, and surfactant mfg.; antifoam in boiler water, latex processing, washing sol'ns., paints, adhesives, salt water evaporators; solvent for inks and dyes; solvent, softener, and plasticizer for textile and paper coatings, NC coatings; mold release agents and dye solvents for vinyl resins; demulsifier; cosmetic emollient

Properties: M.w. 640 (avg.); sp.gr. 0.981 (20/20 C); dens. 8.16 lb/gal (20 C); visc. 135 SUS (38 C); pour pt. -48 C; flash pt. (PMCC) 177 C; ref. index 1.4436 (20 C); nonionic

Ucon® LB-165 [Dow]

Uses: Lubricant for two-stroke cycle engines, rubber, greases, textiles; reactive intermediate for resins, plasticizers, modifiers, and surfactant mfg.; antifoam in boiler water, latex processing, washing sol'ns., paints, adhesives, salt water evaporators; solvent for inks and dyes; solvent, softener, and plasticizer for textile and paper coatings, NC coatings; mold release agents and dye solvents for vinyl resins; demulsifier; cosmetic emollient

Properties: M.w. 740 (avg.); sp.gr. 0.983 (20/20 C); dens. 8.18 lb/gal (20 C); visc. 165 SUS (38 C); pour pt. -46 C; flash pt. (PMCC) 190 C; ref. index 1.4453 (20 C)

Ucon® LB-285 [Dow]

Chem. Descrip.: PPG-15 butyl ether

CAS 9003-13-8

Uses: Lubricant for two-stroke cycle engines, rubber, greases, textiles; reactive intermediate for resins, plasticizers, modifiers, and surfactant mfg.; antifoam in boiler water, latex processing, washing sol'ns., paints, adhesives, salt water evaporators; solvent for inks and dyes; solvent, softener, and plasticizer for textile and paper coatings, NC coatings; mold release agents and dye solvents for vinyl resins; demulsifier; cosmetic emollient; food pkg. adhesives; defoamer for food contact

Regulatory: FDA 21CFR §175.105, 176.210(d)(2) and (3), 178.3120, 178.3910

Properties: M.w. 1020 (avg.); sp.gr. 0.989 (20/20 C); dens. 8.23 lb/gal (20 C); visc. 285 SUS (38 C); pour pt. -40 C; flash pt. (PMCC) 179 C; ref. index 1.4482 (20 C); nonionic

Ucon® LB-385 [Dow]

Chem. Descrip.: PPG-18 butyl ether

CAS 9003-13-8

Uses: Lubricant for two-stroke cycle engines, rubber, greases, textiles; reactive intermediate for resins, plasticizers, modifiers, and surfactant mfg.; antifoam in boiler water, latex processing, washing sol'ns., paints, adhesives, salt water evaporators; solvent for inks and dyes; solvent, softener, and plasticizer for textile and paper coatings, NC coatings; mold release agents and dye solvents for vinyl resins; demulsifier; cosmetic emollient; food pkg. adhesives; defoamer for food contact

Regulatory: FDA 21CFR §175.105, 176.200, 176.210(d)(2) and (3), 177.1200, 178.3120, 178.3910

Properties: Low color and odor; sol. in alcohol; insol. in water; m.w. 1200 avg.; sp.gr. 0.994 (20/20 C); dens. 8.27 lb/gal (20 C); visc. 385 SUS (100 F); pour pt. -37 C; flash pt. (PMCC) 177 C; ref. index 1.4489 (20 C); nonionic

Ucon® LB-525 [Dow]

Chem. Descrip.: PPG-22 butyl ether

CAS 9003-13-8

Uses: Lubricant for two-stroke cycle engines, rubber, greases, textiles; reactive intermediate for resins, plasticizers, modifiers, and surfactant mfg.; antifoam in boiler water, latex processing, washing sol'ns., paints, adhesives, salt water evaporators; solvent for inks and dyes; solvent, softener, and plasticizer for textile and paper coatings, NC coatings; mold release agents and dye solvents for vinyl resins; demulsifier; cosmetic emollient; food pkg. adhesives; defoamer for food contact

Regulatory: FDA 21CFR §175.105, 176.200, 176.210(d)(2) and (3), 177.1200, 178.3120, 178.3910

Properties: M.w. 1420 (avg.); sp.gr. 0.997 (20/20 C); dens. 8.30 lb/gal (20 C); visc. 525 SUS (38 C); pour pt. -34 C; flash pt. (PMCC) 185 C; ref. index 1.4487 (20 C); nonionic

Ucon® LB-625 [Dow]

Chem. Descrip.: PPG-24 butyl ether

CAS 9003-13-8

Uses: Lubricant for two-stroke cycle engines, rubber, greases, textiles; reactive intermediate for resins, plasticizers, modifiers, and surfactant mfg.; antifoam in boiler water, latex processing, washing sol'ns., paints, adhesives, salt water evaporators; solvent for inks and dyes; solvent, softener, and plasticizer for textile and paper coatings, NC coatings; mold release agents and dye solvents for vinyl resins; demulsifier; cosmetic emollient; food pkg. adhesives; defoamer for food contact

Regulatory: FDA 21CFR §175.105, 176.200, 176.210(d)(2) and (3), 177.1200, 178.3120, 178.3570, 178.3910

Properties: Low color and odor; sol. in alcohol; m.w. 1550; sp.gr. 1.00 (20/20 C); dens. 8.32 lb/gal (20 C); visc. 625 SUS (38 C); pour pt. -32 C; flash pt. (PMCC) 168 C; ref. index 1.4491 (20 C); nonionic

Ucon® LB-1145 [Dow]

Chem. Descrip.: PPG-33 butyl ether

CAS 9003-13-8

Uses: Lubricant for two-stroke cycle engines, rubber, greases, textiles; reactive intermediate for resins, plasticizers, modifiers, and surfactant mfg.; antifoam in boiler water, latex processing, washing sol'ns., paints, adhesives, salt water evaporators; solvent for inks and dyes; solvent, softener, and plasticizer for textile and paper coatings, NC coatings; mold release agents and dye solvents for vinyl resins; demulsifier; cosmetic emollient; food pkg. adhesives; defoamer for food contact

Regulatory: FDA 21CFR §175.105, 176.200, 176.210(d)(2) and (3), 177.1200, 178.3120, 178.3570, 178.3910

Properties: Low color and odor; sol. in alcohol; m.w. 2080 (avg.); sp.gr. 1.002 (20/20 C); dens. 8.34 lb/gal (20 C); visc. 1145 SUS (38 C); pour pt. -29 C; flash pt. (PMCC) 190 C; ref. index 1.4499 (20 C); nonionic

Ucon® LB-1715 [Dow]

Chem. Descrip.: PPG-40 butyl ether

CAS 9003-13-8

Uses: Lubricant for two-stroke cycle engines, rubber, greases, textiles; reactive intermediate for resins, plasticizers, modifiers, and surfactant mfg.; antifoam in boiler water, latex processing, washing sol'ns., paints, adhesives, salt water evaporators; solvent for inks and dyes; solvent, softener, and plasticizer for textile and paper coatings, NC coatings; mold release agents and dye solvents for vinyl resins; demulsifier; cosmetic emollient; food pkg. adhesives; defoamer for food contact

Regulatory: FDA 21CFR §175.105, 176.200, 176.210(d)(2) and (3), 177.1200, 178.3120, 178.3570, 178.3910

Properties: Clear liq., low char. odor; sol. in alcohol; m.w. 2490 (avg.); sp.gr. 1.002 (20/20 C); dens. 8.34 lb/gal (20 C); visc. 1715 SUS (38 C); pour pt. -23 C; flash pt. (PMCC) 188 C; ref. index 1.4503 (20 C); pH 5.0-8.5 (10% w/v); nonionic

Ucon® LB-3000 [Dow]

Chem. Descrip.: PPG-53 butyl ether

CAS 9003-13-8

Uses: Lubricant for two-stroke cycle engines, rubber, greases, textiles; reactive intermediate for resins, plasticizers, modifiers, and surfactant mfg.; antifoam in boiler water, latex processing, washing sol'ns., paints, adhesives, salt water evaporators; solvent for inks and dyes; solvent, softener, and plasticizer for textile and paper coatings, NC coatings; mold release agent and dye solv. for vinyl resins; demulsifier; cosmetic emollient; defoamer for food contact; food pkg. cellophane

Regulatory: FDA 21CFR §176.200, 177.1200, 178.3570, 178.3910

Properties: Low color and odor; sol. in alcohol; visc. 3000 SUS (38 C); nonionic

Ukalene® [Atofina SA]

Chem. Descrip.: Methylene chloride

CAS 75-09-2; EINECS/ELINCS 200-838-9

Uses: Solvent

Ukasol® [Atofina SA]

Chem. Descrip.: Methylene chloride

CAS 75-09-2; EINECS/ELINCS 200-838-9

Uses: Solvent

Ultra Low Naphthalene Aromatic 150 Fluid [ExxonMobil]

Chem. Descrip.: Aromatic hydrocarbon fluid

Uses: Solvent for additives, paints/coatings, pesticides, industrial cleaning, mastics, sealants, process fluids

Features: High solvency; controlled evaporation chars.

Properties: I.b.p. 175 C min.; flash pt. > 62 C; > 95 min. vol.% aromatics

Ultra Low Naphthalene Aromatic 200 [ExxonMobil; Exxon Canada]

Chem. Descrip.: Naphtha, heavy aromatic, BHT (17 ppm)

CAS 64742-94-5; EINECS/ELINCS 265-198-5

Uses: Solvent for additives, paints/coatings, pesticides, industrial cleaning, mastics, sealants, process fluids; diluent; chem. feedstock; fuel

Features: High solvency; controlled evaporation chars.

Regulatory: Canada DSL listed; NPRI nonreportable

Properties: Pale yel. bright clear liq.; aromatic hydrocarbon odor; sol. < 0.01% in water; m.w. ≈ 120; sp.gr. 1 (15.6 C); visc. 2.7 cSt; vapor pressure < 0.1 kPa (20 C); f.p. -10 C; b.p. 235-278 C; flash pt. (PMCC) 106 C; autoignition temp. ≈ 484 C; > 97 min. vol.% aromatics; 0.1% max. naphthalene

Toxicology: REL TWA 15 ppm; inh. of high vapor concs. may cause eye/respiratory tract irritation, headaches, dizziness, anesthesia, drowsiness, unconsciousness, other CNS effects, death; sl. eye irritant; frequent/prolonged skin contact may cause irritation, dermatitis; if aspirated, may cause mild to severe pulmonary injury, death

Precaution: Flamm. limits 1.8-11.8 vol.%; low flamm./explosion hazard; electrostatic accumulation hazard; use proper grounding procedures; incompat. with nitric acid, sulfuric acid, strong oxidizing agents

Hazardous decomp. prods.: Combustion: toxic gases, fumes, smoke, CO

Storage: Store in cool, well-ventilated area away from incompat. materials, heat, open flame, ignition sources, direct sunlight; keep container closed

Ultra Low Naphthalene Solvesso 200 Solvent [Exxon Canada]

Chem. Descrip.: Naphtha, heavy aromatic, BHT (17 ppm)

CAS 64742-94-5; EINECS/ELINCS 265-198-5

Uses: Solvent; diluent; chem. feedstock; fuel

Regulatory: Canada DSL listed; NPRI nonreportable

Properties: Pale yel. liq.; aromatic hydrocarbon odor; sol. < 0.01% in water; m.w. ≈ 120; sp.gr. 1. (15.6 C); visc. 2.7 cSt; vapor pressure < 0.1 kPa (20 C); f.p. -10 C; b.p. 235-278 C; flash pt. (PMCC) 106 C; autoignition temp. ≈ 484 C

Toxicology: REL TWA 15 ppm; inh. of high vapor concs. may cause eye/respiratory tract irritation, headaches, dizziness, anesthesia, drowsiness, unconsciousness, other CNS effects, death; sl. eye irritant; frequent/prolonged skin contact may cause irritation, dermatitis; if aspirated, may cause mild to severe pulmonary injury, death

Environmental: Prevent entry into sewers, watercourses, low areas

Precaution: Flamm. limits 1.8-11.8 vol.%; low flamm./explosion hazard; electrostatic accumulation hazard; use proper grounding procedures; incompat. with nitric acid, sulfuric acid, strong oxidizing agents

Hazardous decomp. prods.: Combustion: toxic gases, fumes, smoke, CO

Storage: Store in cool, well-ventilated area away from incompat. materials, heat, open flame, ignition sources, direct sunlight; keep container closed

UltraPure™ Ethylene Carbonate [Huntsman]

Chem. Descrip.: Ethylene carbonate

CAS 96-49-1; EINECS/ELINCS 202-510-0

Uses: Solvent for lithium batteries, condensers, photochromic sol'ns., polymer removal and other cleaning related applics. in microelectronics fabrication

Features: Extremely high purity; very low levels of water, glycol, other contaminants; low m.p.

Properties: Pract. colorless solid; pract. odorless; m.w. 88.06; dens. 11.0 lb/gal (20 C); m.p. 36.4 C; b.p. 248.2 C; flash pt. (PMCC) 160 C; dielec. const. 89.6 esu (40 C); sp. resist. 1-10 x 10^7 ohm-cm (40 C); 99.99% min. assay; 20 ppm max. water

Precaution: LEL 4.5% (200 C); UEL 26.8% (200 C)

UltraPure™ Propylene Carbonate [Huntsman]

Chem. Descrip.: Propylene carbonate

CAS 108-32-7; EINECS/ELINCS 203-572-1

Uses: Solvent for lithium batteries, condensers, photochromic sol'ns., polymer removal for microelectronics, other related electrochemical applics.

Features: Extremely high purity; very low levels of water, glycol, other contaminants

Regulatory: DOT nonregulated; SARA §302/304 extremely hazardous substance (residual propylene oxide, ≤ 0.001%), §311 acute hazard, §313 nonreportable; Australia, Japan listed

Properties: Colorless clear mobile liq.; sl. odor; partly sol. in water; m.w. 102.09; sp.gr. 1.203-1.210 (20/20 C); dens. 10.1 lb/gal (20 C); visc. 2.5 mPa·s; vapor pressure 0.02 mm Hg (20 C); m.p. -49.2 C; b.p. 242 C; flash pt. (PMCC) 135 C; VOC 28.5%; dielec. const. 65 esu; sp. resist. 1-8 x 10^6 ohm-cm; 9.99% min. assay; 20 ppm max. water

Toxicology: LD50 (oral, rat) > 5 g/kg; may cause eye irritation, skin defatting/irritation; inh. of vapors/mist may cause nose/throat irritation, headache, nausea, drowsiness; ing. may cause abdominal discomfort, nausea, diarrhea; aspiration may damage lungsTSCA listed

Precaution: May dec. in presence of water, acids, bases, salts, or metal oxides to cause a pressure buildup in processing/storage vessels; may cause ruptured container

Hazardous Ingredients: May contain residual (< 100 ppm) concs. of propylene oxide

Hazardous decomp. prods.: Toxic levels of CO, CO_2, irritating aldehydes and ketones may be formed on burning; heating in air may produce irritating aldehydes, acids, and ketones

HMIS: Health 1, Flammability 1, Reactivity 0

Storage: Hygroscopic; keep container closed; minimize exposure to high temps.; avoid water contamination

Unichem BZBN [Universal Preserv-A-Chem]

Chem. Descrip.: Benzyl benzoate

CAS 120-51-4; EINECS/ELINCS 204-402-9

Uses: Fixing agent, solvent for musk in perfumes and flavors; external medicine

Uniplex 110 [Unitex]
Chem. Descrip.: Dimethyl phthalate
CAS 131-11-3; EINECS/ELINCS 205-011-6
Uses: Solvent, plasticizer for cellulose acetate butyrate compositions; solvent for org. catalysts; plasticizer, solvent for aerosol hair sprays
Properties: APHA 10 max. clear liq., essentially odorless; sp.gr. 1.188-1.192; visc. 14-15 cps; ref. index 1.512-1.514; 99% min. assay

Uniplex 150 [Unitex]
Chem. Descrip.: Dibutyl phthalate
CAS 84-74-2; EINECS/ELINCS 201-557-4
Uses: Solvent, plasticizer for fingernail polishes, nail polish removers, hair sprays, org. peroxide catalysts, adhesives, coatings; food pkg. adhesives, coatings, paper, cellophane, polymers
Features: Compat. with cellulosics, methacrylate, PS, PVB, vinyl chloride, urea-formaldehyde, melamine-formaldehyde, phenolics
Regulatory: FDA 21CFR §175.105, 175.300, 175.320, 176.170, 177.1200, 177.2430
Properties: APHA 25 max. clear liq., mild odor; pract. insol. in water; m.w. 278; sp.gr. 1.042; dens. 8.72 lb/gal; visc. 59 cps (32 F); f.p. -35 C; b.p. 340 C (760 mm Hg); ref. index 1.4905; 99% min. assay

Uniplex 155 [Unitex]
Chem. Descrip.: Diisobutyl phthalate
CAS 84-69-5; EINECS/ELINCS 201-553-2
Uses: Plasticizer for thermoplastic and thermoset resins; solvent, plasticizer in cellophane, resin-coated sand for foundry casting, org. peroxides; food pkg. adhesives, coatings, paper, cellophane, rubber articles, textiles
Regulatory: FDA 21CFR §175.105, 175.300, 175.320, 176.170, 177.1200, 177.2510, 177.2550, 177.2600, 177.2800, 177.2910
Properties: APHA 25 max. clear liq., mild odor; pract. insol. in water; m.w. 278; sp.gr. 1.038-1.040; dens. 8.65 lb/gal; b.p. 327 C (760 mm Hg); 99% min. assay

Upamide ACMEA [Universal Preserv-A-Chem]
Chem. Descrip.: Acetamide MEA
CAS 142-26-7; EINECS/ELINCS 205-530-8
Uses: Solvent, humectant for cosmetics
Properties: Nonionic

Upiwax 350 [Universal Preserv-A-Chem]
Chem. Descrip.: PEG-6 methyl ether
CAS 9004-74-4
Uses: Intermediate; solvent

Uresolve 411 [Dynaloy]
Chem. Descrip.: Methanol (60-80%) with potassium hydroxide (1-8%)
UN NA 1993
Uses: Solvent for urethane coatings, RTV and silicone oils, for electronic use (conformal coating removal, cable assembly repair, depotting)
Features: Flamm.
Regulatory: SARA §313 reportable (methanol)
Properties: Blue liq.; mild alcoholic odor; sol. in water; sp.gr. .0891; pH 13.7
Toxicology: Toxic; harmful or fatal if swallowed; may cause eye/skin irritation, nausea, abdominal pain, vomiting, headache, dizziness, weakness, confusion, drunken behavior, visual disturbances, drowsiness, coma; liver/kidney/heart damage in massive doses
Precaution: Flamm. liq.; incompat. with alkali metals, acids, aldehydes, acyl chlorides
Hazardous decomp. prods.: CO, CO_2
HMIS: Health 2, Flammability 3, Reactivity 1
Storage: Keep container closed; keep away from heat, spark, flames

Uresolve Blue [Dynaloy]
Chem. Descrip.: Methanol base
Uses: Solvent for electronic conformal coating removal, cable assembly repair, connector repair, depotting, cured urethane removal
Features: For R.T. use only; most selective
Precaution: Flamm.

Uresolve HF [Dynaloy]
Chem. Descrip.: Ethylene glycol ether base
Uses: Solvent for electronic depotting, decapsulation
Features: Stronger than Uresolve 411, but less selective; can be heated (to 37 C) for faster action

Uresolve Plus 500 [Dynaloy]
Chem. Descrip.: Ethylene glycol ether base
Uses: Solvent for electronic depotting, decapsulation, deinking, and surf. prep. applics.
Features: Stronger than Uresolve Plus, but less selective; slow @ R.T., more effective @ 65-100 C

Uresolve Plus SG [Dynaloy]
Chem. Descrip.: Ethylene glycol ether base
Uses: Solvent for depotting, deflashing, decapsulation, surf. prep., photo resist stripping applics.
Features: Semiconductor grade; can be used in contact with aluminum

USP Castor Oil [Süd-Chemie Inc/Rheologicals]
Chem. Descrip.: Castor oil USP
CAS 8001-79-4; EINECS/ELINCS 232-293-8
Uses: Gloss aid, emollient, dye solvent in personal care prods., lipsticks, pharmaceuticals, and urethane applics.
Features: High quality oil
Properties: Gardner 1 color; sp.gr. 0.959; dens. 7.98 lb/gal; visc. 7.5 stokes; acid no. 1; iodine no. 86; sapon. no. 180; hyd. no. 164

V

V-154L [Sasol N. Am.]
Chem. Descrip.: Benzene, mono-C10-13-alkyl derivs., distillation residues
CAS 84961-70-6; EINECS/ELINCS 284-660-7
Uses: Sec. plasticizer for vinyl; wire rope lubricant; heat transfer fluid; solvent for oil-sol. sulfonates such as sacrificial agent corrosion inhibitors; specialty lubricants; sulfonation feedstock; intermediate for agric. emulsifiers, enhanced oil recovery surfactants, dust abatement agents
Features: High flash pt.
Properties: Amber oily liq.; m.w. ≈ 325; sp.gr. 0.87 (60 F); visc. 19-20 cSt (40 C); b.p. 300-500 C; pour pt. -50 to -55 C; flash pt. > 185 C; 50 ppm water
Toxicology: Repeated/prolonged contact may cause skin and eye irritation; prolonged inh. may cause dizziness, headaches

V-159L [Sasol N. Am.]
Chem. Descrip.: C12-14 alkylbenzene (alkylate bottoms produced during mfg. of high m.w. linear alkylbenzene)
CAS 68515-32-2; EINECS/ELINCS 271-073-6
Uses: Sec. plasticizer for vinyl; wire rope lubricant; heat transfer fluid; solvent for oil-sol. sulfonates such as sacrificial agent corrosion inhibitors; specialty lubricants; sulfonation feedstock; intermediate for agric. emulsifiers, enhanced oil recovery surfactants, dust abatement agents
Properties: Amber oily liq.; m.w. ≈ 340; sp.gr. 0.88 (60 F); visc. 20-25 cSt (40 C); b.p. 320-500 C; pour pt. -50 C; flash pt. > 195 C; 50 ppm water
Toxicology: Repeated/prolonged contact may cause skin and eye irritation; prolonged inh. may cause dizziness, headaches

Variquat® CC-42 [Goldschmidt]
Chem. Descrip.: PPG-40 diethylmonium chloride
CAS 9076-43-1
Uses: Detergent, pigment dispersant, particle suspending agent, emulsifier, solvent, conditioner, antistat, lubricant, corrosion inhibitor for toiletries, cosmetics, germicides, syn. fibers and plastics, textiles, industrial processes; ore flotation additive; dry cleaning
Properties: Lt. amber oily liq.; sol. in IPA, acetone, MEK, min. spirits, ethanol; partly sol. in water; sp.gr. 1.01; flash pt. > 200 C (PMCC); pH 6.5; cationic; 100% conc.

Variquat® CC-55 [Goldschmidt]
Chem. Descrip.: Polypropoxy quat. ammonium acetate
Uses: Antistat; conditioner for hair rinse preparations; emulsifier for cosmetics and textile flame retardants; solvent for phenolic-type germicides for cosmetics and toiletries; antistat for syn. fibers and plastics; fabric conditioner; lubricant for textile and industrial formulations; solvent cleaning and scouring agent; corrosion inhibitor in protective coatings; pigment dispersant in nonaq. media; o/w emulsifier
Properties: Lt. amber oily liq.; sol. @ 25% in ethanol, IPA, acetone, MEK; water-disp.; sp.gr. 1.02; flash pt. > 93 C (PMCC); pH 6.5; cationic; 99% solids

Variquat® CC-57 [Goldschmidt]
Chem. Descrip.: Polypropoxy quat. ammonium phosphate
Uses: Antistat, conditioner, emulsifier, solvent, lubricant, solvent cleaning and scouring agent, corrosion inhibitor, and dispersant used in syn. fibers and plastics, dry cleaning, personal care prods., germicides, flame retardants, textile and industrial applic., protective coatings, and pigments
Properties: Lt. amber oily liq.; sol. in water, ethanol, IPA, acetone, MEK; sp.gr. 1.12; flash pt. > 93 C (PMCC); cationic

Varonic® APS [Goldschmidt; Goldschmidt AG]
Chem. Descrip.: PPG-11 stearyl ether
CAS 25231-21-4
Uses: Emollient for skin care; solvent for lipophilic act. ingreds. and perfume oils
Features: Highly polar
Regulatory: JCIC registered
Properties: Liq.

Varonic® DM55 [Goldschmidt]
Chem. Descrip.: Methyl capped glycol ether
Uses: Solvent for surf. cleaners, textile scouring; hair care, skin care
Features: Exc. grease cutting properties; low vapor pressure helps prod. to meet emission stds. for VOCs
Properties: Liq.; water-emulsifiable; 100% solids
Toxicology: Low toxicity

Varsol® 1 [ExxonMobil]
Chem. Descrip.: Stoddard solvent
CAS 8052-41-3; EINECS/ELINCS 232-489-3
UN 1268
Uses: Solvent for solv.-based architectural and industrial coatings, printing inks
Regulatory: SARA §311/312 chronic/acute/fire hazard, §313 reportable [1,2,4-trimethylbenzene (4%)]; contains ethylbenzene
Properties: Clear liq.; min. spirit odor; negligible sol. in water; m.w. 140; sp.gr. 0.80 (15.6 C); dens. 6.65 lb/gal (60 F); visc. 1.0 cP; vapor pressure 1.6 mm Hg (20 C); b.p. 160-205 C; pour pt. -18 C; flash pt. > 38 C; autoignition temp. 254 C; surf. tens. 26.6 dynes/cm; KB value 39; 100% act.; 38% paraffin, 47% cycloparaffin, 15% aromatic
Toxicology: REL TWA (8 h) 100 ppm (vapor); pract. nontoxic by ing., inh., skin contact; skin irritant; prolonged/repeated skin contact may cause dermatitis; high concs. of vapors may cause eye/respiratory system irritation, anesthesia; if swallowed, may enter lungs and cause lung damage; contains ethylbenzene (possible human carcinogen); TSCA listed
Environmental: Expected to be biodeg.; may be harmful to aquatic organisms, toxic to algae, daphnia; prevent entry into sewers, watercourses, low areas
Precaution: Combustible liq.; flamm. limits in air 2.1-13.3 vol.%; liq. can release vapors that readily form flamm. mixts. at or above flash pt.; static accumulation hazard; use proper grounding procedures; avoid buildup of static electricity; incompat. with halogens, strong acids, alkalis, oxidizers
Hazardous decomp. prods.: Combustion: fumes, smoke, CO, SO_x, aldehydes, other decomp. prods.; does not dec. @ ambient temps.
NFPA: Health 1, Flammability 2, Reactivity 0
Storage: Store in cool, well-ventilated area away from heat, open flame

Varsol® 1 Naphtha [ExxonMobil]
Uses: Solvent
Properties: Bright clear liq.; sp.gr. 0.770-0.820 (15.6/15.6 C); i.b.p. 152 C min.; flash pt. > 40.5 C; KB value 29-45; 14-22% aromatics

Varsol® 1 Naphtha Anti-Static [ExxonMobil; Exxon Canada]
Chem. Descrip.: Stoddard solvent with antistatic additive (16 ml/1000 gal)
CAS 8052-41-3; EINECS/ELINCS 232-489-3
UN 1268

Uses: Solvent; diluent; chem. feedstock; fuel
Regulatory: Canada DSL listed; NPRI reportable [1,2,4-trimethylbenzene (4.2%), naphthalene (0.2%)]
Properties: Colorless bright clear liq.; mild petrol. odor; sol. < 0.01% in water; sp.gr. 0.770-0.820 (15.6/15.6 C); visc. 1.14 cSt; vapor pressure < 0.3 kPa (20 C); f.p. -58 C; b.p. 158-197 C; flash pt. (TCC) 43 C; autoignition temp. ≈ 229 C; KB value 29-45; 100 vol.%; 14-22% aromatics
Toxicology: REL TWA 66 ppm; LD50 (oral, rat) > 5 g/kg; inh. of high vapor concs. may cause eye/respiratory tract irritation, headaches, dizziness, anesthesia, drowsiness, unconsciousness, other CNS effects, death; sl. eye irritant; frequent/prolonged skin contact may cause irritation, dermatitis; if aspirated, may cause mild to severe pulmonary injury, death; contains ethylbenzene (possible human carcinogen)
Environmental: Prevent entry into sewers, watercourses, low areas
Precaution: Combustible liq.; flamm. limits 1-13.3 vol.%; volatile; gives off invisible vapors; liq. or vapor may settle in low areas; distant ignition/explosion possible; electrostatic accumulation hazard; use proper grounding procedures; incompat. with strong oxidizing agents
Hazardous decomp. prods.: Combustion: toxic gases, fumes, smoke, CO
Storage: Store in cool, well-ventilated area away from incompat. materials, heat, open flame, ignition sources, direct sunlight; keep container closed

Varsol® 18 [ExxonMobil]
Chem. Descrip.: Stoddard solvent (45-55%), hydrotreated light distillate (45-55%)
UN 1268
Uses: Solvent for solv.-based architectural and industrial coatings, printing inks
Regulatory: SARA §311/312 chronic/acute/fire hazard, §313 reportable [1,2,4-trimethylbenzene (2%)]
Properties: Clear liq.; hydrocarbon odor; negligible sol. in water; m.w. 145; sp.gr. 0.80 (15.6 C); dens. 6.62 lb/gal (60 F); visc. 0.9 cP; vapor pressure 5.6 mm Hg (20 C); b.p. 160-205 C; pour pt. -20 C; aniline pt. 67 C max.; flash pt. > 38 C; surf. tens. 26.3 dynes/cm; KB value 36; 40% paraffin, 53% cycloparaffin, 7% aromatic
Toxicology: REL TWA (8 h) 100 ppm (vapor); pract. nontoxic by ing., inh., skin contact; skin irritant; prolonged/repeated skin contact may cause dermatitis; high concs. of vapors may cause eye/respiratory system irritation, anesthesia; if swallowed, may enter lungs and cause lung damage
Environmental: Expected to be biodeg.; may be harmful to aquatic organisms, toxic to algae, daphnia; prevent entry into sewers, watercourses, low areas
Precaution: Combustible liq.; flamm. limits in air 0.7-5.0 vol.%; liq. can release vapors that readily form flamm. mixts. at or above flash pt.; static accumulation hazard; use proper grounding procedures; avoid buildup of static electricity; incompat. with halogens, strong acids, alkalis, oxidizers
Hazardous decomp. prods.: Combustion: fumes, smoke, CO, SO_x, aldehydes, other decomp. prods.; does not dec. @ ambient temps.
NFPA: Health 1, Flammability 2, Reactivity 0
Storage: Store in cool, well-ventilated area away from heat, open flame

Varsol® 18 Naphtha [ExxonMobil; Exxon Canada]
Chem. Descrip.: Light hydrotreated petroleum distillates (60-100%), Stoddard solvent (15-40%)
UN 1268
Uses: Solvent; diluent; chem. feedstock; fuel
Regulatory: Canada DSL listed; NPRI reportable [1,2,4-trimethylbenzene (1.3%)]
Properties: Colorless bright clear liq.; mild petrol. odor; sol. < 0.01% in water; m.w. 142; sp.gr. 0.770-0.810 (15.6/15.6 C); visc. 1.24 cSt; vapor pressure 0.295 kPa (20 C); f.p. -60 C; b.p. 159-192 C; flash pt. (TCC) 43 C; autoignition temp. ≈ 254 C; KB value 29-45; 4-8% aromatics
Toxicology: REL TWA 97 ppm; inh. of high vapor concs. may cause eye/respiratory tract irritation, headaches, dizziness, anesthesia, drowsiness, unconsciousness, other CNS effects, death; sl. eye irritant; frequent/prolonged skin contact may cause irritation, dermatitis; if aspirated, may cause mild to severe pulmonary injury, death
Environmental: Prevent entry into sewers, watercourses, low areas
Precaution: Combustible liq.; flamm. limits 0.9-7 vol.%; volatile; gives off invisible fumes; liq. or vapor may settle in low areas; distant ignition/explosion possible; electrostatic accumulation hazard; use proper grounding procedures; incompat. with strong oxidizing agents
Hazardous decomp. prods.: Combustion: toxic gases, fumes, smoke, CO
Storage: Store in cool, well-ventilated area away from incompat. materials, heat, open flame, ignition sources, direct sunlight; keep container closed

Varsol® 18 Naphtha Anti-Static [ExxonMobil; Exxon Canada]
Chem. Descrip.: Light hydrotreated petroleum distillates (60-100%), Stoddard solvent (15-40%) with antistatic additive (16 ml/1000 gal)
UN 1268
Uses: Solvent; diluent; chem. feedstock; fuel
Regulatory: Canada DSL listed; NPRI reportable [1,2,4-trimethylbenzene (1.3%)]
Properties: Colorless bright clear liq.; mild petrol. odor; sol. < 0.01% in water; sp.gr. 0.770-0.810 (15.6/15.6 C); visc. 1 cSt; vapor pressure 0.107 kPa (20 C); f.p. -54 C; b.p. 157-197 C; flash pt. (TCC) > 38 C; autoignition temp. ≈ 254 C; KB value 29-45; 4-8% aromatics
Toxicology: REL TWA 132 ppm; inh. of high vapor concs. may cause eye/respiratory tract irritation, headaches, dizziness, anesthesia, drowsiness, unconsciousness, other CNS effects, death; sl. eye irritant; frequent/prolonged skin contact may cause irritation, dermatitis; if aspirated, may cause mild to severe pulmonary injury, death
Environmental: Prevent entry into sewers, watercourses, low areas
Precaution: Combustible liq.; flamm. limits 2.3-14.4 vol.%; volatile; gives off invisible fumes; liq. or vapor may settle in low areas; distant ignition/explosion possible; electrostatic accumulation hazard; use proper grounding procedures; incompat. with strong oxidizing agents
Hazardous decomp. prods.: Combustion: toxic gases, fumes, smoke, CO
Storage: Store in cool, well-ventilated area away from incompat. materials, heat, open flame, ignition sources, direct sunlight; keep container closed

Varsol® 30 [ExxonMobil Europe]
Chem. Descrip.: Nondearomatized hydrocarbon fluid
Uses: Solvent for solv.-based decorative and industrial coatings
Features: Cost effective
Properties: Liq.; dens. 0.776 kg/dm^3 (15 C); visc. 0.71 mPa·s; i.b.p. 142 C; flash pt. 29 C; ref. index 1.433 (20 C); surf. tens. 23.9 mN/m; KB value 39; aniline pt. 52 C; 22% aromatic content

Varsol® 40 [ExxonMobil Europe]
Chem. Descrip.: Nondearomatized hydrocarbon fluid
Uses: Solvent for solv.-based decorative and industrial coatings
Features: Cost effective
Properties: Dens. 0.788 kg/dm^3 (15 C); visc. 0.91 mPa·s; i.b.p. 158 C; flash pt. 41 C; ref. index 1.439 (20 C); surf. tens. 24.8 mN/m; KB value 37; 23% aromatic content

Varsol® 60 [ExxonMobil Europe]
Chem. Descrip.: Nondearomatized hydrocarbon fluid
Uses: Solvent for solv.-based decorative and industrial coatings
Features: Cost effective
Properties: Dens. 0.807 kg/dm^3 (15 C); visc. 1.24 mPa·s; i.b.p. 185 C; flash pt. 64 C; ref. index 1.448 (20 C); surf. tens. 26.1 mN/m; KB value 35; 25% aromatic content

Varsol® 80 [ExxonMobil Europe]
Chem. Descrip.: Nondearomatized hydrocarbon fluid
Uses: Solvent for solv.-based decorative and industrial coatings
Features: Cost effective
Properties: Dens. 0.819 kg/dm^3 (15 C); visc. 1.64 mPa·s; i.b.p. 204 C; flash pt. 78 C; ref. index 1.455 (20 C); surf. tens. 26.8 mN/m; KB value 33; 26% aromatic content

Varsol® 110 [ExxonMobil; ExxonMobil Europe]
Chem. Descrip.: Nondearomatized hydrocarbon fluid
Uses: Solvent for solv.-based architectural and industrial coatings, printing inks
Properties: Dens. 0.847 kg/dm^3 (15 C); visc. 3.53 mm^2/s; i.b.p. 242 C; flash pt. 107 C; ref. index 1.471 (20 C); surf. tens. 28.0 mN/m; KB value 32; aniline pt. 60 C; 32% aromatics

Varsol® 120 [ExxonMobil Europe]
Chem. Descrip.: Nondearomatized hydrocarbon fluid
Uses: Solvent
Properties: Dens. 0.850 kg/dm^3 (15 C); visc. 4.70 mm^2/s; i.b.p. 260 C; flash pt. 122 C; ref. index 1.473 (20 C); surf. tens. 28.5 mN/m; KB value 29; aniline pt. 65 C; 34% aromatics

Varsol® 140 [ExxonMobil Europe]
Chem. Descrip.: Nondearomatized hydrocarbon fluid
Uses: Solvent
Properties: Dens. 0.854 kg/dm^3 (15 C); visc. 6.95 mm^2/s; i.b.p. 290 C; flash pt. 144 C; ref. index 1.476 (20 C); surf. tens. 28.9 mN/m; KB value 26; aniline pt. 72 C; 31% aromatics

Varsol® 140 Naphtha [ExxonMobil]
Chem. Descrip.: Hydrocarbon fluid
Uses: Solvent
Properties: Bright clear liq.; sp.gr. 0.768-0.820 (15.6/15.6 C); i.b.p. 180 C min.; flash pt. > 61.5 C; 14-22% aromatics

Varsol® 3135 Naphtha [ExxonMobil]
Chem. Descrip.: Stoddard solvent
CAS 8052-41-3; EINECS/ELINCS 232-489-3
Uses: Solvent
Properties: Bright clear appearance; sp.gr. 0.754-0.810 (15.6/15.6 C); dens. 6.56 lb/gal (60 F); b.p. 159-176 C; flash pt. (TCC) 42 C; surf. tens. 23.7 dynes/cm; KB value 39; 45% paraffin, 35% cycloparaffin, 20% aromatics
Toxicology: OEL 200 ppm

Varsol® 3139 [Exxon Canada]
Chem. Descrip.: Stoddard solvent
CAS 8052-41-3; EINECS/ELINCS 232-489-3
UN 1268
Uses: Solvent; diluent; chem. feedstock; fuel
Regulatory: Canada DSL listed; NPRI reportable [1,2,4-trimethylbenzene (4.2%), naphthalene 0.2%)]
Properties: Colorless clear liq.; mild petrol. odor; sol. < 0.01% in water; sp.gr. 0.79 (15.5 C); visc. 1.14 cSt; vapor pressure 0.3 kPa (20 C); f.p. -58 c; b.p. 158-195 C; flash pt. (TCC) 43 C; autoignition temp. ≈ 229 C; 100% act.
Toxicology: REL TWA 73 ppm; LD50 (oral, rat) > 5 g/kg; inh. of high vapor concs. may cause eye/respiratory tract irritation, headaches, dizziness, anesthesia, drowsiness, unconsciousness, other CNS effects, death; sl. eye irritant; frequent/prolonged skin contact may cause irritation, dermatitis; if aspirated, may cause mild to severe pulmonary injury, death; contains ethylbenzene, naphthalene (possible human carcinogens)
Environmental: Prevent entry into sewers, watercourses, low areas
Precaution: Combustible liq.; flamm. limits 1-13.3 vol.%; volatile; gives off invisible fumes; liq. or vapor may settle in low areas; distant ignition/explosion possible; electrostatic accumulation hazard; use proper grounding procedures; incompat. with strong oxidizing agents
Hazardous decomp. prods.: Combustion: toxic gases, fumes, smoke, CO
Storage: Store in cool, well-ventilated area away from incompat. materials, heat, open flame, ignition sources, direct sunlight; keep container closed

Varsol® 3139 Anti-Static Solvent [Exxon Canada]
Chem. Descrip.: Stoddard solvent with antistatic additive (16 ml/1000 gal)
CAS 8052-41-3; EINECS/ELINCS 232-489-3
UN 1268
Uses: Solvent; diluent; chem. feedstock; fuel
Regulatory: Canada DSL listed; NPRI reportable [1,2,4-trimethylbenzene (4.2%), naphthalene 0.2%)]
Properties: Colorless clear liq.; mild petrol. odor; sol. < 0.01% in water; sp.gr. 0.79 (15.5 C); visc. 1.14 cSt; vapor pressure 0.3 kPa (20 C); f.p. -58 C; b.p. 158-195 C; flash pt. (TCC) 43 C; autoignition temp. ≈ 229 C; 100% act.
Toxicology: REL TWA 73 ppm; LD50 (oral, rat) > 5 g/kg; inh. of high vapor concs. may cause eye/respiratory tract irritation, headaches, dizziness, anesthesia, drowsiness, unconsciousness, other CNS effects, death; sl. eye irritant; frequent/prolonged skin contact may cause irritation, dermatitis; if aspirated, may cause mild to severe pulmonary injury, death; contains ethylbenzene, naphthalene (possible human carcinogens)
Environmental: Prevent entry into sewers, watercourses, low areas
Precaution: Combustible liq.; flamm. limits 1-13.3 vol.%; volatile; gives off invisible fumes; liq. or vapor may settle in low areas; distant ignition/explosion possible; electrostatic accumulation hazard; use proper grounding procedures; incompat. with strong oxidizing agents
Hazardous decomp. prods.: Combustion: toxic gases, fumes, smoke, CO
Storage: Store in cool, well-ventilated area away from incompat. materials, heat, open flame, ignition sources, direct sunlight; keep container closed

Varsol® DX 140 Naphtha [ExxonMobil]
Uses: Solvent
Properties: Bright clear liq.; i.b.p. 182 C min.; flash pt. > 61.5 C; 4-8% aromatics

Varsol® DX 3139 [Exxon Canada]
Chem. Descrip.: Light hydrotreated petroleum distillates (60-100%), Stoddard solvent (15-40%)
UN 1268
Uses: Solvent; diluent; chem. feedstock; fuel
Regulatory: Canada DSL listed; NPRI reportable [1,2,4-trimethylbenzene (1.3%)]
Properties: Colorless clear liq.; mild petrol. odor; sol. < 0.01% in water; m.w. 142; sp.gr. 0.78 (15.6 C); visc. 1.24 cSt; vapor pressure 0.295 kPa (20 C); f.p. -60 C; b.p. 159-192 C; flash pt. (TCC) 43 C; autoignition temp. ≈ 254 C; 100% act.
Toxicology: REL TWA 97 ppm; inh. of high vapor concs. may cause eye/respiratory tract irritation, headaches, dizziness, anesthesia, drowsiness, unconsciousness, other CNS effects, death; sl. eye irritant; frequent/prolonged skin contact may cause irritation, dermatitis; if aspirated, may cause mild to severe pulmonary injury, death
Environmental: Prevent entry into sewers, watercourses, low areas
Precaution: Combustible liq.; flamm. limits 0.9-7 vol.%; volatile; gives off invisible vapors; liq. or vapor may settle in low areas; distant ignition/explosion possible; electrostatic accumulation hazard; use proper grounding procedures; incompat. with strong oxidizing agents
Hazardous decomp. prods.: Combustion: toxic gases, fumes, smoke, CO
Storage: Store in cool, well-ventilated area away from incompat. materials, heat, open flame, ignition sources, direct sunlight; keep container closed

Varsol® DX 3139 Anti-Static Solvent [Exxon Canada]
Chem. Descrip.: Light hydrotreated petroleum distillates (60-100%), Stoddard solvent (15-40%) with antistatic additive (16 ml/1000 gal)
UN 1268
Uses: Solvent; diluent; chem. feedstock; fuel
Regulatory: Canada DSL listed; NPRI reportable [1,2,4-trimethylbenzene (1.3%)]
Properties: Colorless clear liq.; mild petrol. odor; sol. < 0.01% in water; sp.gr. 0.78 (15.6 C); visc. 1 cSt; vapor pressure 0.107 kPa (20 C); f.p. -54 C; b.p. 157-197 C; flash pt. (TCC) > 38 C; autoignition temp. ≈ 254 C; 100% act.
Toxicology: REL TWA 132 ppm; inh. of high vapor concs. may cause eye/respiratory tract irritation, headaches, dizziness, anesthesia, drowsiness, unconsciousness, other CNS effects, death; sl. eye irritant; frequent/prolonged skin contact may cause irritation, dermatitis; if aspirated, may cause mild to severe pulmonary injury, death
Environmental: Prevent entry into sewers, watercourses, low areas
Precaution: Combustible liq.; flamm. limits 2.3-14.4 vol.%; volatile; gives off invisible vapors; liq. or vapor may settle in low areas; distant ignition/explosion possible; electrostatic accumulation hazard; use proper grounding procedures; incompat. with strong oxidizing agents
Hazardous decomp. prods.: Combustion: toxic gases, fumes, smoke, CO
Storage: Store in cool, well-ventilated area away from incompat. materials, heat, open flame, ignition sources, direct sunlight; keep container closed

Varsol® DX 3641 [Exxon Canada]
Chem. Descrip.: Petroleum distillates, hydrotreated light
CAS 64742-47-8; EINECS/ELINCS 265-149-8
Uses: Solvent; diluent; chem. feedstock; fuel
Regulatory: Canada DSL listed; NPRI reportable [naphthalene (0.3%)]
Properties: Colorless clear liq.; mild petrol. odor; sol. < 0.01% in water; m.w. 154; sp.gr. 0.795 (15.6 C); visc. 1.6 cSt; vapor pressure 0.068 kPa (20 C); f.p. -56 C; b.p. 188-209 C; flash pt. (TCC) 64 C; autoignition temp. ≈ 227 C; 100% act.
Toxicology: REL TWA 93 ppm; LD50 (oral, rat) > 5 g/kg, (skin, rabbit) > 3 g/kg; inh. of high vapor concs. may cause eye/respiratory tract irritation, headaches, dizziness, anesthesia, drowsiness, unconsciousness, other CNS effects, death; sl. eye irritant; frequent/prolonged skin contact may cause irritation, dermatitis; if aspirated, may cause mild to severe pulmonary injury, death; contains naphthalene, possible human carcinogen
Environmental: Prevent entry into sewers, watercourses, low areas

Precaution: Combustible liq.; flamm. limits 2.1-13 vol.%; volatile; gives off invisible vapors; liq. or vapor may settle in low areas; distant ignition/ explosion possible; electrostatic accumulation hazard; use proper grounding procedures; incompat. with strong oxidizing agents
Hazardous decomp. prods.: Combustion: toxic gases, fumes, smoke, CO
Storage: Store in cool, well-ventilated area away from incompat. materials, heat, open flame, ignition sources, direct sunlight; keep container closed

Versadil™ 100 [Neville]
Chem. Descrip.: Petroleum aromatic hydrocarbon resin
Uses: Diluent, alkyd modifier, solvent substitute for meeting low VOC regulations, adhesion promoter, hardener, visc. reducer in high-solids, air-drying coatings
Properties: Gardner 3 resin; sol. in chlorinated hydrocarbons, ketones (except acetone), aromatic, naphthenic, paraffinic, and terpene hydrocarbons; partially sol. in esters and ethers; insol. in alcohols; dens. 7.8 lb/gal; visc. 600 cps; 99% min. NV
Precaution: Incompat. with acrylates

VersaTRANS™ Solvent [PPG Ind.]
Chem. Descrip.: trans-1,2-Dichloroethylene
CAS 156-60-5; EINECS/ELINCS 205-860-2
UN 1150
Uses: Solvent, azeotrope to increase solvency power of weaker solvents for electronic and precision cleaning applics.; hydrochloric acid source in electronics mfg. and etching of silicone chips
Regulatory: SARA §311/312 acute health hazard, §313 nonreportable; Canada DSL, EINECS, Australia AICS, Korea KECI, Japan ENCS listed
Properties: Colorless clear liq.; pleasant odor; sol. 0.063 g/100 g water; m.w. 96.95; sp.gr. 1.27; dens. 10.53 lb/gal (20 C); vapor pressure 336 mm Hg; f.p. -49.4 C; b.p. 47 C; flash pt. (CC) 36 F; autoignition temp. 460 C; pH 6.5-7.2; KB value 117; > 99% act.
Toxicology: ACGIH TWA 200 ppm; LD50 (skin, rabbit) >5000 mg/kg; inh. may cause CNS depression, respiratory tract irritation, dizziness, nausea, headache, loss of coordination, unconsciousness, death; high concs. may produce cardiac sensitization; ing. may cause mouth/GI tract irritation, vomiting; aspiration may cause chem. pneumonia, pulmonary edema; may cause eye pain, irritation; prolonged/repeated contact may cause skin irritation, dermatitis; TSCA listed
Environmental: Prevent contamination of ground and surf. waters
Precaution: Flamm.; flamm. limits in air 6.7-18%; vapors may flash back or collect in low areas; vapor/air mixts. are explosive; incompat. with steam, oxidizers, elevated temps., caustic soda, caustic potash; may form shock-sensitive compds.
Hazardous decomp. prods.: Hydrogen chloride gas; CO_2; liq. or vapor can dec. to form toxic and corrosive acid fumes; emits toxic fumes under fire conditions
Storage: Store in closed, properly labeled containers; keep closed when not in use

Vertrel® MCA [DuPont Fluoroprods.]
Chem. Descrip.: Mixt. of Vertrel® XF (2,3-dihydrodecafluoropentane) and trans-1,2-dichloroethylene
Uses: Solvent for use in vapor degreasing equip. for precision and specialty cleaning in difficult soil situations
Features: Zero ozone depletion potential; low global warming potential; replaces CFC-113, 1,1,1-trichloroethane, HCFCs, and PFCs
Regulatory: EPA approved under SNAP; DOT nonregulated
Properties: Clear, colorless; sp.gr. 1.41; visc. 0.49 cps; vapor pressure 464 mm Hg; f.p. < -50 C; b.p. 39 C; flash pt. none; surf. tens. 15.2 dynes/cm; 10.0 ppm max. NV; < 200 ppm moisture
Toxicology: Very low toxicity; sl. skin and eye irritant, low acute inh. toxicityTSCA listed
Precaution: No vapor flamm. limits in air
Storage: Store in clean, dry area @ < 52 C

Vertrel® MCA Plus [DuPont Fluoroprods.]
Chem. Descrip.: Mixt. of Vertrel® XF (2,3-dihydrodecafluoropentane), trans-1,2-dichloroethylene, and cyclopentane
Uses: Solvent for use in vapor degreasing equip. for precision and specialty cleaning in difficult soil situations
Features: Zero ozone depletion potential; low global warming potential; replaces CFC-113, 1,1,1-trichloroethane, HCFCs, and PFCs
Regulatory: EPA approved under SNAP; DOT nonregulated
Properties: Sp.gr. 1.28; visc. 0.61 cps; vapor pressure 440 mm Hg; f.p. < -50 C; b.p. 37 C; flash pt. none; surf. tens. 15.9 dynes/cm
Toxicology: Very low toxicity; sl. skin and eye irritant, low acute inh. toxicityTSCA listed
Precaution: Contact with highly basic process materials of pH 10 or above not rec.; prod. has vapor flamm. limits in air
Storage: Store in clean, dry area

Vertrel® SMT [DuPont Fluoroprods.]
Chem. Descrip.: Mixt. of Vertrel® XF (2,3-dihydrodecafluoropentane), trans-1,2-dichloroethylene, methanol
Uses: Solvent for use in vapor degreasing equip. for cleaning ionic soils and flux residues from electronic assemblies, precision and industrial cleaning where enhanced solvency is required
Features: Zero ozone depletion potential; low global warming potential; replaces CFC-113, 1,1,1-trichloroethane, HCFCs, and PFCs
Regulatory: EPA approved under SNAP; DOT nonregulated
Properties: Clear, colorless; sp.gr. 1.35; visc. 0.49 cps; vapor pressure 486 mm Hg; f.p. < -50 C; b.p. 37 C; flash pt. none; surf. tens. 15.5 dynes/cm; 10.0 ppm max. NV; < 200 ppm moisture
Toxicology: Very low toxicity; sl. skin and eye irritant, low acute inh. toxicityTSCA listed
Precaution: Contact with highly basic process materials of pH 10 or above not rec.; prod. has vapor flamm. limits in air
Storage: Store in clean, dry area @ < 52 C

Vertrel® X-B3 [DuPont Fluoroprods.]
Chem. Descrip.: Mixt. of Vertrel® XF (2,3-dihydrodecafluoropentane) and 2-butoxyethanol
Uses: Solvent for aircraft and aerospace industry for drilling and machining a variety of high-str. alloys
Features: Assures cooling and lubricant delivery to cutting point; designed to replace Freon® TB-1; nonflamm.; nonozone-depleting
Regulatory: EPA approved under SNAP
Properties: Sp.gr. 1.55; b.p. 61 C; flash pt. none
Toxicology: Avoid skin contactTSCA listed
Storage: Store in clean, dry area @ < 52 C

Vertrel® X-P10 [DuPont Fluoroprods.]
Chem. Descrip.: Mixt. of Vertrel® XF (2,3-dihydroperfluoropentane, 90 ± 0.5%) and isopropanol (10 ± 0.5%)
Uses: Solvent for removal of water from nonporous surfs.
Features: Nonflamm.
Properties: Colorless, clear liq.; sp.gr. 1.42; b.p. 54 C (1 atm); flash pt. none; 500 ppm max. moisture
Toxicology: Very low toxicity; sl. eye and skin irritant, low inh. toxicity
Storage: Store in clean, dry area @ < 52 C; protect from freezing

Vertrel® XE [DuPont Fluoroprods.]
Chem. Descrip.: Mixt. of Vertrel® XF (2,3-dihydrodecafluoropentane) and ethanol
Uses: Solvent for use in vapor degreasing equip. with improved solvency for polar soils and in precision and specialty cleaning and rinsing for removal of particulate, fingerprints, and lt. soils
Features: Zero ozone depletion potential; low global warming potential; replaces CFC-113, 1,1,1-trichloroethane, HCFCs, and PFCs
Regulatory: EPA approved under SNAP; DOT nonregulated
Properties: Clear, colorless; sp.gr. 1.52; visc. 0.73 cps; vapor pressure 250 mm Hg; f.p. < -80 C; b.p. 52 C; flash pt. none; surf. tens. 14.1 dynes/cm; 2.0 ppm max. NV; < 200 ppm moisture
Toxicology: Very low toxicity; sl. skin and eye irritant, low acute inh. toxicityTSCA listed
Precaution: Contact with highly basic process materials of pH 10 or above not rec.; prod. has vapor flamm. limits in air
Storage: Store in clean, dry area @ < 52 C; protect from freezing

Vertrel® XF [DuPont Fluoroprods.]
Chem. Descrip.: 2,3-Dihydroperfluoropentane
Chem. Analysis: Fluoropentanes (99.9%)
Uses: Solvent for industrial aerosol formulations, vapor degreasing equip. for cleaning, rinsing, and drying; carrier fluid for aerosol lubricants; mold release agent
Features: Nonflamm.; zero ozone depletion potential; can replace CFC-113, 1,1,1-trichloroethane, HCFCs, and PFCs
Regulatory: EPA approved under SNAP
Properties: Colorless, clear; sol. 140 ppm in water; completely misc. with most esters, ketones, ethers, ether-alcohol sol'ns., and lower alcohols such as methanol, ethanol, and isopropanol; m.w. 252; sp.gr. 1.58; visc.

0.67 cps; vapor pressure 226 mm Hg; b.p. 55 C; f.p. -80 C; surf. tens. 14.1 dynes/cm; flash pt. none; 2.0 ppm max. NV; < 50 ppm moisture
Toxicology: AEL 200 ppm (8- and 12-h TWA); 400 ppm (ceiling limit); low toxicity; sl. skin and eye irritant, low inh. toxicityTSCA listed
Precaution: Incompat. with strong bases, acrylics, and fluoroelastomers (e.g. Viton® and Kalrez®)
Storage: Store in clean, dry area @ < 52 C

Vertrel® XM [DuPont Fluoroprods.]
Chem. Descrip.: Mixt. of Vertrel® XF (2,3-dihydrodecafluoropentane, 94.0 ± 0.5%) and methanol (6 ± 0.3%)
Uses: Solvent for use in vapor degreasing equip., precision and specialty cleaning and rinsing for removal of particulate, fingerprints, and lt. soils from metal, plastic, and glass
Features: Zero ozone potential; replaces CFC-113, 1,1,1-trichloroethane, HCFCs, and PFCs
Regulatory: EPA approved under SNAP; DOT nonregulated
Properties: Colorless clear; sp.gr. 1.49; visc. 0.63 cps; vapor pressure 298 mm Hg; f.p. < 80 C; b.p. 48 C; flash pt. none; surf. tens. 14.1 dynes/cm; 2.0 ppm max. NV; < 200 ppm moisture
Toxicology: Very low toxicity; sl. skin and eye irritant, low acute inh. toxicityTSCA listed
Precaution: Contact with highly basic process materials of pH 10 or above not rec.; prod. has vapor flamm. limits in air
Storage: Store in clean, dry area @ < 52 C

Vertrel® XMS Plus [DuPont Fluoroprods.]
Chem. Descrip.: Mixt. of Vertrel® XF (2,3-dihydrodecafluoropentane, 50.9%) with trans-1,2-dichloroethylene (43%), cyclopentane (2%), methanol (4%), stabilizer (0.1%)
Uses: Solvent for use in vapor degreasing equip. for cleaning ionic soils and flux residues from electronic assemblies
Features: Zero ozone-depleting potential; low global warming potential; replaces CFC-113, methyl chloroform, HCFCs and PFCs; SNAP compliance; thermally stable; methanol component considered HAP
Regulatory: SARA §313 reportable (methanol); ELINCS, Canada DSL, Japan MITI, Australia NICNAS, Korea TCCL listed
Properties: Colorless clear liq.; m.w. 125; dens. 1.34 g/cc; visc. 0.46 cps; vapor pressure 470 mm Hg; f.p. < -50 C; b.p. 38 C; flash pt. (PMCC) none, (TOC) 20 C; surf. tens. 14.9 dynes/cm; sp.heat 0.29 cal/g°C; VOC 658 g/l
Toxicology: AEL 197 ppm; dihydrodecafluoropentane: overexposure by inh. may cause suffocation, CNS stimulation, irregular heart beat, death; skin contact may cause sl. irritation; repeated/prolonged exposure may cause skin defatting, itching, rash; may cause eye irritation, blurred vision; if aspirated, may cause chem. pneumonia, coughing, choking, shortness of breath, fever, pulmonary edema, coma, seizures; TSCA listed
Environmental: Prevent entry into sewers, waterways, low areas
Precaution: Nonflamm., but potential to ignite; vapor flamm. in air 6-15 vol.%; vapor is heavier than air; incompat. with alkali or alkaline earth metals (powd. Al, Zn, Be, Na, Mg), strong bases (NaOH, KOH)
Hazardous decomp. prods.: Hydrofluoric acids, possibly carbonyl halides
Storage: Store in clean, dry area; protect from freezing temps.; if stored below -10 C, mix prior to use; do not allow stored prod. to exceed 52 C; keep container tightly closed

Vertrel® XSi [DuPont Fluoroprods.]
Chem. Descrip.: Mixt. of Vertrel® XF (2,3-dihydrodecafluoropentane, (57.0 ± 1.0%) and hexamethyldisiloxane (43.0 ± 1.0%)
Uses: Solvent for medical applics. for cleaning or depositing of silicone oil-based lubricants; swelling media for silicone rubber tubing
Features: Nonflamm.; zero ODP; exempted as VOC; can replace hexane, methylchloroform, CFC-113, PFC fluids, and HCFC-141b
Regulatory: EPA approved under SNAP
Properties: Colorless, clear; sp.gr. 1.17; visc. 0.60 cps; vapor pressure 135 mm Hg; b.p. 57 C; f.p. < -50 C; surf. tens. 14.0 dynes/cm; flash pt. none; 10.0 ppm max. NV; < 200 ppm moisture
Toxicology: Very low toxicityTSCA listed
Storage: Store in clean, dry area @ > -10 C to < 52 C

Vinycizer 30 [Kao Corp.]
Chem. Descrip.: Isobutyl oleate
CAS 84988-79-4; EINECS/ELINCS 284-868-8
Uses: Raw material for spin finishes, surfactants; oiling agent for textiles; processing aid for rubber; lubricant, plasticizer for plastics; additive in paints, inks; base for cosmetics; solvent for pharmaceuticals
Properties: Lt. yel. liq.; sol. (@ 10%) in ethanol, hexane, xylene; visc. 9 cp (20 C); solid. pt. -10 max.; acid no 1. max.; iodine no. 78-88; hyd. no. 5 max.

VM&P Naphtha EC [Shell]
Chem. Descrip.: Solvent naphtha (petroleum), light aliphatic (93%), xylene (7%), contains ethylbenzene (1%)
UN 1993
Uses: Solvent for dry cleaning, coatings, automotive, and chemical specialties
Features: Med. evaporating; sl. higher solvency than similar prods. with same aromatic content
Regulatory: SARA §311/312 acute health/chronic health/fire hazard, §313 reportable (ethylbenzene, xylene)
Properties: Lt. colored liq.; hydrocarbon odor; negligible sol. in water; m.w. 117; sp.gr. 0.75 (60 F); dens. 6.24 lb/gal (60 F); vapor pressure 9.8 mm Hg (68 F); b.p. 257-291 F; flash pt. 15 C; VOC 6.3 lb/gal (60 F)
Toxicology: If aspirated, can cause severe lung damage, chem. pneumonitis; may be harmful or fatal by ing.; skin irritant; may cause CNS depression, dizziness, lightheadedness, headache, nausea, loss of coordination, unconsciousness, death; overexposure to high concs. may cause hearing loss; chronic abuse may cause irregular heart rhythm, cardiac arrest; may cause temporary eye irritation; target organs: heart, auditory system; TSCA listed
Environmental: Prevent entry into waterways, sewers
Precaution: Flamm.; flamm. in air 1-7 vol.%; vapors are heavier than air, may flash back; prevent vapor accumulation; floats on water; static electricity may accumulate and create a fire hazard; bond and ground equip. and transfer containers; avoid contact with strong oxidizing agents; prevent entry into basements, confined areas
Hazardous decomp. prods.: Complex mixt. of airborne solids, liqs., and gases incl. CO, CO_2, other org. compds.
HMIS: Health 1, Flammability 3, Reactivity 0
Storage: Keep containers closed when not in use; keep away from heat, sparks, flame

VM&P Naphtha HT [Shell]
Chem. Descrip.: Solvent naphtha (petroleum), light aliphatic, contains ethylbenzene (0.12-0.25%)
CAS 64742-89-8; EINECS/ELINCS 265-192-2
UN 1268
Uses: Solvent for agric., oil/fat extraction, polishes, industrial cleaners, adhesives, inks, paints, lighter fluids, pharmaceuticals, paper/textile coatings, petrol. gas absorp., asphalt, resins, plastics, rubber
Features: Fast/med.-evaporating, high solvency; hydrotreating removes unwanted compds. (e.g., sulfur, nitrogen, oxygen), converts unsat. hydrocarbons (olefins, acetylenes, aromatics), results in relatively low odor; no more than trace amts. of HAPs
Regulatory: SARA §311/312 acute health/fire hazard, §313 nonreportable
Properties: Lt. colored liq.; hydrocarbon odor; negligible sol. in water; sp.gr. 0.743 (60 F); vapor pressure 10 mm Hg (68 F); b.p. 253-282 F; flash pt. (TCC) 13.89 C; KB value 35; VOC 6.2 lb/gal (60 F); 100% act.
Toxicology: LD50 (oral, rat) 5840 mg/kg, (skin, rat) 2920 mg/kg; LC50 (inh., rat) 3400 ppm; if aspirated, can cause severe lung damage, chem. pneumonitis; may be fatal if swallowed; may cause CNS depression, dizziness, lightheadedness, headache, nausea, loss of coordination, unconsciousness, death; may cause temporary eye irritation, sl. skin irritation; TSCA listed
Environmental: Prevent entry into waterways, sewers
Precaution: Flamm.; flamm. in air 1-7 vol.%; vapors are heavier than air, may flash back; prevent vapor accumulation; floats on water; static electricity may accumulate and create a fire hazard; bond and ground equip. and transfer containers; avoid contact with strong oxidizing agents; prevent entry into basements, confined areas
Hazardous decomp. prods.: Complex mixt. of airborne solids, liqs., and gases incl. CO, CO_2, other org. compds.
HMIS: Health 1, Flammability 3, Reactivity 0
Storage: Keep containers closed when not in use; keep away from heat, sparks, flame

Volpo G26 [Croda Oleochems.]
Chem. Descrip.: Glycereth-26
CAS 31694-55-0
Uses: Humectant, visc. control agent, solvent for polar essential oils and perfumery components for hair care, skin care

Properties: Essentially colorless clear liq.

Vorite® 105 [CasChem]

Chem. Descrip.: Polymerized castor oil

Uses: Pigment wetting agent/dispersant; plasticizer for resins, gums, polymers; lubricant; penetrant; coupling agent; solvent; adhesion promoter; for cellulose lacquers, inks, adhesives, industrial lubricants, polishes, caulks, leather dressing; hydraulic fluids; rubber compding., gasket cement

Properties: Gardner 4 color; sp.gr. 0.975; visc. 26 stokes; pour pt. -5 F; acid no. 2; iodine no. 85; sapon. no. 170; hyd. no. 130; pour pt. - 21 C

Vorite® 110 [CasChem]

Chem. Descrip.: Polymerized castor oil

Uses: Pigment wetting agent/dispersant; plasticizer for resins, gums, polymers; lubricant; penetrant; coupling agent; solvent; adhesion promoter; for cellulose lacquers, inks, adhesives, industrial lubricants, polishes, caulks, leather dressing; hydraulic fluids; rubber compding.; gasket cement

Properties: Gardner 3 oil; sp.gr. 0.990; visc. 115 stokes; pour pt. 15 F; acid no. 2; iodine no. 82; sapon. no. 166; hyd. no. 102

Vorite® 115 [CasChem]

Chem. Descrip.: Polymerized castor oil

Uses: Pigment wetting agent/dispersant; plasticizer for resins, gums, polymers; lubricant; penetrant; coupling agent; solvent; adhesion promoter; for cellulose lacquers, inks, adhesives, industrial lubricants, polishes, caulks, leather dressing; hydraulic fluids; rubber compding.; gasket cement

Properties: Gardner 3 oil; sp.gr. 0.995; visc. 192 stokes; pour pt. 20 F; acid no. 2; iodine no. 85; sapon. no. 165; hyd. no. 93

Vorite® 125 [CasChem]

Chem. Descrip.: Polymerized castor oil

Uses: Pigment wetting agent/dispersant; plasticizer for resins, gums, polymers; lubricant; penetrant; coupling agent; solvent; adhesion promoter; for cellulose lacquers, inks, adhesives, industrial lubricants, polishes, caulks, leather dressing; hydraulic fluids; rubber compding.; gasket cement

Properties: Gardner 4 oil; sp.gr. 1.007; visc. 900 stokes; pour pt. 13 C; acid no. 2; iodine no. 84; sapon. no. 157; hyd. no. 72

W

Wickenol® 127 [Alzo]
Chem. Descrip.: Isopropyl stearate
CAS 112-10-7; EINECS/ELINCS 203-934-9
Uses: Emollient; cosolvent; lubricant
Properties: Gardner -1 color; sol. in acetone, min. and veg. oils, ethyl acetate, alcohol; insol. in water; pour pt. 18 C; acid no. 0.1; ref. index 1.438

Wickenol® 136 [Alzo]
Chem. Descrip.: Isopropyl palmitate, isopropyl myristate, isopropyl stearate
Uses: Emollient, conditioner, moisturizer, solubilizer, vehicle, solvent for cosmetics, toiletries
Properties: Gardner -1 liq.; sol. in alcohols, ketones, aromatic and aliphatic hydrocarbons; insol. in water; pour pt. 7 C; acid no. 0.1; ref. index 1.435

Wickenol® 151 [Alzo]
Chem. Descrip.: Isononyl isononanoate
CAS 42131-25-9; EINECS/ELINCS 261-665-2
Uses: Emollient, moisturizer, pigment wetting agent, solvent for hair care
Features: Silky emolliency
Properties: Gardner -1 color; sol. in alcohol, animal, veg., and min. oil; insol. in water; pour pt. < -15 C; acid no. 0.10; ref. index 1.435

Wickenol® 152 [Alzo]
Chem. Descrip.: Isodecyl isononanoate
CAS 41395-89-5; EINECS/ELINCS 261-674-1
Uses: Emollient, solvent, pigment wetting agent for hair care, skin care
Features: Silky emollience
Properties: Gardner -1 color; sol. in alcohol, animal, veg., and min. oils; pour pt. < -15 C; acid no. 0.10; ref. index 1.438

Wickenol® 153 [Alzo]
Chem. Descrip.: Isotridecyl isononanoate
CAS 59231-37-7; EINECS/ELINCS 261-675-7
Uses: Emollient, solvent, pigment wetting agent, moisturizer for hair care, skin care
Features: Silky emollience
Properties: Gardner -1 color; sol. in alcohol, animal, veg., and min. oils; pour pt. < -18 C; acid no. 0.10; ref. index 1.444

Wickenol® 158 [Alzo]
Chem. Descrip.: Dioctyl adipate
CAS 103-23-1; EINECS/ELINCS 203-090-1
Uses: Emollient, moisturizer, pigment wetting agent/dispersant, cosolvent in cosmetics, topical pharmaceuticals
Features: Increases water vapor porosity of fatty components
Properties: Gardner -1 color; sol. in alcohol, animal, veg., and min. oils; pour pt. -20 C; acid no. 0.10; ref. index 1.445

Wickenol® 171 [Alzo]
Chem. Descrip.: Octyl hydroxystearate
CAS 29383-26-4; EINECS/ELINCS 249-793-7
Uses: Emollient, moisturizer, pigment wetting agent/dispersant, refatting agent, counterirritant, cosolvent, solubilizer in cosmetics, topical pharmaceuticals
Features: Increases water vapor porosity of fatty components
Properties: Gardner 2+ color; sol. in acetone, castor, corn, and min. oil, chloroform, ethyl acetate, ethanol; pour pt. 5 C; acid no. 0.30; ref. index 1.456

Witcohol™ 42 [Goldschmidt]
Chem. Descrip.: Cetyl alcohol, stearyl alcohol, oleyl alcohol
Uses: Emulsifier; lubricant; foam control agent; cosolvent; plasticizer; stabilizer; emollient; raw material for detergents, emulsifiers, lube oil additives, textile auxiliaries, germicides, chem. intermediates, org. synthesis, resins, and lubricants
Properties: Wh. soft waxy solid; mild char. fatty odor; sol. in IPA, benzene, trichloroethylene, ethyl ether, acetone, turpentine, VM&P naphtha, kerosene, lt. min. oil; m.w. 250-267; sp.gr. 0.814 (60/25 C); visc. 40 SSU (210 F); b.p. 315-345 C; acid no. 1.5 max.; iodine no. 45-55; sapon. no. 3.5 max.; hyd. no. 210-225; cloud pt. 36 C max.; nonionic

Witconol™ APM [Goldschmidt]
Chem. Descrip.: PPG-3 myristyl ether
CAS 63793-60-2
Uses: Lubricant; emulsifier; wetting agent; penetrant; dye dispersant; scouring aid; antistat; solvent coupling agent; metal processing; textile surfactant; emollient for cosmetics and toiletries, esp. stick prods.; solubilizer for alcohol in oils
Properties: Gardner 1 liq.; sol. in oil; sp.gr. 0.90; pH 7.0; nonionic; 100% act.

Witconol™ APS [Goldschmidt]
Chem. Descrip.: PPG-11 stearyl ether
CAS 25231-21-4
Uses: Lubricant; emulsifier; wetting agent; penetrant; dye dispersant; scouring aid; antistat; solvent coupling agent; for personal care prods.; metal processing; textile surfactant; emollient for cosmetics and toiletries; solubilizer
Properties: Gardner 1 liq.; sol. in oil; sp.gr. 0.94; pH 7.0; nonionic

Witsol® 45 [Crompton/Witco Refined Prods.]
Chem. Descrip.: Aliphatic petroleum solvent manufactured by severe hydrotreating
Uses: Solvent for printing inks
Features: Highly refined
Properties: Saybolt +30 color; sp.gr. 0.801 (15.6/15.6 C); kinematic visc. 2.70 cSt (37.8 C); i.b.p. 232 C; pour pt. -1 C; flash pt. (COC) 104 C

Witsol® 50 [Crompton/Witco Refined Prods.]
Chem. Descrip.: Petroleum solvs.
Uses: Solvent for printing inks

Witsol® 471 [Crompton/Witco Refined Prods.]
Chem. Descrip.: Petroleum hydrocarbon solvent
Uses: Solvent for printing inks
Features: Precise balance of aromatic and aliphatic fractions; better solvency; used alone or with a severely hydrotreated grade
Properties: Saybolt +16 color; sp.gr. 0.806 (15.6/15.6 C); kinematic visc. 2.40 cSt (37.8 C); i.b.p. 235 C; pour pt. -23 C; flash pt. (COC) 107 C

Witsol® 501 [Crompton/Witco Refined Prods.]
Chem. Descrip.: Petroleum hydrocarbon solvent
Uses: Solvent for printing inks
Features: Precise balance of aromatic and aliphatic fractions; better solvency; used alone or with a severely hydrotreated grade
Properties: Saybolt +12 color; sp.gr. 0.817 (15.6/15.6 C); kinematic visc. 1.54 cSt (37.8 C); i.b.p. 268 C; pour pt. -9 C

Witsol® MSO [Crompton/Witco Refined Prods.]
Chem. Descrip.: Petroleum hydrocarbon solvent
Uses: Solvent for printing inks

Features: Precise balance of aromatic and aliphatic fractions; better solvency; used alone or with a severely hydrotreated grade

Properties: Straw color; sp.gr. 0.823 (15.6/15.6 C); kinematic visc. 4.55 cSt (37.8 C); i.b.p. 271 C; pour pt. -4 C; flash pt. (COC) 135 C

X

X-3B Solvent [Shell]
Uses: Solvent for paints
Features: Wide boiling range; med. evaporating
Properties: Immisc. with water; dens. 0.798 kg/l (15 C); i.b.p. 99 C; flash pt. < 0 C; conduct. 450 pS/m (23 C)
Toxicology: Prolonged/repeated skin contact can cause dermatitis, defatting
Precaution: Highly flamm.; can form explosive mixts. with air; electrostatic charge may be generated
Storage: Keep away from heat, sparks, flames

X-55 Solvent [Shell]
Uses: Solvent for rubber
Features: Wide boiling range; rapidly evaporating
Properties: Immisc. with water; dens. 0.685 kg/l (15 C); i.b.p. 45 C; flash pt. < 0 C; conduct. 450 pS/m (23 C)
Toxicology: Prolonged/repeated skin contact can cause dermatitis, defatting
Precaution: Highly flamm.; can form explosive mixts. with air; electrostatic charge may be generated
Storage: Keep away from heat, sparks, flames

X-95 Solvent [Shell; Shell Netherlands]
Chem. Descrip.: Naphtha (petroleum), hydrotreated light, contains n-hexane (1.5%), ethylbenzene (0.60%), benzene (0.20%)
CAS 64742-49-0; EINECS/ELINCS 265-151-9
UN 1268
Uses: Solvent for mfg. of resins
Features: Wide boiling range; med. evaporating
Regulatory: EINECS, Canada DSL, Australia AICS, China IECSC, Japan MITI, Philippines PICCS, Korea TCCL listed
Properties: Colorless liq.; paraffinic odor; insol. with water; dens. 0.754 kg/l (15 C); i.b.p. 94 C; flash pt. (Abel) < 0 C; conduct. 450 pS/m (23 C)
Toxicology: LD50 (oral) > 2000 mg/kg, (skin) > 2000 mg/kg; LC50 (inh.) > 5 mg/l; low toxicity by ing., inh., skin contact; if aspirated into lungs, may cause chem. pneumonitis which can be fatal; narcotic at high vapor concs.; may cause headache, dizziness, nausea, respiratory tract irritation; prolonged/repeated skin contact can cause dermatitis, defatting; may cause cancer; risk of impaired fertility; TSCA listed
Environmental: Readily biodeg.; potential to bioaccumulate; LC50 (fish) ≤ 10 mg/l; toxic to aquatic organisms; may cause long-term adverse effects in aquatic environment; prevent contamination of soil, water
Precaution: Highly flamm.; explosion/flamm. limits in air 0.6-8 vol.%; extreme risk of vapor ignition @ normal handling temps.; can form explosive mixts. with air; vapor is heavier than air, spreads along ground; distant ignition possible; electrostatic charge may be generated; prevent static discharge; ground equip.; floats on water; risk of explosion
Hazardous decomp. prods.: Incomplete combustion: CO
Storage: Store in tightly closed containers @ ambient temps. in well-ventilated area away from direct sunlight, heat, ignition sources

Z

ZELEC® C-48 [Stepan]
Chem. Descrip.: Methyl laurate (57%), methyl myristate (22%), methyl oleate-stearate (12%), methyl palmitate (9%)
Uses: Lubricant for metalworking, textiles, and agric. applics., cutting fluids and drawing compds., drawing lubricants in aluminum beverage cans; solvent, cosolvent, crop oil, or carrier oil in agric. applics.
Features: Min. oil replacement
Regulatory: EPA 40CFR §180.1001 approved
Properties: Gardner 1 max. clear liq.; sol. in oil; emulsifiable in water; sp.gr. 0.86; dens. 7.2 lb/gal; visc. 11 cps; m.p. -6 C; b.p. > 150 C; flash pt. (Seta) > 94 C; 0.2% max. moisture
Toxicology: Pract. nontoxic orally; may cause sl. to mild skin and eye irritation @ 10%
Environmental: Readily biodeg.
Storage: Store in sealed containers @ 90-110 F; storage under nitrogen will prevent degradation

ZELEC® C-49 [Stepan]
Chem. Descrip.: Methyl laurate (49%), methyl myristate (18%), methyl caprylate (10%), methyl palmitate (8%), methyl oleate-stearate (8%), methyl caprate (7%)
Uses: Lubricant for metalworking, textiles, and agric. applics., cutting fluids and drawing compds., drawing lubricants in aluminum beverage cans; solvent, cosolvent, crop oil, or carrier oil in agric. applics.
Features: Min. oil replacement
Regulatory: EPA 40CFR §180.1001 approved
Properties: Gardner 1 max. clear liq.; sol. in oil; emulsifiable in water; sp.gr. 0.86; dens. 7.2 lb/gal; visc. 7 cps; m.p. -8 C; b.p. > 260 C; flash pt. (Seta) > 94 C; 0.2% max. moisture
Toxicology: Practically nontoxic orally; may cause sl. to mild skin and eye irritation @ 0.2%
Environmental: Readily biodeg.
Storage: Store in sealed containers @ 90-110 F; storage under nitrogen will prevent degradation

ZELEC® IPM [Stepan]
Chem. Descrip.: Isopropyl myristate
CAS 110-27-0; EINECS/ELINCS 203-751-4
Uses: Lubricant in textile spin finish, coning oils, carding, dye bath; emollient, solubilizer, vehicle, solvent for makeup, shaving preps., bath oils, hair preps., highly pigmented prods.
Features: Replacement for min. oil
Properties: APHA 20 max. clear liq., pract. odorless; insol. in water; sol. in IPA, min. oil, peanut oil; f.p. -3 C; m.p. -3 C; b.p. 162.5 C; acid no. 1 max.; iodine no. 1 max.; sapon. no. 204-212; flash pt. (COC) 305 F; ref. index 1.433; 100% act.
Environmental: Biodeg.

Part II
Discontinued/Unverified Trade Name Reference

Discontinued/Unverified Trade Name Reference

Acintene® AC-P† [Arizona]
Chem. Descrip.: 32% α-Pinene, 19% dipentene, 13% 1,4-cineole, 9% terpinolene, 8% camphene, 7% 1,8-cineole
Uses: Terpene solvent is nontoxic replacement for petroleum-based hydrocarbon solvents; solvent and antistripping agent for paints, solvent in mfg. of varnishes, component of household and industrial cleaners and polishes; printing ink solvent as flow promoter, solvent in reclaiming rubber, solvent and paraffin wax inhibitor in oil field production chems., replacement for chlorinated solvents, and component of degreasing compds.
Properties: APHA 10 color; mild pine-like odor; misc. in alcohols; insol. in water; sp.gr. 0.860 (15.5 C); dens. 7.16 lb/gal (15 C); i.b.p. 157 C; flash pt. (CC) 35 F; ref. index 1.4680; KB value 94
Toxicology: Nontoxic

Acintene® B Beta Pinene† [Arizona]
Chem. Descrip.: 76.3% β-Pinene, 11% dipentene, 7.5% α-pinene
Uses: Intermediate in mfg. of syn. resins
Properties: APHA 5 color; sp.gr. 0.8708; dens. 7.19 lb/gal; i.b.p. 165-167 C; flash pt. (CC) 100 F; ref. index 1.4774

Acintene® DP 738† [Arizona]
Chem. Descrip.: 24% Dipentene, 17% α-pinene, 16% 1,4-cineole, 12% terpinolene, 11% 1,8-cineole, 7% camphene, 7% α-terpinene
Uses: Solvent and antistripping agent for paints, solvent in mfg. of varnishes, component of household and industrial cleaners and polishes, extender for pine oil disinfectant cleaners; printing ink solvent as flow promoter, solvent in reclaiming rubber, solvent and paraffin wax inhibitor in oil field production chems., replacement for chlorinated solvents, replacement for limonene as solvent, and component of degreasing compds.
Properties: Colorless clear liq.; strong pine-like odor; misc. in alcohols; insol. in water; sp.gr. 0.870 (15.5 C); dens. 7.24 lb/gal (15 C); i.b.p. 170 C; flash pt. (CC) 49 F; ref. index 1.4680; KB value 103

Acintene® LS 160 A† [Arizona]
Uses: Solvent for paints, varnishes; replacement for chlorinated hydrocarbons; substitute for d-limonene; component of household and industrial cleaners and polishes, paint stripping formulations, degreasing compds.; wood penetrant and carrier oil; solvent, flow promoter in printing inks
Properties: Gardner 1 color; citrus-like odor; misc. in alcohols; insol. in water; sp.gr. 0.860 (15.5 C); dens. 7.19 lb/gal (15 C); flash pt. (CC) 35 F; KB value 68

Acintene® LS 160 B† [Arizona]
Uses: Solvent for paints, varnishes; replacement for chlorinated hydrocarbons; substitute for d-limonene; component of household and industrial cleaners and polishes, paint stripping formulations; wood penetrant and carrier oil; solvent, flow promoter for printing inks
Properties: Gardner 1 color; citrus-like odor; misc. in alcohols; insol. in water; sp.gr. 0.860 (15.5 C); dens. 7.16 lb/gal (15 C); flash pt. (CC) 37 F; KB value 83

Acintene® LS 160 D† [Arizona]
Uses: Solvent for paints, varnishes; replacement for chlorinated hydrocarbons; substitute for d-limonene; component of household and industrial cleaners and polishes, paint stripping formulations, degreasing compds.; wood penetrant and carrier oil; solvent, flow promoter for printing inks; extender for pine oil in disinfectant cleaners
Properties: Gardner 1 color; citrus-like top note odor, turpentine-like odor base; misc. in alcohols; insol. in water; sp.gr. 0.880 (15.5 C); dens. 7.33 lb/gal (15 C); flash pt. (CC) 49 F; KB value 100

Acintene® P-AC Alpha-Pinene† [Arizona]
Chem. Descrip.: 81% α-Pinene and 14% camphene
Uses: Solvent intermediate in mfg. of pine oils and camphene; monomer for production of polyterpene resins and terpene-phenolic resins; intermediate in mfg. of wide variety of terpene-based aromatic compds.
Properties: APHA 20 color; mild turpentine-like odor; misc. in alcohols; insol. in water; sp.gr. 0.865 (15.5 C); dens. 7.19 lb/gal (15 C); i.b.p. 157 C; flash pt. (CC) 35 F; ref. index 1.4670; KB value 62; < 0.1% moisture

Adol® 42† [Witco♦]
Chem. Descrip.: Tallow alcohol
EINECS/ELINCS 308-980-4
Uses: Emulsifier, lubricant, foam control agent, cosolvent, plasticizer, stabilizer, emollient, intermediate for metal rolling oils and lubricants, gas scrubbing, printing inks, auto antifreeze, textile dyeing and finishing, emulsion systems, min. processing, oil field chemicals, fabric softeners
Properties: Lovibond 5Y/0.5R color; sol. in fatty alcohols, IPA, benzene, trichlorethylene, acetone, turpentine, VMP naphtha, kerosene, lt. min. oil; m.w. 258; sp.gr. 0.814 (60/25 C); cloud pt. 36 C max.; sapon. no. 3.5 max.; nonionic; 100% conc.

Adol® 52 NF† [Witco♦]
Chem. Descrip.: Cetyl alcohol
CAS 36653-82-4; EINECS/ELINCS 253-149-0
Uses: Coemulsifier, lubricant, foam control agent, cosolvent, plasticizer, stabilizer, emollient, intermediate for metal lubricants, inks, textiles, emulsions, paper, cosmetics, mineral processing, oil field chemicals, fabric softeners
Properties: Lovibond 5Y/0.5R max.; sol. in fatty alcohols, IPA, benzene, trichloroethylene, acetone, turpentine, VMP naphtha, kerosene, lt. min. oil; m.w. 247; sp.gr. 0.815 (60/25 C); m.p. 45-50 C; acid no. 1.0 max.; sapon. no. 3.0 max.; nonionic

Adol® 60† [Witco♦]
Chem. Descrip.: Behenyl alcohol
CAS 661-19-8; EINECS/ELINCS 211-546-6
Uses: Lubricant; foam control agent; cosolvent; plasticizer; emollient
Properties: Lovibond 10Y/1.0R; m.w. 316; sp. gr. 0.79 (100/25 C); acid no. 1.5 max.; iodine no. 6 max.; sapon. no. 3 max.

Adol® 66† [Witco♦]
Chem. Descrip.: Isostearyl alcohol
CAS 27458-93-1; EINECS/ELINCS 248-470-8
Uses: Coemulsifier, lubricant, foam control agent, cosolvent, plasticizer, stabilizer, emollient, intermediate for metal lubricants, inks, textiles, emulsions, paper, mineral processing, oil field chemicals, fabric softeners; emollient for skin care prods.; replacement for oleyl alcohol
Properties: Lovibond 5Y/0.5R max. liq.; sol. in fatty alcohols, IPA, benzene, trichlorethylene, acetone, turpentine, VMP naphtha, kerosene, lt. min. oil; m.w. 295; sp.gr. 0.861; cloud pt. 8 C max.; acid no. 1.0 max.; sapon. no. 2.0 max.; nonionic; 100% conc.

Adol® 80† [Witco♦]
Chem. Descrip.: Oleyl alcohol
CAS 143-28-2; EINECS/ELINCS 205-597-3

Uses: Coemulsifier, lubricant, foam control agent, cosolvent, plasticizer, stabilizer, emollient, intermediate for metal rolling oils and lubricants, gas scrubbing, printing inks, auto antifreeze, textile dyeing and finishing, emulsion systems, paper pulping, cosmetics, cationic surfactants, min. processing, oil field chemicals, fabric softeners
Properties: Lovibond 5Y/0.5R liq.; m.w. 263; sol. in fatty alcohols, IPA, benzene, trichlorethylene, acetone, turpentine, VMP naphtha, kerosene, lt. min. oil; sp.gr. 0.840; cloud pt. 13 C max.; sapon. no. 3.0 max.; nonionic; 100% conc.

Adol® 85 NF† [Witco♦]
Chem. Descrip.: Oleyl alcohol NF
CAS 143-28-2; EINECS/ELINCS 205-597-3
Uses: Emulsifier, lubricant, foam control agent, cosolvent, plasticizer, emollient for personal care prods.
Properties: Lovibond 3Y/0.3R liq.; m.w. 268; sp.gr. 0.84; acid no. 1.0; iodine no. 85-95; sapon no. 2 max.; cloud pt. 10 C max.; nonionic; 100% conc.

Adol® 520 NF† [Witco♦]
Chem. Descrip.: Cetyl alcohol
CAS 36653-82-4; EINECS/ELINCS 253-149-0
Uses: Coemulsifier, lubricant, foam control agent, cosolvent, plasticizer, stabilizer, emollient, intermediate for metal lubricants, inks, textiles, emulsions, paper, cosmetics, mineral processing, oil field chemicals, fabric softeners; emollient, emulsion stabilizer, visc. modifier, conditioner for skin care prods.; opacifier for creams and lotions
Properties: APHA 50 max. flake; sol. in fatty alcohols, IPA, benzene, trichlorethylene, acetone, turpentine, VMP naphtha, kerosene, lt. min. oil; m.w. 246; sp.gr. 0.815 (60/25 C); m.p. 45-50 C; solid pt. 48-53 C; acid no. 1.0 max.; sapon. no. 2.0 max.; nonionic; 100% conc.

Adol® 620 NF† [Witco♦]
Chem. Descrip.: Stearyl alcohol
CAS 112-92-5; EINECS/ELINCS 204-017-6
Uses: Coemulsifier, lubricant, foam control agent, cosolvent, plasticizer, stabilizer, emollient, intermediate for metal lubricants, inks, textiles, emulsions, paper, cosmetics, mineral processing, oil field chemicals, fabric softeners; emollient, emulsion stabilizer, visc. modifier, conditioner for skin care prods.; opacifier for creams and lotions
Properties: APHA 50 max. flake; sol. in fatty alcohols, IPA, benzene, trichlorethylene, acetone, turpentine, VMP naphtha, kerosene, lt. min. oil; m.w. 267; sp.gr. 0.817 (60/25 C); m.p. 55-60 C; acid no. 2.0 max.; sapon. no. 2.0 max.; nonionic; 100% conc.

Adol® 640† [Witco♦]
Chem. Descrip.: Cetearyl alcohol
CAS 8005-44-5; EINECS/ELINCS 267-008-6
Uses: Emulsifier, lubricant, foam control agent, cosolvent, plasticizer, emollient
Properties: APHA 40 max. flake; m.w. 253; sp.gr. 0.82 (50/50 C); m.p. 43-46 C; acid no. 1.0 max.; iodine no. 1.5 max.; sapon. no. 1.0 max.; nonionic; 100% conc.

Agent #5032† [Polymer Research Corp. of Am.♦]
Uses: Antitarnishing agent for pigment roller printing; solvent, lubricant, antistat, conditioner for textile fibers
Properties: Water-wh. clear liq.; dens. 9.16 lb/gal (55 C); 25% act. in water

Antarox® P-400† [Rhodia HPCII]
Chem. Descrip.: PPG
Uses: Component in hydraulic fluids; solvent and cosolvent for cosmetics, pharmaceuticals, industrial applics.; chem. intermediates
Properties: Liq.; pour pt. < -25 C; nonionic; 100% conc.

Antarox® P-700† [Rhodia HPCII]
Chem. Descrip.: PPG
Uses: Component in hydraulic fluids; solvent and cosolvent for cosmetics, pharmaceuticals, industrial applics.; chem. intermediates
Properties: Liq.; pour pt. < -25 C; nonionic; 100% conc.

Arcosolv® DB‡ [Lyondell]
Chem. Descrip.: Diethylene glycol butyl ether
CAS 112-34-5; EINECS/ELINCS 203-961-6
Uses: Solvent
Regulatory: SARA §313 reportable
Properties: Sol. 100% in water; m.w. 162.23; sp.gr. 0.955; dens. 7.94 lb/gal (20 C); visc. 4.7 cps; vapor pressure 0.02 mm Hg (20 C); surf. tens. 30.0 dynes/cm; b.p. 230 C; HLB 7.7; flash pt. (COC) 232 F; ref. index 1.4316
NFPA: Health 1, Flammability 1, Reactivity 0

Arcosolv® DBA† [Lyondell]
Chem. Descrip.: Diethylene glycol butyl ether acetate
CAS 124-17-4; EINECS/ELINCS 204-685-9
Uses: Solvent
Regulatory: SARA §313 reportable
Properties: Sol. 6% in water; m.w. 204.27; sp.gr. 0.980; dens. 8.16 lb/gal (20 C); visc. 3.2 cps; vapor pressure 0.04 mm Hg (20 C); surf. tens. 30 dynes/cm; b.p. 235 C; HLB 7.7; flash pt. (COC) 240 F; ref. index 1.4239
NFPA: Health 1, Flammability 1, Reactivity 0

Arcosolv® DE‡ [Lyondell]
Chem. Descrip.: Diethylene glycol ethyl ether
CAS 111-90-0; EINECS/ELINCS 203-919-7
Uses: Solvent
Regulatory: SARA §313 reportable
Properties: Sol. 100% in water; m.w. 134.17; sp.gr. 0.990; dens. 8.25 lb/gal (20 C); visc. 4.5 cps; vapor pressure 0.12 mm Hg (20 C); surf. tens. 32.2 dynes/cm; b.p. 198 C; HLB 8.6; flash pt. (TCC) 195 F; ref. index 1.4260
NFPA: Health 1, Flammability 1, Reactivity 0

Arcosolv® DEA‡ [Lyondell]
Chem. Descrip.: Diethylene glycol ethyl ether acetate
CAS 112-15-2; EINECS/ELINCS 203-940-1
Uses: Solvent
Regulatory: SARA §313 reportable
Properties: Sol. 100% in water; m.w. 176.21; sp.gr. 1.012; dens. 8.42 lb/gal (20 C); visc. 2.8 cps; vapor pressure 0.05 mm Hg (20 C); surf. tens. 31.7 dynes/cm; b.p. 214 C; HLB 8.7; flash pt. (COC) 225 F; ref. index 1.4220
NFPA: Health 1, Flammability 1, Reactivity 0

Arcosolv® DH† [Lyondell]
Chem. Descrip.: Diethylene glycol hexyl ether
CAS 112-59-4; EINECS/ELINCS 203-988-3
Uses: Solvent
Properties: Sol. 1.7% in water; m.w. 190.28; sp.gr. 0.935; dens. 7.78 lb/gal (20 C); visc. 8.6 cps; vapor pressure < 0.01 mm Hg (20 C); surf. tens. 29.6 dynes/cm; b.p. 259.1 C; HLB 6.8; flash pt. (CC) 271 F; ref. index 1.4381

Arcosolv® DM‡ [Lyondell]
Chem. Descrip.: Diethylene glycol methyl ether
CAS 111-77-3; EINECS/ELINCS 203-906-6
Uses: Solvent
Regulatory: SARA §313 reportable
Properties: Sol. 100% in water; m.w. 120.15; sp.gr. 1.023; dens. 8.51 lb/gal (20 C); visc. 3.9 cps; vapor pressure 0.2 mm Hg (20 C); surf. tens. 34.8 dynes/cm; b.p. 191 C; HLB 9.2; flash pt. (TCC) 191 F; ref. index 1.4268
NFPA: Health 2, Flammability 2, Reactivity 0

Arcosolv® DP† [Lyondell]
Chem. Descrip.: Diethylene glycol propyl ether
CAS 6881-94-3; EINECS/ELINCS 229-985-7
Uses: Solvent
Regulatory: SARA §313 reportable
Properties: Sol. 100% in water; m.w. 148.20; sp.gr. 0.963; dens. 8.04 lb/gal (20 C); visc. 4.1 cps; vapor pressure 0.03 mm Hg (20 C); surf. tens. 32.3 dynes/cm; b.p. 202 C; HLB 8.2; flash pt. (TCC) 200 F; ref. index 1.4290

Arcosolv® DPE† [Lyondell]
Chem. Descrip.: Dipropylene glycol propyl ether
CAS 29911-27-1
Uses: Solvent
Properties: Sol. in water; m.w. 162.2; sp.gr. 0.937 (20/20 C); dens. 7.8 lb/gal (20 C); vapor pressure 0.057 mm Hg (20 C); b.p. 188 C; HLB 7.7; flash pt. (TCC) 188 F; surf. tens. 28.2 dynes/cm; sp. heat 0.53 Cal/g°C

Arcosolv® EB‡ [Lyondell]
Chem. Descrip.: Ethylene glycol butyl ether
CAS 111-76-2; EINECS/ELINCS 203-905-0
Uses: Solvent
Regulatory: SARA §313 reportable
Properties: Sol. 100% in water; m.w. 118.17; sp.gr. 0.902; dens. 7.51 lb/gal (20 C); visc. 6.4 cps; vapor pressure 0.6 mm Hg (20 C); surf. tens.

† Discontinued ‡ Unverified ♦ Defunct Manufacturer or Division

26.6 dynes/cm; b.p. 169 C; HLB 7.4; flash pt. (TCC) 143 F; ref. index 1.4193
NFPA: Health 2, Flammability 2, Reactivity 0

Arcosolv® EBA‡ [Lyondell]
Chem. Descrip.: Ethylene glycol butyl ether acetate
CAS 112-07-2; EINECS/ELINCS 203-933-3
Uses: Solvent
Regulatory: SARA §313 reportable
Properties: Sol. 1% in water; m.w. 106.21; sp.gr. 0.941; dens. 7.84 lb/gal (20 C); visc. 1.8 cps; vapor pressure 0.29 mm Hg (20 C); surf. tens. 30.3 dynes/cm; b.p. 186 C; HLB 7.4; flash pt. (TCC) 160 F; ref. index 1.4142
NFPA: Health 1, Flammability 2, Reactivity 0

Arcosolv® EE‡ [Lyondell]
Chem. Descrip.: Ethylene glycol ethyl ether
CAS 110-80-5; EINECS/ELINCS 203-804-1
Uses: Solvent
Regulatory: SARA §313 reportable
Properties: Sol. 100% in water; m.w. 90.12; sp.gr. 0.931; dens. 7.75 lb/gal (20 C); visc. 2.1 cps; vapor pressure 3.8 mm Hg (20 C); surf. tens. 29.3 dynes/cm; b.p. 134 C; HLB 8.3; flash pt. (TCC) 110 F; ref. index 1.4076
NFPA: Health 2, Flammability 2, Reactivity 0

Arcosolv® EEA‡ [Lyondell]
Chem. Descrip.: Ethylene glycol ethyl ether acetate
CAS 111-15-9; EINECS/ELINCS 203-839-2
Uses: Solvent
Regulatory: SARA §313 reportable
Properties: Sol. 24% in water; m.w. 132.16; sp.gr. 0.973; dens. 8.11 lb/gal (20 C); visc. 1.3 cps; vapor pressure 1.7 mm Hg (20 C); surf. tens. 28.2 dynes/cm; b.p. 150 C; HLB 8.3; flash pt. (TCC) 130 F; ref. index 1.4030
NFPA: Health 1, Flammability 2

Arcosolv® EEH† [Lyondell]
Chem. Descrip.: Ethylene glycol ethyl hexyl ether
CAS 1559-35-9
Uses: Solvent
Properties: Sol. 0.2% in water; m.w. 174.29; sp.gr. 0.892; dens. 7.42 lb/gal (20 C); visc. 7.0 cps; vapor pressure 0.08 mm Hg (20 C); surf. tens. 27.6 dynes/cm; b.p. 224 C; HLB 5.4; flash pt. (Seta) 208 F; ref. index 1.4361

Arcosolv® EH† [Lyondell]
Chem. Descrip.: Ethylene glycol hexyl ether
CAS 112-25-4; EINECS/ELINCS 203-951-1
Uses: Solvent
Regulatory: SARA §313 reportable
Properties: Sol. 1% in water; m.w. 146.23; sp.gr. 0.889; dens. 7.40 lb/gal (20 C); visc. 5.2 cps; vapor pressure < 1.0 mm Hg (20 C); surf. tens. 30.3 dynes/cm; b.p. 208 C; HLB 6.4; flash pt. (TCC) 179 F; ref. index 1.4290

Arcosolv® EM‡ [Lyondell]
Chem. Descrip.: Ethylene glycol methyl ether
CAS 109-86-4; EINECS/ELINCS 203-713-7
Uses: Solvent
Regulatory: SARA §313 reportable
Properties: Sol. 100% in water; m.w. 76.10; sp.gr. 0.966; dens. 8.04 lb/gal (20 C); visc. 1.5 cps; vapor pressure 6.2 mm Hg (20 C); surf. tens. 30.8 dynes/cm; b.p. 124 C; HLB 8.8; flash pt. (TCC) 105 F; ref. index 1.4021
NFPA: Health 2, Flammability 2, Reactivity 0

Arcosolv® EMA† [Lyondell]
Chem. Descrip.: Ethylene glycol methyl ether acetate
CAS 110-49-6; EINECS/ELINCS 203-772-9
Uses: Solvent
Regulatory: SARA §313 reportable
Properties: Sol. 100% in water; m.w. 118.14; sp.gr. 1.006; dens. 8.37 lb/gal (20 C); visc. 1.1 cps; vapor pressure 2 mm Hg (20 C); surf. tens. 34 dynes/cm; b.p. 145 C; HLB 8.8; flash pt. (TCC) 120 F; ref. index 1.4025
NFPA: Health 1, Flammability 2

Arcosolv® EP† [Lyondell]
Chem. Descrip.: Ethylene glycol propyl ether
CAS 2807-30-9; EINECS/ELINCS 220-548-6
Uses: Solvent
Regulatory: SARA §313 reportable
Properties: Sol. 100% in water; m.w. 104.15; sp.gr. 0.913; dens. 7.59 lb/gal (20 C); visc. 2.4 cps; vapor pressure 1.3 mm Hg (20 C); surf. tens. 27.9 dynes/cm; b.p. 149 C; HLB 7.8; flash pt. (TCC) 120 F; ref. index 1.4136

Arlox FE‡ [Aquaness]
Uses: Aq. reactive scale solvent that rapidly dissolves iron sulfide and calcium carbonate scale from wellbore, downhole equip., and formation; evolves carbon dioxide gas, providing agitation, and resulting in faster scale removal; sol. calcium chelate and sol. sulfate salt are reaction prods.; acidizing not necessary following use; may be used as preflush in sour wells ahead of corrosion inhibitor batch treatments
Properties: Liq.; sol. in water, insol. in isopropanol, heavy aromatic solv.; dens. 11.49 lb/gal; pour pt. -20 F; flash pt. (TCC) > 200 F; pH 3.5-4.5
Use Level: 50-100 ppm (initially); 5-15 ppm (continuous)

Arosolve 9D5R† [Arol Chem. Prods.]
Chem. Descrip.: Trichlorobenzene, butyl benzoate
CAS 136-60-7; EINECS/ELINCS 205-252-7
Uses: High exhaustion carrier for use on polyester and blends where extremely effective swelling, penetration and leveling are required
Properties: Lt. yel. clear liq.; sp.gr. 1.25 g/cc; dens. 10.43 lb/gal; pH 8.0 ± 0.3; nonionic

Atomsolve® 1400-1000‡ [Atomergic Chemetals]
Chem. Descrip.: Acetone
CAS 67-64-1; EINECS/ELINCS 200-662-2
Uses: Certified low-particulate HPLC-grade solvent for use in critical chromatography, semiconductor processing, pharmaceutical mfg., spectrophotometry, scintillation counting, spectrofluorescence, high purity org. synthesis, precision cleaning, etc.
Properties: Ref. index 1.3565; ≥ 99.9% assay

Atomsolve® 1400-1025‡ [Atomergic Chemetals]
Chem. Descrip.: Acetonitrile
CAS 75-05-8; EINECS/ELINCS 200-835-2
Uses: Certified low-particulate HPLC-grade solvent for use in critical chromatography, semiconductor processing, pharmaceutical mfg., spectrophotometry, scintillation counting, spectrofluorescence, high purity org. synthesis, precision cleaning, etc.
Properties: Ref. index 1.3415; ≥ 99.9% assay

Atomsolve® 1400-1325‡ [Atomergic Chemetals]
Chem. Descrip.: 1-Butanol
CAS 71-36-3; EINECS/ELINCS 200-751-6
Uses: Certified low-particulate HPLC-grade solvent for use in critical chromatography, semiconductor processing, pharmaceutical mfg., spectrophotometry, scintillation counting, spectrofluorescence, high purity org. synthesis, precision cleaning, etc.
Properties: Ref. index 1.3980; ≥ 99.9% assay

Atomsolve® 1400-1400‡ [Atomergic Chemetals]
Chem. Descrip.: n-Butyl chloride
CAS 109-69-3; EINECS/ELINCS 203-696-6
Uses: Certified low-particulate HPLC-grade solvent for use in critical chromatography, semiconductor processing, pharmaceutical mfg., spectrophotometry, scintillation counting, spectrofluorescence, high purity org. synthesis, precision cleaning, etc.
Properties: Ref. index 1.4010; ≥ 99.9% assay

Atomsolve® 1400-1675‡ [Atomergic Chemetals]
Chem. Descrip.: Chloroform
CAS 67-66-3; EINECS/ELINCS 200-663-8
Uses: Certified low-particulate HPLC-grade solvent for use in critical chromatography, semiconductor processing, pharmaceutical mfg., spectrophotometry, scintillation counting, spectrofluorescence, high purity org. synthesis, precision cleaning, etc.
Properties: Ref. index 1.4430; ≥ 99.9% assay

Atomsolve® 1400-1700‡ [Atomergic Chemetals]
Chem. Descrip.: Cyclohexane
CAS 110-82-7; EINECS/ELINCS 203-806-2
Uses: Certified low-particulate HPLC-grade solvent for use in critical chromatography, semiconductor processing, pharmaceutical mfg., spectrophotometry, scintillation counting, spectrofluorescence, high purity org. synthesis, precision cleaning, etc.
Properties: Ref. index 1.4240; ≥ 99.9% assay

Atomsolve® 1400-2025‡ [Atomergic Chemetals]
Chem. Descrip.: o-Dichlorobenzene
CAS 95-50-1; EINECS/ELINCS 202-425-9
Uses: Certified low-particulate HPLC-grade solvent for use in critical chromatography, semiconductor processing, pharmaceutical mfg., spectro-

photometry, scintillation counting, spectrofluorescence, high purity org. synthesis, precision cleaning, etc.
Properties: Ref. index 1.5490; ≥ 99.9% assay

Atomsolve® 1400-2425‡ [Atomergic Chemetals]
Chem. Descrip.: Ethyl acetate
CAS 141-78-6; EINECS/ELINCS 205-500-4
Uses: Certified low-particulate HPLC-grade solvent for use in critical chromatography, semiconductor processing, pharmaceutical mfg., spectrophotometry, scintillation counting, spectrofluorescence, high purity org. synthesis, precision cleaning, etc.
Properties: Ref. index 1.3700; ≥ 99.9% assay

Atomsolve® 1400-2460‡ [Atomergic Chemetals]
Chem. Descrip.: Reagent alcohol
CAS 64-17-5; EINECS/ELINCS 200-578-6
Uses: Certified low-particulate HPLC-grade solvent for use in critical chromatography, semiconductor processing, pharmaceutical mfg., spectrophotometry, scintillation counting, spectrofluorescence, high purity org. synthesis, precision cleaning, etc.
Properties: Ref. index 1.3580-1.3610; ≥ 89% assay

Atomsolve® 1400-3000‡ [Atomergic Chemetals]
Chem. Descrip.: Heptane
CAS 142-82-5; EINECS/ELINCS 205-563-8
Uses: Certified low-particulate HPLC-grade solvent for use in critical chromatography, semiconductor processing, pharmaceutical mfg., spectrophotometry, scintillation counting, spectrofluorescence, high purity org. synthesis, precision cleaning, etc.
Properties: Ref. index 1.3855; ≥ 97% assay

Atomsolve® 1400-3050‡ [Atomergic Chemetals]
Chem. Descrip.: Hexane
CAS 110-54-3; EINECS/ELINCS 203-777-6
Uses: Certified low-particulate HPLC-grade solvent for use in critical chromatography, semiconductor processing, pharmaceutical mfg., spectrophotometry, scintillation counting, spectrofluorescence, high purity org. synthesis, precision cleaning, etc.
Properties: Ref. index 1.3720; ≥ 95% assay

Atomsolve® 1400-3060‡ [Atomergic Chemetals]
Chem. Descrip.: Hexamethyl disilazane
CAS 999-97-3; EINECS/ELINCS 213-668-5
Uses: Certified low-particulate HPLC-grade solvent for use in critical chromatography, semiconductor processing, pharmaceutical mfg., spectrophotometry, scintillation counting, spectrofluorescence, high purity org. synthesis, precision cleaning, etc.
Properties: Ref. index 1.4055; ≥ 99.9% assay

Atomsolve® 1400-3800‡ [Atomergic Chemetals]
Chem. Descrip.: Methanol
CAS 67-56-1; EINECS/ELINCS 200-659-6
Uses: Certified low-particulate HPLC-grade solvent for use in critical chromatography, semiconductor processing, pharmaceutical mfg., spectrophotometry, scintillation counting, spectrofluorescence, high purity org. synthesis, precision cleaning, etc.
Properties: Ref. index 1.3260; ≥ 99.9% assay

Atomsolve® 1400-3900‡ [Atomergic Chemetals]
Chem. Descrip.: Methyl t-butyl ether
CAS 1634-04-4; EINECS/ELINCS 216-653-1
Uses: Certified low-particulate HPLC-grade solvent for use in critical chromatography, semiconductor processing, pharmaceutical mfg., spectrophotometry, scintillation counting, spectrofluorescence, high purity org. synthesis, precision cleaning, etc.
Properties: Ref. index 1.3660; ≥ 99.9% assay

Atomsolve® 1400-3925‡ [Atomergic Chemetals]
Chem. Descrip.: Methylene chloride
CAS 75-09-2; EINECS/ELINCS 200-838-9
Uses: Certified low-particulate HPLC-grade solvent for use in critical chromatography, semiconductor processing, pharmaceutical mfg., spectrophotometry, scintillation counting, spectrofluorescence, high purity org. synthesis, precision cleaning, etc.
Properties: Ref. index 1.4220; ≥ 99.9% assay

Atomsolve® 1400-4300‡ [Atomergic Chemetals]
Chem. Descrip.: Pentane
CAS 109-66-0; EINECS/ELINCS 203-692-4
Uses: Certified low-particulate HPLC-grade solvent for use in critical chromatography, semiconductor processing, pharmaceutical mfg., spectrophotometry, scintillation counting, spectrofluorescence, high purity org. synthesis, precision cleaning, etc.
Properties: Ref. index 1.3560; ≥ 99.9% assay

Atomsolve® 1400-4400‡ [Atomergic Chemetals]
Chem. Descrip.: 2-Propanol
CAS 67-63-0; EINECS/ELINCS 200-661-7
Uses: Certified low-particulate HPLC-grade solvent for use in critical chromatography, semiconductor processing, pharmaceutical mfg., spectrophotometry, scintillation counting, spectrofluorescence, high purity org. synthesis, precision cleaning, etc.
Properties: Ref. index 1.3740; ≥ 99.9% assay

Atomsolve® 1400-4900‡ [Atomergic Chemetals]
Chem. Descrip.: Tetrahydrofuran
CAS 109-99-9; EINECS/ELINCS 203-726-8
Uses: Certified low-particulate HPLC-grade solvent for use in critical chromatography, semiconductor processing, pharmaceutical mfg., spectrophotometry, scintillation counting, spectrofluorescence, high purity org. synthesis, precision cleaning, etc.
Properties: Ref. index 1.4060; ≥ 99.9% assay

Atomsolve® 1400-4975‡ [Atomergic Chemetals]
Chem. Descrip.: Toluene
CAS 108-88-3; EINECS/ELINCS 203-625-9
Uses: Certified low-particulate HPLC-grade solvent for use in critical chromatography, semiconductor processing, pharmaceutical mfg., spectrophotometry, scintillation counting, spectrofluorescence, high purity org. synthesis, precision cleaning, etc.
Properties: Ref. index 1.4970; ≥ 99.9% assay

Atomsolve® 1400-5000‡ [Atomergic Chemetals]
Chem. Descrip.: 1,2,4-Trichlorobenzene
CAS 120-82-1; EINECS/ELINCS 204-428-0
Uses: Certified low-particulate HPLC-grade solvent for use in critical chromatography, semiconductor processing, pharmaceutical mfg., spectrophotometry, scintillation counting, spectrofluorescence, high purity org. synthesis, precision cleaning, etc.
Properties: Ref. index 1.5707; ≥ 96% assay

Atomsolve® 1400-5075‡ [Atomergic Chemetals]
Chem. Descrip.: 2,2,4-Trimethylpentane
CAS 540-84-1; EINECS/ELINCS 203-892-1
Uses: Certified low-particulate HPLC-grade solvent for use in critical chromatography, semiconductor processing, pharmaceutical mfg., spectrophotometry, scintillation counting, spectrofluorescence, high purity org. synthesis, precision cleaning, etc.
Properties: Ref. index 1.3910; ≥ 99.9% assay

Atomsolve® 1400-5300‡ [Atomergic Chemetals]
Chem. Descrip.: Water
CAS 7732-18-5; EINECS/ELINCS 231-791-2
Uses: Certified low-particulate HPLC-grade solvent for use in critical chromatography, semiconductor processing, pharmaceutical mfg., spectrophotometry, scintillation counting, spectrofluorescence, high purity org. synthesis, precision cleaning, etc.
Properties: ≥ 99.9% assay

Bernel® Ester DOM‡ [Bernel]
Chem. Descrip.: Dicaprylyl maleate
Uses: Emollient, oxybenzone solubilizer, cleanser and wax solvent; noncomedogenic; imparts shine in hair conditioners
Regulatory: FDA, EU, and Japanese compliances
Properties: Colorless liq.; sol. in oil
Use Level: 2-10%

Borothene† [World Wide Chemnet♦]
Chem. Descrip.: Halogenated hydrocarbons
UN 2810
Uses: Precision cleaning solv. and drop-in replacement for 1,1,1 trichloroethane and other chlorofluorocarbons; solv. for high performance applics.
Features: Nonflamm., non-CFC
Properties: Sp.gr. 1.9; visc. 0.42 cps; f.p. -86.5 C; b.p. 153.5 F; flash pt. (TCC) none; ref. index 1.48; KB value 132
Toxicology: May irritate eyes, skin, respiratory tract; inh. may cause headache, dizziness, nausea, vomiting, narcosis in confined or poorly ventilated areas; prolonged skin contact may cause dermatitis; ing. may irritate digestive tractTSCA listed
Precaution: Avoid temps. > 204.4 C, strong acids and oxidizers

Hazardous Ingredients: Halogenated hydrocarbon mixt.
Hazardous decomp. prods.: CO, CO^2, hydrogen halide, and phosgene
HMIS: Health 1, Flammability 0, Reactivity 0
Storage: Store tightly closed; protect from extreme temps.

Butyl Dipropasol† [Dow]
Uses: Solvent, coalescing aid for water-disp. and water-sol. coatings
Properties: Sol. 3% in water (20 C); m.w. 190.30; sp.gr. 0.915 (20/20 C); dens. 7.63 lb/gal (20 C); b.p. 227.9 C; flash pt. (CC) 230 F

Butyl Propasol† [Dow]
Chem. Descrip.: Propylene glycol t-butyl ether
CAS 57018-52-7
Uses: Coalescing aid, solvent for water-disp. and water-sol. coatings
Properties: Sol. 6.4% in water (20 C); m.w. 132.20; sp.gr. 0.880 (20/20 C); dens. 7.37 lb/gal (20 C); b.p. 170.2 C; flash pt. (CC) 138 F

Calgene GMT K† [Lambent Tech.]
Chem. Descrip.: Glycerol monotallate
CAS 61789-12-6
Uses: Emulsifier, pigment dispersant, and lubricant for industrial applics.; rust preventative in oils; anti-icing fuel additive; textile lubricant for syn. fibers and spin finishes; solvent; mold release agent; w/o emulsifier for cutting fluids and drawing compds.
Regulatory: Kosher
Properties: Amber liq.; sol. in ethyl alcohol, propylene glycol, vegetable and min. oils, and org. solvs.; dens. 7.97 lb/gal; visc. 130 cps; HLB 3.4; acid no. 5 max.; iodine no. 73; sapon. no. 165; pour pt. -18 C; flash pt. (COC) > 250 C
Storage: Avoid storage at temps. > 90 F

Captex® 810A† [ABITEC]
Chem. Descrip.: Caprylic/capric/linoleic triglyceride; conforms to hydrogenated veg. oil NF
Uses: Emollient, solvent, fixative, and extender in pharmaceutical, nutritional, and cosmetic applics. incl. lipsticks, lip glosses, makeup bases, bath oils, aftershave lotions, aerosols; carrier for flavors and fragrances; avail. as developmental prod.
Properties: Gardner 2 max. liq.; acid no. 0.1 max.; iodine no. 25; sapon. no. 307-320

Captex® 810B† [ABITEC]
Chem. Descrip.: Caprylic/capric/linoleic triglyceride
Uses: Emollient, solvent, fixative, and extender in pharmaceutical, nutritional, and cosmetic applics. incl. lipsticks, lip glosses, makeup bases, bath oils, aftershave lotions, aerosols; carrier for flavors and fragrances
Properties: Gardner 2 max. color; visc. 30 cps; acid no. 0.1 max.; iodine no. 55; sapon. no. 280-296; cloud pt. 10 C

Captex® 810C† [ABITEC]
Chem. Descrip.: Caprylic/capric/linoleic triglyceride
Uses: Emollient, solvent, fixative, and extender in pharmaceutical, nutritional, and cosmetic applics. incl. lipsticks, lip glosses, makeup bases, bath oils, aftershave lotions, aerosols; carrier for flavors and fragrances
Properties: Gardner 2 max. color; acid no. 0.1 max.; iodine no. 75; sapon. no. 257-275

Captex® 910A† [ABITEC]
Chem. Descrip.: Caprylic/capric/oleic triglyceride
Uses: Emollient, solvent, fixative, and extender in pharmaceutical, nutritional, and cosmetic applics. incl. lipsticks, lip glosses, makeup bases, bath oils, aftershave lotions, aerosols; carrier for flavors and fragrances
Properties: Gardner 2 max. color; acid no. 0.1 max.; iodine no. 19; sapon. no. 304-318

Captex® 910B† [ABITEC]
Chem. Descrip.: Caprylic/capric/oleic triglyceride
Uses: Emollient, solvent, fixative, and extender in pharmaceutical, nutritional, and cosmetic applics. incl. lipsticks, lip glosses, makeup bases, bath oils, aftershave lotions, aerosols; carrier for flavors and fragrances
Properties: Gardner 2 max. color; acid no. 0.1 max.; iodine no. 35; sapon. no. 280-296

Captex® 910C† [ABITEC]
Chem. Descrip.: Caprylic/capric/oleic triglyceride
Uses: Emollient, solvent, fixative, and extender in pharmaceutical, nutritional, and cosmetic applics. incl. lipsticks, lip glosses, makeup bases, bath oils, aftershave lotions, aerosols; carrier for flavors and fragrances
Properties: Gardner 2 max. color; acid no. 0.1 max.; iodine no. 48; sapon. no. 260-275

Captex® 910D† [ABITEC]
Chem. Descrip.: Caprylic/capric/oleic triglyceride
Uses: Emollient, solvent, fixative, and extender in pharmaceutical, nutritional, and cosmetic applics. incl. lipsticks, lip glosses, makeup bases, bath oils, aftershave lotions, aerosols; carrier for flavors and fragrances
Properties: Gardner 2 max. color; acid no. 0.1 max.; iodine no. 60; sapon. no. 238-250

Charlab™ CHT Intensol TH-B† [CHT R. Beitlich Corp.]
Chem. Descrip.: Proprietary
Uses: Dyestuff solvent and machine cleaner
Properties: Straw-colored sl. hazy liq.; sl. odor; infinitely sol. in water; sp.gr. 1.01; b.p. 212 F; flash pt. (TCC) > 200 F
Toxicology: Could irritate or burn eyes; prolonged/repeated contact may cause skin irritation; inh. of very high concs. may cause headache, giddiness, mental confusion, nausea
Environmental: COD 367 mg O^2/g; BOD^5 72 mg O^2/g
Storage: Store in cool area

Cithrol 2DO‡ [Croda Chem. Ltd; Croda Oleochems.]
Chem. Descrip.: PEG-4 dioleate
CAS 9005-07-6
Uses: Dispersant, w/o emulsifier, wetting agent and cosolvent; used in cosmetic and industrial applics.
Properties: Liq.; nonionic; 97% conc.

Cithrol 6DL‡ [Croda Chem. Ltd; Croda Oleochems.]
Chem. Descrip.: PEG-12 dilaurate
CAS 9005-02-1
Uses: Dispersant, w/o emulsifier, wetting agent and cosolvent; used in cosmetic and industrial applics.
Properties: Liq.; HLB 12.0; nonionic; 97% conc.

Cithrol 6DO‡ [Croda Chem. Ltd; Croda Oleochems.]
Chem. Descrip.: PEG-12 dioleate
CAS 9005-07-6; EINECS/ELINCS 288-459-5
Uses: Dispersant, w/o emulsifier, wetting agent and cosolvent; used in cosmetic and industrial applics.
Properties: Liq.; HLB 10.4; nonionic; 97% conc.

Cithrol 10DL‡ [Croda Chem. Ltd; Croda Oleochems.]
Chem. Descrip.: PEG-20 dilaurate
CAS 9005-02-1
Uses: Dispersant, w/o emulsifier, wetting agent and cosolvent; used in cosmetic and industrial applics.
Properties: Paste; HLB 14.7; nonionic; 97% conc.

Cithrol 10DO‡ [Croda Chem. Ltd; Croda Oleochems.]
Chem. Descrip.: PEG-20 dioleate
CAS 9005-07-6
Uses: Dispersant, w/o emulsifier, wetting agent and cosolvent; used in cosmetic and industrial applics.
Properties: Liq.; HLB 13.2; nonionic; 97% conc.

Citroflex® 1† [Morflex]
Chem. Descrip.: Trimethyl citrate
CAS 1587-20-8
Uses: Environmentally friendly solvent offering safety, low volatility, high flash pt.; easily emulsified; compat. with nonaq. systems; used as actives in household, industrial, and institutional cleaners; solvent, film-former, and plasticizer for inks, adhesives, and coatings
Properties: APHA 100 max. solid; sol. 8.2 g/100 ml in water, < 4 g/100 ml in toluene; insol. in heptane; m.w. 234; m.p. 76 C; acid. no. 2.6; flash pt. (COC) 345 F; > 97% assay

Citroflex® 6† [Morflex]
Chem. Descrip.: Trihexyl citrate
CAS 16544-70-0
Uses: Environmentally friendly solvent offering safety, low volatility, high flash pt.; easily emulsified; compat. with nonaq. systems; used as actives in household, industrial, and institutional cleaners; solvent, film-former, and plasticizer for inks, adhesives, and coatings
Properties: APHA 100 max. liq.; sol. < 0.1 g/100 ml in water; infinite sol. in toluene and heptane; m.w. 444; sp.gr. 0.996; visc. 40 cps; vapor pressure 8.3 x 10~-6 mm Hg (20 C); pour pt. -57 C; acid. no. 0.02; flash pt. (COC) 400 F; ref. index 1.449; surf. tens. 29.11 dynes/cm; > 99% assay

Citroflex® A-1† [Morflex]
Chem. Descrip.: Acetyltrimethyl citrate
CAS 146340-22-9
Uses: Environmentally friendly solvent offering safety, low volatility, high flash pt.; easily emulsified; compat. with nonaq. systems; used as actives

in household, industrial, and institutional cleaners; solvent, film-former, and plasticizer for inks, adhesives, and coatings
Properties: APHA 50 max. liq.; sol. 5.5 g/100 ml in water; sol. in toluene; insol. in heptane; m.w. 276; sp.gr. 1.2324; visc. 366 cps; vapor pressure 0.16 mm Hg (20 C); pour pt. -2 C; acid. no. 0.02; flash pt. (COC) 311 F; ref. index 1.4440; surf. tens. 38.01 dynes/cm; > 98.5% assay

CNC Sol BD† [CNC Int'l.]
Uses: Solvent/emulsifier blend for removal of waxes, greases, oils, and greige-mill dirt from cotton, rayon, and blends

CNC Sol UE† [Bayer/Industrial Chems./Textile]
Chem. Descrip.: Aliphatic and aromatic solvs., terpenes, low titer soaps, and nonionic biodeg. detergents
Uses: Detergent/solvent for removal of oil, dirt, waxes, graphite, for textile scouring
Properties: Amber liq.; dissolves in hot or cold water

CNC Sol XN† [CNC Int'l.]
Chem. Descrip.: Solv./emulsifier with alkaline buidlers, nonionic detergents
Uses: Solvent/emulsifier blend for removal of waxes, oils, and greige mill dirt from cotton, rayon, and blends

CNC Sol XNN #11† [CNC Int'l.]
Uses: Solvent/emulsifier blend for removal of waxes, greases, oils, and greige-mill dirt from cotton, rayon, and blends

CNC Solv 809† [Bayer/Industrial Chems./Textile]
Chem. Descrip.: Solv./surfactant blend
Uses: For deinking and repulping resin coated mixed waste furnishes; prevents redeposition of agglomerated inks and binders

Cosmetol® X† [CasChem]
Chem. Descrip.: Unrefined castor oil USP, with antioxidant
CAS 8001-79-4; EINECS/ELINCS 232-293-8
Uses: Emollient, pigment wetting agent, cosolvent, lubricant for cosmetics, makeup, antiperspirant sticks
Features: Deodorized specially refined grade of castor oil containing a food grade antioxidant
Properties: Pour pt. -23 C; acid no. 2; iodine no. 86; sapon. no. 180; hyd. no. 164

Crodamol IPL‡ [Croda Chem. Ltd]
Chem. Descrip.: Isopropyl laurate
CAS 10233-13-3; EINECS/ELINCS 233-560-1
Uses: Emollient, plasticizer, and cosolvent for cosmetics
Properties: Colorless liq.; sol. in castor oil, corn oil, min. oil, oleyl alcohol, ethanol; sp.gr. 0.845-0.852; visc. 3.7 cSt

Custom® DIOL† [Custom Ingreds.]
Chem. Descrip.: 1,2-Propanediol
CAS 57-55-6; EINECS/ELINCS 200-338-0
Uses: Solvent for many essential oils
Properties: Water wh. liq.; mild odor; misc. with water, acetone, chloroform; immisc. with fixed oils; sp.gr. 1.035-1.037; 97.5% min. assay; 0.2% max. water

Cyanex® 925‡ [Cytec Ind.]
Chem. Descrip.: Bis (2,4,4,-trimethylpentyl) octylphosphine oxide
Uses: Solv. extraction reagent used for recovery of tin from chloride sol'ns., arsenic from copper electrolytes, and separation of niobium and tantalum

DB† [Occidental]
Chem. Descrip.: Butoxydiglycol
CAS 112-34-5; EINECS/ELINCS 203-961-6
Uses: Solvent

Dba† [ARCO♦]
Chem. Descrip.: Diethylene glycol butyl ether acetate
CAS 124-17-4; EINECS/ELINCS 204-685-9
Uses: Solvent for coatings and cleaning applics.
Features: HAPs listed
Regulatory: SARA reportable
Properties: 6% sol. in water @ 20 C; m.w. 204.27; sp.gr. 0.980 (20/20 C); dens. 8.16 lb/gal; visc. 3.2 cps; vapor pressure 0.04 mm Hg (20 C); b.p. 235 C; HLB 7.7; flash pt. (COC) 240 F; ref. index 1.4239; surf. tens. 30 dynes/cm

DB Acetate† [Occidental]
Chem. Descrip.: Diethylene glycol butyl ether acetate
CAS 124-17-4; EINECS/ELINCS 204-685-9
Uses: Solvent

DE† [Occidental]
Chem. Descrip.: Ethoxydiglycol
CAS 111-90-0; EINECS/ELINCS 203-919-7
Uses: Solvent

DEA† [ARCO♦]
Chem. Descrip.: Diethylene glycol ethyl ether acetate
CAS 112-15-2; EINECS/ELINCS 203-940-1
Uses: Solvent for coatings and cleaning applics.
Features: HAPs listed
Regulatory: SARA reportable
Properties: 100% sol. in water @ 20 C; m.w. 176.21; sp.gr. 1.012 (20/20 C); dens. 8.42 lb/gal; visc. 2.8 cps; vapor pressure 0.05 mm Hg (20 C); b.p. 214 C; HLB 8.7; flash pt. (TCC) 225 F; ref. index 1.4220; surf. tens. 31.7 dynes/cm

DE Acetate† [Occidental]
Chem. Descrip.: Diethylene glycol ethyl ether acetate
CAS 112-15-2; EINECS/ELINCS 203-940-1
Uses: Solvent

Delios® C‡ [Grünau GmbH]
Chem. Descrip.: Capric acid triglyceride
Uses: Solvent for flavors and oil-sol. food additives; surf. treating agent for dried fruits, confectionery, dietary foodstuffs
Properties: Wh. cryst. mass, neutral odor, sl. to coconut taste; m.p. 30 C; acid no. 0.1 max.; iodine no. 0.5 max.; sapon. no. 300; hyd. no. 5 max.
Storage: Store under cool, dry conditions in tightly closed containers below 25 C; 18 mos storage life in original pkg.

Delios® V‡ [Grünau GmbH]
Chem. Descrip.: Caprylic/capric acid triglyceride
CAS 65381-09-1; EINECS/ELINCS 265-724-3
Uses: Solvent for flavors and oil-sol. food additives for dietetic prods., sport foods; glazing and release agent for dried fruit and confectionery; lubricant for cutting machines; antidusting agent; visc. reducer
Features: Exc. oxidation stability
Properties: Colorless to sl. yel. liq., neutral odor, sl. to coconut taste; misc. with ethanol; visc. 30 mPa·s; cloud pt. 0 C max.; acid no. 0.1 max.; iodine no. 0.5 max.; sapon. no. 330-345; hyd. no. 5 max.; 60% caprylic acid, 40% capric acid.
Storage: Store in tightly closed containers @ 0-30 C; avoid contact with water; 18 mos storage life in original pkg.

DePEG 3350† [DeForest Enterprises]
Chem. Descrip.: PEG-75
CAS 25322-68-3; EINECS/ELINCS 203-989-9
Uses: Surfactant; intermediate for PEG-fatty acid esters; humectant in printing inks; plasticizer for starch pastes; solvent for personal care prods.; binder, lubricant for compressed tablets; plasticizer in PE sheet; food pkg. adhesives; lubricant for rubber molds; humectant, solvent for ink pad dyes; water softener and additive in paper processing; coupling agent in creams and lotions; color stabilizer for fuel oils; leveling agent in electropolishing formulas
Regulatory: FDA 21CFR §172.820, 173.310, 173.340, 175.105, 178.3750, 178.3910; DOT nonregulated
Properties: Wh. waxy flakes; sol. in water; insol. in most solvs. and oils; m.w. 3000-3700; m.p. 52-57 C; flash pt. (PMCC) > 350 F; pH 5.0-8.0 (5% in DW); nonionic; 100% conc.
Toxicology: Nontoxic; LD50 (oral, mouse) 29,500 mg/kg
Precaution: Incompat. with strong oxidizers
Hazardous decomp. prods.: Thermal decomp. prods.: CO^x
HMIS: Health 0, Flammability 1, Reactivity 0

DePEG 6000† [DeForest Enterprises]
Chem. Descrip.: PEG-135
CAS 25322-68-3; EINECS/ELINCS 203-989-9
Uses: Surfactant; intermediate for PEG-fatty acid esters; humectant in printing inks; plasticizer for starch pastes; solvent for personal care prods.; binder, lubricant for compressed tablets; plasticizer in PE sheet; food pkg. adhesives; lubricant for rubber molds; humectant, solvent for ink pad dyes; water softener and additive in paper processing; coupling agent in creams and lotions; color stabilizer for fuel oils; leveling agent in electropolishing formulas
Regulatory: FDA 21CFR §172.820, 173.310, 173.340, 175.105, 178.3750, 178.3910; DOT nonregulated
Properties: Wh. solid; sol. in water; insol. in most solvs. and oils; m.w. 5000-7000; m.p. 59-61 C; flash pt. (COC) > 300 C; pH 4.0-7.0 (5% in

† Discontinued ‡ Unverified ♦ Defunct Manufacturer or Division

DW); nonionic; 100% conc.
Toxicology: Nontoxic; LD50 (oral, mouse) 29,500 mg/kg
Precaution: Incompat. with strong oxidizers
Hazardous Ingredients: DePEG 6000 (100%)
Hazardous decomp. prods.: Thermal decomp. prods.: CO^x
HMIS: Health 1, Flammability 1, Reactivity 0

DePoxy™‡ [Atomergic Chemetals]
Chem. Descrip.: Dimethyl sulfoxide (90%), nitric acid (7%), stabilizers and accelerators (3%)
Uses: Solvent, industrial epoxy remover which breaks down epoxy and PU paints, glues, foams, resins, and finishes, rec. for electronic applics.; cold paint remover
Features: Re-usable; electronic grade materials; most effective when heated
Toxicology: Readily penetrates skin tissue; avoid breathing vapors which may emit fumes of nitric oxide

Dermol EB‡ [Fabriquimica]
Chem. Descrip.: Butyl stearate
CAS 123-95-5; EINECS/ELINCS 204-666-5
Uses: Emollient, dermatophilic additive; plasticizer and solvent for pigments/lipsticks
Properties: Acid no. < 1; iodine no. < 1; sapon. no. 168-176

Detersol HF† [Finetex]
Chem. Descrip.: Hydrocarbon blend, emulsified
Uses: High flash pt. solvent scour for textiles; low foam
Properties: Liq.; nonionic; 100% conc.

Detersol X-66† [Finetex]
Chem. Descrip.: Solvent/surfactant blend
Uses: General-purpose solvent scour
Properties: Liq.; nonionic; 100% conc.

Dional 11, 113† [Hoechst Celanese♦]
Chem. Descrip.: Fluorinated hydrocarbon
Uses: Solvent for dry-cleaning

Dioxitol-High Gravity† [Shell]
Chem. Descrip.: Diethylene glycol monoethyl ether, ethylene glycol
Uses: Solvent for natural and syn. resins, dyes, fats used as component of heavy duty automotive brake fluids; in paint industry in nonaq. stains, and nongrain raising stains for furniture and wood; yarn and cloth conditioner in textile industry; dye solvent where deep penetration and bright shades are obtained; component of industrial cleaners; coupling agent; mfg. of esters; intermediate in mfg. of plasticizers, solvents
Properties: Colorless liq.; water-misc.; sp.gr. 1.021-1.027; 72% Dioxitol-Low Gravity, 28% ethylene glycol.

Dioxitol-Low Gravity† [Shell]
Chem. Descrip.: Diethylene glycol monoethyl ether
CAS 111-90-0; EINECS/ELINCS 203-919-7
Uses: Solvent for resins, dyes, fats, in auto brake fluids, paint industry, industrial cleaners, mfg. of esters; intermediate in mfg. of plasticizers; yarn and cloth conditioner in textiles; dye solvent where deep penetration and bright shades are obtained
Properties: Colorless liq.; water-misc.; sp.gr. 0.986-0.990.

Dizene† [PPG Ind.]
Chem. Descrip.: Orthodichlorobenzene
CAS 95-50-1; EINECS/ELINCS 202-425-9
Uses: Aromatic used in industrial plants, stockyard pens, restaurants, hotels, hospitals, camps, parks, land-fill areas, cesspools, septic tanks, grease traps, dry wells, and soil; solvent for oil and grease
Properties: Ylsh. clear liq.; strong pleasant odor; dens. 10.63 lb/gal; flash pt. 150 F (TCC); 57% min. conc.

DM† [Occidental]
Chem. Descrip.: Methoxydiglycol
CAS 111-77-3; EINECS/ELINCS 203-906-6
Uses: Solvent

Dowanol® TMH† [Dow]
Chem. Descrip.: Triethylene glycol methyl ether/highers
CAS 112-50-5; EINECS/ELINCS 203-978-9
Uses: Solvent
Properties: Infinite sol. in water; m.w. 173 (avg.); sp.gr. 1.054; dens. 8.8 lb/gal; visc. 7.0 cks; vapor pressure < 0.01 mm Hg; b.p. 232 C; flash pt. (PMCC) 255 F; surf. tens. 39.1 dynes/cm (20 C)
Environmental: LC50 (emerald shiner) > 130 mg/l

DP† [ARCO♦]
Chem. Descrip.: Diethylene glycol propyl ether
CAS 6881-94-3; EINECS/ELINCS 229-985-7
Uses: Solvent for coatings and cleaning applics.
Features: HAPs listed
Regulatory: SARA reportable
Properties: 100% sol. in water @ 20 C; m.w. 148.20; sp.gr. 0.963 (20/20 C); dens. 8.04 lb/gal; visc. 4.1 cps; vapor pressure 0.03 mm Hg (20 C); b.p. 202 C; HLB 8.2; flash pt. (TCC) 200 F; ref. index 1.4290; surf. tens. 32.3 dynes/cm; sp. heat 0.55 Cal/g°C

Dynasolve 150‡ [Dynaloy]
Chem. Descrip.: Tetrahydrofuran-based solvent
Uses: Solvent for electronic conformal coating removal
Features: Flamm.; for R.T. use only

Dynasolve 185 SG‡ [Dynaloy]
Chem. Descrip.: N-Methyl pyrrolidone (78-98%) and 2-phenoxy ethanol (5-15%)
UN NA 1993
Uses: Electronic grade solvent for depotting, deflashing, decapsulation, and deinking applics., surf. prep., photo resist stripping
Regulatory: SARA §313 nonreportable
Properties: Ylsh. transparent liq.; mild amine-like odor; completely sol. in water; sp.gr. 1.038; vapor pressure 0.29 mm Hg (20 C); b.p. 202 C; flash pt. (CC) 199 F; pH 9.3; 100% volatile
Toxicology: May cause irritation, inflamm., or burns to eyes, irritation to mouth, nose, throat, mucous membranes, coughing, difficulty breathing, skin irritation, nausea, vomiting, convulsions; high concs. may cause mental confusion; dermatitis (chronic)
Precaution: Combustible liq.; flamm. limits in air 1.3-9.5%; mod. fire hazard exposed to heat or flame; incompat. with strong oxidizing or reducing agents
Hazardous decomp. prods.: Heated to decomp., emits CO and NO^x fumes
HMIS: Health 2, Flammability 2, Reactivity 0
Storage: Keep container closed

Dynasolve 611† [Dynaloy]
Chem. Descrip.: Ethylene glycol ether base
Uses: Solvent for surf. prep. and residual photoresist removal or removal of other org. contaminants; use hot (100-120 C)

Dynasolve 700‡ [Dynaloy]
Chem. Descrip.: Propylene glycol ether base
Uses: Solvent for electronic depotting, decapsulation applics.
Features: For R.T. temp. use only; similar in str. to Uresolve HF
Precaution: Flamm.

Dynasolve 710‡ [Dynaloy]
Chem. Descrip.: Propylene glycol ether base
Uses: Solvent for electronic depotting, decapsulation, deinking, and surf. prep. applics.; suitable for cured epoxies, urethanes, and silicones; stronger than Dynasolve 700, but less selective; similar in str. to Uresolve Plus; slow at R.T.

Dynasolve CU-4† [Dynaloy]
Uses: Solvent similar to Dynaflush, but more effective for cleaning urethane residues and crystallized isocyanates; can be used hot (95 C) for faster results; cleans residues from mixing and metering equip., nozzles, spray guns, RIM molds

Dynasolve DF-1‡ [Dynaloy]
Uses: Solvent; replenisher conc. for Dynaflush; when added to vacuum distilled or spent Dynaflush, will restore its efficiency and loading capacity to original levels

Dynasolve M-30‡ [Dynaloy]
Uses: Nonchlorinated, nonflamm. solvent for general cleaning and degreasing operations; esp. for curing of uncured polymers such as epoxies, urethanes and silicones

Dynasolve MP-710‡ [Dynaloy]
Chem. Descrip.: Propylene glycol methyl ether (40-70%), propylene glycol phenyl ether (10-30%), methanol (2-5%), potassium hydroxide (1-5%)
UN NA 1993
Uses: Solvent for cleaning residues from mixing, metering equip., nozzles, RIM molds; removal of mold releases, adhesives and inks from rollers, general polymer removal, paint stripping; suitable for cured epoxies, silicones, and urethanes; similar in str. to Dynasolve MP-AL GR, but less selective; slow at R.T.; more effective at 65-110 C

Regulatory: SARA §313 reportable (methanol)
Properties: Blue-grn. liq.; sl. pleasant odor; sol. in water; sp.gr. 0.983; flash pt. (CC) 105 F; pH 13.6; 100% volatile by vol.
Toxicology: Irritant to skin, eyes, nose, throat, respiratory tract; may cause dizziness, drowsiness, incoordination, nausea, vomiting, coma, breathing difficulties, headache; may cause anesthesia at higher levels (chronic overexposure)TSCA listed
Precaution: Combustible liq.; flamm. limits n air 1.6-13.8% by vol.; incompat. with oxidizing materials
Hazardous decomp. prods.: CO, CO^2
HMIS: Health 2, Flammability 2, Reactivity 1
Storage: Keep container closed

Dynasolve MP-750‡ [Dynaloy]
Chem. Descrip.: Propylene glycol methyl ether (40-70%), propylene glycol phenyl ether (10-30%), methanol (2-5%), potassium hydroxide (1-5%)
UN NA 1993
Uses: Solvent for cleaning residues from mixing, metering equip., nozzles, RIM molds; removal of mold releases, adhesives and inks from rollers, general polymer removal, paint stripping; suitable for cured epoxies, silicones, and urethanes; stronger than Dynasolve MP-710, but less selective; similar in str. to Dynasolve MP-500; slow at R.T.; more effective at 65-110 C
Regulatory: SARA §313 reportable (methanol)
Properties: Blue-grn. liq.; sl. pleasant odor; sol. in water; sp.gr. 0.996; flash pt. (CC) 105 F; pH 13.7

Dynasolve XD 16-4‡ [Dynaloy]
Chem. Descrip.: Proprietary with 2-phenoxyethanol (35-55%)
Uses: Solvent developed for removal of silicone oils and uncured silicone polymers from molds, molded parts, and processing equip.
Regulatory: DOT nonregulated; SARA §313 reportable (2-phenoxy ethanol)
Properties: Lt. amber liq.; ester odor; sol. in water; sp.gr. 1.032; b.p. > 300 F; flash pt. (CC) 262 F; VOC 1032 g/l
Toxicology: Acute overexposure: irritation of mouth, nose, throat, mucous membranes, coughing, difficulty breathing, nausea, vomiting, skin irritation, burning or watering of eyes, inflamm. of conjunctiva; chronic overexposure: dermatitisTSCA listed
Precaution: Incompat. with strong oxidizing or reducing agents
Hazardous decomp. prods.: Acrid smoke, fumes, CN~–
HMIS: Health 2, Flammability 1, Reactivity 0
Storage: Contains powerful org. solvs., keep away from heat and flame

Dynasolve XD 27-2‡ [Dynaloy]
Uses: Safety solvent for use in cleaning, defluxing of printed circuit boards; usable in commercially available cleaning equip.; superior solvency thoroughly and quickly removes rosin and ionic contamination from circuit boards and other electronic devices

EB† [Occidental]
Chem. Descrip.: Butoxyethanol
CAS 111-76-2; EINECS/ELINCS 203-905-0
Uses: Solvent for coatings and cleaning applics.

EBA† [ARCO♦]
Chem. Descrip.: Ethylene glycol butyl ether acetate
CAS 112-07-2; EINECS/ELINCS 203-933-3
Uses: Solvent for coatings and cleaning applics.
Features: HAPs listed
Regulatory: SARA reportable
Properties: 1% sol. in water @ 20 C; m.w. 160.21; sp.gr. 0.941 (20/20 C); dens. 7.84 lb/gal; visc. 1.8 cps; vapor pressure 0.29 mm Hg (20 C); b.p. 186C; HLB 7.4; flash pt. (TCC) 160 F; ref. index 1.4142; surf. tens. 30.3 dynes/cm

EB Acetate† [Occidental]
Chem. Descrip.: Butoxyethanol acetate
CAS 112-07-2; EINECS/ELINCS 203-933-3
Uses: Solvent

Eccoclean CR-46‡ [Eastern Color & Chem.]
Chem. Descrip.: Self-emulsifying solv. cleaner
Uses: Added solvency for extra cleaning power; useful on difficult cleaning problems
Properties: Water-disp.

Eccoclean RPW-5/WA‡ [Eastern Color & Chem.]
Uses: Solvent cleaner for stubborn cleaning problems
Features: Sl. modified version of Eccoclean RPW

Eccosolv C-14† [Eastern Color & Chem.]
Uses: Self-emulsifying solvent cleaner for cleaning and degreasing of machine parts

Edenol® 302‡ [Cognis/Care Chems.; Cognis Canada]

EEA† [ARCO♦]
Chem. Descrip.: Ethylene glycol ethyl ether acetate
CAS 111-15-9; EINECS/ELINCS 203-839-2
Uses: Solvent for coatings and cleaning applics.
Features: HAPs listed
Regulatory: SARA reportable
Properties: 24% sol. in water @ 20 C; m.w. 132.16; sp.gr. 0.973 (20/20 C); dens. 8.11 lb/gal; visc. 1.3 cps; vapor pressure 1.7 mm Hg (20 C); b.p. 150 C; HLB 8.3; flash pt. (TCC) 130 F; ref. index 1.4030; surf. tens. 28.2 dynes/cm

EEH† [ARCO♦]
Chem. Descrip.: Ethylene glycol ethyl hexyl ether
CAS 1559-35-9
Uses: Solvent for coatings and cleaning applics.
Features: HAPs listed
Properties: 0.2% sol. in water @ 20 C; m.w. 174.29; sp.gr. 0.892 (20/20 C); dens. 7.42 lb/gal; visc. 7.0 cps; vapor pressure 0.08 mm Hg (20 C); b.p. 224 C; HLB 5.4; flash pt. (Seta) 208 F; ref. index 1.4361; surf. tens. 27.6 dynes/cm

EH† [ARCO♦]
Chem. Descrip.: Ethylene glycol hexyl ether
CAS 112-25-4; EINECS/ELINCS 203-951-1
Uses: Solvent for coatings and cleaning applics.
Features: HAPs listed
Regulatory: SARA reportable
Properties: 1% sol. in water @ 20 C; m.w. 146.23; sp.gr. 0.889(20/20 C); dens. 7.40lb/gal; visc. 5.2 cps; vapor pressure < 0.1 mm Hg (20 C); b.p. 208 C; HLB 6.4; flash pt. (TCC) 179 F; ref. index 1.4290; surf. tens. 30.3 dynes/cm

Ektapro® EEP Solvent† [Eastman]
Chem. Descrip.: Ethyl 3-ethoxypropionate
CAS 763-69-9
Uses: High-performance retarder solvent for formulating enamels, lacquers, topcoats, and primers incl. auto OEM topcoats, coil, maintenance, appliance, aerospace, photoresist, marine, and plastic coatings; urethane-grade quality; polymerization solvent
Properties: Pt-Co 15 max. color; moderate odor; 2.9% sol. in water; m.w. 146.19; sp.gr. 0.95; dens. 7.91 lb/gal; visc. 1.0 cP; f.p. < -50 C; b.p. 165 C (760 mm); flash pt. (Seta) 58 C; surf. tens. 27.0 dynes/cm (23 C); elec. resist. 20 megohms

Ektasolve® DB† [Eastman]
Chem. Descrip.: Butoxydiglycol
CAS 112-34-5; EINECS/ELINCS 203-961-6
Uses: Solvent for cosmetic and industrial applics.

Ektasolve® DB Acetate† [Eastman]
Chem. Descrip.: Diethylene glycol butyl ether acetate
CAS 124-17-4; EINECS/ELINCS 204-685-9
Uses: Solvent for cosmetic and industrial applics.

Ektasolve® DE† [Eastman]
Chem. Descrip.: Ethoxydiglycol
CAS 111-90-0; EINECS/ELINCS 203-919-7
Uses: Solvent for cosmetic and industrial applics.

Ektasolve® DE Acetate† [Eastman]
Chem. Descrip.: Ethoxydiglycol acetate
CAS 112-15-2; EINECS/ELINCS 203-940-1
Uses: Solvent for cosmetic and industrial applics.

Ektasolve® DM† [Eastman]
Chem. Descrip.: Diethylene glycol monomethyl ether
CAS 111-77-3; EINECS/ELINCS 203-906-6
Uses: Evaporating solvent used in brushing lacquers and dye stains; useful in wood stains, printing inks, and dye pastes for textiles; coalescing aid for PVAc latex paints; used in stamp pad and stencil inks; diluent for hydraulic brake fluids; also for cosmetics
Properties: Water-wh. liq.; agreeable odor; m.w. 120.15; f.p. -85 C; b.p. 191 C min.; water-sol.; sp.gr. 1.023; dens. 1.02 kg/l; ref. index 1.4263; flash pt. 96 C (COC); fire pt. 96 C; > 99% volatiles by vol.

Ektasolve® EB† [Eastman]

Chem. Descrip.: Butoxyethanol
CAS 111-76-2; EINECS/ELINCS 203-905-0
Uses: Solvent for alkyd, phenolic, maleic, and cellulose nitrate resins; retarder for lacquers, improving gloss and flow-out, blush resistance, and reducing the formation of orange peel, for formulating hot-spray, brushing, flow-coat, and aerosol lacquers
Properties: Colorless liq.; faint odor; m.w. 118.17; f.p. -75 C; b.p. 169.0 C min.; water-sol.; sp.gr. 0.902; dens. 0.90 kg/l; ref. index 1.4193; flash pt. 62 C (TCC); fire pt. 70 C; > 99% volatiles by vol.

Ektasolve® EEH† [Eastman]
Chem. Descrip.: Ethylene glycol 2-ethylhexyl ether
CAS 1559-35-9
Uses: Solvent for cosmetic and industrial applics.

Ektasolve® EP† [Eastman]
Chem. Descrip.: Ethylene glycol monopropyl ether
CAS 2807-30-9; EINECS/ELINCS 220-548-6
Uses: Solvent in coatings, waterborne coating systems, cosmetics; coupling solvent for resin/water systems; controls visc. of waterborne resins; effective for NC, acrylic, epoxy, polyamide, and alkyd resins; retarder in coating systems
Features: Slow evaporating
Properties: Colorless clear liq.; mild odor; m.w. 104.15; f.p. < -90 C; b.p. 149.5 min.; water-sol.; sp.gr. 0.9125; dens. 0.91 kg/l; ref. index 1.4136; flash pt. 49 C (TCC); fire pt. 56 C; 100% volatiles by vol.

Ektasolve® PM Acetate† [Eastman]
Chem. Descrip.: Propylene glycol methyl ether acetate
CAS 108-65-6; EINECS/ELINCS 203-603-9
Uses: Retarder solvent for cosmetic and industrial applics.
Properties: Sol. 25.9% in water; dens. 7.91 lb/gal; b.p. 142-147 C; flash pt. 46 C; surf. tens. 26.4 dynes/cm; elec. resist. 6 megohms.

EM† [Occidental]
Chem. Descrip.: Methoxyethanol
CAS 109-86-4; EINECS/ELINCS 203-713-7
Uses: Solvent

EMA† [ARCO♦]
Chem. Descrip.: Ethylene glycol methyl ether acetate
CAS 110-49-6; EINECS/ELINCS 203-772-9
Uses: Solvent for coatings and cleaning applics.
Features: HAPs listed
Regulatory: SARA reportable
Properties: 100% sol. in water @ 20 C; m.w. 118.14; sp.gr. 1.006 (20/20 C); dens. 8.37 lb/gal; visc. 1.1 cps; vapor pressure 2 mm Hg (20 C); b.p. 145 C; HLB 8.8; flash pt. (TCC) 120 F; ref. index 1.4025; surf. tens. 34.0 dynes/cm

EmCon™ Walnut† [Fanning]
Chem. Descrip.: Walnut oil
CAS 8024-09-7
Uses: Skin conditioning agent, occlusive solvent for suntan gels, creams, lotions, topical pharmaceuticals, ointments
Properties: Clear oil, sl. nutty odor; iodine no. 145-158; sapon. no. 180-200

Emerest® 2419† [Cognis; Cognis Canada]
Chem. Descrip.: Glyceryl dioleate
CAS 25637-84-7; EINECS/ELINCS 247-144-2
Uses: Emulsifier, lubricant, rust preventive additive, mold release agent, solvent for dyes and pigments; used in leathers, lubricants and softeners
Properties: Gardner 8 liq.; sol. in min. oil, toluene, IPA, xylene; dens. 7.7 lb/gal; visc. 55 cSt (100 F); HLB 1.6; pour pt. < 0 C; flash pt. 470 F; nonionic; 100% conc.

Emerest® 5301† [Cognis/Chems. Group]
Chem. Descrip.: Ethyl oleate
CAS 111-62-6; EINECS/ELINCS 203-889-5
Uses: Solv.; best odor for oleate esters, good low temp. props.; appeals to corn farmers

Emery® 2209† [Cognis/Chems. Group]
Chem. Descrip.: Methyl caprylate-caprate
CAS 68937-83-7
Uses: Detergent intermediate; solvent for pesticides and herbicides
Regulatory: EPA exempt
Properties: Gardner 1 color; dens. 7.1 lb/gal; visc. 1.6 cSt (100 C); m.p. -30 C; acid no. 0.5 max.; iodine no. 0.4 max.; sapon. no. 330-336; flash pt. 180 F

Emery® 2224† [Cognis]
Chem. Descrip.: Methyl soyate
CAS 67784-80-9
Uses: Solvent for pesticides and herbicides
Regulatory: EPA exempt
Properties: Gardner 11 color; dens. 7.3 lb/gal; visc. 4.3 cSt (100 C); m.p. 18 C; flash pt. 350 F

Emery® 2231† [Cognis/Chems. Group]
Chem. Descrip.: Dist. methyl canolate
Uses: Solvent and adjuvant for pesticides and herbicides; exc. low temp. props.; used in tank mixes
Regulatory: EPA exempt
Properties: Gardner 1 color; dens. 7.3 lb/gal; visc. 4.6 cSt (100 C); m.p. < -10 C; flash pt. 345 F

Emery® 6686† [Cognis]
Chem. Descrip.: PEG 600
CAS 25322-68-3; EINECS/ELINCS 229-859-1
Uses: Chemical intermediate for coatings, adhesives, lubricants, metalworking, paper mfg., petrol. prod., ceramics, printing, electronics, solvs., cleaners, latex paints, mold release agent, rubber
Properties: Gardner < 1 liq.; sol. in water, xylene; disp. in Stod.; dens. 9.4 lb/gal; visc. 63 cSt (100 F); pour pt. 22 C; flash pt. 475 F

Emery® 6687† [Cognis]
Chem. Descrip.: PEG-300
CAS 25322-68-3; EINECS/ELINCS 220-045-1
Uses: Chemical intermediate for coatings, adhesives, lubricants, metalworking, paper mfg., petrol. prod., ceramics, printing, electronics, solvs., cleaners, latex paints, mold release agent, rubber, textiles
Properties: Gardner < 1 liq.; sol. in water, xylene; disp. in Stod.; dens. 9.4 lb/gal; visc. 33 cSt (100 F); pour pt. -10 C; flash pt. > 400 F

Emery® 6709† [Cognis]
Chem. Descrip.: PEG 400
CAS 25322-68-3; EINECS/ELINCS 225-856-4
Uses: Humectant, binder, solvent, visc. modifier for personal care prods.; chemical intermediate for coatings, adhesives, lubricants, metalworking, paper mfg., petrol. prod., ceramics, printing, electronics, solvs., cleaners, mold release agent
Properties: Gardner < 1 clear, visc. liq.; sol. @ 5% in water, xylene; m.w. 400; dens. 9.4 lb/gal; visc. 45 cSt (100 F); pour pt. 6 C; flash pt. > 350 F; nonionic; 100% act.

Emery® Methyl Lardate† [Cognis]
Chem. Descrip.: Methyl lardate
Uses: Solvent-carrier for agric. spray prods.; defoaming component in metalworking, paper deinking, pharmaceutical fermentation

Emkapon® DAC-500† [Emkay♦]
Uses: Solvent scour for removing fatty soils, oils, and waxes used in sizing, for use on synthetics, nylon, Dacron, Kodel, Arnel, Spandex

Emkapyl 400† [Uniqema; Uniqema Am.]
Chem. Descrip.: Polypropylene glycol
CAS 25322-69-4; EINECS/ELINCS 200-338-0
Uses: Surfactant; modifier for resins; for surfactant mfg., heat transfer fluids, ether/ester mfg., metalworking lubricants, paint and ink solvs., textile lubricants, mutual solvs. (veg. oils)
Features: Good foam control and lubrication props
Properties: Colorless liq.; pour pt. -37 C; hyd. no. 250; pH 6.2 (10 g in 60 ml of 10:1 methanol:water); 0.5% max. water

Emkapyl 600† [Uniqema; Uniqema Am.]
Chem. Descrip.: Polypropylene glycol
CAS 25322-69-4; EINECS/ELINCS 200-338-0
Uses: Surfactant; modifier for resins; for surfactant mfg., heat transfer fluids, ether/ester mfg., metalworking lubricants, paint and ink solvs., textile lubricants, mutual solvs. (veg. oils)
Features: Good foam control and lubrication props.
Properties: Colorless liq.; pour pt. -37 C; hyd. no. 187; pH 6.2 (10 g in 60 ml of 10:1 methanol:water); 0.5% max. water

Emkapyl 4000† [Uniqema; Uniqema Am.]
Chem. Descrip.: Polypropylene glycol
CAS 25322-69-4; EINECS/ELINCS 200-338-0
Uses: Surfactant; modifier for resins; for ceramic antifoams, surfactant and ether/ester mfg., lubricant, softener for leather, base and solv. for cosmetics, metalworking lubricants, petrol. demulsifier; mold release for

rubber
Features: Good foam control and lubrication props.
Properties: Colorless liq.; pour pt. -27 C; hyd. no. 28; pH 6.2 (10 g in 60 ml of 10:1 methanol:water); 0.3% max. water

Emulbon LB-78† [Toho Chem. Ind.]
Chem. Descrip.: Six-membered borate with monoalkyl POE group
Uses: Lubricant; plasticizer; solvent
Properties: Liq.

EP† [ARCO♦]
Chem. Descrip.: Ethylene glycol propyl ether
CAS 2807-30-9; EINECS/ELINCS 220-548-6
Uses: Solvent for coatings and cleaning applics.
Features: HAPs listed
Regulatory: SARA reportable
Properties: 100% sol. in water @ 20 C; m.w. 104.15; sp.gr. 0.913 (20/20 C); dens. 7.59 lb/gal; visc. 2.4 cps; vapor pressure 1.3 mm Hg (20 C); b.p. 149 C; HLB 7.8; flash pt. (TCC) 120 F; ref. index 1.4136; surf. tens. 27.9 dynes/cm

Eposolve® 299-R‡ [Elementis Spec. Perf. Polymers]
Chem. Descrip.: Blend of chlorinated solvs.
Uses: Versatile clean-up solvent (non-red label) for cured and uncured epoxies; designed for attacking and disintegrating fully cured thermosetting plastics such as epoxies, urethanes, and polyesters; clean-up solvent for tools, brushes, automatic dispensing equip., spilled uncured resins; does not contain ODC
Properties: Clear liq.; dens. 11.1 lb/gal; visc. 10 cps; flash pt. 48 C
Toxicology: May be harmful if inhaled; can cause severe eye burns
Storage: Store in cool, dry place; keep containers sealed

Ethana®‡ [Asahi Chem. Ind.]
Chem. Descrip.: 1,1,1-Trichloroethane
CAS 71-55-6; EINECS/ELINCS 200-756-3
Uses: Solvent

Ethylan® HB Series‡ [Akzo Nobel Surf. Chem. AB]
Chem. Descrip.: Aromatic alkoxylates
Uses: Cosolvent with high flash pt. for use in iodophors, rinse aids, detergent sterilizers, glass cleaners, etc.
Properties: Liq.; nonionic

Ethylan® KELD† [Akzo Nobel Surf. Chem. AB]
Chem. Descrip.: Modified fatty alkylolamide
Uses: Emulsifier for wax, kerosene and olein, solvent degreasing
Properties: Clear amber liq.; mild odor; disp. in water; sp.gr. 1.005; visc. 967 cs; flash pt. (COC) > 150 C; pour pt. 5 C; pH 9.0 (1% aq.); nonionic; 100% act.

Ethyl Di-Icinol† [Huntsman]
Chem. Descrip.: 2-(2-Ethoxyethoxy)ethanol
CAS 111-90-0; EINECS/ELINCS 203-919-7
Uses: Solvent for use in protective coatings, inks, cleaning prods., agric. chems.; aids wetting, penetration, and soil removal; coupling solvent

Eumulgin® PWM2† [Cognis/Care Chems.; Cognis Iberica]
Chem. Descrip.: Oleth-2
CAS 9004-98-2
Uses: W/o emulsifier, dispersant, lipophilic cosolvent; fragrance grade; broad pH and electrolyte tolerance
Properties: Liq.; HLB 5.0; hyd. no. 155-165; pH 6.0-7.5 (1%); nonionic; 100% conc.

Exxal® 16† [ExxonMobil]
Chem. Descrip.: Hexyldecanol
CAS 2425-77-6; EINECS/ELINCS 219-370-1
Uses: Extender and solvent for dyes and fragrances

Exx-Print® 314 A‡ [ExxonMobil]
Uses: Solvent for printing inks
Properties: Sp.gr. 1.00 (15.6 C); dens. 8.34 lb/gal (60 F); b.p. 231-276 C; flash pt. (TCC) 219 F; surf. tens. 38.0 dynes/cm; 99.8% aromatic; < 0.2% cycloparaffin

Fancol™ OA-95† [Fanning]
Chem. Descrip.: Oleyl alcohol
CAS 143-28-2; EINECS/ELINCS 205-597-3
Uses: Plasticizer, emulsion stabilizer, antifoam, coupling agent, aerosol lubricant, pigment dispersant, rust preventive, detergent, release agent, cosolvent, softener, tackifier, spreading agent for metalworking, petrochemicals, pulp and paper, paints and coatings, plastics and polymers, inks, food applics., pharmaceuticals, cosmetics; chemical intermediate; petrol. additive
Properties: Liq.; sol. in IPA, acetone, lt. min. oil, trichloroethylene, kerosene, VMP naphtha, benzene, turpentine; acid no. 0.05 max.; iodine no. 90-96; sapon no. 1 max.; hyd. no. 200-212; cloud pt. 5 C max.; nonionic

Filmer EHC† [Dow]
Uses: Solvent, coalescing aid for water-disp. and water-sol. coatings
Properties: Sol. 1.7% in water (20 C); m.w. 190.30; sp.gr. 0.933 (20/20 C); dens. 7.79 lb/gal (20 C); b.p. 259.0 C; flash pt. (CC) 271 F

Flexricin® 9† [CasChem]
Chem. Descrip.: Propylene glycol ricinoleate
CAS 26402-31-3; EINECS/ELINCS 247-669-7
Uses: Wetting agent, dye solvent, wax plasticizer, stabilizer for textile, household, and cosmetic applics., rewetting dried skins
Regulatory: FDA approval
Properties: Gardner 2+ liq.; sol. in toluene, butyl acetate, MEK, ethanol; sp.gr. 0.96; visc. 3 stokes; pour pt. -15 F; acid no. 2; iodine no. 77; sapon. no. 159; hyd. no. 288; nonionic; 100% act.

Formula 111† [Chemtronics]
Uses: Heavy-duty solvent for strong cleaning/degreasing on interfaces, metal contacts, electro-mech. assemblies, PCB connectors

Formula 500-C† [Garratt Callahan]
Uses: Fuel oil tank and piping cleaner; strong solvents dissolve and disperse gunk on screens, fuel lines, pumps, burner guns, and in storage tanks; as a result, burning efficiency and handling ease is increased; high dispersing action which helps penetrate heaviest sludge masses; completely combustible when mixed with fuel oil
Properties: Lt. yel.-brn. clear liq.; dens. 8.6 lb/gal; flash pt. 185 F
Storage: Indefinite storage life when stored in cool, dry area

Frigen 113 TR-N† [Hoechst AG♦]
Chem. Descrip.: Fluorinated hydrocarbon
Uses: Solvent for removing org. and inorg. pollution.

G-100† [Uniqema; Uniqema Am.]
Chem. Descrip.: Dimethylisosorbide
CAS 5306-85-4; EINECS/ELINCS 226-159-8
Uses: Carrier, solvent for cosmetic and pharmaceutical solvs.
Properties: Colorless liq.; 100% act.

G-4972‡ [Uniqema Am.]
Chem. Descrip.: PEG-15 hydroxystearate
Uses: Solvent for injection sol'ns.

Gaftex® HI-1† [ISP]
Chem. Descrip.: N-Methyl-2-pyrrolidone
CAS 872-50-4; EINECS/ELINCS 212-828-1
Uses: Solvent for hard surf., heavy duty, and automobile cleaners; replacement for hazardous solvs.; adds extra cleaning power to remove stubborn stains with much less risk of damage to sensitive surfs.; eliminates need for extra alkalinity; reduces streaking effects; very effective in removing ink stains
Properties: Clear liq.; misc. in water, and most org. solvs. incl. alcohols, ketones, aromatic and chlorinated hydrocarbons; sp.gr. 1.027; visc. 1.65 mPa·s; f.p. -24.4 C; b.p. 202 C; flash pt. (CC) 93 C; pH sl. alkaline (10% aq. sol'n.); 99.8% purity
Toxicology: LD50 (oral) 4200 mg/kg; low toxicity, no skin/eye irritation
Environmental: Biodeg.; environmentally safe

Gaftex® HI-8† [ISP]
Chem. Descrip.: N-Octyl pyrrolidone
CAS 7425-87-8
Uses: Surf. act. solvent for use in hard surf. cleaners, heavy duty concs. and liq., glass cleaners, waxes and polishes; outstanding solvs. with additional surf. act., exc. wetting, and low foaming; low visc.; stable over wide pH range; hydrophilic/lipophilic balance; nonvolatile, unaffected by VOC regulations
Properties: Liq.; sparingly sol. in water, but sol. in other polar and nonpolar solvs.; dens. 0.92 g/cm~3; f.p. -25 C; b.p. 100 C (0.3 mm Hg); HLB 6; flash pt. (TCC) 119 C; nonionic; 97% act.
Toxicology: LD50 (oral) > 2050 mg/kg; low toxicity, no skin/eye irritation
Environmental: Biodeg.

Gaftex® HI-12† [ISP]
Chem. Descrip.: N-Dodecyl pyrrolidone
CAS 2687-96-9; EINECS/ELINCS 403-730-1
Uses: Surf. act. solvent for use in hard surf. cleaners, heavy duty concs. and liq., glass cleaners, waxes and polishes; outstanding solvs. with additional surf. act., exc. wetting, and low foaming; low visc.; stable over

wide pH range; hydrophilic/lipophilic balance; nonvolatile, unaffected by VOC regulations

Properties: Liq.; sparingly sol. in water, but sol. in other polar and nonpolar solvs.; dens. 0.90 g/cm~3; f.p. 10 C; b.p. 145 C (0.2 mm Hg); HLB 3; flash pt. (TCC) 116 C; nonionic; 97% act.

Toxicology: LD50 (oral) > 5000 mg/kg; low toxicity, no skin/eye irritation

Environmental: Biodeg.

Gaftex® HI-20H† [ISP]

Chem. Descrip.: N-2-Hydroxyethyl-2-pyrrolidone

CAS 3445-11-2; EINECS/ELINCS 222-359-4

Uses: Solvent for hard-surf. and heavy-duty cleaners, oven cleaners; replacement for hazardous solvs.; particularly useful in slowly settling or nonsettling sol'ns. and suspensions; provides cleaning power to remove stubborn stains with reduced risk of damage to sensitive surfs.; reduces streaking effects and eliminates need for extra alkalinity; effective in removing ink stains

Properties: Clear liq.; misc. with water, and most org. solvs.; immisc. with aliphatic hydrocarbons, ethers, vegetable and min. oils; sp.gr. 1.139; visc. 53 mPas; f.p. 20 C; b.p. 295 C; flash pt. (CC) > 100 C; 98% purity

Toxicology: LD50 (oral) 14,000 mg/kg; low toxicity, nonirritant

Gaftex® HI-B† [ISP]

Chem. Descrip.: γ-Butyrolactone

CAS 96-48-0; EINECS/ELINCS 202-509-5

Uses: Solvent useful for formulations requiring nonirritating and nonsensitizing ingreds.; replacement for hazardous solvs. in many hard surf. and heavy-duty cleaners; removes ink stains without damage to sensitive surfs.; offers extra stain cleaning power; no extra alkalinity is required; streaking effects reduced; for oven and automobile cleaners

Properties: Clear liq.; misc. with water, methanol, acetone, benzene, carbon tetrachloride, ethyl ether, monochlorobenzene; insol. in aliphatic hydrocarbons; sp.gr. 1.12; visc. 1.7 mPa·s; f.p. -44.4 C; b.p. 204 C; flash pt. (CC) 93 C; pH 4.5 (10% aq. sol'n.); 98% purity

Toxicology: LD50 (oral) 1600 mg/kg; low toxicity, non-irritant

Environmental: Biodeg.

Glidsafe® Avsol 7B† [Millennium Spec. Chems.♦]

Chem. Descrip.: Terpene emulsion

Uses: Solvent cleaning agent

Properties: Flash pt. 142 F

Glidsol® 100† [Millennium Spec. Chems.♦]

Uses: Industrial degreasing solvent

Properties: Flash pt. 106 F

Glidsol® Paint Purge 5125† [Millennium Spec. Chems.♦]

Uses: Paint line purging/cleaning solvent

Properties: Flash pt. 116 F

Glycerine (Pharmaceutical)‡ [Asahi Denka Kogyo]

Chem. Descrip.: Glycerin

CAS 56-81-5; EINECS/ELINCS 200-289-5

Uses: Solvent, emollient, sweetener in cosmetics, soaps, pharmaceuticals

Features: Pharmaceutical grade

Properties: Clear liq.; sp.gr. > 1.2598; > 98% purity

Grindtek AML 60† [Danisco Cultor USA]

Chem. Descrip.: Acetylated palm kernel glycerides

CAS 91744-63-7; EINECS/ELINCS 294-627-9

Uses: Lubricant and plasticizer for plastics and coatings; cosolvent for polar additives in low polarity systems

Properties: Lt. tan liq.; sol. in ethanol, peanut oil, toluene, wh. spirit, paraffin oil; HLB 2.1.

Grindtek AMOS 90† [Danisco Cultor USA]

Chem. Descrip.: Acetylated lard glyceride

CAS 8029-92-3; EINECS/ELINCS 273-609-4

Uses: Lubricant and plasticizer for plastics and coatings; cosolvent for polar additives in low polarity systems

Properties: Yel. liq.; sol. in ethanol, peanut oil, toluene, wh. spirit, paraffin oil; HLB 1.8.

HallStar™ DO† [C.P. Hall]

Chem. Descrip.: Decyl oleate

CAS 3687-46-5; EINECS/ELINCS 222-981-6

Uses: Penetrating emollient, cosolv., lubricant, pigment dispersant for creams, lotions, moisturizers, topical pharmaceuticals, liq. and pressed powd., bath oils, and makeup removers; binder for pressed powds.; carrier for lipid-sol. active ingreds.

Properties: Gardner 1 clear liq.; mild, char. odor; sol. in peanut oil (5%), cyclomethicone, ethanol, min. oil, and IPM; insol. in glycerin and water; m.w. 422; sp.gr. 0.8622; f.p. < 0 C; acid no. 1.0 max.; sapon. no. 136.5; iodine no. 58; cloud pt. < 10 C; ref. index 1.4550

Toxicology: LD50 (oral) > 40 ml/kg; nonirritating to skin and eyes

Halso® AG 125† [Occidental]

Chem. Descrip.: Mixt. of monochlorotoluenes

CAS 25168-05-2

Uses: Solvent

Properties: APHA 25 max. clear liq.; dens. 9 lb/gal; visc. 0.747 cps (100 C); f.p. ≈ -32 C; b.p. 159-162 C; flash pt. (TCC) 123 F; surf. tens. 33 dynes/cm (20 C); 99.5% min. monochlorotoluene

HAN® 857† [ExxonMobil]

Chem. Descrip.: Aromatic

CAS 64742-06-9

Uses: Solvent

Properties: Saybolt +22 color; sp.gr. 0.903 (60/60 F); flash pt. 60 C; KB value 78

Toxicology: TLV/OEL 1000 ppm

Hexcel FO 425A† [Hexcel♦]

Chem. Descrip.: Petrol. solv.

Uses: General purpose cleaning and degreasing solvent for cleaning filled telecommunications cables and tools; noncorrosive

Properties: Clear; odorless; sp.gr. 0.753; flash pt. 200 F (COC)

Hiposcour® NFMS-2† [High Point]

Chem. Descrip.: Complex solv./surfactant blend

Uses: Solvent and detergent scour for removal of oil and grease; stable over broad pH range; can be used in sodium peroxide and sodium hypochlorite bleach baths

Properties: Amber clear liq., mild alcoholic odor; sp.gr. 0.93; dens. 7.75 lb/gal; pH 6.0-6.5; anionic

Hydrosulfite of Soda Conc.‡ [Cognis/Chems. Group]

Chem. Descrip.: Sodium hydrosulfite

CAS 7775-14-6; EINECS/ELINCS 231-890-0

Uses: Reducing agent in redox polymerization; emulsifier in mfg. of food-contact articles; food-contact paper/paperboard

Regulatory: FDA 21CFR §176.170, 178.3400; DOT regulated; SARA §311/312 reportable

Properties: Wh. to off-wh. powd.; odorless; 18% sol. in water (20 C); sp.gr. 0.88; dens. 55 lb/ft~3 (loose), 65 lb/ft~3 (packed); pH 4.2

Toxicology: Causes skin, respiratory tract, and severe eye irritation; may cause respiratory allergic reaction; ing. may cause GI irritation, nausea, vomiting; mod. toxic by ing.TSCA listed

Precaution: Flamm.; heats spontaneously when in contact with moisture; avoid contact with strong oxidizing agents and acids

Hazardous Ingredients: Sodium hydrosulfite (88%); sodium sulfate, sodium thiosulfate, sodium formate, and sodium bisulfite (12%)

Hazardous decomp. prods.: Decomposes in water to sulfur dioxide and other sulfur oxides, sodium sulfite, and sodium thiosulfite; hydrogen sulfide and mercaptans may also be present

NFTA: Health 3, Flammability 2, Reactivity 1

Storage: Store tightly closed in cool, dry place away from low flash point materials; avoid exposure to atmosphere; storage for > 90 days must meet EPA requirements

Icinol BE33† [Huntsman]

Chem. Descrip.: Monobutyl ether of triethylene glycol and tetraethylene glycol

Uses: Solvent for lacquers, enamels, water-borne systems, inks, industrial cleaners, waterless hand cleaners; aids wetting, penetration and soil removal; coupling agent

Properties: Sol. in water, common organic chems.; sp.gr. 0.985-1.00; visc. 9.2 cps; b.p. 260 C; flash pt. (OC) > 250 C; 99% act.

Icinol DPM‡ [Huntsman]

Chem. Descrip.: (2-Methoxy-methylethoxy) propanol

CAS 34590-94-8; EINECS/ELINCS 252-104-2

Uses: Solvent for use in protective coatings, inks, cleaning prods., agric. chems.; aids wetting, penetration, and soil removal; coupling solvent

Icinol EE22† [Huntsman]

Chem. Descrip.: Monoethyl ether of diethylene glycol (80% min.), triethylene glycol, and tetraethylene glycol

Uses: Solvent for protective coatings, inks, cleaning prod., and agric. chems.

Properties: Visc. 4.5 cps; flash pt. (OC) 96 C; 1.0% max.
Storage: Store and handle in stainless or mild steel; do not store in aluminum or aluminum alloys, black iron, or copper and some of its alloys

Icinol H260‡ [Huntsman]
Chem. Descrip.: Polyoxyalkylene glycol ether
Uses: Syn. lubricant, defoamer; high visc. index solvent
Properties: Liq.; water-sol.; visc. 260 SUS (100 F)

Icinol H280X‡ [Huntsman]
Chem. Descrip.: Polyoxyalkylene glycol ether
Uses: Inhibited version of Icinol H260; syn. lubricant, defoamer; high visc. index solvent

Icinol H660‡ [Huntsman]
Chem. Descrip.: Polyoxyalkylene glycol ether
Uses: Syn. lubricant, defoamer; high visc. index solvent
Properties: Liq.; water-sol.; visc. 660 SUS

Icinol H660YA‡ [Huntsman]
Chem. Descrip.: Polyoxyalkylene glycol ether
Uses: Inhibited version of Icinol H660; syn. lubricant, defoamer; high visc. index solvent

Icinol H5100‡ [Huntsman]
Chem. Descrip.: Polyoxyalkylene glycol ether
Uses: Syn. lubricant, defoamer; high visc. index solvent
Properties: Liq.; water-sol.; visc. 5100 SUS

Icinol L65‡ [Huntsman]
Chem. Descrip.: Polyoxyalkylene glycol ether
Uses: Syn. lubricant, defoamer; high visc. index solvent
Properties: Liq.; water-insol.; visc. 65 SUS (100 F)

Icinol L285‡ [Huntsman]
Chem. Descrip.: Polyoxyalkylene glycol ether
Uses: Syn. lubricant, defoamer; high visc. index solvent
Properties: Liq.; water-insol.; visc. 285 SUS

Icinol L300X‡ [Huntsman]
Chem. Descrip.: Polyoxyalkylene glycol ether
Uses: Inhibited version of Icinol L285

Icinol L385‡ [Huntsman]
Chem. Descrip.: Polyoxyalkylene glycol ether
Uses: Syn. lubricant, defoamer; high visc. index solvent
Properties: Liq.; water-insol.; visc. 385 SUS

Icinol L625‡ [Huntsman]
Chem. Descrip.: Polyoxyalkylene glycol ether
Uses: Syn. lubricant, defoamer; high visc. index solvent
Properties: Liq.; water-insol.; visc. 625 SUS

Icinol L1715‡ [Huntsman]
Chem. Descrip.: Polyoxyalkylene glycol ether
Uses: Syn. lubricant, defoamer; high visc. index solvent
Properties: Liq.; water-insol.; visc. 1715 SUS

Icinol ME33† [Huntsman]
Chem. Descrip.: Monomethyl ether of triethylene glycol and tetraethylene glycol
Uses: Solvent for protective coatings, inks, cleaning prod., and agric. chems.
Properties: Sp.gr. 1.058; visc. 7.5 cps; b.p. 258 C; flash pt. (OC) 118 C; 0.3% max.
Storage: Store and handle in stainless or mild steel; do not store in aluminum or aluminum alloys, black iron, or copper and some of its alloys

Icinol ME45† [Huntsman]
Chem. Descrip.: Monomethyl ether of tetraethylene glycol and pentaethylene glycol
Uses: Solvent for protective coatings, inks, cleaning prod., and agric. chems.
Properties: Sp.gr. 1.065-1.075; b.p. 260 C; flash pt. (OC) > 118 C; 0.3% max.
Storage: Store and handle in stainless or mild steel; do not store in aluminum or aluminum alloys, black iron, or copper and some of its alloys

Icinol PM‡ [Huntsman]
Chem. Descrip.: 1-Methoxy 2-propanol
CAS 107-98-2; EINECS/ELINCS 203-539-1
Uses: Solvent for use in protective coatings, inks, cleaning prods., agric. chems.; aids wetting, penetration, and soil removal; coupling solvent

Igepal® Cephene Lave† [Rhodia HPCII France]
Chem. Descrip.: Ethoxylated phenol
Uses: Surfactant; solvent for vinyl, phenolic, polyester, nitrocellulose, cellulose acetate, and other resins; in metal cleaners, paint strippers, and other cleaning compds. where solvs. and solv. boosters are required; vehicle for inks
Properties: Liq.; nonionic; 99% conc.

Igepal® LAVE† [Rhodia HPCII]
Chem. Descrip.: Phenol ethoxylate
Uses: Solvent for vinyl, phenolic, polyester, nitrocellulose, cellulose acetate, and other resins; ingred. for metal cleaners, paint strippers, cleaning compds.; vehicle for inks
Properties: Liq.; nonionic; 99% conc.

Imwitor® 908† [Degussa; Degussa AG]
Chem. Descrip.: Glyceryl caprylate
CAS 26402-26-6; EINECS/ELINCS 247-668-1
Uses: Coemulsifier for o/w emulsions, lipophilic materials; solvent, solubilizer for active substances, surfactant, absorp. promoter; carrier of drugs; bacteriostatic effect; cosmetics ingred.
Properties: Ylsh. oily liq., char. coconut odor, fatty sl. bitter taste; sol. in ether, n-hexane, alcohol; disp. in water; m.p. 20-25 C; HLB 3.0; acid no. 2 max.; iodine no. 1 max.; sapon. no. 250-275; nonionic; 45% min. monoglycerides

Intravon® SOL† [Sensient Colors]
Chem. Descrip.: Aromatic solv./surfactant blend
Uses: Solvent scouring detergent for syn. and natural fibers and their blends, one-bath dyeing/scouring processes; aftersoaping agent, wetting agent, penetrant; alkali stable
Properties: Off-wh. to yel. gel; self-emulsifiable in water; flash pt. (PMCC) 52 C; pH 6.0 ± 0.5; anionic/nonionic
Toxicology: Defatting on prolonged skin contact
Environmental: Biodeg.

Invert™ 1000† [Dow]
Uses: Alternative solvent for aerosol cleaners; carrier solvent for aerosol applics., general metal degreasing, paint strippers, ink and press cleaners, removers for adhesives and graffiti
Features: High solvency; mod. evap. rates; low residues; wide formulation latitude; reduced VOCs; thermodynamically stable
Properties: Lt. straw, clear liq.; sp.gr. 0.8716; visc. 6.2 cps; vapor pressure 34.9 mm Hg (20 C); f.p. 27.7 F; b.p. 183 F; flash pt. (Seta CC) 27 F; surf. tens. 20.8 dynes/cm; VOC 392 g/l; < 5% NV

Invert™ 2000† [Dow]
Uses: Solvent for aerosol cleaners; carrier solvent for aerosol applics., general metal degreasing, paint strippers, ink and press cleaners, removers for adhesives and graffiti
Features: Mod. evap. rates; low residues; wide formulation latitude; reduced VOCs; thermodynamically stable; high solvency
Properties: Sl. yel. hazy liq.; sp.gr. 0.9328; visc. 7.5 cps; vapor pressure 18 mm Hg (20 C); f.p. 27.0 F; b.p. 206 F; flash pt. (Seta CC) 123 F; surf. tens. 28.2 dynes/cm; VOC 419 g/l; < 5% NV

Invert™ 5000† [Dow]
Uses: Solvent for aerosol cleaners; carrier solvent for aerosol applics., general metal degreasing, paint strippers, ink and press cleaners, removers for adhesives and graffiti
Features: Mod. evap. rates; low residues; wide formulation latitude; reduced VOCs; thermodynamically stable; high solvency
Properties: Lt. straw clear liq.; sp.gr. 0.9133; visc. 7.5 cps; vapor pressure 18 mm Hg (20 C); f.p. 28.0 F; b.p. 208 F; flash pt. (Seta CC) 145 F; surf. tens. 25.7 dynes/cm; VOC 410 g/l; < 5% NV

Isanol®† [BASF AG]
Chem. Descrip.: 60% Iso- and 40% n-butanol
CAS 71-36-3; EINECS/ELINCS 200-751-6
Uses: Solvent for prod. of adhesives and surface coating resins.

Isoamyl Alcohol 95%‡ [CPS]
Chem. Descrip.: Isoamyl alcohol
CAS 123-51-3; EINECS/ELINCS 204-633-5
Uses: Used for flavors, fragrances, fine chems., pharmaceuticals, extraction solvent, herbicides, plasticizers, mining chems., solvent for gums, resins, lacquers, oil additives, intermediates
Regulatory: FDA 21CFR §172.515
Properties: APHA 50 color; m.w. 88.15; sp.gr. 0.811 (20/20 C); distillation range 126-132 C (760 mm); flash pt. 47 C; 95% purity, 0.5% water

Isoamyl Alcohol 99%‡ [CPS]
Chem. Descrip.: Isoamyl alcohol
CAS 123-51-3; EINECS/ELINCS 204-633-5

Uses: Used for flavors, fragrances, fine chems., pharmaceuticals, extraction solvent, herbicides, plasticizers, mining chems., solvent for gums, resins, lacquers, oil additives, intermediates
Regulatory: FDA 21CFR §172.515
Properties: APHA 20 color; m.w. 88.15; sp.gr. 0.811 (20/20 C); distillation range 128-132 C (760 mm); flash pt. 47 C; 99% purity, 0.2% water

Jefftreat® MS-100-1‡ [Huntsman]
Uses: Solvent for gas treating

Kemester® 115† [Crompton]
Chem. Descrip.: Methyl oleate
CAS 112-62-9; EINECS/ELINCS 203-992-5
Uses: Intermediate in prod. of superamides, in metalworking lubricants; solvent
Properties: Gardner 4 max.; acid no. 4.0 max.; iodine no. 90 max.; sapon. no. 185-205

Kemester® 9022† [Crompton]
Chem. Descrip.: Methyl behenate
CAS 929-77-1; EINECS/ELINCS 213-207-8
Uses: Intermediate in prod. of superamides, in metalworking lubricants; solvent
Properties: Gardner 2 max.; acid no. 20 max.; iodine no. 1 max.; sapon. no. 150-160

Kodaflex® DBP‡ [Eastman]
Chem. Descrip.: Dibutyl phthalate
CAS 84-74-2; EINECS/ELINCS 201-557-4
Uses: Primary plasticizer-solvent for nitrocellulose lacquers; plasticizer for rubbers and CAB, ethyl cellulose, PVAc, and syn. resins; solvent for oil-sol. dyes, insecticides, peroxides, and org. compds.; antifoamer and fiber lubricant in textile mfg.; cosmetics plasticizer
Properties: Pt-Co 15 ppm liq.; sol. 0.46% in water; m.w. 278; sp.gr. 1.048; dens. 8.72 lb/gal (20 C); visc. 15 cP; f.p. -35 C; b.p. 340 C (760 mm); ref. index 1.4920; flash pt. (COC) 190 C; fire pt. (COC) 202 C; dissip. factor 0.08 x 10~-2 (1 MHz); dielec. const. 5.8 (1 MHz); vol. resist. 3.0 x 10~9 ohm cm; 99% min. assay
Precaution: Combustible

Kodaflex® DEP‡ [Eastman]
Chem. Descrip.: Diethyl phthalate
CAS 84-66-2; EINECS/ELINCS 201-550-6
Uses: Plasticizer; wetting agent in grinding pigments; pigment-disp. medium in cellulose acetate sol'ns. and plastics, and solvent for natural resins and polymers; seldom used in PVC prods. due to relatively high volatility; also for cosmetics
Properties: Pt-Co 10 ppm liq.; sol. 0.12 g/l in water; m.w. 222; sp.gr. 1.120; dens. 9.32 lb/gal (20 C); visc. 9.5 cP; f.p. < -50 C; b.p. 298 C (760 mm); ref. index 1.4990; flash pt. (COC) 161 C; fire pt. (COC) 171 C; dissip. factor 0.10 x 10~-2 (1 MHz); dielec const. 6.7 (1 MHz); vol. resist. 1.45 x 10~9 ohm cm; 99% min. assay

Kodaflex® DMP‡ [Eastman]
Chem. Descrip.: Dimethyl phthalate
CAS 131-11-3; EINECS/ELINCS 205-011-6
Uses: Plasticizer with high solvent power for cellulose acetate extrusion compds.; compatible with ethyl cellulose, CAB, PS, PVAc, polyvinyl butyral, and PVC; used in NC-based printing inks; also for cosmetics and personal care prods.
Properties: Pt-Co 5 ppm liq.; sol. 0.45 g/l in water; m.w. 194; sp.gr. 1.192 (20/20 C); dens. 9.93 lb/gal (20 C); visc. 11.0 cP; f.p. -1 C; b.p. 284 C; ref. index 1.513; flash pt. (COC) 157 C; fire pt. (COC) 168 C; dissip. factor 0.13 x 10~-2 (1 mc); dielec. const. 7.5 (1 mc); vol. resist. 1.07 x 10~9 ohm cm; 99% min. assay

Kodaflex® HS-3‡ [Eastman]
Chem. Descrip.: Butyl octyl phthalate
CAS 84-78-6; EINECS/ELINCS 201-562-1
Uses: High-solvating primary plasticizer for polyvinyl homopolymer and copolymer resins, vinyl plastisols, expanded vinyl foams, and rotational molding and dip coating
Properties: Pt-Co 35 ppm liq.; sol. < 0.01 g/l in water; sp.gr. 0.988-0.996; dens. 8.26 lb/gal (20 C); visc. 42 cP; f.p. -48 C; b.p. 350 C; ref. index 1.4848; flash pt. (COC) 208 C; fire pt. (COC) 224 C

Kodaflex® HS-4‡ [Eastman]
Chem. Descrip.: 60% Dioctyl phthalate, 5% dibutyl phthalate
Uses: High-solvating primary plasticizer; used in formulating vinyl plastisols having processing chars. required to be mechanically frothed; for PVC formulations, rotational molding, slush molding, dip coating, and filament coating applic.
Properties: Pt-Co 35 ppm liq.; sol. < 0.01 g/l in water; sp.gr. 0.993; dens. 8.26 lb/gal (20 C); visc. 31.8 cP; f.p. -47 C; b.p. 337 C; ref. index 1.4798; flash pt. (COC) 182 C; fire pt. (COC) 200 C; dissip. factor 7.9 x 10~-2 (1 MHz); dielec. const. 3.4 (1 MHz); vol. resist. 2.0 x 10~14 ohm-cm

Kronitex® TOF† [Great Lakes]
Chem. Descrip.: Trioctyl phosphate
CAS 78-42-2; EINECS/ELINCS 201-116-6
Uses: Low temp. plasticizer for use in PVC, rubber, paints, coatings; highly efficient solvent

Lanpol 20† [Croda Chem. Ltd; Croda Oleochems.]
Chem. Descrip.: PEG-20 lanolin acids, dist.
CAS 68459-50-7
Uses: O/w emulsifier, solvent for bromo acid dyes in lipsticks, skin care and makeup prods., solubilizer, wetting agent, dispersant
Properties: Solid; HLB 14.1; nonionic; 97% conc.

Lanpol 520† [Croda Chem. Ltd; Croda Oleochems.]
Chem. Descrip.: POE lanolin acid
Uses: O/w emulsifier, solvent for bromo acid dyes in lipsticks, skin care and makeup, solubilizer, wetting agent
Properties: Pale yel. pasty solid; HLB 14.1; sapon. no. 45-60; pH 4-7 (3%); nonionic; 97% conc.

Lowenol 915† [Lowenstein Dyes & Cosmetics]
Chem. Descrip.: Phenylether-based
Uses: Solvent for hair color systems.

Marvanol® Solvent Scour 34† [Marlowe-Van Loan]
Chem. Descrip.: Solvs. and detergents
Uses: Solvent scour for textiles
Properties: Amber liq.; anionic; 100% act.

Methyl Di-Icinol‡ [Huntsman]
Chem. Descrip.: 2-(2-Methoxyethoxy)ethanol
CAS 111-77-3; EINECS/ELINCS 203-906-6
Uses: Solvent for use in protective coatings, as a fuel additive

Methyl Dipropasol† [Dow]
Chem. Descrip.: PPG-2 methyl ether
CAS 34590-94-8; EINECS/ELINCS 252-104-2
Uses: Solvent for acrylic coatings
Properties: Sol. in water (20 C); m.w. 148.20; sp.gr. 0.951 (20/20 C); dens. 7.91 lb/gal (20 C); b.p. 188.3 C; flash pt. (CC) 175 F

Methyl Icinol‡ [Huntsman]
Chem. Descrip.: 2-Methoxyethanol
CAS 109-86-4; EINECS/ELINCS 203-713-7
Uses: Solvent for use as chemical intermediate and fuel additive

Methyl Propasol®† [Dow]
Chem. Descrip.: Methoxypropanol
CAS 107-98-2; EINECS/ELINCS 203-539-1
Uses: Solvent for coatings (NC, butyrate, ethyl cellulose, acrylic, water-disp. and -sol.)
Properties: Sol. in water; m.w. 90.12; sp.gr. 0.923 (20/20 C); dens. 7.69 lb/gal (20 C); b.p. 121 C; flash pt. (CC) 91 F

Methyl Propasol® Acetate† [Dow]
Chem. Descrip.: Propylene glycol methyl ether acetate
CAS 108-65-6; EINECS/ELINCS 203-603-9
Uses: Solvent for coatings (NC, cellulose acetate, ethyl cellulose, butyrate, acrylic, urethane), waterborne coatings
Properties: Sol. 18.5% in water (20 C); m.w. 132.16; sp.gr. 0.969 (20/20 C); dens. 8.09 lb/gal (20 C); b.p. 145.7 C; flash pt. (CC) 114 F

Monestriol GP-60-6%† [Undesa; S. Black]
Chem. Descrip.: Glyceryl stearate SE
CAS 11099-07-3
Uses: Dispersant, solvent for act. ingreds.; vehicle for injectable substances; emulsifier for textiles; emulsifier, opacifier, thickener, dispersant for emulsions; greasing agent in shampoos, foam baths; food lubricant; perfume diluent; intermediate in textiles and tanning industry
Properties: Anionic

Monestriol GP-AE† [Undesa; S. Black]
Chem. Descrip.: Glyceryl stearate SE
CAS 11099-07-3
Uses: Dispersant, solvent for act. ingreds.; vehicle for injectable substances; emulsifier for textiles; emulsifier, opacifier, thickener, dispersant for emulsions; greasing agent in shampoos, foam baths; food lubri-

cant; perfume diluent; intermediate in textiles and tanning industry
Properties: Anionic

Monolan® PPG440, PPG1100, PPG2200† [Akzo Nobel Surf. Chem. AB]
Chem. Descrip.: PPG
Uses: Lubricant, antistat, plasticizer, cosolv. in dyestuffs, inks, resins, rubber, cosmetics; intermediates in surfactants and plastic prod.
Properties: Colorless liq.; faint odor; sol. in water (PPG440), insol. in water (PPG1100, 2200); m.w. 400, 1000, 2000 resp.; sp.gr. 1.010, 1.005, and 1.004; visc. 80, 180, and 450 cs; flash pt. > 450 F (COC); pour pt. < 0 C; nonionic; 100% act.

2429 Naphtha† [ExxonMobil]
Chem. Descrip.: Hydrocarbon solvent
CAS 64742-89-8; EINECS/ELINCS 205-563-8
Uses: Solvent
Properties: Sp.gr. 0.76 (15.6/15.6 C); dens. 6.29 lb/gal (60 F); b.p. 115-143 C; flash pt. (TCC) 10 C; surf. tens. 22.2 dynes/cm; KB value 39; 58% paraffin, 30% cycloparaffin, 12% aromatics
Toxicology: OEL 400 ppm

Naxonac™ 510† [Rütgers Org.]
Chem. Descrip.: Nonyl phenol ether phosphate
Uses: Detergent, wetting agent, emulsifier, lubricant, hydrotrope for heavy-duty and household detergents, waterless hand cleaners, solvent degreasers, emulsion polymerization, paint and wax strippers, electrolytic cleaners
Properties: Clear liq.; sol. and stable in aq. sol'ns. with high alkaline concs.; pH 2; anionic; 99% conc.

NEOBEE® 20† [Stepan]
Chem. Descrip.: Propylene glycol dicaprylate/dicaprate
CAS 68583-51-7; EINECS/ELINCS 271-516-3
Uses: Emollient oil for cosmetics and pharmaceuticals (bath oils, creams, lotions, lipsticks, makeup bases, shaving preps.), carrier/extender for flavors, fragrances, and colors in foods; vehicle for vitamins, antibiotics, and medicinals; solubilizer, cosolvent
Properties: APHA 50 max. liq.; sol. in alcohol, min. oil, acetone; sp.gr. 0.920; visc. 9 cps; acid no. 0.10 max.; sapon. no. 315-335; surf. tens. 31.0 dynes/cm.

NEOBEE® 1054† [Stepan]
Chem. Descrip.: Propylene glycol dicaprylate/dicaprate
CAS 68583-51-7; EINECS/ELINCS 271-516-3
Uses: Cosolvent for cosmetics and pharmaceuticals, carrier for flavors and colors; solubilizer for antibiotics, medicinals; diluent; emollient for creams, lotions, ointments
Regulatory: FDA 21CFR §170.30, GRAS; avail. in kosher grade
Properties: Sl. yel. liq., bland odor; sol. in alcohol, min. oil, acetone; sp.gr. 0.910-0.923; iodine no. 0.1; sapon. no. 325; hyd. no. 1; 0.05% moisture
Storage: Avoid prolonged storage above 90 F

NEOBEE® 1062† [Stepan]
Chem. Descrip.: Coconut oil-derived triglycerides
Uses: Cosolvent, carrier for fat-sol. vitamins, medicinals, and ointments in food, cosmetic and pharmaceutical industries
Regulatory: FDA 21CFR §170.30, GRAS; avail. in kosher grade
Properties: Liq.; sol. in min. oil, acetone; sp.gr. 0.925-0.945; sapon. no. 295-315.

NEOBEE® O† [Stepan]
Chem. Descrip.: Caprylic/capric triglyceride
CAS 65381-09-1; EINECS/ELINCS 265-724-3
Uses: Solubilizer, cosolvent, carrier for fat-sol. vitamins, flavors, medicinals, and ointments in food, cosmetic and pharmaceutical industries; diluent for food colors, essential oils, injectables; clouding agent for beverages; source of nutritional MCT; spreading agent, penetrant
Regulatory: FDA 21CFR §170.30, GRAS; avail. in kosher grade
Properties: Sl. yel. liq., very sl. fatty odor; sol. in min. oil, acetone; sp.gr. 0.938; visc. 30 cps; iodine no. 4.2; sapon. no. 298; hyd. no. 5; surf. tens. 32 dynes/cm; 0.03% moisture
Storage: Avoid prolonged storage above 90 F

Niox KJ-10† [Cognis Iberica]
Chem. Descrip.: Ethoxylated oleyl alcohol
CAS 9004-98-2
Uses: Emulsifier, dispersant, fragrance grade w/o emulsifier; lipophilic cosolvent; solubilizer
Properties: Liq.; oil-misc.

Niox KS Series† [Cognis Iberica]
Chem. Descrip.: PEGs
Uses: Humectant for food, cosmetic, pharmaceutical prods.; lubricant, solvent for dyestuffs and resins; solvent and ointment base for cosmetics and pharmaceuticals
Properties: Sol. in water; partly sol. in some org. solvs.

Nirez® 9007† [Arizona]
Chem. Descrip.: Ester of selected tall oil fatty acids
Uses: Specialty ink solvent for use in high solids, low VOC ink systems and in soybean oil systems to reduce visc. and adjust tack

Nirez® 9011† [Arizona]
Chem. Descrip.: Ester of selected tall oil fatty acids
Uses: Specialty ink solvent for use in high solids, low VOC ink systems and in soybean oil systems to reduce visc. and adjust tack

Nissan Dispanol SN‡ [NOF]
Uses: Auxiliary solvent for dyestuffs
Properties: Liq. (30 C); nonionic

NM-55®‡ [ANGUS]
Chem. Descrip.: Blend of nitromethane
CAS 75-52-5; EINECS/ELINCS 200-876-6
Uses: Fuel for automotive racing and model engines; solvent

Noiox KS-10, -12, -13, -14, -16† [Cognis Iberica]
Chem. Descrip.: PEG
Uses: Lubricant, solvent for dyestuffs and resins; plasticizer for casein, gelatin compositions and printing inks; used in cosmetics and pharmaceuticals; intermediate prods.
Properties: Liq., solid, paste; water-sol.; nonionic; 100% conc.

Nopalcol Series† [Cognis/Chems. Group]
Chem. Descrip.: PEG fatty esters
Uses: Dispersant, wetting agent, emulsifier, plasticizer, lubricant, binder, thickener for cosmetics, distillation defoamers, dry cleaning agents, leather and paper industries, textiles, wall tile mastics, solvents
Properties: Liq. to solid; 100% conc.

Oxitol† [Shell]
Chem. Descrip.: Ethylene glycol monoethyl ether
CAS 110-80-5; EINECS/ELINCS 203-804-1
Uses: Med. boiling solvent with toluene dilution ratio of 4.9; strong solvent action on NC and alkyd resins; also for phenolics, Epon resins, various oils and dyes; used in mfg. of surf. coatings, esp. lacquers where it improves gloss; in printing and dyeing textiles; used in leather finishing; coupling agent for soap-hydrocarbon systems; used in varnish removers and cleaning sol'ns.
Properties: Colorless liq.; mild odor; water-misc.; sp.gr. 0.9256-0.9286; 0.10% max. water

Oxsol® 73† [Makhteshim Agan Ind.]
Chem. Descrip.: Monochlorotoluene (70%) and perchloroethylene (30%) blend
Uses: Non-ozone depleting solvent for industrial solvent cleaning, aerosols, adhesives, coatings, and inks; 1,1,1-trichloroethane alternative; no flash to boiling; aggressive solvency; reduces HAPS and VOCs
Regulatory: Regulated
Properties: Odor; dens. 10.3 lb/gal; visc. 0.91 cps; vapor pressure 8 mm Hg; f.p. -33 C; b.p. 129-158 C; flash pt. none; surf. tens. 32 dynes/cm; KB value 114
Toxicology: May cause mild to severe eye irritation, sl. to severe skin irritation, respiratory tract irritation, CNS depression; contains suspected carcinogen

Oxsol® 253† [Makhteshim Agan Ind.]
Chem. Descrip.: p-Chlorobenzotrifluoride (20%), monochlorotoluene (50%), and perchloroethylene (30%) volumetric blend
Uses: Non-ozone depleting solvent for industrial solvent cleaning, aerosols, adhesives, coatings, and inks; 1,1,1-trichloroethane alternative; reduces VOCs and HAPS
Regulatory: Regulated; partially VOC exempt (58%)
Properties: Odor; dens. 10.72 lb/gal; visc. 0.88 cps; vapor pressure 9 mm Hg; f.p. < -60 C; b.p. 127-159 C; flash pt. none; surf. tens. 32 dynes/cm; KB value 109
Toxicology: May cause mild to severe eye irritation, sl. to severe skin irritation, respiratory tract irritation, CNS depression; contains suspected carcinogen

Oxsol® 325† [Makhteshim Agan Ind.]
Chem. Descrip.: p-Chlorobenzotrifluoride (30%), monochlorotoluene (20%), and perchloroethylene (50%) volumetric blend

Uses: Non-ozone depleting solvent for industrial solvent cleaning, aerosols, adhesives, coatings, and inks; 1,1,1-trichloroethane alternative; reduces VOCs and HAPS
Regulatory: Regulated; partially VOC exempt (85%)
Properties: Odor; dens. 11.8 lb/gal; visc. 0.85 cps; vapor pressure 13 mm Hg; f.p. < -50 C; b.p. 121-155 C; flash pt. none; surf. tens. 32 dynes/cm; KB value 100
Toxicology: May cause mild to severe eye irritation, sl. to severe skin irritation, respiratory tract irritation, CNS depression; contains suspected carcinogen

Palatinol® DBP† [BASF]
Chem. Descrip.: Di-n-butylphthalate
CAS 84-74-2; EINECS/ELINCS 201-557-4
Uses: Plasticizer for PVAc-based coatings and adhesives; used to bond paper and in lamination of vinyl wall coverings to gypsum board; NC lacquers and cellophane coatings; mfg. of smokeless powds. for military use, processing aid for nitrile rubbers; carrier solvent for pigments and dyes, solvent for printing ink vehicles, fixative for perfumes, modifier of phenolic laminates, component of floor waxes
Properties: APHA 35 max.; mild, char. odor; m.w. 278.35; b.p. 182 C; sol. < 0.01% in water; sp.gr. 1.045-1.051; dens. 8.7 lb/gal; visc. 20.1 cps; pour pt. -40 C; acid no. 0.07 max.; ref. index 1.4926; flash pt. (COC) 171 C

Panasol™ Solvents‡ [BP Amoco]
Uses: Solvents providing outstanding flow, control of evaporation rates, and high solvency for use in can and coil coatings

Patogen PD-3† [Yorkshire Am.]
Uses: High flash pt., non 'red label' solvent scour for removal of stubborn sizes; high caustic stability

Pegol® P-1000† [Rhodia HPCII]
Chem. Descrip.: PPG
CAS 25322-69-4; EINECS/ELINCS 200-338-0
Uses: Antifoam for industrial applics.; additive for tire lubricants; carrier and cosolvent for personal care prods.
Properties: Liq.; pour pt. < -25 C; nonionic; 100% conc.

Pegol® P-2000† [Rhodia HPCII]
Chem. Descrip.: PPG
CAS 25322-69-4; EINECS/ELINCS 200-338-0
Uses: Antifoam for industrial applics.; additive for tire lubricants; carrier and cosolvent for personal care prods.
Properties: Liq.; pour pt. < -25 C; nonionic; 100% conc.

PentoXone† [Shell]
Chem. Descrip.: 4-Methoxy-4-methyl-pentanone-2
CAS 107-70-0; EINECS/ELINCS 203-512-4
Uses: High boiling solvent used in preparation of NC, acrylic, vinyl, and cellulose acetate butyrate lacquers, alkyd and epoxy enamels, and urethane and thermosetting acrylic coatings; used in cleaning compds.
Properties: Colorless liq.; pleasant odor; sp.gr. 0.899-0.909; 0.10% max. water

Perclene TG† [Occidental]
Chem. Descrip.: Perchlorethylene
CAS 127-18-4; EINECS/ELINCS 204-825-9
Uses: Transformer grade solvent

Petrosolv®† [Witco♦]
Chem. Descrip.: Aromatic solvent
Uses: Solvent additive for carburetor and soak cleaners.

PF™† [PDM]
Chem. Descrip.: Proprietary
Uses: Solvent/degreaser for various military and governmental applics. incl. naval nuclear approvals; hydraulic, weaponry, and aircraft engine cleaning; QPL solv. remover for inspection materials penetrant; replacement for 1,1,1 trichloroethane, chlorinated solvs., flamm. min. spirits, and other potentially harmful solvs.; does not contain CFCs or HCFCs
Regulatory: SARA §113 nonreportable
Properties: Flash pt. (TCC) 144 F; 100% volatile
Precaution: Combustible
Storage: 3 yrs. shelf life stored in orig. unopened containers @ 10-90 F

Pine Ester† [Millennium Spec. Chems.♦]
Chem. Descrip.: Terpene ester
Uses: Solvent used in cleaning formulations; environmentally safe
Regulatory: DOT nonregulated
Properties: Colorless to yel. liq.; sp.gr. 0.957; dens. 7.97 lb/gal; b.p. 218 C; flash pt. (TCC) 93.3 C
Toxicology: TSCA listed

Pine Oil 50† [Millennium Spec. Chems.♦]
Chem. Descrip.: α-Terpineol, terpene alcohols, and terpene hydrocarbons
UN 1272
Uses: Used in household cleaners and solvs.
Regulatory: DOT nonregulated
Properties: Clear liq.; sp.gr. 0.908; dens. 7.58 lb/gal; b.p. 165 C; flash pt. (TCC) 60 C; ref. index 1.479-1.485
Toxicology: LD50 (oral, rat) 2.65 g/kg, (skin) > 2000 mg/kg; eye and skin irritantTSCA listed

Pluracol® E1500† [BASF/Perf. Chems.]
Chem. Descrip.: PEG-6-32
Uses: Intermediate for preparation of nonionic surfactants; binder, base, coating, stabilizer, solvent, vehicle, extender, and coupling agent for pharmaceutical, cosmetic, and toiletries; lubricant for metal applics., rubber industry; wood treatment; textile conditioning, antistat, and sizing agent, softener; dispersant in food tablets and preparations; plasticizer
Properties: Wh. waxy solid; sol. in water and org. solvs. except aliphatic hydrocarbons; dens. 10.0 lb/gal; sp.gr. 1.20; m.p. 46.0-47.5 C; flash pt. > 490 F; surf. tens. 62.8 dynes/cm (1%); pH 6.7 (5% aq.)

Polylan®† [Amerchol; Amerchol Europe]
Chem. Descrip.: Oleyl linoleate, lanolin linoleate
Uses: Emollient, dispersant, conditioner, penetrant, lubricant, cosolvent, dye solubilizer used in cosmetics and pharmaceuticals
Properties: Amber visc. oily liq.; sol. in min. and castor oil, absolute ethyl alcohol, IPA, alkylolamides, ethyl acetate, and D.C. Silicone Fluid #555; insol. in water; sp.gr. 0.890-0.915; visc. 100-120 cps; HLB 8.0; acid no. 5.0 max.; iodine no. 110-130; sapon. no. 90-105; nonionic; 100% act.

Poly-Solv® DE (High Gravity)† [Arch Chems./Specialty Polyols]
Chem. Descrip.: Diethylene glycol monoethyl ether (75%) and ethylene glycol (25%)
Uses: Solvent for brake fluids, hard-surf. cleaners, leather dyeing, paints, coatings, printing inks, textile vat dyeing and printing, adhesives, antifreeze, floor waxes/polishes, insect repellents; solubilizer for dyes; plasticizer; extraction and crystallization solvent; vinyl chloride dispersant; coupler in cosmetic preparations
Properties: APHA 10 max. liq.; mild, char. odor; f.p. -75 C; b.p. 195 C (760 mm); sol. in water; misc. with many org. solvs.; sp.gr. 1.0253; dens. 8.53 lb/gal; visc. 6.9 cP; ref. index 1.4297; flash pt. (COC) 96 C; sp. heat 2.308 joules/g/°C

Poly-Solv® DE (Low Gravity)† [Arch Chems./Specialty Polyols]
Chem. Descrip.: Diethylene glycol monoethyl ether
CAS 111-90-0; EINECS/ELINCS 203-919-7
Uses: Solvent for brake fluids, hard-surf. cleaners, leather dyeing, paints, coatings, printing inks, textile vat dyeing and printing, adhesives, antifreeze, floor waxes/polishes, insect repellents; solubilizer for dyes; plasticizer
Properties: APHA 10 max. liq.; mild, char. odor; m.w. 134.17; f.p. -76 C; b.p. 202 C (760 mm); sol. in water; misc. with many org. solvs.; sp.gr. 0.989; dens. 8.24 lb/gal; visc. 4.59 cP; ref. index 1.4273; flash pt. (TCC) 85 C; sp. heat 2.308 joules/g/°C; 100% act.

Poly-Solv® EE† [Arch Chems./Specialty Polyols]
Chem. Descrip.: Ethylene glycol monoethyl ether
CAS 110-80-5; EINECS/ELINCS 203-804-1
Uses: Solvent for brake fluids, hard-surf. cleaners, leather dyeing, paints, coatings, printing inks, textile vat dyeing and printing, adhesives, antifreeze, floor waxes/polishes, insect repellents; solubilizer for dyes; plasticizer
Properties: APHA 10 max. liq.; mild char. odor; m.w. 90.12; f.p. -70 C; b.p. 135 C (760 mm); sol. in water; misc. with many org. solvs.; sp.gr. 0.931; dens. 7.76 lb/gal; visc. 2.1 cP; ref. index 1.4076; flash pt. (TCC) 45 C; sp. heat 2.321 joules/g/°C

Polystab RB1483‡ [Akcros Chems. UK]
Chem. Descrip.: Aromatic alkoxyates blend
Uses: Solvent; reduces visc. of polyether polyols; gives improved in-mold flowout for PU mfg.
Properties: Liq.

PrepRite™ Coatings Remover† [ISP]
Chem. Descrip.: N-Methyl-2-pyrrolidone, butyrolactone, and other ingreds.
Uses: Solvent for one-step removal of paints, coatings, and residue from

plant and maintenance equip., tools, removal of adhesives and resins (polyester, acrylic, urethane, vinyl ester, epoxy)
Features: Low VOC emissions; replaces strippers contg. methylene chloride, phenol, alkali
Properties: Thick liq., sweet odor; water-sol.; sp.gr. 1.06; dens. 8.8 lb/gal, visc. 6000 cps (20 C); flash pt. (PMCC) 197 F; pH 4-5 (10 vol.%, 20 C)
Toxicology: TLV 100 ppm; low oral and inhalation toxicity; mod. dermal toxicity; mod. eye irritant; contains org. solvs.; avoid breathing vapors or mist
Precaution: Keep away from excessive heat and open flames

Primarol 1006† [Cognis Canada]
Chem. Descrip.: 2-Hexyl decanol-1
CAS 2425-77-6; EINECS/ELINCS 219-370-1
Uses: Extender and solvent for dyes and fragrance oils; intermediate for prod. of compds. for applics. in lubricants, emulsifiers, metal processing, and textiles
Properties: Water-wh. clear liq.; sol. in solvs.; flash pt. 154 C (COC); cloud pt. < -20 C; sapon. no. < 5.

Primarol 1107† [Cognis Canada]
Chem. Descrip.: C16-C20 2-alkyl alcohols
Uses: Extender and solvent for dyes and fragrance oils; intermediate for prod. of compds. for applics. in lubricants, emulsifiers, metal processing, and textiles
Properties: Water-wh. clear liq.; cloud pt. < 0 C.

Printsolve™ Ink Remover† [ISP]
Chem. Descrip.: N-Methyl-2-pyrrolidone, dipropylene glycol methyl ether
Uses: Solvent cleaner for removal of water-and solv.-based inks, UV-cured inks, and heat-set inks; used in pulp/paper, flexographic/rotogravure, silk-screen printing, and graphics arts industries; safer alternative to xylene, acetates, aromatic hydrocarbons, naphthas, MEK, and alkaline cleaners
Properties: Water-sol.
Toxicology: TLV 100 ppm; low oral and inhalation toxicity; mod. dermal toxicity; mod. eye irritant

Progasol® SN† [Rhodia]
Chem. Descrip.: Solv./surfactant blend
Uses: Solvent scouring agent, detergent
Properties: Liq.; 84% conc.

Propsolv I-2† [Equistar]
Chem. Descrip.: 100 gal SDA 1-2, 1.0 gal rubber hydrocarbon solv., 4.25 gal ethyl acetate
Uses: Solvent for shellac, chemical specialties, latex coagulants

Propyl Dipropasol† [Dow]
Uses: Solvent, coalescing aid for water-disp. and water-sol. coatings
Properties: Sol. 17% in water (20 C); m.w. 176.26; sp.gr. 0.930 (20/20 C); dens. 7.74 lb/gal (20 C); b.p. 211.3 C; flash pt. (CC) 190 F

Propyl Propasol† [Dow]
Uses: Solvent for water-disp. and water-sol. coatings
Properties: Sol. in water (20 C); m.w. 116.18; sp.gr. 0.886 (20/20 C); dens. 7.36 lb/gal (20 C); b.p. 149.8 C; flash pt. (CC) 119 F

Prox-onic MG-020 P‡ [Synthron SA]
Chem. Descrip.: POP (20) methyl glucoside
CAS 61849-72-7
Uses: Cosolvent, perfume fixative
Properties: Liq.

Punctilious® SDA 4‡ [Equistar]
Chem. Descrip.: Specially denatured alcohol 4
CAS 64-17-5; EINECS/ELINCS 200-578-6
Uses: Solvent for hair spray, perfume, surfactants, detergents, pharmaceuticals, agric., cleaning sol'ns., disinfectants, explosives, plastics, polishes, shellac, stains, textiles, adhesives

Punctilious® SDA 25-1† [Equistar]
Chem. Descrip.: Specially denatured ethanol
CAS 64-17-5; EINECS/ELINCS 200-578-6
Uses: Solvent for hair spray, perfume, surfactants, detergents, pharmaceuticals, agric., cleaning sol'ns., disinfectants, explosives, plastics, polishes, shellac, stains, textiles, adhesives

Punctilious® SDA 25-2† [Equistar]
Chem. Descrip.: Specially denatured ethanol
CAS 64-17-5; EINECS/ELINCS 200-578-6
Uses: Solvent for hair spray, perfume, surfactants, detergents, pharmaceuticals, agric., cleaning sol'ns., disinfectants, explosives, plastics, polishes, shellac, stains, textiles, adhesives

Rexoclean®† [Emkay♦]
Chem. Descrip.: Syn. detergents, terpenes, and solvs.
Uses: Scouring agent for stubborn oil and grease stains and for removal of nylon warp size
Properties: 85% act.

Rexoclean® 200N† [Emkay♦]
Chem. Descrip.: Emulsified blend of solvs. and detergents
Uses: Multipurpose cleaner for heavy duty cleaning incl. rubber marks and stains, in soap dispensers

Rexoclean® GA† [Emkay♦]
Uses: Solvent scour, detergent for low or high pH applics, bleaching, scouring, and cleansing obstinate fatty soils

Rexoclean® JA† [Emkay♦]
Chem. Descrip.: Emulsified solvs./detergent blend
Uses: Scouring agent for oil and grease soils and fatty sizing
Properties: Liq.; anionic; 35% conc.

Rexoclean® NB Conc.† [Emkay♦]
Chem. Descrip.: Mixt. of solvs., emulsifiers, and detergents
Uses: Can be cut 50-50 with water to make an efficient blanket cleaner

Rexosolve®† [Emkay♦]
Chem. Descrip.: Blend of petrol. solvs. and terpenes
Uses: Cleaner for oil or grease soils, dirty knots, etc. in textile industry
Properties: Sol. in water

Rexosolve® BCT† [Emkay♦]
Uses: Dewaxing agent; solvent scour; used with caustic for cleaning rubber print blankets and to remove metallic pigment dispersions; leveling agent for overdyed Dacron; stripper on many colors

Rexosolve® CR† [Emkay♦]
Chem. Descrip.: Emulsified blend of solvs. and cresylic acid
Uses: Wool scouring agent for removing min. oil

Rexosolve® EP, EPN† [Emkay♦]
Chem. Descrip.: Blend of aromatic solvs.
Uses: Cleaners for oily and greasy soils; recommended for Burlington machines
Properties: Misc. with hot or cold water to produce highly stable emulsions

Rexosolve® GA, GAX† [Emkay♦]
Uses: Solvent scour for bleach, scour and dye baths to remove fatty soils
Features: Stable in acid and alkaline baths

Rexosolve® HP† [Emkay♦]
Chem. Descrip.: Conc. cleaner containing high flash solvs.
Uses: Cleaner for removing stubborn tars and greases

Rexosolve® MS† [Emkay♦]
Uses: Solvent scour used as general cleaner and washing compd. for back grays; softens and removes dried pigment; usually used with 2.5% caustic at 140 F
Properties: 100% act.

Rexosolve® OA, OAC† [Emkay♦]
Uses: Solvent scours for cleaning stains, rust and grease in one bath; compat. with oxalic acid

Rexosolve® RF-2† [Emkay♦]
Chem. Descrip.: Blend of org. solvs. with syn. emulsifiers and detergents
Uses: Cleaner for removal of stubborn tars, greases and stains
Properties: Lt. free-flowing liq.; disp. readily in water to give stable emulsion

Rexosolve® SL† [Emkay♦]
Uses: Stain and soil remover for use in neutral or alkaline baths; may be used in the dye liquor

Rhodasurf® PEG 300† [Rhodia HPCII]
Chem. Descrip.: PEG-6
CAS 25322-68-3; EINECS/ELINCS 220-045-1
Uses: Intermediate for surfactants; binder, lubricant in pharmaceuticals; plasticizer; paper softener; humectant; solvent; antistat; for cosmetics, textiles, plastics processing, dyes and inks
Properties: Liq.; m.w. 285-315; water-sol.; dens. 1.13 g/ml; pH 5-8 (5% DW); 100% conc.

Rhodasurf® PEG 400† [Rhodia HPCII]
Chem. Descrip.: PEG-9
CAS 25322-68-3; EINECS/ELINCS 222-206-1
Uses: Intermediate, plasticizer, solvent, lubricant, coupling agent in cosmetic lotions

† Discontinued ‡ Unverified ♦ Defunct Manufacturer or Division

Properties: Wh. liq.; water-sol.; m.w. 380-420; sp.gr. 1.13; f.p. 4-8 C; pH 5-8; 100% conc.

Rhodasurf® PEG 4000† [Rhodia HPCII]
Chem. Descrip.: PEG 4000
Uses: Solvent or cosolvent for industrial applics. where lubricity, water sol., low toxicity, and chemical stability are important
Properties: Solid; 100% conc.

Ritalafa® 5† [R.I.T.A.]
Chem. Descrip.: PEG-5 lanolate
CAS 68459-50-7
Uses: Emulsifier, solvent

Ritalafa® 10† [R.I.T.A.]
Chem. Descrip.: PEG-10 lanolate
CAS 68459-50-7
Uses: Emulsifier, solvent

Rotichrom‡ [Atomergic Chemetals]
Uses: Solvents, standards and equip. for chromatography

Saf-T-Sol™‡ [Crowley Chem.]
Chem. Descrip.: Aromatic solvent
Uses: Solvent for paints and varnishes

Schercemol DICA† [Scher]
Chem. Descrip.: Diisocetyl adipate
CAS 57533-90-1; EINECS/ELINCS 260-799-9
Uses: Nonoily emollient and spreading agent for cosmetics; release and anticlogging additive in aerosols due to strong plasticizing action; film spreader for bath oils; solvent for lipstick colors
Properties: Gardner 2.0 max. clear liq.; slight odor; m.w. 594; f.p. < -10.0 C; sol. in hydrophobic solvs.; sp.gr. 0.885 ± 0.005; dens. 7.4 lb/gal; acid no. 2.0 max.; sapon. no. 170-185; ref. index 1.4538 ± 0.001; flash pt. (OC) > 180 C

Schercemol IPM† [Scher]
Chem. Descrip.: IPM
CAS 110-27-0; EINECS/ELINCS 203-751-4
Uses: Emollient and film spreader used in personal care prods.; solvent and solubilizer for additives in cosmetic preparations; nonoily
Properties: Water wh. clear liq.; odorless; m.w. 270; f.p. -4.0 C; sol. in org. solvs., insol. in water; sp.gr. 0.850; acid no. 1.0 max.; sapon. no. 202-212; ref. index 1.4330-1.4350; flash pt. (OC) > 170 C; 99% min. ester conc.

Schercophos NP-9† [Scher]
Chem. Descrip.: Nonoxynol-9 phosphate
CAS 51609-41-7; EINECS/ELINCS 266-231-6
Uses: Low to moderate foam liq. detergent and emulsifier for heavy-duty all-purpose liq. detergents; solvent cleaner for dry-cleaning
Properties: Liq.; anionic; 100% conc.

Scour KSV, KSV Special‡ [CHT R. Beitlich Corp.]
Chem. Descrip.: Emulsified solvent containing stain remover and wetting and scouring agent

SECOSTER® MA 300‡ [Stepan]
Chem. Descrip.: PEG 300 abietate
Uses: Additive for cutting oils; solvent, emulsifier for solvs. and oils, creams, cleansing milks and pesticides; dispersant; for household/industrial cleaners
Properties: Brn. liq.; nonionic; 100% act.

Sepasolv® MPE† [BASF AG]
Uses: For purification of natural gas.

Sesolvan® L† [BASF AG]
Uses: Dye solvent for textile dyeing

Shell TS-28† [Shell]
Chem. Descrip.: Aromatic hydrocarbon
Uses: Solvent for dry-cleaning, coatings, automotive, and chemical specialties
Properties: Sp.gr. 0.852 (60 F); dens. 7.09 lb/gal (60 F); flash pt. (CC) 122 F

Shell Cyclo Sol 42† [Shell]
Chem. Descrip.: Aromatic
Uses: Solvent
Properties: Sp.gr. 0.8532 (60 F); dens. 7.103 lb/gal (60 F); b.p. 321-373 F; flash pt. (TCC) 113 F; KB value 76

Shell Cyclo Sol 63† [Shell]
Chem. Descrip.: C^9 -C^11 aromatic hydrocarbon solv.
Uses: Solvent for dry cleaning, coatings, automotive and chemical specialties, insecticides, wood preservatives, industrial cleaning, inks, petrol. drilling, catalytic and reaction medium, textile printing, asphalt prods.
Properties: Sp.gr. 0.887 (60 F); dens. 7.39 lb/gal (60 F); flash pt. (CC) 128 F; KB value 89; VOC 7.35 lb/gal; 97.5% aromatics

Shellsol® 71 (H), 71 (WR)† [Shell]
Uses: Solvent for dry-cleaning, coatings, automotive, and chemical specialties
Properties: Sp.gr. 0.759 (60 F); dens. 6.32 lb/gal (60 F); flash pt. (CC) 125 F

Shellsol® 72† [Shell]
Uses: Solvent for dry-cleaning, coatings, automotive, and chemical specialties
Properties: Sp.gr. 0.768 (60 F); dens. 6.4 lb/gal (60 F); flash pt. (CC) 137 F

Shellsol® B† [Shell]
Chem. Descrip.: Aliphatic naphtha solvent
CAS 8030-30-6
Uses: Solvent for dry-cleaning, coatings, automotive, and chemical specialties
Features: Low boiling, fast evaporating
Properties: Sp.gr. 0.670 (60 F); dens. 5.58 lb/gal (60 F); flash pt. (CC) < 0 F

Shell Super VM&P Naphtha EC† [Shell]
Chem. Descrip.: Aliphatic naphtha solv.
CAS 8030-30-6
Uses: Solvent for dry-cleaning, coatings, automotive, and chemical specialties; low boiling, fast evaporating
Properties: Sp.gr. 0.754 (60 F); dens. 6.27 lb/gal (60 F); flash pt. (CC) 55 F

Shell Tolu-Sol® 6-EC† [Shell]
Uses: Solvent used in dry-cleaning, coatings, automotive, and chemical specialties; low boiling, fast evaporating
Properties: Sp.gr. 0.752 (60 F); dens. 6.26 lb/gal (60 F); flash pt. (CC) 21 F

Shell Tolu-Sol® 19-EC† [Shell]
Uses: Solvent used in dry-cleaning, coatings, automotive, and chemical specialties; low boiling, fast evaporating
Properties: Sp.gr. 0.730 (60 F); dens. 6.08 lb/gal (60 F); flash pt. (CC) 24 F

Skellite®† [Texaco♦]
Uses: Solvent used as stove and lamp fuel
Properties: Flash pt. -14 F

Solvenon® BB‡ [BASF AG]
Uses: Difficultly volatile solvent for surface coatings and paints; additive in cleaning and pickling agents.

Solvenon® DIP‡ [BASF AG]
Chem. Descrip.: Mixt. of isomers of dipropylene glycol monoisopropyl ethers
Uses: Solvent for resins and dyes, surface coatings industry; leveling agent for water-dilutable baking finishes; film-binding assistant for dispersions; component in cleaning agents and printing inks

Solvenon® I† [BASF AG]
Chem. Descrip.: 85% Isobutyl formate, 15% isobutanol
Uses: Solvent and diluent for adhesives, paints, varnishes

Solvenon® IPP‡ [BASF AG]
Chem. Descrip.: 1-Isopropoxy-2-propanol, 2-isopropoxy-1-propanol
Uses: Solvent for resins and dyes, surface coatings, cleaning agent for printing plates

Solvent APV Spec.† [Degussa AG/Marl]
Chem. Descrip.: Ethoxydiglycol
CAS 111-90-0; EINECS/ELINCS 203-919-7
Uses: High boiling solvent for fragrance materials, essential oils, terpene oils for cosmetics use
Properties: Colorless liq.; dens. 0.986-0.990 kg/l; b.p. 198-203 C; acid no. ≤ 0.05; hyd. no. 420-490

Solvent GC† [BASF AG]
Chem. Descrip.: Mixt. of ethylene glycol acetates
Uses: Solvent for printing inks and rubber stamp inks.

Solvent PM‡ [BASF AG]
Chem. Descrip.: 1- Methoxypropan-2-ol
CAS 107-98-2; EINECS/ELINCS 203-539-1
Uses: Solvent for cosmetic industry

Solvent Scour 25/27† [Huntsman]
Chem. Descrip.: Alkanolamides and alkyl phenol ethoxylate
Uses: Scouring agent and prescour for greasy and oil soiled fabrics
Properties: Yel. liq.; dens. 0.899; nonionic; 100% act.

Soygold® 3000† [Ag Environmental Prods.]
Chem. Descrip.: Methyl soyate with 1-5% diethanolamine and 5-10% poly (oxy-1,2-ethanediyl) (CAS 24939-91-8)
Uses: Blanket and roller wash
Features: Environmentally friendly, low VOC; contains < 10% petrol. distillate; no HAPs; high flash; EPA approved replacement (SNAP)
Regulatory: DOT regulated (Class 65)
Properties: Amber clear microemulsion; sol. in water; sp.gr. 0.9371 (60 F); vapor pressure 7.8 mm Hg; m.p. 19.4 F; b.p. 209 F; flash pt. (TCC) > 90 C; pH 4; nonionic
Toxicology: May cause skin, eye, and respiratory irritation; ing. may cause gastric disturbancesTSCA listed
Precaution: Avoid strong oxidizing and reducing agents
Hazardous Ingredients: Poly (oxy-1,2 ethanediyl) (5-10%); diethanolamine (1-5%)
Hazardous decomp. prods.: CO^2, CO, ammonia, amines, other nitrogen compds.
HMIS: Health 0, Flammability 1, Reactivity 0
Storage: Store in cool, dry, well ventilated location in closed containers

SOYsolv™† [PDM]
Chem. Descrip.: Proprietary
Uses: 100% biodeg./recyclable industrial solvent
Regulatory: SARA nonreportable
Properties: Blue/grn. to clear liq.; negligible sol. in water; sp.gr. 0.87; vapor pressure < 1 mm Hg; b.p. > 400 F; flash pt. (PMCC) > 300 F
Toxicology: Not primary skin irritant
Environmental: Biodeg.
Precaution: Avoid contact with string oxidizing agents
Hazardous decomp. prods.: Decomp. may produce CO and CO^2
HMIS: Health 1, Flammability 0, Reactivity 0

Standamul® G† [Cognis/Care Chems.]
Chem. Descrip.: Octyl dodecanol
CAS 5333-42-6; EINECS/ELINCS 226-242-9
Uses: Carrier for oil-sol. act. ingred.; dissolves dyes; pigment dispersant; coupler for waxes and other materials in personal care prods.
Properties: Water-wh. clear liq.; odorless; cloud pt. -50 C; acid no. 0.5 max.; sapon. no. 5.0 max.; 100% conc.

Standamul® G-16† [Cognis/Care Chems.]
Chem. Descrip.: Isocetyl alcohol
CAS 36311-34-9; EINECS/ELINCS 252-964-9
Uses: Emollient; extender and solvent for pigments and personal care prods.; binder in cosmetics
Properties: Water-wh. clear liq.; visc. 100 cps; cloud pt. -50 C; acid no. 0.5 max.; sapon. no. 5.0 max.; 100% conc.

Stearate 400 WL 817† [Gattefosse SA]
Chem. Descrip.: PEG-8 palmitostearate
Uses: Solvent and emulsifier for cosmetic/pharmaceutical emulsions
Properties: Drop pt. 29-34 C; HLB 11-12; acid no. < 6; iodine no. < 3; sapon. no. 70-95

Stearate 6000 WL 1644† [Gattefosse SA]
Chem. Descrip.: PEG-150 palmitostearate
Uses: Solvent and emulsifier for cosmetic/pharmaceutical emulsions
Properties: Drop pt. 55-60 C; HLB 18; acid no. ≤ 2; iodine no. < 3; sapon. no. 15-25
Toxicology: Nonirritating to skin; sl. irritating to eyes

Striptron Stripper† [Dow]
Chem. Descrip.: Inhibited methylene chloride with additives
CAS 75-09-2; EINECS/ELINCS 200-838-9
Uses: Solvent stripper for dry film photoresist and screen inks
Properties: Sol. in most org. solvs.; low water sol.

Sure Sol®-156† [Koch Chem. Spec.]
Chem. Descrip.: m/p-Diisopropylbenzene
Uses: Solvent for textiles

Sure Sol®-190† [Koch Chem. Spec.]
Chem. Descrip.: Mixed methylnaphthalenes
Uses: Solvent for textiles, agric.
Properties: Flash pt. 85 C min.

Sure Sol®-200† [Koch Chem. Spec.]
Chem. Descrip.: 1,3,5-Triisopropylbenzene
CAS 717-74-8; EINECS/ELINCS 211-941-3
Uses: Solvent for agric. industry
Properties: Flash pt. 85 C min.

Sure Sol®-450† [Koch Chem. Spec.]
Chem. Descrip.: Triisopropylbenzene
CAS 717-74-8; EINECS/ELINCS 211-941-3
Uses: Solvent for carbonless paper and elec. industries

Sylfat® DX† [Arizona]
Chem. Descrip.: Tall oil fatty acid ester
Uses: For plasticizers, extenders, surfactants in grinding and cutting oils, specialty lubricant additives, corrosion inhibitors, specialty solvs. for printing inks, metalworking, and oil well servicing
Properties: Gardner 5+ color; sp.gr. 0.94; visc. 700 cps; acid no. 12; flash pt. (OC) 190 C

T4250† [Degussa]
Chem. Descrip.: Phenyltris(trimethylsiloxy) silane
CAS 2116-84-9; EINECS/ELINCS 218-320-6
Uses: Lubricant
Features: Low visc. silicone fluid with high control over volatility, visc., dens. and ref. index, and unique solvent props.

Tabs† [Bush Boake Allen UK♦]
Chem. Descrip.: Paramenthadiene
Uses: Solvent
Environmental: Biodeg.

Tegosoft® 189† [Goldschmidt]
Chem. Descrip.: Isostearyl isononanoate
Uses: Emollient, superfatting agent, solvent for lipophilic components
Properties: Liq.

Texacar® EC† [Huntsman]
Chem. Descrip.: Ethylene carbonate
CAS 96-49-1; EINECS/ELINCS 202-510-0
Uses: Solvent

Texacar® PC‡ [Huntsman]
Chem. Descrip.: Propylene carbonate
CAS 108-32-7; EINECS/ELINCS 203-572-1
Uses: Solvent

Texsolve B† [Texaco♦]
Chem. Descrip.: Highly refined commercial hexane
CAS 110-54-3; EINECS/ELINCS 203-777-6
Uses: Solvent used for veg. oil and pharmaceutical extraction, compounding rubber cements and sealants, and polyolefin prod.
Properties: Liq.; sp.gr. 0.679 (60 F); dens. 5.65 lb/gal (60 F); i.b.p. 152 F; flash pt. -25 F; 56% act.
Toxicology: TLV 500 ppm (OSHA), 50 ppm (ACGIH)
Precaution: Flamm.

Texsolve C† [Texaco♦]
Chem. Descrip.: Mixed heptane fraction
CAS 142-82-5; EINECS/ELINCS 205-563-8
Uses: Solvent used for compounding rubber cements and sealants, extraction of oils and fats, and recrystallization of org. chemicals
Properties: Liq.; sp.gr. 0.700 (60 F); dens. 5.85 lb/gal (60 F); i.b.p. 185 F; flash pt. (TCC) 13 F
Toxicology: TLV 500 ppm (OSHA), 400 ppm (ACGIH)
Precaution: Flamm.

Texsolve E† [Texaco♦]
Chem. Descrip.: Heptane, hexane, octane, and toluene
Uses: Solvent used in rubber tire mfg., rubber cements, sealants, inks, lacquers, and adhesives, extraction of oils and fats, recrystallization of org. chems.
Properties: Liq.; sp.gr. 0.694 (60/60); i.b.p. 190 F min.; flash pt. -20 F.
Precaution: Flamm.

Texsolve H† [Texaco♦]
Chem. Descrip.: Hexane-heptane fraction
Uses: Solvent used in rubber tire mfg., rubber cements, sealants, inks, lacquers, and adhesives; offers fast evaporation and extremely low evaporation residue
Properties: Liq.; i.b.p. 158 F; flash pt. -20 F
Precaution: Flamm.

Texsolve S† [Texaco♦]
Chem. Descrip.: Mineral spirits
CAS 64475-85-0; EINECS/ELINCS 232-453-7

Uses: Solvent used in paint and protective coatings, dry-cleaning, degreasing, wood treating, and charcoal lighter fluid; meets or exceeds standards for min. spirits and Stod.
Properties: Flash pt. 107 F.

Texsolve S-66† [Texaco♦]
Chem. Descrip.: Exempt mineral spirits with max. aromatic content
CAS 64475-85-0; EINECS/ELINCS 232-453-7
Uses: Solvent used in paint and protective coatings, dry-cleaning, degreasing, wood treating, charcoal lighter fluid; meets or exceeds standards for min. spirits and Stod.
Properties: Flash pt. 107 F.

Texsolve S-LO† [Texaco♦]
Chem. Descrip.: Low aromatic mineral spirits
CAS 64475-85-0; EINECS/ELINCS 232-453-7
Uses: Solvent used in charcoal lighter fluid, floor finishes, paints, coatings, hand cleaners
Properties: Low odor; flash pt. 108 F.

Texsolve V† [Texaco♦]
Chem. Descrip.: VM&P naphtha
CAS 8032-32-4; EINECS/ELINCS 232-453-7
Uses: Rule 66 solvent used in paints, coatings, rubber compounding, sealants, and chemical absorption
Properties: Liq., pleasant odor; sp.gr. 0.751 (60 F); dens. 6.25 lb/gal (60 F); i.b.p. 256 F; flash pt. 58 F
Toxicology: TLV 300 ppm (ACGIH)
Precaution: Flamm.

TF Solvent† [Chemtronics]
Uses: Mild cleaning agent for removal of flux residue, oil, grease and contaminants from magnetic tape heads, thermal print heads, optical equip., contacts, relays, and PCB assemblies

TP-35 Solvent† [Chemtronics]
Uses: Precision cleaner/defluxer; removes residual moisture from electronic equip.; for critical cleaning of PCBs, components, magnetic tape heads, optical lenses, precision instruments; safe for most plastics

Trylon® 6702† [Cognis/Chems. Group; Cognis/Textiles]
Uses: Solv. emulsifier, detergent for metal degreasers, engine block cleaners, solv. scouring, textile dye applics.
Properties: Gardner 7 liq.; sol. in butyl stearate, glycerol trioleate, Stod., xylene; disp. in water; dens. 8.5 lb/gal; visc. 1250 cSt (100 F); flash pt. 530 F; pour pt. -10 C.; anionic

Type CG6-M‡ [Cameron Carbon]
Chem. Descrip.: Activated carbon, coal base
CAS 7440-44-0; EINECS/ELINCS 231-153-3
Uses: For wastewater treatment; adsorbs color bodies and high m.w. organics; solv. purification
Properties: Granules; sp.gr. 0.49; surf. area 1100 B.E.T./N^2; iodine no. 1050; hardness 80; 2% moisture

Ucar® Butyl Acetate‡ [Dow]
Chem. Descrip.: Butyl acetate
CAS 123-86-4; EINECS/ELINCS 204-658-1
Uses: Solvent for higher solids coatings
Properties: M.w. 116.16; dens. 7.34 lb/gal (20 C); visc. 0.7 cps (20 C); vapor pressure 15 mm Hg (20 C); b.p. 126.2 C; flash pt. (CC) 84 F

Ucar® Ethyl Acetate† [Dow]
Chem. Descrip.: Ethyl acetate
CAS 141-78-6; EINECS/ELINCS 205-500-4
Uses: Solvent for higher solids coatings
Properties: M.w. 88.11; dens. 7.51 lb/gal (20 C); visc. 0.5 cps (20 C); vapor pressure 76 mm Hg (20 C); b.p. 77.2 C; flash pt. (CC) 30 F

Ucar® Methyl Ethyl Ketone† [Dow]
Chem. Descrip.: Methylethylketone
CAS 78-93-3; EINECS/ELINCS 201-159-0
Uses: Solvent for higher solids coatings
Properties: M.w. 72.11; dens. 6.71 lb/gal (20 C); visc. 0.4 cps (20 C); vapor pressure 71 mm Hg (20 C); b.p. 79.6 C; flash pt. (CC) 24 F

Ukatronic‡ [Atofina SA]
Chem. Descrip.: Methylene chloride
CAS 75-09-2; EINECS/ELINCS 200-838-9
Uses: Solvent

Ureresolve‡ [Dynaloy]
Chem. Descrip.: Glycol-based solvent
Uses: Solvent used hot (175-205 C) for removing heavily encrusted urethanes; cleans residues from mixing and metering equip., nozzles, spray guns, etc.

Uresolve Plus‡ [Dynaloy]
Chem. Descrip.: Ethylene glycol ether base
Uses: Solvent for electronic depotting, decapsulation, and deinking applics.; surf. prep.; suitable for cured epoxies, urethanes, and silicones; stronger than Uresolve HF, but less selective; slow at R.T.; more effective @ 65-110 C

Urinsteinlöser (Urine Scale Solvent)‡ [Schilling-Chemie]
Uses: Removes urine and water scale from lavatories, urinals, outlets, and pipelines; does not corrode iron, steel, copper or copper alloys; disinfects fitted appliances; removes blockages and stench traps from pipelines; prevents formation of unpleasant odors
Properties: Pale yel. low-visc. liq.; sp.gr. 1.05 ± 0.02 g/cm~3; pH 0.0
Use Level: 3-5 kg (unsealed pipelines); 0.5-1 kg (toilet bowls, siphon traps)
Precaution: Do not use with other cleaners

Vacutec®‡ [Atomergic Chemetals]
Uses: Vacuum evaporation mateorials

Vertrel® XMS† [DuPont Fluoroprods.]
Chem. Descrip.: Mixt. of Vertel® XF (2,3-dihydrodecafluoropentane), trans-1,2-dichloroethylene, cyclopentane, and methanol
Uses: Solvent for use in vapor degreasing equip. for cleaning ionic soils and flux residues from electronic asssemblies; defluxing solvent compat. with polymeric materials (conformal coatings, IC modules, other components of printing wiring board)
Features: Zero ozone depletion potential; low global warming potential; replaces CFC-113, 1,1,1-trichloroethane, HCFCs, and PFCs
Regulatory: EPA approved under SNAP; DOT nonregulated
Properties: Sp.gr. 1.22; visc. 0.65 cps; vapor pressure 465 mm Hg; f.p. < -50 C; b.p. 36 C; flash pt. none; surf. tens. 15.9 dynes/cm
Toxicology: Very low toxicityTSCA listed
Precaution: Contact with highly basic process materials of pH 10 or above not rec.; prod. has vapor flamm. limits in air
Storage: Store in clean, dry area

Witcamide® CMEA† [Witco♦]
Chem. Descrip.: Acetamide MEA
CAS 142-26-7; EINECS/ELINCS 205-530-8
Uses: Solvent, humectant for cosmetics
Properties: Liq.; nonionic; 70-100% conc.

Witcohol™ 320‡ [Goldschmidt]
Chem. Descrip.: Oleyl, cetyl and stearyl alcohols; straight-chain, monohydric, mono-unsat.
Uses: For use in antifoam agents, coupling agents, plasticizers, lubricants, emulsion stabilizers, petrol. additives, chem. intermediates; emulsifier, emollient
Features: Commercial grade
Properties: APHA 100 max. liq.; sol. in warm ethanol, IPA, trichloroethylene, benzene, ethyl ether, acetone, turpentine, VM&P naphtha, kerosene, lt. min. oil; acid no. 1.0 max.; iodine no. 75-85; sapon. no. 3.0 max.; hyd. no. 205-225; cloud pt. 66 F; nonionic

Witcohol™ 330‡ [Goldschmidt]
Chem. Descrip.: Oleyl, cetyl and stearyl alcohols; straight-chain, monohydric, mono-unsat.
Uses: For use in antifoam agents, coupling agents, plasticizers, lubricants, emulsion stabilizers, petrol. additives, chem. intermediates; emulsifier, emollient
Features: Commercial grade
Properties: APHA 100 max. liq.; sol. in warm ethanol, IPA, trichloroethylene, benzene, ethyl ether, acetone, turpentine, VM&P naphtha, kerosene, lt. min. oil; acid no. 1.0 max.; iodine no. 72-82; sapon. no. 3.0 max.; hyd. no. 205-225; cloud pt. 72 F; nonionic

Witcohol™ 340‡ [Goldschmidt]
Chem. Descrip.: Oleyl, cetyl and stearyl alcohols; straight-chain, monohydric, mono-unsat.
Uses: For use in antifoam agents, coupling agents, plasticizers, lubricants, emulsion stabilizers, petrol. additives, chem. intermediates; emulsifier, emollient
Features: Commercial grade
Properties: APHA 100 max. liq.; sol. in warm ethanol, IPA, trichloroethylene, benzene, ethyl ether, acetone, turpentine, VM&P naphtha, kerosene, lt. min. oil; acid no. 1.0 max.; iodine no. 60-70; sapon. no. 3.0

max.; hyd. no. 205-225; cloud pt. 82 F; nonionic

Witcohol™ 350‡ [Goldschmidt]
Chem. Descrip.: Oleyl alcohol
CAS 143-28-2; EINECS/ELINCS 205-597-3
Uses: Chem. intermediate for antifoam agents, coupling agents, plasticizers, lubricants, emulsion stabilizers, petrol. additives
Properties: APHA 100 max. liq.; acid no. 0.2 max.; iodine no. 70-75; sapon. no. 1.0 max.; hyd. no. 210-220; cloud pt. 18-22 C; nonionic

Witconol™ APEM‡ [Goldschmidt]
Chem. Descrip.: PPG-3-myreth-3
CAS 37311-04-9
Uses: Lubricant, emulsifier, wetting agent, penetrant, dye dispersant, scouring aid, antistat, solvent coupling agent; for personal care prods.; metal processing; textile surfactant
Properties: Liq.; sol. in 3A ethanol, min. oil.; nonionic

Part III
Chemical Component Cross-Reference

A

AA. *See* Allyl alcohol
2-AB. *See* s-Butylamine
Abietic acid methyl ester. *See* Methyl abietate
ABS. *See* Dodecylbenzenesulfonic acid
Absolute alcohol; Absolute ethanol. *See* Alcohol
ACAC. *See* Acetylacetone
Aceite de Algodon. *See* Cottonseed (Gossypium) oil
Aceite de ricino. *See* Castor (Ricinus communis) oil

Acenaphthene

CAS 83-32-9; EINECS/ELINCS 201-469-6
Synonyms: Acenaphthylene, 1,2-dihydro-; 5-Aminoacenaphthene; 1,8-Ethylene naphthalene; Naphthalene ethylene; Naphthylene ethylene; Periethylene naphthalene
Classification: Aromatic hydrocarbon
Empirical: $C_{12}H_{10}$
Properties: Cryst. solid; sol. in hot alcohol, chlorinated and aromatic solvs.; insol. in water; m.w. 154.22; dens. 1.024; f.p. 93.6 C; b.p. 277.5 C; flash pt. (CC) 125 C; ref. index 1.6048
Toxicology: LD50 (IP, rat) 600 mg/kg, (IV, mouse) 56 mg/kg; mod. toxic by IP route; irritating to eyes, skin, respiratory system; mutagen; TSCA listed
Precaution: Combustible
Hazardous Decomp. Prods.: Heated to decomp., emits toxic fumes of NO_x, acrid smoke, irritating vapors
Uses: Dye intermediate; pharmaceutical intermediate; insecticide; fungicide; plastics
Manuf./Distrib.: ABCR UK; Acros Org.; Advanced Synthesis Tech.; Aldrich; Alfa Aesar; CBC Europe; ChemPur; Collinda Ltd; Cyclo Int'l.; Fisher Scientific UK; Fluka; Lancaster Synthesis UK; Rochem Int'l.; Sigma; Supelco
Trade Names Containing: Methylnaphthalene Fraction

Acenaphthylene, 1,2-dihydro-. *See* Acenaphthene

Acetal

CAS 105-57-7; EINECS/ELINCS 203-310-6
UN 1088 (DOT); FEMA 2002
Synonyms: Acetaldehyde diethyl acetal; 1,1-Diethoxyacetal; 1,1-Diethoxyethane; Diethyl acetal; Ethylidene diethyl ether
Classification: Diether
Empirical: $C_6H_{14}O_2$
Formula: $CH_3CH(OC_2H_5)_2$
Properties: Colorless volatile liq.; agreeable green woody solvent odor; fruity green flavor; very sol. in ethanol, diethyl ether; sol. in heptane, methylcyclohexane, ethyl acetate, acetone, chloroform, propyl alcohol, butyl alcohol, IPA; sl. sol. in water; m.w. 118.18; dens. 0.831; vapor pressure 10 mm Hg (8 C); m.p. -100 C; b.p. 107-112 C; flash pt. (CC) 36 C
Toxicology: LD50 (oral, rat) 4.57 g/kg, (IP, rat) 900 mg/kg; moderately toxic by ingestion and IP route; skin and eye irritant; narcotic; TSCA listed
Precaution: Highly flamm.; dangerous fire hazard exposed to heat or flames; tends to polymerize on standing; can react vigorously with oxidizers; forms heat-sensitive explosive peroxides on contact with air; readily decomposed by dil. acids
Hazardous Decomp. Prods.: CO, CO_2; heated to decomp., emits acrid smoke and fumes
NFPA: Health 2, Flammability 3, Reactivity 0
Storage: Keep away from ignition sources; light-sensitive, heat-sensitive, moisture-sensitive
Uses: Solvent; cosmetics; organic synthesis; perfumery; synthetic flavoring agent for nonalcoholic beverages, ice cream, pharmaceuticals
Regulatory: FDA §172.515; FEMA GRAS
Manuf./Distrib.: Acros Org.; Aldrich; Ashland; Ashley Polymers; Augustus Oils Ltd; Axxence Aromatic GmbH; Bay Resins; Berje; Degussa AG; Degussa Flavors/Alfrebro; Degussa; Fleurchem; Fluka; Frutarom Ltd; Givaudan Fragrances; LNP; Lambiotte; Lluch Essence; Penta Mfg.; Plastics Color & Compding.; Ticona; Westlake Plastics

Acetaldehyde cyanohydrin. *See* Lactonitrile
Acetaldehyde diethyl acetal. *See* Acetal
Acetaldehyde, trimer. *See* Paraldehyde

Acetamide

CAS 60-35-5; EINECS/ELINCS 200-473-5
Synonyms: Acetic acid amide; Acetic acid amine; Acetimidic acid; Ethanamide; Ethanolamide; Methanecarboxamide
Classification: Nonaromatic amide
Empirical: C_2H_5NO
Formula: CH_3CONH_2
Properties: Wh. cryst.; odorless when pure, but frequently has mousy odor; sol. in water, alcohol, chloroform, glycerol, hot benzene; m.w. 59.07; dens. 1.159; vapor pressure negligible; m.p. 82 C; b.p. 221 C; ref. index 1.4274; neutral reaction
Toxicology: LD50 (oral, rat) 7 g/kg, (IP, rat) 10,300 mg/kg, (IV, rat) 12,500 mg/kg, (subcut., rat) 10 g/kg; eye/skin/respiratory tract irritant; may cause sleep disturbances, muscle weakness, dyspnea, spastic paralysis, lacrimation, change in heart rate; cancer suspect agent; tumorigen; reproductive effector; mutagen; target organ: liver; TSCA listed
Precaution: Incompat. with strong oxidizers, metals, halogenated materials; avoid ignition sources, excess heat, elec. sparks
Hazardous Decomp. Prods.: CO, CO_2, NO_x
Storage: Hygroscopic, deliq.; store under nitrogen in cool, dry, well-ventilated area away from incompat. substances
Uses: Solubilizer; solvent for many org. and inorg. compds., plastic films; dye solvent, dye assistant for textiles; mfg. of methylamine; denaturing alcohol; in organic synthesis; drug intermediate in mfg. of ampicillin, cephaclor, etc.; solder flux ingred.; plasticizer in leather, cloth films, coatings; stabilizer; humectant in paper; accelerator; buffer in lacquers, explosives, cosmetics; prep. of hypnotics; photochemicals; biocides; cryoscopy; vehicle in pharmaceuticals; antidote
Manuf./Distrib.: Aldrich; Alfa Aesar; Fallek; Fluka; Global-Seven; Heico; Integra; J.T. Baker; McIntyre; Otira Pharmaceuticals; Penta Mfg.; Sigma; Spectrum Quality Prods.; Universal Preserv-A-Chem; VWR Int'l.

Acetamide MEA

CAS 142-26-7; EINECS/ELINCS 205-530-8

Synonyms: 2-Acetamidoethanol; 2-Acetylaminoethanol; N-Acetyl ethanolamine; N-Ethanolacetamide; Hydroxyethyl acetamide; β-Hydroxyethylacetamide; N-(2-Hydroxyethyl) acetamide; N-β-Hydroxyethylacetamide
Classification: Aliphatic amide
Empirical: $C_4H_9NO_2$
Formula: $CH_3CONHCH_2CH_2OH$
Properties: Brn. visc. liq.; misc. with water; m.w. 103.14; dens. 1.12 (20/4 C); f.p. 15.8 C; b.p. 195-196 C; flash pt. (OC) 355 F; dec. on heating; nonionic
Toxicology: LD50 (oral, rat) 26,950 mg/kg; mildly toxic by ing.; skin and severe eye irritant; TSCA listed
Precaution: Combustible exposed to heat or flame; can react vigorously with oxidizers
Hazardous Decomp. Prods.: Heated to decomp., emits toxic fumes of NO_x
Uses: Solvent; antistat, humectant in cosmetics; skin and hair conditioner; chemical intermediate; coupling agent; pigment dispersant; clarifier for shampoos; moisturizer; solubilizer, humectant, conditioner, coupling agent, pigment dispersant in topical pharmaceuticals; in food-pkg. adhesives
Regulatory: FDA 21CFR §175.105
Manuf./Distrib.: AXO Chem.; Aldrich; Global-Seven; McIntyre; Scher; Universal Preserv-A-Chem
Trade Names: Foamid AME-70; Foamid AME-75; Foamid AME-100; Incromectant AMEA-70; Upamide ACMEA

Acetamide, N-methyl. *See* N-Methylacetamide
2-Acetamidoethanol. *See* Acetamide MEA
Acetate C-8. *See* Octyl acetate
Acetate ester, C6 alcohol branched. *See* Oxo-hexyl acetate
1-Acetate-1,2,3-propanetriol. *See* Glyceryl acetate
Acetdimethylamide. *See* Dimethyl acetamide

Acetic acid

CAS 64-19-7; EINECS/ELINCS 200-580-7
UN 2789 (DOT); UN 2790 (DOT); FEMA 2006; INS260; E260
Synonyms: Ethanoic acid; Ethyllic acid; Methanecarboxylic acid; Pyroligneus acid; Vinegar acid
Classification: Aliphatic organic acid
Empirical: $C_2H_4O_2$
Formula: CH_3COOH
Properties: Clear colorless liq.; pungent vinegar-like odor; sharply acid taste; misc. with water, alcohol, glycerol, ether; insol. in carbon disulfide; m.w. 60.03; dens. 1.0492 (20/4 C); m.p. 16.63 C; b.p. 118 C (765 mm); visc. 1.22 cps (20 C); vapor pressure 11.4 mm Hg (20 C); flash pt. (OC) 43 C; autoignition temp. 463 C; ref. index 1.3715 (20 C); surf. tens. 27.42 dynes/cm; dielec. const. 61.70
Toxicology: ACGIH TLV/TWA 10 ppm; STEL 15 ppm; LD50 (oral, rat) 3310 mg/kg, (skin, rabbit) 1060 mg/kg; mod. toxic by ingestion, inhalation; corrosive; strong irritant to eyes, skin, and tissue; caustic; can cause burns, lachrymation; human systemic effects by ing.; experimental reproductive effects; mutation data reported; TSCA listed
Environmental: VOC; BOD5 0.65; COD 1.09; ThOD 1.07
Precaution: Flamm.; moderate fire and explosion hazard exposed to heat or flame; can react vigorously with oxidizers; explosive or violent reactions possible; incompat. with many chems.; a common air contaminant
Hazardous Decomp. Prods.: Heated to decomp., emits irritating fumes
NFPA: Health 2, Flammability 3, Reactivity 0
Uses: Mfg. of acetic anhydride, cellulose acetate, vinyl acetate monomer; acetic esters; prod. of plastics, pharmaceuticals, cosmetics, aroma chems., dyes, insecticides, photographic chemicals, food additives; solvent reagent; acidifier; solvent for electronic materials; etchant (semiconductor mfg.); aluminum brightener; laundry sour; boiler water additive; coagulant for latex; buffer in cosmetics; buffer, acidifier in textile baths; preservative, acidity regulator in foods; flavor, solvent, vehicle in foods, pharmaceuticals; color diluent; sanitizing solutions for food contact
Regulatory: FDA 21CFR §73.85, 133.123, 133.124, 133.169, 133.173, 133.178, 133.179, 172.814, 178.1010, 184.1005, GRAS; FEMA GRAS; USDA 9CFR §318.7; CERCLA hazardous substance; Europe listed; UK approved; NF, BP compliance
Manuf./Distrib.: AB R Lundberg; AMC Chems.; AMRESCO; Advanced BioTech; Air Prods.; Aldrich; Alfa Aesar; Alfa Chem; Allchem Ind.; Am. Biorganics; Amyl; Arch Chems.; Ashland; Asiamerica Int'l.; Augustus Oils Ltd; Aventis SA; Axxence Aromatic GmbH; BASF; BP Amoco; BP Chems. Ltd; Brenntag AG; Brown; Captree; Celanese; Chemconnect; Chemical; Coyne; Daicel Chem. Ind.; Degussa AG/Health & Nutrition; Degussa Flavors/Alfrebro; Delta Distributors; DuPont; E&E Ltd; EM Ind.; Eastech; Eastman; Elan; Equistar; Fisher Scientific; Fleurchem; Florida Distillers; Fluka; Frutarom Ltd; General Chem.; George Uhe; Harcros; Integra; J.C. Wilson; J.T. Baker; Keith Harris & Co. Ltd; Lonza Ltd; Mallinckrodt Baker; Millennium; Milsolv; Moore Ingreds.; Noveon Diamalt GmbH; P.B. & S.; PMC Spec.; Penta Mfg.; Pfaltz & Bauer; Prodasynth; Quaker City; Rhodia; Romil Ltd; Ruger; SOCO-Lynch; Seeler Ind.; Showa Denko; Siber Hegner Ltd; Sigma; Southchem; Spectrum Quality Prods.; Sterling Chems.; Textile Chem.; Thomas Scientific; Triple Crown Am.; Universal Preserv-A-Chem; VWR Int'l.; Vam Org. Chems. Ltd; Varsal Instruments; Veckridge; Vopak USA; Vulcan Perf. Chems./GA; Wacker-Chemie GmbH; Whittaker, Clark & Daniels; Xinchem; Zetapharm
Trade Names Containing: Dynasolve 165

Acetic acid, alkyl (C11-C14) esters. *See* Oxo-dodecyl acetate
Acetic acid amide; Acetic acid amine. *See* Acetamide
Acetic acid amyl ester. *See* Amyl acetate
Acetic acid benzyl ester. *See* Benzyl acetate
Acetic acid 2-butoxy ester. *See* s-Butyl acetate
Acetic acid 2-(2-butoxyethoxy) ethyl ester. *See* Diethylene glycol butyl ether acetate
Acetic acid, 2-butoxyethyl ester. *See* Butoxyethanol acetate
Acetic acid, butyl ester. *See* n-Butyl acetate
Acetic acid 2-butyl ester. *See* s-Butyl acetate
Acetic acid n-butyl ester. *See* n-Butyl acetate
Acetic acid, s-butyl ester. *See* s-Butyl acetate
Acetic acid t-butyl ester. *See* t-Butyl acetate
Acetic acid, C7-9-branched alkyl esters, C8-rich. *See* C8 alkyl acetate
Acetic acid, C11-14-branched alkyl esters, C13-rich. *See* Oxo-dodecyl acetate
Acetic acid cyclohexyl ester. *See* Cyclohexyl acetate
Acetic acid, decyl ester, branched. *See* C10 alkyl acetate
Acetic acid, dimethyl-. *See* Isobutyric acid
Acetic acid dimethylamide. *See* Dimethyl acetamide
Acetic acid-1,3-dimethylbutyl ester. *See* Methyl amyl acetate
Acetic acid 1,1-dimethylethyl ester. *See* t-Butyl acetate
Acetic acid, esters with lanolin alcohols. *See* Acetylated lanolin alcohol
Acetic acid esters of mono- and diglycerides; Acetic acid esters of mono- and diglycerides of fatty acids. *See* Acetylated mono- and diglycerides of fatty acids
Acetic acid 2-ethoxyethyl ester. *See* Ethoxyethanol acetate
Acetic acid 2-ethylbutyl ester. *See* 2-Ethylbutyl acetate
Acetic acid, ethyl ester. *See* Ethyl acetate
Acetic acid 2-ethylhexyl ester. *See* 2-Ethylhexyl acetate

Acetic acid, glacial

CAS 64-19-7; EINECS/ELINCS 200-580-7
UN 2789; UN 2790; FEMA 2006; INS260; E260
Synonyms: Concentrated acetic acid; Ethanoic acid; Ethanolic acid; Glacial acetic acid; Vinegar acid
Classification: Aliphatic organic acid
Definition: Pure compd. (99.8% min.) as distinguished from the usual aq. sol'ns.
Empirical: $C_2H_4O_2$
Formula: CH_3COOH
Properties: Clear colorless liq., pungent char. odor, acid taste when dil. with water; misc. with water, alcohol, acetone, glycerol, ether; insol. in chloroform; m.w. 60.05; dens. 1.049 (20/4 C); m.p. 16.2 C; b.p. 118 C; flash pt. 40 C; ref. index 1.3720
Toxicology: TLV/TWA 10 ppm; human poison by unspecified routes; mod. toxic by various routes; corrosive; severe eye and skin irritant; can produce lung obstruction; TSCA listed
Precaution: DOT: Corrosive material
NFPA: Health 3, Flammability 2, Reactivity 0

Uses: Acidifier, buffer, solvent for pharmaceuticals; preservative, pH control agent in foods
Regulatory: NF, BP, EP compliance
Manuf./Distrib.: Aldrich; Am. Biorganics; Ashland; Celanese; Chemical; EM Science; Eastman; Equistar; Fluka; Frutarom Ltd; Galbraith Labs; Houghton Chem.; J.T. Baker; Lluch Essence; Lonza; Penta Mfg.; Ruger; Sal Chem.; Seeler Ind.; Sigma; Spectrum Quality Prods.; Voigt Global Distrib.; Wacker-Chemie GmbH
Trade Names: Eastman® Glacial Acetic Acid; Eastman® Glacial Acetic Acid Food Grade, Kosher

Acetic acid hexyl ester. *See* Hexyl acetate
Acetic acid, hydroxy-, butyl ester. *See* Butyl glycolate
Acetic acid, 2-hydroxyethyl ester. *See* Ethylene glycol acetate
Acetic acid isobutyl ester. *See* Isobutyl acetate
Acetic acid isopentyl ester. *See* Isoamyl acetate
Acetic acid isopropyl ester. *See* Isopropyl acetate
Acetic acid, 3-methoxybutanol ester; Acetic acid, 3-methoxybutyl ester; Acetic acid (3-methoxy-n-butyl) ester. *See* Methoxybutyl acetate
Acetic acid, 2-(2-methoxyethoxy) ethyl ester. *See* Diethylene glycol methyl ether acetate
Acetic acid 2-methoxyethyl ester. *See* Methoxyethanol acetate
Acetic acid, 2-methoxy-1-methylethyl ester. *See* Propylene glycol methyl ether acetate
Acetic acid, 2-methoxypropyl ester. *See* 2-Methoxy-1-propanol acetate
Acetic acid methyl ester. *See* Methyl acetate
Acetic acid 1-methylethyl ester. *See* Isopropyl acetate
Acetic acid 1-methylpropyl ester. *See* s-Butyl acetate
Acetic acid 2-methylpropyl ester. *See* Isobutyl acetate
Acetic acid, monoglyceride. *See* Glyceryl acetate
Acetic acid octyl ester. *See* Octyl acetate
Acetic acid pentyl ester. *See* Amyl acetate
Acetic acid, phenyl ester. *See* Phenyl acetate
Acetic acid phenylmethyl ester. *See* Benzyl acetate
Acetic acid propyl ester; Acetic acid n-propyl ester. *See* Propyl acetate
Acetic acid, trifluoro-. *See* Trifluoroacetic acid
Acetic ester; Acetic ether. *See* Ethyl acetate
Acetic, 1,2,3-propanetriyl ester. *See* Triacetin
Acetimidic acid. *See* Acetamide
Acetin. *See* Glyceryl acetate; Triacetin
Acetoacetic acid ethyl ester. *See* Ethylacetoacetate
Acetoacetic acid methyl ester. *See* Methyl acetoacetate
Acetoacetic ester. *See* Ethylacetoacetate
Acetoacetone. *See* Acetylacetone
Aceto-caustin. *See* Trichloroacetic acid
Acetoglycerides. *See* Acetylated mono- and diglycerides of fatty acids

Acetone

CAS 67-64-1; EINECS/ELINCS 200-662-2
UN 1090 (DOT); UN 1091 (DOT); FEMA 3326
Synonyms: Acetone oils; Dimethyl formaldehyde; Dimethylketal; Dimethylketone; DMK; Ketone, dimethyl; Ketone propane; β-Ketopropane; Methyl ketone; Propanone; 2-Propanone; Pyroacetic acid; Pyroacetic ether
Classification: Aliphatic ketone
Empirical: C_3H_6O
Formula: CH_3COCH_3
Properties: Colorless volatile transparent liq., sweetish odor, pungent sweetish taste; sol. in water, alcohol, chloroform, dimethylformamide, ether, most volatile oils; m.w. 58.09; sp.gr. 0.792 (20/20 C); vapor pressure 400 mm Hg (39.5 C); m.p. -94.3 C; b.p. 56.2 C; flash pt. (CC) -18 C
Toxicology: ACGIH TLV/TWA 750 ppm; STEL 1000 ppm; LD50 (oral, mouse) 3000 mg/kg; narcotic in high conc.; moderately toxic by ingestion and inhalation; lg. doses may cause narcosis; extreme concs. may cause collapse, coma, death; inh. irritates lungs; ing. may cause throat, esophagus, and stomach irritation; peeling and splitting of nails, skin rashes; eye irritant; common air contaminant; TSCA listed
Environmental: BOD5 0.85; COD 1.12-2.07; ThOD 2.21
Precaution: DOT: flamm. liq.; dangerous fire risk; explosive limit in air 2.6-12.8%; reacts violently with oxidizing agents, chlorinated solv./alkali mixts.; reacts vigorously with sulfur dichloride, potassium t-butoxide, hexachloromelamine
Hazardous Decomp. Prods.: Complete combustion yields CO_2, water vapor; incomplete combustion can produce CO
NFPA: Health 1, Flammability 3, Reactivity 0
Storage: Store in cool, dry, well-ventilated area, out of direct sunlight
Uses: Solvent for paints, varnishes, lacquers, resins, fats, waxes, adhesives, printing inks, cellulose resins, plastics, pharmaceuticals, electronic materials; degreasing agent; cleaning and drying precision equip.; delustrant for cellulose acetate fibers; extraction solvent for foods, essential oils; mfg. of acetate rayon, photographic film, paint removers, glues, cements, plastics, rubber, dyes, explosives; denaturant, solvent in cosmetics; color diluent for foods; flavor; in food-pkg. adhesives; adjuvant in resinous/polymeric food-contact coatings; adjuvant in slimicides in food-contact paper/paperboard
Features: Apple-, grape-, pear-, pineapple-like flavor
Regulatory: FDA 21CFR §73.1, 73.30, 73.345, 73.615, 173.210, 175.105, 175.320, 176.180, 176.300, 177.2600, 182.99; 40CFR §180.1001; 30 ppm tolerance in spice oleoresins; FEMA GRAS; SARA §313 reportable; CERCLA hazardous substance; Japan approved with restrictions; NF, BP, EP compliance; VOC-exempt
Manuf./Distrib.: AAE Chemie NV; AMC Chems.; AMRESCO; Aldrich; Allchem Ind.; Alvin Chem.; Amyl; Arch Chems.; Ashland; Atofina; Axxence Aromatic GmbH; BASF; BP Amoco; BP Chems. Ltd; Baychem; Brenntag AG; C.P. Hall; Captree; Chemcentral; Chemconnect; Chemical; ChevronTexaco; Columbus Chem. Ind.; Degussa Flavors/Alfrebro; Dow; E-Chemicals; EM Science; Eastman; Electron Microscopy Sciences; ExxonMobil; Fisher Scientific; Fleurchem; Fluka; GE Plastics; General Chem.; Georgia Gulf; Haltermann Prods. UK; Harcros; Hatco; Herdillia Chems. Ltd; Honeywell Perf. Polymers; Houghton Chem.; Hukill; INEOS Phenol; Integra; J.C. Wilson; J.T. Baker; JLM Marketing; Kyowa Hakko Kogyo; Mallinckrodt Baker; MelChem; Mitsubishi Chem.; Nippon Petrochems.; Penta Mfg.; Primachem; Quaker City; R.E. Carroll; Romil Ltd; Ruger; Sal Chem.; Seeler Ind.; Shell; Sigma-Aldrich; Sigma; Spectrum Quality Prods.; Stanchem; Sumitomo Chem.; Sunnyside; Sunoco; Total Spec. Chems.; Universal Preserv-A-Chem; VWR Int'l.; Veckridge; Vopak USA; Xinchem
Trade Names: Eastman® Acetone, High Purity Sales Grade; Ucar® Acetone
Trade Names Containing: Celtone; Punctilious® SDA 23A 190 Proof; Punctilious® SDA 23A Anhydrous; Punctilious® SDA 23H 190 Proof

Acetone, acetonyl-. *See* 2,5-Hexanedione
Acetone dimethyl acetal. *See* Dimethoxypropane
Acetone glycerol. *See* 2,2-Dimethyl-1,3-dioxolane-4-methanol
Acetone oils. *See* Acetone

Acetone oxime

CAS 127-06-0; EINECS/ELINCS 204-820-1
Synonyms: Acetonoxime; Acetoxime; β-Isonitrosopropane; 2-Propanone oxime
Empirical: C_3H_7NO
Formula: $(CH_3)_2CNOH$
Properties: Colorless cryst., chloral-like odor; sol. in alcohols, ethers, water; m.w. 73.09; dens. 0.97; vapor pressure 3.12 mm Hg; m.p. 61 C; b.p. 136.3 C; flash pt. 60 C
Toxicology: LD (oral, rat) > 500 mg/kg; LD50 (IP, mouse) 4000 mg/kg; mod. toxic by ing. and IP routes; TSCA listed
Precaution: Combustible
Hazardous Decomp. Prods.: Heated to decomp., emits toxic NO_x fumes
Uses: Intermediate for organic synthesis; solvent for cellulose ethers; in sealants and coatings; primer for diesel fuels
Manuf./Distrib.: ABCR UK; Acros Org.; Aldrich; Alfa Aesar; Avocado Research Chems.; Fisher Scientific UK; Fluka; Honeywell Perf. Polymers; ICN Biomed. Research Prods.; Lancaster Synthesis UK; Sigma; TCI Am.; Venchem; Whyte Chems. Ltd

Acetonic acid. *See* Lactic acid

Acetonitrile

CAS 75-05-8; EINECS/ELINCS 200-835-2
UN 1648 (DOT)
Synonyms: ACN; Cyanomethane; Ethanenitrile; Ethyl nitrile; Methanecarbonitrile; Methyl cyanide

Classification: Nitrile/cyano compd.
Empirical: C_2H_3N
Formula: CH_3CN
Properties: Colorless liq., ether-like odor; sol. in water, methanol, ether, acetone, chloroform, CCl_4, ethylene chloride, methyl acetate; m.w. 41.05; sp.gr. 0.783; m.p. -45 C; b.p. 81.6 C; flash pt. (CC) 12.8 C; ref. index 1.3440
Toxicology: ACGIH TLV/TWA 40 ppm; STEL 60 ppm (skin); LD50 (oral, rat) 3800 mg/kg; poisonous; toxic by skin absorp. and inh.; skin irritant; may cause severe eye irritation; lachrymator; inh. of vapors can cause nose/throat irritation; TSCA listed
Precaution: DOT: Flamm. liq.; dangerous fire risk; incompat. with oxidizing agents, acids, reducing agents (violent/explosive reactions possible); reacts slowly with water or steam to release toxic and flamm. vapors (ammonia, hydrogen cyanide)
Hazardous Decomp. Prods.: Acetic acid, ammonia, hydrogen cyanide; heated to decomp., emits highly toxic fumes of CN^-, NO_x
NFPA: Health 2, Flammability 3, Reactivity 0
Storage: Store in cool, dry, well-ventilated area, out of direct sunlight
Uses: Solvent for hydrocarbon extraction processes, esp. for butadiene; intermediate; catalyst; for separation of fatty acids from vegetable oils; intermediate for mfg. of synthetic pharmaceuticals, Vitamin B, perfumes, syn. fibers, other chems.; refining of hydrocarbons, waxes, oils, greases; chem. purification; rubber antioxidant; alcohol denaturant; solvent in pharmaceuticals
Regulatory: SARA reportable
Manuf./Distrib.: Aceto; Aldrich; Alfa Aesar; Alfa Chem; Allchem Ind.; Asahi Glass; BCH Brühl; BP Amoco; BP Chems. Ltd; Burdick & Jackson; Chemconnect; DuPont; EniChem Am.; Filo; Fisher Scientific; Fluka; Honeywell Spec. Chems.; Honeywill & Stein; Integra; J.T. Baker; Mallinckrodt Baker; Mitsui Toatsu; Penta Mfg.; R.W. Greeff; Romil Ltd; Sigma; Spectrum Quality Prods.; Thomas Scientific; Total Spec. Chems.; VWR Int'l.; Whyte Chems. Ltd

Acetonoxime. *See* Acetone oxime
Acetonyl acetone. *See* 2,5-Hexanedione

Acetophenone

CAS 98-86-2; EINECS/ELINCS 202-708-7
UN 1993; FEMA 2009
Synonyms: Acetylbenzene; Benzoyl methide; Hypnone; Ketone methyl phenyl; Methyl phenyl ketone; 1-Phenylethanone; 1-Phenyl-1-ethanone; Phenyl methyl ketone
Classification: Aromatic organic compd.
Empirical: C_8H_8O
Formula: $C_6H_5COCH_3$
Properties: Colorless to lt. yel. liq. or plates; sweet pungent odor and taste; sol. in alcohol, chloroform, ether, fatty oils, glycerol; sl. sol. in water; m.w. 120.15; dens. 1.030 (20/20 C); vapor pressure 1 mm Hg (15 C); m.p. 20.5 C; b.p. 201.7 C; flash pt. 82.2 C; ref. index 1.5339 (20 C)
Toxicology: LD50 (oral, rat) 815 mg/kg, (IP, mouse) 200 mg/kg; poison by IP, subcut. routes; moderate toxicity by ing.; skin and severe eye irritant; narcotic in high concs.; a hypnotic; may cause allergic reaction; mutagenic data; TSCA listed
Precaution: DOT: Combustible liq.; flamm. exposed to heat, flames, or oxidizers
Hazardous Decomp. Prods.: Heated to decomp., emits acrid smoke and fumes
NFPA: Health 1, Flammability 2, Reactivity 0
Uses: Fragrance in cosmetics, perfumes, detergents, tech. prods.; solvent for cellulose ethers and esters; intermediate for pharmaceuticals, resins, paints, etc.; synthetic flavoring agent in foods and pharmaceuticals; polymerization catalyst; organic synthesis; drug (to induce sleep)
Features: Sweet flavor
Regulatory: FDA 21CFR §172.515; FEMA GRAS; Japan approved for flavoring
Manuf./Distrib.: AMC Chems.; Advanced Synthesis Tech.; Aldrich; Alemark; Alfa Chem; Allchem Ind.; Amber Syn.; BP Chems. Ltd; Berje; ChemPur; EniChem Am.; Fluka; Haarmann & Reimer GmbH; Honeywell Spec. Chems.; INEOS Phenol; Integra; J.T. Baker; JLM Marketing; Lluch Essence; Lyondell; Mallinckrodt Baker; MelChem; Mitsui Petrochem. Ind.; Mitsui Toatsu; Penta Mfg.; Polarome Int'l.; R.W. Greeff; Rhodia; Sigma-Aldrich; Sigma; Spectrum Quality Prods.; Synasia Fine Chem.; Takasago Int'l.; Toyotama Perfumery; VWR Int'l.; Whyte Chems. Ltd

Acetoxime. *See* Acetone oxime
Acetoxybenzene. *See* Phenyl acetate
1-Acetoxy-2-butoxyethane. *See* Butoxyethanol acetate
1-Acetoxy-2-t-butylcyclohexane. *See* 2-t-Butylcyclohexyl acetate
Acetoxyethane. *See* Ethyl acetate
1-Acetoxyhexadecane. *See* Cetyl acetate
1-Acetoxy-2-methoxyethane. *See* Methoxyethanol acetate
2-Acetoxy-1-methoxypropane. *See* Propylene glycol methyl ether acetate
Acetoxymethylbenzene. *See* Benzyl acetate
2-Acetoxypentane. *See* s-Amyl acetate
1-Acetoxypropane. *See* Propyl acetate
2-Acetoxypropane. *See* Isopropyl acetate
α-Acetoxytoluene. *See* Benzyl acetate
Acetylacetonate. *See* Acetylacetone

Acetylacetone

CAS 123-54-6; EINECS/ELINCS 204-634-0
UN 2310 (DOT)
Synonyms: ACAC; Acetoacetone; Acetylacetonate; Acetyl 2-propanone; Diacetylmethane; Pentan-2,4-dione; 2,4-Pentanedione; Pentanedione-2,4
Empirical: $C_5H_8O_2$
Formula: $CH_3COCH_2COCH_3$
Properties: Colorless to sl. yel. liq., pleasant odor; misc. with alcohol, benzene, ether, chloroform, acetone, glac. acetic acid, propylene glycol; insol. in water; m.w. 100.13; dens. 0.952-0.962; m.p. -23.2 C; b.p. 139 C (746 mm); flash pt. (TOC) 40.5 C
Toxicology: LD50 (oral, rat) 1000 mg/kg, (IP, mouse) 750 mg/kg; mod. toxic by ingestion, IP, inh. routes; skin and severe eye irritant; narcotic in high doses; TSCA listed
Precaution: Flamm. exposed to heat or flame; incompat. with oxidizing materials
NFPA: Health 2, Flammability 2, Reactivity 0
Storage: Light-sensitive; refrigerate
Uses: Solvent for cellulose acetate; intermediate; chelating agent for metals, paint driers, lubricant additives, pesticides; synthesis of vitamin B_6 and K; intermediate for drugs, pesticides; prod. of corrosion inhibitors; catalyst for esterification and olefin polymerization; resin crosslinking agent for automotive prods., coatings, elastomers, films/paints, graphic arts, plastics; extraction solvent for mineral processing; metal deactivator (lubricants, fuels); drying agent; germicide; insecticide; catalyst promoter; analytical reagent
Manuf./Distrib.: Aceto; Aldrich; Alfa Aesar; Daicel Chem. Ind.; Dow; Filo; Fluka; J.T. Baker; Kowa Am.; Mallinckrodt Baker; Mitsui & Co. USA; Monomer-Polymer & Dajac Labs; Penta Mfg.; Rochem Int'l.; Shandong Haihua Tianhe; Sigma; Spectrum Quality Prods.; Wacker Chems.; Wacker-Chemie GmbH

2-Acetylaminoethanol. *See* Acetamide MEA
Acetylated glycerin monostearate. *See* Acetylated glyceryl stearate

Acetylated glyceryl stearate

Synonyms: Acetylated glycerin monostearate
Uses: Dispersant, solvent for act. ingreds.; vehicle for injectable substances; emulsifier for textiles; emulsifier, opacifier, thickener, dispersant for emulsions; greasing agent in shampoos, foam baths; food lubricant; perfume diluent; intermediate in textiles and tanning industry
Trade Names: Monestriol DM

Acetylated hydrogenated lard glyceride

CAS 8029-91-2; 93572-31-7; EINECS/ELINCS 297-461-5
Synonyms: Glycerides, lard mono-, hydrogenated, acetates
Definition: Acetyl ester of the monoglyceride derived from hydrog. lard
Properties: Nonionic
Uses: Emulsifier, emollient, plasticizer for cosmetics, technical prods.; food additive; lubricant, solvent, plasticizer for foodstuffs; film-former; moisture barrier
Features: Good moisture vapor barrier properties
Regulatory: FDA 21CFR §172.828

Trade Names: Axol® E 61

Acetylated hydrogenated soybean oil glyceride

Synonyms: Glycerides, soybean oil, hydrogenated, acetates
Properties: Nonionic
Uses: Emulsifier; emollient in cosmetics; food-grade lubricant, solvent, emollient; deaerator in some systems; food additive
Regulatory: FDA 21CFR §172.828
Trade Names: Myvacet® 9-45K

Acetylated lanolin alcohol

CAS 61788-49-6; EINECS/ELINCS 262-980-8
Synonyms: Acetic acid, esters with lanolin alcohols; Lanolin, alcohols, acetates
Definition: Acetyl ester of lanolin alcohol
Properties: Pale yel. liq.; HLB 1.0; sapon. no. 180-200; nonionic
Toxicology: TSCA listed
Uses: Surfactant, antistat, emulsifier, emollient, spreading agent for cosmetic and pharmaceutical creams/lotions, skin cleansers, soaps, baby prods., shampoos
Trade Names Containing: Acetol® 1706; Acetulan®; Fancol™ ALA; Ritawax AEO; Ritawax ALA

Acetylated lard glyceride

CAS 8029-92-3; EINECS/ELINCS 273-609-4
Synonyms: Glycerides, lard mono-, acetates
Definition: Acetyl ester of lard glyceride
Uses: Emulsifier; lubricant; emollient in cosmetics; deaerator; food emulsifier; cosolvent for polar additives in low polarity systems; in hot-melt strippable food coatings
Regulatory: FDA 21CFR §172.828, 175.230

Acetylated mono- and diglycerides. *See* Acetylated mono- and diglycerides of fatty acids

Acetylated mono- and diglycerides of fatty acids

CAS 68990-55-6
INS472a; E472a
Synonyms: Acetic acid esters of mono- and diglycerides; Acetic acid esters of mono- and diglycerides of fatty acids; Acetoglycerides; Acetylated mono- and diglycerides; Mono- and diglycerides, acetic acid esters
Definition: Partial or complete esters of glycerin with a mixture of acetic acid and edible fat-forming fatty acids
Properties: Wh. to pale yel. oily to waxy liq., bland taste; sol. in alcohol, acetone; disp. to sol. in edible oils and fats; insol. in water; HLB 2-3; nonionic; HLB 2.0-3.0
Hazardous Decomp. Prods.: Heated to decomp., emits acrid smoke and fumes
Uses: Emulsifier, coating agent, stabilizer, sequestrant, lubricant, solvent, texturizer for foods, pharmaceuticals; plasticizer for fats; emulsifier for cosmetics; plasticizer and slip agent for waxes on paper prods.; solvating plasticizer for PVC and other plastics
Regulatory: FDA 21CFR §172.828; Europe listed; UK approved
Manuf./Distrib.: AB R Lundberg; ABITEC; Adept Sol'ns.; Universal Preserv-A-Chem
Trade Names: Axol® E 66

Acetylated palm kernel glycerides

CAS 91744-63-7; EINECS/ELINCS 294-627-9
Synonyms: Glycerides, palm kernel oil mono-, di-, and tri-, acetates
Definition: Acetyl ester of palm kernel glycerides
Uses: Lubricant, plasticizer for plastics, coatings; emollient, emulsifier in cosmetics; food-grade plasticizer for PVC; cosolvent for polar additives in low polarity systems; food additive

Acetylbenzene. *See* Acetophenone
Acetyl butyl citrate; Acetylcitric acid, tributyl ester. *See* Acetyl tributyl citrate
N-Acetyl-diethyl-L-aspartate. *See* Diethyl acetyl aspartate
Acetyldimethylamine. *See* Dimethyl acetamide

Acetyldimethylamine chloride

CAS 2675-89-0; EINECS/ELINCS 220-224-4
Synonyms: 2-Chloro-N,N-dimethylacetamide; N,N-Dimethyl-2-chloroacetamide
Empirical: C_4H_8ClNO
Formula: $ClCH_2CON(CH_3)_2$
Properties: Colorless clear liq. or solid; sol. in water; m.w. 121.57; sp.gr. 1.18-1.185; m.p. -20 C; b.p. 98-100 C; ref. index 1.478
Toxicology: Irritating to eyes, skin, mucous membranes, upper respiratory tract; may be harmful by ing., inh., skin absorp.
Precaution: Incompat. with strong oxidizing agents
Hazardous Decomp. Prods.: Toxic fumes of CO, CO_2, NO_x, hydrogen chloride gas; emits toxic fumes under fire conditions
NFPA: Health 2, Flammability 2, Reactivity 0
Storage: Store in cool, dry place; keep tightly closed
Uses: Solvent for fibers, coatings, adhesives; reaction medium for prod. of pharmaceuticals, plasticizers
Manuf./Distrib.: Fluka

Acetylene carbinol. *See* Propargyl alcohol
Acetylene dichloride. *See* cis-trans-1,2-Dichloroethylene (mixed isomers)
trans-Acetylene dichloride. *See* trans-1,2-Dichloroethylene
Acetylene tetrabromide. *See* Tetrabromoethane
Acetylene tetrachloride. *See* 1,1,2,2-Tetrachloroethane
Acetylene trichloride. *See* Trichloroethylene
N-Acetyl ethanolamine. *See* Acetamide MEA
Acetylmethylamine. *See* N-Methylacetamide
(Acetyloxy) benzene. *See* Phenyl acetate
2-(Acetyloxy)-1,2,3-propanetricarboxylic acid, tributyl ester. *See* Acetyl tributyl citrate
2-(Acetyloxy)-1,2,3-propanetricarboxylic acid, triethyl ester. *See* Acetyl triethyl citrate
Acetylphenol. *See* Phenyl acetate
2-Acetyl propane. *See* Methyl isopropyl ketone
Acetyl 2-propanone. *See* Acetylacetone

Acetyl tributyl citrate

CAS 77-90-7; EINECS/ELINCS 201-067-0
FEMA 3080
Synonyms: Acetyl butyl citrate; Acetylcitric acid, tributyl ester; 2-(Acetyloxy)-1,2,3-propanetricarboxylic acid, tributyl ester; Acetyl tri-n-butyl citrate; ATBC; Citric acid, tributyl ester, acetate; 1,2,3-Propanetricarboxylic acid, 2-(acetyloxy)-, tributyl ester; Tributyl acetyl citrate; Tri-n-butyl O-acetylcitrate; Tributyl O-acetylcitrate; Tributyl 2-(acetyloxy)-1,2,3-propanetricarboxylate; Tributyl citrate acetate; Tri-n-butyl citrate acetate
Classification: Aliphatic ester
Definition: Ester of citric acid
Empirical: $C_{20}H_{34}O_8$
Formula: $CH_3COOC_3H_4(COOC_4H_9)_3$
Properties: Colorless sl. visc. liq., sweet herbaceous odor; sol. in alcohol; insol. in water; m.w. 402.49; dens. 1.14; b.p. > 300 C; flash pt. 204 C; ref. index 1.4408
Toxicology: LD50 (IP, mouse) > 4 g/kg; low toxicity by ing., skin contact, and IP routes; TSCA listed
Precaution: Combustible
Hazardous Decomp. Prods.: Heated to decomp., emits acrid smoke and irritating fumes
Uses: Plasticizer for vinyl resins, cellulosic resins, rubber, food pkg.; synthetic flavoring agent in foods and pharmaceuticals; solvent in household, I&I cleaners; solvent, film-former, and plasticizer for inks, adhesives, coatings; film-former in cosmetics; plasticizer for aq. coatings, sustained-release drugs; in surf. lubricants for mfg. of food-contact metallic articles
Features: Sweet flavor
Regulatory: FDA 21CFR §172.515, 175.105, 175.300, 175.320, 178.3910, 181.22, 181.27; FEMA GRAS
Manuf./Distrib.: Aldrich; Alfa Chem; Allan; Allchem Ind.; ChemService; Chemial SpA; Jungbunzlauer; Kyowa Hakko Kogyo; Morflex; Pechiney Chems. Div.; Penta Mfg.; Pfizer Int'l.; Reilly Chems. SA; Sanken Chem.; Synasia Fine Chem.; Unitex; Universal Preserv-A-Chem
Trade Names: Citroflex® A-4; Citrofol® BII

Acetyl tri-n-butyl citrate. *See* Acetyl tributyl citrate

Acetyl triethyl citrate

CAS 77-89-4; EINECS/ELINCS 201-066-5

Synonyms: 2-(Acetyloxy)-1,2,3-propanetricarboxylic acid, triethyl ester; ATEC; Tricarballylic acid-β-acetoxytributyl ester; Triethyl acetylcitrate; Triethyl o-acetylcitrate

Classification: Aliphatic ester

Empirical: $C_{14}H_{22}O_8$

Formula: $CH_3COOC_3H_4(COOC_2H_5)_3$

Properties: Colorless liq.; odorless; sl. sol. in water; m.w. 318.36; dens. 1.135 (25 C); m.p. -50 C; flash pt. 187 C

Toxicology: Moderate toxicity by intraperitoneal route; mild toxicity by ingestion; TSCA listed

Precaution: Combustible

Uses: Plasticizer for cellulosics, esp. ethyl cellulose, food pkg.; solvent for household, I&I cleaners, pharmaceuticals; plasticizer for pharmaceutical aq. coatings; film-former in cosmetics; solvent, film-former, plasticizer for inks, adhesives, coatings; in food-pkg. adhesives; in surf. lubricants for mfg. of food-contact metallic articles

Regulatory: FDA 21CFR §175.105, 175.300, 175.320, 178.3910, 181.22, 181.27

Manuf./Distrib.: Aldrich; ChemService; Chemial SpA; Morflex; Penta Mfg.; Pfizer Ltd; Reilly Ind.; Synasia Fine Chem.; Triple Crown Am.; Unitex; Universal Preserv-A-Chem

Trade Names: Citroflex® A-2; Citrofol® AII

Acetyl trimethyl citrate

CAS 146340-22-9

Empirical: $C_{11}H_{16}O_8$

Properties: Sol. in toluene, sol. 5.5 g/100 ml water; insol. in heptane; m.w. 276; sp.gr. 1.2324; vapor pressure 0.16 mm Hg (20 C); flash pt. (COC) 311 F; ref. index 1.4440

Uses: Solvent in household, I&I cleaners; solvent, film-former, plasticizer for inks, adhesives, coatings

Acids, lanolin. *See* Lanolin acid
ACN. *See* Acrylonitrile; Acetonitrile
Acraldehyde. *See* Acrolein

Acrolein

CAS 107-02-8; EINECS/ELINCS 203-453-4
UN 1092 (DOT); UN 2607

Synonyms: Acraldehyde; Acrylaldehyde; Acrylic aldehyde; Allyl aldehyde; Ethylene aldehyde; Prop-2-enal; 2-Propenal; Propylene aldehyde

Classification: Acrylic aldehyde

Empirical: C_3H_4O

Formula: CH_2:CHCHO

Properties: Colorless to yel. liq.; pungent, burnt sweet odor; sol. in water, ethanol, diethyl ether, petrol. ether, acetone, oil; m.w. 56.06; dens. 0.8389 (20 C); m.p. -88 C; b.p. 52.5 C (760 mm); flash pt. (OC) -18 C

Toxicology: LD50 (oral, rat) 46 mg/kg; ACGIH TLV/TWA 0.1 ppm; human poison by inh. and intradermal route; poison experimentally by most routes; vapors cause lacrimation; primary irritant; irritates skin and mucosa; TSCA listed

Environmental: Biodeg.; VOC; BOD5 0; ThOD 2.00

Precaution: Extremely flamm.; unstable; polymerizes in light or presence of alkali or strong acid; can react vigorously with oxidizing materials

Hazardous Decomp. Prods.: Heated to decomp., emits highly toxic fumes, CO, CO_2

NFPA: Health 4, Flammability 3, Reactivity 3

Uses: Bactericide/algicide for control of bacteria in oil fields, weeds in agric. irrigation systems and canals; laboratory reagent; mfg. of perfumes and pharmaceuticals; hydrogen sulfide scavenger for oil and gas prod. systems; intermediate for syn. glycerol, polyurethane and polyester resins, methionine; warning agent in gases; crosslinking agent for cellulose fibers; etherification agent for food starch; slimicide in food-contact paper/paperboard; solvent

Regulatory: FDA 21CFR §172.892, 176.300; SARA reportable; HAP

Manuf./Distrib.: ABCR UK; Advanced Synthesis Tech.; Aldrich; Alfa Aesar; Atofina UK; Baker Petrolite; Daicel Chem. Ind.; Dow; Fisher Scientific UK; Fluka; MDA Chems.; Sigma; Spectrum Quality Prods.

Acrylaldehyde. *See* Acrolein
Acrylic acid, 2-(2-ethoxyethoxy) ethyl ester. *See* 2-(2-Ethoxyethoxy) ethyl acrylate
Acrylic acid, polymer with sucrose-polyallyl ether. *See* Carbomer
Acrylic aldehyde. *See* Acrolein

Acrylonitrile

CAS 107-13-1; EINECS/ELINCS 203-466-5
UN 1093 (DOT)

Synonyms: ACN; Acrylonitrile monomer; AN; Cyanoethylene; Propenenitrile; 2-Propenenitrile; Vinyl cyanide

Classification: Nitrile

Empirical: C_3H_3N

Formula: H_2C:CHCN

Properties: Colorless, mobile liq.; pungent odor similar to garlic or onion; sol. in acetone, alcohol, benzene, ether, org. solvs., partially misc. in water; m.w. 53.06; dens. 0.8004 (25 C); f.p. -83 C; b.p. 77.3-77.4 C; flash pt. (TOC) 32 F

Toxicology: ACGIH TLV/TWA 2 ppm; LD50 (oral, rat) 78 mg/kg; toxic by inh. and skin absorp.; lachrymator; can cause irritation of respiratory tract, lack of oxygen, headache, dizziness, nausea, vomiting, tremors, diarrhea, signs of liver dysfunction; skin contact can cause blisters, toxic and allergic dermatitis; readily absorbed thru skin; vapor or liq. can cause severe eye irritation, corneal damage; ing. may cause sore throat, abdominal pain, possibly death; a carcinogen; TSCA listed

Precaution: Flamm.; dangerous fire risk; potentially explosive; explosive limits in air 3-17%; unstable and polymerizes violently when heated, in absence of oxygen, or on exposure to visible light; normally stable when inhibited

Hazardous Decomp. Prods.: CO, CO_2, NO_x, hydrogen cyanide

NFPA: Health 4, Flammability 3, Reactivity 2

Uses: Monomer for acrylic and modacrylic fibers, paints; in prod. of ABS and SAN copolymers, nitrile rubber, surfactants, adhesives, polyols, barrier resins, carbon fibers; organic synthesis; formerly as pesticide fumigant; solvent

Regulatory: SARA reportable

Manuf./Distrib.: Aldrich; Alfa Aesar; Allchem Ind.; Asahi Chem. Ind.; BP Chems. Ltd; Chemconnect; Cytec Ind.; DSM NV; DuPont; Fluka; Mitsubishi Chem.; Mitsui Toatsu; Monomer-Polymer & Dajac Labs; Monsanto; Showa Denko; Solutia; Sterling Chems.; Sumitomo Chem.; Whyte Chems. Ltd

Acrylonitrile monomer. *See* Acrylonitrile
Active amyl alcohol; Active primary amyl alcohol. *See* 2-Methyl-1-butanol
Adeps lanae. *See* Lanolin
Adipic acid bis (2-ethylhexyl) ester. *See* Dioctyl adipate
Adipic acid, dibutyl ester. *See* Dibutyl adipate
Adipic acid, diisobutyl ester. *See* Diisobutyl adipate
Adipic acid diisopropyl ester. *See* Diisopropyl adipate
Adipic acid dimethyl ester. *See* Dimethyl adipate
Adipic acid dinitrile. *See* Adiponitrile
Adipic acid monomethyl ester. *See* Methyl adipate
Adipic acid nitrile; Adipodinitrile. *See* Adiponitrile

Adiponitrile

CAS 111-69-3; EINECS/ELINCS 203-896-3
UN 2205 (DOT)

Synonyms: Adipic acid dinitrile; Adipic acid nitrile; Adipodinitrile; Adipylnitrile; ADN; 1,4-Dicyanobutane; Hexanedinitrile; Hexanedioic acid dinitrile; Tetramethylene cyanide

Classification: Alkyl nitrile; cyano compd.

Empirical: $C_6H_8N_2$

Formula: $NC(CH_2)_4CN$

Properties: Water-wh. oily liq.; pract. odorless; sol. in methanol, ethanol, chloroalkanes; mod. sol. in water; m.w. 108.1; sp.gr. 0.965 (20/4 C); vapor pressure 0.01 mm Hg (40 C); m.p. 1 C; b.p. 295 C; flash pt. (CC) 159 C

Toxicology: ACGIH TLV/TWA 2 ppm (skin); LD50 (oral, rat) 155 mg/kg, (IP, mouse) 40 mg/kg; poison by inh., ing., subcut., IP routes; may cause skin and eye irritation; inh. of mist can probably cause nose/throat irritation at high concs.; high concs. may cause cyanide poisoning (weakness, headache, dizziness, anxiety, nausea, vomiting, difficulty breathing); severe exposure may cause unconsciousness, convulsions, and death; oxygen deprivation effects possible; TSCA listed

Precaution: Flamm. exposed to heat or flame; incompat. with oxidizing agents (increases fire/explosion risk), strong acids (violent reactions possible releasing toxic and flamm. hydrogen cyanide)
Hazardous Decomp. Prods.: Hydrogen cyanide; heated to decomp., emits toxic fumes of CN^-
NFPA: Health 2, Flammability 2, Reactivity 1
Storage: Store in a cool, dry, well-ventilated area out of direct sunlight
Uses: Intermediate in mfg. of nylon; intermediate for hexamethylenediamine; mfg. of other chems.; adipoguanamine resins; as polar solvent
Manuf./Distrib.: Aldrich; BASF plc; DuPont UK; Fisher Scientific UK; Fluka; Lancaster Synthesis UK; Solutia

Adipylnitrile; ADN. *See* Adiponitrile
Adronol acetate. *See* Cyclohexyl acetate
AEP. *See* Aminoethylpiperazine
Albacol. *See* Propyl alcohol

Alcohol

CAS 64-17-5; EINECS/ELINCS 200-578-6
UN 1170 (DOT); UN 1986 (DOT); UN 1987 (DOT); FEMA 2419
Synonyms: Absolute alcohol; Absolute ethanol; Distilled spirits; Ethanol; Ethanol, undenatured; Ethyl alcohol; Ethyl alcohol, undenatured; Ethyl hydrate; Ethyl hydroxide; EtOH; Grain alcohol; IMS; Industrial methylated spirit; Methylcarbinol; Spirits of wine
Definition: Undenatured ethyl alcohol
Empirical: C_2H_6O
Formula: CH_3CH_2OH
Properties: Colorless limpid, volatile liq., vinous odor, pungent taste; misc. with water, methanol, ether, chloroform, 95% acetone, oxygenated/chlorinated solvs.; m.w. 46.08; dens. 0.816 (15.5 C); vapor pressure 44.6 mm Hg (20 C); f.p. -117.3 C; b.p. 78.3 C; flash pt. (CC) 12.7 C; autoignition temp. 362 C; ref. index 1.365 (15 C); surf. tens. 21.99 dynes/cm
Toxicology: ACGIH TLV/TWA 1000 ppm; LD50 (oral, rat) 7060 mg/kg, (IP, rat) 3750 mg/kg; mod. toxic to humans by ing.; mod. toxic by IV and IP routes; mildly toxic by inh., skin contact; eye, skin, and mucous membrane irritant; severe skin irritant; human systemic effects; CNS depressant; ing. of lg. doses can cause alcohol poisoning; human carcinogen; experimental tumorigen, teratogen; human reproductive effects by ing., IV; human mutagenic data; TSCA listed
Environmental: VOC; BOD5 1.27; COD 1.99; ThOD 2.09
Precaution: DOT: Flamm. liq.; dangerous fire and explosion hazard; can react vigorously with oxidizers; reacts violently with many chemicals
NFPA: Health 0, Flammability 3, Reactivity 0
Storage: Hygroscopic
Uses: Solvent, cosolvent in paints, inks, dyes, lacquer thinners; extraction medium; mfg. of acetaldehyde, denatured alcohol, paints, waterborne coatings, pharmaceuticals (tonics), colognes, perfumery, organic synthesis; octane booster in gasoline; solvent, coupling agent, wetting agent in cleaners, polishes, coatings; solvent in cosmetics; disinfectant; antimicrobial, extraction solvent, vehicle for foods; alcoholic beverages; active ingred. in OTC drug prods.; topical antiseptic; solvent, excipient in oral pharmaceuticals; component of flavored vehicles; in food-pkg. adhesives; defoamer in food-contact coatings and paper/paperboard
Regulatory: FDA 21CFR §73.30, 73.345, 73.615, 169.3, 169.175, 169.176, 169.177, 169.178, 169.180, 169.181, 172.340, 172.560, 175.105, 176.200, 176.210, 177.1200, 177.1440, 177.1650, 178.1010, 184.1293, GRAS; 27CFR §2.5, 2.12; FEMA GRAS; NF compliance (alcohol diluted); SARA §311/312 acute health/chronic health/fire hazard; BP, EP compliance; Japan restricted
Manuf./Distrib.: A.E. Staley Mfg.; ADM Ethanol Sales; AMC Chems.; AMRESCO; AXO Chem.; Aldrich; Alfa Aesar; Amyl; Ashland; BP Amoco; Baychem; Boith China; Brenntag AG; C.P. Hall; Carbery Food Ingreds. Ltd; Chemacon GmbH; Chematch; Chemcentral; Coyne; Croda Chem. Ltd; DSM Food Spec.; Degussa AG/Health & Nutrition; Delta Distributors; Dow; EMCO Chem. Distributors; Eastech; Eastman; Electron Microscopy Sciences; Equistar; Fluka; Galbraith Labs; Georgia-Pacific Resins; Grain Processing; Great Western; Haarmann & Reimer GmbH; Harcros; Houghton Chem.; Hukill; Integra; J.T. Baker; JLM Marketing; Kraft Chem.; Mallinckrodt Baker; Merck KGaA; Midwest Grain Prods.; P.B. & S.; Penta Mfg.; Polysciences; Quaker City; Romil Ltd; Ruger; SOCO-Lynch; Sasol N. Am.; Sigma; Sithean; Southchem; Spectrum Quality Prods.; Stanchem; Sunnyside; Textile Chem.; Thomas Scientific; Universal Preserv-A-Chem; VWR Int'l.; Veltek Assoc.; Vopak USA; Whittaker, Clark & Daniels
Trade Names: CDA 19-3; Punctilious® Ethyl Alcohol 190 Proof Pure USP; Punctilious® Ethyl Alcohol 200 Proof Pure USP; Punctilious® SDA 1-2; Ucar® Ethanol
Trade Names Containing: Eastman® CDA-19; Filmex® 3C-1 190 Proof; Filmex® A-1, 190 Proof; Filmex® A-2 190 Proof; Filmex® A-2 Anhydrous; Filmex® B 190 Proof; Filmex® B Anhydrous; Filmex® C 190 Proof; Filmex® C Anhydrous; Filmex® D-1; Filmex® D-2 190 Proof; Filmex® D-2 Anhydrous; HFE-71DA; Propsolv I-1 190 Proof; Propsolv I-1 Anhydrous; Propsolv III-1 190 Proof; Propsolv III-1 Anhydrous; Propsolv III-2 190 Proof; Propsolv III-2 Anhydrous; Punctilious® CDA 19-3 190 Proof; Punctilious® CDA 19-3 Anhydrous A; Punctilious® SDA 1-1 190 Proof; Punctilious® SDA 1-1 Anhydrous; Punctilious® SDA 1-2 190 Proof; Punctilious® SDA 2B-2 190 Proof; Punctilious® SDA 2B-2 Anhydrous; Punctilious® SDA 2B-3 190 Proof; Punctilious® SDA 2B-4 190 Proof; Punctilious® SDA 2B-4 Anhydrous; Punctilious® SDA 3A 190 Proof; Punctilious® SDA 3A Anhydrous; Punctilious® SDA 3C 190 Proof; Punctilious® SDA 12A-3 190 Proof; Punctilious® SDA 12A-3 Anhydrous; Punctilious® SDA 13A 190 Proof; Punctilious® SDA 19 190 Proof; Punctilious® SDA 20 190 Proof; Punctilious® SDA 20 Anhydrous; Punctilious® SDA 23A 190 Proof; Punctilious® SDA 23A Anhydrous; Punctilious® SDA 23H 190 Proof; Punctilious® SDA 25A-1 190 Proof; Punctilious® SDA 25A-2 190 Proof; Punctilious® SDA 28A Anhydrous; Punctilious® SDA 29-2 190 Proof; Punctilious® SDA 29-2 Anhydrous; Punctilious® SDA 29-3 190 Proof; Punctilious® SDA 29-13 190 Proof; Punctilious® SDA 29-13 Anhydrous; Punctilious® SDA 30 190 Proof; Punctilious® SDA 30 Anhydrous; Punctilious® SDA 32 190 Proof; Punctilious® SDA 32 Anhydrous; Punctilious® SDA 35A-1 190 Proof; Punctilious® SDA 35A-1 Anhydrous; Punctilious® SDA 36-2 190 Proof; Punctilious® SDA 37-1 190 Proof; Punctilious® SDA 38B 190 Proof Prod. Code 13601027; Punctilious® SDA 38B 190 Proof Prod. Code 13602007; Punctilious® SDA 38B 190 Proof Prod. Code 13602027; Punctilious® SDA 38B 190 Proof Prod. Code 13602133; Punctilious® SDA 38B 190 Proof Prod. Code 13603998; Punctilious® SDA 38B 200 Proof Prod. Code 14601023; Punctilious® SDA 38B 200 Proof Prod. Code 14601040; Punctilious® SDA 38B Anhydrous Prod. Code 14680000; Punctilious® SDA 38F 190 Proof; Punctilious® SDA 38F-2 190 Proof Prod. Code 13643020; Punctilious® SDA 38F-2 190 Proof Prod. Code 13645006; Punctilious® SDA 38F-6 190 Proof; Punctilious® SDA 39B 190 Proof; Punctilious® SDA 39C Anhydrous; Punctilious® SDA 40-1 190 Proof; Punctilious® SDA 40-1 Anhydrous; Punctilious® SDA 40-2 Anhydrous; Punctilious® SDA 40A 190 Proof; Punctilious® SDA 40A Anhydrous; Punctilious® SDA 40B Anhydrous; Punctilious® SDA 40B Modified Anhydrous; Punctilious® SDA 40C 190 Proof; Punctilious® SDA Canada DAG-2A Anhydrous; Punctilious® SDA Canada DAG-2A2 Anhydrous; S-327 190 Proof; S-327 Anhydrous; S-333 Anhydrous; S-336 Anhydrous; S-357 Anhydrous; S-358 Anhydrous; S-359 Anhydrous; Tecsol® A, 95%; Tecsol® A Anhydrous; Tecsol® A-2, 95%; Tecsol® A-2 Anhydrous; Tecsol® B; Tecsol® C 95%; Tecsol® C Anhydrous; Tecsol® D, 95%; Tecsol® D-1, 95%; Tecsol® D-2, 95%; Tecsol® D-2 Anydrous; Vertrel® XE

Alcohol C3. *See* Propyl alcohol
Alcohol C-5. *See* n-Amyl alcohol
Alcohol C6. *See* Hexyl alcohol
Alcohol C7. *See* Heptyl alcohol
Alcohol C_8. *See* Isooctyl alcohol; 2-Ethylhexanol
Alcohol C8. *See* Caprylic alcohol
Alcohol C10. *See* Decyl alcohol
Alcohol C12. *See* Lauryl alcohol
Alcohol, C13-15. *See* C13-15 alcohols
Alcohol C16. *See* Cetyl alcohol
Alcohol C20. *See* Arachidyl alcohol
Alcohol C22. *See* Behenyl alcohol

Alcohol denatured

CAS 64-17-5; EINECS/ELINCS 200-578-6
UN 1170 (DOT); UN 1986 (DOT); UN 1987 (DOT)
Synonyms: Denatured alcohol; Denatured ethanol; Denatured ethyl alco-

hol; Denatured spirits; Potato alcohol
Definition: Alcohol denatured with one or more denaturing agents
Empirical: C_2H_6O
Formula: CH_3CH_2OH
Properties: Colorless clear volatile liq., sweet alcohol odor, sol. in water, methanol, diethyl ether, chloroform, acetone; m.w. 46.08; sp.gr. 0.789 (20 C); m.p. -114 C; b.p. 78 C; flash pt. (CC) 13 C
Toxicology: TLV: 1000 ppm in air; potentially poisonous by ingestion
Precaution: Flamm. liq.; fire hazard; mod. explosion hazard
NFPA: Health 0, Flammability 3, Reactivity 0
Storage: Hygroscopic
Uses: Mfg. of acetaldehyde and other chemicals, solvents, antifreeze, brake fluids, fuels; solvent in protective coatings, printing inks, paint and varnish removers, cosmetics
Manuf./Distrib.: Aldrich; Ashland; BP Chems. Ltd; Fluka; General Chem.; Midwest Grain Prods.; Phoenix Nat. Prods.; Sigma; Spectrum Quality Prods.

Alcohol, denatured formula 23A. *See* SD alcohol 23-A
Alcohols, C13-15. *See* C13-15 alcohols
Alcohols, C16-18. *See* Cetearyl alcohol
Alcohols, lanolin. *See* Lanolin alcohol
Alcohols, tallow. *See* Tallow alcohol
Aldehyde C1. *See* Formaldehyde
Aldehyde C3. *See* Propionaldehyde
Aldehyde C14; Aldehyde C14 pure. *See* γ-Undecalactone
Aldehyde C18. *See* γ-Nonalactone
Alginic acid, ester with 1,2-propanediol. *See* Propylene glycol alginate
Aliphatic petroleum distillate. *See* Naphtha, light aliphatic
Aliphatic petroleum distillates. *See* Naphtha, heavy aliphatic
Aliphatic petroleum solvent (naphtha). *See* Naphtha, medium aliphatic
Alkane C4. *See* Butane
Alkane C5. *See* n-Pentane
Alkane C6. *See* Hexane
Alkane C7. *See* Heptane
Alkane C12. *See* n-Dodecane
Alkane C(16). *See* n-Hexadecane
Alkanes, C10-13. *See* C10-13 alkane
Alkanes, C14-16. *See* C14-16 paraffins
Alkanes, C14-17. *See* C14-17 alkane
Alkanes, C7-8-iso-. *See* C7-8 isoparaffin
Alkanes, C7-10, iso-. *See* C7-10 isoparaffin
Alkanes, C8-9-iso-. *See* C8-9 isoparaffin
Alkanes, C9-10-iso-. *See* C9-10 isoparaffin
Alkanes, C9-11-iso-. *See* C9-11 isoparaffin
Alkanes, C9-12-iso-. *See* C9-12 isoparaffin
Alkanes, C9-13-iso-. *See* C9-13 isoparaffin
Alkanes, C9-14-iso-. *See* C9-14 isoparaffin
Alkanes, C9-15-iso-. *See* C9-15 isoparaffin
Alkanes, C10-11-iso-. *See* C10-11 isoparaffin
Alkanes, C10-13-iso-. *See* C10-13 isoparaffin
Alkanes, C11-12-iso-. *See* C11-12 isoparaffin
Alkanes, C11-13-iso-. *See* C11-13 isoparaffin
Alkanes, C11-16, iso-. *See* C11-16 isoparaffin
Alkanes, C12-14-iso-. *See* C12-14 isoparaffin
Alkanes, C13-14-iso-. *See* C13-14 isoparaffin
Alkanes, C13-16-iso-. *See* C13-16 isoparaffin
Alkanes, C13-17 iso-. *See* C13-17 isoparaffin
Alkanes, C20-40-iso-. *See* C20-40 isoparaffin
Alkanes, iso-, C8-9. *See* C8-9 isoparaffin
Alkanes, iso-, C10-11. *See* C10-11 isoparaffin
Alkanes, iso-, C11-12. *See* C11-12 isoparaffin
Alkanes, iso-, C11-13. *See* C11-13 isoparaffin
Alkanes, iso-, C13-14. *See* C13-14 isoparaffin
n-Alkanol (C16-C18). *See* Cetearyl alcohol
Alkylate petroleum distillates. *See* Petroleum distillates, alkylate
Alkylation naphtha, light, C7-10. *See* Naphtha, light alkylate
Alkylbenzenesulfonic acid. *See* Dodecylbenzenesulfonic acid
Alkylbenzol C10-13 distillation residues. *See* Benzene, mono-C10-13-alkyl derivs., distillation residues
Alkyl (C12-15) benzoate. *See* C12-15 alkyl benzoate
Alkyl C16-18 methyl esters. *See* Methyl canolate

Alligator pear oil. *See* Avocado (Persea gratissima) oil

Allyl alcohol

CAS 107-18-6; EINECS/ELINCS 203-470-7
UN 1098 (DOT)
Synonyms: AA; Allylic alcohol; 3-Hydroxypropene; 2-Propene-1-ol; Propenol; Propen-1-ol-3; 1-Propen-3-ol; 2-Propen-1-ol; Propenyl alcohol; 2-Propenyl alcohol; Vinylcarbinol
Classification: Unsat. aliphatic alcohol
Empirical: C_3H_6O
Formula: $H_2C{=}CHCH_2OH$
Properties: Colorless liq.; pungent mustard-like odor; sol. in water, ethanol, chloroform, diethyl ether, petrol. ether; m.w. 58.09; dens. 0.8520 (20/4 C); vapor pressure 10 mm (10.5 C); f.p. -50 C; m.p. -129 C; b.p. 96.9 C; flash pt. (TOC) 21 C; ref. index 1.4131
Toxicology: ACGIH TLVTWA 2 ppm, STEL 4 ppm (skin); LD50 (oral, rat) 64 mg/kg, (IP, rat) 37 mg/kg, (IV, mouse) 78 mg/kg, (skin, rabbit) 45 mg/kg; highly toxic; toxic effects from skin absorption; poison by inh., ing., skin contact, subcut., IP routes; lachrymator; irritates upper respiratory system, eyes, and skin; pulmonary edema; mutagenic data
Precaution: Corrosive; flamm. liq.; dangerous fire and explosion hazard exposed to heat, flame, or oxidizers; explosive and violent reactions possible; avoid contact with strong acids; can polymerize when heated above 100 C or exposed to light
Hazardous Decomp. Prods.: CO, CO_2; heated to decomp., emits acrid smoke and fumes
NFPA: Health 4, Flammability 3, Reactivity 1
Uses: Intermediate for pharmaceuticals and other org. chemicals; mfg. of glycerol, acrolein, herbicides, resins, plasticizers, allyl diglycol carbonate, allyl glycidyl ether, crosslinking agents; solvent
Regulatory: SARA reportable
Manuf./Distrib.: ABCR UK; Aldrich; Alfa Aesar; Avocado Research Chems.; CBC Europe; Daicel Chem. Ind.; Fisher Scientific UK; Fluka; Honeywill & Stein; Indofine; Itochu Spec. Chems.; Lancaster Synthesis UK; Laporte Perf. Chems.; Lyondell; MDA Chems.; Monomer-Polymer & Dajac Labs; Rhodia; SAF Bulk Chems.; Showa Denko; Spectrum Quality Prods.

Allyl aldehyde. *See* Acrolein
Allylguaiacol; 4-Allylguaiacol; 4-Allyl-1-hydroxy-2-methoxybenzene. *See* Eugenol
Allylic alcohol. *See* Allyl alcohol
4-Allyl-2-methoxyphenol. *See* Eugenol
Allyl trichloride. *See* 1,2,3-Trichloropropane
Almond artificial essential oil. *See* Benzaldehyde
Almond oil; Almond oil, sweet. *See* Sweet almond (Prunus amygdalus dulcis) oil
Aluminosilicic acid, sodium salt. *See* Sodium silicoaluminate
Aluminum lithium hydride. *See* Lithium aluminum hydride
Aluminum sodium silicate. *See* Sodium silicoaluminate
American ash. *See* Potassium carbonate
Amide C_1. *See* Formamide
Amides, coco, N,N-bis (2-hydroxyethyl)-. *See* Cocamide DEA
Amides, peanut oil, N-(2-hydroxypropyl). *See* Peanutamide MIPA
Amine C_4. *See* n-Butylamine
Amine C_5. *See* Pentylamine
Aminic acid. *See* Formic acid
5-Aminoacenaphthene. *See* Acenaphthene
Aminobenzene. *See* Aniline
Amino-1,3-bis (β-ethylhexyl)-5-methylhexahydropyrimidine; 5-Amino-1,3-bis (2-ethylhexyl)-5-methyl-hexhydropyrimidine. *See* Hexetidine
1-Aminobutane. *See* n-Butylamine
2-Aminobutane. *See* s-Butylamine

2-Aminobutanol

CAS 96-20-8; 5856-63-3; 13054-87-0; EINECS/ELINCS 202-488-2; 235-940-2
Synonyms: 2-Amino-1-butanol; 2-Aminobutan-1-ol; 2-Amino-n-butyl alcohol; Butanol-2-amine; 1-Butanol-2-amino-; 1-Hydroxy-2-butanamine; 1-Hydroxy-s-butylamine
Classification: Amino alcohol
Empirical: $C_4H_{11}NO$
Formula: $CH_3CH_2CHNH_2CH_2OH$

Properties: Colorless liq.; misc. in water at 20 C, sol. in alcohols; m.w. 89.14; dens. 0.944 (20/20 C); m.p. -2 C; b.p. 178 C (760 mm); flash pt. 73.3 C; ref. index 1.453
Toxicology: LD50 (oral, mouse) 2300 mg/kg, (IV, mouse) 316 mg/kg; poison by IV and IP routes; mod. toxic by ing.; TSCA listed
Precaution: Combustible; mod. flamm. exposed to heat, flame, or oxidizers; can form explosive mixtures with air ≥ 74 C; corrosive to copper, brass, aluminum
Hazardous Decomp. Prods.: CO, CO_2, NO_x
NFPA: Health 2, Flammability 2, Reactivity 0
Uses: Emulsifier for oils, fats, and waxes, cosmetics, polishes, cleaners, soluble oils; coupling solvent for sol. oils, wax emulsions, creams; absorbent for acidic gases, organic synthesis; textile/leather processing; prod. of detergents/surfactants; buffer in cosmetics; corrosion inhibitor; acid-salt catalyst; pH buffer; pharmaceutical intermediate, emulsifier
Manuf./Distrib.: ANGUS; Aldrich; Dow; Fluka; Hampshire; Indofine; Penta Mfg.; Sigma

2-Amino-1-butanol; 2-Aminobutan-1-ol; 2-Amino-n-butyl alcohol. *See* 2-Aminobutanol
4-Aminobutyric acid lactam; γ-Aminobutyric acid lactam; γ-Aminobutyric lactam; γ-Aminobutyrolactam. *See* 2-Pyrrolidone
1-Amino-2-chlorobenzene; 2-Amino-1-chlorobenzene; o-Aminochlorobenzene. *See* o-Chloroaniline
Aminocyclohexane. *See* Cyclohexylamine
Aminoethane; 1-Aminoethane. *See* Ethylamine
2-Aminoethanol. *See* Ethanolamine
2-Aminoethoxyethanol. *See* 2-(2-Aminoethoxy) ethanol

2-(2-Aminoethoxy) ethanol

CAS 929-06-6; EINECS/ELINCS 213-195-4
UN 3055 (DOT)
Synonyms: 2-Aminoethoxyethanol; DGA; Diethylene glycol amine; Diglycolamine; Ethanol, 2-(2-aminoethoxy)-; β-(β-Hydroxyethoxy) ethylamine
Empirical: $C_4H_{11}NO_2$
Formula: $NH_2CH_2CH_2OCH_2CH_2OH$
Properties: Colorless, slightly viscous liq., amine odor; misc. in water, alcohols, and aromatics; immisc. with aliphatic hydrocarbons; m.w. 105.16; dens. 1.0572 (20/20 C); f.p. -12.5 C; b.p. 221 C; flash pt. 126.6 C
Toxicology: LD50 (oral, rat) 5660 mg/kg, (skin, rabbit) 1190 mg/kg; mod. toxic by skin contact; mildly toxic by ing.; primary irritant; strong irritant to tissue; severe eye, skin irritant; anesthetic, somnolent effect; TSCA listed
Precaution: Corrosive
Hazardous Decomp. Prods.: Heated to decomp., emits toxic fumes of NO_x
Uses: Solvent for removal of acid components from gases, recovery of aromatics from refinery streams; hydrogen sulfide/CO_2 extraction solvent; intermediate; prep. of foam stabilizers, wetting agents, emulsifiers, condensation polymers
Manuf./Distrib.: Aldrich; BASF AG; Fluka; Huntsman; Sigma
Trade Names: Diglycolamine® Agent (DGA®)

2-Aminoethyl alcohol; β-Aminoethyl alcohol. *See* Ethanolamine
Aminoethylethandiamine; N-(2-Aminoethyl)-1,2-ethanediamine; N-(2-Aminoethyl) ethylenediamine. *See* Diethylenetriamine

Aminoethylpiperazine

CAS 140-31-8; EINECS/ELINCS 205-411-0
UN 2815 (DOT)
Synonyms: AEP; 1-(2-Aminoethyl) piperazine; N-Aminoethyl piperazine; N-(2-Aminoethyl) piperazine; N-(β-Aminoethyl) piperazine; 1-Piperazineethanamine; 2-Piperazinoethylamine; 1-(1-Piperazinyl)-2-aminoethane; 2-Piperazin-1-yl ethylamine
Classification: Nonaromatic amine
Empirical: $C_6H_{15}N_3$
Formula: $H_2NC_2H_4NCH_2CH_2NHCH_2CH_2$
Properties: Colorless or pale yel liq.; sol. in water; misc. with alcohols, ether, aromatic solvs.; m.w. 129.24; dens. 0.9837; f.p. 17.6 C; b.p. 222.0 C; flash pt. 200 F; ref. index 1.499
Toxicology: LD50 (oral, rat) 214 mg/kg, (IP, mouse) 250 mg/kg, (skin, rabbit) 880 mg/kg; corrosive; strong irritant to tissue; poison by IP route; mod. toxic by ing. and skin contact; readily absorbed thru skin; mutagen; TSCA listed
Precaution: Combustible; corrosive to copper
Hazardous Decomp. Prods.: Heated to decomp., emits toxic fumes of NO_x
Uses: Epoxy curing agent/hardener; intermediate for pharmaceuticals, anthelmintics, surfactants, synthetic fibers, corrosion inhibitors, asphalt additives, emulsion breakers, emulsifiers, textile chems.; used for civil engineering, coatings, adhesives, grouts, casting and elec. encapsulation; corrosion inhibitor in petrol. prod.; solvent
Manuf./Distrib.: Akzo Nobel; Aldrich; Alfa Aesar; Alkyl Amines Chems. Ltd; BASF AG; Dow; Fabrichem; Fluka; Huntsman; Sumitomo Seika; Tosoh

1-(2-Aminoethyl) piperazine; N-Aminoethyl piperazine; N-(2-Aminoethyl) piperazine; N-(β-Aminoethyl) piperazine. *See* Aminoethylpiperazine
Aminohexahydrobenzene. *See* Cyclohexylamine
Aminomethane. *See* Methylamine
1-Amino-3-methoxypropane. *See* Methoxypropylamine
1-Aminopentane. *See* Pentylamine
2-Aminopentane. *See* Diethylamine
Aminophen. *See* Aniline
2-Aminopropane. *See* Isopropylamine

Ammonia

CAS 7664-41-7; EINECS/ELINCS 231-635-3
UN 1005 (anhyd.); UN 2672; UN 2073 (sol'ns.)
Synonyms: Ammonia anhydrous; Ammonia gas; Anhydrous ammonia
Classification: Inorganic nitrogen compd.
Empirical: H_3N
Formula: NH_3
Properties: Colorless gas or liq., sharp, intensely irritating odor; easily liquefied by pressure; sol. in water, alcohol, ether; m.w. 17.03; sp.gr. 0.682 (-33.4 C); f.p. -77 C; b.p. -33.5 C; pH 11.6 (1N aq.)
Toxicology: ACGIH TLV/TWA 25 ppm; STEL 35 ppm; LD50 (inh., mouse, 1 h) 4837 ppm, (oral, rat) 350 mg/kg (ammonium hydroxide); corrosive; inh. of conc. fumes may be fatal; irritating to eyes and mucous membranes; shown to produce cancer of the skin in human doses of 1000 mg/kg of body wt.; TSCA listed
Precaution: Moderate fire risk; caustic
Hazardous Decomp. Prods.: Exposed to heat, emits toxic fumes of NH_3 and NO_x
NFPA: Health 3, Flammability 1, Reactivity 0
Uses: Fertilizer; refrigerant; nitriding of steel; condensation catalyst; neutralizer; extraction solvent; petroleum industry; latex preservative; explosives; taste and odor control agent in water treatment; condensate corrosion inhibitor; buffer in cosmetics; foods (processing aid); pharmaceuticals (alkalizing agent, buffer); microbial fermentation nutrient
Regulatory: FDA 27CFR §21.95; SARA reportable (anhyd.); Japan approved
Manuf./Distrib.: Advanced Synthesis Tech.; Agrium; Air Prods.; Aldrich; Arch Chems.; Ashland; Asiamerica Int'l.; BASF AG; BDH; Brenntag AG; Brown; Chemconnect; Chemical; ChevronPhillips; Columbus Chem. Ind.; Coyne; Cytec Ind.; DPC Ind.; DSM Fine Chems. Austria; Dakota Gasification; Delta Distributors; Dow; DuPont; EM Science; Eastech; Fluka; Galbraith Labs; Harcros; Hill Bros. Chem.; Holland; Honeywell Perf. Polymers; Integra; J.C. Wilson; J.T. Baker; Jilin Chem. Ind.; Kemira BV; Koch Nitrogen; LaRoche Ind.; Lonza; MG Ind.; Mallinckrodt Baker; Miljac; Milsolv; Monsanto; Nissan Chem. Ind.; Nitron; Noah; Norsk Hydro AS; Occidental; P.B. & S.; PPG Ind.; Penta Mfg.; SOCO-Lynch; Sal Chem.; Seeler Ind.; Solutia; Southchem; Spectrum Quality Prods.; Stanchem; Textile Chem.; U.S. Chems.; Universal Preserv-A-Chem; VWR Int'l.; Veckridge; Vopak USA; Whittaker, Clark & Daniels

Ammonia anhydrous. *See* Ammonia
Ammonia aqua; Ammonia aqueous. *See* Ammonium hydroxide
Ammonia gas. *See* Ammonia
Ammonia solution; Ammonia solution, strong; Ammonia water. *See* Ammonium hydroxide
Ammonium, benzyldiethyl ((2,6-xylylcarbamoyl) methyl)-, benzoate. *See*

Denatonium benzoate

Ammonium, benzyltriethyl-, chloride. *See* Benzyltriethyl ammonium chloride

Ammonium hydroxide

CAS 1336-21-6; EINECS/ELINCS 215-647-6
UN NA 2672 (DOT); INS527; E527
Synonyms: Ammonia aqua; Ammonia aqueous; Ammonia solution; Ammonia solution, strong; Ammonia water; Aqua ammonia; Aqua ammonium; Aqueous ammonia; Spirit of Hartshorn; Strong ammonia solution
Classification: Inorganic base
Empirical: H_5NO
Formula: NH_4OH
Properties: Clear colorless liq., very pungent odor, acrid taste; sol. in water; m.w. 35.06; dens. 0.90; m.p. -77 C; flash pt. none; pH 13.6
Toxicology: LD50 (oral, rat) 350 mg/kg; LDLo (IV, rabbit) 10 mg/kg; human poison by ingestion; experimental poison by inh. and ing.; inhalation irritant; severe eye irritant; liq. can inflict burns; mutation data; TSCA listed
Precaution: DOT: Corrosive material; vapor will ignite, but not readily; incompat. with acrolein, nitromethane, acrylic acid, chlorosulfonic acid, dimethyl sulfate, halogens, HCl, HF, HNO_3, propylene oxide, etc.
Hazardous Decomp. Prods.: Heated to decomp., emits NH_3 and NO_x
NFPA: Health 3, Flammability 1, Reactivity 0
Storage: Keep cool in strong glass, plastic, or rubber-stoppered bottles not completely filled; store below 26 C
Uses: Food additive, alkali, leavening agent, pH control agent, processing aid; boiler water additive; corrosion inhibitor for steam generating systems; taste and odor control in water treatment; textiles; mfg. of rayon, rubber, fertilizers; org. synthesis; refrigeration; condensation polymerization; photography; soaps; lubricants; fireproofing wood; ink mfg.; explosives; ceramics; saponifying fats and oils; buffer, denaturant in cosmetics; food coloring diluent/solvent; alkalizing agent for pharmaceuticals
Regulatory: FDA 21CFR §73.85, 163.110, 163.111, 163.112, 177.1600, 182.90, 184.1139, GRAS; Europe listed; UK approved
Manuf./Distrib.: AMRESCO; Acros Org.; Agrium; Aldrich; Alexander; Alfa Aesar; Alpharma; Ashland; Atomergic Chemetals; Brown; Charkit; Chem-Impex Int'l.; Chemical; Coyne; Cytec Ind.; Degussa AG/Health & Nutrition; Electron Microscopy Sciences; Fisher Scientific; Frutarom Ltd; G.J. Chem.; GFS; Honeywell Perf. Polymers; ICN Biomed. Research Prods.; Integra; Isotec; J.C. Wilson; J.T. Baker; Koch Microelectronic; LaRoche Ind.; Lancaster Synthesis; Lonza; Mallinckrodt Baker; Miljac; Penta Mfg.; Pfaltz & Bauer; Pharmco Prods.; Romil Ltd; Ruger; Sal Chem.; Sigma; Spectrum Quality Prods.; Stanchem; Universal Preserv-A-Chem; VWR Int'l.; Voigt Global Distrib.; Vopak USA
Trade Names Containing: Punctilious® SDA 36-2 190 Proof

Ammonium xylenesulfonate

CAS 26447-10-9; EINECS/ELINCS 247-710-9
Synonyms: Benzenesulfonic acid, dimethyl-, ammonium salt
Definition: Ammonium salt of ring sulfonated mixed xylene isomers
Empirical: $C_8H_{10}O_3S \cdot H_3N$
Properties: Sol. in water; m.w. 203.26; anionic
Toxicology: Chronic toxicity or skin effects are not known; narcotic in high doses; TSCA listed
Precaution: Flamm.
Uses: Hydrotrope, coupling agent, stabilizer, solubilizer, processing aid for personal care prods., detergents, inks, electroplating baths, dyestuffs, polymers; surfactant, visc. control agent in cosmetics; lacquer solvent used in nail polishes; antiblocking and anticaking agent in powd. prods.; cloud pt. depressant; textile finishing; aerosols; cutting oils; glues
Trade Names: STEPANATE® AXS

Amyl acetate

CAS 628-63-7; EINECS/ELINCS 211-047-3
UN 1104 (DOT)
Synonyms: Acetic acid amyl ester; Acetic acid pentyl ester; n-Amyl acetate; Amylacetic ester; Amylacetic ether; Pear oil; Pent-acetate; 1-Pentanol acetate; Pentyl acetate; 1-Pentyl acetate; n-Pentyl acetate; Pentyl ethanoate; Primary amyl acetate
Classification: Carboxylic acid ester
Definition: Ester of amyl alcohol and acetic acid
Empirical: $C_7H_{14}O_2$
Formula: $CH_3COOC_5H_{11}$
Properties: Colorless liq. (pure), yel. liq. (commercial); pear or banana-like odor; completely sol. in alcohol, ether; sl. sol. in water; m.w. 130.19; sp.gr. 0.876 (20 C); vapor pressure 3.5 mm Hg; m.p. -70.8 C; b.p. 149 C; flash pt. (CC) 25 C; autoignition temp. 360 C; ref. index 1.4013; surf. tens. 25.13 dynes/cm; dielec. const. 4.475
Toxicology: ACGIH TLV/TWA 100 ppm; LD50 (oral, rat) 6500 mg/kg; mod. toxic by IP route; skin and eye irritant; inh. of vapor can irritate nose/throat; higher exposure may cause breathing difficulty, increased heart rate, CNS depression; severe exposures may cause unconsciousness; low order of chronic toxicity; TSCA listed
Environmental: VOC; BOD5 0.72; ThOD 2.34
Precaution: Flamm.; LEL 1.1%; UEL 7.5%; dangerous fire hazard exposed to heat or flame; mod. explosive as vapor exposed to flame; incompat. with reducing agents, strong acids, strong bases (decomp. can occur), oxidizing agents (increased risk of fire and explosion)
Hazardous Decomp. Prods.: Heated to decomp., emits acrid smoke and irritating fumes
NFPA: Health 1, Flammability 3, Reactivity 0
Storage: Store in cool, dry, well-ventilated area, out of direct sunlight; avoid generating mist
Uses: Solvent for cosmetics, lacquers, paints; extraction of penicillin; mfg. of photographic film, artificial leather, nail polishes; warning odor; printing and finishing fabrics; solvent for phosphors in fluorescent lamps; synthetic flavoring agent in foods and pharmaceuticals; anti-inflammatory agent; in food-pkg. adhesives
Features: Banana-like flavor
Regulatory: FDA 21CFR §175.105; FEMA GRAS; CERCLA hazardous substance
Manuf./Distrib.: A.P. Chems. Ltd; Aceto; Aldrich; Alfa Aesar; Ashland; Augustus Oils Ltd; BP Chems. Ltd; Bell Flavors & Fragrances; Berje; Chemcentral; Coyne; Creative Fragrances; Dow; E-Chemicals; Fluka; Goldensun Mfg.; Grau Aromatics; Houghton Chem.; Hukill; Indofine; Kyowa Hakko Kogyo; Lluch Essence; MelChem; Osaka Org. Chem. Ind.; Penta Mfg.; Polarome Int'l.; Quaker City; Ruger; Sigma; Spectrum Quality Prods.; Stanchem; Toyo Gosei Kogyo; Vopak USA
Trade Names: Ucar® Primary Amyl Acetate

n-Amyl acetate. *See* Amyl acetate

s-Amyl acetate

CAS 626-38-0
UN 1104 (DOT)
Synonyms: 2-Acetoxypentane; 1-Methylbutyl acetate; 2-Pentanol acetate; 2-Pentyl acetate
Classification: Carboxylic acid ester
Empirical: $C_7H_{14}O_2$
Formula: $CH_3COOCH(CH_3)CH_2CH_2CH_3$
Properties: Colorless clear liq.; fruity odor; sol. in alcohol, ether; sl. sol. in water; m.w. 130.19; sp.gr. 0.862-0.866 (20/20 C); m.p. -78.5 C; b.p. 134 C; flash pt. (CC) 32 C
Toxicology: ACGIH TLV/TWA 125 ppm; LCLo (inh., guinea pig, 5 h) 10,000 ppm; mild skin and eye irritant; inh. of vapor may cause nose/throat irritation; higher exposure may cause difficulty breathing, increased heart rate, CNS depression symptoms; ing. may cause mouth/throat irritation, increased heart rate, headache, dizziness, weakness; unconsciousness on severe exposure; TSCA listed
Precaution: Flamm.; mod. explosive as vapor exposed to heat or flame; incompat. with oxidizing materials (increase fire/explosion hazard), strong acids, strong bases, reducing materials (decomp. can occur)
Hazardous Decomp. Prods.: Heated to decomp., emits acrid smoke and irritating fumes
NFPA: Health 1, Flammability 3, Reactivity 0
Storage: Store in cool, dry, well-ventilated area, out of direct sunlight; avoid generating mist
Uses: Solvent for nitrocellulose, ethylcellulose; mfg. of lacquers, varnishes, cements, artificial leather, nail enamels, coated paper, printing compds.; plastic wood; textile sizing

Manuf./Distrib.: E-Chemicals; Penta Mfg.

Amylacetic ester; Amylacetic ether. *See* Amyl acetate

n-Amyl alcohol

CAS 71-41-0; EINECS/ELINCS 200-752-1
UN 1105 (DOT); FEMA 2056
Synonyms: Alcohol C-5; Amyl alcohol, normal; n-Butyl carbinol; 1-Pentanol; Pentanol-1; Pentan-1-ol; n-Pentanol; Pentyl alcohol; n-Pentyl alcohol; Primary amyl alcohol
Classification: Primary aliphatic alcohol
Empirical: $C_5H_{12}O$
Formula: $CH_3(CH_2)_4OH$
Properties: Colorless clear liq., somewhat sweet balsamic odor; sl. sol. in water; misc. with alcohol, ether; m.w. 88.15; sp.gr. 0.812 (20/4 C); vapor pressure 2.2 mm Hg; m.p. -79 C; b.p. 137-139 C; flash pt. (CC) 38 C; autoignition temp. 300 C; ref. index 1.409; surf. tens. 25.6 dynes/cm; dielec. const. 13.9
Toxicology: LD50 (oral, rat) 3030 mg/kg, (skin, rabbit) 4490 mg/kg; highly toxic; corrosive; narcotic; ingestion of 30 mg can cause death in humans; irritating to respiratory tract; severe eye and skin irritant; can be absorbed thru skin in toxic amts.; inh. can cause nose/throat irritation, CNS depression; extreme exposure may cause double vision, deafness, delirium, fatalities; may cause severe lung damage, respiratory/cardiac arrest, or death if aspirated into lungs; TSCA listed
Environmental: VOC; BOD5 1.55; ThOD 2.73
Precaution: DOT: Flamm. liq.; incompat. with oxidizing agents (increases fire/explosion risk)
NFPA: Health 1, Flammability 3, Reactivity 0
Storage: Store in cool, dry, well-ventilated area, out of direct sunlight
Uses: Solvent in organic synthesis, for nail lacquers, pharmaceuticals; extraction solvent for mineral processing; process solvent; raw material for pharmaceutical preps.; plastics processing; synthetic flavoring agent in foods and pharmaceuticals; fragrance in cosmetics; defoamer in food-contact paper/paperboard
Features: Sweet flavor
Regulatory: FDA 21CFR §172.515, 176.180, 176.210; FEMA GRAS
Manuf./Distrib.: A.P. Chems. Ltd; Aceto; Adept Sol'ns.; Advanced Synthesis Tech.; Aldrich; Alfa Aesar; Allchem Ind.; Ashland; BASF; Celanese; Clariant; Dow; E-Chemicals; Fluka; Grau Aromatics; Houghton Chem.; Hukill; Integra; J.T. Baker; Jarchem Ind.; Penta Mfg.; Prodasynth; Ruger; Sasol N. Am.; Sigma; Sithean; Spectrum Quality Prods.; Stanchem; Synasia Fine Chem.; VWR Int'l.; Vopak USA
Trade Names: Ucar® 1-Pentanol
Trade Names Containing: Pentanol 45

s-Amyl alcohol

CAS 6032-29-7; EINECS/ELINCS 227-907-6
UN 1105 (DOT); FEMA 3316
Synonyms: 2-Hydroxypentane; 1-Methyl-1-butanol; Methyl propyl carbinol; Pentan-2-ol; 2-Pentanol; Pentanol-2; s-Pentyl alcohol
Classification: Sec. aliphatic alcohol
Empirical: $C_5H_{12}O$
Formula: $CH_3CH_2CH_2CH(OH)CH_3$
Properties: Colorless liq.; mild winey, ethereal odor; sol. in water, alcohol, ether m.w. 88.15; sp.gr. 0.8098 (20 C); f.p. -50 C; m.p. -75 C; b.p. 118-119 C; flash pt. (CC) 33 C; ref. index 1.403-1.409 (20 C)
Toxicology: LD50 (oral, rat) 1470 mg/kg; mod. toxic by ing., IP route; narcotic; skin and severe eye irritant; can be absorbed thru skin; inh. may cause nose/throat irritation, headache, dizziness, coughing, nausea, vomiting; severe exposure may cause double vision, deafness, delirium, severe nervous symptoms, fatalities; ing. may cause 'alcohol' intoxication symptoms; may cause severe lung damage, respiratory failure, cardiac arrest, death if aspirated into lungs; TSCA listed
Precaution: Flamm. exposed to heat or flame; severe explosion hazard exposed to heat or flame; incompat. with oxidizing agents (increases fire/explosion hazard), hydrogen trisulfide (explosive reaction possible)
Hazardous Decomp. Prods.: Heated to decomp., emits acrid smoke and irritating fumes
NFPA: Health 0, Flammability 2, Reactivity 0
Storage: Store in cool, dry, well-ventilated area out of direct sunlight
Uses: Solvent for paints and lacquers; intermediate for pharmaceuticals; synthetic flavoring agent in foods; fragrance in cosmetics
Regulatory: FEMA GRAS
Manuf./Distrib.: Aceto; Air Prods.; Aldrich; E-Chemicals; Fluka; Penta Mfg.

s-n-Amyl alcohol. *See* 3-Pentanol

t-Amyl alcohol

CAS 75-85-4; EINECS/ELINCS 200-908-9
UN 1105 (DOT)
Synonyms: Amylene hydrate; Dimethyl ethyl carbinol; 2-Methyl-2-butanol; 2-Methyl butanol-2; 3-Methylbutan-3-ol; t-Pentanol; t-Pentyl alcohol
Classification: Tert. aliphatic alcohol
Empirical: $C_5H_{12}O$
Formula: $CH_3CH_2C(CH_3)_2OH$
Properties: Colorless clear volatile liq., char. camphoraceous odor, burning taste; sol. in 8 parts water; misc. with alcohol, ether, benzene, chloroform, glycerin, oils; m.w. 88.15; sp.gr. 0.808 (20/4 C); vapor pressure 10 mm Hg (17.2 C); m.p. -9 C; b.p. 100-103 C; flash pt. (CC) 19 C; ref. index 1.405
Toxicology: LD50 (oral, rat) 1 g/kg; mod. toxic by ing., IP, subcut. routes; mod. irritating to human mucous membranes; inh. of vapors may cause nose/throat irritation, headache, dyspnea, coughing, nausea, vomiting; drying to skin; can be absorbed thru skin; narcotic in high concs.; a hypnotic agent; ing. may cause 'alcohol' intoxication symptoms; may cause severe lung damage, respiratory and cardiac arrest, death if aspirated into lungs; TSCA listed
Precaution: Flamm. exposed to heat, flame, or oxidizing materials; mod. explosive as vapor exposed to heat or flame
Hazardous Decomp. Prods.: Heated to decomp., emits acrid smoke and irritating fumes
NFPA: Health 1, Flammability 3, Reactivity 0
Storage: Store in cool, well-ventilated area; keep tightly closed; protect from light
Uses: Solvent for epoxy and PU coatings; process solvent; solvent for pharmaceuticals; flotation agent; stabilizer and intermediate for organic synthesis; medicine (sedative)
Regulatory: NF compliance
Manuf./Distrib.: Aceto; Aldrich; Alfa Aesar; Allchem Ind.; BASF AG; BASF; Century Multech; E-Chemicals; Fluka; Mallinckrodt Baker; Pechiney Chems. Div.; Penta Mfg.; Sigma; Spectrum Quality Prods.

Amyl alcohol, active. *See* 2-Methyl-1-butanol
Amyl alcohol, commercial. *See* Fusel oil refined
Amyl alcohol, normal. *See* n-Amyl alcohol
Amyl alcohols, mixed. *See* Fusel oil refined
Amylamine; n-Amylamine. *See* Pentylamine

Amyl benzoate

CAS 2049-96-9; EINECS/ELINCS 218-077-6
Synonyms: n-Amyl benzoate; Benzoic acid, pentyl ester; Pentyl benzoate; n-Pentyl benzoate
Definition: Ester of amyl alcohol and benzoic acid
Empirical: $C_{12}H_{16}O_2$
Properties: M.w. 192.28; flash pt. > 100 C
Toxicology: Primary irritant; mild skin irritant; TSCA listed
Uses: Solvent, masking agent in cosmetics; fragrance; synthetic flavoring agent in foods
Use Level: 0.5% max. in finished cosmetics
Manuf./Distrib.: ABCR; Alfa Aesar; Augustus Oils Ltd; Bell Flavors & Fragrances; Goldensun Mfg.; Grau Aromatics; Lancaster Synthesis; TCI Am.

n-Amyl benzoate. *See* Amyl benzoate
Amyl butyrate. *See* Isoamyl butyrate
γ-N-Amylbutyrolactone. *See* γ-Nonalactone
Amylcarbinol. *See* Hexyl alcohol
Amylene hydrate. *See* t-Amyl alcohol
Amyl ether; n-Amyl ether. *See* Diamyl ether
Amyl ethyl ketone; n-Amyl ethyl ketone. *See* 3-Octanone

Amyl formate

CAS 638-49-3; EINECS/ELINCS 211-340-6
UN 1109 (DOT); FEMA 2068
Synonyms: n-Amyl formate; Formic acid, pentyl ester; Pentyl formate; n-Pentyl formate

Empirical: $C_6H_{12}O_2$
Formula: $HCOOC_5H_{11}$
Properties: Colorless liq.; plum-like odor; misc. with oils, hydrocarbons, alcohols, ether, ketones; sl. sol. in water; m.w. 116.16; dens. 0.880-0.885; m.p. -73.5 C; b.p. 123.5 C; flash pt. 26.6 C
Toxicology: LD50 (oral, rat) > 5 g/kg, (skin, rabbit) > 5 g/kg; toxic by inh. and ing.; mod. irritant by skin contact; TSCA listed
Precaution: Flamm.; dangerous fire risk; reacts vigorously with heat, flame, oxidizing materials
NFPA: Health 1, Flammability 3, Reactivity 0
Uses: Solvent for cellulose esters, resins; films and coatings; fragrance in perfumes; leather; synthetic flavoring agent in foods and pharmaceuticals
Features: Ethereal, fruity, earthy flavoring
Regulatory: FDA 21CFR §172.515; FEMA GRAS
Manuf./Distrib.: Aldrich; Degussa Flavors/Alfrebro; Goldensun Mfg.; Grau Aromatics; Lluch Essence; Penta Mfg.

n-Amyl formate. *See* Amyl formate
Amyl hydride. *See* n-Pentane

Amyl lactate

CAS 6382-06-5
Synonyms: n-Amyl lactate; Lactic acid, amyl ester; Lactic acid, pentyl ester; Propanoic acid, 2-hydroxy-, pentyl ester
Classification: Ester
Empirical: $C_8H_{16}O_3$
Properties: Colorless liq.; m.w. 160.13; sp.gr. 0.96; b.p. 112 C; flash pt. 79 C; ref. index 1.425
Toxicology: May cause skin irritation, allergic skin reaction, dermatitis; inh. may cause respiratory tract irritation; may be harmful by ing.; chronic exposure may cause nausea, vomiting; higher exposure causes unconsciousness; may cause headache, dizziness, tiredness; TSCA listed
Environmental: ThOD 2.10; may cause long-term adverse effects in aquatic environment
Precaution: Flamm. IIIA class; incompat. with oxidizing agents, spontaneously flamm. prods.
Hazardous Decomp. Prods.: thermal decomp. may release irritating gases/vapors, e.g., CO_x; combustion: irritating fumes
NFPA: Health 1, Flammability 2, Reactivity 0
Storage: Store @ R.T.; avoid exposure to air or water over prolonged periods
Uses: Solvent
Regulatory: DOT nonregulated; CERCLA, SARA §313 nonreportable; Canada WHMIS noncontrolled
Manuf./Distrib.: Am. Ingreds./Patco; Ashland; Brown; CPS; Coyne; ExxonMobil; Gallard-Schlesinger Ind.; Grau Aromatics; ICN Biomed. Research Prods.; J.T. Baker; Koch Ind.; Occidental; PURAC Am.; Penta Mfg.; Petroquimica Uniao; Pfanstiehl Labs; Shell; Vopak USA; Wilke Int'l.

n-Amyl lactate. *See* Amyl lactate
Amyl methyl alcohol. *See* 2-Methyl-1-pentanol
Amyl methyl carbinol; n-Amyl methyl carbinol. *See* 2-Heptanol

t-Amyl methyl ether

CAS 994-05-8; EINECS/ELINCS 213-611-4
UN 1993
Synonyms: Butane, 2-methoxy-2-methyl-; 1,1-Dimethylpropyl methyl ether; Ether, methyl t-pentyl; 2-Methoxy-2-methylbutane; Methyl t-amyl ether; Methyl 1,1-dimethyl propyl ether; 2-Methyl-2-methoxybutane; Methyl 2-methyl-2-butyl ether; Methyl t-pentyl ether; t-Pentyl methyl ether; TAME
Empirical: $C_6H_{14}O$
Formula: $C_2H_5C(CH_3)_2OCH_3$
Properties: Colorless clear liq.; sol. 12 g/l in water; m.w. 102.18; sp.gr. 0.764; m.p. -80 C; b.p. 85-87 C; flash pt. -7 C; ref. index 1.3896
Toxicology: LD50 (oral, rat) 1602 mg/kg; LC50 (inh., rat, 4 h) > 5400 mg/m^3; irritating to eyes, skin, respiratory system; may be harmful by ing., inh., skin absorp.; vapor/mist irritating to eyes, mucous membranes, upper respiratory tract; may cause ataxia, musculo-skeletal changes, body temp. decrease; TSCA listed
Precaution: Highly flamm.; vapor may flash back; incompat. with strong oxidizing agents
Hazardous Decomp. Prods.: Toxic fumes of CO, CO_2; emits toxic fumes under fire conditions
Storage: Store in cool, dry place away from heat, sparks, open flame, keep tightly closed
Uses: Solvent; extractant; reaction medium; paraffin-removing agent alone or in combination with other solvents
Manuf./Distrib.: Acros Org.; Aldrich; Fluka

Amyl methyl ketone; n-Amyl methyl ketone. *See* Methyl n-amyl ketone
Amyl propanoate. *See* Amyl propionate

Amyl propionate

CAS 624-54-4; EINECS/ELINCS 210-852-7
Synonyms: Amyl propanoate; n-Amyl propionate; Pentyl propanoate; Pentyl propionate; n-Pentyl propionate; Propanoic acid, pentyl ester
Empirical: $C_8H_{16}O_2$
Formula: $CH_3CH_2COOC_5H_{11}$
Properties: Colorless liq., fruity apricot-pineapple odor; sol. in alcohol, most fixed oils; insol. in glycerin, propylene glycol, water @ 160 C; m.w. 144.21; dens. 0.869-0.873; vapor pressure 1.5 mm Hg (20 C); b.p. 156 C; flash pt. 106 F; ref. index 1.405-1.409
Toxicology: LD50 (oral, rat) > 14 g/kg, (skin, rabbit) > 14 g/kg; TSCA listed
Precaution: Flamm.; fire hazard
Hazardous Decomp. Prods.: Heated to decomp., emits acrid smoke and irritating fumes
NFPA: Health 0, Flammability 2, Reactivity 0
Uses: Fragrance in perfumes; lacquers; synthetic flavoring agent in foods; solvent for high-solids coatings, cellulosic and urethane paints
Regulatory: Japan approved as flavoring
Manuf./Distrib.: Acros Org.; Aldrich; Augustus Oils Ltd; Dow; Frutarom Ltd; Frutarom Meer; Grau Aromatics; Lluch Essence; Penta Mfg.; Pfaltz & Bauer
Trade Names: Ucar® n-Pentyl Propionate

n-Amyl propionate. *See* Amyl propionate
AN. *See* Acrylonitrile
Anesthetic ether. *See* Ethyl ether
1,4-Anhydro-D-glucitol, 6-hexadecanoate. *See* Sorbitan palmitate
Anhydrosorbitol monolaurate. *See* Sorbitan laurate
Anhydrosorbitol trioleate. *See* Sorbitan trioleate
Anhydrous ammonia. *See* Ammonia
Anhydrous hydrobromic acid. *See* Hydrobromic acid
Anhydrous lanolin. *See* Lanolin

Aniline

CAS 62-53-3; EINECS/ELINCS 200-539-3
UN 1547 (DOT)
Synonyms: Aminobenzene; Aminophen; Aniline oil; Benzenamine; CI 76000; Phenylamine
Empirical: C_6H_7N
Formula: $C_6H_5NH_2$
Properties: Colorless to pale yel. oily liq., char. odor; darkens on light exposure; misc. with oxygenated and aromatic solvs.; sol. 35 g/kg in water; m.w. 93.13; dens. 1.021 (20/4 C); vapor pressure 1 mm (34.8 C); f.p. -6.2 C; b.p. 184 C; flash pt. (CC) 76 C; ref. index 1.586 (20 C)
Toxicology: ACGIH TLV/TWA 2 ppm (skin); LD50 (oral, rat) 250 mg/kg, (skin, rat) 1400 mg/kg, (IP, rat) 420 mg/kg; toxic absorbed thru skin; severe eye irritant; skin irritant; an allergen; forms methemoglobin in body, causing CNS depression; may cause liver jaundice; suspected carcinogen; experimental reproductive effects; a common air contaminant; TSCA listed
Precaution: DOT: Poisonous material; combustible exposed to heat or flame; can react vigorously with oxidizers; violent or spontaneous explosive reactions possible
Hazardous Decomp. Prods.: Heated to decomp., emits highly toxic fumes of NO_x
NFPA: Health 3, Flammability 2, Reactivity 0
Storage: Photosensitive
Uses: Intermediate for paints; rubber accelerators and antioxidants; dyes and pigments; intermediates; photographic chems.; isocyanates for urethane foams; pharmaceutical intermediate; explosives; petrol. re-

fining; phenolics; diphenylamine; herbicides; fungicides; epoxy curing agents
Manuf./Distrib.: Aldrich; Alfa Aesar; Allchem Ind.; BASF AG; Bayer; ChemFirst; Columbus Chem. Ind.; Crompton/Uniroyal; Dow; DuPont; Fluka; Honeywell Perf. Polymers; Huntsman; Kessler; Mitsui Petrochem. Ind.; Mitsui Toatsu; New Japan Chem.; Nippon Polyurethane Ind.; Rubicon; Sigma; Spectrum Quality Prods.; Sumitomo Chem.; Sunoco; United Min. & Chem.
Trade Names Containing: Punctilious® SDA 29-13 190 Proof; Punctilious® SDA 29-13 Anhydrous

Aniline, N,N-dimethyl. *See* N,N-Dimethylaniline
Aniline oil. *See* Aniline
Anilinomethane. *See* N-Methylaniline

Anisole

CAS 100-66-3; EINECS/ELINCS 202-876-1
UN 2222 (DOT); FEMA 2097
Synonyms: Benzene, methoxy; Ether, methyl phenyl; Methoxybenzene; Methylphenyl ether; Phenol methyl ether; Phenyl methylether
Classification: Ether
Empirical: C_7H_8O
Formula: $CH_3OC_6H_5$
Properties: Colorless or straw liq.; anise-like odor; sol. in alcohol, ether, oxygenated solvs.; insol. in water; m.w. 108.13; dens. 0.992; vapor pressure 10 mm (42.2 C); m.p. -37 C; b.p. 154 C; flash pt. 42 C; autoignition temp. 475 C; ref. index 1.515-1.518; surf. tens. 35 dynes/cm; dielec. const. 4.33
Toxicology: LD50 (oral, rat) 3700 mg/kg; LC50 (inh., mouse, 2 h) 3021 mg/m^3; mod. toxic by ing. and inh.; skin irritant; target organs: CNS, respiratory system; TSCA listed
Environmental: VOC; ThOD 2.52
Precaution: Flamm. liq.; mod. flamm. exposed to heat, flame, or oxidizers
Hazardous Decomp. Prods.: Heated to decomp., emits acrid fumes
NFPA: Health 1, Flammability 2, Reactivity 0
Uses: Pharmaceutical intermediate; perfumes; cosmetics ingred.; synthetic flavoring agent in foods and pharmaceuticals; solvent; vermicide
Regulatory: FDA 21CFR §172.515; FEMA GRAS
Manuf./Distrib.: Advanced ChemTech; Aldrich; Alfa Aesar; Alfa Chem; Allchem Ind.; Armentor; Blackman Uhler; Fluka; Haarmann & Reimer GmbH; Honeywell Spec. Chems.; Indofine; Integra; J.T. Baker; Kessler; Penta Mfg.; Rhodia/Fine Orgs.; Richard's Pacific Chem. Sourcing; Sigma; Spectrum Quality Prods.; VWR Int'l.; Xinchem

Annulene. *See* Benzene
Anthracen. *See* Anthracene

Anthracene

CAS 120-12-7; EINECS/ELINCS 204-371-1
Synonyms: Anthracen; Anthracin; Green oil; Paranpahthalene
Classification: Aromatic hydrocarbon
Empirical: $C_{14}H_{10}$
Formula: $C_6H_4(CH)_2C_6H_4$
Properties: Colorless cryst. (pure), yel. cryst. with blue or green fluorescence (impure); darkens in sunlight; weak aromatic odor; sol. in acetone, benzene; sl. sol. in alcohol, ether; insol. in water; m.w. 178.2; dens. 1.24; m.p. 217 C; b.p. 340 C; sublimes @ 226.5 C; flash pt. (CC) 121 C; semiconducting props.
Toxicology: OSHA PEL/TWA 0.2 mg/m^3; LD50 (IP, mouse) 430 mg/kg; mod. toxic by IP route; ing. may cause sore throat, GI symptoms; fumes can probably cause irritation of nose, throat, respiratory system, and eyes; can be absorbed thru skin; skin irritant and allergen; possible carcinogen, tumorigen; mutagenic data; TSCA listed
Precaution: Combustible; sol'ns. undergo photooxidation and decomp. when exposed to sunlight or UV light; mod. explosive exposed to flame, $Ca(OCl)_2$, chromic acid; explodes on contact with fluorine
Hazardous Decomp. Prods.: CO_2, CO
NFPA: Health 0, Flammability 1
Uses: Refined anthracene in prod. of anthraquinone, a dye intermediate; prod. of alizarin, phenanthrene, carbazole; in calico printing; component of smoke screens; anthracene oil in wood preservatives and as pesticide (except on food crops)
Features: Present in environment due to incomplete combustion of organic matter (e.g., in coal tar volatiles, tobacco smoke, exhaust emissions, charcoal broiled meat smoke, edible oils, surf. waters, lake sediments)
Manuf./Distrib.: Advanced Synthesis Tech.; Alfa Aesar; Allchem Ind.; Kawasaki Steel; Nihon Jyoryu Kogyo; Nippon Steel Chem.; Osaka Gas; Spectrum Quality Prods.
Trade Names Containing: Methylnaphthalene Fraction

Anthracin. *See* Anthracene

Apricot kernel oil PEG-6 esters

CAS 97488-91-0
Definition: Complex mixture formed from the transesterification of apricot kernel oil and PEG-6
Properties: Nonionic
Uses: Emollient, emulsifier, surfactant in cosmetics; solvent, solubilizer, emulsifier, vehicle, excipient for pharmaceuticals
Features: Hydrophilic

Aqua. *See* Water
Aqua ammonia; Aqua ammonium; Aqueous ammonia. *See* Ammonium hydroxide

Arachidyl alcohol

CAS 629-96-9; EINECS/ELINCS 211-119-4
Synonyms: Alcohol C20; Eicosanol; 1-Eicosanol
Classification: Nonaromatic fatty alcohol
Empirical: $C_{20}H_{42}O$
Formula: $CH_3(CH_2)_{18}CH_2OH$
Properties: Wh. wax-like solid; insol. in water; sol. in alcohol, hot benzene; m.w. 298.56; m.p. 66.5 C; b.p. 369 C
Toxicology: TSCA listed
Precaution: Combustible
Uses: Emollient in cosmetics; lubricants; rubber; plastics; textiles; research; pharmaceutical raw material; emollient, lubricant, consistency agent for pharmaceutical topicals
Manuf./Distrib.: Aldrich; Fluka; Indofine; Jarchem Ind.; M. Michel; Sasol N. Am.; Sigma
Trade Names: Cachalot® AR-20

Arachis hypogaea; Arachis oil. *See* Peanut (Arachis hypogaea) oil
Aromatic castor oil. *See* Castor (Ricinus communis) oil
Aromatic naphtha type I. *See* Naphtha, light aromatic
Artificial almond oil. *See* Benzaldehyde
Artificial ant oil; Artificial oil of ants. *See* Furfural
Aspartic acid, N-acetyl-, diethyl ester. *See* Diethyl acetyl aspartate
Asymmetrical trimethyl benzene. *See* 1,2,4-Trimethylbenzene
ATBC. *See* Acetyl tributyl citrate
ATEC. *See* Acetyl triethyl citrate
Automotive gasoline. *See* Gasoline
Avocado oil. *See* Avocado (Persea gratissima) oil

Avocado (Persea gratissima) oil

CAS 8024-32-6; EINECS/ELINCS 232-428-0
Synonyms: Alligator pear oil; Avocado oil; Oils, avocado
Definition: Oil obtained by pressing dehydrated avocado pear *Persea americana*; consists principally of glycerides of fatty acids
Properties: Ylsh.-green to brownish-green oil, faint char. odor; sol. in min. oil, isopropyl esters, ethanol; insol. in water; dens. 0.908-0.925; iodine no. 84-95; sapon. no. 177-198; ref. index 1.460-1.470
Toxicology: Primary irritant; severe skin and eye irritant; TSCA listed
Hazardous Decomp. Prods.: Combustion produces CO_2; heated to decomp., emits acrid smoke and irritating vapors
Uses: Emollient, conditioner, gloss aid for cosmetics, toiletries, pharmaceuticals; food ingred.
Manuf./Distrib.: AXO Chem.; Alban Muller; Aldivia; Alfa Chem; Amend Drug & Chem.; Amerol; Anglia Oils; Arista Ind.; Berje; Centerchem; Charkit; Chart; Cognis/Care Chems.; Cosmetic Supplies USA; Croda Inc; Desert Whale Jojoba; Fabrichem; Int'l. Sourcing; Jeen Int'l.; Lipo; Natural Oils Int'l.; Parchem Trading; Penta Mfg.; Pokonobe Ind.; Provital; Ruger; S. Black; Tri-K Ind.; Welch, Holme & Clark
Trade Names: Phytol™ AVO

Axole. *See* Furan
Azabenzene. *See* Pyridine
Azacyclohexane. *See* Piperidine
1-Azanaphthalene. *See* Quinoline
2-Azanaphthalene. *See* Isoquinoline
3-Azapentane-1,5-diamine. *See* Diethylenetriamine
Azeite. *See* Olive (Olea europaea) oil
Azine. *See* Pyridine

B

2-BA. *See* s-Butylamine
Babassu oil. *See* Babassu (Orbignya oleifera) oil

Babassu (Orbignya oleifera) oil

CAS 91078-92-1; EINECS/ELINCS 293-376-2
Synonyms: Babassu oil; Oils, babassu
Definition: Fixed oil obtained from the nuts of *Orbigyna oleifera*
Properties: Iodine no. 13-18; sapon. no. 247-251
Uses: Emollient for sunscreen prods.; foods, soapmaking
Features: Supply limited
Manuf./Distrib.: ABITEC; Alban Muller; Croda Inc; Fabrichem; Pokonobe Ind.; Thornley
Trade Names: Cropure® Babassu

Banana oil. *See* Isoamyl acetate
BBP. *See* Butyl benzyl phthalate
BC. *See* 1,2-Butylene carbonate
1,4-BD. *See* 1,4-Butanediol
BDCM. *See* Bromodichloromethane
BDO; 1,4-BDO. *See* 1,4-Butanediol

Behenyl alcohol

CAS 661-19-8; EINECS/ELINCS 211-546-6
Synonyms: Alcohol C22; 1-Docosanol
Definition: Mixture of fatty alcohols chiefly of n-docosanol
Empirical: $C_{22}H_{46}O$
Formula: $CH_3(CH_2)_{20}CH_2OH$
Properties: Colorless waxy solid; sol. in oxygenated solvs., ethanol, chloroform; insol. in water, m.w. 326.61; m.p. 71 C; b.p. 180 C (0.22 mm)
Toxicology: Low toxicity; TSCA listed
Uses: Synthetic fibers; evaporation retardant on water surfaces; surfactant for polymerization; emollient in cosmetics; foam control agent; cosolvent; plasticizer; mfg. of household/industrial cleaners, personal care prods., textile auxs., plasticizers, ore flotation, oil well drilling, metal lubricants, agric. additives; raw material, consistency agent, emollient for pharmaceuticals; as antihistamine; in surf. lubricants for mfg. of food-contact metallic articles
Regulatory: FDA 21CFR §178.3910
Manuf./Distrib.: Aldrich; Ashland; Brown; CarboMer; CoKEM Assoc.; Fluka; Jarchem Ind.; M. Michel; Pechiney Chems. Div.; Sasol N. Am.; Schweizerhall; Sea-Land; Sigma; Universal Preserv-A-Chem
Trade Names: Cachalot® BE-22

Behenyl behenate

CAS 17671-27-1; EINECS/ELINCS 241-646-5
Synonyms: Docosanoic acid, docosyl ester; Docosyl docosanoate
Definition: Ester of behenic acid and behenyl alcohol
Empirical: $C_{44}H_{48}O_2$
Formula: $CH_3(CH_2)_{20}COOCH_2(CH_2)_{20}CH_3$
Uses: Emollient, moisturizer, visc. builder for cosmetics, topical pharmaceuticals
Manuf./Distrib.: Sigma
Trade Names: Exceparl BB

Behenyl benzoate

CAS 103403-38-9
Synonyms: 1-Docosanol, benzoate
Properties: M.p. 52-55 C
Uses: Skin conditioner, emollient in cosmetics; structurizing agent, heat stabilizer, solvent in cosmetic stick formulations
Features: High m.p.; compat. with most lipophilic materials
Trade Names: Finsolv® 137

Benne oil. *See* Sesame (Sesamum indicum) oil
Bentanol. *See* Benzyl alcohol
Benylate. *See* Benzyl benzoate
Benzal alcohol. *See* Benzyl alcohol

Benzaldehyde

CAS 100-52-7; EINECS/ELINCS 202-860-4
UN 1989 (DOT); UN 1990; UN NA 1993; FEMA 2127
Synonyms: Almond artificial essential oil; Artificial almond oil; Benzenecarbaldehyde; Benzene carbonal; Benzene carboxaldehyde; Benzene methylal; Benzic aldehyde; Benzoic aldehyde; Phenylmethanal; Synthetic oil of bitter almond
Classification: Aromatic aldehyde
Empirical: C_7H_6O
Formula: C_6H_5CHO
Properties: Colorless to yel. oily liq., bitter almond odor, burning taste; very sol. in alcohol, ether, benzene, acetone, petrol. ether, conc. sulfuric acid, liq. CO_2, liq. ammonia; sl. sol. in water; m.w. 106.13; sp.gr. 1.041; vapor pressure 1 mm Hg (26.2 C); m.p. -26 C; b.p 179 C; flash pt. (CC) 62 C; ref. index 1.544
Toxicology: LD50 (oral, rat) 1300 mg/kg, (IP, mouse) 9 mg/kg, (subcut., rabbit) 5000 mg/kg; highly toxic; poison by ing. and IP routes; allergen; feeble local anesthetic; skin and eye irritant; vapors can irritate nose, throat, lungs; CNS depressant (nausea, vomiting, headache, dizziness, abdominal pain); 1 g/kg may be fatal in humans; mutagenic data; TSCA listed
Precaution: DOT: Combustible liq.; a strong reducing agent; acts violently with oxidizers; incompat. with strong bases, strong reducing agents, alkaline cyanide sol'n., heavy metal salts
NFPA: Health 2, Flammability 2, Reactivity 0
Storage: Light-sensitive; store in cool, dry, well-ventilated area, protected from light; store under nitrogen or other inert atm.; keep tightly closed when not in use
Uses: Chemical intermediate for dyes, odorants, aromatic alcohols, pharmaceuticals, pesticides; paints; solvent for oils, resins, cellulose acetate and nitrate; mfg. of cinnamic acid, benzoic acid, photographic chemicals; plastics additive; solvent in cosmetics; fragrance ingred.; synthetic flavoring agent and fragrance in foods, beverages, cosmetics, pharmaceuticals, perfumes, cleaning prods.
Regulatory: FDA 21CFR §182.60, GRAS; 27CFR §21.65, 21.151, GRAS; FEMA GRAS; Japan approved as flavoring; NF, BP compliance
Manuf./Distrib.: ADA Int'l.; Aceto; Advanced BioTech; Aldrich; Allchem Ind.; Ashland; Atofina SA; Augustus Oils Ltd; Axxence Aromatic GmbH; Bayer; Berje; Biddle Sawyer; Britannia Natural Prods.; Celanese; Chemical; Chemisphere; DSM NV; Delta Distributors; Dragoco Australia; Eastech; Elan; Fleurchem; Fluka; Foote & Jenks; George Uhe; Givaudan Fragrances; H.E. Daniel Ltd; Haarmann & Reimer; Indofine; Integra; J.C. Wilson; J.T. Baker; Kraft Chem.; Lluch Essence; MelChem;

Milsolv; Mitsubishi Chem.; Moore Ingreds.; Naturex; Noveon Kalama; O'Laughlin Ind.; P.B. & S.; Penta Mfg.; Prodasynth; Punda Mercantile; Quimdis; R.W. Greeff; Rit-Chem; Ruger; SOCO-Lynch; Schweizerhall; Sigma; Southchem; Spectrum Quality Prods.; Universal Preserv-A-Chem; Vopak USA, Whittaker, Clark & Daniels; Whyte Chems. Ltd; Xinchem

Benzamide, N,N-diethyl-3-methyl. *See* Diethyl toluamide
Benzanthracene. *See* 1,2-Benzanthracene

1,2-Benzanthracene

CAS 56-55-3; EINECS/ELINCS 200-280-6
Synonyms: Benzanthracene; Benz(a)anthracene; Benzanthrene; 1,2-Benzanthrene; Benzoanthracene; 1,2-Benzoanthracene; 2,3-Benzophenanthrene; Benzo(a)phenanthrene; Benzo(b)phenanthrene; 2,3-Benzphenanthrene; Naphthanthracene; Tetraphene
Classification: Aromatic hydrocarbon, fused ring
Empirical: $C_{18}H_{12}$
Properties: Colorless leaflets or plates; sol. in most org. solvs.; dissolves with difficulty in boiling alcohol; insol. in water; m.w. 228.30; m.p. 160 C; b.p. 400 C; sublimes
Toxicology: TDLo (skin, mouse) 18 mg/kg, (subcut., mouse) 2 mg/kg; LD50 (IV, rat) > 200 mg/kg; poison by IV route; confirmed carcinogen; experimental neoplastigen, tumorigen by skin contact and other routes; human mutagenic data; TSCA listed
Environmental: Very slow biodeg. (close to 1 yr half-life); bioconcentrates in aquatic organisms; in unadsorbed state, degrades by photolysis in hours to days; strongly adsorbed to soil
Hazardous Decomp. Prods.: Heated to decomp., emits acrid smoke and irritating fumes
Uses: Biochemical, biomedical, laboratory, and cancer research
Manuf./Distrib.: Aldrich; Fluka; Sigma; Supelco
Trade Names Containing: Kenplast® G

Benz(a)anthracene; Benzanthrene; 1,2-Benzanthrene. *See* 1,2-Benzanthracene
1-Benzazine. *See* Quinoline
2-Benzazine. *See* Isoquinoline
Benzenamine. *See* Aniline
Benzenamine, N,N-dimethyl-. *See* N,N-Dimethylaniline
Benzenamine, N-methyl-. *See* N-Methylaniline

Benzene

CAS 71-43-2; EINECS/ELINCS 200-753-7
UN 1114 (DOT)
Synonyms: Annulene; Benzol; Benzole; Benzolene; Bicarburet of hydrogen; Carbon oil; Coal naphtha; Cyclohexatriene; Mineral naphtha; Motor benzol; Nitration benzene; Phene; Phenyl hydride; Pyrobenzol; Pyrobenzole
Classification: Aromatic hydrocarbon
Empirical: C_6H_6
Properties: Colorless to light-yel. mobile nonpolar liq., aromatic hydrocarbon odor; misc. with alcohol, ether, chloroform, CCl_4, acetone, oils; sl. sol. in water; m.w. 78.11; sp.gr. 0.8790 (20/4 C); vapor pressure 100 mm Hg (26 C); f.p. 5.5 C; b.p. 80.1 C; flash pt. (CC) -11 C; ref. index 1.50110 (20 C)
Toxicology: ACGIH TLV/TWA 10 ppm; LD50 (oral, rat) 3.8 ml/kg, (skin, mouse) 48 mg/kg; very toxic; narcotic; poison by inh. causing CNS depression, dizziness, headache, nausea, unconsciousness; absorbed thru the skin causing systemic effects; severe eye irritant; mod. skin irritant; chronic exposure can cause bone marrow damage, leukemia; human carcinogen; mutagenic effects; experimental neoplastigen, tumorigen, teratogen; common air contaminant; TSCA listed
Precaution: Extremely flamm.; fire risk; incompat. with oxidizers, peroxides, chlorine, nitric acid, ozone, interhalogens, metal perchlorates, etc.; explosion risk; potential violent reactions
Hazardous Decomp. Prods.: None known
NFPA: Health 2, Flammability 3, Reactivity 0
Storage: Store in cool, dry, well-ventilated area out of direct sunlight, away from heat/ignition sources, away from corrosives
Uses: Mfg. of ethylbenzene, dodecylbenzene, cyclohexane, phenol, nitrobenzene, chlorobenzene, cumene, maleic anhydride, dyes, medicinals, artificial leather, airplane dopes, lacquers; solvent for waxes, resins, oils, paints, pharmaceuticals; very minor component of gasoline; extraction solvent for modified hop extract for flavoring beer; in food-pkg. adhesives
Use Level: 2 ppm (pharmaceuticals)
Regulatory: FDA 21CFR §172.560, 175.105; SARA reportable
Manuf./Distrib.: Aldrich; Alfa Aesar; Ashland; Atofina; BP Amoco; BP Chems. Ltd; ChevronPhillips; Dow; Equistar; ExxonMobil; Fluka; Honeywell Spec. Chems.; Hukill; Huntsman; J.T. Baker; LG Petrochem. Co. Ltd; Lyondell; Marathon Ashland Petrol.; Nova Ltd; Occidental; Petroquimica Uniao; Romil Ltd; Shell; Sigma; Spectrum Quality Prods.; Springfield Scientific Supplies; Sunoco
Trade Names Containing: X-95 Solvent

Benzeneacetic acid, ethyl ester. *See* Ethyl phenylacetate
Benzene, benzyl-. *See* Diphenylmethane
Benzene, 1,4-bis (1-methylethyl)-. *See* 1,4-Diisopropyl benzene
Benzene, bromo-. *See* Bromobenzene
Benzene, C8-16 alkyl derivs. *See* C8-16 alkylbenzene
Benzene, C12-14 alkyl dervis. *See* C12-14 alkylbenzene
Benzenecarbaldehyde. *See* Benzaldehyde
Benzenecarbinol. *See* Benzyl alcohol
Benzene carbonal; Benzene carboxaldehyde. *See* Benzaldehyde
Benzene chloride. *See* Chlorobenzene
Benzene, chloromethyl-. *See* Monochlorotoluene
Benzene, 1-chloro-2-(trifluoromethyl)-. *See* o-Chlorobenzotrifluoride
Benzene, 1-chloro-4 (trifluoromethyl). *See* p-Chlorobenzotrifluoride
1,2-Benzenedicarboxylic acid bi (2-methoxyethyl) ester. *See* Dimethoxyethyl phthalate
1,2-Benzenedicarboxylic acid, butyl octyl ester. *See* Butyl octyl phthalate
1,2-Benzenedicarboxylic acid, butyl phenylmethyl ester. *See* Butyl benzyl phthalate
1,2-Benzenedicarboxylic acid, dibutyl ester; o-Benzenedicarboxylic acid dibutyl ester; Benzene-o-dicarboxylic acid di-n-butyl ester. *See* Dibutyl phthalate
1,2-Benzenedicarboxylic acid diethyl ester. *See* Diethyl phthalate
1,3-Benzenedicarboxylic acid, diethyl ester. *See* Diethyl isophthalate
1,2-Benzenedicarboxylic acid dimethyl ester. *See* Dimethyl phthalate
1,2-Benzenedicarboxylic acid dioctyl ester; o-Benzenedicarboxylic acid dioctyl ester. *See* n-Dioctyl phthalate
Benzene, o-dichloro-. *See* o-Dichlorobenzene
Benzene, 2,4-dichloro-1-methyl-. *See* 2,4-Dichlorotoluene
Benzene, 1,2-dichloro-4-(trifluoromethyl)-. *See* 3,4-Dichlorobenzotrifluoride
Benzene, diethyl-. *See* Diethylbenzene
Benzene-, 1,3-diisocyanatomethyl-. *See* Toluene diisocyanate
Benzene, p-diisopropyl-. *See* 1,4-Diisopropyl benzene
Benzene, 1,2-dimethyl. *See* o-Xylene
1,2-Benzenediol. *See* Pyrocatechol
1,4-Benzenediol. *See* Hydroquinone
o-Benzenediol. *See* Pyrocatechol
p-Benzenediol. *See* Hydroquinone
Benzeneethanol. *See* Phenethyl alcohol
Benzene hexahydride. *See* Cyclohexane
Benzene isopropyl. *See* Cumene
Benzenemethanaminium, N-(2-((2,6-dimethylphenyl) amino)-2-oxoethyl)-N,N-diethyl-, benzoate. *See* Denatonium benzoate
Benzenemethanaminium, N,N,N-triethyl-, chloride. *See* Benzyltriethyl ammonium chloride
Benzenemethanaminium, N,N,N-trimethyl-, chloride. *See* Benzyl trimethyl ammonium chloride
Benzenemethanol. *See* Benzyl alcohol
Benzenemethanol, α-methyl-. *See* α-Methylbenzyl alcohol
Benzene, methoxy. *See* Anisole
Benzene methylal. *See* Benzaldehyde
Benzene, 1,1′-methylenebis-. *See* Diphenylmethane
Benzene, 1-methyl-4-(1-methylethyl)-. *See* p-Cymene
Benzene, (2-methylpropyl)-. *See* Isobutylbenzene
Benzene, mono-C12-14 alkyl derivs. *See* C12-14 alkylbenzene

Benzene, mono-C10-13-alkyl derivs., distillation residues

CAS 84961-70-6; EINECS/ELINCS 284-660-7
Synonyms: Alkylbenzol C10-13 distillation residues
Uses: Plasticizer for vinyl; wire rope lubricant; heat transfer fluid; solvent

for oil-sol. sulfonates; specialty lubricants; sulfonation feedstock; intermediate for agric. emulsifiers, enhanced oil recovery surfactants
Trade Names: V-154L

Benzene monochloride. *See* Chlorobenzene
Benzenenitrile. *See* Benzonitrile
Benzenesulfonic acid, dimethyl-, ammonium salt. *See* Ammonium xylenesulfonate
Benzenesulfonic acid, dodecyl-. *See* Dodecylbenzenesulfonic acid
Benzenesulfonic acid, 4-methyl-, sodium salt. *See* Sodium toluenesulfonate
Benzene, 1,2,4-trichloro-. *See* 1,2,4-Trichlorobenzene
Benzene, 1,3,5-trichloro-. *See* 1,3,5-Trichlorobenzene
Benzene, 1,3,5-trimethyl-. *See* 1,3,5-Trimethylbenzene
Benzene, 1,3,5-tris (1-methylethyl). *See* 1,3,5-Triisopropylbenzene
Benzenol. *See* Phenol
Benzenyl fluoride. *See* Benzotrifluoride
Benzic aldehyde. *See* Benzaldehyde
Benzin; Benzine. *See* Naphtha
1-Benzine. *See* Quinoline
Benzine (light petroleum distillate). *See* VM&P naphtha
Benzinoform. *See* Carbon tetrachloride
Benzoanthracene; 1,2-Benzoanthracene. *See* 1,2-Benzanthracene
1,3-Benzodioxol-5-ol. *See* Sesamol
Benzohydroquinone. *See* Hydroquinone
Benzoic acid, benzyl ester. *See* Benzyl benzoate
Benzoic acid, butyl ester; Benzoic acid n-butyl ester. *See* Butyl benzoate
Benzoic acid, C12-15 alkyl esters. *See* C12-15 alkyl benzoate
Benzoic acid, diester with diethylene glycol. *See* Diethylene glycol dibenzoate
Benzoic acid diester with dipropylene glycol; Benzoic acid-n-dipropylene glycol diester. *See* Dipropylene glycol dibenzoate
Benzoic acid ethyl ester. *See* Ethyl benzoate
Benzoic acid, hexyl ester. *See* Hexyl benzoate
Benzoic acid, 2-hydroxy-, 2-butyloctyl ester. *See* Butyl octyl salicylate
Benzoic acid, 2-hydroxy-, 2-ethylhexyl ester. *See* 2-Ethylhexyl salicylate
Benzoic acid, 2-hydroxy-, isohexadecyl ester. *See* Isocetyl salicylate
Benzoic acid methyl ester. *See* Methyl benzoate
Benzoic acid nitrile. *See* Benzonitrile
Benzoic acid, pentyl ester. *See* Amyl benzoate
Benzoic acid phenylmethyl ester. *See* Benzyl benzoate
Benzoic aldehyde. *See* Benzaldehyde
Benzoic ether. *See* Ethyl benzoate
Benzol; Benzole; Benzolene. *See* Benzene
Benzoline. *See* VM&P naphtha

Benzonitrile

CAS 100-47-0; EINECS/ELINCS 202-855-7
UN 2224 (DOT)
Synonyms: Benzenenitrile; Benzoic acid nitrile; Cyanobenzene; Phenyl cyanide
Empirical: C_7H_5N
Formula: C_6H_5CN
Properties: Colorless oily liq.; almond-like odor; sol. in alcohol, diethyl ether, acetone, benzene; misc. with most org. solvs.; sl. sol. in cold water; m.w. 103.13; dens. 1.246 (20/4 C); m.p. -13 C; b.p. 190.7 C; flash pt. 70 C
Toxicology: LD50 (oral, mouse) 971 mg/kg, (IP, mouse) 400 mg/kg; high toxicity; absorbed by skin; poison by IP and subcut. routes; mod. toxic by ing., inh., skin contact; skin irritant; TSCA listed
Precaution: Combustible; DOT: poisonous material
Hazardous Decomp. Prods.: CO, CO_2, NO_x, hydrogen cyanide; heated to decomp., emits toxic fumes of CN^- and NO_x
Uses: Mfg. of benzoquanamine; intermediate for rubber chemicals; solvent for nitrile rubber, specialty lacquers, resins and polymers, anhydrous metallic salts; cosmetics ingred.
Manuf./Distrib.: Aldrich; Alemark; Alfa Aesar; Amber Syn.; BASF; Biddle Sawyer; Fluka; Honeywell Spec. Chems.; J.T. Baker; Nantong ChangChem; PMC Spec.; Penta Mfg.; SKW Chems.; Sigma; Spectrum Quality Prods.; Springfield Scientific Supplies; Synasia Fine Chem.; Triple Crown Am.

2,3-Benzophenanthrene; Benzo(a)phenanthrene; Benzo(b)phenanthrene. *See* 1,2-Benzanthracene
Benzo[b]pyridine. *See* Quinoline
Benzo(c)pyridine. *See* Isoquinoline
Benzoquinone; p-Benzoquinone. *See* Hydroquinone

Benzotrifluoride

CAS 98-08-8
UN 2338 (DOT)
Synonyms: Benzenyl fluoride; Benzylidyne fluoride; Phenylfluoroform; Toluene trifluoride; Trifluoromethylbenzene; α,α,α-Trifluorotoluene
Empirical: $C_6H_5F_3$
Properties: Water-wh. liq.; aromatic odor; insol. in water; m.w. 146.12; dens. 1.182; vapor pressure 11 mm Hg (0 C); f.p. -29.1 C; b.p. 102.1 C; flash pt. 12.2 C; ref. index 1.4146
Toxicology: LD50 (oral, rat) 15,000 mg/kg, (IP, mouse) 100 mg/kg; TLV (as F) 2.5 mg/m^3 of air; poison by IP route; mod. toxic by subcut. route; highly toxic by inh.; TSCA listed
Precaution: Flammable; dangerous fire risk; incompat. with oxidizers
Hazardous Decomp. Prods.: Heated to decomp., emits toxic fumes of F^-
NFPA: Health 3, Flammability 3, Reactivity 1
Uses: Intermediate for dyes and pharmaceuticals; solvent; dielectric fluid; vulcanizing agent; insecticides
Manuf./Distrib.: Apollo Scientific; Celanese; Nantong ChangChem
Trade Names: Oxsol® 2000

Benzoyl alcohol. *See* Benzyl alcohol
Benzoyl methide. *See* Acetophenone
2,3-Benzphenanthrene. *See* 1,2-Benzanthracene
Benztrimonium chloride. *See* Benzyl trimethyl ammonium chloride

Benzyl acetate

CAS 140-11-4; EINECS/ELINCS 205-399-7
FEMA 2135
Synonyms: Acetic acid benzyl ester; Acetic acid phenylmethyl ester; Acetoxymethylbenzene; α-Acetoxytoluene; Benzyl ethanoate; Phenylmethyl acetate
Classification: Carboxylic acid ester
Definition: Ester of benzyl alcohol and acetic acid
Empirical: $C_9H_{10}O_2$
Formula: $CH_3COOCH_2C_6H_5$
Properties: Colorless liq., sweet floral fruity odor; sol. in alcohol, chloroform, benzene, most fixed oils, propylene glycol; pract. insol. in water, glycerin; m.w. 150.19; sp.gr. 1.06; m.p. -51.5 C; b.p. 213.5 C; flash pt. (CC) 90 C; ref. index 1.501
Toxicology: ACGIH TLV/TWA 10 ppm; LD50 (oral, rat) 2490 mg/kg; LCLo (inh., mouse, 22 h) 1300 mg/m^3; mod. toxic by ingestion and subcutaneous routes; poison by inhalation; may be sl. absorbed thru skin; mild eye irritant; antipsychotic; inh. of vapors can probably cause nose/throat irritation; high concs. can cause CNS depression, drowsiness, poor coordination, unconsciousness; TSCA listed
Precaution: Combustible liq.; incompat. with strong oxidizing agents (increases fire/explosion risk), acids, bases, reducing agents (decomp. can occur)
Hazardous Decomp. Prods.: Benzyl alcohol, benzoic acid, acetic acid; heated to decomp., emits irritating fumes
NFPA: Health 1, Flammability 1, Reactivity 0
Storage: Store in cool, dry, well-ventilated area, out of direct sunlight
Uses: Fragrance in artificial jasmine and other perfumes, soap perfumes; synthetic flavoring agent in foods and pharmaceuticals; solvent and high boiler for cellulose acetate and nitrate, natural and synthetic resins; oils; lacquers; polishes; printing inks; varnish removers; solvent in cosmetics
Features: Apple, apricot, cherry, plum-like flavor
Regulatory: FDA 21CFR §172.515; FEMA GRAS; Japan approved as flavoring
Manuf./Distrib.: Aceto; Advanced BioTech; Akzo Nobel; Aldrich; Alfa Chem; Allan; Ashland; Asiamerica Int'l.; Augustus Oils Ltd; Axxence Aromatic GmbH; BCH Brühl; Berje; Chemisphere; Chr. Hansen Inc; Creative Fragrances; Degussa Flavors/Alfrebro; Elan; Epochem; Fleurchem; Fluka; George Uhe; Goldensun Mfg.; Haarmann & Reimer; Haltermann GmbH; Indofine; Int'l. Chem. Group; J.H. Calo; Koyo; Lluch Essence; MelChem; Moore Ingreds.; Morflex; Noveon Kalama; Penta Mfg.; Polarome Int'l.; Quest Int'l.; R.W. Greeff; Reilly Chems. SA; Rhodia/

Fine Orgs.; Sarcom; Spectrum Quality Prods.; Springfield Scientific Supplies; Whyte Chems. Ltd

Benzyl alcohol

CAS 100-51-6; EINECS/ELINCS 202-859-9
FEMA 2137
Synonyms: Bentanol; Benzal alcohol; Benzenecarbinol; Benzenemethanol; Benzoyl alcohol; (Hydroxymethyl) benzene; Hydroxytoluene; α-Hydroxytoluene; Phenolcarbinol; Phenylcarbinol; Phenylmethanol; Phenylmethyl alcohol; α-Toluenol
Classification: Aromatic alcohol
Empirical: C_7H_8O
Formula: $C_6H_5CH_2OH$
Properties: Water-wh. liq., faint aromatic odor, sharp burning taste; misc. with alcohol, chloroform, ether, water @ 206 C (dec.); m.w. 108.15; sp.gr. 1.042; vapor pressure 1 mm Hg (58 C); m.p. -15.3 C; b.p. 205.7 C; flash pt. (CC) 93 C; ref. index 1.540
Toxicology: LD50 (oral, rat) 1230 mg/kg, (IP, rat) 400 mg/kg, (skin, rabbit) 2000 mg/kg; poison by ing., IP, IV routes; mod. toxic by inh., skin contact; mod. skin and severe eye irritant; may cross-react with balsam Peru in sensitive persons; TSCA listed
Precaution: Combustible liq.; incompat. with oxidizers (increased fire and explosion risk), acids (exothermic polymerization), sulfuric acid
Hazardous Decomp. Prods.: Dec. explosively at 180 C with sulfuric acid; heated to decomp., emits acrid smoke and fumes
NFPA: Health 2, Flammability 1, Reactivity 0
Storage: Store in cool, dry, well-ventilated area, away from heat and ignition sources; light-sensitive
Uses: Perfumes; photographic developer ingred.; dyeing wool, nylon, textiles, leather, sheet plastics; solvent for dyestuffs, cellulose esters, gelatin, casein, waxes, lacquers, shellac, ball-point pen inks; heat-sealing polyethylene films; intermediate for benzyl esters; epoxy curing agent; degreasing agent (rug cleaners); bacteriostat; fragrance/flavoring solvent; preservative, solvent in cosmetics; antimicrobial, preservative, solubilizer, synthetic flavoring agent in foods and pharmaceuticals; astringent; food-pkg. adhesives; in closure-sealing gaskets for food containers
Features: Citrus berry cherry-like flavor
Use Level: 10% (disinfectant, local anesthetic), 1% (additive to injectables), 1% max. (solvs., perfumes, and flavorings in finished prods.)
Regulatory: FDA 21CFR §172.515, 175.105, 175.300, 177.1210; FEMA GRAS; USA EPA registered; Japan approved as flavoring; Europe listed; FDA approved for use in injectables, parenterals, orals, topicals; NF, BP, EP compliance
Manuf./Distrib.: ABCR; AMC Chems.; AXO Chem.; Aceto; Advanced BioTech; Akzo Nobel; Aldrich; Allan; Allchem Ind.; Ashland; Atofina SA; Atofina; Augustus Oils Ltd; Axxence Aromatic GmbH; BCH Brühl; Bayer/Fiber, Addits., Rubber; Brown; Camida Ltd; Charkit; ChemService; DSM Fine Chems.; Degussa Flavors/Alfrebro; EM Ind.; EM Science; Elan; Fleurchem; Fluka; Functional Foods; General Chem.; George Uhe; Givaudan Iberica SA; Goldensun Mfg.; Grau Aromatics; Haarmann & Reimer; Indofine; Int'l. Chem. Group; Integra; J.C. Wilson; J.T. Baker; Kraft Chem.; Lluch Essence; Mallinckrodt Baker; MelChem; Merck KGaA; Miljac; Mutchler; Noveon Kalama; Osaka Org. Chem. Ind.; Parchem Trading; Penta Mfg.; Polysciences; Prodasynth; Quest Int'l.; R.W. Greeff; Richman; Ruger; Sarcom; Schweizerhall; Sigma; Spectrum Quality Prods.; Springfield Scientific Supplies; Takasago Int'l.; Tennants Distrib. Ltd; Tessenderlo Chemie; Tosoh; U.S. Chems.; Ubichem plc; Univar E&E Ltd; Universal Preserv-A-Chem; VWR Int'l.; Varsal Instruments; Velsicol; Voigt Global Distrib.; Vopak USA; Whyte Chems. Ltd; ZenPharm Int'l.

Benzyl alcohol formate. *See* Benzyl formate
Benzylbenzene. *See* Diphenylmethane
Benzyl benzene carboxylate. *See* Benzyl benzoate

Benzyl benzoate

CAS 120-51-4; EINECS/ELINCS 204-402-9
UN 2810; FEMA 2138
Synonyms: Benylate; Benzoic acid, benzyl ester; Benzoic acid phenylmethyl ester; Benzyl benzene carboxylate; Benzyl phenylformate; Phenylmethyl benzoate
Definition: Ester of benzyl alcohol and benzoic acid
Empirical: $C_{14}H_{12}O_2$
Formula: $C_6H_5COOCH_2C_6H_5$
Properties: Colorless oily liq., sl. aromatic odor; sol. in oxygenated/chlorinated solvs.; misc. with alcohol, chloroform, ether; insol. in water, glycerin; m.w. 212.26; sp.gr. 1.116-1.120; m.p. 21 C; b.p. 324 C; flash pt. (CC) 298 F; ref. index 1.568
Toxicology: LD50 (oral, rat) 500 mg/kg, (skin, rabbit) 4000 mg/kg; mod. toxic by ingestion, skin contact; TSCA listed
Precaution: Combustible liq.; reactive with oxidizing materials
Hazardous Decomp. Prods.: Heated to decomp., emits acrid and irritating fumes and smoke
NFPA: Health 1, Flammability 1, Reactivity 0
Storage: Light-sensitive
Uses: Fragrance; fixing agent, solvent for musk in perfumes and flavors; synthetic flavoring agent in foods and pharmaceuticals; external medicine; solvent, vehicle for foods; plasticizer for nitrocellulose and cellulose acetate; miticide; scabicide; insect repellent; dye carrier; solvent in cosmetics; textiles; solvent, preservative for pharmaceuticals; food-pkg. adhesives
Features: Sweet, cherry, pineapple, strawberry-like flavor
Use Level: 0.5% max. (finished cosmetics)
Regulatory: FDA 21CFR §172.515, 175.105; FEMA GRAS; FDA approved for injectables, orals; USP/NF, BP, EP compliance
Manuf./Distrib.: Aceto; Advanced BioTech; Akzo Nobel; Aldrich; Alfa Chem; Allchem Ind.; Asiamerica Int'l.; Augustus Oils Ltd; Aventis Pharmaceuticals; BCH Brühl; Berje; Boehme Filatex; Camida Ltd; DSM Fine Chems.; EM Ind.; Fluka; Haarmann & Reimer; Haltermann GmbH; Indofine; Integra; Lebermuth; Lluch Essence; Magnesia GmbH; Mallinckrodt Baker; MelChem; Merck KGaA; Morflex; Noveon Kalama; Ogawa & Co.; Penta Mfg.; Polarome Int'l.; Polysciences; Rhodia/Fine Orgs.; Ruger; Sarcom; Schweizerhall; Sigma; Spectrum Quality Prods.; Springfield Scientific Supplies; Stason Pharmaceuticals; Univar E&E Ltd; Universal Preserv-A-Chem; Voigt Global Distrib.; ZenPharm Int'l.
Trade Names: Unichem BZBN

Benzyl butyl phthalate; Benzyl n-butyl phthalate. *See* Butyl benzyl phthalate
Benzyl carbinol. *See* Phenethyl alcohol
Benzyl 'Cellosolve'. *See* Ethylene glycol benzyl ether
Benzyldiethyl [(2,6-xylylcarbomoyl) methyl] ammonium benzoate. *See* Denatonium benzoate
Benzyl dodecanoate. *See* Benzyl laurate
Benzyl ethanoate. *See* Benzyl acetate

Benzyl ether

CAS 103-50-4; EINECS/ELINCS 203-118-2
FEMA 2371
Synonyms: Dibenzyl ether; 1,1′-[Oxybis (methylene)] bis [benzene]
Empirical: $C_{14}H_{14}O$
Formula: $(C_6H_5CH_2)_2O$
Properties: Colorless to pale yel. unstable liq.; misc. with ethanol, ether, chloroform, acetone; pract. insol. in water; m.w. 198.28; dens. 1.043 (20/4 C); m.p. 5 C; b.p. 298 C (dec.); flash pt. (CC) 275 F; ref. index 1.557
Toxicology: LD50 (oral, rat) 2500 mg/kg; mod. toxic by ingestion; vapors may be narcotic in high conc.; skin and eye irritant; TSCA listed
Precaution: Combustible exposed to heat or flame; reactive with oxidizing materials; mod. explosion hazard by spontaneous chemical reaction
NFPA: Health 0, Flammability 1, Reactivity 0
Uses: Synthetic flavoring agent in foods and pharmaceuticals; solvent; plasticizer for nitrocellulose; cosmetics ingred.
Features: Sweet chocolate-like, fruity flavor
Regulatory: FDA 21CFR §172.515; FEMA GRAS
Manuf./Distrib.: Aldrich; Fluka

Benzyl formate

CAS 104-57-4; EINECS/ELINCS 203-214-4
FEMA 2145
Synonyms: Benzyl alcohol formate; Benzyl methanoate; Formic acid benzyl ester; Formic acid phenylmethyl ester
Empirical: $C_8H_8O_2$
Formula: $HCOOCH_2C_6H_5$

Properties: Colorless liq.; floral-fruity odor; misc. with alcohols, ketones, oils; pract. insol. in water; m.w. 136.15; dens. 1.081; b.p. 202-203 C; flash pt. 83 C; ref. index 1.5100-1.5120 (20 C)
Toxicology: LD50 (oral, rat) 1400 mg/kg, (skin, rabbit) 2 g/kg; mod. toxic by ing. and skin contact; may be narcotic in high concs.; mutagenic data; TSCA listed
Hazardous Decomp. Prods.: Heated to decomp., emits acrid, irritating fumes
Uses: Synthetic flavoring agent in foods and pharmaceuticals; solvent for cellulose esters; in perfumes; cosmetics ingred.
Regulatory: FDA 21CFR §172.515; FEMA GRAS
Manuf./Distrib.: Augustus Oils Ltd; Berje; Epochem; Fleurchem; Goldensun Mfg.; Penta Mfg.; Salor

Benzyl 2-hydroxybenzoate; Benzyl o-hydroxybenzoate. *See* Benzylsalicylate
Benzylideneacetaldehyde. *See* Cinnamal
Benzylidyne fluoride. *See* Benzotrifluoride

Benzyl laurate

CAS 140-25-0; EINECS/ELINCS 204-405-8
Synonyms: Benzyl dodecanoate; Dodecanoic acid benzyl ester; Dodecanoic acid, phenylmethyl ester; Lauric acid, benzyl ester; Phenylmethyl dodecanoate
Definition: Ester of lauric acid and benzyl alcohol
Empirical: $C_{19}H_{30}O_2$
Properties: M.w. 290.49
Toxicology: Skin irritant; TSCA listed
Hazardous Decomp. Prods.: Heated to decomp., emits acrid smoke and irritating fumes
Uses: Emollient, lubricant in cosmetics, hair and skin care; solvent, degreaser in sunscreens
Manuf./Distrib.: Penta Mfg.

Benzyl methanoate. *See* Benzyl formate
2-Benzyloxyethanol. *See* Ethylene glycol benzyl ether
2-[2-(Benzyloxy) ethoxy] ethanol. *See* Diethylene glycol benzyl ether
Benzyl phenylformate. *See* Benzyl benzoate

Benzylsalicylate

CAS 118-58-1; EINECS/ELINCS 204-262-9
FEMA 2151
Synonyms: Benzyl 2-hydroxybenzoate; Benzyl o-hydroxybenzoate; Phenylmethyl 2-hydroxybenzoate
Definition: Ester of benzyl alcohol and salicylic acid
Empirical: $C_{14}H_{12}O_3$
Properties: Colorless visc. liq., pleasant balsam-like odor; sol. in fixed oils, oxygenated solvs. sl. sol. in water; insol. in glycerin, propylene glycol; m.w. 228.26; dens. 1.175 (20 C); b.p. 208 C (26 mm); ref. index 1.579
Toxicology: LD50 (oral, rat) 2227 mg/kg; mod. toxic by ingestion; TSCA listed
Precaution: Combustible when heated or exposed to flame; incompat. with oxidizing materials
Hazardous Decomp. Prods.: Heated to decomp., emits acrid smoke and irritating fumes
NFPA: Health 1, Flammability 1, Reactivity 0
Uses: Fragrance; perfume fixing agent; solvent for syn. musk; soap odorant; synthetic flavoring agent in foods and pharmaceuticals; UV absorber, sunscreen agent in cosmetics
Regulatory: FDA 21CFR §172.515; FEMA GRAS
Manuf./Distrib.: Aceto; Akzo Nobel; Alfa Aesar; Alfa Chem; Asiamerica Int'l.; Augustus Oils Ltd; Berje; Fleurchem; Goldensun Mfg.; Haarmann & Reimer GmbH; Indofine; Lluch Essence; MelChem; Nantong ChangChem; Noveon Kalama; Penta Mfg.; Quest Int'l. Fragrances; Rhodia/Fine Orgs.; Sarcom; Spectrum Quality Prods.

Benzyltriethyl ammonium chloride

CAS 56-37-1; EINECS/ELINCS 200-270-1
Synonyms: Ammonium, benzyltriethyl-, chloride; Benzenemethanaminium, N,N,N-triethyl-, chloride; BETEC; BTEAC; TEBAC; N,N,N-Triethylbenzenemethanaminium chloride; Triethyl benzyl ammonium chloride
Classification: Quaternary ammonium salt
Empirical: $C_{13}H_{22}N \cdot Cl$
Formula: $C_6H_5CH_2N(Cl)(C_2H_5)_3$
Properties: Solid; m.w. 227.8; m.p. 197-199 C (dec.)
Toxicology: LD50 (IV, mouse) 18 mg/kg; poison by IV route; TSCA listed
Hazardous Decomp. Prods.: Heated to decomp., emits toxic vapors of NO_x and Cl^-
Storage: Hygroscopic
Uses: Solvent for cellulose; gelling inhibitor in polyester resins; intermediate; phase transfer catalyst; antistat in cosmetics; surfactant; catalyst; plasticizer; additive for toners and paper; biocide
Manuf./Distrib.: Accord; Aceto; Aldrich; Alfa Aesar; Amyl; Biddle Sawyer; Chemical; Dishman USA; Fluka; Hawks Chem. Co Ltd; Honeywell Spec. Chems.; Koei Chem.; Lindau Chems; R S A; Raschig; Spectrum Quality Prods.; Triple Crown Am.; Universal Preserv-A-Chem; Zeeland

Benzyl trimethyl ammonium chloride

CAS 56-93-9; EINECS/ELINCS 200-300-3
Synonyms: Benzenemethanaminium, N,N,N-trimethyl-, chloride; Benztrimonium chloride; BTMAC; TMBAC; Trimethylbenzylammonium chloride
Classification: Quaternary ammonium salt
Empirical: $C_{10}H_{16}ClN$
Formula: $C_6H_5CH_2N(CH_3)_3Cl$
Properties: Colorless cryst.; stable to 135 C; readily sol. in water, ethanol, butanol; sl. sol. in butyl phthalate, tributyl phosphate; m.w. 185.7; den. 1.07 (20/20 C, 61% sol'n.); m.p. 234-236 C; b.p. 135 C (some decomp.); cationic
Toxicology: LDLo (oral, rat) 250 mg/kg; poison by ingestion; tumorigen; TSCA listed
Precaution: Combustible
Hazardous Decomp. Prods.: Heated to decomp., emits very toxic fumes of NH_3, NO_x, Cl^-
Storage: Hygroscopic
Uses: Dispersant, dye leveling agent/retarder, emulsifier in textiles; solvent for cellulose; gelling inhibitor in polyester resins; intermediate; agric. chem.; phase transfer catalyst; accelerator in food-contact crosslinked polyesters
Regulatory: FDA 21CFR §177.2420
Manuf./Distrib.: Accord; Aldrich; Alfa Aesar; Amyl; Ashland; Bayer/Industrial Chems./Textile; Biddle Sawyer; Bio-Lab; Charkit; Chemical; Dishman USA; Fluka; Haltermann GmbH; Hawks Chem. Co Ltd; High Point; Koei Chem.; Lindau Chems; Piedmont Chem. Ind.; R S A; RTD Chems.; Rhodia; Spectrum Quality Prods.; Triple Crown Am.; Whyte Chems. Ltd

Bertholetta excelsea nut oil. *See* Brazil (Bertholetta excelsa) nut oil
BETEC. *See* Benzyltriethyl ammonium chloride
Betula; Betula oil. *See* Methyl salicylate

BHA

CAS 25013-16-5; EINECS/ELINCS 246-563-8
FEMA 2183; INS320
Synonyms: Butylated hydroxyanisole; Butylhydroxyanisole; 2(3)-t-Butyl-4-hydroxyanisole; t-Butylhydroxyanisole; t-Butyl-4-hydroxyanisole; t-Butyl-4-methoxyphenol; (1,1-Dimethylethyl)-4-methoxyphenol; Hydroxyanisole, butylated
Definition: Mixture of isomers of tertiary butyl-substituted 4-methoxyphenols
Empirical: $C_{11}H_{16}O_2$
Properties: Wh. or sl. yel. cryst. or waxy solid, faint char. aromatic odor; sol. in ethanol, petrol. ether, 50% or higher alcohol, propylene glycol, chloroform, fats, oils; insol. in water; m.w. 180.27; m.p. 48-55 C; b.p. 264-270 C (733 mm)
Toxicology: LD50 (oral, mouse) 2000 mg/kg; mod. toxicity by ing., IP routes; may cause rashes, hyperactivity; confirmed carcinogen; experimental neoplastigen, tumorigen, reproductive effects; mutagenic data; TSCA listed
Precaution: Combustible
Hazardous Decomp. Prods.: Heated to decomp., emits acrid and irritating fumes
HMIS: Health 1, Flammability 1, Reactivity 0
Uses: Antioxidant for fats and oils; antioxidant, preservative in foods, cosmetics, pharmaceuticals, plastics, food pkg., food-pkg. adhesives, pres-

sure-sensitive adhesives; defoamer in food-contact paper/paperboard; in lubricants for incidental food-contact use; pkg. material for irradiated foods

Use Level: 0.02% max. (preservation of fixed oils, fats, vitamin oil concs.)

Regulatory: FDA 21CFR §166.110, 172.110, 172.515, 172.615, 173.340, 175.105, 175.125, 175.300, 175.380, 175.390, 176.170, 176.210, 177.1010, 177.1210, 177.1350, 178.3120, 178.3570, 179.45; 181.22, 181.24 (0.005% migrating from food pkg.), 182.3169 (0.02% max. of fat or oil), GRAS; FEMA GRAS; USDA 9CFR 318.7, 381.147; Japan approved 0.2-1 g/kg; Europe approved; UK approved; approved for orals, rectals, topicals; NF, BP, EP compliance

Manuf./Distrib.: AMRESCO; Aceto; Aldrich; Allan; Allchem Ind.; Amerol; Ashland; Atomergic Chemetals; Augustus Oils Ltd; Aventis Pharmaceuticals; Blagden Spec. Chems. Ltd; Buckton Page Ltd; Camida Ltd; Eastman; Eggar & Co.; Fluka; Int'l. Chem. Inc; Kraft Chem.; Lavy's; Lluch Essence; Mutchler; Nipa Inc; Penta Mfg.; Ruger; Sigma; Sparkford Chems. Ltd; Spectrum Quality Prods.; UOP; Universal Preserv-A-Chem; Voigt Global Distrib.; Vopak USA

Trade Names Containing: Crystal® Crown

BHT

CAS 128-37-0; EINECS/ELINCS 204-881-4

FEMA 2184; INS321

Synonyms: 2,6-Bis (1,1-dimethylethyl)-4-methylphenol; Butylated hydroxytoluene; DBPC; 2,6-Di-t-butyl-p-cresol; 2,6-Di-t-butyl-4-methylphenol; Hydroxytoluene, butylated; 4-Methyl-2,6-di-t-butylphenol; Methyl di-t-butylphenol

Classification: Substituted toluene

Empirical: $C_{15}H_{24}O$

Formula: $[C(CH_3)_3]_2CH_3C_6H_2OH$

Properties: Wh. cryst. solid, faint char. odor; sol. in toluene, alcohols, MEK, acetone, Cellosolve, petrol. ether, chloroform, benzene, most HC solvs.; insol. in water, propylene glycol; m.w. 220.39; sp.gr. 1.048 (20/4 C); m.p. 68 C; b.p. 265 C; flash pt. (TOC) 260 F

Toxicology: ACGIH TLV 10 mg/m^3; LD50 (oral, rat) 890 mg/kg, (IP, mouse) 138 mg/kg, IV, mouse) 180 mg/kg; mod. toxic by ing.; poison by IP, IV routes; suspected carcinogen; human skin irritant; eye irritant; may cause rashes, hyperactivity; experimental teratogen, reproductive effects; TSCA listed

Precaution: Combustible exposed to heat or flame; reactive with oxidizing materials

Hazardous Decomp. Prods.: Heated to decomp., emits acrid smoke and fumes

HMIS: Health 1, Flammability 0, Reactivity 0

Uses: Antioxidant, preservative for foods, pharmaceuticals, animal feed, petroleum prods., jet fuels, lubricants, solvents, hot-melt adhesives, syn. rubbers, plastics (polyamide, polyolefins, POM, PU, PVC, PS), soaps; antioxidant for fats, oils, soap; antiskinning agent in paints and inks; oxidation/gum inhibitor in gasoline; antioxidant in food-contact coatings, food-pkg. adhesives and pressure-sensitive adhesives; defoamer in food-contact paper/paperboard; antioxidant in food-contact rubber articles for repeated use; in lubricants for incidental food-contact use

Use Level: 0.02% max. (preservation of fixed oils, fats, vitamin oil concs.)

Regulatory: FDA 21CFR §137.350, 166.110, 172.115, 172.615 (0.1% max.), 172.185, 173.340 (0.1% of defoamer), 175.105, 175.125, 175.300, 175.380, 175.390, 176.170, 176.210, 177.1010, 177.1210, 177.1350, 177.2260, 177.2600, 178.2010, 178.3120, 178.3570, 179.45, 181.22, 181.24 (0.005% migrating from food pkg.), 182.3173 (0.02% max. of fat/oil), GRAS; USDA 9CFR §318.7, 381.147; FEMA GRAS; Japan 0.2-1 g/kg; Europe, UK; NF, BP, EP compliance

Manuf./Distrib.: Aceto; Aldrich; Alfa Chem; Allan; Allchem Ind.; Amerol; Ashland; Atomergic Chemetals; Augustus Oils Ltd; Avatar; Aventis Pharmaceuticals; Bayer/Fiber, Addits., Rubber; Blagden Spec. Chems. Ltd; Buckton Page Ltd; Camida Ltd; ChemService; Crompton/Uniroyal; E-Chemicals; Eastman; Eggar & Co.; Equistar; Ferro/Grant; Fluka; GE Spec.; Gallard-Schlesinger Ind.; Great Lakes; Helm NY; Int'l. Chem. Inc; J.C. Wilson; Kraft Chem.; Laporte Perf. Chems.; MelChem; Mutchler; Nipa Inc; PMC Spec.; Penta Mfg.; Peter Whiting Ltd; Polysciences; R.T. Vanderbilt; Raschig; Ruger; Sea-Land; Shell; Sigma; Sparkford Chems. Ltd; Spectrum Quality Prods.; Springfield Scientific Supplies; Triple Crown Am.; UOP; Univar E&E Ltd; Universal Preserv-A-Chem; Varsal Instruments; Voigt Global Distrib.; Vopak USA; Wako Pure Chem. Ind.

Trade Names Containing: Aromatic 200; Bayol 35; Cropure® Almond; Cropure® Babassu; Cropure® Menhaden; Cropure® Olive; Cropure® Sesame; Cropure® Soybean; Eastman® EEP; EcoSolv™ Dry Cleaning Fluid; HAN® 906 Solvent; HC-DCF Low Flash™ Dry Cleaning Solvent; Isopar® V Fluid; Naphthalene Depleted Aromatic 150 Fluid; Naphthalene Depleted Aromatic 200 Fluid; Solvesso™ 100; Solvesso™ 100 Anti-Static Solvent; Solvesso™ 150 Anti-Static Solvent; Solvesso™ 150 ND; Solvesso™ 200; Solvesso™ 200 ND; Ultra Low Naphthalene Aromatic 200; Ultra Low Naphthalene Solvesso 200 Solvent

Bibenzene. *See* Biphenyl
Bicarburet of hydrogen. *See* Benzene
1,2-Bichloroethane. *See* Ethylene dichloride
Bicyclo (4.4.0) decane. *See* Decahydronaphthalene
cis-Bicyclo [4.4.0] decane. *See* cis-Decalin
trans-Bicyclo [4.4.0] decane. *See* trans-Decalin
Bicyclo (2.2.1) heptan-2-ol, 1,7,7-trimethyl-, acetate, endo-. *See* Bornyl acetate
Bicyclo (4.1.0) hept-3-ene, 3,7,7-trimethyl-. *See* 3-Carene
Bihexyl. *See* n-Dodecane
Biisopropyl. *See* 2,3-Dimethylbutane

Biphenyl

CAS 92-52-4; EINECS/ELINCS 202-163-5

FEMA 3129; INS230; E230

Synonyms: Bibenzene; 1,1´-Biphenyl; Diphenyl; Lemonene; Phenyl benzene; PHPH; Xenene

Empirical: $C_{12}H_{10}$

Formula: $C_6H_5C_6H_5$

Properties: Wh. scales or colorless leaflets, pleasant odor; sol. in oxygenated/chlorinated solvs., alcohols, ether; insol. in water; m.w. 154.08; dens. 1.041; m.p. 70 C; b.p. 256 C; flash pt. (CC) 112.7 C; ref. index 1.588

Toxicology: ACGIH TLV/TWA 0.2 ppm; LD50 (oral, rat) 3280 mg/kg, (IV, mouse) 56 mg/kg; irritant; poison by IV route; mod. toxic by ing.; human systemic effects by inh. of very sm. amts.; CNS depression, paralysis, convulsions reported; experimental tumorigen; mutagenic data; TSCA listed

Precaution: Combustible exposed to heat or flame; can react with oxidizers

Hazardous Decomp. Prods.: Heated to decomp., emits acrid smoke and fumes

NFPA: Health 2, Flammability 1, Reactivity 0

Uses: Organic synthesis; heat transfer agent; preservative in foods; fungistat in pkg. citrus fruit; plant disease control; flavoring agent in pharmaceuticals; mfg. of benzidine; dyeing assistant for polyesters; dye carrier

Regulatory: FDA 21CFR §178.2010; FEMA GRAS; SARA reportable; Europe listed; UK approved; Japan approved

Manuf./Distrib.: Aldrich; Bayer/Industrial Chems./Textile; Coalite Chems.; Fluka; Koch Chem. Spec.; Merck S.A.; Monsanto; Sigma; Springfield Scientific Supplies

Trade Names Containing: Methylnaphthalene Fraction

1,1´-Biphenyl. *See* Biphenyl

1,1´-Biphenyl, butenylated

CAS 81846-81-3

Uses: Solvent for aerosols, adhesives, electronic parts mfg. and cleaning, industrial cleaning, inks, paper coatings, metal cleaning, paints, plastics, sealants, tile mfg.

o-Biphenylene methane. *See* Fluorene
Biphenyl, isopropyl. *See* 1,1´-Biphenyl (1-methylethyl)
o-Biphenylmethane. *See* Fluorene

1,1´-Biphenyl (1-methylethyl)

CAS 25640-78-2

Synonyms: Biphenyl, isopropyl; Isopropylbiphenyl; Isopropyl diphenyl; (1-Methylethyl) 1,1´-biphenyl; Monoisopropyl biphenyl

Empirical: $C_{15}H_{16}$

Properties: M.w. 196.31

Toxicology: LD50 (oral, rat) 4570 mg/kg; LDLo (oral, rabbit) 4 g/kg; mod. toxic by ingestion; mutagen; may cause altered sleep time, tremors, ataxia; TSCA listed
Hazardous Decomp. Prods.: Heated to decomp., emits acrid smoke and fumes
Uses: Solvent for aerosols, adhesives, electronic parts mfg. and cleaning, industrial cleaning, inks and paper coatings, metal cleaning, paints, plastics, sealants, tile mfg.

Biphenyl oxide. *See* Diphenyl oxide
Bis (2-aminoethyl) amine; Bis (β-aminoethyl) amine. *See* Diethylenetriamine
Bis (butoxyethyl) ether; Bis (2-butoxyethyl) ether. *See* Diethylene glycol dibutyl ether
3,6-Bis (carboxymethyl)-3,5-diazooctanedioic acid. *See* Edetic acid

Bis (2-chloroethoxy) methane

CAS 111-91-1
UN 2810
Synonyms: Bis (2-chloroethyl) formal; Bis (β-chloroethyl) formal; Dichlorodiethyl formal; Dichloroethyl formal; Ethane, 1,1´-(methylenebis (oxy) bis (2-chloro-; Formaldehyde, bis (2-chloroethyl) acetal; Formaldehyde bis (β-chloroethyl) acetal; 1,1´-[Methylene bis (oxy)] bis (2-chloroethane)
Empirical: $C_5H_{10}Cl_2O_2$
Properties: Colorless clear liq.; sl. sol. in water; m.w. 173.0; sp.gr. 1.2339 (20/20 C); f.p. -32.8 C; b.p. 218.1 C; flash pt. (CC) 110 C; nonflamm.
Toxicology: LD50 (skin, guinea pig) 170 mg/kg; LDLo (oral, rat) 65 mg/kg; LCLo (inh., rat, 4 h) 62 ppm; poison; toxic by inh. and ing.; strong irritant; irritating to eyes/skin; TSCA listed
Precaution: Incompat. with oxidizing agents; unstable; keep away from food
Hazardous Decomp. Prods.: Heated to decomp., emits toxic vapors of HCl
NFPA: Health 2, Flammability 1, Reactivity 0
Storage: Protect from exposure to light; keep container tightly closed under inert atmosphere; store in freezer
Uses: Solvent; intermediate for polysulfide rubber
Manuf./Distrib.: Supelco

Bis (2-chloroethyl) ether; Bis (β-chloroethyl) ether. *See* sym-Dichloroethyl ether
Bis (2-chloroethyl) formal; Bis (β-chloroethyl) formal. *See* Bis (2-chloroethoxy) methane
Bis (2-cyanoethyl) sulfide. *See* Thiodipropionitrile
2,6-Bis (1,1-dimethylethyl)-4-methylphenol. *See* BHT
1,2-Bis (2-ethoxyethoxy) ethane. *See* Triethylene glycol diethyl ether
Bis (2-ethoxyethyl) ether. *See* Diethylene glycol diethyl ether
Bis (2-ethylhexyl) adipate. *See* Dioctyl adipate
Bis (2-ethylhexyl) decanedioate. *See* Dioctyl sebacate
Bis (2-ethyl-1-hexyl)-2-ethyl-1-hexyl phosphonate; Bis (2-ethylhexyl) 2-ethylhexyl phosphonate. *See* Dioctyl octylphosphonate
1,3-Bis (2-ethylhexyl) hexahydro-5-methyl-5-pyrimidiamine. *See* Hexetidine
Bis (2-ethylhexyl) hexanedioate. *See* Dioctyl adipate

Bis (2-ethylhexyl) maleate

CAS 142-16-5; EINECS/ELINCS 205-524-5
Synonyms: 2-Butenedioic acid (Z)-, bis (2-ethylhexyl) ester; DEHM; Diethylhexyl maleate; Di (2-ethylhexyl) maleate; Maleic acid, bis (2-ethylhexyl) ester; Maleic acid, di-(2-ethylhexyl) ester
Definition: Diester of 2-ethylhexyl alcohol and maleic acid
Empirical: $C_{20}H_{36}O_4$
Formula: $C_8H_{17}OCOCH{=}CHCOOC_8H_{17}$
Properties: Liq.; insol. in water; m.w. 340.56; dens. 0.9436 (20/20 C); m.p. -60 C; b.p. 164 C (10 mm); flash pt. 365 F
Toxicology: LD50 (oral, rat) 14 g/kg, (IP, rat) > 5 g/kg, (skin, rabbit) 15 g/kg; mildly toxic by ing. and skin contact; primary skin and eye irritant; TSCA listed
Precaution: Combustible exposed to heat or flame; reactive with oxidizers
Hazardous Decomp. Prods.: Heated to decomp., emits acrid smoke and irritating fumes
Uses: Intermediate; comonomer in copolymerization of PVC, vinyl acetates, vinyl emulsion polymers; flexibilizer; emollient, skin conditioner in cosmetics; plasticizer for PVC
Manuf./Distrib.: Celanese; Salor
Trade Names: HallStar™ DOM

Bis (2-ethylhexyl) sebacate. *See* Dioctyl sebacate
Bis (2-hexyldecyl) adipate. *See* Diisocetyl adipate
Bis (1-hexyl) ether. *See* n-Hexyl ether
1,2-Bis (2-hydroxyethoxy) ethane. *See* Triethylene glycol
Bis [2-(2-hydroxyethoxy) ethyl] ether. *See* PEG-4
Bis (2-hydroxyethyl) amine; N,N-Bis (2-hydroxyethyl) amine. *See* Diethanolamine
N,N-Bis (2-hydroxyethyl) coco amides; N,N-Bis (2-hydroxyethyl) coco fatty acid amide. *See* Cocamide DEA
Bis (2-hydroxyethyl) ether. *See* Diethylene glycol
N,N-Bis (2-hydroxyethyl) ethylamine. *See* N-Ethyldiethanolamine
1,6-Bis (2-hydroxyethyl) hexane. *See* 1,10-Decanediol
Bis (2-hydroxyethyl) sulfide; Bis (β-hydroxyethyl) sulfide. *See* Thiodiglycol
Bis (hydroxymethyl) acetylene. *See* But-2-yne-1,4-diol
Bis (hydroxymethyl)-5,5-dimethylhydantoin; 1,3-Bis (hydroxymethyl)-5,5-dimethylimidazolidinedione; 1,3-Bis (hydroxymethyl)-5,5-dimethylimidazolidine-2,4-dione; Bis (hydroxymethyl)-5,5-dimethyl-2,4-imidazolidinedione. *See* DMDM hydantoin
Bis (2-hydroxypropyl) amine. *See* Diisopropanolamine
Bis (hydroxypropyl) ether; Bis (2-hydroxypropyl) ether. *See* Dipropylene glycol
Bis (isopropyl) naphthalene. *See* Diisopropyl naphthalene
Bis (2-methoxycarbonylethyl) sulfide. *See* Dimethyl thiodipropionate
Bis (2-methoxyethoxyethyl) ether. *See* Dimethoxytetraglycol
Bis (2-methoxyethyl) ether. *See* Diethylene glycol dimethyl ether
Bis (methoxyethyl) phthalate; Bis (2-methoxyethyl) phthalate. *See* Dimethoxyethyl phthalate
Bis (methoxy) methane. *See* Methylal
Bis (methoxypropyl) ether. *See* Dipropylene glycol dimethyl ether
1,3-Bis(1-methylethyl) benzene. *See* 1,3-Diisopropyl benzene
1,4-Bis(1-methylethyl) benzene. *See* 1,4-Diisopropyl benzene

Bis (methylethyl)-1,1´-biphenyl

CAS 69009-90-1
Synonyms: Bis (1-methylethyl)-1,1´-biphenyl; Diisopropyl biphenyl
Empirical: $C_{18}H_{22}$
Properties: M.w. 238.37
Uses: Solvent for aerosols, adhesives, electronic parts mfg. and cleaning, industrial cleaning, inks and paper coatings, metal cleaning, paints, plastics, sealants, tile mfg.
Features: Exc. solvency, chem. and thermal stability, nonvolatility

Bis (1-methylethyl)-1,1´-biphenyl. *See* Bis (methylethyl)-1,1´-biphenyl
Bis (1-methylethyl) decanedioate. *See* Diisopropyl sebacate
Bis (1-methylethyl) hexanedioate. *See* Diisopropyl adipate
Bis (1-methylethyl) naphthalene. *See* Diisopropyl naphthalene
Bis (2-methylpropyl) hexanedioate. *See* Diisobutyl adipate
Bis (1-octyl) maleate. *See* Dioctyl maleate
Bis (2-propanol) amine. *See* Diisopropanolamine

Bis (2,4,4,-trimethylpentyl) dithiophosphinic acid

CAS 107667-02-7
UN 2927
Synonyms: Diisooctyldithiophosphinic acid
Empirical: $C_{16}H_{35}PS_2$
Formula: $[(CH_3)_3CCH_2CHCH_3CH_2]_2PSSH$
Properties: Sp.gr. 0.95; flash pt. 24 C
Toxicology: Very toxic by inh.; harmful by ing., skin absorp.; causes burns
Precaution: Flamm.; incompat. with strong oxidizing agents
Uses: Solvent extractant for metals
Features: Stable
Trade Names: Cyanex® 301

Bis (2,4,4,-trimethylpentyl) phosphinic acid

CAS 83411-71-6; EINECS/ELINCS 280-445-7
Synonyms: Diisooctylphosphinic acid; Phosphinic acid, bis (2,4,4-trimethylpentyl)-

Empirical: $C_{16}H_{35}O_2P$
Formula: $[(CH_3)_3CCH_2CHCH_3CH_2]_2POOH$
Properties: Liq.; m.w. 290.4; sp.gr. 0.916; ref. index 1.460
Toxicology: Irritating to eyes; may cause irritation to skin, mucous membranes, upper respiratory tract; may be harmful by inh., ing., or skin absorp.
Precaution: Incompat. with strong oxidizing agents
Hazardous Decomp. Prods.: Thermal decomp. may produce toxic fumes of phosphorus oxides and/or phosphine, CO, CO_2; emits toxic fumes under fire conditions
Storage: Store in cool, dry place; keep tightly closed
Uses: Reagent, extractant for separating cobalt and nickel from sulfate or chloride media, for separation of heavy rare earths, and selective separation for iron and zinc from cobalt sol'ns.
Regulatory: Australia AICS listed
Manuf./Distrib.: Fluka
Trade Names: Cyanex® 272

Bis (2,4,4,-trimethylpentyl) thiophosphinic acid

CAS 132767-86-3
Synonyms: Diisooctylthiophosphinic acid; Phosphinothioic acid, bis (2,4,4-trimethylpentyl)-
Classification: Phosphine
Empirical: $C_{16}H_{35}OPS$
Formula: $[(CH_3)_3CCH_2CHCH_3CH_2]_2PSOH$
Properties: Dens. 0.93 g/ml; flash pt. 96 C; ref. index 1.48
Toxicology: Corrosive; causes burns; extremely destructive to tissue of mucous membranes, upper respiratory tract, eyes, skin; harmful by ing., inh., skin absorp.; inh. may cause spasm/inflamm./edema of larynx/bronchi, chem. pneumonitis, pulmonary edema; may cause burning sensation, coughing, wheezing, laryngitis, shortness of breath, headache, nausea, vomiting
Precaution: Incompat. with strong oxidizing agents
Hazardous Decomp. Prods.: Thermal decomp. may produce toxic fumes of phosphorus oxides and/or phosphine, CO, CO_2, SO_x; emits toxic fumes under fire conditions
Storage: Store in cool, dry place; keep tightly closed
Uses: Extractant for recovery of cadmium from phosphoric acid; selective extractant for separation of cobalt from manganese, and for impurity cations such as calcium and magnesium; complexing agent for extraction of transition metals
Regulatory: TSCA export notification required
Manuf./Distrib.: Fluka
Trade Names: Cyanex® 302

Bis (trimethylsilyl) amine. *See* Hexamethyldisilazane
Bis (trimethylsilyl) ether; Bis (trimethylsilyl) oxide. *See* Hexamethyldisiloxane
Bisulfate; Bisulfite. *See* Sulfur dioxide
Black copper oxide. *See* Copper oxide (ic)
BLO. *See* Butyrolactone
BM. *See* 1-Methoxy-2-butanol
BME. *See* 2-Mercaptoethanol
BOP. *See* Butyl octyl phthalate
Boracic acid. *See* Boric acid

Boric acid

CAS 10043-35-3; EINECS/ELINCS 233-139-2
E284
Synonyms: Boracic acid; Boron trihydroxide; Orthoboric acid
Classification: Inorganic acid
Empirical: H_3BO_3
Formula: BH_3O_3
Properties: Wh. or colorless cryst. powd., gran.; odorless; almost tasteless or sl. acidic taste; sol. in water, alcohol, glycerol; insol. in ether, benzene; m.w. 61.83; dens. 1.435; m.p. 171 C; b.p. 300 C; loses 1.5 H_2O @ 300 C; nonflamm.
Toxicology: LD50 (oral, rat) 2660 mg/kg, (subcut., rat) 1400 mg/kg, (IV, rat) 1330 mg/kg; LDLo (IP, mouse) 800 mg/kg; poison by inh., subcut. routes; mod. toxic by IP, IV routes; human poison by ing.; mod. toxic to humans by skin contact, subcut. routes; human skin irritant; human systemic effects; acute poisoning causes digestive upsets in man, CNS stimulation, depression, renal damage, circulatory collapse; dose of 5-20 g in adults may be fatal; experimental reproductive effects; mutagenic data; TSCA listed
Precaution: Hygroscopic
Hazardous Decomp. Prods.: Boric anhydride
NFPA: Health 0, Flammability 0, Reactivity 0
Storage: Store @ R.T.
Uses: Heat-resist. glass; porcelain enamels; boron chemicals; metallurgy; flame retardant (cellulosic insulation, mattress batting, cotton textiles, wood, paper); fungistat in citrus fruits, ointments, eyewash; nickel electroplating baths; insecticide; preservative (foods, pharmaceuticals, rubber latex); medicine (protective disinfectant); antiseptic; antimicrobial in cosmetics; oral care agent; photography; leather finishing; buffer (eye drops, chrome plating, aluminum anodizing); oxidation catalyst; welding/brazing fluxes; in food-pkg. adhesives; preservative in coatings for paper/paperboard in contact with dry food; antioxidant/stabilizer for food-grade polymers
Use Level: 0.01-1.0%
Regulatory: FDA 21CFR §175.105, 176.180, 178.2010, 181.22, 181.30; USA not restricted; Europe listed; permitted in Switzerland and Sweden as preservative in some processed seafood; FDA approved for ophthalmics, otics, topicals; NF, BP, EP compliance
Manuf./Distrib.: AMRESCO; Advance Research Chems.; Aldrich; Alfa Aesar; Allchem Ind.; Am. Int'l.; Ashland; Asiamerica Int'l.; BCH Brühl; Brenntag AG; Brown; C.P. Hall; Coyne; Delta Distributors; Dragoco; EM Ind.; EM Science; Eastech; Electron Microscopy Sciences; Fluka; GFS; Galbraith Labs; Hasnil; Howard Hall; Integra; J.C. Wilson; J.T. Baker; Mallinckrodt Baker; Milsolv; Nippon Denko; Occidental; P.B. & S.; Penta Mfg.; Qbiogene; Research Organics; Ruger; SOCO-Lynch; Sal Chem.; San Yuan; Schaefer Tech.; Seeler Ind.; Sigma; Southchem; Spectrum Quality Prods.; Springfield Scientific Supplies; Stanchem; Textile Chem.; Thomas Scientific; U.S. Borax; Universal Preserv-A-Chem; VWR Int'l.; Varsal Instruments; Veckridge; Vopak USA; Whittaker, Clark & Daniels
Trade Names Containing: Punctilious® SDA 38F 190 Proof; Punctilious® SDA 38F-2 190 Proof Prod. Code 13643020; Punctilious® SDA 38F-2 190 Proof Prod. Code 13645006

Boric acid, trimethyl ester. *See* Trimethyl borate
2-Bornanone; Bornan-2-one. *See* Camphor
Borneol acetate. *See* Bornyl acetate

Bornyl acetate

CAS 76-49-3
FEMA 2159
Synonyms: Bicyclo (2.2.1) heptan-2-ol, 1,7,7-trimethyl-, acetate, endo-; Borneol acetate; levo-Bornyl acetate; Bornyl acetic ether; Bornyl ethanoate; 2-Camphanol acetate; endo-2-Camphanyl ethanoate; 1,7,7-Trimethylbicyclo [2.2.1] heptan-2-ol acetate; endo-1,7,7-Trimethylbicyclo [2.2.1] hept-2-yl acetate
Empirical: $C_{12}H_{20}O_2$
Properties: Colorless liq. or wh. cryst. solid; piney odor; fresh, burning taste; sol. in alcohol, most fixed oils; sl. sol. in water; insol. in propyl glycol; m.w. 196.29; dens. 0.981-0.985; m.p. 27.5 C; b.p. 225-226 C; flash pt. 89 C; ref. index 1.462-1.466
Precaution: Combustible liq.
Hazardous Decomp. Prods.: Heated to decomp., emits acrid smoke and fumes
Uses: Synthetic flavoring agent, solvent in foods and pharmaceuticals; fragrance in perfumes, soaps, household prods.; solvent for nitrocellulose; tobacco ingred.
Regulatory: FDA 21CFR §172.515; FEMA GRAS
Manuf./Distrib.: Aldrich; Augustus Oils Ltd; Chr. Hansen Inc; Penta Mfg.

levo-Bornyl acetate; Bornyl acetic ether; Bornyl ethanoate. *See* Bornyl acetate
Boron methoxide. *See* Trimethyl borate
Boron trihydroxide. *See* Boric acid
Bran absolute. *See* Wheat (Triticum vulgare) germ oil
Brassica campestris oil; Brassica oleifera. *See* Rapeseed (Brassica campestris) oil

Brazil (Bertholetta excelsa) nut oil

Synonyms: Bertholetta excelsea nut oil; Brazilian nut oil; Brazil nut oil

Definition: Fixed oil expressed from the nuts of the Brazil nut tree, *Bertholetta excelsa*
Uses: For hair treatment, bath, and massage oils; in cosmetic clay masks; solvent in cosmetics
Features: Contains vitamins A, B_1, B_2, D, E, niacin
Manuf./Distrib.: Aldivia; Pangaea Sciences

Brazilian nut oil; Brazil nut oil. *See* Brazil (Bertholetta excelsa) nut oil
Brevoortia; Brevoortia tyrannus oil. *See* Menhaden oil
Brick oil. *See* Creosote oil
Brimstone. *See* Sulfur
Bromic ether. *See* Ethyl bromide

Bromine

CAS 7726-95-6; EINECS/ELINCS 231-778-1
UN 1744 (DOT)
Synonyms: Dibromine
Classification: Inorganic liq. nonmetallic halogen element
Empirical: Br_2
Properties: Dk. red fuming liq., pungent odor; sol. in CCl_4, chloroform, methylene chloride, CS_2, ether, methanol, conc. HCl; sol. 3.38% in water (20 C); a.w. 159.83; sp.gr. 3.119 (20/4 C); dens. 25.9 lb/gal; vapor pressure 175 mm (21 C); f.p. -7.27 C; b.p. 58.8 C; ref. index 1.6083 (20 C)
Toxicology: LD50 (oral, rat) 2600 ppm; LDLo (inh., human) 1000 ppm; TLV 0.1 ppm in air; toxic by ingestion and inhalation; severe skin, eye and mucous membrane irritant; TSCA listed
Precaution: DOT: Corrosive material; reactive and corrosive; strong oxidizing agent may ignite combustibles on contact; highly dangerous; attacks most metals (Pt, Pd); aluminum reacts vigorously; potassium reacts explosively
Hazardous Decomp. Prods.: When heated, emits highly toxic fumes; will react with water or steam to produce toxic and corrosive fumes
NFPA: Health 3, Flammability 0, Reactivity 0
Uses: Bromination reagent; mfg. of ethylene dibromide; organic synthesis; in water disinfection; swimming pool algicide, disinfectant; bleaching fibers; medicinals; dyestuffs; solvent; intermediate for fumigants; analytical reagent; fire retardant for plastics; photography; shrinkproofing wool; drilling fluids; brominated pesticides; intermediate for drugs, dyes, and fine chems.; rubber additives; weighting agent for flavors and fragrances (foods); agric. chemicals; antiknock additives
Manuf./Distrib.: Advanced Synthesis Tech.; Albemarle; Aldrich; Alfa Aesar; Alfa Chem; AmeriBrom; Assoc. Octel Co Ltd; Atofina; Bilt Chems. Ltd; Bromine & Chems. Ltd; Chemconnect; Columbus Chem. Ind.; Dead Sea Bromine; Dow; Fluka; Great Lakes; Hawks Chem. Co Ltd; Int'l. Chem. Inc; Integra; J.C. Wilson; J.T. Baker; Morre-Tec Ind.; Ocean Chems.; Schumacher; Spectrum Quality Prods.; Springfield Scientific Supplies; VWR Int'l.

Bromobenzene

CAS 108-86-1; EINECS/ELINCS 203-623-8
UN 2514 (DOT)
Synonyms: Benzene, bromo-; Bromobenzol; Monobromobenzene; Phenyl bromide
Classification: Aromatic halogenated hydrocarbon
Empirical: C_6H_5Br
Properties: Colorless clear mobile liq.; aromatic odor; sol. (mg/ml): $\geq$ 100 mg in DMSO, 95% ethanol, acetone (20.5 C); sol. in CCl_4, ether, benzene, misc. with chloroform, petrol. hydrocarbons, most org. solvs.; insol. in water; m.w. 157.01; sp.gr. 1.495 (20/4 C); vapor pressure 5 mm Hg (27.8 C); m.p. -30 C; b.p. 155-156 C; flash pt. 51 C; ref. index 1.5597 (20 C)
Toxicology: LD50 (oral, rat) 2699 mg/kg, (IP, rat) 3882 mg/kg, (subcut., mouse) 2 g/kg; LC50 (inh., rat) 20,411 mg/m^3; mod. toxic by ing., inh., IP routes; irritant to skin, eyes, mucous membranes; can be absorbed thru skin; may cause somnolence, muscle contraction or spasticity, flaccid paralysis, respiratory stimulation, changes in urine composition, wt. loss, enzyme inhibition; mutagen; TSCA listed
Environmental: Toxic to aquatic organisms; may cause long-term adverse effects in aquatic environment
Precaution: DOT: Flamm. liq.; can react with oxidizing agents and sodium bromobutane
Hazardous Decomp. Prods.: Heated to decomp., emits toxic fumes of CO, CO_2, hydrogen bromide gas; high vapor concs. may be anesthetic
NFPA: Health 2, Flammability 2, Reactivity 0
Storage: Light-sensitive; store in tightly closed container in cool, dry, well-ventilated area away from incompat. substances
Uses: Specialty solvent; solvent for heavy liqs.; intermediate for org. synthesis, esp. to make phenyl magnesium bromide, agrochems., pharmaceuticals; motor oil additive; .
Manuf./Distrib.: ABCR; Acros Org.; Albemarle; Aldrich; Alfa Aesar; Diaz; Fisher Scientific; Fluka; ISP; Indofine; J.T. Baker; Morre-Tec Ind.; Ocean Chems.; Sigma; Spectrum Quality Prods.; Synasia Fine Chem.; VWR Int'l.

Bromobenzol. *See* Bromobenzene

Bromochloromethane

CAS 74-97-5; EINECS/ELINCS 200-826-3
UN 1887 (DOT)
Synonyms: CBM; Chlorobromomethane; Chloromethyl bromide; Methane, bromochloro-; Methylene chlorobromide; Monochloromonobromomethane
Classification: Halogenated hydrocarbon
Empirical: CH_2BrCl
Formula: $BrCH_2Cl$
Properties: Colorless to yel. clear volatile liq.; chloroform-like odor; sol. in org. solvs.; insol. in water; m.w. 129.39; dens. 1.93; f.p. -86.5 C; b.p. 67 C; ref. index 1.48; nonflamm.
Toxicology: ACGIH TLV/TWA 200 ppm; LD50 (oral, rat) 5 g/kg, (skin, rabbit) > 20 g/kg; LCLo (inh., rat, 15 min) 28,800 ppm; poisonous substance; mildly toxic by ing. and inh.; mutagen; narcotic action; dangerous; TSCA listed
Hazardous Decomp. Prods.: Heated to decomp., emits highly toxic fumes of Br^- and Cl^-
Uses: Fire extinguishant fluid; intermediate for organic synthesis, agrochems. (biocides, pesticides), pharmaceuticals; process solvent; reaction solvent; flotation agent
Manuf./Distrib.: Albemarle; Alfa Aesar; AmeriBrom; Assoc. Octel Co Ltd; Dead Sea Bromine; Great Lakes; Morre-Tec Ind.; Spectrum Quality Prods.

Bromodichloromethane

CAS 75-27-4; EINECS/ELINCS 200-856-7
UN 8027
Synonyms: BDCM; Dichlorobromomethane; Dichloromethylbromide; Dichloromonobromomethane; Methane, bromodichloro-; Monobromodichloromethane
Empirical: $CHBrCl_2$
Formula: $BrCHCl_2$
Properties: Colorless clear liq.; very sol. in lipids; sol. in DMSO, 95% ethanol, acetone, ether, benzene, chloroform; misc. with org. solvs.; sol. 5-10 mg/ml in water; m.w. 163.83; sp.gr. 1.980 (20/4 C); vapor pressure 50 mm Hg (20 C); m.p. -55 C; b.p. 87 C; flash pt. none; nonflamm.; ref. index 1.4964 (20 C)
Toxicology: LD50 (oral, rat) 430 mg/kg; mod. toxic by ing.; harmful by inh., skin contact; probably narcotic in high concs.; may cause eye/skin/mucous membrane/respiratory tract irritation, nausea, dizziness, headache, liver/kidney damage, CNS effects, somnolence, tremors, etc.; mutagen; cancer suspect agent; TSCA listed
Precaution: Incompat. with strong oxidizing agents, strong bases, and magnesium; sensitive to prolonged exposure to air and light
Hazardous Decomp. Prods.: CO, CO_2, hydrogen chloride gas, hydrogen bromide gas; emits toxic fumes under fire conditions
Storage: Keep container tightly closed under inert atmosphere; store in freezer; protect from light and oxidizing materials
Uses: Fire extinguisher fluid ingred.; solvent for fats, waxes, resins; synthesis intermediate; reagent; heavy liq. for min. and salt separations and in org. synthesis
Regulatory: SARA §313 reportable
Manuf./Distrib.: AccuStandard; Acros Org.; Aldrich; Alfa Aesar; Fluka; Fluorochem USA; Lancaster Synthesis; Oakwood Prods.; Pfaltz & Bauer; Riedel-deHaën; Sigma; Supelco; TCI Am.

Bromoethane. *See* Ethyl bromide

Bromoform

CAS 75-25-2; EINECS/ELINCS 200-854-6
UN 2515 (DOT)
Synonyms: Methenyl tribromide; Methyl tribromide; Tribromomethane
Classification: Halogenated aliphatic hydrocarbon
Empirical: $CHBr_3$
Properties: Colorless to yel. liq. or hexagonal cryst.; sweet chloroform-like odor; sol. in alcohol, ether, chloroform, benzene, fixed and volatile oils; sl. sol. in water; m.w. 252.75; sp.gr. 2.9 (20 C); m.p. 7-9 C; b.p. 148 C; flash pt. none; ref. index 1.6005; surf. tens. 41.53 dynes/cm (20 C)
Toxicology: ACGIH TLV/TWA 0.5 ppm (skin); LD50 (oral, rat) 1147 mg/kg, (IP, rat) 414 mg/kg; very toxic; human poison by ing.; may cause skin irritation; can be absorbed thru skin; lachrymator; mod. eye irritant; vapor may cause nose/throat irritation; anesthetic props.; high concs. may cause CNS depression (sleepiness, incoordination, unconsciousness), liver injury; severe exposure by ing. may cause respiratory failure and death; addiction possible; human mutagenic data; possible carcinogen; TSCA listed
Precaution: Incompat. with strong bases, metals, acetone/potassium hydroxide, cyclic polyethylene oxides/potassium hydroxide; possible violent reactions, explosive hazards
Hazardous Decomp. Prods.: Heated to decomp., emits highly toxic fumes of Br^-
NFPA: Health 3, Flammability 0, Reactivity 0
Storage: Store in cool, dry, well-ventilated area out of direct sunlight, away from heat; limit stored quantities; light-sensitive
Uses: Chem. intermediate in org. synthesis, for agrochems.; solvent for waxes, greases, oils; high-density solvent for min. separation; prod. of high-dens. liqs. for geological assays (petrographic analysis); ingred. in fire-resist. chems., fluid gauges; polymerization agent; medicine (sedative, cough suppressant)
Manuf./Distrib.: Albemarle; Aldrich; Alfa Aesar; Allan; Fluka; Geoliquids; Integra; J.T. Baker; Morre-Tec Ind.; Nipa Labs; Sigma; Spectrum Quality Prods.; Springfield Scientific Supplies; Synasia Fine Chem.; VWR Int'l.

Bromomethane. *See* Methyl bromide
1-Bromopropane. *See* n-Propyl bromide
2-Bromopropane. *See* Isopropyl bromide

Brucine

CAS 357-57-3; EINECS/ELINCS 206-614-7
UN 1570
Synonyms: Brucine alkaloid; Brucine, solid; 2,3-Dimethoxystrychnidin-10-one; Dimethoxystrychnine; 2,3-Dimethoxystrychnine; 10,11-Dimethoxystrychnine; Strychnidin-10-one, 2,3-dimethoxy-; Strychnine, 2,3-dimethoxy-
Empirical: $C_{23}H_{26}N_2O_4$
Properties: Wh. cryst.; very bitter taste; very sol. in ethanol, chloroform, benzene; sl. sol. in water, ether, glycerol, ethyl acetate; m.w. 394.51; m.p. 175-178 C (dec.)
Toxicology: LD50 (oral, rat) 1 mg/kg, (IP, rat) 91 mg/kg; LDLo (IV, rabbit) 30 mg/kg; extremely toxic; deadly poison; highly toxic by ing. and IP routes; irritant; may cause nausea, vomiting, restlessness, excitement, twitching, convulsions, local anesthesia, tremors/stiffness of legs and face, etc.; contraction of muscles of the diaphragm plus spasm of thoracic/abdominal muscles arrests respiration; death from asphyxia or medullary paralysis usually follows the second to fifth seizure; target organ: CNS; TSCA listed
Precaution: Incompat. with strong oxidizing agents
Hazardous Decomp. Prods.: CO, CO_2; emits toxic fumes under fire conditions
Storage: Store refrigerated; keep tightly closed
Uses: Denaturing alcohol and oils; in analytical chemistry; for separating racemic mixts.; has been patented as addition agent for lubricants
Regulatory: SARA §313 reportable
Manuf./Distrib.: Acros Org.; Aldrich; Fluka; ICN Biomed. Research Prods.; Pfaltz & Bauer; Sigma; Spectrum Quality Prods.; Springfield Scientific Supplies; TCI Am.
Trade Names Containing: Punctilious® SDA 40-1 190 Proof; Punctilious® SDA 40-1 Anhydrous

Brucine alkaloid; Brucine, solid. *See* Brucine

Brucine sulfate heptahydrate

CAS 5787-00-8; EINECS/ELINCS 225-432-9
UN 1544; UN 1570; UN 2811
Synonyms: Brucine sulfate (2:1) heptahydrate; 10,11-Dimethoxystrychnine sulfate, heptahydrate
Empirical: $C_{46}H_{68}N_4O_{19}S$
Formula: $(C_{23}H_{26}N_2O_4)_2 \cdot H_2SO_4 \cdot 6H_2O$
Properties: Off-wh. cryst. powd.; m.w. 1013.13; m.p. 180 C (dec.)
Toxicology: Highly toxic; poison; very toxic, possibly fatal by inh., ing., skin contact; ing. may cause nausea, vomiting; inh. may be harmful or fatal; irritating to eyes, skin, mucous membranes, upper respiratory tract; dust may irritate eyes; may cause convulsions; target organs: CNS
Precaution: Closed containers exposed to heat may explode; avoid raising dust; incompat. with strong oxidizing agents
Hazardous Decomp. Prods.: Toxic fumes of CO, CO_2, NO_x, SO_x; combustion: NO_x, SO_x
Storage: Store in cool, dry place in secure poison area; keep tightly closed
Uses: In denatured alcohol
Manuf./Distrib.: Aldrich; Alfa Aesar; ICN Biomed. Research Prods.
Trade Names Containing: Punctilious® SDA 40-2 Anhydrous

Brucine sulfate (2:1) heptahydrate. *See* Brucine sulfate heptahydrate
BTEAC. *See* Benzyltriethyl ammonium chloride
BTMAC. *See* Benzyl trimethyl ammonium chloride
t-BuAc. *See* t-Butyl acetate
Bubulum. *See* Neatsfoot oil
Butal; Butaldehyde; Butalyde; Butanal; n-Butanal. *See* n-Butyraldehyde
1-Butanamine. *See* n-Butylamine
2-Butanamine. *See* s-Butylamine
Butandiol. *See* 2,5-Dihydrofuran

Butane

CAS 106-97-8; EINECS/ELINCS 203-448-7
UN 1011 (DOT); INS943a; E943a
Synonyms: Alkane C4; n-Butane; Butyl hydride; Diethyl; Liquefied petroleum gas; Methylethylmethane
Classification: Alkane; hydrocarbon
Empirical: C_4H_{10}
Formula: $CH_3CH_2CH_2CH_3$
Properties: Colorless gas; faint disagreeable odor; easily liquefied under pressure @ R.T.; sol. in ethanol, ether, chloroform; sl. sol. in water; m.w. 58.12; sp.gr. 0.599; vapor pressure 1620 mm Hg (17 psig, 21 C); f.p. -138 C; b.p. -0.5 C; flash pt. (CC) -60 C
Toxicology: ACGIH TLV/TWA 800 ppm; LC50 (inh., rat, 4 h) 658 g/m^3; mildly toxic by inh.; causes drowsiness; anesthetic and asphyxiant props.; narcotic in high concs.; liq. may cause burns or frostbite to skin and eyes; TSCA listed
Precaution: Flamm. gas; very dangerous fire hazard exposed to heat, flame, oxidizers; highly explosive; explosive limits 1.9-8.5%; incompat. with strong oxidizing agents (increases fire/explosion risk)
Hazardous Decomp. Prods.: Heated to decomp., emits acrid smoke and fumes
NFPA: Health 1, Flammability 4, Reactivity 0
Storage: Store at or above ground level in a cool, dry, well-ventilated area, out of direct sunlight; cylinder temp. should not exceed 51 C; limit quantities
Uses: Propellant; paint aerosols; org. synthesis; fuel for household and industry; raw material for syn. rubber, high octane liq. fuels; mfg. of ethylene; process solvent; refrigerant; pure grades in calibrating instruments; food additive; propellant in cosmetics, pharmaceuticals; propellant, aerating agent, gas for foamed and sprayed food prods.
Regulatory: FDA 21CFR §173.350, 184.1165, GRAS; approved for topicals; NF compliance
Manuf./Distrib.: Aeropres; Air Prods.; Aldrich; BOC Gases; BP Amoco; ChevronPhillips; Degussa AG/Health & Nutrition; MG Ind.; Merck KGaA; Vopak USA

n-Butane. *See* Butane
Butane, 1-chloro-. *See* Butyl chloride
Butane, 2,3-dimethyl-. *See* 2,3-Dimethylbutane

Butanedioic acid, bis (2-ethoxyethyl) ester. *See* Diethoxyethyl succinate
Butanedioic acid diethyl ester. *See* Diethyl succinate
Butanedioic acid, dimethyl ester. *See* Dimethyl succinate
Butanediol. *See* 1,4-Butanediol
1,3-Butanediol; Butane-1,3-diol. *See* Butylene glycol

1,4-Butanediol

CAS 110-63-4; EINECS/ELINCS 203-786-5
Synonyms: 1,4-BD; BDO; 1,4-BDO; Butanediol; Butane-1,4-diol; 1,4-Butylene glycol; 1,4-Dihydroxybutane; Tetramethylene glycol; 1,4-Tetramethylene glycol
Empirical: $C_4H_{10}O_2$
Formula: $HO(CH_2)_4OH$
Properties: Colorless clear visc. liq.; nearly odorless; sol. in alcohol, water; misc. with acetone; sl. sol. in ether; m.w. 90.12; dens. 1.017; m.p. 16 C; b.p. 230, 120-122 C (10 mm); flash pt. > 110 C
Toxicology: LD50 (oral, rat) 1525 mg/kg, (IP, rat) 1370 mg/kg; mod. toxic by ingestion and IP routes; TSCA listed
Precaution: Combustible exposed to heat or flame; incompat. with oxidizers
Hazardous Decomp. Prods.: Heated to decomp., emits acrid smoke and fumes
NFPA: Health 1, Flammability 1, Reactivity 0
Uses: Solvent; humectant; coating resin raw material; in paints; crosslinking agent in polyurethane elastomers; polyurethane chain extender; intermediate for plasticizers, pharmaceuticals; mfg. of tetrahydrofuran, PBT resins, γ-butyrolactone; elastomeric fabric coatings; in food-pkg. adhesives
Regulatory: FDA 21CFR §175.105
Manuf./Distrib.: Aldrich; Allchem Ind.; Ashland; BASF AG; BASF; BP Amoco; Bencorp Int'l.; Browning; C.P. Hall; Chemcentral; DuPont; Fluka; Focus; Honeywell Spec. Chems.; Houghton Chem.; ISP; Lyondell; Spectrum Quality Prods.; Stanchem; Storchem; Tri-Iso; Vopak USA; Whyte Chems. Ltd

Butane-1,4-diol. *See* 1,4-Butanediol
1,2-Butanediol, cyclic carbonate. *See* 1,2-Butylene carbonate
1,2-Butanediol, diesters with octanoic and decanoic acids. *See* Butylene glycol dicaprylate/dicaprate
1,3-Butanediol 3-methyl ether. *See* 3-Methoxybutanol
Butane, 1,4-epoxy-. *See* Tetrahydrofuran
Butane, 2-methoxy-2-methyl-. *See* t-Amyl methyl ether
Butane, 2-methyl-. *See* Isopentane
Butanenitrile; n-Butanenitrile. *See* Butyronitrile
Butanethiol; 1-Butanethiol; n-Butanethiol. *See* n-Butyl mercaptan
Butanoic acid, ethyl ester. *See* Ethyl butyrate
Butanoic acid, 3-methylbutyl ester. *See* Isoamyl butyrate
Butanoic acid methyl ester. *See* Methyl butyrate
Butanoic acid propyl ester. *See* Propyl butyrate
Butanol; 1-Butanol; Butan-1-ol. *See* Butyl alcohol

2-Butanol

CAS 78-92-2; EINECS/ELINCS 201-158-5
UN 1120 (DOT)
Synonyms: Butan-2-ol; s-Butanol; 2-Butyl alcohol; s-Butyl alcohol; Butylene hydrate; Ethylmethyl carbinol; 2-Hydroxybutane; Methyl ethyl carbinol; 1-Methyl propanol; SBA; Secondary butyl alcohol
Classification: Sec. aliphatic alcohol
Empirical: $C_4H_{10}O$
Formula: $CH_3CH_2CH(OH)CH_3$
Properties: Colorless liq.; strong pleasant sweet odor; sol. in water, ethanol, ether, acetone, benzene; m.w. 74.12; sp.gr. 0.807 (20/4 C); vapor pressure 18.3 mm Hg; m.p. -89 C; b.p. 99-100 C; flash pt. 23 C; autoignition temp. 405 C; ref. index 1.3972 (20 C); surf. tens. 23.37 dynes/cm; dielec. const. 16.56
Toxicology: ACGIH TLV/TWA 100 ppm; LD50 (oral, rat) 6480 mg/kg, (IP, rabbit) 277 mg/kg; poison by IV and IP routes; mildly toxic by ing.; skin and eye irritant; excessive inh. may cause nose/throat irritation, headache, nausea, fatigue, dizziness; unconsciousness, coma in extreme cases; if aspirated into lungs, may cause severe lung damage, respiratory and cardiac arrest, possibly death; TSCA listed
Environmental: VOC; BOD5 1.76; COD 2.49; ThOD 2.59
Precaution: Flamm.; incompat. with alkali metals (increases fire/explosion risk, forms flamm. hydrogen gas), acids, acid chlorides, acid anhydrides (vigorous reactions), chlorine, isocyanates, ethylene oxide
Hazardous Decomp. Prods.: Unstable peroxides may form after prolonged storage
NFPA: Health 1, Flammability 3, Reactivity 0
Storage: Store in cool, well-ventilated area out of direct sunlight
Uses: Solvent for coatings, paint removers, adhesives, resins, lacquers, industrial cleaners, pharmaceuticals; prep. of methyl ethyl ketone; petroleum octane booster; extraction of fish protein; ingred. in hydraulic brake fluids; perfumes; cosmetics ingred.; synthetic flavoring agent in foods; in paper/paperboard in contact with dry food
Regulatory: FDA 21CFR §172.515, 176.180; SARA §313 reportable
Manuf./Distrib.: Ashland; Atofina SA; Chemcentral; E-Chemicals; ExxonMobil; Fluka; Houghton Chem.; Hukill; Quaker City; Shell; Spectrum Quality Prods.; Total Spec. Chems.; Vopak USA

Butan-2-ol. *See* 2-Butanol
n-Butanol. *See* Butyl alcohol
s-Butanol. *See* 2-Butanol
t-Butanol. *See* t-Butyl alcohol
2-Butanol acetate. *See* s-Butyl acetate
Butanol-2-amine; 1-Butanol-2-amino-. *See* 2-Aminobutanol
1,2-Butanolide; 1,4-Butanolide. *See* Butyrolactone
1-Butanol, 3-methoxy-. *See* 3-Methoxybutanol
1-Butanol, 4-methoxy-. *See* Butylene glycol methyl ether
1-Butanol, 3-methoxy-3-methyl-. *See* Methoxymethylbutanol
Butanone; 2-Butanone; 3-Butanone. *See* Methyl ethyl ketone
2-Butanone, 3-methyl-. *See* Methyl isopropyl ketone
2-Butenal; 2-Butenaldehyde. *See* Crotonaldehyde
2-Butenedioic acid (Z)-, bis (2-ethylhexyl) ester. *See* Bis (2-ethylhexyl) maleate
2-Butenedioic acid, (Z)-, dioctyl ester. *See* Dioctyl maleate
2-Butenoic acid, ethyl ester; trans-2-Butenoic acid ethyl ester. *See* Ethyl crotonate
2-Butin-1,4-diol. *See* But-2-yne-1,4-diol
Butolether. *See* 3-Methoxybutanol
1-Butoxybutane. *See* Dibutyl ether
Butoxydiethylene glycol. *See* Butoxydiglycol

Butoxydiglycol

CAS 112-34-5; EINECS/ELINCS 203-961-6
Synonyms: Butoxydiethylene glycol; 2-(2-Butoxyethoxy) ethanol; Butyl 'Carbitol'; o-Butyl diethylene glycol; Butyldiglycol; DEGBE; Diethylene glycol butyl ether; Diethylene glycol n-butyl ether; Diethylene glycol monobutyl ether; Diglycol butyl ether; Diglycol monobutyl ether; PEG-2 butyl ether
Classification: Aliphatic ether alcohol; aliphatic glycol ether
Empirical: $C_8H_{18}O_3$
Formula: $C_4H_9OCH_2CH_2OCH_2CH_2OH$
Properties: Colorless liq.; faint butyl odor; very sol. in ether, alcohol, acetone; sol. in oils and water; misc. with most org. solvs.; m.w. 162.26; sp.gr. 0.9553 (20/4 C); vapor pressure 0.02 mm Hg (20 C); m.p. -68.1 C; b.p. 230.6 C; flash pt. (CC) 78 C; ref. index 1.4316 (20 C)
Toxicology: LD50 (oral, rat) 6.56 g/kg, (skin, rabbit) 4120 mg/kg; mod. toxic by ing., IP routes; mildly toxic by skin contact; severe eye irritant; TSCA listed
Environmental: VOC; BOD5 1.18; COD 2.08; ThOD 2.17
Precaution: Flamm. exposed to heat or flame; incompat. with strong oxidizers (increases risk of fire and explosion)
Hazardous Decomp. Prods.: Thermal decomp. prods.: CO, CO_2; heated to decomp., emits acrid smoke and irritating fumes
NFPA: Health 1, Flammability 2, Reactivity 0
Storage: Store in cool, dry area away from heat and ignition sources
Uses: Coupling solvent for liq. cleaners, cutting fluids, textile auxs., nitrocellulose, coatings, oils, dyes, gums, soaps, polymers, stamp pad inks, high-based enamels, household cleaners; solvent for electronic materials, cosmetics; plasticizer intermediate; diluent for hydraulic brake fluids; coalescing agent for latex paints; dispersant for vinyl chloride resins in organosols; wetting-out sol'n. in textile industry; in food-pkg. adhesives; in paper/paperboard in contact with dry foods; surf. lubricant in mfg. of food-contact metallic articles
Regulatory: FDA 21CFR §175.105, 176.180, 178.3910; SARA §313 re-

portable; HAP
Manuf./Distrib.: Aldrich; Allchem Ind.; Ashland; BCH Brühl; Brown; Dow; Eastman; Fluka; Lyondell; Mitsubishi Chem.; Occidental; Ruger; Sal Chem.; Shell; Stanchem
Trade Names: BDGE; Butyl Carbitol®; Butyl Di Icinol; Butyl Dioxitol®; Eastman® DB; Eastman® DB Stabilized

Butoxyethanol

CAS 111-76-2; EINECS/ELINCS 203-905-0
UN 2369 (DOT)
Synonyms: 2-Butoxyethanol; 2-Butoxy-1-ethanol; 2-n-Butoxyethanol; n-Butoxyethanol; o-Butyl ethylene glycol; Butyl glycol; n-Butyl glycol; EGMBE; Ethylene glycol butyl ether; Ethylene glycol n-butyl ether; Ethylene glycol monobutyl ether; Ethylene glycol mono-n-butyl ether; Glycol butyl ether; Glycol monobutyl ether; Monobutyl glycol ether; 3-Oxa-1-heptanol
Classification: Glycol ether
Empirical: $C_6H_{14}O_2$
Formula: $HOCH_2CH_2OC_4H_9$
Properties: Colorless clear mobile liq.; mild pleasant odor; sol. in water, alcohol, ether; misc. wit most org. solvs.; m.w. 118.20; sp.gr. 0.9019 (20/20 C); vapor pressure 300 mm Hg (140 C); m.p. -74.8 C; b.p. 171.2 C; flash pt. 62 C
Toxicology: ACGIH TLV/TWA 25 ppm (skin); LD50 (oral, rat) 1480 mg/kg, (IP, rat) 220 mg/kg, (skin, rabbit) 490 mg/kg; poison by ing., skin contact, IP, IV routes; mod. toxic by inh., subcut.; human systemic effects; skin/eye irritant; TSCA listed
Environmental: BOD5 0.71; COD 2.20; ThOD 2.31
Precaution: DOT: Poisonous material; flamm. exposed to heat or flame; incompat. with oxidizers, strong caustics; attacks some types of rubber, plastics, and coatings
Hazardous Decomp. Prods.: Heated to decomp., emits acrid smoke and irritating fumes
NFPA: Health 2, Flammability 2, Reactivity 0
Storage: Store in tightly closed, grounded containers in cool, ventilated area, away from work areas; use in minimal quantities
Uses: Solvent for nitrocellulose, natural/syn. resins, sol. oils, adhesives, enamels, insecticides, herbicides, dry cleaning compds., textiles, organosols, cosmetics; solvent, coupling agent for printing inks, agric., water-based coatings, paints, varnishes, lacquers, plastics; cosolvent; coupling agent in cleaners; in textile dyeing/printing; leather treatment; in preservatives, disinfectants; prod. of plasticizers; stabilizer in metal and household cleaners, hydraulic fluids, insecticides, herbicides, rust removers; hydraulic brake fluids; gas chromatography; wetting agent; in food-pkg. adhesives; defoamer in food-contact paper/paperboard; solvent in food-contact polysulfide polymer/polyepoxy resins; flume wash water additive for food processing
Regulatory: FDA 21CFR §173.315, 175.105, 176.210, 177.1650, 178.1010, 178.3297, 178.3570; SARA §313 reportable; HAP; VOC
Manuf./Distrib.: Aldrich; Allchem Ind.; Ashland; Brown; Dow; Eastman; Fluka; Lyondell; Mitsubishi Chem.; Occidental; Ruger; Sal Chem.; Shell; Sigma; Spectrum Quality Prods.
Trade Names: BGE; Butyl Cellosolve®; Butyl Icinol; Butyl Oxitol®; Eastman® EB
Trade Names Containing: Ecco Strip LS-W; Ecosoft™ PB Solvent; Vertrel® X-B3

2-Butoxyethanol; 2-Butoxy-1-ethanol; 2-n-Butoxyethanol; n-Butoxyethanol. *See* Butoxyethanol

Butoxyethanol acetate

CAS 112-07-2; EINECS/ELINCS 203-933-3
Synonyms: Acetic acid, 2-butoxyethyl ester; 1-Acetoxy-2-butoxyethane; 2-Butoxyethanol acetate; 2-Butoxyethyl acetate; 2-Butoxyethyl ester acetic acid; Butyl 'Cellosolve' acetate; n-Butyl 'Cellosolve' acetate; Butyl glycol acetate; Ethanol, 2-butoxy-, acetate; Ethylene glycol butyl ether acetate; Ethylene glycol monobutyl ether acetate; Ethylene glycol mono-n-butyl ether acetate; Glycol monobutyl ether acetate
Classification: Ethylene glycol ether acetate
Empirical: $C_8H_{16}O_3$
Formula: $CH_3CH_2CH_2CH_2OCH_2CH_2OCOCH_3$
Properties: Colorless liq., mild fruity odor; sol. 1.1% in water; sol. in hydrocarbons and org. solvs.; m.w. 160.21; dens. 0.941 (20/20 C); f.p. < -63 C; b.p. 186 C; flash pt. (TOC) 74 C; ref. index 1.4142 (20 C)
Toxicology: LD50 (oral, rat) 2400 mg/kg, (skin, rabbit) 1500 mg/kg; mod. toxic by ing., skin contact; mild skin irritant; may cause blood disorders, kidney damage; harmful if inhaled or absorbed thru skin; TSCA listed
Environmental: VOC; ThOD 1.30
Precaution: Combustible liq. and vapor; peroxide former
Hazardous Decomp. Prods.: CO, CO_2; heated to decomp., emits acrid smoke and irritating fumes
NFPA: Health 1, Flammability 2, Reactivity 0
Uses: Solvent for resins and high-solids coatings (NC, butyrate, ethyl cellulose, acrylic), cosmetics; coalescing aid for water-disp. coatings; solvent, coupling agent for printing inks, agric., water-based coatings, paints, varnishes, lacquers, plastics; in cleaning prods., preservatives, rust removers, disinfectants; in food-pkg. adhesives
Regulatory: FDA 21CFR §175.105; SARA §313 reportable
Manuf./Distrib.: A.P. Chems. Ltd; Acros Org.; Aldrich; Alfa Aesar; Allchem Ind.; Ashland; BP Amoco; Chemial SpA; Eastman; Fluka; G.J. Chem.; J.T. Baker; Lancaster Synthesis; Lyondell; Occidental; Oxiteno; Pfaltz & Bauer; Spectrum Quality Prods.; Stanchem; TCI Am.; Whyte Chems. Ltd
Trade Names: Butyl Cellosolve® Acetate; Eastman® EB Acetate

2-Butoxyethanol acetate. *See* Butoxyethanol acetate
2-Butoxyethanol phosphate. *See* Tributoxyethyl phosphate
2-(2-Butoxyethoxy) ethanol. *See* Butoxydiglycol
2-(2-Butoxyethoxy) ethanol acetate. *See* Diethylene glycol butyl ether acetate
2-(2-(2-Butoxyethoxy) ethoxy) ethanol. *See* Butoxytriglycol
2-(2-Butoxyethoxy) ethyl acetate; 2-(2-n-Butoxyethoxy) ethyl acetate. *See* Diethylene glycol butyl ether acetate

1-Butoxyethoxy-2-propanol

CAS 124-16-3
Synonyms: 1-(2-Butoxyethoxy)-2-propanol; 2-Butoxy-1-(2′-hydroxypropoxy) ethane; 4,7-Dioxaundecan-2-ol; Propylene glycol butoxyethyl ether
Classification: Aliphatic ether alcohol
Empirical: $C_9H_{20}O_3$
Formula: $CH_3CH(OH)CH_2OCH_2CH_2OC_4H_9$
Properties: Colorless clear liq.; mild odor; sol. in water, variety of org. solvs.; m.w. 176.29; sp.gr. 0.931 (20/20 C); f.p. -90 C; b.p. 230 C; flash pt. (OC) 121 C
Toxicology: LD50 (oral, rat) 4000 mg/kg, (skin, rabbit) 2830 µl/kg; mod. toxic by ing. and skin contact; low toxicity by inh.; may cause mild skin irritation; absorbed by skin, but health effects unlikely; may cause mod. to severe eye irritation; TSCA listed
Precaution: Combustible exposed to heat or flame; incompat. with oxidizing agents (increases fire risk), strong acids (may cause decomp. of ethers)
Hazardous Decomp. Prods.: Heated to decomp., emits acrid and irritating fumes
NFPA: Health 2, Flammability 1, Reactivity 0
Storage: Store in a cool, dry, well-ventilated area out of direct sunlight
Uses: Solvent for paints, lacquers, resins, inks, dyes, oils, greases, cleaners; component in hydraulic fluids; antistall additive for automotive fuels; plasticizer; intermediate
Manuf./Distrib.: Ashland

1-(2-Butoxyethoxy)-2-propanol. *See* 1-Butoxyethoxy-2-propanol
2-Butoxyethyl acetate; 2-Butoxyethyl ester acetic acid. *See* Butoxyethanol acetate
2-Butoxyethyl ether. *See* Diethylene glycol dibutyl ether
2-Butoxyethyloctadecanoate. *See* Butoxyethyl stearate

Butoxyethyl stearate

CAS 109-38-6; EINECS/ELINCS 203-668-3
Synonyms: 2-Butoxyethyloctadecanoate; Ethylene glycol monobutyl ether stearate; Octadecanoic acid, 2-butoxyethyl ester
Definition: Ester of butoxyethanol and stearic acid
Empirical: $C_{24}H_{48}O_3$
Formula: $C_{17}H_{35}COOC_2H_4OC_4H_9$
Properties: Clear water-wh. to sl. yel. liq., fatty odor; insol. in water; sp.gr.

0.86; vapor pressure < 0.01 mm Hg (20 C); f.p. 14 C; b.p. > 190 C; flash pt. (COC) 381 F
Toxicology: LD50 (acute oral) 10.8 ml/kg; Draize skin irritation 0.42; Draize eye irritation 1.3
Precaution: Combustible; incompat. with oxidizing agents
Storage: Store away from strong oxidizing agents; avoid excessive heat
Uses: Antistat; solvent; emollient in personal care; lubricant, mold release agent, rubber plasticizer for industrial applics., magnetic tapes; resin plasticizer for pigments

2-Butoxy-1-(2´-hydroxypropoxy) ethane. *See* 1-Butoxyethoxy-2-propanol
'Butoxyl'. *See* Methoxybutyl acetate
1-2-[Butoxy-1-methylethoxy)-1-methylethoxy]-2-propanol. *See* Tripropylene glycol n-butyl ether
1-(2-Butoxy-1-methylethoxy)-2-propanol. *See* Dipropylene glycol butyl ether
Butoxypolypropylene glycol. *See* PPG butyl ether

3-Butoxy-1,2-propanediol
CAS 624-52-2
Synonyms: n-Butyl α-glycerol ether; Glyceryl α-mono-n-butyl ether; 1,2-Propanediol, 3-butoxy-
Classification: Glycol ether
Empirical: $C_7H_{16}O_3$
Properties: M.w. 148.13
Environmental: ThOD 2.05
Uses: Solvent
Regulatory: HAP

3-Butoxy-1,2-propanediol
CAS 624-52-2
Synonyms: 1,2-Propanediol, 3-butoxy-
Classification: Glycol ether
Empirical: $C_7H_{16}O_3$
Properties: M.w. 148.13
Environmental: ThOD 2.05
Uses: Solvent
Regulatory: HAP

Butoxypropanol
CAS 5131-66-8; EINECS/ELINCS 225-878-4
Synonyms: 1-Butoxy-2-propanol; n-Butoxypropanol; 2-Hydroxy-3-butyoxypropane; PGnBE; PnB; 2-Propanol, 1-butoxy; Propylene glycol butyl ether; Propylene glycol n-butyl ether; 1,2-Propylene glycol 1-monobutyl ether; Propylene glycol mono-n-butyl ether
Classification: Aliphatic ether alcohol
Empirical: $C_7H_{16}O$
Formula: $CH_3CH_2CH_2CH_2OCH_2CHOHCH_3$
Properties: Colorless clear liq.; mild char. odor; mod. sol. in water; sol. in variety of org. solvs.; m.w. 132.23; sp.gr. 0.884 (20/20 C); f.p. -148 F; b.p. 170.2 C; flash pt. (OC) 65.6 C; ref. index 1.4074 (20 C)
Toxicology: LD50 (oral, rat) 2200 mg/kg, (skin, rabbit) 3100 mg/kg; mod. toxic by ingestion and by skin contact; skin absorption may produce CNS depression; low toxicity by inh.; may cause eye irritation, possible corneal injury; TSCA listed
Precaution: Combustible; incompat. with oxidizing agents (increases risk of fire and explosion)
Hazardous Decomp. Prods.: Heated to decomp., emits acrid smoke and irritating fumes
Storage: Store in cool, dry area out of direct sunlight
Uses: Solvent, coupling agent for coatings, cleaners, electronics, inks, cosmetics, agric., textiles, adhesives, lt.-duty and hard-surf. cleaners, water-reducible polyester and alkyd resin prod.; wetting agent; solubilizer; surf. tens. reducer in water-based cleaners; humectant in cosmetics
Trade Names: Arcosolv® PNB; Dowanol® PnB; Solvenon® PnB

1-Butoxy-2-propanol. *See* Butoxypropanol
1-t-Butoxy-2-propanol. *See* Propylene glycol t-butyl ether
n-Butoxypropanol. *See* Butoxypropanol
Butoxytriethylene glycol. *See* Butoxytriglycol

Butoxytriglycol
CAS 143-22-6; EINECS/ELINCS 205-592-6
Synonyms: 2-(2-(2-Butoxyethoxy) ethoxy) ethanol; Butoxytriethylene glycol; Ethanol, 2-(2-(2-butoxyethoxy) ethoxy)-; PEG-3 butyl ether; Triethylene glycol butyl ether; Triethylene glycol n-butyl ether; Triethylene glycol monobutyl ether; Triglycol butyl ether; Triglycol monobutyl ether
Empirical: $C_{10}H_{22}O_4$
Formula: $C_4H_9O(C_2H_4O)_3H$
Properties: Colorless liq.; sol. in water; m.w. 206.32; dens. 1.0021 (20/20 C); f.p. -47.6 C; b.p. 276 C (dec.); flash pt. 143 C
Toxicology: LD50 (oral, rat) 6730 mg/kg, (skin, rabbit) 3540 mg/kg; mod. toxic by skin contact; mildly toxic by ing.; skin and severe eye irritant; neuromuscular blockage; potential human reproductive effects; TSCA listed
Precaution: Combustible exposed to heat or flames
Hazardous Decomp. Prods.: Heated to decomp., emits acrid smoke and irritating fumes
Uses: Plasticizer; intermediate; coupling solvent for paints, lacquers, detergents, sol. oils
Manuf./Distrib.: Ashland; Dow; Fluka; Mitsubishi Chem.; Oxiteno; Springfield Scientific Supplies

Butter of zinc. *See* Zinc chloride
Butyl acetate; 1-Butyl acetate. *See* n-Butyl acetate
2-Butyl acetate. *See* s-Butyl acetate

n-Butyl acetate
CAS 123-86-4; EINECS/ELINCS 204-658-1
UN 1123 (DOT); FEMA 2174
Synonyms: Acetic acid, butyl ester; Acetic acid n-butyl ester; Butyl acetate; 1-Butyl acetate; Butyl ethanoate
Classification: Carboxylic acid ester
Definition: Ester of butyl alcohol and acetic acid
Empirical: $C_6H_{12}O_2$
Formula: $CH_3COOC_4H_9$
Properties: Colorless liq., fruity odor; sol. in alcohol, ether, benzene, hydrocarbons; sl. sol. in water; m.w. 116.18; sp.gr. 0.8826 (20/20 C); vapor pressure 15 mm Hg; f.p. -75 C; b.p. 126.3 C; flash pt. (TOC) 36.6 C; ref. index 1.2951 (20 C)
Toxicology: ACGIH TLV/TWA 150 ppm; STEL 200 ppm; LD50 (oral, rat) 14.13 g/kg, (IP, mouse) 1230 mg/kg; mildly toxic by inh., ing.; mod. toxic by IP route; skin irritant; severe eye irritant; mild allergen; inh. or ing. can cause CNS depression, headache, nausea, unconsciousness; experimental teratogen; TSCA listed
Environmental: VOC; BOD5 0.52; COD 2.32; ThOD 2.21
Precaution: Flamm.; dangerous fire hazard exposed to flame; mod. explosive when exposed to flame; incompat. with strong oxidizers (increases fire risk), strong acids or strong bases (decomp. can occur), potassium t-butylate (ignition can occur)
Hazardous Decomp. Prods.: Heated to decomp., emits acrid and irritating fumes
NFPA: Health 1, Flammability 3, Reactivity 0
Storage: Store in cool, dry, well-ventilated area, out of direct sunlight; ground drums and bond transfer containers
Uses: Solvent for automotive paints/coatings, thinners, adhesives, cellulosics, cellulose nitrate lacquers/printing inks, cosmetics, pharmaceuticals, perfumes, electronic materials; intermediate for pharmaceuticals, syn. flavorings, org. compds.; mfg. of artificial leather, photographic films, plastics, safety glass; synthetic flavoring agent in foods and pharmaceuticals; defoamer in acute pulmonary edema extractants; in food-pkg. adhesives; adjuvant in resinous/polymeric food-contact coatings; in cellophane for food pkg.
Regulatory: FDA 21CFR §172.515, 175.105, 175.320, 177.1200; FEMA GRAS; CERCLA hazardous substance; Japan approved as flavoring
Manuf./Distrib.: AAE Chemie NV; AMC Chems.; AP Resources; Advanced BioTech; Aldrich; Allchem Ind.; Ashland; Asiamerica Int'l.; Augustus Oils Ltd; Axxence Aromatic GmbH; BASF; BP Amoco; BP Chems. Ltd; Baychem; Brenntag AG; Celanese; Chemcentral; Chisso Am.; Clariant; Coyne; Degussa AG; Degussa Flavors/Alfrebro; Dow; Eastman; Fleurchem; Fluka; Gallard-Schlesinger Ind.; General Chem.; Goldensun Mfg.; Great Western; Harcros; Houghton Chem.; Hukill; Independent Chem.; Integra; J.C. Wilson; J.T. Baker; Lluch Essence; Mallinckrodt Baker; Natural Advantage; Nipa Inc; Nippon Syn. Chem. Ind.; Penta Mfg.; Sigma; Spectrum Quality Prods.; Springfield Scientific Supplies;

Sunnyside; VWR Int'l.; Vopak USA; Xinchem
Trade Names: Eastman® n-Butyl Acetate, Urethane Grade

s-Butyl acetate

CAS 105-46-4; EINECS/ELINCS 203-300-1
UN 1123 (DOT)
Synonyms: Acetic acid 2-butoxy ester; Acetic acid 2-butyl ester; Acetic acid, s-butyl ester; Acetic acid 1-methylpropyl ester; 2-Butanol acetate; 2-Butyl acetate; s-Butyl alcohol acetate; 2-Butyl ethanoate; 1-Methylpropyl acetate
Classification: Carboxylic acid ester
Definition: Ester of butyl alcohol and acetic acid
Empirical: $C_6H_{12}O_2$
Formula: $CH_3COOCH(CH_3)(C_2H_5)$
Properties: Colorless liq.; mild fruity odor; sol. in oxygenated solvs., alcohol, ether, acetone; insol. in water; m.w. 116.16; sp.gr. 0.870 (20/4 C); m.p. -146 C; b.p. 112.2 C; flash pt. (OC) 31 C; ref. index 1.389 (20 C)
Toxicology: ACGIH TLV/TWA 200 ppm; STEL 150 ppm; mild skin and eye irritant; allergen; inh. of vapors can probably irritate nose and throat; ing. may irritate mouth and throat; high concs. may cause CNS depression, headache, dizziness, nausea, unconsciousness; TSCA listed
Environmental: VOC; ThOD 2.21
Precaution: Flamm.; incompat. with strong oxidizers (increases fire/explosion risk), strong acids, strong bases (decomp. can occur)
Hazardous Decomp. Prods.: s-Butyl alcohol, acetic acid; heated to decomp., emits acrid and irritating fumes
NFPA: Health 1, Flammability 3, Reactivity 0
Storage: Store in a cool, dry, well-ventilated area, out of direct sunlight; avoid generating mist
Uses: Solvent for nitrocellulose lacquers, varnishes, thinners, nail enamels, leather finishes, adhesives; textile sizes; paper coatings
Regulatory: CERCLA hazardous substance
Manuf./Distrib.: Aldrich; Coyne; E-Chemicals; Hukill; Schweizerhall; Vopak USA

t-Butyl acetate

CAS 540-88-5; EINECS/ELINCS 208-760-7
UN 1123 (DOT)
Synonyms: Acetic acid t-butyl ester; Acetic acid 1,1-dimethylethyl ester; t-BuAc; 1,1-Dimethylethyl acetate
Classification: Carboxylic acid ester
Definition: Ester of t-butyl alcohol and acetic acid
Empirical: $C_6H_{12}O_2$
Formula: $CH_3COOC(CH_3)_3$
Properties: Colorless clear liq.; fruity odor; pract. insol. in water; sol. in alcohol, ether, acetic acid; m.w. 116.16; sp.gr. 0.896 (20 C); b.p. 96 C; flash pt. 15 C; ref. index 1.3870; flamm.
Toxicology: ACGIH TLV/TWA 200 ppm; poison by inh. and ing.; vapors can probably irritate nose and throat; high concs. may cause CNS depression; mild skin and eye irritant; ing. may cause mouth/throat irritation, CNS depression, unconsciousness @ higher concs.; TSCA listed
Precaution: Incompat. with strong oxidizers (increases risk of fire and explosion), strong bases, strong acids (decomp. can occur)
Hazardous Decomp. Prods.: t-Butyl alcohol, acetic acid; heated to decomp., emits acrid smoke and irritating fumes
Storage: Store in cool, dry, well-ventilated area out of direct sunlight
Uses: As gasoline additive; solvent; building block for org. synthesis in pharmaceuticals, agric. chems.; reagent for fine chems. prod.; intermediate for cholesterol-lowering drugs
Manuf./Distrib.: Aldrich; Celanese; E-Chemicals; Fluka; Itochu Spec. Chems.; Lyondell; Schweizerhall; Sigma; Spectrum Quality Prods.; Wacker-Chemie GmbH

Butylacetone. *See* Methyl n-amyl ketone
Butyl adipate. *See* Dibutyl adipate

Butyl alcohol

CAS 71-36-3; EINECS/ELINCS 200-751-6
UN 1120 (DOT); FEMA 2178
Synonyms: Butanol; 1-Butanol; Butan-1-ol; n-Butanol; n-Butyl alcohol; Butyl hydroxide; Butyric alcohol; 1-Hydroxybutane; Methylolpropane; Propyl carbinol; Propylmethanol
Classification: Primary aliphatic alcohol
Empirical: $C_4H_{10}O$
Formula: $CH_3(CH_2)_2CH_2OH$
Properties: Colorless clear mobile liq., char. penetrating vinous odor; sol. in water, misc. with alcohol, ether, many org. solvs.; m.w. 74.14; sp.gr. 0.8109 (20/20 C); vapor pressure 5.5 mm Hg (20 C); m.p. -90 C; b.p. 117.7 C; flash pt. 35 C; ref. index 1.3993 (20 C)
Toxicology: ACGIH TLV/CL 50 ppm (skin); LD50 (oral, rat) 790 mg/kg, (IV, mouse) 377 mg/kg, (skin, rabbit) 3400 mg/kg; poison by IV route; mod. toxic by skin contact, ing., subcut., IP routes; skin and severe eye irritant; CNS depressant; TSCA listed
Environmental: BOD5 1.66; COD 2.46; ThOD 2.59
Precaution: Flamm.; dangerous fire hazard exposed to heat, flame, oxidizers; mod. explosive exposed to flame; incompat. with oxidizing materials, aluminum, chromium trioxides; increased fire and explosion hazard; reacts at elevated temps. with Al
Hazardous Decomp. Prods.: Heated to decomp., emits acrid smoke and fumes
NFPA: Health 1, Flammability 3, Reactivity 0
Storage: Store in cool, well-ventilated area, out of direct sunlight
Uses: Prep. of esters; solvent for fats, waxes, resins, gums, shellac, varnish, plasticizers, paints, inks, cosmetics; extraction solvent for mineral processing; mfg. of rayon, detergents, other butyl compds.; hydraulic fluids; dehydration agents; denaturant; synthetic flavoring agent in foods and pharmaceuticals; color diluent for confectionery, tableted food supplements, gum; solvent in prod. of hops for beer brewing; solvent in pharmaceutical orals; fragrance; surfactant intermediate; in food-pkg. adhesives; adjuvant in resinous/polymeric food-contact coatings; defoamer in food-contact paper coatings; solvent in food-contact polysulfide polymer/polyepoxy resins
Regulatory: FDA 21CFR §73.1, 172.515, 172.560, 175.105, 175.320, 176.180, 176.200, 176.210, 177.1200, 177.1440, 177.1650, 177.2800; 27CFR §21.99; FEMA GRAS; FDA approved for orals; NF compliance; SARA §313 reportable; CERCLA hazardous substance; VOC
Manuf./Distrib.: AAE Chemie NV; AMC Chems.; Advanced BioTech; Aldrich; Allchem Ind.; Amyl; Ashland; Axxence Aromatic GmbH; BASF; BP Chems. Ltd; Baychem; Brenntag AG; Celanese; Chemcentral; Chisso Am.; Coyne; Degussa Flavors/Alfrebro; Delta Distributors; Dow; Eastech; Eastman; Fisher Scientific; Fleurchem; Fluka; GNC Group; Grau Aromatics; Harcros; Houghton Chem.; Hukill; Integra; J.C. Wilson; J.T. Baker; Jilin Chem. Ind.; Lluch Essence; Mallinckrodt Baker; Milsolv; Nippon Syn. Chem. Ind.; P.B. & S.; Penta Mfg.; Prodasynth; Punda Mercantile; R.E. Carroll; Romil Ltd; Ruger; SOCO-Lynch; Sal Chem.; Sasol N. Am.; Shell; Sigma; Southchem; Spectrum Quality Prods.; Springfield Scientific Supplies; Stanchem; Sunnyside; Synasia Fine Chem.; Textile Chem.; VWR Int'l.; Vopak USA; Whittaker, Clark & Daniels
Trade Names: Alfol® 4; Eastman® n-Butanol; Ucar® Butanol
Trade Names Containing: Formcel®

2-Butyl alcohol. *See* 2-Butanol
n-Butyl alcohol. *See* Butyl alcohol
s-Butyl alcohol. *See* 2-Butanol

t-Butyl alcohol

CAS 75-65-0; EINECS/ELINCS 200-889-7
UN 1120 (DOT)
Synonyms: t-Butanol; t-Butyl hydroxide; Dimethylethanol; 1,1-Dimethylethanol; Methyl-2-propanol; 2-Methyl-2-propanol; 2-Methylpropan-2-ol; 2-Propanol, 2-methyl-; TBA; Trimethyl carbinol; Trimethyl methanol
Classification: Tertiary aliphatic alcohol
Empirical: $C_4H_{10}O$
Formula: $(CH_3)_3COH$
Properties: Colorless liq. or rhombic prisms; unpleasant camphor-like odor; sol. in ethanol, ether, water; m.w. 74.12; sp.gr. 0.786 (20 C); vapor pressure 40 mm Hg (24.5 C); m.p. 25-26 C; b.p. 83 C; flash pt. (CC) 11 C; autoignition temp. 343 C; ref. index 1.3870 (20 C)
Toxicology: ACGIH TLV/TWA 100 ppm; STEL 150 ppm; LD50 (oral, rat) 3500 mg/kg; toxic; irritant to mucous membranes; inh. of 25 ppm causes pulmonary problems in man; CNS depressant; inh. of high concs. can cause headache, drowsiness, unconsciousness; ing. can cause head-

ache, dizziness, drowsiness; may cause severe lung damage, respiratory/cardiac arrest, and death if aspirated into lungs; skin contact can cause contact dermatitis, possible allergic reactions; eye irritant; mutagenic data; TSCA listed

Precaution: Flamm.; dangerous fire hazard exposed to heat, flame; mod. explosive as vapor; incompat. with strong oxidizers, potassium-sodium alloy, hydrogen peroxide and sulfuric acid (fire/explosion risk), strong min. acids (can dec. to flamm. isobutylene gas)

NFPA: Health 1, Flammability 3, Reactivity 0

Storage: Store in cool, dry, well-ventilated area, out of direct sunlight

Uses: Solvent, coupling agent, processing aid for pharmaceuticals, cosmetics, aq. coatings and adhesives, agric. formulations, polymer processing, cleaners/disinfectants; alcohol denaturant; dehydration agent; fragrance/flavoring ingred.; solvent for perfumes; chem. intermediate; paint removers; mfg. of methyl methacrylate, flotation agents; octane improver in unleaded gasoline; denaturant for cosmetics, pharmaceuticals; defoamer in food-contact paper coatings; surf. lubricant in mfg. of food-contact metallic articles

Features: Noncorrosive

Regulatory: FDA 21CFR §176.200, 178.3910, 27CFR §21.100

Manuf./Distrib.: AMC Chems.; Aldrich; Alfa Aesar; Allchem Ind.; Ashland; Bencorp Int'l.; Columbia Sales Int'l.; Degussa; E-Chemicals; Filo; Fluka; Honeywill & Stein; Integra; J.H. Calo; J.T. Baker; Kira; Lyondell; Mallinckrodt Baker; Mitsui & Co. USA; Pechiney Chems. Div.; Penta Mfg.; Quaker City; Romil Ltd; Sigma; Spectrum Quality Prods.; VWR Int'l.; Vopak USA

Trade Names: Tebol®; Tebol® 99

Trade Names Containing: Punctilious® SDA 38B Anhydrous Prod. Code 14680000; Punctilious® SDA 39B 190 Proof; Punctilious® SDA 40-1 190 Proof; Punctilious® SDA 40-1 Anhydrous; Punctilious® SDA 40-2 Anhydrous; Punctilious® SDA 40A 190 Proof; Punctilious® SDA 40A Anhydrous; Punctilious® SDA 40B Anhydrous; Punctilious® SDA 40B Modified Anhydrous; Punctilious® SDA 40C 190 Proof

s-Butyl alcohol acetate. *See* s-Butyl acetate
Butyl alcohol hydrogen phosphite. *See* Dibutyl phosphite
n-Butylaldehyde. *See* n-Butyraldehyde
Butylamine. *See* n-Butylamine

n-Butylamine

CAS 109-73-9; EINECS/ELINCS 203-699-2
UN 1125 (DOT); FEMA 3130

Synonyms: Amine C_4; 1-Aminobutane; 1-Butanamine; Butylamine; MNBA; Monobutylamine; Mono-n-butylamine

Classification: Saturated aliphatic amine

Empirical: $C_4H_{11}N$

Formula: $CH_3(CH_2)_3NH_2$

Properties: Colorless liq., amine odor; very sol. in acetone; sol. in benzene, ethanol, ether, water; misc. with oxygenated solvs.; m.w. 73.14; dens. 0.7327; m.p. -50 C; b.p. 78 C; flash pt. (OC) -1.1 C; ref. index 1.4010

Toxicology: ACGIH TLV/CL 5 ppm; LD50 (oral, rat) 366 mg/kg, (IP, mouse) 629 mg/kg, (IV, mouse) 198 mg/kg, (skin, rabbit) 850 mg/kg; poison by ing., skin contact, IV routes; mod. toxic by inh., IP routes; potent irritant to eyes, mucous membranes; direct skin contact causes severe primary irritation and blistering; experimental tumorigen; mutagenic data; TSCA listed

Environmental: VOC; BOD5 2.24; ThOD 2.63

Precaution: DOT: Flamm. liq.; dangerous fire hazard exposed to heat, flame, or oxidizers; explodes on contact with perchloryl fluoride

Hazardous Decomp. Prods.: CO, CO_2; heated to decomp., emits toxic fumes of NO_x

NFPA: Health 3, Flammability 3, Reactivity 0

Storage: Refrigerate; may be heat- and air-sensitive

Uses: Intermediate for emulsifiers, pharmaceuticals, insecticides, rubber chemicals, dyes, tanning agents; catalyst; corrosion inhibitor; synthetic flavoring agent in foods and pharmaceuticals; solvent

Features: Fishy pungent flavor

Regulatory: FEMA GRAS; CERCLA hazardous substance

Manuf./Distrib.: Air Prods.; Aldrich; Alfa Aesar; Ashland; Atofina; BASF; Celanese; Daicel Chem. Ind.; Fluka; Honeywell Spec. Chems.; Mallinckrodt Baker; Sigma; Sithean; Spectrum Quality Prods.; VWR Int'l.

s-Butylamine

CAS 13952-84-6
UN 1125 (DOT); 1993

Synonyms: 2-AB; 2-Aminobutane; 2-BA; 2-Butanamine; 1-Methylpropylamine

Classification: Sat. aliphatic amine

Empirical: $C_4H_{11}N$

Formula: $CH_3CH_2CH(CH_3)NH_2$

Properties: Colorless liq.; ammonia-like or fish odor; sol. in water, many org. solvs. incl. ethanol, acetone, ether, chloroform; m.w. 73.14; sp.gr. 0.724 (20 C); m.p. -104 C; b.p. 63 C; flash pt. (CC) -9 C

Toxicology: LD50 (oral, rat) 152 mg/kg, (dermal, rabbit) 2500 mg/kg; toxic, corrosive; poison by ingestion; powerful irritant; mod. toxic by skin contact; severe eye irritant; inh. may cause nose/throat/lung irritation, coughing, chest pain, difficulty breathing; high concs. may cause potentially fatal pulmonary edema; ing. may cause burning of the mouth, throat, digestive tract; TSCA listed

Precaution: Extremely flamm.; dangerous fire hazard exposed to heat or flame; incompat. with oxidizing materials, acids, acid chlorides, acid anhydrides, calcium or sodium hypochlorite, nitromethane, nitrosyl perchlorate; violent/explosive reactions possible

Hazardous Decomp. Prods.: Heated to decomp., emits toxic fumes of NO_x

NFPA: Health 3, Flammability 3

Storage: Store in cool, dry, well-ventilated area, out of direct sunlight

Uses: Fungistat for fruits; solvent; used to produce other org. chems.

Manuf./Distrib.: Aldrich; Atofina; BASF AG; BASF

Butylated hydroxyanisole. *See* BHA
Butylated hydroxytoluene. *See* BHT

s-Butylbenzene

CAS 135-98-8

Synonyms: 2-Phenylbutane

Empirical: $C_{10}H_{13}$

Formula: $C_6H_5C(CH_3)C_2H_5$

Properties: Colorless liq.; m.w. 134.24; dens. 0.8618 (204 C); vapor pressure 15 mm Hg (60 C); f.p. -75.68 C; b.p. 170.65 C; flash pt. 62.7 C; ref. index 1.4901 (20 C)

Toxicology: LD50 (oral, rat) 2240 mg/kg; mod. toxic by ing.; skin and eye irritant; TSCA listed

Precaution: Combustible; explosive limits 0.8-6.9%; incompat. with oxidizers

Hazardous Decomp. Prods.: Heated to decomp., emits acrid smoke and fumes

Uses: Solvent for coating compositions, organic synthesis, plasticizer, surface-active agents

Manuf./Distrib.: ChevronPhillips; Degussa

t-Butylbenzene

CAS 98-06-6

Synonyms: (1,1-Dimethylethyl) benzene; 2-Methyl-2-phenylpropane; Pseudobutylbenzene; Trimethylphenylmethane

Empirical: $C_{10}H_{14}$

Formula: $C_6H_5C(CH_3)_3$

Properties: Colorless liq.; sol. in alcohol; misc. with oxygenated and aromatic solvs.; insol. in water; m.w. 134.22; dens. 0.866 (20 C); vapor pressure 1 mm Hg (13.0 C); f.p. -57.8 C; b.p. 169.1 C; flash pt. 60 C; ref. index 1.492 (20 C)

Toxicology: LDLo (oral, rat) 5000 mg/kg; mildly toxic by ing.; TSCA listed

Precaution: Combustible; incompat. with oxidizers

Hazardous Decomp. Prods.: Heated to decomp., emits acrid smoke and fumes

Uses: Organic synthesis; polymerization solvent; polymer linking agent

Manuf./Distrib.: Alfa Aesar; ChevronPhillips; Degussa; Xinchem

Butyl benzoate

CAS 136-60-7; EINECS/ELINCS 205-252-7

Synonyms: Benzoic acid, butyl ester; Benzoic acid n-butyl ester; n-Butyl benzoate

Empirical: $C_{11}H_{14}O_2$
Formula: $C_6H_5COOC_4H_9$
Properties: Colorless oily liq.; sol. in oxygenated solvs.; misc. with alcohol or ether; insol. in water; m.w. 178.23; dens. 1.00 (20 C); vapor pressure < 0.01 mm Hg (20 C); f.p. -22 C; b.p. 247.3 C; flash pt. 107 C
Toxicology: LD50 (oral, rat) 5140 mg/kg, (skin, rabbit) 4 g/kg; mod. toxic by skin contact; mildly toxic by ing.; severe skin/moderate eye irritant; TSCA listed
Environmental: Biodeg.
Precaution: Combustible exposed to heat or flame; can react with oxidizing materials
Hazardous Decomp. Prods.: Heated to decomp., emits acrid and irritating fumes
NFPA: Health 1, Flammability 1, Reactivity 0
Uses: Solvent for cellulose ether, carriers, emulsifiers; plasticizer; perfume ingredient; dye carrier for textiles, polyester; preservative in cosmetics; in food-pkg. adhesives
Use Level: 0.5% max. in finished cosmetics
Regulatory: FDA 21CFR §175.105
Manuf./Distrib.: Aldrich; Alfa Aesar; Allchem Ind.; Bayer/Industrial Chems./Textile; Bell Flavors & Fragrances; Boehme Filatex; Dastech Int'l.; Fluka; Goldensun Mfg.; Haltermann GmbH; Indofine; Penta Mfg.; Piedmont Chem. Ind.; Polarome Int'l.; Stockhausen; Unitex; Wall Chem.

n-Butyl benzoate. *See* Butyl benzoate

Butyl benzyl phthalate

CAS 85-68-7; EINECS/ELINCS 201-622-7
Synonyms: BBP; 1,2-Benzenedicarboxylic acid, butyl phenylmethyl ester; Benzyl butyl phthalate; Benzyl n-butyl phthalate; n-Butyl benzyl phthalate
Classification: Aromatic ester
Empirical: $C_{19}H_{20}O_4$
Formula: $C_4H_9OOCC_6H_4COOC_7H_7$
Properties: Clear oily liq., sl. odor; insol. in water; m.w. 312.39; dens. 1.116 (25/25 C); m.p. < -35 C; b.p. 370 C; flash pt. 390 F; ref. index 1.535-1.540
Toxicology: LD50 (oral, rat) 20,400 mg/kg, (IP, mouse) 3160 mg/kg; moderate toxicity by ingestion, IP routes; experimental carcinogen and reproductive effects; TSCA listed
Precaution: Combustible exposed to heat or flame; can react with oxidizers
Hazardous Decomp. Prods.: Heated to decomp., emits acrid smoke and irritating fumes
Uses: Plasticizer for polyvinyl and cellulosic resins incl. NC lacquers and films, PVC flooring/plastisols, PVAc adhesives/sealants/caulks, acrylic coatings, chlorinated rubber, cellulose propionate, polyamide, castable PU; organic intermediate; film-former, plasticizer in cosmetics; plasticizer in food-contact polymers; in food-pkg. adhesives; in paper/paperboard in contact with aq./fatty/dry foods; solvent for adjuvants in food-contact crosslinked polyesters
Features: High-solvating
Regulatory: FDA 21CFR §175.105, 176.170, 176.180, 177.2420, 178.3740; SARA reportable
Manuf./Distrib.: Aldrich; Allchem Ind.; Ashland; Bayer; Bencorp Int'l.; ChemService; Solutia; St. Lawrence; Westtec Ind.

n-Butyl benzyl phthalate. *See* Butyl benzyl phthalate
N-Butyl-1-butanamine. *See* Dibutylamine
n-Butyl n-butanoate. *See* Butyl butyrate
2-Butylbutanoic acid. *See* 2-Ethylhexoic acid

Butyl butyrate

CAS 109-21-7; EINECS/ELINCS 203-656-8
FEMA 2186
Synonyms: n-Butyl n-butanoate; n-Butyl butyrate; n-Butyl n-butyrate; Butyric acid, butyl ester
Classification: Ester
Empirical: $C_8H_{16}O_2$
Formula: $CH_3CH_2CH_2COO(CH_2)_3CH_3$
Properties: Colorless liq., pineapple odor; misc. with alcohol, ether, veg. oils; sl. sol. in propylene glycol, water; m.w. 144.24; dens. 0.67-0.871; vapor pressure 1.81 mm Hg; b.p. 166 C; flash pt. 53 C; ref. index 1.405
Toxicology: LD50 (oral, rabbit) 9520 mg/kg, (IP, rat) 2300 mg/kg; mod. toxic by IP; mildly toxic by ingestion; mod. irritating to eyes, skin, mucous membranes; narcotic in high concs.; TSCA listed
Environmental: VOC, ThOD 2.44
Precaution: Combustible liq.; incompat. with oxidizers
Hazardous Decomp. Prods.: Heated to decomp., emits acrid smoke and irritating fumes
NFPA: Health 2, Flammability 2, Reactivity 0
Uses: Synthetic flavoring agent in foods and pharmaceuticals; fragrance in cosmetics; solvent
Features: Apple, banana, berry, peach, pear-like flavor
Regulatory: FDA 21CFR §172.515; FEMA GRAS; Japan approved as flavoring
Manuf./Distrib.: Advanced BioTech; Aldrich; Augustus Oils Ltd; Berje; Degussa Flavors/Alfrebro; Elan; Epochem; Fluka; Goldensun Mfg.; Grau Aromatics; Haarmann & Reimer GmbH; Hercules; Indofine; Lluch Essence; Penta Mfg.; Sithean; Xinchem

n-Butyl butyrate; n-Butyl n-butyrate. *See* Butyl butyrate
dl-s-Butylcarbinol. *See* 2-Methyl-1-butanol
n-Butyl carbinol. *See* n-Amyl alcohol
s-Butyl carbinol. *See* 2-Methyl-1-butanol
t-Butyl carbinol. *See* 2,2-Dimethylpropanol
Butyl 'Carbitol'. *See* Butoxydiglycol
Butyl 'Carbitol' acetate; n-Butyl 'Carbitol' acetate. *See* Diethylene glycol butyl ether acetate
Butyl 'Cellosolve' acetate; n-Butyl 'Cellosolve' acetate. *See* Butoxyethanol acetate

Butyl chloride

CAS 109-69-3; EINECS/ELINCS 203-696-6
UN 1127 (DOT)
Synonyms: Butane, 1-chloro-; n-Butyl chloride; 1-Chlorobutane; N-Propylcarbinyl chloride
Classification: Nonaromatic halogenated hydrocarbon
Empirical: C_4H_9Cl
Formula: $CH_3CH_2CH_2CH_2Cl$
Properties: Colorless liq.; misc. with oxygenated solvs., alcohol, ether; insol. in water; m.w. 92.58; dens. 0.8875 (20/20 C); b.p. 78.6 C; flash pt. 15 F; ref. index 1.40223
Toxicology: LD50 (oral, rat) 2670 mg/kg; LCLo (inh., rat, 4 h) 8000 ppm; toxic on prolonged inh.; mod. toxic by ing.; primary skin and eye irritant; mutagen, tumorigen; TSCA listed
Precaution: Highly flamm.; dangerous fire risk; mod. explosive exposed to flame; incompat. with oxidizing materials
Hazardous Decomp. Prods.: Heated to decomp., emits highly toxic fumes of phosgene and Cl^-
NFPA: Health 2, Flammability 3, Reactivity 0
Uses: Solvent for oils, fats, and waxes; alkylating agent in org. synthesis; mfg. of butyl cellulose; high performance liq. chromatography; molecular biology (protein sequencing)
Manuf./Distrib.: Aceto; Aldrich; Alfa Aesar; Allchem Ind.; Boith China; Degussa; Filo; Fluka; Occidental; Romil Ltd; Sigma; Spectrum Quality Prods.

n-Butyl chloride. *See* Butyl chloride
Butyl citrate. *See* Tributyl citrate

Butylcyclohexane

CAS 1678-93-9; EINECS/ELINCS 216-837-1
Synonyms: 1-Cyclohexylbutane
Classification: Sat. alicyclic hydrocarbon
Empirical: $C_{10}H_{20}$
Formula: $C_4H_9C_6H_{11}$
Properties: Colorless liq.; sol. in most org. solvs.; insol. in water; m.w. 140.27; sp.gr. 0.818 (20 C); m.p. -75 C; b.p. 178-180 C; flash pt. 41 C
Toxicology: May cause mild skin irritation; conc. vapor may cause eye irritation, nose/throat irritation, headache, lightheadedness, nausea, drowsiness; probably no significant CNS effects; ing. may result in sore throat, nausea, diarrhea
Precaution: Combustible; incompat. with strong oxidizing agents (increased fire and explosion hazard)

Hazardous Decomp. Prods.: None
NFPA: Health 0, Reactivity 0
Storage: Store in cool, dry, well-ventilated area out of direct sunlight
Uses: Solvent; org. synthesis
Manuf./Distrib.: Aldrich; Fluka

i-Butylcyclohexane. *See* Isobutylcyclohexane

s-Butylcyclohexane

CAS 7058-01-7; 52993-54-1
Synonyms: Cyclohexane, s-butyl-; Cyclohexane, (1-methylpropyl)-; 2-Cyclohexylbutane
Classification: Sat. alicyclic hydrocarbon
Empirical: $C_{10}H_{20}$
Formula: $CH_3CH_2CH(CH_3)C_6H_{11}$
Properties: Colorless liq.; sol. in most org. solvs.; insol. in water; m.w. 140.27; sp.gr. 0.813 (20 C); b.p. 177 C
Toxicology: May cause mild skin irritation; conc. vapor may cause eye irritation, nose/throat irritation, headache, lightheadedness, nausea, drowsiness; probably no significant CNS effects; ing. may result in sore throat, nausea, diarrhea
Precaution: Combustible; incompat. with strong oxidizing agents (increased risk of fire and explosion)
Hazardous Decomp. Prods.: None
NFPA: Health 0, Reactivity 0
Storage: Store in cool, dry, well-ventilated area
Uses: Solvent; org. synthesis
Manuf./Distrib.: ICN Biomed. Research Prods.

t-Butylcyclohexane

CAS 3178-22-1; EINECS/ELINCS 221-652-4
Synonyms: (1,1-Dimethylethyl) cyclohexane
Classification: Sat. alicyclic hydrocarbon
Empirical: $C_{10}H_{20}$
Formula: $C_6H_{11}C(CH_3)_3$
Properties: Colorless liq.; sol. in most org. solvs.; insol. in water; m.w. 140.27; sp.gr. 0.831 (20 C); m.p.-41 C; b.p. 167-169 C; flash pt. 42 C
Toxicology: May cause mild skin irritation; conc. vapor may cause eye irritation, nose/throat irritation, headache, lightheadedness, nausea, drowsiness; probably no significant CNS effects; ing. may result in sore throat, nausea, diarrhea
Precaution: Combustible; incompat. with strong oxidizing agents (increased risk of fire and explosion)
Hazardous Decomp. Prods.: None
NFPA: Health 0, Reactivity 0
Storage: Store in cool, dry, well-ventilated area
Uses: Solvent; org. synthesis
Manuf./Distrib.: Aldrich; Fluka

2-t-Butylcyclohexanol acetate. *See* 2-t-Butylcyclohexyl acetate

2-t-Butylcyclohexyl acetate

CAS 88-41-5; 20298-69-5 (cis); 20298-70-8 (trans); EINECS/ELINCS 201-828-7; 243-718-1 (cis)
Synonyms: 1-Acetoxy-2-t-butylcyclohexane; 2-t-Butylcyclohexanol acetate; Butyl-o-t-cyclohexyl acetate; Cyclohexanol, 2-(1,1-dimethylethyl)-, acetate; 2-(1,1-Dimethylethyl) cyclohexyl acetate
Classification: Organic compd.
Definition: Commercial prod. is mixt. of cis and trans isomers
Empirical: $C_{12}H_{22}O_2$
Properties: Colorless liq.; fruity odor; m.w. 198.31; dens. 0.94 kg/l; flash pt. 90 C
Toxicology: LD50 (oral, rat) 4600 mg/kg, (skin, rabbit) > 5 g/kg; low toxicity by ing.; TSCA listed
Hazardous Decomp. Prods.: Heated to decomp., emits acrid smoke and irritating vapors
Uses: Syn. fragrance for soaps, household prods.; solvent in cosmetics
Features: Fresh, fruity, woody, very tenacious fragrance
Manuf./Distrib.: Degussa; Grau Aromatics; Penta Mfg.

Butyl-o-t-cyclohexyl acetate. *See* 2-t-Butylcyclohexyl acetate

Butylcyclopentane

CAS 2040-95-1
Synonyms: n-Butylcyclopentane; Cyclopentane, butyl-
Classification: Sat. alicyclic hydrocarbon
Empirical: C_9H_8
Formula: $C_4H_9C_5H_9$
Properties: Colorless liq.; sol. in most org. solvs.; insol. in water; m.w. 126.24; sp.gr. 0.785 (20 C); m.p. -108 C; b.p. 157 C; ref. index 1.4316
Toxicology: May cause mild skin irritation; conc. vapor may be irritating to eyes; ing. may cause sore throat, nausea, diarrhea; probably a CNS depressant on inh.; very high concs. may cause unconsciousness and death
Precaution: Combustible; incompat. with strong oxidizing agents (increased fire and explosion hazard)
Hazardous Decomp. Prods.: None
NFPA: Health 0, Reactivity 0
Storage: Store in cool, dry, well-ventilated area
Uses: Solvent; org. synthesis
Manuf./Distrib.: Sigma

n-Butylcyclopentane. *See* Butylcyclopentane

s-Butylcyclopentane

CAS 4850-32-2
Classification: Sat. alicyclic hydrocarbon
Empirical: C_9H_{18}
Formula: $CH_3CH_2CH(CH_3)C_5H_9$
Properties: Colorless liq.; sol. in most org. solvs.; insol. in water; m.w. 125.23
Toxicology: May cause mild skin irritation; conc. vapor may be irritating to eyes; probably a CNS depressant; inh. of high vapor concs. may cause nose/throat irritation, nausea, drowsiness; very high concs. may cause unconsciousness and death; ing. may result in sore throat, nausea, diarrhea
Precaution: Combustible; incompat. with strong oxidizing agents (increased fire and explosion hazard)
Hazardous Decomp. Prods.: None
Storage: Store in cool, dry, well-ventilated area
Uses: Solvent; org. synthesis

t-Butylcyclopentane

CAS 3875-52-3
Classification: Sat. alicyclic hydrocarbon
Empirical: C_9H_{18}
Formula: $C_5H_9C(CH_3)_3$
Properties: Colorless liq.; sol. in most org. solvs.; insol. in water; m.w. 126.23
Toxicology: May cause mild skin irritation; conc. vapor may be irritating to eyes; probably a CNS depressant; inh. of high vapor concs. may cause nose/throat irritation, nausea, drowsiness; very high concs. may cause unconsciousness and death; ing. may result in sore throat, nausea, diarrhea
Precaution: Combustible; incompat. with strong oxidizing agents (increased fire and explosion hazard)
Hazardous Decomp. Prods.: None
Storage: Store in a cool, dry, well-ventilated area
Uses: Solvent; org. synthesis

o-Butyl diethylene glycol; Butyldiglycol. *See* Butoxydiglycol
Butyldiglycol acetate. *See* Diethylene glycol butyl ether acetate

Butyl diglycol carbonate

Synonyms: Carbonic acid, oxydi-2,1-ethanediyl dibutyl ester; Diethylene glycol bis (butyl carbonate); Diethylene glycol bis (n-butylcarbonate)
Classification: Ester
Empirical: $C_{12}H_{22}O_7$
Formula: $(C_4H_9OCO_2CH_2)_2O$
Properties: Colorless liq.; insol. in water; widely sol. in org. solvs.; dens. 1.07; visc. 21cP (20 C); b.p. 164-166 C; flash pt. 188 C; ref. index 1.435 (20 C)
Precaution: Combustible
Uses: Plasticizer; high-boiling solvent; softener; mfg. of pharmaceuticals and lubricants

Butyl diglyme. *See* Diethylene glycol dibutyl ether
Butyl dipropasol solvent. *See* Dipropylene glycol butyl ether

Butylene carbonate. *See* 1,2-Butylene carbonate

1,2-Butylene carbonate

CAS 4437-85-8; EINECS/ELINCS 224-651-7
Synonyms: BC; 1,2-Butanediol, cyclic carbonate; Butylene carbonate; Carbonic acid, cyclic ethylethylene ester; 1,3-Dioxolan-2-one, 4-ethyl; 4-Ethyl-1,3-dioxolan-2-one
Classification: Cyclic carbonate
Empirical: $C_5H_8O_3$
Properties: Clear liq.; m.w. 116.13; sp.gr. 1.141; m.p. -50 C; b.p. 250 C; flash pt. 66 C; pH 6-8
Toxicology: LD50 (oral, rat) > 5 g/kg, (skin, rabbit) > 2 g/kg; low toxicity by ing.; skin and eye irritant; aspiration hazard if swallowed; can enter lungs and cause damage; inh. of high concs. may cause nose/throat irritation, headache, nausea, drowsiness
Environmental: Biodeg.; LC50 (rainbow trout) 480 mg/l; EC50 (daphnia) > 1000 mg/l; not likely to be more than sl. toxic to aquatic organisms
Hazardous Decomp. Prods.: Heated to decomp., emits acrid smoke and irritating fumes
Uses: Solvent for org. and inorg. materials, agric., inks, colors, inorg. salts, textiles, polyacrylonitrile fibers, surf. cleaners, degreasers, dyes, plastics, batteries, aromatic hydrocarbons, nat. gas; gellant for clays; binder for foundry sand molds; accelerant in dyeing; additive in fuel, lube, hydraulic fluids; reactive diluent, plasticizer in PU; prod. of PC, polyester elastomers; component of cleaners and binders; separation of CO_2 and hydrogen sulfide and refinery for aromatics in oil field; component of electrolytes in lithium-ion rechargeable batteries
Trade Names: Jeffsol® 1,2-Butylene Carbonate

Butylene glycol

CAS 107-88-0; EINECS/ELINCS 203-529-7
Synonyms: 1,3-Butanediol; Butane-1,3-diol; 1,3-Butylene glycol; β-Butylene glycol; 1,3-Dihydroxybutane; Methyltrimethylene glycol
Classification: Aliphatic diol; polyhydric alcohol
Empirical: $C_4H_{10}O_2$
Formula: $HOCH_2CH_2CHOHCH_3$
Properties: Pract. colorless visc. liq., sweet flavor with bitter aftertaste; sol. in water, alcohol, oxygenated solvs.; m.w. 90.12; dens. 1.004-1.006 (20/20 C); vapor pressure 0.06 mm Hg (20 C); f.p. < -50 C; b.p. 207.5 C; flash pt. 121 C; ref. index 1.4401 (20 C)
Toxicology: LD50 (oral, rat) 23 g/kg, (subcut., rat) 20 g/kg, (IP, mouse) 10,276 mg/kg, (skin, rabbit) > 20 g/kg; mildly toxic by ing. and subcut. routes; primary eye irritant; temp. nervous system stimulation, depression; may cause gastric upset, drowsiness, respiratory failure, convulsions, kidney damage; reproductive effector; TSCA listed
Environmental: VOC; ThOD 1.07
Precaution: Combustible when exposed to heat or flame; incompat. with oxidizing materials
Hazardous Decomp. Prods.: Heated to decomp., emits acrid smoke and irritating fumes
Storage: Hygroscopic
Uses: Comonomer for oil-free alkyd/unsat. polyester resins; polyurethanes; surfactants; plasticizers; humectant, solubilizer, solvent in cosmetics; coupling agent; solvent for flavors in foods and pharmaceuticals; food additive; formulation aid, processing aid in mfg. of sausage casings; pharmaceutical carrier, vehicle; antioxidant/stabilizer for food-grade polymers; in food-pkg. adhesives; in closure-sealing gaskets for food containers; in cellophane for food pkg.
Regulatory: FDA 21CFR §172.712, 173.220, 175.105, 175.300, 175.320, 177.1200, 177.1210, 177.1680, 177.2420, 178.2010
Manuf./Distrib.: ABCR; Acros Org.; Aldrich; Allchem Ind.; Amend Drug & Chem.; Ashland; Avatar; Bencorp Int'l.; Bruchem; Celanese; Columbia Sales Int'l.; Degussa AG; E-Chemicals; Fluka; Griffin Int'l.; Honeywell Spec. Chems.; Kyowa Hakko USA; Parchem Trading; Penta Mfg.; Quaker City; Ruger; Springfield Scientific Supplies; Stanchem; Universal Preserv-A-Chem; Vopak USA
Trade Names Containing: Dantogard® Plus Liq.

1,3-Butylene glycol. *See* Butylene glycol
1,4-Butylene glycol. *See* 1,4-Butanediol
β-Butylene glycol. *See* Butylene glycol

Butylene glycol dicaprylate/dicaprate

Synonyms: 1,2-Butanediol, diesters with octanoic and decanoic acids
Definition: Mixt. of the butylene glycol diesters of caprylic and capric acids
Uses: Emollient in cosmetics; emollient, dispersant, lubricant, suspending agent, solubilizer, carrier, vehicle, solvent, penetrant, spreading agent for pharmaceuticals
Trade Names: Miglyol® 8810

Butylene glycol methyl ether

CAS 111-32-0
Synonyms: 1-Butanol, 4-methoxy-; 1,3-Butylene glycol methyl ether; Butylene glycol monomethyl ether; 4-Methoxybutanol; 4-Methoxy-1-butanol
Classification: Glycol ether
Empirical: $C_5H_{12}O_2$
Properties: M.w. 104.17
Toxicology: TSCA listed
Environmental: VOC; ThOD 1.23
Uses: Solvent
Regulatory: SARA §313 reportable; HAP
Manuf./Distrib.: Ashland; Coyne; Dow; Ferro/Grant; G.J. Chem.; Gallard-Schlesinger Ind.; Great Western; ICN Biomed. Research Prods.; Marcus Int'l.; Oxiteno; Penta Mfg.; R.E. Carroll; Shell; Vopak USA

1,3-Butylene glycol methyl ether; Butylene glycol monomethyl ether. *See* Butylene glycol methyl ether
1,3-Butylene glycol monomethyl ether. *See* 3-Methoxybutanol
Butylene hydrate. *See* 2-Butanol
Butylene oxide. *See* Tetrahydrofuran
Butyl ethanoate. *See* n-Butyl acetate
2-Butyl ethanoate. *See* s-Butyl acetate
Butyl ether; n-Butyl ether. *See* Dibutyl ether
Butylethylacetic acid. *See* 2-Ethylhexoic acid
Butyl ethyl carbinol. *See* 3-Heptanol
o-Butyl ethylene glycol. *See* Butoxyethanol
Butyl ethylhexyl phthalate. *See* Butyl octyl phthalate
Butyl ethyl ketone; n-Butyl ethyl ketone. *See* 3-Heptanone

Butyl formate

CAS 592-84-7; EINECS/ELINCS 209-772-5
UN 1128 (DOT); FEMA 2196
Synonyms: n-Butyl formate; Formic acid butyl ester
Classification: Nonaromatic ester
Empirical: $C_5H_{10}O_2$
Formula: $HCOO(CH_2)_3CH_3$
Properties: Colorless liq., plum-like odor; sl. sol. in water; misc. with alcohol, ether; m.w. 102.14; dens. 0.892; m.p. -90 C; b.p. 106.8 C; flash pt. (CC) 64 C; ref. index 1.3890-1.3891
Toxicology: LD50 (oral, rabbit) 2656 mg/kg; mod. toxic by ing.; mildly toxic by inh.; human systemic effects by inh.; irritant; narcotic in high concs.; TSCA listed
Precaution: Flamm. liq.; dangerous fire risk; incompat. with oxidizing materials
Hazardous Decomp. Prods.: Heated to decomp., emits acrid and irritating fumes
NFPA: Health 2, Flammability 3, Reactivity 0
Uses: Synthetic flavoring agent in foods and pharmaceuticals; solvent for nitrocellulose, cellulose acetate, many cellulose ethers, many natural and synthetic resins, lacquers, perfumes; intermediate in organic synthesis; fragrance
Features: Plum-like flavor
Regulatory: FDA 21CFR §172.515; FEMA GRAS
Manuf./Distrib.: Aldrich; Alfa Aesar; Fluka; Goldensun Mfg.; Grau Aromatics; Penta Mfg.; Sigma; Springfield Scientific Supplies; Xinchem

n-Butyl formate. *See* Butyl formate
n-Butyl α-glycerol ether. *See* 3-Butoxy-1,2-propanediol
Butyl glycol; n-Butyl glycol. *See* Butoxyethanol
Butyl glycol acetate. *See* Butoxyethanol acetate

Butyl glycolate

CAS 7397-62-8; EINECS/ELINCS 230-991-7

UN 8027
Synonyms: Acetic acid, hydroxy-, butyl ester; n-Butyl glycollate; Butyl hydroxyacetate; n-Butyl hydroxyacetate; Glycolic acid n-butyl ester
Empirical: $C_6H_{12}O_3$
Formula: $CH_3CH_2CH_2CH_2OOCCH_2OH$
Properties: Colorless liq.; very faint ester-like odor; misc. with common org. solvs.; only partly misc. with water; m.w. 132.16; dens. 1.015-1.023 g/cc (20 C); vapor pressure 1.33 mbar (20 C); f.p. ≈ -26 C; b.p. 82-97 C; flash pt. ≈ 75 C; ref. index 1.423-1.426 (20 C)
Toxicology: LD50 (oral, rat) 4595 mg/kg; irritant to mucous membranes; risk of serious eye damage
Environmental: Readily biodeg.
NFPA: Health 0, Flammability 2
Storage: Store in cool, dry, dry place, preventing contact with air
Uses: Paint additive, flow agent for nitrocellulose lacquers, brush-applied paints, alkyd resin and oil paints; leveling agent for varnishes; high-boiling solvent in stoving finishes; mfg. of printing inks
Manuf./Distrib.: Fluka
Trade Names: Polysolvan® O

Butyl glycol ether. *See* Diethylene glycol dibutyl ether
n-Butyl glycollate. *See* Butyl glycolate
Butyl hydride. *See* Butane

t-Butyl hydroquinone

CAS 1948-33-0; EINECS/ELINCS 217-752-2
INS319
Synonyms: 2-(1,1-Dimethylethyl)-1,4-benzenediol; Mono-t-butyl hydroquinone; Mono-tertiarybutylhydroquinone; MTBHQ; TBHQ
Classification: Aromatic organic compd.; phenol; hydroquinone
Empirical: $C_{10}H_{14}O_2$
Properties: Wh. to lt. tan cryst. solid; sol. in ethyl alcohol, ethyl acetate, acetone, ether; sl. sol. in water; m.w. 166.24; m.p. 126.5-128.5 C; flash pt. (COC) 171 C
Toxicology: LD50 (oral, rat) 700 mg/kg, (IP, rat) 300 mg/kg; poison by IP route; mod. toxic by ing.; irritant; tumorigen and mutagen; TSCA listed
Hazardous Decomp. Prods.: Heated to decomp., emits acrid smoke and irritating fumes
Uses: Polymerization inhibitor; antioxidant, preservative, stabilizer for foods, pharmaceuticals, industry, polyesters; antioxidant in cosmetics; inhibitor in food-contact crosslinked polyesters
Regulatory: FDA 21CFR §172.185 (limitation 0.02% of oil), 177.2420; USDA 9CFR §318.7 (limitation 0.003% in dry sausage, 0.006% with BHA/BHT, 0.01% in rendered animal fat, 0.02% with BHA and/or BHT, 0.02% in margarine), 381.147 (limitation 0.01% on fat in poultry
Manuf./Distrib.: AECI; Aceto; Aldrich; Allchem Ind.; Am. Int'l.; Amerol; Charkit; Chugai Boyeki Am.; Eastman; Fallek; Fluka; Mitsui & Co. USA; Pechiney Chems. Div.; Penta Mfg.; Schweizerhall; Showa Denko; Sigma; Skymart Enterprises; Spectrum Quality Prods.; UOP; Universal Preserv-A-Chem; Voigt Global Distrib.
Trade Names Containing: Phytol™ CAN

Butyl hydroxide. *See* Butyl alcohol
t-Butyl hydroxide. *See* t-Butyl alcohol
Butyl hydroxyacetate; n-Butyl hydroxyacetate. *See* Butyl glycolate
Butylhydroxyanisole; 2(3)-t-Butyl-4-hydroxyanisole; t-Butylhydroxyanisole; t-Butyl-4-hydroxyanisole. *See* BHA
Butyl 2-hydroxypropanoate; Butyl α-hydroxypropionate; n-Butyl-S(-)-2-hydroxypropionate. *See* Butyl lactate
Butyl-(S)-(-)-2-hydroxy propionate; Butyl-(S)-2-hydroxy propionate. *See* Butyl-(S)-lactate
Butyl-3-iodo-2-propynylcarbamate. *See* Iodopropynyl butylcarbamate

Butyl lactate

CAS 138-22-7; EINECS/ELINCS 205-316-4
FEMA 2205
Synonyms: Butyl 2-hydroxypropanoate; Butyl α-hydroxypropionate; n-Butyl-S(-)-2-hydroxypropionate; n-Butyl lactate; 2-Hydroxypropanoic acid butyl ester; Lactic acid butyl ester; Propanoic acid, 2-hydroxy-, butyl ester
Classification: Lactate ester
Empirical: $C_7H_{14}O_3$
Formula: $CH_3CHOHCOOC_4H_9$
Properties: Water-wh. liq., mild odor; misc. with many lacquer solvs., diluents, oils, alcohol, ether; sl. sol. in water; hydrolyzed in acids and alkalis; m.w. 146.19; dens. 0.974-0.984 (20/20 C); vapor pressure 0.4 mm Hg (20 C); m.p. -43 C; b.p. 188 C; flash pt. (TOC) 75.5 C
Toxicology: ACGIH TLV/TWA 5 ppm; LD50 (subcut., rat) 12 g/kg; LDLo (IP, mouse) 200 mg/kg; poison by IP route; skin irritant; toxic conc. 4 ppm in air (humans); TSCA listed
Environmental: VOC; COD 1.97; ThOD 1.97
Precaution: Flamm. exposed to heat or flame; can react with oxidizing materials
Hazardous Decomp. Prods.: Heated to decomp., emits acrid smoke and irritating fumes
NFPA: Health 1, Flammability 2, Reactivity 0
Uses: Solvent for nitrocellulose, ethylcellulose, oils, dyes, gums, coatings, specialty finishes, inks, cosmetics; dipped latex prods.; diluents; adhesives; intermediates; photoresist solvents; screen printing of electronic parts; antiskinning agent; dry cleaning fluids; solvent for electronics, coatings, inks, cleaning agents, mold release agents; synthetic flavoring agent and fragrance in foods and pharmaceuticals; pharmaceutical intermediate; in food-pkg. adhesives
Features: Non-ozone depleting; sweet flavor
Regulatory: FDA 21CFR §172.515, 175.105; FEMA GRAS
Manuf./Distrib.: A&E Connock; Aldrich; Alfa Aesar; Bell Flavors & Fragrances; CPS; ChemService; Ciba Spec. Chems./Water Treatment; Degussa Flavors/Alfrebro; Fleurchem; Grau Aromatics; Haltermann Prods. UK; Kaltron/Pettibone; Lluch Essence; PURAC Am.; PURAC biochem; Penta Mfg.; Spectrum Quality Prods.; Universal Preserv-A-Chem; Vopak USA; Xinchem

Butyl L-lactate. *See* Butyl-(S)-lactate
n-Butyl lactate. *See* Butyl lactate
n-Butyl-(S)-lactate. *See* Butyl-(S)-lactate

Butyl-(S)-lactate

CAS 34451-19-9; EINECS/ELINCS 252-036-3
Synonyms: Butyl-(S)-(-)-2-hydroxy propionate; Butyl-(S)-2-hydroxy propionate; Butyl L-lactate; n-Butyl-(S)-lactate; Butyl-(S)-(-)-lactate; Lactic acid butyl ester
Classification: Lactate ester
Empirical: $C_7H_{14}O_3$
Properties: Colorless liq.; sweet fruity fermented odor; sol. in alcohol; insol. in water; m.w. 146.19; sp.gr. 0.97800-0.99200; dens. 8.138-8.254 lb/gal; m.p. -28 C; b.p. 185-187 C; flash pt. 157 F; ref. index 1.4190-1.429 (20 C)
Toxicology: LD50 (oral, rat) > 5 g/kg, (subcut., rat) 12 g/kg, (skin, rabbit) > 5 g/kg; irritating to eyes, skin, respiratory system; may be harmful by ing., inh., skin absorp.; vapor/mist irritating to eyes, mucous membranes, upper respiratory tract; TSCA listed
Precaution: Combustible; incompat. with strong oxidizing agents, strong bases
Hazardous Decomp. Prods.: CO, CO_2; emits toxic fumes under fire conditions
Storage: Store in cool, dry place away from heat, open flame; keep tightly closed
Uses: Solvent for NC, ethyl cellulose, oils, dyes, natural gums, syn. polymers, lacquers, varnishes, inks, stencil pastes, dry-cleaning fluids, adhesives, cosmetics; fragrance in perfumes
Manuf./Distrib.: Aldrich; Lluch Essence
Trade Names: Purasolv® BL

Butyl-(S)-(-)-lactate. *See* Butyl-(S)-lactate
Butyl mercaptan; 1-Butyl mercaptan. *See* n-Butyl mercaptan

n-Butyl mercaptan

CAS 109-79-5; EINECS/ELINCS 203-705-3
UN 2347 (DOT); FEMA 3478
Synonyms: Butanethiol; 1-Butanethiol; n-Butanethiol; Butyl mercaptan; 1-Butyl mercaptan; n-Butyl thioalcohol; 1-Mercaptobutane; Thiobutanol; Thiobutylalcohol
Classification: Alkyl mercaptan
Empirical: $C_4H_{10}S$
Formula: n-C_4H_9SH
Properties: Colorless to pale yel. liq.; strong offensive garlic or skunk-like

odor; very sol. in alcohol, ether, liq. hydrogen sulfide; sl. sol. in oils; pract. insol. in water; m.w. 90.19; dens. 0.8365 (25/4 C); m.p. -116 C; b.p. 98 C; flash pt. (TCC) -7 C

Toxicology: ACGIH TLV/TWA 0.5 ppm; LD50 (oral, rat) 1500 mg/kg, (IP, rat) 399 mg/kg; poison by IP route; mod. toxic by inh. and ing., eye irritant; sl. skin irritant; TSCA listed

Precaution: Flamm.; dangerous fire hazard exposed to heat, flame, sparks, oxidizers; incompat. with strong bases, strong oxidizing agents, reducing agents, strong acids, alkali metals, nitric acid; violent reactions possible

Hazardous Decomp. Prods.: Reaction with strong acids can produce toxic fumes of hydrogen sulfide; heated to decomp., emits toxic SO_x

NFPA: Health 2, Flammability 3, Reactivity 0

Storage: Store in cool, well-ventilated area, out of direct sunlight; ground drums and bond transfer containers

Uses: Intermediate, solvent for prod. of pesticides, acaricides, herbicides, and defoliants; additive in oils, rubber chemicals; polymerization regulator; odorant for natural gas; flavoring agent

Regulatory: FEMA GRAS

Manuf./Distrib.: Aldrich; Alfa Aesar; Atofina; ChevronPhillips; Epochem; Fluka; Natural Advantage

t-Butyl-4-methoxyphenol. *See* BHA
1-t-Butyl-4-methylbenzene. *See* p-t-Butyl toluene
t-Butyl methyl ether. *See* Methyl t-butyl ether
Butyl methyl ketone; N-Butyl methyl ketone. *See* 2-Hexanone
Butyl octadecanoate; n-Butyl octadecanoate. *See* Butyl stearate
Butyl 9-octadecenoate. *See* Butyl oleate

Butyl octyl benzoate

CAS 188038-97-3

Synonyms: 1-Octanol, 2-butyl-, benzoate

Uses: Plasticizer, emollient, skin conditioner, solvent, carrier, pigment wetting agent/dispersant, lubricant in cosmetics, creams/lotions, hair care, skin care; solvent for sunscreen actives

Manuf./Distrib.: C.P. Hall

Butyl octyl phthalate

CAS 84-78-6; EINECS/ELINCS 201-562-1

Synonyms: 1,2-Benzenedicarboxylic acid, butyl octyl ester; BOP; Butyl ethylhexyl phthalate; n-Butyl octylphthalate; Phthalic acid, butyl octyl ester

Empirical: $C_{20}H_{30}O_4$

Formula: $C_4H_9OOCC_6H_4COOC_8H_{17}$

Properties: Water-wh. liq., mild char. odor; misc. with most org. solvs.; insol. in water; m.w. 334.46; dens. 0.988-0.996 (20/20 C); f.p. -48 C; b.p. 350 C; sapon. no. 298-308; flash pt. (COC) 208 C; ref. index 1.4868

Toxicology: LD50 (oral, rat) > 63 ml/kg; TSCA listed

Precaution: Combustible

Uses: High solvating plasticizer for PVC, PS, PVB, PVAc, cellulosics, and rotational molding and dip coating; in food-pkg. adhesives

Regulatory: FDA 21CFR §175.105

Manuf./Distrib.: Ashland; BP Chems. Ltd; Eastman; J.C. Wilson; PolyOne/Elastomers & Perf. Addit.; Sunoco; Unitex

n-Butyl octylphthalate. *See* Butyl octyl phthalate

Butyl octyl salicylate

CAS 190085-41-7; EINECS/ELINCS 431-090-3

Synonyms: Benzoic acid, 2-hydroxy-, 2-butyloctyl ester

Properties: Clear liq.; mild odor; insol. in water; sp.gr. 0.9660; flash pt. (COC) 360 F

Toxicology: LD50 (oral) > 500 mg/kg; not expected to cause skin/eye irritation; may be harmful if swallowed; may irritate mouth, throat, and stomach by ing.; inh. may cause dizziness

Environmental: May cause long-term adverse effects in the aquatic environment; avoid release to environment

Precaution: Incompat with oxidizing materials

HMIS: Health 1, Flammability 1, Reactivity 0

Storage: Store in tightly closed container in well-ventilated area

Uses: Emollient, moisturizer, sunscreen actives stabilizer, solvent for personal care prods.; hair conditioner, skin conditioner in cosmetics

Regulatory: SARA §311/312/313 nonreportable

Trade Names: HallBrite® BHB

Butyl oleate

CAS 142-77-8; EINECS/ELINCS 205-559-6

Synonyms: Butyl 9-octadecenoate; n-Butyl oleate; 9-Octadecenoic acid, butyl ester

Definition: Ester of butyl alcohol and oleic acid

Empirical: $C_{22}H_{42}O_2$

Formula: $CH_3(CH_2)_7CH:CH(CH_2)_7COOC_4H_9$

Properties: Pale yel. oleaginous liq.; mild odor; misc. with alcohol, ether, veg. and min. oils; insol. in water; m.w. 338.58; dens. 0.873; f.p. opaque @ 12 C; solid @ -26.4 C; b.p. 173-227 C (2 mm); iodine no. 76.8; flash pt. 204 C; nonionic

Toxicology: Skin irritant; TSCA listed

Precaution: Combustible exposed to heat or flame; incompat. with oxidizing materials

Hazardous Decomp. Prods.: Heated to decomp., emits acrid smoke and irritating fumes

Uses: Plasticizer for PVC, PVB, PVAc, PS, cellulosics, surface coatings; solvent; lubricant in rolling oils; water-resisting agent; coatings; polishes; waterproofing compds.; emulsifier, bodying agent, emollient for cosmetics, toiletries; textile surf. finisher, softener, thread lubricant and antistat; lubricant in textile spin finishes, coning oils, cording, and dye baths; plasticizer, softener for NR, SR; flexibilizer; heat stabilizer; retards CR cures; plasticizer in food-contact rubber articles for repeated use; defoamer in food-contact paper/paperboard; food-contact textiles

Regulatory: FDA 21CFR §176.210, 177.2600, 177.2800

Manuf./Distrib.: A&E Connock; Anar; Avatar; Bayer/Industrial Chems./Textile; Boehme Filatex; C.P. Hall; Cardinal; ChemService; Chemol; Eastech; Ferro/Keil; Indofine; Inolex; Int'l. Paper; Jarchem Ind.; Mosselman NV; PolyOne/Elastomers & Perf. Addit.; Sea-Land; Sigma; St. Lawrence; Thornley; Uniqema Am.; Velsicol

Trade Names: Tilol B

n-Butyl oleate. *See* Butyl oleate
Butyl oxalate. *See* Dibutyl oxalate
Butyl phosphate; n-Butyl phosphate. *See* Tributyl phosphate
Butyl phthalate; n-Butyl phthalate. *See* Dibutyl phthalate
Butyl propanoate; Butyl propionate. *See* n-Butyl propionate

n-Butyl propionate

CAS 590-01-2; EINECS/ELINCS 209-669-5

UN 1914 (DOT); FEMA 2211

Synonyms: Butyl propanoate; Butyl propionate; Propanoic acid butyl ester

Empirical: $C_7H_{14}O_2$

Formula: $CH_3CH_2COOC_4H_9$

Properties: Colorless liq., earthy faintly sweet odor, apricot-like taste; very sol. in alcohol, ether; sol. in oxygenated solvs.; very sl. sol. in water; m.w. 130.19; dens. 0.8754 (20/4 C); f.p. 32 C; m.p. -89 C; b.p. 145-146 C; flash pt. 90 F; ref. index 1.401 (20 C)

Toxicology: LD50 (oral, rat) 5000 mg/kg; mildly toxic by ingestion; skin and eye irritant; TSCA listed

Precaution: Flamm.; dangerously flamm. exposed to heat or flame; incompat. with oxidizers

NFPA: Health 2, Flammability 3, Reactivity 0

Uses: Solvent for nitrocellulose, paints, lacquers; retarder in lacquer thinner; fragrance in perfumes; synthetic flavoring agent in foods and pharmaceuticals

Features: Banana-like flavor

Regulatory: FDA 21CFR §172.515; FEMA GRAS

Manuf./Distrib.: Advanced Synthesis Tech.; Aldrich; Ashland; Degussa Flavors/Alfrebro; Dow; Grau Aromatics; Haltermann Prods. UK; Penta Mfg.; Shell

Trade Names: Ucar® n-Butyl Propionate

Butyl stearate

CAS 123-95-5; EINECS/ELINCS 204-666-5

FEMA 2214

Synonyms: Butyl octadecanoate; n-Butyl octadecanoate; n-Butyl stearate; Octadecanoic acid butyl ester

Definition: Ester of butyl alcohol and stearic acid

Empirical: $C_{22}H_{44}O_2$
Formula: $CH_3(CH_2)_{16}COO(CH_2)_3CH_3$
Properties: Colorless liq., solidifying @ 19 C; pract. odorless; sol. in alcohol, ether, hydrocarbon oils; sl. sol. in water; m.w. 340.60; dens. 0.86 (20/4 C); m.p. 17-22 C; b.p. 343 C; flash pt. (CC) 160 C; ref. index 1.4430 (20 C)
Toxicology: LD50 (oral, rat) 32 g/kg; low toxicity by ing.; skin irritant; experimental reproductive data; TSCA listed
Precaution: Combustible
Hazardous Decomp. Prods.: Heated to decomp., emits acrid smoke and irritating fumes
NFPA: Health 1, Flammability 1, Reactivity 0
Uses: Solvent, spreading agent, softener in plastics, textiles, cosmetics, rubbers; synthetic flavoring agent in foods and pharmaceuticals; plasticizer for food-pkg. materials; plasticizer, flow aid for plastics, rubber, lacquers; lubricant, slip agent for coatings and inks; polishes; lubricant for metals, textiles, molding; dye solvent in wax polishes; chlorinated rubber; carbon paper and inks; emollient in cosmetics, pharmaceuticals; dampproofer for concrete; pigment wetting agent/dispersant; in food-pkg. adhesives; defoamer in food-contact coatings; plasticizer in food-contact rubber articles for repeated use; in surf. lubricants for mfg. of food-contact metallic articles
Regulatory: FDA 21CFR §172.515, 173.340, 175.105, 175.300, 175.320, 176.200, 176.210, 177.1200, 177.2260, 177.2600, 177.2800, 178.3910, 181.22, 181.27; FDA approved for topicals; FEMA GRAS
Manuf./Distrib.: A&E Connock; Advanced Synthesis Tech.; Aldrich; Alfa Aesar; Alzo; Amerchol; Anar; Ashland; Avatar; C.P. Hall; ChemService; Chemol; CoKEM Assoc.; Cognis/Chems. Group; DelCon; Fluka; Functional Foods; Independent Chem.; Indofine; Inolex; Int'l. Paper; J.T. Baker; Jarchem Ind.; Kenrich Petrochems.; Kraft Chem.; Lipo; MelChem; Mosselman NV; NOF Am.; Norman, Fox; Penta Mfg.; Phoenix; Ruger; Sea-Land; Sigma; Spectrum Quality Prods.; St. Lawrence; Stepan; Thornley; Uniqema Am.; Unitex; Universal Preserv-A-Chem; VWR Int'l.
Trade Names: Exceparl BS; HallStar™ BST; Monestriol B; Norfox® B-54

n-Butyl stearate. *See* Butyl stearate
n-Butyl thioalcohol. *See* n-Butyl mercaptan
4-t-Butyl toluene. *See* p-t-Butyl toluene

p-t-Butyl toluene

CAS 98-51-1; EINECS/ELINCS 202-675-9
UN 2667
Synonyms: 1-t-Butyl-4-methylbenzene; 4-t-Butyl toluene; 1-(1,1-Dimethylethyl)-4-methylbenzene; 1-Methyl-4-t-butylbenzene; p-Methyl-t-butylbenzene; 8-Methylparacymene; PTBT; TBT; Toluene, p-t-butyl-
Classification: Aromatic hydrocarbon
Empirical: $C_{11}H_{16}$
Formula: $(CH_3)_3CC_6H_4CH_3$
Properties: Colorless liq.; aromatic odor; sol. in oxygenated and aromatic solvs.; insol. in water; m.w. 148.25; dens. 0.857-0.863 (20/20 C); m.p. -54 C; b.p. 192.8 C; flash pt. (PMCC) 54 C
Toxicology: ACGIH TLV/TWA 10 ppm; STEL 20 ppm; LD50 (oral, rat) 1500 mg/kg, (skin, rabbit) 16,934 mg/kg; mod. toxic by inh., ing., skin absorption; skin irritant; human eye irritant (conjunctive irritation); inh. causes nausea, vomiting, nose/lung irritation, muscle weakness, labored breathing, CNS depression, possible bone marrow depression, cardiovascular effects; if aspirated, may cause lung inflamm., fluid accumulation; may cause liver/kidney damage on prolonged exposure; TSCA listed
Environmental: Prevent entry into water sources, sewers; marine pollutant
Precaution: Flamm. exposed to heat or flame; LEL 1 vol.%; UEL 6.5 vol.%; highly flamm. vapors which are heavier than air may accumulate in low areas, flash back; incompat. with strong oxidizing agents, oxygen; bond and ground during transfer
Hazardous Decomp. Prods.: Combustion: CO_x, various hydrocarbons; heated to decomp., emits acrid smoke and fumes
NFPA: Health 2, Flammability 2, Reactivity 0
Storage: Store in closed containers in well-ventilated area, away from heat, sparks, flame
Uses: Intermediate for org. synthesis (esp. t-butylbenzoic acid), perfumes, fragrances; fixing agent for fragrances; cosmetics ingred.; solvent for resins; antioxidant
Regulatory: SARA §313 nonreportable; Canada DSL listed
Manuf./Distrib.: Aldrich; ChevronPhillips; Degussa AG; Fluka; Koch Spec. Chem.

Butynediol; 1,4-Butynediol; 2-Butynediol. *See* But-2-yne-1,4-diol

But-2-yne-1,4-diol

CAS 110-65-6; EINECS/ELINCS 203-788-6
UN 2716 (DOT)
Synonyms: Bis (hydroxymethyl) acetylene; 2-Butin-1,4-diol; Butynediol; 1,4-Butynediol; 2-Butynediol; 2-Butyne-1,4-diol; 1,4-Dihydroxy-2-butyne
Empirical: $C_4H_6O_2$
Formula: $HOCH_2{=}CCH_2OH$
Properties: Straw to amber crystals; sol. in water, aq. acids, alcohol, acetone, ethyl acetate; insol. in ether, benzene; m.w. 86.10; sp.gr. 1.2; m.p. 57.5 C; b.p. 194 (100 mm); flash pt. 152 C
Toxicology: LD50 (oral, rat) 104 mg/kg; LCLo (inh., rat, 2h) 150 mg/m^3; poison by ing.; skin sensitizer on long/repeated contact; TSCA listed
Precaution: Mod. explosive; may explode on contamination with mercury salts, strong acids, and alkaline earth hydroxides and halides at high temps.
Hazardous Decomp. Prods.: Heated to decomp., emits acrid smoke and fumes and may explode
NFPA: Health 2, Flammability 1, Reactivity 0
Storage: Hygroscopic
Uses: Corrosion inhibitor in acid pickles and cleaners; pharmaceutical intermediate; electroplating brightener; defoliant; polymerization accelerator; stabilizer for chlorinated hydrocarbons; cosolvent for paint and varnish removal; synthesis of histamine and pyridoxine; alternative route for vitamin B6; emulsifier
Manuf./Distrib.: Advanced Synthesis Tech.; Aldrich; Alfa Aesar; BASF; Fluka; ISP; Sigma; Sithean; Spectrum Quality Prods.; Springfield Scientific Supplies; VWR Int'l.

2-Butyne-1,4-diol. *See* But-2-yne-1,4-diol
1-Butyn-3-ol, 3-methyl-. *See* Methyl butynol
Butyral. *See* n-Butyraldehyde

n-Butyraldehyde

CAS 123-72-8; EINECS/ELINCS 204-646-6
UN 1129 (DOT); FEMA 2219
Synonyms: Butal; Butaldehyde; Butalyde; Butanal; n-Butanal; n-Butylaldehyde; Butyral; Butyric aldehyde
Classification: Aldehyde
Empirical: C_4H_8O
Formula: $CH_3CH_2CH_2CHO$
Properties: Colorless liq.; sol. in water; misc. with ether @ 75 C, org. solvs.; m.w. 72.10; sp.gr. 0.800; m.p. -99 C; f.p. 12 F; b.p. 75-76 C; flash pt. (CC) 20 F; ref. index 1.3843
Toxicology: LD50 (oral, rat) 2490 mg/kg, (IP, rat) 800 mg/kg, (skin, rabbit) 3560 mg/kg; LCLo (inh., rat, 4 h) 8000 ppm; mod. toxic by ing., inh., skin contact, IP, and subcut. routes; severe skin and eye irritant; delayed hypersensitivity; suspected carcinogen; TSCA listed
Precaution: DOT: Flamm. liq.; dangerous fire hazard when exposed to heat and flame; incompat. with oxidizing materials; reacts vigorously with chlorosulfonic acid, HNO_3, H_2SO_4
Hazardous Decomp. Prods.: Heated to decomp., emits acrid smoke and fumes
NFPA: Health 3, Flammability 3, Reactivity 2
Uses: Plasticizers; rubber accelerators; solvent; high polymers; intermediate for paints; synthetic flavoring agent in foods and pharmaceuticals; fragrance intermediate for cosmetics; in food-pkg. adhesives; in paper/paperboard in contact with aq./fatty foods
Features: Fruity flavor
Regulatory: FDA 21CFR §172.515, 175.105, 176.170; FEMA GRAS
Manuf./Distrib.: Advanced BioTech; Aldrich; Alfa Aesar; Axxence Aromatic GmbH; BASF AG; Celanese; Chisso Am.; Degussa Flavors/Alfrebro; Dow; Eastman; Fleurchem; Fluka; Honeywell Perf. Polymers; Penta Mfg.; Spectrum Quality Prods.; Springfield Scientific Supplies; Vopak USA; Xinchem

Butyric acid, butyl ester. *See* Butyl butyrate
Butyric acid, ethyl ester. *See* Ethyl butyrate
Butyric acid isoamyl ester. *See* Isoamyl butyrate
Butyric acid lactone. *See* Butyrolactone
Butyric acid, methyl ester. *See* Methyl butyrate
Butyric acid nitrile. *See* Butyronitrile
Butyric acid, propyl ester. *See* Propyl butyrate
Butyric alcohol. *See* Butyl alcohol
Butyric aldehyde. *See* n-Butyraldehyde
Butyric ether. *See* Ethyl butyrate
Butyrolactam; γ-Butyrolactam. *See* 2-Pyrrolidone

Butyrolactone

CAS 96-48-0; EINECS/ELINCS 202-509-5
FEMA 3291
Synonyms: BLO; 1,2-Butanolide; 1,4-Butanolide; Butyric acid lactone; 4-Butyrolactone; α-Butyrolactone; γ-Butyrolactone; Butyryl lactone; 4-Deoxytetronic acid; Dihydro-2(3H)-furanone; GBL; 4-Hydroxybutanoic acid lactone; γ-Hydroxybutyric acid cyclic ester; 4-Hydroxybutyric acid γ-lactone; γ-Hydroxybutyric acid lactone; γ-Hydroxybutyrolactone; Tetrahydro-2-furanone
Classification: Lactone
Empirical: $C_4H_6O_2$
Properties: Colorless oily liq., mild caramel odor; sol. in methanol, ethanol, acetone, ether, benzene; misc. with water; insol. in aliphatic hydrocarbons; m.w. 86.09; dens. 1.120; m.p. -45 C; b.p. 204-205 C; flash pt. (OC) 98 C; ref. index 1.4348; volatile with steam; hydrolyzed by hot alcohol sol'ns.
Toxicology: LD50 (oral, rat) 1800 mg/kg, (IP, rat) 1000 mg/kg; mod. toxic by ingestion, IV, IP routes; suspected carcinogen, tumorigen; experimental reproductive effects; mutagenic data; TSCA listed
Environmental: VOC; ThOD 1.67
Precaution: Combustible when exposed to heat or flame; reactive with oxidizing materials
Hazardous Decomp. Prods.: Heated to decomp., emits acrid and irritating fumes
NFPA: Health 0, Flammability 1, Reactivity 0
Storage: Hygroscopic
Uses: Intermediate for synthesis of polyvinylpyrrolidone, piperidine, phenylbutyric acid, aliphatic and cyclic compds., prod. of Vitamins B1 and E, alkaloids, analgesics, anesthetics; dyeing of acetate; solvent for polyacrylonitrile, cellulose acetate, methyl methacrylate polymers, polystyrene, paints, fluorinated hydrocarbons, cellulose triacetate, shellac, cosmetics; solvent, stabilizer for pharmaceuticals; pharmaceutical intermediate; reaction/diluent solvent for pesticides; constituent of paint removers, textile aids, drilling oils; flavoring agent for candy, soy milk; in paint removers, petrol. processing, inks; wetting agent for cellulose acetate films, fibers, solvent welding of plastic films in adhesive applics.
Regulatory: FEMA GRAS
Manuf./Distrib.: ADA Int'l.; Advanced Synthesis Tech.; Aldrich; Alfa Aesar; Allchem Ind.; Ashland; BASF; Chemical; Fabrichem; Fluka; G.J. Chem.; Gelest; George Uhe; Great Western; Honeywell Spec. Chems.; Houghton Chem.; ISP; Lluch Essence; Lyondell; Mitsubishi Chem.; Penta Mfg.; Quaker City; Schweizerhall; Sigma; Spectrum Quality Prods.; Synasia Fine Chem.; UCB; VWR Int'l.; Vopak USA; Whyte Chems. Ltd
Trade Names: Agrisynth® BLO; Agsolex® BLO; BLO®; GBL
Trade Names Containing: Dynasolve 699; Dynasolve CU-5; Dynasolve CU-6; FoamFlush® Urethane Remover; Ship Shape® Resin Cleaner

4-Butyrolactone; α-Butyrolactone; γ-Butyrolactone. *See* Butyrolactone
Butyrone. *See* 4-Heptanone

Butyronitrile

CAS 109-74-0; EINECS/ELINCS 203-700-6
UN 2411 (DOT)
Synonyms: Butanenitrile; n-Butanenitrile; Butyric acid nitrile; 1-Cyanopropane; NBN; Nitrile C_4; Propyl cyanide; n-Propyl cyanide
Classification: Aliphatic nitrile
Empirical: C_4H_7N
Formula: $CH_3CH_2CH_2CN$
Properties: Colorless liq.; sharp odor; sol. in alcohol, ether, dimethylformamide; misc. with oxygenated solvs.; sl. sol. in water; m.w. 69.11; dens. 0.794; m.p. -112 C; b.p. 115-117 C; flash pt. 16 C; ref. index 1.3840
Toxicology: LD50 (oral, rat) 0.14 g/kg, (IP, mouse) 38 mg/kg, (skin, rabbit) 500 mg/kg; highly toxic; poison by ing., skin contact, IP and subcut. routes; mod. toxic by inh.; skin irritant; mild eye irritant; inh. of vapor may cause nose/throat irritation at high concs.; breaks down in body to release cyanide ion; may result in cyanide poisoning; experimental reproductive effects; TSCA listed
Precaution: Flamm.; dangerous fire hazard exposed to heat, flame, or oxidizers
Hazardous Decomp. Prods.: CO, CO_2, hydrogen cyanide; heated to decomp., emits toxic fumes of NO_x and CN^-
NFPA: Health 3, Flammability 3, Reactivity 0
Uses: Solvent; basic material in industrial, chemical and pharmaceutical intermediates and products, poultry medicines
Manuf./Distrib.: Advanced Synthesis Tech.; Air Prods.; Aldrich; Eastman; Fluka; Lonza

Butyrospermum parkii; Butyrospermum parkii butter; Butyrospermum parkii fruit. *See* Shea butter (Butyrospermum parkii)
Butyryl lactone. *See* Butyrolactone

C

CAC. *See* Chloracetyl chloride
Cacao butter. *See* Cocoa (Theobroma cacao) butter
Cajeputene. *See* l-Limonene; d-Limonene; dl-Limonene
Cajeputol. *See* Eucalyptol

Calcium chloride

CAS 10043-52-4 (anhyd.); 10035-04-8 (dihydrate); 7774-34-7 (hexahydrate); EINECS/ELINCS 233-140-8
INS509; E509
Synonyms: Calcium chloride anhydrous
Classification: Inorganic salt
Empirical: $CaCl_2$ (anhyd.); $CaCl_2 \cdot 2H_2O$ (dihydrate)
Properties: Grayish-wh. cryst., granules, lumps, or flakes; odorless; freely sol. in water and alcohol; m.w. 110.99 (anhyd.), 147.01 (dihydrate); dens. 2.16 (anhyd.), 1.83 (dihydrate); m.p. 772 C (anhyd.), 176 C (dihydrate); b.p. > 1600 C; pH 4.5-9.2 (5%)
Toxicology: LD50 (oral rat) 1000 mg/kg, (IP, rat) 264 mg/kg; poison by IV, intramuscular, IP, subcut. routes; mod. toxic by ing. causing stomach and heart disturbances; severe eye irritant; human systemic effects; tumorigen; mutagen; questionable carcinogen; TSCA listed
Precaution: Reacts violently with BrF_3, (B_2O_3 + CaO); reaction with zinc releases explosive hydrogen gas; exothermic reaction with water
Hazardous Decomp. Prods.: Heated to decomp., emits toxic fumes of Cl^-
HMIS: Health 1, Flammability 0, Reactivity 0
Storage: Hygroscopic; deliq.; store @ R.T.; keep well closed
Uses: Deicing and dust control of roads; drilling muds; dustproofing, freezeproofing, and thawing coal, coke, stone, sand, ore; concrete conditioning; drying and desiccating agent; water treatment; visc. control agent in cosmetics; newsprint deinking; sequestrant, firming agent, anticaking agent, antimicrobial, curing agent, flavor enhancer, humectant, nutrient supplement, pH control agent, pickling agent, processing aid, stabilizer, surfactant, texturizer, thickener in foods; solvent in prep. of hops for beer brewing; pharmaceuticals (firming agent, antimicrobial, desiccant); antiseptic in eye lotions; diuretic; urinary acidifier
Regulatory: FDA 21CFR §133.102, 133.108, 133.111, 133.113, 133.118, 133.127, 133.136, 133.138, 133.141...145.145, 150.141, 150.161, 172.560, 178.1010, 184.1193, GRAS; USDA 9CFR §318.7, 381.147, 3% max.; Japan approved (1% max.); Europe listed (ADI not specified); UK approved; WHO limitation: 350-800 mg/kg (canned fruits/veg.), 200 mg/kg (preserves, processed cheese); approved for injectables, ophthalmics, orals; USP/NF, BP, EP compliance
Manuf./Distrib.: AAE Chemie NV; AB R Lundberg; AMRESCO; Aktiva Ltd; Akzo Nobel Salt; Aldrich; Alfa Aesar; Allan; Am. Int'l.; Ashland; Atofina; BCH Brühl; Boith China; Brenntag AG; Browning; Charkit; Chemacon GmbH; DSM Food Spec.; Degussa AG/Health & Nutrition; Delta Distributors; Dow; EM Ind.; EM Science; Eastech; FBC Ind.; Fluka; Fortitech; GFS; Gallard-Schlesinger Ind.; Great Lakes; Harcros; Hasnil; Hawkins Chem.; Honeywell Perf. Polymers; Int'l. Sourcing; Integra; J.T. Baker; Jarchem Ind.; Kemira Kemi UK; Lohmann; Mallinckrodt Baker; Milsolv; Morre-Tec Ind.; Mutchler; Occidental; Oriental Chem. Ind.; P.B. & S.; Penta Mfg.; Qbiogene; RIA Int'l.; Reade Advanced Materials; Reheis; Ruger; SOCO-Lynch; Sal Chem.; Sigma; Solvay SA; Southchem; Spectrum Quality Prods.; Stanchem; Textile Chem.; Tomita Pharmaceutical; Universal Preserv-A-Chem; VWR Int'l.; Varsal Instruments; Voigt Global Distrib.; Vopak USA; Vulcan Perf. Chems.; Wako Chem. USA; Whittaker, Clark & Daniels

Calcium chloride anhydrous. *See* Calcium chloride
Calcium dithiocyanate; Calcium rhodanid; Calcium sulfocyanate. *See* Calcium thiocyanate

Calcium thiocyanate

CAS 2092-16-2
Synonyms: Calcium dithiocyanate; Calcium rhodanid; Calcium sulfocyanate; Thiocyan
Empirical: $C_2CaN_2S_2 \cdot 3H_2O$
Formula: $Ca(SCN)_2 \cdot 3H_2O$
Properties: Wh. cryst.; sol. in water and alcohol; m.w. 156.24 (anhyd.)
Toxicology: LDLo (oral, mouse) 120 mg/kg, (IV, rabbit) 250 mg/kg; poison by ing. and IV routes; TSCA listed
Hazardous Decomp. Prods.: Heated to decomp., emits toxic fumes of NO_x, SO_x
Storage: Hygroscopic, deliq.
Uses: Solvent for cellulose and polyacrylate, acrylonitrile polymers; stiffening and swelling of textiles
Manuf./Distrib.: Allchem Ind.

C10 alcohol. *See* Decyl alcohol
C12-15 alcohol benzoate. *See* C12-15 alkyl benzoate
C8 alcohols. *See* Caprylic alcohol

C13-15 alcohols

CAS 90604-31-2; EINECS/ELINCS 292-324-6
Synonyms: Alcohol, C13-15; Alcohols, C13-15; Linear alcohol C13-15
Definition: Mixt. of C13 and C15 primary alcohols
Properties: Dens. 0.834 g/cm^3 (22 C); vapor pressure < 0.1 kPa (100 C); b.p. 281-306 C; hyd. no. 267; flash pt. (PMCC) 140 C
Toxicology: Skin irritant
Environmental: High persistence, bioaccumulation
Uses: Solvent; mfg. of surfactants and base oils for metalworking fluids
Manuf./Distrib.: BASF AG

C16-18 alcohols. *See* Cetearyl alcohol
C12-15 alcohols benzoate. *See* C12-15 alkyl benzoate

C10-13 alkane

CAS 64771-72-8; 129813-66-7
Synonyms: Alkanes, C10-13
Definition: Mixt. of alkanes with 10-13 carbon atoms in the alkyl chain
Uses: Solvent in cosmetics

C14-17 alkane

CAS 64771-72-8; 90622-47-2; EINECS/ELINCS 292-449-6
Synonyms: Alkanes, C14-17
Definition: Mixt. of alkanes with 14-17 carbon atoms in the alkyl chain
Uses: Solvent in cosmetics

C14-16 alkanes. *See* C14-16 paraffins
C6 alkyl acetate. *See* Oxo-hexyl acetate
C7 alkyl acetate. *See* Oxo-heptyl acetate

C8 alkyl acetate

CAS 108419-32-5
Synonyms: Acetic acid, C7-9-branched alkyl esters, C8-rich
Properties: Sp.gr. 0.87; flash pt. (TCC) 171 F
Toxicology: LD50 (oral, rat) 5 g/kg; reproductive effects; TSCA listed
Uses: Solvent for high solids, maintenance, marine, and industrial waterborne coatings, acrylic resin polymerization media, metalworking fluids, urethane foam and industrial cleaners and degreasers, paint strippers, pigment dispersions, pesticide emulsifiable concs., textile scouring aids, NC or screen printing inks, penetrating and lubricating oils, vinyl plastisols; wetting agent; lubricant; visc. modifier; latex coalescer
Features: Good solvency; low surf. tens. and interfacial tens.
Trade Names: Exxate® 800

C9 alkyl acetate

CAS 108419-33-6
Properties: Sp.gr. 0.87; flash pt. (TCC) 194 F
Uses: Solvent for high solids, maintenance, marine, and industrial waterborne coatings, acrylic resin polymerization media, metalworking fluids, urethane foam and industrial cleaners and degreasers, paint strippers, pigment dispersions, pesticide emulsifiable concs., textile scouring aids, NC or screen printing inks, penetrating and lubricating oils, vinyl plastisols; wetting agent; lubricant; visc. modifier; latex coalescer
Features: Good solvency, low surf. tens. and interfacial tens.
Trade Names: Exxate® 900

C10 alkyl acetate

CAS 68478-36-4; 108419-34-7; EINECS/ELINCS 270-817-7
Synonyms: Acetic acid, decyl ester, branched
Properties: Sp.gr. 0.87; flash pt. (TCC) 212 F
Uses: Solvent for high solids, maintenance, marine, and industrial waterborne coatings, acrylic resin polymerization media, metalworking fluids, urethane foam and industrial cleaners and degreasers, paint strippers, pigment dispersions, pesticide emulsifiable concs., textile scouring aids, NC or screen printing inks, penetrating and lubricating oils, vinyl plastisols; wetting agent; lubricant; visc. modifier; latex coalescer; fragrance in cosmetics
Features: Good solvency, low surf. tens. and interfacial tens.
Trade Names: Exxate® 1000

C13 alkyl acetate

CAS 108419-35-8
Properties: Sp.gr. 0.87; flash pt. (TCC) 261 F
Uses: Solvent for high solids, maintenance, marine, and industrial waterborne coatings, acrylic resin polymerization media, metalworking fluids, urethane foam and industrial cleaners and degreasers, paint strippers, pigment dispersions, pesticide emulsifiable concs., textile scouring aids, NC or screen printing inks, penetrating and lubricating oils, vinyl plastisols; wetting agent; lubricant; visc. modifier; latex coalescer
Features: Good solvency, low surf. tens. and interfacial tens.
Trade Names: Exxate® 1300

C8-16 alkylbenzene

CAS 68648-87-3
Synonyms: Benzene, C8-16 alkyl derivs.
Classification: Linear alkylbenzene
Properties: Colorless liq.; mild odor; sol. in water; sp.gr. 1.10-1.12; vapor pressure < 0.01 mm Hg (20 C); b.p. 465-475 F; flash pt. (COC) 298-310 F
Toxicology: May cause sl. transient eye irritation; essentially nonirritating to skin; prolonged/repeated contact may cause skin reddening, itchiness; not expected to cause harmful effects by inh. of vapor
Hazardous Decomp. Prods.: CO_x, NO_x; combustion: NO_x
HMIS: Health 1, Flammability 1, Reactivity 0
Storage: Store in containers resist. to alkaline sol'ns. (pH 9-12); keep containers tightly closed when not in use
Uses: Intermediate for mfg. of anionic surfactants for hand dishwash, hard surf. cleaning, industrial cleaning; solvent; lubricant oil additive intermediate; sec. plasticizer; intermediate for enhanced oil recovery
Trade Names: Alkylate 215; Alkylate 229

C12-14 alkylbenzene

CAS 68515-32-2; EINECS/ELINCS 271-073-6
Synonyms: Benzene, C12-14 alkyl dervis.; Benzene, mono-C12-14 alkyl derivs.
Uses: Sec. plasticizer for vinyl; wire rope lubricant; heat transfer fluid; solvent for oil-sol. sulfonates such as sacrificial agent corrosion inhibitors; specialty lubricants; sulfonation feedstock; intermediate for agric. emulsifiers, enhanced oil recovery surfactants, dust abatement agents
Trade Names: V-159L

C12-15 alkyl benzoate

CAS 68411-27-8; EINECS/ELINCS 270-112-4
Synonyms: Alkyl (C12-15) benzoate; Benzoic acid, C12-15 alkyl esters; C12-15 alcohol benzoate; C12-15 alcohols benzoate
Definition: Ester of benzoic acid and C12-15 alcohols
Empirical: $C_{20}H_{32}O_2$
Properties: M.w. 304; sp.gr. 0.915-0.935; acid no. ≤ 0.5; ref. index 1.483-1.487
Storage: Preserve in tight, light-resistant containers
Uses: Emollient in cosmetics, pharmaceuticals; emollient, solubilizer, sunscreen vehicle, deoiler for creams and oils; vehicle in pharmaceutical topicals; binder in pressed powds.; solvent and perfume fixing agent
Features: Oleaginous
Regulatory: NF listed
Trade Names: Crodamol AB; Finsolv® TN; Finsolv® TN-O; Tegosoft® TN
Trade Names Containing: Finsolv® TPP

2-Camphanol acetate. *See* Bornyl acetate
2-Camphanone. *See* Camphor
endo-2-Camphanyl ethanoate. *See* Bornyl acetate
Camphocean. *See* 1,1,2-Trimethylcyclopentane

Camphor

CAS 76-22-2; 464-49-3 (D-+); 21368-68-3 (DL); EINECS/ELINCS 200-945-0; 207-355-2 (D-+); 244-350-4 (DL)
UN 1325; UN 2717 (DOT); FEMA 2230 (D-)
Synonyms: 2-Bornanone; Bornan-2-one; 2-Camphanone; Camphora; Cinnamomum camphora; Formosa camphor; Gum camphor; Japan camphor; 2-Keto-1,7,7-trimethylnorcamphane; Laurel camphor; Matricaria camphor; 2-Oxobornane; 1,7,7-Trimethylbicyclo (2.2.1)-2-heptanone; 1,7,7-Trimethylbicyclo [2.2.1] heptan-2-one; 1,7,7-Trimethylnorcamphor
Classification: Aliphatic cyclohexyl compd.
Definition: Ketone derived from wood of the camphor tree, *Cinnamomum camphora* or prepared synthetically
Empirical: $C_{10}H_{16}O$
Properties: Colorless or wh. translucent cryst. mass, char. fragrant penetrating odor, sl. bitter cooling taste; sol. in alcohol, chloroform, ether, fixed/volatile oils; sl. sol. in water; m.w. 152.24; dens. 0.992 (25/4 C); m.p. 179.7 C; b.p. 204 C; subl. @ ambient temp./pressure; flash pt. 150 F
Toxicology: LD50 (oral, mouse) 1310 mg/kg, (IP, mouse) 3000 mg/kg; LCLo (inh., mouse, 3 h) 400 mg/m^3; human poison by ing.; experimental poison by inh., subcut., IP routes; local irritant; ingestion by humans may cause nausea, vomiting, vertigo, mental confusion, delirium, convulsions, coma, respiratory failure, death; chronic exposure may cause transient hepatic and renal damage; mutagenic data; TSCA listed
Precaution: Combustible; incompat. with potassium permanganate; can react with oxidizers; vapor is explosive exposed to heat, flame, or CrO_3
HMIS: Health 2, Flammability 2, Reactivity 0
Storage: Keep tightly closed away from heat
Uses: Plasticizer for cellulose nitrate, other explosives, lacquers; insecticides; moth and mildew proofing; embalming; pyrotechnics; intermediate; fragrance ingred. (incense, tech. prods.); denaturant, film-former in cosmetics; in food-pkg. adhesives; natural flavoring agent in foods and pharmaceuticals; tooth powds.; active OTC drug ingred.; medicine (sedative, anodyne, antispasmodic, diaphoretic, anthelmintic, anesthetic, analgesic, topical anti-infective, anti-itching agent)
Features: Medicinal-type flavor
Regulatory: FDA 21CFR §172.510, 172.515, 175.105, 27CFR §21.65, 21.151; FEMA GRAS; Japan approved; banned in U.S. for internal use; BP, EP compliance
Manuf./Distrib.: AMC Chems.; Aldrich; Alfa Chem; Andrea Aromatics; Asiamerica Int'l.; Augustus Oils Ltd; Barnet Prods.; Berje; Bio-Botanica; Boith China; Buckton Scott Ltd; Charkit; Chart; ChemService;

Chemacon GmbH; CoKEM Assoc.; Creative Fragrances; Degussa AG/ Health & Nutrition; Dujodwala Resins & Terpenes; Fanning; Fluka; George Uhe; H.E. Daniel Ltd; Hawkins Chem.; Hunan Xinyu; Integra; J.T. Baker; Lonza; Mallinckrodt Baker; MelChem; Mosselman NV; Mutchler; Pangaea Sciences; Penta Mfg.; Polarome Int'l.; Quest Int'l.; R.W. Greeff; Robeco; Ruger; San Yuan; Sarcom; Schweizerhall; Schütz; Sigma; Spectrum Quality Prods.; U.S. Synthetics; Universal Preserv-A-Chem; Varsal Instruments; Voigt Global Distrib.; Wego Chem. & Min.; ZenPharm Int'l.

Trade Names Containing: Punctilious® SDA 38B 190 Proof Prod. Code 13603998; Punctilious® SDA 38B 200 Proof Prod. Code 14601023

Camphora. *See* Camphor

Camphor (Cinnamomum camphora) oil

CAS 8008-51-3; EINECS/ELINCS 295-980-1
UN 1130 (DOT); FEMA 2231

Synonyms: Camphor, Japanese, white oil; Camphor oil; Camphor oil brown; Camphor oil light; Camphor oil, rectified; Camphor oil white; Camphor oil white Japanese; Camphor oil yellow; Camphor white oil; Cinnamomum camphora oil; Formosa camphor oil; Formose oil of campor; Japanese camphor oil; Light camphor oil; Liquid camphor; White camphor oil

Definition: Oil extracted from the wood of the camphor tree, *Cinnamomum camphora*

Properties: Colorless to yel.-brn. liq.; camphor or safrole odor; sol. in paraffin oil; insol. in water; dens. 0.87 kg/l (white), 0.97 kg/l (yel.), 1.07 kg/l (brown); b.p. > 200 C (yel.); flash pt. (TCC) 52 C; ref. index 1.46700-1.47200 (20 C)

Toxicology: LD50 (oral, rat) 3730 mg/kg; TDLo (oral, human) 29 mg/kg; primary irritant; mild skin irritant; may be harmful if swallowed; may be irritating if inhaled; may cause tremors, convulsions in humans; TSCA listed

Precaution: DOT: Flamm. liq.; avoid contact or contamination with strong acids, alkalis, or oxidizing agents

Hazardous Decomp. Prods.: CO and unidentified org. compds. may be formed during combustion

Uses: Solvent in paints, lacquers; fragrance ingred.; natural flavoring agent in foods; cough medicine ingred.; topical analgesic drug

Regulatory: FDA 21CFR §172.510; FEMA GRAS

Manuf./Distrib.: Aldrich; Alfa Chem; Asiamerica Int'l.; Boith China; Dujodwala Resins & Terpenes; Fleurchem; Fluka; George Uhe; Good Scents; Janousek Industriale Srl; Kira; MelChem; Penta Mfg.; Universal Preserv-A-Chem; Voigt Global Distrib.

Camphor, Japanese, white oil; Camphor oil; Camphor oil brown; Camphor oil light; Camphor oil, rectified; Camphor oil white; Camphor oil white Japanese; Camphor oil yellow. *See* Camphor (Cinnamomum camphora) oil
Camphor tar. *See* Naphthalene
Camphor white oil. *See* Camphor (Cinnamomum camphora) oil
Cananga; Cananga odorata; Cananga odorata oil; Cananga oil. *See* Ylang ylang (Cananga odorata) oil
Canola. *See* Canola oil
Canola methyl ester. *See* Methyl canolate

Canola oil

CAS 8002-13-9 ; 120962-03-0

Synonyms: Canola; Colza oil; Rapeseed; Rapeseed oil

Definition: Low-erucic rapeseed oil

Properties: Yel.-brn. oily liq., odorless, bland flavor; insol. in water; dens. 0.913-0.917; m.p. 17-22 C; iodine no. 110-120; sapon. no. 180-193; flash pt. (CC) 600 F

Toxicology: Can cause acne-like skin eruptions

Uses: Emollient, lubricant for cosmetics, pharmaceuticals; intermediate for mfg. of oleochems. and esters; lubricant (engine lubricants, hydraulic fluids, rolling/cutting oils, textile lubricants, mold release agents, biodeg. greases); leather fatliquoring agent; quenching oil; textile finishing agent; foods (frying shortening, nondairy fat, salad oil)

Features: Natural oil

Manuf./Distrib.: ABITEC; Aarhus Oliefabrik A/S; Alnor Oil; Anar; Anglia Oils; Arista Ind.; Avatar; C&T Refinery; CanAmera Foods; Cargill Ind. Oils & Lubes; Charkit; Chesham Chems. Ltd; Columbus Foods; Desert Whale Jojoba; Jeen Int'l.; KIC Chems.; Karlshamns AB; Lambent Tech.; Parchem Trading; Penta Mfg.; Pokonobe Ind.; Protameen; Sea-Land; Spectrum Naturals; St. Lawrence; Tri-K Ind.; Vopak USA

Trade Names Containing: Phytol™ CAN

See also Rapeseed (Brassica campestris) oil

Canola oil (low erucic acid rapeseed oil). *See* Rapeseed (Brassica campestris) oil
Canola oil, methyl ester. *See* Methyl canolate
Capric acid ethyl ester. *See* Ethyl decanoate
Capric acid methyl ester. *See* Methyl caprate
Capric alcohol. *See* Decyl alcohol

Capric triglyceride

Uses: Food emulsifier; cosmetics ingred.; solvent for flavors and oil-sol. food additives

Features: Exc. fluidity

Caprinic alcohol. *See* Decyl alcohol
Caprolactone; ε-Caprolactone; 6-Caprolactone monomer. *See* ε-Caprolactone monomer

ε-Caprolactone monomer

CAS 502-44-3; EINECS/ELINCS 207-938-1

Synonyms: Caprolactone; ε-Caprolactone; 6-Caprolactone monomer; 6-Hexanolactone; 1,6-Hexanolide; Hexan-6-olide; 6-Hydroxyhexanoic acid lactone; ε-Lactone hexanoic acid; 2-Oxepanone

Empirical: $C_6H_{10}O_2$

Properties: Colorless liq.; misc. with water; m.w. 114.14; dens. 1.030; b.p. 96-97.5 C (10 mm), 98-99 (2 mm); flash pt. 109 C; ref. index 1.4630

Toxicology: LD50 (oral, rat) 4290 mg/kg, (IP, mouse) 1300 mg/kg, (skin, rabbit) 5990 mg/kg; mod. toxic by IP route; mildly toxic by ing.; primary skin irritant; severe eye irritant; mutagen; TSCA listed

Hazardous Decomp. Prods.: Heated to decomp., emits acrid smoke and irritating fumes

Uses: Intermediate in adhesives, urethane coatings, elastomers; solvent; diluent for epoxy resins; synthetic fibers; organic synthesis

Manuf./Distrib.: Advanced Synthesis Tech.; Aldrich; Dow; Fluka

Caproyl alcohol. *See* Hexyl alcohol
Capryl alcohol. *See* 2-Octanol
Caprylic acid, methyl ester. *See* Methyl caprylate
Caprylic acid, monoester with 1,2-propanediol. *See* Propylene glycol caprylate
Caprylic acid, 1,2,3-propanetriyl ester; Caprylic acid triglyceride. *See* Tricaprylin

Caprylic alcohol

CAS 111-87-5; EINECS/ELINCS 203-917-6
FEMA 2800

Synonyms: Alcohol C8; C8 alcohols; Heptyl carbinol; 1-Hydroxyoctane; Octanol; 1-Octanol; Octan-1-ol; n-Octanol; Octyl alcohol; n-Octyl alcohol; Primary octyl alcohol

Classification: Primary aliphatic alcohol

Empirical: $C_8H_{18}O$

Formula: $CH_3(CH_2)_6CH_2OH$

Properties: Colorless oily liq., fresh orange-rose odor, sl. herbaceous taste; sol. in oxygenated solvs.; misc. in alcohol, ether, chloroform; pract. insol. in water; m.w. 130.26; dens. 0.827; vapor pressure 0.0794 mm Hg; m.p. -16.7 C; b.p. 194.5 C; flash pt. 81 C; ref. index 1.429 (20 C); surf. tens. 26.92 dynes/cm; dielec. const. 10.34

Toxicology: LD50 (oral, mouse) 1790 mg/kg, (IV, mouse) 69 mg/kg; poison by IV route; mod. toxic by ing.; skin and eye irritant; does not readily form a vapor; vapors may cause CNS depression; can probably cause 'alcohol' intoxication effects; may cause severe lung damage, respiratory and/or cardiac arrest and death if aspirated into lungs; mutagenic data; TSCA listed

Environmental: VOC; BOD5 1.58; ThOD 2.95

Precaution: Flamm. exposed to heat or flame; can react with oxidizers

Hazardous Decomp. Prods.: Thermal decomp. prods.: CO, CO_2

NFPA: Health 1, Flammability 2, Reactivity 0

Storage: Store in cool, dry area away from heat and ignition sources

Uses: Wetting agent; emulsifier; detergent and surfactant intermediate; raw material for plasticizers; emulsion polymerization; tobacco-offshoots controller; solvent for paints, foods; defoamer; slip agent for plastics; visc. control agent in cosmetics; synthetic flavoring agent in foods and pharmaceuticals; fragrance; mfg. of perfumes; in food-pkg. adhesives; defoamer for cellophane for food pkg.; food-contact textiles

Features: Citrus flavor

Regulatory: FDA 21CFR §172.230, 172.515 (only for encapsulating lemon, dist. lime, orange, peppermint, and spearmint oils), 172.864, 173.280, 175.105, 175.300, 177.1010, 177.1200, 177.1390, 177.2800, 178.3480; FEMA GRAS

Manuf./Distrib.: Advanced BioTech; Advanced Synthesis Tech.; Albemarle; Aldrich; Arizona; Asiamerica Int'l.; Axxence Aromatic GmbH; Cognis/Chems. Group; Fleurchem; Fluka; GFS; Goldensun Mfg.; Grau Aromatics; Haarmann & Reimer GmbH; Int'l. Paper; J.T. Baker; Jarchem Ind.; Kira; M. Michel; Moore Ingreds.; Penta Mfg.; Prodasynth; R.W. Greeff; Sasol Germany; Schweizerhall; Shandong Haihua Tianhe; Sigma; Spectrum Quality Prods.; VWR Int'l.; Vopak USA

Caprylic, capric acid, propylene glycol diester. *See* Propylene glycol dicaprylate/dicaprate

Caprylic/capric glycerides

CAS 85409-09-2; EINECS/ELINCS 287-075-5

Synonyms: Glycerides, C8-10

Definition: Mixture of mono-, di-, and triglycerides of caprylic and capric acids

Properties: Nonionic

Uses: Plasticizer for hard fats; solvent for lipophilic ingreds.; emollient, emulsifier in cosmetics; solvent, coemulsifier, solubilizer, carrier for lipophilic drugs; antifoam for foods

Trade Names: Imwitor® 742

Caprylic/capric/lauric triglyceride

CAS 68991-68-4

Synonyms: Dodecanoic acid, mixed triesters with octanoic acid, decanoic acid, and 1,2,3-propanetriol

Definition: Mixed triester of glycerin with caprylic, capric and lauric acids

Uses: Emollient, solvent, carrier, fixing agent, and extender for pharmaceuticals, cosmetics, nutritional applics.

Trade Names: Captex® 350

Caprylic/capric/linoleic triglyceride

Synonyms: 9,12-Octadecadienoic acid, mixed triesters with octanoic acid, decanoic acid, and 1,2,3-propanetriol

Definition: Mixed triester of glycerin with caprylic, capric, and linoleic acids

Uses: Emollient, solvent, carrier, fixing agent, and extender for pharmaceuticals, cosmetics, nutritional applics.; carrier for flavors and fragrances

Trade Names: Captex® 810D

Caprylic/capric/oleic triglyceride

Uses: Emollient, solvent, extender in pharmaceuticals, nutritional applics.; carrier for flavors and fragrances

Caprylic/capric/stearic triglyceride

Synonyms: Octadecanoic acid, mixed triesters with octanoic acid, decanoic acid, and 1,2,3-propanetriol

Definition: Mixed triester of glycerin with caprylic, capric and stearic acids

Uses: Emollient, solvent in cosmetics; moisturizer; stabilizer; emulsifier; ointment base for the prep. of nonaq. ointments and creams; filler, carrier for gelatin capsules

Features: Good skin compatibility and resorption chars.

Caprylic/capric triglyceride

CAS 526220-27-2; 65381-09-1; 73398-61-5; EINECS/ELINCS 265-724-3

Synonyms: Glycerides, mixed decanoyl and octanoyl; Glyceryl tricaprylate/caprate; Octanoic/decanoic acid triglyceride

Definition: Mixed triester of glycerin and caprylic and capric acids

Properties: Nonionic

Toxicology: Eye irritant

Hazardous Decomp. Prods.: Heated to decomp., emits acrid smoke and irritating fumes

Uses: Solvent for colors and perfumes; emollient, solvent, moisturizer, dispersant, solubilizer, suspending agent in cosmetics, toiletries, pharmaceuticals; plasticizer; lubricity vehicle; food emulsifier; bakery lubricant, release agent, glazing agent for confectionery; solvent carrier for flavors and fragrances

Regulatory: Approved for topicals

Manuf./Distrib.: Avatar; Croda Inc; DelCon; Inolex; Lambent Tech.; Lonza; Mosselman NV; Phoenix Nat. Prods.; Sea-Land; Stepan

Trade Names: Akomed R; Aldo® MCT KFG; Captex® 300; Crodamol GTCC; Crodamol GTCC-PN; Delios® S; Estalan M-5; Estasan™ GT 8-40 3578; Estasan™ GT 8-60 3575; Estasan™ GT 8-60 3580; Estasan™ GT 8-65 3577; Estasan™ GT 8-65 3581; Estasan™ GT 8-70 3579; Labrafac® Lipophile; Lexol® GT-855; Lexol® GT-865; Liponate GC-K; Miglyol® 810; Miglyol® 812; Miglyol® 812 N; Myritol® 318; NEOBEE® M-5; Tegosoft® CT

Caprylic/capric triglyceride PEG-4 esters

Definition: Complex mixture formed from transesterification of caprylic/capric triglyceride and PEG-4

Properties: Oil; nonionic

Uses: Bioavailability enhancer, emulsifier, solvent, solubilizer for cosmetic, pharmaceutical, and veterinary creams, lotions, ointments, shampoos, bath oils

Features: Hydrophilic

Trade Names: Labrafac® Hydrophile

Caprylic ether. *See* Dioctyl ether
Caprylyl acetate. *See* Octyl acetate

Caprylyl pyrrolidone

CAS 2687-94-7

Synonyms: N-Octyl-2-pyrrolidone; Pyrrolidone, N-octyl-

Classification: Substituted heterocyclic organic compd.

Uses: Surfactant in cosmetics; specialty solvent for commercial cleaning, textile processing, lithographic printing prods., coatings, inks

Trade Names: Agsolex® 8; Surfadone® LP-100

Caraway (Carum carvi) oil

CAS 8000-42-8; 85940-31-4; EINECS/ELINCS 288-921-6
FEMA 2238

Synonyms: Caraway oil; Carum carvi; Carum carvi fruit oil; Carum carvi oil

Definition: Volatile oil distilled from the dried ripe fruit of *Carum carvi*, contg. carvone, d-limonene

Properties: Colorless to pale yel. clear oily liq., caraway odor and taste; darkens and thickens with age; sol. in 8 vols. 80% alcohol; pract. insol. in water; dens. 0.900-0.910 (25/25 C); ref. index 1.485-1.497 (20 C)

Toxicology: LD50 (oral, rat) 3500 mg/kg, (skin, rabbit) 1780 mg/kg; mod. toxic by ing., skin contact; skin irritant; mutagenic data; TSCA listed

Hazardous Decomp. Prods.: Heated to decomp., emits acrid smoke and irritating fumes

HMIS: Health 1, Flammability 2, Reactivity 0

Storage: Keep cool, well closed; protect from light

Uses: Natural flavoring agent in foods, drinks, and pharmaceuticals; carminative; stimulant; solvent in cosmetics

Regulatory: FDA 21CFR §182.10, 182.20, GRAS; FEMA GRAS; Japan approved; BP compliance

Manuf./Distrib.: Advanced BioTech; Aldrich; Chart; F.D. Copeland; Fleurchem; George Uhe; H.E. Daniel Ltd; Integra; Janousek Industriale Srl; Liberty Natural Prods.; Penta Mfg.; Ruger; Spectrum Quality Prods.; Treatt USA; Virginia Dare Extract; Voigt Global Distrib.

Caraway oil. *See* Caraway (Carum carvi) oil
Carbamaldehyde. *See* Formamide
Carbamic acid, butyl-3-iodo-2-propynyl ester. *See* Iodopropynyl butylcarbamate
Carbamic acid nitrile. *See* Dimethyl cyanamide
Carbamic acid, 1-phenylethyl-, ethyl ester. *See* trans-1,4-Dimethylcyclohexane
Carbinamine. *See* Methylamine
Carbinol. *See* Methyl alcohol
'Carbitol'. *See* Ethoxydiglycol
'Carbitol' acetate. *See* Ethoxydiglycol acetate

'Carbitol' acrylate. *See* 2-(2-Ethoxyethoxy) ethyl acrylate
'Carbolic acid'. *See* Phenol

Carbomer

CAS 9003-01-4 (generic); 9007-16-3 (generic); 9007-17-4 (generic); 9007-20-9; 9062-04-8; 54182-57-9; 76050-42-5
Synonyms: Acrylic acid, polymer with sucrose-polyallyl ether; Carboxypolymethylene
Definition: Homopolymer of acrylic acid crosslinked with an allyl ether of pentaerythritol or an allyl ether of sucrose
Properties: Wh. fluffy powd., odorless; dens. 1.41; anionic
Toxicology: LD50 (oral, rat) 4100 mg/kg; mod. toxic by ingestion; TSCA listed
Hazardous Decomp. Prods.: Heated to decomp., emits acrid smoke and fumes
Storage: Hygroscopic
Uses: Emulsifier; thickener; stabilizer; suspending agent; for lubricating, quenching and silicone emulsions; graphite; polyethylene; fiber and paper suspensions; textile backcoatings; pharmaceuticals; emulsion stabilizer, visc. control agent in cosmetics; solvent thickener for cosmetics, pharmaceuticals; food-pkg. adhesives, coatings, paper/paperboard, polymers; defoamer in food-contact coatings
Features: Forms gels or aq. ointments at higher concs.
Use Level: 0.4% (suspending agent), 0.5-5% (aq. ointment base)
Regulatory: FDA approved for orals, rectals, topicals; BP, EP compliance
Manuf./Distrib.: Adept Sol'ns.; Aldrich; Cornelius Chem. Co. Ltd; Goldschmidt AG; Noveon; Protameen; R.I.T.A.; Sigma; Spectrum Quality Prods.

Carbomer 940

CAS 9003-01-4 (generic); 9007-17-4 (generic)
Definition: Polymer of acrylic acid crosslinked with allyl ether of pentaerythritol
Properties: Wh. fluffy powd., sl. char. odor; when neutralized with alkali hydroxides or with amines, dissolves in water, alcohol, and glycerin; m.w. nominally 4,000,000; visc. 40,000-60,000 cps (0.5% aq. disp.); pH 3 (1% disp.); anionic
Toxicology: No known toxicity; TSCA listed
Storage: Hygroscopic
Uses: Thickener, suspending agent, dispersant, emulsifier for pharmaceuticals, cosmetics, personal care, industrial applics.; emulsion stabilizer
Features: Reacts with fat particles to form thick stable emulsions of oils in water
Regulatory: FDA approved for ophthalmics, topicals; NF compliance
Manuf./Distrib.: Aldrich; Sigma; Spectrum Quality Prods.
Trade Names: Carbopol® 940; Carbopol® 980

2-Carbomethoxyphenol. *See* Methyl salicylate
Carbona. *See* Carbon tetrachloride
Carbonate of potash. *See* Potassium carbonate
Carbon bichloride. *See* Perchloroethylene
Carbon bisulfide. *See* Carbon disulfide
Carbon bromide. *See* Carbon tetrabromide
Carbon chloride. *See* Carbon tetrachloride
Carbon dichloride. *See* Perchloroethylene

Carbon disulfide

CAS 75-15-0; EINECS/ELINCS 200-843-6
UN 1131 (DOT)
Synonyms: Carbon bisulfide; Carbon sulfide; Dithiocarbonic anhydride
Classification: Inorg. compd. of carbon and sulfur
Empirical: CS_2
Formula: S=C=S
Properties: Colorless clear liq.; mild sweet odor (pure), strong disagreeable odor (impure); sol. 2 g/l in water (20 C); dissolves in ethanol, methanol, benzene, chloroform, CCl_4, ether, oils; m.w. 76.13; vapor pressure 400 mm Hg (28 C); m.p. -112 C; b.p. 46.3 C; flash pt. (CC) -30 C
Toxicology: ACGIH TLV/TWA 10 ppm (skin); LD50 (oral, rat) > 505 mg/kg; LCLo (inh., human, 30 min) 4000 ppm; readily absorbed thru skin; dissolves oils from skin; splashes in eyes cause immediate severe irritation; tumorigen; inh. may cause severe personality disturbances, confusion, insomnia, psychosis, suicide; exposure @ 4800 ppm for 30 min causes coma and may be fatal; ing. may cause vomiting, diarrhea, headache, convulsions, coma, death; toxic to reproduction; TSCA listed
Precaution: Highly flamm.; LFL 1.25%; UFL 50%; incompat. with air, chemically active metals, metal oxides, metal azides, amines, bases, imines, halogens, nitrogen oxides, permanganates and sulfuric acid; violent and explosive reactions possible
Hazardous Decomp. Prods.: Heated to decomp., emits highly toxic fumes of SO_x
NFPA: Health 3, Flammability 4, Reactivity 0
Storage: Store in cool area in isolated buildings; avoid direct sunlight; ground drums and bond transfer containers
Uses: Mfg. of viscose rayon fibers (prod. of rayon filament yarn, rayon tire yarn, rayon stable fiber, Cellophane film), carbon tetrachloride, fluorocarbon chems., thiochemicals, pharmaceuticals, agric., mining, and rubber chems.; flotation agent; solvent for rubber, waxes, plastics, sulfur, selenium, bromine, and iodine; solvent extraction of oil, grease, wax (textiles)
Regulatory: FDA 21CFR §193.225, 193.230 (fumigant); SARA reportable
Manuf./Distrib.: Akzo Nobel; Aldrich; Alfa Chem; Atofina; FMC Foret; Fluka; Integra; J.T. Baker; Mallinckrodt Baker; PPG Ind.; Rhodia; Romil Ltd; Sigma; Spectrum Quality Prods.; VWR Int'l.

Carbonic acid, cyclic ethylene ester. *See* Ethylene carbonate
Carbonic acid, cyclic ethylethylene ester. *See* 1,2-Butylene carbonate
Carbonic acid, cyclic propylene ester. *See* Propylene carbonate
Carbonic acid diethylamide. *See* Dimethyl cyanamide
Carbonic acid, diethyl ester. *See* Diethyl carbonate
Carbonic acid, dimethyl ester. *See* Methyl carbonate
Carbonic acid, diphenyl ester. *See* Diphenyl carbonate
Carbonic acid dipotassium salt. *See* Potassium carbonate
Carbonic acid, oxydi-2,1-ethanediyl dibutyl ester. *See* Butyl diglycol carbonate
Carbonic acid, 1,2-propylene glycol ester. *See* Propylene carbonate
Carbonic ether. *See* Diethyl carbonate
Carbon oil. *See* Benzene
Carbon sulfide. *See* Carbon disulfide
Carbon tet. *See* Carbon tetrachloride

Carbon tetrabromide

CAS 558-13-4; EINECS/ELINCS 209-189-6
UN 2516
Synonyms: Carbon bromide; Methane tetrabromide; Tetrabromomethane
Classification: Brominated hydrocarbon
Empirical: CBr_4
Properties: Colorless powd. or yel.-brn. cryst.; sol. in alcohol, ether, chloroform; sol. 240 mg/l in water (30 C); m.w. 331.63; sp.gr. 1.8-1.9; vapor pressure 0.27 mm Hg; m.p. 88-90 C; b.p. 190 C; nonflamm.
Toxicology: ACGIH TLV/TWA 0.1 ppm, STEL 0.3 ppm; poison; narcotic in high concs.; irritating to eyes, skin; may cause permanent eye damage; inh. may cause nose/throat/lung irritation, coughing, wheezing, shortness of breath; may damage liver, kidneys
NFPA: Health 3, Flammability 0, Reactivity 0
Uses: Solvent for waxes, greases, oils; ingred. in fire-resist. chems., photochem. compds.; mineral ore selecting agent in geological assays; blowing agent, vulcanizing agent in plastics, rubber; polymerization agent; as sedative; intermediate for agrochems., org. synthesis
Manuf./Distrib.: ABCR; Spektr T.T.T.

Carbon tetrachloride

CAS 56-23-5; EINECS/ELINCS 200-262-8
UN 1846 (DOT)
Synonyms: Benzinoform; Carbona; Carbon chloride; Carbon tet; CTC; Methane tetrachloride; Perchloromethane; Tetrachlorocarbon; Tetrachloromethane
Classification: Chlorinated hydrocarbon
Empirical: CCl_4
Properties: Colorless liq., sweetish distinct ether-like odor; misc. with alcohol, ether, chloroform, benzene, solv. naphtha; insol. in water; m.w. 153.84; dens. 1.585 (25/4 C); vapor pressure 115 mm Hg; m.p. -23 C; b.p. 76.74 C; flash pt. none; ref. index 1.457; surf. tens. 26.92 dynes/cm; KB value 104; dielec. const. 2.29

Toxicology: ACGIH TLV/TWA 5 ppm; STEL 30 (skin); LD50 (oral, rat) 2800-2920 mg/kg, (IP, rat) 1500 mg/kg, (skin, rat) 5070 mg/kg; toxic by ing., inh., IV, and skin absorption; human systemic effects (nausea, vomiting, coma, tremors, anorexia, GI effects; narcotic; eye and skin irritant; can cause contact dermatitis; damages liver, kidneys, lungs; confirmed carcinogen; experimental teratogen; mutation data; TSCA listed
Environmental: VOC; BOD5 0; ThOD 0
Precaution: Forms impact-sensitive explosive mixts. with many metal particulates; other explosive mixts. possible
Hazardous Decomp. Prods.: Heated to decomp., emits toxic fumes of Cl^- and phosgene
NFPA: Health 3, Flammability 0, Reactivity 0
Uses: Refrigerants; metal degreasing; agric. fumigant; chlorinating organic compds.; production of semiconductors; solvent for fats, oils, rubber, pharmaceuticals; solvent for chlorination reactions
Use Level: 4 ppm (pharmaceuticals)
Regulatory: SARA §313 reportable; CERCLA hazardous substance; FDA banned from household use; HAP
Manuf./Distrib.: Advanced Synthesis Tech.; Aldrich; Allchem Ind.; Ashland; Asiamerica Int'l.; Atofina; Coyne; Dow; Fluka; ICC Ind.; J.T. Baker; Occidental; Romil Ltd; Sigma; Solvay SA; Spectrum Quality Prods.; Stanchem

Carbon tetrafluoride. *See* Tetrafluoromethane
Carbon trifluoride. *See* Trifluoromethane
1,3-Carbonyldioxyethane. *See* Ethylene carbonate
1,3-Carbonyl dioxypropane. *See* Propylene carbonate
Carboxylic acid C9. *See* Nonanoic acid
Carboxylic acids, naphthenic. *See* Naphthenic acid
Carboxypolymethylene. *See* Carbomer
(+)-3-Carene. *See* 3-Carene

3-Carene

CAS 13466-78-9; EINECS/ELINCS 236-719-3
Synonyms: Bicyclo (4.1.0) hept-3-ene, 3,7,7-trimethyl-; (+)-3-Carene; Δ^3-Carene; S-3-Carene; Isodiprene; 3-Norcarene, 3,7,7-trimethyl-; 3,7,7-Trimethylbicyclo [4.1.0] hept-3-ene; 4,7,7-Trimethyl-3-norcarene
Classification: Nonaromatic hydrocarbon
Definition: Found in many volatiles oils (turpentine oils, galanga root oil, German pine needle oils) and isolated from turpentine fractions
Empirical: $C_{10}H_{16}$
Properties: Colorless mobile liq.; sweet and pungent odor, more agreeable than turpentine odor; misc. with fat solvs. and oils; pract. insol. in water; m.w. 136.24; dens. 0.857; b.p. 168-169 (705 mm); ref. index 1.4740
Toxicology: LD50 (oral, rat) 4800 mg/kg; mildly toxic by ing.; primary irritant; irritating to eyes, skin, respiratory system; TSCA listed
Precaution: Flamm.; readily oxidized on exposure to air
Hazardous Decomp. Prods.: Heated to decomp., emits acrid smoke and fumes
Storage: Handle under nitrogen
Uses: Solvent in cosmetics
Manuf./Distrib.: Aldrich

Δ^3-Carene; S-3-Carene. *See* 3-Carene
C_8 aromatics. *See* Xylene
Carthamus tinctorious oil. *See* Safflower (Carthamus tinctorius) oil
Carthamus tinctorius. *See* Hybrid safflower (Carthamus tinctorius) oil; Safflower (Carthamus tinctorius) oil
Carthamus tinctorius hybrid oil; Carthamus tinctorius oil. *See* Hybrid safflower (Carthamus tinctorius) oil
Carum carvi; Carum carvi fruit oil; Carum carvi oil. *See* Caraway (Carum carvi) oil
Carvene. *See* l-Limonene
(+)-Carvene. *See* d-Limonene
Caryophyllic acid. *See* Eugenol
Caryophyllus oil. *See* Clove (Eugenia caryophyllus) oil
Cashew nutshell liq. *See* Cashew nut shell oil

Cashew nut shell oil

CAS 8007-24-7; EINECS/ELINCS 232-355-4
Synonyms: Cashew nutshell liq.; Cashew shell liq.; CNSL; 3-(n-Penta-8′-decenyl) phenol
Definition: Mixt. of mono- and dihydric phenols contg. cardanol and cardol with long alkyl unsat. chain substitutes
Empirical: $C_{21}H_{34}O$
Formula: $C_6H_4 \cdot OH \cdot C_{15}H_{29}$
Properties: Brn. liq.; dens. 0.943-0.968; visc. 600 cps; b.p. 223-227 C (1.3 kPa)
Toxicology: Strong irritant
Uses: Resin; chemical intermediate for phenolic resin synthesis and modification, varnishes and impregnating materials, plasticizers, germicides, insecticides, coloring material, indelible inks, lubricants, preservatives; raw material for surfactants, antioxidants, anticorrosives; lubricant additive; resin modifier; cosolvent for insecticides
Manuf./Distrib.: Arista Ind.

Cashew shell liq. *See* Cashew nut shell oil
Cassia aldehyde. *See* Cinnamal
Castor oil. *See* Castor (Ricinus communis) oil
Castor oil, acetylated. *See* Glyceryl triacetyl ricinoleate
Castor oil aromatic. *See* Castor (Ricinus communis) oil
Castor oil, hydrogenated. *See* Hydrogenated castor oil

Castor oil, polymerized

Synonyms: Polymerized castor oil
Definition: Rubber-like polymer resulting from combination of castor oil with sulfur or diisocyanates
Uses: Pigment wetting agent/dispersant; plasticizer, processing aid for resins, gums, polymers, rubber; plasticizer for lacquers, inks, adhesives, lubricants, polishes, hydraulic fluids; plasticizer, adhesion promoter, gloss aid, solv. resist. aid in NC lacquers; lubricant; penetrant; coupling solvent
Manuf./Distrib.: John L Seaton Ltd
Trade Names: #15 Oil; #30 Oil; #40 Oil; Pale® 4; Pale® 16; Pale® 170; Pale® 1000; Vorite® 105; Vorite® 110; Vorite® 115; Vorite® 125

Castor (Ricinus communis) oil

CAS 1323-38-2; 8001-79-4; EINECS/ELINCS 232-293-8
FEMA 2263; INS1503
Synonyms: Aceite de ricino; Aromatic castor oil; Castor oil; Castor oil aromatic; Huile de ricini; Oil of Palma Christi; Oleum ricini; Ricini oleum; Ricinus communis oil; Ricinus oil; Tangantangan oil
Classification: Vegetable oil
Definition: Fixed oil obtained from seeds of *Ricinus communis*
Properties: Colorless to pale yel. viscous oily liq., char. odor; sol. in alcohol; misc. with glac. acetic acid, chloroform, ether; negligible sol. in water; dens. 0.961; m.p. -12 C; b.p. 313 C; acid no. < 4; iodine no. 83-88; sapon. no. 176-187; hyd. no. 160-168; flash pt. 230 C; ref. index 1.478; surf. tens. 39 dynes/cm (20 C)
Toxicology: Moderately toxic by ingestion; allergen; eye irritant; mild skin irritant; mist may cause sl. irritation of nose and throat; purgative, laxative in large doses; TSCA listed
Precaution: Combustible when exposed to heat; spontaneous heating may occur; incompat. with acids and bases (hydrolysis can occur), strong oxidizing agents (increased fire hazard)
Hazardous Decomp. Prods.: @ 340-400 C, pyrolysis occurs yielding heptaldehyde and undecylenic acids; further decomp. yields CO, CO_2, possibly acrolein
NFPA: Health 0, Flammability 1, Reactivity 0
Storage: Store in cool, dry, well-ventilated area, out of direct sunlight; avoid generating mist
Uses: Plasticizer in lacquers, nitrocellulose, inks; PU coatings; drying oil; hydraulic fluids; in mfg. of Turkey red oil, soaps; leather treatment; surfactants; lubricant (metal drawing, machine lubrication, engine fuels); solvent, emollient in cosmetics; rubber preservative; foods (emulsifier, stabilizer, foamer, antisticking agent, release agent, drying oil, flavor); component of protective coatings for vitamin/mineral tablets; pharmaceutical vehicle, plasticizer, solvent, cosolvent; embalming fluids; laxative; drying oil in food-contact coatings; plasticizer in food-contact rubber articles for repeated use; in food-contact textiles, paper; in lubricants for incidental food-contact use; in surf. lubricants for mfg. of food-contact metallic articles
Use Level: 0.08-23 mg (solid oral dosage forms); 5-12.5% (topicals)
Regulatory: FDA 21CFR §73.1, 172.510, 172.876, 175.105, 175.300,

176.210, 177.2600, 177.2800, 178.3120, 178.3520, 178.3570, 178.3910, 181.22, 181.26, 181.28; FEMA GRAS; FDA approved for injectables, orals, topicals; USP/NF, BP, EP, JP compliance

Manuf./Distrib.: Acme-Hardesty; Air Prods.; Aldrich; Alfa Chem; Alzo; Am. Oil & Supply; Amber Syn.; Arista Ind.; Ashland; Asiamerica Int'l.; Avatar; Buckton Page Ltd; CasChem; ChemService; Costec; Dastech Int'l.; Degen; EM Science; Fanning; Fluka; Geo. Pfau's Sons; George Uhe; Georgia-Pacific/Actrachem; Harcros; Indofine; Integra; J.H. Calo; J.T. Baker; John L Seaton Ltd; KIC Chems.; Lanaetex Prods.; Lipo; Mosselman NV; Mutchler; Norman, Fox; Penta Mfg.; Pfanstiehl Labs; R.W. Greeff; RTD Chems.; Rhodia HPCII; Ruger; Sea-Land; Sigma; Spectrum Quality Prods.; St. Lawrence; Stevenson Cooper; Süd-Chemie AG; Thornley; Uniqema Am.; Universal Preserv-A-Chem; VWR Int'l.; Vliengenthart BV; Voigt Global Distrib.; Vopak USA; Welch, Holme & Clark; Zeochem

Trade Names: Crystal® O; Crystal® Crown; EmCon™ CO; Krystal Kleer Castor Oil; USP Castor Oil

Castorwax. *See* Hydrogenated castor oil
Catechol. *See* Pyrocatechol
Caustic potash. *See* Potassium hydroxide
Caustic soda. *See* Sodium hydroxide
CBM. *See* Bromochloromethane
C11-14-branched alkyl acetate C13-rich. *See* Oxo-dodecyl acetate
C16-18 and C18 unsaturated alkylcarboxylic acid, methyl ester. *See* Methyl palmitate oleate
CDOL. *See* Cyclododecanol
'Cellosolve'. *See* Ethoxyethanol
'Cellosolve' acetate. *See* Ethoxyethanol acetate
'Cellosolve' solvent. *See* Ethoxyethanol
Cetane; n-Cetane. *See* n-Hexadecane

Cetearyl alcohol

CAS 8005-44-5; 67762-27-0; EINECS/ELINCS 267-008-6; 267-009-1

Synonyms: Alcohols, C16-18; n-Alkanol (C16-C18); C16-18 alcohols; Cetostearyl alcohol; Cetyl-stearyl alcohol; Stearyl-cetyl alcohol

Definition: Mixture of fatty alcohols, predominantly cetyl and stearyl alcohols

Empirical: $C_{18}H_{38}O$

Formula: $CH_3(CH_2)_nOH$, n = 15-17

Properties: Wh. unctuous flakes or gran., faint char. odor, bland mild taste; sol. in alcohol, ether; insol. in water; m.w. 270.50; m.p. 48-55 C; acid no. ≤ 2; iodine no. ≤ 4; hyd. no. 208-228; nonionic

Toxicology: TSCA listed

Storage: Preserve in well-closed containers

Uses: Surfactant intermediate; raw material for ethoxylation, sulfation, etc.; stabilizer in emulsion polymerization; lubricant in rigid PVC; emollient, emulsifier, emulsion stabilizer, opacifier, visc. control agent in cosmetics; cosolvent, emollient, consistency agent for pharmaceutical creams, hand lotions, bath oils, shaving creams; pearlescent shampoo ingred.; antiseptic in topical anti-infective prods.; in food-pkg. adhesives

Regulatory: FDA 21CFR §175.105; FDA approved for orals, otics, topicals; NF, BP, EP compliance

Manuf./Distrib.: A&E Connock; Aarhus Oliefabrik A/S; Ashland; Atofina Bruxelles; Avatar; Chemacon GmbH; Chesham Chems. Ltd; Cognis; Croda Inc; Efkay Chems. Ltd; Eggar & Co.; Mosselman NV; Rhodia; Ruger; Sea-Land; Spectrum Quality Prods.; Surfachem Ltd; Univar E&E Ltd; Westco; Whyte Chems. Ltd

Cetearyl ethylhexanoate. *See* Cetearyl octanoate

Cetearyl isononanoate

CAS 84878-33-1; EINECS/ELINCS 284-424-3

Synonyms: Isononanoic acid, cetearyl ester; Isononanoic acid, cetyl/stearyl ester

Definition: Ester of cetearyl alcohol and a branched chain of nanoic acid

Uses: Emollient, skin softener, moisture retention aid for cosmetics, hair care, skin care, pharmaceutical skin care, massage, and sun protection preps.; solvent for lipophilic ingreds.; solubilizer for act. ingreds.; lubricant for metalworking

Trade Names: Tegosoft® CI

Cetearyl octanoate

CAS 59130-70-7; EINECS/ELINCS 261-620-7

Synonyms: Cetearyl ethylhexanoate; Cetyl-stearyl 2-ethyl hexanoate; Cetyl/stearyl octanoate; 2-Ethylhexanoic acid, cetyl-stearyl ester; Hexanoic acid, 2-ethyl-, C15-18-alkyl esters

Definition: Ester of cetearyl alcohol and octanoic acid

Uses: Emollient, emulsifier, thickener, stabilizer, opacifier, pearlescent, waterproofing agent in cosmetics, skin care preps., pharmaceuticals

Manuf./Distrib.: A&E Connock

Trade Names Containing: Crodamol CAP

Ceteth-16

CAS 9004-95-9 (generic)

Synonyms: PEG-16 cetyl ether; POE (16) cetyl ether

Classification: Nonaromatic ether

Definition: PEG ether of cetyl alcohol

Formula: $CH_3(CH_2)_{14}CH_2(OCH_2CH_2)_nOH$, avg. n = 16

Properties: Nonionic

Toxicology: TSCA listed

Uses: Surfactant, emulsifier, lubricant in cosmetics and pharmaceuticals; defoamer in food-contact paper coatings

Regulatory: FDA 21CFR §176.200

Manuf./Distrib.: Aldrich; Fluka; Sigma

Trade Names Containing: Solulan® 16

Ceteth-24

CAS 9004-95-9 (generic)

Synonyms: PEG-24 cetyl ether; POE (24) cetyl ether

Classification: Nonaromatic ether

Definition: PEG ether of cetyl alcohol

Formula: $CH_3(CH_2)_{14}CH_2(OCH_2CH_2)_nOH$, avg. n = 24

Properties: Nonionic

Toxicology: TSCA listed

Uses: Surfactant, emulsifier, cleaner, dispersant for cosmetics and pharmaceuticals, esp. creams and lotions; defoamer in food-contact paper coatings; food-contact textiles

Regulatory: FDA 21CFR §176.200, 177.2800

Manuf./Distrib.: Aldrich; Fluka; Sigma

Trade Names Containing: Solulan® C-24

Cetin. *See* Cetyl palmitate
Cetostearyl alcohol. *See* Cetearyl alcohol

Cetyl acetate

CAS 629-70-9; EINECS/ELINCS 211-103-7

Synonyms: 1-Acetoxyhexadecane; 1-Hexadecanol, acetate; Hexadecyl acetate; Palmityl acetate

Definition: Ester of cetyl alcohol and acetic acid

Empirical: $C_{18}H_{36}O_2$

Formula: $CH_3COOCH_2(CH_2)_{14}CH_3$

Properties: M.w. 284.54

Toxicology: LD50 (oral, rat) > 5 g/kg, (skin, rabbit) > 5 g/kg; sl. toxic by ing. and skin contact; primary skin irritant; TSCA listed

Hazardous Decomp. Prods.: Heated to decomp., emits acrid smoke and irritating vapors

Uses: Emollient in cosmetics, pharmaceutical topicals

Manuf./Distrib.: Sigma

Trade Names Containing: Acetol® 1706; Acetulan®; Fancol™ ALA; Ritawax AEO; Ritawax ALA

Cetyl alcohol

CAS 124-29-8; 36653-82-4; EINECS/ELINCS 253-149-0
FEMA 2554

Synonyms: Alcohol C16; C16 linear primary alcohol; Ethal; Ethol; 1-Hexadecanol; Hexadecyl alcohol; Palmityl alcohol

Classification: Fatty alcohol

Empirical: $C_{16}H_{34}O$

Formula: $CH_3(CH_2)_{14}CH_2OH$

Properties: Wh. waxy solid; partially sol. in alcohol, chloroform, ether; pract. insol. in water; m.w. 242.27; dens. 0.8176 (49.5 C); m.p. 49.3 C; b.p. 344 C; acid no. ≤ 2; iodine no. ≤ 5; hyd. no. 218-238; flash pt. > 110 C; ref. index 1.4283; surface-active dispersant, wetting agent, and compatibility aid for enhancing color acceptance in high solids coat-

ings (alkyds, epoxies, UV and EB-cured coatings); emulsifier for emulsion polymerization
Toxicology: LD50 (oral, rat) 6400 mg/kg; mod. toxic by ingestion, intraperitoneal routes; eye and human skin irritant; can cause hives; TSCA listed
Precaution: Flamm. when exposed to heat or flame; can react with oxidizing materials
Hazardous Decomp. Prods.: Heated to decomp., emits acrid smoke and fumes
HMIS: Health 1, Flammability 0, Reactivity 1
Uses: Perfumery; emulsifier; emollient; coupling agent; surfactant; dispersant; wetting agent; foam stabilizer in detergents; opacifier; thickener; chemical intermediate; external mold release for acrylics; raw material for antioxidants; emulsion polymerization; color enhancer in high-solids coatings; flavoring agent, color diluent, intermediate for confectionery, food supplements in tablet form, gum; cosolvent; pharmaceutic acid (emulsifier, stiffener); food-contact textiles; emollient, emulsifier, opacifier, visc. control agent in cosmetics; in food-pkg. adhesives; in food-contact coatings; defoamer in food-contact paper coatings; in cellophane for food pkg.; in surf. lubricants for mfg. of food-contact metallic articles
Regulatory: FDA 21CFR §73.1, 73.1001, 172.515, 172.864, 175.105, 175.300, 176.200, 177.1010, 177.1200, 177.2800, 178.3480, 178.3910; FEMA GRAS; FDA approved for ophthalmics, orals, otics, rectals, topicals; NF, BP, EP compliance
Manuf./Distrib.: A&E Connock; AMC Chems.; AMRESCO; Aarhus Oliefabrik A/S; Aceto; Albemarle; Aldrich; Alfa Aesar; Allchem Ind.; Amerchol; Ashland; Avatar; Brown; C.P. Hall; Camida Ltd; Chemacon GmbH; Chemron; Chesham Chems. Ltd; Cognis/Chems. Group; Cornelius Chem. Co. Ltd; Costec; Croda Chem. Ltd; Croda Inc; E-Chemicals; EM Ind.; Efkay Chems. Ltd; Eggar & Co.; Fluka; GMI Prods.; Gattefosse; Indofine; Integra; Jarchem Ind.; Kimpton Bros. Ltd; Kraft Chem.; Lipo; Lonza; M. Michel; Merck KGaA; Mosselman NV; Multi-Kem; Mutchler; Norman, Fox; Pechiney Chems. Div.; Penta Mfg.; Phoenix Nat. Prods.; Procter & Gamble; Protameen; R.I.T.A.; R.W. Greeff; RTD Chems.; Rhodia HPCII; Robeco; Ruger; Salim Oleochems.; Sasol Germany; Sasol N. Am.; Sea-Land; Sigma; Spectrum Quality Prods.; Stepan; Tomen Am.; Univar E&E Ltd; Universal Preserv-A-Chem; Vopak USA; Whyte Chems. Ltd
Trade Names: Cachalot® C-50
Trade Names Containing: Alfol® 14; Witcohol™ 42

Cetyl 2-ethylhexanoate. *See* Cetyl octanoate

Cetyl lactate

CAS 35274-05-6; EINECS/ELINCS 252-478-7
Synonyms: 1-Hexadecanol lactate; n-Hexadecyl-2-hydroxypropanoate; Hexadecyl lactate; n-Hexadecyl lactate; 2-Hydroxypropanoic acid hexadecyl ester
Definition: Ester of cetyl alcohol and lactic acid
Empirical: $C_{19}H_{38}O_3$
Formula: $CH_3CHOHCOO(CH_2)_{15}CH_3$
Properties: Waxy solid; sol. in oil, ethanol, IPM, oleyl alcohol; partly sol. in water, propylene glycol; m.w. 314.49; m.p. 41 C; b.p. 132 C (0.1 mm), 170 C (1 mm); ref. index 1.4410; nonionic
Toxicology: LD50 (oral, rat) > 20 ml/kg; no known toxicity; nonirritating to eyes and skin; TSCA listed
Uses: Emollient, lubricant for cosmetics and pharmaceuticals; binder for pressed powds.; solvent for lipstick dyes; wetting agent; pigment dispersant
Manuf./Distrib.: A&E Connock; Am. Biorganics
Trade Names: Ceraphyl® 28

Cetyl octanoate

CAS 59130-69-7; EINECS/ELINCS 261-619-1
Synonyms: Cetyl 2-ethylhexanoate; 2-Ethylhexanoic acid, hexadecyl ester; Hexadecyl 2-ethylhexanoate; n-Hexadecyl 2-ethylhexanoate; Hexanoic acid, 2-ethyl-, hexadecyl ester; Perceline oil
Definition: Ester of cetyl alcohol and 2-ethylhexanoic acid
Empirical: $C_{24}H_{48}O_2$
Formula: $CH_3(CH_2)_3CH(CH_2CH_3)COOCH_2(CH_2)_{14}CH_3$
Properties: Clear liq.; m.w. 368.72
Toxicology: Primary irritant; TSCA listed
Uses: Emollient, softener, lubricant, gloss aid for cosmetics; solvent for makeup removers; emollient, skin softener, moisture retention aid, emulsifier, thickener, stabilizer, opacifier, pearlescent in creams/lotions, topical pharmaceuticals
Manuf./Distrib.: A&E Connock
Trade Names: Schercemol CO
Trade Names Containing: Super Refined® Crodamol SCO

Cetyl palmitate

CAS 540-10-3; EINECS/ELINCS 208-736-6
Synonyms: Cetin; Hexadecanoic acid, hexadecyl ester; n-Hexadecyl hexadecanoate; Hexadecyl palmitate; Palmitic acid, hexadecyl ester; Palmityl palmitate
Definition: Ester of cetyl alcohol and palmitic acid
Empirical: $C_{32}H_{64}O_2$
Formula: $C_{15}H_{31}COOC_{16}H_{33}$
Properties: Wh. cryst. waxy solid; sol. in abs. alcohol, ether, oxygenated solvs.; pract. insol. in water; m.w. 480.86; dens. 0.832; m.p. 50 C; b.p. 360 C; acid no. max.; iodine no. 5 max.; sapon. no. 110-130; ref. index 1.4398; nonionic
Toxicology: Nontoxic; skin irritant; TSCA listed
Precaution: Combustible
Hazardous Decomp. Prods.: Heated to decomp., emits acrid smoke and irritating vapors
HMIS: Health 1, Flammability 0, Reactivity 0
Uses: Mfg. of candles and soaps; internal processing lubricant for PVC; internal lubricant and binder in pressed powds.; in metalworking lubricant coatings; pharmaceutical ointment base, excipient; emollient, thickener, visc. builder, consistency agent for pharmaceuticals and cosmetics; base for cosmetic stick prods.; w/o emulsifier, lubricant for metalworking fluids
Regulatory: FDA approved for topicals
Manuf./Distrib.: A&E Connock; Aldrich; Alzo; Ashland; CoKEM Assoc.; Cognis/Care Chems.; Croda Inc; Delta Distributors; Eastech; Fanning; Jarchem Ind.; Koster Keunen; Kraft Chem.; Lipo; Milsolv; Mosselman NV; P.B. & S.; Penta Mfg.; RTD Chems.; Robeco; Ruger; SOCO-Lynch; Scher; Sea-Land; Sigma; Southchem; Spectrum Quality Prods.; Stepan; Textile Chem.; Thornley; Trivent; Universal Preserv-A-Chem; Werner G. Smith; Whittaker, Clark & Daniels
Trade Names: Palmil C

Cetyl stearate

CAS 1190-63-2; EINECS/ELINCS 214-724-1
Synonyms: 1-Hexadecyloctadecanoate; Hexadecyl stearate; n-Hexadecyl stearate; Octadecanoic acid, cetyl ester; Octadecanoic acid, 1-hexadecyl ester; Palmityl stearate; Stearic acid, cetyl ester; Stearic acid, hexadecyl ester; Stearic acid, palmityl ester
Definition: Ester of cetyl alcohol and stearic acid
Empirical: $C_{34}H_{68}O_2$
Formula: $CH_3(CH_2)_{16}COO(CH_2)_{15}CH_3$
Properties: Wh. solid; m.w. 508.91; m.p. 56 C
Uses: Emollient, lubricant for cosmetics; conditioner, opacifier, gloss aid for hair care prods.; antistat, processing lubricant for plastics; plasticizer, lubricant in food-contact PS
Regulatory: FDA 21CFR §178.3450
Manuf./Distrib.: A&E Connock; Koster Keunen; Mosselman NV; Pechiney Chems. Div.; Sigma
Trade Names: Monestriol C

Cetyl-stearyl alcohol. *See* Cetearyl alcohol
Cetyl-stearyl 2-ethyl hexanoate; Cetyl/stearyl octanoate. *See* Cetearyl octanoate
CFC 11. *See* Trichlorofluoromethane
CFC 12. *See* Dichlorodifluoromethane
CFC 13. *See* Chlorotrifluoromethane
CFC 22. *See* Chlorodifluoromethane
CFC 112. *See* 1,1,2,2-Tetrachloro-1,2-difluoroethane
CFC 113. *See* Trichlorotrifluoroethane
CFC-113a. *See* 1,1,1-Trichloro-2,2,2-trifluoroethane
CFC 114. *See* 1,2-Dichlorotetrafluoroethane
CFC 115. *See* Chloropentafluoroethane
CHA. *See* Cyclohexylamine
Chinese bean oil. *See* Soybean (Glycine soja) oil

Chinoleine; Chinolin; Chinoline. *See* Quinoline
Chloracetic chlroide. *See* Chloracetyl chloride

Chloracetyl chloride

CAS 79-04-9
UN 1752
Synonyms: CAC; Chloracetic chlroide; Chloroacetic acid chloride; Chloroacetyl chloride; Monochloroacetyl chloride
Empirical: $C_2H_2Cl_2O$
Formula: $ClCH_2COCl$
Properties: Colorless to sl. yel. liq.; strong pungent odor; insol. in water, dec.; m.w. 112.94; sp.gr. 1.41; dens. 11.7 lb/gal; vapor pressure 19 mm Hg (20 C); f.p. -22 C; b.p. 106 C; flash pt. none
Toxicology: NIOSH REL TWA 0.05 ppm; toxic; corrosive; may cause eye irritation/damage, severe skin burns; ing. of lg. amts. can be fatal; inh. may irritate respiratory tract, nasal passages, lungs; may cause coughing, wheezing, dyspnea; lachrymator
Environmental: Toxic to animals incl. fish; keep away from sewers and all water resources
Precaution: Dec. in water to form monochloroacetic acid and hydrochloric acid; incompat. with water, alcohols, bases, metals (corrosive), amines
Storage: Moisture-sensitive
Uses: Chemical intermediate; solvent in chemical processing, synthesis of pharmaceuticals, herbicides, paper modifiers, plastic additives
Manuf./Distrib.: Acros Org.; Alfa Aesar; Bayer; Dow; Lancaster Synthesis; Pfaltz & Bauer; SF Chem; TCI Am.

o-Chloraniine. *See* o-Chloroaniline
Chlorbenzene; Chlorbenzol. *See* Chlorobenzene
Chlorcosane. *See* Paraffin, chlorinated
Chlor-ethamine. *See* Ethylenediamine dihydrochloride
Chlorethyl. *See* Ethyl chloride
Chlorhydrin; α-Chlorhydrin. *See* Glyceryl-α-monochlorohydrin
Chlorinated hydrochloric ether. *See* 1,1-Dichloroethane
Chlorinated paraffin. *See* Paraffin, chlorinated
α-Chlornaphthalene. *See* 1-Chloronaphthalene
Chloroacetic acid chloride; Chloroacetyl chloride. *See* Chloracetyl chloride
2-Chloroaniline. *See* o-Chloroaniline

o-Chloroaniline

CAS 95-51-2; EINECS/ELINCS 202-426-4
UN 2018; UN 2019
Synonyms: 1-Amino-2-chlorobenzene; 2-Amino-1-chlorobenzene; o-Aminochlorobenzene; o-Chloraniine; 2-Chloroaniline; 2-Chlorobenzenamine
Empirical: C_6H_6ClN
Formula: $ClC_6H_4NH_2$
Properties: Liq.; sol. in most org. solvs., aq. acids; pract. insol. in water; m.w. 127.58; dens. 1.212; m.p. 3 C; b.p. 95 C (11 mm); ref. index 1.589
Toxicology: LD50 (oral, mouse) 256 mg/kg, (skin, cat) 222 mg/kg; LDLo (subcut., cat) 310 mg/kg; poison by skin contact, ing., and subcut. routes; tumorigen; mutagen; TSCA listed
Hazardous Decomp. Prods.: Heated to decomp., emits toxic fumes of Cl^- and NO_x
Storage: Photosensitive; sensitive to air; keep under argon
Uses: Dye intermediate; standards for colorimetric apparatus; mfg. of petroleum solvents and fungicides; pharmaceutical intermediate
Manuf./Distrib.: Aceto; Aldrich; DuPont; Fluka; Supelco

Chloroben. *See* o-Dichlorobenzene
2-Chloro-benzenamine. *See* o-Chloroaniline

Chlorobenzene

CAS 108-90-7; EINECS/ELINCS 203-628-5
UN 1134 (DOT)
Synonyms: Benzene chloride; Benzene monochloride; Chlorbenzene; Chlorbenzol; Chlorobenzol; MCB; Monochlorbenzene; Monochlorobenzene; Phenyl chloride
Classification: Halogenated aromatic hydrocarbon
Empirical: C_6H_5Cl
Properties: Clear volatile liq., faint almond-like not unpleasant odor; sol. in alcohol, benzene, chloroform, ether; misc. with most org. solvs.; insol. in water; m.w. 112.50; sp.gr. 1.105; vapor pressure 10 mm Hg (22.2 C); m.p. -45 C; b.p. 131.6 C; flash pt. (CC) 29.4 C; ref. index 1.5
Toxicology: ACGIH TLV/TWA 10 ppm; LD50 (oral, rat) 2910 mg/kg; toxic; poison by ing., mod. toxic by IP route; strong narcotic; inh. causes CNS depression, headaches, nausea, dizziness, drowsiness, confusion; experimental teratogen, reproductive effects; mutagenic data; may cause kidney/liver damage; TSCA listed
Precaution: Flamm.; moderate fire risk; explosive limits 1.8-9.6% in air; incompat. with oxidizers, nitric acid, sodium, dimethyl sulfoxide, silver perchlorate; fire/explosion risk; potential violent reactions
NFPA: Health 2, Flammability 3, Reactivity 0
Storage: Store in cool, dry, well-ventilated area, out of direct sunlight, away from heat/ignition sources, away from corrosives
Uses: Pesticide intermediate; mfg. of phenol, chloronitrobenzene, aniline, DDT, silicone resins, dyes, perfumes, polysulfone; solvent carrier for methylene diisocyanate; solvent for paints, pesticides, degreasing processes, pharmaceuticals; heat transfer medium; adhesive rubbers and adhesives; primer or adhesive additive for adhering elastomeric coatings to syn. fiber fabrics coated with syn. rubber; surf. coating; dry cleaning; intermediate for halogen synthesis; in food-pkg. adhesives; adjuvant for PC food-pkg. resins; solvent in food-contact polysulfone resins
Regulatory: FDA 21CFR §175.105, 177.1580, 177.1655; SARA reportable
Manuf./Distrib.: Aarti Ind. Ltd; Aldrich; Allchem Ind.; Atofina SA; Bayer/Fiber, Addits., Rubber; Esprit; Fluka; Monsanto; PPG Ind.; Sigma; Solutia; Stanchem; Supelco; Tekchem

Chlorobenzol. *See* Chlorobenzene
2-Chlorobenzotrifluoride. *See* o-Chlorobenzotrifluoride
4-Chlorobenzotrifluoride. *See* p-Chlorobenzotrifluoride

o-Chlorobenzotrifluoride

CAS 88-16-4; EINECS/ELINCS 201-805-1
UN 2234
Synonyms: Benzene, 1-chloro-2-(trifluoromethyl)-; 2-Chlorobenzotrifluoride; 1-Chloro-2-(trifluoromethyl) benzene; 2-Chloro (trifluoromethyl) benzene; o-Chlorotrifluoromethylbenzene; 2-Chloro-α,α,α-trifluorotoluene; o-Chloro-α,α,α-trifluorotoluene; Toluene, o-chloro-α,α,α-trifluoro-; o-(Trifluoromethyl) chlorobenzene; o-Trifluoromethylphenyl chloride
Empirical: $C_7H_4ClF_3$
Formula: $ClC_6H_4CF_3$
Properties: Colorless clear liq.; aromatic odor; sol. in DMSO, 95% ethanol, acetone; sol. < 1 mg/ml in water; m.w. 180.56; sp.gr. 1.379 (15.5/15.5 C); vapor pressure 5 mm Hg (24.7 C); m.p. -6 C; b.p. 152 C; flash pt. 58 C; ref. index 1.4550 (20 C)
Toxicology: TLV 2.5 mg(F)/m^3 of air; severe irritant and extremely destructive to tissue of eyes, skin, mucous membranes, upper respiratory tract; harmful by ing., inh., skin absorp.; may cause burning sensation, coughing, wheezing, laryngitis, shortness of breath, headache, nausea, vomiting; prolonged exposure may cause lung irritation, chest pain, edema, death; TSCA listed
Precaution: Combustible; incompat. with strong bases, strong oxidizing agents
Hazardous Decomp. Prods.: Heated to decomp., emits toxic fumes of CO, CO_2, hydrogen fluoride, hydrogen chloride gas; emits toxic fumes under fire conditions
Storage: Store in a cool, dry place; keep tightly closed; keep away from heat and open flame
Uses: Dye intermediate; chemical intermediate; solvent; dielectric fluid
Manuf./Distrib.: Acros Org.; Aldrich; Alfa Aesar; Asahi Glass; Esprit; Fluorochem USA; ICN Biomed. Research Prods.; Lancaster Synthesis; Miteni SpA; Oakwood Prods.; Pfaltz & Bauer; TCI Am.; Wychem Ltd

p-Chlorobenzotrifluoride

CAS 98-56-6; EINECS/ELINCS 202-681-1
UN 2234 (DOT)
Synonyms: Benzene, 1-chloro-4 (trifluoromethyl); 4-Chlorobenzotrifluoride; (p-Chlorophenyl) trifluoromethane; 1-Chloro-4-trifluoromethylbenzene; 4-Chlorotrifluoromethylbenzene; p-Chlorotrifluoromethylbenzene;

4-Chloro-α,α,α-trifluorotoluene; p-Chloro-α,α,α-trifluorotoluene; 1-Chloro-4-(trimethyl)benzene; Parachlorobenzotrifluoride; PCBTF; Toluene, p-chloro-α,α,α,-trifluoro; α,α,α-Trifluoro-4-chlorotoluene; p-(Trifluoromethyl) chlorobenzene; p-Trifluoromethylphenyl chloride
Classification: Aromatic halogenated hydrocarbon
Empirical: $C_7H_4ClF_3$
Properties: Water-wh. liq.; aromatic odor; m.w. 180.56; dens. 1.353; f.p. -36 C; m.p. -33.2 C; b.p. 138.7 C; flash pt. (CC) 116 F; ref. index 1.4460
Toxicology: TLV 2.5 mg(F)/m^3 of air; LD50 (oral, rat) 13 g/kg; LC50 (inh., rat) 22 g/m^3; mildly toxic by inh. or ing.; irritating to eyes, skin, respiratory system; human mutagenic data; TSCA listed
Environmental: Nonozone depleting
Precaution: Flamm.; mod. fire risk; strong exothermic reactions with sodium dimethylsulfinate
Hazardous Decomp. Prods.: Heated to decomp., emits toxic fumes of F^- and Cl^-
Uses: Solvent for industrial solvent cleaning, aerosols, adhesives, coatings, inks, electronic applics.; 1,1,1-trichloroethane alternative; dye intermediate; chemical intermediate; dielec. fluid
Manuf./Distrib.: Aldrich; Fluka; Occidental; Sigma
Trade Names: Oxsol® 100

Chlorobromomethane. *See* Bromochloromethane
1-Chlorobutane. *See* Butyl chloride
1-Chloro-2-(β-chloroethoxy) ethane. *See* sym-Dichloroethyl ether
Chlorocosane. *See* Paraffin, chlorinated
Chloroden. *See* o-Dichlorobenzene
Chlorodeoxyglycerol. *See* Glyceryl-α-monochlorohydrin
1-Chloro-2,2-dichloroethylene. *See* Trichloroethylene
Chlorodifluoroethane. *See* 1,1,1-Chlorodifluoroethane

1,1,1-Chlorodifluoroethane

CAS 75-68-3; EINECS/ELINCS 200-891-8
UN 2517 (DOT)
Synonyms: Chlorodifluoroethane; 1-Chloro-1,1-difluoroethane; Chlorodifluoroethanes; Chloroethylidene fluoride; α-Chloroethylidene fluoride; Chlorofluorocarbon 142b; 1,1,1-Difluorochloroethane; 1,1-Difluoro-1-chloroethane; Difluorochloroethanes; Difluoromonochloroethane; Ethane, 1-chloro-1,1-difluoro-; FC 142b; Fluorocarbon 142b; Freon 142; Freon 142b; HCFC 142b; Hydrochlorofluorocarbon 142B; Propellant 142b; R 142b
Classification: Halocarbon
Empirical: $C_2H_3ClF_2$
Formula: CH_3CClF_2
Properties: Colorless clear liquefied gas; sl. ethereal odor; insol. in water; m.w. 100.50; dens. 1.194 (-9 C); m.p. -131 C; b.p. -9.2 C; sp. heat 1.34 kJ/kg °C (liq., 30 C)
Toxicology: LC50 (inh., rat, 4 h) 2050 g/m^3; very mildly toxic by inh.; mutagenic data; TSCA listed
Precaution: Flamm. gas; very dangerous fire hazard exposed to heat, flame, oxidizers; can react vigorously with oxidizing materials
Hazardous Decomp. Prods.: Heated to decomp., emits toxic fumes of F^- and Cl^-
Uses: Refrigerant; solvent; intermediate; aerosol propellant; blowing agent in PU and styrene foams, PE; working fluid in high-temp. air conditioner, heat pump, and temp. control applics.
Regulatory: FDA 21CFR §701.30
Manuf./Distrib.: Aldrich; Atofina; Zhejiang Sanhuan Chems. Co. Ltd

1-Chloro-1,1-difluoroethane; Chlorodifluoroethanes. *See* 1,1,1-Chlorodifluoroethane

Chlorodifluoromethane

CAS 75-45-6; EINECS/ELINCS 200-871-9
UN 1018 (DOT)
Synonyms: CFC 22; Chlorofluorocarbon 22; Difluorochloromethane; Difluoromonochloromethane; F 22; FC 22; Fluorocarbon 22; Freon 22; HCFC 22; Hydrochlorofluorocarbon 22; Monochlorodifluoromethane; Propellant 22; R 22; Refrigerant 22
Classification: Halogenated aliphatic hydrocarbon
Empirical: $CHClF_2$
Formula: $ClCHF_2$
Properties: Colorless gas or liquefied compressed gas; nearly odorless to sl. sweetish odor; sol. in ether, acetone, chloroform; partly sol. in water; m.w. 86.47; sp.gr. 1.209 (liq., 21 C), 4.82 g/l (gas at b.p.); vapor pressure 4279 mm Hg (4 C); f.p. -160 C; b.p. -40.8 C; ref. index 1.264; surf. tens. 8 dynes/cm; dielec. const. 6.11; noncombustible
Toxicology: ACGIH TLV/TWA 1000 ppm; STEL 1250 ppm; LC50 (inh., mouse, 20 min) 28 pph; LCLo (inh., dog) 70 pph; asphyxiant; low toxicity by inh.; high concs. may cause CNS effects; massive concs. may cause unconsciousness, irregular heartbeat, cardiac arrest; liq. can cause frostbite to skin and eyes; experimental reproductive effects; mutagenic data; TSCA listed
Environmental: ThOD 0.19
Precaution: Compressed gas; 50% mixts. with air are combustible @ elevated pressures, although ignition is difficult; incompat. with reactive metals, e.g., sodium, potassium, powd. aluminum (can react violently or explosively)
Hazardous Decomp. Prods.: Heated to decomp., emits toxic fumes of F^- and Cl^-
Storage: Store cylinders in cool, dry, well-ventilated area away from corrosive atmospheres, out of direct sunlight; leaked gas can accumulate in low areas
Uses: Refrigerant; low-temp. solvent; aerosol propellant; blowing agent; intermediate for mfg. of fluorocarbon resins, esp. tetrafluoroethylene polymers
Regulatory: FDA 21CFR §701.30; SARA §311/312 acute health hazard
Manuf./Distrib.: Air Prods.; Aldrich; Atofina; Fluka; LaRoche Ind.; MG Ind.; Matheson Gas Prods.; Sigma; Tech Spray; UOP

1-Chloro-2,3-dihydroxypropane; 3-Chloro-1,2-dihydroxypropane. *See* Glyceryl-α-monochlorohydrin
2-Chloro-N,N-dimethylacetamide. *See* Acetyldimethylamine chloride
1-Chloro-2,3-epoxypropane; 3-Chloro-1,2-epoxypropane. *See* Epichlorohydrin
Chloroethane. *See* Ethyl chloride
Chloroethanol; 2-Chloroethanol; 2-Chloro-1-ethanol; β-Chloroethanol. *See* Ethylene chlorohydrin
Chloroethene. *See* 1,1,1-Trichloroethane
2-Chloroethyl alcohol; β-Chloroethyl alcohol. *See* Ethylene chlorohydrin
Chloroethyl ether. *See* sym-Dichloroethyl ether
Chloroethylidene fluoride; α-Chloroethylidene fluoride. *See* 1,1,1-Chlorodifluoroethane
Chlorofluorocarbon 22. *See* Chlorodifluoromethane
Chlorofluorocarbon 113. *See* Trichlorotrifluoroethane
Chlorofluorocarbon 142b. *See* 1,1,1-Chlorodifluoroethane

Chloroform

CAS 67-66-3; EINECS/ELINCS 200-663-8
UN 1888 (DOT)
Synonyms: Formyl trichloride; Methane trichloride; Methane, trichloro-; Methenyl chloride; Methenyl trichloride; Methyl trichloride; Trichloroform; Trichloromethane
Classification: Halogenated aliphatic hydrocarbon
Empirical: $CHCl_3$
Properties: Colorless clear mobile liq., ethereal odor, sweet burning taste; misc. with alcohol, benzene, hexane, acetone, diethyl ether, naphtha, fixed and volatile oils; sl. sol. in water; m.w. 119.38; sp.gr. 1.474-1.478; vapor pressure 197 mm Hg; m.p. -63 C; b.p. 61 C; flash pt. none; ref. index 1.4460
Toxicology: ACGIH TLV/TWA 10 ppm; LDLo (oral, human) 2514 mg/kg; human poison by ing., inh.; experimental poison by IV; mod. toxic by IP, subcut. routes; prolonged skin contact may produce irritation, burning; may be absorbed thru skin; vapor may cause stinging of eyes; respiratory and skin allergies; inh. affects CNS, heart, liver, kidneys; poisonous in large doses; very high doses can cause severe narcosis, respiratory or heart failure, liver/kidney damage; suspected carcinogen; TSCA listed
Environmental: VOC; ThOD 0.13
Precaution: DOT: Poisonous material; incompat. with strong bases, reactive metals, alkali metals, strong oxidizers, nitromethane, potassium t-butoxide; vigorous reactions, ignition, explosions possible
Hazardous Decomp. Prods.: Slowly decomp. to produce hydrogen chloride; reacts with oxidizers to form highly toxic phosgene gas, chlorine

NFPA: Health 2, Flammability 0, Reactivity 0
Storage: Store in cool, dry, well-ventilated area out of direct sunlight, away from corrosives; light-sensitive
Uses: Prod. of chlorodifluoromethane (a refrigerant and propellant), fluorocarbon plastics; extraction solvent for pharmaceuticals, nat. oils; solvent for perfume; analytical chemistry; intermediate in prep. of dyes, drugs, pesticides; fumigant; insecticides; ingred. in drugs and cosmetics; preservative, solvent, flavor in pharmaceuticals; in food-pkg. adhesives; adjuvant for PC food-pkg. resins
Features: Sweet pleasant flavor
Regulatory: FDA 21CFR §175.105, 177.1580, 177.1585; SARA §302a extremely hazardous substance, §311/312 acute health/chronic health hazard, §313 reportable; Calif. Prop. 65 reportable; HAP; BP compliance
Manuf./Distrib.: AMRESCO; Aldrich; Allchem Ind.; Atofina; BCH Brühl; Burdick & Jackson; Cross World Sales; Degussa AG; Dow; EM Science; Fisher Scientific; Fluka; Houghton Chem.; ICC Ind.; Integra; J.T. Baker; Mallinckrodt Baker; Occidental; Penta Mfg.; Roche Molecular Biochems.; Rochem Int'l.; Romil Ltd; Ruger; Sigma-Aldrich; Sigma; Solvay SA; Spectrum Quality Prods.; Synasia Fine Chem.; Thomas Scientific; VWR Int'l.; Vulcan Perf. Chems.
Trade Names Containing: Punctilious® SDA 20 190 Proof; Punctilious® SDA 20 Anhydrous

Chloroform, methyl-. *See* 1,1,1-Trichloroethane
Chlorohydrin; α-Chlorohydrin. *See* Glyceryl-α-monochlorohydrin
Chloromethane. *See* Methyl chloride
Chloromethylbenzene. *See* Monochlorotoluene
1-Chloro-2-methylbenzene; 2-Chloro-1-methylbenzene. *See* o-Chlorotoluene
4-Chloro-1-methylbenzene. *See* p-Chlorotoluene
Chloromethyl bromide. *See* Bromochloromethane
1-Chloro-3-methylbutane. *See* Isoamyl chloride
(Chloromethyl) ethylene oxide; Chloromethyloxirane; 2-(Chloromethyl) oxirane. *See* Epichlorohydrin

1-Chloronaphthalene

CAS 90-13-1; EINECS/ELINCS 201-967-3
Synonyms: α-Chlornaphthalene; α-Chloronaphthalene; Chorinated naphthalene; Naphthalene, 1-chloro-; 1-Naphthyl chloride; Naphthylene, 1-chloro-
Empirical: $C_{10}H_7Cl$
Properties: Colorless to amber clear, oily, visc. liq.; sol. in DMSO, 95% ethanol, acetone, ether, benzene, petrol. ether; sol. < 1 mg/ml in water; m.w. 162.62; sp.gr. 1.1938 (20/4 C); m.p. -23 to -20 C; b.p. 263 C; vapor pressure 1 mm Hg (80.6 C); flash pt. 121 C; ref. index 1.6326 (20 C); volatile with steam
Toxicology: LD50 (oral, rat) 1540 mg/kg, (oral, mouse) 1091 mg/kg; mod. toxic by ing.; harmful by inh., ing., skin contact; irritating to eyes, skin, mucous membranes, respiratory system; may cause headache, fatigue, vertigo, anorexia; chronic exposure may cause chloracne; TSCA listed
Precaution: Combustible; incompat. with strong oxidizing agents
Hazardous Decomp. Prods.: Heated to decomp., emits toxic fumes of CO, CO_2, hydrogen chloride gas; emits toxic fumes under fire conditions
NFPA: Health 1, Flammability 1, Reactivity 0
Storage: Store refrigerated; keep tightly closed; keep away from oxidizing materials; store away from ignition sources
Uses: Immersion liq. in microscopic determination of refractive index of crystals; solvent for fats, oils, DDT; in prod. of insulation for electric cables/wires
Manuf./Distrib.: AccuStandard; Aldrich; Alfa Aesar; Fluka; ICN Biomed. Research Prods.; J.T. Baker; Oakwood Prods.; Pfaltz & Bauer; Spectrum Quality Prods.; Supelco; TCI Am.

2-Chloronaphthalene

CAS 91-58-7
UN 8027
Synonyms: β-Chloronaphthalene
Empirical: $C_{10}H_7Cl$
Properties: Off-wh. cryst. powd. or monoclinic plates; odorless; sol. in DMSO, 95% ethanol, acetone, benzene, chloroform, CS_2, ether; sol. < 1 mg/ml in water; m.w. 162.62; sp.gr. 1.1377 (71/4 C); m.p. 59.5 C; b.p. 256 C; ref. index 1.6079 (13 C); volatile with steam
Toxicology: LD50 (oral, rat) 2078 mg/kg, (oral, mouse) 886 mg/kg; mod. toxic by ing.; strong irritant; may be absorbed thru skin; may irritate eyes or skin; may cause dermatitis; chronic exposure to compds. of this class may cause chloracne, cysts, headache, fatigue, vertigo, anorexia, jaundice; TSCA listed
Precaution: Probably combustible; incompat. with strong oxidizing agents; may be sensitive to light
Hazardous Decomp. Prods.: Heated to decomp., emits toxic fumes of chlorides, HCl
Storage: Store refrigerated in sealed container; keep away from oxidizing materials; protect from light
Uses: Prod. of electric condensers; in insulation of electric cables/wires; additive to EP lubricants; supports for storage batteries; coating in foundry use; solvent; immersion liq. in microscopy
Manuf./Distrib.: AccuStandard; Acros Org.; Supelco; TCI Am.

α-Chloronaphthalene. *See* 1-Chloronaphthalene
β-Chloronaphthalene. *See* 2-Chloronaphthalene
Chloronitropropane. *See* 1-Chloro-1-nitropropane

1-Chloro-1-nitropropane

CAS 600-25-9
Synonyms: Chloronitropropane
Classification: Halogenated aliphatic nitro compd.; nitroparaffin
Empirical: $C_3H_6ClNO_2$
Formula: $CH_3CH_2CHClNO_2$
Properties: Colorless liq.; unpleasant odor; sol. in glycols, oils, chloroform, ethanol, diethyl ether; sl. sol. in water; m.w. 123.5; dens. 1.21 (20 C); b.p. 139-143 C; flash pt. (OC) 62 C
Toxicology: ACGIH TLV/TWA 2 ppm; LD50 (oral, mouse) 510 mg/kg, (subcut., mouse) 165 mg/kg; toxic; poison by ing., subcut. routes; mod. toxic by inh.; lachrymatory (vapor irritates eyes, causes tearing); causes injury to liver, kidneys, cardiovascular system
Precaution: Flamm. exposed to heat, flame, oxidizers; mod. explosive exposed to heat; incompat. with strong acids (may form highly toxic chloride fumes) and strong oxidizers (increased risk of fire and explosion)
Hazardous Decomp. Prods.: Highly toxic chloride fumes on contact with acids; heated to decomp., emits toxic fumes of Cl^- and NO_x
NFPA: , Flammability 2, Reactivity 3
Storage: Store in cool, dry, well-ventilated area, away from heat/ignition sources
Uses: Solvent and intermediate in chem. and syn. rubber industries; previously used as fungicide

Chloroparaffin. *See* Paraffin, chlorinated

Chloropentafluoroethane

CAS 76-15-3; EINECS/ELINCS 200-938-2
UN 1020 (DOT); INS945
Synonyms: CFC 115; F 115; Fluorocarbon 115; Freon 115; Monochloropentafluoroethane; R 115
Classification: Nonaromatic halogenated hydrocarbons
Empirical: C_2ClF_5
Formula: $CClF_2CF_3$
Properties: Colorless liquefied gas; odorless; sol. in alcohol, ether; insol. in water; m.w. 154.47; dens. 1.5678 (-42 C); vapor pressure 760 mm Hg; m.p. -106 C; b.p. -39.3 C; ref. index 1.290; surf. tens. 5 dynes/cm; dielec. const. 1.0; nonflamm.
Toxicology: ACGIH TLV/TWA 1000 ppm; mildly toxic by inhalation; TSCA listed
Environmental: ThOD 0.10
Hazardous Decomp. Prods.: Heated to decomp., emits toxic fumes of F^- and Cl^-
Storage: Store away from heat
Uses: Dielectric gas; propellant, aerating agent for foamed or sprayed food prods.; refrigerant gas component; solvent
Regulatory: FDA 21CFR §173.345; SARA §313 reportable
Manuf./Distrib.: Air Prods.; Aldrich; Matheson Gas Prods.

(p-Chlorophenyl) trifluoromethane. *See* p-Chlorobenzotrifluoride

1-Chloro-2,3-propanediol; 3-Chloropropanediol-1,2; 3-Chloropropane-1,2-diol; 3-Chloropropylene glycol. *See* Glyceryl-α-monochlorohydrin
Chloropropylene oxide; 3-Chloro-1,2-propylene oxide; γ-Chloropropylene oxide. *See* Epichlorohydrin
Chlorothene. *See* 1,1,1-Trichloroethane
Chlorotoluene. *See* Monochlorotoluene
2-Chlorotoluene. *See* o-Chlorotoluene
4-Chlorotoluene. *See* p-Chlorotoluene
ar-Chlorotoluene. *See* Monochlorotoluene

o-Chlorotoluene

CAS 95-49-8; EINECS/ELINCS 202-424-3
UN 2238 (DOT)
Synonyms: 1-Chloro-2-methylbenzene; 2-Chloro-1-methylbenzene; 2-Chlorotoluene; 1-Methyl-2-chlorobenzene; 2-Methylchlorobenzene; OCT; Orthochlorotoluene; o-Tolyl chloride
Definition: Major isomer of monochlorotoluene
Empirical: C_7H_7Cl
Properties: Colorless liq.; misc. with most org. solvs., alcohol, acetone, ether, benzene, chloroform, CCl_4, n-heptane; sl. sol. in water; m.w. 126.59; dens. 1.0776 (25/4 C); f.p. -35 C; b.p. 159.2 C; ref. index 1.5268 (20 C); volatile with steam
Toxicology: ACGIH TLV/TWA 50 ppm; LD50 (unreported rat) 5700 mg/kg; mod. toxic by unspecified routes; mildly toxic by inhalation; poison by ingestion; TSCA listed
Precaution: Flamm. liq.
Hazardous Decomp. Prods.: Heated to decomp., emits toxic fumes of Cl^-
Uses: Solvent and intermediate for organic chemicals and dyes
Manuf./Distrib.: Aldrich; Bayer; Fluka; Occidental; Sigma

p-Chlorotoluene

CAS 106-43-3
Synonyms: 4-Chloro-1-methylbenzene; 4-Chlorotoluene; Parachlorotoluene; PCT
Definition: Minor isomer in monochlorotoluene
Empirical: C_7H_7Cl
Formula: $CH_3C_6H_4Cl$
Properties: Colorless liq.; sol. in most org. solvs., alcohol, ether, acetone, benzene, chloroform; sl. sol. in water; m.w. 126.59; dens. 1.065-1.067 (25/15 C); f.p. 6.5 C; b.p. 162-166 C
Toxicology: Strong irritant; avoid inhalation
Uses: Solvent and intermediate for organic chemicals and dyes
Manuf./Distrib.: Austin; Charkit; Occidental; Pechiney Chems. Div.

Chlorotoluenes. *See* Monochlorotoluene

Chlorotrifluoromethane

CAS 75-72-9; EINECS/ELINCS 200-894-4
UN 1022 (DOT)
Synonyms: CFC 13; Freon 13; Monochlorotrifluoromethane; R 13; Trifluorochloromethane; Trifluoromethyl chloride; Trifluoromonochlorocarbon
Empirical: $CClF_3$
Formula: CF_3Cl
Properties: Colorless liquefied gas; ethereal odor; m.w. 104.5; sp.gr. 0.9224; vapor pressure 27,100 mm Hg; f.p. -181 C; b.p. -81.4 C; ref. index 1.199; surf. tens. 0.574 dynes/cm; dielec. const. 2.32; nonflamm.
Toxicology: Toxic by inhalation, sl. irritant; avoid contact and inh.; narcotic in high concs.; target organ: heart; TSCA listed
Environmental: VOC
Precaution: Violent reaction with Al
Hazardous Decomp. Prods.: Heated to decomp., emits highly toxic fumes of F^- and Cl^-
Uses: Ultra low temp. refrigerant; dielectric and aerospace chemical; hardening of metals; pharmaceutical processing; solvent
Manuf./Distrib.: Air Prods.; Aldrich; Archimica Florida; Atofina; Dow; MG Ind.; Matheson Gas Prods.

1-Chloro-2-(trifluoromethyl) benzene. *See* o-Chlorobenzotrifluoride
1-Chloro-4-trifluoromethylbenzene. *See* p-Chlorobenzotrifluoride
2-Chloro (trifluoromethyl) benzene. *See* o-Chlorobenzotrifluoride
4-Chlorotrifluoromethylbenzene. *See* p-Chlorobenzotrifluoride
o-Chlorotrifluoromethylbenzene. *See* o-Chlorobenzotrifluoride
p-Chlorotrifluoromethylbenzene. *See* p-Chlorobenzotrifluoride
2-Chloro-α,α,α-trifluorotoluene. *See* o-Chlorobenzotrifluoride
4-Chloro-α,α,α-trifluorotoluene. *See* p-Chlorobenzotrifluoride
o-Chloro-α,α,α-trifluorotoluene. *See* o-Chlorobenzotrifluoride
p-Chloro-α,α,α-trifluorotoluene; 1-Chloro-4-(trimethyl)benzene. *See* p-Chlorobenzotrifluoride
Chloryl. *See* Ethyl chloride

Choleth-24

CAS 27321-96-6 (generic)
Synonyms: PEG-24 cholesteryl ether; POE (24) cholesteryl ether
Definition: PEG ether of cholesterol with avg. ethoxylation value of 24
Properties: HLB 14.0; nonionic
Toxicology: TSCA listed
Uses: Emulsifier, solubilizer, dispersant, thickener in cosmetics, hair care, creams/lotions, pharmaceutical topicals
Regulatory: Approved for topicals
Trade Names Containing: Solulan® C-24

Chorinated naphthalene. *See* 1-Chloronaphthalene
1-Choropropane-2,3-diol. *See* Glyceryl-α-monochlorohydrin
CI 76000. *See* Aniline
CI 76500. *See* Pyrocatechol
CI 77403. *See* Copper oxide (ic)
Cinene. *See* d-Limonene; dl-Limonene
Cineol; 1,8-Cineol; Cineole; 1,8-Cineole. *See* Eucalyptol

Cinnamal

CAS 104-55-2; 14371-10-9; EINECS/ELINCS 203-213-9
FEMA 2286
Synonyms: Benzylideneacetaldehyde; Cassia aldehyde; Cinnamaldehyde; Cinnamic aldehyde; Cinnamyl aldehyde; Phenylacrolein; 3-Phenylpropenal; 3-Phenyl-2-propenal; 2-Propenal, 3-phenyl-
Classification: Aromatic aldehyde
Definition: A syn. liq. with strong cinnamon odor isolated from a wood-rotting fungus; found in Ceylon and Chinese cinnamon oils
Empirical: C_9H_8O
Formula: $C_6H_5CH{:}CHCHO$
Properties: Ylsh. oily liq., strong cinnamon odor, burning aromatic taste; misc. with oxygenated and chlorinated solvs., alcohol, ether, chloroform, fixed oils; very sl. sol. in water; m.w. 132.16; dens. 1.048-1.052; m.p. -7.5 C; b.p. 246 C; flash pt. 248 F; ref. index 1.619-1.623 (20 C); thickens on exposure to air
Toxicology: LD50 (oral, rat) 2220 mg/kg, (IP, mouse) 200 mg/kg, (IV, mouse) 75 mg/kg; poison by IV and parenteral routes; mod. toxic by ing. and IP routes; severe human skin irritant; one of the most common allergens; mutagenic data; TSCA listed
Precaution: Combustible liq.; may ignite after delay in contact with NaOH; volatile with steam
Hazardous Decomp. Prods.: Heated to decomp., emits acrid smoke and fumes
Uses: Synthetic flavoring agent in foods and pharmaceuticals; fragrance in perfumes; denaturant in cosmetics; topical sunscreens; ointments
Features: Cinnamon-, vanilla-like flavor
Regulatory: FDA 21CFR §182.60, GRAS; 27CFR §21.65, 21.151; FEMA GRAS; Japan approved as flavoring; FDA approved for orals
Manuf./Distrib.: Advanced BioTech; Aldrich; Allan; Augustus Oils Ltd; Axxence Aromatic GmbH; Degussa Flavors/Alfrebro; Doingcom; Elan; Fleurchem; Fluka; George Uhe; Goldensun Mfg.; Haarmann & Reimer GmbH; Indofine; Lluch Essence; MelChem; Moore Ingreds.; Nipa Labs; Noveon Kalama; Penta Mfg.; Prodasynth; Quest Int'l.; Treatt USA; Whyte Chems. Ltd
Trade Names Containing: Punctilious® SDA 38F-2 190 Proof Prod. Code 13645006

Cinnamaldehyde. *See* Cinnamal
Cinnamene; Cinnamenol. *See* Styrene
Cinnamic aldehyde. *See* Cinnamal
Cinnamol. *See* Styrene
Cinnamomum camphora. *See* Camphor
Cinnamomum camphora oil. *See* Camphor (Cinnamomum camphora) oil
Cinnamyl aldehyde. *See* Cinnamal
C7-8 isoalkanes. *See* C7-8 isoparaffin

C7-10 isoalkanes. *See* C7-10 isoparaffin
C8-9 isoalkanes. *See* C8-9 isoparaffin
C9-10 isoalkanes. *See* C9-10 isoparaffin
C9-11 isoalkanes. *See* C9-11 isoparaffin
C9-12 isoalkanes. *See* C9-12 isoparaffin
C9-13 isoalkanes. *See* C9-13 isoparaffin
C9-14 isoalkanes. *See* C9-14 isoparaffin
C9-15 isoalkanes. *See* C9-15 isoparaffin
C10-11 isoalkanes. *See* C10-11 isoparaffin
C10-13 isoalkanes. *See* C10-13 isoparaffin
C11-12 isoalkanes. *See* C11-12 isoparaffin
C11-13 isoalkanes. *See* C11-13 isoparaffin
C11-15 isoalkanes; C11-16 isoalkanes. *See* C11-16 isoparaffin
C12-14 isoalkanes. *See* C12-14 isoparaffin
C13-14 isoalkanes. *See* C13-14 isoparaffin
C13-16 isoalkanes. *See* C13-16 isoparaffin
C13-17 isoalkanes. *See* C13-17 isoparaffin
C20-40 isoalkanes. *See* C20-40 isoparaffin

C isoparaffin

CAS 64742-48-9 (generic)
Synonyms: Isoparaffin solvents
Definition: Mixt. of branched chain aliphatic hydrocarbons with unspecified carbons in the alkyl chain
Formula: C_nH_{2n+2}, n = 7-12
Properties: Dens. 0.70-0.78 kg/l; b.p. 97-249 C; flash pt. (TCC) 5-78 C
Uses: Catalyst carrier; electrostatic photocopying fluid; alkylene oxide entraining solvent; diluent for polyolefin prod.; solvent for aerosols, paints, varnishes
Trade Names: 2024 Naphtha; Shellsol® D60 SC

C7-8 isoparaffin

CAS 64742-48-9 (generic); EINECS/ELINCS 274-273-1
Synonyms: Alkanes, C7-8-iso-; C7-8 isoalkanes; Isoalkanes, C7-8
Definition: Mixt. of branched chain aliphatic hydrocarbons with 7-8 carbons in the alkyl chain
Environmental: Minimizes VOC
Uses: Solvent for coatings, odorless alkyd and acrylic paints, cosmetics; thinner; cleaner; defoamer in processing foods; in coatings on fruits/vegetables/egg shells; in froth flotation cleaning of vegetables; in insecticides for processed foods
Features: Extremely low surf. tens. to improve flow and wetting
Regulatory: FDA 21CFR §172.882, 173.340; virtually HAPs free

C7-10 isoparaffin

CAS 90622-56-3; EINECS/ELINCS 292-458-5
UN 1263
Synonyms: Alkanes, C7-10, iso-; C7-10 isoalkanes; Isoalkanes, C7-10
Classification: Isoparaffinic aliphatic hydrocarbon
Definition: Mixt. of branched chain aliphatic hydrocarbons with 7-10 carbons in alkyl chain
Properties: Colorless clear liq.; solvent-like odor; misc. with aliphatic hydrocarbons; insol. in water; sp.gr. 0.724; vapor pressure 1.6 kPa (20 C); m.p. -20 C; b.p. 113-143 C; flash pt. (Seta CC) 7 C; autoignition temp. > 450 C; VOC 724 g/l
Toxicology: REL 1900 mg/m^3 (vapor); may be harmful by ing., inh.; mod. toxic and narcotic in high concs.; may cause lung damage if swallowed; if aspirated, may cause chem. pneumonia, pulmonary edema; inh. of spray mist may cause severe respiratory tract irritation, coughing, choking, shortness of breath; defatting to skin; sl. eye irritant
Environmental: Readily biodeg.; prevent entry into sewers, low areas, confined areas; not a marine pollutant
Precaution: Flamm. liq.; combustible liq. and vapor; LEL 0.6%; UEL 8%; highly flamm. in presence of heat, sparks, open flames; flamm. in presence of oxidizing materials; sl. flamm. to flamm. in presence of combustible materials, org. materials; vapors may accumulate in low or confined areas, flash back; static charge can accumulate; reactive with oxidizing agents; may attack some plastics, rubber coatings
Storage: Store below 35 C in cool, well-ventilated area away from heat, sparks, flames, ignition sources, direct sunlight; keep container tightly closed
Uses: Solvent for paints, lacquers; diluent; propellant; heating agent; metal cleaning; metal degreaser
Trade Names: Isane IP 130

C8-9 isoparaffin

CAS 64742-48-9 (generic); EINECS/ELINCS 265-150-3
Synonyms: Alkanes, C8-9-iso-; Alkanes, iso-, C8-9; C8-9 isoalkanes; Isoalkanes, C8-9
Definition: Mixt. of branched chain aliphatic hydrocarbons with 8-9 carbons in alkyl chain
Uses: Solvent for coatings, cosmetics; defoamer in processing foods; in coatings on fruits/vegetables/egg shells; in froth flotation cleaning of vegetables; in insecticides for processed foods
Features: Extremely low surf. tens. to improve flow and wetting
Regulatory: FDA 21CFR §172.882, 173.340
Trade Names: Spirdane D30; Spirdane D30L; Spirdane L1

C9-10 isoparaffin

CAS 64742-48-9 (generic)
Synonyms: Alkanes, C9-10-iso-; C9-10 isoalkanes; Isoalkanes, C9-10
Definition: Mixt. of branched chain aliphatic hydrocarbons with 9-10 carbons in alkyl chain
Uses: Solvent
Trade Names: Spirdane D40L

C9-11 isoparaffin

CAS 68551-16-6; 64742-48-9 (generic); EINECS/ELINCS 271-365-3; 265-150-3
Synonyms: Alkanes, C9-11-iso-; C9-11 isoalkanes; Isoalkanes, C9-11
Definition: Mixt. of branched chain aliphatic hydrocarbons with 9-11 carbons in alkyl chain
Uses: Solvent in cosmetics; defoamer in processing foods; in coatings on fruits/vegetables/egg shells; in froth flotation cleaning of vegetables; in insecticides for processed foods
Regulatory: FDA 21CFR §172.882, 173.340
Trade Names: HC-DCF Low Flash™ Dry Cleaning Solvent; Orfom® SX 18; Soltrol® 100; Spirdane BT; Spirdane D40

C9-12 isoparaffin

CAS 90622-57-4; 64742-48-9 (generic); EINECS/ELINCS 292-459-0; 265-150-3
Synonyms: Alkanes, C9-12-iso-; C9-12 isoalkanes; Isoalkanes, C9-12
Classification: Isoparaffinic aliphatic hydrocarbon
Definition: Mixt. of branched chain aliphatic hydrocarbons with 9-12 carbons in alkyl chain
Properties: Colorless clear liq.; insol. in water; sp.gr. 0.765 (16/5 C); flash pt. 56 C
Environmental: Biodeg.
Uses: Solvent, fragrance ingred. in cosmetics
Trade Names: Isane IP 155; Isane IP 165; Isane IP 175; Isane IP 185; Spirdane D60; Spirdane K1; Spirdane K2
Trade Names Containing: Axarel® 2200; Axarel® 4100

C9-13 isoparaffin

CAS 64365-06-6 (generic)
Synonyms: Alkanes, C9-13-iso-; C9-13 isoalkanes; Isoalkanes, C9-13
Definition: Mixt. of branched chain aliphatic hydrocarbons with 9-13 carbons in alkyl chain
Uses: Solvent in cosmetics; defoamer in processing foods; in coatings on fruits/vegetables/egg shells; in froth flotation cleaning of vegetables; in insecticides for processed foods
Regulatory: FDA 21CFR §172.882, 173.340

C9-14 isoparaffin

CAS 246538-73-8
Synonyms: Alkanes, C9-14-iso-; C9-14 isoalkanes; Isoalkanes, C9-14
Definition: Mixt. of branched chain aliphatic hydrocarbons with 9-14 carbons in alkyl chain
Uses: Solvent in cosmetics

C9-15 isoparaffin

Synonyms: Alkanes, C9-15-iso-; C9-15 isoalkanes; Isoalkanes, C9-15
Definition: Mixt. of branched chain aliphatic hydrocarbons with 9-15 carbons in alkyl chain

Uses: Solvent cleaner
Trade Names: Toxfree™

C10-11 isoparaffin

CAS 64742-48-9 (generic)
Synonyms: Alkanes, C10-11-iso-; Alkanes, iso-, C10-11; C10-11 isoalkanes; Isoalkanes, C10-11
Definition: Mixt. of branched chain aliphatic hydrocarbons with 10-11 carbons in the alkyl chain
Environmental: Minimizes VOC
Uses: Solvent in cosmetics; solvent, diluent, carrier for personal care prods. for sprays; solvent for coatings, odorless alkyd/acrylic paints; thinner; cleaner; defoamer in processing foods; in coatings on fruits/vegetables/egg shells; in froth flotation cleaning of vegetables; in insecticides for processed foods
Features: Extremely low surf. tens. to improve flow and wetting
Regulatory: FDA 21CFR §172.882, 173.340; virtually HAPs-free
Trade Names: Spirdane D60L

C10-13 isoparaffin

CAS 68551-17-7; EINECS/ELINCS 271-366-9
Synonyms: Alkanes, C10-13-iso-; C10-13 isoalkanes; Isoalkanes, C10-13
Definition: Mixt. of branched chain aliphatic hydrocarbons with 10-13 carbons in the alkyl chain
Uses: Solvent in cosmetics; in food-contact textiles
Regulatory: FDA 21CFR §177.2800
Trade Names: EcoSolv™ Dry Cleaning Fluid; Soltrol® 130

C11-12 isoparaffin

CAS 64742-48-9 (generic)
Synonyms: Alkanes, C11-12-iso-; Alkanes, iso-, C11-12; C11-12 isoalkanes; Isoalkanes, C11-12
Definition: Mixt. of branched chain aliphatic hydrocarbons with 11-12 carbons in the alkyl chain
Environmental: Minimizes VOC
Uses: Solvent, diluent, carrier for cosmetics, personal care prods. for sprays; solvent for coatings, odorless alkyd/acrylic paints; thinner; cleaner; defoamer in processing foods; in coatings on fruits/vegetables/egg shells; in froth flotation cleaning of vegetables; in insecticides for processed foods
Features: Extremely low surf. tens. to improve flow and wetting
Regulatory: FDA 21CFR §172.882, 173.340; virtually HAPs-free

C11-13 isoparaffin

CAS 64742-48-9 (generic)
Synonyms: Alkanes, C11-13-iso-; Alkanes, iso-, C11-13; C11-13 isoalkanes; Isoalkanes, C11-13
Definition: Mixt. of branched chain aliphatic hydrocarbons with 11-13 carbons in the alkyl chain
Environmental: Minimizes VOC
Uses: Solvent for cosmetics, skin care prods., coatings, odorless alkyd/acrylic paints; thinner; cleaner; defoamer in processing foods; in coatings on fruits/vegetables/egg shells; in froth flotation cleaning of vegetables; in insecticides for processed foods
Features: Extremely low surf. tens. to improve flow and wetting
Regulatory: FDA 21CFR §172.882, 173.340; virtually HAPs-free

C11-16 isoparaffin

CAS 90622-58-5; EINECS/ELINCS 292-460-6
Synonyms: Alkanes, C11-16, iso-; C11-15 isoalkanes; C11-16 isoalkanes; Isoalkanes, C11-16
Classification: Isoparaffinic aliphatic hydrocarbon
Definition: Mixt. of branched chain aliphatic hydrocarbons with 11-16 carbons in the alkyl chain
Properties: Colorless clear liq.; char. odor; insol. in water; sp.gr. 0.77 (15 C); vapor pressure 0.2 kPa (38 C); m.p. < -50 C; b.p. 185-213 C (100 kPa); flash pt. > 62 C
Toxicology: Mild eye/skin irritant; harmful; may cause lung damage if swallowed; if aspirated, may cause severe lung injury, death
Precaution: Combustible; LFL 0.6%; UFL 7.0%; vapors heavier than air, may accumulate in low areas, may form potentially explosive mixts., may flash back; incompat. with strong oxidizing agents
Hazardous Decomp. Prods.: CO_2; incomplete combustion: CO, smoke; combustion may produce toxic mixts. in confined spaces
Storage: Keep containers closed to minimize contamination; keep away from incompat. materials, extreme heat, open flames
Uses: Solvent; fragrance ingred. in cosmetics

C12-14 isoparaffin

CAS 68551-19-9; EINECS/ELINCS 271-369-5
Synonyms: Alkanes, C12-14-iso-; C12-14 isoalkanes; Isoalkanes, C12-14
Definition: Mixt. of branched chain aliphatic hydrocarbons with 12-14 carbons in the alkyl chain
Uses: Solvent in cosmetics; defoamer in processing foods; in coatings on fruits/vegetables/egg shells; in froth flotation cleaning of vegetables; in insecticides for processed foods
Regulatory: FDA 21CFR §172.882, 173.340
Trade Names: Orfom® SX 10; Soltrol® 170

C13-14 isoparaffin

CAS 64742-48-9 (generic)
Synonyms: Alkanes, C13-14-iso-; Alkanes, iso-, C13-14; C13-14 isoalkanes; Isoalkanes, C13-14
Definition: Mixt. of branched chain aliphatic hydrocarbons with 13-14 carbons in the alkyl chain
Environmental: Minimizes VOC
Uses: Solvent for hair sprays, coatings, antiperspirants, laundry prewash, furniture/floor polishes, waterless hand cleaners, agrochems., thinners, cleaners, degreasers, odorless alkyd/acrylic paints, antibiotic gels; emollient, solvent in cosmetics; dispersant; flow control agent; wetting agent; defoamer in processing foods; in coatings on fruits/vegetables/egg shells; in froth flotation cleaning of vegetables; carrier for insecticide actives; in insecticides for processed foods
Regulatory: FDA 21CFR §172.882, 173.340; virtually HAPs-free

C13-16 isoparaffin

CAS 68551-20-2; EINECS/ELINCS 271-370-0
Synonyms: Alkanes, C13-16-iso-; C13-16 isoalkanes; Isoalkanes, C13-16
Classification: Aliphatic hydrocarbon
Definition: Mixt. of branched chain aliphatic hydrocarbons with 13-16 carbons in the alkyl chain
Uses: Solvent in cosmetics; defoamer in processing foods; in coatings on fruits/vegetables/egg shells; in froth flotation cleaning of vegetables; in insecticides for processed foods
Regulatory: FDA 21CFR §172.882, 173.340
Trade Names: Orfom® SX 11; Soltrol® 220

C13-17 isoparaffin

CAS 68551-20-2
Synonyms: Alkanes, C13-17 iso-; C13-17 isoalkanes; Isoalkanes, C13-17
Definition: Mixt. of branched chain aliphatic hydrocarbons with 13-17 carbons in alkyl chain
Uses: Solvent
Trade Names: Soltrol® 300

C20-40 isoparaffin

CAS 64365-06-6
Synonyms: Alkanes, C20-40-iso-; C20-40 isoalkanes; Isoalkanes, C20-40
Definition: Mixt. of branched chain aliphatic hydrocarbons with 20-40 carbons in the alkyl chain
Uses: Emollient, solvent in cosmetics; defoamer in processing foods; in coatings on fruits/vegetables/egg shells; in froth flotation cleaning of vegetables; in insecticides for processed foods
Regulatory: FDA 21CFR §172.882, 173.340

Citric acid, tributyl ester. *See* Tributyl citrate
Citric acid, tributyl ester, acetate. *See* Acetyl tributyl citrate
C18 linear alcohol. *See* Stearyl alcohol
C14 linear alpha olefin. *See* Tetradecene-1
C12 linear primary alcohol. *See* Lauryl alcohol
C13 linear primary alcohol. *See* Tridecyl alcohol

C16 linear primary alcohol. *See* Cetyl alcohol
Cloroben. *See* o-Dichlorobenzene
Clove bud oil. *See* Clove (Eugenia caryophyllus) oil

Clove (Eugenia caryophyllus) oil

CAS 8000-34-8
FEMA 2323
Synonyms: Caryophyllus oil; Clove bud oil; Eugenia caryophyllus; Eugenia caryophyllus oil
Definition: Volatile oil distilled from the dried flower buds of *Eugenia caryophyllus*, contg. 82-87% eugenol, 10% acetyleugenol, caryophyllene, etc.
Properties: Colorless to pale yel. volatile liq.; char. odor and taste of clove; becomes darker and thicker with age; very sol. in strong alcohol, ether, glac. acetic acid; insol. in water; dens. 1.036-1.060; b.p. 250 C; ref. index 1.527-1.538 (20 C)
Toxicology: Strongly irritating to skin; causes allergic skin rashes; may cause hypersensitivity; Bud: LD50 (oral, rat) 2650 mg/kg, (skin, rabbit) 5000 mg/kg; mod. toxic by ing.; mildly toxic by skin contact; mutagenic data; Stem: LD50 (oral, rat) 2020 mg/kg; mod. toxic by ing.; TSCA listed
Hazardous Decomp. Prods.: Heated to decomp., emits acrid smoke and irritating vapors
Storage: Keep cool, well closed; protect from light
Uses: Fragrance in perfumes; natural flavoring agent in foods and pharmaceuticals; medicine; carminative; local anesthetic (toothache); counterirritant; soaps; tonic in cosmetics
Regulatory: FDA 21CFR §184.1257, GRAS; 27CFR §21.65, 21.151; FEMA GRAS; Council of Europe listed; FDA approved for buccals, orals; BP, EP compliance
Manuf./Distrib.: Advanced BioTech; Alfa Chem; Calchauvet; Chart; Fleurchem; George Uhe; H.E. Daniel Ltd; Integra; Penta Mfg.; Ruger; Sigma; Spectrum Quality Prods.; Treatt USA; Voigt Global Distrib.
Trade Names Containing: Punctilious® SDA 38B 190 Proof Prod. Code 13602027; Punctilious® SDA 38F-2 190 Proof Prod. Code 13645006

C18 mixed alcohols. *See* Isostearyl alcohol
C5-20 normal paraffins. *See* Paraffins, normal C5-20
CNSL. *See* Cashew nut shell oil
Coal naphtha. *See* Benzene
Coal oil. *See* Kerosene
Coal tar acids. *See* Cresylic acid
Coal tar creosote. *See* Creosote oil
Coal tar cresols. *See* Cresylic acid
Coal tar naphtha; Coal tar naphtha distillate. *See* Naphtha
Coal tar oil. *See* Creosote oil
Coal tar pitch volatiles: phenanthrene. *See* Phenanthrene

Cocamide DEA

CAS 8051-30-7; 61791-31-9; 68603-42-9; EINECS/ELINCS 263-163-9; 271-657-0
Synonyms: Amides, coco, N,N-bis (2-hydroxyethyl)-; N,N-Bis (2-hydroxyethyl) coco amides; N,N-Bis (2-hydroxyethyl) coco fatty acid amide; Coco DEA; Coconut acid diethanolamide; Coconut acid polydiethanolamide (2:1 type); Coconut acid superamide (1:1 type); Coconut diethanolamide; Cocoyl diethanolamide; Diethanolamine coconut fatty acid condensate
Definition: Ethanolamides of coconut acid
Formula: $RCO–N(CH_2CH_2OH)_2$, RCO– represents the coconut acid radical
Properties: Amber liq.
Toxicology: May produce contact sensitivity; TSCA listed
Uses: Detergent, thickener, emulsifier, foam booster/stabilizer, wetting agent, solubilizer for cosmetics, pharmaceutical topicals, industrial/household cleaners, textiles, etc.; thickener, conditioner, stabilizer in shampoos; surfactant, emulsion stabilizer in cosmetics; plastics antistat; foods (delinting of cottonseed for prod. of cottonseed oil); byprods. for use in animal feeds; in cellophane for food pkg.; defoamer in food-contact paper/paperboard
Features: Mild
Regulatory: FDA 21CFR §172.710, 173.322 (0.2% max.), 175.105, 176.180, 176.210, 177.1200, 177.2260, 177.2800; FDA approved for topicals
Manuf./Distrib.: Cosmetic Supplies USA; Custom Ingreds.; McIntyre; Mitsui & Co. USA; Rhodia HPCII; Stanchem; Surfachem Ltd; Venus Ethoxyethers
Trade Names Containing: Comperlan® COD

Cocoa butter. *See* Cocoa (Theobroma cacao) butter

Cocoa (Theobroma cacao) butter

CAS 8002-31-1; EINECS/ELINCS 310-127-6
Synonyms: Cacao butter; Cocoa butter; Theobroma cacao; Theobroma cacao butter; Theobroma oil
Definition: Ylsh. wh. solid obtained from roasted seeds of *Theobroma cacao*; contains chiefly glycerides of stearic, palmitic, oleic, arachidic, and linoleic acids
Properties: Ylsh.-wh. solid, chocolate-like odor and taste; very sol. in ether, chloroform, benzene, petrol. ether; sol. in boiling abs. alcohol; sl. sol. in alcohol; insol. in water; dens. 0.858-0.864 (100/25 C); m.p. 30-35 C; iodine no. 33-42; sapon. no. 188-195; ref. index 1.4537-1.4585 (40 C)
Toxicology: May cause allergic skin reactions
Precaution: Combustible
HMIS: Health 0, Flammability 1, Reactivity 0
Storage: Preserve in well-closed containers
Uses: Chewing gum base; mfg. of chocolate; toilet soaps; emollient in cosmetics; pharmaceutical emollient, carrier, filler, ointment and suppository base; lubricant in massage; natural flavoring agent in foods; skin conditioner, occlusive solvent, skin protectant for OTC drug prods., makeup, sun preps.
Regulatory: FDA 21CFR §182.20, GRAS; FDA approved for orals, rectals, topicals; NF, BP compliance
Manuf./Distrib.: ABITEC; Adept Sol'ns.; Alban Muller; Alfa Chem; Ashland; CoKEM Assoc.; Fanning; Int'l. Sourcing; Integra; Jarchem Ind.; Jeen Int'l.; Kaufholz & Co.; Kraft Chem.; Mosselman NV; Mutchler; Nat'l. Starch & Chem.; Natra US; Paulaur; Penta Mfg.; Phoenix Nat. Prods.; Protameen; R.W. Greeff; Ruger; Spectrum Quality Prods.; Stevenson Cooper; Thornley; Tri-K Ind.; Triple Crown Am.; Universal Preserv-A-Chem; Whyte Chems. Ltd
Trade Names: Fancol™ CB; Fancol™ CB Extra

Coco DEA; Coconut acid diethanolamide. *See* Cocamide DEA
Coconut acid methyl ester. *See* Methyl cocoate
Coconut acid polydiethanolamide (2:1 type); Coconut acid superamide (1:1 type). *See* Cocamide DEA
Coconut aldehyde. *See* γ-Nonalactone
Coconut butter. *See* Coconut (Cocos nucifera) oil

Coconut (Cocos nucifera) oil

CAS 8001-31-8; EINECS/ELINCS 232-282-8
UN 1363 (DOT)
Synonyms: Coconut butter; Coconut oil; Coconut palm oil; Cocos nucifera; Copra oil
Classification: Saturated fat
Definition: Fixed oil obtained from kernels of seeds of *Cocos nucifera*
Properties: Wh. fatty solid or liq., sweet nutty taste; sol. in oxygenated and chlorinated solvs.; very sol. in chloroform, ether, CS_2; pract. insol. in water; dens. 0.903 (0/4 C); m.p. 21-27 C; acid no. < 6; iodine no. 8-9.5; sapon. no. 255-258; ref. index 1.4485-1.4495; flash pt. (CC) 550 F; surf. tens. 33.4 dynes/cm
Toxicology: May cause allergic skin reaction; TSCA listed
Precaution: Flamm. solid when exposed to heat or flame; may spontaneously heat and ignite if stored wet and hot
Hazardous Decomp. Prods.: CO_2 from combustion
HMIS: Health 0, Flammability 1, Reactivity 0
Uses: Emollient, solvent, superfatting agent, clouding agent, detergent, wetting agent, emulsifier in cosmetics and pharmaceuticals; drying oil for paints; soap ingred.; dyeing cotton; syn. detergents; source of fatty acids and alcohols; base for leather preps.; delivery/absorp. enhancement; diluent; foods (coating agent, emulsifier, formulation aid, texturizer); shortening/margarine ingred.; pharmaceuticals (excipient, ointment base, coating agent); in food-pkg. adhesives; in food-contact coatings and textiles; defoamer in food-contact paper coatings, paper/paperboard
Regulatory: FDA 21CFR §175.105, 175.300, 176.200, 176.210, 177.2800,

182.70; GRAS; FDA approved for orals, topicals; BP, EP compliance
Manuf./Distrib.: ABITEC; Aarhus Oliefabrik A/S; Akzo Nobel; Alban Muller; Aldrich; Alnor Oil; Am. Ingreds.; Amber Syn.; Anglia Oils; Arista Ind.; British Arkady Co. Ltd; Chemron; Croda Singapore; Integra; Jarchem Ind.; Jeen Int'l.; KIC Chems.; L.V. Lomas; Lambent Tech.; Lipo; Mosselman NV; Penta Mfg.; Protameen; Rhodia HPCII; Ruger; Sea-Land; Sigma; Spectrum Naturals; Spectrum Quality Prods.; St. Lawrence; Stanchem; Stevenson Cooper; Thornley; Tri-K Ind.; Universal Preserv-A-Chem; Voigt Global Distrib.; Welch, Holme & Clark

Coconut diethanolamide. *See* Cocamide DEA
Coconut oil; Coconut palm oil; Cocos nucifera. *See* Coconut (Cocos nucifera) oil
Cocoyl diethanolamide. *See* Cocamide DEA

Coenzyme A
CAS 85-61-0; EINECS/ELINCS 201-619-0
Synonyms: Coenzyme A trihydrate
Classification: Heterocyclic compd.
Empirical: $C_{21}H_{36}N_7O_{16}P_3S$
Properties: Wh. powd.; sol. in water; pract. insol. in ethanol, ether, acetone; m.w. 821.59 (trihydrate)
Storage: Sensitive to air and humidity; keep under argon
Uses: Emollient, solvent in cosmetics
Manuf./Distrib.: Fluka; Sigma

Coenzyme A trihydrate. *See* Coenzyme A

Collodion
UN 2059 (DOT)
Synonyms: Pyroxylin solution
Classification: Polymer
Definition: A solution of pyroxylin (chiefly nitrocellulose) (6%) in ether (70%) and alcohol (24%)
Properties: Colorless or pale yel. syrupy liq., ether odor; immiscible with water; sp.gr. 0.765-0.775; b.p. 83 C; flash pt. -45 C
Toxicology: May cause allergic skin reactions; TSCA listed
Precaution: DOT: Flamm. liq.; very dangerous fire hazard exposed to heat, flames, oxidizers
Hazardous Decomp. Prods.: Heated to decomp., emits toxic fumes of NO_x
NFPA: Health 1, Flammability 4, Reactivity 0
Storage: Store in a cool place, below 30 C
Uses: Cements; coating wounds and abrasions; solvent for drugs, corn removers; process engraving; lithography; photography; mfg. lacquers; patent and artificial leathers; artificial pearls; embedding tissues for sectioning; binder, film-former in cosmetics
Manuf./Distrib.: Acros Org.; Aldrich; EM Science; Electron Microscopy Sciences; Fisher Scientific; Fluka; Integra; J.T. Baker; Mallinckrodt Baker; Penta Mfg.; Ruger; Scholle; Sigma; Spectrum Quality Prods.; VWR Int'l.; Voigt Global Distrib.

Colonial spirit; Columbian spirits. *See* Methyl alcohol
Colza oil. *See* Canola oil; Rapeseed (Brassica campestris) oil
Concentrated acetic acid. *See* Acetic acid, glacial
Copper brown; Copper monoxide; Copper oxide; Copper (II) oxide; Copper (2+) oxide; Copper oxide, black. *See* Copper oxide (ic)

Copper oxide (ic)
CAS 1317-38-0; EINECS/ELINCS 215-269-1
Synonyms: Black copper oxide; CI 77403; Copper brown; Copper monoxide; Copper oxide; Copper (II) oxide; Copper (2+) oxide; Copper oxide, black; Cupric oxide; Pigment black 15
Empirical: CuO
Properties: Blk. fine powd. or cryst.; insol. in water, alcohol; m.w. 79.54; dens. 6.4; b.p. dec. @ 1026 C
Toxicology: LDLo (intratracheal, rat) 278 mg/kg; poison by intratracheal route; irritant; TSCA listed
Precaution: Incompat. with metals and reductants; explodes when heated with powd. aluminum, hydrogen, magnesium, etc.; ignites on contact with hydrogen trisulfide; incandescent reactions
HMIS: Health 2, Flammability 0, Reactivity 0
Uses: Ceramic colorant; reagent in analytical chemistry; fungicide; insecticide; catalyst; purification of hydrogen; batteries; electrodes; electroplating; solvent; desulfurizing oils; rayon; metallurgical and welding fluxes; antifouling paints; phosphors; trace mineral in animal feeds; glass/enamel/ceramics/porcelain ingred.
Manuf./Distrib.: Aldrich; Alfa Aesar; Alfa Chem; Am. Chemet; Am. Int'l.; Asiamerica Int'l.; Atlantic Equip. Engrs.; Atomergic Chemetals; Carbochem; Cerac; Chemisphere Ltd; Fluka; Fortitech; Generichem; Hasnil; Integra; J.T. Baker; Mallinckrodt Baker; Noah; Penta Mfg.; Ruger; SCM Metal Prods.; Shance; Sigma; Spectrum Quality Prods.; VWR Int'l.; Vopak USA

Copra oil. *See* Coconut (Cocos nucifera) oil
Corn oil. *See* Corn (Zea mays) oil
Corn sugar syrup. *See* Corn syrup

Corn syrup
CAS 8029-43-4; 977004-12-8; EINECS/ELINCS 232-436-4
Synonyms: Corn sugar syrup; Glucose syrup; Syrups, hydrolyzed starch
Definition: Mixture of D-glucose, maltose, and maltodextrins; obtained by partial hydrolysis of corn starch
Properties: Aq. syrup
Toxicology: May cause allergic reaction
Uses: Nutritive sweetener, dietary supplement, thickener, bodying agent, texturizer, carrier in foods, soft drinks; feedstock in fermentation; sweetener in pharmaceutical orals; in aspirin; humectant, solvent in cosmetics
Regulatory: FDA 21CFR §131.112, 133.124, 133.178, 133.179, 145.3, 145.134, 145.180, 146.3, 146.145, 146.146, 155.200, 169.175, 184.1865, GRAS; cleared by MID to flavor sausage, hamburger, meat loaf, luncheon meat, chopped or pressed ham; for use alone at 2% or in combination with corn syrup solids or glucose syrup, with combination totaling 2% on dry basis; approved for orals
Manuf./Distrib.: A.E. Staley Mfg.; ADM; Ashland; Cerestar USA; DSM Food Spec.; Degussa AG/Health & Nutrition; Jungbunzlauer; Varsal Instruments

Corn (Zea mays) oil
CAS 8001-30-7; EINECS/ELINCS 232-281-2
Synonyms: Corn oil; Maize oil; Zea mays; Zea mays oil
Definition: Refined fixed oil obtained from wet milling of corn, *Zea mays*
Properties: Pale yel. oily liq., faint char. odor and taste; sol. in ether, chloroform, amyl acetate, benzene, CS_2, chlorinated and aromatic solvs.; sl. sol. in alcohol; insol. in water; dens. 0.914-0.921; m.p. -10 C; acid no. 2-6; iodine no. 109-133; sapon. no. 187-193; flash pt. 321 C; ref. index 1.470-1.474
Toxicology: Nontoxic; human skin irritant; experimental teratogen; may be an allergen; TSCA listed
Precaution: Combustible exposed to heat or flame; dangerous spontaneous heating may occur
NFPA: Health 0, Flammability 1, Reactivity 0
Storage: Light-sensitive
Uses: Foods (in preparing foodstuffs, coating agent, emulsifier, formulation aid, texturizer); cooking/salad oil; margarine ingred.; dietary supplement; soap; lubricants; leather finishing; antistat, emollient, solvent in cosmetics; hair dressing; semidrying oil in paints; pharmaceutical solvent, vehicle; in food-pkg. adhesives; drying oil in food-contact coatings; defoamer in food-contact paper coatings, paper/paperboard
Features: Oleaginous
Regulatory: FDA 21CFR §175.105, 175.300, 176.200, 176.210, GRAS; FDA approved for injectables, orals, topicals; NF, JP compliance
Manuf./Distrib.: A&E Connock; A.E. Staley Mfg.; ABITEC; ADM; Aarhus Oliefabrik A/S; Alba Int'l.; Alban Muller; Aldivia; Anglia Oils; Arista Ind.; Ashland; Avatar; British Arkady Co. Ltd; Camida Ltd; Cargill plc; Cascade; Cerestar Int'l.; Cerestar USA; Charkit; ChemService; Corn Prods.; Croda Inc; Eggar & Co.; Fluka; Grain Processing; Karlshamns AB; Lambent Tech.; Lipo; Mosselman NV; Mutchler; Penta Mfg.; Pokonobe Ind.; Roquette; Ruger; Sea-Land; Sigma; Spectrum Naturals; Spectrum Quality Prods.; Tri-K Ind.; Vandemoortele Bakery Prods.; Voigt Global Distrib.; Vopak USA; Welch, Holme & Clark
Trade Names: Corn Oil, Refined; Super Refined® Corn NF

Cotton oil. *See* Cottonseed (Gossypium) oil

Cottonseed (Gossypium) oil

CAS 8001-29-4; EINECS/ELINCS 232-280-7
Synonyms: Aceite de Algodon; Cotton oil; Cottonseed oil; Cottonseed oil, winterized; Deodorized winterized cottonseed oil; Gossypium; Oleum Gossypii seminis
Definition: Refined fixed oil from seeds of various species of *Gossypium*
Properties: Pale yel. oily liq., nearly odorless; sol. in ether, benzene, chloroform; sl. sol. in alcohol; negligible sol. in water; dens. 0.915-0.921; f.p. 0-5 C; solid. pt. 31-33 C; iodine no. 109-120; sapon. no. 190-198; flash pt. (CC) 486 F
Toxicology: Questionable carcinogen; experimental tumorigen and teratogen; allergen; TSCA listed
Precaution: Combustible exposed to heat or flame; may be dangerous hazard due to spontaneous heating
Hazardous Decomp. Prods.: CO_2 from combustion
NFPA: Health 0, Flammability 1, Reactivity 0
Storage: Light-sensitive
Uses: Emollient; foods (coating agent, emulsifier, formulation aid, texturizer); drying oil for paints; leather dressing; soap stock; lubricant; glycerol; cosmetic base; waterproofing compositions; pharmaceutical vehicle, solvent; in food-pkg. adhesives; defoamer in food-contact paper coatings, paper/paperboard; drying oil and surf. lubricant in food-contact coatings; in food-contact textiles
Features: Oleaginous
Use Level: 56-92% (intramuscular preps.), 0.002-402 mg (solid oral dosage forms)
Regulatory: FDA 21CFR §175.105, 175.300, 176.200, 176.210, 177.2800, GRAS; FDA approved for intramuscular injectables, orals; NF compliance
Manuf./Distrib.: A&E Connock; ABITEC; AMRESCO; Aldrich; Allyson Enterprises; Alnor Oil; Amber Syn.; Arista Ind.; Avatar; British Arkady Co. Ltd; Camida Ltd; Charkit; Croda Inc; Fisher Scientific; Forum Chems.; Good Food; Jeen Int'l.; Karlshamns AB; Lambent Tech.; Mutchler; Natural Oils Int'l.; Oilseeds Int'l.; Penta Mfg.; Pokonobe Ind.; Ruger; Sea-Land; Sigma; Spectrum Quality Prods.; Thornley; Tri-K Ind.; Voigt Global Distrib.; Vopak USA; Welch, Holme & Clark
Trade Names: Super Refined® Cottonseed NF

Cottonseed oil; Cottonseed oil, winterized. *See* Cottonseed (Gossypium) oil

C14-16 paraffins

CAS 90622-46-1; EINECS/ELINCS 292-448-0
Synonyms: Alkanes, C14-16; C14-16 alkanes
Definition: Mixt. of tetradecane, pentadecane, hexadecane
Properties: Clear oily liq.; mild hydrocarbon odor; negligible sol. in water; sp.gr. 0.773 (16 C); visc. 2.5-2.7 cSt (40 C); vapor pressure < 0.05 mm Hg (20 C); m.p. 8-10 C; b.p. 254-283 C; flash pt. (PMCC) 121 C; autoignition temp. 216 C
Toxicology: LD50 (oral, rat) > 5 g/kg, (skin, rabbit) > 2 g/kg; LC50 (inh., rat, 4 h) > 5.8 mg/l; may cause eye/skin irritation, dermatitis; inh. of high conc. of vapor may cause headache, stupor, dizziness; prolonged breathing of mist or vomiting following ing. may cause serious lung injury, possibly death; TSCA listed
Environmental: Prevent entry into drains, sewers, waterways
Precaution: LEL 0.5%; UEL 7.0%; incompat. with oxidizers, acids, alkalis; prevent contact with static electricity; empty containers contain residue, may explode and cause injury or death
Hazardous Decomp. Prods.: Combustion: CO_x, other org. compds.
NFPA: Health 1, Flammability 1, Reactivity 0
Storage: Store @ 8-49 C in cool, dry place away from heat, open flame, ignition sources
Uses: Solvent
Regulatory: DOT nonhazardous; SARA §302/313 nonreportable; Calif. Prop. 65 nonreportable
Manuf./Distrib.: Americhem Sales
Trade Names: Linpar® 15-V Normal Paraffin; Linpar® 1416-V Normal Paraffin
Trade Names Containing: AM 230 Solvent; Sasol-47

Creosote; Creosote, from coal tar. *See* Creosote oil

Creosote oil

CAS 8001-58-9
UN NA 1993 (DOT)
Synonyms: Brick oil; Coal tar creosote; Coal tar oil; Creosote; Creosote, from coal tar; Creosotes; Creosotum; Cresylic creosote; Dead oil; Heavy oil; Liquid pitch oil; Naphthalene oil; Tar oil; Wash oil
Definition: Wide-cut coal-tar fraction; contains naphthalene and tar acid fractions
Properties: Ylsh. to dk. grn.-brn. oily liq.; sol. < 0.1 g/100 ml in water (21.5 C); dens. 1.1; b.p. 203-220 C (dec.); flash pt. 74 C
Toxicology: LD50 (oral, rat) 725 mg/kg; TDLo (skin, mouse, 33 wks intermittent) 99 g/kg; may cause somnolence, convulsions, respiratory depression, coma; carcinogenic; tumorigen; reproductive effector; mutagen; TSCA listed
Precaution: DOT: Flamm. liq.
NFPA: Health 2, Flammability 2, Reactivity 0
Uses: Road tar fluxing solvent; fuel oil; wash oil in coal gas purification; wood preservative

Creosotes; Creosotum. *See* Creosote oil
Cresol. *See* Cresylic acid
2-Cresol. *See* o-Cresol

o-Cresol

CAS 95-48-7; EINECS/ELINCS 202-423-8
UN 2076 (DOT); FEMA 3480
Synonyms: 2-Cresol; o-Cresylic acid; 1-Hydroxy-2-methylbenzene; o-Hydroxytoluene; 2-Methylphenol; o-Methylphenol; Orthocresol; o-Oxytoluene; o-Toluol
Classification: Aromatic phenol
Empirical: C_7H_8O
Properties: Cryst. or liq.; darkens on exposure to air or light; m.w. 108.14; dens. 1.048; vapor pressure 1 mm (38.2 C); m.p. 30-34 C; b.p. 191 C; flash pt. 178 F
Toxicology: ACGIH TLV/TWA 5 ppm; LD50 (oral, rat) 121 mg/kg, (skin, rat) 620 mg/kg; highly toxic; poison by ing., inh., subcut., IV, IP routes; readily absorbed thru skin; severe eye and skin irritant; experimental neoplastigen; questionable carcinogen; human mutagenic data; target organs: nerves, lungs, liver, kidneys; TSCA listed
Precaution: Corrosive; flamm. exposed to heat, flame, or oxidizers
NFPA: Health 3, Flammability 2, Reactivity 0
Storage: Air and light-sensitive
Uses: Disinfectant; solvent; phenolic resins; tricresyl phosphate; ore flotation; textile scouring; organic intermediate; mfg. of salicylaldehyde, coumarin, herbicides, antioxidants, dyes; cosmetics ingred.; surfactant
Regulatory: FDA 21CFR §175.300, 177.2410; FEMA GRAS; SARA reportable; BP compliance
Manuf./Distrib.: Aldrich; Allchem Ind.; Atul Ltd; Bayer; Crowley Tar Prods.; Fluka; PMC Spec.; Penta Mfg.; Sigma; Whyte Chems. Ltd

Cresol mixture, m-, p-; Cresols. *See* Cresylic acid

Cresylic acid

CAS 1319-77-3; EINECS/ELINCS 215-293-2
UN 2022 (DOT); UN 2076 (DOT)
Synonyms: Coal tar acids; Coal tar cresols; Cresol; Cresol mixture, m-, p-; Cresols; Cresylol; Methyl phenol; Mixed cresols; Oxytoluene; ar-Toluenol; Tricresol
Classification: Aromatic alcohol
Definition: Commercial mixtures of phenolic materials boiling above the cresol range
Empirical: C_7H_8O
Formula: $CH_3C_6H_4OH$
Properties: Ylsh. to brn.-yel. or pink liq.; turns brn. on exposure to air and light; phenolic odor; sol. in alcohol, glycerol, dil. alkalis, aromatic/oxygenated solvs.; sl. sol. in water, min. oil; m.w. 108.15; dens. 1.030-1.038; vapor pressure 0.11-0.30 mm Hg; m.p. 10.9-33.5 C; b.p. 191-203 C; flash pt. (OC) 79-85 C; pH neutral or sl. acidic (sat. aq. sol'ns.)
Toxicology: ACGIH TLV/TWA 5 ppm (skin); PEL-TWA 5 ppm (skin); LD50 (oral, rat) 1454 mg/kg, (dermal, rabbit) 1782 mg/kg; poison; mod. toxic by ing., skin contact; lachrymator; may be fatal if absorbed thru skin or swallowed; CNS depressant; TSCA listed

Precaution: Flamm. when exposed to heat or flame; combustible liq. and vapor; can form explosive mixts. with air ≥ 79 C; corrosive; incompat. with strong oxidizers, chlorosulfonic acid, nitric acid, oleum, strong bases (explosive reactions possible)
Hazardous Decomp. Prods.: CO, CO_2, toxic aromatic degradation prods., fumes of cresol; heated to decomp., emits highly toxic and irritating fumes
NFPA: Health 3, Flammability 2, Reactivity 0
Storage: Store in a cool, dry area, out of direct sunlight, away from heat and ignition sources; hygroscopic, light-sensitive
Uses: Phosphate esters; wire enamel solvent; plasticizers; oil/gasoline additives; laminates; coating for magnet wire; disinfectant; bacteriostat; fumigant; surfactant; metal cleaning; ore flotation agent (copper extraction); intermediate; PU pitch coating; in photographic developers; solvent refining; pesticides; electrical insulators; starting material in mfg. of antioxidants, dyes, phenolic plastics, paints, agric. chems.; epoxy-pitch coatings (marine paints); engine cleaner ingred.; antimicrobial in cosmetics
Manuf./Distrib.: Allchem Ind.; Atofina; Concord; Crowley Tar Prods.; Dakota Gasification; Honeywell Perf. Polymers; PMC Spec.; Spectrum Quality Prods.

o-Cresylic acid. *See* o-Cresol
Cresylic creosote. *See* Creosote oil
Cresylol. *See* Cresylic acid
Cresyl phosphate. *See* Tricresyl phosphate
Crispmint oil. *See* Spearmint (Mentha viridis) oil
Crotonal. *See* Crotonaldehyde

Crotonaldehyde

CAS 4170-30-3; EINECS/ELINCS 204-647-1
UN 1143
Synonyms: 2-Butenal; 2-Butenaldehyde; Crotonal; Crotonaldehyde stabilized; Crotonic acid aldehyde; Crotonic aldehyde; 1,2-Ethanediol, dipropanoate; Ethylene dipropionate; 1-Formylpropene; Methyl acrolein; 3-Methylacrolein; β-Methyl acrolein; Propylene aldehyde; Topanel
Empirical: C_4H_6O
Formula: $CH_3CH{=}CHCHO$
Properties: Colorless to straw-colored mobile liq.; pungent, suffocating, tar odor; very sol. in ether, benzene; sol. in DMSO, 95% ethanol, acetone, acid; misc. with most org. solvs.; sol. < 0.1 mg/ml in water; m.w. 70.09; sp.gr. 0.495 (25/4 C); vapor pressure 19 mm Hg (20 C); m.p. -76.5 C; b.p. 102 C; flash pt. 13 C; ref. index 1.4355 (20 C)
Toxicology: ACGIH TLV/TWA 2 ppm; LD50 (oral, rat) 206 mg/kg, (skin, guinea pig) 25,590 μg/kg; LC50 (inh., rat, 2 h) 200 mg/m³; poison; highly toxic by ing., inh., dermal routes; extremely corrosive to skin/eyes; lachrymator; very dangerous to eyes; can cause corneal burns; irritating to nose, throat, mucous membranes; may cause nausea, vomiting, unconsciousness, coughing, chest pain, systemic illness, headache, fluid accumulation in lungs, etc.; can cause sensitization; may be fatal; TSCA listed
Environmental: Dangerous for the environment
Precaution: Flamm.; incompat. with caustics, ammonia, org. amines, min. acids, bases, reducing agents, oxidizers, ketones, heat; may polymerize/condense in presence of alkalis, amines, acids; uncontrolled polymerization can cause rapid evolution of heat
Hazardous Decomp. Prods.: Heated to decomp., emits toxic fumes of CO, CO_2
NFPA: Health 3, Flammability 3, Reactivity 2
Storage: Air- and heat-sensitive; may be light-sensitive; keep container tightly closed under inert atmosphere; store in explosion-proof refrigerator; protect from light
Uses: Intermediate in mfg. of n-butanol, crotonic acid, sorbic acids; mfg. of resin and rubber antioxidants; solvent in min. oil purification; warning agent in fuel gas; for locating breaks and leaks in pipes; alcohol denaturant; mfg. of butyraldehyde, quinaldine; minor amts. in mfg. of maleic acid, crotyl alcohol, butyl chloral hydrate, rubber accelerators; in org. synthesis; insecticides; chemical warfare
Manuf./Distrib.: Advanced Synthesis Tech.; Celanese; Fluka; Pfaltz & Bauer; Spectrum Quality Prods.; TCI Am.

Crotonaldehyde stabilized; Crotonic acid aldehyde. *See* Crotonaldehyde
Crotonic acid, ethyl ester. *See* Ethyl crotonate
Crotonic aldehyde. *See* Crotonaldehyde
Cryofluoran; Cryofluorane. *See* 1,2-Dichlorotetrafluoroethane
CTC. *See* Carbon tetrachloride

C10-18 triglycerides

CAS 85665-33-4; EINECS/ELINCS 288-123-8
Synonyms: Glycerides, C10-18; Glyceryl tri-C10-18 acids
Definition: Triester of glycerin and a mixture of normal and branched chain C10-18 fatty acids
Uses: Emollient, solvent in cosmetics

Cumene

CAS 98-82-8
UN 1918 (DOT)
Synonyms: Benzene isopropyl; Isopropylbenzene; Isopropyl benzol; (1-Methylethyl) benzene; 2-Phenylpropane
Empirical: C_9H_{12}
Formula: $C_6H_5CH(CH_3)_2$
Properties: Colorless liq.; misc. with oxygenated and chlorinated solvs.; insol. in water; m.w. 120.21; dens. 0.864 (20/4 C); vapor pressure 10 mm Hg (38.3 C); m.p. -96 C; b.p. 152 C; flash pt. 111 F
Toxicology: ACGIH TLV/TWA 50 ppm (skin); LD50 (oral, rat) 1400 mg/kg, (skin, rabbit) 12,300 mg/kg; TCLo (inh., human) 200 ppm; mod. toxic by ing.; mildly toxic by inh., skin contact; primary eye and skin irritant; human systemic effects; may cause somnolence, antipsychotic effects, irritability; potential narcotic action; CNS depressant; human mutagen; TSCA listed
Precaution: Flamm. exposed to heat or flame; can react with oxidizing materials; violent reaction with HNO_3, chlorosulfonic acid, oleum
NFPA: Health 2, Flammability 3, Reactivity 1
Uses: Production of phenol, acetone and α-methylstyrene; org. solvent; solvent in pharmaceuticals
Manuf./Distrib.: Aldrich; Alvin Chem.; Ashland; ChevronPhillips; Degussa AG; Fluka; Georgia Gulf; JLM Marketing; Shell; Sigma; Spectrum Quality Prods.; Sunoco
Trade Names Containing: Solvent 2439; Solvent 30 823

Cumyl phenol

CAS 599-64-4
Synonyms: 4-Cumylphenol; p-Cumylphenol; p-(α-Cumyl) phenol; p-(α,α-Dimethylbenzyl) phenol; 4-(Dimethylphenylmethyl) phenol; 4-Hydroxydiphenyldimethylmethane; 4-(1-Methyl-1-phenethyl) phenol; Phenol, p-(α,α-dimethylbenzyl); Phenol, 4-(1-methyl-1-phenethyl)-; 4-(2-Phenylisopropyl) phenol
Empirical: $C_{15}H_{16}O$
Formula: $C_6H_5C(CH_3)_2C_6H_4OH$
Properties: Wh. to tan crystals, phenol odor; m.w. 212.29; dens. 1.115 g/ml (25 C); m.p. 74-76 C; b.p. 335 C; f.p. 72 C; flash pt. 320 F
Toxicology: LD50 (oral, frog) 335 mg/kg; poison by ing.; irritant; TSCA listed
Precaution: Combustible
Hazardous Decomp. Prods.: Heated to decomp., emits acrid smoke and irritating vapors
Uses: Intermediate for resins, insecticides, lubricants, surfactants; chain terminator for PC food-pkg. resins; modifier for epoxy, furan, and phenolic resins
Regulatory: FDA 21CFR §177.1580
Manuf./Distrib.: Acros Org.; Aldrich; Chugai Boyeki Am.; Degussa AG; Herdillia Chems. Ltd; Howard Hall; ICN Biomed. Research Prods.; Lancaster Synthesis; PMC Spec.; Pechiney Chems. Div.; Pfaltz & Bauer; Schenectady; TCI Am.; Vilax
Trade Names Containing: Kenplast® ES-2HP

4-Cumylphenol; p-Cumylphenol; p-(α-Cumyl) phenol. *See* Cumyl phenol

Cumylphenyl acetate

CAS 24133-73-1
Synonyms: p-Cumylphenyl acetate; Phenol, p-(α,α-dimethylbenzyl)-, acetate; Phenol, 4-(1-methyl-1-phenylethyl)-, acetate
Properties: Sp.gr. 1.03; m.p. < 0 C; flash pt. (COC) 250 C
Uses: Plasticizer for urethanes; high flash pt. reactive diluent for epoxy, for polyamide-cured epoxy floorings and coal tar pipe coatings; comono-

mer and impact modifier for phenolics
Manuf./Distrib.: Kenrich Petrochems.
Trade Names Containing: Kenplast® ES-2HP

p-Cumylphenyl acetate. *See* Cumylphenyl acetate

Cumylphenyl benzoate

Properties: Sp.gr. 1.10; m.p. 48-52 C; flash pt. (COC) 350 C
Uses: Primary plasticizer for PVC and PVC/nitrile; process aid for semi-rigid PVC extrudates; solvates conductive polyester and acrylic inks to improve impact
Manuf./Distrib.: Kenrich Petrochems.
Trade Names: Kenplast® ESB

Cumylphenyl neodecanoate

Uses: Primary plasticizer for PVC, PVC/nitrile; process aid for semirigid PVC extrusions; solvates and impact modifies conductive polyester and acrylic inks
Trade Names: Kenplast® ESN

Cupric oxide. *See* Copper oxide (ic)
Curled mint oil. *See* Spearmint (Mentha viridis) oil
Cyanobenzene. *See* Benzonitrile
N-Cyanodimethylamine. *See* Dimethyl cyanamide
Cyanoethane. *See* Propionitrile
2-Cyanoethanol. *See* Ethylene cyanohydrin
Cyanoethylene. *See* Acrylonitrile
2-Cyanoethyl thioether. *See* Thiodipropionitrile
Cyanomethane. *See* Acetonitrile
1-Cyano-2-methoxyethane. *See* 3-Methoxypropionitrile
N-Cyano-N-methylmethanamine. *See* Dimethyl cyanamide
1-Cyanopropane. *See* Butyronitrile
Cyclic dimethyl polysiloxane; Cyclic dimethyl polysiloxane with n= 3-6; Cyclic dimethylsiloxane. *See* Cyclomethicone
Cyclic dimethylsiloxane pentamer. *See* Decamethylcyclopentasiloxane
Cyclic dimethylsiloxane tetramer. *See* Octamethylcyclotetrasiloxane
Cyclic ethylene carbonate. *See* Ethylene carbonate
Cyclic methylethylene carbonate. *See* Propylene carbonate
Cyclic pentamer-D5. *See* Decamethylcyclopentasiloxane
Cyclic propylene carbonate; Cyclic 1,2-propylene carbonate. *See* Propylene carbonate
Cyclic tetramethylene sulfone. *See* Sulfolane

Cyclododecanol

CAS 1724-39-6; EINECS/ELINCS 217-031-2
Synonyms: CDOL; Cyclododecyl alcohol
Classification: Nonaromatic alcohol
Empirical: $C_{12}H_{24}O$
Formula: $C_{12}H_{23}OH$
Properties: Wh. cryst. powd. and chunks; insol. in water; m.w. 184.32; dens. 0.91 kg/l (18 C); m.p. 76-79 C; b.p. 273 C; flash pt. 138 C
Toxicology: TSCA listed
Precaution: Incompat. with strong oxidizing agents, acid chlorides, acid anhydrides
Hazardous Decomp. Prods.: CO, CO_2, irritating and toxic fumes and gases
Storage: Store in cool, dry place; keep container closed when not in use
Uses: Leveling agent; specialty solvent
Manuf./Distrib.: Acros Org.; Advanced Synthesis Tech.; Aldrich; Fluka

Cyclododecyl alcohol. *See* Cyclododecanol

Cycloethoxymethicone

Synonyms: Methylethoxysiloxane cyclic polymer
Classification: Cyclic silicone polymer
Uses: Emollient, skin conditioner, solvent in cosmetics

1,4-Cyclohexadiene, 1-methyl-4-(1-methylethyl)-. *See* γ-Terpinene
Cyclohexanamine. *See* Cyclohexylamine
N-Cyclohexanamine. *See* Dicyclohexylamine
Cyclohexanamine, N-methyl-. *See* N-Methylcyclohexylamine

Cyclohexane

CAS 110-82-7; EINECS/ELINCS 203-806-2
UN 1145 (DOT)
Synonyms: Benzene hexahydride; Hexahydrobenzene; Hexamethylene; Hexanaphthene
Classification: Sat. aliphatic hydrocarbon
Empirical: C_6H_{12}
Properties: Colorless mobile liq., pungent gasoline-like odor; sol. in ether, chlorinated/aromatic solvs.; misc. with ethanol, acetone, benzene, ethyl ether, CCl_4; insol. in water; m.w. 84.16; dens. 0.779 (20/4 C); f.p. 4.6 C; m.p. 6.5 C; b.p. 80.7 C; flash pt. (CC) -18.3 C; ref. index 1.4264
Toxicology: ACGIH TLV/TWA 300 ppm; LD50 (oral, rat) 29,820 mg/kg; LDLo (IV, rabbit) 77 mg/kg; poison by IV; mod. toxic by ingestion, inhalation, skin contact; systemic and skin irritant; overexposure may cause eye and respiratory irritation, drowsiness, dermatitis, narcosis, coma; mutagenic data; TSCA listed
Precaution: DOT: Flamm. liq.; dangerous fire hazard exposed to heat or flame; reactive with oxidizers; mod. explosion hazard as vapor exposed to flame; explosive mixed hot with liq. dinitrogen tetraoxide
Hazardous Decomp. Prods.: Heated to decomp., emits acrid smoke, irritating fumes
NFPA: Health 1, Flammability 3, Reactivity 0
Storage: Store in cool, dry, well-ventilated area out of direct sunlight
Uses: Solvent for cellulose ethers, fats, oils, waxes, inks, lacquers, resins, coatings; paint and varnish remover; extraction solvent for essential oils; glass substitutes; laboratory reagent; chemical intermediate; in fungicidal formulations; mfg. of nylon, adipic acid, benzene, cyclohexanone, etc.; mfg. of solid fuel for camp stoves; diluent in food colorants; solvent in cosmetics, pharmaceutical orals; in food-pkg. adhesives; defoamer in food-contact paper coatings
Regulatory: FDA 21CFR §73.1, 175.105, 176.200, 178.3620; FDA approved for orals
Manuf./Distrib.: AAE Chemie NV; Aldrich; Ashland; ChevronPhillips; ExxonMobil Europe; ExxonMobil; Fisher Scientific; Fluka; General Chem.; Hukill; Huntsman; Indofine; Koch Chem. Spec.; Romil Ltd; Sigma; Spectrum Quality Prods.; Sunoco; Triple Crown Am.
Trade Names Containing: Rubber Solvent; Rubber Solvent 332; Shellsol® B HT; Shellsol® B HT60; S&L Naphtha; Solvent 2442; Tolu-Sol 30

Cyclohexane, s-butyl-. *See* s-Butylcyclohexane

1,2-Cyclohexane carbonate

CAS 4389-22-4
Properties: M.w. 142.16; dens. 1.188 g/ml; visc. 10.60 cp (40 C); flash pt. (CC) 305 F; surf. tens. 33.44 dynes/cm (40 C)
Uses: Solvent

Cyclohexane, 1,3-dimethyl-, trans-. *See* trans-1,3-Dimethylcyclohexane
Cyclohexane, 1,4-dimethyl-. *See* 1,4-Dimethylcyclohexane
Cyclohexane, 1,4-dimethyl-, cis-. *See* cis-1,4-Dimethylcyclohexane
Cyclohexane, 1,4-dimethyl-, trans-. *See* trans-1,4-Dimethylcyclohexane
Cyclohexane, ethyl-. *See* Ethylcyclohexane
Cyclohexane, isobutyl-. *See* Isobutylcyclohexane
Cyclohexane, isopropyl-. *See* Isopropylcyclohexane
Cyclohexane, methyl-. *See* Methyl cyclohexane
Cyclohexane, (1-methylethyl)-. *See* Isopropylcyclohexane
Cyclohexane, (1-methylpropyl)-. *See* s-Butylcyclohexane
Cyclohexane, (2-methylpropyl)-. *See* Isobutylcyclohexane

Cyclohexanol

CAS 108-93-0; EINECS/ELINCS 203-630-6
Synonyms: 1-Cyclohexanol; Cyclohexyl alcohol; Hexahydrophenol; Hexalin; Hydrophenol; Hydroxycyclohexane
Classification: Alicyclic alcohol
Empirical: $C_6H_{12}O$
Formula: $CH_2(CH_2)_4CHOH$
Properties: Colorless to yel. oily liq. or cryst. (below 25 C); camphor-like odor; sol. in methanol, ethanol, acetone, ether, benzene, most org. solvs.; mod. sol. in water; m.w. 100.09; sp.gr. 0.937 (37.4 C); vapor pressure 0.8 mm Hg; m.p. 23 C; b.p. 160.9 C; flash pt. 67.7 C; autoignition temp. 300 C; ref. index 1.4656 (20 C); surf. tens. 33.91 dynes/cm; dielec. const. 15
Toxicology: ACGIH TLV/TWA 50 ppm (skin); LD50 (oral, rat) 2.06 g/kg; toxic; poison by IV, mod. toxic by ing., subcut., intramuscular routes; mild skin irritant; lachrymator; mod. to severe eye irritant on direct contact; nose/throat irritant; narcotic-like; high concs. may cause headache, nausea, vomiting, tremors; can be absorbed thru skin; ing. of

large doses can probably cause kidney/liver damage, unconsciousness, death; experimental reproductive effects; human mutagenic data; TSCA listed
Environmental: VOC; BOD5 1.97; ThOD 2.72
Precaution: Combustible; incompat. with strong oxidizing agents (reacts vigorously or explosively generating heat and pressure); ignites on contact with chromium trioxide; violent reaction with HNO_3
Hazardous Decomp. Prods.: Heated to decomp., emits acrid smoke and fumes
NFPA: Health 1, Flammability 2, Reactivity 0
Storage: Hygroscopic; store in cool, dry, well-ventilated area
Uses: Stabilizer, homogenizer in soap making and detergents to incorporate solvents and phenolic insecticides; solvent for alkyd and phenolic resins, paints, lacquers, shellacs, varnishes, rubber, inks, fats, oils; cosmetics ingred.; source of adipic acid for nylon prod., textile finishing; plasticizer prod.; dye solvent and carrier; leather degreasing; polishes; plasticizers; germicides; in food-pkg. adhesives; defoamer in food-contact coatings, paper/paperboard
Regulatory: FDA 21CFR §175.105, 176.200, 176.210
Manuf./Distrib.: Aldrich; Ashland; BASF AG; BASF; Chemcentral; Fluka; Harcros; Honeywell Perf. Polymers; J.T. Baker; Nipa Labs; Spectrum Quality Prods.; Sunnyside; UCB; Vopak USA
Trade Names: Hexalin®; Naxol®

1-Cyclohexanol. *See* Cyclohexanol
Cyclohexanol acetate. *See* Cyclohexyl acetate
Cyclohexanol, 2-(1,1-dimethylethyl)-, acetate. *See* 2-t-Butylcyclohexyl acetate
Cyclohexanol, 2-isopropyl-5-methyl-; Cyclohexanol, 5-methyl-2-(1-methylethyl)-. *See* Menthol
Cyclohexanol, 3,3,5-trimethyl-. *See* Trimethylcyclohexanol

Cyclohexanone

CAS 108-94-1; EINECS/ELINCS 203-631-1
UN 1915 (DOT); FEMA 3909
Synonyms: Cycloyhexyl ketone; Ketohexamethylene; Pimelic ketone; Pimelin ketone
Classification: Cyclic ketone
Empirical: $C_6H_{10}O$
Formula: $CO(CH_2)_4CH_2$
Properties: Water-wh. to pale yel. oily liq., peppermint-acetone odor; sol. in alcohol, ether, common org. solvs.; sl. sol. in water; m.w. 98.08; sp.gr. 0.947 (20/4 C); vapor pressure 10 mm Hg (38.7 C); m.p. -32.1 C; b.p. 156.7 C; flash pt. (CC) 44 C; autoignition temp. 420 C; ref. index 1.4507; surf. tens. 35.05 dynes/cm; dielec. const. 16.10
Toxicology: ACGIH TLV/TWA 25 ppm (skin); LD50 (oral, rat) 1535 mg/kg, (skin, rabbit) 1000 mg/kg; mod. toxic by ing., inh., subcut, IV, IP; corrosive; skin and severe eye irritant; inh. may cause throat irritation; ing. may cause GI tract irritation; extreme exposure by inh. or ing. may cause unconsciousness and death; human mutagenic data; experimental reproductive effects; TSCA listed
Environmental: VOC; BOD5 1.50; ThOD 2.61
Precaution: Flamm. exposed to heat or flame; incompat. with strong oxidizing agents (increases fire and explosion hazard); sl. explosion hazard as vapor exposed to flame; explosive reaction with nitric acid @ 75 C; can form explosive peroxides
Hazardous Decomp. Prods.: Heated to decomp., emits acrid smoke and irritating fumes
NFPA: Health 1, Flammability 2, Reactivity 0
Storage: Store in cool, dry, well-ventilated area, out of direct sunlight, in tightly closed containers; limit quantities in use
Uses: As paint/varnish remover; solvent/thinner for cellulose acetate, nitrocellulose, natural resins, vinyl resins, lacquers, inks, paints, rubber, waxes, fats; in prod. of adipic acid for nylon, cyclohexanone resins, caprolactam, polymers, resins; metal and leather degreasing; textile and silk dyeing, leveling, and processing; lubricating oil additives; mfg. of pharmaceuticals, insecticides, inks, wood stains, polishes, magnetic and video tapes; synthetic flavoring agent in foods; cosmetics ingred.
Regulatory: FEMA GRAS; SARA reportable; CERCLA hazardous substance
Manuf./Distrib.: Aldrich; Allchem Ind.; Ashland; BASF; Chemcentral; Coyne; Dow; Fluka; General Chem.; Harcros; Honeywell Perf. Polymers; Houghton Chem.; J.T. Baker; Mallinckrodt Baker; Penta Mfg.; Romil Ltd; Sigma; Spectrum Quality Prods.; Stanchem; Sunnyside; VWR Int'l.; Vopak USA; Whyte Chems. Ltd
Trade Names: Nadone®

Cyclohexanone, methyl-. *See* Methylcyclohexanone
Cyclohexanyl acetate. *See* Cyclohexyl acetate
Cyclohexasiloxane, dodecamethyl-. *See* Cyclomethicone
Cyclohexatriene. *See* Benzene
3-Cyclohexene-1-methanol, α,α, 4-trimethyl-. *See* α-Terpineol
Cyclohexene, 1-methyl-4-(1-methylethenyl)-. *See* dl-Limonene
Cyclohexene, 1-methyl-4-(1-methylethenyl)-, (S)-. *See* l-Limonene

Cyclohexyl acetate

CAS 622-45-7; EINECS/ELINCS 210-736-6
UN 2243 (DOT); FEMA 2349
Synonyms: Acetic acid cyclohexyl ester; Adronol acetate; Cyclohexanol acetate; Cyclohexanyl acetate; Hexalin acetate; Hexaline acetate
Classification: Acetic acid ester
Empirical: $C_8H_{14}O_2$
Formula: $CH_3COOC_6H_{11}$
Properties: Colorless to pale yel. oily liq.; fruity odor; sol. in alcohol, ether; pract. insol. in water; m.w. 142.20; sp.gr. 0.966; b.p. 172-173 C; flash pt. (CC) 58 C; ref. index 1.4400-1.4410
Toxicology: LD50 (oral, rat) 6730 mg/kg, (skin, rabbit) 10 g/kg; mod. toxic by subcut. route; mildly toxic by ing., skin contact; human systemic effects by inh. (conjunctiva irritation, unspecified respiratory changes); systemic irritant to humans; TSCA listed
Precaution: Flamm. exposed to heat or flame; incompat. with strong oxidizing agents (increases risk of fire and explosion), strong bases (decomp. can occur)
Hazardous Decomp. Prods.: Cyclohexanol, acetic acid; heated to decomp., emits acrid smoke and irritating fumes
NFPA: Health 1, Flammability 2, Reactivity 0
Storage: Store in cool, dry, well-ventilated area out of direct sunlight; avoid generating mist; ground drums and bond transfer containers
Uses: Solvent for nitrocellulose, cellulose ether, bitumens, metallic soaps, basic dyes, brown oils, crude rubber, resins, lacquers; synthetic flavoring agent and fragrance in foods and pharmaceuticals; fragrance in cosmetics
Features: Fruity flavor
Regulatory: FDA 21CFR §172.515; FEMA GRAS; Japan approved as flavoring
Manuf./Distrib.: Advanced Synthesis Tech.; Aldrich; Degussa AG/Health & Nutrition; Elan; Grau Aromatics; Lluch Essence; Prodasynth

Cyclohexyl alcohol. *See* Cyclohexanol

Cyclohexylamine

CAS 108-91-8; EINECS/ELINCS 203-629-0
UN 2357 (DOT)
Synonyms: Aminocyclohexane; Aminohexahydrobenzene; CHA; Cyclohexanamine; Hexahydroaniline; Hexahydrobenzenamine
Classification: Saturated alicyclic amine
Empirical: $C_6H_{13}N$
Formula: $C_6H_{11}NH_2$
Properties: Colorless to yel. liq., strong fishy odor; sol. in water, alcohols, ethers, ketones, esters, min. and veg. oils; m.w. 99.20; dens. 0.865; m.p. -17.7 C; b.p. 134.5 C; flash pt. (CC) 28 C; pH 11.8 (0.1 M)
Toxicology: ACGIH TLV/TWA 10 ppm; LD50 (oral, rat) 156 mg/kg, (subcut., mouse) 1150 mg/kg, (skin, rabbit) 277 mg/kg; poison by ing., skin contact, IP; mod. toxic by subcut.; severe human skin irritant; can cause dermatitis; corrosive; human mutagenic data; experimental teratogen, reproductive effects; TSCA listed
Precaution: Flamm. liq.; dangerous fire hazard exposed to heat, flame, or oxidizers; corrosive to copper and copper alloys, aluminum, zinc, and galvanized steel
Hazardous Decomp. Prods.: Heated to decomp., emits toxic fumes of NO_x
NFPA: Health 3, Flammability 3, Reactivity 0
Uses: Corrosion inhibitor for boiler water treatment (incl. food contact except milk prods.); rubber accelerator; corrosion inhibitor; intermediate in organic synthesis, prod. of plasticizers, dyes, textile chems.;

process solvent; neutralizer
Regulatory: FDA 21CFR §173.310 (10 ppm max. in steam); SARA extremely hazardous substance
Manuf./Distrib.: Air Prods.; Aldrich; Allchem Ind.; Ashland; Atofina; BASF; Bayer; Celanese; Filo; Fluka; Great Western; Huntsman Polyurethanes; PMC Spec.; Sigma; Stanchem; Whyte Chems. Ltd

Cyclohexylbenzene. *See* Phenylcyclohexane
1-Cyclohexylbutane. *See* Butylcyclohexane
2-Cyclohexylbutane. *See* s-Butylcyclohexane
N-Cyclohexylcyclohexanamide. *See* Dicyclohexylamine
Cyclohexylmethane. *See* Methyl cyclohexane
N-Cyclohexylmethylamine. *See* N-Methylcyclohexylamine
1-Cyclohexyl-2-pyrrolidinone; N-Cyclohexylpyrrolidinone; N-Cyclohexyl-2-pyrrolidinone; 1-Cyclohexyl-2-pyrrolidone. *See* N-Cyclohexyl pyrrolidone

N-Cyclohexyl pyrrolidone

CAS 6837-24-7; EINECS/ELINCS 229-919-7
Synonyms: 1-Cyclohexyl-2-pyrrolidinone; N-Cyclohexylpyrrolidinone; N-Cyclohexyl-2-pyrrolidinone; 1-Cyclohexyl-2-pyrrolidone; 2-Pyrrolidinone, 1-cyclohexyl-
Empirical: $C_{10}H_{17}NO$
Properties: Liq.; m.w. 167.25; sp.gr. 1.026; f.p. 12 C; b.p. 284 C; flash pt. (CC) 293 F; ref. index 1.4950
Toxicology: LD50 (oral, rat) 370 mg/kg, (skin, rabbit) 1600 mg/kg; LC50 (inh., rat, 1 h) 120 ppm; poison by ing.; mod. toxic by inh. and skin contact; severe skin and eye irritant; may cause permanent eye damage; TSCA listed
Hazardous Decomp. Prods.: Heated to decomp., emits toxic fumes of NO_x
Uses: Solvent for plating engineering plastics, esp. EMS applics.; dyeing of aromatic amide fibers; photo resists; gas separation; elec. insulating coatings; chem. synthesis; absorp. of sulfur dioxide and nitric acid; aromatic extraction; high temp. polyimide processing aid
Manuf./Distrib.: Aldrich; Fluka
Trade Names: CHP®

Cyclomethicone

CAS 540-97-6; 69430-24-6; EINECS/ELINCS 208-764-9; 209-136-7
Synonyms: Cyclic dimethyl polysiloxane; Cyclic dimethyl polysiloxane with n= 3-6; Cyclic dimethylsiloxane; Cyclohexasiloxane, dodecamethyl-; Cyclopentasiloxane; Cyclopentasiloxane, decamethyl-; Cyclopolydimethylsiloxane; Cyclotrisiloxane, hexamethyl-; Decamethylcyclopentasiloxane; Dodecamethylcyclohexasiloxane; Hexamethylcyclotrisiloxane; Polydimethyl siloxy cyclics
Definition: Fully methylated cyclic dimethyl polysiloxane
Empirical: $(C_2H_6OSi)_n$
Formula: $[—(CH_3)_2SiO—]_n$, avg. n = 3-6
Properties: Insol. in water; completely misc. in lower alcohols and other solvs.
Toxicology: LD50 (skin, rabbit) > 16 ml/kg; TSCA listed
Uses: Antistat, emollient, humectant, solvent, visc. control agent in cosmetics; conditioner, emollient for hair care prods., aerosols, sticks, shaving preps., deodorants, antiperspirants; carrier, lubricant, spreading agent, detackifier for skin cleansers; water repellent in pharmaceuticals; surf. tens. reducer; spreading agent, leveling agent for pigments; thread lubricant
Regulatory: FDA approved for orals, injectables, topicals; NF compliance
Manuf./Distrib.: Ashland; Avatar; Biosil Tech.; C.P. Hall; CoKEM Assoc.; Costec; Crompton/Silanes & Spec. Silicones; Dow Corning; GE Silicones; Surfachem Ltd; Universal Preserv-A-Chem; Vopak USA
Trade Names: Chemsil 995; Chemsil 1585; Chemsil 6535; Chemsil 8515
Trade Names Containing: Fancorsil A
See also Decamethylcyclopentasiloxane; Octamethylcyclotetrasiloxane

Cyclonol. *See* Trimethylcyclohexanol

Cyclooctasiloxane

Uses: Solvent in personal care
Features: Nongreasy feel
Trade Names: Chemsil 994

Cyclopentane

CAS 287-92-3; EINECS/ELINCS 206-016-6
UN 1146 (DOT)
Synonyms: Pentamethylene
Classification: Sat. alicyclic hydrocarbon
Empirical: C_5H_{10}
Properties: Colorless liq.; misc. with most org. solvs.; insol. in water; m.w. 70.13; sp.gr. 0.75 (20 C); vapor pressure 400 mm Hg (31 C); m.p. -93 C; b.p. 49.5 C; flash pt. 19.4 F
Toxicology: ACGIH TLV/TWA 600 ppm; mildly toxic by ing. and inh.; eye and skin irritant; CNS depressant; ing. may cause sore throat, nausea, diarrhea; inh. of high vapor concs. can cause nose and throat irritation, light-headedness; very high concs. can cause unconsciousness and death; TSCA listed
Environmental: Zero ozone depletion
Precaution: Flamm.; very dangerous fire hazard exposed to heat or flame; incompat. with strong oxidizing agents (increased fire and explosion hazard)
Hazardous Decomp. Prods.: Heated to decomp., emits acrid smoke and fumes
NFPA: Health 1, Flammability 3, Reactivity 0
Storage: Store in cool, dry, well-ventilated area out of direct sunlight
Uses: Solvent for cellulose acetate, fat and oil extraction; hydrocarbon propellant; motor fuel; azeotropic distillation agent; org. synthesis; laboratory reagent
Manuf./Distrib.: Aldrich; ChevronPhillips; Fluka; Sigma; Spectrum Quality Prods.; Triple Crown Am.
Trade Names: Exxsol® HP-95; Exxsol® Cyclopentane S
Trade Names Containing: Vertrel® MCA Plus; Vertrel® XMS Plus

Cyclopentane, butyl-. *See* Butylcyclopentane
Cyclopentane, 1,1-dimethyl-. *See* 1,1-Dimethylcyclopentane
Cyclopentane, ethyl-. *See* Ethylcyclopentane
Cyclopentane, isobutyl-. *See* Isobutylcyclopentane
Cyclopentane, isopropyl-; Cyclopentane, (1-methylethyl)-. *See* Isopropylcyclopentane
Cyclopentane, (2-methylpropyl)-. *See* Isobutylcyclopentane
Cyclopentanone, 2-heptyl-. *See* 2-Heptylcyclopentanone
Cyclopentasiloxane; Cyclopentasiloxane, decamethyl-. *See* Cyclomethicone
Cyclopentimine. *See* Piperidine
2-Cyclopentylpropane; Cyclopentyl-2 propane. *See* Isopropylcyclopentane
Cyclopolydimethylsiloxane. *See* Cyclomethicone
Cyclotetramethylene oxide. *See* Tetrahydrofuran
Cyclotetramethylene sulfone. *See* Sulfolane
Cyclotetrasiloxane, octamethyl-. *See* Octamethylcyclotetrasiloxane
Cyclotrisiloxane, hexamethyl-. *See* Cyclomethicone
Cycloyhexyl ketone. *See* Cyclohexanone
Cymene. *See* p-Cymene

p-Cymene

CAS 99-87-6; EINECS/ELINCS 202-796-7
FEMA 2356
Synonyms: Benzene, 1-methyl-4-(1-methylethyl)-; Cymene; Cymol; Dolcymene; 1-Isopropyl-4-methylbenzene; 4-Isopropyl-1-methylbenzene; 4-Isopropyltoluene; p-Isopropyltoluene; p-Methylcumene; 1-Methyl-4-isopropylbenzene; p-Methylisopropyl benzene; 1-Methyl-4-(1-methylethyl) benzene; Paracymene; Paracymol
Classification: Aromatic hydrocarbon
Definition: Obtained chiefly from the wash water of sulfite paper
Empirical: $C_{10}H_{14}$
Formula: $(CH_3)_2CHC_6H_4CH_3$
Properties: Colorless to pale yel. liq., odorless; sol. in alcohol, ether, acetone, benzene, oxygenated and aromatic solvs.; insol. in water; m.w. 134.24; dens. 0.853; m.p. -68 C; b.p. 176 C; flash pt. (CC) 117 F; ref. index 1.489
Toxicology: LD50 (oral, rat) 4750 mg/kg; mildly toxic by ingestion; human CNS effects at low doses; skin irritant; mutagenic data; TSCA listed
Precaution: Flamm. or combustible liq.; sl. explosion hazard in vapor form
Hazardous Decomp. Prods.: Heated to decomp., emits acrid smoke and fumes
Uses: Solvent; mfg. of syn. resins; metal polishes; organic synthesis;

synthetic flavoring agent in foods and pharmaceuticals; masking agent in cosmetics
Features: Citrus flavor
Regulatory: FDA 21CFR §172.515; FEMA GRAS
Manuf./Distrib.: Aldrich; Degussa AG/Health & Nutrition; Fluka; Millennium/F&F; Millennium; Sigma
Trade Names Containing: Dipentene-C

3-p-Cymenol; p-Cymen-3-ol. *See* Thymol
Cymol. *See* p-Cymene

D

DAA. *See* Diacetone alcohol
DAMP. *See* 1,3-Pentanediamine
DANA. *See* N-Nitrosodiethylamine
Dapsone. *See* Sulfolane
DATEM. *See* Diacetyl tartaric acid esters of mono- and diglycerides
DBA. *See* Dibutylamine
DBBP. *See* Dibutyl butyl phosphonate
DBM. *See* Methylene bromide
DBP. *See* Dibutyl phthalate
DBPC. *See* BHT
DCB. *See* o-Dichlorobenzene
DCBTF; 3,4-DCBTF. *See* 3,4-Dichlorobenzotrifluoride
1,1-DCE. *See* Vinylidene chloride monomer
1,2-DCE. *See* trans-1,2-Dichloroethylene; Ethylene dichloride
trans-1,2-DCE. *See* trans-1,2-Dichloroethylene
DCEE. *See* sym-Dichloroethyl ether
DCHA. *See* Dicyclohexylamine
DCM. *See* Methylene chloride
DDBSA. *See* Dodecylbenzenesulfonic acid
DEA. *See* Diethylamine; Diethanolamine
Dead oil. *See* Creosote oil
DEAE. *See* Diethylaminoethanol
Deanol. *See* Dimethylethanolamine
DEC. *See* Diethyl carbonate
Dec. *See* Decahydronaphthalene
1,1,1,2,2,3,4,5,5,5-Decafluoropentane. *See* 2,3-Dihydrodecafluoropentane

Decahydronaphthalene

CAS 91-17-8; EINECS/ELINCS 202-046-9
UN 1147 (DOT)
Synonyms: Bicyclo (4.4.0) decane; Dec; Decalin; Decalin solvent; Dekalin; DHN; Naphthalane; Naphthane; Perhydronaphthalene
Definition: Mixture of cis and trans
Empirical: $C_{10}H_{18}$
Properties: Colorless liq.; aromatic odor; sol. in alcohol, ether; insol. in water; m.w. 138.14; dens. 0.8963; m.p. -43.26 C; b.p. 185.5 C; flash pt. 58 C; ref. index 1.474
Toxicology: LD50 (oral, rat) 4170 mg/kg, (skin, rabbit) 5900 mg/kg; TCLo (inh., human) 100 ppm; mod. toxic by inh. and ing.; mildly toxic by skin contact; primary skin and eye irritant; conjunctive irritant; may cause pulmonary system changes, kidney damage; human tumorigen; questionable carcinogen; mutagenic data; TSCA listed
Precaution: Flamm.; mod. fire risk; can react with oxidizing materials
Hazardous Decomp. Prods.: Heated to decomp., emits acrid smoke and fumes
NFPA: Health 2, Flammability 2, Reactivity 0
Uses: Solvent, stabilizer for shoe creams and floor waxes; solvent in paint and lacquers, for oils, fats, waxes, resins, rubber, and asphalt; substitute for turpentine; cleaning machinery; stain removal; cleaning fluids; lubricants; motor fuel additive
Manuf./Distrib.: Aldrich; Degussa; Fluka; Sigma; Spectrum Quality Prods.
Trade Names: Decalin; Dekalin®

cis-Decahydronaphthalene. *See* cis-Decalin
trans-Decahydronaphthalene. *See* trans-Decalin
(1S-(1α,3aβ,4α,8aβ))-Decahydro-4,8,8-trimethyl-9-methylene-1,4-methanoazulene; Decahydro-4,8,8-trimethyl-9-methylene-1,4-methanoazulene. *See* Longifolene
Decalin. *See* Decahydronaphthalene

cis-Decalin

CAS 493-01-6
Synonyms: cis-Bicyclo [4.4.0] decane; cis-Decahydronaphthalene; Naphthalene, decahydro-, cis-; cis-Perhydronaphthalene
Empirical: $C_{10}H_{18}$
Properties: M.w. 138.25; dens. 0.888 g/ml; visc. 2.99 cp; m.p. -43 C; b.p. 193 C; flash pt. (CC) 137 F; ref. index 1.48; surf. tens. 29.29 dynes/cm
Uses: Solvent
Manuf./Distrib.: ABCR; Alfa Aesar; ICN Biomed. Research Prods.; Lancaster Synthesis; TCI Am.

trans-Decalin

CAS 493-02-7
Synonyms: trans-Bicyclo [4.4.0] decane; trans-Decahydronaphthalene
Empirical: $C_{10}H_{18}$
Properties: M.w. 138.25; dens. 0.866 g/ml; visc. 1.93 cp; b.p. 185 C; flash pt. (CC) 127 F; surf. tens. 26.42 dynes/cm
Uses: Solvent

Decalin solvent. *See* Decahydronaphthalene

Decamethylcyclopentasiloxane

CAS 541-02-6; EINECS/ELINCS 208-764-9
Synonyms: Cyclic dimethylsiloxane pentamer; Cyclic pentamer-D5; Decamethylpentacyclosiloxane; Dimethylsiloxane pentamer
Classification: Cyclic volatile methyl siloxane
Empirical: $C_{10}H_{30}O_5Si_5$
Properties: Oily liq.; m.w. 370.78; dens. 0.9593; vapor pressure 2 mm (50 C); m.p. -38 C; b.p. 210 C; flash pt. 55 C; ref. index 1.3982; dielec. const. 2.50
Toxicology: Skin and eye irritant; TSCA listed
Precaution: Combustible
Hazardous Decomp. Prods.: Heated to decomp., emits acrid smoke and irritating vapors
HMIS: Health 1, Flammability 2, Reactivity 0
Uses: Coupling agent; release agent; lubricant; blocking agent; chemical intermediate; synthesis; hair care; skin care; antiperspirants; deodorants; cleaning solvent
Manuf./Distrib.: Aldrich; Fluka; Gelest
See also Cyclomethicone

Decamethylenediol; 1,10-Decamethylenediol; 1,10-Decamethylene diol; Decamethylene glycol; 1,10-Decamethylene glycol. *See* 1,10-Decanediol
Decamethylpentacyclosiloxane. *See* Decamethylcyclopentasiloxane
Decanal dimethyl acetal. *See* Decyl alcohol
Decanamide, N,N-dimethyl-. *See* Dimethyl caproamide
Decane. *See* n-Decane

n-Decane

CAS 124-18-5; EINECS/ELINCS 204-686-4
UN 2247 (DOT)

Synonyms: Decane
Classification: Sat. aliphatic hydrocarbon
Empirical: $C_{10}H_{22}$
Formula: $CH_3(CH_2)_8CH_3$
Properties: Colorless liq.; gasoline-like odor; sol. in most org. solvs., insol. in water; m.w. 142.29; sp.gr. 0.734; m.p. -29.6 C; b.p. 174 C; flash pt. 46 C; ref. index 1.4114 (20 C)
Toxicology: LC50 (inh., mouse) 2 h) 72,300 mg/m^3; TDLo (skin, mouse, 52 wks intermittent) 25 g/kg; relatively low oral toxicity unless aspirated into lungs; mod. to severe skin irritant; probably mod. to severe eye irritant; narcotic; simple asphyxiant; conc. vapor may cause nose/throat irritation, headache, drowsiness, nausea, difficulty breathing; ing. may cause nausea, vomiting, swelling of abdomen, headache, depression; questionable carcinogen, tumorigen; TSCA listed
Precaution: Combustible; mod. fire risk; incompat. with strong oxidizing agents (increases fire and explosion hazard); mod. explosive as vapor
Hazardous Decomp. Prods.: None
NFPA: Health 0, Flammability 2, Reactivity 0
Storage: Store in cool, dry, well-ventilated area out of direct sunlight
Uses: Component of gasoline, jet fuel, kerosene, petrol. solvents such as wh. spirit; solvent; standardized hydrocarbon; org. synthesis; paper and rubber industries; constituent of polyolefin mfg. wastes; cosmetics ingred.
Manuf./Distrib.: Aldrich; ChevronPhillips; Fluka; Sigma; Triple Crown Am.

1,2-Decane carbonate

CAS 99260-48-7
Properties: M.w. 200.28; dens. 0.982 g/ml; visc. 13.00 cp; b.p. 278 C; flash pt. (CC) 295 F; surf. tens. 31.01 dynes/cm
Uses: Solvent

Decanedioic acid, bis (2-ethylhexyl) ester. *See* Dioctyl sebacate
Decanedioic acid, bis (1-methylethyl) ester. *See* Diisopropyl sebacate
Decanedioic acid, dimethyl ester. *See* Dimethyl sebacate

1,10-Decanediol

CAS 112-47-0; EINECS/ELINCS 203-975-2
Synonyms: 1,6-Bis (2-hydroxyethyl) hexane; Decamethylenediol; 1,10-Decamethylenediol; 1,10-Decamethylene diol; Decamethylene glycol; 1,10-Decamethylene glycol; α,ω-Decanediol; 1,10-Dihydroxydecane
Empirical: $C_{10}H_{22}O_2$
Properties: Wh. chips; m.w. 174.28; m.p. 72-74 C
Toxicology: May cause eye/skin irritation; may be harmful by inh., ing., skin absorp.; may cause GI disturbances, nausea, vomiting, headache; TSCA listed
Precaution: Incompat. with acid chlorides, acid anhydrides, oxidizing agents, chloroformates, reducing agents
Hazardous Decomp. Prods.: CO, CO_2
Storage: Store in cool, dry place; keep tightly closed
Uses: Pharmaceuticals; personal care prods.; flavors; fragrances; polyesters; acrylates; epoxy diluents; plastics; coatings; inks; photographic materials; textiles; leather; lubricants; polyurethanes; agric.; solvents; modifiers in alkyd resins; fine chems.; plasticizers
Manuf./Distrib.: ABCR; Acros Org.; Aldrich; Alfa Aesar; Fluka; ICN Biomed. Research Prods.; Int'l. Chem. Group; Jarchem Ind.; Lancaster Synthesis; Penta Mfg.; Sithean; TCI Am.; Universal Preserv-A-Chem
Trade Names: Speziol C 10/2

α,ω-Decanediol. *See* 1,10-Decanediol
Decanoic acid, 2,2-dimethyl-1,3-propanediol diester. *See* Neopentyl glycol dicaprate
Decanoic acid ethyl ester. *See* Ethyl decanoate
Decanoic acid, methyl ester. *See* Methyl caprate
Decanoic acid, 1-methyl-1,2-ethanediyl ester mixed with 1-methyl-1,2-ethanediyl dioctanoate; Decanoic acid, mixed diesters with octanoic acid and propylene glycol. *See* Propylene glycol dicaprylate/dicaprate
Decanoic acid, monoester with 1,2,3-propanetriol. *See* Glyceryl caprate
Decanol; 1-Decanol; Decan-1-ol; n-Decanol. *See* Decyl alcohol

Deceth-6

CAS 26183-52-8 (generic)
Synonyms: 3,6,9,12,15,18-Hexaoxaoctacosan-1-ol; PEG-6 decyl ether; PEG 300 decyl ether; POE (6) decyl ether
Definition: PEG ether of decyl alcohol
Empirical: $C_{22}H_{46}O_7$
Formula: $CH_3(CH_2)_8CH_2(OCH_2CH_2)_nOH$, avg. n = 6
Properties: Nonionic
Uses: Wetting agent, detergent, penetrant, emulsifier for textiles, clay soils, firefighting; surfactant, detergent raw material and additive in dishwashing agents, in mfg. of tablet soaps; emulsifier in cosmetics
Manuf./Distrib.: DelCon
Trade Names Containing: Ship Shape® Resin Cleaner

Decyl alcohol

CAS 112-30-1; 68526-85-2; EINECS/ELINCS 203-956-9
FEMA 2365
Synonyms: Alcohol C10; C10 alcohol; Capric alcohol; Caprinic alcohol; Decanal dimethyl acetal; Decanol; 1-Decanol; Decan-1-ol; n-Decanol; n-Decyl alcohol; Decylic alcohol; 1-Hydroxydecane; Noncarbinol; Nonylcarbinol; Primary decyl alcohol
Classification: Primary aliphatic alcohol
Empirical: $C_{10}H_{22}O$
Formula: $CH_3(CH_2)_8CH_2OH$
Properties: Colorless to lt. yel. mod. visc. liq., sweet odor; sol. in alcohol, ether, acetone, benzene, chloroform, acetic acid; pract. insol. in water; m.w. 158.32; sp.gr. 0.8297 (20/4 C); m.p. 7 C; b.p. 232.9 C; flash pt. (OC) 82 C; ref. index 1.43587; surf. tens. 28.9 dynes/cm (20 C)
Toxicology: LD50 (oral, rat) 4720 mg/kg; mod. toxic by skin contact; mildly toxic by ing., inh.; irritating to eyes, skin, respiratory system; inh. may cause nose/throat irritation; high concs. may cause CNS depression, headache, nausea, drowsiness; ing. may cause headache, dizziness, nausea, vomiting, and unconsciousness or coma in severe cases; if aspirated into lungs, may cause severe lung damage and possibly death; possible carcinogen, tumorigen; TSCA listed
Precaution: Combustible; flamm. when exposed to heat or flame; incompat. with oxidizing agents (increases fire/explosion risk)
Hazardous Decomp. Prods.: Thermal decomp. prods.: CO, CO_2; heated to decomp., emits acrid smoke and irritating fumes
NFPA: Health 0, Flammability 2, Reactivity 0
Storage: Store in cool, dry area
Uses: In mfg. of plasticizers, syn. lubricants, petroleum additives, herbicides, surfactants, solvents, coatings, cosmetics, laundry softeners, textile/leather additives, lubricating oil additives, plasticizers, plastics additives; emulsion polymerization; synthetic flavoring agent in foods and pharmaceuticals; growth retardant; stabilizer with calcium isostearate for calcium stearate disps. for food-contact paper/paperboard coatings; emollient; visc. control agent; antifoam; solvent; fragrance ingred. in perfumes; in food-contact coatings
Features: Fruity flavor
Regulatory: FDA 21CFR §172.515, 172.864, 175.300, 176.170, 177.1390, 178.3480, 178.3910; FEMA GRAS; Japan approved as flavoring
Manuf./Distrib.: Albemarle; Aldrich; Brown; Cognis/Chems. Group; ExxonMobil; Fleurchem; Fluka; Grau Aromatics; Haarmann & Reimer GmbH; Indofine; Integra; J.T. Baker; Jarchem Ind.; Lluch Essence; M. Michel; Mosselman NV; Penta Mfg.; Sasol Germany; Sasol N. Am.; Schweizerhall; Sigma; Spectrum Quality Prods.; Universal Preserv-A-Chem; VWR Int'l.

n-Decyl alcohol; Decylic alcohol. *See* Decyl alcohol
Decyl 9-octadecenoate. *See* Decyl oleate
Decyl octyl alcohol. *See* Stearyl alcohol

Decyl oleate

CAS 3687-46-5; EINECS/ELINCS 222-981-6
Synonyms: Decyl 9-octadecenoate; 9-Octadecenoic acid decyl ester
Definition: Ester of decyl alcohol and oleic acid
Empirical: $C_{28}H_{54}O_2$
Formula: $CH_3(CH_2)_7CH{=}CH(CH_2)_7COOCH_2(CH_2)_8CH_3$
Properties: Wh. to straw liq.; sol. in min. oil, 95% ethanol, IPM, oleyl alcohol; insol. in water; m.w. 422.74; dens. 0.855-0.865; m.p. 3-6 C; ref. index 1.4530-1.4555
Toxicology: TSCA listed
Uses: Emollient, cosolvent, lubricant, superfatting agent for cosmetics, creams, lotions, hand cleaners, moisturizers, bath oils, pressed powd., pharmaceuticals; pigment dispersant/wetting agent

Regulatory: BP compliance
Manuf./Distrib.: A&E Connock; Mosselman NV
Trade Names: Ceraphyl® 140; Cerette V; Estalan DO; Tegosoft® DO

Decyltetradecanol; 2-Decyltetradecanol. *See* 2-Decyl-1-tetradecanol

2-Decyl-1-tetradecanol

CAS 58670-89-6; EINECS/ELINCS 216-385-0
Synonyms: Decyltetradecanol; 2-Decyltetradecanol; Isolignoceryl alcohol; Isotetracosanol; Isotetracosyl alcohol; 1-Tetradecanol, 2-decyl-
Classification: Aliphatic alcohol
Empirical: $C_{24}H_{50}O$
Properties: Liq.; insol. in water; m.w. 354.66; dens. 0.84 kg/l (17 C)
Uses: Emollient, solvent, solubilizer for cosmetics, toiletries; solubilizer in detergents, lubricants, emulsifiers, solvs., rubber, plastics, polymers, metal processing, paper coating, and textiles; solubilizer for lipid-sol. actives, pharmaceutical creams, ointments, emulsions, sticks
Trade Names: Jarcol™ I-24

DEET. *See* Diethyl toluamide
DEG. *See* Diethylene glycol
DEGBE. *See* Butoxydiglycol
DEGEE. *See* Ethoxydiglycol
DEGHE. *See* Diethylene glycol n-hexyl ether
DEGiBE. *See* Diethylene glycol isobutyl ether
DEGME. *See* Methoxydiglycol
DEGMPE. *See* Diethylene glycol methylpentyl ether
DEHM. *See* Bis (2-ethylhexyl) maleate
Deionized water. *See* Water
DEIP. *See* Diethyl isophthalate
DEK. *See* Diethyl ketone
Dekalin. *See* Decahydronaphthalene
DELA. *See* Diethanolamine
Delphene; m-Delphene. *See* Diethyl toluamide
DEN. *See* N-Nitrosodiethylamine; Diethylamine
DENA. *See* N-Nitrosodiethylamine

Denatonium benzoate

CAS 3734-33-6 (anhyd.); 86398-53-0 (monohydrate); EINECS/ELINCS 223-095-2
Synonyms: Ammonium, benzyldiethyl ((2,6-xylylcarbamoyl) methyl)-, benzoate; Benzenemethanaminium, N-(2-((2,6-dimethylphenyl) amino)-2-oxoethyl)-N,N-diethyl-, benzoate; Benzyldiethyl [(2,6-xylylcarbomoyl) methyl] ammonium benzoate; N-[2-[(2,6-Dimethylphenyl) amino]-2-oxoethyl]-N,N-diethylbenzenemethanaminium benzoate; Lignocaine benzyl benzoate
Classification: Organic compd.
Empirical: $C_{28}H_{34}N_2O_3$; $C_{28}H_{34}N_2O_3 \cdot H_2O$
Formula: $C_{21}H_{29}N_2O \cdot C_7H_5O_2$
Properties: Wh. cryst. powd., odorless, extremely bitter taste; very sol. in chloroform, methanol; sol. in water, alcohol; sparingly sol. in acetone; pract. insol. in ether; m.w. 446.59 (anhyd.); m.p. 174-176 C; pH 6.5-7.5 (3%)
Toxicology: Anhyd.: LD50 (oral, rat) 584 mg/kg; harmful solid; mod. toxic by ing.; irritant; TSCA listed
Hazardous Decomp. Prods.: Heated to decomp., emits toxic vapors of NO_x
Storage: Hygroscopic (anhyd.)
Uses: Aversive bitterant added to toxic substances as a deterrent to ingestion; pharmaceutical aid (alcohol denaturant, flavor); replacement for brucine or quassin as denaturant for ethyl alcohol; denaturant in cosmetics, household prods.
Regulatory: FDA 27CFR §21.151; NF compliance
Manuf./Distrib.: Aldrich; Alfa Chem; Atomergic Chemetals; B.I. Chems.; Barrington; Burlington Bio-Medical; Ceres Chem.; Chart; EM Ind.; Macfarlan Smith; PMC Spec.; Robeco; Sigma; Spectrum Quality Prods.; Universal Preserv-A-Chem
Trade Names Containing: Propsolv I-1 190 Proof; Propsolv I-1 Anhydrous; Propsolv III-1 190 Proof; Propsolv III-1 Anhydrous; Punctilious® SDA 1-1 190 Proof; Punctilious® SDA 1-1 Anhydrous; Punctilious® SDA 40B Anhydrous; Punctilious® SDA 40B Modified Anhydrous

Denatured alcohol; Denatured ethanol; Denatured ethyl alcohol; Denatured spirits. *See* Alcohol denatured

Deodorized kerosene

CAS 8008-20-6; EINECS/ELINCS 232-366-4
UN 1223 (DOT)
Synonyms: Kerosene, deodorized; Kerosene, odorless; Odorless kerosene
Definition: Mixt. of petrol.-derived hydrocarbons contg. 10 to 16 carbon atoms
Toxicology: TSCA listed
Precaution: DOT: Flamm. liq.
HMIS: Health 1, Flammability 2, Reactivity 1
Uses: Solvent for paints/coatings, printing inks, floor polishes, metal degreasing/cleaning; domestic lamp oil; solvent, defoamer, diluent for cosmetic and pharmaceutical creams and lotions; solvent, defoamer in froth flotation of vegetables; dust control agent; quench oil; diluent for pesticides; froth flotation; liq. candles; charcoal lighter fluid; polishes; cleaners; defoamer in food-contact coatings and paper/paperboard
Regulatory: FDA 21CFR §176.200, 176.210
Manuf./Distrib.: Alvin Chem.; Ruger
Trade Names: PD-23

Deodorized winterized cottonseed oil. *See* Cottonseed (Gossypium) oil
2-Deoxyglycerol. *See* Trimethylene glycol
4-Deoxytetronic acid. *See* Butyrolactone
DEP. *See* Diethyl phthalate
DET. *See* Diethyl toluamide
DETA. *See* Diethylenetriamine
DETN. *See* Diethylamine
Dewaxed lanolin. *See* Lanolin oil
Dextro-limonene. *See* d-Limonene
DFBM. *See* Dibromodifluoromethane
DGA. *See* 2-(2-Aminoethoxy) ethanol
DHN. *See* Decahydronaphthalene
Diacetic ether. *See* Ethylacetoacetate
Diacetin. *See* Glyceryl diacetate
Diacetone. *See* Diacetone alcohol

Diacetone alcohol

CAS 123-42-2; EINECS/ELINCS 204-626-7
UN 1148 (DOT)
Synonyms: DAA; Diacetone; Diketone alcohol; 4-Hydroxy-1-keto-4-methylpentane; 4-Hydroxy-4-methyl-2-pentanone; 4-Hydroxy-4-methylpentanone-2; 4-Hydroxy-4-methyl pentan-2-one; 2-Methyl-2-pentanol-4-one; 2-Pentanone, 4-hydroxy-4-methyl-
Classification: Ketone
Empirical: $C_6H_{12}O_2$
Formula: $(CH_3)_2C(OH)CH_2COCH_3$
Properties: Colorless oily liq.; faint minty pleasant odor; sol. in water; m.w. 116.16; sp.gr. 0.940 (20/20 C); vapor pressure 1.1 mm Hg (20 C); f.p. -47 F; b.p. 145.2-172 C; flash pt. (TCC) 126 F; autoignition temp. 603 C; ref. index 1.4234; surf. tens. 31 dynes/cm; dielec. const. 18.2
Toxicology: ACGIH TLV/TWA 50 ppm; LD50 (oral, rat) 4 g/kg, (IP, mouse) 933 mg/kg, (skin, rabbit) 13,500 mg/kg; TCLo (inh., human) 100 ppm; mod. toxic by ing., IP routes; mildly toxic by skin contact; primary irritant; may cause headache, nausea, vomiting; can cause pulmonary changes, anemia, liver/kidney damage; skin, mucous membrane, severe eye irritant; narcotic in high concs.; TSCA listed
Environmental: VOC; BOD5 0.68; COD 2.11; ThOD 2.21
Precaution: Flamm. exposed to heat or flame; LEL 1.8%; UEL 6.9%; can react with oxidizing materials; explosive as vapor exposed to heat or flame
Hazardous Decomp. Prods.: Heated to decomp., emits acrid smoke and irritating fumes
NFPA: Health 1, Flammability 2, Reactivity 0
Uses: Solvent for nitrocellulose, cellulose acetate, oils, resins, waxes, fats, dyes, tars, lacquers, dopes, water-based coatings, printing inks, wood preservatives, rayon, artificial leather, metal cleaning, cosmetics; laboratory reagent; hydraulic fluids; textile stripping agent; in food-pkg. adhesives
Regulatory: FDA 21CFR §175.105
Manuf./Distrib.: Aldrich; Allchem Ind.; Amyl; Ashland; Asiamerica Int'l.;

Atofina SA; Atofina; BP Chems. Ltd; Baychem; Browning; Celanese; Chemcentral; Coyne; Delta Distributors; Dow; Eastech; Fabrichem; Fluka; Great Western; Harcros; Herdillia Chems. Ltd; Houghton Chem.; Indofine; Integra; J.T. Baker; Milsolv; P.B. & S.; Penta Mfg.; R.E. Carroll; Rhodia HPCII; Richard's Pacific Chem. Sourcing; SOCO-Lynch; Shell; Southchem; Spectrum Quality Prods.; Stanchem; Sunnyside; Textile Chem.; VWR Int'l.; Vopak USA; Whittaker, Clark & Daniels

Trade Names: Ucar® Diacetone Alcohol

Diacetonyl. *See* 2,5-Hexanedione
1,2-Diacetoxyethane. *See* Ethylene glycol diacetate
1,2-Diacetoxypropane. *See* Propylene glycol diacetate
Diacetylated monoglyceride. *See* Diacetylated monoglycerides

Diacetylated monoglycerides

Synonyms: Diacetylated monoglyceride
Definition: Glycerin esterified with edible fat-forming fatty acids and acetic acid
Properties: Clear liq.; sol. in 80% aq. alcohol, veg. and min. oils; acid no. 3 max.; sapon. no. 365-395; hyd. no. 15 max.
Storage: Preserve in tight, light-resist. containers
Uses: Excipient, solvent, plasticizer for pharmaceuticals; food additive
Regulatory: FDA approved for orals; NF compliance

Diacetylated tartaric acid ester glycerides. *See* Diacetyl tartaric acid esters of mono- and diglycerides
1,2-Diacetylethane; α,β-Diacetylethane. *See* 2,5-Hexanedione
Diacetyl ether. *See* Ethylacetoacetate
Diacetyl fatty acid monoglyceride tartrate. *See* Diacetyl tartaric acid esters of mono- and diglycerides
Diacetylglycerol. *See* Glyceryl diacetate
Diacetylmethane. *See* Acetylacetone

Diacetyl tartaric acid esters of mono- and diglycerides

CAS 91052-81-2; 977051-29-8
INS472e; E472f
Synonyms: DATEM; Diacetylated tartaric acid ester glycerides; Diacetyl fatty acid monoglyceride tartrate; Diacetyltartaric and fatty acid esters of glycerol; Mono- and diglycerides, diacetyl tartaric acid esters; Mono- and diglycerides of fatty acids esterified with mono- and diacetyltartaric acid
Definition: Esters of glycerol with mono- and diacetyl tartaric acids and fatty acids
Properties: Sticky visc. liqs., amber paste, ivory to yel. waxy flake or powd.; sol. in methanol, ethanol, acetone; disp. in cold and hot water; m.p. 15-43 C; HLB 8.0; Nonionic
Precaution: Hydrolyzes in moist air to liberate acetic acid
Hazardous Decomp. Prods.: Heated to decomp., emits acrid smoke and irritating fumes
Uses: Emulsifier, stabilizer, sequestrant, flavoring agent for foods, beverages; antistaling agent, starch complexing agent; controls starch gelation; dough conditioner for baked goods; carrier or solvent for colors and food antioxidants
Regulatory: FDA 21CFR §136.110, 182.4101, 184.1101, GRAS; USDA 9CFR §318.7, 0.5% in margarine; 381.147; ADI 0-50 mg/kg body wt.
Manuf./Distrib.: AB R Lundberg; Adumim Food Ingreds.; Degussa AG/ Health & Nutrition; Food Ingred. Tech. Ltd

Diacetyltartaric and fatty acid esters of glycerol. *See* Diacetyl tartaric acid esters of mono- and diglycerides
Di-n-alkyl (C7-C9) phthalate. *See* n-Dioctyl phthalate
2,2´-Diaminodiethylamine. *See* Diethylenetriamine
1,2-Diaminoethane. *See* Ethylenediamine
1,2-Diaminoethane dihydrochloride. *See* Ethylenediamine dihydrochloride
1,3-Diaminopentane. *See* 1,3-Pentanediamine
Diamylamine. *See* Di-n-amylamine

Di-n-amylamine

CAS 2050-92-2
UN 2841 (DOT)
Synonyms: Diamylamine; Dipentylamine; Di-n-pentylamine; 1-Pentanamine, N-pentyl; N-Pentyl-1-pentanamine; Pentyl pentylamine
Classification: Sat. aliphatic amine
Empirical: $C_{10}H_{23}N$
Formula: $(CH_3CH_2CH_2CH_2CH_2)_2NH$
Properties: Colorless liq.; ammoniacal odor; sol. in dimethyl sulfoxide, ethanol, ether, acetone; pract. insol. in water; m.w. 157.3; dens. 0.777 (20 C); m.p. -44 C; b.p. 210-211 C; flash pt. (CC) 52 C; strong base
Toxicology: LD50 (oral, rat) 270 mg/kg, (skin, rabbit) 350 mg/kg; poison by inh., ing., skin contact; corrosive; can cause skin irritation, burns; may be absorbed thru skin in toxic amts.; can cause eye irritation, 'halo vision,' possible permanent injury; inh. can cause nose/throat/ lung irritation, coughing, chest pain, difficulty breathing; high concs. may cause potentially fatal pulmonary edema; ing. may cause burning of mouth, throat, digestive tract, headache, nausea, faintness, anxiety; TSCA listed
Precaution: Combustible; mod. fire risk; incompat. with oxidizing agents (fire/explosion risk), strong acids, acid chlorides, acid anhydrides, mercury (violent reactions possible)
Hazardous Decomp. Prods.: Heated to decomp., emits toxic fumes of NO_x
NFPA: Health 3, Flammability 2, Reactivity 0
Storage: Store in a cool, dry, well-ventilated area, out of direct sunlight; limit quantities in storage
Uses: Anticorrosion agent for ferrous metals; solvent for oils, resins, some cellulose esters; raw material for vulcanization accelerators, lubricant additives, antioxidants; flotation reagents; dyestuffs
Manuf./Distrib.: Aldrich; Atofina; Celanese; Fluka

Diamyl ether

CAS 693-65-2; EINECS/ELINCS 211-756-8
Synonyms: Amyl ether; n-Amyl ether; Di-n-amyl ether; Dipentyl ether; Di-n-pentyl ether; 1,1´-Oxybispentane; Pentane, 1,1´-oxybis-; Pentyl ether; 1-(Pentyloxy) pentane; Pentyoxypentane
Classification: Ether
Empirical: $C_{10}H_{22}O$
Properties: Colorless liq.; m.w. 158.32; sp.gr. 0.779; visc. 0.922 cP; vapor pressure 0.6 mm Hg; f.p. -69.4 C; b.p. 186.8 C; flash pt. 57 C; autoignition temp. 170 C; ref. index 1.4119; surf. tens. 24.76 dynes/ cm; dielec. const. 2.77
Toxicology: LD50 (IV, mouse) 164 mg/kg; may be harmful by inh., ing., skin absorp.; may cause eye/skin irritation; TSCA listed
Environmental: ThOD 1.92
Precaution: Combustible; incompat. with strong oxidizing agents, strong acids
Hazardous Decomp. Prods.: Toxic fumes of CO, CO_2
NFPA: Health 1, Flammability 2, Reactivity 0
Storage: Store in cool, dry place away from heat, open flame; keep tightly closed; store under nitrogen; forms explosive peroxides on prolonged storage
Uses: Solvent
Manuf./Distrib.: Acros Org.; Aldrich; Atomergic Chemetals; CTC Orgs.; Eastman; Fluka; Pfaltz & Bauer; TCI Am.

Di-n-amyl ether. *See* Diamyl ether

Diamyl oxalate

CAS 20602-86-2
Synonyms: Dipentyl oxalate; Ethanedioic acid, dipentyl ester
Classification: Ester
Empirical: $C_{12}H_{22}O_4$
Properties: M.w. 230.18
Environmental: ThOD 2.16
NFPA: Health 0, Flammability 1, Reactivity 0
Uses: Solvent

1,4:3,6-Dianhydro-2,5-di-O-methyl-D-glucitol. *See* Dimethyl isosorbide
DIBA. *See* Diisobutyl adipate

Dibenzofuran

CAS 132-34-9; 38178-38-0
Synonyms: Diphenylene oxide
Empirical: $C_{12}H_8O$
Properties: Cryst. solid; sol. in oxygenated and aromatic solvs.; sl. sol. in alcohol, ether, benzene; insol. in water; m.p. 87 C; b.p. 288 C
Uses: Insecticide
Manuf./Distrib.: Chugai Boyeki Am.; Dow

Trade Names Containing: Methylnaphthalene Fraction

Dibenzoyldiethyleneglycol ester. *See* Diethylene glycol dibenzoate
Dibenzoyl dipropylene glycol ester. *See* Dipropylene glycol dibenzoate
Dibenzyl ether. *See* Benzyl ether
DIBK. *See* Diisobutyl ketone
DIBP. *See* Diisobutyl phthalate
Dibromine. *See* Bromine

Dibromodifluoromethane

CAS 75-61-6; EINECS/ELINCS 200-885-5
UN 1941 (DOT)
Synonyms: DFBM; Difluorodibromomethane; R 12B2
Classification: Halogenated aliphatic hydrocarbon
Empirical: CBr_2F_2
Properties: Colorless heavy liq. or gas; odorless or char. odor; sol. in methanol, ether, acetone, benzene; insol. in water; m.w. 209.83; sp.gr. 2.288 (15/4 C); m.p. -141 C; b.p. 24.5 C; ref. index 1.399 (12 C); noncombustible
Toxicology: ACGIH TLV/TWA 100 ppm; LC50 (inh., mouse, 2 h) 140 g/m^3; mildly toxic by inh.; irritant; inh. may cause nose/throat irritation, CNS depression; liq. may cause cooling of skin and eyes due to rapid evaporation; liq. may cause defatting of skin; TSCA listed
Precaution: Reacts with chemically active metals (sodium, potassium, calcium, powd. aluminum, zinc, magnesium)
Hazardous Decomp. Prods.: Heated to decomp., emits very toxic fumes of Br^- and F^-
Storage: Store liq. or gas cylinders in cool, dry, well-ventilated areas; normally handled in closed systems
Uses: Fire-extinguishing agent; solvent for prep. of explosive mixtures; polymer intermediate; synthesis of pharmaceuticals, dyes, quat. ammonium compds.; some use as tracer gas in environmental studies
Manuf./Distrib.: Aldrich; Fluka

Dibromomethane. *See* Methylene bromide
Dibuprol. *See* 1,3-Dibutoxy-2-propanol
Dibutoxy diethylene glycol. *See* Diethylene glycol dibutyl ether
1,2-Dibutoxyethane; 1,2-Di-n-butoxyethane. *See* Ethylene glycol dibutyl ether
2,2´-Dibutoxyethyl ether. *See* Diethylene glycol dibutyl ether

1,3-Dibutoxy-2-propanol

CAS 2216-77-5
Synonyms: Dibuprol; Di-n-butyl-glycerol diether; Glycerine-1,3-dibutylether; 2-Propanol, 1,3-di-n-butoxy-
Empirical: $C_{11}H_{24}O_3$
Properties: M.w. 204.31
Toxicology: LD50 (IP, mouse) 1010 mg/kg; may cause muscle weakness, respiratory depression, flaccid paralysis without anesthesia
Uses: Solvent, intermediate for cosmetics

Dibutyl adipate

CAS 105-99-7; EINECS/ELINCS 203-350-4
Synonyms: Adipic acid, dibutyl ester; Butyl adipate; Di-n-butyl adipate; Dibutyl adipinate; Dibutyl hexanedioate; Hexanedioic acid, dibutyl ester
Definition: Diester of butyl alcohol and adipic acid
Empirical: $C_{14}H_{26}O_4$
Formula: $[-CH_2CH_2CO_2(CH_2)_3CH_3]_2$
Properties: M.w. 258.36; dens. 0.962; b.p. 305 C; flash pt. > 110 C
Toxicology: LD50 (oral, rat) 12,900 mg/kg, (skin, rabbit) 20,000 mg/kg; mildly toxic; primary irritant to skin and eyes; experimental teratogen, reproductive effects; TSCA listed
Hazardous Decomp. Prods.: Heated to decomp., emits acrid smoke and irritating fumes
Uses: Emollient, oily component for day creams, pharmaceutical sun and skin protective oils; plasticizer for cellulosics, PVAc, PVC, VCA; agric. chem.; solvent for lipid-sol. ingreds.; lubricant for metalworking fluids, rolling, stamping and drawing oils, greases, motor oils, extrusion of ceramics; fatting agent in textile and leather auxs.; plasticizer and lubricant for textiles
Manuf./Distrib.: A&E Connock; Aldrich; Boehme Filatex; ChemService; Shandong Haihua Tianhe
Trade Names: Cetiol® B

Di-n-butyl adipate; Dibutyl adipinate. *See* Dibutyl adipate

Dibutylamine

CAS 111-92-2
UN 2248 (DOT)
Synonyms: N-Butyl-1-butanamine; DBA; n-Dibutylamine; Di-n-butylamine; DNBA
Classification: Sat. aliphatic amine
Empirical: $C_8H_{19}N$
Formula: $CH_3CH_2CH_2CH_2NHCH_2CH_2CH_2CH_3$
Properties: Colorless liq.; ammonia-like or fishy odor; sol. in ethanol, acetone, ether, oxygenated and hydrocarbon solvs. incl. benzene; sl. sol. in water; m.w. 129.28; dens. 0.76 kg/l (4 C); vapor pressure 2 mm Hg (20 C); m.p. -60 to -59 C; b.p. 159 C; flash pt. (CC) 41 C
Toxicology: LD50 (oral, rat) 220 mg/kg, (skin rabbit) 1010 mg/kg; corrosive; very toxic; poison by ing. and subcut. routes; may cause serious skin and severe eye irritation and 'halo vision'; inh. of mist can irritate nose/throat/lungs; high concs. may cause potentially fatal pulmonary edema; ing. may cause burning of mouth, throat, and digestive tract; CNS effects possible; mutagenic data; TSCA listed
Precaution: Corrosive; flamm. exposed to heat, flames; incompat. with oxidizing agents (risk of fire/explosion), strong acids, acid chlorides, acid anhydrides (violent reactions possible); exothermic reaction with cellulose nitrate does not proceed to ignition
Hazardous Decomp. Prods.: May react with nitrosating agents to form carcinogenic nitrosamines; heated to decomp., emits toxic fumes of NO_x
NFPA: Health 3, Flammability 2, Reactivity 0
Storage: Store in cool, dry, well-ventilated area, out of direct sunlight; limit quantities in storage; hygroscopic
Uses: Catalyst for formaldehyde polymerization; vulcanization accelerator; corrosion inhibitor for textile lubricants, metalworking fluids; solvent; mfg. of pesticides, polymers, rubber, dyes; intermediate for emulsifiers, dyes, insecticides, flotation agents; inhibitor for butadiene
Manuf./Distrib.: Aldrich; Allchem Ind.; Ashland; Atofina; BASF AG; BASF; Celanese; Fluka; Sigma; Spectrum Quality Prods.

n-Dibutylamine; Di-n-butylamine. *See* Dibutylamine
Dibutyl-1,2-benzene dicarboxylate. *See* Dibutyl phthalate
N,N-Dibutyl-1-butanamine. *See* Tributylamine
Dibutyl butanephosphonate. *See* Dibutyl butyl phosphonate

Dibutyl butyl phosphonate

CAS 78-46-6; EINECS/ELINCS 201-119-2
Synonyms: DBBP; Dibutyl butanephosphonate; Di-n-butyl butylphosphonate; Phosphonic acid, butyl-, dibutyl ester
Classification: Nonaromatic phosphorus compd.
Empirical: $C_{12}H_{27}O_3P$
Formula: $C_4H_9P(O)(OC_4H_9)_2$
Properties: Colorless liq.; mild pleasant odor; misc. with common org. solvs.; insol. in water; m.w. 250.22; dens. 0.946; b.p. 127-128 C; flash pt. 310 F; ref. index 1.4320
Toxicology: LD50 (IV, mouse) 56 mg/kg; LDLo (IP, mouse) 125 mg/kg; poison by IP and IV routes
Precaution: Combustible; can react vigorously with oxidizers
Hazardous Decomp. Prods.: Heated to decomp., emits toxic fumes of PO_x
Uses: Heavy metal extractor; synergist with di (2-ethylhexyl) phosphoric acid in heavy metal extraction; extraction solvent (min. processing); solvent separator; gasoline additive; antifoam; textile conditioner; antistat
Manuf./Distrib.: Aldrich

Di-n-butyl butylphosphonate. *See* Dibutyl butyl phosphonate
Dibutyl 'Carbitol'. *See* Diethylene glycol dibutyl ether
Dibutyl 'Cellosolve'; Di-n-butyl 'Cellosolve'. *See* Ethylene glycol dibutyl ether
2,6-Di-t-butyl-p-cresol. *See* BHT
Dibutyldiglycol. *See* Diethylene glycol dibutyl ether
2,2-Dibutyldihydro-6H-1,3,2-oxathiastannin-6-one. *See* Dibutyltin β-mercaptopropionate
Dibutyl ethanedioate. *See* Dibutyl oxalate

Dibutyl ether

CAS 142-96-1
UN 1149 (DOT)
Synonyms: 1-Butoxybutane; Butyl ether; n-Butyl ether; n-Dibutyl ether; Di-n-butyl ether; Dibutyloxide; 1,1´-Oxybis (butane)
Classification: Ether
Empirical: $C_8H_{18}O$
Formula: $CH_3(CH_2)_3O(CH_2)_3CH_3$
Properties: Colorless clear liq.; mild ether-like odor; stable; sol. in water, alcohol; misc. with most common org. solvs.; m.w. 130.26; sp.gr. 0.768 (20 C); vapor pressure 4.8 mm Hg (20 C); m.p. -95.2 C; b.p. 142 C; flash pt. 77 F; ref. index 1.3992 (20 C)
Toxicology: LD50 (oral, rat) 7400 mg/kg, (skin, rabbit) 10 g/kg; TCLo (inh., human) 200 ppm; mildly toxic by inhalation, ingestion, and skin contact; eye irritant; human systemic effects by inh.; skin irritant; TSCA listed
Environmental: VOC; BOD5 0.48; ThOD 2.95
Precaution: Flamm.; dangerous fire hazard exposed to heat, flame, oxidizers; incompat. with NCl_3, oxidizing materials; may form explosive peroxides, esp. in anhyd. form
Hazardous Decomp. Prods.: Heated to decomp., emits acrid smoke and fumes
NFPA: Health 2, Flammability 3, Reactivity 1
Uses: Solvent for paints, hydrocarbons, Grignard synthesis, fats, oils, org. acids, alkaloids, nat. and syn. resins; extractant solvent for separating metals; solvent purification; constituent of catalysts for copolymerization; mfg. of pesticides; organic synthesis (reaction medium)
Manuf./Distrib.: Filo; Focus; Haltermann Prods. UK; Penta Mfg.; Spectrum Quality Prods.; Wacker Chems.; Wacker-Chemie GmbH

n-Dibutyl ether; Di-n-butyl ether. *See* Dibutyl ether
Di-n-butyl-glycerol diether. *See* 1,3-Dibutoxy-2-propanol
Dibutyl hexanedioate. *See* Dibutyl adipate
Dibutyl hydrogen phosphite; Dibutyl hydrogen phosphonate. *See* Dibutyl phosphite
Dibutyl ketone. *See* Diisobutyl ketone
Dibutyl(3-mercaptopropionato(2-) tin. *See* Dibutyltin β-mercaptopropionate
2,6-Di-t-butyl-4-methylphenol. *See* BHT

Dibutyl oxalate

CAS 2050-60-4; EINECS/ELINCS 218-092-8
Synonyms: Butyl oxalate; Dibutyl ethanedioate; Di-n-butyl oxalate; Ethanedioic acid, dibutyl ester
Classification: Ester
Empirical: $C_{10}H_{18}O_4$
Formula: $(COOC_4H_9)_2$
Properties: Water-wh. liq.; mild odor; m.w. 202.25; sp.gr. 0.986; f.p. -29 C; b.p. 239-240 C; flash pt. 104 C; ref. index 1.423
Toxicology: Causes severe irritation; harmful by ing., inh., skin absorp.; high concs. are extremely destructive to tissues of mucous membranes, upper respiratory tract, eyes, skin; may cause burning sensation, coughing, wheezing, laryngitis, shortness of breath, headache, nausea, vomiting; TSCA listed
Environmental: VOC; ThOD 1.27
Precaution: Incompat. with acids, bases, oxidizing agents, reducing agents
Hazardous Decomp. Prods.: Toxic fumes of CO, CO_2
NFPA: Health 0, Flammability 1, Reactivity 0
Storage: Store in cool, dry place; keep tightly closed
Uses: Solvent; chelating agent in cosmetics
Features: Inert ingred. no longer used in pesticides
Manuf./Distrib.: ABCR; Aceto; Acros Org.; Aldrich; Alfa Aesar; Fluka; ICN Biomed. Research Prods.; TCI Am.

Di-n-butyl oxalate. *See* Dibutyl oxalate
2,2-Dibutyl-1-oxa-2-stanna-3-thiacyclohexan-6-one. *See* Dibutyltin β-mercaptopropionate
Dibutyloxide. *See* Dibutyl ether

Dibutyl phosphite

CAS 1809-19-4; EINECS/ELINCS 217-316-1
Synonyms: Butyl alcohol hydrogen phosphite; Dibutyl hydrogen phosphite; Dibutyl hydrogen phosphonate
Classification: Nonaromatic phosphorus compd.
Empirical: $C_8H_{19}O_3P$
Formula: $(CH_3CH_2CH_2CH_2O)_2POH$
Properties: Water-wh. clear liq.; sol. in common org. solvs.; m.w. 194.21; dens. 0.975-0.995 (20/20 C); vapor pressure < 1 mm Hg (20 C); b.p. 95 C (1 mm); ref. index 1.422-1.424; flash pt. 49 C
Toxicology: LD50 (oral, rat) 3200 mg/kg, (skin, rabbit) 1990 mg/kg; mod. toxic by ing. and skin contact; skin and severe eye irritant; TSCA listed
Precaution: Combustible; mod. fire risk; dangerous; corrosive; vigorous reactions with oxidizers
Hazardous Decomp. Prods.: Heated to decomp., or on contact with acid or acid fumes, emits highly toxic fumes of PO_x
NFPA: Health 3, Flammability 2, Reactivity 0
Storage: Moisture-sensitive
Uses: Intermediate; used as amine salts in min. and syn. base stocks; as load-carrying additives with sec. activity as low-temp. stabilizers and metal deactivators; nonstaining antioxidant for PP; solvent; EP agent, friction modifier for lubricants and greases
Manuf./Distrib.: Akzo Nobel; Aldrich; Fluka; Rhodia/Phosphorus Perf. Derivs.

Dibutyl phthalate

CAS 84-74-2; EINECS/ELINCS 201-557-4
UN 9095 (DOT)
Synonyms: 1,2-Benzenedicarboxylic acid, dibutyl ester; o-Benzenedicarboxylic acid dibutyl ester; Benzene-o-dicarboxylic acid di-n-butyl ester; Butyl phthalate; n-Butyl phthalate; DBP; Dibutyl-1,2-benzene dicarboxylate; Di-n-butyl phthalate
Classification: Ester
Definition: Aromatic diester of butyl alcohol and phthalic acid
Empirical: $C_{16}H_{22}O_4$
Formula: $C_6H_4(COOC_4H_9)_2$
Properties: Colorless to yel. stable oily liq., odorless; misc. with common org. solvs.; insol. in water; m.w. 278.17; dens. 1.0484 (20/20 C); f.p. -35 C; b.p. 340 C; flash pt. (COC) 340 F; autoignition temp. 402 C; ref. index 1.4920; surf. tens. 33.4 dynes/cm; dielec. const. 6.44
Toxicology: ACGIH TLV/TWA 5 mg/m^3; LD50 (oral, rat) 8000 mg/kg, (IP, rat) 3050 mg/kg; mod. toxic by IP, IV routes; mildly toxic by ing.; irritant; if ingested can cause GI upset, human systemic effects (hallucinations, nausea, vomiting); vapor irritating to eyes and mucous membranes; may cause kidney or bladder changes; experimental teratogen, reproductive effects; mutagenic data; TSCA listed
Environmental: BOD5 0.43; ThOD 2.26
Precaution: Combustible exposed to heat or flame; can react with oxidizers; reacts explosively with chlorine
Hazardous Decomp. Prods.: Heated to decomp., emits acrid smoke and fumes
NFPA: Health 0, Flammability 1, Reactivity 0
Uses: Plasticizer, solvent, film-former in nitrocellulose lacquers, elastomers, explosives, nail polishes, solid rocket propellants, adhesives, sealants; solvent for resins, cosmetics, perfume oils; perfume fixing agent; textile lubricant; insect repellent; nonreactive epoxy resin diluent; safety glass; printing inks; paper coatings; oral pharmaceuticals; in food-pkg. adhesives; in paper/paperboard in contact with aq./fatty foods; adjuvant in slimicides in food-contact paper/paperboard; in cellophane for food pkg.; solvent for adjuvants in food-contact crosslinked polyesters; plasticizer in food-contact rubber articles for repeated use
Regulatory: FDA 21CFR §175.105, 175.300, 176.170, 176.300, 177.1200, 177.2420, 177.2600; FDA approved for orals; SARA §313 reportable; CERCLA hazardous substance; HAP; VOC; BP, EP compliance
Manuf./Distrib.: AAE Chemie NV; Akdeniz Kimya; Aldrich; Alfa Chem; Allchem Ind.; Ashland; BCH Brühl; BP Chems. Ltd; C.P. Hall; Celanese; ChemService; Chisso Am.; Coyne; Eastman; Fluka; Great Western; Houghton Chem.; Indofine; Parchem Trading; Ruger; Shandong Haihua Tianhe; Sigma; Spectrum Quality Prods.; St. Lawrence; Sunoco; Unitex; Universal Preserv-A-Chem
Trade Names: Eastman® DBP; Uniplex 150

Di-n-butyl phthalate. *See* Dibutyl phthalate

Dibutyl tartrate

Empirical: $C_{12}H_{22}O_6$
Formula: $C_4H_9OOCCH_2OCH_2OCOOC_4H_9$

Properties: Light tan liq.; m.p. 21 C; b.p. ≈ 204 C (26 mm); flash pt. 90.5 C; ref. index 1.4463
Precaution: Combustible
Uses: Solvent, plasticizer for cellulose esters and ethers, elastomers, lubricants, rubberized fabrics, lacquers, dopes, transfer inks
Manuf./Distrib.: ChemService

Dibutyltin β-mercaptopropionate

CAS 78-06-8
Synonyms: 2,2-Dibutyldihydro-6H-1,3,2-oxathiastannin-6-one; Dibutyl(3-mercaptopropionato(2-) tin; 2,2-Dibutyl-1-oxa-2-stanna-3-thiacyclohexan-6-one; Dibutyltin-O,S-mercaptopropionate; Dibutyltin-S,O-3-mercaptopropionate; Dibutyltin-S,O-β-mercaptopropionate; Mercaptopropionic acid, dibutyltin salt
Definition: Mercaptopropionic acid, dibutyltin salt
Empirical: $C_{11}H_{22}O_2SSn$
Properties: Wh. crystals; m.w. 337.08
Toxicology: LD50 (IV, mouse) 100 mg/kg; poison by IV route; TSCA listed
Hazardous Decomp. Prods.: Heated to decomp., emits toxic fumes of SO_x
Uses: Solvent; stabilizer for PVC
Manuf./Distrib.: ICN Biomed. Research Prods.

Dibutyltin-O,S-mercaptopropionate; Dibutyltin-S,O-3-mercaptopropionate; Dibutyltin-S,O-β-mercaptopropionate. *See* Dibutyltin β-mercaptopropionate
Di-n-butyoxyethane. *See* Ethylene glycol dibutyl ether

Di-C12-15 alkyl fumarate

CAS 142104-11-8
Definition: Diester of C12-15 alcohols and fumaric acid
Formula: ROCOHCCHCOOR, where R rep. C12-15 alkyl group
Uses: Solvent, emollient in cosmetics

Dicapryl ether; Dicaprylyl ether. *See* Dioctyl ether
Dicaprylyl maleate. *See* Dioctyl maleate
β,β-Dicarbomethoxydiethyl sulfide. *See* Dimethyl thiodipropionate
DICHA. *See* Dicyclohexylamine
o-Dichlorbenzene; o-Dichlorbenzol. *See* o-Dichlorobenzene
1,1-Dichlorethane. *See* 1,1-Dichloroethane
2,2´-Dichlorethyl ether. *See* sym-Dichloroethyl ether
1,2-Dichlorobenzene. *See* o-Dichlorobenzene

o-Dichlorobenzene

CAS 95-50-1; EINECS/ELINCS 202-425-9
UN 1591 (DOT)
Synonyms: Benzene, o-dichloro-; Chloroben; Chloroden; Cloroben; DCB; o-Dichlorbenzene; o-Dichlorbenzol; 1,2-Dichlorobenzene; Dizene; ODB; ODCB; Orthodichlorobenzene; Orthodichlorobenzol
Empirical: $C_6H_4Cl_2$
Properties: Colorless clear liq., pungent aromatic odor; sol. in alcohol, ether; misc. with most org. solvs.; insol. in water; m.w. 146.95; dens. 1.2884 (20/4 C); f.p. -17 C; m.p. -17.5 C; b.p. 172-179 C; flash pt. (CC) 72 C; ref. index 1.5459
Toxicology: ACGIH TLV/CL 50 ppm (skin); LD50 (oral, rat) 500 mg/kg, (IP, rat) 840 mg/kg; toxic by inh. and ing.; primary irritant to eyes, skin, mucous membranes; chronic exposure can cause CNS effects; causes liver/kidney injury; suspected carcinogen; tumorigen, mutagen; reproductive effects; TSCA listed
Precaution: Flamm. exposed to heat or flame; can react vigorously with oxidizing materials; slow reaction with Al may cause explosion during storage in sealed aluminum container
Hazardous Decomp. Prods.: Heated to decomp., emits toxic fumes of Cl^-
NFPA: Health 2, Flammability 2, Reactivity 0
Uses: Mfg. of 3,4-dichloroaniline; solvent for wide range of org. materials and for oxides of nonferrous metals; solvent carrier; degreasing hides and wool; industrial odor control; pesticide; herbicide; insecticide; soil fumigant; dye mfg.; metal polishes; heat transfer fluid; mfg. of paints; aircraft/vehicle/industrial engine cleaner ingred.
Regulatory: SARA reportable
Manuf./Distrib.: Aarti Ind. Ltd; Aldrich; Allchem Ind.; Ashland; Bayer/Fiber, Addits., Rubber; Fluka; Monsanto; PPG Ind.; Romil Ltd; Sigma; Spectrum Quality Prods.; Stanchem; Tekchem; Xinchem

3,4-Dichlorobenzotrifluoride

CAS 328-84-7
Synonyms: Benzene, 1,2-dichloro-4-(trifluoromethyl)-; DCBTF; 3,4-DCBTF
Empirical: $C_7H_3Cl_2F_3$
Properties: M.w. 215.00; dens. 1.478; m.p. -13 to -12 C; b.p. 173-174 C; flash pt. 65 C
Toxicology: Irritant
Environmental: Non-ozone depleting
Uses: Solvent for industrial solvent cleaning, aerosols, adhesives, coatings, and inks
Manuf./Distrib.: Aldrich; Occidental
Trade Names: Oxsol® 1000

Dichlorobromomethane. *See* Bromodichloromethane
1,1-Dichloro-2-chloroethylene. *See* Trichloroethylene
1,1-Dichloro-2,2-dichloroethane. *See* 1,1,2,2-Tetrachloroethane
2,2´-Dichlorodiethyl ether; β,β-Dichlorodiethyl ether. *See* sym-Dichloroethyl ether
Dichlorodiethyl formal. *See* Bis (2-chloroethoxy) methane

Dichlorodifluoromethane

CAS 75-71-8; EINECS/ELINCS 200-893-9
UN 1028 (DOT); INS940
Synonyms: CFC 12; Difluorodichloromethane; F 12; FC 12; Fluorocarbon 12; Food freezant 12; Halon; Propellant 12; R 12; Refrigerant 12
Classification: Halogenated aliphatic hydrocarbon
Empirical: CCl_2F_2
Formula: $ClCF_2Cl$
Properties: Colorless gas; may be liquefied under pressure; odorless; noncorrosive; insol. in water; sol. in most org. solvs., alcohol, ether, acetic acid; m.w. 120.9; sp.gr. 1.325 (21 C, liq.); vapor pressure 5 atm (16 C); f.p. -158 C; b.p. -29.8 C; ref. index 1.290; surf. tens. 9 dynes/cm; dielec. const. 2.13; noncombustible
Toxicology: ACGIH TLV/TWA 1000 ppm; LC50 (inh., rat, 30 min) 80 pph; human systemic effects by inh.; narcotic in high concs.; may cause nose/throat/respiratory irritation, lightheadedness, drowsiness, incoordination, liver changes; severe exposure may cause irregular heartbeat, cardiac arrest, death; contact with aerosol or liq. may cause frostbite to skin and eyes; TSCA listed
Precaution: Compressed gas; incompat. with reactive metals, e.g., sodium, potassium, aluminum, magnesium, zinc (can react violently or explosively), water (can cause violent spattering or explosion)
Hazardous Decomp. Prods.: Heated to decomp., emits highly toxic fumes of phosgene, Cl^-, and F^-
Storage: Store cylinders in a cool, dry, well-ventilated area away from corrosive atmospheres, out of direct sunlight; leaked gas can accumulate in low areas
Uses: Refrigerant; air conditioning gas; blowing agent for PVC and PU foams; aerosol propellant; low-temp. solvent; degreasing agent; leak detecting agent; monomer for resins; freezing of foods by direct contact; propellant in foods; prep. of frozen tissue sections
Regulatory: FDA 21CFR §701.30; FDA approved for implants, inhalers, rectals, topicals; NF compliance; SARA §313 reportable; CERCLA hazardous substance
Manuf./Distrib.: Air Prods.; Allchem Ind.; Archimica Florida; Atofina; Aventis Pharmaceuticals; BOC Ltd; Dow; DuPont; Fluorochem Ltd; LaRoche Ind.; Matheson Gas Prods.; Sigma; UOP; Uniqema

Dichloroethane. *See* Ethylene dichloride

1,1-Dichloroethane

CAS 75-34-3
UN 2362
Synonyms: Chlorinated hydrochloric ether; 1,1-Dichlorethane; Dichloroethane, 1,1-; Ethane, 1,1-dichloro-; Ethylidene chloride; Ethylidene dichloride
Empirical: $C_2H_4Cl_2$
Formula: Cl_2CHCH_3
Properties: Colorless clear oily mobile liq.; aromatic ethereal odor; saccharine taste; sol. in DMSO, 95% ethanol, acetone, ether, benzene, fixed and volatile oils; misc. with most common solvs.; sol. < 1 mg/ml

in water; m.w. 98.96; sp.gr. 1.1757 (20/4 C); vapor pressure 234 mm Hg (25 C); m.p. -97 C; b.p. 57.3 C; flash pt. -5 C; ref. index 1.4164 (20 C); surf. tens. 24.75 dynes/cm

Toxicology: ACGIH TLV/TWA 200 ppm; STEL 250 ppm; LD50 (oral, rat) 725 mg/kg; LCLo (inh., rat, 4 h) 16,000 ppm; mod. toxic by ing.; may be harmful by inh., skin absorp.; vapor irritating to eyes, skin, mucous membranes, upper respiratory tract; lachrymator; narcotic, possibly anesthetic in high concs.; may cause liver/kidney damage, unconsciousness, CNS depression, nausea, vomiting, faintness, coughing, cyanosis, etc.; experimental tumorigen, teratogen; suspected carcinogen; reported mutagen; TSCA listed

Precaution: Flamm.; LEL 5.6%; UEL 11.4%; vapor may flash back; incompat. with oxidizing agents, strong bases; contact with strong caustics will emit flamm./toxic acetaldehyde gas; attacks some plastics, rubber, coatings; heat-sensitive

Hazardous Decomp. Prods.: Heated to decomp., emits toxic fumes of CO, CO_2, hydrogen chloride gas, phosgene

NFPA: Health 2, Flammability 3, Reactivity 0

Storage: Store in explosion-proof freezer; keep away from oxidizing materials and ignition sources

Uses: Mfg. of vinyl chloride, H_3CCCl_3, high vacuum rubber, silicon grease; coupling agent in antiknock gasoline; metal degreasing agent; organic synthesis; ore flotation; chemical intermediate; industrial solvent; extraction solvent; chlorinated solvent intermediate; fumigant; dewaxing agent for min. oils; extractant for heat-sensitive substances; paint/varnish/finish remover; formerly as anesthetic

Manuf./Distrib.: AccuStandard; ICN Biomed. Research Prods.; Pfaltz & Bauer; Riedel-deHaën; Spectrum Quality Prods.; Supelco; TCI Am.

Dichloroethane, 1,1-. *See* 1,1-Dichloroethane

1,2-Dichloroethane; Dichloro-1,2-ethane; α,β-Dichloroethane; sym-Dichloroethane. *See* Ethylene dichloride

1,1-Dichloroethene. *See* Vinylidene chloride monomer

1,2-Dichloroethene. *See* cis-trans-1,2-Dichloroethylene (mixed isomers)

1,2-Dichloro-, (E)-ethene. *See* trans-1,2-Dichloroethylene

cis-1,2-Dichloroethene. *See* cis-1,2-Dichloroethylene

(E)-1,2-Dichloroethene; trans-1,2-Dichloroethene. *See* trans-1,2-Dichloroethylene

(Z)-1,2-Dichloroethene. *See* cis-1,2-Dichloroethylene

Dichloroether. *See* sym-Dichloroethyl ether

Dichloroethylene. *See* Ethylene dichloride

1,1-Dichloroethylene. *See* Vinylidene chloride monomer

1,2-Dichloroethylene. *See* cis-trans-1,2-Dichloroethylene (mixed isomers)

1,2-Dichloro-, (E)-ethylene. *See* trans-1,2-Dichloroethylene

asym-Dichloroethylene. *See* Vinylidene chloride monomer

cis-Dichloroethylene. *See* cis-1,2-Dichloroethylene

cis-1,2-Dichloroethylene

CAS 156-59-2; EINECS/ELINCS 205-859-7

UN 1150

Synonyms: cis-1,2-Dichloroethene; (Z)-1,2-Dichloroethene; cis-Dichloroethylene; (Z)-1,2-Dichloroethylene; Ethene, 1,2-dichloro-, (Z)-; Ethylene, 1,2-dichloro-, (Z)-

Empirical: $C_2H_2Cl_2$

Formula: CHCl=CHCl

Properties: Colorless to pale yel. clear liq.; ethereal, sl. acrid odor; sol. in DMSO, 95% ethanol, acetone, alcohol, chloroform, benzene, ether, most org. solvs.; sol. 1-5 mg/ml in water; m.w. 96.94; sp.gr. 1.2837 (20/4 C); vapor pressure 200 mm Hg (25 C); m.p. -80.5 C; b.p. 60 C; flash pt. 3.9 C; ref. index 1.4435; surf. tens. 24 dynes/cm

Toxicology: ACGIH TLV/TWA 200 ppm; LCLo (inh., mouse, 2 h) 65,000 mg/m^3; harmful liq. and fumes; mildly toxic by ing. and inh.; in high concs., it is irritating and narcotic; may cause eye/skin/mucous membrane irritation; may cause CNS depression, dizziness, nausea, vomiting, liver/kidney injury, conjunctivitis, difficult breathing, weakness, epigastric cramps, drowsiness, unconsciousness, etc.; suspected carcinogen; mutation data; target organs: CNS, liver, kidneys; TSCA listed

Precaution: Flamm.; incompat. with oxidizers, bases, copper/alloys; forms explosive compd. with caustic alkalis; dec. by air, light, moisture; corrosive to metals unless inhibited; dec. on contact with hot metals; attacks some plastics, rubber; may flash back

Hazardous Decomp. Prods.: Heated to decomp., emits toxic fumes of hydrogen chloride gas, phosgene, CO, CO_2; emits toxic fumes under fire conditions

NFPA: Health 2, Flammability 3, Reactivity 2

Storage: Air-, light-, and moisture-sensitive; keep container tightly closed under inert atmosphere; store in explosion-proof refrigerator; protect from light

Uses: Solvent for waxes, resins, fats, phenol, camphor, acetyl cellulose, org. materials, and heat-sensitive substances, e.g., caffeine; in rubber mfg.; refrigerant; additive to dye and lacquer sol'ns.; in retarding fermentation; org. synthesis; in medicines; in dye extraction; in chlorination reactions; mfg. of artificial pearls; constituent of perfumes and thermoplastics

Regulatory: SARA §313 reportable

Manuf./Distrib.: AccuStandard; Aldrich; Alfa Aesar; Fluka; Riedel-deHaën; Supelco; TCI Am.

cis-trans-1,2-Dichloroethylene; sym-Dichloroethylene. *See* cis-trans-1,2-Dichloroethylene (mixed isomers)

trans-Dichloroethylene. *See* trans-1,2-Dichloroethylene

trans-1,2-Dichloroethylene

CAS 156-60-5; EINECS/ELINCS 205-860-2

UN 1150

Synonyms: trans-Acetylene dichloride; 1,2-DCE; trans-1,2-DCE; 1,2-Dichloro-, (E)-ethene; (E)-1,2-Dichloroethene; trans-1,2-Dichloroethene; 1,2-Dichloro-, (E)-ethylene; trans-Dichloroethylene; (E)-1,2-Dichlrooethylene; Ethene, 1,2-dichloro-, (E)-; Ethylene, 1,2-dichloro-, (E)-

Empirical: $C_2H_2Cl_2$

Formula: ClCH=CHCl

Properties: Colorless clear visc. liq.; sl. acrid odor; sol. in DMSO, 95% ethanol, acetone, chloroform, ether, benzene, most org. solvs.; sol. < 1 mg/ml in water; m.w. 96.95; sp.gr. 1.2743 (25/4 C); vapor pressure 200 mm Hg (14 C); m.p. -50 C; b.p. 47 C (745 mm Hg); flash pt. 6 C; ref. index 1.4454 (20 C)

Toxicology: ACGIH TLV/TWA 200 ppm; LD50 (oral, rat) 1235 mg/kg, (IP, rat) 7536 mg/kg; LCLo (inh., mouse, 2 h) 75,000 mg/m^3; highly toxic to humans (CNS); mod. toxic by inh., ing., skin contact; low toxicity by IP route; high concs. may be narcotic; irritating to eyes/skin/mucous membranes/respiratory tract; may cause dizziness, nausea, vomiting, CNS intoxication, transient renal effects, sleep disturbances, weakness, tremor, unconsciousness, etc.; target organs: CNS, liver, kidneys; TSCA listed

Precaution: Flamm.; incompat. with oxidizers, bases, org. peroxides; reacts with alkalis, etc.; with solid caustic alkalis, forms chloroacetylene (ignites in air); corrosive to metals unless inhibited; avoid copper/alloys; vapor may flash back

Hazardous Decomp. Prods.: Heated to decomp., emits toxic fumes of HCl gas, CO, CO_2, phosgene; emits toxic fumes under fire conditions

NFPA: Health 2, Flammability 3, Reactivity 2

Storage: Air-, light-, and moisture-sensitive; keep container tightly closed under inert atmosphere; store in explosion-proof refrigerator; protect from light

Uses: Solvent for fats, phenols, camphor, etc.; retards fermentation; rubber mfg.; refrigerants; additive to dyes and lacquers; low-temp. solvent for heat-sensitive substances; constituent of perfumes, thermoplastics; organic synthesis; medicine; cleaning/degreasing agent; rinsing/drying agent

Features: Replaces ODC

Regulatory: SARA §313 reportable

Manuf./Distrib.: AccuStandard; Acros Org.; Aldrich; Fluka; ICN Biomed. Research Prods.; Lancaster Synthesis; Mallinckrodt Baker; Pfaltz & Bauer; Riedel-deHaën; TCI Am.

Trade Names: VersaTRANS™ Solvent

Trade Names Containing: Genesolv® S-TZ; HFE-71DA; HFE-71DE; Vertrel® MCA; Vertrel® MCA Plus; Vertrel® SMT; Vertrel® XMS Plus

(Z)-1,2-Dichloroethylene. *See* cis-1,2-Dichloroethylene

cis-trans-1,2-Dichloroethylene (mixed isomers)

CAS 540-59-0; EINECS/ELINCS 208-750-2

UN 1150

Synonyms: Acetylene dichloride; 1,2-Dichloroethene; 1,2-Dichloroethyl-

ene; cis-trans-1,2-Dichloroethylene; sym-Dichloroethylene; Ethylene, 1,2-dichloro-
Empirical: $C_2H_2Cl_2$
Formula: ClCH=CHCl
Properties: Colorless to lt. yel. clear liq.; ethereal, sl. acrid odor; sol. in DMSO, 95% ethanol, acetone, ether, most org. solvs.; sol. < 1 mg/ml in water; m.w. 96.94; sp.gr. 1.27; vapor pressure 180-265 mm Hg (20 C); m.p.-57 C; b.p. 48-60 C; flash pt. 3.9 C; ref. index 1.4463 (20 C)
Toxicology: ACGIH TLV/TWA 200 ppm; LD50 (oral, rat) 770 mg/kg, (IP, mouse) 2 g/kg; LCLo (inh., frog, 1 h) 117 mg/m^3; poison by inh.; mod. toxic by ing. and other routes; may be harmful by skin contact; may cause irritation of eyes, skin, mucous membranes, upper respiratory tract; prolonged contact can have narcotic effect; may cause drowsiness, unconsciousness, digestive disturbances, CNS depression, nausea, vomiting, weakness, tremors, epigastric cramps, etc.; target organs: CNS, liver, kidneys; TSCA listed
Precaution: Flamm.; reacts violently with Na, NaOH, Cu/alloys; can react with caustic alkalis; forms explosive mixts. with N_2O_4; incompat. with strong oxidizers, bases; corrosive to metals; attacks some plastics, rubber, coatings; vapor can flash back
Hazardous Decomp. Prods.: Heated to decomp., emits highly toxic fumes of hydrogen chloride gas, CO, CO_2, phosgene gas; emits toxic fumes under fire conditions
NFPA: Health 2, Flammability 3, Reactivity 2
Storage: Air-, light-, and moisture-sensitive; keep container tightly closed under inert atmosphere; store in explosion-proof refrigerator; protect from light
Uses: Solvent for organic materials, fats, phenol, camphor, dye extraction, perfumes, pharmaceuticals, thermoplastics; retarding fermentation; additive to dye and lacquer sol'ns.; organic synthesis; medicine; degreasing agent
Regulatory: SARA §313 reportable
Manuf./Distrib.: Aldrich

Dichloroethyl ether; Di-(2-chloroethyl) ether. *See* sym-Dichloroethyl ether

sym-Dichloroethyl ether

CAS 111-44-4; EINECS/ELINCS 203-870-1
UN 1916 (DOT)
Synonyms: Bis (2-chloroethyl) ether; Bis (β-chloroethyl) ether; 1-Chloro-2-(β-chloroethoxy) ethane; Chloroethyl ether; DCEE; 2,2′-Dichlorethyl ether; 2,2′-Dichlorodiethyl ether; β,β-Dichlorodiethyl ether; Dichloroether; Dichloroethyl ether; Di-(2-chloroethyl) ether; Dichloroethyl oxide; 1,1-Oxybis (2-chloro) ethane
Empirical: $C_4H_8Cl_2O$
Formula: $ClCH_2CH_2OCH_2CH_2Cl$
Properties: Colorless liq.; sol. in oxygenated and aromatic solvs.; insol. in water; m.w. 143.02; dens. 1.2220; vapor pressure 0.7 mm Hg (20 C); f.p. -51.8 C; b.p. 178.5 C; flash pt. 55 C; ref. index 1.457 (20 C)
Toxicology: ACGIH TLV/TWA 5 ppm; STEL 10 ppm (skin); LD50 (oral, rat) 75 mg/kg, (skin, rabbit) 720 mg/kg; LC50 (inh., rat, 4 h) 330 mg/m^3; toxic by inh., ing., and skin absorption; strong irritant; lachrymator; exposure to 1000 ppm for 30-60 min may be fatal within days; odor easily detectable at 35 ppm; questionable carcinogen; tumorigen; mutagen; TSCA listed
Precaution: Flamm. exposed to heat, flame, or oxidizers; can react vigorously with oxidizing materials; dangerous explosion hazard
Hazardous Decomp. Prods.: Reacts with H_2O or steam to evolve toxic and corrosive fumes; heated to decomp., emits toxic fumes of Cl^-
NFPA: Health 3, Flammability 2, Reactivity 1
Uses: General solvent, selective solvent for prod. of high-grade lubricating oils, textile scouring and fulling compds., paints, varnishes, lacquers, finish removers, spotting and dry cleaning; soil fumigant; intermediate and crosslinking agent in organic synthesis
Manuf./Distrib.: Aldrich; BASF AG; Fluka; Jarchem Ind.; Sithean; Supelco

Dichloroethyl formal. *See* Bis (2-chloroethoxy) methane
Dichloroethyl oxide. *See* sym-Dichloroethyl ether

Dichlorofluoroethane

CAS 1717-00-6
Synonyms: 1,1-Dichloro-1-fluoroethane; Ethane, 1,1-dichloro-1-fluoro-; HCFC 141b; R 141b
Empirical: $C_2H_3Cl_2F$
Formula: CCl_2FCH_3
Properties: Colorless liq.; faint ethereal odor; insol. in water; m.w. 116.95; dens. 1.20; vapor pressure 520 mm Hg (70 F); f.p. -103 C; b.p. 32 C; flash pt. none
Toxicology: LD50 (oral, rat) > 5000 mg/kg, (inh., rat, 2 h) 240 g/m^3, (skin, rabbit) > 2000 mg/kg; sl. toxic by inh.; mild eye irritant; repeated/prolonged contact may cause skin irritation and defatting; inh. may cause dizziness, CNS effects, cardiac arrhythmia; can cause asphyxiation; TSCA listed
Precaution: No flash pt. but will burn when diluted with air; incompat. with strong oxidizers and active metals; flamm. limits 7.4-15.1%; dec. above 480 F and emits poisonous fumes
Hazardous Decomp. Prods.: HF, HCl, carbonyl halides, CO, CO_2; heated to decomp., emits toxic vapors of F^- and Cl^-
Uses: Blowing agent in rigid board, foam systems, flexible foam; refrigerant gas; solvent; precision cleaning agent
Features: CFC-113 replacement
Regulatory: DOT nonregulated
Manuf./Distrib.: LaRoche Ind.; Spectrum Quality Prods.; Zhejiang Sanhuan Chems. Co. Ltd
Trade Names: Forane 141b DGX; Genesolv® 2000
Trade Names Containing: Forane 141b MGX; Genesolv® 2004

1,1-Dichloro-1-fluoroethane. *See* Dichlorofluoroethane

Dichlorofluoromethane

CAS 75-43-4
UN 1029 (DOT)
Synonyms: Dichloromonofluoromethane; F 21; FC 21; Fluorocarbon 21; Fluorodichloromethane; Freon 21; HCFC 21; HFC 21; R 21; Refrigerant 21
Classification: Halogenated aliphatic hydrocarbon
Empirical: $CHCl_2F$
Formula: Cl_2CHF
Properties: Colorless clear gas or liquefied compressed gas; sl. sweetish ether-like odor; sol. in alcohol, ether, acetic acid, chloroform; sl. sol. in water; m.w. 102.92; dens. 1.48; vapor pressure 2 atm (28.4 C); m.p. -135 C; b.p. 8.9 C; ref. index 1.356; surf. tens. 18 dynes/cm; dielec. const. 5.34; nonflamm.
Toxicology: ACGIH TLV/TWA 10 ppm; LC50 (inh., rat, 4 h) 49,900 ppm; mildly toxic by inh.; may cause cooling of skin or eye tissue due to rapid evaporation; gas unlikely to cause eye irritation; may cause CNS symptoms, mild nose/throat, airway irritation, lightheadedness, drowsiness, incoordination; severe exposure may cause irregular heartbeat, cardiac arrest, death; experimental reproductive effects; TSCA listed
Environmental: VOC; ThOD 0.16; replacement for CFC but characterized as global warming gas by the Kyoto Conference on Climate Change
Precaution: Incompat. with reactive metals (can react violently with sodium, potassium, calcium, powd. aluminum, zinc, magnesium and alloys), strong bases (hydrolysis (dec.) can occur in presence of water)
Hazardous Decomp. Prods.: Heated to decomp., emits very toxic fumes of Cl^- and F^-
Storage: Store cylinders in a cool, dry, well-ventilated area away from flamm. materials and corrosive atmospheres, out of direct sunlight
Uses: Refrigerant; solvent; blowing agent; propellant; heat exchange fluid; pharmaceutical orals
Regulatory: FDA approved for orals; SARA §313 reportable
Manuf./Distrib.: Aldrich

Dichlorohydrin; α-Dichlorohydrin. *See* 1,3-Dichloro-2-propanol
α,β-Dichlorohydrin. *See* 2,3-Dichloro-1-propanol
α,γ-Dichlorohydrin. *See* 1,3-Dichloro-2-propanol
β-Dichlorohydrin; β,γ-Dichlorohydrin. *See* 2,3-Dichloro-1-propanol
1,3-Dichloro-2-hydroxypropane; 1,3-Dichloroisopropanol; sym-Dichloroisopropyl alcohol. *See* 1,3-Dichloro-2-propanol
Dichloromethane. *See* Methylene chloride
2,4-Dichloromethylbenzene; 2,4-Dichloro-1-methylbenzene. *See* 2,4-Dichlorotoluene
Dichloromethylbromide; Dichloromonobromomethane. *See* Bromodichloromethane

Dichloromonofluoromethane. *See* Dichlorofluoromethane
Dichloronitroethane. *See* 1,1-Dichloro-1-nitroethane

1,1-Dichloro-1-nitroethane

CAS 594-72-9; EINECS/ELINCS 209-854-0
UN 2650 (DOT)
Synonyms: Dichloronitroethane
Classification: Halogenated aliphatic nitro compd.
Empirical: $C_2H_3Cl_2NO_2$
Formula: $H_3CC9Cl)_2NO_2$
Properties: Colorless liq.; unpleasant odor; sol. in ethanol, acetone, dimethyl sulfoxide; sl. sol. in water; m.w. 143.96; dens. 1.427 (20 C); b.p. 124 C; flash pt. (CC) 58 C
Toxicology: ACGIH TLV/TWA 2 ppm; LD50 (oral, rat) 410 mg/kg, (IP, mouse) 240 mg/kg; poison by ing. and IP routes; mod. toxic by inh.; strong irritant; aspiration could cause severe lung damage; vapor may cause mod. to severe irritation to nose, throat, and respiratory tract; serious exposure can cause severe lung injury or bronchitis; lachrymator; eye irritant; liq. can cause skin irritation
Precaution: Flamm. exposed to heat, flame, or oxidizers; can react vigorously with oxidizing materials
Hazardous Decomp. Prods.: Can dec. to highly toxic gases such as phosgene, hydrogen chloride, NO_x, CO, CO_2
NFPA: Health 2, Flammability 2, Reactivity 3
Uses: In organic chemical synthesis; in mfg. of highly accelerated rubber cements and insecticides; solvent; grain fumigant

1,2-Dichloropropane; α,β-Dichloropropane. *See* Propylene dichloride
1,2-Dichloro-3-propanol; 1,2-Dichloropropanol-3. *See* 2,3-Dichloro-1-propanol

1,3-Dichloro-2-propanol

CAS 96-23-1; EINECS/ELINCS 202-491-9
UN 2750 (DOT); UN 2844 (DOT)
Synonyms: Dichlorohydrin; α-Dichlorohydrin; α,γ-Dichlorohydrin; 1,3-Dichloro-2-hydroxypropane; 1,3-Dichloroisopropanol; sym-Dichloroisopropyl alcohol; 1,3-Dichloropropan-2-ol; 1,3-Dichloropropanol-2; Glycerol dichlorohydrin; Glycerol 1,3-dichlorohydrin; Glycerol α,γ-dichlorohydrin; sym-Glycerol dichlorohydrin; 2-Propanol, 1,3-dichloro-; Propylene dichlorohydrin
Empirical: $C_3H_6Cl_2O$
Formula: $ClCH_2CH(OH)CH_2Cl$
Properties: Yel. clear, sl. visc. liq.; ethereal odor; sol. in water, DMSO, 95% ethanol, acetone; misc. with alcohol, ether, veg. oils, most org. solvs.; m.w. 128.99; sp.gr. 1.3506 (17/4 C); vapor pressure 1 mm Hg (28 C); m.p. -4 C; f.p. 85 c; b.p. 172-176 C; flash pt. 74 C; ref. index 1.4835
Toxicology: LD50 (oral, rat) 110 mg/kg, (skin, rabbit) 800 mg/kg; LCLo (inh., rat, 4 h) 125 ppm; poison; toxic by ing., inh.; mod. toxic by skin contact; irritating to eyes, skin, mucous membranes, upper respiratory tract; may cause hemorrhagic gastritis, pharyngitis, headache, vertigo, nausea, vomiting, pulmonary edema, liver damage; readily absorbed thru skin; risk of serious eye damage, irreversible effects; may cause cancer; mutagen; human mutagenic data; target organs: kidneys, liver, CNS; TSCA listed
Environmental: May cause long-term adverse effects in aquatic environment
Precaution: Combustible; incompat. with strong oxidizers, strong acids, strong reducing agents, acid chlorides, acid anhydrides; sensitive to heat
Hazardous Decomp. Prods.: Heated to decomp., emits highly toxic fumes of CO, CO_2, hydrogen chloride gas, phosgene gas; emits toxic fumes under fire conditions
NFPA: Health 2, Flammability 2, Reactivity 0
Storage: Refrigerate; store under nitrogen; store in well-ventilated area away from incompat. substances; keep away from heat, sparks, flame
Uses: Chlorinated solvent; solvent for hard resins, nitrocellulose; reactive chloro intermediate for the modification of amines, alcohols, and carboxylic acids; C3-building block; mfg. of photographic and Zapon lacquer; cement for celluloid; binder for water colors; in determination of Vitamin A; intermediate in organic synthesis; in paints, varnishes, lacquers
Manuf./Distrib.: ABCR; Acros Org.; Aldrich; Alfa Aesar; Contract Chems. Ltd; Fluka; G.J. Chem.; ICN Biomed. Research Prods.; Pfaltz & Bauer; Raschig; Sithean; Supelco; Synasia Fine Chem.; TCI Am.
Trade Names: DCP

1,3-Dichloropropan-2-ol; 1,3-Dichloropropanol-2. *See* 1,3-Dichloro-2-propanol
2,3-Dichloropropanol. *See* 2,3-Dichloro-1-propanol

2,3-Dichloro-1-propanol

CAS 616-23-9; EINECS/ELINCS 210-470-0
UN 2810
Synonyms: α,β-Dichlorohydrin; β-Dichlorohydrin; β,γ-Dichlorohydrin; 1,2-Dichloro-3-propanol; 1,2-Dichloropropanol-3; 2,3-Dichloropropanol; 2,3-Dichloropropyl alcohol; Glycerol 1,2-dichlorohydrin; Glycerol-2,3-dichlorohydrin; Glycerol-α,β-dichlorohydrin; 1-Propanol, 2,3-dichloro-
Empirical: $C_3H_6Cl_2O$
Properties: Colorless to amber visc. liq.; ethereal odor; sol. in DMSO, 95% ethanol, acetone, ether, benzene; sol. 1-10 mg/ml in water; m.w. 128.99; sp.gr. 1.3607 (20/4 C); b.p. 183-185 C; flash pt. 91.4 C; ref. index 1.4819 (20 C)
Toxicology: LD50 (oral, rat) 90 mg/kg, (skin, rabbit) 200 mg/kg; LCLo (inh., rat, 4 h) 500 ppm; poison; highly toxic by ing., inh., skin contact; severe eye irritant; irritating to skin, mucous membranes, respiratory tract; may cause abdominal pain, dyspnea, hemolytic anemia, cerebral/GI hemorrhage, toxic hepatitis, nephritis, respiratory edema, headache, nausea, vomiting, somnolence, vertigo, diarrhea, jaundice, coma, CNS depression, liver/kidney damage, death by renal failure, etc.; mutation data; TSCA listed
Precaution: Combustible; incompat. with strong oxidizing agents, oxygen, peroxides; may be sensitive to prolonged exposure to air
Hazardous Decomp. Prods.: Heated to decomp., emits toxic fumes of CO, CO_2, hydrogen chloride gas; emits toxic fumes under fire conditions
Storage: Store in tightly closed container under inert atmosphere; store refrigerated; store away from ignition sources
Uses: Solvent for hard resins, nitrocellulose; mfg. of photographic chems., lacquers; cement for celluloid; binder for water colors
Manuf./Distrib.: Acros Org.; Alfa Aesar; Fluka; J.T. Baker; Pfaltz & Bauer; Spectrum Quality Prods.; TCI Am.

2,3-Dichloropropyl alcohol. *See* 2,3-Dichloro-1-propanol

Dichlorotetrafluoroethane

CAS 1320-37-2
UN 1958 (DOT)
Synonyms: Ethane, dichlorotetrafluoro-; Tetrafluorodichloroethane
Empirical: $C_2Cl_2F_4$
Properties: Colorless gas; m.w. 170.92; b.p. 3.5 C; nonflamm.
Toxicology: LCLo (inh., mouse, 30 min) 700,000 ppm; mildly toxic irritant; narcotic in high concs.; asphyxiant; TSCA listed
Precaution: Reacts violently with alcohol
Hazardous Decomp. Prods.: Heated to decomp., emits toxic fumes of F^- and Cl^-
Uses: Solvent; fire extinguishers; refrigerant; air conditioning fluid; blowing agent; dielectric fluid; aerosol propellant for pharmaceuticals
Manuf./Distrib.: Refron
See also 1,2-Dichlorotetrafluoroethane

1,2-Dichlorotetrafluoroethane

CAS 76-14-2; EINECS/ELINCS 200-937-7
UN 1958
Synonyms: CFC 114; Cryofluoran; Cryofluorane; Dichlorotetrafluoroethane; 1,2-Dichloro-1,1,2,2-tetrafluoroethane; sym-Dichlorotetrafluoroethane; Ethane, 1,2-dichloro-1,1,2,2-tetrafluoro-; F 114; FC 114; Fluorane 114; Fluorocarbon 114; Freon 114; Halocarbon 114; Propellant 114; R 114; Refrigerant 114; Tetrafluorodichloroethane; 1,1,2,2-Tetrafluoro-1,2-dichloroethane
Classification: Halogenated aliphatic hydrocarbon
Empirical: $C_2Cl_2F_4$
Formula: $CClF_2CClF_2$
Properties: Colorless clear gas or liquefied compressed gas; nearly odorless to faint ethereal odor; pract. insol. in water; sol. in alcohol, ether; m.w. 170.92; sp.gr. 1.5312 (liq., 0 C); vapor pressure 1620 mm Hg;

f.p. -94 C; b.p. 3.55 C; ref. index 1.3092; surf. tens. 12 dynes/cm; dielec. const. 2.26; nonflamm.
Toxicology: ACGIH TLV/TWA 1000 ppm; LC50 (inh., rat, 30 min) 72 pph; toxic by inhalation; an asphyxiant; may cause mild irritation of nose, throat, upper airways; high concs. may cause lightheadedness, dizziness, drowsiness; TSCA listed
Environmental: ThOD 0.09
Precaution: Compressed gas; incompat. with reactive metals (sodium, potassium, magnesium, zinc, powd. aluminum); can react violently or explosively
Storage: Store cylinders in cool, dry, well-ventilated area away from corrosive atmospheres, out of direct sunlight; leaked gas can accumulate in low areas
Uses: Refrigerant; air conditioning gas; aerosol propellant for industrial applics., pharmaceuticals; solvent; degreasing agent; fire extinguishing agent; foam blowing agent; dielectric fluid
Regulatory: FDA 21CFR §701.30; FDA approved for inhalants, nasals, rectals, topicals; NF compliance; SARA §313 reportable
Manuf./Distrib.: Aldrich; Allchem Ind.; Aventis Pharmaceuticals; BOC Ltd; DuPont; Fluorochem Ltd; LaRoche Ind.; Matheson Gas Prods.; Uniqema
See also Dichlorotetrafluoroethane

1,2-Dichloro-1,1,2,2-tetrafluoroethane; sym-Dichlorotetrafluoroethane. *See* 1,2-Dichlorotetrafluoroethane

2,4-Dichlorotoluene

CAS 95-73-8; EINECS/ELINCS 202-445-8
Synonyms: Benzene, 2,4-dichloro-1-methyl-; 2,4-Dichloromethylbenzene; 2,4-Dichloro-1-methylbenzene; Toluene, 2,4-dichloro-
Empirical: $C_7H_6Cl_2$
Properties: Colorless clear liq.; m.w. 161.03; sp.gr. 1.2498; m.p. -13.5 C; b.p. 196-200 C; flash pt. 79 C ref. index 1.5454
Toxicology: LD50 (oral, rat) 2400 mg/kg; harmful liq. and fumes; harmful by inh., ing., skin contact; may cause eye/skin irritation; TSCA listed
Precaution: Combustible; incompat. with strong oxidizing agents, strong bases
Hazardous Decomp. Prods.: Toxic fumes of CO, CO_2, hydrogen chloride gas; emits toxic fumes under fire conditions
Storage: Store refrigerated; keep container closed; keep away from heat and open flame
Uses: High-boiling solvent; intermediate for organic synthesis
Manuf./Distrib.: Aldrich; Fluka; Riedel-deHaën

(E)-1,2-Dichlrooethylene. *See* trans-1,2-Dichloroethylene
N,N-Diclohexylamine. *See* Dicyclohexylamine
1,4-Dicyanobutane. *See* Adiponitrile

Dicyclohexylamine

CAS 101-83-7; EINECS/ELINCS 202-980-7
UN 2565 (DOT)
Synonyms: N-Cyclohexanamine; N-Cyclohexylcyclohexanamide; DCHA; DICHA; N,N-Diclohexylamine; Dodecahydrodiphenylamine; Perhydrodiphenylamine
Classification: Nonaromatic amine
Empirical: $C_{12}H_{23}N$
Properties: Yel. clear liq.; amine odor; sol. in oxygenated solvs.; misc. with cyclohexylamine; sl. sol. in water; m.w. 181.32; dens. 0.915; m.p. -1 C; b.p. 256 C; flash pt. 103 C; ref. index 1.4842 (20 C); strong base
Toxicology: LD50 (oral, rat) 373 mg/kg, (subcut., mouse) 135 mg/kg; poison by ing. and subcut. routes; harmful if swallowed; corrosive; causes burns; severe skin and eye irritant; possible sensitizer; possible carcinogen; experimental tumorigen; human mutagenic data; TSCA listed
Environmental: Very toxic to aquatic organisms; may cause long-term adverse effects in the aquatic environment; avoid release to the environment
Precaution: Combustible exposed to heat or flame; can react with oxidizing agents
Hazardous Decomp. Prods.: Heated to decomp., emits toxic fumes of NO_x
NFPA: Health 3, Flammability 1, Reactivity 0
Storage: Store in cool, dry, well-ventilated area under nitrogen, away from incompat. substances
Uses: Corrosion inhibitor in lubricants, steel or tinplate food pkg.; industrial solvent; solvent for antibiotic prod.
Regulatory: FDA 21CFR §178.3300
Manuf./Distrib.: ABCR; Acros Org.; Air Prods.; Aldrich; Allchem Ind.; Bencorp Int'l.; Chemapol Usa; Fluka; Gulf Chem. Int'l.; Haltermann Prods. UK; JLM Marketing; Schweizerhall; Sigma; Whyte Chems. Ltd

Didecyl. *See* Eicosane
N,N-Didodecyl-1-dodecanamine. *See* Trilaurylamine
Diethamine. *See* Diethylamine

Diethanolamine

CAS 111-42-2; EINECS/ELINCS 203-868-0
UN 1760
Synonyms: Bis (2-hydroxyethyl) amine; N,N-Bis (2-hydroxyethyl) amine; DEA; DELA; Diethylolamine; 2,2′-Dihydroxydiethylamine; Di (2-hydroxyethyl) amine; Diolamine; 2-(2-Hydroxyethylamino) ethanol; 2,2′-Iminobisethanol; 2,2′-Iminodiethanol
Classification: Aliphatic amine; alkanolamine
Empirical: $C_4H_{11}NO_2$
Formula: $(HOCH_2CH_2)_2NH$
Properties: Colorless cryst. or liq., mild ammoniacal odor; very sol. in water and alcohol; misc. with acetone, chloroform, glycerin; insol. in ether, benzene; m.w. 105.09; dens. 1.0881 (30/4 C); m.p. 28 C; b.p. 268 C; flash pt. 300 F; ref. index 1.4770
Toxicology: ACGIH TLV/TWA 3 ppm in air; LD50 (oral, rat) 12.76 g/kg, (IP, mouse) 2300 mg/kg, (subcut., mouse) 3553 mg/kg, (skin, rabbit) 12,200 mg/kg; mod. toxic by ing., IP, and subcut. routes; severe eye and skin irritant; mucous membrane irritant; experimental reproductive effects; TSCA listed
Environmental: ThOD 1.52
Precaution: Combustible; DOT: corrosive material; normally stable; slowly oxidized by air; incompat. with oxidizing agents, strong acids, acid anhydrides, nitrosating agents
Hazardous Decomp. Prods.: CO, CO_2, NO_x, ammonia
NFPA: Health 1, Flammability 1, Reactivity 0
Storage: Hygroscopic, deliq.; light-sensitive; store @ R.T. in cool, dry, well-ventilated area away from heat and ignition sources
Uses: Absorbent in gas scrubbing; rubber chemicals intermediate; mfg. of surfactants for textiles, herbicides, petroleum demulsifier; detergent, emulsifier and dispersant in coatings, agric., cosmetics, pharmaceuticals, cutting oils, cleaners, polishes; textile lubricants/processing; humectant; softener; in organic synthesis; crosslinking agent for flexible PU foam slabstock; hydrogen sulfide extraction solvent in gas processing; solubilizer (metalworking fluids, printing inks, textile auxs.); alkalizer; in food-pkg. adhesives; adjuvant to control pulp absorbency and pitch content in mfg. of paper/paperboard in contact with aq./fatty/dry foods; defoamer in food-contact paper/paperboard
Regulatory: FDA 21CFR §175.105, 176.170, 176.180, 176.210, 177.2600; FDA approved for injectables (IV), ophthalmics; NF compliance; SARA §313 reportable
Manuf./Distrib.: AMRESCO; Akzo Nobel; Aldrich; Allchem Ind.; Ashland; BASF AG; BP Amoco; Brenntag AG; Camida Ltd; ChevronTexaco; Coyne; Degussa AG; Dow; Fluka; Great Western; Honeywill & Stein; Houghton Chem.; Huntsman; Mosselman NV; Mutchler; Occidental; Oxiteno; Penta Mfg.; Ruger; Sal Chem.; Sigma; Spectrum Quality Prods.; Stanchem; Surfachem Ltd; Univar E&E Ltd; Varsal Instruments; Vopak USA; Whyte Chems. Ltd
Trade Names Containing: Jefftreat® MS-205; Jefftreat® MS-205 LFG

Diethanolamine coconut fatty acid condensate. *See* Cocamide DEA
Diethanolethylamine. *See* N-Ethyldiethanolamine
1,1-Diethoxyacetal; 1,1-Diethoxyethane. *See* Acetal
1,2-Diethoxyethane. *See* Ethylene glycol diethyl ether

Diethoxyethyl succinate

CAS 26962-29-8
Synonyms: Butanedioic acid, bis (2-ethoxyethyl) ester; Ethanol, 2-ethoxy-, succinate (2:1)
Definition: Diester of ethoxyethanol and succinic acid
Empirical: $C_{12}H_{22}O_6$
Uses: Solvent in cosmetics

Diethoxymethane

CAS 462-95-3; EINECS/ELINCS 207-330-6

Synonyms: 1,1-Diethoxymethane; Diethylformal; 3,5-Dioxapheptene; Ethane, 1,1′-[methylenebis (oxy)] bis-; Ethoxymethyl ethyl ether; Ethylal; Formaldehyde diethyl acetal; Methane, diethoxy-

Empirical: $C_5H_{12}O_2$

Formula: $CH_2(OC_2H_5)_2$

Properties: Clear liq.; mod. sol. in water; m.w. 104.15; sp.gr. 0.832; vapor pressure 60 mm Hg; m.p. -66.5 C; b.p. 85-88 C; flash pt. -6 C; autoignition temp. 174 C; ref. index 1.374

Toxicology: LD50 (oral, rat) 3536 mg/kg; irritating to eyes, skin, mucous membranes, upper respiratory tract; may be harmful by ing., inh, skin absorp.; TSCA listed

Precaution: Flamm. liq.; vapor may flash back; may develop pressure; container explosion may occur under fire conditions; incompat. with strong oxidizing agents, strong acids

Hazardous Decomp. Prods.: CO, CO_2; emits toxic fumes under fire conditions

NFPA: Health 1, Flammability 3, Reactivity 0

Storage: Keep tightly closed away from heat, sparks, open flame, other ignition sources

Uses: Reaction solvent; extraction solvent; ethoxymethylation agent, methylenation agent in org. compd. synthesis; mfg. of pharmaceuticals

Features: Can replace 1,2-dimethoxyethane and methylal

Manuf./Distrib.: Acros Org.; Aldrich; Alfa Aesar; Fluka; Indofine; Lancaster Synthesis; TCI Am.

1,1-Diethoxymethane. *See* Diethoxymethane
Diethyl. *See* Butane
Diethyl acetal. *See* Acetal

Diethyl acetyl aspartate

Synonyms: N-Acetyl-diethyl-L-aspartate; Aspartic acid, N-acetyl-, diethyl ester

Classification: Substituted amino acid

Empirical: $C_{10}H_{17}NO_5$

Uses: Solvent, skin conditioner in cosmetics

Diethylamine

CAS 109-89-7; EINECS/ELINCS 203-716-3

UN 1154 (DOT)

Synonyms: 2-Aminopentane; DEA; DEN; DETN; Diethamine; N,N-Diethylamine; N-Ethylethanamine

Classification: Saturated aliphatic amine

Empirical: $C_4H_{11}N$

Formula: $(C_2H_5)_2NH$

Properties: Colorless mobile liq.; ammoniacal odor; sol. in water, ethanol, acetone, ether, chloroform, dimethyl sulfoxide, benzene; misc. with oxygenated solvs.; m.w. 73.14; dens. 0.707; vapor pressure 400 mm Hg (38 C); m.p. -50 C; b.p. 55 C; flash pt. -28 C; ref. index 1.3850 (20 C)

Toxicology: ACGIH TLV/TWA 5 ppm; STEL 15 ppm; LD50 (oral, rat) 540 mg/kg, (skin, rabbit) 820 mg/kg; LC50 (inh., mammal) 5000 mg/m^3; mod. toxic by ing., inh., skin contact; corrosive; skin and severe eye irritant; inh. of vapor can cause severe cough and chest pains; eye contact with liq. may cause permanent eye damage; skin contact causes necrosis and vesiculation; target organs: eye, skin, heart, liver, kidney, respiratory system; TSCA listed

Environmental: EC50 (algae, 96 h) 20 ppm; LC50 (daphnia magna, 24 h) 161-168 ppm; may be degraded by unacclimated biomass; 17% biodeg. in 31 days

Precaution: Highly flamm.; dangerous fire hazard exposed to heat, flame, oxidizers; explodes on contact with dicyanofurazan; violent reaction with sulfuric acid; ignites on contact with cellulose nitrate; absorbs CO_2 from air to form the carbamate salt; vapor may form explosive mixts. with air; sudden reaction and fire may occur if mixed with an oxidizing agent; incompat. with acids, oxidizers, sodium or calcium hypochlorite; may form nitrosamines with nitrous acid, nitrites, etc.

Hazardous Decomp. Prods.: CO, CO_2, NO_x, ammonia, nitric acid, nitrosamines, irritating or toxic fumes

NFPA: Health 3, Flammability 3, Reactivity 0

Storage: Store in cool, dry, ventilated area in closed containers, away from heat, flames, sparks; do not store in reactive metal containers

Uses: Intermediate in mfg. of rubber accelerators, surfactants, dyes, polymers, DEET pesticides, textile specialties, petroleum chemicals; selective solvent; buffer; flotation agent; polymerization inhibitor; in electroplating sol'ns.; corrosion inhibitor for metals; depilatory for hides; in food-pkg. adhesives; in pharmaceutical injectables; catalyst in food-contact PPO resins; activator in food-contact rubber articles for repeated use

Regulatory: FDA 21CFR §175.105, 177.2460, 177.2600; FDA approved for injectables (IM, IV); SARA §312 immediate health and fire hazard, §313 nonreportable

Manuf./Distrib.: AMRESCO; Air Prods.; Aldrich; Allchem Ind.; Ashland; Atofina; BASF AG; BASF; Celanese; Coyne; Daicel Chem. Ind.; Dow; Fluka; J.T. Baker; Mallinckrodt Baker; Mitsubishi Gas; Pechiney Chems. Div.; Penta Mfg.; Sigma; Spectrum Quality Prods.; Triple Crown Am.; Vinchem; Vopak USA

N,N-Diethylamine. *See* Diethylamine
Diethylamine, N-nitroso-. *See* N-Nitrosodiethylamine
(Diethylamino) ethane. *See* Triethylamine

Diethylaminoethanol

CAS 100-37-8; EINECS/ELINCS 202-845-2

UN 2686 (DOT)

Synonyms: DEAE; 2-Diethylaminoethanol; β-Diethylaminoethanol; N-Diethylaminoethanol; N,N-Diethyl-2-aminoethanol; β-Diethylaminoethyl alcohol; Diethylethanolamine; N,N-Diethylethanolamine; N,N-Diethyl-N-(β-hydroxyethyl) amine; Ethanol, 2-(diethylamino)-; 2-Hydroxytriethylamine

Classification: Aliphatic amino alcohol

Empirical: $C_6H_{15}NO$

Formula: $(C_2H_5)_2NCH_2CH_2OH$

Properties: Colorless liq.; strong ammonia-like odor; sol. in water, alcohol, acetone, ether, benzene, petrol. ether; misc. with most org. solvs.; m.w. 117.19; dens. 0.88-0.89 (20/20 C); vapor pressure 1.4 mm Hg (20 C); f.p. -70 C; b.p. 161 C; flash pt. (OC) 140 F; ref. index 1.4389

Toxicology: ACGIH TLV/TWA 10 ppm (skin); LD50 (oral, rat) 1300 mg/kg, (IP, rat) 1220 mg/kg, (subcut., mouse) 1561 mg/kg, (IV, mouse) 188 mg/kg; poison by IP, IV routes; mod. toxic by ing., skin contact, subcut. routes; human systemic effects by inh. (nausea, vomiting); skin, severe eye irritant; corrosive; TSCA listed

Environmental: VOC; ThOD 2.32

Precaution: DOT: Flamm. liq.; moderate fire risk; reactive with oxidizing materials

Hazardous Decomp. Prods.: Heated to decomp., emits toxic fumes of NO_x

NFPA: Health 3, Flammability 2, Reactivity 0

Storage: Hygroscopic

Uses: Chemical intermediate for prod. of emulsifiers, detergents, solubilizers; water-sol. salts; textile/fabric softeners; cosmetics; pharmaceuticals; emulsifier in acid media; curing agent for resins; surface coatings; pigment dispersant in paints; solvent (gas sweetening); catalyst in PU foams; petrol. and gas processing chems.; boiler water additive for food contact; corrosion inhibitor for boiler water treatment; textile/leather/paper aux.

Regulatory: FDA 21CFR §173.310 (15 ppm max. in steam)

Manuf./Distrib.: Aldrich; Ashland; Atofina; BASF; Dow; Filo; Fluka; Richman; Spectrum Quality Prods.; Vopak USA; Xinchem

2-Diethylaminoethanol; β-Diethylaminoethanol; N-Diethylaminoethanol; N,N-Diethyl-2-aminoethanol; β-Diethylaminoethyl alcohol. *See* Diethylaminoethanol

Diethylbenzene

CAS 25340-17-4; EINECS/ELINCS 246-874-9

UN 2049

Synonyms: Benzene, diethyl-; Diethylbenzene, mixed isomers; Diethylbenzene (mixture); Diethylbenzene, mixture of 1,2-, 1,3- and 1,4-isomers; Diethylbenzene, mixture of o-, m- and p-isomers; Diethylbenzol

Empirical: $C_{10}H_{14}$

Formula: $C_6H_4(C_2H_5)_2$

Properties: Colorless clear liq.; char. aromatic hydrocarbon odor; insol. in water; m.w. 134.22; sp.gr. 0.87; b.p. 180-182 C; flash pt. 56 C; ref. index 1.495 (20 C)

Toxicology: LD50 (oral, rat) 2050 mg/kg, (skin, rabbit) > 5000 mg/kg; vapor/mist irritating to eyes, mucous membranes, upper respiratory tract; skin irritant; may be harmful by inh., ing., or skin absorp.; inh. may cause cough, dizziness, headache; ing. may cause confusion, dizziness, drowsiness, headache, unconsciousness; may cause altered sleep time, somnolence, excitement, fetotoxicity, effects on females; may cause lung injury leading to death; target organs: liver, kidneys, CNS; TSCA listed

Environmental: LC50 (fathead minnow, 96 h) > 1 mg/l; not readily biodeg.; harmful to aquatic organisms; keep out of drains and water courses

Precaution: Flamm. liq. and vapor; LEL 0.8%; UEL 5%; incompat. with strong oxidizing agents; reacts violently with oxidants causing fire, explosion hazard; explosive hazard when vapor mixes with air; attacks rubber

Hazardous Decomp. Prods.: Toxic fumes of CO, CO_2

NFPA: Health 2, Flammability 2, Reactivity 0

Storage: Store in cool, dry place; keep away from heat, open flame; keep tightly closed

Uses: Fragrance in cosmetics; heat transfer fluid

Regulatory: SARA immediate/delayed/fire hazard; EU, Canada DSL, Australia, Japan listed

Manuf./Distrib.: ABCR; Acros Org.; Aldrich; Alfa Aesar; Fluka; ICN Biomed. Research Prods.; Mitsubishi Chem.; TCI Am.

Trade Names: Sure Sol®-150ND

Diethyl 1,2-benzenedicarboxylate. *See* Diethyl phthalate

Diethylbenzene, mixed isomers; Diethylbenzene (mixture); Diethylbenzene, mixture of 1,2-, 1,3- and 1,4-isomers; Diethylbenzene, mixture of o-, m- and p-isomers; Diethylbenzol. *See* Diethylbenzene

Diethyl butanedioate. *See* Diethyl succinate

Diethyl carbinol. *See* 3-Pentanol

Diethyl 'Carbitol'. *See* Diethylene glycol diethyl ether

Diethyl carbonate

CAS 105-58-8; EINECS/ELINCS 203-311-1

UN 2366

Synonyms: Carbonic acid, diethyl ester; Carbonic ether; DEC; Ethoxyformic anhydride; Ethyl carbonate

Empirical: $C_5H_{10}O_3$

Formula: $(C_2H_5O)_2CO$

Properties: Colorless liq.; mild ethereal odor; misc. with most org. solvs., alcohol, ether; insol. in water; m.w. 118.13; dens. 0.975; vapor pressure 10 mm Hg (24 C); m.p. -43 C; b.p. 126-128 C; flash pt. (CC) 25 C; ref. index 1.3837

Toxicology: LD50 (subcut., rat) 8500 mg/kg; LDLo (oral, rat) 15 g/kg; TDLo (IP, mouse) 456 mg/kg; irritating to eyes, skin, respiratory system; tumorigen; reproductive effector; TSCA listed

Precaution: DOT: Flamm. liq.

NFPA: Health 2, Flammability 3, Reactivity 1

Storage: Moisture-sensitive; store in cool, dry place in tightly closed container

Uses: Solvent in resins and lacquers; solvent for nitrocellulose; carboethoxylation reagent; mfg. radio tubes; fixing rare earths to cathode elements

Manuf./Distrib.: ABCR; Acros Org.; Aldrich; Fluka; Sigma

Diethyl 'Cellosolve'. *See* Ethylene glycol diethyl ether

Diethyldiglycol. *See* Diethylene glycol diethyl ether

Diethyl 2,3-dihydroxybutanedioate; Diethyl 2,3-dihydroxysuccinate. *See* Diethyl tartrate

Diethylene dioxide; 1,4-Diethylene dioxide; Diethylene ether. *See* 1,4-Dioxane

Diethylene glycol

CAS 111-46-6; EINECS/ELINCS 203-872-2

Synonyms: Bis (2-hydroxyethyl) ether; DEG; Diglycol; Digol; Dihydroxydiethyl ether; β,β′-Dihydroxydiethyl ether; 2,2′-Dihydroxyethyl ether; Ethanol, 2,2′-oxydi-; Ethylene diglycol; Glycol ether; Glycol ethyl ether; 3-Oxapentane-1,5-diol; 3-Oxa-1,5-pentanediol; 2,2′-Oxybisethanol; 2,2′-Oxydiethanol

Classification: Aliphatic diol; glycol ether

Empirical: $C_4H_{10}O_3$

Formula: $CH_2OHCH_2OCH_2CH_2OH$

Properties: Colorless clear syrupy liq., pract. odorless, sweetish taste; lowers f.p. of water; misc. with water, ethanol, acetone, ether, ethylene glycol; m.w. 106.12; dens. 1.1184 (20/20 C); vapor pressure 1 mm Hg (91.8 C); f.p. -80 C; b.p. 245 C; flash pt. 124 C

Toxicology: LD50 (oral, human) 1000 mg/kg, (oral, rat) 12,565 mg/kg, (skin, rabbit) 11,890 mg/kg; poison by inh.; mod. toxic to humans by ing., IV routes; hazardous for household use in conc. of 10% or more; may cause acidosis and renal failure; experimental carcinogen, tumorigen, teratogen; eye and human skin irritant; TSCA listed

Environmental: VOC; BOD5 0.12; COD 1.51; ThOD 1.51

Precaution: Combustible exposed to heat or flame; reactive with oxidizing materials; dec. exothermically with sodium hydroxide @ 230 C and releases explosive hydrogen gas

Hazardous Decomp. Prods.: CO, CO_2; heated to decomp., emits acrid smoke and irritating fumes

NFPA: Health 1, Flammability 1, Reactivity 0

Storage: Extremely hygroscopic

Uses: Chemical intermediate; prod. of polyurethane and unsat. polyester resins, triethylene glycol, morpholine, paints; textile softener, lubricant; solvent for nitrocellulose, dyes, oils, printing inks, cosmetics; extraction solvent; dehydration of natural gas, plasticizers, surfactants; humectant for tobacco, casein, and syn. sponge; cork compositions; bookbinding adhesives; dyeing assistant; antifreeze ingred.; PU chain extender; processing aid; activator; solvent in pharmaceuticals; in food-pkg. adhesives; solvent for certain accelerators in food-contact crosslinked polyesters

Regulatory: FDA 21CFR §175.105, 177.2420

Manuf./Distrib.: AAE Chemie NV; AMC Chems.; AMRESCO; Adept Sol'ns.; Aldrich; Alfa Chem; Allchem Ind.; Arch Chems.; Ashland; Avatar; BASF; BP Chems. Ltd; Baychem; C.P. Hall; Celanese; Chemcentral; Coyne; Dow; DuPont; Eastman; Equistar; ExxonMobil; Ferro/Grant; Fluka; Harcros; Houghton Chem.; Huntsman; Integra; Itochu Spec. Chems.; J.T. Baker; Occidental; Penta Mfg.; Ruger; Sal Chem.; Shell; Sigma; Spectrum Quality Prods.; Stanchem; Sunnyside; Universal Preserv-A-Chem; Vopak USA

Trade Names Containing: DA-405B

Diethylene glycol amine. *See* 2-(2-Aminoethoxy) ethanol

Diethylene glycol benzyl ether

CAS 2050-25-1

Synonyms: 2-[2-(Benzyloxy) ethoxy] ethanol; Diglycol benzyl ether

Empirical: $C_{10}H_{14}O_3$

Formula: $C_6H_5CH_2O(CH_2)O(CH_2)_2OH$

Properties: Clear liq.; sol. in water; m.w. 196.24; sp.gr. 1.08-1.09; m.p. < -50 C; b.p. 302 C; flash pt. 129 C; ref. index 1.51

Toxicology: May be harmful by ing., inh., skin absorp.; may cause eye/skin/mucous membrane/upper respiratory tract irritation

Precaution: Incompat. with strong oxidizing agents

Hazardous Decomp. Prods.: Toxic fumes of CO, CO_2; emits toxic fumes under fire conditions

NFPA: Health 2, Flammability 1, Reactivity 0

Storage: Store in cool, dry place; keep tightly closed

Uses: Solvent, coupling agent for printing inks, agric., water-based coatings, plastics, lacquers, paints, varnishes, cleaning prods., preservatives, rust removers, disinfectants

Manuf./Distrib.: Aldrich

Diethylene glycol bis (butyl carbonate); Diethylene glycol bis (n-butylcarbonate). *See* Butyl diglycol carbonate

Diethylene glycol butyl ether; Diethylene glycol n-butyl ether. *See* Butoxydiglycol

Diethylene glycol butyl ether acetate

CAS 124-17-4; EINECS/ELINCS 204-685-9

Synonyms: Acetic acid 2-(2-butoxyethoxy) ethyl ester; 2-(2-Butoxyethoxy) ethanol acetate; 2-(2-Butoxyethoxy) ethyl acetate; 2-(2-n-Butoxyethoxy) ethyl acetate; Butyl 'Carbitol' acetate; n-Butyl 'Carbitol' acetate; Butyldiglycol acetate; Diethylene glycol monobutyl ether acetate; Diethylene glycol mono-n-butyl ether acetate; Diglycol monobutyl ether acetate

Classification: Ester
Empirical: $C_{10}H_{20}O_4$
Formula: $CH_3COOCH_2CH_2OCH_2CH_2O(CH_2)_3CH_3$
Properties: Colorless liq.; mild odor; misc. with water and most org. solvs.; m.w. 204.27; dens. 0.977 (20/4 C); vapor pressure 0.01 mm Hg (20 C); f.p. -32.2 C; i.b.p. 235 C; flash pt. 102 C; autoignition temp. 295 C; ref. index 1.43 (20 C); surf. tens. 30 dynes/cm
Toxicology: LD50 (oral, rat) 6500 mg/kg, (skin, rabbit) 14,500 mg/kg; LC50 (inh., rat, 4 h) 8693 ppm; mod. toxic by ing.; mild skin/eye irritant; prolonged/repeated skin contact may cause drying, cracking, or irritation; target organs: CNS, skin, kidney, respiratory system, spleen; TSCA listed
Environmental: VOC; ThOD 1.25
Precaution: Combustible exposed to heat or flame; LEL 0.8%; incompat. with oxidizing materials; peroxide former
Hazardous Decomp. Prods.: Heated to decomp., emits acrid and irritating fumes
NFPA: Health 1, Flammability 1, Reactivity 0
Uses: Solvent for printing inks, water-based paints/coatings, separation processes, electronic materials, cosmetics; coalescing aid; in food-pkg. adhesives
Regulatory: FDA 21CFR §175.105; SARA §313 reportable; HAP
Manuf./Distrib.: A.P. Chems. Ltd; ABCR; Acros Org.; Aldrich; Alfa Aesar; Allchem Ind.; Ashland; Chemial SpA; Dow; Eastman; Fluka; G.J. Chem.; ICN Biomed. Research Prods.; J.T. Baker; Lancaster Synthesis; Lyondell; Occidental; Pfaltz & Bauer; Spectrum Quality Prods.; TCI Am.; Whyte Chems. Ltd
Trade Names: Butyl Carbitol® Acetate; Eastman® DB Acetate

Diethylene glycol dibenzoate

CAS 120-55-8; EINECS/ELINCS 204-407-6
Synonyms: Benzoic acid, diester with diethylene glycol; Dibenzoyldiethyleneglycol ester; PEG-2 dibenzoate; PEG 100 dibenzoate; POE (2) dibenzoate
Classification: Aromatic ester
Definition: PEG diester of benzoic acid
Empirical: $C_{18}H_{18}O_5$
Formula: $C_6H_5COO(CH_2)_2O(CH_2)_2OOCC_6H_5$
Properties: High-boiling liq. or solid; mild ester odor; sol. in aliphatic and aromatic solvs.; insol. in water; m.w. 314.36; dens. 1.1765-1.178 (20 C); m.p. 16 C; b.p. 225-227 C (3 mm); flash pt. (PMCC) 232 C; ref. index 1.5424-1.5449 (20 C)
Toxicology: LD50 (oral, rat) 2830 mg/kg, (skin, rabbit) 20 g/kg; mod. toxic by ing.; primary skin and eye irritant; TSCA listed
Precaution: Combustible; can react with oxidizers
Hazardous Decomp. Prods.: Heated to decomp., emits acrid smoke and fumes
NFPA: Health 0, Flammability 1, Reactivity 0
Uses: Plasticizer for polymers, vinyl acetate resins, adhesives, latex coatings; emollient, plasticizer in cosmetics; in food-pkg. adhesives; plasticizer for PVAc coatings for paper/paperboard in contact with aq./fatty foods; plasticizer for polymers for paper/paperboard in contact with dry food
Regulatory: FDA 21CFR §175.105, 176.170, 176.180
Manuf./Distrib.: Aldrich; Allchem Ind.; ChemService; Haltermann GmbH; Noveon Kalama; Unitex; Velsicol
Trade Names: Benzoflex® 2-45

Diethylene glycol dibutyl ether

CAS 112-73-2; EINECS/ELINCS 204-001-9
Synonyms: Bis (butoxyethyl) ether; Bis (2-butoxyethyl) ether; 2-Butoxyethyl ether; Butyl diglyme; Butyl glycol ether; Dibutoxy diethylene glycol; 2,2′-Dibutoxyethyl ether; Dibutyl 'Carbitol'; Dibutyldiglycol; Diethylene glycol di-n-butyl ether; Ether, bis (2-butoxyethyl); 1,1′-(Oxybis (2,1-ethanediyloxy)) bisbutane; 5,8,11-Trioxapentadecane
Empirical: $C_{12}H_{26}O_3$
Formula: $C_4H_9O(C_2H_4O)_2C_4H_9$
Properties: Almost colorless liq.; char. mild ether odor; sol. in DMSO, 95% ethanol, acetone; misc. with ether, ketones, esters, CCl_4; sol. < 1 mg/ml in water; m.w. 218.38; dens. 0.8853 (20/20 C); vapor pressure 0.02 mm Hg (20 C); f.p. -60.2 C; b.p. 256 C; flash pt. 118 C; ref. index 1.423 (20 C)
Toxicology: LD50 (oral, rat) 3900 mg/kg, (skin, rabbit) 4040 mg/kg; mod. toxic by ing.; mildly toxic by skin contact; skin and eye irritant; experimental reproductive effects; TSCA listed
Environmental: ThOD 1.54
Precaution: Combustible exposed to heat or flame; may decompose violently on contact with nitric acid; can react with strong oxidizers; incompat. with strong acids, copper, and copper alloys; can form peroxides on exposure to air or oxygen
Hazardous Decomp. Prods.: Heated to decomp., emits acrid smoke, irritating fumes, and toxic fumes of CO and CO_2
NFPA: Health 1, Flammability 1, Reactivity 0
Storage: Store in tightly closed container under inert atmosphere at refrigerated temps.
Uses: Solvent in coatings, inks, electrochemistry, polymer and boron chemistry, physical processes such as gas absorption, extraction, stabilization, industrial prods. such as fuels, lubricants, textiles, pharmaceuticals, pesticides; diluent in VC dispersions; extractant for uranium ores; selective solvent for extraction of gold; coupling solvent; solvent, carrier, stabilizer in pharmaceuticals
Features: High-boiling; inert
Regulatory: SARA §313 reportable; HAP
Manuf./Distrib.: Acros Org.; Aldrich; Alfa Aesar; BP Amoco; Brand-Nu Labs; Celanese; ChemService; Ferro/Grant; Filo; Fluka; Great Western; ICN Biomed. Research Prods.; Pfaltz & Bauer; Sigma; TCI Am.
Trade Names: Butyl Diglyme

Diethylene glycol di-n-butyl ether. *See* Diethylene glycol dibutyl ether

Diethylene glycol diethyl ether

CAS 112-36-7; EINECS/ELINCS 203-963-7
Synonyms: Bis (2-ethoxyethyl) ether; Diethyl 'Carbitol'; Diethyldiglycol; Diglycol diethyl ether; 1-Ethoxy-2-(β-ethoxyethoxy) ethane; Ethoxyethyl ether; 2-Ethoxyethyl ether; Ethyl diglyme; 3,6,9-Trioxaundecane
Classification: Glycol ether
Empirical: $C_8H_{18}O_3$
Formula: $C_2H_5O(C_2H_4O)_2C_2H_5$
Properties: Colorless liq.; sol. in hydrocarbons, most org. solvs., and water; m.w. 162.93; dens. 0.9082 (20/20 C); vapor pressure 0.38 mm Hg (20 C); f.p. -44.3 C; b.p. 189 C; flash pt. 180 F; ref. index 1.412 (20 C); surf. tens. 26.68 dynes/cm; dielec. const. 5.70
Toxicology: LD50 (oral, rat) 4970 mg/kg; mod. toxic by ing.; primary irritant; eye irritant; experimental reproductive effects; TSCA listed
Environmental: VOC; ThOD 2.17
Precaution: Flamm. exposed to heat or flame
Hazardous Decomp. Prods.: Heated to decomp., emits acrid smoke and irritating fumes
NFPA: Health 1, Flammability 2, Reactivity 0
Uses: High-boiling solvent for nitrocellulose, resins, lacquers; solubilizer; organic synthesis
Features: Very stable
Regulatory: SARA §313 reportable; HAP
Manuf./Distrib.: Aldrich; Ferro/Grant; Fluka; Sigma
Trade Names: Ethyl Diglyme

Diethylene glycol diethylhexanoate/diisononanoate. *See* Diethylene glycol dioctanoate/diisononanoate

Diethylene glycol dimethyl ether

CAS 111-96-6; EINECS/ELINCS 203-924-4
Synonyms: Bis (2-methoxyethyl) ether; Diethyl glycol dimethyl ether; Diglycol methyl ether; Diglyme; Dimethoxydiglycol; Dimethyldiglycol; 2-Methoxyethyl ether; Methyl diglyme; Trioxanonane
Empirical: $C_6H_{14}O_3$
Formula: $CH_3(OCH_2CH_2)_2OCH_3$
Properties: Colorless liq., faint ether-like odor; misc. with water, alcohol, ether, hydrocarbons, most org. solvs.; m.w. 134.18; dens. 0.9451 (20/20 C); f.p. -58 C; b.p. 162 C; flash pt. (OC) 70 C; ref. index 1.408 (20 C)
Toxicology: TCLo (inh., rat, 6 h) 1100 ppm, (oral, mouse) 5 g/kg; mutagen; reproductive effects; TSCA listed
Environmental: VOC; ThOD 1.07
Precaution: Combustible; readily forms explosive peroxides on exposure to air, light, or heat; other explosive reactions possible; incompat. with

strong oxidizing agents, strong acids, strong bases
Hazardous Decomp. Prods.: Explosive peroxides, CO, CO_2; heated to decomp., emits toxic fumes of NO_x
NFPA: Health 1, Flammability 2, Reactivity 1
Uses: Solvent; solubilizer; anhydrous reaction medium for organo-metallic synthesis; reaction solvent in reduction, isomerization, alkylation condensation, polymerization; extraction and separation solvent; solvent in cosmetics, pharmaceuticals; paints
Regulatory: SARA §313 reportable; HAP
Manuf./Distrib.: AMC Chems.; Aldrich; Brand-Nu Labs; Celanese; Eastman; Ferro/Grant; Filo; Fluka; Great Western; Independent Chem.; J.T. Baker; Sigma
Trade Names: Hisolve MDM

Diethylene glycol dioctanoate/diisononanoate

Synonyms: Diethylene glycol diethylhexanoate/diisononanoate; Oxybis (2-hydroxyethyl), diesters with 3,5,5-trimethylhexanoic acid and 2-ethylhexanoic acid
Definition: Diester of a mixt. of 2-ethylhexanoic and isononanoic acids and diethylene glycol
Uses: Emollient in cosmetics; penetrant, spreading agent, perfume solvent in antiperspirants, deodorants, preshave lotions
Trade Names: Estalan 430

Diethylene glycol ethyl ether. *See* Ethoxydiglycol
Diethylene glycol ethyl ether acetate. *See* Ethoxydiglycol acetate
Diethylene glycol ethyl ether acrylate. *See* 2-(2-Ethoxyethoxy) ethyl acrylate

Diethylene glycol 2-ethylhexyl ether

CAS 1559-36-0
Synonyms: Diethylene glycol mono-2-ethylhexyl ether; 2-[2-(2-Ethylhexyloxy) ethoxy] ethanol
Empirical: $C_{12}H_{25}O_3$
Properties: M.w. 218.3; sp.gr. 0.918; b.p. 272 C; flash pt. > 230 F; ref. index 1.442 (20 C)
Toxicology: LD50 (oral, rat) 7.5 g/kg; irritating to eyes, skin, respiratory system; may be harmful by ing., inh., skin absorp.; target organ: CNS
Precaution: Incompat. with strong oxidizing agents
Hazardous Decomp. Prods.: CO, CO_2; emits toxic fumes under fire conditions
Storage: Store in cool, dry place; keep tightly closed
Uses: Solvent, coalescing agent in coatings
Manuf./Distrib.: Aldrich
Trade Names Containing: Eastman® EEH

Diethylene glycol ethyl methyl ether. *See* Diethylene glycol methyl ethyl ether

Diethylene glycol ethylvinyl ether

CAS 10143-53-0; EINECS/ELINCS 233-411-0
Synonyms: Ethene, [2-(2-ethoxyethoxy) ethoxy]-; 2-[2-(2-Ethoxyethoxy) ethoxy] ethene; 1-(2-Ethoxyethoxy)-2-vinyloxy ethane; 2-Ethoxyethyl 2-(vinyloxy) ethyl ether; Vinyl ether 'Carbitol'
Classification: Aliphatic glycol diether
Empirical: $C_8H_{16}O_3$
Formula: $CH_2=CHOCH_2CH_2OCH_2CH_2OCH_2CH_3$
Properties: Colorless liq.; sol. in water, ethanol, many solvs.; m.w. 160.24; sp.gr. 0.941 (20 C); m.p. -80 C (sets to a glass); b.p. 191.2 C; flash pt. (OC) 32.8 C
Toxicology: LD50 (oral, rat) 11 g/kg, (skin, rabbit) 8410 mg/kg; mildly toxic by ing., skin contact; skin irritant; ing. may cause vomiting, cyanosis, headache, rapid respiration and heart rate, low blood pressure, muscle tenderness, unconsciousness; can be absorbed thru skin, but health effects unlikely; high/repeated exposure may affect kidney; TSCA listed
Precaution: Flamm.; incompat. with oxidizing materials (possible fire risk)
Hazardous Decomp. Prods.: Heated to decomp., emits acrid smoke and fumes
Storage: Store in a cool, dry, well-ventilated area, out of direct sunlight, away from corrosives
Uses: Solvent
Manuf./Distrib.: Aldrich

Diethylene glycol hexyl ether. *See* Diethylene glycol n-hexyl ether

Diethylene glycol n-hexyl ether

CAS 112-59-4; EINECS/ELINCS 203-988-3
Synonyms: DEGHE; Diethylene glycol hexyl ether; Diethylene glycol monohexyl ether; 3,6-Dioxa-1-dodecanol; 3,6-Dioxadodecanol-1; Ethanol, 2-((2-hexyloxy)ethoxy)-; n-Hexoxyethoxyethanol; Hexyl 'Carbitol'; Hexyldiglycol; 2-(2-Hexyloxyethoxy) ethanol
Classification: Aliphatic ether alcohol
Empirical: $C_{10}H_{22}O_3$
Formula: $CH_3(CH_2)_4CH_2OCH_2CH_2OCH_2CH_2OH$
Properties: Water-wh. liq.; sol. 3% in water; sl. to mod. sol. in water; m.w. 190.28; sp.gr. 0.9346 (20/20 C); vapor pressure 0.01 mm Hg (20 C); f.p. -33 C; b.p. 259.1 C; flash pt. 140.5 C; ref. index 1.438 (20 C)
Toxicology: LD50 (oral, rat) 4920 mg/kg, (skin, male rabbit) 2000 mg/kg; mod. toxic by skin contact; mildly toxic by ing.; eye and severe skin irritant; can be absorbed thru skin; possible allergic reactions; mod. to severe eye irritant; ing. may cause vomiting, headache, rapid respiration and heart rate, low blood pressure, muscle tenderness, unconsciousness; high or repeated exposure may affect kidney; TSCA listed
Precaution: Combustible exposed to heat or flame; incompat. with oxidizing materials (increased fire and explosion hazard)
Hazardous Decomp. Prods.: Heated to decomp., emits acrid smoke and fumes
Storage: Store in cool, dry, well-ventilated area, out of direct sunlight
Uses: High-boiling solvent for coatings
Manuf./Distrib.: Aldrich; Dow; Fluka
Trade Names: Hexyl Carbitol®

Diethylene glycol isobutyl ether

CAS 18912-80-6
Synonyms: DEGiBE; Diethylene glycol monoisobutyl ether; Isobutoxy-2-ethoxy-2-ethanol; 2-(2-Isobutoxyethoxy) ethanol; 2-(2-(2-Methoxypropoxy) ethoxy) ethanol
Classification: Aliphatic ether alcohol
Empirical: $C_8H_{18}O_3$
Formula: $CH_3CH(CH_3)CH_2OCH_2CH_2OCH_2CH_2OH$
Properties: Colorless liq.; sol. in water; m.w. 162.26; sp.gr. 0.95-1.0; b.p. 217-225 C; flash pt. (CC) 106 C
Toxicology: LD50 (oral, rat) 4920 mg/kg, (skin, rabbit) 3560 mg/kg; mod. toxic by skin contact; mildly toxic by ing., inh.; nonirritating to skin; can be absorbed thru skin, but health effects unlikely; mod. to severe eye irritant; ing. may cause vomiting, cyanosis, headache, rapid respiration and heart rate, low blood pressure, muscle tenderness, unconsciousness; high/repeated exposure may affect kidney; TSCA listed
Precaution: Incompat. with strong oxidizing agent (increases fire and explosion hazard)
Hazardous Decomp. Prods.: Heated to decomp., emits acrid smoke and irritating fumes
NFPA: Health 1, Flammability 1, Reactivity 0
Storage: Store in cool, dry, well-ventilated area, out of direct sunlight
Uses: Solvent
Manuf./Distrib.: Oxiteno

Diethylene glycol lactamide. *See* PEG-2 lactamide
Diethylene glycol methyl ether. *See* Methoxydiglycol

Diethylene glycol methyl ether acetate

CAS 629-38-9
Synonyms: Acetic acid, 2-(2-methoxyethoxy) ethyl ester; Diethylene glycol monomethyl ether acetate; 2-(2-Methoxyethoxy) ethanol acetate; 2-(2-Methoxyethoxy) ethyl acetate; Methyl 'Carbitol' acetate
Classification: Glycol ether
Empirical: $C_7H_{14}O_4$
Properties: Colorless liq.; m.w. 162.21; sp.gr. 1.0396 (20/20 C); vapor pressure 0.12 mm Hg (20 C); b.p. 209.1 C; flash pt. (OC) 180 F
Toxicology: LD50 (oral, rat) 11,960 mg/kg; mod. toxic by ing.; severe eye irritant; TSCA listed
Environmental: VOC; ThOD 1.68
Precaution: Combustible when exposed to heat or flame; can react with oxidizing materials
Hazardous Decomp. Prods.: Heated to decomp., emits acrid smoke and irritating fumes

NFPA: Health 0, Flammability 2, Reactivity 0
Uses: Solvent
Regulatory: SARA §313 reportable
Manuf./Distrib.: Aldrich; Ashland; Eastman; Harcros; U.S. Chems.

Diethylene glycol methyl ethyl ether

CAS 1002-67-1
Synonyms: Diethylene glycol ethyl methyl ether; Ethane, 1-ethoxy-2-(2-methoxyethoxy)-
Classification: Glycol ether
Empirical: $C_7H_{16}O_3$
Properties: M.w. 148.20
Toxicology: TSCA listed
Uses: Solvent

Diethylene glycol methylpentyl ether

CAS 10143-56-3
Synonyms: DEGMPE; Diethylene glycol monomethylpentyl ether; Diethylene glycol mono-2-methylpentyl ether; 8-Methyl-3,6-dioxa-1-undecanol; 2-Methylpentyl 'Carbitol'; 2-[2-((2-Methylpentyl) oxy) ethoxy] ethanol
Classification: Aliphatic ether alcohol
Empirical: $C_{10}H_{22}O_3$
Formula: $CH_3CH_2CH_2CH(CH_3)CH_2OCH_2CH_2OCH_2CH_2OH$
Properties: Colorless liq.; m.w. 190.32
Toxicology: LD50 (oral, rat) 5660 mg/kg, (skin, rabbit) 1580 mg/kg; mod. toxic by skin contact; mildly toxic by inj.; mod. to severe eye irritant; skin irritant; ing. expected to produce vomiting, cyanosis, headache, rapid respiration and heart rate, low blood pressure, muscle tenderness, unconsciousness; high/repeated exposure may affect kidney
Precaution: Incompat. with oxidizing materials; possible fire risk
Hazardous Decomp. Prods.: Heated to decomp., emits acrid smoke and irritating fumes
Storage: Store in a cool, dry, well-ventilated area, out of direct sunlight
Uses: Solvent

Diethylene glycol monobutyl ether. *See* Butoxydiglycol
Diethylene glycol monobutyl ether acetate; Diethylene glycol mono-n-butyl ether acetate. *See* Diethylene glycol butyl ether acetate
Diethylene glycol, monoester with stearic acid. *See* PEG-2 stearate
Diethylene glycol monoethyl ether. *See* Ethoxydiglycol
Diethylene glycol monoethyl ether acetate. *See* Ethoxydiglycol acetate
Diethylene glycol mono-2-ethylhexyl ether. *See* Diethylene glycol 2-ethylhexyl ether
Diethylene glycol monohexyl ether. *See* Diethylene glycol n-hexyl ether
Diethylene glycol monoisobutyl ether. *See* Diethylene glycol isobutyl ether
Diethylene glycol monomethyl ether. *See* Methoxydiglycol
Diethylene glycol monomethyl ether acetate. *See* Diethylene glycol methyl ether acetate
Diethylene glycol monomethylpentyl ether; Diethylene glycol mono-2-methylpentyl ether. *See* Diethylene glycol methylpentyl ether
Diethylene glycol monophenyl ether. *See* Diethylene glycol phenyl ether
Diethylene glycol monopropyl ether. *See* Diethylene glycol propyl ether
Diethylene glycol monostearate. *See* PEG-2 stearate

Diethylene glycol phenyl ether

CAS 104-68-7
Synonyms: Diethylene glycol monophenyl ether; Ethanol, 2-(2-phenoxyethoxy); Phenoxydiglycol; 2-(2-Phenoxyethoxy) ethanol; Phenyl 'Carbitol'
Empirical: $C_{10}H_{14}O_3$
Properties: Liq.; m.w. 182.24; dens. 1.1158 (20/20 C); vapor pressure < 0.01 mm Hg (20 C); f.p. -50 C; b.p. 207 C (55 mm)
Toxicology: LD50 (oral, rat) 2140 mg/kg, (skin, rabbit) 2120 mg/kg; mod. toxic by ing. and skin contact; skin and severe eye irritant; possible human reproductive effects; TSCA listed
Hazardous Decomp. Prods.: Heated to decomp., emits acrid smoke and irritating fumes
Uses: Solvent for resins, in metal cleaners, paint strippers, cleaning compds., as ink vehicle; in food-pkg. adhesives
Regulatory: FDA 21CFR §175.105
Manuf./Distrib.: Kenrich Petrochems.
Trade Names Containing: Dalpad A; Dowanol® EPh; Igepal® OD-410

Diethylene glycol propyl ether

CAS 6881-94-3; EINECS/ELINCS 229-985-7
Synonyms: Diethylene glycol monopropyl ether; Ethanol, 2-(2-propoxyethoxy)-; 2-(2-Propoxyethoxy) ethanol
Empirical: $C_7H_{16}O_3$
Formula: $CH_2OHCH_2OCH_2CH_2OCH_2CH_2CH_3$
Properties: Liq.; sol. in water; m.w. 148.2; dens. 0.963 (20 C); b.p. 202 C (760 mm); f.p. < -90 C; flash pt. (TCC) 93 C
Toxicology: Causes eye irritation
Precaution: Potential peroxide former
Uses: Coupling solvent for sol'n. and water-dilutable coatings; active solvent for NC, acrylic copolymers, epoxy resins, chlorinated rubber, alkyd resins; solvent for electronic materials; coalescing aid in architectural latex coatings, floor polishes; diluent in hydraulic brake fluids; household and general industrial cleaners
Manuf./Distrib.: Ashland; Eastman; Great Western
Trade Names: Eastman® DP

Diethylene glycol stearate. *See* PEG-2 stearate
Diethyleneimide oxide; Diethylene imidoxide. *See* Morpholine
Diethylene oxide. *See* Tetrahydrofuran
Di (ethylene oxide); 1,4-Diethylene oxide. *See* 1,4-Dioxane
Diethylene oximide. *See* Morpholine

Diethylenetriamine

CAS 111-40-0; EINECS/ELINCS 203-865-4
UN 2079 (DOT)
Synonyms: Aminoethylethandiamine; N-(2-Aminoethyl)-1,2-ethanediamine; N-(2-Aminoethyl) ethylenediamine; 3-Azapentane-1,5-diamine; Bis (2-aminoethyl) amine; Bis (β-aminoethyl) amine; DETA; 2,2´-Diaminodiethylamine; DTA; 2,2´-Iminobisethylamine; 2,2´-Iminodiethylamine
Classification: Aliphatic polyamine
Empirical: $C_4H_{13}N_3$
Formula: $NH_2C_2H_4NHC_2H_4NH_2$
Properties: Colorless to yel. visc. liq.; ammoniacal odor; sol. in water and hydrocarbons; misc. with oxygenated and aromatic solvs.; immisc. with aliphatic solvs.; strongly alkaline; m.w. 103.17; dens. 0.9542 (20/20 C); vapor pressure 0.22 mm Hg (20 C); f.p. -39 C; b.p. 206.7 C; flash pt. (TOC) 99 C
Toxicology: ACGIH TLV/TWA 1 ppm (skin); LD50 (oral, rat) 1080 mg/kg, (IP, rat) 74 mg/kg, (skin, rabbit) 1090 mg/kg; toxic by IP, ing., inh., and skin absorption; strong irritant to eyes and skin; high conc. of vapors causes respiratory tract irritation, nausea, vomiting; repeated exposure may cause asthma, skin sensitization; TSCA listed
Precaution: Combustible exposed to heat or flame; can react with oxidizers; mixt. with nitromethane is a shock-sensitive explosive; ignites on contact with cellulose nitrate; corrosive to copper and its alloys; absorbs CO_2 from air to form carbamate salts
Hazardous Decomp. Prods.: Heated to decomp., emits toxic fumes of NO_x
NFPA: Health 3, Flammability 1, Reactivity 0
Storage: Hygroscopic
Uses: Solvent for sulfur, acid gases, various resins, dyes; saponification agent for acidic materials; fuel component; epoxy curing agent and hardener; intermediate for textile finishes; corrosion inhibitor in petrol. prod.; curing agent for two-pkg. protective coatings; mfg. of chelating agents, surfactants, softeners; ion exchange membrane in prod. of grapefruit juice; modifier for amino resins for paper/paperboard in contact with aq./fatty foods; in paper/paperboard in contact with dry food; defoamer in food-contact paper/paperboard
Regulatory: FDA 21CFR §173.20, 175.105, 175.300, 176.170, 176.180, 176.210, 177.1200
Manuf./Distrib.: Akzo Nobel; Aldrich; Alkyl Amines Chems. Ltd; Allchem Ind.; Ashland; BASF AG; Charkit; Coyne; Dow; Fabrichem; Filo; Filo; Fluka; Houghton Chem.; Lowenstein Dyes & Cosmetics; Rhodia HPCII; Sigma; Tosoh

Diethylenimide oxide. *See* Morpholine
Diethyl ester, oxalic acid. *See* Diethyl oxalate
N,N-Diethylethanamine. *See* Triethylamine

Diethyl ethanedicarboxylate. *See* Diethyl succinate
Diethyl ethanedioate. *See* Diethyl oxalate
2,2-Diethyl ethanol. *See* 2-Ethyl-1-butanol
Diethylethanolamine; N,N-Diethylethanolamine. *See* Diethylaminoethanol
Diethyl ether. *See* Ethyl ether
Diethylformal. *See* Diethoxymethane
Diethyl glycol dimethyl ether. *See* Diethylene glycol dimethyl ether
Di (2-ethylhexyl) adipate. *See* Dioctyl adipate
Di (2-ethylhexyl)-2-ethylhexyl phosphonate. *See* Dioctyl octylphosphonate
Diethylhexyl maleate; Di (2-ethylhexyl) maleate. *See* Bis (2-ethylhexyl) maleate
Di-2-ethylhexyl sebacate. *See* Dioctyl sebacate
N,N-Diethyl-N-(β-hydroxyethyl) amine. *See* Diethylaminoethanol

Diethyl isophthalate

CAS 636-53-3
Synonyms: 1,3-Benzenedicarboxylic acid, diethyl ester; DEIP; Isophthalic acid, diethyl ester
Empirical: $C_{12}H_{14}O_4$
Properties: Colorless oily liq.; m.w. 222.24; sp.gr. 1.122-1.127; m.p. 12 C; b.p. 302 C; ref. index 1.5040-1.5060
Uses: Chemical intermediate; solvent for mfg. of TDI
Manuf./Distrib.: Shandong Haihua Tianhe; TCI Am.
Trade Names: Morflex® DEIP

Diethyl ketone

CAS 96-22-0; EINECS/ELINCS 202-490-3
UN 1156 (DOT)
Synonyms: DEK; Dimethylacetone; Metacetone; Methacetone; 3-Pentanone; Pentanone-3; Propione
Empirical: $C_5H_{10}O$
Formula: $CH_3CH_2COCH_2CH_3$
Properties: Colorless mobile liq.; acetone odor; misc. with alcohol, ether, oxygenated solvs.; sol. in 25 parts water; m.w. 86.14; dens. 0.814; m.p. -42 C; b.p. 99-103 C; flash pt. 55 F; ref. index 1.392
Toxicology: LD50 (oral, rat) 2.1 g/kg, (IV, mouse) 513 mg/kg, (skin, rabbit) 20 g/kg; mod. toxic by ing., IP, IV routes; skin and eye irritant; mutagenic data; TSCA listed
Precaution: Highly flamm.; dangerous fire hazard exposed to heat or flame; can react vigorously with oxidizing materials; reacts with hydrogen peroxide + nitric acid to form a shock- and heat-sensitive explosive peroxide
Hazardous Decomp. Prods.: Heated to decomp., emits acrid smoke and irritating fumes
NFPA: Health 1, Flammability 3, Reactivity 0
Uses: Flavors; medicine; organic synthesis; solvent for paints; cosmetics ingred.
Manuf./Distrib.: Aldrich; BASF AG; BASF; Degussa AG/Health & Nutrition; Dow; Esprit; Fluka; General Chem.; Penta Mfg.; Sigma; Spectrum Quality Prods.

N,N-Diethyl-3-methylbenzamide. *See* Diethyl toluamide
Diethylnitrosamide; Diethylnitrosamine; N,N-Diethylnitrosamine; Diethylnitrosoamine. *See* N-Nitrosodiethylamine
Diethylolamine. *See* Diethanolamine

Diethyl oxalate

CAS 95-92-1; EINECS/ELINCS 202-464-1
UN 2525 (DOT)
Synonyms: Diethyl ester, oxalic acid; Diethyl ethanedioate; Ethanedioic acid diethyl ester; Ethyl oxalate; Oxalic acid, diethyl ester
Classification: Nonaromatic ester
Definition: Diester of ethyl alcohol and oxalic acid
Empirical: $C_6H_{10}O_4$
Formula: $C_2H_5O_2CCO_2C_2H_5$
Properties: Colorless oily liq.; mild odor; misc. in alcohol, ether, ethyl acetate, common org. solvs.; dec. in water; m.w. 146.14; dens. 1.076; vapor pressure 0.414 mm Hg; m.p. -41 C; b.p. 185 C; flash pt. (OC) 76 C; ref. index 1.4100; surf. tens. 32.22 dynes/cm; dielec. const. 1.8
Toxicology: LD50 (oral, rat) 400 mg/kg, (oral, mouse) 2 g/kg; harmful; toxic by ing.; severe irritant; strong irritant to skin and mucous membranes; target organs: respiratory system, CNS, kidney, blood; TSCA listed
Environmental: VOC; ThOD 1.42
Precaution: Flamm. exposed to heat or flame; can react with oxidizing materials
Hazardous Decomp. Prods.: Heated to decomp., emits acrid smoke and fumes
NFPA: Health 2, Flammability 2, Reactivity 0
Storage: Moisture-sensitive
Uses: Solvent for cellulose esters and ethers, many natural and syn. resins, paint stripping; radio-tube cathode fixing lacquers; perfumes; chelating agent; cleaner for polymeric residues; pigment dispersant; intermediate for org. synthesis, pharmaceuticals, barbiturates (CNS depressants), dyes; in food-pkg. adhesives
Regulatory: FDA 21CFR §175.105
Manuf./Distrib.: Aceto; Aldrich; Austin; Biddle Sawyer; Chemapol Usa; Esprit; Fabrichem; Fluka; Hummel Croton; Jonas; Morflex; Penta Mfg.; Rhodia/Fine Orgs.; Sigma; Wall Chem.; Whyte Chems. Ltd

Diethyl oxide. *See* Ethyl ether

Diethyl phthalate

CAS 84-66-2; EINECS/ELINCS 201-550-6
Synonyms: 1,2-Benzenedicarboxylic acid diethyl ester; DEP; Diethyl 1,2-benzenedicarboxylate; Diethyl-o-phthalate; Ethyl phthalate; Phthalic acid, diethyl ester; Phthalol
Definition: Aromatic diester of ethyl alcohol and phthalic acid
Empirical: $C_{12}H_{14}O_4$
Formula: $C_6H_4(CO_2C_2H_5)_2$
Properties: Water-wh. oily liq., odorless, bitter taste; misc. with alcohol, ketones, esters, oxygenated and aromatic solvs.; insol. in water; m.w. 222.24; dens. 1.120 (25/25 C); f.p. -40.5 C; b.p. 298 C; flash pt. (OC) 325 F; visc. 31.3 cs (0 C); ref. index 1.500-1.505; stable
Toxicology: ACGIH TLV/TWA 5 mg/m^3; LD50 (oral, rat) 8600 mg/kg, (IP, rat) 5058 mg/kg; poison by IV route; mod. toxic by ingestion, subcut., IP routes; human systemic effects; strong irritant to eyes and mucous membranes; narcotic in high concs.; experimental teratogen, reproductive effects; TSCA listed
Precaution: Combustible when exposed to heat or flame
Hazardous Decomp. Prods.: Heated to decomp., emits acrid smoke and irritating fumes
NFPA: Health 0, Flammability 1, Reactivity 0
Uses: Solvent for nitrocellulose, cellulose acetate, cosmetics; plasticizer for cellulose acetate, solid rocket propellants, food pkg., pharmaceuticals; wetting agent; pesticides; insecticides; alcohol denaturant; solvent and fixing agent in perfumery; film-former; fragrance diluent; pharmaceutical fragrance; in food-pkg. adhesives; surf. lubricant in mfg. of food-contact metallic articles
Regulatory: FDA 21CFR §175.105, 175.300, 175.320, 178.3910, 181.22, 181.27, 212.177; 27CFR §21.105; FDA approved for orals; NF, BP, EP compliance; SARA reportable
Manuf./Distrib.: Aarti Ind. Ltd; Aldrich; Allan; Allchem Ind.; Ashland; BASF; BP Amoco; Berje; Camida Ltd; ChemService; Chemial SpA; Chesham Chems. Ltd; Creative Fragrances; Eastman; Fluka; GFS; Haarmann & Reimer; Houghton Chem.; Indofine; Morflex; Parchem Trading; Penta Mfg.; Polysciences; Proviron Ind. NV; Rhodia/Fine Orgs.; Ruger; Sigma; Spectrum Quality Prods.; Tennants Distrib. Ltd; Ubichem plc; Unitex; Universal Preserv-A-Chem; Vopak USA; Whyte Chems. Ltd
Trade Names: Eastman® DEP; Eastman® DEP, Fragrance Grade; Eastman® Diethyl Phthalate
Trade Names Containing: Punctilious® SDA 38B Anhydrous Prod. Code 14680000; Punctilious® SDA 39B 190 Proof; Punctilious® SDA 39C Anhydrous

Diethyl-o-phthalate. *See* Diethyl phthalate

Diethyl succinate

CAS 123-25-1; EINECS/ELINCS 204-612-0
FEMA 2377
Synonyms: Butanedioic acid diethyl ester; Diethyl butanedioate; Diethyl ethanedicarboxylate; Ethyl succinate; Succinic acid diethyl ester
Empirical: $C_8H_{14}O_4$
Formula: $C_2H_5OCOCH_2CH_2COOC_2H_5$
Properties: Colorless to lt. yel. clear mobile liq., pleasant odor; sol. in alcohol, fixed oils, ether, oxygenated solvs.; insol. in water; m.w. 174.22;

dens. 1.039 (20/4 C); m.p. -20 C; b.p. 97-99 C (10 mm); flash pt. 230 F; ref. index 1.420

Toxicology: LD50 (oral, rat) 8530 mg/kg; mildly toxic by ing.; skin and eye irritant; TSCA listed

Precaution: Combustible liq.; may cause fire or explosion on reaction with ethyl trifluoroacetate + sodium hydride

Hazardous Decomp. Prods.: Heated to decomp., emits acrid smoke and irritating fumes

NFPA: Health 1, Flammability 1, Reactivity 0

Uses: Chemical intermediate; plasticizer; emollient in cosmetics; synthetic flavoring agent in foods and pharmaceuticals; solvent for candy and bakery flavors

Regulatory: FDA 21CFR §172.515; FEMA GRAS

Manuf./Distrib.: Aldrich; Boith China; ChemService; DSM Fine Chems. Austria; Epochem; Fluka; Morflex; Natural Advantage; Reilly Chems. SA; Wutong Aroma

Diethyl sulfide. *See* Ethyl sulfide

Diethyl tartrate

CAS 87-91-2; EINECS/ELINCS 201-783-3

FEMA 2378

Synonyms: Diethyl 2,3-dihydroxybutanedioate; Diethyl 2,3-dihydroxysuccinate; (+)-Diethyl L-tartrate; (R)-2,3-Dihydroxybutanedioic acid diethyl ester; Ethyl tartrate

Empirical: $C_8H_{14}O_6$

Formula: $C_2H_5OOCCH(OH)CH(OH)COOC_2H_5$

Properties: Colorless thick oily liq.; sl. sol. in water; misc. with alcohol, ether; m.w. 206.19; dens. 1.204 (20/4 C); m.p. 17 C; b.p. 280 C; flash pt. 93 C; ref. index 1.4476 (20 C)

Precaution: Combustible

NFPA: Health 0, Flammability 1, Reactivity 0

Uses: Plasticizer for automobile lacquers; solvent for nitrocellulose, gums, and resins; synthetic flavoring agent in foods and pharmaceuticals

Features: Fruity flavor

Regulatory: FDA 21CFR §172.515; FEMA GRAS

Manuf./Distrib.: Aldrich; Fluka; Natural Advantage; Sigma

(+)-Diethyl L-tartrate. *See* Diethyl tartrate

Diethyl thioether. *See* Ethyl sulfide

Diethyl toluamide

CAS 134-62-3; EINECS/ELINCS 205-149-7

Synonyms: Benzamide, N,N-diethyl-3-methyl; DEET; Delphene; m-Delphene; DET; N,N-Diethyl-3-methylbenzamide; Diethyl-m-toluamide; N,N-Diethyltoluamide; N,N-Diethyl-m-toluamide; Metadelphene; 3-Methyl-N,N-diethylbenzamide; m-Toluic acid diethylamide

Classification: Aromatic amide

Empirical: $C_{12}H_{17}NO$

Formula: $CH_3C_6H_4CON(C_2H_5)_2$

Properties: Colorless liq., mild bland odor; sol. in water, alcohol, ether, benzene, oxygenated and aromatic solvs.; m.w. 191.30; dens. 0.996-1.002 (25/25 C); b.p. 160 C (19 mm); flash pt. > 230 F; ref. index 1.5230 (20 C)

Toxicology: LD50 (oral, rat) 1950 mg/kg, (skin, rabbit) 3180 mg/kg; LDLo (oral, human female) 950 mg/kg (coma), (oral, human male) 679 mg/kg; poison by IV route; mod. toxic by ingestion, skin contact; irritant to eyes, mucous membranes; human systemic effects by skin contact; can cause CNS disturbances; mutagenic data; experimental reproductive effects; TSCA listed

Precaution: Combustible

Hazardous Decomp. Prods.: Heated to decomp., emits toxic fumes of NO_x

Uses: Pesticide; insect repellent; resin solvent; film-former; skin protectant in cosmetics

Manuf./Distrib.: Aldrich; Celanese; DuPont; Fluka; Honeywill & Stein; McLaughlin Gormley King; Morflex; Schütz; Sigma

Diethyl-m-toluamide; N,N-Diethyltoluamide; N,N-Diethyl-m-toluamide. *See* Diethyl toluamide

1,1,1-Difluorochloroethane; 1,1-Difluoro-1-chloroethane; Difluorochloroethanes. *See* 1,1,1-Chlorodifluoroethane

Difluorochloromethane. *See* Chlorodifluoromethane

Difluorodibromomethane. *See* Dibromodifluoromethane

Difluorodichloromethane. *See* Dichlorodifluoromethane

Difluoromonochloroethane. *See* 1,1,1-Chlorodifluoroethane

Difluoromonochloromethane. *See* Chlorodifluoromethane

1,2-Difluoro-1,1,2,2-tetrachloroethane. *See* 1,1,2,2-Tetrachloro-1,2-difluoroethane

Diglycol. *See* Diethylene glycol

Diglycolamine. *See* 2-(2-Aminoethoxy) ethanol

Diglycol benzyl ether. *See* Diethylene glycol benzyl ether

Diglycol butyl ether. *See* Butoxydiglycol

Diglycol diethyl ether. *See* Diethylene glycol diethyl ether

Diglycol methyl ether. *See* Diethylene glycol dimethyl ether

Diglycol monobutyl ether. *See* Butoxydiglycol

Diglycol monobutyl ether acetate. *See* Diethylene glycol butyl ether acetate

Diglycol monoethyl ether. *See* Ethoxydiglycol

Diglycol monoethyl ether acetate. *See* Ethoxydiglycol acetate

Diglycol monostearate; Diglycol stearate. *See* PEG-2 stearate

Diglyme. *See* Diethylene glycol dimethyl ether

Digol. *See* Diethylene glycol

Dihexyl. *See* n-Dodecane

Dihexyl ether; Di-n-hexyl ether. *See* n-Hexyl ether

Dihexylmonooctylphosphine oxide. *See* Dihexyloctylphosphine oxide

Dihexyloctylphosphine oxide

CAS 31160-64-2

Synonyms: Dihexylmonooctylphosphine oxide; Monooctyl dihexylphosphine oxide; Octyldihexylphosphine oxide

Empirical: $C_{20}H_{43}OP$

Properties: M.w. 330.53

Uses: Extractant

Trade Names Containing: Cyanex® 923

Dihydrobutadiene sulfone. *See* Sulfolane

2,3-Dihydrodecafluoropentane

Synonyms: 1,1,1,2,2,3,4,5,5,5-Decafluoropentane; HFC 43-10mee

Classification: Hydrofluorocarbon

Empirical: $C_5H_2F_{10}$

Properties: Liq.

Toxicology: REL 8 h 200 ppm

Environmental: Replacement for CFC but characterized as global warming gas by the Kyoto Conference on Climate Change; VOC exempt

Uses: Vapor degreasing applics.; cleaning/rinsing agent; drying fluid; fluorocarbon lubricant carrier; solvent and disp. media; heat transfer media; CFC replacement

Features: Zero ODP; low global warming potential

Trade Names: Vertrel® XF

Trade Names Containing: Vertrel® MCA; Vertrel® MCA Plus; Vertrel® SMT; Vertrel® X-B3; Vertrel® X-P10; Vertrel® XE; Vertrel® XM; Vertrel® XMS Plus; Vertrel® XSi

Dihydro-1,3-dioxole. *See* 1,3-Dioxolane

2,3-Dihydrofuran

CAS 1191-99-7; EINECS/ELINCS 214-747-7

UN 1993

Synonyms: 4,5-Dihydrofuran; 1-Oxa-2-cyclopentene; 2-Oxolene

Empirical: C_4H_6O

Properties: Clear liq.; sl. sol. in water; m.w. 70.09; sp.gr. 0.927; vapor pressure 3.67 psi (20 C), 14.46 psi (55 C); b.p. 51-54 C; flash pt. -24 C; ref. index 1.415-1.435

Toxicology: LD50 (oral, rat) 1414 mg/kg, (skin, rat) > 2000 mg/kg; LC50 (inh., rat, 6 h) > 3700 ppm; harmful by ing.; eye irritant; may cause skin/mucous membrane/upper respiratory tract irritation; may be harmful by inh.; may cause CNS depression, narcotic effects, coughing, chest pains, difficulty breathing, nausea, dizziness, headache, unconsciousness; target organs: liver, CNS, kidneys

Precaution: Highly flamm. liq.; vapor may flash back; contains oxygen; may form explosive peroxides; container explosion can occur under fire conditions; hazardous polymerization may occur; may undergo autopolymerization; incompat. with oxidizing agents, strong acids, oxygen, polymerizing initiators

Hazardous Decomp. Prods.: CO, CO_2; emits toxic fumes under fire con-

ditions
NFPA: Health 1, Flammability 3
Storage: Store in cool, dry place away from heat, sparks, open flame; keep container closed
Uses: Solvent; intermediate for pharmaceuticals
Manuf./Distrib.: ABCR; Acros Org.; Aldrich; Alfa Aesar; Eastman; Fluka; ICN Biomed. Research Prods.; Lancaster Synthesis; TCI Am.

2,5-Dihydrofuran

CAS 1708-29-8; EINECS/ELINCS 216-957-4
UN 1993
Synonyms: Butandiol; Furan, 2,5-dihydro-; 1-Oxa-3-cyclopentene; 3-Oxolene
Empirical: C_4H_6O
Properties: Yel.-tinted clear liq.; sl. sol. in water; m.w. 70.09; sp.gr. 0.927; b.p. 66-67 C; flash pt. 2 F; ref. index 1.432
Toxicology: May be harmful by ing., inh., skin absorp.; may cause irritation
Environmental: Prevent entry into drains
Precaution: Highly flamm.; forms explosive mixts. in air; vapor may flash back; contains oxygen; may form explosive peroxides; container explosion may occur under fire conditions; prevent static discharges; incompat. with oxidizing agents
Hazardous Decomp. Prods.: CO, CO_2
NFPA: Health 1, Flammability 3
Storage: Store in tightly closed container in cool, well-ventilated area away from heat, sparks, open flame, other ignition sources
Uses: Solvent; intermediate for pharmaceuticals
Manuf./Distrib.: ABCR; Aldrich; TCI Am.

4,5-Dihydrofuran. *See* 2,3-Dihydrofuran
Dihydro-2(3H)-furanone. *See* Butyrolactone
Dihydroisophorol. *See* Trimethylcyclohexanol
4,5-Dihydro-5-methyl-2(3H)-furanone. *See* γ-Valerolactone
Dihydrooxirene. *See* Ethylene oxide
Dihydroxybenzene. *See* Hydroquinone
1,2-Dihydroxybenzene. *See* Pyrocatechol
1,4-Dihydroxybenzene. *See* Hydroquinone
o-Dihydroxybenzene. *See* Pyrocatechol
p-Dihydroxybenzene. *See* Hydroquinone
1,3-Dihydroxybutane. *See* Butylene glycol
1,4-Dihydroxybutane. *See* 1,4-Butanediol
(R)-2,3-Dihydroxybutanedioic acid diethyl ester. *See* Diethyl tartrate
1,4-Dihydroxy-2-butyne. *See* But-2-yne-1,4-diol
1,10-Dihydroxydecane. *See* 1,10-Decanediol
2,2´-Dihydroxydiethylamine. *See* Diethanolamine
Dihydroxydiethyl ether; β,β´-Dihydroxydiethyl ether. *See* Diethylene glycol
β,β´-Dihydroxydiethyl sulfide. *See* Thiodiglycol
2,2´-Dihydroxydipropyl ether. *See* Dipropylene glycol
1,2-Dihydroxyethane. *See* Ethylene glycol
Di-β-hydroxyethoxyethane. *See* Triethylene glycol
Di (2-hydroxyethyl) amine. *See* Diethanolamine
2,2´-Dihydroxyethyl ether. *See* Diethylene glycol
Dihydroxyethyl sulfide; β,β´-Dihydroxyethyl sulfide. *See* Thiodiglycol
1,6-Dihydroxyhexane. *See* Hexamethylene glycol
β,β´-Dihydroxyisopropyl chloride. *See* Glyceryl-α-monochlorohydrin
2,2´-Dihydroxyisopropyl ether. *See* Dipropylene glycol
1,3-Di (hydroxymethyl)-5,5-dimethylhydantoin. *See* DMDM hydantoin
2,4-Dihydroxy-2-methylpentane. *See* Hexylene glycol
1,2-Dihydroxypropane. *See* Propylene glycol
1,3-Dihydroxypropane. *See* Trimethylene glycol
2,3-Dihydroxypropyl acetate. *See* Glyceryl acetate
2,3-Dihydroxypropyl chloride. *See* Glyceryl-α-monochlorohydrin
2,3-Dihydroxypropyl octadecanoate. *See* Glyceryl stearate
Diiodomethane. *See* Methylene iodide

1,3-Diisoamyl glycerol diether

CAS 626-78-8
Synonyms: 1,3-Diisoamyloxy-2-propanol; Glyceryl α,γ-diisoamyl ether; 2-Propanol, 1,3-bis (isopentyloxy)-; 2-Propanol, 1,3-bis (3-methylbutoxy)-; 2-Propanol, 1,3-di (isopentyloxy)-
Classification: Glycol ether
Empirical: $C_{13}H_{28}O_3$
Properties: M.w. 232.22
Environmental: ThOD 2.55
Uses: Solvent
Regulatory: HAP

Diisoamyl ketone

CAS 2050-99-9
Synonyms: 2,8-Dimethyl-5-nonanone; Nonanone, 5-, 2,8-dimethyl-
Empirical: $C_{11}H_{22}O$
Properties: M.w. 170.29
Toxicology: TSCA listed
Uses: Solvent for paints, coil coatings; inert ingred. no long used in pesticides
Regulatory: Calif. Prop. 65 nonreportable
Manuf./Distrib.: Aldrich; Eastman; J.C. Wilson

1,3-Diisoamyloxy-2-propanol. *See* 1,3-Diisoamyl glycerol diether

Diisobutyl adipate

CAS 141-04-8; EINECS/ELINCS 205-450-3
Synonyms: Adipic acid, diisobutyl ester; Bis (2-methylpropyl) hexanedioate; DIBA; Diisobutyl hexanedioate; Hexanedioic acid, bis (2-methylpropyl) ester; Hexanedioic acid, diisobutyl ester; Isobutyl adipate
Definition: Diester of isobutyl alcohol and adipic acid
Empirical: $C_{14}H_{26}O_4$
Formula: $[C_2H_4COOCH_2CH(CH_3)_2]_2$
Properties: Colorless visc. liq., odorless; sol. in most org. solvs.; insol. in water; m.w. 258.40; dens. 0.950 (25 C); f.p. -20 C; b.p. 278-280 C; flash pt. > 110 C
Toxicology: LD50 (IP, rat) 5950 mg/kg, (oral, guinea pig) 12,300 mg/kg; mod. toxic by IP route; mildly toxic by ing.; may be harmful by inh., ing., or skin absorp.; may cause irritation to eyes, skin, mucous membranes, upper respiratory tract; experimental teratogen, reproductive effects; TSCA listed
Precaution: Combustible; incompat. with strong oxidizing agents, strong acids, strong bases
Hazardous Decomp. Prods.: Toxic fumes of CO, CO_2; emits toxic fumes under fire conditions
Storage: Store in cool, dry place; keep tightly closed
Uses: Emollient in cosmetics; plasticizer for some cellulosics, PVAc, PVC, VCA, in food-pkg. adhesives, in food-contact coatings; pigment wetting agent, dispersant; fragrance; solvent
Regulatory: FDA 21CFR §175.105, 175.300, 181.22, 181.27
Manuf./Distrib.: ABCR; Aceto; Aldrich; Alfa Aesar; Bencorp Int'l.; Boehme Filatex; ChemService; Grau Aromatics; Haltermann Prods. UK; Pfaltz & Bauer; TCI Am.; Unitex
Trade Names: HallTress™ DIBA Special
Trade Names Containing: Axarel® 52; Axarel® 56; Axarel® 9100; Coasol™; DBE-IB

Diisobutyl carbinol. *See* 2,6-Dimethyl heptanol-4

Diisobutyl glutarate

CAS 71195-64-7; EINECS/ELINCS 275-257-7
Classification: Oxygenated solvent
Toxicology: OSHA PEL/ACGIH TLV not established
Uses: Solvent
Trade Names Containing: Axarel® 52; Axarel® 56; Axarel® 9100; Coasol™; DBE-IB

Diisobutyl hexanedioate. *See* Diisobutyl adipate

Diisobutyl ketone

CAS 108-83-8; EINECS/ELINCS 203-620-1
UN 1157 (DOT); FEMA 3537
Synonyms: DIBK; Dibutyl ketone; s-Diisopropylacetone; 2,6-Dimethyl-4-heptanone; 2,6-Dimethylheptan-4-one; Isobutyl ketone; Isovalerone; Valerone
Classification: Ketone
Empirical: $C_9H_{18}O$
Formula: $(CH_3)_2CHCH_2COCH_2CH(CH_3)_2$
Properties: Colorless liq.; mild sweet peppermint odor; misc. with most org. liqs.; pract. insol. in water; m.w. 142.24; dens. 0.8076; vapor pres-

sure 1.65 mm Hg; m.p. -46 C; b.p. 165 C; flash pt. (CC) 49 C; autoignition temp. 396 C; ref. index 1.4130 (20 C); pH neutral; surf. tens. 24.54 dynes/cm

Toxicology: ACGIH TLV/TWA 25 ppm; LD50 (oral, rat) 4300 mg/kg, (skin, rabbit) 16 g/kg; mod. toxic by ing. and inh.; irritant to respiratory system; extremely high vapor concs. may cause dizziness, nausea, incoordination, possibly death; severe exposure unlikely; narcotic in high concs.; mild skin and eye irritant; ing. may cause irritation of mouth and stomach; may cause lung injury if aspirated; TSCA listed

Environmental: VOC; BOD5 1.37; COD 2.88; ThOD 2.93

Precaution: Flamm. exposed to heat or flame; LEL 0.81%; UEL 7.1%; incompat. with strong oxidizing agents (increases fire and explosion hazard)

Hazardous Decomp. Prods.: Heated to decomp., emits acrid smoke and fumes

NFPA: Health 1, Flammability 2, Reactivity 0

Storage: Store in cool, dry, well-ventilated area, out of direct sunlight

Uses: Solvent for nitrocellulose, rubber, synthetic resins, lacquers, coatings; organic synthesis; roll-coating inks; stains; flavoring agent in foods and pharmaceuticals; cosmetics ingred.; in food-pkg. adhesives

Features: Fruity flavor

Regulatory: FDA 21CFR §175.105; FEMA GRAS

Manuf./Distrib.: Aldrich; Allchem Ind.; Ashland; Chemcentral; Coyne; Degussa AG; Degussa Flavors/Alfrebro; Dow; Eastman; Fluka; Great Western; Harcros; Houghton Chem.; Shell; Stanchem; Sunnyside; Vopak USA

Trade Names: Ucar® Diisobutyl Ketone

Diisobutyl phthalate

CAS 84-69-5; EINECS/ELINCS 201-553-2

Synonyms: DIBP; Phthalic acid, diisobutyl ester

Classification: Aromatic ester

Empirical: $C_{16}H_{22}O_4$

Properties: Colorless liq.; insol. in water; m.w. 278.35; dens. 1.040 (20 C); f.p. -50 C; m.p. -64 C; b.p. 327 C; flash pt. (PMCC) 185 C; ref. index 1.490

Toxicology: LD50 (oral, rat) 15 g/kg, (IP, rat) 3749 mg/kg, (skin, guinea pig) 10 g/kg; mod. toxic by IP route; mildly toxic by ing. and skin contact; experimental teratogen, reproductive effects; TSCA listed

Precaution: Combustible

Hazardous Decomp. Prods.: Heated to decomp., emits acrid smoke and fumes

Uses: Plasticizer, solvent for PVC, PVAc, PVB, PS, cellulosics, paints; vehicle for pigment dispersions; in food-pkg. adhesives; food pkg. adhesives, coatings, paper, cellophane, rubber articles, textiles

Regulatory: FDA 21CFR §175.105, 177.1200

Manuf./Distrib.: Akdeniz Kimya; Aldrich; Allchem Ind.; Boehme Filatex; ChemService; KIC Chems.; Proviron Ind. NV; Shandong Haihua Tianhe; Unitex

Trade Names: Uniplex 155

Diisobutyl succinate

CAS 925-06-4; EINECS/ELINCS 213-113-7

Classification: Oxygenated solvent

Toxicology: OSHA PEL/ACGIH TLV not established

Uses: Solvent

Trade Names Containing: Axarel® 52; Axarel® 56; Axarel® 9100; Coasol™; DBE-IB

Diisocetyl adipate

CAS 57533-90-1; 58262-41-2; EINECS/ELINCS 260-799-9

Synonyms: Bis (2-hexyldecyl) adipate; Hexanedioic acid, bis (2-hexyl decyl) ester; Hexanedioic acid, bis (1-methylpentadecyl) ester

Definition: Diester of hexadecyl alcohol and adipic acid

Empirical: $C_{38}H_{74}O_4$

Formula: $C_{16}H_{33}OCO(CH_2)_4COOC_{16}H_{33}$

Uses: Emollient, spreading agent for cosmetics; film spreader for bath oils; solvent for lipstick dyes

Manuf./Distrib.: A&E Connock

1,6-Diisocyanatohexane. *See* Hexamethylene diisocyanate

Diisocyanatoluene. *See* Toluene-2,4-diisocyanate

Diisocyanatomethylbenzene; 1,3-Diisocyanatomethylbenzene. *See* Toluene diisocyanate

2,4-Diisocyanato-1-methylbenzene. *See* Toluene-2,4-diisocyanate

Diisocyanatotoluene. *See* Toluene diisocyanate

2,4-Diisocyanatotoluene. *See* Toluene-2,4-diisocyanate

Diisooctyldithiophosphinic acid. *See* Bis (2,4,4,-trimethylpentyl) dithiophosphinic acid

Diisooctylphosphinic acid. *See* Bis (2,4,4,-trimethylpentyl) phosphinic acid

Diisooctylthiophosphinic acid. *See* Bis (2,4,4,-trimethylpentyl) thiophosphinic acid

Diisopropanolamine

CAS 110-97-4; EINECS/ELINCS 203-820-9

Synonyms: Bis (2-hydroxypropyl) amine; Bis (2-propanol) amine; DIPA; Dipropyl-2,2´-dihydroxyamine; 1,1´-Iminobis-2-propanol; 1,1´-Iminobis (propan-2-ol); 1,1´-Iminobis-2-propenol; 1,1´-Iminodi-2-propanol; 1,1´-Iminodipropan-2-ol

Classification: Aliphatic amine alcohol compd.

Empirical: $C_6H_{15}NO_2$

Formula: $HN(CH_2CHOHCH_3)_2$

Properties: Wh. cryst. solid; mild ammoniacal odor; very sol. in water; sol. in ethanol, acetone; sl. sol. in diethyl ether; m.w. 133.19; dens. 1.004; m.p. 44.5-45.5 C; b.p. 249-250 C (745 mm); flash pt. 126 C

Toxicology: LD50 (oral, rat) 6720 mg/kg, (IP, mouse) 96 mg/kg; poison by IP route; mildly toxic by ing.; skin and severe eye irritant; TSCA listed

Precaution: Combustible

Hazardous Decomp. Prods.: Heated to decomp., emits toxic fumes of NO_x

NFPA: Health 2, Flammability 1, Reactivity 0

Storage: Hygroscopic; protect from light

Uses: Emulsifier for polishes, textile specialties, leather compds., insecticides, cutting oils, aq. paints, topical pharmaceuticals; neutralizer for cosmetics and toiletries; solvent in desulfurization and off-gas treatment; solubilizer, corrosion inhibitor in metalworking fluids; defoamer in food-contact paper/paperboard

Regulatory: FDA 21CFR §175.105, 176.210; FDA approved for topicals

Manuf./Distrib.: Aldrich; Ashland; BASF AG; BASF; Coyne; Dow; Fluka; Penta Mfg.; Ruger; Spectrum Quality Prods.; Synasia Fine Chem.; Universal Preserv-A-Chem; Vopak USA

Diisopropyl. *See* 2,3-Dimethylbutane

s-Diisopropylacetone. *See* Diisobutyl ketone

Diisopropyl adipate

CAS 6938-94-9; EINECS/ELINCS 230-072-0

Synonyms: Adipic acid diisopropyl ester; Bis (1-methylethyl) hexanedioate; DIPA; Hexanedioic acid, bis (1-methylethyl) ester; Isopropyl adipate

Definition: Diester of isopropyl alcohol and adipic acid

Empirical: $C_{12}H_{22}O_4$

Formula: $(CH_3)_2CHOCO(CH_2)_4COOCH(CH_3)_2$

Properties: Colorless to pale yel. clear liq.; sol. in min. oil, ethanol, propylene glycol, IPM, oleyl alcohol; insol. in water, glycerin; m.w. 230.31; dens. 0.950-0.962; ref. index 1.4216-1.4245

Toxicology: LD50 (oral, rat) > 20 ± 3 ml/kg, (IV, rat) 640 mg/kg; mod. toxic by IV route; skin irritant; nonirritating to eyes; TSCA listed

Hazardous Decomp. Prods.: Heated to decomp., emits acrid smoke and fumes

Uses: Emollient, cosolvent, coupling agent in aq. alcoholic systems, cosmetics, shave lotions, hair tonics, topical pharmaceuticals; imparts spreadability to bath oils; plasticizer in hair sprays; pigment wetting agent/dispersant; solvent for aromatic oils and perfumes, lipstick dyes

Regulatory: FDA approved for topicals

Manuf./Distrib.: A&E Connock; Alzo; C.P. Hall; ChemService; Haltermann Prods. UK; ISP Van Dyk; Inolex; Int'l. Paper; Jarchem Ind.; Lanaetex Prods.; Phoenix; Scher; St. Lawrence; Universal Preserv-A-Chem

Trade Names: Crodamol DA; Estalan DIA; HallStar™ DIPA; Schercemol DIA; Trivent DIA

Diisopropylamine

CAS 108-18-9; EINECS/ELINCS 203-558-5

UN 1158 (DOT)

Synonyms: DIPA; N-(1-Methylethyl)-2-propanamine; 2-Propanamine, N-(1-methylethyl)-

Classification: Sat. aliphatic amine

Empirical: $C_6H_{15}N$
Formula: $[(CH_3)_2CH]_2NH$
Properties: Colorless volatile liq.; amine, ammonia fishy odor; sol. in most org. solvs. incl. ethanol, acetone, ether, benzene; sl. sol. in water; m.w. 101.19; dens. 0.7162 (20/20 C); vapor pressure 81.157 mm Hg; m.p. -61 C; b.p. 83.9 C; flash pt. (OC) -1.11 C; autoignition temp. 316 C; ref. index 1.3924 (20 C); surf. tens. 19.11 dynes/cm
Toxicology: ACGIH TLV/TWA 5 ppm (skin); LD50 (oral, rat) 770 mg/kg, (oral, mouse) 2120 mg/kg; LC50 (inh., mouse, 2 h) 4200 mg/m^3; toxic by ing., inh., skin absorption; readily absorbed thru skin; primary skin and severe eye irritant; corrosive; lachrymator; inh. of fumes can cause pulmonary edema; mutagen; TSCA listed
Environmental: VOC; BOD5 0.02; COD 1.90; ThOD 1.92
Precaution: Flamm.; dangerous fire risk; LEL 1.1%; UEL 7.1%; can react vigorously with oxidizers; can react with CO_2 in air to form the carbamate salt
Hazardous Decomp. Prods.: Heated to decomp., emits toxic fumes of NO_x
NFPA: Health 3, Flammability 3, Reactivity 0
Uses: Intermediate; catalyst; mfg. of pesticides, pharmaceuticals, rubber; in organic synthesis; solvent
Manuf./Distrib.: AMC Chems.; Air Prods.; Aldrich; Alfa Chem; Ashland; Atofina; BASF; Dow; Fluka; Penta Mfg.; Sigma; Spectrum Quality Prods.; Vopak USA

1,3-Diisopropyl benzene
CAS 99-62-7; EINECS/ELINCS 202-773-1
Synonyms: 1,3-Bis(1-methylethyl) benzene; m-Diisopropylbenzene; m-DIPB; 3-Isopropylcumene
Empirical: $C_{12}H_{18}$
Properties: Colorless liq.; misc. with alcohol, ether, acetone, benzene, CCl_4, most org. solvs.; insol. in water; m.w. 162.30; dens. 0.8559 (20/4 C); f.p. -63 C; b.p. 231 C; flash pt. 76.6 C; ref. index 1.4883 (20 C)
Toxicology: LD50 (oral, rat) 7400 mg/kg, (oral, mouse) 1650 mg/kg, (IP, mouse) 1650 mg/kg; mod. toxic by ing. and IP routes; reproductive effects; TSCA listed
Hazardous Decomp. Prods.: Heated to decomp., emits acrid smoke and fumes
Uses: Solvent; intermediate
Manuf./Distrib.: Aldrich; Fluka; Koch Chem. Spec.

1,4-Diisopropyl benzene
CAS 100-18-5; EINECS/ELINCS 202-826-9
Synonyms: Benzene, 1,4-bis (1-methylethyl)-; Benzene, p-diisopropyl-; 1,4-Bis(1-methylethyl) benzene; p-Diisopropyl benzene; p-Diisopropylbenzol; p-DIPB; 4-Isopropylcumene
Empirical: $C_{12}H_{18}$
Properties: Colorless liq. or wh. solid; sol. in oxygenated and aromatic solvs.; misc. with alcohol, ether, acetone, benzene, CCl_4; insol. in water; m.w. 162.30; dens. 0.8568 (20/4 C); m.p. 64 C; b.p. 210 C; ref. index 1.4898 (20 C)
Toxicology: LD50 (oral, mouse) 3400 mg/kg, (IP, mouse) 1650 mg/kg; mod. toxic by IP and ing.
Precaution: Combustible
Hazardous Decomp. Prods.: Heated to decomp., emits acrid smoke and irritating vapors
Uses: Solvent; intermediate
Manuf./Distrib.: Aldrich; Eastman; Fluka

m-Diisopropylbenzene. *See* 1,3-Diisopropyl benzene
p-Diisopropyl benzene; p-Diisopropylbenzol. *See* 1,4-Diisopropyl benzene
Diisopropyl biphenyl. *See* Bis (methylethyl)-1,1´-biphenyl
Diisopropylene glycol. *See* Dipropylene glycol
Diisopropyl ether. *See* Isopropyl ether

Diisopropyl ketone
CAS 565-80-0; EINECS/ELINCS 209-294-7
UN 1224 (DOT)
Synonyms: 2,4-Dimethylpentan-3-one; 2,4-Dimethyl-3-pentanone; Isobutyrone; Isopropyl ketone; 3-Pentanone, 2,4-dimethyl-
Classification: Nonaromatic ketone
Empirical: $C_7H_{14}O$
Formula: $(CH_3)_2CHCOCH(CH_3)_2$
Properties: Colorless liq.; misc. with oxygenated solvs.; insol. in water; m.w. 114.19; dens. 0.81 kg/l (20 C); vapor pressure 5 mbar (20 C); m.p. -80 C; b.p. 122-126 C; flash pt. (CC) 15 C; ref. index 1.400
Toxicology: LD50 (oral, rat) 3536 mg/kg, (skin, rat) > 16,120 mg/kg; LC50 (inh., rat, 6 h) > 2765 ppm; may cause eye, skin, respiratory and digestive tract irritation, nausea, vomiting, diarrhea; TSCA listed
Precaution: DOT: Flamm. liq.; highly flamm.; avoid incompat. materials, ignition sources, excess heat, strong oxidants, electrical sparks, exposure to flame; incompat. with oxidizing agents, reducing agents, strong bases
Hazardous Decomp. Prods.: CO, CO_2
Storage: Store in tightly closed container in cool, dry, well-ventilated area, away from incompat. substances
Uses: Solvent
Manuf./Distrib.: ABCR; Acros Org.; Aldrich; Fluka

Diisopropylmethane. *See* 2,4-Dimethylpentane

Diisopropyl naphthalene
CAS 38640-62-9
Synonyms: Bis (isopropyl) naphthalene; Bis (1-methylethyl) naphthalene; Naphthalene, bis (1-methylethyl)-; Naphthalene diisopropyl-
Empirical: $C_{16}H_{20}$
Properties: M.w. 212.34
Uses: Solvent for carbonless copy paper
Manuf./Distrib.: Koch Spec. Chem.

1,4-Diisopropyl naphthalene
CAS 24157-79-7
Properties: M.w. 212.34; dens. 0.952 g/ml; visc. 10.50 cp; b.p. 300 C; flash pt. (CC) 302 F; surf. tens. 32.30 dynes/cm
Uses: Solvent

2,6-Diisopropyl naphthalene
CAS 24157-81-1
Properties: M.w. 212.34; dens. 0.952 g/ml; visc. 10.50 cp; b.p. 300 C; flash pt. (CC) 302 F; surf. tens. 32.30 dynes/cm
Uses: Solvent

Diisopropyl oxide. *See* Isopropyl ether

Diisopropyl sebacate
CAS 7491-02-3; EINECS/ELINCS 231-306-4
Synonyms: Bis (1-methylethyl) decanedioate; Decanedioic acid, bis (1-methylethyl) ester; DIPS
Definition: Diester of isopropyl alcohol and sebacic acid
Empirical: $C_{16}H_{30}O_4$
Formula: $(CH_3)_2CHOCO(CH_2)_8COOCH(CH_3)_2$
Properties: Colorless liq.; m.w. 286.41; sp.gr. 0.936; acid no. 1 max.; sapon. no. 380-400; flash pt. (COC) 190 C; ref. index 1.4310
Toxicology: TSCA listed
Uses: Emollient, solubilizer, coupling agent in creams, lotions, bath oils, men's grooming aids, topical pharmaceuticals; plasticizer; fragrance additive; solvent props.; penetrant for topical pharmaceuticals
Features: Liquid feel
Manuf./Distrib.: A&E Connock; Haltermann Prods. UK; Int'l. Paper; Nihon Surfactant; Shandong Haihua Tianhe; St. Lawrence
Trade Names: Estalan-DIPS; HallStar™ DIPS; Schercemol DIS

2,5-Diketohexane. *See* 2,5-Hexanedione
Diketone alcohol. *See* Diacetone alcohol
Dilaurin. *See* Glyceryl dilaurate

Dimethicone
CAS 9006-65-9
Synonyms: Dimethicones; Dimethylpolysiloxane; Dimethyl silicone; Dimeticone; Polydimethylsiloxane; Poly (oxy (dimethylsilylene)), α-(trimethylsilyl)-ω-methyl-; α-(Trimethylsilyl)-ω-methylpoly [oxy (dimethylsilylene)]
Definition: Silicone oil consisting of a mixt. of fully methylated linear siloxane polymers end-blocked with trimethylsiloxy units
Empirical: $(C_2H_6OSi)_xC_4H_{12}Si$
Properties: Colorless clear liq.; misc. with chloroform, ether; immisc. with

water, alcohol; visc. inc. with degree of polymerization
Uses: Ointment and topical drug vehicle; skin protectant; antifoam; defoamer for oil processing, water/waste treatment, foods; surfactant; release agent; adhesives; inks; latexes; soap mfg.; starch mfg.; paint mfg., antifoam, emollient in cosmetics; lubricant in polishes, precision bearings; antiadhesion coatings; dielec. fluids; heat transfer agents; textile softener; textile/paper sizing additive; barrier creams; chewing gum base; prosthetic aid (soft tissue); antiflatulent; in food-pkg. adhesives; release agent in food-contact coatings; in paper/paperboard in contact with aq./fatty foods; defoamer in food-contact coatings and paper/paperboard; in food-contact textiles
Features: Oleaginous
Regulatory: FDA 21CFR §145.180, 145.181, 146.185, 173.340, 175.105, 175.300, 176.170, 176.200, 176.210, 177.2260, 177.2600, 177.2800, 178.3570, 178.3910, 181.22, 181.28; USDA 9CFR §318.7, 381.147; Europe listed; UK approved; FDA approved for topicals; NF, BP, EP compliance
Manuf./Distrib.: Acros Org.; Aldrich; Alfa Chem; Ashland; Avatar; BASF; Biosil Tech.; Chemacon GmbH; Costec; Dow Corning; EMCO Chem. Distributors; Fanning; Functional Foods; GE Silicones; Goldschmidt; Lanaetex Prods.; Norman, Fox; Rhodia HPCII; Sigma; Surfachem Ltd; Universal Preserv-A-Chem; Wacker Silicones
Trade Names Containing: Fancorsil A

Dimethicone copolyol undecylenate

CAS 159100-33-1
Synonyms: Dimethicone copolyol undecylinate; Polydimethylsiloxypropyl polyethoxy undecylenate
Definition: Partial ester of dimethicone copolyol and undecylenic acid
Uses: Emollient, coupling agent, cosolvent in cosmetics, pharmaceuticals
Trade Names: Pecosil® DCU

Dimethicone copolyol undecylinate. *See* Dimethicone copolyol undecylenate
Dimethicones. *See* Dimethicone
Dimethoxydiglycol. *See* Diethylene glycol dimethyl ether
Dimethoxyethane; 1,2-Dimethoxyethane; α,β-Dimethoxyethane. *See* Ethylene glycol dimethyl ether

Dimethoxyethyl phthalate

CAS 117-82-8
Synonyms: 1,2-Benzenedicarboxylic acid bi (2-methoxyethyl) ester; Bis (methoxyethyl) phthalate; Bis (2-methoxyethyl) phthalate; Di (2-methoxyethyl) phthalate; Dimethyl glycol phthalate; DMEP; DMGP; 2-Methoxyethyl phthalate; Phthalic acid bis (2-methoxyethyl) ester; Phthalic acid, di (methoxyethyl) ester
Empirical: $C_{14}H_{18}O_6$
Formula: $C_6H_4(COOCH_2CH_2OCH_3)_2$
Properties: Light-colored clear oily liq.; mild aromatic odor; insol. in water; m.w. 282.32; dens. 1.172 (20/20 C); vapor pressure 0.3 mm Hg (150 C); f.p. -45 C; m.p. -40 C; b.p. 340 C; flash pt. 194 C
Toxicology: LD50 (oral, rat) 2750 mg/kg, (IP, rat) 3735 mg/kg; LCLo (inh., rat, 16 h) 1595 ppm; mod. toxic by ing., inh., and IP routes; primary irritant; mutagen; reproductive effects; TSCA listed
Precaution: Combustible when exposed to heat or flame; can react with oxidizing materials
Hazardous Decomp. Prods.: Heated to decomp., emits acrid smoke and irritating fumes
NFPA: Health 0, Flammability 1, Reactivity 0
Uses: Plasticizer esp. for cellulose acetate; solvent; pesticide
Manuf./Distrib.: Ashland; CTC Orgs.; Fluka; Whyte Chems. Ltd

Di (2-methoxyethyl) phthalate. *See* Dimethoxyethyl phthalate
Dimethoxymethane. *See* Methylal
Dimethoxy polyethylene glycol. *See* PEG dimethyl ether

Dimethoxypropane

CAS 77-76-9; EINECS/ELINCS 201-056-0
UN 1993
Synonyms: Acetone dimethyl acetal; 2,2-Dimethoxypropane
Classification: Aliphatic organic compd.
Definition: Protein precipitant
Empirical: $C_5H_{12}O_2$
Formula: $(CH_3)_2C(OCH_3)_2$
Properties: Colorless liq.; sol. in water; misc. with alcohol, ether; m.w. 104.15; dens. 0.849 (20/4 C); m.p. -47 C; b.p. 81-83 C; flash pt. (CC) -6.7 C; ref. index 1.3780
Toxicology: Irritant; TSCA listed
Precaution: Flamm. liq.; very dangerous fire hazard when exposed to heat, flame or oxidizers; heated to 210 C, burns with cool flame, then explodes; explosive reaction with metal perchlorates above 65 C
Hazardous Decomp. Prods.: Heated to decomp., emits acrid smoke and fumes
Uses: Chemical intermediate for pharmaceuticals; tissue dehydrating agent; pesticides; dyes
Manuf./Distrib.: Aldrich; Fluka; Penta Mfg.; Sigma; Ube Ind.
Trade Names Containing: Celtone

1,2-Dimethoxypropane. *See* Propylene glycol dimethyl ether
2,2-Dimethoxypropane. *See* Dimethoxypropane

1,3-Dimethoxy-2-propanol

CAS 623-69-8
Synonyms: 1,3-DP; Glycerol 1,3-dimethyl ether; Glyceryl α,γ-dimethyl ether; 2-Propanol, 1,3-dimethoxy-
Classification: Glycol ether
Empirical: $C_5H_{12}O_3$
Properties: M.w. 120.15
Environmental: ThOD 1.73
Uses: Solvent
Regulatory: HAP

2,3-Dimethoxystrychnidin-10-one; Dimethoxystrychnine; 2,3-Dimethoxystrychnine; 10,11-Dimethoxystrychnine. *See* Brucine
10,11-Dimethoxystrychnine sulfate, heptahydrate. *See* Brucine sulfate heptahydrate
Dimethoxytetraethylene glycol. *See* Dimethoxytetraglycol

Dimethoxytetraglycol

CAS 143-24-8; EINECS/ELINCS 205-594-7
Synonyms: Bis (2-methoxyethoxyethyl) ether; Dimethoxytetraethylene glycol; Ether, bis[2-(2-methoxyethoxy)ethyl]; Glyme-5; 2,5,8,11,14-Pentaoxapentadecane; Tetraethylene glycol dimethyl ether; Tetraglyme
Empirical: $C_{10}H_{22}O_5$
Formula: $CH_3(OCH_2)_4OCH_3$
Properties: Water-wh., practically odorless liq.; sol. in hydrocarbons, water; m.w. 222.32; dens. 1.0132; bulk dens. 8.4 lb/gal (20 C); visc. 0.0405 poise; vapor pressure < 0.01 mm Hg; f.p. -29.7 C; b.p. 275.8 C; flash pt. 116 C; ref. index 1.4325
Toxicology: LD50 (oral, rat) 5140 mg/kg; mildly toxic by ing.; eye irritant; TSCA listed
Precaution: Combustible
Hazardous Decomp. Prods.: Heated to decomp., emits acrid smoke and irritating fumes
NFPA: Health 1, Flammability 1, Reactivity 0
Storage: Hygroscopic; air-sensitive
Uses: Solvent
Manuf./Distrib.: Acros Org.; Aldrich; Alfa Aesar; Filo; Fluka; Lancaster Synthesis; Pfaltz & Bauer; Sigma; TCI Am.
See also PEG-4 dimethyl ether

Dimethyl acetamide

CAS 127-19-5; EINECS/ELINCS 204-826-4
Synonyms: Acetdimethylamide; Acetic acid dimethylamide; Acetyldimethylamine; N,N-Dimethylacetamide; Dimethylacetone amide; Dimethylamide acetate; DMA; DMAC
Empirical: C_4H_9NO
Formula: $CH_3CON(CH_3)_2$
Properties: Colorless clear oily liq.; weak ammonia fishy odor; misc. with water, oxygenated and aromatic solvs., esters, ketones, ethers; m.w. 87.12; dens. 0.942; vapor pressure 1.3 mm Hg; m.p. -20 C; b.p. 165-167 C; flash pt. 70 C; autoignition temp. 490 C; ref. index 1.437; surf. tens. 32.43 dynes/cm; dielec. const. 37.78
Toxicology: ACGIH TLV/TWA 10 ppm (skin); LD50 (oral, rat) 4300 mg/kg, (IP, rat) 2750 mg/kg, (skin, rabbit) 2240 mg/kg; mod. toxic by skin con-

tact, inh., IV, IP routes; mildly toxic by ing.; primary skin irritant; mutagen; reproductive effects; TSCA listed
Environmental: ThOD 1.84
Precaution: Flamm. exposed to heat or flame; LEL 1.8%; UEL 11.5%; mod. explosion hazard; violent reaction with halogenated compds. above 90 C
Hazardous Decomp. Prods.: Heated to decomp., emits toxic fumes of NO_x
NFPA: Health 2, Flammability 2, Reactivity 0
Uses: Solvent for plastics (vinyl, acrylic, cellulose, polyimide processing), resins, gums, fibers, coatings, adhesives, electrolytes, pharmaceuticals; selective solvent for butadiene extraction; reagent; intermediate; catalyst; paint remover; high-purity solvent for crystallization and purification; reaction medium for prod. of pharmaceuticals, plasticizers
Manuf./Distrib.: Aldrich; Allchem Ind.; BASF AG; BASF; Cross World Sales; Filo; Fluka; Int'l. Chem. Inc; Monomer-Polymer & Dajac Labs; Primachem; Romil Ltd; Sigma; Spectrum Quality Prods.; Whyte Chems. Ltd

N,N-Dimethylacetamide. *See* Dimethyl acetamide
Dimethylacetic acid. *See* Isobutyric acid
Dimethylacetone. *See* Diethyl ketone
Dimethylacetone amide. *See* Dimethyl acetamide
Dimethyl acetylene carbinol; Dimethylacetylenylcarbinol. *See* Methyl butynol

Dimethyl adipate

CAS 627-93-0; EINECS/ELINCS 211-020-6
Synonyms: Adipic acid dimethyl ester; Dimethyl hexanedioate; 1,6-Dimethyl hexanedioate; Hexanedioic acid, dimethyl ester; Methyl adipate
Classification: Dicarboxylic acid ester
Definition: Diester of methyl alcohol and adipic acid
Empirical: $C_8H_{14}O_4$
Formula: $CH_3O_2C(CH_2)_4CO_2CH_3$
Properties: Colorless clear oily liq.; odorless; sol. in oxygenated solvs.; misc. with common org. solvs., alcohols, ether; insol. in water; m.w. 174.22; dens. 1.062 (20/4 C); vapor pressure 0.06 mm Hg (20 C); m.p. 9-11 C; b.p. 115 C (13 mm); flash pt. 107 C; autoignition temp. 359 C; ref. index 1.4285
Toxicology: REL TWA (8 h) 1.5 ppm; LD50 (IP, rat) 1809 mg/kg; may be harmful by inh., ing., or skin absorp.; mod. toxic by IP route; readily absorbed thru skin; may cause irritation; inh. may cause respiratory tract irritation, decreased wt. gain, degeneration of nasal tissue; target organs: eyes, nose; suspected teratogen and reproductive effects; TSCA listed
Precaution: LEL 0.8%; UEL 8.1%; incompat. with acids, bases, oxidizing agents, reducing agents
Hazardous Decomp. Prods.: Toxic fumes of CO, CO_2; acrid smoke and irritating fumes
NFPA: Health 1, Flammability 1, Reactivity 0
Storage: Store in cool, dry place; keep container closed
Uses: Solvent for coatings, cleaners, paint stripping, resins, inks, textile lubricants, urethane prod.; plasticizer for flexible thermoset polyester; polymer intermediate for polyester polyols for urethanes, wet-str. paper resins, polyester resins; cleaner for polymeric residues; pigment dispersant; emollient; chemical intermediate for agrochems., dyes
Manuf./Distrib.: Acros Org.; Aldrich; Alfa Aesar; Ashland; Carboshale AS; ChemService; DuPont; Eastern Chem.; Fluka; Haltermann Prods. UK; ICN Biomed. Research Prods.; Lancaster Synthesis; Morflex; Proviron Ind. NV; Sigma; TCI Am.; UCB
Trade Names: DBE-6; Santosol® DMA
Trade Names Containing: DBE; DBE-2; DBE-3; DBE-9; Estasol™; Santosol® DME; Santosol® DME-1; Santosol® DME-2; Santosol® DME-3; Skykleen® 1000

Dimethylamide acetate. *See* Dimethyl acetamide

Dimethylamine

CAS 124-40-3; EINECS/ELINCS 204-697-4
UN 1032 (DOT); UN 1160 (DOT)
Synonyms: Dimethylamine anhydrous; DMA; Methanamine, N-methyl-; N-Methylmethanamine
Classification: Sat. aliphatic amine
Empirical: C_2H_7N
Formula: CH_3NHCH_3
Properties: Colorless liquefied gas; strong fishy odor (below 100 ppm), ammonia-like odor (above 100 ppm); very sol. in water; sol. in ethanol, ether, oxygenated solvs.; m.w. 45.08; dens. 0.68 (0 C); vapor pressure 1520 mm Hg; m.p. -92.2 C; b.p. 6.9 C; flash pt. 20 C; autoignition temp. 400 C; pH 11.9 (0.1 M aq, sol'n.); surf. tens. 16.33 dynes/cm; dielec. const. 5.26
Toxicology: ACGIH TLV/TWA 5 ppm; STEL 15 ppm; LD50 (oral, rat) 698 mg/kg (33% aq. sol'n.); very toxic; poison on ing.; may cause skin irritation, burns on high exposure; lachrymator; can cause severe eye irritation, 'halo vision'; target organs: liver, skin, testes, respiratory system; TSCA listed
Environmental: VOC; BOD5 1.47; ThOD 2.13
Precaution: Flamm., corrosive, very toxic, compressed gas; LEL 2.8%; UEL 14.4%; incompat. with mercury, acrylaldehyde, acids, acid chlorides, acid anhydrides, nitrosating agents, strong oxidizing agents, fluorine, maleic anhydride; violent/explosive reactions possible
Hazardous Decomp. Prods.: Ammonia; reaction with nitrosating agents can form nitrosodimethylamine, a carcinogen; heated to decomp., emits toxic fumes of NO_x
NFPA: Health 3, Flammability 4, Reactivity 0
Storage: Store in cool, well-ventilated area, out of direct sunlight; protect from extreme temps.; never expose cylinders to temps. > 52 C or < -29 C
Uses: Solvent; antioxidant, stabilizer in natural rubber latex and some resins; prep. of solvents, pesticides, surfactants, soaps, detergents, rubber chems., dyes, pharmaceuticals, textile chems.; acid gas absorbent; reagent for magnesium; mfg. of dimethylformamide, dimethylacetamide; flotation agent; gasoline stabilizer; electroplating; dehairing agent; missile fuels; pesticide propellant; rocket propellant; catalyst accelerator for unsat. polyester resins; ion exchange membrane in prod. of grapefruit juice; water treatment
Regulatory: FDA 21CFR §173.20, 176.170; CERCLA hazardous substance
Manuf./Distrib.: Air Prods.; Aldrich; Allchem Ind.; BASF AG; Celanese; Filo; Fluka; J.T. Baker; Kira; MG Ind.; Penta Mfg.; Richard's Pacific Chem. Sourcing; Sigma; Spectrum Quality Prods.; Triple Crown Am.; VWR Int'l.; Vopak USA

Dimethylamine anhydrous. *See* Dimethylamine
Dimethylamine, N-nitroso-. *See* N-Nitrosodimethylamine
(Dimethylamino) benzene. *See* N,N-Dimethylaniline
2-Dimethylaminoethanol; β-Dimethylaminoethanol; N,N-Dimethylaminoethanol; β-Dimethylaminoethyl alcohol. *See* Dimethylethanolamine
N-Dimethylaniline. *See* N,N-Dimethylaniline

N,N-Dimethylaniline

CAS 121-69-7; EINECS/ELINCS 204-493-5
UN 2253 (DOT)
Synonyms: Aniline, N,N-dimethyl; Benzenamine, N,N-dimethyl-; (Dimethylamino) benzene; N-Dimethylaniline; N,N-Dimethylbenzeneamine; Dimethylphenylamine; N,N-Dimethylphenylamine; DMA; Xylidine
Empirical: $C_8H_{11}N$
Formula: $C_6H_5N(CH_3)_2$
Properties: Ylsh. to brnsh. oily liq.; sol. in alcohol, ether, oxygenated and chlorinated solvs.; insol. in water; m.w. 121.18; dens. 0.956 (20/4 C); m.p. 2-4 C; b.p. 76-78 C (10 mm); flash pt. ≈ 48 C; ref. index 1.558
Toxicology: ACGIH TLV/TWA 5 ppm; STEL 10 ppm (skin); LD50 (oral, rat) 1410 mg/kg; LCLo (inh., rat, 4 h) 250 mg/m^3; toxic by inh., skin contact, ing.; skin irritant; human systemic effects by ing. (nausea, vomiting); CNS depressant; suspected carcinogen; mutagenic data; TSCA listed
Precaution: Combustible; explodes on contact with benzoyl peroxide
Hazardous Decomp. Prods.: Heated to decomp., emits highly toxic fumes of aniline and NO_x
NFPA: Health 3, Flammability 1, Reactivity 0
Storage: Photosensitive
Uses: Curing accelerator for unsat. polyester resins; extraction solvent in sulfur dioxide refining; dyes; intermediates; mfg. of vanillin; stabilizer (acid acceptor); acylation reagent; intermediate for paints; accelerator

in food-contact crosslinked polyesters
Regulatory: FDA 21CFR §177.2420
Manuf./Distrib.: Aarti Ind. Ltd; Aceto; Aldrich; Allchem Ind.; BASF AG; BASF; Buffalo Color; Celanese; DuPont; Fabrichem; Fluka; Kessler; Monomer-Polymer & Dajac Labs; Sigma; Spectrum Quality Prods.; Uniqema Am.

Dimethylbenzene. *See* Xylene
1,2-Dimethylbenzene. *See* o-Xylene
1,3-Dimethylbenzene. *See* m-Xylene
1,4-Dimethylbenzene. *See* p-Xylene
m-Dimethylbenzene. *See* m-Xylene
o-Dimethylbenzene. *See* o-Xylene
p-Dimethylbenzene. *See* p-Xylene
N,N-Dimethylbenzeneamine. *See* N,N-Dimethylaniline
Dimethyl 1,2-benzenedicarboxylate. *See* Dimethyl phthalate
Dimethylbenzene (mixed isomers); Dimethylbenzene, m-, o-, and p-. *See* Xylene
Dimethyl benzeneorthodicarboxylate. *See* Dimethyl phthalate
Dimethylbenzene sulfonic acid, sodium salt. *See* Sodium xylenesulfonate
p-(α,α-Dimethylbenzyl) phenol. *See* Cumyl phenol
1,2-Dimethylbutane. *See* Isohexane
2,2-Dimethylbutane. *See* Neohexane

2,3-Dimethylbutane

CAS 79-29-8; EINECS/ELINCS 201-193-6
UN 2457
Synonyms: Biisopropyl; Butane, 2,3-dimethyl-; Diisopropyl; 1,1,2,2-Tetramethylethane
Empirical: C_6H_{14}
Formula: $(CH_3)_2CHCH(CH_3)_2$
Properties: Colorless clear liq.; mild gasoline-like odor; very sol. in alcohol, hot water; sol. in 95% ethanol, acetone; sol. 10-50 mg/ml in DMSO; sol. > 10% in ether, benzene; sol. < 1 mg/ml in water; m.w. 86.18; sp.gr.0.6616 (20/4 C); vapor pressure 200 mm Hg (21.1 C); m.p. -128.5 C; b.p. 58 C; flash pt. -28.9 C; ref. index 1.3750 (20 C); 100% volatile by vol.
Toxicology: ACGIH TLV/TWA 500 ppm; STEL 1000 ppm; may be harmful by inh., absorp. or ing.; may cause eye/skin/upper respiratory tract irritation, redness/swelling of mucous membranes, headache, fatigue, hypersensitivity, tremors, CNS effects, pulmonary edema, chem. pneumonitis; probably narcotic in high concs.; related compds. have caused dizziness, giddiness, nausea, eye/throat irritation; TSCA listed
Precaution: Flamm.; dangerous fire and explosion risk; LEL 1.2%; UEL 7%; vapors may flash back; container explosion may occur under fire conditions; forms explosive mixts. in air; incompat. with oxidizing agents, oxygen
Hazardous Decomp. Prods.: Heated to decomp., emits acrid smoke and irritating fumes of CO_x and various hydrocarbons
NFPA: Health 1, Flammability 3, Reactivity 0
Storage: Store in explosion-proof refrigerator; keep tightly closed; may develop pressure; keep away from oxidizing materials and ignition sources
Uses: Component of high octane fuel; organic synthesis; solvent for veg. oils, glues, paints, coatings, in gasoline; intermediate for chems., in rubber solvs. and petrol. ether
Manuf./Distrib.: Acros Org.; Aldrich; Alfa Aesar; ChevronPhillips; Fluka; ICN Biomed. Research Prods.; Pfaltz & Bauer; TCI Am.

Dimethyl butanedioate. *See* Dimethyl succinate
1,3-Dimethyl butanol. *See* 2-Methyl-1-pentanol
Dimethylbutyl acetate; 1,3-Dimethylbutyl acetate. *See* Methyl amyl acetate
Dimethyl capramide; N,N-Dimethylcapramide. *See* Dimethyl caproamide

Dimethyl caproamide

CAS 14433-76-2; EINECS/ELINCS 238-405-1
Synonyms: Decanamide, N,N-dimethyl-; Dimethyl capramide; N,N-Dimethylcapramide; N,N-Dimethyldecanamide; N,N-Dimethyldecanoamide
Formula: $C_{12}H_{25}NO$
Properties: M.w. 199.38
Toxicology: Poison by intravenous route; moderately toxic by intraperitoneal route
Hazardous Decomp. Prods.: Heated to decomp., emits toxic NO_x
Uses: Emulsifier, solvent, cosolvent in pesticides
Manuf./Distrib.: ICN Biomed. Research Prods.; Lancaster Synthesis
Trade Names Containing: Hallcomid® M-8-10

N,N-Dimethyl caprylamide. *See* Dimethyl octanamide
Dimethyl carbinol. *See* Isopropyl alcohol
Dimethyl carbonate. *See* Methyl carbonate
N,N-Dimethyl-2-chloroacetamide. *See* Acetyldimethylamine chloride

Dimethyl cyanamide

CAS 1467-79-4; EINECS/ELINCS 215-991-7
UN 2810
Synonyms: Carbamic acid nitrile; Carbonic acid diethylamide; N-Cyanodimethylamine; N-Cyano-N-methylmethanamine; N,N-Dimethylcyanamide
Empirical: $C_3H_6N_2$
Formula: $(CH_3)_2NCN$
Properties: Colorless to yel. mobile liq.; sol. in acetone, ether, 95% ethanol; dec. in water; m.w. 70.11; sp.gr. 0.88; m.p. -41 C; b.p. 161-163 C; flash pt. (TCC) 71.1 C
Toxicology: LD50 (oral, rat) 146 mg/kg, (IP, mouse) 40 mg/kg; (skin, mouse) 125 mg/kg; LC50 (inh., rat) 2500 mg/m^3; poison; toxic by ing., inh., skin contact; severe irritant; readily absorbed thru skin; may be fatal by inh., ing., skin absorp.; high concs. are extremely destructive to tissues of mucous membranes, upper respiratory tract, eyes, skin; may cause burning sensation, coughing, wheezing, laryngitis, shortness of breath, headache, nausea, vomiting, muscle weakness, etc.; TSCA listed
Precaution: Combustible; incompat. with strong acids, strong bases, strong oxidizers, strong reducing agents; reacts with water or steam; may dec. on exposure to moist air or water; keep away from food
Hazardous Decomp. Prods.: Thermal decomp. may produce CO, CO_2, NO_x; reacts with water to produce toxic and flamm. vapors; emits toxic fumes under fire conditions
NFPA: Health 4, Flammability 2, Reactivity 1
Storage: Moisture-sensitive; store refrigerated; keep tightly closed; protect from moisture
Uses: Organic intermediate; solvent
Manuf./Distrib.: Acros Org.; Aldrich; Fluka; Lancaster Synthesis; TCI Am.

N,N-Dimethylcyanamide. *See* Dimethyl cyanamide

1,1-Dimethylcyclohexane

CAS 590-66-9
Synonyms: Dimethyl-1,1-cyclohexane
Classification: Sat. alicyclic hydrocarbon
Empirical: C_8H_{16}
Properties: Colorless liq.; sol. in most org. solvs.; insol. in water; m.w. 122.22; sp.gr. 0.7809 (20 C); m.p. -34 C; b.p. 120 C; flash pt. 7 C
Toxicology: May cause mild skin irritation; vapor may be irritating to eyes; probably a CNS depressant; sl. nose/throat irritant; high vapor concs. may cause headache, lightheadedness, nausea, drowsiness; very high concs. may cause unconsciousness, death; ing. may cause sore throat, nausea, diarrhea
Precaution: Flamm.; incompat. with strong oxidizing agents (increased fire and explosion hazard)
Hazardous Decomp. Prods.: None
Storage: Store in cool, dry, well-ventilated area out of direct sunlight
Uses: Solvent; org. synthesis
Manuf./Distrib.: Aldrich; Fluka

Dimethyl-1,1-cyclohexane. *See* 1,1-Dimethylcyclohexane

1,2-Dimethylcyclohexane

CAS 583-57-3
UN 2263 (DOT)
Synonyms: cis-1,2-Dimethylcyclohexane; o-Dimethylcyclohexane; Hexahydro-o-xylene
Classification: Sat. alicyclic hydrocarbon
Empirical: C_8H_{16}
Formula: $CH_3C_6H_{10}CH_3$

Properties: Colorless liq.; sol. in most org. solvs.; insol. in water; m.w. 112.22; sp.gr. 0.778 (20 C); b.p. 124 C; flash pt. 15 C
Toxicology: May cause mild skin irritation; conc. vapor may be irritating to eyes; probably a CNS depressant; sl. nose/throat irritation; high vapor concs. may cause headache, lightheadedness, nausea, drowsiness; very high concs. may cause unconsciousness, death; ing. may cause sore throat, nausea, diarrhea
Precaution: Flamm.; very dangerous fire hazard exposed to heat, flame, oxidizers; incompat. with strong oxidizing agents (increased fire and explosion hazard)
Hazardous Decomp. Prods.: Heated to decomp., emits acrid smoke and fumes
NFPA: Health 0, Reactivity 0
Storage: Store in cool, dry, well-ventilated area out of direct sunlight
Uses: Solvent; org. synthesis
Manuf./Distrib.: Aldrich; ChevronPhillips; Fluka

1,3-Dimethylcyclohexane

CAS 591-21-9
UN 2263 (DOT)
Synonyms: m-Dimethylcyclohexane; Hexahydroxylene; Hexahydro-m-xylene
Classification: Sat. alicyclic hydrocarbon
Empirical: C_8H_{16}
Formula: $CH_3C_6H_{10}CH_3$
Properties: Colorless liq.; sol. in most org. solvs.; insol. in water; m.w. 112.22; sp.gr. 0.767 (20 C); b.p. 124 C; flash pt. 10 C
Toxicology: May cause mild skin irritation; conc. vapor may be irritating to eyes; probably a CNS depressant; high vapor concs. may cause sl. nose/throat irritation, headache, lightheadedness, nausea, drowsiness; very high concs. may cause unconsciousness, death; ing. may cause sore throat, nausea, diarrhea
Precaution: Flamm.; very dangerous fire hazard exposed to heat, flame, or oxidizers; incompat. with strong oxidizing agents (increased fire and explosion hazard)
Hazardous Decomp. Prods.: Heated to decomp., emits acrid smoke and fumes
NFPA: Health 0, Flammability 3, Reactivity 0
Storage: Store in cool, dry, well-ventilated area out of direct sunlight
Uses: Solvent; org. synthesis
Manuf./Distrib.: Aldrich

1,4-Dimethylcyclohexane

CAS 589-90-2
UN 2263 (DOT)
Synonyms: Cyclohexane, 1,4-dimethyl-; 1,4-Dimethylcyclohexane (cis- and trans-); p-Dimethylcyclohexane; Hexahydroxylene; Hexahydro-p-xylene
Classification: Sat. alicyclic hydrocarbon
Empirical: C_8H_{16}
Formula: $CH_3C_6H_{10}CH_3$
Properties: Colorless liq.; sol. in most org. solvs.; insol. in water; m.w. 112.22; sp.gr. 0.783 (20 C); vapor pressure 10 mm Hg (10.2 C); m.p. -88 C; b.p. 120 C; flash pt. 11 C
Toxicology: May cause mild skin irritation; conc. vapor may be irritating to eyes; probably a CNS depressant; high vapor concs. may cause sl. nose/throat irritation, headache, nausea, drowsiness; very high concs. may cause unconsciousness, death; ing. may cause sore throat, nausea, diarrhea
Precaution: Flamm. liq.; dangerous fire hazard exposed to heat or flame; incompat. with strong oxidizing agents (increased fire and explosion hazard)
Hazardous Decomp. Prods.: None
NFPA: Health 1, Flammability 3, Reactivity 0
Storage: Store in cool, dry, well-ventilated area out of direct sunlight
Uses: Solvent; org. synthesis
Manuf./Distrib.: Aldrich

1,4-Dimethylcyclohexane (cis- and trans-). *See* 1,4-Dimethylcyclohexane

cis-1,2-Dimethylcyclohexane

CAS 2207-01-4; EINECS/ELINCS 218-621-2
UN 2263
Classification: Sat. alicyclic hydrocarbon
Empirical: C_8H_{16}
Formula: $CH_3C_6H_{10}CH_3$
Properties: Colorless liq.; sol. in most org. solvs.; insol. in water; m.w. 112.22; sp.gr. 0.796 (20 C); m.p. -50 C; b.p. 130 C; flash pt. 12 C; ref. index 1.4360
Toxicology: May cause mild skin irritation; conc. vapor may be irritating to eyes; probably a CNS depressant; sl. nose/throat irritation; high vapor concs. may cause headache, lightheadedness, nausea, drowsiness; very high concs. may cause unconsciousness, death; ing. may cause sore throat, nausea, diarrhea; TSCA listed
Precaution: Flamm.; incompat. with strong oxidizing agents (increased fire and explosion hazard)
Hazardous Decomp. Prods.: None
NFPA: Health 0, Reactivity 0
Storage: Store in cool, dry, well-ventilated area out of direct sunlight
Uses: Solvent; org. synthesis
Manuf./Distrib.: Aldrich

cis-1,3-Dimethylcyclohexane

CAS 638-04-0
UN 2263
Classification: Sat. alicyclic hydrocarbon
Empirical: C_8H_{16}
Formula: $CH_3C_6H_{10}CH_3$
Properties: Colorless liq.; sol. in most org. solvs.; insol. in water; m.w. 112.22; sp.gr. 0.784 (20 C); m.p. -76 C; b.p. 120 C; flash pt. 5 C
Toxicology: May cause mild skin irritation; conc. vapor may be irritating to eyes; probably a CNS depressant; sl. nose/throat irritant; high vapor concs. may cause headache, lightheadedness, nausea, drowsiness; very high concs. may cause unconsciousness, death; ing. may cause sore throat, nausea, diarrhea
Precaution: Flamm.; incompat. with strong oxidizing agents (increased fire and explosion hazard)
Hazardous Decomp. Prods.: None
NFPA: Health 0, Flammability 3, Reactivity 0
Storage: Store in cool, dry, well-ventilated area out of direct sunlight
Uses: Solvent; org. synthesis
Manuf./Distrib.: Aldrich

cis-1,4-Dimethylcyclohexane

CAS 624-29-3; EINECS/ELINCS 210-840-1
UN 2263
Synonyms: Cyclohexane, 1,4-dimethyl-, cis-
Classification: Sat. alicyclic hydrocarbon
Empirical: C_8H_{16}
Formula: $CH_3C_6H_{10}CH_3$
Properties: Colorless liq.; sol. in most org. solvs.; insol. in water; m.w. 122.22; sp.gr. 0.783 (20 C); m.p. -87 C; b.p. 124 C; flash pt. 16 C; ref. index 1.4297
Toxicology: May cause mild skin irritation; conc. vapor can irritate eyes; probably a CNS depressant; sl. nose/throat irritant; high vapor concs. may cause headache, lightheadedness, nausea, drowsiness; very high concs. may cause unconsciousness, death; ing. may cause sore throat, nausea, diarrhea
Precaution: Flamm.; incompat. with strong oxidizing agents (increased fire and explosion hazard)
Hazardous Decomp. Prods.: None
NFPA: Health 0, Flammability 3, Reactivity 0
Storage: Store in cool, dry, well-ventilated area out of direct sunlight
Uses: Solvent; org. synthesis
Manuf./Distrib.: Aldrich; Fluka; ICN Biomed. Research Prods.; TCI Am.

m-Dimethylcyclohexane. *See* 1,3-Dimethylcyclohexane
o-Dimethylcyclohexane. *See* 1,2-Dimethylcyclohexane
p-Dimethylcyclohexane. *See* 1,4-Dimethylcyclohexane

trans-1,2-Dimethylcyclohexane

CAS 6876-23-9
UN 2263
Classification: Sat. alicyclic hydrocarbon
Empirical: C_8H_{16}
Formula: $CH_3C_6H_{10}CH_3$

Properties: Colorless liq.; sol. in most org. solvs.; insol. in water; m.w. 112.22; sp.gr. 0.776 (20 C); m.p. -89 C; b.p. 123 C; flash pt. 6 C
Toxicology: May cause mild skin irritation; conc. vapor may be irritating to eyes; probably a CNS depressant; high vapor concs. may cause sl. nose/throat irritation, headache, lightheadedness, nausea, drowsiness; very high concs. may cause unconsciousness, death; ing. may cause sore throat, nausea, diarrhea
Precaution: Flamm.; very dangerous fire hazard exposed to heat, flame, or oxidizers; incompat. with strong oxidizing agents (increased fire and explosion hazard)
Hazardous Decomp. Prods.: Heated to decomp., emits acrid smoke and fumes
NFPA: Health 0, Reactivity 0
Storage: Store in cool, dry, well-ventilated area out of direct sunlight
Uses: Solvent; org. synthesis
Manuf./Distrib.: Aldrich; Fluka

trans-1,3-Dimethylcyclohexane

CAS 2207-03-6
UN 2263
Synonyms: Cyclohexane, 1,3-dimethyl-, trans-
Classification: Sat. alicyclic hydrocarbon
Empirical: C_8H_{16}
Formula: $CH_3C_6H_{10}CH_3$
Properties: Colorless liq.; sol. in most org. solvs.; insol. in water; m.w. 112.22; sp.gr. 0.785 (20 C); m.p. -90 C; b.p. 124 C; ref. index 1.4309
Toxicology: May cause mild skin irritation; conc. vapor may be irritating to eyes; probably a CNS depressant; high vapor concs. may cause sl. nose/throat irritation, headache, lightheadedness, nausea, drowsiness; very high concs. may cause unconsciousness, death; ing. may cause sore throat, nausea, diarrhea
Precaution: Flamm.; incompat. with strong oxidizing agents (increased fire and explosion hazard)
Hazardous Decomp. Prods.: None
NFPA: Health 0, Flammability 3, Reactivity 0
Storage: Store in cool, dry, well-ventilated area out of direct sunlight
Uses: Solvent; org. synthesis
Manuf./Distrib.: Aldrich; ICN Biomed. Research Prods.; TCI Am.

trans-1,4-Dimethylcyclohexane

CAS 2207-04-7
UN 2263
Synonyms: Carbamic acid, 1-phenylethyl-, ethyl ester; Cyclohexane, 1,4-dimethyl-, trans-
Classification: Sat. alicyclic hydrocarbon
Empirical: C_8H_{16}
Formula: $CH_3C_6H_{10}CH_3$
Properties: Colorless liq.; sol. in most org. solvs.; insol. in water; m.w. 122.22; sp.gr. 0.766 (20 C); m.p. -37 C; b.p. 119 C; flash pt. 11 C; ref. index 1.4209
Toxicology: May cause mild skin irritation; conc. vapor can irritate eyes; probably a CNS depressant; high vapor concs. may cause sl. nose/throat irritation, headache, lightheadedness, nausea, drowsiness; very high concs. may cause unconsciousness, death; ing. may cause sore throat, nausea, diarrhea
Precaution: Flamm. gas; incompat. with strong oxidizing agents (increased fire and explosion hazard)
Hazardous Decomp. Prods.: None
NFPA: Health 0, Flammability 3, Reactivity 0
Storage: Store in cool, dry, well-ventilated area out of direct sunlight
Uses: Solvent; org. synthesis
Manuf./Distrib.: Fluka

1,1-Dimethylcyclopentane

CAS 1638-26-2
Synonyms: Cyclopentane, 1,1-dimethyl-
Classification: Sat. alicyclic hydrocarbon
Empirical: C_7H_{14}
Formula: $(CH_3)_2C_5H_8$
Properties: Colorless liq.; sol. in most org. solvs.; insol. in water; m.w. 98.19; sp.gr. 0.749 (20 C); m.p. -70 C; b.p. 88 C; flash pt. < 21 C; ref. index 1.4136
Toxicology: May cause mild skin irritation; conc. vapor may be irritating to eyes; probably a CNS depressant; high vapor concs. may cause nose/ throat irritation, headache, lightheadedness, nausea, dizziness; very high concs. may cause unconsciousness, death; ing. may cause sore throat, nausea, diarrhea
Precaution: Flamm.; incompat. with strong oxidizing agents (increased fire and explosion hazard)
Hazardous Decomp. Prods.: None
Storage: Store in cool, dry, well-ventilated area out of direct sunlight
Uses: Solvent; org. synthesis

1,2-Dimethylcyclopentane

CAS 822-50-4 (trans); 1192-18-3 (cis); 2452-99-5
Classification: Sat. alicyclic hydrocarbon
Empirical: C_7H_{14}
Formula: $CH_3C_5H_8CH_3$
Properties: Colorless liq.; sol. in most org. solvs.; insol. in water; m.w. 98.19; sp.gr. 0.744 (20 C); m.p. -62 C (cis), -120 C (trans); b.p. 99 C (cis), 92 C (trans); flash pt. < 21 C; ref. index 1.4222 (cis), 1.412 (trans)
Toxicology: May cause mild skin irritation; conc. vapor may be irritating to eyes; probably a CNS depressant; high vapor concs. may cause nose/ throat irritation, headache, lightheadedness, nausea, dizziness; very high concs. may cause unconsciousness, death; ing. may cause sore throat, nausea, diarrhea
Precaution: Flamm.; incompat. with strong oxidizing agents (increased fire and explosion hazard)
Hazardous Decomp. Prods.: None
Storage: Store in cool, dry, well-ventilated area out of direct sunlight
Uses: Solvent; org. synthesis

1,3-Dimethylcyclopentane

CAS 2453-00-1; 2532-58-3 (cis)
Synonyms: Dimethyl-1,3 cyclopentane
Classification: Sat. alicyclic hydrocarbon
Empirical: C_7H_{14}
Formula: $(CH_3)_2C_5H_8$
Properties: Colorless liq.; sol. in most org. solvs.; insol. in water; m.w. 98.19; sp.gr. 0.744 (20 C); m.p. -134 C; b.p. 91-92 C; flash pt. < 21 C; ref. index 1.4089 (cis)
Toxicology: May cause mild skin irritation; conc. vapor may be irritating to eyes; probably a CNS depressant; high vapor concs. may cause nose/ throat irritation, headache, lightheadedness, nausea, dizziness; very high concs. may cause unconsciousness, death; ing. may cause sore throat, nausea, diarrhea
Precaution: Flamm.; incompat. with strong oxidizing agents (increased fire and explosion hazard)
Hazardous Decomp. Prods.: None
Storage: Store in cool, dry, well-ventilated area out of direct sunlight
Uses: Solvent; org. synthesis
Manuf./Distrib.: ICN Biomed. Research Prods.; TCI Am.

Dimethyl-1,3 cyclopentane. *See* 1,3-Dimethylcyclopentane
N,N-Dimethyldecanamide. *See* Dimethyl caproamide
Dimethyl decanedioate. *See* Dimethyl sebacate
N,N-Dimethyldecanoamide. *See* Dimethyl caproamide
2,5-Di-o-methyl-1,4:3,6-dianhydro-D-glucitol. *See* Dimethyl isosorbide
Dimethyldiglycol. *See* Diethylene glycol dimethyl ether
3,3-Dimethyl-8,9-dinorbornan-2-ol. *See* Fenchyl alcohol
1,4-Dimethyl-3,6-dioxa-1-heptanol. *See* PPG-2 methyl ether

2,2-Dimethyl-1,3-dioxolane-4-methanol

CAS 100-79-8; EINECS/ELINCS 202-888-7
Synonyms: Acetone glycerol; 2,2-Dimethyl-4-oxymethyl-1,3-dioxolane; Dioxolan; GIE
Empirical: $C_6H_{12}O_3$
Properties: Colorless, odorless liq.; sol. in water and most org. solvs.; m.w. 132.18; dens. 1.065; vapor pressure 70 mm Hg (20 C); m.p. -26.4 C; b.p. 82 C (10 mm); flash pt. 90 C; ref. index 1.43
Toxicology: LD50 (oral, rat) 7 g/kg, (IP, rat) 3 g/kg; poison by IV route; eye irritant; mutagen; TSCA listed
Precaution: Combustible; can react vigorously with oxidizing materials
Hazardous Decomp. Prods.: Heated to decomp., emits acrid smoke and fumes

Uses: Solvent; plasticizer; solubilizer, suspending agent in pharmaceuticals
Manuf./Distrib.: Aldrich; Fluka; Sigma

Dimethyl disulfide. *See* Methyl disulfide
Dimethylenediamine. *See* Ethylenediamine
Dimethylene oxide. *See* Ethylene oxide
Dimethylethanol; 1,1-Dimethylethanol. *See* t-Butyl alcohol

Dimethylethanolamine

CAS 108-01-0; EINECS/ELINCS 203-542-8
UN 2051 (DOT)
Synonyms: Deanol; 2-Dimethylaminoethanol; β-Dimethylaminoethanol; N,N-Dimethylaminoethanol; β-Dimethylaminoethyl alcohol; N,N-Dimethyl ethanolamine; N,N-Dimethyl-2-hydroxyethylamine; DMAE; DMEA; Ethanol, 2-(dimethylamino)-; β-Hydroxyethyldimethylamine
Classification: Amino alcohol; alkanolamine
Empirical: $C_4H_{11}NO$
Formula: $(CH_3)_2NCH_2CH_2OH$
Properties: Colorless to pale yel. liq., amine odor; sol. in water, alcohol, ether; misc. with oxygenated and aromatic solvs.; m.w. 89.14; dens. 0.8866 (20/4 C); visc. 3.8 cps (20 C); vapor pressure 0.42 mm Hg (20 C); f.p. -64 C; b.p. 133 C; flash pt. (TCC) 39 C; autoignition temp. 295 C; ref. index 1.430; surf. tens. 28.3 dynes/cm
Toxicology: LD50 (oral, rat) 2 g/kg, (IP, rat) 1080 mg/kg, (subcut., mouse) 961 mg/kg, (skin, rabbit) 1370 mg/kg; mod. toxic by ing., inh., skin contact, IP, subcut. routes; skin and severe eye irritant; CNS stimulant; TSCA listed
Environmental: VOC; ThOD 1.80
Precaution: Flamm.; corrosive; can form explosive mixts. with air ≥ 39 C; incompat. with oxidizing agents, strong acids, acid anhydrides, acid chlorides, nitrates, nitrous acid
Hazardous Decomp. Prods.: CO, CO_2, NO_x
NFPA: Health 2, Flammability 2, Reactivity 0
Storage: Hygroscopic; store in cool, dry, well-ventilated area away from heat and ignition sources
Uses: Synthesis of dyestuffs, pharmaceuticals, textile and paper auxiliaries; catalyst for PU foam and slabstock; curing agent for epoxy, PU; polymer aggregation agent; emulsifier in paints and coatings; pH control agent; process solvent (offgas treatment); solubilizer for radiation-cured coatings, inks; corrosion inhibitor for ferrous metals, in boiler water and condensate steam systems; corrosion inhibitor, solubilizer in metalworking fluids; medicine (CNS stimulant; antidepressant drug); ion exchange membrane in prod. of grapefruit juice
Regulatory: FDA 21CFR §173.20, 175.105, 175.300
Manuf./Distrib.: Air Prods.; Aldrich; Alfa Chem; Allchem Ind.; Ashland; Atofina; BASF AG; BASF; Dow; Fluka; Great Western; Huntsman; J.T. Baker; Nippon Nyukazai; Pelron; Richman; Sigma; Spectrum Quality Prods.; Stanchem; Triple Crown Am.

N,N-Dimethyl ethanolamine. *See* Dimethylethanolamine

Dimethyl ether

CAS 115-10-6; EINECS/ELINCS 204-065-8
UN 1033 (DOT)
Synonyms: DME; Methane, oxybis-; Methoxymethane; Methyl ether; Methyl oxide; Oxybismethane; Wood ether
Classification: Organic compd.
Empirical: C_2H_6O
Formula: CH_3OCH_3
Properties: Colorless gas or compressed liq.; ether-like odor; sol. in water, alcohol, oxygenated solvs.; m.w. 46.07; f.p. -141.4 C; b.p. -24.5 C; flash pt. -41 C
Toxicology: LC50 (inh., rat) 308 g/m^3; mildly toxic by inh.; irritating to eyes, skin, respiratory tract; may cause CNS depression, difficulty breathing, headache, drowsiness, blurred vision, symptoms of drunkenness; target organ: CNS; TSCA listed
Precaution: Highly flamm.; dangerous fire and explosion hazard; avoid heat, flame; incompat. with oxidizers; violent reaction with AlH_3 and $LiAlH_2$; explosions possible with nitric acid, oxygen, ozone; may form explosive peroxides; may generate electrostatic charge; containers may rupture or explode if exposed to heat
Hazardous Decomp. Prods.: Heated to decomp., emits acrid smoke and irritating fumes
NFPA: Health 2, Flammability 4, Reactivity 1
Storage: Keep in closed container away from heat and flame; grounding and bonding required; protect against static discharges
Uses: Aerosol propellant for paint aerosols; refrigerant; solvent; extraction agent; chemical reaction medium; catalyst and stabilizer in polymerization; propellant, solvent in cosmetics
Regulatory: SARA §311/312 acute, fire, reactive and sudden release hazard, §313 nonreportable
Manuf./Distrib.: Air Prods.; Aldrich; DuPont; Isotec; MG Ind.; Pfaltz & Bauer; TCI Am.
Trade Names: Dymel® A

1-(1,1-Dimethylethoxy-2-methylethoxy)-2-propanol. *See* Dipropylene glycol t-butyl ether
1-(1,1-Dimethylethoxy)-2-propanol. *See* Propylene glycol t-butyl ether
1,1-Dimethylethyl acetate. *See* t-Butyl acetate
(1,1-Dimethylethyl) benzene. *See* t-Butylbenzene
2-(1,1-Dimethylethyl)-1,4-benzenediol. *See* t-Butyl hydroquinone
Dimethyl ethyl carbinol. *See* t-Amyl alcohol
(1,1-Dimethylethyl) cyclohexane. *See* t-Butylcyclohexane
2-(1,1-Dimethylethyl) cyclohexyl acetate. *See* 2-t-Butylcyclohexyl acetate
N,N´-Dimethylethyleneurea. *See* Dimethyl imidazolidinone
(1,1-Dimethylethyl)-4-methoxyphenol. *See* BHA
1-(1,1-Dimethylethyl)-4-methylbenzene. *See* p-t-Butyl toluene
Dimethyl ethynyl carbinol; Dimethylethynylmethanol. *See* Methyl butynol
Dimethyl formal. *See* Methylal
Dimethyl formaldehyde. *See* Acetone

Dimethyl formamide

CAS 68-12-2; EINECS/ELINCS 200-679-5
UN NA 1693 (DOT); UN 2265 (DOT)
Synonyms: N,N-Dimethylformamide; N,N-Dimethylmethanamide; DMF; DMFA; Formic acid, amide, N,N-dimethyl-; N-Formyldimethylamine
Classification: Amide
Empirical: C_3H_7NO
Formula: $CHCON(CH_3)_2$
Properties: Water-wh. clear liq.; fishy unpleasant odor; a dipolar aprotic solv.; sol. in water; misc. with alcohol, ether, acetone, benzene, most org. solvs.; insol. in aliphatic hydrocarbons; m.w. 73.10; sp.gr. 0.953-0.954 (15.6/15.6 C); f.p. -61 C; b.p. 152.8 C; flash pt. 57.7 C; ref. index 1.423-1.431; pH 6.6-8.0 (20 C, 20% aq.)
Toxicology: ACGIH TLV/TWA 10 ppm (skin); LD50 (oral, rat) 2800 mg/kg, (IP, rat) 1400 mg/kg, (IV, rat) 2000 mg/kg; mod. toxic by ing., IV, IP, subcut. routes; mildly toxic by skin contact, inh.; strong irritant to skin and tissue; severe eye irritant; inh. may cause abdominal pains, loss of appetite, nausea, vomiting, constipation, diarrhea, increased blood pressure, liver injury, irritation of mucous membranes of respiratory tract; suspected carcinogen; teratogen; human mutagenic data; TSCA listed
Precaution: Combustible; moderate fire risk; incompat. with strong oxidizers, halogen gases, halogenated compds., 2,4,6-trichloro-1,3,5-triazine; vigorous reactions possible; polymerizes violently on contact with methylene bisphenyl diisocyanate
Hazardous Decomp. Prods.: On decomp., emits toxic fumes of NO_x
NFPA: Health 1, Flammability 2, Reactivity 0
Storage: Store in cool, dry, well-ventilated area, out of direct sunlight
Uses: Solvent for liqs., gases, in vinyl resins, acetylene, butadiene, acid gases, paints, PU coatings (for textile backing materials, imitation leather), paint strippers, cleaners, zinc electroplating, acrylic fiber spinning; extraction solvent for aromatics; process solvent; solvent for welding gas and wire enamel resins, electrolytic capacitors; solvent, reaction medium for pharmaceuticals; organic synthesis; mfg. of polyacrylic fibers, butadiene, pharmaceuticals, dyes, petroleum prods.; carrier for gases; reagent; petroleum anti-icing additive; catalyst in carboxylation reactions; in food-pkg. adhesives; adjuvant in slimicides in food-contact paper/paperboard
Regulatory: FDA 21CFR §175.105, 176.300
Manuf./Distrib.: AMRESCO; Aceto; Advanced Synthesis Tech.; Air Prods.; Aldrich; Allchem Ind.; Ashland; BASF; Baychem; Brown; Browning; Celanese; Chemcentral; Coyne; DuPont; Filo; Fluka; Great Western; Haltermann Prods. UK; Harcros; Houghton Chem.; Integra; J.T. Baker;

Kessler; Mallinckrodt Baker; Monomer-Polymer & Dajac Labs; Nissan Chem. Ind.; Penta Mfg.; Romil Ltd; Ruger; Sigma; Spectrum Quality Prods.; Stanchem; Synasia Fine Chem.; Tulstar Prods.; UCB; VWR Int'l.; Vopak USA; Whyte Chems. Ltd
Trade Names Containing: Dynasolve 100

N,N-Dimethylformamide. *See* Dimethyl formamide

Dimethyl glutarate

CAS 1119-40-0; EINECS/ELINCS 214-277-2
Synonyms: Dimethyl pentanedioate; Pentanedioic acid, dimethyl ester
Empirical: $C_7H_{12}O_4$
Formula: $CH_3OCO(CH_2)_3COOCH_3$
Properties: Colorless liq.; sol. in oxygenated solvs.; m.w. 160.17; dens. 1.087 (20/4 C); m.p. -42 C; b.p. 230 C; flash pt. 97 C; ref. index 1.424 (20 C)
Uses: Solvent for coatings, cleaners, inks, textile lubricants, urethane prod.; plasticizer for flexible thermoset polyester; polymer intermediate for polyester polyols for urethanes, wet-str. paper resins, polyester resins; antistat; specialty chemical intermediate
Manuf./Distrib.: Aldrich; Ashland; Fluka; Sigma
Trade Names: DBE-5; Santosol® DMG
Trade Names Containing: Axarel® 6100; DBE; DBE-2; DBE-3; DBE-9; Estasol™; Santosol® DME; Santosol® DME-1; Santosol® DME-2; Santosol® DME-3; Skykleen® 1000

Dimethylglycol. *See* Ethylene glycol dimethyl ether
Dimethyl glycol phthalate. *See* Dimethoxyethyl phthalate

2,6-Dimethyl heptanol-4

CAS 108-82-7; EINECS/ELINCS 203-619-6
FEMA 3140
Synonyms: Diisobutyl carbinol; 2,6-Dimethyl-4-heptanol; 4-Heptanol, 2,6-dimethyl-; s-Nonyl alcohol
Classification: Nonaromatic alcohol
Empirical: $C_9H_{20}O$
Formula: $[(CH_3)_2CHCH_2]_2CH_2O$
Properties: Colorless liq.; sol. in alcohol and ether; insol. in water; m.w. 144.29; sp.gr. 0.812; f.p. sets to glass at ≈ -65 C; b.p. 177.2-180 C; flash pt. 66 C; ref. index 1.423 (21 C); surf. tens. 26 dynes/cm (20 C)
Toxicology: LD50 (oral, rat) 3560 mg/kg, (IP, rat) 800 mg/kg, (skin, rabbit) 4600 mg/kg; primary irritant; mild skin/eye irritant; TSCA listed
Precaution: Combustible
NFPA: Health 1, Flammability 2, Reactivity 0
Uses: Solvent; surfactants; lubricant additives; rubber chemicals; flotation agents; synthetic flavoring agent in foods; cosmetics ingred.; antifoam; defoamer in food-contact paper/paperboard
Regulatory: FDA 21CFR §176.210; FEMA GRAS
Manuf./Distrib.: Acros Org.; Aldrich; Dow; Pfaltz & Bauer; TCI Am.

2,6-Dimethyl-4-heptanol. *See* 2,6-Dimethyl heptanol-4
2,6-Dimethyl-4-heptanone; 2,6-Dimethylheptan-4-one. *See* Diisobutyl ketone
2,6-Dimethyl-4-heptylphenol (o and p). *See* Nonylphenol
Dimethyl hexanedioate; 1,6-Dimethyl hexanedioate. *See* Dimethyl adipate
3,5-Dimethyl-1-hexyne-3-ol. *See* Dimethyl hexynol

Dimethyl hexynol

CAS 107-54-0; EINECS/ELINCS 203-500-9
UN 1993 (DOT)
Synonyms: 3,5-Dimethyl-1-hexyne-3-ol; 3,5-Dimethyl-1-hexyn-3-ol
Empirical: $C_8H_{14}O$
Formula: $HC{\equiv}CCOH(CH_3)CH_2CH(CH_3)_2$
Properties: Colorless liq., camphor-like odor; sl. sol. in water; m.w. 126.20; dens. 0.8545 (20/20 C); b.p. 150-151 C; flash pt. 56.6 C; nonionic
Toxicology: Toxic
Precaution: Moderate fire risk
NFPA: Health 0, Flammability 2, Reactivity 0
Uses: Stabilizer for chlorinated organic compds.; intermediate; solvent lubricant; glass cleaner additive; surfactant, wetting agent for paper coatings, inks, floor polishes, electroplating, glass cleaning; cleaner in silicon wafer industry; in food-pkg. adhesives
Regulatory: FDA 21CFR §175.105
Manuf./Distrib.: ABCR; Aldrich; GFS; ICN Biomed. Research Prods.; Lancaster Synthesis; Pfaltz & Bauer; TCI Am.

3,5-Dimethyl-1-hexyn-3-ol. *See* Dimethyl hexynol
Dimethylhydantoin; 5,5-Dimethyl hydantoin. *See* DM hydantoin
Dimethylhydroxybenzene. *See* Xylenol
N,N-Dimethyl-2-hydroxyethylamine. *See* Dimethylethanolamine

1,2-Dimethyl imidazole

Synonyms: DMI
Properties: M.w. 96.13; dens. 1.080 g/ml; visc. 1.59 cp (40 C); b.p. 204 C; flash pt. (CC) 198 F; surf. tens. 38.60 dynes/cm (40 C)
Uses: Solvent

5,5-Dimethyl-2,4-imidazolidinedione. *See* DM hydantoin

Dimethyl imidazolidinone

CAS 80-73-9; EINECS/ELINCS 201-304-8
Synonyms: N,N′-Dimethylethyleneurea; 1,3-Dimethyl-2-imidazolidinone; 1,3-Dimethylimidazolidin-2-one; DMI; 2-Imidazolidinone, 1,3-dimethyl-
Classification: Heterocyclic organic compd.
Empirical: $C_5H_{10}N_2O$
Properties: M.w. 114.15; dens. 1.0519 g/ml; m.p. 8.2 C; b.p. 220 C (754 mm); ref. index 1.4720 (20 C); dielec. const. 37.6
Toxicology: LD50 (IP, mouse) 2840 mg/kg
Uses: Humectant, hair conditioner in cosmetics; replacement solvent for HMPA in organic synthesis
Manuf./Distrib.: Aldrich; Fluka

1,3-Dimethyl-2-imidazolidinone; 1,3-Dimethylimidazolidin-2-one. *See* Dimethyl imidazolidinone

Dimethyl isosorbide

CAS 5306-85-4; EINECS/ELINCS 226-159-8
Synonyms: 1,4:3,6-Dianhydro-2,5-di-O-methyl-D-glucitol; 2,5-Di-o-methyl-1,4:3,6-dianhydro-D-glucitol; DMI; D-Glucitol, 1,4:3,6-dianhydro-2,5-di-o-methyl; Isosorbide dimethyl ether
Definition: Dimethyl ether of an anhydride of an isomer of sorbitol
Empirical: $C_8H_{14}O_4$
Properties: Wh. to ylsh. visc. liq.; m.w. 174.20; dens. 1.150; b.p. 93-95 C (0.1 mm); flash pt. 108 C; ref. index 1.4610; nonionic
Uses: Surfactant, carrier, solvent, emollient, solubilizer for cosmetic and pharmaceutical sol'ns.
Manuf./Distrib.: Aldrich; EMS-Dottikon AG; Fluka; Sigma
Trade Names: Arlasolve® DMI

Dimethylketal; Dimethylketone. *See* Acetone
N,N-Dimethylmethanamide. *See* Dimethyl formamide
Dimethylmethane. *See* Propane

Dimethyl methylphosphonate

CAS 756-79-6; EINECS/ELINCS 212-052-3
Synonyms: DMMP; Methylphosphonic acid, dimethyl ester; Phosphonic acid, methyl-, dimethyl ester
Empirical: $C_3H_9O_3P$
Formula: $CH_3P(O)(OCH_3)_2$
Properties: Clear liq.; misc. with water and most org. solvs.; m.w. 124.09; dens. 1.145; f.p. < -50 C; b.p. 181 C
Toxicology: LD50 (oral, rat) 8210 mg/kg, (IV, rat) 1050 mg/kg; mod. toxic by IV route; mutagenic data; suspected carcinogen; experimental reproductive effects; experimental nerve gas stimulant; TSCA listed
Hazardous Decomp. Prods.: Heated to decomp., emits toxic fumes of PO_x
Storage: Hygroscopic
Uses: Flame retardant for rigid urethane foams, unsat. polyester resins, waterborne latexes; visc. depressant and flame retardant in ATH-filled polyester resins; intermediate for other flame retardants; as solvent
Manuf./Distrib.: Aceto; Akzo Nobel; Aldrich; Fluka; Focus; Rhodia/Phosphorus Perf. Derivs.
Trade Names: Amgard® DMMP; Antiblaze® DMMP

Dimethyl monosulfide. *See* Dimethyl sulfide
4,4-Dimethyl-2-neopentyl-1-pentene. *See* 2-(2,2-Dimethylpropyl)-4,4-dimethylpent-1-ene
Dimethylnitromethane. *See* 2-Nitropropane

Dimethylnitrosamine; N,N-Dimethylnitrosamine; Dimethylnitrosoamine. *See* N-Nitrosodimethylamine
2,8-Dimethyl-5-nonanone. *See* Diisoamyl ketone

Dimethyl octanamide

CAS 1118-92-9; EINECS/ELINCS 214-272-5
Synonyms: N,N-Dimethyl caprylamide; N,N-Dimethyl octanamide
Uses: Emulsifier, solvent, cosolvent in pesticides
Trade Names Containing: Hallcomid® M-8-10

N,N-Dimethyl octanamide. *See* Dimethyl octanamide
Dimethyl octane-1,8-dicarboxylate. *See* Dimethyl sebacate
3,6-Dimethyl-octin-4-diol(3,6); 3,6-Dimethyl-octyn-4-diol-(3,6). *See* Dimethyl octynediol

Dimethyl octynediol

CAS 78-66-0; 1321-87-5; EINECS/ELINCS 201-131-8
Synonyms: 3,6-Dimethyl-octin-4-diol(3,6); 3,6-Dimethyl-octyn-4-diol-(3,6); 3,6-Dimethyl-4-octyne-3,6-diol; 3,6-Dimethyloct-4-yne-3,6-diol; 4-Octyn-3,6-diol, 3,6-dimethyl-; Octynediol, dimethyl-; 4-Octyne-3,6-diol, 3,6-dimethyl
Classification: Aliphatic alcohol
Empirical: $C_{10}H_{18}O_2$
Formula: $C_2H_5(CH_3)COHC:CCOH(CH_3)C_2H_5$
Properties: Wh. cryst.; very sol. in acetone, alcohol, benzene; mod. sol. in water; sl. sol. in kerosene; m.w. 170.35; m.p. 55 C; b.p. 222 C; f.p. > 110 C; nonionic
Toxicology: LD50 (oral, mouse) 825 mg/kg; mod. toxic by ing.; TSCA listed
Precaution: Combustible
Hazardous Decomp. Prods.: Heated to decomp., emits acrid smoke and irritating vapors
Uses: Emulsifier, solvent in cosmetics; solubilizer, clarifier in shampoos; developer compds.; visc. reducer; defoamer in electroplating baths; defoamer, wetting agent in pesticides; in food-pkg. adhesives
Regulatory: FDA 21CFR §175.105
Manuf./Distrib.: Aldrich

3,6-Dimethyl-4-octyne-3,6-diol; 3,6-Dimethyloct-4-yne-3,6-diol. *See* Dimethyl octynediol
Dimethylol dimethyl hydantoin; 1,3-Dimethylol-5,5-dimethyl hydantoin; Dimethylol-5,5-dimethylhydantoin. *See* DMDM hydantoin

Dimethyl oleamide

Properties: Nonionic
Uses: Solubilizer, solvent, dispersant, wetting agent, modifier for coatings

2,2-Dimethyl-4-oxymethyl-1,3-dioxolane. *See* 2,2-Dimethyl-1,3-dioxolane-4-methanol
2,2-Dimethylpentane; Dimethyl-2,2-pentane. *See* Neoheptane

2,3-Dimethylpentane

CAS 565-59-3
UN 1206
Synonyms: Pentane, 2,3-dimethyl-
Classification: Sat. aliphatic hydrocarbon
Empirical: C_7H_{16}
Formula: $CH_3CH(CH_3)CH(CH_3)CH_2CH_3$
Properties: Colorless liq.; gasoline-like odor; sol. in most org. solvs.; insol. in water; m.w. 100.2; sp.gr. 0.699 (20 C); m.p. -135 C; b.p. 90 C; flash pt. (CC) -6 C
Toxicology: ACGIH TLV/TWA 400 ppm; STEL 500 ppm; may cause mild skin irritation; conc. vapor may cause sl. eye irritation; conc. vapor may cause sl. nose/throat irritation, lightheadedness, dizziness, nausea, difficulty breathing; very high concs. may cause unconsciousness, death; ing. may cause nausea, vomiting, swelling of abdomen, headache, depression; relatively low oral toxicity unless aspirated into lungs
Precaution: Flamm.; very dangerous fire hazard exposed to heat, flame, oxidizers; incompat. with strong oxidizing agents (increases fire and explosion hazard)
Hazardous Decomp. Prods.: Heated to decomp., emits acrid smoke and fumes
NFPA: Health 0, Flammability 3, Reactivity 0
Storage: Store in cool, dry, well-ventilated area out of direct sunlight
Uses: Ingred. in gasoline, aviation fuels; solvent; org. synthesis
Manuf./Distrib.: Aldrich

2,4-Dimethylpentane

CAS 108-08-7
UN 1206
Synonyms: Diisopropylmethane; Pentane, 2,4-dimethyl-
Classification: Sat. aliphatic hydrocarbon
Empirical: C_7H_{16}
Formula: $(CH_3)_2CHCH_2CH(CH_3)_2$
Properties: Colorless liq.; gasoline-like odor; sol. in most org. solvs. (alcohol); insol. in water; m.w. 100.2; sp.gr. 0.677 (20 C); m.p. -119 C; b.p. 81 C; flash pt. (CC) -12 C; ref. index 1.382 (20 C)
Toxicology: ACGIH TLV/TWA 400 ppm; STEL 500 ppm; may cause mild skin irritation; conc. vapor may cause sl. eye irritation; conc. vapor may cause sl. nose/throat irritation, lightheadedness, dizziness, nausea, difficulty breathing; very high concs. may cause unconsciousness, death; ing. may cause nausea, vomiting, swelling of abdomen, headache, depression; relatively low oral toxicity unless aspirated into lungs
Precaution: Flamm.; dangerous fire risk; incompat. with strong oxidizing agents (increases fire and explosion hazard)
Hazardous Decomp. Prods.: None
NFPA: Health 0, Flammability 3, Reactivity 0
Storage: Store in cool, dry, well-ventilated area out of direct sunlight
Uses: Ingred. in gasoline, aviation fuels; solvent; org. synthesis
Manuf./Distrib.: Aldrich; Fluka; ICN Biomed. Research Prods.; Indofine; Lancaster Synthesis; TCI Am.

3,3-Dimethylpentane

CAS 562-49-2; EINECS/ELINCS 209-230-8
UN 1206
Synonyms: Pentane, 3,3-dimethyl-
Classification: Sat. aliphatic hydrocarbon
Empirical: C_7H_{16}
Formula: $CH_3CH_2C(CH_3)_2CH_2CH_3$
Properties: Colorless liq.; gasoline-like odor; sol. in most org. solvs.; insol. in water; m.w. 100.2; sp.gr. 0.693 (20 C); vapor pressure 2.77 psi (37.7 C); m.p. -135 C; b.p. 86 C; flash pt. (CC) -6 C; ref. index 1.39
Toxicology: ACGIH TLV/TWA 400 ppm; STEL 500 ppm; may cause mild skin irritation; conc. vapor may cause sl. eye irritation; conc. vapor may cause sl. nose/throat irritation, lightheadedness, dizziness, nausea, difficulty breathing; very high concs. may cause unconsciousness, death; ing. may cause nausea, vomiting, swelling of abdomen, headache, depression; relatively low oral toxicity unless aspirated into lungs
Environmental: Very toxic to aquatic organisms; may cause long-term adverse effects in aquatic environment
Precaution: Flamm.; LEL 1%; UEL 7%; incompat. with strong oxidizing agents (increases fire and explosion hazard); forms explosive mixts. in air; vapor may flash back; container explosion may occur under fire conditions; may develop pressure
Hazardous Decomp. Prods.: CO, CO_2
NFPA: Health 0, Flammability 2, Reactivity 0
Storage: Store in cool, dry, well-ventilated area out of direct sunlight, away from heat, sparks, open flame; keep tightly closed
Uses: Ingred. in gasoline, aviation fuels; solvent; org. synthesis
Manuf./Distrib.: Acros Org.; Aldrich; Alfa Aesar; Fluka

Dimethyl pentanedioate. *See* Dimethyl glutarate
2,4-Dimethylpentan-3-one; 2,4-Dimethyl-3-pentanone. *See* Diisopropyl ketone
Dimethylphenol. *See* Xylenol
2,3-Dimethylphenol. *See* 2,3-Xylenol
2,5-Dimethylphenol. *See* 2,5-Xylenol
2,6-Dimethylphenol. *See* 2,6-Xylenol
3,4-Dimethylphenol. *See* 3,4-Xylenol
3,6-Dimethylphenol. *See* 2,5-Xylenol
4,5-Dimethylphenol. *See* 3,4-Xylenol

Dimethylphenylamine; N,N-Dimethylphenylamine. *See* N,N-Dimethylaniline
N-[2-[(2,6-Dimethylphenyl) amino]-2-oxoethyl]-N,N-diethylbenzenemethanaminium benzoate. *See* Denatonium benzoate
4-(Dimethylphenylmethyl) phenol. *See* Cumyl phenol

Dimethyl phthalate

CAS 131-11-3; EINECS/ELINCS 205-011-6
Synonyms: 1,2-Benzenedicarboxylic acid dimethyl ester; Dimethyl 1,2-benzenedicarboxylate; Dimethyl benzeneorthodicarboxylate; DMP; Methyl phthalate; Phthalic acid, dimethyl ester
Definition: Diester of methyl alcohol and phthalic acid
Empirical: $C_{10}H_{10}O_4$
Formula: $C_6H_4(CO_2CH_3)_2$
Properties: Colorless oily liq., sl. aromatic odor; misc. with alcohol, ether, oxygenated and chlorinated solvs.; sl. sol. in min. oil; insol. in water, paraffinic hydrocarbons; m.w. 194.08; dens. 1.189 (25/25 C); visc. 17.2 cps; m.p. 5.5 C; b.p. 282.0 C; flash pt. 149 C; ref. index 1.5138 (25 C)
Toxicology: ACGIH TLV/TWA 5 mg/m^3; LD50 (oral, rat) 6.9 ml/kg, (IP, rat) 3375 mg/kg; LDLo (subcut., mouse) 6500 mg/kg; mod. toxic by ing. and IP routes; mildly toxic by inh.; eye irritant; experimental teratogen, reproductive effects; mutagenic data
Precaution: Combustible; can react with oxidizers
Hazardous Decomp. Prods.: Heated to decomp., emits acrid smoke and irritating fumes
NFPA: Health 0, Flammability 1, Reactivity 0
Uses: Solvent for resins, cosmetics; plasticizer for cellulose acetate and nitrocellulose lacquers, plastics, rubber; coating agents; safety glass; pesticide; insect repellent; peroxide diluent/phlegmatizing agent; film-former; dye carrier for textile dyeing; fixing agent for fragrances; solvent for adjuvants in food-contact crosslinked polyesters; in food-pkg. adhesives; in acrylic food pkg.
Regulatory: FDA 21CFR §175.105, 177.1010, 177.2420; SARA reportable
Manuf./Distrib.: Aarti Ind. Ltd; Aldrich; Alfa Aesar; Alfa Chem; Allan; Allchem Ind.; Ashland; Atomergic Chemetals; BASF; Celanese; ChemService; Chemial SpA; Eastman; Fluka; GFS; Great Western; Houghton Chem.; J.T. Baker; KIC Chems.; Morflex; Parchem Trading; Penta Mfg.; Proviron Ind. NV; Ruger; Spectrum Quality Prods.; Synasia Fine Chem.; UCB; Unitex; VWR Int'l.; Vopak USA; Whyte Chems. Ltd
Trade Names: Uniplex 110

Dimethylpolysiloxane. *See* Dimethicone
2,2-Dimethyl-1,3-propanediyl dioctanoate; 2,2-Dimethyl-1,3-propanediyl 2-ethylhexanoate. *See* Neopentyl glycol dioctanoate

2,2-Dimethylpropanol

CAS 75-84-3
Synonyms: t-Amyl alcohol; t-Butyl carbinol; 2,2-Dimethyl-1-propanol; 2,2-Dimethyl-1,1-propanol; 2,2-Dimethylpropyl alcohol; Neoamyl alcohol; Neopentanol; Neopentyl alcohol; 1-Propanol, 2,2-dimethyl-
Classification: Primary aliphatic alcohol
Empirical: $C_5H_{12}O$
Formula: $CH_3C(CH_3)_2CH_2OH$
Properties: Volatile cryst.; peppermint odor; sol. in ethanol, ether; sl. sol. in water; m.w. 88.15; sp.gr. 0.812 (20 C); m.p. 53 C; b.p. 114 C; flash pt. (CC) 37 C
Toxicology: Toxic; may cause severe eye irritation, sl. skin irritation; can be absorbed thru skin; inh. may cause nose/throat irritation, headache, dizziness, dyspnea, nausea, vomiting; severe exposures may produce double vision, deafness, delirium, severe nervous symptoms, fatalities; can cause severe lung damage, respiratory/cardiac arrest, or death if aspirated into lungs
Precaution: Flamm. solid; incompat. with oxidizing agents (increases risk of fire/explosion), hydrogen trisulfide (explosive reaction possible)
NFPA: Health 2, Flammability 3, Reactivity 0
Storage: Store in cool, dry, well-ventilated area, out of direct sunlight
Uses: Mfg. of flotation agents, pharmaceuticals, lacquers, rubber, plastics, fruit essences, explosives; component of paint strippers, hydraulic fluids; solvent
Manuf./Distrib.: Advanced Synthesis Tech.; Aldrich; Alfa Aesar; Fluka; ICN Biomed. Research Prods.; Indofine; Lancaster Synthesis; TCI Am.

2,2-Dimethyl-1-propanol; 2,2-Dimethyl-1,1-propanol. *See* 2,2-Dimethylpropanol
1,1-Dimethylpropargyl alcohol; α,α-Dimethylpropargyl alcohol. *See* Methyl butynol
2,2-Dimethylpropyl alcohol. *See* 2,2-Dimethylpropanol

2-(2,2-Dimethylpropyl)-4,4-dimethylpent-1-ene

CAS 141-70-8; EINECS/ELINCS 205-495-9
Synonyms: 4,4-Dimethyl-2-neopentyl-1-pentene; 1,1-Dineopentylethylene; Heptane, 2,2,6,6-tetramethyl-4-methylene; 2-Neopentyl-4-4-dimethylpentene-1; 2,2,6,6-Tetramethyl-4-methyleneheptane
Empirical: $C_{12}H_{24}$
Properties: M.w. 168.32
Uses: Emollient, solvent in cosmetics

Dimethylpropylmethane. *See* Isohexane
1,1-Dimethylpropyl methyl ether. *See* t-Amyl methyl ether
1,1-Dimethylpropynol. *See* Methyl butynol
2,6-Dimethylpyridine; α-α′-Dimethylpyridine. *See* 2,6-Lutidine

Dimethyl sebacate

CAS 106-79-6; EINECS/ELINCS 203-431-4
Synonyms: Decanedioic acid, dimethyl ester; Dimethyl decanedioate; Dimethyl octane-1,8-dicarboxylate; DMS; Methyl sebacate; Sebacic acid, dimethyl ester
Empirical: $C_{12}H_{22}O_4$
Formula: $[(CH_2)_4COOCH_3]_2$
Properties: Water-wh. oily liq. or orange liq. and solid; m.w. 230.29; dens. 0.9896 (20/20 C); m.p. 24.5 C; b.p. ≈ 294 C; flash pt. 145 C
Toxicology: May cause eye/skin/mucous membrane/upper respiratory tract irritation; may be harmful by ing., inh., skin absorp.
Precaution: Combustible; incompat. with strong oxidizing agents
Hazardous Decomp. Prods.: CO, CO_2; emits toxic fumes under fire conditions
NFPA: Health 0, Flammability 1, Reactivity 0
Storage: Store in cool, dry place; keep tightly closed
Uses: Solvent, plasticizer, softener for cellulosic resins, nitrocellulose, vinyl resins, PE, syn. rubber; intermediate; raw material for mfg. of light stabilizers
Manuf./Distrib.: Aceto; Acros Org.; Aldrich; ChemService; Chunking; Cognis; Degussa AG; Fluka; Haltermann Prods. UK; ICN Biomed. Research Prods.; Indofine; Int'l. Paper; Lancaster Synthesis; Penta Mfg.; Richman; Shandong Haihua Tianhe; Sigma; TCI Am.; Unitex

Dimethyl silicone. *See* Dimethicone
Dimethylsiloxane pentamer. *See* Decamethylcyclopentasiloxane

Dimethyl succinate

CAS 106-65-0; EINECS/ELINCS 203-419-9
FEMA 2396
Synonyms: Butanedioic acid, dimethyl ester; Dimethyl butanedioate; Methyl succinate; Succinic acid, dimethyl ester
Empirical: $C_6H_{10}O_4$
Formula: $CH_3OCOCH_2CH_2COOCH_3$
Properties: Colorless liq., ethereal winey odor; sol. 8 g/l in water; m.w. 146.14; dens. 1.119 (20/4 C); m.p. 16-19 C; b.p. 190-193 C; flash pt. 90 C; ref. index 1.419
Toxicology: LD50 (oral, rat) > 5 g/kg, (skin, rabbit) > 5 g/kg; sl. toxic by ing. and skin contact; TSCA listed
Hazardous Decomp. Prods.: Heated to decomp., emits acrid smoke and irritating vapors
Uses: Heat/light stabilizer for polyolefins, ABS polymer systems, flexible PVC, food pkg.; emollient; synthetic flavoring agent in foods and pharmaceuticals; solvent for coatings, cleaners, inks, textile lubricants, urethane prod.; plasticizer for flexible thermoset polyester; polymer intermediate for polyester polyols for urethanes, wet-str. paper resins, polyester resins; specialty chemical intermediate
Features: Fruity flavor
Regulatory: FDA 21CFR §172.515; FEMA GRAS
Manuf./Distrib.: Aldrich; Ashland; DSM Chem. N. Am.; DSM Fine Chems. Austria; DuPont; Fluka; Penta Mfg.; Schweizerhall; Sigma
Trade Names: DBE-4; Santosol® DMS
Trade Names Containing: DBE; DBE-2; DBE-3; DBE-9; Estasol™;

Rhodiasolv RPDE; Santosol® DME; Santosol® DME-1; Santosol® DME-2; Skykleen® 1000

Dimethyl sulfide

CAS 75-18-3; EINECS/ELINCS 200-846-2
UN 1164 (DOT); FEMA 2746
Synonyms: Dimethyl monosulfide; DMS; Methyl sulfide; Methyl thioether; Methylthiomethane; 2-Thiapropane; Thiobismethane; Thiopropane; 2-Thiopropane
Empirical: C_2H_6S
Formula: $(CH_3)_2S$
Properties: Colorless to straw-colored liq., disagreeable cabbage-like odor; sol. in alcohol, ether, oxygenated solvs.; insol. in water; m.w. 62.14; dens. 0.8458; m.p. -83.2 C; b.p. 37.5-38 C; flash pt. -38 C; autoignition temp. 206 C; ref. index 1.435 (20 C)
Toxicology: LD50 (oral, rat) 3300 mg/kg, (IP, mouse) 8 g/kg; LC50 (inh., rat) 40,250 ppm; poison by inh.; mod. toxic by ing.; skin and severe eye irritant; TSCA listed
Precaution: Flamm.; dangerous fire hazard; flamm. limits in air 2.2-19.7%; explosive as vapor; vigorous reactions with oxidizers
Hazardous Decomp. Prods.: Heated to decomp., emits highly toxic fumes of SO_x and may explode
NFPA: Health 1, Flammability 4, Reactivity 0
Uses: Odorant for natural gas; coking suppresser, CO formation control agent for ethylene prod.; dust control agent in steel mill furnaces; sulfiding agent; presulfiding agent for catalysts in refinery processes, petrochemicals; solvent for org. and inorg. materials; catalyst impregnator; chem. intermediate for org. synthesis, agrochems.; synthetic flavoring agent in foods and pharmaceuticals
Regulatory: FDA 21CFR §172.515; FEMA GRAS
Manuf./Distrib.: Advanced BioTech; Aldrich; Alfa Chem; Axxence Aromatic GmbH; ChevronPhillips; Degussa Flavors/Alfrebro; Elan; Epochem; Fleurchem; Fluka; Gaylord; Moore Ingreds.; Penta Mfg.; Sigma; Sithean; Spectrum Quality Prods.; Wutong Aroma

Dimethyl sulfone

CAS 67-71-0; EINECS/ELINCS 200-665-9
Synonyms: $DMSO_2$; Methane, sulfonylbis-; Methylsufone; Methylsulfonyl methane; MSM; Sulfonylbismethane
Classification: Nonaromatic sulfur-oxygen compd.
Definition: Org. compd. found in the adrenal cortex of cattle and primitive plants such as *Equisetum arvense L., Equisetaceae*
Empirical: $C_2H_6O_2S$
Properties: Colorless cryst.; freely sol. in water, methanol, ethanol, acetone; sparingly sol. in ether; m.w. 94.13; m.p. 110 C; b.p. 237 C; flash pt. 290 C; sublimes at 90-100 C (13 mm)
Toxicology: TSCA listed
Uses: Color solvent; solvent in cosmetics; pharmaceutical carrier; gives flexibility to skin; thins blood; carries toxins out of systems
Manuf./Distrib.: Aldrich; Alfa Chem; Cornelius Chem. Co. Ltd; Fluka; Gaylord; Parchem Trading; Richman; Sigma; Sithean; Spectrum Quality Prods.; Xinchem

Dimethyl sulfoxide

CAS 67-68-5; EINECS/ELINCS 200-664-3
Synonyms: DMSO; Methyl sulfinylmethane; Methyl sulfoxide; Sulfinylbis (methane)
Classification: Aliphatic organic compd.
Empirical: C_2H_6OS
Formula: $(CH_3)_2SO$
Properties: Colorless clear liq., strong char. odor (garlic-onion); sol. in water; pract. insol. in alcohol, benzene, acetone, chloroform, ether; m.w. 78.13; dens. 1.101; vapor pressure 0.37 mm Hg (20 C); m.p. 18.4 C; b.p. 189 C; acid no. 0.03 max.; flash pt. 95 C; ref. index 1.4790 (20 C)
Toxicology: LD50 (oral, rat) 17,500 mg/kg, (IP, rat) 8200 mg/kg; poison by ing.; mod. toxic by IV, IP routes; human mutagenic data; little systemic toxicity, but a primary irritant when used topically (may cause contact urticaria, shortness of breath, facial swelling, GI disturbances, headache, nausea, diarrhea, photophobia); experimental tumorigen, teratogen, reproductive effect; TSCA listed
Environmental: VOC; ThOD 1.84
Precaution: Combustible exposed to heat or flame; can react with oxidizing materials; violent or explosive reaction with many acyl, aryl, and nonmetal halides, boron compds., etc.; incompat. with metal oxosalts, sulfur trioxide, magnesium perchlorate
Hazardous Decomp. Prods.: Heated to decomp., emits toxic fumes of SO_x
NFPA: Health 1, Flammability 1, Reactivity 0
Storage: Hygroscopic; store @ R.T.
Uses: Extraction solvent, reaction solvent in many industries, agrochems. (esp. pyrethroid insecticides), hydrocarbon refining, paints/coatings, biocides, polymerization; industrial cleaners; pesticides; paint stripping; analytical reagent; organic synthesis; carrier, solubilizer, solvent for pharmaceuticals; medicine (analgesic, anti-inflammatory); veterinary medicine; solvent in food-contact polysulfone and polyethersulfone resins; hydraulic fluids; plant pathology and nutrition; spinning syn. fibers; acrylic fibers
Regulatory: FDA 21CFR §177.1655, 177.2440; BP, EP compliance
Manuf./Distrib.: AMRESCO; Aldrich; Allchem Ind.; Atofina; Fluka; Gaylord; General Chem.; Howard Hall; Itochu Spec. Chems.; KIC Chems.; Mallinckrodt Baker; Monomer-Polymer & Dajac Labs; Parchem Trading; R.W. Greeff; Research Organics; Romil Ltd; Ruger; Sigma; Spectrum Quality Prods.; Toray Fine Chems.; Whyte Chems. Ltd
Trade Names: DMSO
Trade Names Containing: Decap; Strimix® 841

Dimethyl thiodipropionate

CAS 4131-74-2
Synonyms: Bis (2-methoxycarbonylethyl) sulfide; β,β-Dicarbomethoxydiethyl sulfide; Dimethyl 3,3´-thiodipropionate; DMTDP; Propanoic acid, 3,3´-thiobis-, dimethyl ester; Propionic acid, 3,3´-thiodi-, dimethyl ester; 3,3´-Thiobis (methyl propionate); Thiodipropionic acid, dimethyl ester
Empirical: $C_8H_{14}O_4S$
Formula: $S(CH_2CH_2COOCH_3)_2$
Properties: M.w. 206.26; b.p. 130 C (1-2 mm)
Uses: Reaction intermediate in synthesis of thiodipropionic acid and esters; specialty solvent; antioxidant for elastomers
Manuf./Distrib.: Aldrich; Atofina SA; Hampshire; Lancaster Synthesis; TCI Am.

Dimethyl 3,3´-thiodipropionate. *See* Dimethyl thiodipropionate
2,6-Dimethypyridine. *See* 2,6-Lutidine
Dimeticone. *See* Dimethicone
1,1-Dineopentylethylene. *See* 2-(2,2-Dimethylpropyl)-4,4-dimethylpent-1-ene
Dinkum oil. *See* Eucalyptus globulus oil

Dinonyl phenol

CAS 1323-65-5; EINECS/ELINCS 215-356-4
Synonyms: Phenol, dinonyl-
Definition: Mixture of dialkyl substituted phenols
Empirical: $C_{24}H_{42}O$
Formula: $(C_9H_{19})_2C_6H_3OH$
Properties: Colorless liq.; insol. in water; sol. in org. solvs.
Toxicology: TSCA listed
Uses: Solvent; intermediate for mfg. nonionic surfactants, lube oil additives, petroleum demulsifiers, phenolic resins, plasticizers, and insecticides; in food-pkg. adhesives
Regulatory: FDA 21CFR §175.105
Manuf./Distrib.: Allchem Ind.; C & P Sales; Huntsman

Dioctyl adipate

CAS 103-23-1; EINECS/ELINCS 203-090-1
Synonyms: Adipic acid bis (2-ethylhexyl) ester; Bis (2-ethylhexyl) adipate; Bis (2-ethylhexyl) hexanedioate; Di (2-ethylhexyl) adipate; DOA; Hexanedioic acid, bis (2-ethylhexyl) ester
Classification: Aliphatic organic compd.
Definition: Diester of 2-ethylhexyl alcohol and adipic acid
Empirical: $C_{22}H_{42}O_4$
Formula: $[CH_2CH_2COOCH_2CH(C_2H_5)C_4H_9]_2$
Properties: Lt.-colored oily liq.; insol. in water; m.w. 370.64; dens. 0.9268 (20/20 C); b.p. 417 C; flash pt. 196 C; ref. index 1.4472
Toxicology: LD50 (oral, rat) 9110 mg/kg); suspected carcinogen and ter-

atogen; moderately toxic by IV route; mildly toxic by ingestion and skin contact; mutagenic data; eye and skin irritant; TSCA listed
Precaution: Combustible
Hazardous Decomp. Prods.: On decomp., emits acrid smoke and irritating fumes
NFPA: Health 0, Flammability 1, Reactivity 0
Uses: Plasticizer for PVC, rubber, commonly blended with general purpose plasticizers (DOP, DIOP); solvent; aircraft and industrial lubricants; syn. lubricant basestock; vehicle for pigment dispersions in paints; textile spin finishes; pigment wetting agent/dispersant; emollient, moisturizer, pigment dispersant, cosolvent in pharmaceuticals; emollient; film-former; plasticizer in food-contact polymers; in closure-sealing gaskets for food containers; in cellophane for food pkg.; plasticizer in food-contact rubber articles for repeated use
Regulatory: FDA 21CFR §175.105, 175.300, 177.1200, 177.1210, 177.1400, 177.2600, 178.3740; FDA approved for injectables
Manuf./Distrib.: A&E Connock; Aldrich; Allchem Ind.; Ashland; Avatar; BASF; Bencorp Int'l.; Boehme Filatex; ChemService; Chisso Am.; Coyne; DSM Fine Chems. Austria; Degussa AG; Eastech; Eastman; Esprit; Fluka; Haltermann Prods. UK; Houghton Chem.; Inolex; Monsanto; NorChem; Parchem Trading; Shandong Haihua Tianhe; Sigma; Solutia; St. Lawrence; Stanchem; Storchem; Unitex; Velsicol
Trade Names: HallStar™ DOA; Wickenol® 158

Dioctyl-o-benzenedicarboxylate. *See* n-Dioctyl phthalate

Dioctyl ether

CAS 629-82-3; EINECS/ELINCS 211-112-6
Synonyms: Caprylic ether; Dicapryl ether; Dicaprylyl ether; Di-n-octyl ether; Ether, di-n-octyl; Octane, 1,1´-oxybis-; Octyl ether; 1,1´-Oxybisoctane
Empirical: $C_{16}H_{34}O$
Formula: $(C_8H_{17})_2O$
Properties: Colorless liq.; m.w. 242.50; dens. 0.805 (17/4 C); f.p. -7.6 C; b.p. 291.7 C; flash pt. > 100 C; ref. index 1.4329 (24 C)
Toxicology: LD50 (IV, mouse) 1183 mg/kg; mod. toxic by IV route; may be harmful by ing., inh., skin absorp.; may cause eye/skin irritation; TSCA listed
Precaution: Combustible; incompat. with strong oxidizing agents; can react vigorously with oxidizers; forms explosive peroxides on prolonged storage
Hazardous Decomp. Prods.: Toxic fumes of CO, CO_2; heated to decomp., emits acrid smoke and irritating fumes
NFPA: Health 0, Flammability 1, Reactivity 0
Storage: Store in cool, dry place; keep tightly closed
Uses: Electrical insulator; water repellent; mold lubricant; antistat; intermediate; dry feel emollient for cosmetic/pharmaceutical emulsions
Manuf./Distrib.: Aldrich; Fluka
Trade Names: Texaprint® SDCE

Di-n-octyl ether. *See* Dioctyl ether

Dioctylhexylphosphine oxide

CAS 31160-66-4
Synonyms: Dioctylmonohexylphosphine oxide; Hexyldioctylphosphine oxide
Classification: Phosphine
Empirical: $C_{22}H_{47}OP$
Properties: M.w. 358.59
Uses: Extractant
Trade Names Containing: Cyanex® 923

Dioctyl maleate

CAS 2915-53-9; EINECS/ELINCS 220-835-6
Synonyms: Bis (1-octyl) maleate; 2-Butenedioic acid, (Z)-, dioctyl ester; Dicaprylyl maleate; Di-N-octyl maleate; DOM; Maleic acid, dioctyl ester
Definition: Diester of capryl alcohol and maleic acid
Empirical: $C_{20}H_{36}O_4$
Formula: $CH_3(CH_2)_7OCOCH{=}CHCOO(CH_2)_7CH_3$
Properties: Colorless to lt. yel. hazy liq.; mild odor; sol. < 1 g/l in water (20 C); m.w. 340.56; sp.gr. 0.94 (20 C); m.p. -20 C; b.p. 229-239 C; flash pt. (COC) 160 C; ref. index ≈ 1.45
Toxicology: LD50 (oral, rat) 14,200 mg/kg, (skin, rabbit) 14 g/kg; mildly toxic by ing.; TSCA listed
Precaution: Combustible
Hazardous Decomp. Prods.: Heated to decomp., emits acrid smoke and fumes
Uses: Intermediate; emollient in cosmetics; fragrance coupling agent; solubilizer for benzophenone-3; antitackifier in antiperspirants; gloss aid, conditioner for hair care; in copolymerization of PVC and vinyl acetates; modifier, plasticizer for vinyl resins; internal plasticizer in copolymer films; in latex paints, paper and textile coatings, adhesives; surface-active agent; cleanser and wax solvent
Manuf./Distrib.: Allchem Ind.; BASF; Celanese; ChemService; Chemial SpA; DSM Fine Chems. Austria; Filo; Finetex; Haltermann Prods. UK; Monomer-Polymer & Dajac Labs; NOF; Sunoco; Unitex; Velsicol; Whyte Chems. Ltd
Trade Names: Bernel® Ester DCM

Di-N-octyl maleate. *See* Dioctyl maleate
Dioctylmonohexylphosphine oxide. *See* Dioctylhexylphosphine oxide

Dioctyl octylphosphonate

CAS 126-63-6
Synonyms: Bis (2-ethyl-1-hexyl)-2-ethyl-1-hexyl phosphonate; Bis (2-ethylhexyl) 2-ethylhexyl phosphonate; Di (2-ethylhexyl)-2-ethylhexyl phosphonate; Phosphonic acid, (2-ethylhexyl)-, bis (2-ethylhexyl) ester
Empirical: $C_{24}H_{51}O_3P$
Formula: $C_8H_{17}PO(OC_8H_{17})_2$
Properties: Colorless liq.; mild odor; misc. with most common org. solvs.; insol. in water; m.w. 418.64; dens. 0.908 (20/4 C); b.p. 160-161 C (0.26 mm); flash pt. 215 C
Uses: Extraction solvent for mineral processing; solvent separation; gasoline additive; antifoam; plasticizer; stabilizer; textile conditioner and antistat
Manuf./Distrib.: ICN Biomed. Research Prods.

1-(Dioctyl-phosphinoyl)-octane. *See* Trioctylphosphine oxide

Dioctyl phosphite

Synonyms: Dioctyl phosphonate
Empirical: $C_{16}H_{35}O_3P$
Formula: $(C_8H_{17}O)_2P(OH)H$
Properties: Water-wh. liq.; sol. in common org. solvs.; dens. 0.929; b.p. 150-155 C (2-3 mm); ref. index 1.4418
Uses: Antioxidant for PP; solvent; intermediate; lubricant additive; adhesive
Features: Nonstaining
Manuf./Distrib.: GE Spec.; Rhodia/Phosphorus Perf. Derivs.

Dioctyl phosphonate. *See* Dioctyl phosphite
Dioctyl phthalate. *See* n-Dioctyl phthalate

n-Dioctyl phthalate

CAS 117-84-0; EINECS/ELINCS 204-211-0
Synonyms: 1,2-Benzenedicarboxylic acid dioctyl ester; o-Benzenedicarboxylic acid dioctyl ester; Di-n-alkyl (C7-C9) phthalate; Dioctyl-o-benzenedicarboxylate; Dioctyl phthalate; Di-n-octyl phthalate; DNOP; DOP; Octyl phthalate; n-Octyl phthalate; Phthalic acid, dioctyl ester
Empirical: $C_{24}H_{38}O_4$
Formula: $C_6H_4[COOCH_2CH(C_2H_5)C_4H_9]_2$
Properties: Lt.-colored liq., odorless, bitter taste; misc. with min. oil; insol. in water; m.w. 390.62; dens. 0.9861 (20/20 C); b.p. 231 C (5 mm); flash pt. 218 C; ref. index 1.4836
Toxicology: TLV 5 mg/m^3; STEL 10 mg/m^3; LD50 (oral, rat) 47 g/kg, (IP, rat) > 50 ml/kg; mildly toxic by ing.; skin and severe eye irritant; suspected human carcinogen; experimental teratogen; affects human GI tract; experimental teratogen, reproductive effects; TSCA listed
Precaution: Combustible
Hazardous Decomp. Prods.: Heated to decomp., emits acrid smoke and irritating fumes
NFPA: Health 0, Flammability 1, Reactivity 0
Uses: Plasticizer, softener for many resins (esp. PVC), vinyl-coated fabrics, vinyl flooring, vinyl plastisols, elastomers, NC in lacquers, inks (flexographic, screen, aq., gravure), in food-pkg. adhesives; film-former in cosmetics; syn. lubricant basestock; lubricant for melamine-formal-

dehyde resins for food pkg.; lubricant for metalworking fluids, rolling, stamping and drawing oils, greases, motor oils, extrusion of ceramics; fatting agent in textile and leather auxs.; plasticizer in food-contact rubber articles for repeated use; solvent, fixing agent in pharmaceuticals

Regulatory: FDA 21CFR §175.105, 177.1460, 177.2600

Manuf./Distrib.: AAE Chemie NV; Aldrich; Allchem Ind.; Ashland; BASF; BCH Brühl; Bencorp Int'l.; C.P. Hall; Celanese; ChemService; Chemisphere Ltd; Chisso Am.; Coyne; Degussa AG; Eastman; Farabi Petrochem.; Fluka; Great Western; Houghton Chem.; I.C. Trading; KIC Chems.; Parchem Trading; Primachem; Shandong Haihua Tianhe; Sigma; Southern Texas Chem.; Spectrum Quality Prods.; Stanchem; Storchem; Sunoco; UCB; Velsicol; Vopak USA; Westtec Ind.

Di-n-octyl phthalate. *See* n-Dioctyl phthalate

Dioctyl sebacate

CAS 122-62-3; EINECS/ELINCS 204-558-8

Synonyms: Bis (2-ethylhexyl) decanedioate; Bis (2-ethylhexyl) sebacate; Decanedioic acid, bis (2-ethylhexyl) ester; Di-2-ethylhexyl sebacate; DOS; 2-Ethylhexyl sebacate; 1-Hexanol, 2-ethyl-, sebacate; Octyl sebacate; Sebacic acid, bis (2-ethylhexyl) ester

Empirical: $C_{26}H_{50}O_4$

Formula: $(CH_2)_8(CH_2COOC_8H_{17})_2$

Properties: Pale straw-colored oily liq.; mild odor; insol. in water; m.w. 426.68; dens. 0.91 (25 C); m.p. -48 C; f.p. -55 C; b.p. 248 C (4 mm); flash pt. (COC) 210 C; ref. index 1.447 (28 C)

Toxicology: LD50 (oral, rat) 12,800 mg/kg, (oral, mouse) 9500 mg/kg, (IV, rat) 900 mg/kg; mod. toxic by ing. and IV routes; TSCA listed

Precaution: Combustible exposed to heat or flame; can react with oxidizers

Hazardous Decomp. Prods.: Heated to decomp., emits acrid and irritating fumes

Uses: Plasticizer for PVC, PS, cellulosics; lubricant for metalworking fluids, rolling, stamping and drawing oils, greases, motor oils, extrusion of ceramics; solvent; grease; emollient; film-former; pigment wetting agent/dispersant; fatting agent in textile and leather auxs.; plasticizer in food-contact rubber articles for repeated use; surf. lubricant in mfg. of food-contact coatings and metallic articles; in food-pkg. adhesives; in closure-sealing gaskets for food containers

Regulatory: FDA 21CFR §175.105, 175.300, 177.1210, 177.2600, 178.3910

Manuf./Distrib.: A&E Connock; Aldrich; Allchem Ind.; Alzo; Ashland; Avatar; Boehme Filatex; C.P. Hall; ChemService; DSM Fine Chems. Austria; Dastech Int'l.; Fluka; Haltermann Prods. UK; Hatco; Int'l. Paper; Jarchem Ind.; Kyowa Hakko Kogyo; Morflex; New Japan Chem.; Penta Mfg.; Phoenix; PolyOne/Elastomers & Perf. Addit.; Shandong Haihua Tianhe; Sigma; St. Lawrence; Unitex; Velsicol

Trade Names: HallStar™ DOS

Diolamine. *See* Diethanolamine

1,4-Dioxacyclohexane. *See* 1,4-Dioxane

1,3-Dioxacyclopentane. *See* 1,3-Dioxolane

5,8-Dioxadodecane. *See* Ethylene glycol dibutyl ether

3,6-Dioxa-1-dodecanol; 3,6-Dioxadodecanol-1. *See* Diethylene glycol n-hexyl ether

2,5-Dioxahexane. *See* Ethylene glycol dimethyl ether

Dioxan; 1,4-Dioxan; Dioxane. *See* 1,4-Dioxane

1,4-Dioxane

CAS 123-91-1; EINECS/ELINCS 204-661-8

UN 1165 (DOT)

Synonyms: Diethylene dioxide; 1,4-Diethylene dioxide; Diethylene ether; Di (ethylene oxide); 1,4-Diethylene oxide; 1,4-Dioxacyclohexane; Dioxan; 1,4-Dioxan; Dioxane; Dioxane-1,4; p-Dioxane; p-Dioxin, tetrahydro-; Dioxyethylene ether; Ethylene glycol ethylene ether; Glycol ethylene ether; Tetrahydro-1,4-dioxin; Tetrahydro-p-dioxin

Classification: Sat. cyclic aliphatic ether

Empirical: $C_4H_8O_2$

Formula: $OCH_2CH_2OCH_2CH_2$

Properties: Colorless clear liq., ethereal odor; stable; sol. in water, most org. solvs.; m.w. 88.11; sp.gr. 1.0356 (20/20 C); f.p. 10-12 C; b.p. 101.3 C; flash pt. 18.3 C; autoignition temp. 180 C; ref. index 1.4220

Toxicology: ACGIH TLV/TWA 25 ppm (skin); LD50 (oral, mouse) 5700 mg/kg, (skin, rabbit) 7600 mg/kg; poison by IP route; mod. toxic by inh., ing., absorbed by skin; eye and skin irritant; inh. may cause nose/throat irritation, tearing, and burning; coughing, nausea, vomiting, headache, drowsiness, stomach pains; repeated exposure may cause severe injury or death; target organs: liver, kidneys; carcinogen; tumorigen; teratogen; mutagen; TSCA listed

Environmental: VOC; BOD5 0; ThOD 1.09

Precaution: DOT: Flamm. liq.; dangerous fire risk; stable if properly inhibited; may form explosive peroxides on exposure to air/moisture; incompat. with strong oxidizers, catalytic hydrogen, sulfur trioxide, triethynylaluminum

Hazardous Decomp. Prods.: May form peroxides on prolonged contact with air in the absence of inhibitors; heated to decomp., emits acrid smoke and irritating fumes

NFPA: Health 2, Flammability 3, Reactivity 1

Storage: Store in cool, dry, well-ventilated area, out of direct sunlight; hygroscopic

Uses: Stabilizer for chlorinated hydrocarbons, printing inks; solvent for adhesives, sealants, fats, oils, waxes, dyes, cellulosics, resins, lacquers, paints, protective coatings, pharmaceuticals, natural and syn. rubbers, dry cleaning, metal surf. finishes, solvent extraction, transistors; wetting agent, dispersant for textile processing; dyeing, staining, and printing adjuvant; cleaning and detergent preps.; scintillation fluids; agrochems.; in food-pkg. adhesives

Regulatory: FDA 21CFR §175.105; SARA §313 reportable; CERCLA hazardous substance; Calif. Prop. 65 reportable; HAP

Manuf./Distrib.: Aldrich; Alemark; Allchem Ind.; Amber Syn.; Ashland; BASF; Bencorp Int'l.; Brand-Nu Labs; CPS; Columbia Sales Int'l.; Dow; Ferro/Grant; Filo; Fluka; Houghton Chem.; J.T. Baker; Mallinckrodt Baker; Maypro Ind.; Mercian; Nisso Petrochem. Ind.; Pechiney Chems. Div.; Romil Ltd; Sigma; Spectrum Quality Prods.; Toho Chem. Ind.; Vopak USA; Whyte Chems. Ltd

Dioxane-1,4; p-Dioxane. *See* 1,4-Dioxane

3,6-Dioxaoctane-1,8-diol. *See* Triethylene glycol

Dioxapentane. *See* Methylal

3,5-Dioxapheptene. *See* Diethoxymethane

4,7-Dioxaundecan-2-ol. *See* 1-Butoxyethoxy-2-propanol

1,1-Dioxide tetrahydrothiofuran; 1,1-Dioxidetetrahydrothiophene. *See* Sulfolane

p-Dioxin, tetrahydro-. *See* 1,4-Dioxane

Dioxolan. *See* 2,2-Dimethyl-1,3-dioxolane-4-methanol

1,3-Dioxolan; Dioxolane. *See* 1,3-Dioxolane

1,3-Dioxolane

CAS 646-06-0

UN 1166 (DOT)

Synonyms: Dihydro-1,3-dioxole; 1,3-Dioxacyclopentane; 1,3-Dioxolan; Dioxolane; Ethylene glycol formal; Formal glycol; Glycoformal; Glycol formal; Glycol methylene ether

Classification: Sat. cyclic aliphatic ether

Empirical: $C_3H_6O_2$

Formula: $-CH_2OCH_2OCH_2-$ (5-membered ring)

Properties: Water-wh. liq.; ether-like odor; sol. in water, ethanol, ether, acetone, other org. solvs.; m.w. 74.09; dens. 1.067 (20 C); vapor pressure 70 mm Hg (20 C); m.p. -95 C; b.p. 74 C; flash pt. 2 C

Toxicology: LD50 (oral, rat) 3000 mg/kg, (skin, rabbit) 8480 mg/kg; mod. toxic by ing. and IP; mildly to mod. toxic by inh., skin contact; skin and severe eye irritant; may be absorbed thru skin; vapor may be very irritating to eyes; inh. can probably cause nose/throat/respiratory tract irritation, CNS depression (headache, nausea, drowsiness); ing. may cause mouth/throat irritation and CNS effects; mutagenic data; TSCA listed

Precaution: Flamm.; very dangerous fire hazard exposed to heat or flame; incompat. with acids (may form peroxides), oxidizers (may react vigorously and increase risk of fire/explosion); potentially explosive reaction with lithium perchlorate

Hazardous Decomp. Prods.: May form peroxides on prolonged exposure to air, in the absence of inhibitors; heated to decomp., emits acrid smoke and irritating fumes

NFPA: Health 2, Flammability 3, Reactivity 2

Storage: Store in a cool, dry, well-ventilated area out of direct sunlight, away from corrosives
Uses: Low-boiling solvent and extractant for oils, waxes, dyes, cellulose derivs.; crosslinking agent for phenolic novolac resins; in lithium batteries; produces brightening agent for nickel electroplating; mfg. of semipermeable membranes; formulating electrolyte sol'ns. for elec. capacitors; octane booster in gasoline; solubilizer in phenolic resin foams; when nitrated, as plasticizer for propellants and explosives; stabilizer for chlorinated solvs.; absorp. agent for NO_x
Manuf./Distrib.: ABCR; Acros Org.; Aldrich; Alfa Aesar; Amber Syn.; Ferro/Grant; Filo; Fluka; Lancaster Synthesis; Pfaltz & Bauer; Spectrum Quality Prods.; TCI Am.; Whyte Chems. Ltd

1,3-Dioxolane-4-methanol. *See* Glyceryl formal
1,3-Dioxolan-2-one. *See* Ethylene carbonate
1,3-Dioxolan-2-one, 4-ethyl. *See* 1,2-Butylene carbonate
1,3-Dioxolan-2-one, 4-methyl. *See* Propylene carbonate
Dioxolone-2. *See* Ethylene carbonate
Dioxothiolan; 1,1-Dioxothiolan. *See* Sulfolane
o-Dioxybenzene. *See* Pyrocatechol
Dioxyethylene ether. *See* 1,4-Dioxane
DIPA. *See* Diisopropylamine; Diisopropyl adipate; Diisopropanolamine
Dipanol. *See* dl-Limonene
m-DIPB. *See* 1,3-Diisopropyl benzene
p-DIPB. *See* 1,4-Diisopropyl benzene
Dipentene. *See* l-Limonene
Dipentene. *See* dl-Limonene
Dipentylamine; Di-n-pentylamine. *See* Di-n-amylamine
Dipentyl ether; Di-n-pentyl ether. *See* Diamyl ether
Dipentyl oxalate. *See* Diamyl oxalate
o-Diphenol. *See* Pyrocatechol
Diphenyl. *See* Biphenyl

Diphenyl carbonate

CAS 102-09-0; EINECS/ELINCS 203-005-8
Synonyms: Carbonic acid, diphenyl ester; Phenyl carbonate
Classification: Aromatic esters and lactones
Empirical: $C_{13}H_{10}O_3$
Formula: $(C_6H_5O_2)_2CO_3$
Properties: Wh. cryst. solid; sol. in acetone, hot alcohol, benzene, CCl_4, ether, glac. acetic acid, and other org. solvs.; insol. in water; m.w. 214.23; dens. 1.1215; m.p. 78 C; b.p. 302 C; flash pt. 168 C; pH 6.8-7.2
Toxicology: LD50 (oral, rat) 1500 mg/kg; TDLO (oral, mouse, 78 week intermittent) 28 g/kg, (subcut., mouse) 1000 mg/kg; questionable carcinogen; tumorigen; neoplastigenic data; TSCA listed
Precaution: Stable under normal conditions, but easily hydrolyzed
Hazardous Decomp. Prods.: Heated to decomp., emits acrid smoke and irritating fumes
Uses: Plasticizer; solvent; chem. intermediate for synthesis of polycarbonate resins
Manuf./Distrib.: Aldrich; Fluka; Sigma

Diphenylenemethane. *See* Fluorene
Diphenylene oxide. *See* Dibenzofuran
Diphenyl ether. *See* Diphenyl oxide

Diphenylmethane

CAS 101-81-5; EINECS/ELINCS 202-978-6
Synonyms: Benzene, benzyl-; Benzene, 1,1′-methylenebis-; Benzylbenzene; Ditan; Ditane; Methane, diphenyl-; 1,1′-Methylenebisbenzene; Methylene dibenzene; Phenylbenzyl; Toluene, α-phenyl-
Classification: Aromatic hydrocarbon
Empirical: $C_{13}H_{12}$
Formula: $(C_6H_5)_2CH_2$
Properties: Clear sl. yel. liq. or needles; floral-orange odor; sol. in alcohol, ether, chloroform, hexane, benzene, oxygenated, chlorinated, and hydrocarbon solvs.; insol. in water, liq. ammonia; m.w. 168.24; dens. 1.3421 (solid, 10/4 C), 1.0008 (liq., 26/4 C); vapor pressure 1.33 mbar (77 C); m.p. 26 C; b.p. 264 C; flash pt. 127 C; ref. index 1.57683 (20 C)
Toxicology: May cause eye/skin/GI/respiratory tract irritation, nausea, vomiting, diarrhea; TSCA listed
Precaution: Incompat. with oxidizing agents
Hazardous Decomp. Prods.: CO, CO_2
NFPA: Health 1, Flammability 1, Reactivity 0
Storage: Store in cool, dry, well-ventilated area away from incompat. substances; keep container closed when not in use
Uses: Fragrance, solvent in cosmetics
Manuf./Distrib.: ABCR; Acros Org.; Advanced Synthesis Tech.; Aldrich; Atofina; Fluka; Goldensun Mfg.; Kira; Penta Mfg.

Diphenyl oxide

CAS 101-84-8; EINECS/ELINCS 202-981-2
FEMA 3667
Synonyms: Biphenyl oxide; Diphenyl ether; Geranium crystals; 1,1′-Oxybisbenzene; Phenoxybenzene; Phenyl ether
Classification: Aromatic ether
Empirical: $C_{12}H_{10}O$
Formula: $(C_6H_5)_2O$
Properties: Colorless cryst.; geranium odor; sol. in ethanol, benzene, diethyl ether, glac. acetic acid; misc. with oxygenated solvs.; insol. in water; m.w. 170.21; dens. 1.073;' vapor pressure 0.0213 mm Hg; m.p. 26-30 C; b.p. 259 C; flash pt. 96 C; autoignition temp. 618 C; ref. index 1.5790; surf. tens. 38.82 dynes/cm
Toxicology: ACGIH TLV/TWA 1 ppm; STEL 2 ppm (vapor); LD50 (oral, rat) 3370 mg/kg; mod. toxic by ing.; harmful vapor; avoid contact and inh.; primary skin and eye irritant; prolonged exposure causes liver/spleen/kidney/thyroid damage, GI upset; target organs: eyes, skin, respiratory system, liver, kidneys, spleen, thyroid; TSCA listed
Environmental: VOC; ThOD 2.63
Precaution: Combustible exposed to heat or flame; LEL 0.8%; UEL 1.5%; can react with oxidizing materials
Hazardous Decomp. Prods.: Heated to decomp., emits acrid smoke and irritating fumes
NFPA: Health 1, Flammability 1, Reactivity 0
Uses: Perfume ingred., esp. for soaps and detergents; cosmetics ingred.; heat-transfer medium; chemical intermediate for halogenation, acylation, alkylation, etc.; flavoring agent in foods and pharmaceuticals; in org. synthesis; carrier for dyeing; solvent
Regulatory: FEMA GRAS
Manuf./Distrib.: ADA Int'l.; Advanced Synthesis Tech.; Aldrich; Berje; D&O; Fluka; Haltermann GmbH; Herdillia Chems. Ltd; Monsanto; Penta Mfg.; Polarome Int'l.; Stanchem

Diphenyl sulfone

CAS 127-63-9; EINECS/ELINCS 204-853-1
Synonyms: DPS; Phenyl sulfone; Sulfobenzide; 1,1′-Sulfonylbisbenzene
Classification: Aromatic sulfur-oxygen compd.
Empirical: $C_{12}H_{10}O_2S$
Properties: Wh. monoclinic prisms or leaflets; sol. in benzene, hot alcohol, aromatic solvs.; sparingly sol. in hot water; insol. in cold water; m.w. 218.28; m.p. 128-129 C; b.p. 232 C (18 mm)
Toxicology: LD50 (oral, rat) > 2 g/kg, (IV, mouse) 320 mg/kg; LC50 (inh., rat, 8 h) > 1700 mg/m³; poison by IV route; low toxicity by ing.; TSCA listed
Hazardous Decomp. Prods.: Heated to decomp., emits toxic fumes of SO_x
Uses: Diluent for polymers, resins; solvent in food-contact PES; ovicide; acaricide; larvicide
Regulatory: FDA 21CFR §177.2440
Manuf./Distrib.: Aarti Ind. Ltd; Aceto; Aldrich; Chugai Boyeki Am.; Fluka; Proviron Ind. NV; Sloss Ind.

Dipotassium carbonate. *See* Potassium carbonate

Di-PPG-3 myristyl ether adipate

Uses: Emollient for sunscreens, makeup, sticks, creams and lotions; skin care vehicle; sunscreen solvent; pigment dispersant/wetting agent
Trade Names: Cromollient DP3A

Dipropanediol dibenzoate. *See* Dipropylene glycol dibenzoate

Dipropylamine

CAS 142-84-7; EINECS/ELINCS 205-565-9
UN 2383 (DOT)
Synonyms: Di-n-propylamine; n-Dipropylamine; DNPA; N-Propyl-1-

propanamine
Classification: Sat. aliphatic amine
Empirical: $C_6H_{15}N$
Properties: Colorless liq.; ammonia-like odor; sol. in water, org. solvs. incl. ethanol, acetone, ether, benzene; m.w. 101.19; dens. 0.738 (20 C); m.p. -63 C; b.p. 109-110 C; flash pt. (OC) 17 C; pH 12 (0.1M)
Toxicology: LD50 (oral, rat) 930 mg/kg, (skin, rabbit) 1250 mg/kg; mod. toxic by ing., skin contact, inh.; skin irritant; ing. may cause burning of mouth, throat, digestive tract, and CNS effects; vapor can irritate nose, throat, lungs; high concs. may cause potentially fatal pulmonary edema; liq. or mist can cause skin irritation or serious burns; lachrymator; liq. or high vapor concs. can cause severe eye irritation, burns, permanent injury; may cause 'halo vision'; TSCA listed
Precaution: Flamm.; dangerous fire risk; strong base; may absorb CO_2 from air to form the carbamate salt; can react with oxidizers
Hazardous Decomp. Prods.: CO_2, CO, NO_x
NFPA: Health 3, Flammability 3, Reactivity 0
Uses: Intermediate; solvent; corrosion inhibitor; mfg. of pesticides
Manuf./Distrib.: Air Prods.; Aldrich; Asiamerica Int'l.; Celanese; Sithean

Di-n-propylamine; n-Dipropylamine. *See* Dipropylamine

Di-n-propyl carbonate
CAS 623-96-1
Properties: M.w. 146.19; dens. 0.940 g/ml; visc. 1.25 cp; b.p. 165-167 C; flash pt. (CC) 147 F; surf. tens. 26.23 dynes/cm
Uses: Solvent

Dipropyl-2,2′-dihydroxyamine. *See* Diisopropanolamine
Dipropylene carbonate. *See* Propylene carbonate

Dipropylene glycol
CAS 110-98-5; EINECS/ELINCS 203-821-4
Synonyms: Bis (hydroxypropyl) ether; Bis (2-hydroxypropyl) ether; 2,2′-Dihydroxydipropyl ether; 2,2′-Dihydroxyisopropyl ether; Diisopropylene glycol; Di-1,2-propylene glycol; DPG; Methyl-2(methyl-2) oxybispropanol; 1,1′-Oxybis-2-propanol; 1,1′-Oxydi-2-propanol; 1,1′-Oxydipropan-2-ol; 2-Propanol, 1,1′-oxydi-
Classification: Nonaromatic ether; glycol
Empirical: $C_6H_{14}O_3$
Formula: $CH_3CHOHCH_2OCH_2CHOHCH_3$
Properties: Colorless sl. visc. liq., nearly odorless; very sol. in diethyl ether; sol. in ethanol, toluene, water; misc. with oxygenated solvs.; m.w. 134.18; dens. 1.023; vapor pressure 1 mm (73.8 C); b.p. 233 C; flash pt. (OC) 124 C; ref. index 1.4410
Toxicology: LD50 (oral, rat) 14,850 mg/kg, (IP, mouse) 10 g/kg, (IV, rat) 5800 mg/kg, (skin, rabbit) > 20 ml/kg; mildly toxic by ing.; primary skin and eye irritant; may cause somnolence, tremors, kidney/ureter/bladder changes; mutagenic data; TSCA listed
Precaution: Combustible when exposed to heat and flame; can react vigorously with oxidizing materials
Hazardous Decomp. Prods.: Heated to decomp., emits acrid smoke and irritating fumes
NFPA: Health 1, Flammability 1, Reactivity 0
Storage: Hygroscopic
Uses: Solvent in hydraulic brake fluids, cutting oils, textile lubricants, printing inks, cosmetics; aromatics extraction solvent; comonomer in unsat. polyester and alkyd resins; surfactant for unsat. polyester resins, paints; fragrance fixative/diluent; petroleum anti-icing additive; PU chain extender; reinforced plastics; plasticizers; in food-pkg. adhesives; in paper/paperboard in contact with aq./fatty foods; defoamer in food-contact coatings; in surf. lubricants for mfg. of food-contact metallic articles
Regulatory: FDA 21CFR §175.105, 176.170, 176.200, 178.3910
Manuf./Distrib.: ABCR; AMC Chems.; Acros Org.; Adept Sol'ns.; Aldrich; Alfa Aesar; Alfa Chem; Allchem Ind.; Arch Chems./Specialty Polyols; Ashland; Avatar; Bayer; Berje; Brown; C.P. Hall; Chemcentral; Coyne; Dow; Eastman; Fluka; G.J. Chem.; Great Western; Houghton Chem.; Huntsman; KIC Chems.; Lyondell; PMC Spec.; Penta Mfg.; Pharmco Prods.; Ruger; Spectrum Quality Prods.; Stanchem; Sunnyside; TCI Am.; Universal Preserv-A-Chem; Vopak USA; Whyte Chems. Ltd
Trade Names: DPG lo; Eastman® DPG, Fragrance Grade

Di-1,2-propylene glycol. *See* Dipropylene glycol

Dipropylene glycol butoxy ether. *See* Dipropylene glycol butyl ether

Dipropylene glycol butyl ether
CAS 29911-28-2 (α-isomer); 35884-42-5 (mixt.); EINECS/ELINCS 249-951-5; 252-776-7
Synonyms: 1-(2-Butoxy-1-methylethoxy)-2-propanol; Butyl dipropasol solvent; Dipropylene glycol butoxy ether; Dipropylene glycol n-butyl ether; Dipropylene glycol monobutyl ether; Dipropylene glycol mono-n-butyl ether; DPGnBE; DPnB; 2-Propanol, 1-(2-butoxy-1-methylethoxy)-
Empirical: $C_{10}H_{22}O_3$
Properties: Colorless liq.; m.w. 190.32; dens. 0.913 g/ml; b.p. 103 C (10 mm Hg); flash pt. 205 F; ref. index 1.426 (20 C)
Toxicology: LD50 (oral, rat) 1620 μl/kg, (skin, rabbit) 5860 μl/kg; mod. toxic by ingestion; primary eye irritant; may be harmful by inh., skin absorp.; may cause skin/mucous membrane/upper respiratory tract irritation; target organ: liver; TSCA listed
Precaution: Incompat. with strong oxidizing agents, strong acids, strong bases
Hazardous Decomp. Prods.: Toxic fumes of CO, CO_2, aldehydes, ketones, org. acid vapors; heated to decomp., emits acrid smoke and irritating fumes; emits toxic fumes under fire conditions
Storage: Store in cool, dry place; keep tightly closed
Uses: Solvent for coating resins, cleaners, cosmetics, agric., electronics, inks, textiles, and adhesives; wetting agent; solubilizer
Features: Slow evaporating
Manuf./Distrib.: Aldrich
Trade Names: Arcosolv® DPNB; Dowanol® DPnB; Solvenon® DPnB

Dipropylene glycol n-butyl ether. *See* Dipropylene glycol butyl ether

Dipropylene glycol t-butyl ether
CAS 132739-31-2
Synonyms: 1-(1,1-Dimethylethoxy-2-methylethoxy)-2-propanol; DPGtBE; DPtB; Propanol, [2-(1,1-dimethylethoxy) methylethoxy]-
Empirical: $C_{10}H_{22}O_3$
Properties: Colorless liq.; mild odor; m.w. 190.3; dens. 0.9 gm/l; vapor pressure 0.07 mm Hg; b.p. 220-222 C; flash pt. 190 F; ref. index 1.424 (20 C)
Toxicology: Irritating to eyes, skin, respiratory system; may be harmful by ing., inh., skin absorp.
Precaution: Combustible liq.; incompat. with strong oxidizing agents, strong acids, copper, copper alloys
Hazardous Decomp. Prods.: CO, CO_2; emits toxic fumes under fire conditions
Storage: Store in cool, dry place away from heat, open flame; keep tightly closed; protect from moisture, air
Uses: Solvent for coatings, cleaners, agric., cosmetics, inks, textiles, adhesives; coupling agent for greases, oils
Features: Slow evaporating; good solvency
Manuf./Distrib.: Aldrich
Trade Names: Arcosolv® DPTB

Dipropylene glycol dibenzoate
CAS 94-51-9; 27138-31-4; EINECS/ELINCS 202-340-7; 248-258-5
Synonyms: Benzoic acid diester with dipropylene glycol; Benzoic acid-n-dipropylene glycol diester; Dibenzoyl dipropylene glycol ester; Dipropanediol dibenzoate; 3,3′-Oxydi-1-propanol dibenzoate; Oxydipropyl dibenzoate; 3,3′-Oxydyl-1-propanol dibenzoate; POP (2) dibenzoate; PPG-2 dibenzoate; PPG (2) dibenzoate
Definition: Polyoxypropylene glycol diester of benzoic acid
Empirical: $C_{20}H_{22}O_5$
Properties: Lt.-colored liq.; mild ester odor; sol. in aliphatic and aromatic solvs.; insol. in water; m.w. 342.42; dens. 1.1271 (20/20 C); pour pt. -19 C; m.p. 200 C; b.p. 250 C (10 mm)
Toxicology: LD50 (oral, rat) 9800 mg/kg; mildly toxic by ingestion
Precaution: Combustible
Hazardous Decomp. Prods.: Heated to decomp., emits acrid smoke and fumes
Uses: Plasticizer for cellulosics, PVC, plastisols, PS, PVB, PVAc adhesives, VCA, castable PU; latex and lacquer coating applics.; film-former, surf. wetting agent in PVAc homopolymer emulsion adhesives; emollient in cosmetics; plasticizer for PVAc coatings for food-contact pa-

per/paperboard; plasticizer for polymers in paper/paperboard in contact with dry food; in food-pkg. adhesives
Regulatory: FDA 21CFR §175.105, 176.170
Manuf./Distrib.: Aldrich; Allchem Ind.; Ashland; ChemService; Haltermann GmbH; Noveon Kalama; Unitex; Velsicol
Trade Names: Finsolv® PG-22
Trade Names Containing: Finsolv® TPP

Dipropylene glycol dimethyl ether
CAS 111109-77-4; EINECS/ELINCS 404-640-5
Synonyms: Bis (methoxypropyl) ether; DMM; DPGdiMM; Oxygis (methoxy)-propane
Empirical: $C_8H_{18}O_3$
Formula: $CH_3OCH_2CH(CH_3)OCH_2CH(CH_3)OCH_3$
Properties: Colorless liq., low odor; sol. 53 g/100 g water; m.w. 162.23; sp.gr. 0.898; f.p. < -71 C; b.p. 175 C; flash pt. 65 C; ref. index 1.407
Uses: Solvent in coatings incl. PU coatings, agric. formulations, resin prod., cleaning prods.; solvent, coupling agent in cosmetics, toiletries; diluent for thixotrope systems
Manuf./Distrib.: Dow; Fluka
Trade Names: Dowanol® DMM; Proglyde DMM

Dipropylene glycol ethyl ether
CAS 15764-24-6
Synonyms: Dipropylene glycol monoethyl ether; DPGEE; 1-2-(Ethoxy-2-methylethoxy)-2-propanol
Classification: Aliphatic ether alcohol
Empirical: $C_8H_{18}O_3$
Formula: $CH_3CH(OH)CH_2OCH_2CH(CH_3)OCH_2CH_3$
Properties: Colorless liq.; sol. in water, variety of org. solvs.; m.w. 162.26; sp.gr. 0.927; b.p. 193-195 C; flash pt. (COC) 90.5 C
Toxicology: LD50 (oral, rat) 3710 mg/kg; high concs. of vapor expected to cause nose/throat irritation; absorbed thru skin, but health effects unlikely; mild eye irritant; low oral toxicity; TSCA listed
Precaution: Combustible; incompat. with oxidizing agents (increases fire/explosion hazard), strong acids (may cause decomp. of ethers)
Storage: Store in cool, dry, well-ventilated area, out of direct sunlight
Uses: Solvent for paints, lacquers, resins, inks, dyes, oils, greases, cleaners

Dipropylene glycol isobutyl ether
Synonyms: Dipropylene glycol monoisobutyl ether; Isobutoxypropoxypropanol
Empirical: $C_{10}H_{22}O_3$
Formula: $(CH_3)_2CHCH_2OCHCH_3CH_2OCHCH_3CH_2OH$
Properties: Liq.; sol. in water, most org. solvs.; m.w. 190.29; dens. 0.91 kg/l; b.p. 220 C; flash pt. (PMCC) 110 C
Uses: Solvent for paints

Dipropylene glycol methyl ether. *See* PPG-2 methyl ether
Dipropylene glycol methyl ether acetate. *See* PPG-2 methyl ether acetate
Dipropylene glycol monobutyl ether; Dipropylene glycol mono-n-butyl ether. *See* Dipropylene glycol butyl ether
Dipropylene glycol monoethyl ether. *See* Dipropylene glycol ethyl ether
Dipropylene glycol monoisobutyl ether. *See* Dipropylene glycol isobutyl ether
Dipropylene glycol monomethyl ether. *See* PPG-2 methyl ether
Dipropylene glycol monomethyl ether acetate. *See* PPG-2 methyl ether acetate
Dipropylene glycol monopropyl ether; Dipropylene glycol mono-n-propyl ether. *See* Dipropylene glycol n-propyl ether
Dipropylene glycol, myristyl ether, propionate. *See* PPG-2 myristyl ether propionate

Dipropylene glycol phenyl ether
CAS 51730-94-0; EINECS/ELINCS 257-361-4
Synonyms: (Methyl-2-phenoxyethoxy) propanol
Empirical: $C_{12}H_{18}O_3$
Uses: Solvent
Trade Names Containing: Dowanol® PPh

Dipropylene glycol propyl ether. *See* Dipropylene glycol n-propyl ether

Dipropylene glycol n-propyl ether
CAS 29911-27-1; EINECS/ELINCS 249-949-4
Synonyms: Dipropylene glycol monopropyl ether; Dipropylene glycol mono-n-propyl ether; Dipropylene glycol propyl ether; Dipropylene glycol propyl ether, mixt. of isomers; DPnP; 2-Propanol, 1-(2-propoxy-1-methylethoxy)-; 1-(2-Propoxy-1-methylethoxy)-2-propanol; Propyl dipropasol solvent
Empirical: $C_9H_{20}O_3$
Formula: $C_3H_7O[CH_2(CH)CH_3O]_2H$
Properties: Colorless liq., low odor; mod. sol. in water; m.w. 176.2; sp.gr. 0.922; b.p. 212 C; flash pt. 190 F; ref. index 1.424 (20 C); surf. tens. 27.6 dynes/cm
Toxicology: LD50 (oral, rat) 2000-2500 mg/kg, (skin, rabbit) > 2000 mg/kg; irritating to eyes; may be harmful by ing., inh., skin absorp.; may cause skin/mucous membrane/upper respiratory tract irritation; may cause general depressed activity, diarrhea; TSCA listed
Precaution: Combustible liq.; incompat. with strong oxidizing agents
Hazardous Decomp. Prods.: Toxic fumes of CO, CO_2; emits toxic fumes under fire conditions
Storage: Store in cool, dry place away from heat, open flame; keep tightly closed
Uses: Coalescent, solvent for water- or solv.-borne coatings; solvent, coupling agent in cosmetics, toiletries
Manuf./Distrib.: Aldrich; Ashland; Dow
Trade Names: Arcosolv® DPNP; Dowanol® DPnP

Dipropylene glycol propyl ether, mixt. of isomers. *See* Dipropylene glycol n-propyl ether
Dipropyl ketone. *See* 4-Heptanone
Dipropyl methane. *See* Heptane
DIPS. *See* Diisopropyl sebacate
Disiloxane, hexamethyl-. *See* Hexamethyldisiloxane
Distillates, light hydrocracked. *See* Petroleum distillates, light hydrocracked
Distillates (petroleum), acid-treated, medium heavy; Distillates (petroleum), acid-treated, middle. *See* Petroleum distillates, acid-treated middle
Distillates (petroleum), alkylate. *See* Petroleum distillates, alkylate
Distillates (petroleum), catalytic reformer frationator residue, low boiling. *See* Petroleum distillates, catalytic reformer fractionator residue, low boiling
Distillates (petroleum), heavy thermal cracked. *See* Petroleum distillates, heavy thermal cracked
Distillates (petroleum), hydrogen-treated, light naphthene-. *See* Naphthenic oil
Distillates (petroleum), hydroteated (mild) heavy naphthenic; Distillates (petroleum), hydrotreated heavy naphthenic. *See* Petroleum distillates, heavy hydrotreated naphthenic
Distillates (petroleum), hydrotreated light. *See* Petroleum distillates, hydrotreated light
Distillates (petroleum), hydrotreated light paraffinic. *See* Petroleum distillates, hydrotreated light paraffinic
Distillates (petroleum), hydrotreated middle. *See* Petroleum distillates, hydrotreated middle
Distillates (petroleum), hydrotreated (mild) light naphthenic. *See* Naphthenic oil
Distillates (petroleum), light distillate hydrotreating process, low-boiling. *See* Petroleum distillates, hydrotreated light, low-boiling
Distillates (petroleum), light hydrocracked. *See* Petroleum distillates, light hydrocracked
Distillates (petroleum), sweetened middle. *See* Petroleum distillates, sweetened middle
Distilled spirits. *See* Alcohol
Distilled water. *See* Water
Ditan; Ditane. *See* Diphenylmethane
2,3-Dithiobutane. *See* Methyl disulfide
Dithiocarbonic anhydride. *See* Carbon disulfide
Di (trimethylsilyl) amine. *See* Hexamethyldisilazane
Divinylene oxide. *See* Furan
Divinylene sulfide. *See* Thiophene
Dizene. *See* o-Dichlorobenzene
DMA. *See* N,N-Dimethylaniline; Dimethyl acetamide; Dimethylamine
DMAC. *See* Dimethyl acetamide
DMAE. *See* Dimethylethanolamine

DMC. *See* Methyl carbonate
DMDH; DMDMH. *See* DMDM hydantoin

DMDM hydantoin

CAS 6440-58-0; EINECS/ELINCS 229-222-8
Synonyms: Bis (hydroxymethyl)-5,5-dimethylhydantoin; 1,3-Bis (hydroxymethyl)-5,5-dimethylimidazolidinedione; 1,3-Bis (hydroxymethyl)-5,5-dimethylimidazolidine-2,4-dione; Bis (hydroxymethyl)-5,5-dimethyl-2,4-imidazolidinedione; 1,3-Di (hydroxymethyl)-5,5-dimethylhydantoin; Dimethylol dimethyl hydantoin; 1,3-Dimethylol-5,5-dimethyl hydantoin; Dimethylol-5,5-dimethylhydantoin; DMDH; DMDMH; Hydantoin, 1,3-bis (hydroxymethyl)-5,5-dimethyl-; Imidazolidinedione, 1,3-bis (hydroxymethyl)-5,5-dimethyl-; 2,4-Imidazolidinedione, 1,3-bis (hydroxymethyl)-5,5-dimethyl-
Classification: Organic compd.
Empirical: $C_7H_{12}N_2O_4$
Properties: M.w. 188.21
Toxicology: LD50 (oral, rat) 2 g/kg; primary irritant; TSCA listed
Precaution: May release formaldehyde
Uses: Biocide; antimicrobial, preservative for cosmetics, pharmaceutical topicals; intermediate
Regulatory: USA CIR approved, EPA registered; Europe listed 0.6% max.; FDA approved for topicals
Manuf./Distrib.: Acme-Hardesty; High Polymer Labs; Jan Dekker BV
Trade Names Containing: Dantogard® Plus Liq.

DMDS. *See* Methyl disulfide
DME. *See* Ethylene glycol dimethyl ether; Dimethyl ether
DMEA. *See* Dimethylethanolamine
DMEP. *See* Dimethoxyethyl phthalate
DMF; DMFA. *See* Dimethyl formamide
DMGP. *See* Dimethoxyethyl phthalate
DMH. *See* DM hydantoin

DM hydantoin

CAS 77-71-4; EINECS/ELINCS 201-051-3
Synonyms: Dimethylhydantoin; 5,5-Dimethyl hydantoin; 5,5-Dimethyl-2,4-imidazolidinedione; DMH; 2,4-Imidazolidinedione, 5,5-dimethyl-
Classification: Organic compd.
Empirical: $C_5H_8N_2O_2$
Formula: $HNCONHCOC(CH_3)_2$
Properties: Wh. cryst. solid; sol. in water, alcohol, ether, oxygenated solvs.; m.w. 128.13; m.p. 178 C
Toxicology: LD50 (oral, rat) 7800 mg/kg, (subcut., mouse) 2800 mg/kg; mod. toxic by subcut. route; mildly toxic by ing.; CNS depressant; TSCA listed
Hazardous Decomp. Prods.: Heated to decomp., emits toxic fumes of NO_x
Uses: Visc. control agent, hair conditioner in cosmetics; intermediate for textiles and other applics.; synthesis; prep. of water-sol. polymers
Manuf./Distrib.: Advanced Synthesis Tech.; Aldrich; Fluka; Great Lakes; Hampshire; Lonza; Sigma
Trade Names Containing: Dantogard® Plus Liq.

DMI. *See* Dimethyl imidazolidinone; Dimethyl isosorbide; 1,2-Dimethyl imidazole
DMK. *See* Acetone
DMM. *See* Dipropylene glycol dimethyl ether
DMMP. *See* Dimethyl methylphosphonate
DMN; DMNA. *See* N-Nitrosodimethylamine
DMP. *See* Dimethyl phthalate
2,5-DMP. *See* 2,5-Xylenol
2,6-DMP. *See* 2,6-Xylenol
3,4-DMP. *See* 3,4-Xylenol
DMS. *See* Dimethyl sebacate; Dimethyl sulfide
DMSO. *See* Dimethyl sulfoxide
$DMSO_2$. *See* Dimethyl sulfone
DMTDP. *See* Dimethyl thiodipropionate
DNBA. *See* Dibutylamine
DNOP. *See* n-Dioctyl phthalate
DNPA. *See* Dipropylamine
DOA. *See* Dioctyl adipate
Docosanoic acid, docosyl ester. *See* Behenyl behenate
Docosanoic acid, methyl ester. *See* Methyl behenate
Docosanoic acid, 2-octyldodecyl ester. *See* Octyldodecyl behenate
1-Docosanol. *See* Behenyl alcohol
1-Docosanol, benzoate. *See* Behenyl benzoate
Docosenamide; 13-Docosenamide; cis 13-Docosenamide; (Z)-Docos-13-enamide. *See* Erucamide
13-Docosenoic acid, 9-octadecenyl ester. *See* Oleyl erucate
Docosyl docosanoate. *See* Behenyl behenate
Dodecafluoropentane. *See* Perfluoropentane
Dodecahydrodiphenylamine. *See* Dicyclohexylamine
Dodecamethylcyclohexasiloxane. *See* Cyclomethicone
Dodecane. *See* n-Dodecane

n-Dodecane

CAS 112-40-3; EINECS/ELINCS 203-967-9
Synonyms: Alkane C12; Bihexyl; Dihexyl; Dodecane; Duodecane
Classification: Sat. aliphatic hydrocarbon
Empirical: $C_{12}H_{26}$
Formula: $CH_3(CH_2)_{10}CH_3$
Properties: Colorless clear liq.; gasoline-like odor; very sol. in ether; sol. in most org. solvs., acetone, 95% ethanol, chloroform; sol. < 1 mg/ml in water, DMSO; m.w. 170.34; sp.gr. 0.753; vapor pressure 1 mm Hg; m.p. -9.6 C; b.p. 216 C; flash pt. 74 C; autoignition temp. 204 C; ref. index 1.4216 (20 C)
Toxicology: TDLo (skin, mouse, 22 wks intermittent) 11 g/kg; may cause mild skin irritation; conc. vapor may be irritating to eyes; inh. of conc. vapor may cause nose/throat irritation, headache, drowsiness, nausea, difficulty breathing; ing. may cause nausea, vomiting, swelling of abdomen, headache, depression; relatively low oral toxicity unless aspirated into lungs; questionable carcinogen; experimental tumorigen, teratogen; TSCA listed
Precaution: Combustible; LEL 0.6%; incompat. with strong oxidizing agents (increases fire and explosion hazard)
Hazardous Decomp. Prods.: Heated to decomp., emits acrid smoke and irritating fumes
NFPA: Health 0, Flammability 2, Reactivity 0
Storage: Store away from incompat. materials, ignition sources; store refrigerated
Uses: Component of gasoline, jet fuel, kerosene, petrol. solvents such as wh. spirit; solvent; org. synthesis; hydrocarbon standard; copying fluids; jet fuel research; distillation chaser; cosmetics ingred.; mfg. of paraffin prods.; rubber industry; paper processing industry
Manuf./Distrib.: ABCR; Acros Org.; Aldrich; Alfa Aesar; ChevronPhillips; Fisher Scientific; Fluka; ICN Biomed. Research Prods.; Indofine; Lancaster Synthesis; Sigma; TCI Am.

1,2-Dodecane carbonate

Properties: M.w. 228.34 ; dens. 0.963 g/ml; visc. 19.90 cp; b.p. 285 C; flash pt. (CC) 310 F; surf. tens. 32.78 dynes/cm
Uses: Solvent

Dodecanoic acid benzyl ester. *See* Benzyl laurate
Dodecanoic acid, diester with 1,2,3-propanetriol. *See* Glyceryl dilaurate
Dodecanoic acid, 2,3-dihydroxypropyl ester. *See* Glyceryl laurate
Dodecanoic acid, hexyl ester. *See* Hexyl laurate
Dodecanoic acid, 2-hydroxypropyl ester. *See* Propylene glycol laurate
Dodecanoic acid methyl ester. *See* Methyl laurate
Dodecanoic acid, 1-methylethyl ester. *See* Isopropyl laurate
Dodecanoic acid, mixed triesters with octanoic acid, decanoic acid, and 1,2,3-propanetriol. *See* Caprylic/capric/lauric triglyceride
Dodecanoic acid, monoester with 1,2-propanediol. *See* Propylene glycol laurate
Dodecanoic acid, monoester with 1,2,3-propanetriol. *See* Glyceryl laurate
Dodecanoic acid, phenylmethyl ester. *See* Benzyl laurate
Dodecanoic acid, 1,2,3-propanetriyl ester. *See* Trilaurin
1-Dodecanol; Dodecan-1-ol; n-Dodecanol; n-Dodecan-1-ol. *See* Lauryl alcohol
1-Dodecanol, 2-octyl-. *See* Octyldodecanol
2,6,10-Dodecatrienol, 3,7,11-trimethyl-. *See* Farnesol
Dodecyl alcohol; 1-Dodecyl alcohol; n-Dodecyl alcohol. *See* Lauryl alcohol
Dodecyl alcohol, hydrogen sulfate, sodium salt. *See* Sodium lauryl sul-

fate

Dodecylbenzenesulfonic acid

CAS 27176-87-0; 68411-32-5; 68584-22-5; 68608-88-8; 85117-49-3; 85536-14-7; EINECS/ELINCS 248-289-4; 271-807-5
Synonyms: ABS; Alkylbenzenesulfonic acid; Benzenesulfonic acid, dodecyl-; DDBSA; n-Dodecylbenzenesulfonic acid; Laurylbenzenesulfonic acid
Classification: Substituted aromatic acid
Empirical: $C_{18}H_{30}O_3S$
Properties: Wh. to lt. yel. flakes, granules, and powd.; m.w. 326.54; anionic
Toxicology: LD50 (oral, rat) 650 mg/kg; mod. toxic by ing.; TSCA listed
Environmental: Biodeg.
Precaution: DOT: Corrosive
Hazardous Decomp. Prods.: Heated to decomp., emits toxic vapors of SO_x
Uses: Surfactant, detergent raw material in heavy and lt. duty detergents, hand cleaning gels, machine degreasers, tank cleaners; emulsifier for metalworking fluids, pesticides, emulsion polymerization, paints/coatings; catalyst in acid catalyzed reactions; textile scouring agent, wetting agent, bleaching agent, dyeing assistant; defoamer in food-contact paper/paperboard; emulsifier in food pkg.
Regulatory: FDA 21CFR §176.210, 178.1010, 178.3400
Manuf./Distrib.: Allchem Ind.; Ashland; Avatar; Biddle Sawyer; Browning; Chem. Distribution; Expo Chem.; Harcros; Pilot; Rhodia HPCII; Rhodia/Phosphorus Perf. Derivs.; Rütgers Org.; Spectrum Quality Prods.; Stepan; Tulstar Prods.; Vopak USA
Trade Names Containing: Dynasolve 165; Dynasolve 210; Dynasolve 220; Dynasolve 225; Dynasolve 230

n-Dodecylbenzenesulfonic acid. *See* Dodecylbenzenesulfonic acid
Dodecylbenzenesulfonic acid, compd. with isopropylamine (1:1); Dodecylbenzenesulfonic acid, compd. with 2-propanamine (1:1). *See* Isopropylamine dodecylbenzenesulfonate
Dodecyl oleate. *See* Lauryl oleate
1-Dodecyl-2-pyrrolidinone; N-Dodecylpyrrolidinone; N-Dodecyl-2-pyrrolidone. *See* Lauryl pyrrolidone
Dodecyl sodium sulfate; Dodecylsulfate sodium salt. *See* Sodium lauryl sulfate

Dodoxynol-7

CAS 9014-92-0 (generic); 26401-47-8 (generic)
Synonyms: PEG-7 dodecyl phenyl ether; POE (7) dodecyl phenyl ether
Classification: Ethoxylated alkyl phenol
Empirical: $C_{32}H_{58}O_8$
Formula: $C_{12}H_{25}C_6H_4(OCH_2CH_2)_nOH$, avg. n = 7
Properties: Nonionic
Toxicology: TSCA listed
Uses: Emulsifier, surfactant in cosmetics; surfactant for solvents; solvent and emulsion cleaning; surfactant, adjuvant for pesticide use dilutions on growing crops; in food-pkg. adhesives; defoamer in food-contact paper/paperboard; emulsifier in mfg. of food-contact articles
Regulatory: FDA 21CFR §172.710, 175.105, 176.210, 178.3400

Dolcymene. *See* p-Cymene
DOM. *See* Dioctyl maleate
DOP. *See* n-Dioctyl phthalate
DOS. *See* Dioctyl sebacate
1,3-DP. *See* 1,3-Dimethoxy-2-propanol
DPG. *See* Dipropylene glycol
DPGdiMM. *See* Dipropylene glycol dimethyl ether
DPGEE. *See* Dipropylene glycol ethyl ether
DPGME. *See* PPG-2 methyl ether
DPGMEA. *See* PPG-2 methyl ether acetate
DPGnBE. *See* Dipropylene glycol butyl ether
DPGtBE. *See* Dipropylene glycol t-butyl ether
DPM. *See* PPG-2 methyl ether; Propylene glycol dimethyl ether
DPM Acetate. *See* PPG-2 methyl ether acetate
DPMG. *See* PPG-2 methyl ether
DPnB. *See* Dipropylene glycol butyl ether
DPnP. *See* Dipropylene glycol n-propyl ether
DPS. *See* Diphenyl sulfone
DPtB. *See* Dipropylene glycol t-butyl ether
DTA. *See* Diethylenetriamine
Duodecane. *See* n-Dodecane
Duodecyl alcohol. *See* Lauryl alcohol
Dutch liquid; Dutch oil. *See* Ethylene dichloride

E

EA. *See* Ethylamine
EAA. *See* Ethylacetoacetate
EAK. *See* 3-Octanone
Earthnut oil. *See* Peanut (Arachis hypogaea) oil
EB. *See* Ethylbenzene
2EBO. *See* 2-Ethyl-1-butanol
ECH. *See* Epichlorohydrin
EDA. *See* Ethylenediamine
Edathamil. *See* Edetic acid
EDC. *See* Ethylene dichloride
Edetate. *See* Edetic acid

Edetic acid

CAS 60-00-4; EINECS/ELINCS 200-449-4
UN 9117
Synonyms: 3,6-Bis (carboxymethyl)-3,5-diazooctanedioic acid; Edathamil; Edetate; EDTA; EDTA acid; N,N´-1,2-Ethanediylbis[N-(carboxymethyl) glycine]; Ethylenediaminetetraacetic acid; (Ethylenedinitrilo) tetraacetic acid
Classification: Substituted diamine
Empirical: $C_{10}H_{16}N_2O_8$
Formula: $(HOOCCH_2)_2NCH_2CH_2N(CH_2COOH)_2$
Properties: Colorless or wh. cryst. powd.; odorless; sol. in alkali hydroxide sol'ns.; pract insol. in water; insol. in common org. solvs.; m.w. 292.28; sp.gr. 0.721; m.p. dec. @ 240 C
Toxicology: LD50 (oral, mouse) 30 mg/kg, (IP, rat) 397 mg/kg; irritant; poison by IP route; mutagenic data; experimental teratogenic and reproductive effects; TSCA listed
Environmental: Environmentally hazardous substance; VOC; BOD5 0.01; ThOD 0.93
Hazardous Decomp. Prods.: Heated to decomp., emits very toxic fumes of NO_x
Storage: Store @ R.T.
Uses: Chelating agent in detergents, disinfectants, rare earth separation; antioxidant, preservative in cosmetics, foods; stabilizer for cosmetics, pharmaceuticals, water treatment, textiles, soaps, detergents, shampoos; electroless copper plating; solvent; metal cleaning; radioactive decontamination; metal deactivator (lubricants); blood anticoagulant; textile dyeing aid; clarification of liqs.; analytical chemistry; spectrophotometric titration; medicine (to treat lead poisoning); pharmaceutical excipient; in paper/paperboard in contact with aq./fatty foods
Regulatory: FDA 21CFR §175.105, 176.170; FDA approved for otics, rectals, topicals; NF compliance; CERCLA hazardous substance
Manuf./Distrib.: AMRESCO; Akzo Nobel; Aldrich; Allchem Ind.; Am. Int'l.; Ashland; Avatar; Aventis Pharmaceuticals; BASF; Bimal Pharma Chem; Charkit; Chemplex Chems.; Ciba Spec. Chems. Switzerland; Ciba Spec. Chems./Water Treatment; Contract Chems. Ltd; Delta Distributors; Dow; EM Ind.; Eastech; Fluka; Hampshire; Harcros; Hasnil; Independent Chem.; Int'l. Sourcing; Integra; J.T. Baker; Jarchem Ind.; Kraft Chem.; Lowenstein Dyes & Cosmetics; Merck KGaA; Milsolv; Mosselman NV; O.C. Lugo; P.B. & S.; Penta Mfg.; Phoenix Nat. Prods.; Rhodia HPCII; Ruger; SOCO-Lynch; Showa Denko; Sigma; Southchem; Spectrum Quality Prods.; Surfactants Inc; Synthron SA; Textile Chem.; Ubichem plc; Univar E&E Ltd; Universal Preserv-A-Chem; VWR Int'l.; Vopak USA; Whittaker, Clark & Daniels

Edetol. *See* Tetrahydroxypropyl ethylenediamine
EDTA; EDTA acid. *See* Edetic acid
EE; 2-EE. *See* Ethoxyethanol
EEA. *See* Ethoxyethanol acetate; Ethylacetoacetate
2-EEA. *See* Ethoxyethanol acetate
EEEA. *See* 2-(2-Ethoxyethoxy) ethyl acrylate
EEP. *See* Ethyl 3-ethoxypropionate
EG. *See* Ethylene glycol
EGA. *See* Ethoxyethanol acetate
EGDME. *See* Ethylene glycol dimethyl ether
EGDS. *See* Glycol distearate
EGEE. *See* Ethoxyethanol
EGIPE. *See* Ethylene glycol isopropyl ether
EGMBE. *See* Butoxyethanol
EGME. *See* Methoxyethanol
EGMEA. *See* Methoxyethanol acetate
EGMEE. *See* Ethoxyethanol
EGMS. *See* Glycol stearate
2-EH. *See* 2-Ethylhexanol
2-EH acid. *See* 2-Ethylhexoic acid

Eicosane

CAS 112-95-8; EINECS/ELINCS 204-018-1
Synonyms: Didecyl; n-Eicosane; Icosane
Definition: Mixt. of predominantly straight-chain hydrocarbons avg. 20 carbon atoms to the molecule
Empirical: $C_{20}H_{42}$
Properties: Colorless to wh. cryst. solid; sol. in ether, acetone, benzene, petrol. ether; sol. 1-5 mg/ml in 95% ethanol; insol. in water, DMSO; m.w. 282.55; dens. 0.7886 (20/4 C); m.p. 36-38 C; b.p. 342.7 C; flash pt. 100 C; ref. index 1.4348 (20 C)
Toxicology: Irritating to eyes, skin, mucous membranes, upper respiratory tract; may be harmful by inh., ing., or skin absorp.; TSCA listed
Precaution: Combustible; incompat. with strong oxidizing agents
Hazardous Decomp. Prods.: CO, CO_2
NFPA: , Flammability 1, Reactivity 0
Storage: Store refrigerated; keep tightly closed
Uses: Cosmetics ingred.; solvent; lubricants; plasticizers; in closure-sealing gaskets for food containers
Regulatory: FDA 21CFR §177.1210
Manuf./Distrib.: ABCR; Alfa Aesar; Kira; Pfaltz & Bauer; Roper Thermals; Spectrum Quality Prods.; Universal Preserv-A-Chem; VWR Int'l.

n-Eicosane. *See* Eicosane
Eicosanoic acid methyl ester. *See* Methyl eicosanoate
Eicosanol; 1-Eicosanol. *See* Arachidyl alcohol
Elaic acid. *See* Oleic acid
Elaldehyde. *See* Paraldehyde
Elemental iodine. *See* Iodine
Enanthic alcohol; Enanthyl alcohol. *See* Heptyl alcohol
Enzactin. *See* Triacetin
EO. *See* Ethylene oxide
EOEOEA. *See* 2-(2-Ethoxyethoxy) ethyl acrylate
EP. *See* Propylene glycol ethyl ether
EPA. *See* Propylene glycol ethyl ether acetate

Epichlorohydrin

CAS 106-89-8; EINECS/ELINCS 203-439-8
UN 2023 (DOT)
Synonyms: 1-Chloro-2,3-epoxypropane; 3-Chloro-1,2-epoxypropane; (Chloromethyl) ethylene oxide; Chloromethyloxirane; 2-(Chloromethyl) oxirane; Chloropropylene oxide; 3-Chloro-1,2-propylene oxide; γ-Chloropropylene oxide; ECH; α-Epichlorohydrin; 1,2-Epoxy-3-chloropropane; 2,3-Epoxypropyl chloride; Glycerol epichlorhydrin
Classification: Chlorinated epoxide
Empirical: C_3H_5ClO
Properties: Colorless, highly volatile unstable liq., chloroform-like odor; sol. in ethanol, diethyl ether, chloroform, trichloroethylene, CCl_4; misc. with oxygenated solvs.; sl. sol. in water; m.w. 92.53; dens. 1.17 kg/l; f.p. -25 C; b.p. 115.2 C; flash pt. 33.9 C; ref. index 1.4358
Toxicology: ACGIH TLV/TWA 2 ppm (skin); LD50 (oral, rat) 90 mg/kg, (IP, rat) 13 mg/kg, (subcut., rat) 150 mg/kg, (IV, rat) 154 mg/kg; poison by ing., skin contact, IV, and IP routes; mod. toxic by inh.; readily absorbed thru skin; corrosive; sensitizer; primary skin and eye irritant; human systemic effects by inh.; confirmed carcinogen; tumorigen; mutagen; experimental teratogen; human mutagenic data; TSCA listed
Precaution: Flamm.; mod. fire risk; explosive reaction with aniline; violent reaction with sulfuric acid or isopropylamine; exothermic polymerization with strong acids, caustic alkalis, aluminum, zinc, etc.; heating may cause polymerization
Hazardous Decomp. Prods.: CO, CO_2, hydrogen chloride gas, phosgene gas; heated to decomp., emits toxic fumes of Cl^-
NFPA: Health 3, Flammability 3, Reactivity 2
Uses: Intermediate for paints; major raw material for epoxy and phenoxy resins; mfg. of glycerol; solvent for cellulose esters and ethers; high wet-str. resins for paper industry; crosslinking agent for food starch modified; mfg. of copolymers for decolorization, flocculation, clarification of refinery sugar liquors and juices; food pkg.
Regulatory: FDA 21CFR §172.892, 173.60, 175.300, 176.170, 176.180, 177.1440, 177.2280, 177.2550; SARA reportable
Manuf./Distrib.: AP Resources; Aldrich; Alfa Chem; BASF; Dow; Filo; Fluka; ICC Chem.; J.T. Baker; Jarchem Ind.; Kessler; Resolution Perf. Prods.; Richman; Rit-Chem; Sigma; Sithean; Solvay SA; Spectrum Quality Prods.; United Min. & Chem.

α-Epichlorohydrin. *See* Epichlorohydrin
1,4-Epoxy-1,3-butadiene. *See* Furan
1,4-Epoxybutane. *See* Tetrahydrofuran
1,2-Epoxy-3-chloropropane. *See* Epichlorohydrin
Epoxyethane; 1,2-Epoxyethane. *See* Ethylene oxide
1,8-Epoxy-p-menthane. *See* Eucalyptol
Epoxypropane; 1,2-Epoxypropane; 2,3-Epoxypropane. *See* Propylene oxide
2,3-Epoxypropyl chloride. *See* Epichlorohydrin

Erucamide

CAS 112-84-5; EINECS/ELINCS 204-009-2
Synonyms: Docosenamide; 13-Docosenamide; cis 13-Docosenamide; (Z)-Docos-13-enamide; Erucic acid amide; Erucylamide
Classification: Aliphatic amide
Empirical: $C_{22}H_{43}NO$
Formula: $CH_3(CH_2)_7CH=CH(CH_2)_{11}CONH_2$
Properties: Wh. waxy beads; sol. in IPA; sl. sol. in alcohol, acetone, oxygenated solvs.; insol. in water; dens. 0.888; m.p. 75-80 C; iodine no. 72-80
Precaution: Combustible
Uses: Foam stabilizer; solvent for waxes and resins, emulsions; slip agent, antiblocking agent for PE; lubricant, mold release agent for rubber and plastics; slip agent in printing inks; opacifier, visc. control agent in cosmetics; lubricant, slip, antiblock, and mold release agent for plastics, rubber, crayons, petrol. prods., asphalts, inks, metals, textiles; lamination of PE to cellophane and in PE extrusion coatings; defoamer/water repellent in industrial/household applics.; release agent in food-contact polymers and petroleum wax; in food-pkg. adhesives; in paper/paperboard in contact with dry food; in closure-sealing gaskets for food containers
Regulatory: FDA 21CFR §175.105, 176.180, 177.1200, 177.1210, 178.3860
Manuf./Distrib.: Akzo Nobel; Aldrich; Allchem Ind.; Century Multech; ChemService; Chemax; Croda Universal Ltd; Dama Chems.; Eastech; Fallek; Ferro; Lutianhua; Sigma; Uniqema Am.; Vilax

Erucic acid amide. *See* Erucamide
Erucic acid, oleyl ester. *See* Oleyl erucate
Erucylamide. *See* Erucamide
ESE. *See* Ethylsulfonylethyl alcohol
Essence of mirbane; Essence of myrbane. *See* Nitrobenzene
Essence of niobe. *See* Ethyl benzoate; Methyl benzoate
ET. *See* Ethyl toluene
Ethal. *See* Cetyl alcohol
Ethanamide. *See* Acetamide
Ethanamine. *See* Ethylamine
Ethanamine, N,N-diethyl-. *See* Triethylamine
Ethanamine, N-ethyl-N-nitroso-. *See* N-Nitrosodiethylamine
Ethanaminium, 2-hydroxy-N,N,N-tris (2-hydroxyethyl)-, hydroxide. *See* Tetraethanol ammonium hydroxide
Ethane, bromo-. *See* Ethyl bromide
Ethane, 1-chloro-1,1-difluoro-. *See* 1,1,1-Chlorodifluoroethane
1,2-Ethanediamine. *See* Ethylenediamine
1,2-Ethanediamine, N-(2-aminoethyl)-N´-(2-((2-aminoethyl) amino) ethyl)-. *See* Tetraethylenepentamine
1,2-Ethanediamine, dihydrochloride. *See* Ethylenediamine dihydrochloride
Ethane dichloride. *See* Ethylene dichloride
Ethane, 1,1-dichloro-. *See* 1,1-Dichloroethane
Ethane, 1,1-dichloro-1-fluoro-. *See* Dichlorofluoroethane
Ethane, dichlorotetrafluoro-. *See* Dichlorotetrafluoroethane
Ethane, 1,2-dichloro-1,1,2,2-tetrafluoro-. *See* 1,2-Dichlorotetrafluoroethane
Ethanedioic acid, dibutyl ester. *See* Dibutyl oxalate
Ethanedioic acid diethyl ester. *See* Diethyl oxalate
Ethanedioic acid, dipentyl ester. *See* Diamyl oxalate
1,2-Ethanediol; Ethane-1,2-diol. *See* Ethylene glycol
1,2-Ethanediol carbonate. *See* Ethylene carbonate
1,2-Ethanediol diacetate. *See* Ethylene glycol diacetate
1,2-Ethanediol, dipropanoate. *See* Crotonaldehyde
1,2-Ethanediol monoacetate. *See* Ethylene glycol acetate
N,N´-1,2-Ethanediylbis[N-(carboxymethyl) glycine]. *See* Edetic acid
1,1´-(1,2-Ethanediylbis (oxy)) bis-butane. *See* Ethylene glycol dibutyl ether
2,2´-[1,2-Ethanediylbis (oxy)] bisethanol. *See* Triethylene glycol
Ethane, 1-ethoxy-2-(2-methoxyethoxy)-. *See* Diethylene glycol methyl ethyl ether
Ethane, 1,1´-[methylenebis (oxy)] bis-. *See* Diethoxymethane
Ethane, 1,1´-(methylenebis (oxy) bis (2-chloro-. *See* Bis (2-chloroethoxy) methane
Ethanenitrile. *See* Acetonitrile
Ethane, nitro-. *See* Nitroethane
Ethane pentachloride; Ethane, pentachloro-. *See* Pentachloroethane
Ethane, 1,1,2,2-tetrabromo-. *See* Tetrabromoethane
Ethane, 1,1,1,2-tetrachloro-. *See* 1,1,1,2-Tetrachloroethane
Ethane, 1,1´-thiobis-. *See* Ethyl sulfide
Ethane trichloride. *See* 1,1,2-Trichloroethane
Ethane, 1,1,1-trichloro-. *See* 1,1,1-Trichloroethane
Ethane, 1,1,1-trichloro-2,2,2-trifluoro-. *See* 1,1,1-Trichloro-2,2,2-trifluoroethane
Ethane, 1,1,2-trichloro-1,2,2-trifluoro. *See* Trichlorotrifluoroethane
Ethanoic acid. *See* Acetic acid, glacial; Acetic acid
Ethanol. *See* Alcohol
N-Ethanolacetamide. *See* Acetamide MEA
Ethanolamide. *See* Acetamide

Ethanolamine

CAS 141-43-5; EINECS/ELINCS 205-483-3
UN 2491 (DOT)
Synonyms: 2-Aminoethanol; 2-Aminoethyl alcohol; β-Aminoethyl alcohol; β-Ethanolamine; Ethylolamine; Glycinol; 2-Hydroxyethylamine; β-Hydroxyethylamine; MEA; MELA; Monoethanolamine
Classification: Aliphatic amino alcohol; alkanolamine
Empirical: C_2H_7NO
Formula: $NH_2CH_2CH_2OH$
Properties: Colorless clear mod. visc. liq.; ammoniacal odor; misc. with

water, alcohol, acetone, glycerin; sol. in chloroform; sl. sol. in benzene; m.w. 61.10; dens. 1.012; vapor pressure 0.26 mm Hg; m.p. 10.5 C; b.p. 170 C; flash pt. 93 C; autoignition temp. 410 C; ref. index 1.4540; pH 12.1; surf. tens. 48.89 dynes/cm; dielec. const. 37.72

Toxicology: ACGIH TLV/TWA 3 ppm; LD50 (oral, rat) 2140 mg/kg, (skin, rat) 1500 mg/kg; poison by IP route; mod. toxic by ingestion, skin contact, subcut., IV routes; corrosive irritant to eyes, skin, mucous membranes; TSCA listed

Environmental: VOC; BOD5 0.85; COD 1.28; ThOD 1.31

Precaution: DOT: Corrosive material; flamm. exposed to heat or flame; powerful reactive base; incompat. with sulfuric acid, HCl, CO_2 in air, copper and alloys, aluminum, acrylic acid, nitric acid, etc.

Hazardous Decomp. Prods.: CO, CO_2, NO_x

NFPA: Health 3, Flammability 2, Reactivity 0

Storage: Hygroscopic; store in cool, dry well-ventilated area away from heat and ignition sources

Uses: Scrubbing acid gases, esp. in synthesis of ammonia; nonionic detergents for dry cleaning wool treatment, emulsion paints, polishes, agric. sprays; chemical intermediate; emulsifier for pharmaceuticals; corrosion inhibitor; buffer in cosmetics; chem. intermediate for mfg. rubber, vulcanization accelerators; flume wash water additive for food processing; dispersant for agric. chems., polishes, cleansers, paints, cosmetics; solvent; in food-pkg. adhesives; defoamer in food-contact paper/paperboard; adjuvant in slimicides in food-contact paper/paperboard; in food-contact animal glues

Regulatory: FDA 21CFR §173.315, 175.105, 176.180, 176.210, 176.300, 178.3120; not permitted for use in foods intended for babies and young infants in UK; NF, BP compliance

Manuf./Distrib.: AAE Chemie NV; Aldrich; Allchem Ind.; Ashland; BASF; BP Chems. Ltd; Brenntag AG; Camida Ltd; ChevronTexaco; Cosmetic Supplies USA; Degussa AG; Dow; EM Ind.; Fluka; Honeywill & Stein; Houghton Chem.; Huntsman; Int'l. Chem. Inc; J.T. Baker; Mallinckrodt Baker; Merck KGaA; Mosselman NV; Mutchler; Occidental; Oxiteno; Penta Mfg.; Research Organics; Romil Ltd; Ruger; Sal Chem.; Sigma; Spectrum Quality Prods.; Stanchem; Surfachem Ltd; Tennants Distrib. Ltd; Univar E&E Ltd; Universal Preserv-A-Chem; VWR Int'l.; Varsal Instruments; Vopak USA; Whyte Chems. Ltd

β-Ethanolamine. *See* Ethanolamine
Ethanol, 2-(2-aminoethoxy)-. *See* 2-(2-Aminoethoxy) ethanol
Ethanol, 2-butoxy-, acetate. *See* Butoxyethanol acetate
Ethanol, 2-(2-(2-butoxyethoxy) ethoxy)-. *See* Butoxytriglycol
Ethanol, 2-butoxy-, phosphate (3:1). *See* Tributoxyethyl phosphate
Ethanol, 2-(diethylamino)-. *See* Diethylaminoethanol
Ethanol, 2-(dimethylamino)-. *See* Dimethylethanolamine
Ethanol, 2-ethoxy-. *See* Ethoxyethanol
Ethanol, 2-(2-ethoxyethoxy). *See* Ethoxydiglycol
Ethanol, 2-ethoxy-, succinate (2:1). *See* Diethoxyethyl succinate
Ethanol, 2-((2-ethylhexyl)oxy)-. *See* Ethylene glycol 2-ethylhexyl ether
Ethanol, 2,2´-(ethylimino) bis-. *See* N-Ethyldiethanolamine
Ethanol, 2-(ethylphenylamino)-. *See* Phenylethylethanolamine
Ethanol, 2-((2-hexyloxy)ethoxy)-. *See* Diethylene glycol n-hexyl ether
Ethanolic acid. *See* Acetic acid, glacial
Ethanol, 2-isobutyoxy-. *See* Ethylene glycol isobutyl ether
Ethanol, 2-isopropoxy-. *See* Ethylene glycol isopropyl ether
Ethanol, 2-(2-methoxyethoxy)-. *See* Methoxydiglycol
Ethanol, 2(2-(2-methoxyethoxy)ethoxy)-. *See* Triglycol monomethyl ether
Ethanol, 2-methoxy-, stearate. *See* Methoxyethyl stearate
Ethanol, 2-(methylamino)-. *See* N-Methylethanolamine
Ethanol, 2-(1-methylethoxy)-. *See* Ethylene glycol isopropyl ether
Ethanol, 2-(2-methylpropoxy)-. *See* Ethylene glycol isobutyl ether
Ethanol, 2,2´-oxydi-. *See* Diethylene glycol
Ethanol, 2-(2-phenoxyethoxy). *See* Diethylene glycol phenyl ether
Ethanol, 1-phenyl-. *See* α-Methylbenzyl alcohol
Ethanol, 2-phenyl-. *See* Phenethyl alcohol
Ethanol, 2-(phenylmethoxy)-. *See* Ethylene glycol benzyl ether
Ethanol, 2-propoxy. *See* Ethylene glycol propyl ether
Ethanol, 2-(2-propoxyethoxy)-. *See* Diethylene glycol propyl ether
Ethanol, 2,2´-thiodi-. *See* Thiodiglycol
1-Ethanol-2-thiol. *See* 2-Mercaptoethanol
Ethanol, undenatured. *See* Alcohol
Ethene, 1,1-dichloro-. *See* Vinylidene chloride monomer
Ethene, 1,2-dichloro-, (E)-. *See* trans-1,2-Dichloroethylene
Ethene, 1,2-dichloro-, (Z)-. *See* cis-1,2-Dichloroethylene
Ethene, [2-(2-ethoxyethoxy) ethoxy]-. *See* Diethylene glycol ethylvinyl ether
Ethene oxide. *See* Ethylene oxide
Ethenylbenzene. *See* Styrene
Ether. *See* Ethyl ether
Ether, bis (2-butoxyethyl). *See* Diethylene glycol dibutyl ether
Ether, bis[2-(2-methoxyethoxy)ethyl]. *See* Dimethoxytetraglycol
Ether chloratus. *See* Ethyl chloride
Ether cyanatus. *See* Propionitrile
Ether, dihexyl. *See* n-Hexyl ether
Ether, di-n-octyl. *See* Dioctyl ether
Ether, hexyl methyl. *See* Methyl hexyl ether
Ether hydrochloric. *See* Ethyl chloride
Ether, methyl t-pentyl. *See* t-Amyl methyl ether
Ether, methyl phenyl. *See* Anisole
Ether muriatic. *See* Ethyl chloride
2-Ethinylbutanol-2; Ethinylmethylethyl carbinol. *See* Methyl pentynol
Ethinyl trichloride. *See* Trichloroethylene
Ethohexadiol. *See* Ethyl hexanediol
Ethol. *See* Cetyl alcohol
Ethoxy acetate. *See* Ethoxyethanol acetate
1-Ethoxy-2-acetoxypropanol. *See* Propylene glycol ethyl ether acetate

Ethoxydiglycol

CAS 111-90-0; EINECS/ELINCS 203-919-7

Synonyms: 'Carbitol'; DEGEE; Diethylene glycol ethyl ether; Diethylene glycol monoethyl ether; Diglycol monoethyl ether; Ethanol, 2-(2-ethoxyethoxy); 2-(2-Ethoxyethoxy) ethanol; Ethyl diethylene glycol; Ethylene diglycol monoethyl ether

Classification: Aliphatic ether alcohol; glycol ether

Empirical: $C_6H_{14}O_3$

Formula: $CH_2OHCH_2OCH_2CH_2OC_2H_5$

Properties: Colorless liq., mild pleasant odor; sol. in water and org. solvs., e.g., acetone, benzene, chloroform, ethanol, ether; m.w. 134.20; dens. 0.990 (20 C); m.p. -76 C; b.p. 195-202 C; flash pt. (OC) 96.1 C; ref. index 1.425 (25 C)

Toxicology: LD50 (oral, rat) 5500 mg/kg, (IP, rat) 6310 mg/kg, (IV, rat) 2200 mg/kg, (skin, rat) 6000 mg/kg; mod. toxic by ing., IP, IV, and other routes; mildly toxic by skin contact; skin and eye irritant; can be absorbed thru skin, but health effects unlikely; ing. may cause CNS depression, respiratory depression, sl. kidney injury, thirst, metabolic acidosis and cyanosis; high/repeated exposure may affect kidney; experimental reproductive effects; mutation data; TSCA listed

Environmental: VOC; BOD5 0.58; COD 1.85; ThOD 1.91

Precaution: Combustible liq. and vapor; peroxide former; can react with oxidizers

Hazardous Decomp. Prods.: Heated to decomp., emits acrid smoke and irritating fumes

NFPA: Health 1, Flammability 1, Reactivity 0

Storage: Store in cool, dry, well-ventilated area, out of direct sunlight; very hygroscopic

Uses: Solvent for dyes, nitrocellulose, resins, paints, inks, textile printing, lacquers, quick-dry varnishes, enamels, wood stains, thinners, cosmetics; mutual solvent for mineral oil-soap and mineral oil-sulfonated oil mixts.; humectant; soaps; organic synthesis; brake fluid diluent; mfg. of plasticizers; solvent, thinner in nail enamels; ointment base, solvent, solubilizer, cosurfactant for pharmaceuticals; in food-pkg. adhesives; in paper/paperboard in contact with dry food

Regulatory: FDA 21CFR §175.105, 176.180; USP/NF compliance; SARA §313 reportable; HAP

Manuf./Distrib.: AMC Chems.; Adept Sol'ns.; Aldrich; Alfa Aesar; Allchem Ind.; Ashland; Dow; Eastman; Fluka; Gattefosse; Great Western; J.T. Baker; MMP; Occidental; Oxiteno; Penta Mfg.; Ruger; Sigma; Spectrum Quality Prods.; Stanchem; Trivent; Vopak USA

Trade Names: Carbitol®; Carbitol® Low Gravity; Eastman® DE; EDGE; Transcutol® CG; Transcutol® HP; Transcutol® P

Trade Names Containing: Eastman® DE-HG

Ethoxydiglycol acetate

CAS 112-15-2; EINECS/ELINCS 203-940-1

Synonyms: 'Carbitol' acetate; Diethylene glycol ethyl ether acetate; Di-

ethylene glycol monoethyl ether acetate; Diglycol monoethyl ether acetate; 2-(2-Ethoxyethoxy) ethanol acetate; 2-(2-Ethoxyethoxy) ethyl acetate
Empirical: $C_8H_{16}O_4$
Formula: $CH_3COO(CH_2)_2O(CH_2)_2OCH_2CH_3$
Properties: Colorless liq.; sol. in water; m.w. 176.22; sp.gr. 1.011 (20 C); visc. 2.8 cP; vapor pressure 0.1 mm Hg (20 C); m.p. -25 C; b.p. 214 C; flash pt. 107 C; autoignition temp. 360 C; ref. index 1.42 (20 C); surf. tens. 31.4 dynes/cm
Toxicology: LD50 (oral, rat) 11 g/kg, (skin, rabbit) 15,100 μl/kg; mod. toxic by ing.; skin and eye irritant; target organs: lung, kidney; TSCA listed
Environmental: VOC; BOD5 0.25; ThOD 1.09
Precaution: Combustible exposed to heat; can react with oxidizers
Hazardous Decomp. Prods.: Heated to decomp., emits acrid smoke and fumes
NFPA: Health 1, Flammability 1, Reactivity 0
Storage: Hygroscopic
Uses: Solvent, coalescing aid for latex paints; solvent in cosmetics; reagent; textile and silk-screen printing inks; slow-drying lacquers; in food-pkg. adhesives
Features: High-boiling
Regulatory: FDA 21CFR §175.105; SARA §313 reportable
Manuf./Distrib.: Aldrich; Allchem Ind.; Ashland; Delta Distributors; Dow; Eastman; Fluka; Lyondell; Occidental; Oxiteno; Spectrum Quality Prods.
Trade Names: Carbitol® Acetate; Eastman® DE Acetate

Ethoxyethane. *See* Ethyl ether

Ethoxyethanol

CAS 110-80-5; EINECS/ELINCS 203-804-1
UN 1171 (DOT)
Synonyms: 'Cellosolve'; 'Cellosolve' solvent; EE; 2-EE; EGEE; EGMEE; Ethanol, 2-ethoxy-; 2-Ethoxyethanol; Ethyl 'Cellosolve'; Ethylene glycol ethyl ether; Ethylene glycol monoethyl ether; Ethyl glycol; Glycol ether EE; Glycol ethyl ether; Glycol monoethyl ether; Hydroxy ether
Classification: Aliphatic ether alcohol; glycol ether
Empirical: $C_4H_{10}O_2$
Formula: $CH_3CH_2OCH_2CH_2OH$
Properties: Colorless liq.; mild sweet ether odor; misc. with hydrocarbons, alcohol, ether, acetone, liq. esters, water; m.w. 90.14; dens. 0.9311 (20/20 C); vapor pressure 5.31 mm Hg; m.p. -70 C; b.p. 135.6 C; flash pt. 48.9 C; ref. index 1.4060; surf. tens. 28.2 dynes/cm; dielec. const. 29.6
Toxicology: ACGIH TLV/TWA 5 ppm (skin); LD50 (oral, rat) 3 g/kg, (IP, rat) 2800 mg/kg, (IV, rat) 2400 mg/kg, (skin, rabbit) 3300 mg/kg; mod. toxic by ing., skin contact, IV, IP routes; mildly toxic by inhalation, subcut. routes; mild eye/skin irritant; experimental teratogen, reproductive effects; TSCA listed
Environmental: VOC; BOD5 1.03; COD 1.92; ThOD 1.96
Precaution: Flamm.; reacts with oxidizing materials; mod. explosion hazard in form of vapor when exposed to flames; may form explosive peroxides on exposure to light
Hazardous Decomp. Prods.: CO, CO_2
NFPA: Health 2, Flammability 2, Reactivity 0
Uses: Solvent for nitrocellulose, epoxy coatings, alkyd resins, surface coatings, lacquers, lacquer thinners, varnish removers, cleaning solutions, adhesives, cosmetics, pharmaceuticals; intermediate for 2-ethoxyethyl acetate; dyeing and printing textiles; in dye baths; leather treatment; emulsion stabilizer; anti-icing additive for aviation fuels; coupling solvent for sol. oils, agrochems.; wetting agent, penetrant for soil removal; diluent in food colorants; in food-pkg. adhesives; in food-contact rubber articles for repeated use
Regulatory: FDA 21CFR §73.1, 175.105, 177.2600; SARA §313 reportable; CERCLA hazardous substance; Calif. Prop. 65 reportable; HAP
Manuf./Distrib.: Aldrich; Allchem Ind.; Ashland; Brown; Dow; Fluka; Great Western; Lyondell; Occidental; Oxiteno; Romil Ltd; Ruger; Sigma; Spectrum Quality Prods.; Stanchem
Trade Names: Cellosolve®; EGE; Ethyl Icinol
Trade Names Containing: Ecosoft™ PE Solvent

2-Ethoxyethanol. *See* Ethoxyethanol

Ethoxyethanol acetate

CAS 111-15-9; EINECS/ELINCS 203-839-2
UN 1172 (DOT)
Synonyms: Acetic acid 2-ethoxyethyl ester; 'Cellosolve' acetate; EEA; 2-EEA; EGA; Ethoxy acetate; 2-Ethoxyethanol acetate; Ethoxyethyl acetate; 2-Ethoxyethyl acetate; β-Ethoxyethyl acetate; Ethylene glycol ethyl ether acetate; Ethylene glycol monoethyl ether acetate; Ethyl glycol acetate; Glycol monoethyl ether acetate; Oxytol acetate
Classification: Carboxylic acid ester
Definition: Ester of ethoxyethanol and acetic acid
Empirical: $C_6H_{12}O_3$
Formula: $CH_3COOCH_2CH_2OCH_2CH_3$
Properties: Colorless liq.; mild pleasant odor; sol. in ≈ 6 parts water; misc. with aromatic hydrocarbons, most org. solvs.; m.w. 132.16; dens. 0.975 (20/20 C); vapor pressure 2.34 mm Hg; m.p. -62 C; b.p. 156 C; flash pt. (CC) 47 C; ref. index 1.4030
Toxicology: ACGIH TLV/TWA 5 ppm (skin); LD50 (oral, rat) 5.1 g/kg, (IP, mouse) 1420 mg/kg, (skin, rabbit) 10,500 mg/kg; mod. toxic by ing., IP routes; toxic by skin absorp.; skin/eye irritant; inh. may cause mild irritation of nose, throat, and eyes; high concs. may cause CNS depression; ing. may cause GI tract irritation, CNS depression, kidney damage; eye irritant at 1000 ppm; experimental teratogen, reproductive effects; TSCA listed
Environmental: VOC; BOD5 0.65; COD 1.76; ThOD 1.81
Precaution: Combustible; reacts with oxidizing materials; incompat. with acids (can hydrolyze to form acetic acid and alcohol), bases (can hydrolyze to form acetate salt and alcohol); mod. explosion hazard as vapor when heated
Hazardous Decomp. Prods.: Heated to decomp., emits acrid smoke and irritating fumes
NFPA: Health 2, Flammability 2, Reactivity 0
Storage: Store in cool, dry, well-ventilated area, out of direct sunlight; use in minimal quantities; avoid generating mist
Uses: Solvent for paints/coatings (NC, cellulose acetate, ethyl cellulose, butyrate, acrylic, urethane), oils, resins, printing inks, photographic film, lacquers, nail polish, cosmetics; evaporation retarder, gloss aid in automobile lacquers; blush retarder in lacquers; varnish removers; wood stains; textiles; leather finishing; in food-pkg. adhesives
Regulatory: FDA 21CFR §175.105; SARA §313 reportable; Calif. Prop. 65 reportable
Manuf./Distrib.: Aldrich; Allchem Ind.; Ashland; Dow; Fluka; Lyondell; Occidental; Sal Chem.; Spectrum Quality Prods.; Stanchem
Trade Names: Cellosolve® Acetate

2-Ethoxyethanol acetate. *See* Ethoxyethanol acetate
2-(2-Ethoxyethoxy) ethanol. *See* Ethoxydiglycol
2-(2-Ethoxyethoxy) ethanol acetate. *See* Ethoxydiglycol acetate
1-Ethoxy-2-(β-ethoxyethoxy) ethane. *See* Diethylene glycol diethyl ether
2-[2-(2-Ethoxyethoxy) ethoxy] ethanol. *See* Ethoxytriglycol
2-[2-(2-Ethoxyethoxy) ethoxy] ethene. *See* Diethylene glycol ethylvinyl ether
2-(2-Ethoxyethoxy) ethyl acetate. *See* Ethoxydiglycol acetate
Ethoxyethoxyethyl acrylate. *See* 2-(2-Ethoxyethoxy) ethyl acrylate

2-(2-Ethoxyethoxy) ethyl acrylate

CAS 7328-17-8; EINECS/ELINCS 230-811-7
Synonyms: Acrylic acid, 2-(2-ethoxyethoxy) ethyl ester; 'Carbitol' acrylate; Diethylene glycol ethyl ether acrylate; EEEA; EOEOEA; Ethoxyethoxyethyl acrylate; 2-(2-Ethoxyethoxy) ethyl 2-propenoate; Ethoxyethyl acrylate; Ethyl 'Carbitol' acrylate; 2-Propenoic acid, 2-(2-ethoxyethoxy) ethyl ester
Empirical: $C_9H_{16}O_4$
Properties: Colorless liq.; ester-like odor; sol. 140 g/l in water; misc. with other solvs.; m.w. 188.22; dens. 1.0062 g/cm^3 (20 C); vapor pressure 14.4 mbar (50 C); m.p. -66 C; b.p. 250 C; flash pt. > 110 C; ignition temp. 210 C; ref. index 1.439 (20 C)
Toxicology: LD50 (oral, rat) 1100 mg/kg; primary skin irritant; may cause skin sensitization on repeated contact; harmful by ing.; irritating to eyes, mucous membranes, upper respiratory tract; may be harmful by inh., skin absorp.; may cause coughing, chest pains, difficulty breathing; prolonged/repeated skin contact causes defatting, dermatitis
Environmental: Easy to eliminate (> 70%); EC50 (daphnia magna, 48 h) 10.56 mg/l, (algae, 72 h) 21 mg/l; prevent entry into inshore waters

without pretreatment; EC not hazardous to the environment

Precaution: LEL 1 vol.% (50 C); protect against fire and explosion; may form ignitable air mixts. if heated above flash pt., sprayed, or atomized; ground equip.; dangerous reaction with radical formers; incompat. with heat, strong oxidizing agents, direct sunlight, inert gases, free radical initiators; can polymerize if shelf life or storage temp. are greatly exceeded; heat develops during polymerization; closed containers may rupture and explode during runaway polymerization

Hazardous Decomp. Prods.: Toxic fumes of CO, CO_2; emits toxic fumes under fire conditions

Storage: 6 mos max. shelf life; store in cool, dry place under air, below 25 C; protect from light; keep away from heat, sparks, open flame; do not freeze

Uses: Monomer, reactive diluent for radiation-curable ink and coating formulations

Manuf./Distrib.: ABCR; Aldrich; BASF; Cornelius Chem. Co. Ltd; Monomer-Polymer & Dajac Labs; Morton Int'l.

Trade Names: SR 256

2-(2-Ethoxyethoxy) ethyl 2-propenoate. *See* 2-(2-Ethoxyethoxy) ethyl acrylate

1-(2-Ethoxyethoxy)-2-vinyloxy ethane. *See* Diethylene glycol ethylvinyl ether

Ethoxyethyl acetate; 2-Ethoxyethyl acetate; β-Ethoxyethyl acetate. *See* Ethoxyethanol acetate

Ethoxyethyl acrylate. *See* 2-(2-Ethoxyethoxy) ethyl acrylate

Ethoxyethyl ether; 2-Ethoxyethyl ether. *See* Diethylene glycol diethyl ether

2-Ethoxyethyl 2-(vinyloxy) ethyl ether. *See* Diethylene glycol ethylvinyl ether

Ethoxyformic anhydride. *See* Diethyl carbonate

Ethoxylated tridecyl alcohol. *See* Trideceth

1-[2-(2-Ethoxy-1-methylethoxy)-1-methylethoxy]-2-propanol. *See* Tripropylene glycol ethyl ether

1-2-(Ethoxy-2-methylethoxy)-2-propanol. *See* Dipropylene glycol ethyl ether

(2-Ethoxymethylethoxy) propanol

CAS 30025-38-8

Synonyms: Ethoxypropoxypropanol; Propanol, (2-ethoxymethylethoxy)-

Empirical: $C_8H_{18}O_3$

Properties: Dens. 0.942 kg/l (20 C); vapor pressure 0.14 mbar (20 C); b.p. 162-228 C; flash pt. (ACC) 86.5 C; autoignition temp. 190 C; ref. index 1.422 (20 C); vol. resist. 5.7 x 10^4 ohm

Precaution: LFL 0.75 vol.%; UFL 9.6 vol.%

Uses: Solvent

Manuf./Distrib.: BP Amoco plc

Ethoxymethyl ethyl ether. *See* Diethoxymethane

Ethoxynonafluorobutane. *See* Ethyl nonafluorobutyl ether

3-Ethoxypropanoic acid, ethyl ester. *See* Ethyl 3-ethoxypropionate

Ethoxypropanol; 1-Ethoxy-2-propanol. *See* Propylene glycol ethyl ether

3-Ethoxy-1-propanol

CAS 111-35-3; EINECS/ELINCS 203-861-2

UN 1993

Synonyms: 1-Propanol, 3-ethoxy-; Propylene glycol monoethyl ether; β-Propylene glycol monoethyl ether; Propylene glycol β-monoethyl ether

Classification: Glycol ether

Empirical: $C_5H_{12}O_2$

Properties: Colorless clear liq.; sol. (@ 18 C): ≥ 100 mg/ml in water, DMSO, 95% ethanol, acetone; m.w. 104.17; sp.gr. 0.904 (20/4 C); vapor pressure 8.2 mm Hg; b.p. 160-161 C; flash pt. 54 C; ref. index 1.4172 (20 C)

Toxicology: LD50 (oral, rat) 11,752 mg/kg, (skin, rabbit) 2558 mg/kg; mod. toxicity via skin route; mild skin irritant; harmful by inh., ing., skin contact; may cause eye and lung irritation; may be narcotic in high concs.; reproductive/developmental toxicity; target organs: skin, CNS; TSCA listed

Environmental: VOC; ThOD 2.15

Precaution: Flamm. liq.; incompat. with strong oxidizing agents, strong acids, strong bases

Hazardous Decomp. Prods.: CO, CO_2; heated to decomp., emits acrid smoke and fumes

Storage: Store in cool dry place away from heat, open flame, other ignition sources; store refrigerated; keep tightly closed

Uses: Solvent

Regulatory: SARA §313 reportable; HAP

Manuf./Distrib.: Aldrich; ICN Biomed. Research Prods.

Ethoxypropanol acetate; 1(or 2)-Ethoxypropanol acetate; 1-Ethoxy-2-propanol acetate. *See* Propylene glycol ethyl ether acetate

Ethoxypropionic acid ethyl ester; 3-Ethoxypropionic acid ethyl ester. *See* Ethyl 3-ethoxypropionate

Ethoxypropoxypropanol. *See* (2-Ethoxymethylethoxy) propanol

Ethoxypropyl acetate; 1-Ethoxy-2-propyl acetate. *See* Propylene glycol ethyl ether acetate

Ethoxy triethylene glycol. *See* Ethoxytriglycol

Ethoxytriglycol

CAS 112-50-5; EINECS/ELINCS 203-978-9

Synonyms: 2-[2-(2-Ethoxyethoxy) ethoxy] ethanol; Ethoxy triethylene glycol; Ethyl triglycol; PEG-3 ethyl ether; Triethylene glycol ethyl ether; Triethylene glycol monoethyl ether; Triglycol ethyl ether; Triglycol monoethyl ether

Empirical: $C_8H_{18}O_4$

Formula: $C_2H_5O(C_2H_4O)_3H$

Properties: Colorless liq.; sol. in water; misc. with most org. solvs.; m.w. 178.26; dens. 1.0208 (20/20 C); vapor pressure 0.01 mm (20 C); b.p. 255.4 C; f.p. -18.7 C; flash pt. (OC) 135 C; surf. tens. 33.7 dynes/cm (20 C)

Toxicology: LD50 (oral, rat) 10,610 mg/kg, (skin, rabbit) 8 g/kg; mildly toxic by ing. and skin contact; eye irritant; TSCA listed

Precaution: Combustible; can react with oxidizing materials

Hazardous Decomp. Prods.: Heated to decomp., emits acrid smoke and fumes

NFPA: Health 0, Flammability 1, Reactivity 0

Uses: Chemical intermediate; solvent for lacquers and paints

Manuf./Distrib.: Ashland; Dow; Fluka; Oxiteno

Ethyl acetate

CAS 141-78-6; EINECS/ELINCS 205-500-4

UN 1173 (DOT); FEMA 2414

Synonyms: Acetic acid, ethyl ester; Acetic ester; Acetic ether; Acetoxyethane; Ethyl acetic ester; Ethyl ethanoate; Vinegar naphtha

Classification: Sat. aliphatic carboxylic acid ester

Definition: Ester of ethyl alcohol and acetic acid

Empirical: $C_4H_8O_2$

Formula: $CH_3COOC_2H_5$

Properties: Colorless liq., fragrant fruity odor, acetous burning taste; sol. in chloroform, alcohol, ether, acetone, benzene, fixed and volatile oils; sl. sol. in water; m.w. 88.12; dens. 0.902 (20/4 C); bulk dens. 0.8945 g/ml; f.p. -83.6 C; b.p. 77 C; flash pt. -4.4 C; ref. index 1.3723; surf. tens. 24 dynes/cm (20 C)

Toxicology: ACGIH TLV/TWA 400 ppm; LD50 (oral, rat) 5620 mg/kg, (IP, mouse) 709 mg/kg; LC50 (inh., rat, 8 h) 1600 ppm; poison by inh.; mod. toxic by IP and subcut. routes; mildly toxic by ing.; irritant to eyes, skin, mucous membranes; human systemic effects by inh.; mildly narcotic; CNS depressant; can cause dermatitis; high concs. can cause liver/kidney congestion; mutagenic data; TSCA listed

Environmental: VOC; BOD5 1.00; COD 1.69; ThOD 1.82

Precaution: DOT: Flamm. liq.; very dangerous fire hazard exposed to heat or flame; can react vigorously with oxidizers; incompat. with strong oxidizers, strong acids, bases, potassium t-butoxide

Hazardous Decomp. Prods.: Ethanol, acetic acid; heated to decomp., emits acrid smoke and irritating fumes

NFPA: Health 1, Flammability 3, Reactivity 0

Storage: Store in cool, dry, well-ventilated area, out of direct sunlight, away from heat/ignition sources

Uses: Solvent in cellulosics, automotive paints, coatings, plastics, resins, lacquers, printing inks, adhesives, organic synthesis, smokeless powders, artificial leather, photographic films and plates, cosmetics, pharmaceuticals, perfumes; vehicle in cosmetics; synthetic fruit essences; flavoring agent in foods, pharmaceuticals; fragrance ingred.; spotting solvent (dry cleaning); cleaning textiles; nail polish removers; color diluent, extraction solvent for foods; reagent; extractant, purification agent in embalming fluids; recovery, crystallization of drugs; adjuvant in resinous/polymeric food-contact coatings; in cellophane for food pkg.

Features: Pineapple-like flavor

Regulatory: FDA 21CFR §73.1, 172.560, 172.859, 173.228, 175.320, 177.1200, 182.60, GRAS; 27CFR §21.106; FEMA GRAS; FDA approved for ophthalmics, orals, topicals; NF, BP, EP compliance; CERCLA hazardous substance; Japan approved with restrictions
Manuf./Distrib.: AAE Chemie NV; AMC Chems.; AP Resources; Aastrid Int'l.; Advanced BioTech; Aldrich; Allchem Ind.; Ashland; Axxence Aromatic GmbH; BP Amoco; BP Chems. Ltd; Baychem; Berje; Brenntag AG; Brown; Celanese; Chemcentral; Chisso Am.; Chisso Petrochem.; Daicel Chem. Ind.; Degussa AG; Degussa Flavors/Alfrebro; Dow; Eastman; Elan; Fisher Scientific; Fleurchem; Fluka; GFS; General Chem.; Harcros; Houghton Chem.; Hukill; Integra; J.T. Baker; Lluch Essence; Lonza Ltd; Mallinckrodt Baker; Monsanto; Moore Ingreds.; Penta Mfg.; Punda Mercantile; Romil Ltd; Ruger; Sal Chem.; Sigma; Spectrum Quality Prods.; Stanchem; Sunnyside; Synasia Fine Chem.; Thomas Scientific; Tokuyama Petrochem.; Vinchem; Vopak USA; Xinchem
Trade Names: Eastman® Ethyl Acetate, Food Grade
Trade Names Containing: Filmex® C 190 Proof; Filmex® C Anhydrous; Propsolv I-1 190 Proof; Propsolv I-1 Anhydrous; Propsolv III-1 190 Proof; Propsolv III-1 Anhydrous; Propsolv III-2 190 Proof; Propsolv III-2 Anhydrous; Punctilious® SDA 29-3 190 Proof; Punctilious® SDA 35A-1 190 Proof; Punctilious® SDA 35A-1 Anhydrous; Punctilious® SDA Canada DAG-2A Anhydrous; Punctilious® SDA Canada DAG-2A2 Anhydrous; Tecsol® 1; Tecsol® 3; Tecsol® C 95%; Tecsol® C Anhydrous

Ethyl acetic ester. *See* Ethyl acetate

Ethylacetoacetate

CAS 141-97-9; EINECS/ELINCS 205-516-1
UN 1993 (DOT); FEMA 2415
Synonyms: Acetoacetic acid ethyl ester; Acetoacetic ester; Diacetic ether; Diacetyl ether; EAA; EEA; Ethyl acetyl acetate; Ethyl acetylacetonate; Ethyl benzyl acetoacetate; Ethyl 3-oxobutanoate; Ethyl 3-oxobutyrate; 3-Oxobutanoic acid ethyl ester
Classification: Aliphatic organic compd.
Empirical: $C_6H_{10}O_3$
Formula: $CH_3COCH_2COOC_2H_5$
Properties: Colorless liq., fruity odor; sol. in ≈ 35 parts water; misc. with oxygenated, chlorinated, and aromatic solvs.; m.w. 130.14; dens. 1.0213 (25/4 C); m.p. -45 C; b.p. 180.8 C; flash pt. (CC) 184 F; ref. index 1.4180-1.4195
Toxicology: LD50 (oral, rat) 3.98 g/kg; mod. toxic by ingestion; mod. irritating to skin, mucous membranes, eyes; TSCA listed
Precaution: DOT: Flamm. liq.; combustible liq. when exposed to heat or flame; can react with oxidizing materials; explosive reactions possible
Hazardous Decomp. Prods.: Heated to decomp., emits acrid smoke and irritating fumes
NFPA: Health 2, Flammability 2, Reactivity 0
Uses: Catalyst for esterification and olefin polymerization; resin crosslinking agent for automotive prods., coatings, elastomers, films/paints, graphic arts, plastics; synthetic flavoring agent in foods and pharmaceuticals; fragrance; solvent; extractant; solvent for fragrances in personal care; intermediate for vitamins, feed additives, drugs, pesticides; prod. of azo dyes; in food-contact coatings
Features: Sweet flavor
Regulatory: FDA 21CFR §172.515, 175.300; FEMA GRAS; Japan approved as flavoring
Manuf./Distrib.: Aceto; Aldrich; Berje; Classic Flavors & Fragrances; Daicel Chem. Ind.; Eastman; Fluka; Grau Aromatics; Haarmann & Reimer GmbH; Honeywill & Stein; Integra; Lluch Essence; Lonza; Mallinckrodt Baker; Nippon Syn. Chem. Ind.; Penta Mfg.; Prodasynth; S.A. Chems.; Sigma; Spectrum Quality Prods.; VWR Int'l.; Wacker Chems.; Wacker-Chemie GmbH
Trade Names: Eastman® EAA
Trade Names Containing: Punctilious® SDA 29-2 190 Proof; Punctilious® SDA 29-2 Anhydrous

Ethyl acetone. *See* Methyl propyl ketone
Ethyl acetyl acetate; Ethyl acetylacetonate. *See* Ethylacetoacetate
Ethyl 3-acetylpropionate. *See* Ethyl levulinate
Ethylal. *See* Diethoxymethane
Ethyl alcohol; Ethyl alcohol, undenatured. *See* Alcohol

Ethylamine

CAS 75-04-7
UN 1036 (DOT); UN 2270 (DOT)
Synonyms: Aminoethane; 1-Aminoethane; EA; Ethanamine; Ethylamine anhydrous; MEA; Monoethylamine
Classification: Saturated aliphatic amine
Empirical: C_2H_7N
Formula: $CH_3CH_2NH_2$
Properties: Colorless liq. below 16.6 C; colorless gas above 17 C; strong ammonia-like odor; sol. in water, ethanol, acetone, ether, dimethyl sulfoxide; misc. with oxygenated solvs.; m.w. 45.09; dens. 0.689 (15 C); vapor pressure 873 mm Hg (20 C); m.p. -81 C; b.p. 16.6 C; flash pt. (CC) < -18 C; pH 1.9 (0.1 M sol'n.)
Toxicology: ACGIH TLV/TWA 5 ppm (skin); STEL 15 ppm (skin); LD50 (oral, rat) 400 mg/kg, (skin, rabbit) 390 mg/kg; poison by ing., skin contact, IV routes; mod. toxic by inh.; skin irritant; severe eye irritant; lachrymator; inh. of gas can irritate nose, throat, lungs; high concs. may cause pulmonary edema that might be fatal; TSCA listed
Environmental: BOD5 1.11; ThOD 2.13
Precaution: Flamm., corrosive, toxic gas or liq.; dangerous fire hazard; mod. explosion hazard; incompat. with mercury, acids, oxidizing agents, calcium or sodium hypochlorite, nitromethane, nitrosyl perchlorate; violent or explosive reactions possible
Hazardous Decomp. Prods.: Heated to decomp., emits toxic fumes of NO_x
NFPA: Health 3, Flammability 4, Reactivity 0
Storage: Store in cool, well-ventilated area out of direct sunlight; protect from extreme temps.; never expose cylinders to temps. > 52 C or < -29 C
Uses: Intermediate; organic synthesis; mfg. of dyes; pharmaceuticals; emulsifying agents; syn. resins; detergents; pesticides; as solvent; solvent extraction; petroleum refining; stabilizer for rubber latex; curing agent for epoxy resins
Manuf./Distrib.: Air Prods.; Aldrich; Ashland; Atofina; BASF AG; Dow; Fluka; Sigma; Spectrum Quality Prods.

Ethylamine anhydrous. *See* Ethylamine
Ethylamine, N-nitrosodi-. *See* N-Nitrosodiethylamine
Ethyl amyl ketone; Ethyl n-amyl ketone. *See* 3-Octanone
2-(N-Ethylanilino) ethanol. *See* Phenylethylethanolamine

Ethylbenzene

CAS 100-41-4; EINECS/ELINCS 202-849-4
UN 1175 (DOT)
Synonyms: EB; Ethylbenzol; Phenylethane
Classification: Aromatic hydrocarbon
Empirical: C_8H_{10}
Formula: $C_6H_5C_2H_5$
Properties: Colorless liq., aromatic odor; sol. in alcohol, benzene, CCl_4, ether; pract. insol. in water; m.w. 106.17; dens. 0.867 (20 C); vapor pressure 10 mm (25.9 C); f.p. -95 C; b.p. 136.187 C; flash pt. (TCC) 15 C; autoignition temp. 432 C; ref. index 1.495 (20 C); surf. tens. 28.48 dynes/cm; KB value 96; dielec. const. 2.042
Toxicology: ACGIH TLV/TWA 100 ppm; STEL 125 ppm; LD50 (oral, rat) 3500 mg/kg, (dermal, rabbit) 17,800 mg/kg; mod. toxic by ing., IP route; mildly toxic by inh. and skin contact; eye irritant; inh. of vapor can irritate nose and throat, cause CNS depression, headache, vomiting, nausea; can cause severe lung damage or death if liq. is aspirated into lungs; experimental teratogen; human mutagenic data; TSCA listed
Environmental: VOC; BOD5 0.92; ThOD 3.17
Precaution: Highly flamm.; LEL 1%; UEL 6.7%; dangerous fire risk; incompat. with strong oxidizing agents (can react violently; increased risk of fire and explosion)
Hazardous Decomp. Prods.: Heated to decomp., emits acrid smoke and irritating fumes
NFPA: Health 2, Flammability 3, Reactivity 0
Storage: Store in cool, dry, well-ventilated area of out direct sunlight, away from ignition sources
Uses: Solvent and raw material for prod. of cellulose acetate, acetophenone, diethyl benzene, anthraquinones; chem. intermediate in mfg. of styrene, other plastics, syn. rubber, styrene latexes; minor component of gasoline and aviation fuels; reagent; cosmetics ingred.

Regulatory: SARA §313 reportable; CERCLA hazardous substance; HAP
Manuf./Distrib.: Aldrich; Alvin Chem.; Ashland; BP Amoco; ChevronPhillips; Dow; Fluka; Huntsman; Lyondell; Sigma; Spectrum Quality Prods.; Sterling Chems.
Trade Names Containing: ExxonMobil T 500-100; Kristalloel 21; Shellsol® 1495; Shellsol® 1626; Solvent 2216; Solvent 2310; Solvent 2339; Solvent 2439; Solvent 2495; Solvent 30 823; VM&P Naphtha EC; VM&P Naphtha HT; X-95 Solvent

Ethyl benzeneacetate. *See* Ethyl phenylacetate
Ethyl benzenecarboxylate. *See* Ethyl benzoate

Ethyl benzoate

CAS 93-89-0; EINECS/ELINCS 202-284-3
FEMA 2422
Synonyms: Benzoic acid ethyl ester; Benzoic ether; Essence of niobe; Ethyl benzenecarboxylate
Definition: Ester of ethyl alcohol and benzoic acid
Empirical: $C_9H_{10}O_2$
Formula: $C_6H_5COOC_2H_5$
Properties: Colorless liq., aromatic odor; sol. in alcohol, fixed oils, propylene glycol; misc. with chloroform, ether, petroleum ether; insol. in water, glycerin; m.w. 150.18; dens. 1.046 (20/4 C); m.p. -34 C; b.p. 211-214 C; flash pt. 184 F; ref. index 1.505 (20 C)
Toxicology: LD50 (oral, rat) 2100 mg/kg; LDLo (skin, cat) 10 g/kg; mod. toxic by ing.; mildly toxic by skin contact; skin and eye irritant; vapors cause cough; TSCA listed
Precaution: Combustible liq.; can react with oxidizers
Hazardous Decomp. Prods.: Heated to decomp., emits acrid smoke and irritating fumes
NFPA: Health 1, Flammability 1, Reactivity 0
Uses: Preservative in cosmetics; fragrance ingred. in perfumery; synthetic flavoring agent in foods and pharmaceuticals; solvent for many cellulose derivs. and natural and synthetic resins; lacquers
Features: Banana-, cherry-, plum-like flavor
Use Level: 0.5% max. as benzoic acid in finished cosmetics
Regulatory: FDA 21CFR §172.515; FEMA GRAS
Manuf./Distrib.: Advanced BioTech; Advanced Synthesis Tech.; Aldrich; Allan; Axxence Aromatic GmbH; Bell Flavors & Fragrances; Chemial SpA; Degussa Flavors/Alfrebro; Elan; Fleurchem; Fluka; Goldensun Mfg.; Lluch Essence

Ethylbenzol. *See* Ethylbenzene
Ethyl benzyl acetoacetate. *See* Ethylacetoacetate
Ethylbis (2-hydroxyethyl) amine. *See* N-Ethyldiethanolamine

Ethyl bromide

CAS 74-96-4; EINECS/ELINCS 200-825-8
UN 1891
Synonyms: Bromic ether; Bromoethane; Ethane, bromo-; Hydrobromic ether; Monobromoethane
Empirical: C_2H_5Br
Formula: CH_3CH_2Br
Properties: Colorless to lt. yel. clear liq.; turns yel. on exposure to air and light; ethereal odor; burning taste; sol. in DMSO, 95% ethanol, acetone, ether; misc. with alcohol, chloroform, org. solvs.; sol. < 1 mg/ml in water; m.w. 108.98; sp.gr. 1.460 (20/4); vapor pressure 375 mm Hg (20 C); m.p. -119 C; b.p. 38.4 C; flash pt. -23 C; ref. index 1.4235 (20 C)
Toxicology: ACGIH TLV/TWA 200 ppm; STEL 250 ppm; LD50 (oral, rat) 1350 mg/kg, (IP, mouse) 2850 mg/kg; LC50 (inh., rat, 1 h) 26,980 ppm; poison; mod. toxic by ing., IP routes; mildly toxic by inh.; irritating to eyes, skin, mucous membranes, upper respiratory; vapors very irritating to lungs; may cause acute congestion/edema, liver/kidney damage, nausea, dizziness, CNS depression, cardiac arrhythmias/arrest, etc.; anesthetic; narcotic in high concs.; cancer suspect agent; target organ: reproductive; TSCA listed
Precaution: Flamm.; UEL 11.25%; vapor may flash back; incompat. with strong oxidizing agents, strong bases; reacts with chemically act. metals (Na, K, Ca, powd. Al, Zn, Mg); attacks some plastics, rubbers, coatings
Hazardous Decomp. Prods.: Heated to decomp., emits toxic fumes of CO, CO_2, bromine, hydrogen bromide gas; emits toxic fumes under fire conditions; reacts with steam to produce toxic/corrosive fumes
NFPA: Health 2, Flammability 1, Reactivity 0
Storage: Sensitive to heat; protect from light; keep tightly closed under inert atmosphere; store in explosion-proof refrigerator; store away from ignition sources
Uses: Ethylating agent in organic synthesis; anesthetic in medicine; refrigerant; solvent; grain and fruit fumigant; in gasoline; formerly as topical and inhalation anesthetic
Manuf./Distrib.: Acros Org.; Albemarle; Aldrich; Alfa Aesar; Alfa Chem; Am. Radiolabeled Chems.; Dead Sea Bromine; Fluka; Honeywell Spec. Chems.; Isotec; Lancaster Synthesis; Pfaltz & Bauer; Spectrum Quality Prods.; TCI Am.; VWR Int'l.; Yixing Fangqiao East Chemical

Ethyl butanoate. *See* Ethyl butyrate
2-Ethylbutanol. *See* 2-Ethyl-1-butanol

2-Ethyl-1-butanol

CAS 97-95-0; EINECS/ELINCS 202-621-4
UN 2275 (DOT); UN 2282 (DOT)
Synonyms: 2,2-Diethyl ethanol; 2EBO; 2-Ethylbutanol; 2-Ethyl butan-1-ol; 2-Ethylbutanol-1; 2-Ethylbutyl alcohol; s-Hexanol; s-Hexyl alcohol; 3-(Hydroxymethyl)-pentan; 3-Methylolpentane; 3-Pentylcarbinol; s-Pentylcarbinol; Pseudohexyl alcohol
Classification: Primary aliphatic alcohol
Empirical: $C_6H_{14}O$
Formula: $CH_3CH_2CH(CH_2CH_3)CH_2OH$
Properties: Colorless clear liq.; sweet musty odor; sol. in most org. solvs.; sl. sol. in water; m.w. 102.17; sp.gr. 0.8328 (20 C); vapor pressure 0.9 mm Hg (20 C); m.p. -114 C; b.p. 150 C; flash pt. (CC) 58 C; ref. index 1.4229 (20 C)
Toxicology: LD50 (oral, rat) 1850 mg/kg, (skin, rabbit) 1260 mg/kg; mod. toxic by ing. and skin contact; skin and severe eye irritant; inh. may cause nose/throat irritation, CNS effects; ing. may cause 'alcohol' intoxication symptoms; severe exposure may cause unconsciousness and coma; may cause severe lung damage, respiratory arrest, and death if aspirated into lungs; TSCA listed
Precaution: Flamm. exposed to heat or flame; incompat. with oxidizing agents (increase fire/explosion hazard)
Hazardous Decomp. Prods.: Heated to decomp., emits acrid smoke and irritating fumes
NFPA: Health 1, Flammability 2, Reactivity 0
Storage: Store in cool, dry, well-ventilated area
Uses: Solvent for oils, resins, waxes, dyes; synthesis of pharmaceuticals, surfactants, perfumes, syn. lubricants, lubricant additives; diluent; flavoring; cosmetics ingred.; esterification agent for mfg. of org. compds. used as plasticizers, lubricant additives, hydraulic fluids, diffusion pump oils, heat transfer media, pharmaceuticals, crop protection prods., herbicides, fungicides, insecticides, acaricides, wetting agents; penetrating oils; cleaning agents for printed circuit boards; metal extraction and separating agent
Manuf./Distrib.: Aldrich; Celanese; Fluka

2-Ethyl butan-1-ol; 2-Ethylbutanol-1. *See* 2-Ethyl-1-butanol
Ethyl trans-2-butenoate. *See* Ethyl crotonate
3-Ethylbutinol. *See* Methyl pentynol
Ethyl butylacetate. *See* Ethyl caproate

2-Ethylbutyl acetate

CAS 10031-87-5; EINECS/ELINCS 233-095-4
FEMA 2425
Synonyms: Acetic acid 2-ethylbutyl ester
Definition: Obtained by reacting 2-ethylbutanol with acetic anhydride in the presence of sulfuric acid
Empirical: $C_8H_{16}O_2$
Formula: $C_2H_5CH(C_2H_5)CH_2OOCCH_3$
Properties: Colorless liq.; mild odor; m.w. 144.21; dens. 0.876; b.p. 160-163 C; flash pt. 52 C; ref. index 1.4109
Toxicology: TSCA listed
Precaution: Combustible; moderate fire risk
NFPA: Health 1, Flammability 2, Reactivity 0
Uses: Solvent for nitrocellulose lacquers; synthetic flavoring agent in foods and pharmaceuticals
Features: Fruity flavor

Regulatory: FDA 21CFR §172.515; FEMA GRAS
Manuf./Distrib.: Aldrich

2-Ethylbutyl alcohol. *See* 2-Ethyl-1-butanol
Ethyl butyl carbinol. *See* 3-Heptanol
Ethyl butyl ketone; Ethyl-n-butyl ketone. *See* 3-Heptanone
3-Ethylbutynol. *See* Methyl pentynol

Ethyl butyrate

CAS 105-54-4; EINECS/ELINCS 203-306-4
UN 1180 (DOT); FEMA 2427
Synonyms: Butanoic acid, ethyl ester; Butyric acid, ethyl ester; Butyric ether; Ethyl butanoate; Ethyl n-butyrate
Classification: Butanoic acid ester
Definition: Obtained by esterification of n-butyric acid with ethyl alcohol in presence of Twitchell's reagent or $MgCl_2$
Empirical: $C_6H_{12}O_2$
Formula: $CH_3CH_2CH_2COOC_2H_5$
Properties: Colorless liq., banana-pineapple odor; sol. in water, fixed oils, propylene glycol; misc. with alcohol, ether; insol. in glycerin @ 121 C; m.w. 116.18; dens. 0.874; m.p. -100.8 C; b.p. 121.6 C; flash pt. (CC) 24 C; ref. index 1.391
Toxicology: LD50 (oral, rat) 13 g/kg; mildly toxic by ingestion; skin irritant; TSCA listed
Precaution: Flamm. liq.; can react vigorously with oxidizing materials
Hazardous Decomp. Prods.: Heated to decomp., emits acrid smoke and irritating fumes
NFPA: Health 0, Flammability 3, Reactivity 0
Uses: Perfumery; cosmetics ingred.; synthetic flavoring agent and fragrance in foods and pharmaceuticals; flavor improver for foods; solvent mixt. for cellulose ethers and esters
Features: Ethereal, banana-, strawberry-, pineapple-like flavor and fragrance
Regulatory: FDA 21CFR §182.60, GRAS; FEMA GRAS; Japan approved as flavoring
Manuf./Distrib.: Advanced BioTech; Aldrich; Axxence Aromatic GmbH; BASF; Bell Flavors & Fragrances; Boith China; Degussa Flavors/Alfrebro; Elan; Fleurchem; Fluka; Frutarom Ltd; Goldensun Mfg.; Haltermann Prods. UK; Lluch Essence; Moore Ingreds.; Natural Advantage; Sarcom; Sigma; Xinchem

Ethyl n-butyrate. *See* Ethyl butyrate
Ethyl caprate; Ethyl caprinate. *See* Ethyl decanoate

Ethyl caproate

CAS 123-66-0; EINECS/ELINCS 204-640-3
FEMA 2439
Synonyms: Ethyl butylacetate; Ethyl capronate; Ethyl hexanoate; Ethyl hexoate; Ethyl hexylate; Hexanoic acid ethyl ester
Definition: Obtained by esterification of caproic acid with ethyl alcohol in presence of conc. H_2SO_4 or HCl
Empirical: $C_8H_{16}O_2$
Formula: $CH_3(CH_2)_4COOC_2H_5$
Properties: Colorless to ylsh. liq., mild wine odor; sol. in fixed oils, most org. solvs.; sl. sol. in propylene glycol; misc. with alcohol, ether; insol. in water, glycerin; m.w. 144.24; dens. 0.867-0.871; b.p. 163 C; flash pt. (OC) 130 F; ref. index 1.406-1.409
Toxicology: Skin irritant; TSCA listed
Precaution: Flamm. or combustible liq.; can react with oxidizing materials
Hazardous Decomp. Prods.: Heated to decomp., emits acrid smoke and irritating fumes
NFPA: Health 2, Flammability 2, Reactivity 0
Uses: Solvent; organic synthesis; synthetic flavoring agent and fragrance in foods and pharmaceuticals; flavor enhancer for foodstuffs; mfg. of artificial fruit flavors; cosmetics ingred.
Features: Fruity, sl. fermented, apple, banana, pineapple-like flavor and fragrance
Regulatory: FDA 21CFR §172.515; FEMA GRAS; Japan approved as flavoring
Manuf./Distrib.: A&E Connock; Advanced BioTech; Advanced Synthesis Tech.; Aldrich; Axxence Aromatic GmbH; BASF; Bell Flavors & Fragrances; Boith China; Degussa Flavors/Alfrebro; Elan; Fleurchem; Fluka; Frutarom Ltd; Grau Aromatics; Lluch Essence; Moore Ingreds.; Natural Advantage; Sigma

2-Ethylcaproic acid; α-Ethylcaproic acid. *See* 2-Ethylhexoic acid
Ethyl capronate. *See* Ethyl caproate
Ethyl caprylate. *See* Ethyl octanoate
Ethyl carbinol. *See* Propyl alcohol
Ethyl 'Carbitol' acrylate. *See* 2-(2-Ethoxyethoxy) ethyl acrylate
Ethyl carbonate. *See* Diethyl carbonate
Ethyl 'Cellosolve'. *See* Ethoxyethanol

Ethyl chloride

CAS 75-00-3
UN 1037 (DOT)
Synonyms: Chlorethyl; Chloroethane; Chloryl; Ether chloratus; Ether hydrochloric; Ether muriatic; Hydrochloric ether; Monochlorethane; Monochloroethane; Muriatic ether
Classification: Halogenated aliphatic hydrocarbon
Empirical: C_2H_5Cl
Formula: CH_3CH_2Cl
Properties: Colorless gas @ R.T., liq. below 12 C or under pressure; ether-like pungent odor; misc. with methanol, ethanol, diethyl ether, ethyl acetate, methylene chloride, chloroform, CCl_4, benzene; sl. sol. in water; m.w. 64.52; sp.gr. 0.891 (liq.); vapor pressure 1000 mm Hg (20 C); m.p. -139 C; b.p. 12.3 C; flash pt. -50 C
Toxicology: ACGIH TLV/TWA 100 ppm (skin); LC50 (inh., rat, 2 h) 160 g/m^3; mildly toxic by inh.; may cause cooling of skin due to evaporation, allergic reactions; irritating to skin, eyes, mucous membranes; inh. causes CNS depression (headache, dizziness, incoordination); severe exposure can cause narcosis, unconsciousness, death; suspected carcinogen; TSCA listed
Precaution: Flamm.; compressed gas; very dangerous fire hazard exposed to heat or flame; severe explosion hazard exposed to flame; incompat. with alkali metals, powd. metals, strong oxidizers; violent reactions, increased risk of fire or explosion possible
Hazardous Decomp. Prods.: Reacts with water or steam to produce toxic and corrosive fumes; heated to decomp., emits toxic fumes of phosgene and Cl^-
NFPA: Health 1, Flammability 4, Reactivity 0
Storage: Store cylinders in cool, dry, well-ventilated area away from flammables and corrosive atmospheres, out of direct sunlight
Uses: Raw material in prod. of tetraethyl lead and ethyl cellulose; refrigerant; local anesthetic; intermediate in chem. synthesis, mfg. of dyes, drugs, perfumes, insecticides; aerosol propellant; alkylating agent; analytical reagent; ethylation reagent; solvent for phosphorus, sulfur, fats, oils, resins, and waxes; extraction solvent
Manuf./Distrib.: Air Prods.; Aldrich; Alfa Chem; Arch Personal Care Prods.; Bernel; Boith China; Dow; Hatco; Koster Keunen; Lipo; MG Ind.; PPG Ind.; Sigma; Spectrum Quality Prods.; Voigt Global Distrib.

Ethylchlorohydrin. *See* Ethylene chlorohydrin

Ethyl cinnamate

CAS 103-36-6; EINECS/ELINCS 203-104-6
FEMA 2430
Synonyms: Ethyl-trans-cinnamate; Ethyl phenylacrylate; Ethyl-β-phenylacrylate; Ethyl-3-phenylpropenoate; Ethyl 3-phenyl-2-propenoate; Ethyl (E)-3-phenylprop-2-enoate; 3-Phenyl-2-propenoic acid, ethyl ester; 2-Propenoic acid, 3-phenyl-, ethyl ester
Definition: Ester of ethyl alcohol and cinnamic acid; obtained by heating cinnamic acid, alcohol, and sulfuric acid to 100 C in presence of aluminum sulfate
Empirical: $C_{11}H_{12}O_2$
Properties: Nearly colorless oily liq., faint cinnamon odor; misc. with alcohol, ether, fixed oils; insol. in glycerin, water; m.w. 176.23; dens. 1.049 (20/4 C); m.p. 9 C; b.p. 271 C; flash pt. > 212 F; ref. index 1.558-1.561
Toxicology: LD50 (oral, rat) 4000 mg/kg; mod. toxic by ingestion; TSCA listed
Precaution: Combustible liq.
Hazardous Decomp. Prods.: Heated to decomp., emits acrid smoke and irritating fumes
Uses: Solvent; UV absorber in cosmetics; synthetic flavoring agent in foods and pharmaceuticals; perfumery
Features: Sweet cinnamon, plum-like flavor

Regulatory: FDA 21CFR §172.515; FEMA GRAS; Japan approved as flavoring; BP compliance
Manuf./Distrib.: A&E Connock; Advanced BioTech; Advanced Synthesis Tech.; Aldrich; Axxence Aromatic GmbH; Degussa Flavors/Alfrebro; Elan; Fleurchem; Fluka; Goldensun Mfg.; Lluch Essence; Sigma

Ethyl-trans-cinnamate. *See* Ethyl cinnamate
Ethyl citrate. *See* Triethyl citrate

Ethyl crotonate

CAS 623-70-1; 10544-63-5 (trans); EINECS/ELINCS 210-808-7; 234-125-9 (trans)
UN 1862 (DOT); FEMA 3486
Synonyms: 2-Butenoic acid, ethyl ester; trans-2-Butenoic acid ethyl ester; Crotonic acid, ethyl ester; Ethyl trans-2-butenoate; Ethyl α-crotonate; Ethyl β-methylacrylate
Definition: Obtained by esterification of crotonic acid with ethyl alcohol in presence of conc. H_2O_4
Empirical: $C_6H_{10}O_2$
Formula: $CH_3CH{:}CHCOOC_2H_5$
Properties: Water-wh. solid or liq., char. pungent persistent odor; sol. in alcohol, ether, oxygenated solvs.; insol. in water; m.w. 114.15; dens. 0.916 (20/4 C); m.p. 45 C; b.p. 126 C (cis), 145 C (trans); flash pt. 2 C; ref. index 1.425 (20 C)
Toxicology: LD50 (oral, rat) 3 g/kg; corrosive; sl. toxic by ing.; lachrymator; strong irritant to eyes, skin, respiratory system; TSCA listed
Precaution: Highly flamm.; dangerous fire risk
Hazardous Decomp. Prods.: Heated to decomp., emits acrid smoke and irritating fumes
NFPA: Health 2, Flammability 3, Reactivity 0
Storage: Keep away from ignition sources
Uses: Solvent; softener; lacquers; organic synthesis; fragrance in cosmetics; synthetic flavoring agent in foods and pharmaceuticals; mfg. of copolymers with vinyl acetate used in lacquers and paper sizing; mfg. of DL-threonine, vitamin A
Regulatory: FDA 21CFR §172.515; FEMA GRAS
Manuf./Distrib.: Advanced Synthesis Tech.; Aldrich; Bernel; Fluka; Grau Aromatics; Int'l. Chem. Group; Lipo; Penta Mfg.; Sithean

Ethyl α-crotonate. *See* Ethyl crotonate
Ethyl cyanide. *See* Propionitrile

Ethylcyclohexane

CAS 1678-91-7; EINECS/ELINCS 216-835-0
Synonyms: Cyclohexane, ethyl-
Classification: Sat. alicyclic hydrocarbon
Empirical: C_8H_{16}
Properties: Colorless liq.; sol. in most org. solvs.; insol. in water; m.w. 112.21; sp.gr. 0.792 (20 C); m.p. -111 C; b.p. 130-132 C; flash pt. 18 C; ref. index 1.433
Toxicology: May be harmful by ing., inh., skin absorp.; may cause mild skin irritation; vapor may be irritating to eyes; probably a CNS depressant; sl. nose/throat irritant; high vapor concs. may cause headache, lightheadedness, nausea, drowsiness; very high concs. may cause unconsciousness, death; ing. may cause sore throat, nausea, diarrhea
Environmental: Prevent entry into drains
Precaution: Flamm.; mod. fire risk; flamm. limits in air 0.95-6.6%; forms explosive mixts. in air; vapor may flash back; container explosion may occur under fire conditions; take precautions against static discharges; incompat. with strong oxidizing agents (increased fire and explosion hazard)
Hazardous Decomp. Prods.: CO, CO_2
NFPA: Health 1, Flammability 3, Reactivity 0
Storage: Store in cool, dry, well-ventilated area out of direct sunlight, away from heat, sparks, open flame, other ignition sources; keep tightly closed
Uses: Solvent; org. synthesis
Manuf./Distrib.: Acros Org.; Aldrich; ChevronPhillips; Fluka; ICN Biomed. Research Prods.; TCI Am.

Ethylcyclopentane

CAS 1640-89-7; EINECS/ELINCS 216-686-1
Synonyms: Cyclopentane, ethyl-
Classification: Sat. alicyclic hydrocarbon
Empirical: C_7H_{14}
Formula: $C_2H_5C_5H_9$
Properties: Colorless liq.; gasoline-like odor; sol. in most org. solvs.; insol. in water; m.w. 98.19; sp.gr. 0.763 (20 C); vapor pressure 72.8 mm Hg (37.7 C); m.p. -138 C; b.p. 103 C; flash pt. 15 C; autoignition temp. 500 F; ref. index 1.4198 (20 C)
Toxicology: May be harmful by ing., inh., skin absorp.; may cause mild skin irritation; vapor may be irritating to eyes; probably an asphyxiant and CNS depressant; high vapor concs. may cause nose/throat irritation, lightheadedness, nausea, drowsiness; very high concs. may cause unconsciousness, death; ing. may cause sore throat, nausea, diarrhea
Environmental: Prevent entry into drains
Precaution: Flamm.; very dangerous fire and explosion hazard; flamm. limits in air 1.1-6.7%; incompat. with strong oxidizing agents (increased fire and explosion hazard); easily forms explosive mixts. in air; vapor may flash back; container explosion may occur under fire conditions; take precautions against static discharges
Hazardous Decomp. Prods.: Heated to decomp., emits acrid smoke and fumes
NFPA: Health 1, Flammability 3, Reactivity 0
Storage: Store in cool, dry, well-ventilated area out of direct sunlight, away from heat, sparks, open flame, other ignition sources; keep tightly closed
Uses: Extraction solvent for essential oils; org. synthesis; component in gasoline
Manuf./Distrib.: Aldrich; ICN Biomed. Research Prods.; TCI Am.

Ethyl decanoate

CAS 110-38-3; EINECS/ELINCS 203-761-9
FEMA 2432
Synonyms: Capric acid ethyl ester; Decanoic acid ethyl ester; Ethyl caprate; Ethyl caprinate; Ethyl decylate
Definition: Obtained by esterification of decanoic acid and ethyl alcohol in presence of HCl or H_2SO_4
Empirical: $C_{12}H_{24}O_2$
Formula: $CH_3(CH_2)_8COOC_2H_5$
Properties: Colorless liq.; fragrant odor; sol. in fixed oils; misc. with alcohol, chloroform, ether; insol. in water, glycerin, propylene glycol; m.w. 200.32; dens. 0.862 (20 C); m.p. -20 C; b.p. 243-245 C; flash pt. 216 F; ref. index 1.425 (20 C)
Toxicology: Skin irritant; TSCA listed
Precaution: Combustible liq.; reacts with oxidizing materials
Hazardous Decomp. Prods.: Heated to decomp., emits acrid smoke and irritating fumes
NFPA: Health 0, Flammability 1, Reactivity 0
Uses: Solvent; organic synthesis; synthetic flavoring agent in foods and pharmaceuticals; fragrance in cosmetics; mfg. of wine bouquets, cognac essence
Features: Grape, pear-like flavor
Regulatory: FDA 21CFR §172.515; FEMA GRAS; Japan approved as flavoring
Manuf./Distrib.: A&E Connock; ABCR; Advanced BioTech; Aldrich; Axxence Aromatic GmbH; Degussa Flavors/Alfrebro; Fleurchem; Fluka; Grau Aromatics; Lluch Essence; Moore Ingreds.; Sigma

Ethyl decylate. *See* Ethyl decanoate
Ethyldiethanolamine. *See* N-Ethyldiethanolamine

N-Ethyldiethanolamine

CAS 139-87-7; EINECS/ELINCS 205-379-8
Synonyms: N,N-Bis (2-hydroxyethyl) ethylamine; Diethanolethylamine; Ethanol, 2,2′-(ethylimino) bis-; Ethylbis (2-hydroxyethyl) amine; Ethyldiethanolamine; 2-(N-Ethyl-N-2-hydroxyethylamino) ethanol; 2,2′-(Ethylimino) bisethanol; 2,2′-(Ethylimino) diethanol; N-Ethyl-2,2′-iminodiethanol
Classification: Amino alcohol
Empirical: $C_6H_{15}NO_2$
Formula: $CH_3CH_2N(CH_2CH_2OH)_2$
Properties: Water-wh. liq.; amine odor; sol. in water; m.w. 133.22; sp.gr. 1.015 (20 C); m.p. -50 C; sets to glasslike solid below -44 C; b.p. 246-252 C; flash pt. (CC) 123 C; ref. index 1.4665 (20 C)
Toxicology: LD50 (oral, rat) 4570 mg/kg, (IV, rabbit) > 200 mg/kg; harmful

by ing., inh., skin absorp.; mildly toxic by ingestion; skin irritant; severe eye irritant; high concs. extremely destructive to tissues of mucous membranes, upper respiratory tract, eyes, skin; may cause burning sensation, coughing, wheezing, laryngitis, shortness of breath, headache, nausea, vomiting; TSCA listed

Precaution: Combustible; incompat. with oxidizing agents; avoid contact with acid

Hazardous Decomp. Prods.: Thermal decomp. may produce CO, CO_2, NO_x; heated to decomp., emits toxic fumes of NO_x; emits toxic fumes under fire conditions

NFPA: Health 2, Flammability 1, Reactivity 0

Storage: Store in cool, dry place; keep tightly closed

Uses: Solvent; detergents

Manuf./Distrib.: Acros Org.; Aldrich; Atofina; Harcros; ICN Biomed. Research Prods.; TCI Am.

Ethyl diethylene glycol. *See* Ethoxydiglycol
Ethyl diglyme. *See* Diethylene glycol diethyl ether
Ethyldimethyl methane. *See* Isopentane
4-Ethyl-1,3-dioxolan-2-one. *See* 1,2-Butylene carbonate
Ethylene acetate. *See* Ethylene glycol diacetate
Ethylene alcohol. *See* Ethylene glycol
Ethylene aldehyde. *See* Acrolein

Ethylene carbonate

CAS 96-49-1; EINECS/ELINCS 202-510-0

Synonyms: Carbonic acid, cyclic ethylene ester; 1,3-Carbonyldioxyethane; Cyclic ethylene carbonate; 1,3-Dioxolan-2-one; Dioxolone-2; 1,2-Ethanediol carbonate; Ethylene carbonic acid; Ethylene glycol carbonate; Ethylene glycol cyclic carbonate; Glycol carbonate

Classification: Carbonic acid ester

Empirical: $C_3H_4O_3$

Formula: $(-CH_2O)_2CO$

Properties: Colorless solid (below 36.4 C) or liq., odorless; very sol. in water; sol. in ether, n-butanol, CCl_4; misc. with alcohol, ethyl acetate, benzene, chloroform; m.w. 88.07; dens. 1.3218 (39/4 C); vapor pressure 0.01 mm Hg (20 C); m.p. 36.4 C; b.p. 248 C; flash pt. (OC) 143 C; ref. index 1.4158 (50 C)

Toxicology: LD50 (oral, rat) 10 g/kg; LDLo (IP, mouse) 500 mg/kg; moderately toxic by intraperitoneal route; mildly toxic by ingestion; skin and eye irritant; TSCA listed

Environmental: Biodeg.

Precaution: Combustible exposed to heat or flame; incompat. with strong oxidizing agents (increases fire risk), acids, bases, and reducing agents (decomp. can occur)

Hazardous Decomp. Prods.: May produce ethylene oxide and CO_2 if an acid, base, or salt is present in an aq. sol'n. of ethylene carbonate; heated to decomp., emits acrid smoke and irritating fumes

NFPA: Health 2, Flammability 1, Reactivity 1

Storage: Store in cool, dry, well-ventilated area, out of direct sunlight

Uses: Solvent for polymers, resins, polyacrylonitrile polymerization/spinning, chem. synthesis, cosmetics, agric., inks, colors, inorg. salts, textiles, extraction of aromatic hydrocarbons; aprotic extraction solvent; process solvent; intermediate for pharmaceuticals, rubber chemicals, textile finishing agents; hydroxyethylation reagent; accelerant in dyeing; additive in fuel, lube and hydraulic fluids; reactive diluent, plasticizer in PU; prod. of PC, polyester elastomers; component of cleaners, binders; separation of CO_2 and hydrogen sulfide; refining of aromatics in oil field; component of electrolytes in lithium-ion rechargeable batteries

Manuf./Distrib.: Aldrich; BASF; Degussa AG; Ferro; Fluka; Jarchem Ind.; Merck KGaA; Monomer-Polymer & Dajac Labs; New Japan Chem.; Nisso Petrochem. Ind.; Toagosei

Trade Names: Jeffsol® EC; UltraPure™ Ethylene Carbonate

Trade Names Containing: Jeffsol® EC-25; Jeffsol® EC-50; Jeffsol® EC-75

Ethylene carbonic acid. *See* Ethylene carbonate
Ethylene chlorhydrin. *See* Ethylene chlorohydrin
Ethylene chloride. *See* Ethylene dichloride

Ethylene chlorohydrin

CAS 107-07-3; EINECS/ELINCS 203-459-7

UN 1135

Synonyms: Chloroethanol; 2-Chloroethanol; 2-Chloro-1-ethanol; β-Chloroethanol; 2-Chloroethyl alcohol; β-Chloroethyl alcohol; Ethylchlorohydrin; Ethylene chlorhydrin; Ethylene glycol, chlorohydrin; Glycol chlorohydrin; Glycolmonochlorhydrin; Glycol monochlorohydrin; 2-Hydroxyethyl chloride; 2-Monochloroethanol

Empirical: C_2H_5ClO

Formula: $ClCH_2CH_2OH$

Properties: Colorless to lt. yel. clear liq.; faint ether-like odor; sol. (@ 19 C): ≥ 100 mg/ml in water, DMSO, 95% ethanol, acetone; sol. in most org. liqs.; m.w. 80.51; sp.gr. 1.021; vapor pressure 4.9 mm Hg (20 C); m.p. -89 C; b.p. 130 C; dec. > 184 C; acid no. 0.2; flash pt. 55 C; autoignition temp. 425 C; ref. index 1.4410

Toxicology: ACGIH TLV CL 1 ppm (skin); LD50 (oral, rat) 71 mg/kg, (IP, rat) 58 mg/kg, (skin, rabbit) 67 mg/kg; LC50 (inh., rat) 290 mg/m^3; poison by ing., inh., skin contact, IP, IV, subcut., possibly other routes; mod. toxic to humans by inh.; severe eye/mild skin irritant; rapidly absorbed thru skin; may affect liver, spleen, lungs; may cause nausea, vomiting, headache, pulmonary edema, CNS depression; experimental teratogen; mutagenic data; human lethal dose 1-2 ml; TSCA listed

Precaution: Flamm. liq.; LEL 4.9%; UEL 15.9%; reacts violently with chlorosulfonic acid, ethylene diamine, sodium hydroxide; reacts with water or steam to produce toxic and corrosive fumes; may react violently with strong oxidizers; incompat. with strong caustics, bases; attacks some plastics, rubber, coatings; vapors heavier than air, may flash back

Hazardous Decomp. Prods.: Heated to decomp., emits highly toxic fumes of hydrogen chloride gas, chlorine, phosgene, CO

NFPA: Health 4, Flammability 2, Reactivity 0

Storage: Sensitive to moisture, heat; store refrigerated, away from oxidizing materials; protect from moisture; storage under inert atm. rec.

Uses: Solvent for cellulosics, resins, waxes; extraction solvent for separation of butadiene from hydrocarbon mixts., for dewaxing and removing cycloalkanes from mineral oil, for pure lignin, for refining of rosin; cleaning solvent; hydroxyethylating agent in org. synthesis (ethylene oxide, ethylene glycol, amines, carbitols, indigo, malonic acid, novocaine); prod. of dye intermediates, pharmaceuticals, agrochems., pesticides, insecticides, plasticizers; solvent in fabric dyeing; tar spot remover, cleaning agent in dyeing and cleaning industry; treating sweet potatoes before planting

Manuf./Distrib.: Acros Org.; Aldrich; Alfa Aesar; Fluka; Lancaster Synthesis; Riedel-deHaën; Spectrum Labs

Ethylene cyanohydrin

CAS 109-78-4; EINECS/ELINCS 203-704-8

Synonyms: 2-Cyanoethanol; Hydracrylonitrile; 3-Hydroxypropionitrile

Classification: Nonaromatic nitrile

Empirical: C_3H_5NO

Formula: $N\equiv CC_2H_4OH$

Properties: Liq.; m.w. 71.08; dens. 1.040; m.p.-46 C; b.p. 228 C; flash pt. (CC) 235 F

Toxicology: LD50 (oral, rat) 3200 mg/kg; poison by inh.; irritant; harmful liq.; avoid skin contact; TSCA listed

Hazardous Decomp. Prods.: Heated to decomp. or on contact with acid or acid fumes, emits highly toxic fumes of CN^-

NFPA: Health 1, Flammability 2, Reactivity 2

Uses: Solvent for some cellulose esters and many inorganic salts

Manuf./Distrib.: Aldrich; Fluka; Sigma

Ethylene diacetate. *See* Ethylene glycol diacetate

Ethylenediamine

CAS 107-15-3 (anhyd.); 6780-13-8 (monohydrate); EINECS/ELINCS 203-468-6

UN 1604 (DOT)

Synonyms: 1,2-Diaminoethane; Dimethylenediamine; EDA; 1,2-Ethanediamine

Classification: Aliphatic organic compd.; aliphatic polyamine

Empirical: $C_2H_8N_2$

Formula: $NH_2CH_2CH_2NH_2$

Properties: Colorless to sl. yel. clear volatile liq., ammonia-like odor; sol. in acetone, ethanol, dimethylsulfoxide, benzene, water; sl. sol. in diethyl ether, heptane; misc. with water, alcohol, oxygenated and aromatic solvs.; m.w. 60.12; dens. 0.8994 (20/4 C); m.p. 8.5 C; b.p. 117.2 C; flash pt. (CC) 42 C; ref. index 1.4565

Toxicology: TLV/TWA 10 ppm; LD50 (oral, rat) 500 mg/kg; human irritant poison by inh.; mod. toxic by ingestion, skin contact; corrosive; severe skin and eye irritant; lachrymator; allergen, sensitizer; mutagenic data; TSCA listed

Precaution: DOT: Corrosive material; flamm. exposed to heat, flame, oxidizers; can react violently with acetic acid, acetic anhydride, acrylic acid, epichlorohydrin, many others; absorbs CO_2 from air to form the carbamate salt

Hazardous Decomp. Prods.: CO, CO_2, hydrogen cyanide, volatile amines; heated to decomp., emits toxic fumes of NO_x and NH_3

NFPA: Health 3, Flammability 2, Reactivity 0

Storage: Hygroscopic; air-sensitive

Uses: Chemical intermediate in mfg. of fungicides, chelating agents, polyamide resins, corrosion inhibitors, rubber chems., dimethylolethyleneurea resins, syn. waxes, fuel additives; solvent; emulsifier; textile lubricants; antifreeze inhibitor; agric.; ion-exchange resins; epoxy curing agent; thermoplastics lubricants; stabilizer for dyes; solvent for shellac; flume wash water additive for food processing; antimicrobial for cane-sugar/beet-sugar mills; stabilizer, buffer for pharmaceuticals; corrosion inhibitor in petrol. prod.; in food-pkg. adhesives; adjuvant in slimicides in food-contact paper/paperboard; in food-contact animal glues

Regulatory: FDA 21CFR §173.320 (1 ppm max.), 175.105, 175.300, 175.320, 176.180, 176.300, 178.1010, 178.3120, 181.30, 556.270; FDA approved for IV, injectables, orals, rectals, topicals; BP, EP, JP compliance

Manuf./Distrib.: AMC Chems.; Accurate Chem. & Scientific; Advanced Synthesis Tech.; Aldrich; Alfa Aesar; Alkyl Amines Chems. Ltd; Allchem Ind.; BASF AG; BASF; Coyne; Dow; Filo; Filo; Fluka; Houghton Chem.; Integra; Kanto Denka Kogyo; Nova Molecular Tech.; Sigma; Spectrum Quality Prods.; Sumitomo Seika; Tosoh; VWR Int'l.; Voigt Global Distrib.; Vopak USA

Ethylenediamine dihydrochloride

CAS 333-18-6; EINECS/ELINCS 206-369-6

Synonyms: Chlor-ethamine; 1,2-Diaminoethane dihydrochloride; 1,2-Ethanediamine, dihydrochloride; Ethylenediamine hydrochloride; Ethylenediammonium chloride

Classification: Nonaromatic amine

Empirical: $C_2H_8N_2 \cdot 2HCl$

Formula: $H_2NCH_2CH_2NH_2 \cdot 2HCl$

Properties: Prisms; sol. in water; insol. in alcohol and ether; m.w. 133.02; m.p. > 300 C; sublimes

Toxicology: LD50 (oral, mouse) 1620 mg/kg, (IM, rat) 150 mg/kg; harmful solid; poison by IM route; mod. toxic by ing.; irritant; sensitizer; experimental teratogen and reproductive effects; TSCA listed

Hazardous Decomp. Prods.: Heated to decomp., emits very toxic fumes of HCl and NO_x

Storage: Hygroscopic

Uses: Solvent for casein, albumin, shellac, and sulfur; emulsifier; stabilizer for rubber latex; inhibitor in antifreeze sol'ns.; in textile lubricants; pharmaceutical topicals

Regulatory: FDA approved for topicals

Manuf./Distrib.: Aldrich; Fluka; Sigma; Spectrum Quality Prods.

Ethylenediamine hydrochloride. *See* Ethylenediamine dihydrochloride
Ethylenediaminetetraacetic acid. *See* Edetic acid
Ethylenediammonium chloride. *See* Ethylenediamine dihydrochloride

Ethylene dichloride

CAS 107-06-2; EINECS/ELINCS 203-458-1

UN 1184 (DOT)

Synonyms: 1,2-Bichloroethane; 1,2-DCE; Dichloroethane; 1,2-Dichloroethane; Dichloro-1,2-ethane; α,β-Dichloroethane; sym-Dichloroethane; Dichloroethylene; Dutch liquid; Dutch oil; EDC; Ethane dichloride; Ethylene chloride; 1,2-Ethylene dichloride; Glycol dichloride

Classification: Halogenated aliphatic hydrocarbon; chlorinated alkane

Empirical: $C_2H_4Cl_2$

Formula: $Cl \cdot CH_2CH_2 \cdot Cl$

Properties: Colorless oily liq., chloroform-like odor, sweet taste; misc. with ethanol, chloroform, diethyl ether, acetone, benzene, oxygenated and chlorinated solvs.; very sl. sol. in water; m.w. 98.96; dens. 1.2554 (20/4 C); vapor pressure 100 mm Hg (29.4 C); f.p. -35.5 C; b.p. 83.5 C; flash pt. 56 C; ref. index 1.445

Toxicology: ACGIH TLV/TWA 10 ppm; LD50 (oral, rat) 670 mg/kg, (skin, rabbit) 3890 mg/kg; poison by IV, subcut. route; human poison by ing.; mod. toxic by inh., skin contact, IP route; strong narcotic; skin/severe eye irritant; may cause dermatitis; human systemic effects by ing., inh. (somnolence, cough, nausea, vomiting, diarrhea, stomach ulceration, cardiac rate change, cyanosis, coma); carcinogen; tumorigenic data; experimental teratogen, reproductive effects; human mutagenic data; TSCA listed

Precaution: Flamm.; dangerous fire hazard exposed to heat, flame, or oxidizers; mod. explosive as vapor exposed to flame; violent reaction with Al, NH_3; can react vigorously with oxidizers and emit vinyl chloride and HCl; dec. on contact with steam (forms HCl)

Hazardous Decomp. Prods.: Heated to decomp., emits highly toxic fumes of Cl^-, phosgene

NFPA: Health 2, Flammability 3, Reactivity 0

Uses: Prod. of vinyl chloride, trichloroethylene, vinylidene chloride, trichloroethane; lead scavenger in gasoline; paint, varnish remover; metal degreasing; soaps; wetting agent; penetrant; organic synthesis; ore flotation; storage fumigant; pesticides; solvent in paint strippers, adhesives, dewaxing process, pharmaceuticals; extraction solvent for foods; flume wash water additive; intermediate for paints; in food-pkg. adhesives; adjuvant for PC food-pkg. resins

Use Level: 5 ppm (pharmaceuticals)

Regulatory: FDA 21CFR §73.30, 73.345, 73.615, 172.385, 172.560, 172.710, 173.230, 173.315, 175.105, 176.170, 177.1580, 177.1585, 177.2550, 573.440; SARA reportable

Manuf./Distrib.: Albemarle; Aldrich; Ashland; Asiamerica Int'l.; BASF; BP Chems. Ltd; Delta Distributors; Eastech; Fluka; Georgia Gulf; Houghton Chem.; Hukill; Integra; J.T. Baker; Milsolv; Norsk Hydro AS; Occidental; P.B. & S.; PPG Ind.; Rhodia/Phosphorus Perf. Derivs.; Rhodia; Romil Ltd; SOCO-Lynch; Sal Chem.; Sasol N. Am.; Sigma; Southchem; Spectrum Quality Prods.; Textile Chem.; Thomas Scientific; Voigt Global Distrib.; Vopak USA; Vulcan Perf. Chems.; Whittaker, Clark & Daniels; Whyte Chems. Ltd

1,2-Ethylene dichloride. *See* Ethylene dichloride
Ethylene, 1,1-dichloro-. *See* Vinylidene chloride monomer
Ethylene, 1,2-dichloro-. *See* cis-trans-1,2-Dichloroethylene (mixed isomers)
Ethylene, 1,2-dichloro-, (E)-. *See* trans-1,2-Dichloroethylene
Ethylene, 1,2-dichloro-, (Z)-. *See* cis-1,2-Dichloroethylene
Ethylene diethanoate. *See* Ethylene glycol diacetate
Ethylene diglycol. *See* Diethylene glycol
Ethylene diglycol monoethyl ether. *See* Ethoxydiglycol
Ethylene dihydrate. *See* Ethylene glycol
Ethylene dimethyl ether. *See* Ethylene glycol dimethyl ether
(Ethylenedinitrilo) tetraacetic acid. *See* Edetic acid
Ethylenedinitrilotetra-2-propanol; 1,1′,1′′,1′′′-(Ethylenedinitriolo) tetra-2-propanol. *See* Tetrahydroxypropyl ethylenediamine
2,2′-Ethylenedioxydiethanol; 2,2′-Ethylenedioxyethanol. *See* Triethylene glycol
Ethylene dipropionate. *See* Crotonaldehyde

Ethylene glycol

CAS 107-21-1; EINECS/ELINCS 203-473-3

Synonyms: 1,2-Dihydroxyethane; EG; 1,2-Ethanediol; Ethane-1,2-diol; Ethylene alcohol; Ethylene dihydrate; Glycol; Glycol alcohol; MEG; Monoethylene glycol

Classification: Aliphatic dihydric alcohol

Empirical: $C_2H_6O_2$

Formula: $HOCH_2CH_2OH$

Properties: Colorless clear liq., sweet taste (poisonous); sol. in water, lower aliphatic alcohols, glycerol, glycol ethers, acetone, most org. solvs.; poor sol. in toluene, benzene, chloroform; m.w. 62.07; sp.gr. 1.1135 (20/4 C); visc. 17.3 cps; m.p. -13 C; b.p. 197.6 C; flash pt. (OC) 115 C; ref. index 1.43063; surf. tens. 48.49 dynes/cm; dielec. const. 37.7

Toxicology: ACGIH TLV/CL 50 ppm (vapor); LD50 (oral, rat) 4700 mg/kg; toxic by inh.; mod. toxic by ing., subcut., IV, IM routes; human poison by ing.; vapor can cause nose/throat irritation and burning; irritating to eyes, skin, mucous membranes; may be absorbed thru skin; human systemic effects by ing./inh. (lacrimation, anesthesia, headache, cough, nausea, vomiting, pulmonary/kidney/liver changes); lethal dose for humans 100 ml; experimental teratogen; human mutagenic data; TSCA listed

Environmental: VOC; BOD5 0.47; COD 1.29; ThOD 1.29

Precaution: Combustible; incompat. with strong oxidizing agents, perchloric acid, strong bases, phosphorus (V) sulfide, strong acids, silvered copper wires carrying dc current; mod. explosion hazard; violent reactions possible

Hazardous Decomp. Prods.: Heated to decomp., emits acrid smoke and irritating fumes

NFPA: Health 1, Flammability 1, Reactivity 0

Storage: Very hygroscopic; store in cool, dry, well-ventilated area

Uses: Antifreeze (cooling/heating systems); in hydraulic brake fluids; humectant (starch adhesives); solvent in paints, plastics, printing inks; softener for cellophane; stabilizer; in explosives; alkyd resins; elastomers; waxes; asphalt; PU chain extender; wood stains; raw material for prod. of syn. polyester fibers, latex paints, polyester/alkyd resins; plasticizer (regenerated cellulose film); heat transfer agent; low-pressure laminates; textile processing; foam stabilizer; pharmaceutical solvent; solvent, visc. control agent, humectant in cosmetics; in food-pkg. adhesives; adjuvant in slimicides in food-contact paper/paperboard

Regulatory: FDA 21CFR §175.105, 176.300; FDA approved for topicals; SARA §313 reportable; HAP

Manuf./Distrib.: AAE Chemie NV; AMC Chems.; AMRESCO; Adept Sol'ns.; Aldrich; Arch Chems.; Ashland; Asiamerica Int'l.; BASF; Baychem; Brenntag AG; C.P. Hall; CasChem; Celanese; Chemcentral; Chemial SpA; Cognis/Chems. Group; Coyne; Degussa AG; Delta Distributors; Dow; Eastech; Eastman; Equistar; ExxonMobil; Fluka; General Chem.; Haltermann Prods. UK; Harcros; Houghton Chem.; Hukill; Huntsman; Integra; J.T. Baker; Lanaetex Prods.; Laporte Perf. Chems.; Mallinckrodt Baker; Milsolv; Occidental; Oxiteno; P.B. & S.; Penta Mfg.; Reliance Ind. Ltd; Rhodia; Robeco; Ruger; SOCO-Lynch; Sal Chem.; Shell; Sigma; Southchem; Spectrum Quality Prods.; Stanchem; Sunnyside; Textile Chem.; Uniqema Am.; Universal Preserv-A-Chem; Vopak USA; Whittaker, Clark & Daniels

Trade Names Containing: Eastman® DE-HG

Ethylene glycol acetate

CAS 542-59-6; EINECS/ELINCS 208-821-8

Synonyms: Acetic acid, 2-hydroxyethyl ester; 1,2-Ethanediol monoacetate; Ethylene glycol monoacetate; Glycol monoacetate; Glycol monoacetin; 2-Hydroxyethyl acetate

Classification: Glycol ester; carboxylic acid ester

Empirical: $C_4H_8O_3$

Formula: $CH_3COOCH_2CH_2OH$

Properties: Colorless clear liq.; almost odorless; sol. in water, DMSO, alcohol, ether, acetone, benzene, toluene; m.w. 104.22; sp.gr. 1.108; vapor pressure 9.0 mm Hg (23 C); b.p. 181-182 C; flash pt. (OC) 102 C; ref. index 1.422

Toxicology: LD50 (oral, rat) 8250 mg/kg, (IP, mouse) 1310 mg/kg; mod. toxic by ing., IP routes; severe eye irritant; inh. of vapor may irritate nose/throat; ing. expected to cause mild CNS depression; may cause somnolence, GI changes; large doses may cause coma, dehydration, acidosis, fluid accumulation in lungs, kidney damage, respiratory failure, cardiovascular collapse, death; TSCA listed

Precaution: Combustible exposed to heat or flame; incompat. with strong oxidizing agents (increases fire/explosion hazard), strong acids and strong bases (decomp. can occur)

Hazardous Decomp. Prods.: Ethylene glycol, acetic acid; CO, CO_2 (thermal); heated to decomp., emits acrid smoke and irritating fumes

NFPA: Health 0, Flammability 1, Reactivity 0

Storage: Store in cool, dry, well-ventilated area, out of direct sunlight, away from ignition sources, oxidizing materials; limit quantities in use; store refrigerated

Uses: Solvent for camphor, dyes, nitrocellulose, cellulose acetate, lacquers, paints, varnish removers; antifoam for aq. sol'ns.

Manuf./Distrib.: Aldrich; Chemial SpA; Fluka

See also Ethylene glycol diacetate

Ethylene glycol benzyl ether

CAS 622-08-2; EINECS/ELINCS 210-719-3

Synonyms: Benzyl 'Cellosolve'; 2-Benzyloxyethanol; Ethanol, 2-(phenylmethoxy)-; Ethylene glycol monobenzyl ether; Glycol benzyl ether; Glycol monobenzyl ether

Classification: Glycol ether

Empirical: $C_9H_{12}O_2$

Formula: $C_6H_5CH_2OCH_2CH_2OH$

Properties: Clear liq.; sol. 4000 mg/kg in water; m.w. 152.21; sp.gr. 1.065-1.075; vapor pressure 0.02 mm Hg (20 C); m.p. < -50 C; b.p. 265 C; flash pt. 129 C; autoignition temp. 352 C; ref. index 1.521

Toxicology: Harmful by ing.; irritating to eyes, skin, respiratory system; TSCA listed

Environmental: VOC; ThOD 1.68

NFPA: Health 2, Flammability 1, Reactivity 0

Uses: Solvent, coupling agent for printing inks, agric., water-based coatings, plastics, lacquers, paints, varnishes, cleaning prods., preservatives, rust removers, disinfectants

Regulatory: SARA §313 reportable; HAP; Canada WHMIS listed

Manuf./Distrib.: Aldrich; Ashland; Coyne; Dow; Ferro/Grant; G.J. Chem.; Gallard-Schlesinger Ind.; Great Western; Marcus Int'l.; Oxiteno; R.E. Carroll; Shell; Vopak USA

Ethylene glycol bis (2-hydroxyethyl ether). *See* Triethylene glycol

Ethylene glycol butyl ether; Ethylene glycol n-butyl ether. *See* Butoxyethanol

Ethylene glycol butyl ether acetate. *See* Butoxyethanol acetate

Ethylene glycol carbonate. *See* Ethylene carbonate

Ethylene glycol, chlorohydrin. *See* Ethylene chlorohydrin

Ethylene glycol cyclic carbonate. *See* Ethylene carbonate

Ethylene glycol diacetate

CAS 111-55-7; EINECS/ELINCS 203-881-1

Synonyms: 1,2-Diacetoxyethane; 1,2-Ethanediol diacetate; Ethylene acetate; Ethylene diacetate; Ethylene diethanoate; Ethylene glycol acetate; Glycol diacetate

Classification: Glycol ester

Empirical: $C_6H_{10}O_4$

Formula: $CH_3COOCH_2CH_2OOCCH_3$

Properties: Colorless liq. or cryst., faint fruity odor; sol. in alcohol, ether, benzene, acetone, CCl_4, acetic acid, CS_2; mod. sol. in water (10%); insol. in petrol.; m.w. 146.14; sp.gr. 1.1063 (20/20 C); f.p. -31 C; b.p. 190.5 C; flash pt. (OC) 96 C; ref. index 1.416

Toxicology: LD50 (oral, rat) 6850 mg/kg, (skin, rabbit) 8480 mg/kg; mod. toxic by IP route; mildly toxic by ing. and skin contact; eye irritant; inh. may cause nose/throat irritation; TSCA listed

Precaution: Combustible; incompat. with strong oxidizing agents (increases fire/explosion hazard), strong acids, strong reducing agents (decomp. can occur)

Hazardous Decomp. Prods.: Ethylene glycol, acetic acid; heated to decomp., emits acrid smoke and irritating fumes

NFPA: Health 1, Flammability 1, Reactivity 0

Storage: Store in cool, dry, well-ventilated area out of direct sunlight

Uses: Extraction solvent; foundry resins; perfume fixative; solvent for coatings, lacquers, cellulose esters, cellulose ethers, oils, resins; mfg. of printing inks, lacquers; nondiscoloring plasticizer for cellulosics, PVAc; in food-pkg. adhesives

Regulatory: FDA 21CFR §175.105

Manuf./Distrib.: A.P. Chems. Ltd; Aceto; Aldrich; Ashland; BP Amoco; CPS; Chemial SpA; Eastman; Fluka; Haltermann Prods. UK; Houghton Chem.; Sigma

Ethylene glycol dibutyl. *See* Ethylene glycol dibutyl ether

Ethylene glycol dibutyl ether

CAS 112-48-1

Synonyms: 1,2-Dibutoxyethane; 1,2-Di-n-butoxyethane; Dibutyl 'Cellosolve'; Di-n-butyl 'Cellosolve'; Di-n-butyoxyethane; 5,8-Dioxadodecane; 1,1′-(1,2-Ethanediylbis (oxy)) bis-butane; Ethylene glycol dibutyl; Ethylene glycol di-n-butyl ether

Classification: Glycol ether

Empirical: $C_{10}H_{22}O_2$

Formula: $C_4H_9OC_2H_4OC_4H_9$

Properties: Colorless liq.; char. odor; poor sol. in water; m.w. 174.32; sp.gr. 0.83189; visc. 14.04 cP; vapor pressure 10 mm Hg (84 C); f.p. -69.1 C; b.p. 203.3 C; flash pt. 85 C; ref. index 1.411
Toxicology: LD50 (oral) 3250 mg/kg, (skin) 3560 mg/kg; mod. toxic by ing., skin contact; inh. may cause cough, sore throat; ing. may cause abdominal pain, dizziness, vomiting, unconsciousness; eye/skin irritant; may cause dry skin/redness/pain, eye redness/pain; TSCA listed
Environmental: VOC; ThOD 1.65
Precaution: Combustible; explosive vapor/air mixts. may form above 85 C
Hazardous Decomp. Prods.: Heated to decomp., emits acrid smoke and irritating fumes
NFPA: Health 1, Flammability 2, Reactivity 0
Uses: Solvent
Regulatory: SARA §313 reportable; HAP
Manuf./Distrib.: Ashland; Coyne; Dow; Ferro/Grant; G.J. Chem.; Gallard-Schlesinger Ind.; Great Western; Marcus Int'l.; Oxiteno; R.E. Carroll; Shell; TCI Am.; Vopak USA

Ethylene glycol di-n-butyl ether. *See* Ethylene glycol dibutyl ether

Ethylene glycol diethyl ether

CAS 629-14-1; EINECS/ELINCS 211-076-1
UN 1153 (DOT)
Synonyms: 1,2-Diethoxyethane; Diethyl 'Cellosolve'; Ethyl glyme
Classification: Glycol ether
Empirical: $C_6H_{14}O_2$
Formula: $C_2H_5OCH_2CH_2OC_2H_5$
Properties: Colorless liq., sl. odor; sol. in water and most org. solvs.; m.w. 118.20; dens. 0.8417 (20/20 C); vapor pressure 9.4 mm Hg (20 C); f.p. -74 C; b.p. 121.4 C; flash pt. 35 C; autoignition temp. 406 C; ref. index 1.393 (20 C); dielec. const. 5.10
Toxicology: LD50 (oral, rat) 4390 mg/kg; LCLo (inh., rat, 4 h) 8000 ppm; mod. toxic by ing.; mildly toxic by inh.; eye irritant; experimental teratogen, reproductive effects; target organs: CNS, kidney, liver, brain, testes; TSCA listed
Environmental: VOC; ThOD 2.31
Precaution: Flamm.; dangerous fire hazard exposed to heat or flame; can react with oxidizing materials
NFPA: Health 1, Flammability 3, Reactivity 0
Uses: Organic synthesis (reaction medium); solvent, diluent for detergents; solvent for paints; solubilizer
Regulatory: SARA §313 reportable; HAP
Manuf./Distrib.: Aldrich; Brand-Nu Labs; Eastern Chem.; Ferro/Grant; Fluka; Sigma; United-Guardian
Trade Names: Ethyl Glyme

Ethylene glycol dihydroxydiethyl ether. *See* Triethylene glycol

Ethylene glycol dimethyl ether

CAS 110-71-4; EINECS/ELINCS 203-794-9
UN 2252 (DOT)
Synonyms: Dimethoxyethane; 1,2-Dimethoxyethane; α,β-Dimethoxyethane; Dimethylglycol; 2,5-Dioxahexane; DME; EGDME; Ethylene dimethyl ether; Glycol dimethyl ether; Glyme; Monoethylene glycol dimethyl ether; Monoglyme
Empirical: $C_4H_{10}O_2$
Formula: $CH_3OCH_2CH_2OCH_3$
Properties: Water-wh. liq., sharp ethereal odor; sol. in hydrocarbons, most org. solvs.; misc. with water, alcohol; m.w. 90.12; dens. 0.86285 (20/4 C); m.p. -58 C; b.p. 82-83 (760 mm); flash pt. 4.5 C; ref. index 1.3813 (20 C)
Toxicology: Experimental reproductive effects; possible teratogen; target organs: liver, kidneys; TSCA listed
Precaution: DOT: Flamm. liq.; very dangerous fire hazard exposed to heat, flames, oxidizers; readily forms explosive peroxides; mixt. with lithium tetrahydroaluminate may ignite or explode if heated
Hazardous Decomp. Prods.: Heated to decomp., emits acrid smoke and fumes
NFPA: Health 2, Flammability 2, Reactivity 0
Storage: Store under nitrogen
Uses: Solvent for paints, electrochemistry, polymer and boron chemistry, physical processes such as gas absorp., extraction, stabilization, industrial prods. such as fuels, lubricants, textiles, pesticides; process solvent/solubilizer; pharmaceutical solvent, vehicle; in the Reformatsky reaction with methyl bromocrotonate
Manuf./Distrib.: Aldrich; Brand-Nu Labs; Ferro/Grant; Fluka; Sigma; Spectrum Quality Prods.
Trade Names: Monoglyme

Ethylene glycol distearate. *See* Glycol distearate
Ethylene glycol ethylene ether. *See* 1,4-Dioxane
Ethylene glycol ethyl ether. *See* Ethoxyethanol
Ethylene glycol ethyl ether acetate. *See* Ethoxyethanol acetate

Ethylene glycol 2-ethylhexyl ether

CAS 1559-35-9
Synonyms: Ethanol, 2-((2-ethylhexyl)oxy)-; Ethylene glycol mono-2-ethylhexyl ether; 2-((2-Ethylhexyl) oxy) ethanol
Empirical: $C_{10}H_{22}O_2$
Properties: M.w. 174.32
Toxicology: LD50 (oral, rat) 3080 mg/kg, (skin, rabbit) 2120 mg/kg; mod. toxic by ing. and skin contact; primary irritant; severe skin and eye irritant; TSCA listed
Hazardous Decomp. Prods.: Heated to decomp., emits acrid smoke and irritating fumes
NFPA: Health 0, Flammability 1, Reactivity 0
Uses: Solvent for coatings and cleaning applics.
Trade Names Containing: Eastman® EEH

Ethylene glycol formal. *See* 1,3-Dioxolane

Ethylene glycol hexyl ether

CAS 112-25-4; EINECS/ELINCS 203-951-1
Synonyms: Ethylene glycol n-hexyl ether; Ethylene glycol monohexyl ether; n-Hexyl 'Cellosolve'; Hexylglycol; 2-(Hexyloxy) ethanol
Classification: Glycol ether
Empirical: $C_8H_{18}O_2$
Formula: $C_6H_{13}OC_2H_4OH$
Properties: Water-wh. liq.; m.w. 146.26; sp.gr. 0.887 (20/20 C); f.p. -58 F; b.p. 208.1 C; flash pt. 90.5 C; ref. index 1.4290 (20 C)
Toxicology: LD50 (oral, rat) 830 mg/kg, (skin, rabbit) 810 mg/kg; primary irritant; TSCA listed
Precaution: Combustible
Uses: High-boiling solvent, coalescing aid for coatings
Manuf./Distrib.: Ashland; Dow; Eastman; Fluka
Trade Names: Hexyl Cellosolve®
Trade Names Containing: Ecosoft™ PH Solvent

Ethylene glycol n-hexyl ether. *See* Ethylene glycol hexyl ether

Ethylene glycol isobutyl ether

CAS 4439-24-1
Synonyms: Ethanol, 2-isobutyoxy-; Ethanol, 2-(2-methylpropoxy)-; Ethylene glycol monoisobutyl ether; iBG; Isobutyl 'Cellosolve'; Isobutyl glycol; 2-Isobutyoxyethanol; 2-(2-Methylpropoxy) ethanol
Classification: Glycol ether
Empirical: $C_6H_{14}O_2$
Formula: $(CH_3)_2CHCH_2OCH_2CH_2OH$
Properties: Colorless liq.; sol. in water; misc. with alcohol, ether; m.w. 118.18; sp.gr. 0.8925 (20/20 C); vapor pressure 0.473 mm Hg; b.p. 160.5 C
Toxicology: LDLo (oral, rat) 400 mg/kg; LD50 (skin, rabbit) 710 mg/kg; LCLo (inh., rat, 4 h) 1000 ppm; poison by ing.; mod. toxic by skin contact; mildly toxic by inh.; skin irritant; TSCA listed
Hazardous Decomp. Prods.: Heated to decomp., emits acrid smoke and fumes
NFPA: Health 2, Flammability 2
Uses: Solvent
Manuf./Distrib.: TCI Am.

Ethylene glycol isopropyl ether

CAS 109-59-1; EINECS/ELINCS 203-685-6
Synonyms: EGIPE; Ethanol, 2-isopropoxy-; Ethanol, 2-(1-methylethoxy)-; Ethylene glycol methyl-ethyl ether; Ethylene glycol monoisopropyl ether; β-Hydroxyethyl isopropyl ether; IPE; IPG; Isopropoxyethanol; 2-Isopropoxyethanol; Isopropyl 'Cellosolve'; Isopropylethylene glycol

ether; Isopropyl glycol; 2-(1-Methyl ethoxy) ethanol; 4-Methyl-3-oxa-1-pentanol; Monoisopropyl ether of ethylene glycol
Classification: Glycol ether
Empirical: $C_5H_{12}O_2$
Formula: $(CH_3)_2CHOCH_2CH_2OH$
Properties: Colorless liq.; mild ethereal odor; misc. with water; m.w. 104.15; sp.gr. 0.903; vapor pressure 3 mm Hg; b.p. 42-44 C (13 mm Hg); flash pt. 44.4 C; ref. index 1.41 (20 C)
Toxicology: ACGIH TLV/TWA 25 ppm (skin); LD50 (oral, rat) 5660 μl/kg, (skin, rabbit) 1600 μl/kg, (IP, mouse) 1860 mg/kg; LC50 (inh., mouse, 7 h) 1930 ppm; eye/skin irritant; may be harmful by ing., inh., skin absorp.; may cause hematuria, anemia, pulmonary edema, analgesia, dyspnea, somnolence, excitement, birth defects, fetotoxicity, bone marrow damage; target organs: eyes, skin, respiratory system, blood, reproductive system; TSCA listed
Precaution: Flamm. liq.; incompat. with strong oxidizing agents
Hazardous Decomp. Prods.: CO, CO_2; emits toxic fumes under fire conditions
NFPA: Health 1, Flammability 3, Reactivity 0
Storage: Store in cool, dry place away from heat, open flame; keep tightly closed
Uses: Solvent for nat. and syn. resins, surf. coatings, printing inks, household/industrial cleaners, cosmetics, agric., insecticides, herbicides, dyes (textiles, leather, printing); coalescent, flow aid for aq. paints; mutual solvent, coupling agent in sol. oils, metal/glass cleaners, varnish removers; diluent in hydraulic brake fluids
Regulatory: SARA §313 reportable
Manuf./Distrib.: Aldrich; Fluka; Lancaster Synthesis; TCI Am.
Trade Names: i-PGE; Isopropyl Oxitol

Ethylene glycol methyl ether. *See* Methoxyethanol
Ethylene glycol methyl ether acetate. *See* Methoxyethanol acetate
Ethylene glycol methyl-ethyl ether. *See* Ethylene glycol isopropyl ether
Ethylene glycol monoacetate. *See* Ethylene glycol acetate
Ethylene glycol monobenzyl ether. *See* Ethylene glycol benzyl ether
Ethylene glycol monobutyl ether; Ethylene glycol mono-n-butyl ether. *See* Butoxyethanol
Ethylene glycol monobutyl ether acetate; Ethylene glycol mono-n-butyl ether acetate. *See* Butoxyethanol acetate
Ethylene glycol monobutyl ether stearate. *See* Butoxyethyl stearate
Ethylene glycol monoethyl ether. *See* Ethoxyethanol
Ethylene glycol monoethyl ether acetate. *See* Ethoxyethanol acetate
Ethylene glycol mono-2-ethylhexyl ether. *See* Ethylene glycol 2-ethylhexyl ether
Ethylene glycol monohexyl ether. *See* Ethylene glycol hexyl ether
Ethylene glycol monoisobutyl ether. *See* Ethylene glycol isobutyl ether
Ethylene glycol monoisopropyl ether. *See* Ethylene glycol isopropyl ether
Ethylene glycol monomethyl ether. *See* Methoxyethanol
Ethylene glycol monomethyl ether acetate. *See* Methoxyethanol acetate
Ethylene glycol monomethyl ether, stearate. *See* Methoxyethyl stearate
Ethylene glycol monophenyl ether. *See* Phenoxyethanol
Ethylene glycol monopropyl ether; Ethylene glycol mono-n-propyl ether. *See* Ethylene glycol propyl ether
Ethylene glycol monostearate. *See* Glycol stearate
Ethylene glycol phenyl ether. *See* Phenoxyethanol

Ethylene glycol propyl ether

CAS 2807-30-9; EINECS/ELINCS 220-548-6
Synonyms: Ethanol, 2-propoxy; Ethylene glycol monopropyl ether; Ethylene glycol mono-n-propyl ether; Monopropyl ether of ethylene glycol; 2-Propoxyethanol; Propyl 'Cellosolve'
Empirical: $C_5H_{12}O_2$
Formula: $CH_2OHCH_2OCH_2CH_2CH_3$
Properties: Liq.; sol. in water; m.w. 104.17; dens. 0.913 (20/20 C); f.p. -90 C; b.p. 149.5 C; flash pt. (TCC) 49 C
Toxicology: LD50 (oral, rat) 3089 mg/kg, (skin, rabbit) 960 mg/kg; LC50 (inh., mouse, 7 h) 1530 ppm; mod. toxic by ing. and skin contact; mild toxicity by inh.; skin and severe eye irritant; experimental teratogen, reproductive effector; TSCA listed
Precaution: Combustible; can react with oxidizers
Hazardous Decomp. Prods.: Heated to decomp., emits acrid smoke and irritating fumes
Uses: Solvent for coatings, cosmetics, resins; coupling solvent for resin/water systems
Manuf./Distrib.: Aldrich; Ashland; Eastman; Sal Chem.; Stanchem
Trade Names: Eastman® EP; Propyl Cellosolve®

Ethylene glycol stearate. *See* Glycol stearate
1,8-Ethylene naphthalene. *See* Acenaphthene

Ethylene oxide

CAS 75-21-8; EINECS/ELINCS 200-849-9
UN 1040 (DOT); FEMA 2433
Synonyms: Dihydrooxirene; Dimethylene oxide; EO; Epoxyethane; 1,2-Epoxyethane; Ethene oxide; ETO; Oxacyclopropane; Oxane; Oxidoethane; α,β-Oxidoethane; Oxirane
Classification: Ether
Empirical: C_2H_4O
Properties: Colorless gas or liq.; ether-like odor; very sol. in ether; sol. in org. solvs.; misc. with water, alcohol, oxygenated solvs.; m.w. 44.06; dens. 0.8711 (20/20 C); vapor pressure 1310 mm Hg; m.p. -111.3 C; b.p. 10.73 C; flash pt. -29 C; ref. index 1.359; dielec. const. 13.0
Toxicology: ACGIH TLV/TWA 1 ppm; LD50 (oral, rat) 72 mg/kg; poison by ing., IP, subcut., IV routes; mod. toxic by inh.; irritating to eyes, skin, respiratory tract; human systemic effects by inh. (convulsions, nausea, vomiting, pulmonary changes); suspected human carcinogen; experimental tumorigen, neoplastigen, teratogen; mutagenic data; TSCA listed
Environmental: VOC; BOD5 0.06; COD 1.74; ThOD 1.82
Precaution: Flamm.; LEL 3%; UEL 100%; severe explosion hazard exposed to flame; violent polymerization on contact with ammonia, alkali hydroxides, amines, acids, etc.
Hazardous Decomp. Prods.: Heated to decomp., emits acrid smoke and irritating fumes
Uses: Mfg. of ethylene glycol and higher glycols, surfactants, acrylonitrile, ethanolamines, ethoxylates; petroleum demulsifier; rocket propellant; industrial sterilant (medical plastic tubing); fungicide; solvent; fumigant in sizing for paper/paperboard in contact with dry food; mfg. of copolymer condensates for use as solubilizer, stabilizer in flavors, processing aids, wetting agents for foods
Regulatory: FDA 21CFR §172.710, 172.808, 175.105, 176.180, 176.210, 177.2470, 178.3520; FEMA GRAS; FDA approved for ophthalmics, topicals; SARA §302a extremely hazardous, §313 reportable; CERCLA hazardous substance; Calif. Prop. 65 reportable; HAP
Manuf./Distrib.: Aldrich; Asiamerica Int'l.; BASF; BP Amoco; Celanese; Cognis/Chems. Group; Degussa AG; Dow; Eastman; Equistar; Honeywell Perf. Polymers; Huntsman; MG Ind.; Occidental; Rhodia HPCII; Shell; Sunoco; Total Spec. Chems.

Ethylene tetrachloride. *See* Perchloroethylene
Ethylene thioglycol. *See* 2-Mercaptoethanol
Ethylene trichloride. *See* Trichloroethylene
N-Ethylethanamine. *See* Diethylamine
Ethyl ethanoate. *See* Ethyl acetate

Ethyl ether

CAS 60-29-7; EINECS/ELINCS 200-467-2
UN 1155 (DOT)
Synonyms: Anesthetic ether; Diethyl ether; Diethyl oxide; Ether; Ethoxyethane; Ethyl oxide; 1,1′-Oxybisethane; Oxybis-1,1′-ethane; Solvent ether; Sulfuric ether
Classification: Sat. aliphatic ether
Empirical: $C_4H_{10}O$
Formula: $CH_3CH_2OCH_2CH_3$
Properties: Colorless clear volatile liq., sweet pungent odor; sol. in chloroform, benzene, conc. HCl; sol. 85 g/l water (15 C); misc. with alcohol, ether, most org. solvs.; m.w. 74.12; sp.gr. 0.7135 (20/4 C); vapor pressure 537 mm Hg; m.p. -116.2 C; b.p. 34.6 C; flash pt. (CC) -45 C; autoignition temp. 160 C; surf. tens. 16.5 dynes/cm; dielec. const. 4.197
Toxicology: ACGIH TLV/TWA 400 ppm; STEL 500 ppm; LD50 (oral, rat) 1215 mg/kg, (IP, mouse) 2420 mg/kg, (IV, mouse) 996 mg/kg; poison by subcut. route; mod. toxic by ing., IP, IV routes; mildly toxic by inh.; mod. toxic to humans by ing.; human systemic effects by inh.; CNS depressant; anesthetic; mod. to severe skin and eye irritant; extreme exposure may cause respiratory failure and death; mutagenic data;

TSCA listed
Environmental: VOC; BOD20 0.50; ThOD 2.59
Precaution: DOT: Flamm. liq.; LEL 1.9%; UEL 36%; severe fire and explosion hazard when exposed to heat or flame; forms explosive peroxides; incompat. with sulfur compds., halogens, interhalogens, strong oxidizers; violent or explosive reactions possible
Hazardous Decomp. Prods.: Can form peroxides on prolonged exposure to air and light, in absence of inhibitors; heated to decomp., emits acrid smoke and irritating fumes
NFPA: Health 2, Flammability 4, Reactivity 1
Storage: Store in cool, dry, well-ventilated area, out of direct sunlight; hygroscopic
Uses: Organic synthesis; mfg. of monoethanolamine; smokeless powders; analytical chemistry; anesthetic; extraction solvent; solvent for paints, waxes, fats, and oils; solvent in Grignard/Wurtz reactions; diesel fuel ignition primer; ethylation reagent; solvent, antiseptic, analgesic, expectorant in pharmaceuticals
Regulatory: FDA approved for transmucosal pharmaceuticals; CERCLA hazardous substance
Manuf./Distrib.: AMC Chems.; Aldrich; BCH Brühl; Degussa AG; EM Science; Equistar; ExxonMobil; Fluka; GFS; Integra; J.T. Baker; Lipo; Mallinckrodt Baker; Romil Ltd; Ruger; Sigma; Sithean; Spectrum Quality Prods.; Thomas Scientific; Tulstar Prods.; VWR Int'l.; Vopak USA
Trade Names Containing: Punctilious® SDA 13A 190 Proof; Punctilious® SDA 19 190 Proof; Punctilious® SDA 32 190 Proof; Punctilious® SDA 32 Anhydrous

Ethyl 3-ethoxypropionate

CAS 763-69-9
Synonyms: EEP; 3-Ethoxypropanoic acid, ethyl ester; Ethoxypropionic acid ethyl ester; 3-Ethoxypropionic acid ethyl ester; Ethyl-β-ethoxypropionate
Empirical: $C_7H_{14}O_3$
Formula: $C_2H_5OC_3H_5O_2C_2H_5$
Properties: Liq.; sl. sol. in water; m.w. 146.19; dens. 0.9496 (20/20 C); f.p. < -50 C; b.p. 165-172 C; flash pt. (Seta) 58 C; ref. index 1.4074 (20 C)
Toxicology: LD50 (oral, rat) 5000 mg/kg, (skin, rabbit) 10 g/kg; mildly toxic by ing. and skin contact; skin and eye irritant; TSCA listed
Precaution: Flamm. exposed to heat or flame; can react with oxidizing materials
Hazardous Decomp. Prods.: Heated to decomp., emits acrid smoke and irritating fumes
Uses: Solvent for polymerization, esp. high-solids acrylics; retarder solvent for coatings; let-down solvent for thinning synthetic coating resins, esp. moisture-sensitive resins for PU coatings; air-dry and baking finishes; intermediate for vitamin B_1
Manuf./Distrib.: Aldrich; Ashland; Dow; Eastman; General Chem.; Houghton Chem.
Trade Names: Eastman® EEP; Ucar® Ester EEP
Trade Names Containing: Dynasolve CU-5; Dynasolve CU-6

Ethyl-β-ethoxypropionate. *See* Ethyl 3-ethoxypropionate
Ethyl ethynyl methyl carbinol. *See* Methyl pentynol

Ethyl formate

CAS 109-94-4; EINECS/ELINCS 203-721-0
UN 1190 (DOT); FEMA 2434
Synonyms: Ethyl formic ester; Ethyl methanoate; Formic acid ethyl ester; Formic ether
Classification: Aliphatic organic compd.; carboxylic acid ester
Definition: Ester of ethyl alcohol and formic acid
Empirical: $C_3H_6O_2$
Formula: $HCOOC_2H_5$
Properties: Colorless liq., sharp rum-like odor; very sol. in water; sol. in acetone, alcohol, benzene, ether, fixed oils, propylene glycol; sl. sol. in min. oil; m.w. 74.09; sp.gr. 0.9236 (20/20 C); vapor pressure 100 mm Hg (5.4 C); m.p. -79 C; b.p. 54 C; flash pt. (CC) -20 C; ref. index 1.359
Toxicology: ACGIH TLV/TWA 100 ppm; LD50 (oral, rat) 1850 mg/kg, (skin, rabbit) 20 g/kg; mod. toxic by ing., subcut. routes; mildly toxic by skin contact and inh.; skin and eye irritant; vapor can irritate nose, throat, lungs; high concs. may cause CNS depression, unconsciousness; severe exposure may cause potentially fatal pulmonary edema; experimental tumorigen; TSCA listed
Precaution: DOT: Flamm. liq.; dangerous fire and explosion hazard exposed to heat, flame, oxidizers; incompat. with oxidizing materials, acids (decomp. can occur), bases (vigorous decomp. can occur), moisture
Hazardous Decomp. Prods.: Ethanol, formic acid; dec. slowly in water, releasing formic acid; heated to decomp., emits acrid smoke and irritating fumes
NFPA: Health 2, Flammability 3, Reactivity 0
Storage: Store in cool, dry, well-ventilated area, out of direct sunlight; ground drums and bond transfer containers
Uses: Synthetic flavoring agent and fragrance in foods and pharmaceuticals; intermediate for prod. of pharmaceuticals and crop protection agents; solvent for nitrocellulose, pharmaceuticals; insecticide, fungicide, and larvicide for tobacco, cereals, dried; acetone substitute; org. synthesis; disinfectant; preservative; reagent; fragrance in cosmetics
Features: Ethereal, fruity fragrance and flavor, rum-like on dilution
Regulatory: FDA 21CFR §184.1295, GRAS; 193.210, insecticide residue tolerance of 250 ppm in raisins and currants; FEMA GRAS
Manuf./Distrib.: Advanced BioTech; Advanced Synthesis Tech.; Aldrich; Axxence Aromatic GmbH; BASF; Degussa AG/Health & Nutrition; Degussa Flavors/Alfrebro; Fleurchem; Fluka; Integra; J.T. Baker; Kira; Lluch Essence; Ogawa & Co.; Pechiney Chems. Div.; Penta Mfg.; Rhodia/Fine Orgs.; San-Ei Gen F.F.I.; Sigma; Spectrum Quality Prods.; T. Hasegawa; VWR Int'l.; Xinchem

Ethyl formic ester. *See* Ethyl formate
Ethyl glycol. *See* Ethoxyethanol
Ethyl glycol acetate. *See* Ethoxyethanol acetate
Ethyl glyme. *See* Ethylene glycol diethyl ether

Ethyl hexanediol

CAS 94-96-2; EINECS/ELINCS 202-377-9
Synonyms: Ethohexadiol; 2-Ethylhexanediol; 2-Ethyl-1,3-hexanediol; 2-Ethylhexane-1,3-diol; 2-Ethylhexanediol-1,3; Ethyl hexylene glycol; 2-Ethyl-3-propyl-1,3-propanediol; 3-Hydroxymethyl-n-heptan-4-ol; Hydroxymethyl-n-heptan-4-ol; 1,3-Octanediol; Octylene glycol; 1,3-Octylene glycol
Classification: Aliphatic alcohol
Empirical: $C_8H_{18}O_2$
Formula: $CH_3CH_2CH_2CHOHCHCH_2CH_3CH_2OH$
Properties: Colorless, sl. visc. liq., odorless; sol. in alcohol, ether, acetone, toluene, benzene, CCl_4; partly sol. in water; m.w. 146.26; dens. 0.9422 (20/20 C); bulk dens. 7.8 lb/gal (20 C); vapor pressure < 0.01 mm Hg (20 C); f.p. -40 C; b.p. 244 C; flash pt. 126 C; autoignition temp. 225 C; ref. index 1.4465-1.4515
Toxicology: LD50 (oral, rat) 1400 mg/kg, (skin rabbit) 2000 mg/kg; moderately toxic by ingestion and skin contact; skin and severe eye irritant; TSCA listed
Precaution: Combustible exposed to heat or flame; can react with oxidizers
Hazardous Decomp. Prods.: Heated to decomp., emits acrid smoke and irritating fumes
NFPA: Health 2, Flammability 1, Reactivity 0
Uses: Insecticide, insect repellent; vehicle, solvent in printing inks; solvent in cosmetics, urethane industry; chelating agent for boric acid; brake oil; raw material for oily cosmetics, creams/lotions; coupling agent for shampoos, liq. cleansing creams; medicine; solvent for pharmaceutical topical ointments
Regulatory: FDA approved for topicals
Manuf./Distrib.: Aldrich; Degussa AG; Dixie; Dow; Fluka; Kyowa Hakko Kogyo; Ruger

2-Ethylhexanediol; 2-Ethyl-1,3-hexanediol; 2-Ethylhexane-1,3-diol; 2-Ethylhexanediol-1,3. *See* Ethyl hexanediol
Ethyl hexanoate. *See* Ethyl caproate
2-Ethylhexanoic acid; α-Ethylhexanoic acid. *See* 2-Ethylhexoic acid
2-Ethylhexanoic acid, cetyl-stearyl ester. *See* Cetearyl octanoate
2-Ethylhexanoic acid, 2,2-dimethyl-1,3-propanediyl ester. *See* Neopentyl glycol dioctanoate
2-Ethylhexanoic acid, hexadecyl ester. *See* Cetyl octanoate
2-Ethylhexanoic acid, 1-methyl-1,2-ethanediyl ester. *See* Propylene glycol dioctanoate

2-Ethylhexanoic acid, octadecyl ester. *See* Stearyl octanoate

2-Ethylhexanol

CAS 104-76-7; EINECS/ELINCS 203-234-3
FEMA 3151
Synonyms: Alcohol C_8; 2-EH; 2-Ethyl-1-hexanol; 2-Ethylhexan-1-ol; 2-Ethylhexyl alcohol; 1-Hexanol, 2-ethyl
Classification: Alcohol
Empirical: $C_8H_{18}O$
Formula: $CH_3(CH_2)_3CHC_2H_5CH_2OH$
Properties: Colorless to pale yel. clear liq., mild oily sweet sl. rose fragrance; sol. in ethanol, ether, diethyl ether, DMSO, acetone, benzene, chloroform; misc. with most org. solvs.; sl. sol. in water; m.w. 130.26; dens. 0.83 (20 C); vapor pressure 0.05 mm Hg (20 C); f.p. -76 C; b.p. 183.5 C; flash pt. 81.1 C; ref. index 1.4300 (20 C); surf. tens. 27.6 dynes/cm (20 C)
Toxicology: LD50 (oral, rat) 2049 mg/kg, (IP, rat) 500 mg/kg, (subcut., rat) 650 mg/kg, (skin, rabbit) 1970 mg/kg; mod. toxic by ing., skin contact, IP, subcut., parenteral routes; severe eye irritant; moderate skin irritant; experimental teratogen, reproductive effects; mutagenic data; TSCA listed
Precaution: Combustible; dangerous fire hazard exposed to heat or flame; can react vigorously with oxidizing materials; incompat. with strong oxidizing agents and strong acids
Hazardous Decomp. Prods.: CO, CO_2; heated to decomp., emits acrid smoke and fumes
NFPA: Health 2, Flammability 2, Reactivity 0
Storage: Store under ambient temps. away from oxidizing materials and acids
Uses: Organic synthesis; prod. of plasticizers for PVC resins (phthalates, adipates, trimellitates), 2-ethylhexyl acrylate, 2-ethylhexyl nitrate (cetane improver), lube additives, surfactants; defoamer; wetting agent; solvent for nitrocellulose, paints, waterborne coatings, dyes, resins, oils, antifoaming agents; penetrant for plasticizing inks, etc.; disinfectant; flotation frothing agent; mining applics.; in mercerizing textiles; in textile finishing compds.; in paint lacquers, baking finishes, inks, rubber, paper, lubricants, photography, dry cleaning; defoamer in acute pulmonary edema extractants; in food-pkg. adhesives; syn. flavor for foods and pharmaceuticals; cosmetics ingred.; defoamer in food-contact paper/paperboard; in cellophane for food pkg.
Features: Sweet flavor
Regulatory: FDA 21CFR §175.105, 176.180, 176.210, 177.1200, 178.3480; FEMA GRAS
Manuf./Distrib.: ABCR; Aldrich; Allchem Ind.; Ashland; BASF; BP Chems. Ltd; Bencorp Int'l.; Celanese; Chisso Am.; Coyne; Dow; Eastman; Expo Chem.; Fluka; GNC Group; Houghton Chem.; Ivanhoe Ind.; KIC Chems.; Mitsubishi Chem.; Penta Mfg.; Rhodia HPCII; Shell; Stanchem; Sunoco; Total Spec. Chems.; Vopak USA
Trade Names: Ucar® 2-Ethylhexanol

2-Ethyl-1-hexanol; 2-Ethylhexan-1-ol. *See* 2-Ethylhexanol
2-Ethyl-1-hexanol acetate. *See* 2-Ethylhexyl acetate
2-Ethyl-1-hexanol, hydrogen sulfate, sodium salt. *See* Sodium 2-ethylhexyl sulfate
2-Ethylhexanol phosphate; 2-Ethyl-1-hexanol phosphate; 2-Ethylhexanol, phosphate triester. *See* Trioctyl phosphate
2-Ethyl-1-hexanol sulfate sodium salt. *See* Sodium 2-ethylhexyl sulfate
2-Ethylhexanyl acetate. *See* 2-Ethylhexyl acetate
Ethyl hexoate. *See* Ethyl caproate

2-Ethylhexoic acid

CAS 149-57-5; EINECS/ELINCS 205-743-6
Synonyms: 2-Butylbutanoic acid; Butylethylacetic acid; 2-EH acid; 2-Ethylcaproic acid; α-Ethylcaproic acid; 2-Ethylhexanoic acid; α-Ethylhexanoic acid; 3-Heptanecarboxylic acid; Hexanoic acid, 2-ethyl-
Classification: Aliphatic carboxylic acid
Empirical: $C_8H_{16}O_2$
Formula: $C_4H_9CH(C_2H_5)COOH$
Properties: Colorless liq., mild char. odor; sol. in ether; sl. sol. in water, alcohol; m.w. 144.21; dens. 0.9077 (20/20 C); vapor pressure 0.03 mm Hg (20 C0; f.p. -83 C; b.p. 226.9 C; flash pt. 126 C
Toxicology: LD50 (oral, rat) 3000 mg/kg, (skin, rabbit) 126 mg/kg; mild to mod. skin irritant; severe eye irritant; ing. may cause mild irritation, sore throat, abdominal pain, nausea, vomiting
Precaution: Combustible; incompat. with strong bases and strong oxidizers (violent reactions possible), reactive metals (may produce flamm./explosive hydrogen gas)
Hazardous Decomp. Prods.: Thermal decomp. prods.: CO, CO_2
NFPA: Health 1, Flammability 1, Reactivity 0
Storage: Store in cool, dry, well-ventilated area, out of direct sunlight; avoid generating mists; limit quantities in use
Uses: Paint and varnish driers (metallic salts); mfg. of alkyd resins, plasticizers/stabilizers for PVC; metal salts as silicone stabilizers, fungicides; chem. intermediate for lubricants, detergents, corrosion inhibitors; cosolvent and defoamer in pesticides; fragrance intermediate for cosmetics
Manuf./Distrib.: Aldrich; Ashland; BASF; Celanese; Coyne; Dow; Eastman; Filo; Fluka; Houghton Chem.; ICC Chem.; Sigma

2-Ethylhexyl acetate

CAS 103-09-3; EINECS/ELINCS 203-079-1
Synonyms: Acetic acid 2-ethylhexyl ester; 2-Ethyl-1-hexanol acetate; 2-Ethylhexanyl acetate; 2-Ethyl-1-hexyl acetate; β-Ethylhexyl acetate; 2-Ethylhexyl ethanoate
Classification: Carboxylic acid ester
Empirical: $C_{10}H_{20}O_2$
Formula: $CH_3COOCH_2CHC_2H_5C_4H_9$
Properties: Water-wh. stable liq.; sweet odor; sol. in ether, alcohol; very sl. sol. in water; m.w. 172.27; sp.gr. 0.873 (20/20 C); m.p. -93 C; b.p. 199-205 C; flash pt. (CC) 71 C; ref. index 1.4201 (20 C); pH neutral
Toxicology: LD50 (oral, rat) 3000 mg/kg; mod. toxic by ingestion; sl. to mod. skin and eye irritant; may be sl. absorbed thru skin; inh. may cause nose/throat irritation; unlikely to cause significant health effects
Precaution: Combustible; incompat. with strong acids or bases (decomp. can occur), oxidizing materials (increases fire/explosion risk)
Hazardous Decomp. Prods.: CO_2, CO; heated to decomp., emits acrid smoke and fumes
NFPA: Health 2, Flammability 2, Reactivity 0
Storage: Store in cool, dry, well-ventilated area, out of direct sunlight; avoid generating mists
Uses: Solvent for nitrocellulose, resins, baking finishes, waxes, oils; solvent, flow aid, coalescent in paints, lacquers, coatings, printing inks; in fragrances; cosmetics ingred.
Manuf./Distrib.: A.P. Chems. Ltd; Aceto; Aldrich; Allan; Ashland; Chemial SpA; Degussa AG; Dow; Eastman; Haltermann Prods. UK; Houghton Chem.; Penta Mfg.; Vopak USA

2-Ethyl-1-hexyl acetate; β-Ethylhexyl acetate. *See* 2-Ethylhexyl acetate
2-Ethylhexyl alcohol. *See* 2-Ethylhexanol
Ethyl hexylate. *See* Ethyl caproate
Ethyl hexylene glycol. *See* Ethyl hexanediol
2-Ethylhexyl ethanoate. *See* 2-Ethylhexyl acetate
2-Ethylhexyl hexadecanoate. *See* Octyl palmitate
2-Ethylhexyl 2-hydroxybenzoate. *See* 2-Ethylhexyl salicylate
2-Ethylhexyl 12-hydroxyoctadecanoate; 2-Ethylhexylhydroxystearate; 2-Ethylhexyl 12-hydroxystearate. *See* Octyl hydroxystearate
2-Ethylhexyl isononanoate. *See* Octyl isononanoate

2-Ethylhexyl lactate

CAS 6283-86-9
Classification: Lactate ester
Definition: 2-Ethylhexyl ester of lactic acid
Uses: Solvent
Features: Low volatility, high flash pt.

2-Ethylhexyl-S-lactate

CAS 186817-80-1; EINECS/ELINCS 228-503-2
Synonyms: 2-Ethylhexyl lactate; (S)-2-Ethylhexyl lactate; L-Lactic acid, 2-ethylhexyl ester; Propanoic acid, 2-hydroxy-, 2-ethylhexyl ester
Classification: Lactate ester
Empirical: $C_{11}H_{22}O_3$
Properties: Sol. 30 mg/100 ml in water; m.w. 202.3; dens. 0.940 g/ml (20 C); b.p. 246 C; flash pt. 113 C
Uses: Solvent; degreaser
Trade Names: Purasolv® EHL

(S)-2-Ethylhexyl lactate. *See* 2-Ethylhexyl-S-lactate
2-Ethylhexyl octadecanoate. *See* Octyl stearate
2-Ethylhexyl 2-octadecenoate. *See* Octyl oleate
2-((2-Ethylhexyl) oxy) ethanol. *See* Ethylene glycol 2-ethylhexyl ether
2-[2-(2-Ethylhexyloxy) ethoxy] ethanol. *See* Diethylene glycol 2-ethylhexyl ether
2-Ethylhexyl oxystearate. *See* Octyl hydroxystearate
2-Ethylhexyl palmitate. *See* Octyl palmitate

2-Ethylhexyl salicylate

CAS 118-60-5; EINECS/ELINCS 204-263-4
Synonyms: Benzoic acid, 2-hydroxy-, 2-ethylhexyl ester; 2-Ethylhexyl 2-hydroxybenzoate; Salicylic acid-2-ethylhexyl ester
Classification: Aromatic ester
Definition: Ester of 2-ethylhexyl alcohol and salicylic acid
Empirical: $C_{15}H_{22}O_3$
Properties: M.w. 250.34; dens. 1.014; b.p. 189-190 (21 mm); ref. index 1.5020
Toxicology: LD50 (IP, mouse) 200 mg/kg; poison by IP route; primary skin irritant; TSCA listed
Hazardous Decomp. Prods.: Heated to decomp., emits acrid smoke and irritating fumes
Storage: Heat and lt. sensitive
Uses: Fragrance; UV absorber in cosmetics
Manuf./Distrib.: Aldrich
Trade Names: HallBrite® OS

2-Ethylhexyl sebacate. *See* Dioctyl sebacate
2-Ethylhexyl sodium sulfate. *See* Sodium 2-ethylhexyl sulfate
2-Ethylhexyl stearate. *See* Octyl stearate
Ethyl hydrate; Ethyl hydroxide. *See* Alcohol
2-(N-Ethyl-N-2-hydroxyethylamino) ethanol. *See* N-Ethyldiethanolamine
Ethyl (β-hydroxyethyl) aniline; N-Ethyl-N-(2-hydroxyethyl) aniline. *See* Phenylethylethanolamine
Ethyl 2-hydroxypropanoate. *See* Ethyl lactate
Ethyl-(S)-2-hydroxypropanoate. *See* Ethyl-(S)-lactate
Ethyl 2-hydroxypropionate; Ethyl α-hydroxypropionate. *See* Ethyl lactate
Ethyl-S(-)-2-hydroxy propionate; Ethyl-(S)-2-hydroxy propionate. *See* Ethyl-(S)-lactate
Ethylidene chloride; Ethylidene dichloride. *See* 1,1-Dichloroethane
Ethylidene diethyl ether. *See* Acetal
Ethylidenelactic acid. *See* Lactic acid
2,2´-(Ethylimino) bisethanol; 2,2´-(Ethylimino) diethanol; N-Ethyl-2,2´-iminodiethanol. *See* N-Ethyldiethanolamine
Ethylisobutylmethane. *See* Isoheptane
Ethyl ketovalerate; Ethyl 4-ketovalerate; Ethyl γ-ketovalerate. *See* Ethyl levulinate

Ethyl lactate

CAS 97-64-3; EINECS/ELINCS 202-598-0
UN 1192 (DOT); FEMA 2440
Synonyms: Ethyl 2-hydroxypropanoate; Ethyl 2-hydroxypropionate; Ethyl α-hydroxypropionate; 2-Hydroxypropionic acid ethyl ester; Lactic acid ethyl ester
Classification: Lactate ester
Definition: Ethyl ester of lactic acid; commercial prod. is a racemic mixt.
Empirical: $C_5H_{10}O_3$
Formula: $CH_3CHOHCOOC_2H_5$
Properties: Colorless liq.; mild pleasant odor; misc. with water, alcohol, ketones, esters, hydrocarbons, oil, oxygenated solvs.; m.w. 118.13; dens. 1.020-1.036 (20/20 C); vapor pressure 1.2 mm Hg (20 C); m.p. -26 C; b.p. 154 C; flash pt. 46.1 C; autoignition temp. 400 C; ref. index 1.410-1.420; surf. tens. 34 dynes/cm; KB value 1000
Toxicology: LD50 (oral, mouse) 2500 mg/kg, (subcut., mouse) 2500 mg/kg, (IV, mouse) 600 mg/kg, (skin, rabbit) > 5 g/kg; mod. toxic by IP, subcut., IV routes; low toxicity by ing. and skin contact; TSCA listed
Environmental: VOC; BOD5 0.7; COD 1.62; ThOD 1.63
Precaution: Flamm. or combustible; LEL 1.6%; UEL 10.6%; can react with oxidizers; sl. explosion hazard in vapor form exposed to flame
Hazardous Decomp. Prods.: Heated to decomp., emits acrid smoke and irritating fumes
Uses: Dipped latex prods.; coatings; specialty finishes; inks; diluents; adhesives; intermediates; screen printing of electronic parts; solvent resins; safety glass; solvent for electronics, photoresist, lacquers, coatings, cleaning agents, precision cleaning, inks, metal industry, cellulose acetate; synthetic flavoring agent and fragrance in foods and pharmaceuticals; cosmetics ingred.; in food-pkg. adhesives
Features: Fruity flavor; non-ozone depleting
Regulatory: FDA 21CFR §172.515, 175.105; FEMA GRAS
Manuf./Distrib.: A&E Connock; Aceto; Advanced BioTech; Advanced Synthesis Tech.; Aldrich; Alfa Chem; Axxence Aromatic GmbH; Boith China; Brown; CPS; Ciba Spec. Chems./Water Treatment; Degussa Flavors/Alfrebro; Farleyway Chems. Ltd; Fleurchem; Grau Aromatics; Haltermann Prods. UK; Int'l. Chem. Inc; J.F. Henry; J.T. Baker; Jungbunzlauer; Kaltron/Pettibone; Kraft Chem.; Lluch Essence; Moore Ingreds.; PURAC Am.; Penta Mfg.; Spectrum Quality Prods.; Universal Preserv-A-Chem; VWR Int'l.; Vopak USA; Wilke Int'l.; Xinchem
Trade Names: Galaster EL

(-)-Ethyl L-lactate. *See* Ethyl-(S)-lactate

Ethyl-(S)-lactate

CAS 687-47-8; EINECS/ELINCS 211-694-1
UN 1192
Synonyms: Ethyl-(S)-2-hydroxypropanoate; Ethyl-S(-)-2-hydroxy propionate; Ethyl-(S)-2-hydroxy propionate; (-)-Ethyl L-lactate; Ethyl (S)-(-)-lactate; (S)-(-)-2-Hydroxypropionic acid, ethyl ester; Lactic acid ethyl ester; Propanoic acid, 2-hydroxy-, ethyl ester, (S)-
Classification: Lactate ester
Empirical: $C_5H_{10}O_3$
Formula: $CH_3CH(OH)COOC_2H_5$
Properties: Colorless liq.; fruit-like odor; misc. with water, alcohols, ketones, esters, hydrocarbons, ether, oil; m.w. 118.13; sp.gr. 1.042; vapor pressure 2 mm Hg (20 C); m.p. -26 C; b.p. 154 C; flash pt. 46 C; autoignition temp. 400 C; pH 4; ref. index 1.413 (20 C)
Toxicology: LD50 (oral, rat) 8200 mg/kg, (subcut., mouse) 2500 mg/kg, (IV, mouse) 600 mg/kg, (skin, rabbit) > 5 g/kg; LC50 (inh., rat, 4 h) 5400 mg/l; irritant; risk of serious eye damage; may cause skin irritation; may be harmful by ing., inh., skin absorp.; vapor/mist irritating to mucous membranes, upper respiratory tract; may cause coughing, dyspnea, muscle weakness, cyanosis, mouth/pharynx/esophagus/GI tract irritation; TSCA listed
Environmental: Biodeg.; non-ozone depleting; EC50 (daphnia magna, 48 h) 683 mg/l; COD 1.62 g/g; sl. pollutant; prevent entry into sewers, waters, wastewater, soil
Precaution: Flamm. liq.; LEL 1.5 vol.%; UEL 11.4 vol.%; vapors heavier than air; sensitive to air; may form explosive mixts. with air; prevent electrostatic charging; incompat. with strong oxidizing agents, various plastics; hydrolyzes
Hazardous Decomp. Prods.: CO, CO_2; emits toxic fumes under fire conditions
Storage: Store @ 15-25 C in cool, dry, well-ventilated place away from heat, open flame; keep tightly closed
Uses: Solvent for NC, cellulose acetate, cellulose ethers and resins, lacquers, paints, enamels, varnishes, stencil sheets, safety glasses, flavors, cosmetics; degreaser as substitute for trichloroethylene; synthesis
Manuf./Distrib.: Aldrich; Fleurchem; Fluka; Merck KGaA
Trade Names: Purasolv® EL; Purasolv® ELECT; Purasolv® ELS

Ethyl (S)-(-)-lactate. *See* Ethyl-(S)-lactate
Ethyl laevulinate; Ethyl levulate. *See* Ethyl levulinate

Ethyl levulinate

CAS 539-88-8; EINECS/ELINCS 208-728-2
FEMA 2442
Synonyms: Ethyl 3-acetylpropionate; Ethyl ketovalerate; Ethyl 4-ketovalerate; Ethyl γ-ketovalerate; Ethyl laevulinate; Ethyl levulate; Ethyl-4-oxopentanoate; Ethyl-4-oxovalerate; Levulinic acid, ethyl ester; 4-Oxopentanoic acid ethyl ester; Pentanoic acid, 4-oxo-, ethyl ester
Classification: Nonaromatic ester
Definition: Ester of levulinic acid and ethyl alcohol
Empirical: $C_7H_{12}O_3$
Formula: $CH_3COCH_2CH_2COOC_2H_5$

Properties: Colorless liq.; freely sol. in water; misc. with alcohol, oxygenated solvs.; m.w. 144.17; dens. 1.012 (20/4 C); b.p. 203-205 C; flash pt. 94 C; ref. index 1.423 (20 C)
Toxicology: LD50 (oral, rat) > 5 g/kg, (skin, rabbit) > 5 g/kg; primary skin irritant; TSCA listed
Precaution: Combustible
Hazardous Decomp. Prods.: Heated to decomp., emits acrid smoke and irritating fumes
Uses: Solvent for cellulose acetate and starch ethers; synthetic flavoring agent in foods and pharmaceuticals; fragrance in cosmetics
Features: Apple, pineapple-like flavor
Regulatory: FDA 21CFR §172.515; FEMA GRAS
Manuf./Distrib.: Advanced BioTech; Advanced Synthesis Tech.; Aldrich; Axxence Aromatic GmbH; DSM Fine Chems. Austria; Elan; Epochem; Fleurchem; Fluka; Grau Aromatics; Sigma

Ethyllic acid. *See* Acetic acid

Ethyl linoleate

CAS 544-35-4; EINECS/ELINCS 208-868-4
Synonyms: Ethyl (Z,Z)-9,12-octadecadienoate; Linoleic acid ethyl ester; 9,12-Octadecadienoic acid ethyl ester; Vitamin F
Definition: Ester of ethyl alcohol and linoleic acid
Empirical: $C_{20}H_{36}O_2$
Toxicology: Large doses can cause nausea and vomiting; no known skin toxicity
Uses: Emollient, humectant, consistency agent for cosmetics, pharmaceuticals; oil component for cutting oils, release emulsions; solvent for fats; drug delivery; nutrition
Features: Oily
Manuf./Distrib.: A&E Connock; Aldrich; Avrachem AG; Esperis; Fluka; Lluch Essence; Sigma
Trade Names: Arlypon LSE; Crossential® EL90

Ethyl methanoate. *See* Ethyl formate
Ethyl O-(o-(methoxycarbonyl) benzoyl) glycolate; Ethyl o-(methoxycarbonyl) benzoyloxyacetate. *See* Methyl phthalyl ethyl glycolate
Ethyl β-methylacrylate. *See* Ethyl crotonate
Ethylmethylbenzene. *See* Ethyl toluene
Ethylmethyl carbinol. *See* 2-Butanol
Ethyl methyl ketone. *See* Methyl ethyl ketone
Ethyl monosulfide. *See* Ethyl sulfide

Ethyl morpholine

CAS 100-74-3; EINECS/ELINCS 202-885-0
Synonyms: 4-Ethylmorpholine; N-Ethylmorpholine; NEM
Empirical: $C_6H_{13}NO$
Formula: $C_2H_5N(CH_2CH_2)_2O$
Properties: Colorless liq.; char. ammoniacal odor; misc. with water; m.w. 115.20; dens. 0.916 (20/20 C); f.p. -63 C; b.p. 138 C; flash pt. (OC) 89.6 F
Toxicology: ACGIH TLV/TWA 5 ppm (skin); LD50 (oral, rat) 1780 mg/kg, (IV, mouse) 180 mg/kg; LC50 (inh., mouse, 2 h) 18,000 mg/m^3; poison by IV route; mod. toxic by ing.; mildly toxic by inh.; skin and severe eye irritant; absorbed by skin; TSCA listed
Precaution: Flamm.; mod. fire risk; can react vigorously with oxidizers
Hazardous Decomp. Prods.: Heated to decomp., emits toxic fumes of NO_x
Uses: Catalyst for flexible and rigid PU foam, polyester slabstock; intermediate for dyestuffs, pharmaceuticals; rubber accelerators; emulsifying agents; solvent for dyes, resins, oils
Manuf./Distrib.: Aldrich; BASF; Fluka; Sigma; Xinchem

4-Ethylmorpholine; N-Ethylmorpholine. *See* Ethyl morpholine
Ethyl nitrile. *See* Acetonitrile
N-Ethyl-N-nitrosoethanamine. *See* N-Nitrosodiethylamine

Ethyl nonafluorobutyl ether

CAS 163702-05-4
Synonyms: Ethoxynonafluorobutane; Ethyl perfluorobutyl ether
Classification: Hydrofluoroether
Empirical: $C_6H_5F_9O$
Formula: $CF_3CF_2CF_2CF_2OC_2H_5$
Properties: Colorless liq.; low odor; m.w. 264; vapor pressure 109 mm Hg; m.p. -138 C
Toxicology: TWA 200 ppm (as total isomers)
Uses: Solvent cleaner, rinsing agent for precision and electronic cleaning, vapor degreasing, lt. duty cleaning; lubricant carrier for fluorocarbons, hydrocarbons, silicones; cold cleaner; specialty solvent; dispersion medium; reaction medium; spray contact cleaner; heat transfer fluid
Trade Names: HFE-7200
Trade Names Containing: Novec™ HFE-7200; Novec™ HFE-7200DL

Ethyl nonafluoroisobutyl ether

CAS 163702-06-5
Synonyms: Ethyl perfluoroisobutyl ether
Classification: Hydrofluoroether
Empirical: $C_6H_5F_9O$
Formula: $(CF_3)_2CFCF_2OC_2H_5$
Toxicology: TWA 200 ppm (as total isomers)
Uses: Solvent cleaner for precision and electronic cleaning; heat transfer fluid
Trade Names Containing: Novec™ HFE-7200; Novec™ HFE-7200DL

Ethyl (Z,Z)-9,12-octadecadienoate. *See* Ethyl linoleate
Ethyl 9-octadecenoate. *See* Ethyl oleate

Ethyl octanoate

CAS 106-32-1; EINECS/ELINCS 203-385-5
FEMA 2449
Synonyms: Ethyl caprylate; Ethyl octoate; Ethyl octylate; Octanoic acid ethyl ester
Definition: Ester of caprylic acid and ethyl alcohol
Empirical: $C_{10}H_{20}O_2$
Formula: $CH_3(CH_2)_6COOC_2H_5$
Properties: Colorless very mobile liq., pleasant pineapple odor; sol. in fixed oils; misc. with alcohol, ether; sl. sol. in propylene glycol; insol. in water, glycerin; m.w. 172.27; dens. 0.867 (20/4 C); m.p. -47 C; b.p. 207-209; flash pt. 167 F; ref. index 1.418 (20 C)
Toxicology: LD50 (oral, rat) 25,960 mg/kg; mildly toxic by ingestion; skin irritant; TSCA listed
Precaution: Combustible liq.
Hazardous Decomp. Prods.: Heated to decomp., emits acrid smoke and fumes
NFPA: Health 2, Flammability 2, Reactivity 0
Uses: Solvent; synthetic flavoring agent in foods and pharmaceuticals; fruit essences; fragrance in cosmetics
Features: Apricot, pineapple-like flavor
Regulatory: FDA 21CFR §172.515; FEMA GRAS; Japan approved as flavoring
Manuf./Distrib.: Advanced BioTech; Advanced Synthesis Tech.; Aldrich; Axxence Aromatic GmbH; Degussa Flavors/Alfrebro; Fleurchem; Fluka; Goldensun Mfg.; Grau Aromatics; Moore Ingreds.; Sigma

Ethyl octoate; Ethyl octylate. *See* Ethyl octanoate
Ethylolamine. *See* Ethanolamine

Ethyl oleate

CAS 111-62-6; 85049-36-1; EINECS/ELINCS 203-889-5; 285-206-0
FEMA 2450
Synonyms: Ethyl 9-octadecenoate; 9-Octadecenoic acid ethyl ester
Definition: Ester of ethyl alcohol and oleic acid
Empirical: $C_{20}H_{38}O_2$
Formula: $CH_3(CH_2)_7CH:CH(CH_2)_7COOC2H_5$
Properties: Ylsh. oily liq.; misc. with alcohol, ether, veg. oils; insol. in water; m.w. 310.52; dens. 0.870 (20/4 C); visc. 75.15 cp; m.p. -32 C; b.p. 216-218 (15 mm); acid no. 0.5 max.; iodine no. 75-85; sapon. no. 177-188; flash pt. 175 C; ref. index 1.451
Toxicology: LD50 (oral, rat) > 5 g/kg; TSCA listed
Precaution: Combustible
Storage: Air- and light-sensitive; freeze
Uses: Solvent; plasticizer; lubricant; emollient for cosmetics; water-resisting agent; synthetic flavoring agent in foods; vehicle and solvent for pharmaceuticals; diagnostic aid; surfactant for pharmaceuticals, cosmetics, foods, electronics, etc.
Features: Oleaginous

Regulatory: FDA 21CFR §172.225, 172.515; FEMA GRAS; NF, BP, EP compliance
Manuf./Distrib.: A&E Connock; Advanced BioTech; Akzo Nobel; Aldrich; ChemService; Chesham Chems. Ltd; Croda Inc; Degussa Flavors/Alfrebro; Fleurchem; Fluka; Mosselman NV; Penta Mfg.; Pfaltz & Bauer; Sigma; Sithean; Spectrum Quality Prods.; Univar E&E Ltd
Trade Names: Crodamol EO; Tilol ET

Ethyl oxalate. *See* Diethyl oxalate
Ethyl oxide. *See* Ethyl ether
Ethyl 3-oxobutanoate; Ethyl 3-oxobutyrate. *See* Ethylacetoacetate
Ethyl-4-oxopentanoate; Ethyl-4-oxovalerate. *See* Ethyl levulinate
Ethyl pentyl ketone. *See* 3-Octanone
Ethyl perfluorobutyl ether. *See* Ethyl nonafluorobutyl ether
Ethyl perfluoroisobutyl ether. *See* Ethyl nonafluoroisobutyl ether
Ethyl phenacetate. *See* Ethyl phenylacetate

Ethyl phenylacetate

CAS 101-97-3; EINECS/ELINCS 202-993-8
FEMA 2452
Synonyms: Benzeneacetic acid, ethyl ester; Ethyl benzeneacetate; Ethyl phenacetate; Ethyl-2-phenylethanoate; Ethyl-α-toluate; Phenylacetic acid, ethyl ester; α-Toluic acid ethyl ester
Definition: Ester of ethyl alcohol and phenylacetic acid
Empirical: $C_{10}H_{12}O_2$
Formula: $C_6H_5CH_2COOC_2H_5$
Properties: Colorless liq., sweet honey-like odor; sol. in fixed oils, oxygenated solvs.; insol. in glycerin, propylene glycol, water; m.w. 164.22; dens. 1.033 (20 C); b.p. 227 C; flash pt. > 100 C; ref. index 1.496-1.500
Toxicology: LD50 (oral, rat) 3300 mg/kg; mod. toxic by ing.; mutagenic data; TSCA listed
Precaution: Combustible liq.
Hazardous Decomp. Prods.: Heated to decomp., emits acrid smoke and irritating fumes
Uses: Tonic, solvent in cosmetics; fragrance in perfumes; intermediate, esp. for phenobarbital; herbicide; synthetic flavoring agent in foods and pharmaceuticals
Features: Sweet apricot, cherry-like flavor
Regulatory: FDA 21CFR §172.515; FEMA GRAS; Japan approved as flavoring
Manuf./Distrib.: Aceto; Advanced BioTech; Aldrich; Atofina; Axxence Aromatic GmbH; Bell Flavors & Fragrances; Elan; Fleurchem; Fluka; Goldensun Mfg.; Grau Aromatics; Haarmann & Reimer GmbH; Interchem; Lluch Essence; Penta Mfg.; Xinchem

Ethyl phenylacrylate; Ethyl-β-phenylacrylate. *See* Ethyl cinnamate
2-(Ethylphenylamino) ethanol. *See* Phenylethylethanolamine
Ethyl-2-phenylethanoate. *See* Ethyl phenylacetate
Ethyl phenyl ethanolamine; N,N-Ethylphenylethanolamine; N-Ethyl-N-phenylethanolamine. *See* Phenylethylethanolamine
Ethyl-3-phenylpropenoate; Ethyl 3-phenyl-2-propenoate; Ethyl (E)-3-phenylprop-2-enoate. *See* Ethyl cinnamate
Ethyl phosphate. *See* Triethyl phosphate
Ethyl phthalate. *See* Diethyl phthalate
Ethyl propanoate. *See* Ethyl propionate
1-Ethyl-1-propanol. *See* 3-Pentanol

Ethyl propionate

CAS 105-37-3; EINECS/ELINCS 203-291-4
UN 1195 (DOT); FEMA 2456
Synonyms: Ethyl propanoate; Propanoic acid ethyl ester; Propionic acid, ethyl ester; Propionic ether
Classification: Carboxylic acid ester
Definition: Ester of propionic acid and ethyl alcohol
Empirical: $C_5H_{10}O_2$
Formula: $CH_3CH_2COOC_2H_5$
Properties: Colorless liq., fruity rum odor; sol. in ≈ 60 parts water; sol. in fixed oils, most org. solvs.; misc. with alcohol, ether; m.w. 102.14; dens. 0.890 (20/4 C); vapor pressure 40 mm (27.2 C); m.p. -73 C; b.p. 96-99 C; flash pt. (CC) 12 C; autoignition temp. 440 C; ref. index 1.384 (20 C); surf. tens. 24.27 dynes/cm; dielec. const. 5.65
Toxicology: LD50 (oral, rabbit) 3500 mg/kg, (IP, rat) 1200 mg/kg; mod. toxic by ing. and IP routes; skin and eye irritant; probable respiratory tract irritant; high concs. may cause dizziness, headache; extreme concs. may cause unconsciousness, death; ing. may cause gasping breathing, hypothermia, acidosis; TSCA listed
Environmental: VOC; ThOD 2.04
Precaution: Highly flamm. liq.; dangerous fire risk and explosion hazard; LEL 1.9; UEL 11%; reacts vigorously with oxidizing materials
Hazardous Decomp. Prods.: May hydrolyze slowly to ethanol and propionic acid if water is present; heated to decomp., emits acrid smoke and irritating fumes
NFPA: , Flammability 3, Reactivity 0
Storage: Store in tightly closed electrically grounded containers in cool, well-ventilated area, separate from workplace; limit quantities in use
Uses: Solvent for paints, various natural and synthetic resins, cellulose ethers and esters; synthetic flavoring agent in foods, fruit syrups, pharmaceuticals; fragrance in cosmetics; cutting agent for pyroxylin
Features: Pineapple-like flavor
Regulatory: FDA 21CFR §172.515; FEMA GRAS; Japan approved as flavoring
Manuf./Distrib.: Advanced BioTech; Advanced Synthesis Tech.; Aldrich; Ashland; Axxence Aromatic GmbH; Bell Flavors & Fragrances; Berje; Degussa Flavors/Alfrebro; Dow; Elan; Fleurchem; Fluka; Frutarom Ltd; Goldensun Mfg.; Grau Aromatics; Haarmann & Reimer GmbH; Haltermann Prods. UK; Lluch Essence; MelChem; Moore Ingreds.; Northwestern Flavors; Penta Mfg.; Sigma; Spectrum Quality Prods.; VWR Int'l.; Vopak USA; Xinchem

Ethyl propyl ketone. *See* 3-Hexanone
2-Ethyl-3-propyl-1,3-propanediol. *See* Ethyl hexanediol
1-Ethyl-2-pyrrolidinone; N-Ethylpyrrolidinone; N-Ethylpyrrolidone. *See* N-Ethyl-2-pyrrolidone

N-Ethyl-2-pyrrolidone

CAS 2687-91-4
Synonyms: 1-Ethyl-2-pyrrolidinone; N-Ethylpyrrolidinone; N-Ethylpyrrolidone; 2-Pyrrolidinone, 1-ethyl-
Empirical: $C_6H_{11}NO$
Properties: Colorless liq.; m.w. 113.16; dens. 0.992; b.p. 97 C (20 mm); flash pt. 169 F
Toxicology: LD50 (oral, rat) 1350 mg/kg; mod. toxic by ing.; eye irritant; TSCA listed
Precaution: Combustible
Hazardous Decomp. Prods.: Heated to decomp., emits toxic fumes of NO_x
Uses: Reaction solvent for chemical reactions down to -70 C; antibiotic prod.; elec. capacitors; mfg. of polyamide fibers and film; ink; photoresist stripper; polymerization; synthesis
Trade Names: NEP®

Ethyl succinate. *See* Diethyl succinate

Ethyl sulfide

CAS 352-93-2; EINECS/ELINCS 206-526-9
UN 2375 (DOT)
Synonyms: Diethyl sulfide; Diethyl thioether; Ethane, 1,1′-thiobis-; Ethyl monosulfide; Ethylthioethane; Ethyl thioether; 3-Thiapentane; 1,1′-Thiobisethane; Thioethyl ether
Classification: Alkyl sulfide
Empirical: $C_4H_{10}S$
Formula: $(C_2H_5)_2S$
Properties: Colorless oily liq.; garlic-like odor; sol. in alcohol, ether, oxygenated solvs.; sl. sol. in water; m.w. 90.20; dens. 0.837 (20/4 C); m.p. -103.9 C; b.p. 92.1 C; flash pt. -10 C
Toxicology: LD50 (oral, rat) 5930 mg/kg; mildly toxic by ing.; generally low toxicity; due to strong odor, detectable before harmful airborne levels are reached; can be absorbed thru skin; skin and eye irritant; TSCA listed
Precaution: Combustible; very dangerous fire hazard; can react vigorously with oxidizers; reacts with water, steam, acids, or acid fumes to produce toxic and flamm. vapors
Hazardous Decomp. Prods.: CO_2, CO, SO_x, hydrogen sulfide
Uses: Organic synthesis; intermediate in mfg. of chemicals; solvent for min. salts; in electrochemical metal plating; gas odorant

Manuf./Distrib.: Aldrich; Epochem; Occidental; Penta Mfg.; Sithean; Spectrum Quality Prods.

2-(Ethylsulfonyl)ethanol. *See* Ethylsulfonylethyl alcohol

Ethylsulfonylethyl alcohol

CAS 513-12-2
Synonyms: ESE; 2-(Ethylsulfonyl)ethanol
Empirical: $C_4H_{10}O_3S$
Formula: $CH_3CH_2SO_2CH_2CH_2OH$
Properties: Colorless cryst.; sol. in water and common org. solvs.; m.w. 138.20; dens. 1.25; m.p. 45 C; b.p. 153 C (2.5 mm); flash pt. 188 C
Toxicology: LD50 (oral, rat) 18 g/kg; LDLo (oral, mouse) 14 g/kg
Storage: Hygroscopic
Uses: Solvent; organic intermediate; antistat; humectant

Ethyl tartrate. *See* Diethyl tartrate
Ethylthioethane; Ethyl thioether. *See* Ethyl sulfide
Ethyl-α-toluate. *See* Ethyl phenylacetate

Ethyl toluene

CAS 25550-14-5
Synonyms: ET; Ethylmethylbenzene; Ethyl toluene, all isomers; Ethyl toluene, mixed isomers; Ethyl toluenes; Methylethylbenzene; Methylethylbenzene, all isomers; Toluene, ar-ethyl
Empirical: C_9H_{12}
Properties: M.w. 120.19
Toxicology: Irritant
Uses: Solvent
Trade Names Containing: Solvent 2216; Solvent 2439

Ethyl toluene, all isomers; Ethyl toluene, mixed isomers; Ethyl toluenes. *See* Ethyl toluene
Ethyl triglycol. *See* Ethoxytriglycol
2-Ethynyl-2-butanol. *See* Methyl pentynol
Ethynyl carbinol. *See* Propargyl alcohol
Ethynyldimethyl carbinol. *See* Methyl butynol
Ethynyl methanol. *See* Propargyl alcohol
ETO. *See* Ethylene oxide
EtOH. *See* Alcohol

Eucalyptol

CAS 470-82-6; EINECS/ELINCS 207-431-5
FEMA 2465
Synonyms: Cajeputol; Cineol; 1,8-Cineol; Cineole; 1,8-Cineole; 1,8-Epoxy-p-menthane; Eucalyptole; Limonene oxide; p-Menthane, 1,8-epoxy; 2-Oxabicyclo (2.2.2) octane, 1,3,3-trimethyl-; 1,8-Oxido-p-menthane; 1,3,3-Trimethyl-2-oxabicyclo [2.2.2] octane
Definition: Constituent of oil of eucalyptus
Empirical: $C_{10}H_{18}O$
Properties: Colorless to ylsh. liq., camphor-like odor, spicy cooling taste; misc. with alcohol, chloroform, ether, glac. acetic acid, fixed or volatile oils; pract. insol. in water; m.w. 154.25; dens. 0.924 (20/4 C); m.p. > 1.5 C; b.p. 175-179 C; flash pt. 49 C; ref. index 1.4550-1.4600 (20 C)
Toxicology: LD50 (oral, rat) 2480 mg/kg, (subcut., mouse) 1070 mg/kg; poison by subcut. and intramuscular routes; mod. toxic by ing.; experimental carcinogen, reproductive effects; mutagen; TSCA listed
Precaution: Combustible liq.
Hazardous Decomp. Prods.: Heated to decomp., emits acrid smoke and fumes
HMIS: Health 1, Flammability 2, Reactivity 0
Uses: Denaturant, tonic in cosmetics; pharmaceuticals (cough syrups, expectorants); synthetic flavoring agent in foods, beverages, chewing gum, pharmaceuticals; antiseptic in oral rinse prods.; dentals; fragrance in perfumes
Regulatory: FDA 21CFR §172.515; 27CFR §21.65, 21.151; FEMA GRAS; Japan approved as flavoring; FDA approved for dentals
Manuf./Distrib.: AMC Chems.; Aldrich; Barnet Prods.; Berje; Boith China; Chart; Creative Fragrances; Doingcom; Fleurchem; Fluka; George Uhe; Goldensun Mfg.; H.E. Daniel Ltd; Haarmann & Reimer GmbH; Hunan Xinyu; Lluch Essence; MelChem; Millennium; Penta Mfg.; Polarome Int'l.; Quest Int'l.; Ruger; Sarcom; Sigma; Spectrum Quality Prods.; Treatt USA; U.S. Synthetics; Ungerer; Universal Preserv-A-Chem; Wego Chem. & Min.
Trade Names Containing: Punctilious® SDA 37-1 190 Proof; Punctilious® SDA 38B 190 Proof Prod. Code 13602133

Eucalyptole. *See* Eucalyptol
Eucalyptus globulus. *See* Eucalyptus globulus oil

Eucalyptus globulus oil

CAS 8000-48-4
FEMA 2466
Synonyms: Dinkum oil; Eucalyptus globulus; Eucalyptus oil
Definition: Volatile oil obtained from leaves of *Eucalyptus globulus*, contg. eucalyptol (70-80%), plus α-pinene, phellandrene, terpineol, citronellal, geranyl acetate, piperitone, etc.
Properties: Colorless to pale yel. liq., char. camphoraceous odor, pungent spicy cooling taste; sol. in 5 vols 70% alcohol; misc. with abs. alcohol, oils, fats; pract. insol. in water; dens. 0.905-0.925; m.p. -15.4 C; ref. index 1.458-1.470 (20 C)
Toxicology: LD50 (oral, rat) 2480 mg/kg, (skin, rabbit) 2480 mg/kg; human poison and human systemic effects by ing.; skin and eye irritant; can cause allergic reaction; ing. can cause eye spasms, nausea, vomiting, respiratory depression, sweating; TSCA listed
Hazardous Decomp. Prods.: Heated to decomp., emits acrid smoke and irritating fumes
HMIS: Health 1, Flammability 2, Reactivity 0
Storage: Keep cool, well closed; protect from light
Uses: Fat liquor for leather; synthetic sperm oil; isoparaffinic solvent; natural flavoring agent in foods and pharmaceuticals; as local antiseptic, antispasmodic, stimulant
Regulatory: FDA 21CFR §172.510, 27CFR §21.65, 21.151; FEMA GRAS; Japan approved; Europe listed, no restrictions; FDA approved for orals; BP compliance
Manuf./Distrib.: AMC Chems.; Aceto; Advanced BioTech; Aldrich; Alfa Chem; Arista Ind.; Asiamerica Int'l.; Boith China; Chart; Cosmetic Supplies USA; Doingcom; Essence Nat. Prods.; Fleurchem; George Uhe; H.E. Daniel Ltd; Hunan Xinyu; Integra; KIC Chems.; MelChem; Penta Mfg.; Ruger; Sarcom; Spectrum Quality Prods.; Treatt USA; U.S. Synthetics; Universal Preserv-A-Chem; Voigt Global Distrib.

Eucalyptus oil. *See* Eucalyptus globulus oil
Eugenia caryophyllus; Eugenia caryophyllus oil. *See* Clove (Eugenia caryophyllus) oil
Eugenic acid. *See* Eugenol

Eugenol

CAS 97-53-0; EINECS/ELINCS 202-589-1
FEMA 2467
Synonyms: Allylguaiacol; 4-Allylguaiacol; 4-Allyl-1-hydroxy-2-methoxybenzene; 4-Allyl-2-methoxyphenol; Caryophyllic acid; Eugenic acid; 1-Hydroxy-2-methoxy-4-allylbenzene; 4-Hydroxy-3-methoxyallylbenzene; 1-Hydroxy-2-methoxy-4-prop-2-enylbenzene; 2-Methoxy-4-allylphenol; 2-Methoxy-4-(2-propenyl) phenol; 2-Methoxy-4-prop-2-enylphenol; Phenol, 2-methoxy-4-(2-propenyl)-
Classification: Substituted phenol
Empirical: $C_{10}H_{12}O_2$
Properties: Colorless or ylsh. liq., pungent clove odor; sol. in alcohol, chloroform, ether, volatile oils, most org. solvs.; very sl. sol. in water; m.w. 164.22; dens. 1.064-1.070; m.p. 10.3 C; b.p. 253.5 C; flash pt. 219 F; ref. index 1.540
Toxicology: LD50 (oral, rat) 1930 mg/kg, (IP, mouse) 500 mg/kg; LDLo (subcut., rat) 5000 mg/kg; mod. toxic by ing., IP, and subcut. routes; human skin irritant; potential allergen; experimental carcinogen, tumorigen; human mutagenic data; TSCA listed
Precaution: Combustible liq.
Hazardous Decomp. Prods.: Heated to decomp., emits acrid smoke, irritating fumes
HMIS: Health 1, Flammability 1, Reactivity 0
Storage: Light-sensitive
Uses: Denaturant, tonic in cosmetics; fragrance ingred. in perfumes; essential oils; disinfectant; medicine (dental analgesic, anesthetic, antiseptic); synthetic flavoring agent and adjuvant in foods and pharmaceuticals; prod. of isoeugenol for mfg. of vanillin; in food-contact textiles
Regulatory: FDA 21CFR §177.2800, 184.1257, GRAS; 27CFR §21.65,

21.151; FEMA GRAS; Japan approved as flavoring; FDA approved for orals; BP, EP compliance

Manuf./Distrib.: Aldrich; Berje; Charabot USA; Chart; Creative Fragrances; Doingcom; EM Science; Firmenich; Fleurchem; Fluka; Frutarom Ltd; George Uhe; Goldensun Mfg.; H.E. Daniel Ltd; Haarmann & Reimer GmbH; Integra; KIC Chems.; Lluch Essence; Lucta SA; MelChem; Penta Mfg.; Polarome Int'l.; Ruger; Sarcom; Schweizerhall; Sigma; Spectrum Quality Prods.; Treatt USA; Ungerer; VWR Int'l.; Voigt Global Distrib.

Trade Names Containing: Punctilious® SDA 38B 190 Proof Prod. Code 13602133

Extracts (petroleum), heavy paraffinic distillate solvent. *See* Paraffin distillate

F

F 11. *See* Trichlorofluoromethane
F 12. *See* Dichlorodifluoromethane
F 21. *See* Dichlorofluoromethane
F 22. *See* Chlorodifluoromethane
F 112. *See* 1,1,2,2-Tetrachloro-1,2-difluoroethane
F 114. *See* 1,2-Dichlorotetrafluoroethane
F 115. *See* Chloropentafluoroethane

Farnesol

CAS 4602-84-0; EINECS/ELINCS 225-004-1
FEMA 2478
Synonyms: 2,6,10-Dodecatrienol, 3,7,11-trimethyl-; Farnesyl alcohol; Trimethyl dodecatrienol; 3,7,11-Trimethyl-2,6,10-dodecatrien-1-ol
Classification: Organic compd.; unsat. alcohol
Empirical: $C_{15}H_{26}O$
Formula: $(CH_3)_2CCHCH_2CH_2CCH_3CHCH_2CH_2CCH_3CHCH_2OH$
Properties: Colorless to lt. yel. liq.; floral odor; insol. in water; dens. 0.8871 (20/4 C); b.p. 111 C; ref. index 1.487-1.492
Toxicology: LD50 (oral, rat) 6000 mg/kg, (IP, mouse) 4443 mg/kg; mod. toxic by IP route; mildly toxic by ing.; mutagenic data; TSCA listed
Precaution: Combustible
Hazardous Decomp. Prods.: Heated to decomp., emits acrid smoke and irritating fumes
Uses: Solvent, soothing agent, deodorant agent in cosmetics; fragrance in perfumes; synthetic flavoring agent and fragrance in soaps, shampoos, detergents, cosmetics, foods, pharmaceuticals; insect hormone
Regulatory: FDA 21CFR §172.515; FEMA GRAS
Manuf./Distrib.: Advanced Synthesis Tech.; Aldrich; Chemos GmbH; Fluka; Sigma

Farnesyl alcohol. *See* Farnesol
Fatty acid mono/diglycerides. *See* Mono- and diglycerides of fatty acids
Fatty acids, coco, methyl esters. *See* Methyl cocoate
Fatty acids, cottonseed oil, methyl esters. *See* Methyl cottonseed
Fatty acids, lanolin. *See* Lanolin acid
Fatty acids, lanolin, compds. with isopropanolamine. *See* Isopropanolamine lanolate
Fatty acids, lanolin, isopropyl esters. *See* Isopropyl lanolate
Fatty acids, peanut oil, compds. with ethanolamine. *See* Peanutamide MEA
Fatty acids, tallow, methyl esters. *See* Methyl tallowate
FC 11. *See* Trichlorofluoromethane
FC 12. *See* Dichlorodifluoromethane
FC 21. *See* Dichlorofluoromethane
FC 22. *See* Chlorodifluoromethane
FC 112. *See* 1,1,2,2-Tetrachloro-1,2-difluoroethane
FC 113. *See* Trichlorotrifluoroethane
FC 113a. *See* 1,1,1-Trichloro-2,2,2-trifluoroethane
FC 114. *See* 1,2-Dichlorotetrafluoroethane
FC 142b. *See* 1,1,1-Chlorodifluoroethane
2-Fenchanol; Fenchol; (+)-Fenchol. *See* Fenchyl alcohol

Fenchyl alcohol

CAS 1632-73-1; EINECS/ELINCS 216-639-5
FEMA 2480
Synonyms: 3,3-Dimethyl-8,9-dinorbornan-2-ol; 2-Fenchanol; Fenchol; (+)-Fenchol; (1S)-1,3,3-Trimethylbicyclo [2.2.1] heptan-2-ol; 1,3,3-Trimethyl-2-norbornanol
Empirical: $C_{10}H_{18}O$
Properties: Colorless oily liq. (d and l forms) or solid (dl form); camphor-like odor; sol. in alcohol; sl. sol. in water; m.w. 154.25; m.p. 40-43 C; b.p. 201-202 C; flash pt. 165 F; ref. index 1.473
Uses: Solvent; organic intermediate; odorant; synthetic flavoring agent, flavor enhancer, adjuvant for foods and pharmaceuticals; cosmetics ingred.
Features: Lemon-like flavor
Regulatory: FDA 21CFR §172.515; FEMA GRAS
Manuf./Distrib.: Aldrich; Chr. Hansen Inc; Lluch Essence; Natural Advantage

Flour sulfur; Flowers of sulfur. *See* Sulfur
Fluorane 114. *See* 1,2-Dichlorotetrafluoroethane

Fluorene

CAS 86-73-7
Synonyms: o-Biphenylene methane; o-Biphenylmethane; Diphenylenemethane; 9H-Fluorene; 2,2´-Methylenebiphenyl
Classification: Aromatic hydrocarbon
Empirical: $C_{13}H_{10}$
Properties: Wh. cryst. plates; sol. in alcohol, ether, benzene, carbon disulfide, oxygenated and aromatic solvs.; insol. in water; m.w. 166.23; m.p. 116 C; b.p. 295 C (dec.)
Toxicology: LD50 (IP, mouse) 2 g/kg, (parenteral, mouse) > 2 g/kg; sl. toxic by IP and parenteral routes; tumorigen; mutagen; TSCA listed
Precaution: Combustible
Hazardous Decomp. Prods.: Heated to decomp., emits acrid smoke and irritating vapors
Uses: Resinous products; dyestuffs; prod. of fluorenone
Manuf./Distrib.: Aldrich; Biddle Sawyer; Fluka; Sigma; Supelco
Trade Names Containing: Methylnaphthalene Fraction

9H-Fluorene. *See* Fluorene
Fluorocarbon 11. *See* Trichlorofluoromethane
Fluorocarbon 12. *See* Dichlorodifluoromethane
Fluorocarbon 14. *See* Tetrafluoromethane
Fluorocarbon 21. *See* Dichlorofluoromethane
Fluorocarbon 22. *See* Chlorodifluoromethane
Fluorocarbon 112. *See* 1,1,2,2-Tetrachloro-1,2-difluoroethane
Fluorocarbon 113. *See* Trichlorotrifluoroethane
Fluorocarbon 114. *See* 1,2-Dichlorotetrafluoroethane
Fluorocarbon 115. *See* Chloropentafluoroethane
Fluorocarbon 142b. *See* 1,1,1-Chlorodifluoroethane
Fluorodichloromethane. *See* Dichlorofluoromethane
Fluoroform. *See* Trifluoromethane
Fluorotrichloromethane. *See* Trichlorofluoromethane
Food freezant 12. *See* Dichlorodifluoromethane
Formal. *See* Methylal

Formaldehyde

CAS 50-00-0; EINECS/ELINCS 200-001-8
UN 1198 (DOT); UN 2209 (DOT); INS240
Synonyms: Aldehyde C1; Formalin; Formic aldehyde; Formol; HOCH;

Methaldehyde; Methanal; Methyene oxide; Methyl aldehyde; Methylene glycol; Oxymethylene; Paraform; Polyoxymethylene glycols
Classification: Aldehyde
Empirical: CH_2O
Formula: HCHO
Properties: Colorless gas; strong pungent odor; avail. commercially as aq. sol'ns. (37-50% in methanol); sol. in water, alcohol; m.w. 30.03; dens. 1.083; vapor pressure 3284 mm Hg (20 C); f.p. -118 C; b.p. -19 C; flash pt. 56 C; autoignition temp. 424 C; ref. index 1.3765
Toxicology: ACGIH TLV/TWA 1 ppm; LD50 (oral, rat) 800 mg/kg, (subcut., rat) 420 mg/kg, (IV, rat) 87 mg/kg; LC50 (inh., rat) 590 mg/m^3; poison by ing., skin contact, inh., IV, IP, subcut. routes; human poison by ing., systemic effects, skin/eye irritant; vapor intensely irritating to mucous membranes; if ingested, causes violent vomiting and diarrhea and possible collapse; suspected human carcinogen; experimental tumorigen, teratogen; mutagenic data; common air contaminant; TSCA listed
Environmental: VOC; BOD5 0.74; ThOD 1.07
Precaution: Combustible when exposed to heat, or flame; mod. fire risk; mod. explosion hazard; explosive limits 7-73%; can react vigorously with oxidizers; violent reactions possible
Hazardous Decomp. Prods.: Heated to decomp., emits acrid smoke and fumes
HMIS: Health 3, Flammability 2, Reactivity 1
Uses: Urea/melamine resins; polyacetal resins; phenolic resins; fertilizers; preservative for hides/skins, cosmetics; reducing agent; corrosion inhibitor; intermediate for paints; antimicrobial, preservative in pharmaceuticals; solvent; nickel plating brightener; disinfectant; crosslinker (paper waterproofing); latex coagulant; textile finishing agent; embalming fluids; methylation/chloromethylation reagent; preservative in food processing defoamers; animal glue adjuvant; sterilizer for kidney dialyzer membranes; in food-pkg. adhesives; preservative for coatings for paper/paperboard in contact with aq./fatty/dry foods; preservative for defoamers in food-contact coatings; defoamer in food-contact paper/paperboard; preservative in food-contact textiles
Regulatory: FDA 21CFR §173.340 (1% of dimethicone content), 175.105, 175.210, 175.300, 176.170, 176.180, 176.200, 176.210, 177.1200, 177.2410, 177.2800, 178.3120, 573.460; FDA approved for topicals; SARA §302a extremely hazardous, §313 reportable; CERCLA hazardous substance; HAP; BP, EP compliance (sol'n.)
Manuf./Distrib.: AAE Chemie NV; AMRESCO; Aldrich; Ashland; Bayer; Borden; Brenntag AG; C.P. Hall; Celanese; Degussa; DuPont; EM Science; Farleyway Chems. Ltd; Fluka; Georgia-Pacific Resins; Hercules/Aqualon; Houghton Chem.; ISP; Integra; J.T. Baker; Lipo; Mallinckrodt Baker; Merck KGaA; Miljac; Monomer-Polymer & Dajac Labs; Monsanto; Penta Mfg.; Perstorp Polyols; Praxair; Ruger; Sigma; Solutia; Spectrum Quality Prods.; Stanchem; Thomas Scientific; U.S. Chems.; Universal Preserv-A-Chem; VWR Int'l.; Veckridge; Voigt Global Distrib.; Vopak USA; Wright
Trade Names Containing: Eastman® EEP; Formcel®

Formaldehyde, bis (2-chloroethyl) acetal; Formaldehyde bis (β-chloroethyl) acetal. *See* Bis (2-chloroethoxy) methane
Formaldehyde diethyl acetal. *See* Diethoxymethane
Formaldehyde dimethylacetal; Formaldehyde methyl ketal. *See* Methylal
Formal glycol. *See* 1,3-Dioxolane
Formalin. *See* Formaldehyde

Formamide

CAS 75-12-7; EINECS/ELINCS 200-842-0
Synonyms: Amide C_1; Carbamaldehyde; Methanamide; Methanoic acid, amide
Classification: Amide
Empirical: CH_3NO
Formula: $HCONH_2$
Properties: Colorless clear, sl. visc. oily liq.; sol. in water, 95% ethanol, acetone, dimethyl sulfoxide; misc. with methanol, acetic acid, dioxane, phenol, ethylene glycol; very sl. sol. in ether, benzene; m.w. 45.04; dens. 1.134 (20/4 C); vapor pressure 3 mm Hg (30 C); m.p. 2.6 C; b.p. 210 C (dec.); flash pt. (OC) 154 C; ref. index 1.44754 (20 C); pH 7.3
Toxicology: ACGIH TLV/TWA 20 ppm (skin); LD50 (oral, rat) 5577 mg/kg, (IP, rat) 5700 mg/kg, (subcut., rat) > 4 g/kg; poison by skin contact, subcut., IV routes; mod. toxic by ing., IP routes; mildly toxic by inh.; primary irritant; skin and eye irritant; inh. may cause mod. irritation to mucous membranes of respiratory tract; suspected carcinogen; tumorigen; mutagen; experimental reproductive effects; TSCA listed
Precaution: Combustible exposed to heat or flame; incompat. with strong oxidizers, acids, bases, iodine, pyridine, SO_3; reacts explosively with furfuryl alcohol, H_2O_2, nitromethane; attacks copper and brass; incompat. with Karl Fischer reagent (toluene/pyridine/sulfur trioxide); reaction may cause explosion of closed container
Hazardous Decomp. Prods.: Heated to decomp., emits toxic fumes of NO_x
NFPA: Health 2, Flammability 1
Storage: Store in cool, dry, well-ventilated area, out of direct sunlight; inert atm. rec.; restrict access; protect from moisture; avoid generating mists; hygroscopic; light-sensitive
Uses: Extraction solvent for aromatics, acetylene, butadiene; process solvent; ionizing solvent; solvent for pharmaceuticals; softener for paper, gums, animal glues; intermediate in organic synthesis; mfg. of formates and hydrocyanic acid; syn. reagent; in food-pkg. adhesives
Regulatory: FDA 21CFR §175.105
Manuf./Distrib.: AMRESCO; Aldrich; Alfa Aesar; BASF AG; BASF; Electron Microscopy Sciences; Fluka; Integra; J.T. Baker; Mallinckrodt Baker; Nat'l. Diagnostics; Pechiney Chems. Div.; Penta Mfg.; Qbiogene; Research Organics; Roche Molecular Biochems.; Serologicals; Sigma; Sithean; Spectrum Quality Prods.; Thomas Scientific; VWR Int'l.

Formamide, N-methyl-. *See* N-Methylformamide

Formic acid

CAS 64-18-6; EINECS/ELINCS 200-579-1
UN 1779 (DOT); FEMA 2487; INS236
Synonyms: Aminic acid; Formylic acid; Hydrogen carboxylic acid; Methanoic acid
Classification: Organic acid
Empirical: CH_2O_2
Formula: HCOOH
Properties: Colorless fuming liq.; pungent penetrating odor; sol. in water, alcohol, ether; m.w. 46.03; dens. 1.22 (20/4 C); vapor pressure 42.6 mm Hg; m.p. 8.3 C; b.p. 100.8 C; flash pt. (OC) 69 C; autoignition temp. 539 C; ref. index 1.3714; surf. tens. 37.58 dynes/cm; dielec. const. 58.5
Toxicology: ACGIH TLV/TWA 5 ppm; STEL 10 ppm; LD50 (oral, rat) 1100 mg/kg, (IP, mouse) 940 mg/kg; LC50 (inh., rat, 15 min) 15 g/m^3; poison by inh., IV, IP routes; mod. toxic by ing.; corrosive; skin and severe eye irritant; migrates to food from pkg.; mutagenic data; TSCA listed
Environmental: VOC; BOD5 0.20; COD 0.36; ThOD 0.35
Precaution: Flamm.; DOT: corrosive material; can react vigorously with oxidizers; explosive with furfuryl alcohol, nitromethane, etc.
Hazardous Decomp. Prods.: Heated to decomp., emits acrid smoke and irritating fumes
NFPA: Health 3, Flammability 2, Reactivity 0
Uses: Reducing agent, pH control agent in dyeing, electroplating, pickling baths; textile dyeing and finishing; leather treatment; chemicals; mfg. of fumigants, insecticides, refrigerants; solvent for perfumes, lacquers, pharmaceuticals; silvering glass; preservative in cosmetics; ore flotation; reagent; paper mfg. aid; latex coagulant; antibacterial, preservative for foods; disinfectant; synthetic flavoring agent in foods and pharmaceuticals; fragrance; silage fermentation additive; in food pkg. paper/paperboard
Regulatory: FDA 21CFR §172.515, 172.723, 186.1316, GRAS as indirect additive, 573.480; FEMA GRAS; CERCLA hazardous substance; Europe listed; prohibited in UK
Manuf./Distrib.: AMRESCO; Aastrid Int'l.; Advanced BioTech; Aldrich; Alfa Chem; Ashland; Axxence Aromatic GmbH; BASF AG; BASF; BCH Brühl; BP Amoco; Boith China; Brenntag AG; Celanese; Coyne; Degussa Flavors/Alfrebro; Fisher Scientific; Fleurchem; Fluka; Grau Aromatics; Harcros; Hawkins Chem.; Houghton Chem.; Independent Chem.; Integra; J.T. Baker; Lipo; Mallinckrodt Baker; Mutchler; Norsk Hydro AS; Penta Mfg.; Prodasynth; Rochem Int'l.; Romil Ltd; Sal Chem.; Sigma; Spectrum Quality Prods.; Stanchem; Thomas Scientific; VWR Int'l.; Varsal Instruments; Voigt Global Distrib.; Vopak USA

Trade Names Containing: Dynasolve 165

Formic acid, amide, N,N-dimethyl-. *See* Dimethyl formamide
Formic acid benzyl ester. *See* Benzyl formate
Formic acid butyl ester. *See* Butyl formate
Formic acid ethyl ester. *See* Ethyl formate
Formic acid methyl ester. *See* Methyl formate
Formic acid, pentyl ester. *See* Amyl formate
Formic acid phenylmethyl ester. *See* Benzyl formate
Formic aldehyde. *See* Formaldehyde
Formic ether. *See* Ethyl formate
Formol. *See* Formaldehyde
Formosa camphor. *See* Camphor
Formosa camphor oil; Formose oil of campor. *See* Camphor (Cinnamomum camphora) oil
N-Formyldimethylamine. *See* Dimethyl formamide
Formylic acid. *See* Formic acid
N-Formylmethylamine. *See* N-Methylformamide
4-Formylmorpholine. *See* N-Formylmorpholine

N-Formylmorpholine

CAS 4394-85-8; EINECS/ELINCS 224-518-3
Synonyms: 4-Formylmorpholine; 4-Morpholinecarboxaldehyde
Classification: Nonaromatic amide
Empirical: $C_5H_9NO_2$
Properties: Solid or liq.; m.w. 115.13; dens. 1.420; m.p. 23 C; b.p. 240 C; flash pt. 113 C; ref. index 1.4863
Toxicology: LD50 (oral, rat) 6500 μl/kg, (skin, rabbit) > 16 ml/kg; primary irritant; skin, eye, and respiratory system irritant; TSCA listed
Hazardous Decomp. Prods.: Heated to decomp., emits toxic vapors of NO_x
Storage: Hygroscopic
Uses: Selective solvent (aromatics extraction process)
Manuf./Distrib.: Acros Org.; Aldrich; Fluka; Sigma

1-Formylpropene. *See* Crotonaldehyde
Formyl trichloride. *See* Chloroform
Freon 11. *See* Trichlorofluoromethane
Freon 13. *See* Chlorotrifluoromethane
Freon 14. *See* Tetrafluoromethane
Freon 21. *See* Dichlorofluoromethane
Freon 22. *See* Chlorodifluoromethane
Freon 23. *See* Trifluoromethane
Freon 30. *See* Methylene chloride
Freon 112. *See* 1,1,2,2-Tetrachloro-1,2-difluoroethane
Freon 113. *See* Trichlorotrifluoroethane
Freon 114. *See* 1,2-Dichlorotetrafluoroethane
Freon 115. *See* Chloropentafluoroethane
Freon 142; Freon 142b. *See* 1,1,1-Chlorodifluoroethane
Freon FT. *See* 1,1,1-Trichloro-2,2,2-trifluoroethane
Freon HE. *See* Trichlorofluoromethane
β-D-Fructofuranosyl-α-D-glucopyranoside monohexadecanoate. *See* Sucrose palmitate
Fural; 2-Furaldehyde; Furale. *See* Furfural

Furan

CAS 110-00-9; EINECS/ELINCS 203-727-3
UN 2389
Synonyms: Axole; Divinylene oxide; 1,4-Epoxy-1,3-butadiene; Furane; Furfuran; Oxacyclopentadiene; Oxole; Tetrole
Classification: 5-Membered aromatic heterocyclic compd.
Empirical: C_4H_4O
Properties: Colorless volatile liq.; strong ethereal odor; turns brn. on standing; sol. in alcohol, ether; misc. with most org. solvs.; insol. in water; m.w. 68.08; dens. 0.936 (20/4 C); vapor pressure 670 mbar (20 C); f.p. -86 C; b.p. 31-33 C; flash pt. (CC) -35 C; ref. index 1.4216 (20 C)
Toxicology: LD50 (IP, mouse) 7 mg/kg; LDLo (oral, rabbit) 234 mg/kg; LC50 (inh., mouse, 1 h) 120 mg/m^3; highly toxic; poison by inh., IP routes; mod. toxic by ing. and skin contact; readily absorbed thru skin; narcotic; vapors are anesthetic; may cause ptosis, somnolence, dyspnea, acute pulmonary edema, nausea, vomiting, hemorrhage, convulsions, wt. loss, renal failure; carcinogen, tumorigen, mutagen, reproductive effector; target organs: GI system, liver; TSCA listed
Precaution: Extremely flamm.; dangerous fire risk; flamm. limits 2-24%; forms peroxides on exposure to air; can react with oxidizers, acids, peroxides, oxygen; violent exothermic reaction on contact with acids; avoid dust generation, excess heat
Hazardous Decomp. Prods.: CO, CO_2; heated to decomp., emits acrid smoke and irritating fumes
NFPA: Health 1, Flammability 4, Reactivity 1
Storage: Sensitive to heat; photosensitive; store refrigerated; keep container tightly closed under an inert atmosphere; keep away from ignition sources
Uses: Chemical intermediate in mfg. of herbicides, pharmaceuticals, plastics, and fine chemicals (pyrrole, tetrahydrofuran, thiophene); coating asphaltic pavements, foundry sand cores; solvent for resins, lacquers
Manuf./Distrib.: ABCR; Acros Org.; Aldrich; Atlas Mins. & Chems.; Delta Resins & Refractories; Fluka; Great Lakes; Penn Spec. Chems.; Sigma; Spectrum Quality Prods.; Xinchem

2-Furanaldehyde. *See* Furfural
2-Furancarbinol. *See* Furfuryl alcohol
2-Furancarbonal; 2-Furancarboxaldehyde. *See* Furfural
Furan-α-carboxylic acid methyl ester. *See* Methyl 2-furoate
Furan, 2,5-dihydro-. *See* 2,5-Dihydrofuran
Furane. *See* Furan
Furanidine. *See* Tetrahydrofuran
Furanmethanol; 2-Furanmethanol. *See* Furfuryl alcohol
2-Furanmethanol, tetrahydro-, acetate. *See* Tetrahydrofurfuryl acetate
Furan, tetrahydro-. *See* Tetrahydrofuran

Furfural

CAS 98-01-1; EINECS/ELINCS 202-627-7
UN 1199 (DOT); FEMA 2489
Synonyms: Artificial ant oil; Artificial oil of ants; Fural; 2-Furaldehyde; Furale; 2-Furanaldehyde; 2-Furancarbonal; 2-Furancarboxaldehyde; 2-Furfural; Furfuraldehyde; Furfurol; Furfurole; Furole; α-Furole; 2-Furyl methanal; Pyromucic aldehyde
Classification: Cyclic aldehyde
Empirical: $C_5H_4O_2$
Formula: C_4H_3OCHO
Properties: Colorless liq. (pure); reddish-brown (on exposure to air and light); almond-like odor; sol. in alcohol, ether, benzene, 8.3% in water; m.w. 96.08; dens. 1.1598 (20/4 C); f.p. -36.5 C; b.p. 161.7 C; flash pt. 60 C; ref. index 1.5260 (20 C)
Toxicology: ACGIH TLV/TWA 2 ppm (skin); LD50 (oral, rat) 65 mg/kg, (IP, rat) 20 mg/kg, (subcut., rat) 148 mg/kg; LCLo (inh., rat, 4 h) 153 ppm; poison by ing., IP, subcut., IV, IM routes; mod. toxic by inh., skin contact; skin and eye irritant; irritates mucous membranes, acts on CNS; human mutagenic data; TSCA listed
Environmental: VOC; BOD5 0.77; ThOD 1.67
Precaution: Flamm. or combustible; can react with oxidizing materials; mod. explosion hazard exposed to heat or flame or by chem. reaction; incompat. with strong min. acids or alkalis; violent exothermic polymerization on contact with strong min. acids/alkalis
Hazardous Decomp. Prods.: Heated to decomp., emits acrid smoke and irritating fumes
NFPA: Health 3, Flammability 2, Reactivity 0
Storage: Refrigerate; light-sensitive
Uses: Chemical intermediate for mfg. of derivs., furfuryl alcohol, THF, furan and phenolic resins, paints; prod. of lysine; novolac resin comonomer; solvent for petroleum lubes, nitrocellulose, cellulose acetate, shoe dyes, fatty acid/wood rosin separation, cosmetics; selective solvent for butadiene/aromatics extraction; wetting agent; in mfg. of furfural-phenol plastics; refining of rare earth; vulcanization accelerator; insecticide; fungicide; germicide; weed killer; reagent in analytical chemistry; synthetic flavoring agent in foods and pharmaceuticals; in food-pkg. adhesives
Features: Sweet flavor
Regulatory: FDA 21CFR §175.105; FEMA GRAS; SARA reportable; CERCLA hazardous substance; Japan approved as flavoring
Manuf./Distrib.: AAE Chemie NV; Advanced BioTech; Aldrich; Allchem Ind.; Ashland; Axxence Aromatic GmbH; Bencorp Int'l.; Boith China; Chemway Int'l.; Fleurchem; Fluka; Great Lakes; Harborchem; Int'l.

Furan Chems. USA; Lluch Essence; Maypro Ind.; Moore Ingreds.; Penn Spec. Chems.; Penta Mfg.; Sigma; Sinochem Tianjin; Spectrum Quality Prods.; Vanco Trading; Varsal Instruments; Vopak USA; Westpro
Trade Names: QO® Furfural

2-Furfural. *See* Furfural
Furfural alcohol; Furfuralcohol. *See* Furfuryl alcohol
Furfuraldehyde. *See* Furfural
Furfuran. *See* Furan
Furfurol; Furfurole. *See* Furfural

Furfuryl alcohol

CAS 98-00-0; EINECS/ELINCS 202-626-1
UN 2874 (DOT); FEMA 2491
Synonyms: 2-Furancarbinol; Furanmethanol; 2-Furanmethanol; Furfural alcohol; Furfuralcohol; Furfurol; Furyl alcohol; 2-Furylcarbinol; α-Furylcarbinol; (2-Furyl) methanol; 2-Hydroxymethylfuran
Classification: Heterocyclic substituted alcohol
Empirical: $C_5H_6O_2$
Formula: $C_4H_3OCH_2OH$
Properties: Colorless mobile liq., brn.-dk. red (air/lt. exposed), low odor, cooked sugar taste; sol. in alcohol, ether, chloroform, benzene; misc. with water but unstable; misc. with oxygenated solvs.; insol. in petrol. hydrocarbons, most oils; m.w. 98.10; dens. 1.1285 (20/4 C); vapor pressure 1 mm (31.8 C); m.p. -29 C; b.p. 170 C; flash pt. (OC) 75 C; ref. index 1.485
Toxicology: ACGIH TLV/TWA 10 ppm (skin); STEL 15 ppm (skin); LD50 (oral, rat) 275 mg/kg, (dermal, rabbit) 450 mg/kg; poison by ing., skin contact, subcut. route; mod. toxic by inh., IP routes; mild skin irritant; may cause severe eye irritation; high vapor concs. may cause CNS depression; high ing. doses may cause nausea, diarrhea, dizziness, increased urination; unconsciousness and coma in severe cases; mutagenic data; TSCA listed
Environmental: VOC; BOD5 0.53; COD 2.97; ThOD 0.180
Precaution: Combustible; can react with oxidizing materials; mod. explosion hazard when exposed to heat or flame; may react explosively with mineral and some organic acids
Hazardous Decomp. Prods.: Thermal decomp. prods.: CO, CO_2; heated to decomp., emits acrid smoke and fumes
NFPA: Health 1, Flammability 2, Reactivity 1
Storage: Store in cool, dry, well-ventilated area out of direct sunlight
Uses: Mfg. of phenolic and thermosetting resins, furan polymers, urea-formaldehyde resins; wetting agent; foundry sand binders; corrosion-resist. resins; intermediate for esterification and etherification; paints; solvent, plasticizer for phenolic resins; solvent for dyes and resins; nonreactive epoxy resin diluent; visc. reducer, cure promoter, and carrier in amine-cured epoxy resins; gel retarder in casein/protein glues; polymer sealants/cements; synthetic flavoring agent in foods and pharmaceuticals; cosmetics ingred.; in food-pkg. adhesives
Regulatory: FDA 21CFR §175.105; FEMA GRAS
Manuf./Distrib.: Aldrich; Allchem Ind.; Ashland; Axxence Aromatic GmbH; BCH Brühl; Bencorp Int'l.; Browning; CPS Union; F.T.L. Int'l.; Fleurchem; Fluka; Great Lakes; Harborchem; Indofine; Int'l. Furan Chems. USA; Jarchem Ind.; Lluch Essence; Pechiney Chems. Div.; Penn Spec. Chems.; Penta Mfg.; Rhodia; Schweizerhall; Sinochem Tianjin; Spectrum Quality Prods.; Storchem; Vanco Trading; Varsal Instruments; Vopak USA; Westpro; Whyte Chems. Ltd
Trade Names: FA®

2-Furoic acid, methyl ester. *See* Methyl 2-furoate
Furole; α-Furole. *See* Furfural
Furyl alcohol; 2-Furylcarbinol; α-Furylcarbinol. *See* Furfuryl alcohol
2-Furyl methanal. *See* Furfural
(2-Furyl) methanol. *See* Furfuryl alcohol
Fusel oil. *See* Fusel oil refined

Fusel oil refined

CAS 8013-75-0; EINECS/ELINCS 232-395-2
UN 1201 (DOT); FEMA 2497
Synonyms: Amyl alcohol, commercial; Amyl alcohols, mixed; Fusel oil; Whiskey fusel oil
Definition: Mixed amyl alcohols, contg. isoamyl alcohol, methanol, ethanol, acetaldehyde, and other alcohols
Properties: Water-wh. to pale yel. oily liq.; disagreeable odor; sol. in water, alcohol, ether; dens. 0.807-0.813; b.p. 120-140 C; flash pt. 123 F; ref. index 1.405-1.410
Toxicology: Mutagenic data; suspected of containing carcinogens; experimental reproductive effects; TSCA listed
Precaution: Flamm. liq. exposed to heat or flame; reacts with oxidizing materials
Hazardous Decomp. Prods.: Heated to decomp., emits acrid smoke and fumes
Uses: Solvent; synthetic flavoring agent in foods and pharmaceuticals; cosmetics ingred.
Regulatory: FDA 21CFR §172.515; FEMA GRAS
Manuf./Distrib.: Ciba Spec. Chems./Water Treatment; Moore Ingreds.; Penta Mfg.

G

Galam butter. *See* Shea butter (Butyrospermum parkii)
Gas; Gasolene. *See* Gasoline

Gasoline

CAS 8006-61-9; 68425-31-0; 68514-15-8; EINECS/ELINCS 232-349-1
UN 1203 (DOT); UN 1257 (DOT)
Synonyms: Automotive gasoline; Benzin; Gas; Gasolene; Gasoline, natural; Motor fuel; Motor spirit(s); Natural gasoline; Petrol; Unleaded gasoline
Classification: Mixed petroleum hydrocarbons
Definition: Complex combination of hydrocarbons separated from nat. gas by processes such as refrigeration or absorption; predominantly C4-8 sat. aliphatic hydrocarbons; common commercial unleaded automotive gasoline; may contain additives and blending agents
Properties: Colorless volatile liq.; char. odor; insol. in water; sol. in abs. alcohol, ether, chloroform, benzene, ethanol, other petrol. solvs.; dissolves fats, oils, natural resins; m.w. 108 (avg.); sp.gr. 0.72-0.76; m.p. < -60 C; b.p. 32-210 C; flash pt. 40-70 C; pH 9 max.; evaporates quickly
Toxicology: ACGIH TLV/TWA 300 ppm; STEL 500 ppm; LD50 (oral, rat) 18.85 ml/kg; LC50 (inh., rat, 5 min) 300 g/m^3; mod. toxic by ing.; mildly toxic by inh.; human systemic effects; can cause burning, stomach irritation, nausea, cyanosis; human eye irritant; ing./inh. can cause CNS depression, unconsciousness, coma, pulmonary edema, fatalities; prolonged/repeated contact can cause dermatitis, blistering of skin; asphyxiant; suspected carcinogen; TSCA listed
Precaution: Extremely flamm.; very dangerous fire and explosion hazard exposed to heat or flame; incompat. with strong oxidizing agents (can cause fire or explosion); common air contaminant
NFPA: Health 1, Flammability 3, Reactivity 0
Storage: Store in cool, dry, well-ventilated area out of direct sunlight
Uses: Fuel for motor vehicles; as diluent; as industrial solvent; finishing agent
Trade Names Containing: Punctilious® SDA 28A Anhydrous

Gasoline, natural. *See* Gasoline
GBL. *See* Butyrolactone
GDO. *See* Glyceryl dioleate
Geranium crystals. *See* Diphenyl oxide
GIE. *See* 2,2-Dimethyl-1,3-dioxolane-4-methanol
Gingelly oil; Gingilli oil. *See* Sesame (Sesamum indicum) oil
Glacial acetic acid. *See* Acetic acid, glacial
Glicerol. *See* Glycerin
D-Glucitol, 1,4:3,6-dianhydro-2,5-di-o-methyl. *See* Dimethyl isosorbide
α-D-Glucopyranoside, β-D-frutofuranosyl, monohexadecanoate. *See* Sucrose palmitate
α-D-Glucopyranoside, 1,3,4,6-tetra-O-acetyl-β-D-fructofuranosyl-, tetraacetate. *See* Sucrose octaacetate
Glucose syrup. *See* Corn syrup
Glutaric acid, monomethyl ester. *See* Methyl glutarate
Glyceran esters of condensed castor oil fatty acids. *See* Polyglyceryl polyricinoleate

Glycereth-7

CAS 31694-55-0 (generic)
Synonyms: PEG-7 glyceryl ether; POE (7) glyceryl ether
Definition: PEG ether of glycerin with avg. ethoxylation value of 7
Properties: Nonionic
Uses: Humectant, hair conditioner, lubricant, and foam modifier for cosmetics and toiletries, antiperspirants, cosmetic masks, wavesetting lotions, depilatories; solvent in cosmetics; adds lubricity and nongreasy luxurious feel to personal care prods.; pigment dispersant; aluminum brightener and lubricant; fiber antistat; surgical instrument lubricant
Manuf./Distrib.: Aldrich

Glycereth-26

CAS 31694-55-0 (generic)
Synonyms: PEG-26 glyceryl ether; POE (26) glyceryl ether
Definition: PEG ether of glycerin with avg. ethoxylation value of 26
Properties: Nonionic
Toxicology: TSCA listed
Uses: Emulsifier, humectant, lubricant for cosmetics, pharmaceuticals, industrial applics.; solvent in cosmetics; plasticizer for hair resins; pigment dispersant; food-pkg. adhesives
Regulatory: FDA 21CFR §175.105
Manuf./Distrib.: Aldrich
Trade Names: Volpo G26

Glycereth-7 triacetate

CAS 57569-76-3
Synonyms: PEG-7 glyceryl ether triacetate; POE (7) glyceryl ether triacetate
Definition: Triester of acetic acid with a PEG ether of glycerin containing an avg. 7 moles ethylene oxide
Empirical: $C_{23}H_{42}O_{13}$
Properties: Clear pale yel. liq., mild typ. odor; sol. in water; m.w. 526; sp.gr. 1.15; b.p. > 200 C
Toxicology: May cause mild skin and eye irritation on prolonged contact
Precaution: Incompat. with oxidizing agents
Storage: Store away from strong oxidizing agents; avoid excessive heat
Uses: Emollient, humectant, solubilizer in cosmetics; solvent for sunscreen ingreds.
Trade Names: Pelemol® G7A

Glycerides, C8-10. *See* Caprylic/capric glycerides
Glycerides, C10-18. *See* C10-18 triglycerides
Glycerides, lard mono-, acetates. *See* Acetylated lard glyceride
Glycerides, lard mono-, hydrogenated, acetates. *See* Acetylated hydrogenated lard glyceride
Glycerides, mixed decanoyl and octanoyl. *See* Caprylic/capric triglyceride
Glycerides, palm kernel oil mono-, di-, and tri-, acetates. *See* Acetylated palm kernel glycerides
Glycerides, soybean oil, hydrogenated, acetates. *See* Acetylated hydrogenated soybean oil glyceride
Glycerides, tall oil mono-. *See* Glyceryl tallate

Glycerin

CAS 56-81-5; EINECS/ELINCS 200-289-5
FEMA 2525; INS422; E422
Synonyms: Glicerol; Glycerine; Glycerol; Glycyl alcohol; 1,2,3-Propanetriol;

Propane-1,2,3-triol; Trihydroxypropane; 1,2,3-Trihydroxypropane; Trihydroxypropane glycerol

Classification: Polyhydric alcohol

Empirical: $C_3H_8O_3$

Formula: $HOCH_2COHHCH_2OH$

Properties: Clear colorless syrupy liq., odorless, sweet taste; sol. in water, alcohol; insol. in ether, benzene, chloroform; m.w. 92.09; dens. 1.26201 (25/25 C); m.p. 17.8 C; b.p. 290 C (dec.); flash pt. (OC) 176 C; ref. index 1.4730 (25 C)

Toxicology: LD50 (oral, rat) > 20 ml/kg, (IV, rat) 4.4 ml/kg; poison by subcut. route; mildly toxic by ing.; human systemic and GI effects by ing.; skin and eye irritant; nuisance dust; human mutagenic data; TSCA listed

Precaution: Combustible liq. exposed to heat, flame, strong oxidizers; highly explosive with hydrogen peroxide; ignites on contact with calcium hypochlorite; explosive mixts. and violent reactions possible with many chems.

Hazardous Decomp. Prods.: Heated to decomp., emits acrid smoke and fumes

NFPA: Health 1, Flammability 1, Reactivity 0

Storage: Hygroscopic

Uses: Alkyd resins; paints; dynamite; ester gums; perfumery; lubricants; softener; bacteriostat; penetrant; emollient; antifreeze; prod. of antibiotics; plasticizer for regenerated cellulose film, sodium silicate foundry resins; solvent (printing inks); humectant (tobacco, starch adhesives, textile sizes); solvent, humectant, plasticizer, bodying agent for foods, food pkg., cosmetics; plasticizer, solvent, solubilizer, preservative for pharmaceuticals; PU crosslinker; heat costabilizer (PVC); sweetener; preservative; plasticizer in food-contact coatings

Use Level: 0.2-65.7% (topicals), 1-50% (liq. orals), 50% (parenterals), 7-10% (dentifrices), 0.5-3.0% (ophthalmics)

Regulatory: FDA 21CFR §169.175, 169.176, 169.177, 169.178, 169.180, 169.181, 172.811, 172.866, 175.300, 175.320, 176.210, 177.1390, 177.2420, 177.2800, 178.3500, 182.90, 182.1320, GRAS; FEMA GRAS; FDA approved for orals, parenterals, ophthalmics, dentifrices, intramuscular injectables, rectals; Japan approved; Europe listed; UK approved; USP, BP, EP compliance

Manuf./Distrib.: A&E Connock; AAE Chemie NV; AB R Lundberg; ADA Int'l.; AMC Chems.; AMRESCO; Acme-Hardesty; Adept Sol'ns.; Akzo Nobel; Alba Int'l.; Aldrich; Alfa Chem; Allchem Ind.; Alnor Oil; Am. Int'l.; Amalgamet; Anar; Asahi Denka Kogyo; Ashland; Atofina Bruxelles; Avatar; Avrachem AG; BCH Brühl; Baychem; Blagden Spec. Chems. Ltd; Boehle; Brenntag AG; Brown; C.P. Hall; CPI Chems.; Chemacon GmbH; Chemcentral; Chemco France; Chesham Chems. Ltd; Cognis/Chems. Group; Cognis; Croda Chem. Ltd; Croda Universal Ltd; Crompton; Degussa AG/Health & Nutrition; Dial; Dow; E&E Ltd; EM Ind.; EM Science; EMCO Chem. Distributors; Farleyway Chems. Ltd; Fisher Scientific; Fluka; GMI Prods.; Galbraith Labs; General Chem.; Grünau GmbH; Hollister-Stier Labs; Houghton Chem.; ICC Chem.; Int'l. Sourcing; Integra; J.T. Baker; Jeen Int'l.; K & K Greeff; KIC Chems.; Karlshamns AB; Kimpton Bros. Ltd; Kraft Chem.; Lonza; Lutianhua; Magnesia GmbH; Mallinckrodt Baker; Merck KGaA; Miljac; Mosselman NV; Mutchler; NOF Am.; NOF; Norman, Fox; Parchem Trading; Penta Mfg.; Phoenix Nat. Prods.; Polysciences; Procter & Gamble; Protameen; R.I.T.A.; Reliable Biopharmaceutical; Ruger; Schaefer Tech.; Sea-Land; Shandong Haihua Tianhe; Sigma; Sithean; Spectrum Quality Prods.; St. Lawrence; Stan Chem Int'l. Ltd; Tennants Distrib. Ltd; Thew Arnott & Co. Ltd; Thomas Scientific; U.S. Chems.; Ubichem plc; Uniqema Am.; United Min. & Chem.; Univar E&E Ltd; Universal Preserv-A-Chem; VWR Int'l.; Varsal Instruments; Voigt Global Distrib.; Vopak USA; Welch, Holme & Clark; Whyte Chems. Ltd

Trade Names: Dow Glycerine 96% USP; Dow Glycerine 99.5% USP/EP; Emery® 912; Emery® 916; Emery® 917; Kemstrene® 96.0% USP; Kemstrene® 99.7% USP; Kemstrene® Reagent Grade; Optim™ Glycerine 99.7; Pricerine™ 9091; Pricerine™ 9099

Trade Names Containing: Comperlan® COD

Glycerin diacetate. *See* Glyceryl diacetate
Glycerine. *See* Glycerin
Glycerine diacetate. *See* Glyceryl diacetate
Glycerine-1,3-dibutylether. *See* 1,3-Dibutoxy-2-propanol
Glycerine triacetate. *See* Triacetin
Glycerin formal. *See* Glyceryl formal
Glycerin monoacetate. *See* Glyceryl acetate
Glycerin-α-monochlorhydrin. *See* Glyceryl-α-monochlorohydrin
Glycerin monoleate. *See* Glyceryl oleate
Glycerin α-monomethyl ether. *See* 3-Methoxy-1,2-propanediol
Glycerin monostearate. *See* Glyceryl stearate
Glycerol. *See* Glycerin
Glycerol-1-acetate. *See* Glyceryl acetate
Glycerol chlorohydrin. *See* Glyceryl-α-monochlorohydrin
Glycerol diacetate. *See* Glyceryl diacetate
Glycerol dicaprylate/caprate. *See* Glyceryl dicaprylate/caprate
Glycerol dichlorohydrin. *See* 1,3-Dichloro-2-propanol
Glycerol 1,2-dichlorohydrin. *See* 2,3-Dichloro-1-propanol
Glycerol 1,3-dichlorohydrin. *See* 1,3-Dichloro-2-propanol
Glycerol-2,3-dichlorohydrin; Glycerol-α,β-dichlorohydrin. *See* 2,3-Dichloro-1-propanol
Glycerol α,γ-dichlorohydrin; sym-Glycerol dichlorohydrin. *See* 1,3-Dichloro-2-propanol
Glycerol 1,3-dimethyl ether. *See* 1,3-Dimethoxy-2-propanol
Glycerol dioleate. *See* Glyceryl dioleate
Glycerol epichlorhydrin. *See* Epichlorohydrin
Glycerol esters of condensed castor oil fatty acids. *See* Polyglyceryl polyricinoleate
Glycerol formal. *See* Glyceryl formal
Glycerol monoacetate; Glycerol-1-monoacetate; Glycerol-α-monoacetate. *See* Glyceryl acetate
Glycerol-α-monochlorohydrin. *See* Glyceryl-α-monochlorohydrin
Glycerol monolaurate. *See* Glyceryl laurate
Glycerol monoleate. *See* Glyceryl oleate
Glycerol 1-monomethyl ether. *See* 3-Methoxy-1,2-propanediol
Glycerol monooleate. *See* Glyceryl oleate
Glycerol monoricinoleate; Glycerol 1-monoricinoleate. *See* Glyceryl ricinoleate
Glycerol monostearate. *See* Glyceryl stearate
Glycerol oleate. *See* Glyceryl oleate
Glycerol ricinoleate. *See* Glyceryl ricinoleate
Glycerol stearate. *See* Glyceryl stearate
Glycerol triacetate. *See* Triacetin
Glycerol tricaprinate. *See* Tricaprin
Glycerol tricaprylate. *See* Tricaprylin
Glycerol trichlorohydrin. *See* 1,2,3-Trichloropropane
Glycerol trilaurate. *See* Trilaurin
Glycerol trioctanoate. *See* Trioctanoin
Glycerol trioleate. *See* Triolein
Glycerol tripropionate. *See* Glyceryl tripropionate
Glycerol tristearate. *See* Tristearin
Glycerol triundecanoate. *See* Triundecanoin

Glyceryl acetate

CAS 106-61-6; 26446-35-5

Synonyms: 1-Acetate-1,2,3-propanetriol; Acetic acid, monoglyceride; Acetin; 2,3-Dihydroxypropyl acetate; Glycerin monoacetate; Glycerol-1-acetate; Glycerol monoacetate; Glycerol-1-monoacetate; Glycerol-α-monoacetate; Glyceryl monoacetate; Monacetin; 1-Monacetin; α-Monacetin; Monacetyl glycerine; 1,2,3-Propanetriol monoacetate

Empirical: $C_5H_{10}O_4$

Formula: $C_3H_7O_2 \bullet OOCCH_3$

Properties: Colorless or pale yel. (commercial) liq.; char. odor; sol. in water, alcohol; sl. sol. in ether; insol. in benzene; m.w. 134.13; dens. 1.206 (20/4 C); b.p. 158 C (17 mm); ref. index 1.451

Toxicology: LD50 (subcut., rat) 5500 mg/kg, (subcut., mouse) 3500 mg/kg; mod. toxic by subcut. route; mutagen

Hazardous Decomp. Prods.: Heated to decomp., emits acrid smoke and irritating fumes

Storage: Very hygroscopic

Uses: In mfg. of smokeless powd., dynamite; solvent for basic dyes; in tanning leather; plasticizer for cellulose acetate, cellulose nitrate; pigment dispersant; in cellophane for food pkg.

Regulatory: FDA 21CFR §177.1200

Manuf./Distrib.: Aceto; Allan; C.P. Hall; Lanaetex Prods.; Penta Mfg.; Sigma; Universal Preserv-A-Chem

Glyceryl caprate

CAS 26402-22-2; EINECS/ELINCS 247-667-6

Synonyms: Decanoic acid, monoester with 1,2,3-propanetriol; Glyceryl monocaprate

Definition: Monoester of glycerin and capric acid

Empirical: $C_{13}H_{26}O_4$

Formula: $CH_3(CH_2)_8COOCH_2COHHCH_2OH$

Properties: Nonionic

Toxicology: TSCA listed

Uses: Food emulsifier; emollient, solubilizer, absorption promoter, antimicrobial, moisturizer, refatting agent in cosmetics; dispersant, solubilizer, emulsifier, vehicle, carrier, penetrant, bacteriostat for pharmaceuticals (coating, microemulsions, soft gelatin capsules, suppositories); in paper/paperboard in contact with dry food; food-contact textiles

Regulatory: FDA 21CFR §176.180, 176.210, 177.2800

Manuf./Distrib.: A&E Connock; Sigma

Trade Names: Capmul® MCMC10

Glyceryl caprylate

CAS 26402-26-6; EINECS/ELINCS 247-668-1

Synonyms: Glyceryl monocaprylate; Monooctanoin; Octanoic acid, monoester with 1,2,3-propanetriol

Definition: Monoester of glycerin and caprylic acid

Empirical: $C_{11}H_{22}O_4$

Formula: $CH_3(CH_2)_6COOCH_2COHHCH_2OH$

Properties: Crystals; m.p. 39.5-40.5 C; nonionic

Toxicology: TSCA listed

Uses: Solubilizer, emulsifier, emollient for pharmaceuticals, cosmetics; carrier, vehicle for drugs in capsules; solvent, solubilizer for active ingreds.; antistat for syn. materials; softener for chewing gum base; defoamer in food-contact paper/paperboard; in food-contact textiles

Regulatory: FDA 21CFR §176.210, 177.2800

Manuf./Distrib.: A&E Connock; ABITEC

Trade Names: Capmul® MCMC8

Glyceryl caprylate/caprate

Definition: Mixture of monoglycerides of caprylic and capric acids

Properties: Nonionic

Uses: Cosolvent, coupling agent for org. compds.; w/o emulsifier; antistat, lubricant for syn. materials; emulsifier, emollient for cosmetics, foods, pharmaceuticals; solvent, dispersant, solubilizer, vehicle, penetrant for pharmaceuticals; defoamer in food-contact paper/paperboard; in food-contact textiles

Regulatory: FDA 21CFR §176.210, 177.2800

Manuf./Distrib.: A&E Connock

Trade Names: Capmul® MCM; Capmul® MCM-90

Glyceryl diacetate

CAS 25395-31-7; EINECS/ELINCS 246-941-2

Synonyms: Diacetin; Diacetylglycerol; Glycerin diacetate; Glycerine diacetate; Glycerol diacetate; 1,2,3-Propanetriol, diacetate

Classification: Nonaromatic ester

Empirical: $C_7H_{12}O_5$

Properties: Colorless liq.; sol. in water, oxygenated and aromatic solvs.; m.w. 176.17; sp.gr. 1.181; m.p. -35 C; flash pt. (COC) 157 C; ref. index 1.44

Toxicology: LD50 (oral, mouse) 8500 mg/kg, (subcut., rat) 4 g/kg, (IV, dog) 3 g/kg; low toxicity by ing. and other routes; TSCA listed

Hazardous Decomp. Prods.: Heated to decomp., emits acrid smoke and irritating vapors

Storage: Hygroscopic

Uses: Carrier for fragrances and flavors; plasticizer for PVB, cellulosics; solvent in cosmetics; setting agent in sodium silicate foundry mold/soil consolidation binders; in cellophane for food pkg.

Regulatory: FDA 21CFR §177.1200

Manuf./Distrib.: A.P. Chems. Ltd; Aceto; Akzo Nobel; Aldrich; Allan; Bayer; C.P. Hall; ChemService; Chemial SpA; Fluka; General Chem.; Parchem Trading; Uniqema Am.

Glyceryl dicaprylate/caprate

Synonyms: Glycerol dicaprylate/caprate

Formula: $CH_2OCO(CH_2)_nCH_3CHOOC(CH_2)_nCH_3CH_2OH$, n = 8-10

Properties: Liq.; dens. 0.95 kg/l; pour pt. -5 C; flash pt. (COC) 240 C

Uses: Emollient, cosolvent for toiletries

Glyceryl α,γ-diisoamyl ether. *See* 1,3-Diisoamyl glycerol diether

Glyceryl dilaurate

CAS 27638-00-2; EINECS/ELINCS 248-586-9

Synonyms: Dilaurin; Dodecanoic acid, diester with 1,2,3-propanetriol

Definition: Diester of glycerin and lauric acid

Empirical: $C_{27}H_{52}O_5$

Formula: $C_{11}H_{23}COOCH_2CHCH_2OHOCOC_{11}H_{23}$

Properties: Wh. to off-wh. solid; sol. in min. oil, 95% ethanol, IPM, oleyl alcohol, castor oil; insol. in water, glycerin, propylene glycol; sapon. no. 219-229; ref. index 1.4520-1.4560 (35 C); nonionic

Toxicology: LD50 (oral, rat) > 5 g/kg; nonirritating to eyes and skin; TSCA listed

HMIS: Health 0, Flammability 1, Reactivity 0

Uses: Emollient, emulsifier, solubilizer, dispersant in cosmetics, pharmaceuticals; coupling agent; plasticizer; emulsifier, dispersant, antistat for textiles, paper processing, cutting oils, polishes, emulsion cleaners, rubber latexes, wool lubricants; surfactant in foods; in food-pkg. adhesives

Features: Lipid

Regulatory: FDA 21CFR §175.105, 1876.210

Manuf./Distrib.: A&E Connock; Ruger; Sigma

Trade Names: Glicopol 2-G; Protachem™ GDL

Glyceryl α,γ-dimethyl ether. *See* 1,3-Dimethoxy-2-propanol

Glyceryl dioleate

CAS 25637-84-7; EINECS/ELINCS 247-144-2

Synonyms: GDO; Glycerol dioleate; 9-Octadecenoic acid, diester with 1,2,3-propanetriol

Definition: Diester of glycerin and oleic acid

Empirical: $C_{39}H_{72}O_5$

Properties: Pale yel. liq.; HLB 1.8; nonionic

Toxicology: TSCA listed

HMIS: Health 0, Flammability 1, Reactivity 0

Uses: Surfactant for cosmetics, foods, pharmaceuticals, industrial applics.; emulsifier; thickener; stabilizer; emollient for cosmetics; lubricant for pharmaceuticals; internal processing lubricant for PVC; antifog for PVC film; solvent for dyes and pigments; defoamer for paints; emulsifier, lubricant for textile spin finishes, metalworking fluids; superfatting agent for soaps, hand cleaners; emulsifier, stabilizer, wetting agent, lubricant for foods; in food-pkg. adhesives; defoamer in food-contact paper/paperboard; in food-contact textiles

Regulatory: FDA 21CFR §175.105, 176.210, 177.2800

Manuf./Distrib.: A&E Connock; A.P. Chems. Ltd; Aarhus Oliefabrik A/S; Ruger; Sigma; Universal Preserv-A-Chem

Glyceryl formal

CAS 4740-78-7; EINECS/ELINCS 225-248-9

UN 1993

Synonyms: 1,3-Dioxolane-4-methanol; Glycerin formal; Glycerol formal; 1,2-(Methylidene) glycerol; Methylidinoglycerol

Definition: Mixt. of 5-hydroxy-1,3-dioxane and 4-hydroxymethyl-1,3-dioxolane

Empirical: $C_4H_8O_3$

Properties: Colorless clear liq., low odor; misc. with water; m.w. 104.11; dens. 1.218 (20/4 C); b.p. 95-97 C (20 mm); flash pt. 97 C; autoignition temp. 400 C; ref. index 1.451 (20 C); pH 4-7 (10%)

Toxicology: LD50 (oral, rat) 8.6 ml/kg; low toxicity

NFPA: Health 1, Flammability 1, Reactivity 0

Storage: Hygroscopic

Uses: Solvent for pharmaceuticals, cosmetics, water-based inks; slowly evaporating solvent in insecticide delivery systems; vehicle for veterinary injectables (sulphadiazine, trimethoprin, antiparasite, intramuscular injections)

Manuf./Distrib.: DSM Chem. N. Am.; DSM Fine Chems. Austria; Degussa AG; Fluka; Haltermann GmbH; Occidental; Sigma; Xinchem

Glyceryl laurate

CAS 142-18-7; EINECS/ELINCS 205-526-6

Synonyms: Dodecanoic acid, 2,3-dihydroxypropyl ester; Dodecanoic acid,

monoester with 1,2,3-propanetriol; Glycerol monolaurate; Glyceryl monolaurate; 1-Monolaurin
Definition: Monoester of glycerin and lauric acid
Empirical: $C_{15}H_{30}O_4$
Formula: $CH_3(CH_2)_{10}COOCH_2COHHCH_2OH$
Properties: Cream-colored paste, faint odor; disp. in water; sol. in methanol, ethanol, toluene, naphtha, min. oil; dens. 0.98; m.p. 23-27 C; HLB 5.2; pH 8-8.6; nonionic
Toxicology: LD50 (oral, rat) 53 g/kg; mildly toxic by ing.; TSCA listed
Precaution: Combustible
Hazardous Decomp. Prods.: Heated to decomp., emits acrid smoke and irritating fumes
HMIS: Health 0, Flammability 1, Reactivity 0
Uses: Emulsifier, dispersant for foods, oils, waxes, solvents; antifoam; dry-cleaning soap base; plasticizer for PVC, PS, cellulosics; emollient, emulsifier for cosmetics; stabilizer, thickener, opacifier for skin creams/lotions, shampoos; emulsifier, stabilizer, solubilizer, carrier in pharmaceuticals; in food-pkg. adhesives; defoamer in food-contact paper/paperboard; in food-contact textiles
Regulatory: FDA 21CFR §175.105, 176.210, 177.2800, GRAS; Japan approved; Europe listed
Manuf./Distrib.: A&E Connock; ABITEC; Adept Sol'ns.; Ashland; Cognis/Chems. Group; Danisco Ingreds.; Degussa AG; Gattefosse; Global-Seven; Goldschmidt; Inolex; Kraft Chem.; Lanaetex Prods.; Lonza; Mosselman NV; Mutchler; PPG Ind.; Penta Mfg.; Protameen; Ruger; Sasol Germany; Sithean; Stepan; Thornley; U.S. Synthetics; Uniqema Am.; Universal Preserv-A-Chem; VWR Int'l.; Velsicol; Vopak USA
Trade Names: Protachem™ MLD

Glyceryl monoacetate. *See* Glyceryl acetate
Glyceryl α-mono-n-butyl ether. *See* 3-Butoxy-1,2-propanediol
Glyceryl monocaprate. *See* Glyceryl caprate
Glyceryl monocaprylate. *See* Glyceryl caprylate

Glyceryl-α-monochlorohydrin

CAS 96-24-2; EINECS/ELINCS 202-492-4
UN 2689 (DOT)
Synonyms: Chlorhydrin; α-Chlorhydrin; Chlorodeoxyglycerol; 1-Chloro-2,3-dihydroxypropane; 3-Chloro-1,2-dihydroxypropane; Chlorohydrin; α-Chlorohydrin; 1-Chloro-2,3-propanediol; 3-Chloropropanediol-1,2; 3-Chloropropane-1,2-diol; 3-Chloropropylene glycol; 1-Choropropane-2,3-diol; β,β'-Dihydroxyisopropyl chloride; 2,3-Dihydroxypropyl chloride; Glycerin-α-monochlorhydrin; Glycerol chlorohydrin; Glycerol-α-monochlorohydrin; 1,2-Propanediol, 3-chloro-
Classification: Nonaromatic alcohol
Empirical: $C_3H_7ClO_2$
Formula: $ClCH_2CHOHCH_2$
Properties: Colorless or pale yel. liq.; sol. in water, DMSO, 95% ethanol, acetone, ether, polar solvs.; immisc. with oils; m.w. 110.54; sp.gr. 1.326 (18/15 C); vapor pressure 3.75 mm Hg (25 C); m.p. -40 C; b.p. 213 C (dec.); flash pt. 58 C; ref. index 1.4809 (20 C)
Toxicology: LD50 (oral, rat) 26 mg/kg, (IP, mouse) 73 mg/kg; highly toxic; poison by ing. and IP routes; mod. toxic by inh.; can be absorbed thru skin; irritating to eyes, skin, mucous membranes, upper respiratory tract; risk of serious eye damage; may cause nausea, vomiting, giddiness, dizziness, incoordination, coma, abdominal pain, lacrimation, somnolence, ataxia, etc.; may be fatal by inh., ing., skin absorp.; questionable carcinogen, tumorigen; mutagenic data; target organ: CNS; TSCA listed
Precaution: Combustible exposed to heat or flame; forms sensitive explosive prod. on reaction with perchloric acid; incompat. with strong oxidizing agents; may be sensitive to prolonged exposure to air; keep away from food
Hazardous Decomp. Prods.: Heated to decomp., emits toxic fumes of Cl^-, CO, CO_2, hydrogen chloride gas; emits toxic fumes under fire conditions
Storage: Hygroscopic; keep in tightly closed containers under inert atmosphere; store refrigerated; store away from ignition sources
Uses: Intermediate in org. synthesis incl. cosmetics and pharmaceuticals; improves flame resist. in PU; agric. chem.; chemosterilant for rodents; to lower freezing pt. in dynamite; mfg. of dye intermediates; solvent for cellulose acetate, glyceryl phthalate resins; partial solvent for gums
Manuf./Distrib.: AccuStandard; Acros Org.; Aldrich; Alfa Aesar; Borregaard Synthesis US; Dixie; Fluka; Hampshire; ICN Biomed. Research Prods.; Lancaster Synthesis; Pfaltz & Bauer; Raschig AG; Sigma; Solvay SA; TCI Am.
Trade Names: GAM

Glyceryl 1-mono-12-hydroxy-cis-9-octadecenoate. *See* Glyceryl ricinoleate
Glyceryl α-monoisoamyl ether. *See* 1-Isoamyl glycerol ether
Glyceryl monolaurate. *See* Glyceryl laurate
Glyceryl monomyristate. *See* Glyceryl myristate
Glyceryl monooleate. *See* Glyceryl oleate
Glyceryl monoricinoleate; Glyceryl 1-monoricinoleate. *See* Glyceryl ricinoleate
Glyceryl monostearate. *See* Glyceryl stearate
Glyceryl monostearate SE. *See* Glyceryl stearate SE
Glyceryl monotallate. *See* Glyceryl tallate
Glyceryl monotristearate. *See* Tristearin

Glyceryl myristate

CAS 589-68-4; 27214-38-6; EINECS/ELINCS 248-329-0
Synonyms: Glyceryl monomyristate; Monomyristin; Tetradecanoic acid, monoester with 1,2,3-propanetriol
Definition: Monoester of glycerin and myristic acid
Empirical: $C_{17}H_{34}O_4$
Formula: $CH_3(CH_2)_{12}COOCH_2COHHCH_2OH$
Properties: Nonionic
Toxicology: TSCA listed
Uses: Emulsifier, emollient in cosmetics, pharmaceuticals; component in w/o and o/w creams; lubricant, antistat, antifog in plastics; coemulsifier, solubilizer, carrier for lipophilic drugs; in food-pkg. adhesives; defoamer in food-contact paper/paperboard; in food-contact textiles
Regulatory: FDA 21CFR §175.105, 176.210, 177.2800
Manuf./Distrib.: A&E Connock; Mosselman NV
Trade Names: Thomil G

Glyceryl oleate

CAS 111-03-5; 25496-72-4 (generic); 37220-82-9; 68424-61-3 (generic); EINECS/ELINCS 203-827-7; 247-038-6; 253-407-2
FEMA 2526
Synonyms: Glycerin monoleate; Glycerol monoleate; Glycerol monooleate; Glycerol oleate; Glyceryl monooleate; Monoglyceryl oleate; Monoolein; Monooleoylglycerol; 9-Octadecenoic acid, monoester with 1,2,3-propanetriol; Oleic acid glycerol monoester; Oleic acid monoglyceride; Oleoylglycerol; Oleyl monoglyceride
Definition: Monoester of glycerin and oleic acid
Empirical: $C_{21}H_{40}O_4$
Formula: $CH_3(CH_2)_7CH{=}CH(CH_2)_7CHOHOOCH_2CCH_2OH$
Properties: Yel. oil or soft solid; insol. in water; somewhat sol. in alcohol, most org. solvs.; m.w. 356.55; dens. 0.940-0.960; m.p. 149-19 C; HLB 3.4; acid no. 3 max.; iodine no. 65-80; sapon. no. 166-174; nonionic
Toxicology: Skin and eye irritant; TSCA listed
Precaution: Combustible
Hazardous Decomp. Prods.: Heated to decomp., emits acrid smoke and irritating fumes
HMIS: Health 0, Flammability 1, Reactivity 0
Uses: Emulsifier in hydrocarbon oils, textile spin finishes; coemulsifier; wetting agent; antistat; emollient, emulsifier, thickener, opacifier in cosmetics; mold release agent/processing lubricant in plastics; internal antistat for PE, PP, PVC; antifog, cling agent, lubricant for PVC film; visc. stabilizer in paints; rust-preventive oils; softener in textile finishes; flavoring agent/adjuvant, solvent, vehicle, defoamer, dispersant, emulsifier, plasticizer for foods, food pkg.; pharmaceutical excipient, emulsifier, coemulsifier, stabilizer, wetting agent, lubricant; plasticizer in food-contact coatings; defoamer in food-contact paper/paperboard; in food-contact textiles
Regulatory: FDA 21CFR §175.105, 175.300, 175.320, 176.210, 177.2800, 181.22, 181.27, 184.1323, GRAS; FEMA GRAS; FDA approved for orals; BP, EP compliance
Manuf./Distrib.: A&E Connock; A.P. Chems. Ltd; ABITEC; AP Chems.; Accurate Chem. & Scientific; Adept Sol'ns.; Alfa Chem; Allchem Int'l. Ltd; Am. Ingreds.; Ashland; Atofina Bruxelles; Avatar; Camida Ltd;

Cognis/Chems. Group; Cornelius Chem. Co. Ltd; Costec; Croda Oleochems.; Danisco Ingreds.; Degussa AG/Health & Nutrition; DelCon; Dr. W. Kolb AG; Fallek; Ferro/Keil; Fluka; Gallard-Schlesinger Ind.; Gattefosse; Goldschmidt; Honeywill & Stein; Inolex; Kraft Chem.; Lambent Tech.; Lonza; Mosselman NV; Mutchler; PPG Ind.; Penta Mfg.; Ruger; Sigma; Spectrum Quality Prods.; Stepan; Uniqema Am.; Uniqema; Univar E&E Ltd; Universal Preserv-A-Chem; VWR Int'l.; Venus Ethoxyethers; Vopak USA

Trade Names: Lumulse® GMO; Mazol® 300 K; Tilol GP-O

Glyceryl ricinoleate

CAS 141-08-2; EINECS/ELINCS 205-455-0

Synonyms: Glycerol monoricinoleate; Glycerol 1-monoricinoleate; Glycerol ricinoleate; Glyceryl 1-mono-12-hydroxy-cis-9-octadecenoate; Glyceryl monoricinoleate; Glyceryl 1-monoricinoleate; GMRO; 12-Hydroxy-9-octadecenoic acid 2,3-dihydroxypropyl ester; 12-Hydroxy-9-octadecenoic acid, monoester with 1,2,3-propanetriol; Monoricinolein; α-Monoricinolein

Classification: Fatty acid ester

Definition: Monoester of glycerin and ricinoleic acid

Empirical: $C_{21}H_{40}O_5$

Formula: $C_3H_5(OOCC_{16}H_{32}OH)_3$

Properties: Lt. yel. oily liq.; sol. in ethanol, ether, acetone, benzene, cottonseed oil, ethyl acetate; sl. sol. in methanol; insol. in water, min. oil, aliphatic hydrocarbons; m.w. 372.55; dens. 1.028 (20 C); m.p. < -5 C; cloud pt. -10 C; HLB 2.1; flash pt. (COC) 265 C; ref. index 1.470; nonionic

Toxicology: LD50 (oral, mouse) > 25 ml/kg; low oral toxicity; ing. of lg. amts. may cause nausea, vomiting, cramps, diarrhea; unlikely to cause skin irritation; inh. may cause sl. nose/throat irritation; exposure to vapors unlikely to occur; TSCA listed

Precaution: Avoid generating mist; incompat. with acids and bases; decomp. (hydrolysis) can occur

Hazardous Decomp. Prods.: Thermal decomp. prods.: CO, CO_2

Storage: Store away from heat and ignition sources in suitable containers, protected from damage

Uses: Emulsifier; plasticizer for PVB, cellulosics; emollient for cosmetics; emulsifier, solubilizer for cosmetics, pharmaceuticals, household prods., coatings, and printing inks; antifoam; antifog for food pkg.; solvent; lubricant for plastics processing, metalworking, greases, motor oils, extrusion of ceramics; fatting agent in textile and leather auxs.; superfatting agent for hand cleaners, soaps; freeze-thaw stabilizer (latex paint); emulsifier, stabilizer, wetting agent in foods; in food-pkg. adhesives; in paper/paperboard in contact with aq./fatty foods

Regulatory: FDA 21CFR §175.105, 176.170, 176.210, 178.3130; FDA approved for orals, topicals

Manuf./Distrib.: A&E Connock; CasChem; Lambent Tech.; Lonza; Mosselman NV; Pechiney Chems. Div.

Glyceryl stearate

CAS 123-94-4 (pure grade); 11099-07-3 (crude grade); 31566-31-1 (generic); 85666-92-8 (generic); 85251-77-0; EINECS/ELINCS 204-664-4; 234-325-6; 250-705-4; 286-490-9

FEMA 2527; INS471

Synonyms: 2,3-Dihydroxypropyl octadecanoate; Glycerin monostearate; Glycerol monostearate; Glycerol stearate; Glyceryl monostearate; GMS; Monostearin; Octadecanoic acid, monoester with 1,2,3-propanetriol; 1,2,3-Propanetriol octadecanoate; Stearic acid, monoester with glycerol; Stearic monoglyceride

Classification: Aliphatic carboxylic acid ester

Definition: Monoester of glycerin and stearic acid

Empirical: $C_{21}H_{42}O_4$

Formula: $CH_3(CH_2)_{16}COOCH_2COHHCH_2OH$

Properties: Wh. to cream wax-like flakes, sl. fatty odor and taste; sol. in hot org. solvs.; insol. in water, ethanol, glycerin, propylene glycol; disp. in min. oil; m.w. 358.57; sp.gr. 0.97; m.p. 56-59 C; b.p. 238-240 C; HLB 3.8; acid no. 6 max.; iodine no. 3 max.; sapon. no. 162-175; hyd. no. 300-330; flash pt. (OC) 230 C; pH 9.3 (3%); nonionic

Toxicology: ACGIH TLV/TWA 10 $mg/^3$ (stearates); LD50 (IP, mouse) 200 mg/kg; poison by IP route; TSCA listed

Precaution: Combustible dust; incompat. with oxidizing materials (increased risk of fire and explosion)

Hazardous Decomp. Prods.: Heated to decomp., emits acrid smoke and irritating fumes

HMIS: Health 1, Flammability 0, Reactivity 0

Storage: Store in a cool, dry, well-ventilated area, out of direct sunlight, away from heat/ignition sources; light-sensitive

Uses: Emulsifier for cosmetics, toiletries, foods, pharmaceuticals, oils, waxes, solvents; thickener, visc. builder for emulsions; plasticizer for cellulose nitrate; antistat, antifog, lubricant, processing aid for plastics; detackifier; emollient, stabilizer, thickener, opacifier in cosmetics; oil-well drilling mud lubricant; protective coating for hygroscopic powds.; softener in textile finishes; foods (thickener, emulsifier, antisticking, antistalant, solvent); pharmaceutical excipient; surf. lubricant in food-contact coatings; defoamer in food-contact paper coatings, paper/paperboard

Regulatory: FDA 21CFR §139.110, 139.115, 139.117, 139.120, 139.121, 139.122, 139.125, 139.135, 139.138, 139.140, 139.150, 139.155, 139.160, 139.165, 139.180, 175.105, 175.210, 175.300, 176.170, 176.200, 176.210, 177.1200, 177.2800, 184.1324, GRAS; FEMA GRAS; Europe listed; FDA approved for orals, ophthalmics, otics, rectals, topicals; NF, BP compliance

Manuf./Distrib.: A&E Connock; A.C. Hatrick; A.P. Chems. Ltd; AAA Int'l.; AB Tech. Ltd; ABITEC; AP Chems.; Adept Sol'ns.; Akzo Nobel; Allan; Allchem Int'l. Ltd; Am. Ingreds./Patco; Am. Int'l.; Ashland; Atofina Bruxelles; Avatar; Blagden Spec. Chems. Ltd; Boith China; Camida Ltd; Chemacon GmbH; Cognis/Chems. Group; Cognis; Cornelius Chem. Co. Ltd; Costec; Croda Chem. Ltd; Croda Oleochems.; Danisco Ingreds.; Degussa AG/Health & Nutrition; Degussa; DelCon; Dr. W. Kolb AG; Eastman; Eigenmann & Veronelli; Fallek; Functional Foods; GMI Prods.; Gallard-Schlesinger Ind.; Gattefosse; Goldschmidt; Grünau GmbH; Hart Prods.; Honeywill & Stein; ISP Van Dyk; Inolex; Integra; Jarchem Ind.; Karlshamns AB; Koster Keunen; Kraft Chem.; Krishna Org. Pvt. Ltd; Lambent Tech.; Lanaetex Prods.; Lavy's; Lipo; Lonza; Mosselman NV; Mutchler; Nordmann Rassmann GmbH; Norman, Fox; PPG Ind.; Penta Mfg.; Peter Whiting Ltd; Phoenix Nat. Prods.; Protameen; Quest Int'l.; R.I.T.A.; Rhodia HPCII; Riken Vitamin; Ruger; Sasol Germany; Seppic; Sithean; Spectrum Quality Prods.; Stepan; Unipex; Uniqema Am.; Univar E&E Ltd; Universal Preserv-A-Chem; Van Den Burg Eiprodukten BV; Venus Ethoxyethers; Vopak USA

Trade Names: Monestriol GE; Monestriol GP; Monestriol GP-35; Monestriol GP-40; Monestriol GP-40-CF; Monestriol GP-40-DC; Monestriol GP-45-DC; Monestriol GP-60; Monestriol GP-AS; Protachem™ GMS-450; Protachem™ HMS

Trade Names Containing: Protachem™ GMS-165; Protachem™ GMS-AS

Glyceryl stearate SE

CAS 31566-31-1 (generic); 11099-07-3; 85666-92-8 (generic); 977053-96-5; EINECS/ELINCS 234-325-6; 286-490-9

Synonyms: Glyceryl monostearate SE; GMS-SE

Definition: Self-emulsifying grade of glyceryl stearate contg. some sodium and/or potassium stearate

Empirical: $C_{21}H_{42}O_4$

Properties: Wh. to cream flakes; sol. in oleyl alcohol; partly sol. in water, veg. oil, ethanol, propylene glycol; m.w. 358.63; m.p. 57-59 C; HLB 11.0; sapon. no. 150-160; anionic

Toxicology: LD50 (IP, mouse) 200 mg/kg; TSCA listed

HMIS: Health 1, Flammability 0, Reactivity 0

Uses: Aux. emulsifier for soap o/w emulsions, pharmaceutical topicals; emulsifier for polymers; lubricant, antistat, antifog for plastics; plasticizer in syn. fabrics and plastics; emulsifier, emollient in cosmetics, toiletries, pharmaceuticals; mold release compds.; metalworking applics.; thickener, emulsifier, stabilizer, wetting agent, flavoring agent in foods

Regulatory: FDA approved for topicals

Manuf./Distrib.: A&E Connock; Krishna Org. Pvt. Ltd; Mosselman NV; Ruger

Trade Names: Monestriol GP-35-3%; Monestriol GP-35-AE; Protachem™ GMS-D; Protachem™ GMS-T

Glyceryl tallate

CAS 61789-12-6

Synonyms: Glycerides, tall oil mono-; Glyceryl monotallate; Tall oil monoglycerides
Properties: Nonionic
Uses: Emulsifier, pigment dispersant, lubricant, corrosion inhibitor for industrial applics.; anti-icing fuel additive; textile lubricant; cutting fluids; emulsifier in foods, cosmetics
Manuf./Distrib.: St. Lawrence
Trade Names: Lumulse® GMT; Lumulse® GMT-40

Glyceryl triacetate. *See* Triacetin

Glyceryl triacetyl hydroxystearate

CAS 27233-00-7; EINECS/ELINCS 248-351-0
Synonyms: Octadecanoic acid, (acetyloxy)-1,2,3-propanetriyl ester; Octadecanoic acid, 12-hydroxy, 1,2,3-propanetriyl ester
Definition: Triester of glycerin and acetyl hydroxystearic acid
Empirical: $C_{63}H_{116}O_{12}$
Uses: Emollient, solvent, visc. control agent in cosmetics; wetting agent, stabilizer, gloss aid for pigmented prods.; food pkg.
Features: Mild
Regulatory: FDA 21CFR §178.3505

Glyceryl tri-(12-acetylricinolate). *See* Glyceryl triacetyl ricinoleate

Glyceryl triacetyl ricinoleate

CAS 101-34-8; EINECS/ELINCS 202-935-1
Synonyms: Castor oil, acetylated; Glyceryl tri-(12-acetylricinolate); 9-Octadecenoic acid, 12-(acetyloxy)-, 1,2,3-propanetriol ester; 1,2,3-Propanetriyl 12-(acetyloxy)-9-octadecenoate
Definition: Triester of glycerin and acetyl ricinoleic acid
Empirical: $C_{63}H_{110}O_{12}$
Formula: $C_3H_5(OOCC_{17}H_{32}OCOCH_3)_3$
Properties: Clear pale yel. oily liq., mild odor; sol. in most org. liqs.; insol. in water; dens. 0.967 (25/25 C); solidifies @ -40 C; iodine no. 76; sapon. no. 300
Toxicology: TSCA listed
Precaution: Combustible
Uses: Plasticizer for nitrocellulose, ethylcellulose, and PVC; lubricants; protective coatings; solvent, mild emollient in cosmetics; pigment wetting agent; cosolvent; softener for waxes and resins
Manuf./Distrib.: CasChem
Trade Names: Naturechem® GTR

Glyceryl tri-C10-18 acids. *See* C10-18 triglycerides
Glyceryl tricaprate. *See* Tricaprin
Glyceryl tricaprylate. *See* Tricaprylin
Glyceryl tricaprylate/caprate. *See* Caprylic/capric triglyceride
Glyceryl trichlorohydrin. *See* 1,2,3-Trichloropropane
Glyceryl tridecanoate. *See* Tricaprin
Glyceryl tridodecanoate. *See* Trilaurin
Glyceryl tri (2-ethylhexanoate). *See* Trioctanoin
Glyceryl tri (12-hydroxystearate). *See* Trihydroxystearin
Glyceryl triisostearate. *See* Triisostearin
Glyceryl trilaurate. *See* Trilaurin
Glyceryl trilinoleate. *See* Trilinolein
Glyceryl trimyristate. *See* Trimyristin
Glyceryl trioctanoate. *See* Trioctanoin
Glyceryl trioleate. *See* Triolein
Glyceryl tripalmitate. *See* Tripalmitin
Glyceryl tripropanoate. *See* Glyceryl tripropionate

Glyceryl tripropionate

CAS 139-45-7; EINECS/ELINCS 205-365-1
FEMA 3286
Synonyms: Glycerol tripropionate; Glyceryl tripropanoate; Tripropionin
Classification: Nonaromatic ester
Empirical: $C_{12}H_{20}O_6$
Formula: $C_3H_5(OCOC_2H_5)_3$
Properties: Solid; sol. in alcohol, 0.313% in water; m.w. 260.32; dens. 1.078 (20 C); f.p. < -50 C; b.p. 177-182 C (20 mm); ref. index 1.4340
Toxicology: LD50 (IV, mouse) 840 mg/kg; TSCA listed
NFPA: Health 0, Flammability 1, Reactivity 0
Uses: Plasticizer; solvent in cosmetics; flavoring agent in foods and pharmaceuticals
Regulatory: FEMA GRAS
Manuf./Distrib.: Aldrich; Chemial SpA; Haltermann Prods. UK

Glyceryl tris-12-hydroxystearate. *See* Trihydroxystearin
Glyceryl tristearate. *See* Tristearin
Glyceryl triundecanoate. *See* Triundecanoin
Glycine soja; Glycine soja oil. *See* Soybean (Glycine soja) oil
Glycinol. *See* Ethanolamine
Glycoformal. *See* 1,3-Dioxolane

Glycofurol

CAS 9004-76-6; 31692-85-0
Synonyms: Poly (ethylene glycol) tetrahydrofurfuryl ether; Tetraglycol; Tetrahydrofurfuryl polyethylene glycol; Tetrahydrofurfuryl-polyethylene glycol ether
Properties: Colorless clear liq.; sol. in most org. solvs.; completely misc. with water @ 30%; m.w. 190-210; sp.gr. 1.075-1.090 (20 C); ref. index 1.455-1.465 (20 C); pH 4.5-6.5 (10% aq.)
Toxicology: Possibly hazardous
Precaution: Combustible; incompat. with strong oxidizing agents
Uses: Solvent in cosmetics, injectables
Features: Stable
Manuf./Distrib.: Laboratori Alchemia

Glycol; Glycol alcohol. *See* Ethylene glycol
Glycol benzyl ether. *See* Ethylene glycol benzyl ether
Glycol bis (hydroxyethyl) ether. *See* Triethylene glycol
Glycol butyl ether. *See* Butoxyethanol
Glycol carbonate. *See* Ethylene carbonate
Glycol chlorohydrin. *See* Ethylene chlorohydrin
Glycol diacetate. *See* Ethylene glycol diacetate
Glycol dichloride. *See* Ethylene dichloride
Glycol dimethyl ether. *See* Ethylene glycol dimethyl ether

Glycol distearate

CAS 627-83-8; 91031-31-1; EINECS/ELINCS 211-014-3
Synonyms: EGDS; Ethylene glycol distearate; Octadecanoic acid, 1,2-ethanediyl ester
Definition: Diester of ethylene glycol and stearic acid
Empirical: $C_{38}H_{74}O_4$
Formula: $CH_3(CH_2)_{16}COOCH_2CH_2OCO(CH_2)_{16}CH_3$
Properties: Wh. to yel. powd.; m.w. 595.00; dens. 0.97; m.p. 60 C; HLB 1.6; flash pt. (COC) 171 C; nonionic
Toxicology: TSCA listed
Uses: Pearlescent, opacifier, thickener, lubricant, emulsifier, emollient in cosmetics, emulsion shampoos and foam baths; plasticizer, lubricant, antistat for plastics and rubber; surfactant in paints, pharmaceuticals; w/o emulsifier, dispersant, antistat for textiles, paper processing, cutting oils, polishes, emulsion cleaners, rubber latex, wool lubricants; emulsifier, lubricant, antistat, defoamer for metalworking; intermediate; specialized solvent; diluent in food colorants; defoamer in food-contact paper/paperboard
Regulatory: FDA 21CFR §73.1, 176.210
Manuf./Distrib.: A&E Connock; A.P. Chems. Ltd; Ashland; Cedar Concepts; Inolex; Lonza; Mosselman NV; PPG Ind.; Rhodia HPCII; Ruger
Trade Names: Kemester® EGDS; Protachem™ EGDS

Glycol ether. *See* Diethylene glycol
Glycol ether EE. *See* Ethoxyethanol
Glycol ether EM acetate. *See* Methoxyethanol acetate
Glycol ethylene ether. *See* 1,4-Dioxane
Glycol ethyl ether. *See* Diethylene glycol; Ethoxyethanol
Glycol formal. *See* 1,3-Dioxolane
Glycolic acid n-butyl ester. *See* Butyl glycolate
Glycolic acid, ethyl ester, methyl phthalate. *See* Methyl phthalyl ethyl glycolate
Glycol methylene ether. *See* 1,3-Dioxolane
Glycol monoacetate; Glycol monoacetin. *See* Ethylene glycol acetate
Glycol monobenzyl ether. *See* Ethylene glycol benzyl ether
Glycol monobutyl ether. *See* Butoxyethanol
Glycol monobutyl ether acetate. *See* Butoxyethanol acetate
Glycolmonochlorhydrin; Glycol monochlorohydrin. *See* Ethylene chlorohydrin

Glycol monoethyl ether. *See* Ethoxyethanol
Glycol monoethyl ether acetate. *See* Ethoxyethanol acetate
Glycol monomethyl ether. *See* Methoxyethanol
Glycol monomethyl ether acetate. *See* Methoxyethanol acetate
Glycol monophenyl ether. *See* Phenoxyethanol
Glycol monostearate. *See* Glycol stearate
Glycols, polyethylene, dimethyl ether. *See* PEG dimethyl ether
Glycols, polyethylene, mono (nonylphenyl) ether. *See* Nonoxynol

Glycol stearate

CAS 111-60-4; 9004-99-3 (generic); 97281-23-7; EINECS/ELINCS 203-886-9; 306-522-8
Synonyms: EGMS; Ethylene glycol monostearate; Ethylene glycol stearate; Glycol monostearate; 2-Hydroxyethyl ester stearic acid; 2-Hydroxyethyl octadecanoate; PEG-1 stearate; POE (1) stearic acid; Stearic acid, monoester with ethylene glycol
Definition: Ester of ethylene glycol and stearic acid
Empirical: $C_{20}H_{40}O_3$
Formula: $CH_3(CH_2)_{16}COOCH_2CH_2OH$
Properties: Wh. to yel. waxy solid; sol. in alcohol, hot ether, acetone; insol. in water; m.w. 328.60; dens. 0.96 (25 C); m.p. 57-60 C; HLB 2.9; nonionic
Toxicology: LD50 (IP, mouse) 200 mg/kg; poison by IP route; primary skin irritant; TSCA listed
Precaution: Combustible
Hazardous Decomp. Prods.: Heated to decomp., emits acrid smoke and irritating fumes
HMIS: Health 1, Flammability 0, Reactivity 0
Uses: Surfactant, opacifier, pearlescent, emollient, emulsifier, visc. control agent in cosmetics; plasticizer for cellulose nitrate; lubricant for plasticized PVC; surfactant for industrial applics., pulp/paper, pharmaceuticals, metalworking, lubricants, textiles, agric., paints, adhesives, water treatment; specialized solvent; pharmaceutical excipient; defoamer in food-contact paper/paperboard
Regulatory: FDA 21CFR §176.210; FDA approved for topicals
Manuf./Distrib.: A&E Connock; A.P. Chems. Ltd; Adept Sol'ns.; Alfa Chem; Alzo; Ashland; Asiamerica Int'l.; C.P. Hall; CoKEM Assoc.; Costec; Delta Distributors; Eastech; Fanning; Gattefosse; Goldschmidt; ISP Van Dyk; Inolex; Kraft Chem.; Lanaetex Prods.; Lipo; McIntyre; Milsolv; Mosselman NV; Norman, Fox; P.B. & S.; PPG Ind.; Penta Mfg.; Phoenix; Protameen; R.I.T.A.; Rhodia HPCII; Ruger; SOCO-Lynch; Scher; Southchem; Stepan; Textile Chem.; Thornley; Uniqema Am.; Universal Preserv-A-Chem; Vopak USA; Whittaker, Clark & Daniels
Trade Names: Protachem™ EGMS
Trade Names Containing: Protachem™ SAS

Glyco stearin. *See* PEG-2 stearate
Glycyl alcohol. *See* Glycerin
Glyme. *See* Ethylene glycol dimethyl ether
Glyme-3. *See* PEG-3 dimethyl ether
Glyme-5. *See* Dimethoxytetraglycol
GMRO. *See* Glyceryl ricinoleate
GMS. *See* Glyceryl stearate
GMS-SE. *See* Glyceryl stearate SE
Gomenoleo oil. *See* Olive (Olea europaea) oil
Gossypium. *See* Cottonseed (Gossypium) oil
Grain alcohol. *See* Alcohol
Green oil. *See* Anthracene
Groundnut oil. *See* Peanut (Arachis hypogaea) oil
p-Guaiacol. *See* Hydroquinone monomethyl ether
Gum camphor. *See* Camphor
Gum turpentine. *See* Turpentine

H

HA. *See* Hydroquinone monomethyl ether
Halocarbon 11. *See* Trichlorofluoromethane
Halocarbon 23. *See* Trifluoromethane
Halocarbon 112. *See* 1,1,2,2-Tetrachloro-1,2-difluoroethane
Halocarbon 113. *See* Trichlorotrifluoroethane
Halocarbon 114. *See* 1,2-Dichlorotetrafluoroethane
Halon. *See* Dichlorodifluoromethane
Hans solvent. *See* Naphtha, heavy aromatic
Hard paraffin. *See* Paraffin
HCBD. *See* Hexachloro-1,3-butadiene
HCFC 21. *See* Dichlorofluoromethane
HCFC 22. *See* Chlorodifluoromethane
HCFC 23. *See* Trifluoromethane
HCFC 141b. *See* Dichlorofluoroethane
HCFC 142b. *See* 1,1,1-Chlorodifluoroethane
HDI. *See* Hexamethylene diisocyanate
HDO. *See* Hexamethylene glycol
HEAR. *See* Rapeseed (Brassica campestris) oil
Heavy aliphatic petroleum solvent; Heavy aliphatic solvent; Heavy aliphatic solvent naphtha. *See* Naphtha, heavy aliphatic
Heavy aromatic bottoms; Heavy aromatic naphtha; Heavy aromatic naphtha solvent; Heavy aromatic petroleum solvent; Heavy aromatic solvent naphtha; Heavy aromatic solvent naphtha (petroleum). *See* Naphtha, heavy aromatic
Heavy hydrotreated naphtha (petroleum). *See* Naphtha, hydrotreated heavy
Heavy mineral oil. *See* Mineral oil
Heavy naphtha. *See* Naphtha
Heavy normal paraffins concentrate (petroleum); Heavy normal paraffins (petroleum). *See* Paraffins, normal C5-20
Heavy oil. *See* Creosote oil
Heavy paraffinic distillate, solvent extract. *See* Paraffin distillate
Heavy thermal cracked petroleum distillate. *See* Petroleum distillates, heavy thermal cracked
Helianthus annuus; Helianthus annuus oil. *See* Sunflower (Helianthus annuus) seed oil

Hempa

CAS 680-31-9; EINECS/ELINCS 211-653-8
Synonyms: Hexametapol; Hexamethylphosphoramide; Hexamethylphosphoric triamide; HMPA; HMPT; HPT
Classification: Nonaromatic phosphorus compounds
Empirical: $C_6H_{18}N_3OP$
Formula: $[N(CH_3)_2]_3PO$
Properties: Water-wh. liq.; mild amine odor; sol. in water and polar and nonpolar solvs.; m.w. 179.24; dens. 1.021; m.p. 7.20 C; b.p. 230-232 C (739.4 mm); ref. index 1.4572
Toxicology: LD50 (oral, rat) 2650 mg/kg, (oral, mouse) 2400 mg/kg; mod. toxic by ing., IP, and IV routes; toxic by skin contact; suspected human carcinogen; mutagen; tumorigen; TSCA listed
Precaution: Combustible
Hazardous Decomp. Prods.: Heated to decomp., emits very toxic fumes of phosphine, PO_x, and NO_x
Uses: UV inhibitor in PVC; chemosterilant for insects; specialty solvent; aprotic process solvent in chem. synthesis; de-icing additive for jet fuels
Manuf./Distrib.: Aldrich; Fluka; Sigma

Hendecane. *See* n-Undecane
Heptalin acetate. *See* Methylcyclohexanol acetate
2,2,4,4,6,6,8-Heptamethylnonane. *See* Isohexadecane

Heptane

CAS 142-82-5; EINECS/ELINCS 205-563-8
UN 1206 (DOT)
Synonyms: Alkane C7; Dipropyl methane; n-Heptane; Heptyl hydride
Classification: Sat. aliphatic hydrocarbon
Empirical: C_7H_{16}
Formula: $CH_3(CH_2)_5CH_3$
Properties: Volatile colorless liq.; gasoline-like odor; sol. in alcohol, ether, chloroform; misc. with oxygenated, chlorinated, and hydrocarbon solvs.; insol. in water; m.w. 100.21; dens. 0.68368 (20 C); vapor pressure 45.7 mm Hg; m.p. -91 C; b.p. 98.428 C; flash pt. (CC) -3.89 C; autoignition temp. 204 C; ref. index 1.38764 (20 C); surf. tens. 19.7 dynes/cm; KB value 30; dielec. const. 1.925
Toxicology: ACGIH TLV/TWA 400 ppm; STEL 500 ppm; LD50 (IV, mouse) 222 mg/kg; lethal conc. for mice 15,900 ppm in air; irritating to respiratory tract; narcotic in high concs.; poison by IV route; ing. may cause nausea, vomiting, swelling of abdomen, headache, depression; relatively low oral toxicity unless aspirated into lungs; TSCA listed
Environmental: VOC; ThOD 3.52
Precaution: Flamm. when exposed to heat or flame; LEL 1.05%; UEL 6.7%; dangerous fire risk; can react vigorously with oxidizing materials; mod. explosive exposed to heat or flame; violent reaction with phosphorus + chlorine
Hazardous Decomp. Prods.: Heated to decomp., emits acrid smoke and fumes
NFPA: Health 1, Flammability 3, Reactivity 0
Storage: Store in cool, dry, well-ventilated area out of direct sunlight
Uses: Standard for octane rating determinations; anesthetic; solvent for rubber, cements/sealants, inks, fast-drying adhesives, lacquers, fat/oil extraction, polishes, paints, resins, plastics, textile coatings, minerals, agric., pharmaceuticals; organic synthesis; replacement for n-hexane; reagent; major ingred. in gasoline, aviation fuel, petroleum solvents such as petroleum naphtha and rubber solvent; reaction diluent in polyolefin prod.; solvent for protective coatings on citrus fruits; solvent in cosmetics, hair dyeing; in food-pkg. adhesives; adjuvant for PC food-pkg. resins
Regulatory: FDA 21CFR §175.105, 177.1580, 27CFR §21.111
Manuf./Distrib.: Aldrich; Ashland; Burdick & Jackson; ChevronPhillips; Coyne; ExxonMobil; Fisher Scientific; Fluka; Great Western; Houghton Chem.; Hukill; Humphrey; Romil Ltd; Ruger; Sigma; Spectrum Quality Prods.; Stanchem; Triple Crown Am.
Trade Names: Exxsol® Heptane Anti-Static; Hydrosol Heptane; Shellsol® Heptane
Trade Names Containing: Eastman® CDA-19; S&L Naphtha; Solvent 2205; Solvent 2304; Solvent 2340; Tecsol® 1; Tecsol® 3

n-Heptane. *See* Heptane
3-Heptanecarboxylic acid. *See* 2-Ethylhexoic acid
Heptane, 2,2,6,6-tetramethyl-4-methylene. *See* 2-(2,2-Dimethylpropyl)-4,4-dimethylpent-1-ene

Heptanoic acid, ester with 2,2-dimethyl-1,3-propanediol. *See* Neopentyl glycol diheptanoate
Heptanoic acid, octadecyl ester. *See* Stearyl heptanoate
1-Heptanol. *See* Heptyl alcohol

2-Heptanol

CAS 543-49-7; EINECS/ELINCS 208-844-3
UN 1987; FEMA 3288
Synonyms: Amyl methyl carbinol; n-Amyl methyl carbinol; Heptanol-2; Heptan-2-ol; s-Heptyl alcohol; 2-Hydroxyheptane; Methyl amyl carbinol; Methyl-n-amyl carbinol
Classification: Sec. aliphatic alcohol
Empirical: $C_7H_{16}O$
Formula: $CH_3CH_2CH_2CH_2CH_2CH(OH)CH_3$
Properties: Colorless liq.; mild, fresh lemon-like, grass-herbaceous odor; sol. in ethanol, ether, benzene; very sl. sol. in water; m.w. 116.20; sp.gr. 0.8193 (20 C); vapor pressure 1 mm Hg (14.6 C); b.p. 160-162 C; flash pt. (CC) 64 C; ref. index 1.421-1.422
Toxicology: LD50 (oral, rat) 2580 mg/kg, (skin, rabbit) 1780 mg/kg; mod. toxic by ing. and skin contact; severe eye and sl. skin irritant; may be absorbed thru skin; inh. may cause nose/throat irritation; high concs. of vapor may cause CNS effects, depression, headache, nausea, drowsiness; ing. may cause 'alcohol' intoxication symptoms; may cause severe lung damage, respiratory arrest, and death if aspirated into lungs; TSCA listed
Precaution: Flamm. exposed to heat and flame; mod. fire risk; incompat. with oxidizing agents (increases risk of fire and explosion)
NFPA: Health 0, Flammability 2, Reactivity 0
Storage: Store in cool, dry, well-ventilated area
Uses: Solvent for syn. resins; frothing agent in ore flotation; synthetic flavoring agent in foods; cosmetics ingred.
Regulatory: FEMA GRAS
Manuf./Distrib.: Advanced Synthesis Tech.; Aldrich; Degussa Flavors/ Alfrebro; Fluka; Penta Mfg.

Heptanol-2; Heptan-2-ol. *See* 2-Heptanol

3-Heptanol

CAS 589-82-2; EINECS/ELINCS 209-661-1
FEMA 3547
Synonyms: Butyl ethyl carbinol; Ethyl butyl carbinol; Heptan-3-ol; 3-Heptyl alcohol; 3-Hydroxyheptane
Classification: Sec. aliphatic alcohol
Empirical: $C_7H_{16}O$
Formula: $CH_3(CH_2)_3CH(OH)CH_2CH_3$
Properties: Colorless oily liq., herbaceous odor, pungent sl. bitter taste; sol. in alcohol, ether; sl. sol. in water; m.w. 116.20; sp.gr. 0.820 (20/4 C); vapor pressure 0.5 mm Hg (20 C); m.p. -70 C; b.p. 155-157 C; flash pt. (CC) 60 C; ref. index 1.422 (20 C)
Toxicology: LD50 (oral, rat) 1870 mg/kg, (skin, rabbit) 4360 mg/kg; mod. toxic by ing. and skin contact; skin and severe eye irritant; can be absorbed thru skin; inh. may cause nose/throat irritation; high concs. of vapor may cause CNS depression; ing. may cause 'alcohol' intoxication symptoms; may cause severe lung damage, respiratory arrest, possibly death if aspirated into lungs; TSCA listed
Precaution: Flamm. exposed to heat or flame; incompat. with oxidizing agents (increase fire/explosion risk)
Hazardous Decomp. Prods.: Heated to decomp., emits acrid smoke and fumes
NFPA: Health 0, Flammability 2, Reactivity 0
Storage: Store in cool, dry, well-ventilated area
Uses: Flotation frother; solvent, diluent in org. coatings; intermediate in chem. synthesis; synthetic flavoring agent in foods and pharmaceuticals; cosmetics ingred.
Regulatory: FDA 21CFR §172.515; FEMA GRAS
Manuf./Distrib.: Aldrich; Fluka; Penta Mfg.

Heptan-3-ol. *See* 3-Heptanol
n-Heptanol. *See* Heptyl alcohol
4-Heptanol, 2,6-dimethyl-. *See* 2,6-Dimethyl heptanol-4
2-Heptanone; Heptan-2-one. *See* Methyl n-amyl ketone

3-Heptanone

CAS 106-35-4; EINECS/ELINCS 203-388-1
FEMA 2545
Synonyms: Butyl ethyl ketone; n-Butyl ethyl ketone; Ethyl butyl ketone; Ethyl-n-butyl ketone; Heptan-3-one
Classification: Aliphatic ketone
Empirical: $C_7H_{14}O$
Formula: $CH_3(CH_2)_3COC_2H_5$
Properties: Colorless liq.; powerful green fatty fruity odor; melon banana flavor; sol. in alcohol, diethyl ether; misc. with ether; pract. insol. in water; m.w. 114.19; dens. 0.818 (20/4 C); m.p. -39 C; b.p. 145-148 C; flash pt. 38 C; ref. index 1.409 (20 C)
Toxicology: ACGIH TLV/TWA 50 ppm in air; LD50 (oral, rat) 2760 mg/kg; LCLo (inh., rat, 4 h) 2000 ppm; mod. toxic by ing. and inh.; skin and eye irritant; TSCA listed
Precaution: Flamm.; mod. fire risk; can react with oxidizing materials
Hazardous Decomp. Prods.: CO, CO_2
NFPA: Health 1, Flammability 2, Reactivity 0
Uses: Solvent for air-dried and baked finishes, for polyvinyl and nitrocellulose resins; synthetic flavoring agent in foods and pharmaceuticals; cosmetics ingred.
Features: Sweet, fruity flavor
Regulatory: FDA 21CFR §172.515; FEMA GRAS
Manuf./Distrib.: Aldrich; Alfa Chem; Fluka; Penta Mfg.; Sigma; Sithean; Spectrum Quality Prods.

Heptan-3-one. *See* 3-Heptanone

4-Heptanone

CAS 123-19-3; EINECS/ELINCS 204-608-9
UN 2710 (DOT); FEMA 2546
Synonyms: Butyrone; Dipropyl ketone; Heptan-4-one; Propyl ketone
Empirical: $C_7H_{14}O$
Formula: $(CH_3CH_2CH_2)_2CO$
Properties: Colorless liq., penetrating odor, burning taste; misc. with alcohol, ether; insol. in water; m.w. 114.19; dens. 0.814 (20/4 C); vapor pressure 5.2 mm Hg (20 C); m.p. -32.6 C; b.p. 142-144 C; flash pt. 49 C; ref. index 1.4073 (22 C); surf. tens. 25.2 dynes/cm
Toxicology: ACGIH TLV/TWA 50 ppm in air; LD50 (oral, rat) 3730 mg/kg, (skin, rabbit) 5660 mg/kg; LCLo (inh., rat, 4 h) 4000 ppm; mod. toxic by ing., inh., and skin contact; skin and eye irritant; TSCA listed
Precaution: Flamm.; mod. fire risk; can react with oxidizing materials
Hazardous Decomp. Prods.: Heated to decomp., emits acrid smoke and fumes
Uses: Solvent for nitrocellulose, raw and blown oils, resins, and polymers; lacquers; synthetic flavoring agent in foods and pharmaceuticals; pharmaceutical raw material; cosmetics ingred.
Features: Pineapple-like flavor
Regulatory: FDA 21CFR §172.515; FEMA GRAS
Manuf./Distrib.: Aldrich; Degussa Flavors/Alfrebro; Fluka; Moore Ingreds.; Sigma

Heptan-4-one. *See* 4-Heptanone
3,6,9,12,15,18,21-Heptaoxatricosane-1,23-diol. *See* PEG-8

Heptyl alcohol

CAS 111-70-6; EINECS/ELINCS 203-897-9
FEMA 2548
Synonyms: Alcohol C7; Enanthic alcohol; Enanthyl alcohol; 1-Heptanol; n-Heptanol; n-Heptyl alcohol; pri-Heptyl alcohol; Hydroxyheptane; 1-Hydroxyheptane
Classification: Primary aliphatic alcohol
Empirical: $C_7H_{16}O$
Formula: $CH_3(CH_2)_6OH$
Properties: Colorless liq.; pleasant citrus odor; sol. in oxygenated solvs.; misc. with alcohol, fixed oils, ether; sl. sol. in water @ 175 C; m.w. 116.23; dens. 0.824 (20/4 C); vapor pressure 1 mm Hg (42 C); m.p. -34.6 C; b.p. 175.8 C; flash pt. 70 C; ref. index 1.423-.1427; surf. tens. 26.2 dynes/cm
Toxicology: LD50 (oral, rat) 500 mg/kg, (skin, rabbit) 2 g/kg; mod. toxic by ing., skin contact; mildly toxic by inh.; probably an eye irritant; inh. can probably cause nose/throat irritation; high concs. may cause headache, nausea, drowsiness; ing. may cause 'alcohol' intoxication symp-

toms; may cause severe lung damage, respiratory failure, and death if aspirated into lungs; TSCA listed
Environmental: VOC; BOD5 1.63; ThOD 2.90
Precaution: Flamm. exposed to heat, flames; reactive with oxidizing materials
Hazardous Decomp. Prods.: Heated to decomp., emits acrid smoke and fumes
Storage: Store in cool, dry, well-ventilated area
Uses: Organic intermediate; solvent; extraction agent for phosphoric acid purification; cosmetic formulations; mfg. of plastics; synthetic flavoring agent in foods and pharmaceuticals; fragrance in cosmetics
Features: Citrus flavor
Regulatory: FDA 21CFR §172.515; FEMA GRAS
Manuf./Distrib.: Aldrich; Atofina; Fleurchem; Fluka; Grau Aromatics; Lluch Essence; Penta Mfg.; Sigma; Spectrum Quality Prods.; Suchema AG

3-Heptyl alcohol. *See* 3-Heptanol
n-Heptyl alcohol; pri-Heptyl alcohol. *See* Heptyl alcohol
s-Heptyl alcohol. *See* 2-Heptanol
γ-Heptyl butyrolactone. *See* γ-Undecalactone
Heptyl carbinol. *See* Caprylic alcohol

2-Heptylcyclopentanone

CAS 137-03-1; EINECS/ELINCS 205-273-1
Synonyms: Cyclopentanone, 2-heptyl-; 2-n-Heptyl cyclopentanone; α-Heptyl cyclopentanone
Classification: Organic compd.
Empirical: $C_{12}H_{22}O$
Properties: Colorless visc. liq.; jasmine-like odor; m.w. 182.34; dens. 0.89 kg/l (20 C); b.p. 140-143 C (24 mm)
Toxicology: LD50 (oral, rat) > 5 g/kg, (skin, rabbit) 5000 mg/kg; mildly toxic by ing. and skin contact; irritating to skin; TSCA listed
Hazardous Decomp. Prods.: Heated to decomp., emits acrid smoke and fumes
Uses: Fragrance, solvent in cosmetics
Manuf./Distrib.: Advanced Synthesis Tech.

2-n-Heptyl cyclopentanone; α-Heptyl cyclopentanone. *See* 2-Heptylcyclopentanone
5-Heptyldihydrofuran-2-one. *See* γ-Undecalactone
Heptyl hydride. *See* Heptane
Hexachlorbutadiene; Hexachlorobutadiene; 1,1,2,3,4,4-Hexachloro-1,3-butadiene. *See* Hexachloro-1,3-butadiene

Hexachloro-1,3-butadiene

CAS 87-68-3; EINECS/ELINCS 201-765-5
UN 2279
Synonyms: HCBD; Hexachlorbutadiene; Hexachlorobutadiene; 1,1,2,3,4,4-Hexachloro-1,3-butadiene; 1,3-Hexachlorobutadiene; Perchlorobutadiene
Empirical: C_4Cl_6
Formula: $Cl_2C{=}CClCCl{=}CCl_2$
Properties: Colorless to lt. amber heavy clear liq.; faint turpentine odor; sol. in DMSO, 95% ethanol, acetone, ether; sol. < 0.1 mg/ml in water; m.w. 260.76; sp.gr. 1.676 (15.5/15.5 C); vapor pressure 0.15 mm Hg (20 C); m.p. -19 to -22 C; b.p. 210-220 C; nonflamm.; ref. index 1.5550 (20 C)
Toxicology: ACGIH TLV/TWA 0.02 ppm (skin); LD50 (oral, rat) 90 mg/kg, (IP, rat) 175 mg/kg, (skin, rabbit) 1211 mg/kg; LCLo (inh., mouse, 4 h) 235 ppm; poison by ing., IP; mod. toxic by inh., skin contact; irritating to eyes, skin, mucous membranes, lungs; readily absorbed thru skin; corrosive; may cause liver/kidney damage, CNS effects, cardiac disease, chronic bronchitis, etc.; may be fatal; experimental carcinogen, reproductive effector; mutagenic data; target organs: liver, kidneys; TSCA listed
Precaution: Incompat. with strong oxidizing agents; reacts to form an explosive prod. with bromine perchlorate; rapidly dec. rubber on contact; keep away from food
Hazardous Decomp. Prods.: CO, CO_2, hydrogen chloride gas; heated to decomp., emits very toxic fumes of chlorine; emits toxic fumes under fire conditions
NFPA: Health 2, Flammability 1, Reactivity 1
Storage: Store in cool, dry place; keep tightly closed; protect from light
Uses: Solvent for elastomers, nat. rubber, syn. rubber, other polymers; heat transfer fluid; transformer and hydraulic fluid; wash liquor for removing C_4 and higher hydrocarbons; pesticide; insecticide; herbicide; algicide; chem. intermediate; recovery of chorine from 'sniff' gas in chlorine plants and of chlorine fluid for gyroscopes; intermediate in mfg. of rubber compds. and lubricants
Regulatory: SARA §313 reportable
Manuf./Distrib.: AccuStandard; Aldrich; Alfa Aesar; Fluka; Fluorochem USA; ICN Biomed. Research Prods.; Lancaster Synthesis; Oakwood Prods.; Pfaltz & Bauer; Riedel-deHaën; Sigma; Supelco; TCI Am.

1,3-Hexachlorobutadiene. *See* Hexachloro-1,3-butadiene
Hexacid C9. *See* Nonanoic acid
Hexadecafluoroheptane. *See* Perfluoroheptane
Hexadecane. *See* n-Hexadecane

n-Hexadecane

CAS 544-76-3; EINECS/ELINCS 208-878-9
Synonyms: Alkane C(16); Cetane; n-Cetane; Hexadecane
Classification: Nonaromatic hydrocarbon
Empirical: $C_{16}H_{34}$
Formula: $CH_3(CH_2)_{14}CH_3$
Properties: Colorless liq.; sol. in acetone; partly sol. in ether; sl. sol. in alcohol; insol. in water; m.w. 226.45; dens. 0.773 (20/4 C); vapor pressure 1 mm Hg (105 C); f.p. 18 C; b.p. 287 C; flash pt. 135 C; ref. index 1.4335 (20 C)
Toxicology: LDLo (IV, mouse) 9821 mg/kg; possible aspiration hazard; direct aspiration into the lungs may cause chemical pneumonitis, pulmonary edema, hemorrhaging; severe irritant; irritating to skin, eyes, respiratory system; TSCA listed
Environmental: In water, substance is rapidly degraded by marine bacteria
Precaution: Combustible; incompat. with strong oxidizing agents
Hazardous Decomp. Prods.: CO, CO_2
NFPA: Health 0, Flammability 1, Reactivity 0
Storage: Hygroscopic; refrigerate; store in a cool, dry place in tightly closed container; handle and store under nitrogen
Uses: Solvent; organic intermediate; spectrophotometry; gas chromatography; ignition standard for diesel fuels; cosmetics ingred.
Manuf./Distrib.: ABCR; Acros Org.; Aldrich; ChevronPhillips; Fisher Scientific; Fluka; Romil Ltd; Sigma
Trade Names Containing: Sasol C1316

Hexadecanoic acid, 2-ethylhexyl ester. *See* Octyl palmitate
Hexadecanoic acid, hexadecyl ester. *See* Cetyl palmitate
Hexadecanoic acid isopropyl ester. *See* Isopropyl palmitate
Hexadecanoic acid, methyl ester. *See* Methyl palmitate
Hexadecanoic acid, 1-methylethyl ester. *See* Isopropyl palmitate
Hexadecanoic acid, 2-methylpropyl ester. *See* Isobutyl palmitate
Hexadecanoic acid, 1,2,3-propanetriyl ester. *See* Tripalmitin
1-Hexadecanol. *See* Cetyl alcohol
1-Hexadecanol, acetate. *See* Cetyl acetate
1-Hexadecanol lactate. *See* Cetyl lactate
Hexadecyl acetate. *See* Cetyl acetate
Hexadecyl alcohol. *See* Cetyl alcohol
Hexadecyl 2-ethylhexanoate; n-Hexadecyl 2-ethylhexanoate. *See* Cetyl octanoate
n-Hexadecyl hexadecanoate. *See* Cetyl palmitate
n-Hexadecyl-2-hydroxypropanoate; Hexadecyl lactate; n-Hexadecyl lactate. *See* Cetyl lactate
1-Hexadecyloctadecanoate. *See* Cetyl stearate
Hexadecyl palmitate. *See* Cetyl palmitate
Hexadecyl stearate; n-Hexadecyl stearate. *See* Cetyl stearate
Hexaethylene glycol. *See* PEG-6
Hexahydroaniline; Hexahydrobenzenamine. *See* Cyclohexylamine
Hexahydrobenzene. *See* Cyclohexane
Hexahydrocresol. *See* Methylcyclohexanol
Hexahydrocumene. *See* Isopropylcyclohexane
Hexahydromethylphenol. *See* Methylcyclohexanol
Hexahydrophenol. *See* Cyclohexanol
Hexahydropyridine. *See* Piperidine
Hexahydrothymol. *See* Menthol
Hexahydrotoluene. *See* Methyl cyclohexane

Hexahydroxylene; Hexahydro-m-xylene. *See* 1,3-Dimethylcyclohexane
Hexahydro-o-xylene. *See* 1,2-Dimethylcyclohexane
Hexahydro-p-xylene. *See* 1,4-Dimethylcyclohexane
Hexalin. *See* Cyclohexanol
Hexalin acetate; Hexaline acetate. *See* Cyclohexyl acetate
Hexametapol. *See* Hempa
Hexamethylcyclotrisiloxane. *See* Cyclomethicone

Hexamethyldisilazane

CAS 999-97-3; EINECS/ELINCS 213-668-5
Synonyms: Bis (trimethylsilyl) amine; Di (trimethylsilyl) amine; Hexamethylsilazane; HMDS; OAP; 1,1,1-Trimethyl-N-(trimethylsilyl) silanamine
Empirical: $C_6H_{19}NSi_2$
Formula: $(CH_3)_3SiNHSi(CH_3)_3$
Properties: Colorless liq.; ammonia-like odor; sol. in acetone, benzene, ethyl ether, heptane, perchloroethylene; insol. in water, reacts slowly; m.w. 161.44; dens. 0.77; vapor pressure 79 mm (50 C); b.p. 125 C; flash pt. (CC) 27 C; ref. index 1.4057; dielec. const. 2.27 (1000 Hz)
Toxicology: PEL 6 mg/m^3; LD50 (oral, rat) 850 mg/kg, (skin, rabbit) 1350 mg/kg; LDLo (IP, mouse) 650 mg/kg; mod. toxic by ing. and IP route; corrosive; causes severe burns to eyes and irritation of skin; experimental tumorigen; questionable carcinogen; TSCA listed
Environmental: VOC
Precaution: Flamm.; dangerous fire hazard exposed to heat or flame; can react vigorously with oxidizers
Hazardous Decomp. Prods.: Heated to decomp., emits toxic fumes of NO_x
Uses: Chemical intermediate; release agent; lubricant; coupling agent; deactivator for chromatographic support materials; silylating agent in pharmaceuticals; adhesion promoter for photoresists on silicon (electronics); as reagent for temporary protection of reactive sites in mfg. of betalactum antibiotics; org. synthesis of carbohydrates, nucleosides, dyes, perfumes, agrochems.; solubilizer in polar/nonpolar solvs.; solvent
Manuf./Distrib.: AP Resources; Advanced Synthesis Tech.; Aldrich; Archimica Florida; Austin; Degussa AG/Aerosil & Silanes; Degussa AG; Dow Corning; FAR Research; Fluka; GFS; Gelest; General Chem.; Great Western; Integra; Jiangxi Liq. Ind. Chems.; R.W. Greeff; Reliance Silicones; Schweizerhall; Sigma; Sithean; Spectrum Quality Prods.; Wacker-Chemie GmbH

Hexamethyldisiloxane

CAS 107-46-0; EINECS/ELINCS 203-492-7
Synonyms: Bis (trimethylsilyl) ether; Bis (trimethylsilyl) oxide; Disiloxane, hexamethyl-; HMDSO; Oxybis (trimethylsilane); Silane, oxybis (trimethyl-
Classification: Linear siloxane
Empirical: $C_6H_{18}OSi_2$
Formula: $(CH_3)_3SiOSi(CH_3)_3$
Properties: Visc. liq.; sol. 930 ppm in water; m.w. 162.42; dens. 0.764; vapor pressure 55 mm (30 C); m.p. -68 C; b.p. 99-101 C; flash pt. -2 C; ref. index 1.3774; surf. tens. 15.9 dynes/cm; dielec. const. 2.17
Toxicology: LD50 (IP, mouse) 4500 mg/kg, (skin, rabbit) 16 ml/kg; LC50 (inh., rat, 4 h) 15,956 ppm; mildly toxic by ing., IP routes; skin/eye irritant; may cause mucous membrane/upper respiratory tract irritation; may be harmful by inh., ing., skin absorp.; prolonged/repeated exposure causes skin defatting, dermatitis; can cause dizziness; target organs: nerves; TSCA listed
Precaution: Flamm.; vapor may flash back; incompat. with strong acids/bases/oxidizers; container explosion may occur under fire conditions; may dec. to form flamm./explosive mixts. in air; generates formaldehyde in presence of air @ > 150 C
Hazardous Decomp. Prods.: CO, CO_2, silicon oxide; emits toxic fumes under fire conditions
HMIS: Health 1, Flammability 4, Reactivity 0
Storage: Store under nitrogen; hygroscopic; keep tightly closed; keep away from heat, sparks, open flame
Uses: Antifoam, emollient, antistat in cosmetics; silylation of carboxylic acids; endcapper for silicones; low polarity solvent; coupling agent; blocking agent; lubricant; release agent
Manuf./Distrib.: AP Resources; Advanced Synthesis Tech.; Aldrich; Fluka; Gelest; Jiangxi Liq. Ind. Chems.; Sigma
Trade Names: Chemsil DM 0.65
Trade Names Containing: Vertrel® XSi

Hexamethylene. *See* Cyclohexane

Hexamethylene diisocyanate

CAS 822-06-0; EINECS/ELINCS 212-485-8
UN 2281 (DOT)
Synonyms: 1,6-Diisocyanatohexane; HDI; Hexamethylene 1,6-diisocyanate; Isocyanic acid, hexamethylene ester
Empirical: $C_8H_{12}N_2O_2$
Formula: $OCN(CH_2)_6NCO$
Properties: Liq.; sol. in hexane, benzene, chloroform; m.w. 168.20; dens. 1.04 (25/15.5 C); vapor pressure 0.03 mm Hg; f.p. -67 C; b.p. 61 C (0.08 mm); flash pt. (OC)140 C; ref. index 1.453 (20 C); hydrolyzed by water
Toxicology: ACGIH TLV/TWA 0.005 ppm in air; LD50 (oral, rat) 738 mg/kg, (dermal, rabbit) 593 mg/kg; poison by inhalation, intravenous routes; moderately toxic by ingestion, skin contact; target organs: respiratory system, blood, lungs, spleen; TSCA listed
Environmental: VOC; ThOD 1.05
Precaution: Combustible
Uses: Chemical intermediate; solvent; prod. of light-stable, weather-resist. industrial coatings, elastomers, and foams for paints (metal finishing for automotive, aircraft, machines), coatings (wood, plastics, textiles, leathers, flooring), elastomeric casting compds. and mastics, adhesives; prod. of urethane-acrylate resins
Regulatory: HAP
Manuf./Distrib.: Aldrich; Fluka; Sigma

Hexamethylene 1,6-diisocyanate. *See* Hexamethylene diisocyanate
Hexamethylenediol. *See* Hexamethylene glycol

Hexamethylene glycol

CAS 629-11-8; EINECS/ELINCS 211-074-0
Synonyms: 1,6-Dihydroxyhexane; HDO; Hexamethylenediol; 1,6-Hexanediol; Hexane-1,6-diol; α-ω-Hexanediol; ω-Hexanediol
Empirical: $C_6H_{14}O_2$
Formula: $HO(CH_2)_6OH$
Properties: Clear colorless waxy cryst. solid; sol. in water, alcohol, oxygenated solvs.; sparingly sol. in hot ether; m.w. 118.17; dens. 0.953 (50 C); m.p. 42.8 C; b.p. 208 C; flash pt. 147 C; ref. index 1.4579 (25 C)
Toxicology: LD50 (oral, rat) 3.73 g/kg; mod. toxic by ing.; skin and eye irritant; TSCA listed
Precaution: Combustible when exposed to heat or flame; can react with oxidizers
Hazardous Decomp. Prods.: Heated to decomp., emits acrid smoke and fumes
NFPA: Health 1, Flammability 0, Reactivity 0
Uses: Intermediate in the prod. of nylon, hexamethylenediamine, polyesters, polyurethanes, paints, pharmaceuticals; polyurethane chain extender; in gasoline refining; as plasticizer; solvent; coupling agent; coil coating
Manuf./Distrib.: Aldrich; Allchem Ind.; BASF; Eastern Chem.; Fallek; Fluka; Penta Mfg.

1,1,5,5,5-Hexamethyl-3-phenyl-3-[(trimethylsilyl) oxy] trisiloxane. *See* Phenyl trimethicone
Hexamethylphosphoramide; Hexamethylphosphoric triamide. *See* Hempa
Hexamethylsilazane. *See* Hexamethyldisilazane
Hexanaphthene. *See* Cyclohexane

Hexane

CAS 110-54-3; EINECS/ELINCS 203-777-6
UN 1208 (DOT)
Synonyms: Alkane C6; n-Hexane; Hexyl hydride; Normal hexane
Classification: Aliphatic hydrocarbon
Empirical: C_6H_{14}
Formula: $CH_3(CH_2)_4CH_3$
Properties: Colorless volatile liq., faint petrol. odor; sol. in alcohol, ace-

tone, ether; misc. with oxygenated, chlorinated and hydrogenated solvs.; insol. in water; m.w. 86.20; dens. 0.65937 (20/4 C); m.p. -95 C; b.p. 67-70 C; flash pt. (CC) -21.7 C; ref. index 1.37486 (20 C)
Toxicology: ACGIH TLV/TWA 50 ppm; LD50 (oral, rat) 28.710 mg/kg; sl. toxic by ing., inh.; human systemic effects by inh. (nausea, headache, dizziness, drowsiness); irritating to respiratory tract; skin and eye irritant; narcotic in high concs.; massive exposures can cause unconsciousness and death; mutagenic data; TSCA listed
Environmental: VOC; ThOD 3.53
Precaution: Flamm.; very dangerous fire/explosion hazard exposed to heat or flame; can react vigorously with oxidizers
Hazardous Decomp. Prods.: Heated to decomp., emits acrid smoke and fumes
NFPA: Health 1, Flammability 3, Reactivity 0
Storage: Store in cool, dry, well-ventilated area, out of direct sunlight; limit quantities in use; avoid generating mists
Uses: Solvent, thinner for veg. oil and pharmaceutical extraction, compounding rubber cements, cosmetics, shoe industry; alcohol denaturant; paint diluent; polymerization reaction medium for SBR rubber; filling for thermometers; cleaning agent for textiles, furniture, leather; in adhesives; base for rubber cement; mfg. of tires, contact adhesives, paints, inks; laboratory reagent; extraction solvent for hops extract, spice oleoresins; processing desorbent; in food-pkg. adhesives; food processing aid (Japan); adjuvant in resinous/polymeric food-contact coatings; defoamer in food-contact paper coatings
Features: Fast-drying
Regulatory: FDA 21CFR §73.30, 73.295, 73.300, 73.315, 73.345, 73.615, 172.340, 172.560, 173.270, 175.105, 175.320, 176.200, 177.1200; HAP; Japan approved with restrictions
Manuf./Distrib.: AAE Chemie NV; AMC Chems.; Aldrich; Alfa Aesar; Amyl; Ashland; BP Chems. Ltd; Brenntag AG; Burdick & Jackson; ChevronPhillips; Coyne; ExxonMobil; Fisher Scientific; Fluka; GFS; Great Western; Houghton Chem.; Hukill; Humphrey; Integra; J.T. Baker; Mallinckrodt Baker; Penta Mfg.; Richard's Pacific Chem. Sourcing; Romil Ltd; Ruger; Sal Chem.; Shell; Sigma; Spectrum Quality Prods.; Stanchem; VWR Int'l.; Vopak USA
Trade Names: Calumet Hexane; Exxsol® Methylpentane Naphtha; Isohexane LNH
Trade Names Containing: Rubber Solvent; Rubber Solvent 332; SBP 40/65 LNH; SBP 60/95 LNH; SBP 80/95 LNH; SBP 80/110 LNH; Shellsol® BF; Shellsol® B HT; Shellsol® B HT60; S&L Naphtha; Solvent 2254; Solvent 2255; Solvent 2332; Solvent 2430; Solvent 2442; X-95 Solvent

n-Hexane. *See* Hexane

Hexane, branched and linear

CAS 92112-69-1; EINECS/ELINCS 295-570-2
Classification: Hydrocarbon
Definition: Complex mixt. of n-hexane and hexane isomers
Uses: Solvent
Manuf./Distrib.: Chimen Srl
Trade Names: Hydrosol Hexane

1,2-Hexane carbonate

CAS 66675-43-2
Properties: M.w. 144.17; dens. 1.057 g/ml; visc. 4.88 cp; b.p. 257 C; flash pt. (CC) 305 F; surf. tens. 34.19 dynes/cm
Uses: Solvent

Hexanedinitrile. *See* Adiponitrile
Hexanedioic acid, bis (2-ethylhexyl) ester. *See* Dioctyl adipate
Hexanedioic acid, bis (2-hexyl decyl) ester. *See* Diisocetyl adipate
Hexanedioic acid, bis (1-methylethyl) ester. *See* Diisopropyl adipate
Hexanedioic acid, bis (1-methylpentadecyl) ester. *See* Diisocetyl adipate
Hexanedioic acid, bis (2-methylpropyl) ester. *See* Diisobutyl adipate
Hexanedioic acid, dibutyl ester. *See* Dibutyl adipate
Hexanedioic acid, diisobutyl ester. *See* Diisobutyl adipate
Hexanedioic acid, dimethyl ester. *See* Dimethyl adipate
Hexanedioic acid dinitrile. *See* Adiponitrile
1,6-Hexanediol; Hexane-1,6-diol; α-ω-Hexanediol; ω-Hexanediol. *See* Hexamethylene glycol

2,5-Hexanedione

CAS 110-13-4; EINECS/ELINCS 203-738-3
UN 8027
Synonyms: Acetone, acetonyl-; Acetonyl acetone; Diacetonyl; 1,2-Diacetylethane; α,β-Diacetylethane; 2,5-Diketohexane
Classification: Ketone
Empirical: $C_6H_{10}O_2$
Formula: $CH_3COCH_2CH_2COCH_3$
Properties: Colorless to amber liq.; sweet aromatic odor; sol. in water, DMSO, 95% ethanol, acetone, ether, benzene, acid; m.w. 114.14; sp.gr. 0.9737 (20/4 C); vapor pressure 0.43 mm Hg (20 C); m.p. -5.5 C; b.p. 191 C; flash pt. 78 C; autoignition temp. 493 C; ref. index 1.4421 (20 C)
Toxicology: LD50 (oral, rat) 2700 mg/kg, (IP, rat) 800 mg/kg, (skin, guinea pig) 6422 mg/kg; TDLo (oral, male rat, 42 days) 4200 mg/kg; highly toxic by inh.; mod. toxic by ing.; may be harmful by skin absorp.; irritating to eyes, skin, respiratory system; readily absorbed thru skin; high concs. may cause narcosis; neurotoxic; may cause peripheral neuropathy, skin discoloration; chronic exposure may cause severe neurologic damage, testicular germinal cell change, blood changes, CNS impairment; TSCA listed
Environmental: ThOD 2.10
Precaution: Combustible; incompat. with strong oxidizing agents, strong bases, strong reducing agents
Hazardous Decomp. Prods.: Heated to decomp., emits toxic fumes of CO, CO_2
NFPA: Health 1, Flammability 2, Reactivity 0
Storage: Store in cool, dry place; keep tightly closed
Uses: Solvent for cellulose acetate, roll-coating inks, paints, lacquers, stains; intermediate for organic synthesis, pharmaceuticals, photographic chemicals; in electroplating; fragrance in cosmetics
Manuf./Distrib.: Acros Org.; Advanced Synthesis Tech.; Aldrich; Fluka; J.T. Baker; Lancaster Synthesis; Pfaltz & Bauer; Spectrum Quality Prods.; TCI Am.; Wacker Chems.

Hexane, 1-methoxy-. *See* Methyl hexyl ether
Hexane, 1,1´-oxybis-. *See* n-Hexyl ether

1,2,6-Hexanetriol

CAS 106-69-4; EINECS/ELINCS 203-424-6
Synonyms: Hexane-1,2,6-triol; 1,2,6-Trihydroxyhexane
Classification: Aliphatic alcohol
Empirical: $C_6H_{14}O_3$
Formula: $HOCH_2CH(OH)CH_2CH_2CH_2CH_2OH$
Properties: Water-wh. to pale yel. liq.; misc. with water and oxygenated solvs.; m.w. 134.20; dens. 1.1063; vapor pressure < 0.01 mm Hg (20 C); pour pt. -20 C; f.p. 32.8 C; b.p. 178 C (5 mm); flash pt. 380 F; ref. index 1.4760
Toxicology: LD50 (oral, rat) 15,500 mg/kg, (oral, mouse) 11,400 mg/kg; mildly toxic by ing.; eye, skin irritant; TSCA listed
Precaution: Combustible; can react with oxidizing materials
Hazardous Decomp. Prods.: Heated to decomp., emits acrid smoke and fumes
Storage: Hygroscopic liq.
Uses: Alkyd/polyester/polyurethane resin comonomer; softener for tech. papers, printing inks; moistening agent; solvent; humectant, solvent, visc. control agent in cosmetics; ointment base for pharmaceuticals; in food-pkg. adhesives
Regulatory: FDA 21CFR §175.105
Manuf./Distrib.: Aldrich; Fabrichem; Fluka; Indofine; Kira; Maypro Ind.; Penta Mfg.; Ruger; Sigma; Spectrum Quality Prods.; TCI Am.

Hexane-1,2,6-triol. *See* 1,2,6-Hexanetriol
Hexanoic acid, 2-ethyl-. *See* 2-Ethylhexoic acid
Hexanoic acid, 2-ethyl-, C15-18-alkyl esters. *See* Cetearyl octanoate
Hexanoic acid ethyl ester. *See* Ethyl caproate
Hexanoic acid, 2-ethyl-, hexadecyl ester. *See* Cetyl octanoate
Hexanoic acid, 2-ethyl-, octadecyl ester. *See* Stearyl octanoate
Hexanoic acid, methyl ester. *See* Methyl caproate
Hexanoic acid, 3,5,5-trimethyl-, isodecyl ester. *See* Isodecyl isononanoate
Hexanoic acid, 3,5,5-trimethyl-, isooctadecyl ester. *See* Isostearyl isononanoate
Hexanol; 1-Hexanol; Hexan-1-ol; n-Hexanol. *See* Hexyl alcohol

s-Hexanol. *See* 2-Ethyl-1-butanol
6-Hexanolactone. *See* ε-Caprolactone monomer
1-Hexanol, 2-ethyl. *See* 2-Ethylhexanol
1-Hexanol, 2-ethyl-, phosphate. *See* Trioctyl phosphate
1-Hexanol, 2-ethyl-, sebacate. *See* Dioctyl sebacate
1,6-Hexanolide; Hexan-6-olide. *See* ε-Caprolactone monomer
Hexanol, 3,5,5-trimethyl-. *See* 3,5,5-Trimethylhexan-1-ol
Hexanone. *See* Methyl isobutyl ketone

2-Hexanone

CAS 591-78-6; EINECS/ELINCS 209-731-1
UN 1224
Synonyms: Butyl methyl ketone; N-Butyl methyl ketone; Hexanone-2; Ketone, butyl methyl; MBK; Methyl butyl ketone; Methyl N-butyl ketone; MNBK; Propylacetone
Empirical: $C_6H_{12}O$
Formula: $CH_3(CH_2)_3COCH_3$
Properties: Colorless to lt. yel. clear liq.; odor resembling acetone but more pungent; sol. in DMSO, 95% ethanol, acetone, ether, acid; completely misc. with most org. solvs.; sol. 10-50 mg/ml in water; m.w. 100.16; sp.gr. 0.8113 (20/4 C); vapor pressure 2 mm Hg (20 C); m.p. -57 C; b.p. 127 C; flash pt. 23 C; autoignition temp. 423 C; ref. index 1.4005 (20 C); surf. tens. 25.5 dynes/cm; dielec. const. 14.56
Toxicology: ACGIH TLV/TWA 5 ppm (skin); LD50 (oral, rat) 2590 mg/kg, (skin, rabbit) 4800 mg/kg; LC560 (inh., rat, 4 h) 8000 ppm; mod. toxic by ing., inh., IP routes; prolonged inh. may cause serious damage to health; readily absorbed thru skin; irritating to eyes, mucous membranes, upper respiratory tract; narcotic in high concs.; may cause gastric irritation, lethargy, peripheral neuropathy, nausea, dizziness, headache, neurotoxic effects, etc.; risk of irreversible effects; can be fatal; TSCA listed
Environmental: VOC; ThOD 1.76
Precaution: Flamm.; LEL 1.2 in air; UEL 8% in air; forms explosive mixts. in air; vapor may flash back; container explosion may occur under fire conditions; incompat. with oxidizing agents, strong bases, reducing agents
Hazardous Decomp. Prods.: Heated to decomp., emits toxic fumes of CO, CO_2
NFPA: Health 2, Flammability 3, Reactivity 0
Storage: Store in explosion-proof refrigerator; keep tightly closed; keep away from oxidizing materials; store away from ignition sources
Uses: Solvent for lacquers, ink thinners, nitrocellulose, resins, oils, fats, waxes, paints, pharmaceuticals; in printing of plasticized fabrics; medium evaporating solvent for nitrocellulose acrylates, vinyl, and alkyd coatings
Manuf./Distrib.: ABCR; AccuStandard; Acros Org.; Advanced Synthesis Tech.; Aldrich; CU Chemie Uetikon; Degussa; Fluka; ICN Biomed. Research Prods.; Indofine; J.T. Baker; Lancaster Synthesis; Pfaltz & Bauer; Riedel-deHaën; Spectrum Quality Prods.; TCI Am.; Wacker Chems.

Hexanone-2. *See* 2-Hexanone

3-Hexanone

CAS 589-38-8; EINECS/ELINCS 209-645-4
FEMA 3290
Synonyms: Ethyl propyl ketone; Hexan-3-one
Empirical: $C_6H_{12}O$
Formula: $C_2H_5CO(CH_2)_2CH_3$
Properties: Colorless liq.; m.w. 100.16; dens. 0.815; b.p. 123 C; flash pt. (OC) 35 C; ref. index 1.4000
Toxicology: LD50 (oral, rat) 3360 mg/kg, (skin, rabbit) 3170 mg/kg; LCLo (inh., rat, 4 h) 4000 ppm; mod. toxic by ing., skin contact, subcut. routes; mildly toxic by inh.; strong irritant; skin and eye irritant; TSCA listed
Precaution: Flamm.; very dangerous fire hazard exposed to heat or flame; can react vigorously with oxidizers
Hazardous Decomp. Prods.: Heated to decomp., emits acrid smoke and fumes
Uses: Solvent; synthetic flavoring agent in foods, pharmaceuticals; cosmetics ingred.
Features: Grape-like flavor
Regulatory: FEMA GRAS
Manuf./Distrib.: Aldrich; Degussa Flavors/Alfrebro; Fluka

Hexan-3-one. *See* 3-Hexanone
3,6,9,12,15,18-Hexaoxaoctacosan-1-ol. *See* Deceth-6
Hexazane. *See* Piperidine
3-Hexene-1-ol. *See* 3-Hexenol

3-Hexenol

CAS 544-12-7; EINECS/ELINCS 208-860-0
Synonyms: 3-Hexene-1-ol; Hex-3-en-1-ol; 3-Hexen-1-ol; 3-Hexen-1-ol (cis, trans); (3E)-3-Hexen-1-ol
Empirical: $C_6H_{12}O$
Properties: M.w. 100.16
Uses: Solvent, fragrance in cosmetics
Manuf./Distrib.: Advanced BioTech; Fleurchem; Grau Aromatics; Sarcom; Sigma

3-Hexen-1-ol. *See* cis-3-Hexenol
Hex-3-en-1-ol; 3-Hexen-1-ol (cis, trans); (3E)-3-Hexen-1-ol. *See* 3-Hexenol
β-γ-Hexenol. *See* cis-3-Hexenol

cis-3-Hexenol

CAS 928-96-1; EINECS/ELINCS 213-192-8
FEMA 2563
Synonyms: β-γ-Hexenol; cis-3-Hexen-1-ol; cis-Hexen-3-ol-1; Leaf alcohol
Classification: Organic compd.
Definition: Commercial prod. is the cis-isomer
Empirical: $C_6H_{12}O$
Formula: $CH_3CH_2CH{=}CHCH_2CH_2OH$
Properties: Colorless liq., strong grassy-green odor; sol. in alcohol, propylene glycol, fixed oils, oxygenated solvs.; very sl. sol. in water; m.w. 100.18; dens. 0.846-0.850; b.p. 156-157 C; flash pt. 44 C; ref. index 1.43-1.441
Toxicology: LD50 (oral, rat) 4700 mg/kg, (IP, rat) 600 mg/kg; poison by IP route; mildly toxic by ing.; TSCA listed
Precaution: Flamm.
Hazardous Decomp. Prods.: Heated to decomp., emits acrid smoke and fumes
Uses: Solvent in cosmetics; fragrance ingred. in perfumery; synthetic flavoring agent in foods and pharmaceuticals
Regulatory: FDA 21CFR §172.515; FEMA GRAS
Manuf./Distrib.: Advanced Synthesis Tech.; Axxence Aromatic GmbH; Lluch Essence; Moore Ingreds.; Natural Advantage

cis-3-Hexen-1-ol; cis-Hexen-3-ol-1. *See* cis-3-Hexenol

Hexetidine

CAS 141-94-6; EINECS/ELINCS 205-513-5
Synonyms: Amino-1,3-bis (β-ethylhexyl)-5-methylhexahydropyrimidine; 5-Amino-1,3-bis (2-ethylhexyl)-5-methyl-hexhydropyrimidine; 1,3-Bis (2-ethylhexyl) hexahydro-5-methyl-5-pyrimidiamine; Substituted hexa hydropyrimidine
Classification: Heterocyclic organic compd.
Empirical: $C_{21}H_{45}N_3$
Properties: Liq.; sol. in methanol, benzene, petrol. ether; insol. in water; m.w. 339.61; dens. 0.860-0.87; b.p. 172-176 C (1 mm); ref. index 1.460-1.466
Precaution: Combustible
Uses: Antimicrobial; preservative, solvent in cosmetics; bactericide; algicide; antiseptic; antifungal agent for oral hygiene prods., pharmaceutical topicals, feminine hygiene prods.; antistat for synthetics; insect repellent; medicine (antifungal agent)
Use Level: 100-1000 ppm
Regulatory: USA EPA registered; Europe listed 0.1% max.
Manuf./Distrib.: ANGUS; Aldrich; Sigma

Hexone. *See* Methyl isobutyl ketone
n-Hexoxyethoxyethanol. *See* Diethylene glycol n-hexyl ether

Hexyl acetate

CAS 142-92-7; EINECS/ELINCS 205-572-7
FEMA 2565
Synonyms: Acetic acid hexyl ester; Dihexyl ether; 1-Hexyl acetate; n-Hexyl acetate; Hexyl alcohol, acetate; Hexyl ethanoate
Empirical: $C_8H_{16}O_2$

Formula: $CH_3COO(CH_2)_5CH_3$
Properties: Colorless to pale yel. oily liq., pleasant fruity odor, bittersweet taste; sol. in alcohol, ether; insol. in water; m.w. 144.22; dens. 0.873 (20/4 C); vapor pressure 5 mm Hg (45 C); m.p. -81 C; b.p. 167-169 C; flash pt. (PMCC) 57 C; ref. index 1.409 (20 C); surf. tens. 26 dynes/cm
Toxicology: LD50 (oral, rat) 42 g/kg, (skin, rabbit) > 5 g/kg; low toxicity by ing. and skin contact; skin and eye irritant; TSCA listed
Environmental: ThOD 2.44
Precaution: Flamm.
Hazardous Decomp. Prods.: Heated to decomp., emits acrid smoke and fumes
Uses: Solvent for cellulosic and acrylic coatings; synthetic flavoring agent in foods and pharmaceuticals; fragrance in cosmetics
Features: Sweet apple, cherry, pear-like flavor
Regulatory: FDA 21CFR §172.515; FEMA GRAS
Manuf./Distrib.: Advanced BioTech; Advanced Synthesis Tech.; Akzo Nobel; Aldrich; Axxence Aromatic GmbH; Bell Flavors & Fragrances; Berje; Degussa Flavors/Alfrebro; Dow; Elan; Fleurchem; Fluka; Grau Aromatics; Lluch Essence; Moore Ingreds.; Natural Advantage; Penta Mfg.; Prodasynth; Sigma

1-Hexyl acetate; n-Hexyl acetate. *See* Hexyl acetate
s-Hexyl acetate. *See* Methyl amyl acetate

Hexyl alcohol

CAS 111-27-3; 68526-79-4; EINECS/ELINCS 203-852-3
UN 2282 (DOT); FEMA 2567
Synonyms: Alcohol C6; Amylcarbinol; Caproyl alcohol; Hexanol; 1-Hexanol; Hexan-1-ol; n-Hexanol; n-Hexyl alcohol; 1-Hydroxyhexane; Pentylcarbinol
Classification: Primary aliphatic alcohol
Empirical: $C_6H_{14}O$
Formula: $CH_3(CH_2)_4CH_2OH$
Properties: Colorless liq., fruity odor, aromatic flavor; sol. in alcohol, ether, acetone, chloroform, benzene, oxygenated solvs.; sl. sol. in water; m.w. 102.20; dens. 0.8186; vapor pressure 1 mm Hg (24.4 C); f.p. -51.6 C; b.p. 157.2 C; flash pt. (TOC) 65 C; ref. index 1.1469; surf. tens. 25.73 dynes/cm
Toxicology: LD50 (oral, rat) 4.59 g/kg, (skin, rabbit) 3100 mg/kg; poison by IV route; mod. toxic by ing., skin contact; skin and severe eye irritant; can be absorbed thru skin; inh. may cause CNS effects, headache, nausea, dizziness, incoordination; ing. may cause 'alcohol' intoxication symptoms; may cause severe lung damage, respiratory arrest, possibly death if aspirated into lungs; TSCA listed
Environmental: VOC; BOD5 1.50; ThOD 2.82
Precaution: Flamm. or combustible liq.; reactive with oxidizing materials, strong acids
NFPA: Health 1, Flammability 2, Reactivity 0
Storage: Store in cool, dry area away from corrosives
Uses: Solvent for waterborne coatings, cosmetics; plasticizer; emulsion polymerization; intermediate for textile/leather finishing agents; mfg. of food additives, plasticizers, lubricants, biodeg. surfactants, bactericides, fabric softeners, defoamers, rubber chems.; antifoam; flotation frothing agent; fragrance ingred.; perfume esters; synthetic flavoring agent in foods and pharmaceuticals; antiseptics; in food pkg.
Regulatory: FDA 21CFR §172.515, 172.864, 178.3480; FEMA GRAS
Manuf./Distrib.: Advanced BioTech; Albemarle; Aldrich; Ashland; Axxence Aromatic GmbH; Degussa Flavors/Alfrebro; Elan; Fleurchem; Fluka; Grau Aromatics; Jarchem Ind.; Lluch Essence; Moore Ingreds.; Penta Mfg.; Prodasynth; Sasol N. Am.; Sigma; Spectrum Quality Prods.

n-Hexyl alcohol. *See* Hexyl alcohol
s-Hexyl alcohol. *See* 2-Ethyl-1-butanol
Hexyl alcohol, acetate. *See* Hexyl acetate

Hexyl benzoate

CAS 6789-88-4; EINECS/ELINCS 229-856-5
FEMA 3691
Synonyms: Benzoic acid, hexyl ester; 1-Hexyl benzoate; n-Hexyl benzoate
Empirical: $C_{13}H_{18}O_2$
Formula: $C_6H_5COO(CH_2)_5CH_3$
Properties: Colorless clear liq.; fruity odor; wood, balsamic flavor; sol. < 0.01 g/100 ml in water, > 100 mg/ml in 95% ethanol and in acetone; m.w. 206.28; dens. 0.979 g/cm³ (21.5 C); vapor pressure 32 mm Hg (22 C); b.p. 272 C; flash pt. > 93.3 C
Toxicology: LD50 (oral, rat) 12,300 mg/kg, (skin, rabbit) 21,000 mg/kg; mildly toxic by ing. and skin contact; primary irritant; mildly irritating to eyes and skin; TSCA listed
Hazardous Decomp. Prods.: Heated to decomp., emits acrid smoke and fumes
Storage: Store at ambient temps.
Uses: Perfume fixative; dye carrier; plasticizer; solvent for cellulose; fragrance in cosmetics; synthetic flavoring agent in foods
Use Level: 0.5% max. as benzoic acid in finished cosmetics
Regulatory: FEMA GRAS
Manuf./Distrib.: Aldrich; Bell Flavors & Fragrances; Haltermann GmbH; Lluch Essence; Penta Mfg.; Supelco

1-Hexyl benzoate; n-Hexyl benzoate. *See* Hexyl benzoate
Hexyl 'Carbitol'. *See* Diethylene glycol n-hexyl ether
n-Hexyl 'Cellosolve'. *See* Ethylene glycol hexyl ether

Hexyldecanol

CAS 2425-77-6; EINECS/ELINCS 219-370-1
Synonyms: 2-Hexyl-1-decanol
Classification: Aliphatic alcohol
Empirical: $C_{16}H_{34}O$
Formula: $CH_3(CH_2)_7CHCH_2OHCH_3(CH_2)_4CH_2$
Uses: Extender, solvent for dyes, fragrance oils; intermediate for lubricants, emulsifiers, metal processing compds., textiles; solubilizer for lipid-sol. actives for pharmaceutical creams, ointments, emulsions, sticks; humectant, solvent, emollient, carrier, dispersant, spreading agent for cosmetics
Trade Names: Jarcol™ I-16

2-Hexyl-1-decanol. *See* Hexyldecanol; Isocetyl alcohol

Hexyldecyl benzoate

CAS 163883-40-7
Uses: Emollient, lubricant, pigment wetting agent/dispersant, min. oil deoiler/degreaser, solvent, plasticizer for sunscreens, creams/lotions, topical hair/skin preps., bath/shower prods., deodorants, makeup; solubilizer for sunscreen actives
Manuf./Distrib.: C.P. Hall

Hexyldiglycol. *See* Diethylene glycol n-hexyl ether
Hexyldioctylphosphine oxide. *See* Dioctylhexylphosphine oxide

Hexylene glycol

CAS 107-41-5; EINECS/ELINCS 203-489-0
Synonyms: 2,4-Dihydroxy-2-methylpentane; 2-Methyl-2,4-pentanediol; 2-Methylpentane-2,4-diol; 4-Methyl-2,4-pentanediol; 2,4-Pentanediol, 2-methyl; α,α,α'-Trimethyltrimethyleneglycol
Classification: Aliphatic dihydric alcohol
Empirical: $C_6H_{14}O_2$
Formula: $(CH_3)_2COHCH_2CHOHCH_3$
Properties: Colorless liq., nearly odorless; sol. in water, aliphatic and aromatic hydrocarbons, alcohols, ethers, fatty acids; m.w. 118.18; dens. 0.9216 (20/4 C); m.p. -50 C; sets to glass; b.p. 198.3 C; flash pt. (TCC) 98 C; ref. index 1.4276 (20 C)
Toxicology: ACGIH TLV/CL 25 ppm; LD50 (oral, rat) 4.70 g/kg, (dermal, rabbit) 7900-12,000 mg/kg; very low oral toxicity; mod. toxic by IP route; mildly toxic by skin contact; skin and eye irritant; inh. may cause nose/respiratory irritation; very high concs. may cause headache, dizziness, nausea, incoordination; mutagenic data; TSCA listed
Precaution: Combustible; can react vigorously with oxidizing materials; incompat. with strong acids (decomp. occurs)
Hazardous Decomp. Prods.: Thermal decomp. prods.: CO, CO_2; heated to decomp., emits acrid smoke and fumes
NFPA: Health 0, Flammability 1, Reactivity 0
Storage: Store in cool, dry area away from heat and ignition sources; hygroscopic; avoid generating mists
Uses: Coupling agent, penetrant for textiles, cosmetics; ice inhibitor in carburetors; fuel and lubricant emulsifier; solvent for hydraulic fluids, printing inks, textile dyes, in petroleum refining, agrochems., water-based paints, cosmetics; cement additive; component of soaps, cos-

metics; intermediate in chem. prod.; defoamer; pharmaceutic aid, humectant, solvent; comonomer for unsat. polyester resins; in food-pkg. adhesives; in paper/paperboard in contact with dry food; defoamer in food-contact coatings and paper/paperboard; in closure-sealing gaskets for food containers; in food-contact textiles

Regulatory: FDA 21CFR §175.105, 176.180, 176.200, 176.210, 177.1210, 177.2800; FDA approved for topicals; NF compliance

Manuf./Distrib.: AMC Chems.; Aldrich; Allchem Ind.; Altair Ind.; Ashland; Atofina SA; Atofina SA; Avatar; BP Chems. Ltd; Brown; Celanese; Coyne; Dow; Duso; Fluka; G.J. Chem.; Great Western; Herdillia Chems. Ltd; Houghton Chem.; Innovative Processing; Penta Mfg.; Ruger; Sal Chem.; Shell; Sigma; Spectrum Quality Prods.; Stanchem; Universal Preserv-A-Chem; Vopak USA

Hexyl ethanoate. *See* Hexyl acetate
Hexyl ether. *See* n-Hexyl ether

n-Hexyl ether

CAS 112-58-3; EINECS/ELINCS 203-987-8

Synonyms: Bis (1-hexyl) ether; Dihexyl ether; Di-n-hexyl ether; Ether, dihexyl; Hexane, 1,1′-oxybis-; Hexyl ether; 1,1-Oxybishexane

Classification: Ether

Empirical: $C_{12}H_{26}O$

Properties: Colorless; mild odor; sol. in water; m.w. 186.38; sp.gr. 0.794; vapor pressure 0.0461 mm Hg; m.p. -43 C; b.p. 219-228 C; flash pt. 77 C; autoignition temp. 185 C; ref. index 1.402

Toxicology: LD50 (oral, rat) 30,900 mg/kg, (skin, rabbit) 6900 mg/kg; mildly toxic by ing., skin contact; severe skin irritant; irritating to eyes, respiratory system; may be harmful by inh., ing., skin absorp.; vapor/mist irritating to eyes, mucous membranes, upper respiratory tract; may cause altered sleep time; TSCA listed

Environmental: VOC; ThOD 1.98

Precaution: Combustible exposed to heat or flame; incompat. with strong oxidizing agents; can react with oxidizing materials

Hazardous Decomp. Prods.: CO, CO_2; emits toxic fumes under fire conditions

NFPA: Health 2, Flammability 2, Reactivity 0

Storage: Store in cool, dry place away from heat, open flame; keep tightly closed

Uses: Solvent

Manuf./Distrib.: ABCR; Aldrich; Alfa Aesar; Allchem Ind.; Arizona; Fluka; Hercules; J.H. Calo; Lancaster Synthesis; Penta Mfg.; Polarome Int'l.; S & S Chem.; T&R Chem.; TCI Am.; Welch, Holme & Clark

Hexylglycol. *See* Ethylene glycol hexyl ether
Hexyl hydride. *See* Hexane

Hexyl laurate

CAS 34316-64-8; EINECS/ELINCS 251-932-1

Synonyms: Dodecanoic acid, hexyl ester

Definition: Ester of hexyl alcohol and lauric acid

Empirical: $C_{18}H_{36}O_2$

Formula: $CH_3(CH_2)_{10}COO(CH_2)_5CH_3$

Toxicology: TSCA listed

Uses: Emulsifier, emollient, lubricant, vehicle, spreading agent, carrier, base, dispersant in pharmaceuticals and personal care prods.; lubricant for metalworking oils, syn. metalworking fluids, rolling, stamping and drawing oils, greases, motor oils, extrusion of ceramics; fatting agent in textile and leather auxs.; superfatting agent, solvent for hand cleaners

Manuf./Distrib.: A&E Connock

Hexyl methyl carbinol. *See* 2-Octanol
Hexyl methyl ether. *See* Methyl hexyl ether
Hexyl methyl ketone. *See* Methyl hexyl ketone
2-(Hexyloxy) ethanol. *See* Ethylene glycol hexyl ether
2-(2-Hexyloxyethoxy) ethanol. *See* Diethylene glycol n-hexyl ether
HFC 21. *See* Dichlorofluoromethane
HFC 23. *See* Trifluoromethane
HFC 43-10mee. *See* 2,3-Dihydrodecafluoropentane
HFC 245fa. *See* 1,1,1,3,3-Pentafluoropropane
High-erucic acid rapeseed oil. *See* Rapeseed (Brassica campestris) oil
High flash naphtha. *See* Stoddard solvent
High flash white spirit. *See* Naphtha, hydrodesulfurized heavy
HMDS. *See* Hexamethyldisilazane
HMDSO. *See* Hexamethyldisiloxane
HMPA; HMPT. *See* Hempa
HOCH. *See* Formaldehyde
Homomenthol. *See* Trimethylcyclohexanol
HPT. *See* Hempa
HQ. *See* Hydroquinone
HQMEE. *See* Hydroquinone monomethyl ether
Huile de ricini. *See* Castor (Ricinus communis) oil

Hybrid safflower (Carthamus tinctorius) oil

Synonyms: Carthamus tinctorius; Carthamus tinctorius hybrid oil; Carthamus tinctorius oil; Hybrid safflower oil; Safflower oil, hybrid

Definition: Oil derived from safflower seeds of a generic strain (hybrid safflower) which contains predominantly oleic acid triglyceride

Properties: Bland liq.; insol. in water; sol. in oil

Uses: Solubilizer, emollient, conditioner, gloss aid, solvent for cosmetics and pharmaceuticals and in cooking oils

Trade Names: NEOBEE® 18

Hybrid safflower oil. *See* Hybrid safflower (Carthamus tinctorius) oil
Hydantoin, 1,3-bis (hydroxymethyl)-5,5-dimethyl-. *See* DMDM hydantoin
Hydracrylonitrile. *See* Ethylene cyanohydrin
Hydrazine, methyl-; Hydrazomethane. *See* Methylhydrazine

Hydrobromic acid

CAS 10035-10-6; EINECS/ELINCS 233-113-0
UN 1048 (DOT); UN 1788 (DOT)

Synonyms: Anhydrous hydrobromic acid; Hydrobromic acid sol'n.; Hydrogen bromide; Hydrogen bromide anhydrous

Classification: Inorganic gas

Empirical: BrH

Formula: HBr

Properties: Colorless gas or pale yel. liq.; acrid odor (gas); misc. with water, alcohol; m.w. 80.92; dens. 3.50 g/l (0 C); m.p. -87 C; b.p.-66.5 C; nonflamm.

Toxicology: ACGIH TLV/CL 3 ppm; LD50 (IP, rat) 76 mg/kg; LC50 (inh., rat, 1 h) 2858 ppm; poison gas; mildly toxic by inh.; corrosive irritant to eyes, skin, mucous membranes; TSCA listed

Precaution: Corrosive gas; reacts violently with F_2, Fe_2O_3, NH_3, O_3

Hazardous Decomp. Prods.: Heated to decomp. or reacted with water or steam, emits toxic and corrosive fumes of Br^- and HBr

NFPA: Health 1, Flammability 4, Reactivity 4

Storage: Light-sensitive; protect from light

Uses: Analytical chemistry; solvent for ore minerals; mfg. of inorganic and some alkyl bromides; source of bromide; alkylation catalyst; for soldering flux; raw material for pure terephthalic acid, photographic chems., dyes, pharmaceuticals

Manuf./Distrib.: Advance Research Chems.; Advanced Synthesis Tech.; Albemarle; Aldrich; Alfa Aesar; Allan; Allchem Ind.; AmeriBrom; Assoc. Octel Co Ltd; Atofina; Bilt Chems. Ltd; Dastech Int'l.; Dead Sea Bromine; EM Ind.; Fisher Scientific; Fluka; GFS; Great Lakes; Hawks Chem. Co Ltd; Honeywell Perf. Polymers; Integra; J.T. Baker; Morre-Tec Ind.; Noveon Kalama; Ocean Chems.; Sigma; Spectrum Quality Prods.; VWR Int'l.; Vopak USA; Yixing Fangqiao East Chemical

Hydrobromic acid sol'n. *See* Hydrobromic acid
Hydrobromic ether. *See* Ethyl bromide
Hydrochloric ether. *See* Ethyl chloride
Hydrochlorofluorocarbon 22. *See* Chlorodifluoromethane
Hydrochlorofluorocarbon 142B. *See* 1,1,1-Chlorodifluoroethane
Hydrocyanic ether. *See* Propionitrile
Hydrodesulfurized heavy naphtha. *See* Naphtha, hydrodesulfurized heavy
Hydrodesulfurized kerosene. *See* Kerosene, hydrodesulfurized
Hydrodesulfurized naphtha, heavy. *See* Naphtha, hydrodesulfurized heavy
Hydrofluorocarbon 23. *See* Trifluoromethane
Hydrofuran. *See* Tetrahydrofuran

Hydrogenated canola oil

CAS 68334-28-1

Uses: Oil, filler fat, shortening for food prods., frying, coating, and spray applics.

Trade Names: Oleocal® C-105

Hydrogenated castor oil

CAS 8001-78-3; EINECS/ELINCS 232-292-2

Synonyms: Castor oil, hydrogenated; Castorwax; Opalwax

Definition: End prod. of controlled hydrogenation of castor oil, consisting mainly of the triglyceride of hydroxystearic acid

Properties: Wh. hard wax; very insol. in water and in the more common org. solvs.; m.w. ≈ 932; m.p. 86-88 C; iodine no. 5 max.; sapon. no. 176-182; hyd. no. 154-162

Toxicology: LD50 (oral, rat) > 10 g/kg; ingestion of large amts. may cause pelvic congestion; TSCA listed

Uses: Surfactant, emollient, emulsifier, visc. control agent for cosmetics, pharmaceuticals; in water-repellent coatings, candles, polishes, ointments, cosmetics; impregnant for paper, wood, cloth; lubricant, mold release in mfg. of formed plastics and rubber goods; thixotrope for aq. paints; wax for foods, pharmaceutical ointments; solvent for pharmaceutical injectables; tablet coatings; in food-pkg. adhesives; in food-contact coatings; in paper/paperboard in contact with aq./fatty foods; in closure-sealing gaskets for food containers; in cellophane for food pkg.; in food-contact crosslinked polyesters; lubricant for PVC food pkg.; food-contact textiles

Regulatory: FDA 21CFR §175.105, 175.300, 176.170, 176.210, 177.1200, 177.1210, 177.2420, 177.2800, 178.3280; NF compliance

Manuf./Distrib.: A&E Connock; Acme-Hardesty; Akzo Nobel; Allyson Enterprises; Alnor Oil; Amber Syn.; Arista Ind.; Ashland; Avatar; Biddle Sawyer; Brook-Chem; CasChem; Celanese; Charkit; Cognis; Collaborative Labs/Oleochems.; Condor; Cornelius Chem. Co. Ltd; Freund Chem. Co. Ltd; Hansotech; J.H. Calo; Kawaken Fine Chems.; Mutchler; NorChem; Ruger; Sea-Land; Seegott; Seppic; Southern Clay Prods.; Spectrum Quality Prods.; St. Lawrence; Süd-Chemie Inc; Univar E&E Ltd; Universal Preserv-A-Chem; Vilax; Vliengenthart BV; Vopak USA; Welch, Holme & Clark

Hydrogenated menhaden oil

CAS 93572-53-3; 68002-72-2; EINECS/ELINCS 297-485-6

Synonyms: Menhaden oil, hydrogenated

Definition: End prod. of controlled hydrogenation of menhaden oil

Properties: Wh. opaque solid, odorless; iodine no. 10 max.; sapon. no. 180-200

Uses: Emollient, solvent in cosmetics; textile lubricant; pharmaceutical intermediate; emulsifier; mold release agent; buffing compd.; edible fats and oils; in food-pkg. adhesives; in food-contact paper/paperboard; defoamer in food-contact paper/paperboard; in food-contact textiles

Regulatory: FDA 21CFR §175.105, 176.170, 176.210, 177.2800, 184.1472, 186.1551, GRAS

Hydrogenated pyrolysis gasoline. *See* Petroleum distillates, hydrotreated light, low-boiling

Hydrogen bromide; Hydrogen bromide anhydrous. *See* Hydrobromic acid

Hydrogen carboxylic acid. *See* Formic acid

Hydrogen dioxide. *See* Hydrogen peroxide

Hydrogen oxide. *See* Water

Hydrogen peroxide

CAS 7722-84-1; EINECS/ELINCS 231-765-0

UN 2014 (DOT); 2984

Synonyms: Hydrogen dioxide; Hydrogen peroxide solution; Hydroperoxide

Classification: Inorganic oxide

Empirical: H_2O_2

Formula: HOOH

Properties: Colorless visc. liq., cryst. solid at low temp.; sl. pungent irritating odor; bitter taste; sol. in ether, alcohol; misc. with water; dec. by many org. solvs.; m.w. 34.02; dens. (liq.) 1.450 g/cc (20 C); f.p. -0.41 C; b.p. 150.2 C; nonflamm.

Toxicology: ACGIH TLV/TWA 1 ppm; LD50 (oral, mouse) 2 g/kg, (skin, rat) 4060 mg/kg; mod. toxic by inh., ing., skin contact; corrosive irritant to skin, eyes, mucous membranes; tumorigenic; human mutagenic data

Environmental: LC50 (fish, 96 h) 16.4 mg/l; nonbioaccumulative

Precaution: Dangerous fire hazard by chem. reaction with flamm. materials; explosion hazard; strong oxidizer; incompat. with acids, bases, metals, reducing agents, org. materials

Hazardous Decomp. Prods.: Oxygen and heat

HMIS: Health 1, Flammability 0, Reactivity 1

Storage: Store in cool, ventilated, noncombustible area away from incompat. materials

Uses: Oxidizing, bleaching and deodorizing industrial/municipal water/wastewater, textiles, wood pulp, hair, fur, etc.; plasticizers; refining and cleaning metals; visc. control agent for starch and cellulose derivs.; solvent; intermediate for paints; flue gas treatment; sulfite removal in oil refineries, photo finishing, paper mills; removal of hypochlorite, nitrite, cyanide, formaldehyde, phenols, sulfides, SO_2, NO_x, Cl_2; foods; mining; electronics; oxidizing agent, antimicrobial in cosmetics; pharmaceuticals (antiseptic, foamer in dentifrices); resin solubilizer; in food-pkg. adhesives; sterilizer for polymeric food-contact surfaces

Regulatory: FDA 21CFR §133.113, 133.118, 133.133, 133.144, 133.195, 160.105, 160.145, 160.185, 172.723, 172.814, 172.892, 173.315, 175.105, 178.1005, 178.1005 (35% sol'n. max.), 178.1010, 184.1366, GRAS; BATF 27CFR §240.1051 (3 ppm max. in wine), 240.1051a (200 ppm max. in distilling materials); Japan restricted; FDA approved for topicals; BP, EP compliance (sol'n.)

Manuf./Distrib.: AP Resources; Aldrich; Arch Chems.; Ashland; Atofina; Belinka; Brenntag AG; Brown; Browning; Burlington Chem.; C.P. Hall; Coyne; Degussa AG/Health & Nutrition; Degussa; DuPont; EM Science; Eka Chems. AB; FMC Foret; FMC; Farleyway Chems. Ltd; Fluka; GFS; Galbraith Labs; General Chem.; Hansol; Harcros; Honeywell Perf. Polymers; Hutchins & Hutchins; ICC Ind.; Independent Chem.; Integra; J.T. Baker; Mallinckrodt Baker; O.C. Lugo; Oriental Chem. Ind.; Richard's Pacific Chem. Sourcing; Rochem Int'l.; Romil Ltd; Ruger; Sal Chem.; Schaefer Tech.; Seppic; Sigma; Solvay Interox; Solvay SA; Spectrum Quality Prods.; Stanchem; Thomas Scientific; Tri-K Ind.; Universal Preserv-A-Chem; VWR Int'l.; Veckridge; Voigt Global Distrib.; Vopak USA

Trade Names: Oxypure® 50%

Hydrogen peroxide solution. *See* Hydrogen peroxide

α-Hydro-ω-hydroxypoly [oxy (methyl-1,2-ethanediyl)]; α-Hydro-ω-hydroxypoly (oxypropylene). *See* Polypropylene glycol

Hydroperoxide. *See* Hydrogen peroxide

Hydrophenol. *See* Cyclohexanol

Hydroquinol. *See* Hydroquinone

Hydroquinone

CAS 123-31-9; EINECS/ELINCS 204-617-8

UN 2662 (DOT)

Synonyms: 1,4-Benzenediol; p-Benzenediol; Benzohydroquinone; Benzoquinone; p-Benzoquinone; Dihydroxybenzene; 1,4-Dihydroxybenzene; p-Dihydroxybenzene; HQ; Hydroquinol; p-Hydroquinone; 1,4-Hydroxybenzene; p-Hydroxyphenol; Hydroxyquinone; Paradioxybenzene; Pyrogentistic acid; Quinol

Classification: Aromatic organic compd.; aromatic alcohol

Empirical: $C_6H_6O_2$

Formula: $C_6H_4(OH)_2$

Properties: Wh., colorless, lt. tan, or gray cryst. solid; odorless; discolored by light and air; sol. in water, alcohol, ether, CCl_4, oxygenated solvs.; mod. sol. in acetone; sl. sol. in benzene; m.w. 110.11; dens. 1.330; vapor pressure 0.00018 mm Hg; m.p.170 C; b.p. 285 C; flash pt. (TCC) 165 C; autoignition temp. 516 C; pH weakly acidic (aq. sol'ns.)

Toxicology: ACGIH TLV/TWA 2 mg/m³; PEL-TWA 2 mg/m³;LD50 (oral, rat) 320 mg/kg; toxic by ing., inh., IP, IV, subcut. routes; human poison by ing.; harmful if absorbed thru skin or swallowed; human systemic effects by ing. (cyanosis, coma); lachrymator; irritant to eyes, skin, respiratory tract; severe human skin irritant; skin sensitizer; active allergen; human mutagenic data; experimental reproductive data; questionable carcinogen; TSCA listed

Environmental: ThOD 1.89

Precaution: Combustible; dust may be ignited by static discharge; incompat. with strong bases (violent reactions may occur) and strong oxidizers (increased fire/explosion risk); may corrode copper and brass

Hazardous Decomp. Prods.: CO, CO_2

NFPA: Health 2, Flammability 1, Reactivity 0

Storage: Sensitive to air and light; store in cool, dry, well-ventilated area, out of direct sunlight, away from heat and ignition sources, work areas, and incompat. materials

Uses: Photographic developer and reducer (not for color film); dye intermediate; polymerization inhibitor (vinyl acetate, acrylic monomers); polyarylate comonomer; stabilizer in paints and varnishes, motor fuels and oils; hair dyes; bleaching agents; antioxidant for fats and oils, syn. latexes, polyester resins, in suntan lotions; oxygen scavenger in water treatment; reagent for phosphate determination; solvent; in food-pkg. adhesives; monomer inhibitor in paper/paperboard in contact with aq./fatty foods; inhibitor in food-contact crosslinked polyesters

Regulatory: FDA 21CFR §175.105, 176.170, 177.2420; SARA §302a extremely hazardous, §313 reportable; CERCLA hazardous substance; HAP

Manuf./Distrib.: AMC Chems.; AMRESCO; Aceto; Aldrich; Alfa Chem; Allchem Ind.; Atomergic Chemetals; Charkit; Eastman; Fluka; GFS; Goodyear; Hawkins Chem.; Indofine; Integra; Kraeber GmbH; Mallinckrodt Baker; Penta Mfg.; Rhodia/Fine Orgs.; Rochem Int'l.; Ruger; San Yuan; Sigma; Sinochem Liaoning; Spectrum Quality Prods.; Stanchem; Synasia Fine Chem.; Triple Crown Am.; Universal Preserv-A-Chem; VWR Int'l.; Voigt Global Distrib.; Vopak USA; Whyte Chems. Ltd

o-Hydroquinone. *See* Pyrocatechol
p-Hydroquinone. *See* Hydroquinone
Hydroquinone methyl ether. *See* Hydroquinone monomethyl ether

Hydroquinone monomethyl ether

CAS 150-76-5; EINECS/ELINCS 205-769-8

Synonyms: p-Guaiacol; HA; HQMEE; Hydroquinone methyl ether; 4-Hydroxyanisole; p-Hydroxyanisole; MEHQ; 4-Methoxyphenol; p-Methoxyphenol; MME; Monomethyl ether of hydroquinone; Phenol, 4-methoxy-

Classification: Aromatic ether; substituted phenolic compd.

Definition: Monomethyl ether of hydroquinone

Empirical: $C_7H_8O_2$

Formula: $CH_3OC_6H_4OH$

Properties: Colorless to wh. waxy solid or cryst. powd.; caramel/phenol odor; sol. in water, benzene, ether, acetone, ethyl acetate, alcohol, oxygenated solvs.; m.w. 124.14; dens. 1.55 (20/20 C); vapor pressure < 0.01 mm Hg (20 C); m.p. 52.5 C; b.p. 243 C; flash pt. (COC) 132 C; pH 5.6 (aq. sol'n.)

Toxicology: ACGIH TLV/TWA 5 mg/m^3; PEL-TWA 5 mg/m^3; LD50 (oral, rat) 1600 mg/kg, (IP, mouse) 250 mg/kg, (dermal, rabbit) > 1000 mg/kg; poison by IP route; mildly toxic by ing.; severe skin and eye irritant; skin sensitizer; may cause allergic skin reaction; TSCA listed

Precaution: Combustible; may form explosive dust-air mixts.; incompat. with strong oxidizers (increases fire/explosion risk), strong bases, acid chlorides, and acid anhydrides (vigorous exothermic reactions possible)

Hazardous Decomp. Prods.: CO, CO_2; heated to decomp., emits acrid smoke and fumes

NFPA: , Flammability 1, Reactivity 0

Storage: Store in cool, dry, well-ventilated area, out of direct sunlight, away from heat and ignition sources and work areas; hygroscopic

Uses: Mfg. of antioxidants, pharmaceuticals, plasticizers, dyestuffs; stabilizer for chlorinated hydrocarbons, ethylcellulose, styrene monomer; UV inhibitor; inhibitor for acrylic and vinyl monomers and acrylonitrile; antioxidant for fats, oils, vitamins, and cosmetics; monomer inhibitor in mfg. of paper/paperboard in contact with aq./fatty foods; antioxidant in acrylic food pkg.

Regulatory: FDA 21CFR §176.170, 177.1010

Manuf./Distrib.: Aldrich; Alemark; Alfa Chem; Allchem Ind.; Arenol; Borregaard Synthesis US; Charkit; ChemDesign; ChemService; Eastman; Fluka; Kincaid Enterprises; Maypro Ind.; Mitsui & Co. USA; Monomer-Polymer & Dajac Labs; Penta Mfg.; Rhodia/Fine Orgs.; Sigma; Skymart Enterprises; Specialty Chem Prods.; Spectrum Quality Prods.

Trade Names Containing: SR 256

Hydrotreated distillate, light. *See* Petroleum distillates, hydrotreated light, low-boiling
Hydrotreated heavy naphtha. *See* Naphtha, hydrotreated heavy
Hydrotreated heavy naphthenic distillate. *See* Petroleum distillates, heavy hydrotreated naphthenic
Hydrotreated light distillate. *See* Petroleum distillates, hydrotreated light; Petroleum distillates, hydrotreated light, low-boiling
Hydrotreated light distillate (petroleum). *See* Petroleum distillates, hydrotreated light
Hydrotreated light naphtha (petroleum). *See* Naphtha, hydrotreated light
Hydrotreated light naphthenic distillate. *See* Naphthenic oil
Hydrotreated light paraffinic distillate. *See* Petroleum distillates, hydrotreated light paraffinic
Hydrotreated light petroleum distillates. *See* Petroleum distillates, hydrotreated light
Hydrotreated light straight run petroluem. *See* Naphtha, hydrotreated light
Hydrotreated middle distillate; Hydrotreated middle petroleum distillate. *See* Petroleum distillates, hydrotreated middle
Hydrotreated (mild) heavy naphthenic distillate. *See* Petroleum distillates, heavy hydrotreated naphthenic
Hydrotreated (mild) light naphthenic distillate; Hydrotreated (mild) light naphthenic distillates (petroleum). *See* Naphthenic oil
Hydrotreated naphtha. *See* Naphtha
4-Hydroxyanisole; p-Hydroxyanisole. *See* Hydroquinone monomethyl ether
Hydroxyanisole, butylated. *See* BHA
Hydroxybenzene. *See* Phenol
1,4-Hydroxybenzene. *See* Hydroquinone
5-Hydroxy-1,3-benzodioxole. *See* Sesamol
2-Hydroxybenzoic acid, methyl ester; o-Hydroxybenzoic acid, methyl ester. *See* Methyl salicylate
1-Hydroxy-2-butanamine. *See* 2-Aminobutanol
1-Hydroxybutane. *See* Butyl alcohol
2-Hydroxybutane. *See* 2-Butanol
4-Hydroxybutanoic acid lactone. *See* Butyrolactone
1-Hydroxy-s-butylamine. *See* 2-Aminobutanol
2-Hydroxy-3-butyoxypropane. *See* Butoxypropanol
γ-Hydroxybutyric acid cyclic ester; 4-Hydroxybutyric acid γ-lactone; γ-Hydroxybutyric acid lactone; γ-Hydroxybutyrolactone. *See* Butyrolactone
Hydroxycyclohexane. *See* Cyclohexanol
3-Hydroxy-p-cymene. *See* Thymol
1-Hydroxydecane. *See* Decyl alcohol
Hydroxydimethylbenzene. *See* Xylenol
1-Hydroxy-2,3-dimethylbenzene. *See* 2,3-Xylenol
1-Hydroxy-3,4-dimethylbenzene; 4-Hydroxy-1,2-dimethylbenzene. *See* 3,4-Xylenol
N-(2-Hydroxy-1,1-dimethylethyl) octadecanamide. *See* Stearamide AMP
4-Hydroxydiphenyldimethylmethane. *See* Cumyl phenol
1-Hydroxydodecane. *See* Lauryl alcohol
1-Hydroxyethanecarboxylic acid; 1-Hydroxyethane 1-carboxylic acid. *See* Lactic acid
2-Hydroxy-1-ethanethiol. *See* 2-Mercaptoethanol
Hydroxy ether. *See* Ethoxyethanol
2-[2-(2-(2-Hydroxyethoxy) ethoxy) ethoxy] ethanol. *See* PEG-4
β-(β-Hydroxyethoxy) ethylamine. *See* 2-(2-Aminoethoxy) ethanol
2-(2-Hydroxyethoxy) ethyl ester stearic acid; 2-(2-Hydroxyethoxy) ethyl stearate. *See* PEG-2 stearate
Hydroxyethyl acetamide; β-Hydroxyethylacetamide; N-(2-Hydroxyethyl) acetamide; N-β-Hydroxyethylacetamide. *See* Acetamide MEA
2-Hydroxyethyl acetate. *See* Ethylene glycol acetate
2-Hydroxyethylamine; β-Hydroxyethylamine. *See* Ethanolamine
2-(2-Hydroxyethylamino) ethanol. *See* Diethanolamine
β-Hydroxyethylbenzene. *See* Phenethyl alcohol
2-Hydroxyethyl chloride. *See* Ethylene chlorohydrin
β-Hydroxyethyldimethylamine. *See* Dimethylethanolamine
2-Hydroxyethyl ester stearic acid. *See* Glycol stearate
N-(2-Hydroxyethyl)-N-ethylaniline. *See* Phenylethylethanolamine
β-Hydroxyethyl isopropyl ether. *See* Ethylene glycol isopropyl ether
N-2-Hydroxyethyllactamide. *See* Lactamide MEA
Hydroxyethylmercaptan; 2-Hydroxyethylmercaptan. *See* 2-Mercaptoethanol
(2-Hydroxyethyl) methylamine; N-(2-Hydroxyethyl)-N-methylamine. *See* N-Methylethanolamine
2-Hydroxyethyl octadecanoate. *See* Glycol stearate
N-(2-Hydroxyethyl) peanut acid amide. *See* Peanutamide MEA

Hydroxyethyl PEI-1000

Definition: Multihydroxyethylated deriv. of PEI-1000

Uses: Solvent in cosmetics

Hydroxyethyl PEI-1500

Definition: Multihydroxyethylated deriv. of PEI-1500
Uses: Solvent in cosmetics

β-Hydroxyethyl phenyl ether. *See* Phenoxyethanol
1-(2-Hydroxyethyl)-2-pyrrolidinone; 1-(2-Hydroxyethyl) pyrrolidin-2-one. *See* N-Hydroxyethylpyrrolidone

N-Hydroxyethylpyrrolidone

CAS 3445-11-2; EINECS/ELINCS 222-359-4
Synonyms: 1-(2-Hydroxyethyl)-2-pyrrolidinone; 1-(2-Hydroxyethyl) pyrrolidin-2-one; N-(2-Hydroxyethyl)-2-pyrrolidone; 2-Pyrrolidinone, 1-(2-hydroxyethyl)-
Classification: Nonaromatic amide
Empirical: $C_6H_{11}NO_2$
Properties: M.w. 129.18; dens. 1.143; b.p. 140-142 C (3 mm); ref. index 1.4960
Toxicology: LD50 (oral, rat) 14,430 mg/kg; low toxicity by ing.; primary eye irritant; TSCA listed
Hazardous Decomp. Prods.: Heated to decomp., emits toxic vapors of NO_x
Uses: Solvent in coating and cleaning applics., microelectronics, photoresists, to clean silk screens, prep. of polymer membranes for gas scrubbing, anticlogging jet printing ink, aq. veterinary tetracycline antibiotic; reaction solvent for prep. of retinoid esters for skin treatment; mfg. of haloalkyl lactams; alkaline paint stripper; solvent plant growth regulator; intermediate in lube oil additives
Manuf./Distrib.: Aldrich; Fluka
Trade Names: HEP®

N-(2-Hydroxyethyl)-2-pyrrolidone. *See* N-Hydroxyethylpyrrolidone
β-Hydroxyethyl sulfide. *See* Thiodiglycol
Hydroxyheptane; 1-Hydroxyheptane. *See* Heptyl alcohol
2-Hydroxyheptane. *See* 2-Heptanol
3-Hydroxyheptane. *See* 3-Heptanol
1-Hydroxyhexane. *See* Hexyl alcohol
6-Hydroxyhexanoic acid lactone. *See* ε-Caprolactone monomer
2-Hydroxy-N-(2-hydroxyethyl) propanamide. *See* Lactamide MEA
α-Hydroxy-ω-hydroxypoly (oxy-1,2-ethanediyl). *See* Polyethylene glycol
4-Hydroxy-1-keto-4-methylpentane. *See* Diacetone alcohol
1-Hydroxy-2-methoxy-4-allylbenzene; 4-Hydroxy-3-methoxyallylbenzene; 1-Hydroxy-2-methoxy-4-prop-2-enylbenzene. *See* Eugenol
(Hydroxymethyl) benzene. *See* Benzyl alcohol
1-Hydroxy-2-methylbenzene. *See* o-Cresol
2-Hydroxy-2-methyl-3-butyne. *See* Methyl butynol
1-(Hydroxymethyl)-5,5-dimethyl hydantoin; 1-(Hydroxymethyl)-5,5-dimethyl-2,4-imidazolinedione. *See* MDM hydantoin
2-(Hydroxymethyl) ethanol. *See* Trimethylene glycol
2-Hydroxymethylfuran. *See* Furfuryl alcohol
3-Hydroxymethyl-n-heptan-4-ol; Hydroxymethyl-n-heptan-4-ol. *See* Ethyl hexanediol
3-Hydroxy-1-methyl-4-isopropylbenzene. *See* Thymol
3-(Hydroxymethyl)-pentan. *See* 2-Ethyl-1-butanol
4-Hydroxy-4-methyl-2-pentanone; 4-Hydroxy-4-methylpentanone-2; 4-Hydroxy-4-methyl pentan-2-one. *See* Diacetone alcohol
1-Hydroxymethylpropane. *See* Isobutyl alcohol
2-Hydroxymethyltetrahydropyran. *See* Tetrahydropyran-2-methanol
4-Hydroxynonanoic acid lactone; 4-Hydroxynonanoic acid, γ-lactone. *See* γ-Nonalactone
12-Hydroxyoctadecanoic acid, 2-ethylhexyl ester. *See* Octyl hydroxystearate
12-Hydroxyoctadecanoic acid, 1,2,3-propanetriyl ester. *See* Trihydroxystearin
12-Hydroxy-9-octadecenoic acid 2,3-dihydroxypropyl ester. *See* Glyceryl ricinoleate
12-Hydroxy-9-octadecenoic acid, monoester with 1,2-propanediol. *See* Propylene glycol ricinoleate
12-Hydroxy-9-octadecenoic acid, monoester with 1,2,3-propanetriol. *See* Glyceryl ricinoleate
1-Hydroxyoctane. *See* Caprylic alcohol
2-Hydroxyoctane. *See* 2-Octanol
(R)-12-Hydroxyoleic acid, monoester with 1,2-propanediol. *See* Propylene glycol ricinoleate
2-Hydroxypentane. *See* s-Amyl alcohol
4-Hydroxypentanoic acid lactone. *See* γ-Valerolactone
2-Hydroxyphenol; o-Hydroxyphenol. *See* Pyrocatechol
p-Hydroxyphenol. *See* Hydroquinone
1-Hydroxy-2-phenoxyethane. *See* Phenoxyethanol
1-Hydroxypropane. *See* Propyl alcohol
2-Hydroxypropane. *See* Isopropyl alcohol
2-Hydroxy-1,2,3-propanetricarboxylic acid, tributyl ester. *See* Tributyl citrate
2-Hydroxy-1,2,3-propanetricarboxylic acid, triethyl ester. *See* Triethyl citrate
2-Hydroxypropanoic acid. *See* Lactic acid
2-Hydroxypropanoic acid butyl ester. *See* Butyl lactate
2-Hydroxypropanoic acid hexadecyl ester. *See* Cetyl lactate
2-Hydroxypropanoic acid, isostearyl ester. *See* Isostearyl lactate
2-Hydroxypropanoic acid methyl ester. *See* Methyl lactate
2-Hydroxypropanol. *See* Propylene glycol
3-Hydroxypropene. *See* Allyl alcohol
2-Hydroxypropionic acid; α-Hydroxypropionic acid. *See* Lactic acid
2-Hydroxypropionic acid ethyl ester. *See* Ethyl lactate
(S)-(-)-2-Hydroxypropionic acid, ethyl ester. *See* Ethyl-(S)-lactate
2-Hydroxypropionitrile. *See* Lactonitrile
3-Hydroxypropionitrile. *See* Ethylene cyanohydrin
α-Hydroxypropionitrile. *See* Lactonitrile
2-(2-(2-Hydroxypropoxy) propoxy)-1-propanol; 3-[3-(3-Hydroxypropoxy) propoxy] propan-1-ol. *See* Tripropylene glycol
Hydroxypropyl alginate. *See* Propylene glycol alginate
2-Hydroxypropylamine lanolate. *See* Isopropanolamine lanolate
2-Hydroxypropyl dodecanoate; 2-Hydroxypropyl laurate. *See* Propylene glycol laurate
N-(2-Hydroxypropyl) peanut acid amide. *See* Peanutamide MIPA
1-Hydroxy-2-propyne; 3-Hydroxy-1-propyne. *See* Propargyl alcohol
Hydroxyquinone. *See* Hydroquinone

12-Hydroxystearyl alcohol

CAS 2726-73-0; EINECS/ELINCS 220-342-6
Synonyms: 1,12-Octadecanediol
Classification: Aliphatic diol
Empirical: $C_{18}H_{38}O_2$
Properties: Wh. flakes; m.p. 61-65 C; acid no. < 1; hyd. no. 345-360
Toxicology: May cause eye/skin irritation; may be harmful by ing., inh., skin absorp.
Precaution: Incompat. with strong oxidizing agents; avoid raising dust
Hazardous Decomp. Prods.: CO, CO_2; emits toxic fumes under fire conditions
Storage: Store in cool, dry place @ R.T.; keep tightly closed
Uses: Personal care prods.; flavors; fragrances; polyesters; acrylates; epoxy diluents; plastics; coatings; inks; photographic materials; textiles; leather; lubricants; polyurethanes; agric.; solvents; modifiers in alkyd resins; pharmaceuticals; fine chems.; plasticizers
Manuf./Distrib.: Sigma
Trade Names: Speziol C 18/2

14-Hydroxy-3,6,9,12-tetraoxatetradec-1-yl-9-octadecenoic acid. *See* PEG-5 oleate
Hydroxytoluene; α-Hydroxytoluene. *See* Benzyl alcohol
o-Hydroxytoluene. *See* o-Cresol
Hydroxytoluene, butylated. *See* BHT
2-Hydroxytriethylamine. *See* Diethylaminoethanol
4-Hydroxy-2,6,8-trimethylnonane. *See* 2,6,8-Trimethylnonyl-4-alcohol
4-Hydroxyundecanoic acid lactone; 5-Hydroxyundecanoic acid γ-lactone. *See* γ-Undecalactone
4-Hydroxyvaleric acid lactone. *See* γ-Valerolactone
Hydroxyxylene. *See* Xylenol
3-Hydroxy-o-xylene. *See* 2,3-Xylenol
4-Hydroxy-o-xylene. *See* 3,4-Xylenol
Hypnone. *See* Acetophenone

I

IBA. *See* Isobutyl alcohol
iBG. *See* Ethylene glycol isobutyl ether
IBIB. *See* Isobutyl isobutyrate
iBuH. *See* 2-Methylpropanal
Ice. *See* Water
Icosane. *See* Eicosane
IMI. *See* 1-Isopropyl-2-methyl imidazole
Imidazolidinedione, 1,3-bis (hydroxymethyl)-5,5-dimethyl-; 2,4-Imidazolidinedione, 1,3-bis (hydroxymethyl)-5,5-dimethyl-. *See* DMDM hydantoin
2,4-Imidazolidinedione, 5,5-dimethyl-. *See* DM hydantoin
2-Imidazolidinone, 1,3-dimethyl-. *See* Dimethyl imidazolidinone
2,2´-Iminobisethanol. *See* Diethanolamine
2,2´-Iminobisethylamine. *See* Diethylenetriamine
1,1´-Iminobis-2-propanol; 1,1´-Iminobis (propan-2-ol); 1,1´-Iminobis-2-propenol. *See* Diisopropanolamine
2,2´-Iminodiethanol. *See* Diethanolamine
2,2´-Iminodiethylamine. *See* Diethylenetriamine
1,1´-Iminodi-2-propanol; 1,1´-Iminodipropan-2-ol. *See* Diisopropanolamine
IMS. *See* Alcohol
Inactive limonene. *See* dl-Limonene
Indigenous peanut oil. *See* Peanut (Arachis hypogaea) oil
Industrial methylated spirit. *See* Alcohol
IO. *See* 3-Isopropyl-2-oxazolidinone
Iode. *See* Iodine

Iodine

CAS 7553-56-2; EINECS/ELINCS 231-442-4
UN 1759 (DOT)
Synonyms: Elemental iodine; Iode
Classification: Nonmetallic halogen element
Empirical: I_2
Properties: Bluish-black scales or plates, metallic luster, char. odor, sharp acrid taste; sol. (g/100 g): 14.09 benzene, 16.47 CS_2, 21.43 ethanol; sol. in chloroform, CCl_4, glac. acetic acid, alkaline iodide sol'n., oxygenated and chlorinated solvs.; very sl. sol. in water; at.wt. 126.9045; m.w. 253.81; dens. 4.98; vapor pressure 1 mm Hg (38.7 C), 0.030 mm Hg (solid, 0 C); m.p. 113.6 C; b.p. 185 C
Toxicology: LD50 (oral, rat) 14 g/kg, (subcut., rat) 10,500 mg/kg; very toxic material; human poison by ing.; experimental poison by IV and subcut. routes; mod. toxic by inh.; human ing. of large quantities causes abdominal pain, nausea, vomiting, diarrhea, asthma, anaphylactic shock; 2-4 g have been fatal; experimental reproductive effector; mutagenic data; TSCA listed
Precaution: DOT: Corrosive material; incompat. with alkaloids, starch, tannins; reacts vigorously with reducing materials; explosive reactions possible
Hazardous Decomp. Prods.: Heated to decomp., emits toxic fumes of I^- and various iodine compds.
Uses: Dyes; alkylation/condensation/sulfonation/chlorination catalyst; Ziegler catalyst component; iodides; iodates; disinfectant; swimming pool disinfection; antiseptics; germicides; stabilizers; photographic film; water treatment; analytical reagent for fats, oils; ring iodination reagent; dietary supplement; food and feed additive; pharmaceuticals; medicinal soaps; diagnostic, medical, and radiation treatment; x-ray contrast media; in denatured alcohol
Regulatory: BP, EP compliance
Manuf./Distrib.: Aldrich; Alfa Aesar; Alfa Chem; Allan; Asiamerica Int'l.; Atomergic Chemetals; Avrachem AG; Bio-Lab; Buckton Page Ltd; Cerac; Charkit; DSM Andeno BV; EM Science; Fabrichem; Fluka; George Uhe; Hasnil; Hawkins Chem.; Helm NY; Integra; J.T. Baker; Mallinckrodt Baker; Mutchler; Napp Tech.; Nippoh Chems.; RIA Int'l.; Richman; Robeco; Rochem Int'l.; Ruger; Sea-Land; Sigma; Spectrum Quality Prods.; Synasia Fine Chem.; Thomas Scientific; Tomen Am.; Universal Preserv-A-Chem; Yanagishima Pharmaceutical
Trade Names Containing: Punctilious® SDA 25A-1 190 Proof; Punctilious® SDA 25A-2 190 Proof

Iodopropynyl butylcarbamate

CAS 55406-53-6; EINECS/ELINCS 259-627-5
Synonyms: Butyl-3-iodo-2-propynylcarbamate; Carbamic acid, butyl-3-iodo-2-propynyl ester; 3-Iodopropynylbutylcarbamate; 3-Iodo-2-propynyl butyl carbamate; 3-Iodo-2-propynyl-N-butyl carbamate; IPBC
Classification: Organic compd.
Empirical: $C_8H_{12}INO_2$
Formula: $IC_2CH_2OCONH(CH_2)_3CH_3$
Properties: Wh. to off-wh. powd.; m.w. 281.1; m.p. 65-67 C
Toxicology: TSCA listed
Uses: Fungicide, mildewcide, wood preservative in oil-based and latex paints, wood prods., cutting oils, textiles, paper coatings, inks, plastics, adhesives; antifungal preservative in food-pkg. adhesives
Regulatory: FDA 21CFR §175.105; USA EPA registered; Japan approved; Europe listed
Manuf./Distrib.: Arch Chems.; Jan Dekker BV; Troy
Trade Names Containing: Dantogard® Plus Liq.

3-Iodopropynylbutylcarbamate; 3-Iodo-2-propynyl butyl carbamate; 3-Iodo-2-propynyl-N-butyl carbamate. *See* Iodopropynyl butylcarbamate
IPA. *See* Isopropyl alcohol
IPBC. *See* Iodopropynyl butylcarbamate
IPE. *See* Ethylene glycol isopropyl ether
IPG. *See* Isopentyldiol; Ethylene glycol isopropyl ether
IPL. *See* Isopropyl lanolate
IPM. *See* Isopropyl myristate
IPP. *See* Isopropyl palmitate
Isoacetophorone. *See* Isophorone
Isoalkanes, C7-8. *See* C7-8 isoparaffin
Isoalkanes, C7-10. *See* C7-10 isoparaffin
Isoalkanes, C8-9. *See* C8-9 isoparaffin
Isoalkanes, C9-10. *See* C9-10 isoparaffin
Isoalkanes, C9-11. *See* C9-11 isoparaffin
Isoalkanes, C9-12. *See* C9-12 isoparaffin
Isoalkanes, C9-13. *See* C9-13 isoparaffin
Isoalkanes, C9-14. *See* C9-14 isoparaffin
Isoalkanes, C9-15. *See* C9-15 isoparaffin
Isoalkanes, C10-11. *See* C10-11 isoparaffin
Isoalkanes, C10-13. *See* C10-13 isoparaffin
Isoalkanes, C11-12. *See* C11-12 isoparaffin
Isoalkanes, C11-13. *See* C11-13 isoparaffin

Isoalkanes, C11-16. *See* C11-16 isoparaffin
Isoalkanes, C12-14. *See* C12-14 isoparaffin
Isoalkanes, C13-14. *See* C13-14 isoparaffin
Isoalkanes, C13-16. *See* C13-16 isoparaffin
Isoalkanes, C13-17. *See* C13-17 isoparaffin
Isoalkanes, C20-40. *See* C20-40 isoparaffin

Isoamyl acetate

CAS 123-92-2; EINECS/ELINCS 204-662-3
UN 1104; FEMA 2055
Synonyms: Acetic acid isopentyl ester; Banana oil; Isoamyl ethanoate; Isopentyl acetate; Isopentyl alcohol acetate; 3-Methyl-1-butanol acetate; 3-Methylbutyl acetate; 3-Methyl-1-butyl acetate; β-Methylbutyl acetate; 3-Methylbutyl ethanoate; Pear oil
Classification: Carboxylic acid ester
Definition: Ester of isoamyl alcohol and acetic acid
Empirical: $C_7H_{14}O_2$
Formula: $CH_3COOCH_2CH_2CH(CH_3)_2$
Properties: Colorless liq., pear- or banana-like odor and taste; sl. sol. in water; misc. with alcohol, ether, ethyl acetate, fixed oils; insol. in glycerin, propylene glycol; m.w. 130.21; sp.gr. 0.876 (20 C); m.p. -78.5 C; b.p. 142 C; flash pt. (CC) 25 C; ref. index 1.400
Toxicology: ACGIH TLV/TWA 100 ppm; LD50 (oral, rat) 16,600 mg/kg; mildly toxic by ing., inh., subcut. routes; eye and mild skin irritant; inh. can cause nose/throat irritation, CNS depression; ing. may cause mouth/throat irritation, increased heart rate, CNS depression; severe exposure may cause unconsciousness; TSCA listed
Precaution: Highly flamm.; dangerous fire hazard exposed to heat or flame; can react vigorously with reducing materials; also incompat. with oxidizing agents (increases fire/explosion risk), strong acids, strong bases (decomp. can occur)
Hazardous Decomp. Prods.: Heated to decomp., emits acrid smoke and fumes
NFPA: Health 1, Flammability 3, Reactivity 0
Storage: Store in cool, dry, well-ventilated area, out of direct sunlight; ground drums and bond transfer containers
Uses: Synthetic flavoring agent and fragrance in foods, pharmaceuticals; pear flavor (alcohol sol'n.) in min. waters, syrups; fragrance in perfumes; photographic films; celluloid cements; waterproof varnishes; masking agent for undesirable odors; bronzing liqs.; metallic paints; solvent for oil colors, tannins, nitrocellulose, lacquers, celluloid, camphor, resins, waxes, cosmetics; perfuming shoe polish; mfg. artificial silk, leather, pearls; dyeing and finishing textiles
Features: Fresh fruity fragrance and flavoring, pear- or banana-like on dilution
Regulatory: FDA 21CFR §172.515; FEMA GRAS; Japan approved as flavoring
Manuf./Distrib.: Aceto; Advanced BioTech; Advanced Synthesis Tech.; Aldrich; Axxence Aromatic GmbH; Bell Flavors & Fragrances; Berje; Chr. Hansen Inc; Danisco Cultor; Degussa Flavors/Alfrebro; Elan; Epochem; Fleurchem; Fluka; Grau Aromatics; Haltermann GmbH; Haltermann Prods. UK; Lluch Essence; Mallinckrodt Baker; MelChem; Moore Ingreds.; Northwestern Flavors; Oxiteno; Penta Mfg.; Prodasynth; Rhodia/Fine Orgs.; Sarcom; Sigma; Spectrum Quality Prods.; United Pharma Ltd; VWR Int'l.; Xinchem

Isoamyl alcohol

CAS 123-51-3; EINECS/ELINCS 204-633-5
UN 1105; FEMA 2057
Synonyms: Isoamylol; Isobutyl carbinol; Isopentanol; Isopentyl alcohol; 2-Methyl-4-butanol; 3-Methylbutanol; 3-Methyl-1-butanol; 3-Methylbutan-1-ol; Primary isoamyl alcohol
Classification: Primary aliphatic alcohol
Empirical: $C_5H_{12}O$
Formula: $(CH_3)_2CHCH_2CH_2OH$
Properties: Colorless liq.; disagreeable sweet odor; pungent taste; sl. sol. in water; sol. in alcohol, ether, acetone, many org. solvs.; m.w. 88.15; sp.gr. 0.813 (15/4 C); f.p. -117.2 C; b.p. 132 C; flash pt. (CC) 42.7 C; ref. index 1.407 (20 C)
Toxicology: ACGIH TLV/TWA 100 ppm; STEL 125 ppm; LD50 (oral, rat) 1300 mg/kg, (IV, mouse) 234 mg/kg; highly toxic; poison by IP, IV routes; mod. toxic by ing., skin contact; ing. has caused human deaths from respiratory failure; may cause heart, lung, kidney damage; CNS depressant; vapor exposure has caused marked irritation of eyes, nose, throat, and headache; may cause sl. to mild skin irritation; questionable carcinogen; TSCA listed
Precaution: DOT: Flamm. liq.; explosive limits in air 1.2-9%; incompat. with strong oxidizing agents, reducing agents, strong acids, alkalis, hydrogen trisulfide; potential explosive hazard
Hazardous Decomp. Prods.: None known
NFPA: Health 1, Flammability 2, Reactivity 0
Storage: Store in cool, dry, well-ventilated area away from heat/ignition sources
Uses: Synthetic flavoring agent and fragrance in foods and pharmaceuticals; fragrance in cosmetics; fine chemicals; extraction solvent; extraction agent for phosphoric acid purification; solvent for gums, resins, lacquers, pharmaceuticals; oil additives; intermediates; photographic chemicals; herbicides; plasticizers; mining chemicals; organic synthesis; microscopy
Regulatory: FDA 21CFR §172.515; FEMA GRAS
Manuf./Distrib.: AMRESCO; Advanced BioTech; Advanced Synthesis Tech.; Aldrich; Allchem Ind.; Axxence Aromatic GmbH; BASF; Biddle Sawyer; CPS; Celanese; Chr. Hansen Inc; Coyne; Danisco Cultor; Degussa Flavors/Alfrebro; Elan; Fleurchem; Fluka; GFS; Grau Aromatics; Haltermann GmbH; Haltermann Prods. UK; Integra; Jarchem Ind.; Kuraray; Lluch Essence; MelChem; Moore Ingreds.; Osaka Org. Chem. Ind.; Oxiteno; Penta Mfg.; Prodasynth; Sigma; Sithean; Spectrum Quality Prods.; United Pharma Ltd; VWR Int'l.; Wutong Aroma; Xinchem

Isoamyl butanoate; Isoamyl butylate. *See* Isoamyl butyrate

Isoamyl butyrate

CAS 106-27-4; EINECS/ELINCS 203-380-8
FEMA 2060
Synonyms: Amyl butyrate; Butanoic acid, 3-methylbutyl ester; Butyric acid isoamyl ester; Isoamyl butanoate; Isoamyl butylate; Isoamyl n-butyrate; Isopentyl butanoate; Isopentyl butyrate; 3-Methylbutyl butyrate
Definition: Ester of isoamyl alcohol and butyric acid
Empirical: $C_9H_{18}O_2$
Formula: $CH_3(CH_2)_2COOCH_2CH_2CH(CH_3)_2$
Properties: Colorless liq.; strong fruity pear-like odor; sol. in alcohol, fixed oils, most org. solvs.; sl. sol. in water; insol. in glycerin, propylene glycol; m.w. 158.24; dens. 0.860; b.p. 184-185 C; flash pt. 149 F; ref. index 1.409-1.414
Toxicology: LD50 (oral, rat) > 5 g/kg, (skin, rabbit) > 5 g/kg; primary skin irritant; TSCA listed
Precaution: Combustible liq.
Hazardous Decomp. Prods.: Heated to decomp., emits acrid smoke and irritating fumes
Uses: Synthetic flavoring agent and fragrance in foods and pharmaceuticals; mfg. artificial rum and fruit essences; fragrance in cosmetics; solvent and plasticizer for cellulose acetate
Features: Sweet, fruity apricot, banana, pineapple fragrance and flavor
Regulatory: FDA 21CFR §172.515; FEMA GRAS; Japan approved as flavoring
Manuf./Distrib.: A&E Connock; Advanced BioTech; Advanced Synthesis Tech.; Aldrich; Axxence Aromatic GmbH; BASF; Bell Flavors & Fragrances; Degussa Flavors/Alfrebro; Elan; Epochem; Fleurchem; Grau Aromatics; Moore Ingreds.; United Pharma Ltd; Xinchem

Isoamyl n-butyrate. *See* Isoamyl butyrate

Isoamyl chloride

CAS 107-84-6
Synonyms: 1-Chloro-3-methylbutane
Definition: Mixture, usually also containing normal amyl chloride
Empirical: $C_5H_{11}Cl$
Formula: $(CH_3)_2CHCH_2CH_2Cl$
Properties: Colorless or sl. yel. liq.; sl. sol. in water; sol. in alcohol and ether; m.w. 106.60; dens. 0.893; m.p. -104 C; b.p. 99.7 C (758 mm); flash pt. 16 C; ref. index 1.410
Precaution: Combustible; reaction with divalent metals may form reactive products
Hazardous Decomp. Prods.: Heated to decomp. emits toxic fumes of Cl^-

Uses: Solvent (nitrocellulose, varnishes, lacquers, neoprene), rotogravure inks, soil fumigation, organic compds.

Isoamyl ethanoate. *See* Isoamyl acetate

1-Isoamyl glycerol ether

CAS 627-92-9
Synonyms: Glyceryl α-monoisoamyl ether; 3-Isopentyloxy-1,2-propanediol; 3-(3-Methylbutoxy)-1,2-propanediol; 1,2-Propanediol, 3-(3-methylbutoxy)-
Classification: Glycol ether
Empirical: $C_8H_{18}O_3$
Properties: M.w. 162.23
Environmental: ThOD 2.17
Uses: Solvent
Regulatory: HAP

Isoamylhydride. *See* Isopentane
Isoamyl methyl ketone. *See* Methyl isoamyl ketone
Isoamylol. *See* Isoamyl alcohol
Isoarachidyl alcohol. *See* Octyldodecanol
Isobutanal. *See* 2-Methylpropanal

Isobutane

CAS 75-28-5; EINECS/ELINCS 200-857-2
UN 1076 (DOT); UN 1969 (DOT); INS943b; E943b
Synonyms: 2-Methylpropane; Trimethylmethane
Classification: Hydrocarbon gas
Definition: A constituent of natural gas and illuminating gas
Empirical: C_4H_{10}
Formula: $CH(CH_3)_3$
Properties: Colorless gas, odorless; easily liquefied under pressure at R.T.; sol. in oxygenated and hydrocarbon solvs.; insol. in water; m.w. 58.12; dens. 0.5572; vapor pressure 2950 mm Hg (31 psig, 21 C); f.p. -159 C; b.p. -11.73 C; flash pt. -83 C
Toxicology: Asphyxiant; narcotic at high concs.; TSCA listed
Precaution: Highly flamm. gas; explosive; very dangerous fire and explosion hazard exposed to heat, flame, or oxidizers
Hazardous Decomp. Prods.: Heated to decomp., emits acrid smoke and irritating fumes
Uses: Aerosol propellant for paint aerosols, pharmaceutical topicals; propellant in cosmetics; solvent; direct food additive, propellant, aerating agent for foamed and sprayed food prods.
Regulatory: FDA 21CFR §184.1165, GRAS; FDA approved for topicals; NF compliance
Manuf./Distrib.: Aeropres; Air Prods.; Aldrich; BOC Ltd; Calor Gas; ChevronPhillips; Degussa AG/Health & Nutrition; MG Ind.

Isobutanoic acid. *See* Isobutyric acid
Isobutanol. *See* Isobutyl alcohol
Isobutenyl methyl ketone. *See* Mesityl oxide
Isobutoxy-2-ethoxy-2-ethanol; 2-(2-Isobutoxyethoxy) ethanol. *See* Diethylene glycol isobutyl ether
Isobutoxypropanol; 1-Isobutoxy-2-propanol. *See* Propylene glycol isobutyl ether
Isobutoxypropoxypropanol. *See* Dipropylene glycol isobutyl ether

Isobutyl acetate

CAS 110-19-0; EINECS/ELINCS 203-745-1
UN 1123 (DOT); UN 1213 (DOT); FEMA 2175
Synonyms: Acetic acid isobutyl ester; Acetic acid 2-methylpropyl ester; 2-Methylpropyl acetate; 2-Methyl-1-propyl acetate; 2-Methylpropyl ethanoate; β-Methylpropyl ethanoate
Classification: Carboxylic acid ester
Definition: Ester of isobutyl alcohol and acetic acid
Empirical: $C_6H_{12}O_2$
Formula: $CH_3COOCH_2CH(CH_3)_2$
Properties: Colorless clear liq., fruit-like odor; very sol. in alcohol, fixed oils, propylene glycol; sol. in ether, hydrocarbons; sl. sol. in water; m.w. 116.18; sp.gr. 0.8685 (15 C); m.p. -98.9 C; b.p. 118 C; flash pt. (TCC) 18 C; ref. index 1.389
Toxicology: ACGIH TLV/TWA 150 ppm; LD50 (oral, rat) 13,400 mg/kg; mildly toxic by ing. and inh.; vapors can probably irritate nose and throat; high concs. may cause CNS depression; ing. may cause CNS mouth/throat irritation, CNS depression, headache, weakness, dizziness, nausea; higher concs. may cause unconsciousness; skin and eye irritant; TSCA listed
Precaution: Highly flamm.; very dangerous fire and mod. explosion hazard on exposure to heat, flame, oxidizers; incompat. with strong oxidizers, strong bases, strong acids; increased risk of fire/explosion; decomp. (hydrolysis) can occur
Hazardous Decomp. Prods.: Heated to decomp., emits acrid smoke and fumes
NFPA: Health 1, Flammability 3, Reactivity 0
Storage: Store in cool, dry, well-ventilated area, out of direct sunlight; ground drums and bond transfer containers
Uses: Solvent for nitrocellulose, resins, adhesives, thinners, varnishes, sealants, topcoat lacquers, printing inks, cosmetics, pharmaceuticals; intermediate in mfg. of pharmaceuticals, org. compds.; solvent for fragrances in cosmetics; aroma component in perfumes; synthetic flavoring agent in foods and pharmaceuticals; defoamer in acute pulmonary edema extractants
Features: Sweet apple, banana-like flavor
Regulatory: FDA 21CFR §172.515; FEMA GRAS
Manuf./Distrib.: AAE Chemie NV; AMC Chems.; Accurate Chem. & Scientific; Advanced BioTech; Aldrich; Allchem Ind.; Ashland; Axxence Aromatic GmbH; BASF; BP Amoco; Celanese; Coyne; Degussa Flavors/Alfrebro; Dow; Eastman; Elan; Fleurchem; Fluka; Goldensun Mfg.; Houghton Chem.; Hukill; Moore Ingreds.; Novachem Aromatici; Penta Mfg.; Spectrum Quality Prods.; Stanchem; Thomas Scientific; United Pharma Ltd; Vopak USA; Xinchem
Trade Names: Eastman® Isobutyl Acetate; Ucar® Isobutyl Acetate

Isobutylacetone. *See* Methyl isoamyl ketone
Isobutyl adipate. *See* Diisobutyl adipate

Isobutyl alcohol

CAS 78-83-1; EINECS/ELINCS 201-148-0
UN 1212 (DOT); UN 1213; FEMA 2179
Synonyms: 1-Hydroxymethylpropane; IBA; Isobutanol; Isopropylcarbinol; 2-Methylpropanol; 2-Methyl-1-propanol; 2-Methylpropan-1-ol; 2-Methylpropyl alcohol
Classification: Primary aliphatic alcohol
Empirical: $C_4H_{10}O$
Formula: $(CH_3)_2CHCH_2OH$
Properties: Colorless liq.; sweet musty odor; sol. in alcohol, ether; misc. with most org. solvs.; partly sol. in water; m.w. 74.12; dens. 0.806 (15 C); vapor pressure 10.4 mm Hg; f.p. -108 C; b.p. 106-109 C; flash pt. (TCC) 29 C; autoignition temp. 415 C; ref. index 1.396; surf. tens. 22.98 dynes/cm; dielec. const. 17.93
Toxicology: ACGIH TLV/TWA 50 ppm; LD50 (oral, rat) 2460 mg/kg; poison by IV, IP route; mod. toxic by ing., skin contact; severe skin/eye irritant; inh. of high concs. or ing. may cause CNS depression effects; collapse, coma, or death possible at higher doses; liver/kidney damage possible; experimental carcinogen, tumorigen; mutagenic data; TSCA listed
Environmental: VOC; BOD5 0.41; COD 2.46; ThOD 2.59
Precaution: Flamm.; dangerous fire hazard with heat, flame; mod. explosive as vapor with heat, flame, oxidizers; also incompat. with chromium trioxide; explosion potential with various chems.
Hazardous Decomp. Prods.: Heated to decomp., emits acrid smoke and fumes
NFPA: Health 1, Flammability 3, Reactivity 0
Storage: Store in cool, dry, well-ventilated area, out of direct sunlight, away from heat/ignition sources
Uses: Synthetic flavor and fragrance in foods and pharmaceuticals; fragrance in cosmetics; organic synthesis; latent solvent; intermediate; solvent for paints, lacquers, paint removers, cleaners, hydraulic fluids, pharmaceuticals; fluorometric determinations; liq. chromatography; mfg. of isobutyl esters for use in solvents, plasticizers, flavors, and perfumes; extraction agent for phosphoric acid purification; in food-pkg. adhesives; defoamer in food-contact coatings and paper/paperboard; in food-contact textiles
Features: Fruity flavor
Regulatory: FDA 21CFR §73.1, 172.515, 172.859, 175.105, 176.180, 176.200, 176.210, 177.1200, 177.2800; FEMA GRAS; CERCLA haz-

ardous substance
Manuf./Distrib.: AAE Chemie NV; AMC Chems.; Advanced BioTech; Aldrich; Allchem Ind.; Ashland; Axxence Aromatic GmbH; BASF; Brenntag AG; CPS; Celanese; Degussa Flavors/Alfrebro; Dow; Eastman; Fleurchem; Fluka; GFS; GNC Group; Grau Aromatics; Houghton Chem.; Hukill; Integra; J.T. Baker; Mallinckrodt Baker; Moore Ingreds.; Penta Mfg.; Prodasynth; Romil Ltd; Ruger; Shell; Sigma; Spectrum Quality Prods.; Stanchem; Thomas Scientific; United Pharma Ltd; VWR Int'l.; Vopak USA; Wutong Aroma; Xinchem
Trade Names: Ucar® Isobutanol

Isobutyl aldehyde. *See* 2-Methylpropanal

Isobutylbenzene

CAS 538-93-2; EINECS/ELINCS 208-706-2
UN 2709
Synonyms: Benzene, (2-methylpropyl)-; 2-Methyl-1-phenylpropane; (2-Methylpropyl) benzene; 1-Phenyl-2-methylpropane
Classification: Aromatic hydrocarbon
Empirical: $C_{10}H_{14}$
Formula: $(CH_3)_2CHCH_2C_6H_5$
Properties: Colorless to ylsh. liq.; sweet aromatic odor; sol. in ethanol; insol. in water; m.w. 134.22; dens. 0.853; vapor pressure 4.2 mm Hg (37.7 C); m.p. -51 C; b.p. 172-174 C; flash pt. 55 C; autoignition temp. 425 C; ref. index 1.4855 (20 C)
Toxicology: REL 100 ppm; LD50 (ing.) 3321 mg/kg; LC50 (inh.) > 1032 ppm; irritating to eyes; non- to sl. irritating to skin; may be harmful by ing., inh., skin absorp.; sl. to mod. upper airway irritancy possible @ 200 ppm; ing. may cause sluggish, labored respiration, incoordination; if aspirated, may cause inflamm., fluid accumulation in lungs; TSCA listed
Environmental: Marine pollutant; prevent entry into sewers, water sources, wastewater, soil
Precaution: Flamm. liq.; LEL 0.8 vol.%; UEL 6.0 vol.%; floats on water; vapors heavier than air, may accumulate in low areas, spread along ground; bond and ground during transfer; incompat. with oxygen, strong oxidizing agents
Hazardous Decomp. Prods.: combustion: CO_x
Storage: Light-sensitive; store in cool, dry, well-ventilated area away from heat, sparks, flames; keep tightly closed
Uses: High-boiling solvent; intermediate for perfumes; synthesis
Regulatory: SARA §313 nonreportable
Manuf./Distrib.: Acros Org.; ChevronPhillips; Dr. Ehrenstorfer GmbH; ICN Biomed. Research Prods.; Lancaster Synthesis; Merck KGaA; TCI Am.

Isobutyl carbinol. *See* Isoamyl alcohol
Isobutyl 'Cellosolve'. *See* Ethylene glycol isobutyl ether

Isobutylcyclohexane

CAS 1678-98-4
Synonyms: i-Butylcyclohexane; Cyclohexane, isobutyl-; Cyclohexane, (2-methylpropyl)-; (2-Methylpropyl) cyclohexane
Classification: Sat. alicyclic hydrocarbon
Empirical: $C_{10}H_{20}$
Formula: $C_6H_{11}CH_2CH(CH_3)CH_3$
Properties: Colorless liq.; sol. in most org. solvs.; insol. in water; m.w. 140.27; sp.gr. 0.795 (20 C); m.p. -95 C; b.p. 169 C
Toxicology: May cause mild skin irritation; conc. vapor may cause eye irritation, nose/throat irritation, headache, lightheadedness, nausea, drowsiness; probably no significant CNS effects; ing. may result in sore throat, nausea, diarrhea
Precaution: Combustible; incompat. with strong oxidizing agents (increased risk of fire and explosion)
Hazardous Decomp. Prods.: None
NFPA: Health 0, Reactivity 0
Storage: Store in cool, dry, well-ventilated area
Uses: Solvent; org. synthesis
Manuf./Distrib.: Dr. Ehrenstorfer GmbH; TCI Am.

Isobutylcyclopentane

CAS 3788-32-7
Synonyms: Cyclopentane, isobutyl-; Cyclopentane, (2-methylpropyl)-
Classification: Sat. alicyclic hydrocarbon
Empirical: C_9H_{18}
Formula: $C_5H_9CH_2CH(CH_3)CH_3$
Properties: Colorless liq.; sol. in most org. solvs.; insol. in water; m.w. 126.23; sp.gr. 0.7781; m.p. -115 C; b.p. 148 C; ref. index 1.4298
Toxicology: May cause mild skin irritation; conc. vapor may be irritating to eyes; probably a CNS depressant; high vapor concs. may cause nose/throat irritation, lightheadedness, nausea; very high concs. may cause unconsciousness and death; ing. may result in sore throat, nausea, diarrhea
Precaution: Combustible; incompat. with strong oxidizing agents (increased fire and explosion hazard)
Hazardous Decomp. Prods.: None
Storage: Store in cool, dry, well-ventilated area
Uses: Solvent; org. synthesis

Isobutyl glycol. *See* Ethylene glycol isobutyl ether

Isobutyl heptyl ketone

CAS 123-18-2; EINECS/ELINCS 204-607-3
Synonyms: 4-Nonanone, 2,6,8-trimethyl-; 2,6,8-Trimethyl-4-nonanone; Trimethyl 4-nonanone
Empirical: $C_{12}H_{24}O$
Formula: $(CH_3)_2CHCH_2COCH_2CH(CH_3)CH_2CH(CH_3)_2$
Properties: Water-wh. liq., pleasant odor; insol. in water; m.w. 184.32; sp.gr. 0.822 (20/20 C); f.p. -75 C; b.p. 224 C; flash pt. (CC) 90.5 C
Precaution: Combustible
NFPA: Health 2, Flammability 2, Reactivity 0
Uses: Solvent for coatings (NC emulsions, vinyls); dispersant; intermediate; high solvating power for vinyl resins, cellulose esters and ethers; lube oil dewaxing; fragrance in cosmetics
Manuf./Distrib.: ABCR; Alfa Aesar; Dow
Trade Names: Ecosoft™ Solvent IK

Isobutyl isobutyrate

CAS 97-85-8; EINECS/ELINCS 202-612-5
UN 2528 (DOT); FEMA 2189
Synonyms: IBIB; Isobutyric acid, isobutyl ester; 2-Methylpropanoic acid 2-methylpropyl ester; 2-Methylpropyl isobutyrate; 2-Methylpropyl 2-methylpropanoate
Empirical: $C_8H_{16}O_2$
Formula: $(CH_3)_2CHCOOCH_2CH(CH_3)_2$
Properties: Colorless liq., fruity odor; sol. in alcohol, ether; insol. in water; m.w. 144.22; dens. 0.854 (20/4 C); m.p. -81 C; b.p. 149-151 C; flash pt. 40 C; ref. index 1.399 (20 C)
Toxicology: LD50 (oral, rat) 12,800 mg/kg; LC50 (inh., rat, 6 h) 5000 ppm; mildly toxic by ing. and inh.; irritant; TSCA listed
Precaution: Flamm.; can react with oxidizers
Hazardous Decomp. Prods.: Heated to decomp., emits acrid smoke and fumes
Uses: Solvent for paints; nitrocellulose lacquers and thinners; insect repellent; synthetic flavoring agent in foods and pharmaceuticals; fragrance in cosmetics; substitute for methyl amyl acetate
Features: Grape-like flavor
Regulatory: FDA 21CFR §172.515; FEMA GRAS
Manuf./Distrib.: Advanced Synthesis Tech.; Aldrich; Ashland; Bell Flavors & Fragrances; Degussa Flavors/Alfrebro; Eastman; Epochem; Fleurchem; Fluka; Grau Aromatics; Houghton Chem.; Penta Mfg.; United Pharma Ltd; Xinchem
Trade Names: Eastman® IBIB

Isobutyl ketone. *See* Diisobutyl ketone
Isobutyl methyl carbinol. *See* Methyl amyl alcohol
Isobutyl methyl ketone. *See* Methyl isobutyl ketone
Isobutyl methyl methanol. *See* Methyl amyl alcohol

Isobutyl oleate

CAS 84988-79-4; EINECS/ELINCS 284-868-8
Definition: Ester of isobutyl alcohol and oleic acid
Empirical: $C_{22}H_{42}O_2$
Formula: $CH_3(CH_2)_7CH{=}CH(CH_2)_7COOCH_2CH(CH_3)_2$
Properties: Yel. liq.; misc. with paraffin oil; insol. in water; m.w. 338.58; dens. 0.86 kg/l (20 C); pour pt. -21 C; flash pt. (COC) 195 C
Uses: Chemical intermediate; chemical synthesis; carbon source in anti-

biotic culture broths; lubricity improvers in min. oils; lubricant in metalworking fluids, cutting, lamination, and textile oils, rust inhibitors; textile and leather industry; emollient, emulsifier, bodying agent in cosmetics and toiletries; plastics; antiozonant for chloroprene rubber
Manuf./Distrib.: A&E Connock; ChemService; DelCon; Expocell AB; Mosselman NV
Trade Names: Excepar IB-OL; Vinycizer 30

Isobutyl palmitate
CAS 110-34-9; EINECS/ELINCS 203-758-2
Synonyms: Hexadecanoic acid, 2-methylpropyl ester; 2-Methylpropyl hexadecanoate; Palmitic acid, isobutyl ester
Definition: Ester of isobutyl alcohol and palmitic acid
Empirical: $C_{20}H_{40}O_2$
Formula: $CH_3(CH_2)_{14}COOCH_2CH(CH_3)_2$
Properties: M.w. 312.53
Uses: Emollient, lubricant, solvent for cosmetics, textiles, metalworking; for fiber finishing for resin-bonded filters for food contact; defoamer in food-contact paper/paperboard; in food-contact textiles
Regulatory: FDA 21CFR §176.210, 177.2260, 177.2800
Manuf./Distrib.: A&E Connock; ChemService

Isobutyl propionate
CAS 540-42-1; EINECS/ELINCS 208-746-0
UN 2394 (DOT); FEMA 2212
Synonyms: 2-Methylpropyl propanoate; 2-Methylpropyl propionate; Propanoic acid, 2-methylpropyl ester; Propionic acid, isobutyl ester
Empirical: $C_7H_{14}O_2$
Formula: $CH_3CH_2COOCH_2CH(CH_3)_2$
Properties: Water-wh. liq., agreeable ethereal odor; very sol. in alcohol, ether; insol. in water; m.w. 130.18; dens. 0.888 (0/4 C); m.p. -71 C; b.p. 137 C; flash pt. 33 C; ref. index 1.3975 (20 C)
Toxicology: LD50 (oral, rabbit) 5599 mg/kg; mildly toxic by ing.; irritant; TSCA listed
Precaution: Flamm. liq.; vigorous reactions with oxidizing materials
Hazardous Decomp. Prods.: Heated to decomp., emits acrid smoke and irritating fumes
Uses: Solvent; synthetic flavoring agent, fragrance in foods, pharmaceuticals; mfg. of fruit essences; fragrance in cosmetics
Regulatory: FDA 21CFR §172.515; FEMA GRAS
Manuf./Distrib.: Advanced BioTech; Advanced Synthesis Tech.; Aldrich; Axxence Aromatic GmbH; Degussa Flavors/Alfrebro; Elan; Epochem; Fleurchem; Goldensun Mfg.; Grau Aromatics; United Pharma Ltd; Xinchem

Isobutyltrimethylethane. *See* Isooctane
2-Isobutyoxyethanol. *See* Ethylene glycol isobutyl ether
Isobutyraldehyde. *See* 2-Methylpropanal

Isobutyric acid
CAS 79-31-2; EINECS/ELINCS 201-195-7
UN 2529 (DOT); FEMA 2222
Synonyms: Acetic acid, dimethyl-; Dimethylacetic acid; Isobutanoic acid; Isopropylformic acid; 2-Methylpropanoic acid; 2-Methylpropionic acid; α-Methylpropionic acid; Propionic acid, 2-methyl-
Classification: Organic acid
Empirical: $C_4H_8O_2$
Formula: $(CH_3)_2CHCOOH$
Properties: Colorless liq., pungent odor of rancid butter; sol. in 6 parts of water; misc. with alcohol, ether, chloroform, oxygenated and chlorinated solvs.; m.w. 88.11; dens. 0.946-0.950 (20/20 C); m.p. -46 C; b.p. 154.4 C; flash pt. (TOC) 76.6 C; ref. index 1.393 (20 C)
Toxicology: LD50 (oral, rat) 280 mg/kg, (skin, rabbit) 500 mg/kg; poison by ing.; mod. toxic by skin contact; corrosive irritant to eyes and tissue; TSCA listed
Precaution: Flamm., corrosive; reactive with oxidizing materials
Hazardous Decomp. Prods.: Heated to decomp., emits acrid smoke and fumes
Uses: Mfg. of esters for solvents, flavors, perfume bases; disinfectant; preservative; varnish; deliming hides; tanning agent; buffer in cosmetics; synthetic flavoring agent in foods and pharmaceuticals
Features: Pungent, butter-like, fruity on dilution
Regulatory: FDA 21CFR §172.515; FEMA GRAS; SARA reportable
Manuf./Distrib.: Advanced BioTech; Advanced Synthesis Tech.; Aldrich; Axxence Aromatic GmbH; BASF; Celanese; Degussa Flavors/Alfrebro; Eastman; Elan; Fleurchem; Fluka; Grau Aromatics; Moore Ingreds.; Penta Mfg.; Prodasynth; Sigma; United Pharma Ltd; Whyte Chems. Ltd; Xinchem

Isobutyric acid, isobutyl ester. *See* Isobutyl isobutyrate
Isobutyric acid, monoester with 2,2,4-trimethylpentane-1,3-diol. *See* 2,2,4-Trimethyl-1,3-pentanediol isobutyrate
Isobutyric aldehyde. *See* 2-Methylpropanal
Isobutyrone. *See* Diisopropyl ketone

Isocetyl alcohol
CAS 36311-34-9; EINECS/ELINCS 252-964-9
Synonyms: 2-Hexyl-1-decanol; Isohexadecanol; Isohexadecyl alcohol; Isopalmityl alcohol
Definition: Mixture of branched chain C16 aliphatic alcohols
Empirical: $C_{16}H_{34}O$
Formula: $C_8H_{17}CHC_6H_{13}CH_2OH$
Properties: Clear liq.; sol. in min. oil, 95% ethanol, IPM, oleyl alcohol, castor oil, cyclomethicone; insol. in water, glycerin, propylene glycol; m.w. 242; sp.gr. 0.830-0.840; m.p. -30 C; flash pt. (COC) 156 C
Toxicology: LD50 (oral, rat) > 50 g/kg; mildly irritating to skin and eyes; TSCA listed
Uses: Emollient, visc. control agent, solubilizer for cosmetics, emulsions; coupling agent for min. oil and castor oil; carrier, extender for flavor and fragrance oils in foods, pharmaceuticals; pigment dispersant and binder; extender and solvent for pigments, lipsticks dyes; ester and alkoxylate synthesis; in fiber lubricants; in metalworking lubricants
Manuf./Distrib.: Indofine; Jarchem Ind.; Leprino Foods; M. Michel; Mutchler; Penta Mfg.; United Pharma Ltd; Universal Preserv-A-Chem
Trade Names: Ceraphyl® ICA

Isocetyl isostearate
CAS 52006-45-8; EINECS/ELINCS 257-598-3
Definition: Ester of isocetyl alcohol and isostearic acid
Empirical: $C_{34}H_{68}O_2$
Properties: Colorless nonoily liq.; m.w. 508.91; congeal pt. -19.5 C; low viscosity; hydrolysis and oxidation stability
Uses: Emollient, oil phase component for pharmaceutical topicals
Trade Names: Isostriol IC

Isocetyl laurate
Empirical: $C_{28}H_{56}O_2$
Formula: $C_{11}H_{23}COOC_{16}H_{33}$
Properties: Oily liq.; almost odorless; sol. in org. solvs.; insol. in water; m.w. 424.7; f.p. -65 C
Toxicology: May cause adverse or toxic effects
Precaution: Combustible
Uses: Lubricant, fixing agent, and solvent in cosmetics and pharmaceuticals; plasticizer; mold release agent; textile softener
Manuf./Distrib.: Sigma

Isocetyl myristate
CAS 83708-66-1
Synonyms: Myristic acid, isocetyl ester; Tetradecanoic acid, isocetyl ester; Tetradecanoic acid, isohexadecyl ester
Definition: Ester of isocetyl alcohol and myristic acid
Empirical: $C_{30}H_{60}O_2$
Formula: $CH_3(CH_2)_{12}COOC_{16}H_{33}$
Properties: Oily liq.; pract. odorless; sol. in most org. solvs.; insol. in water; dens. 0.857; f.p. -39 C
Precaution: Combustible
Uses: Emollient, lubricant, fixing agent, and solvent in cosmetics and pharmaceuticals; plasticizer; mold release agent; textile softener

Isocetyl salicylate
CAS 138208-68-1
Synonyms: Benzoic acid, 2-hydroxy-, isohexadecyl ester; Salicylic acid, isocetyl ester
Definition: Ester of isocetyl alcohol and salicylic acid
Empirical: $C_{23}H_{38}O_3$
Properties: Water-wh. liq., odorless; insol. in water; sp.gr. 1.02; flash pt.

(PMCC) > 200 F
Toxicology: May cause mild skin and eye irritation on prolonged contact
Precaution: Incompat. with oxidizing agents
Storage: Store away from strong oxidizing agents; avoid excessive heat
Uses: Emollient, solvent in cosmetics; solvent, plasticizer for benzophenone in sunscreen formulations, waxes, resins, polymers
Trade Names: Dermol ICSA

Isocetyl stearate
CAS 25339-09-7; EINECS/ELINCS 246-868-6
Synonyms: Isohexadecyl stearate; Octadecanoic acid, isocetyl ester; Octadecanoic acid, isohexadecyl ester; Stearic acid, isocetyl ester; Stearic acid, isohexadecyl ester
Definition: Ester of isocetyl alcohol and stearic acid
Empirical: $C_{34}H_{68}O_2$
Formula: $CH_3(CH_2)_{16}COOC_{16}H_{33}$
Properties: Oily liq., pract. odorless; insol. in water; sol. in most org. solvs.; dens. 0.862; f.p. 57 C; visc. 29 cp (25 C); ref. index 1.446-1.456
Toxicology: LD50 (oral, rabbit) > 5 g/kg; LDLo (oral, rat) 5 g/kg; low toxicity by ing.; skin and eye irritant; TSCA listed
Precaution: Combustible
Hazardous Decomp. Prods.: Heated to decomp., emits acrid smoke and irritating vapors
Uses: Lubricant, fixing agent, solvent, emollient, thickener, pigment dispersant in cosmetic creams and lotions, pharmaceutical topicals; plasticizer; mold release agent; textile softener, surfactant
Manuf./Distrib.: A&E Connock; ChemService; DelCon; Jarchem Ind.; Mosselman NV
Trade Names: Monestriol IC

Isocumene. *See* Propylbenzene
Isocyanic acid, hexamethylene ester. *See* Hexamethylene diisocyanate
Isocyanic acid, methylphenylene ester. *See* Toluene-2,4-diisocyanate; Toluene diisocyanate
Isocyanic acid, 4-methyl-m-phenylene ester. *See* Toluene-2,4-diisocyanate
Isocyanic acid, methyl-m-phenylene ester. *See* Toluene diisocyanate

Isodecanoic acid
CAS 26403-17-8; EINECS/ELINCS 247-673-9
Synonyms: 8-Methylnonanoic acid
Empirical: $C_{10}H_{20}O_2$
Formula: $C_9H_{19}COOH$
Properties: Liq.; sl. sol. in water; m.w. 172.27; dens. 0.91 kg/l (20 C); pour pt. < -65 C; b.p. 250-280 C
Uses: Emulsifier, coupling solvent for metalworking fluids; modifier for alkyd resins

Isodecanol. *See* Isodecyl alcohol

Isodecyl alcohol
CAS 25339-17-7; EINECS/ELINCS 246-869-1
Synonyms: Isodecanol; 8-Methylnonyl alcohol
Empirical: $C_{10}H_{22}O$
Formula: $C_9H_{19}CH_2OH$
Properties: Colorless liq.; insol. in water; m.w. 158.29; dens. 0.84 kg/l (20 C); b.p. 214-224 C; flash pt. 220 F
Toxicology: LD50 (oral, rat) 6500 μl/kg, (skin, rabbit) 3150 mg/kg; mod. toxic by skin contact; mildly toxic by ing.; primary skin irritant; TSCA listed
Precaution: Combustible exposed to heat or flame; can react vigorously with oxidizers
Hazardous Decomp. Prods.: Heated to decomp., emits acrid smoke and fumes
Uses: Solvent; froth flotation foam modifier; oiliness additive in lubricants; fragrance in cosmetics
Manuf./Distrib.: Bencorp Int'l.; ICC Chem.; Kyowa Hakko USA; United Pharma Ltd

Isodecyl isononanoate
CAS 41395-89-5; 59231-35-5; EINECS/ELINCS 261-674-1
Synonyms: Hexanoic acid, 3,5,5-trimethyl-, isodecyl ester; Isodecyl 3,5,5-trimethylhexanoate; 3,5,5-Trimethylhexanoic acid, isodecyl ester
Definition: Ester of branched chain decyl alcohols and a branched chain nonanoic acid
Empirical: $C_{19}H_{38}O_2$
Formula: $(CH_3)_3CCH_2CCH_3HCH_2COOC_{10}H_{21}$
Properties: Sol. in alcohol, animal and vegetable oils, min. oil; insol. in water, glycerin, propylene glycol
Toxicology: TSCA listed
Uses: Antistat, emollient, skin conditioner, solvent in cosmetics, hair care, skin care; pigment wetting agent
Manuf./Distrib.: A&E Connock
Trade Names: Wickenol® 152

Isodecyl oleate
CAS 59231-34-4; EINECS/ELINCS 261-673-6
Synonyms: 9-Octadecenoic acid, isodecyl ester
Definition: Ester of branched chain decyl alcohols and oleic acid
Empirical: $C_{28}H_{54}O_2$
Formula: $CH_3(CH_2)_7CH{=}CH(CH_2)_7COOC_{10}H_{21}$
Properties: Wh. to straw liq., char. mild odor; sol. in peanut oil, 95% ethanol, IPM, oleyl alcohol; insol. in water, glycerin, propylene glycol; sp.gr. 0.858-0.864; ref. index 1.4540-1.4560; nonionic
Toxicology: LD50 (oral, rat) > 40 ml/kg; nonirritating to eyes; mildly irritating to skin; TSCA listed
Uses: Emollient, cosolvent, solubilizer for cosmetics, pharmaceutical topicals; emulsifier, lubricant, antistat, defoamer for metalworking, textiles, plastics, paper
Manuf./Distrib.: A&E Connock
Trade Names: Ceraphyl® 140-A; Estalan-IDO

Isodecyl 3,5,5-trimethylhexanoate. *See* Isodecyl isononanoate
Isodiprene. *See* 3-Carene

Isododecane
CAS 31807-55-3; 93685-81-5; EINECS/ELINCS 250-816-8; 297-629-8
Synonyms: 2-Methylnonane; 2-Methylundecane
Definition: Branched chain aliphatic hydrocarbon with 12 carbons
Empirical: $C_{12}H_{26}$
Formula: $H_3C(C(CH_3)_2CH_2)_2CH(CH_3)_2$
Properties: M.w. 170.34
Uses: Emollient in cosmetics; rheological additive; cosolubilizer for nonhydrocarbon materials; solvent for debris on skin
Manuf./Distrib.: ABCR; Fluka; United Pharma Ltd
Trade Names: Permethyl® 99A; Ritacane ID

Isoeicosane
CAS 52845-07-5; 93685-79-1; EINECS/ELINCS 297-527-7
Classification: Liquid isoparaffin
Definition: Branched chain aliphatic hydrocarbon with 20 carbons in the alkyl chain
Empirical: $C_{20}H_{42}$
Formula: $H_3C(C(CH_3)_2CH_2)_4CH(CH_3)_2$
Properties: M.w. 282
Uses: Emollient, skin conditioner, solvent, gloss aid, moisture barrier, cosolubilizer, plasticizer in cosmetics, mascara
Manuf./Distrib.: Jarchem Ind.

Isoeicosanol; Isoeicosyl alcohol. *See* Octyldodecanol
Isoforon. *See* Isophorone

Isoheptane
CAS 591-76-4
UN 1206
Synonyms: Ethylisobutylmethane; 2-Methylhexane
Classification: Sat. aliphatic hydrocarbon
Empirical: C_7H_{16}
Formula: $CH_3CHCH_3CH_2CH_2CH_2CH_3$
Properties: Colorless liq.; gasoline-like odor; sol. in most org. solvs.; insol. in water; m.w. 100.2; sp.gr. 0.683 (20 C); vapor pressure 40 mm Hg (14.9 C); m.p. -118 C; b.p. 90 C; flash pt. (CC) < -18 C; ref. index 1.38498 (20 C)
Toxicology: ACGIH TLV/TWA 400 ppm; STEL 500 ppm; may cause mild skin irritation; conc. vapor may cause sl. eye irritation; conc. vapor may cause sl. nose/throat irritation, lightheadedness, dizziness, nausea, difficulty breathing; narcotic @ 0.8-1.1%; very high concs. may cause unconsciousness, death; ing. may cause nausea, vomiting, swelling of abdomen, headache, depression; relatively low oral toxic-

ity unless aspirated into lungs; TSCA listed
Precaution: Flamm.; very dangerous fire hazard exposed to heat, flame; explosive limits in air 1-6%; explosive as vapor exposed to heat, flame; incompat. with strong oxidizing agents (increases fire and explosion hazard)
Hazardous Decomp. Prods.: Heated to decomp., emits acrid smoke and irritating fumes
NFPA: Health 0, Flammability 3, Reactivity 0
Storage: Store in cool, dry, well-ventilated area out of direct sunlight
Uses: Ingred. in gasoline, aviation fuels; solvent; org. synthesis; raw material for prod. of plasticizers; in food-pkg. adhesives
Regulatory: FDA 21CFR §175.105
Manuf./Distrib.: Aldrich; Fluka
Trade Names: Exxsol® Isoheptane

Isoheptanol. *See* Isoheptyl alcohol

Isoheptyl acetate

Empirical: $C_9H_{18}O_2$
Formula: $CH_3COOCH_2C_6H_{13}$
Properties: Colorless liq.; insol. in water; m.w. 158.24; dens. 0.87 kg/l (20 C); b.p. 176-200 C; flash pt. (TCC) 60 C
Uses: Solvent for paints

Isoheptyl alcohol

CAS 70914-20-4
Synonyms: Isoheptanol
Empirical: $C_7H_{16}O$
Properties: Colorless liq.; misc. with oxygenated solvs.; insol. in water; m.w. 116.21; dens. 0.82 kg/l (20 C); b.p. 164-174 C
Uses: Solvent; lubricant/antiwear agent raw material; in coatings

Isohexadecane

CAS 4390-04-9; 60908-77-2; EINECS/ELINCS 224-506-8
Synonyms: 2,2,4,4,6,6,8-Heptamethylnonane
Classification: Liquid isoparaffin
Definition: Branched chain aliphatic hydrocarbon with 16 carbons
Empirical: $C_{16}H_{34}$
Formula: $H_3C(C(CH_3)_2CH_2)_4CH(CH_3)_2$
Properties: M.w. 226
Uses: Emollient, moisture barrier, solvent in cosmetics
Manuf./Distrib.: Aldrich; Jarchem Ind.; Sigma
Trade Names: Ritacane IH

Isohexadecanol; Isohexadecyl alcohol. *See* Isocetyl alcohol
Isohexadecyl stearate. *See* Isocetyl stearate

Isohexane

CAS 107-83-5; EINECS/ELINCS 203-523-4
UN 1208 (DOT); UN 2462 (DOT)
Synonyms: 1,2-Dimethylbutane; Dimethylpropylmethane; Methyl pentane; 2-Methylpentane
Classification: Sat. aliphatic hydrocarbon
Empirical: C_6H_{14}
Formula: $CH_3CH(CH_3)CH_2CH_2CH_3$
Properties: Colorless liq. or oil; faint char. petrol. odor; misc. with most org. solvs.; insol. in water; m.w. 86.18; dens. 0.65 (20 C); vapor pressure 23 kPa (20 C); m.p. -153 C; b.p. 60 C; aniline pt. 63 C; flash pt. (CC) -32 C; autoignition temp. 264 C
Toxicology: ACGIH TLV/TWA 500 ppm; STEL 1000 ppm; may cause skin irritation; human eye irritant; inh. of high concs. can cause respiratory tract irritation, sl. nausea, headache, lightheadedness; very high concs. can cause unconsciousness, death; low oral toxicity unless aspirated into lungs (severe lung damage or death); TSCA listed
Precaution: Highly flamm.; explosive limits in air 1.0-7.0 vol.%; very dangerous fire hazard exposed to heat, flame, or oxidizers; explosive limits in air 1-7%; severe explosion hazard exposed to heat, flame; vapor heavier than air; distant ignition possible; electrostatic charges can be generated; incompat. with strong oxidizing agents (increases risk of fire and explosion); attacks plastic
Hazardous Decomp. Prods.: Heated to decomp., emits acrid smoke and irritating fumes
NFPA: Health 1, Flammability 3, Reactivity 0
Storage: Store in cool, dry, well-ventilated area out of direct sunlight
Uses: Extraction solvent for veg. oils, animal oils, essential oils; solvent for fast-drying adhesives, lacquers; chem. reaction medium; f.p. depressant; reaction diluent for polyolefin prod.; cleaning agent; release agent
Manuf./Distrib.: Aldrich; ChevronPhillips; Fluka; Romil Ltd
Trade Names: Calumet Isohexane; Exxsol® Isohexane

Isohexyl alcohol. *See* 2-Methyl-1-pentanol
Isolignoceryl alcohol. *See* 2-Decyl-1-tetradecanol
Isonitropropane. *See* 2-Nitropropane
β-Isonitrosopropane. *See* Acetone oxime
Isononanoic acid, cetearyl ester; Isononanoic acid, cetyl/stearyl ester. *See* Cetearyl isononanoate
Isononanoic acid, 2-ethylhexyl ester. *See* Octyl isononanoate
Isononanoic acid isononyl ester. *See* Isononyl isononanoate
Isononanoic acid, isotridecyl ester. *See* Isotridecyl isononanoate
Isononanoic acid, 1,2,3-propanetriyl ester. *See* Triisononanoin

Isononyl isononanoate

CAS 42131-25-9; 59219-71-5; EINECS/ELINCS 261-665-2
Synonyms: Isononanoic acid isononyl ester; 3,5,5-Trimethylhexanoic acid, 3,5,5-trimethylhexyl ester; 3,5,5-Trimethylhexyl 3,5,5-trimethylhexanoate
Definition: Ester of a branched chain nonyl alcohol with a branched chain nonanoic acid
Empirical: $C_{18}H_{36}O_2$
Properties: Sol. in alcohol, animal and vegetable oils, min. oil; insol. in water, glycerin, propylene glycol
Toxicology: No known toxicity; TSCA listed
Uses: Antistat, emollient, moisturizer, pigment wetting agent, solvent for cosmetics, hair care prods.; emollient for pharmaceutical topicals, mouthwashes
Manuf./Distrib.: A&E Connock
Trade Names: Salacos® 99; Wickenol® 151

Isooctadecanoic acid, diester with 1,2,3-propanetriol trimer; Isooctadecanoic acid, diester with triglycerol. *See* Polyglyceryl-3 diisostearate
Isooctadecanoic acid, isooctadecyl ester. *See* Isostearyl isostearate
Isooctadecanol; Isooctadecan-1-ol. *See* Isostearyl alcohol
Isooctadecyl 2-hydroxypropanoate. *See* Isostearyl lactate
Isooctadecyl isononanoate. *See* Isostearyl isononanoate
Isooctadecyl isooctadecanoate. *See* Isostearyl isostearate

Isooctane

CAS 540-84-1; EINECS/ELINCS 203-892-1
UN 1262 (DOT)
Synonyms: Isobutyltrimethylethane; 2,4,-TMP; 2,2,4-Trimethylpentane
Classification: Sat. aliphatic hydrocarbon
Empirical: C_8H_{18}
Formula: $CH_3C(CH_3)_2CH_2CH(CH_3)CH_3$
Properties: Colorless clear liq.; gasoline-like odor; completely misc. with most org. solvs.; insol. in water; m.w. 114.26; sp.gr. 0.692 (20 C); vapor pressure 40.6 mm Hg (21 C); m.p. -107 C; b.p. 99 C; flash pt. 4.5 C; ref. index 1.3914 (20 C)
Toxicology: ACGIH TLV/TWA 300 ppm; STEL 375 ppm; mild skin irritant; conc. vapor irritating to eyes; inh. of conc. vapor may cause nose/throat irritation, headache, drowsiness, nausea, difficulty breathing; very high concs. may cause unconsciousness, death; ing. may cause nausea, vomiting, swelling of abdomen, headache, depression; relatively low oral toxicity unless aspirated into lungs (severe lung irritation or damage or death may occur); TSCA listed
Precaution: Flamm.; very dangerous fire hazard exposed to heat, flame, oxidizers; explosive as vapor exposed to heat, flame; can react vigorously with reducing agents; incompat. with strong oxidizing agents (increases fire and explosion hazard)
Hazardous Decomp. Prods.: Heated to decomp., emits acrid smoke and irritating fumes
NFPA: Health 0, Flammability 3, Reactivity 0
Storage: Store in cool, dry, well-ventilated area out of direct sunlight
Uses: Motor fuel; std. reference fuel for octane ratings; petrol. blending component; reagent in spectrophotometric analysis; org. synthesis; solvent; thinner; environmental analysis (e.g., pesticide residues); high performance liq. chromatography; gas chromatography; UV and IR

spectroscopy; solvent, film-former for cosmetics, personal care
Manuf./Distrib.: Aldrich; ChevronPhillips; Fisher Scientific; Fluka; Romil Ltd; Ruger; Sigma; Spectrum Quality Prods.; Triple Crown Am.
Trade Names: Permethyl® 97A
Trade Names Containing: Soltrol® 10

Isooctanol; Isooctan-1-ol. *See* Isooctyl alcohol
Isooctaopherone. *See* Isophorone

Isooctyl alcohol

CAS 26952-21-6; EINECS/ELINCS 248-133-5
Synonyms: Alcohol C_8; Isooctanol; Isooctan-1-ol; 6-Methylheptanol; 6-Methyl-1-heptanol
Classification: Alcohol
Definition: Mixed isomer prod.
Empirical: $C_8H_{18}O$
Formula: $CH_3(CH_2)_3CH(C_2H_5)CH_2OH$
Properties: Colorless clear liq.; mild odor; insol. in water; m.w. 130.26; dens. 0.832 (20/20 C); f.p. -106 C; b.p. 182-185 C; flash pt. (PMCC) 74 C; autoignition temp. 277 C; ref. index 1.431; surf. tens. 40 dynes/cm
Toxicology: ACGIH TLV 50 ppm; LD50 (oral, rat) 1480 mg/kg, (skin, rabbit) 2520 mg/kg; mod. toxic by skin absorption and ing.; severe eye irritant; skin irritant; target organs: respiratory system, CNS; TSCA listed
Environmental: VOC; ThOD 1.84
Precaution: Combustible; LEL 0.9%; UEL 5.7%
Hazardous Decomp. Prods.: Heated to decomp., emits acrid smoke and fumes
NFPA: Health 0, Flammability 2, Reactivity 0
Uses: In plasticizers; intermediate for nonionic detergents and surfactants, syn. drying oils, cutting and lubricating oils, hydraulic fluids; resin solvent; emulsifier; antifoam; in coatings; intermediate to introduce isooctyl group; froth flotation foam modifier; cosmetics ingred.
Manuf./Distrib.: Allchem Ind.; Chemcentral; ExxonMobil; Indofine; Jarchem Ind.; United Pharma Ltd; Universal Preserv-A-Chem

Isooctyl stearate

CAS 91031-48-0; EINECS/ELINCS 292-951-5
Definition: Ester of isooctyl alcohol and stearic acid
Empirical: $C_{26}H_{52}O_2$
Formula: $CH_3(CH_2)_{16}COOCH_2C_7H_{15}$
Properties: Colorless nonvolatile liq.; sol. in hydrocarbon oils; insol. in water; m.w. 396.71; dens. 0.86 kg/l (20 C); sapon. no. 145-150
Uses: Emollient, solvent for cosmetics; plasticizer and lubricant for plastics, paper; lubricant for metalworking fluids, textile spin finishes; plasticizer, flow improver for plastics, rubber, lacquers; surfactant for industrial detergents, cosmetics, toiletries, feed, foods, textiles, agric., lubricants, metal treatment, and petroleum/mining/road construction industries
Manuf./Distrib.: A&E Connock; ChemService

Isopalmityl alcohol. *See* Isocetyl alcohol
Isoparaffin solvents. *See* C isoparaffin

Isopentane

CAS 78-78-4; EINECS/ELINCS 201-142-8
UN 1265 (DOT)
Synonyms: Butane, 2-methyl-; Ethyldimethyl methane; Isoamylhydride; 2-Methylbutane; 1,1,2-Trimethylethane
Classification: Sat. aliphatic hydrocarbon
Empirical: C_5H_{12}
Formula: $CH_3CH_2CH(CH_3)_2$
Properties: Colorless volatile liq.; gasoline-like odor; sol. in hydrocarbons, alcohol, oils, ether; insol. in water; m.w. 72.17; dens. 0.625 (20 C); vapor pressure 595 mm Hg (21 C); f.p. -159.89 C; b.p. 27.85 C; flash pt. -51 C
Toxicology: LCLo (inh., mouse, 2 h) 419 g/m^3; ACGIH TLV/TWA 600 ppm; STEL 750 ppm; mildly toxic and narcotic by inhalation; mild skin irritation on evaporative cooling; may cause temporary eye irritation; TSCA listed
Environmental: Zero ozone depletion
Precaution: Highly flamm.; very dangerous fire and explosion hazard exposed to heat, flame, or oxidizers; incompat. with strong oxidizing agents (increases fire and explosion hazard)
Hazardous Decomp. Prods.: Heated to decomp., emits acrid smoke and irritating fumes
NFPA: Health 1, Flammability 4, Reactivity 0
Storage: Store in cool, dry, well-ventilated area out of direct sunlight
Uses: Solvent for ethylene-propylene rubber prod.; blowing agent for polymers incl. food-grade expandable PS; synthesis of org. chems. (e.g., chlorinated hydrocarbons); component of motor vehicle and aviation fuel; propellant, solvent, visc. control agent in cosmetics
Regulatory: FDA 21CFR §178.3010
Manuf./Distrib.: Aeropres; Aldrich; ChevronPhillips; ExxonMobil Europe; ExxonMobil; Fluka; Int'l. Commodities Export; Shell; Sigma; South Hampton Refining; Spectrum Quality Prods.; Triple Crown Am.
Trade Names Containing: Exxsol® Isopentane; Pentane 60/40

Isopentanol. *See* Isoamyl alcohol
Isopentyl acetate. *See* Isoamyl acetate
Isopentyl alcohol. *See* Isoamyl alcohol
Isopentyl alcohol acetate. *See* Isoamyl acetate
Isopentyl butanoate; Isopentyl butyrate. *See* Isoamyl butyrate

Isopentyldiol

CAS 2568-33-4; 50468-22-9; EINECS/ELINCS 256-597-5
Synonyms: IPG; Isoprene glycol; 3-Methylbutane-1,2-diol; 3-Methyl-1,3-butanediol
Classification: Diol
Empirical: $C_5H_{12}O_2$
Uses: Solvent in cosmetics; humectant, moisturizer, smoothing agent pharmaceuticals, cosmetics, skin care
Manuf./Distrib.: Fluka; Kuraray; Robeco

Isopentyl methyl ketone. *See* Methyl isoamyl ketone
3-Isopentyloxy-1,2-propanediol. *See* 1-Isoamyl glycerol ether
Isophorol, dihydro-. *See* Trimethylcyclohexanol

Isophorone

CAS 78-59-1; EINECS/ELINCS 201-126-0
UN 1638; FEMA 3553
Synonyms: Isoacetophorone; Isoforon; Isooctaopherone; α-Isophorone; 1,1,3-Trimethyl-3-cyclohexene-5-one; 3,3,5-Trimethyl-2-cyclohexene-1-one; 3,3,5-Trimethylcyclohexenone; 3,3,5-Trimethyl-2-cyclohexen-1-one
Classification: Cyclic aliphatic ketone
Empirical: $C_9H_{14}O$
Formula: $OCHC{:}C(CH_3)CH_2C(CH_3)_2CH_2$
Properties: Colorless to pale yel. liq.; peppermint or camphor-like odor; sol. in alcohol, ether, acetone, oxygenated solvs.; misc. with org. solvs.; insol. in water; m.w. 138.20; sp.gr. 0.922 (20/20 C); vapor pressure 1 mm (38 C); m.p. -8 C; b.p. 210-218 C; flash pt. (CC) 84 C; autoignition temp. 380 C; ref. index 1.4781 (20 C)
Toxicology: ACGIH TLV/CL 5 ppm; LD50 (oral, rat) 2330 mg/kg; LCLo (inh., rat, 4 h) 1840 ppm; mod. toxic by ing.; mildly toxic by inh.; irritating to eyes, skin, respiratory system; human systemic effects by inh.; human irritant at 25 ppm; may cause narcosis, lung irritation, death; questionable carcinogen; mutagenic data; target organ: kidneys; TSCA listed
Precaution: Combustible; incompat. with strong oxidizing agents such as peroxides, nitrates, and perchlorates
NFPA: Health 2, Flammability 2, Reactivity 0
Storage: Store in tightly closed containers in cool, dry, well-ventilated area, separate from workplace
Uses: Solvent for lacquers, printing inks, plastics, resins, oils, fats, gums, polyvinyl and nitrocellulose resins; intermediate in mfg. of PU field compds. (isophorone diamine, isophorone diisocyanate, trimethylcyclohexanone); mfg. of pesticides, coatings; leveling agent for textile printing; flavoring agent in foods and pharmaceuticals; fragrance in cosmetics; in food-pkg. adhesives
Features: Sweet flavor
Regulatory: FDA 21CFR §175.105; FEMA GRAS; SARA reportable
Manuf./Distrib.: Aceto; Aldrich; Allchem Ind.; Ashland; Atofina SA; Atofina; BP Amoco; Chemcentral; Coyne; Degussa AG; Degussa; Dow; ExxonMobil; Fabrichem; Fallek; Fluka; Great Western; Herdillia Chems. Ltd; Houghton Chem.; Sal Chem.; Sigma; Stanchem; Sunnyside; Vopak

USA; Whyte Chems. Ltd
Trade Names: Ucar® Isophorone

α-Isophorone. *See* Isophorone
Isophthalic acid, diethyl ester. *See* Diethyl isophthalate
Isoprene glycol. *See* Isopentyldiol
Isopropanol. *See* Isopropyl alcohol

Isopropanolamine lanolate

CAS 85536-46-5; EINECS/ELINCS 287-529-2
Synonyms: Fatty acids, lanolin, compds. with isopropanolamine; 2-Hydroxypropylamine lanolate
Definition: Salt of isopropanolamine and lanolin acid
Uses: Surfactant, cleansing agent, solvent in cosmetics

4-Isopropenyl-1-methylcyclohexene. *See* d-Limonene
4-Isopropenyl-1-methyl-1-cyclohexene. *See* dl-Limonene
(R)-4-Isopropenyl-1-methyl-1-cyclohexene. *See* d-Limonene
Isopropoxyethanol; 2-Isopropoxyethanol. *See* Ethylene glycol isopropyl ether
2-Isopropoxypropane. *See* Isopropyl ether

1-Isopropoxy-2-propanol

Classification: Glycol ether
Empirical: $C_6H_{14}O_2$
Uses: Solvent for resins, dyes, surface coatings

2-Isopropoxy-1-propanol

CAS 3944-37-4
Empirical: $C_6H_{14}O_2$
Properties: M.w. 118.18
Uses: Solvent for resins, dyes, surface coatings

Isopropyl acetate

CAS 108-21-4; EINECS/ELINCS 203-561-1
UN 1220 (DOT); FEMA 2926
Synonyms: Acetic acid isopropyl ester; Acetic acid 1-methylethyl ester; 2-Acetoxypropane; Isopropyl ethanoate; 2Methylethanoate; 1-Methylethyl acetate; 2-Propyl acetate; 2-Propyl ethanoate
Classification: Sat. aliphatic carboxylic acid ester
Empirical: $C_5H_{10}O_2$
Formula: $CH_3COOCH(CH_3)2$
Properties: Colorless clear aromatic liq., fruity odor; sol. in acetone, most org. solvs.; mod. sol. in water; misc. with alcohol, ether, fixed oils; m.w. 102.15; dens. 0.874 (20/20 C); vapor pressure 59.2 mm Hg; f.p. -73 C; b.p. 88.4 C; flash pt. 2 C; autoignition temp. 460 C; ref. index 1.377; surf. tens. 26 dynes/cm (20 C)
Toxicology: ACGIH TLV/TWA 250 ppm; STEL 310 ppm; LD50 (oral, rat) 3000 mg/kg; mod. toxic by ing.; mildly toxic by inh.; high concs. of vapor irritating to nose and throat; human systemic effects on inh.; narcotic in high conc.; chronic exposure can cause liver damage; TSCA listed
Environmental: VOC; BOD5 1.07; COD 1.67; ThOD 2.04
Precaution: Highly flamm.; UEL 8%; dangerous fire hazard with heat, flame, oxidizers; mod. explosive with heat or flame; incompat. with strong oxidizers, strong acids, bases, potassium t-butoxide
Hazardous Decomp. Prods.: Isopropanol, acetic acid
NFPA: Health 1, Flammability 3, Reactivity 0
Storage: Store in cool, dry, well-ventilated area, out of direct sunlight, away from heat/ignition sources
Uses: Solvent for syn. resins (nitrocellulose, ethylcellulose, CAB, some vinyl copolymers, PS, methacrylate resins), gums, plastics, oils and fats, cosmetics, pharmaceuticals; solvent, flow aid, coalescent in paints/coatings, lacquers, printing inks; organic synthesis; odorant in perfumes; recovery of acetic acid from dil. aq. sol'ns.; synthetic flavoring agent in foods and pharmaceuticals; recovery and crystallization of drugs; in food-pkg. adhesives; in cellophane for food pkg.
Features: Sweet banana-like flavor
Regulatory: FDA 21CFR §172.515, 175.105, 177.1200; FEMA GRAS
Manuf./Distrib.: AMC Chems.; Aldrich; Allchem Ind.; Ashland; BP Amoco; BP Chems. Ltd; Brenntag AG; Celanese; Chemcentral; Coyne; Dow; Eastman; Fluka; Goldensun Mfg.; Harcros; Houghton Chem.; Hukill; Indofine; Integra; J.T. Baker; Penta Mfg.; Spectrum Quality Prods.; Stanchem; Sunnyside; Thomas Scientific; Vopak USA
Trade Names: Eastman® Isopropyl Acetate; Ucar® Isopropyl Acetate

Isopropylacetone. *See* Methyl isobutyl ketone
Isopropyl adipate. *See* Diisopropyl adipate

Isopropyl alcohol

CAS 67-63-0; EINECS/ELINCS 200-661-7
UN 1219 (DOT); FEMA 2929
Synonyms: Dimethyl carbinol; 2-Hydroxypropane; IPA; Isopropanol; 1-Methylethanol; 1-Methylethyl alcohol; Petrohol; 2-Propanol; Propan-2-ol; 2-Propyl alcohol; s-Propyl alcohol; Rubbing alcohol; Secondary propyl alcohol
Classification: Aliphatic alcohol
Empirical: C_3H_8O
Formula: $(CH_3)_2CHOH$
Properties: Colorless volatile clear liq.; pleasant alcohol-like odor, sl. bitter taste; sol. in water, alcohol, ether, chloroform, most org. solvs.; insol. in salt sol'ns.; m.w. 60.11; sp.gr. 0.7863 (20/20 C); f.p. -86 C; b.p. 82.4 C; flash pt. (TOC) 11.7 C; ref. index 1.3756 (20 C)
Toxicology: ACGIH TLV/TWA 400 ppm; STEL 500 ppm; LD50 (oral, rat) 5045 mg/kg, (IP, rat) 2735 mg/kg, (IV, rat) 1099 mg/kg; poison by ing., subcut. routes; mod. toxic by IV and IP routes; mildly toxic by skin contact; skin and eye irritant; human systemic effects by ing./inh. (headache, nausea, vomiting, narcosis); 100 ml can be fatal; may cause dry cracking skin; experimental teratogen, reproductive effector; mutagenic data; questionable carcinogen; TSCA listed
Environmental: VOC; BOD5 1.53; COD 2.23; ThOD 2.40
Precaution: DOT: Flamm. liq.; very dangerous fire hazard with heat, flame, oxidizers; reacts with air to form dangerous peroxides; incompat. with strong oxidizers, phosgene, iron salts, hydrogen-palladium, potassium t-butoxide, nitroform
Hazardous Decomp. Prods.: Heated to decomp., emits acrid smoke and fumes
NFPA: Health 1, Flammability 3, Reactivity 0
Storage: Store in tightly closed electrically grounded containers in a cool area separate from workplace; use in minimal quantities
Uses: Solvent for essential oils, alkaloids, gums, resins, cellulose derivs., coatings, lacquers, inks, cosmetics, pharmaceuticals; deicing agent for liq. fuels; dehydrating agent; denaturing ethyl alcohol; in quick-drying oils and inks; preservative in extraction processes; cosolvent for pesticides; mfg. of acetone, glycerol, isopropyl acetate; in antifreezes; in cosmetics, body rubs, after-shaves; synthetic flavoring agent; color diluent for foods and pharmaceuticals; antiseptic in topical anti-infectives; in food-pkg. adhesives; defoamer in food-contact coatings and paper/paperboard; in cellophane for food pkg.; in food-contact textiles; surf. lubricant in mfg. of food-contact metallic articles; food-contact textiles
Regulatory: FDA 21CFR §73.1 (no residue), 73.30, 73.315, 73.345, 73.615, 73.1001, 172.385, 172.515, 172.560, 172.665, 172.695, 172.712, 173.240, 173.340, 175.105, 176.180, 176.200, 176.210, 177.1200, 177.2800, 178.1010, 178.3910; 27CFR §21.112; use in bread is permitted in Ireland and Japan; FEMA GRAS; FDA approved for orals, topicals; USP/NF, BP, EP compliance; SARA §31/312 acute health/chronic health/fire hazard, §313 reportable
Manuf./Distrib.: AAE Chemie NV; AMC Chems.; AMRESCO; AXO Chem.; Aldrich; Alfa Chem; Allchem Ind.; Amyl; Arch Chems.; Ashland; BP Amoco; BP Chems. Ltd; Baychem; Blagden Spec. Chems. Ltd; Brenntag AG; C.P. Hall; Camida Ltd; Chemacon GmbH; Chemcentral; Coyne; Degussa AG; Dow; EM Ind.; EM Science; Eastman; ExxonMobil; Farleyway Chems. Ltd; Fisher Scientific; Fluka; GFS; General Chem.; Haltermann Prods. UK; Harcros; Houghton Chem.; Hukill; Hutchins & Hutchins; Integra; J.T. Baker; Lyondell; Mallinckrodt Baker; Merck KGaA; Mitsui Chems.; Mitsui Toatsu; Mutchler; Nippon Petrochems.; Penta Mfg.; Phoenix Nat. Prods.; Polysciences; R.E. Carroll; Romil Ltd; Ruger; Sal Chem.; Sasol Germany; Schaefer Tech.; Shell; Sigma; Spectrum Quality Prods.; Stanchem; Tech Spray; Tennants Distrib. Ltd; Thomas Scientific; Tokuyama Petrochem.; Ubichem plc; Uniqema; United Pharma Ltd; Univar E&E Ltd; VWR Int'l.; Veckridge; Veltek Assoc.; Voigt Global Distrib.; Vopak USA; Whyte Chems. Ltd
Trade Names: Eastman® Isopropanol; ExxonMobil IPA; Ucar® Isopropanol

Trade Names Containing: Filmex® 3C-1 190 Proof; Filmex® A-1, 190 Proof; Filmex® B 190 Proof; Filmex® B Anhydrous; Filmex® D-1; Punctilious® SDA 3C 190 Proof; S-336 Anhydrous; S-357 Anhydrous; S-358 Anhydrous; S-359 Anhydrous; Tecsol® A, 95%; Tecsol® A Anhydrous; Tecsol® B; Tecsol® D, 95%; Tecsol® D-1, 95%; Tecsol® H; Vertrel® X-P10

Isopropylamine

CAS 75-31-0; EINECS/ELINCS 200-860-9
UN 1221 (DOT)
Synonyms: 2-Aminopropane; 1-Methylethylamine; MIPA; Monoisopropylamine; 2-Propanamine; 2-Propaneamine; 2-Propylamine; s-Propylamine
Classification: Sat. aliphatic amine
Empirical: C_3H_9N
Formula: $(CH_3)_2CHNH_2$
Properties: Colorless liq.; ammonia-like odor; very sol. in acetone; sol. in water, ethanol, ether, benzene, chloroform, many org. solvs.; misc. with oxygenated solvs.; m.w. 59.11; dens. 0.6870 (20/20 C); vapor pressure 472 mm Hg (20 C); m.p. -101 C; b.p. 32. 4 C; flash pt. (CC) -37 C; pH 11.8 (0.1M aq. sol'n.)
Toxicology: ACGIH TLV/TWA 5 ppm; STEL 10 ppm; LD50 (oral, rat) 820 mg/kg, (skin, rabbit) 550 mg/kg; corrosive, very toxic; can be absorbed thru skin in toxic amts.; mod. toxic by ing.; mildly toxic by inh.; severe skin and eye irritant; possible permanent eye injury; inh. can cause nose/throat/lung irritation; narcotic in high concs.; high concs. can cause potentially fatal pulmonary edema; ing. may cause burning of mouth, throat, digestive tract; TSCA listed
Precaution: Flamm.; incompat. with oxidizers, acids, halogenated hydrocarbons, nitroparaffins, nitromethane, calcium or sodium hypochlorite, nitrosyl perchlorate, chloromethyloxirane, mercury, perchloryl fluoride; violent/explosive reactions possible
Hazardous Decomp. Prods.: Heated to decomp., emits toxic fumes of NO_x
NFPA: Health 3, Flammability 4, Reactivity 0
Storage: Store in cool, dry, well-ventilated areas, out of direct sunlight
Uses: Solvent, intermediate in synthesis of rubber accelerators, pharmaceuticals, dyes, insecticides, bactericides, textile specialties, surfactants; corrosion inhibitor; dehairing agent; solubilizer for 2,4-D acid; buffer in cosmetics
Manuf./Distrib.: Air Prods.; Aldrich; Ashland; Atofina; Celanese; Delta Distributors; Dow; Eastech; Fluka; Kira; Milsolv; P.B. & S.; Penta Mfg.; SOCO-Lynch; Sigma; Southchem; Spectrum Quality Prods.; Textile Chem.; United Pharma Ltd; Vopak USA; Whittaker, Clark & Daniels

Isopropylamine dodecylbenzenesulfonate

CAS 26264-05-1; 68584-24-7; EINECS/ELINCS 247-556-2
Synonyms: Dodecylbenzenesulfonic acid, compd. with isopropylamine (1:1); Dodecylbenzenesulfonic acid, compd. with 2-propanamine (1:1); Isopropylamine salt of dodecylbenzene sulfonic acid
Classification: Aromatic compd.
Definition: Salt of isopropylamine and dodecylbenzene sulfonic acid
Empirical: $C_{21}H_{39}NO_3S$
Formula: $C_{18}H_{30}O_3S \cdot C_3H_9N$
Properties: Anionic
Toxicology: TSCA listed
Uses: Emulsifier, solubilizer, solvent, detergent, and wetting agent for oil-based systems; dispersant in oil and water-based systems; in dry-cleaning surfactants; hydrotrope for liq. detergents; latex emulsifier; emulsion polymerization; surfactant, emulsifier, wetting agent for textiles; defoamer in food-contact paper/paperboard
Regulatory: FDA 21CFR §176.210
Trade Names: Rhodacal® 330

Isopropylamine salt of dodecylbenzene sulfonic acid. *See* Isopropylamine dodecylbenzenesulfonate
Isopropylbenzene; Isopropyl benzol. *See* Cumene
Isopropylbiphenyl. *See* 1,1´-Biphenyl (1-methylethyl)

Isopropyl bromide

CAS 75-26-3; EINECS/ELINCS 200-855-1
Synonyms: 2-Bromopropane; Propane, 2-bromo-; s-Propyl bromide
Empirical: C_3H_7Br
Formula: $BrCH(CH_3)_2$
Properties: Colorless to sl. yel. liq.; sol. in ethanol, ether; sl. sol. in water; dens. 1.304; f.p. -90 C; b.p. 58.5-60.5 C; flash pt. none; ref. index 1.4252
Toxicology: LD50 (IP, mouse) 4837 mg/kg; LC50 (inh., unspecified mammal) 36 g/m^3; harmful; may cause hemorrhage, changes to lungs, thorax, or respiration; reproductive effector; TSCA listed
Uses: Intermediate for organic synthesis, dyes, pharmaceuticals; solvent
Manuf./Distrib.: Amyl; Morre-Tec Ind.; Ocean Chems.; Yixing Fangqiao East Chemical

Isopropyl butanoate. *See* Isopropyl butyrate

Isopropyl butyrate

CAS 638-11-9; EINECS/ELINCS 211-320-7
FEMA 2935
Synonyms: Isopropyl butanoate
Empirical: $C_7H_{14}O_2$
Formula: $(CH_3)_2CHOOCC_3H_7$
Properties: Colorless liq., fruity odor; m.w. 130.19; dens. 0.859; b.p. 130-131 C; flash pt. 86 F; ref. index 1.3936
Toxicology: TSCA listed
Uses: Solvent for cellulose ethers; synthetic flavoring agent in foods and pharmaceuticals; fragrance in cosmetics
Features: Fruity flavor
Regulatory: FDA 21CFR §172.515; FEMA GRAS
Manuf./Distrib.: Aldrich; Goldensun Mfg.; Grau Aromatics; United Pharma Ltd

Isopropylcarbinol. *See* Isobutyl alcohol
Isopropyl 'Cellosolve'. *See* Ethylene glycol isopropyl ether
Isopropyl cresol; 6-Isopropyl-m-cresol; Isopropyl-m-cresol. *See* Thymol
3-Isopropylcumene. *See* 1,3-Diisopropyl benzene
4-Isopropylcumene. *See* 1,4-Diisopropyl benzene

Isopropylcyclohexane

CAS 696-29-7; EINECS/ELINCS 211-792-4
Synonyms: Cyclohexane, isopropyl-; Cyclohexane, (1-methylethyl)-; Hexahydrocumene; (1-Methylethyl) cyclohexane; Normanthane
Classification: Sat. alicyclic hydrocarbon
Empirical: C_9H_{18}
Properties: Colorless liq.; sol. in most org. solvs.; insol. in water; m.w. 126.24; sp.gr. 0.802 (20 C); vapor pressure 9.9 mm Hg (37.7 C); m.p. -95 C; b.p. 152-155 C; flash pt. 35 C; ref. index 1.441
Toxicology: May be harmful by ing., inh., skin absorp.; may cause mild skin irritation; vapor may be irritating to eyes; probably a CNS depressant; sl. nose/throat irritant; high vapor concs. may cause headache, lightheadedness, nausea, drowsiness; very high concs. may cause unconsciousness, death; ing. may cause sore throat, nausea, diarrhea
Environmental: Prevent entry into drains
Precaution: Flamm.; LEL 0.8%; forms explosive mixts. in air; vapor may flash back; container explosion may occur under fire conditions; take precautions against static discharges; incompat. with strong oxidizing agents (increased fire and explosion hazard)
Hazardous Decomp. Prods.: Toxic fumes of CO, CO_2
NFPA: Health 1, Reactivity 0
Storage: Store in cool, dry, well-ventilated area out of direct sunlight, away from heat, sparks, open flame, other ignition sources; keep tightly closed
Uses: Solvent; org. synthesis
Manuf./Distrib.: Acros Org.; Aldrich; Fluka; ICN Biomed. Research Prods.; TCI Am.

Isopropylcyclopentane

CAS 3875-51-2
Synonyms: Cyclopentane, isopropyl-; Cyclopentane, (1-methylethyl)-; 2-Cyclopentylpropane; Cyclopentyl-2 propane; (1-Methylethyl) cyclopentane; Propane, 2-cyclopentyl-
Classification: Sat. alicyclic hydrocarbon
Empirical: C_8H_{16}
Formula: $C_5H_9CH(CH_3)CH_3$
Properties: Colorless liq.; sol. in most org. solvs.; insol. in water; m.w. 112.22; sp.gr. 0.7765 (20 C); m.p. -111 C; b.p. 126 C; ref. index 1.4258

Toxicology: May cause mild skin irritation; conc. vapor may be irritating to eyes; probably a CNS depressant; high vapor concs. may cause nose/throat irritation, lightheadedness, nausea, dizziness; very high concs. may cause unconsciousness and death; ing. may cause sore throat, nausea, diarrhea
Precaution: Flamm.; incompat. with strong oxidizing agents (increased fire and explosion hazard)
Hazardous Decomp. Prods.: None
Storage: Store in cool, dry, well-ventilated area out of direct sunlight
Uses: Solvent; org. synthesis

Isopropyl dimethyl carbinol. *See* 2-Methyl-1-pentanol
Isopropyl diphenyl. *See* 1,1´-Biphenyl (1-methylethyl)
Isopropylene glycol. *See* Propylene glycol
Isopropyl ethanoate. *See* Isopropyl acetate

Isopropyl ether

CAS 108-20-3; EINECS/ELINCS 203-560-6
UN 1159 (DOT)
Synonyms: Diisopropyl ether; Diisopropyl oxide; 2-Isopropoxypropane
Classification: Ether
Empirical: $C_6H_{14}O$
Formula: $(CH_3)_2CHOCH(CH_3)_2$
Properties: Colorless volatile liq.; sweet ethereal odor; sol. in oxygenated solvs.; misc. with water, ethanol, ether; m.w. 102.20; dens. 0.719; vapor pressure 0.149 mm Hg; m.p. -60 C; b.p. 68.5 C; flash pt. (CC) -28 C; autoignition temp. 443 C; ref. index 1.368 (20 C); surf. tens. 17.34 dynes/cm; dielec. const. 3.88
Toxicology: ACGIH TLV/TWA 250 ppm, STEL 310 ppm; LD50 (oral, rat) 8470 mg/kg, (dermal, rabbit) 20 g/kg; LC50 (inh., rabbit) 121 g/m^3; mod. toxic by IP route; mildly toxic by ing., inh., skin contact; skin irritant; target organs: respiratory system, skin; TSCA listed
Environmental: VOC; BOD5 0.19; COD 1.75; ThOD 2.82
Precaution: DOT: Flamm. liq.; severe explosion hazard when exposed to heat, flame, sparks, or oxidizers; may form explosive peroxides; shock may cause explosion; violent reaction with chlorosulfonic acid, HNO_3
Hazardous Decomp. Prods.: Heated to decomp., emits acrid smoke and fumes
NFPA: Health 1, Flammability 3, Reactivity 1
Storage: Hygroscopic; refrigerate
Uses: Solvent for paints, stain removers, dyes, animal, vegetable, and min. oils, waxes, and resins; extraction solvent (phosphoric acid, phenol purification); rubber cements; reagent; paint and varnish remover; spotting compositions
Manuf./Distrib.: Aldrich; Allchem Ind.; Ashland; ExxonMobil; Fluka; J.T. Baker; Romil Ltd; Shell; Spectrum Quality Prods.

Isopropylethylene glycol ether. *See* Ethylene glycol isopropyl ether
Isopropylformic acid. *See* Isobutyric acid
Isopropyl glycol. *See* Ethylene glycol isopropyl ether
Isopropyl hexadecanoate; Isopropyl n-hexadecanoate. *See* Isopropyl palmitate
Isopropyl-S-(-)-2-hydroxy propionate. *See* Isopropyl-S-lactate
Isopropylideneacetone. *See* Mesityl oxide
4-Isopropylidene-1-methylcyclohexene. *See* Terpinolene
Isopropyl ketone. *See* Diisopropyl ketone

Isopropyl lactate

CAS 617-51-6
Synonyms: Lactic acid isopropyl ester; 1-Methylethyl 2-hydroxypropanoate; Propanoic acid, 2-hydroxy-, 1-methylethyl ester
Definition: Isopropyl ester of lactic acid
Empirical: $C_6H_{12}O_3$
Formula: $CH_3CH(OH)COOCH(CH_3)_2$
Properties: M.w. 132.18
Toxicology: LDLo (IM, guinea pig) 2500 mg/kg
Uses: Solvent for paints and coatings, electronics, printed circuit boards, metal industry
Manuf./Distrib.: PURAC Am.; Penta Mfg.

(-)-Isopropyl L-lactate. *See* Isopropyl-S-lactate

Isopropyl-S-lactate

CAS 63697-00-7; EINECS/ELINCS 264-417-1
Synonyms: Isopropyl-S-(-)-2-hydroxy propionate; (-)-Isopropyl L-lactate; Isopropyl-S-(-)-lactate; Lactic acid isopropyl ester
Classification: Lactate ester
Empirical: $C_6H_{12}O_3$
Properties: Colorless liq.; m.w. 132.16; sp.gr. 0.988; b.p. 166-168 C; flash pt. 57 C; ref. index 1.41 (20 C)
Toxicology: May be harmful by ing., inh., skin absorp.; may cause eye/skin irritation
Precaution: Combustible; incompat. with strong oxidizing agents, strong acids, strong bases, strong reducing agents
Hazardous Decomp. Prods.: Toxic fumes of CO, CO_2
Storage: Store in cool, dry place away from heat, open flame; keep tightly closed
Uses: Solvent
Manuf./Distrib.: Acros Org.; Aldrich
Trade Names: Purasolv® IPL

Isopropyl-S-(-)-lactate. *See* Isopropyl-S-lactate

Isopropyl lanolate

CAS 63393-93-1; EINECS/ELINCS 264-119-1
Synonyms: Fatty acids, lanolin, isopropyl esters; IPL; Lanolin fatty acids, isopropyl esters
Definition: Ester of isopropyl alcohol and lanolin acid
Properties: Pale yel. liq. or paste; HLB 9.0; nonionic
Toxicology: Eye and skin irritant; may cause skin sensitization; TSCA listed
Hazardous Decomp. Prods.: Heated to decomp., emits acrid smoke and fumes
Uses: Conditioner, penetrant, lubricant, moisturizer, antistat, emollient, w/o emulsifier, stabilizer, opacifier for cosmetics, pharmaceuticals; pigment dispersant; wetting agent and dispersant for solids; plasticizer for wax and pigment systems
Manuf./Distrib.: A&E Connock; Charkit; Costec; Croda/Westbrook; Fanning; Indofine; R.I.T.A.; Thornley; Universal Preserv-A-Chem
Trade Names: Lanesta L

Isopropyl laurate

CAS 10233-13-3; EINECS/ELINCS 233-560-1
Synonyms: Dodecanoic acid, 1-methylethyl ester; 1-Methylethyl dodecanoate
Definition: Ester of isopropyl alcohol and lauric acid
Empirical: $C_{15}H_{30}O_2$
Formula: $CH_3(CH_2)_{10}COOCH(CH_3)_2$
Toxicology: TSCA listed
Uses: Emollient, lubricant, plasticizer, and cosolvent in cosmetics and pharmaceuticals; defoamer in food-contact paper/paperboard; in food-contact textiles
Regulatory: FDA 21CFR §176.210, 177.2800
Manuf./Distrib.: A&E Connock
Trade Names: Lexol® IPL

Isopropyl metacresol. *See* Thymol
1-Isopropyl-4-methylbenzene; 4-Isopropyl-1-methylbenzene. *See* p-Cymene
1-Isopropyl-4-methyl-1,4-cyclohexadiene. *See* γ-Terpinene
2-Isopropyl-5-methylcyclohexanol. *See* Menthol

1-Isopropyl-2-methyl imidazole

CAS 84606-45-1
Synonyms: IMI
Properties: M.w. 124.19; dens. 0.950 g/ml; visc. 2.84 cp; b.p. 217 C; flash pt. (CC) 220 F; surf. tens. 36.19 dynes/cm
Uses: Solvent

Isopropyl methyl ketone. *See* Methyl isopropyl ketone
2-Isopropyl-5-methylphenol. *See* Thymol

Isopropyl myristate

CAS 110-27-0; EINECS/ELINCS 203-751-4
FEMA 3556
Synonyms: IPM; Isopropyl tetradecanoate; 1-Methylethyl tetradecanoate; Myristic acid isopropyl ester; Tetradecanoic acid isopropyl ester; Tetradecanoic acid, 1-methylethyl ester; 1-Tridecanecarboxylic acid,

isopropyl ester
Classification: Synthetic fatty alcohol
Definition: Ester of isopropyl alcohol and myristic acid
Empirical: $C_{17}H_{34}O_2$
Formula: $CH_3(CH_2)_{12}COOCH(CH_3)_2$
Properties: Colorless to pale yel. oily low-visc. liq., odorless; sol. in most org. solvs., acetone, chloroform, ethyl acetate, ethanol, min. oil, veg. oil; dissolves waxes; insol. in water, glycerol; m.w. 270.44; dens. 0.850-0.860; f.p. 3 C; b.p. 192.6 C (20 mm); dec. 208 C; acid no. 1 max.; iodine no. 1 max.; sapon. no. 202-212; cloud pt. -3 to 5 C; ref. index 1.432-1.436 (20 C)
Toxicology: LD50 (oral, mouse) 49,700 mg/kg, (skin, rabbit) 5 g/kg; low toxicity by ing. and skin contact; human skin irritant; causes blackheads; suspected tumorigen; TSCA listed
Hazardous Decomp. Prods.: Heated to decomp., emits toxic smoke and fumes
HMIS: Health 1, Flammability 0, Reactivity 0
Storage: Light-sensitive
Uses: Emulsifier, emollient, lubricant, vehicle, solvent, solubilizer, spreading agent in cosmetic creams, topical medicinals, industrial applics.; plasticizer for cellulosics; pigment dispersant; sewing thread lubricant; binder; synthetic flavoring agent in foods; carrier, solvent for flavors; defoamer in food-contact paper/paperboard; in food-contact textiles
Features: Oleaginous
Regulatory: FDA 21CFR §176.210, 177.2800; FEMA GRAS; FDA approved for topicals; NF, BP, EP compliance
Manuf./Distrib.: A&E Connock; AMC Chems.; AXO Chem.; Akzo Nobel; Aldrich; Allan; Allchem Int'l. Ltd; Alzo; Am. Int'l.; Amerchol; Ashland; Atofina Bruxelles; Berje; Blagden Spec. Chems. Ltd; Burdick & Jackson; C.P. Hall; CasChem; Charkit; ChemService; Chemacon GmbH; Chesham Chems. Ltd; CoKEM Assoc.; Cognis/Chems. Group; Cornelius Chem. Co. Ltd; Creative Fragrances; Croda Inc; Fanning; Fluka; GMI Prods.; Givaudan Fragrances; Goldschmidt; Haltermann GmbH; Haltermann Prods. UK; Inolex; Integra; Jarchem Ind.; KIC Chems.; Kraft Chem.; Lanaetex Prods.; Lipo; MelChem; Mosselman NV; NOF Am.; Norman, Fox; Parchem Trading; Penta Mfg.; Phoenix Nat. Prods.; Phoenix; Protameen; R.I.T.A.; RTD Chems.; Robeco; Ruger; Sea-Land; Sigma; Sithean; Spec. Ingreds. & Perfume; Spectrum Quality Prods.; St. Lawrence; Stepan; Thew Arnott & Co. Ltd; Thornley; U.S. Synthetics; Ubichem plc; Uniqema Am.; United Pharma Ltd; Univar E&E Ltd; Universal Preserv-A-Chem; VWR Int'l.; Vopak USA
Trade Names: Crodamol IPM; Estol IPM; Estol IPM 1509 (BIO-IPM); HallStar™ IPM; Lexol® IPM; Lexol® IPM-NF; Polypax IPM; Promyr; STEPAN® IPM; STEPAN® IPM-NF; Super Refined® Crodamol IPM; Tegosoft® M; Thomil IS; Thomil IS-F; ZELEC® IPM
Trade Names Containing: Crodamol CAP; Lexol® 3975; Wickenol® 136

Isopropyl myristate/palmitate. *See* Isopropyl myristopalmitate

Isopropyl myristopalmitate

Synonyms: Isopropyl myristate/palmitate; Isopropyl palmitate/myristate
Definition: Ester of isopropyl alcohol and mixture of myristic and palmitic acids
Properties: Sp.gr. 0.852; flash pt. (COC) 182 C; ref. index 1.4340
Uses: Emollient, solvent in cosmetics; skin emollient, lubricant, carrier used in lotions, creams, antiperspirants, bath oils; plasticizer for PVC; lubricant for PS
Manuf./Distrib.: Inolex; Int'l. Paper

Isopropyl oleate

CAS 112-11-8; 85116-87-6; EINECS/ELINCS 203-935-4; 285-540-7
Synonyms: 1-Methylethyl-9-octadecenoate; 9-Octadecenoic acid, 1-methylethyl ester
Definition: Ester of isopropyl alcohol and oleic acid
Empirical: $C_{21}H_{40}O_2$
Formula: $CH_3(CH_2)_7CH{=}CH(CH_2)_7COOCH(CH_3)_2$
Properties: Colorless liq.; odorless; misc. with paraffin oil; insol. in water; m.w. 324.55; dens. 0.87 kg/l (20 C); pour pt. -21 C; flash pt. (COC) 185 C
HMIS: Health 1, Flammability 0, Reactivity 0
Uses: Emollient, emulsifier, bodying agent, and spreading agent in cosmetics and toiletries; lubricant for makeup; lubricity additive for lubricants; plasticizer for rubber; defoamer in food-contact paper/paperboard; surf. lubricant in mfg. of food-contact metallic articles; adjuvant to improve lubricity in mineral oil lubricants for incidental food-contact use; food-contact textiles
Regulatory: FDA 21CFR §176.210, 177.2800, 178.3570, 178.3910
Manuf./Distrib.: A&E Connock; ChemService; Inolex; Mosselman NV; Ruger; Uniqema Am.
Trade Names: Tilol IS

3-Isopropyl-2-oxazolidinone

CAS 40482-44-8
Synonyms: IO
Properties: M.w. 129.16; dens. 1.068 g/ml; visc. 3.73 cp; b.p. 151 C (29 mm Hg); surf. tens. 37.93 dynes/cm
Uses: Solvent

Isopropyl palmitate

CAS 142-91-6; EINECS/ELINCS 205-571-1
Synonyms: Hexadecanoic acid isopropyl ester; Hexadecanoic acid, 1-methylethyl ester; IPP; Isopropyl hexadecanoate; Isopropyl n-hexadecanoate; 1-Methylethyl hexandecanoate
Definition: Ester of isopropyl alcohol and palmitic acid
Empirical: $C_{19}H_{38}O_2$
Formula: $CH_3(CH_2)_{14}COOCH(CH_3)_2$
Properties: Colorless mobile liq., very sl. odor; sol. in 4 parts 90% alcohol, min. oil, fixed oils, acetone, castor oil, chloroform, cottonseed oil, ethyl acetate, oxygenated solvs.; insol. in water, glycerin, propylene glycol; m.w. 298.57; dens. 0.850-0.855; m.p. 14 C; acid no. 1 max.; iodine no. 1 max.; sapon. no. 183-193; cloud pt. 12-14 C; flash pt. > 230 F; ref. index 1.4350-1.4390 (20 C)
Toxicology: LD50 (IP, mouse) 100 mg/kg; poison by IP route; human skin irritant; TSCA listed
Precaution: Combustible
Hazardous Decomp. Prods.: Heated to decomp., emits acrid smoke and fumes
HMIS: Health 1, Flammability 0, Reactivity 0
Storage: Light-sensitive
Uses: Antistat, binder, emollient, lubricant, emulsifier, vehicle, solvent, cosolvent, stabilizer in cosmetic creams/lotions, deodorants, industrial applics.; vehicle in topical pharmaceuticals; plasticizer for cellulosics; extender for fragrances; metal stamping oils; defoamer in food-contact paper/paperboard; in food-contact textiles
Features: Oleaginous
Regulatory: FDA 21CFR §176.210, 177.2800; FDA approved for topicals; NF, BP, EP compliance
Manuf./Distrib.: A&E Connock; AMC Chems.; Akzo Nobel; Aldrich; Alfa Chem; Allan; Allchem Int'l. Ltd; Alzo; Am. Int'l.; Amerchol; Ashland; Atofina Bruxelles; Berje; CasChem; Charkit; ChemService; Chesham Chems. Ltd; CoKEM Assoc.; Cognis KGaA/Care Chems.; Cognis/Chems. Group; Croda Inc; Fanning; Fluka; Givaudan Fragrances; Goldschmidt; Haltermann Prods. UK; Hawkins Chem.; Inolex; Integra; Jarchem Ind.; KIC Chems.; Kraft Chem.; Lanaetex Prods.; Lipo; Mosselman NV; NOF Am.; Norman, Fox; Parchem Trading; Penta Mfg.; Phoenix Nat. Prods.; Phoenix; Protameen; R.I.T.A.; RTD Chems.; Robeco; Ruger; Sea-Land; Sithean; Spec. Ingreds. & Perfume; Spectrum Quality Prods.; St. Lawrence; Stepan; Thew Arnott & Co. Ltd; Thornley; U.S. Synthetics; Uniqema Am.; Univar E&E Ltd; Universal Preserv-A-Chem; Vopak USA
Trade Names: Crodamol IPP; Palmil IS; Polypax IPP; Propal NF; STEPAN® IPP; STEPAN® IPP-NF; Tegosoft® P
Trade Names Containing: Lexol® 3975; Wickenol® 136

Isopropyl palmitate/myristate. *See* Isopropyl myristopalmitate

Isopropyl PPG-2-isodeceth-7-carboxylate

Synonyms: PEG-7-PPG-2 isodecyl ether carboxylci acid, isopropyl ester; POE(7) POP(2) isodecyl ether carboxylic acid isopropyl ester
Classification: Organic compd.
Empirical: $C_{33}H_{66}O_{11}$
Formula: $C_{10}H_{21}(OC(CH_3)HCH_2)_x(OCH_2CH_2)_yOCH_2COOCCH_3HCH_3$, avg. x = 2, avg. y = 6
Uses: Solvent, emollient, skin conditioner, solubilizer in cosmetics, skin care, sun care, hair care, bath prods., pigmented prods.

Isopropyl stearate

CAS 112-10-7; EINECS/ELINCS 203-934-9
Synonyms: 1-Methylethyl octadecanoate; Octadecanoic acid, 1-methylethyl ester; Stearic acid, isopropyl ester
Definition: Ester of isopropyl alcohol and stearic acid
Empirical: $C_{21}H_{42}O_2$
Formula: $CH_3(CH_2)_{16}COOCH(CH_3)_2$
Properties: Pale yel. liq.; m.w. 326.63; m.p. 16-20 C
Toxicology: LDLo (oral, rat) 8 g/kg; mildly toxic by ing.; primary skin irritant; TSCA listed
Hazardous Decomp. Prods.: Heated to decomp., emits acrid smoke and irritating vapors
Uses: Binder, emollient, cosolvent, lubricant, refatting agent, skin softener, moisture retention aid for cosmetics, toiletries, hair care, skin care, sunscreens, soaps, topical pharmaceuticals; lubricant for textile processing; solubilizer for essential oils and lipophilic act. ingreds.; defoamer in food-contact paper/paperboard; in food-contact textiles
Regulatory: FDA 21CFR §176.210, 177.2800; FDA approved for topicals
Manuf./Distrib.: A&E Connock; ChemService; Cognis/Chems. Group; Mosselman NV
Trade Names: Monestriol IS-C; Wickenol® 127
Trade Names Containing: Lexol® 3975; Wickenol® 136

Isopropyl tetradecanoate. *See* Isopropyl myristate
4-Isopropyltoluene; p-Isopropyltoluene. *See* p-Cymene
Isopropyltrimethylmethane. *See* 2,2,3-Trimethylbutane

Isoquinoline

CAS 119-65-3; EINECS/ELINCS 204-341-8
FEMA 2978
Synonyms: 2-Azanaphthalene; 2-Benzazine; Benzo(c)pyridine; Leucoline
Empirical: C_9H_7N
Properties: Colorless plates or liq.; pungent odor; sol. in most org. solvs.; misc. with oxygenated solvs.; sl. sol. in water; m.w. 129.16; dens. 1.09 (20/4 C); m.p. 23-25 C; b.p. 243 C; flash pt. 102 C; ref. index 1.615
Toxicology: LD50 (oral, rat) 360 mg/kg, (skin, rabbit) 590 mg/kg; toxic by ing.; mod. toxic by skin contact; severe skin and eye irritant; readily absorbed thru skin; mutagen; TSCA listed
Precaution: Combustible
Hazardous Decomp. Prods.: Heated to decomp., emits toxic fumes of NO_x
Uses: Mfg. of pharmaceuticals (nicotinic acid), dyes, insecticides, rubber accelerators; organic synthesis; synthesis of antimalarials; synthetic flavoring agent in foods and pharmaceuticals
Features: Sweet flavor
Regulatory: FDA 21CFR §172.515; FEMA GRAS
Manuf./Distrib.: Advanced Synthesis Tech.; Aldrich; Fluka; Sigma
Trade Names Containing: Methylnaphthalene Fraction

Isosorbide dimethyl ether. *See* Dimethyl isosorbide

Isostearyl alcohol

CAS 27458-93-1; EINECS/ELINCS 248-470-8
Synonyms: C18 mixed alcohols; Isooctadecanol; Isooctadecan-1-ol; 2-Octyl-1-decanol
Definition: Mixture of branched chain 18 carbon aliphatic alcohols
Empirical: $C_{18}H_{38}O$
Formula: $C_{18}H_{37}OH$
Properties: Insol. in water; m.w. 270.50; flash pt. (COC) 170 C
Toxicology: TSCA listed
Uses: Coemulsifier; lubricant; foam control agent; cosolvent; plasticizer; stabilizer; intermediate; metal lubricants; inks; textiles; emulsions; paper; mineral processing; oil field chemicals; fabric softener; emollient, visc. control agent for cosmetics, skin care emulsions, pharmaceuticals
Manuf./Distrib.: Albemarle; Ashland; Avatar; Celanese; Fluka; Indofine; Jarchem Ind.; M. Michel; Uniqema Am.; Universal Preserv-A-Chem
Trade Names: Diadol 18G-C; Jarcol™ I-18EX

Isostearyl isononanoate

Synonyms: Hexanoic acid, 3,5,5-trimethyl-, isooctadecyl ester; Isooctadecyl isononanoate
Definition: Ester of isostearyl alcohol and isononanoic acid
Empirical: $C_{27}H_{54}O_2$
Formula: $CH_3(CH_2)_5CHCH_3COOC_{18}H_{37}$
Uses: Emollient, moisturizer, hair conditioner in cosmetics; superfatting agent, solvent for lipophilic components

Isostearyl isostearate

CAS 41669-30-1; EINECS/ELINCS 255-485-3
Synonyms: Isooctadecanoic acid, isooctadecyl ester; Isooctadecyl isooctadecanoate; 16-Methylheptadecanoic acid, 16-methylheptadecyl ester; 16-Methylheptadecyl 16-methylheptadecanoate
Definition: Ester of isostearyl alcohol and isostearic acid
Empirical: $C_{36}H_{72}O_2$
Properties: Pale yel. liq.; sol. in hydrocarbons; insol. in water; m.w. 536.98; sapon. no. 95-110; nonionic
Toxicology: No known toxicity; TSCA listed
Uses: Binder, emollient, slip agent, lubricant, gloss aid for cosmetics, toiletries; cosolvent and solubilizer in perfumes; emollient, lubricant for pharmaceuticals
Manuf./Distrib.: A&E Connock; Gattefosse; Mosselman NV
Trade Names: Estalan-ISI; Schercemol 1818

Isostearyl lactate

CAS 42131-28-2; EINECS/ELINCS 255-674-0
Synonyms: 2-Hydroxypropanoic acid, isostearyl ester; Isooctadecyl 2-hydroxypropanoate
Definition: Ester of isostearyl alcohol and lactic acid
Empirical: $C_{21}H_{42}O_3$
Formula: $CH_3CHOHCOOC_{18}H_{37}$
Uses: Emollient, substantivity agent, softener, slip agent, lubricant, cosolvent, solubilizer in cosmetics and pharmaceuticals
Manuf./Distrib.: A&E Connock

Isotetracosanol; Isotetracosyl alcohol. *See* 2-Decyl-1-tetradecanol
Isotridecanol ethoxylates. *See* Trideceth

Isotridecyl isononanoate

CAS 42131-27-1; 59231-37-7; EINECS/ELINCS 255-673-5; 261-675-7
Synonyms: Isononanoic acid, isotridecyl ester; Isotridecyl 3,5,5-trimethylhexanoate; Isotridecyl 3,5,5-trimethylhexanoic acid; 3,5,5-Trimethylhexanoic acid, isotridecyl ester
Definition: Ester of isotridecyl alcohol and isononanoic acid
Empirical: $C_{22}H_{44}O_2$
Formula: $(CH_3)_3CCH_2CCH_3HCH_2COOC_{13}H_{27}$
Properties: Sol. in alcohol, animal and vegetable oils, min. oil; insol. in water, glycerin, propylene glycol
Toxicology: TSCA listed
Uses: Emollient, solvent in cosmetics, hair care, skin care; pigment wetting agent
Manuf./Distrib.: A&E Connock
Trade Names: Wickenol® 153

Isotridecyl octadecanoate. *See* Isotridecyl stearate

Isotridecyl stearate

CAS 31565-37-4; EINECS/ELINCS 250-703-3
Synonyms: Isotridecyl octadecanoate
Definition: Ester of isotridecyl alcohol and stearic acid
Empirical: $C_{31}H_{62}O_2$
Toxicology: TSCA listed
Uses: Lubricant for PVC, metalworking oils, greases, extrusion of ceramics; surfactant for industrial detergents, cosmetics, toiletries, feed, foods, textiles, agric., lubricants, metal treatment, and petroleum/mining/road construction industries
Manuf./Distrib.: A&E Connock; Dasico A/S; Expocell AB; Kao Corp.; Walther AG
Trade Names: Excepari TD-S

Isotridecyl 3,5,5-trimethylhexanoate; Isotridecyl 3,5,5-trimethylhexanoic acid. *See* Isotridecyl isononanoate
Isovalerone. *See* Diisobutyl ketone

JK

Japan camphor. *See* Camphor
Japanese camphor oil. *See* Camphor (Cinnamomum camphora) oil
Juglans regia; Juglans regia oil. *See* Walnut (Juglans regia) oil
Junipen. *See* Longifolene
Junipene. *See* Longifolene

Karite butter. *See* Shea butter (Butyrospermum parkii)
Katchung oil. *See* Peanut (Arachis hypogaea) oil
Kautschin. *See* d-Limonene

Kerosene

CAS 8008-20-6; EINECS/ELINCS 232-366-4
UN 1223 (DOT)
Synonyms: Coal oil; Kerosine; Kerosine (petroleum); Straight-run kerosene; Straight-run kerosine
Classification: Aliphatic petroleum hydrocarbon
Definition: Complex combination of hydrocarbons chiefly of the methane series, produced by distillation of crude oil; consists of hydrocarbons predominantly in C9-16 range, boiling range 150-290 C
Properties: Water-wh. to pale yel. oily liq.; mild char. petrol. odor; sol. in other petrol. solvs.; insol. in water; sp.gr. 0.78; vapor pressure 1.5 mm Hg (20 C); f.p. -45.6 C; b.p. 151-301 C; flash pt. 122 F; ref. index 1.437; KB value 36
Toxicology: LD50 (oral, rabbit) 28 g/kg, (IP, rabbit) 6600 mg/kg, (IV, rabbit) 180 mg/kg; LDLo (oral, human male) 500 mg/kg; poison by IV and intratracheal routes; corrosive; causes burns; extremely destructive to tissue of mucous membranes, upper respiratory tract, eyes, skin; inh. may cause spasm/inflamm./edema of larynx, bronchi, chem. pneumonitis, pulmonary edema; may cause coughing, wheezing, laryngitis, headache, nausea, vomiting, somnolence; tumorigen, mutagen; TSCA listed
Environmental: VOC; BOD5 0.53; ThOD 3.46
Precaution: Combustible exposed to heat or flame; LEL 0.7%; UEL 5%; mod. explosive as vapor exposed to heat or flame; vapor may flash back; risk of fire and explosion; incompat. with strong oxidizing agents, strong acids, strong bases, amines
Hazardous Decomp. Prods.: CO, CO_2, acrid smoke and fumes; emits toxic fumes under fire conditions
NFPA: Health 0, Flammability 2, Reactivity 0
Storage: Store in a cool, dry, well-ventilated area, away from heat, open flame, ignition sources; keep tightly closed
Uses: Fuel for lamps, flares, stoves, rockets, jet engines; diesel fuel for automobiles; domestic heating; vehicle for paints, thinners, enamels, polishes, varnishes; carrier for pesticides; cleaning degreasing solvent; mold release agent for ceramics; solvent in asphalt coatings, cutback bitumens, printing inks; mineral flotation agent; lubricant for plastics processing; in food-pkg. adhesives; defoamer in food-contact coatings and paper/paperboard; in food-contact textiles
Regulatory: FDA 21CFR §175.105, 176.200, 176.210, 177.2800; contains HAPs
Manuf./Distrib.: Aldrich; Fluka; Houghton Chem.; Sigma; Spectrum Quality Prods.; Stanchem
Trade Names Containing: Amsol 230-K Solvent

Kerosene, deodorized. *See* Deodorized kerosene

Kerosene, hydrodesulfurized

CAS 64742-81-0; EINECS/ELINCS 265-184-9
Synonyms: Hydrodesulfurized kerosene; Kerosine (petroleum), hydrodesulfurized
Definition: Complex combination of hydrocarbons obtained from a petrol. stock by treating with hydrogen to convert org. sulfur to hydrogen sulfide which is removed; consists of hydrocarbons in C9-16 range
Properties: B.p. 150-290 C
Toxicology: LD50 (oral, rat) 5000 mg/kg; skin irritant
Precaution: Combustible liq.
Uses: Solvent
Trade Names: Ketrul 212; Ketrul HT

Kerosene, odorless. *See* Deodorized kerosene
Kerosene (petroleum), hydrotreated. *See* Petroleum distillates, hydrotreated light
Kerosine; Kerosine (petroleum). *See* Kerosene
Kerosine (petroleum), hydrodesulfurized. *See* Kerosene, hydrodesulfurized
Ketohexamethylene. *See* Cyclohexanone
Ketone, butyl methyl. *See* 2-Hexanone
Ketone C-7. *See* Methyl n-amyl ketone
Ketone, dimethyl. *See* Acetone
Ketone, methyl isoamyl. *See* Methyl isoamyl ketone
Ketone methyl phenyl. *See* Acetophenone
Ketone propane; β-Ketopropane. *See* Acetone
8-Ketotricyclo (5.2.1.0^{2,6}) decane. *See* Tricyclo [5.2.1.0^{2,6}] decane-8-one
2-Keto-1,7,7-trimethylnorcamphane. *See* Camphor

L

Lactamide DGA

CAS 123792-29-0

Synonyms: Lactamide N-(1-ethoxy-2-hydroxyethyl); Propanamide, 2-hydroxy-N-(1-ethoxy-2-hydroxyethyl)

Definition: Ethoxylated monoethanolamine (DGA) of lactic acid

Properties: Nonionic

Uses: Humectant, emollient, o/w emulsifier, cosolvent for hair care, skin care; gloss aid, conditioner, moisturizer, antistat for hair care

Trade Names: Foamid LM2E

Lactamide N-(1-ethoxy-2-hydroxyethyl). *See* Lactamide DGA

Lactamide MEA

CAS 5422-34-4; EINECS/ELINCS 226-546-1

Synonyms: N-2-Hydroxyethyllactamide; 2-Hydroxy-N-(2-hydroxyethyl) propanamide; Lactic acid monoethanolamide; Monoethanolamine lactic acid amide

Definition: Mixture of ethanolamides of lactic acid

Empirical: $C_5H_{11}NO_3$

Formula: $CH_3COHHCONHCH_2CH_2OH$

Properties: M.w. 133.17; nonionic

Toxicology: LD50 (subcut., rat) 24 g/kg, (IP, mouse) 16,300 mg/kg; mutagen; TSCA listed

Uses: Antistat in cosmetics; surfactant, thickener, foam booster/stabilizer for personal care and industrial applics.; conditioner, clarifier, detangling agent for shampoos and cream rinses; humectant, lubricant in creams and lotions

Manuf./Distrib.: Sigma

Trade Names: Naetex-LAM

Lactic acid

CAS 50-21-5; 598-82-3 (DL); 79-33-4 (L); 10326-41-7 (D); EINECS/ELINCS 200-018-0; 209-954-4 (DL); 201-296-2 (L); 233-713-2 (D)

UN 1760; FEMA 2611; INS270; E270

Synonyms: Acetonic acid; Ethylidenelactic acid; 1-Hydroxyethanecarboxylic acid; 1-Hydroxyethane 1-carboxylic acid; 2-Hydroxypropanoic acid; 2-Hydroxypropionic acid; α-Hydroxypropionic acid; Milk acid; Propanoic acid, 2-hydroxy-; Propionic acid, 2-hydroxy-

Classification: Organic acid

Definition: Prod. of the metabolism of glucose and glycogen

Empirical: $C_3H_6O_3$

Formula: $CH_3CHOHCOOH$

Properties: Colorless to ylsh. cryst. or syrupy liq., nearly odorless; misc. with water, alcohol, glycerol, furfural; insol. in chloroform; m.w. 90.09; dens. 1.249; m.p. 18 C; b.p. 122 C (15 mm); flash pt. 110 C; ref. index 1.4251; anionic

Toxicology: LD50 (oral, rat) 3730 mg/kg, (subcut., mouse) 4500 mg/kg; LDLo (rectal, rabbit) 1200 mg/kg; mod. toxic by ing., rectal routes; corrosive; severe skin and eye irritant; mutagenic data; TSCA listed

Environmental: VOC; ThOD 0.53

Precaution: DOT: Corrosive material; mixts. with nitric acid + hydrofluoric acid may react vigorously

Hazardous Decomp. Prods.: Heated to decomp., emits acrid smoke and irritating fumes

HMIS: Health 3, Flammability 1, Reactivity 0

Storage: Hygroscopic

Uses: Cultured dairy prods.; acidulant; chemicals (salts, plasticizers, adhesives, pharmaceuticals); mordant in wool dyeing; mfg. of lactates; sequestrant in metal plating; cataphoresis paints; leather tanning delimer; electroplating bath additive; acidulant, antimicrobial, preservative, flavoring agent for foods, pharmaceuticals; solvent, vehicle for foods; buffer, humectant, skin conditioner in cosmetics; keratin softener for skin care prods.; treatment of infective skin and vaginal disorders, dermatological problems, e.g., warts; ingred. in IV fluids; prep. of emulsifiers

Use Level: Up to 10%

Regulatory: FDA 21CFR §131.144, 133.123, 133.124, 133.129, 133.169, 133.173, 133.178, 133.179, 150.141, 150.161, 172.814, 178.1010, 184.1061, GRAS; USDA 9CFR §318.7, 381.147; BATF 27CFR §240.1051, GRAS; not for use in infant foods; FEMA GRAS; Japan approved; Europe listed; UK approved; FDA approved for injectables, parenterals, orals, topicals, vaginals; USP/NF, BP, EP, JP compliance

Manuf./Distrib.: A&E Connock; AAA Int'l.; AB R Lundberg; AB Tech. Ltd; ADM; AMC Chems.; AMRESCO; Active Organics; Aldrich; Alfa Chem; Am. Biorganics; Ashland; Asiamerica Int'l.; Axxence Aromatic GmbH; BCH Brühl; Balchem; Barnet Prods.; Boith China; Borculo Domo Ingreds.; Brenntag AG; Brown; Cambridge Research Biochems.; Camida Ltd; CarboMer; Charkit; Chemco France; Chr. Hansen Inc; Cornelius Chem. Co. Ltd; Coyne; Degussa AG/Health & Nutrition; E&E Ltd; EM Ind.; EM Science; FBC Ind.; Farleyway Chems. Ltd; Fisher Scientific; Fleurchem; Fluka; GFS; Gallard-Schlesinger Ind.; Grau Aromatics; Harcros; Hasnil; Honeywill & Stein; ICC Ind.; Int'l. Sourcing; Integra; J.T. Baker; Janssen-Cilag; Jarchem Ind.; Jungbunzlauer; Kraft Chem.; Lluch Essence; Lohmann; Mallinckrodt Baker; Merck KGaA; Mitsubishi Chem.; Moore Ingreds.; Mutchler; PURAC Am.; PURAC biochem; Parchem Trading; Penta Mfg.; Peter Whiting Ltd; Pfanstiehl Europe; Pfanstiehl Labs; Przedsiebiorstwo Przemyslu AKWAWIT; R.I.T.A.; Research Organics; Rhodia/Fine Orgs.; Rona; Ruger; S.A. Chems.; San Yuan; Siber Hegner Ltd; Sigma; Spectrum Quality Prods.; Stanchem; Takeda Vitamin & Food USA; Thomas Scientific; Tri-K Ind.; Ubichem plc; Univar E&E Ltd; Universal Preserv-A-Chem; V.L. Clark; VWR Int'l.; Varsal Instruments; Voigt Global Distrib.; Vopak USA; Wako Chem. USA; Wego Chem. & Min.; Wilke Int'l.

Lactic acid, amyl ester. *See* Amyl lactate

Lactic acid butyl ester. *See* Butyl lactate; Butyl-(S)-lactate

Lactic acid ethyl ester. *See* Ethyl lactate; Ethyl-(S)-lactate

L-Lactic acid, 2-ethylhexyl ester. *See* 2-Ethylhexyl-S-lactate

Lactic acid isopropyl ester. *See* Isopropyl lactate; Isopropyl-S-lactate

Lactic acid methyl ester. *See* Methyl lactate; Methyl-(S)-lactate

Lactic acid monoethanolamide. *See* Lactamide MEA

Lactic acid, pentyl ester. *See* Amyl lactate

Lactol spirits. *See* Naphtha, light aliphatic

ε-Lactone hexanoic acid. *See* ε-Caprolactone monomer

Lactonitrile

CAS 78-97-7

Synonyms: Acetaldehyde cyanohydrin; 2-Hydroxypropionitrile; α-Hydroxypropionitrile; Propionitrile, 2-hydroxy-

Empirical: C_3H_5NO

Formula: CH_3CH_2OCN
Properties: Straw-colored liq.; sol. in water and alcohol; insol. in petroleum ether and carbon disulfide; m.w. 71.09; dens. 0.9919 (18.4 C); f.p. -40 C; b.p. 182-184 C; flash pt. 76.6 C; ref. index 1.4058 (18.4 C)
Toxicology: LD50 (oral, rat) 87 mg/kg, (IP, mouse) 15 mg/kg, (skin, rabbit) 20 mg/kg; toxic by ing., subcut. route, and skin absorption; mod. toxic by inh.; TSCA listed
Precaution: Combustible; can react vigorously with oxidizing materials
Hazardous Decomp. Prods.: Evolves hydrocyanic acid in the presence of alkali; heated to decomp. emits toxic fumes of CN^- and NO_x
Uses: Solvent, intermediate in the production of ethyl lactate and lactic acid

LAH. *See* Lithium aluminum hydride
LAM. *See* 2-Pyrrolidone

Laneth-5

CAS 61790-91-8; 61791-20-6 (generic)
Synonyms: PEG-5 lanolin alcohol; PEG-5 lanolin ether; POE (5) lanolin alcohol; POE (5) lanolin ether
Definition: PEG ether of lanolin alcohol with avg. ethoxylation value of 5
Properties: HLB 7.7; nonionic
Toxicology: TSCA listed
Uses: Emulsifier, emollient, stabilizer, solubilizer, cosolvent, wetting agent, dispersant, conditioner, visc. control agent for cosmetics, pharmaceuticals, creams and lotions; bromo dye solvent
Trade Names: Nikkol BWA-5; Polychol 5

Laneth-10

CAS 61791-20-6 (generic)
Synonyms: PEG-10 lanolin ether; PEG 500 lanolin ether; POE (10) lanolin ether
Definition: PEG ether of lanolin alcohol with avg. ethoxylation value of 10
Properties: Nonionic
Toxicology: TSCA listed
Uses: Emulsifier, emollient, stabilizer, solubilizer, cosolvent, wetting agent, dispersant, conditioner, visc. control agent for cosmetics, creams and lotions, pharmaceuticals
Trade Names: Nikkol BWA-10

Laneth-15

CAS 61791-20-6 (generic); 84650-19-1
Synonyms: PEG-15 lanolin ether; POE (15) lanolin ether
Definition: PEG ether of lanolin alcohol with avg. ethoxylation value of 15
Properties: Nonionic
Toxicology: TSCA listed
Uses: Emulsifier, emollient, stabilizer, solubilizer, wetting agent, dispersant, conditioner, visc. control agent for cosmetics, topical pharmaceuticals, creams and lotions; cosolvent for nonmiscible oils
Trade Names: Aqualose W15; Polychol 15

Laneth-16

CAS 61791-20-6 (generic)
Synonyms: PEG-16 lanolin ether; POE (16) lanolin alcohol; POE (16) lanolin ether
Definition: PEG ether of lanolin alcohol with avg. ethoxylation value of 16
Properties: HLB 13.2-15.0; nonionic
Toxicology: TSCA listed
Uses: Emulsifier for cosmetics and pharmaceuticals
Trade Names Containing: Solulan® 16

Lanolic acids. *See* Lanolin acid

Lanolin

CAS 8006-54-0 (anhyd.); 8020-84-6 (hyd.); EINECS/ELINCS 232-348-6
INS913
Synonyms: Adeps lanae; Anhydrous lanolin; Wool fat; Wool grease; Wool wax
Definition: Deriv. of unctuous fatty sebaceous secretion of sheep, *Ovis aries*, consisting of complex mixt. of esters of high m.w. aliphatic, steroid, or triterpenoid alcohol and fatty acids
Properties: Yel.-wh. semisolid to paste; sl. odor; sol. in chloroform, ether; insol. in water; m.p. 38-42 C; iodine no. 18-36; flash pt. > 230 F; nonionic
Toxicology: Can cause allergic reactions, contact dermatitis; TSCA listed
Hazardous Decomp. Prods.: Heated to decomp., emits acrid smoke and irritating fumes
NFPA: Health 0, Flammability 1, Reactivity 0
Uses: Soaps; face creams; facial tissues; hair set and suntan preps.; plasticizer for rubber; lubricant for textiles, metalworking compds.; EP agent, slip agent for metalworking/rust preventative coatings; in printing inks; waterproofing agent for leather; chewing gum base; food glazing agent; pharmaceutical emollient, ointment base, filler, vehicle; protectant in diaper rash, hemorrhoidal, antibiotic ointments; antistat, emollient, emulsifier, solvent, cosolvent, plasticizer, wetting agent for cosmetics; surf. lubricant in food-contact coatings; in paper/paperboard in contact with aq./fatty foods; defoamer in food-contact paper/paperboard; in cellophane for food pkg.; plasticizer in food-contact rubber articles for repeated use
Regulatory: FDA 21CFR §172.615, 175.300, 176.170, 176.180, 176.210, 177.1200, 177.2600, 178.3910; Japan approved; FDA approved for ophthalmics, topicals; anhyd.: USP/NF, BP compliance
Manuf./Distrib.: AXO Chem.; Aldrich; Alfa Chem; Am. Int'l.; Amerchol; Arch Personal Care Prods.; Arch Personal Care Prods.; Ashland; Asiamerica Int'l.; Camida Ltd; Charkit; Charles Bowman; Chemacon GmbH; Chemetall Chem. Prods.; CoKEM Assoc.; Cognis/Chems. Group; Cornelius Chem. Co. Ltd; Costec; Croda Chem. Ltd; Croda Inc; Croda/Westbrook; Dow; EM Science; Fabrichem; Fanning; Fluka; GMI Prods.; Gallard-Schlesinger Ind.; Integra; J.T. Baker; Kira; Kraft Chem.; Lanaetex Prods.; Mosselman NV; Mutchler; Norman, Fox; O.C. Lugo; Parchem Trading; Penta Mfg.; Protameen; R.I.T.A.; Richard's Pacific Chem. Sourcing; Robeco; Ruger; Sea-Land; Sigma; Spec. Ingreds. & Perfume; Spectrum Quality Prods.; Stevenson Cooper; Takasago Int'l.; Thornley; Uniqema; Universal Preserv-A-Chem; VWR Int'l.; Voigt Global Distrib.
Trade Names: Corona Lanolin; Coronet Lanolin
See also Lanolin alcohol

Lanolin acid

CAS 68424-43-1; EINECS/ELINCS 270-302-7
Synonyms: Acids, lanolin; Fatty acids, lanolin; Lanolic acids; Lanolin fatty acids
Definition: Mixture of organic acids obtained from hydrolysis of lanolin
Properties: Pale yel. to brn. waxy solid
Toxicology: TSCA listed
Uses: Emulsifier, stabilizer, superfatting agent, cosolvent, emollient for fatty acid systems, aerosol shave creams, cream shampoos, wax systems, household prods.; pigment dispersant; tackifier and plasticizer for wax films; in industrial leather treating, polishes, lubricants; corrosion inhibition; protective coatings; wire drawing/metalforming oils; emollient, emulsifier, stabilizer for cosmetics, pharmaceuticals
Manuf./Distrib.: Alfa Chem; Charles Bowman; Chemetall Chem. Prods.; CoKEM Assoc.; Costec; Croda Inc; R.I.T.A.; Richard's Pacific Chem. Sourcing; Universal Preserv-A-Chem

Lanolin alcohol

CAS 8027-33-6; EINECS/ELINCS 232-430-1
Synonyms: Alcohols, lanolin; Lanolin; Lanolin alcohols; Wool alcohol; Wool wax alcohol
Definition: Mixture of organic alcohols obtained from hydrolysis of lanolin; main constituents are cholesterol, lanosterol, dehydrolanosterol, various aliphatic alcohols
Properties: Amber hard waxy solid, char. odor; sol. in ether, chloroform; sl. sol. in alcohol; insol. in water; m.p. 56 C; acid no. 2 max.; sapon. no. 12 max.; nonionic
Toxicology: Less likely to cause allergic reaction than lanolin; TSCA listed
Uses: Antistat, emollient, w/o emulsifier, stabilizer, softener, gellant, thickener, plasticizer, moisturizer for absorption bases, cosmetics, cleansing preps.; emulsifier, solubilizer in pharmaceuticals
Regulatory: FDA approved for ophthalmics, topicals; NF compliance
Manuf./Distrib.: Alfa Chem; Amerchol; Arch Personal Care Prods.; Arch Personal Care Prods.; Ashland; Asiamerica Int'l.; Charkit; Charles Bowman; Chemacon GmbH; Chemetall Chem. Prods.; CoKEM Assoc.; Cognis/Chems. Group; Cornelius Chem. Co. Ltd; Costec; Croda Inc; Gallard-Schlesinger Ind.; Global-Seven; Kraft Chem.; R.I.T.A.; Solvay SA; Universal Preserv-A-Chem; Vopak USA; Westfalia Separator

Trade Names Containing: Argobase 125T; Liquid Absorption Base T

Lanolin alcohols. *See* Lanolin alcohol
Lanolin, alcohols, acetates. *See* Acetylated lanolin alcohol
Lanolin fatty acids. *See* Lanolin acid
Lanolin fatty acids, isopropyl esters. *See* Isopropyl lanolate

Lanolin oil

CAS 8038-43-5; 70321-63-0; EINECS/ELINCS 274-559-6
Synonyms: Dewaxed lanolin; Oils, lanolin
Definition: Liq. fraction of lanolin obtained by physical means from whole lanolin
Properties: Nonionic
Toxicology: LD50 (oral, rat) > 20 g/kg; may cause allergic reactions, contact skin rashes; TSCA listed
Uses: Antistat, emollient, solvent, conditioner, w/o emulsifier for cosmetics; cosolvent, lubricant, moisturizer, penetrant, superfatting agent for lipsticks, baby oils, brilliantines, cleansing lotions; dispersant for pigments; pharmaceutical ointment base
Manuf./Distrib.: Aldrich; Alfa Chem; Arch Personal Care Prods.; Ashland; CoKEM Assoc.; Cognis/Chems. Group; Costec; Croda Inc; Fluka; Kraft Chem.; Lanaetex Prods.; R.I.T.A.; Sigma; Universal Preserv-A-Chem
Trade Names: Lipolan R; Liquid Medilan™; Liquid Medilan™ Ultra

Laurel camphor. *See* Camphor

Laureth-9

CAS 3055-99-0; 9002-92-0 (generic); EINECS/ELINCS 221-284-4
Synonyms: 3,6,9,12,15,18,21,24,27-Nonaoxanonatriacontan-1-ol; PEG-9 lauryl ether; PEG-9 monododecyl ether; POE (9) lauryl ether
Definition: PEG ether of lauryl alcohol
Empirical: $C_{30}H_{62}O_{10}$
Formula: $CH_3(CH_2)_{10}CH_2(OCH_2CH_2)_nOH$, avg. n = 9
Properties: M.w. ≈ 600; sol. in water, ethanol, toluene; misc. with fats, fatty alcohols, and hot min., nat. and syn. oils; nonionic
Toxicology: LD50 (oral, mouse) 1170 mg/kg, (IV, mouse) 125 mg/kg; TSCA listed
Uses: Emulsifier, wetting agent, detergent, dispersant for maintenance and institutional cleaners, textiles, paper, paints, emulsion polymerization; solvent; pharmaceutic aid (surfactant); spermicide; penetrant; coupling agent; in food-contact textiles; antistat in food-pkg. materials
Regulatory: FDA 21CFR §177.2800, 178.3130
Manuf./Distrib.: Aldrich; DelCon; Fluka; Sigma

Lauric acid, benzyl ester. *See* Benzyl laurate
Lauric acid methyl ester. *See* Methyl laurate
Lauric acid, monoester with 1,2-propanediol. *See* Propylene glycol laurate
Lauric acid triglyceride. *See* Trilaurin
Lauric alcohol; Laurinic alcohol. *See* Lauryl alcohol

Lauryl alcohol

CAS 112-53-8; 27342-88-7; EINECS/ELINCS 203-982-0
FEMA 2617
Synonyms: Alcohol C12; C12 linear primary alcohol; 1-Dodecanol; Dodecan-1-ol; n-Dodecanol; n-Dodecan-1-ol; Dodecyl alcohol; 1-Dodecyl alcohol; n-Dodecyl alcohol; Duodecyl alcohol; 1-Hydroxydodecane; Lauric alcohol; Laurinic alcohol; n-Lauryl alcohol; Lauryl alcohol, narrow-cut
Classification: Primary aliphatic alcohol; fatty alcohol
Empirical: $C_{12}H_{26}O$
Formula: $CH_3(CH_2)_{10}CH_2OH$
Properties: Colorless leaflets, liq. above 21 C, floral odor; sol. in alcohol, ether, oxygenated solvs.; insol. in water; m.w. 186.33; dens. 0.8309 (24/4 C); m.p. 24 C; b.p. 259 C; flash pt. (CC) 127 C; ref. index 1.440-1.444
Toxicology: LD50 (oral, rat) 12,800 mg/kg, (dermal, guinea pig) > 8310 mg/kg; mod. toxic by IP route; mildly toxic by ing.; severe human skin irritant; weak to mod. carcinogen; experimental tumorigen; TSCA listed
Precaution: Combustible; reactive with oxidizing materials; incompat. with acetaldehyde, acids, and chlorine (vigorous or violent reactions possible)
Hazardous Decomp. Prods.: CO, CO_2; heated to decomp., emits acrid smoke and irritating fumes
NFPA: Health 0, Flammability 1, Reactivity 0
Storage: Store in suitable labeled containers, protected from damage
Uses: Mfg. of surfactants, wetting agents, emulsifiers, dispersants, antioxidants used in detergents, cosmetics, lube additives, pharmaceuticals; in rubber; textiles; perfumes; synthetic flavoring agent in foods and pharmaceuticals; emollient, emulsion stabilizer, visc. control agent for cosmetics; emulsion polymerization; mfg. of salts used as biocides and disinfectants; foam control; metalworking; froth flotation agent; lubricant in polymer processing; in food-pkg. adhesives; in food-contact coatings; in cellophane for food pkg.
Features: Honey, coconut-like flavor
Regulatory: FDA 21CFR §172.515, 172.864, 175.105, 175.300, 177.1010, 177.1200, 177.1390, 177.2800, 178.3480, 178.3910; FEMA GRAS
Manuf./Distrib.: Albemarle; Aldrich; Allchem Ind.; Ashland; Cognis/Chems. Group; E-Chemicals; Fleurchem; Fluka; Grau Aromatics; Haarmann & Reimer GmbH; Indofine; J.T. Baker; Jarchem Ind.; Kraft Chem.; Lipo; Lluch Essence; M. Michel; Mosselman NV; Multi-Kem; Penta Mfg.; Procter & Gamble; Prodasynth; Protameen; Salim Oleochems.; Sasol Germany; Schweizerhall; Sea-Land; Sigma; Spectrum Quality Prods.; United Coconut Chem.; Universal Preserv-A-Chem; VWR Int'l.; Vopak USA; Whyte Chems. Ltd
Trade Names: Cachalot® L-90
Trade Names Containing: Alfol® 14; Alfol® 1412

n-Lauryl alcohol; Lauryl alcohol, narrow-cut. *See* Lauryl alcohol
Laurylbenzenesulfonic acid. *See* Dodecylbenzenesulfonic acid

Lauryl oleate

CAS 36078-10-1; EINECS/ELINCS 252-862-4
Synonyms: Dodecyl oleate; Oleic acid, lauryl ester
Definition: Ester of lauryl alcohol and oleic acid
Empirical: $C_{30}H_{58}O_2$
Uses: Emollient, skin conditioner in cosmetics
Features: Dermatophilic additive
Manuf./Distrib.: Sigma
Trade Names: Exceparl L-OL

Lauryl pyrrolidone

CAS 2687-96-9; 55257-88-0; EINECS/ELINCS 403-730-1
Synonyms: 1-Dodecyl-2-pyrrolidinone; N-Dodecylpyrrolidinone; N-Dodecyl-2-pyrrolidone; 1-Lauryl-2-pyrrolidone; 2-Pyrrolidinone, 1-dodecyl-
Classification: Substituted heterocyclic compd.; nonaromatic amide
Empirical: $C_{16}H_{31}NO$
Properties: M.w. 253.48; dens. 0.890; b.p. 202-205 (11 mm); flash pt. > 230 F; ref. index 1.4660; nonionic
Toxicology: LDLo (oral, rat) 5 g/kg, (IP, mouse) 1 g/kg; mildly toxic by ing.; primary irritant; severe skin and mod. eye irritant; TSCA listed
Precaution: Combustible
Hazardous Decomp. Prods.: Heated to decomp., emits toxic fumes of NO_x
Uses: Surfactant for cosmetics; conditioner; foam stabilizer; wetting agent; solvent for commercial cleaning, textile processing, lithographic printing prods., waterborne coatings, inks; absorption enhancer in pharmaceuticals; drug
Features: VOC replacement
Manuf./Distrib.: Aldrich
Trade Names: Agsolex® 12; Surfadone® LP-300

1-Lauryl-2-pyrrolidone. *See* Lauryl pyrrolidone
Lauryl sodium sulfate; Lauryl sulfate sodium salt. *See* Sodium lauryl sulfate
Lead, tetraethyl-. *See* Tetraethyl lead
Leaf alcohol. *See* cis-3-Hexenol
LEAR. *See* Rapeseed (Brassica campestris) oil
Lemonene. *See* Biphenyl
Leucol. *See* Quinoline
Leucoline. *See* Isoquinoline; Quinoline
Levulinic acid, ethyl ester. *See* Ethyl levulinate
Light aliphatic naphtha; Light aliphatic solvent naphtha (petroleum). *See* Naphtha, light aliphatic
Light alkylate naphtha; Light alkylation naphtha. *See* Naphtha, light alkylate
Light aromatic naphtha; Light aromatic petroleum naphtha; Light aro-

matic solvent naphtha; Light aromatic solvent naphtha (petroleum). *See* Naphtha, light aromatic
Light camphor oil. *See* Camphor (Cinnamomum camphora) oil
Light hydrocracked distillate. *See* Petroleum distillates, light hydrocracked
Light hydrotreated distillate. *See* Petroleum distillates, hydrotreated light, low-boiling
Light hydrotreated naphtha. *See* Naphtha, hydrotreated light
Lightly hydrocracked distillate. *See* Petroleum distillates, light hydrocracked
Light mineral oil. *See* Mineral oil
Light normal paraffin concentrate (petroleum); Light normal paraffins (petroleum). *See* Paraffins, normal C5-20
Lignin liquor. *See* Tall oil
Lignocaine benzyl benzoate. *See* Denatonium benzoate
Ligroin; Ligroine. *See* VM&P naphtha
(-)-Limonene. *See* l-Limonene
(+)-Limonene. *See* d-Limonene

d-Limonene

CAS 5989-27-5; 68647-72-3; EINECS/ELINCS 227-813-5
UN 1993 (DOT); FEMA 2633
Synonyms: Cajeputene; (+)-Carvene; Cinene; Dextro-limonene; 4-Isopropenyl-1-methylcyclohexene; (R)-4-Isopropenyl-1-methyl-1-cyclohexene; Kautschin; (+)-Limonene; (D)-Limonene; D-(+)-Limonene; R(+)-Limonene; (+)-1,8-p-Menthadiene; d-p-Mentha-1,8-diene; p-Mentha-1,8-diene; (R)-p-Mentha-1,8-diene; (R)-1-Methyl-4-(1-methylethenyl)-cyclohexene
Classification: Unsat. alicyclic hydrocarbon; terpene
Definition: Naturally occurring C10 cycloolefinic hydrocarbon; major component in citrus oils
Empirical: $C_{10}H_{16}$
Formula: $CH_3C_6H_8C(CH_3)=CH_2$
Properties: Colorless liq., citrus odor; sol. in alcohol, ether; insol. in water, propylene glycol; m.w. 136.26; dens. 0.8411 (20 C); vapor pressure 20 mm Hg (68 C); m.p. -74 C; b.p. 176-176.4 C; flash pt. (CC) 48 C; ref. index 1.471; surf. tens. 26.87 dynes/cm; KB value 63 (tech.); VOC 837 g/l; dielec. const. 2.375
Toxicology: LD50 (oral, rat) 4400 mg/kg, (IV, rat) 110 mg/kg; poison by IV route; mod. toxic by IP route; mildly toxic by ing.; skin irritant; inh. of vapor may cause nose/throat irritation; target organs: eyes, skin, respiratory system, kidney; cancer suspect agent; experimental tumorigen, reproductive effects; TSCA listed
Environmental: Fully biodeg.; VOC; ThOD 2.35
Precaution: Combustible; incompat. with strong oxidizers, iodine pentafluoride, tetrafluoroethylene; explosive reactions possible
Hazardous Decomp. Prods.: Thermal decomp. prods.: CO, CO_2; heated to decomp., emits acrid smoke and irritating fumes
NFPA: Health 0, Flammability 2, Reactivity 0
Storage: Store in cool, dry, well-ventilated area out of direct sunlight
Uses: Synthetic flavoring agent in foods and pharmaceuticals; fragrance in cosmetics, perfume; chem. intermediate; wetting agent, dispersant in terpene resin mfg.; industrial solvent for degreasing, in household and I&I cleaning prods.; minor component of turpentine; component of essential oils; inhibitor of tetrafluoroethylene polymerization; gallstone solubilizer; packaging material; pine oil cleaning formulations; adhesives carrier; adhesive removal; mold release compd.; antimicrobial; disinfectant; biodeg. solvent replacement for cleaning/degreasing applics.; odorant for leak detection, paints/coatings; coating carrier; asphalt remover; metalworking cleaner
Features: Citrus flavor
Regulatory: FDA 21CFR §182.60, GRAS; FEMA GRAS
Manuf./Distrib.: Aldrich; Allchem Ind.; Cargill Ind. Oils & Lubes; Charkit; Coyne; Fleurchem; Florida Chem.; Fluka; Int'l. Flavors & Fragrances; Langley-Smith Ltd; Millennium/F&F; Mosselman NV; Parchem Trading; Penta Mfg.; Sarcom; Sea-Land; Sigma; Spectrum Quality Prods.; Vliengenthart BV; Whyte Chems. Ltd
Trade Names: Citrus Stripper Oil; Orange Terpenes
Trade Names Containing: Citrus Terpenes; DA-425

(D)-Limonene; D-(+)-Limonene. *See* d-Limonene

dl-Limonene

CAS 138-86-3; 7705-14-8 (±); EINECS/ELINCS 205-341-0
UN 2052 (DOT)
Synonyms: Cajeputene; Cinene; Cyclohexene, 1-methyl-4-(1-methylethenyl)-; Dipanol; Dipentene; Inactive limonene; 4-Isopropenyl-1-methyl-1-cyclohexene; (±)-Limonene; 1,8(9)-p-Menthadiene; dl-p-Mentha-1,8-diene; p-Mentha-1,8-diene; (±)-1,8-p-Menthadiene; 1-Methyl-4-isopropenylcyclohexene; 1-Methyl-4-isopropenyl-1-cyclohexene; 1-Methyl-4 (1-methylethenyl) cyclohexene; Racemic limonene
Classification: Unsat. aliphatic hydrocarbon
Empirical: $C_{10}H_{16}$
Properties: Colorless liq., pleasant lemon-like odor; sol. in alcohol, ether; pract. insol. in water; m.w. 136.23; dens. 0.847 (15.5/15.5 C); vapor pressure 20 mm Hg (68 C); m.p. -96.9 C; b.p. 175-176 C (763 mm); flash pt. (CC) 45 C; autoignition temp. 237 C; ref. index 1.4744; dielec. const. 2.3
Toxicology: LD50 (oral, rat) 5000 mg/kg; skin irritant and sensitizer; inh. of vapor may cause nose and throat irritation; probable eye irritant; not very toxic by ing.; target organs: CNS, skin, kidney; TSCA listed
Environmental: VOC; ThOD 3.29
Precaution: DOT: Flamm. liq.; can react vigorously with oxidizers; also incompat. with iodine pentafluoride and tetrafluoroethylene
Hazardous Decomp. Prods.: Thermal decomp. prods.: CO, CO_2; heated to decomp., emits acrid smoke and irritating fumes
NFPA: Health 0, Flammability 2, Reactivity 0
Storage: Store in cool, dry, well-ventilated area out of direct sunlight
Uses: Solvent for oleoresinous prods., rosin, ester gum, alkyd and phenolic resins, waxes, rubber, paints, enamels, lacquers, polishes; softener, swelling agent in rubber; antiskinning agent in paints; cosolvent for mineral extraction; pine oil cleaners; minor component in turpentine; wetting agent, dispersant for oils, resins, pigments, driers, paints, enamels, lacquers, printing inks, perfumes, floor waxes, furniture polishes; rubber compding. and reclaiming; solvent in cosmetics; in perfumes for soaps, personal care; intermediate for terpene resins, carvone, terylene, rubber chems.; metal drier; synthetic flavoring agent, fragrance in foods, pharmaceuticals; in food-pkg. adhesives
Regulatory: FDA 21CFR §175.105, 175.300, 175.320, 177.2600, 182.60, GRAS; FDA approved for topicals
Manuf./Distrib.: Aldrich; Arizona; Fluka; Hercules; Millennium; Mosselman NV; Penta Mfg.; Spectrum Quality Prods.; Veitsiluoto Oy
Trade Names: Dipentene No. 122
Trade Names Containing: Dipentene-C

l-Limonene

CAS 5989-54-8; EINECS/ELINCS 227-815-6
UN 2052 (DOT)
Synonyms: Cajeputene; Carvene; Cyclohexene, 1-methyl-4-(1-methylethenyl)-, (S)-; Dipentene; (-)-Limonene; l-p-Mentha-1,8-diene; p-Mentha-1,8-diene, (S)-(-)-; (S)-(-)-p-Mentha-1,8-diene; (S)-p-Mentha-1,8-diene; 1-Methyl-4-(1-methylethenyl)-(S)-cyclohexene
Empirical: $C_{10}H_{16}$
Properties: Liq.; misc. with alcohol; insol. in water; m.w. 136.23; dens. 0.8407 (20.5/4 C); b.p. 175.5-176.5 C (763 mm); flash pt. 168 F; ref. index 1.474 (21 C)
Toxicology: Skin irritant, sensitizer; TSCA listed
Hazardous Decomp. Prods.: Heated to decomp., emits acrid smoke and irritating fumes
NFPA: Health 0, Flammability 2, Reactivity 0
Uses: Fragrance in cosmetics, perfumes; synthetic flavoring agent in foods; solvent; wetting agent; resin mfg.
Regulatory: FDA 21CFR §182.60, GRAS
Manuf./Distrib.: Fleurchem; Millennium; Polarome Int'l.; Spectrum Quality Prods.

R(+)-Limonene. *See* d-Limonene
(±)-Limonene. *See* dl-Limonene
Limonene oxide. *See* Eucalyptol
Linear alcohol C13-15. *See* C13-15 alcohols
Linoleic acid ethyl ester. *See* Ethyl linoleate
Linoleic acid, 1,2,3-propanetriyl ester. *See* Trilinolein

Liquefied petroleum gas

Synonyms: LPG; LP gas; Petroleum gas, liquefied; Propane/butane, mixed
Classification: Alkane
Definition: General term for commercial propane, butane, or mixtures; can be liquefied at normal temp. by mod. pressure increase, or at

normal pressure by cooling; occurs naturally in oil or gas fields or obtained from refining petroleum
Properties: Colorless; odorless (normally odorized for leak detection); vapor pressure 593 kPa (20 C)
Toxicology: Nontoxic but can cause asphyxiation
Precaution: Flamm.; LFL 2.1 vol.%, UFL 9.2 vol.% (15 C, 101 kPa in air); volatile; heavier than air
Uses: Fuel for domestic, industrial, horticultural, agric., cooking, heating, and drying processes; motor vehicle fuel; propellant for aerosols; solvent for solvent deasphalting process; provides lighting through use of pressure lanterns; alternative for diesel and petrol. fuels, fluorocarbon aerosol propellants and refrigerants
Features: Nonreactive; storable in pressure containers; easily transportable; clean burning
See also Butane; Propane

Liquid camphor. *See* Camphor (Cinnamomum camphora) oil
Liquid paraffin; Liquid petrolatum. *See* Mineral oil
Liquid phenol. *See* Phenol
Liquid pitch oil. *See* Creosote oil
Liquid rosin. *See* Tall oil
Lithium alanate; Lithium aluminohydride. *See* Lithium aluminum hydride

Lithium aluminum hydride

CAS 16853-85-3; EINECS/ELINCS 240-877-9
UN 1410 (DOT); UN 1411 (DOT)
Synonyms: Aluminum lithium hydride; LAH; Lithium alanate; Lithium aluminohydride; Lithium aluminum hydride, ethereal; Lithium aluminum tetrahydride; Lithium tetrahydroaluminate
Classification: Lithium compd.
Empirical: H_4AlLi
Formula: $LiAlH_4$
Properties: Wh. (pure) to lt. gray (Al impurity) powd.; odorless; sol. in diethyl ether, tetrahydrofuran, dimethyl Cellosolve; sl. sol. in dioxane; reacts with alcohols; m.w. 37.94; dens. 0.92; m.p. dec. above 125 C without melting
Toxicology: ACGIH TLV/TWA 2 mg(Al)/m^3; corrosive; reacts with moisture in the body to form a corrosive sol'n. (lithium hydroxide) which can cause severe irritation, tissue damage; ing. can cause severe burns, dizziness, nausea, vomiting, diarrhea; severe exposure can cause collapse, death; severe inh. exposure can cause pulmonary edema; can cause severe skin and eye irritation or tissue destruction (chemical burns); TSCA listed
Precaution: Flamm. solid; reacts violently with water, air, acids, alcohols, benzoyl peroxide, etc.; dec. when exposed to moist air or water; incompat. with CO_2, BF_3, ethyl acetate, tetrahydrofuran
Hazardous Decomp. Prods.: Dec. above 125 C to lithium hydride, aluminum metal, and hydrogen gas; fire can produce aluminum oxide
NFPA: Health 3, Flammability 2, Reactivity 2
Storage: Stable under cool, dry conditions
Uses: Reducing agent (converts esters, aldehydes, and ketones to alcohols, and nitriles to amines) in pharmaceuticals, perfumes, and fine organic chemicals; source of hydrogen; a propellant; catalyst in polymerizations; drying solvents
Manuf./Distrib.: Advanced Synthesis Tech.; Aldrich; Alfa Aesar; Asiamerica Int'l.; Atomergic Chemetals; Chemetall Foote; Chemetall GmbH; Clariant; FMC/Lithium; Integra; Sithean; Spectrum Quality Prods.; VWR Int'l.

Lithium aluminum hydride, ethereal; Lithium aluminum tetrahydride; Lithium tetrahydroaluminate. *See* Lithium aluminum hydride
Longifolen. *See* Longifolene

Longifolene

CAS 475-20-7; EINECS/ELINCS 207-491-2
Synonyms: (1S-(1α,3aβ,4α,8aβ))-Decahydro-4,8,8-trimethyl-9-methylene-1,4-methanoazulene; Decahydro-4,8,8-trimethyl-9-methylene-1,4-methanoazulene; Junipen; Junipene; Longifolen; (+)-Longifolene; D-Longifolene; 1,4-Methanoazulene, decahydro-4,8,8-trimethyl-9-methylene-, (1S,3AR,4S,8AS)-(+)-; (1R,2S,7S,9S)-3,3,7-Trimethyl-8-methylenetricyclo [5.4.0.0^{2,9}] undecane;
Empirical: $C_{15}H_{24}$
Properties: Colorless to ylsh. oily liq.; m.w. 204.4; dens. 0.938 g/ml; m.p. 156.2 C; b.p. 258-260 C; flash pt. 98 C; ref. index 1.42
Toxicology: LD50 (oral, rat) > 5 g/kg, (skin, rabbit) > 5 g/kg; may cause eye/skin irritation; may be harmful by inh., ing., or skin absorp.; TSCA listed
Precaution: Incompat. with strong oxidizing agents
Hazardous Decomp. Prods.: CO, CO_2
Storage: Store @ 2-8 C in cool, dry place; keep tightly closed
Uses: Solvent for lacquers, resins, emulsion-bonding adhesives; fragrance in cosmetics; building block for synthesis of hydroazulenes; prep. of dilongifolylborane (a chiral hydroborating agent)
Manuf./Distrib.: Aldrich; Doingcom; Fluka; Lluch Essence; Sigma

(+)-Longifolene; D-Longifolene. *See* Longifolene
Low-erucic acid rapeseed oil. *See* Rapeseed (Brassica campestris) oil
LPG; LP gas. *See* Liquefied petroleum gas

2,6-Lutidine

CAS 108-48-5; EINECS/ELINCS 203-587-3
UN 1993; FEMA 3540
Synonyms: 2,6-Dimethylpyridine; α-α′-Dimethylpyridine; 2,6-Dimethypyridine; α-α′-Lutidine; Pyridine, 2,6-dimethyl-
Empirical: C_7H_9N
Properties: Colorless to pale yel. liq., peppermint odor; sol. in water, alcohol, ether; m.w. 107.16; dens. 0.920; f.p. -6 C; b.p. 143-145 C; flash pt. 33 C; ref. index 1.4970
Toxicology: LD50 (oral, rat) 400 mg/kg, (skin, guinea pig) 2500 mg/kg; LCLo (inh., rat, 1 h) 7500 ppm; poison by ing.; mod. toxic by skin contact; mildly toxic by inh.; mutagenic data; TSCA listed
Precaution: DOT: Flamm. liq.
Hazardous Decomp. Prods.: Heated to decomp., emits toxic vapors of NO_x
Storage: Hygroscopic
Uses: Solvent; synthetic flavor for foods and pharmaceuticals; reagent for pharmaceutical mfg.; resins; dyestuffs; rubber accelerators; insecticides
Regulatory: FEMA GRAS
Manuf./Distrib.: Aldrich; Biddle Sawyer; Fluka; Koei Chem.; Mallinckrodt Baker; Penta Mfg.; Raschig; Reilly Ind.; Sigma; Spectrum Quality Prods.; Synasia Fine Chem.; VWR Int'l.

α-α′-Lutidine. *See* 2,6-Lutidine
Lye. *See* Potassium hydroxide; Sodium hydroxide

M

MAA. *See* Methyl acetoacetate
MAAC. *See* Methyl amyl acetate
Macrogol 200. *See* PEG-4
Macrogol 300. *See* PEG-6
Macrogol 400. *See* PEG-8
Macrogol 600. *See* PEG-12
Macrogol 1000. *See* PEG-20
Macrogol 1540. *See* PEG-32
Macrogol 4000. *See* PEG-75
Macrogol 6000. *See* PEG-150
Maize oil. *See* Corn (Zea mays) oil
MAK. *See* Methyl n-amyl ketone
Maleic acid, bis (2-ethylhexyl) ester; Maleic acid, di-(2-ethylhexyl) ester. *See* Bis (2-ethylhexyl) maleate
Maleic acid, dioctyl ester. *See* Dioctyl maleate
MAOH. *See* Methyl amyl alcohol
Matricaria camphor. *See* Camphor
MB. *See* Methyl bromide
MBK. *See* 2-Hexanone
MBX. *See* Methyl bromide
MBY. *See* Methyl butynol
MCB. *See* Chlorobenzene
MCF. *See* 1,1,1-Trichloroethane
MCT. *See* Monochlorotoluene
MDC. *See* Methylene chloride
MDEA. *See* Methyldiethanolamine
MDG. *See* Mono- and diglycerides of fatty acids
MDMH. *See* MDM hydantoin

MDM hydantoin

CAS 116-25-6; EINECS/ELINCS 204-132-1
Synonyms: 1-(Hydroxymethyl)-5,5-dimethyl hydantoin; 1-(Hydroxymethyl)-5,5-dimethyl-2,4-imidazolinedione; MDMH; Monomethylol dimethyl hydantoin
Classification: Organic compd.
Empirical: $C_6H_{10}N_2O_3$
Toxicology: TSCA listed
Uses: Antimicrobial in cosmetics; hair lacquer resin; intermediate for cosmetics, other applics.; preservative; formaldehyde donor
Regulatory: USA not restricted; Japan not approved
Manuf./Distrib.: Int'l. Sourcing
Trade Names Containing: Dantogard® Plus Liq.

MDS. *See* Methyl disulfide
2-ME. *See* 2-Mercaptoethanol; Methoxyethanol
MEA. *See* Ethylamine; Ethanolamine
2-MEA. *See* Methoxyethanol acetate
MEBR. *See* Methyl bromide
Medium aliphatic solvent naphtha; Medium aliphatic solvent naphtha (petroleum). *See* Naphtha, medium aliphatic
MEG. *See* Ethylene glycol
MEHQ. *See* Hydroquinone monomethyl ether
MEK. *See* Methyl ethyl ketone
MELA. *See* Ethanolamine

Menhaden oil

CAS 8002-50-4; EINECS/ELINCS 232-311-4
Synonyms: Brevoortia; Brevoortia tyrannus oil; Mossbunker oil; Pogy oil
Definition: Oil obtained from the small North Atlantic fish, *Brevoortia tyrannus*
Properties: Ylsh.-brown or reddish-brown oil, char. fishy odor and taste; sol. in ether, benzene, petrol. ether, naphtha, kerosene, CS_2; dens. 0.925-0.933; m.p. 38.5-47.2 C; iodine no. 115-160; sapon. no. 191-200; ref. index 1.480 (20 C)
Toxicology: No known toxicity; TSCA listed
Precaution: Combustible
Uses: Substitute for linseed oil; leather dressing; hydrogenated fats for cooking and industrial use; printing inks; animal feed; lubricants; paint drier; cleansers; lipstick; as tallow substitute in soap mfg.; rubber compounding; emollient, solvent in cosmetics; defoamer in food-contact coatings and paper/paperboard; in resinous/polymeric food-contact coatings; in food-contact textiles; nutritional supplement in pharmaceuticals
Regulatory: FDA 21CFR §175.300, 176.200, 176.210, 177.2800, 184.1472
Manuf./Distrib.: ABITEC; Alnor Oil; Arista Ind.; Croda Oleochems.; Degussa AG/Health & Nutrition; Sigma; Werner G. Smith
Trade Names: Cropure® Menhaden

Menhaden oil, hydrogenated. *See* Hydrogenated menhaden oil
(+)-1,8-p-Menthadiene. *See* d-Limonene
1,4-p-Menthadiene. *See* γ-Terpinene
1,8(9)-p-Menthadiene; dl-p-Mentha-1,8-diene. *See* dl-Limonene
d-p-Mentha-1,8-diene. *See* d-Limonene
l-p-Mentha-1,8-diene. *See* l-Limonene
p-Mentha-1,4-diene. *See* γ-Terpinene
p-Mentha-1,8-diene. *See* d-Limonene; dl-Limonene
p-Mentha-1,8-diene, (S)-(-)-. *See* l-Limonene
(R)-p-Mentha-1,8-diene. *See* d-Limonene
(S)-(-)-p-Mentha-1,8-diene; (S)-p-Mentha-1,8-diene. *See* l-Limonene
(±)-1,8-p-Menthadiene. *See* dl-Limonene
p-Menthane, 1,8-epoxy. *See* Eucalyptol
3-p-Menthanol; p-Menthan-3-ol. *See* Menthol
Mentha piperita; Mentha piperita oil. *See* Peppermint (Mentha piperita) oil
Mentha spicata; Mentha spicata oil; Mentha viridis; Mentha viridis oil. *See* Spearmint (Mentha viridis) oil
p-Menth-1,4(8)-diene. *See* Terpinolene
p-Menthen-1-en-8-ol; 1-p-Menthen-8-ol; p-Menth-1-en-8-ol. *See* α-Terpineol

Menthol

CAS 89-78-1 (DL); 1490-04-6; 2216-51-5 (L, -); 15356-60-2 (+); 15356-70-4 (D, ±); EINECS/ELINCS 201-939-0 (DL); 216-074-4; 218-690-9 (L); 239-388-3 (D)
FEMA 2665
Synonyms: Cyclohexanol, 2-isopropyl-5-methyl-; Cyclohexanol, 5-methyl-2-(1-methylethyl)-; Hexahydrothymol; 2-Isopropyl-5-methylcyclohexanol; 3-p-Menthanol; p-Menthan-3-ol; 5-Methyl-2-isopropylcyclohexanol; 5-Methyl-2-isopropyl hexahydrophenol; 5-Methyl-2-(1-methylethyl) cyclohexanol; Peppermint camphor; Racemic menthol
Classification: Diterpene alcohol

Definition: Alcohol obtained from diverse mint oils, *Mentha spp.* or prepared synthetically; may be levorotatory (l-), from natural or syn. sources, or racemic (dl-)
Empirical: $C_{10}H_{20}O$
Formula: $CH_3C_6H_9(C_3H_7)OH$
Properties: Wh. cryst., cooling peppermint-like odor and taste; very sol. in alcohol, light petrol. solv., glacial acetic acid, min. oil, fixed/volatile oils; sl. sol. in water; m.w. 156.26; dens. 0.89; m.p. 42.5 C; b.p. 215 C; flash pt. 200 F; ref. index 1.461
Toxicology: LD50 (oral, rat) 3180 mg/kg, (IM, rat) 10 g/kg; LDLo (IP, mouse) 1800 mg/kg, (IV, cat) 37 mg/kg; poison by IV route; mod. toxic by ing., IP routes; severe eye irritant; irritant to mucous membranes on inh.; may cause human intolerance reaction; TSCA listed
Precaution: Combustible; incompat. with phenol, β-naphthol, others
Hazardous Decomp. Prods.: Heated to decomp., emits acrid smoke and irritating fumes
HMIS: Health 1, Flammability 0, Reactivity 0
Uses: Perfumery; cigarettes; liqueurs; denaturant; synthetic flavoring agent, fragrance, cooling agent in foods, confectionery, pharmaceuticals, cough drops, nasal inhalers, liniments, chewing gum, veterinary medicated shampoos/ointments; fragrance, cooling agent in chest rubs
Features: Minty flavor
Use Level: < 3%
Regulatory: FDA 21CFR §172.515, 182.20, GRAS; 27CFR §21.65, 21.151; FEMA GRAS; Japan approved as flavoring; FDA approved for buccals, dentals, inhalants, orals, topicals; USP/NF compliance
Manuf./Distrib.: A&E Connock; ADA Int'l.; AMC Chems.; Aastrid Int'l.; Acme-Hardesty; Aldrich; Alfa Chem; Allchem Int'l. Ltd; Am. Fruit Processors; Amalgamet; Anmar Int'l.; Ashland; Asiamerica Int'l.; BCH Brühl; Berje; Berk Chem. Ltd; Biddle Sawyer; Camida Ltd; CarboMer; Charabot USA; Charkit; Chart; Chemacon GmbH; Dastech Int'l.; Doingcom; EM Ind.; EM Science; Essence Nat. Prods.; Fleurchem; Fluka; Forrester Wood & Co. Ltd; Frutarom (UK) Ltd; Fuerst Day Lawson; GMI Prods.; Gallard-Schlesinger Ind.; George Uhe; H.E. Daniel Ltd; Haarmann & Reimer; Hasnil; Helm NY; Hunan Xinyu; Ikeda; Integra; J.T. Baker; Jeen Int'l.; Lebermuth; Lluch Essence; Magnesia GmbH; Mallinckrodt Baker; MelChem; Merck KGaA; Millennium; Mutchler; Noveon Diamalt GmbH; Pangaea Sciences; Parchem Trading; Penta Mfg.; Phoenix Nat. Prods.; Polarome Int'l.; Polysciences; Quest Int'l.; Quimdis; R.C. Treatt & Co. Ltd; R.W. Greeff; RIA Int'l.; Richard's Pacific Chem. Sourcing; Robeco; Robert Bryce; Ruger; Sarcom; Sigma; Spectrum Quality Prods.; Stan Chem Int'l. Ltd; Takasago Int'l. USA; Univar E&E Ltd; Universal Preserv-A-Chem; VWR Int'l.; Voigt Global Distrib.
Trade Names Containing: Punctilious® SDA 37-1 190 Proof; Punctilious® SDA 38B 190 Proof Prod. Code 13602027; Punctilious® SDA 38B 190 Proof Prod. Code 13603998; Punctilious® SDA 38B 200 Proof Prod. Code 14601040; Punctilious® SDA 38F-2 190 Proof Prod. Code 13645006; Punctilious® SDA 38F-6 190 Proof

Meparfynol. *See* Methyl pentynol
1-Mercaptobutane. *See* n-Butyl mercaptan
Mercaptoethanol. *See* 2-Mercaptoethanol

2-Mercaptoethanol

CAS 60-24-2; EINECS/ELINCS 200-464-6
UN 2966 (DOT)
Synonyms: BME; 1-Ethanol-2-thiol; Ethylene thioglycol; 2-Hydroxy-1-ethanethiol; Hydroxyethylmercaptan; 2-Hydroxyethylmercaptan; 2-ME; Mercaptoethanol; β-Mercaptoethanol; Monothioethyleneglycol; 2-Thioethanol; Thioethylene glycol; Thioglycol; Thiomonoglycol
Classification: Aliphatic organic compd.; alkyl mercapto alcohol
Empirical: C_2H_6OS
Formula: $HSCH_2CH_2OH$
Properties: Water-wh. mobile liq.; disagreeable odor; sol. in water, most org. solvs. incl. alcohol, benzene, ether; misc. with oxygenated and aromatic solvs.; m.w. 78.13; dens. 1.114; m.p. < -100 C; b.p. 157 C; flash pt. 74 C; ref. index 1.5006
Toxicology: LD50 (oral, rat) 244 mg/kg, (skin, rabbit) 150 mg/kg; poison by ing., skin contact, IP routes; mod. toxic by IV route; severely irritating to eyes, skin, nose; massive exposures can be fatal; ing. may irritate GI tract; human mutagenic data; TSCA listed
Precaution: Combustible; flamm. exposed to heat, flame, oxidizers; incompat. with oxidizers, reducing agents, strong acids, nitric acid, strong bases, alkali metals, moisture; vigorous or explosive reactions possible; can produce toxic fumes
Hazardous Decomp. Prods.: Heated to decomp., emits highly toxic fumes of SO_x
NFPA: Health 2, Flammability 2
Storage: Store in cool, dry, well-ventilated area, out of direct sunlight
Uses: Solvent for dyestuffs; intermediate for producing dyestuffs, pharmaceuticals, rubber chemicals, insecticides, plasticizers; flotation depressant for copper ores; reducing agent; biochemical reagent; PVC stabilizers; agricultural chemicals; textile auxiliary; chain transfer agent in mfg. of acrylic resins for high-solids automotive topcoats; mfg. of polythioethers for high-solids elastomers, coatings, and adhesives; adjuvant in ABS food pkg.
Regulatory: FDA 21CFR §177.1020, 177.1030
Manuf./Distrib.: AMRESCO; Aldrich; Atofina; BASF AG; BASF; Cardinal; ChevronPhillips; Fluka; Morton Int'l.; Pechiney Chems. Div.; Research Organics; Rhodia HPCII; Sigma; Spectrum Quality Prods.; Toray Thiokol

β-Mercaptoethanol. *See* 2-Mercaptoethanol
Mercaptopropionic acid, dibutyltin salt. *See* Dibutyltin β-mercaptopropionate
Mesitylene. *See* 1,3,5-Trimethylbenzene

Mesityl oxide

CAS 141-79-7; EINECS/ELINCS 205-502-5
UN 1229 (DOT)
Synonyms: Isobutenyl methyl ketone; Isopropylideneacetone; Methyl isobutenyl ketone; 4-Methyl-3-pentene-2-one; 2-Methyl-2-penten-4-one; 4-Methyl-3-penten-2-one
Classification: Ketone
Empirical: $C_6H_{10}O$
Formula: $(CH_3)_2C{:}CHCOCH_3$
Properties: Colorless oily liq.; sweet fruity odor; somewhat water-sol. @ 20 C; misc. with alcohol, ether, most org. liqs., oxygenated solvs.; m.w. 98.15; dens. 0.854 (20/4 C); vapor pressure 8.21 mm Hg; m.p. -59 C; b.p. 128-130 C; flash pt. 31 C; ref. index 1.446 (20 C); surf. tens. 28.09 dynes/cm
Toxicology: ACGIH TLV/TWA 15 ppm; STEL 25 ppm; LD50 (oral, rat) 1120 mg/kg, (skin, rabbit) 5150 mg/kg; LC50 (inh., rat, 4 h) 2242 ppm; poison by IP route; mod. toxic by ing.; mildly toxic by inh., skin contact; human systemic effects; highly irritating to tissue; narcotic in high concs.; target organs: eyes, skin, respiratory system, CNS; TSCA listed
Environmental: VOC; BOD5 1.91; COD 2.40; ThOD 2.61
Precaution: DOT: Flamm. liq.; dangerous fire hazard exposed to heat, sparks, flame; reactive with oxidizers; violent reactions possible
Hazardous Decomp. Prods.: Heated to decomp., emits acrid smoke, irritating fumes
Uses: Solvent for paints, cellulose esters and ethers, vinyl resins, lacquers, inks; ore flotation; paint and varnish removers; insect repellent; mfg. of methyl isobutyl ketone
Manuf./Distrib.: Allchem Ind.; Ashland; Atofina SA; Celanese; Corco; Coyne; Delta Distributors; Dow; Eastman; Fluka; G.J. Chem.; Harcros; Herdillia Chems. Ltd; ICC Ind.; Penta Mfg.; Primachem; Seeler Ind.; Shell; Vopak USA

Metacetone. *See* Diethyl ketone
Metadelphene. *See* Diethyl toluamide
Metaformaldehyde. *See* Trioxane
Methacetone. *See* Diethyl ketone
Methaldehyde. *See* Formaldehyde
Methamine, N-methyl-N-nitroso-. *See* N-Nitrosodimethylamine
Methanal. *See* Formaldehyde
Methanamide. *See* Formamide
Methanamine. *See* Methylamine
Methanamine, N-methyl-. *See* Dimethylamine
Methanamine, N-methyl-N-nitroso-. *See* N-Nitrosodimethylamine
Methane, bromochloro-. *See* Bromochloromethane
Methane, bromodichloro-. *See* Bromodichloromethane
Methanecarbonitrile. *See* Acetonitrile
Methanecarboxamide. *See* Acetamide
Methanecarboxylic acid. *See* Acetic acid

Methane, chloro-. *See* Methyl chloride
Methane dichloride; Methane-, dichloro-. *See* Methylene chloride
Methane, diethoxy-. *See* Diethoxymethane
Methane, diphenyl-. *See* Diphenylmethane
Methane, oxybis-. *See* Dimethyl ether

Methanesulfonic acid

CAS 75-75-2; EINECS/ELINCS 200-898-6
UN 2584; UN 2586
Synonyms: Methylsulfonic acid; MSA; Sulfomethane
Classification: Aliphatic organic compd.
Empirical: CH_4O_3S
Formula: CH_3SO_3H
Properties: Colorless to ylsh. liq.; sol. in water, alcohol, ether, oxygenated solvs.; m.w. 96.10; dens. 1.483 (20/4 C); vapor pressure 1 mm Hg (20 C); m.p. 20 C; b.p. 167 C (10 mm); dec. > 200 C; flash pt. 110 C; ref. index 1.4300
Toxicology: LDLo (oral, rat) 200 mg/kg, (IP, rat) 50 mg/kg; poison by ing., IP routes; corrosive to tissue, skin, eyes, mucous membranes; causes burns; target organs: liver, kidney, stomach, CNS; TSCA listed
Environmental: Good biodegrad.; VOC; ThOD 0.67; do not allow undiluted prod. to reach ground water or sewage systems
Precaution: Corrosive to iron, steel, brass, copper, lead; explosive reaction with ethyl vinyl ether; incompat. with hydrogen fluoride, alkalis, amines, vinyl ethers; reacts violently with water
Hazardous Decomp. Prods.: Heated to decomp., emits toxic fumes of SO_x
Storage: Store container tightly sealed in cool, dry area with sufficient ventilation; keep away from heat, direct sunlight
Uses: Polymerization catalyst; catalyst in esterification, alkylation, olefinic addition, epoxidation, peroxidation reactions; acid catalyst; curing accelerator for coatings; curing agent for thermosetting resins; dehydrating agent; solvent; electroplating bath additive; pesticides; corrosion inhibitors; textile oils; pharmaceutical parenterals
Regulatory: FDA approved for parenterals
Manuf./Distrib.: AMRESCO; Aldrich; Alfa Aesar; Allchem Ind.; Atofina; Atotech USA; Austin; BCH Brühl; Charkit; Chemconserve; Chevron Phillips; Dastech Int'l.; Fluka; J.T. Baker; Kira; Pickering Labs; Richman; Sigma; Toyo Kasei Kogyo; VWR Int'l.; Varsal Instruments; Xinchem

Methane, sulfonylbis-. *See* Dimethyl sulfone
Methane tetrabromide. *See* Carbon tetrabromide
Methane tetrachloride. *See* Carbon tetrachloride
Methane trichloride; Methane, trichloro-. *See* Chloroform
Methane, trichlorofluoro-. *See* Trichlorofluoromethane
Methane, trifluoro. *See* Trifluoromethane
1,4-Methanoazulene, decahydro-4,8,8-trimethyl-9-methylene-, (1S,3AR, 4S,8AS)-(+)-. *See* Longifolene
Methanoic acid. *See* Formic acid
Methanoic acid, amide. *See* Formamide
4,7-Methanoindan-5(4H)-one, tetrahydro-; 4,7-Methano-5H-inden-5-one, octahydro-. *See* Tricyclo [5.2.1.0^2,6] decane-8-one
Methanol. *See* Methyl alcohol
Methanol, benzyl-. *See* Phenethyl alcohol
2-Methanol tetrahydropyran. *See* Tetrahydropyran-2-methanol
Methenyl chloride. *See* Chloroform
Methenyl tribromide. *See* Bromoform
Methenyl trichloride. *See* Chloroform
1-Methoxy-2-acetoxypropane. *See* Propylene glycol methyl ether acetate
2-Methoxy-1-acetoxypropane. *See* 2-Methoxy-1-propanol acetate
2-Methoxy-4-allylphenol. *See* Eugenol
Methoxybenzene. *See* Anisole
1-Methoxybutane-1,3-dione. *See* Methyl acetoacetate

1-Methoxy-2-butanol

CAS 3085-35-6
Synonyms: BM
Properties: M.w. 104.15; dens. 0.90 g/ml; visc. 1.97 cp; b.p. 135 C; flash pt. (CC) 104 F; surf. tens. 26.53 dynes/cm
Uses: Solvent

3-Methoxybutanol

CAS 2517-43-3; EINECS/ELINCS 219-741-8
UN 1987
Synonyms: 1,3-Butanediol 3-methyl ether; 1-Butanol, 3-methoxy-; Butolether; 1,3-Butylene glycol monomethyl ether; 3-Methoxy-n-butanol
Classification: Organic compd.
Empirical: $C_5H_{12}O_2$
Formula: $CH_3CH(OCH_3)CH_2CH_2OH$
Properties: Colorless clear liq.; mild odor; sol. in water; misc. with common org. solvs.; m.w. 104.15; dens. 0.920-0.922 g/cc; vapor pressure 0.9 mm Hg (20 C); sets to glass @ -85 C; b.p. 161 C; flash pt. (CC) 60 C; autoignition temp. 335 C; ref. index 1.415-1.416 (20 C)
Toxicology: LD50 (oral, rat) 4000 mg/kg; toxic by ingestion
Environmental: Good biological eliminability
Precaution: Combustible
NFPA: Health 1, Flammability 2, Reactivity 0
Storage: Store in cool, dark, dry place in tightly closed containers, with air excluded
Uses: Solvent for lacquers, many nat. resins, nitrocellulose, PVB, aldehyde, ketone, indene, phenol-formaldehyde, alkyd, and maleic resins, commonly used plasticizers; coupling agent for brake fluids; hydraulic fluid blending component; intermediate for plasticizers, herbicides; film-former in PVA emulsions; solvent for pharmaceuticals; antiskinning agent in conventional and aq. paints
Features: High-boiling; good dissolving power; low volatility
Manuf./Distrib.: Celanese

3-Methoxy-n-butanol. *See* 3-Methoxybutanol
4-Methoxybutanol; 4-Methoxy-1-butanol. *See* Butylene glycol methyl ether
3-Methoxy-1-butanol acetate; 3-Methoxy butanol, acetic acid ester. *See* Methoxybutyl acetate

Methoxybutyl acetate

CAS 4435-53-4; EINECS/ELINCS 224-644-9
UN 2708 (DOT)
Synonyms: Acetic acid, 3-methoxybutanol ester; Acetic acid, 3-methoxybutyl ester; Acetic acid (3-methoxy-n-butyl) ester; 'Butoxyl'; 3-Methoxy-1-butanol acetate; 3-Methoxy butanol, acetic acid ester; 3-Methoxybutyl acetate; 3-Methoxy-1-butyl acetate; 3-Methoxy-n-butyl acetate; Methyl-1,3-butylene glycol acetate; 3-MOB acetate
Empirical: $C_7H_{14}O_3$
Formula: $CH_3COOCH_2CH_2CH(OCH_3)CH_3$
Properties: Colorless clear liq.; faint acrid ester-like odor; bitter taste; sol. in water, org. solvs.; m.w. 146.21; dens. 0.952-0.958 (20/20 C); f.p. -80 c; b.p. 171 C; flash pt. 62.5 C; autoignition temp. 410 C; ref. index 1.408-1.410 (20 C)
Toxicology: LD50 (oral, rat) 4210 mg/kg; mildly toxic by ingestion; skin and eye irritant; TSCA listed
Precaution: Combustible exposed to heat or flame; can react with oxidizing materials
Hazardous Decomp. Prods.: Heated to decomp., emits acrid smoke and irritating fumes
Uses: Solvent, thinner for coatings, nitrocellulose, acryl, urethane coatings, epoxy resin, aminoalkyd resin
Features: Low volatility; exc. dilutability with ethanol, hydrocarbons
Manuf./Distrib.: Daicel Chem. Ind.
Trade Names: Butoxyl®

3-Methoxybutyl acetate; 3-Methoxy-1-butyl acetate; 3-Methoxy-n-butyl acetate. *See* Methoxybutyl acetate
2-(Methoxycarbonyl) phenol. *See* Methyl salicylate

Methoxydiglycol

CAS 111-77-3; EINECS/ELINCS 203-906-6
Synonyms: DEGME; Diethylene glycol methyl ether; Diethylene glycol monomethyl ether; Ethanol, 2-(2-methoxyethoxy)-; 2-(2-Methoxyethoxy) ethanol; Methyl 'Carbitol'; Methyl digol
Classification: Aliphatic ether alcohol
Empirical: $C_5H_{12}O_3$
Formula: $CH_3OCH_2CH_2OCH_2CH_2OH$
Properties: Water-wh. liq.; mild pleasant odor; sol. in water, alcohols, glycerol, ether, acetone; misc. with most org. solvs.; m.w. 120.17; dens. 1.0211 (20/4 C); vapor pressure 0.2 mm Hg (20 C); m.p. -85 C; b.p. 194 C; flash pt. 93.3 C; ref. index 1.4264 (27 C)

Toxicology: LD50 (oral, rat) 9210 mg/kg; mod. toxic by skin contact, IP route; mildly toxic by ing.; may cause headache, nausea, dizziness, drowsiness, and incoordination; eye irritant; experimental teratogen and reproductive effects; TSCA listed
Environmental: VOC; BOD5 0.12; COD 1.71; ThOD 1.73
Precaution: Flamm. exposed to heat or flame; can react with oxidizers; reacts violently with $Ca(OCl)_2$; chlorosulfonic acid; oleum
Hazardous Decomp. Prods.: Heated to decomp., emits acrid smoke and irritating fumes
NFPA: Health 2, Flammability 2, Reactivity 0
Storage: Hygroscopic
Uses: Solvent for cosmetics, coatings (nongrain-raising wood stains, lacquers), printing inks, resins, waxes, dyes, stamp pad and stencil inks, dye pastes for textiles; diluent for hydraulic brake fluids; intermediate; coalescing agent for PVAc latex paints; solvent, coupling agent, surf. wetting agent, soil penetrant in household, commercial and industrial cleaners; antifreeze; anti-icing additive in aviation fuel; in food-pkg. adhesives
Features: High-boiling solvent
Regulatory: FDA 21CFR §175.105; SARA §313 reportable; HAP
Manuf./Distrib.: Aldrich; Ashland; Brown; Celanese; Dow; Eastman; Fluka; Great Western; Harcros; Occidental; Oxiteno; Sal Chem.; Spectrum Quality Prods.; Stanchem
Trade Names: Eastman® DM; Hisolve DM; MDGE; Methyl Carbitol®; Poly-Solv® DM

Methoxy dipropylene glycol. *See* PPG-2 methyl ether

Methoxyethanol

CAS 109-86-4; EINECS/ELINCS 203-713-7
UN 1188 (DOT)
Synonyms: EGME; Ethylene glycol methyl ether; Ethylene glycol monomethyl ether; Glycol monomethyl ether; 2-ME; 2-Methoxyethanol; β-Methoxyethanol; Methoxyhydroxyethane; Methyl 'Cellosolve'
Classification: Aliphatic ether alcohol
Empirical: $C_3H_8O_2$
Formula: $CH_3OCH_2CH_2OH$
Properties: Colorless liq., mild agreeable ether-like odor.; sol. in water, alcohols, ketones, glycols, most hydrocarbons; misc. with ether, benzene, glycerol, acetone, dimethylformamide; m.w. 76.10; dens. 0.964 (20/4 C); vapor pressure 9.5 mm Hg; f.p. -86.5 C; b.p. 123-124 C; flash pt. (OC) 43 C; autoignition temp. 235 C; ref. index 1.4028 (20 C); surf. tens. 30.84 dynes/cm; pH neutral
Toxicology: ACGIH TLV/TWA 5 ppm (skin); LD50 (oral, rat) 2460 mg/kg, (IP, rat) 2500 mg/kg, (IV, rat) 2140 mg/kg, (skin, rabbit) 1280 mg/kg; mod. toxic by ing., inh., skin contact, IP, IV routes; mod. toxic to humans by ing.; human systemic effects; skin/eye irritant; inh. can cause headache, confusion, weakness, nausea, difficulty breathing, increased heart rate; target organs: blood, brain, kidneys; potential occupational carcinogen; experimental teratogen; mutagenic data; TSCA listed
Environmental: VOC; BOD5 0.12; COD 1.69; ThOD 1.68
Precaution: Flamm. on exposure to heat and flame; mod. explosion hazard; can react with oxidizers to form explosive peroxides; incompat. with acid chlorides, acid anhydrides (forms ester compds.)
Hazardous Decomp. Prods.: Acetaldehyde, methanol; heated to decomp., emits acrid smoke, irritating fumes
NFPA: Health 2, Flammability 2, Reactivity 0
Storage: Store in cool, dry, well-ventilated area, out of direct sunlight
Uses: Solvent for nitrocellulose, cellulose acetate, ethyl cellulose, natural and some synthetic resins (vinyl, phenolic, alkyd, polyester), paints, varnishes, enamels, lacquers, wood stains, printing inks, dyes, leather dyeing, nail polishes, cosmetics, agric., electronic materials; plastics; plasticizers; jet fuel deicing additive; sealing moisture-proof cellophane; textile aux.; fixing agent in perfumes; chem. intermediate; in semiconductor and microfilm industries; antifreeze; in cleaning prods., preservatives, germicides, bactericides, insect repellents, antiseptics; in modified Karl Fischer reagent; solvent in cosmetics, pharmaceuticals; binder, color diluent, extender in foods; in food-pkg. adhesives
Regulatory: FDA 21CFR §73.1 (no residue), 175.105; USDA 9CFR §381.147; SARA §313 reportable; HAP
Manuf./Distrib.: Aldrich; Ashland; Brown; Celanese; Dow; Fluka; Great Western; Harcros; Lyondell; Occidental; Oxiteno; Romil Ltd; Ruger; Sal Chem.; Sigma; Spectrum Quality Prods.
Trade Names: Hisolve MC; Methyl Cellosolve®; MGE; Poly-Solv® EM
Trade Names Containing: Dynasolve MP-500; Dynasolve MP-AL

2-Methoxyethanol; β-Methoxyethanol. *See* Methoxyethanol

Methoxyethanol acetate

CAS 110-49-6; EINECS/ELINCS 203-772-9
UN 1189
Synonyms: Acetic acid 2-methoxyethyl ester; 1-Acetoxy-2-methoxyethane; EGMEA; Ethylene glycol methyl ether acetate; Ethylene glycol monomethyl ether acetate; Glycol ether EM acetate; Glycol monomethyl ether acetate; 2-MEA; 2-Methoxyethanol acetate; Methoxyethyl acetate; 2-Methoxyethyl acetate; β-Methoxyethyl acetate; Methyl 'Cellosolve' acetate; Methyl glycol acetate; Methyl glycol monoacetate
Classification: Aliphatic ether alcohol ester
Empirical: $C_5H_{10}O_3$
Formula: $CH_3COOC(CH_2)_2OCH_3$
Properties: Colorless liq.; mild pleasant sweet ether odor; sol. in water, ethanol, ether; misc. with most org. solvs.; m.w. 118.14; sp.gr. 1.007 (20/20 C); vapor pressure 7 mm Hg (20 C); f.p. -70 C; b.p. 143 C; flash pt. (CC) 49 C; autoignition temp. 392 C; ref. index 1.402 (20 C); surf. tens. 34 dynes/cm
Toxicology: ACGIH TLV/TWA 5 ppm (skin); LD50 (oral, rat) 3390 mg/kg, (dermal, rabbit) 5250 mg/kg; very toxic; mod. toxic by ing., IP, subcut. routes; mildly toxic by inh., skin contact; human systemic effects; eye irritant; inh. irritant; mutagenic data; TSCA listed
Environmental: VOC; BOD5 0.49; COD 1.60; ThOD 1.63
Precaution: DOT: Flamm. liq.; explosive limits 1.7-8.2%; mod. explosion hazard; can react with oxidizing materials; may form peroxides which can be explosive; decomp. (hydrolysis) can occur with strong acids and bases
Hazardous Decomp. Prods.: 2-Methoxyethanol, acetic acid; heated to decomp., emits acrid smoke and irritating fumes
NFPA: Health 1, Flammability 2
Storage: Store in cool, dry, well-ventilated area out of direct sunlight
Uses: Solvent for coatings, waxes, oils, resins, adhesives, various gums, printing inks, cellulose acetate, nitrocellulose, photographic films, lacquers, textile printing, nail polish
Regulatory: SARA §313 reportable; Calif. Prop. 65 reportable
Manuf./Distrib.: Aldrich; Celanese; Cognis; Coyne; Fluka; G.J. Chem.; Hatco; Hercules; Inolex; Joy; Scher; Sigma; Spectrum Quality Prods.; Stanchem; Stepan; Vopak USA; Werner G. Smith
Trade Names: Methyl Cellosolve® Acetate

2-Methoxyethanol acetate. *See* Methoxyethanol acetate
Methoxy ether of propylene glycol. *See* Methoxyisopropanol
2-(2-Methoxyethoxy) ethanol. *See* Methoxydiglycol
2-(2-Methoxyethoxy) ethanol acetate. *See* Diethylene glycol methyl ether acetate
2-(2-(2-Methoxyethoxy) ethoxy) ethanol. *See* Triglycol monomethyl ether
2-(2-Methoxyethoxy) ethyl acetate. *See* Diethylene glycol methyl ether acetate
Methoxyethyl acetate; 2-Methoxyethyl acetate; β-Methoxyethyl acetate. *See* Methoxyethanol acetate
2-Methoxyethyl ether. *See* Diethylene glycol dimethyl ether
2-Methoxyethyl phthalate. *See* Dimethoxyethyl phthalate

Methoxyethyl stearate

CAS 111-09-1
Synonyms: Ethanol, 2-methoxy-, stearate; Ethylene glycol monomethyl ether, stearate; 2-Methoxyethyl stearate; Methyl 'Cellosolve' stearate; Octadecanoic acid, 2-methoxyethyl ester; Stearic acid, 2-methoxyethyl ester
Classification: Glycol ether
Empirical: $C_{21}H_{42}O_3$
Formula: $CH_3OCH_2CH_2OOCC_{17}H_{35}$
Properties: Oily liq.; mild odor; visc. 9 cP; f.p. 19-24 C; b.p. 186-205 C (4 mm); flash pt. 192 C; dielec. const. 3.9 (60 F)
Toxicology: Toxic
Precaution: Combustible
Uses: Plasticizer; solvent
Manuf./Distrib.: ChemService

2-Methoxyethyl stearate. *See* Methoxyethyl stearate
1-Methoxyhexane. *See* Methyl hexyl ether
Methoxyhydroxyethane. *See* Methoxyethanol
1-Methoxy-2-hydroxypropane. *See* Methoxyisopropanol

Methoxyisopropanol

CAS 107-98-2; EINECS/ELINCS 203-539-1
UN 1993; 3092 (DOT)
Synonyms: Methoxy ether of propylene glycol; 1-Methoxy-2-hydroxypropane; 2-Methoxy-1-methylethanol; Methoxypropanol; 1-Methoxy-2-propanol; Monopropylene glycol methyl ether; MP; PGME; PM; Polypropylene glycol methyl ether; Polypropylene glycol monomethyl ether; 2-Propanol, 1-methoxy; Propylene glycol methyl ether; Propylene glycol monomethyl ether; α-Propylene glycol monomethyl ether
Classification: Aliphatic ether alcohol; glycol ether
Empirical: $C_4H_{10}O_2$
Formula: $CH_3OCH_2CH(OH)CH_3$
Properties: Colorless clear liq.; sweet ether-like odor; sol. in water, acetone, benzene, CCl_4, ethyl ether, petrol. ether, methanol; misc. with oxygenated solvs.; m.w. 90.14; dens. 0.919; vapor pressure 12.5 mm Hg; m.p. -96.7 C; b.p. 120 C; flash pt. (CC) 32 C; autoignition temp. 278 C; ref. index 1.402; surf. tens. 28 dynes/cm
Toxicology: ACGIH TLV/TWA 100 ppm; STEL 150 ppm; LD50 (oral, rat) 5660 mg/kg, (IP, rat) 3720 mg/kg, (IV, rat) 4200 mg/kg, (dermal, rabbit) 13 g/kg; mod. toxic by IV route; mildly toxic by ing., inh., skin contact; vapor irritates eyes and causes tearing; skin irritant; human systemic effects by inh. (anesthesia, nausea); experimental teratogen; target organs: respiratory system, eyes, CNS, liver; TSCA listed
Environmental: VOC; ThOD 1.95
Precaution: Flamm. liq.; LEL 3%; UEL 12%; dangerous fire hazard when exposed to heat or flame; incompat. with oxidizing materials, strong acids; may slowly form reactive peroxides during prolonged storage
Hazardous Decomp. Prods.: Heated to decomp., emits acrid smoke and irritating fumes
NFPA: Health 0, Flammability 3, Reactivity 0
Storage: Store in cool, well-ventilated area, out of direct sunlight; hygroscopic
Uses: Antifreeze/coolant for diesel engines; coalescing solvent for paints, lacquers, resins, dyes, oils, greases, cellulosics, acrylics, epoxides, polyesters, PU, cleaners, inks, electronics, textile dyes, agric. prods., cosmetics, floor polish, adhesives, fuel additives; solvent-sealing of cellophane; oil field; mining; epoxy laminates; chem. intermediate; heat transfer fluid; surf. tens. reducer in cleaners; solvent, coupling agent, wetting agent, penetrant for household/industrial cleaners; in food-pkg. adhesives; adjuvant in slimicides in food-contact paper/paperboard
Regulatory: FDA 21CFR §175.105, 176.300, 181.22, 181.30; HAP
Manuf./Distrib.: Aldrich; Allchem Ind.; Ashland; BP Amoco; Coyne; Dow; Ferro/Grant; Fluka; G.J. Chem.; Gallard-Schlesinger Ind.; General Chem.; Great Western; Harcros; Lyondell; Marcus Int'l.; Norman, Fox; Oxiteno; R.E. Carroll; Shell; Spectrum Quality Prods.; Stanchem; Vopak USA; Whyte Chems. Ltd
Trade Names: Arcosolv® PM; Dowanol® PM; Eastman® PM; Methyl Proxitol; Poly-Solv® MPM; Solvenon® PM

1-(2-Methoxyisopropoxy)-2-propanol. *See* PPG-2 methyl ether

Methoxyisopropyl acetamide

CAS 127727-20-2
Properties: M.w. 131.18; dens. 0.981 g/ml; visc. 30.69 cp; b.p. 232 C; flash pt. (CC) 265 F; surf. tens. 33.50 dynes/cm
Uses: Solvent

Methoxyisopropyl formamide

CAS 73579-09-6
Properties: M.w. 117.15; dens. 1.004 g/ml; visc. 8.34 cp; b.p. 231 C; flash pt. (CC) 275 F; surf. tens. 28.10 dynes/cm
Uses: Solvent

Methoxymethane. *See* Dimethyl ether

Methoxymethylbutanol

CAS 56539-66-3; EINECS/ELINCS 260-252-4
Synonyms: 1-Butanol, 3-methoxy-3-methyl-; 3-Methoxy-3-methylbutanol; 3-Methoxy-3-methyl-1-butanol
Classification: Organic compd.
Empirical: $C_6H_{14}O_2$
Formula: $(CH_3)_2COCH_3CH_2CH_2OH$
Properties: Colorless liq.; m.w. 118.18; dens. 0.926; m.p. -50 C; b.p. 173-175 C; flash pt. 71 C; ref. index 1.4280
Toxicology: Harmful by inh., ing., skin contact; irritating to eyes, skin, respiratory system; TSCA listed
Precaution: Combustible; incompat. with strong oxidizing agents, acid chlorides, acid anhydrides
Hazardous Decomp. Prods.: Toxic fumes of CO, CO_2; emits toxic fumes under fire conditions
Storage: Store in cool, dry place; keep tightly closed
Uses: Solvent in cosmetics
Manuf./Distrib.: Acros Org.; Aldrich; Aldrich; Chugai Boyeki Am.; Fluka; Lancaster Synthesis; TCI Am.

2-Methoxy-2-methylbutane. *See* t-Amyl methyl ether
3-Methoxy-3-methylbutanol; 3-Methoxy-3-methyl-1-butanol. *See* Methoxymethylbutanol
2-Methoxy-1-methylethanol. *See* Methoxyisopropanol

1-(2-(2-Methoxy-1-methylethoxy)-1-methylethoxyl)-2-propanol

CAS 20324-33-8 (α-isomer]
Synonyms: 2-Propanol, 1-(2-(2-methoxy-1-methylethoxy)-1-methylethoxy)-; Triglycol monomethyl ether; α-Tripropylene glycol methyl ether; Tripropylene glycol monomethyl ether
Empirical: $C_{10}H_{22}O_4$
Properties: Colorless clear visc. liq.; char. odor; sol. in water; m.w. 206.32; sp.gr. 0.96; vapor pressure < 0.1 mm Hg (20 C); b.p. 457 F; flash pt. (PMCC) 237 F
Toxicology: REL 100 ppm; LD50 (oral, rat) 3300 mg/kg; may cause eye/skin irritation; ing. may cause GI tract irritation; TSCA listed
Environmental: Pollutant; prevent entry into sewers, waterways
Precaution: Flamm. liq.; can release vapors that form flamm. mixts. at or above flash pt.; empty containers retain residue (liq./vapor) that may be dangerous; mists/sprays can ignite below flash pt.; slipping hazard; ground/bond containers during transfer; incompat. with strong oxidizing agents
Hazardous Decomp. Prods.: CO_x, noxious fumes
NFPA: Health 1, Flammability 1, Reactivity 1
Storage: Store unopened containers under cool, dry, ventilated conditions; keep away from heat, sparks, open flames, ignition sources
Uses: Solvent; pesticide inert ingred.
Regulatory: DOT nonregulated; SARA §311/312 fire/reactivity/acute/chronic hazard; CERCLA nonreportable
Manuf./Distrib.: Pharmco Prods.; TCI Am.
See also Triglycol monomethyl ether

(2-(2-Methoxymethylethoxy) methylethoxy) propanol. *See* PPG-3 methyl ether
1-(2-Methoxy-1-methylethoxy)-2-propanol; (2-Methoxymethylethoxy) propanol. *See* PPG-2 methyl ether
2-Methoxy-1-methylethyl acetate. *See* Propylene glycol methyl ether acetate
Methoxymethyl methyl ether. *See* Methylal

4-Methoxy-4-methylpentanone-2

CAS 107-70-0; EINECS/ELINCS 203-512-4
UN 2293 (DOT)
Synonyms: 4-Methoxy-4-methylpentan-2-one; 4-Methoxy-4-methyl-2-pentanone; Methoxy-4-methyl-4-pentanone-2; 4-Methyl-4-methoxy-2-pentanone; 2-Pentanone, 4-methoxy-4-methyl-; Pentoxane
Classification: Aliphatic ketone
Empirical: $C_7H_{14}O_2$
Formula: $CH_3C(CH_3)(OCH_3)CH_2COCH_3$
Properties: Colorless to water-wh. liq.; mod. sol. in water; sol. in ethanol, acetone; m.w. 130.21; sp.gr. 0.899; b.p. 147-163 C; flash pt. (CC) 60.5 C
Toxicology: LD50 (oral, mouse) 2050 mg/kg; LDLo (skin, rabbit) 3 g/kg; LCLo (inh., mouse, 15 h) 2280 ppm; mod. toxic by ing. and skin contact; mildly toxic by inh.; irritant to skin and eyes; TSCA listed
Precaution: Flamm. exposed to heat or flame; moderate fire risk; can

react vigorously with oxidizers
Hazardous Decomp. Prods.: Heated to decomp., emits acrid smoke and irritating fumes
Uses: Solvent for resin coatings

4-Methoxy-4-methylpentan-2-one; 4-Methoxy-4-methyl-2-pentanone; Methoxy-4-methyl-4-pentanone-2. *See* 4-Methoxy-4-methylpentanone-2

2-Methoxy-2-methylpropane. *See* Methyl t-butyl ether

Methoxy PEG

CAS 9004-74-4 (generic)
Synonyms: Methoxy polyethylene glycol; PEG methyl ether; Polyethylene glycol monomethyl ether
Definition: Addition polymer of ethylene oxide and methanol
Formula: $CH_3(CH_2CH_2O)_nOH$, n = 7-16
Properties: Liq. grades: Colorless or pract. colorless clear to sl. hazy sl. hygroscopic visc. liqs., sl. char. odor; sol. in acetone, alcohol, chloroform, ethyl acetate, toluene; misc. with water; sp.gr. 1.09-1.10; Solid grades: wh. waxy plastic, flakes, beads, or powds., pract. odorless and tasteless; sol. in water, acetone, alcohol, chloroform, ethyl acetate, toluene
Uses: Prod. of emulsifiers; ointment base, solvent, plasticizer in pharmaceuticals
Regulatory: NF compliance
Trade Names: Hisolve PM

Methoxy PEG-6

CAS 9004-74-4 (generic)
Synonyms: PEG-6 methyl ether; PEG 300 methyl ether; PEG-6 monomethyl ether; POE (6) methyl ether
Definition: PEG ether of methyl alcohol
Empirical: $C_{13}H_{28}O_4$
Formula: $CH_3(OCH_2CH_2)_nOH$, avg. n = 6
Properties: Nonionic
Toxicology: TSCA listed
Uses: Intermediate, lubricant for adhesives, inks, mining, soaps, detergents; solvent in cosmetics
Features: Provides low odor, low irritation, low toxicity to personal care prods.
Manuf./Distrib.: Aldrich; Fluka; Sigma
Trade Names: Dow Corning® MPEG350; Upiwax 350

Methoxy PEG-10

CAS 9004-74-4 (generic)
Synonyms: PEG 500 methyl ether; PEG 500 monomethyl ether; POE (10) methyl ether; POE (10) monomethyl ether
Classification: Ether
Definition: PEG ether of methyl alcohol
Empirical: $C_{21}H_{44}O_{11}$
Formula: $H(OCH_2CH_2)_nOCH_3$, avg. n = 10
Properties: Nonionic
Uses: Intermediate, lubricant for adhesives, inks, mining, soaps and detergents; humectant, solvent in cosmetics
Features: Provides solvency, low odor, low irritation, low toxicity to personal care prods.
Manuf./Distrib.: Aldrich; Fluka; Sigma
Trade Names: Dow Corning® MPEG550

Methoxy PEG-16

CAS 9004-74-4 (generic)
Synonyms: PEG-16 methyl ether; PEG (16) monomethyl ether; POE (16) methyl ether; POE (16) monomethyl ether
Classification: Ether
Definition: PEG ether of methyl alcohol
Formula: $H(OCH_2CH_2)_nOCH_3$, avg. n = 16
Properties: Nonionic
Uses: Chemical and surfactant intermediate; lubricant for adhesives; humectant, solvent in cosmetics
Features: Provides solvency, low odor, low irritation, low toxicity to personal care prods.
Manuf./Distrib.: Aldrich; Fluka; Sigma
Trade Names: Dow Corning® MPEG750

4-Methoxyphenol; p-Methoxyphenol. *See* Hydroquinone monomethyl ether
Methoxy polyethylene glycol. *See* Methoxy PEG
α,ω-Methoxypoly (ethylene oxide). *See* PEG dimethyl ether

3-Methoxy-1,2-propanediol

CAS 623-39-2; EINECS/ELINCS 210-791-6
Synonyms: Glycerin α-monomethyl ether; Glycerol 1-monomethyl ether; 3-MPD; 1,2-Propanediol, 3-methoxy-
Classification: Glycol ether
Empirical: $C_4H_{10}O_3$
Properties: Colorless liq.; m.w. 106.12; dens. 1.114 gm/l; b.p. 220 C; flash pt. > 230 F; ref. index 1.444 (20 C)
Toxicology: LD50 (subcut., mouse) > 2547 mg/kg; may be harmful by ing., inh., skin absorp.; may cause eye/skin/mucous membrane/upper respiratory tract irritation; TSCA listed
Environmental: ThOD 1.51
Precaution: Incompat. with strong oxidizing agents
Hazardous Decomp. Prods.: Toxic fumes of CO, CO_2
Storage: Store in cool, dry place; keep tightly closed
Uses: Solvent
Regulatory: HAP
Manuf./Distrib.: Aldrich; Porta Ltd; TCI Am.

3-Methoxypropanenitrile. *See* 3-Methoxypropionitrile
Methoxypropanol. *See* Methoxyisopropanol; 3-Methoxy-1-propanol
1-Methoxy-2-propanol. *See* Methoxyisopropanol
3-Methoxypropanol. *See* 3-Methoxy-1-propanol

3-Methoxy-1-propanol

CAS 1320-67-8 (mixt.); 1589-49-7 (β-isomer)
UN 1987
Synonyms: Methoxypropanol; 3-Methoxypropanol; β-PGME; 1,2-Propanediol, monomethyl ether; Propanol, methoxy-; Propanol, 3-methoxy-; β-Propylene glycol monomethyl ether
Empirical: $C_4H_{10}O_2$
Properties: Clear liq.; sol. in water; m.w. 90.12; sp.gr. 0.94-0.945; b.p. 148-152 C; flash pt. 65 C; ref. index 1.413
NFPA: Health 1, Flammability 2, Reactivity 0
Uses: Solvent; flow aid; intermediate for synthesis of proton pump inhibitor (rabeprazole sodium)
Features: Low volatility
Manuf./Distrib.: Fluka

1-Methoxy-2-propanol acetate. *See* Propylene glycol methyl ether acetate

2-Methoxy-1-propanol acetate

CAS 70657-70-4; EINECS/ELINCS 274-724-2
Synonyms: Acetic acid, 2-methoxypropyl ester; 2-Methoxy-1-acetoxypropane; 2-Methoxypropyl acetate; (2-Methoxy-1-propyl) acetate; 2-Methoxypropyl-1-acetate; 2-Methoxypropylacetate-1
Empirical: $C_6H_{12}O_3$
Properties: M.w. 132.16
Toxicology: TCLo (inh., female rat, 6-15 days after conception) 2710 ppm; reproductive effector
Uses: Solvent for paints, lacquers, resins, oils, fats
Trade Names Containing: Dowanol® PMA

2-Methoxy-4-(2-propenyl) phenol; 2-Methoxy-4-prop-2-enylphenol. *See* Eugenol

3-Methoxypropionitrile

CAS 110-67-8; EINECS/ELINCS 203-790-7
Synonyms: 1-Cyano-2-methoxyethane; 3-Methoxypropanenitrile; Methyl β-cyanoethyl ether; Propanenitrile, 3-methoxy-; Propionitrile, 3-methoxy-
Classification: Nonaromatic nitrile
Empirical: C_4H_7NO
Formula: $CH_3OCH_2CH_2C{\equiv}N$
Properties: Liq.; m.w. 85.11; dens. 0.937; m.p. < -20 C; b.p. 164-165 C; flash pt. (OC) 149 F; ref. index 1.4030
Toxicology: LD50 (oral, rat) 4390 mg/kg, (IP, mouse) 2900 mg/kg, (skin, rabbit) > 10 ml/kg; mod. toxic by IP route; mildly toxic by ing.; harmful by ing., inh., skin contact; irritating to eyes, skin, respiratory system; TSCA listed

Precaution: Combustible liq.; flamm. exposed to heat, flame, or oxidizers; reacts with water, steam, or acids to produce toxic/flamm. vapors
Hazardous Decomp. Prods.: Heated to decomp., emits highly toxic fumes of CN^-
Storage: Moisture-sensitive
Uses: Process solvent
Manuf./Distrib.: Aldrich; Fluka

2-(2-(2-Methoxypropoxy) ethoxy) ethanol. *See* Diethylene glycol isobutyl ether
Methoxypropoxypropanol. *See* PPG-2 methyl ether
3-(3-(3-Methoxypropoxy) propoxy) propan-1-ol. *See* PPG-3 methyl ether

Methoxypropyl acetamide

CAS 16339-54-1
Properties: M.w. 131.18; dens. 0.989 g/ml; visc. 15.12 cp; b.p. 256 C; flash pt. (CC) 310 F; surf. tens. 35.50 dynes/cm
Uses: Solvent

Methoxypropyl acetate; 1-Methoxy-2-propyl acetate. *See* Propylene glycol methyl ether acetate
2-Methoxypropyl acetate; (2-Methoxy-1-propyl) acetate; 2-Methoxypropyl-1-acetate; 2-Methoxypropylacetate-1. *See* 2-Methoxy-1-propanol acetate
2-(1-Methoxy) propyl acetate. *See* Propylene glycol methyl ether acetate

Methoxypropylamine

CAS 5332-73-0; EINECS/ELINCS 226-241-3
UN 2924 (DOT)
Synonyms: 1-Amino-3-methoxypropane; 3-Methoxypropylamine; 3-MPA
Empirical: $C_4H_{11}NO$
Formula: $CH_3OCH_2CH_2CH_2NH_2$
Properties: Colorless clear liq.; misc. with water, ethanol, toluene, acetone, CCl_4, hexane, ether, most org. solvs.; m.w. 89.16; dens. 0.873 (20/20 C); vapor pressure 20 mm (30 C); m.p. -75.7 C; b.p. 119 C; flash pt. (TCC) 80 F; ref. index 1.417 (20 C)
Toxicology: LD50 (IV, mouse) 180 mg/kg; poison by IV route; toxic by ing., inh.; irritating to skin, eyes, mucous membranes; causes burns; TSCA listed
Precaution: Flamm.; moderate fire risk; can react with oxidizing materials
Hazardous Decomp. Prods.: Heated to decomp., emits toxic fumes of NO_x
Uses: Solvent; organic intermediate; emulsifier in anionic coatings, wax formulations; emulsifier intermediate for water treating chemicals, agric. chemicals; neutralizer in preboiler system; optical brightener intermediate
Manuf./Distrib.: Aldrich; Atofina; BASF; Fluka; Huntsman

3-Methoxypropylamine. *See* Methoxypropylamine

Methoxypropyl formamide

CAS 71172-42-4
Properties: M.w. 117.15; dens. 1.012 g/ml; visc. 3.24 cp; b.p. 255 C; flash pt. (CC) 255 F; surf. tens. 36.10 dynes/cm
Uses: Solvent

Methoxy triethylene glycol; Methoxytriglycol. *See* Triglycol monomethyl ether
Methyene oxide. *See* Formaldehyde

Methyl abietate

CAS 127-25-3; EINECS/ELINCS 204-832-7
UN 1231 (DOT)
Synonyms: Abietic acid methyl ester; Methyl [1R-(1α,4aβ,4bα,10aα)]-1,2,3,4,41,4b,5,6,10,10a-decahydro-1,4a-dimethyl-7-(1-methylethyl)-1-phenanthrenecarboxylate; Methyl ester of wood rosin; 1-Phenanthrenecarboxylic acid, 1,2,3,4,4a,4b,5,6,10,10a-decahydro-1,4a-dimethyl-7-(1-methylethyl)-, methyl ester, [1R-(1α,4aβ,4bα,10aα)]-; Podocarpa-7,13-dien-15-oic acid, 13-isopropyl-, methyl ester; Wood rosin, methyl ester
Definition: Commercial prod. is a mixt. of methyl esters of the rosin acids
Empirical: $C_{21}H_{32}O_2$
Formula: $C_{19}H_{29}COOCH_3$
Properties: Colorless to yel. thick liq., odorless; misc. with alcohol, ether, common org. solvs., aliphatic hydrocarbons; insol. in water; m.w. 316.47; dens. 1.03; b.p. 360-365 C (dec.); flash pt. (COC) 180 C; ref. index 1.53 (20 C)
Toxicology: LD50 (oral, rat) > 5 g/kg, (skin, rabbit) > 5 g/kg; low toxicity by ing. and skin contact; primary irritant; mod. skin irritant; probably sl. toxic; TSCA listed
Precaution: Combustible exposed to heat or flame; can react with oxidizers
Hazardous Decomp. Prods.: Heated to decomp., emits acrid smoke and irritating fumes
NFPA: Health 0, Flammability 1, Reactivity 0
Uses: Plasticizer, solvent for vinyls, cellulosics, PS, paints, lacquers, varnishes, ester gums, rosin, rubber, adhesives; fragrance in cosmetics
Manuf./Distrib.: ChemService; Hercules; Salor

Methylacetaldehyde. *See* Propionaldehyde
Methyl acetamide. *See* N-Methylacetamide

N-Methylacetamide

CAS 79-16-3; EINECS/ELINCS 201-182-6
Synonyms: Acetamide, N-methyl; Acetylmethylamine; Methyl acetamide; Monomethylacetamide
Classification: Nonaromatic amide
Definition: Substituted amide of acetic acid
Empirical: C_3H_7NO
Formula: $CH_3CONHCH_3$
Properties: Needles; sol. in water, ethanol, benzene; insol. in ligroin; m.w. 73.10; dens. 0.957; m.p. 26-28 C; b.p. 206 C; flash pt. 116 C; ref. index 1.4330
Toxicology: LD50 (oral, rat) 5000 mg/kg, (IP, rat) 2750 mg/kg, (subcut., rat) 3600 mg/kg, (IV, mouse) 4015 mg/kg; mod. toxic by IP and subcut. routes; mildly toxic by ing., IV routes; harmful by ing., inh., skin contact; irritating to eyes, skin, respiratory system; may cause GI, liver, or endocrine changes, wt. loss, body temp. decrease; experimental teratogen, reproductive effects; mutagenic data; TSCA listed
Hazardous Decomp. Prods.: Heated to decomp., emits toxic fumes of NO_x
Storage: Store under nitrogen
Uses: Solvent, visc. control agent in cosmetics
Manuf./Distrib.: ABCR; Acros Org.; Aldrich; BASF; Fluka

Methyl acetate

CAS 79-20-9; EINECS/ELINCS 201-185-2
UN 1231 (DOT); FEMA 2676
Synonyms: Acetic acid methyl ester; Methyl acetic ester; Methyl ethanoate
Classification: Sat. aliphatic carboxylic acid ester
Definition: Ester of methyl alcohol and acetic acid
Empirical: $C_3H_6O_2$
Formula: CH_3COOCH_3
Properties: Colorless volatile liq., fragrant odor, sl. bitter flavor; very sol. in water; sol. in ethanol, diethyl ether, acetone, benzene, chloroform; m.w. 74.09; dens. 0.92438; vapor pressure 100 mm Hg (9.4 C); f.p. -98.05 C; b.p. 54.05 C; flash pt. (CC) -10 C; autoignition temp. 454 C; ref. index 1.3614 (20 C); surf. tens. 24.1 dynes/cm (20 C); dielec. const. 6.68
Toxicology: ACGIH TLV/TWA 200 ppm in air; STEL 250 ppm; LD50 (oral, rabbit) 3705 mg/kg; LCLo (inh., cat, 1 h) 57,000 mg/m^3; mod. toxic by several routes; human systemic irritant by inh.; mod. skin and severe eye irritant; irritant to respiratory tract; narcotic in high concs.; mutagenic data; TSCA listed
Environmental: ThOD 1.51
Precaution: Flamm.; dangerous fire and explosion risk; explosive limits 3.1-16% in air; incompat. with oxidizing agents, strong acids, bases, potassium t-butoxide
Hazardous Decomp. Prods.: Methanol, acetic acid; heated to decomp., emits acrid smoke and fumes
NFPA: Health 1, Flammability 3, Reactivity 0
Storage: Store in cool, dry, well-ventilated area, out of direct sunlight, away from heat/ignition sources
Uses: Solvent for paint removers, paints/coatings, lacquers, resins, oils, cellulosics, cellulose nitrate, nitrocellulose, acetylcellulose, perfumes, cosmetics, pharmaceuticals; odorant; catalyst for biodeg. of org. materials; intermediate for mfg. of acetic anhydride, pharmaceuticals, artificial leather, syn. flavors; formulating cleaners; synthetic flavoring

agent in foods and pharmaceuticals; in food-pkg. adhesives
Regulatory: FDA 21CFR §172.515, 175.105; FEMA GRAS
Manuf./Distrib.: Akzo Nobel; Aldrich; Ashland; Bio-Rad Labs; Celanese; Degussa Flavors/Alfrebro; Eastern Chem.; Eastman; Fleurchem; Fluka; Haltermann Prods. UK; Hampshire; Integra; J.T. Baker; Lyondell; Monsanto; Penta Mfg.; Spectrum Quality Prods.; Synasia Fine Chem.; VWR Int'l.; Vopak USA; Wacker-Chemie GmbH
Trade Names: Eastman® Methyl Acetate, High Purity
Trade Names Containing: Eastman® Methyl Acetate; Eastman® Methyl Acetate, Cosmetic Grade

Methyl acetic ester. *See* Methyl acetate

Methyl acetoacetate

CAS 105-45-3; EINECS/ELINCS 203-299-8
Synonyms: Acetoacetic acid methyl ester; MAA; 1-Methoxybutane-1,3-dione; Methyl acetylacetate; Methyl acetylacetonate; Methyl 3-oxobutanoate; Methyl 3-oxobutyrate; 3-Oxobutanoic acid methyl ester
Classification: Carboxylic acid ester
Empirical: $C_5H_8O_3$
Formula: $CH_3COCH_2COOCH_3$
Properties: Colorless liq.; very sol. in water; sol. in alcohol, ether; misc. with oxygenated solvs.; m.w. 116.1; dens. 1.076 (20/4 C); m.p. -80 C; b.p. 167-170 C; flash pt. 62 C; pH neutral
Toxicology: LD50 (oral, rat) 3228 mg/kg; mod. toxic by ing.; severe eye and mild skin irritant; high concs. of vapor may cause nose/throat irritation, drowsiness, headache, lightheadedness; TSCA listed
Precaution: Flamm. exposed to heat or flame; incompat. with oxidizing agents (increases fire/explosion hazard), bases (decomp. can occur)
Hazardous Decomp. Prods.: Methanol, acetic acid; heated to decomp., emits acrid smoke and irritating fumes
NFPA: Health 2, Flammability 2, Reactivity 0
Storage: Store in cool, dry, well-ventilated area, out of direct sunlight
Uses: Copromoter for cobalt-contg. polyester-accelerator systems; solvent for cellulose ethers; extractant; ingred. of solvent mixts. for cellulose esters; synthesis of nifedipine (a cardiac drug); org. synthesis; reagent for Dane salt formation; intermediate for vitamins, feed additives, pharmaceuticals, pesticides; component for flavors and fragrances; solvent for fragrances in cosmetics; prod. of azo dyes
Manuf./Distrib.: Aceto; Aldrich; Eastman; Fluka; Honeywill & Stein; Lonza; Penta Mfg.; S.A. Chems.; Sigma; Spectrum Quality Prods.; Synasia Fine Chem.; Wacker-Chemie GmbH
Trade Names: Eastman® MAA

Methyl acetone. *See* Methyl ethyl ketone
Methyl 12-acetoxy-9-octadecenoate; Methyl 12-acetoxyoleate. *See* Methyl acetyl ricinoleate
Methyl acetylacetate; Methyl acetylacetonate. *See* Methyl acetoacetate

Methyl acetyl ricinoleate

CAS 140-03-4; EINECS/ELINCS 205-392-9
Synonyms: Methyl 12-acetoxy-9-octadecenoate; Methyl 12-acetoxyoleate; Ricinoleic acid, methyl ester, acetate
Empirical: $C_{21}H_{38}O_4$
Formula: $C_{17}H_{32}(OCOCH_3)COOCH_3$
Properties: Cryst.; sol. in most org. solvs.; insol. in water; pale yel. liquid; mild odor; m.w. 1306.59; sp.gr. 0.938; m.p. -15 C; flash pt. (COC) 196 C; ref. index 1.4545
Toxicology: LD50 (oral, mouse) 34,900 mg/kg; low toxicity by ing.; primary irritant to eyes; probable skin and mucous membrane irritant; TSCA listed
Precaution: Combustible exposed to heat or flame; can react with oxidizing materials
Hazardous Decomp. Prods.: Heated to decomp., emits toxic fumes of NO_x
Uses: Plasticizer; lubricant; protective coatings; synthetic rubbers; vinyl compds.; cellulosics; chewing gum base; solubilizer for benzophenone-3; cosolvent props.; emollient; in food-pkg. adhesives
Regulatory: FDA 21CFR §175.105; Japan approved
Manuf./Distrib.: CasChem; ChemService
Trade Names: Naturechem® MAR

Methyl acrolein; 3-Methylacrolein; β-Methyl acrolein. *See* Crotonaldehyde

Methyl adipate

CAS 627-91-8; EINECS/ELINCS 211-019-0
Synonyms: Adipic acid monomethyl ester; MMAD; Monomethyl adipate
Empirical: $C_7H_{12}O_4$
Properties: Colorless liq. to wh. solid; m.w. 160.2; dens. 1.081 g/ml; m.p. 8-9 C; b.p. 162 C (10 mm Hg); flash pt. > 230 F; ref. index 1.44 (20 C)
Toxicology: May be harmful by inh., ing., skin absorp.; may cause eye/skin irritation; TSCA listed
Precaution: Incompat. with bases, oxidizing agents, reducing agents
Hazardous Decomp. Prods.: Toxic fumes of CO, CO_2
Storage: Store in cool, dry place; keep container closed
Uses: Solvent in paints
Manuf./Distrib.: Aldrich; Fluka; Proviron Ind. NV
Trade Names Containing: Rhodiasolv RPDE
See also Dimethyl adipate

Methylal

CAS 109-87-5; EINECS/ELINCS 203-714-2
UN 1234 (DOT)
Synonyms: Bis (methoxy) methane; Dimethoxymethane; Dimethyl formal; Dioxapentane; Formal; Formaldehyde dimethylacetal; Formaldehyde methyl ketal; Methoxymethyl methyl ether; Methylene dimethyl ether
Empirical: $C_3H_8O_2$
Formula: $CH_3OCH_2OCH_3$
Properties: Colorless clear liq.; chloroform-like odor; pungent taste; mod. sol. in water (20 C); m.w. 76.11; dens. 0.86; vapor pressure 330 mm Hg (20 C); f.p. -105 C; b.p. 42.3 C; flash pt. ≈ -17.7 C; ref. index 1.3589
Toxicology: ACGIH TLV/TWA 1000 ppm; LD50 (oral, rabbit) 5708 mg/kg; LC50 (inh., rat) 15,000 ppm; mod. toxic by subcut. route; mildly toxic by ing. and inh.; target organ: eyes, kidneys, lung, liver, heart; narcotic and anesthetic in high concs.; TSCA listed
Precaution: Flammable, dangerous fire risk exposed to heat, flame, or oxidizers; explosion risk; may ignite or explode when heated with oxygen
Hazardous Decomp. Prods.: Heated to decomp., emits acrid smoke and irritating fumes
NFPA: Health 1, Flammability 3, Reactivity 0
Storage: Refrigerate before opening
Uses: Solvent in cosmetics; reaction solvent, extraction solvent for mfg. pharmaceuticals, aerosols, paints, varnishes, paint strippers, cleaners; resin mfg.; chain length regulator in ion exchange resins, polyacetal; organic synthesis; protective coatings; special fuel; fuel additive for smoke reduction; mfg. of adhesives, glues, fragrances, perfumes, pesticides
Manuf./Distrib.: Aldrich; Fluka; Sigma; Tryline

Methyl alcohol

CAS 67-56-1; EINECS/ELINCS 200-659-6
UN 1230 (DOT)
Synonyms: Carbinol; Colonial spirit; Columbian spirits; Methanol; Methyl hydrate; Methyl hydroxide; Methylol; Monohydroxymethane; Pyroxylic spirit; Wood alcohol; Wood naphtha; Wood spirit
Empirical: CH_4O
Formula: CH_3OH
Properties: Clear colorless liq., alcoholic odor (pure), pungent odor (crude); highly polar; misc. with water, alcohol, ether, benzene, ketones, most org. solvs.; m.w. 32.05; dens. 0.7924; vapor pressure 127 mm Hg; m.p. -97.8 C; b.p. 64.5 C; flash pt.(CC) 6 C; autoignition temp. 385 C; ref. index 1.3292 (20 C); surf. tens. 22.3 dynes/cm; dielec. const. 32.66
Toxicology: ACGIH TLV/TWA 200 ppm; LD50 (oral, rat) 5628 mg/kg; toxic (causes blindness); poisonous by ing., inh., or percutaneous absorp.; eye/skin irritant; narcotic; acute effects: headache, fatigue, nausea, visual impairment, acidosis, convulsions, circulatory collapse, respiratory failure, death; usual fatal dose 100-250 ml; experimental teratogen, reproductive effects; TSCA listed
Environmental: VOC; BOD5 0.85; COD 1.50; ThOD 1.50
Precaution: DOT: Flamm. liq.; dangerous fire risk; explosive limits 6.0-36.5 vol.% in air; reacts vigorously with oxidizers; violent or explosive reactions possible
Hazardous Decomp. Prods.: Heated to decomp., emits acrid smoke and

irritating fumes
NFPA: Health 1, Flammability 3, Reactivity 0
Uses: Industrial solvent; mfg. of formaldehyde, acetic acid, dimethyl terephthalate, methyl esters; chemical synthesis; antifreeze; solvent for nitrocellulose, polyvinyl butyral, shellac, rosin, manila resin, dyes, paints, waterborne coatings; plastics softener; denaturant, solvent in cosmetics; octane booster in gasoline; as fuel for picnic stoves, soldering torches; extractant for animal/veg. oils; extraction solvent for hops extract, spice oleoresins (foods); pharmaceutical solvent, excipient; in food-pkg. adhesives; in paper/paperboard in contact with dry food; defoamer in food-contact coatings and paper/paperboard; solvent in food-contact PPO resins
Regulatory: FDA 21CFR §73.345, 73.615, 172.560, 172.859, 172.867, 173.250, 173.385, 175.105, 175.300, 176.180, 176.200, 176.210, 177.1200, 177.2420, 177.2460, 177.2800, 27CFR §21.115; FDA approved for orals; NF, BP compliance; SARA §313 reportable; CERCLA hazardous substance; HAP
Manuf./Distrib.: AMC Chems.; Air Prods.; Aldrich; Allchem Ind.; Amyl; Ashland; BP Amoco; Baychem; Brenntag AG; Brown; C.P. Hall; CPS; Celanese; Chemcentral; Coastal Chem Inc; Coyne; DuPont; EM Science; Eastman; Equistar; Fisher Scientific; Fluka; General Chem.; Georgia Gulf; Harcros; Hawkins Chem.; Honeywell Perf. Polymers; Houghton Chem.; Hukill; Integra; J.T. Baker; Mallinckrodt Baker; Methanex; Miljac; Nissan Chem. Ind.; Norsk Hydro AS; Penta Mfg.; Rhodia/Phosphorus Perf. Derivs.; Rhodia; Romil Ltd; Ruger; Sal Chem.; Shell; Sigma; Spectrum Quality Prods.; Stanchem; Sterling Chems.; Sunnyside; Thomas Scientific; VWR Int'l.; Varsal Instruments; Veckridge; Vopak USA; Xinchem
Trade Names Containing: Celtone; Dynasolve 190; Eastman® Methyl Acetate; Eastman® Methyl Acetate, Cosmetic Grade; Filmex® A-1, 190 Proof; Filmex® A-2 190 Proof; Filmex® A-2 Anhydrous; Filmex® B 190 Proof; Filmex® B Anhydrous; Filmex® C 190 Proof; Filmex® C Anhydrous; Filmex® D-1; Filmex® D-2 190 Proof; Filmex® D-2 Anhydrous; Forane 141b MGX; Genesolv® 2004; Propsolv I-1 190 Proof; Propsolv I-1 Anhydrous; Propsolv III-1 190 Proof; Propsolv III-1 Anhydrous; Propsolv III-2 190 Proof; Propsolv III-2 Anhydrous; Punctilious® SDA 1-1 190 Proof; Punctilious® SDA 1-1 Anhydrous; Punctilious® SDA 1-2 190 Proof; Punctilious® SDA 3A 190 Proof; Punctilious® SDA 3A Anhydrous; Punctilious® SDA 30 190 Proof; Punctilious® SDA 30 Anhydrous; Punctilious® SDA Canada DAG-2A Anhydrous; Punctilious® SDA Canada DAG-2A2 Anhydrous; S-327 190 Proof; S-327 Anhydrous; S-333 Anhydrous; Tecsol® A, 95%; Tecsol® A Anhydrous; Tecsol® A-2, 95%; Tecsol® A-2 Anhydrous; Tecsol® B; Tecsol® C 95%; Tecsol® C Anhydrous; Tecsol® D, 95%; Tecsol® D-2, 95%; Tecsol® D-2 Anydrous; Uresolve 411; Vertrel® SMT; Vertrel® XM; Vertrel® XMS Plus

Methyl aldehyde. *See* Formaldehyde

Methylamine

CAS 74-89-5; EINECS/ELINCS 200-820-0
UN 1061 (DOT); UN 1235 (DOT)
Synonyms: Aminomethane; Carbinamine; Methanamine; Methylamine, anhydrous; MMA; Monomethylamine
Classification: Sat. aliphatic amine
Empirical: CH_5N
Formula: CH_3NH_2
Properties: Colorless gas @ R.T. and atm. pressure; char. fish odor (low conc.) or ammonia-like odor (high conc.); very sol. in water; sol. in ethanol, acetone, diethyl ether, benzene; sol. in oxygenated solvs.; m.w. 31.07; dens. 0.693 (liq. @ -6.5 C); vapor pressure 2650 mm Hg; f.p. -93.5 C; b.p. -6.3 C; flash pt. 0 C; autoignition temp. 430 C; pH 11.8 (0.1M); surf. tens. 19.19 dynes/cm; dielec. const. 9.4
Toxicology: ACGIH TLV/TWA 5 ppm; LC50 (inh., mouse, 2 h) 2400 mg/m^3; poison by subcut. route; mod. toxic by inh.; severely irritating to nose, throat, lungs; severe exposure can cause potentially fatal pulmonary edema; corrosive to skin; liquefied gas can cause frostbite; irritating to eyes; lachrymator; severe exposure to liquefied gas can cause tissue death and gangrene, permanent eye damage or blindness; may cause 'halo vision'; mutagenic data; TSCA listed
Environmental: VOC; BOD5 0.97; ThOD 1.54
Precaution: Flamm. gas; dangerous fire risk; LEL 4.9%; UEL 20.7%; strong base; forms explosive mixt. with nitromethane
Hazardous Decomp. Prods.: CO, CO_2, NO_x, ammonia, traces of organic compds. incl. hydrogen cyanide, nitriles, isocyanates, nitrosamines, and amines
NFPA: Health 3, Flammability 4, Reactivity 0
Uses: Intermediate in organic synthesis; intermediate for water gel explosives, accelerators, dyes, pharmaceuticals, insecticides, N-methylpyrrolidine, methylalkanolamines, surface active agents, fungicides; in tanning; in photographic developers; in paint removers; fuel additive; polymerization inhibitor; rocket propellant; solvent
Regulatory: CERCLA hazardous substance
Manuf./Distrib.: Air Prods.; Aldrich; Celanese Mexico; Chinook Group; DuPont; MG Ind.; Matheson Gas Prods.; Penta Mfg.; Spectrum Quality Prods.

Methylamine, anhydrous. *See* Methylamine
(Methylamino) benzene; N-Methylaminobenzene. *See* N-Methylaniline
2-(Methylamino) ethanol; β-(Methylamino) ethanol; N-Methylaminoethanol. *See* N-Methylethanolamine

Methyl amyl acetate

CAS 108-84-9; EINECS/ELINCS 203-621-7
UN 1233 (DOT)
Synonyms: Acetic acid-1,3-dimethylbutyl ester; Dimethylbutyl acetate; 1,3-Dimethylbutyl acetate; s-Hexyl acetate; MAAC; Methylisoamyl acetate; Methylisobutyl carbinol acetate; Methylisobutylcarbinyl acetate; 4-Methyl-2-pentanol acetate; 4-Methyl-2-pentyl acetate; 2-Pentanol, 4-methyl-, acetate
Empirical: $C_8H_{16}O_2$
Formula: $CH_3COOCH(CH_3)C_4H_9$
Properties: Colorless liq., mild odor; insol. in water; sol. in alcohol; m.w. 144.21; sp.gr. 0.858 (20/20 C); vapor pressure 3.8 mm Hg (20 C); f.p. -64 C; b.p. 146-150 C; flash pt. (TCC) 96 F; ref. index 1.4008 (20 C)
Toxicology: LD50 (oral, rat) 6160 mg/kg, (skin, rabbit) 20 g/kg; LCLo (inh., rat, 4 h) 2000 ppm; TLV 50 ppm in air; toxic by inhalation; mildly toxic by ing., skin contact; skin and human eye irritant; human systemic effects by inh.; TSCA listed
Precaution: Flamm. exposed to heat or flames; moderate fire risk; can react with oxidizing materials
Uses: Solvent for coatings, nitrocellulose, and other lacquers; cosmetics ingred.
Manuf./Distrib.: CTC Orgs.; Celanese; Eastman; Penta Mfg.

Methyl amyl alcohol

CAS 108-11-2; EINECS/ELINCS 203-551-7
UN 2053 (DOT)
Synonyms: Isobutyl methyl carbinol; Isobutyl methyl methanol; MAOH; Methylisobutyl carbinol; 2-Methyl-4-pentanol; 4-Methylpentanol-2; 4-Methyl-2-pentanol; 4-Methylpentan-2-ol; MIBC; MIC; 2-Pentanol, 4-methyl-
Classification: Alcohol
Empirical: $C_6H_{14}O$
Formula: $(CH_3)_2CHCH_2CH(CH_3)OH$
Properties: Colorless clear liq.; mild odor; misc. with most common org. solvs.; sl. sol. in water; m.w. 102.18; dens. 0.8079 (20/20 C); vapor pressure 2.8 mm Hg (20 C); f.p. < -90 C (sets to a glass); b.p. 131.8 C; flash pt. (OC) 40.5 C; ref. index 1.4089; surf. tens. 22.63 dynes/cm
Toxicology: ACGIH TLV/TWA 25 ppm; STEL 40 ppm (skin); LD50 (oral, rat) 2590 mg/kg, (IP, mouse) 812 mg/kg; mod. toxic by ing., skin contact, IP routes; mildly toxic by inh.; irritating to respiratory system; skin and severe eye irritant; inh. of high concs. can cause anesthesia; TSCA listed
Environmental: VOC; BOD5 2.37; COD 2.60; ThOD 2.82
Precaution: Flamm. exposed to heat or flame; LEL 1%; UEL 5.5%; moderate fire risk; mod. explosion hazard exposed to heat or flame; can react with oxidizers
Hazardous Decomp. Prods.: Heated to decomp., emits acrid smoke and fumes
NFPA: Health 2, Flammability 2, Reactivity 0
Uses: High-boiling solvent for dyestuffs, oils, gums, resins, waxes, lacquers, nitrocellulose, ethylcellulose; latent solvent for coatings; organic synthesis; froth flotation agent; brake fluids; fragrance in cosmetics
Manuf./Distrib.: Aldrich; Allchem Ind.; Ashland; Atofina SA; Celanese; Dow;

Fluka; Houghton Chem.; Hukill; O'Brien Prods.; Shell; Stanchem; Zinkan Enterprises
See also 2-Methyl-1-pentanol

Methyl amyl carbinol; Methyl-n-amyl carbinol. *See* 2-Heptanol
Methyl t-amyl ether. *See* t-Amyl methyl ether
Methyl amyl ketone. *See* Methyl n-amyl ketone

Methyl n-amyl ketone

CAS 110-43-0; EINECS/ELINCS 203-767-1
UN 1110 (DOT); FEMA 2544
Synonyms: Amyl methyl ketone; n-Amyl methyl ketone; Butylacetone; 2-Heptanone; Heptan-2-one; Ketone C-7; MAK; Methyl amyl ketone; Methyl pentyl ketone; n-Pentyl methyl ketone
Classification: Sat. aliphatic ketone
Empirical: $C_7H_{14}O$
Formula: $CH_3CH_2CH_2CH_2CH_2COCH_3$
Properties: Water-wh. clear liq.; penetrating fruity cinnamon-like odor; sol. in ethanol, diethyl ether; misc. with org. solvs.; almost insol. in water; m.w. 114.18; dens. 0.8166 (20/20 C); vapor pressure 2.6 mm Hg (20 C); m.p. -26.9 C; b.p. 150.6 C; acid no. 2; flash pt. 49 C; ref. index 1.4110 (20 C)
Toxicology: ACGIH TLV/TWA 50 ppm; LD50 (oral, rat) 1670 mg/kg, (IP, rat) 800 mg/kg, (skin, rabbit) 12,600 mg/kg; mod. toxic by ing.; mildly toxic by inh., skin contact; irritating to eyes, nose, throat, skin; narcotic in high concs.; inh. can cause dizziness, lightheadedness, unconsciousness; may damage liver/kidneys; may affect nervous system; TSCA listed
Environmental: VOC; BOD5 1.23; COD 2.27; ThOD 2.80; prevent contamination of soil, ground- and surf. waters
Precaution: Flamm. exposed to heat or flames; moderate fire risk; reactive with strong oxidizing materials; incompat. with strong acids, alkalis, hydrogen peroxides; violent reactions possible
Hazardous Decomp. Prods.: CO, CO_2, explosive peroxides; heated to decomp., emits acrid smoke and fumes
NFPA: Health 1, Flammability 2, Reactivity 0
Storage: Store in cool, dry, well-ventilated area, away from heat/ignition sources
Uses: Industrial solvent for nitrocellulose lacquers, syn. resin finishes, high-solids coatings, electronic materials; synthetic flavoring agent in foods and pharmaceuticals; fragrance in cosmetics, creams, lotions, soaps, detergents, perfumes; fortifier for lavender; inert reaction medium
Features: Banana-like flavor
Regulatory: FDA 21CFR §172.515; FEMA GRAS; DOT nonregulated
Manuf./Distrib.: ABCR; Acros Org.; Advanced BioTech; Aldrich; Alfa Aesar; Allchem Ind.; Ashland; Axxence Aromatic GmbH; Bedoukian Research; Bio-Rad Labs; Degussa Flavors/Alfrebro; Dow; Eastman; Fleurchem; Fluka; Houghton Chem.; Hukill; ICN Biomed. Research Prods.; Indofine; Lluch Essence; Moore Ingreds.; Natural Advantage; Shell; Sigma; Stanchem; Vopak USA

Methylaniline. *See* N-Methylaniline

N-Methylaniline

CAS 100-61-8; EINECS/ELINCS 202-870-9
UN 2294 (DOT)
Synonyms: Anilinomethane; Benzenamine, N-methyl-; (Methylamino) benzene; N-Methylaminobenzene; Methylaniline; N-Methylbenzenamine; Methylphenylamine; N-Methylphenylamine; Monomethyl aniline; N-Monomethylaniline; N-Phenylmethylamine
Classification: Aromatic sec. amine
Empirical: C_7H_9N
Formula: $C_6H_5NHCH_3$
Properties: Colorless to sl. yel. oily visc. liq.; char. odor; turns brn. on exposure to air; sol. in ethanol, chloroform, ether, oxygenated solvs.; sl. sol. in water; m.w. 107.16; dens. 0.989 (20 C); m.p. -57 C; b.p. 196 C; flash pt. (CC) 79-80 C
Toxicology: ACGIH TLV/TWA 0.5 ppm (skin); LDLo (oral, rabbit) 280 mg/kg, (IV, rabbit) 24 mg/kg, (subcut, guinea pig) 1200 mg/kg; very toxic; poison by ing. and IV routes; mod. toxic by subcut. route; inh. or ing. may cause oxygen depletion in blood, heart, brain, causing headache, shortness of breath, nausea, vomiting, dry throat, dizziness, weakness, drowsiness, irregular heartbeats; TSCA listed
Precaution: Combustible; incompat. with strong oxidizing materials (violent reaction, risk of fire and explosion), acids (violent reactions to form salts), acid anhydrides, acid chlorides (aniline formation)
Hazardous Decomp. Prods.: CO, CO_2; heated to decomp., emits toxic fumes of NO_x
Storage: Store in cool, dry, well-ventilated area away from heat/ignition sources
Uses: Org. synthesis; solvent; neutralizing agent; acid acceptor
Manuf./Distrib.: Aceto; Allchem Ind.; Bayer; KIC Chems.; Kessler; Richman; Xinchem

Methyl arachidate. *See* Methyl eicosanoate
1-Methylazacyclopentan-2-one. *See* N-Methyl-2-pyrrolidone

Methyl behenate

CAS 929-77-1; EINECS/ELINCS 213-207-8
Synonyms: Docosanoic acid, methyl ester; Methyl docosanoate
Definition: Methyl ester of behenic acid
Empirical: $C_{23}H_{46}O_2$
Formula: $CH_3(CH_2)_{20}COOCH_3$
Properties: Wax-like cryst. solid; insol. in water; sol. in alcohol and ether; m.w. 354.62; m.p. 54 C; b.p. 215.5 C (3.75 mm); ref. index 1.4262 (80 C)
Precaution: Combustible
Uses: Special synthesis; intermediate for pure behenic acid, prod. of superamides; biochemical and medical research; reference standard in gas chromatography; in metalworking lubricants, specialized solvents; opacifier; visc. control agent; as solvent
Manuf./Distrib.: Aldrich; Fluka; Lutianhua; Penn Bio-Organics; Sigma

N-Methylbenzenamine. *See* N-Methylaniline
Methylbenzene. *See* Toluene
Methyl benzenecarboxylate. *See* Methyl benzoate
Methylbenzenesulfonic acid, sodium salt; 4-Methylbenzenesulfonic acid, sodium salt. *See* Sodium toluenesulfonate

Methyl benzoate

CAS 93-58-3; EINECS/ELINCS 202-259-7
UN 2938 (DOT); FEMA 2683
Synonyms: Benzoic acid methyl ester; Essence of niobe; Methyl benzenecarboxylate; Niobe oil; Oil of niobe
Classification: Ester
Empirical: $C_8H_8O_2$
Formula: $C_6H_5COOCH_3$
Properties: Colorless liq., fragrant odor; sol. in alcohol, fixed oils, propylene glycol, water @ 30 C; misc. with alcohol, ether, oxygenated solvs.; insol. in glycerin; m.w. 136.15; dens. 1.082-1.088; vapor pressure 1 mm Hg (39 C); m.p. -12.5 C; b.p. 199.6 C; flash pt. 181 F; ref. index 1.515
Toxicology: LD50 (oral, rat) 1350 mg/kg; LDLo (skin, cat) 10 g/kg; mod. toxic by ing.; mildly toxic by skin contact; skin and eye irritant; TSCA listed
Precaution: Flamm. exposed to heat or flame; reactive with oxidizing materials; keep away from food
Hazardous Decomp. Prods.: Heated to decomp., emits acrid smoke and irritating fumes
NFPA: Health 0, Flammability 2, Reactivity 0
Uses: Preservative, skin conditioner in cosmetics; fragrance in perfumes; dyestuffs intermediate; solvent for cellulose esters and ethers, resins, rubber; synthetic flavoring agent in foods and pharmaceuticals
Features: Fruity flavor
Use Level: 0.5% max. as benzoic acid in finished cosmetics
Regulatory: FDA 21CFR §172.515; FEMA GRAS
Manuf./Distrib.: Advanced BioTech; Aldrich; Alfa Chem; Allan; Bayer/Industrial Chems./Textile; Bell Flavors & Fragrances; Berje; Bio-Rad Labs; Chemial SpA; Degussa AG/Health & Nutrition; Degussa Flavors/Alfrebro; Fleurchem; Fluka; Goldensun Mfg.; Haarmann & Reimer GmbH; Haltermann GmbH; J.T. Baker; Kessler; Lluch Essence; MelChem; Morflex; Penta Mfg.; Reilly Chems. SA; Richard's Pacific Chem. Sourcing; Ruger; Schweizerhall; Sigma; Spectrum Quality Prods.; Synasia Fine Chem.; VWR Int'l.; Whyte Chems. Ltd

Methylbenzol. *See* Toluene

α-Methylbenzyl alcohol

CAS 98-85-1; EINECS/ELINCS 202-707-1
UN 2937 (DOT); FEMA 2685
Synonyms: Benzenemethanol, α-methyl-; Ethanol, 1-phenyl-; Methyl phenylcarbinol; Methylphenylmethanol; α-Phenethyl alcohol; s-Phenethyl alcohol; 1-Phenylethanol; α-Phenylethyl alcohol; Phenylmethylcarbinol; Styrallyl alcohol; Styralyl alcohol
Empirical: $C_8H_{10}O$
Formula: $C_6H_5CH(CH_3)OH$
Properties: Colorless liq.; mild floral odor; sol. in alcohol, glycerol, min. oil, oxygenated and hydrocarbon solvs.; sl. sol. in water; m.w. 122.17; dens. 1.009-1.014; vapor pressure 0.1 mm (20 C); m.p. 20.7 C; b.p. 204 C; flash pt. (COC) 96 C); ref. index 1.525-1.529 (20 C); congeals below R.T.
Toxicology: LD50 (oral, rat) 400 mg/kg, (subcut., mouse) 250 mg/kg, (skin, rabbit) 2500 mg/kg; toxic; poison by ing., subcut. routes; mod. toxic by skin contact; primary skin irritant; severe eye irritant; cancer suspect agent; mutagen; tumorigen; target organ: kidneys; TSCA listed
Precaution: Combustible; can react with oxidizers
Uses: Fragrance in cosmetics, perfumes; solvent; dyes; synthetic flavoring agent in foods and pharmaceuticals; laboratory reagent
Regulatory: FDA 21CFR §172.515; FEMA GRAS
Manuf./Distrib.: Adept Sol'ns.; Advanced Synthesis Tech.; Aldrich; Dyno Spec. Polymers; Lipo; Lluch Essence; Penta Mfg.; Sarcom

Methyl borate. *See* Trimethyl borate

Methyl bromide

CAS 74-83-9
UN 1062 (DOT)
Synonyms: Bromomethane; MB; MBX; MEBR; Monobromomethane
Empirical: CH_3Br
Properties: Colorless odorless gas or liquefied gas; chloroform-like odor; burning taste; readily sol. in lower alcohols, ethers, esters, ketones, halogenated hydrocarbons, aromatic hydrocarbons, carbon disulfide, most org. solvs.; insol. in water; m.w. 94.95; dens. 1.732 (0 C); vapor pressure 1824 mm Hg; f.p. -94 c; b.p. 3.6 C; flash pt. none
Toxicology: ACGIH TLV/TWA 5 ppm (skin); LD50 (oral, rat) 214 mg/kg; LCLo (human, inh., 2 h) 60,000 ppm; human poison by inh.; can cause respiratory distress, cardiac arrest, CNS effects, neurotoxic effects; corrosive to skin; hemotoxic and narcotic with delayed action; cumulative effects; may cause lung, kidney damage, liver changes; acute or chronic poisoning can be fatal; experimental carcinogen; human mutagenic data; TSCA listed
Precaution: 10-15% mixts. with air may be ignited with difficulty; mod. explosive exposed to sparks or flame; incompat. with metals, dimethyl sulfoxide, ethylene oxide
Hazardous Decomp. Prods.: Heated to decomp., emits toxic fumes of Br^-
Uses: Soil/grain/space fumigant; disinfestation of potatoes, tomatoes, other crops; organic synthesis; extraction solvent for vegetable oils
Regulatory: Restricted use; SARA reportable
Manuf./Distrib.: Akzo Nobel; Albemarle; Aldrich; Allchem Ind.; AmeriBrom; Atomergic Chemetals; Dead Sea Bromine; Great Lakes; MG Ind.; Ocean Chems.; Sigma; Sithean; Synasia Fine Chem.

2-Methylbutane. *See* Isopentane
3-Methylbutane-1,2-diol; 3-Methyl-1,3-butanediol. *See* Isopentyldiol
Methyl butanoate; Methyl n-butanoate. *See* Methyl butyrate
1-Methyl-1-butanol. *See* s-Amyl alcohol
2-Methyl butanol. *See* 2-Methyl-1-butanol

2-Methyl-1-butanol

CAS 137-32-6; EINECS/ELINCS 205-289-9
UN 1105; FEMA 3998
Synonyms: Active amyl alcohol; Active primary amyl alcohol; Amyl alcohol, active; dl-s-Butylcarbinol; s-Butyl carbinol; 2-Methyl butanol; 2-Methyl butanol-1; 2-Methyl-n-butanol; (±)-2-Methyl-1-butanol; 2-Methylbutyl alcohol; Primary active amyl alcohol
Classification: Primary aliphatic alcohol
Empirical: $C_5H_{12}O$
Formula: $CH_3CH_2CH(CH_3)CH_2OH$
Properties: Colorless liq.; sour sharp odor; m.w. 88.15; very sol. in acetone; sol. in alcohols, ether; sl. sol. in water; m.w. 88.15; dens. 0.816 (20 C); vapor pressure 2.0 mm Hg (20 C); f.p. < -70 C; b.p. 128 C; flash pt. (TCC) 43 C; surf. tens. 25.7 dynes/cm (20 C)
Toxicology: LD50 (oral, rat) 4920 mg/kg; LDLo (skin, rabbit) 3540 mg/kg; mod. toxic by skin contact, IP; mildly toxic by ing.; can be absorbed thru skin; inh. may cause nose/throat irritation, headache, dizziness, dyspnea, nausea, vomiting; eye, skin, mucous membrane irritant; severe exposures may produce double vision, deafness, delirium, severe nervous symptoms, fatalities; can cause severe lung damage, respiratory/cardiac arrest, or death if aspirated into lungs; TSCA listed
Precaution: Flamm. exposed to heat, flame, oxidizers; explosive as vapor exposed to heat or flame; incompat. with oxidizing agents (increases fire/explosion hazard), hydrogen trisulfide (explosive reaction possible)
Hazardous Decomp. Prods.: Heated to decomp., emits acrid smoke and irritating fumes
NFPA: Health 2, Flammability 2, Reactivity 0
Storage: Store in cool, dry place away from corrosives
Uses: Solvent; org. synthesis (introduction of active amyl group); lubricants; plasticizers; additive for oils and paints; frothing agent; synthetic flavoring agent in foods; fragrance in cosmetics
Regulatory: FEMA GRAS
Manuf./Distrib.: Aldrich; BASF AG; Dow; Wutong Aroma
Trade Names Containing: Pentanol 45

2-Methyl butanol-1. *See* 2-Methyl-1-butanol
2-Methyl-2-butanol; 2-Methyl butanol-2. *See* t-Amyl alcohol
2-Methyl-4-butanol. *See* Isoamyl alcohol
2-Methyl-n-butanol. *See* 2-Methyl-1-butanol
3-Methylbutanol; 3-Methyl-1-butanol; 3-Methylbutan-1-ol. *See* Isoamyl alcohol
3-Methylbutan-3-ol. *See* t-Amyl alcohol
(±)-2-Methyl-1-butanol. *See* 2-Methyl-1-butanol
3-Methyl-1-butanol acetate. *See* Isoamyl acetate
3-Methylbutanone; 3-Methyl-2-butanone; 3-Methylbutan-2-one. *See* Methyl isopropyl ketone
1-Methyl-2-t-butoxyethanol. *See* Propylene glycol t-butyl ether
3-(3-Methylbutoxy)-1,2-propanediol. *See* 1-Isoamyl glycerol ether
1-Methylbutyl acetate. *See* s-Amyl acetate
3-Methylbutyl acetate; 3-Methyl-1-butyl acetate; β-Methylbutyl acetate. *See* Isoamyl acetate
2-Methylbutyl alcohol. *See* 2-Methyl-1-butanol
1-Methyl-4-t-butylbenzene; p-Methyl-t-butylbenzene. *See* p-t-Butyl toluene
3-Methylbutyl butyrate. *See* Isoamyl butyrate
Methyl-1,3-butylene glycol acetate. *See* Methoxybutyl acetate
3-Methylbutyl ethanoate. *See* Isoamyl acetate

Methyl t-butyl ether

CAS 1634-04-4; EINECS/ELINCS 216-653-1
UN 2398 (DOT)
Synonyms: t-Butyl methyl ether; 2-Methoxy-2-methylpropane; Methyl 1,1-dimethylethyl ether; 2-Methyl-2-methoxypropane; Methyl tertiary butyl ether; MTBE; Propane, 2-methoxy-2-methyl-
Classification: Aliphatic ether
Empirical: $C_5H_{12}O$
Formula: $CH_3OC(CH_3)_3$
Properties: Colorless clear liq., terpene-like odor; misc. with all gasoline-type hydrocarbons; sl. sol. in water; m.w. 88.15; dens. 0.7335; vapor pressure 249 mm Hg; f.p. -75 C; b.p. 91.1 C; flash pt. -25.6 C; ref. index 1.369; surf. tens. 19.07 dynes/cm
Toxicology: LD50 (oral, rat) 4 g/kg, (skin, rabbit) > 10,000 mg/kg; LC50 (inh., rat, 4 h) 23,576 ppm; LDLo (IV, rat) 148 mg/kg; poison by IV route; sl. toxic by ing., inh.; sl. skin and eye irritant; target organs: CNS, adrenal, kidney, liver; TSCA listed
Environmental: VOC; ThOD 1.63
Precaution: Flamm. exposed to heat or flame; mod. fire risk
Hazardous Decomp. Prods.: Heated to decomp., emits acrid smoke and irritating fumes
NFPA: Health 2, Flammability 3, Reactivity 0
Uses: Octane booster in gasoline; petrol. blending component; intermediate; solvent for paints; extraction solvent; solvent, reaction medium in pharmaceuticals; for polymerizations; Grignard reactions; mfg. of isobutene

Regulatory: SARA §313 reportable; HAP
Manuf./Distrib.: AMC Chems.; Aldrich; Alfa Chem; Allchem Ind.; Ashland; Equistar; ExxonMobil; Fluka; Honeywill & Stein; Huntsman; Integra; J.T. Baker; Kira; Lyondell; Romil Ltd; Sigma; Spectrum Quality Prods.; Texas Petrochem.; VWR Int'l.
Trade Names: Arcopure® MTBE

Methyl butyl ketone; Methyl N-butyl ketone. *See* 2-Hexanone

Methyl butynol

CAS 115-19-5; EINECS/ELINCS 204-070-5
UN 1993
Synonyms: 1-Butyn-3-ol, 3-methyl-; Dimethyl acetylene carbinol; Dimethylacetylenylcarbinol; Dimethyl ethynyl carbinol; Dimethylethynylmethanol; 1,1-Dimethylpropargyl alcohol; α,α-Dimethylpropargyl alcohol; 1,1-Dimethylpropynol; Ethynyldimethyl carbinol; 2-Hydroxy-2-methyl-3-butyne; MBY; 2-Methyl-3-butyn-2-ol; 2-Methylbutyn-3-ol-2; 3-Methyl-1-butyn-3-ol
Classification: Aliphatic organic compd.
Definition: Tertiary acetylenic alcohol
Empirical: C_5H_8O
Formula: $(CH_3)_2COHCCH$
Properties: Colorless liq., fragrant odor; misc. with water, acetone, benzene, CCl_4; sol. in most org. solvs.; m.w. 84.13; dens. 0.8672 (20/20 C); m.p. 2.6 C; b.p. 104-105 C; flash pt. (TOC) 25 C; ref. index 0.861 (20C)
Toxicology: LD50 (oral, rat) 1950 mg/kg, (IP, mouse) 3600 mg/kg, (subcut., mouse) 2340 mg/kg; LC50 (inh., mouse) 2 g/m^3; mod. toxic by ing., IP, subcut. routes; TSCA listed
Precaution: Flamm.; very dangerous fire hazard exposed to heat or flame; reactive with oxidizing agents
Hazardous Decomp. Prods.: Heated to decomp., emits acrid smoke and irritating fumes
Uses: Corrosion inhibitor for oil field acidizing, steel pickling acids; reactive intermediate in mfg. of pharmaceuticals, plastics, rubbers, fragrances, agric.; solvent, acid inhibitor, viscosity reducer, stabilizer in vinyl plastisols; platinum catalyst blocker for silicones; stabilizer in chlorinated solvents; electroplating brightener
Manuf./Distrib.: Advanced Synthesis Tech.; Air Prods.; Aldrich; BASF; Fluka

2-Methyl-3-butyn-2-ol; 2-Methylbutyn-3-ol-2; 3-Methyl-1-butyn-3-ol. *See* Methyl butynol

Methyl butyrate

CAS 623-42-7; EINECS/ELINCS 210-792-1
UN 1237 (DOT); FEMA 2693
Synonyms: Butanoic acid methyl ester; Butyric acid, methyl ester; Methyl butanoate; Methyl n-butanoate; Methyl-n-butyrate
Empirical: $C_5H_{10}O_2$
Formula: $CH_3CH_2CH_2COOCH_3$
Properties: Colorless liq.; sol. in about 60 parts water; misc. with alcohol, ether; m.w. 102.14; dens. 0.898 (20/4 C); vapor pressure 40 mm (29.6 C); m.p. -95 C; b.p. 99-102 C; flash pt. 14 C; ref. index 1.388 (20 C)
Toxicology: LD50 (oral, rat) > 5 g/kg, (skin, rabbit) 3560 mg/kg; LC50 (inh., mouse, 2 h) 18 g/m^3; mod. toxic by ing., skin contact; primary skin irritant; TSCA listed
Environmental: VOC; ThOD 1.25
Precaution: Highly flamm.; dangerous fire risk exposed to heat, flame, or oxidizers; can react vigorously with oxidizers
Hazardous Decomp. Prods.: Heated to decomp., emits acrid smoke and irritating fumes
Uses: Solvent for ethylcellulose; solvent mixt. for nitrocellulose; synthetic flavoring agent in foods and pharmaceuticals; mfg. of artificial rum and fruit essences; fragrance in cosmetics
Regulatory: FDA 21CFR §172.515; FEMA GRAS
Manuf./Distrib.: Aldrich; Axxence Aromatic GmbH; Berje; Degussa Flavors/Alfrebro; Eastern Chem.; Fluka; Grau Aromatics; Lluch Essence; Moore Ingreds.; Penta Mfg.; Sigma; United-Guardian

Methyl-n-butyrate. *See* Methyl butyrate
3-Methylbutyrolactone; 4-Methyl-γ-butyrolactone; γ-Methyl-γ-butyrolactone. *See* γ-Valerolactone

Methyl canolate

CAS 73891-99-3
Synonyms: Alkyl C16-18 methyl esters; Canola methyl ester; Canola oil, methyl ester; Rape oil, methyl ester; Rapeseed oil, methyl ester
Definition: Methyl esters of canola oil
Properties: Lt. colored liq.; mild fatty odor; insol. in water; sp.gr. 0.876 (20 C); vapor pressure 0.6 mm Hg (68 F); b.p. 425 F (20 mm Hg); flash pt. (TCC) > 93 C; KB value 58
Toxicology: ACGIH TLV none; nonirritating to skin, eyes; noncorrosive; TSCA listed
Environmental: Environmentally friendly; readily biodeg.; low VOC; avoid uncontrolled release into environment
Precaution: Incompat. with strong oxidizing agents, acids, alkalis; spillages may be slippery
Hazardous Decomp. Prods.: CO, CO_2
HMIS: Health 0, Flammability 1, Reactivity 0
Storage: Store away from open flames, excessive heat
Uses: Solvent , adjuvant for pesticides and herbicides
Features: Replaces d-limonene
Regulatory: SARA §311/312 nonhazardous, §313 nonreportable; DOT cleaning compd. n.o.s.; non-HAP; SNAP listed
Manuf./Distrib.: Anar; Parchem Trading
Trade Names: Emery® 2232; Oleocal® ME-112

Methyl caprate

CAS 110-42-9; EINECS/ELINCS 203-766-6
Synonyms: Capric acid methyl ester; Decanoic acid, methyl ester; Methyl n-caprate; Methyl caprinate; Methyl decanoate; Methyl n-decanoate
Empirical: $C_{11}H_{22}O_2$
Formula: $CH_3(CH_2)_8COOCH_3$
Properties: Colorless liq.; pleasant odor; sol. in alcohol, ether, oxygenated solvs.; insol. in water; m.w. 186.17; dens. 0.8733 (20/4 C); f.p. -13.3 C; b.p. 224 C, 130.6 C (30 mm); flash pt. > 100 C; ref. index 1.4250
Toxicology: LDLo (IV, mouse) 48 mg/kg; poison by IV route; TSCA listed
Precaution: Combustible
Hazardous Decomp. Prods.: Heated to decomp., emits acrid smoke and irritating vapors
Uses: Intermediate for detergents, emulsifiers, wetting agents, stabilizers, resins, lubricants, plasticizers; flavoring agent; fragrance
Manuf./Distrib.: Advanced Synthesis Tech.; Aldrich; Fluka; Grau Aromatics; Penta Mfg.; Procter & Gamble; Sigma
Trade Names: Estasol™ MD10
Trade Names Containing: ZELEC® C-49

Methyl n-caprate; Methyl caprinate. *See* Methyl caprate

Methyl caproate

CAS 106-70-7; EINECS/ELINCS 203-425-1
FEMA 2708
Synonyms: Hexanoic acid, methyl ester; Methyl capronate; Methyl hexanoate; Methyl 2-hexanoate; Methyl n-hexanoate; Methyl hexoate; Methyl hexylate
Definition: Ester of methyl alcohol and caproic acid
Empirical: $C_7H_{14}O_2$
Formula: $CH_3(CH_2)_4COOCH_3$
Properties: Colorless liq.; pleasant odor; sol. in alcohol, ether, oxygenated solvs.; insol. in water; m.w. 130.19; dens. 0.884 (20/4 C); f.p. -71 C; b.p. 150-151 C; flash pt. 43 C; ref. index 1.405
Toxicology: LD50 (oral, rat) > 5 g/kg, (skin, guinea pig) > 5 g/kg; LDLo (IV, mouse) 48 mg/kg; poison by IV route; sl. toxic by ing., inh., skin contact; TSCA listed
Precaution: Flamm.
Hazardous Decomp. Prods.: Heated to decomp., emits acrid smoke and irritating vapors
Uses: Solvent; emollient in cosmetics; intermediate for detergents, emulsifiers, wetting agents, stabilizers, resins, lubricants, plasticizers; synthetic flavoring agent in foods and pharmaceuticals; fragrance
Features: Pineapple-like flavor
Regulatory: FDA 21CFR §172.515; FEMA GRAS
Manuf./Distrib.: Aldrich; Degussa Flavors/Alfrebro; Fleurchem; Fluka; Grau Aromatics; Kira; Lluch Essence; Penta Mfg.; Sigma

Methyl capronate. *See* Methyl caproate

Methyl caprylate

CAS 111-11-5; EINECS/ELINCS 203-835-0
FEMA 2728
Synonyms: Caprylic acid, methyl ester; Methyl octanoate; Octanoic acid, methyl ester
Definition: Ester of methyl alcohol and caprylic acid
Empirical: $C_9H_{18}O_2$
Formula: $CH_3(CH_2)_6COOCH_3$
Properties: Colorless liq.; sol. in alcohol, ether; insol. in water; m.w. 158.24; dens. 0.875 (20/4 C); f.p. -37.3 C; b.p. 193-194 C; flash pt. 69 C; ref. index 1.417 (20 C)
Toxicology: LDLo (IV, mouse) 48 mg/kg; poison by IV route; TSCA listed
Precaution: Combustible
Hazardous Decomp. Prods.: Heated to decomp., emits acrid smoke and irritating vapors
Uses: Emollient in cosmetics; intermediate for detergents, emulsifiers, wetting agents, stabilizers, resins, lubricants, plasticizers; synthetic flavoring agent in foods and pharmaceuticals; fragrance; defoamer in food-contact coatings and paper/paperboard; in food-contact textiles
Features: Citrus flavor
Regulatory: FDA 21CFR §172.225, 172.515, 176.200, 176.210, 177.2800; FEMA GRAS
Manuf./Distrib.: A&E Connock; Advanced Synthesis Tech.; Aldrich; Degussa Flavors/Alfrebro; Fluka; Grau Aromatics; Penta Mfg.; Sigma
Trade Names Containing: ZELEC® C-49

Methyl caprylate/caprate

CAS 67762-39-4; 68937-83-7
Synonyms: Mixed decanoic and octanoic acids, methyl esters
Definition: Mixture of esters of methyl alcohol and caprylic and capric acids
Properties: Clear liq.; insol. in water; sp.gr. 0.87; m.p. -30 C; b.p. 100 C; flash pt. 75 C
Uses: Emollient, skin conditioner in cosmetics; in lubricants for automotive, textile, metal rolling operations; food coatings; cosolvent, solvent, oil carrier for agric. pesticides and herbicides; detergent intermediate; defoamer in food-contact coatings and paper/paperboard; in food-contact textiles
Regulatory: FDA 21CFR §172.225, 176.200, 176.210, 177.2800
Manuf./Distrib.: A&E Connock
Trade Names: Edenor® ME C6-10

Methylcarbinol. *See* Alcohol
Methyl 'Carbitol'. *See* Methoxydiglycol
Methyl 'Carbitol' acetate. *See* Diethylene glycol methyl ether acetate

Methyl carbonate

CAS 616-38-6; EINECS/ELINCS 210-478-4
UN 1161 (DOT)
Synonyms: Carbonic acid, dimethyl ester; Dimethyl carbonate; DMC
Classification: Carbonate
Empirical: $C_3H_6O_3$
Formula: $CO(OCH_3)_2$
Properties: Colorless clear liq.; pleasant odor; sol. in most org. solvs.; insol. in water; m.w. 90.09; dens. 1.0718 (20 C); m.p. 0 C; b.p. 90.6 C; flash pt. 18 C; ref. index 1.3687
Toxicology: LD50 (oral, rat) 13 g/kg, (IP, rat) 1600 mg/kg; toxic by inh.; mod. toxic by IP route; mildly toxic by ing.; strong irritant; target organs: kidneys, eyes; TSCA listed
Precaution: Highly flamm.; dangerous fire risk exposed to heat, flames, or oxidizers; violent reaction or ignition on contact with potassium-t-butoxide
Hazardous Decomp. Prods.: Heated to decomp., emits acrid smoke and irritating fumes
NFPA: Health 3, Flammability 3, Reactivity 0
Storage: Moisture-sensitive
Uses: Reaction solvent, extraction solvent in many industries, pharmaceuticals, agrochems., hydrocarbon refining, paints/coatings, fragrances; methylation agent, carbonylation agent, phosgenation reagent in org. synthesis; fuel and lube additive
Manuf./Distrib.: Acros Org.; Aldrich; Alfa Aesar; Am. Radiolabeled Chems.; Biddle Sawyer; Esprit; Fluka; G.J. Chem.; ICN Biomed. Research Prods.; Lancaster Synthesis; Pfaltz & Bauer; SNPE Chimie; Sigma; Spectrum Quality Prods.; TCI Am.; Ube Am.; Whyte Chems. Ltd

Methyl 'Cellosolve'. *See* Methoxyethanol
Methyl 'Cellosolve' acetate. *See* Methoxyethanol acetate
Methyl 'Cellosolve' stearate. *See* Methoxyethyl stearate

Methyl chloride

CAS 74-87-3; EINECS/ELINCS 200-817-4
UN 1063 (DOT)
Synonyms: Chloromethane; Methane, chloro-; Monochloromethane; Refrigerant R40
Classification: Haloalkane
Definition: Obtained by direct reaction of methanol and hydrogen chloride or of methane with chlorine
Empirical: CH_3Cl
Properties: Colorless gas; faint sweet ethereal odor; sweet taste; sol. in alcohol; misc. with chloroform, ether, glac. acetic acid; sl. sol. in water; m.w. 50.49; dens. 0.918 (20/4 C); vapor pressure 4300 mm Hg; m.p. -97 C; b.p. -23.7 C; flash pt. -50 C; autoignition temp. 632 C; ref. index 1.334; surf. tens. 15.19 dynes/cm
Toxicology: ACGIH TLV/TWA 50 ppm; STEL 100 ppm (skin); LD50 (oral, rat) 1800 mg/kg; LCLo (inh., human, 2 h) 20,000 ppm; mildly toxic by inh.; human systemic effects; can cause convulsions, nausea, vomiting; sl. irritant; some narcotic effect; high concs. may cause CNS effects, coma, or death; may affect liver, kidneys, bone marrow; tumorigen, mutagen; reproductive effects; TSCA listed
Environmental: VOC; ThOD 0.95
Precaution: Flamm. gas; LEL 8.1%; UEL 17.4%; very dangerous fire hazard exposed to heat, flame, or powerful oxidizers; mod. explosion hazard; explodes on contact with interhalogens, magnesium, potassium, sodium, zinc, or their alloys
Hazardous Decomp. Prods.: Heated to decomp., emits highly toxic fumes of Cl^-
NFPA: Health 2, Flammability 4, Reactivity 0
Uses: Catalyst carrier in low-temp. polymerization; mfg. of tetramethyl lead (an antiknock agent for gasoline), silicone polymers, methylcellulose (textile sizing agent); methylating agent, chlorinating agent in organic synthesis; refrigerant; fluid for thermometric/thermostatic equipment; extractant for greases, oils, resins; low-temp. solvent in prod. of butyl rubber; propellant, blowing agent in PS foam prod.; as fire extinguisher; as fumigant; as herbicide; intermediate for pharmaceuticals; as topical anesthetic; as food additive
Regulatory: SARA §313 reportable; CERCLA hazardous substance; HAP; permitted as food additive
Manuf./Distrib.: ARC; Air Prods.; Aldrich; Alexander; Alfa Chem; Dow; Electron Microscopy Sciences; MG Ind.; Occidental; PPG Ind.; Richard's Pacific Chem. Sourcing; Sigma; Solvay SA; Stanchem; Uniqema Am.; Vulcan Perf. Chems.

1-Methyl-2-chlorobenzene; 2-Methylchlorobenzene. *See* o-Chlorotoluene
Methylchloroform. *See* 1,1,1-Trichloroethane

Methyl cocoate

CAS 61788-59-8; EINECS/ELINCS 262-988-1
Synonyms: Coconut acid methyl ester; Fatty acids, coco, methyl esters; Methyl ester of coconut oil fatty acids
Definition: Ester of methyl alcohol and coconut fatty acids
Formula: $RCO–OCH_3$, RCO– represents the fatty acids derived from coconut oil
Toxicology: TSCA listed
Uses: Emollient, plasticizer, lubricant for cosmetics, pharmaceuticals, plastics, lubricating oils, textiles, leather, cutting oils; chemical and detergent intermediate; food coatings; solvent for pesticides and herbicides; in food-pkg. adhesives; defoamer in food-contact paper coatings, paper/paperboard; surf. lubricant in mfg. of food-contact metallic articles
Regulatory: FDA 21CFR §172.225, 175.105, 176.200, 176.210, 177.2260, 177.2800, 178.3910
Manuf./Distrib.: A&E Connock; Mosselman NV
Trade Names: Emery® 2253; Emery® 2254; Excepari MC

Methyl cottonseed

CAS 61788-60-1
Synonyms: Fatty acids, cottonseed oil, methyl esters
Definition: Methyl ester of cottonseed oil
Uses: Solvent, cosolvent, penetrant in industrial, food processing, and agric. applics.; in textile processing aids; in printing inks
Trade Names: Norfox® MCS

6-Methyl-m-cresol. *See* 2,5-Xylenol
p-Methylcumene. *See* p-Cymene
Methyl cyanide. *See* Acetonitrile
Methyl β-cyanoethyl ether. *See* 3-Methoxypropionitrile
N-Methylcyclohexanamine. *See* N-Methylcyclohexylamine

Methyl cyclohexane

CAS 108-87-2; EINECS/ELINCS 203-624-3
UN 2296 (DOT)
Synonyms: Cyclohexane, methyl-; Cyclohexylmethane; Hexahydrotoluene; Toluene hexahydride
Classification: Sat. alicyclic hydrocarbon
Empirical: C_7H_{14}
Formula: $C_6H_{11}CH_3$
Properties: Colorless liq.; faint benzene-like odor; sol. in most org. solvs.; insol. in water; m.w. 98.19; sp.gr. 0.77 (20 C); vapor pressure 40 mm Hg (20 C); m.p. -126 C; b.p. 101 C; flash pt.-4 C
Toxicology: ACGIH TLV/TWA 400 ppm; LD50 (oral, mouse) 2250 mg/kg; LC50 (inh., mouse, 2 h) 41,500 mg/m^3; mod. toxic by ing.; may cause mild skin irritation; vapor may irritate eyes; probably a CNS depressant; sl. nose/throat irritant; high vapor concs. may cause headache, lightheadedness, nausea, drowsiness; very high concs. may cause unconsciousness, death; ing. may cause sore throat, nausea, diarrhea; can be aspirated into lungs causing severe damage; TSCA listed
Precaution: Flamm.; dangerous fire risk; lower explosive limit in air 1.2%; incompat. with strong oxidizing agents (increases fire and explosion hazard)
Hazardous Decomp. Prods.: Heated to decomp., emits acrid smoke and fumes
NFPA: Health 2, Flammability 3, Reactivity 0
Storage: Store in cool, dry, well-ventilated area out of direct sunlight
Uses: Solvent for cellulose ethers, pharmaceuticals; organic synthesis; reagent
Manuf./Distrib.: Aldrich; ChevronPhillips; Fluka; Penta Mfg.; Romil Ltd; Spectrum Quality Prods.; Triple Crown Am.
Trade Names Containing: Solvent 2205; Solvent 2241; Solvent 2245; Solvent 2304; Solvent 2310; Solvent 2340; Solvent 2430

Methylcyclohexanol

CAS 25639-42-3; EINECS/ELINCS 247-152-6
UN 2617 (DOT)
Synonyms: Hexahydrocresol; Hexahydromethylphenol; Methylcyclohexyl alcohol
Classification: Alicyclic alcohol
Definition: Mixed isomer prod.
Empirical: $C_7H_{14}O$
Formula: $CH_3C_6H_{10}OH$ (C_6H_{10} = cyclohexane ring)
Properties: Colorless visc. oily liq.; aromatic menthol-like odor; sol. in common org. solvs.; mod. sol. in water; m.w. 114.2; dens. 0.913; m.p. -50 C; b.p. 155-180 C; flash pt. (CC) 68 C
Toxicology: ACGIH TLV/TWA 50 ppm; LD50 (oral, rat) 1660 mg/kg, (subcut., rat) 2900 mg/kg; LDLo (skin, rabbit) 6800 mg/kg; mod. toxic by ing. and subcut. routes; mildly toxic by skin contact; liq. may cause skin irritation; vapor can irritate eyes; liq. splashes may be very irritating; human systemic effects by inh. (liver/kidney effects)
Precaution: Flamm. exposed to heat, flame, or oxidizers; incompat. with strong oxidizing agents (increases fire/explosion hazard)
Hazardous Decomp. Prods.: On heating, emits acrid fumes
NFPA: Health 0, Flammability 2, Reactivity 0
Storage: Store in a cool, dry, well-ventilated area
Uses: Solvent for lacquers, oils, gums, waxes, resins, printing inks; blending agent for soaps and detergents; lubricant antioxidant; degreasing agent
Manuf./Distrib.: Advanced Synthesis Tech.; Fluka; Nipa Labs; Penta Mfg.
Trade Names: Methylhexalin®

Methylcyclohexanol acetate

Synonyms: Heptalin acetate; Methyl cyclohexyl acetate
Empirical: $C_9H_{16}O_2$
Formula: $C_7H_{13}OOCCH_3$
Properties: Colorless liq.; ester-like odor; dens. 0.941; b.p. 176-193 C; flash pt. 64 C
Precaution: Combustible
Uses: Solvent
Manuf./Distrib.: Rhodia/Fine Orgs.

Methylcyclohexanone

CAS 1331-22-2; EINECS/ELINCS 215-556-1
UN 2297 (DOT)
Synonyms: Cyclohexanone, methyl-
Definition: Commercial prod. is a mixt. of isomers
Empirical: $C_7H_{12}O$
Properties: Colorless to pale yel. liq.; strong ketone odor; sol. in alcohol, ether; insol. in water; m.w. 112.18; dens. 0.91 kg/l; b.p. 170 C; flash pt. (CC) 118 F; ref. index 1.446
Toxicology: LDLo (oral, rabbit) 1000 mg/kg, (skin, rabbit) 4900 mg/kg; mod. toxic by ing.; mildly toxic by skin contact; may cause general anesthesia, muscle contractions or spasticity, convulsions, liver and kidney damage; TSCA listed
Precaution: DOT: Flamm. liq.; flamm. exposed to heat, sparks, or flames; violent reactions with HNO_3 and other oxidizers possible
Hazardous Decomp. Prods.: Heated to decomp., emits acrid smoke and irritating fumes
Uses: Solvent for lacquers, printing inks
Manuf./Distrib.: Fluka

Methyl cyclohexyl acetate. *See* Methylcyclohexanol acetate
Methylcyclohexyl alcohol. *See* Methylcyclohexanol
Methylcyclohexylamine. *See* N-Methylcyclohexylamine

N-Methylcyclohexylamine

CAS 100-60-7; EINECS/ELINCS 202-869-3
Synonyms: Cyclohexanamine, N-methyl-; N-Cyclohexylmethylamine; N-Methylcyclohexanamine; Methylcyclohexylamine; NMCHA
Classification: Nonaromatic amines
Empirical: $C_7H_{15}N$
Formula: $C_6H_{11}NHCH_3$
Properties: Water-wh. to amber liq.; sol. in alcohol, ether, oxygenated solvs.; sl. sol. in water; m.w. 113.23; dens. 0.86 (20 C); f.p. -9 C; b.p. 145-147 C; ref. index 1.4560
Toxicology: LD50 (oral, rat) 400 mg/kg; LDLo (skin, rabbit) 2 g/kg; LC50 (inh., guinea pig, 1 h) 7000 mg/m^3; toxic; poison by ing.; mod. toxic by skin contact; mildly toxic by inh.; strong irritant to tissue; can cause severe eye damage; may cause lacrimation, irregular respiration, nose bleeds, prostration, convulsions; TSCA listed
Precaution: Combustible; moderate fire risk
Hazardous Decomp. Prods.: Heated to decomp., emits toxic fumes of NO_x
Storage: Store under nitrogen; absorbs CO_2 from air
Uses: Intermediate; solvent; acid acceptor
Manuf./Distrib.: Aastrid Int'l.; Aldrich; Fluka

Methylcyclopentane

CAS 96-37-7
UN 2298 (DOT)
Classification: Sat. alicyclic hydrocarbon
Empirical: C_6H_{12}
Formula: $C_5H_9CH_3$
Properties: Colorless liq. or solid; sweetish gasoline-like odor; sol. in ether, most org. solvs.; insol. in water; m.w. 84.18; sp.gr. 0.749 (20 C); vapor pressure 100 mm Hg (17.9 C); m.p. -142 C; b.p. 72 C; flash pt. -23.9 C
Toxicology: LCLo (inh., mouse) 9500 mg/m^3; mildly toxic by inh.; may cause mild skin irritation; vapor can temporarily irritate eyes; probably a CNS depressant; high vapor concs. may cause nose/throat irritation, lightheadedness, nausea, drowsiness; very high concs. may cause unconsciousness, death; ing. may cause sore throat, nausea, diarrhea; TSCA listed

Precaution: Flamm.; very dangerous fire hazard exposed to heat, flame, oxidizers; incompat. with strong oxidizing agents (increased fire and explosion hazard)
Hazardous Decomp. Prods.: Heated to decomp., emits acrid smoke and irritating fumes
NFPA: Health 2, Flammability 3, Reactivity 0
Storage: Store in cool, dry, well-ventilated area out of direct sunlight
Uses: Extraction solvent for essential oils; org. synthesis; azeotropic distillation agent
Manuf./Distrib.: Aldrich; Fluka

Methyl [1R-(1α,4aβ,4bα,10aα)]-1,2,3,4,41,4b,5,6,10,10a-decahydro-1,4a-dimethyl-7-(1-methylethyl)-1-phenanthrenecarboxylate. *See* Methyl abietate
Methyl decanoate; Methyl n-decanoate. *See* Methyl caprate
4-Methyl-2,6-di-t-butylphenol; Methyl di-t-butylphenol. *See* BHT

Methyldiethanolamine

CAS 105-59-9; EINECS/ELINCS 203-312-7
Synonyms: MDEA; 2,2´-(Methylimino) diethanol; N-Methyl-2,2´-iminodiethanol
Classification: Nonaromatic amine
Empirical: $C_5H_{13}NO_2$
Formula: $CH_3N(C_2H_4OH)_2$
Properties: Colorless liq.; mild ammoniacal odor; sol. in water, oxygenated solvs.; insol. in aliphatic solvs.; m.w. 119.19; dens. 1.0418; bulk dens. 8.7 lb/gal; vapor pressure < 0.01 mm Hg (20 C); f.p. -21.0 C; pour pt. -55 C; b.p. 247.2 C; flash pt. 126.6 C; ref. index 1.4699; absorbs water and CO_2 from air
Toxicology: LD50 (oral, rat) 4780 mg/kg, (IP, mouse) 500 mg/kg; mod. toxic by IP route; mildly toxic by ing.; irritant; TSCA listed
Hazardous Decomp. Prods.: Heated to decomp., emits toxic fumes of NO_x
Uses: Intermediate; surfactant intermediate for prod. of amphoteric and cationic surfactants for hair care, fabric/textile softeners, antistats; epoxy resin curing agent; absorption of acidic gases; catalyst for PU foams; pH control agent; extraction solvent for hydrogen sulfide gas processing; photoactivator for printing inks, lacquers
Manuf./Distrib.: Aldrich; Dow; Fluka; Huntsman; Richard's Pacific Chem. Sourcing
Trade Names: Jefftreat® M Solvent; Jefftreat® MS-100; Jefftreat® MS-300
Trade Names Containing: Jefftreat® M 45%; Jefftreat® M-504; Jefftreat® M-507; Jefftreat® M-510 45%; Jefftreat® M-510 100%; Jefftreat® MS-200 Series; Jefftreat® MS-205; Jefftreat® MS-205 LFG

3-Methyl-N,N-diethylbenzamide. *See* Diethyl toluamide
Methyl diglyme. *See* Diethylene glycol dimethyl ether
Methyl digol. *See* Methoxydiglycol
Methyl 1,1-dimethylethyl ether. *See* Methyl t-butyl ether
Methyl 2,2´-dimethylpropionate. *See* Methyl pivalate
Methyl 1,1-dimethyl propyl ether. *See* t-Amyl methyl ether
8-Methyl-3,6-dioxa-1-undecanol. *See* Diethylene glycol methylpentyl ether
4-Methyl-1,3-dioxolan-2-one. *See* Propylene carbonate

Methyl disulfide

CAS 624-92-0; EINECS/ELINCS 210-871-0
UN 2381 (DOT); FEMA 3536
Synonyms: Dimethyl disulfide; 2,3-Dithiobutane; DMDS; MDS; Methyldithiomethane
Classification: Alkyl disulfide
Empirical: $C_2H_6S_2$
Formula: CH_3SSCH_3
Properties: Pale yel. liq.; disagreeable onion odor; very sol. in alcohol, ether; sl. sol. in water; m.w. 94.20; sp.gr. 1.062 (20/4 C); vapor pressure 28.6 mm Hg; m.p. -84.7 C; b.p. 108-110 C; flash pt. (CC) 16 C; ref. index 1.527 (20 C)
Toxicology: LD50 (mammal) 138 mg/kg; skin and eye irritant; very toxic by inh. in animal tests; breaks down to methyl mercaptan in body; inh. of vapors probably causes headache, dizziness, nausea, vomiting; massive exposure can be fatal; TSCA listed
Precaution: Highly flamm.; very dangerous fire hazard exposed to heat, flame, oxidizers; incompat. with oxidizing agents (can react violently), reducing materials (can produce methyl mercaptan), copper and alloys (can react corrosively)
Hazardous Decomp. Prods.: Hydrogen sulfide, methyl mercaptan
Storage: Store in cool, well-ventilated area, out of direct sunlight; prevent release of mists into workplace air
Uses: Intermediate in mfg. of chemicals; prep. of catalysts for petrol. refining; solvent for sulfur; synthetic flavoring agent in foods and pharmaceuticals
Regulatory: FDA 21CFR §172.515; FEMA GRAS
Manuf./Distrib.: ADA Int'l.; Aldrich; Atofina; Epochem; Fluka; Sigma

Methyldithiomethane. *See* Methyl disulfide
Methyl docosanoate. *See* Methyl behenate
Methyl dodecanoate; Methyl n-dodecanoate; Methyl dodecylate. *See* Methyl laurate

Methyl eicosanoate

CAS 1120-28-1; EINECS/ELINCS 214-304-8
Synonyms: Eicosanoic acid methyl ester; Methyl arachidate
Definition: Methyl ester of arachidic acid
Empirical: $C_{21}H_{42}O_2$
Formula: $CH_3(CH_2)_{18}COOCH_3$
Properties: Waxy solid; sol. in alcohol, ether; insol. in water; m.w. 326.57; m.p. 45.8 C; b.p. 284 C (100 mm); ref. index 1.4352 (50 C)
Uses: Special synthesis; intermediate in prod. of alkanolamides, for pure arachic acid; in metalworking lubricants; as specialized solvents; defoamer and nutrient in fermentation; reference standard for gas chromatography; medical research
Manuf./Distrib.: Aldrich; Fluka; Sigma
Trade Names: Kemester® 2050

Methylene bichloride. *See* Methylene chloride
2,2´-Methylenebiphenyl. *See* Fluorene
1,1´-Methylenebisbenzene. *See* Diphenylmethane
1,1´-[Methylene bis (oxy)] bis (2-chloroethane). *See* Bis (2-chloroethoxy) methane

Methylene bromide

CAS 74-95-3; EINECS/ELINCS 200-824-2
UN 2664 (DOT)
Synonyms: DBM; Dibromomethane; Methylene dibromide
Classification: Halogenated aliphatic hydrocarbon
Empirical: CH_2Br_2
Formula: $BrCH_2Br$
Properties: Colorless to pale yel. clear liq.; sol. in alcohol, ether, chloroform, acetone; mod. sol. in water; misc. with most org. solvs.; m.w. 173.85; dens. 2.49; m.p. -52 C; b.p. 97 C; nonflamm.
Toxicology: LD50 (oral, rat) 108 mg/kg, (subcut., mouse) 3738 mg/kg; poison; mod. toxic by subcut. route; sl. skin/eye irritant; may be absorbed thru skin; inh. may cause sl. nose/throat irritation, CNS depression, lightheadedness, nausea, fatigue, incoordination; prolonged/repeated high-level exposure may cause liver/kidney damage; very high concs. may cause unconsciousness, death; not very toxic in short-term ing. studies with animals; mutagenic data; TSCA listed
Precaution: Incompat. with strong oxidizers, strong bases, metals (may react vigorously or increase risk of fire and explosion); mixts. with potassium explode on impact
Hazardous Decomp. Prods.: Heated to decomp., emits toxic fumes of Br^-
NFPA: Health 2, Flammability 1, Reactivity 0
Storage: Store in a cool, dry, well-ventilated area out of direct sunlight
Uses: High-density solvent for mineral separation, gauge fluid; process solvent; intermediate in org. synthesis, prod. of herbicides, insecticides, biocides
Manuf./Distrib.: Albemarle; Aldrich; AmeriBrom; Assoc. Octel Co Ltd; Dead Sea Bromine; Fluka; Great Lakes; Sigma

Methylene chloride

CAS 75-09-2; EINECS/ELINCS 200-838-9
UN 1593 (DOT)
Synonyms: DCM; Dichloromethane; Freon 30; MDC; Methane dichloride; Methane-, dichloro-; Methylene bichloride; Methylene dichloride
Classification: Halogenated aliphatic hydrocarbon
Empirical: CH_2Cl_2

Properties: Colorless clear volatile liq., penetrating ether-like odor; sol. in alcohol, ether, phenols, aldehydes, ketones; misc. with fixed and volatile oils, oxygenated and chlorinated solvs.; mod. sol. in water; m.w. 84.93; dens. 1.335 (15/4 C); vapor pressure 380 mm (22 C); f.p. -97 C; b.p. 40.1 C; autoignition temp. 556 C; ref. index 1.4244 (20 C); KB value 136; nonflamm.
Toxicology: ACGIH TLV/TWA 50 ppm; LD50 (oral, rat) 2136 mg/kg, (IP, mouse) 437 mg/kg, (subcut., mouse) 6460 mg/kg; very toxic; poison by IV route; mod. toxic by ing., subcut., IP routes; mildly toxic by inh.; human systemic effects; eye and severe skin irritant; narcotic in high concs.; confirmed carcinogen; experimental tumorigen, teratogen, reproductive effector; human mutagenic data; TSCA listed
Environmental: ThOD 0.38
Precaution: Explosive as vapor exposed to heat or flame; incompat. with methanol, Al powd., amines, quat. ion exchange resins (azide form), DMSO, perchloric acid, alkali metals, nitric acid, etc.
Hazardous Decomp. Prods.: Contact with hot surfaces cause decomp., yielding toxic fumes; heated to decomp., emits highly toxic fumes of phosgene and Cl^-
NFPA: Health 2, Flammability 1, Reactivity 0
Storage: Store in cool, dry, well-ventilated area out of direct sunlight; keep quantities stored as small as possible
Uses: Solvent for vapor degreasing, paint/varnish removers, electronics, resin cleaners, adhesives, cosmetics, tablet coatings; process solvent for cellulose acetate, butyl rubber; cleaning solvent; plastics processing; blowing agent in PU foams; propellant for paint aerosols; refrigerant; aerosol blowing agent/solvent vapor depressant; heat-transfer fluid; extraction solvent for industrial applics., foods (decaffeinated coffee, hops extract, spice oleoresins); color diluent for foods; pharmaceutic aid; pharmaceutical solvent, anesthetic; in food-pkg. adhesives; adjuvant for PC food-pkg. resins
Regulatory: FDA 21CFR §73.1 (no residue), 73.30, 73.345, 73.615, 172.560, 173.255, 175.105, 177.1580, 177.1585; FDA approved for orals; SARA §311/312 acute health/chronic health hazard, §313 reportable; CERCLA hazardous substance; Calif. Prop. 65 reportable; HAP; NF, BP, EP compliance
Manuf./Distrib.: AAE Chemie NV; AMC Chems.; Aldrich; Allchem Ind.; Amyl; Ashland; Atofina; Brenntag AG; C.P. Hall; Chemcentral; Coyne; Dow; Farleyway Chems. Ltd; Filo; Fisher Scientific; Fluka; Haltermann Prods. UK; Harcros; Houghton Chem.; Hukill; ICC Ind.; Indofine; Integra; J.T. Baker; Kessler; Mallinckrodt Baker; Occidental; PPG Ind.; Penta Mfg.; Primachem; R.E. Carroll; Romil Ltd; Sal Chem.; Sigma; Solvay SA; Spectrum Quality Prods.; Stanchem; VWR Int'l.; Voigt Global Distrib.; Vopak USA; Vulcan Perf. Chems.
Trade Names: Aerothene MM; Propulsol®; Solvaclene®; Ukalene®; Ukasol®
Trade Names Containing: Dynasolve 165; Dynasolve 200; Dynasolve 210

Methylene chlorobromide. *See* Bromochloromethane
Methylene dibenzene. *See* Diphenylmethane
Methylene dibromide. *See* Methylene bromide
Methylene dichloride. *See* Methylene chloride
Methylene diiodide. *See* Methylene iodide
Methylene dimethyl ether. *See* Methylal
3,4-Methylenedioxyphenol; Methylene ether of oxyquinone. *See* Sesamol
Methylene glycol. *See* Formaldehyde

Methylene iodide

CAS 75-11-6; EINECS/ELINCS 200-841-5
Synonyms: Diiodomethane; Methylene diiodide
Classification: Halogenated aliphatic hydrocarbon
Empirical: CH_2I_2
Properties: Colorless liq.; darkens (yel./brn.) on exposure to light, air, and moisture; mod. sol. in water; misc. with alcohol, propanol, IPA, hexane, cyclohexane, ether, chloroform, benzene; m.w. 267.87; dens. 3.325; m.p. 5 C; b.p. 180 C (dec.); flash pt. > 110 C
Toxicology: LD50 (IP, rat) 403 mg/kg, (subcut., mouse) 830 mg/kg; mod. toxic by ing., IP, subcut. routes; CNS depressant; inh. can probably cause sl. nose/throat irritation, mild CNS depression; high doses can cause unconsciousness, death; probable skin and sl. eye irritant; may be sl. absorbed thru skin; liq. may cause temporary corneal damage; related chemicals are suspected human carcinogens; mutagen; TSCA listed
Precaution: Explosive reaction possible with diethyl zinc + alkenes; violent reaction with copper-zinc alloys + ether; forms shock-sensitive explosive mixts. with potassium, potassium-sodium alloys, and lithium
Hazardous Decomp. Prods.: CO, CO_2, hydrogen iodide; heated to decomp., emits toxic fumes of I^-
Uses: High-density solvent for mineral separation; in mfg. of x-ray contrast media; determination of sp.gr. of minerals and other substances; organic synthesis
Manuf./Distrib.: Deepwater Chems.; R S A; Spectrum Quality Prods.

3-Methylene-7-methyl-1,6-octadiene. *See* Myrcene
Methyl ester of coconut oil fatty acids. *See* Methyl cocoate
Methyl esters of soybean oil. *See* Methyl soyate
Methyl ester stearic acid. *See* Methyl stearate
Methyl ester of wood rosin. *See* Methyl abietate
((1-Methyl-1,2-ethanediyl) bis (oxy)) bis (propanol). *See* Tripropylene glycol
Methyl ethanoate. *See* Methyl acetate
2Methylethanoate. *See* Isopropyl acetate
1-Methylethanol. *See* Isopropyl alcohol
Methylethanolamine. *See* N-Methylethanolamine

N-Methylethanolamine

CAS 109-83-1; EINECS/ELINCS 203-710-0
UN NA 1993
Synonyms: Ethanol, 2-(methylamino)-; (2-Hydroxyethyl) methylamine; N-(2-Hydroxyethyl)-N-methylamine; 2-(Methylamino) ethanol; β-(Methylamino) ethanol; N-Methylaminoethanol; Methylethanolamine; Methyl (2-hydroxyethyl) amine; Methyl (β-hydroxyethyl) amine; Monomethyl aminoethanol; Monomethylethanolamine
Classification: Organic compd.; amino alcohol
Empirical: C_3H_9NO
Formula: $CH_3NHCH_2CH_2OH$
Properties: Colorless visc. liq.; fishy odor; sol. in water, ethanol, diethyl ether, acetone, benzene, oxygenated solvs.; insol. in aliphatic solvs.; m.w. 75.11; dens. 0.9414 (20 C); vapor pressure 0.5 mm Hg (20 C); m.p. -4.5 C; b.p. 158 C; flash pt. (OC) 74 C; ref. index 1.439; pH strong base; surf. tens. 34.4 mN/m (22 C)
Toxicology: LD50 (oral, rat) 2340 mg/kg, (IP, rat) 130 mg/kg, (subcut., mouse) 1802 mg/kg; poison by IP route; mod. toxic by ing., subcut. routes; corrosive to skin, eyes, mucous membranes; reproductive/developmental toxicity; TSCA listed
Environmental: ThOD 1.71
Precaution: Combustible liq.; can readily absorb moisture and CO_2 from air; corrosive to cork, metals; strong base
Hazardous Decomp. Prods.: CO, CO_2, NO_x
NFPA: Health 2, Flammability 2, Reactivity 0
Uses: Buffer; chemical intermediate for prod. of pharmaceuticals, textile auxs., coatings, optical brighteners, agrochems.; solvent
Manuf./Distrib.: Aldrich; Atofina; Bayer; Dow; Richard's Pacific Chem. Sourcing; Xinchem

Methyl ether. *See* Dimethyl ether
2-(1-Methyl ethoxy) ethanol. *See* Ethylene glycol isopropyl ether
Methylethoxysiloxane cyclic polymer. *See* Cycloethoxymethicone
1-Methylethyl acetate. *See* Isopropyl acetate
Methylethylacetylenyl carbinol. *See* Methyl pentynol
1-Methylethyl alcohol. *See* Isopropyl alcohol
1-Methylethylamine. *See* Isopropylamine
Methylethylbenzene. *See* Ethyl toluene
(1-Methylethyl) benzene. *See* Cumene
Methylethylbenzene, all isomers. *See* Ethyl toluene
(1-Methylethyl) 1,1′-biphenyl. *See* 1,1′-Biphenyl (1-methylethyl)
Methyl ethyl carbinol. *See* 2-Butanol
(1-Methylethyl) cyclohexane. *See* Isopropylcyclohexane
(1-Methylethyl) cyclopentane. *See* Isopropylcyclopentane
1-Methylethyl dodecanoate. *See* Isopropyl laurate
((Methylethylene) bis (oxy)) dipropanol. *See* Tripropylene glycol
1-Methyl ethylene carbonate. *See* Propylene carbonate
1-Methylethylene 2-ethylhexanoate. *See* Propylene glycol dioctanoate
Methyl ethylene glycol. *See* Propylene glycol

Methyl ethylene oxide. *See* Propylene oxide
Methyl ethyl glycol. *See* Propylene glycol
1-Methylethyl hexandecanoate. *See* Isopropyl palmitate
1-Methylethyl 2-hydroxypropanoate. *See* Isopropyl lactate

Methyl ethyl ketone

CAS 78-93-3; EINECS/ELINCS 201-159-0
UN 1193 (DOT); FEMA 2170
Synonyms: Butanone; 2-Butanone; 3-Butanone; Ethyl methyl ketone; MEK; Methyl acetone; Methyl-2-propanone; 2-Oxobutane
Classification: Sat. aliphatic ketone
Empirical: C_4H_8O
Formula: $CH_3COCH_2CH_3$
Properties: Colorless liq., acetone-like odor; sol. in 4 parts water, benzene, alcohol, ether, oxygenated and chlorinated solvs.; misc. with fixed oils; m.w. 72.10; dens. 0.8255 (0/4 C); vapor pressure 71.2 mm (20 C); visc. 0.40 cp; m.p. -86 C; b.p. 79.6 C; flash pt. (TOC) 24 F; ref. index 1.3814 (15 C)
Toxicology: ACGIH TLV/TWA 200 ppm; STEL 300 ppm; LD50 (oral, rat) 2737 mg/kg, (IP, mouse) 616 mg/kg; mod. toxic by ing., skin contact, IP routes; toxic by inh.; strong irritant; mod. to severe eye irritant; human eye irritant @ 350 ppm; affects CNS; ing. of extremely high concs. may cause loss of consciousness or death; experimental teratogen, reproductive effects; TSCA listed
Environmental: VOC; BOD5 2.03; COD 2.31; ThOD 2.44
Precaution: DOT: Flamm. liq; dangerous fire risk; explosive limits in air 2-10%; incompat. with oxidizers, strong acids, mixts. of haloforms and strong bases, mixts. of hydrogen peroxide and nitric acid
Hazardous Decomp. Prods.: Explosive peroxides; heated to decomp., emits acrid smoke and fumes
NFPA: Health 1, Flammability 3, Reactivity 0
Storage: Store in cool, well-ventilated area, out of direct sunlight, away from heat/ignition sources, away from corrosives
Uses: Solvent in nitrocellulose coatings, acrylic coatings, vinyl films, paint removers, paints, lacquers, varnishes, cements, adhesives, magnetic tapes, printing inks, polymerization, organic synthesis, cosmetics, pharmaceuticals; process solvent; metal degreasing; syn. leathers; chem. intermediate; mfg. of smokeless powder; cleaning fluids; lube oil dewaxing; priming, catalyst carrier; extraction solvent in food processing; synthetic flavoring agent in foods and pharmaceuticals; in food-pkg. adhesives; adjuvant in resinous/polymeric food-contact coatings; in cellophane for food pkg.
Regulatory: FDA 21CFR §172.515, 172.859, 175.105, 175.320, 177.1200; FEMA GRAS; SARA §313 reportable; CERCLA hazardous substance
Manuf./Distrib.: AAE Chemie NV; AMC Chems.; Aldrich; Alfa Chem; Allchem Ind.; Alvin Chem.; Amyl; Arch Chems.; Ashland; Asiamerica Int'l.; Atofina SA; Atofina; BP Chems. Ltd; Baychem; Brenntag AG; C.P. Hall; Celanese; Chemcentral; Degussa Flavors/Alfrebro; Delta Distributors; Dow; Eastech; Eastman; ExxonMobil; Fluka; GFS; GNC Group; General Chem.; Harcros; Honeywell Perf. Polymers; Houghton Chem.; Hukill; ICC Ind.; Integra; J.T. Baker; Lluch Essence; Mallinckrodt Baker; Milsolv; Oxiteno; P.B. & S.; Penta Mfg.; Primachem; R.E. Carroll; Romil Ltd; Ruger; SOCO-Lynch; Sal Chem.; Shell; Southchem; Spectrum Quality Prods.; Stanchem; Sunnyside; Textile Chem.; VWR Int'l.; Varsal Instruments; Vopak USA; Whittaker, Clark & Daniels
Trade Names: ExxonMobil MEK
Trade Names Containing: Punctilious® SDA Canada DAG-2A Anhydrous; Punctilious® SDA Canada DAG-2A2 Anhydrous

Methylethylmethane. *See* Butane
1-Methylethyl octadecanoate. *See* Isopropyl stearate
1-Methylethyl-9-octadecenoate. *See* Isopropyl oleate
N-(1-Methylethyl)-2-propanamine. *See* Diisopropylamine
1-Methylethyl tetradecanoate. *See* Isopropyl myristate
Methylformamide. *See* N-Methylformamide

N-Methylformamide

CAS 123-39-7; EINECS/ELINCS 204-624-6
Synonyms: Formamide, N-methyl-; N-Formylmethylamine; Methylformamide; MMF; Monomethylformamide; N-Monomethylformamide; NMF
Classification: Nonaromatic amide
Empirical: C_2H_5NO
Formula: $HCONHCH_3$
Properties: Colorless clear visc. liq.; strong odor; sol. in water, alcohol, acetone; insol. in ether; m.w. 59.07; dens. 0.9961 (25/4 C); m.p. -3 C; b.p. 199-201 C; flash pt. 111 C; ref. index 1.4320
Toxicology: LD50 (oral, rat) 4000 mg/kg, (IP, rat) 3500 mg/kg, (subcut., mouse) 3100 mg/kg, (IV, mouse) 1580 mg/kg; mod. toxic by ing., IP, IV, intramuscular, and subcut. routes; harmful by inh., ing., skin contact; severe irritant; irritating to eyes, skin, respiratory tract; readily absorbed thru skin; may cause respiratory depression, wt. loss, body temp. decrease, somnolence, corneal damage; experimental reproductive effects; mutagen; teratogen; target organs: liver, lungs; TSCA listed
Precaution: Dangerous fire hazard exposed to heat or flame; reacts violently with benzene sulfonyl chloride
Hazardous Decomp. Prods.: Heated to decomp., emits toxic fumes of NO_x
NFPA: Health 1, Flammability 1, Reactivity 0
Uses: Process solvent; solvent for hydrocarbon extraction; selective solvent for aromatics, butadiene, acetylene extraction; intermediate for pharmaceuticals, other org. chems.
Manuf./Distrib.: Acros Org.; Aldrich; Fluka; Sigma

Methyl formate

CAS 107-31-3
UN 1243 (DOT)
Synonyms: Formic acid methyl ester; Methyl methanoate
Classification: Carboxylic acid ester
Empirical: $C_2H_4O_2$
Formula: $HCOOCH_3$
Properties: Colorless volatile liq.; ether-like odor; very sol. in water; sol. in alcohol, ether; m.w. 60.06; sp.gr. 0.975 (20 C); vapor pressure 400 mm Hg (16/0 C); m.p. -100 C; b.p. 32 C; solidifies @ 100 C; flash pt. (CC) -19 C; pH neutral
Toxicology: ACGIH TLV/TWA 100 ppm; STEL 150 ppm; LD50 (oral, rabbit) 1622 mg/kg; LCLo (inh., guinea pig) 10,000 ppm; mod. toxic by ing.; probably skin and mod. to severe eye irritant; can be absorbed thru skin; inh. can cause nose/throat/lung irritation; severe exposure may cause pulmonary edema; ing. can cause mouth/throat irritation, CNS depression; severe exposure may cause vision impairment, unconsciousness, possibly death; TSCA listed
Precaution: Flamm.; very dangerous fire hazard; incompat. with oxidizing materials (increases fire/explosion hazard), acids (decomp. can occur), bases (vigorous decomp. can occur), moisture or water vapor (slowly hydrolyzes to methanol, formic acid)
Hazardous Decomp. Prods.: Methanol, formic acid; breaks down slowly in water releasing formic acid; heated to decomp., emits acrid smoke and irritating fumes
NFPA: Health 2, Flammability 4, Reactivity 0
Storage: Store in cool, dry, well-ventilated area, out of direct sunlight; refrigerated storage rec.
Uses: Solvent for cellulose acetate; fumigant and larvicide for tobacco, dried fruits, cereals, etc.; cold-box foundry mold process curing agent; intermediate for dimethylformamide, other org. synthesis; in prep. of antileukemic agents
Manuf./Distrib.: Aldrich; BASF AG; BASF; Celanese; Fluka; Kira; Penta Mfg.; Richard's Pacific Chem. Sourcing; Sigma; Spectrum Quality Prods.; Xinchem

Methyl 2-furancarboxylate; Methyl furoate. *See* Methyl 2-furoate

Methyl 2-furoate

CAS 611-13-2; EINECS/ELINCS 210-254-6
FEMA 2703
Synonyms: Furan-α-carboxylic acid methyl ester; 2-Furoic acid, methyl ester; Methyl 2-furancarboxylate; Methyl furoate; Methyl pyromucate; Pyromucic acid methyl ester
Empirical: $C_6H_6O_3$
Formula: $C_4H_3OCO_2CH_3$
Properties: Colorless liq.; turns yel. in light; pleasant odor; sol. in alcohol, ether; insol. in water; m.w. 126.11; dens. 1.179; m.p. 164 C; b.p. 181 C; flash pt. 164 F; ref. index 1.4860 (20 C)
Toxicology: LD50 (skin, rabbit) > 1250 mg/kg; LDLo (IP, rat) 75 mg/kg; poison by IP route; skin irritant; lachrymator; TSCA listed
Precaution: Combustible

Hazardous Decomp. Prods.: Heated to decomp., emits acrid smoke and irritating fumes
Storage: Store under nitrogen
Uses: Solvent, fragrance in cosmetics; organic synthesis; pharmaceutical intermediate; synthetic flavoring agent in foods, pharmaceuticals
Features: Fruity flavor
Regulatory: FEMA GRAS
Manuf./Distrib.: Advanced Synthesis Tech.; Aldrich; Bio-Rad Labs; Degussa Flavors/Alfrebro; Fluka; Int'l. Chem. Group; Kira; Penn Spec. Chems.; Penta Mfg.; Spectrum Quality Prods.; Synasia Fine Chem.

Methyl gluceth-10

CAS 68239-42-9
Synonyms: PEG-10 methyl glucose ether; POE (10) methyl glucose ether
Definition: PEG ether of methyl glucose
Formula: $CH_3C_6H_{10}O_5(OCH_2CH_2)_nOH$, avg. n = 10
Properties: Nonionic
Uses: Emulsifier, humectant, moisturizer, emollient for cosmetics, personal care prods.; freezing pt. depressant; foam modifier in detergents and shampoos; solvent and solubilizer for topical pharmaceuticals
Trade Names: Glucam® E-10

Methyl gluceth-20

CAS 68239-42-9; 68239-43-0
Synonyms: PEG-20 methyl glucose ether; POE (20) methyl glucose ether
Definition: PEG ether of methyl glucose
Formula: $CH_3C_6H_{10}O_5(OCH_2CH_2)_nOH$, avg. n = 20
Properties: Nonionic
Uses: Humectant, moisturizer, emollient for cosmetics, personal care prods.; freezing pt. depressant; foam modifier in detergents and shampoos; solvent and solubilizer for topical pharmaceuticals
Trade Names: Glucam® E-20

Methyl glutarate

CAS 1501-27-5; EINECS/ELINCS 216-116-1
Synonyms: Glutaric acid, monomethyl ester; Methyl glutarate, mono; Methyl hydrogen glutarate; Monomethyl ester of glutaric acid; Monomethyl glutarate; Pentanedioic acid, monomethyl ester
Empirical: $C_6H_{10}O_4$
Properties: Colorless liq.; m.w. 146.14; dens. 1.164 g/ml; flash pt. > 110 C; ref. index 1.439
Toxicology: May be harmful by inh., ing., or skin absorp.; may cause irritation
Precaution: Incompat. with strong oxidizing agents, strong acids, strong bases
Hazardous Decomp. Prods.: Toxic fumes of CO, CO_2
Storage: Store in cool, dry place; keep tightly closed
Uses: In solvent blend for cleaning/stripping, paints
Manuf./Distrib.: Aldrich; Fluka; ICN Biomed. Research Prods.; Lancaster Synthesis; TCI Am.
Trade Names Containing: Rhodiasolv RPDE

Methyl glutarate, mono. *See* Methyl glutarate
Methyl glycol. *See* Propylene glycol
Methyl glycol acetate; Methyl glycol monoacetate. *See* Methoxyethanol acetate
16-Methylheptadecanoic acid, 16-methylheptadecyl ester; 16-Methylheptadecyl 16-methylheptadecanoate. *See* Isostearyl isostearate
6-Methylheptanol; 6-Methyl-1-heptanol. *See* Isooctyl alcohol

Methylheptanone

CAS 541-85-5; EINECS/ELINCS 208-793-7
UN 1224 (DOT)
Synonyms: 5-Methylheptan-3-one
Classification: Ketone
Empirical: $C_8H_{16}O$
Properties: Colorless liq.; pungent odor; m.w. 128.24; sp.gr. 0.82; vapor pressure 2 mm Hg; b.p. 157 C; ref. index 1.414
Toxicology: ACGIH TLV 25 ppm; LD50 (oral, rat) 3500 mg/kg, (skin, rabbit) > 16,000 mg/kg; LC50 (inh., mouse, 8 h) > 3483 ppm; target organs: eyes, skin, respiratory system, CNS; TSCA listed
Environmental: VOC; ThOD 1.87
Uses: Fragrance in cosmetics; solvent
Manuf./Distrib.: Aldrich; Atofina; Coyne; Jonas; Penta Mfg.

5-Methyl-3-heptanone. *See* 3-Octanone
5-Methylheptan-3-one. *See* Methylheptanone
Methyl hexadecanoate. *See* Methyl palmitate
2-Methylhexane. *See* Isoheptane
Methyl hexanoate; Methyl 2-hexanoate; Methyl n-hexanoate. *See* Methyl caproate
2-Methyl-5-hexanone; 5-Methyl-2-hexanone; 5-Methylhexan-2-one. *See* Methyl isoamyl ketone
Methyl hexoate; Methyl hexylate. *See* Methyl caproate
Methyl hexyl carbinol. *See* 2-Octanol

Methyl hexyl ether

CAS 4747-07-3; EINECS/ELINCS 225-263-0
Synonyms: Ether, hexyl methyl; Hexane, 1-methoxy-; Hexyl methyl ether; 1-Methoxyhexane
Classification: Aliphatic ether
Empirical: $C_7H_{16}O$
Formula: $CH_3O(CH_2)_5CH_3$
Properties: M.w. 116.20
Uses: Solvent in cosmetics
Manuf./Distrib.: Alfa Aesar; TCI Am.

Methyl hexyl ketone

CAS 111-13-7; EINECS/ELINCS 203-837-1
FEMA 2802
Synonyms: Hexyl methyl ketone; 2-Octanone
Empirical: $C_8H_{16}O$
Formula: $CH_3(CH_2)_5COCH_3$
Properties: Colorless liq., apple odor, camphor taste; sol. in oxygenated solvs.; misc. with alcohol, ether, esters, hydrocarbons; insol. in water; m.w. 128.22; dens. 0.818 (20/4 C); m.p. -16 C; b.p. 170-172 C; flash pt. 56 C; ref. index 1.416 (20 C)
Toxicology: LD50 (mouse) 1600 mg/kg; mod. toxic; skin irritant; TSCA listed
Precaution: Flamm. exposed to heat, flame, oxidizers
Hazardous Decomp. Prods.: Heated to decomp., emits acrid smoke and irritating fumes
Uses: Solvent for epoxy resin coatings, paints, leather finishes; antiblushing agent for nitrocellulose lacquers; fragrance in cosmetics; odorant; perfumes; synthetic flavoring agent in foods and pharmaceuticals
Features: High-boiling; fruity flavor
Regulatory: FDA 21CFR §172.515; FEMA GRAS
Manuf./Distrib.: Advanced BioTech; Aldrich; Degussa Flavors/Alfrebro; Fluka; Int'l. Paper; Lluch Essence; Miljac; Penta Mfg.; U.S. Chems.; UOP

Methyl hydrate. *See* Methyl alcohol

Methylhydrazine

CAS 60-34-4
UN 1244
Synonyms: Hydrazine, methyl-; Hydrazomethane; 1-Methylhydrazine; MMH; Monomethylhydrazine
Empirical: CH_6N_2
Formula: CH_3NHNH_2
Properties: Colorless clear liq.; ammonia-like odor; sol. in hydrocarbons, ether; misc. with hydrazine, low m.w. monohydric alcohols; sol. ≥ 10 mg/ml in DMSO, 95% ethanol; insol. in water; m.w. 46.09; sp.gr. 0.874 (20/4 C); vapor pressure 49.6 mm Hg (25 C); m.p. -20.9 C; b.p. 87.8 C; flash pt. 16.7 C; ref. index 1.4325 (20 C)
Toxicology: ACGIH TLV/CL 0.1 ppm (skin), intent to change to 0.01 ppm; LD50 (oral, rat) 33 mg/kg, (IV, rat) 17 mg/kg, (IP, rat) 21 mg/kg, (subcut., rat) 35 mg/kg; LC50 (inh., rat, 4 h) 74 ppm; highly toxic by inh., dermal, oral, IP, subcut., and IV routes; severe irritant to eyes, skin, respiratory/GI tracts; may cause convulsions, skin/eye burns, vomiting, diarrhea, cyanosis, etc.; may be fatal; experimental teratogen, neoplastigen, carcinogen; suspected human carcinogen; target organ: blood; TSCA listed
Precaution: Flamm.; explosion limits 2.5-97%; strong reducing agent; incompat. with oxidizers (ignites), oxygen, peroxides; unstable in contact with Cu, Fe or their alloys; vapors may flash back; in fire, may dec.

to form flamm./explosive mixts. in air
Hazardous Decomp. Prods.: CO, CO_2, NO_x
NFPA: Health 3, Flammability 3, Reactivity 2
Storage: Hygroscopic; air-sensitive; store refrigerated; keep tightly closed; protect from moisture, oxidizing agents, and metals
Uses: Rocket fuel; missile propellant; intermediate in chemical synthesis; solvent
Regulatory: SARA §313 reportable
Manuf./Distrib.: Acros Org.; Aldrich; Charkit; Fluka; Pfaltz & Bauer; SNPE Chimie; TCI Am.

1-Methylhydrazine. *See* Methylhydrazine
Methyl hydrogen glutarate. *See* Methyl glutarate
Methyl hydroxide. *See* Methyl alcohol
Methyl hydroxybenzoate; Methyl 2-hydroxybenzoate; Methyl o-hydroxybenzoate. *See* Methyl salicylate
Methyl (2-hydroxyethyl) amine; Methyl (β-hydroxyethyl) amine. *See* N-Methylethanolamine
1-Methyl-3-hydroxy-4-isopropylbenzene. *See* Thymol
Methyl 2-hydroxypropanoate. *See* Methyl lactate
Methyl-(S)-2-hydroxy propionate. *See* Methyl-(S)-lactate
1,2-(Methylidene) glycerol; Methylidinoglycerol. *See* Glyceryl formal
2,2′-(Methylimino) diethanol; N-Methyl-2,2′-iminodiethanol. *See* Methyldiethanolamine
Methylisoamyl acetate. *See* Methyl amyl acetate

Methyl isoamyl ketone

CAS 110-12-3; EINECS/ELINCS 203-737-8
UN 2302 (DOT)
Synonyms: Isoamyl methyl ketone; Isobutylacetone; Isopentyl methyl ketone; Ketone, methyl isoamyl; 2-Methyl-5-hexanone; 5-Methyl-2-hexanone; 5-Methylhexan-2-one; Methyl isopentyl ketone; MIAK
Classification: Saturated aliphatic ketone
Empirical: $C_7H_{14}O$
Formula: $CH_3COC_2H_4CH(CH_3)_2$
Properties: Colorless liq., pleasant odor; misc. with oxygenated, chlorinated, and hydrocarbon solvs.; sl. sol. in water; m.w. 114.19; dens. 0.8164; vapor pressure 5.2 mm Hg; f.p. -74 C; b.p. 142-145 C; flash pt. (TCC) 36 C; autoignition temp. 425 C; ref. index 1.4078; surf. tens. 25.33 dynes/cm
Toxicology: ACGIH TLV/TWA 50 ppm; LD50 (oral, rat) 4760 mg/kg, (IP, rat) 400 mg/kg, (skin, rabbit) 10 g/kg; mod. toxic by ing. and IP routes; mildly toxic by inh., skin contact; TSCA listed
Environmental: VOC; ThOD 2.80
Precaution: Flamm. exposed to heat, flame, oxidizers; LEL 1.05%; UEL 8.2%
Hazardous Decomp. Prods.: CO, CO_2; heated to decomp., emits acrid smoke and irritating fumes
NFPA: Health 1, Flammability 2, Reactivity 0
Uses: High-boiling solvent for paints, lacquers, nitrocellulose, cellulose acetate butyrate, acrylics, and vinyl copolymers
Manuf./Distrib.: Aldrich; Allchem Ind.; Ashland; Burdick & Jackson; Chemcentral; Eastman; Fluka; General Chem.; Hukill; Sigma; Vopak USA

Methyl isobutenyl ketone. *See* Mesityl oxide
Methylisobutyl carbinol. *See* Methyl amyl alcohol
Methyl isobutyl carbinol. *See* 2-Methyl-1-pentanol
Methylisobutyl carbinol acetate; Methylisobutylcarbinyl acetate. *See* Methyl amyl acetate

Methyl isobutyl ketone

CAS 108-10-1; EINECS/ELINCS 203-550-1
UN 1245; FEMA 2731
Synonyms: Hexanone; Hexone; Isobutyl methyl ketone; Isopropylacetone; 4-Methyl-2-oxopentane; 2-Methyl-4-pentanone; 4-Methyl-2-pentanone; 4-Methylpentan-2-one; 2-Methylpropyl methyl ketone; MIBK
Classification: Sat. aliphatic ketone
Empirical: $C_6H_{12}O$
Formula: $CH_3COCH_2CH(CH_3)_2$
Properties: Colorless volatile liq., faint ketonic/camphoraceous odor; misc. with alcohol, ether, benzene, most org. solvs.; sl. sol. in water; m.w. 100.18; dens. 0.8042 (20/20 C); vapor pressure 19.9 mm Hg; m.p. -85 C; b.p. 115.8 C; flash pt. (CC) 13 C; autoignition temp. 448 C; ref. index 1.396; surf. tens. 23.29 dynes/cm; dielec. const. 13.11
Toxicology: ACGIH TLV/TWA 50 ppm; STEL 75 ppm; LD50 (oral, rat) 2080 mg/kg; poison by IP route; mod. toxic by ingestion; mildly toxic by inh.; very irritating to skin, eyes, mucous membranes; narcotic in high conc.; TSCA listed
Environmental: VOC; BOD5 1.67; COD 2.16; ThOD 2.72
Precaution: Flamm.; dangerous fire risk; explosive limits 1.4-7.5% in air; incompat. with oxidizing/reducing agents, strong bases, potassium t-butoxide; violent reactions possible
Hazardous Decomp. Prods.: Explosive peroxides (e.g., methyl isobutyl peroxide)
NFPA: Health 2, Flammability 3, Reactivity 1
Storage: Store in well-ventilated area; bond and ground metal containers in storage area
Uses: Solvent for resins, paints, protective coatings, varnishes, nitrocellulose, lacquers, printing inks, adhesives, cosmetics; extraction solvent (mineral processing, metal, uranium from fission prods.); process solvent (lubricant dewaxing); mfg. of methyl amyl alcohol; organic synthesis; raw material in prod. of antioxidants; denaturant for alcohol; prod. of surfactants for inks, paints, pesticides; synthetic flavoring agent in foods, pharmaceuticals; pharmaceutical solvent, extractive agent; in food-pkg. adhesives; in food-contact polysulfide polymer/polyepoxy resins
Features: Fruity flavor
Regulatory: FDA 21CFR §172.515, 172.842, 175.105, 176.180, 176.200, 176.210, 177.1650; FEMA GRAS; NF compliance; SARA §313 reportable; CERCLA hazardous substance; HAP
Manuf./Distrib.: AAE Chemie NV; AMC Chems.; Aldrich; Allchem Ind.; Amyl; Ashland; Asiamerica Int'l.; Atofina SA; Baychem; Brenntag AG; C.P. Hall; Celanese; Chemcentral; Degussa Flavors/Alfrebro; Delta Distributors; Dow; Eastech; Eastman; ExxonMobil; Fluka; General Chem.; Haltermann Prods. UK; Harcros; Hukill; Integra; J.T. Baker; Mallinckrodt Baker; Milsolv; P.B. & S.; Penta Mfg.; Primachem; Romil Ltd; SOCO-Lynch; Sal Chem.; Shell; Sigma; Southchem; Spectrum Quality Prods.; Stanchem; Textile Chem.; Thomas Scientific; VWR Int'l.; Vopak USA; Whittaker, Clark & Daniels
Trade Names: Eastman® MIBK; Ucar® Methyl Isobutyl Ketone
Trade Names Containing: Eastman® CDA-19; Filmex® 3C-1 190 Proof; Filmex® A-1, 190 Proof; Filmex® A-2 190 Proof; Filmex® A-2 Anhydrous; Filmex® B 190 Proof; Filmex® B Anhydrous; Filmex® C 190 Proof; Filmex® C Anhydrous; Filmex® D-1; Filmex® D-2 190 Proof; Filmex® D-2 Anhydrous; Propsolv III-1 190 Proof; Propsolv III-1 Anhydrous; Propsolv III-2 190 Proof; Propsolv III-2 Anhydrous; Punctilious® CDA 19-3 190 Proof; Punctilious® CDA 19-3 Anhydrous A; Punctilious® SDA 1-2 190 Proof; Punctilious® SDA 23H 190 Proof; S-336 Anhydrous; Tecsol® 3; Tecsol® A, 95%; Tecsol® A Anhydrous; Tecsol® A-2, 95%; Tecsol® A-2 Anhydrous; Tecsol® B; Tecsol® C 95%; Tecsol® C Anhydrous; Tecsol® D, 95%; Tecsol® D-1, 95%; Tecsol® D-2, 95%; Tecsol® D-2 Anydrous

Methyl isopentyl ketone. *See* Methyl isoamyl ketone
1-Methyl-4-isopropenylcyclohexene; 1-Methyl-4-isopropenyl-1-cyclohexene. *See* dl-Limonene
1-Methyl-4-isopropylbenzene; p-Methylisopropyl benzene. *See* p-Cymene
1-Methyl-4-isopropylcyclohexadiene-1,4. *See* γ-Terpinene
5-Methyl-2-isopropylcyclohexanol. *See* Menthol
1-Methyl-4-isopropyl-1-cyclohexen-8-ol. *See* α-Terpineol
5-Methyl-2-isopropyl hexahydrophenol. *See* Menthol

Methyl isopropyl ketone

CAS 563-80-4; EINECS/ELINCS 209-264-3
UN 2397 (DOT)
Synonyms: 2-Acetyl propane; 2-Butanone, 3-methyl-; Isopropyl methyl ketone; 3-Methylbutanone; 3-Methyl-2-butanone; 3-Methylbutan-2-one; MIPK
Classification: Ketone
Empirical: $C_5H_{10}O$
Formula: $CH_3COCH(CH_3)_2$
Properties: Colorless liq.; acetone-like odor; sol. in alcohol, ether, oxygenated solvs.; very sl. sol. in water; m.w. 86.15; dens. 0.805 (20/20 C); dens. 6.64 lb/gal; vapor pressure 52.2 mm Hg; f.p. -92 C; b.p. 94.3

C; flash pt. (Seta) 0.5 C; ref. index 1.38788 (16 C); surf. tens. 24.61 dynes/cm
Toxicology: ACGIH TLV/TWA 200 ppm; LD50 (oral, rat) 148 mg/kg, (IP, rat) 800 mg/kg, (skin, rabbit) 6350 mg/kg; LC50 (inh., rat, 4 h) > 5700 mg/kg; primary irritant; poison by ing.; mildly toxic by inh. and skin contact; skin and eye irritant; mutagen; target organ: CNS; TSCA listed
Environmental: VOC; ThOD 1.67
Precaution: Flamm. exposed to heat or flame; can react vigorously with oxidizing materials
Hazardous Decomp. Prods.: Heated to decomp., emits acrid smoke and irritating fumes
Uses: Solvent for nitrocellulose lacquers; dyestuffs intermediate; fragrance in cosmetics
Manuf./Distrib.: Aldrich; Allchem Ind.; Ashland; Chisso Am.; Corco; Coyne; Delta Distributors; Eastern Chem.; Eastman; G.J. Chem.; Great Western; Harcros; ICC Ind.

5-Methyl-2-isopropylphenol. *See* Thymol
Methyl ketone. *See* Acetone

Methyl lactate

CAS 547-64-8; EINECS/ELINCS 208-930-0
UN 1993
Synonyms: 2-Hydroxypropanoic acid methyl ester; Lactic acid methyl ester; Methyl 2-hydroxypropanoate; Methyl DL-lactate; Propanoic acid, 2-hydroxy-, methyl ester
Classification: Lactate ester
Definition: Methyl ester of lactic acid
Empirical: $C_4H_8O_3$
Formula: $CH_3CH(OH)COOCH_3$
Properties: Colorless liq.; mild pleasant odor; sol. in alcohol, ether; dec. by water; m.w. 104.11; sp.gr. 1.085-1.098 (20 C); vapor pressure 2.25 mm Hg (20 C); b.p. 140-150 C; flash pt. 51 C; autoignition temp. 385 C; ref. index 1.410-1.415; surf. tens. 39 dynes/cm; KB value 62
Toxicology: LD50 (oral, rat) 2000 mg/kg; LDLo (IP, rat) > 2 g/kg; sl. toxic by IP route; mild irritant; target organs: respiratory system, skin, eyes, GI tract
Environmental: VOC; COD 1.38; ThOD 1.38
Precaution: Flamm. liq.; LEL 1.1%; UEL 3.6%; mod. fire risk
Hazardous Decomp. Prods.: Heated to decomp., emits acrid smoke and irritating vapors
NFPA: Health 1, Flammability 2, Reactivity 0
Uses: Solvent for cellulose acetate, nitrocellulose, cellulose acetobutyrate, cellulose acetopropionate, paints/coatings, lacquers, stains, electronics, printed circuit boards, metal industry
Manuf./Distrib.: Accurate Chem. & Scientific; Aldrich; Fluka; PURAC Am.; Penta Mfg.; Rhodia; Sithean; Thomas Scientific

Methyl DL-lactate. *See* Methyl lactate
(-)-Methyl L-lactate. *See* Methyl-(S)-lactate

Methyl-(S)-lactate

CAS 27871-49-4; EINECS/ELINCS 248-704-9
UN 1993
Synonyms: Lactic acid methyl ester; Methyl-(S)-2-hydroxy propionate; (-)-Methyl L-lactate; Methyl S-(-)-lactate; Propanoic acid, 2-hydroxy-, methyl ester, (S)-
Classification: Lactate ester
Empirical: $C_4H_8O_3$
Formula: $CH_3CH(OH)COOCH_3$
Properties: Colorless liq.; sol. in water, ethanol, ether; m.w. 104.11; sp.gr. 1.092; b.p. 144-145 C; flash pt. 49 C; autoignition temp. 385 C; ref. index 1.413 (20 C)
Toxicology: May cause eye/skin/mucous membrane/upper respiratory tract irritation; may be harmful by ing., inh., skin absorp.; inh. may cause coughing, dyspnea; ing. may cause irritation to mouth, pharynx, esophagus, GI tract; may cause tiredness, agitation, spasms, narcosis
Environmental: Pollutant; prevent entry into sewers, water, wastewater, soil
Precaution: Flamm. liq.; may dec. on exposure to moist air or water; may form explosive mixts. with air; vapors heavier than air; incompat. with oxidizing agents, bases, acids
Hazardous Decomp. Prods.: CO, CO_2; emits toxic fumes under fire conditions
Storage: Moisture-sensitive; store @ 15-25 C in cool, dry, well-ventilated area away from heat, open flame; keep tightly closed
Uses: Solvent; synthesis
Manuf./Distrib.: Aldrich; Hofmen Int'l.; Interchim; Merck KGaA
Trade Names: Purasolv® ML

Methyl S-(-)-lactate. *See* Methyl-(S)-lactate

Methyl lardate

CAS 112-62-6
Definition: Methyl ester of lard
Uses: Solvent, carrier for agric. sprays; defoamer in metalworking; paper deinking; pharmaceutical fermentation; wetting agent, lubricant for metalworking and lubricating oils
Manuf./Distrib.: Anar; Chemol; Norman, Fox; Sea-Land

Methyl laurate

CAS 111-82-0; EINECS/ELINCS 203-911-3
FEMA 2715
Synonyms: Dodecanoic acid methyl ester; Lauric acid methyl ester; Methyl dodecanoate; Methyl n-dodecanoate; Methyl dodecylate; Methyl laurinate
Definition: Ester of methyl alcohol and lauric acid
Empirical: $C_{13}H_{26}O_2$
Formula: $CH_3(CH_2)_{10}COOCH_3$
Properties: Water-wh. clear liq., fatty floral odor; insol. in water; m.w. 214.35; dens. 0.8702 (20/4 C); m.p. 4-5 C; b.p. 262 C (766 mm); flash pt. 94 C; ref. index 1.4320
Toxicology: LDLo (IV, mouse) 48 mg/kg; poison by IV route; may be harmful by inh., ing., or skin absorption; may cause irritation; TSCA listed
Precaution: Combustible; noncorrosive; incompat. with strong oxidizing agents, strong bases
Hazardous Decomp. Prods.: Heated to decomp., emits acrid smoke and irritating vapors
Storage: Keep tightly closed; store in cool, dry place
Uses: Intermediate for detergents, emulsifiers, wetting agents, stabilizers, lubricants, plasticizers, textiles, plastics, cosmetics; lubricant in textile spin finishes; emollient in cosmetics; solvent, cosolvent, oil carrier for agric. pesticides and herbicides; fragrance in perfumes; synthetic flavoring agent in foods, pharmaceuticals; defoamer in food-contact coatings and paper/paperboard; in resin-bonded filters for food contact; in food-contact textiles
Features: Coconut-like flavor
Regulatory: FDA 21CFR §172.225, 172.515, 176.200, 176.210, 177.2260, 177.2800; FEMA GRAS
Manuf./Distrib.: A&E Connock; ABCR; Acros Org.; Advanced Synthesis Tech.; Aldrich; Chemol; Cognis/Chems. Group; Fleurchem; Fluka; Indofine; Jarchem Ind.; Lluch Essence; NOF Am.; Penta Mfg.; Pfaltz & Bauer; Procter & Gamble; Riedel-deHaën; Sigma; Stepan; Uniqema Am.
Trade Names: Emery® 2270; Emery® 2290; Emery® 2296; Exceparl ML-85
Trade Names Containing: Degreez; ZELEC® C-48; ZELEC® C-49

Methyl laurinate. *See* Methyl laurate
N-Methylmethanamine. *See* Dimethylamine
Methyl methanoate. *See* Methyl formate
2-Methyl-2-methoxybutane. *See* t-Amyl methyl ether
4-Methyl-4-methoxy-2-pentanone. *See* 4-Methoxy-4-methylpentanone-2
2-Methyl-2-methoxypropane. *See* Methyl t-butyl ether
Methyl 2-methyl-2-butyl ether. *See* t-Amyl methyl ether
2-Methyl-6-methylene-2,7-octadiene; 7-Methyl-3-methylene-1,6-octadiene; 7-Methyl-3-methylene-octa-1,6-diene. *See* Myrcene
1-Methyl-4 (1-methylethenyl) cyclohexene. *See* dl-Limonene
1-Methyl-4-(1-methylethenyl)-(S)-cyclohexene. *See* l-Limonene
(R)-1-Methyl-4-(1-methylethenyl)-cyclohexene. *See* d-Limonene
1-Methyl-4-(1-methylethyl) benzene. *See* p-Cymene
1-Methyl-4-(1-methylethyl)-1,4-cyclohexadiene. *See* γ-Terpinene
5-Methyl-2-(1-methylethyl) cyclohexanol. *See* Menthol
1-Methyl-4-(1-methylethylidene) cyclohexene. *See* Terpinolene
5-Methyl-2-(1-methylethyl) phenol. *See* Thymol

Methyl-2(methyl-2) oxybispropanol. *See* Dipropylene glycol
[Methyl-2-(methyl-2-propoxyethoxy) ethoxy]-propanol. *See* Tripropylene glycol n-propyl ether
Methylmorpholine; 4-Methyl morpholine; N-Methyl morpholine. *See* p-Methyl morpholine

p-Methyl morpholine

CAS 109-02-4; EINECS/ELINCS 203-640-0
UN 2535 (DOT)
Synonyms: Methylmorpholine; 4-Methyl morpholine; N-Methyl morpholine; Morpholine, 4-methyl-
Empirical: $C_5H_{11}NO$
Formula: $CH_2CH_2OCH_2CH_2NCH_3$
Properties: Water-wh. liq., ammonia odor; forms constant-boiling mixture with 25% water and boiling at 97 C; misc. with benzene, water, oxygenated solvs.; m.w. 101.17; dens. 0.921; b.p. 115.4 C; f.p. -66 C; flash pt. (TOC) 75 F; ref. index 1.4349
Toxicology: LD50 (oral, rat) 1970 mg/kg, (skin, rabbit) 1242 mg/kg; LC50 (inh., mouse, 2 h) 25,200 mg/m^3; mod. toxic by ing. and skin contact; mildly toxic by inh.; corrosive irritant to skin, eyes, mucous membranes; TSCA listed
Precaution: Flamm. exposed to heat or flames; dangerous fire risk; corrosive; can react vigorously with oxidizing materials
Hazardous Decomp. Prods.: Heated to decomp., emits toxic fumes of NO_x
Uses: Catalyst in polyurethane foams; extraction solvent; stabilizer for chlorinated hydrocarbons; emulsifier, corrosion inhibitor for oils and polishes; self-polishing waxes; pharmaceuticals
Manuf./Distrib.: Aldrich; BASF AG; BASF; Fluka; Sigma; Xinchem

Methyl myristate

CAS 124-10-7; EINECS/ELINCS 204-680-1
FEMA 2722
Synonyms: Methyl tetradecanoate; Tetradecanoic acid, methyl ester
Definition: Ester of methyl alcohol and myristic acid
Empirical: $C_{15}H_{30}O_2$
Formula: $CH_3(CH_2)_{12}COOCH_3$
Properties: Colorless liq., honey and orris-like odor; insol. in water; m.w. 242.40; dens. 0.866 (20/4 C); m.p. 17.8 C; b.p. 186.8 C (30 mm); flash pt. > 112 C; ref. index 1.438 (20 C)
Toxicology: TSCA listed
Precaution: Combustible
Uses: Emollient in cosmetics; intermediate for myristic acid detergents, emulsifiers, wetting agents, stabilizers, resins, lubricants, plasticizers; animal feeds; standard for gas chromatography; fragrance; synthetic flavoring agent in foods, pharmaceuticals; solvent for pesticides and herbicides; defoamer in food-contact coatings and paper/paperboard; in resin-bonded filters for food contact; in food-contact textiles
Features: Honey-like flavor
Regulatory: FDA 21CFR §172.225, 172.515, 176.200, 176.210, 177.2260, 177.2800; FEMA GRAS
Manuf./Distrib.: A&E Connock; Aldrich; Cognis/Chems. Group; Fluka; Grau Aromatics; Indofine; Jarchem Ind.; NOF Am.; Penta Mfg.; RTD Chems.; Sigma; Stepan; Uniqema Am.
Trade Names: Emery® 2214
Trade Names Containing: Degreez; ZELEC® C-48; ZELEC® C-49

Methylnaphthalene

CAS 1321-94-4
Synonyms: Monomethylnaphthalene
Empirical: $C_{11}H_{10}$
Formula: $CH_3C_{10}H_7$
Properties: Yel. to bluish-brn. oily liq.; coal tar or mothball odor; sol. in DMSO, 95% ethanol, acetone, benzene, ether; sol. < 1 mg/ml in water; m.w. 142.20; sp.gr. 1.01 (15.5/15.5 C); vapor pressure 23 mm Hg; b.p. 232-310 C; flash pt. > 93 C
Toxicology: LD50 (oral, rat) 4360 mg/kg; low toxicity by ing.; mod. skin irritant; can cause photosensitization; may be harmful if inh. or absorbed thru skin; may be irritating to eyes; inh. of vapors may cause airway irritation, headache, nausea, weakness, collapse; long-term overexposure may cause corneal changes; ing. may cause systemic poisoning of GI tract, kidneys, blood-forming tissue; TSCA listed
Precaution: Probably combustible; incompat. with strong oxidizers
Hazardous Decomp. Prods.: Heated to decomp., emits acrid smoke and fumes
Storage: Store under ambient temps.
Uses: Organic synthesis; component in slow-release insecticides, mole repellents, and often in combination with naphthalene
Manuf./Distrib.: G.J. Chem.
Trade Names: Arosolve MN-LF
Trade Names Containing: Kenplast® A-450

1-Methylnaphthalene. *See* α-Methylnaphthalene

α-Methylnaphthalene

CAS 90-12-0; EINECS/ELINCS 201-966-8
FEMA 3193
Synonyms: 1-Methylnaphthalene
Empirical: $C_{11}H_{10}$
Formula: $C_{10}H_7CH_3$
Properties: Colorless liq. or oil; sol. in alcohol, ether, oxygenated and aromatic solvs.; insol. in water; m.w. 142.20; dens. 1.020 (20/4 C); m.p. -22 C; b.p. 241-245 C; flash pt. 82 C; ref. index 1.614
Toxicology: LD50 (oral, rat) 1840 mg/kg; mod. toxic by ing.; mutagenic data; TSCA listed
Precaution: Combustible exposed to heat, flame, or oxidizers
Hazardous Decomp. Prods.: Heated to decomp., emits acrid smoke and irritating fumes
Uses: Organic synthesis; surfactant; carrier for polyester/wool blended fabrics; synthetic flavoring agent in foods, pharmaceuticals; solvent for pesticides
Regulatory: FEMA GRAS
Manuf./Distrib.: Aldrich; Allchem Ind.; Coyne; Crowley Tar Prods.; Fluka; Koch Chem. Spec.
Trade Names: MN

Methyl neopentanoate. *See* Methyl pivalate
N-Methyl-N-nitrosomethanamine. *See* N-Nitrosodimethylamine

Methyl nonafluorobutyl ether

CAS 163702-07-6
Classification: Hydrofluoroether
Empirical: $C_5H_3F_9O$
Formula: $CF_3CF_2CF_2CF_2OCH_3$
Properties: Colorless clear liq.
Toxicology: AIHA WEEL 750 ppm; TSCA listed
Environmental: Zero ODP; virtually no VOC emissions
Uses: Solvent cleaner, rinsing agent for precision and electronic cleaning, defluxing of electronic circuit assemblies; heat transfer fluid; lubricant carrier; water drying agent; spray contact cleaner
Trade Names: HFE-7100
Trade Names Containing: HFE-71DA; HFE-71DE; Novec™ HFE-7100; Novec™ HFE-7100DL

Methyl nonafluoroisobutyl ether

CAS 163702-08-7
Classification: Hydrofluoroether
Empirical: $C_5H_3F_9O$
Formula: $(CF_3)_2CFCF_2OCH_3$
Toxicology: AIHA WEEL 750 ppm
Uses: Solvent cleaner for precision and electronic cleaning; heat transfer fluid
Trade Names Containing: Novec™ HFE-7100; Novec™ HFE-7100DL

2-Methylnonane. *See* Isododecane
8-Methylnonanoic acid. *See* Isodecanoic acid
8-Methylnonyl alcohol. *See* Isodecyl alcohol
Methyl octadecanoate; Methyl n-octadecanoate. *See* Methyl stearate
Methyl 9-octadecenoate; Methyl-cis-9-octadecenoate; Methyl (Z)-9-octadecenoate. *See* Methyl oleate
Methyl octanoate. *See* Methyl caprylate
Methylol. *See* Methyl alcohol

Methyl oleate

CAS 112-62-9; 67762-38-3; EINECS/ELINCS 203-992-5; 267-015-4
Synonyms: Methyl 9-octadecenoate; Methyl-cis-9-octadecenoate; Meth-

yl (Z)-9-octadecenoate; 9-Octadecenoic acid, methyl ester; (Z)-9-Octadecenoic acid, methyl ester; Oleic acid, methyl ester, cis-; cis-Oleic acid, methyl ester

Definition: Ester of methyl alcohol and oleic acid

Empirical: $C_{19}H_{36}O_2$

Formula: $CH_3(CH_2)_7CH{=}CH(CH_2)_7COOCH_3$

Properties: Clear to amber liq., faint fatty odor; sol. in alcohols, most org. solvs.; misc. with oxygenated solvs.; insol. in water; m.w. 296.55; dens. 0.8739 (20 C); f.p. -19.9 C; b.p. 218.5 C (20 mm); pour pt. -15 C; ref. index 1.4510 (26 C); nonionic

Toxicology: ACGIH TLV/TWA 10 mg/m^3 (total dust); TDLo (skin, mouse, 45 wk intermittent) 54 g/kg; low oral toxicity; mildly irritating to skin; questionable carcinogen; experimental tumorigen which causes leukemia and skin tumors by skin contact; TSCA listed

Precaution: Combustible

Hazardous Decomp. Prods.: Heated to decomp., emits acrid smoke and irritating fumes

Uses: Emollient, lubricant in cosmetics; intermediate for detergents, emulsifiers, wetting agents, stabilizers, paints, textile treatment, plasticizers for PS, cellulosics, duplicating inks, rubbers, waxes, etc.; chromatographic reference standard; lubricant for textile spin finishes, metalworking fluids, leather; plasticizer, softener for rubbers; extrusion aid; defoamer for pharmaceutical fermentation; solvent, carrier for pesticides and herbicides; food coatings; in food-pkg. adhesives; defoamer in food-contact paper coatings, paper/paperboard

Regulatory: FDA 21CFR §172.225, 175.105, 176.200, 176.210, 177.2260, 177.2800

Manuf./Distrib.: A&E Connock; Aldrich; Anar; Ashland; ChemService; Cognis/Chems. Group; Eastech; Ferro/Polymer Addit.; Fluka; Int'l. Paper; Lambent Tech.; Mosselman NV; NOF Am.; Norman, Fox; Penn Bio-Organics; Sea-Land; Sigma; Stepan; Uniqema Am.

Trade Names: Exceparl M-OL; Kemester® 205

Trade Names Containing: Degreez; Oleocal® ME-70

Methyl oleate/linoleate

Definition: Ester of methyl alcohol and a mixture of oleic and linoleic acids

Uses: Emulsifier, emollient for cosmetics; lubricant for leather; carrier for agric. spray prods.; as solvent

Trade Names: Kemester® 213

Methyl oleate/palmitate. *See* Methyl palmitate oleate

Methyl oleate-stearate

Classification: Fatty acid methyl ester

Uses: Lubricant

Trade Names Containing: ZELEC® C-48; ZELEC® C-49

3-Methylolpentane. *See* 2-Ethyl-1-butanol
Methylolpropane. *See* Butyl alcohol
2-Methyloltetrahydro-1,4-pyran. *See* Tetrahydropyran-2-methanol
Methyloxacyclopropane. *See* Propylene oxide
4-Methyl-3-oxa-1-pentanol. *See* Ethylene glycol isopropyl ether
Methyl oxide. *See* Dimethyl ether
Methyloxirane; S(-)-Methyloxirane. *See* Propylene oxide
Methyl 3-oxobutanoate; Methyl 3-oxobutyrate. *See* Methyl acetoacetate
4-Methyl-2-oxopentane. *See* Methyl isobutyl ketone

Methyl palmitate

CAS 112-39-0; EINECS/ELINCS 203-966-3

Synonyms: Hexadecanoic acid, methyl ester; Methyl hexadecanoate

Definition: Ester of methyl alcohol and palmitic acid

Empirical: $C_{17}H_{34}O_2$

Formula: $CH_3(CH_2)_{14}COOCH_3$

Properties: Colorless liq.; sol. in alcohol, ether, oxygenated, chlorinated, and aromatic solvs.; insol. in water; m.w. 270.46; m.p. 29.5 C; b.p. 211.5 C (30 mm); ref. index 1.4310 (45 C)

Toxicology: TSCA listed

Precaution: Combustible

Uses: Emollient; intermediate for detergents, emulsifiers, wetting agents, stabilizers, resins, lubricants, plasticizers; chemical synthesis; lubricant in min., cutting, lamination, textile oils, and rust inhibitors; animal feeds; flavoring agent; fragrance; textile and leather applics.; solvent for pesticides and herbicides; medical research; in food-pkg. adhesives; defoamer in food-contact paper coatings, paper/paperboard; in food-contact textiles; food coatings

Regulatory: FDA 21CFR §172.225, 175.105, 176.200, 176.210, 177.2260, 177.2800, 178.3910

Manuf./Distrib.: A&E Connock; Aldrich; Cognis/Chems. Group; Fluka; Grau Aromatics; Indofine; Lluch Essence; Norman, Fox; Penta Mfg.; Sigma; Stepan; Uniqema Am.

Trade Names: Emery® 2216

Trade Names Containing: Degreez; ZELEC® C-48; ZELEC® C-49

Methyl palmitate oleate

Synonyms: C16-18 and C18 unsaturated alkylcarboxylic acid, methyl ester; Methyl oleate/palmitate

Classification: Methyl ester

Definition: Ester of methyl alcohol and a mixture of palmitic and oleic acids

Properties: Clear liq.; sp.gr. 0.87; m.p. 14 C; b.p. 260 C; acid no. 2; sapon. no. 199; flash pt. 94 C

Uses: Intermediate for mfg. of detergents, emulsifiers, wetting agents, stabilizers, lubricants, plasticizers, resins, and textile specialties; lubricant, wetting agent for metalworking oils; solvent, cosolvent, oil carrier in agric.; in food-pkg. adhesives

Regulatory: FDA 21CFR §175.105

Methyl palm kernelate

Synonyms: Palm kernel methyl ester; Palm kernel oil methyl ester

Formula: $RCOOOCH_3$, R = palm kernel

Uses: Detergent intermediate; agric. solvent; lubricants; oil components; biodiesel fuel

Trade Names: Emery® 2255

8-Methylparacymene. *See* p-t-Butyl toluene
Methylparafynol. *See* Methyl pentynol
Methyl pentane; 2-Methylpentane. *See* Isohexane

3-Methylpentane

CAS 96-14-0

Properties: M.w. 86.18; dens. 0.659 g/ml; visc. 0.437 cp; b.p. 64 C; flash pt. (CC) 20 F; surf. tens. 17.7 dynes/cm

Uses: Solvent

2-Methyl-2,4-pentanediol; 2-Methylpentane-2,4-diol; 4-Methyl-2,4-pentanediol. *See* Hexylene glycol
2-Methylpentanol. *See* 2-Methyl-1-pentanol

2-Methyl-1-pentanol

CAS 105-30-6

Synonyms: Amyl methyl alcohol; 1,3-Dimethyl butanol; Isohexyl alcohol; Isopropyl dimethyl carbinol; Methylamyl alcohol; Methyl isobutyl carbinol; 2-Methylpentanol; 2-Methylpentan-1-ol; 2-Methylpentanol-1; 2-Methyl-2-propylethanol

Classification: Primary aliphatic alcohol

Empirical: $C_6H_{14}O$

Formula: $CH_3CH_2CH_2CH(CH_3)CH_2OH$

Properties: Colorless liq.; sol. in ethanol, ether, acetone; very sl. sol. in water; m.w. 102.18; dens. 0.8263 (20 C); b.p. 148 C; flash pt. (TCC) 51 C

Toxicology: LD50 (oral, rat) 1410 mg/kg, (skin, rabbit) 3560 mg/kg; toxic; may cause mild skin and severe eye irritation; skin absorp. can occur; inh. of mist may cause nose/throat irritation, headaches, nausea, dizziness, drowsiness, incoordination; ing. may cause 'alcohol' intoxication symptoms; severe exposure may cause unconsciousness, coma; may cause lung damage, respiratory arrest, death if aspirated into lungs; TSCA listed

Precaution: Mod. flamm.; incompat. with oxidizing agents (increases fire/explosion risk)

Hazardous Decomp. Prods.: Heated to decomp., emits smoke and acrid fumes

NFPA: Health 0, Flammability 2, Reactivity 0

Storage: Store in cool, dry, well-ventilated area; avoid generating mists; limit quantities in use; use in area separate from storage area

Uses: Solvent for metallurgy, metal processing, surf. coatings; chem. intermediate for plastics, other org. compds.

Manuf./Distrib.: Aldrich; BASF AG; Dow; Fluka; Penta Mfg.; Sigma

2-Methylpentan-1-ol; 2-Methylpentanol-1. *See* 2-Methyl-1-pentanol
2-Methyl-4-pentanol; 4-Methylpentanol-2; 4-Methyl-2-pentanol; 4-Methylpentan-2-ol. *See* Methyl amyl alcohol
4-Methyl-2-pentanol acetate. *See* Methyl amyl acetate
2-Methyl-2-pentanol-4-one. *See* Diacetone alcohol
2-Methyl-4-pentanone; 4-Methyl-2-pentanone; 4-Methylpentan-2-one. *See* Methyl isobutyl ketone
4-Methyl-3-pentene-2-one; 2-Methyl-2-penten-4-one; 4-Methyl-3-penten-2-one. *See* Mesityl oxide
3-Methylpentin-3-ol. *See* Methyl pentynol
4-Methyl-2-pentyl acetate. *See* Methyl amyl acetate
2-Methylpentyl 'Carbitol'. *See* Diethylene glycol methylpentyl ether
Methyl t-pentyl ether. *See* t-Amyl methyl ether
Methyl pentyl ketone. *See* Methyl n-amyl ketone
2-[2-((2-Methylpentyl) oxy) ethoxy] ethanol. *See* Diethylene glycol methylpentyl ether
3-Methyl-1-pentyne-3-ol. *See* Methyl pentynol

Methyl pentynol

CAS 77-75-8; EINECS/ELINCS 201-055-5
UN 1993
Synonyms: 2-Ethinylbutanol-2; Ethinylmethylethyl carbinol; 3-Ethylbutinol; 3-Ethylbutynol; Ethyl ethynyl methyl carbinol; 2-Ethynyl-2-butanol; Meparfynol; Methylethylacetylenyl carbinol; Methylparafynol; 3-Methylpentin-3-ol; 3-Methyl-1-pentyne-3-ol; 3-Methylpent-1-yn-3-ol; 3-Methyl-1-pentyn-3-ol; 3-Methylpentyn-3-ol; m-Pentynol
Classification: Aliphatic organic compd.
Empirical: $C_6H_{10}O$
Formula: $CH_3CH_2CH_3COHCCH$
Properties: Colorless liq.; acrid odor; burning taste; sol. in water, ether; misc. with acetone, benzene, CCl_4, ethyl acetate, oxygenated, chlorinated, and aromatic solvs.; m.w. 98.16; dens. 0.8721 (20/20 C); f.p. -30.6 C; b.p. 121.4 C; flash pt. (TOC) 38.3 C; ref. index 1.4310
Toxicology: LD50 (oral, rat) 300 mg/kg, (IP, mouse) 625 mg/kg; toxic by ing.; mod. toxic by IP and subcut. routes; irritant; a soporific; may cause dermatitis, psychoses, CNS effects; overdosage may cause coma and death; TSCA listed
Environmental: Dangerous to the environment
Precaution: Flamm. exposed to heat or flame; can react with oxidizing materials
Hazardous Decomp. Prods.: Heated to decomp., emits acrid smoke and fumes
Uses: Acid corrosion inhibitor; reactive intermediate in the mfg. of pharmaceuticals (hypnotics), perfumes, plastics, rubbers, specialty chemicals; as solvent for polyamide resins; visc. reducer; stabilizer; electroplating brightener; stabilizer in chlorinated solvents; halogenated solvent stabilizer; prevention of hydrogen embrittlement; medicine (soporific and anesthetic); fish tranquilizer
Manuf./Distrib.: Advanced Synthesis Tech.; Air Prods.; Aldrich; BASF AG; Fluka

3-Methylpent-1-yn-3-ol; 3-Methyl-1-pentyn-3-ol; 3-Methylpentyn-3-ol. *See* Methyl pentynol
4-(1-Methyl-1-phenethyl) phenol. *See* Cumyl phenol
Methyl phenol. *See* Cresylic acid
2-Methylphenol; o-Methylphenol. *See* o-Cresol
(Methyl-2-phenoxyethoxy) propanol. *See* Dipropylene glycol phenyl ether
Methylphenylamine; N-Methylphenylamine. *See* N-Methylaniline
Methyl phenylcarbinol. *See* α-Methylbenzyl alcohol
4-Methylphenylene diisocyanate. *See* Toluene-2,4-diisocyanate
Methyl-m-phenylene diisocyanate; Methylphenylene isocyanate. *See* Toluene diisocyanate
4-Methylphenylene isocyanate. *See* Toluene-2,4-diisocyanate
Methylphenyl ether. *See* Anisole
Methyl phenyl ketone. *See* Acetophenone
Methylphenylmethanol. *See* α-Methylbenzyl alcohol
Methyl phenyl polysiloxane. *See* Phenyl trimethicone
2-Methyl-1-phenylpropane. *See* Isobutylbenzene
2-Methyl-2-phenylpropane. *See* t-Butylbenzene
Methyl phosphate. *See* Trimethyl phosphate
Methylphosphonic acid, dimethyl ester. *See* Dimethyl methylphosphonate
Methyl phthalate. *See* Dimethyl phthalate

Methyl phthalyl ethyl glycolate

CAS 85-71-2
Synonyms: Ethyl O-(o-(methoxycarbonyl) benzoyl) glycolate; Ethyl o-(methoxycarbonyl) benzoyloxyacetate; Glycolic acid, ethyl ester, methyl phthalate; Phthalic acid, monomethyl ester, ester with ethyl glycolate
Empirical: $C_{13}H_{14}O_6$
Properties: Colorless liq.; sl. char. odor; misc. with most org. solvs.; very sl. sol. in water; m.w. 266.27; dens. 1.217-1.227; flash pt. 190 C
Toxicology: LD50 (skin, guinea pig) > 10 g/kg; mild eye irritant
Precaution: Combustible
Uses: Plasticizer; solvent; in food-pkg. adhesives
Regulatory: FDA 21CFR §175.105
Manuf./Distrib.: ChemService

1-Methylpiperidine

CAS 626-67-5
UN 2399 (DOT)
Synonyms: N-Methylpiperidine
Empirical: $C_6H_{13}N$
Properties: Liq.; m.w. 99.20; dens. 0.820 (20/4 C); b.p. 107 C; flash pt. < 73.4 F
Toxicology: LD50 (IP, mouse) 400 mg/kg; LDLo (subcut., rabbit) 400 mg/kg; poison by IP and subcut. routes; TSCA listed
Precaution: Very dangerous fire hazard exposed to heat or flame
Hazardous Decomp. Prods.: Heated to decomp., emits toxic fumes of NO_x
Uses: Intermediate in the pharmaceutical, agric., photo, polymer and chemical industries; solvent; catalyst
Manuf./Distrib.: Aldrich; Fluka; Raschig; Sigma

N-Methylpiperidine. *See* 1-Methylpiperidine

Methyl pivalate

CAS 598-98-1; EINECS/ELINCS 209-959-1
Synonyms: Methyl 2,2′-dimethylpropionate; Methyl neopentanoate; Methyl trimethylacetate
Definition: Methyl ester of neopentanoic acid
Empirical: $C_6H_{12}O_2$
Properties: Liq.; m.w. 116.17; dens. 0.873 (20/20 C); dens. 7.29 lb/gal (20 C); b.p. 214 C; flash pt. (SCC) 44 F
Toxicology: TSCA listed
Uses: Intermediate for prod. of pesticides, herbicides, solvents, metal extraction agents, syn. lubricants, polymerization aid for acrylic acid esters, industrial fragrances, odorants, cosmetic components, pharmaceuticals, photographic applics.; insect attractants and repellents; fragrance; flavoring agent
Manuf./Distrib.: ExxonMobil; Grau Aromatics

2-Methylpropanal

CAS 78-84-2; EINECS/ELINCS 201-149-6
FEMA 2220
Synonyms: iBuH; Isobutanal; Isobutyl aldehyde; Isobutyraldehyde; Isobutyric aldehyde; 2-Methyl-1-propanal; 2-Methylpropionaldehyde; Valine aldehyde
Empirical: C_4H_8O
Formula: $(CH_3)_2CHCHO$
Properties: Colorless liq., pungent sharp odor; sol. in water; misc. with alcohol, ether, benzene, carbon disulfide, acetone, toluene, chloroform, oxygenated and aromatic solvs.; m.w. 72.12; dens. 0.7938 (20/4 C); m.p. -65 C; b.p. 64 C; flash pt. (CC) -40 F; ref. index 1.374
Toxicology: LD50 (oral, rat) 2810 mg/kg, (skin, rabbit) 7130 mg/kg; LCLo (inh., rat, 4 h) 8000 ppm; mod. toxic by ing.; mildly toxic by skin contact, inh.; severe skin and eye irritant; TSCA listed
Precaution: Flamm.; dangerous fire hazard exposed to heat, flames, oxidizers; can react vigorously with reducing materials; explosive limits 1.6-10.6%
Hazardous Decomp. Prods.: Heated to decomp., emits acrid smoke and fumes
Uses: Intermediate for rubber antioxidants and accelerators; in synthesis of amino acids, pantothenic acid, valine, leucine, cellulose esters, perfumes, flavors, plasticizers, resins, gasoline additives; solvent; synthetic flavoring agent in foods and pharmaceuticals; fragrance inter-

mediate for cosmetics
Features: Banana-like flavor
Regulatory: FDA 21CFR §172.515; FEMA GRAS
Manuf./Distrib.: Advanced BioTech; Aldrich; Axxence Aromatic GmbH; BASF; Celanese; Degussa Flavors/Alfrebro; Dow; Eastman; Fleurchem; Fluka; Moore Ingreds.; Xinchem

2-Methyl-1-propanal. *See* 2-Methylpropanal
2-Methylpropane. *See* Isobutane

Methylpropanediol

CAS 2163-42-0; EINECS/ELINCS 412-350-5
Synonyms: 2-Methyl-1,3-propanediol; MPD; 1,3-Propanediol-2-methyl
Classification: Organic compd.; aliphatic diol
Empirical: $C_4H_{10}O_2$
Formula: $HOCH_2CH(CH_3)CH_2OH$
Properties: Colorless to wh. clear liq.; misc. with water, alcohols; m.w. 90.12; dens. 1.015; m.p. -91 C; b.p. 123-125 C (20 mm); acid no. 0.05; flash pt. > 110 C; autoignition temp. 380 C; ref. index 1.4450 (20 C); pH 6.5
Toxicology: LD50 (oral, rat) > 5000 mg/kg; irritant
Uses: Intermediate in prod. of solvents, urethanes, unsat. polyesters, gel coats, sat. polyester and alkyd coatings, polymeric plasticizers, adhesives, sealants, elastomers (required to keep transparency, weatherability); solvent, compatibilizer for adjuvants, fabric softener carriers, nonionic surfactants, pigment dispersions; solubilizer, emulsifier in air freshener applics.; solvent, emulsifier, humectant, solubilizer, compatibilizer, stabilizer, neutralizer, emollient in cosmetics, personal care; food-contact coatings
Features: Nonvolatile
Manuf./Distrib.: Aldrich; Fluka; Lyondell
Trade Names: MPDiol® Glycol

2-Methyl-1,3-propanediol. *See* Methylpropanediol
Methyl propanoate. *See* Methyl propionate
2-Methylpropanoic acid. *See* Isobutyric acid
2-Methylpropanoic acid 2-methylpropyl ester. *See* Isobutyl isobutyrate
1-Methyl propanol. *See* 2-Butanol
2-Methylpropanol. *See* Isobutyl alcohol
Methyl-2-propanol. *See* t-Butyl alcohol
2-Methyl-1-propanol; 2-Methylpropan-1-ol. *See* Isobutyl alcohol
2-Methyl-2-propanol; 2-Methylpropan-2-ol. *See* t-Butyl alcohol
Methyl-2-propanone. *See* Methyl ethyl ketone
4,4′,4′′-(1-Methyl-1-propanyl-3-ylidene) tris 2-(1,1-dimethylethyl)-5-methylphenol. *See* 1,1,3-Tris (2-methyl-4-hydroxy-5-t-butylphenyl) butane
2-Methylpropionaldehyde. *See* 2-Methylpropanal

Methyl propionate

CAS 554-12-1; EINECS/ELINCS 209-060-4
UN 1248 (DOT); FEMA 2742
Synonyms: Methyl propanoate; Methyl propylate; Propanoic acid methyl ester
Classification: Ester
Empirical: $C_4H_8O_2$
Formula: $CH_3CH_2COOCH_3$
Properties: Colorless liq., fruity rum-like odor; sol. in 16 parts water; misc. with alcohol, ether; m.w. 88.11; dens. 0.915 (20/4 C); vapor pressure 40 mm (11 C); m.p. -87 C; b.p. 78-80 C; flash pt. -2 C; autoignition temp. 469 C; ref. index 1.377 (20 C); dielec. const. 5.5
Toxicology: LD50 (oral, rat) 5 g/kg, (oral, mouse) 3460 mg/kg, (skin, rabbit) > 5000 mg/kg; LC50 (inh., mouse) 27 g/m^3; mod. toxic by ing.; mildly toxic by inh.; primary skin irritant; target organs: CNS, respiratory system; TSCA listed
Environmental: VOC; ThOD 1.09
Precaution: Highly flamm.; explosive limits in air 2.5-13%; dangerous fire hazard exposed to heat, flame, or oxidizers; explosive as vapor exposed to heat or flame
Hazardous Decomp. Prods.: Heated to decomp., emits acrid smoke and irritating fumes
NFPA: Health 1, Flammability 3, Reactivity 0
Uses: Solvent for cellulose nitrate; solvent mixt. for cellulose derivs., lacquers, paints, varnishes, coatings; synthetic flavoring agent in foods and pharmaceuticals; fragrance in cosmetics; in organic synthesis
Features: Apple, banana, strawberry-like flavor
Regulatory: FDA 21CFR §172.515; FEMA GRAS
Manuf./Distrib.: Aldrich; Degussa Flavors/Alfrebro; Fleurchem; Fluka; Goldensun Mfg.; Lluch Essence; Sigma

2-Methylpropionic acid; α-Methylpropionic acid. *See* Isobutyric acid
2-(2-Methylpropoxy) ethanol. *See* Ethylene glycol isobutyl ether
1-(2-Methylpropoxy)-2-propanol. *See* Propylene glycol isobutyl ether
1-Methylpropyl acetate. *See* s-Butyl acetate
2-Methylpropyl acetate; 2-Methyl-1-propyl acetate. *See* Isobutyl acetate
2-Methylpropyl alcohol. *See* Isobutyl alcohol
1-Methylpropylamine. *See* s-Butylamine
Methyl propylate. *See* Methyl propionate
(2-Methylpropyl) benzene. *See* Isobutylbenzene
Methyl propyl carbinol. *See* s-Amyl alcohol
(2-Methylpropyl) cyclohexane. *See* Isobutylcyclohexane
2-Methylpropyl ethanoate; β-Methylpropyl ethanoate. *See* Isobutyl acetate
2-Methyl-2-propylethanol. *See* 2-Methyl-1-pentanol
2-Methylpropyl hexadecanoate. *See* Isobutyl palmitate
4-Methyl-2-propylhexane-1-ol. *See* Propylheptanol
2-Methylpropyl isobutyrate. *See* Isobutyl isobutyrate

Methyl propyl ketone

CAS 107-87-9; EINECS/ELINCS 203-528-1
UN 1249 (DOT); FEMA 2842
Synonyms: Ethyl acetone; Methyl n-propyl ketone; MPK; 2-Pentanone
Classification: Saturated aliphatic ketone
Empirical: $C_5H_{10}O$
Formula: $CH_3COCH_2CH_2CH_3$
Properties: Water-wh. liq.; fruity ethereal odor; sol. in ethanol, diethyl ether, acetone, other ketones, oxygenated solvs.; sl. sol. in water; m.w. 86.14; dens. 0.801-0.806; vapor pressure 35.4 mm Hg; m.p. -78 C; b.p. 102-103 C; flash pt. (CC) 7 C; autoignition temp. 452 C; ref. index 1.388; surf. tens. 33.87 dynes/cm; dielec. const. 13.6
Toxicology: ACGIH TLV/TWA 200 ppm, STEL 250 ppm; LD50 (oral, rat) 3730 mg/kg, (IP, rat) 800 mg/kg; mod. toxic by ingestion, IP; mildly toxic by skin contact, inh.; human systemic effects; skin irritant; mutagenic data; target organs: respiratory system, eyes, skin, CNS; TSCA listed
Environmental: VOC; BOD5 1.18; COD 2.31; ThOD 2.60
Precaution: Highly flamm.; LEL 1.5%; UEL 8.2%; very dangerous fire hazard exposed to heat or flame; reacts vigorously with oxidizers; explosion hazard in vapor form exposed to heat or flame; mixts. with bromine trifluoride may explode
Hazardous Decomp. Prods.: CO, CO_2; heated to decomp., emits acrid smoke and irritating fumes
NFPA: Health 2, Flammability 3, Reactivity 0
Uses: Solvent for paints; substitute for diethyl ketone; synthetic flavoring agent in foods and pharmaceuticals; fragrance in cosmetics; in org. synthesis
Features: Fruity flavor
Regulatory: FDA 21CFR §172.515; FEMA GRAS
Manuf./Distrib.: Aldrich; Ashland; Chisso Am.; Degussa Flavors/Alfrebro; Eastman; Fluka; General Chem.; Houghton Chem.; Moore Ingreds.; Mutchler; Penta Mfg.; S & S Chem.; Spectrum Quality Prods.; VWR Int'l.

Methyl n-propyl ketone. *See* Methyl propyl ketone
2-Methylpropyl methyl ketone. *See* Methyl isobutyl ketone
2-Methylpropyl 2-methylpropanoate. *See* Isobutyl isobutyrate
2-Methylpropyl propanoate; 2-Methylpropyl propionate. *See* Isobutyl propionate
2-Methylpyridine. *See* α-Picoline
3-Methylpyridine. *See* β-Picoline
4-Methylpyridine. *See* γ-Picoline
α-Methylpyridine. *See* α-Picoline
Methyl pyromucate. *See* Methyl 2-furoate
1-Methylpyrrolidinone; 1-Methyl-2-pyrrolidinone; N-Methylpyrrolidinone; N-Methyl-2-pyrrolidinone; Methyl pyrrolidone; 1-Methylpyrrolidone; 1-Methyl-2-pyrrolidone; N-Methylpyrrolidone. *See* N-Methyl-2-pyrrolidone

N-Methyl-2-pyrrolidone

CAS 872-50-4; EINECS/ELINCS 212-828-1
UN NA 1993 (DOT)
Synonyms: 1-Methylazacyclopentan-2-one; 1-Methylpyrrolidinone; 1-Methyl-2-pyrrolidinone; N-Methylpyrrolidinone; N-Methyl-2-pyrrolidinone; Methyl pyrrolidone; 1-Methylpyrrolidone; 1-Methyl-2-pyrrolidone; N-Methylpyrrolidone; NMP; m-Pyrrole
Classification: Sat. nitrogen heterocyclic compd.
Empirical: C_5H_9NO
Properties: Colorless liq., mild amine odor; sol. in water, alcohols, ketones, ethers, ethyl acetate, aromatic hydrocarbons, most org. solvs.; mod. sol. in aliphatic hydrocarbons; misc. with castor oil; m.w. 99.13; dens. 1.032 (20/4 C); vapor pressure 0.3 mm Hg (20 C); f.p. -24 C; b.p. 202 C; flash pt. 95 C; ref. index 1.470 (20 C); pH 7.7-8.0 (10% aq.)
Toxicology: LD50 (oral, rat) 7000 mg/kg, (IP, rat) 2472 mg/kg, (IV, mouse) 54,500 µg/kg, (skin, rabbit) 8000 mg/kg; poison by IV route; mod. toxic by ing. and IP routes; mildly toxic by skin contact; severely irritating to eyes, skin; inh. of high concs. may cause fatigue, difficulty breathing; ing. may cause nausea, vomiting; not a skin absorption hazard; experimental teratogen, reproductive effects; mutagenic data; TSCA listed
Environmental: Readily biodeg.; VOC; ThOD 1.05; EC50 (rainbow trout, 96 h) ≥ 500 mg/l, (algae, 72 h) ≥ 500 mg/l
Precaution: Combustible exposed to heat, open flame, powerful oxidizers; incompat. with oxidizing agents (may spontaneously ignite), strong acids and strong bases (can cause hydrolysis)
Hazardous Decomp. Prods.: May form hydroperoxides on exposure to air and sunlight; heated to decomp., emits CO, CO_2, toxic fumes of NO_x
HMIS: Health 2, Flammability 1, Reactivity 0
Storage: Store in cool, dry, well-ventilated area, out of direct sunlight; hygroscopic; photosensitive
Uses: Solvent for resins, waxes, paints, paint stripping, acetylene, industrial cleaning, agric., mold cleaning; microelectronics industry plastic solvent; solvent for CO_2 removal in ammonia synthesis; aprotic process solvent; selective solvent; cleaner for polymeric residues; extraction of acetylene and butadiene; metal finishing; printed circuit board mfg.; dehydration of natural gas; lube oil processing; SBR latex prod.; petrochem. and petroleum processing; spinning agent for PVC; intermediate; pigment dispersant; veterinary pharmaceuticals; pharmaceutical solvent; synthesis of vitamin E precursor; solubilizer for topicals; bioadhesive; skin penetration enhancer; spray bandagesadjuvant for slimicides in food-contact paper/paperboard; solvent in food-contact polysulfone and PES resins
Regulatory: FDA 21CFR §176.300, 177.1655, 177.2440; SARA §311/312 acute health/chronic health/fire hazard
Manuf./Distrib.: AMC Chems.; Aldrich; Alfa Chem; Allchem Ind.; Arch Chems.; Ashland; BASF; Coyne; Dynaloy; Fluka; General Chem.; Haltermann Prods. UK; ISP; Integra; Kessler; Lyondell; Mitsubishi Chem.; Mutchler; Romil Ltd; S & S Chem.; Sigma; Spectrum Quality Prods.; Stanchem; Thomas Scientific; Tulstar Prods.; VWR Int'l.; Vopak USA; Whyte Chems. Ltd
Trade Names: Agsolex® 1; GAFsolver™ TMP; M-Pyrol®; NMP; NMP-EL; Pharmasolve™
Trade Names Containing: Dynasolve 180; Dynasolve 185; Dynasolve 699; Dynasolve CU-5; Dynasolve CU-6; Ecco Strip LS-W; FoamFlush® Urethane Remover; NMP; Partsprep™ Degreaser; Ship Shape® Resin Cleaner

Methyl rapeseedate

CAS 73891-99-3
Synonyms: Rape oil methyl ester; Rapeseed oil methyl ester
Definition: Methyl ester of rapeseed oil
Uses: Lubricant in metalworking and agric. adjuvants; carrier, solvent for printing inks, leather treating compds.; additive in adhesives for food production or pkg.

Methyl salicylate

CAS 119-36-8; EINECS/ELINCS 204-317-7
UN 3082; FEMA 2745
Synonyms: Betula; Betula oil; 2-Carbomethoxyphenol; 2-Hydroxybenzoic acid, methyl ester; o-Hydroxybenzoic acid, methyl ester; 2-(Methoxycarbonyl) phenol; Methyl hydroxybenzoate; Methyl 2-hydroxybenzoate; Methyl o-hydroxybenzoate; Oil of wintergreen; Sweet birch oil; Synthetic oil of wintergreen; Teaberry oil; Wintergreen oil
Classification: Aromatic organic compd.
Definition: Ester of methyl alcohol and salicylic acid
Empirical: $C_8H_8O_3$
Formula: $C_6H_4OHCOOCH_3$
Properties: Colorless, yel., or red liq., wintergreen odor and taste; sol. in ether, glacial acetic acid, oxygenated and chlorinated solvs.; very sl. sol. in water; m.w. 152.14; dens. 1.180-1.185; m.p. -8.6 C; b.p. 222.2 C; flash pt. (CC) 101 C; ref. index 1.535-1.538 (20 C)
Toxicology: LD50 (oral, rat) 887 mg/kg, (subcut., rabbit) 4250 mg/kg; mod. toxic by ing., IP, IV, subcut. routes; human poison by ing.; lethal dose 30 cc in adults, 10 cc in children; human systemic effects; primary irritant; severe skin/eye irritant; experimental teratogen, reproductive effects; TSCA listed
Precaution: Combustible when exposed to heat or flame; reactive with oxidizing materials; protect from light, alkaline, iron salts
Hazardous Decomp. Prods.: Heated to decomp., emits acrid smoke and irritating fumes
HMIS: Health 2, Flammability 1, Reactivity 0
Uses: Denaturant, soothing agent in cosmetics; synthetic flavoring agent in foods, beverages, chewing gum, pharmaceuticals, dentifrices; odorant; fragrance in perfumes; disinfectant; antiseptic; topical analgesic; local anesthetic; UV absorber in sunburn lotions; dye carrier; light stabilizer; in food-pkg. adhesives; stabilizer in acrylic food pkg.
Features: Sweet minty flavor
Use Level: 0.5 mg/kg/day considered safe for ingestion
Regulatory: FDA 21CFR §175.105, 177.1010; 27CFR §21.65, 21.151; FEMA GRAS; Japan approved as flavoring; FDA approved for buccals, orals, topicals; NF, BP, EP compliance
Manuf./Distrib.: ABCR; ADA Int'l.; AMC Chems.; Aceto; Acme-Hardesty; Advanced BioTech; Aldrich; Alfa Chem; Allan; Allchem Ind.; Am. Int'l.; Andrea Aromatics; Asiamerica Int'l.; Avrachem AG; Barnet Prods.; Berje; CPI Chems.; Chart; Chemacon GmbH; Chr. Hansen Inc; Creative Fragrances; Dastech Int'l.; EMCO Chem. Distributors; Fabrichem; Fleurchem; Fluka; George Uhe; H.E. Daniel Ltd; Haarmann & Reimer GmbH; Indofine; Int'l. Sourcing; Integra; J.H. Calo; J.T. Baker; Kraft Chem.; Lavy's; Lluch Essence; Mallinckrodt Baker; MelChem; Nanjing Meheco; Nantong ChangChem; Nipa Labs; Norman, Fox; Penta Mfg.; Polarome Int'l.; Quest Int'l. Fragrances; R.W. Greeff; Rhodia/Fine Orgs.; Robeco; Ruger; Schweizerhall; Sigma; Spectrum Quality Prods.; Thomas Scientific; Universal Preserv-A-Chem; VWR Int'l.; Vopak USA; Xinchem; ZenPharm Int'l.
Trade Names Containing: Punctilious® SDA 38B 190 Proof Prod. Code 13601027; Punctilious® SDA 38F 190 Proof; Punctilious® SDA 38F-2 190 Proof Prod. Code 13643020

Methyl sebacate. *See* Dimethyl sebacate

Methyl soyate

CAS 67784-80-9
Synonyms: Methyl esters of soybean oil
Definition: Ester of methyl alcohol and soya acid
Uses: Intermediate in prod. of superamides; in metalworking lubricants; as specialized solvent; opacifier; visc. control agent; agric. solvent and adjuvant; additive in adhesives for food production and/or pkg.; food coatings
Trade Names: Agnique® 2237; Kemester® 226; Oleocal® ME-130; Soygold® 1000
Trade Names Containing: Soygold® 2000

Methyl stearate

CAS 112-61-8; 85586-21-6; EINECS/ELINCS 203-990-4; 287-824-6
Synonyms: Methyl ester stearic acid; Methyl octadecanoate; Methyl n-octadecanoate; Octadecanoic acid, methyl ester; Stearic acid, methyl ester
Definition: Ester of methyl alcohol and stearic acid
Empirical: $C_{19}H_{38}O_2$
Formula: $CH_3(CH_2)_{16}COOCH_3$
Properties: Wh. crystals; sol. in ether, alcohol; insol. in water; m.w. 298.57; m.p. 37.8 C; b.p. 234.5 C (30 mm); flash pt. 307 F
Toxicology: TDLo (subcut., mouse, 26 wk intermittent) 5200 mg/kg; ex-

perimental tumorigen; questionable carcinogen; TSCA listed
Precaution: Combustible exposed to heat or flame; can react with oxidizing materials
Hazardous Decomp. Prods.: Heated to decomp., emits acrid smoke and irritating fumes
Uses: Emollient in cosmetics, topical pharmaceuticals; intermediate for stearic acid detergents, emulsifiers, wetting agents, stabilizers, resins, lubricants, plasticizers, paints; defoamer and nutrient in fermentation; specialized solvent; solvent for pesticides and herbicides; defoamer in food-contact coatings and paper/paperboard; in resin-bonded filters for food contact; in food-contact textiles; in surf. lubricants for mfg. of food-contact metallic articles
Regulatory: FDA 21CFR §172.225, 176.200, 176.210, 177.2260, 177.2800, 178.3910; FDA approved for topicals
Manuf./Distrib.: A&E Connock; A.P. Chems. Ltd; Aldrich; Ashland; Charkit; ChemService; Chemol; Cognis/Chems. Group; Ferro/Polymer Addit.; Fluka; George Uhe; Indofine; Int'l. Paper; Jarchem Ind.; Mosselman NV; Norman, Fox; Penn Bio-Organics; Penta Mfg.; Rochem Int'l.; Sea-Land; Sigma; Spec. Ingreds. & Perfume; Uniqema Am.
Trade Names: Edenor® ME C18-90; Exceparl MS
Trade Names Containing: Degreez

Methylstyrene; m-/p-Methylstyrene. *See* Vinyltoluene monomer
Methyl succinate. *See* Dimethyl succinate
Methylsufone. *See* Dimethyl sulfone
Methyl sulfide. *See* Dimethyl sulfide
Methyl sulfinylmethane. *See* Dimethyl sulfoxide
Methylsulfonic acid. *See* Methanesulfonic acid
Methylsulfonyl methane. *See* Dimethyl sulfone
Methyl sulfoxide. *See* Dimethyl sulfoxide

Methyl sunflowerate

Uses: Solvent, adjuvant for pesticides and herbicides
Trade Names: Emery® 2230

Methyl tallowate

CAS 61788-61-2; EINECS/ELINCS 262-989-7
Synonyms: Fatty acids, tallow, methyl esters
Definition: Ester of methyl alcohol and tallow acid
Uses: Wetting agent, oiliness agent for metalworking and lubricating oils; lubricant, plasticizer for cosmetics, leather, rubber prods.; solvent for pesticides and herbicides; in food-pkg. adhesives
Regulatory: FDA 21CFR §175.105
Manuf./Distrib.: Anar; Mosselman NV; Sea-Land
Trade Names: Exceparl MT; Norfox® MTL

Methyl tertiary butyl ether. *See* Methyl t-butyl ether
Methyl tetradecanoate. *See* Methyl myristate
Methyl thioether; Methylthiomethane. *See* Dimethyl sulfide
Methyl toluene. *See* Xylene
3-Methyltoluene. *See* m-Xylene
4-Methyltoluene. *See* p-Xylene
m-Methyltoluene. *See* m-Xylene
o-Methyltoluene. *See* o-Xylene
p-Methyltoluene. *See* p-Xylene
Methyl tribromide. *See* Bromoform
Methyl trichloride. *See* Chloroform
Methyltrichloromethane. *See* 1,1,1-Trichloroethane
Methyl trifluoride. *See* Trifluoromethane
Methyltriglycol; Methyl trigol. *See* Triglycol monomethyl ether
Methyl trimethylacetate. *See* Methyl pivalate
Methyltrimethylene glycol. *See* Butylene glycol
2-Methylundecane. *See* Isododecane
MIAK. *See* Methyl isoamyl ketone
MIBC. *See* Methyl amyl alcohol
MIBK. *See* Methyl isobutyl ketone
MIC. *See* Methyl amyl alcohol
Milk acid. *See* Lactic acid
Mineral grease (petrolatum); Mineral jelly. *See* Petrolatum
Mineral naphtha. *See* Benzene

Mineral oil

CAS 8012-95-1; 8020-83-5 (wh.); 8042-47-5 (wh.); 39355-35-6; 79956-36-8; 83046-05-3; EINECS/ELINCS 232-384-2; 232-455-8 (wh.)
INS905a
Synonyms: Heavy mineral oil; Light mineral oil; Liquid paraffin; Liquid petrolatum; Mineral oil, white; Oil mist, mineral; Paraffin oil; Paraffinum liquidum; Petrolatum liquid; Petroleum oil; White mineral oil; White oil
Definition: Liq. mixture of hydrocarbons obtained from petroleum by intensive treatment with sulfuric acid and oleum, or by hydrogenation, or a combination; consists of predominantly C15-50 sat. hydrocarbons
Properties: Colorless transparent oily liq., odorless, tasteless; insol. in water, alcohol; sol. in benzene, chloroform, ether, petrol. ether, carbon disulfide, volatile oils; dens. 0.83-0.86 (light), 0.875-0.905 (heavy); flash pt. (OC) 444 F; surf. tens. < 35 dynes/cm
Toxicology: ACGIH TLV/TWA 5 mg/m^3; STEL 10 mg/m^3; LD50 (oral, mouse) 22 g/kg; skin and severe eye irritant; inh. of vapor or particulates may cause aspiration pneumonia; human carcinogen and teratogen by inhalation; tumorigen; highly purified food grades are of low toxicity; TSCA listed
Precaution: Combustible
Hazardous Decomp. Prods.: Heated to decomp., emits acrid smoke and fumes
NFPA: Health 0, Flammability 1, Reactivity 0
Uses: Protectant; lubricant; binder; carrier; extender; plasticizer, lubricant for plastics; pharmaceutical vehicle, solvent, lubricant, laxative; cathartic; emollient, antistat, solvent in cosmetics; agric.; paper; textiles; water treatment; floor treatment; foods (mold release agent, formulation aid, sealing agent, glazing agent, binder, defoamer, lubricant, fermentation aid, protective coating for fruits/vegetables); food-pkg. materials; in food-pkg. adhesives; in hot-melt strippable food coatings; surf. lubricant in food-contact coatings; in cellophane for food pkg.; in paper/paperboard in contact with aq./fatty foods; defoamer in food-contact coatings and paper/paperboard; in food-contact textiles; lubricant for incidental food-contact use
Features: Oleaginous
Regulatory: FDA 21CFR §172.842, 172.878, 173.340 (limitation 0.008% in wash water for sliced potatoes, 150 ppm in yeast), 175.105, 175.210, 175.230, 175.300, 176.170, 176.200, 176.210, 177.1200, 177.2260, 177.2600, 177.2800, 178.2010, 178.3570, 178.3620, 178.3740, 178.3910, 179.45, 573.680; ADI not specified (FAO/WHO); FDA approved for ophthalmics, orals, topicals; USP/NF compliance
Manuf./Distrib.: AMC Chems.; AMRESCO; AP Pharma; Adept Sol'ns.; Air-Scent Int'l.; Aldrich; Am. Ingreds.; Ashland; Avatar; BP Amoco; British Wax Refining; Brown; C.P. Hall; Castrol Ind.; Charkit; Chemisphere; ChevronPhillips; Coyne; EM Ind.; EM Science; EMCO Chem. Distributors; Eggar & Co.; ExxonMobil; Fisher Scientific; Fluka; GMI Prods.; Golden Bear Oil Spec.; Harcros; Integra; J.H. Calo; J.T. Baker; Jarchem Ind.; Jeen Int'l.; Magie Bros. Oil; Mallinckrodt Baker; Merck KGaA; Mosselman NV; Mutchler; Norman, Fox; O.C. Lugo; Parchem Trading; Penreco; Penta Mfg.; RTD Chems.; Ruger; San Yuan; Sasol N. Am.; Sea-Land; Sigma; Spec. Ingreds. & Perfume; Spectrum Quality Prods.; St. Lawrence; Stevenson Cooper; Strahl & Pitsch; Surco Prods.; Total Petrol.; U.S. Synthetics; Univar E&E Ltd; Universal Preserv-A-Chem; VWR Int'l.; Voigt Global Distrib.; Vopak USA
Trade Names: Drakeol® 7; Drakesol® 220; Drakesol® 260; Drakesol® 260 AT; Drakesol® 305; MagieSol 47; MagieSol 52; MagieSol 60
Trade Names Containing: Argobase 125T; Liquid Absorption Base T; Magie 4700 Oil; Magie 5200 Oil

Mineral oil, petroleum distillates, hydrotreated. *See* Naphthenic oil
Mineral oil, petroleum distillates, hydrotreated (mild) heavy naphthenic. *See* Petroleum distillates, heavy hydrotreated naphthenic
Mineral oil, petroleum distillates, hydrotreated (mild) light naphthenic. *See* Naphthenic oil
Mineral oil, petroleum extracts, heavy paraffinic distillate solvent. *See* Paraffin distillate
Mineral oil, white. *See* Mineral oil

Mineral spirits

CAS 64475-85-0; EINECS/ELINCS 232-453-7
UN 1271 (DOT)
Synonyms: Petroleum spirits; Turpentine substitutes; White spirits
Classification: Hydrocarbon
Definition: Mixture of hydrocarbons from petroleum with distillation range

of 149-213 C and flash pts. above 38 C; avail. in type I (reg., incl. Stoddard), II (high flash), III (odorless), IV (low dry pt.)
Properties: (I): Clear liq.; dens. 0.754-0.820 (15.6/15.6 C); i.b.p. 149 C; flash pt. 38 C min.; (II): dens. 0.768-0.820 (15.6/15.6 C); i.b.p. 177 C min.; flash pt. 60 C min.; (III): dens. 0.775 max. (15.6/15.6 C); i.b.p. 149 C min.; flash pt. 38 C min.; (IV): dens. 0.754-0.800 (15.6/15.6 C); i.b.p. 149 C; flash pt. 38 C min.
Toxicology: LD50 (oral, rat) 34,600 mg/kg, (skin, rabbit) 15,400 mg/kg; LC50 (inh., rat, 4 h) > 21,400 mg/m^3; LDLo (IP, rat) 8560 mg/kg; target organs: respiratory system, CNS, blood, kidney; TSCA listed
Environmental: VOC
NFPA: Health 0, Flammability 2, Reactivity 0
Uses: Paint and varnish thinner; solvent in cosmetics, dry cleaning, coatings, adhesives, agric., automotive, and chemical specialties, cleaning and polishing compds., pharmaceuticals, inks, textile coatings, asphalt, concrete curing; wood preservative for agric. prods. pkg.
Regulatory: FDA 21CFR §178.3800
Manuf./Distrib.: Aldrich; Alvin Chem.; Ashland; CITGO; ChemService; ChevronTexaco; Houghton Chem.; Hukill; Ruger; Sal Chem.; Sasol N. Am.; Schaefer Tech.; Spectrum Quality Prods.; Stanchem; Sunoco
Trade Names: Calumet Mineral Spirits < 1%; Shell OMS; Shell Mineral Spirits 7 EC; Shell Mineral Spirits 135
See also VM&P naphtha

Mineral spirits Type I. *See* Stoddard solvent
MIPA. *See* Isopropylamine
MIPK. *See* Methyl isopropyl ketone
Mirbane oil. *See* Nitrobenzene
Mixed cresols. *See* Cresylic acid
Mixed decanoic and octanoic acids, methyl esters. *See* Methyl caprylate/caprate

Mixed terpenes

Definition: Mixt. of hydrocarbons distilled from certain plants and trees such as pine and citrus; consists of terpenes, sesquiterpenes, diterpenes, polyterpenes
Uses: Solvent; constituent of industrial cleaners, deodorants; solvent in cosmetics
Trade Names: Solvenol® 2

Mixed xylenes. *See* Xylene
MMA. *See* Methylamine
MMAD. *See* Methyl adipate
MME. *See* Hydroquinone monomethyl ether
MMF. *See* N-Methylformamide
MMH. *See* Methylhydrazine
MNBA. *See* n-Butylamine
MNBK. *See* 2-Hexanone
3-MOB acetate. *See* Methoxybutyl acetate
Monacetin; 1-Monacetin; α-Monacetin; Monacetyl glycerine. *See* Glyceryl acetate
Monoamylamine. *See* Pentylamine
Monobromobenzene. *See* Bromobenzene
Monobromodichloromethane. *See* Bromodichloromethane
Monobromoethane. *See* Ethyl bromide
Monobromomethane. *See* Methyl bromide
Monobutylamine; Mono-n-butylamine. *See* n-Butylamine
Monobutyl glycol ether. *See* Butoxyethanol
Mono-t-butyl hydroquinone. *See* t-Butyl hydroquinone
Monochlorbenzene. *See* Chlorobenzene
Monochlorethane. *See* Ethyl chloride
Monochloroacetyl chloride. *See* Chloracetyl chloride
Monochlorobenzene. *See* Chlorobenzene
Monochlorodifluoromethane. *See* Chlorodifluoromethane
Monochloroethane. *See* Ethyl chloride
2-Monochloroethanol. *See* Ethylene chlorohydrin

α-Monochlorohydrine

Uses: Pharmaceutical intermediate; solvent for agric. chems.; accelerator for hardening epoxy resins
Trade Names: Epiol M-0

Monochloromethane. *See* Methyl chloride
Monochloromonobromomethane. *See* Bromochloromethane
Monochloropentafluoroethane. *See* Chloropentafluoroethane

Monochlorotoluene

CAS 25168-05-2
UN 2238 (DOT)
Synonyms: Benzene, chloromethyl-; Chloromethylbenzene; Chlorotoluene; ar-Chlorotoluene; Chlorotoluenes; MCT; Toluene, ar-chloro-
Empirical: C_7H_7Cl
Properties: Liq.; m.w. 126.59
Toxicology: LC50 (inh., unspecified mammal) 48 g/m^3; low toxicity by inh.; TSCA listed
Precaution: Flamm.
Hazardous Decomp. Prods.: Heated to decomp., emits toxic vapors of Cl^-
Uses: Solvent, extender, diluent for dye carriers in printing and textiles, fuel oils (to dissolve sludge), paint thinners and strippers, metal parts cleaners, adhesives; inert carrier; solvent in heavy-duty industrial and consumer cleaning prods.
Trade Names: Halso® 99

Monochlorotrifluoromethane. *See* Chlorotrifluoromethane
Mono- and diglycerides, acetic acid esters. *See* Acetylated mono- and diglycerides of fatty acids
Mono- and diglycerides, diacetyl tartaric acid esters. *See* Diacetyl tartaric acid esters of mono- and diglycerides

Mono- and diglycerides of fatty acids

CAS 67254-73-3; 67701-32-0; 67701-33-1; 68990-53-4
INS471; E471
Synonyms: Fatty acid mono/diglycerides; MDG
Definition: Mixt. of glycerol mono- and di-esters, with minor amts. of triesters, of fatty acids from edible oils
Properties: Yel. liqs. to ivory plastics to hard solids, bland odor and taste; sol. in alcohol, ethyl acetate, chloroform, other chlorinated hydrocarbons; insol. in water; acid no. 4 max.; nonionic
Hazardous Decomp. Prods.: Heated to decomp., emits acrid smoke and irritating fumes
Uses: Antistat for PE, flexible PVC, plastisols; softener for textile finishes; emulsifier, flavoring agent, formulation aid, lubricant, release agent, softener, solvent, stabilizer, surfactant, surface-finishing agent, texturizer, thickener, vehicle, dough strengthener in foods; emulsifier for pharmaceutical topical ointments, cosmetics; tableting aid
Regulatory: FDA 21CFR §136.110, 172.812, 172.863, 175.105, 184.1505, GRAS; USDA 9CFR §318.7, 381.147; Europe listed; UK approved; NF compliance
Manuf./Distrib.: AB R Lundberg; Adumim Food Ingreds.; Avatar; Browning; Degussa AG/Health & Nutrition; Int'l. Chem. Inc
Trade Names Containing: Hodag® S-35 FGK

Mono- and diglycerides of fatty acids esterified with mono- and diacetyltartaric acid. *See* Diacetyl tartaric acid esters of mono- and diglycerides
Monododecyl sodium sulfate. *See* Sodium lauryl sulfate
Monoethanolamine. *See* Ethanolamine
Monoethanolamine lactic acid amide. *See* Lactamide MEA
Monoethanolamine peanut acid amide. *See* Peanutamide MEA
Monoethylamine. *See* Ethylamine
Monoethylene glycol. *See* Ethylene glycol
Monoethylene glycol dimethyl ether. *See* Ethylene glycol dimethyl ether
Mono (2-ethylhexyl) sulfate sodium salt. *See* Sodium 2-ethylhexyl sulfate
Monofluorotrichloromethane. *See* Trichlorofluoromethane
Monoglyceryl oleate. *See* Glyceryl oleate
Monoglyme. *See* Ethylene glycol dimethyl ether
Monohydroxybenzene. *See* Phenol
Monohydroxymethane. *See* Methyl alcohol
Monoisopropylamine. *See* Isopropylamine
Monoisopropyl biphenyl. *See* 1,1′-Biphenyl (1-methylethyl)
Monoisopropyl ether of ethylene glycol. *See* Ethylene glycol isopropyl ether
1-Monolaurin. *See* Glyceryl laurate
Monomethylacetamide. *See* N-Methylacetamide
Monomethyl adipate. *See* Methyl adipate

Monomethylamine. *See* Methylamine
Monomethyl aminoethanol. *See* N-Methylethanolamine
Monomethyl aniline; N-Monomethylaniline. *See* N-Methylaniline
Monomethyl ester of glutaric acid. *See* Methyl glutarate
Monomethylethanolamine. *See* N-Methylethanolamine
Monomethyl ether of hydroquinone. *See* Hydroquinone monomethyl ether
Monomethylformamide; N-Monomethylformamide. *See* N-Methylformamide
Monomethyl glutarate. *See* Methyl glutarate
Monomethylhydrazine. *See* Methylhydrazine
Monomethylnaphthalene. *See* Methylnaphthalene
Monomethylol dimethyl hydantoin. *See* MDM hydantoin
Monomyristin. *See* Glyceryl myristate
Monooctanoin. *See* Glyceryl caprylate
Monooctyl dihexylphosphine oxide. *See* Dihexyloctylphosphine oxide
Monoolein; Monooleoylglycerol. *See* Glyceryl oleate
Monopalmitate sorbitan. *See* Sorbitan palmitate
Monophenol. *See* Phenol
Monopropylene glycol. *See* Propylene glycol
Monopropylene glycol methyl ether. *See* Methoxyisopropanol
Monopropyl ether of ethylene glycol. *See* Ethylene glycol propyl ether
Monoricinolein; α-Monoricinolein. *See* Glyceryl ricinoleate
Monostearin. *See* Glyceryl stearate
Monosteol. *See* Propylene glycol stearate
Mono-tertiarybutylhydroquinone. *See* t-Butyl hydroquinone
Monothioethyleneglycol. *See* 2-Mercaptoethanol

Morpholine

CAS 110-91-8; EINECS/ELINCS 203-815-1
UN 1760 (DOT); UN 2054 (DOT)
Synonyms: Diethyleneimide oxide; Diethylene imidoxide; Diethylene oximide; Diethylenimide oxide; 1-Oxa-4-azacyclohexane; Tetrahydro-1,4-isoxazine; Tetrahydro-p-isoxazine; Tetrahydro-1,4-oxazine; Tetrahydro-2H-1,4-oxazine; Tetrahydro-4H-1,4-oxazine; Tetrahydro-p-oxazine
Classification: Heterocyclic sec. amine
Empirical: C_4H_9NO
Formula: C_4H_8ONH
Properties: Colorless clear oily liq., amine-like odor; sol. in water, acetone, benzene, ether, ethyl acetate, chloroform, ethylene glycol, castor oil, alcohols; m.w. 87.14; sp.gr. 1.002 (20/20 C); vapor pressure 10 mm Hg (23 C); f.p. -4.9 C; b.p. 128.9 C; flash pt. (OC) 37.7 C; autoignition temp. 590 F; ref. index 1.4540 (20 C)
Toxicology: ACGIH TLV/TWA 20 ppm; STEL 30 ppm (skin); LD50 (oral, rat) 1050 mg/kg, (skin, rabbit) 500 mg/kg; mod. toxic by ing., inh., skin contact, IP; corrosive irritant to skin, eyes, mucous membranes; overexposure may cause visual disturbance, nose and respiratory irritation, coughing, liver and kidney damage; very high concs. may be fatal; TSCA listed
Precaution: DOT: Flamm. liq.; dangerous fire hazard exposed to flame, heat or oxidizers; reactive with oxidizers; explosive with nitromethane; incompat. with strong acids, acid anhydrides, nitrites, cellulose nitrate, isocyanates, org. peroxides, epoxides
Hazardous Decomp. Prods.: Thermal decomp. prods.: CO, CO_2, NO_x, ammonia
NFPA: Health 3, Flammability 3, Reactivity 0
Storage: Store in cool, dry, well-ventilated area, out of direct sunlight; hygroscopic
Uses: Organic intermediate for vulcanization accelerators, textile lubricants, softeners, sizing agents, antioxidants, surfactants, plasticizers, insecticides, dyes, bactericides, analgesics, anesthetics; solvent for resins, waxes, casein, dyes; neutralizer; emulsifier for cosmetics, rubless waxes, polishes; defoamer; optical brightener for detergents; lubricant, coolant for machine tools; solubilizer in printing inks; separating agent for volatile amines; protective food coatings; local anesthetic and antiseptic; analgesics; corrosion inhibitor in steam boiler systems, steel or tinplate food pkg.; in food-pkg. adhesives; defoamer in food-contact paper/paperboard
Regulatory: FDA 21CFR §172.235, 173.310 (10 ppm max. in steam), 175.105, 176.180, 176.210, 178.3300
Manuf./Distrib.: Air Prods.; Aldrich; Allchem Ind.; BASF AG; BASF; Coyne; Fluka; Houghton Chem.; Huntsman; Independent Chem.; J.T. Baker; Mallinckrodt Baker; Nippon Nyukazai; PMC Spec.; Penta Mfg.; Ruger; Sigma; Spectrum Quality Prods.; Thomas Scientific; Universal Preserv-A-Chem; Varsal Instruments; Vopak USA; Whyte Chems. Ltd

4-Morpholinecarboxaldehyde. *See* N-Formylmorpholine
Morpholine, 4-methyl-. *See* p-Methyl morpholine
Mossbunker oil. *See* Menhaden oil
Moth balls; Moth flakes. *See* Naphthalene
Motor benzol. *See* Benzene
Motor fuel; Motor spirit(s). *See* Gasoline
MP. *See* Methoxyisopropanol
MPA. *See* Propylene glycol methyl ether acetate
3-MPA. *See* Methoxypropylamine
MPD. *See* Methylpropanediol
3-MPD. *See* 3-Methoxy-1,2-propanediol
MPK. *See* Methyl propyl ketone
MSA. *See* Methanesulfonic acid
MSM. *See* Dimethyl sulfone
MTBE. *See* Methyl t-butyl ether
MTBHQ. *See* t-Butyl hydroquinone
Muriatic ether. *See* Ethyl chloride
Muthmann's liquid. *See* Tetrabromoethane
MX. *See* m-Xylene

Myrcene

CAS 123-35-3; EINECS/ELINCS 204-622-5
FEMA 2762
Synonyms: 3-Methylene-7-methyl-1,6-octadiene; 2-Methyl-6-methylene-2,7-octadiene; 7-Methyl-3-methylene-1,6-octadiene; 7-Methyl-3-methylene-octa-1,6-diene; β-Myrcene
Empirical: $C_{10}H_{16}$
Formula: $(CH_3)_2C{=}CHCH_2CH_2C(CH_2)CH{=}CH_2$
Properties: Colorless to lt. yel. oily liq., pleasant balsamic odor; sol. in alcohol, chloroform, ether, fixed oils, oxygenated and chlorinated solvs.; insol. in water; m.w. 136.23; dens. 0.791; b.p. 167 C; flash pt. 103 F; ref. index 1.4650
Toxicology: LD50 (oral, rat) > 5 g/kg, (skin, rabbit) > 5 g/kg; low toxicity by ing. and skin contact; primary skin and eye irritant; experimental reproductive effector; TSCA listed
Precaution: Combustible
Hazardous Decomp. Prods.: Heated to decomp., emits acrid smoke and irritating fumes
Uses: Intermediate for flavor and fragrance chems., e.g., citral, citronellol, citronellal, geraniol, nerol, linalool; fragrance in cosmetics, perfumes; synthetic flavoring agent in foods and pharmaceuticals
Features: Sweet flavor
Regulatory: FDA 21CFR §172.515; FEMA GRAS
Manuf./Distrib.: Aldrich; Axxence Aromatic GmbH; Doingcom; Fluka; Millennium/F&F; Millennium; Moore Ingreds.; Sigma
Trade Names Containing: Citrus Terpenes

β-Myrcene. *See* Myrcene
Myristic acid, isocetyl ester. *See* Isocetyl myristate
Myristic acid isopropyl ester. *See* Isopropyl myristate
Myristic alcohol. *See* Myristyl alcohol
Myristin. *See* Trimyristin

Myristyl alcohol

CAS 112-72-1; EINECS/ELINCS 204-000-3
Synonyms: Myristic alcohol; 1-Tetradecanol; N-Tetradecanol-1; Tetradecyl alcohol; N-Tetradecyl alcohol
Empirical: $C_{14}H_{30}O$
Formula: $CH_3(CH_2)_{12}CH_2OH$
Properties: Colorless to wh. waxy solid flakes, waxy odor; sol. in ether; sl. sol. in alcohol; insol. in water; m.w. 214.38; dens. 0.8355 (20/20 C); vapor pressure 0.01 mm Hg (20 C); m.p. 38 C; b.p. 167 C; acid no. 2 max.; iodine no. 1 max.; hyd. no. 250-267; flash pt. 285 F
Toxicology: TDLo (skin, mouse, 24 wk intermittent) 12 g/kg; nontoxic; human skin irritant; mod. eye irritant; questionable carcinogen; experimental tumorigen; TSCA listed
Precaution: Combustible; can react with oxidizers
Hazardous Decomp. Prods.: Heated to decomp., emits acrid smoke and irritating fumes

HMIS: Health 1, Flammability 1, Reactivity 0
Uses: Surfactant intermediate; organic synthesis; antifoam; perfume fixing agent for soaps and cosmetics; specialty cleaning preps.; emollient for cold creams; emulsion polymerization; slip agent for plastics; emollient, emulsion stabilizer, visc. control agent in cosmetics; raw material for antioxidants; synthetic flavoring agent in foods; pharmaceutical oleaginous vehicle; in food-pkg. adhesives; in food-contact coatings; defoamer in food-contact coatings and paper/paperboard; in surf. lubricants for mfg. of food-contact metallic articles; food-contact textiles
Features: Oleaginous
Regulatory: FDA 21CFR §172.864, 175.105, 175.300, 176.180, 176.200, 176.210, 177.1010, 177.1200, 177.1390, 177.2800, 178.3480, 178.3910; FDA approved for orals
Manuf./Distrib.: Albemarle; Aldrich; Ashland; Barnet Prods.; Cognis/Chems. Group; Condor; Costec; Fluka; Indofine; Jarchem Ind.; Kraft Chem.; Lipo; M. Michel; Penta Mfg.; R.W. Greeff; Robeco; Ruger; Sasol Germany; Sasol N. Am.; Schweizerhall; Sea-Land; Sigma; Spectrum Quality Prods.; Universal Preserv-A-Chem; Whyte Chems. Ltd
Trade Names: Alfol® 14; Cachalot® M-43
Trade Names Containing: Alfol® 1412

Myristyl myristate

CAS 3234-85-3; EINECS/ELINCS 221-787-9
Synonyms: Tetradecanoic acid, tetradecyl ester; Tetradecyl tetradecanoate
Definition: Ester of myristyl alcohol and myristic acid
Empirical: $C_{28}H_{56}O_2$
Formula: $CH_3(CH_2)_{12}COO(CH_2)_{13}CH_3$
Properties: Wh. waxy solid; bland char. odor; sol. in min. oil, IPM, oleyl alcohol; insol. in water, glycerin, propylene glycol; m.p. 36-39 C
Toxicology: LD50 (oral, rat) 8.6 g/kg; minimal eye irritation, mild skin irritation; TSCA listed
Uses: Emollient, emulsifier, thickener, lubricant, moisturizer, superfatting agent, opacifier, humectant, solubilizer, pigment dispersant for cosmetics, toiletries, creams, lotions, makeup, topical pharmaceuticals; binder for pressed powds.
Features: Melts at body temp.
Manuf./Distrib.: A&E Connock; Alzo; C.P. Hall; Cascade; Cognis/Care Chems.; Croda Inc; Global-Seven; ISP Van Dyk; Jarchem Ind.; Lanaetex Prods.; Mosselman NV; Phoenix; Rhodia HPCII; Sea-Land; Sigma; Thornley; Universal Preserv-A-Chem
Trade Names: Exceparl MY-M; Thomil 14

Myristyl propionate

CAS 6221-95-0; EINECS/ELINCS 226-300-9
Synonyms: 1-Tetradecanol, propanoate; Tetradecyl propionate
Definition: Ester of myristyl alcohol and propionic acid
Empirical: $C_{17}H_{34}O_2$
Formula: $CH_3CH_2COOCH_2(CH_2)_{12}CH_3$
Uses: Emollient, penetrant, and spreading agent in cosmetics, antiperspirants, pharmaceuticals, creams, lotions; perfume solvent in personal care prods.; humectant
Manuf./Distrib.: A&E Connock
Trade Names: Lonzest® 143-S

N

Naphtha

CAS 8030-30-6; 68920-06-9; 977126-64-9; EINECS/ELINCS 232-443-2
UN 1255 (DOT); UN 1256 (DOT); UN 1270 (DOT); UN 2553 (DOT)
Synonyms: Benzin; Benzine; Coal tar naphtha; Coal tar naphtha distillate; Heavy naphtha; Hydrotreated naphtha; Naphtha coal tar; Naphtha, heavy; Naphtha, hydrotreated; Naphtha, petroleum; Naphtha, solvent; Petroleum benzin; Petroleum-derived naphtha; Petroleum distillates (naphtha); Petroleum ether; Petroleum naphtha; Petroleum oil; Petroleum spirit; VM&P naphtha
Classification: Petroleum distillate; high benzene naphtha
Definition: Refined, partly refined, or unrefined petrol. prods. produced by distillation of nat. gas; consists of predominantly C5-6 hydrocarbons
Properties: Dark straw-colored to colorless liq.; sol. in benzene, toluene, xylene; dens. 0.862-0.892; b.p. 149-216 C; flash pt. (CC) 107 F
Toxicology: ACGIH TLV/TWA 300 ppm; LD50 (oral, rat) > 5 g/kg, (skin, rabbit) 3 g/kg; LCLo (inh., rat, 6 h) 1600 ppm; mildly toxic by inh.; human poison and systemic effects by IV route (dyspnea, respiratory effects); irritating to eyes, skin, respiratory tract; vapor inh. may cause nausea, dizziness, unconsciousness; experimental carcinogen; tumorigen; common air contaminant; TSCA listed
Precaution: Flamm. when exposed to heat or flame; dangerous fire risk; explosive limits in air 1-6%; sl. explosion hazard; can react with oxidizing materials; sl. explosion hazard
NFPA: Health 1, Flammability 3, Reactivity 0
Storage: Keep containers tightly closed
Uses: Source of gasoline, special naphthas, petroleum chems. esp. ethylene; solvent, thinner in paints, dry-cleaning fluids, rubber compounding, sealants, chem. absorption; blending with natural gas; aviation fuel blending component; color diluent, protective coating solvent in foods; in food-pkg. adhesives; defoamer in food-contact coatings and paper/paperboard
Regulatory: FDA 21CFR §73.1, 172.210, 172.250, 175.105, 176.180, 176.200, 176.210
Manuf./Distrib.: Ashland; Baychem; Chemcentral; ChevronPhillips; Crowley Chem.; ExxonMobil; Harcros; Houghton Chem.; J.T. Baker; Kerr-McGee; Monsanto; Norsk Hydro AS; R.E. Carroll; Sal Chem.; Shell; Spectrum Quality Prods.; Stanchem; Sunnyside; Vopak USA
Trade Names: Calumet 142 Flash Naphtha Rule 66; Calumet 180/210 Naphtha; Exxsol® D 40 Naphtha; Exxsol® D 60 Naphtha; Exxsol® D 3135 Naphtha; Exxsol® DSP 115/145 Naphtha; Exxsol® DSP 90/120 Naphtha Anti-Static; Exxsol® Hexane Naphtha; Hi-Sol® 15; Hi-Sol® 70; Kwik Dri; Lacolene™; 1520 Naphtha
See also Naphtha, light aromatic; VM&P naphtha

Naphtha, aromatic, high flash. *See* Naphtha, light aromatic
Naphtha coal tar; Naphtha, heavy. *See* Naphtha

Naphtha, heavy aliphatic

CAS 64742-96-7; EINECS/ELINCS 265-200-4
Synonyms: Aliphatic petroleum distillates; Heavy aliphatic petroleum solvent; Heavy aliphatic solvent; Heavy aliphatic solvent naphtha; Solvent naphtha, heavy aliphatic; Solvent naphtha (petroleum), heavy aliphatic; Solvent naphtha (petroleum), heavy aliphatic C11-16
Classification: Hydrocarbon
Definition: Complex combination of hydrocarbons obtained from distillation of crude oil or nat. gasoline; consists of predominantly C11-16 sat. hydrocarbons
Properties: B.p. 190-290 C
Toxicology: ACGIH TLV/TWA 100 ppm; eye/skin irritant
Uses: Solvent; solvent extraction diluent
Trade Names: Orfom® SX-1

Naphtha, heavy alkylate

CAS 64741-65-7; EINECS/ELINCS 265-067-2
UN 1268
Synonyms: Naphtha (petroleum), heavy alkylate; Petroleum naphtha, heavy alkylate
Classification: Low b.p. modified naphtha
Definition: Complex combination of hydrocarbons produced by distillation of reaction prods. of isobutane with C3-5 monoolefinic hydrocarbons; consists of predominantly C9-12 branched chain sat. hydrocarbons; mixt. of iso- and n-paraffins
Properties: Colorless liq.; insol. in water; dens. 0.75-0.79 g/cm^3; vapor pressure 0.1-0.2 kPa (20 C); m.p. -30 C; b.p. 150-220 C; flash pt. (CC) 44 C; autoignition temp. 355 C
Toxicology: ACGIH TLV/TWA 300 ppm; inh. may cause CNS effects, dizziness, headache, drowsiness, nausea, unconsciousness; ing. may cause cough, diarrhea, sore throat, vomiting; contact may cause dry skin, eye redness; swallowing liq. may cause aspiration into lungs, chem. pneumonitis
Environmental: Toxic to aquatic organisms; prevent from entering sewers
Precaution: Flamm.; explosive limits in air 0.6-8.0 vol.%; explosive vapor/air mixts. may form above 44 C; vapor is heavier than air and may travel along the ground; distant ignition possible; reacts with strong oxidants causing fire and explosion hazard
Storage: Store well closed away from strong oxidants
Uses: Solvent
Trade Names: Shellsol® OMS; Shellsol® T; Shellsol® TC; Shellsol® TD; Shellsol® TK
Trade Names Containing: AM OMS 75 Solvent

Naphtha, heavy aromatic

CAS 64742-94-5; EINECS/ELINCS 265-198-5
Synonyms: Hans solvent; Heavy aromatic bottoms; Heavy aromatic naphtha; Heavy aromatic naphtha solvent; Heavy aromatic petroleum solvent; Heavy aromatic solvent naphtha; Heavy aromatic solvent naphtha (petroleum); Petroleum distillates, intermediate catalytic cracked; (Polyethyl) benzenes; Solvent naphtha, heavy aromatic; Solvent naphtha (petroleum), heavy aromatic
Classification: Petroleum hydrocarbon
Definition: Complex combination of hydrocarbons obtained from distillation of aromatic streams; consists of predominantly C9-16 aromatic hydrocarbons
Properties: B.p. 165-290 C
Toxicology: ACGIH TLV 100 mg/m^3
Uses: Solvent
Trade Names: Aromatic 150; Aromatic 200; HAN® 906 Solvent; Naphthalene Depleted Aromatic 150 Fluid; Naphthalene Depleted Aromatic 200 Fluid; Petrinex® ASB; Shellsol® A200 Solvent; Solv G; Solvarex 10; Solvarex 90/180; Solvaro AFD; Solvesso™ 150 Anti-Static Sol-

vent; Solvesso™ 150 ND; Solvesso™ 200; Solvesso™ 200 ND; Ultra Low Naphthalene Aromatic 200; Ultra Low Naphthalene Solvesso 200 Solvent
Trade Names Containing: Shellsol® A150ND; Shellsol® A150 Solvent

Naphtha, heavy hydrotreated. *See* Naphtha, hydrotreated heavy

Naphtha, hydrodesulfurized heavy

CAS 64742-82-1; EINECS/ELINCS 265-185-4
Synonyms: High flash white spirit; Hydrodesulfurized heavy naphtha; Hydrodesulfurized naphtha, heavy; Naphtha (petroleum), hydrodesulfurized heavy; Petroleum naphtha, hydrodesulfurized heavy; White spirits, hydrodesulfurized; White spirit, White spirit, high flash
Classification: Petroleum distillate
Definition: Complex combination of hydrocarbons obtained from a catalytic hydrodesulfurization process; mixt. of aliphatic and aromatic hydrocarbons, predominantly C7-12
Properties: B.p. 90-230 C
Uses: Solvent
Trade Names: Kerosol; Kristalloel 60; Shellsol® 1495; Shellsol® 1626; Shellsol® 2046; Shellsol® H; Spirdane HT
Trade Names Containing: Kristalloel 21; Kristalloel 30; Solvent 2341

Naphtha, hydrotreated. *See* Naphtha

Naphtha, hydrotreated heavy

CAS 64742-48-9; EINECS/ELINCS 265-150-3
UN 3295
Synonyms: Heavy hydrotreated naphtha (petroleum); Hydrotreated heavy naphtha; Naphtha, heavy hydrotreated; Naphtha (petroleum), hydrotreated heavy
Definition: Complex combination of hydrocarbons obtained by treating a petrol. fraction with hydrogen in the presence of a catalyst; consists predominantly of C6-13 hydrocarbons; mixt. of C9-13 naphthenes, iso- and n-paraffins
Properties: Colorless liq.; insol. in water; dens. 0.76-0.79 g/cm^3; vapor pressure 0.1-0.3 kPa (20 C); m.p. 0 C; b.p. 155-217 C; flash pt. (CC) 40-62 C; autoignition temp. 255-270 C
Toxicology: Inh. may cause dizziness, headache, drowsiness, nausea, unconsciousness; contact may cause dry skin, eye redness; ing. may cause cough, diarrhea, sore throat, vomiting; if aspirated, may cause chem. pneumonitis
Environmental: Toxic to aquatic organisms; prevent entry into sewers
Precaution: Flamm.; explosive limits in air 0.7-6.0 vol.%; may form explosive vapor/air mixts. above 40 C; reacts with strong oxidants, causing fire/explosion hazard
Uses: Solvent; catalytic reformer feed
Features: Low b.p.
Trade Names: Actrel® 3338L; Actrel® 3360L; Isopar® G Fluid; Isopar® H Fluid; Isopar® K Fluid; Isopar® K Naphtha; Isopar® L Fluid; Shellsol® D40; Solvent 2217; Solvent 2324
Trade Names Containing: Dynasolve 220; Dynasolve 225; Dynasolve 230; Solvent 2241; Solvent 2245; Solvent 2310; Solvent 2336; Solvent 2337; Solvent 2339; Solvent 2341; Solvent 2376

Naphtha, hydrotreated light

CAS 64742-49-0; EINECS/ELINCS 265-151-9
UN 1208
Synonyms: Hydrotreated light naphtha (petroleum); Hydrotreated light straight run petroluem; Light hydrotreated naphtha; Naphtha (petroleum), hydrogen-treated, light; Naphtha (petroleum), hydrotreated light; Naphtha (petroleum), light, hydrotreated; SBP; Special boiling point spirit
Classification: High benzene naphtha
Definition: Complex combination of hydrocarbons obtained by treating a petrol. fraction with hydrogen in the presence of a catalyst; consists predominantly of C4-11 hydrocarbons
Properties: B.p. -20 to 190 C
Uses: Solvent; extraction solvent; cleaning agent; diluent; chem. feedstock; fuel
Trade Names: Exxsol® D 75/100; Exxsol® D 75/100 Naphtha; Exxsol® D 115/145; Exxsol® D 115/145 Anti-Static Solvent; Exxsol® D 115/145 Naphtha; Exxsol® Heptane; Exxsol® Hexane; Exxsol® Hexane Anti-Static; Exxsol® Hexane Naphtha Anti-Static; Exxsol® Isohexane Naphtha; Hydrosol 75/95 N; Hydrosol 100/130 N; Hydrosol Essence 60/95; Hydrosol Essence 80/110; Hydrosol Essence 100/140; Hydrosol Essence A; Hydrosol Essence C; Hydrosol Essence E<5; Hydrosol Essence F<5; Hydrosol Isohexane; Iosol RS Solvent; SBP 100/140; SBP 140/165; Shellsol® D25; X-95 Solvent
Trade Names Containing: SBP 40/65 LNH; SBP 60/95 LNH; SBP 80/95 LNH; SBP 80/110 LNH; Shellsol® BF

Naphthalane. *See* Decahydronaphthalene

Naphthalene

CAS 91-20-3; EINECS/ELINCS 202-049-5
UN 1334 (DOT); 2304 (DOT)
Synonyms: Camphor tar; Moth balls; Moth flakes; Naphthalene, crude or refined; Naphthalene, molten; Naphthaline; Naphthene; Tar camphor; White tar
Empirical: $C_{10}H_8$
Properties: Wh. cryst. volatile flakes; strong coal tar odor; sol. in benzene, abs. alcohol, ether, CCl_4, CS_2, fixed/volatile oils, oxygenated and aromatic solvs.; insol. in water; m.w. 128.18; dens. 1.145 (20/4 C); vapor pressure 1 mm (52.6 C); m.p. 76-80 C; b.p. 217.96 C; sublimes @ R.T.; flash pt. 80 C
Toxicology: ACGIH TLV/TWA 10 ppm; STEL 15 ppm; LD50 (oral, rat) 490 mg/kg, (IP, mouse) 150 mg/kg, (IV, mouse) 100 mg/kg, (subcut., mouse) 969 mg/kg; poison by ing., IV, IP; toxic by inh.; human poison by ing.; primary eye and skin irritant; may cause nausea, headache, hematuria, fever, anemia, liver damage, convulsions, coma; questionable carcinogen; tumorigen; experimental teratogen, reproductive effector; mutagen; target organs: eyes, blood, kidneys, lungs; TSCA listed
Precaution: Flamm.; reactive with oxidizers; explosive and violent reactions possible; explosive as vapor or dust exposed to heat or flame
Hazardous Decomp. Prods.: Heated to decomp., emits acrid smoke and irritating fumes
HMIS: Health 2, Flammability 2, Reactivity 0
Uses: Intermediate (phthalic anhydride, naphthol, chlorinated naphthalenes, dyes, surfactants, dispersants, etc.); moth repellent; fungicide; insecticide; smokeless powder; lubricant; synthetic resin; tanning; preservative; textile aux.; emulsion breaker; scintillation counter; antiseptic; solvent
Regulatory: SARA reportable
Manuf./Distrib.: AMRESCO; Aldrich; Alvin Chem.; Cognis/Chems. Group; Crowley Tar Prods.; Fluka; Honeywell Perf. Polymers; Indofine; J.T. Baker; Koppers Ind./Carbon Materials; Penta Mfg.; Recochem; Robeco; Ruger; Sigma; Sithean; Spectrum Quality Prods.; Stan Chem Int'l. Ltd; Sunoco; Synasia Fine Chem.
Trade Names Containing: Kenplast® A-450; Methylnaphthalene Fraction; Shellsol® A150 Solvent; Shellsol® A200 Solvent

Naphthalene, bis (1-methylethyl)-. *See* Diisopropyl naphthalene
Naphthalene, 1-chloro-. *See* 1-Chloronaphthalene
Naphthalene, crude or refined. *See* Naphthalene
Naphthalene, decahydro-, cis-. *See* cis-Decalin
Naphthalene diisopropyl-. *See* Diisopropyl naphthalene
Naphthalene ethylene. *See* Acenaphthene
Naphthalene, molten. *See* Naphthalene
Naphthalene oil. *See* Creosote oil
Naphthalene-1,2,3,4-tetrahydride. *See* Tetrahydronaphthalene

Naphtha, light aliphatic

CAS 64742-89-8; EINECS/ELINCS 265-192-2
Synonyms: Aliphatic petroleum distillate; Lactol spirits; Light aliphatic naphtha; Light aliphatic solvent naphtha (petroleum); Naphtha (petroleum), light aliphatic; Solvent naphtha (petroleum), light aliphatic
Definition: Complex combination of hydrocarbons obtained from distillation of crude oil or nat. gasoline; consists of predominantly C5-10 sat. hydrocarbons
Properties: B.p. 35-160 C
Uses: Solvent
Trade Names: Shellsol® W HT; Special Lactolite; VM&P Naphtha HT
Trade Names Containing: Propsolv I-1 190 Proof; Propsolv I-1 Anhydrous; Propsolv III-1 190 Proof; Propsolv III-1 Anhydrous; Propsolv III-2 190 Proof; Propsolv III-2 Anhydrous; Punctilious® CDA 19-3 190

Proof; Punctilious® CDA 19-3 Anhydrous A; Punctilious® SDA 2B-2 190 Proof; Punctilious® SDA 2B-2 Anhydrous; Punctilious® SDA 2B-4 190 Proof; Punctilious® SDA 2B-4 Anhydrous; Rubber Solvent; Rubber Solvent 332; Shellsol® B HT; Shellsol® B HT60; S&L Naphtha; Solvent 2249; Solvent 2298; Solvent 2299; Solvent 2300; Solvent 2307; Solvent 2313; Solvent 2314; Solvent 2316; Solvent 2318; Solvent 2321; Solvent 2322; Solvent 2326; Solvent 2329; Solvent 2331; Solvent 2335; Solvent 2336; Solvent 2337; Solvent 2338; Solvent 2340; Solvent 2341; Solvent 2376; Tolu-Sol 30; VM&P Naphtha EC

Naphtha, light alkylate

CAS 64741-66-8; EINECS/ELINCS 265-068-8

Synonyms: Alkylation naphtha, light, C7-10; Light alkylate naphtha; Light alkylation naphtha; Naphtha (petroleum), light alkylate

Definition: Complex combination of hydrocarbons produced by distillation of reaction prods. of isobutane with C3-5 monoolefinic hydrocarbons; consists of predominantly C7-10 branched chain sat. hydrocarbons

Properties: B.p. 90-160 C

Uses: Solvent; diluent; chem. feedstock; fuel

Trade Names: Isopar® C; Isopar® E Fluid

Naphtha, light aromatic

CAS 64742-95-6; EINECS/ELINCS 265-199-0

Synonyms: Aromatic naphtha type I; Light aromatic naphtha; Light aromatic petroleum naphtha; Light aromatic solvent naphtha; Light aromatic solvent naphtha (petroleum); Naphtha, aromatic, high flash; Solvent naphtha (petroleum), light aromatic

Classification: Petroleum distillate

Definition: Complex combination of hydrocarbons obtained from distillation of aromatic streams; consists of predominantly C8-10 aromatic hydrocarbons

Properties: B.p. 135-210 C

Toxicology: PEL 5 mg/m^3; LD50 (oral, rat) 4.7 g/kg

Precaution: Flamm.

Uses: Solvent for agrochems., paints

Trade Names: Aromasol® H; Aromatic 100; Hi-Sol® 10; Shellsol® A100 Solvent; Solvarex 9; Solvarex 90/160; Solvesso™ 100; Solvesso™ 100 Anti-Static Solvent; Super Hi-Flash Naphtha

Trade Names Containing: ExxonMobil T 500-100; Shellsol® 3EC; Shellsol® 7EC; Shellsol® 15; Solvent 2216; Solvent 30 823

Naphthaline. *See* Naphthalene

Naphtha, medium aliphatic

CAS 64742-88-7; EINECS/ELINCS 265-191-7

UN 1268; 1993

Synonyms: Aliphatic petroleum solvent (naphtha); Medium aliphatic solvent naphtha; Medium aliphatic solvent naphtha (petroleum); Naphtha (petroleum), medium aliphatic; Solvent naphtha (peroleum), medium heavy, aliphatic hydrocarbons; Solvent naphtha (petroleum), medium aliphatic

Classification: Petroleum distillate solvent

Definition: Mixt. of n- and iso-paraffins, aromatic hydrocarbons and naphthenes, obtained from distillation of crude oil or nat. gasoline; consists of predominantly C9-12 sat. hydrocarbons

Properties: Colorless liq.; char. odor; insol. in water; dens. 0.75-0.82 g/cm^3; vapor pressure 0.2-0.6 kPa (20 C); m.p. -48 to -26 C; b.p. 138-178 C; flash pt. (CC) 38-60 C; autoignition temp. 229-260 C

Toxicology: ACGIH TLV 100 mg/m^3; LD50 (skin, rabbit) 5 ml/kg; inh. may cause dizziness, headache, drowsiness, nausea, cough, unconsciousness; contact may cause dry skin, eye redness; ing. may cause cough, diarrhea, sore throat, vomiting; swallowing liq. may cause aspiration into lungs, chem. pneumonitis; chronic exposure may cause skin defatting, CNS and liver effects

Precaution: Flamm.; explosive limits in air 0.6-6.5 vol.%; explosive vapor/air mixts. may form above 38 C; vapors heavier than air, may travel along ground; distant ignition possible; reacts with strong oxidants; attacks some forms of plastics and rubber

Uses: Solvent, vehicle, thinning agent for paints/coatings, waxes, printing inks, adhesives; solvent in liq. photocopier toners, dry cleaning; degreaser for engine parts in machine and auto repair shops

Manuf./Distrib.: Fisher Scientific

Trade Names: Envirolite® Solvent; Shellsol® 9; Shellsol® 16; Shellsol® D38; Shellsol® D43; Shellsol® D60

Trade Names Containing: Shellsol® 3EC; Shellsol® 7EC; Shellsol® 15; Solvent 30 823

Naphthane. *See* Decahydronaphthalene

Naphthanthracene. *See* 1,2-Benzanthracene

Naphtha, petroleum. *See* VM&P naphtha; Naphtha

Naphtha (petroleum), heavy alkylate. *See* Naphtha, heavy alkylate

Naphtha (petroleum), hydrodesulfurized heavy. *See* Naphtha, hydrodesulfurized heavy

Naphtha (petroleum), hydrogen-treated, light. *See* Naphtha, hydrotreated light

Naphtha (petroleum), hydrotreated heavy. *See* Naphtha, hydrotreated heavy

Naphtha (petroleum), hydrotreated light. *See* Naphtha, hydrotreated light

Naphtha (petroleum), light aliphatic. *See* Naphtha, light aliphatic

Naphtha (petroleum), light alkylate. *See* Naphtha, light alkylate

Naphtha (petroleum), light, hydrotreated. *See* Naphtha, hydrotreated light

Naphtha (petroleum), medium aliphatic. *See* Naphtha, medium aliphatic

Naphtha (petroleum), solvent-refined light. *See* Naphtha, solvent-refined light

Naphtha, solvent. *See* Stoddard solvent; Naphtha

Naphtha, solvent-refined light

CAS 64741-84-0; EINECS/ELINCS 265-086-6

Synonyms: Naphtha (petroleum), solvent-refined light; Petroleum distillates, textile spirits; Solvent-refined light naphtha (petroleum); Textile spirits

Classification: Petroleum hydrocarbon

Definition: Complex combination of hydrocarbons obtained as the raffinate from a solvent extraction process; consists of predominantly C5-11 aliphatic hydrocarbons

Properties: B.p. 35-190 C

Toxicology: OSHA PEL 400 ppm

Uses: Solvent

Trade Names: Textile Spirits

Trade Names Containing: Solvent 2238; Solvent 2249; Solvent 2254; Solvent 2255; Solvent 2307; Solvent 2322; Solvent 2329; Solvent 2331; Solvent 2332; Solvent 2335; Solvent 2336; Solvent 2338; Solvent 2442

Naphtha, VM&P. *See* VM&P naphtha

Naphthene. *See* Naphthalene

Naphthenic acid

CAS 1338-24-5; EINECS/ELINCS 215-662-8

UN 9137 (DOT)

Synonyms: Carboxylic acids, naphthenic; Naphthenic acids

Definition: Any of a group of sat. higher fatty acids derived from gas-oil fraction of petroleum by extraction with caustic acid sol'n. and subsequent acidification

Properties: Yel. visc. liq., mild odor (purified), dk. red visc. liq., strong offensive odor (crude); sol. in most org. solvs.; insol. in water; m.w. 235-280; dens. 0.94-0.96; m.p. 31 C; b.p. 233 C; flash pt. (OC) 149 C

Toxicology: LD50 (oral, rat) 3000 mg/kg, (IP, rat) 640 mg/kg; mod. toxic by ing., IP routes; probable sl. to mod. skin irritant and eye irritant; inh. may cause CNS depression (headache, dizziness, nausea); ing. may cause appetite loss, lethargy, diarrhea, weakness; crude may cause mouth, throat, GI irritation; if aspirated into lungs, may cause severe lung irritation or damage which may be life-threatening; TSCA listed

Precaution: Incompat. with strong oxidizers (increases fire and explosion risk), strong bases (violent reactions possible)

Hazardous Decomp. Prods.: Heated to decomp., emits acrid smoke and irritating fumes

Storage: Store in cool, dry, well-ventilated area out of direct sunlight; avoid release of vapors; crude may require corrosion-resist. materials, lighting

Uses: Paint driers; fungicides; metal catalysts; corrosion inhibitors; lubricants; fracturing fluids; cellulose preservatives; solvent; detergents; rubber reclaiming agent; emulsifier for metalworking fluids; chemical and surfactant intermediate; intermediate for synthesis of metal naphthenates and succinic acid

Manuf./Distrib.: Crowley Chem.; Esprit; Fabrichem; Fluka; Indofine; OM Group; Orange Chems. Ltd; Sea-Land; Sparkford Chems. Ltd

Naphthenic acids. *See* Naphthenic acid
Naphthenic distillates. *See* Naphthenic oil

Naphthenic oil

CAS 64742-53-6; 67254-74-4; EINECS/ELINCS 265-156-6
Synonyms: Distillates (petroleum), hydrogen-treated, light naphthene-; Distillates (petroleum), hydrotreated (mild) light naphthenic; Hydrotreated light naphthenic distillate; Hydrotreated (mild) light naphthenic distillate; Hydrotreated (mild) light naphthenic distillates (petroleum); Mineral oil, petroleum distillates, hydrotreated; Mineral oil, petroleum distillates, hydrotreated (mild) light naphthenic; Naphthenic distillates; Petroleum distillates, hydrogen-treated, light naphthene-
Classification: Petroleum hydrocarbon
Definition: Complex combination of hydrocarbons obtained by treating a petroleum fraction with hydrogen in the presence of a catalyst; consists predominantly of C15-30 hydrocarbons; contains relatively few normal paraffins
Toxicology: LD (oral, rat) > 5 g/kg, (skin, rabbit) > 2 g/kg; low toxicity by ing. and skin contact; primary irritant; severe skin irritant by Draize test; confirmed carcinogen; tumorigen; experimental neoplastigen; TSCA listed
Hazardous Decomp. Prods.: Heated to decomp., emits acrid smoke and irritating fumes
Uses: Extender, process oil, color/heat stabilizer for rubber compounding; plasticizer, extender for resins, PVC, textiles, caulks, food-contact applics.; softener; diluent; dispersion carrier; base for elec. transformer oils, refrigeration oils, hydraulic fluids, compressor oils, transmission fluids, two-cycle- and diesel engine oils, gear oils, chain oils, etc.
Manuf./Distrib.: Avatar; ChemService
Trade Names Containing: Kenflex® N1

Naphthol. *See* Naphthol spirits

Naphthol spirits

CAS 1321-67-1 (generic)
Synonyms: Naphthol
Empirical: $C_{10}H_8O$
Properties: Liq.; monoclinic; sol. in cold and hot water, chloroform; very sol. in alcohol, benzene; m.w. 144.18; dens. 1.217 (4 C); m.p. 122-123 C; b.p. 285-286 C
Toxicology: LDLo (subcut., rabbit) 1000 mg/kg, (subcut., dog) 333 mg/kg; poison by subcutaneous route
Precaution: Combustible
Hazardous Decomp. Prods.: Heated to decomp., emits acrid smoke and irritating fumes
Uses: Solvent
Features: Med.-boiling
Manuf./Distrib.: Aceto; Austin; Brook-Chem; Fanwood; Spectrum Quality Prods.; Stanchem; Wego Chem. & Min.

1-Naphthyl chloride; Naphthylene, 1-chloro-. *See* 1-Chloronaphthalene
Naphthylene ethylene. *See* Acenaphthene
Natural gasoline. *See* Gasoline
NBN. *See* Butyronitrile
NDEA. *See* N-Nitrosodiethylamine
NDMA. *See* N-Nitrosodimethylamine
NE. *See* Nitroethane

Neatsfoot oil

CAS 8002-64-0; 85049-53-2; EINECS/ELINCS 232-314-0
Synonyms: Bubulum
Definition: Fixed oil produced by rendering cattle feet
Properties: Fixed pale yel. oily liq.; peculiar odor; sol. in alcohol, ether, chloroform, kerosene; insol. in water; dens. 0.916; solid. pt. -10 to 0 C; iodine no. 70; sapon. no. 194-199; flash pt. 243 C
Toxicology: TSCA listed
Precaution: Combustible; subject to spontaneous heating
Uses: Fatliquor and softener for leathers; lubricant; oiling wool; emollient, solvent in cosmetics
Manuf./Distrib.: Anar; Mosselman NV; Ruger; Sea-Land; Voigt Global Distrib.

NEM. *See* Ethyl morpholine

Neoamyl alcohol. *See* 2,2-Dimethylpropanol

Neodecanoic acid

CAS 26896-20-8; EINECS/ELINCS 248-093-9
Definition: A mixt. of isomeric 10-carbon saturated monocarboxylic acids
Empirical: $C_{10}H_{20}O_2$
Formula: $C_6H_{13}C(CH_3)_2COOH$
Properties: Colorless to yel. clear liq.; sol. 0.02 g/10 ml water; m.w. 172.30; dens. 0.915 (20/20 C); dens. 7.61 lb/gal (20 C); b.p. 250-257 C; acid no. 322; flash pt. (TCC) 221 F
Toxicology: LD50 (oral, mammal) 3400 mg/kg; mod. toxic by ing.; may cause mod. to marked eye and skin irritation; TSCA listed
Hazardous Decomp. Prods.: Heated to decomp., emits acrid smoke and irritating fumes
Uses: Intermediate for prod. of derivs. used as paint driers, plasticizers, lubricants, and in coatings, cosmetics, agric., pharmaceuticals, etc.; coupling solvent for metalworking fluids; metal extraction agent
Regulatory: DOT nonregulated
Manuf./Distrib.: Biddle Sawyer; ExxonMobil

Neoheptane

CAS 590-35-2
UN 1206
Synonyms: 2,2-Dimethylpentane; Dimethyl-2,2-pentane
Classification: Sat. aliphatic hydrocarbon
Empirical: C_7H_{16}
Formula: $CH_3CH_2CH_2C(CH_3)_3$
Properties: Colorless liq.; gasoline-like odor; sol. in most org. solvs.; insol. in water; m.w. 100.2; sp.gr. 0.674 (20 C); m.p. -124 C; b.p. 79 C; flash pt. (CC) 15 C
Toxicology: ACGIH TLV/TWA 400 ppm; STEL 500 ppm; may cause mild skin irritation; conc. vapor may cause sl. eye irritation; conc. vapor may cause sl. nose/throat irritation, lightheadedness, dizziness, nausea, difficulty breathing; very high concs. may cause unconsciousness, death; ing. may cause nausea, vomiting, swelling of abdomen, headache, depression; relatively low oral toxicity unless aspirated into lungs
Precaution: Flamm.; incompat. with strong oxidizing agents (increases fire and explosion hazard)
Hazardous Decomp. Prods.: None
Storage: Store in cool, dry, well-ventilated area out of direct sunlight
Uses: Ingred. in gasoline, aviation fuels; solvent; org. synthesis
Manuf./Distrib.: Aldrich; Fluka

Neohexane

CAS 75-83-2; EINECS/ELINCS 200-906-8
UN 1208
Synonyms: 2,2-Dimethylbutane
Classification: Sat. aliphatic hydrocarbon
Empirical: C_6H_{14}
Properties: Colorless volatile liq.; faint petrol. odor; misc. with most org. solvs.; insol. in water; m.w. 86.18; dens. 0.65 (20 C); m.p. -100 C; b.p. 50 C; flash pt. (CC) 48 C; ref. index 1.3659
Toxicology: ACGIH TLV/TWA 500 ppm, STEL-TLV 1000 ppm (hexane isomers); low oral toxicity; aspiration can cause severe lung irritation or pulmonary edema; probably a mucous membrane irritant with narcotic props.; very high concs. can cause unconsciousness and death; may cause eye irritation; TSCA listed
Precaution: Highly flamm.; dangerous fire and explosion risk; explosion limits 1.2-7%; vigorous reaction with oxidizers
Hazardous Decomp. Prods.: CO_2, CO; heated to decomp., emits acrid smoke and irritating fumes
NFPA: Health 1, Flammability 3, Reactivity 0
Uses: Solvent; component of high-octane motor and aviation fuels; intermediate for agricultural chemicals
Manuf./Distrib.: ChevronPhillips

Neononane

CAS 3522-94-9
UN 1920
Synonyms: 2,2,5-Trimethylhexane
Classification: Sat. aliphatic hydrocarbon
Empirical: C_9H_{20}

Formula: $CH_3C(CH_3)_2CH_2CH_2CH(CH_3)CH_3$
Properties: Colorless liq.; gasoline-like odor; sol. in most org. solvs.; insol. in water; m.w. 128.26; sp.gr. 0.7174 (20 C); m.p. -106 C; b.p. 124 C; flash pt. 12.8 C
Toxicology: ACGIH TLV/TWA 200 ppm; may cause mild skin irritation; conc. vapor may cause eye irritation; probably a CNS depressant; conc. vapor may cause nose/throat irritation, headache, drowsiness, tremors, difficulty breathing; very high concs. may cause unconsciousness, death; ing. may cause nausea, vomiting, swelling of abdomen, headache, depression; relatively low oral toxicity, unless aspirated into lungs
Precaution: Flamm.; incompat. with strong oxidizers (increases fire and explosion risk)
Hazardous Decomp. Prods.: CO_2, CO
NFPA: Health 2, Flammability 3, Reactivity 0
Storage: Store in cool, dry, well-ventilated area out of direct sunlight
Uses: Ingred. in gasoline and petrol. solvents such as VM&P naphtha and Stoddard solvent
Manuf./Distrib.: Aldrich

Neopentanoic acid, tridecyl ester. *See* Tridecyl neopentanoate
Neopentanol; Neopentyl alcohol. *See* 2,2-Dimethylpropanol
2-Neopentyl-4-4-dimethylpentene-1. *See* 2-(2,2-Dimethylpropyl)-4,4-dimethylpent-1-ene

Neopentyl glycol dicaprate

CAS 27841-06-1; EINECS/ELINCS 248-688-3
Synonyms: Decanoic acid, 2,2-dimethyl-1,3-propanediol diester
Definition: Diester of neopentyl glycol and decanoic acid
Empirical: $C_{25}H_{48}O_4$
Formula: $CH_3(CH_2)_8COOCH_2C(CH_3)_2CH_2OCO(CH_2)_8CH_3$
Uses: Emollient; solvent for make-up removers, topical pharmaceuticals
Trade Names: Estemol N-01; Schercemol NGDC

Neopentyl glycol diheptanoate

CAS 68855-18-5; EINECS/ELINCS 272-469-1
Synonyms: Heptanoic acid, ester with 2,2-dimethyl-1,3-propanediol
Definition: Diester of neopentyl glycol and heptanoic acid
Empirical: $C_{19}H_{36}O_4$
Formula: $CH_3(CH_2)_5COOCH_2C(CH_3)_2CH_2OCO(CH_2)_5CH_3$
Properties: Yel.-amber clear liq.; essentially odorless; insol. in water; sp.gr. 0.943; dens. 7.86 lb/gal; m.p. < 25 C; b.p. > 200 C; flash pt. (COC) 170 C
Toxicology: Relatively low toxicity by single dose oral exposure; swallowing larger amts. may cause injury; nonirritating to skin; may cause minimal eye irritation; TSCA listed
Precaution: Containers may buildup pressure if exposed to heat (fire), may rupture
Hazardous Decomp. Prods.: None
Storage: Keep container closed when not in use
Uses: Emollient, solvent, coupling agent, lubricant for skin care creams/lotions; emollient for antiperspirants
Regulatory: DOT nonregulated; SARA §302 nonreportable, §311/312 nonhazardous, §313 nonreportable; Calif. Prop. 65 nonreportable; Canada DSL listed
Trade Names: LexFeel® 7

Neopentyl glycol dioctanoate

CAS 28510-23-8; 31335-74-7; EINECS/ELINCS 249-060-1; 250-575-9
Synonyms: 2,2-Dimethyl-1,3-propanediyl dioctanoate; 2,2-Dimethyl-1,3-propanediyl 2-ethylhexanoate; 2-Ethylhexanoic acid, 2,2-dimethyl-1,3-propanediyl ester; Octanoic acid, 2,2-dimethyl-1,3-propanediol diester
Definition: Diester of neopentyl glycol and 2-ethylhexanoic acid
Empirical: $C_{21}H_{40}O_4$
Toxicology: TSCA listed
Uses: Emollient, solvent, pigment dispersant for cosmetics; solvent for makeup removers
Features: Low freeze pt.
Trade Names: Cosmol 525; Schercemol NGDO

Niobe oil. *See* Methyl benzoate
Nitration benzene. *See* Benzene
Nitrile C_4. *See* Butyronitrile

2,2′,2′′-Nitrilotriethanol; 2,2′,2′′-Nitrilotris (ethanol). *See* Triethanolamine

Nitrobenzene

CAS 98-95-3
UN 1662 (DOT)
Synonyms: Essence of mirbane; Essence of myrbane; Mirbane oil; Nitrobenzene, liq.; Nitrobenzol; Oil of mirbane
Empirical: $C_6H_5NO_2$
Properties: Greenish-yel. cryst. or yel. oily liq.; volatile almond oil odor; sol. in alcohol, benzene, ether, oils, most org. solvs.; very sl. sol. in water; m.w. 123.12; dens. 1.19867; vapor pressure 1 mm (44.4 C); m.p. 5.7 C; b.p. 210.85 C; flash pt. 87.7 C
Toxicology: ACGIH TLV/TWA 1 ppm (skin); LD50 (oral, rat) 489 mg/kg, (skin rat) 2100 mg/kg; mod. toxic by ing., skin contact, IP route; human poison; human systemic effects by ing.; eye/skin irritant; inh. and skin absorp. result in severe intoxication; reacts with blood causing oxygen depravation, cyanosis, dizziness, headache, nausea, possible unconsciousness; CNS or cardiovascular damage, death is possible; mutagenic data; target organs: blood, nerves, spleen, liver; TSCA listed
Precaution: Combustible; oxidant; mod. explosion hazard exposed to heat or flame; explosive or violent reactions possible; incompat. with potassium hydroxide
Hazardous Decomp. Prods.: Heated to decomp., emits toxic fumes of NO_x
Uses: Mfg. of aniline, benzidine, quinoline, azobenzene, etc.; solvent for cellulose ethers; extraction solvent for lubricating oil refining; process solvent for Friedel-Crafts reactions; modifying esterification of cellulose acetate; in metal and shoe polishes
Manuf./Distrib.: Aarti Ind. Ltd; Aldrich; BCH Brühl; ChemFirst; Fluka; Sigma; Supelco

Nitrobenzene, liq.; Nitrobenzol. *See* Nitrobenzene
Nitrocarbol. *See* Nitromethane

Nitroethane

CAS 79-24-3; EINECS/ELINCS 201-188-9
UN 2842 (DOT)
Synonyms: Ethane, nitro-; NE
Classification: Aliphatic organic compd.; nitroparaffin
Empirical: $C_2H_5NO_2$
Formula: $CH_3CH_2NO_2$
Properties: Colorless oily clear liq.; mild fruity agreeable odor; sol. in water, alcohols, esters, ketones, ethers, aromatic hydrocarbons, chloroform, acid, alkali; misc. with oxygenated solvs.; m.w. 75.07; dens. 1.048 (20/4 C); f.p. -50 C; m.p. -90 C; b.p. 112-116 C; decomp. pt. 335-382 C; flash pt. (TCC) 30 C; autoignition temp. 414 C; ref. index 1.391 (20 C)
Toxicology: ACGIH TLV/TWA 100 ppm; LD50 (oral, rat) 1100 mg/kg, (IP, rat) 310 mg/kg; poison by IP route; mod. toxic by ing.; mildly toxic by inh.; irritating to eyes, mucous membranes; may cause headache, nausea, coughing, sneezing, CNS depression, respiratory obstruction, cyanosis; causes injury to liver and kidneys; mild skin irritant; cancer suspect agent; mutagen; TSCA listed
Precaution: DOT: Flamm. liq.; explodes when heated; incompat. with strong oxidizers, strong acids, reducing agents, alkali metals, inorg. bases, amines, heavy metal oxides, hydrocarbons, activated carbon
Hazardous Decomp. Prods.: Heated to decomp., emits toxic fumes of NO_x
NFPA: Health 1, Flammability 3, Reactivity 3
Storage: Hygroscopic; store in well-ventilated area; bond and ground metal containers in storage area; install pressure and vacuum-relief venting in drums
Uses: Intermediate for Friedel-Crafts synthesis; intermediate for org. synthesis for explosives, medicine, pesticides, chem. fibers, paints, ore dressing; reaction medium; extraction solvent for rosin and some inorg. materials (e.g., aluminum chloride); solvent for coatings, adhesives, inks, cellulose esters, vinyl, alkyd, other resins and waxes; stabilizer for chlorinated solvents; fuel additive; propellant; raw material for pharmaceutical synthesis (α-methyldopa, a hypertensive drug, phenylpropanolamine, in bronchial decongestants and appetite suppressants)
Manuf./Distrib.: ANGUS; Aldrich; Atomergic Chemetals; Fluka; Spectrum Quality Prods.
Trade Names: NE™; NE™ 101

Nitroisopropane. *See* 2-Nitropropane

Nitromethane

CAS 75-52-5; EINECS/ELINCS 200-876-6
UN 1261 (DOT)
Synonyms: Nitrocarbol; NM
Classification: Aliphatic nitro compd.; nitroparaffin
Empirical: CH_3NO_2
Properties: Colorless oily liq., mild fruity to disagreeable odor; sol. in water, alcohol, ether, aromatic hydrocarbons, esters, ketones; misc. with oxygenated and aromatic solvs.; m.w. 61.04; sp.gr. 1.139 (20/20 C); m.p. -29 C; b.p. 100-103 C; flash pt. (TCC) 35-35.6 C; ref. index 1.382 (20 C)
Toxicology: ACGIH TLV/TWA 20 ppm; LD50 (oral, rat) 940 mg/kg, (IP, mouse) 110 mg/kg; LDLo (IV, rabbit) 750 mg/kg; poison by ing., inh., IP routes; mod. toxic by IV route; vapor inh. may cause mild respiratory tract irritation; CNS depression occurs in animals; may cause human anorexia, nausea, vomiting, diarrhea, kidney/liver damage; mild skin irritant; TSCA listed
Precaution: Flamm.; dangerously reactive; dangerous fire/explosion risk; may detonate under high temps. and pressures; incompat. with acids, alkali metals, inorg. bases, amines, strong oxidizers, hydrocarbons, metal oxides, aluminum chloride, etc.
Hazardous Decomp. Prods.: Heated to decomp., emits toxic fumes of NO_x
NFPA: Health 1, Flammability 3, Reactivity 4
Storage: Store in cool, dry, well-ventilated area out of direct sunlight; store separately from incompat. materials; limit quantity in storage
Uses: Stabilizer for chlorinated solvents; chemical intermediate; solvent for cellulosic compds., polymers, waxes, coatings, printing inks, adhesives, pharmaceuticals; process solvent; in chem. synthesis; rocket fuel; gasoline additive; corrosion inhibitor; raw material for pharmaceutical synthesis
Manuf./Distrib.: ANGUS; Aastrid Int'l.; Aldrich; Alfa Chem; Allchem Ind.; Fluka; Integra; J.T. Baker; PCAS; Sigma; Sithean; Spectrum Quality Prods.; VWR Int'l.
Trade Names: NM™
Trade Names Containing: Genesolv® 2004

Nitropropane. *See* 1-Nitropropane; 2-Nitropropane

1-Nitropropane

CAS 108-03-2; EINECS/ELINCS 203-544-9
UN 2608
Synonyms: Nitropropane; n-Nitropropane; 1-NP
Classification: Aliphatic nitro compd.; nitroparaffin
Empirical: $C_3H_7NO_2$
Formula: $CH_3CH_2CH_2NO_2$
Properties: Colorless oily liq.; mild fruity odor; sol. in aromatic hydrocarbons, alcohols, esters, ketones, ethers, lower carboxylic acids; misc. with most org. solvs.; very sl. sol. in water; m.w. 89.09; dens. 1.003 (20/20 C); m.p. -108 C; b.p. 129-133 C; flash pt. (TCC) 35.6 C; autoignition temp. 420 C; ref. index 1.4018 (20 C)
Toxicology: ACGIH TLV/TWA 25 ppm; LD50 (oral, rat) 455 mg/kg, (IP, mouse) 250 mg/kg; poison by IP route and ing.; mildly toxic by inh.; irritating to mucous membranes; inh. may cause eye/nose/throat irritation, headaches, nausea, methemoglobinemia; mild skin irritant; human eye irritant; ing. may produce appetite loss, nausea, vomiting, diarrhea, kidney/liver damage, methemoglobinemia; mutagen; TSCA listed
Precaution: Flamm. liq.; reacts violently; incompat. with strong oxidizers, metal oxides, strong acids, amines, inorg. bases, hydrocarbons, activated carbon; sensitive to heat
Hazardous Decomp. Prods.: Heated to decomp., emits toxic fumes of NO_x
NFPA: Health 1, Flammability 3, Reactivity 2
Storage: Store in cool, well-ventilated area out of direct sunlight
Uses: Solvent for inks, protective coatings, cellulose acetate, vinyl resins, lacquers, synthetic rubbers, fats, oils, dyes; extraction solvent; chemical intermediate for agrochems., pharmaceuticals; rocket propellant; gasoline and fuel additive
Manuf./Distrib.: ANGUS; Aldrich; Fluka
Trade Names: NiPar S-10™

2-Nitropropane

CAS 79-46-9; EINECS/ELINCS 201-209-1
UN 2608 (DOT)
Synonyms: Dimethylnitromethane; Isonitropropane; Nitroisopropane; Nitropropane; β-Nitropropane; s-Nitropropane; 2-NP; Propane, 2-nitro-
Classification: Aliphatic nitro compd.; nitroparaffin
Empirical: $C_3H_7NO_2$
Formula: $CH_3CH(NO_2)CH_3$
Properties: Colorless to very lt. yel. clear oily liq.; mild fruity odor; sol. in most aromatic hydrocarbons, alcohols, esters, ketones, chloroform, ethers, lower carboxylic acids; sl. sol. in water; m.w. 89.09; sp.gr. 0.992 (20/20 C); vapor pressure 12.9 mm Hg (20 C); m.p. -93 C; b.p. 119-122 C; flash pt. (TOC) 24 C; ref. index 1.394 (20 C)
Toxicology: ACGIH TLV/TWA 10 ppm; LD50 (oral, rat) 725 mg/kg, (IP, rat) 75 mg/kg; very toxic; poison by ing., inh., and IP routes (nausea, vomiting, diarrhea, anorexia, severe headaches, cyanosis); eye/skin/mucous membrane/upper respiratory tract irritant; lg. doses may cause respiratory tract/liver damage and CNS effects; acute inh. may cause GI bleeding; may cause methemoglobinemia, depressed appetite, pulmonary edema, etc.; suspected carcinogen; mutagenic; reproductive effects
Precaution: Flamm. when exposed to heat, open flame, oxidizers; LEL 1.6%; UEL 11%; may explode on heating; incompat. with strong oxidizers, strong acids, amines, strong bases, activated carbon, hydrocarbons, chlorosulfonic acid, lead/copper/alloys
Hazardous Decomp. Prods.: Heated to decomp., emits toxic fumes of CO, CO_2, NO_x; emits toxic fumes under fire conditions
NFPA: Health 2, Flammability 3, Reactivity 1
Storage: Heat-sensitive; store in cool, dry, well-ventilated area, out of direct sunlight, away from heat/ignition sources, corrosives, incompat. materials
Uses: Solvent for coatings, resins, varnishes, lacquers, printing inks, polymers, cellulose acetate, vinyl resins, synthetic rubbers, fats, oils, dyes, org. synthesis; chemical intermediate in prod. of pharmaceuticals, dyes, insecticides, textile chems.; rocket propellant; gasoline additive; solvent for fractionating edible fats and oils; in food-pkg. adhesives
Regulatory: FDA 21CFR §175.105; SARA §313 reportable
Manuf./Distrib.: ANGUS; Aldrich; Ashland; Fluka

β-Nitropropane. *See* 2-Nitropropane
n-Nitropropane. *See* 1-Nitropropane
s-Nitropropane. *See* 2-Nitropropane
N-Nitroso-N,N-diethyamine; Nitrosodiethylamine. *See* N-Nitrosodiethylamine

N-Nitrosodiethylamine

CAS 55-18-5
UN 2810
Synonyms: DANA; DEN; DENA; Diethylamine, N-nitroso-; Diethylnitrosamide; Diethylnitrosamine; N,N-Diethylnitrosamine; Diethylnitrosoamine; Ethanamine, N-ethyl-N-nitroso-; Ethylamine, N-nitrosodi-; N-Ethyl-N-nitrosoethanamine; NDEA; N-Nitroso-N,N-diethyamine; Nitrosodiethylamine; Nitrous diethylamide
Empirical: $C_4H_{10}N_2O$
Formula: $(CH_3CH_2)_2NNO$
Properties: Yel. clear liq.; sol. in water, DMSO, 95% ethanol, acetone, alcohol, ether, lipids, org. solvs.; m.w. 102.14; sp.gr. 0.9422 (20/4 C); vapor pressure 5.7 mm Hg (25 C); b.p. 177 C; flash pt. 63 C; ref. index 1.4386 (20 C)
Toxicology: LD50 (oral, rat) 280 mg/kg, (IP, rat) 216 mg/kg, (IV, rat) 280 mg/kg, (subcut., rat) 195 mg/kg; poison by ing., IV, IP, subcut., and possibly other routes; harmful by inh., ing., skin contact; may cause irritation, somnolence, GI ulceration, diarrhea, fatty liver degeneration, bladder inflammation, etc.; experimental carcinogen, neoplastigen, tumorigen, teratogen, reproductive effector; human mutagenic data; a transplacental carcinogen; TSCA listed
Precaution: Combustible; can be oxidized by strong oxidants; incompat. with reducing agents; can be hydrolyzed by hydrogen bromide in acetic acid; sensitive to light, esp. UV light; keep away from food
Hazardous Decomp. Prods.: Heated to decomp., emits toxic fumes of NO_x

Storage: Store under ambient temps.; protect from light
Uses: Gasoline and lubricant additive; antioxidant; stabilizer; fiber industry solvent; copolymer softener; starting material for synthesis of 1,1-diethylhydrazine; to increase dielec. constant in condensers
Regulatory: SARA §313 reportable
Manuf./Distrib.: AccuStandard; Fluka; Pfaltz & Bauer; Sigma; Supelco; TCI Am.

Nitrosodimethylamine. *See* N-Nitrosodimethylamine

N-Nitrosodimethylamine

CAS 62-75-9; EINECS/ELINCS 200-549-8
UN 2810
Synonyms: Dimethylamine, N-nitroso-; Dimethylnitrosamine; N,N-Dimethylnitrosamine; Dimethylnitrosoamine; DMN; DMNA; Methamine, N-methyl-N-nitroso-; Methanamine, N-methyl-N-nitroso-; N-Methyl-N-nitrosomethanamine; NDMA; Nitrosodimethylamine; N-Nitroso-N,N-dimethylamine; Nitrous dimethylamide
Classification: Amine
Empirical: $C_2H_6N_2O$
Formula: $(CH_3)_2NNO$
Properties: Yel. clear oily liq.; char. odor; sol. in water, DMSO, 95% ethanol, acetone, ether, lipids, all common org. solvs.; m.w. 74.08; sp.gr. 1.0048 (20/4 C); low visc.; vapor pressure 5 mm Hg (20 C); b.p. 151-153 C; flash pt. 61 C; ref. index 1.4368 (20 C); can be steam-distilled quantitatively
Toxicology: LD50 (oral, rat) 58 mg/kg, (IP, rat) 26,500 μg/kg, (subcut., rat) 45 mg/kg; LC50 (inh., rat, 4 h) 78 ppm; highly toxic; human poison by ing.; experimental poison by ing., inh., IP, subcut., IV, possibly other routes; may cause nausea, headache, vomiting, abdominal cramps, diarrhea, fever, jaundice, kidney damage, etc.; has caused fatal liver disease in humans; suspected human carcinogen; experimental carcinogen, neoplastigen, tumorigen, teratogen, reproductive effector; human mutagenic data; TSCA listed
Environmental: VOC; ThOD 0.65
Precaution: Combustible; incompat. with strong oxidizers, strong bases, hydrogen bromide in acetic acid; can be reduced by reducing agents; photochemically sensitive; sensitive to exposure to light, esp. UV light
Hazardous Decomp. Prods.: Heated to decomp., emits toxic fumes of CO, CO_2, NO_x; emits toxic fumes under fire conditions
Storage: Keep container tightly closed under inert atmosphere; store refrigerated; protect from light
Uses: In rocket fuels; industrial solvent; solvent for fibers, plastics; rubber accelerator; plasticizer for acrylonitrile polymers, rubber; gasoline and lubricant additive; antioxidant; pesticide; nematicide; inhibition of nitrification in soil; mfg. of 1,1-dimethylhydrazine; in active metal anode-electrolyte systems (high-energy batteries); prep. of thiocarbonyl fluoride polymers; additive in condensers to inc. dielec. const.; lubricant and softener in copolymers
Regulatory: SARA §302a extremely hazardous, §313 reportable; CERCLA hazardous substance; HAP
Manuf./Distrib.: AccuStandard; Acros Org.; Aldrich; Isotec; Sigma; Spectrum Quality Prods.; Supelco; TCI Am.

N-Nitroso-N,N-dimethylamine. *See* N-Nitrosodimethylamine
Nitrous diethylamide. *See* N-Nitrosodiethylamine
Nitrous dimethylamide. *See* N-Nitrosodimethylamine
NM. *See* Nitromethane
NMCHA. *See* N-Methylcyclohexylamine
NMF. *See* N-Methylformamide
NMP. *See* N-Methyl-2-pyrrolidone

γ-Nonalactone

CAS 104-61-0; EINECS/ELINCS 203-219-1
FEMA 2781
Synonyms: Aldehyde C18; γ-N-Amylbutyrolactone; Coconut aldehyde; 4-Hydroxynonanoic acid lactone; 4-Hydroxynonanoic acid, γ-lactone; 1,4-Nonalolide; Nonan-4-olide; Prunolide
Classification: Heterocyclic compd.
Empirical: $C_9H_{16}O_2$
Properties: Colorless to sl. yel. liq., coconut odor; sol. in alcohol, fixed oils, propylene glycol; insol. in water; m.w. 156.25; dens. 0.958-0.966; b.p. 243 C; flash pt. > 212 F; ref. index 1.446-1.450
Toxicology: LD50 (oral, rat) 6600 mg/kg; mod. toxic by ingestion; skin irritant; mutagenic data; TSCA listed
Precaution: Combustible liq.
Hazardous Decomp. Prods.: Heated to decomp., emits acrid smoke and irritating fumes
Uses: Solvent, fragrance in cosmetics; synthetic flavoring agent in foods and pharmaceuticals
Features: Coconut-like flavor
Regulatory: FDA 21CFR §172.515; FEMA GRAS; Japan approved as flavoring
Manuf./Distrib.: Advanced BioTech; Advanced Synthesis Tech.; Aldrich; Alfa Aesar; Elan; J.H. Calo; Penta Mfg.; Polarome Int'l.

1,4-Nonalolide. *See* γ-Nonalactone
Nonane. *See* n-Nonane

n-Nonane

CAS 111-84-2; EINECS/ELINCS 203-913-4
UN 1920 (DOT)
Synonyms: Nonane
Classification: Sat. aliphatic hydrocarbon
Empirical: C_9H_{20}
Formula: $CH_3(CH_2)_7CH_3$
Properties: Colorless liq.; gasoline-like odor; misc. with most org. solvs.; sol. in abs. alcohol, ether; insol. in water; m.w. 128.26; sp.gr. 0.714; vapor pressure 10 mm Hg (38 C); m.p. -53.5 C; b.p. 151 C; flash pt. 31 C
Toxicology: ACGIH TLV/TWA 200 ppm; LD50 (IV, mouse) 218 mg/kg; toxic; poison by IV route; skin irritant; conc. vapor probably irritating to eyes; inh. of conc. vapor may cause nose/throat irritation, headache, drowsiness, nausea, tremors, difficulty breathing; narcotic in high concs.; very high concs. may cause unconsciousness, death; ing. may cause nausea, vomiting, swelling of abdomen, headache, depression; relatively low oral toxicity unless aspirated into lungs; TSCA listed
Precaution: Flamm.; very dangerous fire hazard exposed to heat or flame; explosive as vapor exposed to heat or flame; incompat. with strong oxidizing agents (increases fire and explosion hazard)
Hazardous Decomp. Prods.: Heated to decomp., emits acrid smoke and irritating fumes
NFPA: Health 0, Flammability 3, Reactivity 0
Storage: Store in cool, dry, well-ventilated area out of direct sunlight
Uses: Solvent; org. synthesis; component of gasoline and petrol. solvents such as VM&P naphtha and Stod. solvent; mfg. of paraffin prods.; paper processing; rubber industry; synthesis of biodeg. detergents; jet fuel research; distillation chaser; fragrance in cosmetics
Manuf./Distrib.: Aldrich; Fluka; Romil Ltd; Sigma

Nonanoic acid

CAS 112-05-0; EINECS/ELINCS 203-931-2
FEMA 2784
Synonyms: Carboxylic acid C9; Hexacid C9; n-Nonanoic acid; Nonoic acid; n-Nonoic acid; Nonylic acid; n-Nonylic acid; 1-Octanecarboxylic acid; Pelargic acid; Pelargonic acid
Classification: Aliphatic carboxylic acid
Empirical: $C_9H_{18}O_2$
Formula: $CH_3(CH_2)_7COOH$
Properties: Colorless oily liq., char. fatty odor; cryst. when cooled; sol. in alcohol, chloroform, ether, acetone, benzene, cyclohexane, n-hexane, oxygenated solvs.; pract. insol. in water; m.w. 158.24; dens. 0.907 (20/4 C); m.p. 10-12 C; b.p. 252-253 C (756 mm); acid no. 351; flash pt. 129 C; ref. index 1.433 (20 C); weakly acidic
Toxicology: LD50 (oral, mouse) 15 g/kg, (IV, mouse) 224 ± 4.6 mg/kg; poison by IV route; mod. toxic by ing.; strong irritant to skin and eyes; inh. of mists may cause mild to mod. irritation to nose/throat; ing. of lg. doses may cause mild irritation, sore throat, abdominal pain, nausea, vomiting; TSCA listed
Precaution: Combustible; corrosive
Hazardous Decomp. Prods.: CO, CO_2; heated to decomp., emits acrid smoke and irritating fumes
Uses: Organic synthesis; alkyd resin comonomer; intermediate for paints; mfg. of lacquers, plastics, hydrotropic salts, fragrances, pharmaceuticals, lubricants, vinyl plasticizers, gasoline additives; flotation agent; esters for turbojet lubricants; solvent in cosmetics; synthetic flavoring

agent in foods and pharmaceuticals; butter marker; fragrance
Regulatory: FDA 21CFR §172.515, 173.315, 178.1010; FEMA GRAS
Manuf./Distrib.: Aldrich; Atofina; Celanese; Cognis/Chems. Group; ExxonMobil; Fluka; Grau Aromatics; Lluch Essence; Parchem Trading; Penta Mfg.; Prodasynth; Sea-Land; Sigma; Universal Preserv-A-Chem

n-Nonanoic acid. *See* Nonanoic acid
Nonanoic acid, 1-methyl-1,2-ethanediyl ester. *See* Propylene glycol dipelargonate
Nonanol. *See* 3,5,5-Trimethylhexan-1-ol
Nonan-4-olide. *See* γ-Nonalactone
4-Nonanol, 2,6,8-trimethyl. *See* 2,6,8-Trimethylnonyl-4-alcohol
Nonanone, 5-, 2,8-dimethyl-. *See* Diisoamyl ketone
4-Nonanone, 2,6,8-trimethyl-. *See* Isobutyl heptyl ketone
3,6,9,12,15,18,21,24,27-Nonaoxanonatriacontan-1-ol. *See* Laureth-9
3,6,9,12,15,18,21,24,27-Nonaxanonacosane-1,29-diol. *See* PEG-10
Noncarbinol. *See* Decyl alcohol
Nonoic acid; n-Nonoic acid. *See* Nonanoic acid

Nonoxynol

CAS 9016-45-9 (generic); 37205-87-1 (generic); 68412-54-4 (generic)
Synonyms: Glycols, polyethylene, mono (nonylphenyl) ether; Nonylphenol ethoxylates; Nonylphenol, polyoxyethylene ether; Nonyl phenyl polyethylene glycol; Nonyl phenyl polyethylene glycol ether; Polyethylene glycol nonyl phenyl ether; Polyoxyethylene nonylphenol
Classification: Ethoxylated alkyl phenol
Formula: $(C_2H_4O)_x \cdot C_{15}H_{24}O$
Properties: Pale yel. to wh.; liq. (4-7 EO), paste (13-15 EO), or solid (> 20 EO); oil-sol. (4-7 EO), rest water-sol.; HLB 8.9-18.2; nonionic
Toxicology: LD50 (oral, rat) 1310 mg/kg, (skin, rabbit) 2 ml/kg; primary irritant; mild skin and severe eye irritant; may cause cardiac changes, nausea, vomiting, reproductive system tumors; TSCA listed
Storage: Keep under argon
Uses: Crude oil demulsifier ingred.; defoamer; emulsifier for oils, solvents, waxes, pesticides; emulsifier, protective colloid in emulsion polymerization; spermicides; surfactant for dishwashing detergents, cosmetics; surfactant, wetting agent, dispersant for household/industrial cleaners; textile scouring agent; solubilizer; emulsifier in mfg. of food-contact articles
Trade Names: Berol® 02
Trade Names Containing: Soygold® 2000

Nonoxynol-4

CAS 7311-27-5; 9016-45-9 (generic); 26027-38-3 (generic); 37205-87-1 (generic); 27176-97-2; 68412-54-4 (generic); EINECS/ELINCS 230-770-5
Synonyms: PEG-4 nonyl phenyl ether; PEG 200 nonyl phenyl ether; POE (4) nonyl phenyl ether
Classification: Ethoxylated alkyl phenol
Empirical: $C_{23}H_{40}O_5$
Formula: $C_9H_{19}C_6H_4(OCH_2CH_2)_nOH$, avg. n = 4
Properties: Yel. to almost colorless liq.; sol. in oil; HLB 8.9; nonionic
Toxicology: Moderately toxic by ingestion, skin contact; severe eye and mild skin irritant in humans; TSCA listed
Hazardous Decomp. Prods.: Heated to decomp., emits acrid smoke and fumes
Uses: Surfactant, stabilizer, defoamer, emulsifier, dispersant, detergent, wetting agent, solubilizer, coupling agent for cosmetics, textiles, metalworking, household, industrial, agric., paper, paint, fat liquoring, cutting oils, petrol. oils; intermediate in synthesis of anionic surfactants; pharmaceutic aids; plasticizer, antistat for plastics; solvent cleaner; emulsifier in mfg. of food-contact articles; defoamer for food-contact paper/paperboard; food-pkg. adhesives; in paper/paperboard in contact with dry foods
Regulatory: FDA 21CFR §175.105, 176.180, 176.210, 178.3400; FDA approved for ophthalmics, topicals, vaginals
Manuf./Distrib.: Aldrich; Fluka; Sigma
Trade Names: T-Det® N-4

Nonoxynol-5

CAS 9016-45-9 (generic); 20636-48-0; 26027-38-3 (generic); 26264-02-8; 37205-87-1 (generic); EINECS/ELINCS 247-555-7
Synonyms: 14-(Nonylphenoxy)-3,6,9,12-tetraoxatetradecan-1-ol; PEG-5 nonyl phenyl ether; POE (5) nonylphenol; POE (5) nonyl phenyl ether
Classification: Ethoxylated alkyl phenol
Empirical: $C_{25}H_{44}O_6$
Formula: $C_9H_{19}C_6H_4(OCH_2CH_2)_nOH$, avg. n = 5
Properties: Yel. to almost colorless liq.; sol. in oil; HLB 10.5; nonionic
Toxicology: Moderately toxic by ingestion, skin contact; severe eye and mild skin irritant in humans; TSCA listed
Hazardous Decomp. Prods.: Heated to decomp., emits acrid smoke and fumes
Uses: Surfactant, defoamer, detergent, emulsifier, coemulsifier, wetting agent, penetrant, solubilizer for industrial, household cleaners, agric., textiles, emulsion polymerization, paints, metalworking/cleaning, petrol., cosmetics, latex paints, cutting oils; intermediate in synthesis of anionic surfactants; intermediate for sulfation in mfg. lubricants and antistats; solvent cleaner; antistat; lubricant; latex stabilizer; detergent for petrol. oils; pharmaceutical surfactant; in food-pkg. adhesives; in food-contact paper/paperboard; defoamer in food-contact paper/paperboard; emulsifier in mfg. of food-contact articles
Regulatory: FDA 21CFR §175.105, 176.180, 176.210, 178.3400
Manuf./Distrib.: Aldrich; Fluka; Sigma

Nonoxynol-6

CAS 9016-45-9 (generic); 26027-38-3 (generic); 27177-01-1; 27177-05-5 (generic); 37205-87-1 (generic); 68412-54-4 (generic)
Synonyms: PEG-6 nonyl phenyl ether; PEG 300 nonyl phenyl ether; POE (6) nonyl phenyl ether
Classification: Ethoxylated alkyl phenol
Empirical: $C_{27}H_{48}O_7$
Formula: $C_9H_{19}C_6H_4(OCH_2CH_2)_nOH$, avg. n = 6
Properties: Yel. to almost colorless liq.; HLB 10.9; nonionic
Toxicology: Moderately toxic by ingestion, skin contact; severe eye and mild skin irritant in humans; TSCA listed
Hazardous Decomp. Prods.: Heated to decomp., emits acrid smoke and fumes
Uses: Surfactant, stabilizer, defoamer, emulsifier, dispersant, detergent, wetting agent, solubilizer, coupling agent for cosmetics, textiles, metalworking, household, industrial, agric., paper, paints, foods, pharmaceuticals; intermediate in synthesis of anionic surfactants; solvent cleaner; plasticizer, antistat for plastics; emulsifier in mfg. of food-contact articles; food-pkg. adhesives; in paper/paperboard in contact with dry foods; defoamer in food-contact paper/paperboard
Regulatory: FDA 21CFR §175.105, 176.180, 176.210, 178.3400
Manuf./Distrib.: Aldrich; Fluka; Sigma
Trade Names: T-Det® N-6

Nonoxynol-9 phosphate

CAS 51609-41-7 (generic); 51811-79-1 (generic); 66197-78-2; 68412-53-3 (generic); EINECS/ELINCS 266-231-6
Synonyms: 26-(Nonylphenoxy)-3,6,9,12,15,18,21,24-octaoxahexacosan-1-ol, dihydrogen phosphate; PEG-9 nonyl phenyl ether phosphate; PEG 450 nonyl phenyl ether phosphate; POE (9) nonyl phenyl ether phosphate; Poly(oxy-1,2-ethanediyl), α-(nonylphenyl)-ω-hydroxy-, phosphate
Definition: Complex mixture of esters of phosphoric acid and nonoxynol-9
Empirical: $C_{33}H_{61}O_{13}P$
Properties: Anionic
Uses: Emulsifier, lubricant, antistat, detergent, corrosion inhibitor for agric., industrial use; solvent, antisoil redeposition for dry cleaning; emulsion polymerization; household/industrial detergents; fabric finishes; paints; textile wet processing; metals; surfactant, cleansing agent, hydrotrope in cosmetics; in food-pkg. adhesives
Regulatory: FDA 21CFR §175.105, 178.3400

Nonyl alcohol. *See* 3,5,5-Trimethylhexan-1-ol
s-Nonyl alcohol. *See* 2,6-Dimethyl heptanol-4
Nonylcarbinol. *See* Decyl alcohol
Nonylic acid; n-Nonylic acid. *See* Nonanoic acid
Nonylol. *See* 3,5,5-Trimethylhexan-1-ol

Nonylphenol

CAS 25154-52-3

Synonyms: 2,6-Dimethyl-4-heptylphenol (o and p); Nonyl phenol (mixed isomers)
Definition: A mixture of isomeric monoalkyl phenols
Empirical: $C_{15}H_{24}O$
Formula: $C_9H_{19}C_6H_4OH$
Properties: Straw-colored clear visc. liq.; sl. phenolic odor; sol. in benzene, chlorinated solvs., aniline, heptane, aliphatic alcohol, ethylene glycol, most org. solvs.; insol. in water, dil. aq. NaOH; m.w. 220.36; dens. 0.949 (20/4 C); b.p. 293-297 C; flash pt. > 110 C; ref. index 1.5110 (20 C)
Toxicology: LD50 (oral, rat) 1620 mg/kg, (skin, rabbit) 2140 mg/kg; mod. toxic by ing. and skin contact; severe eye and skin irritant; TSCA listed
Precaution: Combustible exposed to heat or flame
Hazardous Decomp. Prods.: Heated to decomp., emits acrid smoke and irritating fumes
Uses: Nonionic surfactant intermediate (nonbiodegradable); lube oil additives; stabilizers; curing agents; petroleum demulsifiers; plasticizers; dyestuffs; aromatic oils; fungicides; antioxidants for plastics and rubber; solvent; nonreactive epoxy resin diluent; defoamer in food-contact coatings and paper/paperboard
Regulatory: FDA 21CFR §176.200, 176.210
Manuf./Distrib.: Akzo Nobel Surf. Chem. AB; Akzo Nobel; Aldrich; Allchem Ind.; Ashland; Atofina; Bencorp Int'l.; Cognis/Chems. Group; Degussa AG; Dover; Eastech; FRP Services; Filo; GE Spec.; Herdillia Chems. Ltd; Huntsman; Indofine; Noveon Kalama; Rhodia; Schenectady

Nonylphenol ethoxylates. *See* Nonoxynol
Nonyl phenol (mixed isomers). *See* Nonylphenol
Nonylphenol, polyoxyethylene ether. *See* Nonoxynol
26-(Nonylphenoxy)-3,6,9,12,15,18,21,24-octaoxahexacosan-1-ol, dihydrogen phosphate. *See* Nonoxynol-9 phosphate
14-(Nonylphenoxy)-3,6,9,12-tetraoxatetradecan-1-ol. *See* Nonoxynol-5
Nonyl phenyl polyethylene glycol; Nonyl phenyl polyethylene glycol ether. *See* Nonoxynol
3-Norcarene, 3,7,7-trimethyl-. *See* 3-Carene
Normal hexane. *See* Hexane
Normal paraffins. *See* Paraffins, normal C5-20
Normal pentane. *See* n-Pentane
Normal propyl alcohol. *See* Propyl alcohol
Normanthane. *See* Isopropylcyclohexane
1-NP. *See* 1-Nitropropane
2-NP. *See* 2-Nitropropane

O

OAP. *See* Hexamethyldisilazane
Oceol. *See* Oleyl alcohol
OCT. *See* o-Chlorotoluene
Octaacetylsucrose. *See* Sucrose octaacetate
9,12-Octadecadienoic acid ethyl ester. *See* Ethyl linoleate
9,12-Octadecadienoic acid, mixed triesters with octanoic acid, decanoic acid, and 1,2,3-propanetriol. *See* Caprylic/capric/linoleic triglyceride
9,12-Octadecadienoic acid, 1,2,3-propanetriyl ester. *See* Trilinolein
Octadecanamide, N-(2-hydroxy-1,1-dimethylethyl)-. *See* Stearamide AMP
Octadecane. *See* N-Octadecane

N-Octadecane

CAS 593-45-3; EINECS/ELINCS 209-790-3
Synonyms: Octadecane
Empirical: $C_{18}H_{38}$
Properties: Colorless liq. or wh. cryst. solid; sol. in acetone, ether; sl. sol. in alcohol; insol. in water; m.w. 254.51; sp.gr. 0.7768 (28/4 C); m.p. 28-30 C; b.p. 316.1 C; flash pt. > 100 C
Toxicology: Irritating to eyes, skin, mucous membranes, upper respiratory tract; may be harmful by inh., ing., or skin absorp.; TSCA listed
Precaution: Combustible; incompat. with strong oxidizing agents
Hazardous Decomp. Prods.: CO, CO_2
Storage: Store refrigerated; keep tightly closed
Uses: Solvents; organic synthesis; calibration standard; fragrance in cosmetics
Manuf./Distrib.: Acros Org.; Aldrich; Alfa Aesar; Am. Radiolabeled Chems.; Cambrex; ChevronPhillips; Fluka; ICN Biomed. Research Prods.; Indofine; Kira; Lancaster Synthesis; Pfaltz & Bauer; Riedel-deHaën; Sigma; Spectrum Quality Prods.; Supelco; TCI Am.; Universal Preserv-A-Chem

1,12-Octadecanediol. *See* 12-Hydroxystearyl alcohol
Octadecanoic acid, (acetyloxy)-1,2,3-propanetriyl ester. *See* Glyceryl triacetyl hydroxystearate
Octadecanoic acid, 2-butoxyethyl ester. *See* Butoxyethyl stearate
Octadecanoic acid butyl ester. *See* Butyl stearate
Octadecanoic acid, cetyl ester. *See* Cetyl stearate
Octadecanoic acid, 1,2-ethanediyl ester. *See* Glycol distearate
Octadecanoic acid, 2-ethylhexyl ester. *See* Octyl stearate
Octadecanoic acid, 1-hexadecyl ester. *See* Cetyl stearate
Octadecanoic acid, 12-hydroxy, 1,2,3-propanetriyl ester. *See* Glyceryl triacetyl hydroxystearate
Octadecanoic acid, isocetyl ester; Octadecanoic acid, isohexadecyl ester. *See* Isocetyl stearate
Octadecanoic acid, 2-methoxyethyl ester. *See* Methoxyethyl stearate
Octadecanoic acid, methyl ester. *See* Methyl stearate
Octadecanoic acid, 1-methylethyl ester. *See* Isopropyl stearate
Octadecanoic acid, mixed triesters with octanoic acid, decanoic acid, and 1,2,3-propanetriol. *See* Caprylic/capric/stearic triglyceride
Octadecanoic acid, monoester with 1,2-propanediol. *See* Propylene glycol stearate
Octadecanoic acid, monoester with 1,2,3-propanetriol. *See* Glyceryl stearate
Octadecanoic acid, monoester with triglycerol; Octadecanoic acid, monoester with tri-1,2,3-propanetriol diether. *See* Polyglyceryl-3 stearate
Octadecanoic acid, octadecyl ester. *See* Stearyl stearate
Octadecanoic acid, 1,2,3-propanetriyl ester. *See* Tristearin
Octadecanol; 1-Octadecanol; n-Octadecanol. *See* Stearyl alcohol
9-Octadecenoic acid; 9,10-Octadecenoic acid; cis-9-Octadecenoic acid; cis-Octadec-9-enoic acid. *See* Oleic acid
9-Octadecenoic acid, 12-(acetyloxy)-, 1,2,3-propanetriol ester. *See* Glyceryl triacetyl ricinoleate
9-Octadecenoic acid, butyl ester. *See* Butyl oleate
9-Octadecenoic acid decyl ester. *See* Decyl oleate
9-Octadecenoic acid, diester with 1,2,3-propanetriol. *See* Glyceryl dioleate
9-Octadecenoic acid ethyl ester. *See* Ethyl oleate
2-Octadecenoic acid, 2-ethylhexyl ester. *See* Octyl oleate
9-Octadecenoic acid, 2-[2-[2-(2-hydroxyethoxy) ethoxy] ethoxy] ethyl ester. *See* PEG-4 oleate
9-Octadecenoic acid, 17-hydroxy-3,6,9,12,15-pentaoxaheptadec-1-yl ester. *See* PEG-6 oleate
9-Octadecenoic acid, isodecyl ester. *See* Isodecyl oleate
9-Octadecenoic acid, methyl ester; (Z)-9-Octadecenoic acid, methyl ester. *See* Methyl oleate
9-Octadecenoic acid, 1-methylethyl ester. *See* Isopropyl oleate
9-Octadecenoic acid, monoester with 1,2,3-propanetriol. *See* Glyceryl oleate
9-Octadecenoic acid, monoester with tetraglycerol. *See* Polyglyceryl-4 oleate
9-Octadecenoic acid, 9-octadecenyl ester. *See* Oleyl oleate
9-Octadecenoic acid, oxybis (2,1-ethanediyloxy-2,1-ethanediyl) ester. *See* PEG-4 dioleate
9-Octadecenoic acid, 1,2,3-propanetriyl ester. *See* Triolein
Octadecenol; 1-Octadecenol; Octadec-9-en-1-ol; 9-Octadecen-1-ol; cis-9-Octadecen-1-ol; (Z)-Octadec-9-enol; (Z)-9-Octadecen-1-ol. *See* Oleyl alcohol
9-Octadecenyl 13-docosenoate; (Z)-Octadec-9-enyl (Z)-docos-13-enoate. *See* Oleyl erucate
(Z)-α-9-Octadecenyl-ω-hydroxypoly (oxy-1,2-ethanediyl); 2-[2-(9-Octadecenyloxy) ethoxy] ethanol. *See* Oleth-2
Octadecyl alcohol; n-Octadecyl alcohol. *See* Stearyl alcohol
Octadecyl 2-ethylhexanoate. *See* Stearyl octanoate
Octadecyl heptanoate. *See* Stearyl heptanoate
Octadecyl stearate. *See* Stearyl stearate
1H,1H,5H-Octafluoropentanol. *See* 1H,1H,5H-Octafluoro-1-pentanol

1H,1H,5H-Octafluoro-1-pentanol

CAS 355-80-6; EINECS/ELINCS 206-593-4
Synonyms: 1H,1H,5H-Octafluoropentanol; 2,2,3,3,4,4,5,5-Octafluoro-1-pentanol; OFP; 1-Pentanol, 2,2,3,3,4,4,5,5-octafluoro-; 1,1,5-Trihydrofluoropentane-1-ol

Classification: Halogenated alcohol
Empirical: $C_5H_4F_8O$
Formula: $H(CF_2CF_2)_2CH_2OH$
Properties: M.w. 232.07; dens. 1.66 g/cm^3 (20 C); b.p. 139-142 C (1 atm)
Toxicology: Eye/skin irritant
Environmental: Sl. polluting
Uses: Solvent coating for disks; synthesis
Manuf./Distrib.: ABCR; Acros Org.; Alfa Aesar; Daikin Industries Ltd; Fluorochem USA; Indofine; Interchim; Lancaster Synthesis; Merck KGaA; Spektr T.T.T.; TCI Am.; Zhejiang Sanhuan Chems. Co. Ltd

2,2,3,3,4,4,5,5-Octafluoro-1-pentanol. *See* 1H,1H,5H-Octafluoro-1-pentanol

Octamethylcyclotetrasiloxane

CAS 556-67-2; EINECS/ELINCS 209-136-7
Synonyms: Cyclic dimethylsiloxane tetramer; Cyclotetrasiloxane, octamethyl-; Octamethyltetracyclosiloxane
Classification: Silicone oligomer; cyclic siloxane
Empirical: $C_8H_{24}O_4Si_4$
Formula: $[(CH_3)_2SiO]_4$
Properties: Colorless smooth visc. oily liq.; m.w. 296.62; dens. 0.955 (20/4 C); vapor pressure 1 mm (23 C); m.p. 17.5 C; b.p. 175-176 C; flash pt. 60 C; ref. index 1.3968 (20 C); sp. heat 120 cal/g/C; dielec. const. 2.39
Toxicology: LD50 (oral, rat) 1540 mg/kg, (skin, rabbit) 794 µl/kg; LC50 (inh., rat, 4 h) 36 g/m^3; eye irritant; may cause irritation to skin, mucous membranes, upper respiratory tract; harmful by ing., skin contact; may cause tremors, salivary gland changes, liver changes, effects on newborn; TSCA listed
Precaution: Flamm.; incompat. with strong oxidizing agents, acids, bases; may dec. on exposure to moist air or water; may dec. under fire conditions to form flamm./explosive mixts. in air
Hazardous Decomp. Prods.: Toxic fumes of CO, CO_2, silicon oxide; emits toxic fumes under fire conditions
HMIS: Health 1, Flammability 2, Reactivity 0
Storage: Store in cool, dry place; keep tightly closed; keep away from heat, open flame
Uses: Silicone oils; methyl silicone oils; synthesis; reagent; emollient for cosmetics, hair care prods., skin care prods.; antiperspirants and deodorants; cleaning solvent; coupling agent; blocking agent; release agent; lubricant
Manuf./Distrib.: ABCR; AP Resources; Acros Org.; Aldrich; Alfa Aesar; Degussa AG; Fluka; Gelest; ICN Biomed. Research Prods.; Jiangxi Liq. Ind. Chems.; Lancaster Synthesis; Pfaltz & Bauer; Sigma; TCI Am.; United Chem. Tech.

Octamethyltetracyclosiloxane. *See* Octamethylcyclotetrasiloxane

Octamethyltrisiloxane

CAS 107-51-7; EINECS/ELINCS 203-497-4
Classification: Linear siloxane
Empirical: $C_8H_{24}O_2Si_3$
Formula: $[(CH_3)_3SiO]_2Si(CH_3)_2$
Properties: Colorless liq.; sol. in benzene and the lighter hydrocarbons; sl. sol. in alcohol and heavy hydrocarbons; sol. 34 ppb in water; m.w. 236.55; dens. 0.820; vapor pressure 4 mm; m.p. -80 C; b.p. 153 C; flash pt. 29 C; ref. index 1.3848 (20 C); surf. tens. 17.4 dynes/cm; dielec. const. 2.30
Toxicology: Harmful by ing., skin absorp.; may be harmful by inh.; eye irritant; may cause irritation to skin, mucous membranes, upper respiratory tract; TSCA listed
Precaution: Flamm.; vapor may flash back; incompat. with strong oxidizing agents, strong acids; may dec. on exposure to moist air or water; container explosion may occur under fire conditions; protect against static discharges
Hazardous Decomp. Prods.: Emits toxic fumes under fire conditions
HMIS: Health 1, Flammability 3, Reactivity 0
Storage: Store in cool, dry place; keep tightly closed; moisture-sensitive; keep away from heat, sparks, open flame
Uses: As basis for silicone oils for extreme temp. performance; antifoam in petrol. lubricating oils; cleaning solvent; coupling agent; blocking agent; release agent; lubricant; skin conditioner in cosmetics
Manuf./Distrib.: ABCR; AP Resources; Acros Org.; Aldrich; Fluka; Gelest; Pfaltz & Bauer; TCI Am.; United Chem. Tech.

Octane

CAS 111-65-9; EINECS/ELINCS 203-892-1
UN 1262 (DOT)
Synonyms: n-Octane
Classification: Sat. aliphatic hydrocarbon
Empirical: C_8H_{18}
Formula: $CH_3(CH_2)_6CH_3$
Properties: Colorless liq.; gasoline-like odor; completely misc. with most org. solvs.; insol. in water; sl. sol. in alcohol, ether; m.w. 114.23; sp.gr. 0.70 (20 C); vapor pressure 10 mm Hg (19 C); m.p. -56.7 C; b.p. 125-126 C; flash pt. 13 C
Toxicology: ACGIH TLV/TWA 300 ppm; STEL 375 ppm; LDLo (IV, mouse) 428 mg/kg; poison by IV route; skin and eye irritant; mild narcotic at high vapor concs.; may cause nose/throat irritation, headache, drowsiness, nausea, difficulty breathing; very high concs. may cause unconsciousness, death; ing. may cause nausea, vomiting, swelling of abdomen, headache, depression; relatively low oral toxicity unless aspirated into lungs (may cause severe damage, death); asphyxiant; TSCA listed
Precaution: Highly flamm.; severe explosion hazard exposed to heat, flame, oxidizers; incompat. with strong oxidizing agents (increases fire and explosion hazard)
Hazardous Decomp. Prods.: Heated to decomp., emits acrid smoke and irritating fumes
NFPA: Health 0, Flammability 3, Reactivity 0
Storage: Store in cool, dry, well-ventilated area out of direct sunlight
Uses: Organic synthesis; solvent; component of high-octane motor and aviation fuel; ingred. of VM&P naphtha and other petrol. solvents; mfg. of benzene, toluene, xylene aromatics; azeotropic distillations; synthetic flavoring agent in foods; calibrations
Manuf./Distrib.: Aldrich; Alfa Aesar; Asiamerica Int'l.; ChevronPhillips; Fluka; GFS; Integra; J.T. Baker; Romil Ltd; Sigma; Sithean; Spectrum Quality Prods.; Triple Crown Am.; VWR Int'l.

n-Octane. *See* Octane
1-Octanecarboxylic acid. *See* Nonanoic acid
1,3-Octanediol. *See* Ethyl hexanediol
Octane, 1,1´-oxybis-. *See* Dioctyl ether
Octanoic acid, 2,2-dimethyl-1,3-propanediol diester. *See* Neopentyl glycol dioctanoate
Octanoic acid ethyl ester. *See* Ethyl octanoate
Octanoic acid, 2-hydroxypropyl ester. *See* Propylene glycol caprylate
Octanoic acid, methyl ester. *See* Methyl caprylate
Octanoic acid, 1-methyl-1,2-ethanediyl ester. *See* Propylene glycol dicaprylate
Octanoic acid, mixed diesters with decanoic acid and propylene glycol. *See* Propylene glycol dicaprylate/dicaprate
Octanoic acid, monoester with 1,2-propanediol. *See* Propylene glycol caprylate
Octanoic acid, monoester with 1,2,3-propanetriol. *See* Glyceryl caprylate
Octanoic acid, 1,3-propanediyl ester. *See* Propylene glycol dioctanoate
Octanoic acid, 1,2,3-propanetriol ester; Octanoic acid, 1,2,3-propanetriyl ester; Octanoic acid triglyceride. *See* Trioctanoin
Octanoic/decanoic acid triglyceride. *See* Caprylic/capric triglyceride
Octanol; 1-Octanol; Octan-1-ol. *See* Caprylic alcohol

2-Octanol

CAS 123-96-6; EINECS/ELINCS 204-667-0
UN 1987 (DOT); FEMA 2801
Synonyms: Capryl alcohol; Hexyl methyl carbinol; 2-Hydroxyoctane; Methyl hexyl carbinol; Octan-2-ol; 2-n-Octanol; s-n-Octyl alcohol; Secondary caprylic alcohol
Classification: Sec. aliphatic alcohol
Empirical: $C_8H_{18}O$
Formula: $CH_3(CH_2)_5CH(OH)CH_3$
Properties: Colorless clear oily visc. liq.; pungent aromatic odor; sol. in ethanol, ether, acetone, oxygenated solvs., aromatic and aliphatic hydrocarbons; pract. insol. in water; m.w. 130.23; sp.gr. 0.819 (20/4 C); vapor pressure 0.84 mm Hg (40 C); m.p. -39 C; b.p. 178.5 C; flash pt. 76 C; ref. index 1.426 (20 C); surf. tens. 26.38 dynes/cm; dielec. const. 8.173

Toxicology: LD50 (oral, rat) > 3200 mg/kg; toxic; skin and eye irritant; inh. may cause nose/throat irritation, CNS depression symptoms; ing. may cause 'alcohol' intoxication symptoms; may cause severe lung damage, respiratory/cardiac arrest, and death if aspirated into lungs; TSCA listed
Environmental: VOC; ThOD 1.84
Precaution: Combustible liq.; can form explosive mixts. with air ≥ 80 C; incompat. with oxidizing agents (increases risk of fire or explosion)
Hazardous Decomp. Prods.: Thermal decomp. prods.: CO, CO_2, acrid smoke, irritating vapors
NFPA: Health 1, Flammability 2, Reactivity 0
Storage: Store in cool, dry, well-ventilated area away from heat and ignition sources; avoid generating mists and vapors
Uses: Intermediate in mfg. of plasticizers, disinfectant soaps; antifoam; solvent for fats and waxes; mfg. of wetting agents, hydraulic oils, petrol. additives; in lacquers, enamels; masking agent for industrial odors; synthetic flavoring agent and fragrance in foods and pharmaceuticals; fragrance in cosmetics
Regulatory: FDA 21CFR §172.515, 176.180,176.210; FEMA GRAS
Manuf./Distrib.: Advanced Synthesis Tech.; Aldrich; Arizona; Ashland; Austin; Cognis/Chems. Group; Fluka; J.T. Baker; Jarchem Ind.; Penta Mfg.; Sigma; Spectrum Quality Prods.

Octan-2-ol; 2-n-Octanol. *See* 2-Octanol
n-Octanol. *See* Caprylic alcohol
1-Octanol acetate. *See* Octyl acetate
1-Octanol, 2-butyl-, benzoate. *See* Butyl octyl benzoate
2-Octanone. *See* Methyl hexyl ketone

3-Octanone

CAS 106-68-3; EINECS/ELINCS 203-423-0
UN 2271 (DOT); FEMA 2803
Synonyms: Amyl ethyl ketone; n-Amyl ethyl ketone; EAK; Ethyl amyl ketone; Ethyl n-amyl ketone; Ethyl pentyl ketone; 5-Methyl-3-heptanone; Octanone-3; n-Pentyl ethyl ketone
Classification: Nonaromatic ketone
Empirical: $C_8H_{16}O$
Formula: $CH_3(CH_2)_4COC_2H_5$
Properties: Colorless liq.; mild fruity odor; very sol. in ethanol; misc. with oxygenated solvs.; sl. sol. in water; m.w. 128.22; dens. 0.821 (20/4 C); b.p. 166-169 C; flash pt. 51 C; ref. index 1.415 (20 C)
Toxicology: OSHA PEL TWA 25 ppm; LD50 (IP, mouse) 406 mg/kg; poison by IP route; mod. irritating to eyes, skin, respiratory system; narcotic in high concs.; TSCA listed
Precaution: Flamm. exposed to heat, sparks, flame, or oxidizers; mod. fire risk
Hazardous Decomp. Prods.: CO, CO_2; heated to decomp., emits acrid smoke and irritating fumes
Uses: Fragrance in perfumery, cosmetics; synthetic flavoring agent in foods and pharmaceuticals; solvent for nitrocellulose and vinyl resins
Regulatory: FDA 21CFR §172.515; FEMA GRAS
Manuf./Distrib.: Acme-Hardesty; Advanced Synthesis Tech.; Aldrich; Atofina SA; Degussa AG/Health & Nutrition; Degussa Flavors/Alfrebro; Fluka; Lluch Essence

Octanone-3. *See* 3-Octanone
n-Octanyl acetate. *See* Octyl acetate
3,6,9,12,15,18,21,24-Octaoxahexacosane-1,26-diol. *See* PEG-9

Octyl acetate

CAS 112-14-1; EINECS/ELINCS 203-939-6
FEMA 2806
Synonyms: Acetate C-8; Acetic acid octyl ester; Caprylyl acetate; 1-Octanol acetate; n-Octanyl acetate; 1-Octyl acetate; n-Octyl acetate; Octyl alcohol acetate
Classification: Carboxylic acid ester
Empirical: $C_{10}H_{20}O_2$
Formula: $CH_3(CO_2(CH_2)_7CH_3$
Properties: Colorless liq.; floral-fruity odor; sol. in alcohol, ether, fixed oils; sl. sol. in water; m.w. 172.27; dens. 0.868; m.p. -38.5 C; b.p. 199 C; flash pt. (CC) 86 C; ref. index 1.4180 (20 C)
Toxicology: LD50 (oral, rat) 3000 mg/kg, (skin, rabbit) > 5 g/kg; low oral toxicity; eye and sl. skin irritant; inh. of mist may cause nose/throat irritation; high concs. may cause CNS depression, drowsiness, headache, lightheadedness; TSCA listed
Precaution: Combustible; incompat. with strong oxidizing agents (increases fire/explosion hazard), strong bases, strong acids (decomp. can occur)
Hazardous Decomp. Prods.: n-Octyl alcohol, acetic acid; heated to decomp., emits acrid smoke and irritating fumes
Storage: Store in cool, dry, well-ventilated area, out of direct sunlight; keep containers closed; avoid generating mist; limit quantities in use
Uses: Solvent; fragrance in cosmetics, perfumes; synthetic flavoring agent and fragrance in foods and pharmaceuticals
Features: Fruity flavor
Regulatory: FDA 21CFR §172.515; FEMA GRAS
Manuf./Distrib.: Advanced BioTech; Advanced Synthesis Tech.; Aldrich; Axxence Aromatic GmbH; Bell Flavors & Fragrances; Degussa AG; Fleurchem; Goldensun Mfg.; Grau Aromatics; Lluch Essence; Moore Ingreds.; Penta Mfg.; Prodasynth; Sigma

1-Octyl acetate; n-Octyl acetate. *See* Octyl acetate
Octyl alcohol; n-Octyl alcohol. *See* Caprylic alcohol
s-n-Octyl alcohol. *See* 2-Octanol
Octyl alcohol acetate. *See* Octyl acetate
2-Octyl-1-decanol. *See* Isostearyl alcohol
Octyldihexylphosphine oxide. *See* Dihexyloctylphosphine oxide

Octyldodecanol

CAS 5333-42-6; EINECS/ELINCS 226-242-9
Synonyms: 1-Dodecanol, 2-octyl-; Isoarachidyl alcohol; Isoeicosanol; Isoeicosyl alcohol; 2-Octyl dodecanol; 2-Octyl-1-dodecanol; 2-Octyldodecyl alcohol
Classification: Aliphatic alcohol
Empirical: $C_{20}H_{42}O$
Formula: $CH_3(CH_2)_9CHCH_2OHCH_3(CH_2)_6CH_2$
Properties: Water-wh. clear liq.; sol. in alcohol, ether; insol. in water; m.w. 298.62; sp.gr. 0.84-0.85; m.p. -40 C; acid no. 0.5 max.; iodine no. 8 max.; sapon. no. 5 max.; hyd. no. 175-190; flash pt. (COC) 180 C
Toxicology: Eye and severe skin irritant; TSCA listed
Hazardous Decomp. Prods.: Heated to decomp., emits acrid smoke and irritating vapors
Uses: Carrier for oil-sol. prods.; lubricant, emollient, solvent for cosmetics and pharmaceuticals; pigment dispersant, solubilizer, coupling agent; intermediate for ester and alkoxylate synthesis, fiber lubricants, metalworking lubricants, personal care prods., emulsifiers, textiles; pharmaceutical vehicle
Features: Oleaginous
Regulatory: FDA approved for topicals, vaginal creams; NF, BP, EP compliance
Manuf./Distrib.: Mosselman NV
Trade Names Containing: Argobase 125T

2-Octyl dodecanol; 2-Octyl-1-dodecanol; 2-Octyldodecyl alcohol. *See* Octyldodecanol

Octyldodecyl behenate

Synonyms: Docosanoic acid, 2-octyldodecyl ester
Definition: Ester of octyldodecanol and behenic acid
Empirical: $C_{42}H_{84}O_2$
Formula: $CH_3(CH_2)_{20}COOCH_2CH(CH_2)_9CH_3CH_2(CH_2)_6CH_3$
Uses: Emollient in cosmetics
Trade Names: Excepal OD-B

Octyldodecyl isostearate

Classification: Guerbet ester
Definition: Ester of octyldodecanol and isostearic acid
Properties: Lt. colored oily liq.; mild odor; sol. in IPM, min. oil; insol. in water, ethanol; sp.gr. 0.97; b.p. > 100 C; acid no. 9; sapon. no. 95; flash pt. (PMCC) > 93.3 C
Toxicology: Low skin and eye irritation; may cause mild eye irritation; no inh. effects known
Precaution: Spillages may be slippery; incompat. with strong acids, bases, and oxidizing agents
Hazardous Decomp. Prods.: CO, CO_2, hydrocarbons
HMIS: Health 1, Flammability 1, Reactivity 0

Storage: Keep container closed when not in use
Uses: Emollient, moisturizer, gloss aid, application/wear enhancer, pigment dispersant, solvent in personal care, colored cosmetics, hair care, skin care, creams/lotions, ointments, makeup removers, moisturizing cleansers
Features: Color stable

Octylene glycol; 1,3-Octylene glycol. *See* Ethyl hexanediol
Octyl ether. *See* Dioctyl ether

Octyl hydroxystearate

CAS 29383-26-4; 29710-25-6; EINECS/ELINCS 249-793-7
Synonyms: 2-Ethylhexyl 12-hydroxyoctadecanoate; 2-Ethylhexylhydroxystearate; 2-Ethylhexyl 12-hydroxystearate; 2-Ethylhexyl oxystearate; 12-Hydroxyoctadecanoic acid, 2-ethylhexyl ester; Octyl oxystearate
Definition: Ester of 2-ethylhexyl alcohol and 12-hydroxystearic acid
Empirical: $C_{26}H_{52}O_3$
Formula: $CH_3(CH_2)_5COHH(CH_2)_{10}COOCH_2CCH_2CH_3H(CH_2)_3CH_3$
Properties: Amber clear liq.; mild fatty odor; insol. in water; sp.gr. 0.89; f.p. 12 C; b.p. > 225 C; flash pt. (COC) 425 C
Toxicology: Nonhazardous; TSCA listed
Precaution: Incompat. with oxidizing agents
Storage: Store away from strong oxidizing agents; avoid excessive heat
Uses: Emollient, solubilizer, refatting agent, cosolvent, moisturizer, pigment wetting agent/dispersant, counter-irritant for cosmetics, topical pharmaceuticals, cleansers; refatting agent for soaps, cleansers
Manuf./Distrib.: A&E Connock
Trade Names: Wickenol® 171

Octyl hydroxystearate benzoate

CAS 199277-69-5
Uses: Emollient, pigment wetting agent/dispersant in skin care, dry skin and therapeutic prods., bath care; solvent in makeup removers, cleansers; dispersant for org. and inorg. sunscreen actives in sun care; pigment wetting agent, gloss aid, moisturizer, color enhancer in lipsticks, lip glosses
Trade Names: Finsolv® BOHS-111

Octyl isononanoate

CAS 71566-49-9; EINECS/ELINCS 275-637-2
Synonyms: 2-Ethylhexyl isononanoate; Isononanoic acid, 2-ethylhexyl ester
Definition: Ester of 2-ethylhexyl alcohol and a branched chain nonanoic acid
Empirical: $C_{17}H_{34}O_2$
Formula: $CH_3(CH_3)_2CCH_2CH_3CHCH_2COOCH_2(CH_2CH_3)CH(CH_2)_3CH_3$
Properties: Clear water-wh. liq., typ. mild odor; sp.gr. 0.85; f.p. -34 C; b.p. 200 C; flash pt. (COC) 127 C
Toxicology: Nontoxic
Precaution: Incompat. with oxidizing agents
Storage: Store away from strong oxidizing agents
Uses: Emollient, lubricant, solvent for cosmetics, pharmaceutical topicals; emollient for creams/lotions, makeup, lipsticks, nail polish; antitackifier in antiperspirants; plasticizer for hair spray resins; lubricant for textiles, metalworking compds.; slip agent, wetting agent for pigments
Manuf./Distrib.: A&E Connock

Octyl oleate

Synonyms: 2-Ethylhexyl 2-octadecenoate; 2-Octadecenoic acid, 2-ethylhexyl ester
Empirical: $C_{26}H_{50}O_2$
Formula: $CH_3(CH_2)_7CH{=}CH(CH_2)_7COOCH_2CHCH_2CH_3(CH_2)_3CH_3$
Uses: Emollient in cosmetics, creams/lotions
Trade Names: Excepar l O-OL

Octyl oxystearate. *See* Octyl hydroxystearate

Octyl palmitate

CAS 29806-73-3; EINECS/ELINCS 249-862-1
Synonyms: 2-Ethylhexyl hexadecanoate; 2-Ethylhexyl palmitate; Hexadecanoic acid, 2-ethylhexyl ester; Palmitic acid, 2-ethylhexyl ester
Definition: Ester of 2-ethylhexyl alcohol and palmitic acid
Empirical: $C_{24}H_{48}O_2$
Formula: $CH_3(CH_2)_{14}COOCH_2(CH_2CH_3)CH(CH_2)_3CH_3$
Properties: Water-wh. to pale yel. liq.; sol. in min. oil, 95% ethanol, IPM, oleyl alcohol; insol. in water, glycerin, propylene glycol; m.w. 386.72; dens. 0.850-0.856; m.p. 2-6 C; ref. index 1.4445-1.4465
Toxicology: LD50 (oral, rat) > 40 ml/kg, (skin, rabbit) > 5 g/kg; low toxicity by ing. and skin irritant; mild skin irritant; nonirritating to eyes; TSCA listed
Hazardous Decomp. Prods.: Heated to decomp., emits acrid smoke and irritating fumes
Uses: Emollient, spreading agent, gloss aid for sunscreens, antiperspirants, cosmetics, bath oils, make-up, hair grooms, topical pharmaceuticals; binder for pressed powds.; lubricant in textile spin finish, coning oils, carding, dye bath; solubilizer for benzophenone-3; plasticizer; min. oil replacement
Manuf./Distrib.: A&E Connock; DelCon; ISP Van Dyk; Inolex; Int'l. Paper; Phoenix Nat. Prods.; Ruger; St. Lawrence
Trade Names: Excepar l EH-P; Palmil O; Trivent OP

Octyl phosphate. *See* Trioctyl phosphate
Octyl phthalate; n-Octyl phthalate. *See* n-Dioctyl phthalate
1-Octyl-2-pyrrolidinone; N-Octyl pyrrolidinone. *See* N-Octyl pyrrolidone

N-Octyl pyrrolidone

CAS 7425-87-8
Synonyms: 1-Octyl-2-pyrrolidinone; N-Octyl pyrrolidinone; 2-Pyrrolidinone, 1-octyl-
Empirical: $C_{12}H_{23}NO$
Properties: Liq.; m.w. 197.36; nonionic
Toxicology: LD50 (oral, rat) 2050 mg/kg; mod. toxic by ing.; primary irritant; severely irritating to skin and eyes; corrosive; TSCA listed
Precaution: Combustible liq.
Hazardous Decomp. Prods.: Heated to decomp., emits toxic fumes of NO_x
Uses: Surface act. solvent for industrial cleaners, waxes, polishes; wetting agent; foaming agent

N-Octyl-2-pyrrolidone. *See* Caprylyl pyrrolidone
Octyl sebacate. *See* Dioctyl sebacate

Octyl stearate

CAS 22047-49-0; EINECS/ELINCS 244-754-0
Synonyms: 2-Ethylhexyl octadecanoate; 2-Ethylhexyl stearate; Octadecanoic acid, 2-ethylhexyl ester; Stearic acid, 2-ethylhexyl ester
Definition: Ester of 2-ethylhexyl alcohol and stearic acid
Empirical: $C_{26}H_{52}O_2$
Formula: $CH_3(CH_2)_{16}COOCH_2(CH_2CH_3)CH(CH_2)_3CH_3$
Properties: Oily liq.; m.w. 396.71; cloud pt. 10-15 C
Toxicology: Skin and eye irritant; TSCA listed
Hazardous Decomp. Prods.: Heated to decomp., emits acrid smoke and irritating vapors
Uses: EP agent; lubricant for PVC, textile spin finishes, metalworking; plasticizer, solvent for paper coatings, metalworking fluids; emollient, superfatting agent, moisturizer, lubricant, spreading agent, detackifier, solubilizer, gloss aid, pigment wetting agent/dispersant for cosmetics, pharmaceutical topicals; textile surfactant
Manuf./Distrib.: A&E Connock; ChemService; DelCon
Trade Names: Excepar l EH-S

4-Octyn-3,6-diol, 3,6-dimethyl-; Octynediol, dimethyl-; 4-Octyne-3,6-diol, 3,6-dimethyl. *See* Dimethyl octynediol
ODB; ODCB. *See* o-Dichlorobenzene
Odorless kerosene. *See* Deodorized kerosene
OFP. *See* 1H,1H,5H-Octafluoro-1-pentanol
Oil of mirbane. *See* Nitrobenzene
Oil mist, mineral. *See* Mineral oil
Oil of niobe. *See* Methyl benzoate
Oil of Palma Christi. *See* Castor (Ricinus communis) oil
Oils, avocado. *See* Avocado (Persea gratissima) oil
Oils, babassu. *See* Babassu (Orbignya oleifera) oil
Oils, lanolin. *See* Lanolin oil
Oils, rice bran. *See* Rice (Oryza sativa) bran oil
Oil of wintergreen. *See* Methyl salicylate
Olea europaea. *See* Olive (Olea europaea) husk oil; Olive (Olea europaea) oil
Olea europaea husk oil. *See* Olive (Olea europaea) husk oil

Olea europaea oil. *See* Olive (Olea europaea) oil

Oleic acid

CAS 112-80-1; EINECS/ELINCS 204-007-1
FEMA 2815; INS570; E570
Synonyms: Elaic acid; 9-Octadecenoic acid; 9,10-Octadecenoic acid; cis-9-Octadecenoic acid; cis-Octadec-9-enoic acid; Oleinic acid; Red oil
Classification: Aliphatic carboxylic acid; unsaturated fatty acid
Empirical: $C_{18}H_{34}O_2$
Formula: $CH_3(CH_2)_7CH{:}CH(CH_2)_7COOH$
Properties: Colorless liq., odorless; darkens when exposed to oxygen; sol. in alcohol, ether, benzene, chloroform, fixed/volatile oils, oxygenated, chlorinated, and aromatic solvs.; insol. in water; m.w. 282.47; dens. 0.895; vapor pressure 1 mm (176.5 C); m.p. 6 C; b.p. 286 C (100 mm); HLB 1.0; acid no. 196-204; nonionic
Toxicology: LD50 (oral, rat) 74 g/kg, (IV, rat) 2400 μg/kg; poison by IV route; mildly toxic by ing.; irritant to skin, mucous membranes; human eye and skin irritant; questionable carcinogen; experimental tumorigen; mutagenic data; TSCA listed
Precaution: Combustible when exposed to heat or flame; incompat. with Al and perchloric acid
Hazardous Decomp. Prods.: Heated to decomp., emits acrid smoke and irritating fumes
HMIS: Health 1, Flammability 1, Reactivity 0
Storage: Light-sensitive
Uses: Soap base; mfg. of oleates, ointments, polishing compds., lubricants, food-grade additives, Turkey red oil, driers; intermediates for paints; waterproofing textiles; oiling wool; flotation frothing agent; emollient, superfatting agent, emulsifier in cosmetics; emulsifier in metalworking fluids, resins, surf. coatings; activator, plasticizer, softener, lubricant in rubber; emulsifier, solubilizer, stabilizer, foam stabilizer, glazing agent, antifoam in foods; emulsifier, solubilizer, stabilizer, solvent for pharmaceuticals; pharmaceutic aid; food pkg.; in food-pkg. adhesives; defoamer in food-contact paper coatings, paper/paperboard; in food-contact acrylics, cellophane, rubber articles, textiles; in lubricants for incidental food contact; flume wash water additive for food processing
Regulatory: FDA 21CFR §172.210, 172.860, 172.862, 173.315 (0.1 ppm max. in wash water), 173.340, 175.105, 175.320, 176.170, 176.180, 176.200, 176.210, 177.1010, 177.1200, 177.2260, 177.2600, 177.2800, 178.3570, 178.3910, 182.70, 182.90; FEMA GRAS; FDA approved for inhalants, orals, topicals; NF, BP, EP compliance
Manuf./Distrib.: AMRESCO; Adept Sol'ns.; Advanced BioTech; Akzo Nobel; Aldrich; Alfa Chem; Allchem Ind.; Am. Biorganics; Anar; Arista Ind.; Arizona; Ashland; Asiamerica Int'l.; Atofina Bruxelles; Avatar; Axxence Aromatic GmbH; BDH; Baychem; Brown; Browning; C.P. Hall; Chesham Chems. Ltd; Cognis/Chems. Group; Croda Inc; EM Ind.; EM Science; Ferro/Keil; Fisher Scientific; Fleurchem; Fluka; Genzyme; Geo. Pfau's Sons; Georgia-Pacific/Actrachem; Hercules; Int'l. Chem. Inc; Integra; J.T. Baker; KIC Chems.; Kraft Chem.; Lluch Essence; Mallinckrodt Baker; Merck KGaA; Mosselman NV; Mutchler; Parchem Trading; Penn Bio-Organics; Penta Mfg.; Prodasynth; R.E. Carroll; Rhodia HPCII; Rona; Ruger; Schweizerhall; Sea-Land; Sigma; Spectrum Quality Prods.; St. Lawrence; Tennants Distrib. Ltd; Uniqema Am.; Uniqema; Univar E&E Ltd; Universal Preserv-A-Chem; VWR Int'l.; Voigt Global Distrib.; Vopak USA; Welch, Holme & Clark

Oleic acid glycerol monoester. *See* Glyceryl oleate
Oleic acid, lauryl ester. *See* Lauryl oleate
Oleic acid, methyl ester, cis-; cis-Oleic acid, methyl ester. *See* Methyl oleate
Oleic acid monoglyceride. *See* Glyceryl oleate
Olein. *See* Triolein
Oleinic acid. *See* Oleic acid
Oleol; Oleoyl alcohol. *See* Oleyl alcohol
Oleoylglycerol. *See* Glyceryl oleate

Oleth-2

CAS 9004-98-2 (generic); 5274-65-7
Synonyms: (Z)-α-9-Octadecenyl-ω-hydroxypoly (oxy-1,2-ethanediyl); 2-[2-(9-Octadecenyloxy) ethoxy] ethanol; Oleyl alcohol EO (2); PEG-2 oleyl alcohol; PEG-2 oleyl ether; PEG 100 oleyl ether; POE (2) oleyl ether
Definition: PEG ether of oleyl alcohol
Empirical: $C_{22}H_{44}O_3$
Formula: $CH_3(CH_2)_7CH{=}CH(CH_2)_7CH_2(OCH_2CH_2)_nOH$, avg. n = 2
Properties: HLB 4.9; nonionic
Toxicology: LD50 (oral, rat) 25,800 mg/kg; mildly toxic by ing.; TSCA listed
Hazardous Decomp. Prods.: Heated to decomp., emits acrid smoke and irritating fumes
Uses: Solubilizer, emulsifier, surfactant, dispersant, stabilizer, spreading agent, emollient, cosolvent for cosmetics, detergents, fragrances, textiles, leather, paper, pharmaceuticals
Features: Lipophilic
Manuf./Distrib.: Aldrich; DelCon; Fluka; Sigma
Trade Names: Ameroxol® OE-2

Oleth-16

CAS 9004-98-2 (generic); 25190-05-0 (generic)
Synonyms: PEG-16 oleyl ether; POE (16) oleyl ether
Definition: PEG ether of oleyl alcohol
Formula: $CH_3(CH_2)_7CH{=}CH(CH_2)_7CH_2(OCH_2CH_2)_nOH$, avg. n = 16
Properties: Nonionic
Toxicology: TSCA listed
Uses: Surfactant, emulsifier for cosmetics and pharmaceuticals; defoamer in food-contact paper coatings; in food-contact textiles
Regulatory: FDA 21CFR §176.200, 177.2800
Manuf./Distrib.: Aldrich; Fluka; Sigma
Trade Names Containing: Solulan® 16

Oleum abietis. *See* Pine (Pinus palustris) oil
Oleum Gossypii seminis. *See* Cottonseed (Gossypium) oil
Oleum olivae. *See* Olive (Olea europaea) oil
Oleum ricini. *See* Castor (Ricinus communis) oil

Oleyl alcohol

CAS 143-28-2; EINECS/ELINCS 205-597-3
Synonyms: Oceol; Octadecenol; 1-Octadecenol; Octadec-9-en-1-ol; 9-Octadecen-1-ol; cis-9-Octadecen-1-ol; (Z)-Octadec-9-enol; (Z)-9-Octadecen-1-ol; Oleol; Oleoyl alcohol; Oleyl fatty alcohol
Classification: Unsaturated fatty alcohol
Empirical: $C_{18}H_{36}O$
Formula: $CH_3(CH_2)_7CH{=}CH(CH_2)_8OH$
Properties: Colorless to pale yel. oily visc. liq., faint char. odor, bland taste; sol. in alcohol, ether; insol. in water; m.w. 268.49; dens. 0.84; m.p. 13-19 C; b.p. 207 C (13 mm); acid no. 1 max.; iodine no. 85-95; hyd. no. 205-215; cloud pt. ≤ 10 C; flash pt. > 110 C; ref. index 1.4582 (27.5 C); nonionic
Toxicology: Human skin irritant; TSCA listed
Hazardous Decomp. Prods.: Heated to decomp., emits acrid smoke and irritating fumes
Uses: Plasticizer, emulsifier, emulsion stabilizer, lubricant, antifoam, coupling agent, solubilizer, cosolvent, pigment dispersant, detergent in cutting oils, inks, textile finishing, petrochem., pulp/paper, paints, plastics, foods, topical pharmaceuticals, cosmetics; metal cutting lubricant; softener, lubricant in textile processing; mfg. of sulfuric esters used as surfactants; emollient, emulsifier, opacifier, visc. control agent in cosmetics; carrier for medicaments; in paper/paperboard in contact with aq./fatty foods; defoamer in food-contact paper/paperboard; in closure-sealing gaskets for food containers
Regulatory: FDA 21CFR §176.170, 176.210, 177.1010, 177.1210, 177.2800, 178.3910; FDA approved for topicals; NF compliance
Manuf./Distrib.: Aldrich; Ashland; ChemService; Cognis/Chems. Group; Croda Inc; Fluka; Hickson & Welch Ltd; Jarchem Ind.; Lanaetex Prods.; M. Michel; Mosselman NV; Penta Mfg.; R.W. Greeff; Rhodia HPCII; Sea-Land; Sigma; Spectrum Quality Prods.; Universal Preserv-A-Chem; Whyte Chems. Ltd
Trade Names: Novol NF
Trade Names Containing: Witcohol™ 42

Oleyl alcohol EO (2). *See* Oleth-2

Oleyl erucate

CAS 17673-56-2; EINECS/ELINCS 241-654-9
Synonyms: 13-Docosenoic acid, 9-octadecenyl ester; Erucic acid, oleyl ester; 9-Octadecenyl 13-docosenoate; (Z)-Octadec-9-enyl (Z)-docos-

13-enoate
Definition: Ester of oleyl alcohol and erucic acid
Empirical: $C_{40}H_{76}O_2$
Uses: Emollient, lubricant for cosmetics, pharmaceuticals; jojoba oil substitute; diluent for natural jojoba; dispersant, solubilizer, solvent, carrier, vehicle in topical pharmaceuticals
Manuf./Distrib.: A&E Connock
Trade Names: Dynacerin® 660

Oleyl fatty alcohol. *See* Oleyl alcohol
Oleyl monoglyceride. *See* Glyceryl oleate

Oleyl oleate
CAS 3687-45-4; EINECS/ELINCS 222-980-4
Synonyms: 9-Octadecenoic acid, 9-octadecenyl ester
Definition: Ester of oleyl alcohol and oleic acid
Empirical: $C_{36}H_{68}O_2$
Formula: $CH_3(CH_2)_7CH{=}CH(CH_2)_7COO(CH_2)_8CH{=}CH(CH_2)_7CH_3$
Properties: Yel. liq.; insol. in water; m.w. 532.94; m.p. 14-16 C
Toxicology: TSCA listed
Uses: Emollient, lubricant for cosmetics and pharmaceuticals; gloss aid, softener, lubricant in hair and skin care preps.; cosolvent and solubilizer in perfumes; lubricant for metalworking, wire drawing, in textile spin finishes; sperm oil substitute; solubilizer for pharmaceuticals
Regulatory: FDA approved for topicals
Manuf./Distrib.: A&E Connock; Sigma
Trade Names: Arlypon OSO; Exceparl OL-OL; Schercemol OLO

Olive husk oil. *See* Olive (Olea europaea) husk oil
Olive oil. *See* Olive (Olea europaea) oil

Olive (Olea europaea) husk oil
Synonyms: Olea europaea; Olea europaea husk oil; Olive husk oil
Definition: Oil obtained from the solvent extraction of the olive husk, *Olea europaea*
Uses: Emollient, solvent in cosmetics

Olive (Olea europaea) oil
CAS 8001-25-0; EINECS/ELINCS 232-277-0
Synonyms: Azeite; Gomenoleo oil; Olea europaea; Olea europaea oil; Oleum olivae; Olive oil
Definition: Fixed oil obtained from the ripe fruit of *Olea europaea*, contg. glycerides of oleic acid, palmitic acid, linoleic acid, stearic acid, and arachidic acid
Properties: Yel. to lt. grnsh. liq., sl. olive odor, sl. char. taste; sl. sol. in alcohol; misc. with ether, chloroform, carbon disulfide; insol. in water; dens. 0.909-0.915 (25/25 C); m.p. -6 C; solid. pt. 17-26 C; iodine no. 79-88; sapon. no. 187-196; flash pt. (CC) 437 F; ref. index 1.466-1.468; becomes rancid on exposure to air
Toxicology: LD50 (IV, mouse) 1320 mg/kg; mod. toxic by IP route; human skin irritant; devoid of side effects; TSCA listed
Precaution: Combustible exposed to heat or flame; can react with oxidizers; some spontaneous heating
Hazardous Decomp. Prods.: Heated to decomp., emits acrid smoke and irritating fumes
HMIS: Health 0, Flammability 1, Reactivity 0
Storage: Photosensitive
Uses: Lubricant, emollient, solvent, cosolvent, conditioner in cosmetics, pharmaceuticals; vehicle, solvent for pharmaceutical ointments, liniments; castile soaps; textile soaps; sulfonated oils; salad dressing and other foods; softener for ear wax, crusts of eczema and psoriasis; nutrient; in food-pkg. adhesives; defoamer in food-contact paper coatings, paper/paperboard
Features: Oleaginous
Regulatory: FDA 21CFR §175.105, 176.200, 176.210, GRAS; Japan approved (olive); NF, BP, EP, JP compliance
Manuf./Distrib.: ABITEC; Aldivia; Aldrich; Alfa Chem; Alnor Oil; Anglia Oils; Arista Ind.; Avatar; Cascade; Charkit; Croda Chem. Ltd; Croda Inc; Croda Oleochems.; F.H. Taussig; Fanning; Fluka; Hawkins Chem.; Int'l. Chem. Inc; Int'l. Sourcing; Jeen Int'l.; KIC Chems.; Lipo; Mosselman NV; Mutchler; Noveon; Parchem Trading; Penta Mfg.; Pokonobe Ind.; Ruger; Scher; Sigma; Spectrum Quality Prods.; Thornley; Tri-K Ind.; Universal Preserv-A-Chem; Voigt Global Distrib.; Welch, Holme & Clark
Trade Names: Cropure® Olive; EmCon™ Olive; Fanoliv™ Oil; Olive Oil, Refined; Super Refined® Olive NF

Opalwax. *See* Hydrogenated castor oil
Optal. *See* Propyl alcohol
Orange oil. *See* Phenethyl alcohol
Organosiloxane; Organosiloxanes. *See* Silicone
Orthoboric acid. *See* Boric acid
Orthochlorotoluene. *See* o-Chlorotoluene
Orthocresol. *See* o-Cresol
Orthodichlorobenzene; Orthodichlorobenzol. *See* o-Dichlorobenzene
Orthophosphoric acid. *See* Phosphoric acid
Oryza sativa. *See* Rice (Oryza sativa) bran oil
1-Oxa-4-azacyclohexane. *See* Morpholine
2-Oxabicyclo (2.2.2) octane, 1,3,3-trimethyl-. *See* Eucalyptol
Oxacyclopentadiene. *See* Furan
Oxacyclopentane. *See* Tetrahydrofuran
1-Oxa-2-cyclopentene. *See* 2,3-Dihydrofuran
1-Oxa-3-cyclopentene. *See* 2,5-Dihydrofuran
Oxacyclopropane. *See* Ethylene oxide
3-Oxa-1-heptanol. *See* Butoxyethanol
Oxalic acid, diethyl ester. *See* Diethyl oxalate
Oxane. *See* Ethylene oxide
3-Oxapentane-1,5-diol; 3-Oxa-1,5-pentanediol. *See* Diethylene glycol
2-Oxepanone. *See* ε-Caprolactone monomer
Oxidoethane; α,β-Oxidoethane. *See* Ethylene oxide
1,8-Oxido-p-menthane. *See* Eucalyptol
Oxirane. *See* Ethylene oxide
Oxirane, methyl-, homopolymer. *See* Polypropylene glycol
Oxirane, methyl, polymer and oxibane, butyl ether. *See* PPG-2-buteth-3
2-Oxobornane. *See* Camphor
2-Oxobutane. *See* Methyl ethyl ketone
3-Oxobutanoic acid ethyl ester. *See* Ethylacetoacetate
3-Oxobutanoic acid methyl ester. *See* Methyl acetoacetate

Oxo-dodecyl acetate
CAS 108419-35-8
Synonyms: Acetic acid, alkyl (C11-C14) esters; Acetic acid, C11-14-branched alkyl esters, C13-rich; C11-14-branched alkyl acetate C13-rich
Properties: Sp.gr. 0.875; b.p. 240-285 C; flash pt. > 109 C
Toxicology: May be harmful by inh., ing., or skin absorption; may cause eye/skin/mucous membrane/upper respiratory tract irritation; exposure may cause nausea, dizziness, headache
Precaution: Incompat. with strong bases, strong oxidizing agents
Hazardous Decomp. Prods.: Toxic fumes of CO, CO_2; emits toxic fumes under fire conditions
Storage: Store in cool, dry place; keep tightly closed
Uses: Solvent for architectural coatings, industrial coatings, traffic marking, maintenance coatings
Manuf./Distrib.: Aldrich; ExxonMobil
Trade Names: Exxate® 1200

Oxo-heptyl acetate
CAS 90438-79-2
Synonyms: C7 alkyl acetate
Properties: Sp.gr. 0.87; flash pt. (TCC) 150 F
Uses: Solvent for high-solids, maintenance, marine, and industrial waterborne coatings, acrylic resin polymerization media, metalworking fluids, urethane foam and industrial cleaners and degreasers, paint strippers, pigment dispersions, pesticide emulsifiable concs., textile scouring aids, NC or screen printing inks, penetrating and lubricating oils, vinyl plastisols; wetting agent; lubricant; visc. modifier; latex coalescer
Features: Good solvency, low surf. tens. and interfacial tens.
Manuf./Distrib.: Chemcentral; Houghton Chem.
Trade Names: Exxate® 700

Oxo-hexyl acetate
CAS 88230-35-7
Synonyms: Acetate ester, C6 alcohol branched; C6 alkyl acetate
Properties: Sp.gr. 0.87; vapor pressure 1.400 mm Hg; flash pt. (TCC) 57.2 C; autoignition temp. 294 C
Toxicology: REL 50 ppm

Precaution: LEL 1%
Uses: Solvent for high-solids, maintenance, marine, and industrial water-borne coatings, acrylic resin polymerization media, metalworking fluids, urethane foam and industrial cleaners and degreasers, paint strippers, pigment dispersions, pesticide emulsifiable concs., textile scouring aids, NC or screen printing inks, penetrating and lubricating oils, vinyl plastisols; wetting agent; lubricant; visc. modifier; latex coalescer
Features: Good solvency, low surf. tens. and interfacial tens.
Manuf./Distrib.: Houghton Chem.
Trade Names: Exxate® 600

Oxolane. *See* Tetrahydrofuran
Oxole. *See* Furan
2-Oxolene. *See* 2,3-Dihydrofuran
3-Oxolene. *See* 2,5-Dihydrofuran
4-Oxopentanoic acid ethyl ester. *See* Ethyl levulinate
2-Oxopyrrolidine. *See* 2-Pyrrolidone
8-Oxotricyclo (5.2.1.0^{2,6}) decane. *See* Tricyclo [5.2.1.0^{2,6}] decane-8-one
Oxybenzene. *See* Phenol
1,1′-Oxybisbenzene. *See* Diphenyl oxide
1,1′-Oxybis (butane). *See* Dibutyl ether
1,1-Oxybis (2-chloro) ethane. *See* sym-Dichloroethyl ether
1,1′-Oxybisethane; Oxybis-1,1′-ethane. *See* Ethyl ether
1,1′-(Oxybis (2,1-ethanediyloxy)) bisbutane. *See* Diethylene glycol dibutyl ether
2,2′-[Oxybis (2,1-ethanediyloxy)] bisethanol. *See* PEG-4
2,2′-Oxybisethanol. *See* Diethylene glycol
2,2′-(Oxybis (ethyleneoxy)) diethanol. *See* PEG-4
1,1-Oxybishexane. *See* n-Hexyl ether
Oxybis (2-hydroxyethyl), diesters with 3,5,5-trimethylhexanoic acid and 2-ethylhexanoic acid. *See* Diethylene glycol dioctanoate/diisononanoate
Oxybismethane. *See* Dimethyl ether
1,1′-[Oxybis (methylene)] bis [benzene]. *See* Benzyl ether
1,1′-Oxybisoctane. *See* Dioctyl ether
1,1′-Oxybispentane. *See* Diamyl ether
1,1′-Oxybis-2-propanol. *See* Dipropylene glycol
Oxybis (trimethylsilane). *See* Hexamethyldisiloxane
2,2′-Oxydiethanol. *See* Diethylene glycol
1,1′-Oxydi-2-propanol; 1,1′-Oxydipropan-2-ol. *See* Dipropylene glycol
3,3′-Oxydi-1-propanol dibenzoate; Oxydipropyl dibenzoate; 3,3′-Oxydyl-1-propanol dibenzoate. *See* Dipropylene glycol dibenzoate
Oxygis (methoxy)-propane. *See* Dipropylene glycol dimethyl ether
Oxymethylene. *See* Formaldehyde
Oxyphenic acid. *See* Pyrocatechol
Oxytol acetate. *See* Ethoxyethanol acetate
Oxytoluene. *See* Cresylic acid
o-Oxytoluene. *See* o-Cresol

P

Painters naphtha. *See* VM&P naphtha
Palmitic acid, 2-ethylhexyl ester. *See* Octyl palmitate
Palmitic acid, hexadecyl ester. *See* Cetyl palmitate
Palmitic acid, isobutyl ester. *See* Isobutyl palmitate
Palmitic sucrose ester. *See* Sucrose palmitate
Palmitin. *See* Tripalmitin
Palmityl acetate. *See* Cetyl acetate
Palmityl alcohol. *See* Cetyl alcohol
Palmityl palmitate. *See* Cetyl palmitate
Palmityl stearate. *See* Cetyl stearate
Palm kernel methyl ester; Palm kernel oil methyl ester. *See* Methyl palm kernelate

Pancreatin

CAS 8049-47-6; EINECS/ELINCS 232-468-9
Definition: Mixt. of enzymes obtained from the fresh pancreas of animals
Properties: Cream to yel. amorphous powder; sol. in water; insol. in alcohol
Toxicology: LD50 (oral, rat) > 10 g/kg, (IP, rat) 2550 mg/kg, (subcut., mouse) 3870 mg/kg; mod. toxic by subcut. and IP routes; irritant; sensitizer; TSCA listed
HMIS: Health 2, Flammability 1, Reactivity 0
Uses: Food enzyme; nutrient; digestive enzyme; animal feed digestion aid; bating leather; starch and protein sizing removal from textiles; spotting solvent in dry cleaning; hair conditioner, skin conditioner in cosmetics
Features: Changes protein into proteoses and derived substances, and starch into dextrins and sugars
Regulatory: FDA 21CFR §184.1583, GRAS; Canada, Japan approved
Manuf./Distrib.: AMRESCO; Aceto; Aldrich; Alfa Chem; Am. Labs; Asiamerica Int'l.; Barrington; Bioiberica; Degussa AG/Health & Nutrition; EM Science; Fluka; J.T. Baker; Marcor Development; Pangaea Sciences; Penta Mfg.; R.W. Greeff; RIA Int'l.; Ruger; Scientific Protein Labs; Sigma; Spectrum Quality Prods.; Stason Pharmaceuticals; VWR Int'l.; Voigt Global Distrib.; ZenPharm Int'l.

Paraacetaldehyde; Paracetaldehyde. *See* Paraldehyde
Parachlorobenzotrifluoride. *See* p-Chlorobenzotrifluoride
Parachlorotoluene. *See* p-Chlorotoluene
Paracymene; Paracymol. *See* p-Cymene
Paradioxybenzene. *See* Hydroquinone

Paraffin

CAS 8002-74-2; EINECS/ELINCS 232-315-6
UN 1223; FEMA 3216; INS905c(ii)
Synonyms: Hard paraffin; Paraffin wax; Paraffin waxes; Paraffin wax fume; Petroleum wax, crystalline; Poly (methylene) wax
Classification: Aliphatic organic compd.; hydrocarbon
Definition: Solid mixture of hydrocarbons obtained from petroleum fractions by solvent crystallization or by the sweating process; consists predominantly of straight-chain hydrocarbons > C20; characterized by relatively large crystals
Empirical: C_nH_{2n+2}
Properties: Colorless to wh. cryst. solid; odorless; tasteless; greasy feel; sol. in benzene, gasoline, chloroform, ether, carbon disulfide, oils; misc. with fats; insol. in water, alcohol; dens. ≈ 0.9; m.p. 50-57 C; flash pt. (CC) 190 C
Toxicology: ACGIH TLV/TWA 2 mg/m³ (fume); TDLo (implant, rat) 120 mg/kg; primary irritant; skin and eye irritant; anesthetic effect; chronic skin exposure can cause dermatitis, abnormal pigmentation, etc.; fumes cause lung damage; questionable carcinogen; experimental tumorigen by implantation; many paraffin waxes contain carcinogens; TSCA listed
Precaution: Dangerous fire hazard; will burn above 198 C; burns with luminous flame; when heated, produces irritating fumes
NFPA: Health 0, Flammability 1, Reactivity 0
Storage: Store away from heat; store in cool, dry place; refrigerate
Uses: Candles; paper coating; glazing agent, protective sealant for foods; food coatings; plastics lubricant; hot-melt carpetbacking; floor polishes; masticatory substance in chewing gum; raising m.p. of ointments; antiozonant for rubber; mfg. of varnishes; in elec. insulators; perfume extraction (from flowers); paper sizing agent; waterproofing agent for textiles, leather, paper, cork, wood; solvent for inks; emollient, visc. control agent, moisturizer, stabilizer for cosmetics; pharmaceuticals (stiffener, tablet coating); defoamer, surf. lubricant in food-contact coatings; in cellophane for food pkg.; wood preservative for agric. prods. pkg.; pkg. material for irradiated foods
Regulatory: FDA 21CFR §133.150, 133.181, 133.189, 172.275, 172.615, 175.105, 175.210, 175.250, 175.300, 175.320, 176.170, 176.200, 177.1200, 177.2420, 177.2600, 177.2800, 178.3710, 178.3800, 178.3910, 179.45; FEMA GRAS; Canada, Japan approved; FDA approved for implants, orals, topicals; NF, BP, EP compliance
Manuf./Distrib.: Acros Org.; Aldrich; Ashland; Avatar; BP Amoco; BP Global Spec. Prods.; Baker Petrolite; Boith China; British Wax Refining; Chemacon GmbH; Chemcor; ChevronPhillips; Cornelius Chem. Co. Ltd; EM Ind.; Eggar & Co.; Elementis Spec.; ExxonMobil; Fisher Scientific; Frank B. Ross; Honeywell Spec. Wax & Addit.; Humphrey; Int'l. Group; Janssen-Cilag; Koster Keunen; Magnesia GmbH; Merck KGaA; Michelman; Micro Powders; Mutchler; Penreco; Penta Mfg.; Polygon; Polysciences; RTD Chems.; Ruger; Sasol Germany; Sasol N. Am.; Scheel; Sea-Land; Shamrock Tech.; Shell; Spec. Ingreds. & Perfume; Spectrum Quality Prods.; Stevenson Cooper; Strahl & Pitsch; Tryline; U.S. Synthetics; Vopak USA; Whittaker, Clark & Daniels

Paraffin, chlorinated

CAS 63449-39-8
Synonyms: Chlorcosane; Chlorinated paraffin; Chlorocosane; Chloroparaffin; Paraffin waxes and hydrocarbon waxes, chlorinated; Wax, chlorinated
Definition: Paraffin hydrocarbons treated with chlorine; contains about 50% Cl
Properties: Pale to amber, neutral, light viscous oils or soft waxes; sol. in benzene, chloroform, most org. solvs.; sl. sol. in alcohol; insol. in water; dens. 0.900-1.50; nonflamm.
Toxicology: LD50 (oral, rat) > 4000 mg/kg (liq. or resinous, 70% Cl); primary irritant; confirmed carcinogen; TSCA listed
Hazardous Decomp. Prods.: Heated to decomp., emits acrid smoke and irritating fumes
Uses: In mfg. of textiles, draperies, etc.; solvent for dichloramine-T; high-pressure lubricants; flame retardant in plastics, rubber, adhesives, textiles, inks, paper; leather fatliquoring agent; plasticizer and flame retardant in paints; plasticizer for PVC, sealants, PVAc, syn. rubber; EP

agent in lubricants, metalworking fluids; tackifier for rubbers, adhesives, sealants; textile backcoating; extender in rubber and plastic compds.; in detergents; in food-pkg. adhesives; crosslinking agent in food-contact polysulfide polymer/polyepoxy resins
Regulatory: FDA 21CFR §175.105 (for chain lengths of C_{10}-C_{17}, contg. 40-70% Cl by wt.), 177.1650
Manuf./Distrib.: Allchem Ind.; Atofina; C.P. Hall; Celanese; Charlotte Chem.; ChemService; Chemcentral; Dover; Eastech; Ferro/Keil; Morton Int'l.; Mosselman NV; Occidental; Pechiney Chems. Div.; PolyOne/Elastomers & Perf. Addit.; Punda Mercantile; R.E. Carroll; S & S Chem.; Sea-Land; StanChem; Tosoh; Tri-Iso

Paraffin distillate

CAS 64742-04-7
Synonyms: Extracts (petroleum), heavy paraffinic distillate solvent; Heavy paraffinic distillate, solvent extract; Mineral oil, petroleum extracts, heavy paraffinic distillate solvent; Petroleum extracts, heavy paraffinic distillate solvent
Definition: Distilled petroleum fraction which when cooled consists of a mixture of crystalline wax and oil
Toxicology: Confirmed carcinogen; TSCA listed
Hazardous Decomp. Prods.: Heated to decomp., emits acrid smoke and irritating fumes
Uses: Solvent

Paraffinic hydrocarbon solvent, normal. *See* Paraffins, normal C5-20
Paraffin jelly. *See* Petrolatum
Paraffin oil. *See* Mineral oil

Paraffins, normal C5-20

CAS 64771-72-8
Synonyms: C5-20 normal paraffins; Heavy normal paraffins concentrate (petroleum); Heavy normal paraffins (petroleum); Light normal paraffin concentrate (petroleum); Light normal paraffins (petroleum); Normal paraffins; Paraffinic hydrocarbon solvent, normal; Paraffins (petroleum), normal C5-20
Classification: Petroleum distillate
Properties: Negligible sol. in water; m.w. 212; sp.gr. 0.773 (15.6/15.6 C); visc. 2.53 cSt; vapor pressure < 0.5 mm Hg (38 C); b.p. 251-280 C; pour pt. 7 C; flash pt. (PMCC) > 120 C; autoignition temp. ≈ 210 C
Toxicology: ACGIH TLV 5 mg/m^3 (oil mist); LD50 (oral, rat) > 5 g/kg, (skin, rabbit) > 3.16 g/kg; high vapor concs. may cause eye/respiratory tract irritation, headaches, dizziness, anesthesia, drowsiness, unconsciousness, other CNS effects death; prolonged/repeated skin contact may cause irritation, dermatitis; may cause eye irritation; if aspirated, may cause mild to severe pulmonary injury, death; TSCA listed
Environmental: Prevent entry into sewers, watercourses
Precaution: Volatile; invisible vapors; liq. or vapor may settle in low areas; distant ignition/explosion possible; incompat. with strong oxidizing agents (liq. chlorine, conc. oxygen, sodium or calcium hypochlorite)
Hazardous Decomp. Prods.: combustion: fumes, smoke, CO, aldehydes
HMIS: Health 1, Flammability 1, Reactivity 0
Storage: Keep away from heat, sparks, open flames, static electricity; keep containers closed when not in use
Uses: Solvent
Regulatory: DOT nonregulated; SARA 311/312 nonhazardous, §313 nonreportable
Trade Names: Exxpar 451; Exx-Print® 287 D; Jayflex® 215; Norpar® 12 Fluid; Norpar® 13 Fluid; Norpar® 15 Fluid

Paraffins (petroleum), normal C5-20. *See* Paraffins, normal C5-20
Paraffinum liquidum. *See* Mineral oil
Paraffin wax; Paraffin waxes. *See* Paraffin
Paraffin waxes and hydrocarbon waxes, chlorinated. *See* Paraffin, chlorinated
Paraffin wax fume. *See* Paraffin
Paraform. *See* Formaldehyde
Paral. *See* Paraldehyde

Paraldehyde

CAS 123-63-7; EINECS/ELINCS 204-639-8
UN 1264; FEMA 4010
Synonyms: Acetaldehyde, trimer; Elaldehyde; Paraacetaldehyde; Paracetaldehyde; Paral; PCHO; 2,4,6-Trimethyl-1,3,5-trioxacyclohexane; 2,4,6-Trimethyl-1,3,5-trioxane; 2,4,6-Trimethyl-s-trioxane; s-Trimethyltrioxymethylene
Empirical: $C_6H_{12}O_3$
Properties: Colorless to ylsh. clear liq.; pleasant sweet aromatic odor; disagreeable, burning taste; very sol. in chloroform, ether, 95% ethanol; sparingly sol. in water; m.w. 132.16; sp.gr. 0.9943 (20/4 C); m.p. 12.6 C; b.p. 123-128 C; flash pt. 35.5 C; ref. index 1.4049 (20 C)
Toxicology: LD50 (oral, rat) 1530 mg/kg, (skin, rabbit) 14 g/kg; LCLo (inh., rat, 4 h) 2000 ppm; mod. toxic by ing.; low toxicity by skin contact; irritating to eyes, skin, respiratory system; risk of serious eye damage; may cause incoordination, drowsiness; lg. doses may cause dilation of pupils, loss of reflexes, coma; may cause weak pulse followed by cyanosis; death results from respiratory paralysis; psychotropic effects on humans; experimental neoplastigen via dermal route; target organ: CNS; TSCA listed
Precaution: Highly flamm.; incompat. with rubber, plastics, alkalis, hydrocyanic acid, iodides, and oxidizers; dec. by light and air and on prolonged storage; easily depolymerized to acetaldehyde
Hazardous Decomp. Prods.: CO, CO_2
NFPA: Health 2, Flammability 3, Reactivity 1
Storage: Store refrigerated; keep container tightly closed under inert atmosphere; protect from light
Uses: Substitute for acetaldehyde; accelerator and antioxidant for rubber; intermediate for organic chems., dyestuffs, vulcanization accelerators, rubber oxidants; solvent for fats, oils, gums, rubber, resins, leather; solvent mixts. for cellulose derivs.; synthetic flavoring agent in foods; sedative (hypnotic drug)
Regulatory: FEMA GRAS; DEA controlled substance; SARA §313 reportable
Manuf./Distrib.: AccuStandard; Alfa Aesar; Fluka; Lonza; Sigma; Spectrum Quality Prods.; TCI Am.; Wutong Aroma

Paranpahthalene. *See* Anthracene
Paraxylene. *See* p-Xylene
PCBTF. *See* p-Chlorobenzotrifluoride
PCHO. *See* Paraldehyde
PCT. *See* p-Chlorotoluene
PEA; β-PEA. *See* Phenethyl alcohol
Peach aldehyde. *See* γ-Undecalactone

Peanutamide MEA

CAS 93572-05-5; EINECS/ELINCS 297-433-2
Synonyms: Fatty acids, peanut oil, compds. with ethanolamine; N-(2-Hydroxyethyl) peanut acid amide; Monoethanolamine peanut acid amide; Peanut fatty acid amide, N-(2-hydroxyethyl)-; Peanut fatty acid monoethanolamide
Definition: Mixture of ethanolamides of fatty acids derived from peanut oil
Formula: $RCO–NHCH_2CH_2OH$, RCO– represents fatty acids derived from peanut oil
Uses: Emulsifier, emulsion stabilizer, surfactant, visc. control agent in cosmetics; detergent intermediate; solvent

Peanutamide MIPA

Synonyms: Amides, peanut oil, N-(2-hydroxypropyl); N-(2-Hydroxypropyl) peanut acid amide; Peanut fatty acid amide, N-(2-hydroxypropyl)-
Definition: Mixture of isopropanolamides of fatty acids derived from peanut oil
Formula: $RCO–NHCH_2CCH_3HOH$, RCO– represents fatty acids derived from peanut oil
Uses: Emulsifier, emulsion stabilizer, surfactant, visc. control agent in cosmetics; detergent intermediate; solvent

Peanut (Arachis hypogaea) oil

CAS 8002-03-7; EINECS/ELINCS 232-296-4
Synonyms: Arachis hypogaea; Arachis oil; Earthnut oil; Groundnut oil; Indigenous peanut oil; Katchung oil; Peanut oil; Pecan shell powder
Definition: Refined fixed oil obtained from seed kernels of one or more cultivated varieties of *Arachis hypogaea*
Properties: Colorless to pale yel. liq., nutty odor, bland taste; sol. in benzene, ether, chloroform, CCl_4, oils; misc. with carbon disulfide; sl. sol. in alcohol; insol. in alkalis; dens. 0.916-0.922; solid. pt. -5 C; iodine no. 84-100; sapon. no. 185-195; flash pt. 540 F; ref. index 1.466-1.470

Toxicology: TDLo (oral, mouse, 1 yr intermittent) 952 g/kg; human skin irritant, mild allergen; questionable carcinogen; experimental tumorigen; mutagenic data; TSCA listed
Precaution: Combustible exposed to heat or flame; can react with oxidizing materials; sl. spontaneous heating
Hazardous Decomp. Prods.: Heated to decomp., emits acrid smoke and irritating fumes
NFPA: Health 0, Flammability 1, Reactivity 0
Storage: Light-sensitive
Uses: Drying oil for paints; soap ingred.; emollient, solvent in cosmetics; heat transfer medium in laboratory; edible oil, dietary supplement, coating agent, emulsifier, formulation aid, texturizer in foods, margarine, salad oil, shortening; substitute for olive oil; pharmaceutical solvent, vehicle; nutritive emulsions; softener for ear wax, enemas, emollient creams; gall bladder evacuant; in food-pkg. adhesives; defoamer in food-contact paper coatings, paper/paperboard; in food-contact textiles
Features: Oleaginous
Regulatory: FDA 21CFR §164.150, 175.105, 176.200, 176.210, 177.2800, 182.70; GRAS; FDA approved for injectables, orals, vaginals; NF, BP, EP compliance
Manuf./Distrib.: A&E Connock; ABITEC; Aarhus Oliefabrik A/S; Adept Sol'ns.; Aldivia; Aldrich; Allchem Int'l. Ltd; Alnor Oil; Anglia Oils; Arista Ind.; Ashland; Avatar; Berje; Brutanicals; Bunge Foods; CanAmera Foods; Cascade; Charkit; Chesham Chems. Ltd; Croda Inc; Croda Oleochems.; Flavors of N. Am.; H.E. Daniel Ltd; J.T. Baker; Jarchem Ind.; Jeen Int'l.; John L Seaton Ltd; KIC Chems.; Karlshamns AB; Lambent Tech.; MLG Enterprises; Maypro Ind.; Mosselman NV; Mutchler; Natural Oils Int'l.; Penta Mfg.; Pokonobe Ind.; Ruger; S. Black; SKW Chems.; Sea-Land; Sigma; Spectrum Quality Prods.; Tri-K Ind.; Univar E&E Ltd; Universal Preserv-A-Chem; Vamo-Fuji Spec.; Van Den Burg Eiprodukten BV; Voigt Global Distrib.; Welch, Holme & Clark
Trade Names: Super Refined® Peanut NF

Peanut fatty acid amide, N-(2-hydroxyethyl)-. *See* Peanutamide MEA
Peanut fatty acid amide, N-(2-hydroxypropyl)-. *See* Peanutamide MIPA
Peanut fatty acid monoethanolamide. *See* Peanutamide MEA
Peanut oil. *See* Peanut (Arachis hypogaea) oil
Pearl ash. *See* Potassium carbonate
Pear oil. *See* Amyl acetate; Isoamyl acetate
Pecan shell powder. *See* Peanut (Arachis hypogaea) oil
PEEA. *See* Phenylethylethanolamine
PEG. *See* Polyethylene glycol

PEG-4

CAS 112-60-7; 25322-68-3 (generic); EINECS/ELINCS 203-989-9
Synonyms: Bis [2-(2-hydroxyethoxy) ethyl] ether; 2-[2-(2-(2-Hydroxyethoxy) ethoxy) ethoxy] ethanol; Macrogol 200; 2,2´-[Oxybis (2,1-ethanediyloxy)] bisethanol; 2,2´-(Oxybis (ethyleneoxy)) diethanol; PEG 200; POE (4); TEG; Tetraethylene glycol; Tetraglycol
Classification: Polyhydric alcohol
Definition: Polymer of ethylene oxide
Empirical: $C_8H_{18}O_5$
Formula: $H(OCH_2CH_2)_4OH$
Properties: Colorless to pale straw visc. liq., sl. char. odor; sol. in water; misc. with methanol; insol. in benzene, toluene, gasoline; m.w. 194.23; dens. 1.127 (25/25 C); vapor pressure > 0.001 mm Hg (20 C); f.p. -4 C; b.p. 327.3 C; flash pt. (COC) 185 C; ref. index 1.4577 (20 C); surf. tens. 44.13 dynes/cm; supercools on freezing; nonionic
Toxicology: LD50 (oral, rat) 28,900 mg/kg, (IP, mouse) 7500 mg/kg; mildly toxic by ing.; irritating to eyes, skin, respiratory system; reproductive hazard; target organs: CNS, GI, respiratory system; TSCA listed
Environmental: VOC; COD 1.64; ThOD 1.65
Precaution: Combustible; LEL 1%; can react with oxidizers; solvent action on some plastics
Hazardous Decomp. Prods.: Heated to decomp., emits acrid smoke and irritating fumes
NFPA: Health 0, Flammability 1, Reactivity 0
Storage: Hygroscopic
Uses: Lubricant for rubber molds, textile fibers, metalworking fluids; in foods and food pkg.; humectant, solvent in cosmetics; hair preps.; in gas chromatography; in lacquers; paints; paper coatings; polishes; solvent for nitrocellulose; plasticizer; binder in ceramic pastes, printing inks; softener for paper tissue; food coating, binder, plasticizer, lubricant, defoamer; pharmaceutic aid; pharmaceutical solvent, vehicle, bulking agent; in food-pkg. adhesives
Regulatory: FDA 21CFR §73.1, 172.210, 172.770, 172.820, 173.310, 173.340, 175.105, 175.300, 178.3750; FDA approved for topicals; NF compliance
Manuf./Distrib.: Aldrich; Ashland; C.P. Hall; Cognis; Coyne; DelCon; Delta Distributors; Dow; Fluka; Marcus Int'l.; NOF Am.; Oxiteno; PolyOne/Elastomers & Perf. Addit.; Ruger; Sigma; Spectrum Quality Prods.; Uniqema
Trade Names: Carbowax® E200; DePEG 200; PEG 200; Pluracol® E200

PEG-6

CAS 2615-15-8; 25322-68-3 (generic); EINECS/ELINCS 203-989-9; 220-045-1
Synonyms: Hexaethylene glycol; Macrogol 300; PEG 300; 3,6,9,12,15-Pentaoxaheptadecane-1,17-diol; Polyethylene glycol 300
Definition: Polymer of ethylene oxide
Empirical: $C_{12}H_{26}O_7$
Formula: $H(OCH_2CH_2)_nOH$, avg. n = 6
Properties: Sp.gr. 1.124-1.127; visc. 5.4-6.4 cSt (99 C); pour pt. -15 to -8 C; flash pt. (COC) 196 C; ref. index 1.463-1.4641 (20 C); pH 4.5-7.5 (5%); nonionic
Toxicology: LD50 (oral, rat) 27,500 mg/kg, (IP, rat) 17,000 mg/kg; mildly toxic by ing.; TSCA listed
Hazardous Decomp. Prods.: Heated to decomp., emits acrid smoke and irritating fumes
NFPA: Health 0, Flammability 1, Reactivity 0
Uses: Lubricant for rubber molds, textile fibers, metalworking; in food and food pkg.; humectant, solvent in cosmetics; hair preps.; pharmaceutic aid; in gas chromatography; in paints; paper coatings; polishes; ceramics; plasticizer for cellulose nitrate; food coating, binder, plasticizer, lubricant, flavor adjuvant, bodying agent, defoamer; pharmaceutical coating, suppository base, lubricant, solvent; in food-pkg. adhesives
Regulatory: FDA 21CFR §172.210, 172.770, 172.820, 173.310, 173.340, 175.105, 175.300, 178.3750, 178.3910; FDA approved for parenterals, ophthalmics, topicals; NF compliance
Manuf./Distrib.: Aldrich; Ashland; Cognis; DelCon; Dow; Fluka; NOF Am.; PolyOne/Elastomers & Perf. Addit.; Ruger; Sigma; Spectrum Quality Prods.; Uniqema
Trade Names: Carbowax® E300; Carbowax® E300 NF; Carbowax® PEG 300; Carbowax® Sentry® PEG 300; DePEG 300; Lutrol® E 300; PEG 300; Pluracol® E300; Renex® PEG 300
Trade Names Containing: Carbowax® Sentry® PEG 540 Blend

PEG-8

CAS 5117-19-1; 25322-68-3 (generic); EINECS/ELINCS 225-856-4
Synonyms: 3,6,9,12,15,18,21-Heptaoxatricosane-1,23-diol; Macrogol 400; PEG 400; POE (8); Polyethylene glycol 400
Definition: Polymer of ethylene oxide
Empirical: $C_{16}H_{34}O_9$
Formula: $H(OCH_2CH_2)_nOH$, avg. n = 8
Properties: Visc. liq., sl. char. odor, minimal taste; m.w. 380-420; dens. 1.128 (25/25 C); visc. 7.3 cSt (210 F); m.p. 4-8 C; pH 4.5-7.5 (5%); nonionic
Toxicology: LD50 (oral, rat) 30 ml/kg, (IP, rat) 9708 mg/kg, (IV, rat) 7313 mg/kg; low toxicity by ing., IV, IP routes; TSCA listed
Hazardous Decomp. Prods.: Heated to decomp., emits acrid smoke and irritating fumes
NFPA: Health 0, Flammability 1, Reactivity 0
Storage: Sl. hygroscopic
Uses: Lubricant for rubber molds, textile fibers, metalworking; in food and food pkg.; humectant, solvent in cosmetics; hair preps.; pharmaceutic aid; in gas chromatography; in paints; paper coatings; polishes; ceramics; plasticizer for cellulose nitrate; food coating, binder, plasticizer, lubricant, flavor adjuvant, bodying agent, defoamer; pharmaceutical coating, suppository base, bulking agent, solvent; in food-pkg. adhesives
Regulatory: FDA 21CFR §172.210, 172.770, 172.820, 173.310, 173.340, 175.105, 175.300, 178.3750, 178.3910, 181.22, 181.30; NF compliance

Manuf./Distrib.: Aldrich; Ashland; C.P. Hall; Cognis; DelCon; Dow; Fluka; NOF Am.; PolyOne/Elastomers & Perf. Addit.; Ruger; Sigma; Spectrum Quality Prods.; Uniqema
Trade Names: Carbowax® E400; Carbowax® E400 NF; Carbowax® Sentry® PEG 400; DePEG 400; Lutrol® E 400; PEG 400; Pluracol® E400; Pluracol® E400 NF; Pogol® 400; Renex® PEG 400

PEG-9

CAS 3386-18-3; 25322-68-3 (generic); EINECS/ELINCS 222-206-1
Synonyms: 3,6,9,12,15,18,21,24-Octaoxahexacosane-1,26-diol; PEG 450; POE (9)
Definition: Polymer of ethylene oxide
Empirical: $C_{18}H_{38}O_{10}$
Formula: $H(OCH_2CH_2)_nOH$, avg. n = 9
Properties: Nonionic
Toxicology: TSCA listed
Uses: Lubricant for rubber molds, textile fibers, metalworking; in food and food pkg.; humectant, solvent in cosmetics; hair preps.; pharmaceutic aid; in gas chromatography; in paints; paper coatings; polishes; ceramics; food coating, binder, plasticizer, lubricant, flavor adjuvant, bodying agent, defoamer; pharmaceutical bulking agent, solvent; in food-pkg. adhesives
Regulatory: FDA 21CFR §172.210, 172.770, 172.820, 173.310, 173.340, 175.105, 175.300, 178.3750, 178.3910
Manuf./Distrib.: Aldrich; Fluka; Sigma

PEG-10

CAS 5579-66-8; 25322-68-3 (generic); EINECS/ELINCS 226-962-3
Synonyms: 3,6,9,12,15,18,21,24,27-Nonaxanonacosane-1,29-diol; PEG 500; POE (10)
Classification: Polymer of ethylene oxide
Empirical: $C_{20}H_{42}O_{11}$
Formula: $H(OCH_2CH_2)_nOH$, avg. n = 10
Uses: Humectant, solvent in cosmetics; coatings on fresh citrus fruit; food pkg. adhesives; resinous/polymeric food-contact coatings; in surf. lubricants for mfg. of food-contact metallic articles
Regulatory: FDA 21CFR §172.210, 172.770, 172.820, 173.310, 173.340, 175.105, 175.300, 178.3750, 178.3910

PEG-12

CAS 6790-09-6; 25322-68-3 (generic); EINECS/ELINCS 229-859-1
Synonyms: Macrogol 600; PEG 600; POE (12); Polyethylene glycol 600; 3,6,9,12,15,18,21,24,27,3,33-Undecaoxapentatriacontane-1,35-diol
Definition: Polymer of ethylene oxide
Empirical: $C_{24}H_{50}O_{13}$
Formula: $H(OCH_2CH_2)_nOH$, avg. n = 12
Properties: Visc. liq., char. odor; m.w. 570-630; dens. 1.128 (25/25 C); m.p. 20-25 C; visc. 10.5 cSt (210 F); pH 4.5-7.5 (5%); nonionic
Toxicology: LD50 (oral, rat) 38,100 mg/kg; low toxicity by ing.; eye irritant; TSCA listed
Hazardous Decomp. Prods.: Heated to decomp., emits acrid smoke and irritating fumes
NFPA: Health 0, Flammability 1, Reactivity 0
Storage: Sl. hygroscopic
Uses: Lubricant for rubber molds, textile fibers, metalworking; in food and food pkg.; humectant, solvent in cosmetics; hair preps.; pharmaceutic aid; in gas chromatography; in paints; paper coatings; polishes; ceramics; plasticizer for cellulose nitrate; food coating, binder, plasticizer, lubricant, flavor adjuvant, bodying agent, defoamer; pharmaceutical coating, suppository base, bulking agent, solvent; in food-pkg. adhesives
Regulatory: FDA 21CFR §172.210, 172.770, 172.820, 173.310, 173.340, 175.105, 175.300, 178.3750, 178.3910; NF compliance
Manuf./Distrib.: Aldrich; Ashland; C.P. Hall; DelCon; Dow; Fluka; NOF Am.; PolyOne/Elastomers & Perf. Addit.; Ruger; Sigma; Spectrum Quality Prods.; Uniqema
Trade Names: Carbowax® E600; Carbowax® E600 NF; Carbowax® Sentry® PEG 600; DePEG 600; Lutrol® E 600; Pluracol® E600; Pluracol® E600 NF; Renex® PEG 600

PEG-14

CAS 25322-68-3 (generic)
Synonyms: PEG 700; POE (14)
Definition: Polymer of ethylene oxide
Empirical: $C_{28}H_{58}O_{15}$
Formula: $H(OCH_2CH_2)_nOH$, avg. n = 14
Properties: Visc. 11.5-13.0 cSt; nonionic
Toxicology: TSCA listed
Uses: Lubricant for rubber molds, textile fibers, metalworking; in food and food pkg.; humectant, solvent in cosmetics; hair preps.; pharmaceutic aid; in gas chromatography; in paints; paper coatings; polishes; ceramics; food coating, binder, plasticizer, lubricant, flavor adjuvant, bodying agent, defoamer; pharmaceutical bulking agent, solvent; in food-pkg. adhesives
Regulatory: FDA 21CFR §172.210, 172.770, 172.820, 173.310, 173.340, 175.105, 175.300, 178.3750, 178.3910; NF compliance
Manuf./Distrib.: Aldrich; Fluka; Sigma
Trade Names: Rhodasurf® PEG 600

PEG-16

CAS 25322-68-3 (generic)
Synonyms: PEG 800; POE (16)
Definition: Polymer of ethylene oxide
Empirical: $C_{32}H_{66}O_{17}$
Formula: $H(OCH_2CH_2)_nOH$, avg. n = 16
Properties: Visc. 12.5-14.5 cSt; nonionic
Toxicology: TSCA listed
Uses: Lubricant for rubber molds, textile fibers, metalworking; in food and food pkg.; humectant, solvent in cosmetics; hair preps.; pharmaceutic aid; in gas chromatography; in paints; paper coatings; polishes; ceramics; food coating, binder, plasticizer, lubricant, flavor adjuvant, bodying agent, defoamer; pharmaceutical bulking agent, solvent; in food-pkg. adhesives
Regulatory: FDA 21CFR §172.210, 172.770, 172.820, 173.310, 173.340, 175.105, 175.300, 178.3750, 178.3910; NF compliance
Manuf./Distrib.: Aldrich; DelCon; Fluka; Sigma

PEG-18

CAS 25322-68-3 (generic)
Synonyms: PEG (18); POE (18)
Classification: Polymer of ethylene oxide
Formula: $H(OCH_2CH_2)_nOH$, avg. n = 18
Uses: Humectant, solvent in cosmetics; coatings on fresh citrus fruit; food pkg. adhesives; resinous/polymeric food-contact coatings; in surf. lubricants for mfg. of food-contact metallic articles
Regulatory: FDA 21CFR §172.210, 172.770, 172.820, 173.310, 173.340, 175.105, 175.300, 178.3750, 178.3910

PEG (18). *See* PEG-18

PEG-20

CAS 25322-68-3 (generic); EINECS/ELINCS 203-989-9
Synonyms: Macrogol 1000; PEG 1000; POE (20); Polyethylene glycol 1000; Polyglycol 1000
Definition: Polymer of ethylene oxide
Empirical: $C_{40}H_{82}O_{21}$
Formula: $H(OCH_2CH_2)_nOH$, avg. n = 20
Properties: Solid; sp.gr. 1.085; visc. 16-19 cSt (99 C); pour pt. 37-40 C; flash pt. (COC) 265 C; pH 4.5-7.5 (5%); nonionic
Toxicology: LD50 (oral, rat) 32 g/kg, (IP, rat) 15,570 mg/kg; LDLo (IV, dog) 3000 mg/kg; mod. toxic by IP, IV routes; mildly toxic by ing.; questionable carcinogen; experimental tumorigen; TSCA listed
Hazardous Decomp. Prods.: Heated to decomp., emits acrid smoke and irritating fumes
NFPA: Health 0, Flammability 1, Reactivity 0
Uses: Lubricant for rubber molds, textile fibers, metalworking; in food and food pkg.; humectant, solvent in cosmetics; hair preps.; pharmaceutic aid; in gas chromatography; in paints; paper coatings; polishes; ceramics; plasticizer for cellulose nitrate; food coating, binder, plasticizer, lubricant, flavor adjuvant, bodying agent, defoamer; pharmaceutical coating, ointment and suppository base, bulking agent, solvent, plasticizer, lubricant; tablet binder, lubricant; in food-pkg. adhesives
Regulatory: FDA 21CFR §172.210, 172.770, 172.820, 173.310, 173.340, 175.105, 175.300, 178.3750, 178.3910; NF compliance
Manuf./Distrib.: Aldrich; Ashland; DelCon; Dow; Fluka; NOF Am.; PolyOne/Elastomers & Perf. Addit.; Ruger; Sigma; Spectrum Quality Prods.;

Uniqema
Trade Names: Carbowax® E1000; Carbowax® E1000 NF; Carbowax® Sentry® PEG 900; Carbowax® Sentry® PEG 1000

PEG-32

CAS 25322-68-3 (generic); EINECS/ELINCS 203-989-9
Synonyms: Macrogol 1540; PEG 1540; POE (32)
Definition: Polymer of ethylene oxide
Empirical: $C_{64}H_{130}O_{33}$
Formula: $H(OCH_2CH_2)_nOH$, avg. n = 32
Properties: Wh. powd.; nonionic
Toxicology: LD50 (oral, rat) 44,200 mg/kg; mildly toxic by ingestion; human skin irritant; TSCA listed
Hazardous Decomp. Prods.: Heated to decomp., emits acrid smoke and irritating fumes
Uses: Lubricant for rubber molds, textile fibers, metalworking; in food and food pkg.; humectant, solvent in cosmetics; hair preps.; pharmaceutic aid; in gas chromatography; in paints; paper coatings; polishes; ceramics; food coating, binder, plasticizer, lubricant, flavor adjuvant, bodying agent, defoamer; pharmaceutical coating, ointment base, bulking agent; tablet binder, lubricant; solvent; solubilizer; emulsifier; in food-pkg. adhesives
Regulatory: FDA 21CFR §172.210, 172.770, 172.820, 173.310, 173.340, 175.105, 175.300, 178.3750, 178.3910; FDA approved for dentals, orals, rectals, topicals
Manuf./Distrib.: Aldrich; Fluka; NOF Am.; Sigma
Trade Names: Carbowax® E1450; Carbowax® E1450 NF; Carbowax® Sentry® PEG 1450; Lutrol® E 1500; Pluracol® E1450 NF
Trade Names Containing: Carbowax® Sentry® PEG 540 Blend

PEG-40

CAS 25322-68-3 (generic)
Synonyms: PEG 2000; POE (40)
Definition: Polymer of ethylene oxide
Empirical: $C_{80}H_{162}O_{41}$
Formula: $H(OCH_2CH_2)_nOH$, avg. n = 40
Properties: Solid; dens. 1.127; visc. 38-49 cSt; ref. index 1.4590 (20 C); nonionic
Toxicology: TSCA listed
Uses: Lubricant for rubber molds, textile fibers, metalworking; in food and food pkg.; humectant, solvent in cosmetics; hair preps.; pharmaceutic aid; in gas chromatography; in paints; paper coatings; polishes; ceramics; food coating, binder, plasticizer, lubricant, flavor adjuvant, bodying agent, defoamer; pharmaceutical coating, ointment base, bulking agent; tablet binder, lubricant; in food-pkg. adhesives
Regulatory: FDA 21CFR §172.210, 172.770, 172.820, 173.310, 173.340, 175.105, 175.300, 178.3750, 178.3910; FDA approved for parenterals; NF compliance
Manuf./Distrib.: Aldrich; NOF Am.

PEG-75

CAS 25322-68-3 (generic); EINECS/ELINCS 203-989-9
Synonyms: Macrogol 4000; PEG 4000; POE (75); Polyethylene glycol 4000; Polyglycol 4000
Definition: Polymer of ethylene oxide
Formula: $H(OCH_2CH_2)_nOH$, avg. n = 75
Properties: Wh. powd. or creamy-wh. flakes; m.w. 3000-3700; dens. 1.212 (25/25 C); visc. 76-110 cSt (210 F); m.p. 54-58 C; pH 4.5-7.5 (5%); nonionic
Toxicology: LD50 (oral, rat) 50 g/kg, (IP, rat) 11,550 mg/kg, (subcut., mouse) 18 g/kg, (IV, mouse) 16 g/kg; mildly toxic by ing.; skin irritant; TSCA listed
Hazardous Decomp. Prods.: Heated to decomp., emits acrid smoke and irritating fumes
NFPA: Health 0, Flammability 1, Reactivity 0
Uses: Lubricant for rubber molds, textile fibers, metalworking; in food and food pkg.; binder, humectant, solvent in cosmetics; hair preps.; pharmaceutic aid; in gas chromatography; in paints; paper coatings; polishes; ceramics; food coating, binder, plasticizer, lubricant, flavor adjuvant, bodying agent, defoamer; pharmaceutical coating, ointment and suppository base, bulking agent, solvent, plasticizer; in food-pkg. adhesives
Regulatory: FDA 21CFR §172.210, 172.770, 172.820, 173.310, 173.340, 175.105, 175.300, 178.3750, 178.3910; FDA approved for injectables, orals, rectals, topicals, vaginals; NF compliance
Manuf./Distrib.: Aldrich; Fluka; Sigma; Spectrum Quality Prods.; Uniqema
Trade Names: Carbowax® E3350; Carbowax® E3350 NF; Carbowax® Sentry® PEG 3350; Lutrol® E 3350; Pluracol® E4000

PEG-90

CAS 25322-68-3 (generic); EINECS/ELINCS 203-989-9
Synonyms: PEG (90); POE (90)
Definition: Polymer of ethylene oxide
Formula: $H(OCH_2CH_2)_nOH$, avg. n = 90
Properties: Nonionic
Uses: Humectant, solvent in cosmetics, pharmaceuticals; food additive; antistat; chem. intermediate; dye carrier; lubricant; mold release agent; tablet binder/coating; in adhesives, ceramic glazes, creams/lotions, mandrel releases, mining applics., soaps/detergents, food-pkg. adhesives
Regulatory: FDA 21CFR §175.105
Trade Names: Lutrol® E 4000

PEG (90). *See* PEG-90

PEG-100

CAS 25322-68-3 (generic); EINECS/ELINCS 203-989-9
Synonyms: PEG (100); POE (100)
Definition: Polymer of ethylene oxide
Formula: $H(OCH_2CH_2)_nOH$, avg. n = 100
Properties: Nonionic
Uses: Lubricant for rubber molds, textile fibers, metalworking; in food and food pkg.; humectant, solvent in cosmetics; hair preps.; pharmaceutic aid; in gas chromatography; in paints; paper coatings; polishes; ceramics; food coating, binder, plasticizer, lubricant, flavor adjuvant, bodying agent, defoamer; pharmaceutical coating, ointment base, bulking agent, tablet binder; solvent; vehicle; in food-pkg. adhesives
Regulatory: FDA 21CFR §172.210, 172.770, 172.820, 173.310, 173.340, 175.105, 175.300, 178.3750, 178.3910
Manuf./Distrib.: Aldrich; Fluka; Sigma
Trade Names: Carbowax® E4500 NF; Carbowax® Sentry® PEG 4600

PEG (100). *See* PEG-100

PEG-125

CAS 25322-68-3 (generic)
Definition: Polymer of ethylene oxide
Properties: Nonionic
Uses: Solvent, emulsifier, carrier, solubilizer, absorp. enhancer, bioavailability enhancer for drugs that are insol. or sparingly sol. in water, e.g., steroids, fat-sol. vitamins; binder in mfg. of sugar-coated and uncoated tablets; controls release in sustained-release formulations; used with buffers in laxative formulations; antisticking agent , gloss aid for sugar-coated tablets; film-former, plasticizer in tablet coatings; base for ointments and suppositories
Trade Names: Lutrol® E 6000

PEG-135

CAS 25322-68-3 (generic); EINECS/ELINCS 203-989-9
Synonyms: PEG (135); POE (135)
Definition: Polymer of ethylene oxide
Formula: $H(OCH_2CH_2)_nOH$, avg. n = 135
Properties: Nonionic
Uses: Humectant, solvent in cosmetics; surfactant intermediate; humectant for printing inks; plasticizer; binder; lubricant; rubber molds; water softener; paper additive; coupling agent in creams and lotions; in food-pkg. adhesives
Regulatory: FDA 21CFR §175.105
Manuf./Distrib.: Aldrich; Fluka; Sigma

PEG (135). *See* PEG-135

PEG-150

CAS 25322-68-3 (generic); EINECS/ELINCS 203-989-9
Synonyms: Macrogol 6000; PEG 6000; POE (150)
Definition: Polymer of ethylene oxide

Formula: $H(OCH_2CH_2)_nOH$, avg. n = 150
Properties: Creamy-wh. flakes or powd., waxy paraffin-like appearance; very sol. in water, methylene chloride; pract. insol. in alcohol, ether, fatty and min. oils; m.w. 5600-7000; dens. 1.21 (25/25 C); visc. 470-900 cSt (210 F); m.p. 56-63 C; hyd. no. 16-22; flash pt. > 887 F; nonionic
Toxicology: LDLo (oral, rat) 50 g/kg; LD50 (IP, rat) 6790 mg/kg; mildly toxic by ing.; skin irritant; mutagenic data; TSCA listed
Precaution: Combustible exposed to heat or flame
Hazardous Decomp. Prods.: Heated to decomp., emits acrid smoke and irritating fumes
NFPA: Health 0, Flammability 1, Reactivity 0
Uses: Lubricant for rubber molds, textile fibers, metalworking; in food and food pkg.; binder, humectant, solvent in cosmetics; hair preps.; pharmaceutic aid; in gas chromatography; in paints; paper coatings; polishes; ceramics; food coating, binder, plasticizer, lubricant, flavor adjuvant, bodying agent, defoamer; pharmaceutical coating, ointment and suppository base, bulking agent, solvent, plasticizer; in food-pkg. adhesives
Regulatory: FDA 21CFR §172.210, 172.770, 172.820, 173.310, 173.340, 175.300, 177.2420, 178.3750, 178.3910; NF compliance
Manuf./Distrib.: Aldrich; DelCon; Fluka; NOF Am.; Sigma; Spectrum Quality Prods.; Uniqema
Trade Names: Carbowax® E8000; Carbowax® E8000 NF; Carbowax® Sentry® PEG 8000; Pluracol® E6000

PEG-200
CAS 25322-68-3 (generic)
Synonyms: PEG 9000; POE (200)
Definition: Polymer of ethylene oxide
Formula: $H(OCH_2CH_2)_nOH$, avg. n = 200
Properties: Solid; nonionic
Toxicology: TSCA listed
Uses: Lubricant for rubber molds, textile fibers, metalworking; in food and food pkg.; humectant, solvent in cosmetics; hair preps.; pharmaceutic aid; in gas chromatography; in paints; paper coatings; polishes; ceramics; food coating, binder, plasticizer, lubricant, flavor adjuvant, bodying agent, defoamer; pharmaceutical coating, ointment base, bulking agent, tablet binder; in food-pkg. adhesives
Regulatory: FDA 21CFR §172.210, 172.770, 172.820, 173.310, 173.340, 175.105, 175.300, 178.3750, 178.3910; FDA approved for intramuscular injectables, orals, topicals
Manuf./Distrib.: Aldrich; Fluka; Ruger; Sigma

PEG 200. *See* PEG-4
PEG 300. *See* PEG-6

PEG-350
CAS 25322-68-3 (generic)
Synonyms: PEG 20000; POE (350)
Definition: Polymer of ethylene oxide
Formula: $H(OCH_2CH_2)_nOH$, avg. n = 350
Uses: Lubricant for rubber molds, textile fibers, metalworking; binder, solvent in cosmetics; hair preps.; pharmaceutic aid; in gas chromatography; in paints; paper coatings; polishes; ceramics; foam stabilizer in fermented malt beverages; in food-pkg. adhesives
Regulatory: FDA 21CFR §172.770, 173.310, 175.105, 175.300, 178.3910
Manuf./Distrib.: Aldrich; Fluka; NOF Am.; Sigma

PEG 400. *See* PEG-8
PEG 450. *See* PEG-9
PEG 500. *See* PEG-10
PEG 600. *See* PEG-12
PEG 700. *See* PEG-14
PEG 800. *See* PEG-16
PEG 1000. *See* PEG-20
PEG 1540. *See* PEG-32
PEG 2000. *See* PEG-40
PEG 4000. *See* PEG-75
PEG 6000. *See* PEG-150
PEG 9000. *See* PEG-200
PEG 20000. *See* PEG-350

PEG-6 abietate
Synonyms: PEG 300 abietate; POE (6) abietate
Properties: Nonionic
Uses: Additive for cutting oils; solvent, emulsifier for solvents, creams, cleansing milks and pesticides; dispersant

PEG 300 abietate. *See* PEG-6 abietate
PEG-2 butyl ether. *See* Butoxydiglycol
PEG-3 butyl ether. *See* Butoxytriglycol
PEG 400 caprylate/caprate glycerides. *See* PEG-8 caprylic/capric glycerides

PEG-8 caprylic/capric glycerides
CAS 57307-99-0; 85536-07-8
Synonyms: PEG 400 caprylate/caprate glycerides; POE (8) caprylate/caprate glycerides
Definition: PEG deriv. of a mixt. of mono-, di-, and triglycerides of caprylic and capric acids with avg. 8 moles EO
Properties: Oil; nonionic
Uses: Emulsifier in cosmetics; surfactant for drug delivery systems; bioavailability enhancer, emulsifier, solvent, solubilizer, excipient for pharmaceuticals
Features: Hydrophilic
Trade Names: Labrasol®

PEG-30 castor oil
CAS 61791-12-6 (generic)
Synonyms: POE (30) castor oil
Definition: PEG deriv. of castor oil with avg. 30 moles of EO
Properties: Nonionic
Toxicology: TSCA listed
Uses: Lubricant, emulsifier in lubricants for plastics, metals, textiles; clay/pigment dispersant, rewetting agent, softener, dyeing assistant for paints, paper, textiles, and leather; in food-pkg. adhesives; surfactant, emulsifier in cosmetics; emulsifier, wetting agent in pharmaceuticals
Regulatory: FDA 21CFR §175.105, 175.300, 177.2800
Manuf./Distrib.: DelCon; Fluka; Sigma
Trade Names: Incrocas 30

PEG-33 castor oil
CAS 61791-12-6 (generic)
Synonyms: POE (33) castor oil
Definition: PEG deriv. of castor oil with avg. 33 moles of EO
Properties: Nonionic
Uses: Surfactant, emulsifier in cosmetics; solvent, emulsifier for pharmaceuticals; in food-pkg. adhesives; defoamer in food-contact paper/paperboard; in food-contact textiles
Regulatory: FDA 21CFR §175.105, 175.300, 176.210, 177.2800
Manuf./Distrib.: Fluka; Sigma

PEG-35 castor oil
CAS 61791-12-6 (generic)
Synonyms: POE (35) castor oil; Polyoxyl 35 castor oil
Definition: PEG deriv. of castor oil with avg. 35 moles of EO
Properties: Yel. oily liq., faint char. odor, sl. bitter taste; very sol. in water; sol. in alcohol, ethyl acetate; insol. in min. oils; sp.gr. 1.057; vapor pressure 6 2 mm Hg (20 C); acid no. 2 max.; iodine no. 25-35; sapon. no. 60-76; hyd. no. 65-80; nonionic
Toxicology: LD50 (IV, mouse) 6500 mg/kg; may be harmful by ing., inh., or skin absorption; may cause eye/skin/respiratory tract irritation; may cause sensitization by inh. and skin contact; may cause anaphylactic reactions; may cause a hepatitis-like reaction in large doses; may cause GI disturbances on exposure to lg. amts.; TSCA listed
Storage: Store and use with adequate ventilation
Uses: Surfactant, solubilizer, emulsifier, lubricant, emollient, superfatting agent, antistat, softener, detergent, wetting agent for pharmaceuticals, cosmetics, veterinary medicine; solvent for pharmaceuticals (intravenous steroid anesthetic, miconazole); emulsifier for essential oils; in food-pkg. adhesives; food-contact textiles
Regulatory: FDA 21CFR §175.105, 175.300, 177.2800; NF compliance
Manuf./Distrib.: Fluka; Sigma
Trade Names: Cremophor® EL; Cremophor® ELP; Etocas 35 NF

PEG-40 castor oil

CAS 61791-12-6 (generic)
Synonyms: PEG 2000 castor oil; POE (40) castor oil; Polyoxyl 40 castor oil
Definition: PEG deriv. of castor oil with avg. 40 moles of EO
Properties: Nonionic
Toxicology: TSCA listed
Uses: Lubricant, emulsifier in lubricants for plastics, metals, textiles; clay/pigment dispersant, rewetting agent, softener, dyeing assistant for paints, paper, textiles, and leather; surfactant, emulsifier in cosmetics; emulsifier, solubilizer, dispersant for vitamins and pharmaceuticals; in food-pkg. adhesives
Regulatory: FDA 21CFR §175.105, 175.300, 176.170, 176.180, 176.210, 177.2800; FDA approved for parenterals
Manuf./Distrib.: DelCon; Fluka; Sigma
Trade Names: Incrocas 40

PEG 2000 castor oil. *See* PEG-40 castor oil
PEG-16 cetyl ether. *See* Ceteth-16
PEG-24 cetyl ether. *See* Ceteth-24
PEG-24 cholesteryl ether. *See* Choleth-24
PEG-6 decyl ether; PEG 300 decyl ether. *See* Deceth-6
PEG-2 dibenzoate; PEG 100 dibenzoate. *See* Diethylene glycol dibenzoate

PEG-4 dilaurate

CAS 9005-02-1 (generic)
Synonyms: PEG (4) dilaurate; PEG 200 dilaurate; POE (4) dilaurate; Polyoxyl 4 dilaurate
Definition: PEG diester of lauric acid
Empirical: $C_{32}H_{62}O_7$
Formula: $CH_3(CH_2)_{10}CO(OCH_2CH_2)_nOCO(CH_2)_{10}CH_3$, avg. n = 4
Properties: Sol. in alcohol; sol. warm in veg. and min. oils; disp. in water; nonionic
Toxicology: TSCA listed
HMIS: Health 1, Flammability 1, Reactivity 0
Uses: Surfactant, emulsifier, dispersant, solubilizer, visc. control agent, defoamer, lubricant, wetting agent, cosolvent for cosmetics, pharmaceuticals, plastics, textiles, metalworking, agric., foods, water treatment; in food-pkg. adhesives; adjuvant in mfg. of paper/paperboard in contact with aq./fatty/dry foods; defoamer in food-contact coatings
Regulatory: FDA 21CFR §175.105, 175.300, 176.170, 176.180, 176.200, 176.210; FDA approved for topicals
Manuf./Distrib.: A&E Connock; Mosselman NV; Ruger

PEG (4) dilaurate. *See* PEG-4 dilaurate

PEG-8 dilaurate

CAS 9005-02-1 (generic)
Synonyms: PEG 400 dilaurate; POE (8) dilaurate
Definition: PEG diester of lauric acid
Empirical: $C_{40}H_{78}O_{11}$
Formula: $CH_3(CH_2)_{10}CO(OCH_2CH_2)_nOCO(CH_2)_{10}CH_3$, avg. n = 8
Properties: Sol. in alcohol; sol. warm in veg. and min. oils; disp. in water; sp.gr. 1.030; m.p. 15 C; HLB 10.4; flash pt. (COC) 249 C; ref. index 1.459; nonionic
Toxicology: TSCA listed
Uses: Emulsifier, solubilizer, cosolvent, dispersant, wetting agent, lubricant, thickener, emollient, plasticizer, softener, release agent, coupling agent in personal care prods., pharmaceuticals, industrial applics., agric. chemical sprays, industrial and textile lubricants; in food-pkg. adhesives; defoamer in food-contact paper/paperboard; in closures with sealing gaskets for food containers; in resin-bonded filters for food contact; in food-contact textiles
Regulatory: FDA 21CFR §175.105, 175.300, 176.210, 177.1210, 177.2260, 177.2800, 178.3520
Manuf./Distrib.: A&E Connock; C.P. Hall; Inolex; Mosselman NV; Ruger; Velsicol
Trade Names: Cithrol 4DL

PEG-12 dilaurate

CAS 9005-02-1 (generic)
Synonyms: PEG 600 dilaurate; POE (12) dilaurate
Definition: PEG diester of lauric acid
Empirical: $C_{48}H_{94}O_{15}$
Formula: $CH_3(CH_2)_{10}CO(OCH_2CH_2)_nOCO(CH_2)_{10}CH_3$, avg. n = 12
Properties: Nonionic
Toxicology: TSCA listed
HMIS: Health 1, Flammability 1, Reactivity 0
Uses: Dispersant, emulsifier for metalworking lubricants, fiber lubricants and softeners, solubilizers, defoamers, antistats, cosmetics, pharmaceuticals, plastics, agric., and chemical intermediates; cosolvent; emollient; opacifier; spreading agent; thickener; in food-pkg. adhesives; defoamer in food-contact paper/paperboard; in resin-bonded filters for food contact; in food-contact textiles
Regulatory: FDA 21CFR §175.105, 175.300, 176.210, 177.2260, 177.2800
Manuf./Distrib.: A&E Connock; Ruger

PEG-20 dilaurate

CAS 9005-02-1 (generic)
Synonyms: PEG 1000 dilaurate; POE (20) dilaurate
Definition: PEG diester of lauric acid
Empirical: $C_{64}H_{126}O_{23}$
Formula: $CH_3(CH_2)_{10}CO(OCH_2CH_2)_nOCO(CH_2)_{10}CH_3$, avg. n = 20
Properties: Nonionic
Toxicology: TSCA listed
Uses: Dispersant, emulsifier, wetting agent, cosolvent, solubilizer, thickener, emollient, opacifier, spreading agent in cosmetics, pharmaceuticals, foods, agric., plastics, other industries; defoamer in food-contact paper/paperboard; in resinous/polymeric food-contact coatings; in resin-bonded filters for food contact; in food-contact textiles
Regulatory: FDA 21CFR §175.300, 176.210, 177.2260, 177.2800
Manuf./Distrib.: A&E Connock; Ruger

PEG 200 dilaurate. *See* PEG-4 dilaurate
PEG 400 dilaurate. *See* PEG-8 dilaurate
PEG 600 dilaurate. *See* PEG-12 dilaurate
PEG 1000 dilaurate. *See* PEG-20 dilaurate

PEG dimethyl ether

CAS 24991-55-7 (generic)
Synonyms: Dimethoxy polyethylene glycol; Glycols, polyethylene, dimethyl ether; α,ω-Methoxypoly (ethylene oxide); Polyethylene glycol dimethyl ether; Poly (oxy-1,2-ethanediyl), α-methyl-ω-methoxy-
Formula: $(C_2H_4O)_xC_2H_6O$
Toxicology: TDLo (oral, male rabbit, 1 day premating) 2500 mg/kg; reproductive effector; TSCA listed
Uses: Solvent; brake fluid component; solvent for removing CO, CO_2, SO_2, H_2S and other sour-gas components from ammonia gas streams, for absorp. of fluorocarbons and other halogenated liqs./gases; stack gas scrubbing; solubilizer for plastics, textiles, and paper
Trade Names: Polyglyme

PEG-3 dimethyl ether

CAS 112-49-2; 24991-55-7 (generic)
Synonyms: Glyme-3; 2,5,8,11-Tetraoxadodecane; Triethylene glycol dimethyl ether; Triglyme
Classification: Glycol ether
Empirical: $C_8H_{18}O_4$
Formula: $CH_3(OCH_2CH_2)_3OCH_3$
Properties: Water-wh. liq.; mild ether odor; misc. with water, hydrocarbon and most org. solvs.; m.w. 178.26; dens. 0.990 (20/4 C); vapor pressure 0.02 mm Hg (20 C); m.p. -45 C; b.p. 215 C; flash pt. 111 C; autoignition temp. 190 C; ref. index 1.4233; pH 7.0; surf. tens. 29.4 dynes/cm
Toxicology: LD50 (oral, rat) 5000 mg/kg; TDLo (oral, mouse) 7500 mg/kg, (oral, rabbit) 3500 mg/kg; mutagenic; experimental teratogen, reproductive effects; TSCA listed
Environmental: VOC; ThOD 1.08
Precaution: Flamm. exposed to heat or flame
Hazardous Decomp. Prods.: Heated to decomp., emits acrid smoke and irritating fumes
NFPA: Health 1, Flammability 1, Reactivity 0
Uses: Complexing agent; solvent for gases; in electrochem.; polymer/boron chem.; gas absorp.; extraction; stabilization; fuels; lubricants; textiles; solvent for pharmaceuticals; pesticides; coupling agent for immiscible liqs.; solubilizer for organic media

Regulatory: SARA §313 reportable; HAP
Manuf./Distrib.: Ablestik; Aldrich; Brand-Nu Labs; Celanese; Ferro/Grant; Fluka; Sigma
Trade Names: Triglyme

PEG-4 dimethyl ether

CAS 24991-55-7 (generic)
Empirical: $C_{10}H_{22}O_5$
Formula: $CH_3O(CH_2CH_2O)_4CH_3$
Properties: M.w. 222.28
Uses: Complexing agent; solubilizer; selective solvent in Selexol process; solvent in electrochemistry, polymer/boron chemistry, gas absorp., extraction, industrial prods. such as fuels, lubricants, textiles, pesticides; pharmaceutical solvent, drug carrier, stabilizer
Manuf./Distrib.: Aldrich; Fluka
Trade Names: Tetraglyme

PEG-4 dioleate

CAS 9005-07-6 (generic); 52688-97-0 (generic); 134141-38-1
Synonyms: 9-Octadecenoic acid, oxybis (2,1-ethanediyloxy-2,1-ethanediyl) ester; PEG 200 dioleate; POE (4) dioleate
Definition: PEG diester of oleic acid
Empirical: $C_{44}H_{82}O_7$
Properties: Nonionic
Toxicology: LD50 (oral, rat) 1900 mg/kg; TSCA listed
Uses: Surfactant, emulsifier, coemulsifier, thickener, solubilizer, cosolvent, emollient, opacifier, spreading agent, wetting agent, dispersant for cosmetics, pharmaceuticals, foods, agric., plastics, metalworking, fiber lubricants, etc.; defoamer in processing foods; in food-pkg. adhesives; defoamer in food-contact paper/paperboard
Regulatory: FDA 21CFR §173.340, 175.105, 175.300, 176.210
Manuf./Distrib.: A&E Connock; Ruger

PEG-8 dioleate

CAS 9005-07-6 (generic); 52688-97-0 (generic)
Synonyms: PEG 400 dioleate; POE (8) dioleate
Definition: PEG diester of oleic acid
Empirical: $C_{52}H_{98}O_{11}$
Properties: Sol. in veg. and min. oils; partly sol. in alcohol; disp. in water; nonionic
Toxicology: LD50 (oral, rat) 1900 mg/kg; TSCA listed
Uses: Emulsifier, solubilizer, cosolvent, lubricant, dispersant, defoamer, corrosion inhibitor, wetting agent for cosmetics, pharmaceuticals, textiles, metalworking, agric., plastics, foods, water treatment; defoamer for beet sugar and yeast processing; defoamer in food-contact coatings; in food-pkg. adhesives; in paper/paperboard in contact with aq./fatty foods
Regulatory: FDA 21CFR §173.340, 175.105, 175.300, 176.170, 176.200, 176.210, 177.1210, 177.2260, 177.2800
Manuf./Distrib.: A&E Connock; DelCon; Mosselman NV; Ruger
Trade Names: Cithrol 4DO

PEG-12 dioleate

CAS 9005-07-6 (generic); 52688-97-0 (generic); 85736-49-8; EINECS/ELINCS 288-459-5
Synonyms: PEG 600 dioleate; POE (12) dioleate
Definition: PEG diester of oleic acid
Properties: Sol. in alcohol, veg. and min. oils; disp. in water; m.p. 15-20 C; nonionic
Toxicology: LD50 (oral, rat) 1900 mg/kg; TSCA listed
HMIS: Health 0, Flammability 1, Reactivity 0
Uses: Emulsifier; dispersant; additive for cutting oils; component of paper defoamers, softeners, lubricants; emulsifier in cosmetics, pharmaceuticals, foods, agric., plastics, solvent and solventless coatings; lubricant for textile spin finishes; thickener; superfatting agent; leveling agent; defoamer for beet sugar and yeast processing; cosolvent; food-pkg. adhesives; defoamer in food-contact coatings and paper/paperboard; in resinous/polymeric food-contact coatings; in resin-bonded filters for food contact; in food-contact textiles
Regulatory: FDA 21CFR §173.340, 175.105, 175.300, 176.200, 176.210, 177.2260, 177.2800
Manuf./Distrib.: A&E Connock; DelCon; Mosselman NV; Ruger

PEG-20 dioleate

CAS 9005-07-6 (generic); 52688-97-0 (generic)
Synonyms: PEG 1000 dioleate; POE (20) dioleate
Definition: PEG diester of oleic acid
Properties: Nonionic
Toxicology: LD50 (oral, rat) 1900 mg/kg; TSCA listed
Uses: Surfactant, emulsifier, solubilizer, cosolvent, thickener, emollient, opacifier, spreading agent, wetting agent, dispersant for cosmetics, pharmaceuticals, perfumes, foods, agric., plastics, other industries; defoamer in food-contact paper/paperboard; in resinous/polymeric food-contact coatings; in resin-bonded filters for food contact; in food-contact textiles
Regulatory: FDA 21CFR §175.300, 176.210, 177.2260, 177.2800
Manuf./Distrib.: A&E Connock; Ruger

PEG 200 dioleate. *See* PEG-4 dioleate
PEG 400 dioleate. *See* PEG-8 dioleate
PEG 600 dioleate. *See* PEG-12 dioleate
PEG 1000 dioleate. *See* PEG-20 dioleate
PEG-7 dodecyl phenyl ether. *See* Dodoxynol-7
PEG-3 ethyl ether. *See* Ethoxytriglycol
PEG-4 ethyl ether. *See* Tetraethylene glycol ethyl ether

PEG-7 glyceryl cocoate

CAS 66105-29-1; 68201-46-7 (generic)
Synonyms: PEG (7) glyceryl monococoate; POE (7) glyceryl monococoate
Definition: PEG ether of glyceryl cocoate
Formula: $RCO–OCH_2COHHCH_2(OCH_2CH_2)_nOH$, RCO– rep. fatty acids from coconut oil, avg. n = 7
Properties: Nonionic
Uses: Surfactant, emollient, superfatting agent, emulsifier, solubilizer, coupling agent for cosmetics, personal care prods., pharmaceuticals, lubricants, mold release compds.; dispersant for biologically act. ingreds.; mold release agent; plasticizer for syn. fabrics and plastics; food emulsifier; in resinous/polymeric food-contact coatings
Regulatory: FDA 21CFR §175.300
Manuf./Distrib.: A&E Connock; ABITEC
Trade Names: Milopol G-7; Protachem™ GC-7

PEG-30 glyceryl cocoate

CAS 68201-46-7 (generic)
Synonyms: POE (30) glyceryl monococoate
Definition: PEG ether of glyceryl cocoate
Formula: $RCO–OCH_2COHHCH_2(OCH_2CH_2)_nOH$, RCO– rep. fatty acids from coconut oil, avg. n = 30
Properties: Nonionic
Uses: Surfactant, detergent, emollient, emulsifier, solubilizer, thickener, superfatting agent for cosmetics, personal care prods., pharmaceuticals; anti-irritant for surfactants; in resinous/polymeric food-contact coatings; in food-contact textiles
Features: Low irritation
Regulatory: FDA 21CFR §175.300, 177.2800
Manuf./Distrib.: A&E Connock
Trade Names: Protachem™ GC-30

PEG-7 glyceryl ether. *See* Glycereth-7
PEG-26 glyceryl ether. *See* Glycereth-26
PEG-7 glyceryl ether triacetate. *See* Glycereth-7 triacetate
PEG (7) glyceryl monococoate. *See* PEG-7 glyceryl cocoate

PEG-15 hydroxystearate

Synonyms: POE (15) hydroxystearate
Definition: PEG ester of hydroxystearic acid
Formula: $CH_3(CH_2)_5CHOH(CH_2)_{10}CO(OCH_2CH_2)_nOH$, avg. n = 15
Uses: Emulsifier in cosmetics; solvent, solubilizer for injection sol'ns.
Trade Names: Solutol® HS 15

PEG-6 isostearate

CAS 56002-14-3 (generic)
Synonyms: PEG 300 monoisostearate; POE (6) monoisostearate
Definition: PEG ester of isostearic acid
Empirical: $C_{30}H_{60}O_8$
Formula: $C_{17}H_{35}CO(OCH_2CH_2)_nOH$, avg. n = 6

Properties: Nonionic
Toxicology: TSCA listed
Uses: Solvent, emulsifier, emulsion stabilizer, dispersant, solubilizer for cosmetics and pharmaceuticals
Manuf./Distrib.: A&E Connock
Trade Names: Olepal ISO

PEG-2 lactamide

Synonyms: Diethylene glycol lactamide
Classification: Amide
Empirical: $C_7H_{13}NO_4$
Formula: $CH_3CHOHCONH(CH_2CH_2O)_nH$, avg. n = 2
Uses: Coupling agent, lubricant, emollient, humectant for personal care prods.; solvent in cosmetic creams/lotions
Trade Names: Naetex-L

PEG-3 lanolate

CAS 68459-50-7 (generic)
Synonyms: POE (3) lanolate
Definition: PEG ester of lanolin acid
Formula: $RCO-(OCH_2CH_2)_nOH$, RCO– represents fatty acids from lanolin, avg. n = 3
Properties: Nonionic
Uses: Emulsifier in cosmetics; solvent

PEG-5 lanolate

CAS 68459-50-7 (generic)
Synonyms: PEG-5 lanolin acids; POE (5) lanolate
Definition: PEG ester of lanolin acid
Formula: $RCO-(OCH_2CH_2)_nOH$, RCO– rep. fatty acids from lanolin, avg. n = 5
Properties: Nonionic
Toxicology: TSCA listed
Uses: Emulsifier in cosmetics; o/w emulsifier, solvent for bromo acid dyes in lipsticks, skin care prods., makeup; solubilizer; wetting agent; dispersant

PEG-10 lanolate

CAS 68459-50-7 (generic)
Synonyms: PEG 500 lanolate; PEG-10 lanolin acid; POE (10) lanolate
Definition: PEG ester of lanolin acid
Formula: $RCO-(OCH_2CH_2)_nOH$, RCO– rep. fatty acids from lanolin, avg. n = 10
Properties: Nonionic
Toxicology: TSCA listed
Uses: Emulsifier in cosmetics; o/w emulsifier, solvent for bromo acid dyes in lipsticks, skin care and makeup prods.; solubilizer; wetting agent; dispersant

PEG-20 lanolate

CAS 68459-50-7 (generic)
Synonyms: PEG 1000 lanolate; POE (20) lanolate
Definition: PEG ester of lanolin acid
Formula: $RCO-(OCH_2CH_2)_nOH$, RCO– rep. fatty acids from lanolin, avg. n = 20
Properties: HLB 11.0; nonionic
Toxicology: TSCA listed
Uses: Emulsifier in cosmetics; o/w emulsifier, solvent for bromo acid dyes in lipsticks, skin care and makeup prods.; solubilizer; wetting agent; dispersant

PEG 500 lanolate. *See* PEG-10 lanolate
PEG 1000 lanolate. *See* PEG-20 lanolate

PEG-75 lanolin

CAS 8039-09-6; 61790-81-6 (generic)
Synonyms: PEG 4000 lanolin; POE (75) lanolin; Polyoxyl 75 lanolin
Definition: PEG deriv. of lanolin with avg. 75 moles of EO
Properties: Nonionic
Toxicology: TSCA listed
Uses: Surfactant, emulsifier, emollient, dispersant, foam stabilizer, plasticizer for cosmetics (antiperspirants, creams, lotions, makeup, hair prods.), pharmaceutical vehicles, textile processing; solvent, emollient, superfatting agent, conditioner for hair care, skin care, household detergents; solubilizer; wetting agent
Regulatory: FDA approved for topicals
Trade Names: Solulan® 75; Solulan® L-575

PEG 4000 lanolin. *See* PEG-75 lanolin
PEG-10 lanolin acid. *See* PEG-10 lanolate
PEG-5 lanolin acids. *See* PEG-5 lanolate
PEG-5 lanolin alcohol; PEG-5 lanolin ether. *See* Laneth-5
PEG-10 lanolin ether. *See* Laneth-10
PEG-15 lanolin ether. *See* Laneth-15
PEG-16 lanolin ether. *See* Laneth-16
PEG 500 lanolin ether. *See* Laneth-10
PEG-9 lauryl ether. *See* Laureth-9
PEG methyl ether. *See* Methoxy PEG
PEG-3 methyl ether. *See* Triglycol monomethyl ether
PEG-4 methyl ether. *See* Tetraethylene glycol methyl ether
PEG-6 methyl ether. *See* Methoxy PEG-6
PEG-16 methyl ether. *See* Methoxy PEG-16
PEG 300 methyl ether. *See* Methoxy PEG-6
PEG 500 methyl ether. *See* Methoxy PEG-10
PEG-10 methyl glucose ether. *See* Methyl gluceth-10
PEG-20 methyl glucose ether. *See* Methyl gluceth-20
PEG-9 monododecyl ether. *See* Laureth-9
PEG 300 monoisostearate. *See* PEG-6 isostearate
PEG-6 monomethyl ether. *See* Methoxy PEG-6
PEG (16) monomethyl ether. *See* Methoxy PEG-16
PEG 500 monomethyl ether. *See* Methoxy PEG-10
PEG 200 monooleate. *See* PEG-4 oleate
PEG 300 monooleate. *See* PEG-6 oleate
PEG 400 monooleate. *See* PEG-8 oleate
PEG 1000 monooleate. *See* PEG-20 oleate
PEG 100 monostearate. *See* PEG-2 stearate
PEG (100) monostearate. *See* PEG-100 stearate
PEG-4 nonyl phenyl ether. *See* Nonoxynol-4
PEG-5 nonyl phenyl ether. *See* Nonoxynol-5
PEG-6 nonyl phenyl ether. *See* Nonoxynol-6
PEG 200 nonyl phenyl ether. *See* Nonoxynol-4
PEG 300 nonyl phenyl ether. *See* Nonoxynol-6
PEG-9 nonyl phenyl ether phosphate; PEG 450 nonyl phenyl ether phosphate. *See* Nonoxynol-9 phosphate

PEG-4 oleate

CAS 9004-96-0 (generic); 10108-25-5; EINECS/ELINCS 233-293-0
Synonyms: 9-Octadecenoic acid, 2-[2-[2-(2-hydroxyethoxy) ethoxy] ethoxy] ethyl ester; PEG 200 monooleate; PEG 200 oleate; POE (4) monooleate
Definition: PEG ester of oleic acid
Empirical: $C_{26}H_{50}O_6$
Formula: $CH_3(CH_2)_7CHCH(CH_2)_7CO(OCH_2CH_2)_nOH$, avg. n = 4
Properties: Sol. in alcohol; partly sol. in veg. and min. oils; disp. in water; nonionic
Toxicology: TSCA listed
Uses: Wetting agent, penetrant, spreading agent, defoamer, detergent, dispersant, emulsifier, solubilizer, thickener, emollient, softener, lubricant for textiles, cosmetics, metalworking, foods, pharmaceuticals, agric., plastics; textile aux.; solvent in metal cleaning, degreasing; in food-pkg. adhesives; resinous/polymeric food-contact coatings; defoamer in food-contact paper/paperboard
Regulatory: FDA 21CFR §175.105, 175.300, 176.210
Manuf./Distrib.: A&E Connock; Mosselman NV; Ruger

PEG-5 oleate

CAS 9004-96-0 (generic); 23336-36-9
Synonyms: 14-Hydroxy-3,6,9,12-tetraoxatetradec-1-yl-9-octadecenoic acid; POE (5) monooleate
Definition: PEG ester of oleic acid
Empirical: $C_{28}H_{54}O_7$
Formula: $CH_3(CH_2)_7CH=CH(CH_2)_7CO(OCH_2CH_2)_nOH$, avg. n = 5
Properties: Nonionic
Toxicology: TSCA listed
Uses: Surfactant in cutting oils, degreasing solvents, metal cleaners; emulsifier in cosmetics; dyeing assistant for textile processing; lubricant; softener; coupling agent; in food-pkg. adhesives

Regulatory: FDA 21CFR §175.105, 175.300
Manuf./Distrib.: A&E Connock; DelCon
Trade Names: Chemax E-200 MO

PEG-6 oleate
CAS 9004-96-0 (generic); 60344-26-5
Synonyms: 9-Octadecenoic acid, 17-hydroxy-3,6,9,12,15-pentaoxaheptadec-1-yl ester; PEG 300 monooleate; POE (6) monooleate
Definition: PEG ester of oleic acid
Empirical: $C_{30}H_{58}O_8$
Formula: $CH_3(CH_2)_7CHCH(CH_2)_7CO(OCH_2CH_2)_nOH$, avg. n = 6
Properties: Sol. in alcohol, min. oils; disp. in water; m.p. 0 C; nonionic
Toxicology: TSCA listed
Uses: Surfactant, emulsifier, lubricant, antifoam, dispersant, thickener, solubilizer, emollient, spreading agent, wetting agent for cosmetics, foods, pharmaceuticals, agric., plastics, textiles; chemical intermediate; solvent for lipids; in food-pkg. adhesives; resinous/polymeric food-contact coatings; defoamer in food-contact paper/paperboard
Regulatory: FDA 21CFR §175.105, 175.300, 176.210
Manuf./Distrib.: A&E Connock; Mosselman NV

PEG-8 oleate
CAS 9004-96-0 (generic)
Synonyms: PEG 400 monooleate; POE (8) monooleate
Definition: PEG ester of oleic acid
Empirical: $C_{34}H_{66}O_{10}$
Formula: $CH_3(CH_2)_7CHCH(CH_2)_7CO(OCH_2CH_2)_nOH$, avg. n = 8
Properties: Dk. red oil; sol. in alcohol; disp. in water; misc. with cottonseed oil; nonionic
Toxicology: LD50 (IV, mouse) 1080 mg/kg; mod. toxic by IV route; eye irritant; TSCA listed
Hazardous Decomp. Prods.: Heated to decomp., emits acrid smoke and irritating fumes
HMIS: Health 0, Flammability 1, Reactivity 0
Uses: Emulsifier, dispersant, lubricant, solubilizer, solvent, visc. control agent for cosmetics, pharmaceuticals, foods, agric., plastics, coatings; chemical intermediate; in food-pkg. adhesives; in paper/paperboard in contact with aq./fatty foods; defoamer in food-contact coatings; in cellophane for food pkg.; food-contact textiles
Regulatory: FDA 21CFR §175.105, 175.300, 176.170, 176.200, 177.1200, 177.1210, 177.2260, 177.2800
Manuf./Distrib.: A&E Connock; Mosselman NV; Ruger

PEG-20 oleate
CAS 9004-96-0 (generic)
Synonyms: PEG 1000 monooleate; POE (20) monooleate
Definition: PEG ester of oleic acid
Formula: $CH_3(CH_2)_7CHCH(CH_2)_7CO(OCH_2CH_2)_nOH$, avg. n = 20
Properties: HLB 15.4; nonionic
Toxicology: LD50 (IV, mouse) 500 mg/kg; mod. toxic by IV route; skin and eye irritant; TSCA listed
Hazardous Decomp. Prods.: Heated to decomp., emits acrid smoke and irritating fumes
Uses: Raw material for finishing agents in the syn. fiber industry; surfactant, emulsifier, thickener, solubilizer, solvent, emollient, opacifier, spreading agent, wetting agent, dispersant for cosmetics, pharmaceuticals, foods, agric., plastics, paints; defoamer in food-contact paper coatings; in resinous/polymeric food-contact coatings; in resin-bonded filters for food contact; in food-contact textiles
Regulatory: FDA 21CFR §175.300, 176.200, 177.2260, 177.2800
Manuf./Distrib.: A&E Connock; Ruger
Trade Names: Chemax E-1000 MO

PEG 200 oleate. *See* PEG-4 oleate
PEG-2 oleyl alcohol; PEG-2 oleyl ether. *See* Oleth-2
PEG-16 oleyl ether. *See* Oleth-16
PEG 100 oleyl ether. *See* Oleth-2

PEG-8 palmitostearate
Uses: Solvent, emulsifier for cosmetics and pharmaceuticals

PEG-150 palmitostearate
Uses: Solvent, emulsifier for cosmetics and pharmaceuticals

PEG/PPG-17/6 copolymer
CAS 9003-11-6 (generic)
Definition: Random copolymer produced by interaction of 17 moles ethylene oxide with 6 moles of propylene oxide
Toxicology: TSCA listed
Uses: Solvent in cosmetics; lubricant for antiperspirants; emollient; bodying agent; lubricant for textile spin finishes; intermediate for prep. of resins, plasticizers, modifiers, and surfactants; demulsifier; food-pkg. adhesives, animal glues; defoamer in food-contact paper/paperboard
Manuf./Distrib.: Aldrich; Fluka; Sigma
Trade Names: Ucon® 75-H-450

PEG/PPG-18/4 copolymer
CAS 9003-11-6 (generic)
Definition: Random copolymer produced by interaction of 18 moles ethylene oxide with 4 moles of propylene oxide
Uses: Solvent in cosmetics
Manuf./Distrib.: Aldrich; Fluka; Sigma

PEG/PPG-23/50 copolymer
CAS 9003-11-6 (generic)
Definition: Random copolymer produced by interaction of 23 moles ethylene oxide with 50 moles of propylene oxide
Uses: Solvent in cosmetics
Manuf./Distrib.: Aldrich; Fluka; Sigma

PEG/PPG-296/57 copolymer
CAS 9003-11-6 (generic)
Definition: Random copolymer produced by interaction of 296 moles ethylene oxide with 57 moles of propylene oxide
Uses: Solvent in cosmetics

PEG/PPG-24/24 glycerin. *See* PPG-24-glycereth-24
PEG-7-PPG-2 isodecyl ether carboxylci acid, isopropyl ester. *See* Isopropyl PPG-2-isodeceth-7-carboxylate

PEG-20 sorbitan isostearate
CAS 66794-58-9 (generic)
Synonyms: PEG 1000 sorbitan monoisostearate; POE (20) sorbitan monoisostearate; Polysorbate 120
Definition: Ethoxylated sorbitan monoester of isostearic acid with avg. 20 moles EO
Properties: Nonionic
Uses: Surfactant, emulsifier, solubilizer for cosmetics, fragrances, perfumes; emulsifier, solvent, wetting agent for topical pharmaceuticals; o/w emulsifier, surfactant for creams, lotions, ointments, pesticides, polishes, cleaners, textiles
Manuf./Distrib.: A&E Connock
Trade Names: Crillet 6

PEG 1000 sorbitan monoisostearate. *See* PEG-20 sorbitan isostearate
PEG-20 sorbitan oleate. *See* Polysorbate 80

PEG-40 sorbitan peroleate
Synonyms: PEG 2000 sorbitan peroleate; POE (40) sorbitan peroleate; POE (40) sorbitol septaoleate
Definition: Mixture of oleic acid esters of sorbitol condensed with avg. 40 moles EO
Properties: Nonionic
Uses: Surfactant, emulsifier in cosmetics; emulsifier, solubilizer, antistat, lubricant for bath oils, textiles; w/o emulsifier, solvent, solubilizer, antistat, lubricant, spreading agent for pharmaceuticals; defoamer in food-contact paper/paperboard; in resinous/polymeric food-contact coatings
Regulatory: FDA 21CFR §175.300, 176.210
Manuf./Distrib.: A&E Connock
Trade Names: Arlatone® T

PEG 2000 sorbitan peroleate. *See* PEG-40 sorbitan peroleate
PEG-20 sorbitan stearate. *See* Polysorbate 60
PEG-20 sorbitol ether; PEG 1000 sorbitol ether. *See* Sorbeth-20
PEG-1 stearate. *See* Glycol stearate

PEG-2 stearate

CAS 106-11-6; 9004-99-3 (generic); 85116-97-8; EINECS/ELINCS 203-363-5; 285-550-1

Synonyms: Diethylene glycol, monoester with stearic acid; Diethylene glycol monostearate; Diethylene glycol stearate; Diglycol monostearate; Diglycol stearate; Glyco stearin; 2-(2-Hydroxyethoxy) ethyl ester stearic acid; 2-(2-Hydroxyethoxy) ethyl stearate; PEG 100 monostearate; Polyoxyl 2 stearate; Stearic acid, 2-(2-hydroxyethoxy) ethyl ester

Definition: PEG ester of stearic acid

Empirical: $C_{22}H_{44}O_4$

Formula: $CH_3(CH_2)_{16}CO(OCH_2CH_2)_nOH$, avg. n = 2

Properties: Wh. wax-like solid; faint fatty odor; sol. in hot alcohol, oils; m.w. 372.66; m.p. 45-55 C; HLB 4.3; nonionic

Toxicology: LD50 (IP, mouse) 200 mg/kg; poison by IV, IP routes; mildly toxic by ing.; TSCA listed

Precaution: Combustible

Hazardous Decomp. Prods.: Heated to decomp., emits acrid smoke and fumes

Uses: Emulsifier, plasticizer, lubricant, wetting agent, binder, thickener, dispersant, antistat, opacifier, pearlescent, stabilizer, solubilizer in cosmetics, topical pharmaceuticals, dry cleaning, leather, textiles, paper processing, rubber; metal drawing lubricants; protective coating for hygroscopic materials (tablets); mold release lubricant for die casting; binder for ceramics and grinding wheels; defoamer in food-contact coatings

Regulatory: FDA 21CFR §175.300, 176.200, 176.210, 177.2800; FDA approved for topicals

Manuf./Distrib.: A&E Connock; ABITEC; Adept Sol'ns.; Cognis/Chems. Group; Gattefosse; Inolex; Lipo; Lonza; Mosselman NV; Penta Mfg.; Ruger; Sigma; Sithean; Stepan; Universal Preserv-A-Chem

Trade Names: Protachem™ DGS

PEG-100 stearate

CAS 9004-99-3 (generic)

Synonyms: PEG (100) monostearate; POE (100) stearate

Definition: PEG ester of stearic acid

Formula: $CH_3(CH_2)_{16}CO(OCH_2CH_2)_nOH$, avg. n = 100

Properties: Nonionic

Toxicology: Poison by intravenous, intraperitoneal route; mildly toxic by ingestion; TSCA listed

Uses: Surfactant, emulsifier, emollient for cosmetics, hair preps.; solubilizer for bath oils and fragrances; lubricant, softener, antistat, emulsifier, detergent, wetting agent, stabilizer for textiles, tannery, household/industrial cleaners; emulsifier in foods, pharmaceuticals, insecticides, agrochems.; pigment wetting agent; mold release agent; defoamer in food-contact paper/paperboard; in resinous/polymeric food-contact coatings

Regulatory: FDA 21CFR §175.300, 176.210

Manuf./Distrib.: A&E Connock; DelCon; Sigma

Trade Names Containing: Protachem™ GMS-165

PEG-16 stearyl ether. *See* Steareth-16

PEG-3 triethanolamine

CAS 36936-60-4

Uses: Solvent neutralizer for acid dyes

Manuf./Distrib.: Aldrich

Trade Names: Emery® 5794

Pelargic acid; Pelargonic acid. *See* Nonanoic acid

1,4,7,10,13-Pentaazatridecane. *See* Tetraethylenepentamine

Pent-acetate. *See* Amyl acetate

Pentachloroethane

CAS 76-01-7; EINECS/ELINCS 200-925-1

UN 1669

Synonyms: Ethane pentachloride; Ethane, pentachloro-; Pentalin

Empirical: C_2HCl_5

Formula: Cl_2CHCCl_3

Properties: Colorless to ylsh. clear liq.; sweetish chloroform-like odor; sol. in DMSO, 95% ethanol, acetone, ether; sol. < 0.1 mg/ml in water; m.w. 202.3; sp.gr. 1.6796 (20/4 C); vapor pressure 3.4 mm Hg (20 C); m.p. -29 C; b.p. 161-162 C; flash pt. none; ref. index 1.5025 (20 C)

Toxicology: LDLo (oral, dog) 500 mg/kg, (IV, dog) 100 mg/kg, (subcut., rabbit) 700 mg/kg; LC50 (inh., rat, 2 h) 4238 ppm; poison by inh. and IV routes; mod. toxic by ing., subcut. routes; irritating to eyes, lungs, corneas, upper respiratory tract, skin, mucous membranes; strong narcotic effect; may cause drowsiness, giddiness, jaundice, headache, unconsciousness, blood effects, heart/liver/kidney damage, etc.; may be fatal; experimental carcinogen; target organs: nerves, liver, kidneys, lungs; TSCA listed

Precaution: Incompat. with strong oxidizing agents; mixts. with potassium may explode; violent reaction with potassium, alkalis, or metals; reacts violently with NaK alloy + bromoform

Hazardous Decomp. Prods.: Heated to decomp., emits highly toxic fumes of CO, CO_2, HCl; emits toxic fumes under fire conditions

Storage: Store in cool, dry place; keep tightly closed

Uses: Solvent for oil and grease in metal cleaning; separation of coal from impurities by dens. difference; chem. intermediate; mfg. of tetrachloroethylene and dichloroacetic acid; solvent for cellulose acetate, some cellulose ethers, resins, gums; drying agent for timber; in dry cleaning; soil sterilizing

Regulatory: SARA §313 reportable

Manuf./Distrib.: AccuStandard; Aldrich; Alfa Aesar; ICN Biomed. Research Prods.; Lancaster Synthesis; Pfaltz & Bauer; Riedel-deHaën; Supelco; TCI Am.

Pentadecane

CAS 629-62-9; EINECS/ELINCS 211-098-1

Uses: Fragrance in cosmetics; solvent

Trade Names Containing: Sasol C1316

3-(n-Penta-8´-decenyl) phenol. *See* Cashew nut shell oil

3-Pentadecylphenol. *See* 3-(n-Pentadecyl) phenol

3-(n-Pentadecyl) phenol

CAS 501-24-6; 3158-56-3

Synonyms: 3-Pentadecylphenol

Empirical: $C_{21}H_{36}O$

Formula: $CH_3(CH_2)_{14}C_6H_4OH$

Properties: M.w. 304.52; m.p. 50-53 C; b.p. 190-195 C (1 mm); flash pt. > 110 C

Toxicology: Irritant

Uses: Raw material for surfactants, antioxidants, anticorrosives; lubricant additive; cosolvent for insecticides, germicides; resin modifier; coupling agent

Manuf./Distrib.: Aldrich; Sigma

Trade Names: Cardolite® NC-510

Pentaerythrityl oleate

Definition: Ester of pentaerythritol and oleic acid

Uses: Corrosion inhibitor for lubricating oils and greases; lubricant; wetting agent, emulsifier for cosmetics, pharmaceuticals, agric., glass fibers, textiles, and leather applics.

Manuf./Distrib.: A.P. Chems. Ltd

Trade Names: Exceparl PE-MO

Pentaethylene glycol

CAS 4792-15-8; EINECS/ELINCS 225-341-4

Synonyms: Pentaglycol; 3,6,9,12-Tetraoxatetradecane-1,14-diol

Classification: Glycol

Empirical: $C_{10}H_{22}O_6$

Formula: $HO(CH_2CH_2O)_5H$

Properties: Colorless liq.; m.w. 238.28; dens. 1.126 g/ml; b.p. 184 C (2 mm Hg); flash pt. > 230 F; ref. index 1.462 (20 C)

Toxicology: LD50 (oral, guinea pig) 22,500 mg/kg; irritating to eyes, skin, respiratory system; may be harmful by inh., ing., skin absorp.; vapor/mist irritating to eyes, mucous membranes, upper respiratory tract; TSCA listed

Precaution: Incompat. with strong oxidizing agents, acid chlorides, acid anhydrides

Hazardous Decomp. Prods.: Toxic fumes of CO, CO_2; emits toxic fumes under fire conditions

Storage: Store in cool, dry place; keep tightly closed

Uses: Solvent

Manuf./Distrib.: Aldrich; Alfa Aesar; Lancaster Synthesis; TCI Am.

1,1,1,3,3-Pentafluoropropane

CAS 460-73-1; EINECS/ELINCS 419-170-6
Synonyms: HFC 245fa
Empirical: $C_3H_3F_5$
Formula: $CF_3CH_2CHF_2$
Properties: Colorless clear liq.; ethereal sweetish odor; misc. with other fluorocarbons, chlorine solvs., pentanes, hexanes, lower alcohols; sol. 0.16% in water; m.w. 134.05; dens. 1.32 g/cm^3 (20 C); vapor pressure 123 kPa (20 C); f.p. < -160 C; b.p. 15 C; flash pt. none
Toxicology: AIHA WEEL TWA (8 h) 300 ppm; LD50 (skin, rabbit) > 2000 mg/kg; LC50 (inh., rat, 4 h) > 200,000 ppm
Environmental: EC50 (daphnia magna, 48 h) > 97.9 mg/l, (rainbow trout, 96 h) > 81.8 mg/l
Uses: Solvent; blowing agent for PU foaming, appliance insulation, spray, boardstock, and panel
Manuf./Distrib.: ABCR; Fluorochem USA; Interchim; Lancaster Synthesis; Zhejiang Sanhuan Chems. Co. Ltd
Trade Names: Genesolv® SF
Trade Names Containing: Genesolv® S-TZ

Pentafluoro-1-propanol; 1H,1H-Pentafluoropropanol; 2,2,3,3,3-Pentafluoropropanol; 2,2,3,3,3-Pentafluoro-1-propanol. *See* Pentafluoropropyl alcohol

Pentafluoropropyl alcohol

CAS 422-05-9; EINECS/ELINCS 207-012-7
Synonyms: Pentafluoro-1-propanol; 1H,1H-Pentafluoropropanol; 2,2,3,3,3-Pentafluoropropanol; 2,2,3,3,3-Pentafluoro-1-propanol; PFPOH
Classification: Alcohol
Empirical: $C_3H_3F_5O$
Properties: Colorless liq.; sol. in water; m.w. 150.06; sp.gr. 1.51; b.p. 80 C (748 mm Hg); ref. index 1.288
Toxicology: LD50 (oral, mouse) 1000 mg/kg; LC50 (inh., mouse) 1629 ppm; mod. toxic by ing.; may be harmful by inh., skin absorp.; irritating to eyes, skin, mucous membranes, upper respiratory tract; TSCA listed
Environmental: VOC; ThOD 0.27
Precaution: Incompat. with strong oxidizing agents, acid chlorides, acid anhydrides
Hazardous Decomp. Prods.: CO, CO_2, hydrogen fluoride; heated to decomp., emits toxic fumes of F^-; emits toxic fumes under fire conditions
Storage: Store in cool, dry place; keep tightly closed
Uses: Solvent
Manuf./Distrib.: ABCR; Acros Org.; Aldrich; Alfa Aesar; Fluka; Fluorochem USA; ICN Biomed. Research Prods.; Indofine; Johnson Matthey; Lancaster Synthesis; Pfaltz & Bauer; Regis Tech.; Strem Chems.; TCI Am.

Pentaglycol. *See* Pentaethylene glycol
γ-Pentalactone. *See* γ-Valerolactone
Pentalin. *See* Pentachloroethane
Pentamethylene. *See* Cyclopentane
Pentamethyleneimine. *See* Piperidine
1-Pentanamine. *See* Pentylamine
1-Pentanamine, N-pentyl. *See* Di-n-amylamine
Pentan-2,4-dione. *See* Acetylacetone
Pentane. *See* n-Pentane

n-Pentane

CAS 109-66-0; EINECS/ELINCS 203-692-4
UN 1265 (DOT)
Synonyms: Alkane C5; Amyl hydride; Normal pentane; Pentane
Classification: Aliphatic hydrocarbon
Empirical: C_5H_{12}
Formula: $CH_3(CH_2)_3CH_3$
Properties: Colorless liq.; gasoline-like odor; sol. in alcohol, oxygenated, chlorinated, and hydrocarbon solvs.; insol. in water; m.w. 72.17; dens. 0.62624; vapor pressure 400 mm Hg (18.5 C); f.p. -129.7 C; b.p. 36 C; flash pt. (CC) -49 C; autoignition temp. 260 C; surf. tens. 15.48 dynes/cm; KB value 26; dielec. const. 1.841
Toxicology: ACGIH TLV/TWA 600 ppm; STEL 750 ppm; LD50 (IV, mouse) 446 mg/kg; mod. toxic by inh. and IV routes; narcotic in high concs. (dizziness, headache); prolonged skin contact may cause itching, redness, pigmentation, swelling, burning, pain; extreme exposure can cause nausea, confusion, persistent gasoline taste, unconsciousness; lethal @ 130,000 ppm; relatively low oral toxicity unless aspirated into lungs (severe lung irritation or damage can occur); TSCA listed
Environmental: VOC; ThOD 3.56; zero ozone depletion
Precaution: Highly flamm.; LEL 1.5%; UEL 7.8%; severe explosion hazard exposed to heat or flame; shock can shatter metal containers; incompat. with strong oxidizing agents (increases fire and explosion hazard)
Hazardous Decomp. Prods.: Heated to decomp., emits acrid smoke and irritating fumes
NFPA: Health 1, Flammability 4, Reactivity 0
Storage: Store in cool, dry, well-ventilated area, out of direct sunlight; limit quantities in use; avoid generating mists; refrigerated storage rec.
Uses: Solvent; extraction solvent; process solvent in lubricant deasphalting; propellant, solvent in cosmetics; reagent; artificial ice mfg.; thermometer fluid; blowing agent in plastics incl. food-contact expandable PS; pesticides; solvent in pharmaceuticals
Regulatory: FDA 21CFR §178.3010
Manuf./Distrib.: Aeropres; Aldrich; Ashland; Burdick & Jackson; ChevronPhillips; Fisher Scientific; Fluka; GFS; Hukill; Int'l. Commodities Export; Integra; J.T. Baker; Kira; LG Petrochem. Co. Ltd; Mallinckrodt Baker; Penta Mfg.; Romil Ltd; Ruger; Shell; Sigma; Sithean; South Hampton Refining; Spectrum Quality Prods.; Thomas Scientific; Triple Crown Am.; VWR Int'l.
Trade Names: Exxsol® Isopentane S; Exxsol® Pentane 80; Hydrosol Pentane
Trade Names Containing: Exxsol® Isopentane; Pentane 60/40; Pentane 80/20; S&L Naphtha

1,3-Pentanediamine

CAS 589-37-7
UN 2734
Synonyms: DAMP; 1,3-Diaminopentane
Empirical: $C_5H_{14}N_2$
Formula: $H_2NCH_2CH_2CH(C_2H_5)NH_2$
Properties: Liq., sl. odor; misc. with water; m.w. 102; dens. 0.855 g/ml; visc. 1.89 cp; vapor pressure 1 mm Hg (20 C); b.p. 164 C; flash pt. (CC) 59 C; surf. tens. 32.2 dyne/cm
Precaution: Corrosive, flamm.; incompat. with strong oxidants
Hazardous Decomp. Prods.: Thermal decomp. may emit NO_x
Uses: Epoxy curing agent; solvent for gas treatment; surfactant (asphalt emulsifiers, textiles); scale/corrosion inhibitor; additive for fuel, oil, plastic, petrol. and chem. processing; extender, catalyst for PU; in coatings, adhesives, sealants, elastomers
Manuf./Distrib.: Aldrich
Trade Names: Dytek® EP Diamine

Pentane, 2,3-dimethyl-. *See* 2,3-Dimethylpentane
Pentane, 2,4-dimethyl-. *See* 2,4-Dimethylpentane
Pentane, 3,3-dimethyl-. *See* 3,3-Dimethylpentane
Pentanedioic acid, dimethyl ester. *See* Dimethyl glutarate
Pentanedioic acid, monomethyl ester. *See* Methyl glutarate
2,4-Pentanediol, 2-methyl. *See* Hexylene glycol
1,3-Pentanediol, 2,2,4-trimethyl. *See* 2,2,4-Trimethyl-1,3-pentanediol
1,3-Pentanediol, 2,2,4-trimethyl, monoisobutyrate. *See* 2,2,4-Trimethyl-1,3-pentanediol isobutyrate
2,4-Pentanedione; Pentanedione-2,4. *See* Acetylacetone
Pentane, dodecafluoro-. *See* Perfluoropentane
Pentane, 1,1′-oxybis-. *See* Diamyl ether
Pentanoic acid, 4-oxo-, ethyl ester. *See* Ethyl levulinate
1-Pentanol; Pentanol-1; Pentan-1-ol. *See* n-Amyl alcohol
Pentan-2-ol; 2-Pentanol; Pentanol-2. *See* s-Amyl alcohol

3-Pentanol

CAS 584-02-1; EINECS/ELINCS 209-526-7
UN 2706 (DOT)
Synonyms: s-n-Amyl alcohol; Diethyl carbinol; 1-Ethyl-1-propanol; Isoamyl alcohol; Pentanol-3; Pentan-3-ol
Classification: Sec. aliphatic alcohol
Empirical: $C_5H_{12}O$
Formula: $(C_2H_5)_2CHOH$
Properties: Colorless visc. liq.; char. odor; sl. sol. in water; sol. in alcohol,

ether; m.w. 88.15; sp.gr. 0.819 (20/4 C); m.p. < -75 C; b.p. 114-116 C; flash pt. (CC) 41 C; ref. index 1.410

Toxicology: ACGIH TLV/TWA 100 ppm; STEL 125 ppm; LD50 (oral, rat) 1.87 g/kg, (skin, rabbit) 2520 mg/kg; irritant to eyes, nose, and throat; severe eye irritant, sl. skin irritant; inh. may cause dizziness, coughing, nausea, vomiting; severe exposure may cause double vision, deafness, delirium, severe nervous symptoms, fatalities; ing. may produce 'alcohol' intoxication symptoms; may cause severe lung damage, cardiac arrest, death if aspirated into lungs; TSCA listed

Precaution: Flamm.; dangerous fire and explosion hazard exposed to heat, flame, oxidizers; incompat. with oxidizing agents (increases fire/explosion hazard), hydrogen trisulfide (explosive reaction possible)

Hazardous Decomp. Prods.: Heated to decomp., emits acrid smoke and irritating fumes

NFPA: Health 1, Flammability 2, Reactivity 0

Storage: Store in cool, dry, well-ventilated area

Uses: Flotation agent; solvent in organic synthesis, mfg. of pharmaceuticals

Manuf./Distrib.: Aceto; Aldrich; Degussa AG; Fluka; Penta Mfg.

Pentanol-3; Pentan-3-ol. *See* 3-Pentanol
n-Pentanol. *See* n-Amyl alcohol
t-Pentanol. *See* t-Amyl alcohol
1-Pentanol acetate. *See* Amyl acetate
2-Pentanol acetate. *See* s-Amyl acetate
4-Pentanolide. *See* γ-Valerolactone
2-Pentanol, 4-methyl-. *See* Methyl amyl alcohol
2-Pentanol, 4-methyl-, acetate. *See* Methyl amyl acetate
1-Pentanol, 2,2,3,3,4,4,5,5-octafluoro-. *See* 1H,1H,5H-Octafluoro-1-pentanol
2-Pentanone. *See* Methyl propyl ketone
3-Pentanone; Pentanone-3. *See* Diethyl ketone
3-Pentanone, 2,4-dimethyl-. *See* Diisopropyl ketone
2-Pentanone, 4-hydroxy-4-methyl-. *See* Diacetone alcohol
2-Pentanone, 4-methoxy-4-methyl-. *See* 4-Methoxy-4-methylpentanone-2
3,6,9,12,15-Pentaoxaheptadecane-1,17-diol. *See* PEG-6
2,5,8,11,14-Pentaoxapentadecane. *See* Dimethoxytetraglycol
Pentoxane. *See* 4-Methoxy-4-methylpentanone-2
Pentyl acetate; 1-Pentyl acetate. *See* Amyl acetate
2-Pentyl acetate. *See* s-Amyl acetate
n-Pentyl acetate. *See* Amyl acetate
Pentyl alcohol; n-Pentyl alcohol. *See* n-Amyl alcohol
s-Pentyl alcohol. *See* s-Amyl alcohol
t-Pentyl alcohol. *See* t-Amyl alcohol

Pentylamine

CAS 110-58-7; EINECS/ELINCS 203-780-2
UN 1106; 2733 (DOT); 2734 (DOT)

Synonyms: Amine C_5; 1-Aminopentane; Amylamine; n-Amylamine; Monoamylamine; 1-Pentanamine; 1-Pentylamine; n-Pentylamine

Classification: Sat. aliphatic amine

Empirical: $C_5H_{13}N$

Formula: $CH_3(CH_2)_4NH_2$

Properties: Colorless liq.; ammoniacal odor; very sol. in water, ethanol, acetone, ether, benzene; sol. in dimethylsulfoxide; m.w. 87.19; dens. 0.753 (20/4 C); m.p. -55 C; b.p. 102-104 C; flash pt. (CC) -1 C; ref. index 1.412 (20 C); mod. strong base

Toxicology: LDLo (IP, rat) 37,500 μg/kg; poison by IP route; corrosive, toxic; can cause burns of skin; can be absorbed thru skin in toxic amts.; eye irritant; possible permanent eye injury, 'halo vision'; inh. of vapor can cause nose/throat irritation, coughing, chest pain, difficulty breathing; high concs. may cause potentially fatal pulmonary edema; ing. may cause burning of mouth, throat, digestive tract, headache, nausea, anxiety; TSCA listed

Precaution: Highly flamm.; dangerous fire risk; incompat. with oxidizing agents, acids, acid chlorides, acid anhydrides, calcium or sodium hypochlorite, nitromethane, nitrosyl perchlorate; violent/explosive reactions possible

Hazardous Decomp. Prods.: Heated to decomp., emits toxic vapors of NO_x

NFPA: Health 2, Flammability 3, Reactivity 0

Storage: Store in cool, dry, well-ventilated area, out of direct sunlight

Uses: Chemical intermediate; prod. of dyes, rubber chems., topical pharmaceuticals, insecticides, syn. detergents, other chems.; as flotation agent; solvent; corrosion inhibitor; gasoline additive

Regulatory: FDA approved for topicals

Manuf./Distrib.: Aldrich; Atofina; BASF AG; BASF; Fluka; Mallinckrodt Baker; Sigma

1-Pentylamine; n-Pentylamine. *See* Pentylamine
Pentyl benzoate; n-Pentyl benzoate. *See* Amyl benzoate
Pentylcarbinol. *See* Hexyl alcohol
3-Pentylcarbinol; s-Pentylcarbinol. *See* 2-Ethyl-1-butanol
Pentyl ethanoate. *See* Amyl acetate
Pentyl ether. *See* Diamyl ether
n-Pentyl ethyl ketone. *See* 3-Octanone
Pentyl formate; n-Pentyl formate. *See* Amyl formate
t-Pentyl methyl ether. *See* t-Amyl methyl ether
n-Pentyl methyl ketone. *See* Methyl n-amyl ketone
1-(Pentyloxy) pentane. *See* Diamyl ether
N-Pentyl-1-pentanamine; Pentyl pentylamine. *See* Di-n-amylamine
Pentyl propanoate; Pentyl propionate; n-Pentyl propionate. *See* Amyl propionate
m-Pentynol. *See* Methyl pentynol
Pentyoxypentane. *See* Diamyl ether
PEO. *See* Polyethylene glycol
Peppermint camphor. *See* Menthol

Peppermint (Mentha piperita) oil

CAS 8006-90-4
FEMA 2848

Synonyms: Mentha piperita; Mentha piperita oil; Peppermint oil

Definition: Volatile oil from steam distillation of *Mentha piperita*, contg. menthol, and menthyl esters incl. menthyl acetate, and menthone

Properties: Colorless to pale yel. liq., strong penetrating peppermint odor, pungent taste producing sensation of cold when air is drawn into mouth; sol. in 4 vols 90% alcohol; very sl. sol. in water; dens. 0.896-0.908; ref. index 1.460-1.471 (20 C)

Toxicology: LD50 (oral, rat) 2426 mg/kg, (IP, rat) 819 mg/kg; mod. toxic by ing. and IP routes; can cause allergic reactions, hay fever, skin rash; mutagenic data; TSCA listed

Hazardous Decomp. Prods.: Heated to decomp., emits acrid smoke and irritating fumes

HMIS: Health 1, Flammability 2, Reactivity 0

Storage: Light-sensitive

Uses: Source of menthol; pharmaceutic aid; natural flavoring agent and fragrance for foods, liqueurs, pharmaceuticals, toothpaste, mouthwashes, chewing gum; carminative; vehicle in peppermint water; perfume

Regulatory: FDA 21CFR §172.230, 182.20, GRAS; 27CFR §21.65, 21.151; FEMA GRAS; Europe listed (pulegone levels: 25 ppm in food to 350 ppm in mint confectionery); FDA approved for buccals, orals; NF, BP, EP compliance

Manuf./Distrib.: AMC Chems.; Acme-Hardesty; Adept Sol'ns.; Aldrich; Alfa Chem; Arista Ind.; Asiamerica Int'l.; Barnet Prods.; Berje; CarboMer; Chart; Chemacon GmbH; Creative Fragrances; Essence Nat. Prods.; Fleurchem; George Uhe; H.E. Daniel Ltd; Hunan Xinyu; Integra; Janousek Industriale Srl; MelChem; Millennium; Mutchler; Pangaea Sciences; Penta Mfg.; Polarome Int'l.; R.W. Greeff; Robeco; Ruger; Sarcom; Spec. Ingreds. & Perfume; Spectrum Quality Prods.; Treatt USA; U.S. Synthetics; Ultra Int'l.; Virginia Dare Extract; Voigt Global Distrib.; Wm. Leman

Trade Names Containing: Punctilious® SDA 38B 190 Proof Prod. Code 13602007; Punctilious® SDA 38F 190 Proof; Punctilious® SDA 38F-2 190 Proof Prod. Code 13643020; Punctilious® SDA 38F-2 190 Proof Prod. Code 13645006; Punctilious® SDA 38F-6 190 Proof

Peppermint oil. *See* Peppermint (Mentha piperita) oil
Perc. *See* Perchloroethylene
Perceline oil. *See* Cetyl octanoate
Perchlor; Perchlorethylene. *See* Perchloroethylene
Perchlorobutadiene. *See* Hexachloro-1,3-butadiene

Perchloroethylene

CAS 127-18-4; EINECS/ELINCS 204-825-9
UN 1897 (DOT)

Synonyms: Carbon bichloride; Carbon dichloride; Ethylene tetrachloride; Perc; Perchlor; Perchlorethylene; Tetrachlorethylene; Tetrachloroethane; Tetrachloroethene; Tetrachloroethylene; 1,1,2,2-Tetrachloroethylene
Classification: Chlorinated hydrocarbon
Empirical: C_2Cl_4
Formula: $Cl_2C=CCl_2$
Properties: Colorless clear dense liq., strong chloroform-like odor; misc. with alcohol, ether, chloroform, benzene, oil, most org. solvs.; insol. in water; m.w. 165.82; dens. 1.6311 (15/4 C); vapor pressure 18.5 mm Hg; m.p. -23 C; b.p. 121 C; flash pt. none; ref. index 1.5055 (20 C); surf. tens. 31.3 dynes/cm; KB value 90; dielec. const. 2.280; nonflamm.; highly stable
Toxicology: ACGIH TLV/TWA 25 ppm; STEL 100 ppm; LD50 (oral, rat) 2629 mg/kg; poison by IV route; mod. toxic by inh.; narcotic in high concs.; vapor inh. affects CNS; severe acute/chronic exposure causes liver damage; massive overexposure can cause death; prolonged skin contact can cause blistering; vapors may cause mild eye irritation, splashes may cause pain, burning, watering; confirmed carcinogen; tumorigen; reproductive effects; TSCA listed
Environmental: VOC; ThOD 0.19
Precaution: Incompat. with bases, metals (aluminum, barium, beryllium, lithium, magnesium, potassium), liq. oxygen, dinitrogen tetraoxide; oxidized by strong oxidizing agents
Hazardous Decomp. Prods.: Phosgene
NFPA: Health 2, Flammability 0, Reactivity 0
Storage: Store in cool, dry, well-ventilated area out of direct sunlight; avoid generating mist; limit amts. in use
Uses: Solvent for dry cleaning, vapor degreasing, cleaning/degreasing metals, fats, greases, waxes, rubber, gums, decaffeinating coffee; extraction solvent (oils, fat); flotation solvent (coal separation); heat exchange fluid; in copying machines; vermifuge; drying agent; scouring agent, sizing agent, and desizing agent in textiles; intermediate in mfg. of trichloroacetic acid, fluorocarbons, paint removers, printing inks; blowing agent adjuvant in food-contact foamed PS; in food-pkg. adhesives; anthelmintic drug (veterinary medicine)
Regulatory: FDA 21CFR §175.105, 178.3010; SARA §313 reportable; CERCLA hazardous substance; Calif. Prop. 65 reportable; HAP
Manuf./Distrib.: AAE Chemie NV; AMJAY Chems.; Aldrich; Allchem Ind.; Altair Ind.; Amyl; Ashland; Atofina SA; Bencorp Int'l.; Coyne; Cross World Sales; Delta Distributors; Dow; Fluka; General Chem.; Gulf Chem. Int'l.; HCl Chemtech Distribution; Houghton Chem.; Hukill; I.C. Trading; ICC Ind.; J.T. Baker; Janus Resources; Nitron; Occidental; PPG Ind.; Primachem; Romil Ltd; Sal Chem.; Sigma; Spectrum Quality Prods.; Stanchem; Storchem; Uniqema Am.; Universal Preserv-A-Chem; Vulcan Perf. Chems.
Trade Names: Dowper; Isoform; Pertene®; PPG Perchlor
Trade Names Containing: Arosolve TBP

Perchloromethane. *See* Carbon tetrachloride
Perfluoroacetic acid. *See* Trifluoroacetic acid

Perfluoroheptane
CAS 335-57-9; EINECS/ELINCS 206-392-1
Synonyms: Hexadecafluoroheptane; Perfluoro-n-heptane
Empirical: C_7F_{16}
Formula: $CF_3(CF_2)_5CF_3$
Properties: M.w. 388.07; dens. 1.745 (20/4 C); b.p. 82-84 C
Toxicology: LDLo (IP, mouse) 4 g/kg; low toxicity by IP route; irritant; TSCA listed
Environmental: Zero ozone depletion potential
Hazardous Decomp. Prods.: Heated to decomp., emits very toxic fumes of F^-
Uses: Dispersant for lubricants, mold release agents, protective coatings; reaction media for polymerizations, purification, separation processes; solvent for removal of halogenated lubricants, oils, greases; carrier for halogenated material
Manuf./Distrib.: Aldrich; Fluka; Miteni SpA
Trade Names: PF-5070

Perfluoro-n-heptane. *See* Perfluoroheptane

Perfluorohexane
CAS 355-42-0; EINECS/ELINCS 206-585-0
Synonyms: Perfluoro-n-hexane; Tetradecafluorohexane; n-Tetradecafluorohexane
Empirical: C_6F_{14}
Formula: $CF_3(CF_2)_4CF_3$
Properties: Colorless liq.; insol. in water; m.w. 338.04; dens. 1.684 (20/4 C); b.p. 54-58 C; pour pt. -90 C; ref. index 1.2520
Toxicology: LC (inh., rat, 6 h) > 20,000 ppm
Environmental: Zero ozone depletion potential
Uses: Dispersant for lubricants, mold release agents, protective coatings; reaction media for polymerizations, purification, separation processes; solvent for removal of halogenated lubricants, oils, greases; carrier for halogenated material; for quick-cooling or freezing foods; heat-transfer fluid
Regulatory: FDA 21CFR §173.342
Manuf./Distrib.: Aldrich; Fluka; Miteni SpA
Trade Names: PF-5060

Perfluoro-n-hexane. *See* Perfluorohexane
Perfluoromethane. *See* Tetrafluoromethane

Perfluoro-N-methylmorpholine
Environmental: Zero ozone depletion potential
Uses: Dispersant for lubricants, mold release agents, protective coatings; reaction media for polymerizations, purification, separation processes; solvent for removal of halogenated lubricants, oils, greases; carrier for halogenated material; drying agent

Perfluoropentane
CAS 678-26-2
Synonyms: Dodecafluoropentane; Pentane, dodecafluoro-; Perfluoro-n-pentane
Empirical: C_5F_{12}
Formula: $CF_3(CF_2)_3CF_3$
Properties: Colorless liq.; m.w. 288.04; dens. 1.60 kg/l (20 C); b.p. 28-32 C; pour pt. -120 C
Toxicology: LC50 (inh., rat) 120 g/m^3; TSCA listed
Environmental: Zero ozone depletion potential
Uses: Dispersant for lubricants, mold release agents, protective coatings; reaction media for polymerizations, purification, separation processes; solvent for removal of halogenated lubricants, oils, greases; carrier for halogenated material; heat-transfer fluid
Trade Names: PF-5050

Perfluoro-n-pentane. *See* Perfluoropentane
Perfluoropolyalkylether. *See* Perfluoropolyether

Perfluoropolyether
CAS 60164-51-4
Synonyms: Perfluoropolyalkylether; PFPE
Toxicology: Do not breathe gas/fumes/vapor/spray
Uses: Lubricant, EP agent for textiles, aerospace, vacuum pumps, instruments, bearings, surgical equip.; for structural analysis; embedding of crystals to protect them from air and moisture
Manuf./Distrib.: ABCR
Trade Names: Fomblin® Perfluorosolv PFS-1

Perhydrodiphenylamine. *See* Dicyclohexylamine
Perhydronaphthalene. *See* Decahydronaphthalene
cis-Perhydronaphthalene. *See* cis-Decalin
Periethylene naphthalene. *See* Acenaphthene
Petrohol. *See* Isopropyl alcohol
Petrol. *See* Gasoline

Petrolatum
CAS 8009-03-8 (NF); 8027-32-5 (USP); EINECS/ELINCS 232-373-2
INS905b
Synonyms: Mineral grease (petrolatum); Mineral jelly; Paraffin jelly; Petrolatum amber; Petrolatum white; Petroleum jelly; Vaseline; White petrolatum; White soft paraffin
Classification: Petroleum hydrocarbons
Definition: Semisolid mixture of hydrocarbons obtained from dewaxing residual oil; consists predominantly of sat. crystalline and liq. hydro-

carbons with > C25

Properties: Ylsh. to lt. amber or wh. semisolid, unctuous mass; pract. odorless and tasteless; sol. in benzene, chloroform, ether, petrol. ether, oils; pract. insol. in water; insol. in alcohol; dens. 0.820-0.865 (60/25 C); m.p. 38-54 C; soften. pt. 65-70 C; ref. index 1.460-1.474 (60 C)

Toxicology: LD50 (IP, mouse) > 50 g/kg; can cause allergic skin reactions in hypersensitive persons; generally nontoxic; TSCA listed

Hazardous Decomp. Prods.: Heated to decomp., emits acrid smoke and irritating fumes

HMIS: Health 0, Flammability 1, Reactivity 0

Uses: Laxative base; dispersant; leather grease; shoe polish; rust preventives; plastics, rubber, textiles lubricant; cable lubricants; elec. insulation impregnation wax; detackifier in printing inks; glazing agent, sealing agent, release agent in foods; protective food coatings; release agent, lubricant in bakery prods., confectionery; chewing gum ingred.; defoamer in beet sugar and yeast processing; excipient, ointment base, antistat, filler, solvent, emollient, skin protectant in cosmetics, pharmaceuticals, veterinary prods.; adhesive diluent; in food-pkg. adhesives, pressure-sensitive adhesives; release agent and surf. lubricant in food-contact coatings; in paper/paperboard in contact with aq./fatty foods; defoamer in food-contact coatings and paper/paperboard

Features: Protects skin from irritation

Regulatory: FDA 21CFR §172.880, 172.884, 173.340, 175.105, 175.125, 175.300, 176.170, 176.180, 176.200, 176.210, 177.1200, 177.2600, 177.2800, 178.3570, 178.3700, 178.3910, 573.720; FDA approved for ophthalmics, orals, topicals, otics; USP compliance (petrolatum, petrolatum white); USP/NF (petrolatum, hydrophilic); BP compliance (petroleum jelly white, petroleum jelly yellow)

Manuf./Distrib.: AAE Chemie NV; Aldrich; Alfa Chem; Am. Int'l.; Ashland; Avatar; BP Amoco; C.P. Hall; Charkit; Delta Distributors; EM Ind.; Eastech; Eggar & Co.; ExxonMobil; GMI Prods.; Harcros; Integra; Jarchem Ind.; Koster Keunen; Magie Bros. Oil; Magnesia GmbH; Merck KGaA; Milsolv; Mosselman NV; Mutchler; P.B. & S.; Penreco; Penta Mfg.; Phoenix Nat. Prods.; R.I.T.A.; RTD Chems.; Ruger; S & S Chem.; SOCO-Lynch; Sasol Germany; Sea-Land; Sinochem Tianjin; Southchem; Spec. Ingreds. & Perfume; Spectrum Quality Prods.; Stevenson Cooper; Strahl & Pitsch; Textile Chem.; Universal Preserv-A-Chem; VWR Int'l.; Varsal Instruments; Voigt Global Distrib.; Vopak USA; Whittaker, Clark & Daniels

Trade Names: Penreco Snow

Petrolatum amber. *See* Petrolatum
Petrolatum liquid. *See* Mineral oil
Petrolatum white. *See* Petrolatum
Petroleum benzin; Petroleum-derived naphtha. *See* Naphtha
Petroleum distillate. *See* Petroleum distillates
Petroleum distillate, hydrotreated middle. *See* Petroleum distillates, hydrotreated middle
Petroleum distillate light. *See* Petroleum distillates

Petroleum distillates

CAS 8002-05-9; 64742-14-9; 64742-48-9; 64742-56-9; EINECS/ELINCS 232-298-5

UN 1268 (DOT)

Synonyms: Petroleum distillate; Petroleum distillate light

Classification: Petroleum hydrocarbons

Definition: Mixture of volatile hydrocarbons obtained from petroleum

Toxicology: OSHA PEL TWA 400 ppm; TDLo (parenteral, man) 57 mg/kg; human systemic effects by parenteral route (cough, dyspnea, nausea, vomiting); mildly toxic by inh., ing.; mod. skin/eye irritant; skin contact may cause defatting; vapor inh. can cause CNS depression, headache, unconsciousness; TSCA listed

Precaution: Combustible exposed to heat or flame

Hazardous Decomp. Prods.: Heated to decomp., emits acrid smoke and irritating fumes

Uses: Processing solvent, defoamer in waterless hand cleaners, polishes, fruit and veg. processing, cleaning oils; vehicle for pesticides; diluent for agric. sprays; fuel conditioner; antifoam, solvent in cosmetics; solvent, defoamer, diluent for pharmaceutical creams and lotions; food pkg. paper, rubber articles; defoamer in food-contact paper/paperboard

Trade Names: Drakesol® 165AT; PD-25; PD-28

Petroleum distillates, acid-treated, medium heavy. *See* Petroleum distillates, acid-treated middle

Petroleum distillates, acid-treated middle

CAS 64742-13-8; EINECS/ELINCS 265-113-1

Synonyms: Distillates (petroleum), acid-treated, medium heavy; Distillates (petroleum), acid-treated, middle; Petroleum distillates, acid-treated, medium heavy; Severe acid-treated middle distillate

Classification: Petroleum distillate

Definition: Complex combination of hydrocarbons obtained as a raffinate from a sulfuric acid treating process; consists of hydrocarbons in C11-20 range, boiling range 205-345 C

Properties: B.p. 205-345 C

Toxicology: OSHA PEL 5 mg/m^3

Uses: Solvent

Trade Names: Hydroseal G3H; Hydroseal G232H; Hydroseal G240H; Hydroseal G250H; Hydroseal G400H

Petroleum distillates, alkylate

CAS 64741-73-7; EINECS/ELINCS 265-074-0

Synonyms: Alkylate petroleum distillates; Distillates (petroleum), alkylate

Classification: Aliphatic hydrocarbon

Definition: Complex combination of hydrocarbons produced by distillation of reaction prods. of isobutane with C3-5 monoolefinic hydrocarbons; consists of predominantly branched chain sat. HC in C11-17 range, boiling range 205-320 C

Properties: Colorless liq.; pleasant odor; negligible sol. in water; sp.gr. 0.809; vapor pressure 0.2 mm Hg; b.p. 450-550 F; flash pt. (PMCC) 93 C

Uses: Solvent; cable lubricant

Trade Names: Parabase; Shellsol® TM

Petroleum distillates, catalytic reformer fractionator residue, low boiling

CAS 68477-31-6; EINECS/ELINCS 270-722-0

Synonyms: Distillates (petroleum), catalytic reformer frationator residue, low boiling

Classification: Petroleum distillate; aromatic petroleum deriv.

Definition: Complex combination of hydrocarbons from distillation of catalytic reformer fractionator residue

Properties: B.p. < 288 C

Toxicology: Irritant

Uses: Solvent

Trade Names Containing: Kenplast® A-450

Petroleum distillates, heavy hydrotreated naphthenic

CAS 64742-52-5; EINECS/ELINCS 265-155-0

Synonyms: Distillates (petroleum), hydroteated (mild) heavy naphthenic; Distillates (petroleum), hydrotreated heavy naphthenic; Hydrotreated heavy naphthenic distillate; Hydrotreated (mild) heavy naphthenic distillate; Mineral oil, petroleum distillates, hydrotreated (mild) heavy naphthenic; Petroleum distillates, hydrotreated (mild) heavy naphthenic

Classification: Petroleum distillate

Definition: Complex combination of hydrocarbons obtained by treating a petroleum fraction with hydrogen in the presence of a catalyst; consists of predominantly C20-50 hydrocarbons; contains relatively few normal paraffins

Properties: B.p. 350 C

Toxicology: LD (oral, rat) > 5 g/kg, (skin, rabbit) > 5 g/kg; TDLo (skin, mouse, 78 wk intermittent) 402 g/kg; low toxicity by ing., skin contact; severe skin irritant; confirmed carcinogen with experimental tumorigenic data; mutation data reported; TSCA listed

Hazardous Decomp. Prods.: Heated to decomp., emits acrid smoke and irritating fumes

Uses: Solvent; lubricant basestock

Petroleum distillates, hydrogen-treated, light naphthene-. *See* Naphthenic oil

Petroleum distillates, hydrotreated light

CAS 64742-47-8; EINECS/ELINCS 265-149-8

UN 1268
Synonyms: Distillates (petroleum), hydrotreated light; Hydrotreated kerosene; Hydrotreated light distillate; Hydrotreated light distillate (petroleum); Hydrotreated light petroleum distillates; Kerosene (petroleum), hydrotreated; Severely hydrotreated light distillates
Classification: Petroleum distillate
Definition: Complex combination of hydrocarbons obtained by treating a petrol. fraction with hydrogen in presence of a catalyst; consists of hydrocarbons in C9-16 range, boiling range 150-290 C
Properties: Water-wh. oily liq.; hydrocarbon odor; sol. 1.4 ppm in water (38 C); sp.gr. 0.809 (16 C); vapor pressure 1.1 Torr (38 C); m.p. < -70 C; b.p. 150 C; flash pt. (TCC) 65 C
Toxicology: High vapor concs. may cause headaches, stupor, throat/eye irritation, kidney effects; extreme aspiration into lungs may cause pneumonia or death; may cause sl. eye irritation; may cause skin irritation, redness, dermatitis on prolonged/repeated contact; suspected carcinogen; TSCA listed
Environmental: Inherently biodeg.
Precaution: Combustible liq. and vapor; vapors may travel to ignition source and flash back; container may explode in heat of fire; bond and ground equip. during transfer; incompat. with strong oxidizers; avoid high temps.
Hazardous Decomp. Prods.: Heated to decomp., emits acrid smoke and irritating vapors
HMIS: Health 1, Flammability 2, Reactivity 1
Storage: Store @ ambient temps. and pressure
Uses: Solvent for carbonless copy paper for office forms, industrial cleaning solvents, waterless hand cleaners, fragrance bases, automotive and furniture polishes, insecticide bases, lamp oils, defoamers in paper and coatings industries; coating on egg shells; defoamer in foods; froth flotation cleaning of vegetables; electrostatic discharge machine fluids; magnetic particle inspection fluids; food-contact applics.
Regulatory: DOT regulated hazardous material in bulk quantities only; SARA §302/313 nonreportable, §311/312 fire hazard; Canada DSL, Australia listed
Manuf./Distrib.: AccuStandard
Trade Names: Actrel® 1111L; Actrel® 1160L; AM 321 Fluid; Amsol 650; Amsol 3800; Bayol 35; Conosol 215; Conosol C-145; Conosol C-170; Conosol C-200; Drakesol® 165; Escaid® 90; Escaid® 95; Escaid® 110; Exx-Print® 176 D; Exx-Print® 283 D; Exxsol® D 40; Exxsol® D 40 Anti-Static Fluid; Exxsol® D 60; Exxsol® D 60 Anti-Static Fluid; Exxsol® D 60 Naphtha Anti-Static; Exxsol® D 70; Exxsol® D 80; Exxsol® D 90/120; Exxsol® D 90/120 Anti-Static Solvent; Exxsol® D 95; Exxsol® D 110; Exxsol® D 3135; Exxsol® D 3135 Anti-Static Solvent; Exxsol® D 3135 Naphtha Anti-Static; Exxsol® DSP 90/120 Naphtha; Exxsol® S-K Premium Solvent; Isopar® M Fluid; Kerdane D65; Kerdane D75; Ketrul D70; Ketrul D80; Ketrul D85; Ketrul D211; Ketrul D220; MagieSol 44; MagieSol 47LX; Orfom® SX 7; Orfom® SX 12; Penreco 170 Exempt Solvent; Sasol LPA Solvent; Sasol LPA-142 Solvent; Sasol LPA-150 Solvent; Sasol LPA-170 Solvent; Sasol LPA-210 Solvent; Shellsol® D70; Shellsol® D80; Shellsol® D90; Shellsol® D100; Shellsol® D100S; Solvent 170; Varsol® DX 3641
Trade Names Containing: AM 230 Solvent; Amsol 230-K Solvent; Axarel® 52; Axarel® 56; Axarel® 6100; Axarel® 9100; Sasol-47; Solvent 2216; Varsol® 18; Varsol® 18 Naphtha; Varsol® 18 Naphtha Anti-Static; Varsol® DX 3139; Varsol® DX 3139 Anti-Static Solvent

Petroleum distillates, hydrotreated light, low-boiling

CAS 68410-97-9; EINECS/ELINCS 270-093-2
Synonyms: Distillates (petroleum), light distillate hydrotreating process, low-boiling; Hydrogenated pyrolysis gasoline; Hydrotreated distillate, light; Hydrotreated light distillate; Light hydrotreated distillate; Petroleum distillates, light, hydrotreated, low-boiling
Classification: Aliphatic hydrocarbon; high benzene naphtha
Definition: Complex combination of hydrocarbons obtained by the distillation of prods. from the light distillate hydrotreating process; consists predominantly of C6-9 hydrocarbons
Properties: Clear liq.; aromatic odor; negligible sol. in water; b.p. 3-194 C
Toxicology: LD50 (oral, rat) 5.17 g/kg; LC50 (inh., rat) > 12,408 ppm; may cause cancer
Uses: Solvent
Trade Names: Amsol 10

Petroleum distillates, hydrotreated light paraffinic

CAS 64742-55-8; EINECS/ELINCS 265-158-7
Synonyms: Distillates (petroleum), hydrotreated light paraffinic; Hydrotreated light paraffinic distillate
Definition: Distillation fraction between kerosene and gas oil; complex combination of hydrocarbons obtained by treating a petrol. fraction with hydrogen in presence of a catalyst; consists predominantly of C15-30 hydrocarbons
Uses: Solvent in gas absorption processes; lubricant for rolling of metal foil; base oil in many specialty formulations
Trade Names: Shellsol® D120

Petroleum distillates, hydrotreated middle

CAS 64742-46-7; EINECS/ELINCS 265-148-2
Synonyms: Distillates (petroleum), hydrotreated middle; Hydrotreated middle distillate; Hydrotreated middle petroleum distillate; Petroleum distillate, hydrotreated middle
Classification: Sat. aliphatic hydrocarbon; petroleum distillate
Definition: Complex combination of hydrocarbons obtained by treating a petrol. fraction with hydrogen in presence of a catalyst; consists of hydrocarbons in C11-25 range, boiling range 205-400 C
Toxicology: OSHA PEL TLV no limit; TDLo (skin, mouse, 2 yr intermittent) 416 g/kg; questionable carcinogen with tumorigenic data reported; mutation data reported; TSCA listed
Hazardous Decomp. Prods.: Heated to decomp., emits acrid smoke and irritating vapors
Uses: Solvent; component in newsprint formulations, metalworking fluids, lubricants
Trade Names: Amsol 400; Amsol 500; Amsol 600; Conosol 260; Conosol 340; Drakesol® 205; Exx-Print® 588 D; Exxsol® D 130; Hydroseal G2; Isopar® V Fluid; Magie 5200-HTX; MagieSol 55LX; MagieSol 65LX; PCPL Solvent; PCP Solvent

Petroleum distillates, hydrotreated (mild) heavy naphthenic. *See* Petroleum distillates, heavy hydrotreated naphthenic
Petroleum distillates, intermediate catalytic cracked. *See* Naphtha, heavy aromatic

Petroleum distillates, light hydrocracked

CAS 64741-77-1; EINECS/ELINCS 265-078-2
Synonyms: Distillates, light hydrocracked; Distillates (petroleum), light hydrocracked; Light hydrocracked distillate; Lightly hydrocracked distillate
Classification: Petroleum distillate
Definition: Complex combination of hydrocarbons from distillation of the prods. from a hydrocracking process; consists predominantly of sat. hydrocarbons in C10-18 range
Properties: B.p. 160-320 C
Precaution: Combustible
Uses: Solvent; jet fuel; gas oil
Trade Names: Amsol 100

Petroleum distillates, light, hydrotreated, low-boiling. *See* Petroleum distillates, hydrotreated light, low-boiling
Petroleum distillates (naphtha). *See* Naphtha

Petroleum distillates, sweetened middle

CAS 64741-86-2; EINECS/ELINCS 265-088-7
Synonyms: Distillates (petroleum), sweetened middle; Sweetened middle distillate; Sweetened middle petroleum distillates
Classification: Petroleum distillate
Definition: Complex combination of hydrocarbons obtained by subjecting a petrol. distillate to a sweetening process to convert mercaptans or to remove acidic impurities; consists of predominantly C9-20 hydrocarbons
Properties: B.p. 150-345 C
Toxicology: OSHA PEL 5 mg/m^3, STEL 10 mg/m^3
Uses: Solvent for inks
Trade Names: EXK-385-MS0; Magie 470 Oil; Magie 500 Oil
Trade Names Containing: Magie 4700 Oil; Magie 5200 Oil

Petroleum distillates, textile spirits. *See* Naphtha, solvent-refined light
Petroleum ether. *See* VM&P naphtha; Naphtha

Petroleum extracts, heavy paraffinic distillate solvent. *See* Paraffin distillate
Petroleum gases, liquefied. *See* Propane
Petroleum gas, liquefied. *See* Liquefied petroleum gas
Petroleum jelly. *See* Petrolatum
Petroleum naphtha. *See* VM&P naphtha; Naphtha
Petroleum naphtha, heavy alkylate. *See* Naphtha, heavy alkylate
Petroleum naphtha, hydrodesulfurized heavy. *See* Naphtha, hydrodesulfurized heavy
Petroleum oil. *See* Naphtha; Mineral oil
Petroleum raffinate. *See* Petroleum raffinates, sorption process

Petroleum raffinates, sorption process

CAS 64741-85-1; EINECS/ELINCS 265-087-1
Synonyms: Petroleum raffinate; Raffinates (petroleum), sorption process; Sorption raffinate
Uses: Solvent
Trade Names: Penreco 150-B Solvent; Sasol MR Solvent

Petroleum spirit. *See* VM&P naphtha; Naphtha
Petroleum spirits. *See* VM&P naphtha; Mineral spirits
Petroleum thinner. *See* VM&P naphtha
Petroleum wax, crystalline. *See* Paraffin
PFPE. *See* Perfluoropolyether
PFPOH. *See* Pentafluoropropyl alcohol
PG. *See* Propylene glycol
PGDE. *See* Propylene glycol dimethyl ether
PGEE. *See* Propylene glycol ethyl ether
PGEEA. *See* Propylene glycol ethyl ether acetate
PGME. *See* Methoxyisopropanol
β-PGME. *See* 3-Methoxy-1-propanol
PGMEA. *See* Propylene glycol methyl ether acetate
PGML. *See* Propylene glycol laurate
PGMS. *See* Propylene glycol stearate
PGnBE. *See* Butoxypropanol
PGnPE; PGPE. *See* Propylene glycol n-propyl ether
PGPR. *See* Polyglyceryl polyricinoleate

Phenanthrene

CAS 85-01-8; EINECS/ELINCS 201-581-5
Synonyms: Coal tar pitch volatiles: phenanthrene
Empirical: $C_{14}H_{10}$
Properties: Colorless shining cryst. solid; sol. in alcohol, ether, benzene, carbon disulfide, acetic acid, aromatic solvs.; insol. in water; m.w. 178.24; dens. 1.063; vapor pressure 1 mm (118.3 C); m.p. 100.35 C; b.p. 340 C
Toxicology: OSHA PEL/TWA 0.2 mg/m^3; LD50 (oral, mouse) 700 mg/kg, (IP, mouse) 700 mg/kg, (IV, mouse) 56 mg/kg; harmful solid; poison by IV route; mod. toxic by ing.; human skin photosensitizer; irritant; questionable carcinogen; tumorigen; mutagen; TSCA listed
Precaution: Combustible; can react vigorously with oxidizers
Hazardous Decomp. Prods.: Heated to decomp., emits acrid smoke and irritating fumes
Storage: Photosensitizer; handle and store under nitrogen
Uses: Dyestuffs; explosives; synthesis of drugs; biochemical research; mfg. of phenanthrenequinone
Manuf./Distrib.: Advanced Synthesis Tech.; Aldrich; Fluka; Sigma; Supelco; Xinchem
Trade Names Containing: Methylnaphthalene Fraction

1-Phenanthrenecarboxylic acid, 1,2,3,4,4a,4b,5,6,10,10a-decahydro-1,4a-dimethyl-7-(1-methylethyl)-, methyl ester, [1R-(1α,4aβ,4bα,10aα)]-. *See* Methyl abietate
Phene. *See* Benzene
Phenethanol; β-Phenethanol. *See* Phenethyl alcohol

Phenethyl alcohol

CAS 60-12-8; EINECS/ELINCS 200-456-2
FEMA 2858
Synonyms: Benzeneethanol; Benzyl carbinol; Ethanol, 2-phenyl-; β-Hydroxyethylbenzene; Methanol, benzyl-; Orange oil; PEA; β-PEA; Phenethanol; β-Phenethanol; 2-Phenethyl alcohol; β-Phenethyl alcohol; 2-Phenylethanol; β-Phenylethanol; Phenylethyl alcohol; 2-Phenylethyl alcohol; β-Phenylethyl alcohol; Rose oil
Classification: Aromatic alcohol
Empirical: $C_8H_{10}O$
Formula: $C_6H_5CH_2CH_2OH$
Properties: Colorless liq., floral rose odor, burning taste; sol. in fixed oils, glycerin, propylene glycol, oxygenated solvs.; misc. with alcohol, ether; sl. sol. in water; m.w. 122.18; dens. 1.0245 (15 C); vapor pressure 1 mm (58 C); m.p. -27 C; b.p. 220 C; flash pt. 102 C; ref. index 1.532 (20 C)
Toxicology: LD50 (oral, rat) 1790 mg/kg, (skin, rabbit) 790 mg/kg; mod. toxic by ing., skin contact; harmful by inh.; skin, eye, and respiratory system irritant; may cause severe CNS injury; may cause anesthesia, ataxia, coma, gastritis, fetal death; may cause burning sensation, coughing, wheezing, laryngitis, shortness of breath, headache, nausea, vomiting; experimental teratogen, reproductive effector; mutagenic data; target organs: CNS; TSCA listed
Precaution: Combustible when exposed to heat or flame; incompat. with strong oxidizing agents, strong acids; reactive with oxidizing materials
Hazardous Decomp. Prods.: Toxic fumes of CO, CO_2; emits toxic fumes under fire conditions
NFPA: Health 1, Flammability 1, Reactivity 0
Storage: Store in cool, dry place; keep tightly closed
Uses: Organic synthesis; fragrance ingred. and fixing agent; synthetic rose oil; soaps; synthetic flavoring agent and fragrance in foods and pharmaceuticals; deodorant agent in cosmetics; antibacterial; antimicrobial, preservative, and solvent for pharmaceuticals; topical antiseptic; local anesthetic
Features: Mildly floral, rose-like fragrance and flavoring with nuances of hyacinth and hone
Regulatory: FDA 21CFR §172.515, CIR approved, EPA reg.; FEMA GRAS; JSCI listed; FDA approved for ophthalmics, otics, topicals, vaginals
Manuf./Distrib.: ADA Int'l.; AMC Chems.; Aceto; Advanced BioTech; Akzo Nobel; Aldrich; Axxence Aromatic GmbH; BASF; Camida Ltd; Chr. Hansen Inc; Degussa Flavors/Alfrebro; Fleurchem; Fluka; Grau Aromatics; Haarmann & Reimer; Int'l. Flavors & Fragrances; Integra; Lluch Essence; Lyondell; Mallinckrodt Baker; MelChem; Moore Ingreds.; Penta Mfg.; Polarome Int'l.; Prodasynth; Ruger; Sarcom; Sigma; Spectrum Quality Prods.; Takasago Int'l.; Univar E&E Ltd; Voigt Global Distrib.

2-Phenethyl alcohol. *See* Phenethyl alcohol
α-Phenethyl alcohol. *See* α-Methylbenzyl alcohol
β-Phenethyl alcohol. *See* Phenethyl alcohol
s-Phenethyl alcohol. *See* α-Methylbenzyl alcohol
Phenethylene. *See* Styrene
Phenethyltris (trimethylsiloxy) silane. *See* Phenyl trimethicone
Phenic acid. *See* Phenol

Phenol

CAS 108-95-2; EINECS/ELINCS 203-632-7
UN 1671 (solid); UN 2312 (fused); UN 2821 (sol'n.); FEMA 3223
Synonyms: Benzenol; 'Carbolic acid'; Hydroxybenzene; Liquid phenol; Monohydroxybenzene; Monophenol; Oxybenzene; Phenic acid; Phenol alcohol; Phenol, liquefied; Phenyl hydrate; Phenyl hydroxide; Phenylic acid; Phenylic alcohol
Classification: Aromatic organic compd.; carbolic acid
Empirical: C_6H_6O
Formula: C_6H_5OH
Properties: Colorless to lt. pink needle-shaped cryst.; char. odor of coal tar and wood; burning taste; sol. in alcohol, glycerin, chloroform, ether, water; misc. with oxygenated and chlorinated solvs.; sl. sol. in min. oil; m.w. 94.11; dens. 1.07; m.p. 40-42 C; b.p. 182 C; flash pt. 79 C; darkens on exposure to lt., air
Toxicology: ACGIH TLV/TWA 5 ppm (skin); LD50 (oral, rat) 317 mg/kg, (subcut., rat) 460 mg/kg, (IV, mouse) 112 mg/kg, (skin, rabbit) 850 mg/kg; poison by ing., subcut., IV, IP routes; human poison by ing.; mod. toxic by skin contact; severe eye/skin irritant; ing. of even sm. amts. may cause vomiting, circulatory collapse, paralysis, convulsions, coma, grnsh. urine, necrosis of mouth and GI tract; death results from respiratory failure; may cause serious skin burns; possible carcinogen; human mutagen; TSCA listed
Precaution: Combustible; vapor is flamm.; DOT: poisonous material; ex-

plosive and violent reactions possible; polymerizes violently when combined with isocyanates
Hazardous Decomp. Prods.: CO, CO_2; heated to decomp., emits acrid smoke and irritating fumes
NFPA: Health 4, Flammability 2, Reactivity 0
Storage: Hygroscopic; deliq.; moisture- and light-sensitive; store frozen
Uses: Phenolic adhesives and resins; epoxy resins; pesticide intermediates; paints; lubricant additives; chem. deriv. intermediates; as fuel oil sludge inhibitors; as solvent; selective solvent in lubricating oil dearomatization; as rubber chemicals; isocyanate blocking agent; antimicrobial, preservative for pharmaceuticals; antimicrobial, denaturant, deodorant agent in cosmetics; disinfectant, anesthetic for skin; active ingred. in deodorant soaps and mouthwash; flavoring agent in foods; preservative in food-pkg. adhesives; adjuvant for PC food-pkg. resins; in food-contact paper/paperboard; in closures with sealing gaskets for food containers
Use Level: 0.5% (topicals), 0.2-5% (parenterals), 1% max. (soaps, shampoos)
Regulatory: FDA 21CFR §175.105, 175.300, 175.380, 175.390, 176.170, 177.1210, 177.1580, 177.2410, 177.2600, 27CFR §21.65, 21.151; FEMA GRAS; SARA reportable; FDA approved for injectables, parenterals, topicals; USP/NF, BP, EP compliance
Manuf./Distrib.: AMRESCO; Aldrich; Allchem Ind.; Ashland; Asiamerica Int'l.; BP Amoco; Barker Ind.; Brenntag (UK) Ltd; Chance & Hunt; Charkit; Chemical; ChevronTexaco; Croda Chem. Ltd; Dakota Gasification; Dastech Int'l.; Dow; EM Ind.; EM Science; Fisher Scientific; Fluka; GE Plastics; Georgia Gulf; Hawkins Chem.; Herdillia Chems. Ltd; Honeywell Perf. Polymers; INEOS Phenol; Int'l. Chem. Inc; Integra; J.T. Baker; Kessler; Mallinckrodt Baker; MelChem; Merck KGaA; Miljac; Noveon Kalama; PCAS; PMC Spec.; Penta Mfg.; Qbiogene; R.W. Greeff; Research Organics; Rhodia; Roche Molecular Biochems.; Royale Pigments & Chems.; Ruger; Shell; Sigma; Spectrum Quality Prods.; Stanchem; Sunoco; Tennants Distrib. Ltd; U.S. Chems.; Ubichem plc; VWR Int'l.; Voigt Global Distrib.; Vopak USA
Trade Names Containing: Punctilious® SDA 38B 190 Proof Prod. Code 13603998

Phenol acetate. *See* Phenyl acetate
Phenol alcohol. *See* Phenol
Phenolcarbinol. *See* Benzyl alcohol
Phenol, dimethyl-. *See* Xylenol
Phenol, 3,4-dimethyl-. *See* 3,4-Xylenol
Phenol, p-(α,α-dimethylbenzyl). *See* Cumyl phenol
Phenol, p-(α,α-dimethylbenzyl)-, acetate. *See* Cumylphenyl acetate
Phenol, dinonyl-. *See* Dinonyl phenol
Phenol, liquefied. *See* Phenol
Phenol, 4-methoxy-. *See* Hydroquinone monomethyl ether
Phenol, 2-methoxy-4-(2-propenyl)-. *See* Eugenol
Phenol, 3,4-(methylenedioxy). *See* Sesamol
Phenol methyl ether. *See* Anisole
Phenol, 5-methyl-2-(1-methylethyl)-. *See* Thymol
Phenol, 4-(1-methyl-1-phenethyl)-. *See* Cumyl phenol
Phenol, 4-(1-methyl-1-phenylethyl)-, acetate. *See* Cumylphenyl acetate
Phenol, 4,4′,4′′-[1-methyl-1-propanyl-3-ylidene] tris 2-[1,1-dimethylethyl]-5-methyl-. *See* 1,1,3-Tris (2-methyl-4-hydroxy-5-t-butylphenyl) butane
Phenoxethol; Phenoxetol. *See* Phenoxyethanol
Phenoxybenzene. *See* Diphenyl oxide
Phenoxydiglycol. *See* Diethylene glycol phenyl ether

Phenoxyethanol

CAS 122-99-6; EINECS/ELINCS 204-589-7
Synonyms: Ethylene glycol monophenyl ether; Ethylene glycol phenyl ether; Glycol monophenyl ether; β-Hydroxyethyl phenyl ether; 1-Hydroxy-2-phenoxyethane; Phenoxethol; Phenoxetol; 2-Phenoxyethanol; β-Phenoxyethanol; Phenoxyethyl alcohol; Phenoxytol; Phenylglycol; Phenylglycol ether; Phenylmonoglycol ether; Rose ether
Classification: Aromatic ether alcohol
Definition: Phenol polyglycol ether
Empirical: $C_8H_{10}O_2$
Formula: $C_6H_5OCH_2CH_2OH$
Properties: Yel.-brn. or clear liq.; faint aromatic rose-like odor; burning taste; sol. in alcohol, ether, NaOH sol'ns.; sol. 27 g/l in water; m.w. 138.18; dens. 1.1094 (20/20 C); vapor pressure 0.03 mm Hg (20 C); m.p. 14 C; b.p. 242 C; flash pt. 121 C; ref. index 1.534 (20 C); surf. tens. 42 dynes/cm
Toxicology: LD50 (oral, rat) 1.26 g/kg, (skin, rabbit) 5000 mg/kg; mod. toxic by ing. and skin contact; skin and severe eye irritant; mutation data; possible reproductive effector; TSCA listed
Environmental: VOC; ThOD 1.62
Precaution: Combustible; incompat. with strong oxidizers
Hazardous Decomp. Prods.: Heated to decomp., emits acrid smoke and irritating fumes
NFPA: Health 0, Flammability 1, Reactivity 0
Uses: Solvent for resins, dyes, printing inks, in organic synthesis, bactericides, germicides; coalescent solvent in adhesives, water-based architectural and industrial coatings; coalescing agent in water-based latex adhesives; solvent, plasticizer for cellulosics, NC, cellulose acetate, ethyl cellulose, resins (vinyl, phenolic, alkyd, polyester); perfume fixative/solvent; diluent, flexibilizer for epoxies; plasticizer; as insect repellent; antiseptic; carrier solvent, solubilizer for textile dyes; paint removers; ingred. in metal corrosion protection slurries; solvent, emulsifier, coupling agent, wetting agent, soil penetrant in household, commercial and industrial cleaners; antimicrobial, preservative for cosmetics, pharmaceuticals, preservatives; in food-pkg. adhesives
Use Level: 1% max. in finished cosmetics
Regulatory: FDA 21CFR §175.105; CIR approved; EPA reg.; SARA §313 reportable; HAP; JSCI listed 1.0% max.; Europe listed 1% max.; BP, EP compliance
Manuf./Distrib.: Aldrich; Amber Syn.; Arch Personal Care Prods.; Ashland; BASF; Bayer; Camida Ltd; ChemService; Chesham Chems. Ltd; Costec; Degussa AG; Fluka; Haarmann & Reimer; Integra; Jan Dekker BV; Kira; Kowa Am.; Kraft Chem.; Lluch Essence; Morre-Tec Ind.; Napp Tech.; Nipa Inc; Oxiteno; Penta Mfg.; Phoenix Nat. Prods.; Prodasynth; Pyosa S.A. de C.V.; Rhodia HPCII; Seidler; Sigma; Spectrum Quality Prods.; Tri-K Ind.; Univar E&E Ltd; Universal Preserv-A-Chem; Vevy; Yokkaichi Chem.
Trade Names: Emery® 6705; Rewopal® MPG 10; Rewopal® MPG 40
Trade Names Containing: Dalpad A; Dowanol® EPh; Dynasolve 185; Igepal® OD-410

2-Phenoxyethanol; β-Phenoxyethanol. *See* Phenoxyethanol
2-(2-Phenoxyethoxy) ethanol. *See* Diethylene glycol phenyl ether
Phenoxyethyl alcohol. *See* Phenoxyethanol

Phenoxyglycol

Uses: In solvent blend for resins, metal cleaners, paint strippers
Trade Names Containing: Igepal® OD-410

Phenoxyisopropanol

CAS 770-35-4; EINECS/ELINCS 212-222-7
Synonyms: β-Phenoxyisopropanol; 1-Phenoxy-2-propanol; 1-Phenoxypropan-2-ol; 2-Propanol, 1-phenoxy-; α-Propylene glycol 1-phenyl ether; Propylene phenoxetol; Propylene phenoxythol; Propylene phenoxytol
Classification: Aromatic ether alcohol
Empirical: $C_9H_{12}O_2$
Properties: M.w. 152.19; dens. 1.063; b.p. 243 C; ref. index 1.5230
Toxicology: TSCA listed
Uses: Preservative, solvent in cosmetics; antiseptic for skin care prods.; pharmaceutical intermediate
Manuf./Distrib.: Acros Org.; Aldrich; Fluka; Lancaster Synthesis; TCI Am.
Trade Names Containing: Dowanol® PPh; Solvenon® PP

β-Phenoxyisopropanol; 1-Phenoxy-2-propanol; 1-Phenoxypropan-2-ol. *See* Phenoxyisopropanol
2-Phenoxypropanol; 2-Phenoxypropyl alcohol. *See* Propylene glycol phenyl ether
Phenoxytol. *See* Phenoxyethanol

Phenyl acetate

CAS 122-79-2; EINECS/ELINCS 204-575-0
FEMA 3958
Synonyms: Acetic acid, phenyl ester; Acetoxybenzene; (Acetyloxy) benzene; Acetylphenol; Phenol acetate
Classification: Phenyl esters

Empirical: $C_8H_8O_2$
Formula: $C_6H_5OOCCH_3$
Properties: Water-wh. liq.; sol. in alcohol, glac. acetic acid, ether; misc. with oxygenated and chlorinated solvs.; almost insol. in water; m.w. 136.16; sp.gr. 1.073-1.079; b.p. 195-196 C; flash pt. 80 C; ref. index 1.500-1.506
Toxicology: LD50 (oral, rat) 1630 µl/kg, (skin, rabbit) 8000 mg/kg; mod. toxic by ing.; mildly toxic by skin contact; irritating to skin; TSCA listed
Precaution: Combustible exposed to heat, flame, or oxidizers
Hazardous Decomp. Prods.: Heated to decomp., emits acrid smoke and irritating fumes
Uses: High-boiling aprotic solvent; organic synthesis; laboratory reagent; synthetic flavoring agent in foods
Regulatory: FEMA GRAS
Manuf./Distrib.: Aldrich; Alfa Chem; Fluka; Goldensun Mfg.; Kira; Penta Mfg.; Richman; Sigma; Spectrum Quality Prods.

Phenylacetic acid, ethyl ester. *See* Ethyl phenylacetate
Phenylacrolein. *See* Cinnamal
Phenylamine. *See* Aniline
Phenyl benzene. *See* Biphenyl
Phenylbenzyl. *See* Diphenylmethane
Phenyl bromide. *See* Bromobenzene
2-Phenylbutane. *See* s-Butylbenzene
Phenylcarbinol. *See* Benzyl alcohol
Phenyl 'Carbitol'. *See* Diethylene glycol phenyl ether
Phenyl carbonate. *See* Diphenyl carbonate
Phenyl chloride. *See* Chlorobenzene
Phenyl cyanide. *See* Benzonitrile

Phenylcyclohexane

CAS 827-52-1; EINECS/ELINCS 212-572-0
Synonyms: Cyclohexylbenzene
Classification: Aromatic hydrocarbons
Empirical: $C_{12}H_{16}$
Formula: $C_6H_5C_6H_{11}$
Properties: Colorless oily liq.; pleasant odor; insol. in water, glycerol; very sol. in alcohol, acetone, benzene, CCl_4, castor oil, hexane, xylene; m.w. 160.28; dens. 0.938; vapor pressure 1 mm Hg (67.5 C); m.p. 5 C; b.p. 237.5 C; flash pt. 98 C; ref. index 1.523
Toxicology: LD50 (IP, mouse) 248 mg/kg, (IV, mouse) 67 mg/kg; poison by IP and IV routes; mod. toxic by ing.; TSCA listed
Precaution: Combustible; can react with oxidizers
Hazardous Decomp. Prods.: Heated to decomp., emits acrid smoke and irritating fumes
Uses: High-boiling solvent' penetrant; intermediate; laboratory reagent
Manuf./Distrib.: Aldrich; Fluka

o-Phenylenediol. *See* Pyrocatechol
Phenylethane. *See* Ethylbenzene
1-Phenylethanol. *See* α-Methylbenzyl alcohol
2-Phenylethanol; β-Phenylethanol. *See* Phenethyl alcohol
1-Phenylethanone; 1-Phenyl-1-ethanone. *See* Acetophenone
Phenylethene. *See* Styrene
Phenyl ether. *See* Diphenyl oxide
Phenylethyl alcohol; 2-Phenylethyl alcohol. *See* Phenethyl alcohol
α-Phenylethyl alcohol. *See* α-Methylbenzyl alcohol
β-Phenylethyl alcohol. *See* Phenethyl alcohol
Phenylethylene. *See* Styrene

Phenylethylethanolamine

CAS 92-50-2; EINECS/ELINCS 202-160-9
Synonyms: Ethanol, 2-(ethylphenylamino)-; 2-(N-Ethylanilino) ethanol; Ethyl (β-hydroxyethyl) aniline; N-Ethyl-N-(2-hydroxyethyl) aniline; 2-(Ethylphenylamino) ethanol; Ethyl phenyl ethanolamine; N,N-Ethylphenylethanolamine; N-Ethyl-N-phenylethanolamine; N-(2-Hydroxyethyl)-N-ethylaniline; PEEA
Classification: Aromatic amine
Empirical: $C_{10}H_{15}NO$
Formula: $C_6H_5N(C_2H_5)C_2H_4OH$
Properties: Solid; sol. in alcohol, acetone, benzene; very sl. sol. in water; m.w. 165.26; dens. 1.104 (20/20 C); m.p. 37.2 C; b.p. 268 C (740 mm)
Toxicology: LD50 (oral, rat) 1640 mg/kg, (oral, mouse) 2360 mg/kg; harmful; irritating to eyes, skin, respiratory system; TSCA listed
Precaution: Combustible
Uses: Dye intermediate; polyester cure promoter; solvents; chemical intermediates; laboratory reagent; organic synthesis; coupling agent for dyes for syn. fibers; cure promoter for polyester resins; photosensitive chemical for paper coatings
Manuf./Distrib.: Aldrich

Phenylfluoroform. *See* Benzotrifluoride
Phenylglycol; Phenylglycol ether. *See* Phenoxyethanol
Phenyl hydrate. *See* Phenol
Phenyl hydride. *See* Benzene
Phenyl hydroxide; Phenylic acid; Phenylic alcohol. *See* Phenol
4-(2-Phenylisopropyl) phenol. *See* Cumyl phenol
Phenylmethanal. *See* Benzaldehyde
Phenylmethane. *See* Toluene
Phenylmethanol. *See* Benzyl alcohol
Phenylmethyl acetate. *See* Benzyl acetate
Phenylmethyl alcohol. *See* Benzyl alcohol
N-Phenylmethylamine. *See* N-Methylaniline
Phenylmethyl benzoate. *See* Benzyl benzoate
Phenylmethylcarbinol. *See* α-Methylbenzyl alcohol
Phenylmethyl dodecanoate. *See* Benzyl laurate
Phenyl methylether. *See* Anisole
Phenylmethyl 2-hydroxybenzoate. *See* Benzylsalicylate
Phenyl methyl ketone. *See* Acetophenone
1-Phenyl-2-methylpropane. *See* Isobutylbenzene
Phenylmonoglycol ether. *See* Phenoxyethanol
1-Phenylpropane. *See* Propylbenzene
2-Phenylpropane. *See* Cumene

Phenylpropanol

CAS 1335-12-2; EINECS/ELINCS 215-621-4
Synonyms: 1-Phenyl-1-propanol; 1-Propanol, 1-phenyl-
Empirical: $C_9H_{12}O$
Properties: M.w. 136.19
Uses: Solvent, fragrance in cosmetics

1-Phenyl-1-propanol. *See* Phenylpropanol
3-Phenylpropenal; 3-Phenyl-2-propenal. *See* Cinnamal
3-Phenyl-2-propenoic acid, ethyl ester. *See* Ethyl cinnamate
Phenyl sulfone. *See* Diphenyl sulfone

Phenyl trimethicone

CAS 2116-84-9; EINECS/ELINCS 218-320-6
Synonyms: 1,1,5,5,5-Hexamethyl-3-phenyl-3-[(trimethylsilyl) oxy] trisiloxane; Methyl phenyl polysiloxane; Phenethyltris (trimethylsiloxy) silane; Phenyltris (trimethylsiloxyl) silane; Phenyl tristrimethyl siloxysilane; Polyphenylmethyl siloxane; Trisiloxane, 1,1,1,5,5,5-hexamethyl-3-phenyl-3-((trimethylsilyl) oxy)-; Tris (trimethylsiloxy) phenylsilane
Classification: Siloxane polymer
Definition: Phenyl substituted silicone fluid
Empirical: $C_{15}H_{32}O_3Si_4$
Properties: M.w. 372.76; dens. 0.97; vapor pressure 1 mm (105 C); m.p. < -60 C; b.p. 264-266 C; flash pt. 127 C; ref. index 1.459; surf. tens. 27.2 dynes/cm
Toxicology: LDLo (oral, rat) > 34,500 mg/kg; irritating to eyes, skin, respiratory system; TSCA listed
HMIS: Health 1, Flammability 1, Reactivity 0
Storage: Moisture-sensitive; keep under argon
Uses: Conditioner; oil-phase ingred.; hydrophobing agent; lubricant; solvent; antifoam, antistat, emollient, detackifier in cosmetics; emollient, water repellent in skin care prods.; gloss aid in hair care; emollient, conditioner, film-former in pharmaceuticals; lubricant, compatibilizer for silicones and hydrocarbons; silylating agent; cobinder for primers, esp. for facade coatings; antiperspirants; thread lubricant; in food-pkg. adhesives; release agent in food-contact coatings
Regulatory: FDA 21CFR §175.105, 175.300
Manuf./Distrib.: ABCR; AP Resources; Fluka; Gelest; Pfaltz & Bauer; Silar Labs; Surfachem Ltd; United Chem. Tech.

Phenyltris (trimethylsiloxyl) silane; Phenyl tristrimethyl siloxysilane. *See* Phenyl trimethicone

Phosphinic acid, bis (2,4,4-trimethylpentyl)-. *See* Bis (2,4,4,-trimethylpentyl) phosphinic acid
Phosphinothioic acid, bis (2,4,4-trimethylpentyl)-. *See* Bis (2,4,4,-trimethylpentyl) thiophosphinic acid
Phosphonic acid, butyl-, dibutyl ester. *See* Dibutyl butyl phosphonate
Phosphonic acid, (2-ethylhexyl)-, bis (2-ethylhexyl) ester. *See* Dioctyl octylphosphonate
Phosphonic acid, methyl-, dimethyl ester. *See* Dimethyl methylphosphonate

Phosphoric acid

CAS 7664-38-2; EINECS/ELINCS 231-633-2
UN 1805 (DOT); FEMA 2900; INS338
Synonyms: Orthophosphoric acid; Phosphoric acid, pure
Classification: Inorganic mineral acid
Empirical: H_3O_4P
Formula: H_3PO_4
Properties: Colorless liq. or rhombic crystals, odorless; sol. in water, alcohol; misc. with many org. solvs.; m.w. 97.99; dens. 1.70 (20/4 C); vapor pressure 0.0285 mm (20 C); m.p. 42.4 C; b.p. 158 C; starts to dec. above 200 C forming polyphosphoric acids
Toxicology: ACGIH TLV/TWA 1 mg/m^3; STEL 3 mg/m^3; LD50 (oral, rat) 1530 mg/kg, (skin, rabbit) 2740 mg/kg; mod. toxic by skin contact; human poison by ing.; corrosive irritant to eyes, skin, mucous membranes; systemic irritant by inh.; common air contaminant; TSCA listed
Environmental: Environmentally hazardous
Precaution: DOT: Corrosive material; strong acid; mixts. with nitromethane are explosive; incompat. with alkalis; corrosive to many metals; violent reaction possible with sodium tetrahydroborate
Hazardous Decomp. Prods.: Heated to decomp., emits toxic fumes of PO_x
NFPA: Health 3, Flammability 0, Reactivity 0
Uses: In mfg. of inorganic phosphates, fertilizers, detergents, chemical polishing, paints, priming metals, petroleum refining; acid catalyst; acidulant, flavoring agent, antioxidant synergist, sequestrant in foods; pharmaceutic acid; buffer in cosmetics; in dental cements; in analytical chem.; corrosion/scale inhibitor in drinking water treatment; bonding agent (ceramics); metal cleaning; electropolishing; dil. as tonic for treatment of nausea/vomiting; in resin-bonded filters for food contact; in resinous/polymeric food-contact coatings
Regulatory: FDA 21CFR §73.85, 73.275, 131.144, 133.123, 133.124, 133.129, 133.169, 133.173, 133.178, 133.179, 163.110, 163.111, 163.112, 175.300, 177.2260, 178.1010, 178.3520, 182.1073, GRAS; USDA 9CFR §318.7, 381.147 (0.01% max. in lard, shortening, poultry fat); FEMA GRAS; SARA reportable; Japan approved; Europe listed; UK approved; FDA approved for parenterals, intramuscular injectables, orals, topicals, vaginals; NF, BP, EP compliance
Manuf./Distrib.: AB R Lundberg; AMRESCO; Agrium; Aldrich; Amyl; Arch Chems.; Ashland; Astaris; Bilt Chems. Ltd; Boith China; Brown; Browning; Coyne; E + E (US); EM Ind.; EM Science; EMCO Chem. Distributors; FMC Foret; FMC; Farleyway Chems. Ltd; Fisher Scientific; Fluka; Fuerst Day Lawson; GFS; General Chem.; Haco Ltd; Houghton Chem.; Hydrite; Integra; J.T. Baker; Mallinckrodt Baker; Monsanto; Occidental; Oriental Chem. Ind.; PCS Purified Phosphates; Penta Mfg.; Prayon Inc; Rasa Ind.; Rhodia Food; Rhodia HPCII; Rhodia UK/Phosphorus Perf. Derivs.; Rhodia/Phosphorus Perf. Derivs.; Rhodia; Ruger; Schaefer Tech.; Sigma; Sinochem Liaoning; Spectrum Quality Prods.; Stanchem; Tekchem; Thomas Scientific; Universal Preserv-A-Chem; VWR Int'l.; Varsal Instruments; Voigt Global Distrib.; Vopak USA
Trade Names Containing: Surphos 515; Surphos 620; Surphos 645

Phosphoric acid, pure. *See* Phosphoric acid
Phosphoric acid tributyl ester; Phosphoric acid, tri-n-butyl ester. *See* Tributyl phosphate
o-Phosphoric acid, triethyl ester. *See* Triethyl phosphate
Phosphoric acid, trimethyl ester. *See* Trimethyl phosphate
Phosphoric acid, triphenyl ester. *See* Triphenyl phosphate
Phosphoric acid, tris (2-ethylhexyl) ester. *See* Trioctyl phosphate
Phosphoric acid, tris (methylphenyl) ester. *See* Tricresyl phosphate
Phosphoric acid, tris (4-methylphenyl) ester. *See* Tri-p-tolyl phosphate
Phosphoric acid, tritolyl ester. *See* Tricresyl phosphate
Phosphorous acid, triisooctyl ester. *See* Triisooctyl phosphite
Phosphorous acid, triisopropyl ester; Phosphorous acid, tris (1-methylethyl) ester. *See* Triisopropyl phosphite
PHPH. *See* Biphenyl
Phthalic acid bis (2-methoxyethyl) ester. *See* Dimethoxyethyl phthalate
Phthalic acid, butyl octyl ester. *See* Butyl octyl phthalate
Phthalic acid, diethyl ester. *See* Diethyl phthalate
Phthalic acid, diisobutyl ester. *See* Diisobutyl phthalate
Phthalic acid, di (methoxyethyl) ester. *See* Dimethoxyethyl phthalate
Phthalic acid, dimethyl ester. *See* Dimethyl phthalate
Phthalic acid, dioctyl ester. *See* n-Dioctyl phthalate
Phthalic acid, monomethyl ester, ester with ethyl glycolate. *See* Methyl phthalyl ethyl glycolate
Phthalol. *See* Diethyl phthalate
2-Picoline. *See* α-Picoline
3-Picoline. *See* β-Picoline
4-Picoline. *See* γ-Picoline

α-Picoline

CAS 109-06-8; EINECS/ELINCS 203-643-7
UN 2313 (DOT)
Synonyms: 2-Methylpyridine; α-Methylpyridine; 2-Picoline; o-Picoline
Empirical: C_6H_7N
Formula: $C_5H_4N(CH_3)$
Properties: Colorless liq., strong unpleasant odor; very sol. in water; misc. with alcohol, ether, oxygenated solvs.; m.w. 93.13; dens. 0.943; vapor pressure 10 mm Hg (24.4 C); m.p. -70 C; b.p. 128-129 C; flash pt. 26 C; ref. index 1.500
Toxicology: LD50 (oral, rat) 790 mg/kg, (IP, rat) 200 mg/kg, (skin, rabbit) 410 mg/kg; poison by IP route; mod. toxic by ing. and skin contact; primary irritant; skin and severe eye irritant; mutagen; TSCA listed
Precaution: Flamm. exposed to heat and flame; mixts. with hydrogen peroxide + iron (II) sulfate + sulfuric acid may ignite and explode
Hazardous Decomp. Prods.: Heated to decomp., emits toxic fumes of NO_x
Uses: Organic intermediate for pharmaceuticals, dyes, rubber chemicals; solvent; source for vinyl pyridine; laboratory reagent
Manuf./Distrib.: Aldrich; Fluka; Lonza; Nepera; Raschig; Reilly Ind.; Schweizerhall

β-Picoline

CAS 108-99-6; EINECS/ELINCS 203-636-9
UN 2313 (DOT)
Synonyms: 3-Methylpyridine; 3-Picoline; m-Picoline; Pyridine, 3-methyl-
Empirical: C_6H_7N
Properties: Colorless liq.; sweetish, not unpleasant odor; sol. in water, alcohol, ether; m.w. 93.13; dens. 0.957; f.p. -18 C; b.p. 144 C; flash pt. 36 C; ref. index 1.5050
Toxicology: LCLo (inh., rat, 2 h) 8700 ppm; LD50 (oral, rat) 400 mg/kg, (IP, rat) 150 mg/kg; poison by IV and IP routes; mod. toxic by ing.; irritant; TSCA listed
Precaution: Flamm. exposed to heat or flame; moderate fire risk; can react vigorously with oxidizers
Hazardous Decomp. Prods.: Heated to decomp., emits toxic fumes of NO_x
Uses: Process solvent; solvent in synthesis of pharmaceuticals, resins, dyestuffs, rubber accelerators, insecticides; prep. of nicotinic acid, nicotinic acid amide, waterproofing agents; laboratory reagent
Manuf./Distrib.: Aldrich; Alfa Chem; Fabrichem; Fluka; Nepera; Raschig; Reilly Ind.; Schweizerhall; Sithean; Synasia Fine Chem.; Whyte Chems. Ltd

γ-Picoline

CAS 108-89-4; EINECS/ELINCS 203-626-4
UN 2313 (DOT)
Synonyms: 4-Methylpyridine; 4-Picoline; p-Picoline; Pyridine, 4-methyl-
Empirical: C_6H_7N
Formula: $NCHCHC(CH_3)CHCH$
Properties: Colorless, mod. volatile liq.; disagreeable odor; sol. in water, alcohol, ether, oxygenated solvs.; m.w. 93.13; dens. 0.957; m.p. 2.4 C; b.p. 145 C; flash pt. 56 C; ref. index 1.5050
Toxicology: LD50 (oral, rat) 1290 mg/kg, (IP, rat) 163 mg/kg; poison by skin contact, IP routes; mod. toxic by ing.; mildly toxic by inh.; severe primary irritant to eyes and skin; TSCA listed

Precaution: Flamm. exposed to heat, flames, oxidizers
Hazardous Decomp. Prods.: Heated to decomp., emits toxic fumes of NO_x
Uses: Solvent in synthesis of pharmaceuticals, resins, dyestuffs, rubber accelerators, pesticides, waterproofing agents; laboratory reagent; mfg. of isoniazid; catalyst; curing agent
Manuf./Distrib.: Aldrich; Fluka; Lonza; Raschig; Reilly Ind.; Schweizerhall

m-Picoline. *See* β-Picoline
o-Picoline. *See* α-Picoline
p-Picoline. *See* γ-Picoline
Pigment black 15. *See* Copper oxide (ic)
Pimelic ketone; Pimelin ketone. *See* Cyclohexanone
Pine balsam; Pine gum. *See* Turpentine
2-Pinene. *See* α-Pinene

α-Pinene

CAS 80-56-8; 7785-26-4 (L, -); 7785-70-8 (D, +); EINECS/ELINCS 201-291-9; 232-077-3 (L,-); 232-087-8 (D,+)
UN 2368 (DOT); FEMA 2902
Synonyms: 2-Pinene; 4,6,6-Trimethylbicyclo (3,1,1) hept-3-en; 2,6,6-Trimethylbicyclo (3.1.1)-2-heptene; 2,6,6-Trimethylbicyclo (3.1.1) hept-2-ene; 2,6,6-Trimethylbicyclo (3.1.1)-2-hept-2-ene
Classification: Terpene hydrocarbon
Empirical: $C_{10}H_{16}$
Properties: Colorless liq., turpentine odor; sol. in alcohol, fixed oils, chloroform, ether, glac. acetic acid, oxygenated and chlorinated solvs.; insol. in water; m.w. 136.26; dens. 0.8592 (20/4 C); vapor pressure 10 mm (37.3 C); m.p. -55 C; b.p. 155 C; flash pt. 91 F; ref. index 1.464-1.468
Toxicology: LD50 (oral, rat) 3700 mg/kg; LCLo (inh., rat) 625 μg/m^3; deadly poison by inh.; mod. toxic by ing.; eye, mucous membrane, and severe skin irritant; toxic effects similar to turpentine
Precaution: DOT: Flamm. liq.; dangerous fire hazard exposed to heat, flame, oxidizers; explodes on contact with nitrosyl perchlorate
NFPA: Health 1, Flammability 3, Reactivity 0
Uses: Solvent for protective coatings, polishes, waxes; synthesis of camphene, camphor, pine oil; odorant; insecticides; additive for lubricants; odorant; synthetic flavoring agent in foods and pharmaceuticals; fragrance in personal care; in food-contact coatings
Regulatory: FDA 21CFR §172.515, 175.300, 175.320; FEMA GRAS
Manuf./Distrib.: Aldrich; Arizona; Berje; Doingcom; Fluka; Hercules; J.H. Calo; J.H. Hinz; Lluch Essence; Millennium; PDM; Penta Mfg.; Polarome Int'l.; Prodasynth; Sigma; Spectrum Quality Prods.; Technichem; Tecnal; Veitsiluoto Oy
Trade Names Containing: Dipentene-C

Pine oil. *See* Pine (Pinus palustris) oil
Pine oil synthetic. *See* α-Terpineol

Pine (Pinus palustris) oil

CAS 8002-09-3
UN 1272 (DOT)
Synonyms: Oleum abietis; Pine oil; Pine, white, oil; Pinus palustris; Pinus palustris oil; White pine oil; Yarmor; Yarmor pine oil
Definition: Volatile oil obtained from distillation of *Pinus palustris*
Properties: Colorless to pale yel. liq., penetrating pine odor; sol. in org. solvs.; insol. in water; dens. 0.86; b.p. 200-220 C; flash pt. (CC) 172 F
Toxicology: LD50 (oral, rat) 3200 mg/kg, (skin, rabbit) 5 g/kg; mod. toxic by ing.; mildly toxic by skin contact; irritating to eyes, skin, mucous membranes; weak allergen; human systemic effects by ing. (excitement, ataxia, headache); large doses may cause CNS depression; TSCA listed
Precaution: Flamm. exposed to heat or flame; can react with oxidizers; mod. spontaneous heating
NFPA: Health 0, Flammability 2, Reactivity 0
Uses: Odorant; disinfectant; sanitizer ingred.; penetrant; wetting agent; solvent; flotation frothing agent; preservative (textiles, papers); nonreactive epoxy resin diluent; laboratory reagent; fragrance ingred.; natural flavoring agent in foods; in food-pkg. adhesives; in paper/paperboard in contact with dry food; defoamer in food-contact coatings and paper/paperboard; in food-contact textiles
Regulatory: FDA 21CFR §172.510, 175.105, 176.180, 176.200, 176.210, 177.2800; 27CFR §21.65, 21.151
Manuf./Distrib.: Alfa Chem; Allchem Ind.; Andrea Aromatics; Arizona; Ashland; Berje; Charabot USA; Chr. Hansen Inc; Dujodwala Resins & Terpenes; George Uhe; Givaudan Fragrances; H.E. Daniel Ltd; Harcros; Hercules; J.H. Calo; Langley-Smith Ltd; Lebermuth; MelChem; Millennium; Pangaea Sciences; Penta Mfg.; Polarome Int'l.; Ruger; Schaefer Tech.; Sea-Land; Spec. Ingreds. & Perfume; Spectrum Quality Prods.; T&R Chem.; Technichem; U.S. Synthetics; Ultra Int'l.; Voigt Global Distrib.; Wego Chem. & Min.; Welch, Holme & Clark
Trade Names: Herco® Pine Oil

Pine, white, oil; Pinus palustris; Pinus palustris oil. *See* Pine (Pinus palustris) oil
1-Piperazineethanamine; 2-Piperazinoethylamine; 1-(1-Piperazinyl)-2-aminoethane; 2-Piperazin-1-yl ethylamine. *See* Aminoethylpiperazine

Piperidine

CAS 110-89-4; EINECS/ELINCS 203-813-0
UN 2401 (DOT); FEMA 2908
Synonyms: Azacyclohexane; Cyclopentimine; Hexahydropyridine; Hexazane; Pentamethyleneimine
Classification: Heterocyclic nitrogen compd.
Empirical: $C_5H_{11}N$
Formula: $NH(CH_2)_4CH_2$
Properties: Colorless clear liq.; amine-like odor; sol. in alcohol, acetone, diethyl ether, benzene, chloroform, oxygenated and aromatic solvs.; misc. with water; m.w. 85.15; dens. 0.862 (20/4 C); vapor pressure 40 mm Hg (29.2 C); m.p. -11 to -9 C; b.p. 104-106 C; flash pt. 16 C; ref. index 1.453 (20 C)
Toxicology: LD50 (oral, rat) 0.52 ml/kg, (skin, rabbit) 320 mg/kg, (inh., mammal) 6500 mg/m^3; toxic by ing., skin contact, IP routes; mod. toxic by subcut. route; mildly toxic by inh.; skin irritant; causes burns; experimental carcinogen, teratogen, reproductive effects; mutagenic data; TSCA listed
Precaution: Highly flamm.; very dangerous fire hazard exposed to heat, flame, or oxidizers; can react vigorously with oxidizers; explodes on contact with 1-perchloryl piperidine, dicyanofurazan, N-nitrosoacetanilide
Hazardous Decomp. Prods.: Heated to decomp., emits highly toxic fumes of NO_x
NFPA: Health 3, Flammability 3, Reactivity 0
Storage: Hygroscopic
Uses: Solvent; intermediate; curing agent for rubber and epoxy resins; catalyst for condensation reactions; ingred. in oils and fuels; complexing agent; intermediate for rubber accelerators; in agric.; synthetic flavoring agent in foods and pharmaceuticals
Features: Sweet flavor
Regulatory: FDA 21CFR §133, 172.515; FEMA GRAS
Manuf./Distrib.: Advanced ChemTech; Aldrich; Alfa Aesar; Am. Biorganics; BASF AG; Fluka; Nepera; Penta Mfg.; Raschig; Reilly Ind.; Schweizerhall; Sigma; Sithean; Spectrum Quality Prods.

PM. *See* Methoxyisopropanol
PMA; PM acetate. *See* Propylene glycol methyl ether acetate
PnB. *See* Butoxypropanol
Podocarpa-7,13-dien-15-oic acid, 13-isopropyl-, methyl ester. *See* Methyl abietate
POE (4). *See* PEG-4
POE (8). *See* PEG-8
POE (9). *See* PEG-9
POE (10). *See* PEG-10
POE (12). *See* PEG-12
POE (14). *See* PEG-14
POE (16). *See* PEG-16
POE (18). *See* PEG-18
POE (20). *See* PEG-20
POE (32). *See* PEG-32
POE (40). *See* PEG-40
POE (75). *See* PEG-75
POE (90). *See* PEG-90
POE (100). *See* PEG-100
POE (135). *See* PEG-135
POE (150). *See* PEG-150

POE (200). *See* PEG-200
POE (350). *See* PEG-350
POE (6) abietate. *See* PEG-6 abietate
POE (8) caprylate/caprate glycerides. *See* PEG-8 caprylic/capric glycerides
POE (30) castor oil. *See* PEG-30 castor oil
POE (33) castor oil. *See* PEG-33 castor oil
POE (35) castor oil. *See* PEG-35 castor oil
POE (40) castor oil. *See* PEG-40 castor oil
POE (16) cetyl ether. *See* Ceteth-16
POE (24) cetyl ether. *See* Ceteth-24
POE (24) cholesteryl ether. *See* Choleth-24
POE (6) decyl ether. *See* Deceth-6
POE (2) dibenzoate. *See* Diethylene glycol dibenzoate
POE (4) dilaurate. *See* PEG-4 dilaurate
POE (8) dilaurate. *See* PEG-8 dilaurate
POE (12) dilaurate. *See* PEG-12 dilaurate
POE (20) dilaurate. *See* PEG-20 dilaurate
POE (4) dioleate. *See* PEG-4 dioleate
POE (8) dioleate. *See* PEG-8 dioleate
POE (12) dioleate. *See* PEG-12 dioleate
POE (20) dioleate. *See* PEG-20 dioleate
POE (7) dodecyl phenyl ether. *See* Dodoxynol-7
POE (7) glyceryl ether. *See* Glycereth-7
POE (26) glyceryl ether. *See* Glycereth-26
POE (7) glyceryl ether triacetate. *See* Glycereth-7 triacetate
POE (7) glyceryl monococoate. *See* PEG-7 glyceryl cocoate
POE (30) glyceryl monococoate. *See* PEG-30 glyceryl cocoate
POE (15) hydroxystearate. *See* PEG-15 hydroxystearate
POE (3) lanolate. *See* PEG-3 lanolate
POE (5) lanolate. *See* PEG-5 lanolate
POE (10) lanolate. *See* PEG-10 lanolate
POE (20) lanolate. *See* PEG-20 lanolate
POE (75) lanolin. *See* PEG-75 lanolin
POE (5) lanolin alcohol. *See* Laneth-5
POE (16) lanolin alcohol. *See* Laneth-16
POE (5) lanolin ether. *See* Laneth-5
POE (10) lanolin ether. *See* Laneth-10
POE (15) lanolin ether. *See* Laneth-15
POE (16) lanolin ether. *See* Laneth-16
POE (9) lauryl ether. *See* Laureth-9
POE (6) methyl ether. *See* Methoxy PEG-6
POE (10) methyl ether. *See* Methoxy PEG-10
POE (16) methyl ether. *See* Methoxy PEG-16
POE (10) methyl glucose ether. *See* Methyl gluceth-10
POE (20) methyl glucose ether. *See* Methyl gluceth-20
POE (6) monoisostearate. *See* PEG-6 isostearate
POE (10) monomethyl ether. *See* Methoxy PEG-10
POE (16) monomethyl ether. *See* Methoxy PEG-16
POE (4) monooleate. *See* PEG-4 oleate
POE (5) monooleate. *See* PEG-5 oleate
POE (6) monooleate. *See* PEG-6 oleate
POE (8) monooleate. *See* PEG-8 oleate
POE (20) monooleate. *See* PEG-20 oleate
POE (5) nonylphenol. *See* Nonoxynol-5
POE (4) nonyl phenyl ether. *See* Nonoxynol-4
POE (5) nonyl phenyl ether. *See* Nonoxynol-5
POE (6) nonyl phenyl ether. *See* Nonoxynol-6
POE (9) nonyl phenyl ether phosphate. *See* Nonoxynol-9 phosphate
POE (2) oleyl ether. *See* Oleth-2
POE (16) oleyl ether. *See* Oleth-16
POE (20) POP (5) cetyl ether. *See* PPG-5-ceteth-20
POE (24) POP (24) glyceryl ether. *See* PPG-24-glycereth-24
POE(7) POP(2) isodecyl ether carboxylic acid isopropyl ester. *See* Isopropyl PPG-2-isodeceth-7-carboxylate
POE (3) POP (2) monobutyl ether. *See* PPG-2-buteth-3
POE (5) POP (3) monobutyl ether. *See* PPG-3-buteth-5
POE (7) POP (5) monobutyl ether. *See* PPG-5-buteth-7
POE (10) POP (7) monobutyl ether. *See* PPG-7-buteth-10
POE (12) POP (9) monobutyl ether. *See* PPG-9-buteth-12
POE (16) POP (12) monobutyl ether. *See* PPG-12-buteth-16
POE (30) POP (20) monobutyl ether. *See* PPG-20-buteth-30
POE (35) POP (28) monobutyl ether. *See* PPG-28-buteth-35
POE (45) POP (33) monobutyl ether. *See* PPG-33-buteth-45
POE (3) POP (3) myristyl ether. *See* PPG-3-myreth-3
POE (20) sorbitan monoisostearate. *See* PEG-20 sorbitan isostearate
POE (20) sorbitan monooleate. *See* Polysorbate 80
POE (20) sorbitan monostearate. *See* Polysorbate 60
POE (40) sorbitan peroleate. *See* PEG-40 sorbitan peroleate
POE (20) sorbitol ether. *See* Sorbeth-20
POE (40) sorbitol septaoleate. *See* PEG-40 sorbitan peroleate
POE (100) stearate. *See* PEG-100 stearate
POE (1) stearic acid. *See* Glycol stearate
POE (16) stearyl ether. *See* Steareth-16
Pogy oil. *See* Menhaden oil

Poloxamer 181

CAS 9003-11-6 (generic); 53637-25-5
Classification: Polyoxyethylene, polyoxypropylene block polymer
Formula: $HO(CH_2CH_2O)_x(CCH_3HCH_2O)_y(CH_2CH_2O)_zH$, avg. x = 3, y = 30, z = 3
Properties: Nonionic
Toxicology: LD50 (oral, rat) 2300 mg/kg; moderately toxic by ingestion, intraperitoneal route; TSCA listed
Uses: Surfactant, emulsifier in cosmetics, paints, paper, agric., foods, pharmaceuticals, water treatment; defoamer; antistat; demulsifier; detergent; wetting agent; gellant; dispersant; dye leveling agents; food additive; emulsifier, solubilizer, wetting agent, stabilizer in flavor concs.; defoamer in food-contact coatings and paper/paperboard; food-pkg. adhesives; in paper/paperboard in contact with dry foods; in cellophane for food pkg.
Features: Provides low odor, low irritation, low toxicity to personal care prods.
Regulatory: FDA 21CFR §172.808, 173.340, 175.105, 176.180, 176.200, 176.210, 177.1200
Manuf./Distrib.: Aldrich; Fluka; Sigma
Trade Names: Dow EP530

Polydimethylsiloxane. *See* Dimethicone
Polydimethyl siloxy cyclics. *See* Cyclomethicone
Polydimethylsiloxypropyl polyethoxy undecylenate. *See* Dimethicone copolyol undecylenate
Polyether glycol. *See* Polyethylene glycol
(Polyethyl) benzenes. *See* Naphtha, heavy aromatic

Polyethylene glycol

CAS 25322-68-3 (generic); EINECS/ELINCS 203-473-3
INS1521
Synonyms: α-Hydroxy-ω-hydroxypoly (oxy-1,2-ethanediyl); PEG; PEO; Polyether glycol; Poly (ethylene oxide); Polyglycol; Polyox; Poly (oxy-1,2-ethanediyl), α-hydro-ω-hydroxy-
Classification: Aliphatic organic compd.
Definition: Condensation polymer of ethylene glycol
Empirical: $(C_2H_4O)_n \cdot H_2O$
Formula: $H(OC_2H_4)_nOH$, $n \geq 4$
Properties: Clear visc. liq. or wh. solid; sol. in water, org. solvs., aromatic hydrocarbons; dens. 1.110-1.140 (20 C); m.p. 4-10 C; flash pt. 471 F; ref. index 1.4590
Toxicology: LD50 (oral, rat) 33,750 mg/kg; LDLo (IV, rat) 22 g/kg; mod. toxic by IV route; sl. toxic by ing.; skin and eye irritant; contact dermatitis sensitization; may cause hives; TSCA listed
Precaution: Combustible exposed to heat or flames
Hazardous Decomp. Prods.: Heated to decomp., emits acrid smoke and irritating fumes
Uses: Chemical intermediates; plasticizers; softeners; humectants; ointments; polishes; paper coating; mold lubricants; bases for cosmetics and pharmaceuticals; solvents; binders; metal and rubber processing; laboratory reagent; antistat; nickel plating brightening agent; crude oil demulsifier component; paints; textile fibers; boiler water additive for food processing; pharmaceutical ointment/suppository base, plasticizer, solvent; food additives; antifoam in foods; coating for tablets; dispersant, antisettling agent in calamine lotion; lubricant, evaporation inhibitor for rubbing alcohol, controlled-release drugs, contact lens fluid; food pkg.; in food-contact coatings
Regulatory: FDA 21CFR §172.210, 172.820, 173.310, 173.340, 175.300,

176.180; FDA approved for orals, topicals; NF compliance
Manuf./Distrib.: AMRESCO; Aldrich; Arch Chems.; Ashland; Avatar; Aventis Pharmaceuticals; BASF; BCH Brühl; BP Chems. Ltd; Brenntag (UK) Ltd; CarboMer; Celanese; Chemax; Cognis/Chems. Group; Degussa AG; Dow; DuPont; Eastman; Ferro/Grant; Fluka; Harcros; Honeywill & Stein; Huntsman; Inolex; Lambent Tech.; Mosselman NV; Mutchler; Norman, Fox; PPG Ind.; Penta Mfg.; Polysciences; Protameen; Rhodia HPCII; Ruger; Sal Chem.; Sigma; Spec. Ingreds. & Perfume; Spectrum Quality Prods.; Surfactants Inc; Thornley; Univar E&E Ltd; Universal Preserv-A-Chem; Varsal Instruments; Vopak USA; Whyte Chems. Ltd

Polyethylene glycol 300. *See* PEG-6
Polyethylene glycol 400. *See* PEG-8
Polyethylene glycol 600. *See* PEG-12
Polyethylene glycol 1000. *See* PEG-20
Polyethylene glycol 4000. *See* PEG-75
Polyethylene glycol dimethyl ether. *See* PEG dimethyl ether
Polyethylene glycol monomethyl ether. *See* Methoxy PEG
Polyethylene glycol nonyl phenyl ether. *See* Nonoxynol
Poly (ethylene glycol) tetrahydrofurfuryl ether. *See* Glycofurol
Poly (ethylene oxide). *See* Polyethylene glycol
Polyglycerol esters of interesterified ricinoleic acid; Polyglycerol esters of polycondensed fatty acids from castor oil. *See* Polyglyceryl polyricinoleate

Polyglyceryl-3 diisostearate

CAS 63705-03-3 (generic); 66082-42-6; 85404-84-8; EINECS/ELINCS 291-548-1
Synonyms: Isooctadecanoic acid, diester with 1,2,3-propanetriol trimer; Isooctadecanoic acid, diester with triglycerol; Triglyceryl diisostearate
Definition: Diester of isostearic acid and polyglycerin-3
Empirical: $C_{45}H_{88}O_9$
Properties: Nonionic
Toxicology: TSCA listed
Uses: Emulsifier, emollient, thickener, solvent in cosmetics, creams, lotions, lip prods., pharmaceuticals; dye and pigment wetting agent; food emulsifier; pharmaceutical excipient
Manuf./Distrib.: A&E Connock
Trade Names: Emerest® 2452

Polyglyceryl-4 oleate

CAS 9007-48-1 (generic); 71012-10-7
Synonyms: 9-Octadecenoic acid, monoester with tetraglycerol; Tetraglyceryl monooleate
Definition: Ester of oleic acid and polyglycerin-4
Empirical: $C_{30}H_{58}O_{10}$
Properties: Nonionic
Toxicology: TSCA listed
Uses: W/o emulsifier, lubricant, and antifoam for industrial applics.; emulsifier for cosmetics, foods and pharmaceutical ointments
Regulatory: FDA 21CFR §172.854
Manuf./Distrib.: A&E Connock
Trade Names: Protachem™ 100

Polyglyceryl polyricinoleate

INS476; E476
Synonyms: Glyceran esters of condensed castor oil fatty acids; Glycerol esters of condensed castor oil fatty acids; PGPR; Polyglycerol esters of interesterified ricinoleic acid; Polyglycerol esters of polycondensed fatty acids from castor oil
Definition: Polyglyceryl esters of interesterified ricinoleic acid
Properties: Lt. brn. highly visc. liq.; sol. in edible oils and fats; insol. in cold and hot water; does not crystallize @ 0 C; nonionic
Uses: Emulsifier, stabilizer, visc. modifier, emulsion stabilizer, pan release agent in foods; emulsifier, stabilizer in cosmetics, pharmaceuticals, emulsions; release agent
Regulatory: Europe listed; UK approved; ADI 0-7.5 mg/kg body wt. (FAO/WHO)
Trade Names: Tilol GCT-R

Polyglyceryl-3 stearate

CAS 37349-34-1 (generic); 27321-72-8; 26855-43-6; 61790-95-2; EINECS/ELINCS 248-403-2
Synonyms: Octadecanoic acid, monoester with triglycerol; Octadecanoic acid, monoester with tri-1,2,3-propanetriol diether; Stearic acid, monoester with triglycerol; Triglyceryl stearate
Definition: Ester of stearic acid and polyglycerin-3
Empirical: $C_{27}H_{54}O_8$
Properties: Nonionic
Toxicology: TSCA listed
Uses: Emulsifier in cosmetics; food emulsifier, stabilizer, whipping agent; emulsifier, antifog for plastics; bodying agent, solubilizer, dispersant, emulsifier for pharmaceutical delivery/absorp. enhancement, dermatologicals, suppositories
Regulatory: FDA 21CFR §172.854
Manuf./Distrib.: A&E Connock; ABITEC
Trade Names: Protachem™ G-5566

Polyglyceryl-9 stearate

Properties: Nonionic
Uses: Emulsifier, emollient, lubricant, solvent, stabilizer, emulsion thickener, opacifier in personal care prods.
Trade Names: Protachem™ G-5509

Polyglycol. *See* Polyethylene glycol
Polyglycol 1000. *See* PEG-20
Polyglycol 4000. *See* PEG-75
Polymerized castor oil. *See* Castor oil, polymerized
Polymerized siloxane. *See* Silicone
Poly (methylene) wax. *See* Paraffin

Petroleum distillates, heavy thermal cracked

CAS 64741-81-7; EINECS/ELINCS 265-082-4
Synonyms: Distillates (petroleum), heavy thermal cracked; Heavy thermal cracked petroleum distillate; Polynuclear aromatic hydrocarbon
Classification: Aromatic
Definition: Complex combination of hydrocarbons from distillation of prods. from thermal cracking process; consists predominantly of unsat. hydrocarbons in C15-36 range; likely to contain ≥ 5% 4-6 membered condensed ring aromatic hydrocarbons
Properties: B.p. 260-480 C
Toxicology: May cause cancer
Uses: Sec. vinyl plasticizer; solvent
Trade Names Containing: Kenplast® G

Polynuclear aromatic hydrocarbon. *See* Petroleum distillates, heavy thermal cracked
Polyorganosiloxanes. *See* Silicone
Polyox. *See* Polyethylene glycol
Poly (oxy (dimethylsilylene)), α-(trimethylsilyl)-ω-methyl-. *See* Dimethicone
Poly (oxy-1,2-ethanediyl), α-hydro-ω-hydroxy-. *See* Polyethylene glycol
Poly (oxy-1,2-ethanediyl), α-methyl-ω-methoxy-. *See* PEG dimethyl ether
Poly(oxy-1,2-ethanediyl), α-(nonylphenyl)-ω-hydroxy-, phosphate. *See* Nonoxynol-9 phosphate
Polyoxyethylene nonylphenol. *See* Nonoxynol
Polyoxyl 35 castor oil. *See* PEG-35 castor oil
Polyoxyl 40 castor oil. *See* PEG-40 castor oil
Polyoxyl 4 dilaurate. *See* PEG-4 dilaurate
Polyoxyl 75 lanolin. *See* PEG-75 lanolin
Polyoxyl 2 stearate. *See* PEG-2 stearate
Polyoxymethylene. *See* Trioxane
Polyoxymethylene glycols. *See* Formaldehyde
Poly (oxy (methyl-1,2-ethanediyl), α-butyl-ω-hydroxy. *See* PPG butyl ether
Polyoxypropylene (9). *See* PPG-9
Polyoxypropylene (17). *See* PPG-17
Polyoxypropylene (20). *See* PPG-20
Polyoxypropylene (26). *See* PPG-26
Polyoxypropylene (30). *See* PPG-30
Polyoxypropylene (3) methyl ether. *See* PPG-3 methyl ether
Polyphenylmethyl siloxane. *See* Phenyl trimethicone

Polypropylene glycol

CAS 25322-69-4 (generic); EINECS/ELINCS 200-338-0
Synonyms: α-Hydro-ω-hydroxypoly [oxy (methyl-1,2-ethanediyl)]; α-Hy-

dro-ω-hydroxypoly (oxypropylene); Oxirane, methyl-, homopolymer; Poly (propylene oxide); PPG; Propylene oxide homopolymer
Classification: Aliphatic organic compd.
Empirical: $(C_3H_8O_2)_n$
Formula: $HO(C_3H_6O)_nH$, n = 4-70
Properties: Colorless clear visc. liq., sl. bitter taste; sol. in water, aliphatic ketones, alcohol; insol. in ether, aliphatic hydrocarbons; m.w. 400-2000; dens. 1.001-1.007; m.p. does not cryst.; flash pt. > 390 F
Toxicology: LD50 (oral, rat) 4190 mg/kg; mildly toxic by ingestion; skin and eye irritant; linked to sensitive reactions; TSCA listed
Precaution: Combustible exposed to heat or flame; reactive with oxidizers
Hazardous Decomp. Prods.: Heated to decomp., emits acrid smoke and irritating fumes
Uses: Aq. fire-resist. hydraulic fluids; rubber processing lubricant; high-temp. lubricants for furnaces; antifoam agents; PU chain extender; intermediates for urethane foams, adhesives, coatings, elastomers, plasticizers, paint formulations; lab reagent; flotation frothing agent; defoamer for beet sugar and yeast processing, foods; pharmaceutical solvent for fats, oils, waxes, resins; plasticizer in food-contact polymers; in food-pkg. adhesives; in side seam cements for food-contact containers
Regulatory: FDA 21CFR §173.310, 173.340, 175.105, 175.300, 177.1680, 178.3740; FDA approved for ophthalmics, orals, topicals
Manuf./Distrib.: Aldrich; Arch Chems.; Ashland; BASF; BP Chems. Ltd; Bayer; Coyne; Degussa AG; Dow; Eastman; Fluka; Harcros; Lambent Tech.; Lyondell; Monomer-Polymer & Dajac Labs; Mosselman NV; PPG Ind.; Rhodia HPCII; Sal Chem.

Polypropylene glycol (9). *See* PPG-9
Polypropylene glycol (17). *See* PPG-17
Polypropylene glycol (20). *See* PPG-20
Polypropylene glycol (26). *See* PPG-26
Polypropylene glycol (30). *See* PPG-30
Polypropylene glycol 1200. *See* PPG-20
Polypropylene glycol methyl ether; Polypropylene glycol monomethyl ether. *See* Methoxyisopropanol
Poly (propylene oxide). *See* Polypropylene glycol

Polysorbate 60
CAS 9005-67-8 (generic)
FEMA 2916; INS435
Synonyms: PEG-20 sorbitan stearate; POE (20) sorbitan monostearate; Sorbimacrogol stearate 300; Sorbitan, monooctadecanoate, poly (oxy-1,2-ethanediyl) derivs.
Definition: Mixture of stearate esters of sorbitol and sorbitol anhydrides, with ≈ 20 moles ethylene oxide
Empirical: $C_{64}H_{126}O_{26}$
Properties: Lemon to orange oily liq., faint char. odor, bitter taste; sol. in water, alcohol, aniline, ethyl acetate, toluene; insol. in min. and veg. oils; m.w. 1311.90; HLB 14.9; acid no. 2 max.; sapon. no. 45-55; hyd. no. 81-96; nonionic
Toxicology: LD50 (IV, rat) 1220 mg/kg; mod. toxic by IV route; questionable carcinogen; experimental tumorigen, reproductive effects; TSCA listed
Hazardous Decomp. Prods.: Heated to decomp., emits acrid smoke and irritating fumes
Uses: Surfactant, emulsifier, solvent, solubilizer for polymerization, cosmetics, pharmaceuticals, industrial chemicals, pesticides, textiles, plastics, paints, pulp/paper, polishes, dyes; solubilizer; textile lubricant/antistat (spin finishes); dye penetrant/leveling aid; lubricant; softener; emulsifier, dispersant, protective coating, dough conditioner, stabilizer in foods; foaming agent in beverage mixes; pharmaceutical and veterinary drug; emulsifier in food pkg.; in food-pkg. adhesives
Regulatory: FDA 21CFR §73.1001, 172.515, 172.836, 173.310, 173.340, 175.105, 175.300, 178.3400; USDA CFR9 §318.7, 381.147 (limitation 1% max., 1% total combined with polysorbate 80, 0.0175% in scald water); FEMA GRAS; Europe listed; UK approved; FDA approved for orals, rectals, topicals, vaginals; NF, BP, EP compliance
Manuf./Distrib.: Aldrich; Am. Ingreds.; DelCon; Fluka; Mosselman NV; Phoenix Nat. Prods.; Rhodia HPCII; Sigma; Spectrum Quality Prods.; Voigt Global Distrib.

Polysorbate 80
CAS 9005-65-6 (generic); 37200-49-0; 61790-86-1
FEMA 2917; INS433
Synonyms: PEG-20 sorbitan oleate; POE (20) sorbitan monooleate; Sorbimacrogol oleate 300
Definition: Mixture of oleate esters of sorbitol and sorbitol anhydrides, with ≈ 20 moles EO
Properties: Amber visc. liq.; faint odor; bitter taste; nonionic; very sol. in water; sol. in alcohol, fixed oils, cottonseed oil, corn oil, ethyl acetate, methanol, toluene; insol. in min. oil; dens. 1.06-1.10; visc. 270-430 cSt; HLB 15.0; acid no. 2 max.; sapon. no. 45-55; hyd. no. 65-80; pH 5-7 (5% aq.)
Toxicology: LD50 (oral, mouse) 25 g/kg, (IP, rats) 6.3 ml/kg, (IV, rat) 179 mg/kg; mod. toxic by IV route; mildly toxic by ing.; eye irritant; questionable carcinogen; experimental tumorigen, reproductive effects; human mutagenic data; TSCA listed
Hazardous Decomp. Prods.: Heated to decomp., emits acrid smoke and irritating fumes
Uses: Surfactant, emulsifier, solubilizer, wetting agent, dispersant in cosmetics, pharmaceuticals, polymerization, paints; emulsifier, dispersant in medicinals; adjuvant in herbicides; diluent in food colorants; flavor; antistat, lubricant for textile fibers; emulsifier for textile lubricants, spin finishes; lubricant, antistat and antifog for plastics; defoamer, dispersant, emulsifier in foods; emulsifier in food pkg.; in food-pkg. adhesives
Regulatory: FDA 21CFR §73.1, 73.1001, 172.515, 172.840, 172.842, 173.340, 175.105, 175.300, 176.180, 178.3400, 573.860; USDA 9CFR §318.7, 381.147 (limitation 1% alone, 1% total combined with polysorbate 60); FEMA GRAS; Europe listed; UK approved; FDA approved for buccals, intramuscular injectables, intravenous, parenterals, ophthalmics, orals, otics, rectals, topicals, vaginals; NF, BP, EP compliance
Manuf./Distrib.: Aldrich; CarboMer; Chemacon GmbH; Croda Chem. Ltd; DelCon; Fluka; J.T. Baker; Mosselman NV; Sigma; Spectrum Quality Prods.; Voigt Global Distrib.
Trade Names: Capmul® POE-O
Trade Names Containing: Punctilious® SDA 38F-6 190 Proof; Ritawax AEO

Polysorbate 120. *See* PEG-20 sorbitan isostearate
Polyterpene resin. *See* Terpene resin
POP (5) butyl ether. *See* PPG-5 butyl ether
POP (9) butyl ether. *See* PPG-9 butyl ether
POP (14) butyl ether. *See* PPG-14 butyl ether
POP (15) butyl ether. *See* PPG-15 butyl ether
POP (18) butyl ether. *See* PPG-18 butyl ether
POP (22) butyl ether. *See* PPG-22 butyl ether
POP (24) butyl ether. *See* PPG-24 butyl ether
POP (33) butyl ether. *See* PPG-33 butyl ether
POP (40) butyl ether. *See* PPG-40 butyl ether
POP (53) butyl ether. *See* PPG-53 butyl ether
POP (10) cetyl ether. *See* PPG-10 cetyl ether
POP (20) cetyl ether. *See* PPG-20 cetyl ether
POP (30) cetyl ether. *See* PPG-30 cetyl ether
POP (50) cetyl ether. *See* PPG-50 cetyl ether
POP (2) dibenzoate. *See* Dipropylene glycol dibenzoate
POP (55) glyceryl ether. *See* PPG-55 glyceryl ether
POP (40) methyl diethyl ammonium chloride. *See* PPG-40 diethylmonium chloride
POP (2) methyl ether. *See* PPG-2 methyl ether
POP (10) methyl glucose ether. *See* PPG-10 methyl glucose ether
POP (20) methyl glucose ether. *See* PPG-20 methyl glucose ether
POP (3) myristyl ether. *See* PPG-3 myristyl ether
POP (2) myristyl ether propionate. *See* PPG-2 myristyl ether propionate
POP (23) oleyl ether. *See* PPG-23 oleyl ether
POP (37) oleyl ether. *See* PPG-37 oleyl ether
POP (5) POE (20) cetyl ether. *See* PPG-5-ceteth-20
POP (24) POE (24) glyceryl ether. *See* PPG-24-glycereth-24
POP (2) POE (3) monobutyl ether. *See* PPG-2-buteth-3
POP (3) POE (5) monobutyl ether. *See* PPG-3-buteth-5
POP (5) POE (6) monobutyl ether. *See* PPG-5-buteth-7
POP (7) POE (10) monobutyl ether. *See* PPG-7-buteth-10

POP (9) POE (12) monobutyl ether. *See* PPG-9-buteth-12
POP (12) POE (16) monobutyl ether. *See* PPG-12-buteth-16
POP (20) POE (30) monobutyl ether. *See* PPG-20-buteth-30
POP (28) POE (35) monobutyl ether. *See* PPG-28-buteth-35
POP (33) POE (45) monobutyl ether. *See* PPG-33-buteth-45
POP (3) POE (3) myristyl ether. *See* PPG-3-myreth-3
POP (11) stearyl ether. *See* PPG-11 stearyl ether
POP (15) stearyl ether. *See* PPG-15 stearyl ether
POP (15) stearyl ether benzoate. *See* PPG-15 stearyl ether benzoate
Potash. *See* Potassium carbonate
Potash lye; Potassa. *See* Potassium hydroxide

Potassium carbonate

CAS 584-08-7; EINECS/ELINCS 209-529-3
INS501(i); E501
Synonyms: American ash; Carbonate of potash; Carbonic acid dipotassium salt; Dipotassium carbonate; Pearl ash; Potash; Potassium carbonate (2:1); Salt of tartar
Classification: Inorganic salt
Empirical: CK_2O_3
Formula: K_2CO_3
Properties: Wh. gran., translucent powd., odorless; alkaline taste; sol. in water; insol. in alcohol, acetone, glycerol; m.w. 138.20; dens. 2.428 (19 C); m.p. 891 C; b.p. dec.; noncombustible
Toxicology: LD50 (oral, rat) 1870 mg/kg, (oral, mouse) 2570 mg/kg; poison by ing.; sol'ns. irritating to tissue; strong caustic; TSCA listed
Precaution: Strong caustic; incompat. with KCO, magnesium, chlorine trifluoride
Hazardous Decomp. Prods.: Heated to decomp., emits toxic fumes of K_2O
HMIS: Health 2, Flammability 0, Reactivity 0
Storage: Deliq., hygroscopic
Uses: Special glasses (optical, TV tubes); potassium silicate; fertilizer mfg.; dehydrating agent; pigments; printing inks; lab reagent; textile dyeing, raw wool washing; detergents/soft soap ingred.; buffer in cosmetics, pharmaceuticals; photography; enamels; process solvent; agrochemicals; prod. of org. chems.; potassium ion source; rubber; gas treatment; cement; corrosion/scale control in drinking water treatment; boiler water additive for food processing; flavoring agent, adjuvant, nutrient supplement, pH control agent, alkali, processing aid, yeast nutrient, stabilizer in foods; chocolate ingred.
Regulatory: FDA 21CFR §73.85, 163.110, 163.111, 163.112, 172.560, 173.310, 184.1619, GRAS; USDA 9CFR §318.7; BATF 27CFR §240.1051; Japan approved; Europe listed; UK approved; FDA approved for orals, topicals; USP/NF compliance
Manuf./Distrib.: AMRESCO; ASHTA; Albemarle; Aldrich; Alfa Aesar; Alfa Chem; Allan; Armand Prods.; Ashland; Asiamerica Int'l.; BCH Brühl; BassTech Int'l.; Boith China; Browning; Charkit; Church & Dwight; Degussa AG/Health & Nutrition; EM Ind.; EM Science; Fluka; Functional Foods; GFS; Gallard-Schlesinger Ind.; Harcros; Hasnil; Hays Chems. Ltd; ICC Ind.; Independent Chem.; Integra; J.T. Baker; Lavy's; Lohmann; Mallinckrodt Baker; Mutchler; Occidental; Penta Mfg.; Peter Whiting Ltd; Potash Corp./Saskatchewan Sales; Ruger; Schaefer Tech.; Sigma; Spectrum Quality Prods.; Stanchem; Technichem; Thomas Scientific; Universal Preserv-A-Chem; VWR Int'l.; Varsal Instruments; Voigt Global Distrib.; Vopak USA

Potassium carbonate (2:1). *See* Potassium carbonate
Potassium hydrate. *See* Potassium hydroxide

Potassium hydroxide

CAS 1310-58-3; EINECS/ELINCS 215-181-3
UN 1813 (DOT); UN 1814 (DOT); INS525; E525
Synonyms: Caustic potash; Lye; Potash lye; Potassa; Potassium hydrate
Classification: Inorganic base
Empirical: HKO
Formula: KOH
Properties: Wh. flakes, lumps or pellets; sol. in water, alcohol, glycerol; sl. sol. in ether; m.w. 56.11; dens. 2.044; m.p. 405 C; b.p. 1320 C
Toxicology: ACGIH TLV/CL 2 mg/m^3; LD50 (oral, rat) 365 mg/kg; toxic by ing., inh.; strong caustic; eye and severe human skin irritant; ing. may cause violent throat pain and stricture of esophagus; mutagenic data; TSCA listed
Precaution: DOT: Corrosive material; releases heat in contact with water and acids; corrodes many metals when wet; violent exothermic reactions with water; potentially explosive reactions with chlorine dioxide, nitrobenzene, nitromethane
Hazardous Decomp. Prods.: Heated to decomp., emits toxic fumes of K_2O; above 84 C reacts with reducing sugars to form CO
NFPA: Health 3, Flammability 0, Reactivity 1
Storage: Highly deliq.; hygroscopic
Uses: Soap raw material; bleaching; mfg. of potassium chemicals; battery electrolyte; absorbent; liq. fertilizer ingred.; herbicides; electroplating reagent; photographic reagent; mercerizing; paint removers; corrosion/scale inhibitor in water treatment; buffer in cosmetics; pharmaceutical alkalizer, cuticle solvent, caustic for wart removal; reagent for CO_2 absorption processes; syn. rubber; direct food additive (formulation aid, pH control agent, alkali, processing aid, stabilizer, thickener, oxidizing agent; poultry scald agent); in acrylate ester copolymer food-contact coatings
Regulatory: FDA 21CFR §73.85, 163.110, 163.111, 163.112, 172.841, 175.210, 176.180, 176.210, 177.1600, 177.2800, 184.1631, GRAS; USDA 9CFR §381.147; Europe listed; UK approved; FDA approved for intravenous, parenterals, orals, topicals; NF, BP, EP, JP compliance
Manuf./Distrib.: AMRESCO; ASHTA; Albemarle; Aldrich; Alfa Aesar; Alfa Chem; Amyl; Arch Chems.; Ashland; Asiamerica Int'l.; Avrachem AG; BCH Brühl; Borden & Remington; Charkit; Degussa AG/Health & Nutrition; E + E (US); EM Ind.; EM Science; Fluka; GFS; General Chem.; Hasnil; Hydrite; Int'l. Chem. Inc; Integra; J.T. Baker; K3; Kuehne Chem.; Mallinckrodt Baker; Occidental; P.B. & S.; Penta Mfg.; Phoenix Nat. Prods.; Ruger; Sal Chem.; San Yuan; Schaefer Tech.; Sigma; Spectrum Quality Prods.; Stanchem; Thomas Scientific; Universal Preserv-A-Chem; VWR Int'l.; Voigt Global Distrib.; Vopak USA; Vulcan Perf. Chems.
Trade Names Containing: Uresolve 411

Potassium iodide

CAS 7681-11-0; EINECS/ELINCS 231-659-4
Classification: Inorganic salt
Definition: Potassium salt of hydriodic acid
Empirical: IK
Formula: KI
Properties: Colorless to wh. cryst. granules or powd., strong bitter saline taste; sol. in water, alcohol, acetone, glycerol, oxygenated solvs.; m.w. 166.02; dens. 3.123; m.p. 723 C; b.p. 1420 C
Toxicology: LDLo (oral, mouse) 1862 mg/kg, (IV, rat) 167 mg/kg; irritant; poison by IV route; mod. toxic by ing. and IP routes; allergic reactions; human teratogenic effects; mutagenic data; TSCA listed
Precaution: Explosive reaction with charcoal + ozone; incompat. with oxidants, BrF_3, FClO, metallic salts
Hazardous Decomp. Prods.: Heated to decomp., emits very toxic fumes of K_2O and I^-
HMIS: Health 2, Flammability 0, Reactivity 1
Storage: Sl. hygroscopic; moisture- and light-sensitive; store @ R.T.
Uses: Reagent in analytical chemistry; photographic emulsions; animal feed additive; dietary supplement; source of iodine in table salt and pharmaceuticals; dye remover; antiseptic; nylon stabilizer; photoengraving; syn. raw materials; ring iodination reagent; antimicrobial in cosmetics; as sanitizing solution for food contact; in denatured alcohol
Regulatory: FDA 21CFR §100.155, 172.375, 178.1010, 184.1634, 582.80, GRAS; BP, EP compliance
Manuf./Distrib.: AMRESCO; Aldrich; Alfa Aesar; Alfa Chem; Allan; Am. Int'l.; Amyl; Ashland; Asiamerica Int'l.; Atomergic Chemetals; Avrachem AG; Calabrian; Charkit; Clariant; Cornelius Chem. Co. Ltd; Crystran Ltd; Deepwater Chems.; Degussa AG/Health & Nutrition; EM Ind.; EM Science; Fabrichem; Filo; Fluka; GFS; Gallard-Schlesinger Ind.; Hasnil; Hawkins Chem.; Helm NY; Independent Chem.; Integra; Interchem; J.T. Baker; Mallinckrodt Baker; Miljac; Morre-Tec Ind.; Mutchler; Penta Mfg.; R.W. Greeff; RIA Int'l.; Robeco; Ruger; Seltzer; Sigma; Spectrum Quality Prods.; Takeda Vitamin & Food USA; Thomas Scientific; Tomen Am.; Universal Preserv-A-Chem; VWR Int'l.; Voigt Global Distrib.; Vopak USA; Wego Chem. & Min.; West Agro; Whyte Chems. Ltd; Wm. Blythe Ltd; Xinchem
Trade Names Containing: Punctilious® SDA 25A-1 190 Proof

Potato alcohol. *See* Alcohol denatured
PPG. *See* Polypropylene glycol
PPG-3. *See* Tripropylene glycol

PPG-9

CAS 25322-69-4 (generic)
Synonyms: Polyoxypropylene (9); Polypropylene glycol (9); PPG 400
Definition: Polymer of propylene oxide
Formula: $H(OCH_2CHCH_3)_nOH$, avg. n = 9
Properties: Nonionic
Toxicology: TSCA listed
Uses: Chemical intermediate; intermediate yielding esters useful as lubricants, defoamers in rubber, pharmaceuticals; antifoam in fermentation and paints; antiblooming agent for pentachlorophenol-treated wood; binder and lubricant for ceramics; plasticizer of resin-treated papers; mold release applics.; emollient, solvent in cosmetics, pharmaceuticals; boiler water additive (food contact); in food-pkg. adhesives
Features: Provides low odor, low irritation, low toxicity to personal care prods.
Regulatory: FDA 21CFR §173.310, 175.105, 176.200, 176.210
Manuf./Distrib.: Aldrich; Fluka
Trade Names: Jeffox® PPG-400; Polyglycol P425

PPG-17

CAS 25322-69-4 (generic)
Synonyms: Polyoxypropylene (17); Polypropylene glycol (17); PPG (1000)
Definition: Polymer of propylene oxide
Formula: $H(OCH_2CHCH_3)_nOH$, avg. n = 17
Properties: Nonionic
Toxicology: TSCA listed
Uses: Chemical intermediate for surfactants, ethers, esters; antifoam in fermentation, paints, ceramics, rubber; binder and lubricant for ceramics; metalworking lubricants; textile lubricant; plasticizer for plastics, resin-treated papers; antiblooming agent for pentachlorophenol-treated wood; hydraulic fluids; dyeing; latex coagulant; emollient, solvent in cosmetics; defoamer for food-grade coatings and paper prods.; boiler water additive (food contact); in food-pkg. adhesives
Features: Provides low odor, low irritation, low toxicity to personal care prods.
Regulatory: FDA 21CFR §173.310, 175.105, 176.170, 176.200, 176.210
Manuf./Distrib.: Aldrich; Fluka
Trade Names: Polyglycol P1000TB

PPG-20

CAS 25322-69-4 (generic)
Synonyms: Polyoxypropylene (20); Polypropylene glycol (20); Polypropylene glycol 1200; PPG 1200
Definition: Polymer of propylene oxide
Formula: $H(OCH_2CHCH_3)_nOH$, avg. n = 20
Properties: Nonionic
Toxicology: LD50 (IP, mouse) 113 mg/kg, (skin, rabbit) > 10 g/kg; LDLo (IV, dog) 20 mg/kg; poison by IV and IP routes; irritating to skin and eyes; TSCA listed
Hazardous Decomp. Prods.: Heated to decomp., emits acrid smoke and irritating fumes
Uses: Defoamer; mold release applics.; chemical intermediates for fatty acid esters; components for urethane resins; textile lubricants; metalworking compds.; paints; urethane foams; brake and hydraulic fluids; plasticizers; release agents; lubricating oils/greases; rubber processing; emollient, solvent in cosmetics; defoamer for beet sugar and yeast processing; boiler water additive (food contact); in food-pkg. adhesives; defoamer in food-contact paper coatings, paper/paperboard; plasticizer in food-contact polymers
Features: Provides low odor, low irritation, low toxicity to personal care prods.
Regulatory: FDA 21CFR §173.310, 173.340, 175.105, 176.170, 176.200, 176.210, 178.3740
Manuf./Distrib.: Aldrich; Fluka
Trade Names: Polyglycol P1200

PPG-26

CAS 25322-69-4 (generic)
Synonyms: Polyoxypropylene (26); Polypropylene glycol (26); PPG 2000
Definition: Polymer of propylene oxide
Formula: $H(OCH_2CHCH_3)_nOH$, avg. n = 26
Properties: Nonionic
Toxicology: TSCA listed
Uses: Defoamer; mold release applics.; intermediate for esters useful as lubricants, defoamers in rubber, pharmaceuticals; components for urethane resins; boiler water additive for food contact; defoamer for beet sugar and yeast processing; emollient, solvent in cosmetics, pharmaceuticals; lubricant base for textiles, paper, metalworking; food-pkg. adhesives; in paper/paperboard in contact with aq./fatty foods; plasticizer in food-contact polymers; defoamer in food-contact coatings and paper/paperboard
Features: Provides low odor, low irritation, low toxicity to personal care prods.
Regulatory: FDA 21CFR §173.310, 173.340, 175.105, 176.170, 176.200, 176.210, 178.3740
Manuf./Distrib.: Aldrich; Fluka
Trade Names: Jeffox® PPG-2000; Polyglycol P2000

PPG-30

CAS 25322-69-4 (generic)
Synonyms: Polyoxypropylene (30); Polypropylene glycol (30); PPG 4000
Definition: Polymer of propylene oxide
Formula: $H(OCH_2CHCH_3)_nOH$, avg. n = 30
Properties: Nonionic
Toxicology: TSCA listed
Uses: Defoamer; mold release applics.; chemical intermediates for fatty acid esters; components for urethane resins; antifoam in fermentation and paints, antiblooming agent for pentachlorophenol-treated wood; binder and lubricant for ceramics; plasticizer of resin-treated papers; emollient, solvent in cosmetics; boiler water additive (food contact); in food-pkg. adhesives
Features: Provides low odor, low irritation, low toxicity to personal care prods.
Regulatory: FDA 21CFR §173.310, 173.340, 175.105, 176.170, 176.200, 178.3740
Manuf./Distrib.: Aldrich; Fluka
Trade Names: Polyglycol P4000

PPG-400

Properties: Nonionic
Uses: Intermediate yielding esters useful as lubricants, defoaming agents in rubber and pharmaceuticals, solvents and humectant modifiers for inks, plasticizers, and functional fluids

PPG 400. *See* PPG-9
PPG (1000). *See* PPG-17
PPG 1200. *See* PPG-20

PPG-2000

Properties: Nonionic
Uses: Intermediate yielding esters useful as lubricants, defoaming agents in rubber and pharmaceuticals, solvents and humectant modifiers for inks, plasticizers, and functional fluids

PPG 2000. *See* PPG-26
PPG 4000. *See* PPG-30

PPG-10 butanediol

Definition: Polyoxypropylene ether of butanediol
Empirical: $C_{34}H_{70}O_{12}$
Formula: $HO(CH_2CHCH_3O)_xCH_2CH_2CH_2CH_2(OCHCH_3CH_2)_yOH$, avg. (x+y)=10
Properties: Nonionic
Uses: Emollient, solubilizer, coupling agent for cosmetics, pharmaceuticals, antiperspirant sticks, makeup removers, lotions, foundations; solvent for sunscreen actives; crystal growth inhibitor
Trade Names: Luvitol® BD 10 P; Macol® 57

PPG-2-buteth-3

CAS 9038-95-3 (generic); 9065-63-8 (generic)
Synonyms: Oxirane, methyl, polymer and oxibane, butyl ether; POE (3) POP (2) monobutyl ether; POP (2) POE (3) monobutyl ether

Definition: Polyoxypropylene, polyoxyethylene ether of butyl alcohol
Empirical: $C_{16}H_{34}O_6$
Formula: $C_4H_9(OCHCH_3CH_2)_x(OCH_2CH_2)_yOH$, avg. x = 2, avg. y = 3
Properties: Nonionic
Toxicology: LD50 (oral, rat) 8530 mg/kg; mildly toxic by ing.; skin and severe eye irritant; TSCA listed
Hazardous Decomp. Prods.: Heated to decomp., emits acrid smoke and irritating fumes
Uses: Antistat, emollient, emulsifier in cosmetics; lubricant for compressor, rubber, greases, textiles; intermediate for prep. of resins, plasticizers, surfactants; solvent for inks and dyes; antifoam in boiler water, latex processing, washing sol'ns., paints, adhesives, salt water evaporators; food-pkg. adhesives; defoamer in food-contact paper/paperboard; in lubricants for incidental food contact; boiler water additive (food contact)
Regulatory: FDA 21CFR §173.310, 173.340, 175.105, 176.210, 178.3570
Manuf./Distrib.: Aldrich; Fluka; Sigma
Trade Names: Ucon® 50-HB-55

PPG-3-buteth-5

CAS 9038-95-3 (generic); 9065-63-8 (generic)
Synonyms: POE (5) POP (3) monobutyl ether; POP (3) POE (5) monobutyl ether
Definition: Polyoxypropylene, polyoxyethylene ether of butyl alcohol
Empirical: $C_{23}H_{48}O_9$
Formula: $C_4H_9(OCHCH_3CH_2)_x(OCH_2CH_2)_yOH$, avg. x = 3, avg. y = 5
Properties: Nonionic
Toxicology: LD50 (oral, rat) 9170 mg/kg, (skin, rabbit) 14 g/kg; mildly toxic by ing. and skin contact; skin irritant; TSCA listed
Hazardous Decomp. Prods.: Heated to decomp., emits acrid smoke and irritating fumes
Uses: Antistat, emulsifier, emollient in cosmetics; lubricant for compressor, rubber, greases, textiles; intermediate for prep. of resins, plasticizers, surfactants; solvent for inks and dyes; antifoam in boiler water, latex processing, washing sol'ns., paints, adhesives, salt water evaporators; food-pkg. adhesives; defoamer in food-contact paper/paperboard; in lubricants for incidental food contact; boiler water additive (food contact)
Regulatory: FDA 21CFR §173.310, 175.105, 176.210, 178.3570
Manuf./Distrib.: Aldrich; Fluka; Sigma
Trade Names: Ucon® 50-HB-100

PPG-5-buteth-7

CAS 9038-95-3 (generic); 9065-63-8 (generic); 74623-31-7 (generic)
Synonyms: POE (7) POP (5) monobutyl ether; POP (5) POE (6) monobutyl ether
Definition: Polyoxypropylene, polyoxyethylene ether of butyl alcohol
Empirical: $C_{33}H_{68}O_{13}$
Formula: $C_4H_9(OCH_3CHCH_2)_x(OCH_2CH_2)_yOH$, avg. x = 5, avg. y = 7
Properties: Nonionic
Toxicology: LD50 (oral, rat) 5370 mg/kg; TSCA listed
Uses: Component in demulsifying and wetting formulations; brake and metalworking fluids; rubber and fiber lubricant; textile applics.; defoamer for hot and cold applics., foods, chemical processing; antistat, emulsifier, skin conditioner, emollient in cosmetics; boiler water additive (food contact); solvent for inks and dyes; antifoam in boiler water, latex processing, washing sol'ns., paints, adhesives, salt water evaporators; food-pkg. adhesives; defoamer in food-contact paper/paperboard; in lubricants for incidental food contact
Regulatory: FDA 21CFR §173.310, 175.105, 176.210, 178.3570
Manuf./Distrib.: Aldrich; Fluka; Sigma

PPG-7-buteth-10

CAS 9038-95-3 (generic); 9065-63-8 (generic)
Synonyms: POE (10) POP (7) monobutyl ether; POP (7) POE (10) monobutyl ether
Definition: Polyoxypropylene, polyoxyethylene ether of butyl alcohol
Empirical: $C_{45}H_{92}O_{18}$
Formula: $C_4H_9(OCHCH_3CH_2)_x(OCH_2CH_2)_yOH$, avg. x = 7, avg. y = 10
Properties: Nonionic
Toxicology: LD50 (oral, rat) 4000 mg/kg; mod. toxic by ing. and skin contact; skin irritant; TSCA listed
Uses: Detergent for toilet cleaners, laundry detergents, emulsion polymerization, metalworking fluids, hydraulic fluids; antifoam in boiler water, latex processing, washing sol'ns., paints, adhesives, salt water evaporators; antistat, emulsifier, emollient for cosmetics; lubricant for compressor, rubber, greases, textiles; intermediate for prep. of resins, plasticizers, surfactants; solvent for inks and dyes; boiler water additive (food contact); food-pkg. adhesives; defoamer in food-contact paper/paperboard; in lubricants for incidental food contact
Regulatory: FDA 21CFR §173.310, 175.105, 176.210, 178.3570
Manuf./Distrib.: Aldrich; Fluka; Sigma
Trade Names: Ucon® 50-HB-260

PPG-9-buteth-12

CAS 9038-95-3 (generic); 9065-63-8 (generic)
Synonyms: POE (12) POP (9) monobutyl ether; POP (9) POE (12) monobutyl ether
Definition: Polyoxypropylene, polyoxyethylene ether of butyl alcohol
Empirical: $(C_7H_{14}O_2 \cdot C_6H_{12}O_2)_x$
Formula: $C_4H_9(OCH_3CHCH_2)_x(OCH_2CH_2)_yOH$, avg. x = 9, avg. y = 12
Properties: Nonionic
Toxicology: LD50 (oral, rat) 5370 mg/kg; mildly toxic by ing.; skin irritant; TSCA listed
Uses: Antistat, emulsifier, emollient in cosmetics; lubricant for compressor, rubber, greases, textiles; intermediate for prep. of resins, plasticizers, surfactants; solvent for inks and dyes; antifoam in boiler water, latex processing, washing sol'ns., paints, adhesives, salt water evaporators; food-pkg. adhesives; defoamer in food-contact paper/paperboard; in lubricants for incidental food contact; boiler water additive (food contact)
Regulatory: FDA 21CFR §173.310, 175.105, 176.210, 178.3570
Manuf./Distrib.: Aldrich; Fluka; Sigma
Trade Names: Ucon® 50-HB-400

PPG-12-buteth-16

CAS 9038-95-3 (generic); 9065-63-8 (generic); 74623-31-7 (generic)
Synonyms: POE (16) POP (12) monobutyl ether; POP (12) POE (16) monobutyl ether
Definition: Polyoxypropylene, polyoxyethylene ether of butyl alcohol
Empirical: $(C_7H_{14}O_2 \cdot C_6H_{12}O_2)_x$
Formula: $C_4H_9(OCH_3CHCH_2)_x(OCH_2CH_2)_yOH$, avg. x = 12, avg. y = 16
Properties: Nonionic
Toxicology: LD50 (oral, rat) 18 g/kg; mildly toxic by ing.; skin irritant; TSCA listed
Uses: Defoamer; lubricant; in demulsifying and wetting formulations; brake fluids; metalworking; rubber lubricant; textiles; fiber lubricant; chemical processing; antistat, emulsifier, skin conditioner, emollient in cosmetics; boiler water additive (food contact); solvent for inks and dyes; antifoam in boiler water, latex processing, washing sol'ns., paints, adhesives, salt water evaporators; food-pkg. adhesives; defoamer in food-contact paper/paperboard; in lubricants for incidental food contact
Regulatory: FDA 21CFR §173.310, 175.105, 176.210, 178.3570
Manuf./Distrib.: Aldrich; Fluka; Sigma
Trade Names: Ucon® 50-HB-660

PPG-20-buteth-30

CAS 9038-95-3 (generic); 9065-63-8 (generic); 74623-31-7 (generic)
Synonyms: POE (30) POP (20) monobutyl ether; POP (20) POE (30) monobutyl ether
Definition: Polyoxypropylene, polyoxyethylene ether of butyl alcohol
Empirical: $(C_7H_{14}O_2 \cdot C_6H_{12}O_2)_x$
Formula: $C_4H_9(OCH_3CHCH_2)_x(OCH_2CH_2)_yOH$, avg. x = 20, avg. y = 30
Properties: Nonionic
Toxicology: LD50 (oral, rat) 21 g/kg; mildly toxic by ing.; skin irritant; TSCA listed
Uses: Antistat, emulsifier, skin conditioner, emollient in cosmetics; defoamer for food processing; antifoam in boiler water, latex processing, washing sol'ns., paints, adhesives, salt water evaporators; lubricant for compressor, rubber, greases, textiles; intermediate for prep. of resins, plasticizers, surfactants; solvent for inks and dyes; food-pkg. adhesives; defoamer in food-contact paper/paperboard; in lubricants for incidental food contact; boiler water additive (food contact)
Regulatory: FDA 21CFR §173.310, 175.105, 176.210, 178.3570

Manuf./Distrib.: Aldrich; Fluka; Sigma
Trade Names: Ucon® 50-HB-2000

PPG-28-buteth-35

CAS 9038-95-3 (generic); 9065-63-8 (generic)
Synonyms: POE (35) POP (28) monobutyl ether; POP (28) POE (35) monobutyl ether
Definition: Polyoxypropylene, polyoxyethylene ether of butyl alcohol
Empirical: $(C_7H_{14}O_2 \cdot C_6H_{12}O_2)_x$
Formula: $C_4H_9(OCH_3CHCH_2)_x(OCH_2CH_2)_yOH$, avg. x = 28, avg. y = 35
Properties: Nonionic
Toxicology: LD50 (oral, rat) 38 g/kg; mildly toxic by ing.; skin irritant; TSCA listed
Uses: Detergent; defoamer; metalworking and hydraulic fluids; mold release agent; emulsion polymerization; defoamer for food processing; antifoam in boiler water, latex processing, washing sol'ns., paints, adhesives, salt water evaporators; lubricant for compressor, rubber, greases, textiles; intermediate for prep. of resins, plasticizers, surfactants; solvent for inks and dyes; antistat, emulsifier, skin conditioner, emollient in cosmetics; boiler water additive (food contact); food-pkg. adhesives; defoamer in food-contact paper/paperboard; in lubricants for incidental food contact
Regulatory: FDA 21CFR §173.310, 175.105, 176.210, 178.3570
Manuf./Distrib.: Aldrich; Fluka; Sigma
Trade Names: Pluracol® W3520N; Ucon® 50-HB-3520

PPG-33-buteth-45

CAS 9038-95-3 (generic); 9065-63-8 (generic); 74623-31-7 (generic)
Synonyms: POE (45) POP (33) monobutyl ether; POP (33) POE (45) monobutyl ether
Definition: Polyoxypropylene, polyoxyethylene ether of butyl alcohol
Empirical: $(C_7H_{14}O_2 \cdot C_6H_{12}O_2)_x$
Formula: $C_4H_9(OCH_3CHCH_2)_x(OCH_2CH_2)_yOH$, avg. x = 33, avg. y = 45
Properties: Nonionic
Toxicology: LD50 (oral, rat) 49 g/kg; mildly toxic by ing.; skin irritant; TSCA listed
Uses: Defoamer; lubricant; emollient; in demulsifying and wetting formulations; brake fluids; metalworking; rubber lubricant; textiles; fiber lubricant; defoamer in food processing; chemical processing; antistat, emulsifier in cosmetics; boiler water additive (food contact); solvent for inks and dyes; antifoam in boiler water, latex processing, washing sol'ns., paints, adhesives, salt water evaporators; food-pkg. adhesives; defoamer in food-contact paper/paperboard; in lubricants for incidental food contact
Regulatory: FDA 21CFR §173.310, 173.340, 175.105, 176.210, 178.3570
Manuf./Distrib.: Aldrich; Fluka; Sigma
Trade Names: Ucon® 50-HB-5100

PPG butyl ether

CAS 9003-13-8 (generic)
Synonyms: Butoxypolypropylene glycol; Poly (oxy (methyl-1,2-ethanediyl), α-butyl-ω-hydroxy
Definition: Polypropylene glycol ether of butyl alcohol
Toxicology: Irritating to eyes, skin, respiratory system
Storage: Store @ R.T.
Uses: Additive in cosmetics; in veterinary fly sprays and repellents
Trade Names Containing: Dowanol® TPnB

PPG-3 butyl ether. *See* Tripropylene glycol n-butyl ether

PPG-5 butyl ether

CAS 9003-13-8 (generic)
Synonyms: POP (5) butyl ether
Definition: Polypropylene glycol ether of butyl alcohol
Formula: $C_4H_9(OCH_3CHCH_2)_nOH$, avg. n = 5
Properties: Nonionic
Toxicology: LD50 (oral, rat) 9160 mg/kg; mildly toxic by ingestion and skin contact; skin irritant; TSCA listed
Hazardous Decomp. Prods.: Heated to decomp., emits acrid smoke and irritating fumes
Uses: Lubricant for two-stroke cycle engines, rubber, greases, textiles; reactive intermediate for resins, plasticizers, surfactants; antifoam in boiler water, latex processing; solvent for inks and dyes; insect repellent; antistat, skin conditioner in cosmetics
Manuf./Distrib.: Aldrich
Trade Names: Ucon® LB-65

PPG-9 butyl ether

CAS 9003-13-8 (generic)
Synonyms: POP (9) butyl ether
Definition: Polypropylene glycol ether of butyl alcohol
Formula: $C_4H_9(OCH_3CHCH_2)_nOH$, avg. n = 9
Properties: Nonionic
Toxicology: LD50 (oral, rat) 9160 mg/kg; mildly toxic by ingestion and skin contact; skin irritant; TSCA listed
Hazardous Decomp. Prods.: Heated to decomp., emits acrid smoke and irritating fumes
Uses: Lubricant for two-stroke cycle engines, rubber, greases, textiles; reactive intermediate for resins, plasticizers, surfactants; antifoam in boiler water, latex processing; solvent for inks and dyes; antistat in cosmetics
Manuf./Distrib.: Aldrich
Trade Names: Ucon® LB-135

PPG-14 butyl ether

CAS 9003-13-8 (generic)
Synonyms: POP (14) butyl ether
Definition: Polypropylene glycol ether of butyl alcohol
Formula: $C_4H_9(OCH_3CHCH_2)_nOH$, avg. n = 14
Properties: Nonionic
Toxicology: LD50 (oral, rat) 9160 mg/kg; mildly toxic by ingestion and skin contact; skin irritant; TSCA listed
Hazardous Decomp. Prods.: Heated to decomp., emits acrid smoke and irritating fumes
Uses: Antistat in cosmetics; emollient, solvent, bodying agent for personal care prods.; solubilizer, coupling agent, emollient, lubricant for antiperspirants; dispersant in antiperspirant sticks; detackifier in creams/lotions; conditioner in shaving and hair prods.
Manuf./Distrib.: Aldrich
Trade Names: Dow Corning® L910; Ucon® Fluid AP

PPG-15 butyl ether

CAS 9003-13-8 (generic)
Synonyms: POP (15) butyl ether
Definition: Polypropylene glycol ether of butyl alcohol
Formula: $C_4H_9(OCH_3CHCH_2)_nOH$, avg. n = 15
Properties: Nonionic
Toxicology: LD50 (oral, rat) 9160 mg/kg; mildly toxic by ingestion and skin contact; skin irritant; TSCA listed
Hazardous Decomp. Prods.: Heated to decomp., emits acrid smoke and irritating fumes
Uses: Lubricant for two-stroke cycle engines, rubber, greases, textiles; reactive intermediate for resins, plasticizers, surfactants; antifoam in boiler water, latex processing; solvent for inks and dyes; antistat in cosmetics; food-pkg. adhesives; defoamer for food contact
Manuf./Distrib.: Aldrich
Trade Names: Ucon® LB-285

PPG-18 butyl ether

CAS 9003-13-8 (generic)
Synonyms: POP (18) butyl ether; PPG (18) butyl ether
Definition: Polypropylene glycol ether of butyl alcohol
Formula: $C_4H_9(OCH_3CHCH_2)_nOH$, avg. n = 18
Properties: Liq.; insol. in water; sol. in alcohol; nonionic
Toxicology: LD50 (oral, rat) 9160 mg/kg; mildly toxic by ingestion and skin contact; skin irritant; TSCA listed
Hazardous Decomp. Prods.: Heated to decomp., emits acrid smoke and irritating fumes
Uses: Lubricant for two-stroke cycle engines, rubber, greases, textiles; reactive intermediate for resins, plasticizers, surfactants; antifoam in boiler water, latex processing; solvent for inks and dyes; antistat, skin conditioner, emollient in cosmetics; food-pkg. adhesives; defoamer in food-contact paper/paperboard
Regulatory: FDA 21CFR §175.105, 176.210
Manuf./Distrib.: Aldrich
Trade Names: Dow Corning® L1150; Ucon® LB-385

PPG (18) butyl ether. *See* PPG-18 butyl ether

PPG-22 butyl ether

CAS 9003-13-8 (generic)
Synonyms: POP (22) butyl ether
Definition: Polypropylene glycol ether of butyl alcohol
Formula: $C_4H_9(OCH_3CHCH_2)_nOH$, avg. n = 22
Properties: Nonionic
Toxicology: LD50 (oral, rat) 9160 mg/kg; mildly toxic by ingestion and skin contact; skin irritant; TSCA listed
Hazardous Decomp. Prods.: Heated to decomp., emits acrid smoke and irritating fumes
Uses: Lubricant for two-stroke cycle engines, rubber, greases, textiles; reactive intermediate for resins, plasticizers, surfactants; antifoam in boiler water, latex processing; solvent for inks and dyes; antistat in cosmetics; food pkg. adhesives; defoamer in food-contact paper coatings, paper/paperboard
Regulatory: FDA 21CFR §175.105, 176.200, 176.210
Manuf./Distrib.: Aldrich
Trade Names: Ucon® LB-525

PPG-24 butyl ether

CAS 9003-13-8 (generic)
Synonyms: POP (24) butyl ether
Definition: Polypropylene glycol ether of butyl alcohol
Formula: $C_4H_9(OCH_3CHCH_2)_nOH$, avg. n = 24
Properties: Liq.; insol. in water; sol. in alcohol; nonionic
Toxicology: LD50 (oral, rat) 9160 mg/kg; mildly toxic by ingestion and skin contact; skin irritant; TSCA listed
Hazardous Decomp. Prods.: Heated to decomp., emits acrid smoke and irritating fumes
Uses: Lubricant for two-stroke cycle engines, rubber, greases, textiles; reactive intermediate for resins, plasticizers, surfactants; antifoam in boiler water, latex processing; solvent for inks and dyes; surfactant, emollient in cosmetics; defoamer in food-contact coatings and paper/paperboard
Regulatory: FDA 21CFR §175.105, 176.200, 176.210, 178.3570
Manuf./Distrib.: Aldrich
Trade Names: Ucon® LB-625

PPG-33 butyl ether

CAS 9003-13-8 (generic)
Synonyms: POP (33) butyl ether; PPG (33) butyl ether
Definition: Polypropylene glycol ether of butyl alcohol
Formula: $C_4H_9(OCH_3CHCH_2)_nOH$, avg. n = 33
Properties: Liq.; insol. in water; sol. in alcohol; nonionic
Toxicology: LD50 (oral, rat) 9160 mg/kg; mildly toxic by ingestion and skin contact; skin irritant; TSCA listed
Hazardous Decomp. Prods.: Heated to decomp., emits acrid smoke and irritating fumes
Uses: Lubricant for two-stroke cycle engines, rubber, greases, textiles; reactive intermediate for resins, plasticizers, surfactants; antifoam in boiler water, latex processing; solvent for inks and dyes; antistat, emollient in cosmetics; chewing gum base; defoamer in food-contact coatings and paper/paperboard
Regulatory: FDA 21CFR §166.110, 172.615, 175.105, 175.125, 175.300, 175.380, 175.390, 176.170, 176.200, 176.210, 177.1010, 177.1210, 177.1350, 178.3570
Manuf./Distrib.: Aldrich
Trade Names: Ucon® LB-1145

PPG (33) butyl ether. *See* PPG-33 butyl ether

PPG-40 butyl ether

CAS 9003-13-8 (generic)
Synonyms: POP (40) butyl ether; PPG (40) butyl ether
Definition: Polypropylene glycol ether of butyl alcohol
Formula: $C_4H_9(OCH_3CHCH_2)_nOH$, avg. n = 40
Properties: Liq.; insol. in water; sol. in alcohol; nonionic
Toxicology: LD50 (oral, rat) 9160 mg/kg; mildly toxic by ingestion and skin contact; skin irritant; TSCA listed
Hazardous Decomp. Prods.: Heated to decomp., emits acrid smoke and irritating fumes
Uses: Lubricant for two-stroke cycle engines, rubber, greases, textiles; reactive intermediate for resins, plasticizers, surfactants; antifoam in boiler water, latex processing; solvent for inks and dyes; antistat, emollient in cosmetics; in food-pkg. adhesives; defoamer in food-contact coatings and paper/paperboard
Regulatory: FDA 21CFR §175.105, 176.200, 176.210, 178.3570
Manuf./Distrib.: Aldrich
Trade Names: Ucon® LB-1715

PPG (40) butyl ether. *See* PPG-40 butyl ether

PPG-53 butyl ether

CAS 9003-13-8 (generic)
Synonyms: POP (53) butyl ether; PPG (53) butyl ether
Definition: Polypropylene glycol ether of butyl alcohol
Formula: $C_4H_9(OCH_3CHCH_2)_nOH$, avg. n = 53
Properties: Liq.; insol. in water; sol. in alcohol; nonionic
Toxicology: LD50 (oral, rat) 9160 mg/kg; mildly toxic by ingestion and skin contact; skin irritant; TSCA listed
Hazardous Decomp. Prods.: Heated to decomp., emits acrid smoke and irritating fumes
Uses: Lubricant for two-stroke cycle engines, rubber, greases, textiles; reactive intermediate for resins, plasticizers, surfactants; antifoam in boiler water, latex processing; solvent for inks and dyes; antistat, emollient in cosmetics; defoamer in food-contact paper coatings; in lubricants for incidental food contact
Regulatory: FDA 21CFR §176.200, 178.3570
Manuf./Distrib.: Aldrich
Trade Names: Ucon® LB-3000

PPG (53) butyl ether. *See* PPG-53 butyl ether

PPG-5-ceteth-20

CAS 9087-53-0 (generic); 37311-01-6 (generic)
Synonyms: POE (20) POP (5) cetyl ether; POP (5) POE (20) cetyl ether
Definition: Polyoxypropylene, polyoxyethylene ether of cetyl alcohol
Formula: $CH_3(CH_2)_{14}CH_2(OCH_3CHCH_2)_x(OCH_2CH_2)_yOH$, avg. x = 5, avg. y = 20
Properties: Liq.; sol. in water, alcohol; nonionic
Toxicology: TSCA listed
Uses: Surfactant, emollient, emulsifier, lubricant, solubilizer, coupling agent, humectant, dispersant, plasticizer for cosmetics, perfumes, antiperspirants, aq. and aq. alcoholic systems; solvent, emollient, and humectant in topical pharmaceuticals
Trade Names: Procetyl AWS; Protachem™ AWS-100

PPG-10 cetyl ether

CAS 9035-85-2 (generic)
Synonyms: POP (10) cetyl ether; PPG (10) cetyl ether
Definition: Polypropylene glycol ether of cetyl alcohol
Formula: $CH_3(CH_2)_{14}CH_2(OCH_3CHCH_2)_nOH$, avg. n = 10
Properties: Nonionic
Toxicology: TSCA listed
Uses: Emollient, coupling agent, cosolvent, plasticizer, superfatting agent, wetting agent, spreading agent, penetrant, surfactant, emulsifier, lubricant in cosmetics, personal care prods.
Trade Names: Procetyl 10

PPG (10) cetyl ether. *See* PPG-10 cetyl ether

PPG-20 cetyl ether

CAS 9035-85-2 (generic)
Synonyms: POP (20) cetyl ether; PPG (20) cetyl ether
Definition: Polypropylene glycol ether of cetyl alcohol
Formula: $CH_3(CH_2)_{14}CH_2(OCH_3CHCH_2)_nOH$, avg. n = 20
Uses: Emollient, lubricant, cosolvent, plasticizer, superfatting agent for personal care prods.

PPG (20) cetyl ether. *See* PPG-20 cetyl ether

PPG-30 cetyl ether

CAS 9035-85-2 (generic)
Synonyms: POP (30) cetyl ether; PPG (30) cetyl ether
Definition: Polypropylene glycol ether of cetyl alcohol
Formula: $CH_3(CH_2)_{14}CH_2(OCH_3CHCH_2)_nOH$, avg. n = 30

Properties: Liq.; sol. in oils, alcohol; insol. in water; nonionic
Toxicology: No known toxicity; TSCA listed
Uses: Emollient, moisturizer, surfactant, coupling agent, cosolvent, plasticizer, wetting agent, spreading agent, penetrant, lubricant, superfatting agent in cosmetics, pharmaceuticals

PPG (30) cetyl ether. *See* PPG-30 cetyl ether

PPG-50 cetyl ether

CAS 9035-85-2 (generic)
Synonyms: POP (50) cetyl ether; PPG (50) cetyl ether
Definition: Polypropylene glycol ether of cetyl alcohol
Formula: $CH_3(CH_2)_{14}CH_2(OCH_3CHCH_2)_nOH$, avg. n = 50
Properties: Liq.; sol. in lanolin oil, alcohol; nonionic
Toxicology: TSCA listed
Uses: Surfactant, emollient, coupling agent, cosolvent, plasticizer, superfatting agent, wetting agent, spreading agent, penetrant, lubricant in cosmetics

PPG (50) cetyl ether. *See* PPG-50 cetyl ether
PPG-2 dibenzoate; PPG (2) dibenzoate. *See* Dipropylene glycol dibenzoate

PPG-40 diethylmonium chloride

CAS 9076-43-1
Synonyms: POP (40) methyl diethyl ammonium chloride; Quaternium-21
Classification: Quaternary ammonium salt
Formula: $[CH_3N(CH_2CH_3)_2(CH_2CH_3CHO)_nH]^+Cl^-$, avg. n = 40
Properties: Cationic
Uses: Surfactant, pigment dispersant, particle suspending agent, emulsifier, solvent, conditioner, antistat, lubricant, corrosion inhibitor for toiletries, cosmetics, germicides, syn. fibers and plastics, textiles, industrial processes; ore flotation additive
Trade Names: Variquat® CC-42

PPG-66-glycereth-12

CAS 9082-00-2 (generic); 51258-15-2 (generic)
Definition: Polyoxypropylene, polyoxyethylene ether of glycerin
Formula: $C_3H_7O_2(OCHCH_3CH_2)_x(OCH_2CH_2)_yOH$, avg. x = 66, avg. y = 12
Properties: Nonionic
Toxicology: TSCA listed
Uses: Emulsifier, solvent in cosmetics; defoamer for cutting and drilling fluids, fermentation, chem. mfg., acid gas treatment, pulp and paper, and textiles
Features: Provides low odor, low irritation, low toxicity to personal care prods
Manuf./Distrib.: Aldrich
Trade Names: Polyglycol P112-2

PPG-24-glycereth-24

CAS 9082-00-2 (generic); 51258-15-2 (generic)
Synonyms: PEG/PPG-24/24 glycerin; POE (24) POP (24) glyceryl ether; POP (24) POE (24) glyceryl ether
Definition: Polyoxypropylene, polyoxyethylene ether of glycerin
Formula: $C_3H_7O_2(OCHCH_3CH_2)_x(OCH_2CH_2)_yOH$, avg. x = 24, avg. y = 24
Properties: Nonionic
Toxicology: TSCA listed
Uses: Emollient, emulsifier, solvent in cosmetics; defoamer for aq. systems
Features: Provides low odor, low irritation, low toxicity to personal care prods.
Manuf./Distrib.: Aldrich
Trade Names: Polyglycol P15-200

PPG-10 glyceryl ether

CAS 25791-96-2 (generic)
Definition: Polypropylene glycol ether of glycerin
Empirical: $C_{33}H_{68}O_{13}$
Formula: $C_3H_7O_2(OCHCH_3CH_2)_nOH$, avg. n = 10
Uses: Emulsifier, solvent in cosmetics
Features: Provides low odor, low irritation, low toxicity to personal care prods.
Manuf./Distrib.: Aldrich
Trade Names: Polyglycol PT700

PPG-55 glyceryl ether

CAS 25791-96-2 (generic)
Synonyms: POP (55) glyceryl ether
Definition: Polypropylene glycol ether of glycerin
Formula: $C_3H_7O_2(OCHCH_3CH_2)_nOH$, avg. n = 55
Properties: Nonionic
Toxicology: TSCA listed
Uses: Lubricant, emulsifier, antistat, solvent for cosmetics
Features: Provides low odor, low irritation, low toxicity to personal care prods.
Manuf./Distrib.: Aldrich
Trade Names: Polyglycol PT3000

PPG-2 methyl ether

CAS 13429-07-7; 20324-32-7 (α-isomer); 34590-94-8 (mixt.); EINECS/ELINCS 236-547-9; 252-104-2
Synonyms: 1,4-Dimethyl-3,6-dioxa-1-heptanol; Dipropylene glycol methyl ether; Dipropylene glycol monomethyl ether; DPGME; DPM; DPMG; Methoxy dipropylene glycol; 1-(2-Methoxyisopropoxy)-2-propanol; 1-(2-Methoxy-1-methylethoxy)-2-propanol; (2-Methoxymethylethoxy) propanol; Methoxypropoxypropanol; POP (2) methyl ether; 2-Propanol, 1-(2-methoxy-1-methylethoxy)-
Classification: Aliphatic ether alcohol; glycol ether
Definition: Polypropylene glycol ether of methyl alcohol
Empirical: $C_7H_{16}O_3$
Formula: $CH_3(OCHCH_3CH_2)_nOH$, avg. n = 2
Properties: Colorless clear liq.; mild ethereal sweet odor; sol. in water, acetone, ethanol, benzene, CCl_4, ether, methanol, monochlorobenzene, petrol, ether, VM&P naphtha; misc. with most org. solvs.; m.w. 148.23; dens. 0.951; vapor pressure 0.55 mm Hg; f.p. -83 C; b.p. 190 C; flash pt. (CC) 86 C; ref. index 1.4205; surf. tens. 28.8 dynes/cm
Toxicology: ACGIH TLV/TWA 100 ppm (skin); STEL 150 ppm (skin); LD50 (oral, rat) 5660 mg/kg, (skin, rat) 9500 mg/kg; mildly toxic by ing. and skin contact; skin and eye irritant; human eye irritant; causes tears; mild allergen; may cause nose/throat irritation, weakness, exhaustion, dizziness, headache; target organs: eyes, respiratory system, CNS; TSCA listed
Environmental: VOC; ThOD 1.19
Precaution: Combustible exposed to heat or flame; LEL 1.1% (392 F); UEL 3%; can react with oxidizing materials
Hazardous Decomp. Prods.: Heated to decomp., emits acrid smoke, irritating fumes
NFPA: Health 0, Flammability 2, Reactivity 0
Storage: Store in cool, dry, well-ventilated area, out of direct sunlight
Uses: Solvent for paints/coatings, lacquers, resins, dyes, oils/greases, cellulose, cleaners, inks, cosmetics, agric., epoxy laminates, adhesives, floor polish, fuel additives, antifreeze, oil field, mining, electronic chems., textile vat dyeing/printing; chem. intermediate (reacts with acids to form esters, oxidizing agents to form aldehydes or carboxylic acids, alkali metals to form alcoholates, aldehydes to form acetals); mfg. of esters used as plasticizers; heat transfer fluid; wetting agent, soil penetrant in cleaners; paint remover additive; component in hydraulic fluids; in food-pkg. adhesives; adjuvant in slimicides in food-contact paper/paperboard
Regulatory: FDA 21CFR §175.105, 176.300, 181.22, 181.30; SARA §311/312 fire hazard, §313 reportable; HAP
Manuf./Distrib.: Aldrich; Arch Chems.; Ashland; Celanese; Fluka; Lyondell; Oxiteno; Stanchem; TCI Am.; Vopak USA
Trade Names: Arcosolv® DPM; Dowanol® DPM; Methyl Diproxitol; Poly-Solv® DPM

PPG-3 methyl ether

CAS 25498-49-1; EINECS/ELINCS 243-734-9; 247-045-4
Synonyms: (2-(2-Methoxymethylethoxy) methylethoxy) propanol; 3-(3-(3-Methoxypropoxy) propoxy) propan-1-ol; Polyoxypropylene (3) methyl ether; PPG (3) methyl ether; Propanol, (2-(2-methoxymethylethoxy) methylethoxy)-; Propanol, 3-(3-(3-methoxypropoxy) propoxy)-; TPGME; TPM; Tripropylene glycol methyl ether; Tripropylene glycol monomethyl ether
Classification: Aliphatic ether alcohol
Definition: Polypropylene glycol ether of methyl alcohol
Empirical: $C_{10}H_{22}O_4$

Formula: $CH_3(OCHCH_3CH_2)_nOH$, avg. n = 3
Properties: Colorless clear liq.; sl. sweetish ether odor; sol. in water, acetone, ethanol, benzene, CCl_4, ether, methanol, monochlorobenzene, VM&P naphtha; misc. with most org. solvs.; m.w. 206.32; dens. 0.967; f.p. -60 C; b.p. 243 C, pour pt. -78 C; flash pt. (CC) 121 C; ref. index 1.4300
Toxicology: LD50 (oral, rat) 3200 mg/kg; LDLo (skin, rabbit) 14,505 mg/kg; mod. toxic by ing.; mild eye irritant; vapor inh. nonirritating; high concs. of mists can irritate nose/throat, cause CNS depression, nausea, headache, weakness, fatigue; may cause wt. loss, sleepiness; TSCA listed
Precaution: Incompat. with oxidizing agents; increases risk of fire and explosion
Hazardous Decomp. Prods.: Heated to decomp., emits acrid smoke and irritating fumes
NFPA: Health 0, Flammability 1, Reactivity 0
Storage: Store in a cool, dry area, out of direct sunlight
Uses: Solvent, coupling agent for paints/coatings, lacquers, resins, dyes, oils, greases, cellulosics, cleaners, printing inks, functional fluids, agric., cosmetics, epoxy laminates, adhesives, floor polish, fuel additives, antifreeze, oil field, mining, electronic chems., brake fluids, leather dyeing, textile vat dyeing/printing, insect repellents; in hard surf. cleaners, penetrating oils; chemical intermediate; flotation frothing agent; wetting agent, soil penetrant in cleaners; solubilizer for dyes; plasticizer; in food-pkg. adhesives; adjuvant in slimicides in food-contact paper/paperboard
Features: High-boiling; good solvency; low volatility; slow evaporation
Regulatory: FDA 21CFR §175.105, 176.300, 181.22, 181.30
Manuf./Distrib.: Aldrich; Oxiteno; Stanchem
Trade Names: Arcosolv® TPM; Dowanol® TPM; Poly-Solv® TPM
Trade Names Containing: Dynasolve CU-6

PPG (3) methyl ether. *See* PPG-3 methyl ether

PPG-2 methyl ether acetate

CAS 88917-22-0
UN NA 1993
Synonyms: Dipropylene glycol methyl ether acetate; Dipropylene glycol monomethyl ether acetate; DPGMEA; DPM Acetate; Propanol, 1 (or 2)-(2-methoxymethylethoxy)-, acetate
Empirical: $C_9H_{18}O_4$
Properties: Liq.; mod. sol. in water; m.w. 190.24; sp.gr. 0.97; visc. 2 cP; vapor pressure 5.17 mm Hg (22 C); b.p. 200 C; flash pt. 85 C; surf. tens. 28.6 dynes/cm
Toxicology: Target organs: eyes, skin, CNS, kidneys; TSCA listed
Environmental: VOC; ThOD 1.77
Precaution: Combustible liq.
HMIS: Health 1, Flammability 1, Reactivity 1
Uses: Solvent for paints, inks, cleaners, epoxy laminates, agric., adhesives, floor polishes, oil field, mining, and electronic chems.; solvent, coupling agent, wetting agent, soil penetrant in cleaners; coalescent for architectural water-borne coatings
Regulatory: SARA §311/312 fire hazard; HAPs-free
Manuf./Distrib.: Aldrich; Ashland; G.J. Chem.; Lyondell; Spectrum Quality Prods.
Trade Names: Arcosolv® DPMA; Dowanol® DPMA

PPG-10 methyl glucose ether

CAS 61849-72-7 (generic)
Synonyms: POP (10) methyl glucose ether
Definition: Polypropylene glycol ether of methyl glucose
Empirical: $C_{37}H_{74}O_{16}$
Formula: $CH_3(CH_6H_{10}O_5)(OCHCH_3CH_2)_nOH$, avg. n = 10
Properties: Nonionic
Uses: Emollient in cosmetics; humectant for personal care prods.; freezing pt. depressant; emollient, moisturizer, foam modifier in detergents and shampoos; solvent and solubilizer for topical pharmaceuticals
Trade Names: Glucam® P-10

PPG-20 methyl glucose ether

CAS 61849-72-7 (generic)
Synonyms: POP (20) methyl glucose ether; PPG (20) methyl glucose ether
Definition: Polypropylene glycol ether of methyl glucose
Formula: $CH_3(CH_6H_{10}O_5)(OCHCH_3CH_2)_nOH$, avg. n = 20
Properties: Nonionic
Uses: Antistat, binder, humectant in cosmetics; emollient, moisturizer, foam modifier in detergents and shampoos; solvent and solubilizer for topical pharmaceuticals; freezing pt. depressant
Trade Names: Glucam® P-20

PPG (20) methyl glucose ether. *See* PPG-20 methyl glucose ether

PPG-3-myreth-3

CAS 37311-04-9 (generic)
Synonyms: POE (3) POP (3) myristyl ether; POP (3) POE (3) myristyl ether
Definition: Polyoxypropylene, polyoxyethylene ether of myristyl alcohol
Formula: $CH_3(CH_2)_{12}CH_2(OCHCH_3CH_2)_x(OCH_2CH_2)_yOH$, avg. x = 3, avg. y = 3
Properties: Nonionic
Toxicology: TSCA listed
Uses: Emulsifier in cosmetics; lubricant; wetting agent; penetrant; dye dispersant; scouring aid; antistat; solvent coupling agent; metal processing; textile surfactant

PPG-3 myristyl ether

CAS 63793-60-2 (generic)
Synonyms: POP (3) myristyl ether; Tripropylene glycol myristyl ether
Definition: Polypropylene glycol ether of myristyl alcohol
Empirical: $C_{23}H_{48}O_4$
Formula: $CH_3(CH_2)_{12}CH_2(OCHCH_3CH_2)_nOH$, avg. n = 3
Properties: Liq.; insol. in water; sol. in alcohol, min. oil; nonionic
Uses: Emollient, emulsifier for cosmetics, toiletries; lubricant, emulsifier, wetting agent, penetrant, dye dispersant, scouring aid for metal processing, textile use; solvent coupling agent
Trade Names: Promyristyl PM-3; Witconol™ APM

PPG-1 myristyl ether acetate. *See* Propylene glycol myristyl ether acetate

PPG-2 myristyl ether propionate

CAS 111497-87-1
Synonyms: Dipropylene glycol, myristyl ether, propionate; POP (2) myristyl ether propionate
Definition: Ester of propionic acid and the polypropylene glycol ether of myristyl alcohol
Empirical: $C_{20}H_{52}O_4$
Formula: $CH_3(CH_2)_{12}CH_2(OCHCH_3CH_2)_2OCOCH_2CH_3$
Uses: Emollient, lubricant, emulsion stabilizer, spreading agent, antitackifier, coupling agent in cosmetics, bath oils, creams, moisturizers, emulsions, pharmaceuticals; solvent for sunscreen actives; surfactant, pigment dispersant, antistat, emulsifier for creams/lotions, antiperspirants, depilatories
Trade Names: Crodamol PMP

PPG-23 oleyl ether

CAS 52581-71-2 (generic)
Synonyms: POP (23) oleyl ether
Definition: Polypropylene glycol ether of oleyl alcohol
Formula: $CH_3(CH_2)_7CH{=}CH(CH_2)_7CH_2(OCHCH_3CH_2)_nOH$, avg. n = 23
Toxicology: TSCA listed
Uses: Emollient in cosmetics; cosolvent

PPG-37 oleyl ether

CAS 52581-71-2 (generic)
Synonyms: POP (37) oleyl ether
Definition: Polypropylene glycol ether of oleyl alcohol
Formula: $CH_3(CH_2)_7CH{=}CH(CH_2)_7CH_2(OCHCH_3CH_2)_nOH$, avg. n = 37
Toxicology: TSCA listed
Uses: Emollient in cosmetics; cosolvent

PPG-11 stearyl ether

CAS 25231-21-4 (generic)
Synonyms: POP (11) stearyl ether; PPG (11) stearyl ether
Definition: Polypropylene glycol ether of stearyl alcohol
Formula: $CH_3(CH_2)_{16}CH_2(OCHCH_3CH_2)_nOH$, avg. n = 11
Properties: Liq.; insol. in water; sol. in alcohol, min. oil; nonionic

Toxicology: TSCA listed
Uses: Emollient, lubricant, emulsifier, wetting agent, penetrant, dye dispersant, scouring aid, antistat in cosmetics, toiletries, metal processing; textile surfactant; solvent coupling agent
Trade Names: Varonic® APS; Witconol™ APS

PPG (11) stearyl ether. *See* PPG-11 stearyl ether

PPG-15 stearyl ether

CAS 25231-21-4 (generic)
Synonyms: POP (15) stearyl ether; PPG (15) stearyl ether
Definition: Polypropylene glycol ether of stearyl alcohol
Formula: $CH_3(CH_2)_{16}CH_2(OCHCH_3CH_2)_nOH$, avg. n = 15
Properties: Nonionic
Toxicology: TSCA listed
Uses: Surfactant, emulsifier, dispersant, solubilizer, visc. control agent, emollient, solvent, lubricant for cosmetics, pharmaceuticals, and industrial applics.; coupling agent for fragrances
Regulatory: FDA approved for topicals
Manuf./Distrib.: ABITEC
Trade Names: Arlamol® E; Lipocol P-15; Prostearyl 15

PPG (15) stearyl ether. *See* PPG-15 stearyl ether

PPG-15 stearyl ether benzoate

Synonyms: POP (15) stearyl ether benzoate
Definition: Ester of PPG-15 stearyl ether and benzoic acid
Uses: Emollient, solubilizer, foam modifier for soap and soap/syndet systems, cosmetics
Trade Names Containing: Finsolv® TPP

Precipitated sulfur. *See* Sulfur
Primary active amyl alcohol. *See* 2-Methyl-1-butanol
Primary amyl acetate. *See* Amyl acetate
Primary amyl alcohol. *See* n-Amyl alcohol
Primary decyl alcohol. *See* Decyl alcohol
Primary isoamyl alcohol. *See* Isoamyl alcohol
Primary octyl alcohol. *See* Caprylic alcohol
Propanal. *See* Propionaldehyde
Propanamide, 2-hydroxy-N-(1-ethoxy-2-hydroxyethyl). *See* Lactamide DGA
2-Propanamine. *See* Isopropylamine
2-Propanamine, N-(1-methylethyl)-. *See* Diisopropylamine

Propane

CAS 74-98-6; EINECS/ELINCS 200-827-9
UN 1075 (DOT); UN 1978 (DOT); INS944; E944
Synonyms: Dimethylmethane; Liquefied petroleum gas; Petroleum gases, liquefied; n-Propane; Propane liquefied; Propyl hydride
Classification: Sat. aliphatic hydrocarbon
Empirical: C_3H_8
Formula: $CH_3CH_2CH_3$
Properties: Colorless gas, nat. gas odor; easily liquefied under pressure at R.T.; noncorrosive; sol. in ether, alcohol, chloroform, org. solvs.; sl. sol. in water, acetone; m.w. 44.09; sp.gr. 0.513 (0 C, as liq.), 1.56 (0 C, as vapor); f.p. -189.9 C; b.p. -42.5 C; flash pt. -104 C
Toxicology: OSHA PEL/TWA 1000 ppm; simple asphyxiant; narcotic in high concs.; overexposure may cause dizziness, disorientation, excitation; liq. propane may cause frostbite to skin and eyes; TSCA listed
Precaution: Flamm.; dangerous fire hazard exposed to heat, flame; explosive limits in air 2.4-9.5%; vigorously reactive with oxidizers; explosive as vapor exposed to heat or flame; explosive with ClO_2; violent exothermic reaction with barium peroxide, heat
Hazardous Decomp. Prods.: Heated to decomp., emits acrid smoke and irritating fumes
NFPA: Health 1, Flammability 4, Reactivity 0
Storage: Store in cool, dry, well-ventilated area out of direct sunlight
Uses: Aerosol propellant for paint aerosols; organic synthesis; mfg. of ethylene; extractant; solvent; process solvent for deasphalting processes; refrigerant; household and industrial fuel; as fuel gas, sometimes mixed with butane; gas enricher; propellant in cosmetics; direct food additive, propellant, aerating agent; pharmaceutical propellant
Regulatory: FDA 21CFR §173.350, 184.1655, GRAS; FDA approved for topicals; NF compliance
Manuf./Distrib.: Aeropres; Air Prods.; Aldrich; Atofina Bruxelles; BOC Ltd; BP Amoco; ChevronPhillips; Degussa AG/Health & Nutrition; ExxonMobil; Fluka; MG Ind.; Marathon Ashland Petrol.; Supelco; Vopak USA

n-Propane. *See* Propane
2-Propaneamine. *See* Isopropylamine
Propane, 1-bromo-. *See* n-Propyl bromide
Propane, 2-bromo-. *See* Isopropyl bromide
Propane/butane, mixed. *See* Liquefied petroleum gas
Propane, 2-cyclopentyl-. *See* Isopropylcyclopentane
Propane, 1,2-dichloro. *See* Propylene dichloride
Propane, 1,2-dimethoxy-. *See* Propylene glycol dimethyl ether
1,2-Propanediol; Propane-1,2-diol. *See* Propylene glycol
1,3-Propanediol; Propane-1,3-diol. *See* Trimethylene glycol
Propane-1,2-diol alginate. *See* Propylene glycol alginate
1,2-Propanediol, 3-butoxy-. *See* 3-Butoxy-1,2-propanediol
1,2-Propanediol carbonate. *See* Propylene carbonate
1,2-Propanediol, 3-chloro-. *See* Glyceryl-α-monochlorohydrin
1,2-Propanediol cyclic carbonate. *See* Propylene carbonate
1,2-Propanediol, 3-methoxy-. *See* 3-Methoxy-1,2-propanediol
1,3-Propanediol-2-methyl. *See* Methylpropanediol
1,2-Propanediol, 3-(3-methylbutoxy)-. *See* 1-Isoamyl glycerol ether
1,2-Propanediol monododecanoate. *See* Propylene glycol laurate
1,2-Propanediol, monomethyl ether. *See* 3-Methoxy-1-propanol
1,2-Propanediol monomethyl ether acetate. *See* Propylene glycol methyl ether acetate
1,2-Propanediol monostearate. *See* Propylene glycol stearate
1,2-Propanediyl carbonate. *See* Propylene carbonate
Propane, 1,2-epoxy-. *See* Propylene oxide
Propane liquefied. *See* Propane
Propane, 2-methoxy-2-methyl-. *See* Methyl t-butyl ether
Propanenitrile. *See* Propionitrile
Propanenitrile, 3-methoxy-. *See* 3-Methoxypropionitrile
Propane, 2-nitro-. *See* 2-Nitropropane
1,2,3-Propanetricarboxylic acid, 2-(acetyloxy)-, tributyl ester. *See* Acetyl tributyl citrate
1,2,3-Propanetricarboxylic acid, 2-hydroxy-, tributyl ester. *See* Tributyl citrate
1,2,3-Propanetriol; Propane-1,2,3-triol. *See* Glycerin
1,2,3-Propanetriol, diacetate. *See* Glyceryl diacetate
1,2,3-Propanetriol monoacetate. *See* Glyceryl acetate
1,2,3-Propanetriol octadecanoate. *See* Glyceryl stearate
1,2,3-Propanetriol triacetate. *See* Triacetin
1,2,3-Propanetriol tridodecanoate. *See* Trilaurin
1,2,3-Propanetriol tri (12-hydroxystearate). *See* Trihydroxystearin
1,2,3-Propanetriol triisononoate. *See* Triisononanoin
1,2,3-Propanetriol trilinoleate. *See* Trilinolein
1,2,3-Propanetriol trioctadecanoate. *See* Tristearin
1,2,3-Propanetriol trioctanoate. *See* Trioctanoin
1,2,3-Propanetriol tritetradecanoate. *See* Trimyristin
1,2,3-Propanetriyl 12-(acetyloxy)-9-octadecenoate. *See* Glyceryl triacetyl ricinoleate
1,2,3-Propanetriyl triisooctadecanoate. *See* Triisostearin
Propane-1,2,3-triyl 3,5,5-trimethylhexanoate. *See* Triisononanoin
1,2,3-Propanetriyl tris (12-hydroxyoctadecanoate). *See* Trihydroxystearin
Propanoic acid butyl ester. *See* n-Butyl propionate
Propanoic acid ethyl ester. *See* Ethyl propionate
Propanoic acid, 2-hydroxy-. *See* Lactic acid
Propanoic acid, 2-hydroxy-, butyl ester. *See* Butyl lactate
Propanoic acid, 2-hydroxy-, ethyl ester, (S)-. *See* Ethyl-(S)-lactate
Propanoic acid, 2-hydroxy-, 2-ethylhexyl ester. *See* 2-Ethylhexyl-S-lactate
Propanoic acid, 2-hydroxy-, methyl ester. *See* Methyl lactate
Propanoic acid, 2-hydroxy-, methyl ester, (S)-. *See* Methyl-(S)-lactate
Propanoic acid, 2-hydroxy-, 1-methylethyl ester. *See* Isopropyl lactate
Propanoic acid, 2-hydroxy-, pentyl ester. *See* Amyl lactate
Propanoic acid methyl ester. *See* Methyl propionate
Propanoic acid, 2-methyl-, monoester with 2,2,4-trimethyl-1,3-pentanediol. *See* 2,2,4-Trimethyl-1,3-pentanediol isobutyrate
Propanoic acid, 2-methylpropyl ester. *See* Isobutyl propionate
Propanoic acid, pentyl ester. *See* Amyl propionate

Propanoic acid, propyl ester. *See* Propyl propionate
Propanoic acid, 3,3′-thiobis-, dimethyl ester. *See* Dimethyl thiodipropionate
1-Propanol; Propanol-1; Propan-1-ol. *See* Propyl alcohol
2-Propanol; Propan-2-ol. *See* Isopropyl alcohol
n-Propanol. *See* Propyl alcohol
2-Propanol, 1,3-bis (isopentyloxy)-; 2-Propanol, 1,3-bis (3-methylbutoxy)-. *See* 1,3-Diisoamyl glycerol diether
2-Propanol, 1-butoxy. *See* Butoxypropanol
2-Propanol, 1-(2-butoxy-1-methylethoxy)-. *See* Dipropylene glycol butyl ether
Propanol, [2-(2-butoxymethylethoxy) methylethoxy]-. *See* Tripropylene glycol n-butyl ether
2-Propanol, 1,3-di-n-butoxy-. *See* 1,3-Dibutoxy-2-propanol
1-Propanol, 2,3-dichloro-. *See* 2,3-Dichloro-1-propanol
2-Propanol, 1,3-dichloro-. *See* 1,3-Dichloro-2-propanol
2-Propanol, 1,3-di (isopentyloxy)-. *See* 1,3-Diisoamyl glycerol diether
2-Propanol, 1,3-dimethoxy-. *See* 1,3-Dimethoxy-2-propanol
1-Propanol, 2,2-dimethyl-. *See* 2,2-Dimethylpropanol
2-Propanol, (1,1-dimethylethoxy)-. *See* Propylene glycol t-butyl ether
Propanol, [2-(1,1-dimethylethoxy) methylethoxy]-. *See* Dipropylene glycol t-butyl ether
1-Propanol, 3-ethoxy-. *See* 3-Ethoxy-1-propanol
2-Propanol, 1-ethoxy-. *See* Propylene glycol ethyl ether
Propanol, 1(or 2)-ethoxy-, acetate. *See* Propylene glycol ethyl ether acetate
Propanol, (2-ethoxymethylethoxy)-. *See* (2-Ethoxymethylethoxy) propanol
2-Propanol, 1-isobutoxy-. *See* Propylene glycol isobutyl ether
Propanol, methoxy-. *See* 3-Methoxy-1-propanol
2-Propanol, 1-methoxy. *See* Methoxyisopropanol
Propanol, 3-methoxy-. *See* 3-Methoxy-1-propanol
2-Propanol, 1-(2-methoxy-1-methylethoxy)-. *See* PPG-2 methyl ether
Propanol, 1 (or 2)-(2-methoxymethylethoxy)-, acetate. *See* PPG-2 methyl ether acetate
2-Propanol, 1-(2-(2-methoxy-1-methylethoxy)-1-methylethoxy)-. *See* 1-(2-(2-Methoxy-1-methylethoxy)-1-methylethoxyl)-2-propanol
Propanol, (2-(2-methoxymethylethoxy) methylethoxy)-; Propanol, 3-(3-(3-methoxypropoxy) propoxy)-. *See* PPG-3 methyl ether
2-Propanol, 2-methyl-. *See* t-Butyl alcohol
Propanol, ((1-methyl-1,2-ethanediyl) bis (oxy)) bis-. *See* Tripropylene glycol
2-Propanol, 1-(2-methylpropoxy)-. *See* Propylene glycol isobutyl ether
2-Propanol, 1,1′-oxydi-. *See* Dipropylene glycol
1-Propanol, 2-phenoxy-. *See* Propylene glycol phenyl ether
2-Propanol, 1-phenoxy-. *See* Phenoxyisopropanol
1-Propanol, 1-phenyl-. *See* Phenylpropanol
2-Propanol, 1-propoxy-. *See* Propylene glycol n-propyl ether
2-Propanol, 1-(2-propoxy-1-methylethoxy)-. *See* Dipropylene glycol n-propyl ether
1-Propanol, 2-(tetradecyloxy)-, acetate. *See* Propylene glycol myristyl ether acetate
1,2,3-Propanol tridecanoate. *See* Tricaprin
Propanone; 2-Propanone. *See* Acetone
2-Propanone oxime. *See* Acetone oxime

Propargyl alcohol

CAS 107-19-7; EINECS/ELINCS 203-471-2
UN 2929
Synonyms: Acetylene carbinol; Ethynyl carbinol; Ethynyl methanol; 1-Hydroxy-2-propyne; 3-Hydroxy-1-propyne; Propiolic alcohol; 1-Propyne-3-ol; Propyn-1-ol; 1-Propyn-3-ol; 2-Propynol; 2-Propyn-1-ol; 3-Propynol; 2-Propynyl alcohol
Classification: Unsat. aliphatic alcohol
Empirical: C_3H_4O
Formula: $HC{\equiv}CCH_2OH$
Properties: Colorless to straw-colored clear liq.; mild geranium-like odor; sol. in water, ethanol, acetone, diethyl ether, benzene, chloroform, dioxane, THF; misc. with oxygenated, chlorinated, and aromatic solvs.; insol. in aliphatic hydrocarbons; m.w. 56.06; dens. 0.9715 (20/4 C); vapor pressure 11.6 mm (20 C); m.p. -53 C; b.p. 114-115 C; flash pt. 33 C; autoignition temp. 400 C
Toxicology: ACGIH TLV/TWA 1 ppm (skin); LD50 (oral, rat) 55 mg/kg, (skin, rabbit) 88 mg/kg; LC50 (inh., mouse, 2 h) 2000 mg/m^3; very toxic by inhalation; poison by ing., skin contact, subcut. routes; corrosive; CNS depressant; skin and mucous membrane irritant; vapor irritates eyes and causes tearing; very short exposure may cause death or major residual injury; mutagenic data; TSCA listed
Precaution: Flamm.; reacts violently with oxidants; incompat. with oxidizing agents, strong acids, alkalis, phosphorus pentoxide, metallic halide salts, heavy metal compds.; risk of fire and explosion
Hazardous Decomp. Prods.: Heated to decomp., emits acrid smoke and irritating fumes
NFPA: Health 4, Flammability 3, Reactivity 3
Storage: Store in cool, dry, well-ventilated area away from heat/ignition sources
Uses: Corrosion inhibitor for oil-well acidizing, electroplating baths; inhibits attack of min. acids on steel; prevents hydrogen embrittlement of steel; metal pickling/plating; electroplating brightener additive; stabilizer for solvents, chlorinated hydrocarbon formulations; polishing agent in galvanotechnics; soil fumigant; solvent for cellulose acetate; chem. intermediate in org. synthesis, prod. of alkaloids, antibiotics, vitamins, pharmaceuticals, pesticides, biocides; laboratory reagent
Manuf./Distrib.: Advanced Synthesis Tech.; Aldrich; BASF; GFS; ISP; Int'l. Chem. Group; Spectrum Quality Prods.; Universal Preserv-A-Chem

Propasol solvent P. *See* Propylene glycol n-propyl ether
Propellant 11. *See* Trichlorofluoromethane
Propellant 12. *See* Dichlorodifluoromethane
Propellant 22. *See* Chlorodifluoromethane
Propellant 23. *See* Trifluoromethane
Propellant 114. *See* 1,2-Dichlorotetrafluoroethane
Propellant 142b. *See* 1,1,1-Chlorodifluoroethane
Prop-2-enal; 2-Propenal. *See* Acrolein
2-Propenal, 3-phenyl-. *See* Cinnamal
Propenenitrile; 2-Propenenitrile. *See* Acrylonitrile
2-Propene-1-ol. *See* Allyl alcohol
Propene oxide. *See* Propylene oxide
1,2,3-Propenetriol triisooctadecanoate. *See* Triisostearin
1,2,3-Propenetriol tri (9,12-octadecadienoate). *See* Trilinolein
2-Propenoic acid, 2-(2-ethoxyethoxy) ethyl ester. *See* 2-(2-Ethoxyethoxy) ethyl acrylate
2-Propenoic acid, 3-phenyl-, ethyl ester. *See* Ethyl cinnamate
Propenol; Propen-1-ol-3; 1-Propen-3-ol; 2-Propen-1-ol; Propenyl alcohol; 2-Propenyl alcohol. *See* Allyl alcohol
Propiolic alcohol. *See* Propargyl alcohol

Propionaldehyde

CAS 123-38-6; EINECS/ELINCS 204-623-0
UN 1275 (DOT); FEMA 2923
Synonyms: Aldehyde C3; Methylacetaldehyde; Propanal; Propionic aldehyde; Propylaldehyde
Empirical: C_3H_6O
Formula: CH_3CH_2CHO
Properties: Colorless mobile liq., suffocating odor; sol. in 5 vols water; misc. with alcohol, ether, oxygenated solvs.; m.w. 58.08; dens. 0.807 (20/4 C); m.p. -81 C; b.p. 47-49 C; flash pt. -40 C; ref. index 1.362 (20 C)
Toxicology: LD50 (oral, rat) 1.4 g/kg, (subcut., rat) 820 mg/kg, (skin, rabbit) 5040 mg/kg; LCLo (inh., rat, 4 h) 8000 ppm; mod. toxic by skin contact, ing., subcut. routes; mildly toxic by inh.; skin and severe eye irritant; TSCA listed
Precaution: Highly flamm.; dangerous fire hazard exposed to heat or flame; reacts vigorously with oxidizers; explosive limits 2.9-17%
Hazardous Decomp. Prods.: Heated to decomp., emits acrid smoke and irritating fumes
Storage: Store refrigerated
Uses: Mfg. of propionic acid, polyvinyl and other plastics, rubber chemicals; intermediate for paints; disinfectant; preservative; synthetic flavoring agent in foods and pharmaceuticals; fragrance in cosmetics; solvent
Regulatory: FDA 21CFR §172.515; FEMA GRAS; SARA reportable
Manuf./Distrib.: Advanced BioTech; Aldrich; BASF AG; Celanese; Dow; Eastman; Fluka; Penta Mfg.; Sigma; Spectrum Quality Prods.

Propione. *See* Diethyl ketone

Propionic acid, ethyl ester. *See* Ethyl propionate
Propionic acid, 2-hydroxy-. *See* Lactic acid
Propionic acid, isobutyl ester. *See* Isobutyl propionate
Propionic acid, 2-methyl-. *See* Isobutyric acid
Propionic acid, 2-methyl-, monoester with 2,2,4-trimethyl-1,3-pentanediol. *See* 2,2,4-Trimethyl-1,3-pentanediol isobutyrate
Propionic acid, propyl ester. *See* Propyl propionate
Propionic acid, 3,3´-thiodi-, dimethyl ester. *See* Dimethyl thiodipropionate
Propionic aldehyde. *See* Propionaldehyde
Propionic ether. *See* Ethyl propionate
Propionic nitrile. *See* Propionitrile

Propionitrile

CAS 107-12-0
UN 2404 (DOT)
Synonyms: Cyanoethane; Ether cyanatus; Ethyl cyanide; Hydrocyanic ether; Propanenitrile; Propionic nitrile; Propylnitrile
Classification: Nitrile/cyano compd.
Empirical: C_3H_5N
Formula: CH_3CH_2CN
Properties: Colorless mobile liq.; sweet ether-like odor; sol. in water, dimethylformamide, ether, alcohol; misc. with oxygenated solvs.; m.w. 55.08; dens. 0.783 (20 C); m.p. -93 C; b.p. 97 C; flash pt. (CC) 2 C; ref. index 1.3664 (20 C)
Toxicology: LD50 (oral, rat) 39 mg/kg, (IP, mouse) 28 mg/kg, (IV, rabbit) 50 mg/kg, (skin, rabbit) 210 mg/kg; poison by ing., skin contact, IV, IP routes; mod. toxic by inh.; mild skin irritant; can be absorbed thru skin; mod. eye irritant; vapor can cause nose/throat irritation; high concs. can result in cyanide poisoning (weakness, headache, dizziness, anxiety, nausea, vomiting); experimental teratogen, reproductive effector; TSCA listed
Precaution: Flamm.; dangerous fire hazard exposed to heat, flame, oxidizers; incompat. with oxidizing agents (can react explosively), strong acids or bases (can react violently); explosive with N-bromosuccinimide and heat
Hazardous Decomp. Prods.: Hydrogen cyanide; heated to decomp., emits toxic fumes of NO_x and CN^-
NFPA: Health 4, Flammability 3, Reactivity 1
Storage: Store in cool, dry, well-ventilated area out of direct sunlight
Uses: Solvent in petrol. refining; dielectric fluid; chem. intermediate; raw material for drug mfg.; high performance liq. chromatography; UV spectroscopy
Manuf./Distrib.: Advanced Synthesis Tech.; Aldrich; BASF AG; Fluka; Kira; Lonza; Penta Mfg.; Romil Ltd; Spectrum Quality Prods.

Propionitrile, 2-hydroxy-. *See* Lactonitrile
Propionitrile, 3-methoxy-. *See* 3-Methoxypropionitrile
2-Propoxyethanol. *See* Ethylene glycol propyl ether
2-(2-Propoxyethoxy) ethanol. *See* Diethylene glycol propyl ether
1-(2-Propoxy-1-methylethoxy)-2-propanol. *See* Dipropylene glycol n-propyl ether

Propoxypropanol

CAS 30136-13-1
Synonyms: n-Propoxypropanol; n-Propoxypropanol mixed isomers; Propylene glycol monopropyl ether
Classification: Alcohol/ether
Empirical: $C_6H_{14}O_2$
Properties: Liq.; sol. in water; m.w. 118.18; dens. 0.8865 (20/20 C); f.p. -80 C (sets to glass below f.p.); b.p. 149.8 C; flash pt. 53.3 C
Toxicology: LD50 (oral, rat) 3250 mg/kg, (skin, rabbit) 4000 mg/kg; mod. toxic by ing. and skin contact; skin and severe eye irritant
Precaution: Flamm. liq.
Hazardous Decomp. Prods.: Heated to decomp., emits acrid smoke and irritating fumes
Uses: Solvent; pesticides
Regulatory: Calif. Prop. 65 nonreportable; non-HAP

1-Propoxy-2-propanol. *See* Propylene glycol n-propyl ether
n-Propoxypropanol. *See* Propoxypropanol; Propylene glycol n-propyl ether
n-Propoxypropanol mixed isomers. *See* Propoxypropanol

Propyl acetate

CAS 109-60-4; EINECS/ELINCS 203-686-1
UN 1276 (DOT); FEMA 2925
Synonyms: Acetic acid propyl ester; Acetic acid n-propyl ester; 1-Acetoxypropane; 1-Propyl acetate; n-Propyl acetate; n-Propyl ethanoate
Classification: Sat. aliphatic carboxylic acid ester
Definition: Ester of propyl alcohol and acetic acid
Empirical: $C_5H_{10}O_2$
Formula: $CH_3COOCH_2CH_2CH_3$
Properties: Colorless liq., pear-like odor; sol. in alcohols, ketones, esters, hydrocarbons, diethyl ether, castor oil, linseed oil; mod. sol. in water; m.w. 102.14; dens. 0.887 (20/20 C); vapor pressure 40 mm (28.8 C); m.p. -92 C; b.p. 99-102 C; flash pt. (CC) 14 C; autoignition temp. 450 C; ref. index 1.384 (20 C); surf. tens. 24.3 dynes/cm (20 C); dielec. const. 6.002
Toxicology: ACGIH TLV/TWA 200 ppm; STEL 250 ppm; LD50 (oral, rat) 9370 mg/kg, (IP, mouse) 1420 mg/kg; mod. toxic by IP and subcut. routes; mildly toxic by ing., inh.; may be irritating to skin, mucous membranes; narcotic in high concs.; human systemic effects by inh. (lacrimation, cough); TSCA listed
Environmental: VOC; BOD5 1.26; COD 2.04; ThOD 2.04
Precaution: Highly flamm.; UEL 8%; dangerous fire hazard exposed to heat, flame, oxidizers; explosive as vapor exposed to heat, flame; incompat. with oxidizing agents, strong acids, potassium t-butoxide, reactive nitrogen compds.
Hazardous Decomp. Prods.: n-Propanol, acetic acid; heated to decomp., emits acrid smoke and irritating fumes
NFPA: Health 1, Flammability 3, Reactivity 0
Storage: Store in cool, dry, well-ventilated area, out of direct sunlight and away from heat and ignition sources
Uses: Synthetic flavoring agent in foods and pharmaceuticals; solvent for nitrocellulose and other cellulose derivs., natural and syn. resins, lacquers, plastics, printing inks, chlorinated rubber, heat-reactive phenolics, cosmetics, pharmaceuticals; fragrance ingred. in perfumes; organic synthesis; lab reagent; in recovery of acetic acid from dil. aq. sol'ns.; in cellophane for food pkg.
Regulatory: FDA 21CFR §172.515, 177.1200; FEMA GRAS
Manuf./Distrib.: A.P. Chems. Ltd; Advanced BioTech; Aldrich; Allchem Ind.; Ashland; Axxence Aromatic GmbH; BASF; BP Chems. Ltd; Bencorp Int'l.; Berje; Celanese; Coyne; Degussa Flavors/Alfrebro; Dow; Eastman; Fluka; G.J. Chem.; General Chem.; Grau Aromatics; HCH Marketing; HCI Chemtech Distribution; Houghton Chem.; Hukill; JLM Marketing; Lluch Essence; Moore Ingreds.; Penta Mfg.; Quaker City; Vopak USA
Trade Names: Eastman® n-Propyl Acetate; Ucar® n-Propyl Acetate
Trade Names Containing: S-327 190 Proof; S-327 Anhydrous; S-333 Anhydrous; S-357 Anhydrous; S-358 Anhydrous; S-359 Anhydrous

1-Propyl acetate. *See* Propyl acetate
2-Propyl acetate. *See* Isopropyl acetate
n-Propyl acetate. *See* Propyl acetate
Propylacetone. *See* 2-Hexanone

Propyl alcohol

CAS 71-23-8; EINECS/ELINCS 200-746-9
UN 1274 (DOT); FEMA 2928
Synonyms: Albacol; Alcohol C3; Ethyl carbinol; 1-Hydroxypropane; Normal propyl alcohol; Optal; 1-Propanol; Propanol-1; Propan-1-ol; n-Propanol; 1-Propyl alcohol; n-Propyl alcohol; Propylic alcohol
Classification: Aliphatic alcohol
Empirical: C_3H_8O
Formula: $CH_3CH_2CH_2OH$
Properties: Colorless clear liq., alcoholic and sl. stupefying odor; very sol. in benzene; sol. in water, ethyl alcohol, ether, acetone, propylene glycol; misc. with oxygenated solvs.; dissolves fat; m.w. 60.10; sp.gr. 0.804 (20/4 C); vapor pressure 10 mm Hg (14.7 C); m.p. -127 C; b.p. 97-98 C; flash pt. (TCC) 23 C; ref. index 1.385 (20 C)
Toxicology: ACGIH TLV/TWA 200 ppm; STEL 250 ppm (skin); LD50 (oral, rat) 1.87 g/kg, (IP, rat) 2164 mg/kg, (IV, rat) 590 mg/kg; poison by subcut. route; mod. toxic by inh., ing., IP, IV routes; toxic by skin absorption; skin and severe eye irritant; drying effect on skin may lead to

cracking, fissuring, and infections; mildly irritating to mucous membranes; questionable carcinogen; mutagenic data; TSCA listed
Environmental: VOC; BOD5 1.50; COD 2.18; ThOD 2.40
Precaution: Highly flamm.; dangerous fire hazard exposed to heat, flame, oxidizers; explosive limits in air 2-13%; incompat. with oxidizing agents (increases risk of fire and explosion); ignites on contact with potassium t-butoxide
Hazardous Decomp. Prods.: Heated to decomp., emits acrid smoke and irritating fumes
NFPA: Health 1, Flammability 3, Reactivity 0
Storage: Store in tightly closed containers in cool, well-ventilated area separate from work place; limit quantities in use
Uses: Solvent for natural and syn. resins, cellulose esters and ethers, paints, printing inks, waxes, veg. oils, pharmaceuticals; organic synthesis; chemical intermediate; in polishes; brake fluids; degreasing solvent; antiseptic; antifoam, solvent in cosmetics; synthetic flavoring agent in foods and pharmaceuticals; in food-pkg. adhesives; in cellophane for food pkg.
Features: Sweet flavor
Regulatory: FDA 21CFR §172.515, 175.105, 176.180, 176.210, 177.1200, 573.880; FEMA GRAS; FDA approved for topicals
Manuf./Distrib.: AMC Chems.; Advanced BioTech; Aldrich; Allchem Ind.; Amyl; Ashland; Axxence Aromatic GmbH; BASF; Baychem; Burdick & Jackson; Celanese; Chemcentral; Coyne; Degussa Flavors/Alfrebro; Dow; EM Science; Eastman; Fluka; General Chem.; Grau Aromatics; Houghton Chem.; Hukill; Integra; J.T. Baker; Lluch Essence; Lluch Essence; Lyondell; Mallinckrodt Baker; Penta Mfg.; Romil Ltd; Ruger; Sigma; Spectrum Quality Prods.; Stanchem; Sunnyside; VWR Int'l.; Vopak USA; Wutong Aroma
Trade Names: Eastman® n-Propanol; Ucar® 1-Propanol
Trade Names Containing: Punctilious® SDA 40B Modified Anhydrous

1-Propyl alcohol. *See* Propyl alcohol
2-Propyl alcohol. *See* Isopropyl alcohol
n-Propyl alcohol. *See* Propyl alcohol
s-Propyl alcohol. *See* Isopropyl alcohol
Propylaldehyde. *See* Propionaldehyde
2-Propylamine; s-Propylamine. *See* Isopropylamine

Propylbenzene

CAS 103-65-1; EINECS/ELINCS 203-132-9
UN 2364 (DOT)
Synonyms: Isocumene; 1-Phenylpropane; 1-Propylbenzene; n-Propylbenzene
Classification: Aromatic hydrocarbon
Empirical: C_9H_{12}
Formula: $C_6H_5CH_2CH_2CH_3$
Properties: Colorless clear liq.; mild odor; sol. in acetone, benzene, petrol. ether; pract. insol. in water; m.w. 120.19; sp.gr. 0.852 (20 C); vapor pressure 10 mm Hg (43.4 C); m.p. -99 C; b.p. 159 C; flash pt. (TCC) 30 C
Toxicology: REL 100 ppm; LD50 (oral, rat) 6040 mg/kg; LCLo (inh., mouse) 20 g/m^3; mildly toxic by ing. and inh.; vapor may be eye irritant; liq. may be mod. eye/skin irritant; vapor inh. may cause irritation to mucous membranes, nose, throat, respiratory tract; prolonged inh. exposure may cause fatigue, vertigo, chest constriction, CNS effects; ing. may cause depression, CNS effects; if aspirated, may cause lung inflamm., fluid accumulation; kidney toxin; TSCA listed
Environmental: Prevent entry into water sources, sewers; marine pollutant
Precaution: Flamm.; very dangerous fire hazard exposed to heat, flame, oxidizers; reactive with oxidizers; incompat. with oxygen; mod. explosive hazard as vapor exposed to heat or flame; bond and ground during transfer
Hazardous Decomp. Prods.: Combustion: CO_x; heated to decomp., emits acrid smoke and fumes
NFPA: Health 2, Flammability 3, Reactivity 0
Storage: Store in closed container in well-ventilated area away from heat, sparks, flames
Uses: Solvent for cellulose acetate; in textile printing and dyeing; mfg. of methylstyrene; constituent of asphalt and naphtha
Manuf./Distrib.: Aldrich; ChevronPhillips; Fluka; Sigma
Trade Names Containing: Solvent 2216; Solvent 2439

1-Propylbenzene; n-Propylbenzene. *See* Propylbenzene

n-Propyl bromide

CAS 106-94-5; EINECS/ELINCS 203-445-0
UN 2344
Synonyms: 1-Bromopropane; Propane, 1-bromo-
Classification: Alkyl bromide
Empirical: C_3H_7Br
Formula: $CH_3CH_2CH_2Br$
Properties: Colorless to sl. yel. liq.; sol. in alcohol, acetone, ether; sol. 0.25 g/100 ml water; m.w. 122.99; sp.gr. 1.3537; vapor pressure 143 mm Hg; m.p. -110 C; b.p. 71 C; flash pt. 25 C; ref. index 1.4343; pH 6-8
Toxicology: LD50 (oral, rat) 4260 mg/kg; LC50 (inh., rat, 0.5 h) 253,000 mg/m^3; irritating to eyes, skin, mucous membranes; harmful by inh.; may affect respiratory tract, liver and CNS; may cause sleepiness, muscular incoordination, respiratory depression; TSCA listed
Precaution: Flamm.; incompat. with strong oxidizers, strong bases; forms flamm. mixts. with air in a narrow flammability range; containers may explode in fire
Hazardous Decomp. Prods.: Heated to decomp., may release poisonous and corrosive fumes of HBr, CO, CO_2
NFPA: Health 2, Flammability 3, Reactivity 0
Storage: Store in cool, dry, well-ventilated area away from incompat. materials, heat, sparks, flame; keep containers tightly closed
Uses: Intermediate for pharmaceuticals, insecticides, quaternary ammonium compds., flavors, fragrances; solvent for industrial cleaning (degreasing, metal processing/finishing, electronics, aerospace, aviation), aerosols, textiles, adhesives, inks, coatings
Features: Possible substitute for ODPs
Regulatory: DOT nonregulated
Manuf./Distrib.: Contract Chems. Ltd; Dead Sea Bromine; Ocean Chems.; Tulstar Prods.; Xinchem; Yixing Fangqiao East Chemical
Trade Names: AG 101; Leksol™; Leksol™ AL

s-Propyl bromide. *See* Isopropyl bromide
Propyl butanoate. *See* Propyl butyrate

Propyl butyrate

CAS 105-66-8; EINECS/ELINCS 203-320-0
FEMA 2934
Synonyms: Butanoic acid propyl ester; Butyric acid, propyl ester; Propyl butanoate; n-Propyl butyrate
Empirical: $C_7H_{14}O_2$
Formula: $CH_3CH_2CH_2COOCH_2CH_2CH_3$
Properties: Colorless liq.; sol. in oxygenated solvs.; misc. with alcohol, ether; sl. sol. in water; m.w. 130.19; dens. 0.87 (20/4 C); m.p. -95 C; b.p. 142-144 C; flash pt. 37 C; ref. index 1.400 (20 C)
Toxicology: LD50 (oral, rat) 15,000 mg/kg; sl. toxic by ing.; irritant to mucous membranes; narcotic in high concs.; TSCA listed
Precaution: Flamm.
Hazardous Decomp. Prods.: Heated to decomp., emits acrid smoke and irritating vapors
Uses: Solvent for cellulose ethers; synthetic flavoring agent in foods and pharmaceuticals; fragrance in cosmetics
Regulatory: FDA 21CFR §172.515; FEMA GRAS
Manuf./Distrib.: Advanced BioTech; Aldrich; Axxence Aromatic GmbH; Degussa Flavors/Alfrebro; Fluka; Grau Aromatics; Lluch Essence

n-Propyl butyrate. *See* Propyl butyrate
Propyl carbinol. *See* Butyl alcohol
N-Propylcarbinyl chloride. *See* Butyl chloride
Propyl 'Cellosolve'. *See* Ethylene glycol propyl ether
Propyl cyanide; n-Propyl cyanide. *See* Butyronitrile

Propylcyclohexane

CAS 1678-92-8
Classification: Sat. alicyclic hydrocarbon
Empirical: C_9H_{18}
Formula: $C_3H_7C_6H_{11}$
Properties: Colorless liq.; sol. in most org. solvs.; insol. in water; m.w. 126.24; sp.gr. 0.794 (20 C); m.p. -95 C; b.p. 156-157 C; flash pt. 35 C

Toxicology: May cause mild skin irritation; conc. vapor can irritate eyes; probably a CNS depressant; high vapor concs. may cause sl. nose/throat irritation, headache, lightheadedness, nausea, drowsiness; very high concs. may cause unconsciousness, death; ing. may cause sore throat, nausea, diarrhea
Precaution: Flamm.; incompat. with strong oxidizing agents (increased fire and explosion hazard)
Hazardous Decomp. Prods.: None
NFPA: Health 0, Reactivity 0
Storage: Store in cool, dry, well-ventilated area out of direct sunlight
Uses: Solvent; org. synthesis
Manuf./Distrib.: Aldrich

Propylcyclopentane

CAS 2040-96-2
Classification: Sat. alicyclic hydrocarbon
Empirical: C_8H_{16}
Formula: $C_3H_7C_5H_9$
Properties: Colorless liq.; sol. in most org. solvs.; insol. in water; m.w. 112.22; sp.gr. 0.776 (20 C); m.p. -117 C; b.p. 131 C
Toxicology: May cause mild skin irritation; conc. vapor may be irritating to eyes; probably a CNS depressant; high vapor concs. may cause nose/throat irritation, lightheadedness, nausea, dizziness; very high concs. may cause unconsciousness, death; ing. may result in sore throat, nausea, diarrhea
Precaution: Flamm.; incompat. with strong oxidizing agents (increased risk of fire and explosion)
Hazardous Decomp. Prods.: None
NFPA: Health 0, Reactivity 0
Storage: Store in cool, dry, well-ventilated area
Uses: Solvent; org. synthesis
Manuf./Distrib.: Fluka; Sigma

Propyl dipropasol solvent. *See* Dipropylene glycol n-propyl ether
Propylene aldehyde. *See* Crotonaldehyde; Acrolein

Propylene carbonate

CAS 108-32-7; EINECS/ELINCS 203-572-1
Synonyms: Carbonic acid, cyclic propylene ester; Carbonic acid, 1,2-propylene glycol ester; 1,3-Carbonyl dioxypropane; Cyclic methylethylene carbonate; Cyclic propylene carbonate; Cyclic 1,2-propylene carbonate; 1,3-Dioxolan-2-one, 4-methyl; Dipropylene carbonate; 4-Methyl-1,3-dioxolan-2-one; 1-Methyl ethylene carbonate; 1,2-Propanediol carbonate; 1,2-Propanediol cyclic carbonate; 1,2-Propanediyl carbonate; 1,2-Propylene carbonate; Propylene glycol cyclic carbonate
Classification: Organic compd.; carbonic acid ester
Empirical: $C_4H_6O_3$
Properties: Colorless clear liq.; odorless; very sol. in water; misc. with acetone, benzene, chloroform, ether, ethyl acetate; m.w. 102.10; dens. 1.2069 (20/20 C); vapor pressure 0.03 mm Hg (20 C); m.p. -48.8 C; b.p. 242.1 C; flash pt. (OC) 135 C; ref. index 1.422 (20 C); pH 8.8; can remain liq. (supercools) below its f.p.
Toxicology: LD50 (oral, rat) 29 g/kg, (skin, rabbit) 20 ml/kg; mildly toxic by ingestion; human skin irritant; eye irritant; TSCA listed
Environmental: Biodeg.; VOC; ThOD 1.25
Precaution: Combustible exposed to heat or flame; incompat. with strong oxidizing agents (increases fire risk), acids, bases, reducing agents (decomp. can occur)
Hazardous Decomp. Prods.: May produce propylene oxide if an acid, base, or salt is present in an aq. sol'n. of propylene carbonate; heated to decomp., emits acrid smoke and irritating fumes
NFPA: Health 1, Flammability 1, Reactivity 0
Storage: Store in cool, dry, well-ventilated area, out of direct sunlight; avoid generating mist
Uses: Solvent for inorg. salts, plasticizers, syn. fibers, polymers, pigments, dyes, paints, agric. inks, colors, textiles, prod. of acrylonitrile fibers; extraction solvent; polar process solvent; intermediate for organic syntheses; prod. of PC, polyester elastomers; component of cleaners, binders; reactive diluent, plasticizer for urethane foams and coatings; foundry sand binders; natural gas purification; bentonite/montmorillonite gellant (greases, cosmetics); separation of CO_2 and hydrogen sulfide; refining of aromatics in oil field; component of electrolytes in lithium-ion rechargeable batteries; hydraulic fluid, lube, and fuel additive; plasticizer; accelerant in dyeing; lubricant, solvent in cosmetics; flavor for pharmaceuticals; antifoam for antifreeze; in food-pkg. adhesives
Features: Medicinal, minty flavor; exc. solvency and reactivity
Regulatory: FDA 21CFR §175.105; FDA approved for topicals; NF compliance
Manuf./Distrib.: Aceto; Aldrich; Allchem Ind.; Ashland; BASF AG; Bencorp Int'l.; Camida Ltd; ChemService; ChevronTexaco; Degussa AG; EM Ind.; Ferro/Grant; Fluka; Great Western; Honeywill & Stein; Huntsman; Itochu Spec. Chems.; Loos & Dilworth; Lyondell; Merck KGaA; Penta Mfg.; Romil Ltd; Ruger; Sigma; Spectrum Quality Prods.; Stanchem; Univar E&E Ltd; Vopak USA
Trade Names: Arconate® PC-1000; Arconate® PC-HP; Jeffsol® PC; PCHP; Solvenon® PC; UltraPure™ Propylene Carbonate
Trade Names Containing: Jeffsol® EC-25; Jeffsol® EC-50; Jeffsol® EC-75

1,2-Propylene carbonate. *See* Propylene carbonate
Propylene chloride. *See* Propylene dichloride

Propylene dichloride

CAS 78-87-5; EINECS/ELINCS 201-152-2
UN 1279 (DOT)
Synonyms: 1,2-Dichloropropane; α,β-Dichloropropane; Propane, 1,2-dichloro; Propylene chloride; α,β-Propylene dichloride
Empirical: $C_3H_6Cl_2$
Formula: $CH_3CHClCH_2Cl$
Properties: Colorless liq.; sol. in oxygenated solvs.; sl. sol. in water; m.w. 112.99; dens. 1.1593 (20/20 C); vapor pressure 40 mm Hg (19.4 C); f.p -100 C; b.p. 96.8 C; flash pt. (TCC) 60 C
Toxicology: ACGIH TLV/TWA 75 ppm; STEL 110 ppm; LD50 (oral, rat) 1947 mg/kg, (skin, rabbit) 8750 mg/kg, (IP, rat) 1100 mg/kg; mod. toxic by inh. and ing.; mildly toxic by skin contact; primary eye irritant; can cause dermatitis; can cause liver, kidney, and heart damage; suspected carcinogen; tumorigen; mutagen; TSCA listed
Precaution: Very dangerous fire hazard exposed to heat or flame; reacts with aluminum to form aluminum chloride (reaction can be explosive); can react vigorously with oxidizers
Hazardous Decomp. Prods.: Heated to decomp., emits toxic fumes of Cl^-
Uses: Intermediate for perchloroethylene, CCl_4; lead scavenger for antiknock fluids; solvent for fats, oils, waxes, gums, resins, cellulose esters and ethers; scouring compounds; metal degreasers; soil fumigant for nematodes
Manuf./Distrib.: Aldrich; Fluka; Hukill; Sigma

α,β-Propylene dichloride. *See* Propylene dichloride
Propylene dichlorohydrin. *See* 1,3-Dichloro-2-propanol
Propylene epoxide. *See* Propylene oxide

Propylene glycol

CAS 57-55-6; 4254-15-3 (+); 4254-14-2 (-); 4254-16-4 (±); EINECS/ELINCS 200-338-0
FEMA 2940; INS1520; E1520
Synonyms: 1,2-Dihydroxypropane; 2-Hydroxypropanol; Isopropylene glycol; Methyl ethylene glycol; Methyl ethyl glycol; Methyl glycol; Monopropylene glycol; PG; 1,2-Propanediol; Propane-1,2-diol; 1,2-Propylene glycol; α-Propylene glycol; Trimethyl glycol
Classification: Aliphatic dihydric alcohol
Empirical: $C_3H_8O_2$
Formula: $CH_3CHOHCH_2OH$
Properties: Colorless clear visc. liq., odorless, sl. acrid taste; sol. in water, essential oils, acetone, chloroform, ether, ethanol; misc. with oxygenated solvs.; m.w. 76.11; dens. 1.0362; vapor pressure 0.08 mm (20 C); m.p. -60 C; b.p. 188.2 C; flash pt. (OC) 99 C; autoignition temp. 510 C
Toxicology: LD50 (oral, rat) 25 ml/kg, (IP, rat) 6660 mg/kg, (IV, rat) 6423 mg/kg, (skin, rabbit) 20,800 mg/kg; sl. toxic by ing., skin contact, IP,IV, subcut. routes; eye and human skin irritant; human systemic effects by ing. (anesthesia, convulsions, EEG changes); experimental teratogen, reproductive effector; mutagenic data; TSCA listed
Environmental: VOC; BOD5 1.08; COD 1.63; ThOD 1.68

Precaution: Combustible exposed to heat or flame; explosive as vapor exposed to heat or flames; explosive limits 2.6-12.6%; reactive with oxidizers; can increase risk of fire and explosion
Hazardous Decomp. Prods.: Heated to decomp., emits acrid smoke and irritating fumes
NFPA: Health 0, Flammability 1, Reactivity 0
Storage: Store in cool, dry area; hygroscopic
Uses: Intermediate in synthesis of org. compds., polyester and alkyd resins, polypropylene glycol; solvent, emulsifier in paints, foods, pharmaceuticals; substitute for ethylene glycol and glycerol; hygroscopic agent; energy source in animal feed; plasticizer for adhesives, cork, paper prods.; mold growth and fermentation inhibitor; bactericide (air sterilization); component of antifreeze, hydraulic and brake fluids; humectant, solvent in cosmetics; textile softener; pharmaceutical vehicle, humectant, preservative, plasticizer; stabilizer in vitamin preps.; protectant in hemorrhoidal prods.; solvent, vehicle, emulsifier, wetting agent, dispersant, humectant, anticaking agent, antioxidant, dough strengthener, flavoring agent, formulation aid, processing aid, stabilizer, thickener, surfactant, texturizer, clarifier in foods; plasticizer in food-contact coatings and rubber articles for repeated use
Use Level: 10-25% (oral sol'ns.), 10-80% (parenterals), 5-80% (topicals)
Regulatory: FDA 21CFR §73.30, 169.175, 169.176, 169.177, 169.178, 169.180, 169.181, 175.300, 175.320, 176.180, 176.210, 177.1390, 177.1680, 177.2420, 177.2600, 177.2800, 178.3300, 184.1666, 582.4666, GRAS; USDA 9CFR §318.7, 381.147; BATF 27CFR §240.1051; EPA reg., approved for some drugs; Japan approved with limitations; Europe listed; FEMA GRAS; FDA approved for orals, parenterals, topicals; NF, BP, EP, JP compliance
Manuf./Distrib.: AAE Chemie NV; ADA Int'l.; AMC Chems.; AMRESCO; AXO Chem.; Adept Sol'ns.; Aldrich; Allchem Ind.; Arch Chems./Specialty Polyols; Asahi Denka Kogyo; Ashland; Avatar; BASF; BP Amoco; Baychem; Bayer; Blagden Spec. Chems. Ltd; Brenntag AG; Brown; C.P. Hall; Chemacon GmbH; Chemcentral; Coyne; Crompton; Degussa AG/Health & Nutrition; Degussa AG; Dow; E&E Ltd; EM Ind.; EM Science; EMCO Chem. Distributors; Eastman; Farleyway Chems. Ltd; Fisher Scientific; Fluka; Focus; Frutarom Meer; GMI Prods.; General Chem.; Haarmann & Reimer; Haltermann Prods. UK; Harcros; Hawkins Chem.; Hays Chems. Ltd; Honeywill & Stein; Houghton Chem.; Hukill; Huntsman; Integra; J.T. Baker; K & K Greeff; KIC Chems.; Kraft Chem.; Lebermuth; Lluch Essence; Lyondell; Magnesia GmbH; Mallinckrodt Baker; Merck KGaA; Miljac; Mutchler; Nisso Petrochem. Ind.; Norman, Fox; PPG Ind.; Parchem Trading; Penta Mfg.; Phoenix Nat. Prods.; Polysciences; Primachem; R.C. Treatt & Co. Ltd; R.I.T.A.; Romil Ltd; Ruger; Sal Chem.; Schaefer Tech.; Seeler Ind.; Shell; Showa Denko; Sigma; Spectrum Quality Prods.; Stanchem; Sunnyside; Thomas Scientific; Univar E&E Ltd; Universal Preserv-A-Chem; VWR Int'l.; Veckridge; Voigt Global Distrib.; Vopak USA; Westco; Whyte Chems. Ltd
Trade Names: Propylene Glycol USP, FCC; Propylene Glycol USP/FCC Ultra Grade
Trade Names Containing: Hodag® S-35 FGK

1,2-Propylene glycol. *See* Propylene glycol
1,3-Propylene glycol. *See* Trimethylene glycol
α-Propylene glycol. *See* Propylene glycol
β-Propylene glycol. *See* Trimethylene glycol

Propylene glycol alginate

CAS 9005-37-2
FEMA 2941; INS405
Synonyms: Alginic acid, ester with 1,2-propanediol; Hydroxypropyl alginate; Propane-1,2-diol alginate
Definition: Mixture of propylene glycol esters of alginic acid
Empirical: $C_{72}H_{112}O_{56}$
Formula: $(C_9H_{14}O_7)_8$
Properties: Wh. to ylsh. fibrous or gran. powd., pract. odorless and tasteless; sol. in water, dil. organic acids, hydroalcoholic mixts.; m.w. 1873.6; DE 0.75-0.85
Toxicology: LD50 (oral, rat) 7200 mg/kg; mildly toxic by ing.; TSCA listed
Hazardous Decomp. Prods.: Heated to decomp., emits acrid smoke and irritating fumes
Uses: Foods (emulsifier, flavoring adjuvant, formulation aid, stabilizer, surfactant, thickener, solvent); gellant; film-former; emulsifier; suspending agent; binder, visc. control agent in cosmetics, pharmaceuticals, lotions, vitamin suspensions; industrial (paper coatings, adhesives, textile printing, dyeing, explosives); pharmaceutical solvent, flavoring adjuvant, suspending agent, thickener, gellant, film-former, emulsifier, formulation aid, stabilizer, defoamer; in paper/paperboard in contact with aq./fatty foods
Features: Reactive with milk
Regulatory: FDA 21CFR §133.133, 133.134, 133.162, 133.178. 133.179, 172.210, 172.820, 172.858, 173.340, 176.170, GRAS; FEMA GRAS; Japan approved (1% max.); Europe listed; UK approved; FDA approved for orals; NF compliance
Manuf./Distrib.: AB R Lundberg; Adept Sol'ns.; Ashland; B2E; CP Kelco; CarboMer; FMC Biopolymer; Frutarom Meer; Functional Foods; Hydro Pronova a.s.; Monsanto; Spectrum Quality Prods.; Voigt Global Distrib.; Vopak USA

Propylene glycol butoxyethyl ether. *See* 1-Butoxyethoxy-2-propanol
Propylene glycol butyl ether. *See* Butoxypropanol
Propylene glycol 1-t-butyl ether. *See* Propylene glycol t-butyl ether
Propylene glycol n-butyl ether. *See* Butoxypropanol

Propylene glycol t-butyl ether

CAS 57018-52-7
UN 1993
Synonyms: 1-t-Butoxy-2-propanol; 1-(1,1-Dimethylethoxy)-2-propanol; 1-Methyl-2-t-butoxyethanol; 2-Propanol, (1,1-dimethylethoxy)-; Propylene glycol 1-t-butyl ether; Propylene glycol mono-t-butyl ether; PTB
Empirical: $C_7H_{16}O_2$
Formula: $C_4H_9OCH_2CHOHCH_3$
Properties: Colorless clear liq.; ethereal odor; sol. (@ 19C): ≥ 100 mg/ml in water, DMSO, 95% ethanol, acetone; m.w. 132.21; dens. 0.874 (20/4 C); vapor pressure 4.8 mm Hg (20 C); f.p. -69 F; b.p. 151 C; flash pt. (TCC) 44.4 C; ref. index 1.4116
Toxicology: LD50 (oral, rat) 3.7 g/kg, (skin, rabbit) > 2 g/kg; LC50 (inh., rat, 4 h) > 2680 mg/m^3; skin and eye irritant; may be absorbed thru skin; tumorigen; TSCA listed
Precaution: Flamm. liq.; incompat. with acids, oxidizers, oxygen (prolonged contact), water, dehydrating agents
Hazardous Decomp. Prods.: Heated to decomp., emits toxic fumes of CO, CO_2, and unidentified org. compds. in black smoke
NFPA: Health 2, Flammability 2, Reactivity 0
Storage: Sensitive to heat and prolonged exposure to air; store in tightly closed container under inert atmosphere; store refrigerated
Uses: Coupling solvent, coalescing aid for water-based paints, cleaners, electronics, inks, cutting fluids, lt.-duty and hard-surf. cleaners, polyester and alkyd resins, agric., cosmetics, textiles, adhesives; wetting agent; solubilizer; chemical intermediate in the synthesis of monomeric and polymeric prods.
Manuf./Distrib.: Aldrich; Eastman; Fluka
Trade Names: Arcosolv® PTB

Propylene glycol caprylate

CAS 31565-12-5
Synonyms: Caprylic acid, monoester with 1,2-propanediol; Octanoic acid, 2-hydroxypropyl ester; Octanoic acid, monoester with 1,2-propanediol
Definition: Ester of caprylic acid and propylene glycol
Empirical: $C_{11}H_{22}O_3$
Formula: $CH_3(CH_2)_6COOCH_2CHCH_3OH$
Uses: Emollient, solubilizer, solvent for cosmetics, pharmaceuticals; solubilizer, stabilizer for foods
Trade Names: Imwitor® 408; Nikkol Sefsol 218

Propylene glycol, caprylate caprate diester. *See* Propylene glycol dicaprylate/dicaprate
Propylene glycol cyclic carbonate. *See* Propylene carbonate

Propylene glycol diacetate

CAS 623-84-7; EINECS/ELINCS 210-817-6
Synonyms: 1,2-Diacetoxypropane; α-Propylene glycol diacetate
Empirical: $C_7H_{12}O_4$
Properties: Transparent liq., fruit-like odor; sol. in water; m.w. 160.19; dens. 1.040-1.060; m.p. -31 C; b.p. 190.2 C; flash pt. 87 C; ref. index

1.413-1.415 (20 C); pH 4-6 (5%)
Toxicology: LD50 (oral, rat 13,530 mg/kg, (oral, guinea pig) 3420 mg/kg; mod. toxic by ing.; primary irritant; eye and human skin irritant; experimental teratogen; TSCA listed
Hazardous Decomp. Prods.: Heated to decomp., emits acrid smoke and irritating fumes
Uses: Emulsifier, solubilizer in pharmaceutical otics
Regulatory: FDA approved for otics
Manuf./Distrib.: A.P. Chems. Ltd; Fluka; Haltermann GmbH
Trade Names: Dowanol® PGDA

α-Propylene glycol diacetate. *See* Propylene glycol diacetate

Propylene glycol dibenzoate
CAS 19224-26-1
FEMA 3419
Properties: Sp.gr. 1.146; flash pt. (COC) 199 C; ref. index 1.544
Uses: Solvating plasticizer for PVC applic., latex caulk formulations, PVAc adhesives, and castable PU; plasticizer in food-pkg. adhesives; coalescent in latex paints; flavoring agent in beverages and pharmaceuticals
Regulatory: FDA 21CFR §175.105; FEMA GRAS
Manuf./Distrib.: Unitex; Universal Preserv-A-Chem; Velsicol
Trade Names: Benzoflex® 284

Propylene glycol dicaprylate
CAS 7384-98-7; EINECS/ELINCS 230-962-9
Synonyms: Octanoic acid, 1-methyl-1,2-ethanediyl ester
Definition: Diester of propylene glycol and caprylic acid
Empirical: $C_{19}H_{36}O_4$
Formula: $CH_3(CH_2)_6COOCH_2CHCH_3OCO(CH_2)_6CH_3$
Properties: Colorless liq.; m.w. 328.491
Toxicology: TSCA listed
Uses: Emollient, visc. control agent, solvent, solubilizer for cosmetics, pharmaceuticals; solvent for sl. sol. act. substances; in resinous/polymeric food-contact coatings; in paper/paperboard in contact with aq./fatty foods; defoamer in food-contact paper/paperboard; in food-contact textiles
Features: Good compat. with oily materials
Regulatory: FDA 21CFR §172.856, 173.340, 175.300, 176.170, 176.210, 177.2800
Manuf./Distrib.: A&E Connock; Nihon Surfactant
Trade Names: Nikkol Sefsol 228

Propylene glycol dicaprylate-caprate. *See* Propylene glycol dicaprylate/dicaprate

Propylene glycol dicaprylate/dicaprate
CAS 58748-27-9; 68583-51-7; 68988-72-7; EINECS/ELINCS 271-516-3
Synonyms: Caprylic, capric acid, propylene glycol diester; Decanoic acid, 1-methyl-1,2-ethanediyl ester mixed with 1-methyl-1,2-ethanediyl dioctanoate; Decanoic acid, mixed diesters with octanoic acid and propylene glycol; Octanoic acid, mixed diesters with decanoic acid and propylene glycol; Propylene glycol, caprylate caprate diester; Propylene glycol dicaprylate-caprate
Definition: Mixture of the propylene glycol diesters of caprylic and capric acids
Formula: $C_{10}H_{20}O_2 \bullet C_8H_{16}O_2 \bullet C_3H_8O_2$
Properties: Liq.; insol. in water; dens. 0.92 kg/l; pour pt. -40 C; flash pt. (COC) 184 C; nonionic
Toxicology: TSCA listed
Uses: Coupling agent, solvent, vehicle, diluent, carrier for flavors, fragrance oils, sol. colorants, vitamins, medicinals, cosmetics; emollient for cosmetics, pharmaceuticals, creams/lotions, makeup; food additive, emulsifier, stabilizer, defoamer; cosolvent for toiletries; in resinous/polymeric food-contact coatings; in paper/paperboard in contact with aq./fatty foods; defoamer in food-contact paper/paperboard; in food-contact textiles
Regulatory: FDA 21CFR §172.856, 173.340, 175.300, 176.170, 176.210, 177.2800
Manuf./Distrib.: A&E Connock; Ashland; MMP; Mosselman NV
Trade Names: Captex® 200; Crodamol PC-PN; Lexol® PG-865; Liponate PC; Miglyol® 840; Myritol® PC; NEOBEE® M-20

Propylene glycol dimethyl ether
CAS 7778-85-0; EINECS/ELINCS 404-630-0
Synonyms: 1,2-Dimethoxypropane; DPM; PGDE; Propane, 1,2-dimethoxy-
Empirical: $C_5H_{12}O_2$
Properties: Colorless visc. liq.; pleasant odor; completely sol. in water; misc. with alcohol, org. solvs.; m.w. 104.15; sp.gr. 0.95-0.96; vapor pressure 84.6 mm Hg; b.p. 180-195 C; flash pt. 33 F; ref. index 1.3835 (20 C)
Toxicology: Irritating to eyes; may be harmful by ing., inh., skin absorp.; vapor/mist irritating to eyes, mucous membranes, upper respiratory tract
Precaution: Highly flamm.; vapor may flash back; incompat. with strong oxidizing agents, strong acids
Hazardous Decomp. Prods.: Toxic fumes of CO, CO_2; emits toxic fumes under fire conditions
Storage: Store in cool, dry, well-ventilated place away from heat, sparks, open flame, other ignition sources; keep tightly closed
Uses: Solvent for cellulosics, acrylics, epoxides, alkyds, polyester, PU, paints/coatings, industrial applics.; solvent-sealing of cellophane; polyester and alkyd resins; antifreeze coolant; heat transfer fluid; deicing fluid; plasticizer; detergents and surfactants; bactericide; surf. tens. reducer in cleaners; chem. intermediate; in foods, pharmaceuticals, personal care
Manuf./Distrib.: Aldrich

Propylene glycol dinonanoate. *See* Propylene glycol dipelargonate

Propylene glycol dioctanoate
CAS 56519-71-2; EINECS/ELINCS 301-185-3
Synonyms: 2-Ethylhexanoic acid, 1-methyl-1,2-ethanediyl ester; 1-Methylethylene 2-ethylhexanoate; Octanoic acid, 1,3-propanediyl ester
Definition: Diester of propylene glycol and 2-ethylhexanoic acid
Empirical: $C_{19}H_{36}O_4$
Formula: $CH_3(CH_2)_3CHCH_3CH_2COOCH_2CHCH_3OCOCHCH_2CH_3(CH_2)_3CH_3$
Properties: Sol. in alcohol, min. oil; sp.gr. 0.921; flash pt. (COC) 159 C; ref. index 1.4350
Uses: Emollient for cosmetics, creams/lotions, topicals, makeup, bath preps., after-shave lotions; plasticizer; solvent, carrier, vehicle in flavors, fragrances, vitamins, antibiotics, medicinals; vehicle for nutritional prods.; emollient, moisturizer, lubricant, penetrant, extender, carrier, vehicle, solvent, solubilizer, fixing agent, humectant in pharmaceuticals, dermatologicals, dry skin prods., suppositories
Manuf./Distrib.: A&E Connock; Inolex
Trade Names: Captex® 355EP; Lexol® PG-800

Propylene glycol dipelargonate
CAS 41395-83-9; EINECS/ELINCS 255-350-9
Synonyms: Nonanoic acid, 1-methyl-1,2-ethanediyl ester; Propylene glycol dinonanoate
Definition: Diester of propylene glycol and pelargonic acid
Empirical: $C_{21}H_{40}O_4$
Formula: $CH_3(CH_2)_7COOCH_2CHCH_3OCO(CH_2)_7CH_3$
Properties: Liq.; m.w. 356.55; dens. 0.918; m.p. -21 C; flash pt. (COC) 197 C; ref. index 1.4404
Toxicology: TSCA listed
Uses: Emollient, visc. control agent, lubricant, penetrant, spreading agent, cosolvent for cosmetics, bath oils, preshave lotions, aerosol systems, lipsticks, glosses, makeup bases, pharmaceuticals; carrier for fragrances; plasticizer; lubricant for textile processing
Manuf./Distrib.: A&E Connock; Cognis; Inolex; Mosselman NV
Trade Names: Estalan-PGDP; Schercemol PGDP

Propylene glycol ethyl ether
CAS 1569-02-4 (α-isomer); 52125-53-8 (mixt.)
Synonyms: EP; Ethoxypropanol; 1-Ethoxy-2-propanol; PGEE; 2-Propanol, 1-ethoxy-; Propylene glycol monoethyl ether
Classification: Aliphatic ether alcohol
Empirical: $C_5H_{12}O_2$
Formula: $CH_3CH(OH)CH_2OCH_2CH_3$
Properties: Colorless clear liq.; sol. in water, ethanol, ether, other org. solvs.; m.w. 104.17; dens. 0.896; m.p. -100 C; b.p. 131 C; flash pt. (COC) 43 C
Toxicology: LD50 (oral, rat) 4400 mg/kg, (skin, rabbit) 8100 mg/kg; mildly

toxic by ing. and skin contact; vapor irritating to eyes, nose, and throat; absorbed thru skin, but health effects unlikely; mod. to severe eye irritant; probable lachrymator; low oral toxicity
Precaution: Combustible exposed to heat and flame; incompat. with oxidizing agents (increases fire/explosion hazard), strong acids (may cause decomp. of ethers)
Hazardous Decomp. Prods.: Heated to decomp., emits acrid smoke and irritating fumes
Storage: Store in cool, dry, well-ventilated area, out of direct sunlight
Uses: Coalescing solvent for paints/coatings, lacquers, resins, dyes, oils, greases, cleaners, cosmetics, agric., electronics, printing inks, textiles, adhesives; wetting agent; solubilizer; surf. tens. reducer in cleaning formulations
Features: Fast evaporating
Manuf./Distrib.: Acros Org.; Alfa Aesar; BP Amoco; Parchem Trading; TCI Am.; Whyte Chems. Ltd
Trade Names: Arcosolv® PE; Ethyl Proxitol

Propylene glycol ethyl ether acetate

CAS 54839-24-6 (α-isomer); 98516-30-4
Synonyms: EPA; 1-Ethoxy-2-acetoxypropanol; Ethoxypropanol acetate; 1(or 2)-Ethoxypropanol acetate; 1-Ethoxy-2-propanol acetate; Ethoxypropyl acetate; 1-Ethoxy-2-propyl acetate; PEA; PGEEA; Propanol, 1(or 2)-ethoxy-, acetate; Propylene glycol monoethyl ether acetate
Empirical: $C_7H_{14}O_3$
Formula: $C_2H_5OC(CH_3)HCH_2OC(O)CH_3$
Properties: Colorless liq.; low residual odor; sol. in water; misc. with most org. solvs.; m.w. 146.19; dens. 0.9409 kg/l (20 C); vapor pressure 2.27 mbar (20 C); b.p. 158 C; flash pt. (PMCC) 54 C; ref. index 1.4042 (20 C)
Toxicology: OSHA PEL 100 ppm
Uses: Solvent for paints/coatings, cleaners, agric., cosmetics, electronics, printing inks, textiles, adhesives
Features: Highly misc.; broad solvency; low surf. tens. improves substrate wetting, flow out; med. evaporation rate; high flash pt.
Manuf./Distrib.: ; Ashland; BP Amoco; Harcros
Trade Names: Arcosolv® PEA

Propylene glycol isobutyl ether

CAS 23436-19-3; EINECS/ELINCS 245-663-9
Synonyms: Isobutoxypropanol; 1-Isobutoxy-2-propanol; 1-(2-Methylpropoxy)-2-propanol; 2-Propanol, 1-isobutoxy-; 2-Propanol, 1-(2-methylpropoxy)-; Propylene glycol monoisobutyl ether
Classification: Aliphatic ether alcohol
Empirical: $C_7H_{16}O_2$
Formula: $CH_3CHCH_3CH_2OC_3H_6OH$
Properties: Liq.; sol. in water and most org. solvs.; m.w. 132.21; dens. 0.87 kg/l; b.p. 161 C; flash pt. (PMCC) 60 C
Toxicology: LD50 (oral, rat) 4290 mg/kg, (skin, rabbit) 8 g/kg; TSCA listed
Uses: Solvent for paints, cosmetics; coalescing solvent for water-based paints

Propylene glycol laurate

CAS 142-55-2; 10108-22-2; 27194-74-7; EINECS/ELINCS 205-542-3
Synonyms: Dodecanoic acid, 2-hydroxypropyl ester; Dodecanoic acid, monoester with 1,2-propanediol; 2-Hydroxypropyl dodecanoate; 2-Hydroxypropyl laurate; Lauric acid, monoester with 1,2-propanediol; PGML; 1,2-Propanediol monododecanoate; Propylene glycol monolaurate
Definition: Ester of propylene glycol and lauric acid
Empirical: $C_{15}H_{30}O_3$
Formula: $CH_3(CH_2)_{10}COOCH_2CHCH_3OH$
Properties: Yel to pale amber oily liq.; m.w. 258.40; dens. 0.911; m.p. 0-12 C; HLB 4.5; flash pt. (COC) 188 C; nonionic
Toxicology: Nontoxic but can cause allergic reactions in hypersensitive persons; TSCA listed
Uses: Surfactant, emulsifier, coemulsifier, stabilizer, wetting agent, lubricant, plasticizer, emollient, solvent, and antistat in cosmetics, pharmaceuticals, foods, industrial applics., pulp/paper, metalworking, lubricants, textiles, agric., paints, adhesives; solvent for paints/coatings, industrial applics.; polyester and alkyd resins; antifreeze coolant; heat transfer fluid; deicing fluid; plasticizer; detergents and surfactants; bactericide; food additive, defoamer, emulsifier, emulsion stabilizer; pharmaceutical emulsifier, stabilizer, lubricant, solvent, excipient, permeation enhancer; in food-pkg. adhesives
Features: Lipophilic
Regulatory: FDA 21CFR §172.856, 173.340, 175.105, 175.300, 176.170, 176.210, 177.2800
Manuf./Distrib.: A&E Connock; ChemService; Inolex; Lambent Tech.; Ruger; Stepan; Velsicol
Trade Names: Imwitor® 412; Schercemol PGML

Propylene glycol methyl ether. *See* Methoxyisopropanol

Propylene glycol methyl ether acetate

CAS 108-65-6; EINECS/ELINCS 203-603-9
UN 1993 (DOT)
Synonyms: Acetic acid, 2-methoxy-1-methylethyl ester; 2-Acetoxy-1-methoxypropane; 1-Methoxy-2-acetoxypropane; 2-Methoxy-1-methylethyl acetate; 1-Methoxy-2-propanol acetate; Methoxypropyl acetate; 1-Methoxy-2-propyl acetate; 2-(1-Methoxy) propyl acetate; MPA; PGMEA; PMA; PM acetate; 1,2-Propanediol monomethyl ether acetate; Propylene glycol monomethyl ether acetate; Propylene glycol monomethyl ether-1,2-acetate
Classification: Aliphatic ether alcohol; ester
Empirical: $C_6H_{12}O_3$
Formula: $CH_3COOCHCH_3CH_2OCH_3$
Properties: Colorless liq.; mild fruity odor; sol. 20% in water; misc. with most common org. solvs.; m.w. 132.17; dens. 0.97 (20/20 C); vapor pressure 3.7 mm Hg (20 C); m.p. < -67 C; b.p. 146 C; flash pt. (Seta) 45 C; autoignition temp. 354 C; ref. index 1.402 (20 C); surf. tens. 26 dynes/cm
Toxicology: LD50 (oral, rat) 8532 mg/kg, (skin, rabbit) 5 g/kg, (IP, mouse) 750 mg/kg; mod. toxic by IP route; sl. toxic by ing. and skin contact; inh. of vapor may cause nose/throat irritation; very high exposures unlikely but may cause CNS depression; low toxicity by skin absorption; may cause eye irritation on direct contact; TSCA listed
Environmental: VOC; BOD5 0.36; COD 1.11; ThOD 1.82
Precaution: Flamm.; LEL 1.3%; UEL 13.1%; incompat. with oxidizing agents (increases fire/explosion hazard)
Hazardous Decomp. Prods.: Heated to decomp., emits acrid smoke and irritating vapors
NFPA: Health 0, Flammability 2, Reactivity 0
Storage: Store in cool, dry, well-ventilated area, out of direct sunlight; hygroscopic
Uses: Solvent for electronics, acrylics, nitrocellulose, paints, water-based paints, urethane coatings, lacquers, oils, gums, cosmetics; solvent, flow aid, coalescent in coatings, printing inks; solvent, coupling agent, wetting agent, soil penetrant in cleaners
Regulatory: SARA §311/312 fire hazard
Manuf./Distrib.: Acros Org.; Aldrich; Allchem Ind.; Ashland; BP Amoco; Coyne; Fluka; G.J. Chem.; General Chem.; Harcros; Hatco; Hercules; Inolex; Joy; Lyondell; Oxiteno; Scher; Spectrum Quality Prods.; Stanchem; Stepan; Vopak USA; Werner G. Smith; Whyte Chems. Ltd
Trade Names: Arcosolv® PMA; Dowanol® PMA; Eastman® PM Acetate; Methyl Proxitol Acetate

1,2-Propylene glycol 1-monobutyl ether; Propylene glycol mono-n-butyl ether. *See* Butoxypropanol
Propylene glycol mono-t-butyl ether. *See* Propylene glycol t-butyl ether
Propylene glycol monoethyl ether; β-Propylene glycol monoethyl ether; Propylene glycol β-monoethyl ether. *See* 3-Ethoxy-1-propanol
Propylene glycol monoethyl ether acetate. *See* Propylene glycol ethyl ether acetate
Propylene glycol monoisobutyl ether. *See* Propylene glycol isobutyl ether
Propylene glycol monolaurate. *See* Propylene glycol laurate
Propylene glycol monomethyl ether; α-Propylene glycol monomethyl ether. *See* Methoxyisopropanol
β-Propylene glycol monomethyl ether. *See* 3-Methoxy-1-propanol
Propylene glycol monomethyl ether acetate; Propylene glycol monomethyl ether-1,2-acetate. *See* Propylene glycol methyl ether acetate
Propylene glycol monophenyl ether. *See* Propylene glycol phenyl ether
Propylene glycol monopropyl ether. *See* Propoxypropanol; Propylene glycol n-propyl ether

Propylene glycol monoricinoleate. *See* Propylene glycol ricinoleate
Propylene glycol monostearate. *See* Propylene glycol stearate

Propylene glycol myristyl ether acetate

CAS 93385-02-5
Synonyms: PPG-1 myristyl ether acetate; 1-Propanol, 2-(tetradecyloxy)-, acetate
Definition: Ester of propylene glycol myristyl ether and acetic acid
Uses: Emollient, solvent, plasticizer for anhyd. oil systems, emulsions; emollient in cosmetics
Trade Names: Hetester PMA

Propylene glycol octadecanoate. *See* Propylene glycol stearate

Propylene glycol phenyl ether

CAS 4169-04-4 (β-isomer)
Synonyms: 2-Phenoxypropanol; 2-Phenoxypropyl alcohol; 1-Propanol, 2-phenoxy-; Propylene glycol monophenyl ether
Classification: Glycol ether
Empirical: $C_9H_{12}O_2$
Formula: $C_6H_5OC_3H_6OH$
Properties: Colorless clear liq.; sl. 'rotten orange' odor; sol. in a variety of org. solvs.; mod. sol. in water; m.w. 152.20; sp.gr. 1.063; vapor pressure 0.1 mm Hg; f.p. 11.4 C; b.p. 242.7 C; flash pt. (TCC) 116 C; ref. index 1.522; surf. tens. 38.1 dynes/cm
Toxicology: LD50 (oral, rat) 2830 mg/kg, (skin, rabbit) > 2 g/kg; mod. toxic by ing., skin contact; inh. may cause nose/throat irritation; low toxicity by inh. unless material is heated or a mist formed; repeated skin contact may cause mild irritation; direct contact may cause sl. eye irritation and corneal injury; ing. of large amts. can cause CNS depression, but such ing. is unlikely in normal handling; target organs: CNS, respiratory system; TSCA listed
Environmental: VOC; ThOD 1.68
Precaution: Combustible; incompat. with oxidizing agents (increases fire risk), strong acids (may cause decomp. of ethers)
Hazardous Decomp. Prods.: Heated to decomp., emits acrid smoke and irritating vapors
Storage: Store in cool, dry area
Uses: Solvent for coatings, paints, lacquers, resins, inks, dyes, oils, greases, cleaners, textile printing pastes; bactericide; fixing agent for soaps and perfumes; intermediate for plasticizers; coalescent in latex adhesives and coatings; carrier solvent for textile dyes; paint removers; solvent, coupling agent in cosmetics, toiletries
Regulatory: HAP
Manuf./Distrib.: Aldrich; Fluka; TCI Am.; Yokkaichi Chem.
Trade Names Containing: Solvenon® PP

α-Propylene glycol 1-phenyl ether. *See* Phenoxyisopropanol
Propylene glycol propyl ether. *See* Propylene glycol n-propyl ether

Propylene glycol n-propyl ether

CAS 1569-01-3; EINECS/ELINCS 250-069-8
Synonyms: PGnPE; PGPE; 2-Propanol, 1-propoxy-; Propasol solvent P; 1-Propoxy-2-propanol; n-Propoxypropanol; Propylene glycol monopropyl ether; Propylene glycol propyl ether; Propyl propasol
Classification: Aliphatic ether alcohol
Empirical: $C_6H_{14}O_2$
Formula: $C_3H_7OCH_2CH(CH_3)OH$
Properties: Colorless clear liq.; mild char. odor; very sol. in water; sol. in a variety of org. solvs.; m.w. 118.18; sp.gr. 0.886 (20/20 C); vapor pressure 1.7 mm Hg (20 C); f.p. -80 C; b.p. 149.8 C; flash pt. (TCC) 48 C; autoignition temp. 252 C; ref. index 1.4121
Toxicology: ACGIH TLV 25 ppm; LD50 (oral, rat) 3250 mg/kg, (skin, rabbit) 3560 mg/kg; toxic; low toxicity absorbed thru skin; low toxicity by ing. and short-term inh.; mild to mod. skin irritant; mod. to severe eye irritant; may cause corneal injury; vapors may cause nose/throat irritation; high concs. may cause CNS depression; TSCA listed
Precaution: Flamm.; incompat. with oxidizing agents (increases fire/explosion hazard), strong acids (may cause decomp. of ethers)
Hazardous Decomp. Prods.: Heated to decomp., emits acrid smoke and irritating fumes
Storage: Store in cool, dry area, away from corrosives
Uses: Coalescing solvent for coatings, cleaners, cosmetics, agric., electronics, inks, textiles, adhesives; solvent, coupling agent for household/industrial cleaners, hard-surf. cleaners, water-reducible aerosol paints, grease/paint removers, metal cleaners; wetting agent; solubilizer
Features: Fast evaporating
Manuf./Distrib.: Aldrich
Trade Names: Arcosolv® PNP; Dowanol® PnP
Trade Names Containing: Axarel® 2200; Axarel® 4100; Prothene

Propylene glycol ricinoleate

CAS 26402-31-3; EINECS/ELINCS 247-669-7
Synonyms: 12-Hydroxy-9-octadecenoic acid, monoester with 1,2-propanediol; (R)-12-Hydroxyoleic acid, monoester with 1,2-propanediol; Propylene glycol monoricinoleate
Definition: Ester of propylene glycol and ricinoleic acid
Empirical: $C_{21}H_{40}O_4$
Formula: $C_{17}H_{32}(OH)COOCH_2CH_2OCH_3$
Properties: Pale yel. mod. visc. oily liq.; mild odor; sol. in most org. solvs.; insol. in water; sp.gr. 0.960; m.p. < -16 C; hyd. no. 285; sapon. no. 160; flash pt. (COC) 221 C; ref. index 1.469; nonionic
Toxicology: TSCA listed
Precaution: Combustible
Uses: Emulsifier, coemulsifier, stabilizer, wetting agent, lubricant, emollient, and antistat in cosmetics, pharmaceuticals, industrial applics., textiles, foods; dye solvent; plasticizer for cellulosics; urethane polymers; hydraulic fluids
Manuf./Distrib.: A&E Connock; CasChem

Propylene glycol stearate

CAS 1323-39-3; EINECS/ELINCS 215-354-3
FEMA 2942
Synonyms: Monosteol; Octadecanoic acid, monoester with 1,2-propanediol; PGMS; 1,2-Propanediol monostearate; Propylene glycol monostearate; Propylene glycol octadecanoate; Prostearin; Stearic acid, monoester with 1,2-propanediol
Definition: Ester of propylene glycol and stearic acid
Empirical: $C_{21}H_{42}O_3$
Formula: $CH_3(CH_2)_{16}COOCH_2CHCH_3OH$
Properties: Wh. to cream flakes, bland typ. fatty odor and taste; sol. in min. oil, IPM, oleyl alcohol; insol. in water, glycerin, propylene glycol; m.w. 342.63; m.p. 35-38 C; HLB 3.4; acid no. 4 max.; iodine no. 3 max.; sapon. no. 155-165; hyd. no. 160-175; nonionic
Toxicology: LD50 (IP, mouse) 200 mg/kg; poison by IP route; TSCA listed
Hazardous Decomp. Prods.: Heated to decomp., emits acrid smoke and irritating fumes
Uses: Emulsifier for lotions, soft creams, makeup, paints, adhesives; emollient, emulsifier, visc. builder for cosmetics; humectant in dry skin prods.; plasticizer for cellulose nitrate; solvent for paints/coatings, industrial applics.; polyester and alkyd resins; antifreeze coolant; heat transfer fluid; deicing fluid; detergents and surfactants; bactericide; food additive, emulsifier, stabilizer, emulsion stabilizer, defoamer, visc. builder; pharmaceutical surfactant, dispersant, emulsifier, solubilizer, lubricant, humectant, excipient, thickener, vehicle; in food-pkg. adhesives; in resinous/polymeric food-contact coatings; defoamer in food-contact paper/paperboard; in food-contact textiles
Features: Lipophilic
Regulatory: FDA 21CFR §172.856, 172.860, 172.862, 173.340, 175.105, 175.300, 176.170, 176.210, 177.2800; USDA 9CFR §318.7, 381.147; FEMA GRAS; FDA approved for rectals, vaginals; NF compliance
Manuf./Distrib.: A&E Connock; A.P. Chems. Ltd; ABITEC; Acros Org.; Adept Sol'ns.; Ashland; Bernel; CoKEM Assoc.; Danisco Ingreds.; Eastman; Gattefosse; ICN Biomed. Research Prods.; ISP Van Dyk; Inolex; Kraft Chem.; Lambent Tech.; Lipo; Lonza; Mosselman NV; Mutchler; Nihon Surfactant; Penta Mfg.; Pfaltz & Bauer; Ruger; Sea-Land; Sithean; Spectrum Quality Prods.; Stepan; TCI Am.; Universal Preserv-A-Chem; Vopak USA
Trade Names: Protachem™ PGMS

Propylene oxide

CAS 75-56-9; EINECS/ELINCS 200-879-2
UN 1280 (DOT)
Synonyms: Epoxypropane; 1,2-Epoxypropane; 2,3-Epoxypropane; Methyl ethylene oxide; Methyloxacyclopropane; Methyloxirane; S(-)-

Methyloxirane; Propane, 1,2-epoxy-; Propene oxide; Propylene epoxide; 1,2-Propylene oxide
Classification: Ether
Empirical: C_3H_6O
Properties: Colorless liq., ethereal odor; sol. in alcohol and ether; misc. with oxygenated solvs.; sl. sol. in water; m.w. 58.08; dens. 0.8304 (20/20 C); vapor pressure 400 mm Hg (17.8 C); f.p. -104.4 C; b.p. 33.9 C; flash pt. (TCC) < -25 C; ref. index 1.366 (20C)
Toxicology: ACGIH TLV/TWA 20 ppm; LD50 (oral, rat) 380 mg/kg, (IP, rat) 150 mg/kg, (skin, rabbit) 1245 mg/kg; LC50 (inh., mouse, 4 h) 1740 ppm; poison by ing. and IP routes; mod. toxic by inh., skin contact; severe skin and eye irritant; can cause somnolence, headache, nausea, vomiting; suspected human carcinogen; experimental neoplastigen, teratogen, reproductive effector, tumorigen; human mutagenic data; TSCA listed
Environmental: VOC; BOD5 0.20; ThOD 2.21
Precaution: Highly flamm.; explosive limits in air 2-22%; very dangerous fire and explosion hazard exposed to heat or flame; explosive reaction with epoxy resin and sodium hydroxide; forms explosive mixts. with oxygen; reacts vigorously with oxidizers
Hazardous Decomp. Prods.: Heated to decomp., emits acrid smoke and fumes
NFPA: Health 3, Flammability 4, Reactivity 2
Storage: Refrigerate
Uses: Intermediate; polyols for urethane foams, propylene glycols, surfactants, detergents, isopropanolamines, fumigant, synthetic lubricants, synthetic elastomers, solvent; food starch modified; food pkg. adhesives; defoamer in food-contact paper/paperboard
Regulatory: FDA 21CFR §172.892, 175.105, 176.210; SARA §302a extremely hazardous, §313 reportable; CERCLA hazardous substance; Calif. Prop. 65 reportable; HAP
Manuf./Distrib.: ARC; Aldrich; Ashland; Degussa AG; Dow; Electron Microscopy Sciences; Fluka; Huntsman; Lyondell; Romil Ltd; Spectrum Quality Prods.
Trade Names Containing: Aerothene MM

1,2-Propylene oxide. *See* Propylene oxide
Propylene oxide homopolymer. *See* Polypropylene glycol
Propylene phenoxetol; Propylene phenoxythol; Propylene phenoxytol. *See* Phenoxyisopropanol
2-Propyl ethanoate. *See* Isopropyl acetate
n-Propyl ethanoate. *See* Propyl acetate
2-Propylheptane-1-ol. *See* Propylheptanol

Propylheptanol

CAS 10042-59-8; EINECS/ELINCS 233-126-1
Synonyms: 4-Methyl-2-propylhexane-1-ol; 2-Propylheptane-1-ol; 2-Propyl-1-heptanol
Empirical: $C_{10}H_{22}O$
Properties: Clear liq.; mild odor; misc. with most common org. solvs.; pract. insol. in water; m.w. 158.28; dens. 0.8322 (20 C); visc. 15.2 mPa•s (20 C); vapor pressure 0.02 hPa (20 C); m.p. < -50 C; b.p. 207-212 C (1013 mbar); acid no. 0.1 max.; ref. index 1.4371 (20 C); surf. tens. 27.5 mN/m (20 C)
Storage: Store tightly closed under dry nitrogen blanket
Uses: Solvent; raw material for prod. of plasticizers and auxiliaries in the chemical industry; intermediate in prod. of esters
Features: High boiling
Manuf./Distrib.: BASF AG; ICN Biomed. Research Prods.

2-Propyl-1-heptanol. *See* Propylheptanol
Propyl hydride. *See* Propane
Propylic alcohol. *See* Propyl alcohol
Propyl ketone. *See* 4-Heptanone
Propylmethanol. *See* Butyl alcohol
Propylnitrile. *See* Propionitrile
N-Propyl-1-propanamine. *See* Dipropylamine
Propyl propanoate. *See* Propyl propionate
Propyl propasol. *See* Propylene glycol n-propyl ether

Propyl propionate

CAS 106-36-5; EINECS/ELINCS 203-389-7
FEMA 2958
Synonyms: Propanoic acid, propyl ester; Propionic acid, propyl ester; Propyl propanoate; n-Propyl propionate
Empirical: $C_6H_{12}O_2$
Formula: $CH_3CH_2COOCH_2CH_2CH_3$
Properties: Clear liq., complex fruity odor; sol. in 200 parts water; misc. with alcohol, ether; m.w. 116.16; dens. 0.881 (20/4 C); vapor pressure 10 mm Hg (19.4 C); m.p. -76 C; b.p. 120-122 C; flash pt. 22 C; ref. index 1.3935 (20 C)
Toxicology: LD50 (oral, rat) 10,331 mg/kg; LC50 (inh., unspecified mammal) 16,900 mg/m^3; mod. toxic by ing.; very mildly toxic by inh.; skin irritant; TSCA listed
Precaution: Flamm. exposed to heat or flame; can react with oxidizers
Hazardous Decomp. Prods.: Heated to decomp., emits acrid smoke and irritating fumes
Uses: Solvent for paints, varnishes, lacquers, coating agents, nitrocellulose; synthetic flavoring agent in foods and pharmaceuticals; fragrance in cosmetics
Features: Fruity flavor
Regulatory: FDA 21CFR §172.515; FEMA GRAS
Manuf./Distrib.: Adept Sol'ns.; Advanced Synthesis Tech.; Aldrich; Ashland; Axxence Aromatic GmbH; Degussa Flavors/Alfrebro; Dow; Fluka; Grau Aromatics; Moore Ingreds.; Penta Mfg.
Trade Names: Ucar® n-Propyl Propionate

n-Propyl propionate. *See* Propyl propionate
1-Propyne-3-ol; Propyn-1-ol; 1-Propyn-3-ol; 2-Propynol; 2-Propyn-1-ol; 3-Propynol; 2-Propynyl alcohol. *See* Propargyl alcohol
Prostearin. *See* Propylene glycol stearate
Prunolide. *See* γ-Nonalactone
Prunus dulcis. *See* Sweet almond (Prunus amygdalus dulcis) oil
Pseudobutylbenzene. *See* t-Butylbenzene
Pseudocumene; Pseudocumol. *See* 1,2,4-Trimethylbenzene
Pseudohexyl alcohol. *See* 2-Ethyl-1-butanol
PTB. *See* Propylene glycol t-butyl ether
PTBT. *See* p-t-Butyl toluene
Purified gum spirits. *See* Turpentine
Purified water. *See* Water
PX. *See* p-Xylene
Pyran-2-methanol, tetrahydro-; 2H-Pyran-2-methanol, tetrahydro-. *See* Tetrahydropyran-2-methanol

Pyridine

CAS 110-86-1; EINECS/ELINCS 203-809-9
UN 1282 (DOT); FEMA 2966
Synonyms: Azabenzene; Azine
Classification: Heterocyclic aromatic compd.
Empirical: C_5H_5N
Formula: NCHCHCHCHCH
Properties: Colorless to pale yel. liq., char. disagreeable odor, sharp burning taste; sol. in water, alcohol, ether, chloroform, petrol. ether, oils, other org. liqs.; misc. with oxygenated and aromatic solvs.; m.w. 79.10; dens. 0.98272 (20/4 C); vapor pressure 10 mm Hg (13.2 C); m.p. -41.6 C; b.p. 115-116 C; flash pt. (CC) 20 C; ref. index 1.510 (20 C); pH 8.5 (0.2 molar sol'n. in water)
Toxicology: ACGIH TLV/TWA 5 ppm; LD50 (oral, rat) 1.58 g/kg, (skin, rabbit) 1121 mg/kg; poison by IP route; mod. toxic by ing., IV, subcut.; may cause human CNS depression, irritation of skin and respiratory tract; may cause severe corrosive damage to eyes; ing. may cause mouth/throat irritation, nausea, vomiting, diarrhea; large doses may produce GI disturbances, kidney/liver damage, convulsions, unconsciousness, death; mutagenic data; TSCA listed
Precaution: Highly flamm.; dangerous fire hazard exposed to heat, flame, oxidizers; volatile with steam; incompat. with oxidizing materials, strong acids or bases, formamide, iodine, maleic anhydride, silver perchlorate, perchromates, etc.
Hazardous Decomp. Prods.: Heated to decomp., emits highly toxic fumes of NO_x
NFPA: Health 2, Flammability 3, Reactivity 0
Storage: Store in cool, dry, well-ventilated area out of direct sunlight
Uses: Chemical intermediate; acid scavenger; in agrochemicals; fungicides; surfactants; photographic materials; coatings; curing agents; rubber chemicals; vulcanization accelerator; plastics; antidandruff

shampoos; waterproofing; denaturant for alcohols; solvent in pharmaceuticals; synthesis of vitamins, drugs; sedative; synthetic flavoring agent in foods and pharmaceuticals; processing aid in textile dyeing; adjuvant for PC food-pkg. resins

Regulatory: FDA 21CFR §172.515, 177.1580, 177.1585; FEMA GRAS; BP, EP compliance

Manuf./Distrib.: Aldrich; Alfa Aesar; Alfa Chem; Allchem Ind.; BCH Brühl; Bencorp Int'l.; Berje; Daicel Chem. Ind.; Fabrichem; Fluka; Global Chemsources; H&A (Canada) Ind.; Honeywell Perf. Polymers; Integra; J.T. Baker; Koei Chem.; Kowa Am.; Nepera; Pechiney Chems. Div.; Penta Mfg.; Pharmrite N. Am.; Raschig; Reagents; Reilly Ind.; Romil Ltd; Ruger; Schweizerhall; Sigma; Spectrum Quality Prods.; Total Spec. Chems.; VWR Int'l.; Whyte Chems. Ltd

Pyridine, 2,6-dimethyl-. *See* 2,6-Lutidine
Pyridine, 3-methyl-. *See* β-Picoline
Pyridine, 4-methyl-. *See* γ-Picoline
Pyroacetic acid; Pyroacetic ether. *See* Acetone
Pyrobenzol; Pyrobenzole. *See* Benzene
Pyrocatechin; Pyrocatechinic acid. *See* Pyrocatechol

Pyrocatechol

CAS 120-80-9; EINECS/ELINCS 204-427-5

Synonyms: 1,2-Benzenediol; o-Benzenediol; Catechol; CI 76500; 1,2-Dihydroxybenzene; o-Dihydroxybenzene; o-Dioxybenzene; o-Diphenol; o-Hydroquinone; 2-Hydroxyphenol; o-Hydroxyphenol; Oxyphenic acid; o-Phenylenediol; Pyrocatechin; Pyrocatechinic acid; Pyrocatechuic acid

Classification: Aromatic hydrocarbon; phenol

Empirical: $C_6H_6O_2$

Formula: $C_6H_4(OH)_2$

Properties: Colorless monocl. leaflets; discolored by lt. and air; sol. in water, oxygenated solvs., chloroform, alcohol, benzene, ether; m.w. 110.11; dens. 1.371 (15 C); vapor pressure 10 mm (118.3 C); m.p. 105 C; b.p. 240 C; flash pt. 127 C

Toxicology: LD50 (oral, rat) 260 mg/kg, (subcut., mouse) 247 mg/kg, (skin, rabbit) 800 mg/kg; toxic; poison by ing., subcut., IP, IV, parental routes; mod. toxic by skin contact; corrosive; skin irritant; allergen; readily absorbed thru skin; tumorigen; experimental reproductive effects; human mutagenic data; target organs: liver, kidneys, nerves; TSCA listed

Precaution: Oxygen-sensitive; combustible exposed to heat or flame; can react vigorously with oxidizers; hypergolic reaction with conc. nitric acid

Hazardous Decomp. Prods.: Heated to decomp., emits acrid smoke and irritating fumes

Storage: Light- and air-sensitive; store under nitrogen

Uses: Reagent; synthesis of rubber chems., resins, antioxidants, adhesives, insecticides; electroplating; flavors and fragrances, pharmaceuticals; fax papers; dyes; photographic chems.; fur/leather dyeing reagent; antiseptic; specialty inks; antioxidant for perfumes, essential oils; solvent

Regulatory: SARA §313 reportable; HAP

Manuf./Distrib.: Aldrich; Coalite Chems.; EM Ind.; EM Science; Filo; Fluka; Howard Hall; Int'l. Fiber; Lipo; Penta Mfg.; Schweizerhall; Sigma; Spectrum Quality Prods.

Pyrocatechuic acid. *See* Pyrocatechol
Pyrogentistic acid. *See* Hydroquinone
2-Pyrol. *See* 2-Pyrrolidone
Pyroligneus acid. *See* Acetic acid
Pyromucic acid methyl ester. *See* Methyl 2-furoate
Pyromucic aldehyde. *See* Furfural
Pyroxylic spirit. *See* Methyl alcohol
Pyroxylin solution. *See* Collodion
m-Pyrrole. *See* N-Methyl-2-pyrrolidone
2-Pyrrolidinone; α-Pyrrolidinone. *See* 2-Pyrrolidone
2-Pyrrolidinone, 1-cyclohexyl-. *See* N-Cyclohexyl pyrrolidone
2-Pyrrolidinone, 1-dodecyl-. *See* Lauryl pyrrolidone
2-Pyrrolidinone, 1-ethyl-. *See* N-Ethyl-2-pyrrolidone
2-Pyrrolidinone, 1-(2-hydroxyethyl)-. *See* N-Hydroxyethylpyrrolidone
2-Pyrrolidinone, 1-octyl-. *See* N-Octyl pyrrolidone
Pyrrolidone. *See* 2-Pyrrolidone

2-Pyrrolidone

CAS 616-45-5; EINECS/ELINCS 204-648-7

Synonyms: 4-Aminobutyric acid lactam; γ-Aminobutyric acid lactam; γ-Aminobutyric lactam; γ-Aminobutyrolactam; Butyrolactam; γ-Butyrolactam; LAM; 2-Oxopyrrolidine; 2-Pyrol; 2-Pyrrolidinone; α-Pyrrolidinone; Pyrrolidone; Pyrrolidone-2; α-Pyrrolidone

Empirical: C_4H_7NO

Formula: $CH_2CH_2CH_2C(O)NH^*$

Properties: Colorless to lt. yel. liq. or solid; sol. in water, ethanol, ethyl ether, chloroform, benzene, ethyl acetate, carbon disulfide; misc. with oxygenated, chlorinated, and aromatic solvs.; insol. in aliphatic solvs.; m.w. 85.12; dens. 1.11 kg/l (20 C); m.p. 25 C; b.p. 245 C; flash pt. 265 F

Toxicology: LD50 (oral, rat) 6500 mg/kg, (IP, rat) 160 mg/kg; LDLo (subcut., mouse) 10 g/kg; mildly toxic by ing., subcut. route; mutagen; TSCA listed

Hazardous Decomp. Prods.: Heated to decomp., emits toxic fumes of NO_x

Uses: Plasticizer, emulsifier, and coalescent for acrylic latexes, styrene-acrylic floor polishes; solvent for polymers, insecticides, veterinary medicine; selective solvent (acetylene, acrylonitrile prod.); process solvent (pharmaceuticals); specialty inks; monomer for nylon-4; pharmaceutical solubilizer, intermediate; synthesis of piracetam; treatment of cerebral distress

Manuf./Distrib.: Aldrich; Allchem Ind.; BASF; Fluka; ISP; Kraft Chem.; Mitsubishi Chem.; Sithean; Spectrum Quality Prods.; UCB; Unitex

Trade Names: 2-Pyrol®; Soluphor® P

Pyrrolidone-2; α-Pyrrolidone. *See* 2-Pyrrolidone
Pyrrolidone, N-octyl-. *See* Caprylyl pyrrolidone

Q

Quaternium-21. *See* PPG-40 diethylmonium chloride
Quinol. *See* Hydroquinone

Quinoline

CAS 91-22-5; EINECS/ELINCS 202-051-6
UN 2656 (DOT); FEMA 3470
Synonyms: 1-Azanaphthalene; 1-Benzazine; 1-Benzine; Benzo[b]pyridine; Chinoleine; Chinolin; Chinoline; Leucol; Leucoline
Empirical: C_9H_7N
Formula: $C_6H_4N{=}CHCH{=}CH$
Properties: Colorless liq., penetrating unpleasant odor; sol. in hot water; misc. with alcohol, ether, CS_2, oxygenated solvs.; difficultly sol. in cold water; m.w. 129.16; dens. 1.093 (20/4 C); vapor pressure 1 mm (59.7 C); m.p. -17 to -13 C; b.p. 236-238 C; flash pt. 92 C; ref. index 1.625 (20 C)
Toxicology: LD50 (oral, rat) 331 mg/kg, (skin, rabbit) 540 mg/kg; LDLo (IP, mouse) 64 mg/kg; toxic; poison by ing., subcut., IP routes; mod. toxic by skin contact; primary skin irritant; severe eye irritant; readily absorbed thru skin; may cause retinitis; questionable carcinogen; experimental neoplastigen, tumorigen; mutagen; target organ: liver; TSCA listed
Precaution: Combustible exposed to heat or flames; explosive reaction possible with hydrogen peroxide; violent reaction with dinitrogen tetraoxide, perchromates; incompat. with linseed oil + thionyl chloride; maleic anhydride; unpredictably violent
Hazardous Decomp. Prods.: Heated to decomp., emits toxic fumes of NO_x
Storage: Hygroscopic; photosensitive; protect from light and moisture
Uses: Preserving anatomical specimens; mfg. of quinolinol sulfate, niacin and copper-8-quinolinolate; synthetic flavoring agent in foods and pharmaceuticals; fragrance in cosmetics; antimalarial; mfg. of dyes; solvent for resins, terpenes
Regulatory: FEMA GRAS
Manuf./Distrib.: Aldrich; Allchem Ind.; Biddle Sawyer; Expansia SA; Fluka; Jarchem Ind.; Koppers Ind.; Sigma
Trade Names Containing: Methylnaphthalene Fraction

Quinoline, 1,2,3,4-tetrahydro-. *See* 1,2,3,4-Tetrahydroquinoline

R

R 11. *See* Trichlorofluoromethane
R 12. *See* Dichlorodifluoromethane
R 12B2. *See* Dibromodifluoromethane
R 13. *See* Chlorotrifluoromethane
R 21. *See* Dichlorofluoromethane
R 22. *See* Chlorodifluoromethane
R 23. *See* Trifluoromethane
R 112. *See* 1,1,2,2-Tetrachloro-1,2-difluoroethane
R 113. *See* Trichlorotrifluoroethane
R 114. *See* 1,2-Dichlorotetrafluoroethane
R 115. *See* Chloropentafluoroethane
R 141b. *See* Dichlorofluoroethane
R 142b. *See* 1,1,1-Chlorodifluoroethane
Racemic limonene. *See* dl-Limonene
Racemic menthol. *See* Menthol
Raffinates (petroleum), sorption process. *See* Petroleum raffinates, sorption process
Rape oil, methyl ester. *See* Methyl canolate
Rape oil methyl ester. *See* Methyl rapeseedate
Rapeseed. *See* Canola oil

Rapeseed (Brassica campestris) oil

CAS 8002-13-9; 120962-03-0; EINECS/ELINCS 232-299-0
Synonyms: Brassica campestris oil; Brassica oleifera; Canola oil (low erucic acid rapeseed oil); Colza oil; HEAR; High-erucic acid rapeseed oil; LEAR; Low-erucic acid rapeseed oil; Rapeseed oil; Rapeseed oil, blown
Definition: Vegetable oil expressed from seeds of *Brassica campestris*
Properties: Brn. viscous liq., yel. when refined, noxious odor; sol. in chloroform, ether, CS_2; dens. 0.913-0.916; m.p. 17-22 C; solidifies at 0 C; iodine no. 97-105; sapon. no. 170-177 C; flash pt. 162 C; ref. index 1.4720-1.4752
Toxicology: Toxic; allergic potential; can cause acne-like skin eruptions
Precaution: Subject to spontaneous heating
NFPA: Health 0, Flammability 1, Reactivity 0
Uses: Lubricant and slip agent for plastics, polymers, rubber, metal lubricants; quenching oil; blown oils; soft soaps; textile finishing agent; drying oil for paints; leather fat-liquoring agent; lubricant, wetting agent for continuous films; emollient in cosmetics; edible oil for salad dressings, margarine; pharmaceutical excipient; in food-pkg. adhesives; defoamer in food-contact paper/paperboard; in cellophane for food pkg.; in food-contact textiles
Regulatory: FDA 21CFR §175.105, 176.210, 177.1200, 177.2800, 184.1555; Japan approved (extract); FDA approved for orals; BP, EP compliance (refined)
Manuf./Distrib.: ABITEC; Aldivia; Alfa Aesar; Alnor Oil; Anar; Anglia Oils; Arista Ind.; Degen; Degussa AG/Health & Nutrition; Freund Chem. Co. Ltd; Georgia-Pacific/Actrachem; Industrial Oil Prods.; Jarchem Ind.; John L Seaton Ltd; Lambent Tech.; Mosselman NV; Noveon; Penta Mfg.; Sea-Land; St. Lawrence; Stevenson Cooper; Vliengenthart BV; Welch, Holme & Clark; Werner G. Smith
Trade Names: Oleocal® C-102; Oleocal® C-102 A
See also Canola oil

Rapeseed oil; Rapeseed oil, blown. *See* Rapeseed (Brassica campestris) oil
Rapeseed oil, methyl ester. *See* Methyl canolate
Rapeseed oil methyl ester. *See* Methyl rapeseedate
Red oil. *See* Oleic acid
Refined solvent naphtha. *See* VM&P naphtha
Refrigerant 12. *See* Dichlorodifluoromethane
Refrigerant 21. *See* Dichlorofluoromethane
Refrigerant 22. *See* Chlorodifluoromethane
Refrigerant 23. *See* Trifluoromethane
Refrigerant 112. *See* 1,1,2,2-Tetrachloro-1,2-difluoroethane
Refrigerant 113. *See* Trichlorotrifluoroethane
Refrigerant 114. *See* 1,2-Dichlorotetrafluoroethane
Refrigerant R40. *See* Methyl chloride
Rice bran oil; Rice oil. *See* Rice (Oryza sativa) bran oil

Rice (Oryza sativa) bran oil

CAS 68553-81-1; 84696-37-7; EINECS/ELINCS 271-397-8
Synonyms: Oils, rice bran; Oryza sativa; Rice bran oil; Rice oil
Definition: Oil expressed from rice bran, *Oryza sativa*
Properties: Golden yel. oil; misc. with hexane and other fat solvs.; negligible sol. in water; dens. 0.916-0.921; vapor pressure very low; cloud pt. < -7 C; acid no. < 2; iodine no. 92-115; sapon. no. 180-190; ref. index 1.470-1.473
Hazardous Decomp. Prods.: CO_2 from combustion
Uses: Soap raw material; hydrogenated shortening ingred.; edible oil in foods and pharmaceuticals; skin conditioner, solvent, emollient for cosmetics; pharmaceutical emollient; ointment solvent; in food-pkg. adhesives; defoamer in food-contact paper coatings, paper/paperboard; for fiber finishing for resin-bonded filters for food contact; in food-contact textiles
Regulatory: FDA 21CFR §175.105, 176.200, 176.210, 177.2260, 177.2800
Manuf./Distrib.: Arista Ind.; Ashland; B&D Nutritional Ingreds.; Barnet Prods.; Chugai Boyeki Am.; Fabrichem; Freeman Ind.; Int'l. Sourcing; Maypro Ind.; Natural Oils Int'l.; Oilseeds Int'l.; Penta Mfg.; Pokonobe Ind.; Seltzer; Tri-K Ind.; Welch, Holme & Clark
Trade Names: EmCon™ Rice Bran

Ricini oleum. *See* Castor (Ricinus communis) oil
Ricinoleic acid, methyl ester, acetate. *See* Methyl acetyl ricinoleate
Ricinus communis oil; Ricinus oil. *See* Castor (Ricinus communis) oil
Rose ether. *See* Phenoxyethanol
Rose oil. *See* Phenethyl alcohol
Rubbing alcohol. *See* Isopropyl alcohol

S

Saccharose palmitate. *See* Sucrose palmitate
Safety solvent naphtha. *See* Stoddard solvent

Safflower (Carthamus tinctorius) oil

CAS 8001-23-8; EINECS/ELINCS 232-276-5
Synonyms: Carthamus tinctorious oil; Carthamus tinctorius; Safflower oil; Safflower oil (unhydrogenated); Safflower seed oil
Definition: Oily liq. obtained from seeds of *Carthanus tinctorius* consisting principally of triglycerides of linoleic acid
Properties: Lt. yel. oily liq.; sl. veg. odor; sol. in oil and fat solvs.; misc. with ether, chloroform; insol. in water; dens. 0.9211-0.9215 (25/25 C); iodine no. 135-150; sapon. no. 188-194; ref. index 1.472-1.475
Toxicology: Human skin and eye irritant; ing. in large volumes produces vomiting; TSCA listed
Precaution: Combustible
Hazardous Decomp. Prods.: CO_2 from combustion; heated to decomp., emits acrid smoke and irritating fumes
HMIS: Health 0, Flammability 1, Reactivity 0
Storage: Light-sensitive; becomes rancid on exposure to air
Uses: Drying oil for paints, varnishes, alkyd resins; emollient in cosmetics; edible oil in salad oil blends, margarine, shortening; food coatings; food emulsifier, texturizer, flavoring agent; medicine (laxative, diaphoretic); pharmaceutical emollient, vehicle, solvent; in food-pkg. adhesives; drying oil in food-contact coatings; in resinous/polymeric food-contact coatings; defoamer in food-contact paper coatings, paper/paperboard
Features: Oleaginous
Regulatory: FDA 21CFR §175.105, 175.300, 176.200, 176.210, GRAS; Japan approved (safflower); USP compliance
Manuf./Distrib.: ABITEC; Alban Muller; Aldivia; Aldrich; Alfa Chem; Alnor Oil; Anglia Oils; Arista Ind.; Avatar; Biosil Tech.; Cascade; Charkit; Croda Chem. Ltd; Croda Inc; Desert Whale Jojoba; Int'l. Sourcing; Jeen Int'l.; John L Seaton Ltd; KIC Chems.; Lambent Tech.; Lipo; Mosselman NV; Penta Mfg.; Protameen; Ruger; Sea-Land; Sigma; Spec. Ingreds. & Perfume; Spectrum Quality Prods.; Thornley; Tri-K Ind.; Vliengenthart BV; Voigt Global Distrib.; Welch, Holme & Clark
Trade Names: Super Refined® Safflower USP

Safflower oil. *See* Safflower (Carthamus tinctorius) oil
Safflower oil, hybrid. *See* Hybrid safflower (Carthamus tinctorius) oil
Safflower oil (unhydrogenated); Safflower seed oil. *See* Safflower (Carthamus tinctorius) oil
Salicylic acid-2-ethylhexyl ester. *See* 2-Ethylhexyl salicylate
Salicylic acid, isocetyl ester. *See* Isocetyl salicylate
Salt of tartar. *See* Potassium carbonate
SBA. *See* 2-Butanol
SBP. *See* Naphtha, hydrotreated light

SD alcohol 1

Classification: Specially denatured alcohol
Definition: Ethyl alcohol (100 gal) denatured with methyl alcohol (4 gal) and one of the following: denatonium benzoate NF (1/8 oz), MIBK (1 gal), mixed isomers of nitropropane (1 gal), or methyl n-butyl ketone (1 gal)
Uses: Solvent for coatings (cellulose, syn. resin, shellac, nat. resin), plastics, photographic film, explosives, soldering flux, adhesives, binders, polishes, inks, stains, shampoos, soaps, bath preps., petrol. prods., disinfectants, insecticides, dehydration (cellulosics, sodium hydrosulfite), embalming fluids, sterilizing/preserving sol'ns., industrial/household detergents; thinner; solvent, astringent, diluent for cosmetics; process solvent for pectin, food prods., crude drugs, vitamins, hormones, yeasts, antibiotics, vaccines, blood prods., dyes, intermediates, perfumes, photographic chems., rosin, rubber latex; fuel (automobiles, airplanes, rocket, jets, heating); fluid for scientific instruments, brake fluids, cutting oils, refrigeration; raw material for ethyl acetate, ethyl chloride, ethylamines, dyes, intermediates, etc.
Regulatory: FDA 27CFR §20.11, 21.32
Trade Names: SDA-1
Trade Names Containing: Tecsol® 1; Tecsol® 3

SD alcohol 2-B

Classification: Specially denatured alcohol
Definition: Ethyl alcohol (100 gal) denatured with benzene (0.5 gal), rubber hydrocarbon solvent (0.5 gal), toluene (0.5 gal), or heptane (0.5 gal)
Uses: Solvent for plastics, resins, photographic film, transparent sheeting, explosives, petrol. prods., dehydration (cellulosics, sodium hydrosulfite); process solvent for pectin, food prods., crude drugs, vitamins, hormones, yeasts, antibiotics, vaccines, medicinal chems., dyes, intermediates, perfumes; raw material for ethyl acetate, ethyl chloride, ethylamines, dyes, intermediates, etc.
Regulatory: FDA 27CFR §21.33
Trade Names: SDA-2B

SD alcohol 2-C

Classification: Specially denatured alcohol
Definition: Ethyl alcohol (100 gal) denatured with metallic sodium (≥ 33 lb) and either benzene (0.5 gal), toluene (0.5 gal), or rubber hydrocarbon solvent (0.5 gal)
Uses: Process solvent for medicinal chems. incl. alkaloids, other chems.; raw material for ethyl esters, ethylamines, dyes, intermediates, drugs, medicinal chems.
Regulatory: FDA 27CFR §21.34

SD alcohol 3-A

CAS 977021-59-2
Classification: Specially denatured alcohol
Definition: Ethyl alcohol (100 gal) denatured with methyl alcohol (5 gal)
Toxicology: Methanol may modify and increase toxic symptoms caused by ing. and exposure to fumes
Uses: Solvent in coatings (cellulose, syn. resin), plastics, photographic film, explosives, industrial collodions, soldering flux, adhesives, binders, polishes, inks, stains, shampoos, soaps, bath preps., dehydration (cellulosics, sodium hydrosulfite), petrol. prods., disinfectants, insecticides, fungicides, embalming fluids, sterilizing/preserving sol'ns., industrial/household detergents, room deodrants, photoengraving; process solvent for pectin, food prods., crude drugs, vitamins, hormones, yeasts, antibiotics, vaccines, medicinal chems., blood prods., dyes, intermediates, perfumes, rosin, rubber latex; raw material for ethylamines, dyes, intermediates, medicinal chems., organo-silicone prods., syn. resins; solvent in pharmaceutical tablet

coatings; solvent, astringent, diluent in cosmetics; diluent in colorants (marking food); fuel (automobile, airplane, rocket, jet, heating); fluid (scientific instruments, brake fluids, cutting oils, refrigration)
Regulatory: FDA 21CFR §73.1, 27CFR §20.11, 21.35
Manuf./Distrib.: Eastman
Trade Names: SDA-3A
Trade Names Containing: Tecsol® H

SD alcohol 3-B

Classification: Specially denatured alcohol
Definition: Ethyl alcohol (100 gal) denatured with pine tar USP (1 gal)
Uses: Solvent for cosmetics, hair/scalp preps., shampoos, soaps, bath preps., disinfectants, insecticides, fungicides, other biocides
Regulatory: FDA 27CFR §20.11, 21.36

SD alcohol 3-C

Classification: Specially denatured alcohol
Definition: Ethyl alcohol (100 gal) denatured with isopropyl alcohol (5 gal)
Uses: Solvent for coatings (cellulose, syn. resin), plastics, photographic film, explosives, industrial collodions, soldering flux, adhesives, binders, polishes, inks, meat branding inks, stains, shampoos, soaps, bath preps., dehydration (cellulosics, sodium hydrosulfite), petrol. prods., disinfectants, insecticides, fungicides, embalming fluids, sterilizing/preserving sol'ns., household/industrial detergents, room deodorants, photoengraving dyes; process solvent for pectin, food prods., crude drugs, vitamins, hormones, yeasts, antibiotics, vaccines, medicinal chems., blood prods., dyes, intermediates, perfumes, rosin, rubber latex; solvent, astringent, diluent for cosmetics; raw material for ethylamines, dyes, intermediates, drugs, organo-silicone prods.; fuel (automobile, airplane, rocket, jet, heating); fluid (scientific instruments, brake fluids, cutting oils, refrigeration)
Regulatory: FDA 27CFR §20.37, 21.161
Manuf./Distrib.: Eastman
Trade Names: SDA-3C

SD alcohol 4

Classification: Specially denatured alcohol
Definition: Ethyl alcohol (100 gal) or rum not less than 150 proof (100 gal) denatured with 1 gal of the following sol'n.: 5 gal of aq. sol'n. contg. 40% nicotine, 3.6 oz of methylene blue USP, and water sufficient to make 100 gal
Uses: Solvent for tobacco sprays and flavors
Regulatory: FDA 27CFR §21.38

SD alcohol 12-A

Classification: Specially denatured alcohol
Definition: Ethyl alcohol (100 gal) denatured with benzene (5 gal) or toluene (5 gal)
Uses: Solvent for plastics, resins, cellulose plastics, adhesives, binders, sterilizing/preserving sol'ns., pharmaceuticals; process solvent for vitamins, hormones, yeasts, antibiotics, vaccines, medicinal chems. incl. alkaloids, blood prods., dyes, intermediates, perfumes/fixatives, rosin, other chems.; raw material for ethyl esters, ethylamines, dyes, intermediates, drugs, medicinal chems.
Regulatory: FDA 27CFR §21.40
Trade Names: SDA-12A

SD alcohol 13-A

Classification: Specially denatured alcohol
Definition: Ethyl alcohol (100 gal) denatured with ethyl ether (10 gal)
Uses: Solvent for candy glazes, plastics, resins, cellulose plastics, photographic film, transparent sheeting, industrial collodions and collodion USP, inks, sterilizing/preserving sol'ns.; process solvent for pectin, food prods., vitamins, hormones, yeasts, antibiotics, vaccines, medicinal chems. incl. alkaloids, blood prods., pharmaceuticals, perfumes/fixatives, photographic chems., other chems.; raw material for ethyl esters, ethyl ether, other ethers, drugs, medicinal chems., other chems.
Regulatory: FDA 27CFR §21.41

SD alcohol 17

Classification: Specially denatured alcohol
Definition: Ethyl alcohol (100 gal) denatured with bone oil (Dipple's oil) (6.4 fl. oz)
Uses: Process solvent for medicinal chems. incl. alkaloids, other chems.; raw material for drugs, medicinal chems., other chems.
Regulatory: FDA 27CFR §21.42

SD alcohol 19

Classification: Specially denatured alcohol
Definition: Ethyl alcohol (100 gal) denatured with ethyl ether (100 gal)
Uses: Solvent for photographic film, cellulose intermediates, industrial collodions, collodion USP
Regulatory: FDA 27CFR §21.44

SD alcohol 22

Classification: Specially denatured alcohol
Definition: Ethyl alcohol (100 gal) denatured with formaldehyde sol'n. USP (10 gal)
Uses: Solvent for embalming fluids, sterilizing/preserving sol'ns., theater sprays, incense, room deodorants
Regulatory: FDA 27CFR §21.46

SD alcohol 23-A

CAS 977021-64-9
Synonyms: Alcohol, denatured formula 23A
Classification: Specially denatured alcohol
Definition: Ethyl alcohol (100 gal) denatured with acetone USP (8 gal)
Uses: Solvent for coatings (cellulose, syn. resin, shellac, other nat. resin), candy glazes, transparent sheeting, cellulose intermediates, industrial collodions, soldering flux, adhesives, binders, inks, meat branding inks, stains, bay rum, hair/scalp preps., hand/face/body creams/ lotions, deodorants, shampoos, soaps, bath preps., external pharmaceuticals, antiseptics (USP or NF), disinfectants, insecticides, fungicides, other biocides, embalming fluids, sterilizing/preserving sol'ns., household/industrial detergents, dyes; process solvent for pectin, food prods., crude drugs, vitamins, hormones, yeasts, antibiotics, vaccines, medicinal chems. incl. alkaloids, blood prods., other chems.; thinner; solvent, astringent, diluent for cosmetics; diluent in colorants for coloring shell eggs; refrigeration fluid
Regulatory: FDA 21CFR §73.1, 27CFR §20.11, 21.47
Manuf./Distrib.: Eastman
Trade Names: SDA-23A

SD alcohol 23-F

Classification: Specially denatured alcohol
Definition: Ethyl alcohol (100 gal) denatured with salicylic acid USP (1 lb), resorcinol USP (1 lb), and 1 gal of bergamot oil NF or bay oil (myrica oil) NF
Uses: Solvent for cosmetics, hair/scalp preps., external pharmaceuticals (not USP or NF)
Regulatory: FDA 27CFR §20.11, 21.48

SD alcohol 23-H

Classification: Specially denatured alcohol
Definition: Ethyl alcohol (100 gal) denatured with acetone USP (8 gal) and MIBK (1.5 gal)
Uses: Solvent for cosmetics, hair/scalp preps., hand/face/body creams/ lotions, external pharmaceuticals (not USP or NF), rubbing alcohols, disinfectants, insecticides, fungicides, other biocides, household detergents
Regulatory: FDA 27CFR §20.11, 21.49
Trade Names: SDA-23H

SD alcohol 25

Classification: Specially denatured alcohol
Definition: Ethyl alcohol (100 gal) denatured with iodine USP (20 lb) and 15 lb of potassium iodide USP or sodium iodide USP
Uses: Solvent for tinctures of iodine, external pharmaceuticals (USP or NF)
Regulatory: FDA 27CFR §21.50

SD alcohol 25-A

Classification: Specially denatured alcohol
Definition: Ethyl alcohol (100 gal) denatured with sol'n. of 20 lb of iodine

USP, 15 lb of potassium iodide USP or sodium iodide USP, and 15 lb of water
Uses: Solvent for tinctures of iodine, external pharmaceuticals (USP or NF)
Regulatory: FDA 27CFR §21.51

SD alcohol 27

Classification: Specially denatured alcohol
Definition: Ethyl alcohol (100 gal) denatured with rosemary oil NF (1 gal) and camphor USP (30 lb)
Uses: Solvent for external pharmaceuticals (not USP or NF), liniments (USP or NF)
Regulatory: FDA 27CFR §21.52

SD alcohol 27-A

Classification: Specially denatured alcohol
Definition: Ethyl alcohol (100 gal) denatured with camphor USP (35 lb) and clove oil NF (1 gal)
Uses: Solvent for external pharmaceuticals (not USP or NF), disinfectants, insecticides, fungicides, other biocides
Regulatory: FDA 27CFR §21.53

SD alcohol 27-B

Classification: Specially denatured alcohol
Definition: Ethyl alcohol (100 gal) denatured with lavender oil NF (1 gal) and medicinal soft soap (green soap USP) (100 lb)
Uses: Solvent for cosmetics, shampoos, external pharmaceuticals (not USP or NF), liniments (USP or NF), disinfectants, insecticides, fungicides, other biocides
Regulatory: FDA 27CFR §20.11, 21.54

SD alcohol 29-3

Classification: Specially denatured alcohol
Definition: Ethyl alcohol denatured with ethyl acetate
Uses: Solvent
Trade Names: SDA-29-3

SD alcohol 29-4

Classification: Specially denatured alcohol
Uses: Solvent
Trade Names: SDA-29-4

SD alcohol 29-5

Classification: Specially denatured alcohol
Definition: Ethyl alcohol denatured with ethyl acetate
Uses: Solvent
Trade Names: SDA-29-5

SD alcohol 29-8

Classification: Specially denatured alcohol
Definition: Ethyl alcohol denatured with ethyl acrylate
Uses: Solvent; solvent for mfg. of pharmaceuticals
Trade Names: SDA-29-8

SD alcohol 29-9

Classification: Specially denatured alcohol
Definition: Ethyl alcohol denatured with diethylamine
Uses: Solvent
Trade Names: SDA-29-9

SD alcohol 29-11

Classification: Specially denatured alcohol
Uses: Solvent
Trade Names: SDA-29-11

SD alcohol 30

Classification: Specially denatured alcohol
Definition: Ethyl alcohol (100 gal) denatured with methyl alcohol (10 gal)
Toxicology: Methanol may modify and increase toxic symptoms caused by ing. and exposure to fumes
Uses: Solvent for coatings (cellulose, syn. resin), plastics, resins, cellulose plastics, photographic film, soldering flux, adhesives, binders, polishes, inks, stains, soaps, bath preps., disinfectants, insecticides, fungicides, sterilizing/preserving sol'ns., household/industrial detergents, photoengraving dyes, duplicating fluids; process solvent for pectin, food prods., crude drugs, vitamins, hormones, yeasts, antibiotics, vaccines, medicinal chems. incl. alkaloids, blood prods., drugs, perfumes/fixatives, photographic chems., other chems.; thinner; solvent, astringent, diluent for cosmetics; raw material for drugs, organo-silicone prods., other chems., syn. resins; refrigeration fluid
Regulatory: FDA 27CFR §20.11, 21.57
Manuf./Distrib.: Eastman
Trade Names: SDA-30

SD alcohol 31-A

Classification: Specially denatured alcohol
Definition: Ethyl alcohol (100 gal) denatured with glycerin USP (100 lb) and hard soap NF (20 lb)
Uses: Solvent for cosmetics, hand/face/body creams/lotions, toothpaste, tooth powd., shampoos
Regulatory: FDA 27CFR §20.11, 21.58

SD alcohol 32

Classification: Specially denatured alcohol
Definition: Ethyl alcohol (100 gal) denatured with ethyl ether (5 gal)
Uses: Solvent for photographic film/emulsions, cellulose intermediates, industrial collodions, collodion USP, inks, dehydration (ethyl cellulose compds.), sterilizing/preserving sol'ns., photoengraving/rotogravure dyes; process solvent for food prods., vitamins, hormones, yeasts, antibiotics, vaccines, medicinal chems. incl. alkaloids; raw material for ethyl chloride, ethyl esters, ethyl ether, ethylene gas, drugs, syn. rubber, etc.
Regulatory: FDA 27CFR 21.59
Trade Names: SDA-32

SD alcohol 33

Classification: Specially denatured alcohol
Definition: Ethyl alcohol (100 gal) denatured with gentian violet or gentian violet USP (30 lb)
Uses: Solvent for inks, external pharmaceuticals (USP or NF)
Regulatory: FDA 27CFR §21.60

SD alcohol 35

Classification: Specially denatured alcohol
Definition: Ethyl alcohol (100 gal) denatured with ethyl acetate 100% ester content (29.75 gal) or equivalent
Uses: Solvent for candy glazes
Regulatory: FDA 27CFR §21.61

SD alcohol 35-A

Classification: Specially denatured alcohol
Definition: Ethyl alcohol (100 gal) denatured with ethyl acetate 100% ester content (4.25 gal) or equivalent
Uses: Solvent for candy glazes; process solvent for pectin, food prods., vitamins, hormones, yeasts, antibiotics, vaccines, medicinal chems. incl. alkaloids, pharmaceuticals, pill mfg., other chems. and prods.; raw material for vinegar, acetic acid, ethyl acetate, other ethyl esters, syn. resins, animal feed supplements
Regulatory: FDA 27CFR §21.62
Trade Names: SDA-35A

SD alcohol 36

Classification: Specially denatured alcohol
Definition: Ethyl alcohol denatured with ammonium hydroxide and sodium hydroxide
Uses: Solvent for cosmetics, shampoos, soaps, bath preps., external pharmaceuticals (not USP or NF), household detergents; raw material for ethylamines, dyes, intermediates
Regulatory: FDA 27CFR §20.11, 21.63
Trade Names: SDA-36

SD alcohol 37

Classification: Specially denatured alcohol
Definition: Ethyl alcohol (100 gal) denatured with eucalyptol NF (45 fl. oz), thymol NF (30 oz), and menthol USP (20 oz)

Uses: Solvent for cosmetics, hair/scalp preps., hand/face/body creams/lotions, mouthwashes, dentifrices, bay rum, external pharmaceuticals (not USP or NF), antiseptics (USP or NF), disinfectants, insecticides, fungicides, other biocides, sterilizing/preserving sol'ns., theater sprays, incense, room deodorants
Regulatory: FDA 27CFR §20.11, 21.64

SD alcohol 38-B

Classification: Specially denatured alcohol
Definition: Ethyl alcohol denatured with one or more of: anethol, anise oil, bay oil (myrica oil), benzaldehyde, bergamot oil, bitter almond oil, camphor, cedar leaf oil, chlorothymol, cinnamal, clove oil, coal tar, eucalyptol, menthol, etc.
Uses: Solvent for cosmetics, shampoos, hair/scalp preps., hand/face/body creams/lotions, body deodorants, perfumes, toilet waters, colognes, soaps, bath preps., bay rum, dentifrices, toothpaste, mouthwashes, external pharmaceuticals (USP or NF), liniments (USP or NF), antiseptics (USP or NF), drug processing, mfg. of pills, disinfectants, insecticides, fungicides, other biocides, sterilizing/preserving sol'ns., theater sprays, incense, room deodorants
Regulatory: FDA 27CFR §20.11, 21.65

SD alcohol 38-C

Classification: Specially denatured alcohol
Definition: Ethyl alcohol (100 gal) denatured with menthol USP (10 lb) and formaldehyde sol'n. USP (1.25 gal)
Uses: Solvent for cosmetics, dentifrices, mouthwashes
Regulatory: FDA 27CFR §20.11, 21.66

SD alcohol 38-D

Classification: Specially denatured alcohol
Definition: Ethyl alcohol (100 gal) denatured with menthol USP (2.5 lb) and formaldehyde sol'n. USP (2.5 gal)
Uses: Solvent for cosmetics, dentifrices, mouthwashes
Regulatory: FDA 27CFR §20.11, 21.67

SD alcohol 38-F

Classification: Specially denatured alcohol
Definition: Ethyl alcohol denatured with boric acid, polysorbate 80, zinc chloride or hydrochloric acid, and at least two of the denaturants listed in SD Alcohol 38-B
Uses: Solvent for cosmetics, mouthwashes, external pharmaceuticals (not USP or NF), antiseptics (USP or NF)
Regulatory: FDA 27CFR §20.11, 21.68

SD alcohol 39

Classification: Specially denatured alcohol
Definition: Ethyl alcohol (100 gal) denatured with 9 lb of sodium salicylate USP or salicylic acid USP, 1.25 gal of fluid extract of quassia NF, and 1/8 gal of t-butyl alcohol
Uses: Solvent for cosmetics, hair/scalp preps., hand/face/body creams/lotions, bay rum, perfumes, toilet waters, colognes
Regulatory: FDA 27CFR §20.11, 21.69

SD alcohol 39-A

Classification: Specially denatured alcohol
Definition: Ethyl alcohol (100 gal) denatured with t-butyl alcohol (1/8 gal) and 60 oz of one of the following: quinine NF, quinine bisulfate NF, quinine dihydrochloride NF, cinchonidine, or cinchonidine sulfate NF or their salts
Uses: Solvent for cosmetics, hair/scalp preps., toilet waters, colognes, shampoos
Regulatory: FDA 27CFR §20.11, 21.70

SD alcohol 39-B

Classification: Specially denatured alcohol
Definition: Ethyl alcohol (100 gal) denatured with t-butyl alcohol (1/8 gal) and diethyl phthalate (2.5 gal)
Uses: Solvent for cosmetics, hair/scalp preps., hand/face/body creams/lotions, body deodorants, perfumes, toilet waters, colognes, shampoos, soaps, bath preps., bay rum, external pharmaceuticals (not USP or NF), disinfectants, insecticides, fungicides, other biocides, household detergents, theater sprays, incense, room deodorants
Regulatory: FDA 27CFR §20.11, 21.71

SD alcohol 39-C

Classification: Specially denatured alcohol
Definition: Ethyl alcohol (100 gal) denatured with diethyl phthalate (1 gal)
Uses: Solvent for hair/scalp preps., hand/face/body creams/lotions, body deodorants, perfumes, toilet waters, colognes, soaps, bath preps., theater sprays, incense, room deodorants; solvent, astringent, diluent for cosmetics
Regulatory: FDA 27CFR §20.11, 21.72
Manuf./Distrib.: Eastman
Trade Names: SDA-39C

SD alcohol 39-D

Classification: Specially denatured alcohol
Definition: Ethyl alcohol (100 gal) denatured with bay oil (myrica oil) NF (1 gal) and either 50 oz of quinine sulfate USP or sodium salicylate USP
Uses: Solvent for cosmetics, hair/scalp preps., bay rum
Regulatory: FDA 27CFR §20.11, 21.73

SD alcohol 40

Classification: Specially denatured alcohol
Definition: Ethyl alcohol (100 gal) denatured with t-butyl alcohol (1/8 gal) and 1.5 oz of either brucine alkaloid, brucine sulfate NF, quassin, or combination thereof
Uses: Mfg. of acetaldehyde and other chems.; solvent for hair/scalp preps., hand/face/body creams/lotions, body deodorants, perfumes, toilet waters, colognes, shampoos, soaps, bath preps., external pharmaceuticals (not USP or NF), disinfectants, insecticides, fungicides, other biocides, household detergents, theater sprays, incense, room deodorants; solvent, astringent, diluent in cosmetics; thickener, solidifier, liquefier in pharmaceuticals; antifreeze; brake fluids; fuels
Regulatory: FDA 27CFR §20.11, 21.74
Manuf./Distrib.: Eastman
Trade Names: SDA-40

SD alcohol 40-A

Classification: Specially denatured alcohol
Definition: Ethyl alcohol (100 gal) denatured with t-butyl alcohol (1/8 gal) and sucrose octaacetate (1 lb)
Uses: Solvent for hair/scalp preps., hand/face/body creams/lotions, body deodorants, perfumes, toilet waters, colognes, soaps, bath preps., bay rum, external pharmaceuticals (not USP or NF), disinfectants, insecticides, fungicides, other biocides, household detergents, theater sprays, incense, room deodorants; solvent, astringent, diluent for cosmetics
Regulatory: FDA 27CFR §20.11, 21.75
Manuf./Distrib.: Eastman
Trade Names: SDA-40A

SD alcohol 40-B

Classification: Specially denatured alcohol
Definition: Ethyl alcohol (100 gal) denatured with denatonium benzoate (1/16 oz) and t-butyl alcohol (1/8 gal)
Uses: Mfg. of acetaldehyde and other chems.; solvent for hair/scalp preps., hand/face/body creams/lotions, body deodorants, perfumes, toilet waters, colognes, shampoos, soaps, bath preps., external pharmaceuticals (not USP or NF), bay rum, disinfectants, insecticides, fungicides, other biocides, household detergents, theater sprays, incense, room deodorants; thickener, solidifier, liquefier in pharmaceuticals; solvent, astringent, diluent in cosmetics; antifreeze; brake fluids; fuels
Regulatory: FDA 27CFR §20.11, 21.76
Manuf./Distrib.: Eastman
Trade Names: SDA-40B

SD alcohol 40-C

Classification: Specially denatured alcohol
Definition: Ethyl alcohol (100 gal) denatured with t-butyl alcohol (3 gal)
Uses: Solvent for hair/scalp preps., hand/face/body creams/lotions, body

deodorants, perfumes, toilet waters, colognes, shampoos, soaps, bath preps., external pharmaceuticals (not USP or NF), disinfectants, insecticides, fungicides, other biocides, household detergents, bay rum, theater sprays, incense, room deodorants; solvent, astringent, diluent for cosmetics
Regulatory: FDA 27CFR §20.11, 21.77
Manuf./Distrib.: Eastman
Trade Names: SDA-40C

SD alcohol 42

Classification: Specially denatured alcohol
Definition: Ethyl alcohol (100 gal) denatured with (1) potassium iodide USP (80 g) and red mercuric iodide NF (109 g), or (2) thimerosal USP (95 g) or (3) 76 g of any of: phenyl mercuric nitrate NF, phenyl mercury chloride NF, or phenyl mercury benzoate
Uses: Solvent for sterilizing/preserving sol'ns.
Regulatory: FDA 27CFR §21.78

SD alcohol 44

Classification: Specially denatured alcohol
Definition: Ethyl alcohol (100 gal) denatured with n-butyl alcohol (10 gal)
Uses: Solvent for sterilizing/preserving sol'ns.
Regulatory: FDA 27CFR §21.79

SD alcohol 45

Classification: Specially denatured alcohol
Definition: Ethyl alcohol (100 gal) denatured with refined white or orange shellac (300 lb)
Uses: Solvent for candy glazes
Regulatory: FDA 27CFR §21.80

SD alcohol 46

Classification: Specially denatured alcohol
Definition: Ethyl alcohol (100 gal) denatured with phenol USP (25 fl. oz) and wintergreen oil (methyl salicylate NF) (4 fl. oz)
Uses: Solvent for cosmetics, in antiseptic, sterilizing, and bath sol'ns. having restricted use
Regulatory: FDA 27CFR §20.11, 21.81

SDS. *See* Sodium lauryl sulfate
Sebacic acid, bis (2-ethylhexyl) ester. *See* Dioctyl sebacate
Sebacic acid, dimethyl ester. *See* Dimethyl sebacate
Secondary butyl alcohol. *See* 2-Butanol
Secondary caprylic alcohol. *See* 2-Octanol
Secondary propyl alcohol. *See* Isopropyl alcohol
Sesame oil; Sesame seed oil. *See* Sesame (Sesamum indicum) oil

Sesame (Sesamum indicum) oil

CAS 8008-74-0; EINECS/ELINCS 232-370-6
Synonyms: Benne oil; Gingelly oil; Gingilli oil; Sesame oil; Sesame seed oil; Sesamum indicum; Sesamum indicum oil; Teel oil
Definition: Refined fixed oil obtained from the seeds of *Sesamum indicum*
Properties: Bland ylsh. oily liq., pract. odorless, bland taste; sl. sol. in alcohol; misc. with ether, chloroform, hexane, carbon disulfide; negligible sol. in water; dens. 0.916-0.921; vapor pressure very low; iodine no. 103-116; sapon. no. 188-195; flash pt. 491 F; ref. index 1.4575-1.4598 (60 C)
Toxicology: LD50 (IP, mouse) > 50 g/kg, (IV, rabbit) 678 μg/kg; poison by IV route; primary irritant; human skin irritant; may cause allergic reactions, primarily contact dermatitis; questionable carcinogen; experimental tumorigen; TSCA listed
Precaution: Combustible exposed to heat or flame
Hazardous Decomp. Prods.: CO_2 from combustion; heated to decomp., emits acrid smoke and irritating fumes
NFPA: Health 0, Flammability 1, Reactivity 0
Storage: Light-sensitive
Uses: Emollient in cosmetics, lotions, injectables; natural flavoring agent in foods; mfg. of oleomargarine, iodized oil; pharmaceutical solvent, vehicle, keratin softener; solvent for steroids; in food-pkg. adhesives; drying oil in food-contact coatings; in resinous/polymeric food-contact coatings; defoamer in food-contact paper coatings, paper/paperboard
Features: Oleaginous
Regulatory: FDA 21CFR §175.105, 175.300, 176.200, 176.210; FDA approved for parenterals, orals, topicals; NF, JP compliance; BP, EP compliance (refined)
Manuf./Distrib.: A&E Connock; Aarhus Oliefabrik A/S; Adept Sol'ns.; Alban Muller; Aldivia; Aldrich; Alfa Chem; Allchem Int'l. Ltd; Anglia Oils; Arista Ind.; Asiamerica Int'l.; Avatar; Barrington; Blagden Spec. Chems. Ltd; Cascade; Charkit; Chart; Chesham Chems. Ltd; CoKEM Assoc.; Cornelius Chem. Co. Ltd; Cosmetic Supplies USA; Croda Inc; Desert Whale Jojoba; Efkay Chems. Ltd; Eggar & Co.; Fanning; Fluka; Hawkins Chem.; Helm NY; Int'l. Sourcing; Jarchem Ind.; Jeen Int'l.; KIC Chems.; Lipo; Mosselman NV; Mutchler; Parchem Trading; Penta Mfg.; Pokonobe Ind.; Protameen; Quest Int'l.; Ruger; Seltzer; Sigma; Spectrum Quality Prods.; Takasago Int'l.; Thornley; Tri-K Ind.; Univar E&E Ltd; Universal Preserv-A-Chem; Vitamins, Inc; Voigt Global Distrib.; Vopak USA; Welch, Holme & Clark
Trade Names: Cropure® Sesame; EmCon™ SES; Lipovol SES; Sesame Oil USP/NF 16; Super Refined® Sesame NF

Sesamol

CAS 533-31-3; EINECS/ELINCS 208-561-5
Synonyms: 1,3-Benzodioxol-5-ol; 5-Hydroxy-1,3-benzodioxole; 3,4-Methylenedioxyphenol; Methylene ether of oxyquinone; Phenol, 3,4-(methylenedioxy); 1,2,4-Trihydroxybenzene-1,2-methylene ether
Definition: Natural component of sesame seed oil
Empirical: $C_7H_6O_3$
Properties: Colorless to brn. cryst. solid; may be sol. in water, ethanol, MEK; m.p. 63-65 C
Toxicology: LD50 (IP, mouse) 470 mg/kg; may be harmful by inh., ing., or skin absorp.; irritating to eyes, skin, mucous membranes, upper respiratory tract; prolonged contact can cause eye damage, severe irritation or burns; was found to reduce activity in the hexobarbital sleeping time assay; TSCA listed
Precaution: Incompat. with strong oxidizing agents
Hazardous Decomp. Prods.: Toxic fumes of CO, CO_2; emits toxic fumes under fire conditions
Storage: Store in cool, dry place; keep tightly closed
Uses: Solvent, vehicle for fat-sol. substances in pharmaceuticals, foods, cosmetics, iodized oils, liniments, ointments, oleomargarine; antioxidant in sesame oil
Manuf./Distrib.: Acros Org.; Advanced Synthesis Tech.; Aldrich; Alfa Aesar; Davos Chem.; Fluka; Hokko Chem. Ind.; Indofine; Lancaster Synthesis; Pfaltz & Bauer; Sigma; Spectrum Quality Prods.; Syntai Chems. & Pharm.; TCI Am.; Ube Ind. Am./Fine Chems.

Sesamum indicum; Sesamum indicum oil. *See* Sesame (Sesamum indicum) oil
Severe acid-treated middle distillate. *See* Petroleum distillates, acid-treated middle
Severely hydrotreated light distillates. *See* Petroleum distillates, hydrotreated light
Shea butter. *See* Shea butter (Butyrospermum parkii)

Shea butter (Butyrospermum parkii)

CAS 68424-60-2; 977026-99-5; EINECS/ELINCS 293-515-7
Synonyms: Butyrospermum parkii; Butyrospermum parkii butter; Butyrospermum parkii fruit; Galam butter; Karite butter; Shea butter; Shea butter fruit
Definition: Natural fat obtained from fruit of the Karite tree, *Butyrospermum parkii*
Properties: Gray-wh. solid; dens. 0.9175; iodine no. 53-65; sapon. no.178-190
Uses: Emollient, consistency agent, lubricant, moisturizer, vehicle, carrier, visc. modifier, skin protectant, fatting agent for lotions, pharmaceutical dermatologicals, suppositories, o/w and w/o creams and emulsions; skin conditioner, occlusive agent, solvent for suntan preps., body lotions, soaps, shampoos
Manuf./Distrib.: ABITEC; Alban Muller; Desert Whale Jojoba; Fanning; Int'l. Sourcing; Jarchem Ind.; Mosselman NV; Phoenix Nat. Prods.; Protameen; R.I.T.A.; Spec. Ingreds. & Perfume; Terry Labs; Thornley; Tri-K Ind.; Universal Preserv-A-Chem
Trade Names: Fancol™ Karite Butter

Shea butter fruit. *See* Shea butter (Butyrospermum parkii)

Silane, oxybis (trimethyl-. *See* Hexamethyldisiloxane
Silane, tetramethyl-. *See* Tetramethylsilane
Silicic acid, aluminum sodium salt. *See* Sodium silicoaluminate

Silicone

Synonyms: Organosiloxane; Organosiloxanes; Polymerized siloxane; Polyorganosiloxanes; Silicone polymer; Siloxane; Siloxanes
Classification: Organosilicon compd.; thermosetting siloxane polymer
Definition: Family of syn. polymers contg. a repeating silicon-oxygen backbone with org. side groups attached via carbon-silicon bonds; classified as fluids, resins, and elastomers
Formula: $[—OSiR_2—]_n$, R not equal to H
Properties: Liq., semisolid, or solid; cis 1 to >1,000,000 cs; water repellent; sol. in most organic solvents
Toxicology: Silicone-related diseases can occur, e.g., silicone-induced synovitis and lymphadenopathy, acute and chronic pneumonitis from 'bleeding' from ruptured bag-gel breast implants, pulmonary lesions, granulomatous reactions
Precaution: Unhalogenated types combustible
Uses: Mold release for plastics and rubber; defoamer for mining, latex, ink, soaps, agric., food processing, pulp/paper, effluent, antifreeze, textiles; wetting agent; surfactant; hydraulic and dielec. oils; plasticizer; textile surf. treatment; detergents; adhesives; caulking compds.; water repellents for textile, paper, masonry, concrete; damping fluids; lubricant, conditioner, and emollient in personal care prods.; molding compds.; encapsulants; medical implants; dental molds; contact lenses; release agent in food-contact coatings; food pkg.; food processing
Features: Thermal and oxidative stability; chemically inert; resist. to weathering; good dielec. str.; low surf. tens.
Regulatory: FDA 21CFR §175.300
Manuf./Distrib.: Ashland; Bayer; Chemcentral; Clariant; Crucible; D.N. Lukens; Degussa AG/Aerosil & Silanes; Dow Corning; Dow; GE Silicones; Genesee Polymers; Goldschmidt; Harcros; Reichhold; Rhodia HPCII; Ruger; Seegott; Shin-Etsu; Stanchem; Tego GmbH; Transene; Veckridge; Vopak USA; Wacker Silicones; Wacker-Chemie GmbH
Trade Names Containing: Hedislide RS
See also Cyclomethicone; Dimethicone

Silicone polymer; Siloxane; Siloxanes. *See* Silicone
SLS. *See* Sodium lauryl sulfate
Soda lye. *See* Sodium hydroxide
Sodium aluminosilicate; Sodium aluminum silicate. *See* Sodium silicoaluminate
Sodium dimethylbenzenesulfonate. *See* Sodium xylenesulfonate
Sodium dodecyl sulfate; Sodium n-dodecyl sulfate. *See* Sodium lauryl sulfate
Sodium etasulfate; Sodium (2-ethylhexyl) alcohol sulfate. *See* Sodium 2-ethylhexyl sulfate

Sodium 2-ethylhexyl sulfate

CAS 126-92-1; EINECS/ELINCS 204-812-8
Synonyms: 2-Ethyl-1-hexanol, hydrogen sulfate, sodium salt; 2-Ethyl-1-hexanol sulfate sodium salt; 2-Ethylhexyl sodium sulfate; Mono (2-ethylhexyl) sulfate sodium salt; Sodium etasulfate; Sodium (2-ethylhexyl) alcohol sulfate; Sulfuric acid, mono (2-ethylhexyl) ester sodium salt
Definition: Sodium salt of 2-ethylhexyl sulfate
Empirical: $C_8H_{17}O_4S \cdot Na$
Properties: M.w. 232.28; anionic
Toxicology: LD50 (oral, rat) 4 g/kg, (oral, mouse) 1550 mg/kg; poison by IP route; mod. toxic by ing., skin contact; skin and eye irritant; TSCA listed
Hazardous Decomp. Prods.: Heated to decomp., emits very toxic fumes of SO_x and Na_2O
Uses: Wetting agent for electroplating baths, alkaline textile processing aids, industrial cleaners; coemulsifier for polymerization; surfactant, detergent, wetting agent, emulsifier, penetrant, stabilizer for cosmetics, pharmaceuticals, textiles, household and industrial cleaners, metal cleaning, paints, plastics, rubber, food pkg. and processing, adhesives; food-pkg. adhesives; in paper/paperboard in contact with aq./fatty foods; washing/lye peeling of fruits and vegetables
Features: Low-foaming
Regulatory: FDA 21CFR §173.315, 175.105, 176.170
Manuf./Distrib.: Chemron; Dai-ichi Karkaria; Niacet; Rhodia HPCII; Rhodia/ Phosphorus Perf. Derivs.; Sigma
Trade Names Containing: DA-425

Sodium feldspar. *See* Sodium silicoaluminate
Sodium hydrate. *See* Sodium hydroxide

Sodium hydroxide

CAS 1310-73-2; EINECS/ELINCS 215-185-5
UN 1823 (DOT); 1824 (DOT); INS524; E524
Synonyms: Caustic soda; Lye; Soda lye; Sodium hydrate; White caustic
Classification: Alkali hydroxide
Empirical: HNaO
Formula: NaOH
Properties: Wh. solid beads or pellets; absorbs water and CO_2 from air; sol. in water, alcohol, methanol, glycerin; m.w. 40.00; dens. 2.12 (20/4 C); vapor pressure 1 mm (739 C); m.p. 318 C; b.p. 1390 C
Toxicology: ACGIH TLV/CL 2 mg/m^3 of air; LD50 (IP, mouse) 40 mg/kg; LDLo (oral, rabbit) 500 mg/kg; poison by IP route; mod. toxic by ing.; corrosive irritant to eyes, skin, mucous membrane; mists and dusts cause small burns; inh. can damage upper respiratory tract and lungs; mutagenic data; TSCA listed
Precaution: DOT: Corrosive material; strong base; incompat. with acids and water; may ignite or react violently with many org. compds.; dangerous material to handle
Hazardous Decomp. Prods.: Heated to decomp., emits toxic fumes of Na_2O
NFPA: Health 3, Flammability 0, Reactivity 1
Storage: Deliq.
Uses: Chemical mfg.; rayon; cellophane; neutralizing agent in petrol. refining; textile processing; pulp and paper bleaching; deinking and water treatment at pulp mills; soaps; vegetable oil refining reagent; reclaiming rubber; solvent for colorants; processing aid in petrol. mfg.; prod. of alumina; etching and electroplating; boiler water treatment, coagulation, corrosion/scale control, pH control, softening; industrial cleaner ingred.; ion-exchange resin regeneration; dyes; pharmaceutical alkalizer, buffer; food additive; buffer, denaturant in cosmetics; pH control agent in foods, cosmetics; dispersant for suspended material such as silt and metal oxides
Regulatory: FDA 21CFR §73.85, 155.191, 155.194, 163.110, 163.111, 163.112, 172.560, 172.814, 172.892, 173.310, 176.170, 176.180, 176.210, 177.1600, 177.2800, 184.1763, GRAS; USDA 9CFR §318.7, 381.147; BATF 27CFR §21.101, 240.1051a; Japan restricted; Europe listed; UK approved; FDA approved for injectables, dentals, parenterals, inhalants, ophthalmics, orals, rectals, topicals, vaginals; NF, BP, EP compliance; SARA reportable
Manuf./Distrib.: AAE Chemie NV; AMRESCO; Advance Research Chems.; Akzo Nobel; Aldrich; Alfa Chem; Allied Universal; Altivia; Am. Biorganics; Arch Chems.; Asahi Chem. Ind.; Asahi Denka Kogyo; Ashland; Asiamerica Int'l.; Atofina; Avrachem AG; BCH Brühl; Bay Chem. & Supply; Bayer; Bilt Chems. Ltd; Boith China; Brenntag AG; CXY Chems.; Chem. Formulators; DPC Ind.; Degussa AG/Health & Nutrition; Delta Distributors; Dow; E + E (US); EM Ind.; EM Science; Eastech; FMC; Fluka; Formosa Plastics; GE Plastics; GFS; GMI Prods.; Galbraith Labs; General Chem.; George Mann; Georgia Gulf; Georgia-Pacific Resins; HCl Ind. Chem. & Supply; HCl Worth; HVC; Harcros; Hawkins Chem.; HoltraChem; Hydrite; Hydro Plus; Integra; Interstate Chem.; J.T. Baker; Jiangxi Liq. Ind. Chems.; K.A. Steel; Kuehne Chem.; LaRoche Ind.; Mallinckrodt Baker; Milsolv; Nissan Chem. Ind.; Norsk Hydro AS; Occidental; Olin/Chlor Alkali; P.B. & S.; PCI Chems. Canada; PPG Ind.; PVS; Phoenix Nat. Prods.; Pioneer Am.; Primachem; Rasa Ind.; Ruger; Russ Chem.; SOCO-Lynch; Sal Chem.; Schaefer Tech.; Sierra; Sigma; Solvay Mins.; Southchem; Southern Wisconsin Chems.; Spectrum Quality Prods.; Stanchem; Textile Chem.; Thomas Scientific; Universal Preserv-A-Chem; VWR Int'l.; Varsal Instruments; Vertex; Voigt Global Distrib.; Vopak USA; Vulcan Perf. Chems.; Westlake CA & O; Weyerhaeuser; Whittaker, Clark & Daniels

Sodium iodide

CAS 7681-82-5; EINECS/ELINCS 231-679-3
Synonyms: Sodium iodine; Sodium monoiodide
Classification: Inorganic salt
Empirical: INa

Formula: NaI
Properties: Wh. or colorless cryst. or powd.; discolored by air; very sol. in water; sol. in alcohol and glycerin; m.w. 149.89; dens. 3.667; vapor pressure 1 mm Hg (767 C); m.p. 651 C; b.p. 1300 C
Toxicology: LD50 (oral, rat) 4340 mg/kg, (IV, rat) 1060 mg/kg, (IP, mouse) 430 mg/kg; mod. toxic by ing., IV, IP routes; primary skin and eye irritant; human teratogenic and reproductive effects by ing.; TSCA listed
Precaution: Reacts violently with BrF_3, $HClO_4$, oxidants
Hazardous Decomp. Prods.: Heated to decomp., emits toxic fumes of I^- and Na_2O
HMIS: Health 1, Flammability 0, Reactivity 1
Storage: Hygroscopic, deliq.; store @ R.T.
Uses: Photographic emulsions; solvent for iodine; in denatured alcohol; organic chemicals; iodination reagent; animal feed supplement; cloud seeding; scintillation; antimicrobial in cosmetics; pharmaceuticals (expectorant; antiseptic in topical anti-infective prods.)
Regulatory: FDA approved for parenterals, topicals, vaginals; BP, EP compliance
Manuf./Distrib.: AMRESCO; Advance Research Chems.; Ajay; Aldrich; Alfa Chem; Allan; Am. Int'l.; Asiamerica Int'l.; Atomergic Chemetals; Barrington; Calabrian; Charkit; Chemetall Chem. Prods.; Clariant; EM Ind.; EM Science; Fluka; GFS; Helm NY; Integra; Interchem; J.T. Baker; Mallinckrodt Baker; Miljac; Mutchler; Penta Mfg.; R.W. Greeff; Research Organics; Robeco; Ruger; Sigma; Spectrum Quality Prods.; Thomas Scientific; Tomen Am.; Universal Preserv-A-Chem; VWR Int'l.; Voigt Global Distrib.; Vopak USA; Wego Chem. & Min.; West Agro; Wm. Blythe Ltd; Xinchem
Trade Names Containing: Punctilious® SDA 25A-2 190 Proof

Sodium iodine. *See* Sodium iodide
Sodium isothiocyanate. *See* Sodium thiocyanate

Sodium lauryl sulfate

CAS 151-21-3; 68585-47-7; 68955-19-1; EINECS/ELINCS 205-788-1; 271-557-7; 273-257-1
INS487
Synonyms: Dodecyl alcohol, hydrogen sulfate, sodium salt; Dodecyl sodium sulfate; Dodecylsulfate sodium salt; Lauryl sodium sulfate; Lauryl sulfate sodium salt; Monododecyl sodium sulfate; SDS; SLS; Sodium dodecyl sulfate; Sodium n-dodecyl sulfate; Sodium monododecyl sulfate; Sulfuric acid monododecyl ester sodium salt
Classification: Alkyl sulfate salt
Definition: Sodium salt of lauryl sulfate
Empirical: $C_{12}H_{25}O_4S \cdot Na$
Formula: $CH_3(CH_2)_{10}CH_2OSO_3Na$
Properties: Wh. to cream crystals, flakes, or powd., faint fatty odor; mod. sol. in water; m.w. 288.38; sp.gr. > 1.1 (20 C); m.p. 204-207 C; anionic
Toxicology: LD50 (oral, rat) 1288 mg/kg, (IP, rat) 210 mg/kg, (IV, rat) 118 mg/kg; poison by IV, IP routes; mod. toxic by ing.; human skin irritant; experimental eye, severe skin irritant; mild allergen; experimental teratogen, reproductive effects; mutagenic data; TSCA listed
Precaution: Burns above 93.3 C; incompat. with strong oxidizers (increases fire risk)
Hazardous Decomp. Prods.: Thermal decomp. prods.: CO, CO_2, SO_2, H_2S; heated to decomp., emits toxic fumes of SO_x and Na_2O
HMIS: Health 1, Flammability 0, Reactivity 0
Storage: Store in cool, dry, well-ventilated area, out of direct sunlight, away from incompat. materials
Uses: Detergent; surface tension depressant; wetting agent in textiles, toothpaste; emulsifier for emulsion polymerization (vinyl chloride, S/B, acrylic), paints, coatings, printing inks, fats; flotation collector; denaturant, emulsifier, surfactant in cosmetics; lubricant; pearlescent; visc. builder; penetrant; food additive (thickener, emulsifier, wetting agent, whipping agent, hog/poultry scald agent); dispersant in creams, lotions, medical preps.; foaming agent (foamed rubber/carpetbacking latex); electrophoretic separation of proteins and lipids; surfactant in food-contact coatings; in food-pkg. adhesives; in paper/paperboard in contact with aq./fatty foods; defoamer in food-contact paper/paperboard; in cellophane for food pkg.; in closure-sealing gaskets for food containers
Use Level: 0.004-0.6 mg (oral solids); 0.01-0.02% (oral liqs.); 0.1-12.7% (topical pharmaceuticals)
Regulatory: FDA 21CFR §172.210, 172.822, 175.105, 175.300, 175.320, 176.170, 176.180, 176.210, 177.1200, 177.1210, 177.1630, 177.2600, 177.2800, 178.1010, 178.3400, 179.45; USDA 9CFR §318.7, 381.147; FDA approved for dentals, orals, topicals, vaginals; NF, BP, EP compliance
Manuf./Distrib.: AMRESCO; Accurate Chem. & Scientific; Aceto; Acme-Hardesty; Akzo Nobel; Aldrich; Alfa Chem; Am. Biorganics; Ashland; Blagden Spec. Chems. Ltd; Chemacon GmbH; Chemron; Clariant; Cognis/Chems. Group; Costec; DeForest Enterprises; Degussa AG; DuPont; EM Ind.; EM Science; Fisher Scientific; Fluka; GMI Prods.; Integra; J.T. Baker; Kraft Chem.; Lonza; Magnesia GmbH; Merck KGaA; Monomer-Polymer & Dajac Labs; Mutchler; Norman, Fox; Patco Chem. Ind.; Penta Mfg.; Phoenix Nat. Prods.; Pilot; Reliable Biopharmaceutical; Rhodia HPCII; Rhodia UK; Rochem Int'l.; Ruger; Sea-Land; Sigma; Spectrum Quality Prods.; Stanchem; Stepan; Thornley; U.S. Synthetics; Ubichem plc; Unger Fabrikker AS; Universal Preserv-A-Chem; VWR Int'l.; Vivion; Voigt Global Distrib.; Vopak USA
Trade Names Containing: Protachem™ GMS-AS

Sodium p-methylbenzenesulfonate. *See* Sodium toluenesulfonate
Sodium monododecyl sulfate. *See* Sodium lauryl sulfate
Sodium monoiodide. *See* Sodium iodide
Sodium paratoluene sulfonate. *See* Sodium toluenesulfonate
Sodium rhodanate; Sodium rhodanide. *See* Sodium thiocyanate

Sodium silicoaluminate

CAS 1344-00-9; EINECS/ELINCS 215-684-8
INS554
Synonyms: Aluminosilicic acid, sodium salt; Aluminum sodium silicate; Silicic acid, aluminum sodium salt; Sodium aluminosilicate; Sodium aluminum silicate; Sodium feldspar; Zeolite; Zeolites
Definition: Series of hydrated sodium aluminum silicates; produced by reaction of sodium silicate and kaolinite clay
Formula: $Na_2O : Al_2O_3 : SiO_2$ with mole ratio ≈ 1:1:13.2
Properties: Wh. fine amorphous powd. or beads, odorless and tasteless; insol. in water, alcohol, org. solvs.; partly sol. in strong acids and alkali hydroxides @ 80-100 C; pH 6.5-10.5 (20% slurry)
Toxicology: Irritant to skin, eyes, mucous membranes; mutagen; TSCA listed
Precaution: Noncombustible
Hazardous Decomp. Prods.: Heated to decomp., emits toxic fumes of Na_2O
Uses: Anticaking agent in detergents, desiccants, foods; reinforcing filler for rubbers; extender for paints; solvent; detergent builder; gas separation; white pigment in paper, paints, plastics; ion exchange and selective absorp./adsorp.; removes hardness ions from wash water; abrasive, visc. control agent in cosmetics, pharmaceuticals; in coatings for paper/paperboard in contact with aq./fatty foods
Regulatory: FDA 21CFR §133.146, 160.105, 160.185, 176.170, 182.2727 (2% max.), 582.2727, GRAS; Europe listed; UK approved; FDA approved for orals
Manuf./Distrib.: Albemarle; Am. Colloid; Am. Int'l.; Fluka; INEOS Silicas Am.; J.M. Huber/Chems.; Kraft Chem.; PPG Ind.; PQ; Punda Mercantile; Sigma; Spectrum Quality Prods.; Stanchem; Voigt Global Distrib.
Trade Names: EZA®

Sodium sulfocyanate; Sodium sulfocyanide. *See* Sodium thiocyanate

Sodium thiocyanate

CAS 540-72-7; EINECS/ELINCS 208-754-4
Synonyms: Sodium isothiocyanate; Sodium rhodanate; Sodium rhodanide; Sodium sulfocyanate; Sodium sulfocyanide; Sodium thiocyanide; Thiocyanate sodium; Thiocyanic acid, sodium salt
Classification: Inorganic compd.
Empirical: CNNaS
Formula: NaSCN
Properties: Colorless cryst.; sol. in water, alcohol, acetone; m.w. 81.07; m.p. 287 C
Toxicology: LD50 (oral, rat) 764 mg/kg, (IP, rat), 540 mg/kg, (IV, mouse) 484 mg/kg; poison by ing., IV, subcut. routes; mod. toxic by IP route; irritant; ing. of large doses may cause vomiting, convulsions; chronic exposure causes weakness, confusion, diarrhea, skin rashes; TSCA listed

Hazardous Decomp. Prods.: Heated to decomp., emits very toxic fumes of NO_x, SO_x, Na_2O
Storage: Hygroscopic, deliq.; light-sensitive
Uses: Intermediate for agric. chems., slimicides for paper, pharmaceuticals; brightener in electroplating; analytical reagent; photographic emulsion stabilizer; textile dye and printing assistant; fiber swelling agent; in food-pkg. adhesives; acrylic fiber spinning solvent additive; corrosion inhibitor for water treatment; solvent for polyacrylates, textile baths; extraction solvent for metals; mfg. of other thiocyanate salts and artificial mustard oil; medicine (antihypertensive); hair conditioner in cosmetics
Regulatory: FDA 21CFR §175.105
Manuf./Distrib.: AMRESCO; Akzo Nobel bv; Aldrich; Alemark; Alfa Chem; Allan; Allchem Ind.; Am. Int'l.; Amber Syn.; Austin; Barker Ind.; Contract Chems. Ltd; Degussa; EM Ind.; EM Science; Fluka; GFS; Hasnil; Integra; J.T. Baker; Mallinckrodt Baker; Pacific West; Pechiney Chems. Div.; Penta Mfg.; Richman; Sanko Kagaku; Sigma; Spectrum Quality Prods.; Thomas Scientific; Toyo Kasei Kogyo; VWR Int'l.; Vopak USA; Westco

Sodium thiocyanide. *See* Sodium thiocyanate

Sodium toluenesulfonate

CAS 657-84-1; 12068-03-0; EINECS/ELINCS 235-088-1
Synonyms: Benzenesulfonic acid, 4-methyl-, sodium salt; Methylbenzenesulfonic acid, sodium salt; 4-Methylbenzenesulfonic acid, sodium salt; Sodium p-methylbenzenesulfonate; Sodium paratoluene sulfonate; Sodium toluene-4-sulfonate; Sodium p-toluenesulfonate; Sodium-p-tolyl sulfonate; Sodium tosylate; STS; 4-Toluenesulfonic acid sodium salt; p-Toluenesulfonic acid, sodium salt
Classification: Substituted aromatic compd.
Definition: Mixed prod. contg. typically 80% toluene-4-sulfonate, 15% toluene-2-sulfonate, and 5% toluene-3-sulfonate
Empirical: $C_7H_7O_3S \cdot Na$
Formula: $CH_3C_6H_4SO_3Na$
Properties: Crystals; sol. in water forming acidic sol'ns.; m.w. 194.19; anionic
Toxicology: LD50 (IV, mouse) 1700 mg/kg; mod. toxic by IV route; TSCA listed
Hazardous Decomp. Prods.: Heated to decomp., emits toxic fumes of SO_x and Na_2O
Uses: Hydrotrope, solubilizer, coupling agent, detergent, cloud pt. depressant, processing aid for detergents, industrial processes, shampoos, aerosols, cutting oils, glues, textile finishing; surfactant in cosmetics; solvent; visc. modifier; antiblocking agent, anticaking agent for powds.; dye chemistry; hydrotrope, stabilizer, solubilizer for inks, electroplating baths, dyestuffs, polymers
Regulatory: FDA 21CFR §178.1010
Manuf./Distrib.: Alchemie USA; Aldrich; Ashland; Charkit; Degussa AG; Fabrichem; Fluka; H & C Ind.; Howard Hall; Intertrade Holdings; Maypro Ind.; Monomer-Polymer & Dajac Labs; Rit-Chem; Rütgers Org.; TCI Am.

Sodium toluene-4-sulfonate; Sodium p-toluenesulfonate; Sodium-p-tolyl sulfonate; Sodium tosylate. *See* Sodium toluenesulfonate

Sodium xylenesulfonate

CAS 1300-72-7; EINECS/ELINCS 215-090-9
Synonyms: Dimethylbenzene sulfonic acid, sodium salt; Sodium dimethylbenzenesulfonate; SXS; Xylenesulfonic acid, sodium salt
Definition: Sodium salt of ring sulfonated mixed xylene isomers
Empirical: $C_8H_9O_3S \cdot Na$
Formula: $(CH_3)_2C_6H_3SO_3Na$
Properties: Lt. yel. liq.; m.w. 208.21; sp.gr. 1.23; m.p. 27 C; b.p. 157 C; pH 7-10 (3%); anionic
Toxicology: No known toxicity; irritant; TSCA listed
NFPA: Health 1, Flammability 0, Reactivity 0
Uses: Hydrotrope, solubilizer, coupling agent, processing aid, solvent in heavy-duty detergents, shampoos, degreasing compds., industrial processes, aerosols, cutting oils, glues, dyestuffs, textile finishing, petrol. industry, pulping, printing pastes; extract solvent for pentosans and lignin in paper industry; additive for glues in leather industry; surfactant in cosmetics; stabilizer; dispersant; visc. reducer; antiblocking agent, anticaking agent in powd. prods.; pharmaceutical solubilizer; in food-pkg. adhesives; in paper/paperboard in contact with dry food
Regulatory: FDA 21CFR §175.105, 176.180; FDA approved for topicals
Manuf./Distrib.: Aldrich; Ashland; Browning; Burlington Chem.; Fluka; Howard Hall; Huntsman; J.F. Henry; Pilot; R.W. Greeff; Rütgers Org.; Stanchem; Stepan
Trade Names: Pilot® SXS-40
Trade Names Containing: DA-425

Soluble sulfur. *See* Sulfur
Solvent ether. *See* Ethyl ether
Solvent naphtha. *See* VM&P naphtha; Stoddard solvent
Solvent naphtha, heavy aliphatic. *See* Naphtha, heavy aliphatic
Solvent naphtha, heavy aromatic. *See* Naphtha, heavy aromatic
Solvent naphtha (peroleum), medium heavy, aliphatic hydrocarbons. *See* Naphtha, medium aliphatic
Solvent naphtha (petroleum), heavy aliphatic; Solvent naphtha (petroleum), heavy aliphatic C11-16. *See* Naphtha, heavy aliphatic
Solvent naphtha (petroleum), heavy aromatic. *See* Naphtha, heavy aromatic
Solvent naphtha (petroleum), light aliphatic. *See* Naphtha, light aliphatic
Solvent naphtha (petroleum), light aromatic. *See* Naphtha, light aromatic
Solvent naphtha (petroleum), medium aliphatic. *See* Naphtha, medium aliphatic
Solvent-refined light naphtha (petroleum). *See* Naphtha, solvent-refined light

Sorbeth-20

CAS 53694-15-8 (generic)
Synonyms: PEG-20 sorbitol ether; PEG 1000 sorbitol ether; POE (20) sorbitol ether
Definition: PEG ether of sorbitol with an avg. 20 moles EO
Properties: Nonionic
Uses: Solvent, emulsifier, humectant in cosmetics, pharmaceuticals, industrial applics.; tackifier in adhesives

Sorbimacrogol oleate 300. *See* Polysorbate 80
Sorbimacrogol stearate 300. *See* Polysorbate 60
Sorbitan, esters, monododecanoate. *See* Sorbitan laurate
Sorbitan, esters, monohexadecanoate. *See* Sorbitan palmitate

Sorbitan laurate

CAS 1338-39-2; 5959-89-7; EINECS/ELINCS 215-663-3; 227-729-9
INS493; E493
Synonyms: Anhydrosorbitol monolaurate; Sorbitan, esters, monododecanoate; Sorbitan monododecanoate; Sorbitan monolaurate
Definition: Monoester of lauric acid and hexitol anhydrides derived from sorbitol
Empirical: $C_{18}H_{34}O_6$
Properties: Yel. to amber oily liq., bland char. odor; sol. in methanol, alcohol, min. oil; sl. sol. in cottonseed oil, ethyl acetate; insol. in water; m.w. 346.47; dens. 1.032; m.p. 18 C; HLB 8.6; acid no. 8 max.; sapon. no. 158-170; hyd. no. 330-358; flash pt. > 230 F; ref. index 1.4740; nonionic
Toxicology: LD50 (oral, rat) 33,600 mg/kg; TDLo (skin, mouse, 24 wk intermittent) 1350 mg/kg; sl. toxic by ing.; questionable carcinogen; experimental neoplastigen, reproductive effects; TSCA listed
Hazardous Decomp. Prods.: Heated to decomp., emits acrid smoke and irritating fumes
Uses: Emulsifier, solubilizer for foods, cosmetics, household prods., industrial applics., coatings, household (polishes, cleaners), pharmaceuticals; lubricant, antifog, antistat for PVC and other plastics; antifog for food-pkg. films; emulsifier for thermoplastics polymerization; emulsifier in food pkg.; emollient for skin care prods.; lubricant, emulsifier in textile spin finishes; cosolvent; foods (crystallization retarder, dough improver)
Regulatory: FDA 21CFR §175.320, 178.3400; Europe listed; UK approved; FDA approved for ophthalmics, orals; NF, BP, EP compliance
Manuf./Distrib.: A&E Connock; A.P. Chems. Ltd; Aldrich; Ashland; DelCon; Fluka; Jeen Int'l.; Lonza; Mosselman NV; Protameen; Rhodia HPCII; Sigma; Spectrum Quality Prods.; Universal Preserv-A-Chem
Trade Names: Crill 1

Sorbitan monododecanoate; Sorbitan monolaurate. *See* Sorbitan laurate

Sorbitan, monooctadecanoate, poly (oxy-1,2-ethanediyl) derivs. *See* Polysorbate 60

Sorbitan monopalmitate. *See* Sorbitan palmitate

Sorbitan palmitate

CAS 26266-57-9; EINECS/ELINCS 247-568-8
INS495; E495
Synonyms: 1,4-Anhydro-D-glucitol, 6-hexadecanoate; Monopalmitate sorbitan; Sorbitan, esters, monohexadecanoate; Sorbitan monopalmitate
Definition: Partial ester of palmitic acid with sorbitol mono- and dianhydrides
Empirical: $C_{22}H_{42}O_6$
Properties: Tan gran. waxy solid, faint tallow-like odor; sol. in ethyl acetate, warm abs. alcohol; sol. hazy in warm peanut or min. oil; insol. in water; m.w. 402.57; dens. 0.989; HLB 6.7; acid no. 8 max.; sapon. no. 140-150; hyd. no. 275-305; flash pt. > 230 F; ref. index 1.4780; nonionic
Toxicology: No known toxicity; skin and eye irritant; TSCA listed
Hazardous Decomp. Prods.: Heated to decomp., emits acrid smoke and irritating vapors
Uses: Surfactant, emulsifier for foods, cosmetics, pharmaceuticals, household prods., industrial applics., polymerization, pulp/paper, metalworking, lubricants, textiles, agric., paints, adhesives; emulsifier in food pkg.; oil additive; corrosion inhibitor; cosolvent; stabilizer for foods; antimicrobial in cane and beet sugar mills
Regulatory: FDA 21CFR §175.320, 178.3400; Europe listed; UK approved; FDA approved for intramuscular injectables; NF, BP, EP compliance
Manuf./Distrib.: A&E Connock; Aldrich; Ashland; DelCon; Fluka; Jeen Int'l.; Lonza; Mosselman NV; Protameen; Sigma; Spectrum Quality Prods.; Universal Preserv-A-Chem
Trade Names: Crill 2

Sorbitan tri-9-octadecenoate. *See* Sorbitan trioleate

Sorbitan trioleate

CAS 26266-58-0; 85186-88-5; EINECS/ELINCS 247-569-3; 286-074-7
INS496
Synonyms: Anhydrosorbitol trioleate; Sorbitan tri-9-octadecenoate; STO
Definition: Triester of oleic acid and hexitol anhydrides derived from sorbitol
Empirical: $C_{60}H_{108}O_8$
Properties: Yel. to amber oily liq.; sol. in alcohol, veg. oil, min. oil; insol. in water, ethylene glycol, propylene glycol; m.w. 957.52; dens. 0.956; HLB 1.8; acid no. 17 max.; iodine no. 77-85; sapon. no. 170-190; hyd. no. 50-75; flash pt. > 230 F; ref. index 1.4760; nonionic
Toxicology: Irritant; TSCA listed
Uses: Emulsifier, stabilizer in foods; emulsifier for pharmaceuticals, cosmetics, household prods., industrial applics., polymerization; antifog, lubricant, cling additive for plastic film; emollient; solvent; oil phase regulator; antirust; coupling agent, emulsifier, lubricant, softener for textiles, fibers, leather, agric., petrol. oils and waxes, natural fats and waxes; coemulsifier for min. oil; emulsifier in food pkg.
Regulatory: FDA 21CFR §175.320, 178.3400; FDA approved for inhalants, orals, topicals; NF, BP, EP compliance
Manuf./Distrib.: A&E Connock; Aldrich; Ashland; DelCon; Fluka; Jeen Int'l.; Lonza; Mosselman NV; Nihon Surfactant; Pfaltz & Bauer; Protameen; Sigma; Spectrum Quality Prods.; TCI Am.
Trade Names: Nikkol SO-30; Nikkol SO-30R

Sorption raffinate. *See* Petroleum raffinates, sorption process

Soya bean oil; Soya oil. *See* Soybean (Glycine soja) oil

Soybean (Glycine soja) oil

CAS 8001-22-7; EINECS/ELINCS 232-274-4
Synonyms: Chinese bean oil; Glycine soja; Glycine soja oil; Soya bean oil; Soya oil; Soybean oil; Soy oil
Classification: Fixed oil
Definition: Oil obtained from seeds of soya plant, *Glycine soja*, by extraction or expression; consists of triglycerides of oleic, linoleic, linolenic, and saturated acids
Properties: Pale yel. to brnsh. yel. oil, sl. char. odor and taste; sol. in alcohol, ether, chloroform, carbon disulfide; insol. in water; dens. 0.924-0.929; visc. 50.09 cps; vapor pressure very low; m.p. 22-31 C; iodine no. 120-141; sapon. no. 189-195; flash pt. 540 F; ref. index 1.471
Toxicology: May cause allergic reactions incl. hair damage and acne-like pimples; TSCA listed
Precaution: Combustible
Hazardous Decomp. Prods.: CO_2 from combustion; heated to decomp., emits acrid smoke and irritating fumes
HMIS: Health 0, Flammability 1, Reactivity 0
Storage: Light-sensitive
Uses: Soap mfg.; plasticizer (epoxidized); plasticizer for rubber; high-protein foods; edible cooking oil; drying oil for inks, paints, varnishes, in food-contact coatings; alkyd resins; cattle feeds; pesticide carrier; dietary supplement; food coatings, emulsifier, formulation aid, texturizer, lubricant; microbial fermentation nutrient; pharmaceutical vehicle, solvent; in food-pkg. adhesives; in resinous/polymeric food-contact coatings; defoamer in food-contact paper coatings, paper/paperboard; in food-contact textiles
Features: Oleaginous
Regulatory: FDA 21CFR §175.105, 175.300, 176.200, 176.210, 177.2800, GRAS; FDA approved for orals, topicals; USP/NF, BP, EP compliance
Manuf./Distrib.: A&E Connock; ABITEC; ADM; Aarhus Oliefabrik A/S; Akzo Nobel; Alba Int'l.; Alban Muller; Aldivia; Aldrich; Alnor Oil; Am. Ingreds.; Amcan Ingreds.; Amerol; Anglia Oils; Arista Ind.; Ashland; Atofina; Avatar; Blagden Spec. Chems. Ltd; Camida Ltd; CanAmera Foods; Cargill Ind. Oils & Lubes; Central Soya; Charkit; ChemService; Chesham Chems. Ltd; Croda Chem. Ltd; Croda Inc; Degen; Eastman; Efkay Chems. Ltd; Jeen Int'l.; John L Seaton Ltd; KIC Chems.; Karlshamns AB; L.V. Lomas; Lambent Tech.; Lipo; Mosselman NV; Mutchler; Penta Mfg.; Pokonobe Ind.; Reichhold; Riceland Foods; Ruger; S. Black; Sea-Land; Seltzer; Sigma; Soya Mainz; Spec. Ingreds. & Perfume; Spectrum Naturals; Spectrum Quality Prods.; St. Lawrence; Stevenson Cooper; Thornley; Tri-K Ind.; U.S. Synthetics; Univar E&E Ltd; Vamo-Fuji Spec.; Vivion; Vliengenthart BV; Voigt Global Distrib.; Vopak USA; Welch, Holme & Clark; Werner G. Smith
Trade Names: Cropure® Soybean; Oleocal® IVO-114; Super Refined® Soybean USP

Soybean oil; Soy oil. *See* Soybean (Glycine soja) oil

Spearmint (Mentha viridis) oil

CAS 8008-79-5
FEMA 3032
Synonyms: Crispmint oil; Curled mint oil; Mentha spicata; Mentha spicata oil; Mentha viridis; Mentha viridis oil; Spearmint oil
Definition: Volatile oil obtained from the dried tops and leaves of *Mentha viridis*, contg. chiefly carvone
Properties: Colorless or greenish-yel. liq., spearmint odor and taste; sol. in equal vol 80% alcohol; very sl. sol. in water; dens. 0.917-0.934 (25/25 C); ref. index 1.4820-1.4900 (20 C)
Toxicology: LD50 (oral, rat) 5 g/kg; mildly toxic by ing.; skin irritant; allergen causing skin rash; mutagenic data; TSCA listed
Hazardous Decomp. Prods.: Heated to decomp., emits acrid smoke and irritating fumes
HMIS: Health 0, Flammability 1, Reactivity 0
Storage: Keep cool, well closed; protect from light
Uses: Source of carvone; natural flavoring agent and fragrance in foods, confectionery, chewing gum, toothpaste, and pharmaceuticals; stimulant; carminative; antispasmodic; cosmetic ingred.
Regulatory: FDA 21CFR §172.230, 182.20, GRAS; 27CFR §21.65, 12.128, 21.151; FEMA GRAS; Japan approved (spearmint); Europe listed, no restrictions; FDA approved for buccals, orals; BP, EP compliance
Manuf./Distrib.: A.M. Todd; AMC Chems.; Acme-Hardesty; Adept Sol'ns.; Aldrich; Alfa Chem; Arista Ind.; Asiamerica Int'l.; Barnet Prods.; Berje; Chart; Creative Fragrances; Essence Nat. Prods.; Fabrichem; Fleurchem; Foote & Jenks; Frutarom Meer; George Uhe; H.E. Daniel Ltd; Hunan Xinyu; Integra; MelChem; Millennium; Penta Mfg.; Polarome Int'l.; PureWorld Botanicals; R.W. Greeff; Ruger; Sarcom; Spectrum Quality Prods.; Treatt USA; U.S. Synthetics; Virginia Dare Extract; Voigt Global Distrib.; Wm. Leman
Trade Names Containing: Punctilious® SDA 38B 190 Proof Prod. Code 13602007

Spearmint oil. *See* Spearmint (Mentha viridis) oil

Special boiling point spirit. *See* Naphtha, hydrotreated light

Spirit of Hartshorn. *See* Ammonium hydroxide
Spirits of turpentine. *See* Turpentine
Spirits of wine. *See* Alcohol
Spirit of turpentine. *See* Turpentine
Steam. *See* Water

Stearamide AMP

CAS 36284-86-3
Synonyms: N-(2-Hydroxy-1,1-dimethylethyl) octadecanamide; Octadecanamide, N-(2-hydroxy-1,1-dimethylethyl)-
Classification: Organic compd.
Empirical: $C_{22}H_{45}NO_2$
Formula: $CH_3(CH_2)_{16}CONHC(CH_3)_2CH_2OH$
Uses: Visc. control agent in cosmetics, pharmaceuticals
Trade Names Containing: Protachem™ SAS

Steareth-16

CAS 9005-00-9 (generic)
Synonyms: PEG-16 stearyl ether; POE (16) stearyl ether
Definition: PEG ether of stearyl alcohol
Formula: $CH_3(CH_2)_{16}CH_2(OCH_2CH_2)_nOH$, avg. n = 16
Properties: Nonionic
Uses: Surfactant, emulsifier in cosmetics, pharmaceuticals; in food-contact textiles
Regulatory: FDA 21CFR §177.2800
Manuf./Distrib.: Aldrich; Fluka; Sigma
Trade Names Containing: Solulan® 16

Stearic acid, cetyl ester. *See* Cetyl stearate
Stearic acid, 2-ethylhexyl ester. *See* Octyl stearate
Stearic acid, hexadecyl ester. *See* Cetyl stearate
Stearic acid, 2-(2-hydroxyethoxy) ethyl ester. *See* PEG-2 stearate
Stearic acid, isocetyl ester; Stearic acid, isohexadecyl ester. *See* Isocetyl stearate
Stearic acid, isopropyl ester. *See* Isopropyl stearate
Stearic acid, 2-methoxyethyl ester. *See* Methoxyethyl stearate
Stearic acid, methyl ester. *See* Methyl stearate
Stearic acid, monoester with ethylene glycol. *See* Glycol stearate
Stearic acid, monoester with glycerol. *See* Glyceryl stearate
Stearic acid, monoester with 1,2-propanediol. *See* Propylene glycol stearate
Stearic acid, monoester with triglycerol. *See* Polyglyceryl-3 stearate
Stearic acid, palmityl ester. *See* Cetyl stearate
Stearic monoglyceride. *See* Glyceryl stearate
Stearin. *See* Tristearin
Stearol. *See* Stearyl alcohol

Stearyl alcohol

CAS 112-92-5; EINECS/ELINCS 204-017-6
Synonyms: C18 linear alcohol; Decyl octyl alcohol; Octadecanol; 1-Octadecanol; n-Octadecanol; Octadecyl alcohol; n-Octadecyl alcohol; Stearol
Classification: Fatty alcohol
Empirical: $C_{18}H_{38}O$
Formula: $CH_3(CH_2)_{16}CH_2OH$
Properties: Wh. unctuous flakes or gran., faint odor, bland taste; sol. in alcohol, acetone, ether; insol. in water; m.w. 270.56; dens. 0.8124 (59/4 C); m.p. 55-60 C; b.p. 210.5 C (15 mm); acid no. 2 max.; iodine no. 2 max.; hyd. no. 195-220
Toxicology: LD50 (oral, rat) 20 g/kg; TDLo (implant, mouse) 1000 mg/kg; mildly toxic by ing.; irritating to skin and eyes; may cause skin sensitivity; nonallergenic; questionable carcinogen; experimental neoplastigen; TSCA listed
Precaution: Flamm. when exposed to heat or flame; can react with oxidizers
Hazardous Decomp. Prods.: Heated to decomp., emits acrid smoke and irritating fumes
HMIS: Health 1, Flammability 1, Reactivity 0
Uses: Perfumery; surfactants; lubricant for plastics processing; resins; antifoam; food additive; intermediate in mfg. of food additives; surfactant, emulsion stabilizer, emulsifier, emollient, stiffener, astringent in pharmaceuticals; cosolvent; emulsifier, thickener, emollient, emulsion stabilizer, opacifier, visc. control agent in cosmetics; in food-pkg. adhesives; in food-contact coatings; defoamer in food-contact coatings and paper/paperboard; in cellophane for food pkg.; food-contact textiles
Regulatory: FDA 21CFR §172.755, 172.864, 175.105, 175.300, 176.180, 176.200, 176.210, 177.1010, 177.1200, 177.1390, 177.2800, 178.3480, 178.3910; FDA approved for orals, topicals, vaginals; NF, BP, EP compliance
Manuf./Distrib.: A&E Connock; AMC Chems.; Aarhus Oliefabrik A/S; Albemarle; Aldrich; Alfa Chem; Allchem Ind.; Amerchol; Ashland; Avatar; Brown; Chemacon GmbH; Chemron; Chesham Chems. Ltd; CoKEM Assoc.; Cognis/Chems. Group; Costec; Croda Chem. Ltd; Croda Inc; Efkay Chems. Ltd; Eggar & Co.; Filo; Fluka; Integra; Jarchem Ind.; Kimpton Bros. Ltd; Kraft Chem.; Lipo; Lonza; M. Michel; Mosselman NV; Mutchler; Pechiney Chems. Div.; Penta Mfg.; Procter & Gamble; Protameen; R.I.T.A.; R.W. Greeff; RTD Chems.; Ruger; Sasol Germany; Sea-Land; Sigma; Spectrum Quality Prods.; Universal Preserv-A-Chem; Vopak USA; Whyte Chems. Ltd
Trade Names: Cachalot® S-56
Trade Names Containing: Witcohol™ 42

Stearyl-cetyl alcohol. *See* Cetearyl alcohol
Stearyl 2-ethylhexanoate. *See* Stearyl octanoate

Stearyl heptanoate

CAS 66009-41-4; EINECS/ELINCS 266-065-4
Synonyms: Heptanoic acid, octadecyl ester; Octadecyl heptanoate
Definition: Ester of stearyl alcohol and heptanoic acid
Empirical: $C_{25}H_{50}O_2$
Formula: $CH_3(CH_2)_5COOCH_2(CH_2)_{16}CH_3$
Uses: Emollient and water repellent for cosmetics, toiletries, and pharmaceuticals
Manuf./Distrib.: A&E Connock
Trade Names: Crodamol W

Stearyl octanoate

CAS 59130-70-0; EINECS/ELINCS 261-620-7
Synonyms: 2-Ethylhexanoic acid, octadecyl ester; Hexanoic acid, 2-ethyl-, octadecyl ester; Octadecyl 2-ethylhexanoate; Stearyl 2-ethylhexanoate
Definition: Ester of stearyl alcohol and 2-ethylhexanoic acid
Empirical: $C_{26}H_{52}O_2$
Formula: $CH_3(CH_2)_3CHCH_2CH_3COOCH_2(CH_2)_{16}CH_3$
Uses: Emollient, emulsifier, thickener, stabilizer, opacifier, pearlescent in cosmetics, creams, shaving creams, cream shampoos, topical pharmaceuticals
Manuf./Distrib.: A&E Connock; AXO Chem.
Trade Names Containing: Super Refined® Crodamol SCO

Stearyl stearate

CAS 2778-96-3; 85536-04-5; EINECS/ELINCS 220-476-5; 287-484-9
Synonyms: Octadecanoic acid, octadecyl ester; Octadecyl stearate
Definition: Ester of stearyl alcohol and stearic acid
Empirical: $C_{36}H_{72}O_2$
Formula: $CH_3(CH_2)_{16}COOCH_2(CH_2)_{16}CH_3$
Properties: M.w. 536.98; m.p. 56 C; flash pt. (COC) 242 C
Toxicology: TSCA listed
Uses: Emollient, thickener, visc. control agent, pigment dispersant, spreading agent, solubilizer for cosmetics, bath oils, pharmaceuticals, creams, lotions; plasticizer; processing lubricant for plastics; lubricant for metalworking fluids, rolling, stamping and drawing oils, greases, motor oils, extrusion of ceramics; fatting agent in textile and leather auxs.; textile lubricant; in surf. lubricants for mfg. of food-contact metallic articles
Regulatory: FDA 21CFR §178.3910
Manuf./Distrib.: A&E Connock; Alfa Chem; Ashland; Bruchem; Chem. Group; Fluka; Inolex; Jarchem Ind.; Lipo; R.I.T.A.; RTD Chems.; Rhodia HPCII; Sigma; Sithean; Stason Pharmaceuticals; Universal Preserv-A-Chem
Trade Names: Exceparl SS

STO. *See* Sorbitan trioleate

Stoddard solvent

CAS 8052-41-3; EINECS/ELINCS 232-489-3
UN 1256 (DOT)

Synonyms: High flash naphtha; Mineral spirits Type I; Naphtha safety solvent; Naphtha, solvent; Safety solvent naphtha; Solvent naphtha; White spirits
Classification: Mixed petroleum hydrocarbons
Definition: Mixed isomer contg. 85% nonane and 15% trimethylbenzene
Properties: Colorless clear liq.; petrol. kerosene-like odor; misc. with abs. alcohol, benzene, ether, chloroform, CCl_4, CS_2, most org. solvs.; insol. in water; m.w. ≈ 150; sp.gr. 0.78; vapor pressure 2 mm Hg (20 C); f.p. -70 C; b.p. 157-196 C; flash pt. (CC) 40 C; ref. index 1.43 (20 C); KB value 32-37
Toxicology: ACGIH TLV/TWA 100 ppm; LD50 (oral, rat) > 5 g/kg, (skin, rabbit) > 3 g/kg; LCLo (inh., cat, 2.5 h) 10 g/m^3; mildly toxic by inh. (irritation, CNS effects, headache, dizziness, confusion, fatigue); human eye irritant; probably a mild to mod. skin irritation; target organs: skin, eyes, respiratory system, CNS, kidney; TSCA listed
Environmental: VOC
Precaution: Combustible; flamm. exposed to heat or flame; explosive in vapor form when exposed to heat or flame; reactive with oxidizing materials
Hazardous Decomp. Prods.: Heated to decomp., may explode; heated to decomp., emits acrid fumes
NFPA: Health 0, Flammability 2, Reactivity 0
Storage: Store in cool, dry, well-ventilated area away from incompat. materials
Uses: Solvent and thinner for paints, varnishes, other coatings; dry-cleaning solvent; spot and stain removal; solvent in textile industry; mfg. of pesticides, cleaners, aerosol prods.; metal and leather cleaning/degreasing; extraction of fats and oils; in rubber cements and polishes; tackifying agent for rubber
Manuf./Distrib.: Ashland; Coyne; Hukill; R.E. Carroll; Schaefer Tech.; Spectrum Quality Prods.; Stanchem; Vopak USA
Trade Names: 0-Aromatic 142 Solvent; 0-Aromatic Mineral Spirits; Mineral Spirits 066/3; Mineral Spirits 075; Mineral Spirits 150; Mineral Spirits 150 66/3; Naphthol Spirits 66/3; 142 Solvent 66/3; Solvent 2215; Solvent 2222; Solvent 2231; Solvent 2256; Solvent 2257; Solvent 2284; Solvent 2292; Solvent 2302; Solvent 2350; Solvent 2351; Solvent 2352; Solvent 2359; Solvent 2361; Solvent 2363; Solvent 2364; Solvent 2379; Solvent 2380; Solvent 2383; Solvent 2388; Solvent 2417; Varsol® 1; Varsol® 1 Naphtha Anti-Static; Varsol® 3135 Naphtha; Varsol® 3139; Varsol® 3139 Anti-Static Solvent
Trade Names Containing: AM OMS 75 Solvent; Solvent 2461; Varsol® 18; Varsol® 18 Naphtha; Varsol® 18 Naphtha Anti-Static; Varsol® DX 3139; Varsol® DX 3139 Anti-Static Solvent

Straight-run kerosene; Straight-run kerosine. *See* Kerosene
Strong ammonia solution. *See* Ammonium hydroxide
Strychnidin-10-one, 2,3-dimethoxy-; Strychnine, 2,3-dimethoxy-. *See* Brucine
STS. *See* Sodium toluenesulfonate
Styrallyl alcohol; Styralyl alcohol. *See* α-Methylbenzyl alcohol

Styrene

CAS 100-42-5; EINECS/ELINCS 202-851-5
UN 2055 (DOT); FEMA 3233
Synonyms: Cinnamene; Cinnamenol; Cinnamol; Ethenylbenzene; Phenethylene; Phenylethene; Phenylethylene; Styrene monomer; Styrene monomer, inhibited; Styrol; Styrole; Styrolene; Styron; Styropor; Vinylbenzene; Vinylbenzol
Classification: Aromatic hydrocarbon; alkenylbenzene
Empirical: C_8H_8
Formula: $C_6H_5CH{:}CH_2$
Properties: Colorless to ylsh. oily liq., penetrating odor; sol. in alcohol, ether, methanol, acetone, CS_2; sparingly sol. in water; m.w. 104.15; dens. 0.906 (20/4 C); m.p. -30.6 C; b.p. 145-146 C; flash pt. (CC) 31 C; ref. index 1.546 (20 C)
Toxicology: ACGIH TLV/TWA 50 ppm; STEL 100 ppm (skin); LD50 (oral, rat) 5000 mg/kg, (IV, mouse) 90 mg/kg, (IP, mouse) 660 mg/kg; poison by ing., inh., IV routes; mod. toxic by IP; mildly toxic by inh. to humans; human systemic effects by inh. (eye/olfactory changes); skin, eye, mucous membrane irritant; human skin irritant; narcotic in high concs.; carcinogen; experimental teratogen, reproductive effector; human mutagenic data; TSCA listed
Precaution: Flamm.; dangerous fire hazard exposed to flame, heat, oxidizers; vapor explosive; slowly undergoes polymerization and oxidation on exposure to light and air, yielding peroxides; violent/explosive polymerizations possible; reacts with oxidizers
Hazardous Decomp. Prods.: Heated to decomp., emits acrid smoke and irritating fumes
NFPA: Health 2, Flammability 3, Reactivity 2
Storage: Storage hazard above 32 C
Uses: Monomer, coreactant, diluent for prod. of paints, plastics, and resins (polystyrene resins, styrene-acrylonitrile, styrene-divinylbenzene copolymers, styrene-butadiene rubber, acrylonitrile-butadiene-styrene, unsat. polyesters); for crosslinking reactive coatings; intermediate; synthetic flavoring agent in foods and pharmaceuticals; fragrance in cosmetics; modifier for oils/alkyds in food-contact coatings; solvent for adjuvants in food-contact crosslinked polyesters; in ion-exchange membranes for prod. of grapefruit juice; in ion-exchange resins for purification of foods; in food-contact rubber articles for repeated use
Regulatory: FDA 21CFR §172.515, 173.20, 173.25, 175.300, 175.320, 175.380, 175.390, 176.170, 176.180, 177.1010, 177.1020, 177.1030, 177.1040, 177.1050, 177.1210, 177.1630, 177.1640, 177.1810, 177.1820, 177.1830, 177.2260, 177.2420, 177.2600, 177.2710, 177.2800, 178.3790; FEMA GRAS
Manuf./Distrib.: Advanced BioTech; Aldrich; Allchem Ind.; Ashland; Atofina; BP Amoco; Bencorp Int'l.; Brøste; Chemcentral; Chemical; ChevronPhillips; Dow Plastics; Fluka; GFI Chems. USA; GFI; Gulf Chem. Int'l.; HCI Chemtech Distribution; Houghton Chem.; Huntsman; JLM Marketing; Janus Resources; KIC Chems.; Lyondell; Magna-Kron; Mays; Mitsui & Co. USA; Nova Ltd; Reach Group; Sigma; Sparkford Chems. Ltd; Sterling Chems.; Total Spec. Chems.; U.S. Chems.; Vopak USA

Styrene monomer; Styrene monomer, inhibited; Styrol; Styrole; Styrolene; Styron; Styropor. *See* Styrene
Sublimed sulfur. *See* Sulfur
Substituted hexa hydropyrimidine. *See* Hexetidine
Succinic acid diethyl ester. *See* Diethyl succinate
Succinic acid, dimethyl ester. *See* Dimethyl succinate

Sucroglycerides

INS474; E474
Synonyms: Sugar glycerides
Definition: Obtained by reaction of sucrose with an edible fat or oil with or without the presence of a solvent; mixt. of mono- and di-esters of sucrose and fatty acids, with mono-, di-, and tri-glycerides from the fat or oil
Properties: Wh. to off-wh. powd., soft solid mass, or stiff gel; odorless; sol. in ethanol; insol. in cold water
Uses: Emulsifier, aerating agent, starch complexing agent in foods; dispersant, stabilizer for fats; carrier, solvent for colors and fat-sol. antioxidants; emulsifier in cosmetic and pharmaceutical creams, lotions, ointments
Regulatory: ADI 0-30 mg/kg body wt.
Manuf./Distrib.: Alfa Chems. Ltd; Rhodia Ltd

Sucrose esters; Sucrose esters of fatty acids. *See* Sucrose fatty acid esters

Sucrose fatty acid esters

CAS 977019-37-6
INS473; E473
Synonyms: Sucrose esters; Sucrose esters of fatty acids; Sugar esters
Definition: Mono-, di-, and triesters of sucrose with fatty acids
Properties: Wh. to sl. grayish powd. (based on sat. fatty acids); ylsh. pasty to waxy substance (based on unsat. fatty acids); sol. in ethanol; sparingly sol. in water; insol. in veg. oils; nonionic
Hazardous Decomp. Prods.: Heated to decomp., emits acrid smoke and irritating fumes
Uses: Emulsifier, stabilizer, texturizer for foods, beverages, chewing gum; protective coating on fruits; aerating agent; starch complexing agent; dispersant, stabilizer for fats; carrier, solvent for colors and fat-sol. antioxidants; emulsifier in cosmetic and pharmaceutical creams, lotions, ointments; emulsion stabilizer, dispersant, solubilizer for pharmaceutical tablets

Regulatory: FDA 21CFR §172.859; ADI 0-30 mg/kg body wt.; Europe listed; UK approved; FAO/WHO 8.446
Manuf./Distrib.: Dai-ichi Kogyo Seiyaku; Mitsubishi Chem.; Multi-Kem

Sucrose monopalmitate. *See* Sucrose palmitate

Sucrose octaacetate

CAS 126-14-7; EINECS/ELINCS 204-772-1
FEMA 3038
Synonyms: α-D-Glucopyranoside, 1,3,4,6-tetra-O-acetyl-β-D-fructofuranosyl-, tetraacetate; Octaacetylsucrose; D-(+)-Sucrose octaacetate
Empirical: $C_{28}H_{38}O_{19}$
Properties: Wh. need., pract. odorless, intensely bitter taste; sol. in ether, methanol, chloroform, 1100 parts water, 11 parts alcohol, 22 parts CCl_4, oxygenated and aromatic solvs.; m.w. 678.60; m.p. 82-85 C; b.p. 260 C (1 mm); dec. above 285 C; ref. index 1.4660
Toxicology: LD50 (oral, rat) > 5 g/kg, (skin, rabbit) > 5 g/kg; sl. toxic by ing. and skin contact; irritating to skin; TSCA listed
Precaution: Combustible
Hazardous Decomp. Prods.: Heated to decomp., emits acrid smoke and irritating vapors
HMIS: Health 1, Flammability 0, Reactivity 0
Storage: Hygroscopic
Uses: Plasticizer for cellulose esters and plastics; adhesives; coatings; lacquers; insecticide; termite repellent; synthetic flavoring agent in foods; alcohol denaturant in pharmaceuticals; denaturant in cosmetics; in preps. intended to deter nail biting; in food-pkg. adhesives
Regulatory: FDA 21CFR §172.515, 175.105, 27CFR §21.130; FEMA GRAS; NF compliance
Manuf./Distrib.: Adept Sol'ns.; Aldrich; Atomergic Chemetals; B2E; Barrington; Burlington Bio-Medical; CarboMer; Fluka; Jarchem Ind.; M.W. Hardy; Mentha & Allied Prods.; Morre-Tec Ind.; Paulaur; Penta Mfg.; Ruger; Sigma; Spectrum Quality Prods.; Synasia Fine Chem.; Universal Preserv-A-Chem
Trade Names Containing: Punctilious® SDA 40A 190 Proof; Punctilious® SDA 40A Anhydrous

D-(+)-Sucrose octaacetate. *See* Sucrose octaacetate

Sucrose palmitate

CAS 26446-38-8; EINECS/ELINCS 247-706-7
Synonyms: β-D-Fructofuranosyl-α-D-glucopyranoside monohexadecanoate; α-D-Glucopyranoside, β-D-frutofuranosyl, monohexadecanoate; Palmitic sucrose ester; Saccharose palmitate; Sucrose monopalmitate; Sucrose palmitic acid ester
Definition: Monoester of palmitic acid and sucrose
Empirical: $C_{28}H_{52}O_{12}$
Properties: M.w. 580.71; nonionic
Uses: O/w and w/o emulsifier, softener, conditioner, texturizer, and aerating agent in foods; protective fruit coatings; emulsifier, solubilizer, stabilizer, tablet lubricant, excipient, diluent for pharmaceuticals; emulsifier, surfactant, solvent in cosmetics; polymerization emulsifier for food-pkg. film
Regulatory: FDA 21CFR §172.859

Sucrose palmitic acid ester. *See* Sucrose palmitate
Sugar esters. *See* Sucrose fatty acid esters
Sugar glycerides. *See* Sucroglycerides
Sulfalone. *See* Sulfolane
Sulfinylbis (methane). *See* Dimethyl sulfoxide
Sulfobenzide. *See* Diphenyl sulfone
Sulfolan. *See* Sulfolane

Sulfolane

CAS 126-33-0; EINECS/ELINCS 204-783-1
Synonyms: Cyclic tetramethylene sulfone; Cyclotetramethylene sulfone; Dapsone; Dihydrobutadiene sulfone; 1,1-Dioxide tetrahydrothiofuran; 1,1-Dioxidetetrahydrothiophene; Dioxothiolan; 1,1-Dioxothiolan; Sulfalone; Sulfolan; Tetrahydrothiophene dioxide; Tetrahydrothiophene-1,1-dioxide; 2,3,4,5-Tetrahydrothiophene-1,1-dioxide; Tetramethylene sulfone; Thiacyclopentane dioxide; Thiocyclopentane-1,1-dioxide; Thiolane-1,1-dioxide; Thiophane dioxide; Thiophan sulfone
Empirical: $C_4H_8O_2S$
Formula: $CH_2CH_2CH_2CH_2SO_2$
Properties: Wh. or creamy wh. cryst. powd., odorless, sl. bitter taste; sol. in alcohol, acetone, dilute min. acids; misc. with acetone, toluene, oxygenated/aromatic solvs.; sl. misc. with octanes, olefins, naphthenes, aliphatic hydrocarbons; very sl. sol. in water; m.w. 120.16; dens. 1.27 (20 C); vapor pressure 0.0062; m.p. 25-28 C; b.p. 285 C; flash pt. 177 C; ref. index 1.485; surf. tens. 35.5 dynes/cm; dielec. const. 43.26
Toxicology: REL 5 ppm; LD50 (oral, rat) 1941 mg/kg, (IP, rat) 1600 mg/kg, (IV, mouse) 1080 mg/kg, (skin, rabbit) 4009 mg/kg; LC50 (inh., rat, 24 h) > 956 ppm; mod. toxic by ing., IP, IV routes; mildly toxic by inh., skin contact; eye irritant; mutagenic; target organs: skin, CNS, GI, lungs, liver; TSCA listed
Environmental: VOC; BOD5 0.0; COD 1.75; ThOD 1.73
Precaution: Combustible
Hazardous Decomp. Prods.: Heated to decomp., emits toxic fumes of SO_x
NFPA: Health 2, Flammability 1, Reactivity 0
Uses: Curing agent for epoxy resins; extraction solvent for extraction of benzene, toluene, xylene, other aromatic hydrocarbons from oil refinery streams; cosolvent in Sulfinol gas desulfurization process; aprotic process solvent (chem. synthesis); solvent in surf. coatings, pharmaceuticals, polymerization; reaction solvent for prep. of aromatic sulfonic acids, pyridines, isocyanates, pharmaceuticals; selective solvent; in halogen exchange process; solvent for separation of low boiling alcohols (propyl alcohol, s-butyl alcohol), mineral oils, tars; to purify natural gas streams; for fractionation of fatty acids into sat. and unsat. components; plasticizer; dielec. in elec. equipment; component in hydraulic fluids; medicine (antibacterial); adjuvant in prod. of polyarylsulfone resins for food pkg.
Regulatory: FDA 21CFR §177.1560
Manuf./Distrib.: Aldrich; Allchem Ind.; Alpha Chem; ChevronPhillips; Fluka; Laporte Perf. Chems.; Shell; Sigma; Sumitomo Seika
Trade Names: Sulfolane W

Sulfomethane. *See* Methanesulfonic acid
1,1´-Sulfonylbisbenzene. *See* Diphenyl sulfone
Sulfonylbismethane. *See* Dimethyl sulfone

Sulfur

CAS 7704-34-9; EINECS/ELINCS 231-722-6
UN NA 1350 (DOT); NA 2448 (DOT)
Synonyms: Brimstone; Flour sulfur; Flowers of sulfur; Precipitated sulfur; Soluble sulfur; Sublimed sulfur; Sulfur flower; Sulfur flowers; Sulfur soluble; Sulfur sublimed; Sulphur; Wettable powder sulfur
Classification: Nonmetallic element
Empirical: S
Properties: Sol. in carbon disulfide, CCl_4, benzene, chlorinated and aromatic solvs.; sl. sol. in alcohol, ether; insol. in water; at.wt. 32.064; α: rhombic yel. crystals; dens. 2.06; β: monoclinic pale yel. crystals; dens. 1.96; vapor pressure 1 mm Hg (183.8 C); m.p. 119 C; b.p. 444.6 C; flash pt. 405 F
Toxicology: LD (oral, rat) > 8437 mg/kg; LDLo (IV, rat) 8 mg/kg, (IP, guinea pig) 55 mg/kg; poison by ing., IV, and IP routes; skin, eye, mucous membrane irritant; human eye irritant; TSCA listed
Environmental: Nontoxic to fish and bees
Precaution: DOT: Flamm. solid; explosive as dust exposed to flame; reactive with oxidizers; violent reactions possible with halogens, carbides, zinc, uranium, tin, Na, lithium, nickel, phosphorus, potassium, aluminum, ammonia, etc.
Hazardous Decomp. Prods.: When heated, it burns and emits highly toxic fumes of SO_x
NFPA: Health 2, Flammability 1, Reactivity 0
Uses: Sulfuric acid mfg.; petroleum refining; dyes; chemicals; fungicide; acaricide; insecticide; explosives; gunpowder; matches; detergents; plastics; rubber vulcanizer/processing aid; pulp and paper mfg.; soil conditioner; bleaching wood pulp, textiles; nucleating agent for photographic film; enamels; cement sealant, binder; asphalt extender; mortars; precipitant for heavy metal removal in wastewater treatment; keratolytic in dermatitis and psoriasis preps.; pharmaceuticals; antistat, antidandruff agent in cosmetics; in food-pkg. adhesives; vulcanizing agent in vulcanized nat. or syn. rubber closure-sealing gaskets for food containers
Regulatory: FDA 21CFR §175.105, 177.1210, 177.2600

Manuf./Distrib.: ABCR; Akrochem; Aldrich; Allan; Ashland; Atofina; BP Amoco; Chemacon GmbH; ChevronPhillips; Crompton/Uniroyal; Dow; Drexel; EM Science; ExxonMobil; Fluka; H.M. Royal; Helm AG; Integra; J.T. Baker; Kraft Chem.; Lohmann; Lyondell/Tech.; Mallinckrodt Baker; Mutchler; Norsk Hydro AS; Penta Mfg.; Reade Advanced Materials; Rhodia; Riedel-deHaën; Ruger; S & S Chem.; San Yuan; Shell; Skylighter; Solvay GmbH; Spectrum Quality Prods.; Universal Preserv-A-Chem; VWR Int'l.; Voigt Global Distrib.; Vopak USA; Wilbur-Ellis
Trade Names Containing: Kenplast® G

Sulfur dioxide

CAS 7446-09-5; EINECS/ELINCS 231-195-2
UN 1079 (DOT); FEMA 3039; INS220; E220
Synonyms: Bisulfate; Bisulfite; Sulfurous acid anhydride; Sulfurous anhydride; Sulfurous oxide; Sulfur oxide; Sulphur dioxide, liquefied
Classification: Inorganic sulfur compd.
Empirical: O_2S
Formula: O=S=O
Properties: Colorless gas, strong suffocating odor; condenses @ -10 C to colorless liq.; sol. 8.5% in water, 25% in alcohol, 32% in methanol; sol. in ether, chloroform, oxygenated solvs.; m.w. 64.06; dens. 1.5 (liq.); vapor pressure 2538 mm Hg (21.1 C); m.p. -72 C; b.p. -10 C (liq.); nonflamm.
Toxicology: ACGIH TLV/TWA 2 ppm; STEL 5 ppm; LCLo (inh., human, 5 min) 3000 ppm; poison gas; mildly toxic to humans by inh.; corrosive irritant to eyes, respiratory tract; gas causes skin irritation, liq. may cause burns due to freezing; human systemic effects by inh. (pulmonary changes); causes upper respiratory/bronchi effects; can cause respiratory paralysis; questionable carcinogen; experimental tumorigen, teratogen, reproductive effects; human mutagenic data; TSCA listed
Precaution: Incompat. with bases, chlorates, fluorine, interhalogens, powd. metals, metal oxides, metal acetylides, sodium hydroxide, diethyl zinc; violent reactions, ignition, explosions possible; common air contaminant
Hazardous Decomp. Prods.: Forms an acid sol'n. (sulfurous acid) on contact with moisture; heated to decomp., emits toxic fumes of SO_x
NFPA: Health 3, Flammability 0, Reactivity 0
Storage: Store cylinders in cool, dry, well-ventilated, fireproof area away from flamm. materials and corrosive atmospheres, out of direct sunlight
Uses: Chem. intermediate; bleaching of paper pulp; ore and metal refining; soybean protein; extraction solvent; catalyst; bleaching oils and starch; sulfonation of oils; reducing agent; antioxidant; fungicide; oxygen scavenger; disinfectant in breweries; dechlorination and silica activation in water treatment (potable, sewage, effluent); antioxidant, bleaching agent, antimicrobial, fumigant, preservative for foods; corn processing; antioxidant, preservative for pharmaceuticals; air pollutant, major contributor to acid rain
Regulatory: FDA 21CFR §73.170, 172.892, 182.3862, GRAS; BATF 27CFR §240.1051; FEMA GRAS; Japan approved (0.03-5 g/kg); Europe listed; UK approved; NF compliance; SARA reportable
Manuf./Distrib.: Air Prods.; Aldrich; Alexander; BASF; Boliden Intertrade; Calabrian; Celanese; Clariant; Coyne; Danisco Cultor; Grillo-Werke AG; Kuehne Chem.; MG Ind.; Olin/Chlor Alkali; Outokumpu Mintec Oy; PCAS; PVS; Peridot; Rhodia; Scott Spec. Gases; Stanchem; Thatcher

Sulfur flower; Sulfur flowers. *See* Sulfur
Sulfuric acid monododecyl ester sodium salt. *See* Sodium lauryl sulfate
Sulfuric acid, mono (2-ethylhexyl) ester sodium salt. *See* Sodium 2-ethylhexyl sulfate
Sulfuric ether. *See* Ethyl ether
Sulfurous acid anhydride; Sulfurous anhydride; Sulfurous oxide; Sulfur oxide. *See* Sulfur dioxide
Sulfur soluble; Sulfur sublimed; Sulphur. *See* Sulfur
Sulphur dioxide, liquefied. *See* Sulfur dioxide

Sunflower (Helianthus annuus) seed oil

CAS 8001-21-6; EINECS/ELINCS 232-273-9
Synonyms: Helianthus annuus; Helianthus annuus oil; Sunflower oil; Sunflower seed oil
Definition: Oil expressed from seeds of the sunflower, *Helianthus annuus*
Properties: Amber liq., pleasant odor, mild taste; sol. in alcohol, ether, chloroform, CS_2; negligible sol. in water; dens. 0.924-0.926; vapor pressure very low; m.p. -18 C; iodine no. 125-140; sapon. no.188-194; ref. index 1.4611
Toxicology: No known toxicity; TSCA listed
Precaution: Combustible
Hazardous Decomp. Prods.: CO_2 from combustion; heated to decomp., emits acrid smoke and irritating fumes
HMIS: Health 0, Flammability 1, Reactivity 0
Uses: Raw material in plastics, polymers, alkyd resins, rubbers, soaps, pharmaceuticals; drying oil for paints, alkyd resins; emollient for cosmetics; edible oil; in prep. of margarine; food coating agent, emulsifier, formulation aid, texturizer; dietary supplement; pharmaceuticals (diluent, carrier, emulsifier, emollient, tablet binder; diuretic; expectorant); soybean oil replacement; drying oil, defoamer in food-contact paper coatings; in resinous/polymeric food-contact coatings
Regulatory: FDA 21CFR §175.300, 176.200, GRAS; BP, EP compliance (refined)
Manuf./Distrib.: ABITEC; ADM; Aldivia; Alnor Oil; Anglia Oils; Arista Ind.; Avatar; Cargill Ind. Oils & Lubes; Charkit; Desert Whale Jojoba; Int'l. Chem. Inc; Int'l. Flora Tech.; Jeen Int'l.; John L Seaton Ltd; Lipo; Mosselman NV; Penta Mfg.; Protameen; Ruger; Sea-Land; Sigma; St. Lawrence; Tri-K Ind.; Welch, Holme & Clark; Werner G. Smith
Trade Names: EmCon™ SUN

Sunflower oil; Sunflower seed oil. *See* Sunflower (Helianthus annuus) seed oil
Sweet almond oil. *See* Sweet almond (Prunus amygdalus dulcis) oil

Sweet almond (Prunus amygdalus dulcis) oil

CAS 8007-69-0
Synonyms: Almond oil; Almond oil, sweet; Prunus dulcis; Sweet almond oil
Definition: Fixed oil obtained from the ripe seed kernel of *Prunus amygdalus dulcis*, contg. chiefly glyceryl oleate
Properties: Colorless or pale yel. oily liq., almost odorless, bland taste; sl. sol. in alcohol; misc. with benzene, chloroform, ether, petrol. ether; insol. in water; dens. 0.910-0.915; iodine no. 93-100; sapon. no. 191-200; ref. index 1.4593-1.4646 (40 C)
Toxicology: LD50 (oral, rat) > 5 g/kg, (skin, rabbit) > 5 g/kg; TSCA listed
HMIS: Health 1, Flammability 0, Reactivity 0
Storage: Keep cool, well closed; protect from light
Uses: Emollient, conditioner, gloss aid in cosmetics; food ingred.; vehicle, solvent, flavor, emollient, nutritive, demulcent in pharmaceuticals; in rose water ointment USP; softener for ear wax
Features: Oleaginous
Regulatory: NF, BP, EP compliance
Manuf./Distrib.: AXO Chem.; Adept Sol'ns.; Aldivia; Alfa Chem; Anglia Oils; Arista Ind.; Ashland; Berje; CoKEM Assoc.; Costec; Croda Inc; Desert Whale Jojoba; Fleurchem; George Uhe; H.E. Daniel Ltd; Integra; Jeen Int'l.; Mosselman NV; Mutchler; Pangaea Sciences; Penta Mfg.; Pokonobe Ind.; Ruger; Seltzer; Sigma; Spectrum Quality Prods.; Total Spec. Chems.; Tri-K Ind.; Triple Crown Am.; U.S. Chems.; Universal Preserv-A-Chem; Voigt Global Distrib.; Welch, Holme & Clark
Trade Names: Cropure® Almond; Super Refined® Almond NF

Sweet birch oil. *See* Methyl salicylate
Sweetened middle distillate; Sweetened middle petroleum distillates. *See* Petroleum distillates, sweetened middle
SXS. *See* Sodium xylenesulfonate
Synthetic oil of bitter almond. *See* Benzaldehyde
Synthetic oil of wintergreen. *See* Methyl salicylate
Syrups, hydrolyzed starch. *See* Corn syrup

T

Talleol. *See* Tall oil

Tall oil

CAS 8002-26-4; EINECS/ELINCS 232-304-6
Synonyms: Lignin liquor; Liquid rosin; Talleol; Tallol
Definition: Byprod. of wood pulp contg. rosin acids, oleic and linoleic acids, and long chain alcohols
Properties: Dk. brn. liq., acrid odor; dens. 0.95; acid no. 178 min.; flash pt. 182 C
Toxicology: LD50 (unreported, mouse) 7300 mg/kg; mild allergen; may cause wt. loss, blood changes; migrates to food from pkg. materials
Precaution: Combustible when exposed to heat or flame; can react with oxidizing materials
Hazardous Decomp. Prods.: Heated to decomp., emits acrid smoke and irritating fumes
Uses: Drying oil for paints, in food-contact coatings; paint vehicle; source of rosin; alkyd resins; soaps; cutting oils; flotation agents; core oils; lubricants and greases; asphalt derivs.; rubber reclaiming; chemical intermediate; emulsifier in oil-based drilling fluids; fungicide for topical pharmaceuticals; synthesis of cortisone and sex hormones; emollient, solvent in cosmetics; in food-pkg. adhesives; in food-contact rubber articles for repeated use; in food-contact textiles; defoamer in food-contact paper coatings, paper/paperboard
Regulatory: FDA 21CFR §175.105, 175.300, 175.320, 176.200, 176.210, 177.2600, 177.2800, 181.22, 181.26, 182.70, 186.1557, GRAS as indirect food additive; FDA approved for topicals
Manuf./Distrib.: Arakawa USA; Arizona; Avatar; Bencorp Int'l.; Chemcentral; Chemical; Dow; Eastech; Geo. Pfau's Sons; Georgia-Pacific Resins; Georgia-Pacific/Actrachem; Harcros; Int'l. Paper; L.V. Lomas; MeadWestvaco; Nat'l. Purity; Norman, Fox; Nottingham; Sea-Land; Spectrum Quality Prods.; St. Lawrence; Total Spec. Chems.; Voigt Global Distrib.; Vopak USA; Welch, Holme & Clark

Tall oil monoglycerides. *See* Glyceryl tallate
Tallol. *See* Tall oil

Tallow alcohol

CAS 99561-04-3; EINECS/ELINCS 308-980-4
Synonyms: Alcohols, tallow; Tallow fatty alcohols
Definition: Mixt. of fatty alcohols derived from tallow
Formula: $CH_3(CH_2)_nOH$, n = 16, 18
Properties: Waxy solid; m.p. 46 C
Uses: Emollient, emulsifier, consistency agent, cosolvent for cosmetics, skin creams/lotions; defoamer; in paper/paperboard in contact with aq./fatty foods
Regulatory: FDA 21CFR §176.170, 176.210
Manuf./Distrib.: Ashland; Cognis/Care Chems.; Goldschmidt; Indofine; Jarchem Ind.; Lipo; Mosselman NV; Procter & Gamble; RTD Chems.; Sasol N. Am.

Tallow fatty alcohols. *See* Tallow alcohol
TAME. *See* t-Amyl methyl ether
Tangantangan oil. *See* Castor (Ricinus communis) oil
Tar camphor. *See* Naphthalene
Tar oil. *See* Creosote oil
TBA. *See* t-Butyl alcohol
TBC. *See* Tributyl citrate
TBE. *See* Tetrabromoethane
TBEP. *See* Tributoxyethyl phosphate
TBHQ. *See* t-Butyl hydroquinone
TBP. *See* Tributyl phosphate
TBT. *See* p-t-Butyl toluene
TCA. *See* Trichloroacetic acid
TCE. *See* 1,1,2,2-Tetrachloroethane; 1,1,1-Trichloroethane; Trichloroethylene
1,1,1-TCE. *See* 1,1,1-Trichloroethane
1,1,2-TCE. *See* 1,1,2-Trichloroethane
TCP. *See* Tricresyl phosphate
TCTFE. *See* Trichlorotrifluoroethane
TDG. *See* Thiodiglycol
TDI. *See* Toluene-2,4-diisocyanate; Toluene diisocyanate
2,4-TDI. *See* Toluene-2,4-diisocyanate
TDI 80-20. *See* Toluene diisocyanate
TEA. *See* Triethylamine; Triethanolamine
Teaberry oil. *See* Methyl salicylate
TEBAC. *See* Benzyltriethyl ammonium chloride
TEC. *See* Triethyl citrate
Teel oil. *See* Sesame (Sesamum indicum) oil
TEG. *See* Triethylene glycol; PEG-4
TEL. *See* Tetraethyl lead
TELA. *See* Triethanolamine
TEN. *See* Triethylamine
TEP. *See* Triethyl phosphate
TEPA. *See* Tetraethylenepentamine
Terebenthine. *See* Turpentine
1,4(8)-Terpadiene. *See* Terpinolene
Terpene polymer resin. *See* Terpene resin

Terpene resin

CAS 9003-74-1
Synonyms: Polyterpene resin; Terpene polymer resin
Classification: Unsat. hydrocarbon; thermoplastic resin
Definition: Thermoplastic resin obtained by polymerization of turpentine in presence of cataysts
Properties: Sol. in most org. solvs.
Toxicology: LD50 (oral, rat) > 15 g/kg; somnolent effect; general depressed activity; TSCA listed
Hazardous Decomp. Prods.: Heated to decomp., emits acrid smoke and irritating fumes
Uses: Tackifier in hot-melt and pressure-sensitive adhesives; modifier resin in rubber compounding, coatings, laminations, waxes; extender for paints; binder in printing inks; impregnating paper; reinforcing agent; tires; tackifier, adhesion promoter in elastomers, polymers, adhesives, coatings, concrete-curing compds.; foods (color diluent, moisture barrier; as masticatory agents and solvent for elastomers in chewing gum); antiseptic in oral pharmaceuticals; in food-pkg. adhesives and pressure-sensitive adhesives; plasticizer in food-contact rubber articles for repeated use; in food-contact PP film; in food-contact coatings; in cellophane for food pkg.; defoamer in food-contact paper/paperboard
Use Level: Limitation 0.07% (of wt. of capsule) 7% (of ascorbic acid and salts)

Regulatory: FDA 21CFR §73.1, 172.280, 172.615, 175.105, 175.125, 175.300, 175.320, 177.1200, 177.2600, 178.3930; FDA approved for orals
Manuf./Distrib.: Allchem Ind.; Arizona; Cardolite; Cary; Chemcentral; Dujodwala Resins & Terpenes; Hercules; J.H. Calo; Langley-Smith Ltd; Monomer-Polymer & Dajac Labs
See also d-Limonene; dl-Limonene; l-Limonene; α-Pinene

γ-Terpinene

CAS 99-85-4; EINECS/ELINCS 202-794-6
FEMA 3559
Synonyms: 1,4-Cyclohexadiene, 1-methyl-4-(1-methylethyl)-; 1-Isopropyl-4-methyl-1,4-cyclohexadiene; 1,4-p-Menthadiene; p-Mentha-1,4-diene; 1-Methyl-4-isopropylcyclohexadiene-1,4; 1-Methyl-4-(1-methylethyl)-1,4-cyclohexadiene
Empirical: $C_{10}H_{16}$
Properties: Colorless oil or liq.; herbaceous citrus odor; sol. in alcohol, fixed oils; insol. in water; m.w. 136.24; dens. 0.848 (20/4 C); b.p. 183-186 C; flash pt. 50 C; ref. index 1.474 (20 C)
Toxicology: LD50 (oral, rat) 3650 mg/kg; mod. toxic by ing.; primary skin irritant; TSCA listed
Precaution: Flamm.
Hazardous Decomp. Prods.: Heated to decomp., emits acrid smoke and irritating fumes
Uses: Synthetic flavoring agent and fragrance in foods and pharmaceuticals; fragrance in cosmetics, perfumes
Features: Citrus flavor
Regulatory: FDA 21CFR §172.515; FEMA GRAS
Manuf./Distrib.: Aldrich; Fleurchem; Fluka; Millennium; Moore Ingreds.; Sigma; Treatt USA
Trade Names Containing: Citrus Terpenes

Terpineol. *See* α-Terpineol

α-Terpineol

CAS 98-55-5 (mixt. of α,β,γ); 8000-41-7 (mixed isomers); 10482-56-1; EINECS/ELINCS 202-680-6; 232-268-1 (mixed isomers)
UN NA 1993; FEMA 3045
Synonyms: 3-Cyclohexene-1-methanol, α,α, 4-trimethyl-; p-Menthen-1-en-8-ol; 1-p-Menthen-8-ol; p-Menth-1-en-8-ol; 1-Methyl-4-isopropyl-1-cyclohexen-8-ol; Pine oil synthetic; Terpineol; α,α,4-Trimethyl-3-cyclohexene-1-methanol
Definition: Several grades avail.; lower grades are mixed isomer prods.; higher grades are used in perfumes
Empirical: $C_{10}H_{18}O$
Formula: $C_{10}H_{17}OH$
Properties: Colorless visc. liq. or solid; lilac-like odor; very sol. in alcohol, ether; sol. in propylene glycol; very sl. sol. in water; insol. in min. oil; m.w. 154.24; dens. 0.935 (20/20 C); vapor pressure 5 mm Hg (80 C); m.p. 40-41 C; b.p. 214-224 C; flash pt. 89 C; ref. index 1.482; surf. tens. 31.6 dynes/cm
Toxicology: LD50 (oral, rat) 5170 mg/kg, (intramuscular, mouse) 2000 mg/kg; mod. toxic by intramuscular route; harmful by ing.; may be harmful by inh., skin absorp.; vapor/mist may cause eye/mucous membrane/upper respiratory tract irritation; skin irritant; can be a sensitizer; TSCA listed
Environmental: VOC; ThOD 1.97
Precaution: Combustible; incompat. with strong oxidizing agents, acid chlorides, acid anhydrides
Hazardous Decomp. Prods.: CO, CO_2; combustion: toxic fumes; heated to decomp., emits acrid smoke and irritating fumes
NFPA: Health 0, Flammability 2, Reactivity 0
Storage: Store in cool, dry place away from heat, open flame; keep tightly closed
Uses: Solvent for hydrocarbon materials; mutual solvent for resins, cellulose esters and ethers; fragrance in cosmetics, perfumes, soaps, detergents; sanitizer, fragrance in cleaners, disinfectants; antioxidant; denaturant, solvent in cosmetics; flotation frothing agent; reodorant in paints; synthetic flavoring agent in foods, pharmaceuticals; antiseptic; pharmaceutical solvent; in food-pkg. adhesives
Regulatory: FDA 21CFR §172.515, 175.105. 178.1010; FEMA GRAS; FDA approved for topicals; BP compliance
Manuf./Distrib.: ADA Int'l.; AMC Chems.; Advanced BioTech; Aldrich; Arakawa Chem. Ind.; Axxence Aromatic GmbH; Berje; Boith China; Creative Fragrances; Fleurchem; Givaudan Fragrances; Haarmann & Reimer GmbH; Hercules; J.H. Calo; Lluch Essence; MelChem; Millennium/F&F; Millennium; Penta Mfg.; Polarome Int'l.; Prodasynth; Quest Int'l.; Sarcom; Sigma; Soda Aromatic; Spectrum Quality Prods.; T&R Chem.; Toyotama Perfumery
Trade Names Containing: Dipentene-C; Pine Oil 150

Terpinolene

CAS 586-62-9; 68956-56-9; EINECS/ELINCS 209-578-0; 273-309-3
UN 2541 (DOT); FEMA 3046
Synonyms: 4-Isopropylidene-1-methylcyclohexene; p-Menth-1,4(8)-diene; 1-Methyl-4-(1-methylethylidene) cyclohexene; 1,4(8)-Terpadiene
Classification: Nonaromatic hydrocarbon
Empirical: $C_{10}H_{16}$
Properties: Water-wh. to pale amber liq. or oil; sol. in alcohol, ether, glycol; insol. in water; m.w. 136.24; dens. 0.864 (15.5/15.5 C); b.p. 183-185 C; flash pt. (CC) 37.2 C); ref. index 1.4890
Toxicology: LD50 (oral, rat) 4390 mg/kg; mildly toxic by ing.; TSCA listed
Precaution: Combustible; very dangerous fire hazard exposed to heat or flames; can react with oxidizing materials
Hazardous Decomp. Prods.: Heated to decomp., emits acrid smoke and irritating fumes
Uses: Solvent for resins, essential oils; mfg. of synthetic resins; synthetic flavoring agent in foods and pharmaceuticals; chem. intermediate, solvent, fragrance in personal care prods.
Regulatory: FDA 21CFR §172.515; FEMA GRAS
Manuf./Distrib.: Aldrich; Fluka; J.H. Calo; Lluch Essence; Millennium; Mosselman NV; Penta Mfg.
Trade Names: Dipentene Extra
Trade Names Containing: Citrus Terpenes; Dipentene-C

Terpinyl ethylene glycol ether

Classification: Glycol ether
Empirical: $C_{12}H_{22}O_2$
Properties: M.w. 174.18
Environmental: VOC; ThOD 3.03
Uses: Solvent
Regulatory: SARA §313 reportable; HAP

Tetrabromoacetylene. *See* Tetrabromoethane

Tetrabromoethane

CAS 79-27-6; EINECS/ELINCS 201-191-5
UN 2504
Synonyms: Acetylene tetrabromide; Ethane, 1,1,2,2-tetrabromo-; Muthmann's liquid; TBE; Tetrabromoacetylene; 1,1,2,2-Tetrabromoethane; s-Tetrabromoethane
Classification: Nonaromatic halogenated hydrocarbon
Empirical: $C_2H_2Br_4$
Formula: $CHBr_2CHBr_2$
Properties: Colorless to pale yel. heavy liq.; odor of camphor and iodoform; misc. with alcohol, chloroform, ether, aniline, glac. acetic acid, oxygenated and chlorinated solvs.; insol. in water; m.w. 345.66; dens. 2.96 kg/l (20 C); f.p. 0 C; b.p. 246 C (dec.); autoignition temp. 335 C; ref. index 1.6370; noncombustible
Toxicology: ACGIH TLV/TWA 1 ppm; LD50 (oral, rat) 1200 mg/kg, (IP, mouse) 443 mg/kg, (skin, rat) 5250 mg/kg; poison by inh., ing., and IP routes; eye/skin irritant; narcotic; may cause anorexia, nausea, severe headaches, abdominal pain, jaundice, monocytosis; may cause anesthesia, liver degeneration, hemorrhage, wt. loss, changes in liver wt.; questionable carcinogen; tumorigen; experimental neoplastigen; mutagenic data; target organs: nerves, liver; TSCA listed
Environmental: Harmful to aquatic organisms; may cause long-term adverse effects in aquatic environment; avoid release to environment
Hazardous Decomp. Prods.: Heated to decomp., emits highly toxic fumes of carbonyl bromide and Br^-
NFPA: Health 3, Flammability 0, Reactivity 1
Uses: Flame retardant for fibers; polymer/polyester fibers; solvent for fats, oils, greases, microscopy; high-density solvent for mineral separation, gauge fluids
Manuf./Distrib.: ABCR; Acros Org.; Aldrich; Dead Sea Bromine; Fluka; Ocean Chems.; Yixing Fangqiao East Chemical

1,1,2,2-Tetrabromoethane; s-Tetrabromoethane. *See* Tetrabromoethane
Tetrabromomethane. *See* Carbon tetrabromide
Tetrachlorethane. *See* 1,1,2,2-Tetrachloroethane
Tetrachlorethylene. *See* Perchloroethylene
Tetrachlorocarbon. *See* Carbon tetrachloride
Tetrachlorodifluoroethane; 1,1,2,2-Tetrachlorodifluoroethane. *See* 1,1,2,2-Tetrachloro-1,2-difluoroethane

1,1,2,2-Tetrachloro-1,2-difluoroethane

CAS 76-12-0
Synonyms: CFC 112; 1,2-Difluoro-1,1,2,2-tetrachloroethane; F 112; FC 112; Fluorocarbon 112; Freon 112; Halocarbon 112; R 112; Refrigerant 112; Tetrachlorodifluoroethane; 1,1,2,2-Tetrachlorodifluoroethane; sym-Tetrachlorodifluoroethane
Classification: Halogenated aliphatic hydrocarbon
Empirical: $C_2Cl_4F_2$
Formula: Cl_2FCCCl_2F
Properties: Colorless liq. or wh. cryst. solid; sl. ether-like odor; sol. in alcohol, ether, chloroform; pract. insol. in water; m.w. 203.83; sp.gr. 1.6447; vapor pressure 22,000 mm Hg; m.p. 25 C; b.p. 93 C; ref. index 1.412; surf. tens. 22.73 dynes/cm; dielec. const. 2.52; noncombustible
Toxicology: ACGIH TLV/TWA 500 ppm; LD50 (oral, mouse) 800 mg/kg; LC50 (inh., mouse, 2 h) 123 g/m^3; poison by inh.; mod. toxic by ing.; may cause drowsiness and difficulty breathing; skin and eye irritant; direct skin contact may cause drying or cracking of skin; may cause sl. eye irritation; TSCA listed
Environmental: VOC; ThOD 0.08
Precaution: Incompat. with reactive metals (sodium, potassium, magnesium, zinc, powd. aluminum); can react violently or explosively
Hazardous Decomp. Prods.: Heated to decomp., emits toxic fumes of F^- and Cl^-
Storage: Store in cool, dry, well-ventilated area; avoid handling @ temps. > 260 C (dec.)
Uses: Degreasing solvent in veterinary medicine to destroy parasitic worms
Manuf./Distrib.: Air Prods.; Halocarbon Prods.; Scott Spec. Gases

sym-Tetrachlorodifluoroethane. *See* 1,1,2,2-Tetrachloro-1,2-difluoroethane
Tetrachloroethane. *See* Perchloroethylene

1,1,1,2-Tetrachloroethane

CAS 630-20-6; EINECS/ELINCS 211-135-1
UN 1702
Synonyms: Ethane, 1,1,1,2-tetrachloro-; asym-Tetrachloroethane
Empirical: $C_2H_2Cl_4$
Formula: $ClCH_2CCl_3$
Properties: Colorless clear liq.; sol. in DMSO, 95% ethanol, acetone, chloroform, benzene; sol. > 10% in acid; misc. with ether, alcohol; sol. < 1 mg/ml in water; m.w. 167.85; sp.gr. 1.5406 (20/4 C); vapor pressure 10 mm Hg (19.3 C); m.p. -70.2 C; b.p. 130.5 C; nonflamm.; ref. index 1.4819 (20 C)
Toxicology: LD50 (oral, rat) 670 mg/kg, (skin, rabbit) 20 g/kg; poison; toxic by ing., inh., skin absorp.; skin/severe eye irritant; irritating to mucous membranes, upper respiratory tract; corrosive; lachrymator; readily absorbed thru skin; risk of irreversible effects; may cause CNS depression, nausea, vomiting, laryngitis, shortness of breath, headache, albuminuria, dizziness, somnolence, hematuria, jaundice, coma, etc.; experimental carcinogen; target organs: nerves, liver; TSCA listed
Environmental: Classified as hazardous waste and priority toxic pollutant by EPA
Precaution: Incompat. with strong oxidizers, strong bases, nitrogen tetraoxide, potassium, potassium hydroxide, sodium/sodium potassium alloy, etc.; may react with chemically act. metals, strong caustics, hot iron, Al, Zn in presence of steam
Hazardous Decomp. Prods.: Heated to decomp., emits very toxic fumes of CO, CO_2, hydrogen chloride gas; emits toxic fumes under fire conditions
Storage: Air-, moisture- and light-sensitive; keep container tightly closed under inert atmosphere; store refrigerated; protect from light
Uses: Chemical intermediate; solvent
Regulatory: SARA §313 reportable
Manuf./Distrib.: AccuStandard; Aldrich; Fluka; Riedel-deHaën; Supelco; TCI Am.

1,1,2,2-Tetrachloroethane

CAS 79-34-5; EINECS/ELINCS 211-135-1
UN 1702 (DOT)
Synonyms: Acetylene tetrachloride; 1,1-Dichloro-2,2-dichloroethane; TCE; Tetrachlorethane; sym-Tetrachloroethane
Classification: Halogenated aliphatic hydrocarbon
Empirical: $C_2H_2Cl_4$
Formula: $CHCl_2CHCl_2$
Properties: Colorless liq.; sickly sweet chloroform-like odor; sol. in alcohol, ether; misc. with most org. solvs.; sl. sol. in water; m.w. 167.84; dens. 1.596; m.p. -36 C; b.p. 146 C; flash pt. none; ref. index 1.4918
Toxicology: ACGIH TLV/TWA 1 ppm (skin); highly toxic; mild skin irritant; absorption thru skin may cause acute poisoning; may cause severe injury or death by inh., skin absorption, and ing.; inh. causes nausea, vomiting, weight loss, anemia, liver injury; peripheral nervous system damage; CNS depression, pulmonary edema, kidney damage; ing. may cause jaundice and severe liver injury, dizziness, unconsciousness, death; 5 ml can be fatal; experimental carcinogen; mutagen; TSCA listed
Precaution: Incompat. with strong bases (may react explosively), reactive metals (may react explosively with sodium, magnesium, aluminum)
Hazardous Decomp. Prods.: Hydrogen chloride, phosgene
Storage: Store in cool, dry, well-ventilated area out of direct sunlight; limit quantity stored
Uses: Intermediate in prod. of trichloroethylene; solvent; insecticide; cleansing and degreasing metals; paint removers; varnishes; lacquers; photographic film; resins; waxes; extraction of oils and fats; alcohol denaturant; org. synthesis; weed killer; fumigant
Manuf./Distrib.: Aldrich; Fluka; Pechiney Chems. Div.; Sigma; Spectrum Quality Prods.

asym-Tetrachloroethane. *See* 1,1,1,2-Tetrachloroethane
sym-Tetrachloroethane. *See* 1,1,2,2-Tetrachloroethane
Tetrachloroethene; Tetrachloroethylene; 1,1,2,2-Tetrachloroethylene. *See* Perchloroethylene
Tetrachloromethane. *See* Carbon tetrachloride
Tetradecafluorohexane; n-Tetradecafluorohexane. *See* Perfluorohexane
Tetradecane. *See* n-Tetradecane

n-Tetradecane

CAS 629-59-4; EINECS/ELINCS 211-096-0
UN 8027
Synonyms: Tetradecane
Empirical: $C_{14}H_{30}$
Formula: $CH_3(CH_2)_{12}CH_3$
Properties: Colorless liq. or oil; very sol. in alcohol, ether; insol. in water; m.w. 198.40; sp.gr. 0.7628 (20/4 C); vapor pressure 1 mm Hg (76.4 C); m.p. 5.89 C; b.p. 252-254 C; flash pt. 100 C; ref. index 1.4302 (20 C)
Toxicology: LDLo (IV, mouse) 5800 mg/kg; low toxicity by IV route; vapor or mist is irritating to eyes, skin, mucous membranes, upper respiratory tract; may be harmful by inh., ing., or skin absorp.; probably irritant and narcotic in high concs.; questionable carcinogen; experimental tumorigen; TSCA listed
Precaution: Combustible exposed to heat or flame; must be preheated before ignition can occur; LEL 0.5% in air; mod. explosion hazard in form of vapor when exposed to heat or flame; incompat. with strong oxidizing agents
Hazardous Decomp. Prods.: CO, CO_2; heated to decomp., emits acrid smoke and irritating fumes
Storage: Store refrigerated; keep tightly closed
Uses: Organic synthesis; solvent; standardized hydrocarbon; distillation chaser; fragrance in cosmetics
Manuf./Distrib.: AccuStandard; Acros Org.; Aldrich; Alfa Aesar; ChevronPhillips; Fisher Scientific; Fluka; ICN Biomed. Research Prods.; Indofine; Lancaster Synthesis; Pfaltz & Bauer; Riedel-deHaën; Sigma; Spectrum Quality Prods.; Supelco; TCI Am.
Trade Names Containing: Sasol C1316

Tetradecanoic acid, isocetyl ester; Tetradecanoic acid, isohexadecyl ester. *See* Isocetyl myristate
Tetradecanoic acid isopropyl ester. *See* Isopropyl myristate

Tetradecanoic acid, methyl ester. *See* Methyl myristate
Tetradecanoic acid, 1-methylethyl ester. *See* Isopropyl myristate
Tetradecanoic acid, monoester with 1,2,3-propanetriol. *See* Glyceryl myristate
Tetradecanoic acid, tetradecyl ester. *See* Myristyl myristate
1-Tetradecanol; N-Tetradecanol-1. *See* Myristyl alcohol
1-Tetradecanol, 2-decyl-. *See* 2-Decyl-1-tetradecanol
1-Tetradecanol, propanoate. *See* Myristyl propionate
Tetradecene. *See* Tetradecene-1

Tetradecene-1

CAS 1120-36-1; EINECS/ELINCS 272-493-2
Synonyms: C14 linear alpha olefin; Tetradecene; 1-Tetradecene; α-Tetradecylene
Empirical: $C_{14}H_{28}$
Formula: $CH_3(CH_2)_{11}CH:CH_2$
Properties: Colorless liq., mild hydrocarbon odor; sl. sol. in alcohol, ether; insol. in water; m.w. 196.38; dens. 0.775; m.p. -13 C; pour pt. -18 C; b.p. 251 C; flash pt. (Seta) 226 F; ref. index 1.4360
Toxicology: Irritating to eyes and skin; low acute inhalation toxicity; low acute ingestion toxicity but ingestion may cause vomiting, aspiration of vomitus
Precaution: Combustible; avoid contact with air or oxygen; explosion danger from peroxide formation
Storage: Store under nitrogen blanket
Uses: Intermediate for surfactants and specialty industrial chemicals (polymers, plasticizers, lubricants, gasoline additives, paper sizing, PVC lubricants); prod. of amines, amine oxides, oxo alcohols, alkylated aromatics, tanning oils, alpha olefin sulfonates, epoxides, syn. fatty acids; solvent in perfumes, flavors, medicines, dyes, oils, resins
Manuf./Distrib.: Advanced Synthesis Tech.; Albemarle; Aldrich; BP Amoco; ChevronPhillips; Fluka; Shell; Sigma; Spectrum Quality Prods.

1-Tetradecene. *See* Tetradecene-1
Tetradecyl alcohol; N-Tetradecyl alcohol. *See* Myristyl alcohol
α-Tetradecylene. *See* Tetradecene-1
Tetradecyl propionate. *See* Myristyl propionate
Tetradecyl tetradecanoate. *See* Myristyl myristate

Tetraethanol ammonium hydroxide

CAS 631-41-4
Synonyms: Ethanaminium, 2-hydroxy-N,N,N-tris (2-hydroxyethyl)-, hydroxide
Definition: Commercial grade is 40% aq. sol'n.
Empirical: $C_8H_{21}NO_5$
Formula: $(HOCH_2CH_2)_4NOH$
Properties: Wh. cryst. solid; completely sol. in water; m.w. 211.30; vapor pressure < 0.01 mm (20 C); strong base; aq. sol'ns. stable at R.T., but dec. on heating
Toxicology: LD50 (oral, rat) 2250 mg/kg; mod. toxic by ing.; severe eye irritant; strong irritant to skin and tissue; TSCA listed
Hazardous Decomp. Prods.: Heated to decomp., emits toxic fumes of NH_3 and NO_x
Uses: Alkaline catalyst; solvent for some dyes, metal-plating sol'ns.
Manuf./Distrib.: R S A

Tetraethylene glycol. *See* PEG-4
Tetraethylene glycol dimethyl ether. *See* Dimethoxytetraglycol

Tetraethylene glycol ethyl ether

CAS 5650-20-4
Synonyms: PEG-4 ethyl ether; Tetraethylene glycol monoethyl ether
Uses: In solvent blends for protective coatings, inks, cleaners, agric. chems.

Tetraethylene glycol methyl ether

CAS 23783-42-8
Synonyms: PEG-4 methyl ether; Tetraethylene glycol monomethyl ether
Empirical: $C_9H_{20}O_5$
Uses: In solvent blends for protective coatings, inks, cleaners, agric. chems.

Tetraethylene glycol monoethyl ether. *See* Tetraethylene glycol ethyl ether
Tetraethylene glycol monomethyl ether. *See* Tetraethylene glycol methyl ether

Tetraethylenepentamine

CAS 112-57-2; EINECS/ELINCS 203-986-2
UN 2320 (DOT)
Synonyms: 1,2-Ethanediamine, N-(2-aminoethyl)-N′-(2-((2-aminoethyl) amino) ethyl)-; 1,4,7,10,13-Pentaazatridecane; TEPA
Classification: Amine
Empirical: $C_8H_{23}N_5$
Formula: $NH_2(CH_2CH_2NH)_3CH_2CH_2NH_2$
Properties: Amber visc. liq.; ammonia odor; sol. in most org. solvs. and water; m.w. 189.31; dens. 0.9980 (20/20 C); vapor pressure < 0.01 mm Hg (20 C); m.p. -30 C; b.p. 340 C (dec.); flash pt. (OC) 163 C; autoignition temp. 321 C; ref. index 1.5055
Toxicology: LD50 (oral, rat) 3990 mg/kg, (dermal, rabbit) 660 mg/kg, (IV, mouse) 320 mg./kg; poison by ing. and IV route; mod. toxic by skin contact; corrosive; strong primary irritant to skin, eyes, mucous membranes; causes burns; mutagen; TSCA listed
Environmental: VOC; ThOD 0.72
Precaution: Combustible exposed to heat or flame; LEL 0.8%; UEL 4.6%; corrosive; can react with oxidizing materials
Hazardous Decomp. Prods.: Heated to decomp., emits toxic fumes of NO_x
Storage: Hygroscopic
Uses: Solvent for sulfur, acid gases, various resins and dyes; saponifying agent for acidic materials; mfg. of syn. rubber; dispersant in motor oils; intermediate for oil additives; ion-exchange resins; surfactants; R.T. curing agent in two-pkg. protective coating systems; nickel plating brightening agent; corrosion inhibitor in petrol. prod.; catalyst activator in acrylic food pkg.; in food-pkg. adhesives; amino resin modifier in paper/paperboard in contact with aq./fatty foods; polymerization crosslinking agent in paper/paperboard in contact with dry food
Regulatory: FDA 21CFR §175.105, 176.170, 176.180, 177.1010
Manuf./Distrib.: Aldrich; Alkyl Amines Chems. Ltd; Allchem Ind.; Ashland; Brook-Chem; Coyne; Dow; Filo; Filo; Fluka; Great Western; Houghton Chem.; J.H. Calo; Loos & Dilworth; Tosoh

Tetraethyl lead

CAS 78-00-2; EINECS/ELINCS 201-075-4
UN NA 1649 (DOT)
Synonyms: Lead, tetraethyl-; TEL; Tetraethyl lead, liquid; Tetraethyl plumbane; Tetraethylplumbium
Classification: Organometallic compounds
Empirical: $C_8H_{20}Pb$
Formula: $Pb(C_2H_5)_4$
Properties: Colorless, oily liq.; pleasant odor; sol. in all org. solvs.; insol. in water and dil. acids or alkalis; m.w. 323.47; dens. 1.65; vapor pressure 1 mm Hg (38.4 C); f.p. -136 C; b.p. 198-202 C, 75-85 C (13-14 mm); flash pt. 200 F; ref. index 1.5198
Toxicology: ACGIH TLV/TWA 0.1 mg(Pb)/m^3 (skin); LD50 (oral, rat) 12,300 μg/kg, (IP, rat) 15 mg/kg, (IV, rat) 14,400 μg/kg; poison by ing., IV, subcut., parenteral routes; mod. toxic by inh., skin contact; human poison; toxic to CNS; can cause elemental lead intoxication by skin contact; questionable carcinogen; experimental teratogen, reproductive effects; tumorigen; mutagen; target organs: nerves, blood, kidneys, male reproductive; TSCA listed
Precaution: Combustible; can react vigorously with oxidizing materials; dec. exposed to sunlight or allowed to evaporate; on decomp., forms triethyl lead (a poison)
Hazardous Decomp. Prods.: Heated to decomp., emits toxic fumes of Pb
Storage: Freeze
Uses: Antiknock agent; petrol. octane booster; solvent for fatty materials; some solvent action on rubber
Manuf./Distrib.: Aldrich; Sigma

Tetraethyl lead, liquid; Tetraethyl plumbane; Tetraethylplumbium. *See* Tetraethyl lead
Tetrafluorodichloroethane; 1,1,2,2-Tetrafluoro-1,2-dichloroethane. *See* 1,2-Dichlorotetrafluoroethane

Tetrafluoromethane

CAS 75-73-0; EINECS/ELINCS 200-896-5
UN 1982 (DOT)

Synonyms: Carbon tetrafluoride; Fluorocarbon 14; Freon 14; Perfluoromethane
Classification: Nonaromatic halogenated hydrocarbon
Empirical: CF_4
Properties: Colorless inert gas; sl. sol. in water; m.w. 88.01; dens. 1.96 (-184 C); f.p. -184 C; b.p. -128 C; ref. index 1.151; surf. tens. 4 dynes/cm; dielec. const. 1.001
Toxicology: Toxic by inh.; target organ: heart; TSCA listed
Environmental: ThOD 0
Hazardous Decomp. Prods.: Heated to decomp., emits toxic fumes of F^-
Storage: Store away from heat
Uses: Refrigerant; gaseous insulator; inert filler gas in elec. switchgear; solvent
Manuf./Distrib.: Air Prods.; Aldrich; Dow; Matheson Gas Prods.

1H,1H,3H-Tetrafluoro-1-propanol; 2,2,3,3-Tetrafluoropropanol. *See* 2,2,3,3-Tetrafluoro-1-propanol

2,2,3,3-Tetrafluoro-1-propanol

CAS 76-37-9; EINECS/ELINCS 200-955-5
Synonyms: 1H,1H,3H-Tetrafluoro-1-propanol; 2,2,3,3-Tetrafluoropropanol; TFP
Empirical: $C_3H_4G_4O$
Formula: $CHF_2CF_2CH_2OH$
Properties: Colorless clear liq.; m.w. 132.06; dens. 1.471 g/cm^3 (20 C); m.p. -15 C; b.p. 106 C (1 atm); flash pt. 49 C; ref. index 1.3210; pH 6
Toxicology: LDLo (rat) 3400 mg/kg; may be harmful by ing., inh.; irritating to eyes, skin, respiratory system
Precaution: Flamm.; incompat. with strong oxidizing agents, strong bases, sodium, potassium, acid chlorides, acid anhydrides
Uses: Intermediate for pharmaceuticals, pesticides; cleaning agent; solvent coating for disks
Manuf./Distrib.: Interchim; Peakchem; Wuzhou Int'l.; Zhejiang Grand Chembay; Zhejiang Sanhuan Chems. Co. Ltd

Tetraglyceryl monooleate. *See* Polyglyceryl-4 oleate
Tetraglycol. *See* PEG-4; Glycofurol
Tetraglyme. *See* Dimethoxytetraglycol
1,2,3,4-Tetrahydrochinolin. *See* 1,2,3,4-Tetrahydroquinoline
Tetrahydro-1,4-dioxin; Tetrahydro-p-dioxin. *See* 1,4-Dioxane

Tetrahydrofuran

CAS 109-99-9; EINECS/ELINCS 203-726-8
UN 2056 (DOT)
Synonyms: Butane, 1,4-epoxy-; Butylene oxide; Cyclotetramethylene oxide; Diethylene oxide; 1,4-Epoxybutane; Furanidine; Furan, tetrahydro-; Hydrofuran; Oxacyclopentane; Oxolane; Tetramethylene oxide; THF
Classification: Sat. cyclic aliphatic ether
Empirical: C_4H_8O
Formula: $CH_2CH_2CH_2CH_2O$
Properties: Water-wh. volatile liq.; ethereal odor; very sol. in water; sol. in alcohols, ketones, esters, ethers, hydrocarbons, oxygenated solvs.; m.w. 72.11; dens. 0.888 (20/4 C); vapor pressure 114 mm (15 C); f.p. -65 C; m.p. -108.5 C; b.p. 66 C; flash pt. (OC) -15 C; ref. index 1.4070
Toxicology: ACGIH TLV/TWA 200 ppm; STEL 250 ppm; LD50 (oral, rat) 1650 mg/kg, (IP, rat) 2900 mg/kg; LC50 (inh., rat, 3 h) 21,000 ppm; mod. toxic by ing., IP; mildly toxic by inh.; eye/mucous membrane irritant; essentially nonirritating to skin; CNS depressant; narcotic, anesthetic at high concs.; may injure liver and kidneys; 25,000 ppm reported lethal in humans; tumorigen; mutagen; reproductive effects; TSCA listed
Environmental: VOC; ThOD 2.44
Precaution: Extremely flamm.; flamm. limits in air 2-11.8%; incompat. with strong oxidizing agents (increases fire/explosion risk), bromine (reacts vigorously with gas evolution), caustic alkalis (violent reactions possible)
Hazardous Decomp. Prods.: Peroxides may form and accumulate on exposure to air and light, in the absence of inhibitors; heated to decomp., emits acrid smoke and irritating fumes
NFPA: Health 2, Flammability 3, Reactivity 1
Storage: Store in cool, dry, well-ventilated area, out of direct sunlight
Uses: Solvent for Grignard reactions, reductions, polymerizations, sol'n. S/B prod., resins, paints, adhesives, sealants, magnetic tapes, printing inks; solvent welding agent for PVC; extractant; reaction medium; chemical intermediate; polyacetal resin comonomer; surf. coating of films and textiles with PVC or PU sol'ns.; mfg. of magnetizable information carriers; intermediate, extraction solvent for pharmaceuticals; steroid hormone prod. for use in birth control pills; preservative for histological samples; in food-pkg. adhesives; adjuvant in resinous/polymeric food-contact coatings; adjuvant in slimicides in food-contact paper; in cellophane for food pkg.; solvent in casting of food-contact film
Regulatory: FDA 21CFR §175.105, 175.320, 176.300, 177.1200, 178.3950; CERCLA hazardous substance
Manuf./Distrib.: AAE Chemie NV; AMC Chems.; Aldrich; Alfa Chem; Allchem Ind.; Ashland; BASF; Brenntag AG; Burdick & Jackson; Chemcentral; Coyne; DuPont; Fisher Scientific; Fluka; Great Lakes; Haltermann Prods. UK; Harcros; Honeywill & Stein; Houghton Chem.; Hukill; ISP; Integra; J.T. Baker; KIC Chems.; Kessler; Lyondell Europe; Lyondell; Mallinckrodt Baker; Mitsubishi Chem.; Parchem Trading; Richman; Romil Ltd; Sigma; Spectrum Quality Prods.; Stanchem; VWR Int'l.; Vopak USA; Whyte Chems. Ltd
Trade Names: Lyondell Tetrahydrofuran (THF)

Tetrahydro-2-furancarbinol; Tetrahydro-2-furanmethanol. *See* Tetrahydrofurfuryl alcohol
Tetrahydro-2-furanmethanol acetate. *See* Tetrahydrofurfuryl acetate
Tetrahydro-2-furanone. *See* Butyrolactone

Tetrahydrofurfuryl acetate

CAS 637-64-9; EINECS/ELINCS 211-296-8
FEMA 3055
Synonyms: 2-Furanmethanol, tetrahydro-, acetate; Tetrahydro-2-furanmethanol acetate
Definition: Ester of tetrahydrofurfuryl alcohol and acetic acid
Empirical: $C_7H_{12}O_3$
Formula: $C_4H_7OCH_2OOCCH_3$
Properties: Colorless liq.; sol. in water, alcohol, ether, chloroform; dens. 1.061 (20/0 C); b.p. 194-195 C (753 mm)
Precaution: Combustible
Uses: Synthetic flavoring agent in foods and pharmaceuticals; solvent in cosmetics
Regulatory: FDA 21CFR §172.515; FEMA GRAS
Manuf./Distrib.: Aldrich

Tetrahydrofurfuryl alcohol

CAS 97-99-4; EINECS/ELINCS 202-625-6
FEMA 3056
Synonyms: Tetrahydro-2-furancarbinol; Tetrahydro-2-furanmethanol; Tetrahydro-2-furylmethanol; THFA
Classification: Cyclic alcohol
Empirical: $C_5H_{10}O_2$
Formula: $C_4H_7OCH_2OH$
Properties: Colorless liq., mild odor; misc. with water, alcohol, ether, acetone, chloroform, benzene, oxygenated solvs.; m.w. 102.14; dens. 1.053 (20/4 C); visc. 6.24 cp (20 C); vapor pressure 0.801 mm Hg; m.p. < -80 C; b.p. 173-177 C; flash pt. (TOC) 84 C; autoignition temp. 282 C; ref. index 1.453 (20 C); surf. tens. 37 dyne/cm; dielec. const. 13.61
Toxicology: LD50 (oral, rat) 2500 mg/kg, (IV, rabbit) 725 mg/kg; mod. toxic by ing., IV, IP routes; severe eye irritant; mod. irritating to skin, mucous membranes; TSCA listed
Environmental: VOC; BOD5 1.50; ThOD 2.04
Precaution: Combustible exposed to heat, flame; explosive limit 1.5-9.7 vol.%; explosive as vapor exposed to heat or flame; incompat. with oxidizing materials; violent or explosive reaction with 3-nitro-N-bromophthalimide
Hazardous Decomp. Prods.: Heated to decomp., emits acrid smoke and irritating fumes
NFPA: Health 2, Flammability 2, Reactivity 0
Storage: Hygroscopic
Uses: Solvent for vinyl resins, paints, paint strippers, cleaners, dyes for leather, chlorinated rubber, cellulose esters, cosmetics; coupling agent; coupling solvent for pesticides, textile auxs.; solvent, softener for nylon; synthetic flavoring agent in foods and pharmaceuticals; in food-

pkg. adhesives; defoamer in food-contact paper/paperboard
Regulatory: FDA 21CFR §172.515, 175.105, 176.180, 176.210; FEMA GRAS
Manuf./Distrib.: Aldrich; Allchem Ind.; Ashland; Bencorp Int'l.; Continental Ind. Group; Fluka; General Chem.; Great Lakes; Great Western; Int'l. Furan Chems. USA; Interchem; Penn Spec. Chems.; Penta Mfg.; Schweizerhall; Sigma; Vopak USA; Westpro
Trade Names: THFA®

Tetrahydrofurfuryl benzoate

Empirical: $C_{12}H_{14}O_3$
Properties: M.w. 206.24
Uses: Solvent for agric. formulations
Manuf./Distrib.: Penn Spec. Chems.

Tetrahydrofurfuryl polyethylene glycol; Tetrahydrofurfuryl-polyethylene glycol ether. *See* Glycofurol
Tetrahydro-2-furylmethanol. *See* Tetrahydrofurfuryl alcohol
Tetrahydro-1,4-isoxazine; Tetrahydro-p-isoxazine. *See* Morpholine

Tetrahydronaphthalene

CAS 119-64-2; EINECS/ELINCS 204-340-2
Synonyms: Naphthalene-1,2,3,4-tetrahydride; 1,2,3,4-Tetrahydronaphthalene; Tetralin; Tetranap; THN
Empirical: $C_{10}H_{12}$
Properties: Colorless liq., pungent menthol odor; misc. with most solvs., ethanol, butanol, acetone, ether, benzene; insol. in water; m.w. 132.21; dens. 0.981 (13 C); vapor pressure 1 mm (38 C); m.p. -25 C; b.p. 206 C; flash pt. 160 F; ref. index 1.5410
Toxicology: LD50 (oral, rat) 2860 mg/kg, (dermal, rabbit) 17 g/kg; mod. toxic by ing.; mildly toxic by skin contact; primary irritant to eyes, mucous membranes, respiratory system; severe skin irritant; narcotic in high concs.; may cause cataracts, kidney damage; tumorigen; TSCA listed
Precaution: Combustible exposed to heat or flame; can react with oxidizers; explosive as vapor; explosive limits 0.8% (212 F) to 5% (302 F); may be explosive on prolonged close contact with air
Hazardous Decomp. Prods.: Heated to decomp., emits acrid smoke and irritating fumes
Uses: Chemical intermediate; solvent for greases, fats, oils, waxes, rubber, asphalt, coatings, pharmaceuticals; heat-transfer fluid; substitute for turpentine
Manuf./Distrib.: Aldrich; Degussa AG; Degussa; DuPont; Fluka; J.T. Baker; Penta Mfg.; Sigma; Sithean; Spectrum Quality Prods.
Trade Names: Tetralin®; Tetralin

1,2,3,4-Tetrahydronaphthalene. *See* Tetrahydronaphthalene
Tetrahydro-1,4-oxazine; Tetrahydro-2H-1,4-oxazine; Tetrahydro-4H-1,4-oxazine; Tetrahydro-p-oxazine. *See* Morpholine
Tetrahydropyran-2-carbinol; 2-Tetrahydropyranilcarbinol. *See* Tetrahydropyran-2-methanol

Tetrahydropyran-2-methanol

CAS 100-72-1; EINECS/ELINCS 202-882-4
Synonyms: 2-Hydroxymethyltetrahydropyran; 2-Methanol tetrahydropyran; 2-Methyloltetrahydro-1,4-pyran; Pyran-2-methanol, tetrahydro-; 2H-Pyran-2-methanol, tetrahydro-; Tetrahydropyran-2-carbinol; 2-Tetrahydropyranilcarbinol; Tetrahydropyranyl-2-methanol
Classification: Ether
Empirical: $C_6H_{12}O_2$
Properties: Colorless liq.; sol. in water; m.w. 116.18; sp.gr. 1.0272 (20/20 C); visc. 11 cP; vapor pressure 0.1 mm Hg (20 C); f.p. -70 C; b.p. 187 C; flash pt. 93 C; ref. index 1.458; surf. tens. 34.1 dynes/cm
Toxicology: LD50 (oral, rat) 3730 mg/kg, (skin, rabbit) 4000 mg/kg; mod. toxic by ing., skin contact; severe skin and eye irritant; harmful by inh., skin absorp.; high concs. extremely destructive to tissues of mucous membranes, upper respiratory tract, eyes, skin; may cause burning sensation, coughing, wheezing, laryngitis, shortness of breath, headache, nausea, vomiting; TSCA listed
Environmental: VOC; ThOD 2.21
Precaution: Combustible exposed to heat or flame; incompat. with strong oxidizing agents; can react with oxidizing materials
Hazardous Decomp. Prods.: Toxic fumes of CO, CO_2; heated to decomp., emits acrid smoke and irritating fumes; emits toxic fumes under fire conditions
NFPA: Health 1, Flammability 2, Reactivity 0
Storage: Store in cool, dry place; keep tightly closed
Uses: Solvent
Manuf./Distrib.: Aldrich

Tetrahydropyranyl-2-methanol. *See* Tetrahydropyran-2-methanol

1,2,3,4-Tetrahydroquinoline

CAS 635-46-1; EINECS/ELINCS 211-237-6
Synonyms: Quinoline, 1,2,3,4-tetrahydro-; 1,2,3,4-Tetrahydrochinolin
Classification: Aromatic amine
Empirical: $C_9H_{11}N$
Properties: Yel. clear liq.; m.w. 133.19; dens. 1.061; vapor pressure 0.1 mbar (20 C); m.p. 15-17 C; b.p. 249 C; flash pt. 100 C; ref. index 1.5945
Toxicology: Irritating to eyes, skin, respiratory system; TSCA listed
Storage: Store in tightly closed container in cool, dry, well-ventilated area away from incompat. substances; keep container closed when not in use
Uses: Intermediate in the chemical, polymer, agric., dye, and pharmaceutical industries; basic solvent with high b.p.
Manuf./Distrib.: Acros Org.; Advanced Synthesis Tech.; Aldrich; Fluka; Raschig

Tetrahydrothiophene

CAS 110-01-0; EINECS/ELINCS 203-728-9
UN 2412 (DOT)
Synonyms: Tetramethylene sulfide; Thiacyclopentane; Thiolane; Thiophane; Thiophene, tetrahydro-; THT
Empirical: C_4H_8S
Formula: $S(CH_2)_3CH_2$
Properties: Water-wh. liq.; penetrating odor; sol. in oxygenated solvs.; misc. with most solvs. except water; m.w. 88.18; dens. 1.00 (13.6/15.6 C); m.p. -96 C; b.p. 115-124.4 C; flash pt. 55 F; ref. index 1.5030
Toxicology: LC50 (inh., mouse, 2 h) 27 g/m^3; mildly toxic by inh.; TSCA listed
Precaution: Flamm.; potentially explosive reaction with hydrogen peroxide
Hazardous Decomp. Prods.: Heated to decomp., emits toxic fumes of SO_x
Uses: Solvent; intermediate; fuel gas odorant
Manuf./Distrib.: Aldrich; Atofina; ChevronPhillips; Fluka; Natural Gas Odorizing; Occidental

Tetrahydro-3,4-thiophenediol. *See* Thiolanediol
Tetrahydrothiophene dioxide; Tetrahydrothiophene-1,1-dioxide; 2,3,4,5-Tetrahydrothiophene-1,1-dioxide. *See* Sulfolane
N,N,N´,N´-Tetra-(2-hydroxypropyl)-1,2-diaminoethane. *See* Tetrahydroxypropyl ethylenediamine

Tetrahydroxypropyl ethylenediamine

CAS 102-60-3; EINECS/ELINCS 203-041-4
Synonyms: Edetol; Ethylenedinitrilotetra-2-propanol; 1,1´,1´´,1´´´-(Ethylenedinitriolo) tetra-2-propanol; N,N,N´,N´-Tetra-(2-hydroxypropyl)-1,2-diaminoethane; N,N,N´,N´-Tetrakis (2-hydroxypropyl) ethylenediamine
Classification: Substituted amine
Empirical: $C_{14}H_{32}N_2O_4$
Formula: $(HOC_3H_6)_2NCH_2CH_2N(C_3H_6OH)$
Properties: Water-wh. visc. liq.; sol. in ethanol, toluene, ethylene glycol; misc. with water; m.w. 292.42; dens. 1.013; b.p. 175-181 C (0.8 mm); flash pt. > 230 F; ref. index 1.4812; cationic
Toxicology: LD50 (mammal) 3900 mg/kg; mod. toxic by unspecified route; may be irritating to skin and mucous membranes; may cause skin sensitization; TSCA listed
Precaution: Combustible
Hazardous Decomp. Prods.: Heated to decomp., emits toxic fumes of NO_x
Uses: Chelating agent in cosmetics; emulsifier; antistat; rewetting agent; grease additive; textile lubricant; herbicides; fungicides; insecticides; adhesives; latexes; resins; plasticizers; inks; resins; pharmaceutical

solvent, preservative, chelating agent, emulsifier; propoxylated/ ethoxylated surfactant intermediate; curing agent in food-contact PU resins
Regulatory: FDA 21CFR §177.1680
Manuf./Distrib.: Aldrich; Fluka; Sigma

N,N,N´,N´-Tetrakis (2-hydroxypropyl) ethylenediamine. *See* Tetrahydroxypropyl ethylenediamine
Tetralin. *See* Tetrahydronaphthalene
Tetramethylene cyanide. *See* Adiponitrile
Tetramethylene glycol; 1,4-Tetramethylene glycol. *See* 1,4-Butanediol
Tetramethylene oxide. *See* Tetrahydrofuran
Tetramethylene sulfide. *See* Tetrahydrothiophene
Tetramethylene sulfone. *See* Sulfolane
1,1,2,2-Tetramethylethane. *See* 2,3-Dimethylbutane
2,2,6,6-Tetramethyl-4-methyleneheptane. *See* 2-(2,2-Dimethylpropyl)-4,4-dimethylpent-1-ene

Tetramethylsilane

CAS 75-76-3; EINECS/ELINCS 200-899-1
UN 2749
Synonyms: Silane, tetramethyl-; Tetramethyl silicane; TMS
Classification: Organo-silicon compd.
Empirical: $C_4H_{12}Si$
Formula: $(CH_3)_2Si(CH_3)_2$
Properties: Colorless volatile liq.; insol. in water, cold conc. sulfuric acid; sol. in most org. solvs.; m.w. 88.22; dens. 0.6411; vapor pressure 589 mm (20 C); m.p. -99.5 C; b.p. 26.6-26.7 C; flash pt. -28 C; ref. index 1.3588 (20 C)
Toxicology: May be irritating to eyes, skin, mucous membranes, upper respiratory tract; may be harmful by inh., ing., skin absorp.; TSCA listed
Precaution: Extremely flamm.; vapor may flash back; spontaneously flamm. in air; incompat. with strong acids/bases/oxidizers; explosive reaction with chlorine + antimony trichloride (above 100 C); protect against static discharges; container explosion may occur
Hazardous Decomp. Prods.: CO, CO_2, silicon oxide; emits toxic fumes under fire conditions
HMIS: Health 1, Flammability 4, Reactivity 0
Storage: Keep container tightly closed in cool place; store under nitrogen; refrigerate; moisture-sensitive; may develop pressure; keep away from heat, sparks, open flame
Uses: Fuel mixture component (aviation fuel); internal standard for NMR analytical instruments; coupling agent; blocking agent; release agent; lubricant; solvent for tissue drying
Features: Chemically inert
Manuf./Distrib.: ABCR; AP Resources; Acros Org.; Advanced Synthesis Tech.; Aldrich; Alfa Aesar; EM Science; Electron Microscopy Sciences; Fluka; Gelest; ICN Biomed. Research Prods.; J.T. Baker; Lancaster Synthesis; Mallinckrodt Baker; Pfaltz & Bauer; Sigma; Spectrum Quality Prods.; Strem Chems.; TCI Am.; United Chem. Tech.

Tetramethyl silicane. *See* Tetramethylsilane

Tetramethylurea

CAS 632-22-4; EINECS/ELINCS 211-173-9
Synonyms: 1,1,3,3-Tetramethylurea; TMU; Urea, 1,1,3,3-tetramethyl-
Classification: Nonaromatic amide
Empirical: $C_5H_{12}N_2O$
Formula: $[(CH_3)_2N]_2CO$
Properties: Liq.; faint pleasant odor; very sol. in alcohol, ether; misc. with water, petrol. ether, most common org. solvs.; m.w. 116.16; dens. 0.97 kg/l (20 C); m.p. -1.2 C; b.p. 176-178 C; flash pt. 167 F; ref. index 1.4493
Toxicology: LD50 (oral, mouse) 2920 mg/kg, (IV, rat) 1100 mg/kg; LDLo (IP, mouse) 2900 mg/kg; mod. toxic by ing. and IV routes; may cause somnolence, anesthesia; experimental teratogen, reproductive effector; human mutagenic data; TSCA listed
Precaution: Combustible exposed to heat, flame, or oxidizers
Hazardous Decomp. Prods.: Heated to decomp., emits toxic fumes of NO_x
Storage: Hygroscopic
Uses: High-boiling aprotic solvent and reagent
Manuf./Distrib.: Aceto; Acros Org.; Aldrich; Fluka; Sigma

1,1,3,3-Tetramethylurea. *See* Tetramethylurea
Tetranap. *See* Tetrahydronaphthalene
2,5,8,11-Tetraoxadodecane. *See* PEG-3 dimethyl ether
3,6,9,12-Tetraoxatetradecane. *See* Triethylene glycol diethyl ether
3,6,9,12-Tetraoxatetradecane-1,14-diol. *See* Pentaethylene glycol
Tetraphene. *See* 1,2-Benzanthracene
Tetrole. *See* Furan
Textile spirits. *See* Naphtha, solvent-refined light
TFE. *See* 2,2,2-Trifluoroethanol
TFP. *See* 2,2,3,3-Tetrafluoro-1-propanol
Theobroma cacao; Theobroma cacao butter; Theobroma oil. *See* Cocoa (Theobroma cacao) butter
THF. *See* Tetrahydrofuran
THFA. *See* Tetrahydrofurfuryl alcohol
Thiacyclopentadiene. *See* Thiophene
Thiacyclopentane. *See* Tetrahydrothiophene
Thiacyclopentane dioxide. *See* Sulfolane
3-Thiapentane. *See* Ethyl sulfide
Thiaphene. *See* Thiophene
2-Thiapropane. *See* Dimethyl sulfide
1,1´-Thiobisethane. *See* Ethyl sulfide
2,2´-Thiobisethanol. *See* Thiodiglycol
Thiobismethane. *See* Dimethyl sulfide
3,3´-Thiobis (methyl propionate). *See* Dimethyl thiodipropionate
Thiobutanol; Thiobutylalcohol. *See* n-Butyl mercaptan
Thiocyan. *See* Calcium thiocyanate
Thiocyanate sodium; Thiocyanic acid, sodium salt. *See* Sodium thiocyanate
Thiocyclopentane-1,1-dioxide. *See* Sulfolane
2,2´-Thiodiethanol; Thiodiethylene glycol. *See* Thiodiglycol

Thiodiglycol

CAS 111-48-8; EINECS/ELINCS 203-874-3
Synonyms: Bis (2-hydroxyethyl) sulfide; Bis (β-hydroxyethyl) sulfide; β,β´-Dihydroxydiethyl sulfide; Dihydroxyethyl sulfide; β,β´-Dihydroxyethyl sulfide; Ethanol, 2,2´-thiodi-; β-Hydroxyethyl sulfide; TDG; 2,2´-Thiobisethanol; 2,2´-Thiodiethanol; Thiodiethylene glycol; β-Thiodiglycol
Classification: Aliphatic alcohol
Empirical: $C_4H_{10}O_2S$
Formula: $HOCH_2CH_2 \cdot S \cdot CH_2CH_2OH$
Properties: Colorless visc. liq., char. odor; sol. in acetone, alcohol, chloroform, water; sl. sol. in benzene, CCl_4, ether; m.w. 122.18; dens. 1.1852 (20/20 C); m.p. -16 C; b.p. 283 C (dec.); flash pt. (OC) 320 F
Toxicology: LD50 (oral, rat) 6610 mg/kg, (IV, rabbit) 3000 mg/kg, (skin, rabbit) 20 ml/kg, (subcut., rat) 4 g/kg; mod. toxic by ing., IV, subcut. routes; skin and eye irritant; TSCA listed
Precaution: Combustible exposed to heat or flame; reacts violently with acetone and sodium peroxide; can react with oxidizers
Hazardous Decomp. Prods.: Heated to decomp., emits highly toxic fumes of SO_x
Uses: Intermediate for elastomers, antioxidants, warfare chemicals; solvent for printing inks, dyes in textile printing; electroplating bath ingred.; photographic developer ingred.; antioxidant in cosmetics
Manuf./Distrib.: Aldrich; Fluka; Morton Int'l.; Sigma

β-Thiodiglycol. *See* Thiodiglycol
Thiodipropionic acid, dimethyl ester. *See* Dimethyl thiodipropionate

Thiodipropionitrile

CAS 111-97-7; EINECS/ELINCS 203-926-5
Synonyms: Bis (2-cyanoethyl) sulfide; 2-Cyanoethyl thioether; 3,3´-Thiodipropionitrile; β,β-Thiodipropionitrile
Empirical: $C_6H_8N_2S$
Formula: $S(CH_2CH_2CN)_2$
Properties: Pale yel. liq.; malodorous; sol. in acetone, chloroform, benzene, oxygenated and chlorinated solvs.; sl. sol. in water, alcohol; m.w. 140.21; dens. 1.1095; m.p. 28.65 C; b.p. 200 C (4 mm)
Toxicology: LD50 (oral, rat) 4210 mg/kg, (IP, mouse) 300 mg/kg; primary irritant; TSCA listed
Uses: Preservative; selective solvent; chromatography; nickel plating brightening agent
Manuf./Distrib.: Fluka; Hampshire; Salor

3,3´-Thiodipropionitrile; β,β-Thiodipropionitrile. *See* Thiodipropionitrile
2-Thioethanol; Thioethylene glycol. *See* 2-Mercaptoethanol
Thioethyl ether. *See* Ethyl sulfide
Thiofuram; Thiofuran; Thiofurfuran. *See* Thiophene
Thioglycol. *See* 2-Mercaptoethanol
Thiolane. *See* Tetrahydrothiophene

Thiolanediol

CAS 3976-75-8
Synonyms: Tetrahydro-3,4-thiophenediol; 3,4-Thiolanediol; 3,4-Thiophenediol, tetrahydro-
Classification: Org. compd.
Empirical: $C_4H_8O_2S$
Uses: Solvent in cosmetics

3,4-Thiolanediol. *See* Thiolanediol
Thiolane-1,1-dioxide. *See* Sulfolane
Thiole. *See* Thiophene
Thiomonoglycol. *See* 2-Mercaptoethanol
Thiophane. *See* Tetrahydrothiophene
Thiophane dioxide; Thiophan sulfone. *See* Sulfolane
Thiophen. *See* Thiophene

Thiophene

CAS 110-02-1; EINECS/ELINCS 203-729-4
UN 2414 (DOT)
Synonyms: Divinylene sulfide; Thiacyclopentadiene; Thiaphene; Thiofuram; Thiofuran; Thiofurfuran; Thiole; Thiophen; Thiotetrole
Empirical: C_4H_4S
Properties: Colorless clear liq.; sl. aromatic odor similar to benzene; misc. with most org. solvs.; sol. in alcohol, ether; insol. in water; m.w. 84.14; dens. 1.051; vapor pressure 40 mm Hg (12.5 C); m.p. -38 C; b.p. 84 C; flash pt. 21.2 F; ref. index 1.5270; may be heated to 850 C without decomp.
Toxicology: LD50 (oral, rat) 1400 mg/kg, (subcut., rabbit) 830 mg/kg, (IP, mouse) 100 mg/kg; LC50 (inh., mouse, 2 h) 9500 mg/m^3; poison by ing. and IP routes; mildly toxic by inh. and subcut. routes; TSCA listed
Precaution: DOT: Flamm. liq.; dangerous fire hazard exposed to heat or flame; explosive reaction with N-nitrosoacetanilide; violent or explosive reaction with nitric acid; incompat. with oxidizers
Hazardous Decomp. Prods.: Heated to decomp., emits highly toxic fumes of SO_x
NFPA: Health 2, Flammability 3, Reactivity 0
Uses: Organic synthesis; phenol-formaldehyde resin comonomer; copolymerizes with maleic anhydride; solvent; dye; pharmaceutical mfg.
Manuf./Distrib.: Aldrich; Atofina; Biddle Sawyer; ChevronPhillips; Fluka; Kira; Laporte Perf. Chems.; Penta Mfg.; Sigma; Spectrum Quality Prods.

3,4-Thiophenediol, tetrahydro-. *See* Thiolanediol
Thiophene, tetrahydro-. *See* Tetrahydrothiophene
Thiopropane; 2-Thiopropane. *See* Dimethyl sulfide
Thiotetrole. *See* Thiophene
THN. *See* Tetrahydronaphthalene
THT. *See* Tetrahydrothiophene
Thyme camphor; Thymic acid. *See* Thymol

Thymol

CAS 89-83-8; 3228-03-3 (m-); EINECS/ELINCS 201-944-8
FEMA 3066
Synonyms: 3-p-Cymenol; p-Cymen-3-ol; 3-Hydroxy-p-cymene; 3-Hydroxy-1-methyl-4-isopropylbenzene; Isopropyl cresol; 6-Isopropyl-m-cresol; Isopropyl-m-cresol; Isopropyl metacresol; 2-Isopropyl-5-methylphenol; 1-Methyl-3-hydroxy-4-isopropylbenzene; 5-Methyl-2-isopropylphenol; 5-Methyl-2-(1-methylethyl) phenol; Phenol, 5-methyl-2-(1-methylethyl)-; Thyme camphor; Thymic acid; m-Thymol
Classification: Aromatic organic compd.; substituted phenol
Empirical: $C_{10}H_{14}O$
Formula: $(CH_3)_2CHC_6H_3(CH_3)OH$
Properties: Colorless translucent cryst.; herbal, medicinal odor; pungent caustic taste; very sol. in alcohol, ether, chloroform, olive oil; sol. in water, alkali, glac. acetic acid, fixed/volatile oils, oxygenated and chlorinated solvs.; m.w. 150.24; dens. 0.972; vapor pressure 1 mm Hg (64 C); m.p. 51 C; b.p. 233 C; flash pt. 216 F; ref. index 1.523
Toxicology: LD50 (oral, rat) 980 mg/kg, (IV, mouse) 100 mg/kg; LDLo (subcut., rat) 1600 mg/kg; poison by ing., IV, IP; mod. toxic by subcut. route; local irritant; allergen; accumulation of toxic levels causes pulmonary edema; experimental reproductive effector; mutagenic data; TSCA listed
Precaution: Combustible; affected by light; incompat. with acetanilide
Hazardous Decomp. Prods.: Heated to decomp., emits acrid smoke and irritating fumes
HMIS: Health 2, Flammability 1, Reactivity 0
Storage: Light-sensitive
Uses: Antibacterial; antifungal agent; mold and mildew preventive; antimicrobial preservative; antiseptic; antioxidant; laboratory reagent; microscopy; preservative in food-pkg. adhesives; hair dye; denaturant, masking agent in cosmetics; synthetic menthol; fragrance ingred. in perfumes; synthetic flavoring agent in foods and pharmaceuticals; pharmaceuticals (disinfectant, deodorant, antimicrobial, antiseptic, preservative, anesthetic, cooling agent); FDA OTC drug; anthelmintic drug
Features: Medicinal flavor
Regulatory: FDA 21CFR §172.515, 175.105; 27CFR §21.65, 21.151; FEMA GRAS; FDA approved for buccals, inhalants; NF, BP, EP compliance
Manuf./Distrib.: Aldrich; Alfa Chem; Barnet Prods.; Berje; Creative Fragrances; Doingcom; EM Science; Fabrichem; Fleurchem; Fluka; Frutarom Ltd; George Uhe; H&R Florasynth; H. Interdonati; Haarmann & Reimer; Integra; J.T. Baker; Kishida Chem.; Lluch Essence; Mallinckrodt Baker; Penta Mfg.; Polarome Int'l.; Quest Int'l.; R.W. Greeff; Ruger; Schütz; Sigma; Spectrum Quality Prods.; Thomas Scientific; Universal Preserv-A-Chem; VWR Int'l.
Trade Names Containing: Punctilious® SDA 37-1 190 Proof

m-Thymol. *See* Thymol
1,3,5-TIBP. *See* 1,3,5-Triisopropylbenzene
TMB. *See* 1,3,5-Trimethylbenzene
TMBAC. *See* Benzyl trimethyl ammonium chloride
2,4,-TMP. *See* Isooctane
TMPD. *See* 2,2,4-Trimethyl-1,3-pentanediol
TMS. *See* Tetramethylsilane
TMU. *See* Tetramethylurea
TNBA. *See* Tributylamine
TOF. *See* Trioctyl phosphate

Toluene

CAS 108-88-3; EINECS/ELINCS 203-625-9
UN 1294 (DOT)
Synonyms: Methylbenzene; Methylbenzol; Phenylmethane; Toluol
Classification: Aromatic hydrocarbon
Empirical: C_7H_8
Formula: $C_6H_5CH_3$
Properties: Colorless clear liq., benzene odor; sol. in alcohol, benzene, ether, acetone, chloroform, petrol. ether, ethyl acetate; misc. with most org. solvs.; very sl. sol. in water; m.w. 92.13; dens. 0.866 (20/4 C); vapor pressure 36.7 mm Hg (30 C); m.p. -94.5 C; b.p. 110.7 C; flash pt. (CC) 4.4 C; ref. index 1.4967 (20 C); KB value 105
Toxicology: ACGIH TLV/TWA 100 ppm; STEL 150 ppm; LD50 (oral, rat) 7.53 g/kg, (skin, rabbit) 12,124 mg/kg; toxic by ing., inh., IP, and skin absorption; irritant to eyes, skin, respiratory tract; severe dermatitis on direct contact; ing. may cause lung aspiration; overexposure by inh. may cause CNS excitation/depression; high concs. may cause paresthesia, vision disturbances, dizziness, nausea, headache, narcosis, death; experimental teratogen, reproductive effector; mutagenic data; TSCA listed
Environmental: VOC; BOD5 2.15; COD 2.52; ThOD 3.13
Precaution: Highly flamm.; incompat. with strong oxidizers, nitric acid, sulfuric acid, nitrogen tetroxide, silver perchlorate, sulfur dichloride, etc.; fire and explosion risk; potential violent reactions; common air contaminant
Hazardous Decomp. Prods.: Heated to decomp., emits acrid smoke and irritating fumes
NFPA: Health 2, Flammability 3, Reactivity 0
Storage: Store in cool well-ventilated area out of direct sunlight, away

from heat/ignition sources, away from corrosives
Uses: Aviation gasoline additive; octane booster in gasoline; solvent for paints, gums, resins, rubber, inks, cleaners, adhesives, cosmetics, pharmaceuticals; diluent, thinner in nitrocellulose lacquers; thinner for perfumes, dyes; alcohol denaturant; process solvent; adhesive solvent in plastic toys; in extraction of various principles from plants; mfg. of benzene, benzoic acid, benzaldehyde, toluene diisocyanate for PU resins, dyes, TNT for explosives, toluene sulfonates for detergents and dyestuffs; adjuvant in resinous/polymeric food-contact coatings; in paper/paperboard in contact with dry food; in acrylic food pkg.; in cellophane for food pkg.; adjuvant for PC food-pkg. resins; solvent in food-contact PPO resins; in food-pkg. adhesives
Regulatory: FDA 21CFR §175.105, 175.320, 176.180, 177.1010, 177.1200, 177.1440, 177.1580, 177.1650, 177.2460, 178.3010, 27CFR §21.131; SARA reportable; SARA §313 reportable; CERCLA hazardous substance; Calif. Prop. 65 reportable; HAP
Manuf./Distrib.: AAE Chemie NV; AMC Chems.; Aldrich; Alfa Aesar; Alvin Chem.; Amyl; Arch Chems.; Ashland; Atofina Bruxelles; BASF AG; BP Amoco; BP Chems. Ltd; Baychem; Brenntag AG; Burdick & Jackson; Chemcentral; ChevronPhillips; Coyne; Delta Distributors; Dow; Eastech; EniChem Elastomeri SpA; Equistar; ExxonMobil; Ferro/Grant; Fisher Scientific; Fluka; GFS; General Chem.; Harcros; Houghton Chem.; Hukill; Huntsman; Idemitsu Petrochem.; Integra; J.T. Baker; Koch Ind.; LG Petrochem. Co. Ltd; Lyondell; Mallinckrodt Baker; Marathon Ashland Petrol.; Maruzen Petrochem.; Miljac; Milsolv; Mitsubishi Oil; Mitsui Chems.; Nova Ltd; P.B. & S.; R.E. Carroll; Romil Ltd; Ruger; SOCO-Lynch; Sal Chem.; Sasol Germany; Shell; Sigma; Southchem; Spectrum Quality Prods.; Stanchem; Sunoco; Textile Chem.; Thomas Scientific; Tonen Chem.; Triple Crown Am.; VWR Int'l.; Vopak USA; Whittaker, Clark & Daniels
Trade Names: Shell Toluene Solvent
Trade Names Containing: Punctilious® SDA 2B-3 190 Proof; Punctilious® SDA 12A-3 190 Proof; Punctilious® SDA 12A-3 Anhydrous; Rubber Solvent; Shellsol® X7B; S&L Naphtha; Solvent 2205; Solvent 2238; Solvent 2249; Solvent 2254; Solvent 2298; Solvent 2299; Solvent 2300; Solvent 2304; Solvent 2307; Solvent 2313; Solvent 2314; Solvent 2316; Solvent 2318; Solvent 2321; Solvent 2322; Solvent 2326; Solvent 2331; Solvent 2336; Solvent 2337; Solvent 2338; Solvent 2341; Solvent 2442; Solvent 2495; Tolu-Sol 30

Toluene, p-t-butyl-. *See* p-t-Butyl toluene
Toluene, ar-chloro-. *See* Monochlorotoluene
Toluene, o-chloro-α,α,α-trifluoro-. *See* o-Chlorobenzotrifluoride
Toluene, p-chloro-α,α,α,-trifluoro. *See* p-Chlorobenzotrifluoride
Toluene, 2,4-dichloro-. *See* 2,4-Dichlorotoluene

Toluene diisocyanate

CAS 26471-62-5
UN 2078
Synonyms: Benzene-, 1,3-diisocyanatomethyl-; Diisocyanatomethylbenzene; 1,3-Diisocyanatomethylbenzene; Diisocyanatotoluene; Isocyanic acid, methylphenylene ester; Isocyanic acid, methyl-m-phenylene ester; Methyl-m-phenylene diisocyanate; Methylphenylene isocyanate; TDI; TDI 80-20; Toluene diisocyanate mixed isomers; Tolylene diisocyanate; Tolylene isocyanate
Definition: Mixt. of the 2,4-isomer (89.3%) and the 2,6-isomer (10.7%)
Empirical: $C_9H_6N_2O_2$
Properties: Water-wh. to pale yel. liq.; sharp pungent odor; sol. in DMSO, acetone, org. solvs.; misc. with CCl_4, chlorobenzenes, olive oil, kerosene, ether, benzene, diglycol methyl ether; reacts with water; m.w. 174.17; sp.gr. 1.22; vapor pressure 0.025 mm Hg (25 C); m.p. 19.5-21.5 C; b.p. 251 C; flash pt. 121 C; ref. index 1.5689 (20 C)
Toxicology: ACGIH TLV/TWA 0.005 ppm; STEL 0.02 ppm; LD50 (oral, rat) 4130 mg/kg; LCLo (inh., rat, 6 h) 600 ppm; poison by ing., inh., IV routes; highly toxic by skin contact; irritating to eyes, skin, mucous membranes, respiratory tract; lachrymator; may cause bronchial asthma, pulmonary edema, nausea, vomiting, abdominal pain, dyspnea, cough, diarrhea, chem. pneumonia, etc.; possibly fatal; experimental carcinogen; possible human carcinogen; mutagenic data; a common air contaminant; TSCA listed
Precaution: Combustible; incompat. with strong oxidizers, acids, amines, alcohols; may polymerize vigorously with bases, acyl chlorides; reacts with water to release CO_2; can react vigorously with compds. contng. active H; corrosive to Cu alloys, Al
Hazardous Decomp. Prods.: Heated to decomp., emits very toxic fumes of CO, CO_2, hydrogen cyanide, NO_x
NFPA: Health 3, Flammability 1, Reactivity 1
Storage: Sensitive to light, moisture and heat; store refrigerated in tightly closed containers under inert atmosphere; protect from lt.; storage in PE containers may evolve gas
Uses: Mfg. of PU foams used in furniture, pkg., insulation, boat bldg.; mfg. of PU coatings used on leather, wire, tank linings, masonry; mfg. of other elastomers used in adhesives, films, linings, abrasive wheels, other mech. items; solvent
Manuf./Distrib.: BASF; Bayer; Dow; Lyondell; Pfaltz & Bauer; Stanchem; TCI Am.
See also Toluene-2,4-diisocyanate

Toluene-2,4-diisocyanate

CAS 584-84-9; EINECS/ELINCS 209-544-5
UN 2206 (DOT); UN 2207 (DOT); UN 2478 (DOT); UN 3080 (DOT)
Synonyms: Diisocyanatoluene; 2,4-Diisocyanato-1-methylbenzene; 2,4-Diisocyanatotoluene; Isocyanic acid, methylphenylene ester; Isocyanic acid, 4-methyl-m-phenylene ester; 4-Methylphenylene diisocyanate; 4-Methylphenylene isocyanate; TDI; 2,4-TDI; Toluene diisocyanate; 2,4-Toluenediisocyanate; Toluylene-2,4-diisocyanate; Tolylene diisocyanate; 2,4-Tolylene diisocyanate; m-Tolylene diisocyanate
Classification: Isocyanic acid ester
Empirical: $C_9H_6N_2O_2$
Formula: $CH_3C_6H_3(NCO)_2$
Properties: Water-wh. to pale yel. liq., sharp pungent odor; misc. with alcohol (dec.), ether, acetone, CCl_4, benzene, chlorobenzene, kerosene, olive oil; m.w. 174.15; dens. 1.22 (25/15.5 C); vapor pressure 0.01 mm (20 C); m.p. 19.4-21.5 C; b.p. 251 C; flash pt. (OC) 132 C; ref. index 1.5680; hydrolyzed by water
Toxicology: ACGIH TLV/TWA 0.005 ppm; STEL 0.02 ppm; LD50 (oral, rat) 5800 mg/kg; LC50 (inh., rat, 4 h) 14 ppm; poison by ing., inh., IV routes; strong irritant to eyes, skin, and tissue; can cause severe dermatitis and bronchial spasm; human systemic effects by inh. (pulmonary and GI changes); carcinogen; mutagenic data; common air contaminant; target organs: respiratory system, skin; TSCA listed
Environmental: VOC; ThOD 1.20
Precaution: Combustible; explosive as vapor; explosive limits 0.9-9.5%; reacts with water producing CO_2; violent reactions possible with compds. contg. act. hydrogen; violent polymerizations possible with bases or acyl chlorides
Hazardous Decomp. Prods.: Heated to decomp., emits highly toxic fumes of NO_x
NFPA: Health 3, Flammability 1, Reactivity 1
Uses: In mfg. of polyurethane foams, elastomers, coatings, adhesives, sealants; crosslinking agent for nylon 6; solvent; in food-pkg. adhesives
Regulatory: FDA 21CFR §175.105; SARA §302a extremely hazardous, §313 reportable; CERCLA hazardous substance; Calif. Prop. 65 reportable; HAP
Manuf./Distrib.: Aldrich; Allchem Ind.; Arch Chems.; BASF; Bayer Hispania Industrial SA; Bayer; Continental Ind. Group; Cross World Sales; D.B. Becker; Dow; Fluka; Gulf Chem. Int'l.; Huntsman Polyurethanes; Lyondell; Neuchem; Rhodia; Rubicon; Sigma; White Cross Labs

2,4-Toluenediisocyanate. *See* Toluene-2,4-diisocyanate
Toluene diisocyanate mixed isomers. *See* Toluene diisocyanate
Toluene, ar-ethyl. *See* Ethyl toluene
Toluene hexahydride. *See* Methyl cyclohexane
Toluene, α-phenyl-. *See* Diphenylmethane
4-Toluenesulfonic acid sodium salt; p-Toluenesulfonic acid, sodium salt. *See* Sodium toluenesulfonate
Toluene trifluoride. *See* Benzotrifluoride
Toluene, vinyl- (mixed isomers). *See* Vinyltoluene monomer
α-Toluenol. *See* Benzyl alcohol
ar-Toluenol. *See* Cresylic acid
m-Toluic acid diethylamide. *See* Diethyl toluamide
α-Toluic acid ethyl ester. *See* Ethyl phenylacetate
Toluol. *See* Toluene

o-Toluol. *See* o-Cresol
Toluylene-2,4-diisocyanate. *See* Toluene-2,4-diisocyanate
o-Tolyl chloride. *See* o-Chlorotoluene
Tolylene diisocyanate. *See* Toluene-2,4-diisocyanate; Toluene diisocyanate
2,4-Tolylene diisocyanate; m-Tolylene diisocyanate. *See* Toluene-2,4-diisocyanate
Tolylene isocyanate. *See* Toluene diisocyanate
TOP. *See* Trioctyl phosphate
Topanel. *See* Crotonaldehyde
TOPO. *See* Trioctylphosphine oxide
TPCP. *See* Tri-p-tolyl phosphate
TPG. *See* Tripropylene glycol
TPGEE. *See* Tripropylene glycol ethyl ether
TPGME. *See* PPG-3 methyl ether
TPGnBE. *See* Tripropylene glycol n-butyl ether
TPGnPE. *See* Tripropylene glycol n-propyl ether
TPM. *See* PPG-3 methyl ether
TPnB. *See* Tripropylene glycol n-butyl ether
TPNC. *See* 1,1,3-Tris (2-methyl-4-hydroxy-5-t-butylphenyl) butane
TPnP. *See* Tripropylene glycol n-propyl ether
TPP. *See* Triphenyl phosphate
Tri. *See* Trichloroethylene

Triacetin

CAS 102-76-1; EINECS/ELINCS 203-051-9
FEMA 2007; INS1518; E1518
Synonyms: Acetic, 1,2,3-propanetriyl ester; Acetin; Enzactin; Glycerine triacetate; Glycerol triacetate; Glyceryl triacetate; 1,2,3-Propanetriol triacetate; Triacetyl glycerin; Triacetyl glycerol
Definition: Triester of glycerin and acetic acid
Empirical: $C_9H_{14}O_6$
Formula: $C_3H_5(OCOCH_3)_3$
Properties: Colorless or pale yel. oily liq., sl. fatty odor, bitter taste; sol. in water, alcohol, ether, other org. solvs.; sl. sol. in CS_2; m.w. 218.20; dens. 1.160 (20 C); m.p. -78 C; b.p. 258-260 C; sapon. no. 765-775; flash pt. 300 F; ref. index 1.4307 (20 C)
Toxicology: LD50 (oral, rat) 3000 mg/kg, (IV, mouse) 1600 ± 81 mg/kg, (IP, rat) 2100 mg/kg, (subcut., rat) 2800 mg/kg; poison by ing.; mod. toxic by IP, IV, subcut. routes; eye irritant; TSCA listed
Precaution: Combustible exposed to heat, flame, or powerful oxidizers
Hazardous Decomp. Prods.: Heated to decomp., emits acrid smoke and irritating fumes
NFPA: Health 1, Flammability 1, Reactivity 0
Uses: Plasticizer for vinyl, cellulosic, and acrylic resins, in food-contact coatings; fixing agent in perfumes; specialty solvent; solvent for dyes; mfg. of celluloid, photographic films; adhesives; coatings; paper; setting agent for sodium silicate foundry mold/soil consolidation binders; antimicrobial, film-former, solvent in cosmetics; emollient, solvent, fixing agent, solubilizer, extender, carrier in pharmaceuticals; synthetic flavoring agent in foods, pharmaceuticals; food additive, nutrient, solvent, vehicle, humectant, formulation aid; OTC drug; topical antifungal
Regulatory: FDA 21CFR §175.300, 175.320, 181.22, 181.27, 184.1901, GRAS; FEMA GRAS; FDA approved for orals; USP/NF, BP, EP compliance
Manuf./Distrib.: A.P. Chems. Ltd; ABITEC; AMRESCO; Aastrid Int'l.; Aceto; Akzo Nobel; Aldrich; Allan; Ashland; Avatar; B&D Nutritional Ingreds.; Bayer AG; Blagden Spec. Chems. Ltd; Britannia Natural Prods.; Brown; CTC Orgs.; ChemUSA; Chemial SpA; Croda Inc; Degussa AG/Health & Nutrition; Eastman; Eggar & Co.; Fluka; Haarmann & Reimer; Honeywill & Stein; Industrias Monfel; Integra; J.T. Baker; K3; Lluch Essence; Mallinckrodt Baker; Parchem Trading; Penta Mfg.; Ruger; Sigma; Spectrum Quality Prods.; Stan Chem Int'l. Ltd; Unipex; Uniqema Am.; Universal Preserv-A-Chem; VWR Int'l.; Voigt Global Distrib.; Vopak USA
Trade Names: Captex® 500; Captex® 500P; Eastman® Triacetin, Food Grade; Triacetina

Triacetyl glycerin; Triacetyl glycerol. *See* Triacetin
Tribromomethane. *See* Bromoform

Tributoxyethyl phosphate

CAS 78-51-3; EINECS/ELINCS 201-122-9
Synonyms: 2-Butoxyethanol phosphate; Ethanol, 2-butoxy-, phosphate (3:1); TBEP; Tris (2-butoxyethyl) phosphate
Empirical: $C_{18}H_{39}O_7P$
Formula: $[CH_3(CH_2)_3O(CH_2)_2O]_3PO$
Properties: Sl. yel. oily liq.; sol. in most org. liqs.; insol. or limited sol. in glycerol, glycols, certain amines; m.w. 398.48; dens. 1.020 (20 C); vapor pressure 0.03 mm Hg (150 C); m.p. -70 C; b.p. 215-228 C (4 mm); flash pt. 438 F; ref. index 1.4380
Toxicology: LD50 (oral, rat) 3000 mg/kg, (IV, mouse) 180 mg/kg, (skin, rabbit) > 16 ml/kg; mod. toxic by ing.; poison by IV route; irritant; TSCA listed
Precaution: Combustible; can react with oxidizing materials
Hazardous Decomp. Prods.: Heated to decomp., emits toxic fumes of PO_x
Uses: Primary plasticizer for most resins and elastomers; coalescing solvent, plasticizer for acrylic-based polishes, gloss paints, adhesives; leveling agent in floor finishes and waxes; flame retardant for plastics; lubricant; antiwear additive; defoamer for drilling muds, cements, fracturing fluids, plasters, paper coatings, pulp bleaching, aq. emulsion paints, adhesives, textiles, mercerizing liquors/dye baths, antifreeze, fermentation, detergents; in food-pkg. adhesives; defoamer in food-contact paper/paperboard
Regulatory: FDA 21CFR §175.105, 176.210
Manuf./Distrib.: Akzo Nobel; Aldrich; Ashland; C.P. Hall; Celanese; ChemService; Chemcor; FMC; Houghton Chem.; PolyOne/Elastomers & Perf. Addit.; Rhodia HPCII; Rhodia/Phosphorus Perf. Derivs.; Surfactants Inc; Velsicol

Tributyl acetyl citrate; Tri-n-butyl O-acetylcitrate; Tributyl O-acetylcitrate; Tributyl 2-(acetyloxy)-1,2,3-propanetricarboxylate. *See* Acetyl tributyl citrate

Tributylamine

CAS 102-82-9
UN 2542 (DOT)
Synonyms: N,N-Dibutyl-1-butanamine; TNBA; Tri-n-butylamine; Tris-n-butylamine
Classification: Sat. aliphatic amine
Empirical: $C_{12}H_{27}N$
Formula: $N(CH_2CH_2CH_2CH_3)_3$
Properties: Colorless to pale yel. liq.; ammonia-like odor; sol. in many org. solvs. incl. ethanol, acetone, ether, benzene; sl. sol. in water; m.w. 185.36; dens. 0.78 (20 C); m.p. -70 C; b.p. 216-27 C; flash pt. (CC) 63 C
Toxicology: LD50 (oral, rat) 540 mg/kg, (skin, rabbit) 250 mg/kg; poison by ing., inh., skin contact, subcut. routes; corrosive irritant to skin, eyes, mucous membranes; can be absorbed thru skin in toxic amts.; may cause 'halo vision'; inh. can cause nose/throat/lung irritation, chest pain, difficulty breathing; higher concs. may cause potentially fatal pulmonary edema; ing. may cause burning of mouth, throat, digestive tract, headache, nausea, faintness, anxiety; CNS effects; TSCA listed
Precaution: Flamm. exposed to heat, flame, oxidizers; incompat. with oxidizing agents, strong acids, mercury; violent reactions and fire/explosion hazards possible
Hazardous Decomp. Prods.: Heated to decomp., emits toxic fumes of NO_x
NFPA: Health 3, Flammability 2, Reactivity 0
Storage: Store in cool, dry, well-ventilated area, out of direct sunlight; hygroscopic; limit stored quantities
Uses: Inhibitor in hydraulic fluids; solvent; intermediate in mfg. of pesticides, other org. chems.
Manuf./Distrib.: Air Prods.; Aldrich; Ashland; Atofina; BASF AG; BASF; Celanese; Fluka; J.T. Baker; Sithean; Spectrum Quality Prods.

Tri-n-butylamine. *See* Tributylamine

Tributyl citrate

CAS 77-94-1; EINECS/ELINCS 201-071-2
Synonyms: Butyl citrate; Citric acid, tributyl ester; 2-Hydroxy-1,2,3-propanetricarboxylic acid, tributyl ester; 1,2,3-Propanetricarboxylic acid, 2-hydroxy-, tributyl ester; TBC; Tri-n-butyl citrate; Tributyl 2-hydroxy-1,2,3-propanetricarboxylate
Definition: Triester of butyl alcohol and citric acid

Empirical: $C_{18}H_{32}O_7$
Formula: $C_3H_5O(COOC_4H_9)_3$
Properties: Colorless or pale yel., odorless; insol. in water; m.w. 360.45; dens. 1.042 (25/25 C); m.p. -20 C; pour pt. -85 C; b.p. 233.5 C (22.5 mm); flash pt. (COC) 315 F; ref. index 1.4453
Toxicology: LD50 (IP, mouse) 2900 mg/kg; mod. toxic by IP route; TSCA listed
Precaution: Combustible
Hazardous Decomp. Prods.: Heated to decomp., emits acrid smoke and irritating vapors
Uses: Plasticizer for lacquers, printing inks; antifoam agent; solvent for nitrocellulose; film-former, solvent for cosmetics, pharmaceuticals; plasticizer for aq. pharmaceutical coatings, sustained-release drugs; in food-pkg. adhesives
Regulatory: FDA 21CFR §175.105; NF compliance
Manuf./Distrib.: Alfa Chem; ChemService; Chemial SpA; J.T. Baker; Morflex; Penta Mfg.; Reilly Ind.; Spectrum Quality Prods.; Unitex; Universal Preserv-A-Chem
Trade Names: Citroflex® 4; Citrofol® BI

Tri-n-butyl citrate. *See* Tributyl citrate
Tributyl citrate acetate; Tri-n-butyl citrate acetate. *See* Acetyl tributyl citrate
Tributylcresyl butane. *See* 1,1,3-Tris (2-methyl-4-hydroxy-5-t-butylphenyl) butane
Tributyl 2-hydroxy-1,2,3-propanetricarboxylate. *See* Tributyl citrate

Tributyl phosphate

CAS 126-73-8; EINECS/ELINCS 204-800-2
Synonyms: Butyl phosphate; n-Butyl phosphate; Phosphoric acid tributyl ester; Phosphoric acid, tri-n-butyl ester; TBP; Tri-n-butyl phosphate
Classification: Organophosphate
Empirical: $C_{12}H_{27}O_4P$
Formula: $(C_4H_9)_3PO_4$
Properties: Stable colorless liq., odorless; misc. with alcohol, ether, most org. solvs.; insol. in water; m.w. 266.32; dens. 0.978 (20/20 C); m.p. < -80 C; b.p. 292 C (dec.); flash pt. (COC) 146 C; ref. index 1.4215 (25 C)
Toxicology: ACGIH TLV/TWA 0.2 ppm; LD50 (oral, rat) 1400 mg/kg, (IP, rat) 251 mg/kg; LDLo (IV, rat) 100 mg/kg; LC50 (inh., mouse) 1300 mg/m^3; poison by IP, IV routes; mod. toxic by ing., inh., subcut. routes; irritant to skin, mucous membranes, eyes; experimental reproductive effects; TSCA listed
Precaution: Combustible exposed to heat or flame
Hazardous Decomp. Prods.: Heated to decomp., emits toxic fumes of PO_x
NFPA: Health 2, Flammability 1, Reactivity 0
Uses: Heat-exchange medium; extraction agent for separation of rare earth metals, uranium, plutonium, phosphoric acid; dielectric; flame retardant for plastics, cellulosics, coatings; chem. reagent; fire-resist. hydraulic fluid; plasticizer for lacquers, plastics, NC, cellulose esters, chlorinated rubber, vinyl resins, PVC; lubricant; antiwear additive; paint additive; adhesive; printing ink solvent; organic intermediate; defoamer for drilling muds, cements, paper coatings, pulp bleaching, mold coating, aq. paints, adhesives, mercerizing liquors/dye baths, antifreeze, fermentation, detergents; in food-pkg. adhesives; defoamer in food-contact paper/paperboard
Regulatory: FDA 21CFR §175.105, 176.210
Manuf./Distrib.: ABCR; Acros Org.; Akzo Nobel; Aldrich; Allchem Ind.; Ashland; Bayer; C.P. Hall; ChemService; Chemron; EM Ind.; EM Science; F.T.L. Int'l.; FMC; Fluka; Houghton Chem.; Independent Chem.; Pechiney Chems. Div.; PolyOne/Elastomers & Perf. Addit.; Rhodia/ Phosphorus Perf. Derivs.; Sigma; Sinochem Jiangsu Wuxi; Spectrum Quality Prods.; Springfield Scientific Supplies; Velsicol; Vopak USA
Trade Names: Albrite® TBP; Albrite® Tributyl Phosphate; Amgard® $TBPO_4$; Phosflex® 4; Reomol® TBP; TBP

Tri-n-butyl phosphate. *See* Tributyl phosphate
Tric. *See* Trichloroethylene

Tricaprin

CAS 621-71-6; EINECS/ELINCS 210-702-0
Synonyms: Glycerol tricaprinate; Glyceryl tricaprate; Glyceryl tridecanoate; 1,2,3-Propanol tridecanoate; Tridecanoin
Definition: Triester of glycerin and capric acid
Empirical: $C_{33}H_{62}O_6$
Formula: $[CH_3(CH_2)_8COOCH_2]_2CHOCO(CH_2)_8CH_3$
Properties: M.w. 554.86; m.p. 31-32 C
Toxicology: LD50 (IV, mouse) > 10 g/kg; TSCA listed
Uses: Emollient, lubricant, emulsifier, moisturizer, solvent, solubilizer, carrier, visc. modifier for cosmetics, sticks, creams, lotions, powds.; carrier for flavors, fragrances, colors, and vitamins; diluent, emollient, lubricant, moisturizer, solubilizer, vehicle, carrier, visc. modifier for aerosols, clinical nutrition, coating, delivery/absorption enhancement, dermatologicals, infant formulas, nutritional/sports supplements, gelatin capsules
Manuf./Distrib.: ABITEC; Fluka; Sigma
Trade Names Containing: Triol 91

Tricaprylic glyceride. *See* Tricaprylin

Tricaprylin

CAS 538-23-8; EINECS/ELINCS 208-686-5
Synonyms: Caprylic acid, 1,2,3-propanetriyl ester; Caprylic acid triglyceride; Glycerol tricaprylate; Glyceryl tricaprylate; Tricaprylic glyceride
Classification: Triester
Definition: Triester of glycerin and caprylic acid
Empirical: $C_{27}H_{50}O_6$
Formula: $[CH_3(CH_2)_6COOCH_2]_2CHOCO(CH_2)_6CH_3$
Properties: Cryst.; m.w. 470.70; dens. 0.954 (20/4 C); m.p. 9-10 C; b.p. 233 C (1 mm); ref. index 1.447
Toxicology: LD50 (oral, rat) 33,300 mg/kg, (IP, rat) 50 mg/kg, (IV, mouse) 3700 mg/kg; poison by IP route; mod. toxic by IV route; mildly toxic by ing.; experimental reproductive effects; tumorigen; mutagen; TSCA listed
Hazardous Decomp. Prods.: Heated to decomp., emits acrid smoke and irritating fumes
Uses: Emollient, lubricant, spreading agent, penetrant for cosmetics and pharmaceuticals; carrier for essential oils, flavors; vehicle for vitamins, medicinals, nutritional prods.; prod. of dietary foods
Features: Nonoily
Manuf./Distrib.: Fluka; Sigma
Trade Names Containing: Triol 91

Tricarballylic acid-β-acetoxytributyl ester. *See* Acetyl triethyl citrate
Trichlor. *See* Trichloroethylene
Trichloracetic acid. *See* Trichloroacetic acid
Trichloran; Trichlorethylene. *See* Trichloroethylene

Trichloroacetic acid

CAS 76-03-9
UN 1839 (DOT); UN 2564 (DOT)
Synonyms: Aceto-caustin; TCA; Trichloracetic acid; Trichloroacetic acid, solution; Trichloroethanoic acid
Classification: Halogenated carboxylic acid
Empirical: $C_2HCl_3O_2$
Formula: CCl_3COOH
Properties: Colorless cryst.; sharp, pungent odor; sol. in water producing highly acidic sol'ns.; very sol. in alcohol, ether; m.w. 163.39; sp.gr. 1.62; vapor pressure 1 mm Hg (51 C); m.p. 58 C; b.p. 197 C; flash pt. none
Toxicology: ACGIH TLV/TWA 1 ppm; LD50 (oral, rat) 400 mg/kg, (subcut., mouse) 270 mg/kg; poison by ing., subcut. route; mod. toxic by IP; corrosive irritant to skin, eyes, mucous membranes; inh. may cause tissue damage, sore throat, nasal irritation, coughing; extreme exposure may cause difficulty breathing, pulmonary edema; ing. can produce severe burns to lips, mouth, throat, vomiting, diarrhea, pain, permanent digestive damage; fatalities possible; questionable carcinogen; mutagenic data; TSCA listed
Precaution: DOT: Corrosive; incompat. with dimethylsulfoxide, copper wool, reactive metals, bases, strong oxidizing agents; violent/explosive reactions possible
Hazardous Decomp. Prods.: May produce flamm. and explosive hydrogen gas with reactive metals; heated to decomp., emits toxic fumes of Cl^- and Na_2O
HMIS: Health 3, Flammability 0, Reactivity 1

Storage: Store in cool, dry, well-ventilated area, out of direct sunlight; deliq.; avoid generating mists; limit quantities in use; use corrosion-resist. materials; hygroscopic, deliq.
Uses: Intermediate in chem. synthesis of esters; organic synthesis; prod. of its sodium salt; as herbicide; pickling or etching agent in metal surf. finishing; solvent for plastics; protein precipitant in laboratories; additive in min. lubricating oils; reagent for detection of albumin; medical agent in treating skin disorders
Manuf./Distrib.: AMRESCO; Aldrich; Alemark; Alfa Aesar; Celanese; Fluka; GFS; Garuda Chems.; Integra; J.T. Baker; Mallinckrodt Baker; Penta Mfg.; Rhodia; Ruger; Sigma; Spectrum Quality Prods.; VWR Int'l.

Trichloroacetic acid, solution. *See* Trichloroacetic acid
Trichlorobenzene. *See* 1,2,4-Trichlorobenzene

1,2,4-Trichlorobenzene

CAS 120-82-1; EINECS/ELINCS 204-428-0
UN 2321 (DOT)
Synonyms: Benzene, 1,2,4-trichloro-; Trichlorobenzene; 1,2,5-Trichlorobenzene; 1,3,4-Trichlorobenzene; unsym-Trichlorobenzene
Empirical: $C_6H_3Cl_3$
Properties: Colorless stable liq.; misc. with most org. solvs., ether, benzene, petrol. ether, carbon disulfide; insol. in water; m.w. 181.45; dens. 1.4634; vapor pressure 1 mm Hg (38.4 C); m.p. 17 C; b.p. 213 C; flash pt. 98.9 C; ref. index 1.5710
Toxicology: ACGIH TLV/CL 5 ppm; LD50 (oral, rat) 756 mg/kg, (IP, mouse) 1223 mg/kg; poison by ing.; toxic by inh.; mod. toxic by IP route; primary skin irritant; experimental teratogen, reproductive effects; mutagenic data; TSCA listed
Precaution: Combustible exposed to heat or flame; can react vigorously with oxidizing materials
Hazardous Decomp. Prods.: Heated to decomp., emits toxic fumes of Cl^-
NFPA: Health 2, Flammability 1, Reactivity 0
Uses: Solvent in chemical mfg., dyes, intermediates, dielectric fluid, synthetic transformer oils, lubricants, heat-transfer media, insecticides; herbicide for aquatic weed control in irrigation canals, lakes, ponds; gel permeation chromatography
Manuf./Distrib.: Aldrich; Allchem Ind.; Ashland; Bayer/Fiber, Addits., Rubber; Fabrichem; Fluka; PPG Ind.; Romil Ltd; Sigma; Spectrum Quality Prods.; Stan Chem Int'l. Ltd

1,2,5-Trichlorobenzene; 1,3,4-Trichlorobenzene. *See* 1,2,4-Trichlorobenzene

1,3,5-Trichlorobenzene

CAS 108-70-3; EINECS/ELINCS 203-608-6
UN 2321
Synonyms: Benzene, 1,3,5-trichloro-; Trichloro-1,3,5-benzene; s-Trichlorobenzene; sym-Trichlorobenzene
Classification: Aromatic halogenated hydrocarbon
Empirical: $C_6H_3Cl_3$
Formula: $ClC_6H_3Cl_2$
Properties: Wh. to colorless needle-like cryst.; aromatic odor; freely sol. in diethyl ether, benzene, petrol. ether, glac. acetic acid; sparingly sol. in ethanol; pract. insol. in water; m.w. 181.45; m.p. 63-65 C; b.p. 208.5 C; flash pt. 107 C
Toxicology: LD50 (oral, rat) 800 mg/kg, (IP, mouse) 2260 mg/kg; harmful solid; mod. toxic by ing. and IP routes; irritant; mutagen; target organs: liver, kidneys, thyroid; TSCA listed
Hazardous Decomp. Prods.: Heated to decomp., emits toxic vapors of Cl^-
Uses: Component in trichlorobenzene mixt. as dye solvent, degreasing solvent, dielectric fluid, lubricant additive; termite control; org. synthesis
Manuf./Distrib.: Aldrich; Fluka; Ishihara USA

Trichloro-1,3,5-benzene; s-Trichlorobenzene; sym-Trichlorobenzene. *See* 1,3,5-Trichlorobenzene
unsym-Trichlorobenzene. *See* 1,2,4-Trichlorobenzene
Trichloroethane. *See* 1,1,1-Trichloroethane

1,1,1-Trichloroethane

CAS 71-55-6; EINECS/ELINCS 200-756-3
UN 2831 (DOT)
Synonyms: Chloroethene; Chloroform, methyl-; Chlorothene; Ethane, 1,1,1-trichloro-; MCF; Methylchloroform; Methyltrichloromethane; TCE; 1,1,1-TCE; Trichloroethane; α-Trichloroethane; Trichloromethylmethane
Classification: Halogenated aliphatic hydrocarbon
Empirical: $C_2H_3Cl_3$
Formula: CH_3CCl_3
Properties: Colorless liq.; char. chloroform-like odor; sol. in alcohol, ether, acetone, benzene, chloroform, methanol, CCl_4; pract. insol. in water; m.w. 133.42; dens. 1.3376 (20/4 C); vapor pressure 100 mm Hg (20 C); m.p. -32.5 C; b.p. 74.1 C; flash pt. none; autoignition temp. 537 C; ref. index. 1.43838 (20 C); surf. tens. 25.56 dynes/cm; KB value 124
Toxicology: ACGIH TLV/TWA 350 ppm; STEL 450 ppm; LD50 (oral, rat) 10,300 mg/kg, (IP, rat) 3593 mg/kg; poison by IV; mod. toxic by ing., inh., skin contact; primary irritant to eyes, tissue, skin, respiratory tract; vapor inh. produces CNS effects; overexposure can cause headache, lassitude, poor equilibrium, eye irritation, dermatitis, cardiac arrhythmias; narcotic in high concs. causing unconsciousness, coma; tumorigen; mutagen; reproductive effects; TSCA listed
Environmental: Ozone-depleting; ThOD 0.48
Precaution: Nonflamm., but vapor may burn in presence of oxidizers or ignition source; LEL 7.5%; UEL 12.5%; incompat. with aluminum/magnesium/potassium and alloys, sodium, NaOH, KOH, nitrogen dioxide, oxygen (gas or liq.)
Hazardous Decomp. Prods.: Hydrogen chloride, hydrochloric acid, phosgene; heated to decomp., emits toxic fumes of Cl^-
NFPA: Health 2, Flammability 1, Reactivity 0
Storage: Store in cool, dry, well-ventilated, clearly identified area, out of direct sunlight, away from heat/ignition sources
Uses: Solvent for cleaning precision instruments, vapor degreasing of metals, cold-cleaning, dip cleaning, plastic molds, elec. equip., motors, electronic components, dry-cleaning, resins, oils, waxes, adhesives, insecticides, pesticides, textile processing, cutting fluids, surf. coatings, printing inks, shoe polishes, spot cleaners, cosmetics, pharmaceuticals; extraction solvent; diluent for paints; chem. intermediate for prod. of vinylidene chloride; catalyst for PU foams; printed circuit board fabrication; motion picture film cleaning; in stain repellents for upholstery fabrics; for on-site cleaning of printing presses, food pkg. machinery, molds; carrier, vapor pressure depressant for aerosols; as aerosol propellant; reagent; in food-pkg. adhesives; crosslinking agent in food-contact polysulfide polymer/polyepoxy resins
Use Level: 1500 ppm (pharmaceuticals)
Regulatory: FDA 21CFR §175.105, 177.1650; SARA §311/312 acute health/reactivity hazard, §313 reportable; CERCLA hazardous substance; HAP; covered under Montreal Protocol
Manuf./Distrib.: Aldrich; Allchem Ind.; Amyl; Arch Chems.; Ashland; Atofina; Dow; Fabrichem; Fluka; Haltermann Prods. UK; Houghton Chem.; Hukill; ICC Ind.; O.C. Lugo; PPG Ind.; Romil Ltd; Sal Chem.; Schumacher; Sigma; Solvay SA; Spectrum Quality Prods.; Stanchem; Uniqema Am.; Universal Preserv-A-Chem; Vulcan Perf. Chems.
Trade Names: Balmix®; Baltane®; Mecloran®

1,1,2-Trichloroethane

CAS 79-00-5; EINECS/ELINCS 201-166-9
Synonyms: Ethane trichloride; 1,1,2-TCE; 1,2,2-Trichloroethane; β-Trichloroethane; Vinyl trichloride
Classification: Halogenated aliphatic hydrocarbon
Empirical: $C_2H_3Cl_3$
Formula: Cl_2CHCH_2Cl
Properties: Colorless clear volatile liq.; char. sweet but sl. irritating odor; sol. in ethanol, chloroform, diethyl ether, esters, ketones; misc. with oxygenated and chlorinated solvs.; poorly sol. in water; m.w. 133.41; sp.gr. 1.44 (20 C); m.p. -36.5 C; b.p. 113.8 C; surf. tens. 32.5 dynes/cm (20 C)
Toxicology: ACGIH TLV/TWA 10 ppm (skin); LD50 (oral, rat) 580 mg/kg, (skin, rabbit) 3730 mg/kg; very toxic; poison by ing., IV, subcut. routes; mod. toxic by inh., skin contact, IP routes; eye and severe skin irritant; narcotic; may cause liver/kidney injury; suspected carcinogen; experimental reproductive effects; mutagenic data; TSCA listed
Precaution: Nonflamm., but vapor can be ignited by high-energy source; incompat. with strong oxidizers, strong alkalis, chemically active met-

als, potassium and alloys, sodium amide; fire and explosion risk
Hazardous Decomp. Prods.: Hydrogen chloride, hydrochloric acid; heated to decomp., emits toxic fumes of Cl^-
NFPA: Health 2, Flammability 1, Reactivity 0
Storage: Store in clearly identified area accessible to authorized personnel only
Uses: Intermediate in prod. of vinylidene chloride; solvent for fats, oils, waxes, resins, chlorinated rubbers, pharmaceuticals; mfg. of electronic components; in chem. synthesis; in food-pkg. adhesives
Regulatory: FDA 21CFR §175.105
Manuf./Distrib.: Aldrich; Fluka; Sigma

1,2,2-Trichloroethane. *See* 1,1,2-Trichloroethane
α-Trichloroethane. *See* 1,1,1-Trichloroethane
β-Trichloroethane. *See* 1,1,2-Trichloroethane
Trichloroethanoic acid. *See* Trichloroacetic acid
Trichloroethene; 1,1,2-Trichloroethene. *See* Trichloroethylene

Trichloroethylene

CAS 79-01-6; EINECS/ELINCS 201-167-4
UN 1710 (DOT)
Synonyms: Acetylene trichloride; 1-Chloro-2,2-dichloroethylene; 1,1-Dichloro-2-chloroethylene; Ethinyl trichloride; Ethylene trichloride; TCE; Tri; Tric; Trichlor; Trichloran; Trichlorethylene; Trichloroethene; 1,1,2-Trichloroethene; 1,1,2-Trichloroethylene
Classification: Aliphatic organic compd.
Empirical: C_2HCl_3
Formula: $CHCl{:}CCl_2$
Properties: Colorless, photoreactive liq., chloroform odor; misc. with common org. solvs.; sl. sol. in water; m.w. 131.40; dens. 1.456-1.462; vapor pressure 100 mm Hg (32 C); f.p. -86.8 C; m.p. -73 C; b.p. 86.7 C; flash pt. 89.6 F; autoignition temp. 420 C; ref. index 1.45560 (25 C); surf. tens. 28.8 dynes/cm; KB value 130; dielec. const. 3.42
Toxicology: ACGIH TLV/TWA 50 ppm; STEL 200 ppm; LD50 (oral, rat) 4.92 ml/kg, (IV, mouse) 34 mg/kg; poison by IV, subcut. route; mod. toxic by ing. and IP; mildly toxic by inh.; mildly toxic to humans by ing., inh.; human systemic effects; can cause somnolence, GI changes, jaundice, narcosis, anesthesia; eye, severe skin irritant; chronic exposure causes liver damage; suspected carcinogen; experimental tumorigen, teratogen, reproductive effects; human mutagenic data; TSCA listed
Environmental: VOC; ThOD 0.37
Precaution: Very dangerous fire hazard exposed to heat or flame; LEL 8%; UEL 10.5%; high concs. of vapor in high-temp. air can be made to burn mildly under strong flame; reacts with alkali, epoxides; violent reactions possible; can form HCl gas with water
Hazardous Decomp. Prods.: Heated to decomp., emits toxic fumes of Cl^-
NFPA: Health 2, Flammability 1, Reactivity 0
Storage: Light-sensitive
Uses: Extraction solvent for oils, fats, waxes, mfg. of decaffeinated coffee, spice oleoresins; solvent for cellulose esters and ethers, dry cleaning, textile scouring, metal vapor degreasing, surf. coatings, corrosion preventatives, pharmaceuticals; solvent dyeing; flame retardant for plastics; chain transfer agent (PVC); refrigerant; heat-transfer fluid; fumigant; diluent in paints, adhesives; textile processing; intermediate for organic chemicals, pharmaceuticals, insecticides; analgesic, anesthetic; in food-pkg. adhesives; mfg. of modified hop extract for beer making; food pkg. adhesives, polymers
Regulatory: FDA 21CFR §73.345, 73.615, 172.560, 173.290, 175.105, 177.1960; SARA §311/312 acute health/chronic health hazard, §313 reportable; CERCLA hazardous substance; HAP
Manuf./Distrib.: AAE Chemie NV; Aldrich; Allchem Ind.; Amyl; Arch Chems.; Ashland; Atofina; Coyne; Dow; Fluka; General Chem.; Harcros; Houghton Chem.; Hukill; ICC Ind.; Independent Chem.; Integra; Kanto; Kimson; PPG Ind.; Primachem; Romil Ltd; Ruger; Sal Chem.; Sigma; Spectrum Quality Prods.; Stanchem; Uniqema Am.; Universal Preserv-A-Chem; Varsal Instruments; Vulcan Perf. Chems.
Trade Names: Altene®; Hi-Tri; Neu-Tri E; PPG Trichlor; Tavoxene®; Tristabil®

1,1,2-Trichloroethylene. *See* Trichloroethylene

Trichlorofluoromethane

CAS 75-69-4; EINECS/ELINCS 200-892-3
Synonyms: CFC 11; F 11; FC 11; Fluorocarbon 11; Fluorotrichloromethane; Freon 11; Freon HE; Halocarbon 11; Methane, trichlorofluoro-; Monofluorotrichloromethane; Propellant 11; R 11; Trichloromonofluoromethane
Classification: Halogenated aliphatic hydrocarbon
Empirical: CCl_3F
Formula: Cl_3CF
Properties: Colorless clear volatile liq. or gas; nearly odorless @ < 20% conc., faint sweet odor at high concs.; sol. in alcohol, ether; insol. in water; m.w. 137.4; sp.gr. 1.494 (liq., 17.2 C); f.p. -111 C; b.p. 23.7 C; ref. index 1.379; surf. tens. 18 dynes/cm; dielec. const. 2.303; noncombustible
Toxicology: ACGIH TLV/CL 1000 ppm; LD50 (IP, mouse) 1743 mg/kg; TDLo (inh., human, 30 min) 50,000 ppm; poison by inh.; mod. toxic by IP route; human systemic effects; CNS effects @ > 1% conc.; can cause narcosis and anesthesia in humans; severe exposure can cause nausea, vomiting, irregular heartbeat, cardiac arrest, death; TSCA listed
Precaution: Incompat. with reactive metals, e.g., powd. aluminum, sodium, potassium (can react violently), lithium (can detonate)
Hazardous Decomp. Prods.: Heated to decomp., emits highly toxic fumes of F^- and Cl^-
Storage: Store liq. in cool, dry, well-ventilated area
Uses: Refrigerant; air conditioning gas; solvent; fire extinguishing agent; chemical intermediate; blowing agent for PU foams; aerosol propellant; degreasing agent; dry-cleaning solvent; heat transfer fluid; pharmaceutical propellant
Regulatory: FDA 21CFR §710.3; FDA approved for inhalants; NF compliance; SARA §313 reportable; CERCLA hazardous substance
Manuf./Distrib.: Aldrich; Allchem Ind.; Atofina SA; Atofina; Aventis Pharmaceuticals; BOC Ltd; DuPont; Fluka; Fluorochem Ltd; Matheson Gas Prods.; Sigma; Uniqema

Trichloroform. *See* Chloroform
Trichlorohydrin. *See* 1,2,3-Trichloropropane
Trichloromethane. *See* Chloroform
Trichloromethylmethane. *See* 1,1,1-Trichloroethane
Trichloromonofluoromethane. *See* Trichlorofluoromethane

1,2,3-Trichloropropane

CAS 96-18-4; EINECS/ELINCS 202-486-1
Synonyms: Allyl trichloride; Glycerol trichlorohydrin; Glyceryl trichlorohydrin; Trichlorohydrin
Classification: Nonaromatic halogenated hydrocarbons
Empirical: $C_3H_5Cl_3$
Formula: $CH_2ClCHClCH_2Cl$
Properties: Colorless liq.; sol. in oxygenated solvs.; sl. sol. in water; m.w. 147.43; dens. 1.3888; f.p. -15 C; b.p. 156.17 C; flash pt. 82.2 C; ref. index 1.4822 (20 C)
Toxicology: ACGIH TLV/TWA 10 ppm (skin); LD50 (oral, rat) 505 mg/kg, (skin, rabbit) 1770 mg/kg; LC50 (inh., mouse, 2 h) 3400 mg/m^3; poison by ing.; mod. toxic by inh. and skin absorption; skin and severe eye irritant; carcinogen; tumorigen; experimental reproductive effects; mutagenic data; target organ: liver, kidneys, pancreas; TSCA listed
Precaution: Combustible exposed to heat, flames, or powerful oxidizers
Hazardous Decomp. Prods.: Heated to decomp., yields highly toxic Cl^-
Uses: Polysulfide rubber comonomer; paint/varnish remover, solvent, degreasing agent; crosslinking agent in food-contact polysulfide polymer/polyepoxy resins
Regulatory: FDA 21CFR §177.1650
Manuf./Distrib.: Aldrich; Fluka; Sigma; Supelco

Trichlorotrifluoroethane

CAS 76-13-1; EINECS/ELINCS 200-936-1
Synonyms: CFC 113; Chlorofluorocarbon 113; Ethane, 1,1,2-trichloro-1,2,2-trifluoro; FC 113; Fluorocarbon 113; Freon 113; Halocarbon 113; R 113; Refrigerant 113; TCTFE; 1,1,2-Trichloro-1,2,2-trifluoroethane; TTE
Classification: Halogenated aliphatic hydrocarbon
Empirical: $C_2Cl_3F_3$
Formula: $ClCF_2CCl_2F$

Properties: Colorless volatile gas; nearly odorless to sweet ether-like odor at high concs.; sol. in alcohol, ether, benzene; very sl. sol. in water; m.w. 187.37; dens. 1.42; vapor pressure 362 mm Hg; f.p. -35 C; m.p. 13.2 C; b.p. 47.6 C; ref. index 1.355; surf. tens. 17.75 dynes/cm; dielec. const. 2.41; noncombustible
Toxicology: ACGIH TLV/TWA 1000 ppm; STEL 1250 ppm; LD50 (oral, rat) 43 g/kg; LCLo (inh., rat, 6 h) 87,000 ppm; mildly toxic by ing. and inh.; primary skin irritant; may cause eye irritation, drying/cracking of skin; high concs. may cause CNS symptoms, dizziness, headache; TSCA listed
Environmental: ThOD 0.09
Precaution: Combustible exposed to heat or flame; incompat. with chemically active metals (calcium, powd. aluminum, zinc, magnesium, beryllium, titanium, samarium, lithium, barium); may react explosively
NFPA: Health 2, Flammability 0, Reactivity 0
Storage: Store in cool, dry, well-ventilated area; avoid handling near 260 C (dec.)
Uses: Refrigerant; dry-cleaning solvent; fire extinguishing agent; mfg. of chlorotrifluoroethylene; foaming/blowing agent; polymer intermediate for dechlorination of chems.; solvent drying; drying electronic parts and precision equipment; degreasing agent; vapor degreasing; metal/plastic cleaning solvent; quick-cooling or freezing of foods; aerosol propellant
Regulatory: FDA 21CFR §173.342; SARA §311/312 acute health/chronic health hazard, §313 reportable
Manuf./Distrib.: Air Prods.; Aldrich; Allchem Ind.; Archimica Florida; Atofina; BCH Brühl; Burdick & Jackson; Fluka; Honeywell Perf. Polymers; LaRoche Ind.; Matheson Gas Prods.; Romil Ltd; Sigma; Spectrum Quality Prods.
Trade Names: Genesolv® D
See also 1,1,1-Trichloro-2,2,2-trifluoroethane

1,1,1-Trichlorotrifluoroethane. *See* 1,1,1-Trichloro-2,2,2-trifluoroethane

1,1,1-Trichloro-2,2,2-trifluoroethane

CAS 354-58-5; EINECS/ELINCS 206-564-6
Synonyms: CFC-113a; Ethane, 1,1,1-trichloro-2,2,2-trifluoro-; FC 113a; Freon FT; Trichlorotrifluoroethane; 1,1,1-Trichlorotrifluoroethane; 1,1,1-Trifluorotrichloroethane
Classification: Halogenated aliphatic
Empirical: $C_2Cl_3F_3$
Formula: CCl_3CF_3
Properties: Colorless clear volatile liq.; faint sweet odor; sol. in alcohol, ether, chloroform; insol. in water; m.w. 187.38; dens. 1.579 g/cm^3; m.p. 14 C (1 atm); b.p. 45.7 C (1 atm)
Toxicology: OEL none; LC50 (inh., rat, 15 min) 13 pph; relatively nontoxic by inh.; may cause dizziness, symptoms of drunkenness; may aggravate heart problems; TSCA listed
Environmental: Do not discharge into atmosphere
Precaution: Sl. fire hazard; incompat. with metals, combustible materials
Hazardous Decomp. Prods.: thermal decomp.: halogenated compds., CO_x
NFPA: Health 0, Flammability 1, Reactivity 0
Storage: Store in cool, dry, well-ventilated area in tightly closed container away from heat, flames, sparks, ignition sources, incompat. materials
Uses: Inert solvent; intermediate for specialty chems.; raw material for mfg. of fine chems., insecticides, pharmaceuticals
Regulatory: DOT nonregulated; SARA §302/304 nonregulated, §311/312 nonhazardous, §313 nonreportable; Calif. Prop. 65 nonreportable; TSCA 12b export notification not required
Manuf./Distrib.: ABCR; Acros Org.; Alfa Aesar; Atofina UK; Fluorochem Ltd; Halocarbon Prods.; ICN Biomed. Research Prods.; Molekular Sarl; TCI Am.; Zhejiang Sanhuan Chems. Co. Ltd
See also Trichlorotrifluoroethane

1,1,2-Trichloro-1,2,2-trifluoroethane. *See* Trichlorotrifluoroethane
Tricresol. *See* Cresylic acid

Tricresyl phosphate

CAS 1330-78-5; EINECS/ELINCS 215-548-8
UN 2574 (DOT)
Synonyms: Cresyl phosphate; Phosphoric acid, tris (methylphenyl) ester; Phosphoric acid, tritolyl ester; TCP; Tricresyl phosphate, with > 3% ortho isomer; Tris (methylphenyl) phosphate; Tris (tolyloxy) phosphine oxide; Tritolyl phosphate
Classification: Aromatic phosphorus compd.
Definition: Mixed isomer prod., usually excluding the very toxic ortho isomer
Empirical: $C_{21}H_{21}O_4P$
Properties: Colorless oily liq.; odorless; misc. with common org. solvs. and thinners, linseed oil, chinawood oil, castor oil; insol. in water; m.w. 368.39; dens. 1.143; pour pt. -28 C; b.p. 265 C (0.09 mm); flash pt. (CC) 410 F; ref. index 1.5550; sp. heat 0.38; flame-resist.
Toxicology: LD50 (oral, rat) 5190 mg/kg, (skin, cat) 1500 mg/kg; highly toxic; mod. toxic by ing. and skin contact; readily absorbed thru skin; eye and skin irritant; human systemic effects by ing. (motor activity changes, muscle weakness); may cause cyanosis, somnolence, ataxia, GI effects, blood/liver changes, weight loss; avoid contact and inh.; tumorigen; experimental teratogen, reproductive effects; TSCA listed
Precaution: Combustible; DOT: poison
Hazardous Decomp. Prods.: Heated to decomp., emits toxic fumes of PO_x
Uses: Clarifier in casein polymer prod.; waterproofing agent in EP lubricants; fire-resist. hydraulic fluids; fire retardant for plastics, coatings, rubber; plasticizer for adhesives, lacquers, PVC, PS, nitrocellulose, rubber; plasticizer and coalescing agent for coatings; petrol./diesel antipreignition additive; heat-exchange medium; air filter medium; solvent for nitrocellulose; lead scavenger in gasoline; sterilizer for surgical instruments; emulsifier in cosmetics
Manuf./Distrib.: Akzo Nobel; Aldrich; ChemService; Chemron; FMC; Fluka; Great Western; Houghton Chem.; Sinochem Jiangsu Wuxi

Tri-p-cresyl phosphate. *See* Tri-p-tolyl phosphate
Tricresyl phosphate, with > 3% ortho isomer. *See* Tricresyl phosphate

Tricyclo [5.2.1.0^{2,6}] decane-8-one

CAS 13380-94-4; EINECS/ELINCS 236-458-5
Synonyms: 8-Ketotricyclo (5.2.1.0^{2,6}) decane; 4,7-Methanoindan-5(4H)-one, tetrahydro-; 4,7-Methano-5H-inden-5-one, octahydro-; 8-Oxotricyclo (5.2.1.0^{2,6}) decane; Tricyclo [5.2.1.0(2,6)] decan-8-one
Classification: Ketone
Empirical: $C_{10}H_{14}O$
Properties: Colorless liq.; char. ketone odor; m.w. 150.22
Uses: Fragrance; solvent for inks
Manuf./Distrib.: TCI Am.
Trade Names: TCD Ketone A

Tricyclo [5.2.1.0(2,6)] decan-8-one. *See* Tricyclo [5.2.1.0^{2,6}] decane-8-one
Tridecane. *See* n-Tridecane

n-Tridecane

CAS 629-50-5; EINECS/ELINCS 211-093-4
Synonyms: Tridecane
Empirical: $C_{13}H_{28}$
Formula: $CH_3(CH_2)_{11}CH_3$
Properties: Colorless liq.; very sol. in alcohol, ether; insol. in water; m.w. 184.37; sp.gr. 0.7564 (20/4 C); vapor pressure 1 mm Hg (59.4 C); m.p. -5 to -4 C; b.p. 235.4 C; flash pt. 79.4 C; ref. index 1.4250 (20 C)
Toxicology: LD50 (IV, mouse) 1161 mg/kg; may be harmful by inh., ing., or skin absorp.; irritating to eyes, skin, respiratory system; vapor/mist is irritating to eyes, mucous membranes, upper respiratory tract; TSCA listed
Precaution: Combustible; incompat. with strong oxidizing agents
Hazardous Decomp. Prods.: Toxic fumes of CO, CO_2; emits toxic fumes under fire conditions
Storage: Store refrigerated; keep tightly closed
Uses: Organic synthesis; distillation chaser; fragrance in cosmetics
Manuf./Distrib.: AccuStandard; Acros Org.; Aldrich; Alfa Aesar; Cambrex; ChevronPhillips; Fluka; ICN Biomed. Research Prods.; Indofine; Isotec; Lancaster Synthesis; Pfaltz & Bauer; Riedel-deHaën; Sigma; Spectrum Quality Prods.; Supelco; TCI Am.
Trade Names Containing: Sasol C1316

1-Tridecanecarboxylic acid, isopropyl ester. *See* Isopropyl myristate
Tridecanoin. *See* Tricaprin
Tridecanol; 1-Tridecanol; n-Tridecanol. *See* Tridecyl alcohol

Trideceth

CAS 24938-91-8 (generic)
Synonyms: Ethoxylated tridecyl alcohol; Isotridecanol ethoxylates
Definition: PEG ether of tridecyl alcohol; commercial prods. contain 3-20 moles EO
Formula: $(C_2H_4O)_xC_{13}H_{28}O$, x = 3-20
Properties: Liq. to paste; HLB 8-16.3 (3-20 moles EO); nonionic; HLB 8.0-16.3
Toxicology: TDLo (oral, rat, 13 wk continuous) 45 g/kg; primary irritant; TSCA listed
Uses: Emulsifier for olein, oils, solvs., waxes; foam builder for detergents; low-temp. scouring agent, wetting agent for household/industrial cleaners; corrosion inhibitor, wetting agent for metal pickling baths; detergent, wetting agent for textiles, hard surf. cleaning, paper, metal processing; solubilizer; dispersant and suspending agent for solids; wetting agent, rewetting agent for dust settling; coemulsifier, wetting agent, dispersant for cutting oils, low temp. wool scouring; emulsifier, wetting agent for cosmetics, pharmaceuticals
Trade Names Containing: DA-425

Tridecyl alcohol

CAS 112-70-9; EINECS/ELINCS 203-998-8
Synonyms: C13 linear primary alcohol; Tridecanol; 1-Tridecanol; n-Tridecanol; n-Tridecyl alcohol
Classification: Synthetic alcohol; commercial mixt. of isomers
Empirical: $C_{13}H_{28}O$
Formula: $C_{12}H_{25}CH_2OH$
Properties: Low melting wh. solid, pleasant odor; insol. in water; m.w. 200.41; dens. 0.845 (20/20 C); m.p. 31 C; b.p. 274 C; flash pt. (PMCC) 118 C
Toxicology: LD50 (oral, rat) 17,200 mg/kg, (skin, rabbit) 5600 mg/kg; mildly toxic by ingestion and skin contact; skin irritant; TSCA listed
Precaution: Combustible exposed to heat or flame
Hazardous Decomp. Prods.: Heated to decomp., emits acrid smoke and irritating fumes
Uses: Esters for synthetic lubricants, detergents, antifoam agents, other tridecyl compds.; perfumery; solvent for paints; metal extraction solvent; oiliness additive for lubricants; emollient, emulsion stabilizer, visc. control agent in cosmetics; processing aid in PVC food-contact coatings and paper/paperboard; in food-pkg. adhesives
Regulatory: FDA 21CFR §175.105, 175.300, 176.210
Manuf./Distrib.: Aldrich; Allchem Ind.; Ashland; Bencorp Int'l.; Berje; Browning; Celanese; Chemcentral; ExxonMobil; Fluka; ICC Ind.; Indofine; Jarchem Ind.; Kyowa Hakko USA; Penta Mfg.; Sasol Germany; Sigma; Sithean

n-Tridecyl alcohol. *See* Tridecyl alcohol

Tridecyl neopentanoate

CAS 106436-39-9
Synonyms: Neopentanoic acid, tridecyl ester
Definition: Ester of tridecyl alcohol and neopentanoic acid
Empirical: $C_{18}H_{36}O_2$
Formula: $(CH_3)_3CCOO(CH_2)_{11}CH_3$
Properties: Liq., char. mild odor; sol. in min. oil, ethanol (> 95%), IPM, cyclomethicone; insol. in water; m.w. 284; sp.gr. 0.850-0.860; ref. index 1.4345-1.4365
Toxicology: LD50 (oral, rat) > 5 g/kg; minimal eye irritation; mild skin irritation
Uses: Binder, emollient, lubricant, gloss aid, spreading agent for cosmetics, creams/lotions, pharmaceuticals; cosolvent for oils and waxes in cosmetics sticks; SPF booster for sunscreens; pigment binder and dispersant; plasticizer for hair setting polymers; wetting agent for Carbopol
Trade Names: Trivent NP-13

Tridodecylamine. *See* Trilaurylamine

Triethanolamine

CAS 102-71-6; EINECS/ELINCS 203-049-8
Synonyms: 2,2′,2′′-Nitrilotriethanol; 2,2′,2′′-Nitrilotris (ethanol); TEA; TELA; Triethylolamine; Tri (2-hydroxyethyl)amine; Trihydroxytriethylamine; 2,2′,2′′-Trihydroxytriethylamine; Tris (2-hydroxyethyl) amine; Trolamine
Classification: Aliphatic amino alcohol; alkanolamine
Empirical: $C_6H_{15}NO_3$
Formula: $N(CH_2CH_2OH)_3$
Properties: Colorless to pale yel. visc. liq., sl. ammoniacal odor; sol. in water, alcohol, acetone, chloroform; sl. sol. in benzene, ether; m.w. 149.19; dens. 1.126; m.p. 21.2 C; b.p. 335 C (dec.); flash pt. 179 C; ref. index 1.4835; pH 10.5 (0.1N aq.); surf. tens. 45.24 dynes/cm
Toxicology: ACGIH TLV/TWA 5 mg/m^3; LD50 (oral, rat) 8 g/kg; mod. toxic by IP; mildly toxic by ing. (irritation of mouth, abdominal pain, vomiting, diarrhea); human skin irritant; eye irritant; skin sensitizer; TSCA listed
Environmental: VOC; BOD5 0.84; COD 1.53; ThOD 1.61
Precaution: Combustible when exposed to heat or flame; LEL 1.2%; incompat. with oxidizers, strong acids; can react vigorously with oxidizing materials; increased risk of fire or explosion
Hazardous Decomp. Prods.: Heated to decomp., emits toxic fumes of NO_x and CN^-
NFPA: Health 2, Flammability 1, Reactivity 1
Storage: Very hygroscopic; light-sensitive; store in cool, dry, well-ventilated area away from heat/ignition sources
Uses: Intermediate in mfg. of surfactants, textile specialties, waxes, polishes, toiletries, herbicides, cutting oils, fatty acid soaps (dry-cleaning), cosmetics, household detergents, emulsions; solvent for casein, shellac, dyes; water repellent; electroplating; corrosion inhibitor; dispersant for dyes, casein, shellac, rubber latex; activator; neutralizer; grinding aid; sequestrant; rubber accelerator; flume wash water additive (foods); food coatings; buffer in cosmetics; buffer, alkalizer, emulsifier, solubilizer for pharmaceuticals; in food-pkg. adhesives; pH adjuster for amino resins in mfg. of food-contact paper/paperboard; defoamer in food-contact paper coatings/paper/paperboard; curing agent in food-contact PU resins; surf. lubricants for food-contact metallic articles
Use Level: 2.5% max. in non-rinse-off finished cosmetics
Regulatory: FDA 21CFR §173.315, 175.105, 175.300, 175.380, 175.390, 176.170, 176.180, 176.200, 176.210, 177.1200, 177.1210, 177.1680, 177.2260, 177.2600, 177.2800, 178.3120, 178.3910; FDA approved for orals, topicals, vaginals; NF, BP compliance
Manuf./Distrib.: AMC Chems.; Aldrich; Allchem Ind.; Ashland; BASF AG; Bencorp Int'l.; Brown; Bruchem; Coyne; Degussa AG; Dow; Duso; EM Science; Expo Chem.; Fluka; GMI Prods.; HCI Chemtech Distribution; Harcros; Houghton Chem.; Huntsman; Int'l. Chem. Inc; Integra; J.T. Baker; Kramer; Mays; Mosselman NV; Mutchler; Occidental; Oxiteno; Penta Mfg.; Phoenix Nat. Prods.; PolyOne/Elastomers & Perf. Addit.; Research Organics; Ruger; Sal Chem.; Schweizerhall; Sigma; Spectrum Quality Prods.; Stanchem; Surfachem Ltd; Thomas Scientific; Universal Preserv-A-Chem; VWR Int'l.; Varsal Instruments; Venus Ethoxyethers; Vopak USA; Whyte Chems. Ltd

Triethyl acetylcitrate; Triethyl o-acetylcitrate. *See* Acetyl triethyl citrate

Triethylamine

CAS 121-44-8; EINECS/ELINCS 204-469-4
UN 1296 (DOT)
Synonyms: (Diethylamino) ethane; N,N-Diethylethanamine; Ethanamine, N,N-diethyl-; TEA; TEN
Classification: Sat. aliphatic amine
Empirical: $C_6H_{15}N$
Formula: $(C_2H_5)_3N$
Properties: Colorless liq., strong ammoniacal odor; very sol. in chloroform, acetone, benzene; sol. in water, alcohol, ether, other org. solvs.; m.w. 101.22; sp.gr. 0.7293 (20/20 C); vapor pressure 53.5 mm Hg; f.p. -115.3 C; b.p. 89.7 C; flash pt. (CC) -9 C; autoignition temp. 249 C; ref. index 1.4000; surf. tens. 20.66 dynes/cm; dielec. const. 2.423
Toxicology: ACGIH TLV/TWA 10 ppm; STEL 15 ppm; LD50 (oral, rat) 460 mg/kg; LDLo (inh., rat, 4 h) 1000 ppm; very toxic, corrosive; mod. toxic by ing., inh., skin contact; strong irritant to tissue; lachrymator; can injure kidneys, liver; can be absorbed thru skin in toxic amts. on prolonged contact; inh. of high concs. may cause potentially fatal pulmonary edema; ing. may cause burning of the mouth, throat, digestive tract; experimental reproductive effects; mutagenic data; TSCA listed
Environmental: VOC; BOD5 0.30; ThOD 2.85; ecotoxicity: sl. toxic, TLM

10-100 ppm
Precaution: Flamm.; dangerous fire risk; explosive limits in air 1.2-8%; incompat. with oxidizers, strong acids, maleic anhydride, nitrosating agents, nitrogen dioxide, nitroparaffins; violent reactions possible
Hazardous Decomp. Prods.: Heated to decomp., emits toxic fumes of NO_x; can react with nitrosating agents to form carcinogenic nitrosamines; dec. maleic anhydride @ temps. > 150 C generating heat and CO_2
NFPA: Health 3, Flammability 3, Reactivity 0
Storage: Store in a cool, dry, well-ventilated area out of direct sunlight in closed containers; store away from ignition sources
Uses: Propellant; corrosion inhibitor; catalytic solvent in chemical synthesis; accelerator activator for rubber; wetting agent, penetrant, waterproofing agents of quat. ammonium types; curing/hardening agent for polymers; solvent for paints; catalyst for PU foams, epoxy resins, phenolic resins, foundry resins; cold box foundry mold process curing agent; intermediate for mfg. of pesticides, pharmaceuticals, polymers, paints, rubber, dyes; stabilizer for chlorinated solvents; flux for copper soldering; adjuvant for PC food-pkg. resins
Regulatory: FDA 21CFR §177.1580; SARA §312 immediate health and fire hazard, §313 reportable; CERCLA hazardous substance; HAP
Manuf./Distrib.: ABCR; Acros Org.; Air Prods.; Aldrich; Allchem Ind.; Ashland; Atofina; Austin; BASF AG; BASF; Benco; Brown; Coyne; Dow; FRP Services; Fluka; Focus; Houghton Chem.; Integra; J.T. Baker; JLM Marketing; Mays; Penta Mfg.; Romil Ltd; Sigma; Spectrum Quality Prods.; Stanchem; Storchem; Vopak USA

N,N,N-Triethylbenzenemethanaminium chloride; Triethyl benzyl ammonium chloride. *See* Benzyltriethyl ammonium chloride

Triethyl citrate

CAS 77-93-0; EINECS/ELINCS 201-070-7
FEMA 3083; INS1505
Synonyms: Ethyl citrate; 2-Hydroxy-1,2,3-propanetricarboxylic acid, triethyl ester; TEC; Triethyl-2-hydroxypropan-1,2,3-tricarboxylate
Definition: Triester of ethyl alcohol and citric acid
Empirical: $C_{12}H_{20}O_7$
Formula: $C_3H_5O(COOC_2H_5)_3$
Properties: Colorless mobile oily liq., odorless, bitter taste; sol. 65 g/100 cc water; sol. 0.8 g/100 cc oil; sol. in oxygenated solvs.; misc. with alcohol, ether; m.w. 276.32; dens. 1.136 (25 C); vapor pressure 1 mm Hg (107 C); b.p. 294 C; flash pt. (COC) 303 F; ref. index 1.4420
Toxicology: LD50 (oral, rat) 5900 mg/kg, (IP, mouse) 1750 mg/kg, (subcut., rat) 6600 mg/kg; LC50 (inh., rat, 6 h) 1300 ppm; mod. toxic by IP route; mildly toxic by ing., inh.; no known skin toxicity; TSCA listed
Precaution: Combustible exposed to heat or flame
Hazardous Decomp. Prods.: Heated to decomp., emits acrid smoke and irritating fumes
Uses: Solvent, plasticizer for nitrocellulose and natural resins; plasticizer for lacquers, printing inks, cigarette filters; softener; paint removers; agglutinant; perfume base/fixative; food additive, sequestrant, foam stabilizer, flavoring agent; flavoring agent, plasticizer in pharmaceuticals; antioxidant, deodorant, solvent in cosmetics; plasticizer in food-contact coatings
Features: Sweet plum-like flavor
Regulatory: FDA 21CFR §175.300, 175.320, 181.22, 181.27; FEMA GRAS; FDA approved for orals; NF, BP, EP compliance
Manuf./Distrib.: Advanced BioTech; Aldrich; Camida Ltd; Charkit; ChemService; Chemial SpA; Cognis; Degussa AG/Health & Nutrition; Fluka; Great Western; H&R Florasynth; H.E. Daniel Ltd; Int'l. Chem. Inc; Jungbunzlauer; Moore Ingreds.; Morflex; Parchem Trading; Penta Mfg.; Sharon Labs; Spectrum Quality Prods.; Ubichem plc; Unitex; Universal Preserv-A-Chem; Voigt Global Distrib.; Vopak USA; Wilshire
Trade Names: Citroflex® 2; Citrofol® AI

Triethylene glycol

CAS 112-27-6; EINECS/ELINCS 203-953-2
Synonyms: 1,2-Bis (2-hydroxyethoxy) ethane; Di-β-hydroxyethoxyethane; 3,6-Dioxaoctane-1,8-diol; 2,2′-[1,2-Ethanediylbis (oxy)] bisethanol; 2,2′-Ethylenedioxydiethanol; 2,2′-Ethylenedioxyethanol; Ethylene glycol bis (2-hydroxyethyl ether); Ethylene glycol dihydroxydiethyl ether; Glycol bis (hydroxyethyl) ether; TEG; Triglycol
Classification: Aliphatic alcohol; glycol ether
Empirical: $C_6H_{14}O_4$
Formula: $HOCH_2CH_2OCH_2CH_2OCH_2CH_2OH$
Properties: Colorless clear liq., odorless; sol. in water, ethanol, toluene, benzene, oxygenated solvs.; sparingly sol. in diethyl ether; immisc. with gasoline; m.w. 150.17; sp.gr. 1.12 (20 C); visc. 49 cP; vapor pressure 1 mm Hg (114 C); f.p. -7.2 C; b.p. 278 C; flash pt. (PMCC) 166 C; autoignition temp. 371 C; ref. index 1.4550; surf. tens. 45.2 dynes/cm
Toxicology: LD50 (oral, rat) 17 g/kg, (IV, rat) 11,700 μg/kg, (IP, mouse) 8141 mg/kg; poison by IV route; mildly toxic by ing.; skin and eye irritant; experimental reproductive effects; TSCA listed
Environmental: VOC; BOD5 0.03; COD 1.57; ThOD 1.60
Precaution: Combustible exposed to heat or flame; LEL 0.9%; UEL 9.2%; reactive with oxidizing materials; explosive as vapor exposed to heat, flame
Hazardous Decomp. Prods.: CO, CO_2; heated to decomp., emits acrid smoke and irritating fumes
NFPA: Health 1, Flammability 1, Reactivity 0
Storage: Hygroscopic
Uses: Unsat. polyester resin comonomer; solvent, plasticizer in vinyl, polyester, and polyurethane resins; plasticizer for polyvinyl alcohol; dehydrating agent for natural gas; solvent, lubricant for textile dyeing/printing; solvent for printing inks, dyes, cosmetics, pharmaceuticals; extraction solvent; humectant for tobacco, printing inks, cork; chem. intermediate; plasticizer in food-contact polymers and coatings; in food-pkg. adhesives; in cellophane for food pkg.; in surf. lubricants for mfg. of food-contact metallic articles
Regulatory: FDA 21CFR §175.105, 175.300, 177.1200, 178.3740, 178.3910, 179.45
Manuf./Distrib.: AMC Chems.; Adept Sol'ns.; Aldrich; Allchem Ind.; Ashland; Avatar; BASF; Celanese; Coyne; Dow; Eastman; Equistar; Fluka; Haltermann Prods. UK; Houghton Chem.; Huntsman; Integra; J.T. Baker; Marcus Int'l.; Occidental; Oxiteno; Penta Mfg.; Rambach; Ruger; Sal Chem.; Shell; Spectrum Quality Prods.; Stanchem; Universal Preserv-A-Chem; VWR Int'l.; Vopak USA

Triethylene glycol butyl ether; Triethylene glycol n-butyl ether. *See* Butoxytriglycol

Triethylene glycol diethyl ether

CAS 4499-99-4
Synonyms: 1,2-Bis (2-ethoxyethoxy) ethane; 3,6,9,12-Tetraoxatetradecane
Classification: Glycol ether
Empirical: $C_{10}H_{22}O_4$
Properties: M.w. 206.18
Environmental: ThOD 2.10
Uses: Solvent
Regulatory: SARA §313 reportable

Triethylene glycol dimethyl ether. *See* PEG-3 dimethyl ether
Triethylene glycol ethyl ether. *See* Ethoxytriglycol
Triethylene glycol monobutyl ether. *See* Butoxytriglycol
Triethylene glycol monoethyl ether. *See* Ethoxytriglycol
Triethylene glycol monomethyl ether. *See* Triglycol monomethyl ether
Triethylhexyl phosphate; Tri (2-ethylhexyl) phosphate. *See* Trioctyl phosphate
Triethyl-2-hydroxypropan-1,2,3-tricarboxylate. *See* Triethyl citrate
Triethylolamine. *See* Triethanolamine

Triethyl phosphate

CAS 78-40-0; EINECS/ELINCS 201-114-5
Synonyms: Ethyl phosphate; o-Phosphoric acid, triethyl ester; TEP
Empirical: $C_6H_{15}O_4P$
Formula: $(C_2H_5O)_3PO$
Properties: Colorless clear liq., mild odor; sol. in oxygenated solvs.; misc. with water, ethanol, ethyl acetate, acetone, chloroform, benzene, xylene, castor oil, isobutanol; m.w. 182.16; dens. 1.067-1.072 (20/20 C); vapor pressure 1 mm Hg (39.6 C); m.p. -56 C; b.p. 215 C; flash pt. (PMCC) 99 C; ref. index 1.4019
Toxicology: LD50 (oral, rat) 1311 mg/kg, (dermal, guinea pig) > 20 ml/kg; LDLo (oral, rat) 1600 mg/kg; mod. toxic by ing., IP, IV routes; causes eye irritation; may cause nerve damage; cholinesterase inhibitor; experimental reproductive effects; mutagenic data; TSCA listed

Precaution: Combustible exposed to heat or flames; can react vigorously with oxidizing agents
Hazardous Decomp. Prods.: Heated to decomp., emits toxic fumes of PO_x
Uses: Intermediate for agric. insecticides; floor polishes; lubricants; hydraulic fluids; aprotic solvent; flame retardant plasticizer in cellulosics, polyester resins, PU; visc. depressant in polyester laminates, cellulosics; catalyst for synthesizing ketene in prod. of acetic anhydride; lacquer remover; solvating and desensitizing agent for org. peroxides; solvent for textiles; dyeing assistant; in sizes; in food-pkg. adhesives
Regulatory: FDA 21CFR §175.105
Manuf./Distrib.: Aldrich; Allchem Ind.; Ashland; Bayer; ChemService; Eastman; F.T.L. Int'l.; Fluka; Focus; J.T. Baker; Penta Mfg.; Sinochem Jiangsu Wuxi; Sithean; Spectrum Quality Prods.; Xinchem
Trade Names: Eastman® TEP; Eastman® Triethyl Phosphate

Trifluoracetic acid. *See* Trifluoroacetic acid

Trifluoroacetic acid

CAS 76-05-1; EINECS/ELINCS 200-929-3
UN 2699 (DOT)
Synonyms: Acetic acid, trifluoro-; Perfluoroacetic acid; Trifluoracetic acid; Trifluoroethanoic acid
Empirical: $C_2HF_3O_2$
Formula: CF_3COOH
Properties: Colorless liq.; strong pungent odor; sol. in oxygenated solvs.; misc. with water; m.w. 114.03; dens. 1.535 (0 C); m.p.-15.25 C; b.p. 71.1 C (734 mm); strongly acidic
Toxicology: LD50 (oral, rat) 200 mg/kg, (IV, mouse) 1200 mg/kg; LDLo (IP, mouse) 150 mg/kg; LC50 (inh., rat) 10 g/m^3; poison by ing. and IP routes; mod. toxic by IV route; mildly toxic by inh.; corrosive irritant to skin, eyes, mucous membranes; mutagen; TSCA listed
Precaution: Corrosive
Hazardous Decomp. Prods.: Heated to decomp., emits toxic fumes of F^-
Storage: Hygroscopic
Uses: Strong nonoxidizing acid; laboratory reagent; solvent; alkylation/acylation catalyst; high performance liq. chromatography; molecular biology
Manuf./Distrib.: Advance Research Chems.; Advanced Synthesis Tech.; Aldrich UK; Aldrich; Allchem Ind.; Archimica Florida; Atomergic Chemetals; Biddle Sawyer; Filo; Fluka; Halocarbon Prods.; Honeywell Perf. Polymers; Monomer-Polymer & Dajac Labs; Rhodia/Fine Orgs.; Romil Ltd; Sigma; Solvay Fluorides; Solvay GmbH; Spectrum Quality Prods.; Xinchem

Trifluorochloromethane. *See* Chlorotrifluoromethane
α,α,α-Trifluoro-4-chlorotoluene. *See* p-Chlorobenzotrifluoride
Trifluoroethanoic acid. *See* Trifluoroacetic acid
Trifluoroethanol. *See* 2,2,2-Trifluoroethanol

2,2,2-Trifluoroethanol

CAS 75-89-8; EINECS/ELINCS 200-913-6
Synonyms: TFE; Trifluoroethanol; Trifluoroethyl alcohol
Empirical: $C_2H_3F_3O$
Formula: CF_3CH_2OH
Properties: Colorless liq.; alcoholic odor; misc. with water; m.w. 100.04; dens. 1.390 (20 C); m.p. -45 C; b.p. 72-74 C; ref. index < 1.3000
Toxicology: LD50 (oral, rat) 240 mg/kg, (IP, rat) 210 mg/kg, (IV, mouse) 250 mg/kg, (skin, rat) 1680 mg/kg; LC50 (inh., mouse) 2900 mg/m^3; poison by ing., IV, IP routes; mod. toxic by inh., skin contact; severe skin and eye irritant; irritating to respiratory system; experimental reproductive effects; TSCA listed
Precaution: Flamm.
Hazardous Decomp. Prods.: Heated to decomp., emits toxic fumes of F^-
Uses: Solvent for nylon; UV spectroscopy; working fluid in Rankine cycle engines
Manuf./Distrib.: Aldrich; Fluka; Halocarbon Prods.; Romil Ltd; Schweizerhall; Sigma; Solvay GmbH

Trifluoroethyl alcohol. *See* 2,2,2-Trifluoroethanol

Trifluoromethane

CAS 75-46-7; EINECS/ELINCS 200-872-4
UN 1984 (DOT); 3136 (DOT)
Synonyms: Carbon trifluoride; Fluoroform; Freon 23; Halocarbon 23; HCFC 23; HFC 23; Hydrofluorocarbon 23; Methane, trifluoro; Methyl trifluoride; Propellant 23; R 23; Refrigerant 23
Empirical: CHF_3
Properties: Colorless liquefied gas; odorless; sl. sol. in water; m.w. 70.02; sp.gr. 0.8781; dens. 1.152 (liq., -100 C); m.p. -160 C; b.p. -84 C; ref. index 1.215; surf. tens. 0.03 dynes/cm; dielec. const. 1.007; nonflamm.
Toxicology: Mild respiratory irritant; avoid inhalation; narcotic in high conc.; target organ: heart; mutagen; TSCA listed
Environmental: ThOD 0.23; replacement for CFC but characterized as global warming gas by the Kyoto Conference on Climate Change
Hazardous Decomp. Prods.: Heated to decomp., emits toxic fumes of F^-
Storage: Store away from heat
Uses: Low-temp. refrigerant; intermediate in organic synthesis; direct coolant for infrared detector cells; blowing agent for urethane foams; solvent
Manuf./Distrib.: Aldrich

Trifluoromethylbenzene. *See* Benzotrifluoride
Trifluoromethyl chloride. *See* Chlorotrifluoromethane
o-(Trifluoromethyl) chlorobenzene. *See* o-Chlorobenzotrifluoride
p-(Trifluoromethyl) chlorobenzene. *See* p-Chlorobenzotrifluoride
o-Trifluoromethylphenyl chloride. *See* o-Chlorobenzotrifluoride
p-Trifluoromethylphenyl chloride. *See* p-Chlorobenzotrifluoride
Trifluoromonochlorocarbon. *See* Chlorotrifluoromethane
α,α,α-Trifluorotoluene. *See* Benzotrifluoride
1,1,1-Trifluorotrichloroethane. *See* 1,1,1-Trichloro-2,2,2-trifluoroethane
Triformol. *See* Trioxane
Triglyceryl diisostearate. *See* Polyglyceryl-3 diisostearate
Triglyceryl stearate. *See* Polyglyceryl-3 stearate
Triglycol. *See* Triethylene glycol
Triglycol butyl ether. *See* Butoxytriglycol
Triglycol ethyl ether. *See* Ethoxytriglycol
Triglycol methyl ether. *See* Triglycol monomethyl ether
Triglycol monobutyl ether. *See* Butoxytriglycol
Triglycol monoethyl ether. *See* Ethoxytriglycol

Triglycol monomethyl ether

CAS 112-35-6
Synonyms: Ethanol, 2(2-(2-methoxyethoxy)ethoxy)-; 2-(2-(2-Methoxyethoxy) ethoxy) ethanol; Methoxy triethylene glycol; Methoxytriglycol; Methyltriglycol; Methyl trigol; PEG-3 methyl ether; Triethylene glycol monomethyl ether; Triglycol methyl ether
Empirical: $C_7H_{16}O_4$
Formula: $CH_3O(CH_2CH_2O)_2CH_2CH_2OH$
Properties: Colorless liq.; misc. with water, most org. solvs.; m.w. 164.20; dens. 1.048 (20/4 C); b.p. 249 C; flash pt. 118 C; ref. index 1.439 (20 C)
Toxicology: LD50 (oral, rat) 11,300 mg/kg, (dermal, rabbit) 7100 mg/kg; mildly toxic by ing. and skin contact; skin and eye irritant; possible human reproductive effector; TSCA listed
Precaution: Combustible exposed to heat or flame
Hazardous Decomp. Prods.: Heated to decomp., emits acrid smoke and irritating fumes
Uses: Solvent for brake fluids, hard-surf. cleaners, leather dyeing, paints/coatings, printing inks, textile vat dyeing and printing, adhesives, antifreeze, floor waxes/polishes, insect repellents; solubilizer for dyes; plasticizer, wetting agent for resins, paints; coupling solvent
Manuf./Distrib.: Aldrich; Ashland; Dow; Fluka; Oxiteno; Sigma
Trade Names: Hisolve TM; Poly-Solv® TM
See also 1-(2-(2-Methoxy-1-methylethoxy)-1-methylethoxyl)-2-propanol

Triglyme. *See* PEG-3 dimethyl ether

Trihexyl citrate

CAS 16544-70-0
Uses: Solvent for household, I&I cleaners; solvent, film-former, and plasticizer for inks, adhesives, and coatings

Trihexylphosphine oxide

CAS 3084-48-8
Classification: Phosphine
Empirical: $C_{18}H_{39}OP$

Properties: M.w. 308.48
Uses: Extractant
Regulatory: DSL listed
Trade Names Containing: Cyanex® 923

1,1,5-Trihydrofluoropentane-1-ol. *See* 1H,1H,5H-Octafluoro-1-pentanol
1,2,4-Trihydroxybenzene-1,2-methylene ether. *See* Sesamol
Tri (2-hydroxyethyl)amine. *See* Triethanolamine
1,2,6-Trihydroxyhexane. *See* 1,2,6-Hexanetriol
Trihydroxypropane; 1,2,3-Trihydroxypropane; Trihydroxypropane glycerol. *See* Glycerin

Trihydroxystearin

CAS 139-44-6; EINECS/ELINCS 205-364-6
Synonyms: Glyceryl tri (12-hydroxystearate); Glyceryl tris-12-hydroxystearate; 12-Hydroxyoctadecanoic acid, 1,2,3-propanetriyl ester; 1,2,3-Propanetriol tri (12-hydroxystearate); 1,2,3-Propanetriyl tris (12-hydroxyoctadecanoate)
Definition: Triester of glycerin and hydroxystearic acid
Empirical: $C_{57}H_{110}O_9$
Properties: Ylsh. to milky wh. hard brittle wax-like solid; insol. in water; m.w. 939.50; dens. 0.899 (100/25 C); m.p. 86-88 C
Uses: Thixotrope, gellant for coatings, aliphatic solvent systems; lubricant for plastics processing; heavy-metal soaps; waxes; plasticizers; binder, lubricant for tablets, pressed powds.; water resist. aid in cosmetics; emollient, solvent, visc. control agent for cosmetics, topical pharmaceuticals; chemical intermediate; suspending agent for solv.-borne systems, polyester patch and fillers, powd. coatings; stabilizer; thickener; rheology control agent, thixotrope, pigment/filler suspending agent/antisettling agent, flow control agent, and leveling agent in paints/coatings, stains, mastics, inks, textile backings, solv.-free epoxy systems
Regulatory: FDA approved for topicals

Trihydroxytriethylamine; 2,2′,2′′-Trihydroxytriethylamine. *See* Triethanolamine

Triisobutylphosphine sulfide

CAS 3982-87-4; EINECS/ELINCS 223-620-5
Empirical: $C_{12}H_{27}PS$
Formula: $(CH_3CHCH_3CH_2)_3PS$
Properties: M.w. 234.38; m.p. 60 C
Uses: Extractant for metals which are soft Lewis acids (i.e., Ag, Pd, Au, Hg); selective extractant for silver in presence of metals, e.g., copper, zinc, iron (3+), nickel, in Pd-Pt separation, and in the recovery of mercury from effluents
Manuf./Distrib.: Fluka
Trade Names: Cyanex® 471X

Triisononanoin

CAS 56554-53-1; EINECS/ELINCS 260-257-1
Synonyms: Isononanoic acid, 1,2,3-propanetriyl ester; 1,2,3-Propanetriol triisononoate; Propane-1,2,3-triyl 3,5,5-trimethylhexanoate
Definition: Triester of glycerin and a branched chain nonanoic acid
Empirical: $C_{30}H_{56}O_6$
Formula: $C(CH_2OCOC_8H_{17})_2HOCOC_8H_{17}$
Uses: Emollient, solvent, visc. control agent in cosmetics
Trade Names: Isodragol 2/050300

Triisooctyl phosphite

CAS 25103-12-2; EINECS/ELINCS 246-614-4
Synonyms: Phosphorous acid, triisooctyl ester
Empirical: $C_{24}H_{51}O_3P$
Formula: $(C_8H_{17}O)_3P$
Properties: Colorless liq.; char. odor; misc. with most common org. solvs.; insol. in water; m.w. 418.64; dens. 0.891 (20/4 C); b.p. 161-164 C (0.3 mm); flash pt. (COC) 385 F
Toxicology: LD50 (oral, rat) 9200 mg/kg, (dermal, rabbit) 3970 mg/kg; mod. toxic by skin contact; mildly toxic by ing.; skin irritant; TSCA listed
Precaution: Combustible; hydrolyzes very slowly; minimize exposure to air
Hazardous Decomp. Prods.: Heated to decomp., emits toxic fumes of PO_x
Uses: Intermediate for insecticides; lubricant additive; specialty solvent; stabilizer and antioxidant for acrylics, nylon, unsat. polyester, PVC; improves antiwear and antifriction props.; stabilizer in coatings
Manuf./Distrib.: Ashland; GE Spec.; Rhodia/Phosphorus Perf. Derivs.

Triisopropylbenzene. *See* 1,3,5-Triisopropylbenzene

1,3,5-Triisopropylbenzene

CAS 717-74-8; EINECS/ELINCS 211-941-3
Synonyms: Benzene, 1,3,5-tris (1-methylethyl); 1,3,5-TIBP; Triisopropylbenzene
Classification: Aromatic hydrocarbon
Empirical: $C_{15}H_{24}$
Properties: Clear liq.; m.w. 204.36; dens. 0.845; b.p. 232-236 C; flash pt. (CC) 188 F; ref. index 1.4880
Toxicology: TSCA listed
Precaution: Combustible
Uses: Solvent; carbonless paper; elec. industry; agric.; swelling agent
Manuf./Distrib.: Aldrich; Fluka

Triisopropyl phosphite

CAS 116-17-6; EINECS/ELINCS 204-130-0
Synonyms: Phosphorous acid, triisopropyl ester; Phosphorous acid, tris (1-methylethyl) ester
Classification: Nonaromatic phosphorus compd.
Empirical: $C_9H_{21}O_3P$
Formula: $[(CH_3)_2CH]_3PO_3$
Properties: Colorless liq.; char. odor; misc. with most common org. solvs.; insol. in water; m.w. 208.27; dens. 0.914 (20/4 C); b.p. 94-96 C (50 mm); flash pt. (COC) 73.9 C; ref. index 1.4110
Toxicology: LD50 (oral, rat) 167 mg/kg, (IP, mouse) 500 mg/kg, (skin, rabbit) > 2 g/kg; poison by ing.; mod. toxic by IP route; irritant; mutagenic data; TSCA listed
Precaution: Combustible; hydrolyzes slowly in water; minimize exposure to air
Hazardous Decomp. Prods.: Heated to decomp., emits toxic fumes of PO_x
Storage: Moisture-sensitive
Uses: Intermediate for insecticides, pharmaceuticals, mfg. of phosphonate esters, acids, and salts; component of vinyl stabilizers; lubricant additive; specialty solvent; heat and light stabilizer, antioxidant, metal chelator for polyester
Manuf./Distrib.: Aldrich; Fluka; Rhodia/Phosphorus Perf. Derivs.

Triisostearin

CAS 26942-95-0; EINECS/ELINCS 248-122-5
Synonyms: Glyceryl triisostearate; 1,2,3-Propanetriyl triisooctadecanoate; 1,2,3-Propenetriol triisooctadecanoate
Definition: Triester of glycerin and isostearic acid
Empirical: $C_{57}H_{110}O_6$
Toxicology: TSCA listed
Uses: Emollient, solvent, visc. control agent in cosmetics; emollient, moisturizer, lubricant, conditioner for hair care, skin care; visc. builder, gellant, stiffener for makeup, deodorants

Triisostearin PEG-6 esters

Definition: Mixt. formed from the transesterification of triisostearin and PEG-6
Uses: Emollient for cosmetics; emollient, excipient, solvent for pharmaceuticals; amphiphilic agent improving drug delivery
Trade Names: Labrafil® Isostearique

Trilauramine. *See* Trilaurylamine

Trilaurin

CAS 538-24-9; EINECS/ELINCS 208-687-0
Synonyms: Dodecanoic acid, 1,2,3-propanetriyl ester; Glycerol trilaurate; Glyceryl tridodecanoate; Glyceryl trilaurate; Lauric acid triglyceride; 1,2,3-Propanetriol tridodecanoate
Definition: Triester of glycerin and lauric acid
Empirical: $C_{39}H_{74}O_6$
Properties: Wh. solid, sl. bay odor; sol. in ether, chloroform, petrol. ether; m.w. 639.01; m.p. 45-47 C
Toxicology: Mild irritant; TSCA listed
Uses: Emollient, solvent, visc. control agent, consistency agent, foaming

agent for cosmetics, pharmaceuticals; surfactant; aux. in prod. of compressed sweets; tablet lubricant, binder, retarding agent; nucleating agent in sticks; in food-contact textiles
Regulatory: FDA 21CFR §177.2800
Manuf./Distrib.: Fluka; Sigma; TCI Am.

Trilaurylamine

CAS 102-87-4; EINECS/ELINCS 203-063-4
Synonyms: N,N-Didodecyl-1-dodecanamine; Tridodecylamine; Trilauramine
Classification: Tertiary amine
Empirical: $C_{36}H_{75}N$
Formula: $N[CH_2(CH_2)_{10}CH_3]_3$
Properties: Colorless liq.; sol. in org. solvs.; insol. in water; m.w. 522.00; dens. 0.82; m.p. 14 C; b.p. 230 C (0.01 mm); cationic
Toxicology: TSCA listed
Precaution: Combustible
Uses: Chemical and surfactant intermediate, metal complexes; extraction solvent for citric acid (foods); carrier for mfg. of citric acid and oil; antistat in cosmetics
Regulatory: FDA 21CFR §173.280
Manuf./Distrib.: Aldrich; Fluka; Sigma

Trilinolein

CAS 537-40-6; EINECS/ELINCS 208-666-6
Synonyms: Glyceryl trilinoleate; Linoleic acid, 1,2,3-propanetriyl ester; 9,12-Octadecadienoic acid, 1,2,3-propanetriyl ester; 1,2,3-Propanetriol trilinoleate; 1,2,3-Propenetriol tri (9,12-octadecadienoate)
Definition: Triester of glycerin and linoleic acid
Empirical: $C_{57}H_{98}O_6$
Uses: Emollient, solvent, visc. control agent in cosmetics
Manuf./Distrib.: Fluka; Sigma

Trimethoxyborane; Trimethoxyborine; Trimethoxyboron. *See* Trimethyl borate

Trimethylbenzene

CAS 25551-13-7
UN 1993; UN 2325
Synonyms: Trimethylbenzene, mixed isomers
Empirical: C_9H_{12}
Formula: $(CH_3)_3C_6H_3$
Properties: Colorless liq.; char. peculiar odor; sl. sol. in water; m.w. 120.19; sp.gr. 0.86-0.89; vapor pressure 0.18-0.25 kPa (20 C); m.p. -25 to 45 C; b.p. 165-176 C; flash pt. (CC) 44-53 C; autoignition temp. 470-550 C
Toxicology: ACGIH TLV TWA (8 h) 25 ppm; may cause eye/skin redness, dry skin, eye pain; inh. may cause confusion, cough, dizziness, lightheadedness, drowsiness, headache, sore throat, vomiting; may cause nervousness, tension, anxiety, lung irritation, phlegm, shortness of breath, asthmatic bronchitis, blood changes, liver changes
Environmental: Harmful to aquatic organisms; bioaccumulation may occur in fish; prevent entry into sewers and the environment
Precaution: Flamm. liq.; may form explosive vapor/air mixts. above 44 C; reacts violently with strong oxidants causing fire and explosion hazard; avoid contact with oxidizers (e.g., perchlorates, peroxides, permanganates, chlorates, nitrates), nitric acid, strong oxidizers (chlorine, bromine, fluorine)
Hazardous Decomp. Prods.: Dec. on burning, producing toxic and irritating fumes; combustion: poisonous gases
NFPA: Health 0, Flammability 2, Reactivity 0
Uses: Solvent
Trade Names Containing: Solvent 2216; Solvent 2439
See also 1,3,5-Trimethylbenzene

1,2,4-Trimethylbenzene

CAS 95-63-6; EINECS/ELINCS 202-436-9
Synonyms: Asymmetrical trimethyl benzene; Pseudocumene; Pseudocumol; Trimethyl-1,2,4-benzene; 1,2,5-Trimethylbenzene; as-Trimethylbenzene
Classification: Aromatic hydrocarbon
Empirical: C_9H_{12}
Formula: $C_6H_3(CH_3)_3$
Properties: Colorless liq.; char. aromatic odor; sol. in alcohol, benzene, ether; insol. in water; m.w. 120.21; sp.gr. 0.889; m.p. -44 C; b.p. 168 C; flash pt. (CC) 44 C; ref. index 1.5040
Toxicology: ACGIH TLV/TWA 25 ppm; LD50 (oral, rat) 5 g/kg; LDLo (IP, rat) 1752 mg/kg; LC50 (inh., rat, 4 h) 18 g/m^3; mod. toxic by IP route; mildly toxic by inh.; irritant to mucous membranes; mild skin and eye irritant; CNS depressant; may cause anemia, bronchitis; very large dosages can probably cause convulsions and death; TSCA listed
Precaution: Combustible; mod. fire risk; incompat. with oxidizing materials (increased fire and explosion hazard)
Hazardous Decomp. Prods.: Heated to decomp., emits acrid smoke and irritating fumes
NFPA: Health 0, Flammability 2, Reactivity 0
Storage: Store in cool, dry, well-ventilated area out of direct sunlight
Uses: Chem. intermediate in mfg. of trimellitic anhydride, dyes, pharmaceuticals, pseudocumidine; dye carrier solvent; component of some paint solvents and high octane fuels
Manuf./Distrib.: Aldrich; Fluka; Koch Chem. Spec.; Sigma
Trade Names Containing: Kristalloel 21; Kristalloel 30; Shellsol® A150ND; Shellsol® A150 Solvent; Solvent 2216; Solvent 2439; Solvent 30 823

Trimethyl-1,2,4-benzene; 1,2,5-Trimethylbenzene. *See* 1,2,4-Trimethylbenzene

1,3,5-Trimethylbenzene

CAS 108-67-8; EINECS/ELINCS 203-604-4
UN 2325 (DOT)
Synonyms: Benzene, 1,3,5-trimethyl-; Mesitylene; TMB; Trimethylbenzene; Trimethyl-1,3,5-benzene; sym-Trimethylbenzene; Trimethylbenzol
Classification: Aromatic hydrocarbon
Empirical: C_9H_{12}
Formula: $C_6H_3(CH_3)_3$
Properties: Colorless clear liq.; char. aromatic odor; sol. in alcohol, ether, benzene, acetone, CCl_4, petrol. ether; misc. with oxygenated and aromatic solvs.; insol. in water; m.w. 120.21; dens. 0.863; f.p. -52.7 C; b.p. 164.6 C; flash pt. 50 C
Toxicology: ACGIH TLV/TWA 25 ppm; LC50 (inh., rat, 4 h) 24 mg/m^3; toxic by inh. (can cause nose, throat, and lung irritation, headache, dizziness, nausea, CNS depression); mod. toxic by IP route; mild eye and skin irritant; ing. of very lg. amts. can probably cause convulsions and death; TSCA listed
Precaution: Combustible; mod. fire hazard; incompat. with strong oxidizing agents (increased risk of fire and explosion); reacts violently with nitric acid
Hazardous Decomp. Prods.: Heated to decomp., emits acrid smoke and irritating fumes
NFPA: Health 0, Flammability 2, Reactivity 0
Storage: Store in cool, dry, well-ventilated area out of direct sunlight
Uses: Intermediate in prod. of dyestuffs and UV oxidation stabilizers for plastics; solvent and paint thinner
Manuf./Distrib.: Aldrich; B.I. Chems.; Fluka; Koch Chem. Spec.; Sigma
Trade Names Containing: Kristalloel 30; Shellsol® A150ND

Trimethyl-1,3,5-benzene. *See* 1,3,5-Trimethylbenzene
as-Trimethylbenzene. *See* 1,2,4-Trimethylbenzene
sym-Trimethylbenzene. *See* 1,3,5-Trimethylbenzene
Trimethylbenzene, mixed isomers. *See* Trimethylbenzene
Trimethylbenzol. *See* 1,3,5-Trimethylbenzene
Trimethylbenzylammonium chloride. *See* Benzyl trimethyl ammonium chloride
(1S)-1,3,3-Trimethylbicyclo [2.2.1] heptan-2-ol. *See* Fenchyl alcohol
1,7,7-Trimethylbicyclo [2.2.1] heptan-2-ol acetate. *See* Bornyl acetate
1,7,7-Trimethylbicyclo (2.2.1)-2-heptanone; 1,7,7-Trimethylbicyclo [2.2.1] heptan-2-one. *See* Camphor
4,6,6-Trimethylbicyclo (3,1,1) hept-3-en; 2,6,6-Trimethylbicyclo (3.1.1)-2-heptene; 2,6,6-Trimethylbicyclo (3.1.1) hept-2-ene; 2,6,6-Trimethylbicyclo (3.1.1)-2-hept-2-ene. *See* α-Pinene
3,7,7-Trimethylbicyclo [4.1.0] hept-3-ene. *See* 3-Carene
endo-1,7,7-Trimethylbicyclo [2.2.1] hept-2-yl acetate. *See* Bornyl acetate

Trimethyl borate

CAS 121-43-7; EINECS/ELINCS 204-468-9

UN 2416 (DOT)
Synonyms: Boric acid, trimethyl ester; Boron methoxide; Methyl borate; Trimethoxyborane; Trimethoxyborine; Trimethoxyboron
Empirical: $C_3H_9BO_3$
Formula: $(CH_3O)_3B$
Properties: Wh. fuming liq.; sol. in polar nonhydroxylic solvs.; misc. with alcohol, ether, hexane, tetrahydrofuran; dec. in water; m.w. 103.93; dens. 0.915; f.p. -29 C; b.p. 67-68 C; flash pt. < 73 F; ref. index 1.3570
Toxicology: LD50 (oral, rat) 6140 mg/kg, (skin, rabbit) 1980 mg/kg; LDLo (IP, rat) 1600 mg/kg; mod. toxic by ing., skin contact, IP routes; eye irritant; TSCA listed
Precaution: Flamm. liq.; very dangerous fire hazard exposed to heat, flame, oxidizers; mod. explosive exposed to flame; reacts with water, oxidizing agents; reactive with water or steam to produce toxic and flamm. vapors
Hazardous Decomp. Prods.: Heated to decomp., emits acrid smoke and irritating fumes
NFPA: Health 2, Flammability 3, Reactivity 1
Storage: Moisture-sensitive; store under nitrogen; store @ 2-8 C
Uses: Solvent, catalyst for prod. of resins, waxes, paints, varnishes; dehydrating agent; fungicide for citrus fruit; neutron scintillation counters; gaseous brazing/welding flux; boron compds.; brazing flux; methylation agent in org. reactions
Manuf./Distrib.: Aldrich; Fluka; Haltermann Prods. UK; Sigma

2,2,3-Trimethylbutane

CAS 464-06-2
UN 1206
Synonyms: Isopropyltrimethylmethane; Triptane
Classification: Sat. aliphatic hydrocarbon
Empirical: C_7H_{16}
Formula: $CH_3C(CH_3)_2CH(CH_3)CH_3$
Properties: Colorless liq.; gasoline-like odor; sol. in most org. solvs.; insol. in water; m.w. 100.20; sp.gr. 0.69 (20 C); m.p. -25 C; b.p. 81 C; flash pt. -6 C; ref. index 1.3895 (20 C)
Toxicology: ACGIH TLV/TWA 400 ppm; STEL 500 ppm; may cause mild skin irritation; conc. vapor may cause sl. eye irritation; probably a CNS depressant; conc. vapor may cause sl. nose/throat irritation, lightheadedness, dizziness, nausea, difficulty breathing; very high concs. may cause unconsciousness, death; ing. may cause nausea, vomiting, swelling of abdomen, headache, depression; relatively low oral toxicity, unless aspirated into lungs
Precaution: Flamm.; mod. fire risk; incompat. with strong oxidizers (increases fire and explosion risk)
Hazardous Decomp. Prods.: None
NFPA: Health 0, Flammability 3, Reactivity 0
Storage: Store in cool, dry, well-ventilated area out of direct sunlight
Uses: Ingred. in gasoline and aviation fuel; org. synthesis; solvent
Manuf./Distrib.: Aldrich; Fluka

Trimethyl carbinol. *See* t-Butyl alcohol

Trimethyl citrate

CAS 1587-20-8
Empirical: $C_9H_{14}O_7$
Properties: Solid; sol. 8.2 g/100 ml in water; insol. in heptane; m.p. 76 C
Uses: Solvent for household, I&I cleaners; solvent, film-former, and plasticizer in inks, adhesives, and coatings

Trimethylcyclohexanol

CAS 116-02-9; EINECS/ELINCS 204-122-7
FEMA 3962
Synonyms: Cyclohexanol, 3,3,5-trimethyl-; Cyclonol; Dihydroisophorol; Homomenthol; Isophorol, dihydro-; 3,3,5-Trimethylcyclohexanol; 3,3,5-Trimethyl-1-cyclohexanol; 3,5,5-Trimethylcyclohexanol
Empirical: $C_9H_{18}O$
Formula: $CH_3(C_6H_8)(CH_3)_2OH$
Properties: Wh. cryst. or liq.; sol. in most organic solvents, hydrocarbons, oils; insol. in water; m.w. 142.24; dens. 0.878 (40/20 C); vapor pressure 0.1 mm (20 C); f.p. 37 C; b.p. 198 C; flash pt. (OC) 190 F
Toxicology: LD50 (oral, rat) 3250 mg/kg, (skin, rabbit) 2800 mg/kg; toxic by inh.; mod. toxic by ing., skin contact; harmful if ing., inh., or absorbed thru skin; skin and severe eye irritant; high concs. are extremely destructive to tissues of mucous membranes/upper respiratory tract, eyes and skin; may cause burning sensation, coughing, wheezing, laryngitis, headache, nausea, vomiting; TSCA listed
Precaution: Combustible exposed to heat, flame, oxidizers; reactive with oxidizers
Hazardous Decomp. Prods.: CO, CO_2; heated to decomp., emits acrid smoke and irritating fumes; emits toxic fumes under fire conditions
NFPA: Health 2, Flammability 2, Reactivity 0
Storage: Store in cool, dry place; keep container closed when not in use; keep away from heat and open flame
Uses: Dye leveling agent; solvent; menthol and camphor substitute; antifoam; hydraulic fluids; textile soaps; odor masking; esterification agent; pharmaceuticals; wax additive; printing inks; pesticide; synthetic flavoring agent in foods; fragrance in cosmetics
Regulatory: FEMA GRAS
Manuf./Distrib.: Acros Org.; Salor

3,3,5-Trimethylcyclohexanol; 3,3,5-Trimethyl-1-cyclohexanol; 3,5,5-Trimethylcyclohexanol. *See* Trimethylcyclohexanol

3,3,5-Trimethyl cyclohexanone

CAS 873-94-9
Classification: Nonaromatic ketone
Empirical: $C_9H_{16}O$
Properties: Liq.; m.w. 140.23; dens. 0.887; m.p. -10 C; b.p. 188-192 C; ref. index 1.4450
Toxicology: TSCA listed
Precaution: Combustible
Uses: Specialty solvent; leveling agent, flow control agent for paints/coatings; shelf life extender for vinyl coatings for metal or plastic; thinner for PVC pastes and plasticizers; raw material
Manuf./Distrib.: Aldrich; Atofina SA; Degussa; Fluka; Nipa Labs; Salor

α,α,4-Trimethyl-3-cyclohexene-1-methanol. *See* α-Terpineol
1,1,3-Trimethyl-3-cyclohexene-5-one; 3,3,5-Trimethyl-2-cyclohexene-1-one; 3,3,5-Trimethylcyclohexenone; 3,3,5-Trimethyl-2-cyclohexen-1-one. *See* Isophorone

1,1,2-Trimethylcyclopentane

CAS 4259-00-1
Synonyms: Camphocean
Classification: Sat. alicyclic hydrocarbon
Empirical: C_8H_{16}
Formula: $(CH_3)_3C_5H_7$
Properties: Colorless liq.; sol. in most org. solvs.; insol. in water; m.w. 112.21; sp.gr. 0.7661 (20 C); b.p. 113-114 C
Toxicology: May cause mild skin irritation; conc. vapor may be irritating to eyes; probably a CNS depressant; high vapor concs. may cause nose/throat irritation, headache, lightheadedness, nausea, dizziness; very high concs. may cause unconsciousness, death; ing. may cause sore throat, nausea, diarrhea
Precaution: Flamm.; incompat. with strong oxidizing agents (increased fire and explosion hazard)
Hazardous Decomp. Prods.: None
Storage: Store in cool, dry, well-ventilated area out of direct sunlight
Uses: Solvent; org. synthesis

1,1,3-Trimethylcyclopentane

CAS 4516-69-2
Classification: Sat. alicyclic hydrocarbon
Empirical: C_8H_{16}
Formula: $(CH_3)_3C_5H_7$
Properties: Colorless liq.; sol. in most org. solvs.; insol. in water; m.w. 112.21; sp.gr. 0.7703 (20 C); b.p. 115-116 C
Toxicology: May cause mild skin irritation; conc. vapor may be irritating to eyes; probably a CNS depressant; high vapor concs. may cause nose/throat irritation, headache, lightheadedness, nausea, dizziness; very high concs. may cause unconsciousness, death; ing. may cause sore throat, nausea, diarrhea
Precaution: Flamm.; incompat. with strong oxidizing agents (increased fire and explosion hazard)
Hazardous Decomp. Prods.: None
Storage: Store in cool, dry, well-ventilated area out of direct sunlight

Uses: Solvent; org. synthesis

Trimethyl dodecatrienol; 3,7,11-Trimethyl-2,6,10-dodecatrien-1-ol. *See* Farnesol

Trimethylene glycol

CAS 504-63-2; EINECS/ELINCS 207-997-3

Synonyms: 2-Deoxyglycerol; 1,3-Dihydroxypropane; 2-(Hydroxymethyl) ethanol; 1,3-Propanediol; Propane-1,3-diol; 1,3-Propylene glycol; β-Propylene glycol

Classification: Aliphatic dihydric alcohol

Empirical: $C_3H_8O_2$

Formula: $CH_2OHCH_2CH_2OH$

Properties: Colorless to pale yel. visc. liq.; odorless; sol. in water, ethanol, ether; misc. with oxygenated solvs.; m.w. 76.11; dens. 1.0597 (20 C); vapor pressure 0.0441 mm Hg; m.p. -27 C; b.p. 210-212 C; flash pt. (CC) 79 C; autoignition temp. 400 C; pH neutral; ref. index 1.439; dielec. const. 35.0

Toxicology: LD50 (oral, mouse) 4773 mg/kg, (skin, rabbit) > 20,000 mg/kg; mildly toxic by ing.; vapor unlikely to cause health effects on inh.; liq. may cause mild skin and eye irritation; mutagenic; target organs: skin, CNS, liver; TSCA listed

Environmental: VOC; BOD5 0.18; COD 1.64; ThOD 1.68

Hazardous Decomp. Prods.: CO, CO_2; heated to decomp., emits acrid smoke and irritating fumes

NFPA: Health 1, Reactivity 0

Storage: Hygroscopic

Uses: Intermediate in mfg. of polyesters; polyurethane polyester polyol comonomer; by-product in the mfg. of glycerin thru the saponification of fat; used to lower the f.p. of water (e.g., antifreeze); solvent

Manuf./Distrib.: Aldrich; Degussa; Eastern Chem.; Pfaltz & Bauer; United-Guardian

1,1,2-Trimethylethane. *See* Isopentane

Trimethyl glycol. *See* Propylene glycol

2,2,5-Trimethylhexane. *See* Neononane

3,5,5-Trimethylhexanoic acid, isodecyl ester. *See* Isodecyl isononanoate

3,5,5-Trimethylhexanoic acid, isotridecyl ester. *See* Isotridecyl isononanoate

3,5,5-Trimethylhexanoic acid, 3,5,5-trimethylhexyl ester. *See* Isononyl isononanoate

Trimethylhexanol; 3,5,5-Trimethylhexanol. *See* 3,5,5-Trimethylhexan-1-ol

3,5,5-Trimethylhexan-1-ol

CAS 3452-97-9; EINECS/ELINCS 222-376-7

FEMA 3324

Synonyms: Hexanol, 3,5,5-trimethyl-; Nonanol; Nonyl alcohol; Nonylol; Trimethylhexanol; 3,5,5-Trimethylhexanol; 3,5,5-Trimethyl-1-hexanol; Trimethylhexyl alcohol; 3,5,5-Trimethylhexyl alcohol

Classification: Nonaromatic alcohol

Empirical: $C_9H_{20}O$

Formula: $(CH_3)_3CCH_2CHCH_3CH_2CH_2OH$

Properties: Colorless liq.; mild odor; insol. in water; m.w. 144.26; dens. 0.8236 kg/l (20 C); bulk dens. 6.86 lb/gal; b.p. 194 C; flash pt. 200 F; ref. index 1.4300; b.p. 193-194 (760 mm); flash pt. 178 F; ref. index 1.4320

Toxicology: TSCA listed

Precaution: Combustible liq.; incompat. with strong oxidizing agents

Hazardous Decomp. Prods.: Toxic fumes of CO, CO_2; emits toxic fumes under fire conditions

NFPA: Health 2, Flammability 2, Reactivity 0

Storage: Store in cool, dry place; keep tightly closed; keep away from heat and open flame

Uses: Synthetic lubricants; additives to lubricating oils; wetting agent; softener in mfg. of various plastics; disinfectants; germicides; froth flotation foam modifier; antifoam for pesticides; process solvent; fragrance; synthetic flavoring agent in foods; solvent in cosmetics

Regulatory: FEMA GRAS

Manuf./Distrib.: Aldrich; ICN Biomed. Research Prods.; Penta Mfg.; TCI Am.

3,5,5-Trimethyl-1-hexanol; Trimethylhexyl alcohol; 3,5,5-Trimethylhexyl alcohol. *See* 3,5,5-Trimethylhexan-1-ol

3,5,5-Trimethylhexyl 3,5,5-trimethylhexanoate. *See* Isononyl isononanoate

Trimethylmethane. *See* Isobutane

Trimethyl methanol. *See* t-Butyl alcohol

(1R,2S,7S,9S)-3,3,7-Trimethyl-8-methylenetricyclo [5.4.0.0^{2,9}] undecane. *See* Longifolene

Trimethylnonanol; 2,4,8-Trimethyl-6-nonanol; 2,6,8-Trimethyl-4-nonanol; 2,6,8-Trimethylnonanol-4. *See* 2,6,8-Trimethylnonyl-4-alcohol

2,6,8-Trimethyl-4-nonanone; Trimethyl 4-nonanone. *See* Isobutyl heptyl ketone

2,6,8-Trimethylnonyl-4-alcohol

CAS 123-17-1

Synonyms: 4-Hydroxy-2,6,8-trimethylnonane; 4-Nonanol, 2,6,8-trimethyl; Trimethylnonanol; 2,4,8-Trimethyl-6-nonanol; 2,6,8-Trimethyl-4-nonanol; 2,6,8-Trimethylnonanol-4

Empirical: $C_{12}H_{26}O$

Formula: $(CH_3)_2CHCH_2CH_2OCH_2CH(CH_3)CH_2CH(CH_3)_2$

Properties: Colorless liq.; char. odor; insol. in water; m.w. 186.38; dens. 0.8913; bulk dens. 6.9 lb/gal; visc. 21.4 cP; vapor pressure < 0.01 mm (20 C); f.p. -60 C; b.p. 225.2 C; flash pt 200 F (93.3 C)

Toxicology: LD50 (oral, rat) 17 g/kg, (skin, rabbit) 11,220 µl/kg; mildly toxic by ing.; primary irritant; skin and eye irritant; TSCA listed

Precaution: Combustible

Hazardous Decomp. Prods.: Heated to decomp., emits acrid smoke and irritating fumes

Uses: Surfactant; flotation agent; lube additive; rubber chemical; solvent

Manuf./Distrib.: Dow; Pfaltz & Bauer

1,3,3-Trimethyl-2-norbornanol. *See* Fenchyl alcohol

1,7,7-Trimethylnorcamphor. *See* Camphor

4,7,7-Trimethyl-3-norcarene. *See* 3-Carene

1,3,3-Trimethyl-2-oxabicyclo [2.2.2] octane. *See* Eucalyptol

2,2,4-Trimethylpentane. *See* Isooctane

Trimethylpentanediol. *See* 2,2,4-Trimethyl-1,3-pentanediol

2,2,4-Trimethyl-1,3-pentanediol

CAS 144-19-4; EINECS/ELINCS 205-619-1

Synonyms: 1,3-Pentanediol, 2,2,4-trimethyl; TMPD; Trimethylpentanediol

Empirical: $C_8H_{18}O_2$

Formula: $(CH_3)_2CHCH(OH)C(CH_3)_2CH_2OH$

Properties: Wh. cryst. solid; sol. in alcohol, acetone, ether, benzene; sl. sol. in water; m.w. 146.23; sp.gr. 0.93 (55/15 C); m.p. 50 C; b.p. 225 C; flash pt. (COC) 113 C

Toxicology: LDLo (oral, rat) 2000 mg/kg, (IP, rat) 800 mg/kg, (IV, rat) 145 mg/kg; poison by IV; mod. toxic by ing. and IP routes; skin irritant; TSCA listed

Precaution: Combustible; can react with oxidizers

Hazardous Decomp. Prods.: Heated to decomp., emits acrid smoke and irritating fumes

Uses: Resin intermediate; coalescing solvent for water-based paints; unsat. polyester resin comonomer for surf. coatings; plasticizers; lubricants; printing inks; insect repellent

Manuf./Distrib.: Aldrich; Eastman; Fluka

2,2,4-Trimethyl-1,3-pentanediol isobutyrate

CAS 25265-77-4; EINECS/ELINCS 246-771-9

Synonyms: Isobutyric acid, monoester with 2,2,4-trimethylpentane-1,3-diol; 1,3-Pentanediol, 2,2,4-trimethyl, monoisobutyrate; Propanoic acid, 2-methyl-, monoester with 2,2,4-trimethyl-1,3-pentanediol; Propionic acid, 2-methyl-, monoester with 2,2,4-trimethyl-1,3-pentanediol; 2,2,4-Trimethyl-1,3-pentanediol monoisobutyrate; 2,2,4-Trimethyl-1,3-pentanediolmono (2-methylpropanoate)

Classification: Ester alcohol

Empirical: $C_{12}H_{24}O_3$

Formula: $(CH_3)_2CHCH(OH)C(CH_3)_2CH_2OOCCH(CH_3)_2$

Properties: Colorless liq.; sol. in benzene, alcohol, acetone, Cl_4; insol. in water; m.w. 216.32; dens. 0.945-0.955 (20/20 C); vapor pressure 0.012 mm Hg; m.p. -50 C; b.p. 244 C; flash pt. (COC) 245 F; ref. index 1.4423

Toxicology: LD50 (oral, mouse) 3200 mg/kg, (oral, rat) 3200 mg/kg; mod. toxic by ing.; may cause digestive tract irritation; TSCA listed

Precaution: Combustible

Hazardous Decomp. Prods.: CO, CO_2; heated to decomp., emits acrid smoke and irritating fumes

NFPA: Health 0, Flammability 1, Reactivity 0

Storage: Store in cool, dry place in a tightly closed container
Uses: Solvent; chemical intermediate for prod. of esters used as plasticizers, lubricant basestocks, solvents, surfactants, syn. detergents, herbicides, pesticides, resins, inks, latexes, drilling muds; slow-evaporating solvent used as coalescing agent in latex finishes and water-base inks, PVAc latexes, PVAc-acrylic latexes, and acrylic, EVA, and B/S latexes; solvent in electrodeposition coatings, lacquer coatings, high-bake enamels, cosmetics; defoamer in waterborne systems and drilling muds for oil industry
Manuf./Distrib.: Acros Org.; Aldrich; Ashland; Chisso Am.; Eastman
Trade Names: Texanol® Ester-Alcohol; Ucar® Filmer IBT

2,2,4-Trimethyl-1,3-pentanediol monoisobutyrate; 2,2,4-Trimethyl-1,3-pentanediolmono (2-methylpropanoate). *See* 2,2,4-Trimethyl-1,3-pentanediol isobutyrate
Trimethylphenylmethane. *See* t-Butylbenzene

Trimethyl phosphate

CAS 512-56-1
Synonyms: Methyl phosphate; Phosphoric acid, trimethyl ester; O,O,O-Trimethyl phosphate
Empirical: $C_3H_9O_4P$
Formula: $(CH_3O)_3PO$
Properties: Colorless liq.; sol. in water, alcohol, ether, gasoline; m.w. 140.09; dens. 1.210; b.p. 193 C; pour pt. -46 C; flash pt. > 148 C; ref. index 1.397 (20 C)
Toxicology: LD50 (oral, rat) 840 mg/kg, (skin, rabbit) 3388 mg/kg; LDLo (IP, rat) 800 mg/kg, (IV, rat) 2400 mg/kg; mod. toxic by ing., skin contact, IP, IV routes; strong irritant to skin and eyes; suspected carcinogen; experimental neoplastigen, tumorigen, teratogen, reproductive effects; human mutagenic data; TSCA listed
Precaution: Combustible; explodes when heat distilled
Hazardous Decomp. Prods.: Heated to decomp., emits toxic fumes of PO_x
Uses: Flame retardant; controls spark plug fouling, surface ignition, and rumble in gasoline engines; agric. chemical; solvent
Manuf./Distrib.: Beckmann Chemikalien

O,O,O-Trimethyl phosphate. *See* Trimethyl phosphate
α-(Trimethylsilyl)-ω-methylpoly [oxy (dimethylsilylene)]. *See* Dimethicone
α,α,α´-Trimethyltrimethyleneglycol. *See* Hexylene glycol
1,1,1-Trimethyl-N-(trimethylsilyl) silanamine. *See* Hexamethyldisilazane
2,4,6-Trimethyl-1,3,5-trioxacyclohexane; 2,4,6-Trimethyl-1,3,5-trioxane; 2,4,6-Trimethyl-s-trioxane; s-Trimethyltrioxymethylene. *See* Paraldehyde

Trimyristin

CAS 555-45-3; EINECS/ELINCS 209-099-7
Synonyms: Glyceryl trimyristate; Myristin; 1,2,3-Propanetriol tritetradecanoate
Definition: Triester of glycerin and myristic acid
Empirical: $C_{45}H_{86}O_6$
Properties: Wh. to ylsh.-gray solid; insol. in water; sol. in alcohol, benzene, chloroform, ether; m.w. 723.14; dens. 0.885 (60/4 C); m.p. 56-57 C; ref. index 1.4429 (60 C)
Toxicology: TSCA listed
Uses: Emollient, solvent, visc. control agent, consistency agent for cosmetics, pharmaceuticals; binder, lubricant, mold release agent, retarding agent for tablets and compressed confectioneries; emulsifier; in food-contact textiles
Regulatory: FDA 21CFR §177.2800
Manuf./Distrib.: Fluka; Mosselman NV; Sigma

Trioctanoin

CAS 7360-38-5; EINECS/ELINCS 230-896-0
Synonyms: Glycerol trioctanoate; Glyceryl tri (2-ethylhexanoate); Glyceryl trioctanoate; Octanoic acid, 1,2,3-propanetriol ester; Octanoic acid, 1,2,3-propanetriyl ester; Octanoic acid triglyceride; 1,2,3-Propanetriol trioctanoate; Trioctanoylglycerol; 1,2,3-Trioctanoylglycerol
Definition: Triester of glycerin and 2-ethylhexanoic acid
Empirical: $C_{27}H_{50}O_6$
Formula: $(CH_2OCOCHCH_2CH_3(CH_2)_3CH_3)_2HCOCOCHCH_2CH_3(CH_2)_3CH_3$
Properties: M.w. 470.70
Toxicology: TSCA listed
Uses: Antistat, emollient, solvent, lubricant for cosmetics and pharmaceuticals; pigment wetting agent/dispersant/binder in liq. makeup, pressed powds., lip prods; color enhancer for pressed powds.; carrier for essential oils, flavors; vehicle for vitamins, medicinals, nutritional prods.
Manuf./Distrib.: A&E Connock

Trioctanoylglycerol; 1,2,3-Trioctanoylglycerol. *See* Trioctanoin

Trioctyl phosphate

CAS 78-42-2; EINECS/ELINCS 201-116-6
Synonyms: 2-Ethylhexanol phosphate; 2-Ethyl-1-hexanol phosphate; 2-Ethylhexanol, phosphate triester; 1-Hexanol, 2-ethyl-, phosphate; Octyl phosphate; Phosphoric acid, tris (2-ethylhexyl) ester; TOF; TOP; Triethylhexyl phosphate; Tri (2-ethylhexyl) phosphate; Tris (2-ethylhexyl) phosphate
Classification: Phosphoric acid ester
Empirical: $C_{24}H_{51}O_4P$
Formula: $[CH_3(CH_2)_3CH(C_2H_5)CH_2O]_3PO$
Properties: Colorless to pale yel. clear liq.; sl. sharp odor; sol. in alcohol, acetone, ether, min. oil, gasoline; pract. insol. in water; m.w. 434.72; dens. 0.924 (26 C); vapor pressure 1.9 mm Hg (20 C); f.p. -90 C (sets to glass); pour pt. -74 C; b.p. 190-233 C (dec.); flash pt.(OC) 207 C; ref. index 1.4440; anionic
Toxicology: LD50 (oral, rat) 37 g/kg, (dermal, rabbit) 20 g/kg; mildly toxic by ing., skin contact; irritating to eyes, skin, respiratory system; suspected carcinogen; tumorigen; TSCA listed
Precaution: Combustible exposed to heat or flame; can react with oxidizing materials; may soften or deteriorate certain plastics/elastomers; incompat. with cellulose acetate, cellulose acetate butyrate
Hazardous Decomp. Prods.: Heated to decomp., emits toxic fumes of PO_x, CO, CO_2, phosphoric acid
Storage: Store under ambient temps.
Uses: Solvent, antifoam for paints; flame retardant, plasticizer for vinyl resins, rubbers, nitrocellulose; fungicide; cosolvent for hydrogen peroxide prod.; lubricant, EP agent for aq. metalworking systems; emulsifier, wetting agent, corrosion inhibitor, lubricant, and antistat for textiles, metalworking, leather, household/I&I, agrochems., emulsion polymerization, and polymers and plastics; hydrotrope in mercerizing baths; defoamer in food-contact paper/paperboard; in food pkg. adhesives
Regulatory: FDA 21CFR §175.105, 176.210
Manuf./Distrib.: Akzo Nobel; Aldrich; Ashland; Bayer; ChemService; FMC; Ferro/Keil; Fluka; Rhodia HPCII; Rhodia/Phosphorus Perf. Derivs.; Sigma
Trade Names: Reomol® TOP

Tri-n-octyl phosphate

CAS 1806-54-8
Empirical: $C_{24}H_{51}O_4P$
Properties: Colorless clear liq.; sol. in ether; sol. > 10 mg/ml in 95% ethanol; sl. sol. in DMSO; insol. in water; m.w. 434.64; m.p. -74 C; b.p. 220-230 C (8 mm); flash pt. > 93.3 C
Toxicology: Cholinesterase inhibitor
Precaution: Probably combustible; reacts with oxidizing materials
Storage: Store refrigerated; keep away from oxidizing materials; store away from ignition sources
Uses: Solvent; antifoam; plasticizer

Trioctylphosphine oxide

CAS 78-50-2; EINECS/ELINCS 201-121-3
Synonyms: 1-(Dioctyl-phosphinoyl)-octane; TOPO; Tri-n-octylphosphine oxide; Tri-n-octylphosphinic oxide
Classification: Nonaromatic phosphorus compd.
Empirical: $C_{24}H_{51}OP$
Formula: $[CH_3(CH_2)_7]_3PO$
Properties: Off-wh. waxy solid; sol. in hydrocarbons; insol. in water; m.w. 386.65; sp.gr. 0.88; m.p. 50-54 C; b.p. 201-202 C (2 mm); flash pt. (CC) 235 F
Toxicology: Severe irritant; TSCA listed
Uses: Extraction agent for recovering uranium ore from wet process phosphoric acid, and for other metals and some organics, such as acetic

acid and phenol
Manuf./Distrib.: Aldrich; Fluka; Sigma
Trade Names: Cyanex® 921
Trade Names Containing: Cyanex® 923

Tri-n-octylphosphine oxide; Tri-n-octylphosphinic oxide. *See* Trioctylphosphine oxide

Triolein

CAS 122-32-7; 67701-30-8; EINECS/ELINCS 204-534-7; 266-948-4
Synonyms: Glycerol trioleate; Glyceryl trioleate; 9-Octadecenoic acid, 1,2,3-propanetriyl ester; Olein
Definition: Triester of glycerin and oleic acid
Empirical: $C_{57}H_{104}O_6$
Formula: $(CH_2O)2CHO[CO(CH_2)_7CH=CH(CH_2)_7CH_3]_3$
Properties: Colorless to ylsh. oily liq., tasteless, odorless; sol. in chloroform, ether, CCl_4, chlorinated solvs.; sl. sol. in alcohol; pract. insol. in water; m.w. 885.40; dens. 0.915 (15/4 C); m.p. -4 to -5 C; b.p. 235-240 C (15 mm); HLB 0.8; sapon. no. 188-195; cloud pt. -10 C; ref. index 1.4676 (20 C); nonionic
Toxicology: No known toxicity; TSCA listed
Precaution: Combustible
Uses: Emollient, solvent, visc. control agent, lubricant, emulsifier for cosmetics; lubricant, emulsifier for metals, leather, textiles; carbon source in antibiotic culture broths; raw material for engineering plastics; impact modifier; plasticizer; stabilizer; solvent for carbon paper, printing ink dyes; food emulsifier, stabilizer; solubilizer for flavors, vitamin oils; pharmaceutical lubricant, emollient, solubilizer, stabilizer; in food-contact textiles
Regulatory: FDA 21CFR §177.2800
Manuf./Distrib.: A.P. Chems. Ltd; ABITEC; Ashland; Fluka; Mosselman NV; Sigma; Uniqema Am.
Trade Names: Arlypon SO; Trioleina

1,3,5-Trioxacyclohexane; Trioxan. *See* Trioxane

Trioxane

CAS 110-88-3; EINECS/ELINCS 203-812-5
UN 1325
Synonyms: Metaformaldehyde; Polyoxymethylene; Triformol; 1,3,5-Trioxacyclohexane; Trioxan; 1,3,5-Trioxane; s-Trioxane; sym-Trioxane; Trioxymethylene
Classification: Nonaromatic ether
Empirical: $C_3H_6O_3$
Formula: $OCH_2OCH_2OCH_2$
Properties: Cryst. solid; mild pleasant char. ethanol/chloroform-like odor; very sol. in water, alcohol, ketones, ether, acetone, chlorinated and aromatic hydrocarbons, org. solvs.; sl. sol. in pentane, petrol. ether, lower paraffins; m.w. 90.08; dens. 1.17 kg/l (65 C); vapor pressure 13 mm Hg; m.p. 61-62 C; b.p. 115 C; sublimes; flash pt. (OC) 45 C; autoignition temp. 414 C
Toxicology: LD50 (oral, rat) > 3200 mg/kg, (skin, rabbit) > 10,000 mg/kg; harmful if swallowed; mutagenic data; target organ: skin; TSCA listed
Environmental: VOC; ThOD 1.07
Precaution: Flamm. solid exposed to heat, flame, or oxidizers; LEL 3.6%; UEL 29%; may explode when heated or as vapor; explodes on impact; explosive limits 3.6-28.7%; incompat. with oxidizers; mixts. with hydrogen peroxide or liq. oxygen are explosive
Hazardous Decomp. Prods.: Heated to decomp., emits acrid smoke and irritating fumes; can evolve toxic formaldehyde fumes when strongly heated or in contact with strong acids or acid fumes
NFPA: Health 2, Flammability 2, Reactivity 0
Uses: Stabilizer for chlorinated solvs.; solvent
Manuf./Distrib.: ABCR; Acros Org.; Aldrich; BASF; Fluka; Sigma

1,3,5-Trioxane; s-Trioxane; sym-Trioxane. *See* Trioxane
Trioxanonane. *See* Diethylene glycol dimethyl ether
5,8,11-Trioxapentadecane. *See* Diethylene glycol dibutyl ether
3,6,9-Trioxaundecane. *See* Diethylene glycol diethyl ether
Trioxymethylene. *See* Trioxane

Tripalmitin

CAS 555-44-2; EINECS/ELINCS 209-098-1
Synonyms: Glyceryl tripalmitate; Hexadecanoic acid, 1,2,3-propanetriyl ester; Palmitin
Definition: Triester of glycerin and palmitic acid
Empirical: $C_{51}H_{98}O_6$
Formula: $C_3H_5(OOCC_{15}H_{31})_3$
Properties: Wh. crystalline powd. or needles; sol. in ether, chloroform, benzene; pract. insol. in alcohol; insol. in water; m.w. 807.29; dens. 0.886 (80/4 C); m.p. 65.5 C; b.p. 310-320 C; ref. index 1.43807 (80 C); sapon. no. 208.5
Toxicology: LD (IP, mouse) > 500 mg/kg; TSCA listed
Precaution: Combustible
Storage: Refrigerate
Uses: Soaps; leather dressing; aux. in prep. of compressed sweets; emollient, solvent, visc. control agent, lubricant, consistency agent in cosmetics and pharmaceuticals; nucleating agent in sticks; lubricant, mold release agent, binder, retarding agent for tablets; in food-contact textiles
Regulatory: FDA 21CFR §177.2800
Manuf./Distrib.: Aldrich; Fluka; Sigma

Triphenyl phosphate

CAS 115-86-6; EINECS/ELINCS 204-112-2
Synonyms: Phosphoric acid, triphenyl ester; TPP
Classification: Phosphate
Empirical: $C_{18}H_{15}O_4P$
Formula: $PO(OC_6H_5)_3$
Properties: Colorless cryst. powd. or needles, odorless; sol. in benzene, chloroform, ether, acetone, lacquers, solvent, thinners, oil; insol. in water; m.w. 326.28; dens. 1.268 (60 C); m.p. 50 C; b.p. 245 C (11 mm); flash pt. (CC) 220 C; ref. index 1.550
Toxicology: ACGIH TLV/TWA 3 mg/m^3; LD50 (oral, rat) 3500 mg/kg, (subcut., cat) 100 mg/kg, (skin, rabbit) > 7900 mg/kg; poison by subcut. route; mod. toxic by ing.; toxic by inhalation; absorbed slowly, esp. by skin contact; not a potent cholinesterase inhibitor; TSCA listed
Precaution: Combustible
Hazardous Decomp. Prods.: Heated to decomp., emits toxic fumes of PO_x
NFPA: Health 2, Flammability 1, Reactivity 0
Uses: Flame retardant plasticizer for cellulosics, coatings, triacetate film and sheet, engineering thermoplastics (PPO, HIPS, ABS-PC); noncombustible substitute for camphor in celluloid; EP agent in lubricants; in lacquers; varnishes; impregnating roofing paper; in food-pkg. adhesives; solvent for adjuvants in food-contact crosslinked polyesters
Regulatory: FDA 21CFR §175.105, 177.2420
Manuf./Distrib.: Akzo Nobel; Alchemie USA; Aldrich; Ashland; Bayer; ChemService; FMC; Fluka; Monsanto; Pechiney Chems. Div.; PolyOne/ Elastomers & Perf. Addit.; Sinochem Jiangsu Wuxi

Tripropionin. *See* Glyceryl tripropionate

Tripropylene glycol

CAS 24800-44-0; EINECS/ELINCS 246-466-0
Synonyms: 2-(2-(2-Hydroxypropoxy) propoxy)-1-propanol; 3-[3-(3-Hydroxypropoxy) propoxy] propan-1-ol; ((1-Methyl-1,2-ethanediyl) bis (oxy)) bis (propanol); ((Methylethylene) bis (oxy)) dipropanol; PPG-3; Propanol, ((1-methyl-1,2-ethanediyl) bis (oxy)) bis-; TPG
Empirical: $C_9H_{20}O_4$
Formula: $HO(C_3H_6O)_2C_3H_6OH$
Properties: Colorless clear liq.; sol. in water, methanol, ether; misc. with oxygenated solvs.; m.w. 192.26; dens. 1.019; vapor pressure 1 mm (96 C); b.p. 268 C; flash pt. 235 F; ref. index 1.442
Toxicology: LD50 (oral, rat) 3 g/kg; mod. toxic by ing.; TSCA listed
Precaution: Combustible; reactive with oxidizers
Hazardous Decomp. Prods.: Heated to decomp., emits acrid smoke and irritating fumes
NFPA: Health 0, Flammability 1, Reactivity 0
Storage: Hygroscopic
Uses: Intermediate in emulsifiers, resins, plasticizers, pharmaceuticals, insecticides, dyestuffs, mold lubricants; brake/hydraulic fluid component; comonomer for alkyd resins and unsat. polyester resins; chain extender for PU; thickener in cosmetics; solvent, homogenizer for inks, other coloring materials; initiator for urethane polyols; in varnish removers, hard surf. cleaners, penetrating oils; lubricant, antifreeze for carburetor fluids; cooling media; disinfectant; in foods, soaps, personal

care
Manuf./Distrib.: Aldrich; Arch Chems./Specialty Polyols; Ashland; Coyne; Dow; Fluka; Lyondell

Tripropylene glycol butyl ether. *See* Tripropylene glycol n-butyl ether

Tripropylene glycol n-butyl ether

CAS 55934-93-5; EINECS/ELINCS 259-910-3
Synonyms: 1-2-[Butoxy-1-methylethoxy)-1-methylethoxy]-2-propanol; PPG-3 butyl ether; Propanol, [2-(2-butoxymethylethoxy) methylethoxy]-; TPGnBE; TPnB; Tripropylene glycol butyl ether; Tripropylene glycol monobutyl ether; Tripropylene glycol mono-n-butyl ether
Classification: Nonaromatic ether
Empirical: $C_{13}H_{28}O_4$
Formula: $C_4H_9OC_3H_6OC_3H_6OCH_2CHOHCH_3$
Properties: M.w. 248.37; dens. 0.932; b.p. 276 C; ref. index 1.4320; nonionic
Toxicology: LD50 (oral, rat) 9160 mg/kg; mildly toxic by ingestion and skin contact; irritating to eyes, skin, respiratory system; TSCA listed
Hazardous Decomp. Prods.: Heated to decomp., emits acrid smoke and irritating fumes
Uses: Solvent for heavy-duty cleaners, floor wax strippers, coatings, inks; plasticizer for resin systems; coupling agent, grease penetrant in oven cleaners; syn. lubricating oil; cosmetics raw material; material for surfactants; mold lubricant for ceramics; textile printing pastes
Manuf./Distrib.: Aldrich; Lyondell
Trade Names: Arcosolv® TPNB; Solvenon® TPnB
Trade Names Containing: Dowanol® TPnB

Tripropylene glycol ethyl ether

CAS 20178-34-1
Synonyms: 1-[2-(2-Ethoxy-1-methylethoxy)-1-methylethoxy]-2-propanol; TPGEE; Tripropylene glycol monoethyl ether
Classification: Aliphatic ether alcohol
Empirical: $C_{11}H_{24}O_4$
Formula: $CH_3CH(OH)CH_2OCH(CH_3)CH_2OCH(CH_3)CH_2O$
Properties: Colorless clear liq.; sl. pleasant ether-like odor; sol. in water, variety of org. solvs.; m.w. 220.3; sp.gr. 0.948; b.p. 252 C; flash pt. (COC) 132 C
Toxicology: Low oral toxicity; unlikely to irritate skin; may cause mild eye irritation; readily absorbed thru skin, but health effects unlikely; extreme/prolonged contact may cause CNS depression, possible unconsciousness
Precaution: Incompat. with oxidizing agents (increases fire/explosion hazard)
Storage: Store in a cool, dry area, out of direct sunlight
Uses: Solvent for paints, lacquers, resins, inks, dyes, oils, greases, cleaners

Tripropylene glycol methyl ether. *See* PPG-3 methyl ether
α-Tripropylene glycol methyl ether. *See* 1-(2-(2-Methoxy-1-methylethoxy)-1-methylethoxyl)-2-propanol
Tripropylene glycol monobutyl ether; Tripropylene glycol mono-n-butyl ether. *See* Tripropylene glycol n-butyl ether
Tripropylene glycol monoethyl ether. *See* Tripropylene glycol ethyl ether
Tripropylene glycol monomethyl ether. *See* 1-(2-(2-Methoxy-1-methylethoxy)-1-methylethoxyl)-2-propanol; PPG-3 methyl ether
Tripropylene glycol monopropyl ether; Tripropylene glycol mono-n-propyl ether. *See* Tripropylene glycol n-propyl ether
Tripropylene glycol myristyl ether. *See* PPG-3 myristyl ether
Tripropylene glycol propyl ether. *See* Tripropylene glycol n-propyl ether

Tripropylene glycol n-propyl ether

CAS 96077-04-2
Synonyms: [Methyl-2-(methyl-2-propoxyethoxy) ethoxy]-propanol; TPGnPE; TPnP; Tripropylene glycol monopropyl ether; Tripropylene glycol mono-n-propyl ether; Tripropylene glycol propyl ether
Empirical: $C_{12}H_{26}O_4$
Formula: $C_3H_7O(C_3H_6O)_3H$
Properties: Sol. 11% in water; m.w. 234.3; dens. 0.938; f.p. -112 C; b.p. 261 C; flash pt. 130 C
Uses: Solvent for water-based coatings and cleaners; coupling agent
Features: Slow-evaporating
Trade Names: Dowanol® TPnP

Triptane. *See* 2,2,3-Trimethylbutane
Tris (2-butoxyethyl) phosphate. *See* Tributoxyethyl phosphate
Tris-n-butylamine. *See* Tributylamine
1,1,3-Tris (5-t-butyl-4-hydroxy-2-methylphenyl) butane. *See* 1,1,3-Tris (2-methyl-4-hydroxy-5-t-butylphenyl) butane
Tris (2-ethylhexyl) phosphate. *See* Trioctyl phosphate
Tris (2-hydroxyethyl) amine. *See* Triethanolamine
Trisiloxane, 1,1,1,5,5,5-hexamethyl-3-phenyl-3-((trimethylsilyl) oxy)-. *See* Phenyl trimethicone

1,1,3-Tris (2-methyl-4-hydroxy-5-t-butylphenyl) butane

CAS 1843-03-4; EINECS/ELINCS 217-420-7
Synonyms: 4,4′,4′′-(1-Methyl-1-propanyl-3-ylidene) tris 2-(1,1-dimethylethyl)-5-methylphenol; Phenol, 4,4′,4′′-[1-methyl-1-propanyl-3-ylidene] tris 2-[1,1-dimethylethyl]-5-methyl-; TPNC; Tributylcresyl butane; 1,1,3-Tris (5-t-butyl-4-hydroxy-2-methylphenyl) butane; Tris (2-methyl-4-hydroxy-5-t-butylphenyl) butane
Classification: Organic compd.
Definition: Condensation prod. of 3-methyl-6-t-butylphenol and crotonaldehyde
Empirical: $C_{37}H_{52}O_3$
Properties: Wh. powd.; sol. in acetone, ethanol, diethyl ether, ethyl acetate; m.w. 544.83; m.p. 181 C; flash pt. 225 F
Toxicology: LD50 (mammal) 16,100 mg/kg; irritant; TSCA listed
Uses: Antioxidant for polyolefins, polyamides, polyesters, POM, PU, PVC, PS, cellulosics, rubbers, food-grade polymers and adhesives; film-former, solvent in cosmetics
Regulatory: FDA 21CFR §175.105, 178.2010
Manuf./Distrib.: Aldrich; Fairmount; Uniqema Am.

Tris (2-methyl-4-hydroxy-5-t-butylphenyl) butane. *See* 1,1,3-Tris (2-methyl-4-hydroxy-5-t-butylphenyl) butane
Tris (methylphenyl) phosphate. *See* Tricresyl phosphate

Tristearin

CAS 555-43-1; EINECS/ELINCS 209-097-6
Synonyms: Glycerol tristearate; Glyceryl monotristearate; Glyceryl tristearate; Octadecanoic acid, 1,2,3-propanetriyl ester; 1,2,3-Propanetriol trioctadecanoate; Stearin
Definition: Triester of glycerin and stearic acid
Empirical: $C_{57}H_{110}O_6$
Formula: $[CH_3(CH_2)_{16}COOCH_2]_2CHOCO(CH_2)_{16}CH_3$
Properties: Colorless to wh. crystals or powd., odorless, tasteless; sol. in hot alcohol, benzene, chloroform, carbon disulfide, chlorinated and aromatic solvs.; pract. insol. in cold alcohol, ether, petrol. ether; insol. in water; m.w. 891.45; dens. 0.943 (65 C); m.p. 71.6 C; ref. index 1.4385 (80 C); nonionic
Toxicology: No known toxicity; TSCA listed
Precaution: Combustible
Hazardous Decomp. Prods.: Heated to decomp., emits acrid smoke and irritating fumes
Uses: Soap; candles; adhesive pastes; metal polishes; waterproofing paper; textile sizes; leather stuffing; mfg. of stearic acid; processing aid for EPS; dispersant, lubricant for colorants; lubricant for PVC, textile processing; plastics mold release agent; polishing/grinding pastes; crystallization accelerator, fermentation aid, formulation aid, lubricant, release agent, emulsifier, surface-finishing agent for foods; emollient, solvent, visc. control agent in cosmetics; pharmaceuticals (lubricant, mold release agent, binder, retarding agent in tablets); in food-contact textiles
Regulatory: FDA 21CFR §172.811, 177.2800; FDA approved for orals
Manuf./Distrib.: A&E Connock; Fluka; Sigma; Spectrum Quality Prods.

Tris (tolyloxy) phosphine oxide. *See* Tricresyl phosphate
Tris (trimethylsiloxy) phenylsilane. *See* Phenyl trimethicone
Triticum aestivum germ oil; Triticum vulgare; Triticum vulgare oil. *See* Wheat (Triticum vulgare) germ oil
Tritolyl phosphate. *See* Tricresyl phosphate

Tri-p-tolyl phosphate

CAS 78-32-0
Synonyms: Phosphoric acid, tris (4-methylphenyl) ester; TPCP; Tri-p-cresyl

phosphate
Empirical: $C_{21}H_{21}O_4P$
Properties: Solid, need. or tablet; sol. in 95% ethanol, chloroform, ether, benzene; insol. in water; m.w. 368.37; sp.gr. 1.247 g/ml; m.p. 77-78 C; b.p. 224 C (3.5 mm)
Toxicology: LD50 (oral, rat) > 12.8 g/kg, (subcut., cat) > 1 g/kg; may cause nausea, vomiting, abdominal pain, diarrhea; after 3-28 days incubation: nontender enlargement of parotid gland, coldness/sweating/muscular pain/cramping of hands/legs, paresthesia of extremities, muscular tremor, symmetrical polyneuritis
Precaution: Probably combustible
Storage: Store refrigerated
Uses: Plasticizer for PVC, PS, nitrocellulose; fire retardant for plastics; air filter medium; solvent mixtures; waterproofing; additive to EP lubricants; hydraulic fluid; heat exchange medium
Manuf./Distrib.: Acros Org.; ICN Biomed. Research Prods.; Pfaltz & Bauer

Triundecanoin
CAS 13552-80-2; EINECS/ELINCS 236-935-8
Synonyms: Glycerol triundecanoate; Glyceryl triundecanoate; Undecanoic acid, 1,2,3-propanetriyl ester
Definition: Triester of glycerin and undecanoic acid
Empirical: $C_{36}H_{68}O_6$
Formula: $HCO(CH_2OCO(CH_2)_9CH_3)_2CO(CH_2)_9CH_3$
Toxicology: TSCA listed
Uses: Antistat, emollient, solvent for cosmetics; emollient, solvent, fixing agent, and extender in nutritional applics.; carrier for flavors and fragrances; pharmaceutical emollient, lubricant, moisturizer, visc. control agent
Manuf./Distrib.: ABITEC; Fluka; Sigma
Trade Names: Captex® 8227

Trolamine. *See* Triethanolamine
TTE. *See* Trichlorotrifluoroethane

Turpentine
CAS 8006-64-2 (steam distilled); 8052-14-0; 9005-90-7; 977022-00-6 (rectified); EINECS/ELINCS 232-350-7; 232-688-5
UN 1299 (DOT); 1300 (DOT); FEMA 3088
Synonyms: Gum turpentine; Pine balsam; Pine gum; Purified gum spirits; Spirits of turpentine; Spirit of turpentine; Terebenthine; Turpentine gum; Turpentine oil; Turpentine oil, rectified; Turpentine oil, rectifier; Turpentine, purified; Turpentine, rectified; Turpentine, steam distilled; Wood turpentine
Classification: Unsat. alicyclic hydrocarbon
Definition: Volatile essential oil obtained by distillation and rectification from turpentine, an oleoresin obtained from *Pinus* spp., contg. pinene and diterpene
Formula: $\approx C_{10}H_{16}$
Properties: Colorless sticky visc. balsamic liq.; char. sharp penetrating piney odor; sol. in alcohol, benzene, CS_2, CS_2, petrol. ether, ether, chloroform, glac. acetic acid; insol. in water; m.w. ≈ 136; dens. 0.854-0.868; m.p. -50 to -60 C; b.p. 154-170 C; flash pt. (CC) 32-46 C; autoignition temp. 253 C; ref. index 1.463-1.483 (20 C); KB value 56
Toxicology: ACGIH TLV/TWA 100 ppm; LD50 (oral, rat) 5760 mg/kg, (IV, mouse) 1180 μg/kg; LC50 (inh., mouse, 2 h) 29 g/m^3; poison by IV route; mildly toxic by ing., inh.; mod. toxic to humans by ing.; irritating to skin, eyes, respiratory tract; allergen; skin absorp. and inh. can cause CNS depression, headaches, confusion, delirium, convulsions, and death due to respiratory failure; serious kidney irritant; questionable carcinogen; experimental tumorigen; common air contaminant; TSCA listed
Environmental: VOC
Precaution: Highly flamm.; mod. explosive hazard as vapor; lower explosive limit 0.8%; incompat. with oxidizers, halogens, stannic chloride, hexachloromelamine, diatomaceous earth; violent reactions possible; avoid impregnation of combustibles with turpentine
Hazardous Decomp. Prods.: Heated to decomp., emits acrid smoke and irritating fumes
NFPA: Health 1, Flammability 3, Reactivity 0
Storage: Store in cool, dry, well-ventilated area away from heat/ignition sources; protect from light
Uses: Source of turpentine oil and gum rosin; natural flavoring agent in foods, pharmaceuticals; solvent in shoe polishes, printing inks, cleaning compds., waxes, paper prods., cosmetics; solvent, thinner for paints, lacquers; solvent, reclaiming agent for rubber; synthesis of camphor and menthol; medicines (liniments); perfumery; pesticide mfg.; pharmaceutical solvent; rubefacient; diuretic; preps. for respiratory tract disorders; in food-pkg. adhesives
Regulatory: FDA 21CFR §172.510, 175.105; FDA approved for inhalants; FEMA GRAS; BP compliance
Manuf./Distrib.: Ascona; Bencorp Int'l.; Berje; Buckton Page Ltd; Doingcom; Dujodwala Resins & Terpenes; George Uhe; J.H. Calo; J.H. Hinz; Lluch Essence; Mosselman NV; Mutchler; PDM; Pangaea Sciences; Penta Mfg.; Polarome Int'l.; Quest Int'l. Fragrances; Ruger; Sea-Land; Spectrum Quality Prods.; Technichem; Vliengenthart BV
Trade Names: Brazilian Gum Turpentine-NT

Turpentine gum; Turpentine oil; Turpentine oil, rectified; Turpentine oil, rectifier; Turpentine, purified; Turpentine, rectified; Turpentine, steam distilled. *See* Turpentine
Turpentine substitutes. *See* Mineral spirits

U

γ-Undecalactone

CAS 104-67-6; EINECS/ELINCS 203-225-4
FEMA 3091
Synonyms: Aldehyde C14; Aldehyde C14 pure; γ-Heptyl butyrolactone; 5-Heptyldihydrofuran-2-one; 4-Hydroxyundecanoic acid lactone; 5-Hydroxyundecanoic acid γ-lactone; Peach aldehyde; Undecanoic γ-lactone; Undecan-4-olide; γ-Undecyl lactone
Classification: Heterocyclic compd.; nonaromatic lactone
Empirical: $C_{11}H_{20}O_2$
Formula: $CH_3(CH_2)_6CHCH_2CH_2COO$
Properties: Colorless to sl. yel. liq., creamy peach-like odor; sol. in alcohol, most fixed oils, propylene glycol; insol. in glycerin, water; m.w. 184.28; dens. 0.943; b.p. 164-166 C (13 mm); flash pt. > 230 F; ref. index 1.430
Toxicology: Irritating to eyes, skin, respiratory system; TSCA listed
Precaution: Combustible liq.
Hazardous Decomp. Prods.: Heated to decomp., emits acrid smoke and irritating fumes
Uses: Solvent, masking agent in cosmetics; fragrance in perfumes; synthetic flavoring agent in foods and pharmaceuticals
Features: Peach-like flavor
Regulatory: FDA 21CFR §172.515; FEMA GRAS
Manuf./Distrib.: Acme-Hardesty; Advanced BioTech; Advanced Synthesis Tech.; Aldrich; Axxence Aromatic GmbH; Axxence Aromatic GmbH; Degussa Flavors/Alfrebro; Elan; Fleurchem; GuiZhou Essence; Moore Ingreds.

Undecane. *See* n-Undecane

n-Undecane

CAS 1120-21-4
UN 2330 (DOT)
Synonyms: Hendecane; Undecane
Classification: Sat. aliphatic hydrocarbon
Empirical: $C_{11}H_{24}$
Properties: Colorless liq.; gasoline-like odor; sol. in most org. solvs.; insol. in water; m.w. 156.31; dens. 0.744 (20 C); m.p. -26 C; b.p. 196 C; flash pt. (CC) 60 C
Toxicology: LD50 (IV, mouse) 517 mg/kg; mod. toxic by IV route; conc. vapor may be irritating to eyes; conc. vapor can probably cause nose/throat irritation, headache, drowsiness, nausea, difficulty breathing, however no significant narcotic effects likely; ing. may cause nausea, vomiting, swelling of abdomen, headache, depression; relatively low oral toxicity unless aspirated into lungs; may cause mild skin irritation; TSCA listed
Precaution: Flamm. exposed to heat, flame, oxidizers; incompat. with strong oxidizing agents (increases fire and explosion hazard)
Hazardous Decomp. Prods.: Heated to decomp., emits acrid smoke and irritating fumes
NFPA: Health 0, Flammability 2, Reactivity 0
Storage: Store in cool, dry, well-ventilated area
Uses: Component of gasoline, jet fuel, kerosene, photocopying fluids, petrol. solvents such as wh. spirit; solvent; org. synthesis; petrol. research; distillation chaser
Manuf./Distrib.: Aldrich; ChevronPhillips; Fluka; Sigma; Triple Crown Am.

Undecanoic acid, 1,2,3-propanetriyl ester. *See* Triundecanoin
Undecanoic γ-lactone; Undecan-4-olide. *See* γ-Undecalactone
3,6,9,12,15,18,21,24,27,3,33-Undecaoxapentatriacontane-1,35-diol. *See* PEG-12
γ-Undecyl lactone. *See* γ-Undecalactone
Unleaded gasoline. *See* Gasoline
Urea, 1,1,3,3-tetramethyl-. *See* Tetramethylurea

V

4-Valerolactone. *See* γ-Valerolactone

γ-Valerolactone

CAS 108-29-2; EINECS/ELINCS 203-569-5
FEMA 3103
Synonyms: 4,5-Dihydro-5-methyl-2(3H)-furanone; 4-Hydroxypentanoic acid lactone; 4-Hydroxyvaleric acid lactone; 3-Methylbutyrolactone; 4-Methyl-γ-butyrolactone; γ-Methyl-γ-butyrolactone; γ-Pentalactone; 4-Pentanolide; 4-Valerolactone
Empirical: $C_5H_8O_2$
Properties: Colorless mobile liq., sweet herbaceous odor; misc. with alcohol, fixed oils, water; m.w. 100.13; dens. 1.047-1.054; m.p. -31 C; b.p. 205-206.5 C; flash pt. (COC) 205 F; ref. index 1.433
Toxicology: LD50 (oral, rat) 8800 mg/kg; mod. toxic by ing.; skin irritant; mutagenic data; TSCA listed
Precaution: Combustible liq. exposed to heat or flame; reactive with oxidizers
Hazardous Decomp. Prods.: Heated to decomp., emits acrid smoke and irritating fumes
Uses: Coupling agent in dyebaths; brake fluids; cutting oils; solvent for adhesives, insecticides, and lacquers; synthetic flavoring agent in foods and pharmaceuticals; fragrance in cosmetics
Regulatory: FDA 21CFR §172.515; FEMA GRAS
Manuf./Distrib.: Advanced Synthesis Tech.; Aldrich; Fluka; Int'l. Chem. Group; J.T. Baker; Kira; Lluch Essence; Penta Mfg.; Sigma; Spectrum Quality Prods.; VWR Int'l.

Valerone. *See* Diisobutyl ketone
Valine aldehyde. *See* 2-Methylpropanal
Varnish Makers naphtha; Varnish Makers and Painters naphtha. *See* VM&P naphtha
Vaseline. *See* Petrolatum
VC; VDC. *See* Vinylidene chloride monomer
Vinegar acid. *See* Acetic acid, glacial; Acetic acid
Vinegar naphtha. *See* Ethyl acetate
Vinylbenzene; Vinylbenzol. *See* Styrene
Vinylcarbinol. *See* Allyl alcohol
Vinyl cyanide. *See* Acrylonitrile
Vinyl ether 'Carbitol'. *See* Diethylene glycol ethylvinyl ether
Vinylidene chloride; Vinylidene chloride (II); Vinylidene chloride, inhibited. *See* Vinylidene chloride monomer

Vinylidene chloride monomer

CAS 75-35-4; EINECS/ELINCS 200-864-0
UN 1303
Synonyms: 1,1-DCE; 1,1-Dichloroethene; 1,1-Dichloroethylene; asym-Dichloroethylene; Ethene, 1,1-dichloro-; Ethylene, 1,1-dichloro-; VC; VDC; Vinylidene chloride; Vinylidene chloride (II); Vinylidene chloride, inhibited; Vinylidene dichloride; Vinylidine chloride
Empirical: $C_2H_2Cl_2$
Formula: $CH_2=CCl_2$
Properties: Colorless mobile liq., mild sweet odor resembling that of chloroform; pract. insol. in water; sol. in org. solvs.; m.w. 96.95; dens. 1.2129 (20/4 C); m.p. -122.5 C; b.p. 31.7 C; flash pt. -15 C; ref. index 1.4249 (20 C)
Toxicology: ACGIH TLV/TWA 5 ppm; STEL 20 ppm; LD50 (oral, rat) 200 mg/kg; LC50 (inh., rat, 4 h) 6350 ppm; LDLo (IV, dog) 225 mg/kg; poison by inh., ing., IV routes; mod. toxic by subcut.; irritant to skin, mucous membranes; narcotic in high concs.; human systemic effects by inh. (CNS, lung, liver, and kidney changes); suspected human carcinogen; experimental neoplastigen, tumorigen, teratogen, reproductive effects; mutagenic data; TSCA listed
Environmental: Evaporates very quickly from water and soil; breaks down in about 4 days in air; breaks down very slowly in water; does not accumulate very much in fish or birds
Precaution: Extremely flamm.; dangerous fire risk; explosive limits in air 5.6-11.4%; uncontrollable polymerization may lead to explosive reaction prods. with O_2 or ozone; can explode spontaneously; vigorous reaction with oxidizers; violent reactions possible
Hazardous Decomp. Prods.: Heated to decomp., emits toxic fumes of Cl^-
NFPA: Health 2, Flammability 3, Reactivity 0
Storage: Keep in tightly closed containers; photosensitive; keep away from ignition sources
Uses: Intermediate in prod. of vinylidene polymer plastics such as flexible film, food wrap, pkg. materials, Saran; prod. of flame-retardant coatings for fiber/carpet backings; piping; coating for steel pipes; comonomer for nitrile rubber; paints; adhesives; syn. fibers; solvent in pharmaceuticals
Use Level: 8 ppm (pharmaceuticals)
Manuf./Distrib.: Aldrich; Fluka; PPG Ind.; Sigma; Spectrum Quality Prods.; Total Spec. Chems.

Vinylidene dichloride; Vinylidine chloride. *See* Vinylidene chloride monomer
Vinyl toluene; Vinyl toluene, inhibited (mixed isomers); 3- and 4-Vinyl toluene (mixed isomers). *See* Vinyltoluene monomer

Vinyltoluene monomer

CAS 25013-15-4; EINECS/ELINCS 246-562-2
UN 2618 (DOT)
Synonyms: Methylstyrene; m-/p-Methylstyrene; Toluene, vinyl- (mixed isomers); Vinyl toluene; Vinyl toluene, inhibited (mixed isomers); 3- and 4-Vinyl toluene (mixed isomers); VT
Empirical: C_9H_{10}
Formula: $CH_3C_6H_4CH:CH_2$
Properties: Colorless liq.; sol. in methanol, ether; misc. with oxygenated solvs.; very sl. sol. in water; m.w. 118.19; dens. 0.896 (20/4 C); f.p. -77 C; b.p. 169-172 C; flash pt. (TCC) 51 C; ref. index 1.542
Toxicology: ACGIH TLV/TWA 50 ppm; STEL 100 ppm; LD50 (oral, rat) 4000 mg/kg, (IP, rat) 2324 mg/kg; mod. toxic by inh., ing.; skin and eye irritant; human systemic effects by inh. (eye and olfactory effects); experimental teratogen, reproductive effects; mutagenic data; TSCA listed
Precaution: Flamm.; mod. fire risk; vigorous reactions with oxidizers
Hazardous Decomp. Prods.: Heated to decomp., emits acrid smoke and irritating fumes
Uses: Comonomer for polystyrene, alkyd, and epoxy ester resins; monomer for prod. of paints; solvent; intermediate; reactive diluent for unsat. polyester resins; modifier for oils/alkyds in food-contact coatings
Regulatory: FDA 21CFR §175.300
Manuf./Distrib.: Ashland; Chemcentral; Deltech; Dow; Fluka; Monomer-

Polymer & Dajac Labs; Vopak USA

Vinyl trichloride. *See* 1,1,2-Trichloroethane
Vitamin F. *See* Ethyl linoleate

VM&P naphtha

CAS 8032-32-4; EINECS/ELINCS 232-453-7
UN 1255 (DOT); 1271 (DOT)
Synonyms: Benzin; Benzine (light petroleum distillate); Benzoline; Ligroin; Ligroine; Mineral spirits; Naphtha, petroleum; Naphtha, VM&P; Painters naphtha; Petroleum ether; Petroleum naphtha; Petroleum spirit; Petroleum spirits; Petroleum thinner; Refined solvent naphtha; Solvent naphtha; Varnish Makers naphtha; Varnish Makers and Painters naphtha; White spirit
Classification: Hydrocarbon
Definition: Complex combination of hydrocarbons obtained by fractional distillation of petroleum; narrow boiling fraction of petroleum with distillation range 119-143 C
Properties: Colorless clear liq.; gasoline odor; nonfluorescent; volatile; insol. in water; dens. 0.850-0.870 (15.6/15.6 C); vapor pressure 40 mm Hg (20 C); pour pt. -56 C; b.p. 93-204 C; flash pt. (TCC) -6.6 C; autoignition temp. 232 C; ref. index 1.3630
Toxicology: ACGIH TLV/TWA 300 ppm; LD50 (oral, rat) > 5000 mg/kg, (skin, rabbit) > 3000 mg/kg, (IV, mouse) 40 mg/kg; LC50 (inh., rat, 4 h) 3400 ppm; poison by IV; mildly toxic by inh.; eye, skin, and respiratory irritant; ing. can cause vomiting, diarrhea, drowsiness, pulmonary edema in severe cases; inh. of vapor can cause intoxication, headache; target organs: respiratory system, eyes, skin, CNS, blood, PNS; TSCA listed
Environmental: VOC
Precaution: Flamm.; LEL 1%, UEL 6%, dangerous fire hazard exposed to heat, flame, sparks, or oxidizers; explosive as vapor exposed to heat or flame
Hazardous Decomp. Prods.: Heated to decomp., emits acrid smoke and irritating fumes
NFPA: Health 1, Flammability 3, Reactivity 0
Uses: Solvent for agric., oil/fat extraction, polishes, industrial cleaners, adhesives, inks, paints, lighter fluids, pharmaceuticals, paper/textile coatings, petrol. gas absorp., asphalt, resins, plastics, rubber; thinner in paints and varnish; in wood preservatives for food contact
Regulatory: FDA 21CFR §178.3800; SARA §311/312 acute health/chronic health/fire hazard
Manuf./Distrib.: Aldrich; Amyl; Ashland; ChevronTexaco; Coyne; Delta Distributors; Harcros; Romil Ltd; Shell; Sigma; Vopak USA
Trade Names: Calumet VM&P Naphtha < 1%; RotoSolv; Solvent 2218; Solvent 2268; Solvent 2391; Solvent 2409; Solvent 2423; Solvent 2472; Special Naphtholite 66/3
Trade Names Containing: Solvent 2461
See also Mineral spirits; Naphtha

VT. *See* Vinyltoluene monomer

W

Walnut (Juglans regia) oil

CAS 8024-09-7; 84604-00-2
Synonyms: Juglans regia; Juglans regia oil; Walnut oil
Definition: Oil derived from the nut meats of walnuts, *Juglans regia*
Properties: Brn. oil; insol. in water; dens. 0.93 kg/l; iodine no. 145-155; sapon. no. 190; ref. index 1.4691 (40 C)
Toxicology: TSCA listed
Uses: Emollient for makeup, hair care, skin care; drying oil for paints, alkyd resins, in food-contact coatings; occlusive solvent for suntan gels, topical pharmaceuticals, ointments; natural flavoring agent in foods; cooking/salad oil; pharmaceutical lubricant
Regulatory: FDA 21CFR §172.510, 175.300
Manuf./Distrib.: Alban Muller; Aldivia; Alfa Chem; Anglia Oils; Arista Ind.; Charkit; Desert Whale Jojoba; Int'l. Sourcing; Natural Oils Int'l.; Penta Mfg.; Pokonobe Ind.; Thornley; Tri-K Ind.; Welch, Holme & Clark

Walnut oil. *See* Walnut (Juglans regia) oil
Wash oil. *See* Creosote oil

Water

CAS 7732-18-5; EINECS/ELINCS 231-791-2
Synonyms: Aqua; Deionized water; Distilled water; Hydrogen oxide; Ice; Purified water; Steam
Classification: Polar liq.
Empirical: H_2O
Properties: Colorless liq.; odorless; tasteless; m.w. 18.02; dens. 1.00 (4 C); vapor pressure 760 mm Hg (100 C); f.p. 0 C; b.p. 100 C; ref. index 1.333; surf. tens. 73 dynes/cm (20 C); high dielec. const.; weak electrolyte
Toxicology: LD50 (IP, mouse) 190 g/kg, (IV, mouse) 25 g/kg; human systemic effects by ing. of very large amts. (body temp. inc., convulsions, diarrhea, fever, hypermotility, muscle contraction/spasticity, nausea, vomiting, tremors); human and experimental death reported at sufficiently large doses; TSCA listed
Uses: Suspending agent (papermaking, coal slurries); extraction solvent; diluent; hydration of lime; paper coatings; textile processing; industrial coolant; filtration; washing/scouring; sulfur mining; hydrolysis; Portland cement; hydraulic systems; power source; steam generation; hydrogen source; solvent in critical chromatography, semiconductor processing, pharmaceutical mfg., spectrophotometry, scintillation counting, spectrofluorescence, high purity org. synthesis, precision cleaning, etc.
Regulatory: FDA 21CFR §103.35
Manuf./Distrib.: Ruger

Wax, chlorinated. *See* Paraffin, chlorinated
Wettable powder sulfur. *See* Sulfur
Wheat germ oil; Wheat husk oil. *See* Wheat (Triticum vulgare) germ oil

Wheat (Triticum vulgare) germ oil

CAS 8006-95-9; 68917-73-7
Synonyms: Bran absolute; Triticum aestivum germ oil; Triticum vulgare; Triticum vulgare oil; Wheat germ oil; Wheat husk oil
Definition: Oil obtained by expression or extraction of wheat germ
Properties: Lt. yel. oily liq.; bland; fat-sol.; misc. with chloroform, ether, benzene, petrol. ether; sl. sol. in alcohol; insol. in water; dens. 0.93-0.94; acid no. 6-20; iodine no. 120-140; sapon. no. 179-194; hyd. no. 10-48; ref. index 1.469-1.479
Toxicology: Skin and eye irritant; TSCA listed
Hazardous Decomp. Prods.: CO_2; heated to decomp., emits acrid smoke and irritating fumes
Uses: Emollient for hair care, skin care; gloss aid in makeup; moisturizer; antioxidant; food ingred., nutritive additive, dietary supplement; pharmaceutical nutritional source, topical moisturizer, lubricant; occlusive solvent; source of natural vitamin E
Regulatory: BP, EP compliance (refined and virgin)
Manuf./Distrib.: ABITEC; Aldivia; Alfa Chem; Amerol; Anglia Oils; Arista Ind.; Asiamerica Int'l.; Avatar; British Arkady Co. Ltd; Charkit; Cornelius Chem. Co. Ltd; Cosmetic Supplies USA; Croda Chem. Ltd; Croda Inc; Desert Whale Jojoba; Henry Lamotte; Jeen Int'l.; KIC Chems.; Lipo; Marine Bio-Prods.; Mosselman NV; Natural Oils Int'l.; Parchem Trading; Penta Mfg.; Pokonobe Ind.; Provital; Quimdis; R.I.T.A.; Ruger; S. Black; Spectrum Naturals; Spectrum Quality Prods.; Tri-K Ind.; Universal Preserv-A-Chem; Viobin; Vitamins, Inc; Voigt Global Distrib.; Westhove
Trade Names: EmCon™ W

Whiskey fusel oil. *See* Fusel oil refined
White camphor oil. *See* Camphor (Cinnamomum camphora) oil
White caustic. *See* Sodium hydroxide
White mineral oil; White oil. *See* Mineral oil
White petrolatum. *See* Petrolatum
White pine oil. *See* Pine (Pinus palustris) oil
White soft paraffin. *See* Petrolatum
White spirit. *See* VM&P naphtha
White spirits. *See* Stoddard solvent; Mineral spirits
White spirits, hydrodesulfurized; White spirit, White spirit, high flash. *See* Naphtha, hydrodesulfurized heavy
White tar. *See* Naphthalene
Wintergreen oil. *See* Methyl salicylate
Wood alcohol. *See* Methyl alcohol
Wood ether. *See* Dimethyl ether
Wood naphtha. *See* Methyl alcohol
Wood rosin, methyl ester. *See* Methyl abietate
Wood spirit. *See* Methyl alcohol
Wood turpentine. *See* Turpentine
Wool alcohol. *See* Lanolin alcohol
Wool fat; Wool grease; Wool wax. *See* Lanolin
Wool wax alcohol. *See* Lanolin alcohol

X

Xenene. *See* Biphenyl

Xylene

CAS 1330-20-7; EINECS/ELINCS 215-535-7
UN 1307 (DOT)
Synonyms: C_8 aromatics; Dimethylbenzene; Dimethylbenzene (mixed isomers); Dimethylbenzene, m-, o-, and p-; Methyl toluene; Mixed xylenes; Xylene, mixed; Xylenes, mixed isomers; Xylol
Classification: Aromatic hydrocarbon
Definition: Commercial mixture of 3 isomers: o-, m-, and p-xylene (1,3-dimethylbenzene, 1,2-dimethylbenzene, 1,4-dimethylbenzene)
Empirical: C_8H_{10}
Formula: $C_6H_4(CH_3)_2$
Properties: Colorless clear liq., char. sweet odor; sol. in abs. alcohol, ethanol, diethyl ether, many org. liqs.; pract. insol. in water; m.w. 106.16; dens. 0.86; vapor pressure 6.72 mm Hg (21 C); b.p. 137-144 C; flash pt. (CC) 29 C; ref. index 1.4970; KB value 98
Toxicology: ACGIH TLV/TWA 100 ppm; STEL 150 ppm; LD50 (oral, rat) 4300 mg/kg, (IP, rat) 2459 mg/kg; mod. toxic by IP and subcut. routes; mildly toxic by ing., inh.; irritant to eyes (severe), skin, respiratory tract; human systemic effects; may be narcotic in high concs. causing CNS depression, dizziness, nausea, unconsciousness; repeated skin contact causes defatting, dermatitis; possible liver damage; experimental teratogen, reproductive effects; mutagenic data; TSCA listed
Environmental: VOC; ThOD 2.41
Precaution: Highly flamm.; moderate fire risk; incompat. with strong oxidizing agents, nitric acid, dichlorohydantoin; increased fire and explosion risk
Hazardous Decomp. Prods.: Heated to decomp., emits acrid smoke and irritating fumes
NFPA: Health 2, Flammability 3, Reactivity 0
Storage: Store in cool well-ventilated area out of direct sunlight, away from heat/ignition sources
Uses: Solvent for alkyd resins, lacquers, enamels, rubber cements, paints, printing inks, leather, gums, acrylic resins, rubber, adhesives, castor/linseed oils, cosmetics, pharmaceuticals; raw material for prod. of benzoic acid, phthalic anhydride, dyes, paints, other organics; insecticide carrier; aviation gasoline; protective coating; degreaser; cleaning agent; carrier in epoxy resin prod.; chem. synthesis; mfg. of quartz crystal oscillators, perfumes, insect repellents; liq. scintillation; in food-pkg. adhesives; in paper/paperboard in contact with dry food; in acrylic food pkg.
Regulatory: FDA 21CFR §175.105, 176.180, 177.1010, 177.1650; SARA §313 reportable; CERCLA hazardous substance; HAP
Manuf./Distrib.: AAE Chemie NV; AMC Chems.; Aldrich; Alvin Chem.; Amyl; Arch Chems.; Ashland; Atofina Bruxelles; BP Chems. Ltd; Baychem; Chemcentral; ChevronPhillips; Coyne; Crowley Chem.; Delta Distributors; Eastech; ExxonMobil; Fluka; GFS; General Chem.; Harcros; Honeywell Perf. Polymers; Houghton Chem.; Hukill; Integra; J.T. Baker; LG Petrochem. Co. Ltd; Lyondell; Mallinckrodt Baker; Marathon Ashland Petrol.; Milsolv; Nova Ltd; P.B. & S.; R.E. Carroll; Romil Ltd; Ruger; SOCO-Lynch; Sal Chem.; Shell; Sigma; Southchem; Spectrum Quality Prods.; Stanchem; Sunoco; Textile Chem.; Thomas Scientific; VWR Int'l.; Vopak USA; Whittaker, Clark & Daniels
Trade Names: Shell Xylene
Trade Names Containing: ExxonMobil T 500-100; Kristalloel 21; Kristalloel 30; Shellsol® X7B; Solvent 2216; Solvent 2310; Solvent 2339; Solvent 2439; Solvent 2495; Solvent 30 823; VM&P Naphtha EC

1,2-Xylene. *See* o-Xylene
1,3-Xylene. *See* m-Xylene
1,4-Xylene. *See* p-Xylene
3-Xylene. *See* m-Xylene
4-Xylene. *See* p-Xylene

m-Xylene

CAS 108-38-3
UN 1307 (DOT)
Synonyms: 1,3-Dimethylbenzene; m-Dimethylbenzene; 3-Methyltoluene; m-Methyltoluene; MX; 1,3-Xylene; 3-Xylene; m-Xylol
Classification: Aromatic hydrocarbon
Empirical: C_8H_{10}
Formula: $CH_3C_6H_4CH_3$
Properties: Colorless clear liq.; char. aromatic odor; sol. in ethanol, diethyl ether, acetone, benzene, other org. solvs.; misc. with ether, hydrocarbon solvs.; pract. insol. in water; m.w. 106.16; sp.gr. 0.8642 (20 C); vapor pressure 10 mm Hg (28.3 C); m.p. -47 C; b.p. 139 C; flash pt. (CC) 25 C; autoignition temp. 527 C; ref. index 1.495; surf. tens. 28.1 dynes/cm; dielec. const. 2.374
Toxicology: ACGIH TLV/TWA 100 ppm; STEL 150 ppm; LD50 (IP, mouse) 1739 mg/kg, (skin, rabbit) 14,100 mg/kg; mod. toxic by IP; mildly toxic by ing., skin contact, inh.; severe skin irritant; human systemic effects; ing. of large amts. may cause CNS depression; inh. may cause nose/throat irritation, CNS depression, headache, dizziness, nausea, vomiting; experimental teratogen, reproductive effects; common air contaminant; TSCA listed
Environmental: VOC; BOD5 2.53; COD 2.62; ThOD 3.17
Precaution: Flamm.; very dangerous fire hazard exposed to heat, flame; incompat. with oxidizers, nitric acid, sulfuric acid, sulfur dichloride (increases fire and explosion hazard); common air contaminant
Hazardous Decomp. Prods.: Heated to decomp., emits acrid smoke and irritating fumes
NFPA: Health 2, Flammability 3, Reactivity 0
Storage: Store in cool well-ventilated area out of direct sunlight, away from heat/ignition sources, away from corrosives
Uses: Chem. intermediate in mfg. of isophthalic acid for polyester resins, mfg. of alkyd resins; chem. intermediate for dyes, insecticides, org. synthesis; as solvent; in aviation fuel
Regulatory: SARA §313 reportable; CERCLA hazardous substance; HAP
Manuf./Distrib.: Advanced Synthesis Tech.; Aldrich; ChevronPhillips; EMS-Dottikon AG; Fluka; Mitsubishi Gas; Sigma; Spectrum Quality Prods.

o-Xylene

CAS 95-47-6; EINECS/ELINCS 202-422-2
UN 1307 (DOT)
Synonyms: Benzene, 1,2-dimethyl; 1,2-Dimethylbenzene; o-Dimethylbenzene; o-Methyltoluene; 1,2-Xylene; o-Xylol
Classification: Aromatic hydrocarbon
Empirical: C_8H_{10}
Formula: $C_6H_4(CH_3)_2$
Properties: Colorless liq.; pungent sweet aromatic odor; misc. with abs.

alcohol, ether, most org. solvs.; insol. in water; m.w. 106.17; dens. 0.8802 (20 C); vapor pressure 0.26-0.34 psia; m.p. -25 C; b.p. 141.5-144.5 C; flash pt. (TOC) 30 C; autoignition temp. 463 C; ref. index 1.503; surf. tens. 29.49 dynes/cm; dielec. const. 2.568

Toxicology: ACGIH TLV/TWA 100 ppm; STEL 150 ppm; LD50 (oral, rat) 3.5-8.6 g/kg, (IP, mouse) 1364 mg/kg; LC50 (inh., rat, 4 h) 6350-6700 ppm; mod. toxic by IP route; mildly toxic by ing., inh.; inh. may cause respiratory tract irritation; eye/skin irritant; may cause CNS depression, auditory system damage; may cause birth defects; experimental teratogen; common air contaminant; TSCA listed

Environmental: Expected to be readily biodeg.; VOC; BOD5 1.64; COD 2.91; ThOD 3.17; toxic to aquatic organisms; prevent contamination of soil, surf. water, ground water; prevent entry into sewers

Precaution: Flamm. liq. and vapor; LEL 1.1 vol.%; UEL 6.4 vol.%; dangerous fire hazard exposed to heat or flame; explosive as vapor exposed to heat or flame; electrostatic charge may accumulate; bond and ground; incompat. with strong oxidizers (e.g., chlorates, nitrates, peroxides); reacts with nitric acid and sulfuric acid with release of energy

Hazardous Decomp. Prods.: Combustion: CO. CO_2; heated to decomp., emits acrid smoke and irritating fumes

NFPA: Health 2, Flammability 3, Reactivity 0

Storage: Store in closed container away from heat, sparks, open flame

Uses: Solvent for paints; mfg. of phthalic anhydride, vitamins, pharmaceuticals, dyes, insecticides; motor fuels; prod. of plasticizers

Regulatory: SARA §311/312 acute health/chronic health/fire hazard, §313 reportable; CERCLA hazardous substance; HAP; Canada DSL, Australia AICS, China, Japan ENCS, Korea ECL, Philippines PICCS listed

Manuf./Distrib.: Aldrich; ChevronPhillips; Equistar; ExxonMobil; G.J. Chem.; Koch Ind.; Mitsubishi Gas; Petroquimica Uniao; Shell; Spectrum Quality Prods.; Triple Crown Am.

p-Xylene

CAS 106-42-3; EINECS/ELINCS 203-396-5

UN 1307 (DOT)

Synonyms: 1,4-Dimethylbenzene; p-Dimethylbenzene; 4-Methyltoluene; p-Methyltoluene; Paraxylene; PX; 1,4-Xylene; 4-Xylene; p-Xylol

Classification: Aromatic hydrocarbon

Empirical: C_8H_{10}

Formula: $CH_3C_6H_4CH_3$

Properties: Colorless clear liq.; char. aromatic odor; colorless cryst. solid below 13 C; sol. in ethanol, diethyl ether, acetone, benzene, other org. solvs.; pract. insol. in water; m.w. 106.16; sp.gr. 0.861 (20 C); vapor pressure 10 mm Hg (27.3 C); m.p. 13.3 C; b.p. 138.3 C; flash pt. (CC) 25 C; autoignition temp. 528 C; surf. tens. 27.76 dynes/cm; KB value 93; dielec. const. 2.27

Toxicology: ACGIH TLV/TWA 100 ppm; STEL 150 ppm; LD50 (oral, rat) 5 g/kg, (IP, rat) 3810 mg/kg; toxic; mod. toxic by IP route; sl. toxic by ing., inh.; may be narcotic in high concs.; ing. of large amts. causes CNS effects; inh. causes nose/throat irritation, CNS depression, headache, dizziness, nausea, vomiting; experimental teratogen, reproductive effects; TSCA listed

Environmental: VOC; BOD5 1.40; COD 2.56; ThOD 3.17

Precaution: Flamm.; LEL 1.1%; UEL 7%; very dangerous fire hazard exposed to heat or flame; explosive as vapor exposed to heat or flame; incompat. with oxidizers, nitric acid, sulfuric acid, sulfur dichloride (increases fire and explosion hazard)

Hazardous Decomp. Prods.: Heated to decomp., emits acrid smoke and irritating fumes

NFPA: Health 2, Flammability 3, Reactivity 0

Storage: Store in cool well-ventilated area out of direct sunlight, away from heat/ignition sources and corrosives

Uses: Chem. intermediate for synthesis of terephthalic acid for polyester resins and fibers, in mfg. of plastics, vitamins, pharmaceuticals, insecticides; solvent for adhesives and coatings; mfg. of di-p-xylene and herbicides

Regulatory: SARA §313 reportable; CERCLA hazardous substance; HAP

Manuf./Distrib.: Aldrich; BP Amoco; ChevronPhillips; Equistar; ExxonMobil; Fluka; Koch Chem. Spec.; Mitsubishi Gas; Sigma; Spectrum Quality Prods.

Xylene, mixed; Xylenes, mixed isomers. *See* Xylene

Xylenesulfonic acid, sodium salt. *See* Sodium xylenesulfonate

Xylenol

CAS 1300-71-6

UN 2261 (DOT)

Synonyms: Dimethylhydroxybenzene; Dimethylphenol; Hydroxydimethylbenzene; Hydroxyxylene; Phenol, dimethyl-; Xylenol, mixed; Xylenols

Definition: Mixed tar acid fraction contg. various xylenol isomers, ethylphenol, and higher boiling phenols

Empirical: $C_8H_{10}O$

Formula: $(CH_3)_2C_6H_3OH$

Properties: M.w. 122.17; Commercial mixt.: wh. cryst. solid; sol. in aq. alkali, oxygenated and aromatic solvs., most org. solvs., alcohol, chloroform, ether, benzene; sl. sol. in water; dens. 1.02-1.03 (15 C); m.p. 20-76 C; b.p. 203-225 C

Toxicology: LDLo (oral, man) 5 g/kg; poison; toxic by ing., skin absorption; TSCA listed

Precaution: Combustible

Hazardous Decomp. Prods.: Heated to decomp., emits acrid smoke and irritating fumes

Uses: Disinfectant; solvent for wire enamels; pharmaceuticals; insecticides; fungicides; herbicides; plasticizers; rubber chemicals; paints; additives to lubricants and gasoline; mfg. of polyphenylene oxide; wetting agents; dyestuffs; synthetic resin; comonomer for alkali-resist. phenolic resins; froth flotation agent; engine cleaner ingred.

Manuf./Distrib.: Aldrich; Coalite Chems.; Concord; Crowley Chem.; Crowley Tar Prods.; Fabrichem; J.T. Baker; Merichem; Penta Mfg.; Spectrum Quality Prods.; Thomas Scientific

1,2,5-Xylenol. *See* 2,5-Xylenol

1,3,4-Xylenol. *See* 3,4-Xylenol

2,3-Xylenol

CAS 526-75-0; EINECS/ELINCS 208-395-3

UN 2261

Synonyms: 2,3-Dimethylphenol; 1-Hydroxy-2,3-dimethylbenzene; 3-Hydroxy-o-xylene; o-Xylenol; o-3-Xylenol

Empirical: $C_8H_{10}O$

Formula: $(CH_3)_2C_6H_3OH$

Properties: Colorless cryst. solid or brn. cryst. chunks; sol. in DMSO, 95% ethanol, acetone, ether, chloroform, sodium hydroxide sol'n., benzene, caustic soda sol'ns., alkalis, most org. solvs.; sol. < 1 mg/ml in water; m.w. 122.17; vapor pressure 1 mm Hg (56 C); m.p. 73-75 C; b.p. 218 C; ref. index 1.5420 (20 C)

Toxicology: LD50 (IV, mouse) 56 mg/kg; poison; highly toxic by inh., ing., skin absorp., IV route; corrosive and extremely destructive to tissue of mucous membranes, upper respiratory tract, eyes, and skin; may cause burning sensation, coughing, wheezing, laryngitis, shortness of breath, dizziness, stomach pain, exhaustion, coma, headaches, nausea, vomiting, kidney/liver damage; inh. may be fatal due to spasm/inflamm. of larynx/bronchi, chem. pneumonitis, pulmonary edema; readily absorbed thru skin; TSCA listed

Precaution: Probably combustible; incompat. with bases, acid chlorides, acid anhydrides, oxidizing agents; corrodes steel, brass, copper, copper alloys; may dec. under fire conditions to form flamm./explosive mixts. in air

Hazardous Decomp. Prods.: Heated to decomp., emits acrid smoke and toxic fumes of CO, CO_2

Storage: Hygroscopic; store in cool, dry place; keep tightly closed; protect from moisture and oxidizing materials

Uses: Prep. of coal tar disinfectants; mfg. of artificial resins, solvents, pharmaceuticals, insecticides, fungicides, plasticizers, and rubber chemicals; lubricant additive; gasoline additive; wetting agents; dyestuffs; fragrance in cosmetics

Manuf./Distrib.: Acros Org.; Advanced Synthesis Tech.; Aldrich; Chugai Kasei; Fluka; ICN Biomed. Research Prods.; Indofine; Lancaster Synthesis; Pfaltz & Bauer; Riedel-deHaën; Supelco; TCI Am.

2,5-Xylenol

CAS 95-87-4; EINECS/ELINCS 202-461-5

FEMA 3595

Synonyms: 2,5-Dimethylphenol; 3,6-Dimethylphenol; 2,5-DMP; 6-Methyl-m-cresol; 1,2,5-Xylenol; p-Xylenol

Empirical: $C_8H_{10}O$

Properties: Cryst.; m.w. 122.17; dens. 0.971; m.p. 72-76 C; b.p. 212 C
Toxicology: LD50 (oral, rat) 444 mg/kg; toxic; poison by ing.; mod. toxic by unspecified route; corrosive; readily absorbed thru skin; questionable carcinogen; experimental tumorigen; TSCA listed
Uses: Flavoring agent for medicines, foods; fragrance in cosmetics; prep. of coal tar disinfectants; mfg. of artificial resins; in solvents, pharmaceuticals, plasticizers, and wetting agents
Regulatory: FEMA GRAS
Manuf./Distrib.: Advanced Synthesis Tech.; Aldrich; Biddle Sawyer; Fluka; Int'l. Chem. Group; Kira; Penta Mfg.; Richman

2,6-Xylenol

CAS 576-26-1; EINECS/ELINCS 209-400-1
FEMA 3249
Synonyms: 2,6-Dimethylphenol; 2,6-DMP
Classification: Aromatic phenol
Empirical: $C_8H_{10}O$
Properties: Needles; sol. in hot water and alcohol; m.w. 122.17; m.p. 45-48 C; b.p. 203 C; flash pt. 173 F
Toxicology: LD50 (oral, rat) 296 mg/kg, (IP, mouse) 150 mg/kg, (IV, mouse) 80 mg/kg, (skin, rabbit) 1000 mg/kg; toxic; poison by ing., IV, IP routes; mod. toxic by skin contact; corrosive; readily absorbed thru skin; eye irritant; questionable carcinogen; experimental tumorigen; TSCA listed
Precaution: Flamm.
Hazardous Decomp. Prods.: Heated to decomp., emits acrid smoke and irritating fumes
Storage: Store under nitrogen
Uses: Synthetic flavoring agent in medicines, foods; fragrance in cosmetics; in disinfectants, solvents, pharmaceuticals; as antioxidant in gas, oils, elastomers; mfg. of polyphenylene oxide, food-contact resins
Regulatory: FDA 21CFR §177.2460; FEMA GRAS
Manuf./Distrib.: Advanced Synthesis Tech.; Aldrich; Fluka; Int'l. Chem. Group; Kira; Sigma; Spectrum Quality Prods.

3,4-Xylenol

CAS 95-65-8; EINECS/ELINCS 202-439-5
UN 2261; FEMA 3596
Synonyms: 3,4-Dimethylphenol; 4,5-Dimethylphenol; 3,4-DMP; 1-Hydroxy-3,4-dimethylbenzene; 4-Hydroxy-1,2-dimethylbenzene; 4-Hydroxy-o-xylene; Phenol, 3,4-dimethyl-; 1,3,4-Xylenol; as-o-Xylenol; o-4-Xylenol
Empirical: $C_8H_{10}O$
Formula: $(CH_3)_2C_6H_3OH$
Properties: Colorless to lt. tan cryst. powd.; sol. in DMSO, 95% ethanol, acetone, ether, chloroform, benzene, sodium hydroxide sol'n., caustic soda sol'n., alkalis, most org. solvs.; sol. < 1 mg/ml in water; m.w. 122.17; sp.gr. 0.9830 (20/4 C); vapor pressure 1 mm Hg (66.2 C); m.p. 62.5 C; b.p. 225 C; flash pt. 110 C
Toxicology: LD50 (oral, rat) 727 mg/kg, (IP, rat) 200 mg/kg; poison; highly toxic by inh., ing., skin absorp.; corrosive and extremely destructive to tissue of mucous membranes, upper respiratory tract, eyes, and skin; may cause burning sensation, coughing, wheezing, laryngitis, stomach pain, exhaustion, coma, headache, nausea, vomiting, dyspnea, liver/kidney damage, etc.; inh. can be fatal due to spasm/inflamm. of larynx/bronchi, chem. pneumonitis, pulmonary edema; experimental equivocal tumorigen; TSCA listed
Precaution: Probably combustible; incompat. with bases, acid chlorides, acid anhydrides, oxidizing agents; corrosive to steel, brass, Cu, Cu alloys; under fire conditions, may dec. to form flamm./explosive mixts. in air
Hazardous Decomp. Prods.: Heated to decomp., emits acrid smoke and toxic fumes of CO, CO_2
Storage: Hygroscopic; light-sensitive; store in cool, dry place; keep tightly closed; protect from moisture and oxidizing materials
Uses: Prep. of coal tar disinfectants; mfg. of artificial resins, solvents, pharmaceuticals, insecticides, fungicides, plasticizers, rubber chems.; lubricant and gasoline additive; wetting agents; dyestuffs; flavor; fragrance
Regulatory: FEMA GRAS
Manuf./Distrib.: Acros Org.; Advanced Synthesis Tech.; Aldrich; Alfa Aesar; Fluka; Lancaster Synthesis; Pfaltz & Bauer; Riedel-deHaën; Spectrum Quality Prods.; TCI Am.

as-o-Xylenol. *See* 3,4-Xylenol
o-Xylenol; o-3-Xylenol. *See* 2,3-Xylenol
o-4-Xylenol. *See* 3,4-Xylenol
p-Xylenol. *See* 2,5-Xylenol
Xylenol, mixed; Xylenols. *See* Xylenol
Xylidine. *See* N,N-Dimethylaniline
Xylol. *See* Xylene
m-Xylol. *See* m-Xylene
o-Xylol. *See* o-Xylene
p-Xylol. *See* p-Xylene

YZ

Yarmor; Yarmor pine oil. *See* Pine (Pinus palustris) oil
Ylang ylang absolute. *See* Ylang ylang (Cananga odorata) oil

Ylang ylang (Cananga odorata) oil

CAS 8006-81-3; 68606-83-7; EINECS/ELINCS 281-092-1
FEMA 3119
Synonyms: Cananga; Cananga odorata; Cananga odorata oil; Cananga oil; Ylang ylang absolute; Ylang ylang oil
Definition: Oil from steam distillation of flowers of the ylang ylang, *Cananga odorata*
Properties: Lt. yel. oily liq.; sl. woody floral odor; burning taste; insol. in water; dens. 0.930-0.950 (20/20 C); flash pt. (TCC) 75 C; ref. index 1.495-1.503 (20 C)
Toxicology: LD50 (oral, rat) > 5 g/kg, (skin, rabbit) > 5 g/kg; low toxicity by ing. and skin contact; primary irritant; may be irritating to eyes, skin, and if inhaled; sensitizer; TSCA listed
Precaution: Avoid contact or contamination with strong acids, alkalis, or oxidizing agents
Hazardous Decomp. Prods.: CO and unidentified org. compds. may be formed during combustion; heated to decomp., emits acrid smoke and irritating vapors
Uses: Fragrance in soaps, perfumes; natural flavoring agent in foods and pharmaceuticals; solvent in cosmetics
Regulatory: FDA 21CFR §182.20, GRAS; FEMA GRAS; Japan approved (ylang-ylang)
Manuf./Distrib.: AMC Chems.; Calchauvet; Fleurchem; George Uhe; Good Scents; Hunan Xinyu; Janousek Industriale Srl; Penta Mfg.; Sarcom; Spectrum Quality Prods.; Treatt USA; Voigt Global Distrib.

Ylang ylang oil. *See* Ylang ylang (Cananga odorata) oil

Zea mays; Zea mays oil. *See* Corn (Zea mays) oil
Zeolite; Zeolites. *See* Sodium silicoaluminate
Zinc butter. *See* Zinc chloride

Zinc chloride

CAS 7646-85-7; EINECS/ELINCS 231-592-0
UN 1840 (DOT); 2331 (DOT)
Synonyms: Butter of zinc; Zinc butter; Zinc chloride anhydrous; Zinc chloride fume; Zinc chloride, solution; Zinc dichloride; Zinc muriate, solution
Classification: Inorganic zinc compd.
Empirical: Cl_2Zn
Formula: $ZnCl_2$
Properties: Wh. cubic cryst., odorless; very sol. in water, alcohol, glycerol, ether; sol. in acetone; m.w. 136.27; dens. 2.91; vapor pressure 1 mm Hg (428 C); m.p. 290 C; b.p. 732 C; pH 4.0 (10% aq.); noncombustible
Toxicology: ACGIH TLV/TWA 1 mg/m^3 (fume); STEL 2 mg/m^3 (fume); LD50 (oral, rat) 350 mg/kg; poison by ing., IV, subcut., IP routes; corrosive irritant to skin, eyes, mucous membranes; causes burns; ing. may cause vomiting, abdominal pain; human systemic effects by inh. (pulmonary changes); can cause dermatitis, boils, conjunctivitis, GI upset on exposure to fumes or dusts; questionable carcinogen; experimental tumorigen, teratogen, reproductive effects; human mutagenic data
Precaution: DOT: Corrosive material; incompat. with cyanides, sulfides, potassium; mixts. with powd. zinc are flamm.
Hazardous Decomp. Prods.: Heated to decomp., emits toxic fumes of Cl^- and ZnO
HMIS: Health 2, Flammability 0, Reactivity 2
Storage: Store in cool, dry, well-ventilated area, out of direct sunlight; hygroscopic; deliq.; use dust-tight containers; avoid generating dust
Uses: Catalyst; dehydrating agent, condensing agent in org. synthesis; fireproofing; preservative in foods; electroplating; metal processing; galvanization; antiseptic denaturant for alcohol; corrosion inhibitor for water treatment; solvent for cellulose; battery electrolyte; deodorant; disinfectant preservative for medical specimens and embalming; lumber treatment; fiber and textile processing/dyeing; reverse emulsion breaker for oil refining; therapeutic uses; pharmaceutical colorant; trace mineral in food supplements, animal feeds; oral care agent, soothing agent in cosmetics
Regulatory: FDA 21CFR §182.70, 182.8985, 582.80, GRAS; SARA reportable; FDA approved for parenterals; BP, EP compliance
Manuf./Distrib.: AMRESCO; Aldrich; Alfa Aesar; Allan; Am. Int'l.; Asiamerica Int'l.; Atofina; Boith China; Coyne; EM Ind.; EM Science; Filo; Fluka; GFS; Hasnil; Hawkins Chem.; Honeywell Perf. Polymers; Hosoi Chem. Ind.; Int'l. Chem. Inc; Integra; J.T. Baker; Kraft Chem.; Madison Ind.; Mallinckrodt Baker; Miljac; Min. R&D; Nagai Chem. Ind.; Penta Mfg.; RIA Int'l.; Ruger; SAF Bulk Chems.; Sal Chem.; San Yuan; Sigma; Sinochem Liaoning; Spectrum Quality Prods.; Stanchem; Tetra; Thomas Scientific; VWR Int'l.; Voigt Global Distrib.; Vopak USA; Wm. Blythe Ltd; Zaclon

Zinc chloride anhydrous; Zinc chloride fume; Zinc chloride, solution; Zinc dichloride; Zinc muriate, solution. *See* Zinc chloride

Part IV
Functional/Application Index

Functional/Application Index

Hexamethyldisilazane

adhesion promoter, textiles
Filmex® A-2 190 Proof; Filmex® A-2 Anhydrous; Filmex® B 190 Proof; Filmex® B Anhydrous; Filmex® C 190 Proof; Filmex® C Anhydrous; Filmex® D-1; Filmex® D-2 190 Proof; Filmex® D-2 Anhydrous

adhesion promoter, vinyl plastisols coating metals
Kenplast® ES-2HP

adhesive
Acetamide MEA; Acetone; Acetyl triethyl citrate; Acrylonitrile; Alcohol; Aminoethylpiperazine; Amyl acetate; Benzene; Benzyl alcohol; Benzyl benzoate; Boric acid; 1,4-Butanediol; Butoxydiglycol; Butoxyethanol; Butoxyethanol acetate; n-Butyl acetate; Butyl alcohol; Butyl benzoate; Butyl benzyl phthalate; Butylene glycol; Butyl lactate; Butyl octyl phthalate; Butyl stearate; n-Butyraldehyde; Camphor; Caprylic alcohol; Carbomer; Cetearyl alcohol; Cetyl alcohol; Chlorobenzene; Chloroform; Coconut (Cocos nucifera) oil; Corn (Zea mays) oil; Cottonseed (Gossypium) oil; Cyclohexane; Cyclohexanol; Diacetone alcohol; Dibutyl phthalate; Diethanolamine; Diethylamine; Diethylene glycol; Diethylene glycol butyl ether acetate; Diethylene glycol dibenzoate; Diethylene glycol phenyl ether; Diethyl oxalate; Diethyl phthalate; Diisobutyl ketone; Diisobutyl phthalate; Dimethicone; Dimethyl formamide; Dimethyl hexynol; Dimethyl octynediol; Dimethyl phthalate; Dinonyl phenol; Dioctyl maleate; Dioctyl phosphite; Dioctyl sebacate; 1,4-Dioxane; Dipropylene glycol; Dipropylene glycol dibenzoate; Dodoxynol-7; Erucamide; Ethanolamine; Ethoxydiglycol; Ethoxydiglycol acetate; Ethoxyethanol; Ethoxyethanol acetate; Ethylenediamine; Ethylene dichloride; Ethylene glycol; Ethylene glycol diacetate; 2-Ethylhexanol; Ethyl lactate; Formaldehyde; Formamide; Furfural; Furfuryl alcohol; Glycereth-26; Glyceryl dilaurate; Glyceryl dioleate; Glyceryl laurate; Glyceryl myristate; Glyceryl ricinoleate; Heptane; Hexamethylene diisocyanate; Hexane; 1,2,6-Hexanetriol; Hexylene glycol; Hydrogenated castor oil; Hydrogenated menhaden oil; Hydrogen peroxide; Hydroquinone; Isobutyl alcohol; Isoheptane; Isophorone; Isopropyl acetate; Isopropyl alcohol; Kerosene; Lactic acid; Lauryl alcohol; dl-Limonene; 2-Mercaptoethanol; Methoxydiglycol; Methoxyethanol; Methoxyisopropanol; Methyl acetate; Methyl acetyl ricinoleate; Methylal; Methyl alcohol; Methyl cocoate; Methylene chloride; Methyl ethyl ketone; Methyl isobutyl ketone; Methyl oleate; Methyl palmitate; Methyl palmitate oleate; Methyl phthalyl ethyl glycolate; Methyl rapeseedate; Methyl salicylate; Methyl soyate; Methyl tallowate; Mineral oil; Morpholine; Myristyl alcohol; Naphtha; 2-Nitropropane; Nonoxynol-4; Nonoxynol-5; Nonoxynol-6; Nonoxynol-9 phosphate; Oleic acid; Olive (Olea europaea) oil; Paraffin, chlorinated; Peanut (Arachis hypogaea) oil; PEG-4; PEG-6; PEG-8; PEG-9; PEG-10; PEG-12; PEG-14; PEG-16; PEG-18; PEG-20; PEG-32; PEG-40; PEG-75; PEG-90; PEG-100; PEG-135; PEG-150; PEG-200; PEG-350; PEG-30 castor oil; PEG-33 castor oil; PEG-35 castor oil; PEG-40 castor oil; PEG-4 dilaurate; PEG-8 dilaurate; PEG-12 dilaurate; PEG-4 dioleate; PEG-8 dioleate; PEG-12 dioleate; PEG-4 oleate; PEG-5 oleate; PEG-6 oleate; PEG-8 oleate; PEG/PPG-17/6 copolymer; 1,3-Pentanediamine; Perchloroethylene; Petrolatum; Phenoxyethanol; Phenyl trimethicone; Pine (Pinus palustris) oil; Poloxamer 181; Polypropylene glycol; Polysorbate 60; Polysorbate 80; PPG-9; PPG-17; PPG-20; PPG-26; PPG-30; PPG-2-buteth-3; PPG-3-buteth-5; PPG-5-buteth-7; PPG-7-buteth-10; PPG-9-buteth-12; PPG-12-buteth-16; PPG-20-buteth-30; PPG-28-buteth-35; PPG-33-buteth-45; PPG-15 butyl ether; PPG-18 butyl ether; PPG-22 butyl ether; PPG-40 butyl ether; PPG-2 methyl ether; PPG-3 methyl ether; Propyl alcohol; Propylene carbonate; Propylene glycol alginate; Propylene glycol laurate; Propylene glycol stearate; Propylene oxide; Pyrocatechol; Rapeseed (Brassica campestris) oil; Rice (Oryza sativa) bran oil; Safflower (Carthamus tinctorius) oil; Sesame (Sesamum indicum) oil; Silicone; Sodium 2-ethylhexyl sulfate; Sodium lauryl sulfate; Sodium thiocyanate; Sodium xylenesulfonate; Soybean (Glycine soja) oil; Stearyl alcohol; Sucrose octaacetate; Sulfur; Tall oil; Terpene resin; α-Terpineol; Tetraethylenepentamine; Tetrahydrofuran; Tetrahydrofurfuryl alcohol; Tetrahydroxypropyl ethylenediamine; Toluene; Toluene diisocyanate; Toluene-2,4-diisocyanate; Triacetin; Tributoxyethyl phosphate; Tributyl citrate; Tributyl phosphate; 1,1,1-Trichloroethane; 1,1,2-Trichloroethane; Trichloroethylene; Tridecyl alcohol; Triethanolamine; Triethylene glycol; Triethyl phosphate; Trioctyl phosphate; Triphenyl phosphate; Turpentine; Vinylidene chloride monomer; Xylene

adhesive removal agent
d-Limonene

adhesive, pastes
Tristearin

adhesive, phenolic
Phenol

adipic acid source, nylon prod.
Cyclohexanol

adipic acid source, textile finishing
Cyclohexanol

adjuvant, ABS food pkg.
2-Mercaptoethanol

adjuvant, agric.
Methyl soyate

adjuvant, agric.: tank mixes
Agnique® 2237

adjuvant, animal glue
Formaldehyde

adjuvant, blowing agent: foamed PS, food-contact
Perchloroethylene

adjuvant, dyeing
1,4-Dioxane

adjuvant, flavor: pharmaceuticals
Propylene glycol alginate

adjuvant, food
Eugenol; Fenchyl alcohol; Potassium carbonate

adjuvant, herbicides
Emery® 2230; *Methyl canolate; Methyl sunflowerate; Polysorbate 80*

adjuvant, lubricants: mineral oil, incidental food-contact
Isopropyl oleate

adjuvant, paper/paperboard: aq./fatty food/dry food
PEG-4 dilaurate

adjuvant, PC food-pkg. resins
Chlorobenzene; Chloroform; Ethylene dichloride; Heptane; Methylene chloride; Phenol; Pyridine; Toluene; Triethylamine

adjuvant, pesticides
Emery® 2230; *Methyl canolate; Methyl sunflowerate*

adjuvant, pesticides: growing crops
Dodoxynol-7

adjuvant, pharmaceuticals
Eugenol; Fenchyl alcohol

adjuvant, pitch content: paper/ paperboard, aq./fatty/dry food contact
Diethanolamine

adjuvant, polyarylsulfone resin prod.: food pkg.
Sulfolane

adjuvant, printing
1,4-Dioxane

adjuvant, pulp absorbency: paper/ paperboard, aq./fatty/dry food contact
Diethanolamine

adjuvant, resinous/polymeric food-contact coatings
Acetone; n-Butyl acetate; Butyl alcohol; Ethyl acetate; Hexane; Methyl ethyl ketone; Tetrahydrofuran; Toluene

adjuvant, slimicides: food-contact paper
Tetrahydrofuran

adjuvant, slimicides: food-contact paper/paperboard
Acetone; Dibutyl phthalate; Dimethyl formamide; Ethanolamine; Ethylenediamine; Ethylene glycol; Methoxyisopropanol; N-Methyl-2-pyrrolidone; PPG-2 methyl ether; PPG-3 methyl ether

adjuvant, staining
1,4-Dioxane

adsorbent, ion exchange
Sodium silicoaluminate

aerating agent
Sucrose fatty acid esters

aerating agent, foamed food
Butane; Chloropentafluoroethane; Isobutane

aerating agent, food
Propane; Sucroglycerides; Sucrose palmitate

aerating agent, sprayed food
Butane; Chloropentafluoroethane; Isobutane

aerosol propellant
Dymel® A

agglutinant
Triethyl citrate

aggregation agent, polymer
Dimethylethanolamine

air conditioning fluid
Dichlorotetrafluoroethane

air conditioning gas
Dichlorodifluoromethane; 1,2-Dichlorotetrafluoroethane; Trichlorofluoromethane

air filter medium
Tricresyl phosphate; Tri-p-tolyl phosphate

alcohol source
Coconut (Cocos nucifera) oil

algicide
Hexachloro-1,3-butadiene; Hexetidine

algicide, swimming pools
Bromine

alkalizer
Diethanolamine

alkalizer, food
Ammonium hydroxide; Potassium carbonate; Potassium hydroxide

alkalizer, pharmaceuticals
Ammonia; Ammonium hydroxide; Potassium hydroxide; Sodium hydroxide; Triethanolamine

alkylating agent
Ethyl chloride

alkylating agent, organic synthesis
Butyl chloride

amphiphilic agent, drug delivery
Labrafil® Isostearique; *Triisostearin PEG-6 esters*

analgesic, pharmaceuticals
Camphor; Dimethyl sulfoxide; Ethyl ether

analgesic, topical
Methyl salicylate

analgesic, topical pharmaceuticals
Camphor (Cinnamomum camphora) oil

anesthetic
1,1-Dichloroethane; Ethyl bromide; Ethyl ether; Heptane; Trichloroethylene

anesthetic, local
Ethyl chloride; Methyl salicylate; Morpholine; Phenethyl alcohol

anesthetic, local: toothache
Clove (Eugenia caryophyllus) oil

anesthetic, pharmaceuticals
Ethyl bromide; Eugenol; Methylene chloride; Methyl pentynol; Thymol

anesthetic, skin care
Phenol

anesthetic, topical
Methyl chloride

animal feed supplement
Sodium iodide

anodyne, pharmaceuticals
Camphor

anthelmintic drug
Thymol

anthelmintic drug, veterinary medicine
Perchloroethylene

anthelmintic, pharmaceuticals
Camphor

antibacterial
Phenethyl alcohol; Thymol

antibacterial, food
Formic acid

antibacterial, pharmaceuticals
Sulfolane

antiblistering agent, high temp. curing

Aromatic 200
antiblocking agent
Crill 2
antiblocking agent, adhesives
Crill 1
antiblocking agent, asphalt
Erucamide
antiblocking agent, cosmetics
Crill 1
antiblocking agent, crayons
Erucamide
antiblocking agent, food
Crill 1
antiblocking agent, food pkg.
Crill 1
antiblocking agent, food pkg. plastics
Lumulse® GMO
antiblocking agent, herbicides
Crill 1
antiblocking agent, inks
Crill 1; *Erucamide*
antiblocking agent, insecticides
Crill 1
antiblocking agent, leather treatment
Crill 1
antiblocking agent, makeup
Crodamol CAP
antiblocking agent, metals
Erucamide
antiblocking agent, metalworking fluids
Crill 1
antiblocking agent, oil slick dispersing
Crill 1
antiblocking agent, paints
Crill 1
antiblocking agent, paper
Crill 1
antiblocking agent, PE
Erucamide
antiblocking agent, petroleum prods.
Erucamide
antiblocking agent, pharmaceuticals
Crill 1
antiblocking agent, plastics
Crill 1; *Erucamide*
antiblocking agent, polishes
Crill 1
antiblocking agent, powder prods.
Ammonium xylenesulfonate; *Sodium xylenesulfonate*
antiblocking agent, powders
Sodium toluenesulfonate
antiblocking agent, rubber
Erucamide
antiblocking agent, skin care
Crodamol CAP
antiblocking agent, textiles
Crill 1; *Erucamide*
antiblooming agent, pentachlorophenol-treated wood
PPG-9; *PPG-17*; *PPG-30*
antiblush agent, nitrocellulose lacquers
Methyl hexyl ketone
antiblush agent, paints
Butoxyl®
antiblush agent, spray paints
Butoxyl®
anticaking agent, cosmetics
Miglyol® 812
anticaking agent, desiccants
EZA®; *Sodium silicoaluminate*
anticaking agent, detergents
EZA®; *Sodium silicoaluminate*
anticaking agent, food
Calcium chloride; *Propylene glycol*; *Sodium silicoaluminate*
anticaking agent, pharmaceuticals
Miglyol® 812
anticaking agent, powder prods.
Ammonium xylenesulfonate; *Sodium xylenesulfonate*
anticaking agent, powders
Sodium toluenesulfonate
anticoagulant, blood
Edetic acid
antidandruff agent, cosmetics
Sulfur
antidepressant
Dimethylethanolamine
antidusting agent
Delios® S
antiflatulent
Dimethicone
antifoam
Albrite® TBP; Cachalot® AR-20; Cachalot® BE-22; Cachalot® C-50; Cachalot® L-90; Cachalot® M-43; Cachalot® S-56; Crill 2; *Decyl alcohol*; *Dibutyl butyl phosphonate*; *Dimethicone*; *2,6-Dimethyl heptanol-4*; *Dioctyl octylphosphonate*; *Glyceryl laurate*; *Glyceryl ricinoleate*; Herco® Pine Oil; *Hexyl alcohol*; *Isooctyl alcohol*; *Myristyl alcohol*; *2-Octanol*; *Polypropylene glycol*; *Stearyl alcohol*; *Tributyl citrate*; *Trimethylcyclohexanol*; *Tri-n-octyl phosphate*
antifoam, adhesives
Crill 1; *PPG-2-buteth-3*; *PPG-3-buteth-5*; *PPG-5-buteth-7*; *PPG-7-buteth-10*; *PPG-9-buteth-12*; *PPG-12-buteth-16*; *PPG-20-buteth-30*; *PPG-28-buteth-35*; *PPG-33-buteth-45*; Ucon® 50-HB-55; Ucon® 50-HB-100; Ucon® 50-HB-260; Ucon® 50-HB-400; Ucon® 50-HB-660; Ucon® 50-HB-2000; Ucon® 50-HB-3520; Ucon® 50-HB-5100; Ucon® LB-65; Ucon® LB-135; Ucon® LB-165; Ucon® LB-285; Ucon® LB-385; Ucon® LB-525; Ucon® LB-625; Ucon® LB-1145; Ucon® LB-1715; Ucon® LB-3000
antifoam, agric.
PEG-6 oleate
antifoam, antifreeze
Jeffsol® PC; *Propylene carbonate*
antifoam, aq. adhesives
Phosflex® 4; TBP
antifoam, aq. solutions
Ethylene glycol acetate
antifoam, boiler water
PPG-2-buteth-3; *PPG-3-buteth-5*; *PPG-5-buteth-7*; *PPG-7-buteth-10*; *PPG-9-buteth-12*; *PPG-12-buteth-16*; *PPG-20-buteth-30*; *PPG-28-buteth-35*; *PPG-33-buteth-45*; *PPG-5 butyl ether*; *PPG-9 butyl ether*; *PPG-15 butyl ether*; *PPG-18 butyl ether*; *PPG-22 butyl ether*; *PPG-24 butyl ether*; *PPG-33 butyl ether*; *PPG-40 butyl ether*; *PPG-53 butyl ether*; Ucon® 50-HB-55; Ucon® 50-HB-100; Ucon® 50-HB-260; Ucon® 50-HB-400; Ucon® 50-HB-660; Ucon® 50-HB-2000; Ucon® 50-HB-3520; Ucon® 50-HB-5100; Ucon® LB-65; Ucon® LB-135; Ucon® LB-165; Ucon® LB-285; Ucon® LB-385; Ucon® LB-525; Ucon® LB-625; Ucon® LB-1145; Ucon® LB-1715; Ucon® LB-3000
antifoam, calf starter
Crill 1; Crill 2
antifoam, casein solutions
TBP
antifoam, ceramics
PPG 1650; *PPG-17*
antifoam, cosmetics
Crill 1; *Dimethicone*; *Hexamethyldisiloxane*; *Oleyl alcohol*; *PEG-6 oleate*; *Petroleum distillates*; *Phenyl trimethicone*; *Propyl alcohol*
antifoam, cutting oils
Oleyl alcohol
antifoam, detergent solutions
Phosflex® 4; TBP
antifoam, fermentation
Kemester® 2050; *PPG-9*; *PPG-17*; *PPG-30*
antifoam, food
Caprylic/capric glycerides; Crill 1; *Oleic acid*; *Oleyl alcohol*; *PEG-6 oleate*; *Polyethylene glycol*
antifoam, food pkg.
Crill 1
antifoam, food-contact
Ucon® 50-HB-660; Ucon® 50-HB-2000; Ucon® 50-HB-3520; Ucon® 50-HB-5100
antifoam, food-contact paper
Hodag® S-35 FGK
antifoam, gas prod.
Prochinor® SOL
antifoam, herbicides
Crill 1
antifoam, hydraulic fluids
Albrite® Tributyl Phosphate
antifoam, industrial
Polyglyceryl-4 oleate
antifoam, inks
Crill 1; *Oleyl alcohol*; Phosflex® 4; TBP
antifoam, insecticides
Crill 1
antifoam, latex processing
PPG-2-buteth-3; *PPG-3-buteth-5*; *PPG-5-buteth-7*; *PPG-7-buteth-10*; *PPG-9-buteth-12*; *PPG-12-buteth-16*; *PPG-20-buteth-30*; *PPG-28-buteth-35*; *PPG-33-buteth-45*; *PPG-5 butyl ether*; *PPG-9 butyl ether*; *PPG-15 butyl ether*; *PPG-18 butyl ether*; *PPG-22 butyl ether*; *PPG-24 butyl ether*; *PPG-33 butyl ether*; *PPG-40 butyl ether*; *PPG-53 butyl ether*; Ucon® 50-HB-55; Ucon® 50-HB-100; Ucon® 50-HB-260; Ucon® 50-HB-400; Ucon® 50-HB-660; Ucon® 50-HB-2000; Ucon® 50-HB-3520; Ucon® 50-HB-5100; Ucon® LB-65; Ucon® LB-135; Ucon® LB-165; Ucon® LB-285; Ucon® LB-385; Ucon® LB-525; Ucon® LB-625; Ucon® LB-1145; Ucon® LB-1715; Ucon® LB-3000
antifoam, leather treatment
Crill 1
antifoam, metalworking fluids
Crill 1
antifoam, oil prod.
Prochinor® SOL
antifoam, oil slick dispersing
Crill 1
antifoam, paints
Crill 1; *Oleyl alcohol*; *PPG-9*; *PPG-17*; *PPG-30*; *PPG-2-buteth-3*; *PPG-3-buteth-5*; *PPG-5-buteth-7*; *PPG-7-buteth-10*; *PPG-9-buteth-12*; *PPG-12-buteth-16*; *PPG-20-buteth-30*; *PPG-28-buteth-35*; *PPG-33-buteth-45*; *Trioctyl phosphate*; Ucon® 50-HB-55; Ucon® 50-HB-100; Ucon® 50-HB-260; Ucon® 50-HB-400; Ucon® 50-HB-660; Ucon® 50-HB-2000; Ucon® 50-HB-3520; Ucon® 50-HB-5100; Ucon® LB-65; Ucon® LB-135; Ucon® LB-165; Ucon® LB-285; Ucon® LB-385; Ucon® LB-525; Ucon® LB-625; Ucon® LB-1145; Ucon® LB-1715; Ucon® LB-3000
antifoam, paper
Crill 1
antifoam, paper coatings
Phosflex® 4; TBP
antifoam, pesticides
3,5,5-Trimethylhexan-1-ol
antifoam, petrochemicals
Oleyl alcohol
antifoam, pharmaceuticals
Crill 1; *PEG-6 oleate*
antifoam, plastics
Crill 1; *Oleyl alcohol*; *PEG-6 oleate*
antifoam, polishes
Crill 1
antifoam, polymer systems
Albrite® Tributyl Phosphate
antifoam, protein processing
Mazol® 300 K
antifoam, pulp/paper
Oleyl alcohol
antifoam, rubber
PPG 1650; *PPG-17*
antifoam, salt water evaporators
PPG-2-buteth-3; *PPG-3-buteth-5*; *PPG-5-buteth-7*; *PPG-7-buteth-10*; *PPG-9-buteth-12*; *PPG-12-buteth-16*; *PPG-20-buteth-30*; *PPG-28-buteth-35*; *PPG-33-buteth-45*; Ucon® 50-HB-55; Ucon® 50-HB-100; Ucon® 50-HB-260; Ucon® 50-HB-400; Ucon® LB-65; Ucon® LB-135; Ucon® LB-165; Ucon® LB-285; Ucon® LB-385; Ucon® LB-525; Ucon® LB-625; Ucon® LB-1145; Ucon® LB-1715; Ucon® LB-3000
antifoam, sugar processing
Mazol® 300 K
antifoam, syrups
Crill 1; Crill 2
antifoam, textile finishing
Oleyl alcohol
antifoam, textile sizes
Phosflex® 4; TBP
antifoam, textiles
Crill 1; Eastman® DBP; *PEG-6 oleate*
antifoam, topical pharmaceuticals
Oleyl alcohol
antifoam, washing solutions
PPG-2-buteth-3; *PPG-3-buteth-5*; *PPG-5-buteth-7*; *PPG-7-buteth-10*; *PPG-9-buteth-12*; *PPG-12-buteth-16*; *PPG-20-buteth-30*; *PPG-28-buteth-35*; *PPG-33-buteth-45*; Ucon® 50-HB-55; Ucon® 50-HB-100; Ucon® 50-HB-260; Ucon® 50-HB-400; Ucon® 50-HB-660; Ucon® 50-HB-2000; Ucon® 50-HB-3520; Ucon® 50-HB-5100; Ucon® LB-65; Ucon® LB-135; Ucon® LB-165; Ucon® LB-285; Ucon® LB-385; Ucon® LB-525; Ucon® LB-625; Ucon® LB-1145; Ucon® LB-1715; Ucon® LB-3000
antifog, food pkg.
Glyceryl ricinoleate
antifog, food-pkg. film
Sorbitan laurate
antifog, plastic films
Sorbitan trioleate
antifog, plastics
Glyceryl myristate; *Glyceryl stearate*; *Glyceryl stearate SE*; *Polyglyceryl-3 stearate*; *Polysorbate 80*; *Sorbitan laurate*
antifog, PVC

Sorbitan laurate
antifog, PVC film
Glyceryl dioleate; *Glyceryl oleate*
antifreeze
Alcohol denatured; *Glycerin*; *Methoxydiglycol*; *Methoxyethanol*; *Methyl alcohol*; *Propylene glycol*; Propylene Glycol USP/FCC Ultra Grade; Reomol® TBP; *SD alcohol 40*; *SD alcohol 40-B*; *Trimethylene glycol*
antifreeze ingred.
Diethylene glycol
antifreeze inhibitor
Ethylenediamine; *Ethylenediamine dihydrochloride*
antifreeze, carburetor fluids
Tripropylene glycol
antifreeze, cooling/heating systems
Ethylene glycol
antifreeze, diesel engines
Methoxyisopropanol
antifreeze, refrigeration units
Propylene Glycol USP, FCC
antifriction agent
Triisooctyl phosphite
antifungal agent
Thymol
antifungal agent, feminine hygiene prods.
Hexetidine
antifungal agent, oral hygiene prods.
Hexetidine
antifungal agent, pharmaceuticals
Hexetidine
antifungal agent, topical pharmaceuticals
Hexetidine
antifungal, topical
Triacetin
antihistamine
Behenyl alcohol
antihypertensive, pharmaceuticals
Sodium thiocyanate
anti-icing additive, aviation fuel
Ethoxyethanol; *Methoxydiglycol*
anti-icing additive, fuel
Glyceryl tallate; Lumulse® GMO; Lumulse® GMT; Lumulse® GMT-40
anti-icing agent, petroleum
Dimethyl formamide; *Dipropylene glycol*
anti-infective, topical: pharmaceuticals
Camphor
anti-inflammatory
Amyl acetate
anti-inflammatory, pharmaceuticals
Dimethyl sulfoxide
anti-irritant, skin
Fancol™ Karite Butter
anti-irritant, surfactants
PEG-30 glyceryl cocoate
anti-itching agent, pharmaceuticals
Camphor
antiknock agent
Bromine; *Methyl chloride*; *Tetraethyl lead*
antimalarial
Quinoline
antimicrobial
Hexetidine; *d-Limonene*; *Thymol*
antimicrobial agent: sugar mills
Ethylenediamine
antimicrobial, beet sugar mills
Sorbitan palmitate
antimicrobial, cane sugar mills
Sorbitan palmitate
antimicrobial, cosmetics
Boric acid; *Cresylic acid*; Dalpad A; *DMDM hydantoin*; Dowanol® EPh; *Glyceryl caprate*; *Hydrogen peroxide*; *MDM hydantoin*; *Phenol*; *Potassium iodide*; *Sodium iodide*; *Triacetin*
antimicrobial, food
Alcohol; *Benzyl alcohol*; *Calcium chloride*; *Lactic acid*; *Sulfur dioxide*
antimicrobial, fragrances
Dalpad A; Dowanol® EPh
antimicrobial, oral pharmaceuticals
Soluphor® P
antimicrobial, parenteral pharmaceuticals
Soluphor® P
antimicrobial, pharmaceutical injectables
Soluphor® P
antimicrobial, pharmaceuticals
Benzyl alcohol; *Calcium chloride*; *Formaldehyde*; *Lactic acid*; *Phenethyl alcohol*; *Phenol*; *Phenoxyethanol*; *Thymol*
antimicrobial, preservatives
Phenoxyethanol
antimicrobial, topical pharmaceuticals
DMDM hydantoin
antimicrobial, veterinary medicine
Soluphor® P
antioxidant
p-t-Butyl toluene; *N-Nitrosodiethylamine*; *N-Nitrosodimethylamine*; *Sulfur dioxide*; *α-Terpineol*; *Thymol*; *Wheat (Triticum vulgare) germ oil*
antioxidant, acrylic food pkg.
Hydroquinone monomethyl ether
antioxidant, acrylics
Triisooctyl phosphite
antioxidant, animal feed
BHT
antioxidant, cellulosics
Albrite® TBP; *1,1,3-Tris (2-methyl-4-hydroxy-5-t-butylphenyl) butane*
antioxidant, cosmetics
BHA; *t-Butyl hydroquinone*; *Edetic acid*; *Hydroquinone monomethyl ether*; *Thiodiglycol*; *Triethyl citrate*
antioxidant, elastomers
Dimethyl thiodipropionate; *2,6-Xylenol*
antioxidant, essential oils
Pyrocatechol
antioxidant, fats
BHT; *Hydroquinone monomethyl ether*
antioxidant, fats/oils
BHA; *Hydroquinone*
antioxidant, food
BHA; *BHT*; *t-Butyl hydroquinone*; *Edetic acid*; Mazol® 300 K; *Propylene glycol*; *Sulfur dioxide*
antioxidant, food pkg.
BHA
antioxidant, food-contact coating
BHT
antioxidant, food-grade adhesives
1,1,3-Tris (2-methyl-4-hydroxy-5-t-butylphenyl) butane
antioxidant, food-grade polymers
Boric acid; *Butylene glycol*; *1,1,3-Tris (2-methyl-4-hydroxy-5-t-butylphenyl) butane*
antioxidant, food-pkg. adhesives
BHA; *BHT*
antioxidant, gas
2,6-Xylenol
antioxidant, gasoline
BHT
antioxidant, hot-melt adhesives
BHT
antioxidant, jet fuels
BHT
antioxidant, lubricants
BHT; *Methylcyclohexanol*
antioxidant, nonstaining: PP
Dibutyl phosphite
antioxidant, NR latexes
Dimethylamine
antioxidant, nylon
Triisooctyl phosphite
antioxidant, oils
BHT; *Hydroquinone monomethyl ether*; *2,6-Xylenol*
antioxidant, perfumes
Pyrocatechol
antioxidant, petroleum
BHT
antioxidant, pharmaceuticals
BHA; *BHT*; *t-Butyl hydroquinone*; *Sulfur dioxide*
antioxidant, plastics
BHA; *BHT*; *Nonylphenol*
antioxidant, polyamides
1,1,3-Tris (2-methyl-4-hydroxy-5-t-butylphenyl) butane
antioxidant, polyester
Albrite® TBP; *Triisopropyl phosphite*
antioxidant, polyester resins
Hydroquinone
antioxidant, polyesters
t-Butyl hydroquinone; *1,1,3-Tris (2-methyl-4-hydroxy-5-t-butylphenyl) butane*
antioxidant, polyolefins
1,1,3-Tris (2-methyl-4-hydroxy-5-t-butylphenyl) butane
antioxidant, POM
1,1,3-Tris (2-methyl-4-hydroxy-5-t-butylphenyl) butane
antioxidant, PP
Albrite® TBP; *Dioctyl phosphite*
antioxidant, pressure-sensitive adhesives
BHA; *BHT*
antioxidant, PS
1,1,3-Tris (2-methyl-4-hydroxy-5-t-butylphenyl) butane
antioxidant, PU
1,1,3-Tris (2-methyl-4-hydroxy-5-t-butylphenyl) butane
antioxidant, PVC
Albrite® TBP; *Triisooctyl phosphite*; *1,1,3-Tris (2-methyl-4-hydroxy-5-t-butylphenyl) butane*
antioxidant, rubber
Acetonitrile; *Aniline*; *Crotonaldehyde*; *Nonylphenol*; *Paraldehyde*; *1,1,3-Tris (2-methyl-4-hydroxy-5-t-butylphenyl) butane*
antioxidant, rubber articles for repeated use: food-contact
BHT
antioxidant, sesame oil
Sesamol
antioxidant, soaps
BHT
antioxidant, solvents
BHT
antioxidant, SR
BHT
antioxidant, suntan lotions
Hydroquinone
antioxidant, synthetic latexes
Hydroquinone
antioxidant, vitamins
Hydroquinone monomethyl ether
antioxidant: unsat. polyesters
Triisooctyl phosphite
antiozonant, chloroprene rubber
Isobutyl oleate
antiozonant, rubber
Paraffin
antipreignition additive, petroleum/diesel
Tricresyl phosphate
antiredeposition agent, scour/dye operations
Stetlasolv SES-CONC.
antiredeposition agent, soil and sizings
Arosolve FX-6
antiredeposition agent, soil: dry cleaning
Nonoxynol-9 phosphate
antiredeposition agent, spotted goods repair
Stetlasolv SES-CONC.
antisagging agent, high temp. curing
Aromatic 200
antiseptic
Boric acid; *Hexetidine*; *Hexyl alcohol*; *Iodine*; *Methoxyethanol*; *Methyl salicylate*; *Naphthalene*; *Phenoxyethanol*; *Potassium iodide*; *Propyl alcohol*; *Pyrocatechol*; *α-Terpineol*; *Thymol*
antiseptic denaturant, alcohol
Zinc chloride
antiseptic, dentifrices
Hydrogen peroxide
antiseptic, eye lotions
Calcium chloride
antiseptic, local
Eucalyptus globulus oil; *Morpholine*
antiseptic, oral pharmaceuticals
Terpene resin
antiseptic, oral rinses
Eucalyptol
antiseptic, pharmaceuticals
Ethyl ether; *Eugenol*; *Thymol*
antiseptic, skin care
Phenoxyisopropanol
antiseptic, topical anti-infectives
Cetearyl alcohol; *Isopropyl alcohol*; *Sodium iodide*
antiseptic, topical pharmaceuticals
Alcohol; *Phenethyl alcohol*
antisettling agent, fillers: inks
Trihydroxystearin
antisettling agent, fillers: mastics
Trihydroxystearin
antisettling agent, fillers: paints/coatings
Trihydroxystearin
antisettling agent, fillers: solvent-free epoxies
Trihydroxystearin
antisettling agent, fillers: stains
Trihydroxystearin
antisettling agent, fillers: textile backings
Trihydroxystearin
antisettling agent, pigment: inks
Trihydroxystearin
antisettling agent, pigment: mastics
Trihydroxystearin
antisettling agent, pigment: paints/coatings
Trihydroxystearin
antisettling agent, pigment: solvent-free epoxy systems
Trihydroxystearin
antisettling agent, pigment: stains
Trihydroxystearin
antisettling agent, pigment: textile backings
Trihydroxystearin
antishrinking agent, wool
Bromine
antiskinning agent
Butyl lactate

antitackifier, moisturizers
PPG-2 myristyl ether propionate
antitackifier, ointments
Crodamol PMP
antitackifier, pharmaceuticals
PPG-2 myristyl ether propionate
antitackifier, resins
Citrofol® AII
antiwear agent
Tributoxyethyl phosphate; Tributyl phosphate; Triisooctyl phosphite
antiwear agent, metalworking compds.
Norfox® MTL
application enhancer, colored cosmetics
Octyldodecyl isostearate
application enhancer, creams/lotions
Octyldodecyl isostearate
application enhancer, hair care
Octyldodecyl isostearate
application enhancer, makeup removers
Octyldodecyl isostearate
application enhancer, moisturizing cleansers
Octyldodecyl isostearate
application enhancer, ointments
Octyldodecyl isostearate
application enhancer, personal care
Octyldodecyl isostearate
application enhancer, skin care
Octyldodecyl isostearate
aprotic reaction medium, polymerization reactions
Proglyde DMM
aprotic solvent, high-boiling
Phenyl acetate
aroma component, perfumes
Isobutyl acetate
aromatic compd. substitute
Estasol™ MD10
asphalt component
Propylbenzene
asphalt remover
d-Limonene
astringent
Benzyl alcohol
astringent, cosmetics
SDA-1; SDA-3A; SDA-3C; SDA-23A; SDA-30; SDA-39C; SDA-40; SDA-40A; SDA-40B; SDA-40C; *SD alcohol 1; SD alcohol 3-A; SD alcohol 3-C; SD alcohol 23-A; SD alcohol 30; SD alcohol 39-C; SD alcohol 40; SD alcohol 40-A; SD alcohol 40-B; SD alcohol 40-C*
astringent, personal care
SDA-1; SDA-3A; SDA-3C; SDA-23A; SDA-30; SDA-39C; SDA-40; SDA-40A; SDA-40B; SDA-40C
astringent, pharmaceuticals
Stearyl alcohol
aviation fuel blending component
Naphtha
aviation gasoline agent
Toluene
azeotrope
Aromatic 100; Aromatic 150; Aromatic 200
azeotrope, electronic cleaning
VersaTRANS™ Solvent
azeotrope, precision cleaning
VersaTRANS™ Solvent
backcoatings, textiles
Carbomer
bactericide
Cachalot® AR-20; Cachalot® BE-22; Cachalot® C-50; Cachalot® L-90; Cachalot® M-43; Cachalot® S-56; *Hexetidine; Propylene glycol dimethyl ether; Propylene glycol laurate; Propylene glycol phenyl ether; Propylene glycol stearate*
bactericide, air sterilization
Propylene glycol
bactericide, gas prod.
Prochinor® SOL
bactericide, oil prod.
Prochinor® SOL
bactericide/algicide, oil fields
Acrolein
bactericide/algicide, weeds in agric. irrigation systems, canals
Acrolein
bacteriostat
Benzyl alcohol; Cresylic acid; Glycerin; Imwitor® 742
bacteriostat, clinical nutrition
Capmul® MCM
bacteriostat, delivery/absorption enhancement
Capmul® MCM; Capmul® MCMC8
bacteriostat, dermatologicals
Capmul® MCM; Capmul® MCMC8
bacteriostat, nutritional/sports supplements
Capmul® MCM
bacteriostat, pharmaceutical aerosols
Capmul® MCM
bacteriostat, pharmaceutical coatings
Capmul® MCM; Capmul® MCMC10; *Glyceryl caprate*
bacteriostat, pharmaceutical microemulsions
Capmul® MCM; Capmul® MCMC8; Capmul® MCMC10; *Glyceryl caprate*
bacteriostat, pharmaceutical soft gelatin capsules
Glyceryl caprate
bacteriostat, pharmaceuticals
Capmul® MCM; Capmul® MCMC8; *Glyceryl caprate*
bacteriostat, suppositories
Capmul® MCM; Capmul® MCMC8; *Glyceryl caprate*
base fluid, environmentally friendly cleaners
Oleocal ME-70; Oleocal ME-112
base fluid, environmentally friendly solvents
Oleocal ME-70; Oleocal ME-112
base oil, drilling fluids
Sasol MR Solvent
base oil, lubricant: bar lubricants
Oleocal C-102; Oleocal C-102 A; Oleocal C-105; Oleocal IVO-114
base oil, lubricant: chain lubricants
Oleocal C-102; Oleocal C-102 A; Oleocal C-105; Oleocal IVO-114
base oil, lubricant: environmentally friendly lubricants
Oleocal C-102; Oleocal C-102 A; Oleocal C-105; Oleocal IVO-114
base oil, lubricant: environmentally sensitive forestry applics.
Oleocal C-102; Oleocal C-102 A; Oleocal C-105
base oil, lubricant: gear oils
Oleocal C-102; Oleocal C-102 A; Oleocal C-105; Oleocal IVO-114
base oil, lubricant: greases
Oleocal C-102; Oleocal C-102 A; Oleocal C-105; Oleocal IVO-114
base oil, lubricant: hydraulic fluids
Oleocal C-102; Oleocal C-102 A; Oleocal C-105; Oleocal IVO-114
base oil, lubricant: marine lubricants
Oleocal C-102; Oleocal C-102 A; Oleocal C-105; Oleocal IVO-114
base oil, lubricant: mining
Oleocal C-102; Oleocal C-102 A; Oleocal C-105
base, absorption: cosmetics
Argobase 125T
base, absorption: hair care
Argobase 125T
base, absorption: pharmaceuticals
Argobase 125T
base, absorption: skin care
Argobase 125T
base, adhesives
Carbowax® PEG 300
base, agric.
Carbowax® PEG 300
base, animal feed supplement
Penreco Snow
base, antiperspirants
Carbowax® Sentry® PEG 300; Carbowax® Sentry® PEG 540 Blend; Carbowax® Sentry® PEG 600; Carbowax® Sentry® PEG 900; Carbowax® Sentry® PEG 1000; Carbowax® Sentry® PEG 1450; Carbowax® Sentry® PEG 3350; Carbowax® Sentry® PEG 4600; Carbowax® Sentry® PEG 8000
base, baby oils
Drakeol® 7
base, bath oils
Carbowax® Sentry® PEG 400
base, bath powders
Carbowax® Sentry® PEG 3350
base, ceramics
Carbowax® PEG 300
base, chain oils
Naphthenic oil
base, chewing gum
Cocoa (Theobroma cacao) butter; Dimethicone; Lanolin; Methyl acetyl ricinoleate; PPG-33 butyl ether
base, cleaners
Oleocal ME-130
base, colognes
Carbowax® Sentry® PEG 400; Carbowax® Sentry® PEG 600
base, compressor oils
Naphthenic oil
base, cosmetic sticks
Cetyl palmitate
base, cosmetics
Carbowax® PEG 300; Carbowax® Sentry® PEG 300; Carbowax® Sentry® PEG 400; Carbowax® Sentry® PEG 540 Blend; Carbowax® Sentry® PEG 600; Carbowax® Sentry® PEG 900; Carbowax® Sentry® PEG 1000; Carbowax® Sentry® PEG 1450; Carbowax® Sentry® PEG 3350; Carbowax® Sentry® PEG 4600; Carbowax® Sentry® PEG 8000; *Cottonseed (Gossypium) oil;* Exceparl BB; Exceparl BS; Exceparl EH-P; Exceparl EH-S; Exceparl IB-OL; Exceparl L-OL; Exceparl MC; Exceparl ML-85; Exceparl M-OL; Exceparl MS; Exceparl MT; Exceparl MY-M; Exceparl OD-B; Exceparl OL-OL; Exceparl O-OL; Exceparl SS; Exceparl TD-S; Penreco Snow; Pluracol® E200; Pluracol® E300; Pluracol® E400; Pluracol® E400 NF; Pluracol® E600; Pluracol® E600 NF; Pluracol® E1450 NF; Pluracol® E4000; Pluracol® E6000; Vinycizer 30
base, creams
Caprylic/capric/stearic triglyceride
base, creams/lotions
Carbowax® Sentry® PEG 300; Carbowax® Sentry® PEG 400; Carbowax® Sentry® PEG 540 Blend; Carbowax® Sentry® PEG 600; Carbowax® Sentry® PEG 900; Carbowax® Sentry® PEG 1000; Carbowax® Sentry® PEG 1450; Carbowax® Sentry® PEG 3350; Carbowax® Sentry® PEG 4600; Carbowax® Sentry® PEG 8000; Penreco Snow
base, dental adhesives
Penreco Snow
base, dentifrices
Carbowax® Sentry® PEG 600
base, diesel engine oils
Naphthenic oil
base, elastomers
Carbowax® PEG 300
base, elec. transformer oils
Naphthenic oil
base, electroplating
Carbowax® PEG 300
base, electropolishing
Carbowax® PEG 300
base, emollient: cosmetics
NEOBEE® 18
base, emollient: pharmaceuticals
NEOBEE® 18
base, food
Penreco Snow
base, food processing
Carbowax® PEG 300
base, fruit/veg. coatings
Penreco Snow
base, gear oils
Naphthenic oil
base, hair dressing
Carbowax® Sentry® PEG 400; Carbowax® Sentry® PEG 540 Blend; Carbowax® Sentry® PEG 3350; Carbowax® Sentry® PEG 8000
base, hand cleaners
Carbowax® Sentry® PEG 300; Carbowax® Sentry® PEG 540 Blend
base, high-foaming: acid aluminum cleaners
Rhodacal® 330
base, high-foaming: agric. pesticides
Rhodacal® 330
base, high-foaming: cleaners
Rhodacal® 330
base, high-foaming: dry cleaning
Rhodacal® 330
base, high-foaming: emulsion polymerization
Rhodacal® 330
base, high-foaming: latex paints
Rhodacal® 330
base, high-foaming: metal cleaning
Rhodacal® 330
base, high-foaming: metalworking fluids
Rhodacal® 330
base, high-foaming: shampoos
Rhodacal® 330
base, household prods.
Carbowax® PEG 300
base, hydraulic fluids
Naphthenic oil
base, inks
Pluracol® E400 NF; Pluracol® E600 NF; Pluracol® E1450 NF
base, intermediate
Butyronitrile
base, laxatives
Petrolatum

Carbowax® Sentry® PEG 400; Carbowax® Sentry® PEG 600; Carbowax® Sentry® PEG 3350
binder, makeup removers
Carbowax® Sentry® PEG 400
binder, metal fabrication
Carbowax® PEG 300
binder, nail polish removers
Carbowax® Sentry® PEG 400; Carbowax® Sentry® PEG 3350; Carbowax® Sentry® PEG 8000
binder, oil: pressed powder cosmetics
Myritol® 318
binder, ointments
Carbowax® Sentry® PEG 540 Blend; Carbowax® Sentry® PEG 1000; Carbowax® Sentry® PEG 1450
binder, paints
Carbowax® PEG 300
binder, paper
Carbowax® PEG 300; *PEG-2 stearate*
binder, personal care
Carbowax® PEG 300; Carbowax® Sentry® PEG 300; Carbowax® Sentry® PEG 400; Carbowax® Sentry® PEG 540 Blend; Carbowax® Sentry® PEG 600; Carbowax® Sentry® PEG 900; Carbowax® Sentry® PEG 1000; Carbowax® Sentry® PEG 1450; Carbowax® Sentry® PEG 3350; Carbowax® Sentry® PEG 4600; Carbowax® Sentry® PEG 8000; *Isopropyl stearate*; *Isostearyl isostearate*; Pluracol® E200; Pluracol® E300; Pluracol® E400; Pluracol® E600; Pluracol® E4000; Pluracol® E6000
binder, pharmaceutical tablets
Carbowax® Sentry® PEG 300; *PEG-100*; *PEG-200*; *Sunflower (Helianthus annuus) seed oil*
binder, pharmaceuticals
Carbowax® PEG 300; Pluracol® E300; Pluracol® E400; Pluracol® E600; Pluracol® E4000; Pluracol® E6000; *Propylene glycol alginate*; Rhodasurf® PEG 600; *Tridecyl neopentanoate*; *Tristearin*
binder, pigment
Isocetyl alcohol
binder, pigment: bath oils
Fancol™ ALA
binder, pigment: body lotions
Fancol™ ALA
binder, pigment: cosmetics
Fancol™ ALA
binder, pigment: hand lotions
Fancol™ ALA
binder, pigment: ointments
Fancol™ ALA
binder, pigment: personal care
Fancol™ ALA
binder, pigment: skin care
Fancol™ ALA
binder, pigment: suntan oils
Fancol™ ALA
binder, pigments
Tridecyl neopentanoate
binder, pressed powders
Acetulan®; *C12-15 alkyl benzoate*; Ceraphyl® 28; Ceraphyl® 140; Ceraphyl® 140-A; Ceraphyl® ICA; *Cetyl lactate*; *Cetyl palmitate*; Estalan-PGDP; Finsolv® TN; *Myristyl myristate*; *Octyl palmitate*; *Trihydroxystearin*; Trivent OP
binder, printing
Carbowax® PEG 300
binder, printing inks
PEG-4; *Terpene resin*
binder, rubber
Carbowax® PEG 300; *PEG-2 stearate*
binder, shampoos
Carbowax® Sentry® PEG 300; Carbowax® Sentry® PEG 400; Carbowax® Sentry® PEG 600
binder, shaving creams
Carbowax® Sentry® PEG 300; Carbowax® Sentry® PEG 400; Carbowax® Sentry® PEG 600; Carbowax® Sentry® PEG 900; Carbowax® Sentry® PEG 1000; Carbowax® Sentry® PEG 1450
binder, shaving preps.
Carbowax® Sentry® PEG 540 Blend
binder, skin care
Isopropyl stearate
binder, soaps
Carbowax® Sentry® PEG 400; Carbowax® Sentry® PEG 3350; *Isopropyl stearate*
binder, sugar-coated tablets
PEG-125
binder, sunscreens
Carbowax® Sentry® PEG 300; Carbowax® Sentry® PEG 400; Carbowax® Sentry® PEG 540 Blend; Carbowax® Sentry® PEG 600; *Isopropyl stearate*
binder, tablets
Lutrol® E 1500; Lutrol® E 3350; Lutrol® E 4000; Lutrol® E 6000; *PEG-20*; *PEG-32*; *PEG-40*; *PEG-90*; *Trihydroxystearin*; *Trilaurin*; *Trimyristin*; *Tripalmitin*
binder, textiles
Carbowax® PEG 300; *PEG-2 stearate*
binder, toothpaste
Carbowax® Sentry® PEG 300; Carbowax® Sentry® PEG 400; Carbowax® Sentry® PEG 3350
binder, topical pharmaceuticals
Isopropyl stearate; *PEG-2 stearate*
binder, uncoated tablets
PEG-125
binder, vitamin suspensions
Propylene glycol alginate
binder, water colors
1,3-Dichloro-2-propanol; *2,3-Dichloro-1-propanol*
binder, wood processing
Carbowax® PEG 300
bioadhesive
N-Methyl-2-pyrrolidone
bioavailability enhancer, bath oils
Caprylic/capric triglyceride PEG-4 esters
bioavailability enhancer, cosmetic creams/lotions/shampoos/bath oils
Labrafac® Hydrophile
bioavailability enhancer, cosmetics
Caprylic/capric triglyceride PEG-4 esters; Captex® 200; Labrasol®
bioavailability enhancer, fat-soluble vitamins
Lutrol® E 1500; Lutrol® E 3350; Lutrol® E 4000; Lutrol® E 6000; *PEG-125*
bioavailability enhancer, flavors
Captex® 200
bioavailability enhancer, food
Captex® 200
bioavailability enhancer, fragrance oils
Captex® 200
bioavailability enhancer, medicinals
Captex® 200
bioavailability enhancer, nasal pharmaceuticals
Labrafac® Hydrophile
bioavailability enhancer, ointments
Caprylic/capric triglyceride PEG-4 esters
bioavailability enhancer, oral pharmaceuticals
Labrafac® Hydrophile
bioavailability enhancer, pharmaceuticals
Caprylic/capric triglyceride PEG-4 esters; Imwitor® 742; Labrasol®; *PEG-8 caprylic/capric glycerides*
bioavailability enhancer, rectal pharmaceuticals
Labrafac® Hydrophile
bioavailability enhancer, shampoos
Caprylic/capric triglyceride PEG-4 esters
bioavailability enhancer, sl. water-soluble pharmaceuticals
PEG-125
bioavailability enhancer, soluble colorants
Captex® 200
bioavailability enhancer, steroids
Lutrol® E 1500; Lutrol® E 3350; Lutrol® E 4000; Lutrol® E 6000; *PEG-125*
bioavailability enhancer, topical pharmaceuticals
Labrafac® Hydrophile; Pharmasolve™
bioavailability enhancer, vaginal pharmaceuticals
Labrafac® Hydrophile
bioavailability enhancer, veterinary creams/lotions
Caprylic/capric triglyceride PEG-4 esters
bioavailability enhancer, veterinary preps.
Labrafac® Hydrophile; Labrasol®
bioavailability enhancer, vitamins
Captex® 200
bioavailability enhancer, vitamins veterinary medicine
Cremophor® EL
bioavailability enhancer, vitamins: animal feed
Cremophor® EL
bioavailability enhancer, water-insoluble pharmaceuticals
PEG-125
bioavailability enhancer, water-insoluble/poorly soluble pharmaceuticals
Lutrol® E 1500; Lutrol® E 3350; Lutrol® E 4000; Lutrol® E 6000
biochemical/medical research
Methyl behenate
biocide
Acetamide; *Benzyltriethyl ammonium chloride*; *DMDM hydantoin*; *Lauryl alcohol*
biocide, pharmaceuticals
Dantogard® Plus Liq.
bitterant, aversive: ingestion deterrent
Denatonium benzoate
bitumen modifier, road construction
Shellsol® D100; Shellsol® D100S
bleaching agent
Hydroquinone; *Potassium hydroxide*
bleaching agent, fibers
Bromine
bleaching agent, food
Sulfur dioxide
bleaching agent, fur
Hydrogen peroxide
bleaching agent, hair
Hydrogen peroxide
bleaching agent, industrial water
Hydrogen peroxide
bleaching agent, municipal water
Hydrogen peroxide
bleaching agent, oils
Sulfur dioxide
bleaching agent, pulp/paper
Sulfur dioxide
bleaching agent, textiles
Dodecylbenzenesulfonic acid; *Hydrogen peroxide*; *Sulfur*
bleaching agent, wastewater
Hydrogen peroxide
bleaching agent, wood pulp
Hydrogen peroxide; *Sulfur*
bleaching, paper
Sodium hydroxide
bleaching, pulp
Sodium hydroxide
bleaching, starch
Sulfur dioxide
blending agent, bath oils
STEPAN® IPM-NF
blending agent, cleansers
STEPAN® IPM-NF
blending agent, creams/lotions
STEPAN® IPM-NF
blending agent, dusting powders
STEPAN® IPM-NF
blending agent, eye shadows
STEPAN® IPM-NF
blending agent, lipstick
STEPAN® IPM-NF
blending agent, makeup
STEPAN® IPM-NF
blending agent, personal care
STEPAN® IPM-NF
blending agent, petroleum
Isooctane; *Methyl t-butyl ether*
blending agent, talcum powders
STEPAN® IPM-NF
blending agent: detergents
Methylcyclohexanol
blending component, hydraulic fluid
3-Methoxybutanol
blocking agent
Decamethylcyclopentasiloxane; *Hexamethyldisiloxane*; *Octamethylcyclotetrasiloxane*; *Octamethyltrisiloxane*; *Tetramethylsilane*
blocking agent, isocyanate
Phenol
blood thinner
Dimethyl sulfone
blowing agent
Chlorodifluoromethane; *Dichlorofluoromethane*; *Dichlorotetrafluoroethane*; *Trichlorotrifluoroethane*
blowing agent, aerosols
Methylene chloride
blowing agent, appliance insulation
1,1,1,3,3-Pentafluoropropane
blowing agent, boardstocks
1,1,1,3,3-Pentafluoropropane
blowing agent, expandable PS
Pentane 60/40; Pentane 80/20
blowing agent, expandable PS: food-contact
n-Pentane
blowing agent, flexible foam
Dichlorofluoroethane
blowing agent, foam
1,2-Dichlorotetrafluoroethane
blowing agent, foam systems
Dichlorofluoroethane
blowing agent, food-grade PS
Isopentane

Capmul® MCM; Capmul® MCMC8; Captex® 300; Captex® 350; *Propylene glycol dioctanoate*; *Shea butter (Butyrospermum parkii)*; *Tricaprin*

carrier, dietetic prods.
Miglyol® 840

carrier, dispersions
Naphthenic oil

carrier, dissolved pharmaceuticals: soft gelatin capsules
Lutrol® E 300; Lutrol® E 400; Lutrol® E 600

carrier, dry skin care
Propylene glycol dioctanoate

carrier, dye
Benzyl benzoate; *Biphenyl*; *Cyclohexanol*; *Diphenyl oxide*; *Hexyl benzoate*; *Methyl salicylate*; *PEG-90*; Pogol® 400

carrier, dye: polyester
Butyl benzoate

carrier, dye: textiles
Butyl benzoate; *Dimethyl phthalate*

carrier, epoxies
Xylene

carrier, essential oils
NEOBEE® M-5; *Tricaprylin*; *Trioctanoin*

carrier, fats/oils
Eastman® Glacial Acetic Acid Food Grade, Kosher

carrier, fat-soluble antioxidants
Sucroglycerides; *Sucrose fatty acid esters*

carrier, fat-soluble vitamins
Lutrol® E 1500; Lutrol® E 3350; Lutrol® E 4000; Lutrol® E 6000; *PEG-125*

carrier, film
PEG 300

carrier, flavor oils: food
Isocetyl alcohol

carrier, flavor oils: pharmaceuticals
Isocetyl alcohol

carrier, flavors
Capmul® MCM; Capmul® MCMC8; Capmul® MCMC10; *Caprylic/capric/linoleic triglyceride*; *Caprylic/capric/oleic triglyceride*; Captex® 200; Captex® 810D; Captex® 8227; Ceraphyl® ICA; Citrofol® AI; Estasan™ GT 8-40 3578; Estasan™ GT 8-60 3575; Estasan™ GT 8-60 3580; Estasan™ GT 8-65 3577; Estasan™ GT 8-65 3581; Estasan™ GT 8-70 3579; *Glyceryl diacetate*; *Isopropyl myristate*; Lexol® GT-865; Lexol® PG-800; NEOBEE® M-5; *Propylene glycol dicaprylate/dicaprate*; *Propylene glycol dioctanoate*; *Tricaprin*; *Tricaprylin*; *Trioctanoin*; *Triundecanoin*

carrier, fluorocarbon lubricants
2,3-Dihydrodecafluoropentane

carrier, foam control agents
Carbowax® E200; Carbowax® E300; Carbowax® E400; Carbowax® E600; Carbowax® E900; Carbowax® E1000; Carbowax® E1450; Carbowax® E3350; Carbowax® E8000

carrier, food
Captex® 200; *Corn syrup*; Estasan™ GT 8-40 3578; Estasan™ GT 8-60 3575; Estasan™ GT 8-60 3580; Estasan™ GT 8-65 3577; Estasan™ GT 8-65 3581; Estasan™ GT 8-70 3579; Penreco Snow; Polarin® Range

carrier, food processing
Eastman® Glacial Acetic Acid Food Grade, Kosher

carrier, foundation
Capmul® MCM; Capmul® MCMC8

carrier, fragrance oils
Captex® 200; *Propylene glycol dicaprylate/dicaprate*

carrier, fragrance oils: food
Isocetyl alcohol

carrier, fragrance oils: pharmaceuticals
Isocetyl alcohol

carrier, fragrances
Capmul® MCM; Capmul® MCMC8; Capmul® MCMC10; *Caprylic/capric/linoleic triglyceride*; *Caprylic/capric/oleic triglyceride*; Captex® 810D; Captex® 8227; Ceraphyl® ICA; Citrofol® AI; Estasan™ GT 8-40 3578; Estasan™ GT 8-60 3575; Estasan™ GT 8-60 3580; Estasan™ GT 8-65 3577; Estasan™ GT 8-65 3581; Estasan™ GT 8-70 3579; *Glyceryl diacetate*; Lexol® PG-800; *Propylene glycol dioctanoate*; *Propylene glycol dipelargonate*; *Tricaprin*; *Triundecanoin*

carrier, fragrances: deodorants
Citroflex® 2

carrier, fragrances: detergents
Citroflex® 2

carrier, fragrances: perfumes
Citroflex® 2

carrier, fragrances: shampoos
Citroflex® 2

carrier, frozen dairy prods.
Eastman® Glacial Acetic Acid Food Grade, Kosher

carrier, fruit/veg. coatings
Penreco Snow

carrier, gases
Dimethyl formamide

carrier, gelatin
Eastman® Glacial Acetic Acid Food Grade, Kosher

carrier, gelatin capsules
Caprylic/capric/stearic triglyceride; *Tricaprin*

carrier, hair care
Butyl octyl benzoate; Captex® 300; HallBrite® BHB

carrier, hair conditioners
HallBrite® BHB

carrier, halogenated material
Perfluoroheptane; *Perfluorohexane*; *Perfluoro-N-methylmorpholine*; *Perfluoropentane*; PF-5050; PF-5060; PF-5070

carrier, herbicides
Methyl oleate

carrier, inert
Monochlorotoluene

carrier, infant formulas
Captex® 300; Captex® 350; *Tricaprin*

carrier, inj. prods.
Miglyol® 810; Miglyol® 840; Miglyol® 8810

carrier, inks
Oleocal ME-70; Oleocal ME-112; Oleocal ME-130; Pluracol® E1450 NF

carrier, insecticide
Xylene

carrier, insecticide actives
C13-14 isoparaffin; Isopar® M Fluid

carrier, intermediates
Carbowax® E300; Carbowax® E400; Carbowax® E600; Carbowax® E900; Carbowax® E1000; Carbowax® E1450; Carbowax® E3350; Carbowax® E8000

carrier, leather treatment
Methyl rapeseedate; Oleocal ME-70; Oleocal ME-112; Oleocal ME-130

carrier, lip balms
HallBrite® BHB

carrier, lip gloss
Captex® 810D

carrier, lip preps.
Capmul® MCM; Capmul® MCMC8; Captex® 300; Captex® 350

carrier, lipophilic oral pharmaceuticals
Imwitor® 742

carrier, lipophilic pharmaceuticals
Caprylic/capric glycerides; *Glyceryl myristate*; Imwitor® 742

carrier, lipophilic rectal pharmaceuticals
Imwitor® 742

carrier, lipophilic topicals
Imwitor® 742

carrier, lipstick
Captex® 810D; Crystal® Crown; HallBrite® BHB

carrier, liq. pharmaceuticals
Carbowax® E300 NF; Carbowax® E400 NF; Carbowax® E600 NF; Carbowax® E900 NF; Carbowax® E1000 NF; Carbowax® E1450 NF; Carbowax® E3350 NF; Carbowax® E4500 NF; Carbowax® E8000 NF; PEG 300

carrier, lotions
Drakeol® 7; *Shea butter (Butyrospermum parkii)*

carrier, lubricant
Methyl nonafluorobutyl ether

carrier, lubricant: fluorocarbons
HFE-7200

carrier, lubricant: hydrocarbons
HFE-7200

carrier, lubricant: silicones
HFE-7200

carrier, lubricants
HFE-7100; Novec™ HFE-7100; Novec™ HFE-7200

carrier, lubricants: fluorocarbons
Ethyl nonafluorobutyl ether

carrier, lubricants: hydrocarbons
Ethyl nonafluorobutyl ether

carrier, lubricants: silicones
Ethyl nonafluorobutyl ether

carrier, makeup
Carbowax® E200; Carbowax® E300; Carbowax® E400; Carbowax® E600; Carbowax® E900; Carbowax® E1000; Carbowax® E1450; Carbowax® E3350; Carbowax® E8000; Crystal® Crown; Drakeol® 7

carrier, makeup bases
Captex® 810D

carrier, mascara
Captex® 300; Captex® 350

carrier, metal cleaners
Pluracol® E1450 NF

carrier, metal polishes
Pluracol® E1450 NF

carrier, metalworking fluids
Carbowax® E200; Carbowax® E300; Carbowax® E400; Carbowax® E600; Carbowax® E900; Carbowax® E1000; Carbowax® E1450; Carbowax® E3350; Carbowax® E8000

carrier, metalworking lubricants
PEG 300

carrier, milk prods.
Eastman® Glacial Acetic Acid Food Grade, Kosher

carrier, moisturizers
HallBrite® BHB

carrier, nutritional fluids
NEOBEE® M-5

carrier, nutritional preps.
Caprylic/capric/lauric triglyceride; *Caprylic/capric/linoleic triglyceride*; Captex® 810D

carrier, nutritional supplements
Tricaprin

carrier, nutritional/sports supplements
Capmul® MCM; Captex® 350

carrier, o/w creams
Shea butter (Butyrospermum parkii)

carrier, o/w emulsions
Shea butter (Butyrospermum parkii)

carrier, oils
Miglyol® 840

carrier, oils: agric.
Methyl palmitate oleate

carrier, oils: agric. herbicides
Methyl caprylate/caprate; *Methyl laurate*

carrier, oils: agric. pesticides
Methyl caprylate/caprate; *Methyl laurate*

carrier, oil-soluble prods.
Octyldodecanol

carrier, ointment bases
PEG 300

carrier, ointments
Drakeol® 7; Dynacerin® 660; Miglyol® 810; Miglyol® 840; Miglyol® 8810; Penreco Snow

carrier, oral preps.
Miglyol® 840; Pluracol® E1450 NF

carrier, paints
PEG 300

carrier, paints/coatings
Oleocal C-102; Oleocal C-102 A; Oleocal C-105; Oleocal IVO-114

carrier, paper
PEG 300; Pluracol® E1450 NF

carrier, parenteral pharmaceuticals
Miglyol® 840

carrier, pentachlorophenol wood preservatives
PCP Solvent

carrier, personal care
Carbowax® E200; Carbowax® E300; Carbowax® E400; Carbowax® E600; Carbowax® E900; Carbowax® E1000; Carbowax® E1450; Carbowax® E3350; Carbowax® E8000; Crystal® Crown; *Hexyl laurate*; Isopar® G Fluid; Isopar® H Fluid; Isopar® K Fluid

carrier, personal care sprays
C10-11 isoparaffin; *C11-12 isoparaffin*

carrier, pesticides
Condea Vista LPA; Condea Vista LPA-210; *Kerosene*; *Methyl oleate*; Oleocal ME-70; Oleocal ME-112; PEG 300; PEG 400; Sasol LPA Solvent; Sasol LPA-210 Solvent; *Soybean (Glycine soja) oil*; THFA®

carrier, petroleum jellies
Penreco Snow

carrier, pharmaceutical
Dimethyl sulfone

carrier, pharmaceutical aerosols
Capmul® MCM; Polarin® Range

carrier, pharmaceutical capsules
Glyceryl caprylate; Polarin® Range

carrier, pharmaceutical coatings
Capmul® MCM; Capmul® MCMC10;

Captex® 350; *Glyceryl caprate*
carrier, pharmaceutical lipophilic active ingreds.
Delios®
carrier, pharmaceutical microemulsions
Capmul® MCM; Capmul® MCMC8; Capmul® MCMC10; *Glyceryl caprate*
carrier, pharmaceutical soft gelatin capsules
Glyceryl caprate
carrier, pharmaceutical solutions
Dimethyl isosorbide
carrier, pharmaceutical syrups
Polarin® Range
carrier, pharmaceuticals
Butyl Diglyme; *Butylene glycol*; *Butylene glycol dicaprylate/dicaprate*; Capmul® MCM; Capmul® MCMC8; *Caprylic/capric/lauric triglyceride*; *Caprylic/capric/linoleic triglyceride*; Captex® 200; Captex® 300; Captex® 350; Captex® 355EP; Captex® 500P; Captex® 810D; Carbowax® E300 NF; Carbowax® E400 NF; Carbowax® E600 NF; Carbowax® E900 NF; Carbowax® E1000 NF; Carbowax® E1450 NF; Carbowax® E3350 NF; Carbowax® E4500 NF; Carbowax® E8000 NF; *Cocoa (Theobroma cacao) butter*; Crystal® Crown; *Diethylene glycol dibutyl ether*; *Dimethyl sulfoxide*; Estasan™ GT 8-40 3578; Estasan™ GT 8-60 3575; Estasan™ GT 8-60 3580; Estasan™ GT 8-65 3577; Estasan™ GT 8-65 3581; Estasan™ GT 8-70 3579; *Glyceryl caprate*; *Glyceryl laurate*; *Hexyl laurate*; Lexol® PG-800; Miglyol® 812; Miglyol® 840; NEOBEE® M-5; *Oleyl alcohol*; *PEG-4 dimethyl ether*; Penreco Snow; Pluracol® E400 NF; Pluracol® E600 NF; Pluracol® E1450 NF; Polarin® Range; *Propylene glycol dicaprylate/dicaprate*; *Propylene glycol dioctanoate*; *Sunflower (Helianthus annuus) seed oil*; Tetraglyme; *Triacetin*
carrier, plasticizing waxes
Oleocal ME-112
carrier, plastics
Carbowax® E200; Carbowax® E300; Carbowax® E400; Carbowax® E600; Carbowax® E900; Carbowax® E1000; Carbowax® E1450; Carbowax® E3350; Carbowax® E8000
carrier, polyester disperse dyes: synthetic fabrics
Arosolve VNO-LF
carrier, polyester dyes: high pressure dyeing equipment
Arosolve MN; Arosolve MN-LF
carrier, polyester dyes: pressure jet operations
Arosolve MN; Arosolve MN-LF; Arosolve TMB
carrier, polyester/wool blended fabrics
α-Methylnaphthalene
carrier, powders
Tricaprin
carrier, priming catalyst
Methyl ethyl ketone
carrier, printing inks
Methyl rapeseedate; PEG 300
carrier, processed fruit
Eastman® Glacial Acetic Acid Food Grade, Kosher
carrier, puddings
Eastman® Glacial Acetic Acid Food Grade, Kosher
carrier, rectal pharmaceuticals
Miglyol® 840
carrier, relishes
Eastman® Glacial Acetic Acid Food Grade, Kosher
carrier, resins
PEG 300
carrier, rubber
Carbowax® E200; Carbowax® E300; Carbowax® E400; Carbowax® E600; Carbowax® E900; Carbowax® E1000; Carbowax® E1450; Carbowax® E3350; Carbowax® E8000; Pluracol® E1450 NF
carrier, skin care
Butyl octyl benzoate; HallBrite® BHB
carrier, skin cleansers
Cyclomethicone
carrier, skin creams/lotions
Isopropyl myristopalmitate
carrier, sl. water-soluble pharmaceuticals
PEG-125
carrier, snack food
Eastman® Glacial Acetic Acid Food Grade, Kosher
carrier, soaps
Crystal® Crown
carrier, soft beverages
Eastman® Glacial Acetic Acid Food Grade, Kosher
carrier, soft candy
Eastman® Glacial Acetic Acid Food Grade, Kosher
carrier, soft drinks
Corn syrup
carrier, soft gelatin capsules
Captex® 300
carrier, soluble colorants
Captex® 200; *Propylene glycol dicaprylate/dicaprate*
carrier, solvent
o-Dichlorobenzene
carrier, solvent dyes
Arosolve H99B
carrier, solvent extraction
Orfom® SX 7; Orfom® SX 10; Orfom® SX 11; Orfom® SX 12
carrier, solvent: flavors
Caprylic/capric triglyceride; Crodamol GTCC
carrier, solvent: fragrances
Caprylic/capric triglyceride; Crodamol GTCC
carrier, solvents: synthetic fabrics/blends
Arosolve KJ; Arosolve RCB; Arosolve TUV
carrier, sports supplements
Tricaprin
carrier, spray coatings
Aldo® MCT KFG
carrier, steroids
Lutrol® E 1500; Lutrol® E 3350; Lutrol® E 4000; Lutrol® E 6000; *PEG-125*
carrier, sticks
Tricaprin
carrier, sun care
Capmul® MCM; Capmul® MCMC8; Captex® 300; Captex® 350
carrier, sun tan lotions
Drakeol® 7
carrier, sunscreens
HallBrite® BHB
carrier, suppositories
Capmul® MCM; Capmul® MCMC8; *Glyceryl caprate*; Miglyol® 810; Miglyol® 840; *Propylene glycol dioctanoate*; *Shea butter (Butyrospermum parkii)*
carrier, suspended drugs: soft gelatin capsules
Lutrol® E 300; Lutrol® E 400; Lutrol® E 600
carrier, suspensions
Captex® 300
carrier, sweet sauces
Eastman® Glacial Acetic Acid Food Grade, Kosher
carrier, synthetic lubricants
Carbowax® E200; Carbowax® E300; Carbowax® E400; Carbowax® E600; Carbowax® E900; Carbowax® E1000; Carbowax® E1450; Carbowax® E3350; Carbowax® E8000
carrier, textiles
Pluracol® E1450 NF
carrier, toothpaste
Carbowax® E200; Carbowax® E300; Carbowax® E400; Carbowax® E600; Carbowax® E900; Carbowax® E1000; Carbowax® E1450; Carbowax® E3350; Carbowax® E8000
carrier, topical pharmaceuticals
Oleyl erucate
carrier, transparency film
Citrofol® AII
carrier, vinyl stabilizer
THFA®
carrier, vitamins
Captex® 200; Lexol® PG-800; NEOBEE® M-5; *Propylene glycol dicaprylate/dicaprate*; *Propylene glycol dioctanoate*; *Tricaprin*
carrier, w/o creams
Shea butter (Butyrospermum parkii)
carrier, w/o emulsions
Shea butter (Butyrospermum parkii)
carrier, water-insoluble pharmaceuticals
PEG-125
carrier, water-insoluble/poorly soluble pharmaceuticals
Lutrol® E 1500; Lutrol® E 3350; Lutrol® E 4000; Lutrol® E 6000
carrier, waterless hand cleaners
Condea Vista LPA; Sasol LPA Solvent
carrier, wood prods.
Pluracol® E1450 NF
carvone source
Spearmint (Mentha viridis) oil
catalyst
Acetonitrile; *Benzyltriethyl ammonium chloride*; *n-Butylamine*; *Copper oxide (ic)*; *Diisopropylamine*; *Dimethyl acetamide*; Eastman® TEP; Eastman® Triethyl Phosphate; Isoform; *1-Methylpiperidine*; *γ-Picoline*; *Sulfur dioxide*; *Zinc chloride*
catalyst blocker, platinum: silicones
Methyl butynol
catalyst component, copolymerization
Dibutyl ether
catalyst component, Ziegler
Iodine
catalyst impregnator
Dimethyl sulfide
catalyst promoter
Acetylacetone
catalyst, acid
Methanesulfonic acid; *Phosphoric acid*
catalyst, acid catalyzed reactions
Dodecylbenzenesulfonic acid
catalyst, acid-salt
2-Aminobutanol
catalyst, acrylic food pkg.
Tetraethylenepentamine
catalyst, alkaline
Tetraethanol ammonium hydroxide
catalyst, alkylation
Hydrobromic acid; *Methanesulfonic acid*
catalyst, alkylation/acylation
Trifluoroacetic acid
catalyst, alkylation/condensation/sulfonation/chlorination
Iodine
catalyst, biodeg., organic materials
Methyl acetate
catalyst, carboxylation reactions
Dimethyl formamide
catalyst, condensation
Ammonia
catalyst, condensation reactions
Piperidine
catalyst, epoxidation
Methanesulfonic acid
catalyst, epoxies
Triethylamine
catalyst, esterification
Acetylacetone; *Ethylacetoacetate*; *Methanesulfonic acid*
catalyst, flexible PU foam
Ethyl morpholine
catalyst, formaldehyde polymerization
Dibutylamine
catalyst, foundry resins
Triethylamine
catalyst, metal
Naphthenic acid
catalyst, olefin polymerization
Acetylacetone; *Ethylacetoacetate*
catalyst, olefinic addition
Methanesulfonic acid
catalyst, oxidation
Boric acid
catalyst, paints
Trimethyl borate
catalyst, peroxidation reactions
Methanesulfonic acid
catalyst, phase transfer
Benzyltriethyl ammonium chloride; *Benzyl trimethyl ammonium chloride*
catalyst, phenolic resins
Triethylamine
catalyst, polyester slabstock
Ethyl morpholine
catalyst, polymerization
Acetophenone; *Dimethyl ether*; *Lithium aluminum hydride*; *Methanesulfonic acid*
catalyst, PPO resins: food
Diethylamine
catalyst, prod. of acetic anhydride: synthesizing ketene
Triethyl phosphate
catalyst, PU
Dytek® EP Diamine; *1,3-Pentanediamine*
catalyst, PU foam
Diethylaminoethanol; *Methyldiethanolamine*; *p-Methyl morpholine*; *1,1,1-Trichloroethane*; *Triethylamine*
catalyst, PU foam slabstock
Dimethylethanolamine
catalyst, resins
Trimethyl borate

catalyst, rigid PU foam
Ethyl morpholine
catalyst, varnishes
Trimethyl borate
catalyst, waxes
Trimethyl borate
catalytic solvent, chemical synthesis
Triethylamine
catalytic/reaction medium
Pharmasolve™; Shellsol® D38
cathartic
Mineral oil
caustic, wart removal
Potassium hydroxide
cellophane, sealing
Methoxyethanol
CFC replacement
2,3-Dihydrodecafluoropentane
chain extender, PU
1,4-Butanediol; *Diethylene glycol*; *Dipropylene glycol*; *Ethylene glycol*; *Hexamethylene glycol*; *Polypropylene glycol*; *Tripropylene glycol*
chain length regulator, ion exchange resins
Methylal
chain length regulator, polyacetal
Methylal
chain terminator, PC food-pkg. resins
Cumyl phenol
chain transfer agent, acrylic resin prod.: automotive topcoats, high-solids
2-Mercaptoethanol
chain transfer agent, PVC
Trichloroethylene
charcoal lighter fluid
Deodorized kerosene
chelating agent
Diethylenetriamine; *Diethyl oxalate*
chelating agent, boric acid
Ethyl hexanediol
chelating agent, cosmetics
Dibutyl oxalate; *Tetrahydroxypropyl ethylenediamine*
chelating agent, detergents
Edetic acid
chelating agent, disinfectants
Edetic acid
chelating agent, lubricant additives
Acetylacetone
chelating agent, metal: polyester
Triisopropyl phosphite
chelating agent, metals
Acetylacetone
chelating agent, paint driers
Acetylacetone
chelating agent, pesticides
Acetylacetone
chelating agent, pharmaceuticals
Tetrahydroxypropyl ethylenediamine
chelating agent, rare earth separation
Edetic acid
chemosterilant, insects
Hempa
chemosterilant, rodents
Glyceryl-α-monochlorohydrin
chlorinating agent, organic compds.
Carbon tetrachloride
chlorinating agent, organic synthesis
Methyl chloride
chorine fluid recovery, chlorine plants
Hexachloro-1,3-butadiene
chorine recovery, chlorine plants
Hexachloro-1,3-butadiene
chromatographic reference standard
Methyl oleate
Cl_2, removal
Hydrogen peroxide
clarifier, casein polymers
Tricresyl phosphate
clarifier, cream rinses
Lactamide MEA
clarifier, food
Propylene glycol
clarifier, food pkg. plastics
Lumulse® GMO
clarifier, liqs.
Edetic acid
clarifier, shampoos
Acetamide MEA; *Dimethyl octynediol*; Incromectant AMEA-70; *Lactamide MEA*; Surfadone® LP-300
cleaner, solvent
Berol® 02; *C9-15 isoparaffin*; Eccoclean RPW; HFE-7100; *Nonoxynol-4*; *Nonoxynol-5*; *Nonoxynol-6*; Surphos 515; Surphos 620; Surphos 630; Surphos 645; T-Det® N-6
cleaner, solvent precision: fines
Axarel® 2200
cleaner, solvent precision: general purpose cleaning
Axarel® 2200
cleaner, solvent precision: lubricating fluids
Axarel® 2200
cleaner, solvent precision: metal cleaning
Axarel® 2200
cleaner, solvent precision: metal cutting
Axarel® 2200
cleaner, solvent precision: oil removal
Axarel® 2200
cleaner, solvent: acrylic removal
Ship Shape® Resin Cleaner
cleaner, solvent: adhesive pretreatment cleaners/degreasers
Toxfree™
cleaner, solvent: adhesive removal
Marthene; Ship Shape® Resin Cleaner
cleaner, solvent: adhesive removal: metal surfaces
Ecco Cleaner HR-LO
cleaner, solvent: adhesive removal: nonmetallic surfaces
Ecco Cleaner HR-LO
cleaner, solvent: adhesive removal: textile equip.
Ecco Cleaner HR-LO
cleaner, solvent: adhesives: foam cutting, fabrication
FoamFlush® Urethane Remover
cleaner, solvent: aerospace
Marthene; Partsprep™ Degreaser
cleaner, solvent: aircraft
Ship Shape® Resin Cleaner
cleaner, solvent: artist supplies
Brazilian Gum Turpentine-NT
cleaner, solvent: automotive
Ship Shape® Resin Cleaner
cleaner, solvent: buffing compds.
Actrel® 1111L; Actrel® 1138L; Actrel® 1160L; Actrel® 3338L; Actrel® 3349L; Actrel® 3360L; Actrel® 4493L
cleaner, solvent: carbonized oils/residues
Actrel® 1111L; Actrel® 1138L; Actrel® 1160L; Actrel® 3338L; Actrel® 3349L; Actrel® 3360L; Actrel® 4493L
cleaner, solvent: coating deposition applics.
HFE-71DE
cleaner, solvent: coating excess resin remover
Prothene
cleaner, solvent: coatings
Marthene
cleaner, solvent: coatings: foam cutting, fabrication
FoamFlush® Urethane Remover
cleaner, solvent: degreasing
Axarel® 9100
cleaner, solvent: drawing/stamping oils
Actrel® 1111L; Actrel® 1138L; Actrel® 1160L; Actrel® 3338L; Actrel® 3349L; Actrel® 3360L; Actrel® 4493L
cleaner, solvent: dye penetrants
Actrel® 1111L; Actrel® 1138L; Actrel® 1160L; Actrel® 3338L; Actrel® 3349L; Actrel® 3360L; Actrel® 4493L
cleaner, solvent: elec./electronic cleaning
Toxfree™
cleaner, solvent: electro-discharge machining oils
Actrel® 1111L; Actrel® 1138L; Actrel® 1160L; Actrel® 3338L; Actrel® 3349L; Actrel® 3360L; Actrel® 4493L
cleaner, solvent: electronic cleaning
Ethyl nonafluoroisobutyl ether, *Methyl nonafluoroisobutyl ether*
cleaner, solvent: electronic components
Marthene
cleaner, solvent: engineering
Marthene
cleaner, solvent: engineering maintenance degreasers
Toxfree™
cleaner, solvent: epoxy removal
Ship Shape® Resin Cleaner
cleaner, solvent: equip.
Ecco Cleaner AK; Eccoflex 090; Ecco Strip LS-W
cleaner, solvent: fabric spot cleaners
Toxfree™
cleaner, solvent: fabrics
Ecco Cleaner 930 Conc.
cleaner, solvent: fiberglass fabrication
Ship Shape® Resin Cleaner
cleaner, solvent: fibers
Ecco Cleaner 930 Conc.
cleaner, solvent: flux removal
Marthene
cleaner, solvent: furniture refinishing
Ship Shape® Resin Cleaner
cleaner, solvent: general purpose cleaning
Axarel® 4100; Axarel® 6100
cleaner, solvent: glass
Axarel® 52; Axarel® 56
cleaner, solvent: greases
HFE-71DA
cleaner, solvent: handling oils
HFE-71DA
cleaner, solvent: honing/lapping oils
Actrel® 1111L; Actrel® 1138L; Actrel® 1160L; Actrel® 3338L; Actrel® 3349L; Actrel® 3360L; Actrel® 4493L
cleaner, solvent: industrial degreasing
Toxfree™
cleaner, solvent: ink removal
Marthene
cleaner, solvent: ink removal: textile equip.
Ecco Cleaner HR-LO
cleaner, solvent: jewelry
Ship Shape® Resin Cleaner
cleaner, solvent: leather
Eccobrite Cleaner AK; Ecco Cleaner AK; Eccoflex 090; Ecco Strip LS-W
cleaner, solvent: leather finishing
Eccobrite Cleaner AK
cleaner, solvent: light duty cleaning
HFE-7200
cleaner, solvent: lubricant deposition applics.
HFE-71DE
cleaner, solvent: lubricating fluid removal
Axarel® 4100; Axarel® 6100
cleaner, solvent: lubricating greases
Actrel® 1111L; Actrel® 1138L; Actrel® 1160L; Actrel® 3338L; Actrel® 3349L; Actrel® 3360L; Actrel® 4493L
cleaner, solvent: machinery
Ecco Cleaner AK; Eccoflex 090; Ecco Strip LS-W
cleaner, solvent: maintenance applics.
Actrel® 1111L; Actrel® 1138L; Actrel® 1160L; Actrel® 3338L; Actrel® 3349L; Actrel® 3360L; Actrel® 4493L
cleaner, solvent: manual cold degasser
Marthene
cleaner, solvent: metal cleaning
Axarel® 9100
cleaner, solvent: metal cuttings/fines removal
Axarel® 4100; Axarel® 6100
cleaner, solvent: metal surface coating removal
Ecco Cleaner HR-LO
cleaner, solvent: metal surface ink removal
Ecco Cleaner HR-LO
cleaner, solvent: metals
Axarel® 52; Axarel® 56; Partsprep™ Degreaser; T-Det® N-4
cleaner, solvent: nonmetallic surface oil/grease removal
Ecco Cleaner HR-LO
cleaner, solvent: nonmetallic surfaces ink removal
Ecco Cleaner HR-LO
cleaner, solvent: OEM applics.
Actrel® 1111L; Actrel® 1138L; Actrel® 1160L; Actrel® 3338L; Actrel® 3349L; Actrel® 3360L; Actrel® 4493L
cleaner, solvent: oil removal
Axarel® 4100; Axarel® 6100
cleaner, solvent: oils
HFE-71DA
cleaner, solvent: oils/greases
Actrel® 1111L; Actrel® 1138L; Actrel® 1160L; Actrel® 3338L; Actrel® 3349L; Actrel® 3360L; Actrel® 4493L
cleaner, solvent: oil-soluble metalworking cutting oils
Actrel® 1111L; Actrel® 1138L; Actrel® 1160L; Actrel® 3338L; Actrel® 3349L; Actrel® 3360L; Actrel® 4493L
cleaner, solvent: paint pretreatment cleaners/degreasers
Toxfree™
cleaner, solvent: paint removal
Brazilian Gum Turpentine-NT; Ship Shape® Resin Cleaner
cleaner, solvent: paints
Ship Shape® Resin Cleaner

cleansing agent, skin cleansers
HallStar™ DOM
cleansing agent, sunscreens
HallStar™ DOM
cleansing/degreasing metals
1,1,2,2-Tetrachloroethane
cling additive, plastic films
Sorbitan trioleate
cling agent, PVC film
Glyceryl oleate
cloud point depressant
Ammonium xylenesulfonate
cloud point depressant, aerosols
Sodium toluenesulfonate
cloud point depressant, cutting oils
Sodium toluenesulfonate
cloud point depressant, detergents
Sodium toluenesulfonate
cloud point depressant, glues
Sodium toluenesulfonate
cloud point depressant, industrial processes
Sodium toluenesulfonate
cloud point depressant, personal care
Lexol® IPM
cloud point depressant, shampoos
Sodium toluenesulfonate
cloud point depressant, textile finishing
Sodium toluenesulfonate
clouding agent, beverages
NEOBEE® M-5
clouding agent, cosmetics
Coconut (Cocos nucifera) oil
clouding agent, pharmaceuticals
Coconut (Cocos nucifera) oil
clouding agent, vitamins
Liponate GC-K; Liponate PC
CNS stimulant, pharmaceuticals
Dimethylethanolamine
CO formation control agent, ethylene prod.
Dimethyl sulfide
coagulant, latexes
Acetic acid; Filmex® A-2 190 Proof; Filmex® A-2 Anhydrous; Filmex® B 190 Proof; Filmex® B Anhydrous; Filmex® C 190 Proof; Filmex® C Anhydrous; Filmex® D-1; Filmex® D-2 190 Proof; Filmex® D-2 Anhydrous; *Formaldehyde*; *Formic acid*; PPG 2250; *PPG-17*
coagulant, water treatment
Prochinor® SOL
coalescing agent
Diethylene glycol butyl ether acetate; Eastman® DE-HG; Eastman® EB; Purasolv® EL; Purasolv® IPL; Purasolv® ML
coalescing agent, acrylic coatings
Arcosolv® DPNB
coalescing agent, acrylic emulsions
Aromatic 100; Aromatic 150; Aromatic 200
coalescing agent, acrylic latex: emulsion-type floor coatings
2-Pyrol®
coalescing agent, acrylic latexes
2-Pyrrolidone; Texanol® Ester-Alcohol; *2,2,4-Trimethyl-1,3-pentanediol isobutyrate*
coalescing agent, acrylic-styrene copolymers: emulsion-type floor coatings
2-Pyrol®
coalescing agent, adhesives
Arcosolv® PNP; Dalpad A; Dowanol® EPh; *Propylene glycol t-butyl ether*
coalescing agent, agric.
Arcosolv® PNP; Proglyde DMM; *Propylene glycol t-butyl ether*
coalescing agent, alkyds
Propylene glycol t-butyl ether
coalescing agent, aq. architectural paints/coatings
Solvenon® TPnB
coalescing agent, aq. coatings
Coasol™; Dalpad A; Dowanol® EPh
coalescing agent, aq. emulsion systems
Arcosolv® DPMA
coalescing agent, aq. formulations
Purasolv® BL
coalescing agent, aq. industrial paints/coatings
Solvenon® TPnB
coalescing agent, aq. inks
Texanol® Ester-Alcohol; *2,2,4-Trimethyl-1,3-pentanediol isobutyrate*
coalescing agent, aq. latex coatings
Dowanol® DPnB; Dowanol® DPnP; Dowanol® PnP
coalescing agent, aq. paints
Ethylene glycol isopropyl ether; Isopropyl Oxitol; *Propylene glycol t-butyl ether*
coalescing agent, architectural aq. coatings
Dowanol® DPMA; *PPG-2 methyl ether acetate*
coalescing agent, architectural coatings
Lusolvan® FBH
coalescing agent, architectural emulsions/disps.
Eastman® EEH
coalescing agent, architectural latex coatings
Eastman® DP
coalescing agent, architectural latex paints
Eastman® DB
coalescing agent, B/S latexes
Texanol® Ester-Alcohol; *2,2,4-Trimethyl-1,3-pentanediol isobutyrate*
coalescing agent, carpet shampoos
Coasol™
coalescing agent, cleaners
Arcosolv® PNP; Proglyde DMM; *Propylene glycol t-butyl ether*
coalescing agent, coatings
Albrite® Tributyl Phosphate; Arcosolv® PNP; *Diethylene glycol 2-ethylhexyl ether*; Dowanol® PnB; Eastman® EEH; *Ethylene glycol hexyl ether*; Proglyde DMM; *Propylene glycol methyl ether acetate*; *Tricresyl phosphate*
coalescing agent, cosmetics
Arcosolv® PNP; *Propylene glycol t-butyl ether*
coalescing agent, cutting fluids
Propylene glycol t-butyl ether
coalescing agent, electronics
Arcosolv® PNP; *Propylene glycol t-butyl ether*
coalescing agent, electronics cleaners
Proglyde DMM
coalescing agent, EVA latexes
Texanol® Ester-Alcohol; *2,2,4-Trimethyl-1,3-pentanediol isobutyrate*
coalescing agent, felt pens
Ethyl Proxitol
coalescing agent, floor polishes
Eastman® DM; Eastman® DP
coalescing agent, floor waxes
Coasol™
coalescing agent, furniture polishes
Ethyl Proxitol
coalescing agent, hand cleaners
Coasol™
coalescing agent, hard-surface cleaners
Propylene glycol t-butyl ether
coalescing agent, high-solids coatings
Butyl Carbitol®
coalescing agent, industrial aq. coatings
Ecosoft™ PB Solvent; Ecosoft™ PE Solvent; Ecosoft™ PH Solvent
coalescing agent, industrial emulsions/disps.
Eastman® EEH
coalescing agent, industrial latex paints
Eastman® DB
coalescing agent, ink cleaners
Coasol™
coalescing agent, inks
Arcosolv® PNP; *Propylene glycol t-butyl ether*
coalescing agent, interior paints
Lusolvan® FBH
coalescing agent, lacquers
2-Ethylhexyl acetate; *Isopropyl acetate*
coalescing agent, latex adhesives
Dowanol® PPh; *Propylene glycol phenyl ether*
coalescing agent, latex coatings
Diethylene glycol propyl ether; *Propylene glycol phenyl ether*
coalescing agent, latex finishes
Texanol® Ester-Alcohol; *2,2,4-Trimethyl-1,3-pentanediol isobutyrate*
coalescing agent, latex paints
Benzoflex® 284; *Butoxydiglycol*; Butyl Dioxitol®; Eastman® DE; Eastman® DE Acetate; Eastman® EB Acetate; *Ethoxydiglycol acetate*; *Propylene glycol dibenzoate*
coalescing agent, latexes
C8 alkyl acetate; *C9 alkyl acetate*; *C10 alkyl acetate*; *C13 alkyl acetate*; Exxate® 600; Exxate® 700; Exxate® 800; Exxate® 900; Exxate® 1000; Exxate® 1300; *Oxo-heptyl acetate*; *Oxo-hexyl acetate*; Proglyde DMM
coalescing agent, latexes: aq. architectural coatings
Dowanol® PPh
coalescing agent, latexes: industrial coatings
Dowanol® PPh
coalescing agent, leather finishes
Ethyl Proxitol
coalescing agent, light-duty cleaners
Propylene glycol t-butyl ether
coalescing agent, liq. cleaners
Eastman® DP
coalescing agent, low-pH latex paints
Eastman® DB Acetate
coalescing agent, paint removers
Proglyde DMM
coalescing agent, paints/coatings
Coasol™; *2-Ethylhexyl acetate*; *Isopropyl acetate*
coalescing agent, polishes
Coasol™
coalescing agent, polyesters
Propylene glycol t-butyl ether
coalescing agent, printing inks
2-Ethylhexyl acetate; Ethyl Proxitol; *Isopropyl acetate*; *Propylene glycol methyl ether acetate*
coalescing agent, PU coatings
Proglyde DMM
coalescing agent, PVA emulsions
Eastman® DE
coalescing agent, PVAc coatings
Arcosolv® DPNB
coalescing agent, PVAc latex paints
Eastman® DM; *Methoxydiglycol*
coalescing agent, PVAc latexes
Texanol® Ester-Alcohol; *2,2,4-Trimethyl-1,3-pentanediol isobutyrate*
coalescing agent, PVAc-acrylic latexes
Texanol® Ester-Alcohol; *2,2,4-Trimethyl-1,3-pentanediol isobutyrate*
coalescing agent, resin cleaners
Coasol™
coalescing agent, resins
Proglyde DMM
coalescing agent, solvent degreasers
Coasol™
coalescing agent, solvent-based coatings
Proglyde DMM
coalescing agent, stamp pad inks
Ethyl Proxitol
coalescing agent, styrene-acrylic coatings
Arcosolv® DPNB
coalescing agent, styrene-acrylic floor polishes
2-Pyrrolidone
coalescing agent, surface coatings
Ethylan® HB1; Ethyl Proxitol
coalescing agent, textile chemicals
Coasol™
coalescing agent, textiles
Arcosolv® PNP; *Propylene glycol t-butyl ether*
coalescing agent, water-/solvent-borne coatings
Dipropylene glycol n-propyl ether
coalescing agent, water-disp
Butyl Carbitol®; Carbitol® PM-600
coalescing agent, water-disp.
Methyl Carbitol®; Ucar® Filmer IBT
coalescing agent, water-disp. coatings
Butoxyethanol acetate; Butyl Carbitol® Acetate; Butyl Cellosolve® Acetate; Hexyl Cellosolve®
coalescing agent, water-disp./water-soluble coatings
Butyl Cellosolve®
coalescing agent, water-sol
Butyl Carbitol®
coalescing agent, water-soluble coatings
Carbitol® PM-600; Methyl Carbitol®; Ucar® Filmer IBT
coalescing agent, wood paints
Lusolvan® FBH
coalescing agent, wood stains
Ethyl Proxitol; Lusolvan® FBH
coalescing agent,aq. latex adhesives
Phenoxyethanol
coating agent
Dimethyl phthalate
coating agent, asphaltic pavements
Furan
coating agent, citrus fruits
PEG-10; *PEG-18*
coating agent, egg shells
Drakesol® 165; Drakesol® 220; Drakesol® 260; Isopar® V Fluid;

Zinc chloride
conditioner
Lauryl pyrrolidone; *Phenyl trimethicone*
conditioner, acne preps.
Lipolan R
conditioner, aerosols
Cyclomethicone
conditioner, antiperspirants
Carbowax® Sentry® PEG 300; *Cyclomethicone*; *Glycereth-7*; Ritacane ID
conditioner, bath oils
Fancol™ ALA
conditioner, blushes
EmCon™ CO
conditioner, body lotions
Fancol™ ALA
conditioner, cooking oils
Hybrid safflower (Carthamus tinctorius) oil
conditioner, cosmetic creams
Finsolv® TN
conditioner, cosmetic masks
Glycereth-7
conditioner, cosmetic oils
Finsolv® TN
conditioner, cosmetics
Avocado (Persea gratissima) oil; Carbowax® Sentry® PEG 300; Comperlan® COD; Corona Lanolin; Coronet Lanolin; Fancol™ ALA; *Glycereth-7*; *Hybrid safflower (Carthamus tinctorius) oil*; *Isodecyl isononanoate*; *Isopropyl lanolate*; *Laneth-5*; *Laneth-10*; *Laneth-15*; *Lanolin oil*; Lipolan R; *Olive (Olea europaea) oil*; Olive Oil, Refined; Pluracol® E400 NF; Pluracol® E600 NF; Pluracol® E1450 NF; Polypax IPM; Polypax IPP; *PPG-40 diethylmonium chloride*; *Sweet almond (Prunus amygdalus dulcis) oil*; Variquat® CC-42; Wickenol® 136
conditioner, cream rinses
Lactamide MEA
conditioner, creams/lotions
Carbowax® Sentry® PEG 300; *Laneth-5*; *Laneth-10*; *Laneth-15*
conditioner, creme rinses
Incromectant AMEA-70
conditioner, deodorants
Cyclomethicone
conditioner, depilatories
Glycereth-7
conditioner, dermatology prods.
Solulan® 16; Solulan® 75; Solulan® L-575
conditioner, dry cleaning
Variquat® CC-57
conditioner, emulsions
Glucam® E-10
conditioner, eye gels
Glucam® E-10
conditioner, eye makeup
Ritacane IH
conditioner, eye shadow
EmCon™ CO
conditioner, eyeliner
Ritacane ID
conditioner, fabric
Variquat® CC-55
conditioner, face makeup
Ritacane IH
conditioner, face powders
EmCon™ Rice Bran
conditioner, facial cleansers
Glucam® E-10
conditioner, flame retardants
Variquat® CC-57
conditioner, food
Sucrose palmitate
conditioner, fuel
Petroleum distillates
conditioner, germicides
PPG-40 diethylmonium chloride; Variquat® CC-42; Variquat® CC-57
conditioner, hair
Acetamide MEA; *Butyl octyl salicylate*; Ceraphyl® ICA; *Cetyl stearate*; *Cyclomethicone*; *Dimethyl imidazolidinone*; *Dioctyl maleate*; *DM hydantoin*; EmCon™ Olive; EmCon™ SES; EmCon™ SUN; EmCon™ W; Fancorsil A; Fanoliv™ Oil; *Isodecyl isononanoate*; *Isostearyl isononanoate*; *Lactamide DGA*; *Pancreatin*; *PEG-75 lanolin*; *PPG-14 butyl ether*; Ritawax AEO; Ritawax ALA; *Sodium thiocyanate*; STEPAN® IPM-NF; *Triisostearin*
conditioner, hair rinses
Variquat® CC-55
conditioner, hair sprays
Estol IPM 1509 (BIO-IPM)
conditioner, hand cleaners
Carbowax® Sentry® PEG 300
conditioner, hand lotions
Fancol™ ALA
conditioner, household detergents
PEG-75 lanolin
conditioner, industrial
PPG-40 diethylmonium chloride; Variquat® CC-42; Variquat® CC-57
conditioner, inks
Pluracol® E400 NF; Pluracol® E600 NF; Pluracol® E1450 NF
conditioner, lipstick
EmCon™ CO; Lipolan R
conditioner, makeup
Carbowax® Sentry® PEG 300; EmCon™ CO; Fancol™ CB; Fancol™ CB Extra; Lipolan R
conditioner, mascara
Ritacane ID
conditioner, metal cleaners
Pluracol® E400 NF; Pluracol® E600 NF; Pluracol® E1450 NF
conditioner, metal polishes
Pluracol® E400 NF; Pluracol® E600 NF; Pluracol® E1450 NF
conditioner, moisturizing creams/ lotions
EmCon™ Rice Bran; Fancol™ CB; Fancol™ CB Extra
conditioner, nail polishes
EmCon™ CO
conditioner, ointments
EmCon™ Rice Bran; Fancol™ ALA; Lipolan R
conditioner, oral care
Pluracol® E400 NF; Pluracol® E600 NF; Pluracol® E1450 NF
conditioner, OTC pharmaceuticals
Fancol™ CB; Fancol™ CB Extra
conditioner, overtreated hair preps.
Foamid LM2E
conditioner, paper
Pluracol® E400 NF; Pluracol® E600 NF; Pluracol® E1450 NF
conditioner, personal care
Avocado (Persea gratissima) oil; Carbowax® Sentry® PEG 300; Fancol™ ALA; Glucam® E-10; *Glycereth-7*; *PPG-40 diethylmonium chloride*; *Silicone*; Solulan® 16; Solulan® 75; Solulan® C-24; Solulan® L-575; Variquat® CC-42; Variquat® CC-57; Wickenol® 136
conditioner, pharmaceuticals
Avocado (Persea gratissima) oil; Comperlan® COD; Corona Lanolin; Coronet Lanolin; EmCon™ CO; *Hybrid safflower (Carthamus tinctorius) oil*; *Isopropyl lanolate*; *Laneth-5*; *Laneth-10*; Lipolan R; *Olive (Olea europaea) oil*; *Phenyl trimethicone*; Pluracol® E1450 NF
conditioner, pigments
Variquat® CC-57
conditioner, plastics
PPG-40 diethylmonium chloride; Variquat® CC-42; Variquat® CC-57
conditioner, protective coatings
Variquat® CC-57
conditioner, rinses
Lipolan R
conditioner, rubber
Pluracol® E400 NF; Pluracol® E600 NF; Pluracol® E1450 NF
conditioner, shampoos
Carbowax® Sentry® PEG 300; *Cocamide DEA*; Comperlan® COD; Glucam® E-10; *Lactamide MEA*; Lipolan R
conditioner, shaving creams
Carbowax® Sentry® PEG 300; Glucam® E-10
conditioner, shaving preps.
Cyclomethicone; Lipolan R; *PPG-14 butyl ether*
conditioner, shower gels
Glucam® E-10
conditioner, skin
Acetamide MEA; *Bis (2-ethylhexyl) maleate*; *Butyl octyl benzoate*; Ceraphyl® ICA; *Cycloethoxymethicone*; EmCon™ Olive; EmCon™ SES; EmCon™ SUN; EmCon™ W; Fancol™ ALA; Fancol™ Karite Butter; Fancorsil A; Fanoliv™ Oil; Finsolv® PG-22; Finsolv® TPP; *Isodecyl isononanoate*; *Isopropyl PPG-2-isodeceth-7-carboxylate*; *Lauryl oleate*; Lipolan R; *Methyl caprylate/caprate*; *PEG-75 lanolin*; Ritacane ID; Ritawax AEO; Ritawax ALA
conditioner, skin cleansers
EmCon™ Rice Bran
conditioner, skin creams/lotions
Butyl octyl benzoate
conditioner, skin: bath care
Isopropyl PPG-2-isodeceth-7-carboxylate
conditioner, skin: body lotions
Shea butter (Butyrospermum parkii)
conditioner, skin: cleansers
EmCon™ W
conditioner, skin: cosmetics
Behenyl benzoate; *Butyl octyl benzoate*; *Butyl octyl salicylate*; *Diethyl acetyl aspartate*; *Isoeicosane*; *Isopropyl PPG-2-isodeceth-7-carboxylate*; *Lactic acid*; *Methyl benzoate*; *Octamethyltrisiloxane*; *Pancreatin*; *PPG-5-buteth-7*; *PPG-12-buteth-16*; *PPG-20-buteth-30*; *PPG-28-buteth-35*; *PPG-5 butyl ether*; *PPG-18 butyl ether*; *Rice (Oryza sativa) bran oil*
conditioner, skin: lipstick
EmCon™ W
conditioner, skin: makeup
Cocoa (Theobroma cacao) butter
conditioner, skin: moisturizing creams/lotions
EmCon™ W
conditioner, skin: OTC pharmaceuticals
Cocoa (Theobroma cacao) butter
conditioner, skin: pigmented prods.
Isopropyl PPG-2-isodeceth-7-carboxylate
conditioner, skin: shampoos
Shea butter (Butyrospermum parkii)
conditioner, skin: soaps
Shea butter (Butyrospermum parkii)
conditioner, skin: sun care
Isopropyl PPG-2-isodeceth-7-carboxylate
conditioner, skin: sun preps.
Cocoa (Theobroma cacao) butter
conditioner, skin: suntan preps.
Shea butter (Butyrospermum parkii)
conditioner, soap/bath prods.
Lipolan R
conditioner, soil
Sulfur
conditioner, sticks
Cyclomethicone
conditioner, sun care
Lipolan R; Ritacane ID
conditioner, sunscreens
Carbowax® Sentry® PEG 300
conditioner, suntan oils
Fancol™ ALA
conditioner, suntan preps.
Fancol™ CB; Fancol™ CB Extra
conditioner, synthetic fibers
PPG-40 diethylmonium chloride; Variquat® CC-42; Variquat® CC-57
conditioner, textiles
Dibutyl butyl phosphonate; *Dioctyl octylphosphonate*; Pluracol® E200; Pluracol® E300; Pluracol® E400; Pluracol® E400 NF; Pluracol® E600; Pluracol® E600 NF; Pluracol® E1450 NF; Pluracol® E4000; Pluracol® E6000; *PPG-40 diethylmonium chloride*; Variquat® CC-42; Variquat® CC-57
conditioner, toothpaste
Carbowax® Sentry® PEG 300
conditioner, topical pharmaceuticals
Laneth-15; Solulan® 16; Solulan® 75; Solulan® L-575
conditioner, topicals
Acetamide MEA
conditioner, veterinary prods.
Lipolan R
conditioner, wave-setting lotions
Glycereth-7
conditioner, wood prods.
Pluracol® E400 NF; Pluracol® E600 NF; Pluracol® E1450 NF
consistency agent
Arlypon® DV; Arlypon LSE; Arlypon OSO; Arlypon SO; Arlypon SV
consistency agent, bath oils
Cetearyl alcohol
consistency agent, controlled-release solid pharmaceuticals
Renex® PEG 300; Renex® PEG 400; Renex® PEG 600
consistency agent, cosmetics
Cetyl palmitate; *Ethyl linoleate*; *Tallow alcohol*; *Trilaurin*; *Trimyristin*; *Tripalmitin*
consistency agent, dermatologicals
Shea butter (Butyrospermum parkii)
consistency agent, hand lotions
Cetearyl alcohol
consistency agent, laxatives
Renex® PEG 300; Renex® PEG 400; Renex® PEG 600
consistency agent, liq. makeup
Estalan DO
consistency agent, lotions

Methyl caprylate/caprate; *Methyl laurate*
cosolvent, agrochemical toxicants
Agrisol PX401; Agrisol PX413
cosolvent, alcoholic compositions
Procetyl 10
cosolvent, antiperspirant sticks
Crystal® O
cosolvent, antiperspirants
Myritol® PC
cosolvent, aq. alcoholic compositions
Procetyl 10
cosolvent, aq. alcoholic systems
Diisopropyl adipate
cosolvent, baby care
Acetol® 1706
cosolvent, baby oils
Lanolin oil
cosolvent, bath care
Acetol® 1706
cosolvent, bath oils
Cetearyl alcohol; *Decyl oleate*; Estalan-PGDP; Myritol® PC; *Propylene glycol dipelargonate*
cosolvent, brilliantines
Lanolin oil
cosolvent, cleaners
Arcosolv® PTB
cosolvent, cleansers
Octyl hydroxystearate
cosolvent, cleansing lotions
Lanolin oil
cosolvent, coatings
Arcosolv® PTB; NMP
cosolvent, cosmetic creams/lotions
Isopropyl palmitate
cosolvent, cosmetic raw materials
Crodamol IPM
cosolvent, cosmetics
Arcosolv® PTB; Ceraphyl® 140; Ceraphyl® 140-A; Cithrol 4DL; Cithrol 4DO; Corona Lanolin; Coronet Lanolin; Crill 1; Crodamol IPP; Crystal® O; *Decyl oleate*; *Diisopropyl adipate*; *Dimethicone copolyol undecylenate*; *Isodecyl oleate*; *Isopropyl laurate*; *Isopropyl stearate*; *Isostearyl lactate*; *Laneth-5*; *Laneth-10*; *Lanolin*; Myritol® PC; *Octyl hydroxystearate*; *Oleth-2*; *Oleyl alcohol*; *Olive (Olea europaea) oil*; Olive Oil, Refined; *PEG-4 dilaurate*; *PEG-20 dilaurate*; *PEG-4 dioleate*; *PEG-8 dioleate*; *PEG-20 dioleate*; *PPG-10 cetyl ether*; *PPG-30 cetyl ether*; *PPG-50 cetyl ether*; Procetyl 10; *Propylene glycol dipelargonate*; *Tallow alcohol*; Wickenol® 158; Wickenol® 171
cosolvent, cosmetics sticks
Tridecyl neopentanoate
cosolvent, cream shampoos
Lanolin acid
cosolvent, creams
Nikkol BWA-10
cosolvent, creams/lotions
Acetol® 1706; Ameroxol® OE-2; *Decyl oleate*; Estalan-PGDP; *Laneth-5*; *Laneth-10*; Naturechem® MAR; Schercemol PGDP
cosolvent, cutting oils
Oleyl alcohol
cosolvent, defoamers
2-Ethylhexoic acid
cosolvent, deodorants
Isopropyl palmitate
cosolvent, detergent sterilizers
Ethylan® HB4
cosolvent, detergents
Ameroxol® OE-2; *Oleth-2*
cosolvent, d-limonene
FAME
cosolvent, dyes
Alcohol
cosolvent, electronics
Arcosolv® PTB
cosolvent, essential oils
Prostearyl 15
cosolvent, fatty acid systems
Lanolin acid
cosolvent, fatty compds.
Promyristyl PM-3
cosolvent, fiber lubricants
PEG-4 dioleate
cosolvent, flavors
NEOBEE® M-5
cosolvent, food
Crill 1; *Oleyl alcohol*; *PEG-4 dilaurate*; *PEG-20 dilaurate*; *PEG-4 dioleate*; *PEG-8 dioleate*; *PEG-20 dioleate*
cosolvent, food colors
NEOBEE® M-5
cosolvent, food pkg.
Crill 1
cosolvent, food processing
Methyl cottonseed; Norfox® MCS
cosolvent, fragrances
Ameroxol® OE-2; NEOBEE® M-5; *Oleth-2*
cosolvent, germicides
Cardolite® NC-510; *3-(n-Pentadecyl) phenol*
cosolvent, glass cleaners
Ethylan® HB4
cosolvent, glosses
Propylene glycol dipelargonate
cosolvent, hair care
Acetol® 1706; *Isopropyl stearate*; *Lactamide DGA*
cosolvent, hair creams/lotions
Foamid LM2E
cosolvent, hair tonics
Diisopropyl adipate
cosolvent, hand cleaners
Decyl oleate
cosolvent, hand lotions
Cetearyl alcohol
cosolvent, hard-surface cleaners
Arcosolv® PTB
cosolvent, herbicides
Crill 1
cosolvent, high-solids inks
Sylfat® 9012
cosolvent, household prods.
Lanolin acid
cosolvent, hydrogen peroxide prod.
Trioctyl phosphate
cosolvent, industrial
Cithrol 4DL; Cithrol 4DO; *Isopropyl palmitate*; Norfox® MCS; *PEG-8 dilaurate*
cosolvent, industrial cleaning
NMP
cosolvent, industrial lubricants
PEG-8 dilaurate
cosolvent, industrial processing
Methyl cottonseed
cosolvent, inks
Alcohol; Arcosolv® PTB; Crill 1; *Oleyl alcohol*
cosolvent, insecticides
Cardolite® NC-510; *Cashew nut shell oil*; Crill 1; *3-(n-Pentadecyl) phenol*
cosolvent, iodophors
Ethylan® HB4
cosolvent, lacquer thinners
Alcohol
cosolvent, latent: NC
Pentanol 45
cosolvent, leather
Oleth-2
cosolvent, leather treatment
Crill 1
cosolvent, light-duty cleaners
Arcosolv® PTB
cosolvent, lip balm
Naturechem® MAR
cosolvent, lipophilic components
HallStar™ DIPA; HallStar™ DOA
cosolvent, lipstick
Lanolin oil; Naturechem® MAR; *Propylene glycol dipelargonate*
cosolvent, liq. makeup
Ceraphyl® 140-A
cosolvent, low-VOC lithographic inks
Sylfat® 9012
cosolvent, makeup
Crystal® O
cosolvent, makeup bases
Propylene glycol dipelargonate
cosolvent, makeup remover
Estalan-PGDP
cosolvent, medicinals
NEOBEE® M-5
cosolvent, metalworking
PEG-4 dilaurate; *PEG-4 dioleate*; *PEG-8 dioleate*
cosolvent, metalworking fluids
Crill 1
cosolvent, milk lotions
Nikkol BWA-10
cosolvent, mineral extraction
dl-Limonene
cosolvent, moisturizers
Decyl oleate
cosolvent, mold cleaning
NMP
cosolvent, nonmiscible oils
Laneth-15; Polychol 5; Polychol 15
cosolvent, oil slick dispersing
Crill 1
cosolvent, oils
Tridecyl neopentanoate
cosolvent, oils/waxes in anhyd. sticks
Trivent NP-13
cosolvent, organic compds.
Glyceryl caprylate/caprate
cosolvent, paint removal
But-2-yne-1,4-diol
cosolvent, paint/varnish cleaning
NMP
cosolvent, paint/varnish stripping
NMP
cosolvent, paints
Alcohol; Crill 1; *Oleyl alcohol*
cosolvent, paper
Crill 1; *Oleth-2*
cosolvent, perfumed bath oils
Schercemol PGDP
cosolvent, perfumes
Estalan-ISI; Estalan-PGDP; *Isostearyl isostearate*; *Oleyl oleate*; *PEG-20 dioleate*; Prostearyl 15; Schercemol 1818; Schercemol OLO
cosolvent, personal care
Ceraphyl® 140; *Glyceryl dicaprylate/caprate*; *Isopropyl stearate*; Lipolan R; Naturechem® MAR; *PEG-8 dilaurate*; *PPG-10 cetyl ether*; *PPG-20 cetyl ether*; Procetyl 10; *Propylene glycol dicaprylate/dicaprate*; Ritawax AEO; Ritawax ALA
cosolvent, pesticides
Dimethyl caproamide; *Dimethyl octanamide*; *Isopropyl alcohol*; *2-Ethylhexoic acid*
cosolvent, petrochemical processing
NMP; *Oleyl alcohol*
cosolvent, pharmaceutical creams
Cetearyl alcohol
cosolvent, pharmaceuticals
Ameroxol® OE-2; Ceraphyl® 140; Corona Lanolin; Coronet Lanolin; Crill 1; Crystal® O; *Decyl oleate*; *Dimethicone copolyol undecylenate*; *Dioctyl adipate*; *Isopropyl laurate*; *Isostearyl lactate*; *Laneth-5*; *Laneth-10*; Myritol® PC; *Oleth-2*; *Olive (Olea europaea) oil*; *PEG-4 dilaurate*; *PEG-8 dilaurate*; *PEG-20 dilaurate*; *PEG-4 dioleate*; *PEG-8 dioleate*; *PEG-20 dioleate*; *PPG-30 cetyl ether*; *Propylene glycol dipelargonate*; Ritawax AEO; Ritawax ALA
cosolvent, pigment: antiperspirant sticks
Crystal® Crown
cosolvent, pigment: cosmetics
Crystal® Crown
cosolvent, pigment: creams/lotions
Naturechem® GTR
cosolvent, pigment: lip balm
Naturechem® GTR
cosolvent, pigment: lipstick
Crystal® Crown; Naturechem® GTR
cosolvent, pigment: makeup
Crystal® Crown
cosolvent, pigment: personal care
Crystal® Crown
cosolvent, pigment: pharmaceuticals
Crystal® Crown
cosolvent, pigment: soaps
Crystal® Crown; Naturechem® GTR
cosolvent, pigment: sunscreens
Naturechem® GTR
cosolvent, plastics
Crill 1; *Oleyl alcohol*; *PEG-4 dilaurate*; *PEG-20 dilaurate*; *PEG-4 dioleate*; *PEG-8 dioleate*; *PEG-20 dioleate*
cosolvent, polar additives: low polarity systems
Acetylated lard glyceride; *Acetylated palm kernel glycerides*
cosolvent, polishes
Crill 1
cosolvent, preshave lotions
Propylene glycol dipelargonate
cosolvent, pressed powder
Decyl oleate
cosolvent, printing inks
Norfox® MCS
cosolvent, pulp/paper
Oleyl alcohol
cosolvent, rinse aids
Ethylan® HB4
cosolvent, shampoos
Ameroxol® OE-2; Nikkol BWA-5
cosolvent, shaving creams
Cetearyl alcohol
cosolvent, shaving lotions
Diisopropyl adipate
cosolvent, skin care
Isopropyl stearate; *Lactamide DGA*; Myritol® PC
cosolvent, skin creams/lotions
Foamid LM2E; *Tallow alcohol*
cosolvent, soaps
Crystal® O; *Isopropyl stearate*; Naturechem® MAR
cosolvent, soybean oil inks
Sylfat® 9012
cosolvent, Sulfinol gas desulfurization process
Sulfolane
cosolvent, sunscreens
Isopropyl stearate; Naturechem® MAR

PPG-3 methyl ether
coupling agent, food
Captex® 200; *Nonoxynol-6*; *Oleyl alcohol*
coupling agent, food processing
Carbowax® PEG 300
coupling agent, foundations
PPG-10 butanediol
coupling agent, fragrance oils
Captex® 200; *Propylene glycol dicaprylate/dicaprate*
coupling agent, fragrance sticks
Promyristyl PM-3
coupling agent, fragrances
Dioctyl maleate; HallStar™ DIPA; Lexol® PG-865; *PPG-15 stearyl ether*; Prostearyl 15
coupling agent, fresheners
HallStar™ DIPA
coupling agent, fuel additives
PPG-3 methyl ether
coupling agent, functional fluids
PPG-3 methyl ether
coupling agent, glass cleaners
Ethylan® HB4; *Ethylene glycol isopropyl ether*
coupling agent, glues
Sodium toluenesulfonate; *Sodium xylenesulfonate*
coupling agent, grease/paint removers
Dowanol® DPM; Dowanol® DPnB; Dowanol® PnB; *Propylene glycol n-propyl ether*
coupling agent, greases
Dipropylene glycol t-butyl ether, *PPG-3 methyl ether*
coupling agent, hair care
Pelemol® G7A; Promyristyl PM-3
coupling agent, hair dressing
Carbowax® Sentry® PEG 400; Carbowax® Sentry® PEG 540 Blend; Carbowax® Sentry® PEG 3350; Carbowax® Sentry® PEG 8000
coupling agent, hair tonics
Diisopropyl adipate
coupling agent, hand cleaners
Carbowax® Sentry® PEG 300; Carbowax® Sentry® PEG 540 Blend
coupling agent, hard surface cleaners
Dowanol® DPnB; Dowanol® PnB; Ecosoft™ PB Solvent; Ecosoft™ PE Solvent; Ecosoft™ PH Solvent
coupling agent, hard-surface cleaners
Arcosolv® PTB; *Butoxypropanol*; *Propylene glycol n-propyl ether*
coupling agent, heavy-duty cleaners
STEPANATE® AXS
coupling agent, heavy-duty detergent
Sodium xylenesulfonate
coupling agent, household cleaners
Dowanol® TPM; *Methoxydiglycol*; *Methoxyisopropanol*; *Phenoxyethanol*
coupling agent, household cleaners,
Propylene glycol n-propyl ether
coupling agent, household prods.
Carbowax® PEG 300; *Nonoxynol-4*; *Nonoxynol-6*
coupling agent, household/I&I cleaners
Dowanol® DPM; Dowanol® DPMA; Dowanol® DPnB; Dowanol® PM; Dowanol® PMA; Dowanol® PnB
coupling agent, household/I&I cleaning
Dalpad A; Dowanol® EPh
coupling agent, household/industrial cleaners
Ecosoft™ PB Solvent; Ecosoft™ PE Solvent; Ecosoft™ PH Solvent
coupling agent, hydroalcoholic: aftershaves
HallStar™ DIPS
coupling agent, hydroalcoholic: bath care
HallTress™ DIBA Special
coupling agent, hydroalcoholic: cleansers
HallStar™ DIPS
coupling agent, hydroalcoholic: clear sticks/gels
HallStar™ DIPS
coupling agent, hydroalcoholic: colognes
HallStar™ DIPS; HallTress™ DIBA Special
coupling agent, hydroalcoholic: fragrances
HallStar™ DOA
coupling agent, hydroalcoholic: fresheners
HallStar™ DIPS
coupling agent, hydroalcoholic: hair sprays
HallTress™ DIBA Special
coupling agent, hydroalcoholic: makeup
HallStar™ DIPS
coupling agent, hydroalcoholic: nail polish/enamels
HallTress™ DIBA Special
coupling agent, hydroalcoholic: perfumes
HallStar™ DIPS
coupling agent, hydroalcoholic: skin creams/lotions
HallStar™ DIPS
coupling agent, hydroalcoholic: skin lotions
HallTress™ DIBA Special
coupling agent, hydroalcoholic: sunscreens
HallStar™ DIPS
coupling agent, hydroalcoholic: toilet water
HallStar™ DIPS; HallTress™ DIBA Special
coupling agent, hydroalcoholic: tonics
HallStar™ DIPS
coupling agent, immiscible liqs.
PEG-3 dimethyl ether
coupling agent, industrial
Nonoxynol-4; *Nonoxynol-6*; *PEG-8 dilaurate*
coupling agent, industrial cleaners
Dowanol® TPM; *Methoxydiglycol*; *Methoxyisopropanol*; *Phenoxyethanol*; *Propylene glycol n-propyl ether*
coupling agent, industrial lubricants
PEG-8 dilaurate
coupling agent, industrial processes
Sodium toluenesulfonate; *Sodium xylenesulfonate*
coupling agent, inks
Ammonium xylenesulfonate; Arcosolv® PNB; Arcosolv® PTB; *Butoxypropanol*; Dowanol® DPM; Jeffox® WL-440; Jeffox® WL-660; Jeffox® WL-1400; Jeffox® WL-5000; *Oleyl alcohol*; Pluracol® E400 NF; Pluracol® E600 NF; Pluracol® E1450 NF
coupling agent, insect repellents
PPG-3 methyl ether; Promyristyl PM-3
coupling agent, iodophors
Ethylan® HB4
coupling agent, lacquers
Butoxyethanol; *Butoxyethanol acetate*; *Diethylene glycol benzyl ether*; *Ethylene glycol benzyl ether*; *PPG-3 methyl ether*
coupling agent, latexes
Proglyde DMM
coupling agent, leather
Sorbitan trioleate
coupling agent, leather dyeing
PPG-3 methyl ether
coupling agent, light-duty cleaners
Arcosolv® PTB; *Butoxypropanol*
coupling agent, lipstick
Carbowax® Sentry® PEG 400; Carbowax® Sentry® PEG 600; HallStar™ DOA
coupling agent, liq. cleansing creams
Ethyl hexanediol
coupling agent, lotions
PPG-10 butanediol
coupling agent, lubricants
Carbowax® PEG 300; *PEG-7 glyceryl cocoate*
coupling agent, makeup
Carbowax® Sentry® PEG 300; Carbowax® Sentry® PEG 400; Carbowax® Sentry® PEG 600; Carbowax® Sentry® PEG 3350; HallStar™ DIPA; HallStar™ DOA; Promyristyl PM-3
coupling agent, makeup removers
Carbowax® Sentry® PEG 400; *PPG-10 butanediol*
coupling agent, medicinals
Captex® 200; Lexol® PG-865; *Propylene glycol dicaprylate/dicaprate*
coupling agent, men's grooming aids
Diisopropyl sebacate
coupling agent, metal cleaners
Dowanol® DPM; Dowanol® DPnB; Dowanol® PnB; *Ethylene glycol isopropyl ether*; Pluracol® E400 NF; Pluracol® E600 NF; Pluracol® E1450 NF; *Propylene glycol n-propyl ether*
coupling agent, metal fabrication
Carbowax® PEG 300
coupling agent, metal polishes
Pluracol® E400 NF; Pluracol® E600 NF; Pluracol® E1450 NF
coupling agent, metal/glass cleaners
BDGE; BGE; EDGE; EGE; i-PGE; MDGE; MGE
coupling agent, metalworking
Nonoxynol-4; *Nonoxynol-6*; Oleocal ME-70; Oleocal ME-112; Oleocal ME-130
coupling agent, mining
PPG-3 methyl ether
coupling agent, moisturizers
PPG-2 myristyl ether propionate
coupling agent, mold release compds.
PEG-7 glyceryl cocoate
coupling agent, nail polish removers
Carbowax® Sentry® PEG 400; Carbowax® Sentry® PEG 3350; Carbowax® Sentry® PEG 8000; HallStar™ DOA
coupling agent, nail polishes
HallStar™ DOA
coupling agent, natural fats
Sorbitan trioleate
coupling agent, natural waxes
Sorbitan trioleate
coupling agent, NC
Eastman® DE
coupling agent, oil field
PPG-3 methyl ether
coupling agent, oil into water
Eastman® EB
coupling agent, oil: bath oils
Fancol™ ALA
coupling agent, oil: body lotions
Fancol™ ALA
coupling agent, oil: cosmetics
Fancol™ ALA
coupling agent, oil: hand lotions
Fancol™ ALA
coupling agent, oil: ointments
Fancol™ ALA
coupling agent, oil: personal care
Fancol™ ALA
coupling agent, oil: skin care
Fancol™ ALA
coupling agent, oil: suntan oils
Fancol™ ALA
coupling agent, 'oil-free' pigmented prods.
Foamid LM2E
coupling agent, oils
Dipropylene glycol t-butyl ether, *PPG-3 methyl ether*
coupling agent, ointments
Carbowax® Sentry® PEG 540 Blend; Carbowax® Sentry® PEG 1000; Carbowax® Sentry® PEG 1450
coupling agent, oral care
Pluracol® E400 NF; Pluracol® E600 NF; Pluracol® E1450 NF
coupling agent, ore flotation
Jeffox® WL-440; Jeffox® WL-660; Jeffox® WL-1400; Jeffox® WL-5000
coupling agent, organic compds.
Glyceryl caprylate/caprate
coupling agent, organic dyestuffs
Pilot® SXS-40
coupling agent, organic polymers
Pilot® SXS-40
coupling agent, oven cleaners
Dowanol® TPnB; *Tripropylene glycol n-butyl ether*
coupling agent, paint removers
Dowanol® TPM; Proglyde DMM
coupling agent, paints
Butoxyethanol; *Butoxyethanol acetate*; Carbowax® PEG 300; *Diethylene glycol benzyl ether*; *Ethylene glycol benzyl ether*; *Nonoxynol-4*; *Nonoxynol-6*; *Oleyl alcohol*
coupling agent, paints/coatings
PPG-3 methyl ether
coupling agent, paper
Carbowax® PEG 300; *Nonoxynol-4*; *Nonoxynol-6*; Pluracol® E400 NF; Pluracol® E600 NF; Pluracol® E1450 NF
coupling agent, penetrating oils
Dowanol® TPM
coupling agent, perfumes
HallStar™ DIPA; *PPG-5-ceteth-20*; Schercemol PGML
coupling agent, personal care
Ammonium xylenesulfonate; Carbowax® PEG 300; Carbowax® Sentry® PEG 300; Carbowax® Sentry® PEG 400; Carbowax® Sentry® PEG 540 Blend; Carbowax® Sentry® PEG 600; Carbowax® Sentry® PEG 900; Carbowax® Sentry® PEG 1000; Carbowax® Sentry® PEG 1450; Carbowax® Sentry® PEG 3350; Carbowax® Sentry® PEG 4600; Carbowax® Sentry® PEG 8000; *Dipropylene glycol dimethyl ether*, *Dipropylene glycol n-propyl ether*,

curing agent, epoxies: polymer concrete
Jeffamine® D-230
curing agent, epoxies: pottingcasting
Jeffamine® D-230
curing agent, epoxies: reinforced composites
Jeffamine® D-230
curing agent, epoxies: surface coatings
Jeffamine® D-230
curing agent, epoxies: syntactic foam
Jeffamine® D-230
curing agent, fats/oils
Eastman® Glacial Acetic Acid Food Grade, Kosher
curing agent, food
Calcium chloride
curing agent, food processing
Eastman® Glacial Acetic Acid Food Grade, Kosher
curing agent, frozen dairy prods.
Eastman® Glacial Acetic Acid Food Grade, Kosher
curing agent, gelatin
Eastman® Glacial Acetic Acid Food Grade, Kosher
curing agent, milk prods.
Eastman® Glacial Acetic Acid Food Grade, Kosher
curing agent, polyesters
Phenylethylethanolamine
curing agent, polyestes
Phenylethylethanolamine
curing agent, polymers
Triethylamine
curing agent, processed fruit
Eastman® Glacial Acetic Acid Food Grade, Kosher
curing agent, PU
Dimethylethanolamine
curing agent, PU resins: food-contact
Tetrahydroxypropyl ethylenediamine; *Triethanolamine*
curing agent, puddings
Eastman® Glacial Acetic Acid Food Grade, Kosher
curing agent, R.T.: protective coating systems, two-pkg.
Tetraethylenepentamine
curing agent, relishes
Eastman® Glacial Acetic Acid Food Grade, Kosher
curing agent, resins
Diethylaminoethanol
curing agent, rubber
Piperidine
curing agent, snack food
Eastman® Glacial Acetic Acid Food Grade, Kosher
curing agent, soft beverages
Eastman® Glacial Acetic Acid Food Grade, Kosher
curing agent, soft candy
Eastman® Glacial Acetic Acid Food Grade, Kosher
curing agent, sweet sauces
Eastman® Glacial Acetic Acid Food Grade, Kosher
curing agent, thermosetting resins
Methanesulfonic acid
curing agent, two-pkg. protective coatings
Diethylenetriamine
curing catalyst, foundry cores
Mod Acid
cutting agent, pyroxylin
Ethyl propionate
cutting oil fluid
SD alcohol 1; *SD alcohol 3-A*; *SD alcohol 3-C*
dampproofer, concrete
Butyl stearate
deactivator, chromatographic support materials
Hexamethyldisilazane
deactivator, pesticides
Dowanol® PM; Eastman® DB; Eastman® DE; Eastman® DM
deaerating agent
Acetylated hydrogenated soybean oil glyceride; *Acetylated lard glyceride*
dechlorination agent, effluent
Sulfur dioxide
dechlorination agent, potable water
Sulfur dioxide
dechlorination agent, sewage
Sulfur dioxide
dechlorination agent, water treatment
Sulfur dioxide
decolorizing agent, wood rosin
QO® Furfural
dedusting agent
Surphos 620; Surphos 645
defoamer
Alfol® 1412; *Caprylic alcohol*; Carbowax® Sentry® PEG 300; *2-Ethylhexanol*; *Hexylene glycol*; Hydrol H660; Hydrol H660YA; Hydrol H5100; Hydrol L385; Isopar® I Fluid; Isopar® J Fluid; Jeffox® PPG-400; Jeffox® PPG-2000; *Morpholine*; *Nonoxynol*; *Poloxamer 181*; *PPG-20*; *PPG-26*; *PPG-30*; *PPG-12-buteth-16*; *PPG-28-buteth-35*; *PPG-33-buteth-45*; *Tallow alcohol*
defoamer, acid gas treatment
Polyglycol P112-2; *PPG-66-glycereth-12*
defoamer, acute pulmonary edema extractants
n-Butyl acetate; *2-Ethylhexanol*; *Isobutyl acetate*
defoamer, adhesives
Albrite® Tributyl Phosphate; Reomol® TBP; *Tributoxyethyl phosphate*; *Tributyl phosphate*
defoamer, agric.
Nonoxynol-4; *Nonoxynol-5*; *Nonoxynol-6*; *PEG-4 dilaurate*; *PEG-8 dioleate*; *PEG-4 oleate*; *Silicone*
defoamer, antifreeze
Silicone; *Tributoxyethyl phosphate*; *Tributyl phosphate*
defoamer, aq. emulsion paints
Reomol® TBP; *Tributoxyethyl phosphate*
defoamer, aq. paints
Tributyl phosphate
defoamer, aq. systems
Polyglycol P15-200; *PPG-24-glycereth-24*; Texanol® Ester-Alcohol; *2,2,4-Trimethyl-1,3-pentanediol isobutyrate*
defoamer, beet sugar processing
PEG-8 dioleate; *PEG-12 dioleate*; *Petrolatum*; *Polypropylene glycol*; *PPG-20*; *PPG-26*
defoamer, boiler defoaming
Jeffox® WL-440; Jeffox® WL-660; Jeffox® WL-1400; Jeffox® WL-5000
defoamer, cellophane: food pkg.
Caprylic alcohol
defoamer, cements
Reomol® TBP; *Tributoxyethyl phosphate*; *Tributyl phosphate*
defoamer, charcoal lighter fluids
PD-23; PD-25; PD-28
defoamer, chemical mfg.
Polyglycol P112-2
defoamer, chemical processing
PPG-5-buteth-7
defoamer, chemicals
PPG-66-glycereth-12
defoamer, cleaners
PD-23; PD-25; PD-28
defoamer, cleaning oils
Petroleum distillates
defoamer, coatings
Drakesol® 165; Drakesol® 165AT; Drakesol® 205; Drakesol® 220; Drakesol® 260; Drakesol® 260 AT; Drakesol® 305
defoamer, coatings: food grade
Dow EP530
defoamer, coatings: food-contact
Butyl stearate; *Carbomer*; Carbopol® 940; Carbopol® 980; Condea Vista LPA; Condea Vista LPA-142; Condea Vista LPA-170; *Cyclohexanol*; *Dipropylene glycol*; *PEG-4 dilaurate*; *PEG-8 dioleate*; *PEG-8 oleate*; *PEG-2 stearate*
defoamer, coatings: food-grade
PPG-17
defoamer, cold applics.
PPG-5-buteth-7
defoamer, cosmetic creams/lotions
Deodorized kerosene; PD-23
defoamer, cosmetics
Nonoxynol-4; *Nonoxynol-5*; *Nonoxynol-6*; *PEG-4 dilaurate*; *PEG-8 dioleate*; *PEG-4 oleate*; Pluracol® E400 NF; Pluracol® E600 NF
defoamer, creams/lotion
Petroleum distillates
defoamer, cutting fluids
Polyglycol P112-2; *PPG-66-glycereth-12*
defoamer, cutting oils
Nonoxynol-4; *Nonoxynol-5*
defoamer, detergents
Tributoxyethyl phosphate; *Tributyl phosphate*
defoamer, drilling fluids
Polyglycol P112-2; *PPG-66-glycereth-12*
defoamer, drilling muds
Reomol® TBP; Texanol® Ester-Alcohol; *Tributoxyethyl phosphate*; *Tributyl phosphate*; *2,2,4-Trimethyl-1,3-pentanediol isobutyrate*
defoamer, dye baths
Reomol® TBP; *Tributoxyethyl phosphate*; *Tributyl phosphate*
defoamer, dyes
Jeffox® WL-440; Jeffox® WL-660; Jeffox® WL-1400; Jeffox® WL-5000; Pluracol® SRF-2; Pluracol® W3520N; Pluracol® WI 285; Pluracol® WI 625; Pluracol® WI 1715; Pluracol® WS 170; Pluracol® WS 260; Pluracol® WS 660; Pluracol® WS 2000; Pluracol® WS 3520; Pluracol® WS 5100; Pluracol® WT 1400; Pluracol® WT 9150; Pluracol® WT 90,000
defoamer, effluents
Silicone
defoamer, electroplating baths
Dimethyl octynediol
defoamer, emulsion polymerization
Nonoxynol-5
defoamer, fat liquoring
Nonoxynol-4
defoamer, fermentation
Carbowax® E300 NF; Carbowax® E400 NF; Carbowax® E600 NF; Carbowax® E900 NF; Carbowax® E1000 NF; Carbowax® E1450 NF; Carbowax® E3350 NF; Carbowax® E4500 NF; Carbowax® E8000 NF; *Methyl eicosanoate*; *Methyl stearate*; Polyglycol P112-2; *PPG-66-glycereth-12*; *Tributoxyethyl phosphate*; *Tributyl phosphate*
defoamer, fiber lubricants
PD-23; PD-25; PD-28
defoamer, food
Carbowax® E300 NF; Carbowax® E400 NF; Carbowax® E600 NF; Carbowax® E900 NF; Carbowax® E1000 NF; Carbowax® E1450 NF; Carbowax® E3350 NF; Carbowax® E4500 NF; Carbowax® E8000 NF; Condea Vista 47; *Dimethicone*; Drakesol® 165; Drakesol® 220; Drakesol® 260; *Glyceryl oleate*; *Mineral oil*; *Nonoxynol-6*; *PEG-4*; *PEG-6*; *PEG-8*; *PEG-9*; *PEG-12*; *PEG-14*; *PEG-16*; *PEG-20*; *PEG-32*; *PEG-40*; *PEG-75*; *PEG-100*; *PEG-150*; *PEG-200*; *PEG-4 dilaurate*; *PEG-8 dioleate*; *PEG-4 oleate*; *Petroleum distillates, hydrotreated light*; *Polypropylene glycol*; *Polysorbate 80*; *PPG-5-buteth-7*; *PPG-20-buteth-30*; *PPG-28-buteth-35*; *Propylene glycol dicaprylate/dicaprate*; *Propylene glycol laurate*; *Propylene glycol stearate*
defoamer, food pkg.
Glyceryl oleate
defoamer, food processing
Carbowax® Sentry® PEG 400; Carbowax® Sentry® PEG 540 Blend; Carbowax® Sentry® PEG 600; Carbowax® Sentry® PEG 900; Carbowax® Sentry® PEG 1000; Carbowax® Sentry® PEG 1450; Carbowax® Sentry® PEG 3350; Carbowax® Sentry® PEG 4600; Carbowax® Sentry® PEG 8000; *C7-8 isoparaffin*; *C8-9 isoparaffin*; *C9-11 isoparaffin*; *C9-13 isoparaffin*; *C10-11 isoparaffin*; *C11-12 isoparaffin*; *C11-13 isoparaffin*; *C12-14 isoparaffin*; *C13-14 isoparaffin*; *C13-16 isoparaffin*; *C20-40 isoparaffin*; Dow EP530; *PEG-4 dioleate*; Polyglycol P1200; Polyglycol P2000; *PPG-33-buteth-45*; *Silicone*
defoamer, food-contact
PD-23; PD-25; PD-28; *PPG-15 butyl ether*; Ucon® LB-285; Ucon® LB-385; Ucon® LB-525; Ucon® LB-625; Ucon® LB-1145; Ucon® LB-1715; Ucon® LB-3000
defoamer, food-contact coatings
Alcohol; *Deodorized kerosene*; *Dimethicone*; *Hexylene glycol*; *Isobutyl alcohol*; *Isopropyl alcohol*; *Kerosene*; *Menhaden oil*; *Methyl alcohol*; *Methyl caprylate*; *Methyl caprylate/caprate*; *Methyl laurate*; *Methyl myristate*; *Methyl stearate*; *Mineral oil*; *Myristyl alcohol*; *Naphtha*; *Nonylphenol*; *Paraffin*; *PEG-12 dioleate*; *Petrolatum*; *Pine (Pinus palustris) oil*; *Poloxamer 181*; *PPG-26*; *PPG-24 butyl ether*; *PPG-33 butyl ether*; *PPG-40 butyl ether*; *Stearyl alcohol*; *Triethanolamine*
defoamer, food-contact paper
Hodag® S-35 FGK
defoamer, food-contact paper coatings

Butyl alcohol; *t-Butyl alcohol*; *Ceteth-16*; *Ceteth-24*; *Cetyl alcohol*; *Coconut (Cocos nucifera) oil*; *Corn (Zea mays) oil*; *Cottonseed (Gossypium) oil*; *Cyclohexane*; *Glyceryl stearate*; *Hexane*; *Methyl cocoate*; *Methyl oleate*; *Methyl palmitate*; *Oleic acid*; *Oleth-16*; *Olive (Olea europaea) oil*; *Peanut (Arachis hypogaea) oil*; *PEG-20 oleate*; *PPG-20*; *PPG-22 butyl ether*; *PPG-53 butyl ether*; *Rice (Oryza sativa) bran oil*; *Safflower (Carthamus tinctorius) oil*; *Sesame (Sesamum indicum) oil*; *Soybean (Glycine soja) oil*; Soygold® 1000; Soygold® 2000; *Tall oil*

defoamer, food-contact paper/paperboard
n-Amyl alcohol; *BHA*; *BHT*; *Butoxyethanol*; *Butyl oleate*; *Cocamide DEA*; *Diethanolamine*; *Diethylenetriamine*; *Diisopropanolamine*; *2,6-Dimethyl heptanol-4*; *Dodecylbenzenesulfonic acid*; *Dodoxynol-7*; *Ethanolamine*; *2-Ethylhexanol*; *Formaldehyde*; *Glyceryl caprylate*; *Glyceryl caprylate/caprate*; *Glyceryl dioleate*; *Glyceryl laurate*; *Glyceryl myristate*; *Glyceryl oleate*; *Glycol distearate*; *Glycol stearate*; *Hydrogenated menhaden oil*; *Isobutyl palmitate*; *Isopropylamine dodecylbenzenesulfonate*; *Isopropyl laurate*; *Isopropyl myristate*; *Isopropyl oleate*; *Isopropyl palmitate*; *Isopropyl stearate*; *Lanolin*; *Morpholine*; *Nonoxynol-5*; *Nonoxynol-6*; *Oleyl alcohol*; *PEG-33 castor oil*; *PEG-8 dilaurate*; *PEG-12 dilaurate*; *PEG-20 dilaurate*; *PEG-4 dioleate*; *PEG-20 dioleate*; *PEG-4 oleate*; *PEG-6 oleate*; *PEG/PPG-17/6 copolymer*; *PEG-40 sorbitan peroleate*; *PEG-100 stearate*; *Petroleum distillates*; *PPG-2-buteth-3*; *PPG-3-buteth-5*; *PPG-5-buteth-7*; *PPG-7-buteth-10*; *PPG-9-buteth-12*; *PPG-12-buteth-16*; *PPG-20-buteth-30*; *PPG-28-buteth-35*; *PPG-33-buteth-45*; *PPG-18 butyl ether*; *Propylene glycol dicaprylate*; *Propylene glycol dicaprylate/dicaprate*; *Propylene glycol stearate*; *Propylene oxide*; *Rapeseed (Brassica campestris) oil*; *Sodium lauryl sulfate*; *Terpene resin*; *Tetrahydrofurfuryl alcohol*; *Tributoxyethyl phosphate*; *Tributyl phosphate*; *Triethanolamine*; *Trioctyl phosphate*

defoamer, food-grade coatings
Polyglycol P1000TB; Polyglycol P1200; Polyglycol P2000

defoamer, food-grade paper coatings
Polyglycol P4000

defoamer, food-grade paper prods.
Polyglycol P1000TB; Polyglycol P1200; Polyglycol P2000

defoamer, fracturing fluids
Reomol® TBP; *Tributoxyethyl phosphate*

defoamer, fruit processing
Petroleum distillates

defoamer, general purpose
Macol® 85

defoamer, glass luminescent coatings
Reomol® TBP

defoamer, hot applics.
PPG-5-buteth-7

defoamer, household cleaners
Nonoxynol-5

defoamer, household prods.
Erucamide; *Nonoxynol-4*; *Nonoxynol-6*

defoamer, industrial
Erucamide; *Nonoxynol-4*; *Nonoxynol-6*

defoamer, industrial cleaners
Nonoxynol-5

defoamer, inks
Jeffox® WL-440; Jeffox® WL-660; Jeffox® WL-1400; Jeffox® WL-5000; Pluracol® E400 NF; Pluracol® E600 NF; Pluracol® E1450 NF; Pluracol® SRF-2; Pluracol® W3520N; Pluracol® WI 285; Pluracol® WI 625; Pluracol® WI 1715; Pluracol® WS 170; Pluracol® WS 260; Pluracol® WS 660; Pluracol® WS 2000; Pluracol® WS 3520; Pluracol® WS 5100; Pluracol® WT 1400; Pluracol® WT 9150; Pluracol® WT 90,000; *Silicone*

defoamer, latex
Silicone

defoamer, latex paints
Nonoxynol-5

defoamer, liq. candles
PD-23; PD-25; PD-28

defoamer, mercerizing liquors
Reomol® TBP; *Tributoxyethyl phosphate*; *Tributyl phosphate*

defoamer, metal cleaners
Pluracol® E400 NF; Pluracol® E600 NF

defoamer, metal foil rolling oils
PD-23; PD-25; PD-28

defoamer, metal polishing
Pluracol® E400 NF; Pluracol® E600 NF

defoamer, metal sheet rolling oils
PD-23; PD-25; PD-28

defoamer, metalworking
Glycol distearate; *Isodecyl oleate*; *Methyl lardate*; *Nonoxynol-4*; *Nonoxynol-6*; *PEG-4 dilaurate*; *PEG-8 dioleate*; *PEG-4 oleate*

defoamer, metalworking/cleaning
Nonoxynol-5

defoamer, mining
Silicone

defoamer, mold coating
Reomol® TBP; *Tributyl phosphate*

defoamer, oil processing
Dimethicone

defoamer, oral care
Pluracol® E400 NF; Pluracol® E600 NF

defoamer, ore flotation
Jeffox® WL-440; Jeffox® WL-660; Jeffox® WL-1400; Jeffox® WL-5000

defoamer, paints
Glyceryl dioleate; *Nonoxynol-4*; *Nonoxynol-5*; *Nonoxynol-6*

defoamer, paper
Drakesol® 165; Drakesol® 165AT; Drakesol® 205; Drakesol® 220; Drakesol® 260; Drakesol® 260 AT; Drakesol® 305; *Isodecyl oleate*; *Nonoxynol-4*; *Nonoxynol-6*; Pluracol® E400 NF; Pluracol® E600 NF

defoamer, paper coatings
Reomol® TBP; *Tributoxyethyl phosphate*; *Tributyl phosphate*

defoamer, paper coatings: food-contact
Coconut (Cocos nucifera) oil; *Corn (Zea mays) oil*; *Cottonseed (Gossypium) oil*; *Glyceryl stearate*; *Methyl cocoate*; *Methyl oleate*; *Methyl palmitate*; *Oleic acid*; *Olive (Olea europaea) oil*; *Peanut (Arachis hypogaea) oil*; *PPG-20*; *PPG-22 butyl ether*; *Rice (Oryza sativa) bran oil*; *Safflower (Carthamus tinctorius) oil*; *Sesame (Sesamum indicum) oil*; *Soybean (Glycine soja) oil*; *Sunflower (Helianthus annuus) seed oil*; *Tall oil*

defoamer, paper/paperboard
Alcohol; *Coconut (Cocos nucifera) oil*; Condea Vista LPA; Condea Vista LPA-142; Condea Vista LPA-170; *Corn (Zea mays) oil*; *Cottonseed (Gossypium) oil*; *Cyclohexanol*; *Deodorized kerosene*; *Dimethicone*; *Glyceryl stearate*; *Hexylene glycol*; *Isobutyl alcohol*; *Isopropyl alcohol*; *Kerosene*; *Menhaden oil*; *Methyl alcohol*; *Methyl caprylate*; *Methyl caprylate/caprate*; *Methyl cocoate*; *Methyl laurate*; *Methyl myristate*; *Methyl oleate*; *Methyl palmitate*; *Methyl stearate*; *Mineral oil*; *Myristyl alcohol*; *Naphtha*; *Nonylphenol*; *Oleic acid*; *Olive (Olea europaea) oil*; PD-23; PD-25; PD-28; *Peanut (Arachis hypogaea) oil*; *PEG-12 dioleate*; *Petrolatum*; *Pine (Pinus palustris) oil*; *Poloxamer 181*; *PPG-20*; *PPG-26*; *PPG-22 butyl ether*; *PPG-24 butyl ether*; *PPG-33 butyl ether*; *PPG-40 butyl ether*; *Rice (Oryza sativa) bran oil*; *Safflower (Carthamus tinctorius) oil*; *Sesame (Sesamum indicum) oil*; *Soybean (Glycine soja) oil*; *Stearyl alcohol*; *Tall oil*

defoamer, paper/paperboard: food-contact
Coconut (Cocos nucifera) oil; *Corn (Zea mays) oil*; *Cottonseed (Gossypium) oil*; Crill 2; Exxsol® D 60; Exxsol® D 80; Exxsol® D 110; Exxsol® D 130; Exxsol® D 3135; *Glyceryl stearate*; Isopar® G Fluid; Isopar® H Fluid; Isopar® K Fluid; Isopar® L Fluid; Isopar® M Fluid; Isopar® V Fluid; Lumulse® GMO; *Methyl cocoate*; *Methyl oleate*; *Methyl palmitate*; *Nonoxynol-4*; Norpar® 12 Fluid; Norpar® 13 Fluid; Norpar® 15 Fluid; *Oleic acid*; *Olive (Olea europaea) oil*; *Peanut (Arachis hypogaea) oil*; *PPG-20*; *PPG-22 butyl ether*; *Rice (Oryza sativa) bran oil*; *Safflower (Carthamus tinctorius) oil*; *Sesame (Sesamum indicum) oil*; *Soybean (Glycine soja) oil*; *Tall oil*; Ucon® 50-HB-260; Ucon® 50-HB-400; Ucon® 75-H-450

defoamer, paper: food-grade
Dow EP530; *PPG-17*

defoamer, pesticides
Dimethyl octynediol; *2-Ethylhexoic acid*

defoamer, petroleum
Nonoxynol-5

defoamer, petroleum lubricants
Octamethyltrisiloxane

defoamer, petroleum oils
Nonoxynol-4

defoamer, pharmaceutical creams/lotions
Deodorized kerosene; PD-23; PD-25; PD-28

defoamer, pharmaceutical fermentation
Methyl oleate

defoamer, pharmaceuticals
Nonoxynol-6; *PEG-4 dilaurate*; *PEG-8 dioleate*; *PEG-4 oleate*; *Propylene glycol alginate*

defoamer, plaster compositions
Reomol® TBP

defoamer, plasters
Tributoxyethyl phosphate

defoamer, plastics
Isodecyl oleate; *PEG-4 dilaurate*; *PEG-8 dioleate*; *PEG-4 oleate*

defoamer, polishes
PD-23; PD-25; PD-28; *Petroleum distillates*

defoamer, proteinaceous solutions
Citroflex® 4

defoamer, pulp bleaching
Reomol® TBP; *Tributoxyethyl phosphate*; *Tributyl phosphate*

defoamer, pulp/paper
Polyglycol P112-2; *PPG-66-glycereth-12*; *Silicone*

defoamer, PVAc emulsions
Schercemol PGML

defoamer, rubber
Pluracol® E400 NF; Pluracol® E600 NF

defoamer, sanitizers: food processing
Dow EP530

defoamer, soaps
Silicone

defoamer, solder reflow applics.
Jeffox® WL-440; Jeffox® WL-660; Jeffox® WL-1400; Jeffox® WL-5000

defoamer, textile sizes
Albrite® Tributyl Phosphate

defoamer, textiles
Isodecyl oleate; *Nonoxynol-4*; *Nonoxynol-5*; *Nonoxynol-6*; *PEG-4 dilaurate*; *PEG-8 dioleate*; *PEG-4 oleate*; Pluracol® E400 NF; Pluracol® E600 NF; Polyglycol P112-2; *PPG-66-glycereth-12*; *Silicone*; *Tributoxyethyl phosphate*

defoamer, vat dyeing
Albrite® Tributyl Phosphate

defoamer, veg. processing
Petroleum distillates

defoamer, vegetable froth flotation
Deodorized kerosene; PD-23; PD-25; PD-28

defoamer, water treatment
PD-23; PD-25; PD-28; *PEG-4 dilaurate*; *PEG-8 dioleate*

defoamer, water/waste treatment
Dimethicone

defoamer, waterless hand cleaners
PD-23; PD-25; PD-28; *Petroleum distillates*

defoamer, wood prods.
Pluracol® E400 NF; Pluracol® E600 NF

defoamer, yeast processing
PEG-8 dioleate; *PEG-12 dioleate*; *Petrolatum*; *Polypropylene glycol*; *PPG-20*; *PPG-26*

defoliant
But-2-yne-1,4-diol

degreaser
Acetone; Arosolve B-950; Arosolve XNF-1; Chemax E-1000 MO; Condea Vista LPA; Crodamol AB; *Dichlorodifluoromethane*; *cis-trans-1,2-Dichloroethylene (mixed isomers)*; *trans-1,2-Dichloroethylene*; *1,2-Dichlorotetrafluoroethane*; *2-Ethylhexyl-S-lactate*; *Methylcyclo-*

hexanol; Safety-Solv 140 Plus; Sasol LPA Solvent; Sasol LPA-142 Solvent; Sasol LPA-150 Solvent; Sasol LPA-170 Solvent; Sasol LPA-210 Solvent; *Trichlorofluoromethane*; *1,2,3-Trichloropropane*; *Trichlorotrifluoroethane*; *Xylene*

degreaser, coating deposition applics.
HFE-71DE

degreaser, cosmetics
Chemax E-200 MO

degreaser, elec. components/motors
Ecco Cleaner AK; Eccoflex 090; Ecco Strip LS-W

degreaser, emulsions
Ritawax AEO; Ritawax ALA

degreaser, engine parts: auto repair shops
Naphtha, medium aliphatic

degreaser, engine parts: machine repair shops
Naphtha, medium aliphatic

degreaser, hides
o-Dichlorobenzene

degreaser, leather
Cyclohexanol; *Cyclohexanone*; *Stoddard solvent*

degreaser, lubricant deposition applics.
HFE-71DE

degreaser, metal
C7-10 isoparaffin; *Cyclohexanone*; *Propylene dichloride*; *Stoddard solvent*

degreaser, metal cleaning
Degreez

degreaser, metals
Chemax E-200 MO; *1,1-Dichloroethane*

degreaser, mineral oil: bath care
Hexyldecyl benzoate

degreaser, mineral oil: creams/lotions
Hexyldecyl benzoate

degreaser, mineral oil: deodorants
Hexyldecyl benzoate

degreaser, mineral oil: hair care
Hexyldecyl benzoate

degreaser, mineral oil: makeup
Hexyldecyl benzoate

degreaser, mineral oil: shower prods.
Hexyldecyl benzoate

degreaser, mineral oil: skin care
Hexyldecyl benzoate

degreaser, mineral oil: sunscreens
Hexyldecyl benzoate

degreaser, plastisols
Chemax E-200 MO

degreaser, rug cleaners
Benzyl alcohol

degreaser, solvent
T-Mulz® 426

degreaser, solvent: dry cleaning solvents
BIO-SOFT® N-411

degreaser, solvent: emulsion degreasers
BIO-SOFT® N-411

degreaser, solvent: laundry prespotters
BIO-SOFT® N-411

degreaser, solvent: paint strippers
BIO-SOFT® N-411

degreaser, solvent: waterless hand cleaners
BIO-SOFT® N-411

degreaser, specialty solvents
HFE-71DE

degreaser, sunscreens
Benzyl laurate

degreaser, textile cleaning
Degreez

degreaser, textiles
Chemax E-200 MO

degreaser, wool
o-Dichlorobenzene

degreaser/dewaxer
Shellsol® D120

dehairing agent
Dimethylamine; *Isopropylamine*

dehydrating agent
Butyl alcohol; *t-Butyl alcohol*; *Isopropyl alcohol*; *Methanesulfonic acid*; *Potassium carbonate*; *Trimethyl borate*

dehydrating agent, natural gas
Diethylene glycol; *N-Methyl-2-pyrrolidone*; *Triethylene glycol*

dehydrating agent, organic synthesis
Zinc chloride

dehydrating agent, plasticizers
Diethylene glycol

dehydrating agent, surfactants
Diethylene glycol

dehydration preventive, dried fruit
Crodamol GTCC

de-icing agent
Eastman® DM

de-icing agent, diesel fuels
Ecosoft™ PB Solvent; Ecosoft™ PE Solvent; Ecosoft™ PH Solvent

de-icing agent, jet fuels
Hempa; *Methoxyethanol*

de-icing agent, liq. fuels
Isopropyl alcohol

deicing fluid
Propylene glycol dimethyl ether; *Propylene glycol laurate*; *Propylene glycol stearate*

de-icing, roads
Calcium chloride

de-inking agent, paper
Methyl lardate

de-inking agent, pulp/paper
T-Det® N-6

de-inking, pulp mills
Sodium hydroxide

deliming agent, leather tanning
Lactic acid

deliming hides
Isobutyric acid

delinting cottonseed, cottonseed oil prod. (food)
Cocamide DEA

delivery/absorp. enhancer
Coconut (Cocos nucifera) oil

delustering agent, cellulose acetate fibers
Acetone

demulcent, pharmaceuticals
Sweet almond (Prunus amygdalus dulcis) oil

demulsifier
PEG/PPG-17/6 copolymer; *Poloxamer 181*; Ucon® 50-HB-55; Ucon® 50-HB-100; Ucon® 50-HB-260; Ucon® 50-HB-400; Ucon® 50-HB-660; Ucon® 50-HB-2000; Ucon® 50-HB-3520; Ucon® 50-HB-5100; Ucon® 75-H-450; Ucon® LB-65; Ucon® LB-135; Ucon® LB-165; Ucon® LB-285; Ucon® LB-385; Ucon® LB-525; Ucon® LB-625; Ucon® LB-1145; Ucon® LB-1715; Ucon® LB-3000

demulsifier, crude oil
Nonoxynol; *Polyethylene glycol*

demulsifier, gas prod.
Prochinor® SOL

demulsifier, oil prod.
Prochinor® SOL

demulsifier, petroleum
Diethanolamine; *Ethylene oxide*; *Nonylphenol*; PPG 1650; PPG 2250

denaturant
Butyl alcohol; *Menthol*

denaturant, alcohol
Acetamide; *Acetonitrile*; *Brucine*; *t-Butyl alcohol*; *Crotonaldehyde*; *Diethyl phthalate*; *Hexane*; *Methyl isobutyl ketone*; *Pyridine*; *1,1,2,2-Tetrachloroethane*; *Toluene*

denaturant, alcohol: pharmaceuticals
Denatonium benzoate; *Sucrose octaacetate*

denaturant, cosmetics
Acetone; *Ammonium hydroxide*; *t-Butyl alcohol*; *Camphor*; *Cinnamal*; *Denatonium benzoate*; *Eucalyptol*; *Eugenol*; *Methyl alcohol*; *Methyl salicylate*; *Phenol*; *Sodium hydroxide*; *Sodium lauryl sulfate*; *Sucrose octaacetate*; *α-Terpineol*; *Thymol*

denaturant, ethyl alcohol
Denatonium benzoate; *Isopropyl alcohol*

denaturant, household prods.
Denatonium benzoate

denaturant, oils
Brucine

denaturant, pharmaceuticals
t-Butyl alcohol

denaturant, specially denatured alcohols: alcohol 39-B
Eastman® Diethyl Phthalate

denaturant, specially denatured alcohols: alcohol 39-C
Eastman® Diethyl Phthalate

dental analgesic, pharmaceuticals
Eugenol

deodorant
Zinc chloride

deodorant active
Citrofol® AI

deodorant component
Mixed terpenes

deodorant, cosmetics
Farnesol; *Phenethyl alcohol*; *Phenol*; *Triethyl citrate*

deodorant, pharmaceuticals
Thymol

deodorizer, fur
Hydrogen peroxide

deodorizer, hair
Hydrogen peroxide

deodorizer, industrial water
Hydrogen peroxide

deodorizer, municipal water
Hydrogen peroxide

deodorizer, textiles
Hydrogen peroxide

deodorizer, wastewater
Hydrogen peroxide

deodorizer, wood pulp
Hydrogen peroxide

deoiler, cosmetic creams
Finsolv® TN

deoiler, cosmetic oils
Finsolv® TN

deoiler, creams
C12-15 alkyl benzoate

deoiler, mineral oil: bath care
Hexyldecyl benzoate

deoiler, mineral oil: creams/lotions
Hexyldecyl benzoate

deoiler, mineral oil: deodorants
Hexyldecyl benzoate

deoiler, mineral oil: hair care
Hexyldecyl benzoate

deoiler, mineral oil: makeup
Hexyldecyl benzoate

deoiler, mineral oil: shower prods.
Hexyldecyl benzoate

deoiler, mineral oil: skin care
Hexyldecyl benzoate

deoiler, mineral oil: sunscreens
Hexyldecyl benzoate

deoiler, oils
C12-15 alkyl benzoate

depilatory, hides
Diethylamine

deposition agent, skin
NEOBEE® M-20

depressant, flotation: copper ores
2-Mercaptoethanol

desensitizing agent, organic peroxides
Triethyl phosphate

desiccant
Calcium chloride

desiccant, pharmaceuticals
Calcium chloride

desizing agent, one-bath: blended fabrics
Arosolve FX-6

desizing agent, one-bath: natural fabrics
Arosolve FX-6

desizing agent, one-bath: synthetic fabrics
Arosolve FX-6

desizing agent, textiles
Perchloroethylene

desorbent, processing
Hexane

detackifier
Glyceryl stearate

detackifier, antiperspirants
Trivent DIA

detackifier, cosmetics
Octyl stearate; *Phenyl trimethicone*

detackifier, creams/lotions
PPG-14 butyl ether

detackifier, deodorant sticks
Ceraphyl® 28; Ceraphyl® ICA

detackifier, emulsions
Ritawax AEO; Ritawax ALA

detackifier, high-solids inks
Sylfat® 9012

detackifier, low-VOC lithographic inks
Sylfat® 9012

detackifier, mineral oils
Finsolv® TN

detackifier, printing inks
Petrolatum

detackifier, skin cleansers
Cyclomethicone

detackifier, soybean oil ink systems
Sylfat® 9012

detackifier, topical pharmaceuticals
Octyl stearate

detackifier, web-offset heat-set inks
Sylfat® 9012

detangling agent, cream rinses
Lactamide MEA

detangling agent, shampoos
Lactamide MEA

detergent
2-Aminobutanol; *Butyl alcohol*; Cachalot® AR-20; Cachalot® BE-22; Cachalot® C-50; Cachalot® L-90; Cachalot® M-43; Cachalot® S-56; Chemax E-1000 MO; *Coconut (Cocos nucifera) oil*; *Dimethylamine*; *1,4-Dioxane*; *Diphenyl oxide*; *Ethylamine*; *N-Ethyldiethanolamine*; *Lauryl alcohol*; *Naphthenic acid*; *n-Nonane*; *Nonoxynol-9 phosphate*; *Paraffin, chlorinated*; *PEG-90*; *Pentylamine*; *Phosphoric acid*;

D; Exx-Print® 588 D; Exxsol® D 40; Exxsol® D 40 Anti-Static Fluid; Exxsol® D 60; Exxsol® D 60 Anti-Static Fluid; Exxsol® D 60 Naphtha Anti-Static; Exxsol® D 70; Exxsol® D 75/100; Exxsol® D 80; Exxsol® D 90/120; Exxsol® D 90/120 Anti-Static Solvent; Exxsol® D 95; Exxsol® D 110; Exxsol® D 115/145; Exxsol® D 115/145 Anti-Static Solvent; Exxsol® D 130; Exxsol® D 3135; Exxsol® D 3135 Anti-Static Solvent; Exxsol® DSP 90/120 Naphtha; Exxsol® S-K Premium Solvent; Exxsol® Heptane; Exxsol® Hexane Anti-Static; Exxsol® Hexane Naphtha Anti-Static; Exxsol® Isohexane Naphtha; Exxsol® Isopentane; Exxsol® Methylpentane Naphtha; *Gasoline*; HAN® 906 Solvent; Iosol RS Solvent; Isopar® C; Isopar® E Fluid; Isopar® K Naphtha; Isopar® M Fluid; Isopar® V Fluid; Jayflex® 215; *Naphtha, hydrotreated light*; Naphthalene Depleted Aromatic 150 Fluid; Naphthalene Depleted Aromatic 200 Fluid; *Naphtha, light alkylate*; *Naphthenic oil*; Solvarex 9; Solvarex 10; Solvarex 90/160; Solvarex 90/180; Solvesso™ 100 Anti-Static Solvent; Solvesso™ 150 Anti-Static Solvent; Solvesso™ 150 ND; Solvesso™ 200; Solvesso™ 200 ND; Ultra Low Naphthalene Aromatic 200; Ultra Low Naphthalene Solvesso 200 Solvent; Varsol® 1 Naphtha Anti-Static; Varsol® 18 Naphtha; Varsol® 18 Naphtha Anti-Static; Varsol® 3139; Varsol® 3139 Anti-Static Solvent; Varsol® DX 3139; Varsol® DX 3139 Anti-Static Solvent; Varsol® DX 3641; *Water*

diluent, ABS
Styrene

diluent, adhesives
Monochlorotoluene; *Petrolatum*; Shellsol® X7B; *Trichloroethylene*

diluent, aerosols
Captex® 300; Captex® 810D; *Tricaprin*

diluent, aftershave lotions
Captex® 810D

diluent, agric. chemicals
Aromatic 100; Aromatic 150; Aromatic 200

diluent, agric. sprays
Petroleum distillates

diluent, air-drying coatings
Versadil™ 100

diluent, antibiotics
NEOBEE® M-5

diluent, bath oils
Captex® 810D

diluent, brake fluids
Eastman® DE; *Ethoxydiglycol*

diluent, cellulosics
Kenplast® A-450

diluent, charcoal lighter fluids
PD-23; PD-25; PD-28

diluent, cleaners
Fomblin® Perfluorosolv PFS-1; PD-23; PD-25; PD-28

diluent, cleaning agents
Ketrul 212; Ketrul HT

diluent, clinical nutrition
Captex® 300; Captex® 350; *Tricaprin*

diluent, coatings
Captex® 300; *Tricaprin*

diluent, color
Acetic acid; NEOBEE® M-5

diluent, color: confectionery
Butyl alcohol; *Cetyl alcohol*

diluent, color: food
Acetone; *Ethyl acetate*; *Isopropyl alcohol*; *Methylene chloride*; *Naphtha*; *Terpene resin*

diluent, color: food supplements (tablet form)
Cetyl alcohol

diluent, color: gum
Butyl alcohol; *Cetyl alcohol*

diluent, color: pharmaceuticals
Isopropyl alcohol

diluent, color: tableted food supplements
Butyl alcohol

diluent, cosmetic active ingreds.
Crodamol GTCC

diluent, cosmetic creams/lotions
Deodorized kerosene; PD-23

diluent, cosmetics
Captex® 810D; *C11-12 isoparaffin*; Eastman® DE Acetate; *Propylene glycol dicaprylate/dicaprate*; SDA-1; SDA-3A; SDA-3C; SDA-23A; SDA-30; SDA-39C; SDA-40; SDA-40A; SDA-40B; SDA-40C; *SD alcohol 1*; *SD alcohol 3-A*; *SD alcohol 3-C*; *SD alcohol 23-A*; *SD alcohol 30*; *SD alcohol 39-C*; *SD alcohol 40*; *SD alcohol 40-A*; *SD alcohol 40-B*; *SD alcohol 40-C*

diluent, creams/lotion
Petroleum distillates

diluent, decorative paints
Exxsol® D 180/200 SP; Isopar® I Fluid; Isopar® J Fluid

diluent, degreasing
Fomblin® Perfluorosolv PFS-1; Ketrul 212; Ketrul HT; Spirdane BT; Spirdane HT; Spirdane K1

diluent, delivery/absorption enhancement
Captex® 300; Captex® 350; *Tricaprin*

diluent, dermatologicals
Captex® 300; Captex® 350; *Tricaprin*

diluent, detergents
Ethylene glycol diethyl ether

diluent, dewatering
Fomblin® Perfluorosolv PFS-1

diluent, dewaxing
Ketrul 212; Ketrul HT

diluent, dip coatings
Shellsol® X7B

diluent, egg shell colorants
SD alcohol 23-A

diluent, enamels
Shellsol® X7B

diluent, epoxies
ε-Caprolactone monomer; *1,10-Decanediol*; *12-Hydroxystearyl alcohol*; Kenplast® A-450; Kenplast® G; *Phenoxyethanol*; Speziol C 10/2; Speziol C 18/2

diluent, essential oils
NEOBEE® M-5; NEOBEE® M-20

diluent, essential oils: food
NEOBEE® M-20

diluent, fiber lubricants
PD-23; PD-25; PD-28

diluent, flash point reduction
Fomblin® Perfluorosolv PFS-1

diluent, flavors
Aldo® MCT KFG; NEOBEE® M-5; *Propylene glycol dicaprylate/dicaprate*

diluent, fluorochemical acrylate polymer coatings: electronics
Novec™ HFE-7100DL

diluent, fluorochemical acrylate polymer coatings: medical
Novec™ HFE-7100DL

diluent, food colorants
Ammonium hydroxide; *Cyclohexane*; *Ethoxyethanol*; *Glycol distearate*; *Polysorbate 80*

diluent, food marking colorants
SD alcohol 3-A

diluent, food-contact applics.
PD-23; PD-25; PD-28

diluent, fragrance oils
Propylene glycol dicaprylate/dicaprate

diluent, fragrances
Diethyl phthalate; *Dipropylene glycol*

diluent, fuel oils
Monochlorotoluene

diluent, gelatin capsules
Tricaprin

diluent, high flash point reactive: coal tar pipe coatings
Cumylphenyl acetate

diluent, high flash point reactive: epoxies
Cumylphenyl acetate

diluent, high flash point reactive: polyamide-cured epoxy floorings
Cumylphenyl acetate

diluent, high-solids coatings
Versadil™ 100

diluent, household prods.
Spirdane BT; Spirdane HT

diluent, hydraulic brake fluids
BDGE; BGE; *Butoxydiglycol*; *Diethylene glycol propyl ether*; Eastman® DB; Eastman® DM; Eastman® DP; EDGE; EGE; *Ethylene glycol isopropyl ether*; i-PGE; MDGE; *Methoxydiglycol*; MGE

diluent, industrial cleaners
Spirdane BT; Spirdane HT

diluent, infant formulas
Captex® 300; Captex® 350; *Tricaprin*

diluent, injectables
NEOBEE® M-20

diluent, lacquer thinners
Shellsol® X7B

diluent, lacquers
Shellsol® BF; Tolu-Sol 20; Tolu-Sol 30

diluent, lip gloss
Captex® 810D

diluent, lipstick
Captex® 810D

diluent, liq. candles
PD-23; PD-25; PD-28

diluent, liq. printing inks
Shellsol® X7B

diluent, lubricants
Ketrul 212; Ketrul HT; Spirdane BT; Spirdane HT; Spirdane K2

diluent, makeup bases
Captex® 810D

diluent, medicinals
NEOBEE® M-5; *Propylene glycol dicaprylate/dicaprate*

diluent, metal foil rolling oils
PD-23; PD-25; PD-28

diluent, metal parts cleaners
Monochlorotoluene

diluent, metal sheet rolling oils
PD-23; PD-25; PD-28

diluent, natural jojoba
Oleyl erucate

diluent, nitrocellulose lacquers
Toluene

diluent, nonreactive epoxies
Dibutyl phthalate; *Furfuryl alcohol*; *Nonylphenol*; *Pine (Pinus palustris) oil*

diluent, nutritional applics.
Captex® 810D

diluent, nutritional fluids
NEOBEE® M-5

diluent, nutritional supplements
Tricaprin

diluent, nutritional/sports supplements
Captex® 350

diluent, offset printing inks
Texaprint® SDCE; Texaprint® SDM; Texaprint® SJEH; Texaprint® SKEH; Texaprint® SLIB EXP; Texaprint® SLIP; Texaprint® SRM; Texaprint® SSEH

diluent, organic coatings
3-Heptanol

diluent, paint
Hexane

diluent, paint strippers
Monochlorotoluene

diluent, paint thinners
Monochlorotoluene

diluent, paints
Isane IP 130; Isane IP 155; Isane IP 165; Isane IP 175; Isane IP 185; Ketrul 212; Ketrul HT; Shellsol® BF; Spirdane BT; Spirdane HT; Spirdane K1; Spirdane K2; *Styrene*; *1,1,1-Trichloroethane*; *Trichloroethylene*

diluent, paints/coatings
Aromatic 100; Aromatic 150; Aromatic 200

diluent, paper/paperboard
PD-23; PD-25; PD-28

diluent, pentachlorophenol wood preservatives
PCP Solvent

diluent, perfumes
Acetylated glyceryl stearate; Citrofol® AI; Glicopol 2-G; Milopol G-7; Monestriol DM; Monestriol GCT; Monestriol GE; Monestriol GP; Monestriol GP-35; Monestriol GP-35-3%; Monestriol GP-35-AE; Monestriol GP-40; Monestriol GP-40-CF; Monestriol GP-40-DC; Monestriol GP-45; Monestriol GP-45-DC; Monestriol GP-60; Monestriol GP-AS; Thomil G; Tilol GCT-R; Tilol GP-O; Triacetina; Triol 91; Trioleina

diluent, peroxide
Dimethyl phthalate

diluent, personal care
Eastman® DE Acetate; Isopar® G Fluid; Isopar® H Fluid; Isopar® K Fluid; SDA-1; SDA-3A; SDA-3C; SDA-23A; SDA-30; SDA-39C; SDA-40; SDA-40A; SDA-40B; SDA-40C

diluent, personal care sprays
C10-11 isoparaffin; *C11-12 isoparaffin*

diluent, pesticides
Agrisynth® BLO; BLO®; *Deodorized kerosene*; PD-23; PD-25; PD-28

diluent, pharmaceutical active ingreds.
Crodamol GTCC

diluent, pharmaceutical coatings
Captex® 350

diluent, pharmaceutical creams/lotions
Deodorized kerosene; PD-23; PD-25; PD-28

diluent, pharmaceuticals
Captex® 300; Captex® 350; Captex® 810D; *Sucrose palmitate*; *Sunflower (Helianthus annuus) seed oil*

diluent, polishes
PD-23; PD-25; PD-28

diluent, polymers

dl-Limonene
dispersant, fat liquoring
Nonoxynol-4
dispersant, fats
Sucroglycerides; *Sucrose fatty acid esters*
dispersant, fiber lubricants
PEG-12 dilaurate; *PEG-4 dioleate*
dispersant, fiber softeners
PEG-12 dilaurate
dispersant, flame retardants
Variquat® CC-57
dispersant, floor waxes
dl-Limonene
dispersant, food
Glyceryl laurate; *Glyceryl oleate*; *Nonoxynol-6*; *PEG-4 dilaurate*; *PEG-20 dilaurate*; *PEG-4 dioleate*; *PEG-8 dioleate*; *PEG-20 dioleate*; *PEG-4 oleate*; *PEG-6 oleate*; *PEG-8 oleate*; *PEG-20 oleate*; Pluracol® E300; Pluracol® E400; Pluracol® E600; Pluracol® E4000; *Polysorbate 60*; *Polysorbate 80*; *Propylene glycol*
dispersant, food pkg.
Glyceryl oleate
dispersant, food processing
Carbowax® Sentry® PEG 400; Carbowax® Sentry® PEG 540 Blend; Carbowax® Sentry® PEG 600; Carbowax® Sentry® PEG 900; Carbowax® Sentry® PEG 1000; Carbowax® Sentry® PEG 1450; Carbowax® Sentry® PEG 3350; Carbowax® Sentry® PEG 4600; Carbowax® Sentry® PEG 8000
dispersant, food tablets
Pluracol® E300; Pluracol® E400; Pluracol® E600; Pluracol® E4000
dispersant, foundation
Capmul® MCM; Capmul® MCMC8
dispersant, fragrances
Ameroxol® OE-2; *Oleth-2*
dispersant, fuel oils
T-Det® N-6
dispersant, furniture polishes
dl-Limonene
dispersant, general purpose
Macol® 85
dispersant, germicides
Variquat® CC-57
dispersant, grease: acid aluminum cleaners
Rhodacal® 330
dispersant, grease: agric. pesticides
Rhodacal® 330
dispersant, grease: cleaners
Rhodacal® 330
dispersant, grease: dry cleaning
Rhodacal® 330
dispersant, grease: emulsion polymerization
Rhodacal® 330
dispersant, grease: latex paints
Rhodacal® 330
dispersant, grease: metal cleaning
Rhodacal® 330
dispersant, grease: metalworking fluids
Rhodacal® 330
dispersant, grease: shampoos
Rhodacal® 330
dispersant, hair care
Choleth-24; *PEG-75 lanolin*
dispersant, hand cleaners
Carbowax® Sentry® PEG 300
dispersant, household cleaners
Nonoxynol; SECOSTER® A
dispersant, household prods.
Nonoxynol-4; *Nonoxynol-6*
dispersant, industrial
Carbomer 940; Cithrol 4DL; Cithrol 4DO; *Nonoxynol-4*; *Nonoxynol-6*; *PEG-8 dilaurate*; *PPG-15 stearyl ether*; Variquat® CC-57
dispersant, industrial cleaners
Nonoxynol; SECOSTER® A
dispersant, industrial lubricants
PEG-8 dilaurate
dispersant, inks
Pluracol® E400 NF; Pluracol® E600 NF; Pluracol® E1450 NF
dispersant, inorganic sunscreen actives: sun care
Finsolv® BOHS-111
dispersant, institutional cleaners
Laureth-9
dispersant, lacquers
dl-Limonene
dispersant, leather
Oleth-2; *PEG-2 stearate*
dispersant, lime soap: alkaline scouring systems
Incrocas 30
dispersant, lip preps.
Capmul® MCM; Capmul® MCMC8
dispersant, lubricants
Perfluoroheptane; *Perfluorohexane*; *Perfluoro-N-methylmorpholine*; *Perfluoropentane*; PF-5050; PF-5060; PF-5070
dispersant, maintenance cleaners
Laureth-9
dispersant, makeup
Carbowax® Sentry® PEG 300; *PEG-75 lanolin*
dispersant, medical preps.
Sodium lauryl sulfate
dispersant, medicinals
Polysorbate 80
dispersant, metal cleaners
Pluracol® E400 NF; Pluracol® E600 NF; Pluracol® E1450 NF
dispersant, metal oxides
Sodium hydroxide
dispersant, metal polishes
Pluracol® E1450 NF
dispersant, metal polishing
Pluracol® E400 NF; Pluracol® E600 NF
dispersant, metals
Chemax E-200 MO
dispersant, metalworking
Nonoxynol-4; *Nonoxynol-6*; *PEG-4 dilaurate*; *PEG-4 dioleate*; *PEG-8 dioleate*; *PEG-4 oleate*
dispersant, metalworking lubricants
PEG-12 dilaurate
dispersant, mold release agents
Perfluoroheptane; *Perfluorohexane*; *Perfluoro-N-methylmorpholine*; *Perfluoropentane*; PF-5050; PF-5060; PF-5070
dispersant, motor oils
Tetraethylenepentamine
dispersant, nutritional/sports supplements
Capmul® MCM
dispersant, oil drilling
Dipentene-C
dispersant, oil field
SECOSTER® A
dispersant, oil systems
Isopropylamine dodecylbenzenesulfonate
dispersant, oils
Glyceryl laurate; *dl-Limonene*; Mazol® 300 K
dispersant, oral care
Pluracol® E400 NF; Pluracol® E600 NF; Pluracol® E1450 NF
dispersant, oral pharmaceuticals
Miglyol® 810
dispersant, organic sunscreen actives: sun care
Finsolv® BOHS-111
dispersant, organosol
Eastman® DB
dispersant, paints
Dipentene-C; *Ethanolamine*; *Laureth-9*; *dl-Limonene*; *Nonoxynol-4*; *Nonoxynol-6*; *PEG-20 oleate*; *Polysorbate 80*
dispersant, paper
Glyceryl dilaurate; *Glycol distearate*; *Laureth-9*; *Nonoxynol-4*; *Nonoxynol-6*; *Oleth-2*; *PEG-2 stearate*; Pluracol® E400 NF; Pluracol® E600 NF; Pluracol® E1450 NF
dispersant, parenteral pharmaceuticals
Miglyol® 810
dispersant, perfumes
dl-Limonene; *PEG-20 dioleate*; *PPG-5-ceteth-20*
dispersant, personal care
Akomed R; *Caprylic/capric triglyceride*; *Carbomer 940*; Carbowax® Sentry® PEG 300; Eastman® DE Acetate; *Hexyl laurate*; *PEG-8 dilaurate*; Polychol 5; Procetyl AWS; Ritawax AEO; Solulan® 16; Solulan® 75; Solulan® C-24; Solulan® L-575; Variquat® CC-57
dispersant, petroleum oils
Nonoxynol-4; T-Det® N-6
dispersant, pharmaceutical absorption enhancement
Polyglyceryl-3 stearate
dispersant, pharmaceutical aerosols
Capmul® MCM
dispersant, pharmaceutical coatings
Capmul® MCM; Capmul® MCMC10; *Glyceryl caprate*
dispersant, pharmaceutical microemulsions
Capmul® MCM; Capmul® MCMC8; Capmul® MCMC10; *Glyceryl caprate*
dispersant, pharmaceutical soft gelatin capsules
Glyceryl caprate
dispersant, pharmaceutical tablets
Sucrose fatty acid esters
dispersant, pharmaceutical vehicles
PEG-75 lanolin
dispersant, pharmaceuticals
Akomed R; Ameroxol® OE-2; *Butylene glycol dicaprylate/dicaprate*; Capmul® MCM; Capmul® MCMC8; *Caprylic/capric triglyceride*; *Carbomer 940*; *Ceteth-24*; Corona Lanolin; Coronet Lanolin; Dynacerin® 660; *Glyceryl caprate*; *Glyceryl caprylate/caprate*; *Glyceryl dilaurate*; *Hexyl laurate*; *Laneth-5*; *Laneth-10*; Lipocol P-15; Miglyol® 812; Miglyol® 840; Miglyol® 8810; *Nonoxynol-6*; *Oleth-2*; *PEG-40 castor oil*; *PEG-4 dilaurate*; *PEG-8 dilaurate*; *PEG-12 dilaurate*; *PEG-20 dilaurate*; *PEG-4 dioleate*; *PEG-8 dioleate*; *PEG-20 dioleate*; *PEG-6 isostearate*; *PEG-4 oleate*; *PEG-6 oleate*; *PEG-8 oleate*; *PEG-20 oleate*; Pluracol® E1450 NF; *Polysorbate 80*; *PPG-15 stearyl ether*; *Propylene glycol stearate*; Ritawax AEO
dispersant, phaxrmaceuticals
Diethanolamine
dispersant, pigment
Acetamide MEA; *Butyl stearate*; *Castor oil, polymerized*; Ceraphyl® 140; *Cetyl lactate*; Crill 2; Crillet 6; Cromollient DP3A; *Decyl oleate*; *Diethyl oxalate*; *Diisobutyl adipate*; *Diisopropyl adipate*; *Dimethyl adipate*; *Dioctyl adipate*; *Dioctyl sebacate*; *Di-PPG-3 myristyl ether adipate*; *Glycereth-7*; *Glycereth-26*; *Glyceryl acetate*; *Isocetyl alcohol*; *Isopropyl lanolate*; *Isopropyl myristate*; *Lanolin acid*; *Lanolin oil*; *dl-Limonene*; Lipolan R; *N-Methyl-2-pyrrolidone*; *Octyldodecanol*; #15 Oil; #30 Oil; #40 Oil; Pale 4; Pale 16; Pale 170; Pale 1000; Reomol® TBP; Surfadone® LP-100; Surfadone® LP-300; *Tridecyl neopentanoate*; Variquat® CC-57; Vorite 105; Vorite 110; Vorite 115; Vorite 125
dispersant, pigment: acid aluminum cleaners
Rhodacal® 330
dispersant, pigment: adhesives
Crill 1
dispersant, pigment: aftershaves
HallStar™ DIPA; HallStar™ DIPS
dispersant, pigment: agric. pesticides
Rhodacal® 330
dispersant, pigment: antiperspirant sticks
Crystal® O; Crystal® Crown
dispersant, pigment: antiperspirants
PPG-2 myristyl ether propionate
dispersant, pigment: bath care
Finsolv® BOHS-111; HallStar™ DOA; HallTress™ DIBA Special; *Hexyldecyl benzoate*; Liponate GC-K; Liponate PC; *Octyl hydroxystearate benzoate*
dispersant, pigment: bath oils
HallStar™ DIPA; *Stearyl stearate*
dispersant, pigment: cleaners
Rhodacal® 330
dispersant, pigment: cleansers
HallStar™ DIPA; HallStar™ DIPS; HallStar™ DOS; *Octyl hydroxystearate*
dispersant, pigment: clear sticks/gels
HallStar™ DIPS
dispersant, pigment: coatings
Morcryl® 535
dispersant, pigment: colognes
HallStar™ DIPA; HallStar™ DIPS; HallStar™ DOA; HallTress™ DIBA Special
dispersant, pigment: colored cosmetics
Octyldodecyl isostearate
dispersant, pigment: cosmetic creams/lotions
Isocetyl stearate; Liponate GC-K; Liponate PC
dispersant, pigment: cosmetics
Butyl octyl benzoate; Cosmol 525; Crill 1; Crystal® O; Crystal® Crown; *Myristyl myristate*; *Neopentyl glycol dioctanoate*; *Octyl hydroxystearate*; *Octyl stearate*; *Oleyl alcohol*; *PPG-40 diethylmonium chloride*; *Stearyl stearate*; Variquat® CC-42; Wickenol® 158; Wickenol® 171
dispersant, pigment: creams/lotions
Butyl octyl benzoate; Estalan-IDO; HallStar™ BST; *Hexyldecyl*

Tetradecane; *n-Tridecane*; *n-Undecane*
diuretic
Calcium chloride; *Sunflower (Helianthus annuus) seed oil*; *Turpentine*
divider oil
Drakeol® 7
d-limonene substitute
Soygold® 1000; Soygold® 2000
dough conditioner, baked goods
Diacetyl tartaric acid esters of mono- and diglycerides
dough conditioner, food
Polysorbate 60
dough conditioner, tacos/tortillas
Lumulse® GMO
dough improver, food
Sorbitan laurate
dough strengthener, baked goods
Hodag® S-35 FGK
dough strengthener, coffee whiteners
Hodag® S-35 FGK
dough strengthener, food
Hodag® S-35 FGK; *Mono- and diglycerides of fatty acids*; *Propylene glycol*
dough strengthener, frozen desserts
Hodag® S-35 FGK
dough strengthener, shortening
Hodag® S-35 FGK
dough strengthener, whipped toppings
Hodag® S-35 FGK
drier, electronic parts
Trichlorotrifluoroethane
drier, heavy oily systems
Trivent DIA
drier, metals
dl-Limonene
drier, paints
2-Ethylhexoic acid; *Menhaden oil*; *Naphthenic acid*
drier, precision equipment
Trichlorotrifluoroethane
drier, varnishes
2-Ethylhexoic acid
dry cleaning fluid
Butyl lactate; DF-2000™; EcoSolv™ Dry Cleaning Fluid
drying agent
Acetylacetone; *Calcium chloride*; *trans-1,2-Dichloroethylene*; *Perchloroethylene*; *Perfluoro-N-methylmorpholine*
drying agent, coating deposition applics.
HFE-71DE
drying agent, greases
HFE-71DA
drying agent, handling oils
HFE-71DA
drying agent, lubricant deposition applics.
HFE-71DE
drying agent, oils
HFE-71DA
drying agent, solder flux residue
HFE-71DA
drying agent, specialty solvent applics.
HFE-71DE
drying agent, spot-free water
HFE-7100
drying agent, timber
Pentachloroethane
drying agent, water
Methyl nonafluorobutyl ether, Novec™ HFE-7100
drying agent, waxes
HFE-71DA
drying fluid
2,3-Dihydrodecafluoropentane
drying oil
Castor (Ricinus communis) oil; *Isopropyl alcohol*
drying oil, alkyd resins
Safflower (Carthamus tinctorius) oil; *Sunflower (Helianthus annuus) seed oil*; *Walnut (Juglans regia) oil*
drying oil, food
Castor (Ricinus communis) oil
drying oil, food-contact coatings
Castor (Ricinus communis) oil; *Corn (Zea mays) oil*; *Cottonseed (Gossypium) oil*; *Safflower (Carthamus tinctorius) oil*; *Sesame (Sesamum indicum) oil*; *Soybean (Glycine soja) oil*; *Tall oil*; *Walnut (Juglans regia) oil*
drying oil, inks
Oleocal IVO-114; *Soybean (Glycine soja) oil*
drying oil, paints
Coconut (Cocos nucifera) oil; *Cottonseed (Gossypium) oil*; Oleocal IVO-114; *Peanut (Arachis hypogaea) oil*; *Rapeseed (Brassica campestris) oil*; *Safflower (Carthamus tinctorius) oil*; *Soybean (Glycine soja) oil*; *Sunflower (Helianthus annuus) seed oil*; *Tall oil*; *Walnut (Juglans regia) oil*
drying oil, paper coatings: food-contact
Sunflower (Helianthus annuus) seed oil
drying oil, varnishes
Oleocal IVO-114; *Safflower (Carthamus tinctorius) oil*; *Soybean (Glycine soja) oil*
dust control agent
Deodorized kerosene
dust control agent, charcoal lighter fluids
PD-23; PD-25; PD-28
dust control agent, cleaners
PD-23; PD-25; PD-28
dust control agent, cosmetic creams/lotions
PD-23
dust control agent, fiber lubricants
PD-23; PD-25; PD-28
dust control agent, food-contact applics.
PD-23; PD-25; PD-28
dust control agent, liq. candles
PD-23; PD-25; PD-28
dust control agent, metal foil rolling oils
PD-23; PD-25; PD-28
dust control agent, metal sheet rolling oils
PD-23; PD-25; PD-28
dust control agent, paper/paperboard
PD-23; PD-25; PD-28
dust control agent, pharmaceutical creams/lotions
PD-23; PD-25; PD-28
dust control agent, polishes
PD-23; PD-25; PD-28
dust control agent, steel mill furnaces
Dimethyl sulfide
dust control agent, vegetable froth flotation
PD-23; PD-25; PD-28
dust control agent, water treatment
PD-23; PD-25; PD-28
dust control agent, waterless hand cleaners
PD-23; PD-25; PD-28
dust control, roads
Calcium chloride
dustproofing agent, coal
Calcium chloride
dustproofing agent, coke
Calcium chloride
dustproofing agent, ore
Calcium chloride
dustproofing agent, sand
Calcium chloride
dustproofing agent, stone
Calcium chloride
dye
Acetic acid; *Acetone*; *Aniline*; *Benzene*; *Bromine*; *Chlorobenzene*; *o-Cresol*; *Di-n-amylamine*; *Dibromodifluoromethane*; *Dibutylamine*; *o-Dichlorobenzene*; *cis-1,2-Dichloroethylene*; *trans-1,2-Dichloroethylene*; *Dimethoxypropane*; *Dimethylamine*; *N,N-Dimethylaniline*; *Dimethylethanolamine*; *Dimethyl formamide*; *Ethylacetoacetate*; *Ethylamine*; *Fluorene*; *Hexamethyldisilazane*; *Hydroquinone monomethyl ether*; *Iodine*; *Isoquinoline*; *2,6-Lutidine*; *Methyl acetoacetate*; *α-Methylbenzyl alcohol*; *Nonylphenol*; *Pentylamine*; *Phenanthrene*; *Pyrocatechol*; *Quinoline*; *Sodium hydroxide*; *Sulfur*; *Thiophene*; *Toluene*; *o-Xylene*; *Xylenol*; *2,3-Xylenol*; *3,4-Xylenol*
dye remover
Potassium iodide
dye, hair
Hydroquinone; *Thymol*
dyeing assistant
Diethylene glycol; Eastman® TEP; *Triethyl phosphate*
dyeing assistant, aromatic amide fibers
N-Cyclohexyl pyrrolidone
dyeing assistant, blended fabrics
Arosolve FX-6
dyeing assistant, cotton
Coconut (Cocos nucifera) oil
dyeing assistant, disperse dyes: polyesters
Burcolev DLS
dyeing assistant, leather
PEG-30 castor oil; *PEG-40 castor oil*
dyeing assistant, natural fabrics
Arosolve FX-6
dyeing assistant, paints
PEG-30 castor oil; *PEG-40 castor oil*
dyeing assistant, paper
PEG-30 castor oil; *PEG-40 castor oil*
dyeing assistant, polyesters
Biphenyl
dyeing assistant, synthetic fabrics
Arosolve FX-6
dyeing assistant, textiles
Acetamide; *Dodecylbenzenesulfonic acid*; *Edetic acid*; *PEG-30 castor oil*; *PEG-40 castor oil*; *PEG-5 oleate*; *Sodium thiocyanate*
edible oil, food
Peanut (Arachis hypogaea) oil
edible oil, margarine
Peanut (Arachis hypogaea) oil
edible oil, salad oils
Peanut (Arachis hypogaea) oil
edible oil, shortening
Peanut (Arachis hypogaea) oil
electrical insulation impregnation wax
Petrolatum
electrostatic discharge machine fluid
Drakesol® 165; Drakesol® 165AT; Drakesol® 205; Drakesol® 220; Drakesol® 260; Drakesol® 260 AT; Drakesol® 305
embedding agent, crystals
Perfluoropolyether
emollient
Cetyl alcohol; *Cottonseed (Gossypium) oil*; *Decyl alcohol*; Dermol ICSA; *Dimethyl adipate*; *Dimethyl succinate*; *Dioctyl adipate*; *Dioctyl sebacate*; *Glycerin*; *Methyl acetyl ricinoleate*; *Methyl palmitate*; MPDiol® Glycol; *Neopentyl glycol dicaprate*; *PEG-12 dilaurate*; *PEG/PPG-17/6 copolymer*; *PPG-33-buteth-45*; Schercemol PGDP; *Sorbitan trioleate*; Ucon® 75-H-450; Ucon® Fluid AP; Wickenol® 127; Witcohol™ 42
emollient, absorption bases
Lanolin alcohol
emollient, acne creams/lotions
Crodamol CAP; Crodamol PMP; Crodamol W
emollient, acne preps.
Lipolan R
emollient, adhesives
Propylene glycol laurate
emollient, aerosol personal care
HallStar™ IPM
emollient, aerosol shaving creams
Lanolin acid
emollient, aerosols
Captex® 300; Captex® 810D; Captex® 8227; *Cyclomethicone*; Estol IPM 1509 (BIO-IPM); Lexol® PG-865; Myritol® PC; Pecosil® DCU; *Propylene glycol dipelargonate*; *Tricaprin*
emollient, after-bath prods.
Finsolv® TN-O
emollient, after-bath splashes
Trivent DIA
emollient, aftershave lotions
Captex® 810D; Captex® 8227; Estalan-DIPS; *Propylene glycol dioctanoate*; Trivent DIA
emollient, aftershaves
Glucam® P-20; HallStar™ DIPA; HallStar™ DIPS; Lexol® PG-800; Macol® 57
emollient, after-sun prods.
Finsolv® TN-O; Trivent OP
emollient, agric.
Incrocas 30; Incrocas 40; *PEG-20 dilaurate*; *PEG-4 dioleate*; *PEG-20 dioleate*; *PEG-4 oleate*; *PEG-6 oleate*; *PEG-20 oleate*; *Propylene glycol laurate*
emollient, agric. chemical sprays
PEG-8 dilaurate
emollient, alcoholic compositions
Procetyl 10
emollient, analgesics
Promyristyl PM-3
emollient, anhyd. oil systems
Hetester PMA; *Propylene glycol myristyl ether acetate*
emollient, animal feed supplements
Penreco Snow
emollient, antibiotic ointments
Crodamol CAP; Crodamol PMP; Crodamol W
emollient, antiperspirant sticks
Crystal® O; Crystal® Crown; *PPG-10 butanediol*
emollient, antiperspirant/deodorant gels

2452; Estalan-ISI; Estalan M-5; Estalan-PGDP; Estasan™ GT 8-60 3575; Estasan™ GT 8-65 3577; Estasan™ GT 8-70 3579; HallBrite® BHB; HallStar™ BST; HallStar™ DOM; HallStar™ IPM; *Hexyldecyl benzoate*; Isostriol IC; *Laneth-5*; *Laneth-10*; *Laneth-15*; Lexol® GT-855; Lexol® PG-865; Luvitol® BD 10 P; Monestriol B; Monestriol C; Monestriol IC; Monestriol IS-C; *Myristyl myristate*; *Myristyl propionate*; Naetex-LAM; Naturechem® GTR; Naturechem® MAR; NEOBEE® M-5; NEOBEE® M-20; *Octyldodecyl isostearate*; *Octyl isononanoate*; *Octyl oleate*; Palmil C; Palmil IS; Palmil O; Pecosil® DCU; *PEG-75 lanolin*; Pelemol® G7A; Penreco Snow; *Polyglyceryl-3 diisostearate*; *Propylene glycol dicaprylate/dicaprate*; *Propylene glycol dioctanoate*; *Stearyl stearate*; STEPAN® IPM-NF; Super Refined® Almond NF; Tegosoft® CI; Tegosoft® CT; Tegosoft® DO; Tegosoft® M; Tegosoft® P; Thomil 14; Thomil IS; Thomil IS-F; Tilol B; Tilol ET; Tilol IS; *Tricaprin*; *Tridecyl neopentanoate*

emollient, cutting oils
Methyl cocoate

emollient, day creams
Cetiol® B; *Dibutyl adipate*

emollient, decorative cosmetics
Finsolv® TN-O

emollient, delivery/absorption enhancement
Captex® 300; Captex® 350; *Tricaprin*

emollient, dental adhesives
Penreco Snow

emollient, deodorants
Cerette V; *Cyclomethicone*; Estalan 430; HallStar™ BST; HallStar™ DOA; *Hexyldecyl benzoate*; *Isopropyl palmitate*; Isostriol IC; Monestriol B; Monestriol C; Monestriol IC; Monestriol IS-C; Palmil C; Palmil IS; Palmil O; Promyristyl PM-3; Thomil 14; Thomil IS; Thomil IS-F; Tilol B; Tilol ET; Tilol IS

emollient, dermatologicals
Captex® 300; Captex® 350; *Propylene glycol dioctanoate*; Solulan® 16; Solulan® 75; Solulan® L-575; *Tricaprin*

emollient, detergents
Incrocas 30; Incrocas 40; *Oleth-2*; *PPG-10 methyl glucose ether*; *PPG-20 methyl glucose ether*

emollient, dry skin care
Propylene glycol dioctanoate

emollient, dry skin prods.
Finsolv® BOHS-111; *Octyl hydroxystearate benzoate*

emollient, dry skin treatment
Tegosoft® DO

emollient, dusting powders
STEPAN® IPM-NF

emollient, electric preshaves
Lexol® IPM-NF

emollient, emulsion shampoos
Glycol distearate

emollient, emulsions
Crodamol PMP; Hetester PMA; *Isocetyl alcohol*; *PPG-2 myristyl ether propionate*; *Propylene glycol myristyl ether acetate*

emollient, eye makeup
Ritacane IH

emollient, eye makeup removers
Glucam® P-10

emollient, eye shadows
STEPAN® IPM-NF

emollient, eyeliner
Ritacane ID

emollient, face cleansing toners
Glucam® P-20

emollient, face makeup
Ritacane IH

emollient, face preps.
Estalan-DIPS

emollient, facial cleansers
Pelemol® G7A; Promyristyl PM-3

emollient, fatty acid systems
Lanolin acid

emollient, fiber lubricants
PEG-4 dioleate

emollient, fine fragrances
Finsolv® TN-O

emollient, foam baths
Glycol distearate

emollient, foam baths/gels
Lanesta L

emollient, food
Glyceryl caprylate/caprate; Lipovol SES; *PEG-20 dilaurate*; *PEG-4 dioleate*; *PEG-20 dioleate*; *PEG-4 oleate*; *PEG-6 oleate*; *PEG-20 oleate*; Penreco Snow; *Propylene glycol laurate*; *Propylene glycol ricinoleate*

emollient, foundations
Captex® 200; *PPG-10 butanediol*

emollient, fragrance sticks
Promyristyl PM-3

emollient, fragrance-free prods.
Finsolv® TN-O

emollient, fragrances
Glucam® P-20; *Oleth-2*; Pecosil® DCU

emollient, fresheners
HallStar™ DIPA; HallStar™ DIPS; HallStar™ DOS

emollient, fruit/veg. coatings
Penreco Snow

emollient, gelatin capsules
Tricaprin

emollient, glosses
Lexol® PG-800; *Propylene glycol dipelargonate*

emollient, grooming prods.
Estalan DIA

emollient, hair care
Acetol® 1706; *Benzyl laurate*; *Butyl octyl benzoate*; Captex® 200; Captex® 300; *Cetearyl isononanoate*; Cropure® Olive; *Cyclomethicone*; HallBrite® BHB; HallStar™ IPM; *Hexyldecyl benzoate*; *Isodecyl isononanoate*; *Isononyl isononanoate*; *Isopropyl PPG-2-isodeceth-7-carboxylate*; *Isopropyl stearate*; *Isotridecyl isononanoate*; *Lactamide DGA*; Lonzest® 143-S; *Octamethylcyclotetrasiloxane*; *Octyldodecyl isostearate*; *PEG-75 lanolin*; *PEG-100 stearate*; Pelemol® G7A; Phytol™ CAN; Promyristyl PM-3; Schercemol OLO; Tegosoft® CI; Tegosoft® CT; Tegosoft® DO; Tegosoft® M; Tegosoft® P; *Triisostearin*; *Walnut (Juglans regia) oil*; *Wheat (Triticum vulgare) germ oil*; Wickenol® 151; Wickenol® 152; Wickenol® 153

emollient, hair conditioners
Fancorsil A; Glucam® P-10; HallBrite® BHB; HallStar™ DOM; Tegosoft® CI; Tegosoft® CT; Tegosoft® M; Tegosoft® P

emollient, hair conditioners/glosses
Tegosoft® DO

emollient, hair creams/lotions
Foamid LM2E

emollient, hair fixatives
Finsolv® TN-O

emollient, hair glosses
Tegosoft® CI; Tegosoft® CT; Tegosoft® M; Tegosoft® P

emollient, hair grooms
Octyl palmitate

emollient, hair sprays
Finsolv® TN-O; HallStar™ DOM; HallTress™ DIBA Special

emollient, hair tonics
Diisopropyl adipate

emollient, hand cleaners
Decyl oleate

emollient, hand creams
Trivent DIA

emollient, hand creams/lotions
Trivent OP

emollient, hand lotions
Cetearyl alcohol; Estalan DIA; Fancol™ ALA; Finsolv® TN-O; Schercemol DIA

emollient, household detergents
PEG-75 lanolin

emollient, household prods.
Incrocas 30; Incrocas 40; *Lanolin acid*

emollient, hydroalcoholic cosmetics
Schercemol DIA

emollient, hydroalcoholic personal care
Schercemol DIS

emollient, hydroalcoholic prods.
Glucam® E-10; Glucam® E-20; Glucam® P-10; Glucam® P-20

emollient, industrial
Isopropyl myristate; *Isopropyl palmitate*; *PEG-8 dilaurate*; *PPG-15 stearyl ether*; *Propylene glycol laurate*; *Propylene glycol ricinoleate*

emollient, industrial lubricants
PEG-8 dilaurate

emollient, infant formulas
Captex® 300; Captex® 350; *Tricaprin*

emollient, injectables
Sesame (Sesamum indicum) oil

emollient, insect repellents
Cerette V; Isostriol IC; Monestriol B; Monestriol C; Monestriol IC; Monestriol IS-C; Palmil C; Palmil IS; Palmil O; Promyristyl PM-3; Thomil 14; Thomil IS; Thomil IS-F; Tilol B; Tilol ET; Tilol IS

emollient, IV emulsions
Super Refined® Cottonseed NF; Super Refined® Olive NF; Super Refined® Safflower USP; Super Refined® Sesame NF

emollient, leather
Methyl cocoate; *Oleth-2*

emollient, liniments
Super Refined® Cottonseed NF; Super Refined® Olive NF; Super Refined® Safflower USP; Super Refined® Sesame NF

emollient, lip balm
HallBrite® BHB; Naturechem® GTR; Naturechem® MAR

emollient, lip care
Captex® 200; Captex® 300; Captex® 350; Emerest® 2452; *Polyglyceryl-3 diisostearate*

emollient, lip glosses
Captex® 810D; Captex® 8227

emollient, lip prods.
Pecosil® DCU

emollient, lipstick
Captex® 810D; Captex® 8227; Cropure® Babassu; Crystal® Crown; Estalan DIA; HallBrite® BHB; HallStar™ DOA; HallStar™ IPM; Krystal Kleer Castor Oil; Lexol® GT-855; Lexol® GT-865; Lexol® PG-800; Lipolan R; Naturechem® GTR; Naturechem® MAR; NEOBEE® M-5; Novol NF; *Octyl isononanoate*; *Propylene glycol dipelargonate*; Schercemol DIA; STEPAN® IPM-NF; USP Castor Oil

emollient, liq. emulsions
Cetiol® B

emollient, liq. makeup
Ceraphyl® 140-A; Trivent OP

emollient, liq. preps.
Lutrol® E 300; Lutrol® E 400

emollient, liqs.
Estalan 430

emollient, lotions
Myritol® 318; *PPG-10 butanediol*; *Sesame (Sesamum indicum) oil*; *Shea butter (Butyrospermum parkii)*

emollient, lubricants
PEG-7 glyceryl cocoate; *Propylene glycol laurate*

emollient, lubricating oils
Methyl cocoate

emollient, make-up
Octyl palmitate

emollient, makeup
Captex® 200; Crodamol AB; Crodamol CAP; Cromollient DP3A; Crystal® O; Crystal® Crown; *Di-PPG-3 myristyl ether adipate*; Fancorsil A; Glucam® P-10; HallStar™ BST; HallStar™ DIPA; HallStar™ DIPS; HallStar™ DOA; HallStar™ DOS; *Hexyldecyl benzoate*; Lexol® GT-855; Lexol® GT-865; Lexol® PG-865; Lipolan R; *Myristyl myristate*; *Octyl isononanoate*; *PEG-75 lanolin*; Promyr; Promyristyl PM-3; Propal NF; *Propylene glycol dicaprylate/dicaprate*; *Propylene glycol dioctanoate*; STEPAN® IPM-NF; *Walnut (Juglans regia) oil*

emollient, makeup bases
Captex® 810D; Captex® 8227; Lexol® PG-800; NEOBEE® M-5; *Propylene glycol dipelargonate*

emollient, makeup removers
Estalan-PGDP; HallStar™ BST; *Octyldodecyl isostearate*; *PPG-10 butanediol*; Schercemol NGDO

emollient, mascara
Captex® 200; Captex® 300; Captex® 350; *Isoeicosane*; Ritacane ID

emollient, massage oils
Cetiol® B

emollient, massage preps.
Cetearyl isononanoate

emollient, men's grooming aids
Diisopropyl sebacate; Schercemol DIA

emollient, metal processing
PPG-11 stearyl ether

emollient, metalworking
Isobutyl palmitate; *PEG-4 dioleate*; *PEG-4 oleate*; *Propylene glycol laurate*

emollient, metalworking fluids
Incrocas 30; Incrocas 40

emollient, milk lotions

Peanut NF; Super Refined® Sesame NF; Super Refined® Soybean USP; Tegosoft® CI; Tegosoft® CT; Tegosoft® DO; Tegosoft® M; Tegosoft® P; Tegosoft® TN; *Triisostearin*; Trivent DIA; Trivent OP; Varonic® APS; *Walnut (Juglans regia) oil*; *Wheat (Triticum vulgare) germ oil*; Wickenol® 152; Wickenol® 153

emollient, skin cleansers
Acetylated lanolin alcohol; HallStar™ DOM

emollient, skin cosmetics
Lonzest® 143-S

emollient, skin creams
Estol IPM 1509 (BIO-IPM)

emollient, skin creams/lotions
Cromollient DP3A; Cropure® Babassu; Foamid LM2E; HallStar™ DIPA; HallStar™ DIPS; HallStar™ DOS; *Isopropyl myristopalmitate*; LexFeel® 7; Lexol® IPL; *Neopentyl glycol diheptanoate*; *Tallow alcohol*

emollient, skin emulsions
Isostearyl alcohol

emollient, skin fresheners
Macol® 57

emollient, skin lotions
HallTress™ DIBA Special

emollient, skin oils
Cetiol® B

emollient, skin protective oils
Dibutyl adipate

emollient, soap
PPG-15 stearyl ether benzoate

emollient, soap/bath prods.
Lipolan R

emollient, soap/syndet systems
PPG-15 stearyl ether benzoate

emollient, soaps
Acetylated lanolin alcohol; Cachalot® AR-20; Cachalot® BE-22; Cachalot® C-50; Cachalot® L-90; Cachalot® M-43; Cachalot® S-56; Crystal® O; Crystal® Crown; *Isopropyl stearate*; Naturechem® GTR; Naturechem® MAR; Tegosoft® CI; Tegosoft® CT; Tegosoft® DO; Tegosoft® M; Tegosoft® P

emollient, soft gelatin capsules
Captex® 300

emollient, splash-on prods.
Pelemol® G7A

emollient, sports supplements
Tricaprin

emollient, stick formulations
Crodamol W

emollient, stick preps.
Cromollient DP3A; *Cyclomethicone*; *Di-PPG-3 myristyl ether adipate*; Luvitol® BD 10 P; *Tricaprin*; Witconol™ APM

emollient, styling prods.
Glucam® P-10

emollient, substantive: cosmetics
Schercemol 1818

emollient, sun care
Captex® 200; Captex® 300; Captex® 350; Estol IPM 1509 (BIO-IPM); *Isopropyl PPG-2-isodeceth-7-carboxylate*; Lipolan R; Ritacane ID; Tegosoft® TN

emollient, sun creams
Cerette V; Isostriol IC; Monestriol B; Monestriol C; Monestriol IC; Monestriol IS-C; Palmil C; Palmil IS; Palmil O; Thomil 14; Thomil IS; Thomil IS-F; Tilol B; Tilol ET; Tilol IS

emollient, sun preps.
Cropure® Babassu

emollient, sun protection oils
Cetiol® B

emollient, sun protection preps.
Cetearyl isononanoate

emollient, sunscreen oils
Luvitol® BD 10 P

emollient, sunscreens
Babassu (Orbignya oleifera) oil; Crodamol AB; Cromollient DP3A; *Di-PPG-3 myristyl ether adipate*; Finsolv® TN-O; HallBrite® BHB; HallStar™ DIPA; HallStar™ DIPS; HallStar™ DOA; HallStar™ DOM; HallStar™ DOS; *Hexyldecyl benzoate*; *Isopropyl stearate*; Lanesta L; Naturechem® GTR; Naturechem® MAR; Novol NF; *Octyl palmitate*; Pecosil® DCU; Promyristyl PM-3; Tegosoft® CI; Tegosoft® CT; Tegosoft® DO; Tegosoft® M; Tegosoft® P

emollient, suntan oils
Fancol™ ALA

emollient, suntan preps.
Acetol® 1706; Estalan DIA

emollient, suppositories
Propylene glycol dioctanoate; *Shea butter (Butyrospermum parkii)*

emollient, suspensions
Captex® 300

emollient, technical prods.
Acetylated hydrogenated lard glyceride

emollient, textile lubricants
PEG-8 dilaurate

emollient, textiles
Isobutyl palmitate; *Methyl cocoate*; *Oleth-2*; *PEG-75 lanolin*; *PEG-4 oleate*; *PEG-6 oleate*; *Propylene glycol laurate*; *Propylene glycol ricinoleate*

emollient, therapeutic prods.
Finsolv® BOHS-111; *Octyl hydroxystearate benzoate*

emollient, toilet water
HallTress™ DIBA Special; HallStar™ DIPA; HallStar™ DIPS; HallStar™ DOA

emollient, tonics
HallStar™ DIPA; HallStar™ DIPS; HallStar™ DOS

emollient, topical medicinals
Estol IPM 1509 (BIO-IPM); *Isopropyl myristate*

emollient, topical ointments
Fancorsil A

emollient, topical pharmaceuticals
Arachidyl alcohol; *Behenyl behenate*; Ceraphyl® 28; Ceraphyl® 140-A; Ceraphyl® ICA; *Cetyl acetate*; *Cetyl octanoate*; Crodamol PMP; Crodamol W; *Diisopropyl adipate*; *Diisopropyl sebacate*; Estasan™ GT 8-60 3575; Estasan™ GT 8-60 3580; Estasan™ GT 8-65 3577; Estasan™ GT 8-70 3579; Estol IPM; Etocas 35 NF; Incrocas 30; Incrocas 40; *Isocetyl isostearate*; *Isocetyl stearate*; *Isodecyl oleate*; *Isononyl isononanoate*; *Isopropyl stearate*; *Laneth-15*; Lexol® 3975; *Methyl stearate*; Miglyol® 810; *Myristyl myristate*; *Octyl hydroxystearate*; *Octyl isononanoate*; *Octyl palmitate*; *Octyl stearate*; Pecosil® DCU; Pelemol® G7A; Polychol 5; Polychol 15; Procetyl AWS; Promyr; Schercemol NGDO; Schercemol OLO; Solulan® 16; Solulan® 75; Solulan® L-575; *Stearyl octanoate*; Super Refined® Crodamol SCO; Tegosoft® CT; Tegosoft® DO; Tegosoft® M; *Trihydroxystearin*; Wickenol® 158; Wickenol® 171

emollient, topicals
Crodamol GTCC-PN; Lexol® PG-800; *PPG-5-ceteth-20*; *Propylene glycol dioctanoate*; Schercemol PGML; Super Refined® Cottonseed NF; Super Refined® Olive NF; Super Refined® Safflower USP; Super Refined® Sesame NF; Tegosoft® CI; Tegosoft® P

emollient, urethane applics.
Krystal Kleer Castor Oil; USP Castor Oil

emollient, valve: lotions
Promyristyl PM-3

emollient, valve: sprays
Promyristyl PM-3

emollient, veterinary medicine
PEG-35 castor oil

emollient, veterinary preps.
Petrolatum

emollient, w/o emulsions
Shea butter (Butyrospermum parkii)

emollient, waterless hand cleaners
Drakeol® 7; Lonzest® 143-S

emollient, wax systems
Lanolin acid

emulsifier
Acetylated hydrogenated soybean oil glyceride; *Acetylated lard glyceride*; Arosolve TUV; Berol® 02; *But-2-yne-1,4-diol*; Cachalot® AR-20; Cachalot® BE-22; Cachalot® C-50; Cachalot® L-90; Cachalot® M-43; Cachalot® S-56; *Caprylic alcohol*; *Caprylic/capric/stearic triglyceride*; *Carbomer*; *Cetyl alcohol*; Crill 2; *Ethylamine*; *Ethylenediamine*; *Ethylenediamine dihydrochloride*; *Ethyl morpholine*; *Glyceryl dioleate*; *Glyceryl ricinoleate*; *Hydrogenated menhaden oil*; *Isooctyl alcohol*; *PEG-32*; *PEG-12 dioleate*; *Propylene glycol alginate*; Surfadone® LP-100; Surfadone® LP-300; *Tetrahydroxypropyl ethylenediamine*; *Trimyristin*; Witcohol™ 42; Witconol™ APM; Witconol™ APS

emulsifier, acid aluminum cleaners
Rhodacal® 330

emulsifier, acid media
Diethylaminoethanol

emulsifier, acrylic latexes
2-Pyrrolidone

emulsifier, adhesives
Crill 1; *Propylene glycol laurate*; *Propylene glycol stearate*; *Sodium 2-ethylhexyl sulfate*; *Sorbitan palmitate*

emulsifier, aerosol shaving creams
Lanolin acid

emulsifier, aerosols
Ameroxol® OE-2

emulsifier, agric.
Diethanolamine; Incrocas 30; Incrocas 40; *Nonoxynol-4*; *Nonoxynol-5*; *Nonoxynol-6*; *Nonoxynol-9 phosphate*; *PEG-4 dilaurate*; *PEG-12 dilaurate*; *PEG-20 dilaurate*; *PEG-4 dioleate*; *PEG-8 dioleate*; *PEG-12 dioleate*; *PEG-20 dioleate*; *PEG-4 oleate*; *PEG-6 oleate*; *PEG-8 oleate*; *PEG-20 oleate*; *Pentaerythrityl oleate*; *Poloxamer 181*; *Propylene glycol laurate*; *Sorbitan trioleate*

emulsifier, agric. chemical sprays
PEG-8 dilaurate

emulsifier, agric. chemicals
PEG-100 stearate; *Trioctyl phosphate*

emulsifier, agric. pesticides
Rhodacal® 330

emulsifier, agric. toxicants
Surphos 515; Surphos 620; Surphos 630; Surphos 645

emulsifier, air fresheners
Methylpropanediol; MPDiol® Glycol

emulsifier, antiperspirant sticks
Crystal® O; Crystal® Crown

emulsifier, antiperspirants
PEG-75 lanolin; *PPG-5-ceteth-20*; *PPG-2 myristyl ether propionate*

emulsifier, antistats
PEG-12 dilaurate

emulsifier, aq. alcoholic systems
PPG-5-ceteth-20; Procetyl AWS

emulsifier, aq. paints
Diisopropanolamine

emulsifier, aq. systems
Procetyl AWS

emulsifier, auxiliary: personal care
Protachem™ 100; Protachem™ G-5509; Protachem™ G-5566; Protachem™ GC-7; Protachem™ GC-30; Protachem™ GDL; Protachem™ GMS-165; Protachem™ GMS-450; Protachem™ GMS-AS; Protachem™ GMS-D; Protachem™ GMS-T; Protachem™ HMS; Protachem™ MLD

emulsifier, auxiliary: soap o/w emulsions
Glyceryl stearate SE

emulsifier, auxiliary: topical pharmaceuticals
Glyceryl stearate SE

emulsifier, baby care
Acetylated lanolin alcohol

emulsifier, baked goods
Hodag® S-35 FGK

emulsifier, bath oils
Caprylic/capric triglyceride PEG-4 esters; *PEG-40 sorbitan peroleate*

emulsifier, beverages
Diacetyl tartaric acid esters of mono- and diglycerides; *Sucrose fatty acid esters*

emulsifier, chemical industry
Sovatex C Series

emulsifier, chemical intermediates
PEG-12 dilaurate

emulsifier, chewing gum
Sucrose fatty acid esters

emulsifier, chlorinated insecticides
Surphos 515; Surphos 620; Surphos 630; Surphos 645

emulsifier, clay soils
Deceth-6

emulsifier, cleaners
2-Aminobutanol; *Diethanolamine*; Rhodacal® 330

emulsifier, cleansing milks
PEG-6 abietate

emulsifier, clinical nutrition
Capmul® MCM

emulsifier, coatings
Diethanolamine; *Dimethylethanolamine*; *Glyceryl ricinoleate*; *Methoxypropylamine*; *PEG-8 oleate*

emulsifier, coffee whiteners
Hodag® S-35 FGK

emulsifier, commercial cleaners

Sorbitan trioleate; Variquat® CC-42; Variquat® CC-57
emulsifier, industrial chemicals
Polysorbate 60
emulsifier, industrial cleaners
Cocamide DEA; *Nonoxynol-5*; *PEG-100 stearate*; *Phenoxyethanol*; *Sodium 2-ethylhexyl sulfate*
emulsifier, industrial lubricants
PEG-8 dilaurate
emulsifier, industrial processes
PPG-40 diethylmonium chloride
emulsifier, inks
Crill 1; *Oleyl alcohol*
emulsifier, insecticides
Crill 1; *Diisopropanolamine*; *PEG-100 stearate*
emulsifier, institutional cleaners
Laureth-9
emulsifier, iodophors
Ethylan® HB4
emulsifier, latex paints
Nonoxynol-5; Rhodacal® 330; T-Det® N-4
emulsifier, latexes
Isopropylamine dodecylbenzenesulfonate
emulsifier, laundry
Sovatex C Series
emulsifier, laundry prespotters
BIO-SOFT® N-411
emulsifier, leather
Diisopropanolamine; *Oleth-2*; *PEG-2 stearate*; *Pentaerythrityl oleate*; *Sorbitan trioleate*; Sovatex C Series; *Trioctyl phosphate*; *Triolein*
emulsifier, leather treatment
Crill 1
emulsifier, lip balm
Naturechem® GTR
emulsifier, lip care
Capmul® MCM; Capmul® MCMC8; Emerest® 2452; *Polyglyceryl-3 diisostearate*
emulsifier, lipstick
Crystal® Crown; Naturechem® GTR
emulsifier, liq. preps.
Lutrol® E 300; Lutrol® E 400
emulsifier, lotions
Propylene glycol stearate
emulsifier, lubricant removal
Stetlasolv SES-CONC.
emulsifier, lubricants
Hexylene glycol; *PEG-7 glyceryl cocoate*; *Propylene glycol laurate*
emulsifier, maintenance cleaners
Laureth-9
emulsifier, makeup
Crystal® O; Crystal® Crown; *Myristyl myristate*; *PEG-75 lanolin*; *Propylene glycol stearate*
emulsifier, margarine
Peanut (Arachis hypogaea) oil
emulsifier, medicinals
Polysorbate 80
emulsifier, metal cleaning
Rhodacal® 330; *Sodium 2-ethylhexyl sulfate*
emulsifier, metals
PPG-3 myristyl ether; *PPG-11 stearyl ether*; *Triolein*
emulsifier, metalworking
Glycol distearate; *Isodecyl oleate*; *Nonoxynol-4*; *Nonoxynol-6*; *PEG-4 dilaurate*; *PEG-4 dioleate*; *PEG-8 dioleate*; *PEG-4 oleate*; *Propylene glycol laurate*; *Trioctyl phosphate*
emulsifier, metalworking fluids
Crill 1; *Dodecylbenzenesulfonic acid*; *Glyceryl dioleate*; Incrocas 30; Incrocas 40; *Isodecanoic acid*; *Naphthenic acid*; *Oleic acid*; Rhodacal® 330
emulsifier, metalworking lubricants
PEG-12 dilaurate
emulsifier, metalworking/cleaning
Nonoxynol-5
emulsifier, methylene chloride
Surphos 620; Surphos 645
emulsifier, mineral oils
Chemax E-200 MO; Chemax E-1000 MO
emulsifier, mold release compds.
PEG-7 glyceryl cocoate
emulsifier, nasal pharmaceuticals
Labrafac® Hydrophile
emulsifier, natural fats
Sorbitan trioleate
emulsifier, natural waxes
Sorbitan trioleate
emulsifier, nutritional/sports supplements
Capmul® MCM
emulsifier, o/w
Aqualose W15; Variquat® CC-55
emulsifier, o/w: cleaners
PEG-20 sorbitan isostearate
emulsifier, o/w: cosmetics
Polychol 15
emulsifier, o/w: creams/lotions
PEG-20 sorbitan isostearate
emulsifier, o/w: food
Sucrose palmitate
emulsifier, o/w: fragrances/perfumes
Crillet 6
emulsifier, o/w: hair care
Lactamide DGA
emulsifier, o/w: hair creams/lotions
Foamid LM2E
emulsifier, o/w: lipstick
PEG-10 lanolate; *PEG-20 lanolate*
emulsifier, o/w: lipstick bromo acid dyes
PEG-5 lanolate
emulsifier, o/w: makeup
PEG-5 lanolate; *PEG-10 lanolate*; *PEG-20 lanolate*
emulsifier, o/w: ointments
PEG-20 sorbitan isostearate
emulsifier, o/w: personal care
Polychol 5
emulsifier, o/w: pesticides
PEG-20 sorbitan isostearate
emulsifier, o/w: polishes
PEG-20 sorbitan isostearate
emulsifier, o/w: skin care
Lactamide DGA; *PEG-5 lanolate*; *PEG-10 lanolate*; *PEG-20 lanolate*
emulsifier, o/w: skin creams/lotions
Foamid LM2E
emulsifier, o/w: textiles
PEG-20 sorbitan isostearate
emulsifier, o/w: topical pharmaceuticals
Crillet 6; Polychol 5; Polychol 15
emulsifier, oil slick
Rhodacal® 330
emulsifier, oil slick dispersing
Crill 1
emulsifier, oil-based systems
Isopropylamine dodecylbenzenesulfonate
emulsifier, oils
2-Aminobutanol; *Glyceryl laurate*; *Glyceryl stearate*; *p-Methyl morpholine*; *Nonoxynol*; *Trideceth*
emulsifier, oils: household cleaners
SECOSTER® A
emulsifier, oils: industrial cleaners
SECOSTER® A
emulsifier, oils: oil field prod.
SECOSTER® A
emulsifier, ointments
Caprylic/capric triglyceride PEG-4 esters; *Polyglyceryl-4 oleate*; *Sucroglycerides*; *Sucrose fatty acid esters*
emulsifier, olein
Trideceth
emulsifier, one-bath scouring
Burcolev DLS
emulsifier, oral pharmaceuticals
Etocas 35 NF; Labrafac® Hydrophile
emulsifier, paint strippers
BIO-SOFT® N-411
emulsifier, paints
Crill 1; *Dimethylethanolamine*; *Laureth-9*; *Nonoxynol-4*; *Nonoxynol-5*; *Nonoxynol-6*; *Oleyl alcohol*; *PEG-20 oleate*; *Poloxamer 181*; *Polysorbate 60*; *Polysorbate 80*; *Propylene glycol*; *Propylene glycol laurate*; *Propylene glycol stearate*; *Sodium 2-ethylhexyl sulfate*
emulsifier, paints/coatings
Dodecylbenzenesulfonic acid; *Sodium lauryl sulfate*
emulsifier, paper
Crill 1; *Glyceryl dilaurate*; *Isodecyl oleate*; *Laureth-9*; *Nonoxynol-4*; *Nonoxynol-6*; *Oleth-2*; *PEG-2 stearate*; *Poloxamer 181*
emulsifier, parenteral pharmaceuticals
Etocas 35 NF; Incrocas 40
emulsifier, parenterally administered cyclosporine
Cremophor® EL
emulsifier, perfumes
PEG-20 dioleate; *PEG-20 sorbitan isostearate*; *PPG-5-ceteth-20*
emulsifier, personal care
Butyl oleate; *Carbomer 940*; Crystal® Crown; *Glyceryl stearate*; *Glyceryl stearate SE*; *Hexyl laurate*; Incrocas 30; Incrocas 40; *Isobutyl oleate*; *Isopropyl oleate*; *Methyl gluceth-10*; *Methylpropanediol*; MPDiol® Glycol; *Myristyl myristate*; *PEG-8 dilaurate*; *PEG-7 glyceryl cocoate*; *PEG-30 glyceryl cocoate*; *Polyglyceryl-9 stearate*; *PPG-10 cetyl ether*; *PPG-40 diethylmonium chloride*; *PPG-3 myristyl ether*; *PPG-11 stearyl ether*; Procetyl AWS; Ritawax AEO; Solulan® 16; Solulan® 75; Solulan® C-24; Solulan® L-575; Variquat® CC-42; Variquat® CC-57
emulsifier, pesticides
Dimethyl caproamide; *Dimethyl octanamide*; *Dodecylbenzenesulfonic acid*; *Nonoxynol*; *PEG-6 abietate*; *Polysorbate 60*
emulsifier, petrochemicals
Oleyl alcohol
emulsifier, petroleum
Nonoxynol-5; Trax K-40
emulsifier, petroleum oils
Nonoxynol-4
emulsifier, petroleum oils/waxes
Sorbitan trioleate
emulsifier, pharmaceutical absorption enhancement
Polyglyceryl-3 stearate
emulsifier, pharmaceutical aerosols
Capmul® MCM
emulsifier, pharmaceutical coatings
Capmul® MCM; Capmul® MCMC10; *Glyceryl caprate*
emulsifier, pharmaceutical creams/lotions
Acetylated lanolin alcohol; *Sucroglycerides*; *Sucrose fatty acid esters*
emulsifier, pharmaceutical emulsions
Olepal ISO
emulsifier, pharmaceutical inj. solutions
Cremophor® ELP
emulsifier, pharmaceutical microemulsions
Capmul® MCM; Capmul® MCMC8; Capmul® MCMC10; *Glyceryl caprate*; Olepal ISO
emulsifier, pharmaceutical otics
Propylene glycol diacetate
emulsifier, pharmaceutical soft gelatin capsules
Glyceryl caprate
emulsifier, pharmaceutical topical ointments
Mono- and diglycerides of fatty acids
emulsifier, pharmaceutical vehicles
PEG-75 lanolin
emulsifier, pharmaceuticals
Acetylated mono- and diglycerides of fatty acids; Ameroxol® OE-2; *2-Aminobutanol*; *Apricot kernel oil PEG-6 esters*; Capmul® MCM; Capmul® MCMC8; *Caprylic/capric triglyceride PEG-4 esters*; *Carbomer 940*; *Cetearyl octanoate*; *Ceteth-16*; *Ceteth-24*; *Cetyl alcohol*; *Coconut (Cocos nucifera) oil*; Comperlan® COD; Crill 1; Crystal® O; Crystal® Crown; *Diethanolamine*; *Ethanolamine*; *Glycereth-26*; *Glyceryl caprate*; *Glyceryl caprylate*; *Glyceryl caprylate/caprate*; *Glyceryl dilaurate*; *Glyceryl laurate*; *Glyceryl myristate*; *Glyceryl oleate*; *Glyceryl ricinoleate*; *Glyceryl stearate*; *Glyceryl stearate SE*; *Hexyl laurate*; *Hydrogenated castor oil*; *Laneth-5*; *Laneth-10*; *Laneth-16*; *Lanolin acid*; *Lanolin alcohol*; Lipocol P-15; Lumulse® GMO; *Nonoxynol-6*; *Oleic acid*; *Oleth-2*; *Oleth-16*; *PEG-8 caprylic/capric glycerides*; *PEG-30 castor oil*; *PEG-33 castor oil*; *PEG-35 castor oil*; *PEG-40 castor oil*; *PEG-4 dilaurate*; *PEG-8 dilaurate*; *PEG-12 dilaurate*; *PEG-20 dilaurate*; *PEG-4 dioleate*; *PEG-8 dioleate*; *PEG-12 dioleate*; *PEG-20 dioleate*; *PEG-7 glyceryl cocoate*; *PEG-30 glyceryl cocoate*; *PEG-6 isostearate*; *PEG-4 oleate*; *PEG-6 oleate*; *PEG-8 oleate*; *PEG-20 oleate*; *PEG-8 palmitostearate*; *PEG-150 palmitostearate*; *PEG-100 stearate*; *Pentaerythrityl oleate*; *Poloxamer 181*; *Polyglyceryl-3 diisostearate*; *Polyglyceryl polyricinoleate*; *Polysorbate 60*; *Polysorbate 80*; *PPG-15 stearyl ether*; *Propylene glycol*; *Propylene glycol alginate*; *Propylene glycol laurate*; *Propylene glycol ricinoleate*; *Propylene glycol stearate*; Ritawax AEO; *Sodium 2-ethylhexyl sulfate*; *Sorbeth-20*; *Sorbitan laurate*; *Sorbitan trioleate*; Stearate 1500; *Stearyl alcohol*; *Sucrose palmitate*; *Sunflower (Helianthus annuus) seed oil*; *Tetrahydroxypropyl ethylenediamine*; *Trideceth*; *Triethanolamine*
emulsifier, pigments
Variquat® CC-57
emulsifier, plastics

EP agent, bearings
Perfluoropolyether
EP agent, greases
Dibutyl phosphite
EP agent, instruments
Perfluoropolyether
EP agent, lubricants
Dibutyl phosphite; Paraffin, chlorinated
EP agent, metalworking coatings
Lanolin
EP agent, metalworking fluids
Paraffin, chlorinated
EP agent, rust preventative coatings
Lanolin
EP agent, surgical equip.
Perfluoropolyether
EP agent, textiles
Perfluoropolyether
EP agent, vacuum pumps
Perfluoropolyether
essence, fruit
2,2-Dimethylpropanol; Ethyl octanoate; Isoamyl butyrate; Isobutyl propionate; Methyl butyrate
essential oil
Citrus Stripper Oil; Citrus Terpenes
essential oil, food
Orange Terpenes
essential oil, industrial
Orange Terpenes
esterification agent, organic compds.: acaricides
2-Ethyl-1-butanol
esterification agent, organic compds.: crop protection prods.
2-Ethyl-1-butanol
esterification agent, organic compds.: diffusion pump oils
2-Ethyl-1-butanol
esterification agent, organic compds.: fungicides
2-Ethyl-1-butanol
esterification agent, organic compds.: heat transfer media
2-Ethyl-1-butanol
esterification agent, organic compds.: herbicides
2-Ethyl-1-butanol
esterification agent, organic compds.: hydraulic fluids
2-Ethyl-1-butanol
esterification agent, organic compds.: insecticides
2-Ethyl-1-butanol
esterification agent, organic compds.: lubricant additives
2-Ethyl-1-butanol
esterification agent, organic compds.: pharmaceuticals
2-Ethyl-1-butanol
esterification agent, organic compds.: plasticizers
2-Ethyl-1-butanol
esterification agent, organic compds.: wetting agents
2-Ethyl-1-butanol
esterifying agent
Trimethylcyclohexanol
etchant, semiconductor prod.
Acetic acid
etching agent, metal surface finishing
Trichloroacetic acid
etherification agent, food starches
Acrolein
ethoxymethylation agent, organic compd. synthesis
Diethoxymethane
ethylating agent, organic synthesis
Ethyl bromide
ethylation reagent
Ethyl chloride; Ethyl ether
ethylene glycol substitute
Propylene glycol
ethylene source
Naphtha
evacuant, gall bladder
Peanut (Arachis hypogaea) oil
evaporation control agent
Cachalot® AR-20; Cachalot® BE-22; Cachalot® C-50; Cachalot® L-90; Cachalot® M-43; Cachalot® S-56
evaporation inhibitor, controlled-releases
Polyethylene glycol
evaporation inhibitor, rubbing alcohol
Polyethylene glycol
evaporation rate control aid, aq. coatings
Tecsol® C 95%
evaporation rate control aid, industrial coatings
Tecsol® C 95%
evaporation rate control aid, trade sales coatings
Tecsol® C 95%
evaporation retarder, automobile lacquers
Ethoxyethanol acetate
excipient, aerosols
Labrafil® Isostearique
excipient, cosmetics
Labrafil® Isostearique; Labrasol®; *Petrolatum*; Pluracol® E1450 NF
excipient, creams
Labrafil® Isostearique
excipient, injectable substances
Cerette V; Isostriol IC; Monestriol B; Monestriol C; Monestriol IC; Monestriol IS-C; Palmil C; Palmil IS; Palmil O; Thomil 14; Thomil IS; Thomil IS-F; Tilol B; Tilol ET; Tilol IS
excipient, inks
Pluracol® E1450 NF
excipient, massage oils
Labrafil® Isostearique
excipient, metal cleaners
Pluracol® E1450 NF
excipient, metal polishes
Pluracol® E1450 NF
excipient, nasal preps.
Labrafil® Isostearique
excipient, oral care
Pluracol® E1450 NF
excipient, oral pharmaceuticals
Alcohol
excipient, paper
Pluracol® E1450 NF
excipient, pharmaceutical gelatin capsules
Labrafac® Lipophile
excipient, pharmaceuticals
Apricot kernel oil PEG-6 esters; Cetyl palmitate; Coconut (Cocos nucifera) oil; Diacetylated monoglycerides; Edetic acid; Glyceryl oleate; Glyceryl stearate; Glycol stearate; Labrasol®; *Methyl alcohol; PEG-8 caprylic/capric glycerides; Petrolatum*; Pluracol® E400 NF; Pluracol® E600 NF; Pluracol® E1450 NF; *Polyglyceryl-3 diisostearate; Propylene glycol laurate; Propylene glycol stearate; Rapeseed (Brassica campestris) oil; Sucrose palmitate; Triisostearin PEG-6 esters*
excipient, rubber
Pluracol® E1450 NF
excipient, textiles
Pluracol® E1450 NF
excipient, veterinary preps.
Labrafil® Isostearique; Labrasol®; *Petrolatum*
excipient, wood prods.
Pluracol® E1450 NF
expectorant
Sunflower (Helianthus annuus) seed oil
expectorant, pharmaceuticals
Ethyl ether; Eucalyptol; Sodium iodide
expedient, cosmetics
Cachalot® AR-20; Cachalot® BE-22; Cachalot® C-50; Cachalot® L-90
extender
Mineral oil; Promyristyl PM-3
extender, adhesives
Halso® 99; *Monochlorotoluene*
extender, aerosols
Captex® 810D; Captex® 8227
extender, aftershave lotions
Captex® 810D; Captex® 8227
extender, asphalt
Sulfur
extender, bath oils
Captex® 810D; Captex® 8227
extender, caulks
Naphthenic oil
extender, clinical nutrition
Captex® 350
extender, coatings
Morcryl® 535
extender, cosmetics
Caprylic/capric/lauric triglyceride; Caprylic/capric/linoleic triglyceride; Captex® 810D; Captex® 8227; Jarcol™ I-16; Pluracol® E200; Pluracol® E300; Pluracol® E400; Pluracol® E600; Pluracol® E4000; Pluracol® E6000
extender, delivery/absorption enhancement
Captex® 350
extender, dermatologicals
Captex® 350; *Propylene glycol dioctanoate*
extender, detergents
Jarcol™ I-16
extender, dry skin care
Propylene glycol dioctanoate
extender, dye carriers: printing
Monochlorotoluene
extender, dye carriers: textiles
Monochlorotoluene
extender, dyes
Hexyldecanol
extender, emulsifiers
Jarcol™ I-16
extender, flavor oils: food
Isocetyl alcohol
extender, flavor oils: pharmaceuticals
Isocetyl alcohol
extender, flavors
Ceraphyl® ICA
extender, food
Methoxyethanol
extender, food pkg. adhesives
Morcryl® 535
extender, food-contact applics.
Naphthenic oil
extender, fragrance oils
Hexyldecanol
extender, fragrance oils: food
Isocetyl alcohol
extender, fragrance oils: pharmaceuticals
Isocetyl alcohol
extender, fragrances
Ceraphyl® ICA; *Isopropyl palmitate*
extender, fuel oils
Monochlorotoluene
extender, infant formulas
Captex® 350
extender, inks
Morcryl® 535
extender, lip gloss
Captex® 810D; Captex® 8227
extender, lipstick
Captex® 810D; Captex® 8227; *Isocetyl alcohol*
extender, lubricants
Jarcol™ I-16
extender, makeup bases
Captex® 810D; Captex® 8227
extender, metal parts cleaners
Halso® 99; *Monochlorotoluene*
extender, metals
Jarcol™ I-16
extender, nutritional applics.
Caprylic/capric/lauric triglyceride; Caprylic/capric/linoleic triglyceride; Caprylic/capric/oleic triglyceride; Captex® 810D; Captex® 8227; *Triundecanoin*
extender, nutritional/sports supplements
Captex® 350
extender, overprints
Morcryl® 535
extender, paint strippers
Monochlorotoluene
extender, paint thinners
Halso® 99; *Monochlorotoluene*
extender, paints
Sodium silicoaluminate; Terpene resin
extender, paper
Morcryl® 535
extender, paper coating
Jarcol™ I-16
extender, personal care
Pluracol® E200; Pluracol® E300; Pluracol® E400; Pluracol® E600; Pluracol® E4000; Pluracol® E6000
extender, pharmaceutical coatings
Captex® 350
extender, pharmaceuticals
Caprylic/capric/lauric triglyceride; Caprylic/capric/linoleic triglyceride; Caprylic/capric/oleic triglyceride; Captex® 350; Captex® 355EP; Captex® 500P; Captex® 810D; Captex® 8227; Jarcol™ I-16; Pluracol® E300; Pluracol® E400; Pluracol® E600; Pluracol® E4000; Pluracol® E6000; *Propylene glycol dioctanoate; Triacetin*
extender, pigments
Isocetyl alcohol
extender, plastics
Jarcol™ I-16; *Paraffin, chlorinated*
extender, polymers
Jarcol™ I-16
extender, PU
Dytek® EP Diamine; *1,3-Pentanediamine*
extender, PVC
Naphthenic oil
extender, resins
Naphthenic oil
extender, rubber
Jarcol™ I-16; *Naphthenic oil; Paraffin, chlorinated*
extender, solvents
Jarcol™ I-16
extender, suppositories
Propylene glycol dioctanoate
extender, textiles
Jarcol™ I-16; *Naphthenic oil*
extractant, solvents
Amgard® TBPO$_4$
extracting agent, cellulose derivs.
1,3-Dioxolane

Acetyl tributyl citrate; *Acetyl triethyl citrate*; *Acetyl trimethyl citrate*; Citroflex® 4; Citroflex® A-2; Citroflex® A-4
film-former, lipstick
Crystal® Crown
film-former, makeup
Crystal® O; Crystal® Crown
film-former, nail polishes
Dibutyl phthalate
film-former, nitrocellulose lacquers
Dibutyl phthalate
film-former, paper/paperboard: food-pkg.
Citroflex® A-2
film-former, personal care
Crystal® Crown; *Isooctane*; Permethyl® 97A
film-former, pharmaceuticals
Crystal® O; Crystal® Crown; *Phenyl trimethicone*; *Propylene glycol alginate*; *Tributyl citrate*
film-former, PVA emulsions
3-Methoxybutanol
film-former, PVAc homopolymer emulsion adhesives
Dipropylene glycol dibenzoate
film-former, soaps
Crystal® O; Crystal® Crown
film-former, solid rocket propellants
Dibutyl phthalate
film-former, tablet coatings
Lutrol® E 1500; Lutrol® E 3350; Lutrol® E 4000; Lutrol® E 6000; *PEG-125*
film-former, topical pharmaceuticals
Super Refined® Crodamol SCO
finishing agent
Gasoline
finishing agent, textiles
Canola oil; *Formaldehyde*; *Rapeseed (Brassica campestris) oil*
fire extinguishing agent
Bromochloromethane; *Dibromodifluoromethane*; *Dichlorotetrafluoroethane*; *1,2-Dichlorotetrafluoroethane*; *Methyl chloride*; *Trichlorofluoromethane*; *Trichlorotrifluoroethane*
fire extinguishing fluid
Bromodichloromethane
fire resistant hydraulic fluids
Polypropylene glycol; *Tricresyl phosphate*
fire retardant, coatings
Tricresyl phosphate
fire retardant, plastics
Bromine; *Tricresyl phosphate*; *Tri-p-tolyl phosphate*
fire retardant, rubber
Tricresyl phosphate
fireproofing agent
Zinc chloride
fireproofing agent, wood
Ammonium hydroxide
firming agent, food
Calcium chloride
firming agent, pharmaceuticals
Calcium chloride
fixing agent
Citrofol® BII; *Phenethyl alcohol*
fixing agent, aerosols
Captex® 810D; Captex® 8227
fixing agent, aftershave lotions
Captex® 810D; Captex® 8227
fixing agent, bath oils
Captex® 810D; Captex® 8227
fixing agent, can coatings
Citrofol® AII
fixing agent, clinical nutrition
Captex® 350
fixing agent, colognes
Eastman® Diethyl Phthalate
fixing agent, cosmetic creams/lotions
Isocetyl stearate
fixing agent, cosmetics
Caprylic/capric/lauric triglyceride; *Caprylic/capric/linoleic triglyceride*; Captex® 810D; Captex® 8227; Eastman® DEP, Fragrance Grade; *Isocetyl laurate*; *Isocetyl myristate*
fixing agent, delivery/absorption enhancement
Captex® 350
fixing agent, dermatologicals
Captex® 350; *Propylene glycol dioctanoate*
fixing agent, dry skin care
Propylene glycol dioctanoate
fixing agent, flavors
Aldo® MCT KFG; Miglyol® 840; Unichem BZBN
fixing agent, flavors: food
Captex® 500
fixing agent, flavors: pharmaceuticals
Captex® 500
fixing agent, fragrances
p-t-Butyl toluene; *Dimethyl phthalate*; *Dipropylene glycol*; Eastman® DBP; Eastman® DEP; Eastman® DEP, Fragrance Grade; Eastman® DPG, Fragrance Grade; Eastman® Diethyl Phthalate
fixing agent, fragrances: food
Captex® 500
fixing agent, fragrances: pharmaceuticals
Captex® 500
fixing agent, infant formulas
Captex® 350
fixing agent, lip gloss
Captex® 810D; Captex® 8227
fixing agent, lipstick
Captex® 810D; Captex® 8227
fixing agent, makeup bases
Captex® 810D; Captex® 8227
fixing agent, musk: flavors
Benzyl benzoate
fixing agent, musk: perfumes
Benzyl benzoate; Unichem BZBN
fixing agent, nail care
Eastman® DEP, Fragrance Grade
fixing agent, nutritional applics.
Caprylic/capric/lauric triglyceride; *Caprylic/capric/linoleic triglyceride*; Captex® 810D; Captex® 8227; *Triundecanoin*
fixing agent, nutritional/sports supplements
Captex® 350
fixing agent, perfumes
Benzylsalicylate; *C12-15 alkyl benzoate*; Citrofol® AI; Crodamol AB; *Dibutyl phthalate*; *Diethyl phthalate*; Eastman® Diethyl Phthalate; *Ethylene glycol diacetate*; *Hexyl benzoate*; *Methoxyethanol*; *Phenoxyethanol*; *Propylene glycol phenyl ether*; *Triacetin*; *Triethyl citrate*
fixing agent, perfumes: cosmetics
Myristyl alcohol
fixing agent, perfumes: soaps
Myristyl alcohol
fixing agent, pharmaceutical coatings
Captex® 350
fixing agent, pharmaceuticals
Caprylic/capric/lauric triglyceride; *Caprylic/capric/linoleic triglyceride*; Captex® 350; Captex® 355EP; Captex® 500P; Captex® 810D; Captex® 8227; *n-Dioctyl phthalate*; *Isocetyl laurate*; *Isocetyl myristate*; *Propylene glycol dioctanoate*; *Triacetin*
fixing agent, rare earths—>cathode elements
Diethyl carbonate
fixing agent, soaps
Propylene glycol phenyl ether
fixing agent, suppositories
Propylene glycol dioctanoate
fixing agent, textile finishing
Glyezin® A
fixing agent, toilet waters
Eastman® Diethyl Phthalate
fixing agent, topical pharmaceuticals
Isocetyl stearate
fixing agent, transparency film
Citrofol® AII
flame retardant
Trimethyl phosphate
flame retardant, ABS-PC
Triphenyl phosphate
flame retardant, adhesives
Paraffin, chlorinated
flame retardant, aq. latexes
Amgard® DMMP; *Dimethyl methylphosphonate*
flame retardant, ATH-filled polyester resins
Dimethyl methylphosphonate
flame retardant, cellulose acetate
Phosflex® 4
flame retardant, cellulosic insulation
Boric acid
flame retardant, cellulosics
TBP; *Tributyl phosphate*; *Triethyl phosphate*; *Triphenyl phosphate*
flame retardant, coatings
Amgard® $TBPO_4$; *Tributyl phosphate*; *Triphenyl phosphate*
flame retardant, cotton textiles
Boric acid
flame retardant, CR
Phosflex® 4
flame retardant, engineering thermoplastics
Triphenyl phosphate
flame retardant, epoxies
TBP
flame retardant, fibers
Tetrabromoethane
flame retardant, foams: rigid urethane
Dimethyl methylphosphonate
flame retardant, HIPS
Triphenyl phosphate
flame retardant, hydraulic fluids
Amgard® $TBPO_4$
flame retardant, inks
Paraffin, chlorinated
flame retardant, latexes
Antiblaze® DMMP
flame retardant, mattress batting
Boric acid
flame retardant, nitrocellulose
Phosflex® 4; *Trioctyl phosphate*
flame retardant, paints
Paraffin, chlorinated
flame retardant, paper
Boric acid; *Paraffin, chlorinated*
flame retardant, plastics
Paraffin, chlorinated; *Tributoxyethyl phosphate*; *Tributyl phosphate*; *Trichloroethylene*
flame retardant, polyesters
Triethyl phosphate
flame retardant, PPO
Triphenyl phosphate
flame retardant, PS
TBP
flame retardant, PU
Glyceryl-α-monochlorohydrin; *Triethyl phosphate*
flame retardant, PVAc
TBP
flame retardant, PVC
TBP
flame retardant, rigid PU foam
Amgard® DMMP; Antiblaze® DMMP
flame retardant, rubber
Paraffin, chlorinated; *Trioctyl phosphate*
flame retardant, textiles
Paraffin, chlorinated
flame retardant, triacetate film/sheet
Triphenyl phosphate
flame retardant, unsat. polyester resins
Amgard® DMMP
flame retardant, unsat. polyesters
Antiblaze® DMMP; *Dimethyl methylphosphonate*
flame retardant, vinyls
Phosflex® 4; *Trioctyl phosphate*
flame retardant, wood
Boric acid
flame-retardant coating prod., carpet backings
Vinylidene chloride monomer
flatting agent, paints
Cachalot® AR-20; Cachalot® BE-22; Cachalot® C-50; Cachalot® L-90; Cachalot® M-43; Cachalot® S-56
flavor adjuvant, food
Glyceryl oleate; *PEG-6*; *PEG-8*; *PEG-9*; *PEG-12*; *PEG-14*; *PEG-16*; *PEG-20*; *PEG-32*; *PEG-40*; *PEG-75*; *PEG-100*; *PEG-150*; *PEG-200*; *Propylene glycol alginate*
flavor adjuvant, food pkg.
Glyceryl oleate
flavor enhancer, baked goods
Eastman® Glacial Acetic Acid Food Grade, Kosher
flavor enhancer, condiments
Eastman® Glacial Acetic Acid Food Grade, Kosher
flavor enhancer, fats/oils
Eastman® Glacial Acetic Acid Food Grade, Kosher
flavor enhancer, food
Calcium chloride; Emery® 912; Emery® 916; Emery® 917; *Ethyl butyrate*; *Fenchyl alcohol*
flavor enhancer, food processing
Eastman® Glacial Acetic Acid Food Grade, Kosher
flavor enhancer, foodstuffs
Ethyl caproate
flavor enhancer, frozen dairy prods.
Eastman® Glacial Acetic Acid Food Grade, Kosher
flavor enhancer, gelatin
Eastman® Glacial Acetic Acid Food Grade, Kosher
flavor enhancer, milk prods.
Eastman® Glacial Acetic Acid Food Grade, Kosher
flavor enhancer, pharmaceuticals
Fenchyl alcohol
flavor enhancer, processed fruit
Eastman® Glacial Acetic Acid Food Grade, Kosher
flavor enhancer, puddings
Eastman® Glacial Acetic Acid Food Grade, Kosher
flavor enhancer, relishes
Eastman® Glacial Acetic Acid Food Grade, Kosher

flavor enhancer, snack food
Eastman® Glacial Acetic Acid Food Grade, Kosher
flavor enhancer, soft beverages
Eastman® Glacial Acetic Acid Food Grade, Kosher
flavor enhancer, soft candy
Eastman® Glacial Acetic Acid Food Grade, Kosher
flavor enhancer, sweet sauces
Eastman® Glacial Acetic Acid Food Grade, Kosher
flavoring agent
Acetone; t-Butyl alcohol; n-Butyl mercaptan; 1,10-Decanediol; Diethyl ketone; Eastman® Triacetin, Food Grade; *2-Ethyl-1-butanol; Ethyl caproate; Ethylene oxide; 12-Hydroxystearyl alcohol; Isobutyl alcohol; Isobutyric acid; Methyl acetoacetate; Methyl caprate; Methyl palmitate; Methyl pivalate; 2-Methylpropanal; Polysorbate 80;* Purasolv® BL; Purasolv® EL; *Pyrocatechol;* Speziol C 10/2; Speziol C 18/2; *3,4-Xylenol*
flavoring agent, baked goods
Hodag® S-35 FGK
flavoring agent, beverages
Diacetyl tartaric acid esters of mono- and diglycerides; Propylene glycol dibenzoate
flavoring agent, candy
Butyrolactone
flavoring agent, coffee whiteners
Hodag® S-35 FGK
flavoring agent, confectionery
Cetyl alcohol
flavoring agent, food
Acetic acid; Castor (Ricinus communis) oil; Diacetyl tartaric acid esters of mono- and diglycerides; Diisobutyl ketone; Diphenyl oxide; Ethyl acetate; Glyceryl oleate; Glyceryl stearate SE; Glyceryl tripropionate; Hodag® S-35 FGK; *Isophorone; Lactic acid; Mono- and diglycerides of fatty acids; Phenol; Phosphoric acid; Potassium carbonate; Propylene glycol; Safflower (Carthamus tinctorius) oil; Triethyl citrate; 2,5-Xylenol*
flavoring agent, food pkg.
Glyceryl oleate
flavoring agent, frozen desserts
Hodag® S-35 FGK
flavoring agent, gum
Cetyl alcohol
flavoring agent, natural: beverages
Caraway (Carum carvi) oil
flavoring agent, natural: chewing gum
Peppermint (Mentha piperita) oil; Spearmint (Mentha viridis) oil
flavoring agent, natural: confectionery
Spearmint (Mentha viridis) oil
flavoring agent, natural: food
Camphor; Camphor (Cinnamomum camphora) oil; Caraway (Carum carvi) oil; Clove (Eugenia caryophyllus) oil; Cocoa (Theobroma cacao) butter; Eucalyptus globulus oil; Peppermint (Mentha piperita) oil; Pine (Pinus palustris) oil; Sesame (Sesamum indicum) oil; Spearmint (Mentha viridis) oil; Turpentine; Walnut (Juglans regia) oil; Ylang ylang (Cananga odorata) oil
flavoring agent, natural: liqueurs
Peppermint (Mentha piperita) oil
flavoring agent, natural: mouthwashes
Peppermint (Mentha piperita) oil
flavoring agent, natural: pharmaceuticals
Camphor; Caraway (Carum carvi) oil; Clove (Eugenia caryophyllus) oil; Eucalyptus globulus oil; Peppermint (Mentha piperita) oil; Spearmint (Mentha viridis) oil; Turpentine; Ylang ylang (Cananga odorata) oil
flavoring agent, natural: toothpaste
Peppermint (Mentha piperita) oil; Spearmint (Mentha viridis) oil
flavoring agent, pear: mineral waters
Isoamyl acetate
flavoring agent, pear: syrups
Isoamyl acetate
flavoring agent, pharmaceuticals
Acetic acid; Biphenyl; Chloroform; Denatonium benzoate; Diisobutyl ketone; Diphenyl oxide; Ethyl acetate; Glyceryl tripropionate; Isophorone; Lactic acid; Propylene carbonate; Propylene glycol dibenzoate; Sweet almond (Prunus amygdalus dulcis) oil; Triethyl citrate; 2,5-Xylenol
flavoring agent, shortenings
Hodag® S-35 FGK
flavoring agent, soy milk
Butyrolactone
flavoring agent, synthetic
Isopropyl alcohol
flavoring agent, synthetic: beverages
Benzaldehyde; Eucalyptol; Methyl salicylate
flavoring agent, synthetic: chewing gum
Eucalyptol; Menthol; Methyl salicylate
flavoring agent, synthetic: cleaners
Benzaldehyde
flavoring agent, synthetic: confectionery
Menthol
flavoring agent, synthetic: cosmetics
Benzaldehyde
flavoring agent, synthetic: cough drops
Menthol
flavoring agent, synthetic: dentifrices
Methyl salicylate
flavoring agent, synthetic: food
Acetophenone; Acetyl tributyl citrate; Amyl acetate; n-Amyl alcohol; s-Amyl alcohol; Amyl benzoate; Amyl formate; Amyl propionate; Anisole; Benzaldehyde; Benzyl acetate; Benzyl alcohol; Benzyl benzoate; Benzyl ether; Benzyl formate; Benzylsalicylate; Bornyl acetate; 2-Butanol; n-Butyl acetate; Butyl alcohol; n-Butylamine; Butyl butyrate; Butyl formate; Butyl lactate; n-Butyl propionate; Butyl stearate; n-Butyraldehyde; Caprylic alcohol; Cinnamal; Cyclohexanone; Cyclohexyl acetate; p-Cymene; Decyl alcohol; Diethyl succinate; Diethyl tartrate; 2,6-Dimethyl heptanol-4; Dimethyl succinate; Dimethyl sulfide; Ethylacetoacetate; Ethyl benzoate; 2-Ethylbutyl acetate; Ethyl butyrate; Ethyl caproate; Ethyl cinnamate; Ethyl crotonate; Ethyl decanoate; Ethyl formate; 2-Ethylhexanol; Ethyl lactate; Ethyl levulinate; Ethyl octanoate; Ethyl oleate; Ethyl phenylacetate; Ethyl propionate; Eucalyptol; Eugenol; Farnesol; Fenchyl alcohol; Formic acid; Furfural; Furfuryl alcohol; Fusel oil refined; 2-Heptanol; 3-Heptanol; 3-Heptanone; 4-Heptanone; Heptyl alcohol; 3-Hexanone; cis-3-Hexenol; Hexyl acetate; Hexyl alcohol; Hexyl benzoate; Isoamyl acetate; Isoamyl alcohol; Isoamyl butyrate; Isobutyl acetate; Isobutyl alcohol; Isobutyl isobutyrate; Isobutyl propionate; Isobutyric acid; Isopropyl acetate; Isopropyl butyrate; Isopropyl myristate; Isoquinoline; Lauryl alcohol; d-Limonene; dl-Limonene; l-Limonene; 2,6-Lutidine; Menthol; Methyl acetate; Methyl n-amyl ketone; Methyl benzoate; α-Methylbenzyl alcohol; 2-Methyl-1-butanol; Methyl butyrate; Methyl caproate; Methyl caprylate; Methyl disulfide; Methyl ethyl ketone; Methyl 2-furoate; Methyl hexyl ketone; Methyl isobutyl ketone; Methyl laurate; Methyl myristate; α-Methylnaphthalene; 2-Methylpropanal; Methyl propionate; Methyl propyl ketone; Methyl salicylate; Myrcene; Myristyl alcohol; γ-Nonalactone; Nonanoic acid; Octane; 2-Octanol; 3-Octanone; Octyl acetate; Paraldehyde; Phenethyl alcohol; Phenyl acetate; α-Pinene; Piperidine; Propionaldehyde; Propyl acetate; Propyl alcohol; Propyl butyrate; Propyl propionate; Pyridine; Quinoline; Styrene; Sucrose octaacetate; γ-Terpinene; α-Terpineol; Terpinolene; Tetrahydrofurfuryl acetate; Tetrahydrofurfuryl alcohol; Thymol; Triacetin; Trimethylcyclohexanol; 3,5,5-Trimethylhexan-1-ol; γ-Undecalactone; γ-Valerolactone; 2,6-Xylenol
flavoring agent, synthetic: fruit syrups
Ethyl propionate
flavoring agent, synthetic: ice cream
Acetal
flavoring agent, synthetic: liniments
Menthol
flavoring agent, synthetic: nasal inhalers
Menthol
flavoring agent, synthetic: nonalcoholic beverages
Acetal
flavoring agent, synthetic: perfumes
Benzaldehyde
flavoring agent, synthetic: pharmaceuticals
Acetal; Acetophenone; Acetyl tributyl citrate; Amyl acetate; n-Amyl alcohol; Amyl formate; Anisole; Benzaldehyde; Benzyl acetate; Benzyl alcohol; Benzyl benzoate; Benzyl ether; Benzyl formate; Benzylsalicylate; Bornyl acetate; n-Butyl acetate; Butyl alcohol; n-Butylamine; Butyl butyrate; Butyl formate; Butyl lactate; n-Butyl propionate; Butyl stearate; n-Butyraldehyde; Caprylic alcohol; Cinnamal; Cyclohexyl acetate; p-Cymene; Decyl alcohol; Diethyl succinate; Diethyl tartrate; Dimethyl succinate; Dimethyl sulfide; Ethylacetoacetate; Ethyl benzoate; 2-Ethylbutyl acetate; Ethyl butyrate; Ethyl caproate; Ethyl cinnamate; Ethyl crotonate; Ethyl decanoate; Ethyl formate; 2-Ethylhexanol; Ethyl lactate; Ethyl levulinate; Ethyl octanoate; Ethyl phenylacetate; Ethyl propionate; Eucalyptol; Eugenol; Farnesol; Fenchyl alcohol; Formic acid; Furfural; Furfuryl alcohol; Fusel oil refined; 3-Heptanol; 3-Heptanone; 4-Heptanone; Heptyl alcohol; 3-Hexanone; cis-3-Hexenol; Hexyl acetate; Hexyl alcohol; Isoamyl acetate; Isoamyl alcohol; Isoamyl butyrate; Isobutyl acetate; Isobutyl alcohol; Isobutyl isobutyrate; Isobutyl propionate; Isobutyric acid; Isopropyl acetate; Isopropyl butyrate; Isoquinoline; Lauryl alcohol; d-Limonene; dl-Limonene; 2,6-Lutidine; Menthol; Methyl acetate; Methyl n-amyl ketone; Methyl benzoate; α-Methylbenzyl alcohol; Methyl butyrate; Methyl caproate; Methyl caprylate; Methyl disulfide; Methyl ethyl ketone; Methyl 2-furoate; Methyl hexyl ketone; Methyl isobutyl ketone; Methyl laurate; Methyl myristate; α-Methylnaphthalene; 2-Methylpropanal; Methyl propionate; Methyl propyl ketone; Methyl salicylate; Myrcene; γ-Nonalactone; Nonanoic acid; 2-Octanol; 3-Octanone; Octyl acetate; Phenethyl alcohol; α-Pinene; Piperidine; Propionaldehyde; Propyl acetate; Propyl alcohol; Propyl butyrate; Propyl propionate; Pyridine; Quinoline; Styrene; γ-Terpinene; α-Terpineol; Terpinolene; Tetrahydrofurfuryl acetate; Tetrahydrofurfuryl alcohol; Thymol; Triacetin; γ-Undecalactone; γ-Valerolactone; 2,6-Xylenol
flavoring agent, synthetic: veterinary medicated shampoos
Menthol
flavoring agent, synthetic: veterinary ointments
Menthol
flavoring agent, tablet food supplements
Cetyl alcohol
flavoring agent, whipped toppings
Hodag® S-35 FGK
flexibilizer
Bis (2-ethylhexyl) maleate; Butyl oleate; Macol® 85
flexibilizer, epoxies
Kenplast® G; *Phenoxyethanol*
flexibilizer, low temp. resins
Citrofol® BII
flexibilizer, thermoplastics
Carbowax® E200; Carbowax® E300; Carbowax® E400; Carbowax® E600; Carbowax® E900; Carbowax® E1000; Carbowax® E1450; Carbowax® E3350; Carbowax® E8000
fllexibilizer, skin
Dimethyl sulfone
flocculant, water treatment
Prochinor® SOL
flotation agent
t-Amyl alcohol; Bromochloromethane; t-Butyl alcohol; Cachalot® AR-20; Cachalot® BE-22; Cachalot® C-50;

Cachalot® L-90; Cachalot® M-43; Cachalot® S-56; *Carbon disulfide*; *Diethylamine*; *Dimethylamine*; *2,6-Dimethyl heptanol-4*; *2,2-Dimethylpropanol*; *Nonanoic acid*; *3-Pentanol*; *Pentylamine*; *Tall oil*; *2,6,8-Trimethylnonyl-4-alcohol*

flotation agent, froth
Lauryl alcohol; *Methyl amyl alcohol*; *Xylenol*

flotation agent, mineral
Kerosene

flotation collector
Sodium lauryl sulfate

flow agent, alkyd resin/oil paints
Butyl glycolate

flow agent, brush-applied paints
Butyl glycolate

flow agent, nitrocellulose lacquers
Butyl glycolate

flow aid
Dekalin®; Isopar® G Fluid; Isopar® K Fluid; Isopar® M Fluid; *3-Methoxy-1-propanol*

flow aid, agrochemical toxicants
Agrisol PX401; Agrisol PX413

flow aid, air-dry coatings
Eastman® EB Acetate

flow aid, alkyd oil paints
Polysolvan® O

flow aid, alkyd resin paints
Polysolvan® O

flow aid, alkyds
Butoxyl®

flow aid, aq. paints
Ethylene glycol isopropyl ether; Isopropyl Oxitol

flow aid, brush-applied paints
Polysolvan® O

flow aid, coatings
Hexalin®; Isopar® H Fluid; Methylhexalin®; *Propylene glycol methyl ether acetate*; Tetralin®

flow aid, deposition applics.
Novec™ HFE-7200

flow aid, epoxies
Eastman® EB Acetate

flow aid, high-bake coatings
Eastman® EB Acetate

flow aid, insecticides
Pentanol 45

flow aid, lacquers
BDGE; BGE; *Butyl stearate*; EDGE; EGE; *2-Ethylhexyl acetate*; i-PGE; *Isopropyl acetate*; MDGE; MGE

flow aid, nitrocellulose lacquers
Polysolvan® O

flow aid, paints/coatings
2-Ethylhexyl acetate; *Isopropyl acetate*

flow aid, PFPE lubricants
Novec™ HFE-7100; Novec™ HFE-7100DL; Novec™ HFE-7200DL

flow aid, plastics
Butyl stearate

flow aid, printing inks
2-Ethylhexyl acetate; *Isopropyl acetate*; *Propylene glycol methyl ether acetate*

flow aid, rubber
Butyl stearate

flow aid, stoving enamels
Butoxyl®

flow control agent
C13-14 isoparaffin

flow control agent, butyl rubber
Kenflex® N1

flow control agent, chlorosulfonated PE
Kenflex® N1

flow control agent, inks
Trihydroxystearin

flow control agent, mastics
Trihydroxystearin

flow control agent, neoprene
Kenflex® N1

flow control agent, paints/coatings
Trihydroxystearin; *3,3,5-Trimethyl cyclohexanone*

flow control agent, polar thermoplastics
Kenflex® N1

flow control agent, polysulfide
Kenflex® N1

flow control agent, solvent-free epoxy systems
Trihydroxystearin

flow control agent, stains
Trihydroxystearin

flow control agent, textile backings
Trihydroxystearin

flow control aid, felt pens
Ethyl Proxitol

flow control aid, furniture polishes
Ethyl Proxitol

flow control aid, leather finishes
Ethyl Proxitol

flow control aid, printing inks
Ethyl Proxitol

flow control aid, stamp pad inks
Ethyl Proxitol

flow control aid, surface coatings
Ethyl Proxitol

flow control aid, wood stains
Ethyl Proxitol

flow improver, lacquers
Isooctyl stearate

flow improver, plastics
Isooctyl stearate

flow improver, rubber
Isooctyl stearate

flow-out improver, aq. latex disps.
Eastman® EB

flow-out improver, aq. latex emulsions
Eastman® EB

flow-out improver, resins
Eastman® EB

flow-out improver, water-dilutable coatings
Eastman® EB

flue gas treatment
Hydrogen peroxide

fluidizer
STEPANATE® AXS

flume wash water additive
Ethylene dichloride

flume wash water additive, food processing
Butoxyethanol; *Ethanolamine*; *Ethylenediamine*; *Oleic acid*; *Triethanolamine*

fluorometric determinations
Isobutyl alcohol

flux, copper soldering
Triethylamine

foam booster/stabilizer, shampoos
Surfadone® LP-300

foam builder
N-Octyl pyrrolidone; *Trichlorotrifluoroethane*

foam builder, beverage mixes
Polysorbate 60

foam builder, carpetbacking latexes
Sodium lauryl sulfate

foam builder, cosmetics
Comperlan® COD; *Trilaurin*

foam builder, dentifrices
Hydrogen peroxide

foam builder, detergents
Trideceth

foam builder, foamed rubber
Sodium lauryl sulfate

foam builder, food
Castor (Ricinus communis) oil

foam builder, industrial applics.
Lactamide MEA

foam builder, personal care
Lactamide MEA

foam builder, pharmaceuticals
Comperlan® COD; *Trilaurin*

foam builder, shampoos
Comperlan® COD

foam builder/stabilizer, cosmetics
Cocamide DEA

foam builder/stabilizer, household cleaners
Cocamide DEA

foam builder/stabilizer, industrial cleaners
Cocamide DEA

foam builder/stabilizer, textiles
Cocamide DEA

foam builder/stabilizer, topical pharmaceuticals
Cocamide DEA

foam control agent
Behenyl alcohol; *Isostearyl alcohol*; *Lauryl alcohol*; Witcohol™ 42

foam control agent, detergents
Glucam® E-20

foam control agent, shampoos
Glucam® E-20

foam control modifier
Carbowax® E200; Carbowax® E300; Carbowax® E400; Carbowax® E600; Carbowax® E900; Carbowax® E1000; Carbowax® E1450; Carbowax® E3350; Carbowax® E8000

foam modifier, antiperspirants
Glycereth-7

foam modifier, cosmetic masks
Glycereth-7

foam modifier, cosmetics
Glycereth-7; *PPG-15 stearyl ether benzoate*

foam modifier, depilatories
Glycereth-7

foam modifier, detergents
Glucam® E-10; Glucam® P-10; Glucam® P-20; *PPG-10 methyl glucose ether*; *PPG-20 methyl glucose ether*

foam modifier, detergents foam modifier, shampoos
Methyl gluceth-10; *Methyl gluceth-20*

foam modifier, froth flotation
Isodecyl alcohol; *Isooctyl alcohol*; *3,5,5-Trimethylhexan-1-ol*

foam modifier, personal care
Glycereth-7

foam modifier, shampoos
Glucam® E-10; Glucam® P-10; Glucam® P-20; *PPG-10 methyl glucose ether*; *PPG-20 methyl glucose ether*

foam modifier, soap
PPG-15 stearyl ether benzoate

foam modifier, soap/syndet systems
PPG-15 stearyl ether benzoate

foam modifier, wavesetting lotions
Glycereth-7

formaldehyde donor
MDM hydantoin

formaldehyde scavenger
Eastman® EAA

formulation aid
Santosol® DME-1

formulation aid, baked goods
Eastman® Glacial Acetic Acid Food Grade, Kosher; Hodag® S-35 FGK

formulation aid, coffee whiteners
Hodag® S-35 FGK

formulation aid, condiments
Eastman® Glacial Acetic Acid Food Grade, Kosher

formulation aid, fats/oils
Eastman® Glacial Acetic Acid Food Grade, Kosher

formulation aid, food
Coconut (Cocos nucifera) oil; *Corn (Zea mays) oil*; *Cottonseed (Gossypium) oil*; Hodag® S-35 FGK; *Mineral oil*; *Mono- and diglycerides of fatty acids*; *Peanut (Arachis hypogaea) oil*; *Propylene glycol*; *Propylene glycol alginate*; *Soybean (Glycine soja) oil*; *Sunflower (Helianthus annuus) seed oil*; *Triacetin*; *Tristearin*

formulation aid, food processing
Eastman® Glacial Acetic Acid Food Grade, Kosher

formulation aid, foods
Potassium hydroxide

formulation aid, frozen dairy prods.
Eastman® Glacial Acetic Acid Food Grade, Kosher

formulation aid, frozen desserts
Hodag® S-35 FGK

formulation aid, gelatin
Eastman® Glacial Acetic Acid Food Grade, Kosher

formulation aid, margarine
Peanut (Arachis hypogaea) oil

formulation aid, milk prods.
Eastman® Glacial Acetic Acid Food Grade, Kosher

formulation aid, pharmaceuticals
Propylene glycol alginate

formulation aid, processed fruit
Eastman® Glacial Acetic Acid Food Grade, Kosher

formulation aid, puddings
Eastman® Glacial Acetic Acid Food Grade, Kosher

formulation aid, relishes
Eastman® Glacial Acetic Acid Food Grade, Kosher

formulation aid, salad oils
Peanut (Arachis hypogaea) oil

formulation aid, sausage casings
Butylene glycol

formulation aid, shortening
Hodag® S-35 FGK; *Peanut (Arachis hypogaea) oil*

formulation aid, snack food
Eastman® Glacial Acetic Acid Food Grade, Kosher

formulation aid, soft beverages
Eastman® Glacial Acetic Acid Food Grade, Kosher

formulation aid, soft candy
Eastman® Glacial Acetic Acid Food Grade, Kosher

formulation aid, sweet sauces
Eastman® Glacial Acetic Acid Food Grade, Kosher

formulation aid, whipped toppings
Hodag® S-35 FGK

fortifier, lavender
Methyl n-amyl ketone

fragrance
Amyl benzoate; *Benzaldehyde*; *Benzyl benzoate*; *Benzylsalicylate*; *Butyl alcohol*; *t-Butyl alcohol*; *Butyl formate*; *Camphor (Cinnamomum camphora) oil*; *Caprylic alcohol*; *1,10-Decanediol*; *Diisobutyl adipate*;

Kerosene
fumigant
Acrylonitrile; *Chloroform*; *Cresylic acid*; *1,1-Dichloroethane*; *Methyl chloride*; *Propylene oxide*; *1,1,2,2-Tetrachloroethane*; *Trichloroethylene*
fumigant, agric.
Carbon tetrachloride
fumigant, cereals
Methyl formate
fumigant, dried fruits
Methyl formate
fumigant, food
Sulfur dioxide
fumigant, fruit
Ethyl bromide
fumigant, grain
1,1-Dichloro-1-nitroethane; *Ethyl bromide*; *Methyl bromide*
fumigant, paper/paperboard sizing: dry food contact
Ethylene oxide
fumigant, soil
o-Dichlorobenzene; *sym-Dichloroethyl ether*; *Methyl bromide*; *Propargyl alcohol*
fumigant, soil: nematodes
Propylene dichloride
fumigant, space
Methyl bromide
fumigant, storage
Ethylene dichloride
fumigant, tobacco
Methyl formate
functional fluid
Jeffox® PPG-400; Jeffox® PPG-2000; Jeffox® WL-440; Jeffox® WL-660; Jeffox® WL-1400; Jeffox® WL-5000
fungicide
Acenaphthene; *Aniline*; Cachalot® AR-20; Cachalot® BE-22; Cachalot® C-50; Cachalot® L-90; Cachalot® M-43; Cachalot® S-56; *o-Chloroaniline*; *1-Chloro-1-nitropropane*; *Copper oxide (ic)*; *Ethylene oxide*; *2-Ethylhexoic acid*; *Furfural*; *Naphthalene*; *Naphthenic acid*; *Nonylphenol*; *Pyridine*; *Sulfur*; *Sulfur dioxide*; *Tetrahydroxypropyl ethylenediamine*; *Trioctyl phosphate*; *Xylenol*; *2,3-Xylenol*; *3,4-Xylenol*
fungicide, adhesives
Iodopropynyl butylcarbamate
fungicide, cereals
Ethyl formate
fungicide, citrus fruit
Trimethyl borate
fungicide, cutting oils
Iodopropynyl butylcarbamate
fungicide, inks
Iodopropynyl butylcarbamate
fungicide, latex paints
Iodopropynyl butylcarbamate
fungicide, oil-based paints
Iodopropynyl butylcarbamate
fungicide, paper coatings
Iodopropynyl butylcarbamate
fungicide, pharmaceuticals
Dantogard® Plus Liq.
fungicide, plastics
Iodopropynyl butylcarbamate
fungicide, textiles
Iodopropynyl butylcarbamate
fungicide, tobacco
Ethyl formate
fungicide, topical pharmaceuticals
Tall oil
fungicide, wood prods.
Iodopropynyl butylcarbamate
fungistat, citrus fruit
Boric acid
fungistat, citrus fruit pkg.
Biphenyl
fungistat, eyewashes
Boric acid
fungistat, fruit
s-Butylamine
fungistat, ointments
Boric acid
galvanization
Zinc chloride
gas chromatography standard
Methyl myristate
gas chromatography, reference standard
Methyl behenate; *Methyl eicosanoate*
gas enricher
Propane
gasoline additive
Cresylic acid; *N-Nitrosodiethylamine*; *N-Nitrosodimethylamine*; *3,4-Xylenol*
gasoline source
Naphtha
gel control agent, starches
Diacetyl tartaric acid esters of mono- and diglycerides
gellant
Poloxamer 181; *Propylene glycol alginate*
gellant, absorption bases
Lanolin alcohol
gellant, aliphatic solvent systems
Trihydroxystearin
gellant, bentonite
Propylene carbonate
gellant, clays
1,2-Butylene carbonate; Jeffsol® 1,2-Butylene Carbonate; Jeffsol® EC-25; Jeffsol® EC-50; Jeffsol® EC-75
gellant, cleansing preps.
Lanolin alcohol
gellant, coatings
Trihydroxystearin
gellant, cosmetics
Lanolin alcohol; Polychol 15; *Propylene carbonate*
gellant, deodorants
Triisostearin
gellant, general purpose
Macol® 85
gellant, greases
Propylene carbonate
gellant, makeup
Triisostearin
gellant, montmorillonite
Propylene carbonate
gellant, personal care
Polychol 5
gellant, pharmaceuticals
Propylene glycol alginate
gellant, topical pharmaceuticals
Polychol 5; Polychol 15
gelling inhibitor, polyester resins
Benzyltriethyl ammonium chloride; *Benzyl trimethyl ammonium chloride*
germicide
Acetylacetone; *Cyclohexanol*; *Furfural*; *Iodine*; *Methoxyethanol*; *3,5,5-Trimethylhexan-1-ol*
glass cleaner additive
Dimethyl hexynol
glazing agent, confectionery
Caprylic/capric triglyceride; Delios®
glazing agent, flour confectionery
Crodamol GTCC
glazing agent, food
Delios®; Delios® S; *Lanolin*; *Mineral oil*; *Oleic acid*; *Paraffin*; *Petrolatum*
glazing agent, sugar confectionery
Crodamol GTCC
gloss aid,
Cropure® Babassu
gloss aid, alkyds
Butoxyl®
gloss aid, antiperspirants
Octyl palmitate
gloss aid, aq. latex disps.
Eastman® EB
gloss aid, aq. latex emulsions
Eastman® EB
gloss aid, automobile lacquers
Ethoxyethanol acetate
gloss aid, baby oils
Lanesta L
gloss aid, baking enamels
Eastman® DB
gloss aid, balsams
Fancol™ Karite Butter
gloss aid, bath oils
Octyl palmitate
gloss aid, body lotions
Fancol™ Karite Butter
gloss aid, cleansers
Lanesta L
gloss aid, coatings
Morcryl® 535
gloss aid, colored cosmetics
Octyldodecyl isostearate
gloss aid, cooking oils
Hybrid safflower (Carthamus tinctorius) oil
gloss aid, cosmetics
Avocado (Persea gratissima) oil; *Cetyl octanoate*; *Hybrid safflower (Carthamus tinctorius) oil*; *Isoeicosane*; *Isostearyl isostearate*; *Octyl palmitate*; *Octyl stearate*; Schercemol 1818; Schercemol OLO; *Sweet almond (Prunus amygdalus dulcis) oil*; *Tridecyl neopentanoate*
gloss aid, creams/lotions
HallStar™ BST; HallStar™ DOM; *Octyldodecyl isostearate*; *Tridecyl neopentanoate*
gloss aid, deodorants
HallStar™ BST
gloss aid, emulsions
Glucam® E-10
gloss aid, enamels
Butyl Oxitol®
gloss aid, eye gels
Glucam® E-10
gloss aid, facial cleansers
Glucam® E-10
gloss aid, floor polishes
Eastman® DE
gloss aid, foam baths/gels
Lanesta L
gloss aid, food pkg. adhesives
Morcryl® 535
gloss aid, hair care
Cetyl stearate; *Dioctyl maleate*; Finsolv® PG-22; *Lactamide DGA*; Macol® 57; *Octyldodecyl isostearate*; *Oleyl oleate*; *Phenyl trimethicone*; Schercemol OLO
gloss aid, hair conditioners
Bernel® Ester DCM; Fancorsil A; HallStar™ DOM
gloss aid, hair grooms
Octyl palmitate
gloss aid, hair sprays
Estol IPM 1509 (BIO-IPM); HallBrite® OS; HallStar™ BST; HallStar™ DOM
gloss aid, hairsprays
Fancol™ ALA
gloss aid, inks
Morcryl® 535
gloss aid, lacquers
BDGE; BGE; Butyl Oxitol®; EDGE; EGE; i-PGE; MDGE; MGE
gloss aid, lip balms
HallBrite® OS
gloss aid, lip gloss
Finsolv® BOHS-111; *Octyl hydroxystearate benzoate*
gloss aid, lipstick
Finsolv® BOHS-111; Krystal Kleer Castor Oil; *Octyl hydroxystearate benzoate*; USP Castor Oil
gloss aid, lotions
HallBrite® OS
gloss aid, make-up
Octyl palmitate
gloss aid, makeup
Fancorsil A; HallStar™ BST; *Wheat (Triticum vulgare) germ oil*
gloss aid, makeup removers
HallStar™ BST; *Octyldodecyl isostearate*
gloss aid, mascara
Isoeicosane
gloss aid, moisturizing cleansers
Octyldodecyl isostearate
gloss aid, moisturizing creams/ lotions
Fancorsil A
gloss aid, nail varnishes
HallStar™ BST
gloss aid, NC lacquers
Castor oil, polymerized
gloss aid, ointments
Fancorsil A; *Octyldodecyl isostearate*
gloss aid, overprints
Morcryl® 535
gloss aid, overtreated hair preps.
Foamid LM2E
gloss aid, paints
Butoxyl®
gloss aid, paper
Morcryl® 535
gloss aid, personal care
Avocado (Persea gratissima) oil; Glucam® E-10; HallBrite® BHB; HallBrite® OS; *Isostearyl isostearate*; Krystal Kleer Castor Oil; *Octyldodecyl isostearate*; USP Castor Oil
gloss aid, pharmaceuticals
Avocado (Persea gratissima) oil; *Hybrid safflower (Carthamus tinctorius) oil*; Krystal Kleer Castor Oil; *Tridecyl neopentanoate*; USP Castor Oil
gloss aid, pigmented prods.
Glyceryl triacetyl hydroxystearate
gloss aid, pigmented sticks
Lipovol SES
gloss aid, preshave preps.
HallStar™ BST
gloss aid, resins
Eastman® EB
gloss aid, shampoos
Fancol™ Karite Butter; Glucam® E-10; Surfadone® LP-300
gloss aid, shaving creams
Glucam® E-10
gloss aid, shaving foams
Fancol™ Karite Butter
gloss aid, shower gels
Glucam® E-10
gloss aid, skin care
Octyldodecyl isostearate; *Oleyl oleate*; Schercemol OLO
gloss aid, skin cleansers
HallStar™ DOM
gloss aid, soaps
Fancol™ Karite Butter

Blend
humectant, household prods.
Carbowax® PEG 300
humectant, industrial
Glycereth-26; *Sorbeth-20*
humectant, ink pad dyes
DePEG 200; DePEG 300; DePEG 400; DePEG 600
humectant, inks
Kemstrene® 96.0% USP; Kemstrene® 99.7% USP; Kemstrene® Reagent Grade
humectant, lipstick
Carbowax® Sentry® PEG 400; Carbowax® Sentry® PEG 600
humectant, liq. soaps
Glucam® E-20; Kemstrene® 96.0% USP; Kemstrene® 99.7% USP; Kemstrene® Reagent Grade
humectant, lubricants
Carbowax® PEG 300; Kemstrene® 96.0% USP; Kemstrene® 99.7% USP; Kemstrene® Reagent Grade
humectant, makeup
Carbowax® Sentry® PEG 300; Carbowax® Sentry® PEG 400; Carbowax® Sentry® PEG 600; Carbowax® Sentry® PEG 3350; Glucam® P-10; *Myristyl myristate*
humectant, makeup removers
Carbowax® Sentry® PEG 400
humectant, metal fabrication
Carbowax® PEG 300
humectant, metalworking
Pogol® 400
humectant, moisturizing creams
Glucam® P-10
humectant, nail conditioners
Glucam® P-20
humectant, nail polish removers
Carbowax® Sentry® PEG 400; Carbowax® Sentry® PEG 3350; Carbowax® Sentry® PEG 8000
humectant, ointments
Carbowax® Sentry® PEG 540 Blend; Carbowax® Sentry® PEG 1000; Carbowax® Sentry® PEG 1450
humectant, paints
Carbowax® PEG 300
humectant, paper
Acetamide; Carbowax® PEG 300
humectant, perfumery components: hair care
Volpo G26
humectant, perfumery components: skin care
Volpo G26
humectant, perfumes
Eastman® DPG, Fragrance Grade; *PPG-5-ceteth-20*
humectant, personal care
Carbowax® PEG 300; Carbowax® Sentry® PEG 300; Carbowax® Sentry® PEG 400; Carbowax® Sentry® PEG 540 Blend; Carbowax® Sentry® PEG 600; Carbowax® Sentry® PEG 900; Carbowax® Sentry® PEG 1000; Carbowax® Sentry® PEG 1450; Carbowax® Sentry® PEG 3350; Carbowax® Sentry® PEG 4600; Carbowax® Sentry® PEG 8000
humectant, personal care
Dow Glycerine 96% USP; Dow Glycerine 99.5% USP/EP
humectant, personal care
Glucam® E-10; Glucam® E-20; Glucam® P-10; Glucam® P-20; *Glycereth-7*; *Methyl gluceth-10*; *Methyl gluceth-20*; *Methylpropanediol*; *Myristyl myristate*; Naetex-L
humectant, personal care
Optim™ Glycerine 99.7
humectant, personal care
PEG-2 lactamide; *PPG-10 methyl glucose ether*; Procetyl AWS
humectant, pet food
Propylene Glycol USP, FCC
humectant, pharmaceutical aerosols
Polarin® Range
humectant, pharmaceutical capsules
Polarin® Range
humectant, pharmaceutical syrups
Polarin® Range
humectant, pharmaceuticals
Carbowax® PEG 300; Dow Glycerine 96% USP; Dow Glycerine 99.5% USP/EP; *Ethyl linoleate*; *Glycereth-26*; *Hexylene glycol*; *Isopentyldiol*; Kemstrene® 96.0% USP; Kemstrene® 99.7% USP; Optim™ Glycerine 99.7; *PEG-90*; Polarin® Range; Pricerine™ 9091; Pricerine™ 9099; *Propylene glycol*; Propylene Glycol USP, FCC; Propylene Glycol USP/FCC Ultra Grade; *Propylene glycol stearate*; *Sorbeth-20*
humectant, polar essential oils: hair care
Volpo G26
humectant, polar essential oils: skin care
Volpo G26
humectant, printing
Carbowax® PEG 300
humectant, printing inks
DePEG 200; DePEG 300; DePEG 400; DePEG 600; *PEG-135*; *Triethylene glycol*
humectant, PU
Pricerine™ 9091; Pricerine™ 9099
humectant, rubber
Carbowax® PEG 300
humectant, shampoos
Carbowax® Sentry® PEG 300; Carbowax® Sentry® PEG 400; Carbowax® Sentry® PEG 600; Glucam® E-10; Glucam® E-20
humectant, shaving creams
Carbowax® Sentry® PEG 300; Carbowax® Sentry® PEG 400; Carbowax® Sentry® PEG 600; Carbowax® Sentry® PEG 900; Carbowax® Sentry® PEG 1000; Carbowax® Sentry® PEG 1450; Glucam® E-10
humectant, shaving preps.
Carbowax® Sentry® PEG 540 Blend; Pelemol® G7A
humectant, shower gels
Glucam® E-10
humectant, skin care
Finsolv® PG-22; *Isopentyldiol*; *Lactamide DGA*
humectant, skin creams/lotions
Foamid LM2E
humectant, snacks
Propylene Glycol USP, FCC
humectant, soap
Carbowax® Sentry® PEG 400; Carbowax® Sentry® PEG 3350
humectant, soap bars
Glucam® E-20
humectant, splash-on prods.
Pelemol® G7A
humectant, sponges
Diethylene glycol
humectant, starch adhesives
Ethylene glycol; *Glycerin*
humectant, styling preps.
Glucam® P-10
humectant, sunscreens
Carbowax® Sentry® PEG 300; Carbowax® Sentry® PEG 400; Carbowax® Sentry® PEG 540 Blend; Carbowax® Sentry® PEG 600; Glucam® E-20
humectant, textile sizes
Glycerin
humectant, textiles
Carbowax® PEG 300; Pogol® 400
humectant, tobacco
Diethylene glycol; *Glycerin*; *Triethylene glycol*
humectant, toilet articles
Glucam® E-10
humectant, toothpaste
Carbowax® Sentry® PEG 300; Carbowax® Sentry® PEG 400; Carbowax® Sentry® PEG 3350
humectant, topical pharmaceuticals
Glucam® E-10; Glucam® E-20; *Myristyl myristate*; Pelemol® G7A; Procetyl AWS
humectant, topicals
Acetamide MEA; *PPG-5-ceteth-20*
humectant, wavesetting lotions
Glycereth-7
humectant, wood processing
Carbowax® PEG 300
hydraulic fluid
1-Butoxyethoxy-2-propanol; *Butyl alcohol*; *1,2-Butylene carbonate*; *Castor (Ricinus communis) oil*; *Diacetone alcohol*; *2,2-Dimethylpropanol*; *Dimethyl sulfoxide*; *Ethylene carbonate*; *Hexachloro-1,3-butadiene*; *PPG-17*; *PPG-20*; *PPG-28-buteth-35*; *PPG-2 methyl ether*; *Propylene carbonate*; *Propylene glycol ricinoleate*; *Sulfolane*; *Triethyl phosphate*; *Trimethylcyclohexanol*; *Tri-p-tolyl phosphate*
hydraulic fluid inhibitor
Tributylamine
hydraulic fluid, boundary lubrication
Pluracol® SRF-2; Pluracol® W3520N; Pluracol® WI 285; Pluracol® WI 625; Pluracol® WI 1715; Pluracol® WS 170; Pluracol® WS 260; Pluracol® WS 660; Pluracol® WS 2000; Pluracol® WS 3520; Pluracol® WS 5100; Pluracol® WT 1400; Pluracol® WT 9150; Pluracol® WT 90,000
hydraulic fluid, fire resistant
Tributyl phosphate
hydraulic fluid, low-temp. applics.
Pluracol® SRF-2; Pluracol® W3520N; Pluracol® WI 285; Pluracol® WI 625; Pluracol® WI 1715; Pluracol® WS 170; Pluracol® WS 260; Pluracol® WS 660; Pluracol® WS 2000; Pluracol® WS 3520; Pluracol® WS 5100; Pluracol® WT 1400; Pluracol® WT 9150; Pluracol® WT 90,000
hydraulic oil
Silicone
hydrochloric acid source, electronics
VersaTRANS™ Solvent
hydrochloric acid source, silicone chip etching
VersaTRANS™ Solvent
hydrogen embrittlement inhibitor
Methyl pentynol
hydrogen embrittlement preventer, steel
Propargyl alcohol
hydrogen source
Lithium aluminum hydride; *Water*
hydrophobing agent
Phenyl trimethicone
hydrotrope
Mod Acid
hydrotrope, aerosols
Sodium toluenesulfonate; *Sodium xylenesulfonate*
hydrotrope, animal glues
Pilot® SXS-40
hydrotrope, cosmetics
Nonoxynol-9 phosphate
hydrotrope, cutting oils
Sodium toluenesulfonate; *Sodium xylenesulfonate*
hydrotrope, degreasers
Sodium xylenesulfonate
hydrotrope, detergent sterilizers
Ethylan® HB4
hydrotrope, detergents
Ammonium xylenesulfonate; *Sodium toluenesulfonate*; STEPANATE® AXS
hydrotrope, dishwashing detergents
STEPANATE® AXS
hydrotrope, dyes
Ammonium xylenesulfonate; *Sodium toluenesulfonate*; *Sodium xylenesulfonate*
hydrotrope, electroplating baths
Ammonium xylenesulfonate; *Sodium toluenesulfonate*
hydrotrope, glass cleaners
Ethylan® HB4
hydrotrope, glues
Sodium toluenesulfonate; *Sodium xylenesulfonate*
hydrotrope, heavy-duty cleaners
STEPANATE® AXS
hydrotrope, heavy-duty detergents
Sodium xylenesulfonate
hydrotrope, industrial processes
Sodium toluenesulfonate; *Sodium xylenesulfonate*
hydrotrope, inks
Ammonium xylenesulfonate; *Sodium toluenesulfonate*
hydrotrope, iodophors
Ethylan® HB4
hydrotrope, liq. detergents
Isopropylamine dodecylbenzenesulfonate
hydrotrope, mercerizing baths
Trioctyl phosphate
hydrotrope, organic dyestuffs
Pilot® SXS-40
hydrotrope, organic polymers
Pilot® SXS-40
hydrotrope, personal care
Ammonium xylenesulfonate
hydrotrope, petroleum industry
Pilot® SXS-40; *Sodium xylenesulfonate*
hydrotrope, polymers
Ammonium xylenesulfonate; *Sodium toluenesulfonate*
hydrotrope, printing pastes
Sodium xylenesulfonate
hydrotrope, pulping
Pilot® SXS-40; *Sodium xylenesulfonate*
hydrotrope, rinse aids
Ethylan® HB4
hydrotrope, shampoos
Sodium toluenesulfonate; *Sodium xylenesulfonate*
hydrotrope, textile finishing

intermediate, chlorinated solvents
1,1-Dichloroethane
intermediate, chlorination
Linpar® 1416-V Normal Paraffin
intermediate, cholesterol-lowering drugs
t-Butyl acetate
intermediate, citral
Myrcene
intermediate, citronellal
Myrcene
intermediate, citronellol
Myrcene
intermediate, CNS depressants
Diethyl oxalate
intermediate, coatings
N-Methylethanolamine; *Neodecanoic acid*; NiPar S-10™; *Polypropylene glycol*
intermediate, coloring materials
Cashew nut shell oil
intermediate, condensation polymers
Diglycolamine® Agent (DGA®)
intermediate, confectionery
Cetyl alcohol
intermediate, corrosion inhibitors
Aminoethylpiperazine; *Ethylenediamine*; *2-Ethylhexoic acid*
intermediate, cosmetics
1,3-Dibutoxy-2-propanol; *Glyceryl-α-monochlorohydrin*; *MDM hydantoin*; *Methyl laurate*; *Methyl pivalate*; *Neodecanoic acid*; Pluracol® E1450 NF; *Triethanolamine*
intermediate, crop protection agents
Ethyl formate
intermediate, crotonic acid
Crotonaldehyde
intermediate, cutting oils
Isooctyl alcohol; *Triethanolamine*
intermediate, cyclic compds.
Agrisynth® BLO; BLO®; *Butyrolactone*
intermediate, dechlorination polymers
Trichlorotrifluoroethane
intermediate, DEET pesticides
Diethylamine
intermediate, defoliant prod.
n-Butyl mercaptan
intermediate, detergents
Alfol® 4; Alfol® 14; *Caprylic alcohol*; *Diethylaminoethanol*; Edenor® ME C18-90; Emery® 2214; Emery® 2216; Emery® 2253; Emery® 2254; Emery® 2255; Emery® 2270; Emery® 2290; Emery® 2296; Estasol™ MD10; *2-Ethylhexoic acid*; *Methoxy PEG-10*; *Methyl caprate*; *Methyl caproate*; *Methyl caprylate*; *Methyl caprylate/caprate*; *Methyl cocoate*; *Methyl laurate*; *Methyl oleate*; *Methyl palmitate*; *Methyl palmitate oleate*; *Methyl palm kernelate*; *Peanutamide MEA*; *Peanutamide MIPA*
intermediate, dimethylformamide
Methyl formate
intermediate, dimethylolethylene-urea resins
Ethylenediamine
intermediate, disinfectant soaps
2-Octanol
intermediate, dispersants
Naphthalene
intermediate, dry cleaning prod.
Triethanolamine
intermediate, duplicating inks
Methyl oleate
intermediate, dust abatement agents
C12-14 alkylbenzene; V-154L; V-159L
intermediate, dyes
Acenaphthene; *Anthracene*; *Benzaldehyde*; *Benzotrifluoride*; *Bromine*; *n-Butylamine*; *o-Chloroaniline*; *o-Chlorobenzotrifluoride*; *p-Chlorobenzotrifluoride*; *Chloroform*; *Cyclohexylamine*; *Dibutylamine*; *Diethylamine*; *Diethyl oxalate*; *Dimethyl adipate*; *Ethyl chloride*; *Ethylene chlorohydrin*; *Ethyl morpholine*; *Glyceryl-α-monochlorohydrin*; *Hydroquinone*; *Isopropylamine*; *Isopropyl bromide*; Jeffox® WL-440; Jeffox® WL-660; Jeffox® WL-1400; Jeffox® WL-5000; *2-Mercaptoethanol*; *Methylamine*; *Methyl benzoate*; *Methyl isopropyl ketone*; *Naphthalene*; *2-Nitropropane*; *Paraldehyde*; *Phenylethylethanolamine*; *1,2,3,4-Tetrahydroquinoline*; *Triethylamine*; *1,2,4-Trimethylbenzene*; *1,3,5-Trimethylbenzene*; *Tripropylene glycol*; *m-Xylene*
intermediate, elastomers
ε-Caprolactone monomer; *Methylpropanediol*; *Polypropylene glycol*; *Thiodiglycol*
intermediate, emulsifiers
Aminoethylpiperazine; *n-Butylamine*; *Dibutylamine*; *Diethylaminoethanol*; Estasol™ MD10; *Hexyldecanol*; *Methoxypropylamine*; *Methyl caprate*; *Methyl caproate*; *Methyl caprylate*; *Methyl laurate*; *Methyl myristate*; *Methyl oleate*; *Methyl palmitate*; *Methyl palmitate oleate*; *Methyl stearate*; *Octyldodecanol*; *Tripropylene glycol*
intermediate, emulsifiers: agric. chemicals
Methoxypropylamine
intermediate, emulsifiers: water treating chemicals
Methoxypropylamine
intermediate, emulsion breakers
Aminoethylpiperazine
intermediate, emulsions
Triethanolamine
intermediate, enhanced oil recovery
C8-16 alkylbenzene
intermediate, enhanced oil recovery surfactants
Benzene, mono-C10-13-alkyl derivs., distillation residues; *C12-14 alkylbenzene*; V-154L; V-159L
intermediate, EP additives
Linpar® 1416-V Normal Paraffin; Oleocal ME-70; Oleocal ME-112; Oleocal ME-130
intermediate, epoxides
Dowanol® DPnB; Dowanol® PnB
intermediate, esterification
FA®; *Furfuryl alcohol*
intermediate, esters
Canola oil; *Octyldodecanol*; *PPG-17*; *PPG-2 methyl ether*; *Propylheptanol*; *Trichloroacetic acid*
intermediate, esters: drilling muds
2,2,4-Trimethyl-1,3-pentanediol isobutyrate
intermediate, esters: functional fluids
PPG-400; *PPG-2000*
intermediate, esters: herbicides
2,2,4-Trimethyl-1,3-pentanediol isobutyrate
intermediate, esters: humectant modifiers for inks
PPG-400; *PPG-2000*
intermediate, esters: inks
2,2,4-Trimethyl-1,3-pentanediol isobutyrate
intermediate, esters: latexes
2,2,4-Trimethyl-1,3-pentanediol isobutyrate
intermediate, esters: lubricant basestocks
2,2,4-Trimethyl-1,3-pentanediol isobutyrate
intermediate, esters: lubricants
Alfol® 1412; *PPG-9*; *PPG-26*; *PPG-400*; *PPG-2000*
intermediate, esters: metalworking
Alfol® 1412
intermediate, esters: pesticides
2,2,4-Trimethyl-1,3-pentanediol isobutyrate
intermediate, esters: pharmaceutical defoaming agents
PPG-400; *PPG-2000*
intermediate, esters: pharmaceuticals
PPG-9; *PPG-26*
intermediate, esters: plasticizers
PPG-400; *PPG-2000*; *2,2,4-Trimethyl-1,3-pentanediol isobutyrate*
intermediate, esters: plastics
Alfol® 1412
intermediate, esters: resins
2,2,4-Trimethyl-1,3-pentanediol isobutyrate
intermediate, esters: rubber
Alfol® 1412
intermediate, esters: rubber defoamers
PPG-9; *PPG-26*
intermediate, esters: rubber defoaming agents
PPG-400; *PPG-2000*
intermediate, esters: solvents
PPG-400; *PPG-2000*; *2,2,4-Trimethyl-1,3-pentanediol isobutyrate*
intermediate, esters: surfactants
2,2,4-Trimethyl-1,3-pentanediol isobutyrate
intermediate, esters: synthetic detergents
2,2,4-Trimethyl-1,3-pentanediol isobutyrate
intermediate, etherification
FA®; *Furfuryl alcohol*
intermediate, ethers
PPG-17
intermediate, ethoxylated surfactant
Tetrahydroxypropyl ethylenediamine
intermediate, ethyl lactates
Lactonitrile
intermediate, fabric/textile softeners
Methyldiethanolamine
intermediate, fatty acid esters
Alfol® 4; *PPG-20*; *PPG-30*
intermediate, fatty acid soap prod.
Triethanolamine
intermediate, feed additives
Ethylacetoacetate; *Methyl acetoacetate*
intermediate, fiber lubricants
Octyldodecanol
intermediate, fine chemicals
Bromine; *Furan*
intermediate, flame retardants
Dimethyl methylphosphonate; Linpar® 1416-V Normal Paraffin
intermediate, flavor chemicals
Myrcene
intermediate, flavors
n-Propyl bromide
intermediate, flexible film
Vinylidene chloride monomer
intermediate, flotation agents
Dibutylamine
intermediate, fluorocarbon resins
Chlorodifluoromethane
intermediate, fluorocarbons
Perchloroethylene
intermediate, foam stabilizers
Diglycolamine® Agent (DGA®)
intermediate, food additives
Stearyl alcohol
intermediate, food supplements (tablet form)
Cetyl alcohol
intermediate, food wrap
Vinylidene chloride monomer
intermediate, foundry resins
Formcel®
intermediate, fragrances
p-t-Butyl toluene; *Myrcene*; *n-Propyl bromide*
intermediate, fragrances: cosmetics
n-Butyraldehyde; *2-Ethylhexoic acid*; *2-Methylpropanal*
intermediate, Friedel-Crafts synthesis
Nitroethane
intermediate, fuel additives
Ethylenediamine
intermediate, fumigants
Bromine
intermediate, fungicides
Ethylenediamine; *Methylamine*
intermediate, furan
QO® Furfural
intermediate, furan resins
Furfural
intermediate, furfural alcohol
Formcel®
intermediate, furfuryl alcohol
Furfural
intermediate, gasoline additives
Tetradecene-1
intermediate, gel coats
Methylpropanediol
intermediate, geraniol
Myrcene
intermediate, germicides
Cashew nut shell oil
intermediate, gum
Cetyl alcohol
intermediate, halogen synthesis
Chlorobenzene
intermediate, halogenated solvents
NE™; NM™
intermediate, halogenation
Diphenyl oxide
intermediate, hand dishwashing anionic surfactants
Alkylate 215
intermediate, hard surface cleaning
Alkylate 215
intermediate, herbicides
n-Butyl mercaptan; *Furan*; *3-Methoxybutanol*; *Methylene bromide*; *Methyl pivalate*; Texanol® Ester-Alcohol; *Triethanolamine*
intermediate, hexamethylenediamine
Adiponitrile; *Hexamethylene glycol*
intermediate, household detergents
T-Det® N-4; *Triethanolamine*
intermediate, hydraulic fluids
Isooctyl alcohol
intermediate, impregnating materials
Cashew nut shell oil
intermediate, indelible inks
Cashew nut shell oil
intermediate, industrial cleaning
Alkylate 215
intermediate, industrial coatings
Formcel®
intermediate, industrial fragrances

intermediate, photographic chemicals
2,5-Hexanedione; 1-Methylpiperidine; Methyl pivalate
intermediate, phthalic anhydride
Naphthalene
intermediate, piperidine
Butyrolactone
intermediate, pkg. materials
Vinylidene chloride monomer
intermediate, plasticizers
Alfol® 4; 1,4-Butanediol; Butoxydiglycol; Cashew nut shell oil; Cyclohexylamine; Dinonyl phenol; Dowanol® DPnB; Dowanol® PnB; Estasol™ MD10; Linpar® 1416-V Normal Paraffin; 2-Mercaptoethanol; 3-Methoxybutanol; Methyl caprate; Methyl caproate; Methyl caprylate; Methyl laurate; Methyl myristate; Methyl palmitate; Methyl palmitate oleate; Methyl stearate; Neodecanoic acid; 2-Octanol; PEG/PPG-17/6 copolymer; Polypropylene glycol; PPG-2-buteth-3; PPG-3-buteth-5; PPG-7-buteth-10; PPG-9-buteth-12; PPG-20-buteth-30; PPG-28-buteth-35; Propylene glycol phenyl ether; Tetradecene-1; Tripropylene glycol; Ucon® 50-HB-55; Ucon® 50-HB-100; Ucon® 50-HB-260; Ucon® 50-HB-400; Ucon® 50-HB-660; Ucon® 50-HB-2000; Ucon® 50-HB-3520; Ucon® 50-HB-5100; Ucon® 75-H-450
intermediate, plastics
Ethylbenzene; Furan; Methyl laurate; 2-Methyl-1-pentanol; Vinylidene chloride monomer; p-Xylene
intermediate, polish prod.
Triethanolamine
intermediate, polyamide resins
Ethylenediamine
intermediate, polycarbonate resin synthesis
Diphenyl carbonate
intermediate, polyester polyols
DBE; DBE-2; DBE-3; DBE-4; DBE-5; DBE-6; DBE-9; Dimethyl adipate; Dimethyl glutarate; Dimethyl succinate
intermediate, polyester polyols/resins
DBE; DBE-2; DBE-3; DBE-4; DBE-5; DBE-6; DBE-9; Dimethyl adipate; Dimethyl glutarate; Dimethyl succinate
intermediate, polyester polyols: wet-str. paper resins
DBE; DBE-2; DBE-3; DBE-4; DBE-5; DBE-6; DBE-9; Dimethyl adipate; Dimethyl glutarate
intermediate, polyesters
Acrolein; Hexamethylene glycol; Kemstrene® 96.0% USP; Kemstrene® 99.7% USP; Propylene glycol; Trimethylene glycol
intermediate, polymer: polyester polyols, wet-str. paper resins
Dimethyl succinate
intermediate, polymeric plasticizers
Methylpropanediol
intermediate, polymers
Dibromodifluoromethane; Diethylamine; 1-Methylpiperidine; Propylene glycol t-butyl ether; Tetradecene-1; 1,2,3,4-Tetrahydroquinoline; Triethylamine
intermediate, polypropylene glycol
Propylene glycol
intermediate, polysulfide rubber
Bis (2-chloroethoxy) methane
intermediate, polyvinylpyrrolidone
Butyrolactone
intermediate, preservatives
Cashew nut shell oil
intermediate, printing inks
Perchloroethylene
intermediate, propoxylated surfactant
Tetrahydroxypropyl ethylenediamine
intermediate, proton pump inhibitor synthesis
3-Methoxy-1-propanol
intermediate, PS plasticizers
Methyl oleate
intermediate, pseudocumidine
1,2,4-Trimethylbenzene
intermediate, PU
Acrolein; Hexamethylene glycol; Kemstrene® 96.0% USP; Kemstrene® 99.7% USP
intermediate, PU field compds.
Isophorone
intermediate, PU foam
PEG 200; PEG 300; PEG 400
intermediate, pure behenic acid
Methyl behenate
intermediate, PVC lubricants
Tetradecene-1
intermediate, pyrrole
Furan
intermediate, quaternaries: fabric softeners
Alfol® 1412
intermediate, quaternaries: germicides
Alfol® 1412
intermediate, quaternary ammonium compds.
n-Propyl bromide
intermediate, rabeprazole sodium synthesis
3-Methoxy-1-propanol
intermediate, reactive chloro: alcohols
1,3-Dichloro-2-propanol
intermediate, reactive chloro: amines
1,3-Dichloro-2-propanol
intermediate, reactive chloro: carboxylic acids
1,3-Dichloro-2-propanol
intermediate, reactive: agric. prod.
Methyl butynol
intermediate, reactive: alcohol modification
DCP; GAM
intermediate, reactive: amine modification
DCP; GAM
intermediate, reactive: carboxylic acid modification
DCP; GAM
intermediate, reactive: fragrances
Methyl butynol
intermediate, reactive: hypnotic preps.
Methyl pentynol
intermediate, reactive: modifiers
Ucon® LB-65; Ucon® LB-135; Ucon® LB-165; Ucon® LB-285; Ucon® LB-385; Ucon® LB-525; Ucon® LB-625; Ucon® LB-1145; Ucon® LB-1715; Ucon® LB-3000
intermediate, reactive: perfumes
Methyl pentynol
intermediate, reactive: pharmaceuticals
Methyl butynol
intermediate, reactive: plasticizers
PPG-5 butyl ether; PPG-9 butyl ether; PPG-15 butyl ether; PPG-18 butyl ether; PPG-22 butyl ether; PPG-24 butyl ether; PPG-33 butyl ether; PPG-40 butyl ether; PPG-53 butyl ether; Ucon® LB-65; Ucon® LB-135; Ucon® LB-165; Ucon® LB-285; Ucon® LB-385; Ucon® LB-525; Ucon® LB-625; Ucon® LB-1145; Ucon® LB-1715; Ucon® LB-3000
intermediate, reactive: plastics
Methyl butynol; Methyl pentynol
intermediate, reactive: resins
PPG-5 butyl ether; PPG-9 butyl ether; PPG-15 butyl ether; PPG-18 butyl ether; PPG-22 butyl ether; PPG-24 butyl ether; PPG-33 butyl ether; PPG-40 butyl ether; PPG-53 butyl ether; Ucon® LB-65; Ucon® LB-135; Ucon® LB-165; Ucon® LB-285; Ucon® LB-385; Ucon® LB-525; Ucon® LB-625; Ucon® LB-1145; Ucon® LB-1715; Ucon® LB-3000
intermediate, reactive: rubber
Methyl butynol; Methyl pentynol
intermediate, reactive: specialty chemicals
Methyl pentynol
intermediate, reactive: surfactants
PPG-5 butyl ether; PPG-9 butyl ether; PPG-15 butyl ether; PPG-18 butyl ether; PPG-22 butyl ether; PPG-24 butyl ether; PPG-33 butyl ether; PPG-40 butyl ether; PPG-53 butyl ether; Ucon® LB-65; Ucon® LB-135; Ucon® LB-165; Ucon® LB-285; Ucon® LB-385; Ucon® LB-525; Ucon® LB-625; Ucon® LB-1145; Ucon® LB-1715; Ucon® LB-3000
intermediate, resins
Acetophenone; Cumyl phenol; Methyl caprate; Methyl caproate; Methyl caprylate; Methyl myristate; Methyl palmitate; Methyl palmitate oleate; Methyl stearate; PEG/PPG-17/6 copolymer; PPG-2-buteth-3; PPG-3-buteth-5; PPG-7-buteth-10; PPG-9-buteth-12; PPG-20-buteth-30; PPG-28-buteth-35; Tripropylene glycol; Ucon® 50-HB-55; Ucon® 50-HB-100; Ucon® 50-HB-260; Ucon® 50-HB-400; Ucon® 50-HB-660; Ucon® 50-HB-2000; Ucon® 50-HB-3520; Ucon® 50-HB-5100; Ucon® 75-H-450
intermediate, resorcinol
Formcel®
intermediate, rubber
Ethanolamine; Methyl oleate; Pluracol® E1450 NF; Triethylamine
intermediate, rubber accelerators
Diethylamine; Isopropylamine; 2-Methylpropanal; Piperidine
intermediate, rubber antioxidants
2-Methylpropanal
intermediate, rubber chemicals
Benzonitrile; n-Butylamine; Diethanolamine; Ethylene carbonate; Ethylenediamine; dl-Limonene; 2-Mercaptoethanol
intermediate, rubber compds.
Hexachloro-1,3-butadiene
intermediate, rubber oxidants
Paraldehyde
intermediate, rubber solvents
2,3-Dimethylbutane
intermediate, Saran
Vinylidene chloride monomer
intermediate, sat. polyester coatings
Methylpropanediol
intermediate, sealants
Methylpropanediol
intermediate, soap/perfume fixatives
Dalpad A; Dowanol® EPh
intermediate, soaps
Methoxy PEG-10
intermediate, solder reflow applics.
Jeffox® WL-440; Jeffox® WL-660; Jeffox® WL-1400; Jeffox® WL-5000
intermediate, solubilizers
Diethylaminoethanol
intermediate, solvents
Dowanol® DPnB; Dowanol® PnB; Methyl pivalate; Methylpropanediol
intermediate, sorbic acids
Crotonaldehyde
intermediate, specialty
Dimethyl glutarate; Dimethyl succinate
intermediate, specialty chemicals
DBE; DBE-2; DBE-3; DBE-4; DBE-5; DBE-6; DBE-9; Tetradecene-1; 1,1,1-Trichloro-2,2,2-trifluoroethane
intermediate, SR
Ethylbenzene
intermediate, stabilizers
Methyl caprate; Methyl caproate; Methyl caprylate; Methyl laurate; Methyl myristate; Methyl oleate; Methyl palmitate; Methyl palmitate oleate; Methyl stearate
intermediate, stain-resistance plasticizers
Texanol® Ester-Alcohol
intermediate, stearic acid detergents
Methyl stearate
intermediate, styrene
Ethylbenzene
intermediate, styrene latexes
Ethylbenzene
intermediate, succinic acid
Naphthenic acid
intermediate, sulfation: antistats
Nonoxynol-5
intermediate, sulfation: lubricants
Nonoxynol-5
intermediate, sulfochlorination: plasticizers
Linpar® 1416-V Normal Paraffin
intermediate, sulfochlorination: surfactants
Linpar® 1416-V Normal Paraffin
intermediate, sulfoxidation: plasticizers
Linpar® 1416-V Normal Paraffin
intermediate, sulfoxidation: surfactants
Linpar® 1416-V Normal Paraffin
intermediate, superamides
Methyl behenate; Methyl soyate
intermediate, surfactants
Alkylate L210-L; Aminoethylpiperazine; Butyl alcohol; Caprylic alcohol; Cetearyl alcohol; Cumyl phenol; Diethylamine; Isopropylamine; Methoxy PEG-16; Methylamine; Myristyl alcohol; Naphthalene; Naphthenic acid; Oleocal ME-70; Oleocal ME-112; Oleocal ME-130; PEG-135; PEG/PPG-17/6 copolymer; PPG-17; PPG-2-buteth-3; PPG-3-buteth-5; PPG-7-buteth-10; PPG-9-buteth-12; PPG-20-buteth-30; PPG-28-buteth-35; Rhodasurf® PEG 600; Tetradecene-1; Triethanolamine; Trilaurylamine; Ucon® 50-HB-55; Ucon® 50-HB-100; Ucon® 50-HB-260; Ucon® 50-HB-400; Ucon® 50-HB-660; Ucon® 50-HB-2000; Ucon® 50-HB-3520; Ucon® 50-HB-5100; Ucon® 75-H-450
intermediate, surfactants: cosmetics
Alfol® 1412

intermediate, surfactants: hair care
Methyldiethanolamine
intermediate, surfactants: household prods.
Alfol® 1412
intermediate, surfactants: industrial
Alfol® 1412
intermediate, synthetic detergents
Texanol® Ester-Alcohol
intermediate, synthetic drying oils
Isooctyl alcohol
intermediate, synthetic fibers
Acetonitrile; *Aminoethylpiperazine*
intermediate, synthetic flavorings
n-Butyl acetate
intermediate, synthetic flavors
Methyl acetate
intermediate, synthetic glycerol
Acrolein
intermediate, synthetic lubricants
Methyl pivalate
intermediate, synthetic pharmaceuticals
Acetonitrile
intermediate, synthetic waxes
Ethylenediamine
intermediate, tanning
Acetylated glyceryl stearate
intermediate, tanning agents
n-Butylamine; Glicopol 2-G; Milopol G-7; Monestriol DM; Monestriol GCT; Monestriol GE; Monestriol GP; Monestriol GP-35; Monestriol GP-35-3%; Monestriol GP-35-AE; Monestriol GP-40; Monestriol GP-40-CF; Monestriol GP-40-DC; Monestriol GP-45; Monestriol GP-45-DC; Monestriol GP-60; Monestriol GP-AS; Thomil G; Tilol GCT-R; Tilol GP-O; Triacetina; Triol 91; Trioleina
intermediate, t-butylbenzoic acid
p-t-Butyl toluene
intermediate, terephthalic acid synthesis: polyester fibers
p-Xylene
intermediate, terephthalic acid synthesis: polyester resins
p-Xylene
intermediate, terpene resins
dl-Limonene
intermediate, terylene
dl-Limonene
intermediate, tetrafluoroethylene polymers
Chlorodifluoromethane
intermediate, tetrahydrofuran
Furan
intermediate, textile auxs.
N-Methylethanolamine
intermediate, textile chemicals
Aminoethylpiperazine; *Cyclohexylamine*; *2-Nitropropane*
intermediate, textile finishes
Diethylenetriamine
intermediate, textile finishing agents
Ethylene carbonate; *Hexyl alcohol*
intermediate, textile specialties
Diethylamine; *Isopropylamine*; *Methyl palmitate oleate*; *Triethanolamine*
intermediate, textile treatment
Methyl oleate
intermediate, textiles
Acetylated glyceryl stearate; *DM hydantoin*; Formcel®; Glicopol 2-G; *Hexyldecanol*; *Methyl laurate*; Milopol G-7; Monestriol DM; Monestriol GCT; Monestriol GE; Monestriol GP; Monestriol GP-35; Monestriol GP-35-3%; Monestriol GP-35-AE; Monestriol GP-40; Monestriol GP-40-CF; Monestriol GP-40-DC; Monestriol GP-45; Monestriol GP-45-DC; Monestriol GP-60; Monestriol GP-AS; *Octyldodecanol*; Pluracol® E1450 NF; Thomil G; Tilol GCT-R; Tilol GP-O; Triacetina; Triol 91; Trioleina
intermediate, THF
Furfural; QO® Furfural
intermediate, thiophene
Furan
intermediate, trichloroacetic acid
Perchloroethylene
intermediate, trichloroethylene
1,1,2,2-Tetrachloroethane
intermediate, trimellitic anhydride
1,2,4-Trimethylbenzene
intermediate, trimethylcyclohexanone
Isophorone
intermediate, unsat. polyesters
Methylpropanediol
intermediate, urea
Formcel®
intermediate, urethane coatings
ε-Caprolactone monomer
intermediate, urethane foams
Polypropylene glycol
intermediate, urethanes
DBE; DBE-2; DBE-3; DBE-4; DBE-5; DBE-6; DBE-9; *Dimethyl adipate*; *Dimethyl glutarate*; *Dimethyl succinate*; *Methylpropanediol*
intermediate, uv oxidation stabilizers: plastics
1,3,5-Trimethylbenzene
intermediate, varnishes
Cashew nut shell oil
intermediate, vinylidene chloride prod.
1,1,2-Trichloroethane
intermediate, vinylidene chlorides
1,1,1-Trichloroethane
intermediate, Vitamin B
Acetonitrile
intermediate, Vitamin B_1
BLO®; *Butyrolactone*
intermediate, vitamin B_1
Ethyl 3-ethoxypropionate
intermediate, Vitamin E
BLO®; *Butyrolactone*
intermediate, vitamins
Ethylacetoacetate; *Methyl acetoacetate*; *Propargyl alcohol*; *p-Xylene*
intermediate, vulcanization accelerators
Ethanolamine; *Paraldehyde*
intermediate, warfare chemicals
Thiodiglycol
intermediate, water gel explosives
Methylamine
intermediate, waxes
Methyl oleate; *Triethanolamine*
intermediate, wetting agents
Diglycolamine® Agent (DGA®); Estasol™ MD10; *Methyl caprate*; *Methyl caproate*; *Methyl caprylate*; *Methyl laurate*; *Methyl myristate*; *Methyl oleate*; *Methyl palmitate*; *Methyl palmitate oleate*; *Methyl stearate*
intermediate, wood prods.
Formcel®; Pluracol® E1450 NF
intermediate: rubber industry, synthetic
1-Chloro-1-nitropropane
iodine source, pharmaceuticals
Potassium iodide
iodine source, table salt
Potassium iodide
isoparaffin substitute
Estasol™ MD10
jojoba oil substitute
Oleyl erucate
keratolytic, dermatitis
Sulfur
keratolytic, psoriasis preps.
Sulfur
ketone substitute
Estasol™ MD10
lacquer remover
Triethyl phosphate
larvicide
Diphenyl sulfone
larvicide, cereals
Ethyl formate; *Methyl formate*
larvicide, dried fruits
Methyl formate
larvicide, tobacco
Ethyl formate; *Methyl formate*
laxative
Castor (Ricinus communis) oil; *PEG-125*
laxative, pharmaceuticals
Mineral oil; *Safflower (Carthamus tinctorius) oil*
lead poisoning treatment, pharmaceuticals
Edetic acid
lead scavenger, antiknock fluids
Propylene dichloride
lead scavenger, gasoline
Ethylene dichloride; *Tricresyl phosphate*
leak detecting agent
Dichlorodifluoromethane
leak detection aid, pipes
Crotonaldehyde
leather auxiliary
Diethylaminoethanol
leavening agent, food
Ammonium hydroxide
leveling agent
Arosolve TUV; *Cyclododecanol*; *PEG-12 dioleate*; *Polysorbate 60*; Surfadone® LP-100; Surfadone® LP-300
leveling agent, air-dry coatings
Eastman® EB Acetate
leveling agent, disperse dyes: polyesters
Burcolev DLS
leveling agent, dye
Poloxamer 181; *Trimethylcyclohexanol*
leveling agent, dye: textiles
Benzyl trimethyl ammonium chloride; Incrocas 30; Incrocas 40
leveling agent, electropolishing
DePEG 600
leveling agent, enamels
Butyl Oxitol®
leveling agent, epoxies
Eastman® EB Acetate
leveling agent, felt pens
Ethyl Proxitol
leveling agent, floor finishes
Tributoxyethyl phosphate
leveling agent, furniture polishes
Ethyl Proxitol
leveling agent, high-bake coatings
Eastman® EB Acetate
leveling agent, inks
Trihydroxystearin
leveling agent, lacquers
Butyl Oxitol®
leveling agent, leather finishes
Ethyl Proxitol
leveling agent, mastics
Trihydroxystearin
leveling agent, paints/coatings
Trihydroxystearin; *3,3,5-Trimethyl cyclohexanone*
leveling agent, pigments
Cyclomethicone
leveling agent, printing inks
Ethyl Proxitol
leveling agent, pulp/paper
Paxsolv PC
leveling agent, solvent-free epoxy systems
Trihydroxystearin
leveling agent, stains
Trihydroxystearin
leveling agent, stamp pad inks
Ethyl Proxitol
leveling agent, surface coatings
Butyl Oxitol®; Ethyl Proxitol
leveling agent, textile backings
Trihydroxystearin
leveling agent, textile printing
Isophorone
leveling agent, textiles
Incrocas 30; Incrocas 40
leveling agent, varnishes
Butyl glycolate
leveling agent, waxes
Tributoxyethyl phosphate
leveling agent, wood stains
Ethyl Proxitol
leveling, silk
Cyclohexanone
leveling, textiles
Cyclohexanone
limonene extender
Glidsol® 5140
liniment, pharmaceuticals
Turpentine
linseed oil substitute
Menhaden oil
liq. fuels, high octane: raw material
Butane
liquefier, pharmaceuticals
SD alcohol 40; *SD alcohol 40-B*
load-carrying additive
Dibutyl phosphite
lubricant
Acetylated lard glyceride; *Acetyl tributyl citrate*; *Acetyl triethyl citrate*; Albrite® TBP; *Ammonium hydroxide*; *Arachidyl alcohol*; *Behenyl alcohol*; *BHA*; *BHT*; *Butyl diglycol carbonate*; *Butyl stearate*; Cachalot® AR-20; Cachalot® BE-22; Cachalot® C-50; Cachalot® L-90; Cachalot® M-43; Cachalot® S-56; *C8 alkyl acetate*; *C9 alkyl acetate*; *C10 alkyl acetate*; *C13 alkyl acetate*; *Cashew nut shell oil*; *Castor (Ricinus communis) oil*; *Castor oil, polymerized*; *Cetyl alcohol*; *Cetyl palmitate*; Chemax E-1000 MO; *2-Chloronaphthalene*; *Corn (Zea mays) oil*; *Cottonseed (Gossypium) oil*; Crill 2; *Decahydronaphthalene*; *Decamethylcyclopentasiloxane*; *1,10-Decanediol*; *Decyl alcohol*; Diadol 18G-C; *2,6-Dimethyl heptanol-4*; *Dioctyl phosphite*; *Dipropylene glycol*; Edenor® ME C6-10; Edenor® ME C18-90; *Eicosane*; Emery® 2214; Emery® 2254; Emery® 2255; Emery® 2270; Emery® 2296; *2-Ethyl-1-butanol*; *2-Ethylhexanol*; *Ethyl oleate*; Exxate® 600; Exxate® 700; Exxate® 800; Exxate® 900; Exxate® 1000; Exxate® 1300; *Glycerin*; *Glyceryl triacetyl ricinoleate*; *Hexamethyldisilazane*; *Hexamethyl-*

disiloxane; *Hexyl alcohol*; Hydrol H660; Hydrol H660YA; Hydrol H5100; Hydrol L385; *12-Hydroxystearyl alcohol*; *Isostearyl alcohol*; Jarcol™ I-16; Jeffox® PPG-400; Jeffox® PPG-2000; Kemstrene® 99.7% USP; *Lanolin acid*; *Lauryl alcohol*; *Menhaden oil*; *Methyl acetyl ricinoleate*; *2-Methyl-1-butanol*; *Methyl caprylate/caprate*; *Methyl carbonate*; *Methyl oleate-stearate*; *Methyl palm kernelate*; *Methyl stearate*; *Mineral oil*; *Myristyl alcohol*; *Naphthalene*; *Naphthenic acid*; *Neatsfoot oil*; NEOBEE® 18; NEOBEE® M-20; *N-Nitrosodiethylamine*; *N-Nitrosodimethylamine*; *Nonanoic acid*; *Nonoxynol-5*; *Octamethylcyclotetrasiloxane*; *Octamethyltrisiloxane*; #15 Oil; *Oleic acid*; *Oxo-heptyl acetate*; *Oxo-hexyl acetate*; Pale 16; Pale 170; Pale 1000; *PEG-10*; *PEG-18*; *PEG-90*; *PEG-135*; *PEG-3 dimethyl ether*; *PEG-12 dioleate*; *PEG-5 oleate*; *3-(n-Pentadecyl) phenol*; *Pentaerythrityl oleate*; *Petroleum distillates, hydrotreated middle*; *Phenol*; *Phenyl trimethicone*; *α-Pinene*; *Polysorbate 60*; *PPG-2-buteth-3*; *PPG-3-buteth-5*; *PPG-5-buteth-7*; *PPG-7-buteth-10*; *PPG-9-buteth-12*; *PPG-12-buteth-16*; *PPG-20-buteth-30*; *PPG-28-buteth-35*; *PPG-33-buteth-45*; *PPG-53 butyl ether*; *PPG-3-myreth-3*; *Sodium lauryl sulfate*; Soygold® 1000; Soygold® 2000; Speziol C 10/2; Speziol C 18/2; *Stearyl stearate*; *Tall oil*; *Tetramethylsilane*; *Tributoxyethyl phosphate*; *Tributyl phosphate*; *1,3,5-Trichlorobenzene*; *Triethylene glycol*; *Triethyl phosphate*; *Triisooctyl phosphite*; *Triisopropyl phosphite*; *2,6,8-Trimethylnonyl-4-alcohol*; *2,2,4-Trimethyl-1,3-pentanediol*; *Tri-p-tolyl phosphate*; Vorite 105; Vorite 110; Vorite 115; Vorite 125; Wickenol® 127; Witcohol™ 42; Witconol™ APM; Witconol™ APS; *Xylenol*; *2,3-Xylenol*; *3,4-Xylenol*; ZELEC® C-48; ZELEC® C-49

lubricant basestock
Petroleum distillates, heavy hydrotreated naphthenic

lubricant enhancer, mineral oils
Isobutyl oleate

lubricant, acid aluminum cleaners
Rhodacal® 330

lubricant, acne creams/lotions
Crodamol PMP

lubricant, adhesives
Carbowax® PEG 300; Crill 1; *Methoxy PEG-6*; *Methoxy PEG-10*; *Methoxy PEG-16*; #30 Oil; #40 Oil; Pale 4; *Propylene glycol laurate*

lubricant, aerosols
Captex® 300; Captex® 8227; Lexol® PG-865; *Tricaprin*

lubricant, aerospace
Perfluoropolyether

lubricant, afterbath splashes
Trivent DIA

lubricant, aftershave lotions
Captex® 8227; Estalan-DIPS; Trivent DIA

lubricant, aftershaves
HallStar™ DIPA; HallStar™ DIPS

lubricant, agric.
Carbowax® PEG 300; Incrocas 30; Incrocas 40; *Nonoxynol-9 phosphate*; *PEG-4 dilaurate*; *PEG-4 oleate*; *PEG-6 oleate*; *PEG-8 oleate*; *Propylene glycol laurate*; *Sorbitan trioleate*; ZELEC® C-48; ZELEC® C-49

lubricant, agric. adjuvants
Methyl rapeseedate

lubricant, agric. chemical sprays
PEG-8 dilaurate

lubricant, agric. chemicals
Trioctyl phosphate

lubricant, agric. pesticides
Rhodacal® 330

lubricant, aircraft
Dioctyl adipate

lubricant, alcoholic compositions
Procetyl 10

lubricant, aluminum
Glycereth-7

lubricant, antibiotic ointments
Crodamol PMP

lubricant, antiperspirant sticks
Crystal® O; Crystal® Crown

lubricant, antiperspirants
Carbowax® Sentry® PEG 300; Carbowax® Sentry® PEG 540 Blend; Carbowax® Sentry® PEG 600; Carbowax® Sentry® PEG 900; Carbowax® Sentry® PEG 1000; Carbowax® Sentry® PEG 1450; Carbowax® Sentry® PEG 3350; Carbowax® Sentry® PEG 4600; Carbowax® Sentry® PEG 8000; *Glycereth-7*; *Isopropyl myristopalmitate*; Lexol® IPL; *PEG/PPG-17/6 copolymer*; *PPG-14 butyl ether*; *PPG-5-ceteth-20*; Trivent DIA; Ucon® 75-H-450; Ucon® Fluid AP

lubricant, aq. alcoholic systems
PPG-5-ceteth-20; Procetyl 10

lubricant, aq. metalworking
Trioctyl phosphate

lubricant, asphalt
Erucamide

lubricant, baby care
Acetol® 1706

lubricant, baby oils
Lanesta L; *Lanolin oil*

lubricant, bakery prods.
Caprylic/capric triglyceride; Crodamol GTCC; Delios® S; Hodag® S-35 FGK; *Petrolatum*

lubricant, bath care
Acetol® 1706; Carbowax® E200; Carbowax® E300; Carbowax® E400; Carbowax® E600; Carbowax® E900; Carbowax® E1000; Carbowax® E1450; Carbowax® E3350; Carbowax® E8000; *Hexyldecyl benzoate*

lubricant, bath oils
Arlatone® T; Captex® 8227; Carbowax® Sentry® PEG 400; Crodamol PMP; Cropure® Almond; *Decyl oleate*; Estalan DIA; HallStar™ DIPA; *Isopropyl myristopalmitate*; Lexol® PG-865; *PEG-40 sorbitan peroleate*; *PPG-2 myristyl ether propionate*; Schercemol DIA

lubricant, bath powders
Carbowax® Sentry® PEG 3350

lubricant, bath/shower
Captex® 300

lubricant, bearings
Perfluoropolyether

lubricant, beverage machinery
Propylene Glycol USP, FCC

lubricant, biodeg. greases
Canola oil

lubricant, boiler defoaming
Jeffox® WL-440; Jeffox® WL-660; Jeffox® WL-1400; Jeffox® WL-5000

lubricant, brilliantines
Lanolin oil

lubricant, burn creams
Crodamol PMP

lubricant, cable
Petrolatum; *Petroleum distillates, alkylate*

lubricant, calendering
Pluracol® SRF-2; Pluracol® W3520N; Pluracol® WI 285; Pluracol® WI 625; Pluracol® WI 1715; Pluracol® WS 170; Pluracol® WS 260; Pluracol® WS 660; Pluracol® WS 2000; Pluracol® WS 3520; Pluracol® WS 5100; Pluracol® WT 1400; Pluracol® WT 9150; Pluracol® WT 90,000

lubricant, carburetor fluids
Tripropylene glycol

lubricant, carding
Octyl palmitate; STEPAN® IPM; STEPAN® IPP; ZELEC® IPM

lubricant, caulks
#30 Oil; #40 Oil; Pale 4

lubricant, cellulose lacquers
#30 Oil; #40 Oil; Pale 4

lubricant, ceramic extrusion
Dibutyl adipate; *n-Dioctyl phthalate*; *Dioctyl sebacate*; *Glyceryl ricinoleate*; *Hexyl laurate*; *Stearyl stearate*

lubricant, ceramics
Carbowax® PEG 300; *PPG-9*; *PPG-17*; *PPG-30*

lubricant, ceramics extrusion
Isotridecyl stearate

lubricant, clay
Acetulan®

lubricant, cleaners
Rhodacal® 330

lubricant, cleansers
HallStar™ DIPA; HallStar™ DIPS; HallStar™ DOS; Lanesta L

lubricant, cleansing lotions
Lanolin oil

lubricant, clear sticks/gels
HallStar™ DIPS

lubricant, clinical nutrition
Captex® 300; *Tricaprin*

lubricant, coatings
Acetylated palm kernel glycerides; *Butyl stearate*; Captex® 300; *PEG-8 oleate*; *Tricaprin*

lubricant, coffee whiteners
Hodag® S-35 FGK

lubricant, colognes
Carbowax® Sentry® PEG 400; Carbowax® Sentry® PEG 600; HallStar™ DIPA; HallStar™ DIPS

lubricant, colorants
Tristearin

lubricant, compressed confectionery
Trimyristin

lubricant, compressed tablets
DePEG 200; DePEG 300; DePEG 400; DePEG 600

lubricant, compressors
PPG-2-buteth-3; *PPG-3-buteth-5*; *PPG-7-buteth-10*; *PPG-9-buteth-12*; *PPG-20-buteth-30*; *PPG-28-buteth-35*

lubricant, confectionery
Petrolatum

lubricant, confectionery coatings
Aldo® MCT KFG

lubricant, coning oils
Butyl oleate; *Octyl palmitate*; STEPAN® IPM; STEPAN® IPP; ZELEC® IPM

lubricant, contact lens fluid
Polyethylene glycol

lubricant, continuous film
Rapeseed (Brassica campestris) oil

lubricant, controlled-releases
Polyethylene glycol

lubricant, copolymers
N-Nitrosodimethylamine

lubricant, cording
Butyl oleate

lubricant, cosmetic creams
Isopropyl myristate

lubricant, cosmetic creams/lotions
Estalan DIA; *Isocetyl stearate*; *Isopropyl palmitate*

lubricant, cosmetic masks
Glycereth-7

lubricant, cosmetics
Alfol® 1412; Arconate® PC-1000; Arconate® PC-HP; Axol® E 61; Axol® E 66; *Benzyl laurate*; *Butyl octyl benzoate*; *Canola oil*; Captex® 300; Captex® 8227; Carbowax® PEG 300; Carbowax® Sentry® PEG 300; Carbowax® Sentry® PEG 400; Carbowax® Sentry® PEG 540 Blend; Carbowax® Sentry® PEG 600; Carbowax® Sentry® PEG 900; Carbowax® Sentry® PEG 1000; Carbowax® Sentry® PEG 1450; Carbowax® Sentry® PEG 3350; Carbowax® Sentry® PEG 4600; Carbowax® Sentry® PEG 8000; Ceraphyl® 28; Ceraphyl® 140; *Ceteth-16*; *Cetyl octanoate*; *Cetyl stearate*; Chemax E-200 MO; Crill 1; Crystal® O; Crystal® Crown; *Decyl oleate*; Dynacerin® 660; *Glycereth-7*; *Glycereth-26*; *Glycol distearate*; *Isobutyl palmitate*; *Isocetyl laurate*; *Isocetyl myristate*; *Isopropyl lanolate*; *Isopropyl laurate*; *Isopropyl stearate*; *Isostearyl isostearate*; *Isostearyl lactate*; Lipocol P-15; *Methyl cocoate*; *Methyl oleate*; *Methyl tallowate*; Miglyol® 810; Miglyol® 812; Miglyol® 840; Miglyol® 8810; *Myristyl myristate*; *Octyldodecanol*; *Octyl isononanoate*; *Octyl stearate*; *Oleyl alcohol*; *Oleyl erucate*; *Olive (Olea europaea) oil*; Olive Oil, Refined; *PEG-35 castor oil*; *PEG-4 dilaurate*; *PEG-4 oleate*; *PEG-6 oleate*; *PEG-8 oleate*; *PEG-2 stearate*; Pluracol® E400 NF; Pluracol® E600 NF; Pluracol® E1450 NF; Polypax IPM; Polypax IPP; *PPG-5-ceteth-20*; *PPG-10 cetyl ether*; *PPG-30 cetyl ether*; *PPG-50 cetyl ether*; *PPG-40 diethylmonium chloride*; *PPG-55 glyceryl ether*; *PPG-2 myristyl ether propionate*; *PPG-11 stearyl ether*; *PPG-15 stearyl ether*; Procetyl 10; *Propylene carbonate*; *Propylene glycol laurate*; *Propylene glycol ricinoleate*; Prostearyl 15; Schercemol 1818; Schercemol OLO; *Tricaprin*; *Tricaprylin*; *Tridecyl neopentanoate*; *Trioctanoin*; *Triolein*; *Tripalmitin*; Variquat® CC-42

lubricant, crayons
Erucamide

lubricant, creams
Crodamol PMP; *PPG-2 myristyl ether propionate*

lubricant, creams/lotions

phosphate; *Triolein*
lubricant, leather dressing
#30 Oil; #40 Oil; Pale 4
lubricant, leather treatment
Crill 1
lubricant, liniments
Super Refined® Olive NF; Super Refined® Sesame NF; Super Refined® Soybean USP
lubricant, lip balm
Naturechem® GTR
lubricant, lip care
Captex® 300
lubricant, lip glosses
Captex® 8227
lubricant, lipstick
Captex® 8227; Carbowax® Sentry® PEG 400; Carbowax® Sentry® PEG 600; Crystal® Crown; Estalan DIA; Isodragol 2/050300; *Lanolin oil*; Naturechem® GTR; Schercemol DIA
lubricant, liq. preps.
Lutrol® E 300; Lutrol® E 400
lubricant, lotions
Shea butter (Butyrospermum parkii)
lubricant, lubricants
Carbowax® PEG 300
lubricant, lubricating oils
Methyl cocoate; *Methyl lardate*
lubricant, machine lubrication
Castor (Ricinus communis) oil
lubricant, machine part cleaning
Hedislide RS
lubricant, machine tools
Morpholine
lubricant, magnetic tapes
Butoxyethyl stearate
lubricant, makeup
Carbowax® E200; Carbowax® E300; Carbowax® E400; Carbowax® E600; Carbowax® E900; Carbowax® E1000; Carbowax® E1450; Carbowax® E3350; Carbowax® E8000; Carbowax® Sentry® PEG 300; Carbowax® Sentry® PEG 400; Carbowax® Sentry® PEG 600; Carbowax® Sentry® PEG 3350; Crystal® O; Crystal® Crown; Fancorsil A; HallStar™ BST; HallStar™ DIPA; HallStar™ DIPS; HallStar™ DOS; *Hexyldecyl benzoate*; *Isopropyl oleate*; Lexol® PG-865; *Myristyl myristate*
lubricant, makeup bases
Captex® 8227
lubricant, makeup removers
Carbowax® Sentry® PEG 400; HallStar™ BST
lubricant, mascara
Captex® 300
lubricant, massage
Cocoa (Theobroma cacao) butter
lubricant, melamine-formaldehyde resins: food pkg.
n-Dioctyl phthalate
lubricant, men's grooming aids
Schercemol DIA
lubricant, metal cleaners
Pluracol® E400 NF; Pluracol® E600 NF; Pluracol® E1450 NF; Rhodacal® 330
lubricant, metal cutting
Methyl palmitate; *Oleyl alcohol*; Soygold® 1000; Soygold® 2000
lubricant, metal drawing
Castor (Ricinus communis) oil; *PEG-2 stearate*
lubricant, metal fabrication
Carbowax® PEG 300
lubricant, metal foil rolling
Petroleum distillates, hydrotreated light paraffinic
lubricant, metal lubricants
Rapeseed (Brassica campestris) oil
lubricant, metal polishes
Pluracol® E1450 NF
lubricant, metal polishing
Pluracol® E400 NF; Pluracol® E600 NF
lubricant, metal processing
PPG-3 myristyl ether, *PPG-11 stearyl ether*
lubricant, metal rolling/forming
Alfol® 1412
lubricant, metals
Behenyl alcohol; *Butyl stearate*; Chemax E-200 MO; *Erucamide*; *Isostearyl alcohol*; *PEG-30 castor oil*; *PEG-40 castor oil*; Pluracol® E200; Pluracol® E300; Pluracol® E400; Pluracol® E600; Pluracol® E4000; Pluracol® E6000; Sasol LPA-210 Solvent; *Triolein*
lubricant, metalworking
Cetearyl isononanoate; *Glyceryl ricinoleate*; *Glycol distearate*; *Isobutyl palmitate*; *Isocetyl alcohol*; *Isodecyl oleate*; Kemester® 205; Kemester® 213; Kemester® 226; Kemester® EGDS; *Lanolin*; *Methyl behenate*; *Methyl eicosanoate*; *Methyl rapeseedate*; *Methyl soyate*; Norfox® MTL; *Octyl isononanoate*; *Octyl stearate*; Oleocal ME-70; Oleocal ME-112; Oleocal ME-130; *Oleyl oleate*; PEG 200; PEG 400; *PEG-6*; *PEG-8*; *PEG-9*; *PEG-12*; *PEG-14*; *PEG-16*; *PEG-20*; *PEG-32*; *PEG-40*; *PEG-75*; *PEG-100*; *PEG-150*; *PEG-200*; *PEG-350*; *PEG-4 dilaurate*; *PEG-4 oleate*; Pogol® 400; PPG 400; PPG 1650; PPG 2250; *PPG-17*; *Propylene glycol laurate*; Schercemol DIA; Schercemol OLO; *Trioctyl phosphate*; ZELEC® C-48; ZELEC® C-49
lubricant, metalworking fluids
Carbowax® E200; Carbowax® E300; Carbowax® E400; Carbowax® E600; Carbowax® E900; Carbowax® E1000; Carbowax® E1450; Carbowax® E3350; Carbowax® E8000; *Cetyl palmitate*; Crill 1; *Dibutyl adipate*; *n-Dioctyl phthalate*; *Dioctyl sebacate*; Incrocas 30; Incrocas 40; *Isobutyl oleate*; *Isooctyl stearate*; *Methyl oleate*; *PEG-4*; Pluracol® SRF-2; Pluracol® W3520N; Pluracol® WI 285; Pluracol® WI 625; Pluracol® WI 1715; Pluracol® WS 170; Pluracol® WS 260; Pluracol® WS 660; Pluracol® WS 2000; Pluracol® WS 3520; Pluracol® WS 5100; Pluracol® WT 1400; Pluracol® WT 9150; Pluracol® WT 90,000; Rhodacal® 330; *Stearyl stearate*
lubricant, metalworking oils
Hexyl laurate; *Isotridecyl stearate*; *Methyl lardate*; *Methyl palmitate oleate*
lubricant, mills
Pluracol® SRF-2; Pluracol® W3520N; Pluracol® WI 285; Pluracol® WI 625; Pluracol® WI 1715; Pluracol® WS 170; Pluracol® WS 260; Pluracol® WS 660; Pluracol® WS 2000; Pluracol® WS 3520; Pluracol® WS 5100; Pluracol® WT 1400; Pluracol® WT 9150; Pluracol® WT 90,000
lubricant, mineral
Methyl palmitate
lubricant, mining
Methoxy PEG-6; *Methoxy PEG-10*
lubricant, moisturizers
Crodamol PMP; *Decyl oleate*; *PPG-2 myristyl ether propionate*
lubricant, moisturizing creams/lotions
Fancorsil A; Trivent DIA
lubricant, mold
Butyl stearate; *Dioctyl ether*, *Polyethylene glycol*
lubricant, mold release agents
Canola oil
lubricant, mold release: die casting
PEG-2 stearate
lubricant, mold release: food
Drakeol® 7
lubricant, mold release: food pkg.
Drakeol® 7
lubricant, mold release: meat packing
Drakeol® 7
lubricant, mold: ceramics
Tripropylene glycol n-butyl ether
lubricant, motor oils
Dibutyl adipate; *n-Dioctyl phthalate*; *Dioctyl sebacate*; *Glyceryl ricinoleate*; *Hexyl laurate*; *Stearyl stearate*
lubricant, motor vehicle brake fluids
Pluracol® SRF-2; Pluracol® W3520N; Pluracol® WI 285; Pluracol® WI 625; Pluracol® WI 1715; Pluracol® WS 170; Pluracol® WS 260; Pluracol® WS 660; Pluracol® WS 2000; Pluracol® WS 3520; Pluracol® WS 5100; Pluracol® WT 1400; Pluracol® WT 9150; Pluracol® WT 90,000
lubricant, nail polish removers
Carbowax® Sentry® PEG 400; Carbowax® Sentry® PEG 3350; Carbowax® Sentry® PEG 8000
lubricant, natural fats
Sorbitan trioleate
lubricant, natural products
Pogol® 400
lubricant, natural waxes
Sorbitan trioleate
lubricant, nutritional applics.
Captex® 8227
lubricant, nutritional supplements
Super Refined® Olive NF; Super Refined® Sesame NF; Super Refined® Soybean USP; *Tricaprin*
lubricant, o/w creams
Shea butter (Butyrospermum parkii)
lubricant, o/w emulsions
Shea butter (Butyrospermum parkii)
lubricant, oil slick dispersing
Crill 1
lubricant, oil well drilling muds
Glyceryl stearate
lubricant, ointments
Carbowax® Sentry® PEG 540 Blend; Carbowax® Sentry® PEG 1000; Carbowax® Sentry® PEG 1450; Super Refined® Olive NF; Super Refined® Sesame NF; Super Refined® Soybean USP
lubricant, oral care
Pluracol® E400 NF; Pluracol® E600 NF; Pluracol® E1450 NF
lubricant, oral pharmaceuticals
Etocas 35 NF; Miglyol® 810; Super Refined® Olive NF; Super Refined® Sesame NF; Super Refined® Soybean USP
lubricant, ore flotation
Jeffox® WL-440; Jeffox® WL-660; Jeffox® WL-1400; Jeffox® WL-5000
lubricant, paints
Carbowax® PEG 300; Crill 1; *Oleyl alcohol*; *Propylene glycol laurate*
lubricant, paper
Carbowax® PEG 300; Crill 1; *Isodecyl oleate*; *Isooctyl stearate*; *PEG-2 stearate*; Penreco Snow; Pluracol® E400 NF; Pluracol® E600 NF; Pluracol® E1450 NF
lubricant, parenteral pharmaceuticals
Etocas 35 NF; Incrocas 40; Miglyol® 810; Super Refined® Olive NF; Super Refined® Sesame NF; Super Refined® Soybean USP
lubricant, perfumes
HallStar™ DIPA; HallStar™ DIPS; *PPG-5-ceteth-20*
lubricant, personal care
Carbowax® E200; Carbowax® E300; Carbowax® E400; Carbowax® E600; Carbowax® E900; Carbowax® E1000; Carbowax® E1450; Carbowax® E3350; Carbowax® E8000; Carbowax® PEG 300; Carbowax® Sentry® PEG 300; Carbowax® Sentry® PEG 400; Carbowax® Sentry® PEG 540 Blend; Carbowax® Sentry® PEG 600; Carbowax® Sentry® PEG 900; Carbowax® Sentry® PEG 1000; Carbowax® Sentry® PEG 1450; Carbowax® Sentry® PEG 3350; Carbowax® Sentry® PEG 4600; Carbowax® Sentry® PEG 8000; Ceraphyl® 140; Crystal® Crown; *Glycereth-7*; *Hexyl laurate*; Incrocas 30; Incrocas 40; *Isopropyl stearate*; *Isostearyl isostearate*; Lipolan R; *Myristyl myristate*; Naetex-L; *PEG-8 dilaurate*; *PEG-2 lactamide*; *Polyglyceryl-9 stearate*; *PPG-10 cetyl ether*; *PPG-20 cetyl ether*, *PPG-40 diethylmonium chloride*; *PPG-11 stearyl ether*, Procetyl 10; Prostearyl 15; Protachem™ 100; Protachem™ AWS-100; Protachem™ DGS; Protachem™ EGDS; Protachem™ EGMS; Protachem™ G-5509; Protachem™ G-5566; Protachem™ GC-7; Protachem™ GC-30; Protachem™ GDL; Protachem™ GMS-165; Protachem™ GMS-450; Protachem™ GMS-AS; Protachem™ GMS-D; Protachem™ GMS-T; Protachem™ HMS; Protachem™ MLD; Protachem™ PGMS; Protachem™ SAS; Ritawax AEO; Ritawax ALA; *Silicone*; Solulan® 16; Solulan® 75; Solulan® C-24; Solulan® L-575; Variquat® CC-42; Variquat® CC-57
lubricant, petrochemicals
Oleyl alcohol
lubricant, petroleum industry
Schercemol DIA
lubricant, petroleum oils/waxes
Sorbitan trioleate
lubricant, petroleum prods.
Erucamide
lubricant, pharmaceutical machinery
Propylene Glycol USP, FCC
lubricant, pharmaceutical tablets
Carbowax® Sentry® PEG 300; *Sucrose palmitate*

Trimyristin; *Tripalmitin*
lubricant, talc
Acetulan®
lubricant, tanning
PEG-100 stearate
lubricant, textile dyeing/printing
Triethylene glycol
lubricant, textile fibers
PEG-4; *PEG-6*; *PEG-8*; *PEG-9*; *PEG-12*; *PEG-14*; *PEG-16*; *PEG-20*; *PEG-32*; *PEG-40*; *PEG-75*; *PEG-100*; *PEG-150*; *PEG-200*; *PEG-350*; Pluracol® SRF-2; Pluracol® W3520N; Pluracol® WI 285; Pluracol® WI 625; Pluracol® WI 1715; Pluracol® WS 170; Pluracol® WS 260; Pluracol® WS 660; Pluracol® WS 2000; Pluracol® WS 3520; Pluracol® WS 5100; Pluracol® WT 1400; Pluracol® WT 9150; Pluracol® WT 90,000; *Polysorbate 80*
lubricant, textile finishing
Oleyl alcohol
lubricant, textile lubricants
PEG-8 dilaurate
lubricant, textile oils
Isobutyl oleate; *Methyl palmitate*
lubricant, textile spin finishes
Butyl oleate; *Glyceryl dioleate*; *Isooctyl stearate*; *Methyl laurate*; *Methyl oleate*; *Octyl palmitate*; *Octyl stearate*; *Oleyl oleate*; *PEG-12 dioleate*; *PEG/PPG-17/6 copolymer*; *Polysorbate 60*; *Sorbitan laurate*; STEPAN® IPM; STEPAN® IPP; Ucon® 75-H-450; ZELEC® IPM
lubricant, textiles
Arlatone® T; *Butyl stearate*; *Canola oil*; Carbowax® PEG 300; Chemax E-200 MO; Crill 1; DBE-IB; *Dibutyl adipate*; *Dibutyl phthalate*; *Diethylene glycol*; *Erucamide*; *Ethylenediamine*; *Ethylenediamine dihydrochloride*; *Glyceryl tallate*; *Hydrogenated menhaden oil*; *Isobutyl palmitate*; *Isodecyl oleate*; *Isopropyl stearate*; *Lanolin*; Lumulse® GMO; *Methyl cocoate*; *Octyl isononanoate*; *Oleyl alcohol*; *PEG-4 dilaurate*; *PEG-4 oleate*; *PEG-6 oleate*; *PEG-40 sorbitan peroleate*; *PEG-2 stearate*; *PEG-100 stearate*; Penreco Snow; *Perfluoropolyether*, *Petrolatum*; Pluracol® E400 NF; Pluracol® E600 NF; Pluracol® E1450 NF; Pogol® 400; PPG 400; *PPG-17*; *PPG-20*; *PPG-2-buteth-3*; *PPG-3-buteth-5*; *PPG-7-buteth-10*; *PPG-9-buteth-12*; *PPG-20-buteth-30*; *PPG-28-buteth-35*; *PPG-5 butyl ether*, *PPG-9 butyl ether*, *PPG-15 butyl ether*, *PPG-18 butyl ether*, *PPG-22 butyl ether*, *PPG-24 butyl ether*, *PPG-33 butyl ether*, *PPG-40 butyl ether*, *PPG-53 butyl ether*, *PPG-40 diethylmonium chloride*; *PPG-3 myristyl ether*, *Propylene glycol dipelargonate*; *Propylene glycol laurate*; *Propylene glycol ricinoleate*; *Sorbitan trioleate*; *Stearyl stearate*; Surphos 515; *Tetrahydroxypropyl ethylenediamine*; T-Mulz® 426; *Trioctyl phosphate*; *Triolein*; *Tristearin*; Ucon® 50-HB-55; Ucon® 50-HB-100; Ucon® 50-HB-260; Ucon® 50-HB-400; Ucon® 50-HB-660; Ucon® 50-HB-2000; Ucon® 50-HB-3520; Ucon® 50-HB-5100; Ucon® LB-65; Ucon® LB-135; Ucon® LB-165; Ucon® LB-285; Ucon® LB-385; Ucon® LB-525; Ucon® LB-625; Ucon® LB-1145; Ucon® LB-1715; Ucon® LB-3000; Variquat® CC-42; Variquat® CC-55; Variquat® CC-57; ZELEC® C-48; ZELEC® C-49
lubricant, textiles: spin finishes
Lumulse® GMT; Lumulse® GMT-40
lubricant, textiles: synthetic fibers
Lumulse® GMT; Lumulse® GMT-40
lubricant, thermoplastics
Ethylenediamine
lubricant, thread
Butyl oleate; *Cyclomethicone*; *Phenyl trimethicone*
lubricant, toilet waters
HallStar™ DIPA; HallStar™ DIPS
lubricant, tonics
HallStar™ DIPA; HallStar™ DIPS; HallStar™ DOS
lubricant, toothpaste
Carbowax® E200; Carbowax® E300; Carbowax® E400; Carbowax® E600; Carbowax® E900; Carbowax® E1000; Carbowax® E1450; Carbowax® E3350; Carbowax® E8000; Carbowax® Sentry® PEG 300; Carbowax® Sentry® PEG 400; Carbowax® Sentry® PEG 3350
lubricant, topical ointments
EmCon™ W; Fancol™ ALA; Fancorsil A
lubricant, topical pharmaceuticals
Arachidyl alcohol; Ceraphyl® 28; Crodamol PMP; Etocas 35 NF; Incrocas 30; Incrocas 40; *Isocetyl stearate*; *Isopropyl myristate*; *Isopropyl stearate*; Miglyol® 810; *Myristyl myristate*; *Octyl isononanoate*; *Octyl stearate*; *Oleyl alcohol*; *PEG-2 stearate*; Schercemol OLO; Solulan® 16; Solulan® 75; Solulan® L-575; Super Refined® Olive NF; Super Refined® Sesame NF; Super Refined® Soybean USP
lubricant, two-stroke cycle engines
PPG-5 butyl ether, *PPG-9 butyl ether*, *PPG-15 butyl ether*, *PPG-18 butyl ether*, *PPG-22 butyl ether*, *PPG-24 butyl ether*, *PPG-33 butyl ether*, *PPG-40 butyl ether*, *PPG-53 butyl ether*
lubricant, vacuum pumps
Perfluoropolyether
lubricant, valve: lotions
Promyristyl PM-3
lubricant, valve: sprays
Promyristyl PM-3
lubricant, veterinary medicine
PEG-35 castor oil
lubricant, w/o creams
Shea butter (Butyrospermum parkii)
lubricant, w/o emulsions
Shea butter (Butyrospermum parkii)
lubricant, water treatment
PEG-4 dilaurate; *PEG-8 dioleate*
lubricant, wave-setting lotions
Glycereth-7
lubricant, whipped toppings
Hodag® S-35 FGK
lubricant, wire drawing
Oleyl oleate; Schercemol OLO
lubricant, wire rope
Benzene, mono-C10-13-alkyl derivs., distillation residues; *C12-14 alkylbenzene*; V-154L; V-159L
lubricant, wood processing
Carbowax® PEG 300
lubricant, wood prods.
Pluracol® E400 NF; Pluracol® E600 NF; Pluracol® E1450 NF
lubricant/antioxidant
Methylcyclohexanol
lubricant: PVC, plasticized
Glycol stearate
lubricating oil
Condea Vista LPA; *Cyclohexanone*; *Decyl alcohol*; Sasol LPA Solvent; Sasol LPA-142 Solvent; Sasol LPA-150 Solvent; Sasol LPA-170 Solvent
lubricating oil, synthetic
Tripropylene glycol n-butyl ether
lubricating oils/greases
PPG-20
lubricity additive, lubricants
Isopropyl oleate
magnetic particle inspection fluid
Drakesol® 165; Drakesol® 165AT; Drakesol® 205; Drakesol® 220; Drakesol® 260; Drakesol® 260 AT; Drakesol® 305
masking agent, cosmetics
Amyl benzoate; *p-Cymene*; *Thymol*; *γ-Undecalactone*
masking agent, industrial odors
2-Octanol
masking agent, odors
Isoamyl acetate
massage oil
Brazil (Bertholetta excelsa) nut oil
masticatory agent, chewing gum
Paraffin
masticatory agent, elastomers: chewing gum
Terpene resin
medical agent, skin disorders
Trichloroacetic acid
medium, pigment-disp.: plastics
Eastman® DEP, Fragrance Grade
melting point booster, ointments
Paraffin
menthol source
Peppermint (Mentha piperita) oil
menthol substitute
Trimethylcyclohexanol
metal deactivator
Dibutyl phosphite
metal deactivator, fuels
Acetylacetone
metal deactivator, lubricants
Acetylacetone; *Edetic acid*
methyl amyl acetate substitute
Isobutyl isobutyrate
methylating agent, organic synthesis
Methyl chloride
methylation agent, organic reactions
Trimethyl borate
methylation agent, organic synthesis
Methyl carbonate
methylenation agent, organic compd. synthesis
Diethoxymethane
migrating agent, disperse dyes: polyesters
Burcolev DLS
mildew preventive
Thymol
mildewcide, adhesives
Iodopropynyl butylcarbamate
mildewcide, cutting oils
Iodopropynyl butylcarbamate
mildewcide, inks
Iodopropynyl butylcarbamate
mildewcide, latex paints
Iodopropynyl butylcarbamate
mildewcide, oil-based paints
Iodopropynyl butylcarbamate
mildewcide, paper coatings
Iodopropynyl butylcarbamate
mildewcide, plastics
Iodopropynyl butylcarbamate
mildewcide, textiles
Iodopropynyl butylcarbamate
mildewcide, wood prods.
Iodopropynyl butylcarbamate
mineral acid inhibitor, steel
Propargyl alcohol
mineral oil substitute
Octyl palmitate
mineral spirits substitute
Soygold® 1000; Soygold® 2000
mineral, trace: animal feeds
Copper oxide (ic); *Zinc chloride*
mineral, trace: food supplements
Zinc chloride
missile fuels
Dimethylamine
missile propellant
Methylhydrazine
miticide
Benzyl benzoate
modifier, alkyd resins
1,10-Decanediol; *12-Hydroxystearyl alcohol*; Speziol C 10/2; Speziol C 18/2
modifier, alkyd: air-drying coatings
Versadil™ 100
modifier, alkyd: high-solids coatings
Versadil™ 100
modifier, alkydss
Isodecanoic acid
modifier, amino resin: paper/ paperboard, aq./fatty food contact
Tetraethylenepentamine
modifier, amino resins: paper/ paperboard, food contact
Diethylenetriamine
modifier, bath care
Carbowax® E200; Carbowax® E300; Carbowax® E400; Carbowax® E600; Carbowax® E900; Carbowax® E1000; Carbowax® E1450; Carbowax® E3350; Carbowax® E8000
modifier, coatings
Dimethyl oleamide
modifier, epoxies
Cumyl phenol
modifier, esterification: cellulose acetate
Nitrobenzene
modifier, furan
Cumyl phenol
modifier, intermediates
Carbowax® E200; Carbowax® E300; Carbowax® E400; Carbowax® E600; Carbowax® E900; Carbowax® E1000; Carbowax® E1450; Carbowax® E3350; Carbowax® E8000
modifier, makeup
Carbowax® E200; Carbowax® E300; Carbowax® E400; Carbowax® E600; Carbowax® E900; Carbowax® E1000; Carbowax® E1450; Carbowax® E3350; Carbowax® E8000
modifier, metalworking fluids
Carbowax® E200; Carbowax® E300; Carbowax® E400; Carbowax® E600; Carbowax® E900; Carbowax® E1000; Carbowax® E1450; Carbowax® E3350; Carbowax® E8000
modifier, oils/alkyds: food-contact coatings
Styrene; *Vinyltoluene monomer*

moisturizer, pharmaceuticals
Caprylic/capric triglyceride; Captex® 300; Captex® 350; Captex® 810D; Captex® 8227; Corona Lanolin; Coronet Lanolin; *Dioctyl adipate*; *Isopentyldiol*; *Isopropyl lanolate*; Lipolan R; *PPG-30 cetyl ether*; *Propylene glycol dioctanoate*; Ritawax AEO; Ritawax ALA; *Triundecanoin*
moisturizer, powders
Tricaprin
moisturizer, pre-electric shaving lotions
Lexol® PG-865
moisturizer, preshave lotions
Lexol® GT-865
moisturizer, rinses
Lipolan R
moisturizer, shampoos
Glucam® E-10; Glucam® E-20; Lipolan R; *PPG-10 methyl glucose ether*, *PPG-20 methyl glucose ether*
moisturizer, shaving creams
Glucam® E-10
moisturizer, shaving preps.
Lipolan R
moisturizer, shower gels
Glucam® E-10
moisturizer, skin care
Cropure® Menhaden; Cropure® Olive; Finsolv® TPP; HallBrite® BHB; *Isopentyldiol*; Lipolan R; *Octyldodecyl isostearate*; *Triisostearin*; Wickenol® 153
moisturizer, soap bars
Glucam® E-20
moisturizer, soaps
Naturechem® GTR
moisturizer, soaps/bath care
Lipolan R
moisturizer, soft gelatin capsules
Captex® 300
moisturizer, sports supplements
Tricaprin
moisturizer, stick preps.
Tricaprin
moisturizer, styling preps.
Glucam® P-10
moisturizer, sun care
Captex® 200; Captex® 300; Captex® 350; Lipolan R
moisturizer, sunscreens
Glucam® E-20; HallBrite® BHB; HallStar™ DOA; Naturechem® GTR
moisturizer, suppositories
Propylene glycol dioctanoate; *Shea butter (Butyrospermum parkii)*
moisturizer, suspensions
Captex® 300
moisturizer, toilet articles
Glucam® E-10
moisturizer, toilet waters
HallStar™ DOA
moisturizer, topical pharmaceuticals
Behenyl behenate; *Myristyl myristate*; *Octyl hydroxystearate*; *Octyl stearate*; Wickenol® 158; Wickenol® 171
moisturizer, topical: pharmaceuticals
Wheat (Triticum vulgare) germ oil
moisturizer, topicals
Lexol® PG-800
moisturizer, veterinary prods.
Lipolan R
moisturizer, w/o creams
Shea butter (Butyrospermum parkii)
moisturizer, w/o emulsions
Shea butter (Butyrospermum parkii)
mold growth inhibitor
Propylene glycol
mold preventive
Thymol
monomer inhibitor, paper/paperboard: aq./fatty food contact
Hydroquinone; *Hydroquinone monomethyl ether*
mordant, wool dyeing
Lactic acid
moth repellent
Naphthalene
motor vehicle fuel
Liquefied petroleum gas
nail polish removers
Ethyl acetate
naphtha component
Propylbenzene
naphtha source
Naphtha
naphtha substitute
Soygold® 1000; Soygold® 2000
nematicide
N-Nitrosodimethylamine
neutralizer
Ammonia; *Cyclohexylamine*; *N-Methylaniline*; *Morpholine*; MPDiol® Glycol; *Triethanolamine*
neutralizer, cosmetics
Diisopropanolamine; *Methylpropanediol*
neutralizer, personal care
Diisopropanolamine; *Methylpropanediol*
neutralizer, petroleum refining
Sodium hydroxide
neutralizer, preboiler system
Methoxypropylamine
neutralizer, solvent: acid dyes
Emery® 5794; *PEG-3 triethanolamine*
n-hexane replacement
Heptane
nitrite remover
Hydrogen peroxide
nonoxidizing acid, strong
Trifluoroacetic acid
nonwhitening aid, aftershaves
Macol® 57
nonwhitening aid, antiperspirants
Macol® 57
nonwhitening aid, colognes
Macol® 57
nonwhitening aid, nail enamel removers
Macol® 57
nonwhitening aid, skin fresheners
Macol® 57
NO_x, removal
Hydrogen peroxide
nucleating agent, photographic film
Sulfur
nucleating agent, sticks
Trilaurin; *Tripalmitin*
nutrient
Olive (Olea europaea) oil; *Pancreatin*; *Wheat (Triticum vulgare) germ oil*
nutrient supplement, food
Calcium chloride; *Potassium carbonate*
nutrient, adult nutrition
Estasan™ GT 8-60 3575; Estasan™ GT 8-65 3577; Estasan™ GT 8-70 3579
nutrient, dietetic prods.
Estasan™ GT 8-60 3575; Estasan™ GT 8-65 3577; Estasan™ GT 8-70 3579
nutrient, enteral feeding formulas
Estasan™ GT 8-60 3575; Estasan™ GT 8-65 3577; Estasan™ GT 8-70 3579
nutrient, fat metabolization: dietetic food
Delios®
nutrient, fat rapid absorption: dietetic food
Delios®
nutrient, fermentation
Kemester® 2050; *Methyl eicosanoate*; *Methyl stearate*
nutrient, food
Triacetin
nutrient, infant nutrition
Estasan™ GT 8-60 3575; Estasan™ GT 8-65 3577; Estasan™ GT 8-70 3579
nutrient, microbial fermentation
Ammonia; *Soybean (Glycine soja) oil*
nutrient, parenteral feeding formulas
Estasan™ GT 8-60 3575; Estasan™ GT 8-65 3577; Estasan™ GT 8-70 3579
nutritional fluid
NEOBEE® 18
nutritional source, IV emulsions
Super Refined® Soybean USP
nutritional source, lipids
Crodamol GTCC; Crodamol GTCC-PN
nutritional source, oral pharmaceuticals
Super Refined® Soybean USP
nutritional source, parenteral pharmaceuticals
Super Refined® Soybean USP
nutritional source, pharmaceuticals
Super Refined® Soybean USP; *Sweet almond (Prunus amygdalus dulcis) oil*; *Wheat (Triticum vulgare) germ oil*
nutritional source, supplements
Super Refined® Soybean USP
nutritional source, topicals
Super Refined® Soybean USP
nutritional supplement, pharmaceuticals
Menhaden oil
occlusive agent, body lotions
Shea butter (Butyrospermum parkii)
occlusive agent, shampoos
Shea butter (Butyrospermum parkii)
occlusive agent, soaps
Shea butter (Butyrospermum parkii)
occlusive agent, suntan preps.
Shea butter (Butyrospermum parkii)
octane booster, gasoline
Alcohol; *1,3-Dioxolane*; *Methyl alcohol*; *Methyl t-butyl ether*; *Toluene*
octane booster, petroleum
2-Butanol; *Tetraethyl lead*
octane enhancer, gasoline
Arcopure® MTBE
octane improver, unleaded gasoline
t-Butyl alcohol
octane rating determination standard
Heptane
octane/cetane booster, fuels
Ecosoft™ PB Solvent; Ecosoft™ PE Solvent; Ecosoft™ PH Solvent
odor control agent, industrial
o-Dichlorobenzene
odor control agent, water treatment
Ammonia; *Ammonium hydroxide*
odorant
Fenchyl alcohol; *Methyl acetate*; *Methyl hexyl ketone*; *Methyl salicylate*; *α-Pinene*; *Pine (Pinus palustris) oil*
odorant, fuel gas
Tetrahydrothiophene
odorant, gas
Ethyl sulfide
odorant, leak detection
d-Limonene
odorant, natural gas
n-Butyl mercaptan; *Dimethyl sulfide*
odorant, paints/coatings
d-Limonene
odorant, perfumes
Isopropyl acetate
odorant, shoe polishes
Isoamyl acetate
odorant, soaps
Benzylsalicylate
offshoot inhibitor, tobacco
Caprylic alcohol
oil component, cutting oils
Arlypon® DV; Arlypon LSE; Arlypon OSO; Arlypon SO; Arlypon SV; *Ethyl linoleate*
oil component, release emulsions
Arlypon® DV; Arlypon LSE; Arlypon OSO; Arlypon SO; Arlypon SV; *Ethyl linoleate*
oil phase component, topical pharmaceuticals
Isocetyl isostearate
oil phase ingred., primary: o/w emulsions
Liquid Absorption Base T
oil phase regulator
Nikkol SO-30; Nikkol SO-30R; *Sorbitan trioleate*
oil solubilizer
Intravon® SOL-W
oil, cooking: edible
Soybean (Glycine soja) oil
oil, edible
Hydrogenated menhaden oil; *Sunflower (Helianthus annuus) seed oil*
oil, edible: food
Rice (Oryza sativa) bran oil
oil, edible: margarine
Rapeseed (Brassica campestris) oil; *Safflower (Carthamus tinctorius) oil*
oil, edible: pharmaceuticals
Rice (Oryza sativa) bran oil
oil, edible: salad dressings
Rapeseed (Brassica campestris) oil
oil, edible: salad oil blends
Safflower (Carthamus tinctorius) oil
oil, edible: shortening
Safflower (Carthamus tinctorius) oil
oil, food
Hydrogenated canola oil; Oleocal C-102; Oleocal C-102 A
oil, food coatings
Hydrogenated canola oil
oil, food sprays
Hydrogenated canola oil
oil, frying
Hydrogenated canola oil
oiliness additive, lubricants
Isodecyl alcohol; *Tridecyl alcohol*
oiliness agent, lubricating oils
Methyl tallowate
oiliness agent, metalworking oils
Methyl tallowate
oiling agent, textiles
Exceparl BB; Exceparl BS; Exceparl EH-P; Exceparl EH-S; Exceparl IB-OL; Exceparl L-OL; Exceparl MC; Exceparl ML-85; Exceparl M-OL; Exceparl MS; Exceparl MT; Exceparl MY-M; Exceparl OD-B; Exceparl OL-OL; Exceparl O-OL; Exceparl SS; Exceparl TD-S; Vinycizer 30
oiling wool
Neatsfoot oil; *Oleic acid*
oily component, bath care

penetrant, cosmetics
Akomed R; Estasan™ GT 8-40 3578; Estasan™ GT 8-60 3575; Estasan™ GT 8-60 3580; Estasan™ GT 8-65 3577; Estasan™ GT 8-65 3581; Estasan™ GT 8-70 3579; Estol IPM 1509 (BIO-IPM); *Hexylene glycol*; *Isopropyl lanolate*; Lexol® IPM-NF; *Myristyl propionate*; *Nonoxynol-5*; Novol NF; *PEG-4 oleate*; *PPG-10 cetyl ether*; *PPG-30 cetyl ether*; *PPG-50 cetyl ether*; *PPG-11 stearyl ether*; Procetyl 10; *Propylene glycol dipelargonate*; *Sodium 2-ethylhexyl sulfate*; *Tricaprylin*

penetrant, creams/lotions
Acetol® 1706; Estasan™ GT 8-40 3578; Estasan™ GT 8-60 3575; Estasan™ GT 8-60 3580; Estasan™ GT 8-65 3577; Estasan™ GT 8-65 3581; Estasan™ GT 8-70 3579; HallStar™ IPM; *Myristyl propionate*

penetrant, cutting oils
Nonoxynol-5

penetrant, delivery/absorption enhancement
Capmul® MCM; Capmul® MCMC8

penetrant, deodorants
Diethylene glycol dioctanoate/ diisononanoate; Estalan 430

penetrant, dermatologicals
Capmul® MCM; Capmul® MCMC8; *Propylene glycol dioctanoate*

penetrant, detergent sterilizers
Ethylan® HB4

penetrant, direct dyes: heavy-weight cotton fabrics
Stetlasolv VJS

penetrant, dry cleaning
Rhodacal® 330

penetrant, dry skin preps.
Propylene glycol dioctanoate

penetrant, dye
Polysorbate 60

penetrant, dye: aramid fibers
CHP®

penetrant, dyes: textiles
Filmex® A-2 190 Proof; Filmex® A-2 Anhydrous; Filmex® B 190 Proof; Filmex® B Anhydrous; Filmex® C 190 Proof; Filmex® C Anhydrous; Filmex® D-1; Filmex® D-2 190 Proof; Filmex® D-2 Anhydrous

penetrant, elec. preshaves
Lexol® IPM-NF

penetrant, emulsion polymerization
Nonoxynol-5; Rhodacal® 330

penetrant, fibers/blends: one-bath dyeing/scouring processes
Intravon® SOL-W

penetrant, firefighting
Deceth-6

penetrant, food
PEG-4 oleate

penetrant, food pkg.
Sodium 2-ethylhexyl sulfate

penetrant, food processing
Methyl cottonseed; Norfox® MCS; *Sodium 2-ethylhexyl sulfate*

penetrant, gasket cement
Pale 4

penetrant, glass cleaners
Ethylan® HB4

penetrant, glosses
Propylene glycol dipelargonate

penetrant, grease: oven cleaners
Dowanol® TPnB; *Tripropylene glycol n-butyl ether*

penetrant, grooming prods.
Estalan DIA

penetrant, hair care
Acetol® 1706; HallStar™ IPM; Lonzest® 143-S

penetrant, hair sprays
HallTress™ DIBA Special

penetrant, hand lotions
Estalan DIA; Schercemol DIA

penetrant, high-boiling solvents
Phenylcyclohexane

penetrant, household cleaners
Methoxyisopropanol; *Nonoxynol-5*; *Sodium 2-ethylhexyl sulfate*

penetrant, hydraulic fluids
#30 Oil; #40 Oil; Pale 4

penetrant, hydroalcoholic cosmetics
Schercemol DIA

penetrant, industrial
Norfox® MCS

penetrant, industrial cleaners
Methoxyisopropanol; *Nonoxynol-5*; *Sodium 2-ethylhexyl sulfate*

penetrant, industrial lubricants
#30 Oil; #40 Oil; Pale 4

penetrant, industrial processing
Methyl cottonseed

penetrant, inks
#30 Oil; #40 Oil; Pale 4

penetrant, iodophors
Ethylan® HB4

penetrant, latex paints
Nonoxynol-5; Rhodacal® 330

penetrant, leather dressing
#30 Oil; #40 Oil; Pale 4

penetrant, lipstick
Estalan DIA; HallStar™ IPM; *Lanolin oil*; Novol NF; *Propylene glycol dipelargonate*; Schercemol DIA

penetrant, liqs.
Estalan 430

penetrant, makeup bases
Propylene glycol dipelargonate

penetrant, men's grooming aids
Schercemol DIA

penetrant, metal cleaning
Rhodacal® 330; *Sodium 2-ethylhexyl sulfate*

penetrant, metal processing
PPG-3 myristyl ether; *PPG-11 stearyl ether*

penetrant, metalworking
PEG-4 oleate

penetrant, metalworking fluids
Rhodacal® 330

penetrant, metalworking/cleaning
Nonoxynol-5

penetrant, nail polish/enamels
HallTress™ DIBA Special

penetrant, nutritional/sports supplements
Capmul® MCM

penetrant, nylon
Consoscour TKB

penetrant, ointments
HallStar™ IPM

penetrant, oral pharmaceuticals
Labrasol®

penetrant, paints
Nonoxynol-5; *Sodium 2-ethylhexyl sulfate*

penetrant, perfumes
HallStar™ IPM

penetrant, personal care
Akomed R; Estasan™ GT 8-40 3578; Estasan™ GT 8-60 3575; Estasan™ GT 8-60 3580; Estasan™ GT 8-65 3577; Estasan™ GT 8-65 3581; Estasan™ GT 8-70 3579; HallStar™ IPM; Lexol® IPM; Novol NF; *PPG-10 cetyl ether*; *PPG-11 stearyl ether*; Procetyl 10; Ritawax AEO; Ritawax ALA

penetrant, petroleum
Nonoxynol-5

penetrant, pharmaceutical aerosols
Capmul® MCM

penetrant, pharmaceutical coatings
Capmul® MCM; Capmul® MCMC10; *Glyceryl caprate*

penetrant, pharmaceutical micro-emulsions
Capmul® MCM; Capmul® MCMC8; Capmul® MCMC10; *Glyceryl caprate*

penetrant, pharmaceutical soft gelatin capsules
Glyceryl caprate

penetrant, pharmaceuticals
Akomed R; *Butylene glycol dicaprylate/dicaprate*; Capmul® MCM; Capmul® MCMC8; *Glyceryl caprate*; *Glyceryl caprylate/caprate*; *Isopropyl lanolate*; Lexol® IPM-NF; *Myristyl propionate*; Novol NF; *PEG-4 oleate*; *PPG-30 cetyl ether*; *Propylene glycol dioctanoate*; *Propylene glycol dipelargonate*; Ritawax AEO; Ritawax ALA; *Sodium 2-ethylhexyl sulfate*; *Tricaprylin*

penetrant, plasticizing inks
2-Ethylhexanol

penetrant, plastics
PEG-4 oleate; *Sodium 2-ethylhexyl sulfate*

penetrant, polishes
#30 Oil; #40 Oil; Pale 4

penetrant, preshave lotions
Diethylene glycol dioctanoate/ diisononanoate; Estalan 430; Estol IPM 1509 (BIO-IPM); *Propylene glycol dipelargonate*

penetrant, printing inks
Norfox® MCS

penetrant, quat. ammonium compds.
Triethylamine

penetrant, reactive dyes: heavy-weight cotton fabrics
Stetlasolv VJS

penetrant, rectal pharmaceuticals
Labrasol®

penetrant, rinse aids
Ethylan® HB4

penetrant, rubber
#30 Oil; #40 Oil; Pale 4; *Sodium 2-ethylhexyl sulfate*

penetrant, shampoos
Rhodacal® 330

penetrant, shaving lotions
Estalan DIA; HallStar™ IPM

penetrant, skin care
Cetiol® B; Miglyol® 810; Miglyol® 812; Miglyol® 812 N; Miglyol® 840; Miglyol® 8810; Pharmasolve™

penetrant, skin cosmetics
Lonzest® 143-S

penetrant, skin creams
Estol IPM 1509 (BIO-IPM)

penetrant, skin lotions
HallTress™ DIBA Special

penetrant, soil removal
Ethoxyethanol

penetrant, soil: cleaners
PPG-2 methyl ether; *PPG-3 methyl ether*; *PPG-2 methyl ether acetate*; *Propylene glycol methyl ether acetate*

penetrant, soil: commercial cleaners
Methoxydiglycol; *Phenoxyethanol*

penetrant, soil: grease/paint removers
Dowanol® DPM

penetrant, soil: household cleaners
Methoxydiglycol; *Phenoxyethanol*

penetrant, soil: household/I&I cleaners
Dowanol® DPM; Dowanol® DPMA; Dowanol® DPnB; Dowanol® PM; Dowanol® PMA; Dowanol® PnB

penetrant, soil: household/I&I cleaning
Dalpad A; Dowanol® EPh

penetrant, soil: industrial cleaners
Methoxydiglycol; *Phenoxyethanol*

penetrant, soil: metal cleaners
Dowanol® DPM

penetrant, soil: rust removers
Dowanol® PM

penetrant, spandex
Consoscour TKB

penetrant, sun care
Estol IPM 1509 (BIO-IPM)

penetrant, sunscreens
Novol NF

penetrant, suntan preps.
Acetol® 1706; Estalan DIA

penetrant, suppositories
Capmul® MCM; Capmul® MCMC8; *Glyceryl caprate*; *Propylene glycol dioctanoate*

penetrant, synthetic fiber
Consoscour TKB

penetrant, textile processing aids
Norfox® MCS

penetrant, textiles
Deceth-6; *Hexylene glycol*; *Nonoxynol-5*; *PEG-4 oleate*; *PPG-3 myristyl ether*; *Sodium 2-ethylhexyl sulfate*

penetrant, toilet water
HallTress™ DIBA Special

penetrant, topical emulsions
Transcutol® P

penetrant, topical gels
Transcutol® P

penetrant, topical medicinals
Estol IPM 1509 (BIO-IPM)

penetrant, topical pharmaceuticals
Diisopropyl sebacate; Estasan™ GT 8-60 3575; Estasan™ GT 8-65 3577; Estasan™ GT 8-70 3579; Labrasol®

penetrant, topical solutions
Transcutol® P

penetrant, trans-skin emulsions
Transcutol® P

penetrant, trans-skin gels
Transcutol® P

penetrant, trans-skin solutions
Transcutol® P

penetration aid, dye: leather
BDGE; BGE; EDGE; EGE; i-PGE; MDGE; MGE

penetration aid, dye: printing
BDGE; BGE; EDGE; EGE; i-PGE; MDGE; MGE

penetration aid, dye: textiles
BDGE; BGE; EDGE; EGE; i-PGE; MDGE; MGE

penetration enhancer, skin
N-Methyl-2-pyrrolidone

perfume
Butyl benzoate; *Diphenyl oxide*; *Hexyl alcohol*; *Isobutyric acid*; *Peppermint (Mentha piperita) oil*

perfume constituent
cis-1,2-Dichloroethylene; *trans-1,2-Dichloroethylene*

perfume extraction, from flowers
Paraffin

permeation enhancer, pharmaceuticals

myristopalmitate; Kenplast® A-450; *Naphthenic oil*; *Paraffin, chlorinated*; Reomol® TOP; Rhodiasolv RPDE; *Tributyl phosphate*; *Tricresyl phosphate*; *Tri-p-tolyl phosphate*

plasticizer, PVC copolymers
Citrofol® BII

plasticizer, PVC flooring
Butyl benzyl phthalate

plasticizer, PVC plastisols
Butyl benzyl phthalate

plasticizer, PVC-PVA copolymers
Citrofol® BII

plasticizer, PVdC
Citroflex® 2; Citroflex® 4; Citroflex® A-2; Citroflex® A-4

plasticizer, PVP
Citroflex® 2; Citroflex® 4; Citroflex® A-2; Citroflex® A-4

plasticizer, regenerated cellulose film
Glycerin

plasticizer, resin systems
Tripropylene glycol n-butyl ether

plasticizer, resin-coated sand: foundry casting
Uniplex 155

plasticizer, resins
Castor oil, polymerized; Dermol ICSA; *n-Dioctyl phthalate*; Dowanol® TPnB; *Isocetyl salicylate*; Jeffsol® EC; Jeffsol® PC; *Naphthenic oil*; #15 Oil; #30 Oil; #40 Oil; Pale 4; Pale 16; Pale 170; Pale 1000; PEG 300; *Phenoxyethanol*; *Triglycol monomethyl ether*; Vorite 105; Vorite 110; Vorite 115; Vorite 125

plasticizer, resins: pigments
Butoxyethyl stearate

plasticizer, resin-treated paper
PPG-9; *PPG-17*; *PPG-30*

plasticizer, rosin
Methyl abietate

plasticizer, rubber
Acetyl tributyl citrate; *Butyl stearate*; *Castor oil, polymerized*; *Dimethyl phthalate*; *Dioctyl adipate*; Eastman® DBP; *Glycol distearate*; *Isooctyl stearate*; *Isopropyl oleate*; *Lanolin*; *Methyl abietate*; *Methyl oleate*; *Methyl tallowate*; *N-Nitrosodimethylamine*; *Oleic acid*; *PEG-2 stearate*; *Soybean (Glycine soja) oil*; *Trioctyl phosphate*

plasticizer, rubber: industrial
Butoxyethyl stearate

plasticizer, rubber: magnetic tapes
Butoxyethyl stearate

plasticizer, rubberized fabrics
Dibutyl tartrate

plasticizer, safety glass
Citrofol® BII

plasticizer, sausage cellulose coatings
Propylene Glycol USP/FCC Ultra Grade

plasticizer, screen inks
n-Dioctyl phthalate

plasticizer, sealants
Paraffin, chlorinated

plasticizer, secondary
C8-16 alkylbenzene

plasticizer, secondary: vinyls
C12-14 alkylbenzene; *Petroleum distillates, heavy thermal cracked*

plasticizer, shower prods.
Hexyldecyl benzoate

plasticizer, skin care
Butyl octyl benzoate; *Hexyldecyl benzoate*

plasticizer, skin creams/lotions
HallStar™ DIPA; HallStar™ DIPS

plasticizer, soaps
Crystal® O; Crystal® Crown

plasticizer, sodium silicate foundry resins
Glycerin

plasticizer, solder reflow applics.
Jeffox® WL-440; Jeffox® WL-660; Jeffox® WL-1400; Jeffox® WL-5000

plasticizer, solid rocket propellants
Dibutyl phthalate; *Diethyl phthalate*

plasticizer, solvating: castable PU
Benzoflex® 284; *Propylene glycol dibenzoate*

plasticizer, solvating: latex caulks
Benzoflex® 284; *Propylene glycol dibenzoate*

plasticizer, solvating: plastics
Acetylated mono- and diglycerides of fatty acids

plasticizer, solvating: PVAc adhesives
Propylene glycol dibenzoate

plasticizer, solvating: PVAc emulsion adhesives
Benzoflex® 284

plasticizer, solvating: PVC
Acetylated mono- and diglycerides of fatty acids; Benzoflex® 284; *Propylene glycol dibenzoate*

plasticizer, solvating: resilient vinyl flooring top coats
Benzoflex® 284

plasticizer, solvating: stain-resist. flooring
Benzoflex® 284

plasticizer, SR
Butyl oleate; *Dimethyl sebacate*; Kemstrene® 99.7% USP; *Paraffin, chlorinated*

plasticizer, starch pastes
DePEG 200; DePEG 300; DePEG 400; DePEG 600

plasticizer, styrene-acrylic floor polishes
2-Pyrrolidone

plasticizer, sunscreens
HallStar™ DIPA; HallStar™ DIPS; HallStar™ DOA; HallStar™ DOS; *Hexyldecyl benzoate*

plasticizer, surface coatings
Butyl oleate; Jeffsol® PC

plasticizer, sustained-release pharmaceuticals
Acetyl tributyl citrate; *Tributyl citrate*

plasticizer, synthetic fabrics
Glyceryl stearate SE; *PEG-7 glyceryl cocoate*

plasticizer, synthetic resins
Eastman® DBP

plasticizer, tablet coatings
Lutrol® E 1500; Lutrol® E 3350; Lutrol® E 4000; Lutrol® E 6000; *PEG-125*

plasticizer, technical prods.
Acetylated hydrogenated lard glyceride

plasticizer, textile coatings
Ucon® 50-HB-55; Ucon® 50-HB-100; Ucon® 50-HB-260; Ucon® 50-HB-400; Ucon® 50-HB-660; Ucon® 50-HB-2000; Ucon® 50-HB-3520; Ucon® 50-HB-5100; Ucon® 75-H-450; Ucon® LB-65; Ucon® LB-135; Ucon® LB-165; Ucon® LB-285; Ucon® LB-385; Ucon® LB-525; Ucon® LB-625; Ucon® LB-1145; Ucon® LB-1715; Ucon® LB-3000

plasticizer, textile finishing
Oleyl alcohol

plasticizer, textile lubricants
PEG-8 dilaurate

plasticizer, textiles
Dibutyl adipate; Jeffsol® EC; Jeffsol® PC; *Methyl cocoate*; *Naphthenic oil*; *PEG-75 lanolin*; *PEG-2 stearate*; *Propylene glycol laurate*

plasticizer, thermoplastics
Uniplex 155

plasticizer, thermosets
Uniplex 155

plasticizer, toilet waters
HallStar™ DIPA; HallStar™ DIPS; HallStar™ DOA

plasticizer, tonics
HallStar™ DIPA; HallStar™ DIPS

plasticizer, topical pharmaceuticals
Oleyl alcohol; *PEG-2 stearate*; Solulan® 16; Solulan® 75; Solulan® L-575

plasticizer, transfer inks
Dibutyl tartrate

plasticizer, transparency film
Citrofol® AII; Citrofol® BI

plasticizer, triacetate film/sheet
Triphenyl phosphate

plasticizer, urethane
Cumylphenyl acetate; Kenplast® A-450; Kenplast® ES-2HP

plasticizer, urethane coatings
Propylene carbonate

plasticizer, urethane foam
Propylene carbonate

plasticizer, varnishes
Jeffsol® PC; *Methyl abietate*

plasticizer, VCA
Dibutyl adipate; *Diisobutyl adipate*; *Dipropylene glycol dibenzoate*

plasticizer, vinyl
Benzene, mono-C10-13-alkyl derivs., distillation residues

plasticizer, vinyl acetate resins
Diethylene glycol dibenzoate

plasticizer, vinyl copolymers
Schercemol PGML

plasticizer, vinyl flooring
n-Dioctyl phthalate

plasticizer, vinyl plastisols
n-Dioctyl phthalate

plasticizer, vinyl-coated fabrics
n-Dioctyl phthalate

plasticizer, vinyls
Acetyl tributyl citrate; Albrite® Tributyl Phosphate; *Dimethyl sebacate*; *Dioctyl maleate*; Eastman® DBP; *Methyl abietate*; *Phenoxyethanol*; Phosflex® 4; TBP; *Triacetin*; *Tributyl phosphate*; *Triethylene glycol*; *Trioctyl phosphate*

plasticizer, water-repelling: creams/ lotions
HallStar™ BST

plasticizer, water-repelling: deodorants
HallStar™ BST

plasticizer, water-repelling: makeup
HallStar™ BST

plasticizer, water-repelling: makeup removers
HallStar™ BST

plasticizer, water-repelling: preshave preps.
HallStar™ BST

plasticizer, wax
Isopropyl lanolate

plasticizer, wax films
Lanolin acid

plasticizer, waxes
Dermol ICSA; *Isocetyl salicylate*

plasticizer: flexible thermoset polyesters
DBE; DBE-2; DBE-3; DBE-4; DBE-5; DBE-6; *Dimethyl adipate*; *Dimethyl glutarate*; *Dimethyl succinate*

plasticizer: lacquers, automobile
Diethyl tartrate

polar additive, bentonite clay gellants
Arconate® PC-HP; Jeffsol® PC

polar additive, montmorillonite clay gellants
Arconate® PC-HP; Jeffsol® PC

polishing agent
NEOBEE® M-5

polishing agent, galvanotechnics
Propargyl alcohol

polyester inhibitor, crosslinked: food-contact
t-Butyl hydroquinone; *Hydroquinone*

polyester resin comonomer, unsat.: surface coatings
2,2,4-Trimethyl-1,3-pentanediol

polyester resins raw material
Ethylene glycol

polymer linking agent
t-Butylbenzene

polymerization agent
Bromoform; *Carbon tetrabromide*

polymerization inhibitor
t-Butyl hydroquinone; *Diethylamine*; *Methylamine*

polymerization inhibitor, acrylic monomers)
Hydroquinone

polymerization inhibitor, vinyl acetate
Hydroquinone

polymerization reaction medium, SBR rubber
Hexane

polymerization regulator
n-Butyl mercaptan

polymerization, condensation
Ammonium hydroxide

polymerizations
Methyl t-butyl ether

polymers, raw material
Sunflower (Helianthus annuus) seed oil

potassium ion source
Potassium carbonate

pour point depressant, gas prod.
Prochinor® SOL

pour point depressant, oil prod.
Prochinor® SOL

power source
Water

precipitant, heavy metals: wastewater treatment
Sulfur

precipitant, protein: laboratories
Trichloroacetic acid

preservative
Ethyl formate; *Glycerin*; *Isobutyric acid*; *MDM hydantoin*; *Naphthalene*; *Propionaldehyde*; Rewopal® MPG 10; *Thiodipropionitrile*; *Thymol*

preservative, anatomical specimens
Quinoline

preservative, animal feed
BHT

preservative, antifungal: food-pkg. adhesives
Iodopropynyl butylcarbamate

preservative, baked goods
Eastman® Glacial Acetic Acid Food Grade, Kosher

preservative, cellulose
Naphthenic acid

preservative, condiments
Eastman® Glacial Acetic Acid Food Grade, Kosher

romethane; Trichlorotrifluoroethane
propellant, aerosols: pharmaceuticals
Dichlorotetrafluoroethane; 1,2-Dichlorotetrafluoroethane
propellant, aerosols: topical pharmaceuticals
Isobutane
propellant, chlorodifluoromethane
Chloroform
propellant, cosmetics
Butane; Dimethyl ether; Isobutane; Isopentane; n-Pentane; Propane
propellant, fluorocarbon aerosols
Liquefied petroleum gas
propellant, foamed food
Butane; Chloropentafluoroethane; Isobutane
propellant, food
Dichlorodifluoromethane; Propane
propellant, hydrocarbon
Cyclopentane
propellant, industrial aerosols
1,2-Dichlorotetrafluoroethane
propellant, paint aerosols
Dimethyl ether; Isobutane; Methylene chloride; Propane
propellant, pesticides
Dimethylamine
propellant, pharmaceuticals
Butane; Propane; Trichlorofluoromethane
propellant, PS foam
Methyl chloride
propellant, rocket
Dimethylamine; Ethylene oxide; Methylamine; 1-Nitropropane; 2-Nitropropane
propellant, sprayed food
Butane; Chloropentafluoroethane; Isobutane
prosthetic aid, soft tissue
Dimethicone
protectant
Mineral oil
protectant, antibiotic ointments
Lanolin
protectant, diaper rash preps.
Lanolin
protectant, hemorrhoidal preps.
Lanolin; Propylene glycol
protectant, makeup
Fancol™ CB; Fancol™ CB Extra
protectant, moisturizing creams/lotions
Fancol™ CB; Fancol™ CB Extra
protectant, OTC pharmaceuticals
Fancol™ CB; Fancol™ CB Extra
protectant, skin care
Cropure® Menhaden; Cropure® Olive; *Dimethicone*; Naetex-LAM
protectant, skin: bath oils
Cropure® Almond
protectant, skin: cosmetics
Diethyl toluamide; Petrolatum
protectant, skin: creams/lotions
Cropure® Almond
protectant, skin: lotions
Shea butter (Butyrospermum parkii)
protectant, skin: makeup
Cocoa (Theobroma cacao) butter
protectant, skin: o/w creams
Shea butter (Butyrospermum parkii)
protectant, skin: o/w emulsions
Shea butter (Butyrospermum parkii)
protectant, skin: OTC pharmaceuticals
Cocoa (Theobroma cacao) butter
protectant, skin: pharmaceutical dermatologicals
Shea butter (Butyrospermum parkii)
protectant, skin: pharmaceuticals
Petrolatum
protectant, skin: sun preps.
Cocoa (Theobroma cacao) butter
protectant, skin: suppositories
Shea butter (Butyrospermum parkii)
protectant, skin: veterinary preps.
Petrolatum
protectant, skin: w/o creams
Shea butter (Butyrospermum parkii)
protectant, skin: w/o emulsions
Shea butter (Butyrospermum parkii)
protectant, suntan preps.
Fancol™ CB; Fancol™ CB Extra
protective colloid, emulsion polymerization
Nonoxynol
purification agent, embalming fluids
Ethyl acetate
purification agent, natural gas streams
Sulfolane
quench oil
Canola oil; Deodorized kerosene; Rapeseed (Brassica campestris) oil
quench oil, charcoal lighter fluids
PD-23; PD-25; PD-28
quench oil, cleaners
PD-23; PD-25; PD-28
quench oil, cosmetic creams/lotions
PD-23
quench oil, fiber lubricants
PD-23; PD-25; PD-28
quench oil, food-contact applics.
PD-23; PD-25; PD-28
quench oil, liq. candles
PD-23; PD-25; PD-28
quench oil, metal foil rolling oils
PD-23; PD-25; PD-28
quench oil, metal sheet rolling oils
PD-23; PD-25; PD-28
quench oil, paper/paperboard
PD-23; PD-25; PD-28
quench oil, pharmaceutical creams/lotions
PD-23
quench oil, polishes
PD-23; PD-25; PD-28
quench oil, vegetable froth flotation
PD-23; PD-25; PD-28
quench oil, water treatment
PD-23; PD-25; PD-28
quench oil, waterless hand cleaners
PD-23; PD-25; PD-28
raw material, acetic acid
SD alcohol 35-A
raw material, acetophenone
Ethylbenzene
raw material, active pharmaceutical compds.
NM™
raw material, active pharmaceutical compound synthesis
NE™
raw material, α-methyldopa
NE™
raw material, amyl xanthate flotation collectors
Pentanol 45
raw material, animal feed supplements
SD alcohol 35-A
raw material, anthraquinone
Ethylbenzene
raw material, anticorrosives
Cashew nut shell oil; 3-(n-Pentadecyl) phenol
raw material, antioxidants
Cardolite® NC-510; *Cashew nut shell oil; Cetyl alcohol; Di-n-amylamine; Methyl isobutyl ketone; Myristyl alcohol; 3-(n-Pentadecyl) phenol*
raw material, cellulose acetate
Ethylbenzene
raw material, chemical industry auxiliaries
Propylheptanol
raw material, chemical intermediates
Witcohol™ 42
raw material, chemicals
SD alcohol 13-A; SD alcohol 17; SD alcohol 30
raw material, coating resins
1,4-Butanediol
raw material, cosmetics
NM™; *Tripropylene glycol n-butyl ether*
raw material, crankcase oil
Pentanol 45
raw material, creams/lotions
Ethyl hexanediol
raw material, detergent: dishwashing compds.
Deceth-6
raw material, detergent: hand cleaning gels
Dodecylbenzenesulfonic acid
raw material, detergent: heavy duty detergents
Dodecylbenzenesulfonic acid
raw material, detergent: light duty detergents
Dodecylbenzenesulfonic acid
raw material, detergent: machine degreasers
Dodecylbenzenesulfonic acid
raw material, detergent: tablet soap mfg.
Deceth-6
raw material, detergent: tank cleaners
Dodecylbenzenesulfonic acid
raw material, detergents
Witcohol™ 42
raw material, diethyl benzene
Ethylbenzene
raw material, dyes
Hydrobromic acid; SD alcohol 1; SD alcohol 2-B; SD alcohol 2-C; SD alcohol 3-A; SD alcohol 3-C; SD alcohol 12-A; SD alcohol 36; Xylene
raw material, emulsifiers
Witcohol™ 42
raw material, epoxies
Epichlorohydrin
raw material, ethers
SD alcohol 13-A
raw material, ethoxylation
Cetearyl alcohol
raw material, ethyl acetate
SD alcohol 1; SD alcohol 2-B; SD alcohol 35-A
raw material, ethyl cellulose
Ethyl chloride
raw material, ethyl chloride
SD alcohol 1; SD alcohol 2-B; SD alcohol 32
raw material, ethyl esters
SD alcohol 2-C; SD alcohol 12-A; SD alcohol 13-A; SD alcohol 32; SD alcohol 35-A
raw material, ethyl ethers
SD alcohol 13-A; SD alcohol 32
raw material, ethylamines
SD alcohol 1; SD alcohol 2-B; SD alcohol 2-C; SD alcohol 3-A; SD alcohol 3-C; SD alcohol 12-A; SD alcohol 36
raw material, ethylene gas
SD alcohol 32
raw material, fermentation
Linpar® 1416-V Normal Paraffin
raw material, fine chemicals
1,1,1-Trichloro-2,2,2-trifluoroethane
raw material, finishing agents: synthetic fibers
PEG-20 oleate
raw material, flavors
Eastman® DEP
raw material, foundry resins
Formcel®
raw material, fragrances
Eastman® DEP; TCD Ketone A
raw material, furfural alcohol
Formcel®
raw material, gear lubricant corrosion inhibitors
Pentanol 45
raw material, germicides
Witcohol™ 42
raw material, hypertensive drugs
NE™
raw material, industrial coatings
Formcel®
raw material, insecticides
1,1,1-Trichloro-2,2,2-trifluoroethane
raw material, iopamidol
NM™
raw material, latex paints
Ethylene glycol
raw material, light stabilizers
Dimethyl sebacate
raw material, lube oil additives
Witcohol™ 42
raw material, lubricant additives
Di-n-amylamine
raw material, lubricants
Exceparl PE-MO; Witcohol™ 42
raw material, lubricants/antiwear agents
Isoheptyl alcohol
raw material, medicinal chemicals
SD alcohol 2-C; SD alcohol 3-A; SD alcohol 12-A; SD alcohol 13-A; SD alcohol 17
raw material, melamine formaldehyde resins
Formcel®
raw material, microbiological oxidation
Linpar® 1416-V Normal Paraffin
raw material, oiling agents
Excepárl PE-MO
raw material, oily cosmetics
Ethyl hexanediol
raw material, organic synthesis
Witcohol™ 42
raw material, organics
Xylene
raw material, organo-silicone prods.
SD alcohol 3-A; SD alcohol 3-C; SD alcohol 30
raw material, paint stabilizers
Pentanol 45
raw material, paints
Xylene
raw material, personal care
NM™
raw material, pesticides
Pentanol 45
raw material, pharmaceutical synthesis: α-methyldopa
Nitroethane
raw material, pharmaceutical synthesis: appetite suppressants
Nitroethane
raw material, pharmaceutical synthesis: bronchial decongestants
Nitroethane
raw material, pharmaceutical synthesis: hypertensive drug

Cyanex® 272
reagent, solvent extraction: organics recovery
Cyanex® 921
reagent, solvent extraction: palladium separation from platinum
Cyanex® 471X
reagent, solvent extraction: phenol recovery
Cyanex® 921
reagent, solvent extraction: phenol recovery from effluent streams
Cyanex® 923
reagent, solvent extraction: silver selective recovery
Cyanex® 471X
reagent, solvent extraction: uranium recovery
Cyanex® 921
reagent, spectrophotometric analysis
Isooctane
reagent, synthetic
Formamide
reagent, zinc separation: from cobalt solutions
Bis (2,4,4,-trimethylpentyl) phosphinic acid
reclaiming agent, rubber
Naphthenic acid; *Turpentine*
reducer, photography
Hydroquinone
reducing agent
Formaldehyde; *2-Mercaptoethanol*; *Sulfur dioxide*
reducing agent, dyeing
Formic acid
reducing agent, electroplating
Formic acid
reducing agent, fine organic chemicals
Lithium aluminum hydride
reducing agent, perfumes
Lithium aluminum hydride
reducing agent, pharmaceuticals
Lithium aluminum hydride
reducing agent, pickling baths
Formic acid
refatting agent, cleansers
Octyl hydroxystearate
refatting agent, cosmetics
Glyceryl caprate; *Isopropyl stearate*; *Octyl hydroxystearate*; Wickenol® 171
refatting agent, creams/lotions
Tegosoft® DO; Tegosoft® P
refatting agent, dry skin treatment
Tegosoft® DO
refatting agent, hair care
Isopropyl stearate; Tegosoft® DO; Tegosoft® P
refatting agent, hair conditioners
Tegosoft® DO; Tegosoft® P
refatting agent, hair glosses
Tegosoft® DO; Tegosoft® P
refatting agent, personal care
Isopropyl stearate
refatting agent, skin care
Isopropyl stearate; Tegosoft® DO; Tegosoft® P
refatting agent, soaps
Isopropyl stearate; *Octyl hydroxystearate*; Tegosoft® DO; Tegosoft® P
refatting agent, sunscreens
Isopropyl stearate; Tegosoft® DO; Tegosoft® P
refatting agent, topical pharmaceuticals
Isopropyl stearate; *Octyl hydroxystearate*; Tegosoft® DO; Tegosoft® P; Wickenol® 171
refrigerant
Ammonia; *Butane*; *Carbon tetrachloride*; *1,1,1-Chlorodifluoroethane*; *Chlorodifluoromethane*; *Chloropentafluoroethane*; *Dichlorodifluoromethane*; *cis-1,2-Dichloroethylene*; *trans-1,2-Dichloroethylene*; *Dichlorofluoroethane*; *Dichlorofluoromethane*; *Dichlorotetrafluoroethane*; *1,2-Dichlorotetrafluoroethane*; *Dimethyl ether*; *Ethyl bromide*; *Ethyl chloride*; *Formic acid*; *Methyl chloride*; *Methylene chloride*; *Propane*; *Tetrafluoromethane*; *Trichloroethylene*; *Trichlorofluoromethane*; *Trichlorotrifluoroethane*
refrigerant, chlorodifluoromethane
Chloroform
refrigerant, fluorocarbon aerosol
Liquefied petroleum gas
refrigerant, low-temp.
Trifluoromethane
refrigerant, ultra low temp.
Chlorotrifluoromethane
refrigeration fluid
SD alcohol 1; *SD alcohol 3-A*; *SD alcohol 3-C*; *SD alcohol 23-A*; *SD alcohol 30*
reinforcing agent
Terpene resin
release agent
Arlypon® DV; Arlypon LSE; Arlypon OSO; Arlypon SO; Arlypon SV; *Decamethylcyclopentasiloxane*; *Dimethicone*; *Hexamethyldisilazane*; *Hexamethyldisiloxane*; *Isohexane*; NEOBEE® M-5; *Octamethylcyclotetrasiloxane*; *Octamethyltrisiloxane*; *Polyglyceryl polyricinoleate*; *PPG-20*; *Tetramethylsilane*
release agent, agric. chemical sprays
PEG-8 dilaurate
release agent, asphalt
Soygold® 1000; Soygold® 2000
release agent, baked goods
Hodag® S-35 FGK
release agent, baked prods.
Delios® S
release agent, bakery
Crodamol GTCC
release agent, bakery prods.
Petrolatum
release agent, coffee whiteners
Hodag® S-35 FGK
release agent, concrete form oils
Norfox® MTL
release agent, confectionery
Caprylic/capric triglyceride; *Petrolatum*
release agent, external mold: acrylics
Cetyl alcohol
release agent, food
Castor (Ricinus communis) oil; Delios® S; Hodag® S-35 FGK; *Mono- and diglycerides of fatty acids*; *Petrolatum*; *Tristearin*
release agent, food-contact coatings
Dimethicone; *Petrolatum*; *Phenyl trimethicone*; *Silicone*
release agent, food-contact polymers
Erucamide
release agent, frozen desserts
Hodag® S-35 FGK
release agent, industrial
PEG-8 dilaurate
release agent, industrial lubricants
PEG-8 dilaurate
release agent, metal rolling oils
Norfox® MTL
release agent, mold
Carbowax® E200; Carbowax® E300; Carbowax® E400; Carbowax® E600; Carbowax® E900; Carbowax® E1000; Carbowax® E1450; Carbowax® E3350; Carbowax® E8000; Crill 2; *Hydrogenated menhaden oil*; *Isocetyl laurate*; *Isocetyl myristate*; *Isocetyl stearate*; Kemstrene® 99.7% USP; Lumulse® GMO; Lumulse® GMT; Lumulse® GMT-40; *PEG-90*; *PEG-7 glyceryl cocoate*; *PEG-100 stearate*; *PPG-28-buteth-35*; Vertrel® XF
release agent, mold: adhesives
Crill 1
release agent, mold: asphalt
Erucamide
release agent, mold: butyl rubber
Kenflex® N1
release agent, mold: ceramics
Kerosene
release agent, mold: chlorosulfonated PE
Kenflex® N1
release agent, mold: compressed confectionery
Trimyristin
release agent, mold: cosmetics
Crill 1; Pluracol® E400 NF; Pluracol® E600 NF
release agent, mold: crayons
Erucamide
release agent, mold: food
Crill 1; Estasan™ GT 8-40 3578; Estasan™ GT 8-60 3575; Estasan™ GT 8-60 3580; Estasan™ GT 8-65 3577; Estasan™ GT 8-65 3581; Estasan™ GT 8-70 3579; *Mineral oil*
release agent, mold: food pkg.
Crill 1
release agent, mold: formed plastics mfg.
Hydrogenated castor oil
release agent, mold: herbicides
Crill 1
release agent, mold: industrial
Butoxyethyl stearate
release agent, mold: inks
Crill 1; *Erucamide*; Pluracol® E400 NF; Pluracol® E600 NF
release agent, mold: insecticides
Crill 1
release agent, mold: leather treatment
Crill 1
release agent, mold: magnetic tapes
Butoxyethyl stearate
release agent, mold: metals
Erucamide
release agent, mold: metalworking fluids
Crill 1
release agent, mold: natural products
Pogol® 400
release agent, mold: neoprene
Kenflex® N1
release agent, mold: oil slick dispersing
Crill 1
release agent, mold: oral care
Pluracol® E400 NF; Pluracol® E600 NF
release agent, mold: paints
Crill 1
release agent, mold: paper
Crill 1; Pluracol® E400 NF; Pluracol® E600 NF
release agent, mold: petroleum prods.
Erucamide
release agent, mold: pharmaceuticals
Crill 1; *Tristearin*
release agent, mold: plastics
Crill 1; *Erucamide*; *Glyceryl oleate*; *Silicone*; *Tristearin*
release agent, mold: polar thermoplastics
Kenflex® N1
release agent, mold: polishes
Crill 1
release agent, mold: polysulfide
Kenflex® N1
release agent, mold: rubber
Erucamide; *Hydrogenated castor oil*; Pluracol® E400 NF; Pluracol® E600 NF; *Silicone*
release agent, mold: synthetic products
Pogol® 400
release agent, mold: tablets
Tripalmitin
release agent, mold: textiles
Crill 1; *Erucamide*; Pluracol® E400 NF; Pluracol® E600 NF
release agent, mold: vinyl resins
Ucon® 50-HB-55; Ucon® 50-HB-100; Ucon® 50-HB-260; Ucon® 50-HB-400; Ucon® 50-HB-660; Ucon® 50-HB-2000; Ucon® 50-HB-3520; Ucon® 50-HB-5100; Ucon® 75-H-450; Ucon® LB-65; Ucon® LB-135; Ucon® LB-165; Ucon® LB-285; Ucon® LB-385; Ucon® LB-525; Ucon® LB-625; Ucon® LB-1145; Ucon® LB-1715; Ucon® LB-3000
release agent, mold: wood prods.
Pluracol® E400 NF; Pluracol® E600 NF
release agent, pan: food
Polyglyceryl polyricinoleate
release agent, personal care
PEG-8 dilaurate
release agent, petroleum wax
Erucamide
release agent, pharmaceuticals
Imwitor® 742; *PEG-8 dilaurate*
release agent, rubber
PEG 300; PPG 2250
release agent, shortenings
Hodag® S-35 FGK
release agent, textile lubricants
PEG-8 dilaurate
release agent, whipped toppings
Hodag® S-35 FGK
removal agent, cyanide
Hydrogen peroxide
removal aid, finishes
1,1-Dichloroethane
removal aid, oils/greases/contaminants
Arosolve TBP
removal aid, paints
1,1-Dichloroethane
removal aid, phenols
Hydrogen peroxide
removal aid, tar
Arosolve 570-HF; Arosolve 570 Special; Arosolve B-950; Arosolve SH-180 LF; Arosolve XNF-1
removal aid, varnishes
1,1-Dichloroethane
reodorant, paints
α-Terpineol
resin intermediate
2,2,4-Trimethyl-1,3-pentanediol
resin modifier, coatings
Terpene resin
resin modifier, laminations
Terpene resin

scouring agent, solvent: protective coatings
Variquat® CC-57
scouring agent, solvent: rubber rollers
Arosolve SH-180 LF
scouring agent, solvent: screen clear applics.
Scour NFP
scouring agent, solvent: synthetic fabrics
Arosolve FX-6
scouring agent, solvent: synthetic fabrics/blends
Arosolve IWS
scouring agent, solvent: synthetic fibers
Arosolve B-950; Variquat® CC-57
scouring agent, solvent: synthetics/ blends
Arosolve TBP
scouring agent, solvent: textile jet machines
Stetlasolv SES-CONC.
scouring agent, solvent: textile machinery
Cekasol BOD
scouring agent, solvent: textile operations
Hipochem MS-BW
scouring agent, solvent: textile roller cleaners
Scour NFP
scouring agent, solvent: textiles
Arosolve 570-HF; Arosolve 570 Special; Arosolve DMI-F; Hiposcour® MS-BW; Marlasol EX-1; Stetlasolv SES-CONC.; Supersol CSS; Variquat® CC-57
scouring agent, spandex
Consoscour TKB
scouring agent, synthetic fibers
Consoscour TKB
scouring agent, textiles
Dodecylbenzenesulfonic acid; Eccobrite Cleaner AK; *Nonoxynol*; *Perchloroethylene*; *PPG-3 myristyl ether*; Sovatex C Series
scouring agent, wool
Trax K-40
scouring agent, woolen fabrics
Ecco Cleaner 930 Conc.
scrubbing agent, acid gases: ammonia synthesis
Ethanolamine
sealing agent, food
Mineral oil; *Petrolatum*
sedative
Carbon tetrabromide; *Paraldehyde*; *Pyridine*
sedative, pharmaceuticals
t-Amyl alcohol; *Bromoform*; *Camphor*
selecting agent, mineral ore: geological assays
Carbon tetrabromide
semidrying oil, paints
Corn (Zea mays) oil
separating agent, metals
2-Ethyl-1-butanol
separating agent, racemic mixtures
Brucine
separating agent, volatile amines
Morpholine
separator, solvent
Dibutyl butyl phosphonate
sequestrant
Triethanolamine
sequestrant, beverages
Diacetyl tartaric acid esters of mono- and diglycerides
sequestrant, food
Acetylated mono- and diglycerides of fatty acids; *Calcium chloride*; *Diacetyl tartaric acid esters of mono- and diglycerides*; *Phosphoric acid*; *Triethyl citrate*
sequestrant, metal plating
Lactic acid
sequestrant, pharmaceuticals
Acetylated mono- and diglycerides of fatty acids
setting agent, consolidation binders: sodium silicate foundry mold/soil
Glyceryl diacetate
setting agent, foundry mold/soil consolidation binders
Triacetin
shelf life extender, vinyl coatings: metals
3,3,5-Trimethyl cyclohexanone
shelf life extender, vinyl coatings: plastics
3,3,5-Trimethyl cyclohexanone
shortening, food
Hydrogenated canola oil
shortening, food coatings
Hydrogenated canola oil
shortening, food sprays
Hydrogenated canola oil
shortening, frying
Hydrogenated canola oil
silvering glass
Formic acid
silylating agent
Phenyl trimethicone
silylating agent, carboxylic acids
Hexamethyldisiloxane
silylating agent, pharmaceuticals
Hexamethyldisilazane
sizing additive, textile
Dimethicone
sizing agent, paper
Paraffin; Pluracol® E400 NF; Pluracol® E600 NF; Pluracol® E1450 NF
sizing agent, textiles
s-Amyl acetate; *s-Butyl acetate*; *Methyl chloride*; *Perchloroethylene*; Pluracol® E200; Pluracol® E300; Pluracol® E400; Pluracol® E400 NF; Pluracol® E600; Pluracol® E600 NF; Pluracol® E1450 NF; Pluracol® E4000; Pluracol® E6000; *Tristearin*
sizing agent, wood prods.
Pluracol® E400 NF; Pluracol® E600 NF; Pluracol® E1450 NF
sizing removal aid, protein: textiles
Pancreatin
sizing removal aid, starch: textiles
Pancreatin
skin protectant
Dimethicone
skinfeel enhancer, decorative cosmetics
Isodragol 2/050300
skinfeel enhancer, skin care
Isodragol 2/050300
slimicide, paper
Sodium thiocyanate
slimicide, paper/paperboard: food-contact
Acrolein
slimicide, pharmaceuticals
Sodium thiocyanate
slip agent, asphalt
Erucamide
slip agent, coatings
Butyl stearate
slip agent, cosmetics
Ceraphyl® 140; *Isostearyl isostearate*; *Isostearyl lactate*; Schercemol 1818
slip agent, crayons
Erucamide
slip agent, creams/lotions
HallStar™ BST
slip agent, deodorants
HallStar™ BST
slip agent, inks
Butyl stearate; *Erucamide*
slip agent, makeup
HallStar™ BST
slip agent, makeup removers
HallStar™ BST
slip agent, metal lubricants
Rapeseed (Brassica campestris) oil
slip agent, metals
Erucamide
slip agent, metalworking coatings
Lanolin
slip agent, paper product waxes
Acetylated mono- and diglycerides of fatty acids
slip agent, PE
Erucamide
slip agent, personal care
Ceraphyl® 140; *Isostearyl isostearate*
slip agent, petroleum prods.
Erucamide
slip agent, pharmaceuticals
Ceraphyl® 140; *Isostearyl lactate*
slip agent, pigments
Octyl isononanoate
slip agent, plastics
Caprylic alcohol; *Erucamide*; *Myristyl alcohol*; *Rapeseed (Brassica campestris) oil*
slip agent, polymers
Rapeseed (Brassica campestris) oil
slip agent, preshave preps.
HallStar™ BST
slip agent, printing inks
Erucamide
slip agent, rubber
Erucamide; *Rapeseed (Brassica campestris) oil*
slip agent, rust preventative coatings
Lanolin
slip agent, textiles
Erucamide
sludge inhibitor, fuel oil
Phenol
smooth rub-out enhancer, cosmetics
Estalan DO
smoothing agent, cosmetics
Isopentyldiol
smoothing agent, pharmaceuticals
Isopentyldiol
smoothing agent, skin care
Isopentyldiol
SO_2, removal
Hydrogen peroxide
softener
Butyl diglycol carbonate; *Decyl alcohol*; *Diethanolamine*; *Diethylenetriamine*; *Ethyl crotonate*; *Glycerin*; *Naphthenic oil*; *PEG-12 dioleate*; *PEG-5 oleate*; *Polyethylene glycol*; *Polysorbate 60*; *Triethyl citrate*
softener, absorption bases
Lanolin alcohol
softener, agric.
PEG-4 oleate; *Sorbitan trioleate*
softener, agric. chemical sprays
PEG-8 dilaurate
softener, animal glues
Formamide
softener, aq. inks
n-Dioctyl phthalate
softener, butyl rubber
Kenflex® N1
softener, cellophane
Ethylene glycol
softener, cellulosic resins
Dimethyl sebacate
softener, chewing gum base
Glyceryl caprylate
softener, chlorosulfonated PE
Kenflex® N1
softener, cleansers
Lanolin alcohol
softener, copolymers
N-Nitrosodiethylamine; *N-Nitrosodimethylamine*
softener, cosmetics
Butyl stearate; *Cetyl octanoate*; Emery® 912; Emery® 916; Emery® 917; *Isostearyl lactate*; *Lanolin alcohol*; Norfox® B-54; *PEG-35 castor oil*; *PEG-4 oleate*; Pluracol® E400 NF; Pluracol® E600 NF; Pluracol® E1450 NF; Propylene Glycol USP/FCC Ultra Grade; Schercemol OLO; Tegosoft® CT
softener, creams
Cerette V; Isostriol IC; Monestriol B; Monestriol C; Monestriol IC; Monestriol IS-C; Palmil C; Palmil IS; Palmil O; Thomil 14; Thomil IS; Thomil IS-F; Tilol B; Tilol ET; Tilol IS
softener, creams/lotions
Tegosoft® CT; Tegosoft® DO
softener, dry skin treatment
Tegosoft® DO
softener, ear wax
Olive (Olea europaea) oil; *Peanut (Arachis hypogaea) oil*; *Sweet almond (Prunus amygdalus dulcis) oil*
softener, eczema crusts
Olive (Olea europaea) oil
softener, elastomers
n-Dioctyl phthalate
softener, emollient creams
Peanut (Arachis hypogaea) oil
softener, enemas
Peanut (Arachis hypogaea) oil
softener, fabrics
Diethylaminoethanol; *Hexyl alcohol*; *Isostearyl alcohol*
softener, fibers
Sorbitan trioleate
softener, fibers: textiles
Filmex® A-2 190 Proof; Filmex® A-2 Anhydrous; Filmex® B 190 Proof; Filmex® B Anhydrous; Filmex® C 190 Proof; Filmex® C Anhydrous; Filmex® D-1; Filmex® D-2 190 Proof; Filmex® D-2 Anhydrous
softener, flexographic inks
n-Dioctyl phthalate
softener, food
Mono- and diglycerides of fatty acids; *PEG-4 oleate*; *Sucrose palmitate*
softener, food-pkg. adhesives
n-Dioctyl phthalate
softener, gravure inks
n-Dioctyl phthalate
softener, gums
Formamide
softener, hair care
Cerette V; Isostriol IC; Monestriol B; Monestriol C; Monestriol IC; Monestriol IS-C; Naetex-L; *Oleyl oleate*; Palmil C; Palmil IS; Palmil O; Ritawax AEO; Ritawax ALA; Schercemol OLO; Tegosoft® CT; Tegosoft® DO; Thomil 14; Thomil IS; Thomil IS-F; Tilol B; Tilol ET; Tilol

solubilizer, printing inks
Diethanolamine; *Glyceryl ricinoleate*; *Morpholine*; PEG 300
solubilizer, printing pastes
Sodium xylenesulfonate
solubilizer, pulp/paper
Oleyl alcohol; Paxsolv PC; *Polysorbate 60*
solubilizer, pulping
Pilot® SXS-40; *Sodium xylenesulfonate*
solubilizer, rectal pharmaceuticals
Imwitor® 412; Labrafac® Hydrophile; Miglyol® 810; Miglyol® 812 N
solubilizer, resins
Hydrogen peroxide; PEG 300
solubilizer, rinse aids
Ethylan® HB4
solubilizer, rinses
Lipolan R
solubilizer, rubber
2-Decyl-1-tetradecanol; Jarcol™ I-24; *PEG-2 stearate*
solubilizer, shampoos
Ameroxol® OE-2; *Caprylic/capric triglyceride PEG-4 esters*; *Dimethyl octynediol*; Lipolan R; Nikkol BWA-5; Rhodacal® 330; *Sodium toluenesulfonate*; *Sodium xylenesulfonate*
solubilizer, shaving preps.
Lipolan R; Pelemol® G7A
solubilizer, skin care
Isopropyl PPG-2-isodeceth-7-carboxylate; Lipolan R; Promyristyl PM-3
solubilizer, sl. water-soluble pharmaceuticals
PEG-125
solubilizer, soaps
Lipolan R; *PPG-15 stearyl ether benzoate*
solubilizer, soft gelatin capsules
Captex® 300
solubilizer, soluble colorants
Captex® 200
solubilizer, soluble ingreds.: cosmetics
Nikkol Sefsol 218; Nikkol Sefsol 228
solubilizer, solvents
2-Decyl-1-tetradecanol; Jarcol™ I-24
solubilizer, spice oils
Axol® E 66
solubilizer, splash-on prods.
Pelemol® G7A
solubilizer, sports supplements
Tricaprin
solubilizer, steroids
Cremophor® EL; Lutrol® E 1500; Lutrol® E 3350; Lutrol® E 4000; Lutrol® E 6000; *PEG-125*
solubilizer, sticks
Tricaprin
solubilizer, sun care
Capmul® MCM; Capmul® MCMC8; Captex® 200; Captex® 300; Finsolv® TPP; *Isopropyl PPG-2-isodeceth-7-carboxylate*; Lipolan R
solubilizer, sunscreen actives
Finsolv® TPP; HallBrite® BHB; *Hexyldecyl benzoate*
solubilizer, sunscreens
Novol NF; Promyristyl PM-3; Schercemol PGML
solubilizer, suppositories
Capmul® MCM; Capmul® MCMC8; *Glyceryl caprate*; *Polyglyceryl-3 stearate*; *Propylene glycol dioctanoate*
solubilizer, suspensions
Captex® 300
solubilizer, syndets
PPG-15 stearyl ether benzoate
solubilizer, talcum powders
STEPAN® IPM-NF
solubilizer, textile auxiliaries
Diethanolamine
solubilizer, textile dyes
Phenoxyethanol
solubilizer, textile finishing
Oleyl alcohol; *Sodium toluenesulfonate*; *Sodium xylenesulfonate*
solubilizer, textile lubricants
PEG-8 dilaurate
solubilizer, textiles
Arlatone® T; *Cocamide DEA*; *2-Decyl-1-tetradecanol*; Jarcol™ I-24; *Nonoxynol-4*; *Nonoxynol-5*; *Nonoxynol-6*; *Oleth-2*; *PEG-4 dilaurate*; *PEG dimethyl ether*; *PEG-8 dioleate*; *PEG-4 oleate*; *PEG-6 oleate*; *PEG-40 sorbitan peroleate*; *PEG-2 stearate*; Pilot® SXS-40; Polyglyme; *Polysorbate 60*
solubilizer, topical pharmaceuticals
Acetamide MEA; Ceraphyl® 140-A; *Choleth-24*; *Cocamide DEA*; *2-Decyl-1-tetradecanol*; *Diisopropyl sebacate*; Glucam® E-10; Glucam® E-20; Glucam® P-10; Glucam® P-20; Imwitor® 408; *Isodecyl oleate*; *Isopropyl myristate*; Labrafac® Hydrophile; *Laneth-15*; *Methyl gluceth-10*; *Methyl gluceth-20*; *N-Methyl-2-pyrrolidone*; Miglyol® 810; Miglyol® 812 N; *Myristyl myristate*; *Octyl hydroxystearate*; *Octyl stearate*; *Oleyl alcohol*; *Oleyl erucate*; *PEG-2 stearate*; Pelemol® G7A; Pharmasolve™; Polychol 5; Polychol 15; *PPG-10 methyl glucose ether*; *PPG-20 methyl glucose ether*; Wickenol® 171
solubilizer, transparent emulsions
Ameroxol® OE-2
solubilizer, vaginal pharmaceuticals
Labrafac® Hydrophile
solubilizer, veterinary creams/lotions
Caprylic/capric triglyceride PEG-4 esters
solubilizer, veterinary inj. solutions
Solutol® HS 15
solubilizer, veterinary medicine
Cremophor® EL; *PEG-35 castor oil*
solubilizer, veterinary preps.
Labrafac® Hydrophile; Labrasol®; Lipolan R
solubilizer, vitamin A
Solutol® HS 15
solubilizer, vitamin D
Solutol® HS 15
solubilizer, vitamin E
Solutol® HS 15
solubilizer, vitamin K
Solutol® HS 15
solubilizer, vitamin oils
Cerette V; Isostriol IC; Monestriol B; Monestriol C; Monestriol IC; Monestriol IS-C; Palmil C; Palmil IS; Palmil O; Thomil 14; Thomil IS; Thomil IS-F; Tilol B; Tilol ET; Tilol IS; *Triolein*
solubilizer, vitamins
Capmul® POE-O; Captex® 200; Lexol® PG-865; *PEG-40 castor oil*
solubilizer, water treatment
PEG-4 dilaurate; *PEG-8 dioleate*
solubilizer, water-insoluble pharmaceuticals
PEG-125
solubilizer, water-insoluble/poorly soluble pharmaceuticals
Lutrol® E 1500; Lutrol® E 3350; Lutrol® E 4000; Lutrol® E 6000
solubilizer, water-misc. cutting fluids
Jarcol™ I-16
solubilizer, wax strippers
STEPANATE® AXS
solvating agent
Cumylphenyl neodecanoate
solvating agent, acrylic inks: impact improvement
Cumylphenyl benzoate; Kenplast® ESB
solvating agent, conductive polyester/acrylic inks
Kenplast® ESN
solvating agent, conductive polyester: impact improvement
Cumylphenyl benzoate; Kenplast® ESB
solvating agent, organic peroxides
Triethyl phosphate
solvator, PVC
Benzoflex® 2-45
solven, cosmetics
PEG-90
solvent
Acetal; *Acetamide MEA*; *Acrolein*; *Acrylonitrile*; *Alcohol denatured*; Alkylate 215; Alkylate 229; *Allyl alcohol*; Altene®; AM 230 Solvent; AM 321 Fluid; AM OMS 75 Solvent; Amgard® DMMP; *Aminoethylpiperazine*; Amsol 10; Amsol 100; Amsol 230-K Solvent; Amsol 400; Amsol 500; Amsol 600; Amsol 650; Amsol 3800; *Amyl lactate*; *t-Amyl methyl ether*; *Anisole*; Antiblaze® DMMP; Arconate® PC-1000; 0-Aromatic 142 Solvent; 0-Aromatic Mineral Spirits; Arosolve TUV; Balmix®; Baltane®; Bayol 35; *Benzotrifluoride*; *Benzyl ether*; *Bis (2-chloroethoxy) methane*; *Bromine*; *1,4-Butanediol*; *Butoxyethyl stearate*; *3-Butoxy-1,2-propanediol*; *t-Butyl acetate*; *n-Butylamine*; *s-Butylamine*; *Butyl butyrate*; *Butylcyclohexane*; *s-Butylcyclohexane*; *t-Butylcyclohexane*; *Butylcyclopentane*; *s-Butylcyclopentane*; *t-Butylcyclopentane*; *Butylene glycol methyl ether*; *Butyl oleate*; *n-Butyraldehyde*; *Butyronitrile*; *C13-15 alcohols*; *C8-16 alkylbenzene*; *C12-15 alkyl benzoate*; Calprint 35; Calprint 38; Calumet 142 Flash Naphtha Rule 66; Calumet 180/210 Naphtha; Calumet 420/460; Calumet Hexane; Calumet Isohexane; Calumet Mineral Spirits < 1%; Calumet VM&P Naphtha < 1%; Capmul® POE-O; *ε-Caprolactone monomer*; Carbitol® Acetate; *Carbon tetrachloride*; Carbopol® 940; Carbopol® 980; CDA 19-3; Chemax E-200 MO; Chemax E-1000 MO; *o-Chloroaniline*; *o-Chlorobenzotrifluoride*; *1,1,1-Chlorodifluoroethane*; *2-Chloronaphthalene*; *Chloropentafluoroethane*; *Chlorotrifluoromethane*; *C9-10 isoparaffin*; *C11-16 isoparaffin*; *C13-17 isoparaffin*; Citrofol® AI; Citrus Stripper Oil; Citrus Terpenes; *Copper oxide (ic)*; *C14-16 paraffins*; *o-Cresol*; Crodamol AB; *1,2-Cyclohexane carbonate*; *p-Cymene*; DCP; *cis-Decalin*; *trans-Decalin*; *1,2-Decane carbonate*; *1,10-Decanediol*; *n-Decane*; *Decyl alcohol*; Dekalin®; Diadol 18G-C; *Diamyl ether*; *Diamyl oxalate*; *Dibutylamine*; *Dibutyl oxalate*; *Dibutyl phosphite*; *Dibutyltin β-mercaptopropionate*; *Dichlorofluoroethane*; *Dichlorofluoromethane*; *1,1-Dichloro-1-nitroethane*; *Dichlorotetrafluoroethane*; *1,2-Dichlorotetrafluoroethane*; *Diethylene glycol dimethyl ether*; *Diethylene glycol ethylvinyl ether*; *Diethylene glycol isobutyl ether*; *Diethylene glycol methyl ether acetate*; *Diethylene glycol methyl ethyl ether*; *Diethylene glycol methylpentyl ether*; *2,3-Dihydrodecafluoropentane*; *2,3-Dihydrofuran*; *2,5-Dihydrofuran*; *1,3-Diisoamyl glycerol diether*; *Diisobutyl adipate*; *Diisobutyl glutarate*; *Diisobutyl succinate*; *Diisopropylamine*; *1,3-Diisopropyl benzene*; *1,4-Diisopropyl benzene*; *Diisopropyl ketone*; *1,4-Diisopropyl naphthalene*; *2,6-Diisopropyl naphthalene*; *Diisopropyl sebacate*; *Dimethoxyethyl phthalate*; *1,3-Dimethoxy-2-propanol*; *Dimethoxytetraglycol*; *Dimethylamine*; *Dimethyl cyanamide*; *1,1-Dimethylcyclohexane*; *1,2-Dimethylcyclohexane*; *cis-1,2-Dimethylcyclohexane*; *trans-1,2-Dimethylcyclohexane*; *1,3-Dimethylcyclohexane*; *cis-1,3-Dimethylcyclohexane*; *trans-1,3-Dimethylcyclohexane*; *1,4-Dimethylcyclohexane*; *cis-1,4-Dimethylcyclohexane*; *trans-1,4-Dimethylcyclohexane*; *1,1-Dimethylcyclopentane*; *1,2-Dimethylcyclopentane*; *1,3-Dimethylcyclopentane*; *2,2-Dimethyl-1,3-dioxolane-4-methanol*; *Dimethyl ether*; *2,6-Dimethyl heptanol-4*; *1,2-Dimethyl imidazole*; *Dimethyl methylphosphonate*; *2,3-Dimethylpentane*; *2,4-Dimethylpentane*; *3,3-Dimethylpentane*; *2,2-Dimethylpropanol*; *Dinonyl phenol*; *Dioctyl adipate*; *Dioctyl phosphite*; *Dioctyl sebacate*; Dipentene No. 122; *Diphenyl carbonate*; *Diphenyl oxide*; *Dipropylamine*; *Di-n-propyl carbonate*; *Dipropylene glycol phenyl ether*; *1,2-Dodecane carbonate*; *n-Dodecane*; Dowanol® PGDA; Dymel® A; Eastman® C-7 Ketone; Eastman® DEP; Eastman® PM Acetate; Eastman® Glacial Acetic Acid Food Grade, Kosher; Eastman® Triacetin, Food Grade; Eastman® Triethyl Phosphate; Ecosoft™ Solvent IK; *Edetic acid*; *Eicosane*; Escaid® 90; Escaid® 95; Escaid® 110; *Ethanolamine*; *(2-Ethoxymethylethoxy) propanol*; *3-Ethoxy-1-propanol*; *Ethylacetoacetate*; *Ethylamine*; *Ethyl bromide*; *Ethyl caproate*; *Ethyl cinnamate*; *Ethyl crotonate*; *Ethylcyclohexane*; *Ethyl decanoate*; *N-Ethyldiethanolamine*; *Ethylenediamine*; *Ethylene glycol dibutyl ether*;

Sovatex C Series
solvent carrier, methylene diisocyanate
Chlorobenzene
solvent replacement, biodegradable: cleaning/degreasing applics.
d-Limonene
solvent resins
Ethyl lactate
solvent resistance aid, NC lacquers
Castor oil, polymerized
solvent stabilizer, halogenated
Methyl pentynol
solvent substitute, cleaners/hand cleaners
Rilanit® EHK
solvent substitute, low VOC: air-drying coatings
Versadil™ 100
solvent substitute, low VOC: high-solids coatings
Versadil™ 100
solvent substitute, organic: dye carriers
Halso® 99
solvent system, mold cleaning
Clean Wiz-9
solvent welding: plastic films: adhesivess
Agrisynth® BLO
solvent, abrasive wheels
FA®
solvent, acaricides
n-Butyl mercaptan
solvent, acaroids
Eastman® IBIB
solvent, accelerators: food-contact crosslinked polyesters
Diethylene glycol
solvent, acetophenone
Ethylbenzene
solvent, acetyl cellulose
cis-1,2-Dichloroethylene
solvent, acetylcellulose
Methyl acetate
solvent, acetylene
Dimethyl formamide; *N-Methyl-2-pyrrolidone*
solvent, acid aluminum cleaners
Rhodacal® 330
solvent, acid gases
Diethylenetriamine; *Dimethyl formamide*; *Tetraethylenepentamine*
solvent, acid rust removers
Solvenon® DPM
solvent, acidic materials
Syn Fac® TEA-97
solvent, acne creams/lotions
Crodamol CAP; Crodamol PMP
solvent, acryl
Methoxybutyl acetate
solvent, acrylates
Eastman® n-Butanol
solvent, acrylic coating resins
Arcosolv® PMA
solvent, acrylic coatings
Hexyl acetate; *Methyl ethyl ketone*
solvent, acrylic copolymers
Diethylene glycol propyl ether; Eastman® PM Acetate; Ethylan® HB1
solvent, acrylic fiber spinning
Dimethyl formamide; *Sodium thiocyanate*
solvent, acrylic high-solids coatings
Butoxyethanol acetate
solvent, acrylic lacquers
Rhodiasolv RPDE; Santosol® DMA; Santosol® DME; Santosol® DME-2; Santosol® DME-3; Santosol® DMG; Santosol® DMS
solvent, acrylic latexes
Texanol® Ester-Alcohol
solvent, acrylic paints
C7-8 isoparaffin
solvent, acrylic paints/coatings
Ethoxyethanol acetate
solvent, acrylic polymerization
Aromatic 100; Aromatic 150; Aromatic 200
solvent, acrylic resin polymerization
Exxate® 600; Exxate® 700; Exxate® 800; Exxate® 900; Exxate® 1000; Exxate® 1300
solvent, acrylic resin polymerization media
C8 alkyl acetate; *C9 alkyl acetate*; *C10 alkyl acetate*; *C13 alkyl acetate*; *Oxo-heptyl acetate*; *Oxo-hexyl acetate*
solvent, acrylic resins
Xylene
solvent, acrylics
BDGE; BGE; Butyl Dioxitol®; Butyl Oxitol®; Cellosolve®; Cellosolve® Acetate; *Dimethyl acetamide*; Eastman® Methyl Acetate, High Purity; EDGE; EGE; i-PGE; MDGE; Methyl Cellosolve®; Methyl Cellosolve® Acetate; MGE; M-Pyrol®; NMP; Propyl Cellosolve®; *Propylene glycol dimethyl ether*; *Propylene glycol methyl ether acetate*; Ucar® Ester EEP
solvent, acrylonitrile
Calcium thiocyanate
solvent, acrylonitrile fibers
Propylene carbonate
solvent, active ingreds.
Acetylated glyceryl stearate; Akomed R; Glicopol 2-G; *Glyceryl caprylate*; Imwitor® 742; Milopol G-7; Monestriol DM; Monestriol GCT; Monestriol GE; Monestriol GP; Monestriol GP-35; Monestriol GP-35-3%; Monestriol GP-35-AE; Monestriol GP-40; Monestriol GP-40-CF; Monestriol GP-40-DC; Monestriol GP-45; Monestriol GP-45-DC; Monestriol GP-60; Monestriol GP-AS; Myritol® PC; Tegosoft® M; Thomil G; Tilol GCT-R; Tilol GP-O; Triacetina; Triol 91; Trioleina
solvent, active ingreds.: cosmetic anhyd. microemulsions
Transcutol® CG
solvent, active ingreds.: cosmetics
Crodamol GTCC
solvent, active ingreds.: liq. pharmaceuticals
Lutrol® E 300; Lutrol® E 400; Lutrol® E 600
solvent, active ingreds.: pharmaceuticals
Crodamol GTCC; Liquid Medilan™; Liquid Medilan™ Ultra
solvent, active ingreds.: semisolid pharmaceuticals
Lutrol® E 300; Lutrol® E 400; Lutrol® E 600
solvent, active tail: polyester coil coatings
Aromatic 200
solvent, active: acrylic copolymers
Eastman® DP
solvent, active: alkyd resins
Eastman® DP
solvent, active: CR
Eastman® DP
solvent, active: epoxies
Eastman® DP
solvent, active: NC
Eastman® DP
solvent, active: solvent-borne coatings
Dowanol® DPnP; Dowanol® PnB
solvent, additives
Solvesso™ 150 ND; Sunoco 100; Sunoco 150
solvent, adhesive removal
Dynasolve 100; Dynasolve 699; Dynasolve M-10; Dynasolve MP-500; Soygold® 1000; Soygold® 2000
solvent, adhesive/ink removal: rollers
Dynasolve 180; Dynasolve 185; NMP
solvent, adhesive/polymer removal
Decap
solvent, adhesives
Acetyldimethylamine chloride; *Acetyl tributyl citrate*; *Acetyl triethyl citrate*; *Acetyl trimethyl citrate*; Arcosolv® DPM; Arcosolv® DPMA; Arcosolv® DPNB; Arcosolv® PE; Arcosolv® PEA; Arcosolv® PM; Arcosolv® PMA; Arcosolv® PNB; Arcosolv® PNP; Arcosolv® PTB; Arcosolv® TPM; *1,1′-Biphenyl, butenylated*; *1,1′-Biphenyl (1-methylethyl)*; *Bis (methylethyl)-1,1′-biphenyl*; *2-Butanol*; *Butoxyethanol*; *Butoxypropanol*; *n-Butyl acetate*; *s-Butyl acetate*; *t-Butyl alcohol*; *Butyl-(S)-lactate*; Carbowax® PEG 300; *p-Chlorobenzotrifluoride*; Citroflex® 4; Citroflex® A-2; Citroflex® A-4; Coasol B®; Coasol C; *3,4-Dichlorobenzotrifluoride*; *Dimethyl acetamide*; *Dipropylene glycol butyl ether*; *Dipropylene glycol t-butyl ether*; Dynasolve 200; Eastman® DBP; Eccoflex 090; Ecco Strip LS-W; Emery® 912; Emery® 916; Emery® 917; *Ethoxyethanol*; *Ethyl acetate*; *Ethylene dichloride*; Exxsol® D 95; GAFsolver™ TMP; Halso® 99; Hydrosol 75/95 N; Hydrosol Essence 60/95; Hydrosol Essence 80/110; Hydrosol Essence C; Hydrosol Heptane; Hydrosol Hexane; Hydrosol Isohexane; *Isobutyl acetate*; Isohexane LNH; Jeffsol® PC; Leksol™; *Methoxyethanol acetate*; *Methyl abietate*; *Methylene chloride*; *Methyl ethyl ketone*; *Methyl isobutyl ketone*; *Mineral spirits*; *Monochlorotoluene*; Nadone®; *Naphtha, medium aliphatic*; NE™ 101; Neosolv™ 2; Neosolv™ 4; Neosolv™ 6; Neosolv™ 8; Neosolv™ 24; Neosolv™ 150; *Nitroethane*; *Nitromethane*; NMP; N-Terpinal RS-50; Oxsol® 100; Oxsol® 1000; Poly-Solv® DM; Poly-Solv® DPM; Poly-Solv® EM; Poly-Solv® MPM; Poly-Solv® TM; Poly-Solv® TPM; *PPG-2 methyl ether*; *PPG-3 methyl ether*; *PPG-2 methyl ether acetate*; *n-Propyl bromide*; *Propylene glycol ethyl ether acetate*; *Propylene glycol laurate*; Punctilious® Ethyl Alcohol 190 Proof Pure USP; Punctilious® Ethyl Alcohol 200 Proof Pure USP; Punctilious® SDA 1-2 190 Proof; Punctilious® SDA 1-2; Punctilious® SDA 2B-2 190 Proof; Punctilious® SDA 2B-2 Anhydrous; Punctilious® SDA 2B-3 190 Proof; Punctilious® SDA 2B-4 190 Proof; Punctilious® SDA 2B-4 Anhydrous; Punctilious® SDA 3A 190 Proof; Punctilious® SDA 3A Anhydrous; Punctilious® SDA 3C 190 Proof; Punctilious® SDA 12A-3 190 Proof; Punctilious® SDA 12A-3 Anhydrous; Punctilious® SDA 13A 190 Proof; Punctilious® SDA 19 190 Proof; Punctilious® SDA 20 190 Proof; Punctilious® SDA 20 Anhydrous; Punctilious® SDA 23A 190 Proof; Punctilious® SDA 23A Anhydrous; Punctilious® SDA 23H 190 Proof; Punctilious® SDA 25A-1 190 Proof; Punctilious® SDA 25A-2 190 Proof; Punctilious® SDA 29-2 190 Proof; Punctilious® SDA 29-2 Anhydrous; Punctilious® SDA 29-3 190 Proof; Punctilious® SDA 30 190 Proof; Punctilious® SDA 30 Anhydrous; Punctilious® SDA 32 190 Proof; Punctilious® SDA 32 Anhydrous; Rhodiasolv RPDE; Rubber Solvent 332; SBP 40/65 LNH; SBP 60/95 LNH; SBP 80/95 LNH; SBP 80/110 LNH; SBP 100/140; SBP 140/165; *SD alcohol 1*; *SD alcohol 3-A*; *SD alcohol 3-C*; *SD alcohol 12-A*; *SD alcohol 23-A*; *SD alcohol 30*; Shell Mineral Spirits 135; Shellsol® 426; Shellsol® A100 Solvent; Shellsol® D25; Shellsol® D40; Shellsol® D43; Shellsol® D60 SC; Shellsol® D80; Shellsol® H; Shellsol® Heptane; Shellsol® TC; Shellsol® W HT; Shellsol® X7B; Surfadone® LP-300; *Tetrahydrofuran*; *Toluene*; *1,1,1-Trichloroethane*; *Triglycol monomethyl ether*; *Trihexyl citrate*; *Trimethyl citrate*; Uniplex 150; *γ-Valerolactone*; *VM&P naphtha*; VM&P Naphtha HT; *Xylene*; *p-Xylene*
solvent, adhesives/sealants
AG 101; *Dibutyl phthalate*; *1,4-Dioxane*
solvent, adhesives: fast-drying
Isohexane
solvent, adhesives: food-pkg.
Citroflex® A-2
solvent, adhesivesadhesives
Acetone
solvent, adjuvants
Methylpropanediol
solvent, adjuvants: food-contact crosslinked polyesters
Butyl benzyl phthalate; *Dibutyl phthalate*; *Dimethyl phthalate*; *Styrene*; *Triphenyl phosphate*
solvent, aerosol coatings
Arcosolv® TPNB
solvent, aerosol electronic contact cleaners
Genesolv® D
solvent, aerosol hair sprays
Uniplex 110
solvent, aerosol pkg.
Aerothene MM
solvent, aerosol sprays
Shellsol® OMS
solvent, aerosol wood furniture coatings
ExxonMobil MEK
solvent, aerosols
Argobase 125T; *1,1′-Biphenyl, butenylated*; *1,1′-Biphenyl (1-methylethyl)*; *Bis (methylethyl)-1,1′-biphenyl*; Captex® 810D; Captex® 8227; *p-Chlorobenzotrifluoride*; *C isoparaffin*; *3,4-*

Crodamol DA; *Diisopropyl adipate*
solvent, aromatic polyamides
NMP
solvent, aromatic polyimides
NMP
solvent, aromatic recovery: refinery streams
2-(2-Aminoethoxy) ethanol; Diglycolamine® Agent (DGA®)
solvent, aromatics extraction
Dipropylene glycol
solvent, artificial leather
Diacetone alcohol; *Ethyl acetate*
solvent, asphalt
Decahydronaphthalene; Decalin; *Mineral spirits*; Shell Mineral Spirits 135; Shellsol® D40; Shellsol® D43; *Tetrahydronaphthalene*; Tetralin; *VM&P naphtha*; VM&P Naphtha HT
solvent, asphalt cement extraction: from bituminous mixes
Safety-Solv 140 Plus
solvent, asphalt coatings
Kerosene
solvent, auto refinish coatings
Exxsol® D 130; Exxsol® D 3135
solvent, automatic temp. control valve mfg.
Genesolv® D
solvent, automobile undercoatings
Shellsol® Heptane
solvent, automobile/furniture polishes
Drakesol® 165AT; Drakesol® 205
solvent, automotive brakes
Neosolv™ 2; Neosolv™ 4; Neosolv™ 6; Neosolv™ 8; Neosolv™ 24; Neosolv™ 150
solvent, automotive cleaners
Exxate® 1200
solvent, automotive coatings
Arcosolv® TPNB; Eastman® IBIB; Eastman® Acetone, High Purity Sales Grade; Eastman® n-Butanol; Eastman® Isopropanol; Eastman® Isopropyl Acetate; Eastman® n-Propanol; Eastman® n-Propyl Acetate; Santosol® DMA; Santosol® DME; Santosol® DME-2; Santosol® DME-3; Santosol® DMG; Santosol® DMS
solvent, automotive degreasers
Exxate® 1200
solvent, automotive engine part cleaning
GAFsolver™ TMP
solvent, automotive lubricants
Norpar® 14 Fluid
solvent, automotive OEM coatings
Eastman® EEP; Isopar® C; Solvesso™ 100
solvent, automotive paints
Eastman® DE Acetate; Eastman® DE-HG; Eastman® EB Acetate; Eastman® PM; Eastman® PM Acetate; *Ethyl acetate*; Rhodiasolv RPDE
solvent, automotive paints/coatings
n-Butyl acetate
solvent, automotive penetrants
Norpar® 14 Fluid
solvent, automotive polishes
Drakesol® 165; Drakesol® 220; Drakesol® 260; Drakesol® 260 AT; Drakesol® 305; *Petroleum distillates, hydrotreated light*
solvent, automotive prod.
Soygold® 1000; Soygold® 2000
solvent, automotive prods.
Conosol 260; Conosol 340; Conosol C-145; Conosol C-200; Exxsol® D 95; *Mineral spirits*; Shell Mineral Spirits 135; Shell Toluene Solvent
solvent, automotive refinish coatings
Solvesso™ 100
solvent, automotive refinishes
Eastman® EEP; ExxonMobil MEK; Ucar® n-Propyl Propionate
solvent, automotive specialties
Shell OMS; Shell Mineral Spirits 7 EC; Shell Xylene; VM&P Naphtha EC
solvent, automotive undercoatings
Shellsol® W HT
solvent, aux.: dyestuffs
Trax K-40
solvent, aviation
n-Propyl bromide
solvent, aviation: adhesive removal
Skykleen® 1000; Skykleen® 2000
solvent, aviation: painting/bonding surface prep.
Skykleen® 2000
solvent, aviation: painting/bonding surface preparation
Skykleen® 1000
solvent, aviation: sealant removal
Skykleen® 1000; Skykleen® 2000
solvent, aviation: surface cleaning
Skykleen® 1000; Skykleen® 2000
solvent, aviation: surface degreasing
Skykleen® 1000; Skykleen® 2000
solvent, avobenzone
Finsolv® TPP
solvent, B/S latexes
Texanol® Ester-Alcohol
solvent, baby food
Delios®
solvent, baby skin care
Miglyol® 8810
solvent, bactericides
Isopropylamine; *Phenoxyethanol*
solvent, baked finishes
3-Heptanone
solvent, baked goods
Hodag® S-35 FGK
solvent, bakery flavors
Diethyl succinate
solvent, baking enamels
Eastman® DB; Eastman® PM Acetate
solvent, baking finishes
Eastman® EEP; *2-Ethylhexyl acetate*
solvent, ball-point pen inks
Benzyl alcohol
solvent, balsams
Fancol™ Karite Butter
solvent, barbecue
Shellsol® D60 SC
solvent, barbecue lighter fluids
Shellsol® D60 SC; Shellsol® D70; Shellsol® D-LAWS
solvent, base: specialty cleaners
GAFsolver™ TMP
solvent, basic dyes
Cyclohexyl acetate; *Glyceryl acetate*
solvent, bath care
Carbowax® E200; Carbowax® E300; Carbowax® E400; Carbowax® E600; Carbowax® E900; Carbowax® E1000; Carbowax® E1450; Carbowax® E3350; Carbowax® E8000; Finsolv® TN-O; HallStar™ DOA; HallTress™ DIBA Special; *Hexyldecyl benzoate*; *Isopropyl PPG-2-isodeceth-7-carboxylate*; *SD alcohol 1*; *SD alcohol 3-A*; *SD alcohol 3-B*; *SD alcohol 3-C*; *SD alcohol 23-A*; *SD alcohol 30*; *SD alcohol 36*; *SD alcohol 38-B*; *SD alcohol 39-B*; *SD alcohol 39-C*; *SD alcohol 40*; *SD alcohol 40-A*; *SD alcohol 40-B*; *SD alcohol 40-C*
solvent, bath oils
Arlatone® T; *Caprylic/capric triglyceride PEG-4 esters*; Captex® 810D; Captex® 8227; Carbowax® Sentry® PEG 400; Crodamol PMP; Cropure® Almond; Estalan DIA; Estol IPM 1509 (BIO-IPM); Fancol™ ALA; HallStar™ DIPA; Labrafac® Hydrophile; Lexol® IPM-NF; Myritol® 318; Novol NF; Schercemol DIA; STEPAN® IPM-NF
solvent, bath solutions
SD alcohol 46
solvent, bathroom cleaners
Neosolv™ 2; Neosolv™ 4; Neosolv™ 6; Neosolv™ 8; Neosolv™ 24; Neosolv™ 150
solvent, batteries
1,2-Butylene carbonate; Jeffsol® 1,2-Butylene Carbonate; Jeffsol® EC-25; Jeffsol® EC-50; Jeffsol® EC-75
solvent, bay rum
SD alcohol 23-A; *SD alcohol 37*; *SD alcohol 38-B*; *SD alcohol 39*; *SD alcohol 39-B*; *SD alcohol 39-D*; *SD alcohol 40-A*; *SD alcohol 40-B*; *SD alcohol 40-C*
solvent, benzophenone: sunscreens
Dermol ICSA; *Isocetyl salicylate*
solvent, benzophenone-3
Finsolv® TPP
solvent, binders
SD alcohol 1; *SD alcohol 3-A*; *SD alcohol 3-C*; *SD alcohol 12-A*; *SD alcohol 23-A*; *SD alcohol 30*
solvent, biocides
SD alcohol 3-B; *SD alcohol 23-A*; *SD alcohol 23-H*; *SD alcohol 27-A*; *SD alcohol 27-B*; *SD alcohol 37*; *SD alcohol 38-B*; *SD alcohol 39-B*; *SD alcohol 40*; *SD alcohol 40-A*; *SD alcohol 40-B*; *SD alcohol 40-C*
solvent, biomedical parts
Genesolv® D
solvent, bitumens
Cyclohexyl acetate
solvent, blending
Hi-Tri
solvent, blushes
EmCon™ CO
solvent, body creams/lotions
SD alcohol 23-A; *SD alcohol 23-H*; *SD alcohol 31-A*; *SD alcohol 37*; *SD alcohol 38-B*; *SD alcohol 39*; *SD alcohol 39-B*; *SD alcohol 39-C*; *SD alcohol 40*; *SD alcohol 40-A*; *SD alcohol 40-B*; *SD alcohol 40-C*
solvent, body deodorants
SD alcohol 38-B; *SD alcohol 39-B*; *SD alcohol 39-C*; *SD alcohol 40*; *SD alcohol 40-A*; *SD alcohol 40-B*; *SD alcohol 40-C*
solvent, body lotions
EmCon™ Olive; Fancol™ ALA; Fancol™ Karite Butter; Fanoliv™ Oil; Finsolv® TN-O; *Shea butter (Butyrospermum parkii)*
solvent, boiler defoaming
Jeffox® WL-440; Jeffox® WL-660; Jeffox® WL-1400; Jeffox® WL-5000
solvent, boron chemistry
Butyl Diglyme; *Diethylene glycol dibutyl ether*; Ethyl Diglyme; *Ethylene glycol dimethyl ether*; Ethyl Glyme; Monoglyme; *PEG-4 dimethyl ether*; Tetraglyme; Triglyme
solvent, brake cleaners
Exxsol® D 110
solvent, brake fluids
Hisolve DM; Hisolve PM; Hisolve TM; Neosolvent NSG; Poly-Solv® DM; Poly-Solv® DPM; Poly-Solv® EM; Poly-Solv® MPM; Poly-Solv® TM; Poly-Solv® TPM; *PPG-3 methyl ether*; *Triglycol monomethyl ether*
solvent, bromine
Carbon disulfide
solvent, bromo acid dye: lipstick
PEG-10 lanolate; *PEG-20 lanolate*
solvent, bromo acid dye: makeup
PEG-10 lanolate; *PEG-20 lanolate*
solvent, bromo acid dye: skin care
PEG-10 lanolate; *PEG-20 lanolate*
solvent, bromo acid dyes
Aqualose W15
solvent, bromo dye
Laneth-5; Polychol 5
solvent, brown oils
Cyclohexyl acetate
solvent, buffer coating removal: optical fibers
Dynasolve 210
solvent, burn creams
Crodamol CAP; Crodamol PMP
solvent, butadiene
Dimethyl formamide
solvent, butyrate
Butyl Cellosolve®; Butyl Cellosolve® Acetate; Cellosolve®; Cellosolve® Acetate; Methyl Cellosolve®; Methyl Cellosolve® Acetate; Propyl Cellosolve®; Ucar® Ester EEP
solvent, butyrate high-solids coatings
Butoxyethanol acetate
solvent, butyrate paints/coatings
Ethoxyethanol acetate
solvent, CAB
Eastman® DE; Eastman® PM Acetate; *Isopropyl acetate*
solvent, cable assembly repair
Dynasolve 190; Uresolve Blue
solvent, caffeine
cis-1,2-Dichloroethylene
solvent, camper fuel
Rubber Solvent 332
solvent, camphor
cis-1,2-Dichloroethylene; *cis-trans-1,2-Dichloroethylene (mixed isomers)*; *trans-1,2-Dichloroethylene*; *Ethylene glycol acetate*; *Isoamyl acetate*
solvent, can coatings
Citrofol® AI; Rhodiasolv RPDE
solvent, candy flavors
Diethyl succinate
solvent, candy glazes
SD alcohol 13-A; *SD alcohol 23-A*; *SD alcohol 35*; *SD alcohol 35-A*; *SD alcohol 45*
solvent, capsules
Estasan™ GT 8-40 3578; Estasan™ GT 8-60 3575; Estasan™ GT 8-60 3580; Estasan™ GT 8-65 3577; Estasan™ GT 8-65 3581; Estasan™ GT 8-70 3579; Labrafac® Lipophile
solvent, car dewaxing
Exxsol® D 80; Shellsol® D60 SC
solvent, carbon deposit removal
GAFsolver™ TMP
solvent, carbon paper
Triolein
solvent, carbonless copy paper
Diisopropyl naphthalene; Drakesol® 165AT; Drakesol® 205; Shellsol® D120
solvent, carbonless copy paper office forms
Drakesol® 165; Drakesol® 220; Drakesol® 260; Drakesol® 260 AT;

Punctilious® SDA 23H 190 Proof; Punctilious® SDA 25A-1 190 Proof; Punctilious® SDA 25A-2 190 Proof; Punctilious® SDA 29-2 190 Proof; Punctilious® SDA 29-2 Anhydrous; Punctilious® SDA 29-3 190 Proof; Punctilious® SDA 30 190 Proof; Punctilious® SDA 30 Anhydrous; Punctilious® SDA 32 190 Proof; Punctilious® SDA 32 Anhydrous

solvent, cleaning/degreasing circuit boards
Arcosolv® PMA

solvent, cleaning/depositing silicone oil-based lubricants: medical applics.
Vertrel® XSi

solvent, cleaning: cold cleaning
Novec™ HFE-7200

solvent, cleaning: electronic equip. cleaning
Novec™ HFE-7100; Novec™ HFE-7100DL

solvent, cleaning: electronics
Abzol® EG; Abzol® JEG

solvent, cleaning: electronics industry cleaning
Novec™ HFE-7200DL

solvent, cleaning: ionics removal
Abzol® EG; Abzol® JEG

solvent, cleaning: metals
FAME; *Trichlorotrifluoroethane*

solvent, cleaning: PC boards
Abzol® JEG

solvent, cleaning: PCs
Abzol® EG

solvent, cleaning: plastic
Trichlorotrifluoroethane

solvent, cleaning: precision cleaning
Abzol® VG; FAME

solvent, cleaning: printing
SBP 60/95 LNH

solvent, cleaning: textile blends
Abzol® LS

solvent, cleaning: textiles
Abzol® LS

solvent, cleaning: ultrasonic baths
Abzol® VG

solvent, cleaning: vapor degreasers
Abzol® VG

solvent, cleaning: vapor degreasing
Novec™ HFE-7100; Novec™ HFE-7100DL; Novec™ HFE-7200; Novec™ HFE-7200DL

solvent, cleaning: wet cleaning
Novec™ HFE-7100; Novec™ HFE-7100DL; Novec™ HFE-7200; Novec™ HFE-7200DL

solvent, cleansers
EmCon™ Olive; EmCon™ Rice Bran; EmCon™ W; Fanoliv™ Oil; Finsolv® BOHS-111; HallStar™ DIPA; HallStar™ DIPS; HallStar™ DOS; *Octyl hydroxystearate benzoate*; STEPAN® IPM-NF

solvent, cleansing milks
PEG-6 abietate

solvent, clear cosmetic prods.
Finsolv® PG-22

solvent, clear sticks/gels
HallStar™ DIPS

solvent, clear water topical gels
Carbopol® 980

solvent, clinical nutrition
Capmul® MCM; Captex® 350

solvent, CO removal: from ammonia gas streams
PEG dimethyl ether

solvent, CO_2 removal: ammonia mfg.
Jefftreat® M 45%; Jefftreat® M-510 45%; Jefftreat® M-510 100%; *N-Methyl-2-pyrrolidone*

solvent, CO_2 removal: Claus off-gas treating
Jefftreat® MS-100

solvent, CO_2 removal: cryogenic gas processing
Jefftreat® M 45%; Jefftreat® M-510 45%; Jefftreat® M-510 100%

solvent, CO_2 removal: from ammonia gas streams
PEG dimethyl ether

solvent, CO_2 removal: from gases
Diglycolamine® Agent (DGA®)

solvent, CO_2 removal: gas processing
Jefftreat® M 45%; Jefftreat® M-510 45%; Jefftreat® M-510 100%

solvent, CO_2 removal: gas treatment
Jefftreat® MS-200 Series

solvent, CO_2 removal: high pressure natural gas
Jefftreat® MS-100

solvent, CO_2 removal: hydrogen prod.
Jefftreat® M 45%; Jefftreat® M-510 45%; Jefftreat® M-510 100%

solvent, CO_2 removal: hydrogen sulfide enrichment
Jefftreat® M Solvent

solvent, CO_2 removal: LNG prod.
Jefftreat® M 45%; Jefftreat® M-510 45%; Jefftreat® M-510 100%

solvent, CO_2 removal: natural gas plants
Jefftreat® M Solvent

solvent, CO_2 removal: natural gas upgrading
Jefftreat® M 45%; Jefftreat® M-510 45%; Jefftreat® M-510 100%

solvent, CO_2 removal: refinery gas
Jefftreat® MS-100

solvent, CO_2 removal: tail gas treatment
Jefftreat® M Solvent

solvent, coagulating: textiles
GBL

solvent, coalescing: acrylic-based polishes
Tributoxyethyl phosphate

solvent, coalescing: acrylics
Methoxyisopropanol

solvent, coalescing: adhesives
Methoxyisopropanol; *Phenoxyethanol*; *Propylene glycol ethyl ether*; *Propylene glycol n-propyl ether*; *Tributoxyethyl phosphate*

solvent, coalescing: agric.
Methoxyisopropanol; *Propylene glycol ethyl ether*; *Propylene glycol n-propyl ether*

solvent, coalescing: alkyds
Macol® 85

solvent, coalescing: aq. architectural/industrial coatings
Phenoxyethanol

solvent, coalescing: aq. paints
Propylene glycol isobutyl ether; *2,2,4-Trimethyl-1,3-pentanediol*

solvent, coalescing: cellulosics
Methoxyisopropanol

solvent, coalescing: cleaners
Methoxyisopropanol; *Propylene glycol ethyl ether*; *Propylene glycol n-propyl ether*

solvent, coalescing: coatings
Propylene glycol n-propyl ether

solvent, coalescing: cosmetics
Methoxyisopropanol; *Propylene glycol ethyl ether*; *Propylene glycol n-propyl ether*

solvent, coalescing: dyes
Methoxyisopropanol; *Propylene glycol ethyl ether*

solvent, coalescing: electronics
Methoxyisopropanol; *Propylene glycol ethyl ether*; *Propylene glycol n-propyl ether*

solvent, coalescing: epoxides
Methoxyisopropanol

solvent, coalescing: epoxies
Macol® 85

solvent, coalescing: floor polishes
Methoxyisopropanol

solvent, coalescing: fuel additives
Methoxyisopropanol

solvent, coalescing: gloss paints
Tributoxyethyl phosphate

solvent, coalescing: greases
Methoxyisopropanol; *Propylene glycol ethyl ether*

solvent, coalescing: inks
Methoxyisopropanol; *Propylene glycol n-propyl ether*

solvent, coalescing: lacquers
Methoxyisopropanol; *Propylene glycol ethyl ether*

solvent, coalescing: latex caulks
Macol® 85

solvent, coalescing: oils
Methoxyisopropanol; *Propylene glycol ethyl ether*

solvent, coalescing: paints
Methoxyisopropanol

solvent, coalescing: paints/coatings
Propylene glycol ethyl ether

solvent, coalescing: phenolics
Macol® 85

solvent, coalescing: plasticizing acrylic emulsion polymers
Macol® 85

solvent, coalescing: polyesters
Macol® 85; *Methoxyisopropanol*

solvent, coalescing: printing inks
Propylene glycol ethyl ether

solvent, coalescing: PU
Methoxyisopropanol

solvent, coalescing: resins
Methoxyisopropanol; *Propylene glycol ethyl ether*

solvent, coalescing: textile dyes
Methoxyisopropanol

solvent, coalescing: textiles
Propylene glycol ethyl ether; *Propylene glycol n-propyl ether*

solvent, coating defoamers
Petroleum distillates, hydrotreated light

solvent, coating polymers
Eastman® PM Acetate

solvent, coating resins
Arcosolv® DPNB; Arcosolv® DPTB; Arcosolv® PNB; *Dipropylene glycol butyl ether*; Eastman® Methyl Acetate, High Purity; Nadone®

solvent, coatings
Acetyldimethylamine chloride; *Acetyl tributyl citrate*; *Acetyl triethyl citrate*; *Acetyl trimethyl citrate*; *Alcohol*; Arcosolv® DPM; Arcosolv® DPMA; Arcosolv® PE; Arcosolv® PEA; Arcosolv® PM; Arcosolv® PNP; Arcosolv® PTB; Arcosolv® TPM; Arcosolv® TPNB; *2-Butanol*; *Butoxypropanol*; *s-Butylbenzene*; Butyl Carbitol® Acetate; Butyl Icinol; *Butyl lactate*; Cellosolve®; Cellosolve® Acetate; Celtone; *p-Chlorobenzotrifluoride*; *C7-8 isoparaffin*; *C8-9 isoparaffin*; *C10-11 isoparaffin*; *C11-12 isoparaffin*; *C11-13 isoparaffin*; *C13-14 isoparaffin*; Citroflex® 2; Citroflex® 4; Citroflex® A-2; Citroflex® A-4; Coasol B®; Coasol C; Condea Vista LPA-142; Condea Vista LPA-210; *Cyclohexane*; DBE; DBE-2; DBE-3; DBE 4; DBE-5; DBE-6; DBE-9; *3,4-Dichlorobenzotrifluoride*; *Diethylene glycol dibutyl ether*; *Diethylene glycol 2-ethylhexyl ether*; *Diisobutyl ketone*; *Dimethyl acetamide*; *Dimethyl adipate*; *2,3-Dimethylbutane*; *Dimethyl glutarate*; *Dimethyl oleamide*; *Dimethyl succinate*; *Dipropylene glycol t-butyl ether*; *Dipropylene glycol dimethyl ether*; Dowanol® DPM; Dowanol® DPMA; Dowanol® DPnB; Dowanol® PMA; Dowanol® PnB; Eastman® C-11 Ketone; Eastman® DE; Eastman® DE Acetate; Eastman® DE-HG; Eastman® DM; Eastman® EAA; Eastman® EEH; Eastman® EEP; Eastman® EP; Eastman® IBIB; Eastman® MIBK; Eastman® PM; Eastman® PM Acetate; Eastman® Acetone, High Purity Sales Grade; Eastman® n-Butanol; Eastman® Isobutyl Acetate; Eastman® Isopropanol; Eastman® Isopropyl Acetate; Eastman® Methyl Acetate; Eastman® Methyl Acetate, High Purity; Eastman® n-Propanol; Eastman® n-Propyl Acetate; Ecco Cleaner AK; Eccoflex 090; Ecco Strip LS-W; Estasol™; *Ethyl acetate*; *Ethylene glycol diacetate*; *Ethylene glycol 2-ethylhexyl ether*; *Ethylene glycol propyl ether*; *Ethyl lactate*; Exxsol® D 75/100 Naphtha; Exxsol® D 80; Exxsol® D 110; Exxsol® D 115/145 Naphtha; Hetoxide MPC; Hexalin®; *N-Hydroxyethylpyrrolidone*; Isopar® E Fluid; Isopar® G Fluid; Isopar® H Fluid; Isopar® K Fluid; Isopar® L Fluid; Isopar® M Fluid; *Isopropyl alcohol*; *Isopropyl lactate*; Leksol™; *Methoxybutyl acetate*; *Methoxydiglycol*; *Methoxyethanol acetate*; *Methyl amyl acetate*; Methyl Cellosolve®; Methyl Cellosolve® Acetate; Methylhexalin®; *Methyl propionate*; *Mineral spirits*; Morcryl® 535; NE™; NE™ 101; NiPar S-10™; *Nitroethane*; *Nitromethane*; *2-Nitropropane*; NM™; NMP; NMP-EL; Oxsol® 100; Oxsol® 1000; *PEG-8 oleate*; Poly-Solv® DM; Poly-Solv® DPM; Poly-Solv® EM; Poly-Solv® MPM; Poly-Solv® TM; Poly-Solv® TPM; Proglyde DMM; *n-Propyl bromide*; *Propylene glycol methyl ether acetate*; *Propyl propionate*; Rubber Solvent; Rubber Solvent 332; *SD alcohol 1*; *SD alcohol 3-A*; *SD alcohol 3-C*; *SD alcohol 23-A*; *SD alcohol 30*; Shell OMS; Shell Mineral Spirits 7 EC; Shell Mineral Spirits 135; Shellsol® D70; Shell Xylene; Solvenol® 2; Solvenon® PM; SR 256; *Stoddard solvent*; Sure Sol®-150ND; Sure Sol®-202; Sure Sol®-205; Surfadone® LP-100; Tecsol® A, 95%; Tecsol® A Anhydrous; Tecsol® A-2, 95%; Tecsol® A-2 Anhydrous; Tecsol® B; Tecsol® C 95%; Tecsol® C Anhydrous; Tecsol® D, 95%; Tecsol® D-1, 95%; Tecsol® D-2, 95%; Tecsol® D-2 Anydrous;

Polyglycol P15-200; Polyglycol P112-2; Polyglycol P425; Polyglycol P1000TB; Polyglycol P1200; Polyglycol P2000; Polyglycol P4000; Polyglycol PT250; Polyglycol PT700; Polyglycol PT3000; Polypax IPM; Polypax IPP; *Polysorbate 60*; PPG 1650; *PPG-9*; *PPG-17*; *PPG-20*; *PPG-26*; *PPG-30*; *PPG-40 diethylmonium chloride*; *PPG-24-glycereth-24*; *PPG-66-glycereth-12*; *PPG-10 glyceryl ether*; *PPG-55 glyceryl ether*; *PPG-2 methyl ether*; *PPG-3 methyl ether*; *PPG-15 stearyl ether*; Pricerine™ 9091; Pricerine™ 9099; Promyr; Propal NF; *Propyl acetate*; *Propyl alcohol*; *Propylene carbonate*; *Propylene glycol*; Propylene Glycol USP, FCC; *Propylene glycol caprylate*; *Propylene glycol dicaprylate*; *Propylene glycol dicaprylate/dicaprate*; *Propylene glycol ethyl ether acetate*; *Propylene glycol isobutyl ether*; *Propylene glycol laurate*; *Propylene glycol methyl ether acetate*; *Propylene glycol phenyl ether*; Punctilious® SDA 1-1 190 Proof; Punctilious® SDA 1-1 Anhydrous; *Rice (Oryza sativa) bran oil*; Salacos® 99; Schercemol PGML; SDA-1; SDA-3A; SDA-3C; SDA-12A; SDA-23A; SDA-23H; SDA-29-8; SDA-30; SDA-39C; SDA-40; SDA-40A; SDA-40B; SDA-40C; *SD alcohol 1*; *SD alcohol 3-A*; *SD alcohol 3-B*; *SD alcohol 3-C*; *SD alcohol 23-A*; *SD alcohol 23-F*; *SD alcohol 23-H*; *SD alcohol 27-B*; *SD alcohol 30*; *SD alcohol 31-A*; *SD alcohol 36*; *SD alcohol 37*; *SD alcohol 38-B*; *SD alcohol 38-C*; *SD alcohol 38-D*; *SD alcohol 38-F*; *SD alcohol 39*; *SD alcohol 39-A*; *SD alcohol 39-B*; *SD alcohol 39-C*; *SD alcohol 39-D*; *SD alcohol 40*; *SD alcohol 40-A*; *SD alcohol 40-B*; *SD alcohol 40-C*; *SD alcohol 46*; *Sesamol*; Shellsol® D60 SC; Shellsol® D70; Shellsol® TD; Shellsol® TM; *Sorbeth-20*; Stearate 1500; STEPAN® IPM; STEPAN® IPP; *Sucrose palmitate*; *Tall oil*; *α-Terpineol*; *Tetrahydrofurfuryl acetate*; *Tetrahydrofurfuryl alcohol*; Texanol® Ester-Alcohol; *Thiolanediol*; *Toluene*; *Triacetin*; *Tributyl citrate*; *Tricaprin*; *1,1,1-Trichloroethane*; *Triethyl citrate*; *Trihydroxystearin*; *Triisononanoin*; *Triisostearin*; *Trilaurin*; *Trilinolein*; *3,5,5-Trimethylhexan-1-ol*; *2,2,4-Trimethyl-1,3-pentanediol isobutyrate*; *Trimyristin*; *Trioctanoin*; *Triolein*; *Tripalmitin*; *1,1,3-Tris (2-methyl-4-hydroxy-5-t-butylphenyl) butane*; *Tristearin*; *Triundecanoin*; *Turpentine*; *γ-Undecalactone*; Upamide ACMEA; Variquat® CC-42; Wickenol® 136; *Xylene*; *Ylang ylang (Cananga odorata) oil*; ZELEC® IPM

solvent, coupling
Butyl Di-Icinol; *Castor oil, polymerized*; *Diethylene glycol dibutyl ether*; Ethyl Icinol; #15 Oil; Pale 16; Pale 170; Pale 1000; *Triglycol monomethyl ether*

solvent, coupling: adhesives
#30 Oil; #40 Oil; Pale 4; *Propylene glycol t-butyl ether*

solvent, coupling: agric.
Propylene glycol t-butyl ether

solvent, coupling: agric. chemicals
Ethoxyethanol

solvent, coupling: alkyds
Propylene glycol t-butyl ether

solvent, coupling: aq. coatings
Eastman® EP

solvent, coupling: aq. paints
Propylene glycol t-butyl ether

solvent, coupling: caulks
#30 Oil; #40 Oil; Pale 4

solvent, coupling: cellulose lacquers
#30 Oil; #40 Oil; Pale 4

solvent, coupling: cleaners
Propylene glycol t-butyl ether

solvent, coupling: cleaning solutions
Butyl Dioxitol®

solvent, coupling: coatings
Butoxydiglycol

solvent, coupling: cosmetics
Propylene glycol t-butyl ether

solvent, coupling: creams
2-Aminobutanol

solvent, coupling: cutting fluids
Butoxydiglycol; *Propylene glycol t-butyl ether*

solvent, coupling: cutting oils
Butyl Dioxitol®

solvent, coupling: detergents
Butoxytriglycol

solvent, coupling: dyes
Butoxydiglycol

solvent, coupling: electronics
Propylene glycol t-butyl ether

solvent, coupling: gasket cement
Pale 4

solvent, coupling: gums
Butoxydiglycol

solvent, coupling: hard-surface cleaners
Propylene glycol t-butyl ether

solvent, coupling: high-based enamels
Butoxydiglycol

solvent, coupling: household cleaners
Butoxydiglycol

solvent, coupling: hydraulic fluids
#30 Oil; #40 Oil; Pale 4

solvent, coupling: industrial lubricants
#30 Oil; #40 Oil; Pale 4

solvent, coupling: inks
#30 Oil; #40 Oil; Pale 4; *Propylene glycol t-butyl ether*

solvent, coupling: lacquers
Butoxytriglycol

solvent, coupling: leather dressing
#30 Oil; #40 Oil; Pale 4

solvent, coupling: light-duty cleaners
Propylene glycol t-butyl ether

solvent, coupling: liq. cleaners
Butoxydiglycol

solvent, coupling: metalworking fluids
Isodecanoic acid; *Neodecanoic acid*

solvent, coupling: nitrocellulose
Butoxydiglycol

solvent, coupling: oils
Butoxydiglycol

solvent, coupling: paints
Butoxytriglycol

solvent, coupling: pesticides
Tetrahydrofurfuryl alcohol

solvent, coupling: polishes
#30 Oil; #40 Oil; Pale 4

solvent, coupling: polyesters
Propylene glycol t-butyl ether

solvent, coupling: polymers
Butoxydiglycol

solvent, coupling: resin/water systems
Eastman® EP; *Ethylene glycol propyl ether*

solvent, coupling: rubber
#30 Oil; #40 Oil; Pale 4

solvent, coupling: soaps
Butoxydiglycol

solvent, coupling: sol'n. coatings
Diethylene glycol propyl ether

solvent, coupling: soluble oils
2-Aminobutanol; *Butoxytriglycol*; *Ethoxyethanol*

solvent, coupling: stamp pad inks
Butoxydiglycol

solvent, coupling: textile auxiliaries
Butoxydiglycol; *Tetrahydrofurfuryl alcohol*

solvent, coupling: textiles
Propylene glycol t-butyl ether

solvent, coupling: water-dilutable coatings
Diethylene glycol propyl ether

solvent, coupling: wax emulsions
2-Aminobutanol

solvent, CR
Diethylene glycol propyl ether; *Propyl acetate*; TBP; *Tetrahydrofurfuryl alcohol*; *1,1,2-Trichloroethane*

solvent, creams
Crodamol PMP; Drakeol® 7; *PEG-6 abietate*

solvent, creams/lotions
Butyl octyl benzoate; Carbowax® Sentry® PEG 300; Carbowax® Sentry® PEG 400; Carbowax® Sentry® PEG 540 Blend; Carbowax® Sentry® PEG 600; Carbowax® Sentry® PEG 900; Carbowax® Sentry® PEG 1000; Carbowax® Sentry® PEG 1450; Carbowax® Sentry® PEG 4600; Carbowax® Sentry® PEG 8000; Cropure® Almond; Dynacerin® 660; EmCon™ Olive; Emerest® 2452; Estasan™ GT 8-40 3578; Estasan™ GT 8-60 3575; Estasan™ GT 8-60 3580; Estasan™ GT 8-65 3577; Estasan™ GT 8-65 3581; Estasan™ GT 8-70 3579; HallBrite® BHB; HallStar™ BST; HallStar™ DOM; *Hexyldecyl benzoate*; Labrafac® Lipophile; Miglyol® 810; Miglyol® 812 N; Miglyol® 840; Miglyol® 8810; Naetex-LAM; *Octyldodecyl isostearate*; Penreco Snow; *Petroleum distillates*; *Polyglyceryl-3 diisostearate*; STEPAN® IPM-NF; *Tricaprin*

solvent, creosote
Aromasol® 12; Petrinex® ASB

solvent, cresol removal: skin burns
Lutrol® E 300; Lutrol® E 400; Lutrol® E 600

solvent, critical chromatography
Water

solvent, crop protection compds.
Shellsol® D70

solvent, crude rubber
Cyclohexyl acetate

solvent, cured cyanoacrylate instant glue dissolving
Dynasolve M-35

solvent, cured epoxies
Dynasolve 711; Dynasolve 750; Dynasolve MP-500

solvent, cured epoxy removal
Dynasolve 180; Dynasolve 185; NMP

solvent, cured urethane dissolving
Dynasolve MP-AL

solvent, cured urethane removal
Dynasolve 190; Uresolve Blue

solvent, cutback bitumens
Kerosene

solvent, cuticles
Potassium hydroxide

solvent, cutting fluids
Isane IP 130; Isane IP 155; Isane IP 165; Isane IP 175; Isane IP 185; *1,1,1-Trichloroethane*

solvent, cutting oils
Dipropylene glycol; *Sodium xylenesulfonate*

solvent, dammar
Isopropyl Oxitol

solvent, dark colors
Arosolve CZU

solvent, DDT
1-Chloronaphthalene

solvent, decaffeinating coffee
Perchloroethylene

solvent, decapsulation
Dynasolve 180; Dynasolve 185; Dynasolve 210; Dynasolve 220; Dynasolve 225; Dynasolve 230; Dynasolve 711; Dynasolve 750; NMP; Uresolve Plus 500; Uresolve Plus SG

solvent, decapsulation: adhesive removal
Dynasolve 165

solvent, decapsulation: conformal coating removal
Dynasolve 165

solvent, decapsulation: cured acrylics
Dynasolve 100

solvent, decapsulation: deflashing
Dynasolve 165

solvent, decapsulation: depotting
Dynasolve 165

solvent, decapsulation: dissolving cast epoxies
Dynasolve 165

solvent, decapsulation: epoxies
Dynasolve 100

solvent, decapsulation: paint stripping
Dynasolve 165

solvent, decapsulation: polymer removal
Dynasolve 165

solvent, decapsulation: urethanes
Dynasolve 100

solvent, decorative cosmetics
Finsolv® TN-O

solvent, deflashing
Dynasolve 180; Dynasolve 185; NMP; Uresolve Plus SG

solvent, defoliants
n-Butyl mercaptan

solvent, degreasers
1,2-Butylene carbonate; *C13-14 isoparaffin*; Conosol 215; Conosol 260; Conosol 340; Conosol C-145; Conosol C-170; Conosol C-200; Dynasolve XD 22-1; Exxsol® D 40; Exxsol® D 60; Exxsol® D 95; Exxsol® D 110; Exxsol® D 130; Exxsol® D 3135; Fomblin® Perfluorosolv PFS-1; GAFsolver™ TMP; Hydroseal G232H; Hydrosol 75/95 N; Hydrosol 100/130 N; Hydrosol Essence F<5; Isane IP 155; Isane IP 165; Isane IP 175; Isane IP 185; Isopar® G Fluid; Isopar® H Fluid; Isopar® K Fluid; Isopar® L Fluid; Isopar® M Fluid; Isopar® V Fluid; Jeffsol® 1,2-

Butylene Carbonate; Jeffsol® EC-25; Jeffsol® EC-50; Jeffsol® EC-75; Ketrul 212; Ketrul D80; Ketrul D211; Ketrul D220; Ketrul HT; NMP-EL; Norpar® 12 Fluid; Norpar® 13 Fluid; Norpar® 14 Fluid; Norpar® 15 Fluid; N-Terpinal RS-50; *PEG-4 oleate*; Penreco 150-B Solvent; Penreco 170 Exempt Solvent; *Propyl alcohol*; SBP 140/165; Shellsol® D25; *Sodium xylenesulfonate*; Solvarex 10; Solvarex 90/160; Solvaro AFD; Spirdane BT; Spirdane D30; Spirdane D30L; Spirdane D40; Spirdane D40L; Spirdane D60; Spirdane D60L; Spirdane HT; Spirdane K1; Spirdane L1; *1,3,5-Trichlorobenzene*

solvent, degreasing processes
Chlorobenzene

solvent, degreasing: adhesives/ sealants
AG 101

solvent, degreasing: dry cleaning
PPG Perchlor

solvent, degreasing: high speed cleaning
Protene® SC

solvent, degreasing: hot zone cleaning
Protene® SC

solvent, degreasing: industrial
SBP 100/140

solvent, degreasing: inks
AG 101

solvent, degreasing: manual precision cleaning
Protene® SC

solvent, degreasing: metals
Perchloroethylene

solvent, degreasing: paints
AG 101

solvent, degreasing: textile grease spot removal
AG 101

solvent, dehydration: cellulosics
SD alcohol 1; *SD alcohol 2-B*; *SD alcohol 3-A*; *SD alcohol 3-C*

solvent, dehydration: ethyl cellulose compds.
SD alcohol 32

solvent, dehydration: sodium hydrosulfite
SD alcohol 1; *SD alcohol 2-B*; *SD alcohol 3-A*; *SD alcohol 3-C*

solvent, deinking
Dynasolve 180; Dynasolve 185; Dynasolve 711; Dynasolve 750; NMP; Uresolve Plus 500

solvent, delacquering
Polyglycol DME 200

solvent, delivery/absorption enhancement
Capmul® MCM; Captex® 350

solvent, dental adhesives
Penreco Snow

solvent, dentifrices
Carbowax® Sentry® PEG 600; *SD alcohol 37*; *SD alcohol 38-B*; *SD alcohol 38-C*; *SD alcohol 38-D*

solvent, deodorants
DPG lo; Finsolv® PG-22; HallStar™ BST; HallStar™ DOA; *Hexyldecyl benzoate*; *Isopropyl palmitate*; Renex® PEG 300; Renex® PEG 400; Renex® PEG 600; *SD alcohol 23-A*

solvent, deposition
Genesolv® SF; Genesolv® S-TZ; Novec™ HFE-7200

solvent, depotting
Dynasolve 210; Uresolve Blue; Uresolve Plus SG

solvent, dermatologicals
Capmul® MCM; Captex® 350; *Propylene glycol dioctanoate*; Solulan® 16; Solulan® 75; Solulan® L-575

solvent, desulfurization
Diisopropanolamine

solvent, detergents
Ethylene glycol diethyl ether; GBL; Jarcol™ I-16; Jarcol™ I-24; Naxol®; Punctilious® Ethyl Alcohol 190 Proof Pure USP; Punctilious® Ethyl Alcohol 200 Proof Pure USP; Punctilious® SDA 1-1 190 Proof; Punctilious® SDA 1-1 Anhydrous; Punctilious® SDA 1-2 190 Proof; Punctilious® SDA 1-2; Punctilious® SDA 2B-2 190 Proof; Punctilious® SDA 2B-2 Anhydrous; Punctilious® SDA 2B-3 190 Proof; Punctilious® SDA 2B-4 190 Proof; Punctilious® SDA 2B-4 Anhydrous; Punctilious® SDA 3A 190 Proof; Punctilious® SDA 3A Anhydrous; Punctilious® SDA 3C 190 Proof; Punctilious® SDA 12A-3 190 Proof; Punctilious® SDA 12A-3 Anhydrous; Punctilious® SDA 13A 190 Proof; Punctilious® SDA 19 190 Proof; Punctilious® SDA 20 190 Proof; Punctilious® SDA 20 Anhydrous; Punctilious® SDA 23A 190 Proof; Punctilious® SDA 23A Anhydrous; Punctilious® SDA 23H 190 Proof; Punctilious® SDA 25A-1 190 Proof; Punctilious® SDA 25A-2 190 Proof; Punctilious® SDA 29-2 190 Proof; Punctilious® SDA 29-2 Anhydrous; Punctilious® SDA 29-3 190 Proof; Punctilious® SDA 30 190 Proof; Punctilious® SDA 30 Anhydrous; Punctilious® SDA 32 190 Proof; Punctilious® SDA 32 Anhydrous

solvent, dewatering
Fomblin® Perfluorosolv PFS-1

solvent, dewatering fluids
Shellsol® D60 SC

solvent, dewaxing
Ketrul 212; Ketrul D80; Ketrul D211; Ketrul D220; Ketrul HT

solvent, dewaxing process
Ethylene dichloride

solvent, dichloramine-T
Paraffin, chlorinated

solvent, dielectric fluid
1,2,4-Trichlorobenzene

solvent, diesel fuel
QO® Furfural

solvent, dietetic prods.
Delios®; Liponate GC-K; Liponate PC; Miglyol® 840

solvent, diethyl benzene
Ethylbenzene

solvent, diluent: carbonless copy paper
Norpar® 12 Fluid; Norpar® 13 Fluid; Norpar® 15 Fluid

solvent, diluent: spark erosion machining
Norpar® 12 Fluid; Norpar® 13 Fluid; Norpar® 15 Fluid

solvent, dip cleaning
1,1,1-Trichloroethane

solvent, dip coatings
Shellsol® X7B

solvent, disinfectants
t-Butyl alcohol; *Diethylene glycol benzyl ether*; Ecosoft™ PB Solvent; Ecosoft™ PE Solvent; Ecosoft™ PH Solvent; *Ethylene glycol benzyl ether*; Parabase; Punctilious® Ethyl Alcohol 190 Proof Pure USP; Punctilious® Ethyl Alcohol 200 Proof Pure USP; Punctilious® SDA 1-2 190 Proof; Punctilious® SDA 1-2; Punctilious® SDA 2B-2 190 Proof; Punctilious® SDA 2B-2 Anhydrous; Punctilious® SDA 2B-3 190 Proof; Punctilious® SDA 2B-4 190 Proof; Punctilious® SDA 2B-4 Anhydrous; Punctilious® SDA 3A 190 Proof; Punctilious® SDA 3A Anhydrous; Punctilious® SDA 3C 190 Proof; Punctilious® SDA 12A-3 190 Proof; Punctilious® SDA 12A-3 Anhydrous; Punctilious® SDA 13A 190 Proof; Punctilious® SDA 19 190 Proof; Punctilious® SDA 20 190 Proof; Punctilious® SDA 20 Anhydrous; Punctilious® SDA 23A 190 Proof; Punctilious® SDA 23A Anhydrous; Punctilious® SDA 23H 190 Proof; Punctilious® SDA 25A-1 190 Proof; Punctilious® SDA 25A-2 190 Proof; Punctilious® SDA 29-2 190 Proof; Punctilious® SDA 29-2 Anhydrous; Punctilious® SDA 29-3 190 Proof; Punctilious® SDA 30 190 Proof; Punctilious® SDA 30 Anhydrous; Punctilious® SDA 32 190 Proof; Punctilious® SDA 32 Anhydrous; *SD alcohol 1*; *SD alcohol 3-A*; *SD alcohol 3-B*; *SD alcohol 3-C*; *SD alcohol 23-A*; *SD alcohol 23-H*; *SD alcohol 27-A*; *SD alcohol 27-B*; *SD alcohol 30*; *SD alcohol 37*; *SD alcohol 38-B*; *SD alcohol 39-B*; *SD alcohol 40*; *SD alcohol 40-A*; *SD alcohol 40-B*; *SD alcohol 40-C*

solvent, disk lube deposition applics.
Novec™ HFE-7100DL; Novec™ HFE-7200DL

solvent, disk precleaning: pre-lubrication deposition
Novec™ HFE-7100DL

solvent, dopes
Diacetone alcohol; *Dibutyl tartrate*

solvent, drilling
Shellsol® D80

solvent, drilling muds
Shellsol® D70

solvent, drug delivery systems
Crodamol EO; Labrafil® Isostearique

solvent, dry cleaning
Butoxyethanol; Condea Vista LPA; *1,4-Dioxane*; Dowper; HC-DCF Low Flash™ Dry Cleaning Solvent; Hydrosol Essence F<5; *Mineral spirits*; *Naphtha, medium aliphatic*; Naxol®; *Nonoxynol-9 phosphate*; *Perchloroethylene*; PPG Perchlor; Rhodacal® 330; Shell OMS; Shell Mineral Spirits 7 EC; Shell Mineral Spirits 135; Shellsol® D38; Shellsol® D40; Shellsol® D43; Shellsol® D60; Shellsol® D70; Shellsol® D80; Shellsol® T; Shell Toluene Solvent; Shell Xylene; *Stoddard solvent*; *1,1,1-Trichloroethane*; *Trichloroethylene*; *Trichlorofluoromethane*; *Trichlorotrifluoroethane*; Variquat® CC-57; VM&P Naphtha EC

solvent, dry cleaning fluids
Butyl-(S)-lactate; *Naphtha*

solvent, dry skin care
Propylene glycol dioctanoate

solvent, drying
Forane 141b DGX; *Lithium aluminum hydride*

solvent, duplicating fluids
SD alcohol 30

solvent, dusting aids
Neosolv™ 2; Neosolv™ 4; Neosolv™ 6; Neosolv™ 8; Neosolv™ 24; Neosolv™ 150

solvent, dusting powders
STEPAN® IPM-NF

solvent, dye carriers
1,2,4-Trimethylbenzene

solvent, dye carriers: printing
Monochlorotoluene

solvent, dye carriers: textiles
Monochlorotoluene

solvent, dye extraction
cis-trans-1,2-Dichloroethylene (mixed isomers)

solvent, dyes
Alcohol; Alkylate L210-L; BDGE; *Benzyl alcohol*; BGE; *1-Butoxyethoxy-2-propanol*; *1,2-Butylene carbonate*; *Butyl lactate*; *Butyl-(S)-lactate*; Butyl Oxitol®; *Cyclohexanol*; *Diacetone alcohol*; *Diethylene glycol*; *Diethylenetriamine*; *1,4-Dioxane*; *Dipropylene glycol ethyl ether*; EDGE; EGE; *Ethoxydiglycol*; *2-Ethyl-1-butanol*; *Ethylene glycol acetate*; *2-Ethylhexanol*; *Ethyl morpholine*; *Furfuryl alcohol*; GBL; *Glyceryl dioleate*; *Hexyldecanol*; Hisolve DM; i-PGE; *1-Isopropoxy-2-propanol*; *2-Isopropoxy-1-propanol*; *Isopropylamine*; *Isopropyl ether*; Jeffox® WL-440; Jeffox® WL-660; Jeffox® WL-1400; Jeffox® WL-5000; Jeffsol® 1,2-Butylene Carbonate; Jeffsol® EC; Jeffsol® EC-25; Jeffsol® EC-50; Jeffsol® EC-75; Jeffsol® PC; MDGE; *2-Mercaptoethanol*; *Methoxydiglycol*; *Methoxyethanol*; *Methyl alcohol*; Methyl Diproxitol; MGE; *Morpholine*; *1-Nitropropane*; *2-Nitropropane*; NMP; *Phenoxyethanol*; *β-Picoline*; *γ-Picoline*; Pluracol® SRF-2; Pluracol® W3520N; Pluracol® WI 285; Pluracol® WI 625; Pluracol® WI 1715; Pluracol® WS 170; Pluracol® WS 260; Pluracol® WS 660; Pluracol® WS 2000; Pluracol® WS 3520; Pluracol® WS 5100; Pluracol® WT 1400; Pluracol® WT 9150; Pluracol® WT 90,000; *Polysorbate 60*; *PPG-2-buteth-3*; *PPG-3-buteth-5*; *PPG-5-buteth-7*; *PPG-7-buteth-10*; *PPG-9-buteth-12*; *PPG-12-buteth-16*; *PPG-20-buteth-30*; *PPG-28-buteth-35*; *PPG-33-buteth-45*; *PPG-5 butyl ether*; *PPG-9 butyl ether*; *PPG-15 butyl ether*; *PPG-18 butyl ether*; *PPG-22 butyl ether*; *PPG-24 butyl ether*; *PPG-33 butyl ether*; *PPG-40 butyl ether*; *PPG-53 butyl ether*; *PPG-2 methyl ether*; *PPG-3 methyl ether*; *Propylene carbonate*; Propylene Glycol USP/FCC Ultra Grade; *Propylene glycol phenyl ether*; *Propylene glycol ricinoleate*; *SD alcohol 23-A*; *Sodium xylenesulfonate*; *Tetradecene-1*; *Tetraethanol ammonium hydroxide*; *Tetraethylenepentamine*; *Triacetin*; *1,2,4-Trichlorobenzene*; *1,3,5-Trichlorobenzene*; *Triethanolamine*;

Triethylene glycol; *Tripropylene glycol ethyl ether*; Ucon® 50-HB-55; Ucon® 50-HB-100; Ucon® 50-HB-260; Ucon® 50-HB-400; Ucon® 50-HB-660; Ucon® 50-HB-2000; Ucon® 50-HB-3520; Ucon® 50-HB-5100; Ucon® 75-H-450; Ucon® LB-65; Ucon® LB-135; Ucon® LB-165; Ucon® LB-285; Ucon® LB-385; Ucon® LB-525; Ucon® LB-625; Ucon® LB-1145; Ucon® LB-1715; Ucon® LB-3000

solvent, dyes/pigments
Solvenon® PC

solvent, dyes: creams/lotions
HallStar™ BST

solvent, dyes: deodorants
HallStar™ BST

solvent, dyes: leather
BDGE; BGE; EDGE; EGE; i-PGE; MDGE; MGE

solvent, dyes: lipstick
Ceraphyl® 28; Crodamol DA; Krystal Kleer Castor Oil; USP Castor Oil

solvent, dyes: makeup
HallStar™ BST

solvent, dyes: makeup removers
HallStar™ BST

solvent, dyes: personal care
Krystal Kleer Castor Oil; USP Castor Oil

solvent, dyes: pharmaceuticals
Krystal Kleer Castor Oil; USP Castor Oil

solvent, dyes: preshave preps.
HallStar™ BST

solvent, dyes: printing
BDGE; BGE; EDGE; EGE; i-PGE; MDGE; MGE

solvent, dyes: textile finishing
Glyezin® A

solvent, dyes: textiles
Acetamide; BDGE; BGE; EDGE; EGE; i-PGE; MDGE; MGE

solvent, dyes: urethane applics.
Krystal Kleer Castor Oil; USP Castor Oil

solvent, dyes: vinyl resins
Ucon® 50-HB-55; Ucon® 50-HB-100; Ucon® 50-HB-260; Ucon® 50-HB-400; Ucon® 50-HB-660; Ucon® 50-HB-2000; Ucon® 50-HB-3520; Ucon® 50-HB-5100; Ucon® 75-H-450; Ucon® LB-65; Ucon® LB-135; Ucon® LB-165; Ucon® LB-285; Ucon® LB-385; Ucon® LB-525; Ucon® LB-625; Ucon® LB-1145; Ucon® LB-1715; Ucon® LB-3000

solvent, dyes: wax polishes
Butyl stearate

solvent, e-coat systems
N-Terpinal RS-75

solvent, elastomer-based adhesives
Isopar® V Fluid

solvent, elastomers
Dibutyl phthalate; *Dibutyl tartrate*; *Hexachloro-1,3-butadiene*

solvent, elastomers: chewing gum
Terpene resin

solvent, elec. connector repair
Dynasolve 190

solvent, elec. equip.
HC-250; HC-370; HC-FX50; HC-FX70; *1,1,1-Trichloroethane*

solvent, elec. part cleaning
N-Terpinal RS-GEL

solvent, elec. preshaves
Lexol® IPM-NF

solvent, elec. varnishes
Rhodiasolv RPDE

solvent, electric-discharge machining
Hydroseal G250H

solvent, electrochemistry
Butyl Diglyme; *Diethylene glycol dibutyl ether*; Ethyl Diglyme; *Ethylene glycol dimethyl ether*; Ethyl Glyme; Monoglyme; *PEG-4 dimethyl ether*; Tetraglyme; Triglyme; UltraPure™ Propylene Carbonate

solvent, electrodeposition coatings
Texanol® Ester-Alcohol; *2,2,4-Trimethyl-1,3-pentanediol isobutyrate*

solvent, electrodipping varnishes
Solvenon® PP

solvent, electro-discharge machining
Hydroseal G232H

solvent, electrolytes
Dimethyl acetamide; Jeffsol® EC; Jeffsol® PC

solvent, electrolytic capacitors
Dimethyl formamide

solvent, electronic chemicals
Arcosolv® DPM; Arcosolv® PM; Arcosolv® TPM; *PPG-2 methyl ether*; *PPG-3 methyl ether*; *PPG-2 methyl ether acetate*

solvent, electronic cleaning
1,1´-Biphenyl, butenylated; VersaTRANS™ Solvent

solvent, electronic component cleaning
Genesolv® D; PPG Trichlor

solvent, electronic components
1,1,1-Trichloroethane

solvent, electronic conformal coating removal
Uresolve Blue

solvent, electronic depotting
Dynasolve 100; Dynasolve 180; Dynasolve 185; Dynasolve 190; Dynasolve 220; Dynasolve 225; Dynasolve 230; Dynasolve 699; Dynasolve 711; Dynasolve 750; NMP; Uresolve Plus 500

solvent, electronic depotting/ decapsulation
Decap; Uresolve HF

solvent, electronic materials
Acetic acid; *Acetone*; *Butoxydiglycol*; *n-Butyl acetate*; *Diethylene glycol butyl ether acetate*; *Diethylene glycol propyl ether*; *Methoxyethanol*

solvent, electronic parts
1,1´-Biphenyl, butenylated; *1,1´-Biphenyl (1-methylethyl)*; *Bis (methylethyl)-1,1´-biphenyl*

solvent, electronic parts cleaning
1,1´-Biphenyl (1-methylethyl); *Bis (methylethyl)-1,1´-biphenyl*

solvent, electronics
Arcosolv® DPMA; Arcosolv® DPNB; Arcosolv® PE; Arcosolv® PEA; Arcosolv® PMA; Arcosolv® PNB; Arcosolv® PNP; Arcosolv® PTB; *Butoxypropanol*; *Butyl lactate*; *p-Chlorobenzotrifluoride*; *Dipropylene glycol butyl ether*; DMSO; *Ethyl lactate*; GBL; HC-250; HC-370; HC-FX50; HC-FX70; *Isopropyl lactate*; *Methylene chloride*; *Methyl lactate*; NMP-EL; Oxsol® 100; *n-Propyl bromide*; *Propylene glycol ethyl ether acetate*; *Propylene glycol methyl ether acetate*; Purasolv® BL; Purasolv® EL; Purasolv® ELECT; Purasolv® IPL; Purasolv® ML

solvent, electronics cleaners
Proglyde DMM

solvent, electroplating
Carbowax® PEG 300; Polyglycol DME 500; Polyglycol DME 1000; Polyglycol DME 2000

solvent, electropolishing
Carbowax® PEG 300

solvent, electrostatic applics.
Solvesso™ 100; Solvesso™ 150; Solvesso™ 150 ND; Solvesso™ 200

solvent, electrostatically sprayed coatings
Ucar® n-Propyl Propionate

solvent, embalming fluids
SD alcohol 1; *SD alcohol 3-A*; *SD alcohol 3-C*; *SD alcohol 22*; *SD alcohol 23-A*

solvent, EMS
CHP®; *N-Cyclohexyl pyrrolidone*

solvent, emulsifiable concs.
Agsolex® BLO

solvent, emulsifiers
Butyl benzoate; Jarcol™ I-16; Jarcol™ I-24

solvent, emulsion polymerization
Rhodacal® 330

solvent, emulsion-bonding adhesives
Longifolene

solvent, emulsions
Crodamol PMP; *Erucamide*; Hetester PMA; *Propylene glycol myristyl ether acetate*

solvent, emulsions, vinyl coatings
Isobutyl heptyl ketone

solvent, enamel finishing coats
Rhodiasolv RPDE

solvent, enamels
BDGE; BGE; *Butoxyethanol*; Butyl Dioxitol®; Eastman® EEP; Eastman® EP; EDGE; EGE; *Ethoxydiglycol*; *Ethyl-(S)-lactate*; i-PGE; *dl-Limonene*; MDGE; *Methoxyethanol*; MGE; Santosol® DMA; Santosol® DME; Santosol® DME-2; Santosol® DME-3; Santosol® DMG; Santosol® DMS; Shellsol® X7B; *Xylene*

solvent, engine degreasers
Exxsol® D 80; Neosolv™ 2; Neosolv™ 4; Neosolv™ 6; Neosolv™ 8; Neosolv™ 24; Neosolv™ 150

solvent, Epon resins
Butyl Dioxitol®

solvent, epoxides
Propylene glycol dimethyl ether

solvent, epoxies
Diethylene glycol propyl ether; Eastman® DE; Eastman® EB; Eastman® PM Acetate; Eastman® Methyl Acetate, High Purity; Isopropyl Oxitol; Jeffsol® PC; Kenplast® A-450; *Methoxybutyl acetate*

solvent, epoxy cleaning
Dynasolve M-35

solvent, epoxy coatings
t-Amyl alcohol; BDGE; BGE; EDGE; EGE; *Ethoxyethanol*; i-PGE; MDGE; *Methyl hexyl ketone*; MGE

solvent, epoxy laminates
Arcosolv® DPM; Arcosolv® DPMA; Arcosolv® PM; Arcosolv® TPM; *PPG-2 methyl ether*; *PPG-3 methyl ether*; *PPG-2 methyl ether acetate*

solvent, epoxy powder coatings
Dynasolve 699

solvent, epoxy removal
N-Terpinal RS-50

solvent, epoxy sealing coats
Rhodiasolv RPDE

solvent, epoxy surface primers
Santosol® DMA; Santosol® DME; Santosol® DME-2; Santosol® DME-3; Santosol® DMG; Santosol® DMS

solvent, epoxy-contg. systems
Butoxyl®

solvent, equip. cleaning
N-Terpinal RS-25; N-Terpinal RS-GEL

solvent, equip. resin removal
GAFsolver™ TMP

solvent, equip./parts
N-Terpinal RS-75

solvent, essential oil extraction: perfumes
Hydrosol Hexane

solvent, essential oils
Cremophor® EL; *Isopropyl alcohol*; *Terpinolene*

solvent, ester gums
dl-Limonene; *Methyl abietate*

solvent, esters: printing inks
Emery® 6705

solvent, ethyl acetate
Cellosolve®

solvent, ethyl cellulose high-solids coatings
Butoxyethanol acetate

solvent, ethyl cellulose paints/ coatings
Ethoxyethanol acetate

solvent, ethyl lactate
Lactonitrile

solvent, ethylcellulose
s-Amyl acetate; Butyl Cellosolve®; *Butyl lactate*; *Butyl-(S)-lactate*; Cellosolve® Acetate; Emery® 6705; *Isopropyl acetate*; Isopropyl Oxitol; *Methoxyethanol*; *Methyl butyrate*; Methyl Cellosolve®; Methyl Cellosolve® Acetate; *Phenoxyethanol*; Propyl Cellosolve®; Ucar® Ester EEP

solvent, ethylcellulose acrylic coatings
Butyl Cellosolve® Acetate

solvent, EVA latexes
Texanol® Ester-Alcohol

solvent, excipients: liq. pharmaceuticals
Lutrol® E 300; Lutrol® E 400; Lutrol® E 600

solvent, excipients: semisolid pharmaceuticals
Lutrol® E 300; Lutrol® E 400; Lutrol® E 600

solvent, expandable PS
Hydrosol Pentane

solvent, explosive mixture prod.
Dibromodifluoromethane

solvent, explosives
Dibutyl phthalate; Punctilious® Ethyl Alcohol 190 Proof Pure USP; Punctilious® Ethyl Alcohol 200 Proof Pure USP; Punctilious® SDA 1-2 190 Proof; Punctilious® SDA 1-2; Punctilious® SDA 2B-2 190 Proof; Punctilious® SDA 2B-2 Anhydrous; Punctilious® SDA 2B-3 190 Proof; Punctilious® SDA 2B-4 190 Proof; Punctilious® SDA 2B-4 Anhydrous; Punctilious® SDA 3A 190 Proof; Punctilious® SDA 3A Anhydrous; Punctilious® SDA 3C 190 Proof; Punctilious® SDA 12A-3 190 Proof; Punctilious® SDA 12A-3 Anhydrous; Punctilious® SDA 13A 190 Proof; Punctilious® SDA 19 190 Proof; Punctilious® SDA 20 190 Proof; Punctilious® SDA 20 Anhydrous; Punctilious® SDA 23A 190 Proof;

Acetyldimethylamine chloride; *Dimethyl acetamide*; Jeffsol® 1,2-Butylene Carbonate; Jeffsol® EC; Jeffsol® EC-25; Jeffsol® EC-50; Jeffsol® EC-75; Jeffsol® PC; M-Pyrol®; *N-Nitrosodiethylamine*; *N-Nitrosodimethylamine*

solvent, film
PEG 300; Shellsol® D90

solvent, fine fragrances
Finsolv® TN-O

solvent, fingernail polishes
Uniplex 150

solvent, fixture coatings
Arcosolv® TPNB

solvent, flame retardants
Variquat® CC-57

solvent, flash point reduction
Fomblin® Perfluorosolv PFS-1

solvent, flavors
Aldo® MCT KFG; *Benzyl alcohol*; Capmul® MCM; Capmul® MCMC8; Capmul® MCMC10; *Capric triglyceride*; Captex® 200; Delios® S; Eastman® Ethyl Acetate, Food Grade; Estasan™ GT 8-40 3578; Estasan™ GT 8-60 3575; Estasan™ GT 8-60 3580; Estasan™ GT 8-65 3577; Estasan™ GT 8-65 3581; Estasan™ GT 8-70 3579; *Ethyl-(S)-lactate*; *Isopropyl myristate*; Lexol® GT-855; Lexol® PG-800; Lexol® PG-865; Liponate GC-K; Liponate PC; *Propylene glycol dicaprylate/dicaprate*; *Propylene glycol dioctanoate*; *SD alcohol 4*; *Tetradecene-1*; Unichem BZBN

solvent, flavors: food
Butylene glycol; Captex® 500

solvent, flavors: pharmaceuticals
Butylene glycol; Captex® 500

solvent, floor cleaners
Norpar® 15 Fluid; N-Terpinal RS-25; N-Terpinal RS-98

solvent, floor polishes
Arcosolv® DPM; Arcosolv® DPMA; Arcosolv® PM; Arcosolv® TPM; *Deodorized kerosene*; Exxsol® D 95; Isopar® M Fluid; Neosolv™ 2; Neosolv™ 4; Neosolv™ 6; Neosolv™ 8; Neosolv™ 24; Neosolv™ 150; Norpar® 15 Fluid; *PPG-2 methyl ether*; *PPG-3 methyl ether*; *PPG-2 methyl ether acetate*; *Triglycol monomethyl ether*

solvent, floor wax strippers
Tripropylene glycol n-butyl ether

solvent, floor waxes
Coasol™; *Decahydronaphthalene*; Decalin; Neosolv™ 2; Neosolv™ 4; Neosolv™ 6; Neosolv™ 8; Neosolv™ 24; Neosolv™ 150; *Triglycol monomethyl ether*

solvent, floor waxes/polishes
Poly-Solv® DM; Poly-Solv® DPM; Poly-Solv® EM; Poly-Solv® MPM; Poly-Solv® TM; Poly-Solv® TPM

solvent, flotation: coal separation
Perchloroethylene

solvent, flue gas cleaning
Polyglycol DME 200

solvent, fluorinated hydrocarbons
Agrisynth® BLO; BLO®; *Butyrolactone*

solvent, fluorocarbon absorption
PEG dimethyl ether, Polyglyme

solvent, flushing
Genesolv® SF; Genesolv® S-TZ

solvent, flux removal
Forane 141b DGX; Forane 141b MGX

solvent, foam control agents
Carbowax® E200; Carbowax® E300; Carbowax® E400; Carbowax® E600; Carbowax® E900; Carbowax® E1000; Carbowax® E1450; Carbowax® E3350; Carbowax® E8000

solvent, food
Acetic acid; *Acetylated mono- and diglycerides of fatty acids*; Axol® E 61; *Benzyl benzoate*; *Bornyl acetate*; *Caprylic alcohol*; Captex® 200; Condea Vista LPA; Condea Vista LPA-142; Crodamol PC-PN; Dow Glycerine 96% USP; Dow Glycerine 99.5% USP/EP; Eastman® Ethyl Acetate, Food Grade; Emery® 912; Emery® 916; Emery® 917; Estasan™ GT 8-40 3578; Estasan™ GT 8-60 3575; Estasan™ GT 8-60 3580; Estasan™ GT 8-65 3577; Estasan™ GT 8-65 3581; Estasan™ GT 8-70 3579; *Glycerin*; *Glyceryl oleate*; *Glyceryl stearate*; Hodag® S-35 FGK; *Lactic acid*; Lipovol SES; Mazol® 300 K; *Mono- and diglycerides of fatty acids*; Myvacet® 9-45K; Optim™ Glycerine 99.7; Orange Terpenes; *PEG-8 oleate*; *PEG-20 oleate*; Penreco Snow; Polarin® Range; Pricerine™ 9091; Pricerine™ 9099; *Propylene glycol*; Propylene Glycol USP, FCC; Propylene Glycol USP/FCC Ultra Grade; *Propylene glycol alginate*; *Propylene glycol laurate*; *Sesamol*; Shellsol® D100S; *Triacetin*

solvent, food coloring
Ammonium hydroxide

solvent, food pkg.
Glycerin; *Glyceryl oleate*; Isopar® C

solvent, food pkg. adhesives
Morcryl® 535

solvent, food processing
Carbowax® PEG 300; Drakesol® 165AT; Drakesol® 205; *Methyl cottonseed*; Norfox® MCS; Shellsol® D70

solvent, food-contact adhesives
Sasol LPA Solvent; Sasol LPA-142 Solvent; Sasol LPA-150 Solvent; Sasol LPA-170 Solvent; Sasol LPA-210 Solvent

solvent, food-contact applics.
Eastman® PM Acetate; PD-23; PD-25; PD-28

solvent, food-contact coatings
Sasol LPA Solvent; Sasol LPA-142 Solvent; Sasol LPA-150 Solvent; Sasol LPA-170 Solvent; Sasol LPA-210 Solvent

solvent, food-contact paper
Sasol LPA Solvent; Sasol LPA-142 Solvent; Sasol LPA-150 Solvent; Sasol LPA-170 Solvent; Sasol LPA-210 Solvent

solvent, food-contact plastic materials
Citrofol® BI; Citrofol® BII

solvent, food-contact plastics
Citrofol® AII

solvent, food-contact rubber
Sasol LPA Solvent; Sasol LPA-142 Solvent; Sasol LPA-150 Solvent; Sasol LPA-170 Solvent; Sasol LPA-210 Solvent

solvent, food-contact textiles
Sasol LPA Solvent; Sasol LPA-142 Solvent; Sasol LPA-150 Solvent; Sasol LPA-170 Solvent; Sasol LPA-210 Solvent

solvent, food-pkg. coatings
Citroflex® A-2

solvent, food-pkg. inks
Propylene Glycol USP/FCC Ultra Grade

solvent, foodstuffs
Acetylated hydrogenated lard glyceride; Axol® E 66

solvent, foundry core binders
Estasol™; Santosol® DMA; Santosol® DME; Santosol® DME-2; Santosol® DME-3; Santosol® DMG; Santosol® DMS

solvent, foundry core resins
Rhodiasolv RPDE

solvent, foundry sand binders
Jeffsol® PC

solvent, fractionating edible fats/oils
2-Nitropropane

solvent, fragrance bases
Drakesol® 165; Drakesol® 165AT; Drakesol® 205; Drakesol® 220; Drakesol® 260; Drakesol® 260 AT; Drakesol® 305; *Petroleum distillates, hydrotreated light*

solvent, fragrance oils
Captex® 200; *Hexyldecanol*; *Propylene glycol dicaprylate/dicaprate*

solvent, fragrance-free prods.
Finsolv® TN-O

solvent, fragrances
Benzyl alcohol; Capmul® MCM; Capmul® MCMC8; Capmul® MCMC10; DPG lo; Eastman® DBP; Eastman® DE; Eastman® DEP, Fragrance Grade; Eastman® DPG, Fragrance Grade; Eastman® EAA; Eastman® MAA; Eastman® n-Butanol; Eastman® n-Butyl Acetate, Urethane Grade; Eastman® Ethyl Acetate, Food Grade; Eastman® Isobutyl Acetate; Eastman® Isopropanol; Eastman® Isopropyl Acetate; Eastman® Methyl Acetate, High Purity; Eastman® n-Propyl Acetate; Estasan™ GT 8-40 3578; Estasan™ GT 8-60 3575; Estasan™ GT 8-60 3580; Estasan™ GT 8-65 3577; Estasan™ GT 8-65 3581; Estasan™ GT 8-70 3579; HallStar™ DIPA; HallStar™ DOA; Lexol® PG-800; Lexol® PG-865; *Propylene glycol dioctanoate*

solvent, fragrances: cosmetics
Isobutyl acetate; *Methyl acetoacetate*

solvent, fragrances: food
Captex® 500

solvent, fragrances: personal care
Ethylacetoacetate

solvent, fragrances: pharmaceuticals
Captex® 500

solvent, fragrant polishes
Shellsol® T

solvent, frame drying
Arosolve KJ

solvent, freeze point depressants
Sasol MR Solvent

solvent, fresheners
HallStar™ DIPA; HallStar™ DIPS; HallStar™ DOS

solvent, frozen desserts
Hodag® S-35 FGK

solvent, fruit/veg. coatings
Penreco Snow

solvent, fuel additives
Arcosolv® DPM; Arcosolv® DPMA; Arcosolv® PM; Arcosolv® TPM; Eastman® EB; *PPG-2 methyl ether*, *PPG-3 methyl ether*

solvent, fuel oils
Monochlorotoluene

solvent, fuels
Butyl Diglyme; *Diethylene glycol dibutyl ether*, Ethyl Diglyme; *Ethylene glycol dimethyl ether*, Ethyl Glyme; Monoglyme; *PEG-4 dimethyl ether*, Tetraglyme; Triglyme

solvent, functional fluids
PPG-3 methyl ether

solvent, fungicides
Eastman® MIBK; *SD alcohol 3-A*; *SD alcohol 3-B*; *SD alcohol 3-C*; *SD alcohol 23-A*; *SD alcohol 23-H*; *SD alcohol 27-A*; *SD alcohol 27-B*; *SD alcohol 30*; *SD alcohol 37*; *SD alcohol 38-B*; *SD alcohol 39-B*; *SD alcohol 40*; *SD alcohol 40-A*; *SD alcohol 40-B*; *SD alcohol 40-C*

solvent, furniture polishes
Drakesol® 165; Drakesol® 220; Drakesol® 260; Drakesol® 260 AT; Drakesol® 305; Ethyl Proxitol; Isopar® M Fluid; Neosolv™ 2; Neosolv™ 4; Neosolv™ 6; Neosolv™ 8; Neosolv™ 24; Neosolv™ 150; Norpar® 13 Fluid; *Petroleum distillates, hydrotreated light*

solvent, furniture waxes
Neosolv™ 2; Neosolv™ 4; Neosolv™ 6; Neosolv™ 8; Neosolv™ 24; Neosolv™ 150

solvent, furniture/floor polishes
C13-14 isoparaffin

solvent, gas absorption
Butyl Diglyme; *Diethylene glycol dibutyl ether*, Ethyl Diglyme; *Ethylene glycol dimethyl ether*, Ethyl Glyme; Monoglyme; *PEG-4 dimethyl ether*, *Petroleum distillates, hydrotreated light paraffinic*; Tetraglyme; Triglyme

solvent, gas oil
QO® Furfural

solvent, gas sweetening
Diethylaminoethanol

solvent, gas treating
Dytek® EP Diamine; Jeffsol® PC; Jefftreat® M-504; Jefftreat® M-505; Jefftreat® M-507; Jefftreat® MS-205; Jefftreat® MS-205 LFG; *1,3-Pentanediamine*

solvent, gas treating: natural gases
Selexol®

solvent, gas treating: refinery gases
Selexol®

solvent, gases
Dimethyl formamide; *PEG-3 dimethyl ether*

solvent, gasoline
2,3-Dimethylbutane

solvent, gelatin
Benzyl alcohol

solvent, gels
Labrafac® Lipophile

solvent, general-purpose cleaners
Exxsol® D 95

solvent, general-purpose degreasers
Exxsol® D 95

solvent, general-purpose industrial applics.
Filmex® A-1, 190 Proof

solvent, general-purpose: dry cleaning
sym-Dichloroethyl ether

solvent, general-purpose: finish removers

Diethylene glycol diethyl ether; *Methyl amyl alcohol*; *Methyl isoamyl ketone*

solvent, high-boiling: nitrocellulose
Diethylene glycol diethyl ether; *Methyl amyl alcohol*; *Methyl isoamyl ketone*

solvent, high-boiling: oils
Methyl amyl alcohol

solvent, high-boiling: paints
Methyl isoamyl ketone

solvent, high-boiling: resins
Diethylene glycol diethyl ether; *Methyl amyl alcohol*

solvent, high-boiling: stoving finishes
Butyl glycolate

solvent, high-boiling: vinyl copolymers
Methyl isoamyl ketone

solvent, high-boiling: waxes
Methyl amyl alcohol

solvent, high-density: gauge fluids
Methylene bromide; *Tetrabromoethane*

solvent, high-density: mineral separation
Bromoform; *Methylene bromide*; *Methylene iodide*; *Tetrabromoethane*

solvent, high-purity: crystallization
Dimethyl acetamide

solvent, high-purity: purification
Dimethyl acetamide

solvent, high-solids
Dowanol® TPM

solvent, high-solids acrylic polymerization
Ethyl 3-ethoxypropionate

solvent, high-solids coatings
Amyl propionate; *Butoxyethanol acetate*; Butyl Carbitol®; Butyl Cellosolve®; Butyl Cellosolve® Acetate; *C8 alkyl acetate*; *C9 alkyl acetate*; *C10 alkyl acetate*; *C13 alkyl acetate*; Carbitol®; Exxate® 600; Exxate® 700; Exxate® 800; Exxate® 900; Exxate® 1000; Exxate® 1300; *Oxo-heptyl acetate*; *Oxo-hexyl acetate*; Propyl Cellosolve®; Ucar® Acetone; Ucar® Butanol; Ucar® Diacetone Alcohol; Ucar® Diisobutyl Ketone; Ucar® Ethanol; Ucar® 2-Ethylhexanol; Ucar® Isobutanol; Ucar® Isobutyl Acetate; Ucar® Isophorone; Ucar® Isopropanol; Ucar® Isopropyl Acetate; Ucar® Methyl Isobutyl Ketone; Ucar® 1-Pentanol; Ucar® Primary Amyl Acetate; Ucar® 1-Propanol; Ucar® n-Propyl Acetate; Ucar® n-Propyl Propionate

solvent, high-solids paints
Butoxyl®

solvent, high-solids systems
Purasolv® BL; Purasolv® ELECT

solvent, hops: beer brewing
Butyl alcohol; *Calcium chloride*

solvent, hot caustic paint strippers
Alkasurf® OD-410

solvent, hot degreasing equip.
Tavoxene®

solvent, household cleaners
Acetyl tributyl citrate; *Acetyl triethyl citrate*; *Acetyl trimethyl citrate*; Condea Vista LPA; Conosol 215; Conosol 260; Conosol C-145; Conosol C-170; Dowanol® TPM; Eastman® EB; Eastman® EP; *Ethylene glycol isopropyl ether*; Exxate® 1200; GBL; Herco® Pine Oil; Hydrosol Essence C; Isane IP 130; Isane IP 155; Isane IP 165; Isane IP 175; Isane IP 185; Isopropyl Oxitol; *Methoxydiglycol*; *Methoxyisopropanol*; Penreco 150-B Solvent; Penreco 170 Exempt Solvent; *Phenoxyethanol*; *Propylene glycol n-propyl ether*; Sasol LPA Solvent; Sasol LPA-142 Solvent; Sasol LPA-150 Solvent; Sasol LPA-170 Solvent; Sasol LPA-210 Solvent; SBP 80/95 LNH; SBP 80/110 LNH; SBP 100/140; SECOSTER® A; Shellsol® 426; Solvenon® DPM; Solvenon® DPnB; Solvenon® PnB

solvent, household degreasers
Exxate® 1200

solvent, household detergents
PEG-75 lanolin; *SD alcohol 1*; *SD alcohol 3-A*; *SD alcohol 3-C*; *SD alcohol 23-A*; *SD alcohol 23-H*; *SD alcohol 30*; *SD alcohol 36*; *SD alcohol 39-B*; *SD alcohol 40*; *SD alcohol 40-A*; *SD alcohol 40-B*; *SD alcohol 40-C*

solvent, household prods.
Arlatone® T; Carbowax® PEG 300; Condea Vista LPA-142; Condea Vista LPA-170; Condea Vista LPA-210; Crodamol W; Exxsol® D 95; Hydrosol Essence 60/95; Hydrosol Essence A; Hydrosol Essence E<5; Hydrosol Essence F<5; Hydrosol Heptane; Isopar® V Fluid; Norpar® 14 Fluid; SBP 140/165; Shellsol® A150ND; Shellsol® D25; Shellsol® D60 SC; Shellsol® D70; Shellsol® H; Shellsol® TD; Shellsol® TM; Spirdane BT; Spirdane D40; Spirdane D60; Spirdane HT; Spirdane L1

solvent, household specialty prods.
Shellsol® D80

solvent, household sprays
Shellsol® T

solvent, household/I&I cleaners
Citroflex® 2; Citroflex® 4; Citroflex® A-2; Citroflex® A-4; Dalpad A; Dowanol® DPM; Dowanol® DPMA; Dowanol® DPnB; Dowanol® EPh; Dowanol® PM; Dowanol® PMA; Dowanol® PnB; *Trihexyl citrate*; *Trimethyl citrate*

solvent, household/industrial cleaners
Conosol 340; Conosol C-200; Ecosoft™ PB Solvent; Ecosoft™ PE Solvent; Ecosoft™ PH Solvent

solvent, human inj. solutions
Solutol® HS 15

solvent, hydraulic brake fluids
Dipropylene glycol

solvent, hydraulic fluids
Hexylene glycol; *Isobutyl alcohol*

solvent, hydroalcoholic cosmetics
Schercemol DIA

solvent, hydroalcoholic topical gels
Carbopol® 980

solvent, hydrocarbon extraction
N-Methylformamide

solvent, hydrocarbons
Dibutyl ether; *α-Terpineol*

solvent, hydrogen sulfide extraction: gas processing
Diethanolamine

solvent, I&I cleaners
Acetyl tributyl citrate; *Acetyl triethyl citrate*; *Acetyl trimethyl citrate*; Arcosolv® DPNP

solvent, imitation leather
Dimethyl formamide

solvent, incense
SD alcohol 22; *SD alcohol 37*; *SD alcohol 38-B*; *SD alcohol 39-B*; *SD alcohol 40*; *SD alcohol 40-A*; *SD alcohol 40-B*; *SD alcohol 40-C*

solvent, indenes
3-Methoxybutanol

solvent, industrial
Butyl Diglyme; Crodamol W; *1,1-Dichloroethane*; *Dicyclohexylamine*; Ethyl Diglyme; *Ethylene glycol dimethyl ether*; Ethyl Glyme; *Gasoline*; *Isopropyl myristate*; *Isopropyl palmitate*; Lyondell Tetrahydrofuran (THF); *Methyl alcohol*; *Methyl cottonseed*; Monoglyme; NE™; *N-Nitrosodimethylamine*; NM™; Norfox® MCS; Orange Terpenes; *PEG-4 dimethyl ether*; *PPG-40 diethylmonium chloride*; *PPG-15 stearyl ether*; *Propylene glycol dimethyl ether*; *Propylene glycol laurate*; *Propylene glycol stearate*; Renex® PEG 300; Renex® PEG 400; Renex® PEG 600; *Sodium xylenesulfonate*; *Sorbeth-20*; Tetraglyme; Triglyme; Variquat® CC-42; Variquat® CC-57

solvent, industrial aerosols
Genesolv® SF; Genesolv® S-TZ; Vertrel® XF

solvent, industrial aq. coatings
C8 alkyl acetate

solvent, industrial aq.coatings
C9 alkyl acetate; *C10 alkyl acetate*; *C13 alkyl acetate*; Exxate® 600; Exxate® 700; Exxate® 800; Exxate® 900; Exxate® 1000; Exxate® 1300; *Oxo-heptyl acetate*; *Oxo-hexyl acetate*

solvent, industrial chemicals
Polysorbate 60

solvent, industrial cleaners
Aromatic 100; Aromatic 100 Anti-Static; Aromatic 150; Aromatic 200; *1,1´-Biphenyl, butenylated*; *1,1´-Biphenyl (1-methylethyl)*; *Bis (methylethyl)-1,1´-biphenyl*; *2-Butanol*; Butyl Icinol; *C8 alkyl acetate*; *C9 alkyl acetate*; *C10 alkyl acetate*; *C13 alkyl acetate*; *p-Chlorobenzotrifluoride*; Condea Vista LPA; Condea Vista LPA-142; Conosol 215; Conosol 260; Conosol C-145; Conosol C-170; *3,4-Dichlorobenzotrifluoride*; Dowanol® TPM; Drakesol® 165; Drakesol® 165AT; Drakesol® 205; Drakesol® 220; Drakesol® 260; Drakesol® 260 AT; Drakesol® 305; Eastman® EB; Eastman® EP; *Ethylene glycol isopropyl ether*; Exxate® 600; Exxate® 700; Exxate® 800; Exxate® 900; Exxate® 1000; Exxate® 1300; Herco® Pine Oil; Hydrosol Essence 60/95; Hydrosol Essence A; Hydrosol Essence E<5; Hydrosol Essence F<5; Isopropyl Oxitol; *Methoxydiglycol*; *Methoxyisopropanol*; *N-Methyl-2-pyrrolidone*; Naphthalene Depleted Aromatic 150 Fluid; Naphthalene Depleted Aromatic 200 Fluid; NMP; *Oxo-heptyl acetate*; *Oxo-hexyl acetate*; Oxsol® 100; Oxsol® 1000; Penreco 150-B Solvent; Penreco 170 Exempt Solvent; *Petroleum distillates, hydrotreated light*; *Phenoxyethanol*; *n-Propyl bromide*; *Propylene glycol n-propyl ether*; Santosol® DMA; Santosol® DME; Santosol® DME-2; Santosol® DME-3; Santosol® DMG; Santosol® DMS; Sasol LPA Solvent; Sasol LPA-142 Solvent; Sasol LPA-150 Solvent; Sasol LPA-170 Solvent; Sasol LPA-210 Solvent; SECOSTER® A; Shellsol® BF; Shellsol® D38; Shellsol® D40; Shellsol® D43; Shellsol® D60; Shellsol® TD; Shellsol® TM; Solvarex 9; Solvarex 90/160; Solvarex 90/180; Solvenon® DPnB; Solvenon® PnB; Solvesso™ 100; Solvesso™ 150; Solvesso™ 150 ND; Solvesso™ 200; Solvesso™ 200 ND; Solvesso™ 200 S; Spirdane BT; Spirdane D40; Spirdane D60; Spirdane HT; Spirdane L1; Sunoco 100; Sunoco 150; Ultra Low Naphthalene Aromatic 150 Fluid; Ultra Low Naphthalene Aromatic 200; Vertrel® SMT; *VM&P naphtha*; VM&P Naphtha HT

solvent, industrial cleaning
Hydrosol Heptane

solvent, industrial coating removal
Emery® 6705

solvent, industrial coatings
Arcosolv® TPNB; Exxate® 1200; ExxonMobil MEK; Exxsol® D 30; Exxsol® DSP 80/110; Exxsol® DSP 100/120; Exxsol® DSP 100/140; Exxsol® DSP 100/160; Exxsol® DSP 145/160; Exxsol® Heptane; *Oxo-dodecyl acetate*

solvent, industrial collodions
SD alcohol 3-A; *SD alcohol 3-C*; *SD alcohol 13-A*; *SD alcohol 19*; *SD alcohol 23-A*; *SD alcohol 32*

solvent, industrial degreasers
C8 alkyl acetate; *C9 alkyl acetate*; *C10 alkyl acetate*; *C13 alkyl acetate*; Exxate® 600; Exxate® 700; Exxate® 800; Exxate® 900; Exxate® 1000; Exxate® 1300; *Oxo-heptyl acetate*; *Oxo-hexyl acetate*; *n-Propyl bromide*; Shellsol® BF

solvent, industrial detergents
SD alcohol 1; *SD alcohol 3-A*; *SD alcohol 3-C*; *SD alcohol 23-A*; *SD alcohol 30*

solvent, industrial emulsions/disps.
Eastman® EEH

solvent, industrial finish removal
Emery® 6705

solvent, industrial metal coatings
Exxsol® D 130; Exxsol® D 3135

solvent, industrial paints
Rhodiasolv RPDE

solvent, industrial tank washing
N-Terpinal RS-98

solvent, industrial wood coatings
Exxsol® D 130; Exxsol® D 3135

solvent, industrial: degreasing
d-Limonene

solvent, industrial: electronic materials
Methyl n-amyl ketone

solvent, industrial: high-solids coatings
Methyl n-amyl ketone

solvent, industrial: household cleaners
d-Limonene

solvent, industrial: I&I cleaners
d-Limonene

solvent, industrial: inks
Filmex® A-2 190 Proof; Filmex® A-2 Anhydrous; Filmex® B 190 Proof;

Filmex® B Anhydrous; Filmex® C 190 Proof; Filmex® C Anhydrous; Filmex® D-1; Filmex® D-2 190 Proof; Filmex® D-2 Anhydrous

solvent, industrial: nitrocellulose lacquers
Methyl n-amyl ketone

solvent, industrial: synthetic resin finishes
Methyl n-amyl ketone

solvent, inert
1,1,1-Trichloro-2,2,2-trifluoroethane

solvent, infant formulas
Captex® 350

solvent, inj. prods.
Miglyol® 810; Miglyol® 840; Miglyol® 8810

solvent, inj. solutions
PEG-15 hydroxystearate

solvent, injectables
Glycofurol

solvent, ink cleaners
Coasol™; N-Terpinal RS-50

solvent, ink oils
Condea Vista LPA-142; Condea Vista LPA-170

solvent, ink pad dyes
DePEG 200; DePEG 300; DePEG 400; DePEG 600

solvent, ink removal: rollers
Dynasolve MP-500

solvent, ink resins
Eastman® Methyl Acetate, High Purity

solvent, ink thinners
2-Hexanone

solvent, inks
Acetyl tributyl citrate; *Acetyl triethyl citrate*; *Acetyl trimethyl citrate*; AG 101; Albrite® TBP; *Alcohol*; Alkasurf® OD-410; Arcosolv® DPM; Arcosolv® DPMA; Arcosolv® DPNB; Arcosolv® PE; Arcosolv® PEA; Arcosolv® PM; Arcosolv® PMA; Arcosolv® PNB; Arcosolv® PNP; Arcosolv® PTB; Arcosolv® TPM; *1,1´-Biphenyl, butenylated*; *1,1´-Biphenyl (1-methylethyl)*; *Bis (methylethyl)-1,1´-biphenyl*; *1-Butoxyethoxy-2-propanol*; *Butoxypropanol*; *Butyl alcohol*; Butyl Di-Icinol; *1,2-Butylene carbonate*; Butyl Icinol; *Butyl lactate*; *Butyl-(S)-lactate*; Cachalot® AR-20; Cachalot® BE-22; Cachalot® C-50; Cachalot® L-90; Cachalot® M-43; Cachalot® S-56; *p-Chlorobenzotrifluoride*; Citroflex® 4; Citroflex® A-2; Citroflex® A-4; Coasol B®; Coasol C; *Cyclohexane*; *Cyclohexanol*; *Cyclohexanone*; DBE; DBE-2; DBE-3; DBE-4; DBE-5; DBE-6; DBE-9; DBE-IB; *3,4-Dichlorobenzotrifluoride*; *Diethylene glycol dibutyl ether*; *Dimethyl adipate*; *Dimethyl glutarate*; *Dimethyl succinate*; *Dipropylene glycol butyl ether*; *Dipropylene glycol t-butyl ether*; *Dipropylene glycol ethyl ether*; Dowanol® DPM; Dowanol® DPMA; Dowanol® PMA; Dowanol® PPh; Dynasolve 200; Eastman® DM; Eccoflex 090; Ecco Strip LS-W; Estasol™; *Ethoxydiglycol*; *Ethylene carbonate*; Ethyl Icinol; *Ethyl lactate*; *Heptane*; Hisolve DM; Isane IP 130; Isane IP 155; Isane IP 165; Isane IP 175; Isane IP 185; Isopar® V Fluid; *Isopropyl alcohol*; Jeffox® WL-440; Jeffox® WL-660; Jeffox® WL-1400; Jeffox® WL-5000; Kemstrene® 96.0% USP; Kemstrene® 99.7% USP; Kemstrene® Reagent Grade; *Lauryl pyrrolidone*; Leosolb 703B; *Mesityl oxide*; *Mineral spirits*; Morcryl® 535; NE™ 101; *Nitroethane*; *1-Nitropropane*; NMP; Oleocal ME-70; Oleocal ME-112; Oleocal ME-130; Oxsol® 100; Oxsol® 1000; *Paraffin*; *Petroleum distillates, sweetened middle*; Pluracol® E1450 NF; Pluracol® SRF-2; Pluracol® W3520N; Pluracol® WI 285; Pluracol® WI 625; Pluracol® WI 1715; Pluracol® WS 170; Pluracol® WS 260; Pluracol® WS 660; Pluracol® WS 2000; Pluracol® WS 3520; Pluracol® WS 5100; Pluracol® WT 1400; Pluracol® WT 9150; Pluracol® WT 90,000; *PPG-2-buteth-3*; *PPG-3-buteth-5*; *PPG-5-buteth-7*; *PPG-7-buteth-10*; *PPG-9-buteth-12*; *PPG-12-buteth-16*; *PPG-20-buteth-30*; *PPG-28-buteth-35*; *PPG-33-buteth-45*; *PPG-5 butyl ether*; *PPG-9 butyl ether*; *PPG-15 butyl ether*; *PPG-18 butyl ether*; *PPG-22 butyl ether*; *PPG-24 butyl ether*; *PPG-33 butyl ether*; *PPG-40 butyl ether*; *PPG-53 butyl ether*; *PPG-2 methyl ether*; *PPG-2 methyl ether acetate*; *n-Propyl bromide*; *Propylene glycol phenyl ether*; *SD alcohol 1*; *SD alcohol 3-A*; *SD alcohol 3-C*; *SD alcohol 13-A*; *SD alcohol 23-A*; *SD alcohol 30*; *SD alcohol 32*; *SD alcohol 33*; Shell Mineral Spirits 135; Shellsol® D40; Shellsol® D43; Shellsol® Heptane; Shellsol® W HT; Solvenon® DPM; Solvenon® PM; Sunoco 100; Sunoco 150; Surfadone® LP-100; TCD Ketone A; *Toluene*; *Tricyclo [5.2.1.0^{2,6}] decane-8-one*; *Trihexyl citrate*; *Trimethyl citrate*; *Tripropylene glycol*; *Tripropylene glycol n-butyl ether*; *Tripropylene glycol ethyl ether*; Ucon® 50-HB-55; Ucon® 50-HB-100; Ucon® 50-HB-260; Ucon® 50-HB-400; Ucon® 50-HB-660; Ucon® 50-HB-2000; Ucon® 50-HB-3520; Ucon® 50-HB-5100; Ucon® 75-H-450; Ucon® LB-65; Ucon® LB-135; Ucon® LB-165; Ucon® LB-285; Ucon® LB-385; Ucon® LB-525; Ucon® LB-625; Ucon® LB-1145; Ucon® LB-1715; Ucon® LB-3000; *VM&P naphtha*; VM&P Naphtha HT

solvent, inks: ball-point pens
Dowanol® TPnB

solvent, inks: felt tip pens
Dowanol® TPnB

solvent, inorganic compds.
Acetamide; *1,2-Butylene carbonate*; *Dimethyl sulfide*

solvent, inorganic salts
1,2-Butylene carbonate; *Ethylene carbonate*; *Ethylene cyanohydrin*; *Propylene carbonate*

solvent, insect repellents
Neosolv™ 2; Neosolv™ 4; Neosolv™ 6; Neosolv™ 8; Neosolv™ 24; Neosolv™ 150; Poly-Solv® DM; Poly-Solv® DPM; Poly-Solv® EM; Poly-Solv® MPM; Poly-Solv® TM; Poly-Solv® TPM; *PPG-3 methyl ether*; Renex® PEG 300; Renex® PEG 600; *Triglycol monomethyl ether*

solvent, insecticide bases
Drakesol® 165; Drakesol® 165AT; Drakesol® 205; Drakesol® 220; Drakesol® 260; Drakesol® 260 AT; Drakesol® 305; *Petroleum distillates, hydrotreated light*

solvent, insecticides
BDGE; BGE; *Butoxyethanol*; Conosol 215; Conosol 260; Conosol 340; Conosol C-145; Conosol C-170; Conosol C-200; Eastman® DBP; EDGE; EGE; Estasol™; *Ethylene glycol isopropyl ether*; Exxate® 1200; Exxsol® D 95; Exxsol® D 110; i-PGE; Isopar® V Fluid; *Isopropylamine*; MDGE; MGE; Neosolv™ 2; Neosolv™ 4; Neosolv™ 6; Neosolv™ 8; Neosolv™ 24; Neosolv™ 150; Penreco 150-B Solvent; Penreco 170 Exempt Solvent; *β-Picoline*; 2-Pyrol®; *2-Pyrrolidone*; Rhodiasolv RPDE; *SD alcohol 1*; *SD alcohol 3-A*; *SD alcohol 3-B*; *SD alcohol 3-C*; *SD alcohol 23-A*; *SD alcohol 23-H*; *SD alcohol 27-A*; *SD alcohol 27-B*; *SD alcohol 30*; *SD alcohol 37*; *SD alcohol 38-B*; *SD alcohol 39-B*; *SD alcohol 40*; *SD alcohol 40-A*; *SD alcohol 40-B*; *SD alcohol 40-C*; Shellsol® D38; Shellsol® D40; Shellsol® D43; Shellsol® D60; Shellsol® D-LAWS; Sunoco 100; Sunoco 150; *1,2,4-Trichlorobenzene*; *1,1,1-Trichloroethane*; *γ-Valerolactone*

solvent, integrated circuit cleaners
Genesolv® D

solvent, interior paints
Lusolvan® FBH

solvent, intermediates
Arcosolv® DPM; Arcosolv® PM; Arcosolv® TPM; Carbowax® E200; Carbowax® E300; Carbowax® E400; Carbowax® E600; Carbowax® E900; Carbowax® E1000; Carbowax® E1450; Carbowax® E3350; Carbowax® E8000; Santosol® DMA; Santosol® DME; Santosol® DME-2; Santosol® DME-3; Santosol® DMG; Santosol® DMS; *1,2,4-Trichlorobenzene*

solvent, iodine
Carbon disulfide; 2-Pyrol®; *Sodium iodide*

solvent, iodized oils
Sesamol

solvent, ionizing
Formamide

solvent, isocyanate
Eastman® Methyl Acetate, High Purity

solvent, isocyanate-contg. systems
Butoxyl®

solvent, isoparaffinic
Eucalyptus globulus oil

solvent, IV emulsions
Super Refined® Almond NF; Super Refined® Corn NF; Super Refined® Cottonseed NF; Super Refined® Olive NF; Super Refined® Peanut NF; Super Refined® Safflower USP; Super Refined® Sesame NF; Super Refined® Soybean USP

solvent, IV steroid anesthetics
PEG-35 castor oil

solvent, jet engine part cleaning
GAFsolver™ TMP

solvent, ketones
3-Methoxybutanol

solvent, label adhesive removal
N-Terpinal RS-25

solvent, lacquer coatings
Texanol® Ester-Alcohol; *2,2,4-Trimethyl-1,3-pentanediol isobutyrate*

solvent, lacquer thinners
Alcohol; *Ethoxyethanol*; Hisolve DM; Shellsol® X7B

solvent, lacquer: nail polishes
Ammonium xylenesulfonate

solvent, lacquers
Acetone; *Amyl acetate*; *s-Amyl alcohol*; Arcosolv® PMA; BDGE; *Benzyl alcohol*; BGE; *2-Butanol*; *Butoxyethanol*; *Butoxyethanol acetate*; *1-Butoxyethoxy-2-propanol*; Butyl Dioxitol®; *Butyl formate*; *Butyl-(S)-lactate*; *n-Butyl propionate*; *Camphor (Cinnamomum camphora) oil*; *C7-10 isoparaffin*; *Cyclohexane*; *Cyclohexanol*; *Cyclohexanone*; *Cyclohexyl acetate*; *Decahydronaphthalene*; Decalin; *Diacetone alcohol*; *Dibutyl tartrate*; *Diethyl carbonate*; *Diethylene glycol benzyl ether*; *Diisobutyl ketone*; *1,4-Dioxane*; *Dipropylene glycol ethyl ether*; Eastman® C-11 Ketone; Eastman® EB; Eastman® EP; Eastman® PM Acetate; EDGE; EGE; *Ethoxydiglycol*; *Ethoxyethanol*; *Ethoxyethanol acetate*; *Ethoxytriglycol*; *Ethyl acetate*; *Ethylene glycol acetate*; *Ethylene glycol benzyl ether*; *Ethylene glycol diacetate*; *2-Ethylhexyl acetate*; *Ethyl lactate*; *Ethyl-(S)-lactate*; *Formic acid*; *Furan*; Galaster EL; *Heptane*; Hetoxide MPC; *2,5-Hexanedione*; *2-Hexanone*; i-PGE; *Isoamyl acetate*; *Isoamyl alcohol*; *Isoamyl chloride*; *Isobutyl alcohol*; *Isohexane*; *Isophorone*; *Isopropyl acetate*; *Isopropyl alcohol*; *dl-Limonene*; *Longifolene*; MDGE; *Mesityl oxide*; *3-Methoxybutanol*; *Methoxydiglycol*; *Methoxyethanol*; *Methoxyethanol acetate*; *2-Methoxy-1-propanol acetate*; *Methyl abietate*; *Methyl acetate*; *Methyl amyl acetate*; *Methylcyclohexanol*; *Methylcyclohexanone*; *Methyl ethyl ketone*; *Methyl isobutyl ketone*; *Methyl lactate*; *Methyl propionate*; MGE; *1-Nitropropane*; *2-Nitropropane*; *PPG-2 methyl ether*; *PPG-3 methyl ether*; *Propyl acetate*; *Propylene glycol methyl ether acetate*; *Propylene glycol phenyl ether*; *Propyl propionate*; Tecsol® 1; Tecsol® 3; *Tripropylene glycol ethyl ether*; *Turpentine*; *γ-Valerolactone*; *Xylene*

solvent, lactic acid
Lactonitrile

solvent, lamp oil
Conosol C-145; Drakesol® 165; Drakesol® 165AT; Drakesol® 205; Drakesol® 220; Drakesol® 260; Drakesol® 260 AT; Drakesol® 305; *Petroleum distillates, hydrotreated light*; Shellsol® D60 SC; Shellsol® D70; Shellsol® D80

solvent, latent
Isobutyl alcohol; *Methyl amyl alcohol*

solvent, latent: cellulosic coatings
Tecsol® A Anhydrous; Tecsol® A-2, 95%; Tecsol® B; Tecsol® C 95%; Tecsol® D, 95%; Tecsol® D-2, 95%;

Exxate® 600; Exxate® 700; Exxate® 800; Exxate® 900
solvent, NC paints/coatings
Ethoxyethanol acetate
solvent, NC printing inks
C8 alkyl acetate; *C9 alkyl acetate*; *C10 alkyl acetate*; *C13 alkyl acetate*; *Oxo-heptyl acetate*; *Oxo-hexyl acetate*
solvent, neoprene
Isoamyl chloride
solvent, nitrile rubber
Benzonitrile
solvent, nitrocellulose
Alkasurf® OD-410; *s-Amyl acetate*; *Bornyl acetate*; *Butoxyethanol*; Butyl Dioxitol®; *Butyl formate*; *Butyl lactate*; *n-Butyl propionate*; *Cyclohexyl acetate*; *Diacetone alcohol*; *1,3-Dichloro-2-propanol*; *2,3-Dichloro-1-propanol*; *Diethyl carbonate*; *Diethylene glycol*; *Diethyl phthalate*; *Diethyl tartrate*; *Diisobutyl ketone*; *Dimethyl sebacate*; *Ethoxydiglycol*; *Ethoxyethanol*; *Ethylene glycol acetate*; *Ethyl formate*; *2-Ethylhexanol*; *2-Ethylhexyl acetate*; *Furfural*; *4-Heptanone*; *2-Hexanone*; *Isoamyl acetate*; *Isoamyl chloride*; *Isobutyl acetate*; *Isopropyl acetate*; *3-Methoxybutanol*; *Methoxybutyl acetate*; *Methoxyethanol*; *Methoxyethanol acetate*; *Methyl acetate*; *Methyl alcohol*; *Methyl amyl acetate*; *Methyl butyrate*; *Methyl isobutyl ketone*; *Methyl lactate*; *3-Octanone*; *PEG-4*; *Propyl acetate*; *Propylene glycol methyl ether acetate*; *Propyl propionate*; *Tributyl citrate*; *Tricresyl phosphate*
solvent, nitrocellulose coatings
Methyl ethyl ketone
solvent, nitrocellulose lacquers
s-Butyl acetate; *Dibutyl phthalate*; Eastman® DBP; *2-Ethylbutyl acetate*; *Methyl isopropyl ketone*
solvent, nitrocellulose resins
3-Heptanone; *Isophorone*; *Triethyl citrate*
solvent, nonferrous metal oxides
o-Dichlorobenzene
solvent, nonferrous metals
HC-250; HC-370; HC-FX50; HC-FX70
solvent, nongrain-raising wood stains
BDGE; BGE; Eastman® DM; EDGE; EGE; i-PGE; MDGE; *Methoxydiglycol*; MGE
solvent, nonionic surfactants
Methylpropanediol; MPDiol® Glycol
solvent, NR
Butyl Cellosolve®; Butyl Oxitol®
solvent, nutritional applics.
Caprylic/capric/lauric triglyceride; *Caprylic/capric/linoleic triglyceride*; *Caprylic/capric/oleic triglyceride*; Captex® 810D; Captex® 8227; *Triundecanoin*
solvent, nutritional supplements
Super Refined® Almond NF; Super Refined® Corn NF; Super Refined® Cottonseed NF; Super Refined® Olive NF; Super Refined® Peanut NF; Super Refined® Safflower USP; Super Refined® Sesame NF; Super Refined® Soybean USP
solvent, nutritional/sports supplements
Capmul® MCM; Captex® 350
solvent, nylon
Tetrahydrofurfuryl alcohol; *2,2,2-Trifluoroethanol*
solvent, o/w based systems
N-Terpinal RS-50
solvent, occlusive
Wheat (Triticum vulgare) germ oil
solvent, occlusive: makeup
Cocoa (Theobroma cacao) butter
solvent, occlusive: ointments
Walnut (Juglans regia) oil
solvent, occlusive: OTC pharmaceuticals
Cocoa (Theobroma cacao) butter
solvent, occlusive: sun preps.
Cocoa (Theobroma cacao) butter
solvent, occlusive: suntan gels
Walnut (Juglans regia) oil
solvent, occlusive: topical pharmaceuticals
Walnut (Juglans regia) oil
solvent, odor-free paints
Shellsol® T
solvent, odorless acrylic paints
Isopar® C; Isopar® E Fluid
solvent, odorless alkyd paints
Isopar® C; Isopar® E Fluid
solvent, odorless alkyd/acrylic paints
C11-13 isoparaffin; *C13-14 isoparaffin*; Isopar® G Fluid; Isopar® H Fluid; Isopar® K Fluid; Isopar® L Fluid; Isopar® M Fluid
solvent, odorless/low odor solvent-based decorative coatings
Isopar® I Fluid; Isopar® J Fluid
solvent, OEM coatings
Exxsol® D 130; Exxsol® D 3135; Ucar® n-Propyl Propionate
solvent, off-gas treatment
Diisopropanolamine
solvent, office form carbonless copy paper
Petroleum distillates, hydrotreated light
solvent, offset printing inks
Texaprint® SDCE; Texaprint® SDM; Texaprint® SJEH; Texaprint® SKEH; Texaprint® SLIB EXP; Texaprint® SLIP; Texaprint® SRM; Texaprint® SSEH
solvent, oil burners
Kerdane D65; Kerdane D75
solvent, oil colors
Isoamyl acetate
solvent, oil drilling
Dipentene-C
solvent, oil extraction
Cyclopentane; Petrinex® ASB; VM&P Naphtha HT
solvent, oil field
Arcosolv® DPM; Arcosolv® DPMA; Arcosolv® PM; Arcosolv® TPM; *PPG-2 methyl ether*; *PPG-3 methyl ether*; *PPG-2 methyl ether acetate*; SECOSTER® A
solvent, oil field drilling
Sasol C1316
solvent, oil preps.
Miglyol® 812 N
solvent, oil seed extraction
Hydrosol Hexane
solvent, oil/fat extraction
VM&P naphtha
solvent, oil/grease removal
Perfluoroheptane; *Perfluorohexane*; *Perfluoro-N-methylmorpholine*; *Perfluoropentane*
solvent, oil/particulate matter removal/suspension: vapor degreasing
Genesolv® D
solvent, oil-based systems
Isopropylamine dodecylbenzenesulfonate
solvent, oil-dispersant materials
Liponate GC-K; Liponate PC
solvent, 'oil-free' pigmented prods.
Foamid LM2E
solvent, oils/greases removal
PF-5050; PF-5060; PF-5070
solvent, oils/waxes
Butyl Oxitol®; Tetralin
solvent, oil-soluble dyes
Eastman® DBP; Liquid Absorption Base T
solvent, oil-soluble dyes: makeup
Lanesta L
solvent, oil-soluble food additives
Capric triglyceride
solvent, oil-soluble food additives: dietetic prods.
Delios® S
solvent, oil-soluble food additives: sport food
Delios® S
solvent, oil-soluble pigments
Tegosoft® M
solvent, oil-soluble sulfonates
Benzene, mono-C10-13-alkyl derivs., distillation residues; *C12-14 alkylbenzene*; V-154L; V-159L
solvent, oil-soluble UV filters
Tegosoft® M
solvent, oils
BDGE; *Benzaldehyde*; *Benzene*; BGE; *Bromoform*; *1-Butoxyethoxy-2-propanol*; *Butyl chloride*; *Butyl lactate*; *Butyl-(S)-lactate*; *Carbon tetrabromide*; *Cyclohexane*; *Cyclohexanol*; Decalin; *Diethylene glycol*; *Dipropylene glycol ethyl ether*; EDGE; EGE; *Ethoxyethanol acetate*; *Ethylene glycol diacetate*; *2-Ethylhexanol*; *2-Ethylhexyl acetate*; *Ethyl morpholine*; i-PGE; MDGE; *Methoxyethanol acetate*; *Methyl acetate*; *Methylcyclohexanol*; MGE; Miglyol® 840; *2-Nitropropane*; *PPG-2 methyl ether*; *PPG-3 methyl ether*; *Propylene glycol methyl ether acetate*; *Propylene glycol phenyl ether*; *Tetradecene-1*; *1,1,1-Trichloroethane*; *Tripropylene glycol ethyl ether*
solvent, ointment bases
PEG 300
solvent, ointments
Caprylic/capric triglyceride PEG-4 esters; Carbowax® Sentry® PEG 540 Blend; Carbowax® Sentry® PEG 1000; Carbowax® Sentry® PEG 1450; EmCon™ Olive; EmCon™ Rice Bran; Fancol™ ALA; Labrafac® Lipophile; *Octyldodecyl isostearate*; *Olive (Olea europaea) oil*; Penreco Snow; *Rice (Oryza sativa) bran oil*; *Sesamol*; Super Refined® Almond NF; Super Refined® Corn NF; Super Refined® Cottonseed NF; Super Refined® Olive NF; Super Refined® Peanut NF; Super Refined® Safflower USP; Super Refined® Sesame NF; Super Refined® Soybean USP
solvent, oleomargarine
Sesamol
solvent, oleoresinous prods.
dl-Limonene
solvent, one-bath scouring
Burcolev DLS
solvent, oral care
Pluracol® E1450 NF
solvent, oral pharmaceuticals
Alcohol; Crodamol GTCC-PN; Crodamol PC-PN; *Cyclohexane*; Etocas 35 NF; Labrafac® Hydrophile; Miglyol® 812 N; Miglyol® 840; Soluphor® P; Super Refined® Cottonseed NF; Super Refined® Olive NF; Super Refined® Safflower USP; Super Refined® Sesame NF; Super Refined® Soybean USP
solvent, ore flotation
Glidsol® 140; Jeffox® WL-440; Jeffox® WL-660; Jeffox® WL-1400; Jeffox® WL-5000
solvent, ore minerals
Hydrobromic acid
solvent, organic
Cumene
solvent, organic acids
Dibutyl ether
solvent, organic chemicals
o-Chlorotoluene; *p-Chlorotoluene*
solvent, organic coatings
3-Heptanol
solvent, organic compds.
Acetamide; *1,2-Butylene carbonate*; *o-Dichlorobenzene*; *Dimethyl sulfide*; Eastman® DBP
solvent, organic compounds
Isoamyl chloride
solvent, organic dyes
o-Chlorotoluene; *p-Chlorotoluene*
solvent, organic dyestuffs
Pilot® SXS-40
solvent, organic materials
cis-1,2-Dichloroethylene; *cis-trans-1,2-Dichloroethylene (mixed isomers)*
solvent, organic peroxide catalysts
Uniplex 150
solvent, organic peroxides
Uniplex 155
solvent, organic polymers
Pilot® SXS-40
solvent, organic synthesis
n-Amyl alcohol; *s-Butylbenzene*; *Methyl ethyl ketone*; *2-Nitropropane*; *3-Pentanol*; *Phenoxyethanol*
solvent, organic synthesis,
Ethyl acetate
solvent, organic/inorganic materials
Foamid AME-70; Foamid AME-75; Foamid AME-100
solvent, organosols
Butoxyethanol
solvent, OTC pharmaceuticals
Fancol™ CB; Fancol™ CB Extra
solvent, oven cleaners
Neosolv™ 2; Neosolv™ 4; Neosolv™ 6; Neosolv™ 8; Neosolv™ 24; Neosolv™ 150
solvent, oven-dried industrial paints
Rhodiasolv RPDE
solvent, overprints
Morcryl® 535
solvent, paint cleaners
Jeffstrip™ FST
solvent, paint removers
Agrisynth® BLO; *2-Butanol*; Dowanol® TPM; Eastman® CDA-19; GBL; Hisolve MC; *Isobutyl alcohol*; *Methyl acetate*; *Methyl ethyl ketone*; Proglyde DMM; Shell Toluene Solvent
solvent, paint strippers
C8 alkyl acetate; *C9 alkyl acetate*; *C10 alkyl acetate*; *C13 alkyl acetate*; *Diethylene glycol phenyl ether*; *Dimethyl formamide*; Estasol™;

300; DePEG 400; DePEG 600; Dipentene Extra; *Dipropylene glycol dimethyl ether*; *Dipropylene glycol n-propyl ether*; Dow B100-4800; Dow EP530; Dowanol® DMM; Dowanol® DPM; Dowanol® DPnB; Dowanol® DPnP; Dowanol® PM; Dowanol® PnP; Dowanol® PPh; Dowanol® TPM; Dowanol® TPnB; Dow Corning® L910; Dow Corning® L1150; Dow Corning® MPEG350; Dow Corning® MPEG550; Dow Corning® MPEG750; Dow Glycerine 96% USP; Dow Glycerine 99.5% USP/EP; Eastman® DB Stabilized; Eastman® DE Acetate; Eastman® DM; Eastman® EAA; Eastman® MAA; Eastman® MIBK; Eastman® PM; Eastman® PM Acetate; Eastman® Acetone, High Purity Sales Grade; Eastman® n-Butanol; Eastman® n-Butyl Acetate, Urethane Grade; Eastman® Ethyl Acetate, Food Grade; Eastman® Isobutyl Acetate; Eastman® Isopropanol; Eastman® Isopropyl Acetate; Eastman® Methyl Acetate; Eastman® Methyl Acetate, Cosmetic Grade; Eastman® Methyl Acetate, High Purity; Eastman® n-Propanol; Eastman® n-Propyl Acetate; Estasan™ GT 8-40 3578; Estasan™ GT 8-60 3575; Estasan™ GT 8-60 3580; Estasan™ GT 8-65 3577; Estasan™ GT 8-65 3581; Estasan™ GT 8-70 3579; Fancol™ ALA; HallBrite® OS; *Isooctane*; Isopar® G Fluid; Isopar® H Fluid; Isopar® K Fluid; Lexol® IPM; Lipovol SES; Lonzest® 143-S; *Methylpropanediol*; Novol NF; *Octyldodecyl isostearate*; Optim™ Glycerine 99.7; PCHP; Permethyl® 97A; Phytol™ AVO; Pluracol® E200; Pluracol® E300; Pluracol® E400; Pluracol® E600; Pluracol® E4000; Pluracol® E6000; *Polyglyceryl-9 stearate*; Polyglycol P15-200; Polyglycol P112-2; Polyglycol P425; Polyglycol P1000TB; Polyglycol P1200; Polyglycol P2000; Polyglycol P4000; Polyglycol PT250; Polyglycol PT700; Polyglycol PT3000; *PPG-14 butyl ether*; *PPG-40 diethylmonium chloride*; Promyr; Propal NF; *Propylene glycol phenyl ether*; Protachem™ 100; Protachem™ AWS-100; Protachem™ DGS; Protachem™ EGDS; Protachem™ EGMS; Protachem™ G-5509; Protachem™ G-5566; Protachem™ GC-7; Protachem™ GC-30; Protachem™ GDL; Protachem™ GMS-165; Protachem™ GMS-450; Protachem™ GMS-AS; Protachem™ GMS-D; Protachem™ GMS-T; Protachem™ HMS; Protachem™ MLD; Protachem™ PGMS; Protachem™ SAS; Renex® PEG 300; Renex® PEG 400; Renex® PEG 600; SDA-1; SDA-3A; SDA-3C; SDA-12A; SDA-23A; SDA-23H; SDA-29-8; SDA-30; SDA-39C; SDA-40; SDA-40A; SDA-40B; SDA-40C; Solulan® 16; Solulan® 75; Solulan® C-24; Solulan® L-575; STEPAN® IPM; STEPAN® IPM-NF; STEPAN® IPP; *Terpinolene*; Variquat® CC-42; Variquat® CC-57; Wickenol® 136; ZELEC® IPM

solvent, personal care sprays
C10-11 isoparaffin; *C11-12 isoparaffin*

solvent, PES: food-contact
Dimethyl sulfoxide; *Diphenyl sulfone*; *N-Methyl-2-pyrrolidone*

solvent, pesticide emulsifiable concs.
C8 alkyl acetate; *C9 alkyl acetate*; *C10 alkyl acetate*; *C13 alkyl acetate*; Exxate® 600; Exxate® 700; Exxate® 800; Exxate® 900; Exxate® 1000; Exxate® 1300; *Oxo-heptyl acetate*; *Oxo-hexyl acetate*

solvent, pesticides
Agsolex® BLO; Aromatic 100; Aromatic 100 Anti-Static; Aromatic 150; Aromatic 200; Butyl Diglyme; *n-Butyl mercaptan*; *Chlorobenzene*; Condea Vista LPA-142; Condea Vista LPA-170; *Diethylene glycol dibutyl ether*; *Dimethyl caproamide*; *Dimethyl octanamide*; Dowanol® PM; Eastman® EAA; Eastman® MAA; Edenor® ME C18-90; Emery® 2214; Emery® 2216; Emery® 2230; Emery® 2232; Emery® 2253; Emery® 2254; Emery® 2270; Emery® 2290; Emery® 2296; Estasol™; Ethyl Diglyme; *Ethylene glycol dimethyl ether*; Exxsol® D 100; Exxsol® D 100 S; Exxsol® D 120; Exxsol® D 130; Exxsol® D 140; Isopar® P; *Methyl canolate*; *Methyl cocoate*; *Methyl myristate*; *α-Methylnaphthalene*; *Methyl oleate*; *Methyl palmitate*; *Methyl stearate*; *Methyl sunflowerate*; *Methyl tallowate*; MN; Monoglyme; Naphthalene Depleted Aromatic 150 Fluid; Naphthalene Depleted Aromatic 200 Fluid; Norpar® 14 Fluid; Norpar® 15 Fluid; Oleocal ME-70; Oleocal ME-112; PEG 300; *PEG-6 abietate*; *PEG-4 dimethyl ether*; *γ-Picoline*; *Polysorbate 60*; Rhodiasolv RPDE; Sasol LPA Solvent; Sasol LPA-142 Solvent; Sasol LPA-150 Solvent; Sasol LPA-170 Solvent; Sasol LPA-210 Solvent; SBP 100/140; SBP 140/165; Schercemol PGML; Shellsol® A150ND; Shellsol® D25; Shellsol® H; Shellsol® TD; Shellsol® TM; Solvarex 9; Solvarex 10; Solvaro AFD; Solvesso™ 100; Solvesso™ 150; Solvesso™ 150 ND; Solvesso™ 200; Solvesso™ 200 ND; Solvesso™ 200 S; Tetraglyme; THFA®; *1,1,1-Trichloroethane*; Triglyme; Ultra Low Naphthalene Aromatic 150 Fluid; Ultra Low Naphthalene Aromatic 200

solvent, pet food
Propylene Glycol USP, FCC

solvent, petrochemical processing
NMP

solvent, petroleum degreasing
Soygold® 1000; Soygold® 2000

solvent, petroleum gas absorption
Shellsol® D38; Shellsol® Heptane; Shellsol® W HT; *VM&P naphtha*; VM&P Naphtha HT

solvent, petroleum industry
GBL; Pilot® SXS-40; *Sodium xylenesulfonate*

solvent, petroleum jellies
Penreco Snow

solvent, petroleum lubes
Furfural

solvent, petroleum lubricants
QO® Furfural

solvent, petroleum processing
Agrisynth® BLO

solvent, petroleum prods.
SD alcohol 1; *SD alcohol 2-B*; *SD alcohol 3-A*; *SD alcohol 3-C*

solvent, petroleum refining
Hexylene glycol; *Propionitrile*

solvent, PFPE lubricants
Novec™ HFE-7100; Novec™ HFE-7100DL

solvent, pharmaceutical aerosols
Capmul® MCM; Polarin® Range

solvent, pharmaceutical capsules
Polarin® Range

solvent, pharmaceutical coatings
Capmul® MCM; Captex® 350

solvent, pharmaceutical creams/lotions
Deodorized kerosene; PD-23; PD-25; PD-28

solvent, pharmaceutical emulsions
Labrafac® Lipophile; Olepal ISO

solvent, pharmaceutical extraction
Hexane

solvent, pharmaceutical inj. solutions
Cremophor® ELP

solvent, pharmaceutical injectables
Hydrogenated castor oil; Soluphor® P

solvent, pharmaceutical lipophilic actives
Solutol® HS 15

solvent, pharmaceutical microemulsions
Capmul® MCM; Olepal ISO

solvent, pharmaceutical syrups
Polarin® Range

solvent, pharmaceutical tablet coatings
SD alcohol 3-A

solvent, pharmaceutical vehicles
Emery® 912; Emery® 916; Emery® 917

solvent, pharmaceuticals
Acetic acid; *Acetic acid, glacial*; *Acetone*; *Acetonitrile*; *Acetylated mono- and diglycerides of fatty acids*; *Acetyl triethyl citrate*; *n-Amyl alcohol*; *t-Amyl alcohol*; *Apricot kernel oil PEG-6 esters*; Arlasolve® DMI; Arlatone® T; *Benzene*; *Benzyl benzoate*; *Bornyl acetate*; *2-Butanol*; *n-Butyl acetate*; *t-Butyl alcohol*; Butyl Diglyme; *Butylene glycol dicaprylate/dicaprate*; *Butyrolactone*; Capmul® MCM; Capmul® MCM-90; *Caprylic/capric/lauric triglyceride*; *Caprylic/capric/linoleic triglyceride*; *Caprylic/capric/oleic triglyceride*; *Caprylic/capric triglyceride*; *Caprylic/capric triglyceride PEG-4 esters*; Captex® 200; Captex® 350; Captex® 355EP; Captex® 500P; Captex® 810D; Captex® 8227; Carbowax® PEG 300; Celtone; *Chloracetyl chloride*; *Chlorobenzene*; *Chloroform*; *Coconut (Cocos nucifera) oil*; *Collodion*; Comperlan® COD; *Corn (Zea mays) oil*; *Cottonseed (Gossypium) oil*; Crodamol CAP; Crystal® O; Crystal® Crown; *Cumene*; Dantogard® Plus Liq.; *Diacetylated monoglycerides*; *cis-trans-1,2-Dichloroethylene (mixed isomers)*; *Diethylene glycol*; *Diethylene glycol dibutyl ether*; *Diethylene glycol dimethyl ether*; *Dimethyl acetamide*; *Dimethyl formamide*; *Dimethyl isosorbide*; *Dimethyl sulfoxide*; *n-Dioctyl phthalate*; *1,4-Dioxane*; Dow Glycerine 96% USP; Dow Glycerine 99.5% USP/EP; Eastman® EAA; Eastman® MAA; Eastman® Isopropanol; EmCon™ CO; EmCon™ W; Estasan™ GT 8-40 3578; Estasan™ GT 8-60 3575; Estasan™ GT 8-60 3580; Estasan™ GT 8-65 3577; Estasan™ GT 8-65 3581; Estasan™ GT 8-70 3579; Estasan™ GT 8-99 3596; Estasan™ GT 10-99 3599; *Ethoxydiglycol*; *Ethoxyethanol*; *Ethyl acetate*; Ethyl Diglyme; *Ethylene dichloride*; *Ethylene glycol*; *Ethylene glycol dimethyl ether*; *Ethyl ether*; *Ethyl formate*; Ethyl Glyme; *Ethyl oleate*; Exceparl BB; Exceparl BS; Exceparl EH-P; Exceparl EH-S; Exceparl IB-OL; Exceparl L-OL; Exceparl MC; Exceparl ML-85; Exceparl M-OL; Exceparl MS; Exceparl MT; Exceparl MY-M; Exceparl OD-B; Exceparl OL-OL; Exceparl O-OL; Exceparl SS; Exceparl TD-S; Filmex® A-2 190 Proof; Filmex® A-2 Anhydrous; Filmex® B 190 Proof; Filmex® B Anhydrous; Filmex® C 190 Proof; Filmex® C Anhydrous; Filmex® D-1; Filmex® D-2 190 Proof; Filmex® D-2 Anhydrous; *Formamide*; *Formic acid*; *Glycerin*; *Glyceryl caprylate/caprate*; *Glyceryl formal*; *Heptane*; *2-Hexanone*; *Hexylene glycol*; *Hybrid safflower (Carthamus tinctorius) oil*; *Isoamyl alcohol*; *Isobutyl acetate*; *Isobutyl alcohol*; *Isocetyl laurate*; *Isocetyl myristate*; *Isopropyl acetate*; *Isopropyl alcohol*; *Isopropylamine*; Jarcol™ I-16; Jarcol™ I-24; Kemstrene® 96.0% USP; Kemstrene® 99.7% USP; Labrasol®; Lexol® IPM-NF; Lexol® PG-800; Lexol® PG-865; Lipocol P-15; Lipovol SES; *3-Methoxybutanol*; *Methoxyethanol*; *Methoxy PEG*; *Methyl acetate*; *Methyl alcohol*; *Methyl t-butyl ether*; *Methyl cyclohexane*; *Methylene chloride*; *Methyl ethyl ketone*; *Methyl isobutyl ketone*; *N-Methyl-2-pyrrolidone*; Miglyol® 812; Miglyol® 812 N; Miglyol® 840; *Mineral oil*; *Mineral spirits*; Monoglyme; Myritol® 318; NEOBEE® M-20; *Nitromethane*; Novol NF; *Octyldodecanol*; *Oleic acid*; *Olive (Olea europaea) oil*; Optim™ Glycerine 99.7; *Peanut (Arachis hypogaea) oil*; *PEG-4*; *PEG-6*; *PEG-8*; *PEG-9*; *PEG-12*; *PEG-14*; *PEG-16*; *PEG-20*; *PEG-75*; *PEG-90*; *PEG-150*; *PEG-8 caprylic/capric glycerides*; *PEG-33 castor oil*; *PEG-35 castor oil*; *PEG-3 dimethyl ether*; *PEG-4 dimethyl ether*; *PEG-6 isostearate*; *PEG-8 oleate*; *PEG-20 oleate*; *PEG-8 palmitostearate*; *PEG-150 palmitostearate*; *PEG-40 sorbitan peroleate*; Penreco Snow; *n-Pentane*; *3-Pentanol*; *Petrolatum*; Pharmasolve™; *Phenethyl alcohol*; *β-Picoline*; *γ-Picoline*; Pluracol® E300; Pluracol® E400; Pluracol® E400 NF; Pluracol® E600; Pluracol® E600 NF; Pluracol® E1450 NF; Pluracol® E4000; Pluracol® E6000; Polarin® Range; *Polyethylene glycol*; *Polyglyceryl-3 diisostearate*; Polyglycol DME 250; *Polysorbate 60*; *PPG-9*; *PPG-26*;

PPG-15 stearyl ether; Pricerine™ 9091; Pricerine™ 9099; Propulsol®; *Propyl acetate*; *Propyl alcohol*; *Propylene glycol*; Propylene Glycol USP, FCC; Propylene Glycol USP/FCC Ultra Grade; *Propylene glycol alginate*; *Propylene glycol caprylate*; *Propylene glycol dicaprylate*; *Propylene glycol dicaprylate/dicaprate*; *Propylene glycol dioctanoate*; *Propylene glycol laurate*; Punctilious® Ethyl Alcohol 190 Proof Pure USP; Punctilious® Ethyl Alcohol 200 Proof Pure USP; Punctilious® SDA 1-1 190 Proof; Punctilious® SDA 1-1 Anhydrous; Punctilious® SDA 1-2 190 Proof; Punctilious® SDA 1-2; Punctilious® SDA 2B-2 190 Proof; Punctilious® SDA 2B-2 Anhydrous; Punctilious® SDA 2B-3 190 Proof; Punctilious® SDA 2B-4 190 Proof; Punctilious® SDA 2B-4 Anhydrous; Punctilious® SDA 3A 190 Proof; Punctilious® SDA 3A Anhydrous; Punctilious® SDA 3C 190 Proof; Punctilious® SDA 12A-3 190 Proof; Punctilious® SDA 12A-3 Anhydrous; Punctilious® SDA 13A 190 Proof; Punctilious® SDA 19 190 Proof; Punctilious® SDA 20 190 Proof; Punctilious® SDA 20 Anhydrous; Punctilious® SDA 23A 190 Proof; Punctilious® SDA 23A Anhydrous; Punctilious® SDA 23H 190 Proof; Punctilious® SDA 25A-1 190 Proof; Punctilious® SDA 25A-2 190 Proof; Punctilious® SDA 29-2 190 Proof; Punctilious® SDA 29-2 Anhydrous; Punctilious® SDA 29-3 190 Proof; Punctilious® SDA 30 190 Proof; Punctilious® SDA 30 Anhydrous; Punctilious® SDA 32 190 Proof; Punctilious® SDA 32 Anhydrous; *Pyridine*; 2-Pyrol®; *Safflower (Carthamus tinctorius) oil*; SDA-12A; SDA-23H; SDA-29-8; *SD alcohol 12-A*; *SD alcohol 29-8*; *SD alcohol 38-B*; Sesame Oil USP/NF 16; *Sesame (Sesamum indicum) oil*; Shell Mineral Spirits 135; Shellsol® D40; Shellsol® D43; Solvaclene®; *Sorbeth-20*; *Soybean (Glycine soja) oil*; Stearate 1500; *Sulfolane*; Super Refined® Almond NF; Super Refined® Corn NF; Super Refined® Cottonseed NF; Super Refined® Olive NF; Super Refined® Peanut NF; Super Refined® Safflower USP; Super Refined® Sesame NF; Super Refined® Soybean USP; *Sweet almond (Prunus amygdalus dulcis) oil*; *α-Terpineol*; *Tetradecene-1*; Tetraglyme; *Tetrahydronaphthalene*; *Tetrahydroxypropyl ethylenediamine*; *Toluene*; *Triacetin*; *Tributyl citrate*; *1,1,1-Trichloroethane*; *1,1,2-Trichloroethane*; *Trichloroethylene*; *Triethylene glycol*; Triglyme; *Triisostearin PEG-6 esters*; *Trilaurin*; *Trimyristin*; *Trioctanoin*; *Tripalmitin*; *Turpentine*; Vinycizer 30; *Vinylidene chloride monomer*; *VM&P naphtha*; VM&P Naphtha HT; *Water*; *Xylene*

solvent, pharmaceuticals: fats
Polypropylene glycol

solvent, pharmaceuticals: oils
Polypropylene glycol

solvent, pharmaceuticals: resins
Polypropylene glycol

solvent, pharmaceuticals: waxes
Polypropylene glycol

solvent, phenol
cis-1,2-Dichloroethylene; *cis-trans-1,2-Dichloroethylene (mixed isomers)*

solvent, phenol removal: skin burns
Lutrol® E 300; Lutrol® E 400; Lutrol® E 600

solvent, phenol-formaldehyde
3-Methoxybutanol

solvent, phenolic
Eastman® EB

solvent, phenolic resins
Cyclohexanol; FA®; *Furfuryl alcohol*; Isopropyl Oxitol; *dl-Limonene*; *Methoxyethanol*

solvent, phenolic varnishes
BDGE; BGE; EDGE; EGE; i-PGE; MDGE; MGE

solvent, phenolic-type germicides: cosmetics
Variquat® CC-55

solvent, phenolics
Alkasurf® OD-410; Butyl Dioxitol®; Butyl Oxitol®; Eastman® Methyl Acetate, High Purity; Igepal® OD-410; *Phenoxyethanol*

solvent, phenols
trans-1,2-Dichloroethylene

solvent, phenoxy resins
Eastman® PM Acetate

solvent, phosphors: fluorescent lamps
Amyl acetate

solvent, phosphorus
Ethyl chloride

solvent, photo resist stripping
Dynasolve 711; Uresolve Plus SG

solvent, photochromic solutions
UltraPure™ Ethylene Carbonate; UltraPure™ Propylene Carbonate

solvent, photocopier fluids
Shellsol® D60 SC; Shellsol® D70

solvent, photocopier prods.
Isane IP 130; Isane IP 155; Isane IP 165; Isane IP 175; Isane IP 185; Shellsol® T

solvent, photocopier toner dispersants
Shellsol® TD; Shellsol® TM

solvent, photoengraving
SD alcohol 3-A

solvent, photoengraving dyes
SD alcohol 3-C; *SD alcohol 30*; *SD alcohol 32*

solvent, photographic emulsions
SD alcohol 32

solvent, photographic film
Ethoxyethanol acetate; *Ethyl acetate*; *Methoxyethanol acetate*; *SD alcohol 1*; *SD alcohol 2-B*; *SD alcohol 3-A*; *SD alcohol 3-C*; *SD alcohol 13-A*; *SD alcohol 19*; *SD alcohol 30*; *SD alcohol 32*

solvent, photographic plates
Ethyl acetate

solvent, photoresist
Butyl lactate

solvent, photoresist coatings
Eastman® EEP

solvent, photoresist stripping
Dynasolve 180; Dynasolve 185; NMP

solvent, photoresists
Arcosolv® PMA; *Ethyl lactate*; Galaster EL; HEP®; *N-Hydroxyethylpyrrolidone*

solvent, pigment dispersions
C8 alkyl acetate; *C9 alkyl acetate*; *C10 alkyl acetate*; *C13 alkyl acetate*; Exxate® 600; Exxate® 700; Exxate® 800; Exxate® 900; Exxate® 1000; Exxate® 1300; *Methylpropanediol*; MPDiol® Glycol; *Oxo-heptyl acetate*; *Oxo-hexyl acetate*

solvent, pigmented cosmetics
Lexol® PG-865

solvent, pigmented prods.
Isopropyl PPG-2-isodeceth-7-carboxylate

solvent, pigments
Eastman® EAA; *Glyceryl dioleate*; *Isocetyl alcohol*; *Propylene carbonate*; Variquat® CC-57

solvent, pill mfg.
SD alcohol 38-B

solvent, plastic additives
Chloracetyl chloride

solvent, plastic coatings
Arcosolv® TPNB; Eastman® DE Acetate; Eastman® DE-HG; Eastman® EP; Eastman® IBIB; Eastman® MIBK; Eastman® PM; Eastman® PM Acetate; Eastman® Acetone, High Purity Sales Grade; Eastman® n-Butanol; Eastman® Isobutyl Acetate; Eastman® Isopropanol; Eastman® Isopropyl Acetate; Eastman® Methyl Acetate; Eastman® Methyl Acetate, High Purity; Eastman® n-Propanol; Eastman® n-Propyl Acetate; Tecsol® A, 95%; Tecsol® A-2, 95%; Tecsol® A-2 Anhydrous; Tecsol® B; Tecsol® C 95%; Tecsol® C Anhydrous; Tecsol® D, 95%; Tecsol® D-1, 95%; Tecsol® D-2, 95%; Tecsol® D-2 Anydrous

solvent, plastic films
Acetamide

solvent, plastic molds
1,1,1-Trichloroethane

solvent, plastic toy adhesives
Toluene

solvent, plasticizers
Butyl alcohol; *s-Butylbenzene*; Eastman® n-Butanol; *3-Methoxybutanol*; *Propylene carbonate*

solvent, plasticizing waxes
Oleocal ME-112

solvent, plastics
Acetone; *1,1′-Biphenyl, butenylated*; *1,1′-Biphenyl (1-methylethyl)*; *Bis (methylethyl)-1,1′-biphenyl*; *Butoxyethanol*; *Butoxyethanol acetate*; *1,2-Butylene carbonate*; *Butyl stearate*; *Carbon disulfide*; Carbowax® E200; Carbowax® E300; Carbowax® E400; Carbowax® E600; Carbowax® E900; Carbowax® E1000; Carbowax® E1450; Carbowax® E3350; Carbowax® E8000; Coasol B®; *Diethylene glycol benzyl ether*; *Dimethyl acetamide*; *Ethyl acetate*; *Ethylene glycol*; *Ethylene glycol benzyl ether*; *Heptane*; Isopar® C; Isopar® I Fluid; Isopar® J Fluid; *Isophorone*; *Isopropyl acetate*; Jarcol™ I-16; Jarcol™ I-24; Jeffsol® 1,2-Butylene Carbonate; Jeffsol® EC; Jeffsol® EC-25; Jeffsol® EC-50; Jeffsol® EC-75; Jeffsol® PC; *N-Nitrosodimethylamine*; *PEG-8 oleate*; *PEG-20 oleate*; *Polysorbate 60*; *PPG-40 diethylmonium chloride*; *Propyl acetate*; Punctilious® Ethyl Alcohol 190 Proof Pure USP; Punctilious® Ethyl Alcohol 200 Proof Pure USP; Punctilious® SDA 1-2 190 Proof; Punctilious® SDA 1-2; Punctilious® SDA 2B-2 190 Proof; Punctilious® SDA 2B-2 Anhydrous; Punctilious® SDA 2B-3 190 Proof; Punctilious® SDA 2B-4 190 Proof; Punctilious® SDA 2B-4 Anhydrous; Punctilious® SDA 3A 190 Proof; Punctilious® SDA 3A Anhydrous; Punctilious® SDA 3C 190 Proof; Punctilious® SDA 12A-3 190 Proof; Punctilious® SDA 12A-3 Anhydrous; Punctilious® SDA 13A 190 Proof; Punctilious® SDA 19 190 Proof; Punctilious® SDA 20 190 Proof; Punctilious® SDA 20 Anhydrous; Punctilious® SDA 23A 190 Proof; Punctilious® SDA 23A Anhydrous; Punctilious® SDA 23H 190 Proof; Punctilious® SDA 25A-1 190 Proof; Punctilious® SDA 25A-2 190 Proof; Punctilious® SDA 29-2 190 Proof; Punctilious® SDA 29-2 Anhydrous; Punctilious® SDA 29-3 190 Proof; Punctilious® SDA 30 190 Proof; Punctilious® SDA 30 Anhydrous; Punctilious® SDA 32 190 Proof; Punctilious® SDA 32 Anhydrous; *SD alcohol 1*; *SD alcohol 2-B*; *SD alcohol 3-A*; *SD alcohol 3-C*; *SD alcohol 12-A*; *SD alcohol 13-A*; *SD alcohol 30*; Shellsol® D38; Shellsol® D40; Shellsol® D43; Shellsol® D60; Shellsol® Heptane; Shellsol® W HT; *Trichloroacetic acid*; Variquat® CC-42; Variquat® CC-57; *VM&P naphtha*; VM&P Naphtha HT

solvent, plastics: microelectronics
N-Methyl-2-pyrrolidone

solvent, plastics: plating engineering
CHP®; *N-Cyclohexyl pyrrolidone*

solvent, polar
Adiponitrile

solvent, polar essential oils: hair care
Volpo G26

solvent, polar essential oils: skin care
Volpo G26

solvent, polar process
Propylene carbonate

solvent, polar: elastomers
Hallcomid® M-8-10

solvent, polar: resins
Hallcomid® M-8-10

solvent, polishes
Alcohol; Coasol™; Dipentene-C; Exxsol® D 95; *Heptane*; Isopar® V Fluid; *dl-Limonene*; *Mineral spirits*; PD-23; PD-25; PD-28; *α-Pinene*; *Polysorbate 60*; Punctilious® Ethyl Alcohol 190 Proof Pure USP; Punctilious® Ethyl Alcohol 200 Proof Pure USP; Punctilious® SDA 1-2 190 Proof; Punctilious® SDA 1-2; Punctilious® SDA 2B-2 190 Proof; Punctilious® SDA 2B-2 Anhydrous; Punctilious® SDA 2B-3 190 Proof; Punctilious® SDA 2B-4 190 Proof; Punctilious® SDA 2B-4 Anhydrous; Punctilious® SDA 3A 190 Proof; Punctilious® SDA 3A Anhydrous; Punctilious® SDA 3C 190 Proof; Punctilious® SDA 12A-3 190 Proof; Punctilious® SDA 12A-3 Anhydrous; Punctilious® SDA 13A 190 Proof; Punctilious® SDA 19 190 Proof; Punctilious® SDA 20 190 Proof; Punctilious® SDA 20 Anhydrous; Punctilious® SDA 23A 190 Proof; Punctilious® SDA 23A Anhydrous;

Punctilious® SDA 23H 190 Proof; Punctilious® SDA 25A-1 190 Proof; Punctilious® SDA 25A-2 190 Proof; Punctilious® SDA 29-2 190 Proof; Punctilious® SDA 29-2 Anhydrous; Punctilious® SDA 29-3 190 Proof; Punctilious® SDA 30 190 Proof; Punctilious® SDA 30 Anhydrous; Punctilious® SDA 32 190 Proof; Punctilious® SDA 32 Anhydrous; SBP 140/165; *SD alcohol 1*; *SD alcohol 3-A*; *SD alcohol 3-C*; *SD alcohol 30*; Shell Mineral Spirits 135; Shellsol® D25; Shellsol® D38; Shellsol® D40; Shellsol® D43; Shellsol® D60; Shellsol® D60 SC; Shellsol® D70; Shellsol® D-LAWS; Shellsol® Heptane; Shellsol® TD; Shellsol® TM; *VM&P naphtha*; VM&P Naphtha HT

solvent, polyacrylates
Calcium thiocyanate; *Sodium thiocyanate*

solvent, polyacrylonitrile fibers
1,2-Butylene carbonate

solvent, polyacrylonitrile polymerization/spinning
Ethylene carbonate

solvent, polyacrylonitriles
Butyrolactone

solvent, polyamides
BDGE; BGE; Eastman® DE; Eastman® EEH; Eastman® Methyl Acetate, High Purity; EDGE; EGE; i-PGE; MDGE; *Methyl pentynol*; MGE

solvent, polyester dissolving
Dynasolve MP-AL

solvent, polyester sealing coats
Rhodiasolv RPDE

solvent, polyester surface primers
Santosol® DMA; Santosol® DME; Santosol® DME-2; Santosol® DME-3; Santosol® DMG; Santosol® DMS

solvent, polyesters
Alkasurf® OD-410; Eastman® EEH; Eastman® Methyl Acetate, High Purity; Igepal® OD-410; *Methoxyethanol*; NMP; *Phenoxyethanol*; *Propylene glycol dimethyl ether*; *Triethylene glycol*

solvent, polyether urethane dissolving
Dynasolve MP-AL

solvent, polyhydroxylic alcohols
2-Pyrol®

solvent, polyimides
Dimethyl acetamide

solvent, polymer chemistry
Butyl Diglyme; *Diethylene glycol dibutyl ether*; Ethyl Diglyme; *Ethylene glycol dimethyl ether*; Ethyl Glyme; Monoglyme; *PEG-4 dimethyl ether*; Tetraglyme; Triglyme

solvent, polymer emulsions
Condea Vista LPA-170; Sasol LPA-170 Solvent

solvent, polymer membranes: gas scrubbing
HEP®; *N-Hydroxyethylpyrrolidone*

solvent, polymer removal
Dynasolve 100; Dynasolve 180; Dynasolve 185; Dynasolve 699; Dynasolve M-10; Dynasolve MP-500; NMP

solvent, polymerization
t-Butylbenzene; *Methyl ethyl ketone*; *Polysorbate 60*; *Sulfolane*; Sulfolane W; *Tetrahydrofuran*; Ucar® n-Propyl Propionate

solvent, polymerization: photoresist inks
Eastman® EEP

solvent, polymers
Albrite® Tributyl Phosphate; *Benzonitrile*; *t-Butyl alcohol*; Dermol ICSA; DMSO; Dynasolve 200; Dynasolve 220; Dynasolve 225; Dynasolve 230; *Ethylene carbonate*; GBL; *4-Heptanone*; *Hexachloro-1,3-butadiene*; *Isocetyl salicylate*; Jarcol™ I-16; Jarcol™ I-24; *Nitromethane*; *2-Nitropropane*; *Propylene carbonate*; 2-Pyrol®; *2-Pyrrolidone*

solvent, polymers: food-pkg.
Citroflex® A-2

solvent, polystyrene
Butyrolactone

solvent, polysulfide polymer/polyepoxies: food-contact
Butoxyethanol; *Butyl alcohol*

solvent, polysulfide removal: aircraft fuel tanks
Dynasolve 200

solvent, polysulfone: food-contact
Chlorobenzene; *Dimethyl sulfoxide*; *N-Methyl-2-pyrrolidone*

solvent, polyvinyl butyral
Methyl alcohol

solvent, polyvinyl resins
3-Heptanone; *Isophorone*

solvent, poorly soluble actives: liq. pharmaceuticals
Transcutol® P

solvent, poorly soluble actives: pharmaceutical capsules
Transcutol® HP

solvent, poorly soluble drug actives: hard gelatin capsules
Transcutol® P

solvent, poorly soluble drug actives: liq. oral pharmaceuticals
Transcutol® HP

solvent, poorly soluble drug actives: soft gelatin capsules
Transcutol® P

solvent, powder coat systems
N-Terpinal RS-75

solvent, powders
Tricaprin

solvent, PPO: food-contact
Methyl alcohol; *Toluene*

solvent, precision cleaning
Purasolv® ELECT; Purasolv® ELS; Soygold® 1000; Soygold® 2000; VersaTRANS™ Solvent

solvent, precision cleaning
Vertrel® SMT

solvent, precision cleaning
Vertrel® XE; Vertrel® XM; *Water*

solvent, precision cleaning: aerospace parts/equip.
Leksol™ AL

solvent, precision cleaning: aq. paints/coatings
Purasolv® ELS

solvent, precision cleaning: circuit boards
Hydrothene™; Leksol™

solvent, precision cleaning: cold cleaning
Leksol™ AL

solvent, precision cleaning: elec. equip.
Leksol™ AL

solvent, precision cleaning: electronic components
Hydrothene™; Leksol™; Leksol™ AL

solvent, precision cleaning: electronics
Purasolv® ELS

solvent, precision cleaning: high-solids paints/coatings
Purasolv® ELS

solvent, precision cleaning: instruments
Leksol™ AL

solvent, precision cleaning: metal components
Leksol™ AL

solvent, precision cleaning: mold release agents
Purasolv® ELS

solvent, precision cleaning: motors
Leksol™

solvent, precision cleaning: optical lenses
Leksol™ AL

solvent, precision cleaning: paints/coatings
Purasolv® ELS

solvent, precision cleaning: PCBs
Leksol™ AL

solvent, precision cleaning: printing inks
Purasolv® ELS

solvent, precision cleaning: temp.-sensitive parts
Leksol™

solvent, precision cleaning: ultrasonic cleaning
Leksol™ AL

solvent, precision cleaning: vapor degreasing
Leksol™ AL

solvent, precision equip.
HC-250; HC-370; HC-FX50; HC-FX70

solvent, prescouring
Burcolev DLS

solvent, preservatives
Diethylene glycol benzyl ether; *Ethylene glycol benzyl ether*

solvent, preserving solutions
SD alcohol 1; *SD alcohol 3-A*; *SD alcohol 3-C*; *SD alcohol 12-A*; *SD alcohol 13-A*; *SD alcohol 22*; *SD alcohol 23-A*; *SD alcohol 30*; *SD alcohol 32*; *SD alcohol 37*; *SD alcohol 38-B*; *SD alcohol 42*; *SD alcohol 44*

solvent, preshave lotions
Estalan-DIPS; Estol IPM 1509 (BIO-IPM)

solvent, preshave preps.
HallStar™ BST

solvent, pressure washers
N-Terpinal RS-98

solvent, pressure-sensitive tapes
Shellsol® BF

solvent, primary
Reomol® TBP

solvent, primers
Conap® S-1; Conap® S-13

solvent, printed circuit boards
Isopropyl lactate; *Methyl lactate*; Purasolv® EL; Purasolv® IPL; Purasolv® ML

solvent, printing
Carbowax® PEG 300; Hydrosol Essence 80/110; Hydrosol Essence F<5; RotoSolv; Solvarex 9; Solvarex 10; Solvaro AFD; Soygold® 1000; Soygold® 2000

solvent, printing dyes
Ethylene glycol isopropyl ether

solvent, printing ink dyes
Triolein

solvent, printing ink oils
Condea Vista LPA-210; Conosol 215; Conosol 260; Conosol 340; Conosol C-200

solvent, printing inks
Acetone; *Alcohol denatured*; Aromatic 100; Aromatic 150; *Butoxyethanol*; *Butoxyethanol acetate*; Condea Vista 47; Dalpad A; *Deodorized kerosene*; *Diacetone alcohol*; *Diethylene glycol*; *Diethylene glycol benzyl ether*; *Diethylene glycol butyl ether acetate*; *Dipropylene glycol*; Dowanol® EPh; Dowanol® PM; Eastman® CDA-19; Eastman® DB Acetate; Eastman® DM; Eastman® MIBK; Eastman® PM; Eastman® Isopropyl Acetate; Eastman® Methyl Acetate; Eastman® Methyl Acetate, High Purity; Eastman® n-Propanol; Eastman® n-Propyl Acetate; *Ethoxyethanol acetate*; *Ethyl acetate*; *Ethylene glycol*; *Ethylene glycol benzyl ether*; *Ethylene glycol isopropyl ether*; *Ethyl hexanediol*; *2-Ethylhexyl acetate*; Ethyl Proxitol; Exx-Print® 176 D; Exx-Print® 274 A; Exx-Print® 281 A; Exx-Print® 283 D; Exx-Print® 287 D; Exx-Print® 392 D; Exx-Print® 493 D; Exx-Print® 588 D; Exxsol® D 40; Exxsol® D 60; Exxsol® Heptane; *Glycerin*; *Hexylene glycol*; *Isobutyl acetate*; *Isophorone*; *Isopropyl acetate*; Isopropyl Oxitol; *Kerosene*; *Methoxydiglycol*; *Methoxyethanol*; *Methoxyethanol acetate*; *Methylcyclohexanol*; *Methylcyclohexanone*; *Methyl ethyl ketone*; *Methyl isobutyl ketone*; *Methyl rapeseedate*; Nadone®; *Naphtha, medium aliphatic*; *Nitromethane*; *2-Nitropropane*; Norfox® MCS; PEG 300; *Phenoxyethanol*; Poly-Solv® DM; Poly-Solv® DPM; Poly-Solv® EM; Poly-Solv® MPM; Poly-Solv® TM; Poly-Solv® TPM; PPG 400; *PPG-3 methyl ether*; *Propyl acetate*; *Propyl alcohol*; *Propylene glycol ethyl ether acetate*; *Propylene glycol methyl ether acetate*; Purasolv® BL; Purasolv® ELECT; Rhodiasolv RPDE; Rubber Solvent; Rubber Solvent 332; Santosol® DMA; Santosol® DME; Santosol® DME-2; Santosol® DME-3; Santosol® DMG; Santosol® DMS; Sasol-47; SBP 60/95 LNH; SBP 80/95 LNH; SBP 80/110 LNH; SBP 100/140; SBP 140/165; Shellsol® 426; Shellsol® A100 Solvent; Shellsol® A150ND; Shellsol® BF; Shellsol® D25; Shellsol® D80; Shellsol® D90; Shellsol® D100; Shellsol® D100S; Shellsol® T; Shellsol® TC; Shellsol® TD; Shellsol® TM; Tecsol® A, 95%; Tecsol® A Anhydrous; Tecsol® A-2, 95%; Tecsol® A-2 Anhydrous; Tecsol® B; Tecsol® C 95%; Tecsol® C Anhydrous; Tecsol® D, 95%; Tecsol® D-1, 95%; Tecsol® D-2, 95%; Tecsol® D-2 Anydrous; Tecsol® H; *Tetrahydrofuran*; *Thiodiglycol*; Tolu-Sol 20; Tolu-Sol 30; *Tributyl phosphate*; *1,1,1-Trichloroethane*; *Triethylene glycol*; *Triglycol monomethyl ether*; *Turpentine*; Ucar® n-Propyl Propionate; Varsol® 1; Varsol® 18; Varsol® 110; Witsol® 45; Witsol® 50; Witsol® 471; Witsol® 501; Witsol® MSO; *Xylene*

GAM
solvent, rectal pharmaceuticals
Labrafac® Hydrophile; Miglyol® 812 N; Miglyol® 840
solvent, reductions
Tetrahydrofuran
solvent, refinery cleaning chemicals
Petrinex® ASB
solvent, refining
DMSO
solvent, refinish coatings
Isopar® C
solvent, removers
Dowanol® TPM
solvent, residue cleaning: PVC
M-Pyrol®
solvent, residue cleaning: urethane processing equipment
M-Pyrol®
solvent, residue removal: urethane processing equip.
Dynaflush
solvent, resin cleaners
Coasol™; *Methylene chloride*; N-Terpinal RS-50
solvent, resin clean-up
Santosol® DMA; Santosol® DME; Santosol® DME-2; Santosol® DME-3; Santosol® DMG; Santosol® DMS
solvent, resin coatings
4-Methoxy-4-methylpentanone-2; *SD alcohol 1*; *SD alcohol 23-A*
solvent, resin removal
Tavoxene®
solvent, resin-coated sand: foundry casting
Uniplex 155
solvent, resins
Acetone; Alkasurf® OD-410; *Amyl formate*; BDGE; *Benzaldehyde*; *Benzene*; *Benzonitrile*; *Benzyl acetate*; BGE; *Bromodichloromethane*; *2-Butanol*; *Butoxyethanol*; *Butoxyethanol acetate*; *1-Butoxyethoxy-2-propanol*; *Butyl alcohol*; *Butyl formate*; Butyl Oxitol®; *p-t-Butyl toluene*; *Cyclohexane*; *Cyclohexyl acetate*; *Decahydronaphthalene*; Decalin; Dermol ICSA; *Dibutyl ether*; *Dibutyl phthalate*; *cis-1,2-Dichloroethylene*; *Diethyl carbonate*; *Diethylene glycol phenyl ether*; *Diethylenetriamine*; *Diethyl oxalate*; *Diethyl tartrate*; *Diethyl toluamide*; *Dimethyl acetamide*; *Dimethyl adipate*; *Dimethyl phthalate*; *1,4-Dioxane*; *Dipropylene glycol dimethyl ether*; *Dipropylene glycol ethyl ether*; Eastman® EB; EDGE; EGE; *Erucamide*; *Ethoxydiglycol*; *Ethoxyethanol acetate*; *Ethyl acetate*; *Ethyl benzoate*; *2-Ethyl-1-butanol*; *Ethyl chloride*; *Ethylene carbonate*; *Ethylene chlorohydrin*; *Ethylene glycol diacetate*; *Ethylene glycol isopropyl ether*; *Ethylene glycol propyl ether*; *2-Ethylhexanol*; *2-Ethylhexyl acetate*; *Ethyl morpholine*; *Ethyl propionate*; *Furan*; *Furfuryl alcohol*; *Heptane*; *4-Heptanone*; *2-Hexanone*; i-PGE; *Isoamyl acetate*; *Isoamyl alcohol*; *Isobutyl acetate*; *Isocetyl salicylate*; *Isooctyl alcohol*; *Isophorone*; *1-Isopropoxy-2-propanol*; *2-Isopropoxy-1-propanol*; *Isopropyl alcohol*; *Isopropyl ether*; Isopropyl Oxitol; Jeffsol® EC; Jeffsol® PC; Ketrul 212; Ketrul HT; *Longifolene*; MDGE; *Methoxydiglycol*; *Methoxyethanol acetate*; *2-Methoxy-1-propanol acetate*; *Methyl acetate*; *Methylcyclohexanol*; *Methyl isobutyl ketone*; *N-Methyl-2-pyrrolidone*; MGE; *Morpholine*; M-Pyrol®; Naxol®; *Nitroethane*; *2-Nitropropane*; *Paraldehyde*; PEG 300; *Pentachloroethane*; *Phenoxyethanol*; *β-Picoline*; *γ-Picoline*; *PPG-2 methyl ether*; *PPG-3 methyl ether*; Proglyde DMM; *Propyl acetate*; *Propyl alcohol*; *Propylene dichloride*; *Propylene glycol phenyl ether*; *Quinoline*; *SD alcohol 2-B*; *SD alcohol 12-A*; *SD alcohol 13-A*; *SD alcohol 30*; Shellsol® D38; Shellsol® D40; Shellsol® D43; Shellsol® D60; Shellsol® Heptane; Shellsol® W HT; TBP; *Terpinolene*; *Tetradecene-1*; *Tetraethylenepentamine*; *Tetrahydrofuran*; Tetralin; *Toluene*; *1,1,1-Trichloroethane*; *1,1,2-Trichloroethane*; *Trimethyl borate*; *Tripropylene glycol ethyl ether*; *VM&P naphtha*; VM&P Naphtha HT; X-95 Solvent
solvent, retarder: acrylic coatings
Eastman® EP
solvent, retarder: alkyd resin coatings
Eastman® EP
solvent, retarder: epoxy coatings
Eastman® EP
solvent, retarder: NC coatings
Eastman® EP
solvent, retarder: polyamide coatings
Eastman® EP
solvent, retarding: coatings
Ethyl 3-ethoxypropionate
solvent, road tar fluxing
Creosote oil
solvent, roll-coating inks
2,5-Hexanedione
solvent, room deodorants
SD alcohol 3-A; *SD alcohol 3-C*; *SD alcohol 22*; *SD alcohol 37*; *SD alcohol 38-B*; *SD alcohol 39-B*; *SD alcohol 39-C*; *SD alcohol 40*; *SD alcohol 40-A*; *SD alcohol 40-B*; *SD alcohol 40-C*
solvent, rosin
Eastman® IBIB; *dl-Limonene*; *Methyl abietate*; *Methyl alcohol*
solvent, rotogravure dyes
SD alcohol 32
solvent, rotogravure inks
Isoamyl chloride
solvent, RTV removal: CRT tubes
Dynasolve 210; Dynasolve 220; Dynasolve 225; Dynasolve 230
solvent, RTV rubbers
Dynasolve MP-AL
solvent, RTV silicones
Dynasolve 210; Dynasolve 220; Dynasolve 225; Dynasolve 230
solvent, rubber
Butyl stearate; *Carbon disulfide*; *Carbon tetrachloride*; Carbowax® E200; Carbowax® E300; Carbowax® E400; Carbowax® E600; Carbowax® E900; Carbowax® E1000; Carbowax® E1450; Carbowax® E3350; Carbowax® E8000; Carbowax® PEG 300; *Cyclohexanol*; *Cyclohexanone*; *Decahydronaphthalene*; Decalin; *Diisobutyl ketone*; Dipentene-C; *Heptane*; *Hexachloro-1,3-butadiene*; Hi-Sol® 10; Hi-Sol® 15; Hydrosol 100/130 N; Hydrosol Essence 80/110; Hydrosol Essence 100/140; Hydrosol Essence C; Hydrosol Essence E<5; Hydrosol Essence F<5; Hydrosol Heptane; Isohexane LNH; Jarcol™ I-16; Jarcol™ I-24; Kwik Dri; Lacolene™; *dl-Limonene*; *Methyl abietate*; *Methyl benzoate*; *Paraldehyde*; *Perchloroethylene*; Pluracol® E1450 NF; SBP 40/65 LNH; SBP 60/95 LNH; SBP 80/95 LNH; SBP 80/110 LNH; SBP 100/140; SBP 140/165; SDA-2B; Shellsol® 426; Shellsol® A150ND; Shellsol® D25; Shellsol® H; Shellsol® Heptane; Shellsol® W HT; *Tetraethyl lead*; *Tetrahydronaphthalene*; Tetralin; *Toluene*; Tolu-Sol 20; Tolu-Sol 30; *Turpentine*; *VM&P naphtha*; VM&P Naphtha HT; X-55 Solvent; *Xylene*
solvent, rubber accelerators
Isopropylamine; *β-Picoline*; *γ-Picoline*
solvent, rubber cements
Shellsol® BF; *Xylene*
solvent, rubber compounding
Naphtha
solvent, rubber processing: ethylene-propylene
Isopentane
solvent, rubber reclaiming
Dipentene-C
solvent, rubber solutions
Shellsol® BF
solvent, rubberized fabrics
Dibutyl tartrate
solvent, rubberproofing
Shellsol® BF
solvent, rubbing alcohols
SD alcohol 23-H
solvent, rust removers
Diethylene glycol benzyl ether; Dowanol® PM; Dowanol® TPM; Ecosoft™ PB Solvent; Ecosoft™ PE Solvent; Ecosoft™ PH Solvent; *Ethylene glycol benzyl ether*
solvent, S/B latex finishes
Genesolv® D
solvent, S/B solution prods.
Tetrahydrofuran
solvent, sacrificial agent corrosion inhibitors
C12-14 alkylbenzene; V-154L; V-159L
solvent, safety glasses
Ethyl-(S)-lactate
solvent, SBR
Butyl Cellosolve®
solvent, scintillation counting
Water
solvent, screen inks
Eastman® C-11 Ketone; Eastman® DB Acetate; Eastman® EB Acetate; Eastman® PM Acetate; Estasol™ MD10
solvent, screen printing inks
C8 alkyl acetate; *C9 alkyl acetate*; *C10 alkyl acetate*; *C13 alkyl acetate*; Exxate® 600; Exxate® 700; Exxate® 800; Exxate® 900; Exxate® 1000; Exxate® 1300; *Oxo-heptyl acetate*; *Oxo-hexyl acetate*
solvent, sealants
Aromatic 100; Aromatic 100 Anti-Static; Aromatic 150; Aromatic 200; *1,1′-Biphenyl, butenylated*; *1,1′-Biphenyl (1-methylethyl)*; *Bis (methylethyl)-1,1′-biphenyl*
solvent, sealants
Estasol™
solvent, sealants
Estasol™; Exxsol® D 95; *Heptane*; *Isobutyl acetate*; *Naphtha*; Naphthalene Depleted Aromatic 150 Fluid; Naphthalene Depleted Aromatic 200 Fluid; SBP 80/95 LNH; SBP 80/110 LNH; Shellsol® 426; Solvesso™ 100; Solvesso™ 150; Solvesso™ 150 ND; Solvesso™ 200; Solvesso™ 200 ND; Solvesso™ 200 S; *Tetrahydrofuran*; Ultra Low Naphthalene Aromatic 150 Fluid; Ultra Low Naphthalene Aromatic 200
solvent, selective
Diethylamine; *N-Methyl-2-pyrrolidone*; *Sulfolane*; *Thiodipropionitrile*
solvent, selective extraction: acetylene
N-Methylformamide
solvent, selective extraction: butadiene
Dimethyl acetamide; *Furfural*; *N-Methylformamide*
solvent, selective extraction: metals
Cyanex® 272
solvent, selective: acetylene
2-Pyrrolidone
solvent, selective: acrylonitrile
2-Pyrrolidone
solvent, selective: alcohol separation
Eastman® DB Acetate
solvent, selective: aromatics
N-Methylformamide
solvent, selective: aromatics extraction
Furfural
solvent, selective: aromatics extraction process
N-Formylmorpholine
solvent, selective: dry cleaning
sym-Dichloroethyl ether
solvent, selective: finish removers
sym-Dichloroethyl ether
solvent, selective: gold extraction
Diethylene glycol dibutyl ether
solvent, selective: ketone separation
Eastman® DB Acetate
solvent, selective: lacquers
sym-Dichloroethyl ether
solvent, selective: lubricating oil dearomatization
Phenol
solvent, selective: lubricating oils
sym-Dichloroethyl ether
solvent, selective: paints
sym-Dichloroethyl ether
solvent, selective: Selexol process
PEG-4 dimethyl ether
solvent, selective: spot cleaning
sym-Dichloroethyl ether
solvent, selective: textile fulling compds.
sym-Dichloroethyl ether
solvent, selective: textile scouring compds.
sym-Dichloroethyl ether
solvent, selective: varnishes
sym-Dichloroethyl ether
solvent, selenium
Carbon disulfide
solvent, semiaq. defluxing
NMP-EL
solvent, semiconductor processing
Arcosolv® PMA; *Water*
solvent, semipermanent liq. hair color systems
Lowenol Solvent 402
solvent, separation

solvent, spotting: dry cleaning
Pancreatin
solvent, spray gun clean-up
N-Terpinal RS-25
solvent, spray paints
Butoxyl®
solvent, spray washers
DA-405B
solvent, sprays
Isane IP 130; Isane IP 155; Isane IP 165; Isane IP 175; Isane IP 185
solvent, SR
Butyl Oxitol®; *1-Chloro-1-nitropropane*; *Dimethyl sebacate*; *1,4-Dioxane*; *1-Nitropropane*; *2-Nitropropane*
solvent, stabilization
Butyl Diglyme; *Diethylene glycol dibutyl ether*; Ethyl Diglyme; *Ethylene glycol dimethyl ether*; Ethyl Glyme; Monoglyme; Tetraglyme; Triglyme
solvent, stain removers
Hydrosol Essence A; Isopar® V Fluid; *Isopropyl ether*
solvent, stains
2,5-Hexanedione; *Methyl lactate*; Punctilious® Ethyl Alcohol 190 Proof Pure USP; Punctilious® Ethyl Alcohol 200 Proof Pure USP; Punctilious® SDA 1-2 190 Proof; Punctilious® SDA 1-2; Punctilious® SDA 2B-2 190 Proof; Punctilious® SDA 2B-2 Anhydrous; Punctilious® SDA 2B-3 190 Proof; Punctilious® SDA 2B-4 190 Proof; Punctilious® SDA 2B-4 Anhydrous; Punctilious® SDA 3A 190 Proof; Punctilious® SDA 3A Anhydrous; Punctilious® SDA 3C 190 Proof; Punctilious® SDA 12A-3 190 Proof; Punctilious® SDA 12A-3 Anhydrous; Punctilious® SDA 13A 190 Proof; Punctilious® SDA 19 190 Proof; Punctilious® SDA 20 190 Proof; Punctilious® SDA 20 Anhydrous; Punctilious® SDA 23A 190 Proof; Punctilious® SDA 23A Anhydrous; Punctilious® SDA 23H 190 Proof; Punctilious® SDA 25A-1 190 Proof; Punctilious® SDA 25A-2 190 Proof; Punctilious® SDA 29-2 190 Proof; Punctilious® SDA 29-2 Anhydrous; Punctilious® SDA 29-3 190 Proof; Punctilious® SDA 30 190 Proof; Punctilious® SDA 30 Anhydrous; Punctilious® SDA 32 190 Proof; Punctilious® SDA 32 Anhydrous; Sasol LPA Solvent; Sasol LPA-142 Solvent; Sasol LPA-150 Solvent; Sasol LPA-170 Solvent; Sasol LPA-210 Solvent; *SD alcohol 1*; *SD alcohol 3-A*; *SD alcohol 3-C*; *SD alcohol 23-A*; *SD alcohol 30*
solvent, stamp pad inks
Eastman® EB Acetate; Ethyl Proxitol; *Methoxydiglycol*
solvent, stamp pad/pen inks
Dowanol® TPM
solvent, stamp pads
Dowanol® TPnB
solvent, starch ethers
Ethyl levulinate
solvent, steel
HC-250; HC-370; HC-FX50; HC-FX70
solvent, stencil inks
Methoxydiglycol
solvent, stencil pastes
Butyl-(S)-lactate
solvent, stencil sheets
Ethyl-(S)-lactate
solvent, sterilizing
SD alcohol 46
solvent, sterilizing sol'ns
SD alcohol 38-B
solvent, sterilizing solutions
SD alcohol 1; *SD alcohol 3-A*; *SD alcohol 3-C*; *SD alcohol 12-A*; *SD alcohol 13-A*; *SD alcohol 22*; *SD alcohol 23-A*; *SD alcohol 30*; *SD alcohol 32*; *SD alcohol 37*; *SD alcohol 42*; *SD alcohol 44*
solvent, steroids
Cremophor® EL; Lutrol® E 1500; Lutrol® E 3350; Lutrol® E 4000; Lutrol® E 6000; *PEG-125*; *Sesame (Sesamum indicum) oil*
solvent, sticks
Tricaprin
solvent, stoving enamels
Butoxyl®
solvent, strippers
Eastman® DE-HG
solvent, stripping
GAFsolver™ TMP; Rhodiasolv RPDE
solvent, stripping prods.
Purasolv® EL; Purasolv® IPL; Purasolv® ML
solvent, sugars
2-Pyrol®
solvent, Sulfinol process: acid gas removal
Sulfolane W
solvent, sulfur
Carbon disulfide; *Diethylenetriamine*; *Ethyl chloride*; *Ethylenediamine dihydrochloride*; *Methyl disulfide*; *Tetraethylenepentamine*
solvent, sulfur: chemical plant flowlines
Sulfa-Hitech® 0382
solvent, sulfur: refinery flowlines
Sulfa-Hitech® 0382
solvent, sulfur: sour gas wells
Sulfa-Hitech® 0382
solvent, sulfur: sour-gas pipelines
Sulfa-Hitech® 0382
solvent, sun care
Estol IPM 1509 (BIO-IPM); Finsolv® TPP; *Isopropyl PPG-2-isodeceth-7-carboxylate*; Ritacane ID
solvent, sun oils
Labrafac® Lipophile
solvent, sunscreen actives
Butyl octyl benzoate; Crodamol PMP; Finsolv® TPP; Luvitol® BD 10 P; Macol® 57; *PPG-10 butanediol*; *PPG-2 myristyl ether propionate*
solvent, sunscreens
Akomed R; *Benzyl laurate*; Carbowax® Sentry® PEG 300; Carbowax® Sentry® PEG 400; Carbowax® Sentry® PEG 540 Blend; Carbowax® Sentry® PEG 600; Cromollient DP3A; Cropure® Babassu; *Di-PPG-3 myristyl ether adipate*; Finsolv® TN-O; GBL; *Glycereth-7 triacetate*; HallBrite® BHB; HallBrite® OS; HallStar™ DIPA; HallStar™ DIPS; HallStar™ DOA; HallStar™ DOM; HallStar™ DOS; *Hexyldecyl benzoate*; Novol NF; Pelemol® G7A; Trivent OP
solvent, suntan creams/lotions
Fanoliv™ Oil
solvent, suntan gels
EmCon™ Olive; Fanoliv™ Oil; Renex® PEG 300
solvent, suntan oils
Fancol™ ALA
solvent, suntan preps.
Estalan DIA; Fancol™ CB; Fancol™ CB Extra; Fancol™ Karite Butter; *Shea butter (Butyrospermum parkii)*
solvent, suppositories
Capmul® MCM; Miglyol® 810; Miglyol® 840; *Propylene glycol dioctanoate*
solvent, suppository bases
Renex® PEG 400; Renex® PEG 600
solvent, surface active: industrial cleaners
N-Octyl pyrrolidone
solvent, surface active: polishes
N-Octyl pyrrolidone
solvent, surface active: waxes
N-Octyl pyrrolidone
solvent, surface cleaners
1,2-Butylene carbonate; Jeffsol® 1,2-Butylene Carbonate; Jeffsol® EC-25; Jeffsol® EC-50; Jeffsol® EC-75; Varonic® DM55
solvent, surface coatings
BDGE; BGE; EDGE; EGE; *Ethoxyethanol*; *Ethylene glycol isopropyl ether*; Ethyl Proxitol; i-PGE; *1-Isopropoxy-2-propanol*; *2-Isopropoxy-1-propanol*; Isopropyl Oxitol; Jeffsol® PC; MDGE; *2-Methyl-1-pentanol*; MGE; Parabase; SBP 80/95 LNH; SBP 80/110 LNH; SBP 100/140; SBP 140/165; Shellsol® 426; Shellsol® 664; Shellsol® A150ND; Shellsol® D25; Shellsol® D-LAWS; Shellsol® H; Shellsol® OMS; Shellsol® TC; Shellsol® TD; Shellsol® TM; Shell Toluene Solvent; Solvenon® PP; *Sulfolane*; Sulfolane W; *1,1,1-Trichloroethane*; *Trichloroethylene*
solvent, surface preps.
Dynasolve 180; Dynasolve 185; Dynasolve 711; Dynasolve 750; NMP; Uresolve Plus 500; Uresolve Plus SG
solvent, surfactants
s-Butylbenzene; *Isopropylamine*; Punctilious® Ethyl Alcohol 190 Proof Pure USP; Punctilious® Ethyl Alcohol 200 Proof Pure USP; Punctilious® SDA 1-2 190 Proof; Punctilious® SDA 1-2; Punctilious® SDA 2B-2 190 Proof; Punctilious® SDA 2B-2 Anhydrous; Punctilious® SDA 2B-3 190 Proof; Punctilious® SDA 2B-4 190 Proof; Punctilious® SDA 2B-4 Anhydrous; Punctilious® SDA 3A 190 Proof; Punctilious® SDA 3A Anhydrous; Punctilious® SDA 3C 190 Proof; Punctilious® SDA 12A-3 190 Proof; Punctilious® SDA 12A-3 Anhydrous; Punctilious® SDA 13A 190 Proof; Punctilious® SDA 19 190 Proof; Punctilious® SDA 20 190 Proof; Punctilious® SDA 20 Anhydrous; Punctilious® SDA 23A 190 Proof; Punctilious® SDA 23A Anhydrous; Punctilious® SDA 23H 190 Proof; Punctilious® SDA 25A-1 190 Proof; Punctilious® SDA 25A-2 190 Proof; Punctilious® SDA 29-2 190 Proof; Punctilious® SDA 29-2 Anhydrous; Punctilious® SDA 29-3 190 Proof; Punctilious® SDA 30 190 Proof; Punctilious® SDA 30 Anhydrous; Punctilious® SDA 32 190 Proof; Punctilious® SDA 32 Anhydrous
solvent, synthetic coating resin thinning
Eastman® EEP
solvent, synthetic elastomers
NMP
solvent, synthetic fibers
PPG-40 diethylmonium chloride; *Propylene carbonate*; Variquat® CC-42; Variquat® CC-57
solvent, synthetic lubricants
Carbowax® E200; Carbowax® E300; Carbowax® E400; Carbowax® E600; Carbowax® E900; Carbowax® E1000; Carbowax® E1450; Carbowax® E3350; Carbowax® E8000
solvent, synthetic musk
Benzylsalicylate
solvent, synthetic polymers
Butyl-(S)-lactate
solvent, synthetic resin coatings
SD alcohol 3-A; *SD alcohol 3-C*; *SD alcohol 30*
solvent, synthetic resins
Diisobutyl ketone; Eastman® IBIB; *2-Heptanol*; *Isopropyl acetate*
solvent, synthetic transformer oils
1,2,4-Trichlorobenzene
solvent, tablet coatings
Methylene chloride
solvent, tail-gas-treating units
Jefftreat® MS-300
solvent, talcum powders
STEPAN® IPM-NF
solvent, tank reconditioning
N-Terpinal RS-25
solvent, tannins
Isoamyl acetate
solvent, tar separation
Sulfolane
solvent, tars
Diacetone alcohol
solvent, TDI
Diethyl isophthalate
solvent, technical concs.
Agsolex® BLO
solvent, terpenes
Quinoline
solvent, textile backing materials
Dimethyl formamide
solvent, textile baths
Sodium thiocyanate
solvent, textile chemicals
Coasol™
solvent, textile cleaning
Shellsol® BF; Shellsol® TD; Shellsol® TM
solvent, textile coatings
Heptane; *Mineral spirits*; Shell Mineral Spirits 135; Shellsol® D40; Shellsol® D43; Shellsol® D60; Shellsol® Heptane; Ucon® 50-HB-55; Ucon® 50-HB-100; Ucon® 50-HB-260; Ucon® 50-HB-400; Ucon® 50-HB-660; Ucon® 50-HB-2000; Ucon® 50-HB-3520; Ucon® 50-HB-5100; Ucon® 75-H-450; Ucon® LB-65; Ucon® LB-135; Ucon® LB-165; Ucon® LB-285; Ucon® LB-385; Ucon® LB-525; Ucon® LB-625; Ucon® LB-1145; Ucon® LB-1715; Ucon® LB-3000; *VM&P naphtha*; VM&P Naphtha HT
solvent, textile dry cleaning
Shellsol® H
solvent, textile dye pastes
Eastman® DM; *Methoxydiglycol*
solvent, textile dyeing/printing
Triethylene glycol
solvent, textile dyes
Ethylene glycol isopropyl ether; *Hexylene glycol*

Vertrel® XE; Vertrel® XF; Vertrel® XM

solvent, vapor degreasing equip.: electronic assemblies
Vertrel® XMS Plus

solvent, vapor degreasing equip.: electronic assembly cleaning
Vertrel® SMT

solvent, vapor degreasing equip.: precision cleaning
Vertrel® MCA; Vertrel® MCA Plus

solvent, vapor degreasing equip.: specialty cleaning
Vertrel® MCA; Vertrel® MCA Plus

solvent, varnish removers
Eastman® CDA-19; *Ethoxyethanol*; *Ethylene glycol acetate*

solvent, varnishes
Acetone; *Butoxyethanol*; *Butoxyethanol acetate*; *s-Butyl acetate*; *Butyl alcohol*; *Butyl-(S)-lactate*; *C isoparaffin*; *Cyclohexanol*; *Diethylene glycol benzyl ether*; Dipentene-C; Dowanol® TPM; Dynasolve MP-AL; *Ethylene glycol benzyl ether*; *Ethyl-(S)-lactate*; Galaster EL; Hydrosol 100/130 N; Hydrosol Essence E<5; Hydrosol Essence F<5; *Isoamyl chloride*; *Isobutyl acetate*; Jeffsol® PC; Ketrul 212; Ketrul D70; Ketrul D85; Ketrul D211; Ketrul D220; Ketrul HT; *Methoxyethanol*; *Methyl abietate*; *Methyl ethyl ketone*; *Methyl isobutyl ketone*; *Methyl propionate*; *2-Nitropropane*; N-Terpinal RS-50; PPG 400; *Propyl propionate*; Solvarex 9; Solvarex 10; Solvarex 90/160; Solvarex 90/180; Solvaro AFD; Spirdane BT; Spirdane D30; Spirdane D40; Spirdane D40L; Spirdane D60; Spirdane D60L; Spirdane HT; Spirdane K1; Spirdane K2; Spirdane L1; *Stoddard solvent*; Tecsol® 1; Tecsol® 3; *Trimethyl borate*

solvent, vegetable froth flotation
Deodorized kerosene; PD-23; PD-25; PD-28

solvent, vegetable oil extraction
Hexane

solvent, vegetable oils
2,3-Dimethylbutane; *Isopropyl ether*; PPG 400; PPG 2250; *Propyl alcohol*

solvent, vehicle coatings
Sasol MR Solvent

solvent, vehicle underbody coatings
Shellsol® D100; Shellsol® D100S

solvent, veterinary creams/lotions
Caprylic/capric triglyceride PEG-4 esters

solvent, veterinary emulsions
Labrafac® Lipophile

solvent, veterinary inj. solutions
Solutol® HS 15

solvent, veterinary medicine
Cremophor® EL; *2-Pyrrolidone*; Soluphor® P

solvent, veterinary microemulsions
Labrafac® Lipophile

solvent, veterinary parasitic worm destruction
1,1,2,2-Tetrachloro-1,2-difluoroethane

solvent, veterinary preps.
Labrafac® Hydrophile; Labrasol®; *Petrolatum*

solvent, vinyl acetate copolymers
Ethylan® HB1

solvent, vinyl butyral
Eastman® Methyl Acetate, High Purity

solvent, vinyl chloride acetate
Eastman® Methyl Acetate, High Purity

solvent, vinyl copolymers
Isopropyl acetate; NMP

solvent, vinyl film
Methyl ethyl ketone

solvent, vinyl plastisols
C8 alkyl acetate; *C9 alkyl acetate*; *C10 alkyl acetate*; *C13 alkyl acetate*; Exxate® 600; Exxate® 700; Exxate® 800; Exxate® 900; Exxate® 1000; Exxate® 1300; *Methyl butynol*; *Oxo-heptyl acetate*; *Oxo-hexyl acetate*

solvent, vinyls
Alkasurf® OD-410; *Cyclohexanone*; *Dimethyl acetamide*; *Dimethyl formamide*; *Dimethyl sebacate*; Emery® 6705; Igepal® OD-410; *Mesityl oxide*; *Methoxyethanol*; *Methyl abietate*; M-Pyrol®; *Nitroethane*; *1-Nitropropane*; *2-Nitropropane*; *3-Octanone*; *Phenoxyethanol*; Propyl Cellosolve®; TBP; *Tetrahydrofurfuryl alcohol*; *Triethylene glycol*

solvent, vitamin A
Solutol® HS 15

solvent, vitamin D
Solutol® HS 15

solvent, vitamin E
Solutol® HS 15

solvent, vitamin K
Solutol® HS 15

solvent, vitamins
Akomed R; Captex® 200; Delios®; GBL; Lexol® PG-800; Lexol® PG-865; Propylene Glycol USP/FCC Ultra Grade; *Propylene glycol dicaprylate/dicaprate*; *Propylene glycol dioctanoate*

solvent, VOC-free formulations
Paxsolv CS

solvent, wallpaper
Shellsol® T; Shellsol® TD; Shellsol® TM

solvent, water removal: nonporous surfs.
Vertrel® X-P10

solvent, water treatment
Condea Vista LPA-142; Condea Vista LPA-170; Condea Vista LPA-210; Eastman® Glacial Acetic Acid; PD-23; PD-25; PD-28

solvent, water treatment chemicals
Condea Vista LPA; Conosol 215; Conosol 260; Conosol 340; Conosol C-145; Conosol C-170; Conosol C-200; Sasol LPA Solvent; Sasol LPA-142 Solvent; Sasol LPA-150 Solvent; Sasol LPA-170 Solvent; Sasol LPA-210 Solvent

solvent, water-/solvent-borne coatings
Dipropylene glycol n-propyl ether

solvent, water-dilutable coatings
Eastman® DP; Eastman® EB

solvent, water-disp.
Butyl Carbitol®; Carbitol® PM-600; Methyl Carbitol®; Ucar® Filmer IBT

solvent, water-disp. coatings
Hexyl Cellosolve®

solvent, water-insoluble pharmaceuticals
PEG-125

solvent, water-insoluble/poorly soluble pharmaceuticals
Lutrol® E 1500; Lutrol® E 3350; Lutrol® E 4000; Lutrol® E 6000

solvent, waterless hand cleaners
Butyl Icinol; *C13-14 isoparaffin*; Conosol C-145; Drakeol® 7; Drakesol® 165; Drakesol® 165AT; Drakesol® 205; Drakesol® 220; Drakesol® 260; Drakesol® 260 AT; Drakesol® 305; Isopar® M Fluid; Isopar® V Fluid; Lonzest® 143-S; Neosolv™ 2; Neosolv™ 4; Neosolv™ 6; Neosolv™ 8; Neosolv™ 24; Neosolv™ 150; Parabase; PD-23; PD-25; PD-28; *Petroleum distillates, hydrotreated light*; Sasol LPA Solvent; Sasol LPA-142 Solvent; Sasol LPA-150 Solvent; Sasol LPA-170 Solvent; Sasol LPA-210 Solvent

solvent, waterproofing agents
γ-Picoline

solvent, water-reducible aerosol paints
Propylene glycol n-propyl ether

solvent, water-reducible alkyds
Arcosolv® PTB

solvent, water-reducible polyesters
Arcosolv® PTB; *Butoxypropanol*

solvent, water-soluble
Butyl Carbitol®; Cellosolve®; Propyl Cellosolve®

solvent, water-soluble coatings
Carbitol® PM-600; Methyl Carbitol®; Ucar® Filmer IBT

solvent, wax removal
Tavoxene®

solvent, wax/oil removal: fabrics
Stetlasolv VJS

solvent, waxes
Acetone; BDGE; *Benzene*; *Benzyl alcohol*; BGE; *Bromodichloromethane*; *Bromoform*; *Butyl alcohol*; *Butyl chloride*; *Carbon disulfide*; *Carbon tetrabromide*; *Cyclohexane*; *Cyclohexanone*; *Decahydronaphthalene*; Dermol ICSA; *cis-1,2-Dichloroethylene*; *Dioctyl maleate*; *1,4-Dioxane*; Dipentene-C; EDGE; EGE; *Erucamide*; Estasol™; *2-Ethyl-1-butanol*; *Ethyl chloride*; *Ethylene chlorohydrin*; *Ethyl ether*; *2-Ethylhexyl acetate*; Exxsol® D 95; HallStar™ DOM; *2-Hexanone*; i-PGE; *Isoamyl acetate*; *Isocetyl salicylate*; Isodragol 2/050300; *Isopropyl ether*; *dl-Limonene*; MDGE; *Methoxydiglycol*; *Methoxyethanol acetate*; *Methylcyclohexanol*; *N-Methyl-2-pyrrolidone*; MGE; *Morpholine*; M-Pyrol®; Nadone®; *Naphtha, medium aliphatic*; Naxol®; *Nitroethane*; *Nitromethane*; NMP; Norpar® 15 Fluid; *2-Octanol*; *Perchloroethylene*; *α-Pinene*; *Propyl alcohol*; *Propylene dichloride*; SBP 140/165; Shellsol® D25; Shellsol® D60 SC; Shellsol® D70; Shellsol® D-LAWS; Shellsol® TD; Shellsol® TM; *Tetrahydronaphthalene*; *1,1,1-Trichloroethane*; *1,1,2-Trichloroethane*; *Trimethyl borate*; *Turpentine*

solvent, waxes: cosmetics
Bernel® Ester DCM

solvent, welding gas
Dimethyl formamide

solvent, well drilling aq. muds
Ecosoft™ PB Solvent; Ecosoft™ PE Solvent; Ecosoft™ PH Solvent

solvent, wet felt continuous cleaning: paper machine press sections
Qemifeltwash 8020A; Qemifeltwash 8020N

solvent, wet process duplicator liqs.
Hisolve TM

solvent, whipped toppings
Hodag® S-35 FGK

solvent, winter sports prods.
Fancol™ Karite Butter

solvent, wire enamel resins
Dimethyl formamide; Santosol® DMA; Santosol® DME; Santosol® DME-2; Santosol® DME-3; Santosol® DMG; Santosol® DMS

solvent, wire enamels
Cresylic acid; *Xylenol*

solvent, wood furniture coatings
Arcosolv® TPNB

solvent, wood paints
Lusolvan® FBH

solvent, wood preservatives
Diacetone alcohol; Isane IP 165; Isane IP 175; Isane IP 185; Ketrul 212; Ketrul D70; Ketrul D80; Ketrul D211; Ketrul D220; Ketrul HT; Shellsol® A150ND; Shellsol® D38; Shellsol® D60 SC; Shellsol® D70; Shellsol® D-LAWS; Solvarex 9; Spirdane BT; Spirdane D40; Spirdane D60; Spirdane HT; Spirdane K1; Spirdane K2; Sunoco 100; Sunoco 150

solvent, wood processing
Carbowax® PEG 300

solvent, wood prods.
Pluracol® E1450 NF

solvent, wood stains
Ethoxydiglycol; Ethyl Proxitol; Lusolvan® FBH; *Methoxyethanol*

solvent, wood treatment
Shellsol® D80; Shellsol® H

solvent, wrinkle creams
Fancol™ Karite Butter

solvent, zinc electroplating
Dimethyl formamide

solvent/carrier
Eastman® Diethyl Phthalate

solvent/cosolvent, adhesives
Tebol®; Tebol® 99

solvent/cosolvent, agric.
Tebol®; Tebol® 99

solvent/cosolvent, cleaners/ disinfectants
Tebol®; Tebol® 99

solvent/cosolvent, coatings
Tebol®; Tebol® 99

solvent/cosolvent, personal care
Acetulan®; Tebol®; Tebol® 99

solvent/cosolvent, pharmaceuticals
Castor (Ricinus communis) oil; Tebol®; Tebol® 99

solvent/cosolvent, polymer processing
Tebol®; Tebol® 99

solvent: acid component removal, gases
2-(2-Aminoethoxy) ethanol

soothing agent, cosmetics
Farnesol; *Methyl salicylate*; *Zinc chloride*

soporific, pharmaceuticals
Methyl pentynol

sour, laundry
Acetic acid

soybean oil substitute
Sunflower (Helianthus annuus) seed oil

sperm oil substitute
Oleyl oleate

sperm oil, synthetic
Eucalyptus globulus oil

spermicide
Laureth-9; *Nonoxynol*

esters

stabilizer, calcium stearate disps.: paper/paperboard coatings, food-contact
Decyl alcohol

stabilizer, chewing gum
Sucrose fatty acid esters

stabilizer, chlorinated hydrocarbons
But-2-yne-1,4-diol; *1,4-Dioxane*; *Hydroquinone monomethyl ether*; *p-Methyl morpholine*; *Propargyl alcohol*; Tebol®; Tebol® 99

stabilizer, chlorinated organic compds.
Dimethyl hexynol

stabilizer, chlorinated solvents
1,3-Dioxolane; *Methyl butynol*; *Methyl pentynol*; *Nitroethane*; *Nitromethane*; *Triethylamine*; *Trioxane*

stabilizer, cleansers
Lanolin alcohol

stabilizer, cloud point lowering
Pilot® SXS-40

stabilizer, coatings
Triisooctyl phosphite

stabilizer, coffee whiteners
Hodag® S-35 FGK

stabilizer, color: fuel oils
DePEG 600; Rhodasurf® PEG 600

stabilizer, color: rubber compounding
Naphthenic oil

stabilizer, cosmetic creams/lotions
Isopropyl palmitate

stabilizer, cosmetics
Argobase 125T; Carbopol® 940; *Cetearyl octanoate*; *Edetic acid*; *Glyceryl stearate*; *Isopropyl lanolate*; *Laneth-5*; *Laneth-10*; *Laneth-15*; *Lanolin acid*; *Lanolin alcohol*; Lipolan R; Lumulse® GMO; *Methylpropanediol*; *Nonoxynol-4*; *Nonoxynol-6*; *Oleth-2*; *Paraffin*; *PEG-2 stearate*; Pluracol® E200; Pluracol® E300; Pluracol® E400; Pluracol® E600; Pluracol® E4000; Pluracol® E6000; *Polyglyceryl polyricinoleate*; *Propylene glycol laurate*; *Propylene glycol ricinoleate*; *Sodium 2-ethylhexyl sulfate*; *Stearyl octanoate*

stabilizer, cream shampoos
Lanolin acid; *Stearyl octanoate*

stabilizer, creams
Stearyl octanoate

stabilizer, creams/lotions
Ameroxol® OE-2; *Cetyl octanoate*; *Laneth-5*; *Laneth-10*; *Laneth-15*

stabilizer, cutting oils
Nonoxynol-4

stabilizer, deodorants
Isopropyl palmitate

stabilizer, detergents
Ameroxol® OE-2; *Ammonium xylenesulfonate*; *Cyclohexanol*; *Edetic acid*; *Oleth-2*

stabilizer, die-casting/forging lubricants
Carbopol® 940

stabilizer, dry cleaning
PEG-2 stearate

stabilizer, dye
Ammonium xylenesulfonate; *Ethylenediamine*; *Sodium toluenesulfonate*

stabilizer, dye: leather
BDGE; BGE; EDGE; EGE; i-PGE; MDGE; MGE

stabilizer, dye: printing
BDGE; BGE; EDGE; EGE; i-PGE; MDGE; MGE

stabilizer, dye: textiles
BDGE; BGE; EDGE; EGE; i-PGE; MDGE; MGE

stabilizer, electroplating baths
Ammonium xylenesulfonate; *Sodium toluenesulfonate*

stabilizer, emulsion
Carbomer 940; *Ethoxyethanol*

stabilizer, emulsion polymerization
Cetearyl alcohol

stabilizer, emulsion: acne creams/lotions
Crodamol PMP

stabilizer, emulsion: antibiotic ointments
Crodamol PMP

stabilizer, emulsion: antiperspirants
Novol NF

stabilizer, emulsion: bath oils
Crodamol PMP; Lexol® GT-865; Novol NF; *PPG-2 myristyl ether propionate*

stabilizer, emulsion: burn creams
Crodamol PMP

stabilizer, emulsion: cosmetic creams/lotions
Lexol® GT-865

stabilizer, emulsion: cosmetics
Carbomer; *Cetearyl alcohol*; *Cocamide DEA*; *Lauryl alcohol*; *Myristyl alcohol*; Novol NF; *Oleyl alcohol*; *Peanutamide MEA*; *Peanutamide MIPA*; *PEG-6 isostearate*; *PPG-2 myristyl ether propionate*; Schercemol PGML; *Stearyl alcohol*; *Tridecyl alcohol*

stabilizer, emulsion: creams
Crodamol PMP; *PPG-2 myristyl ether propionate*

stabilizer, emulsion: cutting oils
Oleyl alcohol

stabilizer, emulsion: emulsions
PPG-2 myristyl ether propionate

stabilizer, emulsion: food
Oleyl alcohol; *Polyglyceryl polyricinoleate*; *Propylene glycol laurate*; *Propylene glycol stearate*

stabilizer, emulsion: inks
Oleyl alcohol

stabilizer, emulsion: lipstick
Lexol® GT-865; Novol NF

stabilizer, emulsion: makeup
Lexol® GT-865

stabilizer, emulsion: moisturizers
Crodamol PMP; *PPG-2 myristyl ether propionate*

stabilizer, emulsion: paints
Oleyl alcohol

stabilizer, emulsion: personal care
Novol NF

stabilizer, emulsion: petrochemicals
Oleyl alcohol

stabilizer, emulsion: pharmaceutical tablets
Sucrose fatty acid esters

stabilizer, emulsion: pharmaceuticals
Novol NF; *PEG-6 isostearate*; *PPG-2 myristyl ether propionate*; *Stearyl alcohol*

stabilizer, emulsion: plastics
Oleyl alcohol

stabilizer, emulsion: preshave lotions
Lexol® GT-865

stabilizer, emulsion: pulp/paper
Oleyl alcohol

stabilizer, emulsion: sunscreens
Novol NF

stabilizer, emulsion: textile finishing
Oleyl alcohol

stabilizer, emulsion: topical pharmaceuticals
Crodamol PMP; *Oleyl alcohol*; Schercemol PGML

stabilizer, emulsions
Carbomer 940; Crodamol PMP; *Ethoxyethanol*; *Polyglyceryl polyricinoleate*

stabilizer, ethylcellulose
Hydroquinone monomethyl ether

stabilizer, fat liquoring
Nonoxynol-4

stabilizer, fats
Sucroglycerides; *Sucrose fatty acid esters*

stabilizer, fatty acid systems
Lanolin acid

stabilizer, fertilizers
Agsolex® BLO

stabilizer, flavor concs.
Poloxamer 181

stabilizer, floor waxes
Decahydronaphthalene; Decalin

stabilizer, foam
2-(2-Aminoethoxy) ethanol; *Erucamide*; *Ethylene glycol*; *Lauryl pyrrolidone*

stabilizer, foam: antiperspirants
PEG-75 lanolin

stabilizer, foam: cosmetics
PEG-75 lanolin

stabilizer, foam: creams/lotions
PEG-75 lanolin

stabilizer, foam: detergents
Aqualose W15; *Cetyl alcohol*; Solulan® 16; Solulan® 75; Solulan® C-24; Solulan® L-575

stabilizer, foam: food
Oleic acid

stabilizer, foam: hair care
PEG-75 lanolin

stabilizer, foam: industrial applics.
Lactamide MEA

stabilizer, foam: makeup
PEG-75 lanolin

stabilizer, foam: malt beverages
PEG-350

stabilizer, foam: ood
Triethyl citrate

stabilizer, foam: personal care
Lactamide MEA

stabilizer, foam: pharmaceutical vehicles
PEG-75 lanolin

stabilizer, foam: textiles
PEG-75 lanolin

stabilizer, food
Acetylated mono- and diglycerides of fatty acids; *t-Butyl hydroquinone*; *Calcium chloride*; *Castor (Ricinus communis) oil*; *Diacetyl tartaric acid esters of mono- and diglycerides*; *Glyceryl dioleate*; *Glyceryl ricinoleate*; *Glyceryl stearate SE*; Hodag® S-35 FGK; Lumulse® GMO; *Mono- and diglycerides of fatty acids*; *Nonoxynol-6*; *Oleic acid*; *Polyglyceryl polyricinoleate*; *Polyglyceryl-3 stearate*; *Polysorbate 60*; *Potassium carbonate*; *Potassium hydroxide*; *Propylene glycol*; *Propylene glycol alginate*; *Propylene glycol caprylate*; *Propylene glycol dicaprylate/dicaprate*; *Propylene glycol laurate*; *Propylene glycol ricinoleate*; *Propylene glycol stearate*; *Sorbitan palmitate*; *Sorbitan trioleate*; *Sucrose fatty acid esters*; *Triolein*

stabilizer, food pkg.
Sodium 2-ethylhexyl sulfate

stabilizer, food processing
Sodium 2-ethylhexyl sulfate

stabilizer, food-grade polymers
Boric acid; *Butylene glycol*

stabilizer, fragrances
Ameroxol® OE-2; *Oleth-2*

stabilizer, freeze/thaw
STEPANATE® AXS

stabilizer, frozen desserts
Hodag® S-35 FGK

stabilizer, gasoline
Dimethylamine

stabilizer, halogenated solvents
NE™; NM™

stabilizer, heat
Butyl oleate

stabilizer, heat: ABS polymers
Dimethyl succinate

stabilizer, heat: cosmetic sticks
Behenyl benzoate; Finsolv® 137

stabilizer, heat: flexible PVC
Dimethyl succinate

stabilizer, heat: food pkg.
Dimethyl succinate

stabilizer, heat: polyesters
Triisopropyl phosphite

stabilizer, heat: polyolefins
Dimethyl succinate

stabilizer, heat: resins
Citrofol® BII

stabilizer, heat: rubber compounding
Naphthenic oil

stabilizer, herbicides
Butoxyethanol; Dowanol® DPM

stabilizer, household cleaners
Butoxyethanol; *PEG-100 stearate*; *Sodium 2-ethylhexyl sulfate*

stabilizer, household prods.
Lanolin acid; *Nonoxynol-4*; *Nonoxynol-6*

stabilizer, hydraulic fluids
Butoxyethanol

stabilizer, industrial
Isopropyl palmitate; *Nonoxynol-4*; *Nonoxynol-6*; *Propylene glycol laurate*; *Propylene glycol ricinoleate*

stabilizer, industrial cleaners
PEG-100 stearate; *Sodium 2-ethylhexyl sulfate*

stabilizer, inks
Ammonium xylenesulfonate; *Sodium toluenesulfonate*

stabilizer, insecticides
Butoxyethanol

stabilizer, lanolin
Acetulan®

stabilizer, latex paints
T-Det® N-4

stabilizer, latexes
Nonoxynol-5

stabilizer, leather
Oleth-2; *PEG-2 stearate*

stabilizer, light
Methyl salicylate

stabilizer, light: ABS polymers
Dimethyl succinate

stabilizer, light: flexible PVC
Dimethyl succinate

stabilizer, light: food pkg.
Dimethyl succinate

stabilizer, light: polyester
Triisopropyl phosphite

stabilizer, light: polyolefins
Dimethyl succinate

stabilizer, lipstick
Isodragol 2/050300; Lipolan R

stabilizer, lubricants
Propylene glycol laurate; *Propylene glycol ricinoleate*

superfatting agent, cosmetics
Coconut (Cocos nucifera) oil; Corona Lanolin; Coronet Lanolin; *Decyl oleate*; *Myristyl myristate*; Novol NF; *Octyl stearate*; *Oleic acid*; *PEG-35 castor oil*; *PEG-7 glyceryl cocoate*; *PEG-30 glyceryl cocoate*; *PPG-10 cetyl ether*; *PPG-30 cetyl ether*; *PPG-50 cetyl ether*; Procetyl 10
superfatting agent, cream shampoos
Lanolin acid
superfatting agent, creams/lotions
Decyl oleate; Labrafac® Lipophile; *Myristyl myristate*
superfatting agent, detergents
Incrocas 30; Incrocas 40
superfatting agent, fatty acid systems
Lanolin acid
superfatting agent, foam baths
Prostearyl 15
superfatting agent, gels
Labrafac® Lipophile
superfatting agent, hair care
Cetiol® B; *PEG-75 lanolin*
superfatting agent, hair shampoos
Prostearyl 15
superfatting agent, hair sprays
HallTress™ DIBA Special
superfatting agent, hand cleaners
Decyl oleate; *Glyceryl dioleate*; *Glyceryl ricinoleate*; *Hexyl laurate*
superfatting agent, household detergents
PEG-75 lanolin
superfatting agent, household prods.
Incrocas 30; Incrocas 40; *Lanolin acid*
superfatting agent, lipophilic components
Isostearyl isononanoate
superfatting agent, lipstick
Lanolin oil; Novol NF
superfatting agent, lubricants
PEG-7 glyceryl cocoate
superfatting agent, makeup
Myristyl myristate
superfatting agent, massage oils
Labrafac® Lipophile
superfatting agent, metalworking fluids
Incrocas 30; Incrocas 40
superfatting agent, moisturizers
Decyl oleate
superfatting agent, mold release compds.
PEG-7 glyceryl cocoate
superfatting agent, nail polishes/enamels
HallTress™ DIBA Special
superfatting agent, ointments
Labrafac® Lipophile
superfatting agent, personal care
Incrocas 30; Incrocas 40; Labrasol®; *Myristyl myristate*; Novol NF; *PEG-7 glyceryl cocoate*; *PEG-30 glyceryl cocoate*; *PPG-10 cetyl ether*; *PPG-20 cetyl ether*; Procetyl 10
superfatting agent, pharmaceutical emulsions
Labrafac® Lipophile
superfatting agent, pharmaceuticals
Coconut (Cocos nucifera) oil; Corona Lanolin; Coronet Lanolin; *Decyl oleate*; Novol NF; *PEG-35 castor oil*; *PEG-7 glyceryl cocoate*; *PEG-30 glyceryl cocoate*; *PPG-30 cetyl ether*
superfatting agent, pressed powder
Decyl oleate
superfatting agent, shampoos
Fancol™ Karite Butter; Labrafac® Lipophile
superfatting agent, shaving foams
Fancol™ Karite Butter
superfatting agent, skin care
PEG-75 lanolin
superfatting agent, skin fresheners
Crodamol DA
superfatting agent, skin lotions
HallTress™ DIBA Special
superfatting agent, soaps
Fancol™ Karite Butter; *Glyceryl dioleate*; *Glyceryl ricinoleate*
superfatting agent, sun oils
Labrafac® Lipophile
superfatting agent, sunscreens
Novol NF
superfatting agent, suntan preps.
Fancol™ Karite Butter
superfatting agent, toilet water
HallTress™ DIBA Special
superfatting agent, topical pharmaceuticals
Myristyl myristate; *Octyl stearate*
superfatting agent, veterinary emulsions
Labrafac® Lipophile
superfatting agent, veterinary medicine
PEG-35 castor oil
superfatting agent, veterinary microemulsions
Labrafac® Lipophile
superfatting agent, wax systems
Lanolin acid
superfatting agent, winter sports preps.
Fancol™ Karite Butter
superfatting agent, wrinkle creams
Fancol™ Karite Butter
surface finishing agent, textiles
Butyl oleate
surface tens. reducer, aq. coatings
Tecsol® C 95%
surface tens. reducer, cleaners
Methoxyisopropanol; *Propylene glycol dimethyl ether*
surface tens. reducer, industrial coatings
Tecsol® C 95%
surface tens. reducer, trade sales coatings
Tecsol® C 95%
surface tension depressant
Sodium lauryl sulfate
surface tension reducer
Cyclomethicone
surface tension reducer, aq. cleaners
Arcosolv® PNB; *Butoxypropanol*
surface tension reducer, cleaners
Arcosolv® PE; Arcosolv® PM; Arcosolv® TPM; *Propylene glycol ethyl ether*
surface-finishing agent, baked goods
Hodag® S-35 FGK
surface-finishing agent, coffee whiteners
Hodag® S-35 FGK
surface-finishing agent, food
Hodag® S-35 FGK; *Mono- and diglycerides of fatty acids*; *Tristearin*
surface-finishing agent, frozen desserts
Hodag® S-35 FGK
surface-finishing agent, shortening
Hodag® S-35 FGK
surface-finishing agent, whipped toppings
Hodag® S-35 FGK
surfactant
Acrylonitrile; *2-Aminobutanol*; *Benzyltriethyl ammonium chloride*; *Butylene glycol*; *C13-15 alcohols*; *Castor (Ricinus communis) oil*; *Cetyl alcohol*; *o-Cresol*; *Cresylic acid*; *Decyl alcohol*; DePEG 200; DePEG 300; DePEG 400; DePEG 600; Diadol 18G-C; *Diethanolamine*; *Diethylenetriamine*; *Dimethicone*; *Dimethylamine*; *2,6-Dimethyl heptanol-4*; *Dioctyl maleate*; Edenor® ME C6-10; *2-Ethyl-1-butanol*; *Ethylene oxide*; *2-Ethylhexanol*; *Hexyl alcohol*; Imwitor® 742; *Isopropylamine dodecylbenzenesulfonate*; *Lauryl alcohol*; Liquid Absorption Base T; *Methyl isobutyl ketone*; *α-Methylnaphthalene*; *Oleyl alcohol*; Pluracol® E200; Pluracol® E300; Pluracol® E400; Pluracol® E600; Pluracol® E4000; Pluracol® E6000; PPG 400; PPG 1650; PPG 2250; *Propylene glycol dimethyl ether*; *Propylene glycol laurate*; *Propylene glycol stearate*; *Propylene oxide*; *Pyridine*; *Silicone*; *Stearyl alcohol*; *Tetraethylenepentamine*; *Trilaurin*; *2,6,8-Trimethylnonyl-4-alcohol*; *Tripropylene glycol n-butyl ether*
surfactant, adhesives
Glycol stearate; *Propylene glycol laurate*; *Sodium 2-ethylhexyl sulfate*; *Sorbitan palmitate*
surfactant, aftershaves
Renex® PEG 600
surfactant, agric.
Glycol stearate; *Isooctyl stearate*; *Isotridecyl stearate*; *Nonoxynol-4*; *Nonoxynol-5*; *Nonoxynol-6*; *PEG-4 dilaurate*; *PEG-4 dioleate*; *PEG-20 dioleate*; *PEG-6 oleate*; *PEG-20 oleate*; *Poloxamer 181*; *Propylene glycol laurate*
surfactant, agric. chemicals
Renex® PEG 300; Renex® PEG 400; Renex® PEG 600
surfactant, antiperspirant sticks
Crystal® Crown
surfactant, antiperspirants
PEG-75 lanolin; *PPG-5-ceteth-20*; *PPG-2 myristyl ether propionate*
surfactant, aq. alcoholic systems
PPG-5-ceteth-20
surfactant, asphalt emulsifiers
Dytek® EP Diamine; *1,3-Pentanediamine*
surfactant, baby care
Acetylated lanolin alcohol
surfactant, baked goods
Hodag® S-35 FGK
surfactant, cleaners
Crillet 6; *PEG-20 sorbitan isostearate*
surfactant, coffee whiteners
Hodag® S-35 FGK
surfactant, cosmetic creams/lotions
Acetylated lanolin alcohol
surfactant, cosmetic solutions
Dimethyl isosorbide
surfactant, cosmetics
Ammonium xylenesulfonate; *Apricot kernel oil PEG-6 esters*; *Caprylyl pyrrolidone*; *Ceteth-16*; *Ceteth-24*; *Cocamide DEA*; Crystal® Crown; *Dodoxynol-7*; *Ethyl oleate*; *Glyceryl dioleate*; *Glycol stearate*; *Hydrogenated castor oil*; *Isooctyl stearate*; *Isopropanolamine lanolate*; *Isotridecyl stearate*; *Lauryl pyrrolidone*; Lipocol P-15; *Nonoxynol*; *Nonoxynol-4*; *Nonoxynol-5*; *Nonoxynol-6*; *Nonoxynol-9 phosphate*; *Oleth-2*; *Oleth-16*; *Peanutamide MEA*; *Peanutamide MIPA*; *PEG-30 castor oil*; *PEG-33 castor oil*; *PEG-35 castor oil*; *PEG-40 castor oil*; *PEG-4 dilaurate*; *PEG-4 dioleate*; *PEG-20 dioleate*; *PEG-7 glyceryl cocoate*; *PEG-30 glyceryl cocoate*; *PEG-75 lanolin*; *PEG-6 oleate*; *PEG-20 oleate*; *PEG-20 sorbitan isostearate*; *PEG-40 sorbitan peroleate*; *PEG-100 stearate*; *Poloxamer 181*; *Polysorbate 60*; *Polysorbate 80*; *PPG-24 butyl ether*; *PPG-5-ceteth-20*; *PPG-10 cetyl ether*; *PPG-30 cetyl ether*; *PPG-50 cetyl ether*; *PPG-40 diethylmonium chloride*; *PPG-15 stearyl ether*; *Propylene glycol laurate*; *Sodium 2-ethylhexyl sulfate*; *Sodium lauryl sulfate*; *Sodium toluenesulfonate*; *Sodium xylenesulfonate*; *Steareth-16*; *Sucrose palmitate*
surfactant, creams/lotions
Ceteth-24; Crillet 6; *PEG-75 lanolin*; *PEG-20 sorbitan isostearate*; *PPG-2 myristyl ether propionate*
surfactant, cutting oils
Nonoxynol-4; *Nonoxynol-5*; *PEG-5 oleate*
surfactant, degreasing solvents
PEG-5 oleate
surfactant, deodorants
Renex® PEG 300; Renex® PEG 400; Renex® PEG 600
surfactant, depilatories
PPG-2 myristyl ether propionate
surfactant, detergents
Oleth-2
surfactant, dishwashing detergents
Nonoxynol
surfactant, dishwashing prods.
Deceth-6
surfactant, drug delivery systems
PEG-8 caprylic/capric glycerides
surfactant, dry cleaning solvents
BIO-SOFT® N-411
surfactant, dyes
Polysorbate 60
surfactant, electronics
Ethyl oleate
surfactant, electroplating
Dimethyl hexynol
surfactant, emulsion degreasers
BIO-SOFT® N-411
surfactant, emulsion polymerization
Nonoxynol-5
surfactant, fat liquoring
Nonoxynol-4
surfactant, feed
Isooctyl stearate; *Isotridecyl stearate*
surfactant, fiber lubricants
PEG-4 dioleate
surfactant, floor polishes
Dimethyl hexynol
surfactant, food
Calcium chloride; *Ethyl oleate*; *Glyceryl dilaurate*; *Glyceryl dioleate*; Hodag® S-35 FGK; *Isooctyl stearate*; *Isotridecyl stearate*; *Mono- and diglycerides of fatty acids*; *Nonoxynol-6*; *PEG-4 dilaurate*; *PEG-4 dioleate*; *PEG-20 dioleate*; *PEG-6 oleate*; *PEG-20 oleate*; *Poloxamer 181*; *Propylene glycol*; *Propylene glycol alginate*; *Propylene glycol laurate*
surfactant, food pkg.
Sodium 2-ethylhexyl sulfate
surfactant, food processing

suspending agent, personal care
Caprylic/capric triglyceride; *Carbomer 940*

suspending agent, pharmaceuticals
Butylene glycol dicaprylate/dicaprate; *Caprylic/capric triglyceride*; *Carbomer 940*; *2,2-Dimethyl-1,3-dioxolane-4-methanol*; Miglyol® 812; Miglyol® 840; Miglyol® 8810; *Propylene glycol alginate*

suspending agent, pigment: antiperspirants
Novol NF

suspending agent, pigment: bath oils
Novol NF

suspending agent, pigment: cosmetics
Novol NF

suspending agent, pigment: inks
Trihydroxystearin

suspending agent, pigment: lipstick
Novol NF

suspending agent, pigment: mastics
Trihydroxystearin

suspending agent, pigment: paints/coatings
Trihydroxystearin

suspending agent, pigment: personal care
Novol NF

suspending agent, pigment: pharmaceuticals
Novol NF

suspending agent, pigment: solvent-free epoxy systems
Trihydroxystearin

suspending agent, pigment: stains
Trihydroxystearin

suspending agent, pigment: sunscreens
Novol NF

suspending agent, pigment: textile backings
Trihydroxystearin

suspending agent, polyester patch/fillers
Trihydroxystearin

suspending agent, powder coatings
Trihydroxystearin

suspending agent, rectal pharmaceuticals
Miglyol® 810

suspending agent, solids
Trideceth

suspending agent, solvent-borne systems
Trihydroxystearin

suspending agent, thixotropic paints
Carbopol® 940

suspending agent, topical pharmaceuticals
Carbopol® 940; Miglyol® 810

suspension medium, pharmaceuticals
Estasan™ GT 8-99 3596; Estasan™ GT 10-99 3599

sweetener
Glycerin

sweetener, nutritive: food
Corn syrup

sweetener, nutritive: soft drinks
Corn syrup

sweetener, oral pharmaceuticals
Corn syrup

swelling agent
1,3,5-Triisopropylbenzene

swelling agent, fibers
Sodium thiocyanate

swelling agent, rubber
dl-Limonene

swelling media, silicone rubber tubing
Vertrel® XSi

synergist, antioxidant: food
Phosphoric acid

synergist, flame retardant: unsat. polyesters
Eastman® TEP

synergist, with di (2-ethylhexyl) phosphoric acid: heavy metal extraction
Dibutyl butyl phosphonate

synthesis intermediate
Bromodichloromethane

synthesis medium
Agsolex® 1

synthetic lubricating oil
Tripropylene glycol n-butyl ether

tableting agent
Mono- and diglycerides of fatty acids

tackifier
Kemstrene® 99.7% USP

tackifier, adhesives
Cachalot® AR-20; Cachalot® BE-22; Cachalot® C-50; Cachalot® L-90; Cachalot® M-43; Cachalot® S-56; *Paraffin, chlorinated*; *Sorbeth-20*; *Terpene resin*

tackifier, coatings
Terpene resin

tackifier, concrete-curing compds.
Terpene resin

tackifier, elastomers
Terpene resin

tackifier, hot-melt adhesives
Terpene resin

tackifier, polymers
Terpene resin

tackifier, pressure-sensitive adhesives
Terpene resin

tackifier, rubber
Paraffin, chlorinated; *Stoddard solvent*

tackifier, sealants
Paraffin, chlorinated

tackifier, wax films
Lanolin acid

tallow substitute, soaps
Menhaden oil

tanning agent
Isobutyric acid

tar spot remover
Ethylene chlorohydrin

taste control agent, water treatment
Ammonia; *Ammonium hydroxide*

termite control
1,3,5-Trichlorobenzene

termite repellent
Sucrose octaacetate

tetrafluoroethylene polymerization inhibitor
d-Limonene

textile aid
Butyrolactone

textile auxiliary
Diethylaminoethanol; *2-Mercaptoethanol*; *Methoxyethanol*; *Naphthalene*; PEG 300; *PEG-4 oleate*; Shellsol® D120; Surphos 620; Surphos 630; Surphos 645

texturizer, baked goods
Hodag® S-35 FGK

texturizer, beverages
Sucrose fatty acid esters

texturizer, chewing gum
Sucrose fatty acid esters

texturizer, coffee whiteners
Hodag® S-35 FGK

texturizer, food
Acetylated mono- and diglycerides of fatty acids; *Calcium chloride*; *Coconut (Cocos nucifera) oil*; *Corn (Zea mays) oil*; *Corn syrup*; *Cottonseed (Gossypium) oil*; Hodag® S-35 FGK; *Mono- and diglycerides of fatty acids*; *Peanut (Arachis hypogaea) oil*; *Propylene glycol*; *Safflower (Carthamus tinctorius) oil*; *Soybean (Glycine soja) oil*; *Sucrose fatty acid esters*; *Sucrose palmitate*; *Sunflower (Helianthus annuus) seed oil*

texturizer, frozen desserts
Hodag® S-35 FGK

texturizer, margarine
Peanut (Arachis hypogaea) oil

texturizer, pharmaceuticals
Acetylated mono- and diglycerides of fatty acids

texturizer, salad oils
Peanut (Arachis hypogaea) oil

texturizer, shortening
Hodag® S-35 FGK; *Peanut (Arachis hypogaea) oil*

texturizer, soft drinks
Corn syrup

texturizer, whipped toppings
Hodag® S-35 FGK

thawing agent, coal
Calcium chloride

thawing agent, coke
Calcium chloride

thawing agent, ore
Calcium chloride

thawing agent, sand
Calcium chloride

thawing agent, stone
Calcium chloride

thermometer fluid
n-Pentane

thickener
Carbomer; Carbopol® 940; Carbowax® E300 NF; Carbowax® E400 NF; Carbowax® E600 NF; Carbowax® E900 NF; Carbowax® E1000 NF; Carbowax® E1450 NF; Carbowax® E3350 NF; Carbowax® E4500 NF; Carbowax® E8000 NF; *Cetyl alcohol*; *Glyceryl dioleate*; *PEG-12 dilaurate*; *PEG-12 dioleate*; *Trihydroxystearin*

thickener, absorption bases
Lanolin alcohol

thickener, adhesives
Emery® 912; Emery® 916; Emery® 917

thickener, agric.
Emery® 912; Emery® 916; Emery® 917; *PEG-20 dilaurate*; *PEG-4 dioleate*; *PEG-20 dioleate*; *PEG-4 oleate*; *PEG-6 oleate*; *PEG-20 oleate*

thickener, agric. chemical sprays
PEG-8 dilaurate

thickener, aq. systems
Carbopol® 980

thickener, baked goods
Hodag® S-35 FGK

thickener, bath care
Carbowax® E300; Carbowax® E400; Carbowax® E600; Carbowax® E900; Carbowax® E1000; Carbowax® E1450; Carbowax® E3350; Carbowax® E8000; Liponate GC-K; Liponate PC

thickener, bath oils
Stearyl stearate

thickener, ceramics
Emery® 912; Emery® 916; Emery® 917

thickener, cleansers
Lanolin alcohol

thickener, clear water topical gels
Carbopol® 980

thickener, coffee whiteners
Hodag® S-35 FGK

thickener, cosmetic creams/lotions
Isocetyl stearate; Liponate GC-K; Liponate PC

thickener, cosmetics
Carbomer 940; Carbopol® 940; Carbopol® 980; *Cetearyl octanoate*; *Cetyl palmitate*; *Choleth-24*; *Cocamide DEA*; Emery® 912; Emery® 916; Emery® 917; *Glyceryl oleate*; *Glyceryl stearate*; *Glycol distearate*; Kemester® EGDS; *Lanolin alcohol*; *Myristyl myristate*; *PEG-20 dilaurate*; *PEG-4 dioleate*; *PEG-20 dioleate*; *PEG-30 glyceryl cocoate*; *PEG-4 oleate*; *PEG-6 oleate*; *PEG-20 oleate*; *PEG-2 stearate*; Pluracol® E400 NF; Pluracol® E600 NF; Pluracol® E1450 NF; Polychol 15; *Polyglyceryl-3 diisostearate*; *Stearyl alcohol*; *Stearyl octanoate*; *Stearyl stearate*; *Tripropylene glycol*

thickener, cream shampoos
Stearyl octanoate

thickener, creams
Stearyl octanoate

thickener, creams/lotions
Cetyl octanoate; *Choleth-24*; Emerest® 2452; *Myristyl myristate*; *Polyglyceryl-3 diisostearate*; *Stearyl stearate*

thickener, die-casting/forging lubricants
Carbopol® 940

thickener, dry cleaning
PEG-2 stearate

thickener, emulsion shampoos
Glycol distearate

thickener, emulsions
Acetylated glyceryl stearate; Glicopol 2-G; *Glyceryl stearate*; Milopol G-7; Monestriol DM; Monestriol GCT; Monestriol GE; Monestriol GP; Monestriol GP-35; Monestriol GP-35-3%; Monestriol GP-35-AE; Monestriol GP-40; Monestriol GP-40-CF; Monestriol GP-40-DC; Monestriol GP-45; Monestriol GP-45-DC; Monestriol GP-60; Monestriol GP-AS; Thomil G; Tilol GCT-R; Tilol GP-O; Triacetina; Triol 91; Trioleina

thickener, fiber lubricants
PEG-4 dioleate

thickener, foam baths
Glycol distearate

thickener, foam control agents
Carbowax® E300; Carbowax® E400; Carbowax® E600; Carbowax® E900; Carbowax® E1000; Carbowax® E1450; Carbowax® E3350; Carbowax® E8000

thickener, food
Calcium chloride; *Corn syrup*; Emery® 912; Emery® 916; Emery® 917; *Glyceryl stearate*; *Glyceryl stearate SE*; Hodag® S-35 FGK; *Mono- and diglycerides of fatty acids*; *PEG-20 dilaurate*; *PEG-4 dioleate*; *PEG-20 dioleate*; *PEG-4 oleate*; *PEG-6 oleate*; *PEG-20 oleate*; *Potassium hydroxide*; *Propylene glycol*; *Propylene glycol alginate*; *Sodium lauryl sulfate*

Dimethoxypropane
tonic, cosmetics
Clove (Eugenia caryophyllus) oil; *Ethyl phenylacetate*; *Eucalyptol*; *Eugenol*
tonic, nausea treatment
Phosphoric acid
tonic, vomiting treatment
Phosphoric acid
toxin eliminator
Dimethyl sulfone
tracer gas, environmental studies
Dibromodifluoromethane
tranquilizer, fish
Methyl pentynol
transformer fluid
Hexachloro-1,3-butadiene
translocation enhancer, adjuvants
Agsolex® 1
trichloroethane substitute
p-Chlorobenzotrifluoride
trichloroethylene substitute, degreasers
Ethyl-(S)-lactate
turpentine component
d-Limonene
turpentine oil source
Turpentine
turpentine substitute
Decahydronaphthalene; *Tetrahydronaphthalene*
urethane polyol initiator
Tripropylene glycol
UV absorber, balsams
Fancol™ Karite Butter
UV absorber, body lotions
Fancol™ Karite Butter
UV absorber, cosmetics
Benzylsalicylate; *Ethyl cinnamate*; *2-Ethylhexyl salicylate*
UV absorber, hair sprays
HallBrite® OS
UV absorber, lip balms
HallBrite® OS
UV absorber, lotions
HallBrite® OS
UV absorber, shampoos
Fancol™ Karite Butter
UV absorber, shaving foams
Fancol™ Karite Butter
UV absorber, soaps
Fancol™ Karite Butter
uv absorber, sunburn lotions
Methyl salicylate
UV absorber, sunscreens
HallBrite® OS
UV absorber, suntan preps.
Fancol™ Karite Butter
UV absorber, winter sports prods.
Fancol™ Karite Butter
UV absorber, wrinkle creams
Fancol™ Karite Butter
UV inhibitor
Hydroquinone monomethyl ether
UV inhibitor, PVC
Hempa
UV resistance aid, resins
Citrofol® BII
vapor depressant, solvents
Methylene chloride
vapor pressure depressant, aerosols
1,1,1-Trichloroethane
varnish remover
Benzyl acetate; *Cyclohexane*; *Cyclohexanone*; *Ethoxyethanol acetate*; *Ethylene dichloride*; *Isopropyl ether*; *Mesityl oxide*; *1,2,3-Trichloropropane*; *Tripropylene glycol*
vehicle
Isopentane; *PEG-100*; Promyristyl PM-3
vehicle,
Sesamol
vehicle, active ingreds.: cosmetics
Crodamol GTCC
vehicle, active ingreds.: pharmaceuticals
Crodamol GTCC
vehicle, adhesives
Carbowax® PEG 300; *Naphtha, medium aliphatic*
vehicle, aerosols
Captex® 300; Captex® 810D; *Tricaprin*
vehicle, aftershave lotions
Captex® 810D
vehicle, agric.
Carbowax® PEG 300
vehicle, antibiotics
Lexol® GT-855; Lexol® GT-865; Lexol® PG-800; Lexol® PG-865; NEOBEE® M-5; NEOBEE® M-20; *Propylene glycol dioctanoate*
vehicle, antiperspirants
Carbowax® Sentry® PEG 300; Carbowax® Sentry® PEG 540 Blend; Carbowax® Sentry® PEG 600; Carbowax® Sentry® PEG 900; Carbowax® Sentry® PEG 1000; Carbowax® Sentry® PEG 1450; Carbowax® Sentry® PEG 3350; Carbowax® Sentry® PEG 4600; Carbowax® Sentry® PEG 8000; Crodamol AB
vehicle, baby skin care
Miglyol® 8810
vehicle, baked goods
Hodag® S-35 FGK
vehicle, bath oils
Captex® 810D; Carbowax® Sentry® PEG 400; Lexol® GT-865
vehicle, bath powders
Carbowax® Sentry® PEG 3350
vehicle, ceramics
Carbowax® PEG 300
vehicle, clinical nutrition
Capmul® MCM; Captex® 300; Captex® 350; *Tricaprin*
vehicle, coatings
Captex® 300; *Tricaprin*
vehicle, coffee whiteners
Hodag® S-35 FGK
vehicle, colognes
Carbowax® Sentry® PEG 600
vehicle, colors
NEOBEE® M-5
vehicle, cosmetic creams
Isopropyl myristate
vehicle, cosmetic creams/lotions
Isopropyl palmitate; Lexol® GT-865
vehicle, cosmetic raw materials
Crodamol IPM
vehicle, cosmetics
Captex® 200; Captex® 810D; Carbowax® PEG 300; Carbowax® Sentry® PEG 300; Carbowax® Sentry® PEG 400; Carbowax® Sentry® PEG 540 Blend; Carbowax® Sentry® PEG 600; Carbowax® Sentry® PEG 900; Carbowax® Sentry® PEG 1000; Carbowax® Sentry® PEG 1450; Carbowax® Sentry® PEG 3350; Carbowax® Sentry® PEG 4600; Carbowax® Sentry® PEG 8000; Crodamol IPP; *Ethyl acetate*; Lipovol SES; Pluracol® E200; Pluracol® E300; Pluracol® E400; Pluracol® E600; Pluracol® E4000; Pluracol® E6000; *Propylene glycol dicaprylate/dicaprate*; STEPAN® IPM; STEPAN® IPP; Wickenol® 136; ZELEC® IPM
vehicle, creams/lotions
Carbowax® Sentry® PEG 300; Carbowax® Sentry® PEG 400; Carbowax® Sentry® PEG 540 Blend; Carbowax® Sentry® PEG 600; Carbowax® Sentry® PEG 900; Carbowax® Sentry® PEG 1000; Carbowax® Sentry® PEG 1450; Carbowax® Sentry® PEG 3350; Carbowax® Sentry® PEG 4600; Carbowax® Sentry® PEG 8000; Dynacerin® 660; Miglyol® 810; Miglyol® 840; Miglyol® 8810
vehicle, delivery/absorption enhancement
Capmul® MCM; Capmul® MCMC8; Captex® 300; Captex® 350; *Tricaprin*
vehicle, dentifrices
Carbowax® Sentry® PEG 600
vehicle, deodorants
Isopropyl palmitate
vehicle, dermatologicals
Capmul® MCM; Capmul® MCMC8; Captex® 300; Captex® 350; *Propylene glycol dioctanoate*; *Tricaprin*
vehicle, dietetic prods.
Miglyol® 840
vehicle, drug delivery systems
Crodamol EO
vehicle, dry skin care
Propylene glycol dioctanoate
vehicle, elastomers
Carbowax® PEG 300
vehicle, electroplating
Carbowax® PEG 300
vehicle, electropolishing
Carbowax® PEG 300
vehicle, enamels
Kerosene
vehicle, essential oils
NEOBEE® M-5
vehicle, flavors
Captex® 200; Lexol® PG-800; Lexol® PG-865; NEOBEE® M-5; *Propylene glycol dicaprylate/dicaprate*; *Propylene glycol dioctanoate*
vehicle, food
Acetic acid; *Alcohol*; *Benzyl benzoate*; Captex® 200; Crodamol PC-PN; *Glyceryl oleate*; Hodag® S-35 FGK; *Lactic acid*; Lipovol SES; *Mono- and diglycerides of fatty acids*; *Propylene glycol*; *Triacetin*
vehicle, food pkg.
Glyceryl oleate
vehicle, food processing
Carbowax® PEG 300
vehicle, fragrance oils
Captex® 200; *Propylene glycol dicaprylate/dicaprate*
vehicle, fragrances
Lexol® PG-800; Lexol® PG-865; *Propylene glycol dioctanoate*
vehicle, frozen desserts
Hodag® S-35 FGK
vehicle, gelatin capsules
Carbowax® E300 NF; Carbowax® E400 NF; Carbowax® E600 NF; Carbowax® E900 NF; Carbowax® E1000 NF; Carbowax® E1450 NF; Carbowax® E3350 NF; Carbowax® E4500 NF; Carbowax® E8000 NF; *Tricaprin*
vehicle, hair dressing
Carbowax® Sentry® PEG 400; Carbowax® Sentry® PEG 540 Blend; Carbowax® Sentry® PEG 3350; Carbowax® Sentry® PEG 8000
vehicle, hand cleaners
Carbowax® Sentry® PEG 300; Carbowax® Sentry® PEG 540 Blend
vehicle, household prods.
Carbowax® PEG 300
vehicle, industrial
Isopropyl myristate; *Isopropyl palmitate*
vehicle, infant formulas
Captex® 300; Captex® 350; *Tricaprin*
vehicle, injectable substances
Acetylated glyceryl stearate; Glicopol 2-G; Miglyol® 810; Miglyol® 840; Miglyol® 8810; Milopol G-7; Monestriol DM; Monestriol GCT; Monestriol GE; Monestriol GP; Monestriol GP-35; Monestriol GP-35-3%; Monestriol GP-35-AE; Monestriol GP-40; Monestriol GP-40-CF; Monestriol GP-40-DC; Monestriol GP-45; Monestriol GP-45-DC; Monestriol GP-60; Monestriol GP-AS; Thomil G; Tilol GCT-R; Tilol GP-O; Triacetina; Triol 91; Trioleina
vehicle, inks
Diethylene glycol phenyl ether; Igepal® OD-410
vehicle, IV emulsions
Super Refined® Almond NF; Super Refined® Corn NF; Super Refined® Cottonseed NF; Super Refined® Olive NF; Super Refined® Peanut NF; Super Refined® Safflower USP; Super Refined® Sesame NF; Super Refined® Soybean USP
vehicle, liniments
Olive (Olea europaea) oil; Super Refined® Almond NF; Super Refined® Corn NF; Super Refined® Cottonseed NF; Super Refined® Olive NF; Super Refined® Peanut NF; Super Refined® Safflower USP; Super Refined® Sesame NF; Super Refined® Soybean USP
vehicle, lip gloss
Captex® 810D
vehicle, lipstick
Captex® 810D; Carbowax® Sentry® PEG 400; Carbowax® Sentry® PEG 600; Cerette V; Isostriol IC; Lexol® GT-865; Monestriol B; Monestriol C; Monestriol IC; Monestriol IS-C; Palmil C; Palmil IS; Palmil O; Thomil 14; Thomil IS; Thomil IS-F; Tilol B; Tilol ET; Tilol IS
vehicle, lotions
Shea butter (Butyrospermum parkii)
vehicle, lubricants
Caprylic/capric triglyceride; Carbowax® PEG 300
vehicle, makeup
Carbowax® Sentry® PEG 300; Carbowax® Sentry® PEG 400; Carbowax® Sentry® PEG 600; Carbowax® Sentry® PEG 3350; Crodamol AB; Lexol® GT-865
vehicle, makeup bases
Captex® 810D
vehicle, makeup removers
Carbowax® Sentry® PEG 400
vehicle, metal fabrication
Carbowax® PEG 300
vehicle, nail polish removers
Carbowax® Sentry® PEG 400;

Triisostearin
viscosity builder, pharmaceuticals
Cetyl palmitate; Comperlan® COD
viscosity builder, semipermanent liq. hair color systems
Lowenol Solvent 402
viscosity builder, shampoos
Comperlan® COD; Surfadone® LP-300
viscosity builder, soaps
Naturechem® GTR
viscosity builder, sunscreen
Naturechem® GTR
viscosity builder, topical pharmaceuticals
Behenyl behenate
viscosity control agent
Decyl alcohol; Kemester® 205; Kemester® 226; *Methyl behenate*; *Methyl soyate*
viscosity control agent, aerosols
Propylene glycol dipelargonate
viscosity control agent, agric.
PEG-4 dilaurate; *PEG-8 oleate*
viscosity control agent, aq. resins
Eastman® EP
viscosity control agent, bath care
Liponate GC-K; Liponate PC
viscosity control agent, bath oils
Propylene glycol dipelargonate; *Stearyl stearate*
viscosity control agent, cellulose derivs.
Hydrogen peroxide
viscosity control agent, coatings
PEG-8 oleate
viscosity control agent, controlled-release solid pharmaceuticals
Renex® PEG 300; Renex® PEG 400; Renex® PEG 600
viscosity control agent, cosmetic creams/lotions
Liponate GC-K; Liponate PC
viscosity control agent, cosmetics
Ammonium xylenesulfonate; *Calcium chloride*; *Caprylic alcohol*; *Carbomer*; *Cetearyl alcohol*; *Cetyl alcohol*; *Cyclomethicone*; *DM hydantoin*; *Erucamide*; *Ethylene glycol*; *Glyceryl triacetyl hydroxystearate*; *Glycol stearate*; *1,2,6-Hexanetriol*; *Hydrogenated castor oil*; *Isocetyl alcohol*; *Isopentane*; *Isostearyl alcohol*; *Laneth-5*; *Laneth-10*; *Laneth-15*; *Lauryl alcohol*; Lipocol P-15; *N-Methylacetamide*; *Myristyl alcohol*; *Oleyl alcohol*; *Paraffin*; *Peanutamide MEA*; *Peanutamide MIPA*; *PEG-4 dilaurate*; *PEG-8 oleate*; *PPG-15 stearyl ether*; *Propylene glycol alginate*; *Propylene glycol dicaprylate*; *Propylene glycol dipelargonate*; *Sodium silicoaluminate*; *Stearamide AMP*; *Stearyl alcohol*; *Stearyl stearate*; *Tridecyl alcohol*; *Trihydroxystearin*; *Triisononanoin*; *Triisostearin*; *Trilaurin*; *Trilinolein*; *Trimyristin*; *Triolein*; *Tripalmitin*; *Tristearin*
viscosity control agent, creams/ lotions
Laneth-5; *Laneth-10*; *Laneth-15*; *Stearyl stearate*
viscosity control agent, emulsions
Isocetyl alcohol
viscosity control agent, food
PEG-4 dilaurate; *PEG-8 oleate*
viscosity control agent, glosses
Propylene glycol dipelargonate
viscosity control agent, industrial
PPG-15 stearyl ether
viscosity control agent, laxatives
Renex® PEG 300; Renex® PEG 400; Renex® PEG 600
viscosity control agent, lipstick
Propylene glycol dipelargonate
viscosity control agent, lotions
Propylene glycol alginate
viscosity control agent, makeup bases
Propylene glycol dipelargonate
viscosity control agent, metalworking
PEG-4 dilaurate
viscosity control agent, perfumes: hair care
Volpo G26
viscosity control agent, perfumes: skin care
Volpo G26
viscosity control agent, pharmaceuticals
Hydrogenated castor oil; *Isostearyl alcohol*; *Laneth-5*; *Laneth-10*; Lipocol P-15; *PEG-4 dilaurate*; *PEG-8 oleate*; *PPG-15 stearyl ether*; *Propylene glycol alginate*; *Propylene glycol dicaprylate*; *Sodium silicoaluminate*; *Stearamide AMP*; *Stearyl stearate*; *Trilaurin*; *Trimyristin*; *Tripalmitin*; *Triundecanoin*
viscosity control agent, plastics
PEG-4 dilaurate; *PEG-8 oleate*
viscosity control agent, polar essential oils: hair care
Volpo G26
viscosity control agent, polar essential oils: skin care
Volpo G26
viscosity control agent, preshave lotions
Propylene glycol dipelargonate
viscosity control agent, skin care emulsions
Isostearyl alcohol
viscosity control agent, starch derivs.
Hydrogen peroxide
viscosity control agent, textiles
PEG-4 dilaurate
viscosity control agent, topical pharmaceuticals
Laneth-15; *Trihydroxystearin*
viscosity control agent, vitamin suspensions
Propylene glycol alginate
viscosity control agent, water treatment
PEG-4 dilaurate
viscosity depressant, ATH-filled polyester resins
Dimethyl methylphosphonate
viscosity depressant, cellulosics
Triethyl phosphate
viscosity depressant, general polymers
Eastman® TEP
viscosity depressant, polyester laminates
Triethyl phosphate
viscosity index improver
Cachalot® AR-20; Cachalot® BE-22; Cachalot® C-50; Cachalot® L-90; Cachalot® M-43; Cachalot® S-56
viscosity modifier
C8 alkyl acetate; *C9 alkyl acetate*; *C10 alkyl acetate*; *C13 alkyl acetate*; Exxate® 600; Exxate® 700; Exxate® 800; Exxate® 900; Exxate® 1000; Exxate® 1300; *Oxo-heptyl acetate*; *Oxo-hexyl acetate*; *Sodium toluenesulfonate*
viscosity modifier, adhesives
Emery® 912; Emery® 916; Emery® 917
viscosity modifier, aerosols
Captex® 300; Captex® 8227; *Tricaprin*
viscosity modifier, aftershave lotions
Captex® 8227
viscosity modifier, agric.
Emery® 912; Emery® 916; Emery® 917
viscosity modifier, bath oils
Captex® 8227
viscosity modifier, bath/shower preps.
Captex® 200; Captex® 300
viscosity modifier, brake fluids
PEG 300
viscosity modifier, ceramics
Emery® 912; Emery® 916; Emery® 917
viscosity modifier, clinical nutrition
Captex® 300; *Tricaprin*
viscosity modifier, coatings
Captex® 300; *Tricaprin*
viscosity modifier, cosmetics
Capmul® MCMC10; Captex® 200; Captex® 300; Captex® 8227; Emery® 912; Emery® 916; Emery® 917; *Tricaprin*
viscosity modifier, creams/lotions
Captex® 200; Captex® 300; *Tricaprin*
viscosity modifier, delivery/ absorption enhancement
Captex® 300; *Tricaprin*
viscosity modifier, dermatologicals
Captex® 300; *Tricaprin*
viscosity modifier, drug vehicles
Emery® 912; Emery® 916; Emery® 917
viscosity modifier, flavors
Capmul® MCM; Capmul® MCMC8; Capmul® MCMC10
viscosity modifier, food
Emery® 912; Emery® 916; Emery® 917; *Polyglyceryl polyricinoleate*
viscosity modifier, foundation
Captex® 200
viscosity modifier, fragrances
Capmul® MCM; Capmul® MCMC8; Capmul® MCMC10
viscosity modifier, gelatin capsules
Tricaprin
viscosity modifier, glass
Emery® 912; Emery® 916; Emery® 917
viscosity modifier, hair care
Captex® 200; Captex® 300
viscosity modifier, infant formulas
Captex® 300; *Tricaprin*
viscosity modifier, lip care
Captex® 200; Captex® 300
viscosity modifier, lip glosses
Captex® 8227
viscosity modifier, lipstick
Captex® 8227
viscosity modifier, lotions
Shea butter (Butyrospermum parkii)
viscosity modifier, makeup
Captex® 200
viscosity modifier, makeup bases
Captex® 8227
viscosity modifier, mascara
Captex® 200; Captex® 300
viscosity modifier, nutritional applics.
Captex® 8227
viscosity modifier, nutritional supplements
Tricaprin
viscosity modifier, o/w creams
Shea butter (Butyrospermum parkii)
viscosity modifier, o/w emulsions
Shea butter (Butyrospermum parkii)
viscosity modifier, pharmaceutical dermatologicals
Shea butter (Butyrospermum parkii)
viscosity modifier, pharmaceuticals
Captex® 300; Captex® 8227
viscosity modifier, powders
Tricaprin
viscosity modifier, soft gelatin capsules
Captex® 300
viscosity modifier, sticks
Tricaprin
viscosity modifier, sun care
Captex® 200; Captex® 300
viscosity modifier, suppositories
Shea butter (Butyrospermum parkii)
viscosity modifier, suspensions
Captex® 300
viscosity modifier, w/o creams
Shea butter (Butyrospermum parkii)
viscosity modifier, w/o emulsions
Shea butter (Butyrospermum parkii)
viscosity reducer
Crill 2; Delios® S; *Dimethyl octynediol*; *Methyl pentynol*; *Sodium xylenesulfonate*
viscosity reducer, adhesives
Crill 1
viscosity reducer, air-drying coatings
Versadil™ 100
viscosity reducer, amine-cured epoxies
FA®
viscosity reducer, appliance coatings
Ucar® n-Propyl Propionate
viscosity reducer, automotive refinishes
Ucar® n-Propyl Propionate
viscosity reducer, coatings
Ucar® n-Propyl Propionate
viscosity reducer, cosmetics
Crill 1; SDA-1; SDA-3A; SDA-3C; SDA-23A; SDA-30; SDA-39C; SDA-40; SDA-40A; SDA-40B; SDA-40C
viscosity reducer, drug delivery systems
Crodamol EO
viscosity reducer, electrostatically sprayed coatings
Ucar® n-Propyl Propionate
viscosity reducer, food
Crill 1
viscosity reducer, food pkg.
Crill 1
viscosity reducer, herbicides
Crill 1
viscosity reducer, highly filled systems
Kenplast® ES-2HP
viscosity reducer, high-solids coatings
Ucar® n-Propyl Propionate; Versadil™ 100
viscosity reducer, high-solids inks
Sylfat® 9012
viscosity reducer, high-solids paints
Butoxyl®
viscosity reducer, inks
Crill 1
viscosity reducer, insecticides
Crill 1
viscosity reducer, leather treatment
Crill 1
viscosity reducer, low-VOC lithographic inks

wetting agent, burn creams
Crodamol CAP
wetting agent, Carbopol
Tridecyl neopentanoate; Trivent NP-13
wetting agent, cellulose acetate film
Agrisynth® BLO; BLO®; *Butyrolactone*
wetting agent, cellulose acetate film/fibers
Agrisynth® BLO
wetting agent, chemical industry
Sovatex C Series
wetting agent, clay soils
Deceth-6
wetting agent, cleaners
Alcohol; Dipentene-C; *PPG-2 methyl ether*; *PPG-3 methyl ether*; *PPG-2 methyl ether acetate*; *Propylene glycol methyl ether acetate*; Rhodacal® 330
wetting agent, coatings
Alcohol; *Dimethyl oleamide*; Isopar® H Fluid
wetting agent, commercial cleaners
Phenoxyethanol
wetting agent, cosmetics
Cithrol 4DL; Cithrol 4DO; *Cocamide DEA*; *Coconut (Cocos nucifera) oil*; Corona Lanolin; Coronet Lanolin; Crill 1; Crystal® O; Foamid AME-70; Foamid AME-75; Foamid AME-100; *Laneth-5*; *Laneth-10*; *Laneth-15*; *Lanolin*; *Nonoxynol-4*; *Nonoxynol-5*; *Nonoxynol-6*; *PEG-35 castor oil*; *PEG-4 dilaurate*; *PEG-20 dilaurate*; *PEG-4 dioleate*; *PEG-8 dioleate*; *PEG-20 dioleate*; *PEG-4 oleate*; *PEG-6 oleate*; *PEG-20 oleate*; *PEG-2 stearate*; *Pentaerythrityl oleate*; *Polysorbate 80*; *PPG-10 cetyl ether*; *PPG-30 cetyl ether*; *PPG-50 cetyl ether*; *PPG-11 stearyl ether*; Procetyl 10; *Propylene glycol laurate*; *Propylene glycol ricinoleate*; *Sodium 2-ethylhexyl sulfate*; *Trideceth*
wetting agent, creams/lotions
Laneth-5; *Laneth-10*; *Laneth-15*
wetting agent, cutting oils
Nonoxynol-4; *Nonoxynol-5*; *Trideceth*
wetting agent, deodorants
Renex® PEG 600
wetting agent, dermatologicals
Solulan® 16; Solulan® 75; Solulan® L-575
wetting agent, detergent sterilizers
Ethylan® HB4
wetting agent, driers
dl-Limonene
wetting agent, dry cleaning
PEG-2 stearate; Rhodacal® 330; Sovatex C Series
wetting agent, dust settling
Trideceth
wetting agent, dye
Polyglyceryl-3 diisostearate
wetting agent, dye: creams/lotions
Emerest® 2452
wetting agent, dye: lip care
Emerest® 2452
wetting agent, electroplating
Dimethyl hexynol
wetting agent, electroplating baths
Sodium 2-ethylhexyl sulfate
wetting agent, emulsion polymerization
Laureth-9; *Nonoxynol-5*; Rhodacal® 330; *Trioctyl phosphate*
wetting agent, enamels
dl-Limonene
wetting agent, fat liquoring
Nonoxynol-4
wetting agent, fiber lubricants
PEG-4 dioleate
wetting agent, fibers
Agrisynth® BLO; *Butyrolactone*
wetting agent, fibers/blends: one-bath dyeing/scouring processes
Intravon® SOL-W
wetting agent, firefighting prods.
Deceth-6
wetting agent, flavor concs.
Poloxamer 181
wetting agent, floor polishes
Dimethyl hexynol
wetting agent, floor waxes
dl-Limonene
wetting agent, food
Crill 1; *Glyceryl dioleate*; *Glyceryl ricinoleate*; *Glyceryl stearate SE*; *Nonoxynol-6*; *PEG-4 dilaurate*; *PEG-20 dilaurate*; *PEG-4 dioleate*; *PEG-8 dioleate*; *PEG-20 dioleate*; *PEG-4 oleate*; *PEG-6 oleate*; *PEG-20 oleate*; *Propylene glycol*; *Propylene glycol laurate*; *Propylene glycol ricinoleate*; *Sodium lauryl sulfate*
wetting agent, food pkg.
Crill 1; *Sodium 2-ethylhexyl sulfate*
wetting agent, food processing
Sodium 2-ethylhexyl sulfate
wetting agent, furniture polishes
dl-Limonene
wetting agent, general purpose
Macol® 85
wetting agent, glass cleaners
Ethylan® HB4
wetting agent, glass cleaning
Dimethyl hexynol
wetting agent, glass fibers
Pentaerythrityl oleate
wetting agent, grease/paint removers
Dowanol® DPM
wetting agent, grinding pigments
Eastman® DEP, Fragrance Grade
wetting agent, hair straighteners
Renex® PEG 600
wetting agent, hard surface cleaning
Trideceth
wetting agent, health care
Foamid AME-70; Foamid AME-75; Foamid AME-100
wetting agent, herbicides
Crill 1
wetting agent, household cleaners
Cocamide DEA; *Methoxyisopropanol*; *Nonoxynol*; *Nonoxynol-5*; *PEG-100 stearate*; *Phenoxyethanol*; *Sodium 2-ethylhexyl sulfate*
wetting agent, household prods.
Nonoxynol-4; *Nonoxynol-6*; *Trioctyl phosphate*
wetting agent, household/I&I cleaners
Dowanol® DPM; Dowanol® DPMA; Dowanol® DPnB; Dowanol® PM; Dowanol® PMA; Dowanol® PnB
wetting agent, household/I&I cleaning
Dalpad A; Dowanol® EPh
wetting agent, I&I
Trioctyl phosphate
wetting agent, industrial
Cithrol 4DL; Cithrol 4DO; *Nonoxynol-4*; *Nonoxynol-6*; *PEG-8 dilaurate*; *Propylene glycol laurate*; *Propylene glycol ricinoleate*; Renex® PEG 600
wetting agent, industrial cleaners
Cocamide DEA; *Methoxyisopropanol*; *Nonoxynol*; *Nonoxynol-5*; *PEG-100 stearate*; *Phenoxyethanol*; *Sodium 2-ethylhexyl sulfate*; *Trideceth*
wetting agent, industrial lubricants
PEG-8 dilaurate
wetting agent, inks
Crill 1; *Dimethyl hexynol*
wetting agent, insect repellents
Renex® PEG 600
wetting agent, insecticides
Crill 1
wetting agent, institutional cleaners
Laureth-9
wetting agent, iodophors
Ethylan® HB4
wetting agent, iron oxides
Ceraphyl® 140-A
wetting agent, lacquers
dl-Limonene
wetting agent, latex paints
Nonoxynol-5; Rhodacal® 330
wetting agent, laundry
Sovatex C Series
wetting agent, leather
PEG-2 stearate; *Pentaerythrityl oleate*; Sovatex C Series; *Trioctyl phosphate*
wetting agent, leather treatment
Crill 1
wetting agent, lipstick dyestuffs
HallStar™ BST
wetting agent, low-temp. wool scouring
Trideceth
wetting agent, lubricants
Propylene glycol laurate
wetting agent, lubricating oils
Methyl lardate; *Methyl tallowate*
wetting agent, maintenance cleaners
Laureth-9
wetting agent, makeup
Crystal® O
wetting agent, metal cleaners
Dowanol® DPM
wetting agent, metal cleaning
Rhodacal® 330; *Sodium 2-ethylhexyl sulfate*
wetting agent, metal pickling/plating baths
Surphos 630
wetting agent, metal processing
PPG-3 myristyl ether; *PPG-11 stearyl ether*
wetting agent, metals
Trideceth
wetting agent, metalworking
Nonoxynol-4; *Nonoxynol-6*; *PEG-4 dilaurate*; *PEG-4 dioleate*; *PEG-8 dioleate*; *PEG-4 oleate*; *Propylene glycol laurate*; *Trioctyl phosphate*
wetting agent, metalworking fluids
Crill 1; Pluracol® SRF-2; Pluracol® W3520N; Pluracol® WI 285; Pluracol® WI 625; Pluracol® WI 1715; Pluracol® WS 170; Pluracol® WS 260; Pluracol® WS 660; Pluracol® WS 2000; Pluracol® WS 3520; Pluracol® WS 5100; Pluracol® WT 1400; Pluracol® WT 9150; Pluracol® WT 90,000; Rhodacal® 330
wetting agent, metalworking oils
Methyl lardate; *Methyl palmitate oleate*; *Methyl tallowate*
wetting agent, metalworking/cleaning
Nonoxynol-5
wetting agent, oil drilling
Dipentene-C
wetting agent, oil-based systems
Isopropylamine dodecylbenzenesulfonate
wetting agent, 'oil-free' pigmented prods.
Foamid LM2E
wetting agent, oil-slick dispersing
Crill 1
wetting agent, oils
dl-Limonene
wetting agent, oral pharmaceuticals
Etocas 35 NF; Labrasol®
wetting agent, paints
Crill 1; Dipentene-C; *Laureth-9*; *dl-Limonene*; *Nonoxynol-4*; *Nonoxynol-5*; *Nonoxynol-6*; *PEG-20 oleate*; *Polysorbate 80*; *Propylene glycol laurate*; *Sodium 2-ethylhexyl sulfate*; *Triglycol monomethyl ether*
wetting agent, paper
Crill 1; *Laureth-9*; *Nonoxynol-4*; *Nonoxynol-6*; *PEG-2 stearate*; *Trideceth*
wetting agent, paper coatings
Dimethyl hexynol
wetting agent, parenteral pharmaceuticals
Etocas 35 NF; Incrocas 40
wetting agent, perfumes
dl-Limonene; *PEG-20 dioleate*
wetting agent, personal care
PEG-8 dilaurate; *PPG-10 cetyl ether*; *PPG-11 stearyl ether*; Procetyl 10; Renex® PEG 600; Solulan® 16; Solulan® 75; Solulan® C-24; Solulan® L-575
wetting agent, pesticides
Dimethyl octynediol
wetting agent, petroleum
Nonoxynol-5
wetting agent, petroleum oils
Nonoxynol-4
wetting agent, pharmaceuticals
Coconut (Cocos nucifera) oil; Corona Lanolin; Coronet Lanolin; Crill 1; Crodamol CAP; Crystal® O; *Glyceryl oleate*; *Laneth-5*; *Laneth-10*; *Nonoxynol-6*; *PEG-30 castor oil*; *PEG-35 castor oil*; *PEG-4 dilaurate*; *PEG-8 dilaurate*; *PEG-20 dilaurate*; *PEG-4 dioleate*; *PEG-8 dioleate*; *PEG-20 dioleate*; *PEG-4 oleate*; *PEG-6 oleate*; *PEG-20 oleate*; *Pentaerythrityl oleate*; *Polysorbate 80*; *PPG-30 cetyl ether*; *Propylene glycol laurate*; *Propylene glycol ricinoleate*; *Sodium 2-ethylhexyl sulfate*; *Trideceth*
wetting agent, pigment
Butyl stearate; *Castor oil, polymerized*; Cerette V; Cromollient DP3A; *Decyl oleate*; *Diisobutyl adipate*; *Diisopropyl adipate*; *Dioctyl adipate*; *Dioctyl sebacate*; *Di-PPG-3 myristyl ether adipate*; *Glyceryl triacetyl ricinoleate*; *Isodecyl isononanoate*; Isodragol 2/050300; Isostriol IC; *Isotridecyl isononanoate*; *dl-Limonene*; Monestriol B; Monestriol C; Monestriol IC; Monestriol IS-C; *Octyl isononanoate*; #15 Oil; #30 Oil; #40 Oil; Pale 4; Pale 16; Pale 170; Pale 1000; Palmil C; Palmil IS; Palmil O; *PEG-100 stearate*; *Polyglyceryl-3 diisostearate*; Thomil 14; Thomil IS; Thomil IS-F; Tilol B; Tilol ET; Tilol IS; Vorite 105; Vorite 110; Vorite 115; Vorite 125
wetting agent, pigment: aftershaves
HallStar™ DIPA; HallStar™ DIPS
wetting agent, pigment: antiperspirant sticks
Crystal® Crown
wetting agent, pigment: bath care
Finsolv® BOHS-111; HallStar™ DOA;

Part V
Manufacturers Directory

Manufacturers Directory

AAA International Corp.

515 Ogden Ave., Professional Bldg. Suite 317, Downers Grove, IL, 60515, USA (Tel: 630-964-1249; FAX 630-964-1649; E-mail: sales@aaainternational.com; Internet: www.aaainternational.com)

AAE Chemie NV

Verbindingsdok-Oostkaai 13, 2000 Antwerp, Belgium (Tel: 32 3 568 1166; FAX 32 3 568 0597; E-mail: info@aaechemie.com; Internet: www.aaechemie.com)

Aarhus

Aarhus Oliefabrik A/S, M.P. Bruuns Gade 27, PO Box 50, 8100 Aarhus C, Denmark (Tel: 45 87 30 60 00; FAX 45 87 30 60 12; E-mail: aarhus@aarhus.com; Internet: www.aarhus.com)

Aarhus Inc., 131 Marsh St., PO Box 4240, Port Newark, NJ, 07114, USA (Tel: 973-344-1300; FAX 973-344-9049; E-mail: aar@aarhususa.com; Internet: www.aarhususa.com)

Aarhus Australia Pty. Ltd., 4 Endeavour Close, Castle Hill, NSW, 2154, Australia (Tel: 61 2 885 03 522; FAX 61 2 885 03 422; E-mail: brazelp@aol.com)

Aarti Industries Ltd.

201, Udyog Kshetra, Mulund Goregaon Link Rd., Mulund (W), Mumbai, Maharashtra, 400 080, India (Tel: 91 22 5918195; FAX 91 22 5653185; E-mail: info@aartigroup.com; Internet: www.aartigroup.com)

Aastrid International

238, Udyog Bhavan, Sonawala Rd., Goregaon (E), Mumbai, 400 063, India (Tel: 91 22 871 7238; FAX 91 22 876 0845; E-mail: aastrid@bom2.vsnl.net.in; Internet: www.aastrid.com)

ABCR

ABCR GmbH & Co. KG, Postfach 21 01 35, 76151 Karlsuhe, Germany (Tel: 49-721-95061-0; FAX 49-721-95061-80; E-mail: info@abcr.de; Internet: www.abcr.de)

ABCR (UK) Ltd., Unit 60, The Greenhays Business Centre, 10 Pencroft Way, Manchester, M15 6JJ, UK (Tel: 44 161 226 9843; FAX 44 161 226 9850; E-mail: arrowsmith@abcr.de; Internet: www.abcr.de)

ABITEC

ABITEC Corp., Subsid. of Associated British Foods (ABF), 501 W. First Ave., PO Box 569, Columbus, OH, 43216-0569, USA (Tel: 614-429-6464; 800-848-1340; FAX 800-300-0818; 614-299-2445; E-mail: sales@abiteccorp.com; Internet: www.abiteccorp.com)

AB Technology Ltd., Div. of ABITEC, Salthouse Rd, Brackhills Business Park, Northampton, Northamptonshire, NN4 7EX, UK (Tel: 44 1604 768999; FAX 44 1604 701503; E-mail: abitec@global.co.uk)

Ablestik

Ablestik Laboratories, Subsid. of National Starch and Chemical Co., an ICI Co., 20021 Susana Road, Rancho Dominguez, CA, 90221, USA (Tel: 310-764-4600; FAX 310-764-2545; E-mail: ablestik.customerservice@ablestik.com; Internet: www.ablestik.com)

Ablestik Laboratories/EMCA-REMEX Products, Station Rd., Linton, Cambridge, CB1 6NW, UK (Tel: 44 1223 893771; FAX 44 1223 893546)

Accord Corporation

PO Box 1288, Highland Park, NJ, 08904, USA (Tel: 732-819-8738; FAX 732-819-7431; Internet: www.accordchem.com)

Accurate Chemical & Scientific Corp.

300 Shames Dr., Westbury, NY, 11590, USA (Tel: 516-333-2221; 800-645-6264; FAX 516-997-4948; E-mail: info@accuratechemical.com; Internet: www.accuratechemical.com)

AccuStandard Inc.

25 Science Park, New Haven, CT, 06511, USA (Tel: 203-786-5290; 800-442-5290; FAX 203-786-5287; E-mail: usa@accustandard.com; Internet: www.accustandard.com)

Aceto Corporation

1 Hollow Lane, Suite 201, Lake Success, NY, 11042-1215, USA (Tel: 516-627-6000; FAX 516-627-6093; E-mail: aceto@aceto.com; Internet: www.aceto.com)

Acme-Hardesty Co.

Div. of Jacob Stern & Sons, Inc., 1787 Sentry Parkway West, Suite 18-460, Blue Bell, PA, 19422, USA (Tel: 215-591-3610; 800-223-7054; FAX 215-591-3620; E-mail: info@acme-hardesty.com; Internet: www.acme-hardesty.com)

Acros Organics N.V.

Div. of Fisher Scientific, 1 Reagent Lane, Fairlawn, NJ, 07410, USA (Tel: 201-703-3163; 800-ACROS-01; FAX 800-248-3079; E-mail: TDAiken@aol.com; Internet: www.acros.be)

Acti-Chem Specialties Inc.

56 Quarry Rd., Trumbull, CT, 06611, USA (Tel: 203-365-6530; FAX 203-365-6537; E-mail: actichem@snet.net)

Active Organics, Inc.

1097 Yates St., Lewisville, TX, 76051, USA (Tel: 972-253-7676; 800-541-1478; FAX 972-221-3324; E-mail: sales@activeorganics.com; Internet: www.activeorganics.com)

ADA International

Room 415, Huikai Bldg., Bldg. No. 35, Erlizhuang Xiaoqu, Haidian District, Beijing, 100083, P.R. China (Tel: 86 10 6226 2026; FAX 86 10 6226 2025; E-mail: ghwchem@public3.bta.net.cn; Internet: www.joinme.net/ada/index.htm)

Adept Solutions Inc.

331 Capitola Ave., Suite L, Capitola, CA, 95010, USA (Tel: 831-477-1344; 888-477-6644; FAX 831-477-1348; E-mail: russ@adepthq.com)

ADM

ADM, Archer Daniels Midland Co., 4666 Faries Pkwy., Decatur, IL, 62526, USA (Tel: 217-362-3949; 800-637-5843; FAX 217-424-4119; Internet: www.admworld.com)

ADM Ethanol Sales, Div. Archer Daniels Midland Co., Box 1470, Decatur, IL, 62525, USA (Tel: 217-424-5200; FAX 217-424-5978)

Adumim Food Ingredients

Mishor Adumim 90610, Israel (Tel: 972 2 535 3565; FAX 972 2 535 4187; E-mail: info@adumim.co.il; Internet: www.adumim.co.il)

Advanced BioTech

85 Fifth Ave., Bldg. 5, Paterson, NJ, 07524, USA (Tel: 973-357-0577; FAX 973-357-0644; E-mail: info@adv-bio.com; Internet: www.adv-bio.com)

Advanced ChemTech, Inc.

5609 Fern Valley Rd., Louisville, KY, 40228-1075, USA (Tel: 502-969-0000; 800-456-1403; FAX 502-962-5368; E-mail: info@peptide.com; Internet: www.peptide.com)

Advanced Synthesis Technologies, S.A.

PO Box 437920, San Ysidro, CA, 92173, USA (Tel: 619-423-7821; FAX 619-423-7793; E-mail: sales@advancedsynthesis.com; Internet: www.advancedsynthesis.com)

Advance Research Chemicals Inc.

Tanumal Chem. Complex Bldg., 1110 W. Keystone Ave., Catossa, OK, 74015, USA (Tel: 918-266-6789; FAX 918-266-6789; E-mail: sales@fluoridearc.com)

AECI Aroma & Fine Chemicals (Pty.) Ltd.

Subsid. of AECI Ltd., PO Box 1606, Richards Bay, KZN, 3900, Rep. of S. Africa (Tel: 27 35 797 6000; FAX 27 35 797 3336; E-mail: stevens@aecifinechem.co.za; Internet: www.aecifinechem.co.za/)

Aeropres

1324 N. Hearne, Suite 200, PO Box 78588, Shreveport, LA, 71137-8588, USA (Tel: 318-221-6282; FAX 318-213-1270; E-mail: mrivers@aeropres.com; Internet: www.aeropres.com)

Ag Environmental Products L.L.C. (AEP)

9804 Pflumm Rd., Lenexa, KS, 66215, USA (Tel: 913-599-6911; 800-599-9209; FAX 913-599-2121; Internet: www.soygold.com)

Agrium

Agrium Inc., 13131 Lake Fraser Drive, Calgary, Alberta, T2J 7E8, Canada (Tel: 403-225-7000; 800-490-4555; FAX 403-225-7609; E-mail: indprod@agrium.com; Internet: www.agrium.com)

Agrium U.S., Inc., 4582 South Ulster St., Suite 1400, Denver, CO, 80237, USA (Tel: 303-804-4400; FAX 303-804-4476)

Air Products

Air Products and Chemicals, Inc., 7201 Hamilton Blvd., Allentown, PA, 18195-1501, USA (Tel: 610-481-4911; 800-345-3148; FAX 610-481-5900; E-mail: info@apci.com; Internet: www.airproducts.com)

Air Products and Chemicals de México, S.A. de C.V., Suite 26, Piso 12 World Trade Center, Avenida de las Naciones No. 1, Col. Napoles, Mexico D.F., 03810, Mexico (Tel: 52 5 488 0790; FAX 52 5 488 0798)

Air Products Gases Industriais Ltda., Praca Radialista Manoel de Nobrega, 65, Casa Verde, Sao Paulo SP, 02517-160, Brazil (Tel: 55 11 856 1600; FAX 55 11 856 1783)

Air Products Nederland B.V., Kanaalweg 15, PO Box 3193, 3502 GD Utrecht, The Netherlands (Tel: 31 302 857100; FAX 31 302 857111)

Air Products (Chemicals) PLC, Clayton Lane, Clayton, Manchester, Lancashire, M11 4SR, UK (Tel: 44 161 230 4265; FAX 44 161 223 4753)

Air Products S.A., 78 Rue Championnet, 75881 Paris Cedex 18, France (Tel: 33 1 44 92 51 00; FAX 33 1 44 92 51 01)

Air Products S.A., Chaussée de Wavre 1789, 1160 Brussels, Belgium (Tel: 32 2 674 9411; FAX 32 2 674 9462)

Air Products c/o Anchor Italiana SPA, 33 Via Scaldasole, 27020 Pavia Dorno, Italy (Tel: 39 382 812281; FAX 39 382 84017)

Air Products GmbH, Hauptverwaltung Hattingen, Huettenstrasse 50, 45527 Hattingen/Ruhr, Germany (Tel: 49 2324 6890; FAX 49 2324 689100)

Air Products Sweden, World Trade Centre, Maessans Gata 10, Box 5264, 402 25 Göteborg, Sweden (Tel: 46 31 357511; FAX 46 31 164200)

Air Products South Africa (Pty) Ltd., 3 Protea Place, Sandton, Private Bag 784090, Sandton, 2146, Rep. of S. Africa (Tel: 27 11 322 0000; FAX 27 11 322 0222)

Air Products Asia, Inc., Suite 6505-7, Central Plaza, 18 Harbour Rd., Wanchai, Hong Kong (Tel: 852 2527 0515; FAX 852 2527 1957)

Air Products Japan, Inc., Shuwa No. 2 Kamiyacho Bldg., 11 Fl., 3-18-19, Toranomon, Minato-ku, Tokyo, 105, Japan (Tel: 81 3 3432 7031; FAX 81 3 3432 7052)

Air Products China, Inc., Rm. 611 Beijing Silver Tower, No. 2 North Rd. Dong San Huan, Chaoyang District, Beijing, 100027, P.R. China (Tel: 86 10 6410 6151; FAX 86 10 6410 6153)

Air Products Singapore Pte. Ltd., 9 Temasek Blvd., #28-01, Suntec Tower 2, Singapore, 038989, Singapore (Tel: 65 332 1610; FAX 65 332 1600)

Air Products Australia Pty Ltd., 2/20 Hunter St., PO Box 4068, Paramatta, NSW, 2124, Australia (Tel: 61 2 9687 1944; FAX 61 2 9687 1950)

Air-Scent International

290 Alpha Dr., RIDC Industrial Park, Pittsburgh, PA, 15238, USA (Tel: 412-252-2000; 800-247-0770; FAX 412-252-1010; E-mail: info@airscent.com; Internet: www.airscent.com)

Ajay North America

Ajay North America, LLC, PO Box 127, 1400 Industry Rd., Powder Springs, GA, 30127, USA (Tel: 770-943-6202; FAX 770-439-0369; E-mail: ajayna@bellsouth.net; Internet: www.ajayna.com)

Ajay Europe, Sarl, Z.I. Du Grand Verger, B.P. 227, 53602 Evron Cedex, France (Tel: 33 2 43 01 35 35; FAX 33 2 43 01 76 18; E-mail: ajay@wanadoo.fr)

Ajay Pacific, Ltd., 1002 10-F Taurus Bldg., 21 A/B Granville Rd., Tsim Sha Tsui, Kowloon, Hong Kong (Tel: 852 2366 1533; FAX 852 2366 1513; E-mail: eson@netvigator.com)

Akcros Chemicals UK

Business Unit of Akzo Nobel, PO Box 1, Lankro Way, Eccles, Manchester, Lancashire, M30 0BH, UK (Tel: 44 161 785 1111; FAX 44 161 788 7886; E-mail: info.eccles@akcros.com; Internet: www.akcros.com)

Akdeniz Kimya A.S.

PO Box 342, 35212 Izmir, Turkey (Tel: 90 232 877 01 44; FAX 90 232 877 01 50)

Akrochem Chemical Co.

255 Fountain St., Akron, OH, 44304-1991, USA (Tel: 330-535-2100; 800-321-2260; FAX 330-535-8947; Internet: www.akrochem.com)

Aktiva Biochemical Plant Ltd.

Kaznejov 125, 331 51 Kaznejov, Czech Republic (Tel: 42 0 182 932532; FAX 42 0 182 932618)

Akzo Nobel

Akzo Nobel nv, Velperweg 76, 6824 BM Arnhem, The Netherlands (Tel: 31 26 3664433; FAX 31 26 3663250; E-mail: acc@akzonobel.com; Internet: www.akzonobel.com)

Akzo Nobel Chemicals bv, Postbus 247, Stationsplein 4, 3800 AE Amersfoort, The Netherlands (Tel: 31 33 467 62 67; FAX 31 33 467 61 52; E-mail: jan.hasselman@akzonobel.com; Internet: www.akzonobel.com; www.functionalchemicals.com)

Akzo Nobel Surface Chemistry AB, 444 85 Stenungsund, Sweden (Tel: 46 303 85000; FAX 46 303 84659; Internet: www.surface.akzonobel.com)

Akzo Nobel Chemicals Ltd., 1-5 Queens Road, Hersham, Walton-on-Thames, Surrey, KT12 5NL, UK (Tel: 44 1932 247891; FAX 44 1932 231204)

Akzo Nobel Chemicals GmbH, Phillippstrasse 27, 52349 Düren, Germany (Tel: 49 2421 49201; FAX 49 2421 492240)

Akzo Nobel Chemicals S.A., Parc Industriel de Ghlin Zone A, 7011 Ghlin/Mons, Belgium (Tel: 32 65 642381; FAX 32 65 642385)

Akzo Nobel Chemicals SpA, 80 Via E. Vismara, 20020 Arese (MI), Italy (Tel: 39 2 935 251; FAX 39 2 935 89002)

Akzo Nobel Chemicals SA, BP 217, 60202 Compiègne Cedex, France (Tel: 33 3 44 90 30 00; FAX 33 3 44 90 30 44)

Akzo Nobel Representative Office (Russia), Vavliova St. 69, Rooms 406-429, 117846 Moscow, Russia (Tel: 7 502 220 3052; FAX 7 095 134 0425)

Akzo Nobel Chemicals Inc., 300 S. Riverside Plaza, Chicago, IL, 60606-6697, USA (Tel: 312-906-7500; 800-227-7070; FAX 312-906-7680; Internet: www.akzonobel.com)

Akzo Nobel Chemicals Inc., 5 Livingstone Ave., Dobbs Ferry, NY, 10522-3401, USA (Tel: 914-674-5000; 800-666-1212; FAX 914-693-3417)

Akzo Nobel Surface Chemistry LLC, 300 South Riverside Plaza, Chicago, IL, 60606, USA (Tel: 312-906-7500; FAX 312-906-7680; Internet: www.surface.akzonobelusa.com)

Akzo Nobel Surface Chemistry, Inc., 99 Hawley Lane, Merritt 8 Corporate Park, Stratford, CT, 06614, USA (Tel: 203-378-0500; FAX 203-378-5960; E-mail: Susan.E.Booth@akzo-nobel.com; Internet: www.surface.akzonobelusa.com)

Akzo Nobel Chemicals Ltd., 1 City Center Dr., Suite 318, Mississauga, Ontario, L5B 1M2, Canada (Tel: 905-273-5959; FAX 905-273-7339; Internet: www.surface.akzonobelusa.com)

Akzo Nobel Chemicals SA de CV, Calle Morelos No. 49, Col. Tecamachalco, Los Reyes La Paz, Edo. de Mexico, 56500, Mexico (Tel: 52 5 855 1295; FAX 52 5 857 0772)

Akzo Nobel Chemicals SA, Tucuman 359, piso 9, 1049 Buenos Aries, Argentina (Tel: 54 1 313 5040)

Akzo Nobel Quimica Ltda., Estrada Municpal Iva. 188 no. 707, Bairro Sao Roque da Chave, Itupeva, Sao Paulo, 13225-000, Brazil (Tel: 55 11 7801 1130; FAX 55 11 7801 1495)

Akzo Nobel Chemicals C.A., C.C.C.T. Torre-C, Piso S. Oficina 504-C Chaus, Caracas, 1060, Venezuela (Tel: 58 2 959 0952; FAX 58 2 959 0154)

Akzo Nobel Chemicals Ltd., 19 Norwoot Crescent, Suite 10, Monnee Ponds, Victoria (Melbourne), 3039, Australia (Tel: 61 3 9391 3633; FAX 61 3 9391 4007)

Akzo Nobel K.K., Bancho Kaikan, 12-1, Gobancho, Chiyoda-ku, Tokyo, 102, Japan (Tel: 81 3 5275 6273; FAX 81 3 3263 0713)

Akzo Nobel China Beijing, Representative Office, Corporate East Ocean Center Rm 1901-04, Jian Guo Men Wai Dajie, Beijing, 100004, P.R. China (Tel: 86 10 65155736; FAX 86 10 65155737)

Akzo Nobel Hong Kong, 6/F., Allied Capital Resources Bldg., 32-38 Ice House St., Central Hong Kong, Hong Kong (Tel: 852 5217828; FAX 852 5258414)

Akzo Nobel Korea Ltd., Chongdundang Bldg., Rm. 805, 368, 3-ka, Chungchong-ro, Seodaemun-ku, Seoul, 120-756, Korea (Tel: 82 2 393 0083; FAX 82 2 392 5188)

Akzo Nobel Chemicals Pte. Ltd., 41 Science Park Rd., #03-03 The Gemini, Singapore Science Park II, Singapore, 117610, Singapore (Tel: 65 773 8488; FAX 65 773 8484)

Akzo Nobel Chemicals Pte. Ltd., #7A02, No. 3, Hsin YI Roac, Sec. 5, PO Box 109-822, Taipei, 105, Taiwan R.O.C. (Tel: 886 2 345 5200; FAX 886 2 723 7456)

Akzo Nobel Salt

Akzo Nobel Salt America, Inc., PO Box 938, Georgetown, SC, 29440, USA (Tel: 843-546-1925; 800-377-1017; FAX 843-527-1580; E-mail: saltusa@sccoast.net; Internet: www.akzonobelsalt.com)

Akzo Nobel Salz GmbH, Am Sandtorkai 71, 20457 Hamburg, Germany (Tel: 49 40 36 97 70 42; FAX 49 40 36 97 70 20)

Akzo Nobel Salt bv, PO Box 247, 3800 AE Amersfoort, The Netherlands (Tel: 31 33 4 67 68 56; FAX 31 33 4 67 67 61 32; E-mail: angelique.terweij@akzonobel.com; Internet: www.akzonobelsalt.com)

Akzo Nobel nv/sa Salt and Chemicals, Rederskaai 3 bus 1, 1000 Brussels, Belgium (Tel: 32 2 27 45 527; FAX 32 2 27 45 501; E-mail: Magda.Huenaerts@AkzoNobel.com)

Alba International Inc.

PO Box 201, N. Aurora, IL, 60542, USA (Tel: 858-272-0444; 800-669-9333; FAX 858-272-0400; E-mail: albaintl@cari.net; Internet: www.alba-intl.com)

Alban Muller

Alban Muller Int'l., 8 Rue Charles Pathe, 96300 Vincennes, France (Tel: 33 1 48 08 81 00; FAX 33 1 48 08 81 01; E-mail: info@albanmuller.com; Internet: www.albanmuller.com)

Tri-K Industries, Distributor for Alban Muller, 151 Veterans Drive, Northvale, NJ, 07647, USA (Tel: 201-750-1055; 800-526-0372; FAX 201-750-9785; E-mail: info@tri-k.com; Internet: www.tri-k.com)

Albemarle

Albemarle Corp., 451 Florida Blvd., Baton Rouge, LA, 70801, USA (Tel: 225-388-7402; 800-535-3030; FAX 225-388-7848; Internet: www.albemarle.com)

Albemarle S.A., Div. of Albemarle Corp., Parc Scientifique Einstein, Rue du Bosquet 9, 1348 Louvain-la-Neuve Sud, Belgium (Tel: 32 10 48 1711; FAX 32 10 48 1717)

Albemarle Asia Pacific Co., 111Somerset Rd. #13-03, Singapore, 238164, Singapore (Tel: 65 732 6286; FAX 65 737 4155)

Albemarle Asano Corp., 16th Fl., Fukoku Seimei Bldg., 2-2 Uchisaiwaicho, 2-chome, Tokyo, 100, Japan (Tel: 81 3 5251 0791; FAX 81 3 3500 5623)

Albemarle China Corp., China World Tower, Rm. 1317, No. 1 Jian Guo Men Wai Ave., Beijing, 100004, P.R. China (Tel: 86 10 6505 4153; FAX 86 10 6505 4150)

Alchemie USA Inc.

790 S. Main St., Suite 2C, Plantville, CT, 06479, USA (Tel: 860-621-2470; FAX 860-621-9570; E-mail: allchemieus@aol.com)

Aldivia

49, rue des Sources, BP 1, 69561 St-Genis-Laval Cedex, France (Tel: 33 4 72 66 66 90; FAX 33 4 72 66 66 99; E-mail: info@aldivia.com; Internet: www.aldivia.com)

Aldrich. *See under* Sigma-Aldrich

Alemark Chemicals, Amsyn Inc.

1011 High Ridge Rd., Stamford, CT, 06905, USA (Tel: 203-966-7410; FAX 203-966-4276; E-mail: mail@amsyn.com; Internet: www.amsyn.com)

Alexander Chemical Corp.

One Corporate Lakes, 2525 Cabot Drive, Suite 201, Lisle, IL, 60532, USA (Tel: 630-955-6050; 800-445-9458; FAX 630-955-6065)

Alfa Aesar

Alfa Aesar, A Johnson Matthey Co., 30 Bond St., Ward Hill, MA, 01835, USA (Tel: 978-521-6300; 800-343-0660; FAX 800-322-4757; E-mail: info@alfa.com; Internet: www.alfa.com)

Alfa Aesar (Germany), A Johnson Matthey Co., Postfach 110765, 76057 Karlsruhe, Germany (Tel: 49 721 84007 260; FAX 49 721 84 96 74)

Alfa Aesar (Switzerland), A Johnson Matthey Co., Glattalstrasse 18, 8052 Zurich, Switzerland (Tel: 41 1 302 4488; FAX 41 1 302 4701; E-mail: chcat@alfa.com)

Alfa (Japan), A Johnson Matthey Co., Nittobo Bldg., 2-8-1 Yaesu Chuo-ku, Tokyo, 104, Japan (Tel: 81 3 3278 3414; FAX 81 3 3278 3401; E-mail: jmjcatalog@matthey.com)

Alfa Chem

98 Kings Pt. Rd., King's Point, NY, 11024-1303, USA (Tel: 516-504-0059; 800-375-6869; FAX 516-504-0039; E-mail: alfachem1@aol.com; Internet: www.alfachem.homepage.com)

Alfa Chemicals, Ltd

Arc House, Terrace Rd., S. Binfield, Bracknell, Berkshire, RG42 4PZ, UK (Tel: 44 1344 861 800; FAX 44 1344 862 010)

Alkyl Amines Chemicals Limited

401-407, Nirman Vyapar, Kendra, Plot No. 10, Sector 17, Vashi, Navi Mumbai, 400 703, India (Tel: 91 22 7890632; FAX 91 22 7890631; E-mail: customercare@alkylamines.com; Internet: www.alkylamines.com)

Allan Chemical Corp.

PO Box 1837, Fort Lee, NJ, 07024-8337, USA (Tel: 201-592-8122; 800-633-0274; FAX 201-592-9298; E-mail: allanchem@allanchem.com; Internet: www.allanchem.com)

Allchem Industries

Allchem Industries Inc., 6010 NW First Place, Gainesville, FL, 32607, USA (Tel: 352-378-9696; 800-523-0450; FAX 352-338-0400; E-mail: info@allchem.com; Internet: www.allchem.com)

AllChem Industries B.V., PO Box 4061, 2980 GB Ridderkerk, The Netherlands (Tel: 31 180 470 470; FAX 31 180 470 479)

AllChem Industries (Singapore) Pte., 101 Thompson Rd., 26-02 United Square, Singapore, 307591, Singapore (Tel: 65 253 6711; FAX 65 253 6733)

Allchem International Ltd.

Broadway House, 21 Broadway, Maidenhead, Berkshire, SL6 1NJ, UK (Tel: 44 1753 443322; FAX 44 1753 443356; E-mail: info@allchem.co.uk; Internet: www.allchem.co.uk)

Allied Universal Corp.

8350 NW 93 St., Miami, FL, 33166, USA (Tel: 305-888-2623; 800-437-8715; FAX 305-885-4671)

Allyson Enterprises Inc.

133 E. 58th St., Suite 703, New York, NY, 10022, USA (Tel: 212-223-8770; FAX 212-223-8755; E-mail: allysonent@cs.com)

Alnor Oil Company, Inc.

70 E. Sunrise Hwy., Suite 418, Valley Stream, NY, 11581-1221, USA (Tel: 516-561-6146; FAX 516-561-6123; E-mail: sales@alnoroil.com; Internet: www.alnoroil.com)

Alpha Chem

3 Wallis Ct, Lexington, MA, 02421, USA (Tel: 781-863-7290; FAX 781-863-0147; E-mail: info@alphachem.com; Internet: www.alphachem.com)

Alpharma Inc.

One Executive Dr., Fort Lee, NJ, 07024, USA (Tel: 201-947-7774; FAX 201-947-4879)

Altair Industries, Inc.

343 Millburn Ave., Millburn, NJ, 07041, USA (Tel: 973-564-6400; FAX 973-564-6401; E-mail: altairind@aol.com)

Altivia

1100 Louisiana, Suite 3160, Houston, TX, 77002-5217, USA (Tel: 713-658-9000; FAX 713-658-0102; E-mail: cpcsales@phoenix.net; Internet: www.altivia.com)

Alvin Chemical, Inc.

PO Box 94, Alvin, TX, 77512, USA (Tel: 281-331-5258; FAX 281-585-8738; E-mail: mike@alvinchemical.com; Internet: www.alvinchemical.com)

Alzo International Inc.

650 Jernee Mill Rd., Sayreville, NJ, 08872, USA (Tel: 732-254-1901; FAX 732-254-4423; E-mail: alzoinc@worldnet.att.net; Internet: www.alzointernational.com)

Amalgamet

Amalgamet Canada, Div. of Amalgamated Metal Corp. PLC, 111 Richmond St. W, Suite 418, Toronto, Ontario, M5H 2G4, Canada (Tel: 416-366-3954; FAX 416-366-0586; E-mail: info@amalgamet.com; Internet: www.amalgamet.com)

Amalgamet Inc., 50 Main St., 12th Fl., White Plains, NY, 10606, USA (Tel: 914-286-2300; FAX 914-286-2330; E-mail: bsilverstein@amctin.com; Internet: www.amcgroup.com)

Amber Synthetics, Amsyn Inc.

1011 High Ridge Rd., Stamford, CT, 06905, USA (Tel: 203-329-6500; FAX 203-329-6600; E-mail: mail@amsyn.com; Internet: www.amsyn.com)

Amcan Ingredients

6 Ave. Charles de Gaulle, 78510 Le Chesnay, France (Tel: 33 1 39 55 52 60; FAX 33 1 39 54 72 68; E-mail: amcaneurop@aol.com; Internet: www.amcaningredients.fr.st)

AMC Chemicals

AMC Chemicals USA, 1480 Rt. 9 North, Suite 207, Woodbridge, NJ, 07095, USA (Tel: 732-634-6034; FAX 734-634-6140; E-mail: robert@amcchem.com)

AMC Chemicals (UK), 10 Charterhouse Sq., London, EC1M 6EH, UK (Tel: 44 1202 620710; FAX 44 1202 620698; E-mail: amcoils@amcgroup.com; Internet: www.amcchemicals.co.uk)

Amend Drug & Chemical Co., Inc.

83 Cordier St., Irvington, NJ, 07111, USA (Tel: 973-926-0331; 800-274-2636; FAX 973-926-4921; E-mail: rugeramend@worldnet.att.net; Internet: www.amenddrug.com)

Amerchol

Amerchol Corp., Subsid. of Dow Chemical Co., PO Box 4051, 136 Talmadge Rd., Edison, NJ, 08818-4051, USA (Tel: 800-367-3534; FAX 989-832-1465; Internet: www.dow.com)

Union Carbide Benelux N.V./Amerchol Div., A Subsid. of The Dow Chemical Co., Havenstraat 86, 1800 Vilvoorde, Belgium (Tel: 32 3 450 2240; FAX 32 3 450 2815)

AmeriBrom, Inc.

Member of the Dead Sea Bromine Group, 2115 Linwood Ave., Fort Lee, NJ, 07024-5004, USA (Tel: 201-242-6560; FAX 201-242-6561; E-mail: info@ameribrom.dsbg.com; Internet: www.ameribrom.com)

American Biorganics, Inc.

Subsid. of ANGUS Chemical Co., 2236 Liberty Dr, Niagara Falls, NY, 14304, USA (Tel: 716-283-1434; 800-648-6689; FAX 716-283-1570; E-mail: ali@data.abi.angus.com)

American Chemet Corp.

400 Lake Cook Rd., PO Box 437, Deerfield, IL, 60015, USA (Tel: 847-948-0800; FAX 847-948-0811)

American Colloid Co.

Wholly owned subsid. of AMCOL International Corp., 1500 W. Shure Dr., Arlington Hts., IL, 60004-7803, USA (Tel: 847-392-4600; 800-426-5564; FAX 847-506-6199; Internet: www.colloid.com)

American Emulsions Co., Inc.

An RPM Company, 1202 Dozier St., PO Box 3787, Dalton, GA, 30721, USA (Tel: 706-226-7028; FAX 706-278-5183; E-mail: info@americanemulsions.com; Internet: www.americanemulsions.com)

American Fruit Processors

10725 Sutter Ave., Pacoima, CA, 91333, USA (Tel: 818-899-9574; FAX 818-899-6042)

American Ingredients

American Ingredients Co., Wholly owned subsid. of CSM NV, 3947 Broadway, Kansas City, MO, 64111, USA (Tel: 816-561-9050; 800-669-4092; FAX 816-561-9909; Internet: www.americaningredients.com)

American Ingredients Co./Patco Additives Div., Wholly owned subsid. of CSM NV, 3947 Broadway, Kansas City, MO, 64111, USA (Tel: 816-561-9050; 800-669-2250; FAX 816-561-7778; E-mail: info@patco-additives.com; Internet: www.patco-additives.com)

American International Chemical, Inc.

17 Strathmore Rd, Natick, MA, 01760, USA (Tel: 508-655-5805; 800-238-0001; FAX 508-655-0927; E-mail: info@aicma.com; Internet: www.aicma.com)

American Laboratories Inc.

4410 South 102 St., Omaha, NE, 68127, USA (Tel: 402-339-2494; 800-445-5989; FAX 402-339-0801; E-mail: sales@americanlaboratories.com; Internet: www.americanlaboratories.com)

American Oil & Supply Co./AOS Thermal Compds.

238 Wilson Ave., Newark, NJ, 07105, USA (Tel: 201-589-0250; 800-582-3267; FAX 201-465-5083; E-mail: ZZZAOS1895@aol.com; Internet: www.aoils.com)

American Radiolabeled Chemicals, Inc. (ARC)

11624 Bowling Green Drive, St. Louis, MO, 63146, USA (Tel: 314-991-4545; 800-331-6661; FAX 800-999-9925; E-mail: arcinc@arc-inc.com; Internet: www.arc-inc.com)

Americhem Sales Corp.

340 North St., Mason, MI, 48854, USA (Tel: 517-676-9363; Internet: www.americhemsales.com)

Amerol Corp.

71 Carolyn Blvd., Farmingdale, NY, 11735-1527, USA (Tel: 631-694-9090; FAX 631-694-9177; E-mail: amerolcorp@aol.com; Internet: www.amerolcorp.com)

Amity

Amity UK Ltd., Friendship House, Dodworth Business Park, Barnsley, S. Yorkshire, S75 3SP, UK (Tel: 44 1226 770787; FAX 44 1226 770757; Internet: www.amityinternational.com)

Amity International, Inc., 30285 Bruce Industrial, Suite B, Cleveland, OH, 44139, USA (Tel: 440-248-7895; FAX 440-248-7917; E-mail: sales@amityinternational.com)

AMJAY Chemicals

PO Box 218786, 16350 Park 10 Place, Suite 233, Houston, TX, 77218-8786, USA (Tel: 281-492-2000; FAX 281-953-1282; E-mail: amjay@worldnet.att.net)

AMRESCO

30175 Solon Ind. Pkwy., Solon, OH, 44139, USA (Tel: 440-349-1313; 800-448-4442; FAX 440-349-1182; E-mail: amresco@worldnet.att.net; Internet: www.amresco-inc.com)

Amyl, Inc.

13111 Westheimer, Suite 315, Houston, TX, 77077-5520, USA (Tel: 281-584-9994; FAX 281-584-9996; E-mail: info@amyl.com; Internet: www.amyl.com)

Anar Chemical Co.

1765F Cortland Court, Addison, IL, 60101, USA (Tel: 708-953-1660; 800-344-1660; FAX 708-953-1698; E-mail: anarchem@aol.com)

Andrea Aromatics

PO Box 3091, Princeton, NJ, 08543-3091, USA (Tel: 609-695-7710; FAX 609-392-8914)

Anglia Oils Ltd

Subsid. of Aarhus Olie, King George Dock, Kingston-upon-Hull, North Humberside, HU9 5PX, UK (Tel: 44 1482 701 271; FAX 44 1482 709 447; E-mail: info@angliaoils.co.uk; Internet: www.angliaoils.co.uk)

ANGUS Chemical Co.

Subsid. of Dow Chemical Co., 2825 Willard Dow Center, Midland, MI, 48642-2825, USA (Tel: 800-447-4369; FAX 517-832-1465; Internet: www.dow.com/angus/)

Anmar International Ltd.

540 Barnum Ave., Bridgeport, CT, 06608, USA (Tel: 203-336-8330; FAX 203-336-5508; E-mail: cheldianmar@snet.net; Internet: www.anmarinternational.com)

AP Chemicals

223 Westhampton Ave., Capitol Hts., MD, 20743, USA (Tel: 888-336-9303; FAX 301-336-6597; Internet: www.apchem.com)

A.P. Chemicals Ltd.

A Member of the BTP plc Group of Companies, Station Road, Cheddleton, Leek, Staffordshire, ST13 7EF, UK (Tel: 44 1538 369000; FAX 44 1538 361330; E-mail: sales@apchem.u.net.com)

Apollo Scientific

Unit 1A, Bingswood Industrial Estate, Whaley Bridge, Derbyshire, SK23 7LY, UK (Tel: 44 1663 719222; FAX 44 1663 719333; E-mail: sales@apolloscientific.co.uk; Internet: www.apolloscientific.co.uk)

AP Pharma

123 Saginaw Drive, Redwood City, CA, 94063, USA (Tel: 650-366-2626; FAX 650-365-6490; E-mail: jbarr@appharma.com; Internet: www.advancedpolymer.com)

AP Resources Corp.

1712 Windstone Officetel, 275-2, Yangjae-Dong, Seocho-Gu, Seoul, 137-130, Korea (Tel: 82 2 3463 1163; FAX 82 2 3463 1162; E-mail: aprc@netsgo.com; Internet: www.apr.co.kr)

Aquaness Chemical

Div. of Baker Petrolite, 12645 West Airport, Sugarland, TX, 77478, USA (Tel: 281-276-5400; 800-531-2025; E-mail: Larry.Graham@BakerPetrolite.com)

Arakawa

Arakawa Chemical Industries Ltd., 1-3-7, Hirano-machi, Chuo-ku, Osaka, 541-046, Japan (Tel: 81 6 6209 8581; FAX 81 6 6209 8542; E-mail: info@arakawachem.co.jp; Internet: www.arakawachem.co.jp/e)

Arakawa Chemical (USA) Inc., 625 N. Michigan Ave., Suite 1700, Chicago, IL, 60611, USA (Tel: 312-642-1750; FAX 312-642-0089; E-mail: arakawa@arakawa-usa.com; Internet: www.arakawa-usa.com)

ARC Specialty Products Div.

Div. of Balchem Corp., PO Box 175, Slate Hill, NY, 10973, USA (Tel: 845-355-5399; 800-431-5641; FAX 914-355-6314; E-mail: arc@balchem.com; Internet: www.balchem.com)

Arch Chemicals

Arch Chemicals, Inc., 501 Merritt Seven, PO Box 4500, Norwalk, CT, 06856-4500, USA (Tel: 203-229-2900; 800-654-6018; FAX 203-229-3652; E-mail: afnask@archchemicals.com; Internet: www.archchemicals.com)

Arch Chemicals, Inc./Performance Chems./Specialty Polyols, PO Box 547, Highway 933, Brandenburg, KY, 40108-0547, USA (Tel: 800-370-9674)

Arch Chemicals, Inc./Personal Care Products, 70 Tyler Place, South Plainfield, NJ, 07080, USA (Tel: 908-561-5200; FAX 908-561-9174; E-mail: ArchPC@archchemicals.com; Internet: www.archchemicals.com)

Archimica (Florida) Inc.

9250 Baymeadows Rd., Suite 2000, Jacksonville, FL, 32256, USA (Tel: 904-732-2100; FAX 904-732-4799; E-mail: sales@archimica.com; Internet: www.archimica.com)

ARCO Chemical. *See* Lyondell/Performance Chemicals

Arenol Chemical Corp.

189 Meister Ave., Somerville, NJ, 08876, USA (Tel: 908-526-5900; FAX 908-526-9688)

Arista Industries, Inc.

557 Danbury Rd., Wilton, CT, 06897, USA (Tel: 203-761-1009; 800-255-6457; FAX 203-761-4980; E-mail: info@aristaindustries.com; Internet: www.aristaindustries.com)

Arizona Chemical

Arizona Chemical Co., Wholly owned subsid. of International Paper, PO Box 550850, Jacksonville, FL, 32225, USA (Tel: 904-928-8700; 800-526-5294; FAX 904-928-8779; E-mail: info.arizona@ipaper.com; Internet: www.arizonachemical.com)

Arizona Chemical Ltd., Manor House, 1 The Crescent, Leatherhead, Surrey, KT22 8DH, UK (Tel: 44 1372 363433; FAX 44-1372-360025; E-mail: ianmelluish@ipaper.com)

Arizona Chemical BV, Almere, The Netherlands (Tel: 31 36 5462 800; FAX 31 36 5462 811; E-mail: arizona-chemica@export.nl)

Armand Products Co.

469 North Harrison st., Princeton, NJ, 08543-5297, USA (Tel: 609-683-7090; 800-522-0540; FAX 609-497-7176; E-mail: armand@armandproducts.com; Internet: www.armandproducts.com)

Armentor Chemical Marketing

1240 Blalock, Suite 240, Houston, TX, 77055, USA (Tel: 713-932-8920; FAX 713-932-8921; Internet: www.armentorchem.com)

Arol Chemical Products Co.

649-T Ferry St., Newark, NJ, 07105, USA (Tel: 718-824-2162; FAX 718-319-9089)

Asahi Chemical Industry Co., Ltd.

Hibiya Mitsui Bldg., 1-1-2, Yuraku-cho, Chiyoda-ku, Tokyo, 100-8440, Japan (Tel: 81 3 3507 2525; FAX 81 3 3507 2496)

Asahi Denka Kogyo K.K.

Furukawa Bldg., 2-3-14, Nihonbashi Muro-machi, Chuo-ku, Tokyo, 103, Japan (Tel: 81 3 5255 9002; FAX 81 3 3246 2090; Internet: www.adk.co.jp)

Asahi Glass Co. Ltd.

Chiyoda Bldg, 2-1-2, Marunouchi, Chiyoda-ku, Tokyo, 100-8305, Japan (Tel: 81 3 3218 5462; FAX 81 3 3211 7672; Internet: www.agc.co.jp)

Ascona

Div. of International Pine Products S.A., 5 Cold Hill Rd., Suite 9, Mendham, NJ, 07945, USA (Tel: 973-543-9666; FAX 973-543-9614; E-mail: jlsamsurco@msn.com)

Ashland Distribution Co./Industrial Chemicals & Solvents

Div. of Ashland Inc., PO Box 2219, Columbus, OH, 43216, USA (Tel: 614-790-3333; 888-ASHCHEM; FAX 614-889-3465; E-mail: csr@ashchem.com; Internet: www.ashchem.com)

Ashley Polymers Inc.

5114 Fort Hamilton Parkway, Brooklyn, NY, 11219, USA (Tel: 718-851-8111; FAX 718-972-3256; E-mail: mail@ashleypoly.com; Internet: www.ashleypoly.com)

ASHTA Chemicals, Inc.

PO Box 858, 3509 Middle Rd., Ashtabula, OH, 44005-0858, USA (Tel: 440-997-5221; 800-492-5082; FAX 440-992-0151; Internet: www.ashtachemicals.com)

Asiamerica International Inc.

38 Flint Terr., Harrington Park, NJ, 07640-1058, USA (Tel: 201-750-2730; FAX 201-750-2734)

The Associated Octel Co. Ltd.

PO Box 17, Oil Sites Rd., Ellesmere Port, S. Wirral, Cheshire, L65 4HF, UK (Tel: 44 151 355 3611; FAX 44 151 356 2349; Internet: www.octel-corp.com)

Astaris LLC

Joint venture of FMC Corp. and Solutia, Creve Coeur IV, 622 Emerson Rd., St. Louis, MO, 63141, USA (Tel: 314-983-7500; FAX 314-983-7642; E-mail: info@astaris.com; Internet: www.astaris.com)

Atlantic Equip. Engrs.

Div. of Micron Metals Inc., 13 Foster St., PO Box 181, Bergenfield, NJ, 07621, USA (Tel: 201-384-5606; 800-486-2436; FAX 201-387-0291; E-mail: aee@micronmetals.com; Internet: www.micronmetals.com)

Atlas Minerals & Chemicals, Inc.

1227 Valley Rd., PO Box 38, Mertztown, PA, 19539-0038, USA (Tel: 610-682-7171; 800-523-8269; FAX 610-682-9200; E-mail: sales@atlasmin.com; Internet: www.atlasmin.com)

Atofina

Atofina S.A., Branch of TotalFinaElf, 4-8, cours Michelet, La Défense 10, 92091 Paris La Defense Cedex, France (Tel: 33 1 49 00 80 80; FAX 33 1 49 00 83 96; E-mail: info@atofina.com; Internet: www.atofina.com)

Atofina Bruxelles, Div. of Total Fina, 52, Rue de l'Industrie, 1040 Brussels, Belgium (Tel: 32 2 288 9111; FAX 32 2 288 33 22; Internet: www.petrochemicals.atofina.com)

Atofina UK Ltd., Globe House, Bayley St., Stalybridge, Cheshire, SK15 1PY, UK (Tel: 44 161 338 4411; FAX 44 161 303 1908; Internet: www.atofina.co.uk)

Atofina Deutschland GmbH, Tersteegenstrasse 28, Postfach 30 01 52, 40474 Düsseldorf, Germany (Tel: 49 211 4552 0; FAX 49 211 50 66 78 0; E-mail: info.deutschland@atofina.com; Internet: www.atofina.de)

Atofina Italia S.r.l., Via Degli Artigianelli, 10, 20159 Milan (MI), Italy (Tel: 39 2 668111; FAX 39 2 66803607; E-mail: info.italy@atofina.com)

Atofina España SA, Avenida de Burgos 12-7°, 28036 Madrid, Spain (Tel: 34 91 334 34 34; FAX 34 91 334 34 20)

Atofina Norden, Herlev Hovedgade 195, 2730 Herlev, Denmark (Tel: 45 44 50 3939; FAX 45 44 50 3940)

Atofina Nederland B.V., Ottho Heldringstraat 41, 1066 XT Amsterdam, The Netherlands (Tel: 31 20 408 94 00; FAX 31 20 669 19 09)

Atofina Chemicals, Inc, 2000 Market St., Philadelphia, PA, 19103-3222, USA (Tel: 215-419-7000; 800-225-7788; FAX 215-419-7591; E-mail: webmaster@ato.com; Internet: www.atofinachemicals.com)

Atofina North America Inc./Organic Chemicals Div., 2000 Market St., Philadelpha, PA, 19103-3222, USA (Tel: 800-225-7788; FAX 215-419-7591)

Atofina Canada, Subsid. of ATOFINA Chemicals, Inc., 700 Third Line, Oakville, Ontario, L6J 5A3, Canada (Tel: 905-827-9841; 800-567-5726; FAX 905-827-7913; E-mail: lkwasnik@ato.com; Internet: www.atofinacanada.com)

Atofina (Australia), Bldg. 10, 658 Church St., Richmond, Victoria, 3121, Australia (Tel: 61 3 9425 7777; FAX 61 3 9425 7700; E-mail: info@atofina.com.au)

Atofina Japan K.K., Bungei Shunju Annex, 3-23 Kioicho, Chiyoda-ku, Tokyo, 102-0094, Japan (Tel: 81 3 3288 7110; FAX 81 3 3288 7130; E-mail: info.japan@atofina.com)

Atofina China Shanghai, Unit 01-06, 28/F, Hong Kong Plaza 283, Huai Hai Road (M), Shanghai, 200021, P.R. China (Tel: 86 21 6386 3028; FAX 86 21 6386 0668; E-mail: atofina@public8.stat.net.cn)

Atofina Hong Kong Ltd., Suite 3207-3211, Tower 1 The Gateway, Harbour City, 25 Canton Rd., Tsimshatsui, Kowloon, Hong Kong (Tel: 852 2110 1188; FAX 852 2110 1122; E-mail: atohk@netvigator.com; Internet: www.atofina.com/cn)

Atofina South East Asia Pte. Ltd., 53 Tuas Crescent, Singapore, 638732, Singapore (Tel: 65 862 1228; FAX 65 862 3167; E-mail: atochem.SEA@sing.elf-atomchem.fr)

Atofina Korea Ltd., Rm. 1401, Leema Bldg., 146-1, Susong Dong, Chongro-ku, Seoul, 110-755, Korea (Tel: 82 2 732 9622; FAX 82 2 736 7174; E-mail: sohhyun.lee@atofina.com)

Atomergic Chemetals Corp.

71 Carolyn Blvd., Farmingdale, NY, 11735, USA (Tel: 631-694-9000; FAX 631-694-9177; E-mail: info@atomergic.com; Internet: www.atomergic.com)

Atotech USA Inc.

Subsid. of Atofina, 1750 Overview Dr., PO Box 12000, Rock Hill, SC, 29731-2000, USA (Tel: 803-817-3500; 800-843-4460; FAX 803-817-3666; E-mail: whoward@cetlink.net; Internet: www.atotechusa.com)

Atul Ltd./Aromatics Div.

Atul, Gujarat, 396 020, India (Tel: 91 2632 33261; FAX 91 2632 33633; E-mail: wk.ar@atul.co.in; Internet: www.atul.co.in)

Augustus Oils Ltd.

64 Woolmer Trading Estate, Bordon, Hampshire, GU35 9QF, UK (Tel: 44 1420 488555; FAX 44 1420 476777; E-mail: sales@augustus-oils.ltd.uk; Internet: www.augustus-oils.ltd.uk)

Austin Chemical Co. Inc.

1565 Barclay Blvd., Buffalo Grove, IL, 60089, USA (Tel: 847-520-9600; FAX 847-520-9160; E-mail: info@austinchemical.com; Internet: www.austinchemical.com)

Avatar Corp.

500 Central Ave., University Park, IL, 60466, USA (Tel: 708-534-5511; 800-255-3181; FAX 708-534-0123; E-mail: sales@avatarcorp.com; Internet: www.avatarcorp.com)

Aventis S.A.

Member of the Hoechst Group, Espace European de l'Enterprise, 16, Ave. de l'Europe, 67917 Strasbourg, France (Tel: 33 3 88 99 11 00; Internet: www.aventis.com)

Aventis Pharmaceuticals Inc.

300, 400 Somerset Corporate Blvd., Bridgewater, NJ, 08807-2854, USA (Tel: 908-243-6000; 800-981-2491)

Avocado Research Chemicals Ltd.

Shore Rd., Port of Heysham Ind. Park, Heysham, Lancs, LA3 2XY, UK (Tel: 44 1524 850506; FAX 44 1524 850608)

Avrachem AG

Westrasse 119, 8036 Zurich, Switzerland (Tel: 41 1 451 3132; FAX 41 1 451 3222)

Axel Plastics Research Laboratories Inc.

58-20 Broadway, PO Box 770855, Woodside, NY, 11377-0855, USA (Tel: 718-672-8300; 800-332-2935; FAX 718-565-7447; E-mail: info@axelplast.com; Internet: www.axelplast.com)

AXO Chemical, Inc.

PO Box 557973, Miami, FL, 33255, USA (Tel: 305-663-1820; FAX 305-663-1830; Internet: www.axochem.com)

Axxence Aromatic GmbH

Tackenweide 28, 46446 Emmerich, Germany (Tel: 49 28 22 68 561 0; FAX 49 28 22 68 561 39; E-mail: info@axxence.com; Internet: www.axxence.com)

B2E Corporation

500 Mamaroneck Ave., Harrison, NY, 10528, USA (Tel: 914-777-1111; FAX 914-777-0781; E-mail: info@protectivechemicals.com; Internet: www.protective chemicals.com)

J.T. Baker, Inc.

Div. of Mallinckrodt Baker, 222 Red School Lane, Phillipsburg, NJ, 08865, USA (Tel: 908-859-2151; 800-582-2537; FAX 908-859-6905; E-mail: infombi@mkg.com; Internet: www.jtbaker.com)

Baker Petrolite/World Headquarters

A Baker Hughes Co., 12645 W. Airport Blvd., Sugar Land, TX, 77478, USA (Tel: 281-276-5400; 800-231-3606; FAX 281-276-7395; Internet: www.bakerhughes.com/bakerpetrolite)

Balchem Corp.

PO Box 175, Slate Hill, NY, 10973, USA (Tel: 845-355-5300; FAX 845-355-6314; Internet: www.balchem.com)

Barker Industries, Inc.

2841 Old Steele Creek Rd., Charlotte, NC, 28208, USA (Tel: 704-391-1023; FAX 704-393-0464; E-mail: barkerind@msn.com; Internet: www.barkerind.com)

Barnet Products Corp.

140 Sylvan Ave., Englewood Cliffs, NJ, 07632, USA (Tel: 201-346-4620; FAX 201-346-4333)

Barrington Chemical Corp.

500 Mamaroneck Ave., Harrison, NY, 10528, USA (Tel: 914-381-3500; 800-684-CHEM; FAX 914-381-2232; E-mail: barrchem@worldnet.att.net; Internet: www.barringtonchem.com)

BASF

BASF AG, Carl-Bosch Str. 38, 67056 Ludwigshafen, Germany (Tel: 49 621 60 0; FAX 49-621-60 42525; E-mail: info.service@basf.de; Internet: www.basf.de)

BASF Corp., 3000 Continental Dr. North, Mount Olive, NJ, 07828-1234, USA (Tel: 973-426-2600; 800-669-2273; FAX 973-426-2610; Internet: www.basf.com)

BASF Corp./Performance Chemicals, 3000 Continental Dr. North, Mount Olive, NJ, 07828-1234, USA (Tel: 800-367-9861; FAX 973-426-4999; E-mail: perfchem@basf.com; Internet: www.basf.com/businesses/chemicals/performance)

BASF plc, PO Box 4, Earl Road, Cheadle Hulme, Cheadle, Cheshire, SK8 6QG, UK (Tel: 44 161 485-6222; FAX 44 161 486 0891)

BASF Belgium S.A./N.V., Ave. Hamoir-laan 14, 1180 Brussels, Belgium (Tel: 32 2 373 2111; FAX 32 2 375 1042)

BASF Española S.A., Paseo de Gracia 99, 08008 Barcelona, Spain (Tel: 34 93 4 96 40 00; FAX 34 93 4 96 41 00)

BASF France & Co. Cie, 49 ave. Georges Pompidou, 92593 Levallois-Perret Cedex, France (Tel: 33 1 49 64 50 00; FAX 33 1 49 64 50 50)

BASF Italia SpA, Via Marconato 8, 20031 Cesano Maderno (MI), Italy (Tel: 39 3 62 5 12 1; FAX 39 3 62 5 12 2 10)

BASF Nederland BV, Postbus 1019, 6801 MC Arnhem, The Netherlands (Tel: 31 26 3 71 71 71; FAX 31 26 3 71 72 46)

BASF Denmark A/S, Postboks 1734, Ved Stadsgraven 15, 2300 Copenhagen S, Denmark (Tel: 45 32 66 07 00; FAX 45 32 57 22 02)

BASF Österreich GmbH, Hietzinger Hauptstrasse 119, 1131 Vienna, Austria (Tel: 43 1 8 78 90 0; FAX 43 1 8 78 90 110)

BASF (Schweiz) AG, Appital, Postfach 99, 8820 Wädenswil, Switzerland (Tel: 41 1 781 9111; FAX 41 1 781 9388)

BASF Svenska AB, Haraldsgatan 5, 413 14 Göteborg, Sweden (Tel: 46 31 639800; FAX 46 31 639900)

BASF India, Ltd., Rhone-Poulenc House, Sudam Kalu Ahire Marg, Worli, Mumbai, 400 025, India (Tel: 91 22 493 0703; FAX 91 22 495 0512)

BASF Australia Ltd., 500 Princes Highway, Nobel Park, Victoria, 3174, Australia (Tel: 61 3 92 12 15 00; FAX 61 3 92 12 15 11)

BASF Japan Ltd., Nanbu Bldg., 3-3, Kioi-cho, Chiyoda-ku, Tokyo, 102-8570, Japan (Tel: 81 3 3238 2300; FAX 81 3 3238 2290)

BASF (China) Co. Ltd., Beijing Sunflower Tower, 15/F, No. 37, Maizidian Street, Chaoyang District, Beijing, 100026, P.R. China (Tel: 86 10 65 91 88 99; FAX 86 10 85 27 55 99)

BASF China Ltd., 7/F, Tower 1, South Seas Centre, 75 Mody Rd., Tsim Sha Tsui East, Kowloon, Hong Kong (Tel: 852 2731 1222; FAX 852 2734 9631)

BassTech International LLC

300 Grand Ave., Englewood, NJ, 07631, USA (Tel: 201-569-8686; FAX 201-569-7511)

Baychem, Inc.

3200 Moon Station Rd., Kennesaw, GA, 30144, USA (Tel: 770-429-1405; 800-422-1405; FAX 770-425-9141)

Bay Chemical & Supply Co.

PO Box 1160, Odem, TX, 78370, USA (Tel: 361-883-0422; FAX 361-368-2711)

Bayer

Bayer AG, Werk Leverkusen, 51368 Leverkusen, Germany (Tel: 49 214 301; FAX 49 214 30 66328; Internet: www.bayer-ag.de; www.bayer.com)

Bayer plc, Bayer House, Strawberry Hill, Newbury, Berkshire, RG14 1JA, UK (Tel: 44 1635 563000; FAX 44 1635 563393; E-mail: corporate.communications@bayer.co.uk; Internet: www.bayer.co.uk)

Bayer s.a.-n.v., Div. of Bayer AG, 143, Ave. Louise, 1050 Brussels, Belgium (Tel: 32 2 535 6311; FAX 32 2 537 4996; Internet: www.bayer.be)

Bayer Hispania Industrial S.A., Calle Pau Claris, 196, 08037 Barcelona, Spain (Tel: 34 3 218 45 50; FAX 34 3 217 41 49)

Bayer Austria GmbH, Postfach 10, am Heumarket 10, 1030 Vienna, Austria (Tel: 43 1 71146-0; FAX 43 1 71146-14)

Bayer Danmark A/S, Postboks 2090, Noergaardsvej 32, 2800 Lyngby, Denmark (Tel: 45 38 17 7100; FAX 45 38346833; Internet: www.bayer.dk)

Bayer S.A., 49-51 quai de Dion Bouton, 92815 Puteaux Cedex, France (Tel: 33 1 49 06 54 31; FAX 33 1 49 06 52 19)

Bayer Nederland BV, Energieweg 1, 3641 RT Mijdrecht, The Netherlands (Tel: 31 297 280666; FAX 31 297 284165)

Bayer Portugal SA, Apartado 666, 2795 Carnaxide, Portugal

Bayer (Schweiz) AG, Postfach 8045, Grubenstrasse 6, 8045 Zurich, Switzerland (Tel: 41 1 465 811; FAX 41 1 462 0754)

Bayer Corp./Corporate Headquarters, 100 Bayer Rd., Pittsburgh, PA, 15205-9741, USA (Tel: 412-777-2000; FAX 412-778-4430; Internet: www.bayerus.com)

Bayer Corp./Fibers, Additives & Rubber (FAR) Div., 2603 W. Market St., Akron, OH, 44313, USA (Tel: 330-836-0451; 800-662-2927; FAX 330-836-3200; Internet: www.bayerus.com)

Bayer Corp./Industrial Chemicals Div./Textile Processing Chems., 100 Bayer Rd., Bldg. 14, Pittsburgh, PA, 15205-9741, USA (Tel: 412-777-2781; 800-662-2927 x 2781; FAX 412-777-4078; E-mail: cheryl-r.wylie.b@bayer.com; Internet: www.bayerus.com)

Bayer Inc., S. Vidal Street, Sarnia, Ontario, N7V 4H4, Canada (Tel: 800-321-0997)

Bay Resins, Inc.

Div. of Clariant Corp., PO Box 630, 3011 Millington Rd., Millington, MD, 21651, USA (Tel: 410-928-3083; FAX 410-928-5412)

BCH Brühl-Chemikalien Handel GmbH

Postfach 12 48, 50302 Brühl, Germany (Tel: 49 2232 933070; FAX 49 2232 93307 9; E-mail: info@bch-bruehl.de; Internet: www.bch-bruehl.de)

BDH

350 Evans Ave., Toronto, ON, M8Z 1K5, Canada (Tel: 416-255-8521)

B&D Nutritional Ingredients, Inc.

2794 Loker Ave., Suite 103, Carlsbad, CA, 92008, USA (Tel: 619-931-0900; FAX 619-931-9670; E-mail: bvdofc@bndni.com)

D.B. Becker Co., Inc.

23 West Main St., Suite L12, Clinton, NJ, 08809, USA (Tel: 908-730-6010; 800-394-3991; FAX 908-730-9118; E-mail: beckerdbb@aol.com)

Beckmann

Beckmann Chemikalien KG, Auf Dem Scheunenbrink 8-10, 27211 Bassum, Germany (Tel: 49 42 41 93 08 0; FAX 49 42 41 93 08 30; E-mail: beckmannchemkg@t-online.de)

Beckmann Chemical Corp., Inc., 120-A Pleasant Ave., Upper Saddle River, NJ, 07458, USA (Tel: 201-785-1950; FAX 201-785-1955; E-mail: BeckChemUS@aol.com)

Bedoukian Research, Inc.

21 Finance Dr., Danbury, CT, 06810-4192, USA (Tel: 203-830-4000; FAX 203-830-4010; E-mail: customerservice@bedoukian.com; Internet: www.bedoukian.com)

Belinka Perkemija Chemical Industries d.o.o.

Zasavska cesta 95, POB 4973, 1001 Ljubljana, Slovenia (Tel: 386 1 5886 299; FAX 386 1 5886 303; E-mail: info-BL@belinka.si; Internet: www.belinka.si/)

Bell Flavors & Fragrances

Bell Flavors & Fragrances, Inc., 500 Academy Dr., Northbrook, IL, 60062, USA (Tel: 847-291-8300; 800-323-4387; FAX 847-291-1217; E-mail: info@bellff.com; Internet: www.bellff.com)

Bell Flavors & Fragrances Canada, 3800-H Isabelle, Brossard (Montreal), Quebec, J4Y 2R3, Canada (Tel: 450-444-3819; 800-444-3819; FAX 450-444-3974; E-mail: info@bellff.com)

Bell Flavors & Fragrances Mexico, Gabino Barreda 1122, Col. San Carlos, Guadalajara, Jal., 44460, Mexico (Tel: 52 33 3650 1212; FAX 52 33 3619 0017; E-mail: info@bellff.com)

Bell Flavors & Fragrances GmbH, Schimmelstrasse 1, 04205 Miltitz Bei Leipzig, Germany (Tel: 49 341 945 10; FAX 49 341 941 1669; E-mail: info@bell-europe.com; Internet: www.bell-europe.com)

Bell Flavors & Fragrances Philippines, Matrinco Bldg., 4th Fl., Rm. 43, 2178 Pasong Tamo Street, Makati City, Philippines (Tel: 632 810 1341; FAX 632 819 8534)

Bell Flavors & Fragrances China, 3rd Fl., No. 9 Bldg., Dayuan Industrial Zone, Pinghsan, Xili, Nanshan District, Shenzhen, P.R. China (Tel: 86 755 698 5507; FAX 86 755 698 5473)

Benco

PO Box 7207, 11616 Wilmar Blvd., Charlotte, NC, 28273, USA (Tel: 704-588-6297; FAX 704-588-1306; E-mail: benco@lhtech.com)

Bencorp International Ltd.

393 West End Ave., Ste. 2A, New York, NY, 10024, USA (Tel: 212-496-2850; FAX 212-721-4265; E-mail: mailbox@bencorpintl.com)

Berjé Inc.

5 Lawrence St., Bloomfield, NJ, 07003-4691, USA (Tel: 973-748-8980; FAX 973-680-9618; E-mail: berjeinc@aol.com)

Berk Chemicals Ltd.

Div. of Berk Ltd., PO Box 56, Priestley Rd., Basingstoke, Hampshire, RG24 9QB, UK (Tel: 44 1256 312000; FAX 44 1256 814163)

Bernel Chemical Co., Inc.

174 Grand Ave., Englewood, NJ, 07631, USA (Tel: 201-569-8934; FAX 201-569-1741; E-mail: mail@bernel-chemical.com; Internet: www.bernel-chemical.com)

B.I. Chemicals, Inc.

3330 S. Crater Rd., Petersburg, VA, 23805, USA (Tel: 804-863-0098; 800-BICHEM6; FAX 804-863-0998; E-mail: smitchard@bichemicals.com)

Biddle Sawyer Corp.

2 Penn Plaza, New York, NY, 10121-0034, USA (Tel: 212-736-1580; FAX 212-239-1089; E-mail: bsc@biddlesawyer.com; Internet: www.biddlesawyer.com)

Bilt Chemicals Ltd.

Part of the Thapar Group, 31-A, Saronjini Devi Rd., Secunderabad, 500003, India (Tel: 91 40 7707712; FAX 91 40 7707717; Internet: www.biltchemicals.com)

Bimal Pharma Chem

10, Indraprastha, 1-C, Raheja Township, Jitendra Road, Malad (East), Mumbai, 400 097, India (Tel: 91 22 8778563; FAX 91 22 8775755; E-mail: jp2@vsnl.com)

Bio-Botanica, Inc.

75 Commerce Dr., Hauppauge, NY, 11788-3942, USA (Tel: 631-231-5522; 800-645-5720; FAX 631-231-7332; E-mail: gene@bio-botanica.com; Internet: www.bio-botanica.com)

Bioiberica S.A.

Poligono Industrial de Palafolls, Ctra. Nacional II, Km. 680.6, 08389 Palafolls (Barcelona), Spain (Tel: 34 937 65 03 90; FAX 34 937 65 01 02; E-mail: info@bioiberica.com; Internet: www.bioiberica.com)

BioLab Inc.

Subsid. of Great Lakes Chemical Corp., PO Box 1489, Decatur, GA, 30031-1489, USA (Tel: 404-378-8585; 404-687-1028)

Bio-Rad

Bio-Rad Laboratories, 1000 Alfred Nobel Dr., Hercules, CA, 94547, USA (Tel: 510-724-7000; 800-4-BIORAD; FAX 510-741-5817, 800-879-2289; Internet: www.bio-rad.com)

Bio-Rad Laboratories Ltd., Bio-Rad House, Maylands Ave., Hemel Hempstead, Hertfordshire, HP2 7TD, UK (Tel: 44 208 328 2555; FAX 44 1442 232356; E-mail: sadtler.europe@bio-rad.com; Internet: www.bio-rad.com)

Biosil Technologies, Inc.

510 E. 31st St., Paterson, NJ, 07504, USA (Tel: 973-684-2000; FAX 973-742-9048; E-mail: info@biosiltech.com; Internet: www.biosiltech.com)

Blackman Uhler Chemical Co.

Div. of Synalloy Corp., PO Box 5627, Spartanburg, SC, 29304-5627, USA (Tel: 864-585-3661; FAX 864-596-1501; E-mail: sales@buchemical.com; Internet: www.buchemical.com)

S. Black

S. Black (Import & Export) Ltd., Foxholes Business Park, John Tate Road, Hertford, Hertfordshire, SG13 7YH, UK (Tel: 44 1992 825555; FAX 44 1992 825566; E-mail: enquires@sblack.com; Internet: www.sblack.com)

S. Black GmbH, Baumstrasse 41, 47198 Duisburg, Germany (Tel: 49 2066 20380; FAX 49 2066 203815; E-mail: infode@sblack.com)

Blagden Specialty Chemicals Ltd.

Osprey House, Black Eagle Square, Westerham, Kent, TN16 1PA, UK (Tel: 44 1959 562 000; FAX 44 1959 565 111; E-mail: sales@blagdenspecchem.co.uk; Internet: www.blagdenspecchem.co.uk)

William Blythe Ltd.

Subsid. of Holliday Chemical Holdings plc, Holland Bank Works, Bridge St., Church, Accrington, Lancashire, BB5 4PD, UK (Tel: 44 1254 320000; FAX 44 1254 320001; E-mail: info@wm-blythe.co.uk; Internet: www.wm-blythe.co.uk)

BOC

BOC Ltd., Div. of BOC Group plc, The Priestley Centre, 10 Priestley Rd., The Surrey Research Park, Guildford, Surrey, GU2 5XY, UK (Tel: 44 1483 244420; FAX 44 1483 244175; Internet: www.boc.com)

BOC Gases, The Priestley Centre, 10 Priestley Rd., The Surrey Research Park, Guildford, Surrey, GU2 5XY, UK (Tel: 44 1483 579857; FAX 44-1483 244175; Internet: www.boc.com/gases/bocgases.htm)

BOC NV/SA, Member of the BOC Group, Excelsiorlaan 41, 1930 Zaventem, Belgium (Tel: 32 2 719 7111; FAX 32 2 725 3282)

BOC Gases GmbH, Loffelstrasse 3, 7000 Stuttgart 70, Germany (Tel: 49 711 766041; FAX 49 711 7657530)

BOC Gases, 575 Mountain Ave., Murray Hill, NJ, 07974, USA (Tel: 908-464-8100; 800-742-4726; FAX 908-771-1764; E-mail: bruce.bolduc@us.gases.boc.com; Internet: www.boc.com/gases)

BOC Gases, Div. of BOC Canada Ltd., 5975 Falbourne St., Unit 2, Mississauga, Ontario, L5R 3V8, Canada (Tel: 905-501-1700; FAX 905-501-1717)

BOC Gases Australia Limited, A.C.N. 000 029 729 The CIG Bldg., 15th Fl., 799 Pacific Highway, Chatswood, NSW, 2057, Australia (Tel: 61 2 936 6666; FAX 61 2 415 1777)

BOC Technologies Ltd, Shanghai, Rm. 1909, Shanghai Int'l. Trade Center, 2200 Yan An Rd. (W), Shanghai, 200335, P.R. China (Tel: 21 320 9220; FAX 21 320 9267)

BOC Japan Ltd., Shuwa Shiba Park Bldg., A-3F, 2-4-1 Shiba-koen, Minato-ku, Tokyo, 105, Japan (Tel: 81 3 3434 8261; FAX 81 3 5472 8728)

Boehle Chemicals, Inc.

PO Box 2001, 19306 W. Ten Mile Rd., Southfield, MI, 48037, USA (Tel: 313-255-2210; FAX 313-255-4474; E-mail: sales@boehlechem.com)

Böhme

Dr. Th. Böhme KG Chemische Fabrik GmbH & Co., Isardamm 79-83, 82538 Geretsried, Germany (Tel: 49 8171 628 0; FAX 49 8171 628 388; E-mail: contact@boehme-chemie.com; Internet: www.boehme-chemie.com)

Boehme Filatex Inc., 209 Watlington Industrial Dr., Reidsville, NC, 27320-8147, USA (Tel: 336-342-6631; 800-443-8814; FAX 336-634-3355; E-mail: info.us@boehmefilatex.com; Internet: www.boehmefilatex.com)

Boehme Filatex Canada Inc., Wholly owned subsid. of Boehme Filatex Inc., 178 St. Pierre, St-Jean-sur-Richelieu, Quebec, J3B 5W4, Canada (Tel: 450-346-6848; FAX 450-346-7263; E-mail: info@boehme.ca; Internet: www.boehme.ca)

Boehme Filatex Mexico S.A. de C.V., San Andres Atoto No. 5-A, Fracc. Alce Blanco, 53370 Naucalpan, Estado de Mexico, Mexico (Tel: 52 5 357 0169; FAX 52 5 358 6551; E-mail: rroph@boehmefilatex.com)

Boehme Argentina S.R.L., Comandante Piedrabuena 4729, 1605 Munro, Provincia de Buenos Aires, Argentina (Tel: 54 11 4721 0400; FAX 54 11 4721 0444; E-mail: boehmearg@house.com.ar)

Boehme Blumenau Industrial Ltda., Rodovia BR 470, s/n-km 96,7, Bairro Ribeirão São Luiz, Apiuna - SC, 89135-000, Brazil (Tel: 55 47 353 9900; FAX 55 47 353 9901; E-mail: diretoria@boehmesouthamerica.com.br)

Dr. Th. Boehme (UK) Limited, 31, Charles St., Hucknall, Nottingham, NG15 7FL, UK (Tel: 44 115 963 9029; FAX 44 115 964 0210; E-mail: drboehmeukltd@aol.com)

Dr. Th. Böhme Belgium N.V., Keizer Karelstraat, 62, 8930 Menen, Belgium (Tel: 32 56 51 40 92; FAX 32 56 51 68 15; E-mail: vercruysse.alain@yucom.be)

Dr. Th. Böhme Austria Chem. Fabrik GmbH, Josef-Holzinger-Str. 2, 2602 Blumau, Austria (Tel: 43 2628 48205; FAX 43 2628 48816; E-mail: boehmeaustria@aon.at)

Böhme France, Siège Social, 105, Rue Sadi Carnot, 59280 Armentières, France (Tel: 33 3 20 77 65 76; FAX 33 3 20 77 57 53; E-mail: bohme.france@free.fr)

Dr. Bohme Iberica, s.l., Turó de L'Home 21, Apartado 247, 08480 L'Ametlla del Vallées (Barcelona), Spain (Tel: 34 93 8 43 17 56; FAX 34 93 8 43 17 58; E-mail: bohmeiberica@versin.com)

Boehme Asia Ltd., Flat A1, 6/F, Block A, Yee Lim Industrial Center, 2-28 Kwai Lok St., Kwai Chung, N.T., Hong Kong (Tel: 852 2 7701866; FAX 852 2 7805555; E-mail: boehmehk@netvigator.com)

Boith

Boith Limited (UK), 34 Church St., Ashbourne, Derbyshire, DE6 1AE, UK (Tel: 44 1335 343 051; FAX 44 1335 342 053; E-mail: mail@boith.co.uk; Internet: www.boith.co.uk; www.boith.com)

Boith Limited (China), 19/F, HuaMin Bldg., No. 9 HuBin East Rd., XiaMen, P.R. China (Tel: 86 592 5144016; FAX 86 592 5034960; E-mail: boith@21cn.com; Internet: www.boith.com)

Boliden Intertrade Inc.

3379 Peachtree Rd. NE, Suite 300, Atlanta, GA, 30326, USA (Tel: 404-239-6700; 800-241-1912; FAX 404-239-6701)

Borculo Domo Ingredients

Part of Friesland Coberco Dairy Foods, PO Box 449, 8000 AK Zwolle, The Netherlands (Tel: 31 38 46 77 444; FAX 31 38 46 77 555; Internet: www.borculodomo.com)

Borden

Borden Chemical, Inc., 180 E. Broad St., Columbus, OH, 43215, USA (Tel: 614-225-4000; FAX 614-225-4465)

Borden Chemical (UK) Ltd., Div. of Borden Inc, St. Christopher Works, Rownhams Rd., North Baddesley, Southampton, Hamsphire, SO52 9ZB, UK (Tel: 44 1703 732131; FAX 44 1703 738656)

Borden & Remington Corp.

106 Ferry St., PO Box 2573, Fall River, MA, 02722, USA (Tel: 508-675-0096)

Borregaard

Borregaard Synthesis, PO Box 162, 1701 Sarpsborg, Norway (Tel: 47 69 11 8948; FAX 47 69 11 8640; E-mail: pal.romberg@borregaard.com; Internet: www.borregaard.com)

Borregaard France, 86, ave. de Saint-Ouen, 75018 Paris, France (Tel: 33 1 53 06 60 40; FAX 33 1 53 06 60 45)

Borregaard Deutschland, Hansa Allee 201, Haus 2, 40549 Dusseldorf, Germany (Tel: 49 211 59 51 90; FAX 49 211 59 51 922)

Borregaard Italia, Via Carso n. 73, 24040 Madone (BG), Italy (Tel: 39 035 4 99 81 11; FAX 39 035 4 99 83 52)

Borregaard Synthesis (US), 9 Opportunity Way, Newburyport, MA, 01950, USA (Tel: 978-462-5555; 800-992-8506; FAX 978-465-2057)

Borregaard Shanghai Rep. Office, Room 2005, 20/F Westgate Mall, 1038, West Nanjing Rd., Shanghai, 200041, P.R. China (Tel: 86 21 62 18 11 90; FAX 86 21 62 18 37 31)

Borregaard Japan, Nikko Sanno Bldg., #402, 5-3, 2-chome, Akasaka, Minato-ku, Tokyo, 107-0052, Japan (Tel: 81 3 35 84 17 05; FAX 81 3 35 84 64 25)

Charles Bowman & Co.

3328 John F. Donnelly Dr., Holland, MI, 49422-2427, USA (Tel: 616-786-4000; FAX 616-786-2864; E-mail: cbc@macatawa.com; Internet: www.charlesbowman.com)

BP Amoco

BP Amoco plc, Britannic House, 1 Finsbury Circus, London, EC2M 7BA, UK (Tel: 44 20 7496 4000; FAX 44 20 7496 4630; Internet: www.bp.com)

BP Chemicals Ltd., Bldg. A, Chertsey Rd., Sunbury-on-Thames, Middlesex, TW16 7LL, UK (Tel: 44 1932 774 321; FAX 44 1932 774 372; E-mail: chemical_enquiries_eu@bp.com; Internet: www.bpamocochemicals.com)

BP France, 8 rues des Gemeaux, Cergy St-Christophe, 95866 Cergy-Pontoise Cedex, France (Tel: 33 1 34 22 40 00; FAX 33 1 34 22 44 17; E-mail: web_dircom@bp.com)

BP Belgium NV/SA, Uitbreidingstraat 60, 2600 Berchem, Belgium (Tel: 32 3 286 8250; FAX 32 3 286 8269)

BP (Switzerland), Grafenauweg 4, 6301 Zug, Switzerland (Tel: 41 41 724 6111; FAX 41 41 724 6440; E-mail: bpmbs@bp.com)

Deutsche BP Aktiengesellschaft, Ueberseering 2, 22297 Hamburg, Germany (Tel: 49 40 6395 0; FAX 49 40 6395 2224)

BP Italia SpA, BP Chemicals, Centro Direzionale e Commerciale Milanofiori, Srada 6 - Palazzo E/5, 20090 Assago (MI), Italy (Tel: 39 2 82274 1; FAX 39 2 5751 1959)

BP Amoco Chemicals, 200 E. Randolph Dr., MC 1204, Chicago, IL, 60601-7125, USA (Tel: 312-856-6111; FAX 312-856-4151; E-mail: chem_americas@bp.com; Internet: www.bpamoco.com; www.bpamocochemicals.com)

Amoco Canada Petroleum Co., PO Box 200-Station M, Calgary, Alberta, T2P 2H8, Canada (Tel: 403 233 1425; FAX 403 233 1476)

BP Mexico, Paseo de los Tamarindos No. 400, Torre A piso 17, Col. Bosques de las Lomas, 05120, Mexico (Tel: 52 5 081 2100; FAX 52 5 081 2124)

BP Brasil Ltda., Avenida Rio Branco 1, 10 Andar, Rio de Janeiro RJ, 20090-003, Brazil (Tel: 55 21 550 8710; FAX 55 21 518 1244; E-mail: bracelc@bp.com)

BP Australia Ltd., G.P.O Box 5222 BB, Melbourne, 3001, Australia (Tel: 61 3 9268 4111; FAX 61 3 9268 3321; E-mail: bpawebmaster@az1.bp.com)

BP Japan K.K., 3F Petro House, 4-5-21 Kojimachi, Chiyoda-ku, Tokyo, 102-0083, Japan (Tel: 81 3 3238 7310; FAX 81 3 3238 7302)

BP Hong Kong Limited, 22nd Fl., Devon House, 979 King's Rd., Taikoo Place, Hong Kong (Tel: 852 2586 8899; FAX 852 2586 8981)

BP Global Specialty Products

BP Global Special Products, 3650 Touhy Ave., PO Box 549, Skokie, IL, 60076, USA (Tel: 847-679-6300; 800-628-9299; FAX 800-679-6312; Internet: www.dussekwax.com)

BP Global Special Products UK, Witan Gate House, 5000-600 Witan Gate, Central Milton Keynes, MK9 1ES, UK (Tel: 44 1908 853344; FAX 44 1908 853896)

BP Global Special Products Australia, 144-152 Fitzgerald Rd., Laverton North, Milbourne, Victoria, Australia (Tel: 61 3 9368 0088; FAX 61 3 9369 6070)

Brand-Nu Laboratories, Inc.

PO Box 895, 30 Maynard St., Meriden, CT, 06450, USA (Tel: 203-235-7989; 800-243-3768; FAX 203-235-7163; E-mail: contact@brandnu.com; Internet: www.brandnu.com)

Brenntag

Brenntag AG, Humboldtring 15, 45472 Mülheim an der Ruhr, Germany (Tel: 49 208 7828 0; FAX 49 208 7828 698; E-mail: info@brenntag.de; Internet: www.brenntag.de)

Brenntag N.V.-DBI, Mechelsesteenweg 34 bus 12, 2018 Antwerp, Belgium (Tel: 32 3 238 49 66; FAX 32 3 238 07 63; E-mail: dbi@brenntag.be; Internet: www.brenntag.be)

Brenntag (UK) Ltd., Brenntag House, 45 High St., Hampton Wick, Kingston-upon-Thames, Surrey, KT1 4DG, UK (Tel: 44 20 0977 3200; FAX 44 20 8943 4350)

Brenntag Portugal - Produtos Quimicos, Lda., Estrada de Albarraque, Linho - Manique de Cima, Apartado 1, 2646-901 Alcabideche, Portugal (Tel: 351 219248800; FAX 351219248845; E-mail: komer@brenntag.pt; Internet: www.brenntag.pt)

Brenntag, Inc., PO Box 13786, Reading, PA, 19612, USA (Tel: 610-926-6100; FAX 610-926-0420; Internet: www.brenntag.com)

Britannia Natural Proucts Ltd.

5 Woodlands Business Park, Rougham Ind. Estate, Rougham, Suffolk, IP30 9ND, UK (Tel: 44 1359 271461; FAX 44 1359 271672; E-mail: sales@bnpl.co.uk; Internet: www.bnpl.co.uk)

British Arkady Co. Ltd.

Skerton Rd., Old Trafford, Manchester, Lancashire, M16 0NJ, UK (Tel: 44 161 872 7161; FAX 44 161 873 8083)

British Traders & Shippers Ltd.

Div. of Linton Park plc, 429-31 Rainham Road South, Dagenham, Essex, RM10 8RN, UK (Tel: 44 20 8596 7500; FAX 44 20 8596 7509; E-mail: sales@britishtraders.co.uk; Internet: www.britishtraders.co.uk)

British Wax Refining Co. Ltd.

62 Holmethorpe Ave., Holmethorpe Industrial Estate, Redhill, Surrey, RH1 2NL, UK (Tel: 44 1737 761242; FAX 44 1737 761472; E-mail: wax@british-wax.com; Internet: www.british-wax.com)

Bromine & Chemicals Ltd.

Div. of Dead Sea Bromine, 201 Haverstock Hill, Hampstead, London, NW3 4QG, UK (Tel: 44 20 7431 7707; FAX 44 20 7431 7797; E-mail: bromineUK@compuserve.com)

Brook-Chem Inc.

1770 North Park Place, Ste. 109, Naperville, IL, 60563, USA (Tel: 630-369-7300; FAX 630-369-7325; E-mail: brchem1992@aol.com)

Brøste

P. Brøste A/S, Lundtoftegardsvej 95, 2800 Lyngby, Denmark (Tel: 45 45 26 33 33; FAX 45 4593 1334; E-mail: broste@broste.com; Internet: www.broste.com)

Broste Ltd, No. 8 North Lynn Business Village-Bergen Way, North Lynn Industrial, King Lynn, Norfolk, PE302JU, UK (Tel: 44 1553 776066; FAX 44 1553 767319)

Brown Chemical Co., Inc./Industrial and Fine Chems.

302 W. Oakland Ave., PO Box 440, Oakland, NJ, 07436-0440, USA (Tel: 201-337-0900; FAX 201-337-9026; Internet: www.brownchem.com)

Browning Chemical Corp.

Subsid. of JLM Industries, Inc., 707 Westchester Ave., White Plains, NY, 10604, USA (Tel: 914-686-0300; 800-771-0008; FAX 914-686-0310; E-mail: vivan.stone@jlm.industries.com; Internet: www.jlmgroup.com)

Bruchem, Inc.

100 Oak St., Norwood, NJ, 07648, USA (Tel: 201-750-1400; FAX 201-750-1414; E-mail: bgrobman@bellatlantic.net)

Brutanicals, Inc.

100 Oak St., Norwood, NJ, 07648, USA (Tel: 201-750-1200; FAX 201-750-1414; E-mail: bgrobman@bellatlantic.net)

Robert Bryce & Co.

Block 6, 1B Unwin St., Rosehill, NSW, 2142, Australia (Tel: 61 2 9897 6225; FAX 61 2 9897 6299; E-mail: smills@im.aust.com)

Buckton Page Limited

Unit 5, Prospect Business Park, Langston Rd., Loughton, Essex, IG10 3TR, UK (Tel: 44 20 8532 1022; FAX 44 20 8532 1023; E-mail: sales@bucktonpage.com; Internet: www.bucktonpage.com)

Buckton Scott

Buckton Scott Chemicals Ltd., B.S. Chem., Canewdon House, Locks Hill, Rochford, Essex, SS4 1BB, UK (Tel: 44 1702 560600; FAX 44 1702 560606; E-mail: info@buckton.co.uk; Internet: www.buckton.com)

Buckton Scott USA, Inc., 330 N. Harrison St., Princeton, NJ, 08540, USA (Tel: 609-279-9609; FAX 609-279-1909; E-mail: buckton@ix.netcom.com; Internet: www.buckton.com)

Buffalo Color Corp.

100 Lee St., Buffalo, NY, 14210, USA (Tel: 716-827-4500; 800-631-0171; FAX 716-827-4718; E-mail: info@buffalocolor.com; Internet: www.buffalocolor.com)

Bunge

Bunge Limited, 50 Main St., 6th Fl., White Plains, NY, 10606, USA (Tel: 924-684-2800; Internet: www.bunge.com)

Bunge North America Inc./Bunge Foods, 11720 Borman Dr., St. Louis, MO, 63146, USA (Tel: 314-872-3030; 800-828-0800; FAX 314-872-0159; Internet: www.bunge.com)

Bunge Corporation Limited, London Fruit Exchange, Suite 17, Brushfield St., London, E1 6HB, UK (Tel: 44 207 247 4444)

Bunge Global Markets S.P.A., Via Paolo Frisi, 23, 00197 Roma, Italy (Tel: 39 06 80 21 81)

Bunge Trade Service, PO Box 1882, 3000 BW Rotterdam, The Netherlands (Tel: 31 10 43 30 700)

Bunge Handelsgessellschaft GmbH, PO Box 11353, 20434 Hamburg, Germany (Tel: 49 40 8222 17100)

Bunge Australia Holdings Pty. Ltd., Redlands Rd., PO Box 78, Corowa, NSW, 2646, Australia (Tel: 61 2 6033 1200)

Burdick & Jackson

Div. of Honeywell, 1953 S. Harvey St., Muskegon, MI, 49442-6184, USA (Tel: 231-726-3171; 800-368-0050; FAX 231-725-6216; E-mail: sandy.giddings@honeywell.com; Internet: www.bandj.com)

Burlington Bio-Medical & Scientific Corp.

71 Carolyn Blvd., Farmingdale, NY, 11735-1527, USA (Tel: 631-694-4700; FAX 631-694-9177; E-mail: BSCCHEMNY@aol.com; Internet: www.burlington scientific.com)

Burlington Chemical Co., Inc.

PO Box 111, 615 Huffman Mill Rd., Burlington, NC, 27216, USA (Tel: 336-584-0111; 800-672-5888; FAX 336-584-3548; E-mail: info@burco.com; Internet: www.burco.com)

Bush Boake Allen Inc. *See* International Flavors & Fragrances

Calabrian Corp.

1521 Green Oak Pl., Suite 200, Kingwood, TX, 77339, USA (Tel: 281-348-2303; 888-910-4567; FAX 281-348-2310; E-mail: hoppermann@calbriancorp.com)

Calchauvet

Calchauvet, Route de la Parfumerie, 83440 Seillans, Fayence, France (Tel: 33 4 94 39 00 00; FAX 33 4 94 39 00 79; E-mail: infoFR@calchauvet.com; Internet: www.calchauvet.com)

Calchauvet (US), 1 Reuten Dr., Closter, NJ, 07624, USA (Tel: 201-784-9300; FAX 201-784-0604; E-mail: infoUS@calchauvet.com; Internet: www.calchauvet.com)

J.H. Calo Company Inc.

1600 Stewart Ave., Suite 303, PO Box 647, Westbury, NY, 11590, USA (Tel: 516-832-7750; 800-626-2256; FAX 516-832-7758; E-mail: sales@jhcalo.com; Internet: www.jhcalo.com)

Calor Gas Northern Ireland Limited

Airport Rd. West, Sydenham, Belfast, BT3 9EE, Ireland (Tel: 44 28 9045 5588; FAX 44 28 9045 8072; E-mail: info@calorgas.ie; Internet: www.calorgas.ie)

Calumet Lubricants Co.

2780 Waterfront Pkwy. E. Dr., Suite 200, Indianapolis, IN, 46214, USA (Tel: 317-328-5660; 800-437-3188; FAX 317-328-5668; Internet: www.calumetlubricants.com)

Cambrex Corp.

One Meadowlands Plaza, E. Rutherford, NJ, 07073, USA (Tel: 201-804-3000; FAX 201-804-9852; E-mail: mike.scigliano@cambrex.com; Internet: www.cambrex.com)

Cambridge Research Biochemicals (CRB)

10 Belasis Court, Belasis Hall Technology Park, Billingham, Cleveland, TS23 4AZ, UK (Tel: 44 1642 567182; FAX 44 1642 567181; E-mail: crbinc@crb.gb.com; Internet: www.crb.gb.com)

Cameron Carbon Inc.

PO Box 18810, Baltimore, MD, 21206, USA (Tel: 410-254-6070; 800-394-6844; FAX 410-254-8254; E-mail: camcarb@ix.netcom.com; Internet: www.cameron.thomasregister.com/olc/cameron/home.htm)

Camida

Camida Ltd, Tower House, New Quay, Clonmel, Co. Tipperary, Ireland (Tel: 353 52 25455; FAX 53 52 25466; E-mail: info@camida.ie; Internet: www.camida.com)

Camida Ltd (UK), Willow Mill, Caton, Lancaster, LA2 9RA, UK (Tel: 44 1524 771881; FAX 44 1524 771882; E-mail: info@camida.co.uk; Internet: www.camida.com)

Camida LLC, One Gateway Centre, Suite 2600, Newark, NJ, 07102, USA (Tel: 973-645-9227; FAX 973-645-9484; E-mail: infocamida@cs.com; Internet: www.camida.com)

CanAmera Foods

Part of Central Soya, 14711-128th Avenue, Edmonton, Alberta, T5L 3H3, Canada (E-mail: gwmorhun@canamera.ca; Internet: www.canamera.ca)

Captree Chemical Corp.

605 Albany Ave., Amityville, NY, 11702, USA (Tel: 516-841-0200; 800-899-2725; FAX 516-841-5370)

Carbery Food Ingredients Ltd.

Ballineen, Co Cork, Ireland (Tel: 353 23 47222; FAX 353 23 47541; E-mail: info@carbery.com; Internet: www.carbery.com)

Carbochem, Inc.

326 W. Lancaster Ave., Ardmore, PA, 19003, USA (Tel: 610-645-9200; FAX 610-645-5501; E-mail: hq@carbochem.com; Internet: www.carbochem.com)

CarboMer, Inc.

PO Box 721, Westborough, MA, 01581, USA (Tel: 508-870-0806; 800-239-7129; FAX 508-898-0432; E-mail: sales@carbomer.com; Internet: www.carbomer.com)

Carboshale AS

Akadeemia tee 23A, 12618 Tallinn, Estonia (Tel: 372 670 3645; FAX 372 656 0159; E-mail: eke@rnk.ee; Internet: www.carboshale.ee; www.rnk.ee)

Cardinal Companies, L.P.

2010 S. Beltline, Columbia, SC, 29201, USA (Tel: 803-799-7190; 800-788-7274; FAX 803-799-8742; Internet: www.cardinallp.com)

Cardolite

Cardolite Corp., 500 Doremus Ave., Newark, NJ, 07105-4805, USA (Tel: 973-344-5015; 800-322-7365; FAX 973-344-1197; E-mail: csr@cardolite.com; Internet: www.cardolite.com)

Cardanol Chemicals N.V., Offerlaan 116-118, 9000 Ghent, Belgium (Tel: 32 9 265 88 20; FAX 32 9 265 88 29; E-mail: nic.baetsle@cardanol.be; Internet: www.cardanol.be)

Cardanol Far East Pte. Ltd., 371 Beach Rd. #20-08, Key Point, Singapore, 199597, Singapore (Tel: 65 291 9851; FAX 65 298 5798; E-mail: admin99@cardanol.com.sg)

Cardanol China, 28F Bldg. No. 28, Cao Xi Rd., Xu Jia Jui, Shanghai, 200030, P.R. China (Tel: 86 21 64879631; FAX 86 21 64860684; E-mail: j.dai@cardolite.com)

Cargill

Cargill, PO Box 9300, Minneapolis, MN, 55440-9300, USA (Tel: 952-742-7575; Internet: www.cargill.com)

Cargill Industrial Oils & Lubricants, PO Box 5700, MS 66, Minneapolis, MN, 55440, USA (Tel: 612-742-6286; 800-842-3631; FAX 612-742-6722; E-mail: techoils@cargill.com; lubricants@cargill.com; Internet: www.techoils.cargill.com)

Cargill plc, Witham St. Hughs, Lincoln, LN6 9TN, UK (Tel: 44 1522 556100; FAX 44 1522 868244)

Deutsche Cargill GmbH, Ruedekenstrasse 51/ Am Hafen, 38239 Salzgitter-Beddingen, Germany (Tel: 49 5341 224 0; FAX 49 5341 224 888)

Cargill NV, Prins Albertlaan 12, 8870 Izegem, Belgium (Tel: 32 5 133 2811; FAX 32 5 133 2671)

Cargill Srl, Via Rombon 11, 20134 Milano, Italy (Tel: 39 2 215 64200; FAX 39 2 215 64209)

Cargill Espana, S.A., Apartado de Correos 314, 08190 Sant Cugat del Valles (Barcelona), Spain (Tel: 34 3 581 91 00; FAX 34 3 581 92 31)

Cargill International S.A., 14 chemin de Normandie, 1206 Geneva, Switzerland (Tel: 41 22 703 21 11; FAX 41 22 703 25 55)

R.E. Carroll, Inc.

1570 N. Olden Ave., PO Box 5806, Trenton, NJ, 08638-0806, USA (Tel: 609-695-6211; 800-257-9365; FAX 609-695-0102; E-mail: info@recarroll.com; Internet: www.recarroll.com)

The Cary Co.

1195 W. Fullerton Ave., PO Box 403, Addison, IL, 60101-0403, USA (Tel: 630-629-6600; FAX 630-629-3690; Internet: www.thecarycompany.com)

Cascade Chemical Co.

111 Westland Rd., Cedar Grove, NJ, 07009-2219, USA (Tel: 973-239-2523; FAX 973-239-2501)

CasChem Inc.

A Cambrex Co., 40 Ave. A, Bayonne, NJ, 07002, USA (Tel: 201-858-7900; 800-CASCHEM; FAX 201-437-2728; E-mail: annmarie.hansen@cambrex.com; Internet: www.caschem.com)

Castrol Industrial North America Inc.

1001 W. 31st St., Downers Grove, IL, 60515, USA (Tel: 630-241-4000; 800-621-2661; FAX 630-241-1957; Internet: www.castrolindustrial.com)

CBC (Europe) Ltd., Chemical Div.

7/8 Garrick Industrial Centre, Irving Way, London, NW9 6AQ, UK (Tel: 44 20 8732 3303; FAX 44 20 8202 3387; E-mail: kfurukubo@cbcuk.com)

Ceca

Ceca S.A., Subsid. of Atofina S.A., Immeuble Iris, La Défense 2, 92062 paris La Defense Cedex, France (Tel: 33 1 47 96 90 90; FAX 33 1 47 96 91 91; E-mail: info@atofina.com; Internet: www.ceca.fr)

Irish Ceca, Allenwood Enterprise Prk, Allenwood, Naas, Co. Kildare, Ireland (Tel: 353 45 870855; FAX 353 45 870866)

Ceca Chemical Specialities Belgium, Rue de Stalle, 63, Bte 1, 1180 Brussels, Belgium (Tel: 32 2 370 20 42; FAX 32 2 332 01 91)

Ceca Italiana SpA, Via Galileo Galilei 51-53, 20096 Pioltello (MI), Italy (Tel: 39 2 92 91 91; FAX 39 2 92 69 507)

Chesham Chemicals Ltd., Distributor for Ceca, Cunningham House, Westfield Lane, Kenton, Harrow, Middlesex, HA3 9ED, UK (Tel: 44 20 8907 7779; FAX 44 20 8909 1053; E-mail: sales@cheshamchemicals.co.uk; Internet: www.cheshamchemicals.co.uk)

Cedar Concepts Corp.

4342 S. Wolcott Ave., Chicago, IL, 60609, USA (Tel: 773-890-5790; FAX 773-890-1606; E-mail: cedartr@aol.com)

Cekal Specialties, Inc.

801 W. Central Ave., PO Box 788, Mt. Holly, NC, 28120, USA (Tel: 704-822-6206; FAX 704-822-6641)

Celanese

Celanese Chemicals, A Business of Celanese AG, 1601 West LBJ Freeway, Dallas, TX, 75234-6034, USA (Tel: 972-443-4000; 800-344-2497; FAX 972-443-8538; Internet: www.celanesechemicals.com; www.chemvip.com)

Celanese Mexicana, S.A. de C.V., Tecoyotitla No. 412, Col. Ex-Hacienda de Guadalupe Chimalistac, Mexico D.F., 01050, Mexico (Tel: 52 5 480 9100; FAX 52 5 480 9145; E-mail: jagarcia@celanese.com.mx; Internet: www.celanese.com.mx)

Celanese Canada Inc., PO Box 99, Stn. Main, Edmonton, Alberta, T5J 2H7, Canada (Tel: 780-471-0511; FAX 780-471-0113; E-mail: EDMinformation@celanese.com)

Celanese Chemicals UK Ltd., Vista Office Centre, 50 Salisbury Rd., Hounslow, Middlesex, TW4 6JH, UK (Tel: 44 20 8814 1034; FAX 44 20 8577 3194)

Celanese Chemicals Europe GmbH, A Business of Celanese AG, Frankfurterstrasse 111, 61476 Kronberg, Germany (Tel: 49 69 305 16000)

Celanese Tarragona, CN 340, Km 1155, Apartado 18, 43080 Tarragona, Spain (Tel: 34 977 307700; FAX 34-977-307790)

Celanese Nantong, 27 East Zhong Xiu Rd., Nantong City, Lianesu Province, 226008, P.R. China (Tel: 86 513 356 0512; FAX (011) 86-513-356-0793)

Celanese Singapore/Celanese Acetyls, PO Box 219, Jurong Pt. Post Office, Singapore, Singapore (Tel: 65 867 6005; FAX 65 867 5925)

Centerchem Inc.

225 High Ridge Rd, Stamford, CT, 06905-3036, USA (Tel: 203-975-9800; FAX 203-975-8777)

Central Soya

Central Soya Co., Inc., A Company of Cereol, PO Box 1400, 1946 West Cook Rd., Fort Wayne, IN, 46801-1400, USA (Tel: 260-425-5432; 800-348-0960; FAX 260-425-5301; E-mail: NALecithin@centralsoya.com; Internet: www.centralsoya.com)

Central Soya European Proteins A/S, Sydhavnsgade 7, PO Box 380, 8100 Aarhus C, Denmark (Tel: 45 89 31 21 11; FAX 45 89 31 21 12; E-mail: EuProtein@centralsoya.com)

Central Soya European Lecithins GmbH & Co. KG, Div. of Central Soya, Lange Reihe 2, 20099 Hamburg, Germany (Tel: 49 40 248 3001; FAX 49 40 280 30155; E-mail: EuLecithin@centralsoya.com)

Century Multech, Inc.

33-70 Prince St., Suite 605, Flushing, NY, 11354, USA (Tel: 718-353-9885; FAX 718-353-9878; E-mail: rylchu@aol.com)

Cerac, Inc.

PO Box 1178, 407 N. 13th St., Milwaukee, WI, 53201, USA (Tel: 414-289-9800; FAX 414-289-9805; E-mail: marketing@cerac.com; Internet: www.cerac.com)

Ceres Chemical Co., Inc.

707 Westchester Ave., White Plains, NY, 10604, USA (Tel: 914-422-8000; FAX 914-422-8000; E-mail: sales@cereschemical.com)

Cerestar

Cerestar International Sales, A company of Eridania Beghin-Say, 14, Blvd. du General Leclerc, 92572 Neuilly-sur-Seine Cedex, France (Tel: 33 1 41 43 19 24; FAX 33 1 41 43 19 74; Internet: www.cerestar.com)

Cerestar Benelux BV, Nijverheidsstraat 1, 4551 LA Sas Van Gent, The Netherlands (Tel: 31 115 459 000; FAX 31 115 451 123)

Cerestar UK Ltd., Trafford Park Rd, Trafford Park, Manchester, Lancashire, M17 1PA, UK (Tel: 44 161 872 5959; FAX 44 161-848 9034; Internet: www.cerestar.com)

Cerestar Deutschland GmbH, Cerestarstrasse 2, 47809 Krefeld, Germany (Tel: 49 2151 575 01; FAX 49 2151 575 775)

Cerestar Austria, Handelsgesselschaft mbH, Schmidtgasse 18, 4600 Wels, Austria (Tel: 43 72 42 663 160; FAX 43 72 42 663 1618)

Cerestar Scandinavia A/S, Skovlytoften 33, 2840 Holte, Denmark (Tel: 45 45 46 90 00; FAX 45 45 46 90 40)

Cerestar Finland Oy, Piispansilta 9, Toimisto D, III krs, 02230 Espoo, Finland (Tel: 358 9 417 67 100; FAX 358 9 417 67 159)

Cerestar Italia, Via del Lavoro 17, 44100 Ferrara, Italy (Tel: 39 532 590 111; FAX 39 532 590 191)

Cerestar Ibérica SA, Avda. Montserrat s/n, Apartado de Coreos 54, 08760 Martorell (Barcelona), Spain (Tel: 34 93 773 38 00; FAX 34 93 775 59 50)

Cerestar USA Inc., 1100 Indianapolis Blvd., Hammond, IN, 46320-1094, USA (Tel: 219-659-2000; 800-348-9896; FAX 219-473-6601)

Cerestar Jiliang Maize Industry Co. Ltd. (CJMI), Songyuan, Jiangnan Industry Development Zone, Jilin Province, CM 138000, Songyuan City, P.R. China (Tel: 86 438 218 11 01; FAX 86 438 218 11 03; E-mail: cjmicom@mail.jl.cn)

Cerestar Japan Ltd., Eiko Bldg., 6F, 2-6-1 Nihonbashi Honcho, Chuo-ku, Tokyo, 103-0023, Japan (Tel: 81 3 3664 7321; FAX 81 3 3664 7323)

Chance & Hunt

Alexander House, Crown Gate, Runcorn, Cheshire, WA7 2UP, UK (Tel: 44 1928 793000; FAX 44 1928 714351; E-mail: webmaster@chance-hunt.com; Internet: www.chance-hunt.com)

Charabot

Charabot & Co. Inc. USA, 30 Corporate Dr., Orangeburg, NY, 10962-2622, USA (Tel: 914-398-1200; FAX 914-398-1440)

Charabot Hong Kong Limited, 19th Fl., Nine Queen's Road, Central, Hong Kong

Charkit Chemical Corp.

9 Old Kings Way South, PO Box 1725, Darien, CT, 06820-1725, USA (Tel: 203-655-3400; FAX 203-655-8643; E-mail: sales@charkit.com; Internet: www.charkit.com)

Charlotte Chemical Inc.

2500 Wilcrest, Suite 334, Houston, TX, 77042, USA (Tel: 281-955-1162; FAX 281-955-9632; E-mail: maggie@charlotte.com.mx; Internet: www.charlotte.com.mx)

Chart Corporation Inc.

787 E. 27th St., Paterson, NJ, 07504, USA (Tel: 973-345-5554; FAX 973-345-2139; E-mail: chartcorp@aol.com; Internet: www.chartcorp.com)

Chemacon GmbH

Hochstrasse 45 c, 47506 Neukirchen-Vluyn, Germany (Tel: 49 2845 9838 0; FAX 49 2845 9838 28; E-mail: info@chemacon.de; Internet: www.chemacon.de)

Chemapol Usa, Inc.

560 Sylvan Ave., Englewood Cliffs, NJ, 07632, USA (Tel: 201-816-1382; FAX 201-816-1384)

Chematch

1281 Main St., Stamford, CT, 06902, USA (Tel: 203-961-1686; FAX 203-357-7011; E-mail: tfoster@petrochem.net; Internet: www.chematch.com)

Chemax Performance Products

Member of Rütgers Organics Corp., PO Box 6067, Greenville, SC, 29606, USA (Tel: 864-277-7000; 800-334-6234; FAX 864-277-7807; E-mail: info@chemaxinc.com; Internet: www.chemaxinc.com)

Chemcentral Corp.

7050 W. 71 St., PO Box 730, Bedford Park, IL, 60499-0730, USA (Tel: 708-594-7000; 800-331-6174; FAX 708-594-6328)

Chemco France

10 av. Maurice Berteaux, 78300 Poissy, France (Tel: 33 1 30 65 75 00; FAX 33 1 30 65 74 94; E-mail: chemco@club-internet.fr; Internet: www.chemco-france.fr)

Chemconnect

44 Montgomery St., Suite 250, San Francisco, CA, 94104, USA (Tel: 415-364-3300; FAX 415-646-0010; Internet: www.chemconnect.com)

Chemconserve B.V.

PO Box 566, 2280 AN Rijswijk, The Netherlands (Tel: 31 70 394 9512; FAX 31 70 396 0811; E-mail: chemconserve@wxs.nl; Internet: http://home.wxs.nl/~chemconserve/)

Chemcor

48 Leone Lane, Chester, NY, 10918, USA (Tel: 914-469-5800; 888-243-6267; FAX 914-469-5855)

ChemDesign Corp.

A Bayer Company, 310 Authority Dr., Fitchburg, MA, 01420, USA (Tel: 978-345-9999; FAX 978-342-9620; Internet: www.bayerus.com)

Chemetall

Chemetall GmbH, Subsid. of Dynamit Nobel AG, Troisdorf, Trakehner Strasse 3, 60487 Frankfurt Main, Germany (Tel: 49 69 7165 0; FAX 49 69 7165 3018; E-mail: info@chemetall.com; Internet: www.chemetall.com)

Chemetall S.A., 51, rue Pierre, 92111 Clichy, France (FAX 33 1 47 15 38 00)

Chemetall Chemical Products, Inc., 50 Valley Rd., Berkeley Heights, NJ, 07922, USA (Tel: 908-665-2002; 800-654-8059; FAX 908-665-8543; Internet: www.chemetall.com)

Chemetall Foote Corp., 348 Holiday Inn Dr., Kings Mountain, NC, 28086, USA (Tel: 704-739-2501; 800-523-7116; FAX 704-734-2670; E-mail: foote.lithium@chemetall.com)

Chemetall Japan K.K., 1-7-10, Kanda Surugadai, YK Surugadai Bldg., Chiyoda-ku, Tokyo, 101, Japan (Tel: 81 3 3259 2061; FAX 81 3 3259 2065)

ChemFirst Fine Chemicals

PO Box 1249, Jackson, MS, 39215-1249, USA (Tel: 601-948-7550; 800-828-7940; FAX 601-949-0264; Internet: www.chemfirst.com)

Chemial SpA

Via Aresenale 27/E, 10121 Torino, Italy (Tel: 39 11 562 03 31; FAX 39 11 562 09 10; E-mail: info@lachemial.com; Internet: www.lachemial.com)

The Chemical Co.

PO Box 436, 19B Narragansett Ave, Jamestown, RI, 02835, USA (Tel: 401-423-3100; FAX 401-423-3102; Internet: www.thechemco.com)

Chemical Distribution, Inc.

240 Fost Ave., Bensenville, IL, 60106, USA (Tel: 630-521-1291; 800-533-3287; FAX 630-521-1353; Internet: www.chemdis.com)

Chemical Formulators, Inc.

6215 W. Tyson Ave., Tampa, FL, 33611, USA (Tel: 813-832-4868)

Chemical Group Inc.

134 Main St., Eatontown, NJ, 07724, USA (Tel: 215-860-0488; FAX 215-860-2256)

Chem-Impex International, Inc.

935 Dillon Dr., Wood Dale, IL, 61091, USA (Tel: 630-766-2112; 800-869-9290; FAX 630-766-2218; Internet: www.chemimpex.com)

Chemisphere Corp.

2101 Clifton Ave., St. Louis, MO, 63139, USA (Tel: 314-644-1300; 800-844-1301; FAX 314-644-7194; E-mail: mpclote@chemispherecorp.com; Internet: www.chemispherecorp.com)

Chemisphere Ltd.

38 King St., Chester, Cheshire, CH1 2AH, UK (Tel: 44 1244 320878; FAX 44 1244 320858; E-mail: sales@chemisphere.co.uk; Internet: www.chemisphere.co.uk)

Chemol Co.

2300 Randolph Ave., PO Drawer 16286, Greensboro, NC, 27406, USA (Tel: 910-333-3060; 800-849-3050; FAX 910-273-4645)

Chemos GmbH

Werner von Siemens Str. 3, 93128 Regenstauf, Germany (Tel: 49 9402 9336 0; FAX 49 9402 9336 13; E-mail: sales@chemos-group.com; Internet: www.chemos-group.com)

Chemplex Chemicals, Inc.

Crossroads Corporate Center, 1 International Blvd., Suite 400, Mahwah, NJ, 07495-0400, USA (Tel: 201-512-8788; FAX 201-512-8791; E-mail: chemplx@aol.com)

ChemPur Feinchemikalien und Forschungsbedarf GmbH

Postfach 41 04 50, 76204 Karlsruhe, Germany (Tel: 49 721 933 81 40; FAX 49 721 47 20 01; E-mail: info@ChemPur.de; Internet: www.ChemPur.de)

Chemron Corp.

PO Box 2299, Paso Robles, CA, 93446, USA (Tel: 805-239-1550; 800-491-1550; FAX 805-239-8551; E-mail: info@Chemron.com; Internet: www.chemron.com)

ChemService, Inc.

660 Tower Lane, PO Box 599, West Chester, PA, 19381-0599, USA (Tel: 610-692-3026; 800-452-9994; FAX 610-692-8729; E-mail: info@chemservice.com; Internet: www.chemservice.com)

Chemsil Silicones, Inc.

9445 De Soto Ave., Chatsworth, CA, 91311, USA (Tel: 818-700-0302; FAX 818-700-0507; E-mail: chemsilsilicones@aol.com; Internet: www.chemsil.com)

Chemtronics Inc.

An Illinois Tool Works Company, 8125 Cobb Center Dr., Kennesaw, GA, 30152, USA (Tel: 770-424-4888; 800-645-5244; FAX 770-424-4267; E-mail: msauer@chemtronics.com; Internet: www.chemtronics.com)

ChemUSA, Corp.

13575 East Oil Dock Rd., Brownsville, TX, 78521, USA (Tel: 956-831-5100; FAX 956-831-9309; E-mail: chemusa@worldnet.att.net)

Chemway International Inc.

375 Sylvan Ave., Suite 111, Englewood Cliffs, NJ, 07632, USA (Tel: 201-541-0118; FAX 201-541-0120; E-mail: chemway@aol.com)

Chesham Chemicals Ltd.

Cunningham House, Westfield Lane, Harrow, Middlesex, HA3 9ED, UK (Tel: 44 20 8907 7779; FAX 44 20 8909 1053; E-mail: sales@cheshamchemicals.co.uk; Internet: www.cheshamchemicals.co.uk)

Chevron Phillips

Chevron Phillips Chemical Co. LP, Joint venture with ChevronTexaco, PO Box 4910, The Woodlands, TX, 77387-4910, USA (Tel: 832-813-4100; 800-231-1212; E-mail: specialtychemicals@ppco.com; Internet: www.cpchem.com)

Chevron Phillips Chemicals International NV, Haandorpweg 1, Haven 1227, 9130 Kallo-Beveren, Belgium (Tel: 32 3 570 26 11; FAX 32 3 570 26 26)

Chevron Phillips Chemical (China) Co. Ltd., East of Shizi Port Bridge, Zhangjiagang Free Trade Zone, Jiangsu Province, 215634, PR China (Tel: 86 520 838 6688; FAX 86 520 838 6605)

Chevron Phillips Chemicals Asia Pte. Ltd., 21 Tuas Ave. 3, Singapore, 639417, Singapore

ChevronTexaco

ChevronTexaco Corp., 6001 Bollinger Canyo Rd., San Ramon, CA, 94583, USA (Tel: 925-842-1000; E-mail: texacoweb@chevrontexaco.com; Internet: www.chevrontexaco.com)

Texaco Products Inc., 6975-A Pacific Circle, Mississauga, Ontario, L5T 2H3, Canada (Tel: 905-670-6470; FAX 905-670-6476; E-mail: mcwhish@texaco.com)

Productos ChevronTexaco, S.A. de C.V., Blvd. Manuel Avila Camacho #36/20th Fl., Lomas de Chapultepec, Delegacion Miguel Hidalgo, Mexico City, 11000, Mexico (Tel: 52 5 55747 4900; FAX 52 5 5520 93 58)

Texaco Brasil S.A., Av. Republica do Chile, 230, 18th to 25th Floors, Rio de Janeiro - Centro, 20031-170, Brazil (Tel: 55 21 2271 9140; FAX 55 21 21609670)

ChevronTexaco Global Lubricants, 43/45 The Promenade, Cheltenham, Gloucestershire, GL50 1LE, UK (Tel: 44 1242 266700; FAX 44 1242 266749)

Texaco (Ireland) Ltd., Texaco House, Ballsbridge, Dublin, 4, Ireland (Tel: 353 1 668 6822)

SA Texaco Belgium NV/SA, Ave. Arnaud Fraituer 25, 1050 Brussels, Belgium (Tel: 32 2 639 9111)

Texaco France S.A., Batiment Platon, Parc Les Algorithmes, 141-145, rue Michel Carre, 95815 Argentuil Cedex, France (Tel: 33 1 34 34 13 73; FAX 33 1 34 34 13 70)

Texaco Deutschland GmbH, Grafenberger Allee 125, 40237 Dusseldorf, Germany (Tel: 49 211 968 45 0; FAX 49 211 968 45 45)

Texaco Italiana SpA, Via Laurentina, 456, 00142 Rome, Italy (Tel: 39 06 549351; FAX 39 06 5924137)

Chimen S.r.l.

Via Kennedy, 67, 30027 San Dona di Piave (VE), Italy (Tel: 32 421 41442; FAX 32 421 43898; E-mail: info@chimen.it; Internet: www.chimen.it)

Chinook Group

Suite 2750, 25 King St. West, Toronto, ON, M5L 1E2, Canada (Tel: 416-363-6046; FAX 416-363-6337)

Chisso

Chisso Corp., Tokyo Bldg., 2-7-3, Marunouchi, Chiyoda-ku, Tokyo, 100, Japan (Tel: 81 3 3284 8411; FAX 81 3 3284 8412; E-mail: t.taji@chisso.co.jp; Internet: www.chisso.co.jp)

Chisso Petrochemical Corp., Subsid. of Chisso Corp., Tokyo Bldg., 2-7-3 Marunouchi, Chiyoda-ku, Tokyo, 100, Japan (Tel: 81 3 3284 8411; FAX 81 3 3284 8412)

Chisso America Inc., 1185 Ave of the Americas, 32nd Floor, New York, NY, 10036, USA (Tel: 212-302-0500; FAX 212-302-0643; E-mail: kwatabe@email.msn.com; Internet: www.chisso.co.jp)

Christianson S.A. de C.V.

Part of Clariant, Acatl No. 381, Fracc. Industrial San Antonio, Mexico D.F., 02760, Mexico (Tel: 52 5 328 4700; FAX 52 5 352 4981; E-mail: chsavent@mail.internet.com.mx; Internet: www.cosmos.com/mx/chem/christianson)

CHT

CHT Group/CHT R. Beitlich GmbH, Bismarckstrasse 102, 72072 Tübingen, Germany (Tel: 49 70 71 154 0; FAX 49 70 71 154 290; E-mail: info@cht-group.de; Internet: www.cht-group.de)

CHT Austria R. Beitlich GmbH, Aueleweg 3, 6812 Meiningen, Austria (Tel: 43 55 22 3 12 23; FAX 43 55 22 3 12 29; E-mail: chtaustria@cht-group.at)

CHT Belgium N.V., Pres. Kennedypark 39, 8500 Kortrijk, Belgium (Tel: 32 56 20 31 14; FAX 32 56 20 01 19; E-mail: chtbelgium@cht-group.de)

CHT R. Beitlich Danmark, Postbox 58, Hovedgaden 26-28, 1. sal, 2970 Hørsholm, Denmark (Tel: 45 45 76 92 80; FAX 45 45 76 06 60; E-mail: cdanmark@cht-group.de)

CHT France, 141 Ave. d'Echenilly, BP 43, 10122 St. Andre Les Vergers Cedex, France (Tel: 33 3 25 71 34 60; FAX 33 3 25 75 25 83; E-mail: cdanmark@cht-group.de)

CHT Italia S.R.L. Prodotti Chimici Tessili, Via Liguria, 2, 24052 Azzano S. Paolo (Bg), Italy (Tel: 39 35 32 50 11; FAX 39 35 33 01 19; E-mail: cht@chtitalia.it; Internet: www.chtitalia.it)

CHT R. Beitlich GmbH Vertretungsbüro, Simferopolskij Bulvar, dom 7A, kv. 15, 16, 113556 Moscow, Russia (Tel: 7 95 3 17 95 13; FAX 7 95 1 13 97 11; E-mail: cht@co.ru)

CHT R. Beitlich Corp., Subsid. of CHT Group, PO Box 240497, Charlotte, NC, 28217, USA (Tel: 704-523-4242; 800-277-4941; FAX 704-522-8142; E-mail: sales@chtusa.com; Internet: www.chtusa.com)

CHT-Avcomex S.A. de C.V., Emiliano Zapato No. 159-A, Col. Zona Industrial Lerma, Lerma Edo. de Mexico, 52000, Mexico (Tel: 52 72 82 02 05; FAX 52 72 82 11 51; E-mail: textil@chtavco.com.mx)

CHT Brasil Quimica Ltda., Av. Antonio Candido Machado, 1779, Cx. Postal 533, Cajamar - SP, 07760-000, Brazil (Tel: 55 11 44 47 89 00; FAX 55 11 44 47 89 09; E-mail: chtbelgium@cht-group.de)

CHT-Wide Corp., 12th Fl., 137, Nanking E. Rd. Sec. 2, Taipei, Taiwan R.O.C. (Tel: 886 2 25 07 11 58; FAX 886 2 25 06 08 11; E-mail: chtaustria@cht-group.at)

CHT (India) Pvt. Ltd., 405-48 Navbharat Estates, B Wing, 172 Zakaria Bunder Rd., Sweri (Est), Mumbai, 400 015, India (Tel: 91 22 4 15 88 60; FAX 91 22 4 15 88 62; E-mail: chtindia@agmgroup.com)

CHT Australia Pty. Ltd., 33 Elliott Rd., Dandenong, Victoria, 3175, Australia (Tel: 61 3 97 06 74 00; FAX 61 3 97 06 74 11)

CHT Textile Chemicals Mktg., 49 T. Evangelista, B.F. Homes, Paranaque, Metro Manila, Philippines (Tel: 63 2 8 07 40 28; FAX 63 2 8 07 74 60; E-mail: kmadamba@info.com.ph)

CHT South Africa, 5, Daytona Pl., Mahogany Ridge, PO Box 15792, 3608 Westmead, Rep. of S. Africa (Tel: 27 31 7 00 84 36; FAX 27 31 7 00 84 45; E-mail: chtsa@dbn.lia.net)

Chugai Boyeki (America) Corp.

500 5th Ave., Suite 1730, New York, NY, 10110, USA (Tel: 212-869-4860; FAX 212-391-3817)

Chugai Kasei Co., Ltd.

Kamirenjaku 6-6-1, Mitaka, Tokyo, 181, Japan (Tel: 81 4 2249 5131; FAX 81 4 2248 1761)

Chunking Chemical Corp.

No. 58, Hubing Rd., Jiangnin Industry Development Zone, Nanjing, P.R. China (Tel: 86 25 2101240; FAX 86 25 2101533; Internet: www.chunkingchem.com)

Church & Dwight Co. Inc./Performance Prods. Group

Box CN5297, 469 N. Harrison St., Princeton, NJ, 08543-5297, USA (Tel: 609-497-7113; 800-221-0453; FAX 609-497-7176; E-mail: KIRSCLA@churchdwight.com; Internet: www.ahperformance.com)

Ciba Specialty Chemicals

Ciba Specialty Chemicals Inc., Klybeckstrasse 141, 4002 Basel, Switzerland (Tel: 41 61 636 11 11; FAX 41 61 636 12 12; Internet: www.cibasc.com)

Ciba Specialty Chemicals plc, Charter Way, Macclesfield, Cheshire, SK10 2NX, UK (Tel: 44 1625 421933; FAX 44 1625 619637)

Ciba Spécialités Chimiques SA, 8, rue Lionel Terray, 92563 Rueil-Malmaison Cedex, France (Tel: 33 1 47 52 40 00; FAX 33 1 47 52 40 05)

Ciba Specialty Chemicals Sweden AB, Box 605, 421 26 Västra Frölunda, Sweden (Tel: 46 31 892300; FAX 46 31 892350)

Ciba Specialty Chemicals S.p.A., Via Pila 6/3, 40044 Sasso Marconi (BO), Italy (Tel: 39 51 6786.111; FAX 39 51 845516; E-mail: additivi.italia@cibasc.com)

Ciba Specialty Chemicals GmbH/Additives, 68619 Lampertheim, Germany (Tel: 49 6206 15 2745; FAX 49 6206 15 2742)

Ciba Specialitatenchemie GmbH, Postfach 319, 1231 Vienna, Austria (Tel: 43 1 801110; FAX 43 1 80111421)

Ciba Specialty Chemicals N.V., Noordkustlaan 18, 1702 Groot-Bijgaarden, Belgium (Tel: 32 2 467 0611; FAX 32 2 4670774)

Ciba Specialty Chemicals A/S Denmark, Teknikerbyen 38, 8230 Virum, Denmark (Tel: 45 45 950600; FAX 45 45 950610)

Ciba Specialty Chemicals Finland, OY PL 16, 00241 Helsinki, Finland (Tel: 358 9 476911; FAX 358 9 47691260)

Ciba Especialidades Quimicas S.L., Apartado 744, 08080 Barcelona, Spain (Tel: 34 3 4040300; FAX 34 3 4040301)

Ciba Specialty Chemicals Corp., 540 White Plains Rd., PO Box 2005, Tarrytown, NY, 10591-9005, USA (Tel: 914-785-2000; 800-431-2360; FAX 914-785-2183; Internet: www.cibasc.com)

Ciba Specialty Chemicals Corp./Water Treatment Div., PO Box 2005, Tarrytown, NY, 10591-9005, USA (Tel: 800-795-6245; FAX 914-785-4244)

Ciba Specialty Chemicals Canada, Inc., 7030 Century Ave., Mississauga, Ontario, L5N 2V8, Canada (Tel: 905-812 7280; 888-371-3422; FAX 905-821-8466; E-mail: additives.canada@cibasc.com)

Ciba Specialty Chemicals Mexico, Calzada de Tlalpan 3058, Apartado Postal 22-115, Mexico D.F., 04850, Mexico (Tel: 52 5 422 6119; FAX 52 5 677 1342; E-mail: sergio.pulido@dibasc.com)

Ciba Specialty Chemicals Pty. Ltd., PO Box 332, 235 Settlement Rd., Thomastown, Victoria, 3074, Australia (Tel: 61 3 9282 06 00; FAX 61 3 9465 90 70)

Ciba Specialty Chemicals K.K., Kansai Head Office, 10-66, Miyuki-cho, Takarazuka, 665-8666, Japan (Tel: 81 797 73 90 00; FAX 81 797 73 90 25)

Ciba Specialty Chemicals (China) Ltd., 16/F Golden Land Bldg., No. 32 Liang Ma Bridge Rd., Beijing, 100016, P.R. China (Tel: 86 10 6464 37 00; FAX 86 10 6464 37 43)

CITGO Petroleum Corp.

1701 Golf Rd., Suite 1-1101, Rolling Meadows, IL, 60008-4295, USA (Tel: 918-495-5583; 800-967-7601)

Clariant

Clariant Corp., 4000 Monroe Rd., Charlotte, NC, 28205, USA (Tel: 704-331-7000; 800-631-8077; FAX 704-377-1063; E-mail: info@clariant.com; Internet: www.clariant.com; www.clariant-northamerica.com)

Clariant (Mexico) S.A. de C.V., Blvd. Toluca #46, Col. El Conde, Naucalpan de Juarez, Edo de Mexico, 53500, Mexico (Tel: 52 5 329 1800; FAX 52 5 576 3259; Internet: www.clariant-latinamerica.com)

Clariant S.A. (Brazil), Av. de Nacoes Unidas, 18001, Sao Paulo, SP, 04795-900, Brazil (Tel: 55 11 5683 7233; FAX 55 11 5642 1654)

Clariant UK Ltd., Calverley Lane, Horsforth, Leeds, W. Yorkshire, LS18 4RP, UK (Tel: 44 113 2584 646; FAX 44 113 2398 473)

Clariant Benelux SA, Parc Scientifique Fleming, rue Fond Jean Paques 1, 1348 Louvain-la-Neuve, Belgium (Tel: 32 10 480 511; FAX 32 10 480 666)

Clariant (Deutschland) GmbH, Stresemannallee 61, 60596 Frankfurt/Main, Germany (Tel: 49 69 6 33 09 603; FAX 49 69 6 33 09 605; Internet: www.clariant.de)

Clariant (France) SA, 70, Avenue du Général de Gaulle, 92800 Puteaux, France (Tel: 33 1 46 96 96 00; FAX 33 1 46 96 96 01)

Clariant (Italia) S.p.A., Via A. Manzoni, 37, 20030 Palazzolo Milanese (MI), Italy (Tel: 39 2 9918 1; FAX 39 2 9918 8789)

Clariant (Switzerland) Ltd., Rothausstrasse 61, 4132 Muttenz 1, Switzerland (Tel: 41 61 469 7885; FAX 41 61 469 7080; E-mail: BU.Textile.Chemicals@clariant.com; Internet: http://ppp.clariant.com)

Clariant Iberica S.A., Via Augusta, 252-260, 08017 Barcelona, Spain (Tel: 34 93 306 81 21; FAX 34 93 201 13 22)

Clariant (Japan) K.K., Bunkyo Green Court Center Office 9F, 2-28-8 Honkomagome, Bunkyo-ku, Tokyo, 113-8662, Japan (Tel: 81 3 5977 7880; FAX 81 3 5977 7886; Internet: www.clariant.co.jp)

Clariant (Australia) Pty. Ltd., 675-685 Warrigal Rd., Chadstone, Victoria, 3148, Australia (Tel: 61 3 9254 1000; FAX 61 3 9254 1010)

V.L. Clark Chemical Co., Inc.

60 Hiline Dr., PO Box 87, Union, MO, 63084, USA (Tel: 636-583-4304; 888-VLC-CHEM; FAX 636-583-5218; E-mail: sales@vlclark.com; Internet: www.vlclark.com)

Classic Flavors & Fragrances Inc.

125 E. 23rd St., Suite 400, New York, NY, 10010-4511, USA (Tel: 212-777-0004; FAX 212-353-0404)

CNC International, L.P. DEFUNCT

LAST KNOWN: POB 3000, 20 Privilege St., Woonsock, RI, 02895, USA (Tel: 401-769-6100; FAX 401-769-4509)

Coalite Chemicals

Subsid. of Anglo United plc, PO Box 152, Buttermilk Lane, Bolsover, Chesterfield, Derbyshire, S44 6AZ, UK (Tel: 44 1246 826816; FAX 44 1246 240309; E-mail: chemicals@coalitechemicals.com; Internet: www.coalitechemicals.com)

Coastal Chem Inc.

6955 S. Union Park Center, Suite 550, Midvale, UT, 84047, USA (Tel: 801-568-8000; FAX 801-568-8099)

Cognis

Cognis Deutschland GmbH, Postfach 130164, 40551 Dusseldorf, Germany (Tel: 49 211 7940 0; FAX 49 211 798 4008; Internet: www.cognis.de)

Cognis KGaA/Care Chemicals, Postfach 130164, 40551 Düsseldorf, Germany (Tel: 49 211 794 00; FAX 49 211 798 7696; E-mail: care.chemicals@cognis.de)

Cognis KGaA/Coatings & Inks, 40191 Düsseldorf, Germany (Tel: 49 211 7940 9390; FAX 49 211 798 2261; E-mail: coatings.inks@Henkel.de)

Cognis UK Ltd., Eleanor House, 3rd Fl., 33-35 Elanor Cross Rd., Waltham Cross, Hertfordshire, EN8 7LE, UK (Tel: 44 1992 765008; FAX 44 1992 765040; E-mail: chris.bawden@cognis.co.uk; Internet: www.cognis.com)

Cognis France, 185 Av. de Fontainebleau, Boite Postale 4, 77981 St.Fargeau Ponthierry, France (Tel: 33 1 60 65 21 39; FAX 33 1 60 65 21 07)

Cognis Iberica S.L., Affiliate of Cognis Corp., Poligono Industrial San Vicente, 08755 Castellbisbal (Barcelona), Spain (Tel: 34 93 7730 000; FAX 34 93 7730 150; Internet: www.es.cognis.com)

Cognis S.p.A., Via Scalabrini, 34, 22073 Fino Mornasco (CO), Italy (Tel: 39 31 884266; FAX 39 31 880360)

Cognis/Care Chemicals, 300 Brookside Ave., Ambler, PA, 19002, USA (Tel: 215-628-1000; 800-531-0815; FAX 215-628-1450)

Cognis/Chemicals Group, 5051 Estecreek Rd., Cincinnati, OH, 45232-1446, USA (Tel: 513-482-3000; 800-543-7370; FAX 513-482-5501; Internet: www.cognis-us.com)

Cognis/Chemicals Group, 300 Brookside Ave., Ambler, PA, 19002, USA (Tel: 215-628-1000; 800-654-7588; FAX 215-628-1200)

Cognis/Textile Technology, 3304 Westinghouse Blvd., Charlotte, NC, 28273, USA (Tel: 704-587-3800; 800-634-2436; FAX 704-945-8804; E-mail: jim.mcbride@cognis-us.com; Internet: www.cognis.com)

Cognis Canada Ltd., 2290 Argentia Rd., Mississauga, Ontario, L5N 6H9, Canada (Tel: 905-542-7554; 800-668-6023; FAX 905-542-7588; E-mail: alan.mcdonald@cognis.ca)

Cognis Brasil Ltda., Av. Das Nacoes Unidas, 10.989-40 Andar, Vila Olimpia, Sao Paulo SP, 04578-000, Brazil (Tel: 55 11 3848 2463; FAX 55 11 3848 2489; E-mail: Customer-Service.LS@cognis.com)

Cognis Venezolania, S.A., Urbanizacion Maturin, Centro Industrial del Este, Pirmera Ave., Manzana 3, Parcela F3, Edif. Corinta, Caracas, Venezuela (Tel: 58 2 362 3131; FAX 58 2 360 3336)

Cognis South Africa (Pty) Ltd., Potgieter & Bosworth Sts., Alrode Gauteng, 1407, Rep. of S. Africa (Tel: 27 11 864 4950; FAX 27 11 908 4029)

Cognis Australia (Pty) Ltd., Locked Bag 15, Ermington, NSW, 5115, Australia (Tel: 61 2 974 84222; FAX 61 2 974 83863; E-mail: glen.wells@cognis.com)

Cognis New Zealand Ltd., 2 Allens Rd., Greenmount, East Tamaki, Auckland, New Zealand (Tel: 64 9 274 3164; FAX 64 9 274 3172)

Cognis Japan Ltd., 14F Sphere Tower Tennoze, 2-2-8 Hinashi-Shinagawa, Shinagawa-ku, Tokyo, 140-0002, Japan (Tel: 81 3 5769 6444; FAX 81 3 5769 6438)

Cognis Korea Ltd., 13 Fl. Kumbo Electric Bldg., 418 Mapo-Dong, Mapo-ku, Seoul, 121-734, Korea (Tel: 82 2 706 9701; FAX 82 2 706 6091)

Cognis Taiwan Ltd., 16th Fl., 3 Min Shen Rd., Sec. 1, Pan Chiao, Cheng Loong Plaza, Taipei Hsien, 220, Taiwan R.O.C. (Tel: 886 2 2 958 9766; FAX 886 2 2 958 1238)

CoKEM Associates, Inc.

PO Box 1338, Mountainside, NJ, 07092-1338, USA (Tel: 908-654-5550; FAX 908-654-5552; E-mail: cokem@worldnet.att.net)

Collaborative Laboratories

Collaborative Laboratories, Inc., Subsid. of Collaborative Group, Ltd., 50 E. Loop Rd., Stony Brook, NY, 11790, USA (Tel: 631-689-0200; FAX 631-689-6880; E-mail: sales@collabo.com; Internet: www.collabo.com)

Collaborative Laboratories, Inc./Oleochemicals, Subsid. of Collaborative Group, Ltd., 50 E. Loop Rd., Stony Brook, NY, 11790, USA (Tel: 631-689-0200; FAX 631-689-0205; E-mail: sales@collabo.com; Internet: www.collabo.com)

Collinda Ltd.

Collinda House, 25 Ottways Lane, Ashstead, Surrey, KT21 2PZ, UK (Tel: 44 1372 278416; FAX 44 1372-278559; E-mail: info@collinda.co.uk; Internet: www.collinda.co.uk)

Columbia Sales International, Inc.

8640 Guilford Rd., Suite 213, Columbia, MD, 21046, USA (Tel: 410-312-7979; FAX 410-312-7421; E-mail: columbia-sales@connext.net)

Columbus Chemical Industries, Inc.

N4335 Temkin Rd., PO Box 8, Columbus, WI, 53925-0008, USA (Tel: 920-623-2140; FAX 920-623-2577)

Columbus Foods Co.

730 N. Albany Ave., Chicago, IL, 60612, USA (Tel: 773-265-6500; 800-322-6457; FAX 773-265-6985; E-mail: sales@columbusfoods.net; Internet: www.columbusfoods.com)

Concord Chemical Co. Inc.

1700 Federal St., Camden, NJ, 08105, USA (Tel: 609-966-1526; 800-282-2436; FAX 609-963-0246)

Condor Corp.

Executive Center, 560 Sylvan Ave., Englewood Cliffs, NJ, 07632-3193, USA (Tel: 201-567-3337; 800-321-3005; FAX 201-567-6489)

A&E Connock (Perfumery & Cosmetics) Ltd.

Alderholt Mill House, Fordingbridge, Hampshire, SP6 1PU, UK (Tel: 44 1425 653367; FAX 44 1425 656041; E-mail: sales@connock.co.uk; Internet: www.connock.co.uk)

Consos, Inc.

PO Box 34186, Charlotte, NC, 28234, USA (Tel: 704-596-2813; 800-849-2980; FAX 704-596-4861; E-mail: consos@consos.com; Internet: www.consos.com)

Continental Industries Group Inc. (CIG)

245 E. 58th St., 25th Fl., New York, NY, 10022, USA (Tel: 212-752-2020; FAX 212-821-0020; E-mail: cigny@cigusa.com; Internet: www.cigusa.com)

Contract Chemicals Ltd.

Penrhyn Road, Knowsley Business Park, Prescot, Merseyside, L34 9HY, UK (Tel: 44 151 548 8840; FAX 44 151 548 6548; Internet: www.contract-chemicals.com)

F.D. Copeland & Sons Ltd.

Claanol House, 5 Westfield St., Woolwich, London, SE18 5TL, UK (Tel: 44 20 8854 8101; FAX 44 20 8854 1077; E-mail: sales@copelandoil.co.uk; Internet: www.copelandoil.co.uk)

Corco Chemical Corp.

451 Tyburn Rd., Fairless Hills, PA, 19030, USA (Tel: 215-295-5006; FAX 215-295-0781)

Cornelius Chemical Co. Ltd.

Cornelius House, Woodside, Bishop's Stortford, Hertfordshire, CM23 5RG, UK (Tel: 44 1279 714300; FAX 44 1279 714320; E-mail: sales.dept@cornelius.co.uk; Internet: www.cornelius.co.uk)

Corn Products International, Inc.

Unit of CPC Int'l., 6500 S. Archer Ave., Bedford Park, IL, 60501-1933, USA (Tel: 708-563-2400; 800-538-0394; FAX 708-563-6878; E-mail: corpcomm@cornproducts.com; Internet: www.cornproducts.com)

Cosmetic Supplies USA, Inc.

PO Box 470535, Celebration, FL, 34747, USA (Tel: 407-390-9542; 877-9-COS-USA; FAX 407-390-9542; E-mail: info@cosmeticsuppliesusa.com; Internet: www.cosmeticsuppliesusa.com)

Costec, Inc.

800 N. Oakwood Rd., Lake Zurich, IL, 60047-1522, USA (Tel: 847-540-8931; 800-450-8111; FAX 800-5COSTEC; E-mail: personalcare@costec.com; Internet: www.costec.com)

Coyne Chemical Co., Inc., George S.

3015 State Rd., Croydon, PA, 19021-6997, USA (Tel: 215-785-3000; FAX 215-785-1585; Internet: www.coynechemical.com)

CPI Chemicals

Rudolfsplatz 7, 1010 Vienna, Austria (Tel: 43 535 26 12 0; FAX 43 535 26 12 12; E-mail: office@cpichem.com; Internet: www.cpichem.com)

CP Kelco

CP Kelco U.S., Inc., 311 South Wacker Dr., Suite 3700, Chicago, IL, 60606-6627, USA (Tel: 312-554-7800; 800-535-2687; 312-554-7810; Internet: www.cpkelco.com)

CP Kelco U.S., Inc., 1313 North Market St., Wilmington, DE, 19894-0001, USA (Tel: 302-594-7200; 888-FOODGUM; FAX 302-594-7201; Internet: www.cpkelco.com)

CP Kelco U.S., Inc., 8355 Aero Dr., San Diego, CA, 92123, USA (Tel: 858-292-4900; 800-535-2656; FAX 858-292-4901)

CP Kelco Mexico, World Trade Center, Av. de las Naciones No. 1, Piso 19, Oficinas 27-28 Col. Napoles, Mexico, D.F., 03810, Mexico (Tel: 52 5 488 0055; FAX 52 5 488 0899)

CP Kelco Brazil S.A., Rua Teixeira Marques 845, Chacara Sao Jose, Limeira SP, 13485-127, Brazil (Tel: 55 19 440 4600; FAX 55 19 451 1948)

CP Kelco ApS, Div. of CP Kelco Inc., Ved Banen 16, 4623 Lille Skensved, Denmark (Tel: 45 56 165 616; FAX 45 56 169 446; E-mail: solutions@cpkelco.com; Internet: www.cpkelco.com)

CP Kelco UK Limited, Cleeve Court, Cleeve Rd., Leatherhead, Surrey, KT22 7UD, UK (Tel: 44 1372 369 400; FAX 44 1372 369 401)

CP Kelco Germany GmbH, Neuer Wall 63, 20354 Hamburg, Germany (Tel: 49 40 37 49 92 0; FAX 49 40 36 57 47)

CP Kelco France SARL, Tour Neptune, 20 place de Seine, 92086 Paris La Defense Cedex, France (Tel: 33 1 49 03 78 00; FAX 33 1 49 03 78 29)

CP Kelco Belgium B.V.B.A., Industrieweg PO Box 1, 3583 Beringen, Belgium (Tel: 32 1145 8656; FAX 32 1142 0361)

CP Kelco Japan ApS, Izumi Kamiyacho Bldg., 8-1, Higashi-Azabu 1-chome, Minato-ku, Tokyo, 106-00445, Japan (Tel: 81 3 3560 7313; FAX 81 33 560 7316)

CP Kelco Singapore Pte. Ltd., 151 Lorong Chuan, #06-07 New Tech Park, Singapore, 556741, Singapore (Tel: 65 383 4440; FAX 65 383 4441)

S. Black (Import & Export) Ltd., Distrib. for CP Kelco, Foxholes Business Park, John Tate Road, Hertford, Hertfordshire, SG13 7YH, UK (Tel: 44 1992 825555; FAX 44 1992 825566; E-mail: enquires@sblack.com; Internet: www.sblack.com)

C & P Sales Inc.

4917 Clearview Dr., PO Box 781, Williamsville, NY, 14221, USA (Tel: 716-634-3061; FAX 716-631-5180; E-mail: crs3@juno.com; Internet: www.candpsales.com)

CPS Chemical Co. Inc.

Member Allied Colloids Group plc, PO Box 162, 123 White Oak Lane, Old Bridge, NJ, 08857, USA (Tel: 732-607-2700; FAX 732-607-2562)

CPS Union Corp.

PO Box 827, New City, NY, 10956, USA (Tel: 914-639-6983; FAX 914-639-9903; E-mail: cpsunion@tco.com; Internet: www.tco.com/cpsunion/)

Creative Fragrances Mfg. Inc.

10890 Alder Circle, Dallas, TX, 75238, USA (Tel: 214-341-3666; FAX 214-341-1423)

Croda

Croda International plc, Cowick Hall, Snaith Goole, N. Humberside, E. Yorkshire, DN14 9AA, UK (Tel: 44 1405 860551; FAX 44 1405 861767; E-mail: enquires@croda.co.uk; Internet: www.croda.co.uk)

Croda Chemicals Ltd., Div. of Croda International plc, Cowick Hall, Snaith Goole, Humberside, DN14 9AA, UK (Tel: 44 1405 860 551; FAX 44 1405 860 205; Internet: www.croda.co.uk)

Croda Food Services Ltd., Div. of Croda International plc, Falcon St., Oldham, Lancashire, OL8 1JU, UK (Tel: 44 161 652 6311; FAX 44 161 627 2346; Internet: www.croda-foods.co.uk)

Croda Oleochemicals, Cowick Hall, Snaith Goole, Humberside, E. Yorkshire, DN14 9AA, UK (Tel: 44 1405 860551; FAX 44 1405 860205)

Croda Universal Ltd., Div. of Croda International plc, Oak Road, off Clough Road, Hull, E. Yorkshire, HU6 7PH, UK (Tel: 44 1482 443181; FAX 44 1482 341792)

Croda/Westbrook, Div. of Croda International plc, Cowick Hall, Snaith Goole, N. Humberside, E. Yorkshire, DN14 9AA, UK (Tel: 44 1405 860551; FAX 44 1405 860205)

Croda Food Services (Ireland) Ltd, Sovereign Bldgs., Malahide Rd. Ind. Estate, Dublin, 17, Ireland (Tel: 3 53 1 847 4666; FAX 3 53 1 847 4143)

Croda GmbH, Herrenpfad Süd 33, 41334 Nettetal, Germany (Tel: 49 2157 81730; FAX 49 2157 132350; Internet: www.croda.de)

Croda Italiana SpA, Via Grocco, 917/19, 27036 Mortara (PV), Italy (Tel: 39 38 4 205011; FAX 39 38 4 91973)

Croda France SA, BP 49, 78193 Trappes Cedex, France (Tel: 33 1 30 13 94 30; FAX 33 1 30 62 91 07)

Croda Nordica AB, Tegnérgatan 74, 216 14 Limhamn, Sweden (Tel: 46 40 368550; FAX 46 40 150570)

Croda Poland Sp.z.o.o., 30-042 Krakow ul, Kujawska 13, Poland (Tel: 48 12 633 1529; FAX 48 12 634 3742)

Croda Middle East, Branch of Croda International plc, PO Box 61146, Jebel Ali Free Zone, Dubai, United Arab Emirates (Tel: 971 4 817610; FAX 971 4 817 614)

Croda Chemicals (SA) (Pty) Ltd., PO Box 1641, Kempton Park, Gauteng, 1620, Rep. of S. Africa (Tel: 27 11 397 2380; FAX 27 11 397 2442)

Croda Zimbabwe (Pvt) Ltd., PO Box ST510, Southerton Harare, Zimbabwe (Tel: 263 46 62701; FAX 263 46 62704)

Croda Inc., 7 Century Dr., Parsippany, NJ, 07054-4698, USA (Tel: 973-644-4900; FAX 973-644-9222; E-mail: marketing@croda.com; Internet: www.croda.com)

Croda Canada Ltd., 78 Tisdale Ave., Toronto, Ontario, M4A 1Y7, Canada (Tel: 416-751-3571; FAX 416-751-9611; E-mail: fkhory@croda.ca; Internet: www.croda.ca)

Croda Argentina SA, Avenida Blanco, Encalada 193, Oficina 71, 1642 San Isidro, Buenos Aires, Argentina (Tel: 54 114 763 1006; FAX 54 114 763 9555)

Croda do Brasil Ltda., Rua Croda, n°580, Cx Postal 1098, Campinas SP, 13001-970, Brazil (Tel: 55 19 225 6000; FAX 55 19 225 6001; Internet: www.croda.com.br)

Quimica Croda S.A. de C.V., Circuito Medicos 47 Apartado Postal 71-A, Cuidad Satelite, Naucalpan, Edo de Mexico, CP531000, Mexico (Tel: 52 5 393 9111; FAX 52 5 562 4297)

Croda Chemicals (India) Pvt Ltd, 6 Fort Vijay Society, 1/3 Gola Ln Off D N Rd., Mumbai, 400 001, India (Tel: 91 22 265 8248; FAX 91 22 262 2754)

Croda Oleochemicals (New Zealand), PO Box 22122, Otahuhu, Auckland, 1133, New Zealand (Tel: 64 9 573 5187; FAX 64 9 573 5188)

Croda Australia Pty. Ltd., PO Box 6585, Wetherill Park, NSW, 2164, Australia (Tel: 61 2 756 4400; FAX 61 2 756 4414)

Croda Singapore Pte. Ltd., 30 Seraya Ave., Singapore, 627884, Singapore (Tel: 65 551 9600; FAX 65 551 9550; Internet: www.croda.com.sg)

Croda Japan K.K., Aceman Bldg. 5F, 1-1-10, Tokui-cho, Chuo-ku, Osaka, 540, Japan (Tel: 81 6 6942 1791; FAX 81 6 6942 1790)

Croda Hong Kong, Branch of Croda Singapore Pte Ltd., Room 1706, Hong Kong Plaza, 186-191 Connaught Road West, Hong Kong (Tel: 852 2559 3856; FAX 852 2858 6047)

Crompton

Crompton Corp., 199 Benson Rd., Middlebury, CT, 06749, USA (Tel: 203-573-2000; 800-336-9312; Internet: www.cromptoncorp.com)

Crompton Corp./Silanes & Specialty Silicones, 318-24 Fourth Ave., Plaza IV, S. Charleston, WV, 25303-3802, USA (Tel: 800-523-5862; E-mail: osina@cromptoncorp.com)

Crompton Corp./Uniroyal Chemical Co., Inc., 199 Benson Road, PO Box 846, Middlebury, CT, 06749, USA (Tel: 203-573-2000; 800-243-3027; Internet: www.cromptoncorp.com)

Crompton Corp./Witco Refined Products, 100 Sonneborn Lane, Petrolia, PA, 16050, USA (Tel: 877-948-2688; E-mail: rpinfo@cromptoncorp.com)

Crompton Corp. (Canada), PO Box 250, 25 Erb St., Elmira, Ontario, N3B 3A3, Canada (Tel: 519-669-1671; 800-265-2157; FAX 519-669-1679)

Uniroyal Quimica Ltda., Av. Juscelino Kubitscheck, 1830, 10° Andar Torre 1, Sao Paulo - SP - CEP 04543 900, Brazil (Tel: 55 11 3896 1531)

Crompton Europe Ltd., Kennet House, 4 Langley Quay, Waterside Drive, Slough, Berkshire, SL3 6EH, UK (Tel: 44 1753 603000; FAX 44 1753 603078)

Crompton SA, 7 rue du Pre-Bouvier, 1217 Meyrin, Switzerland (Tel: 41 22 989 2111; E-mail: osa@cromptoncorp.com)

Uniroyal Chimica S.r.l., Viale delle Industrie 40, 04013 Latina-Scalo (LT), Italy (Tel: 39 773 6151; FAX 39 773 615334)

Uniroyal Chemical Taiwan Ltd., Bank Tower, Suite 1302, 205, Tun Hwa North Rd., Taipei, Taiwan R.O.C. (Tel: 886 2 2713 1790)

Uniroyal Chemical Co. (Japan), Central Kyobashi Bldg. 2 F, 1-4-13, Kyobashi, Chuo-ku, Tokyo, 104-0031, Japan (Tel: 81 3 3517 5771; FAX 81 3 3281 5701)

Cross World Sales Corp.

1891 Main St., Mohegan Lake, NY, 10547, USA (Tel: 914-528-9142; FAX 914-528-9205; E-mail: crossws@aol.com)

Crowley

Crowley Chemical Co., 261 Madison Ave., New York, NY, 10016, USA (Tel: 212-682-1200; FAX 212-953-3487; E-mail: crowleychemical@msn.com)

Crowley Tar Products Co., Inc., 261 Madison Ave., New York, NY, 10016, USA (Tel: 212-682-1200; FAX 212-953-3487; E-mail: crowleychemical@msn.com)

Crucible Chemical Co. Inc.

Donaldson Center, Bldg. 69, PO Box 6786, Greenville, SC, 29606, USA (Tel: 864-277-1284; 800-845-8873; FAX 864-299-1192)

Crystran Ltd.

27 Factor Rd., Poole, Dorset, BH16 5SL, UK (Tel: 44 1202 632833; FAX 44 1202 632832; E-mail: sales@crystran.co.uk; Internet: www.crystran.co.uk)

CTC Organics

792 Windsor St. SW, Atlanta, GA, 30315, USA (Tel: 404-524-6744; FAX 404-577-1651)

C&T Refinery LLC

7110 Forest Ave., Suite 200, Richmond, VA, 23226, USA (Tel: 804-359-5786; 800-284-6457; FAX 800-285-9170)

CU Chemie Uetikon GmbH

Raiffeisenstrasse 4, PO Box 1248, 77933 Lahr, Germany (Tel: 49 7821 585 208; FAX 49 7821 585 230; E-mail: info@uetikon.com; Internet: www.uetikon.com)

Custom Ingredients

Custom Ingredients, Inc., PO Box 772, 712 Wilson St., Chester, SC, 29706-0772, USA (Tel: 803-581-5800; 888-PHYTOLS; FAX 803-581-5802; Internet: www.chembuyersguide.com/partners/customingredients.html)

CA Specialties, Ltd., Distrib. for Custom Ingredients, 712 Wilson St., PO Box 9, Chester, SC, 29706, USA (Tel: 803-581-5800; 888-227-7324; FAX 803-581-5802; E-mail: specialities@infoave.net; Internet: www.ca-specialties.com)

CXY Chemicals

CXY Chemicals U.S.A. Inc., Wholly owned by Canadian Occidental Petroleum Ltd., 12450 Greenspoint Dr., Suite 1350, Houston, TX, 77060, USA (Tel: 281-876-2267; FAX 281-872-3798; Internet: www.cxychem.com)

CXY Chemicals Canada Limited Partnership, Wholly owned by Canadian Occidental Petroleum Ltd., 635 8th Ave. SW, Calgary, Alberta, T2P 3Z1, Canada (Tel: 403-234-6842; FAX 403-234-1091; Internet: www.cxychem.com)

Cyclo International

PO Box 438000, San Ysidro, CA, 92143, USA (Tel: 619-441-9108; FAX 619-441-2739)

Cytec Industries

Cytec Industries Inc., Five Garret Mountain Plaza, West Paterson, NJ, 07424, USA (Tel: 973-357-3100; 800-652-6013; FAX 973-357-3065; E-mail: info@gm.cytec.com; Internet: www.cytec.com)

Cytec Olean Inc., 1405 Buffalo St., Olean, NY, 14760, USA (Tel: 716-372-9650; 800-836-3666; FAX 716-372-1594; E-mail: bill_hettrick@cnp.cytec.com; Internet: www.conap.com)

Cytec do Brasil Ltda., Rua Quintana 753, 5 Andar, Brooklin Novo, Sao Paulo SP, 04569-011, Brazil (Tel: 55 11 5505 4588; FAX 55 11 5505 4565; E-mail: cytec@cytec.com.br)

Cytec Industries UK Ltd., Bowling Park Dr., Bradford, West Yorkshire, BD4 7TT, UK (Tel: 44 1274 73 3891; FAX 44 1274 73 4770; Internet: www.cytec.com)

Cytec Industries B.V., Botlekweg 175, 3197 KA Botlek-Rotterdam, The Netherlands (Tel: 31 181 295 400; FAX 31 181 295 401)

Cytec Industries Italia s.r.l., Via Sporting Mirasole, 4, 20090 Noverasco Di Opera Milano, Italy (Tel: 39 2 576 06175; FAX 39 2 576 0664)

Mitsui Cytec Ltd., Torii Nihonbashi Bldg., 6th Fl., 3-4-1 Nihonbashi-Honcho, Chuo-ku, Tokyo, 103, Japan (Tel: 81 3 3231 6072; FAX 81 3 3231 6070)

Cytec Industries Pte. Ltd. (Asia/Pacific), 350 Orchrard Rd., #17-01/02 Shaw House, Singapore, 238868, Singapore (Tel: 65 733 6788)

Cytec Taiwan Corp., 6F-1 380, Sec. 1, Fu-Hsing South Rd., Taipei, 106, Taiwan R.O.C. (Tel: 886 2 705 3969; FAX 886 2 705 3967)

Cytec Australia Ltd., Suite 1, 7-11 Railway St., PO Box 384, Baulkham Hills, NSW, 2153, Australia (Tel: 61 2 686 4188; FAX 61 2 639 4248)

Cytec Korea Inc., 7th Fl., Wooduk Bldg., #832-2 Yoksam-dong, Kangnam-ku, Seoul, Korea (Tel: 82 2 3452 7331; FAX 82 2 3452 1202)

Daicel Chemical

Daicel Chemical Industries, Ltd., Toranomon Mitsui Bldg., 3-8-1 Kasumigaseki, Chiyoda-ku, Tokyo, 100, Japan (Tel: 81 3 3507 3111; FAX 81 3 3507 3139; Internet: www.daicel.co.jp)

Daicel Chemical (Asia) Pte. Ltd., 15A Senoko Way, Singapore, 2775, Singapore (Tel: 65 756 2811)

Daicel (Hong Kong) Ltd., Suite 2014, 20F, CMG Asia Tower, The Gateway, Harbour City, Kowloon, Hong Kong (Tel: 852 2730 2129)

Daicel (Europa) GmbH, Oststr. 22, 40211 Dusseldorf, Germany (Tel: 49 211 369848)

Daicel (USA) Inc., Subsid. of Diacel Chem. Industries, Ltd., One Parker Plaza, 9th Fl., 400 Kelby St., Fort Lee, NJ, 07024, USA (Tel: 201-461-4466; FAX 201-461-2776; E-mail: funato@daicel.com)

Dai-ichi Karkaria Ltd. (DIK)

Liberty Bldg., SVT Marg, Mumbai, 400 020, India (Tel: 91 22 201 5895; FAX 91 22 209 6976; E-mail: diksales@vsnl.net; Internet: www.dikexports.com)

Dai-ichi Kogyo Seiyaku Co., Ltd. (DKS)

55, Nishi-Shichijo, Higashikubo-cho, Shimogyo-ku, Kyoto, 600-8873, Japan (Tel: 81 75 321 1550; FAX 81 75 321 1599; Internet: www.dks-web.co.jp/english/)

Daikin

Daikin Industries. Ltd., Umeda Center Bldg, 2-4-12, Nakazaki-nishi, Kita-ku, Osaka, 530-8323, Japan (Tel: 81 6 6374 9300; FAX 81 6 6373 4389; Internet: www.daikin.co.jp)

Daikin Chemical Europe GmbH, Immermannstr. 65D, 40210 Dusseldorf, Germany (Tel: 49 211 1640834; FAX 49 211 1640732)

Daikin America Inc., 20 Olympic Dr., Orangeburg, NY, 10962, USA (Tel: 845-365-9500; 800-365-9570; FAX 845-365-9598; E-mail: customerservice@daikin-america.com; Internet: www.daikin-america.com)

Dakota Gasification Co.

44 Inverness Dr. East, Bldg. E, Engelwood, CO, 80112, USA (Tel: 303-799-1531; 800-736-6868; FAX 303-799-0929; E-mail: dgcmkt@dakotagas.com; Internet: www.dakotagas.com)

Dama Chemicals

4420 Trade Center Blvd., I.T.C. Park, Lared, TX, USA (Tel: 877-329-2476; FAX 877-329-2476; E-mail: damachemicals@infosel.net.mx)

H.E. Daniel Ltd.

Belasis Ave., Billingham, Cleveland, TS23 1LQ, UK (Tel: 44 1642 379900; FAX 44 1642 379901; E-mail: hedinfo@hedaniel.com; Internet: www.hedaniel.com)

Danisco Cultor

Danisco Cultor, Ingredients Sector of Danisco A/S, Langebrogade 1, PO Box 17, 1001 Copenhagen, Denmark (Tel: 45 32 66 20 00; FAX 45 32 66 21 88; Internet: www.daniscocultor.com)

Danisco Cultor (UK) Ltd., Denington Rd., Wellingborough, Northants, NN8 2QJ, UK (Tel: 44 1933 304200; FAX 44 1933 304224; E-mail: wellingborough.ingredients@danisco.com)

Danisco Cultor Deutschland GmbH, Roberts-Bosch Strasse 20-24, 25451 Quickborn, Germany (Tel: 49 4106 70 96 0; FAX 49 4106 70 96 66; E-mail: quickborn.ingredients@danisco.com)

Danisco Cultor France S.A.R.L., Parc D'Activités de Pissaloup, 5, Ave. Jean D'Alembert, 78190 Trappes, France (Tel: 33 1 30 66 08 08; FAX 33 1 30 66 75 08)

Danisco Cultor USA, Inc., 201 New Century Pkwy., PO Box 26, New Century, KS, 66031-0026, USA (Tel: 913-764-8100; 800-255-6837; FAX 913-764-5407; E-mail: sdp1@iname.com; Internet: www.danisco.com)

Danisco Cultor Canada Ltd., 10 Carlson Court, Suite 215, Rexdale, Ontario, M9W 6L2, Canada (Tel: 416-674-7340; FAX 416-674-7378)

Danisco Cultor Mexicana, S.A. de C.V., Benjamin Hill No. 188, Col. Condesa, Mexico D.F., 06100, Mexico (Tel: 52 5 272 0774; FAX 52 5 273 7917)

Danisco Cultor Brasil Ltda., Rodovia Raposo Tavares, Km 27,2, Cotia SP, 06700-000, Brazil (Tel: 55 11 4613 3800; FAX 55 11 4612 1101; E-mail: info.brasil.ingredients@danisco.com)

Danisco Ingredients

Danisco Ingredients, Div. of Danisco A/S, Edwin Rarhs Vej 38, 8220 Brabrand, Denmark (Tel: 45 89 43 50 00; FAX 45 86 25 10 77; E-mail: info@danisco.com; Internet: www.danisco.nl)

Danisco Ingredients (UK) Ltd., Northern Way, Bury St Edmunds, Suffolk, IP32 6NP, UK (Tel: 44 1284 769631; FAX 44 1284 760839)

Dasico A/S

Molledamsvej 10, 3460 Birkerod, Denmark (Tel: 45 45 81 44 88; FAX 45 45 81 23 40; E-mail: dasico@inet.uni2.dk; Internet: www.dasico.dk)

Dastech Int'l. Inc.

10 Cutter Mill Rd., Great Neck, NY, 11021-3201, USA (Tel: 516-466-7676; FAX 516-466-7699; E-mail: dastech@worldnet.att.net)

Davos Chemical Corp.

464 Hudson Terrace, Englewood Cliffs, NJ, 07632, USA (Tel: 201-569-2200; 800-229-3285; FAX 201-569-2201; E-mail: info@davos.com; Internet: www.davos.com)

Dead Sea Bromine

Dead Sea Bromine Group, Flame Retardants Marketing Div., Member of Israel Chemicals Ltd (ICL) Group, Makleff House, 12 Kroitzer St., PO Box 180, Beer-Sheva, 84101, Israel (Tel: 972 8 629 7209; FAX 972 8 629 7848; E-mail: golanb@dsbg.com; Internet: www.deadseabromine.com)

Dead Sea Bromine Group International, London Office, 201 Haverstock Hill, London, NW3 4QG, UK (Tel: 44 20 7431 7707; FAX 44 20 7431 7797; E-mail: BromineUK@compuserve.com)

Eurobrom B.V., Verrijn Stuartlaan 1, 2288 EE Rijswijk, The Netherlands (Tel: 31 70 340 8408; FAX 31 70 399 9035; E-mail: elhananyy@eurobrom.dsbg.com)

Eurobrom B.V. Espana, Rambla Cataluna 10, 08807 Barcelona, Spain (Tel: 34 3 317 9910; FAX 34 3 317 6403; E-mail: eurobrom.spain@deinfo.es)

Dead Sea Bromine Co. Ltd. China, 14 Fl., Bldg. A, Fuhua Mansion, No. 8 North Chaoyangmen St., Dong Cheng District, Beijing, 100027, P.R. China (Tel: 86 10 6554 1972; FAX 86 10 6554 3200; E-mail: dsbbj@public.bta.net.cn)

Deepwater Chemicals, Inc.

Subsid. of Tomen America, 1210 Airpark Rd., Woodward, OK, 73801, USA (Tel: 580-256-0500; 800-854-4064; FAX 580-256-0575; Internet: www.tomenamerica.com/deepwater)

DeForest Enterprises, Inc.

Amtec Center, 6421 Congress Ave. #121, Boca Raton, FL, 33487, USA (Tel: 561-994-9696; 800-388-2301; FAX 561-994-9995; E-mail: dforest@gate.net; Internet: www.deforest.net)

Degen Oil & Chemical Co.

200 Kellogg St., PO Box 5240, Jersey City, NJ, 07305, USA (Tel: 201-432-1192; FAX 201-432-8483; E-mail: degen@eclipse.net; Internet: www.eclipse.net/~degen)

Degussa

Degussa AG, Bennigsenplatz 1, 40474 Dusseldorf, Germany (Tel: 49 211 65041 0; FAX 49 211 65041 555; Internet: www.degussa.com)

Degussa AG/Aerosil & Silanes, Weissfrauenstrasse 9, 60287 Frankfurt am Main, Germany (Tel: 49 69 218 01; FAX 49 69 218 3218; E-mail: silanes@degussa.com; Internet: www.degussa.de)

Degussa AG/Health & Nutrition, Dr.-Albert-Frank-Str. 32, 83308 Trostberg, Germany (Tel: 49 8621 86 0; Internet: www.degussa-health-nutrition.de)

Degussa AG/Marl, Marl Chemical Park, Paul-Baummanstrasse 1, 45764 Marl, Germany (Tel: 49 2365 49 02; FAX 49 2365 49 19020)

Degussa Ltd., Winterton House, Winterton Way, Macclesfield, Cheshire, SK11 0LP, UK (Tel: 44 1625 503050; FAX 44 1625 502096)

Degussa Ireland, Branch of Degussa Ltd., Unit 3A, Ballymount Cross Industrial Estate, Dublin, 24, Ireland (Tel: 353 1 460 3344; FAX 353 1 460 3450; E-mail: moeve.reidy@degussa-heuls.co.uk)

Degussa Benelux BV, Postbus 326, Rokin 81-83, 1000 AH Amsterdam, The Netherlands (Tel: 31 20 626 2050; FAX 31 20 627 3835)

Degussa (Switzerland) AG, Postfach 2160, 6342 Baar, Switzerland (Tel: 41 41 768 3800; FAX 41 41 760 1360)

Degussa Norden AB, PO Box 50559, 20215 Malmo, Sweden (Tel: 46 40 288 280; FAX 46 40 187 277)

Degussa France SNC, 37-39 Ave. Marceau, B.P. 317, 92402 Courbevoie, France (Tel: 33 1 43 34 72 00; FAX 33 1 43 34 73 00; E-mail: klaus.hermann@degussa.com; Internet: www.degussa.fr)

Degussa CEE GmbH, Liesinger Flurgasse 2, Postfach 162, 1235 Vienna, Austria (Tel: 43 1 8 63 91 0; FAX 43 1 8 63 91 70 29; E-mail: office@degussa.at)

Degussa Antwerp NV, Tijsmanstunnel West, 2040 Antwerp 4, Belgium (Tel: 32 35603 211; FAX 32 35603 311)

Degussa Benelux N.V/S.A., Ave. Louise 489, Bote 23, 1050 Brussels, Belgium (Tel: 32 2 6270030; FAX 32 2 6480745; E-mail: info.brussels@degussa.com)

Degussa Italia S.p.A., Via Carlo Pisacane 7/B, 20016 Pero (MI), Italy (Tel: 39 2 339 261; FAX 39 2 339 26 274; E-mail: pero@degussa.it)

Degussa Iberia S.A., Paseo San Juan, 75, 8°, 08009 Barcelona, Spain (Tel: 34 93 4 76 16 00; FAX 34 93 4 59 30 26; E-mail: general.general@degussa.com)

Degussa Corp., 379 Interpace Pkwy., Bldg. C, PO Box 677, Parsippany, NJ, 07054-0677, USA (Tel: 973-541-8000; FAX 973-541-8013; E-mail: rick.rosner@degussa.com; Internet: www.degussa-huls.com)

Degussa Corp./Mobile, 4201 Degussa Rd., Theodore, AL, 36582, USA (Tel: 251-443-4000; FAX 251-443-1620; E-mail: wolfgang.buder@degussa.com)

Degussa/Aerosil & Silanes, 379 Interpace Pkwy., PO Box 677, Parsippany, NJ, 07054, USA (Tel: 973-541-8000; 877-748-3686; FAX 973-541-8013; E-mail: aerosil@degussa.com; silanes@degussa.com; Internet: www.aerosil.com; www.sivento-silanes.com)

Degussa Canada Inc., 4261 Mainway Dr., PO Box 5097, Burlington, Ontario, L7R 3Y8, Canada (Tel: 514-337-2421; FAX 514-337-0057)

Degussa Mexico S.A. de C.V., Calzada Mexico-Xochimilco No. 5149, Col. Arenal, Del. Tlalpan, Apartado Postal 22-252, Mexico D.F., 14610, Mexico (Tel: 52 5 483 1000; FAX 52 5 673 1016; E-mail: christel.saupe@degussa.com)

Degussa Brasil Ltda., Alameda Compinas 579-3 13 andar, Sao Paulo / SP, 01404-000, Brazil (Tel: 55 11 3146 4100; FAX 55 11 3146 4031; Internet: www.degussa.com.br)

Degussa Argentina S.A., Murguiondo 3176, 1439 Buenos Aires, Argentina (Tel: 54 11 46 38 33 55; FAX 54 11 46 38 13 72; E-mail: info@degussa.com.ar; Internet: www.degussa.com.ar)

Degussa Southern Africa (Pty) Ltd., PO Box 3404, Halfway House 1685, Midrand, Rep. of S. Africa (Tel: 27 11 315 6880; FAX 27 11 315 6870)

Degussa India Pvt. Ltd., 12 A/A (13th Fl.), Bhaktawar, 229, Nariman Point, Mumbai, 400 021, India (Tel: 91 22 230 7070; FAX 91 22 230 7071)

Degussa Australia Pty. Ltd., 30, Commercial Dr., PO Box 996, Dandenong, Victoria, 3175, Australia (Tel: 61 3 9703 8888; FAX 61 3 9703 8890)

Degussa China Ltd., 13. F./Manulife Tower, 169, Electric Rd., G.P.O. Box 10005, Northpoint, Hong Kong (Tel: 852 2510 2510; FAX 852 2510 7454)

Degussa Pacific Ltd., Unit 1029, Landmark Tower, 8 N. Dongsanhuan Rd., Chao Yang Distr., Beijing, 100004, P.R. China (Tel: 86 10 65 90 09 65; FAX 86 10 65 90 09 67)

Degussa Japan Co., Ltd., Shinjuku Monolith 12F, 2-3-1, Nishi-shinjuku, Shinjuku-ku, Tokyo, 163-0938, Japan (Tel: 81 3 5323 7300; FAX 81 3 5323 7399)

Degussa Korea Limited, PO Box 66 Buk-Inchon Postoffice, Incheon, 403 600, Korea (Tel: 82 32 510 24 00; FAX 82 32 505 25 10)

Degussa (SEA) Pte. Ltd., 16 International Business Park, #03-08 Mannesmann Centre, Singapore, 609929, Singapore (Tel: 65 899 0080; FAX 65 899 0090)

Degussa Taiwan Ltd., Artist Contstruction Bldg., 9F, No. 133, Min Sheng East Rd. Sec. 3, Taipei, Taiwan R.O.C. (Tel: 886 2 2717 1242; FAX 886 2 2717 2106)

Degussa Flavors & Fruit Systems

Degussa Flavors & Fruit Systems, Member of Degussa Health & Nutrition Div., Z.I. du Plan, BP 82067, 06131 Grasse Cedex, France (Tel: 33 4 93 09 78 00; FAX 33 4 93 09 78 01; Internet: www. flavors-fruit-systems.com)

Degussa Flavors & Fruit Systems Inc./Alfrebro, Inc., 1055 Reed Rd., Monroe, OH, 45050, USA (Tel: 513-539-7373; 800-669-9263; FAX 513-539-7372)

Jan Dekker

Jan Dekker International, PO Box 10, 1520 AA Wormerveer, The Netherlands (Tel: 31 75 647 9999; FAX 31 75 640 3830; E-mail: jandekkernl@jandekker.com; Internet: www.jandekker.com)

Jan Dekker (UK) Ltd., Suite 3, Hillbrow House, Linden Dr., Liss, Hampshire, GU33 7RJ, UK (Tel: 44 1730 895511; FAX 44 1730 89 5111; E-mail: sales@jan-dekker.co.uk)

Jan Dekker France S.A.R.L., 20 Quater, rue Schnapper, B.P. 5205, 78175 St. Germain en Laye Cedex, France (Tel: 33 1 39 04 10 00; FAX 33 1 39 04 10 20; E-mail: france.jan.dekker@wanadoo.fr)

Jan Dekker (Deutschland) GmbH, Hans-Böckler-Strasse 21, 40764 Langenfield, Germany (Tel: 49 2173 90 41 10; FAX 49 2173 81 90 0; E-mail: jan dekker@debitel.net)

DelCon Inc.

108 W. Franklin Ave., Suite I-6A, Pennington, NJ, 08534, USA (Tel: 609-818-1090; FAX 609-818-1117; E-mail: info@seabrook-usa.com; Internet: www.seabrook-usa.com)

Delta Distributors, Inc.

A member of the Brenntag group, 610 Fisher Rd., Longview, TX, 75604, USA (Tel: 903-759-7151; FAX 903-759-7548; E-mail: rpoole@deltadist.com; Internet: www.deltadist.com)

Delta Resins & Refractories Inc.

17350 Ryan Rd., Detroit, MI, 48212, USA (Tel: 313-368-7000)

Deltech Corp.

PO Box 97875, 11911 Scenic Hwy., Baton Rouge, LA, 70874-7875, USA (Tel: 225-775-0150; 800-535-9945; FAX 504-358-3149; E-mail: melduhon@deltechcorp.com; Internet: www.deltechcorp.com)

Desert Whale Jojoba Co. Inc.

PO Box 41594, 2101 E. Beverly St., Tucson, AZ, 85717-1594, USA (Tel: 520-882-4195; FAX 520-882-7821; E-mail: soraya@desertwhale.com; Internet: www.desertwhale.com)

The Dial Corp.

15501 N. Dial Blvd., Scottsdale, AZ, 85260-1619, USA (Tel: 480-754-3425; 800-323-5385; Internet: www.dialcorp.com)

Diaz Chemical Corp.

PO Box 194, Holley, NY, 14470, USA (Tel: 716-638-6321; FAX 716-638-8356)

Dishman USA

550 Union Ave., Suite 9, Middlesex, NJ, 08846, USA (Tel: 732-560-4300; FAX 732-560-4343; E-mail: dishmanusa@att.net; Internet: www.dishman-usa.com)

Dixie Chemical Co., Inc.

PO Box 130410, 300 Jackson Hill, Houston, TX, 77219-0410, USA (Tel: 713-863-1947; FAX 713-863-8316; E-mail: mail@dixiechemical.com; Internet: www.dixiechemical.com)

D&O Chemicals, Inc.

401 South Van Brunt St., Englewood, NJ, 07631, USA (Tel: 201-871-8500; 800-722-3686; FAX 201-871-8505; E-mail: info@dochem.com; Internet: www.dochem.com)

Doingcom Chemical Co., Ltd.

9E, Bonded Market Bldg., Siamen, 361006, China (Tel: 86 592 5626669; FAX 86 592 5626659; E-mail: market@doingcom.com; Internet: www.doingcom.com)

Dover Chemical Corp.

Subsid. of ICC Industries Inc., 3676 Davis Rd. N.W., PO Box 40, Dover, OH, 44622-0040, USA (Tel: 330-343-7711; 800-321-8805; FAX 330-364-9626; E-mail: jmacneil@doverchem.com; Internet: www.doverchem.com)

Dow

Dow Chemical North America, 2030 Willard H. Dow Center, Midland, MI, 48642, USA (Tel: 989-636-1000; 800-441-4DOW; FAX 989-636-3518; Internet: www.dow.com)

Dow Plastics, Business Group of The Dow Chemical Co., Dow Ashman Center, 4520 Ashman St., PO Box 1206, Midland, MI, 48642, USA (Tel: 989-832-1426; 800-441-4DOW; FAX 989-832-1465; E-mail: dowcig@dow.com; Internet: www.dow.com)

Dow Chemical Canada Inc., 250 6th Ave. S.W., Suite 2200, Calgary, Alberta, T2P 3H7, Canada (Tel: 403-267-3500; 800-447-4369; FAX 403-267-3597)

Dow Quimica Mexicana, S.A. de C.V., Av. de las Palmas No. 405 - Mezzanine, Torre Optima, Lomas de Chapultepec, Mexico D.F., 11000, Mexico (Tel: 52 5 201 4700; FAX 52 5 201 4723)

Dow Europe S.A., Bachtobelstrasse 3, 8810 Horgen, Switzerland (Tel: 41 1 728 2111; FAX 41 1 728 2935)

Dow Chemical Co. Ltd., Div. of The Dow Chemical Co., No. 2 Heathrow Boulevard, 284 Bath Road, West Drayton, Middlesex, UB7 0DQ, UK (Tel: 44 181 917 5000; FAX 44 181 917 5400)

Dow Benelux NV, Prins Boudewijnlaan 41, 2650 Edegem, Belgium (Tel: 32 34 50 2011; FAX 32 34 50 2913)

Dow Chemical Ibérica SA, PO Box 36208, c/Cerro del Castanar 72B, Plantas 4a y 5a, 28034 Madrid, Spain (Tel: 34 91 740 7700; FAX 34 91 740 7701)

Dow Deutschland Inc., Am Kronberger Hang 4, 65824 Schwalbach/Ts., Germany (Tel: 49 6196 566 0; FAX 49 6196 566 444)

Dow France S.A., 21, rue Saint Denis, B.P. 110, 92106 Boulogne Billancourt Cedex, France (Tel: 33 1 49 09 78 78; FAX 33 1 49 09 06 19)

Dow Italia SpA, Palazzo Smeraldo, Via Patroclo, 21, 20151 Milan (MI), Italy (Tel: 39 2 4822 1; FAX 39 2 4822 4366)

Dow Sverige AB, Kungsgatan 38, Box 7038, 103 86 Stockholm, Sweden (Tel: 46 8 676 57 00; FAX 46 8 676 57 80)

Dow Chemical (Australia) Ltd., 541-583 Kororoit Creek Rd., Altona, Victoria, 3018, Australia (Tel: 61 3 9226 3500; FAX 61 3 9226 3681)

Dow Chemical Japan Ltd./Dow Polyurethanes Japan Ltd., Tennoz Central Tower, 2-2-24, Higashi-shinagawa, Shinagawa-ku, Tokyo, 140-8617, Japan (Tel: 81 3 5460 2100; FAX 81 3 5460 2231)

Dow Chemical Pacific Ltd., 47/F Sun Hung Kai Centre, 30 Harbour Rd., Wanchai, Hong Kong (Tel: 852 2879 7333; FAX 852 2827 6019)

Dow Chemical Taiwan Ltd., Rm 501, 5th Fl., Bank Tower, 205, Tun Hua North Rd., Taipei, 10592, Taiwan R.O.C. (Tel: 886 2 271 53385; FAX 886 2 271 74119)

Dow Chemical Pacific (Singapore) Pte. Ltd., 260 Orchard Rd., #18-01 The Heeren, Singapore, 238855, Singapore (Tel: 65 835 3773; FAX 65 834 0315)

S. Black (Import & Export) Ltd., Distrib. for Dow, Foxholes Business Park, John Tate Road, Hertford, Hertfordshire, SG13 7YH, UK (Tel: 44 1992 825555; FAX 44 1992 825566; E-mail: enquires@sblack.com; Internet: www.sblack.com)

Dow Corning

Dow Corning Corp./Corporate Headquarters, 2200 W. Salsburg Rd., Auburn, MI, 48611, USA (Tel: 989-496-4000; 800-346-9882; FAX 989-496-4393; E-mail: usdccnn3@ibmmail.com; Internet: www.dowcorning.com/)

Dow Corning do Brasil Ltda., Rodovia Campinas-Monte Mor, Km 9, Hortolandia SP, Brazil (Tel: 55 19 3887 9600; FAX 55 19 3887 1681)

Dow Corning S.A., Parc Industriel, Zone C, 7180 Seneffe, Belgium (Tel: 32 64 88 80 00; FAX 32 64 88 82 72)

Dow Corning Ltd., Div. of Dow Corning Corp., Cardiff Rd., Barry, Vale of Glamorgan, CF63 2YL, Wales UK (Tel: 44 1446 732350; FAX 44 1446 747944)

Dow Corning GmbH, Postfach 13 03 32, Rheingaustrasse 34, 65201 Wiesbaden, Germany (Tel: 49 611 23 71; FAX 49 611 237 620)

Dow Corning France S.A., Div. of Dow Corning Corp., BP 203, European Health Care Center, 300 route des Cretes, 06904 Sophio Antipoles Cedex, France (Tel: 33 92 94 40 00; FAX 33 78 62 78 98)

Dow Corning SpA, Via Lombardia, 31/33, 20098 Sesto, Ulteriano (MI), Italy (Tel: 39 2 98 832 1; FAX 39 2 98 807 10)

Dow Corning Spain, Pol. Ind. Riera de Caldes, C/La Jorja 38, 08184 Palau de Plagamans (Barcelona), Spain (Tel: 34 93 864 0055; FAX 34 93 864 0059)

Dow Corning Asia Ltd., 507-1 Kishi, Yamakita-machi, Ashigarakami-gun, Kanagawa, 258-0112, Japan (Tel: 81 465 76 3108; FAX 81 465 76 3308)

Dow Corning (Shanghai) Co. Ltd., 448 Eastern Ave., Songjiang Industrial Zone, Shanghai, 201613, P.R. China (Tel: 86 21 5774 1161; FAX 86 21 5774 1162)

Dow Corning Korea Ltd., 690-1 Kwanghaewon-Ri, Kwanghyewon-Myon, Jin Chon Kun, Chungchongbuk-do, 365-830, Korea (Tel: 82 43 539 1114; FAX 82 43 535 0131)

S. Black (Import & Export) Ltd., Distrib. for Dow Corning, Foxholes Business Park, John Tate Road, Hertford, Hertfordshire, SG13 7YH, UK (Tel: 44 1992 825555; FAX 44 1992 825566; E-mail: enquires@sblack.com; Internet: www.sblack.com)

DPC Industries, Inc.

300 Jackson Hill, Houston, TX, 77007, USA (Tel: 281-457-4835)

Dragoco

Dragoco Gerberding & Co. AG, 37601 Holzminden, Germany (Tel: 49 5531 970; FAX 49 5531 971391)

Dragoco (Great Britain) Ltd., Unit 4 Alpha Way, Thorpe Industrial Park, Thorpe, Surrey, TW20 8RZ, UK (Tel: 44 1784 222 500; FAX 44 1784 222 556)

Dragoco, Fabrik für Riech- und Geschmackstoffe, GmbH, Mellergasse 4, 1235 Vienna, Austria (Tel: 43 1 86682; FAX 43 1 8691225)

Dragoco S.A.R.L., 51, ave. Flachat, 92600 Asnières, France (Tel: 33 1 41 11 75 00; FAX 33 1 40 86 99 32)

Dragoco Benelux B.V., Stationsplein 3 c, Postbus 156, 5240 AD Rosmalen, The Netherlands (Tel: 31 73 5235151; FAX 31 73 5218443)

Dragoco S.A., Calle Dr. Pedro i Pons, 9-11, Planta 13a, 08034 Barcelona, Spain (Tel: 34 93 252 1700; FAX 34 93 205 5355)

Dragoco Italia S.p.A., Piazza Duomo, 22, (Galleria Pattari, 2), 20122 Milan, Italy (Tel: 39 2 8067 121; FAX 39 2 8067 1280)

Dragoco AG, Industriestrasse 19, 8304 Wallisellen, Switzerland (Tel: 41 1 8779030; FAX 41 1 8779040)

Dragoco Inc., 10 Gordon Dr., Totowa, NJ, 07512, USA (Tel: 973-256-3850; FAX 973-256-6420; Internet: www.dragoco.com)

Dragoco Perfumes e Aromas Ltda., Rua Tavares Carbral, 102- 8 andar, 05423-030 Pinheiros, Sao Paulo SP, Brazil (Tel: 55 11 30 94 2828; FAX 55 11 30 94 2836)

Dragoco S.A., Calz, Union No. 7, Compl. Industrial Cuamatla, Cuautitlan-Izcalli, Edo. de Mexico, Mexico (Tel: 52 5 873 1476; FAX 52 5 873 1062)

Dragoco Argentina S.A., Barsil 2278, 1667 Tortuguitas, Buenos Aires, Argentina (Tel: 54 2320 494001; FAX 54 2320 499658)

Dragoco South Africa (Pty.) Ltd., 144 Boeing Rd. East, Elma Park Ext. 11, Edenvale, 1610, Rep. of S. Africa (Tel: 27 11 453 7760; FAX 27 11 453 1164)

Dragoco Australia Pty. Ltd., PO Box 643, 168 South Creek Rd., Dee Why West, Sydney, NSW, 2099, Australia (Tel: 61 2 9982 7800; FAX 61 2 9981 2536)

Dragoco (Far East) Ltd., 2 B Eton Bldg., 288 Dee Voeux Rd., Central, Hong Kong (Tel: 852 2 854 4173; FAX 852 2 542 0792)

Dragoco (Shanghai) Ltd., No. 356 Guoshoujing Rd., Zhangjiang High-Tech Park, Pudong New Area, Shanghai, 201203, P.R. China (Tel: 86 21 5080 0789; FAX 86 21 5080 0906)

Dragoco Japan Ltd., Root Takanawa Bldg. 3F, 1-5-21, Takanawa, Minato-ku, Tokyo, 108-0074, Japan (Tel: 81 3 3280 6601; FAX 81 3 3280 6606)

Dragoco India Ltd., 52, Dev. Plots for Electric, Seevaram, Perungudi, Chennai, 600 096, India (Tel: 91 44 4960186; FAX 91 44 4960773)

Dragoco (Thailand) Ltd., 350/67 Rama III Rd., Chongnonsee, Yannawa, Bangkok, 10120, Thailand (Tel: 66 285 3682; FAX 66 285 3683)

Dragoco (Malaysia) SDN.BHD., 11, Ground Fl, Jalan SS 25/24, Taman Mayang, 47301, Petaling Jaya, Selangor, Kuala Lumpur, Malaysia (Tel: 60 3 7030950; FAX 60 3 7030955)

Dragoco (Philippines) Inc., PO Box 7577, Domestic Airport PO, Lock Box, 1300 Domestic Rd., Pasay City, Metro Manila, Philippines (Tel: 63 2 7265961; FAX 63 2 7247175)

Dragoco Asia Pacific Pte. Ltd., 226 Pandan Loop, Singapore, 128412, Singapore (Tel: 65 779 4551; FAX 65 779 5843)

Drexel Chemical Co.

PO Box 13327, Memphis, TN, 38113-0327, USA (Tel: 901-774-4370; FAX 901-774-4666; E-mail: DrexChem@bellsouth.net; Internet: www.DrexChem.com)

DSM

DSM N.V., Postbus 6500, 6401 JH Heerlen, The Netherlands (Tel: 31 45 578 32 73; FAX 31 45 578 33 11; Internet: www.dsm.nl)

DSM United Kingdom Ltd., Div. of DSM N.V., DSM House, Papermill Dr., Redditch, Worcestershire, B98 8QJ, UK (Tel: 44 1527 590590; FAX 44 1527 590555; E-mail: contact.dsm-uk@dsm-group.com)

DSM France S.A., Tour Atlantique, 9, Place de la Pyramide, La Défense 9, 92911 Paris La Défense Cedex, France (Tel: 33 1 41 25 05 05; FAX 33 1 47 76 01 00)

DSM Deutschland GmbH & Co., Postfach 300239, Tersteegenstrasse 77, 40402 Düsseldorf (Golzheim), Germany (Tel: 49 211 45 57 600; FAX 49 211 45 57 999)

DSM España S.A., Edificio Euro-3, C/ Frederic Monpou 5, Planta 7a Puerta 3a y 4a, 08960 Sant just Desvern, Barcelona, Spain (Tel: 34 93 470 30 30; FAX 34 93 473 63 73)

DSM Italia s.r.l., Via Silvio Pellico 12, PO Box 321, 22100 Como, Italy (Tel: 39 31 236 700; FAX 39 31 510 716)

DSM Belgium N.V., Bus 4, Woluwellan 140 B, 1831 Diegem, Belgium (Tel: 32 2 712 05 30; FAX 32 2 712 05 55)

DSM South America, Rua Dr. Renato Paes de Barros 717, 9th Fl., 93/94, 04530-001 Itaim Bibi, Sao Paulo SP, Brazil (Tel: 55 11 30 46 33 50; FAX 55 11 30 46 33 52)

DSM Japan K.K., 3rd Fl., Kokuryu Shiba Koen Bldg., 2-6-15, Shiba Koen, Minato-ku, Tokyo, 105-0011, Japan (Tel: 81 3 34 37 76 70; FAX 81 3 34 37 76 80)

DSM China, 10A China Overseas Bldg., No. 25, Chongqing Zhong Rd., 200020 Shanghai, P.R. China (Tel: 86 21 63 86 30 80; FAX 86 21 63 86 29 28)

DSM Taipei Ltd. Corp., 3rd Fl., 90 Chung Shan North Rd., Section 6, Taipei, Taiwan R.O.C. (Tel: 886 22 833 00 01; FAX 886 22 831 30 95; Internet: www.dsmtaipei.com)

DSM Andeno BV, Norderpoort 9, 5900 AB Venlo, The Netherlands (Tel: 31 773 899 555; FAX 31 773 899 300)

DSM Chemicals North America, Inc., 1 Columbia Nitrogen Rd., PO Box 2451, Augusta, GA, 30903, USA (Tel: 706-849-6600; 800-825-4376; FAX 706-849-6999; Internet: www.dsmna.com)

DSM Fine Chemicals Netherlands, Noorderpoort 9, PO Box 81, 5900 AB Venlo, The Netherlands (Tel: 31 77 389 95 55; FAX 31 77 389 93 00)

DSM Fine Chemicals Austria, St.-Peter-Strasse 25, Postfach 933, 4021 Linz, Austria (Tel: 43 70-6916-3619; FAX 43-70-6916-63619; E-mail: guenter.stadlmair @dsm.at)

DSM Fine Chemicals, Inc., Park 80 West, Plaza Two, Saddle Brook, NJ, 07663-5817, USA (Tel: 201-845-4404; FAX 201-845-4406)

DSM Food Specialties, PO Box 1, 2600 MA Delft, The Netherlands (Tel: 31 15 279 3474; FAX 31 15 279 3540; E-mail: jos.haastrecht-van@dsm.com)

Dujodwala Resins & Terpenes Ltd.

812/813 Tulsiani Chambers, 212 Nariman Point, Mumbai, 400 021, India (Tel: 282 4089; FAX 91-22-284 1284; E-mail: power@dujodwala.com; Internet: www.dujodwala.com)

DuPont

DuPont Chemicals, 1007 Market St., Wilmington, DE, 19898, USA (Tel: 302-774-1000; 800-441-7515; FAX 302-774-7321; E-mail: info@dupont.com; Internet: www.dupont.com)

DuPont Co./Fluoroproducts, 1007 Market St., Wilmington, DE, 19898, USA (Tel: 302-774-1000; 800-441-7515; Internet: www.dupont.com/dymel)

DuPont Nylon Intermediates & Specialties, PO Box 1089, Orange, TX, 77631-1089, USA (Tel: 409-882-3720; 800-231-0998; FAX 409-886-9100; E-mail: NISP@usa.dupont.com; Internet: www.dupont.com/intermediates)

DuPont Canada Inc., PO Box 2200, Streetsville Postal Station, Mississauga, Ontario, L5M 2H3, Canada (Tel: 905-821-5953; 800-268-3943; FAX 905-821-5230; Internet: www.dupont.ca)

DuPont S.A. de C.V., Homero 206 Anexo Planta Alta, Col. Chapultepec Morales, Mexico D.F., 11570, Mexico (Tel: 52 5 722 1150; FAX 52 5 722 1149; E-mail: gonzalt@a1.mxvax.umc.dupont.com; Internet: www.dupont.com.mx)

DuPont Argentina S.A., Av. Eduardo Madero 1020 3°, Capital Federal, Buenos Aires, Argentina (Tel: 54 11 4319 4287; FAX 54 11 4319 4287; E-mail: info.argentina@arg.dupont.com; Internet: www.dupont.com.ar)

DuPont do Brasil S.A., Alameda Itapecuru, 506, Alphaville, Barueri, Sao Paulo, 06454-080, Brazil (Tel: 55 11 7266 8542; FAX 55 11 7266 8720; E-mail: produtos.brasil@bra.dupont.com; Internet: www.dupont.com.br)

DuPont de Nemours International S.A., 2, Chemin du Pavillon, PO Box 50, 1218 Le Grand-Saconnex Geneva, Switzerland (Tel: 41 22 717 5111; FAX 41 22 717 5500; Internet: www.dupont.com/enggpolymers/europe)

DuPont (UK) Ltd., Div. of E.I. Du Pont de Nemours & Co., Wedgewood Way, Stevenage, Hertsfordshire, SG1 4QN, UK (Tel: 44 1438 734000; FAX 44 1438 734836)

DuPont de Nemours (France) SA, 137 rue de L'Université, 75334 Paris Cedex 07, France (Tel: 33 1 45 50 65 50; FAX 33 1 47 53 09 67)

DuPont de Nemours (Deutschland) GmbH, DuPont Strasse 1, 61343 Bad Hamburg, Germany (Tel: 49 6172 870; FAX 49 6172 872 701)

DuPont de Nemours (Belgium) SA, Antoon Spinoystraat 6, Industriezone, 2800 Mechelen, Belgium (Tel: 32 15 44 1411; FAX 32 15 44 1409)

DuPont de Nemours Italiana SpA, Via Volta 16, 20063 Cologno Monzese, Italy (Tel: 39 2 25 30 21; FAX 39 2 25 30 23 06)

DuPont (Australia) Ltd., PO Box 930, North Sydney, NSW, 2060, Australia (Tel: 61 2 923 6165)

DuPont Kabushiki Kaisha, Arco Tower, 8-1, Shimomeguor 1-chome, Meguro-ku, Tokyo, 153-0064, Japan (Tel: 81 3 5434 6992; FAX 81 3 5434 6980)

DuPont Mitsui Fluorochemicals Co., Ltd., Chiyoda Honsha Bldg., 1-5-18 Saragaku-cho, Chiyoda-ku, Tokyo, 101, Japan (Tel: 81 3 5281 5805; FAX 81 3 5281 5881)

DuPont Asia Pacific Ltd., PO box TST 99851, Tsim Sha Tsui, Kowloon, Hong Kong (Tel: 852 734 5345)

DuPont China Holding Co. Ltd., Unit 1101, 11th Fl., China World Tower 2, No. 1 Jian Guo Men Wai Ave., Beijing, 100004, P.R. China (Tel: 86 10 6505 8000; FAX 86 10 6505 8008)

DuPont Singapore Pte., Ltd., 1 Maritime Square #07-01, World Trade Center, Singapore, 0409, Singapore (Tel: 65 273 2244; FAX 65 272 7494)

DuPont Taiwan Ltd., PO Box 81, Tapei, 777, Taiwan R.O.C. (Tel: 886 2 514 4400)

Duso Chemical Co., Inc.

173 Smith St., PO Box 665, Poughkeepsie, NY, 12602-0665, USA (Tel: 845-454-6500; FAX 845-454-0188)

Dynaloy, Inc.

PO Box 33609, 1535 E. Naomi St., Indianapolis, IN, 46203, USA (Tel: 317-788-5694; 800-669-5709; FAX 800-671-9583; E-mail: info@dynaloy.com; Internet: www.dynaloy.com)

Dyno Specialty Polymers AS

Box 160, Lillestrom, 2001, Norway (Tel: 47 63 89 71 00; FAX 47 63 89 74 72)

Eastech Chemical Inc.

A member of the Brenntag group, 5700 Tacony St., Philadelphia, PA, 19135, USA (Tel: 215-537-1000; 800-423-7423; FAX 215-537-8575; E-mail: mail@eastechchemical.com; Internet: www.eastechchemical.com)

Eastern Chemical

Subsid. of United Guardian, PO Box 18050, 230 Marcus Blvd., Hauppauge, NY, 11788, USA (Tel: 516-273-0900; 800-645-5566; FAX 516-273-0858; E-mail: evp@u-g.com; Internet: www.u-g.com)

Eastern Color & Chemical Co.

35 Livingston St., PO Box 6161, Providence, RI, 02904, USA (Tel: 401-331-9000; FAX 401-331-2155; E-mail: tech@ecco-lenox.com; Internet: www.ecco-lenox.com)

Eastman

Eastman Chemical Co., 100 North Eastman Rd., PO Box 511, Kingsport, TN, 37662-5075, USA (Tel: 423-229-2000; 800-EASTMAN; FAX 423-229-1673; E-mail: gquillan@eastman.com; Internet: www.eastman.com)

Eastman Chemical Mexicana, S.A. de C.V., Insurgentes Sur No. 1605, Piso 23, Col. San Jose Insurgentes, Mexico D.F., 03900, Mexico (Tel: 52 5 662 9962; FAX 52 5 662 9953)

Eastman Chemical (UK) Limited, Brindley House, Corner Hall, Lawn Lane, Hemel Hempstead, Hertsfordshire, HP3 9YT, UK (Tel: 44 1442 241171; FAX 44 144 224 1177)

Eastman Chemical SA, 65, rue de Bercy, 75012 Paris, France (Tel: 33 1 44 67 88 88; FAX 33 1 44 67 88 99)

Eastman Chemical, Europe, Middle East, & Africa Ltd., Tobias Asserlaan 5, 2517 KC The Hague, The Netherlands (Tel: 31 70 370 1711; FAX 31 70 370 1704)

Eastman Chemical B.V., PO Box 3263, 6300 Zug 3, Switzerland (Tel: 41 41 726 61 00; FAX 41 41 726 62 00)

Eastman Chemical B.V., Charlottenstrasse 61, 51149 Koln, Germany (Tel: 49 2203 1705 24; FAX 49 2203 1705 0)

Eastman Chemical B.V. Filiale Italiana, Via Rosellini 12, 20124 Milan, Italy (Tel: 39 2 6991 151; FAX 39 2 6991 1532)

Eastman Chemical España, S.A., Poligono Industrial Guadarranque, manzana 3, 11369 San Roque, Cadiz, Spain (Tel: 34 956 585 200; FAX 34 956 585 222)

Eastman Chemical Ltd., Level 9, 123 Epping Rd., North Ryde, Sydney, NSW, 2113, Australia (Tel: 61 2 8875 7933; FAX 61 2 8875 7934)

E-Chemicals

505 E. Huron St., Suite 208, Ann Arbor, MI, 48104, USA (Tel: 732-214-0550; FAX 732-214-0558; Internet: www.e-chemicals.com)

E&E Ltd.

A Royal Vopak Co., Lakeside, Cheadle Royal Business Park, Cheadle, Cheshire, SK8 3GR, UK (Tel: 44 161 741 7000; FAX 44 161 741 7047; E-mail: cheadle@univareurope.com; Internet: www.univar.co.uk)

E + E (US) Inc.

E. Emaus St., PO Box 280, Middletown, PA, 17057, USA (Tel: 717-948-5389)

Efkay Chemicals Ltd.

204 Banderway House, 156-162 Kilburn High Rd, London, NW6 4JD, UK (Tel: 44 20 7625 4445; FAX 44 20 7328 9101; E-mail: ekfachem@aol.com)

Eggar & Co. (Chemicals) Ltd.

High St., Theale, Reading, Berkshire, RG7 5AR, UK (Tel: 44 118 930 2379; FAX 44 118 932 3224; E-mail: info@eggar.co.uk; Internet: www.eggar.co.uk)

Dr. Ehrenstorfer GmbH

Bgm.-Schlosser-Str. 6A, 86199 Augsburg, Germany (Tel: 49 821 90 60 80; FAX 49 821 90 60 888; E-mail: info@analytical-standards.com; Internet: www.analytical-standards.com)

Eigenmann & Veronelli SpA

Via della Mosa 6, 20017 Rho (MI), Italy (Tel: 39 2 93539 1; FAX 39 2 93539 300; E-mail: info@eigver.it; Internet: www.eigver.it)

Eka Chemicals

Eka Chemicals Inc., Div. of Akzo Nobel, 1775 West Oak Commons Court, Marietta, GA, 30062, USA (Tel: 770-578-0858; 800-241-3900; FAX 770-321-5865; Internet: www.ekachem.com)

Eka Chemicals Inc./Int'l. Resins Group, 12 Beeches Lane, Woodstock, CT, 06281, USA (Tel: 860-963-2061; FAX 860-963-2164)

Eka Chemicals Ltd., 304 Pkwy., Summer Lane, Worle, Weston-Super-Mare, Avon, BS22 6WA, UK (Tel: 44 1934 522244; FAX 44 1934 522577)

Eka Chemicals AB, Business Unit of Akzo Nobel, 445 80 Bohus, Sweden (Tel: 46 31 58 70 00; FAX 46 31 58 74 00; E-mail: ew@bhs.ekachemicals.com)

Eka Chemicals Iberica SA, Badajoz 57, 08005 Barcelona, Spain (Tel: 34 93 3096050; FAX 34 93 3096378)

Eka Chemicals Oy, Kaivokatu 10A, 00100 Helsinki, Finland (Tel: 358 9 415 2520; FAX 358 9 415 25250; Internet: www.akzonobel.fi)

Eka Chemicals bv, Postbus 3045, 6202 NA Maastricht, The Netherlands (Tel: 31 43 352 5354; FAX 31 43 363 4755; E-mail: av@mst.ekachemicals.com)

Eka Chemicals (Australia) Pty Ltd., Div. of Nobel Industries, 15 Conquest Way, Hallam, Victoria, 3803, Australia (Tel: 61-3-9702-3422; FAX 61-3-9703-1859)

Eka Chemicals (NZ) Ltd., 211 Totara St., PO Box 5136, Mount Maunganui, 3030, New Zealand (Tel: 64 7 575 7089; FAX 64 7 575 2484; E-mail: eka@mni.ekachemicals.co.nz)

Elan Chemical Co. Inc.

268 Doremus Ave., Newark, NJ, 07105, USA (Tel: 973-344-8014; FAX 973-344-1948; E-mail: sales@elan-chemical.com; Internet: www.elan-chemcal.com)

Electron Microscopy Sciences Corp.

PO Box 251, 321 Morris Rd., Ft. Washington, PA, 19034, USA (Tel: 215-646-1566; 800-523-5874; FAX 215-646-8931; E-mail: sgkcck@aol.com; Internet: www.emsdiasum.com)

Elementis

Elementis plc/Group Headquarters, One Great Tower St., London, EC3R 5AH, UK (Tel: 44 207 398 1400; FAX 44 207 398 1401; E-mail: elementis.info@elementis-eu.com; Internet: www.elementis.com)

Elementis Specialties, Durham, Birtley, Chester-le-Street, Co. Durham, DH3 1QX, UK (Tel: 44 191 410 2361; FAX 44 191 410 6005; E-mail: specinfo.uk@elementis.com; Internet: www.elementis-specialties.com)

Elementis Specialties Europe, Zwijnaardseseteenweg 316a-blok E, 9000 Gent, Belgium (Tel: 32 9 242 96 42; FAX 32 9 242 96 58)

Elementis Specialties (U.S.), PO Box 700, Wyckoffs Mill Rd., Hightstown, NJ, 08520, USA (Tel: 609-443-2500; 800-866-6800; E-mail: specialinfo.usa@elementis-na.com; rheox.marketing@elementis-na.com; Internet: www.elementis-specialties.com)

Elementis Specialties Performance Polymers, 600 Cortlandt St., Belleville, NJ, 07109, USA (Tel: 973-751-3000; 888-442-7362; FAX 973-751-8407; E-mail: polymers.info@elementis-na.com; Internet: www.elementis-specialties.com)

Elementis Specialties (Singapore), Blk 16 Science Park Dr., #04-03 The Pasteur, Singapore Science Park 1, Singapore, 118227, Singapore (Tel: 65 775 1922; FAX 65 775 1972)

S. Black (Import & Export) Ltd., Distrib. for Elementis Specialities (UK), Foxholes Business Park, John Tate Road, Hertford, Hertfordshire, SG13 7YH, UK (Tel: 44 1992 825555; FAX 44 1992 825566; E-mail: enquires@sblack.com; Internet: www.sblack.com)

EMCO Chemical Distributors, Inc.

PO Box 1030, 2100 Commonwealth Ave., N. Chicago, IL, 60064, USA (Tel: 847-689-2200; FAX 847-689-8470; E-mail: info@emcochem.com; Internet: www.emcochem.com)

EM Industries

EM Industries, Inc., Affiliate of Merck KGaA, 7 Skyline Dr., Hawthorne, NY, 10532, USA (Tel: 914-592-4660; 888-367-3275; FAX 914-592-9469; Internet: www.emindustries.com)

EM Industries Canada Inc., 5409 Eglinton Ave., West, Suite 209, Toronto, Ontario, M9C 5K6, Canada

Emkay Chemical Co. DEFUNCT

LAST KNOWN: PO Box 6537, 319-325 Second St., Elizabeth, NJ, 07206-6537, USA (Tel: 732-650-9779; 888-EMKAYCHEM; FAX 908-352-6398; E-mail: sales@emkaychemical.com; Internet: www.arjayonline.com/multimedia/flash_emkay.htm)

EMS

EMS-Chemie Holding AG, 7013 Domat/Ems, Switzerland (Tel: 41 81 632 6111; FAX 41 81 632 7401; Internet: www.emschem.com)

EMS-Dottikon AG, PO Box, 5605 Dottikon, Switzerland (Tel: 41 56 616 811 11; FAX 41-56 616 81 20; E-mail: info@ems-dottikon.ch; Internet: www.ems-dottikon.ch)

EMS-CHEMIE (North America) Inc., 2060, Corporate Way, PO Box 1717, Sumter, SC, 29151-1717, USA (Tel: 803-481-9173; FAX 803-481-3820; E-mail: welcome@us.emsgrivory.com)

EM Science

Div. of EM Industries, Inc., 480 S. Democrat Rd., Gibbstown, NJ, 08027, USA (Tel: 856-423-6300; 800-222-0342; FAX 856-423-4389)

EniChem

EniChem SpA, An Eni Group Company, Piazza Boldrini, 1, 20097 San Donato Milanese (MI), Italy (Tel: 39 2 520 1)

EniChem Elastomeri SpA, Via Baiona, 107, 48100 Ravenna, Italy (Tel: 39 544 513111; FAX 39 544 513323)

EniChem Elastomers Ltd., Div. of EniChem Elastomeri Srl, Cadland Rd., Hardley, Hythe, Southampton, Hamsphire, SO45 3YY, UK (Tel: 44 2380 894919; FAX 44 2380 883306)

EniChem France SA, 1, Place Firmin Gautier, 38000 Grenoble, France (Tel: 33 4 76 70 95 00; FAX 33 4 76 70 95 59)

EniChem Deutschland GmbH, Koelner Strasse 3a, D-65760 Eschborn, Germany (Tel: 49 6196 492 0; FAX 49 6196 492 218; E-mail: enichem.deutschland@enichem.it)

EniChem Deutschland GmbH, Zweigniederlassung Wien, Marokkanergasse 22/9, A-1030 Wien, Austria (Tel: 43 1 5037 390; FAX 43 1 5037 390 20; E-mail: enichem.deutschland@enichem.it)

EniChem Benelux SA, 8 Rue de l'Industrie, B-1400 Nivelles, Belgium (Tel: 32 67 880 611; FAX 32 67 210 587; E-mail: enichem.benelux@enichem.it)

EniChem Iberica S.A., Avda. Diagonal 652-656, Edificio B 3°, 08034 Barcelona, Spain (Tel: 34 93 40123 00; FAX 34 93 2803050; E-mail: enichem.iberica@enichem.it)

EniChem Norden A/S, Borups Allè 177 D, DK-2400 Copenhagen NV, Denmark (Tel: 45 38 381080; FAX 45 38 381099; E-mail: enichem.norden@enichem.dk)

EniChem Americas, Inc., 2000 West Loop South, Suite 2010, Houston, TX, 77027, USA (Tel: 713-940-0700; 800-441-3646; FAX 713-850-8606; E-mail: Gmaniscalco@enichem-americas.com)

EniChem Elastomers America, Inc., 4803 Decker Dr., Baytown, TX, 77520, USA (Tel: 281-424-6100; 800-882-7080; FAX 281-424-6171)

EniChem Singapore Pte. Ltd., 79, Anson Rd. 15-01, Singapore 079906, Singapore (Tel: 65 2243132; FAX 65 2279112; E-mail: enichem@singnet.com.sg)

EniChem Pacific Ltd.-Taiwan, Rm 1501, 15th Fl., 142 Min-Chuan E. Rd., Sec. 3, Taipei, 105, Taiwan R.O.C. (Tel: 886 2 27176697; FAX 886 2 27177579; E-mail: enichem@ms12.hinet.net)

EniChem Pacific Ltd.-Beijing, 2504 Citic Bldg., #19 Jianguomenwai St., Beijing, 100004, P.R. China (Tel: 86 10 6500 3441; FAX 86 10 65126128; E-mail: enichem@public3.bta.net.cn)

Epochem Co., Ltd./International Marketing Div.

168 Changdong Rd., Suite 679, Shanghai, 201612, P.R. China (Tel: 86 21 6764 3383; FAX 86 21 6764 3350; E-mail: info@epochem.com; Internet: www.epochem.com)

Equistar Chemicals

Equistar Chemicals, LP, Partnership of Lyondell, Millennium Chem., & Occidental Petroleum, 1221 McKinney St., Suite 700, PO Box 2583, Houston, TX, 77252-2583, USA (Tel: 713-652-7300; 888-777-0232; FAX 713-652-6119; Internet: www.equistarchem.com)

Equistar Chemicals, LP, 11530 Northlake Dr., Cincinnati, OH, 45249, USA (Tel: 513-530-4000; 800-323-4905; FAX 513-530-4208)

Esperis SpA

Via Ambrogio Binda, 29, 20143 Milan, Italy (Tel: 39 2 891 22219; FAX 39 2 891 22257; Internet: www.esperis.it)

Esprit Chemical Co.

Esprit Plaza, 7680 Matoaka Rd., Sarasota, FL, 34236, USA (Tel: 941-355-5100; 800-237-7748; FAX 941-358-1339; E-mail: loucohen@espritchem.com; Internet: www.espritchem.com)

Essence Natural Products

Essence Natural Products Pvt. Ltd., 5-4-187/3&4, III Floor, Mahatma Gandhi Rd., Secundrabrad, Andhra Pradesh, 500 003, India (Tel: 91 40 7543658; FAX 91 40 7541450; E-mail: info@ea-menthol.com; Internet: www.ea-menthol.com)

Essence Natural Products, 910 S. El Camino Real, San Clemente, CA, 92672, USA (Tel: 949-481-6848; FAX 949-492-8517; E-mail: usa@ea-menthol.com; Internet: www.ea-menthol.com)

Expansia S.A.

30390 Aramon, France (Tel: 33 4 66 57 01 01; FAX 33 4 66 57 01 48; E-mail: contact.expansia@expansia.com; Internet: www.expansia.com)

Expocell AB

Gustavsgatan 23, 21611 Limhamn, Sweden (Tel: 46 40 21 10 20; FAX 46 40 21 10 40; E-mail: info@expocell.se; Internet: www.expocell.se)

Expo Chemical Co.

12602 Manorwood Lane, Cypress, TX, 77429, USA (Tel: 281-890-4553; 877-258-EXPO; FAX 281-890-4576; E-mail: expo@expochem.com; Internet: www.expochem.com)

ExxonMobil

ExxonMobil Chemical Co., 13501 Katy Freeway, Houston, TX, 77079-1398, USA (Tel: 281-870-6000; 800-231-6633; FAX 281-870-6661; Internet: www.exxonmobilchemical.com)

Exxon Chemical Canada, 111 St. Clair Ave. West, 7th Fl., PO Box 4029, Station A, Toronto, Ontario, M5W 1K3, Canada (Tel: 416-968-4751; 800-567-ESSO; FAX 416-968-4850; Internet: www.imperialoil.com)

ExxonMobil Mexico S.A. de C.V., Aristoteles No. 77-101, Col. Chapultepec Polanco, Mexico D.F., 11560, Mexico (Tel: 52 5 279 4800; FAX 52 5 280 0070)

Exxon Quimica, Ltda, Rua Barao de Monte Santo, 700, Parque da Mooca, Sao Paulo-SP, 03123-020, Brazil (Tel: 55 11 6914 2022; FAX 55 11 273 2106)

Exxon Chemical Argentina SAIC, Avda. Leandro N. Alem 928 - 2° Piso, 1001 Buenos Aires, Argentina (Tel: 54 11 4319 0761; FAX 54 11 4319 0721)

ExxonMobil Chemical Europe Inc., Hermeslaan 2, 1831 Machelen, Belgium (Tel: 32 2 722 21 11; FAX 32 2 722 27 80)

ExxonMobil Chemical Ltd., 4600 Parkway, Solent Business Park, Whiteley, Fareham, Hampshire, PO15 7AP, UK (Tel: 44 1489 884400; FAX 44 1489 884411; Internet: www.exxonchemical.com)

ExxonMobil Chemical France, 2, rue des Martinets, BP 270, 92505 Rueil Malmaison Cedex, France (Tel: 33 1 47 10 60 00; FAX 33 1 47 10 55 11; Internet: www.esso.com/eaff/exxonchemical_france/english/index.html)

ExxonMobil Chemical Mediterranea SpA, Via Paleocapa 7, 20121 Milan (MI), Italy (Tel: 39 2 880 31; FAX 39 2 880 32 31)

ExxonMobil Chemical Holland BV, PO Box 1, 4803 AA Breda, The Netherlands (Tel: 31 76 529 26 00; FAX 31 76 529 27 00)

ExxonMobil Chemical Central Europe GmbH, Neusser Landstrasse 16, 50735 Cologne, Germany (Tel: 49 221 770 31; FAX 49 221 770 3320)

ExxonMobil Chemical Iberia, Avenida del Partenon 4-3a planta, 28042 Madrid, Spain (Tel: 34 91 300 92 00; FAX 34 91 300 92 40)

ExxonMobil Denmark Holdings International ApS, Frederiksberggade 24, 1459 Copenhagen, Denmark (Tel: 45 33 69 17 10; FAX 45 33 12 47 46)

ExxonMobil Chemical Norden AB, Box 1035, 40522 Göteborg, Sweden (Tel: 46 31 63 82 00; FAX 46 31 15 50 08)

ExxonMobil Chemical Central Europe GmbH (Moscow), 5 Nikitsky Pereulok, 103009 Moscow, Russia (Tel: 7 095 564 89 71; FAX 7 095 564 89 94)

Exxon Chemical Australia Ltd., Level 2, 12 Riverside Quay, Southbank, Melbourne, 3006, Australia (Tel: 61 3 9205 4445; FAX 61 3 9205 4433)

ExxonMobil Chemical New Zealand Ltd., 164-188 Beaumont St., PO Box 8789, Symonds St., Auckland, New Zealand (Tel: 64 9 379 6204; FAX 64 9 377 5324)

ExxonMobil Chemical Singapore Pte. Ltd., 1 Raffles Place, OUB Centre #27-00, Singapore, 048616, Singapore (Tel: 65 6230 3556; FAX 65 6230 3529)

ExxonMobil Chemical Japan Pte. Ltd., New Pier Takeshiba South Tower, 16-1, Kaigan 1-chome, Minato-ku, Tokyo, 105-8616, Japan (Tel: 81 3 5425 9730; FAX 81 3 5425 9771)

Exxon Chemical International Services Ltd., 22nd Fl., Central Plaza, 18 Harbour Road, Wanchai, Hong Kong (Tel: 852 582 0888; FAX 852 2582 0389)

ExxonMobil Company India Pvt. Ltd., 92 Maker Chambers VI, Nariman Pt., Mumbai, 400 021, India (Tel: 91 22 282 9726; FAX 91 22 288 5400)

ExxonMobil Petroleum & Chemical Holdings Inc., Taiwan, Suite 805, 131 Min Sheng East Rd., Section 3, Taipei, 10446, Taiwan R.O.C. (Tel: 886 2 2715 4420; FAX 886 2 2715 5582)

Exxon China Inc., 8th Fl., 818 China World Tower, China World Trade Center, No. 1, Jian Guo Men Wai Ave., Beijing, 100004, P.R. China (Tel: 86 10 6505 3880; FAX 86 10 6505 3889)

Mobil Chemical Products International, Inc., 101 Thomson Rd., United Square #12-04, Singapore, 307591, Singapore (Tel: 65 254 0929; FAX 65 254 1305)

Fabrichem Inc.

2226 Black Rock Tpke., Suite 206, Fairfield, CT, 06430, USA (Tel: 203-366-1820; FAX 203-366-1850; E-mail: FabriChem@Worldnet.att.net)

Fabriquimica S.R.L.

Calle 32 No. 3313, San Martin, 1650, Argentina (Tel: 54 1 753 0894; FAX 54-1-755-7290)

Fairmount Chemical Co., Inc.

117 Blanchard St., Newark, NJ, 07105, USA (Tel: 973-344-5790; 800-872-9999; FAX 973-690-5298)

Fallek Chemical Co.

Div. of ICC Chemical Corp., 460 Park Ave., New York, NY, 10022, USA (Tel: 212-521-1700; FAX 212-521-1946; E-mail: trading@iccchem.com; Internet: www.iccchem.com/fallchem.htm)

The Fanning Corp.

2450 W. Hubbard St., Chicago, IL, 60612-1408, USA (Tel: 312-563-1234; FAX 312-563-0087; E-mail: CustomerService@fanncorp.com; Internet: www.fanncorp.com)

Fanwood Chemical, Inc.

219 Martine Ave., North, PO Box 159, Fanwood, NJ, 07023-0159, USA (Tel: 908-322-8440; FAX 908-322-8494; E-mail: info@fanwoodchemical.com; Internet: www.fanwoodchemical.com)

Farabi Petrochemical Co. (FPC)

35.2 Sanei St., Gandi Ave., Tehran, 19699, Iran (Tel: 98 21 8783040; FAX 98 21 8779174)

Farleyway Chemicals Ltd.

Ham Lane, Kingswinford, West Midlands, DY6 7JU, UK (Tel: 44 1384 400 222; FAX 44 1384 400 020; E-mail: farleyway@btinternet.com)

FAR Research, Inc.

2210 Wilitelming Ct., N.E., Palm Bay, FL, 32905, USA (Tel: 407-723-6160; FAX 407-723-8753)

FBC Industries, Inc.

500 E. Remington Rd., Suite 300, Schaumburg, IL, 60173, USA (Tel: 847-839-0880; 888-FBC-INDS; FAX 847-839-0884; E-mail: info@fbcindustries.com; Internet: www.fbcindustries.com)

Ferro

Ferro Corp./World Headquarters, 1000 Lakeside Ave., PO Box 147000, Cleveland, OH, 44114-7000, USA (Tel: 216-641-8580; FAX 216-696-6958; Internet: www.ferro.com)

Ferro Corp./Grant Chemical Div., 111 W. Irene Rd., Zachary, LA, 70791-9738, USA (Tel: 225-654-6801; 800-325-3578; FAX 225-654-3268; E-mail: 103021.2370@compuserve.com; Internet: www.ferro.com)

Ferro Corp./Keil Chemical Div., 3000 Sheffield Ave., Hammond, IN, 46327-1099, USA (Tel: 219-931-2630; 800-628-9079; FAX 219-931-6318; Internet: www.ferro.com)

Ferro Corp./Polymer Additives Div., Route 130 South, PO Box 309, Bridgeport, NJ, 08014, USA (Tel: 856-467-3000; FAX 856-467-8308; Internet: www.ferro.com)

Ferro Mexicana, S.A. de C.V., Oriente 171 #450, Col. Prol. Aragon-Inguaran, Del. G.A. Madero, 07490 Mexico D.F., Mexico (Tel: 52 5 760 6100; FAX 52 5 760 5154)

Ferro (Great Britain) Limited/Colour & Enamel Divs., Ounsdale Rd., Wombourne, Wolverhampton, WV5 8DA, UK (Tel: 44 1902 324144; FAX 44 1902 324265)

Ferro Chemicals SA, BP 28, Etang de la Gafette, 13521 Port-de-Bouc Cedex, France (Tel: 33 4 42 40 73 00; FAX 33 4 42 40 73 33)

Ferro (Italia) Srl, Via Trentino, 9, 41049 Sassuolo (MO), Italy (Tel: 39 536 819111; FAX 39 536 811641)

Ferro Enamel Espanola S.A., Apartado 232, 12080 Castellon de la Plana, Spain (Tel: 34 64 50 44 50; FAX 34 64 50 44 21)

Ferro (Deutschland) GmbH, Langenbergstrasse 10, 67657 Kaiserslautern, Germany (Tel: 49 631 4164 0; FAX 49 631 4164 147)

Ferro (Holland) BV, PO Box 6088, 3002 AB Rotterdam, The Netherlands (Tel: 31 10 478 4911; FAX 31 10 477 8202)

Ferro Corp. (Aust.) Pty. Ltd., 469 Somerville Rd., West Footscray, Victoria, 3012, Australia (Tel: 61 3 9314 6977; FAX 61 3 9314 5045)

Ferro Enamels (Japan) Ltd., 2-1-27, Oyodo Kita, Kita-ku, Osaka, 531, Japan (Tel: 81 6 6458 3551; FAX 81 6 6458 8911)

Ferro (Ningbo) Powder Coatings Ltd., Ningchuan Rd., Wuxiang Town, Ningbo, Zheijiang, 31511, P.R. China (Tel: 86 574 848 5682; FAX 86 574 848 5029)

Ferro Industrial Products Limited (Taiwan), No. 239, Hsia Chung Tzu Rd., Yuang An Village, Hsin Wu Hsiang, Taoyuan Hsien, 32707, Taiwan R.O.C. (Tel: 886 3 476 1251; FAX 886 3 476 1216)

Filo Chemical Inc.

50 Broadway, New York, NY, 10004-1698, USA (Tel: 212-514-9330; 800-232-7436; FAX 212-514-9085; E-mail: filo@filochemical.com; Internet: www.filochemical.com)

Finetex Inc.

418 Falmouth Ave., PO Box 216, Elmwood Park, NJ, 07407, USA (Tel: 201-797-4686; FAX 201-797-6558; E-mail: information@finetexinc.com; Internet: www.finetexinc.com)

Firmenich

Firmenich Inc., PO Box 5880, Princeton, NJ, 08543, USA (Tel: 609-452-1000; 800-257-9591; FAX 609-520-9780)

Firmenich UK Ltd., Hayes Rd., Southall, Middlesex, UB2 5NN, UK (Tel: 44 208 843 6111; FAX 44 208 843 6471)

Firmenich S.A., PO Box 239, 1211 Geneva 8, Switzerland (Tel: 41 22 780 2211; FAX 41 22 343 7322)

Nihon Fermenich K.K., PO Box 237, Kyobashi, Tokyo, 104-8601, Japan (Tel: 81 3 35 43 52 61; FAX 81 3 35 43 52 89)

Fisher Scientific

Fisher Scientific USA, 2000 Park Lane Drive, Pittsburgh, PA, 15275-1126, USA (Tel: 800-766-7000; FAX 800-926-1166; E-mail: John.Fusco@plpit.fishersci.com; Internet: www.fisher1.com)

Fisher Scientific UK, Bishop Meadow Rd., Loughborough, Leicestershire, LE11 5RG, UK (Tel: 44 1509 231166; FAX 44 1509 231893; E-mail: info@fisher.co.uk; Internet: www.fisher.co.uk)

Flavors of North America Inc.

525 Randy Rd., Carol Stream, IL, 60188, USA (Tel: 630-462-1414; FAX 630-462-8855)

Fleurchem

33 Sprague Ave., Middletown, NY, 10940, USA (Tel: 845-341-2100; FAX 845-341-2121; E-mail: info@fleurchem.com; Internet: www.fleurchem.com)

Floratech Americas

1151 N. Fiesta Blvd., Gilbert, AZ, 85233-2238, USA (Tel: 480-545-7000; FAX 480-892-3000; E-mail: whatsnew@floratech.com; Internet: www.floratech.com)

Florida Chemical Co., Inc.

351 Winter Haven Blvd. NE, Winter Haven, FL, 33881-9432, USA (Tel: 863-294-8483; FAX 863-294-7783; E-mail: info@ floridachemical.com; Internet: www.floridachemical.com)

Florida Distillers Co.

PO Box 4057, West Palm Beach, FL, 33402, USA (Tel: 561-655-8977; FAX 561-655-9718)

Fluka

Fluka Chemical Corp., PO Box 355, Milwaukee, WI, 53201, USA (Tel: 414-273-3850; 800-558-9160; FAX 800-962-9591; Internet: www.sigma-aldrich.com)

Fluka Sarl, L'isle D'Abeau Chesnes, BP 705, 38297 St. Quentin Fallavier Cedex, France (Tel: 33 4 74 82 28 88; FAX 33 4 74 95 68 08)

Fluka Chemical LLC, Industriestrasse 25, 9471 Buchs, Switzerland (Tel: 41 81 755 2828; FAX 41 81 756 5449; E-mail: fluka@sial.com)

Fluka Chemical Ltd., The Old Brickyard, New Road, Gillingham, Dorset, SP8 4XT, UK (Tel: 44 1747 822211; FAX 44 1747 823779)

Fluorochem

Fluorochem Ltd., Wesley St., Old Glossop, Derbyshire, SK13 7RY, UK (Tel: 44 1457 868921; FAX 44 1457 869360; E-mail: enquiries@fluorochem.net; Internet: www.fluorochem.net)

Fluorochem USA, 1741 Old Dunbar Road, West Columbia, SC, 29169, USA (Tel: 803-739-8800; 800-467-3386; FAX 803-739-6957; E-mail: sales@fluorochemusa.com; Internet: www.fluorochemusa.com)

FMC

FMC Corp./Chemical Products Group, 1735 Market St., Philadelphia, PA, 19103, USA (Tel: 215-299-6000; 800-468-3853; FAX 215-299-5999; E-mail: indchem-info@fmc.com; Internet: www.fmc.com)

FMC Biopolymer, 1735 Market St., Philadelphia, PA, 19103, USA (Tel: 215-299-6000; 800-526-3649 x3; FAX 215-299-5809)

FMC Corp./Lithium Div., 449 N. Cox Road, PO Box 3925, Gastonia, NC, 28054-0020, USA (Tel: 704-868-5300; 800-362-2548; FAX 704-868-5370; E-mail: lithium-info@fmc.com)

FMC Corp. Canada, 625 Howe St., Suite 1200, Vancouver, British Columbia, V6C 2T6, Canada (Tel: 604-685-6508; FAX 604-689-5917)

FMC do Brasil Ind. e Com. Ltda., Av. Marg. Eng. Camilo Dinucci, 4605, Araraquara, Sao Paulo, 14808-900, Brazil (Tel: 55 16 201 2000; FAX 55 16 232 0565)

FMC Argentina S.A., Oncativo 954, B° General Paz, 5000 Cordoba, Argentina (Tel: 54 351 421 8094; FAX 54 351 424 8607)

FMC Corp. (UK) Ltd., Balfour House, Churchfield Rd., Walton-on-Thames, Surrey, KT12 2TD, UK (Tel: 44 1932 256 000; FAX 44 1932 253 227)

FMC Europe N.V., Ave. Louise 480-B9, 1050 Brussels, Belgium (Tel: 32 2 645 9211; FAX 32 2 645 9434)

FMC Europe S.A., Route des Clerimois, B.P. 705, 89107 Sens Cedex, France (Tel: 33 3 86 95 87 00; FAX 33 3 86 95 19 16)

FMC A/S, Risingvej 1, 2665 Vallensbaek Strand, Denmark (Tel: 45 43 56 7300; FAX 45 43 56 7341)

FMC Industrial Chemicals (Netherlands) B.V., Oosterhorn 14, Box 153, 9930 AD Delfzijl, The Netherlands (Tel: 31 596 642 333; FAX 31 596 642 300)

FMC International A.G., Rigistrasse 184, 6340 Baar, Switzerland (Tel: 41 41 760 6616; FAX 41 41 760 6616)

FMC Foret SA, Còrsega, 293, 08008 Barcelona, Spain (Tel: 34 93 4 167 400; FAX 34 93 4 167 403; E-mail: fmc_foret@fmc.com)

FMC Italia S.p.A., Via Mantova 63/A, Box 333, 43100 Parma, Italy (Tel: 39 521 908 411; FAX 39 521 487 960)

FMC Corp. (Russia), 3d Samotechny, Pereulok 11, 6th Fl., 103473 Moscow, Russia (Tel: 7 95 288 9666; FAX 7 95 221 3226)

FMC (Australia) Ltd., 82 Biloela St., Villawood, 2163, Australia (Tel: 61 2 9723 2000; FAX 61 2 9723 2085)

FMC Asia-Pacific, Inc., 1/F, Kirslon House, Saki Vihar Rd., Saki Naka, Mumbai, 400 072, India (Tel: 91 22 850 0395; FAX 91 22 850 0398)

FMC Asia-Pacific, Inc., Room 604 CITIC Bldg., 19, Jianguomen Wai Dajie, Beijing, 100004, P.R. China (Tel: 86 10 6500 2255 ext. 3640; FAX 86 10 6500 2251)

FMC Asia Pacific, Inc., 7F Assend Kanda Bldg., 10-2 Kanda Tomiyamacho, Chiyoda-ku, Tokyo, 101-0043, Japan (Tel: 81 3 3252 9520; FAX 81 3 3252 9605)

FMC Asia-Pacific Inc., Suite 2401-02, Central Plaza, 18 Harbour Rd., Wanchai, Hong Kong (Tel: 852 2839 6600; FAX 852 2576 3770)

FMC Southeast Asia Pte. Ltd., 149 Gul Circle, Singapore, 629605, Singapore (Tel: 65 861 3011; FAX 65 862 0527)

FMC BioPolymer Asia, 85 Science Park Dr., #02-08 The Cavendish, Singapore, 118259, Singapore (Tel: 65 872 2920; FAX 65 872 2927)

FMC Korea Ltd., Seo Woo Bldg., 7/F, 837-12, Yeoksam-dong, Kangnam-ku, Seoul, 135-080, Korea (Tel: 82 2 539 6411; FAX 82 2 567 4662)

Focus Chemical Corp.

875 Greenland Rd., Orchard Park, Suite B9, Portsmouth, NH, 03801-4122, USA (Tel: 603-430-9802; FAX 603-431-5390; E-mail: focuschm@nh.ultranet.com)

Food Ingredient Technology Ltd.

HiTec House, Sand Rd. Industrial Estate, Great Gransden, Bedfordshire, SG19 3AH, UK (Tel: 44 1767 677666; FAX 44 1767 677966; E-mail: sales@fit-ltd.co.uk; Internet: www.fit-ltd.com)

The Foote & Jenks Corp.

1420 Crestmont Ave., Camden, NJ, 08103, USA (Tel: 856-966-0700; FAX 856-966-6137)

Formosa Plastics Corp. (FPC USA)

9 Peach Tree Hill Rd., Livingston, NJ, 07039-5702, USA (Tel: 973-992-2090; 888-372-8723; FAX 973-992-9627; E-mail: communications@fpcusa.com; Internet: www.fpcusa.com)

Forrester Wood & Co. Ltd.

Hawksley Ind. Estate, Heron St., Hollinwood, Oldham, Lancashire, OL8 4UJ, UK (Tel: 44 161 620 4124; FAX 44 161 627 1050)

Fortitech

Fortitech Inc., Riverside Technology Park, 2105 Technology Dr., Schenectady, NY, 12308, USA (Tel: 518-372-5155; 800-950-5156; FAX 518-372-5599; E-mail: info@fortitech.com; Internet: www.fortitech.com)

Fortitech UK Ltd., Monton House, Monton Green, Manchester, M30 9LE, UK (Tel: 44 161 281 6003; FAX 44 161 281 6004; E-mail: phillipsi@fortitech.com; Internet: www.fortitech.com)

Forum Chemicals Ltd.

Part of Ajinomoto Co. Inc., Forum House, 41-51 Brighton Rd., Redhill, Surrey, RH1 6YS, UK (Tel: 44 1737 773711; FAX 44 1737 773116; E-mail: enquiries@forum.co.uk; Internet: www.forum.co.uk)

Freeman Industries, L.L.C.

100 Marbledale Rd., PO Box 415, Tuckhoe, NY, 10707-0415, USA (Tel: 914-961-2100; 800-666-6454; FAX 914-961-5793; E-mail: freeman@cyburban.com; Internet: www.freemanllc.com)

Freund

Freund Chemical Co., Ltd., 2-2-10 Numakage, Saitama City, Saitama Prf., 336-0027, Japan (E-mail: freund@freund.co.jp; Internet: www.freund.co.jp)

Freund International Ltd., 100 W. Tenth St., Wilmington, DE, USA

FRP Services & Co. (America) Inc.

10 Bank St., Suite 450, White Plains, NY, 10606, USA (Tel: 914-686-4090; 888-686-4090; FAX 914-686-3988; E-mail: info@frpservices.com; Internet: www.frpservices.com)

Frutarom

Frutarom Ltd., 25 Hashaish St., PO Box 10067, Haifa, 26110, Israel (Tel: 972 4 846 2462; FAX 972 4 872 2517; E-mail: info@frutarom.com; Internet: www.frutarom.com)

Frutarom (UK) Ltd., Northbridge Works, Northbridge Rd., Berkhamsted, Hertsfordshire, HP4 1EF, UK (Tel: 44 1442 876611; FAX 44 1442 876204; E-mail: sales@frutarom.co.uk)

Frutarom Europe, Acticlub BATA, 7, rue des Champs, Z.I. de la Pilaterie, 59650 Villeneuve d'Ascq, France (Tel: 33 3 20 89 70 20; FAX 33 3 20 89 74 00; E-mail: europe@frutarom.com)

Frutarom Russia, Sadovoya-Samotiochnaya St., 24/27, 103051 Moscow, Russia (Tel: 7 095 258 7683; FAX 7 095 732 7205; E-mail: russia@frutarom.com)

Frutarom Turkey Ltd., Imes Sanayi Sitesi, A-Block 101 Sokak No. 11, Y. Dudullu, Umraniye, Istanbul, Turkey (Tel: 216 466 6090; FAX 216 466 6091; E-mail: turkey@frutarom.com)

Frutarom Meer Corp., Member of the Frutarom Group, 9500 Railroad Ave., N. Bergen, NJ, 07047-1206, USA (Tel: 201-861-9500; FAX 201-861-9267; E-mail: usa@frutarom.com; Internet: www.frutarom.com)

Frutarom Flavors (Kunshan) Co. Ltd., 115 Xiang Shi Rd., Dian Shan Hu Town, Kunshan City, Jiangsu, P.R. China (Tel: 86 520 748 1537; FAX 86 520 748 1561; E-mail: china@frutarom.com)

F.T.L. International, Inc.

68-08 174th St., Fresh Meadows, NY, 11365, USA (Tel: 718-939-5505; 800-758-0066; FAX 718-939-4995; E-mail: ftl@chem2china.com; Internet: www.chem2china.com)

Fuerst Day Lawson

Devon House, 58-60 St Katharine's Way, London, E1W 1JP, UK (Tel: 44 20 7488 0777; FAX 44 20 7265 5285; E-mail: fdlgroup@fdl.co.uk; Internet: www.fdl.co.uk)

Functional Foods Corp.

470 Rt. 9, Englishtown, NJ, 07726, USA (Tel: 732-972-2232; FAX 732-536-9179; E-mail: functionalfoods@aol.com)

Galactic

Chaussee de St. Job 12, 1180 Brussels, Belgium (Tel: 32 2 332 1400; FAX 32 2 332 1611; E-mail: galactic@lactic.com; Internet: www.lactic.com)

Galbraith Laboratories, Inc.

2323 Sycamore Dr., PO Box 51610, Knoxville, TN, 37950-1610, USA (Tel: 865-546-1335; 877-449-8797; FAX 865-546-7209; E-mail: labinfo@galbraith.com; Internet: www.galbraith.com)

Gallard-Schlesinger Industries, Inc.

777 Zeckendorf Blvd., Garden City, NY, 11530, USA (Tel: 516-683-6900; 800-645-3044; FAX 516-683-6990; E-mail: info@gallard.com; Internet: www.gallard-schlesinger.com)

Garratt Callahan Company

111 Rollins Rd., Millbrae, CA, 94030-3114, USA (Tel: 650-697-5811; FAX 650-692-6098; E-mail: info@g-c.com; Internet: www.g-c.com)

Garuda Chemicals

15/605-606, Mayuresh c.h.s., Majaswadi-Mhada Colony, Off Mahakali Caves Rd., Andheri (East), Mumbai, 400 093, India (Tel: 91 22 2820 1198; FAX 91 22 2820 1201; E-mail: sales@garudachem.com; Internet: www.garudachem.com)

Gattefossé

Gattefossé S.A., 36 Chemin de Genas, BP 603, 69804 Saint Priest, Cedex, France (Tel: 33 4 72 22 98 00; FAX 33 4 78 90 45 67; Internet: www.gattefosse.com)

Gattefossé (Deutschland) GmbH, Rheincenter, Hauptstrasse 435, 79576 Weil-Am-Rhein, Germany (Tel: 49 76 21 720 07; FAX 49 76 21 79 22 93; E-mail: tmagnet@gattefosse.de)

Gattefossé A.G., Haldenstrasse 11, 6006 Luzern, Switzerland (Tel: 41 41 410 49 66; FAX 41 41 410 35 53)

Gattefosse (UK) Ltd., Arc House, Terrace Rd. South, Binfield, Bracknell, Berkshire, RG42 4PZ, UK (Tel: 44 1344 86 18 00; FAX 44 1344 45 14 00)

Gattefossé Italia s.r.l., Via Derganino 20, 20158 Milan, Italy (Tel: 39 2 39 31 40 73; FAX 39 2 66 200 440; E-mail: cperego@gattefosse.it)

Gattefossé España, S.A., Avda. Diagonal 460, 6°A, 08006 Barcelona, Spain (Tel: 34 93 416 05 20; FAX 34 93 415 35 46)

Gattefosse Corp., 650 From Rd., Paramus, NJ, 07652, USA (Tel: 201-265-4800; FAX 201-265-4853; E-mail: gcsamples@gattefossecorp.com; Internet: www.gattefosse.com)

Gattefosse Canada, Inc., 711 Devon Pl., Baie d'Urfe, Quebec, H9X 2T3, Canada (Tel: 514-457-0921; FAX 514-457-0922; E-mail: jasminep@gattefosseca.com)

Gattefossé Korea, # 1403 Sung Ji Heights III 642-6, Yuksamdong, Kangnamgu, Seoul, 135-080, Korea (Tel: 82 2 508 3877; FAX 82 2 568 0425)

Gaylord Chemical Co.

PO Box 1209, 106 Galeria Blvd., Slidell, LA, 70459-1209, USA (Tel: 504-649-5464; 800-426-6620; FAX 504-649-0068; E-mail: dmso@gaylordchem.com; Internet: www.gaylordchem.com)

Gelest Inc.

612 William Leigh Dr., Tullytown, PA, 19007-6308, USA (Tel: 215-547-1015; FAX 215-547-2484; Internet: www.gelest.com)

General Chemical

General Chemical Corp., A GenTek Company, 90 East Halsey Rd., Parsippany, NJ, 07054-0397, USA (Tel: 973-515-0900; 800-631-8050; FAX 973-515-3232; E-mail: info@genchemcorp.com; Internet: www.genchemcorp.com)

General Chemical Canada Ltd., 201 City Centre Dr., Mississauga, Ontario, L5B 3A3, Canada (Tel: 416-896-9595; 800-668-0433)

General Electric

GE Plastics, One Plastics Ave., Pittsfield, MA, 01201, USA (Tel: 413-448-5800; 800-845-0600; FAX 800-433-2925; Internet: www.geplastics.com)

General Electric Plastics B.V., Plasticslaan 1, PO Box 117, 4600 AC Bergen op Zoom, The Netherlands (Tel: 31 164 29 2391; FAX 31 164 29 1725)

GE Plastics Ltd., Old Hall Rd., Sale, Manchester, Cheshire, M33 2HG, UK (Tel: 44 161 905 5000; FAX 44 161 905 5119)

General Electric Plastics France S.a.r.L., Boite Postale No. 67, 91002 Evry/Cedex, France (Tel: 33 1 60 79 69 00; FAX 33 1 60 77 56 53)

General Electric Plastics GmbH, Storchengasse 1, 1150 Vienna, Austria (Tel: 43 1815 72030; FAX 43 1815 720320)

GE Plastics Japan Ltd., Nihonbashi Hamacho Park Bldg., 2-35-4, Nihonbashi Hama-cho, Chuo-ku, Tokyo, 103, Japan (Tel: 81 3 5695 4888; FAX 81 3 5695 4860)

GE Plastics Pacific Pte. Ltd., 240 Tanjong Pagar Rd., GE Towers #09-00, Singapore, 088540, Singapore (Tel: 65 220 7022; FAX 65 326 3290)

General Electric Co./Silicone Products Div., Subsid. of GE Plastics, 260 Hudson River Rd., Waterford, NY, 12188, USA (Tel: 518-233-3330; 800-255-8886; FAX 518-233-2308; E-mail: ges.help@gep.ge.com; Internet: www.gesilicones.com)

GE Silicones Canada, Inc., 2300 Meadowvale Blvd., Mississauga, Ontario, L5N 5P9, Canada (Tel: 416-858-5741; 800-332-3390)

GE Silicones Mexico, Av. Prolongacion Reforma, #490-Piso 4, Col. Santa Fe, Mexico D.F., 01207, Mexico (Tel: 52 5 257 6095; FAX 52 5 257 6094)

GE Silicones (Argentina), Ave. Leandro N. Alem 619, Piso 4, Buenos Aires, Argentina (Tel: 54 11 43178703)

GE Silicones, c/o General Electric do Brasil S.A., 10th Floor, Av. Nove do Julho, 5229, Sao Paulo SP, 01407-907, Brazil (Tel: 55 11 897 8144; FAX 55 11 897 8197; Internet: www.ge.com)

GE Bayer Silicones bv, Postbus 117, Plasticslaan 1, 4600 AC Bergen op Zoom, The Netherlands (Tel: 31 164 29 3882; FAX 31 164 29 1819)

GE Silicones (Australia), 175 Hammond Rd., Dandenong, Victoria, 3175, Australia (Tel: 61 3 9703 7200; FAX 61 3 9706 8727)

GE Silicones (Japan), Tonichi Bldg. 4F, Minato-ku, Tokyo, 106, Japan (Tel: 81 3 3479 5361; FAX 81 3 3479 5391)

GE Toshiba Silicones Hong Kong, The Gateway, 25 Canton Rd., Tsimshatsui, Kowloon, Hong Kong (Tel: 852 2629 0888; FAX 852 2629 0803)

GE Toshiba Silicones Shanghai, 317 Meigui (N) Rd., Waigaoqiao FTZ, Shanghai, 200131, P.R. China (Tel: 86 21 50460005)

GE Toshiba Silicones Singapore, 240 Tanjong Pagar Rd. #05-00, GE Tower, Singapore, 088540, Singapore (Tel: 65 326 3900)

GE Specialty Chemicals, 1503 23rd St., Parkersburg, WV, USA (Tel: 304-422-2832; 800-872-0022; FAX 304-422-2942; E-mail: geinfo@www.ge.com; Internet: www.ge.com/specialtychemicals)

GE Specialty Chemicals, Postbus 117, 4600 AC Bergen op Zoom, The Netherlands (Tel: 31 164 29 3187; FAX 31 164 29 2425; E-mail: Jean-Pierre.Vissers@gep.ge.com)

GE Specialty Chemicals-Pacific, 240 Tanjong Pagar Rd. #05-00, GE Tower, Singapore, 088540, Singapore (Tel: 65 326 3675; FAX 65 326 3946; E-mail: Jean-Pierre.Vissers@gep.ge.com)

Generichem Corp.

755 Union Blvd., PO Box 457, Totowa, NJ, 07511-0457, USA (Tel: 973-256-9266; FAX 973-256-0069; E-mail: info@generichem.com; Internet: www.generichem.com)

Genesee Polymers Corp.

G-5251 Fenton Rd., PO Box 7047, Flint, MI, 48507-4036, USA (Tel: 810-238-4966; FAX 810-767-3016; E-mail: customerservice@gpcsilicones.com; Internet: www.gpcsilicones.com)

Genzyme

Genzyme Corp., One Kendall Sq., Cambridge, MA, 02139-1562, USA (Tel: 617-252-7500; 800-868-8208; FAX 617-252-7600; E-mail: pharmaceuticals@genzyme.com; Internet: www.genzyme.com)

Genzyme Pharmaceuticals Europe, Eichenweg 1, 4410 Liestal, Switzerland (Tel: 41 61 906 5959; FAX 41 61 906 5958; Internet: www.genzyme.com/pharmaceuticals)

Genzyme NV/SA (Benelux), Interleuvenlaan, 5, 3001 Leuven, Belgium (Tel: 32 16 40 1550; FAX 32 16 40 0391)

Genzyme Japan, Izumiyamabuki-cho Bldg., 333 Yamabuki-cho, Shinjuku-ku, Tokyo, 162-0801, Japan (Tel: 81 3 5228 6464; FAX 81 3 5228 5933)

Geoliquids, Inc.

15 E. Palatine Rd., Suite 109, Prospect Heights, IL, 60070, USA (Tel: 847-215-0938; 800-827-2411; FAX 847-215-9821)

Georgia Gulf Corp.

400 Perimeter Center Terrace, Suite 595, Atlanta, GA, 30346, USA (Tel: 770-395-4591; FAX 770-671-8260)

Georgia-Pacific

Georgia-Pacific Resins Inc., Subsid. of Georgia-Pacific Corp., 1 Atlanta Center, Ste. 1200, 1201 West Peachtree St., Atlanta, GA, 30309, USA (Tel: 866-494-2178; 800-765-7374; FAX 404-817-7743; E-mail: ceking2@gapac.com; Internet: www.gp.com/chemical)

Georgia-Pacific Resins/Actrachem Operations, 7201 W. 65th St., Bedford Park, IL, 60638, USA (Tel: 708-458-8450; 800-323-3231; FAX 708-458-0286)

GFI

GFI, Inc., 111 Boston Post Rd., PO Box 777, Sudbury, MA, 01776, USA (Tel: 508-443-3674; 800-343-1179; FAX 508-443-5714)

GFI Chemicals USA, Northbrook Corporate Ctr., 1210 Northbrook Dr., Suite 100, Trevose, PA, 19053, USA (Tel: 215-942-2100; FAX 215-942-2113)

GFI UK Ltd., Marie Place, Brenchley, Tonbridge, Kent, TN12 7HS, UK (Tel: 44 1892 724077; FAX 44 1892 724099)

GFI Intertrade (Asia) Ltd., 1207-09, Uppo Sun Plaza, 28 Canton Rd., Tsimshatsui, Kowloon, Hong Kong (Tel: 862 27219319; FAX 862 23663194)

GFI (Australia), PO Box 304, Northbridge, Sydney, 2063, Australia (Tel: 61 2 9958 0130; FAX 61 2 9958 0292)

GFS Chemicals, Inc.

PO Box 245, 3041 Home Rd., Powell, OH, 43065, USA (Tel: 740-881-5501; 800-858-9682; FAX 740-881-5989; E-mail: gfschem@gfschemicals.com; Internet: www.gfschemicals.com)

Givaudan

Givaudan SA, 5, Chemin de la Parfumerie, 1214 Vernier, Switzerland (Tel: 41 22 780 9111; FAX 41 22 780 9150)

Givaudan Fragrances Corp., 1775 Windsor Rd., Teaneck, NJ, 07666, USA (Tel: 201-833-7500; FAX 201-833-8165; Internet: www.givaudan.com)

Givaudan Canada Inc., 2400 Matheson Blvd. East, Mississauga, Ontario, L4W 5G9, Canada (Tel: 905-282-9808; FAX 905-282-9972)

Givaudan Dubendorf Ltd., Ueberlandstrasse 138, 8600 Dubendorf, Switzerland (Tel: 41 1 824 2424; FAX 41 1 821 4478)

Givaudan France SA, 19-23 Voie des Bans, B.P. 98, 95102 Argenteuil, France (Tel: 33 1 39 98 15 15; FAX 33 1 39 98 17 13)

Givaudan UK Ltd./Flavors, Chippenham Dr., Kingston, Milton Keynes, MK10 0AE, UK (Tel: 44 1908 242 424; FAX 44 1908 282 232)

Givaudan UK Ltd./Fragrances, Magna House, 76-80 Church St., Staines, Middx., TW18 4XR, UK (Tel: 44 1784 417 700; FAX 44 1784 417 730)

Givaudan Iberica S.A., Plá d'en Batlle s/n, 08470 Sant Celoni, Barcelona, Spain (Tel: 34 93 864 1000; FAX 34 93 867 0319; Internet: www.givaudan.com)

Givaudan Japan K.K., 2-3-23 Shimomeguro, Meguro-ku, Tokyo, 153-0064, Japan (Tel: 81 3 3493 3781; FAX 81 3 3493 3786)

G.J. Chemical Co. Inc.

370-376 Adams St., Newark, NJ, 07114, USA (Tel: 973-589-1450; FAX 973-589-5786; E-mail: gjchemical@aol.com)

Global Chemsources, Inc.

PO Box 69, Cedar Grove, NJ, 07009, USA (Tel: 973-857-1076; FAX 973-857-4611; E-mail: global@webquill.com)

Global-Seven Inc.

303 Vreeland Ave., Paterson, NJ, 07513, USA (Tel: 973-345-3201; FAX 973-278-6809; E-mail: sales@global-seven.com; Internet: www.global-seven.com)

GMI Products Inc.

1301 Sawgrass Corp. Pkwy., Sunrise, FL, 33323, USA (Tel: 954-233-2500; 800-999-9373; FAX 954-233-1681; E-mail: info@gmi-originates.com; Internet: www.gmi-originates.com)

GNC Group

51 E. 42nd St., Suite 309, New York, NY, 10017, USA (Tel: 212-692-9463; FAX 212-692-9479; E-mail: gncgrpinc@aol.com)

Golden Bear Oil Specialties

10100 Santa Monica Blvd., Suite 1470, Los Angeles, CA, 90067, USA (Tel: 310-277-4511; 888-WE-DO-OIL; FAX 310-201-0383; E-mail: gbos@lacorp.goldenbearoil.com; Internet: www.goldenbearoil.com)

Goldensun Manufacturing Co.

Ranvir Bldg., 70 Princess St., Mumbai, 400 002, India (Tel: 91 22 205 2745; FAX 91 22 201 8054; E-mail: goldensun@vsnl.com; Internet: www.bicserve.com/htm/golden/index.htm)

Goldschmidt

Goldschmidt AG, Part of Degussa's Performance Chems. Div., Goldschmidtstrasse 100, 45127 Essen, Germany (Tel: 49 201 173 01; FAX 49 201 173 3000; E-mail: surfactants@de.goldschmidt.com; Internet: www.goldschmidtsurfactants.com)

Goldschmidt Rewo GmbH & Co. KG, Max-Wolf-Strasse 7, Postfach 1160, 36392 Steinau an der Strasse, Germany (Tel: 49 6663 54 0; FAX 49 6663 54 129)

Goldschmidt UK Ltd., Tego House, Chippenham Dr., Kingston, Milton Keynes, Bucks, MK10 0AF, UK (Tel: 44 1908 58 22 50; FAX 44 1908 58 22 54; E-mail: angus.smith@uk.goldschmidt.com)

Goldschmidt Benelux, Kapucijnenlaan 1, 1030 Brussels, Belgium (Tel: 32 2 241 8750; FAX 32 2 241 5452)

Goldschmidt France S.A., 3, ave. des Chaumes, B.P. 47, 78184 St. Quentin Yvelines Cedex, France (Tel: 33 1 30 12 01 69; FAX 33 1 30 12 01 70)

Goldschmidt Austria GmbH, Th., Liesinger-Flurgasse 2c, 1230 Vienna, Austria (Tel: 43 1 4857 3300; FAX 43 1 4857 33026)

Goldschmidt España S.A., Av. Sant Julia, 156-166, Pol. Ind. El Congost, 08400 Granollers (Barcelona), Spain (Tel: 34 93 8 642 100; FAX 34 93 8 642 112)

Goldschmidt Italia S.r.l., Via Valconera, 7, 26025 Pandino (CR), Italy (Tel: 39 373 975 222; FAX 39 373 975 260)

Goldschmidt Chemical Corp., 710 South 6th Ave., Hopewell, VA, 23860, USA (Tel: 804-452-5650; 800-446-1809; FAX 804-541-2783; E-mail: general@goldschmidtusa.com; Internet: www.goldschmidt.com)

Goldschmidt Chemical Canada, 3380 South Service Rd., Burlington, Ontario, L7R 3Y8, Canada (Tel: 905-336 3423; FAX 905-633-7553; E-mail: dtaylor@goldschmidtusa.com)

Goldschmidt Quimica de Mexico, S.A. de C.V., PO Box 366, San Luis Potosi, S.L.P., 78000, Mexico (Tel: 52 2 254 9896; FAX 52 2 531 5102; E-mail: gqm@goldschmidtmex.com.mx)

Goldschmidt Industrias Quimicas Ltda., Th., POB Caixa Postal, 106 Guarulhos Sao Paulo, Brazil (Tel: 55 11 6402 6917; FAX 55 11 6402 0888; E-mail: thgbra@uol.com.br)

Goldschmidt AG Japan Representative Office, Shinjuku Monolith 12F, 2-3-1, Nishi-Shinjuku, Shinuku-ku, Tokyo, 163-0912, Japan (Tel: 81 3 3288 9761; FAX 81 3 3288 7572)

Goldschmidt AG South East Asia Pte. Ltd., Th., 3 Int'l. Business Park, #07-18 Nordic European Centre, Singapore, 609927, Singapore (Tel: 65 5 6890 6056; FAX 65 5 6890 6058; E-mail: gchemsea@mbox3.singnet.com.sg)

Th. Goldschmidt Pacific Ltd., Xinzhuang Industry Park, 55, Chundong Rd., Shanghai, 201108, P.R. China (Tel: 86 21 5385 2045; FAX 86 21 6501 3005)

Th. Goldschmidt Pacific Ltd., Taiwan Branch, 9 F-2, No. 204, Min Sheng East Rd. Sec. 5, Taipei, Taiwan R.O.C. (Tel: 886 2 8787 7079; FAX 886 2 8787 5147)

Good Food Inc.

4960 Horseshoe Pike, PO Box 160, Honey Brook, PA, 19344, USA (Tel: 610-273-3776; FAX 610-273-2087; E-mail: goodfood@goldenbarrel.com)

The Good Scents Co.

USA (Tel: 414-764-2659; FAX 414-764-6035; E-mail: goodscnt@execpc.com; Internet: www.execpc.com/~goodscnt/)

Goodyear

Goodyear Tire & Rubber Co./Specialty Chemical Div., 1117 Perimeter Center West, Suite E-400, Atlanta, GA, 30338-5417, USA (Tel: 770-698-5406; FAX 770-698-5445; E-mail: bnelson@goodyear.com; Internet: www.goodyearchemical.com)

Goodyear do Brazil, Rua Da Intendencia, 91, PO Box 1424, Sao Paulo, 03015-010, Brazil (Tel: 55 11 608 7487 1; FAX 55 11 608 7600)

Goodyear Chemicals Europe, 14 Ave. des Tropiques, Z.A. de Courtaboeuf 2, 91955 Les Ulis Cedex, France (Tel: 33 1 69 29 27 00; FAX 33 1 69 29 27 01; E-mail: evelyne_dufour@goodyear.com; Internet: www.goodyear.com)

Goodyear Chemical, BTS House, 69/73 Manor Rd., Wallington, Surrey, SM6 0DD, UK (Tel: 44 208 6470737; FAX 44 208 6690460)

Goodyear Chemical Italy, 14 Via Frua, 20146 Milano, Italy (Tel: 39 2 48 02 20 42; FAX 39 2 48 00 82 66)

Goodyear Chemical c/o Deutsche Goodyear GmbH, Xantener Str. 105, 50733 Koln, Germany (Tel: 49 221 97 666265; FAX 49 221 97 666586)

Goodyear International Corp., 9 Penang Rd., #08-06 Park Mall, Singapore 238459, Singapore (Tel: 65 332 0508; FAX 65 337 1156)

Goodyear International Corp., 12C-2 Jiu Shi Fu Xin Mansion, No. 918 Huai Hai Road (M), Shanghai, 200020, P.R. China (Tel: 86 21 6415 9083; FAX 86 21 6415 3455)

Goodyear Chemicals India, Rm no. 308, Gokul Arcade (B), Subhash Rd., Vile Parle (East), Mumbai, 400 057, India (Tel: 91 22 82 36 171; FAX 91 22 82 36 175)

Grain Processing Corp.

1600 Oregon St., Muscatine, IA, 52761-1494, USA (Tel: 563-264-4265; FAX 563-264-4289; E-mail: sales@grainprocessing.com; Internet: www.grainprocessing.com)

Grau Aromatics

Grau Aromatics GmbH & Co. KG, Bismarckstrasse 4, 73525 Schwabisch Gmünd, Germany (Tel: 49 7171 9114 0; FAX 49 7171 9114 10; E-mail: contact@grau-aromatics.de; Internet: www.grau-aromatics.de)

J. Manheimer Inc., Distrib. for Grau Aromatics, 100 Hollister Rd., Teterboro, NJ, 07608, USA (Tel: 201-373-1111; FAX 201-288-8389; E-mail: sales@manheimer.com; Internet: www.manheimer.com)

Great Lakes

Great Lakes Chemical Corp., PO Box 2200, One Great Lakes Blvd., W. Lafayette, IN, 47906-0200, USA (Tel: 765-497-6100; 800-428-7947; FAX 765-497-6123; E-mail: hjames@glcc.com; Internet: www.greatlakeschem.com)

Great Lakes Chemical (Europe) Ltd., Sycamore House, Lloyd Drive, Ellesmere Port, South Wirral, Merseyside, L65 9HQ, UK (Tel: 44 151 357 8100; FAX 44 151 357 8120)

Great Western Chemical Co.

808 SW 15th Ave., Portland, OR, 97205, USA (Tel: 503-228-2644 x 208; 800-547-1400; FAX 503-221-5752; E-mail: ScottWhllhite@gwchem.com; Internet: www.gwchem.com)

K & K Greeff Ltd.

Div. of Beijer Industries AB, Suffolk House, George St., Croydon, Surrey, CR9 3QL, UK (Tel: 44 181 686 0544; FAX 44 181 686 4799)

R.W. Greeff. *See* Pechiney Chems./R.W. Greeff Chems. Div.

Griffin International Inc.

302 Clark Ave., Billings, MT, 59101, USA (Tel: 406-248-7275; FAX 406-248-4890; E-mail: information@griffininternational.net)

Grillo-Werke AG

Postfach 11 02 65, Weseler Strasse 1, 47142 Duisburg, Germany (Tel: 49 203 5557 1; FAX 49 203 5557 440; Internet: www.grillo.de)

Grünau Illertissen GmbH

A Cognis Group Co., Robert-Hansen-Strasse 1, 89357 Illertissen, Germany (Tel: 49 7303 13 0; FAX 49 7303 13 210; Internet: www.gruenau-illertissen.de/com/index.html)

GuiZhou Essence Fine Chemicals Ltd. Co.

No. 1 ShaiTianBa Rd., GuiYang, 550002, China (Tel: 86 851 5926830; FAX 86 851 5620341; E-mail: contact@eschemical.com; Internet: www.eschemical.com)

Gulf Chemical International Corp.

2925 Briarpark, Ste. 130, Houston, TX, 77042-3715, USA (Tel: 713-783-7081; FAX 713-783-2479; E-mail: jr@gciusa.com)

Haarmann & Reimer

Haarmann & Reimer GmbH, Subsid. of Bayer Corp., Posftach 12 53, 37601 Holzminden, Germany (Tel: 49 5531 90 0; FAX 49 5531 90 16 49; Internet: www.haarmann-reimer.com)

Haarmann & Reimer Ltd., Fieldhouse Lane, Marlow, Buckinghamshire, SL7 1NA, UK (Tel: 44 1635 562020; FAX 44 1635 562030)

Haarmann & Reimer S.R.L., 13 Via Ludovico di Breme, 20156 Milan, Italy (Tel: 39 2 38 09 81 1; FAX 39 2 38 09 81 42)

Haarmann & Reimer S.A., 13, Rue Jean jaures, 92807 Puteaux, France (Tel: 33 1 49 06 55 00; FAX 33 1 49 06 81 50)

Bayer A/S/Div. Haarmann & Reimer Norden, Norgaardsvej 32, 2800 Lyngby, Denmark (Tel: 45 45 23 52 36; FAX 45 45 23 52 58)

Haarmann & Reimer S.A.E., Subsid. of Haarman & Reimer GmbH, Apartado de Correos 48, 08150 Parets del Valles (Barcelona), Spain (Tel: 34 93 573 83 40; FAX 34 93 573 12 35)

Haarmann & Reimer Corp., Subsid. of Bayer Corp., 300 North St., Teterboro, NJ, 07608, USA (Tel: 201-288-3200; 800-HRFLORA; FAX 201-462-2200)

H&R Florasynth, 300 North St., Teterboro, NJ, 07608, USA (Tel: 973-912-5898; FAX 973-912-5731)

Haarmann & Reimer Canada, Div. of Bayer Canada Inc., 7600 Autoroute Transcanadienne, Pointe Claire, Quebec, H9R 1C8, Canada (Tel: 514-697-5550; FAX 514-697-4334)

Haarmann & Reimer S.A., Apartado Postal 1292, Monterrey N.L., 64000, Mexico (Tel: 52 83 29 6700; FAX (8) 3 29 67 01)

Haarmann & Reimer S.A., Valparaiso 4125, 1667 Tortuguitas, Prov. de Buenos Aires, Argentina (Tel: 54 33 27 45 95 08; FAX 54 33 27 45 46 50)

Haarmann & Reimer de Venezuela S.A., Torre Bayer, Piso 7, Av. Tamanaco, El Rosal, Caracas, 1060, Venezuela (Tel: 58 2 905 2700; FAX 58 2 951 1438)

Haarmann & Reimer Ltda., Rua Alexandre de Gusmao 568, Socorro, CEP 04760-020 Sao Paulo-SP, Brazil (Tel: 55 11 56 94 60 00; FAX 55 11 5 23 52 29)

Haarmann & Reimer (Chile) S.A., Lira 2338, Communa San Joaquin, Casilla 870, Santiago de Chile, Chile (Tel: 56 2 555 85 42; FAX 56 2 556 56 48)

Haarmann & Reimer de Colombia Ltda., Carrera 58 No. 10-80, Santa Fe de Bogota, Colombia (Tel: 57 1 425 22 00; FAX 57 1 425 22 10)

Haarmann & Reimer Australia Pty. Ltd., PO Box 6005, Delivery Centre, Blacktown, NSW, 2148, Australia (Tel: 61 2 96 71 34 44; FAX 61 2 96 21 80 86)

Haarmann & Reimer New Zealand, Div. of Bayer New Zealand Ltd., 3 Argus Place, Glenfield, Auckland, 10, New Zealand (Tel: 64 9 443 3093; FAX 64 9 443 3094)

Haarmann & Reimer S.A. (Pty.) Ltd., 27 Wrench Rd., Isando, 1601, Rep. of S. Africa (Tel: 27 11 921 5911; FAX 27 11 921 5453)

Haarmann & Reimer K.K., 3-32-3, Shiba, Minato-ku, Tokyo, 105-0014, Japan (Tel: 81 3 3798 4211; FAX 81 3 3798 4223)

Haarmann & Reimer Hong Kong, Div. Bayer China Co. Ltd., 48F Caroline Centre, 28 Yunn Ping Rd., Causeway, Hong Kong (Tel: 852 28 95 58 88; FAX 852 28 08 07 32)

Haarmann & Reimer (SEA) Pte. Ltd., No. 190 Pandan Loop, Singapore, 128379, Singapore (Tel: 65 774 5880; FAX 65 774 5233)

Haarmann & Reimer (Thailand) Co., Ltd., Bayer Center, 2nd Fl., 130/1 North Sathorn Rd., Silom, Bangrak, Bangkok, 10500, Thailand (Tel: 66 2 23 54 44 03; FAX 66 2 2 67 27 96)

Bayer Taiwan Co. Ltd./Div. Haarmaan & Reimer, 237 Sung Chian Rd., 8th Fl., PO Box 55-35, Taipei, Taiwan, R.O.C. (Tel: 886 2 25 03 91 23; FAX 886 2 25 16 05 40)

Bayer Korea Co. Ltd./Div. Haarmaan & Reimer, 395-62, Shindae bang-dong, Dongjak-ku, Seoul, Korea (Tel: 82 2 829 6689; FAX 82 2 849 8744)

Bayer India Ltd./Div. Haarmaan & Reimer, Bldg. C-565, Kolshet Rd., Thane, 40 06 07, India (Tel: 91 22 534 2851; FAX 91 22 533 7924)

H&A (Canada) Industrial Inc.

11260 Tapscott Rd., Toronto, Ontario, M1X 1E9, Canada (Tel: 416-412-9518; FAX 416-293-9066; E-mail: sales@hacanada.com; Internet: www.hacanada.com)

Haco Ltd.

Worbstrasse 262, 3073 Gumligen, Switzerland (Tel: 41 31 950 1111; FAX 41 31 950 1515; E-mail: fritz.aeberharcl@haco.ch; Internet: www.haco.ch)

C.P. Hall Co.

311 South Wacker Dr., Suite 4700, Chicago, IL, 60606-6622, USA (Tel: 312-554-7400; 877-427-4255; FAX 312-554-7499; E-mail: info@cphall.com; Internet: www.cphall.com)

Howard Hall

Div. of R.W. Greeff & Co. Inc., 777 West Putnam Ave., Greenwich, CT, 06830, USA (Tel: 203-532-2900; FAX 203-532-2980)

Halocarbon Products Corp.

PO Box 661, 887 Kinderkamack Rd., River Edge, NJ, 07661, USA (Tel: 201-262-8899; FAX 201-262-0019; E-mail: info@halocarbon.com; Internet: www.halocarbon.com)

Haltermann

Haltermann GmbH, Div. of Ascot Specialty Chemicals, Part of Dow Chemical Co., Schopenstehl 15, 20095 Hamburg, Germany (Tel: 49 40 333 18 0; FAX 49 40 333 18 214; E-mail: info@hg.haltermann.de; Internet: www.haltermann.com)

Haltermann Products UK, Div. of Ascot plc, All Saints Refinery, Cargo Fleet Rd, Middlesbrough, Cleveland, TS3 6AF, UK (Tel: 44 1642 248 555; FAX 44 1642 250 369; E-mail: marketing@haltermann.co.uk; Internet: www.haltermann.com)

Haltermann Ltd., PO Box 25, Barnet, Herts, EN5 1TX, UK (Tel: 44 20 81441 1449; FAX 44 20 8441 1388; E-mail: haltermannuk@lineone.net)

Haltermann N.V., Haven 1972, Kentenislaan 3, 9130 Kallo (Kieldrecht), Belgium (Tel: 32 3 570 0211; FAX 32 3 575 0535; E-mail: products@haltermann.be)

Haltermann S.A.R.L., 29/31 Rue Raspail, 92300 Levallois/Frankreich, France (Tel: 33 1 42 70 78 59; FAX 33 1 47 56 99 19; E-mail: haltermannfrance@wanadoo.fr)

Haltermann AB, Petroleumgatan 7, 21124 Malmo, Sweden (Tel: 46 40 680 32 00; FAX 46 40 93 84 85; E-mail: info@haltermann.se)

Haltermann A/S, Sondre Molevej 14-16, 4600 Koge, Denmark (Tel: 45 56 65 33 70; FAX 45 56 65 70 09; E-mail: info@haltermann.dk)

Haltermann AS, PO 86, Lysacer Torg 8, 1325 Lysacer, Norway (Tel: 47 67 516 950; FAX 47 67 516 951; E-mail: info@haltermann.no)

Haltermann Products, 1201 Sheldon Rd., PO Box 429, Channelview, TX, 77530-0429, USA (Tel: 281-457-2768; FAX 281-457-1469; E-mail: info@halterman-usa.com; Internet: www.haltermann-ltd.com)

Hampshire Chemical

Hampshire Chemical Corp., Subsid. of Dow Chemical Co., 55 Hayden Ave., Lexington, MA, 02421, USA (Tel: 781-869-3400; FAX 781-860-5290; Internet: www.dow.com/hampshire)

Hampshire Chemical Ltd., Craven House, 40 Uxbridge Rd., London, W5 2BS, UK (Tel: 44 1208 579 5443; FAX 44 1208 579 3494)

Chr. Hansen

Chr. Hansen's Holding A/S, 10-12 Bøge Allé, 2970 Hørsholm, Denmark (Tel: 45 45 74 74 74; FAX 45 45 74 88 88; Internet: www.chr-hansen.com)

Chr. Hansen (UK) Ltd., 2 Tealgate, Charnham Park, Hungerford, Berkshire, RG17 0YT, UK (Tel: 44 1488 689800; FAX 44 1488 685436)

Chr. Hansen France S.A., PO 64, 91292 Arpajon Cedex, France (Tel: 33 1 69 88 36 36; FAX 33 1 0 84 15 94)

Chr. Hansen GmbH, Giessener STr. 94, 35414 Pohlheim, Germany (Tel: 49 64 03 95 01 0; FAX 49 64 03 95 01 30)

Chr. Hansen S.p.A., Viale della Liberazione 1/5, 20094 Corsico (MI), Italy (Tel: 39 2 44 87 341; FAX 39 2 44 72 497)

Chr. Hansen, Inc., 9015 W. Maple St., Milwaukee, WI, 53214-4298, USA (Tel: 414-607-5700; 888-828-6600; FAX 414-607-5959)

Chr. Hansen Ltd., 1146 Aerowood Dr., Mississauga, Ontario, L4W 1Y5, Canada (Tel: 905-625-2560; FAX 905-625-8157)

Chr. Hansen Industria e Comercio Ltda., Caixa Postal 371, CEP 13276-970, Valinhos - SP, Brazil (Tel: 55 19 3881 8300; FAX 55 19 3881 8253)

Chr. Hansen Pty. Ltd., PO Box 591, Bayswater, Victoria, 3153, Australia (Tel: 61 3 9762 9600; FAX 61 3 9762 9700)

Chr. Hansen Japan, Yokohama Landmark Tower, 2-2-1-1-, Minato-Mirai, Nishi-ku, Yokohama, 220-8130, Japan (Tel: 81 45 224 1690; FAX 81 45 224 1692)

Hansol Chemical

18th Fl., Hansol Bldg., 736-1 Yoksam-dong Kangnam-ku, Seoul, 135-080, Korea (Tel: 82 2 3287-6714; FAX 82 2 3287 6712; Internet: www.hansolchem.co.kr)

Hansotech, Inc.

Part of Petroferm Inc., 144 Woodbury Rd., Woodbury, NY, 11797, USA (Tel: 516-692-6890; FAX 516-692-6865)

Harborchem

186 North Ave. E., PO Box 630, Cranford, NJ, 07016, USA (Tel: 908-272-7070; FAX 908-272-8966; E-mail: inquiries@harborchem.com; Internet: www.harborchem.com)

Harcros Chemicals Inc.

Part of HCI Acquisitions Inc., PO Box 2930, 5200 Speaker Rd., Kansas City, KS, 66110-2930, USA (Tel: 913-321-3131; FAX 913-621-7718; E-mail: custserv@harcroschem.com; Internet: www.harcroschem.com)

M.W. Hardy & Co., Inc.

111 Broadway, New York, NY, 10006, USA (Tel: 212-964-1550; FAX 212-964-1553)

Keith Harris & Co. Ltd.

7 Sefton Rd., Pennant Hills, PO Box 147, Thornleigh, NSW, 2120, Australia (Tel: 61 2 9484 1341; FAX 61 2 9481 8145; E-mail: enquiries@keithharris.com.au; Internet: www.keithharris.com.au)

Hart Products Corp.

173 Sussex St., Jersey City, NJ, 07302, USA (Tel: 201-433-6632; FAX 201-435-7268)

Harwick Standard Distribution Corp.

60 S. Seiberling St., PO Box 9360, Akron, OH, 44305-0360, USA (Tel: 330-798-9300; 800-899-4412; FAX 330-798-0214; E-mail: smcgovern@harwickstandard.com; Internet: www.harwickstandard.com)

T. Hasegawa Co., Ltd.

4-4-14, Nihonbashi Hon-cho, Chuo-ku, Tokyo, 103, Japan (Tel: 81 3 3241 1151; FAX 81 3 3278 8075)

Hasnil Chemicals Inc.

911 Campbell Ave., West Haven, CT, 06516, USA (Tel: 203-937-5786; FAX 201-931-0382)

Hatco Corp.

1020 King George Post Rd., Fords, NJ, 08863-0601, USA (Tel: 732-738-1000; FAX 732-738-9385; E-mail: info@hatcocorporation.com; Internet: www.hatcocorporation.com)

A.C. Hatrick Chemicals Pty. Ltd.

49-61 Stephen Road, Botany Bay, Botany, NSW, 2019, Australia (Tel: 61 2 9666-0331; FAX 61 2 9666 3872)

Hawkins Chemical, Inc.

3100 E. Hennepin Ave., Minneapolis, MN, 55413, USA (Tel: 612-331-6910; 800-328-5460; FAX 612-331-5304; E-mail: tom.babineau@hawkinsinc.com; Internet: www.hawkinschemical.com)

Hawks Chemical Co. Ltd.

2 Tower St, Hyde, Cheshire, SK14 1JW, UK (Tel: 44 161 367 9441; FAX 44 161 367 9443; Internet: www.hawks-chem.com)

Hays Chemical Distribution Ltd.

Part of Albion, Rawdon House, Green Lane, Yeadon, Leeds, West Yorkshire, LS19 7XX, UK (Tel: 44 1132 505811; FAX 44 1132 500113; E-mail: enquiries@hayschem.co.uk; Internet: www.albionchemicals.co.uk/haysweb)

HCH Marketing Inc.

PO Box 159, Manvel, TX, 77578, USA (Tel: 281-331-7766; 800-645-1551; FAX 281-585-3685)

Holland Chemical International

Holland Chemical International (HCI), Chesterfield, MO, USA (Tel: 314-205-2266; Internet: www.hollandchemical.com)

Holland Chemical Int'l. N.V., Postbus 12910, 1100 AX Amsterdam, The Netherlands (Tel: 31 20 697 0666; FAX 31 20 697 3869; E-mail: info@hcinl.com)

HCI Chemtech Distribution Inc., 424 S. Woods Mill Road, Suite 325, Chesterfield, MO, 63017-3428, USA (Tel: 314-832-5010; FAX 314-481-4733)

HCI Industrial Chemical & Supply Co.

3520 Adamo Dr., Tampa, FL, 33605, USA (Tel: 813-247-7354; 800-968-1386)

H & C Industries, Inc.

1311 Crenshaw Blvd., Torrance, CA, 90501, USA (Tel: 310-328-3683; FAX 310-533-8297; E-mail: handcind@ix.netcom.com)

HCI Worth Chemical Corp.

2 Segal Blvd., PO Box 20725, Greensboro, NC, 27420, USA (Tel: 336-292-5166)

Heico Chemicals, Inc.

A Cambrex Co., Route 611, PO Box 730, Delaware Water Gap, PA, 18327-0160, USA (Tel: 570-420-3900; 800-344-3426; FAX 570-421-9012; E-mail: jdoherty@cambrex.com; Internet: www.heicochemicals.com)

Helm

Helm AG, Nordkanalstrasse 28, 20097 Hamburg, Germany (Tel: 49 40 2375 0; FAX 49 40 2375 1845; E-mail: info@helmag.com; Internet: www.helmag.com)

Helm Great Britain Ltd., Wimbledon Bridge House, 1 Hartfield Rd, London, N3 2TL, UK (Tel: 44 20 8544 9000; FAX 44 20 8544 1011; E-mail: chemicals@helm-gb.co.uk)

Helm New York Inc., A Helm Company, 1110 Centennial Ave., Piscataway, NJ, 08854-4169, USA (Tel: 732-981-1160; FAX 732-981-0965; E-mail: helm_ny@compuserve.com; Internet: www.helmnewyork.com; www.helmny.com)

J.F. Henry Chemical Co., Inc.

PO Box 3848, 1620 Rte. 22, Union, NJ, 07083, USA (Tel: 908-688-6777; FAX 908-688-8338; E-mail: rhenry@henrychem.com)

Hercules

Hercules Inc., 1313 N. Market St., Wilmington, DE, 19894-0001, USA (Tel: 302-594-5000; 800-247-4372; FAX 302-594-5400; E-mail: resins@herc.com; Internet: www.herc.com)

Hercules Inc./Aqualon Div., 1313 N. Market St., Wilmington, DE, 19894-0001, USA (Tel: 302-594-5000; 800-345-0447; FAX 302-995-4776; E-mail: aqualon-us@herc.com; Internet: www.aqualon.com)

S.A. Hercules Europe, Interleuvenlaan 25, 3001 Heverlee, Belgium (Tel: 32 16 40 2000; FAX 32 16 40 00 87)

Hercules Ltd., Div. of Hercules Inc., 31 London Road, Reigate, Surrey, RH2 9YA, UK (Tel: 44 1737 242434; FAX 44 1737 224288)

Hercules Ltd./Aqualon Div., Foundry Lane, Widnes, Cheshire, WA8 8UD, UK (Tel: 44 151 422 5897; FAX 44 151 423 2722; E-mail: aqualon-eur@herc.com)

Hercules Int'l. Ltd. LLC, 8 Veraartlaan, 2288 GM Rijswijk, The Netherlands (Tel: 31 70 4134 200; FAX 31 70 3368 116)

Hercules Int'l. Ltd. LLC/Aqualon Div., Postbus 5822, 2280 HV Rijswijk, The Netherlands (Tel: 31 70 413 4200; FAX 31 70 398 9893)

Aqualon France S.A., Parc du St Laurent, Immeuble Le Toronto, 54, route de Sartrouville, 78232 Le Pecq Cedex (St. Germain), France (Tel: 33 1 30 09 40 00; FAX 33 1 39 76 02 07; E-mail: aqualon-eur@herc.com)

Hercules GmbH/Aqualon Div., Paul-Thomas-Strasse 56/58, PO Box 130125, 40551 Dusseldorf, Germany (Tel: 49 211 749 10; FAX 49 211 749 1200)

Hercules Italia SpA/Aqualon Div., Via Bondanello 15, PO Box 1726, 40013 Castelmaggiore (Bologna), Italy (Tel: 39 051 715 333; FAX 39 051 714 734)

Hercules Quimica S.A./Aqualon Div., Carretera de Valencia, 5/9, PO Box 32, 43006 Tarragona, Spain (Tel: 34 9 77 558 800; FAX 34 9 77 558 814)

Hercules Asia Pacific/Hercules China Limited, 1907-8 Harcourt House, 39 Gloucester Rd., Wanchai, Hong Kong (Tel: 852 2527 2638; FAX 852 2528 1598; E-mail: aqualon-asiapac@herc.com)

Hercules Chemicals Singapore Pte. Ltd./Aqualon Div., 200 Pandan Loop, #07-01 Pantech 21, Singapore, 128388, Singapore (Tel: 6775 5366; FAX 6775 5433)

Herdillia Chemicals Limited

Earnest House, 13th Fl., Nariman Point, Mumbai, 400 021, India (Tel: 91 22 281 6470; FAX 91 22 281 6469; E-mail: export@herdillia.com; Internet: www.herdillia.com)

Hexcel Corp./Chemical Products Div. DEFUNCT DIV.

215 N. Centennial St., Zeeland, MI, 49464, USA (Tel: 269-772-2193; FAX 269-772-7344)

Hickson & Welch Ltd.

Div. of Arch Chemicals, Wheldon Rd., Castleford, Yorkshire, WF10 2JT, UK (Tel: 44 1977 556 565; FAX 44 1977 712 222; E-mail: sales@hickson.co.uk; Internet: www.hicksonandwelch.co.uk)

High Point Textile Auxiliaries LLC (HPTA)

Part of Boehme Filatex Inc., PO Box 2208, 1620 Belmar Street, High Point, NC, 27261, USA (Tel: 336-887-6410; 877-678-6410; FAX 336-884-8786; E-mail: information@hptallc.com; Internet: www.hptallc.com)

High Polymer Labs

High Polymer Labs Ltd., 803 Vishal Bhawan, 95 Nehru Place, New Delhi, 110 019, India (Tel: 91 11 643 1379; FAX 91 11 647 4350; E-mail: hpll@hpl-group.com; Internet: www.highpolymerlabs.com)

High Polymer Labs Inc., 4 Fairhaven Mall, Suite E15, Mineola, NY, 11501, USA (Tel: 516-294-3387; FAX 516-294-1034; E-mail: adshpl@aol.com; Internet: www.highpolymerlabs.com)

Hill Brothers Chem. Co.

1675 N. Main St., Orange, CA, 92667, USA (Tel: 714-998-8800; 800-821-7234; FAX 714-998-6310; Internet: www.hillbrothers.com)

J.H. Hinz Co.

2001 Crocker Rd., Suite 200, Westlake, OH, 44145, USA (Tel: 440-871-3333; 800-929-5552; FAX 440-871-4475; E-mail: jhhinz@msn.com; Internet: www.jhhinz.com)

Hoechst

Hoechst AG. *See* Aventis SA

Hoechst Celanese/Int'l. Headquarters. DEFUNCT, LAST KNOWN: PO Box 2500, Rt. 202-206 North, Somerville, NJ, 08876-1258, USA (Tel: 908-231-2000; Internet: www.hoechst.com)

Hofmen International Inc.

8216 Princeton-Glendale Rd., Suite 176, West Chester, OH, 45069-1675, USA (Tel: 973-882-2898; 877-882-2898; FAX 973-556-1214; E-mail: CustomerService@HofmenInternational.com; Internet: www.hofmeninternational.com)

Hokko Chemical Industry Co., Ltd.

Mitsui Bldg., No. 2, 4-4-20, Nihonbashi Hongoku-cho, Chuo-ku, Tokyo, 103, Japan (Tel: 81 3 3279 5142; FAX 81 3 3279 3857)

Holland Company Inc.

153 Howland Ave., Adams, MA, 01220, USA (Tel: 413-743-1292; 800-639-9602; FAX 413-743-1298; E-mail: info@hollandcompany.com; Internet: www.hollandcompany.com)

Hollister-Stier Laboratories

3524 N. Regal St., Spokane, WA, 99207, USA (Tel: 509-489-5656; 800-541-3782; FAX 509-482-3543)

HoltraChem, Inc.

159 Boden Lane, Natick, MA, 01760, USA (Tel: 617-655-2510; 800-343-6470; FAX 617-653-2682)

Honeywell

Honeywell Specialty Chemicals, 101 Columbia Rd., Morristown, NJ, 07962, USA (Tel: 973-455-2000; 800-222-0094; FAX 973-455-4807; E-mail: info@corp.honeywell.com; Internet: www.honeywell.com; www.specialtychem.com)

Honeywell Performance Polymers & Chemicals, 101 Columbia Rd., Morristown, NJ, 07962, USA (Tel: 973-455-2000; FAX 973-455-4807; Internet: www.honeywellppc.com; www.honeywell-plastics.com)

Honeywell Specialty Wax & Additives, 3079 Premiere Pkwy., Suite 100, Duluth, GA, 30097, USA (Tel: 800-421-4929; FAX 770-689-0522)

Honeywell Europe N.V., Haasrode Research Park, Grauwmeer 1, 3001 Heverlee (Leuven), Belgium (Tel: 32-16-39 12 11; FAX 32-16-39 13 71)

Honeywell Specialty Chemicals (Singapore) Pte. Ltd., 16A Science Park, 03-01 Pascal Drive, Singapore, 118228, Singapore (Tel: 65 775 2133; FAX 65 775 2311)

Honeywill & Stein Ltd.

An IMCD Company, Times House, Throwley Way, Sutton, Surrey, SM1 4AF, UK (Tel: 44 20 8770 7090; FAX 44 20 8770 7295; E-mail: info@honeywill.co.uk; Internet: www.honeywill.co.uk)

Hosoi Chemical Industry Co., Ltd.

2-3-16, Nihonbashi, Muro-machi, Tokyo, 103, Japan (Tel: 81 3 3270 3601; FAX 81 3 3279 5863)

Houghton Chemical Corp.

PO Box 307, 52 Cambridge St., Allston, MA, 02134, USA (Tel: 617-254-1010; 800-777-2466; FAX 617-254-2713)

J.M. Huber

J.M. Huber Corp., 333 Thornall St., Edison, NJ, 08837-2220, USA (Tel: 732-549-8600; E-mail: webmaster@huber.com; Internet: www.huber.com)

J.M. Huber Corp./Chemicals Div., PO Box 310, 907 Revolution St., Havre de Grace, MD, 21078, USA (Tel: 410-939-3500; FAX 410-939-7301; E-mail: hgtcv@huber.com; Internet: www.huberchemicals.com)

Hukill Chemical Corp.

7013 Krick Rd., Bedford, OH, 44146, USA (Tel: 440-232-9400; FAX 440-232-9477; Internet: www.hukill.com)

Hummel Croton

10 Harmich Rd., South Plainfield, NJ, 07080-4899, USA (Tel: 908-754-1800; FAX 908-754-1815; E-mail: humchem@aol.com; Internet: www.hummelcroton.com)

The Humphrey Chemical Co. Inc.

A Cambrex Co., 45 Devine St., North Haven, CT, 06473-0325, USA (Tel: 203-230-4940; 800-652-3456; FAX 203-287-9197; E-mail: jdoherty@cambrex.com; Internet: www.cambrex.com)

Hunan Xinyu Native Produce & Animal By-Products Import & Export Co., Ltd.

183 Yuandayi Rd., Changsha, Hunan, China (Tel: 86 731 4421482; FAX 86 731 4427356; E-mail: xinyu@hunanxinyu.com; Internet: www.hunanxinyu.com)

Huntsman

Huntsman Corp., 500 Huntsman Way, Salt Lake City, UT, 84108, USA (Tel: 801-584-5700; 800-421-2411; FAX 801-584-5781; E-mail: kelly_mckibbin@huntsman.com; Internet: www.huntsman.com)

Huntsman Petrochemical Corp., PO Box 27707, Houston, TX, 77227-7707, USA (Tel: 713-235-6000; FAX 713-235-6416)

Huntsman Polyurethanes Group, 286 Mantua Grove Rd., West Deptford, NJ, 08066-1732, USA (Tel: 856-423-8300; 800-257-5547; FAX 856-423-8501; Internet: www.icipu.com; www.rubinate.com)

Huntsman Surface Sciences, PO Box 27707, Houston, TX, 77227-7707, USA (Tel: 713-235-6000; FAX 713-235-6416; E-mail: surface_sciences_americas@huntsman.com; Internet: www.huntsman.com/surface_sciences)

Huntsman Corp. Canada Inc., PO Box 450, Guelph, Ontario, N1H 6K8, Canada (Tel: 519-824-3280; FAX 519-824-0755)

Huntsman Polyurethanes (Canada), 2795 Slough St., Mississauga, Ontario, L4T 1G2, Canada (Tel: 905-678-9150; FAX 905-678-9350)

Huntsman De Mexico, S.A. de C.V., Angel Urraza Num. 303, Col. Insurgentes San Borja, Mexico City, 03100, Mexico (Tel: 52 5 687 2558; FAX 52 5 687 0991)

Huntsman de Brasil Participacoes Ltda., Rua André de Leão , 210, Capela do Socorro, Sao Paulo SP, 04762-030, Brazil (Tel: 55 11 5696 9312; FAX 55 11 5521 9933)

Huntsman Polyuretanos (Argentina), Reconquista 2780, 1617 El Talar de Pacheco, Pcia, Buenos Aires BA, Argentina (Tel: 54 114 736 6000; FAX 54 114 736 6051)

Huntsman Corp. Belgium N.V., Everslaan 45, 3078 Everberg, Belgium (Tel: 32 2 758 9211; FAX 32 2 758 9068)

Huntsman Polyurethanes (Belgium), Everslaan 45, 3078 Everberg, Belgium (Tel: 32-2-758 92 11; FAX 32-2-758 90 18)

Huntsman Petrochemicals (UK) Ltd., PO Box 99, Wilton Centre, Redcar, TS10 4YA, UK (Tel: 44 1642 453366; FAX 44 1642 834608)

Huntsman Polyurethanes (UK), PO Box 99, Wilton Int'l., Redcar, TS10 4YA, UK (Tel: 44 1642 834700; FAX 44 1642 834742)

Huntsman Surface Sciences UK, PO box 9792, 210-222 Hagley Rd. West, Oldbury, West Midlands, B68 0WA, UK (Tel: 44 121 429 6700; FAX 44 121 420 5870; E-mail: surface_sciences_europe@huntsman.com)

Huntsman (Germany) GmbH/Performance Chemicals, Baumwall 5, 20459 Hamburg, Germany (Tel: 49 40 37670 0; FAX 49 40 37670 150)

Huntsman Polyurethanes (Germany), Land-Au 30, 94469 Deggendorf, Germany (Tel: 49 991 27040; FAX 49 991 24884)

Huntsman International Trading Corp., 150 Beach Rd., #14-05/08 Gateway West, Singapore, Singapore (Tel: 65 297 3363; FAX 65 298 8606)

Huntsman Japan, 2-7-6, Sogo Kudan Minami Bldg., 2-7-6, Kudan Minani, Chiyoda-ku, Tokyo, 104-0074, Japan (Tel: 81 3 5215 0771; FAX 81 3 5215 0777)

Huntsman Polyurethanes (Taiwan), 19 Industrial Third Rd., Kuan Yin, Taoyuan County, Taiwan R.O.C. (Tel: 886 3 483 8616; FAX 886 3 483 9324)

Huntsman Surface Sciences (Australia), Gate 2 Newsom St., Ascot Vale, Victoria, 3032, Australia (Tel: 61 1300 301 529; FAX 61 392 836 207; E-mail: surface_sciences_asiapacific@huntsman.com)

Hutchins & Hutchins Inc.

PO Box 40, Waynesboro, VA, 22980, USA (Tel: 540-949-6663; 800-554-4736; FAX 540-943-9803)

HVC, Inc.

An Ellis & Everard Subsid., 4600 Dues Dr., Cincinnati, OH, 45246, USA (Tel: 513-714-5264; 800-947-7373; FAX 513-874-3107)

Hydrite Chemical Co.

Drawer 0948, 300 N. Patrick Blvd., Brookfield, WI, 53008-0948, USA (Tel: 414-792-1450; 800-346-7960 x 741; FAX 414-792-7862; E-mail: deb.mayersak@hydrite.com; Internet: www.hydrite.com)

Hydro Plus, Inc.

3465 Curry Lane, Abilene, TX, 79606, USA (Tel: 915-695-4642)

Hydro Pronova a.s.

Wholly owned subsid. of Norsk Hydro ASA, 0240 Oslo, Norway (Tel: 47 22 53 81 00; FAX 47 22 53 74 90; E-mail: age.wik@hydro.com; Internet: www.pronova.hydro.com)

ICC

ICC Industries Inc., 460 Park Ave., New York, NY, 10022, USA (Tel: 212-521-1700; FAX 212-521-1970; E-mail: icc@iccchem.com; Internet: www.iccchem.com)

ICC Chemical Corp., An affiliate of ICC Industries, Inc., 460 Park Ave., New York, NY, 10022, USA (Tel: 212-521-1700; 800-422-1720; FAX 212-521-1970; E-mail: trading@iccchem.com; Internet: www.iccchem.com)

ICC Hong Kong, Rm. #1110, New World Office Bldg., West Wing, 20 Salisbury Rd., Tsimshatsui East, Kowloon, Hong Kong (Tel: 852-2366-1678; FAX 852-2367-1377; E-mail: hngkng@iccchem.com)

ICC Chemical Corp., Gold Name Tower, 21st Fl., 143 Zhongshan Rd., No. 68, Renmin Rd., Zhongshan District, Dalian, 116001, P.R. China (Tel: 86 411 272 1755; FAX 86 411 272 1955; E-mail: dalchi@iccchem.com)

ICN Pharmaceuticals

ICN Pharmaceuticals Inc., ICN Plaza, 3300 Hyland Ave, Costa Mesa, CA, 92626, USA (Tel: 714-545-0100; 800-556-1937; Internet: www.icnpharm.com)

ICN Biomedical Research Products, ICN Plaza, 3300 Hyland Ave, Costa Mesa, CA, 92646, USA (Tel: 714-545-0100; 800-854-0530; FAX 800-334-6999; E-mail: sales@icnbiomed.com; Internet: www.icnbiomed.com)

ICN Pharmaceuticals Ltd., Cedarwood, Chineham Business Park, Crockford Lane, Basingstoke, Hampshire, RG24 8WD, UK (Tel: 44 1256 707744; FAX 44 1256 707334)

I.C. Trading Co., Inc.

249 Walnut Rd., Glen Cove, NY, 11542, USA (Tel: 516-609-9191; FAX 516-609-9199; E-mail: sales@ic-trading.com; Internet: www.ic-trading.com)

Idemitsu Petrochemical

Idemitsu Petrochemical Co., Ltd., Idemitsu Mita Bldg., 5-6-1, Shiba, Minato-ku, Tokyo, 108, Japan (Tel: 81 3 3457 8610; FAX 81 3 3457 8657)

British Traders & Shippers Ltd., Distrib. for Idemitsu, 429-431 Rainham Road South, Dagenham, Essex, RM10 8RN, UK (Tel: 44 181 596 7500; FAX 44 181 596 7509; E-mail: pvax@britishtraders.co.uk; Internet: www.britishtraders.co.uk)

Ikeda

Ikeda Corp., New Tokyo Bldg., 3-3-1-Marunouchi, Chiyoda-ku, Tokyo, 100-0005, Japan (Tel: 81 3 3212 8791; FAX 81 3 3215 5069; E-mail: b-dept@ikeda-corp.jp; Internet: www.ikeda-corp.co.jp)

S. Black (Import & Export) Ltd., Distrib. for Ikeda, Foxholes Business Park, John Tate Road, Hertford, Hertfordshire, SG13 7YH, UK (Tel: 44 1992 825555; FAX 44 1992 825566; E-mail: enquires@sblack.com; Internet: www.sblack.com)

Independent Chemical Corp.

79-51 Cooper Ave., Glendale, NY, 11385, USA (Tel: 718-894-0700; FAX 718-894-9224)

Indofine Chemical Co., Inc.

PO Box 473, Somerville, NJ, 08876, USA (Tel: 908-359-6778; 888-463-6346; FAX 908-359-1179; E-mail: chemical@indofinechemical.com; Internet: www.indofinechemical.com)

Industrial Oil Products

144 Woodbury Rd., Woodbury, NY, 11797, USA (Tel: 516-692-6862; FAX 516-692-3140; E-mail: blakehanson@tungoil.com)

Industrias Monfel, S.A. de C.V.

Barracuda #4, Apartado Postal 98, Cuautitlan, Edo. de Mexico, 54800, Mexico (Tel: 52 5 899 0029; FAX 52 5 899 0010; E-mail: ricardo.medina@monfel.com; Internet: www.monfel.com)

INEOS

INEOS PLC, Hawkslease, Chapel Lane, Lyndhurst, SO43 7FG, UK (Tel: 44 2380 287063; FAX 44 2380 287069; E-mail: info@ineos.com; Internet: www.ineos.com)

INEOS Belgium nv, Haven 1053, Nieuwe Weg 1, 2070 Zwijndrecht, Belgium (Tel. 32 3 250 90 38; FAX 32 3 250 94 75; E-mail: jean.pierre.etienne@ineos.com)

INEOS Italia, Piazza Emilia 1, 20129 Milano, Italy (Tel: 39 2 7611 5407; FAX 39 2 7611 8189; E-mail: stefano.bazzano@ineos.com)

INEOS Iberica, C/Provenza 288, Pral., 8008 Barcelona, Spain (Tel: 34 934 678301; FAX 34 934 879676; E-mail: jaime_valls@ineos-a.com)

INEOS Oxide, Nieuwe Weg 1, 2070 Zwijndrecht, Belgium (Tel: 32 3 250 9111; FAX 32 3 252 8453)

INEOS Phenol, Dechenstrasse 3, 45966 Gladbeck, Germany (Tel: 49 2043 958 0; FAX 49 2043 9589 00; E-mail: info@ineosphenol.de; Internet: www.phenolchemie.de)

INEOS Silicas Limited, PO Box 26, 4 Liverpool Rd., Warrington, Cheshire, WA5 1AB, UK (Tel: 44 1925 416100; FAX 44 1925 416116; E-mail: steven.steggerda@crogb.sprint.com)

INEOS Silicas Netherlands BV, PO Box 1, 6245 ZG Eijsden, The Netherlands (Tel: 31 43 409 9333; FAX 31 43 409 4865)

INEOS Silicas Italia SpA, Via dei Cipressi, 10, 37033 Montorio (Verona), Italy (Tel: 39 45 895 0611; FAX 39 45 895 0620)

INEOS Silicas Americas LLC, 111 Ingalls Ave., Joliet, IL, 60435, USA (Tel: 815-727-3651; 800-727-3651; FAX 815-727-3667; E-mail: techsupport@ineossilicas.com; Internet: www.ineossilicas.com)

INEOS Silicas Brasil Ltda., Rua Luis Stamatis, 1100, Jacana, Sao Paulo, 02260-001, Brazil (Tel: 55 11 36241 1444; FAX 55 11 6241 9247)

INEOS Silicas South Africa, 169 Tedstone Rd., PO Box 14016, Wadeville, Ganteng, 1422, Rep. of S. Africa (Tel: 27 11 820 7111; FAX 27 11 827 6922)

PT INEOS Silicas Indonesia, Jl. Rembang Industri Raya no. 24, Bangil 0 Pasuruan km 50, Jawa Timur, 67152, Indonesia (Tel: 62 34 3740 114; FAX 62 34 3740 113)

INEOS Silicas Asia Pacific Pte. Ltd., No. 3 Shenton Way, #15-09 Shenton House, Singapore, 068805, Singapore (Tel: 65 325 0898; FAX 65 324 2243)

Innovative Processing, Inc.

50 Abele Rd., Suite 1005, Bridgeville, PA, 15017, USA (Tel: 412-221-3700; FAX 412-221-7470; E-mail: info@rammglobal.com; Internet: www.rammglobal.com)

Inolex

Inolex Chemical Co., Jackson & Swanson Sts., Philadelphia, PA, 19148-3497, USA (Tel: 215-271-0800; 800-521-9891; FAX 215-271-2621; E-mail: cheminfo@inolex.com; Internet: www.inolex.com)

S. Black (Import & Export) Ltd., Distrib. for Inolex, Foxholes Business Park, John Tate Road, Hertford, Hertfordshire, SG13 7YH, UK (Tel: 44 1992 825555; FAX 44 1992 825566; E-mail: enquires@sblack.com; Internet: www.sblack.com)

Integra Chemical Co.

710 Thomas Ave. SW, Renton, WA, 98055, USA (Tel: 425-277-9244; 800-322-6646; FAX 425-277-9246; E-mail: chemicals@integrachem.com; Internet: www.integrachem.com)

Interchem Corp.

120 Rt. 17 North, Suite 115, PO Box 1579, Paramus, NJ, 07652, USA (Tel: 201-261-7333; FAX 201-261-7339; Internet: www.interchem.com)

Interchim

213 Av. J.F. Kennedy, BP 1140, 03103 Montlucon Cedex, France (Tel: 33 4 70 03 88 55; FAX 33 4 70 03 82 60; E-mail: interchim@interchim.com; Internet: www.interchim.com)

H. Interdonati, Inc.

PO Box 262, Cold Spring Harbor, NY, 11724-0262, USA (Tel: 631-367-6611; FAX 631-367-6626; E-mail: flavorplus@aol.com)

International Chemical Group

1147 Merritt Dr., El Cajon, CA, 92020, USA (Tel: 619-441-9108; FAX 619-441-2739)

International Chemical, Inc.

50 West Ferris St., E. Brunswick, NJ, 08816, USA (Tel: 732-238-5160; 800-914-2436; FAX 732-238-5970)

International Commodities Export Co.

2975 Westchester Ave., Purchase, NY, 10577, USA (Tel: 914-642-8900; FAX 914-694-4711)

International Fiber

International Fiber Corp., 50 Bridge St., N. Tonawanda, NY, 14120, USA (Tel: 716-693-4040; FAX 716-693-3528; E-mail: ifc@ifcfiber.com; Internet: www.ifcfiber.com)

International Fiber Europe N.V., Eurolaan 5 - Zone B, Industriezone T.T.S., 9140 Temse, Belgium (Tel: 32 3 711 1636; FAX 32 3 771 3399; E-mail: info@ife-fiber.be; Internet: www.ifefiber.com)

International Flavors & Fragrances

International Flavors & Fragrances Inc. (IFF), 650 Hwy. 36, Hazlet, NJ, 07730, USA (Tel: 732-264-4500; 800-765-9433; FAX 908-335-2627; E-mail: iffusaflavors@iff.com; iffusafragrances@iff.com; Internet: www.iff.com)

International Flavours & Fragrances (GB) Ltd., Duddery Hill, Haverhill, Suffolk, CB9 8LG, UK (Tel: 44 1440 7 15000; FAX 44 1440 7 62199; E-mail: iffuk@iff.com)

International Flavours & Fragrances (Australia), 310 Dandenong Valley Hwy., Dandenong, Victoria, 3175, Australia (Tel: 61 3 9788 9200; FAX 61 3 9768 3652; E-mail: iffaustralia@iff.com)

International Flavours & Fragrances (Japan) Ltd., IFF Bldg., 1-21-4, Higashi-ohi, Shinagawa-ku, Tokyo, 140-0011, Japan (Tel: 81 3 5460 0835; FAX 81 3 5460 0769; E-mail: iffjapan@iff.com)

International Flora Technologies, Inc./FloraTech Am. *See* Floratech Americas

International Furan Chemicals

International Furan Chemicals, Airportplein 7, 3045 AP Rotterdam, The Netherlands (Tel: 31 10 238 0555)

International Furan Chemicals USA, Inc., 120 Allens Creek Rd., Rochester, NY, 14618, USA (Tel: 716-271-2235; FAX 716-271-8457; E-mail: ifc.u@world.att.net)

International Group

International Group Inc. (IGI), 85 Old Eagle School Rd., PO Box 383, Wayne, PA, 19087, USA (Tel: 610-687-9030; 800-852-6537; FAX 610-254-8548; E-mail: igiwax@msn.com; Internet: www.igiwax.com/)

International Group Inc. (IGI), 5625 FM 1960 W., Suite 400, Houston, TX, 77069, USA (Tel: 281-537-7434; FAX 281-537-7835)

International Group Inc., Canada, 50 Salome Dr., Toronto, Ontario, M1S 2A8, Canada (Tel: 416-293-4151; FAX 416-293-0617)

IGI Europe, S.A., Ave. Louise 349 Bte. 4, 1050 Brussels, Belgium (Tel: 32 2 644 0060; FAX 32 2 647 7585)

International Paper

International Paper, 400 Atlantic St., Stamford, CT, 06921, USA (Tel: 203-541-8000; 800-223-1268; E-mail: comm@ipaper.com; Internet: www.internationalpaper.com)

International Paper Ltd-Canada, 1210 Sheppard Ave. East, Suite 405, Willowdale, Ontario, M2K 1E3, Canada (Tel: 416-499-0277)

International Paper do Brasil, Rua Francisco Matarazzo, 1700-18th Fl., Sao Paulo SP, 05001-400, Brazil (Tel: 55 11 3670 5250; Internet: www.internationalpaper.com.br)

International Paper Europe, Chaussee De La Hulpe, 166, 1170 Brussels, Belgium (Tel: 32 2 774 1211)

International Paper (Leeds) Ltd., Unit 7a, Leeds Bradford Airport Ind. Estate, Harrogate Rd., West Yorkshire, LS19 7WP, UK (Tel: 44 113 230 1810)

International Paper (Europe) Ltd., Talacker 41, PO Box, 8039 Zurich, Switzerland (Tel: 41 1 225 3131)

International Paper (Asia) Ltd., 1201-1203 Central Plaza, 18 Harbour Rd., Wanchai, Hong Kong (Tel: 852 2824 3000; FAX 852 2824 3035)

International Paper Co. (Japan) Ltd., The Imperial Tower, 1-1, Uchisaiwaicho 1-chome, Chiyoda-ku, Tokyo, 100-0011, Japan (Tel: 81 3 3593 2490; Internet: www.ipinet.co.jp)

International Sourcing Inc.

121 Pleasant Ave., Upper Saddle River, NJ, 07458, USA (Tel: 201-934-8900; 800-772-7672; FAX 201-934-8291)

Interstate Chemical Co., Inc.

2797 Freedland Rd., Hermitage, PA, 16148, USA (Tel: 724-981-3771; 800-422-2436)

Intertrade Holdings Inc.

State Highway 68, Copperhill, TN, 37317, USA (Tel: 423-496-1339; FAX 423-496-1378; E-mail: klong@ellijay.com)

Ishihara Corp. (USA)

600 Montgomery St., 38th Floor, San Francisco, CA, 94111-2702, USA (Tel: 415-421-8207; 800-477-2833; FAX 415-397-5403; E-mail: sales@ishihara.com; Internet: www.ishihara.com)

Isotec Inc.

A Matheson Tri-Gas Company, 3858 Benner Rd., Miamisburg, OH, 45342-4304, USA (Tel: 937-859-1808; 800-448-9760; FAX 937-859-4878; E-mail: isosales@isotec.com; Internet: www.isotec.com)

ISP

ISP, International Specialty Products, World Headquarters, 1361 Alps Rd., Wayne, NJ, 07470-3688, USA (Tel: 800-622-4423; FAX 973-628-4001; E-mail: info@ispcorp.com; Internet: www.ispcorp.com)

ISP Van Dyk, Inc., Member of the ISP Group, 11 William St., Belleville, NJ, 07109, USA (Tel: 973-450-7722; FAX 973-751-2047)

ISP (Canada) Inc., 4055 Sladeview Crescent #7, Mississauga, Ontario, L5L 5Y1, Canada (Tel: 905-607-2392; 800-465-5094; FAX 905-607-9086)

ISP (Mexico) S.A. de C.V., Fuente de Trevi, No. 23, Col. Techamachalco, Mexico City, 53950, Mexico (Tel: 52 5 589 0721; FAX 52 5 589 7345)

ISP Argentina S.A., Av. Cordoba 679, Piso 5A, 1365 Buenos Aires, Argentina (Tel: 54 11 4314 8971; FAX 54 11 4314 8976)

ISP do Brasil Ltda., Rua General Jardim, 633-1° Andar, Sao Paulo SP, 01271-900, Brazil (Tel: 55 11 3034 4866; FAX 55 11 3034 5707)

ISP Venezuela, Torre Las Mercedes, Piso 4, Oficina 4-07, Av. La Estancia, Urbanizacion Chuao, Caracas, 1060, Venezuela (Tel: 58 2 92 9703; FAX 58 2 9 9705)

ISP Europe, Watefield, Tadworth, Surrey, KT20 5HQ, UK (Tel: 44 207 519 5054; FAX 44 207 519 5056)

ISP Global Technologies Deutschland GmbH, Emil-Hoffmann-Strasse 1A, 50996 Cologne-Rodenkirchen, Germany (Tel: 49 2236 9649 260; FAX 49 2236 9649 295)

ISP (France) SA, ZAC Paris Nord II, BP 50007, 95945 Roissy CDG Cedex, France (Tel: 33 1 49 93 21 58; FAX 33 1 49 93 21 62)

ISP (Belgium) NV, Hoogkamerstraat 42, 9100 Sint-Niklaas, Belgium (Tel: 32 3 626 49 30; FAX 32 3 626 49 32)

ISP (Italia) Srl, Via Dei Gracchi, 30, 20146 Milano, Italy (Tel: 39 2 485340 1; FAX 39 2 43912274)

ISP (Switzerland) AG, Zugerstrasse 77, Baar Postfach 4354, 6340 Baar, Switzerland (Tel: 41 1 439 53 66; FAX 41 1 439 53 68)

ISP International Corp. Sucursal en España, Calle Moianés, 16-Entresuelo 2ª izquierda, 08014 Barcelona, Spain (Tel: 34 91 375 3026; FAX 34 91 375 3028)

ISP (Polska) Sp.z.o., Ul.Tamka 38 pok. 508, 00355 Warsaw, Poland (Tel: 48 22 828 1230; FAX 48 22 827 4371)

ISP Russia, Leninsky Prospekt, 95A, Room 503, 117313 Moscow, Russia (Tel: 7 95 232 0214; FAX 7 95 232 3385)

ISP Hungaria, Kozvetlen Kepviselet, 1113 Budapest, XI, Hungary (Tel: 36 1 385 8288; FAX 36 1 466 2550)

ISP (Australasia) Pty. Ltd., 73-75 Derby St., Silverwater, NSW, 2141, Australia (Tel: 61 2 9648 5177; FAX 61 2 9647 1608)

ISP Asia Pacific Pte. Ltd., 200 Cantonment Rd., #06-05 Southpoint, Singapore, 089763, Singapore (Tel: 65 223 3778; FAX 65 226 0853)

ISP (Hong Kong) Ltd., Suite 1102, Ming An Plaza, No. 8 Sunning Rd., Causeway Bay, Hong Kong (Tel: 852 2 881 6108; FAX 852 2 895 1250)

ISP (Thailand) Co., Ltd., Rm. 502 Panavongs Bldg., #104 Suriwong Rd., Bangkok, 10500, Thailand (Tel: 66 2 267 8103; FAX 66 2 236 0041)

ISP (Japan) Ltd., Across Shinkawa Bldg., Annex 10F, 16-14, Shinkawa 1-chome, Chuo-ku, Tokyo, 104-0033, Japan (Tel: 81 3 5566 8661; FAX 81 3 5566 8682)

ISP (Hong Kong) Ltd., Beijing Rep. Office, Room 1607 Jing Tai Mansion, No. 24 St. Jai Guo Men Wai, Beijing, 100022, P.R. China (Tel: 86 10 6515 6265; FAX 86 10 6515 6267)

Itochu Specialty Chemicals Inc.

Subsid. of Itochu International Inc., 50 Main St., 3rd Floor, White Plains, NY, 10606, USA (Tel: 914-448-7700; FAX 9124-448-7748)

Ivanhoe Industries, Inc.

818 William Leigh Dr., Bldg. H, Tullytown, PA, 19007, USA (Tel: 215-547-1200; FAX 215-946-7411; E-mail: ivanhoeinds@cs.com; Internet: www.ivanhoe industries.com)

Janousek Industriale Srl

Strada per i Laghetti 3, ZI delle Noghere, 340115 Muggia, Trieste, Italy (Tel: 39 40 232691; FAX 39 40 232698; E-mail: janousek@janousek.com; Internet: www.janousek.com)

Janssen-Cilag

Dr. Paul Janssenweg 150, 5026 RH Tilburg, The Netherlands (Tel: 31 13 583 7373; FAX 31 13 583 7300; E-mail: jcinfo@jacnl.jnj.com; Internet: www.janssen-cilag.nl)

Janus Resources Inc.

1501 Broadway, Suite 1304, New York, NY, 10036, USA (Tel: 212-840-8787; FAX 212-382-1639; E-mail: janusresources@worldnet.att.net)

Jarchem Industries Inc.

414 Wilson Ave., Newark, NJ, 07105, USA (Tel: 973-344-0600; FAX 973-344-5743; E-mail: info@jarchem.com; Internet: www.jarchem.com)

Jeen International Corp.

24 Madison Rd., Fairfield, NJ, 07004, USA (Tel: 800-771-5336; FAX 973-439-1402; E-mail: info@jeen.com; Internet: www.jeen.com)

Jiangxi Liquid Industrial Chemicals Co. (JLIC)

274, Railway Station West Ave., Nanchang, Jiangxi, 330003, P.R. China (Tel: 86 791 623 8607; FAX 86 791 621 0984; E-mail: office@jlic.com.cn; Internet: www.jlic.com.cn)

Jilin Chemical Industrial Company Limited

No. 31 East Zunyi Rd., Jilin City CI, P.R. China (Internet: www.jcic.com)

JLM Marketing, Inc.

8675 Hidden River Pkwy., Tampa, FL, 33637, USA (Tel: 813-632-3300; FAX 813-632-3301; E-mail: info@jlmindustries.com; Internet: www.jlmindustries.com; www.jlmgroup.com)

Johnson Matthey

Johnson Matthey plc, 2-4 Cockspur St., Trafalgar Square, London, SW1Y 5BQ, UK (Tel: 44 20 7269 8400; FAX 44 20 7269 8433; Internet: www.matthey.com)

Johnson Matthey S.A., Div. of Johnson Matthey plc, BP 50240, 13 rue de la Perdrix, ZI Paris Nord II, 95956 Roissy CDG Cedex, France (Tel: 33 1 48 17 21 99; FAX 33 1 48 63 27 02)

Johnson Matthey Inc., 2001 Nolte Dr., West Deptford, NJ, 08066, USA (Tel: 856-384-7000; 800-444-1411; FAX 856-384-7282)

Johnson Matthey Japan Inc., Nishishimbashi Tokyo Bldg., 3-16-11 Nishishimbashi, Minato-ku, Tokyo, 105-0003, Japan (Tel: 81 3 5776 6161; FAX 81 3 5776 6162)

Jonas Chemical Corp./Specialty Chemical Div.

1682 59th St., Brooklyn, NY, 11204, USA (Tel: 718-236-1666; FAX 718-236-2248)

Joy Chemical Co.

35 Livingston St., Providence, RI, 02904, USA (Tel: 401-331-9000; FAX 401-331-2155)

Jungbunzlauer

Jungbunzlauer International AG, St. Alban-Vorstadt 90, Postfach, 4002 Basel, Switzerland (Tel: 41 61 295 51 00; FAX 41 61 295 51 08; E-mail: headquarters@jungbunzlauer.com; Internet: www.jungbunzlauer.com)

Jungbunzlauer GmbH, Schwarzenbergplatz 16, 1011 Vienna, Austria (Tel: 43 1 502 00 0; FAX 43 1 502 00 8; E-mail: office.vie@jungbunzlauer.at)

Jungbunzlauer Ladenburg GmbH, Dr. Albert-Reimann-Strasse 18, 68526 Ladenburg, Germany (Tel: 49 6203 104 0; FAX 49 6203 104 210; E-mail: jungbunzlauer@jungbunzlauer.de)

Jungbunzlauer B.V., De Overtoom 8, 3353 HC Papendrecht, The Netherlands (Tel: 31 78 615 55 66; FAX 31 78 641 06 79; E-mail: jungbunzlauer.netherlands@jungbunzlauer.ch)

Jungbunzlauer S.A., Z.I. et Portuaire, B.P. 32, 67390 Marckolsheim, France (Tel: 33 3 88 58 60 60; FAX 33 3 88 58 60 61; E-mail: jungbunzlauer.sales@jungbunzlauer.fr)

Jungbunzlauer Inc., 7 Wells Ave., Newton Centre, MA, 02159-3214, USA (Tel: 617-969-0900; 800-828-0062; FAX 617-964-2921; E-mail: info@jungbunzlauer-inc.com; Internet: www.jungbunzlauer.com)

Jungbunzlauer Canada Inc., 1555 Elm St., Port Colborne, Ontario, Canada (Tel: 905-835-5444; FAX 905-835-0061; E-mail: howard.roberts@jungbunzlauer.ca)

Jungbunzlauer Singapore Pte. Ltd., 99 Bukit Timah Rd., #04-04 Alfa Centre, Singapore, 229835, Singapore (Tel: 65 633 72 300; FAX 65 633 75 900; E-mail: jungbunzlauer.singapore@jungbunzlauer.ch)

K3 Corp.

10115 Walker Lake Dr., Great Falls, VA, 22066, USA (Tel: 703-759-9731; FAX 703-759-3799; E-mail: K3CZECHEM@aol.com)

Kaltron/Pettibone

1241 Ellis St., Bensenville, IL, 60106, USA (Tel: 630-350-1116; FAX 630-350-1606; E-mail: info@concentric.net)

Kanto

Kanto Chemical Co., Inc., 3-2-8, Nihonbashi Hon-cho, Chuo-ku, Tokyo, 103, Japan (Tel: 81 3 3279 1751; FAX 81 3 3279 5560; E-mail: info@gms.kanto.co.jp; Internet: www.kanto.co.jp)

Kanto Corp., Wholly owned subsid. of Kanto Chemical Co., Inc., 12905 NE Airport Way, Portland, OR, 97230, USA (Tel: 503-283-0405; FAX 503-254-4450; E-mail: sales@kantocorp.com; Internet: www.kantocorp.com)

Kanto Denka Kogyo Co., Ltd.

New Tokyo Kaijo Bldg., 1-2-1, Marunouchi, Chiyoda-ku, Tokyo, 100, Japan (Tel: 81 3 3216 4561; FAX 81 3 3216 4581)

Kao

Kao Corp./Consumer Information Center, 1-3, Bunka 2-chome, Sumida-ku, Tokyo, 131-8501, Japan (Tel: 81 3 5630 7700; FAX 81 3 5630 7889; Internet: www.kao.co.jp)

Kao Chemical Corp. Shanghai, No. 2 Bridge, Bei Song Road, Min Hang District, Shanghai, 201111, P.R. China (Tel: 86 21 6409 2880; FAX 86 21 6409 2881)

Kao (Southeast Asia) Pte. Ltd., 247 Alexandra Rd., Singapore, 159934, Singapore (Tel: 65 471 2811; FAX 65 473 0485)

Kao Corp. S.A., Puig dels Tudons, 10, 08210 Barberá Del Vallés (Barcelona), Spain (Tel: 34 93 739 9300; FAX 34 93 7399 333; E-mail: marketing@kao.es; Internet: www.kao.es)

Kao Corp. (France) SARL, 2 rue Eugene et Armand Peugeot, 92500 Rueil Malmaison, France (Tel: 33 1 47 16 21 40; FAX 33 1 47 49 40 67)

Kao Chemicals GmbH, Kupferstrasse 1, 46446 Emmerich, Germany (Tel: 49 2822 7110; FAX 49 2822 711201)

Kao Specialties Americas LLC, 243 Woodbine St., PO Box 2316, High Point, NC, 27261, USA (Tel: 336-884-2214; FAX 336-884-8786)

Karlshamns AB

374 82 Karlshamn, Sweden (Tel: 46 454 82000; FAX 46 454 82810; E-mail: info@karlshamns.se; Internet: www.karlshamns.com)

K.A. Steel Chemicals, Inc.

15185 Main St., PO Box 729, Lemont, IL, 60439, USA (Tel: 630-257-3900; 800-677-8335)

Kaufholz & Co., Inc.

PO Box 99, Lansdowne, PA, 19050-0099, USA (Tel: 610-622-0170; FAX 610-622-0171)

Kawaken Fine Chemicals Co., Ltd.

Horidome Bldg., 2-3-3, Nihombashi Horidome-cho, Chuo-ku, Tokyo, 103-0012, Japan (Tel: 81 3 3663 9521; FAX 81 3 3661 5630; E-mail: hg@kawakenfc.co.jp; Internet: www.kawakenfc.co.jp)

Kawasaki Steel Corp.

Hibiya Kokusai Bldg., 2-2-3 Uchisaiwai-cho, Chiyoda-ku, Tokyo, 100, Japan (Tel: 81 3 3597 4929; FAX 81 3 3595 4868)

Kemira

Kemira Oy, PO Box 330, Porkkalankatu 3, 00101 Helsinki, Finland (Tel: 358 10 861 211; FAX 358 10 862 1124; Internet: www.kemira.com)

Kemira Kemi AB, Div. of Kemira Oy, Strandbadsvägen 15, PO Box 902, 251 09 Helsingborg, Sweden (Tel: 46 42 17 10 00; FAX 46 42 14 06 35)

Kemira Chemicals AS, Postboks 1177, 1631 Gamle Fredrikstad, Norway (Tel: 47 69 358 585; FAX 47 69 358 595)

Kemira Danmark AS, Kongensgade 107, 7000 Fredericia, Denmark (Tel: 45 7922 3366; FAX 45 7922 3322)

Kemira Chemicals BV, PO Box 1015, 3180 AA Rozenburg, The Netherlands (Tel: 31 10 295 2540; FAX 31 10 295 2536)

Kemira Chemicals (UK) Ltd., Kemwater (UK) Ltd., Div. of Kemira Oy, New Potter Grange Rd., MG2 Trading Estate, Goole, DN14 6BZ, UK (Tel: 44 1405 765 131; FAX 44 1405 765 147)

Kemira Chimie SA, BP 5, 67630 Lauterbourg, France (Tel: 33 3 88 54 98 50; FAX 33 3 88 54 98 51)

Kemira SA NV, Tertre Plant, Rue de la Carbo, 7333 Tertre, Belgium (Tel: 32 65 712 211; FAX 32 65 641 917)

Kemira Deutschland GmbH, Postfach 51 07 20, 30637 Hannover, Germany (Tel: 49 511 647 460; FAX 49 511 647 4625)

Kemira Iberica SA, Gran Via Corts Catalanas 641, 08010 Barcelona, Spain (Tel: 34 93 412 3050; FAX 34 93 412 0884)

Kemira Paper Chemicals, Inc., One Kemira Rd., Box 368, Savannah, GA, 31402, USA (Tel: 912-688-5992; 800-4-KEMIRA; FAX 912-652-1160)

Kenrich Petrochemicals, Inc.

140 E. 22nd St., PO Box 32, Bayonne, NJ, 07002-0032, USA (Tel: 201-823-9000; 866-536-7424; FAX 201-823-0691; E-mail: customerservice@4kenrich.com; Internet: www.westnet.com/~kpiemm/)

Kerr-McGee Chemical Corp.

Kerr-McGee Ctr., PO Box 25861, MT-1603, Oklahoma City, OK, 73125, USA (Tel: 405-270-1313; 800-654-3911; FAX 405-270-3955)

Kessler Chemical, Inc.

77 West Broad St., Unit 21A, Bethlehem, PA, 18018, USA (Tel: 610-758-9602; FAX 610-758-9615; E-mail: info@kesslerchemical.com; Internet: www.kesslerchemical.com)

KIC Chemicals, Inc.

451 Main St., PO Box 437, Armonk, NY, 10504-0437, USA (Tel: 914-273-6555; FAX 914-273-6760; E-mail: chem@kicgroup.com; Internet: www.kicchemicals.com; www.kicgroup.com)

Kimpton Brothers Ltd.

10-14 Hewett St., London, EC2A 3RL, UK (Tel: 44 20 7456 9999; FAX 44 20 7247 2784; E-mail: info@kimpton.co.uk; Internet: www.kimpton.co.uk)

Kimson Chemical Inc.

77 Summer St., Boston, MA, 02110, USA (Tel: 617-292-7773; FAX 617-292-7766)

Kincaid Enterprises, Inc.

PO Box 549, Viscose Rd., Nitro, WV, 25143-0549, USA (Tel: 304-755-3377; FAX 304-755-4547)

Kira Corp.

630 N. Craycroft Rd., Tucson, AZ, 85711-1440, USA (Tel: 520-514-2374; FAX 520-514-2375)

Kishida Chemical Co., Ltd.

1-6-9, Dosho-machi, Chuo-ku, Osaka, 541, Japan (Tel: 81 6 6202 0451; FAX 81 6 6226 0670)

Koch

Koch Industries, Inc., 4111 E 37 St N., Wichita, KS, 672206, USA (Tel: 316-828-5500; 800-FOR-ACID; FAX 316-828-6726; E-mail: info@kochind.com; Internet: www.kochind.com)

Koch Chemical Specialties, Div. of Koch Refining Co., Specialties Group, PO Box 2608, Corpus Christi, TX, 78403, USA (FAX 512-242-8353)

Koch Specialty Chemical Co., Div. of Koch Refining Co., 20 Greenway Plaza, Houston, TX, 77046, USA (Tel: 713-544-6015; FAX 713-544-4350; E-mail: specheminfo@kochind.com; Internet: www.kochind.com)

Koch Chemical Co. NV, Tiensesteenweg 80, 3800 Sint-Truiden, Belgium (Tel: 32-11-69 36 06; FAX 32-11-69 36 08)

Koch Industries International Ltd., 7th Fl., 2 George Yard, Lombard St., London, EC3V 9DH, UK (Tel: 44 171 648-6300; FAX 44 171 648-6301)

Koch Microelectronic Service Co., Inc.

20 E. Greenway Plaza, Houston, TX, 77046, USA (Tel: 713-544-5243; FAX 713-544-5708; E-mail: byrdzb@kochind.com; Internet: www.kochmicroelectronic.com)

Koch Nitrogen Co.

4111 E. 37th St. N., Wichita, KS, 67220, USA (Tel: 316-828-3364; FAX 316-828-4084; E-mail: moulda@kochind.com; Internet: www.kochind.com)

Koei Chemical Co., Ltd.

Subsid. of Sumitomo Chem. Co. Ltd., Sumika Fudosan Yokobori Bldg., 4-6-17, Koraibashi, Chuo-ku, Osaka, 541-0043, Japan (Tel: 81 6 6204 1515; FAX 81 6 6204 1510; Internet: www.koeichem.com)

Dr. W. Kolb

Dr. W. Kolb AG, Maienbrunnenstrasse 1, 8908 Hedingen, Switzerland (Tel: 41 1 762 4646; FAX 41-1 762 46 00; E-mail: tensinfo@kolb.ch; ppcinfo@kolb.ch; Internet: www.kolb.ch)

Dr. W. Kolb Nederland B.V., PO Box 123, 4790 AC Klundert, The Netherlands (Tel: 31 168 38 70 80; FAX 31 168 33 04 81)

Kolb (Asia Support Office Surfactants), 40, Jalan Intan 1/5, Taman Puchong Intan, 47100 Selangor, Malaysia (Tel: 60 3 572 5286; FAX 60 3 572 5293; E-mail: kian@pd.jaring.my)

Koppers Industries

Koppers Industries, Inc./Carbon Materials and Chemicals Div., 436 Seventh Ave., Room 1600, Koppers Bldg, Pittsburgh, PA, 15219-1800, USA (Tel: 412-227-2249; 800-321-9876; FAX 412-227-2202; E-mail: sara_baughman@koppers.com; Internet: www.koppers.com)

Koppers Industries/Carbon Materials & Chemicals, 100 Koppers Rd., Follansbee, WV, 26037, USA

Koster Keunen

Koster Keunen, Inc., 1021 Echo Lake Rd., PO Box 69, Watertown, CT, 06795-0069, USA (Tel: 860-945-3333; FAX 860-945-0330; E-mail: info@kosterkeunen.com; Internet: www.kosterkeunen.com)

Koster Keunen Holland B.V., Postbus 53, 5530 AB Bladel, The Netherlands (Tel: 31 4973 61010; FAX 31 4973 87902; E-mail: kosterwx@iae.nl)

Kowa American Corp.

Subsid. of Kowa Co. Ltd., Japan, 55 E. 59th St., 19th Floor, New York, NY, 10022, USA (Tel: 212-303-7800; 800-221-2076; FAX 212-310-0101; E-mail: eva@kowa.com; Internet: www.kowa.com/chemical/)

Koyo Chemical Co., Ltd.

Iidabashi, Hitown-Bldg., 2-28, Shimomiyabi-cho, Shinjuku-ku, Tokyo, 162, Japan (Tel: 81 3 3268 1717; FAX 81 3 3268 1723)

Kraeber GmbH & Co.

Pharmazeutische Rohstoffe, Waldhotstrasse 14, 25474 Ellerbek, Germany (Tel: 49 4101 3053 0; FAX 49 4101 3053 90; E-mail: info@kraeber.de; Internet: www.kraeber.de)

Kraft Chemical Co.

1975 N. Hawthorne, Melrose Park, IL, 60160, USA (Tel: 708-345-5200; 800-345-5200; FAX 708-345-4005; E-mail: kraftchem@aol.com; Internet: www.kraftchemical.com)

Kramer Chemicals Inc.

An Ellis & Everard Subsid., 266 Harristown Rd., PO Box 1118, Glen Rock, NJ, 07452-1118, USA (Tel: 201-670-1600; FAX 201-670-0085)

Krishna Organic Pvt. Ltd.

21, Sugar Market Bldg., 104/114, P'Demello Rd., Mumbai, 400009, India (Tel: 91 22 343 5494; FAX 91 22 342 5679; E-mail: sales@atulorganics.com; Internet: www.atulorganics.com)

Kuehne Chemical Co. Inc.

86 Hackensack Ave., South Kearny, NJ, 07032, USA (Tel: 201-589-0700; FAX 201-589-4866; E-mail: kocarola@kuehnechemical.com; Internet: www.kuehnechemical.com)

Kuraray

Kuraray Co. Ltd., Shin Hankyu Bldg., 1-12-39, Umeda, Kita-ku, Osaka, 530, Japan (Tel: 81 6 6348 2111; FAX 81 6 6348 2189; E-mail: koho@kuraray.co.jp; Internet: www.kuraray.co.jp)

Kuraray America Inc., 101 E 52nd St., 26th Fl., New York, NY, 10022, USA (Tel: 212-986-2230; FAX 212-867-3543)

Kuraray Europe GmbH, Schiess-Strasse 68, 40549 Dusseldorf, Germany (Tel: 49 211 538880; FAX 49 211 538848; Internet: www.kuraray.de)

Kuraray Hong Kong Co. Ltd., Unit 314, 3/F Lippo Sun Plaza, 28 Canton Rd., Tsimshatsui, Kowloon, Hong Kong (Tel: 852 2377 3781; FAX 852 2377 9703)

Kyowa Hakko

Kyowa Hakko Kogyo Co. Ltd., Ohtemachi Bldg., 1-6-1, Ohte-machi, Chiyoda-ku, Tokyo, 100-8185, Japan (Tel: 81 3 3282 0979; FAX 81 3 3284 1839; Internet: www.kyowa.co.jp)

Kyowa Hakko USA Inc., 85 Enterprise, Suite 430, Aliso Viejo, CA, 92656, USA (Tel: 949-425-0707; FAX 949-425-0708; Internet: www.kyowa-usa.com)

Laboratori Alchemia

Via San Faustino 68, Milan, Italy (Tel: 39 2 2640136; E-mail: alchemia@netitalia.it; Internet: www.laboratorialchemia.it)

Lambent Technologies

Lambent Technologies Corp., A Petroferm Co., 7247 North Central Park Ave., Skokie, IL, 60076-4093, USA (Tel: 847-675-3950; 800-432-7187; FAX 847-675-3013; E-mail: lambent@petroferm.com; Internet: www.petroferm.com)

Lambent Technologies Inc., 4437 Park Dr., Norcross, GA, 30093, USA (Tel: 770-279-8601; 888-LAMBENT; FAX 770-279-8535; Internet: www.petroferm.com)

Lambiotte & Cie

Ave. des Aubepines 18, 1180 Brussels, Belgium (Tel: 32 2 374 44 65; FAX 32 2 375 31 55; E-mail: mail@lambiotte.com; Internet: www.lambiotte.com)

Henry Lamotte GmbH

PO Box 103849, Merkurstrasse 47, 28197 Bremen, Germany (Tel: 49 421 52390; FAX 49 421 5239199; E-mail: admin@lamotte.de; Internet: www.lamotte.de)

Lanaetex Products, Inc.

151-157 Third Ave., Elizabeth, NJ, 07206, USA (Tel: 908-351-9700; FAX 908-351-8753)

Lancaster Synthesis

Lancaster Synthesis Ltd., Part of the Clariant Organization, Newgate, White Lund, Morecambe, Lancashire, LA3 3BN, UK (Tel: 44 1524 36101; FAX 44 1524 39727; E-mail: sales@lancastersynthesis.com; Internet: www.lancastersynthesis.com)

Lancaster Synthesis Ltd (France), 15 Rue de l'Atome, Zone Industrielle, 67800 Bischheim, Strasbourg, France (Tel: 33 3 88 62 26 90; 0800 03 51 47; 33 3 88 62 26 81)

Lancaster Synthesis GmbH, Postfach 15 18, 63155 Mühlheim am Main, Germany (Tel: 49 06108 7 30 19; 0800 0006562; 49 06108 7 48 14; E-mail: verkauf@lancaster-gmbh.com)

Lancaster Synthesis, Inc., PO Box 1000, Windham, NH, 03087-9977, USA (Tel: 603-889-3306; 800-238-2324; FAX 603-889-3326; E-mail: andyb@lancaster-us.com)

Langley-Smith & Co. Ltd.

36 Spital Square, London, E1 6DY, UK (Tel: 44 207 247 7473; FAX 44 207 375 1470; Internet: www.langley-smith.co.uk)

Laporte Performance

Laporte Performance Chemicals UK Ltd., Part of Laporte plc, Charleston Ind. Estate, Hardley, Hythe, Southampton, SO45 3ZG, UK (Tel: 44 2380 894666; FAX 44 2380 243113; E-mail: charrison.hythe@laporteplc.com; Internet: www.laporteperform.com)

Laporte Speciality Organics S.A.R.L., Chateau Rouge, 276/6 Ave. de la Marne, 59700 Marcq En Baroeul, Lille, France (Tel: 33 3 20 81 97 40; FAX 33 3 20 81 97 45)

Laporte Speciality Organics (Italy), Via Correggio 19, 20149 Milan, Italy (Tel: 39 2 480 22439; FAX 39 2 480 22465)

Laporte Speciality Organics Austria, Löwengasse 47/Top 16, 1030 Wien, Austria (Tel: 43 1 7144514; FAX 43 1 71445149)

Laporte Speciality Organics Denmark, International House, Center Boulevard, 2300 Copenhagen S, Denmark (Tel: 45 32 52 02 11; FAX 45 3252 5720)

Laporte Speciality Organics, c/o Quimidroga SA, Tuset 26, 08006 Barcelona, Spain (Tel: 34 9 3236 3636; FAX 34 9 2328 0919)

Laporte Speciality Organics, Part of Laporte plc, 500 International Dr., Mt. Olive, NJ, 07828, USA (Tel: 973-426-0123; FAX 973-426-0423)

Laporte Speciality Organics Brazil, Rua Joaquim Floriano 101, cj804/806 Itaim Bibi, Sao Paulo-SP CEP 04534-0010, Brazil (Tel: 55 11 3078 5373)

Laporte Speciality Organics (Korea), 7th Fl., Son Won Bldg. 59-4, Banpo-4, Dong, Seocho-ku, Seoul, Korea (Tel: 82 2 725 8822; FAX 82 2 725 8829)

Laporte Speciality Organics Singapore, 101 Cecil St., #11-04 Tong Eng Bldg., Singapore, 069533, Singapore (Tel: 65 6429 0707; FAX 65 6429 0609; E-mail: G1hill@cyberway.com.sg)

LaRoche Industries Inc.

1100 Johnson Ferry Rd. NE, Atlanta, GA, 30342, USA (Tel: 404-851-0300; 800-226-4572; FAX 404-851-0389; E-mail: larocheinfo@larocheind.com; Internet: www.larocheind.com)

Lavy's Inc.

PO Box 6956, Piscataway, NJ, 08854, USA (Tel: 732-536-0042; FAX 732-536-9179)

The Lebermuth Co. Inc.

PO Box 4103, South Bend, IN, 46634, USA (Tel: 219-259-7000; 800-648-1123; FAX 800-852-4722; E-mail: info@lebermuth.com; Internet: www.lebermuth.com)

Wm. Leman, Inc.

114 N. Center St., Bremen, IN, 46506, USA (Tel: 219-546-2371; FAX 219-546-5762; E-mail: info@wmleman.com; Internet: www.wmleman.com)

Lenmar Chemical Corp.

PO Box 571, 2474 Lakeland Rd., Dalton, GA, 30722-0571, USA (Tel: 706-277-9505; FAX 706-277-9287; Internet: www.lenmarchemical.com)

Leprino Foods

1830 W. 38 Ave., Denver, CO, 80211, USA (Tel: 303-480-2623; 800-LEPRINO; FAX 303-480-2605)

LG Petrochemical Co., Ltd.

LG Twin Tower 20, Youido-dong, Youngdungpo-gu, Seoul, 150-721, Korea (Tel: 82 2 3773 1114; FAX 82 2 3773 3929; E-mail: webmaster@petro.lg.co.kr; Internet: www.lgpetro.com; www.lgpetro.co.kr)

Liberty Natural Products Inc.

8120 SE Stark St., Portland, OR, 97215, USA (Tel: 503-256-1227; 800-289-8427; FAX 503-256-1182; Internet: www.libertynatural.com)

Lindau Chems, Inc.

731 Rosewood Dr., Columbia, SC, 29201, USA (Tel: 803-799-6863; FAX 803-256-3639; E-mail: lci@lindau.net)

Lion Corp.

1-3-7, Honjo, Sumida-ku, Tokyo, 130-8644, Japan (Tel: 81 3 3621 6211; FAX 81 3 3621 6137; Internet: www.lion.co.jp)

Lipo Chemicals Inc.

207 19th Ave., Paterson, NJ, 07504, USA (Tel: 973-345-8600; FAX 973-345-8365; E-mail: salesandmarketing@lipochemicals.com; Internet: www.lipochemicals.com)

Lluch Essence

C/.Lo Gaiter, 160, 08820 Prat de Uobregat, Spain (Tel: 34 93 379 38 49; FAX 34 93 370 65 04; E-mail: buzon@lluch-essence.com; Internet: www.lluch-essence.com)

LNP Engineering Plastics

LNP Engineering Plastics Inc., 475 Creamery Way, Exton, PA, 19341-2537, USA (Tel: 610-363-4500; 800-532-2LNP; FAX 610-363-4749; Internet: www.lnp.com)

LNP Engineering Plastics Europe B.V., Ottergeerde 24, 4941 VM Raamsdonksveer, The Netherlands (Tel: 31 1621 87600; FAX 31 1621-21924)

LNP Engineering Plastics (Asia) Pte. Ltd., 171 Chin Swee Rd. #03-07, San Centre, Singapore, 0316, Singapore (Tel: 65 535 2022; FAX 65-535-5022)

Lohmann Chemicals, Dr. Paul Lohmann GmbH KG

PO Box 1220, Hauptstrasse 2, 31857 Emmerthal, Germany (Tel: 49 5155 63 0; FAX 49 5155 63 118; E-mail: service@lohmann-chemikalien.de; Internet: www.lohmann-chemikalien.de)

L.V. Lomas

99 Summerlea Rd., Brampton, ON, L6T 4V2, Canada (Tel: 905-458-1555; FAX 905-458-0722; Internet: www.lvlomas.com)

Lonza

Lonza Ltd., Member of the Alusuisse-Lonza Group, Münchensteinerstrasse 38, 4002 Basel, Switzerland (Tel: 41 61 316 8111; FAX 41 61 316 91 11; E-mail: contact.bs@lonzagroup.com; Internet: www.lonza.com)

Lonza SpA, Via Vittor Pisani, 31, 20124 Milan, Italy (Tel: 39 2 66 99 91; FAX 39 2 66 98 76 30; E-mail: contact.it@lonza.com)

Lonza UK Ltd., Imperial House, Lypiatt Road, Cheltenham, Gloucestershire, GL50 2QJ, UK (Tel: 44 1242 513211; FAX 44 1242 222294; E-mail: contact.uk@lonza.com)

Lonza France SARL, Div. of Alusuisse, 55, rue Aristide Briand, 92309 Levallois-Perret Cedex, France (Tel: 33 1 40 89 99 00; FAX 33 1 40 89 99 01; E-mail: contact.fr@lonzagroup.com)

Lonza GmbH, Morianstrasse 32, Postfach 13 14 53, 42041 Wuppertal, Germany (Tel: 49 202 245380; FAX 49 202 2453820; E-mail: contact.de@lonzagroup.com)

Lonza Ibérica SA, Via Augusta, n° 18, 2° Izq., 08006 Barcelona, Spain (Tel: 34 93 238 54 60; FAX 34 93 416 02 28; E-mail: contact.es@lonzagroup.com)

Lonza Inc., Member of the Alusuisse-Lonza Group, 17-17 Route 208, Fair Lawn, NJ, 07410, USA (Tel: 201-794-2400; 800-777-1875; FAX 201-794-2597; E-mail: contact@lonza-us.com; Internet: www.lonza.com)

Lonza do Brasil Ltda, Av. das Nações Unídas 1800, Sao Paulo, 04795-900, Brazil (Tel: 55 11 5683 7682; FAX 55 11 5683 7419; E-mail: werner.grauwiler@lonzagroup.com)

Lonza Ltd. India Liaison Office, Sandoz House, Dr. Annie Besant Rd., Worli, Bombay, Maharashtra, 400 018, India (Tel: 91 22 497 5893; FAX 91 22 497 5885; E-mail: contact.in@lonzagroup.com)

Lonza Hongkong, Ltd., 393 Jaffe Rd., Wanchai, Hong Kong (Tel: 852 2545 6868; FAX 852 2854 1881; E-mail: contact.hk@lonzagroup.com)

Lonza Japan Ltd., Kyowa Shinkawa Bldg., 8th Fl., 2-20-8, Shinkawa, Chuo-ku, Tokyo, 104-0033, Japan (Tel: 81 3 5566 0612; FAX 81 3 5566 0619; E-mail: contact.jp@lonzagroup.com)

Lonza Guangzhou Ltd., 821, Gongye Rd. S., Guangzhou, Guangdong, 510280, P.R. China (Tel: 86 20 8433 8998; FAX 86 20 8433 1998; E-mail: gzlonza@public.guangzhou.gd.cn)

Lonza Singapore Pte. Ltd., 12 Sakra Rd., Jurong Island, Singapore, 627892, Singapore (Tel: 65 867 6393; FAX 65 867 6204; E-mail: gzlonza@public.guangzhou.gd.cn)

Loos & Dilworth, Inc.

61 E. Green Lane, PO Box 25, Bristol, PA, 19007, USA (Tel: 215-785-3591; 800-229-5667; FAX 215-785-3597; E-mail: lschepise@loosanddilworth.com; Internet: www.loosanddilworth.com)

Lowenstein Dyes & Cosmetics Inc.

420 Morgan Ave., Brooklyn, NY, 11222, USA (Tel: 718-388-5410; FAX 718-387-3806; E-mail: dye@jhlowenstein.com; Internet: www.jhlowenstein.com)

Lucta S.A.

Carretera Masnou-Granollers km 12.4, 08170 Montornés del Vallés (Barcelona), Spain (Tel: 34 93 845 8888; FAX 34 93 845 9812; E-mail: sp@lucta.com; Internet: www.lucta.com)

The O.C. Lugo Co., Inc.

321 N. Main St., New City, NY, 10956, USA (Tel: 914-708-7080; FAX 914-708-7081; E-mail: dbm9@aol.com)

D.N. Lukens Inc.

15 Old Flanders Rd., Westborough, MA, 01581, USA (Tel: 508-366-1300; 800-3-LUKENS; FAX 508-366-0771; E-mail: sales@dnlukens.com; Internet: www.dnlukens.com)

AB R Lundberg

Box 104, Frihamnsallen 7, 201 21 Malmo, Sweden (Tel: 46 40 698 1100; FAX 46 40 698 1101)

Lutianhua Group Inc.

No. 202, Unit 5, Bldg. 16 Fahua Nanli, Chongwen District, Beijing, 100061, P.R. China (Tel: 86 10 6713 1427; FAX 86 10 6713 1428; Internet: www.lth-inc.com)

Lyondell

Lyondell Chemical Co., PO Box 3646, Houston, TX, 77253-3646, USA (Tel: 713-652-7200; 888-777-0232; Internet: www.lyondell.com)

Lyondell Chemical Co./Technical Center, 3801 West Chester Pike, Newtown Sq., PA, 19073-2387, USA (Tel: 610-359-2000; 800-345-0252; FAX 610-359-2722)

Lyondell Chemical Europe, Inc., Bridge Ave., Maidenhead, Berkshire, SL6 1YP, UK (Tel: 44 1628 775000; FAX 44 1628 775180)

Lyondell Chemical Products Europe, Inc., Z.I. du Pre Sarrazin, Boite Postale No. 34, 60870 Villers Saint Paul, France (Tel: 33 3 44 29 63 00; FAX 33 3 44 71 76 32)

Lyondell Chemical Nederland, Ltd., Postbus 2416, 3000 CK Rotterdam, The Netherlands (Tel: 31 10 275 5500; FAX 31 10 275 5599)

Lyondell Asia Pacific, Ltd., 41st Fl., The Lee Gardens, 33 Hysan Ave., Causeway Bay, Hong Kong (Tel: 852 2822 2668; FAX 852 2840 1690)

Lyondell Asia Pacific Service Center, 37th Fl., 152 Beach Rd., #37-00 The Gateway East, Singapore, 0718, Singapore (Tel: 65 291 2011; FAX 65 297 0517)

Macfarlan Smith Ltd.

10 Wheatfield Road, Midlothian, Edinburgh, EH11 2QA, Scotland UK (Tel: 44 131 337 2434; FAX 44 131 337 9813; E-mail: enquiry@macsmith.com; Internet: www.macsmith.com)

Madison Industries Inc.

Old Waterworks, Old Bridge, NJ, 08857, USA (Tel: 732-727-2225; 800-275-3924; FAX 732-727-2653)

Magie Bros. Oil Co.

Div. of Penreco, 9101 Fullerton Ave., PO Box 1089, Franklin Park, IL, 60131-1089, USA (Tel: 847-455-4500; 800-MAGIE 47; FAX 847-455-0383)

Magna-Kron Corp.

640 Herman Rd., Suite 1, Jackson, NJ, 08527, USA (Tel: 732-928-5800; FAX 732-928-9227; E-mail: magnak@worldnet.att.net)

Magnesia GmbH

Postfach 2168, Kurt-Höbold-Strasse 6, 21311 Lüneburg, Germany (Tel: 49 4131 8710 0; FAX 49 4131 8710 55; E-mail: info@magnesia.de; Internet: www.magnesia.de)

Makhteshim Agan

Makhteshim Agan Industries Ltd., Subsid. of Koor Industries, 9 Omarim St., POB 1646, Omer, 84965, Israel (Tel: 972 8 620 0700; FAX 972 8 646 9846; E-mail: main@main.co.il; Internet: www.main.co.il)

Makhteshim Chemical Works Ltd., Wholly owned subsid. of Makhteshim Agan Industries Ltd., PO Box 60, Industrial Zone, Beer-Sheva, 84100, Israel (Tel: 972 8 629 6611; FAX 972 8 628 0304; E-mail: mcw@makhteshim.co.il; Internet: www.makhteshim.co.il)

Makhteshim Agan International Coordination Centre, Ave. Louise 283, Box No. 7, 1050 Brussels, Belgium (Tel: 32 2 646 8606; FAX 32 2 646 9152; E-mail: maicc@maicc.be)

Makhteshim Agan of North America Inc., 551 Fifth Ave., Suite 1100, New York, NY, 10176, USA (Tel: 212-661-9800; FAX 212-661-9038; E-mail: fay@manainc.com)

Mallinckrodt Baker

Mallinckrodt Baker Inc., Div. of Mallinckrodt Inc., 222 Red School Lane, Phillipsburg, NJ, 08865, USA (Tel: 908-859-2151; 800-582-2537; FAX 908-859-6905; E-mail: infombi@mkg.com; Internet: www.mallbaker.com; www.mallchem.com)

Mallinckrodt Baker S.A. de C.V., Apartado Postal #75595, Col. Lindavista, Deleg. Gustavo A. Madero, Mexico D.F., 07300, Mexico (Tel: 52 5 569 1100; FAX 52 5 755 2585; E-mail: mbakerinf@imfosel.net.mx)

Mallinckrodt Baker do Brasil Ltda., Av. das Nacoes Unidas, 23.013 Santo Amavo, Sao Paulo SP, 04795-100, Brazil (Tel: 55 11 548 3550; FAX 55 11 548 3549; E-mail: mantonio@regra.net)

Mallinckrodt Baker B.V., Rijsterborgherweg 20, PO Box 1, 7400 AA Deventer, The Netherlands (Tel: 31 570 687500; FAX 31 570 611342; E-mail: service@mbe.nl)

Mallinckrodt Baker UK, 107/112 Leadenhall St., London, EC3A 4AH, UK (Tel: 44 1908 506000; FAX 44 1908 503290; E-mail: service.uk@mkg.com)

Mallinckrodt Baker France, 33 Ave. Philippe Auguste, 75011 Paris, France (Tel: 33 1 44 64 89 71; FAX 33 1 44 64 89 73; E-mail: service.fr@mkg.com)

Mallinckrodt Baker Deutschland, Im Leuschnerpark 4, 64347 Griesheim, Germany (Tel: 49 6155 6095 18; FAX 49 6155 6095 25; E-mail: service.de@mkg.com)

Mallinckrodt Baker Italia, Via P. Portaluppi 11/2, 20138 Milan, Italy (Tel: 39 2 580891; FAX 39 2 58019264; E-mail: service.it@mkg.com)

Mallinckrodt Baker, Inc. c/o Mallinckrodt Asia Pacific Pte. Ltd., 83 Clemenceau Ave., #16-03, UE Square, Singapore, 239920, Singapore (Tel: 65 738 4333; FAX 65 738 4334; E-mail: ben.lam@mkg.com)

J. Manheimer, Inc.

100 Hillister Rd., Teterboro, NJ, 07608, USA (Tel: 201-373-1111; FAX 201-288-8389; E-mail: sales@manheimer.com; Internet: www.manheimer.com)

George Mann & Co., Inc.

A Royal Vopak Co., PO Box 9066, Harborside Blvd. & Terminal Rd., Providence, RI, 02940-9066, USA (Tel: 401-781-5600; 800-556-2426; FAX 401-785-1070; Internet: www.univarusa.com)

Marathon Ashland Petroleum, LLC

555 San Felipe, Rm. 2451, Houston, TX, 77056-2723, USA (Tel: 713-646-4441; FAX 713-646-4483)

Marcor Development Corp.

341 Michele Pl., Carlstadt, NJ, 07072, USA (Tel: 201-935-2111; FAX 201-935-5223; E-mail: marcordev@aol.com)

Marcus International, Inc.

616 F.M. 1960 West, Suite 210, Houston, TX, 77090, USA (Tel: 281-893-8970; 800-487-8970; FAX 281-893-2502; E-mail: sag@marcusintl.com; Internet: www.marcusintl.com)

Marine Bio-Products Inc.

39 Broad St., Quincy, MA, 02169, USA (Tel: 617-472-9300; FAX 617-472-9359)

Marlowe-Van Loan Corp.

PO Box 1851, 1511 Joshua Circle, High Point, NC, 27261, USA (Tel: 336-886-7126; 800-422-4MVL; FAX 336-886-3388; E-mail: info@marlowevanloan.com; Internet: www.marlowevanloan.com)

Maruzen Petrochemical Co., Ltd.

Tokyo Sanshin Bldg., 2-25-10, Hatchobori, Chuo-ku, Tokyo, 104, Japan (Tel: 81 3 3552 9361; FAX 81 3 3552 9443)

Matheson Gas Products

30 Seaview Dr., Secaucus, NJ, 07096, USA (Tel: 201-867-4100; FAX 201-867-4572)

Maypro Industries, Inc.

2700 Westchester Ave., Purchase, NY, 10577, USA (Tel: 914-251-0701; FAX 914-251-0746; E-mail: maypro@worldnet.att.net; Internet: www.maypro.com)

Mays Chemical Co., Inc.

5611 E. 71st St., Indianapolis, IN, 46220, USA (Tel: 317-842-8722; FAX 317-842-4604; E-mail: gloriay@fuse.net)

McIntyre

McIntyre Group Ltd., 24601 Governors Hwy., University Park, IL, 60466-4127, USA (Tel: 708-534-6200; 800-645-6457; FAX 708-534-6216; E-mail: info@mcintyregroup.com; Internet: www.mcintyregroup.com)

McIntyre Chem. Co. Ltd., Holywell Green, Halifax, HX4 9DL, UK (Tel: 44 1422 312 200; FAX 44 1422 312 222; E-mail: mick.holland@mcintyre-ltd.co.uk)

McIntyre Group Singapore Pte. Ltd., Blk 513, Bishan Town Centre Street 13 #01-500, Singapore, 570513, Singapore (Tel: 65 354 3547; FAX 65 353 9068; E-mail: mcintyre@swiftech.com.sg)

McLaughlin Gormley King Co., Inc.

8810 10th Ave. N., Golden Valley, MN, 55427, USA (Tel: 612-544-0341; 800-645-6466; FAX 612-544-6437; E-mail: sales@mgk.com; Internet: www.mgk.com)

MDA Chemicals Ltd.

Willow Mill, Caton, Lancaster, Lancs, LA2 9RA, UK (Tel: 44 1524 771881; FAX 44-1524-771882; E-mail: info@mda-chem.co.uk; Internet: www.mda-aci.com/)

MeadWestvaco Corp./World Headquarters

One High Ridge Park, Stamford, CT, 06905, USA (Tel: 203-461-7400)

MelChem Inc.

317 Godwin Ave., Midland Park, NJ, 07432, USA (Tel: 201-445-5695; FAX 201-445-8612; E-mail: henrymellett@melchem.com; Internet: www.melchem.com)

Mentha & Allied Products Ltd.

1/10 Asaf Ali Rd., New Delhi, 110002, India (Tel: 91 11 731355; FAX 91 11 752 1189; E-mail: mentha@del2.vsnl.net.in)

Mercian Corp.

1-5-8, Kyobashi, Chuo-ku, Tokyo, 104, Japan (Tel: 81 3 3231 3917; FAX 81 3 3276 0151)

Merck

Merck KGaA, Frankfurter Str. 250, 64271 Darmstadt, Germany (Tel: 49 6151 720; FAX 49 6151 722000; E-mail: service@merck.de; Internet: www.merck.de)

Merck S.A., 5-9 Rue Anquetil, 94736 Nogent-sur-Marne Cedex, France (Tel: 33 1 43 94 54 00; FAX 33 1 48 76 58 15)

Merck Ltd., Div. of Merck AG, Merck House, Poole, Dorset, BH15 1TD, UK (Tel: 44 1202 669700; FAX 44 1202 665599; E-mail: info@merck-ltd.co.uk; Internet: www.merck-ltd.co.uk)

Merck Nederland BV, Basisweg 34, 1005 AD Amsterdam, The Netherlands (Tel: 31 20 4808 400; FAX 31 20 4808 480; E-mail: office@merck.nl; Internet: www.merck.nl)

Merck AG, Postfach, Rüchligstrasse 20, 8953 Dietikon, Switzerland (Tel: 41 17 45 1111; FAX 41 17 451100)

Merck Pty. Ltd., 207 Colchester Rd., Kilsyth, Victoria, 3137, Australia (Tel: 61 39 728 5855; FAX 61 39 728 1351; E-mail: merck@merck.com.au; Internet: www.merck.com.au)

S. Black (Import & Export) Ltd., Distrib. for Merck KGaA, Foxholes Business Park, John Tate Road, Hertford, Hertfordshire, SG13 7YH, UK (Tel: 44 1992 825555; FAX 44 1992 825566; E-mail: enquires@sblack.com; Internet: www.sblack.com)

Merichem Co.

5450 Old Spanish Trail, Houston, TX, 77023, USA (Tel: 713-428-5000; FAX 713-921-4604)

Methanex

1800 Waterfront Centre, 200 Burrard St., Vancouver, BC, V6C 3M1, Canada (Tel: 604-661-2600; 800-661-8851; FAX 604-661-2676; E-mail: sales@methanex.com; Internet: www.methanex.com)

MG Industries

3 Great Valley Pkwy., Malvern, PA, 19355, USA (Tel: 610-695-7400; 800-759-9899; FAX 610-695-7596; E-mail: mkting_svcs@mgind.com; Internet: www.mgindustries.com)

M. Michel & Co., Inc.

PO Box 788 Planetarium Station, New York, NY, 10024-0545, USA (Tel: 212-344-3878; FAX 212-344-3880; E-mail: corblok@aol.com; Internet: www.mmichel.com)

Michelman

Michelman, Inc., 9080 Shell Rd., Cincinnati, OH, 45236-1299, USA (Tel: 513-793-7766; 877-ADDITIV; FAX 877-464-2436; 513-793-2504; E-mail: info@michem.com; coatings@michem.com; Internet: www.michem.com; www.expandingpaper.com)

Michelman Int'l. & Co., SNC, Zoning Industriel, 6790 Aubange, Belgium (Tel: 32 63 38 18 00; FAX 32 63 38 96 92; E-mail: marie-annettemahy@michelmaninc.com)

Michelman Asia-Pacific Pte. Ltd., 1 Tuas Ave. 16, Singapore, 638924, Singapore (Tel: 65 861 2822; FAX 65 861 2262)

Micro Powders

Micro Powders, Inc., 580 White Plains Rd., Tarrytown, NY, 10591, USA (Tel: 914-793-4058; FAX 914-472-7098; E-mail: mpi@micropowders.com; Internet: www.micropowders.com)

Chesham Chemicals Ltd., Distributor for Micro Powders Inc., Cunningham House, Westfield Lane, Harrow, Middlesex, HA3 9ED, UK (Tel: 44 20 8907 7779; FAX 44 20 8909 1053; E-mail: sales@cheshamchemicals.co.uk; Internet: www.cheshamchemicals.co.uk)

Midwest Grain Products Inc.

1300 Main St., PO Box 130, Atchison, KS, 66002-0130, USA (Tel: 913-367-1480; 800-255-0302; FAX 913-367-1838; E-mail: natec@midwestgrain.com; Internet: www.midwestgrain.com)

Miljac Inc.

280 Elm St., New Canaan, CT, 06840, USA (Tel: 203-966-8777; FAX 203-966-3577; E-mail: sales@miljac.com; Internet: www.miljac.com)

Millennium

Millennium Chemicals, Inc., 230 Half Mile Road, Red Bank, NJ, 07701, USA (Tel: 732-933-5000; FAX 732-933-5200)

Millennium Chemicals/Fragrance & Flavor Div., 601 Crestwood St., Jacksonville, FL, 32208, USA (Tel: 904-768-5800; 800-231-6728; FAX 904-768-2200; E-mail: inquire@aromachem.com; Internet: www.aromachem.com)

Millennium Canada, 468 White Dr., Milton, Ontario, L9T 4G4, Canada (Tel: 905-876-3595; FAX 905-876-3582)

Millennium Europe, PO Box 26, Grimsby, N.E. Lincolnshire, DN40 2PR, UK (Tel: 44 1469 57 1000; FAX 44 1469 57 1234)

Millennium Belgium, SPRL, Ave. Ariane 5, 2nd Floor, 1200 Brussels, Belgium (Tel: 32 2 774 2511; FAX 32 2 774 2555)

Millennium Netherlands, Lange Bunder 7, 4854 MB Bavel, The Netherlands (Tel: 31 161 436600)

Millennium Germany, Neuer Markt 1, 42781 Haan, Dusseldorf, Germany (Tel: 49 2129 93010; FAX 49 2139 31584)

Millennium France, 85 ave. Victor Hugo, 92563 Rueil Malmaison Cedex, France (Tel: 33 1 55 47 22 50; FAX 33 1 55 47 22 51)

Millennium Australia, 15 Charles St., Mt. Waverley, Victoria, 3149, Australia (Tel: 61 3 9807 5028; FAX 61 3 9807 5994)

Millennium New Zealand, 49 C Nesdale Ave., Wiri, Manukaru City, Auckland, New Zealand (Tel: 64 9 570 9999; FAX 64 9 279 7105)

Millennium Singapore, 1 Fifth Ave., #03-04 Guthrie House, Singapore, 268802, Singapore (Tel: 65 465 0710; FAX 65 465 0716)

Millennium Shanghai Chemicals Trading Ltd., 9th Floor, Shanghai HSBC Tower, 101 Yin Cheng East Rd., Shanghai, Pudong, 200120, P.R. China (Tel: 86 21 6841 1090; FAX 86 21 6841 0428)

Millennium Hong Kong, China Hong Kong Tower, 22nd Floor, 8-12 Hennessy Rd., Wanchai, Hong Kong (Tel: 852 2528 4667; FAX 852 2865 1246)

Millennium Korea, 6/F Samdo Bldg., 1-170 Soonhwa-Dong, Chung-ku, Seoul, Korea (Tel: 82 2 779 4844; FAX 822 779 4883)

Millennium Specialty Chemicals. *See* Millennium Chemicals/Fragrance & Flavor Div.

Milliken

Milliken Chemical, Div. of Milliken & Co., PO Box 1927, M-400, Spartanburg, SC, 29304-1927, USA (Tel: 864-503-2200; 800-345-0372; FAX 864-503-2430; E-mail: millichem@milliken.com; Internet: www.millikenchemical.com)

Milliken Servicos, Av. das Nacoes Unidas, 12.551-22° floor/2205, Sao Paulo-SP, 04578-903, Brazil (Tel: 55 11 3043 7170; FAX 55 11 3043 7096)

Milliken Europe N.V., 18-24 Ham, 9000 Ghent, Belgium (Tel: 32 9 265 10 84; FAX 32 9 265 11 95; E-mail: eurochem@milliken.com)

Milliken Asia Pte. Ltd., 438B Alexandra Rd., #08-04 Alexandra Technopark Blk B, Singapore, 119968, Singapore (Tel: 65 377 0770; FAX 65 377 0990; E-mail: asiachem@milliken.com)

Milsolv Corp.

14765 W. Boblink Ave., Menomonee Falls, WI, 53051, USA (Tel: 262-252-3550; 800-558-8501; FAX 262-252-6155)

Mineral Research & Development Corp.

One Woodlawn Green, Suite 250, 200 East Woodlawn Rd., Charlotte, NC, 28217, USA (Tel: 704-525-2771; 800-334-0417; FAX 704-527-8232; E-mail: mrdc@mrdc.com; Internet: www.mrdc.com)

Miteni SpA

Via Mecenate, 90, 20138 Milan, Italy (Tel: 39 2 580 399 1; FAX 39 2 52 580 399 36; Internet: www.miteni.it)

Mitsubishi Chemical

Mitsubishi Chemical Corp., Mitsubishi Bldg., 2-5-2, Marunouchi, Chiyoda-ku, Tokyo, 100-0005, Japan (Tel: 81 3 3283 6111; FAX 81 3 3283 6287; E-mail: mccpr@cc.m-kagaku.co.jp; Internet: www.m-kagaku.co.jp/index_en.htm)

Mitsubishi Chemical Hong Kong Ltd., Unit 1901 & 06, 19th Fl., Peragrine Tower, Lippo Centre, 89 Queensway, Hong Kong (Tel: 852 2522 7031; FAX 852 2868 1470)

Mitsubishi Chemical Singapore Pte. Ltd., 79 Anson Rd., #12-01, Singapore, 079906, Singapore (Tel: 226 3707; FAX 226 1676)

Mitsubishi Chemical America, Inc., One North Lexington Ave., White Plains, NY, 10601, USA (Tel: 914-286-3697; FAX 914-681-0995)

Mitsubishi Chemical do Brasil Ltda., Av. Rio Branco 89, S/1001, Edificio Manhattan Tower, Centro, Rio de Janeiro RJ, 20040-004, Brazil (Tel: 55 21 233 3236; FAX 55 21 516 9890)

Mitsubishi Chemical Europe GmbH, Niederkasseler Lohweg 8, 40547 Dusseldorf, Germany (Tel: 49 211 523 9230; FAX 49 211 523 9281)

Mitsubishi Chemical (UK) plc, Jupiter House, Triton Court, 14 Finsbury Sq., London, EC2A 1BR, UK (Tel: 44 171 588 1810; FAX 44 171 588 1814)

Mitsubishi Gas

Mitsubishi Gas Chemical Co., Inc., Mitsubishi Bldg., 2-5-2, Marunouchi, Chiyoda-ku, Tokyo, 100, Japan (Tel: 81 3 3283 4799; FAX 81 3 3214 0938; Internet: www.mgc-a.com)

Mitsubishi Gas Chemical America, 520 Madison Ave., 17th Floor, New York, NY, 10022, USA (Tel: 212-752-4620; FAX 212-758-4012; E-mail: mike@mgc-a.com)

Mitsubishi Gas Chemical Europe GmbH, Immermannstrasse 43, 40210 Dusseldorf 1, Germany (Tel: 49 211 363088; FAX 49 211 354457; E-mail: tnb-mgce@kddnet.de)

Mitsubishi Oil Co., Ltd.

1-6-41, Konan, Shinagawa-ku, Tokyo, 105, Japan (Tel: 81 3 3472 7500; FAX 81 3 3450 0329)

Mitsui

Mitsui & Co., Ltd., 2-1, Ohtemachi 1-chome, Chiyoda-ku, Tokyo, Japan (Tel: 81 3 3285 1111; E-mail: InfoTKABZ@tkab.xm.mitsui.co.jp)

Mitsui & Co. France S.A., 19 r Alphonse de Neuville, 75017 Paris, France (Tel: 33 1 44 01 70 00; FAX 01 44 01 70 70; Internet: www.mitsui-france.com/)

Mitsui and Co. UK plc, 20 Old Bailey, London, EC4M 7QQ, UK (Tel: 44 20 7822 0582; FAX 44 20 7822 0698)

Mitsui & Co. (USA), Inc., 200 Park Ave., New York, NY, 10166, USA (Tel: 212-878-4000; FAX 212-878-4800; Internet: www.mitsui.com)

Mitsui & Co. (Australia) Ltd., Level 46, Gateway, 1 Macquarie Pl., Sydney, NSW, 2000, Australia (Tel: 61 2 9256 9500; FAX 61 2 9251 1788; Internet: www.mitsui.com.au/)

Mitsui Chemicals

Mitsui Chemicals, Inc., Japan (Tel: 81 3 3592 4466; FAX 81 3 3592 4253)

Mitsui Chemicals America, Inc., 2500 Westchester Ave., Suite 110, Purchase, NY, 10577, USA (Tel: 914-253-0777; 800-682-2377; FAX 914-253-0790; Internet: www.mitsuichemicals.com)

Mitsui Chemicals, Inc., Oststrasse 10, 40211 Dusseldorf, Germany (Tel: 49 211 173 3245; FAX 49 211 171 9961; Internet: www.mitsui.de/en/aktiv/chemie.htm)

Mitsui Petrochemical Industries, Ltd.

Kasumigaseki Bldg., 3-2-5, Kasumigaseki, Chiyoda-ku, Tokyo, 100, Japan (Tel: 81 3 3580 3616; FAX 81 3 3593 0028)

Mitsui Toatsu Chemicals, Inc.

Kasumigaseki Bldg., 3-2-5, Kasumigaseki, Chiyoda-ku, Tokyo, 100, Japan (Tel: 81 3 3592 4111; FAX 81 3 3592 4267)

MLG Enterprises Ltd.

PO Box 52568, Turtle Creek PO, 1801 Lakeshore Road West, Mississauga, Ontario, L5J 4S6, Canada (Tel: 905-569-3330; FAX 905-569-2133; E-mail: mlgent@worldchat.com)

MMP, Inc.

3470 S. Clinton Ave., South Plainfield, NJ, 07080, USA (Tel: 908-561-4435; FAX 908-561-4780; E-mail: inquiry@mmpinc.com; Internet: www.mmpinc.com)

Molekular

Molekular S.a.r.l., Z.I. du Chapelier, B.P. 83 St. Jean-de-Soudain, 38353 La Tour du Pin Cedex, France (Tel: 33 4 37 05 00 50; FAX 33 4 37 05 00 51; E-mail: info@molekula.fr; Internet: www.molekula.co.uk)

Molekular Ltd., Technology Rd., Poole, Dorset, BH17 7DA, UK (Tel: 44 1202 330066; FAX 44 1202 330055; E-mail: info@molekula.co.uk; Internet: www.molekula.co.uk)

Molekular USA, 9648 Olive Blvd., Suite 239, St. Louis, MO, 63132-3002, USA (Tel: 314-503-2245; FAX 314-569-0893; E-mail: info@molekula.com)

Monomer-Polymer & Dajac Labs, Inc.

1675 Bustleton Pike, Feasterville, PA, 19053, USA (Tel: 215-364-1155; FAX 215-364-1583)

Monsanto

Monsanto Co., Wholly owned subsid. of Pharmacia, 800 N. Lindbergh Blvd., St. Louis, MO, 63167, USA (Tel: 314-694-1000; 800-325-4330; FAX 314-674-2211; Internet: www.monsanto.com)

Monsanto Canada, 2330 Argentia Rd., PO Box 667, Steetsville PO, Mississauga, Ontario, L5M 2C2, Canada (Tel: 905-826-2180; FAX 905-826-9979)

Monsanto Europe S.A., Ave. de Tervuren 270-272, Boite 1, 1150 Brussels, Belgium (Tel: 32 2 776 4111; FAX 32 2 776 4040)

Monsanto France SA, Europarc de Chene, 11 rue Pascal, 69673 Bron Cedex, France (Tel: 33 4 72 14 40 40)

Monsanto Danmark A/S, Banemarksvej 50 A, 2605 Brondby, Denmark (Tel: 45 43 22 75 00; FAX 45 4322 7599)

Monsanto (Deutschland) GmbH, Vogelsanger Weg 91, 40470 Dusseldorf, Germany (Tel: 49 211 36750)

Monsanto Australia Ltd., 600 St. Kilda Rd., 12th Floor, Melbourne, Victoria, 3004, Australia (Tel: 61-3-9522-7122; FAX 61-3-9525-2253)

Monsanto Japan, Nihonbashi Daini Bldg., 41-12, Hakozaki-cho, Nihonbashi, Chuo-ku, Tokyo, 103, Japan (Tel: 81 3 5644 1662; FAX 81 3 5644 1661)

Monsanto Singapore Co. (Pte) Ltd., 101 Thomson Rd., #19-00 United Square, Singapore, 307591, Singapore (Tel: 65 250 2000)

Moore Ingredients

9047 Sutton Pl., Hamilton, OH, 45011, USA (Tel: 513-881-7144; FAX 513-881-7145; E-mail: mail@moorelab.com; Internet: www.moorelab.com)

Morflex

Morflex, Inc., Subsid. of Reilly Industries, 2110 High Point Rd., Greensboro, NC, 27403, USA (Tel: 336-292-1781; FAX 336-854-4058; E-mail: skennedy@morflex.com; Internet: www.morflex.com)

Reilly Chemicals S.A., 154, Chaussèe de la Hulpe, 1170 Brussels, Belgium (Tel: 32 2 663 4400; FAX 32 2 673 8104; E-mail: rcsa@reillyind.com)

Morre-Tec Industries, Inc.

One Gary Rd., Union, NJ, 07083, USA (Tel: 908-688-9009; FAX 908-688-9005; E-mail: info@morretec.com; Internet: www.morretec.com)

Morton

Morton International Inc., Wholly owned subsid. of Rohm & Haas, 123 North Wacker Dr., Chicago, IL, 60606, USA (Tel: 312-807-2000; FAX 312-807-2040; Internet: www.morton.com; www.rohmhaas.com)

Morton International Ltd., Westward House, 155-157 Staines Rd., Hounslow, Middlesex, TW3 3JB, UK (Tel: 44 20 8570 7766; FAX 44 20 8572 8891)

Mosselman N.V./S.A.

Route de Wallonie 4, 7011 Ghlin, Belgium (Tel: 32 65 395610; FAX 32 65 395612; E-mail: sales@mosselman.be; Internet: www.mosselman.be)

Multi-Kem Corp.

553 Broad Ave., PO Box 538, Ridgefield, NJ, 07657-0538, USA (Tel: 201-941-4520; 800-462-4425; FAX 201-941-5239; E-mail: multikem@csnet.net; Internet: www.multikem.com)

Mutchler Chemical Co., Inc.

99 Kinderkermack Rd., Westwood, NJ, 07675, USA (Tel: 201-666-7002; FAX 201-666-3652; E-mail: info@mutchlerchem.com; Internet: www.mutchlerchem.com)

Nagai Chemical Industrial Co., Ltd. (Nagai Seiyakusho K.K.)

3-27-15, Chikko-shinmachi, Sakai-shi, Osaka, 592, Japan (Tel: 81 722 44 2321; FAX 81 722 44 2818)

Nanjing Meheco

199 Jianye Rd., Nanjing, P.R. China (Tel: 86 25 2265666; FAX 86 25 2260231; E-mail: meheco@njmco.com)

Nantong ChangChem

Rm. 1208, Nantong International Mansion, 38 Yue Long Rd., Nantong, 226001, P.R. China (Tel: 86 513 515 5236; FAX 86 513 515 2292; E-mail: info@changchem.com; Internet: www.changchem.com)

Napp Technologies, Inc.

299 Market Street, PO Box 893, Saddle Brook, NJ, 07663, USA (Tel: 201-843-4664; FAX 201-843-4737; Internet: www.napptech.com)

National Diagnostics, Inc.

305 Patton Dr., Atlanta, GA, 30336, USA (Tel: 404-699-2121; 800-526-3867; FAX 404-699-2077; E-mail: natldiagnos@mindspring.com; Internet: www.nationaldiagnostics.com)

National Purity Inc.

434 Lakeside Ave. N., Minneapolis, MN, 55405, USA (Tel: 612-672-0022; FAX 612-672-0027)

National Starch & Chemical

National Starch & Chemical Corp., Member of the ICI Group, 10 Finderne Ave., Bridgewater, NJ, 08807-3300, USA (Tel: 908-685-5000; 800-797-4992; FAX 908-685-5005; E-mail: nsta171596464b@adh.com; Internet: www.nationalstarch.com)

National Starch & Chemical, S.A. de C.V., Blvd. Miguel De Cervantes Saavedra 71, Col. Granada, Mexico D.F., 11520, Mexico (Tel: 52 5 545 3787; FAX 52 5 531 0417)

National Starch & Chemical Ltd., Prestbury Court, Greencourts Business Park, 333 Styal Rd., Heald Green, Manchester, Lancashire, M22 5LW, UK (Tel: 44 161 435 3200; FAX 44 161 435 3300)

National Starch & Chemical GmbH, Gruener Diech 110, 20097 Hamburg, Germany (Tel: 49 40 23915 0; FAX 49 40 23915 170)

National Starch & Chemical B.V., Henri Dunantweg 2N, 7201 EV Zutphen, The Netherlands (Tel: 31 575 582500; FAX 31 575 582690)

National Starch & Chemical AB, Box 12033, Torsgatan 12, 600 12 Norrköping, Sweden (Tel: 46 11 219700; FAX 46 11 125451)

National Starch & Chemical SA, BP 438, 299 rue Grange Morin, 69655 Villefranche sur Saone, France (Tel: 33 4 74 02 38 00; FAX 33 4 74 02 39 00)

National Starch & Chemical SpA, Via Roma 29, 20050 Mezzago (MI), Italy (Tel: 39 39 60921; FAX 39 39 6092201)

National Starch & Chemical Pty. Ltd., 7-9 Stanton Rd., Seven Hills, Sydney, NSW, 2147, Australia (Tel: 61 2 9624 6022; FAX 61 2 9624 1468)

National Starch & Chemical (Asia) Pte., Ltd., 10 Science Park Rd., #04-21 The Alpha, Singapore Science Park II, Singapore, 11784, Singapore (Tel: 65 872 5335; FAX 65 872-5988)

S. Black (Import & Export) Ltd., Distrib. for National Starch & Chem., Foxholes Business Park, John Tate Road, Hertford, Hertfordshire, SG13 7YH, UK (Tel: 44 1992 825555; FAX 44 1992 825566; E-mail: enquires@sblack.com; Internet: www.sblack.com)

Natra US Inc.

1059 Tierra Del Ray, Chula Vista, FL, 91910, USA (Tel: 619-397-4120; FAX 619-397-4121; E-mail: pete@natra-group.com)

Natural Advantage

3587 Highway 9 #320, Freehold, NJ, 07728, USA (Tel: 732-274-9683; 877-NAT-FLAV; FAX 732-274-9217; E-mail: customerservice@natural-advantage.net; Internet: www.natural-advantage.net)

Natural Gas Odorizing, Inc.

3601 Decker Dr., PO Box 1429, Baytown, TX, 77522-1429, USA (Tel: 281-424-5568; FAX 281-424-3681)

Natural Oils International, Inc.

9243 Cranford Ave., Arleta, CA, 91331-4309, USA (Tel: 818-504-7383; FAX 818-504-7386; E-mail: info@naturaloils.com; Internet: www.naturaloils.com)

Naturex

Site d'Agroparc, 84911 Avignon Cedex 09, France (Tel: 33 4 90 23 96 89; FAX 33 4 90 23 96 84)

Nepera, Inc.

A Cambrex Co., 41 Arden House Rd., Harriman, NY, 10926, USA (Tel: 845-782-1200; 800-637-3721; FAX 845-783-9713; E-mail: gwillcox-jones@cambrex.com; Internet: www.cambrex.com)

Neuchem Inc.

20 Alteza, Santa Fe, NM, 87505, USA (Tel: 505-466-2221; FAX 505-466-1528; E-mail: neucheminc@aol.com)

Neville Chemical Co.

2800 Neville Rd., Pittsburgh, PA, 15225-1496, USA (Tel: 412-331-4200; 877-704-4200; FAX 412-777-4234; E-mail: neville@usaor.net; Internet: www.nevchem.com)

New Japan Chemical Co., Ltd.

2-1-8, Bingo-machi, Chuo-ku, Osaka, 541, Japan (Tel: 81 6 6202 0621; FAX 81 6 6222 0062)

Niacet Corp.

47 St. & Niagara Falls Blvd., Niagara Falls, NY, 14304, USA (Tel: 716-285-1474; 800-828-1207; FAX 716-285-1497)

Nihon Jyoryu Kogyo K.K.

9-1, Koya-shinmachi, Ichikawa-shi, Chiba, 272, Japan (Tel: 81 473 28 1185; FAX 81 473 28 1189)

Nihon Surfactant Kogyo K.K.

Part of Nikko Chemicals Group, 3-24-3 Hasune, Itabashi-Ku, Tokyo, 174, Japan (Tel: 81 3 3966 7331; FAX 81 3-3966-7307; E-mail: info@ns-cosmos.co.jp; Internet: www.ns-cosmos.co.jp/e/)

Nikko Chemicals

Nikko Chemicals Co., Ltd., 1-4-8, Nihonbashi, Bakuro-cho, Chuo-ku, Tokyo, 103-0002, Japan (Tel: 81 3 3661 1677; FAX 81 3 3664 8620; E-mail: info@nikkol.co.jp; Internet: www.nikkol.co.jp/index.html)

Barnet Products Corp., Distrib. for Nikko Chemicals Co., Ltd., 140 Sylvan Ave., Englewood Cliffs, NJ, 07632, USA (Tel: 201-346-4620; FAX 201-346-4333)

Nipa Laboratories

Nipa Laboratories Ltd., Part of Clariant Group, Llantwit Fardre, Pontypridd, Mid Glamorgan, CF38 2SN, UK (Tel: 44 1443 205311; FAX 44 1443-207746; E-mail: lyndon_payne@compuserve.com; Internet: www.nipa.com)

Nipa Laboratorien GmbH, Hans-Bockler Ring 9, 22851 Norderstedt, Germany (Tel: 49 40 529 54623; FAX 49 40 529 54630)

Nipa SA, 70 Ave. du General de Gaulle, 92058 Paris La Defense Cedex, France (Tel: 33 1 46 96 97 5; FAX 33 1 46 96 99 08)

Nipa Inc./Clariant Corp., 625 E. Catawba Ave., Mount Holly, NC, 28210, USA (FAX 704-822-2241; E-mail: Dave.Andrews@na.clariant.com; Internet: www.nipa.com)

Nippoh Chemicals Co., Ltd.

3-3-3, Nihonbashi Muro-machi, Chuo-ku, Tokyo, 103, Japan (Tel: 81 3 3270 5341; FAX 81 3 3246 0346)

Nippon Denko Co., Ltd.

2-11-8, Ginza, Chuo-ku, Tokyo, 104, Japan (Tel: 81 3 3546 9331; FAX 81 3 3542 3690)

Nippon Nyukazai Co., Ltd.

Subsid. of Sankyo Co., Ltd., Yoshizawa Bldg., 3-9-19, Ginza, Chuo-ku, Tokyo, 104, Japan (Tel: 81 3 3543 8571; FAX 81 3 3546 3174; Internet: www.sankyo.co.jp)

Nippon Petrochemicals Co., Ltd.

1-21-18, Shinbashi, Minato-ku, Tokyo, 105, Japan (Tel: 81 3 3501 7313; FAX 81 3 3501 7630)

Nippon Polyurethane Industry Co., Ltd.

Kotohira Kaikan Bldg., 1-2-8, Toranomon, Minato-ku, Tokyo, 105, Japan (Tel: 81 3 3508 0624; FAX 81 3 3508 0617)

Nippon Steel Chemical Co., Ltd.

7-21-11, Gotanda, Shinagawa-ku, Tokyo, 141-0031, Japan (Tel: 81 3 5759 2741; FAX 81 3 5759 2777)

Nippon Synthetic Chemical Industry Co., Ltd.

Umeda Sky Bldg. Tower East, 1-88, Oyodonaka 1-chome, Kita-ku, Osaka, 531-0076, Japan (Tel: 81 6 6440 5300; FAX 81 6 6440 5330)

Nissan Chemical

Nissan Chemical Industries, Ltd., Kowa-Hitotsubashi Bldg., 3-7-1, Kanda, Nishiki-cho, Chiyoda-ku, Tokyo, 101-0054, Japan (Tel: 81 3 3296-8065; FAX 81 3 3296-8360; Internet: www.snowtex.com; www.nissanchem.co.jp)

Nissan Chemical America Corp., 10777 Westheimer, Suite 150, Houston, TX, 77042, USA (Tel: 713-532-4745; FAX 713-532-0363; E-mail: snowtex@aol.com; Internet: www.snowtex.com)

Nissan Chemical Europe GmbH, Deutsch-Japanisches Center, Immermann-strasse 45, 40210 Dusseldorf, Germany (Tel: 49 211 17 2270; FAX 49 211 16 2243)

Nissan Chemical Industries, Ltd./Singapore Representative Office, 101 Cecil Street, #16-09 Tong Eng Bldg., Singapore, 069533, Singapore (Tel: 220 9296; FAX 224 9624)

British Traders & Shippers Ltd., Distrib. for Nissan Chem. Ind., 429-431 Rainham Road South, Dagenham, Essex, RM10 8RN, UK (Tel: 44 181 596 7500; FAX 44 181 596 7509; E-mail: pvax@britishtraders.co.uk; Internet: www.british traders.co.uk)

Nisshin Oillio Group Ltd.

1-23-1 Shinkawa, Chuo-Ku, Tokyo, 104-8285, Japan (Tel: 81 3 3206 5025; FAX 81 3 3206 6452; Internet: www.oilliogroup.com)

Nisso Petrochemical Industries Co., Ltd.

Joint venture of Nippon Soda Co., Ltd. and Teijin Ltd., Seisho Nihonbashi Bldg., 12-2, Nihonbashi Ohdenma-cho, Chuo-ku, Tokyo, 103, Japan (Tel: 81 3 3664 1991; FAX 81 3 3664 1990)

Nitron Chem Corp.

PO Box 978, 22 W. Putnam Ave., Greenwich, CT, 06830, USA (Tel: 203-622-9200; FAX 203-622-0457)

Noah Chemical

Div. of Noah Technologies Corp., 1 Noah Park, San Antonio, TX, 78249, USA (Tel: 210-691-2000; FAX 210-691-2600; E-mail: sales@noahtech.com; Internet: www.noahtech.com)

NOF

NOF Corp., Yebisu Garden Place Tower, 4-20-3, Ebisu, Sibuya-ku, Tokyo, 150-6019, Japan (Tel: 81 3 5424 6600; FAX 81 3 5424 6800; Internet: www.nof.co.jp)

NOF America Corp., Subsid. of NOF Corp., 11 Martine Ave., Sutie 1170, White Plains, NY, 10606, USA (Tel: 914-681-9790; FAX 914-681-9791; Internet: www.nof.co.jp)

NOF Europe N.V., Subsid. of NOF Corp., Bouwelven 1, Industriezoen Klein-Gent, 2280 Grobbendonk, Belgium (Tel: 32 14 230 001; FAX 32-14-224 563)

NOF Asia Pacific Pte., Ltd., Subsid. of NOF Corp., 10 Anson Rd., #19-10 International Plaza, Singapore, 079903, Singapore (Tel: 65 220 4881; FAX 65 225 9718)

NorChem Corporation

PO Box 6022, Lawrenceville, NJ, 08648, USA (Tel: 609-585-3627; FAX 609-585-3629)

Nordmann Rassmann & Co. GmbH

Kajen 2, 20459 Hamburg, Germany (Tel: 49 40 3687 0; FAX 49 40 3687 249; E-mail: info@nrc.de; Internet: www.nrc.de)

Norman, Fox & Co.

5511 S. Boyle Ave., PO Box 58727, Vernon, CA, 90058-0727, USA (Tel: 323-583-0016; 800-632-1777; FAX 323-583-9769; E-mail: norfox@worldnet.att.net; Internet: www.norfoxx.com)

Norsk Hydro AS

Bygdoy Alle 2, 0240 Oslo, Norway (Tel: 47 22 53 81 00; FAX 47 22 53 27 25; E-mail: corporate@hydro.com)

Northwestern Flavors, Inc.

Subsid. of Wm. Wrigley Jr. Co., 120 N. Aurora St., W. Chicago, IL, 60185, USA (Tel: 708-231-6111; FAX 708-231-1302; Internet: www.wrigley.com)

The Nottingham Co.

A PPI Company, 1303 Boyd Ave. NW, Atlanta, GA, 30318, USA (Tel: 404-351-3501; 800-868-6260; FAX 404-351-7731; E-mail: ppi@ppiinc.com; Internet: www.ppiinc.com)

Novachem Aromatici S.R.I.

Via Carlo Noe 7, 21013 Gallarate (VA), Italy (Tel: 39 33 179 9878; FAX 39 33 179 6531; E-mail: info@novachemaromatici.com; Internet: www.novachemaromatici.com)

NOVA Chemicals

NOVA Chemicals Ltd., Wholly owned subsid. of NOVA Corp., PO Box 2518, Station M, 645 Seventh Ave. SW, Calgary, Alberta, T2P 5C6, Canada (Tel: 403-750-3600; 800-661-1548; FAX 403-269-7410; E-mail: public@novachem.com; Internet: www.nova.ca; www.novachem.com)

NOVA Chemicals Inc., 1550 Coraopolis Hts. Rd., Moon Township, PA, 15108, USA (Tel: 412-490-4000; 800-222-7213; FAX 412-494-4861; E-mail: kwokjc@novachem.com; Internet: www.novachem.com)

NOVA Chemicals Int'l. S.A., Chemin des Mazots 2, 1700 Fribourg, Switzerland (Tel: 41 26 426 5656; FAX 41 26 426 5657)

NOVA France Sarl, 704 Rue P et M Curie, BP 30215, 60772 Ribecourt Cedex, France (Tel: 33 0344 75 7000)

NOVA Chemicals UK, Urmston, Isherwood Rd., Manchester, M31 4TJ, UK (Tel: 44 161 776 5300)

NOVA Chemicals Vertriebs GmbH, Mergenthalerallee 79-81, 65760 Eschborn, Germany (Tel: 49 6196 927530; FAX 49 6196 9275333)

NOVA Chemicals Japan Ltd., Parkwest Bldg., 5F, 6-12-1 Nishi-shinjku, Shinjuku-ku, Tokyo, 160-0023, Japan (Tel: 81 3 3344-5871; FAX 81-3-3344-5876)

NOVA Chemicals (International) S.A., 80 Anson Rd., #31-01, IBM Towers, Singapore, 079909, Singapore (Tel: 65 224 8807; FAX 65-224-1877)

NOVA Chemicals (International) S.A., Room 1603 Beijing Guang Ming Hotel, Liang Ma Qiao Rd. Chao Yang District, Beijing, 100016, P.R. China (Tel: 86 10 6460 0582; FAX 86-10-6460-0581)

NOVA Molecular Technologies, Inc.

1 Parker Place, Suite 725, Janesville, WI, 53545, USA (Tel: 608-754-NOVA; FAX 608-754-6878; Internet: www.novamolecular.com)

Noveon

Noveon Inc., 9911 Brecksville Rd., Cleveland, OH, 44141-3247, USA (Tel: 216-447-5000; 800-379-5389; FAX 216-447-5669; Internet: www.noveoninc.com; www.carbopol.com; www.noveoncoatings.com)

Noveon Canada, 100 Regina St., South, Suite 36D, Waterloo, Ontario, N2J 4P9, Canada (Tel: 519-888-3330; FAX 519-888-3337)

Noveon Performance Material Europe, Chaussèe de Wavre, 1945, 1160 Brussels, Belgium (Tel: 32 2 678 1911; FAX 32 2 678 1990)

Noveon (Germany), Rosenheimer Strasse 43, 83064 Raubling, Germany (Tel: 49 80 35880)

Noveon (Italia) Srl, Viale Gian Galeazzo 25, 20136 Milan, Italy (Tel: 39 2 5830 6961; FAX 39 2 5831 0390)

Noveon UK, Unit 5, Cherrywood, Chineham Business Park, Basingstoke, Hampshire, RG24 8WF, UK (Tel: 44 1256 707546)

Noveon Performance Materials Asia Pacific Ltd., 1107-1110 Shui On Centre, 6-8 Harbour Rd., Wanchai, Hong Kong (Tel: 852 2508 1021; FAX 852 2512 2241; Internet: www.asiapacific.noveoninc.com)

Noveon Diamalt GmbH

Georg-Reismüller-Strasse 32, 80999 Munich, Germany (Tel: 49 89 8106 335; FAX 49 89 8106 513)

Noveon Kalama Inc.

Suite 1110, Bank of California Center, Seattle, WA, 98164, USA (Tel: 206-682-7890; 800-742-6147; FAX 206-682-1907)

Oakwood Products

1741 Old Dunbar Rd., West Columbia, SC, 29172, USA (Tel: 803-739-8800; 800-467-3386; FAX 803-739-6957; E-mail: sales@oakwoodchemical.com; Internet: www.oakwoodchemical.com)

O'Brien Products, Inc.

A Zinkan Enterprises Co., 10574 Ravenna Rd., Twinsburg, OH, 44087, USA (Tel: 330-487-1500; FAX 330-425-8202; E-mail: sales@zinkan.com; Internet: www.zinkan.com)

Occidental Chemical

Occidental Chemical Corp., Wholly owned subsid. of Occidental Petroleum Corp., 5005 LBJ Freeway, Dallas, TX, 75244-6119, USA (Tel: 972-404-3800; 800-578-8880; FAX 972-404-3669; E-mail: info@oxychem.com; Internet: www.oxychem.com)

OxyChem, An Occidental Chemical Corp., Occidental Tower, 5005 LBJ Freeway, Dallas, TX, 75244, USA (Tel: 972-404-3800; 800-752-5151; FAX 972-404-3669; E-mail: info@oxychem.com; Internet: www.oxychem.com)

Occidental Quimica de Mexico, S.A. de C.V., Periferico Sur 3343, Piso 5, Col. San Jeronimo Lidice, Mexico D.F., 10200, Mexico (Tel: 52 5 668 0554; FAX 52 5 595 9923; E-mail: latin_america_north.sales@oxychem.com)

Occidental Chemical Europe S.A., Excelsiorlaan 19, 1930 Zaventem, Belgium (Tel: 32 2 715 6600; FAX 32 2 725 4676; E-mail: emea.sales@oxychem.com)

Occidental Chemical (UK) Ltd., Park House, 354 Broadway, Horsforth, Leeds, West Yorkshire, LS18 4RE, UK (Tel: 44 1132 585702; FAX 44 113 258 1163)

Occidental Chemical Asia Ltd., Kouyo Toranomon Bldg., 5th Fl., 17-2 Nishi-Shinbashi 2-chome, Minato-ku, Tokyo, 105-003, Japan (Tel: 81 3 3431 9681; FAX 81 3 5776 2765; E-mail: far_east_north.sales@oxychem.com)

Occidental Chemical Far East, 152 Beach Rd., #08-06 Gateway East, Singapore, 189721, Singapore (Tel: 65 299 9962; FAX 65 299 2771; E-mail: far_east_cs.sales@oxychem.com)

Ocean Chemicals Group

Div. of Fluorochem Ltd., Wesley St., Old Glossop, Derbyshire, SK13 7RY, UK (Tel: 44 1457 867708; FAX 44 1457 860927; E-mail: enquiries@oceanchemicals.co.uk; Internet: www.oceanchemicals.co.uk)

Ogawa & Co., Ltd.

4-1-11, Nihonbashi Honcho, Chuo-ku, Tokyo, 103-0023, Japan (Tel: 81 3 3270 1720; FAX 81 3 3279 2469)

Oilseeds International, Ltd.

855 Sansome St., #100, San Francisco, CA, 94111, USA (Tel: 415-956-7251; FAX 415-394-9023; E-mail: info@oilseedsf.com)

O'Laughlin Industries Co. Ltd.

4036 Maxanne Dr., Kennesaw, GA, 30144, USA (Tel: 770-517-7837; FAX 770-517-7843)

Olin Chlor Alkali Div.

Div. of the Olin Corp., 490 Stuart Rd. NE, Cleveland, TN, 37312, USA (Tel: 423-336-4830; 800-636-3756; FAX 423-336-4830; E-mail: plperkins@corp.olin.com; Internet: www.olinchloralkali.com)

OM Group (OMG)

OM Group, Inc. (OMG), World Headquarters, 50 Public Square, 3500 Terminal Tower, Cleveland, OH, 44113, USA (Tel: 216-781-0083; 800-519-0083; FAX 216-781-1502; E-mail: bob.strenk@omgi.com; Internet: www.omgi.com)

OMG Americas, Inc., 2601 Weck Dr., Research Triangle Park, NC, 27709-2166, USA (Tel: 919-544-8090; 800-321-9696; FAX 919-544-7996; Internet: www.omgi.com)

OMG Europe GmbH, Munsterstrsse 248, 40470 Dusseldorf, Germany (Tel: 49 211 961 880; FAX 49 211 614 629)

OMG Kokkola Chemicals Oy, PO Box 286, 67101 Kokkola, Finland (Tel: 358 6 8280 111; FAX 358 6 8280 307)

OMG Asia Pacific C., Ltd., Suite B, 18F-1, No. 156, Sec. 3, Min Sheng Eazst Road, Taipei, Taiwan R.O.C. (Tel: 886 2 2712 8566; FAX 886 2 2719 1915)

ONDEO Nalco

ONDEO Nalco Co., Subsid. of Suez, ONDEO Nalco Center, 1601 W. Diehl Road, Naperville, IL, 60563-1198, USA (Tel: 630-305-1000; FAX 630-305-2900; E-mail: cdugina@ondeo-nalco.com; Internet: www.ondeo-nalco.com)

ONDEO Nalco Latin America Operations, Av. Das Nacoes Unidas, 17.891, 11°, Andar, Santo Amaro, Sao Paulo SP, 04795-100, Brazil (Tel: 55 11 5644 6500; FAX 55 11 5641 7687)

ONDEO Nalco European Operations, PO Box 627, 2300 AP Leiden, The Netherlands (Tel: 31 71 524 1100; FAX 31 71 524 1197)

ONDEO Nalco Pacific Pte. Ltd., 21 Gul Lane, Jurong Town, Singapore, 2262, Singapore (Tel: 65 861 4011; FAX 65 862 0850)

Orange Chemicals Ltd.

34 St Thomas Street, Winchester, Hampshire, SO23 9HJ, UK (Tel: 44 1962 842525; FAX 44 1962 841101; E-mail: brianorange@orangechem.demon.co.uk; Internet: www.orangechem.demon.co.uk)

Oriental Chemical Industries

50, Sokong-dong, Chung-ku, Seoul, 100-718, Korea (Tel: 82 2 727 9500; FAX 82 2 757 4111; E-mail: webmaster@oci.co.kr; Internet: www.oci.co.kr)

Osaka Gas Co., Ltd.

4-1-2, Hirano-machi, Chuo-ku, Osaka, 541, Japan (Tel: 81 6 6202 2221; FAX 81 6 6226 1681)

Osaka Organic Chemical Ind. Co., Ltd.

Shin Toyama Bldg., 1-7-20, Azuchi-machi, Chuo-ku, Osaka, 541, Japan (Tel: 81 6 6264 5071; FAX 81 6 6264 1675)

Otira Pharmaceuticals Pvt. Ltd.

401, Lakshmi Nivas, Greenlands, Ameerpet, Hyderabad, 500 016 AP, India (Tel: 91 40 341 5667; FAX 91 40 341 5669; E-mail: otira@otira.com; Internet: www.otira.co.in; www.otira.com)

Outokumpu Mintec Oy

Riihitontuntie, 02201 Espoo, Finland (Tel: 358 9 4211; FAX 358 9 421 3156)

Oxiteno S.A. Industria e Comercio

Av. Brigedeiro Luiz Antonio 1343, 10o andar, Sao Paulo SP, 01350-900, Brazil (Tel: 55 11 3177 6043; FAX 55 11 285 0327; E-mail: oxiteno@oxiteno.com.br; Internet: www.oxiteno.com.br/ingles/)

Pacific West Ind. Co., Ltd.

PO Box 26-1134, Taipei, Taiwan R.O.C. (Tel: 886 2 2358 7111; FAX 886 2 2325 8985; E-mail: sanyuan2@tpts6.seed.net.tw)

Pangaea Sciences

2962 Saint Malo Circle, Mississauga, Ontario, L5N 1S9, Canada (Tel: 905-814-0798; 888-PANGAEA; FAX 905-814-0858; E-mail: sales@pangaeasciences.com; Internet: www.pangaeasciences.com)

Parchem Trading, Ltd.

2 William St., Suite 303, White Plains, NY, 10601, USA (Tel: 914-997-1050; FAX 914-997-1025; E-mail: info@par_chem.com; Internet: www.par_chem.com)

Patco Chemical Ind. Inc.

125 Sanford Ave., Kearny, NJ, 07032, USA (Tel: 201-955-0590; FAX 201-955-0717)

Paulaur Corp.

105 Melrich Rd., Cranbury, NJ, 08512, USA (Tel: 609-395-8844; FAX 609-395-8850)

Pax Group

Pax Group of Companies, Shreyas, Plot No. 72, Mogal Lane, Mahim, Mumbai, 400 016, India (Tel: 91 22 5660 2400; FAX 91 22 5660 5090; E-mail: sales@paxchem.com; Internet: www.paxgroup.com; www.paxchem.com)

PaxChem Ltd., Part of PAX Group, Shreyas, Plot No. 72, Mogal Lane, Mahim, Mumbai, 400 016, India (Tel: 91 22 5660 2400; FAX 91 22 5660 5090; E-mail: sales@paxchem.com; Internet: www.paxchem.com)

P.B. & S. Chemical Co., Inc.

A member of the Brenntag Group, 1405 Highway 136 West, PO Box 20, Henderson, KY, 42419-0020, USA (Tel: 270-827-4509; 800-950-7267; FAX 270-827-3990; E-mail: info@pbschemical.com; Internet: www.pbschemical.com)

PCAS (Produits Chimiques Auxiliaires et de Synthese)

23, Rue Bossuet, Z.I. de la vigne aux Loups, B.P. 181, 91160 Lonjumeau, France (Tel: 33 1 69 09 77 85; FAX 33 1 64 48 23 19; E-mail: pcaschemicals@compuserve.com)

PCI Chemicals Canada Inc.

630 Rene-Levesque Blvd West, 31st Floor, Montreal, Quebec, H3B 1S6, Canada (Tel: 514-397-6184)

PCS Purified Phosphates

1101 Skokie Blvd., Suite 400, PO Box 3320, Northbrook, IL, 60062, USA (Tel: 847-849-4301)

PDM Inc. (Pine Derivatives Marketing, Inc.)

104 Webster Building, 3411 Silverside Rd., Wilmington, DE, 19810-4806, USA (Tel: 302-478-0768; 800-288-0768; FAX 302-478-0299; E-mail: don@pdmchemicals.com; Internet: www.pdmchemicals.com)

Peakchem, Fine Chems. Dept. of City Pride Co. Ltd.

Suite 502, Unit 3, Bldg. 3 Tucuiyuan, Lakeside Garden, No. 176 Wenhua Rd., Hangzhou, P.R. China (Tel: 86 571 8823 3070; FAX 86 5718823 3092; E-mail: peakchem@mail.hz.zj.cn; peakchem@hotmail.com; Internet: www.peakchem.com)

Pechiney

Pechiney Groupe, 7, place du Chancelier Adenauer, 75218 Paris Cedex 16, France (Tel: 33 1 56 28 20 00; E-mail: pechiney.webmaster@pechiney.com; Internet: www.pechiney.com)

Pechiney Chemicals Div., Stamford Harbor Park, 6th Fl., North Tower, 333 Ludlow St., Stamford, CT, 06902, USA (Tel: 203-541-9190; 800-736-7893; FAX 203-541-9191; E-mail: chemicals@pechiney.com; Internet: www.pechiney-chemicals.com)

Pechiney World Trade USA/R.W. Greeff Chemical Div., Stamford Harbor Park, North Tower, 333 Ludlow St., Stamford, CT, 06902, USA (Tel: 203-541-9190; 800-736-7893; FAX 203-541-9191; E-mail: chemicals@pechiney.com; Internet: www.pechiney-chemicals.com)

Pelron Corp.

7847 W. 47 St., Lyons, IL, 60534, USA (Tel: 708-442-9100; FAX 708-442-0213)

Penn Bio-Organics, Inc.

B 212, One University Pl., Rensselaer, NY, 12144, USA (Tel: 518-591-8328; 866-PENNBIO; FAX 518-525-2799; E-mail: info@pennbio.com; Internet: www.pennbio.com)

Penn Specialty Chemicals

Penn Specialty Chemicals, Inc., 1000 Germantown Pike #H-3, Plymouth Greene Office Campus, Plymouth Meeting, PA, 19462, USA (Tel: 610-292-3051; 877-895-PENN; FAX 610-292-3055; Internet: www.pschem.com)

Penn Specialty Chemicals, Ltd., Belgian Branch, Oosterweelsesteenweg 53, Port 269, B-2030 Antwerp, Belgium (Tel: 32 3 2 27 19 55; FAX 32 3 27 19 75; E-mail: pschem@skynet.be)

Penn Specialty Chemicals GmbH, Hahlweg 2, 36093 Kunzell, Germany (Tel: 49 66 58 17 60; FAX 49 66 58 16 77; E-mail: penn-spec-chem@ecos.net)

Penreco

Penreco, Div. of Pennzoil Prods. Co., PO Box 4274, 910 Louisiana, Suite 400, Houston, TX, 77210, USA (Tel: 713-236-6950; FAX 713-236-6990; Internet: www.penreco.com)

Penreco, Div. of Pennzoil Prods. Co., 138 Petrolia St., Karns City, PA, 16041, USA (Tel: 724-756-0110; 800-245-3952; FAX 724-756-1050; Internet: www.penreco.com)

Penreco, European Office, Park House, 116 Park St., London, W1K 6NN, UK (Tel: 44 207 408 6266; FAX 44 207 408 6968)

Penreco Australia, Suite 3, 2-6 New St., Richmond, Victoria, 3121, Australia (Tel: 61 3 9429 6222; FAX 61 3 9421 0666)

Penta Manufacturing Co.

Div. of Penta International Corp., 50 Okner Pkwy., Livingston, NJ, 07039-1604, USA (Tel: 973-740-2300; FAX 973-740-1839)

Peridot Chemicals, Inc.

90 East Halsey Rd., Parsippany, NJ, 07054, USA (Tel: 800-631-8050; FAX 973-515-3241)

Perstorp Polyols, Inc.

Div. of Perstorp Chemicals AB, 600 Matzinger Rd, Toledo, OH, 43612, USA (Tel: 419-729-5448; 800-537-0280; FAX 419-729-3291; Internet: www.perstorppolyols.com)

Petroferm Inc.

2416 Lynndale Rd., Fernandina Beach, FL, 32034, USA (Tel: 904-261-8286; FAX 904-261-6994; Internet: www.petroferm.com)

Petroquimica Uniao S.A.

Av. Presidente Costa e Silva, 1.178, 09270-901 Santo Andre-SP, Brazil (Tel: 55 11 714-1515; FAX 55 11-714-1502; E-mail: pquiniao@ruralsp.com.br; Internet: www.pqu.com.br)

Pfaltz & Bauer Inc.

172 E. Aurora St., Waterbury, CT, 06708, USA (Tel: 203-574-0075; 800-225-5172; FAX 203-574-3181; E-mail: sales@pfaltzandbuaer.com; Internet: www.pfaltzandbauer.com)

Pfanstiehl

Pfanstiehl Laboratories, Inc., Part of Ferro, 1219 Glen Rock Ave., PO Box 439, Waukegan, IL, 60085, USA (Tel: 847-623-0370; 800-383-0126; FAX 847-623-9173; E-mail: pfanstiehl-info@ferro.com; Internet: www.pfanstiehl.com)

Pfanstiehl (Europe) Ltd., Unit 27, Meridian House, Road One, Winsford Industrial Estate, Winsford, Cheshire, CW7 3QG, UK (Tel: 44 1606 559163; FAX 44 1606 559641; E-mail: custserv@pfaneur.u-net.com)

George Pfau's Sons Co., Inc.

824 Wall St., PO Box 7, Jeffersonville, IN, 47130, USA (Tel: 812-283-6697; 800-PFAU-OIL; FAX 812-283-0765)

Pfizer

Pfizer International, 235 East 42nd Street, New York, NY, 10017, USA (Tel: 212-573-7851; FAX 212-573-2273; Internet: www.pfizer.com)

Pfizer Canada, 17300 Trans-Canada Hwy., Kirkland, Quebec, H9J 2M5, Canada (Tel: 514-695-0500; 877-633-2001; FAX 514-426-6831)

Pfizer Ltd., 10 Dover Rd., Sandwich, Kent, CT13 0BN, UK (Tel: 44 1304 616161; FAX 44 1304 616221)

Pfizer Europe, Hoge Wei, 10, 1930 Zaventem, Belgium (Tel: 32 2 722 02 11)

Pharmco Products Inc.

58 Vale Rd., Brookfield, CT, 06804-3967, USA (Internet: www.pharmco-prod.com)

Pharmrite North America Corp.

114 St. Josephs Dr., Stirling, NJ, 07980, USA (Tel: 416-412-9281; FAX 416-293-5798)

Phoenix Chemical, Inc.

60 Fourth St., Somerville, NJ, 08876, USA (Tel: 908-707-0232; FAX 908-707-0186; Internet: www.phoenix-chem.com)

Phoenix Natural Products Ltd.

Unit 15 Witley Industrial Estate, Southall, Middx, UB2 4EZ, UK (Tel: 44 20 8574 7308; FAX 44 20 8574 0536; Internet: www.phoenixuk.com)

Pickering Laboratories Inc.

1280 Space Park Way, Mountain View, CA, 94043-0749, USA (Tel: 650-694-6700; FAX 650-968-0749; E-mail: marketing@pickeringlabs.com; Internet: www.pickeringlabs.com)

Piedmont Chem. Industries

PO Box 2728, High Point, NC, 27261, USA (Tel: 336-885-5131; FAX 336-887-5563; Internet: www.piedmontchemical.com)

Pilot Chemical Co.

11756 Burke St., Santa Fe Springs, CA, 90670, USA (Tel: 213-723-0036; 800-707-4568; FAX 562-945-1877; E-mail: info@pilotchemical.com; Internet: www.pilotchemical.com)

Pioneer Americas, Inc.

700 Louisiana St., Suite 4300, Houston, TX, 77002, USA (Tel: 514-397-6184)

Plastics Color & Compounding, Inc.

Part of Plastics Color Corp., 349 Lake Rd., Dayville, CT, 06241, USA (Tel: 888-549-7822; FAX 860-779-7320; Internet: www.plasticscolor.com)

PMC Specialties

PMC Specialties Group, Inc., Div. of PMC, Inc., 26032 Detroit Rd. Ste # 1, Westlake, OH, 44145-2427, USA (Tel: 800-543-2466; FAX 440-356-2787; Internet: www.pmcsg.com)

PMC Specialties Group, Inc., Div. of PMC, Inc., 501 Murray Rd., Cincinnati, OH, 45217, USA (Tel: 513-242-3300; 800-543-2466; FAX 513-482-7377; Internet: www.pmcsg.com)

PMC Specialities International Ltd., Div. of PMC Specialities Group, 63 Wigmore Street, London, W1H 9LG, UK (Tel: 44 171 935 4058; FAX 44 171 935 9895)

Pokonobe Industries Inc.

Box 1756, Santa Monica, CA, 90406, USA (Tel: 310-392-1259; 877-POKONOBE; FAX 310-392-3659; E-mail: sales@pokonobe.com; Internet: www.pokonobe.com)

Polarome International Inc.

200 Theodore Conrad Dr., Jersey City, NJ, 07305, USA (Tel: 201-333-8700; FAX 201-433-0638; E-mail: plepore@polarome.com; Internet: www.polarome.com)

Polygon Corp.

200 W. Second St., S. Boston, MA, 02127, USA (Tel: 617-268-4455; FAX 617-268-9636)

Polymer Research Corp. of America (PRCA). NO LONGER SELLING CHEMICALS

2186 Mill Ave., Brooklyn, NY, 11234, USA (Tel: 718-444-4300; FAX 718-241-3930; E-mail: abarton@polymer-rd.com; Internet: www.polymer-rd.com)

PolyOne

PolyOne Corp., Suite 36-5000, 200 Public Square, Cleveland, OH, 44114-2304, USA (Tel: 216-589-4000; 800-321-1954; FAX 216-589-4200)

PolyOne/Elastomers & Perf. Additives Group, 2020 Front St., Suite 106, Cuyahoga Falls, OH, 44221, USA (Tel: 330-916-6300; 888-427-6341; FAX 330-916-2351; Internet: www.polyone.com)

Polysciences

Polysciences Inc., 400 Valley Road, Warrington, PA, 18976-2590, USA (Tel: 215-343-6484; 800-523-2575; FAX 800-343-3291; E-mail: info@polysciences.com; Internet: www.polysciences.com)

Polysciences Europe GmbH, Handelsstr. 3, 69214 Eppelheim, Germany (Tel: 49 6221 765767; FAX 49 6221 764620)

Porta Ltd.

Zeleny Pruh 99, Praha 4, 14002, Czech Republic (Tel: 420 2 410 912 11; FAX 420 2 414 427 34; E-mail: porta@porta.cz; Internet: www.porta.cz)

Potash Corp. of Saskatchewan Sales Ltd. (PCS)

122 1st Ave. South, Suite 500, Saskatoon, Saskatchewan, S7K 7G3, Canada (Tel: 306-933-8500; 800-667-3930 (U.S.); FAX 306-652-2699; Internet: www.potashcorp.com)

PPG Industries

PPG Industries, Inc., One PPG Place, Pittsburgh, PA, 15272, USA (Tel: 412-434-3131; 800-CHEM-PPG; FAX 412-434-2891; Internet: www.ppg.com; www.ppgchloralkali.com)

PPG Canada Inc., 2301 Royal Windsor Dr., Mississauga, Ontario, L5J 1K5, Canada (Tel: 905-823-1100; FAX 905-855-6840)

PPG Industrial do Brazil Ltda., Av. Moema, 94-11o, Suite 111/114, CEP, Sao Paulo SP, 04077-020, Brazil (Tel: 55 11 539 7989)

PPG Mexico, S.A. de C.V., Av. Presidente Juarez No. 1978, Col. San Jeronimo Tepetlacalco, Tlalnepantla, Edo de Mexico, 54090, Mexico (Tel: 52 5 397 8222; FAX 52 5 398 3090)

PPG Industries (UK) Ltd./Specialty Chems., Carrington Business Park, Carrington, Urmston, Manchester, M31 4DD, UK (Tel: 44 161 777 9203; FAX 44 161 777 9064)

PPG Industries Sale Inc., Immeubel SCOR, 1 Ave. du President Wilson, 92800, Puteaux, France (Tel: 33 1 46 98 81 33; FAX 33 1 46 98 82 63)

PPG Industries-Asia/Pacific Ltd., Takanawa Court, 5th Fl., 3-13-1 Takanawa, Minato-ku, Tokyo, 105, Japan (Tel: 81 3 3280 2861; FAX 81 3 3280 2920)

PQ

PQ Corp., PO Box 840, Valley Forge, PA, 19482-0840, USA (Tel: 610-651-4200; 800-944-7411; FAX 610-651-4463; Internet: www.pqcorp.com)

Silicatos y Derivados S.A. de C.V., Fracc. Ind. San Nicolas, Rio Lerma 55, Tlalnepantla, Edo. de Mexico, 54030, Mexico (Tel: 52 5 227 6800; FAX 52 5 227 6898)

National Silicates, 429 Kipling Ave., Etobicoke, Ontario, M8Z 5C7, Canada (Tel: 416-255-7771; FAX 416-201-4343)

PQ Hollow Spheres Ltd., St. Edmunds House, Lower Baxter St., Bury St. Edmunds, Suffolk, IP33 1ET, UK (Tel: 44 1284 749200; FAX 44 1284 715402)

Akzo-PQ Silica VOF, Stationsplein 4, PO box 247, 3800 AE Amersfoort, The Netherlands (Tel: 31 33 467 677; FAX 31 33 467 6169)

PQ Chemicals (Thailand) Ltd., 246 Sukhumvit Rd., Times Sq. Bldg., 26th fl., Between Soi 12-14, Bangkok, 10110, Thailand (Tel: 66 2229 4694; FAX 66 2229 4689)

PQ Silicates Ltd., 105 Run Hwa South Rd., Section 2, 4th Fl., Unit 1, Taipei, Taiwan R.O.C. (Tel: 886 227 849593; FAX 886 220 03038)

PQ Australia Pty. Ltd., 13 Rhur St., PO Box 6599, Dandenong, Victoria, 3175, Australia (Tel: 61 39 708 9200; FAX 61 39 791 6511)

Praxair

Praxair Inc., 39 Old Ridgebury Rd., Danbury, CT, 06810-5113, USA (Tel: 716-879-4077; 800-PRAXAIR; FAX 800-772-9985; E-mail: info@praxair.com; Internet: www.praxair.com)

Praxair Canada Inc., 1 City Centre Dr., Suite 1200, Mississauga, ON, L5B 1M2, Canada (Tel: 905-803-1600; FAX 905-803-1696; E-mail: info@praxair.com; Internet: www.praxair.com)

Praxair India Private, Ltd., Praxair House, No. 8, Ulsoor Rd., Bangalore, 560 042, India (Tel: 91 80555 9841; FAX 91 80559 5925)

Prayon

Prayon-Rupel SA, Rue J. Watuers, 144, 4480 Engis, Belgium (Tel: 32-42 739 646; FAX 32-42 739 644; Internet: http://www.prayon.com)

Prayon Inc., PO Box 669, Augusta, GA, 30903, USA (Tel: 706-771-3460; Internet: http://www.prayon.com)

Quadra Chemicals Inc., Distrib. for Prayon-Rupel, 13801 Reese Blvd. W., Suite 190, Huntersville, NC, 28078, USA (Tel: 704-875-9968; FAX 704-875-9847; E-mail: quadra1@perigee.net; Internet: www.quadrachemicals.com)

Presperse

Presperse Inc., 141 Ethel Road West, Piscataway, NJ, 08854, USA (Tel: 732-819-8009; FAX 732-819-7175; Internet: www.presperse.com)

Chesham Chemicals Ltd., Distributor for Presperse, Cunningham House, Westfield Lane, Harrow, Middlesex, HA3 9ED, UK (Tel: 44 20 8907 7779; FAX 44 20 8909 1053; E-mail: sales@cheshamchemicals.co.uk; Internet: www.cheshamchemicals.co.uk)

Primachem Inc.

12 Greenwoods Rd., Old Tappan, NJ, 07675, USA (Tel: 201-784-3434; FAX 201-784-7997; E-mail: primachm@prodigy.net)

Process Chemicals LLC

777 S. Harbour Island Bl, Suite 250, Tampa, FL, 33602, USA (Tel: 813-229-1050; FAX 813-229-1123; Internet: www.process-chemicals.com)

Procter & Gamble

Procter & Gamble Co./Chemicals Div., Rm C2N09A, 11530 Reed Hartman Hwy., Cincinnati, OH, 45140, USA (Tel: 513-626-6882; 800-477-8899; FAX 513-626-3145; E-mail: taylor.rw@pg.com; Internet: www.pg.com/chemicals)

Procter & Gamble Inc. Canada, 4711 Yonge St., North York, Ontario, M2N 6K8, Canada (Tel: 416-730-4064; FAX 416-730-4122)

Procter & Gamble Ltd./Europe Div., Cobalt 3, Silver Fox Way, Cobalt Business Park, Newcastle upon Tyne, NE27 0QN, UK (Tel: 44 191 297 6000; FAX 44 191 297 6296)

Procter & Gamble Far East Inc., Japan Branch, 1-17 Koyo-cho Naka, Higashinada-ku, Kobe, 658-0032, Japan (Tel: 81 78 845 5000; FAX 81 78 845 6912)

Prodasynth

Parc Industriel des Bois de Grasse, Ave. Louison Bobet, 06130 Grasse, France (Tel: 33 4 93 09 00 11; FAX 33 4 93 09 00 88; E-mail: support@prodasynth.com; Internet: www.prodasynth.com)

Protameen Chemicals

Protameen Chemicals, Inc., 375 Minnisink Rd., PO Box 166, Totowa, NJ, 07511, USA (Tel: 973-256-4374; FAX 973-256-6764; E-mail: info@protameen.com; Internet: www.protameen.com)

S. Black (Import & Export) Ltd., Distrib. for Protameen, Foxholes Business Park, John Tate Road, Hertford, Hertfordshire, SG13 7YH, UK (Tel: 44 1992 825555; FAX 44 1992 825566; E-mail: enquires@sblack.com; Internet: www.sblack.com)

Proviron Industries

Proviron Industries NV, G. Gilliotstraat 60, 2620 Hemiksem, Belgium (Tel: 32 3 870 88 20; FAX 32 3 877 23 33; E-mail: info@proviron.com; Internet: www.proviron.com)

Proviron America Inc., 3807 South State Rd. 2, Friendly, WV, 26146, USA (Tel: 304-652-6932; FAX 304-652-2586; E-mail: info@proviron.com)

Provital

Provital S.A., Poligono Industrial Can Salvatella, Gorgs Lladó, PO Box 78, 08210 Barbera Del Valles (Barcelona), Spain (Tel: 34 937 192350; FAX 34 937 190294; E-mail: marketing@provital.org)

Centerchem Inc., Distrib. for Provital, 225 High Ridge Rd, Stamford, CT, 06905-3036, USA (Tel: 203-975-9800; FAX 203-975-8777)

S. Black (Import & Export) Ltd., Distrib. for Provital, Foxholes Business Park, John Tate Road, Hertford, Hertfordshire, SG13 7YH, UK (Tel: 44 1992 825555; FAX 44 1992 825566; E-mail: enquires@sblack.com; Internet: www.sblack.com)

Przedsiebiorstwo Przemyslu Fermentacyjnego AKWAWIT Spólka Akcyjna

ul. Swieciechowska 2, 64-100 Leszno, Poland (Tel: 48 65 529 40 40; FAX 48 65 529 90 55; E-mail: akwawit@akwawit.com.pl; Internet: www.akwawit.com.pl)

Punda Mercantile Inc.

4115 Sherbrooke St. West, Suite 610, Montreal, Quebec, H3Z 1K9, Canada (Tel: 514-931-7278; FAX 415-931-7200; E-mail: host@pundagroup.com)

PURAC

PURAC biochem BV, Arkelsedijk 46, Postbus 21, 4200 AA Gorinchem, The Netherlands (Tel: 31 183 695 695; FAX 31 183 695 600; E-mail: pnl@purac.com; Internet: www.purac.com)

PURAC Biochem (UK), 50-54 St. Paul's Square, Birmingham, West Midlands, B3 1QS, UK (Tel: 44 121 236 1828; FAX 44 121 236 1401; E-mail: puk@purac.com)

PURAC France, 10 Place Charles Beraudier, L'Orient-Gare Part Dieu, 69428 Lyon Cedex 03, France (Tel: 33 4 72 11 4667; FAX 33 4 72 11 4669; E-mail: pfr@purac.com)

PURAC bioquímica, Gran Vial 19-25, 08160 Montmeló (Barcelona), Spain (Tel: 34 93 568 6300; FAX 34 93 568 3955; E-mail: psp@purac.com)

PURAC Germany, Mainzer Strasse 160, 55411 Bingen, Germany (Tel: 49 6721 181740; FAX 49 6721 181750; E-mail: pge@purac.com)

PURAC Poland, ul. Saska 105 m. 5, 03-914 Warszawa, Poland (Tel: 48 22 616 18 52; FAX 48 22 616 09 50; E-mail: ppl@purac.com)

PURAC Russia, Bolshaya Yakimanka St. 31, office 315, 109180 Moscow, Russia (Tel: 7 095 974 1521; FAX 7 095 974 1521; E-mail: pru@purac.com)

PURAC America, Inc., 111 Barclay Blvd., Suite 100, Lincolnshire Corporate Center, Lincolnshire, IL, 60069, USA (Tel: 847-634-6330; FAX 847-634-1992; E-mail: pam@purac.com; Internet: www.purac.com)

PURAC sínteses, Rua Augusta, 1939, Conj. 123-Jardins, Sao Paulo-SP, 01413-000, Brazil (Tel: 55 11 3062 1535; FAX 55 11 3062-4011; E-mail: pbr@purac.com)

PURAC Asia Pacific Pte. Ltd., 138 Robinson Rd. #18-01, The Corporate Office, Singapore, 068906, Singapore (Tel: 65 6349 1350; FAX 65 6222 1707; E-mail: pap@purac.com)

PURAC China, No. 660 Shang Cheng Rd., Rm. 1605, Lucky Mansion, Pudong New Area, Shanghai, P.C., 200120, P.R. China (Tel: 86 21 6875 4755; FAX 86 21 6875 6974; E-mail: pcn@purac.com)

PURAC Japan, Ebisu Sunny Bldg. 3F, 2-23-12 Ebisu-minami, Shibuya-ku, Tokyo, 150-0022, Japan (Tel: 81 3 5720 7531; FAX 81 3 5720 7565; E-mail: pjk@purac.com)

PureWorld Botanicals

Subsid. of Pure World, Inc., 375 Huyler St., South Hackensack, NJ, 07606, USA (Tel: 201-440-5000; FAX 201-342-8000; E-mail: pureworld@pureworld.com; Internet: www.pureworld.com)

PVS Chemicals, Inc.

12260 S. Carondolet Ave., Chicago, IL, 60633, USA (Tel: 773-933-8800; 800-932-8860; FAX 773-933-0957)

Pyosa S.A. de C.V.

Ave. Industrias 1200 Pte. Col., Bella Vista, Monterrey N.L., 64410, Mexico (Tel: 52 8 331 2600; 800-PYOSAPY; FAX 52 8 376 9398; E-mail: hugoh@pyosa.com; Internet: www.pyosa.com)

Qbiogene

2251 Rutherford Rd., Carlsbad, CA, 92008, USA (Tel: 800-424-6101; FAX 760-918-9313; E-mail: technical@qbiogene.com; Internet: www.qbiogene.com)

Qemi International, Inc.

800 Rockmead Dr., Suite 120, Kingwood, TX, 77339, USA (Tel: 281-359-8188; FAX 281-359-8192; Internet: www.qemi.com)

Quadra Corp.

13420 Reese Blvd. W., Ste. 190, Huntersville, NC, 28078, USA (Tel: 704-875-9968; FAX 704-875-9847; E-mail: quadra1@perigee.net)

Quaker City Chemicals, Inc.

7360 Milnor St., Philadelphia, PA, USA (Tel: 215-333-2000; 800-887-CHEM; FAX 215-333-4408; E-mail: custserv@chemical.net; Internet: www.chemical.net)

Quest International

Quest International BV, An ICI company, 28 Huizerstraatweg, 1411 GP Naarden, The Netherlands (Tel: 31 35 699 9111; FAX 31 35 694 6067; Internet: www.quest.nl)

Quest International (UK) Ltd., Bromborough Port, Wirral, Merseyside, Liverpool, L62 4SU, UK (Tel: 44 151 645 2060; FAX 44 151 645 6975)

Quest International UK Ltd./Fragrance Div., Willesborough Rd., Ashford, Kent, TN24 0LT, UK (Tel: 44 1233 644444; FAX 44 1233 644146)

Quest International Deutschland GmbH/Fragrance Div., Holzdamm 28-32 (Pacific Haus), 20099 Hamburg, Germany (Tel: 49 40 2840550; FAX 49 40 28405510)

Quest International Italia, Via Gargano 7, 20139 Milan, Italy (Tel: 39 2 574661; FAX 39 2 57466248)

Quest International France S.A., 43 Rue de Villiers, 92523 Neuilly sur Seine, France (Tel: 33 1 41 05 66 66; FAX 33 1 41 05 66 99)

Quest International Espana S.A., Edificio Conata I, Fructuos Gelabert, 2-4, 08970 Sant Joan Despi (Barcelona), Spain (Tel: 34 93 4808910; FAX 34 93 4770322)

Quest International Flavors & Food Ingreds. USA Inc., 5115 Sedge Blvd., Hoffman Estates, IL, 60192, USA (Tel: 847-645-7000; 800-833-8308; FAX 847-645-7070)

Quest International Fragrances USA, Inc., Member of the ICI Group, 400 International Dr., Mt. Olive, NJ, 07828, USA (Tel: 973-691-7100; 800-598-5986; FAX 973-691-6871; Internet: www.questintl.com)

Quest International de Mexico S.A. de C.V., Blvd. Miguel de Cervantes, Saavedra 71, Col. Granada, Mexico D.E., 11520, Mexico (Tel: 52 5 262 7300; FAX 52 5 254 5793)

Quest International Argentina S.A., Ruta 9 Panamericanca Km 36, Colectora Oeste, Buenos Aires, Argentina (Tel: 54 3327 456800; FAX 54 3327 456910)

Quest International Australia Pty. Ltd., 6 Britton St., Smithfield, NSW, 2164, Australia (Tel: 61 2 9604 1177; FAX 61 2 9604 1988)

Quest International New Zealand Ltd., 12 Averton Pl., East Tamaki, Auckland, New Zealand (Tel: 64 9 2745393; FAX 64 9 2744734)

Quest International India Ltd., 6 Saki Vihar Rd., Saki Naka, PO Box 8920, Mumbai, 400 072, India (Tel: 91 22 8522305; FAX 91 22 8521787)

Quest International Singapore Pte. Ltd., 152 Beach Rd., #15-00 Gateway East, Singapore, 189721, Singapore (Tel: 65 299 0032; FAX 65 299 0031)

Quest International Japan K.K., 5-13 Koraibashi 2-chome, Chuo-ku, Osaka, 541, Japan (Tel: 81 6 2037608; FAX 81 6 2032918)

Quest International China Ltd. (Hong Kong), Suite 2208, Island Place Tower, 510 King's Rd., North Point, Hong Kong (Tel: 852 25600133; FAX 852 25359335)

Quest International China Ltd., 7/F Victory Mansion, 2200 Kai Xuan Rd., Shanghai, 200030, P.R. China (Tel: 86 21 6481 2223; FAX 86 21 6481 2248)

Quimdis S.A.

85 Rue Edouard Vaillant, B.P. 329, 92307 Levalois-Perret Cedex, France (Tel: 33 1 47 39 91 91; FAX 33 1 47 39 9190; E-mail: lifquarre@worldnet.fr)

The Rambach Corp.

52 Vesey St., PO Box 5187, Newark, NJ, 07105, USA (Tel: 973-589-7774; FAX 973-589-2615)

Rasa Industries, Ltd.

Yaesu Dai Bldg., 1-1-1, Kyobashi, Chuo-ku, Tokyo, 104, Japan (Tel: 81 3 3278 3801; FAX 81 3 3281 6697)

Raschig

Raschig AG, Mundenheimer Strasse 100, 67061 Ludwigshafen/Rhein, Germany (Tel: 49 621 56 18 0; FAX 49 621 58 28 85; E-mail: info@raschig.de; Internet: www.raschig.de)

Raschig Corp., 129 S. Scoville Ave., Oak Park, IL, 60302, USA (Tel: 708-386-2799; FAX 708-386-1021; E-mail: ReinwaldB@aol.com)

Raschig UK Ltd., Dock Office, Trafford Rd., Salford Quays, Salford, Lancashire, M5 2XB, UK (Tel: 44 161 877 3933; FAX 44 161 877 3944)

Raschig France S.A.R.L., 1, rue Jean Carrasso, 95870 Bezons, France (Tel: 33 1 34 23 38 78; FAX 33 1 34 23 38 18)

Reach Group

PO Box 33429, Las Vegas, NV, 89133-3429, USA (Tel: 702-363-7391; 800-882-6315; FAX 702-363-7393; E-mail: info@reachgroup.com; Internet: www.reachgroup.com)

Reade Advanced Materials

PO Box 15039, Riverside, RI, 02915-0039, USA (Tel: 401-433-7000; FAX 401-433-7001; E-mail: sales@reade.com; Internet: www.reade.com)

Reagents, Inc.

PO Box 240746, 4746 Sweden Rd., Charlotte, NC, 28224-0746, USA (Tel: 704-554-7474; 800-732-8484; FAX 704-552-7594; E-mail: info@reagents.com; Internet: www.reagents.com)

Recochem, Inc.

850 Montée de Liesse Rd., Montreal, Quebec, H4T 1P4, Canada (Tel: 514-341-3550; FAX 514-341-1292/6553)

Refron, Inc.

38-18 33rd St., Long Island City, NY, 11101, USA (Tel: 718-392-8002; 800-4-REFRON; FAX 718-392-8006; E-mail: custserv@refron.com; Internet: www.refron.com)

Regis Technologies, Inc.

8210 Austin Ave., Morton Grove, IL, 60053, USA (Tel: 847-967-6000; 800-323-8144; FAX 847-967-1214; E-mail: sales@registech.com; Internet: www.registech.com)

Reheis

Reheis Inc., A GenTek company, PO Box 609, 235 Snyder Ave., Berkeley Heights, NJ, 07922, USA (Tel: 908-464-1500; FAX 908-464-7726; E-mail: info@reheis.com; Internet: www.reheis.com)

Reheis Inc.-Canada, 15 rue Du Lipizzan, Blainville, Quebec, J7C 4X4, Canada (Tel: 450-433-9922; FAX 450-433-5953; E-mail: wzacard@reheis.com)

Reheis Ireland, Div. of Reheis Inc., Kilbarrack Rd., Dublin, 5, Ireland (Tel: 353 1 8322621; FAX 353-1-8392205; E-mail: d.mcfarland@reheis.ie)

Reheis Inc.-United Kingdom, Willowbank House, 97 Oxford Rd., Highbridge Estate, Uxbridge, Middlesex, UB8 1LU, UK (Tel: 44 1895 819316; FAX 44 1895 819333)

Reheis Inc.-Netherlands, Sonsbeekweg 12, 6814 BA Arnhem, The Netherlands (Tel: 31 26 4424245; FAX 31 26 4459968)

Reichhold

Reichhold Chemicals, Inc./Corporate Headquarters, Wholly owned subsid. of Dainippon Ink & Chemicals, 2400 Ellis Rd., Durham, NC, 27703, USA (Tel: 919-990-7500; 800-431-1920; FAX 919-990-7711; 800-554-0116; E-mail: question.box@reichhold.com; Internet: www.reichhold.com)

Reichhold Quimica de Mexico, S.A. de C.V., Norte 45 No. 731, Col Industrial Vallejo, Deleg. Azcapotzalco, Vallejo D.F., 02300, Mexico (Tel: 52 5 729 8600; FAX 52 5 368 4524)

Reichhold Pte. Claire, 108 Hymus Blvd., Pointe Claire, Quebec, Canada (Tel: 514-697-5920; FAX 514-697-8803)

Reichhold UK Ltd., 54 Willow Lane, Mitcham, Surrey, UK (Tel: 44 20 8648 4684; FAX 44 20 8640 6432)

Reichhold Chimie SA, 105 rue des Campanules, Parc d'Affaires Silic, 77185 Lognes, France (Tel: 33 1 64 11 55 60; FAX 33 1 64 11 55 70)

Reichhold SRL, Via Romagnoli, 23, 43056 San Polo di Torrile, Parma, Italy (Tel: 39 521 812 811; FAX 39 521 813 445)

Reichhold Chemie GmbH, Winsbergring 25, 22525 Hamburg, Germany (Tel: 49 40 853 992 0; FAX 49 40 857 369)

Reichhold BV, Postbus 20, 2670 Naaldwijk, The Netherlands (Tel: 31 174 612 002; FAX 31 174 612 016)

Reichhold Chemie GmbH, Breitenleer Strasse 97-99, 1222 Wien, Austria (Tel: 43 1 201 10; FAX 43 1 201 10 255)

Dainippon Ink & Chemicals, 7-20 Nihonbasho, 3-chome, Chuo-ku, Tokyo, Japan (Tel: 81 3 3272 4511; FAX 81 3 3272 0436)

Reilly Industries

Reilly Industries Inc., 300 N. Meridian St., Suite 1500, Indianapolis, IN, 46204-1763, USA (Tel: 317-247-8141; 800-777-3536; FAX 317-248-6534; E-mail: corpinfo@reillyind.com; Internet: www.reillyind.com)

Reilly Chemicals S.A., 154, Chaussèe de la Hulpe, 1170 Brussels, Belgium (Tel: 32-2-663-4400; FAX 32-2-673-8104; E-mail: rcsa@reillyind.com)

Reilly Industries Inc., Pacific Rim, 801 Bank of America Tower, 12 Harcourt Rd., Central Hong Kong, Hong Kong (Tel: 852-2522-6361; FAX 852-2522-6665; E-mail: jwhite@reillyind.com)

Reliable Biopharmaceutical Corp.

1945 Walton Rd., St. Louis, MO, 63114, USA (Tel: 314-429-7700; FAX 314-429-0937; E-mail: info@reliablebiopharm.com; Internet: www.reliablebiopharm.com)

Reliance Industries Limited

Maker Chabers IV, Nariman Point, Mumbai, 400 021, India (Tel: 91 22 242384; FAX 91 22 285228)

Reliance Silicones

Reliance Silicones (India) Ltd., 95, Mittal Tower, 210 Nariman Point, Mumbai, 400 021, India (Tel: 91 22 28724808; FAX 91 22 2047455; Internet: www.reliancesilicones.com)

Reliance Silicones International, 20 Alan Turing Rd., Surrey Research Park, Guildford, Surrey, GU2 5YF, UK (Tel: 44 1483 457878; FAX 44 1483 455042; E-mail: 100765.641@compuserve.com; Internet: www.reliancesilicones.com)

Research Organics Inc.

4353 E. 49th St., Cleveland, OH, 44125, USA (Tel: 216-883-8025; 800-321-0570; FAX 216-883-1576; E-mail: info@resorg.com; Internet: www.resorg.com)

Resolution Performance

Resolution Performance Products, 1600 Smith St., 24th Fl., PO Box 4500, Houston, TX, 77210-4500, USA (Tel: 832-366-2300; 800-TEC-EPON; 877-859; -2800; Internet: www.resins.com; www.resins-versatics.com)

Resolution Nederland BV, PO Box 606, 3190 AN Hoogvliet Rt, The Netherlands (FAX 31 10 431 4649; Internet: www.resins-versatics.com)

Rhodia

Rhodia USA, 259 Prospect Plains Rd., CN 7500, Cranbury, NJ, 08512-7500, USA (Tel: 609-860-4000; 800-922-2189; FAX 609-860-0459; E-mail: roliver@us.rhodia.com; Internet: www.rhodia.com)

Rhodia Inc./Fine Organics, 259 Prospect Plains Rd., CN 7500, Cranbury, NJ, 08512-7500, USA (Tel: 609-860-4001; FAX 609-860-0344; E-mail: Rhodiac@us.rhodia.com; Internet: www.rhodia.com)

Rhodia Food, 259 Prospect Plains Rd., CN 7500, Cranbury, NJ, 08512-7500, USA (Tel: 609-860-4600; 800-243-5052; FAX 609-860-0138; E-mail: food@us.rhodia.com; Internet: www.food.us.rhodia.com)

Rhodia Home, Personal Care, Industrial Ingreds. (HPCII), 259 Prospect Plains Rd., CN 7500, Cranbury, NJ, 08512-7500, USA (Tel: 609-860-4000; 888-776-7337; FAX 609-860-0458; E-mail: hpcii@us.rhodia.com; Internet: www.rhodia-hpcii.com)

Rhodia Phosphorus Performance Derivs., 259 Prospect Plains Rd., CN 7500, Cranbury, NJ, 08512-7500, USA (Tel: 609-860-3919; 888-776-7337; FAX 609-860-0076; E-mail: Ann.Mitchell@us.rhodia.com; Internet: www.rhodia-ppd.com)

Rhodia Canada, 2000 Argentia Rd., Plaza 3, Suite 400, Mississauga, Ontario, L5N 1V9, Canada (Tel: 905-821-4450; 800-998-5512; FAX 905-821-8825; Internet: www.rpsurfactants.com)

Rhodia (Mexico), Av. Vasco de Quiroga 3000, 2° piso, Lomas de Santa Fe, Mexico D.F., 01210, Mexico (Tel: 52 5 261 1758; FAX 52 5 261 1702)

Rhodia (Argentina), Av. L.N. Alem 1050 p10°, Buenos Aires, 1001, Argentina (Tel: 54 11 4510 1602; FAX 54 11 4510 1637)

Rhodia Brazil Ltda., Centro Empresarial Av., Maria Coelho Aquiar, 215, BL B 2 andar, Sao Paulo SP, 05804-902, Brazil (Tel: 55 11 3741 8238; FAX 55 11 3741 7542)

Rhodia Chemicals Ltd., Poleacre Lane, Woodley, Stockport, Cheshire, SK6 1PQ, UK (Tel: 44 161 43 04 391; FAX 44 161 910 15 54)

Rhodia Ltd./Staveley Chemicals, Staveley Works, Chesterfield, Derbyshire, S43 2PB, UK (Tel: 44 1246 277251; FAX 44 1246 280090; E-mail: debbie.hayward@rhone-poulenc.com)

Rhodia Consumer Specialties Ltd./Phosphorus & Perf. Derivs., PO Box 3, 210-222 Hagley Road West, Oldbury, Warley, W. Midlands, B68 0NN, UK (Tel: 44 121 420 5324; FAX 44 121 420 5365)

Rhodia SA, 26 quai Alphonse Le Gallo, 92512 Boulogne-Billancourt Cedex, France (Tel: 33 1 55 38 40 00)

Rhodia Home, Personal Care, Industrial Ingredients (France), 40, rue de la Haie-Coq, 93306 Aubervilliers, France (Tel: 33 1 53 56 53 89; Internet: www.hpcii.rhodia.com)

Rhodia Chimie, 243 Blvd. Sylvain Dupuis, 1070 Bruxelles, Belgium (Tel: 32 2 529 4568; FAX 32 252 945 98)

Rhodia Geronazzo SpA, Via Milano 78-80, 20021 Ospiate Di Bollate (MI), Italy (Tel: 39 2 383 341; FAX 39 2 383 343 03)

Rhodia Quimica SA, Diagonal, 640, 2nd Planta, 08017 Barcelona, Spain (Tel: 34 93 430 49 00; FAX 34 93 419 73 54)

Rhodia GmbH, Staedelstrasse 10, 60596 Frankfurt, Germany (Tel: 49 69 60 93 222; FAX 49 69 60 93 409)

Rhodia (Netherlands), Postbus 418, 1180 AK Amstel Veen, The Netherlands (Tel: 31 20 547 3911; FAX 31 20 647 2479)

Rhodia Sveridge AB, Folkungagatan 121, 102 62 Stockholm, Sweden (Tel: 46 8 23 74 80; FAX 46 864 18 863)

Rhodia (Australia), 100 York St., PO Box 401, South Melbourne, 3205, Australia (Tel: 61 3 928 1000; FAX 61 3 96 46 38 17)

Rhodia (Japan), Rappongi First Bldg., 9-9 Rappongi 1-chome, Minato ku, Tokyo, 106-8540, Japan (Tel: 81 3 3585 4231; FAX 81 3 3582 6255)

Rhodia SA Asia Pacific PTE Ltd., #27-06/07 The Concourse, 300 Beach Rd., Singapore, 199555, Singapore (Tel: 65 291 1921; FAX 65 296 6044)

RIA International LLC

123 Columbia Tpke., Suite 104A, Florham Park, NJ, 07932, USA (Tel: 973-301-2011; FAX 973-301-2012; E-mail: nj@riausa.com; Internet: www.riausa.com)

Riceland Foods, Inc./Lecithin Div.

10800 Financial Centre Pkwy., Suite 380, Little Rock, AR, 72211, USA (Tel: 501-225-0936; 888-LECITHIN; FAX 501-225-9179; E-mail: lecithin@riceland.com; Internet: www.lecithin.com)

Richard's Pacific Chemical Sourcing

986 Wildwood Ave., Daly City, CA, 94015, USA (Tel: 650-757-3638; FAX 650-755-8349)

Richman Chemical, Inc.

768 N. Bethlehem Pike, Lower Gwynedd, PA, 19002, USA (Tel: 215-628-2946; FAX 215-628-4262; E-mail: kac@richmanchemical.com; Internet: www.richmanchemical.com)

Riedel-deHaën

Riedel-deHaën Laborchemikalien GmbH & Co. KG, Member of the Sigma-Aldrich Group, PO Box 10 02 62, 30918 Seelze, Germany (Tel: 49 5137 8238 0; FAX 49 5137 8238 120; E-mail: riedel@sial.com; Internet: www.rdh-lab.de)

Riedel-deHaën, Member of the Sigma-Aldrich Group, 3050 Spruce St., St. Louis, MO, 63178, USA (Tel: 314-289-6000)

Riken Vitamin Co., Ltd.

TDC Bldg., 2-9-18, Misaki-cho, Chiyoda-Ku, Tokyo, 101-8370, Japan (Tel: 81 3 5275 5130; FAX 81 3 5275 2905)

R.I.T.A. Corp.

1725 Kilkenny Court, PO Box 1487, Woodstock, IL, 60098, USA (Tel: 815-337-2500; 800-426-7759; FAX 815-337-2522; E-mail: cservice@ritacorp.com; Internet: www.ritacorp.com)

Rit-Chem Co. Inc.

109 Wheeler Ave., PO Box 435, Pleasantville, NY, 10570-0435, USA (Tel: 914-769-9110; FAX 914-769-1408; E-mail: ritchem@compuserve.com; Internet: www.ritchem.com)

Robeco Chemicals Inc.

99 Park Ave., New York, NY, 10016, USA (Tel: 212-986-6410; 800-992-1067; FAX 212-986-6419; E-mail: robecoinc@aol.com)

Rochem International, Inc.

305 Northern Blvd., Great Neck, NY, 11021, USA (Tel: 516-829-4060; FAX 516-829-4059; E-mail: rochemNY@aol.com)

Roche Molecular Biochemicals, Roche Diagnostics

9115 Hague Rd., PO Box 50414, Indianapolis, IN, 46250-0414, USA (Tel: 800-428-5433; FAX 800-428-2883)

Rohm and Haas

Rohm and Haas Co., 100 Independence Mall West, Philadelphia, PA, 19106-2399, USA (Tel: 215-592-3000; 800-922-8765; FAX 215-592-3377; E-mail: acumer@rohmhaas.com; Internet: www.rohmhaas.com; www.acusol.com)

Rohm and Haas Canada Inc., 2 Manse Rd., West Hill, Ontario, M1E 3T9, Canada (Tel: 416-284-4711; 800-268-4201; FAX 888-654-0599)

Rohm and Haas France SAS, 371 rue L. v. Beethoven, Sophia Antipolis, 06565 Valbonne-Cedex, France (Tel: 33 4 93 95 53 53; FAX 33 4 93 65 30 16)

Rohm and Haas Austria GmbH, Diefenbachgasse 35-41, 1150 Vienna, Austria (Tel: 43 1 894 17600; FAX 43 1 894 1869)

Rohm and Haas Nordiska AB, Ringragen 163, 261 22 Landskrona, Sweden (Tel: 46 418 450 400; FAX 46 418 450 499)

Rohm and Haas Belgium N.V., Ankerrui 9, 2000 Antwerpen, Belgium (Tel: 32 3 543 6180; FAX 32 3 541 7769)

Rohm and Haas Italia Srl, Via della Filanda, 2, 20060 Gessate Milano, Italy (Tel: 39 2 952 501; FAX 39 2 952 59399)

Rohm and Haas B.V., PO Box 32, 3800 AA Amersfoort, The Netherlands (Tel: 31 33 479 2600; FAX 31 33 479 2650)

Rohm and Haas Espana S.A., Provenza, 216 3rd Fl., 08036 Barcelona, Spain (Tel: 34 93 227 1900; FAX 34 93 323 4043)

Rohm and Haas Deutschland GmbH, Postfach 940322, 60461 Frankfurt/Main, Germany (Tel: 49 69 78 996 0; FAX 49 69 789 53 56)

Rohm and Haas (UK) Ltd., Tyneside Works, Ellison St., Jarrow, Tyne and Wear, NE32 3DJ, UK (Tel: 44 191 489 8181; FAX 44 191 489 8520)

Rohm and Haas Scotland Ltd., Grangemouth Works, Wholeflats Rd., Grangemouth, Stirlingshire, FK3 9UY, Scotland UK (Tel: 44 1324 473361; FAX 44 1324 47303)

Rohm and Haas China, Inc., Suite 819-825, Beijing Golden Land Bldg., No. 32 Liang Ma Qiao Rd., Chaoyang District, Beijing, P.R. China (Tel: 86 10 6464 3450)

Rohm and Haas Singapore (Pte.) Ltd., No. 11 Tuas Ave. 12, Singapore, 639034, Singapore (Tel: 65 861 1077; FAX 65 861 6865)

Chesham Chemicals Ltd., Distributor for Rohm & Haas (UK) Ltd., Cunningham House, Westfield Lane, Harrow, Middlesex, HA3 9ED, UK (Tel: 44 20 8907 7779; FAX 44 20 8909 1053; E-mail: sales@cheshamchemicals.co.uk; Internet: www.cheshamchemicals.co.uk)

Romil Ltd.

The Source, Convent Dr., Waterbeach, Cambridge, CB5 9QT, UK (Tel: 44 1223 863 873; FAX 44 1223 862 700; E-mail: pure.chemistry@romil.com; Internet: www.romil.com)

Rona

Rona, Div. of EM Industries, Inc., 7 Skyline Dr., Hawthorne, NY, 10532, USA (Tel: 914-592-4660; FAX 914-785-5804; E-mail: wehmsen@emindustries.com)

S. Black (Import & Export) Ltd., Distrib. for Rona, Foxholes Business Park, John Tate Road, Hertford, Hertfordshire, SG13 7YH, UK (Tel: 44 1992 825555; FAX 44 1992 825566; E-mail: enquires@sblack.com; Internet: www.sblack.com)

Roper Thermals

26B Nod Pl., Clinton, CT, 06413, USA (Tel: 860-669-6108; FAX 860-669-6233)

Roquette

Roquette Freres, 62080 Lestrem Cedex, France (Tel: 33 3 21 63 36 00; FAX 33 3 21 63 38 50; Internet: www.roquette.fr/eng)

Roquette America, Inc., 1417 Exchange St., PO Box 6647, Keokuk, IA, 52632-6647, USA (Tel: 319-524-5757; 800-553-7035; FAX 319-526-2345; E-mail: rasales@roquetteamerica.com; Internet: www.roquette.fr)

Roquette (UK) Ltd., Sallow Rd., Corby, Northants, NN17 5JX, UK (Tel: 44 1536 273 000; FAX 44 1536 263 873)

Frank B. Ross Co., Inc.

22 Halladay St., PO Box 4085, Jersey City, NJ, 07304-0085, USA (Tel: 201-433-4512; FAX 201-332-3555; E-mail: techinfo@rosswaxes.com)

H.M. Royal, Inc.

PO Box 28, 689 Pennington Ave., Trenton, NJ, 08601-0028, USA (Tel: 609-396-9176; 800-257-9452; FAX 609-396-3185; Internet: www.hmroyal.com)

Royale Pigments & Chemicals Inc.

12 Rte. 17 North, Suite 309, Paramus, NJ, 07652, USA (Tel: 201-845-4666; FAX 201-845-0719)

Royal Vopak

Royal Vopak, Westerlaan 10, 3016 CK Rotterdam, The Netherlands (Tel: 31 10 400 29 11; FAX 31 10 413 98 29; E-mail: www.vopak.com/general/contact_1628.html; Internet: www.vopak.com; www.vopak.nl)

Vopak USA Inc., PO Box 34325, Seattle, WA, 98124-1325, USA (Tel: 425-889-3400; 800-234-4588; FAX 425-889-4100; Internet: www.vopakusa.com)

R S A Corp.

36 Old Sherman Tpke., Danbury, CT, 06810, USA (Tel: 203-790-8100; FAX 203-790-1709; E-mail: RSASteph@cs.com; Internet: www.rsa-corporation.com)

RTD Chemicals Corp.

1500 Rt. 517, Suite 305, Hackettstown, NJ, 07840, USA (Tel: 908-852-6128; FAX 908-852-1335)

Rubicon, Inc.

Joint venture of Huntsman Int'l. LLC and Uniroyal Chemical Co., 9156 Highway 75, PO Box 517, Geismar, LA, 70734, USA (Tel: 225-673-6141; FAX 225-673-6442)

Ruger Chemical Co. Inc.

83 Cordier St., Irvington, NJ, 07111, USA (Tel: 973-926-0331; 800-274-RUGER; FAX 973-926-4921; E-mail: ruger.amend@worldnet.att.net; Internet: www.rugerchemical.com)

Russ Chemical Co., Inc.

220 Proctor, PO Box 2688, Odessa, TX, 79760, USA (Tel: 915-366-5911)

Rütgers

Rütgers AG, Rellinghauser Strasse 3a, 45128 Essen, Germany (Tel: 49 201 177 02; FAX 49 201 177 2050; E-mail: connect@ruetgers.de; Internet: www.ruetgers.de)

Rütgers Organics Corp., Subsid. of Rütgers AG, 201 Struble Rd., State College, PA, 16801, USA (Tel: 814-238-2424; 800-458-3434; FAX 814-238-1567; E-mail: info@ruetgers-organics-corp.com; Internet: www.ruetgers-organics-corp.com)

S.A. Chemicals

220, Udyog Bhavan, Sonawala Lane, Goregaon (East), Mumbai, 400 063, India (Tel: 91 22 876 9531; FAX 91 22 883 5347; E-mail: sac@sachemicals.com; Internet: www.sachemicals.com)

SAF Bulk Chemicals

3050 Spruce St., St. Louis, MO, 63103, USA (Tel: 314-534-4900; 800-336-9719; FAX 800-368-4661)

Sal Chemical Co., Inc.

Half Moon Industrial Park, 3036 Birch Dr., Weirton, WV, 26062, USA (Tel: 304-748-8200; 800-879-1725; FAX 304-797-8751; Internet: www.salchem.com)

Salim Oleochemicals

Salim Oleochemicals (Singapore) Pte. Ltd., 20 Tuas St., Singapore, 638457, Singapore (Tel: 8619688; FAX 8619050; E-mail: info@salimoleo.com.sg)

Salim Oleochemicals GmbH, Benrather Schlossalle 10, 40597 Dusseldorf, Germany (Tel: 49 211 996 9826; FAX 49 211 996 9822; E-mail: info@salim-oleochem.de)

Salim Oleochemicals USA, 63B Corbins Mill Dr., Dublin, OH, 43017, USA (Tel: 614-889-2009; FAX 614-889-4844)

Salor

Div. of Aldrich Chemical Co., 1001 W. St. Paul Ave., Milwaukee, WI, 53233, USA (Tel: 800-558-9160)

San-Ei Gen

San-Ei Gen F.F.I., Inc., 1-4-9, Hirano-machi, Chuo-ku, Osaka, 541, Japan (Tel: 81 6 6202 3751; FAX 81 6 6202 3770)

San-Ei Gen F.F.I. (U.S.A.), Inc., 489 Fifth Ave., New York, NY, 10017, USA (Tel: 800-252-1144; FAX 212-557-5257)

Sanken Chemical Co., Ltd.

4-6-17, Koraibashi, Chuo-ku, Osaka, 541, Japan (Tel: 81 6 6223 1407; FAX 81 6 6222 2973)

Sanko Kagaku K.K.

2-18-29, Jyuhachijyo, Yodogawa-ku, Osaka, 532, Japan (Tel: 81 6 6391 1971)

San Yuan Chemical Co., Ltd.

3rd Fl., 3 Alley 6 Lane, 329 Tun Hwa S. Rd., Sec. 1, Taipei, Taiwan R.O.C. (Tel: 886 2 702 8048; FAX 886 2 325 8985)

Sarcom, Inc.

18709 S.E. Federal Hwy., Tequesta, FL, 33469, USA (Tel: 561-744-8808; 800-752-5983; FAX 561-744-8807; E-mail: sarcom@sarcominc.com; Internet: www.sarcominc.com)

Sartomer

Sartomer Co., Part of TotalFinaElf, Oaklands Corp. Center, 502 Thomas Jones Way, Exton, PA, 19341, USA (Tel: 610-363-4100; 800-SARTOMER; FAX 610-363-4140; Internet: www.sartomer.com)

Cray Valley, European sister company to Sartomer, 4-8, cours Michelet, La Défense 10, 92091 Paris La Defense Cedex, France (Tel: 33 1 49 00 80 21; FAX 33 1 49 00 80 76; E-mail: norsolene@crayvalley.com; Internet: www.crayvalley.com)

Sasol

Sasol Chemical Industries Limited (SCI), 1 Sturdee Avenue, Rosebank, Johannesburg, Gauteng, 2196, Rep. of S. Africa (Tel: 27 11 441 3422; FAX 27 11 441 3288; E-mail: trevor.munday@sasol.com; Internet: www.sasol.com)

Sasol Germany, Überseering 40, Postfach 60 04 49, 22297 Hamburg, Germany (Tel: 49 40 63684 1645; FAX 49 40 63684 3588; E-mail: harry.hyatt@sasol.com; Internet: www.sasol.com; www.sasololefinssurfactants.com)

Sasol Chemicals Europe, 1 Hockley Court, 2401, Stratford Rd., Hockley Heath, Solihull, West Midlands, B94 6NW, UK (Tel: 44 1564 783060; FAX 44 1564 784088; E-mail: peter.norton@sasol.com)

Sasol Servo BV, Langestraat 197, 7491 AE Delden, The Netherlands (Tel: 31 74 3775 000; FAX 31 74 3775 075; E-mail: info@servo.nl; Internet: www.sasolservo.com)

Sasol Italy S.p.A., Via Medici del Vascello 26, 20138 Milan, Italy (Tel: 39 2 58453 1; FAX 39 2 58453 205; E-mail: sasol.italy@it.sasol.com)

Sasol Chemicals North America LLC, Affiliate of Sasol North America Inc., 900 Threadneedle, Houston, TX, 77079-2990, USA (Tel: 281-588-3000; 800-231-8216; FAX 281-588-3800; E-mail: info@us.sasol.com; Internet: www.sasolnorthamerica.com)

Sasol Chemicals Pacific Limited, Rm. 1005, 10/F, Olympia Plaza, 255 King's Rd., North Point, Hong Kong (Tel: 852 2827 6600; FAX 852 2827 8023; E-mail: bill.bryant@sasol.com)

Säurefabrik

Säurefabrik Schweizerhall (SF-Chem), PO Box 1964, 4133 Pratteln 1, Switzerland (Tel: 41 61 825 3111; FAX 41 61 821 8027; E-mail: contact@sf-chem.com; Internet: www.sf-chem.com)

SF Chem, Inc., 2125 Center Ave., Suite 500, Fort Lee, NJ, 07024, USA (Tel: 201-302-6072; FAX 201-302-6062; E-mail: contact-usa@sf-chem.com)

Schaefer Technologies Inc.

3000 Carrollton Road, Saginaw, MI, 48604, USA (Tel: 517-753-1877; 800-444-9034; FAX 517-753-3346)

Scheel Corp.

PO Box 567, Cranbury, NJ, 08512, USA (Tel: 609-395-1100; FAX 609-395-5525)

Schenectady

Schenectady Chemicals, Inc., PO Box 1046, Schenectady, NY, 12301, USA (Tel: 518-347-4200; FAX 518-346-3111; E-mail: product.info@siigroup.com; Internet: www.siigroup.com)

Schenectady Canada Ltd., 319 Comstock Rd., Scarborough, Ontario, M1L 2H3, Canada (Tel: 416-615-4700; FAX 416-757-2468)

Schnectady Europe GmbH, PO Box 280 180, 20514 Hamburg, Germany (Tel: 49 40 78946 0; FAX 49 40 78946 276; E-mail: sales@schnectady.de)

Schenectady Europe Ltd., Div. of Schenectady Chemicals Inc., Four Ashes, Wolverhampton, Staffs, WV10 7BT, UK (Tel: 44 1902 790555; FAX 44 1902 791640)

Scher Chemicals

Scher Chemicals, Inc., Industrial West & Styertowne Rd., PO Box 4317, Clifton, NJ, 07012-0997, USA (Tel: 973-471-1300; FAX 973-471-3783; E-mail: scherchem@scherchem.com; Internet: www.scherchem.com)

Chesham Chemicals Ltd., Distributor for Scher Chemicals Inc., Cunningham House, Westfield Lane, Harrow, Middlesex, HA3 9ED, UK (Tel: 44 20 8907 7779; FAX 44 20 8909 1053; E-mail: sales@cheshamchemicals.co.uk; Internet: www.cheshamchemicals.co.uk)

Schilling-Chemie GmbH und Produktions KG

Steinbeisstrasse 20-22, 71691 Freiberg, Germany (Tel: 49 7141 703 0; FAX 49 7141 703 178)

Scholle Chemical

Part of Scholle Corp., 2300 W. Point Ave., College Park, GA, 30337, USA (Tel: 404-761-3134; FAX 404-559-8892; Internet: www.scholle.com)

Schumacher

Unit of Air Products & Chemicals, 1969 Palomar Oaks Way, Carlsbad, CA, 92009, USA (Tel: 760-931-9555; 800-545-9242; FAX 760-931-7819; E-mail: brookscn@apci.com; Internet: www.schumacher.com)

Schütz

Schütz & Co. (GmbH & Co.), Kattrepelsbrücke 1, 20095 Hamburg, Germany (Tel: 49 40 300 838 0; FAX 49 40 300 838 99; E-mail: schuetz@sch.de)

S. Black (Import & Export) Ltd., Distrib. for Schütz & Co., Foxholes Business Park, John Tate Road, Hertford, Hertfordshire, SG13 7YH, UK (Tel: 44 1992 825555; FAX 44 1992 825566; E-mail: enquires@sblack.com; Internet: www.sblack.com)

Schweizerhall Inc.

10 Corporate Place South, Piscataway, NJ, 08854, USA (Tel: 732-981-8200; 800-243-6564; FAX 732-981-8282; E-mail: info@susinc.com; Internet: www.susinc.com)

Scientific Protein Laboratories, Inc.

Div. of American Home Prods. Corp., PO Box 158, 700 E. Main St., Waunakee, WI, 53597-0158, USA (Tel: 608-849-5944; 800-334-4775; FAX 608-849-5939; E-mail: splcs@md.ahp.com; Internet: www.spl-pharma.com)

SCM Metal Products, Inc.

Subsid. of OM Group Inc., 2601 Weck Dr., PO Box 12166, Research Triangle Park, NC, 27709-2166, USA (Tel: 919-544-8090; FAX 919-544-7996; E-mail: scminfo@scm-metals.com)

Scott Specialty Gases

6141 Easton Rd., PO Box 310, Plumsteadville, PA, 18949-0310, USA (Tel: 215-766-8861; 800-21-SCOTT; FAX 215-766-2476; Internet: www.scottgas.com)

Sea-Land Chemical Co.

902 Westpoint Pkwy., Suite 350, Westlake, OH, 44145, USA (Tel: 440-871-7887; FAX 440-871-7949; E-mail: sealand@sealandchem.com; Internet: www.sealandchem.com)

John L. Seaton & Co. Ltd.

Bankside, Hull, North Humberside, HU5 1RR, UK (Tel: 44 1482 341345; FAX 44 1482 447157; E-mail: enquiries@seatons-uk.co.uk; Internet: www.seatons-uk.co.uk)

Seegott Inc.

10040 Aurora Hudson Rd., Streetsboro, OH, 44241, USA (Tel: 330-528-0808; 800-321-2865; FAX 330-528-0675; E-mail: jweiss@seegott.com; Internet: www.seegott.com)

Seeler Industries Inc.

1 Genstar Rd., Joliet, IL, 60435, USA (Tel: 815-740-2640; 800-336-2422; FAX 815-740-6469)

Seidler Chemical Co., Inc.

537 Raymond Blvd., Newark, NJ, 07105, USA (Tel: 973-465-1122; FAX 973-465-4469; E-mail: sales@seidlerchem.com; Internet: www.seidlerchem.com)

Seltzer Chemicals, Inc.

5927 Geiger Court, Carlsbad, CA, 92008, USA (Tel: 760-438-0089; 800-735-8137; FAX 760-438-0336; E-mail: sci@seltzerchemicals.com; Internet: www.seltzerchemicals.com)

Sensient

Sensient Technologies, 777 E. Wisconsion Ave., Milwaukee, WI, 53202-5304, USA (Tel: 800-558-9892; E-mail: info@sensient-tech.com; Internet: www.sensient-tech.com)

Sensient Color Group, Part of Sensient Technologies, 2526 Baldwin St., St. Louis, MO, 63106-1903, USA (Tel: 314-889-7600; 800-325-8110; FAX 314-658-7318; E-mail: wjinfo@sensient-tech.com)

Seppic

Seppic, Div. of Air Liquide Group, 75 Quai d'Orsay, 75007 Paris Cedex, France (Tel: 33 1 40 62 55 55; FAX 33 1 40 62 52 53; Internet: www.airliquide.com)

Seppic UK Ltd., 50 Salisbury Rd., PO Box 338, Hounslow, Middlesex, TA1 6JH, UK (Tel: 44 181 477 8800; FAX 44 181 570 2106)

Seppic Italia Srl, Via Quarenghi 27, 20151 Milano, Italy (Tel: 39 2 3800 9110; FAX 39 2 3800 9140)

Seppic Inc., Subsid. of Seppic France, 30 Two Bridges Rd., Suite 225, Fairfield, NJ, 07004-1530, USA (Tel: 973-882-5597; FAX 973-882-5178; E-mail: info@seppicinc.com; Internet: www.seppicinc.com)

Seppic Shanghai, 201 Qin Jiang Rd., Shanghai, 200233, P.R. China (Tel: 86 21 64 85 02 02; FAX 86 21 64 95 75 42)

Serologicals Corp.

5655 Spalding Dr., Norcross, GA, 30092, USA (Tel: 678-728-2000; E-mail: info@serologicals.co; Internet: www.serologicals.com)

Shamrock Technologies

Shamrock Technologies Inc., 117 Docks Corner Rd., Dayton, NJ, 08810-1531, USA (Tel: 732-355-4500; 800-349-1822; FAX 732-355-1995; E-mail: marketing@shamrocktechnologies.com; Internet: www.shamrocktechnologies.com)

Shamrock Technologies European Headquarters, Heesterveldweg 21, 3700 Tongeren, Belgium (Tel: 32 12 45 83 30; FAX 32 12 45 83 40)

Shance Chemical Corp.

PO Box 3200, 170 Division St., Sag Harbor, NY, 11963, USA (Tel: 516-725-2704; FAX 516-725-2704)

Shandong Haihua Tianhe Organic Chemical Co., Ltd.

919 Yuanfei Rd., Kuiwen, Weifang, Shandong, P.R. China (Tel: 86 536 8663484; FAX 86 536 8666048; E-mail: tianhe@ec.com.cn; Internet: www.tianhe-chemical.com)

Sharon Laboratories

Sharon Laboratories Israel, Industrial Zone Ad Halom, PO Box 832, 77106 Ashdod, Israel (Tel: 972 8 8549800; FAX 972 8 8544494)

S. Black (Import & Export) Ltd., Distrib. for Sharon Laboratories, Foxholes Business Park, John Tate Road, Hertford, Hertfordshire, SG13 7YH, UK (Tel: 44 1992 825555; FAX 44 1992 825566; E-mail: enquires@sblack.com; Internet: www.sblack.com)

Shell

Shell Chemical Co., Div. of Shell Oil Co., One Shell Plaza, PO Box 2463, Houston, TX, 77252-2463, USA (Tel: 713-241-6161; 800-872-7435; FAX 713-241-4043; E-mail: CustomerCentreUS@shellus.com; Internet: www.shellchemicals.com)

Shell Chemicals Canada Ltd., 400-4th Ave. South West, PO Box 4280, Station C, Calgary, Alberta, T2T 525, Canada (Tel: 800-567-8717; FAX 800-567-8862; E-mail: SCCACustomerService@shellchem.ca)

Shell Mexico, S.A. de C.V., Insurgentes Sur No. 954 50 Piso, Col del Valle, 03100, Mexico (Tel: 52 5 687 4088; FAX 52 5 536 7920)

Shell Chemicals UK Ltd., Kessler Place, Bldg. 301, Chester Innovation Park, PO Box 1, Chester, Cheshire, CH1 3SH, UK (Tel: 44 1244 685000; FAX 44 1244 685010; E-mail: CustomerCentreUKIreland@OPC.shell.com; Internet: www.shellchemicals.com)

Shell Chimie S.A., B.P. 215, 13131 Berre l'Etang Cedex, France (Tel: 33 4 42 34 05 50; FAX 33 4 42 34 05 55; E-mail: CustomerCentreSouthEurope@opc.shell.com)

Deutsche Shell Chemie GmbH, Kölner Strasse 6, 65760 Eschborn, Germany (Tel: 49 6196 4740; FAX 49 6196 474502)

Shell Nederland Chemie B.V., Posbus 8610, 3009 AP Rotterdam, The Netherlands (Tel: 31 10 231 7104; FAX 31 10 231 7115)

Shell Chemicals Japan Ltd., Daiba Frontier Bldg., 2-3-2 Daiba, Minato-ku, Tokyo, 135-8074, Japan (Tel: 81 3 5500 3002; FAX 81 3 5500 3031; E-mail: Hideo.H.Kimura@japan.shell.com)

Shell Eastern Chemicals (Singapore), Shell House, UE Square, 83 Clemenceau Ave., Singapore, 239920, Singapore (Tel: 65 6384 8180; FAX 65 6384 8499; E-mail: yean-sung.wang@shell.com)

Shell South Africa (Pty) Ltd., Wholly owned subsid. of Shell, PO Box 2231, Cape Town, 8000, Rep. of S. Africa (Tel: 27 21 408 4911; FAX 27 21 419 2864; E-mail: cuscare@shell.co.za; Internet: www.shell.co.za)

Shin-Etsu

Shin-Etsu Chemical Co., Ltd./Silicone Div., Asahi Tokai Bldg., 2-6-1, Ohte-machi, Chiyoda-ku, Tokyo, 100-0004, Japan (Tel: 81 3 3246 5171; FAX 81 3 3246 5363)

Shin-Etsu Silicone Taiwan Co. Ltd., 7F, No. 102, Civil Blvd., Sec. 4, Taipei, Taiwan R.O.C. (Tel: 886 2 2751 6999; FAX 886 2 2751 6769)

Shin-Etsu Silicone Korea Co. Ltd., Int'l. Insurance Bldg. 904, 120, 5-ka, Namdaemoon-ro, Chung-ku, Seoul, Korea (Tel: 82 2 775 9691; FAX 82 2 775 9690; Internet: www.shinetsu.net)

Shin-Etsu Silicone Singapore Pte. Ltd., 150 Ubi Ave. 4, #03-00, Singapore, 408825, Singapore (Tel: 65 6743 7277; FAX 65 6743 7477)

Shin-Etsu Chem. Co., Ltd./Shanghai Laison Office, 2211 Shui On Plaza, 333 Huaihai Zhong Rd., Shanghai, China (Tel: 86 21 5306 9966; FAX 86 21 5382 3970)

Shin-Etsu Silicones of America, Inc., 1150 Damar Dr., Akron, OH, 44305, USA (Tel: 330-630-9860; FAX 330-630-9855; Internet: www.shinetsusilicones.com)

Shin-Etsu Silicones Europe B.V., Bolderweg 32, 1332 AV Almere, The Netherlands (Tel: 31 36 5493170; FAX 31 36 5326459)

Showa Denko K.K.

1-13-9, Shiba-Daimon, Minato-ku, Tokyo, 105, Japan (Tel: 81 3 5470-3462; FAX (03) 3436-4668; E-mail: spec_chem@hq.sdk.co.jp; Internet: www.sdk.co.jp)

Siber Hegner Ltd.

Div. of Siber Hegner & Co. Ltd., County House, 221-241 Beckenham Road, Beckenham, Kent, BR3 4UF, UK (Tel: 44 20 8676 6800; FAX 44 20 8659 1292; E-mail: siber@siberhegner.co.uk; Internet: www.siberhegner.com)

Sierra Chemical Co.

2302 Larkin Circle, Sparks, NV, 89431, USA (Tel: 775-358-0888; 800-777-8965; FAX 775-358-0987)

Sigma-Aldrich

Aldrich Chemical Co., Inc., PO Box 355, 1101 West St. Paul Ave., Milwaukee, WI, 53201, USA (Tel: 414-273-3850; 800-558-9160; FAX 800-962-9591; E-mail: aldrich@sial.com; Internet: www.sigma-aldrich.com)

Sigma-Aldrich Fine Chems., 3050 Spruce St., St. Louis, MO, 63103, USA (Tel: 314-771-5765; FAX 314-771-5757; Internet: www.sigma-aldrich.com)

Sigma-Aldrich N.V./S.A., K. Cardijnplein 8, 2880 Bornem, Belgium (Tel: 32 3 899 1301; FAX 32 3 899 1311; E-mail: sigmaldr@sial.be)

Sigma-Aldrich Co. Ltd., PO Box 1033, Fancy Rd, Poole, Dorset, BH12 4ZL, UK (Tel: 44 1202 712222; FAX 44 1202 740171; E-mail: safinechem@sial.com; Internet: www.sigma-aldrich.com/uk)

Aldrich Chemical Co., Div. of Sigma Aldrich Co. Ltd., The Old Brickyard, New Road, Gillingham, Dorset, SP8 4XT, UK (Tel: 44 1747 822211; 0800-71 71 81; FAX 44 1747 823779; E-mail: ukcustsv@eurnotes.sial.com; Internet: www.sigma-aldrich.com/uk)

Sigma-Aldrich Pty. Ltd., PO Box 970, Castle Hill, NSW, 2154, Australia (Tel: 61 2 9841 0555; 1-800-800-097; FAX 61-2-98410555; E-mail: sigmaa@ibm.net; Internet: www.aldrich.sial.com)

Sigma-Aldrich Japan K.K., 1-10-15, Nihonbashi Horidome-cho, Chuo-ku, Tokyo, 103, Japan (Tel: 81 3 5640 8850; FAX 81 3 5640 8855)

Silar Laboratories

PO Box 9009, Wilmington, NC, 28402, USA (Tel: 910-762-5800; FAX 910-762-9923)

Sinochem Jiangsu Wuxi Import & Export Corp.

17/F, 258 Zhong Shan Rd., Wuxi, P.R. China (Tel: 86 510 2724440; FAX 86 510 2700472; E-mail: wxchem@puglic1.wx.js.cn; Internet: www.sinochemwuxi.com)

Sinochem Liaoning Import & Export Corp.

No. 50 Youhao Rd., Dalian, Liaoning, 116001, P.R. China (Tel: 86 411 282 0619; FAX 86 411 263 4689; E-mail: songyz@sinochemliaoning.com; Internet: www.sinochemliaoning.com)

Sinochem Tianjin Import & Export Corp.

58 Nanjing Rd., Tianjin, P.R. China (Tel: 86 22 23301755; FAX 86-22-23393176; E-mail: sncmtjtw@public.tjuc.com.cn)

Sithean Corp.

630 N. Craycroft Rd., Tucson, AZ, 85711-1440, USA (Tel: 520-514-2374; FAX 520-514-2375)

SKW Chemicals Inc.

Member of the VIAG Group, 1509 Johnson Ferry Rd., Suite 150, Marietta, GA, 30062, USA (Tel: 770-971-1317; FAX 770-971-4306; E-mail: info@skwchem.com; Internet: www.skwchem.com)

Skylighter, Inc.

PO Box 480-W, Round Hill, VA, 20142-0480, USA (Tel: 540-554-4543; FAX 540-554-2849; E-mail: custservice@skylighter.com; Internet: www.skylighter.com)

Skymart Enterprises, Inc.

PO Box 637, San Gabriel, CA, 91778, USA (Tel: 626-286-3742; FAX 626-285-6842; E-mail: genechen@pacbell.net)

Sloss Industries Corp.

PO Box 5327, Birmingham, AL, 35207, USA (Tel: 205-808-7915; FAX 205-808-7885; Internet: www.sloss.com)

Werner G. Smith, Inc.

1730 Train Ave., Cleveland, OH, 44113, USA (Tel: 216-861-3676; 800-535-8343; FAX 216-861-3680)

SNPE

SNPE Chimie, Chemin de la Loge, 31078 Toulouse Cedex 04, France (Tel: 33 5 62 25 73 00; FAX 33 5 61 25 73 23; E-mail: l.pichon@snpe.com; Internet: www.snpe.com)

SNPE Ltd., Div. of SNPE, Suffolk House, George St., Croydon, Surrey, CR0 1PE, UK (Tel: 44 20 8649 7474; FAX 44 20 8760 0420; E-mail: jc.lecoq@snpe.com)

SNPE Deutschland, Bolongarostrasse 82, 65929 Frankfurt/Main, Germany (Tel: 49 6931 6031; FAX 49 6931 6466; E-mail: e.muller@snpe.com)

SNPE Italia, Via Barbarra Melzi 91/93, 20027 Rescaldina (MI), Italy (Tel: 39 33 1579 820; FAX 39 33 1579 520; E-mail: m.montini@snpe.com)

SNPE North America, 103 College Rd. East, 3rd Fl., Princeton, NJ, 08540, USA (Tel: 609-987-9424; FAX 609-987-2767; E-mail: dslick@snpe.com; Internet: www.snpe.com)

SNPE Japan K.K., Onarimon Yusen Bldg. 13F, 3-23-5 Nishi, Shinbashi, Minato-ku, Tokyo, 105, Japan (Tel: 81 3 3433 2880; FAX 81 3 3343 6652; E-mail: s.ikeda@snpe.com)

SOCO-Lynch Corp.

10747 Patterson Pl., Santa Fe Springs, CA, 90670, USA (Tel: 562-903-9626; FAX 562-903-9622)

Soda Aromatic Co., Ltd.

4-15-9, Nihonbashi Hon-cho, Chuo-ku, Tokyo, 103, Japan (Tel: 81 3 3666 0553; FAX 81 3 3666 2913)

Solutia

Solutia Inc., PO Box 66760, St. Louis, MO, 63166-6760, USA (Tel: 413-730-3241; 800-964-5224; FAX 314-674-2490; E-mail: jmfern@solutia.com; Internet: www.solutia.com; www.coatings-solutia.com)

Solutia Canada Inc., 6800 St. Patrick St., LaSalle, Quebec, H8N 2H3, Canada (Tel: 514-366-4855)

Solutia Mexico, S. de R.L. de C.V., Edificio Torre Esmeralda, Blvd. Manuel Avila Camacho, No. 40 Piso 12 Colonia Lomas de Chapultepec, Mexico D.F., 11000, Mexico (Tel: 52 5 202 5600; FAX 52 5 202 5232)

Solutia Argentina S.R.L., Avenida Alicia Moreau de Justo 1960, 2nd Fl., Office 203, Buenos Aires, Argentina (Tel: 54 11 4515 0709; FAX 54 11 4515 0728)

Solutia Brasil Ltda., Caixa Postal 21338, Sao Paulo SP, 04602-970, Brazil (Tel: 55 11 3365 1811; FAX 55 11 3365 1818)

Solutia Venezuela S.R.L., Avenida Francisco de Miranda, Edificio Parque Cristal, Torre Este, Piso 13 Ofc 13-14, Los Palos Grandes, Caracas, 1062, Venezuela (Tel: 58 212 286 6144; FAX 58 212 285 7113)

Solutia Europe N.V./S.A., Rue Laid Burniat, 3, Parc Scientifique-Fleming, 1348 Louvain-la-Neuve Sud, Belgium (Tel: 32 10 48 12 11; FAX 32 10 48 12 12; E-mail: michael.a.scharf@solutia.com; Internet: www.solutia.com)

Solutia UK Ltd., Corporation Rd., Newport, South Wales, Gwent, NP19 UT, UK (Tel: 44 1633 27 8221; FAX 44 1633 27 6317)

Solutia Europe N.V./S.A., Trekhpurdniy lane 9, office 501, 103001 Moscow, Russia (Tel: 7 95 933 5911; FAX 7 95 933 5910)

Solutia Italy S.r.l. (Resins & Additives), Piazzale Stefano Turr, 5, 20149 Milan, Italy (Tel: 39 2 3 45 49 51; FAX 39 2 3 45 49 550)

Solutia Europe NV (Belgium), Ottergemsesteenweg, 707, 9000 Gent, Belgium (Tel: 32 9 243 6211; FAX 32 9 243 6212)

Solutia Germany GmbH & Co. KG (Resins & Additives), Boelckestr. 26, 55252 Mainz-Kastel, Germany (Tel: 49 611 962 02; FAX 49 611 962 94 83)

Solutia Australia Pty. Ltd., 1/437 Canterbury Rd., Surrey Hills, Victoria, 3127, Australia (Tel: 61 3 9888 4590; FAX 61 3 9888 4562)

Solutia Chemicals India Private Ltd., 205-207, Midas Sahar Plaza Complex, Andheri-Kurla Rd., Andheri (East), Mumbai, 400 059, India (Tel: 91 22 830 2860; FAX 91 22 830 2859)

Solutia Japan Ltd., Shinkawa Sano Bldg., 2nd Fl., 1-3-17 Shinkawa, Chuo-ku, Tokyo, 104-0033, Japan (Tel: 81 3 3523 2080; FAX 81 3 3523 2070)

Solutia Hong Kong Ltd., Unit 1706, 17th Fl., Central Plaza, 18 Harbour Road, Wanchai, Hong Kong (Tel: 852 2828 3232; FAX 852 2828 3282)

Solutia Int'l. Trading (Shanghai) Co. Ltd., Unit 18-06/07 Harbour Ring Plaza, 18 Xi Zang Zhong Rd., Shanghai, 200001, P.R. China (Tel: 86 21 6386 7500; FAX 86 21 6385 1882)

Solutia Korea Ltd., 3rd Fl., Anglican Church Bldg., 3-7, Jeong-dong, Joong-gu, Seoul, 100-120, Korea (Tel: 82 2 736 7112; FAX 82 2 739 5049)

Solutia Singapore Pte. Ltd., 101 Thomson Rd., #19-04 United Square, Singapore, 307591, Singapore (Tel: 65 357 6100; FAX 65 357 6201)

Solutia Taiwan Inc., 2/F No. 124 Chung Cheng Rd., Shinlin District, Taipei, Taiwan R.O.C. (Tel: 886 2 2835 1666; FAX 886 2 8866 2703)

Solutia Thailand Ltd., 193/80 Lake Rajada Bldg., 19th Fl., Ratchadapisek Rd., Klongtoey, Bangkok, 10110, Thailand (Tel: 66 2 264 0505; FAX 66 2 264 0499)

Solvay

Solvay S.A, 33 rue du Prince Albert, 1050 Brussels, Belgium (Tel: 32 2 509 6111; FAX 32 2 509 6617; E-mail: solvay.international@solvay.com; Internet: www.solvay.com)

Solvay S.A. (France), 12, Cours Albert 1er, 75008 Paris Cedex 08, France (Tel: 33 1 40 75 80 00; FAX 33 1 45 63 57 28)

Solvay Chemicals Ltd., Grovelands Business Centre, Boundary Way, Hemel Hempstead, Herts, HP2 7TE, UK (Tel: 44 1442 236555; FAX 44 1442 238770)

Solvay Deutschland GmbH, Postfach 220, Hans-Böckler-Allee 20, 30173 Hannover, Germany (Tel: 49 511 857 0; FAX 49 511 282126; Internet: www.solvay.com/de)

Solvay Chimica Italia S.p.A., Via Filippo Turati, 12, 20121 Milano (MI), Italy (Tel: 39 2 290921; FAX 39 2 6570581)

Solvay Duphar BV, C.J. van Houtenlaan 36, Postbus 900, 1380 DA Weesp, The Netherlands (Tel: 31 294 477000; FAX 31 294 480253; E-mail: info@solvayvitamins.nl; Internet: www.solvayvitamins.com)

Solvay Portugal, Avenida Marechal Gomes da Costa 33, 1800 Lisbon, Portugal (Tel: 351 2183 16100; FAX 351 2183 16250)

Solvay Quimica S.L., Calle Mallorca 269, 08008 Barcelona, Spain (Tel: 34 93 484 74 00; FAX 34 93 484 76 56; Internet: www.solvay.com/es)

Solvay Vienna GmbH, Staettermayergasse 28-30, 1150 Wien, Austria (Tel: 43 1 716 880; FAX 43 1 710 24 26; E-mail: solvay.vienna@solvay.com)

Solvay America Inc., Member of Solvay Group, 3333 Richmond Ave., Houston, TX, 77098, USA (Tel: 713-525-6000; FAX 713-525-7887; Internet: www.solvay.com/na/sii)

Solvay Fluorides, Inc., 1630 Des Peres Rd., Suite 305, St. Louis, MO, 63131, USA (Tel: 314-965-7100; FAX 314-966-2907; E-mail: ken.neugebauer@solvay.com; Internet: www.solvay.com)

Solvay Interox, Member of Solvay Group, 3333 Richmond Ave., Houston, TX, 77098-3009, USA (Tel: 713-525-6500; 800-INTEROX; FAX 713-524-9032; E-mail: David.Henry@Solvay.com; Internet: www.solvay-interox.com)

Solvay Minerals, Inc., Subsid. of Solvay America Inc., 3333 Richmond Ave., Houston, TX, 77098-3099, USA (Tel: 713-525-6800; 800-443-2785; FAX 713-525-7804)

Solvay do Brasil Ltda., Rua Urussuí, 300 3°Andar - Itaim Bibi, 04542-903 Sao Paulo-SP, Brazil (Tel: 55 11 3708 5100; FAX 55 11 3708 5283; E-mail: quimicos.br@solvay.com)

Solvay Interox Pty Ltd., 20-22 McPherson St., Banksmeadow, Botany, 2019, Australia (Tel: 61 2 9316 8000; FAX 61 2 9316 6445)

Nippon Solvay K.K., Tabata Asuka Tower 16 F, 6-1-1, Tabata, Kita-ku, Tokyo, 114, Japan (Tel: 81 3 5814 0851; FAX 81 3 5814 0855)

Solvay Asia Pacific Pte. Ltd., 8 Cross St. #24-01, PWC Bldg., Singapore, 048424, Singapore (Tel: 65 438 8886; FAX 65 438 8222)

Solvay China Ltd., Suite 1011, Chinachem Golden Plaza, 77 Mody Rd., Tsimshatsui East, Kowloon, Hong Kong (Tel: 852 23127113; FAX 852 23127099; E-mail: maureen.chou@solvay.com.hk)

Solvay Solexis

Solvay Solexis S.p.A., Viale Lombardia, 20, 20021 Bollate (MI), Italy (Tel: 39 2 38351; FAX 39 2 38352110; E-mail: solvaysolexis.ita@solvay.com; Internet: www.solvaysolexis.com)

Solvay Solexis Ltda., Rua Urussui, 300 - 4° andar, Sao Paulo SP, 04542-903, Brazil (Tel: 55 11 3708 5023; FAX 55 11 3708 5115)

Solvay Solexis Ltd., Erico House, 93-99 Upper Richmond Rd., Putney, London, SW15 2TG, UK (Tel: 44 208 7804000; FAX 44 208 7802871; E-mail: brianh@ausimont.telme.com)

Solvay Solexis (Sweden), Skårs Led 3, 412 63 Göteborg, Sweden (Tel: 46 31 703 71 42; FAX 46 31 703 71 01; E-mail: hasse.andersson@ausimont.se)

Solvay Solexis S.A.S., 12, Cours Albert 1er, 753383 Paris Cedex 08, France (Tel: 33 1 40 75 80 00; FAX 33 1 45 63 82 26)

Solvay Solexis, Inc., 10 Leonard Lane, Thorofare, NJ, 08086, USA (Tel: 856-853-8119; FAX 856-853-6405; E-mail: solvaysolexisinfo@solvay.com; Internet: www.solvaysolexis.com)

Solvay Solexis K.K., Izumi Akasaka Bldg., 3rd Fl., 2-22-24, Akasaka, Minato-ku, Tokyo, 107, Japan (Tel: 81 3 3224 7226; FAX 81 3 3224 7218)

Solvay Solexis Singapore Pte. Ltd., 96 Robinson Rd., SIF Bldg. #04-05, Singapore, 068899, Singapore (Tel: 65 6223 3581; FAX 65 6223 2127)

Solvay Beijing Representative Office, 2515, China World Tower 1, n.1 Jian Guo Men Wai Ave., Beijing, 100004, P.R. China (Tel: 86 10 6505 2280; FAX 86 10 6505 9389)

Southchem, Inc.

Member of Brenntag Group, 2000 E. Pettigrew St., Durham, NC, 27703, USA (Tel: 919-596-0681; FAX 919-596-6438)

Southern Clay Products, Inc.

Subsid. of Laporte, 1212 Church St., PO Box 44, Gonzales, TX, 78629, USA (Tel: 830-672-2891; 800-324-2891; FAX 830-672-1903; E-mail: jrichardson.scprod@laporteplc.com; Internet: www.scprod.com)

Southern Texas Chemical Corp.

6374 Creekbend, Houston, TX, 77096, USA (Tel: 713-988-6089; FAX 713-988-8936)

Southern Wisconsin Chemicals, Inc.

PO Box 42, Springfield, WI, 53176, USA (Tel: 888-293-5857)

South Hampton Refining Co.

PO Box 1636, Silsbee, TX, 77656, USA (Tel: 409-385-1400; FAX 409-385-2453; E-mail: sales@southhamptonrefining.com; Internet: www.southhamptonrefining.com)

Soya Mainz GmbH & Co. KG

Dammweg 2, PO Box 3767, 55020 Mainz, Germany (Tel: 49 6131 895 0; FAX 49 6131 834104)

Sparkford Chemicals Ltd.

72 Portswood Road, Southampton, Hamsphire, SO17 2FW, UK (Tel: 44 23 8228747; FAX 44 23 80586682; E-mail: info@sparkford.co.uk; Internet: www.sparkford.co.uk)

Specialty Chem Products Corp.

A Bayer Company, Two Stanton St., Marinette, WI, 54143, USA (Tel: 715-735-9033; FAX 715-735-5304; Internet: www.bayerus.com)

Specialty Ingredients & Perfume Co.

260 Ryan St., S. Plainfield, NJ, 07080, USA (Tel: 908-412-9292; FAX 908-412-9339; E-mail: specingred@aol.com)

Spectrum Laboratories, Inc.

1460 West McNab Rd., Ft. Lauderdale, FL, 33309, USA (Tel: 954-978-6400; FAX 954-978-2233; E-mail: contact@speclab.com; Internet: www.speclab.com)

Spectrum Naturals Inc.

1304 S. Point Blvd., Suite 280, Petaluma, CA, 94954, USA (Tel: 707-778-8900; FAX 707-765-8470; E-mail: info@spectrumorganic.com; Internet: www.spectrum naturals.com)

Spectrum Quality Products

14422 S. San Pedro St., Gardena, CA, 90248-2027, USA (Tel: 310-516-8000; 800-772-8786; FAX 800-525-2299; E-mail: sales@spectrumchemical.com; Internet: www.spectrumchemical.com)

Spektr T.T.T.

22, Shepkin St., Moscow, 129090, Russia (Tel: 7 95 742 3075; FAX 7 095 288 1652; E-mail: spektr@ivk.ru; Internet: www.spektr-ttt.ru)

Springfield Scientific Supplies

2600 Main St., Springfield, OR, 97477, USA (Tel: 800-344-2047; FAX 541-747-1875)

S & S Chemical Co., Inc.

188 Main St., PO Box 659, Northport, NY, 11768, USA (Tel: 631-261-5920; FAX 631-261-5934; E-mail: rdssands@aol.com)

A.E. Staley Manufacturing Co.

Subsid. of Tate & Lyle PLC, 2200 E. Eldorado St., Decatur, IL, 62525, USA (Tel: 217-423-4411; 800-526-5728; FAX 217-421-3167; Internet: www.aestaley.com)

Stanchem Inc.

Business Unit of Holland Chemical International (HCl), 2900, J.-B. Deschamps Blvd., Lachine, Quebec, H8T 1C8, Canada (Tel: 514-636-9230; FAX 514-636-0877; E-mail: stanchem@adicq.qc.ca; Internet: www.adicq.qc.ca/stanchem/)

StanChem, Inc.

401 Berlin St., East Berlin, CT, 06023, USA (Tel: 860-828-0571; FAX 860-828-3297; E-mail: info@albi.com; Internet: www.albi.com)

Stan Chem International Ltd.

4 Kings Rd, Reading, Berkshire, RG1 3AA, UK (Tel: 44 118 958 0247; FAX 44 118 958 9580; E-mail: sales@stanchem.co.uk; Internet: www.stanchem.co.uk)

Stason Pharmaceuticals Inc.

11 Morgan, Irvine, CA, 92618-2005, USA (Tel: 949-380-4327; FAX 949-380-4345)

Stepan

Stepan Co., 22 West Frontage Rd., Northfield, IL, 60093, USA (Tel: 847-446-7500; 800-745-7837; FAX 847-501-2100; E-mail: techserv@stepan.com; Internet: www.stepan.com)

Stepan Canada, 2000 Argentia Rd., Plaza 5, Suite 102, Mississauga, Ontario, L5N 2R7, Canada (Tel: 905-814-7231; FAX 905-814-7233)

Stepan Mexico, S.A. de C.V., Reforma 403-1004, Mexico, D.F., 06500, Mexico (Tel: 52 5 533 1697; FAX 52 5 514 5448)

Stepan Quimica Ltda., Rua Pelotas, 499 Vila Mariana, 04012-001 Sao Paulo, Brazil (Tel: 55 11 573 0120; FAX 55 11 539 4587)

Stepan Colombiana de Quimicos, Ltda., Calle 114A, No. 19-12, Santa Barbara, Santa Fe de Bogota, Colombia (Tel: 571 522 2523; FAX 571 637 2790)

Stepan Europe, Chemin Jongkind, BP127, 38340 Voreppe (Grenoble), France (Tel: 33 4 76 50 51 00; FAX 33 4 76 56 71 65)

Stepan UK Ltd., Bridge St., Stalybridge, Cheshire, SK15 1PH, UK (Tel: 44 161 338 5511; FAX 44 161 303 2991; E-mail: sales@manro-pc.com; Internet: www.manro-pc.com)

Stepan Philippines, Inc. (J.V.), 7th Fl, NOL Tower, Commerce Ave., Madrigal Business Park, Ayala Alabang, Muntinlupa City, Philippines (Tel: 632 842 9751; FAX 632 842 1236)

Sterling Chemicals, Inc.

1200 Smith St. #1900, Houston, TX, 77002-4312, USA (Tel: 713-650-3700; FAX 713-654-9551; Internet: www.sterlingchemicals.com)

Stevenson Cooper, Inc.

PO Box 46345, 1039 West Venango St., Philadelphia, PA, 19160, USA (Tel: 215-223-2600; 888-420-1663; FAX 215-223-3597; E-mail: waxcooper@aol.com; Internet: www.stevensoncooper.com)

Stev-Tex Laboratories

Stev-Tex Laboratories, 12123 Clark St., Santa Fe Springs, CA, 90670, USA (Tel: 562-906-5090; 800-279-3965; FAX 562-906-5094; E-mail: info@stev-tex.com; Internet: www.stev-tex.com)

Stev-Tex Laboratorios de Mexico S. de R.L. de C.V., Ave. Independencia #183, Col. Chapultepec, Puebla, Pue., 72320, Mexico (Tel: 52 2 253 0002; FAX 52 2 253 0987)

Stev-Tex Laboratories Australia Pty. Ltd., 57 Keys Rd., Moorabbin, Victoria, 3189, Australia (Tel: 61 3 9555 2688; FAX 61 3 9555 2699; E-mail: stevtex@smartchat.net.au)

St. Lawrence Chemicals

St. Lawrence Chemicals, 1920 Ave. Clark Graham, Baie D'Urfe, Quebec, H9X 3P5, Canada (Tel: 514-457-3628; FAX 514-457-9773; E-mail: info@stlawrence chem.com; Internet: www.stlawrencechem.com)

St. Lawrence Chemicals, 35 Vulcan St., Rexdale, Ontario, M9W 1L3, Canada (Tel: 416-243-9615; FAX 416-243-9731)

Stockhausen

Stockhausen GmbH & Co. KG, A Company of the Degussa Group, Bakerpfad 25, 47805 Krefeld, Germany (Tel: 49 2151 38 01; FAX 49 2151 38 1647; E-mail: mail@stockhausen.com; Internet: www.stockhausen.com; www.stoko.de)

Stockhausen, Inc., A Company of the Degussa Group, 2401 Doyle St., Greensboro, NC, 27406, USA (Tel: 336-333-3500; 800-334-0242; FAX 336-333-3545; E-mail: h2otreat@stockhausen-inc.com; Internet: www.stockhausen-inc.com)

Storchem Inc.

3425 Harvester Rd., Unit #213, Burlington, Ontario, L7N 3N1, Canada (Tel: 905-639-9700; FAX 905-639-5244; E-mail: sales@storchem.on.ca; Internet: www.storchem.on.ca)

Strahl & Pitsch, Inc.

PO Box 1098, 230 Great E. Neck Rd., W. Babylon, NY, 11704, USA (Tel: 631-587-9000; FAX 631-587-9120; E-mail: info@strahlpitsch.com; Internet: www.strahlpitsch.com)

Strem Chemicals, Inc.

7 Mulliken Way, Newburyport, MA, 01950-4098, USA (Tel: 978-462-3191; 800-647-8736; FAX 800-517-8736; E-mail: info@strem.com; Internet: www.strem.com)

Suchema AG

Haupstrasse 15, 8259 Kaltenbach, Switzerland (Tel: 41 54 41 12 65; FAX 41 54 41 32 34)

Süd-Chemie

Süd-Chemie AG, Lenbachplatz 6, 80333 München, Germany (Tel: 49 89 5110 0; FAX 49 89 5110 375; E-mail: info@sued-chemie.de; Internet: www.sued-chemie.de)

Sud Chemie MT S.r.l., Via G. Fauser 50, 28100 Novara, Italy (Tel: 39 321 676300; FAX 39 321 676490)

Süd-Chemie Inc., PO Box 32370, Louisville, KY, 40232, USA (Tel: 502-634-7200; FAX 502-637-3732; Internet: www.sud-chemieinc.com)

Süd-Chemie Inc./Rheologicals and Performance Additives, 1600 West Hill St., Louisville, KY, 40210, USA (Tel: 502-634-7500; FAX 502-634-7727; E-mail: rheoinfous@sud-chemieinc.com; Internet: www.rheological.com)

Sumitomo Chemical

Sumitomo Chemical Co., Ltd., New Sumitomo Bldg., 4-5-33, Kitahama, Chuo-ku, Osaka, 541-8550, Japan (Tel: 81 6 6220 3272; FAX 81 6 6220 3345; Internet: www.sumitomo-chem.co.jp)

Sumitomo Chemical America Inc., 345 Park Ave., New York, NY, 10154, USA (Tel: 212-572-8200; FAX 212-572-8234; E-mail: bunji1103@msn.com)

British Traders & Shippers Ltd., Distrib. for Sumitomo Chem., 429-431 Rainham Road South, Dagenham, Essex, RM10 8RN, UK (Tel: 44 181 596 7500; FAX 44 181 596 7509; E-mail: pvax@britishtraders.co.uk; Internet: www.britishtraders.co.uk)

Sumitomo Seika Chemicals

Sumitomo Seika Chemicals Co., Ltd., The Sumitomo Bldg., No. 2, 4-7-28, Kitahama, Chuo-ku, Osaka, 541-0041, Japan (Tel: 81 6 6220 8508; FAX 81 6 6220 8541)

Sumitomo Seika Chemicals (US), 335 Madison Ave., Suite 830, New York, NY, 10017, USA (Tel: 212-572-8245; FAX 212-421-1179; E-mail: 79047goda@msn.com)

Sunnyside Corp.

225 Carpenter Ave., Wheeling, IL, 60090-6095, USA (Tel: 847-541-5700; 800-323-8611; FAX 847-541-9043; E-mail: rlueders@sunnysidecorp.com; Internet: www.sunnysidecorp.com)

Sunoco Chemicals

Business Unit of Sun Company, Inc., Ten Penn Center, 1801 Market Street, Philadelphia, PA, 19103-1699, USA (Tel: 215-977-3321; 877-481-7840; FAX 215-977-3470; E-mail: info@sunocochem.com; Internet: www.sunocochem.com)

Supelco Inc.

Supelco Park, Bellefonte, PA, 16823, USA (Tel: 814-359-3441; 800-247-6628; FAX 800-447-3044; E-mail: supelco@sial.com; Internet: www.sigma-aldrich.com)

Surco Products, Inc.

292 Alpha Dr., RIDC Industrial Park, Pittsburgh, PA, 15238-2903, USA (Tel: 412-252-7000; 800-556-0111; FAX 800-417-8726; E-mail: odorstop@iearthlink.com; Internet: www.surco.com)

Surfachem Ltd.

Div. of Surfachem Group plc, Wellington Park House, Thirsk Row, Leeds, West Yorkshire, LS1 4DP, UK (Tel: 44 113 234 2636; FAX 44 113 244 5910; E-mail: enquiries@surfachem.com; Internet: www.surfachem.com)

Surfactants, Inc.

260 Ryan St., S. Plainfield, NJ, 07080, USA (Tel: 908-755-3300; FAX 755-0592; E-mail: r.coakley@worldnet.att.net)

Synasia Fine Chemical, Inc.

68-19 Clyde St., Forest Hills, NY, 11375, USA (Tel: 718-544-8988; FAX 718-544-5112; E-mail: synasiany@aol.com; Internet: www.synasia.com)

Syntai Chemicals & Pharmaceuticals, Ltd.

23-1, Haccobori 2-chome, Chuo-ku, Tokyo, Japan (FAX 81 3 3523 3490)

Synthron

Synthron S.A., Subsid. of Protex International Group, B.P. 177, 6 rue Barbès, 92305 Levallois-Paris Cedex, France (Tel: 33 1 41 34 14 00; FAX 33 1 41 34 14 16; E-mail: postmaster@protex-international.com; Internet: www.protex-international.com)

Synthron Inc., 305 Amherst Rd., Morganton, NC, 28655-9362, USA (Tel: 828-437-8611; FAX 828-437-4126)

Protex Chemicals Ltd., Astley Lane Industrial Estate, Astley Way, Swillington, Leeds, LS26 8XT, UK (Tel: 44 1132 876002; FAX 44 1132 875003)

Protex Nederland, Kerkstraat 21 a, 5095 AE Hooge Mierde, The Netherlands (Tel: 31 13 509 22 62; FAX 31 13 509 17 19)

Protex Sverige AB, Landex-Kemi AB, Box 320, 261 23 Landskrona, Sweden (Tel: 46 418 58470; FAX 46 418 58788)

Protex Chemie Basel AG, Thannerstrasse 72, 4054 Basel, Switzerland (Tel: 41 61 302 0066; FAX 41 61 302 0076)

Protex Warenhandel GmbH, Hainburger Strasse 19A, 1030 Vienna III, Austria (Tel: 43 222 73 93 864; FAX 43 222 7137 552)

Protex Korea Co. Ltd., PO Box 1193, Seoul, 1351010, Korea (Tel: 82 2 3453 8213; FAX 82 2 3453 8082)

Taiwan Surfactant Corp.

8 Fl., No. 11, Sec. 1, Chung Shan North Rd., Taipei, Taiwan R.O.C. (Tel: 886 2 2541 0022; FAX 886 2 2542 3773; E-mail: twsa@ms2.hinet.net; Internet: www.taiwansurfactant.com.tw)

Takasago International

Takasago International Corp., 17/F, 5-37-1, Kamata, Ohta-ku, Tokyo, 144-8721, Japan (Tel: 81 3 5744 0511; FAX 81 3 5744 0512; Internet: www.takasago.com)

Takasago International Corp., 4 Volvo Dr., Rockleigh, NJ, 07647-0932, USA (Tel: 201-767-9001; FAX 201-784-7277)

Takasago Limited, 27 Victoria St., Windsor, Berkshire, SL4 1HE, UK (Tel: 44 1751 620 032; FAX 44 1753 621 903)

Takasago International (Deutschland) GmbH, Inselstrasse 29, 4047 Dusseldorf, Germany (Tel: 49 211 4976 6660; FAX 49 211 4976 6666)

Takasago International (Australia) Pte. Ltd., 275 Alfred St., Level 4, North Sydney, NSW, 2060, Australia (Tel: 61 2 9959 1000; FAX 61 2 9959 3225)

Takeda Vitamin & Food USA Inc.

101 Takeda Dr., Wilmington, NC, 28401, USA (FAX 800-825-0333; Internet: www.takeda-usa.com)

F.H. Taussig, Inc.

111 Brook St., Scarsdale, NY, 10583-5192, USA (Tel: 914-472-6464; FAX 914-472-1846; E-mail: fhtaussig@aol.com)

TCI America

U.S. Subsid. of Tokyo Kasei Kogyo, 9211 North Harborgate St., Portland, OR, 97203, USA (Tel: 503-283-1681; 800-423-8616; FAX 503-283-1987; E-mail: sales@tciamerica.com; Internet: www.tciamerica.com)

Technichem Inc.

5 West Park Ave., Merchantville, NJ, 08109, USA (Tel: 609-486-0060; 800-510-5086; FAX 609-486-0490; E-mail: 71702.72@compuserve.com)

Tech Spray, L.P.

PO Box 949, Amarillo, TX, 79105-0949, USA (Tel: 806-372-8523; 800-858-4043; FAX 806-372-8750; E-mail: tsales@techspray.com; Internet: www.techspray.com)

Tecnal Corporation

9056 N. Texas Rd., Anacortes, WA, 98221, USA (Tel: 360-293-3200; FAX 360-293-3131; E-mail: cstecnal@aol.com)

Tego

Tego Chemie Service GmbH, A Goldschmidt Group Company, Goldschmidtrasse 100, 45127 Essen, Germany (Tel: 49 201 173 06; FAX 49 201 173 1939; E-mail: info@tego.de; Internet: www.tego.de)

Tego Chemie Service Italia, Via Falconera, 7, 26025 Pandino (CF), Italy (Tel: 39 0373 975 11; FAX 39 0373 975 260; E-mail: franco.franzetti@it.goldschmidt.com)

Degussa Tego Coating & Ink Additives, 710 South 6th Ave., PO Box 1111, Hopewell, VA, 23860-1111, USA (Tel: 804-541-8658; 800-446-1809; FAX 804-541-6290; E-mail: jim.yosh@degussa.com; Internet: www.additiweb.com)

Goldschmidt Pacific Ltd., 13/F, Manulife, Tower 169 Elect, North Point, Hong Kong (Tel: 852 2519 8758; FAX 852 2519 0960)

Tekchem, S.A. de C.V.

Ave. Jalisco 180-3, Col. Tacubaya, Delegacion Miguel Hidalgo, Mexico D.F., 11870, Mexico (Tel: 52 5 272 2221; FAX 52 5 272 2950; E-mail: irivero@tekchem.com.mx; Internet: www.tekchem.com.mx)

Tennants Distribution Limited

Grosvenor Street, London, W1K 3BP, UK (Tel: 44 207 493 5451; FAX 44 207 495 1269; Internet: www.tennantsdistribution.com)

Terry Laboratories

Terry Laboratories, Inc., 390 N. Wickham Rd., Suite F, Melbourne, FL, 32935-8647, USA (Tel: 321-259-1630; 800-367-2563; FAX 321-242-0625; E-mail: aloe@terrylabs.com; Internet: www.terrylabs.com)

S. Black (Import & Export) Ltd., Distrib. for Terry Labs, Foxholes Business Park, John Tate Road, Hertford, Hertfordshire, SG13 7YH, UK (Tel: 44 1992 825555; FAX 44 1992 825566; E-mail: enquires@sblack.com; Internet: www.sblack.com)

Tessenderlo

Tessenderlo Chemie, Rue du Trone 130, 1050 Brussels, Belgium (Tel: 32 2 639 18 11; FAX 32 2 639 19 99; Internet: www.tessenderlo.com)

Tessenderlo Kerley, Inc., 2255 North 44th St., Suite 300, Phoenix, AZ, 85008-3279, USA (Tel: 602-889-8300; 800-669-0559; FAX 602-889-8430; E-mail: im-info@tkinet.com)

Tetra Chemicals

A Div. of TETRA Technologies Inc., 25025 I-45 North, The Woodlands, TX, 77380, USA (Tel: 281-367-1983; 800-327-7817; FAX 281-367-6471; E-mail: chemsales@tetratec.com; Internet: www.tetratec.com)

Texaco Inc. *See* ChevronTexaco

Texas Petrochemicals Corp.

8707 Katy Pkwy., Suite 300, Houston, TX, 77024, USA (Tel: 713-461-3322; FAX 713-461-1029)

Textile Chemical Company

Member of Brenntag Group, Pottsville Pike & Huller Ln., PO Box 13788, Reading, PA, 19612-3788, USA (E-mail: customerservice@textilechem.com; Internet: www.textilechem.com)

Thatcher Co.

1900 Fortune Rd., PO Box 27407, Salt Lake City, UT, 84127, USA (Tel: 801-972-4587; FAX 801-972-4606)

Thew Arnott & Co. Ltd.

Newman Works, 270 London Rd., Wallington, Surrey, SM6 7DJ, UK (Tel: 44 20 8669 3131; FAX 44 20 8669 7747; E-mail: sales@thewarnott.co.uk; Internet: www.thewarnott.co.uk)

Thomas Scientific

99 High Hill Rd., PO Box 99, Swedesboro, NJ, 08085, USA (Tel: 856-467-2000; 800-345-2100; FAX 856-467-3087)

Thor

Thor Specialities (UK) Ltd., Wincham Ave., Wincham, Northwich, Cheshire, CW9 6GB, UK (Tel: 44 1606 818800; FAX 44 1606 818801; E-mail: info@thor.uk.com; Internet: www.thor.com; www.thor-chemicals.com)

Thor GmbH, Landwehrstrasse 1, 67346 Speyer, Germany (Tel: 49 6232 636 0; FAX 49 6232 636 111; E-mail: info@thor-chemie.de)

Thor Specialities SRL, Via Puccini 30, 21010 Besnate (VA), Italy (Tel: 39 331 279311; FAX 39 331 272915; E-mail: thor@thoritaly.it)

Thor SARL, 325 Rue des Balmes, 38150 Salaise-sur-Sanne, France (Tel: 33 4 74 11 20 00; FAX 33 4 74 29 65 35; E-mail: thorfr@thor-france.fr)

Thor Nordic, Molndalsvagen 24, 41263 Goteborg, Sweden (Tel: 46 31 404 146; FAX 46 31 404 125; E-mail: biocid@thornordic.com)

Thor Especialidades, SA, Poligono Industrial El Pla, Avda. de la Industria 1, 08297 Castellgali (Barcelona), Spain (Tel: 34 93 833 28 00; FAX 34 93 833 37 13; E-mail: thor@thor-spain.com)

Thor Specialties Pty. Ltd., 67 Newton Rd., Wetherill Park, NSW, 2164, Australia (Tel: 61 2 9725 1177; FAX 61 2 9725 5677; E-mail: info@thorchem.com.au)

Thor Specialties Pty. Ltd., 15 Kalmia St., Ellerslie, Auckland, New Zealand (Tel: 64 9 579 5037; FAX 64 9 579 4673; E-mail: thor.nz@xtra.co.nz)

Acti-Chem Specialties Inc., 56 Quarry Rd., Trumbull, CT, 06611, USA (Tel: 203-365-6530; FAX 203-365-6537; E-mail: actichem@snet.net)

Thornley Company Inc.

1500 East Newport Pike, Suite 204, Wilmington, DE, 19804-2346, USA (Tel: 302-993-8300; FAX 302-993-8308; E-mail: thornleyco@aol.com; Internet: www.thornleycompany.com)

3M

3M, Product Information Center, 3M Center, Bldg. 304-1-01, St. Paul, MN, 55144-1000, USA (Tel: 651-737-6501; 800-364-3577; FAX 800-713-6329; E-mail: innovation@mmm.com; Internet: www.mmm.com; www.3m.com)

3M Co./Performance Materials Div., 3M Center, Bldg. 223-6S-04, St. Paul, MN, 55144-1000, USA (Tel: 800-360-6777; FAX 651-736-8643)

3M Co./Specialty Materials, 3M Center Bldg. 223-6S-04, St. Paul, MN, 55144-1000, USA (Tel: 612-733-3112; 800-833-5045; FAX 612-736-8643; Internet: www.3m.com/fluids)

3M Puerto Rico, Inc., Puerto Rico Industrial Park, 65 Infanteria, Km. 12.16 Lote #6, 7, Carolina, PR, 00986-0100, USA (FAX 787-257-2745)

3M Canada Company, PO Box 5757, London, Ontario, N6A 4T1, Canada (Tel: 800-364-3477; FAX 800-479-4453)

3M Mexico, Avenida Santa Fe #55, Col. Sante Fe, Delegacion Alvaro Obregon, Mexico D.F., 01210, Mexico

3M Manufacturera Venezuela, S.A., Puente Brion a Luis Razetti Edif, 3M, La Candelaria, Caracas, 1010-A, Venezuela (FAX 58 2 572 9276)

3M Chile, S.A., Santa Isabel 1001 Providencia, Santiago, 6641431, Chile (FAX 56 2 204 8900)

3M United Kingdom plc, 3M House, PO Box 1, Market Pl., Bracknell, Berkshire, RG12 1JU, UK (Tel: 44 1344 858000; FAX 44 1344 858479)

3M Deutschland GmbH, Carl-Schurz-Strasse 1, 41453 Neuss, Germany (Tel: 49 2131 140)

3M Belgium NV/SA, Hermeslaan 7, 1831 Diegem, Belgium (Tel: 32 2 722 5111; FAX 32 2 720 0225)

3M Belgium NV/Specialty Materials, Haven 1005, Canadastraat 11, 2070 Zwijndrecht, Belgium (Tel: 32 3 250 7521)

3M Nederland Holding BV, Industrieweg 24, 2382 NW Zoeterwoude, The Netherlands

3M España SA, Juan Ignacio Luca de Tena, 19-25, 28027 Madrid, Spain (FAX 34 91 321 6305)

3M France, Blvd de l'Oise, 95006 Cergy-Pontoise Cedex, France (Tel: 33 1 30 31 61 61; FAX 33 1 30 31 74 26)

3M Italia S.p.a., Via S. Bovio, 20090 Segrate (MI), Italy (Tel: 39 2 70351)

3M South Africa (Pty) Ltd., 181 Barbara Rd., Elandsfontein, 1406, Rep. of S. Africa (Tel: 27 11 922 2016)

Birla 3M Ltd., Jubilee Bldg. 45, Museum Rd., Bangalore, 560025, India (Tel: 91 80 558 8881)

3M Australia, 950 Pacific Hwy., PO Box 99, Pymble, NSW, 2073, Australia (Tel: 61 2 9498 9333; FAX 61 2 9498 9666)

Sumitomo 3M Ltd., Joint venture of Sumitomo Elec. Ind., Nippon Electric Co., Ltd. and 3M USA, 2-33-1, Tamagawadai, Setagaya-ku, Tokyo, 158-8583, Japan (Tel: 81 3 3709 8250; FAX 81 3 3709 8752)

3M China Ltd., /F, New Town Mansion, 55 Lou Shan Guan Rd., Shanghai, 200336, P.R. China (FAX 86 21 6275 2343)

3M Hong Kong Ltd., Victoria Centre, 5th Fl., 15 Watson Rd., Causeway Bay, Hong Kong (Tel: 852 2806 6111)

Ticona

Ticona GmbH, A Business of Celanese AG, Industriepark Hoechst, 65926 Frankfurt, Germany (Tel: 49 69 30 58 47 32; FAX 49 69 30 58 47 35; E-mail: junker@ticona.de)

Ticona Austria GmbH, Altmannsdorfer Strasse 104, 1121 Vienna, Austria (Tel: 43 1 8 01 01 23 61; FAX 43 1 8 01 01 28 80; E-mail: kalab@ticona.de)

Ticona Benelux SA, Chaussee de Charleroi 111-113, 1060 Bruxelles, Belgium (Tel: 32 2 5 33 49 31; FAX 32 2 5 33 49 33)

Ticona Benelux N.V., Diemerhof 36, 1112 XN Diemen, The Netherlands (Tel: 31 20 3 98 98 91; FAX 31 20 3 98 98 95; E-mail: hullekes@ticona.nl)

Ticona UK Limited/Technical Polymers Div., Michigan Dr., Tongwell, Milton Keynes, MK15 8HQ, UK (Tel: 44 1908 513424; FAX 44 1908 513411; E-mail: michael.oberste-wilms@ticona.co.uk)

Ticona Norden Danmark A/S, Naverland 8, 2600 Glostrup, Denmark (Tel: 45-43 27 12 12; FAX 45-43 27 12 13; E-mail: petersen@ticona.dk)

Ticona Norden Finland Oy, PL 318, 00101 Helsinki, Finland (Tel: 358 9 8 70 98 82; FAX 3 58 9 8 70 97 46; E-mail: liimatainen@ticona.fi)

Ticona Norden Sverige AB, Box 5415, 401 23 Göteborg, Sweden (Tel: 46 31 67 86 01; FAX 46 31 67 86 10)

Ticona France S.a.r.l., 1, Terrasse Bellini, 92910 Paris La Defense Cedex, France (Tel: 33 1 40 81 45 37; FAX 33 1 40 81 45 38; E-mail: brigitte.brinster@hiserv.com)

Ticona Italia S.r.L., Piazza S.T. rr 5, 20149 Milan, Italy (Tel: 39 02 34 54 91 21; FAX 39 02 3 31 13 05; E-mail: siro.guerini@hiserv.com)

Ticona Iberia S.L., Via Augsuta, 252-260, 08017 Barcelona, Spain (Tel: 34 93 3 06 81 71; FAX 34 93 3 06 82 71; E-mail: cendra@msmbcnhub.hoechst.com)

Ticona LLC, A Celanese AG company, 90 Morris Ave., Summit, NJ, 07901, USA (Tel: 908-598-4000; 800-833-4882; FAX 908-598-4306; Internet: www.ticona.com)

Toagosei Co., Ltd.

1-14-1, Nishi-shimbashi, Minato-ku, Tokyo, 105, Japan (Tel: 81 3 3597 7215; FAX 81 3 3597 7217)

A.M. Todd Botanicals

Div. of Zink & Triest, a member of the A.M. Todd Group of Companies, 150 Domorah Dr., Montgomeryville, PA, 18936, USA (Tel: 215-628-8895; 800-634-4418; FAX 215-628-8651; E-mail: info@amtbotanicals.com; Internet: www.amtbotanicals.com)

Toho Chemical Industry Co., Ltd.

6-4, Akashi-cho, Chuo-ku, Tokyo, 104-0044, Japan (Tel: 81 3 5550 3721; FAX 81 3 5550 1988; E-mail: overseas@toho-chem.co.jp; Internet: www.toho-chem.co.jp)

Tokuyama Petrochemical Co., Ltd.

4980, Kasei-cho, Shin-nanyo-shi, Yamaguchi, 746, Japan (Tel: 81 834 62 4121; FAX 81 834 62 3545)

Tomen

Tomen Chemical Co., Ltd., 8-1 Marunouchi 3-Chome, Chiyoda-ku, Tokyo, 100-8623, Japan

Tomen America Inc./Kawaguchi, Subsid. of Tomen Corp., 1285 Ave. of the Americas, New York, NY, 10019, USA (Tel: 212-397-4600; 800-781-0037; FAX 212-582-2007; Internet: www.tomenamerica.com)

Tomita Pharmaceutical Co., Ltd.

85-1, Maruyama Akinokami, Seto-cho, Naruto-shi, Tokushima, 771-0360, Japan (Tel: 81 886 88 0511; FAX 81 886 88 0565)

Tonen Chemical Corp.

Subsid. of Tonen Corp., Togeki Bldg., 4-1-1, Tsukiji, Chuo-ku, Tokyo, 104, Japan (Tel: 81 3 3542 7361; FAX 81 3 5565 1402)

Toray Fine Chemicals Co., Ltd.

273, Katsube-cho, Moriyama-shi, Shiga, 524, Japan (Tel: 81 775 83 2570; FAX 81 775 83 3312)

Toray Thiokol Co., Ltd.

Toray Bldg., 1-8-1, Mihama, Urayasu-shi, Chiba, 279, Japan (Tel: 81 473 50 6151; FAX 81 473 50 6091)

Tosoh

Tosoh Corp., Shiba-koen First Bldg., 3-8-2, Shiba, Minato-ku, Tokyo, 105-8623, Japan (Tel: 81 3 5427 5118; FAX 81 3 5427 5298; E-mail: info@tosoh.co.jp; Internet: www.tosoh.co.jp)

Tosoh Europe B.V., Crown Bldg.-South, Hullenbergweg 359, 1101 CP Amsterdam Z.O., The Netherlands (Tel: 31 20 565 0010; FAX 31 20 691 5458; E-mail: tosoh@tosoheur.com)

Tosoh USA Inc., 1100 Circle 75 Pkwy., Suite 600, Atlanta, GA, 30339-3097, USA (Tel: 770-956-1100; 800-221-4801; FAX 770-956-7368; E-mail: info@tosohusa.com)

TotalFina-Elf Group

2 place de la Coupole, 92400 Courbevoie La Defense 6, France (Tel: 33 1 47 44 45 46; FAX 33 1 47 44 78 78; Internet: www.totalfinaelf.com)

Total Petroleum Inc.

East Superior St., Alma, MI, 48801, USA (Tel: 517-463-9630; 800-292-9033; FAX 517-463-9623)

Total Specialty Chemicals, Inc.

47 Elm St., New Canaan, CT, 06840, USA (Tel: 203-966-1525; FAX 203-966-1505; E-mail: totalsc@aol.com; Internet: www.totalspecialtychemicals.com)

Toyo Gosei Kogyo Co., Ltd.

1603, Kamimyoden, Ichikawa-shi, Chiba, 272-01, Japan (Tel: 81 473 27 8080; FAX 81 3 3275 2238)

Toyo Kasei Kogyo Co., Ltd.

Shindai Bldg., 1-2-6, Dojimahama, Kita-ku, Osaka, 530, Japan (Tel: 81 6 6346 6707; FAX 81 6 6341 6715)

Toyotama Perfumery Co., Ltd.

1-14-5, Nihonbashi, Kakigara-cho, Chuo-ku, Tokyo, 103, Japan (Tel: 81 3 5640 5511; FAX 81 3 5640 5523)

Transene Co. Inc.

10 Electronics Ave., Danvers Industrial Park, Danvers, MA, 01923, USA (Tel: 978-777-7860; FAX 978-739-5640; Internet: www.transene.com)

T&R Chemicals, Inc.

700 Celum Rd., Box 330, Clint, TX, 79836, USA (Tel: 915-851-2761; FAX 915-851-2961)

Treatt

R.C. Treatt & Co. Ltd., Northern Way, Bury St Edmunds, Suffolk, IP32 6NL, UK (Tel: 44 1284 702500; FAX 44 1284 752888; E-mail: enquires@rctreatt.com; Internet: www.rctreatt.com)

Treatt USA, 4900 Lakeland Commerce Pkwy., Lakeland, FL, 33805, USA (Tel: 863-668-9500; 800-866-7704; E-mail: enquires@treattUSA.com; Internet: www.rctreatt.com)

Tri-Iso, Inc.

480 N. Indian Hill Blvd., Suite 2A, Claremont, CA, 91711, USA (Tel: 909-626-4855; FAX 909-621-9119; E-mail: service@tri-iso.com; Internet: www.tri-iso.com)

Tri-K Industries, Inc.

151 Veterans Drive, Northvale, NJ, 07647, USA (Tel: 201-750-1055; 800-526-0372; FAX 201-750-9785; E-mail: info@tri-k.com; Internet: www.tri-k.com)

Triple Crown America, Inc.

13 N. Seventh St., Perkasie, PA, 18944, USA (Tel: 215-453-2500; FAX 215-453-2508; E-mail: info@triplecrownamerica.com; Internet: www.triplecrownamerica.com)

Trivent Chemical Co., Inc.

4266 U.S. Rt. 1, PO Box 194, Monmouth Junction, NJ, 08852, USA (Tel: 732-438-0995; FAX 732-438-1020)

Troy

Troy Corp., PO Box 955, 8 Vreeland Rd., Florham Park, NJ, 07932-0955, USA (Tel: 973-443-4200; FAX 973-443-0258; E-mail: marketing@troycorp.com; Internet: www.troycorp.com)

Troy Chemical Co., Ltd., 3 Director Court, Unit 104, Woodbridge, Ontario, L4L 4S5, Canada (Tel: 905-264-2264; FAX 905-264-2266)

Troy Productos Quimicos Ltda., Av. Jorge Joao Saad, 547/82-c, Morumbi-Sao Paulo-SP, 05618-000, Brazil (Tel: 55 11 844 8317; FAX 55 11 3771 2675)

Troy Chemical Co. UK, 3rd Floor, Adelphia Mill, Grimshaw Lane, Bollington, Macclesfield, Cheshire, SK10 5JB, UK (Tel: 44 1625 575597; FAX 44 12625 575009)

Troy Chemical Co. B.V., Uiverlaan 12e, 3145 XN Maassluis, The Netherlands (Tel: 31 10 592 7494; FAX 31-10 592 8877)

Troy Chemie GmbH, Wunstorferstrasse 40, 30926 Seelze, Germany (Tel: 49 5137 8236 316; FAX 49 5137 8236 106)

Troy France S.A.R.L., 196 Rue Houdan, 92330 Sceaux, France (Tel: 33 1 41 13 88 99; FAX 33 1 41 13 60 40)

Troy Chemical Company B.V., Via Lorenteggio, 35, 20146 Milan, Italy (Tel: 390 2 475054; FAX 390 2 471712)

Troy Chemical Company B.V., Sucursal en Espana, C/ Santiago Rusinyol, 1, 08470 Sant Celoi (Barcelona), Spain (Tel: 34 93 867 5690; FAX 34 93 867 5584)

Troy Chemical Company B.V., Stokbrovej 9, 8520 Lystrup/Arhus, Denmark (Tel: 45 86 232477; FAX 44 86 232479)

Troy Southeast Asia, 35-B Polaris St., Bel-Air Makati, Metro-Manila, Philippines (Tel: 63 2 896 6964; FAX 63 2 895-5551)

The Tryline Co., Inc.

375 118th Ave. SE, Wilburton Ridge, Sutie 200, Bldg. D, Bellevue, WA, 98005, USA (Tel: 425-450-8822; FAX 425-637-7200; E-mail: DavidBruck@TryLine.com; Internet: www.tryline.com)

Tulstar Products Inc.

5510 S. Lewis Ave., Tulsa, OK, 74105, USA (Tel: 918-749-9060; 800-988-5782; FAX 918-747-1444; E-mail: tulstar@tulstar.com; Internet: www.tulstar.com)

Ube Industries

Ube Industries, Ltd., UBE Bldg., 2-3-11, Higashi-shinagawa, Shinagawa-ku, Tokyo, 140, Japan (Tel: 81 3 5460 3311; FAX 81 3 5460 3388; Internet: www.ube-ind.co.jp)

Ube America Inc., 55 E. 59th St., 18th Fl., New York, NY, 10022, USA (Tel: 212-813-8300; FAX 212-826-0454; Internet: www.ube-ind.co.jp)

Ube Industries Ltd., Fine Chems. Div., Chicago Technology Park, 2201 W. Campbell Dr., Chicago, IL, 60612, USA (Tel: 312-82-2380; 888-858-2380; FAX 312-829-2381)

Ubichem plc

Mayflower Close, Chandlers Ford Industrial Estate, Eastleigh, Hampshire, SO53 4AR, UK (Tel: 44 23 8026 3030; FAX 44 23 8026 3012; E-mail: sales@ubichem.com; Internet: www.ubichem.com)

UCB

UCB S.A./Chemical Sector, Allee de la Recherche, 60, 1070 Brussels, Belgium (Tel: 32 2 559 9999; FAX 32 2 559 9900; E-mail: Arnaud.Denis@ucb-group.com; Internet: www.ucb.be; www.chemicals.ucb-group.com)

UCB (Chem) Ltd., Div. of UCB S.A., UCB House, 3 George St., Watford, Hertfordshire, WD18 0UH, UK (Tel: 44 1923 248011; FAX 44 1923 250225; E-mail: richard.scutt@ucb-group.com)

UCB Chemicals Nordic A/S, Tobaksvejen 25, 2, 2860 Soborg, Denmark (Tel: 45 39 66 90 66; FAX 45 3966 69065; E-mail: per.gundersen@ucb-group.com)

Societe Commerciale UCB s.a., Rue Dideerot, 3, BP 314, 92003 Nanterre Cedex, France (Tel: 33 1 47 29 44 68; FAX 33 1 47 25 46 93; E-mail: claudio.eleuteri@ucb-group.com)

UCB GmbH, Huttenstrasse 205, 50170 Kerpen (Sindorf), Germany (Tel: 49 2273 5630; FAX 49 2273 563239; E-mail: Birgit.Offermanns@ucb-group.com)

UCB Quimica Iberica s.a., Plaza del Marques de Salamanca 11, Bajo Izqda, 28006 Madrid, Spain (Tel: 34 91 4264782; FAX 34 91 5766021; E-mail: jose.mendonca@ucb-group.com)

UCB Chemicals Italia S.p.A., Via Monte di Pieta, 1/A, 20121 Milan, Italy (Tel: 39 2 723391; FAX 39 2 8646 0997; E-mail: claudio.eleuteri@ucb-group.com)

UCB Chemicals Corp., Wholly owned subsid. of UCB Group, 2000 Lake Park Dr., Smyrna, GA, 30080, USA (Tel: 770-434-6188; 800-362-1986; FAX 770-434-8314; E-mail: carolyn.mayfield@ucb-group.com)

UCB De Mexico, S.A. de C.V., Antonio M. Rivera No. 1 altos, Centro Industrial Tlalnepantla, Tlalnepantla, Edo de Mexico, 54030, Mexico (Tel: 52 5 565 6798; FAX 52 5 565 8150)

UCB do Brasil Ltda., Av. Magalhaes de Castro 974, Sao Paulo - SP, 05502-001, Brazil (Tel: 55 11 3038 0800; FAX 55 11 3816 2391; E-mail: laura.blanco@ucb-group.com)

UCB Chemicals Australia, 19 Potter St., PO Box 341, Craigieburn, Victoria, 3064, Australia (Tel: 61 3 9303 0900; FAX 61 3 9303 0905; E-mail: Ted.Pearce@ucb-group.com)

Shanghai UCB Speciality Chemicals Co., Ltd., Suite 1508, United Plaza, 1468 Nanjing Xi Road, Shanghai, 200040, P.R. China (Tel: 86 21 62470318; FAX 86 21 62894459; E-mail: jean-yves.dumousseau@ucb-group.com)

Daicel-UCB Co. Ltd., Ochanomizu Kyoun Bldg., 2-2, Kanda-Surugadai, Chiyoda-ku, Tokyo, 101-0062, Japan (Tel: 81 3 5282 4101; FAX 81 3 5282 4102; E-mail: Michio81@mb.infoweb.ne.jp)

UCB Chemicals (Australia), 19 Potter St., PO Box 341, Craigieburn, Victoria, 3064, Australia (Tel: 61 3 9303 0900; FAX 61 3 9303 0905; E-mail: ted.pearce@ucb-group.com)

George Uhe Co., Inc.

12 Rt. 17 North, PO Box 970, Paramus, NJ, 07653-0970, USA (Tel: 201-843-4000; 800-850-4075; FAX 201-843-7517; E-mail: global@uhe.com; Internet: www.uhe.com)

Ultra International Ltd.

304, AVG Bhawan, M-3, Conaught Circus, New Delhi, 110 001, India (Tel: 91 11 332 4100; FAX 91 11 331 776; E-mail: ultra@del2.vsnl.net.in; Internet: www.ultrainternational.com)

Undesa

Undesa (Unión Deriván S.A.), Avda. Geneeralitat 175-179, 08840 Viladecans (Barcelona), Spain (Tel: 34 93 637 35 37; FAX 34 93 659 19 02; E-mail: sales@undesa.com; Internet: www.undesa.com)

Undesa Italia, Via Garibaldi, 10, 40012 Calderara di Reno (BO), Italy (Tel: 39 51 6467200; FAX 39 51 6467298)

Ungerer & Co.

4 Bridgewater Lane, PO Box U, Lincoln Park, NJ, 07035, USA (Tel: 201-628-0600; FAX 201-628-0251)

Unger Fabrikker AS

Postboks 254, 1601 Fredrikstad, Norway (Tel: 47 69 70 82 00; FAX 47 69 32 37 75; E-mail: marunger@unger.no; Internet: www.unger.no/)

Unipex

30 Rue du Fort, B.P. 150, 92500 Rueil Malmaison Cedex, France (Tel: 33 1 47 32 81 30; FAX 33 1 41 96 22 00)

Uniqema

Uniqema, Part of ICI Group of Companies, Buurtje 1, 2802 BE Gouda, The Netherlands (Tel: 31 182 542 911; FAX 31 182 542 255)

Uniqema Chemicals Ltd., Pool Lane, Bebington, Wirral, Merseyside, CH62 4UF, UK (Tel: 44 151 643 3200; FAX 44 151 645 9197)

Uniqema UK, PO Box 90, Wilton Centre, Middlesbrough, Cleveland, TS90 8JE, UK (Tel: 44 1642 432438; FAX 44 1642 437431)

Uniqema, Part of ICI Group of Companies, 30 Queen Anne's Gate, London, SW1H 9AB, UK (Tel: 44 171 798 5101; FAX 44 171 798 5813; Internet: www.uniqema.com)

Uniqema Belgium, Everslaan 45, 3078 Everberg, Belgium (Tel: 32 2 758 9675; FAX 32 2 758 9045)

Uniqema France S.A., 84-88 Blvd. de la Maison Marchand, 92411 Courbevoie Cedex, France (Tel: 33 1 55 70 22 08; FAX 33 1 55 70 22 01)

Uniqema Chemie GmbH, Steintor 9, 46446 Emmerich, Germany (Tel: 49 2822 72456; FAX 49 2822 72289)

Uniqema Italiana S.p.A., Via Bergamo 66, Casella Postale 204, 26100 Cremona, Italy (Tel: 39 372 487267; FAX 39 372 412123)

Uniqema, 1000 Uniqema Blvd., New Castle, DE, 19720-2790, USA (Tel: 302-574-5000; 800-424-2024; FAX 302-574-3525; Internet: www.uniqema.com)

Uniqema, 4650 S. Racine Ave., Chicago, IL, 60609, USA (Tel: 773-376-9000; 800-833-2864; FAX 773-376-0095)

Uniqema, PO Box 425, 76 E. 24th St., Paterson, NJ, 07544, USA (Tel: 973-345-8220; 800-553-6662; FAX 973-345-3527; E-mail: info@monaweb.com)

Uniqema Canada, A Uniqema Business Unit, 26 Easton Rd., Unit 6, Brantford, Ontario, N3T 5T2, Canada (Tel: 519-756-6181; FAX 519-758-8140)

Uniqema Japan, Sin-San Bldg., 5F, 5-10, Shinbashi 3-chome, Minato-ku, Tokyo, 105-0004, Japan (Tel: 81 3 3504 9670; FAX 81 3 3504 9679)

Uniqema (Asia), Level 13A, Menara John Hancock, No. 6 Jalan Gelenggang, Damansara Hts., 50490 Kuala Lumpur, Malaysia (Tel: 60 3 254 7878; FAX 60 3 250 3340)

Uniqema (Korea), 18th Fl., Sam Boo Bldg., 676 Yeoksam-Dong, Kangnam-Ku, Seoul, 135-080, Korea (Tel: 82 2 527 5646; FAX 82 2 527 5650)

Uniqema (Taiwan), 9F-14, No. 189,, 189 Keelung Rd. Sec. 2, Taipei, Taiwan R.O.C. (Tel: 886 2 2377 5222; FAX 886 2 2377 5136)

Uniroyal. *See* Crompton Corp./Uniroyal

United Chemical Technologies, Inc.

2731 Bartram Rd., Bristol, PA, 19007, USA (Tel: 215-781-9255; 800-541-0559; FAX 215-785-1226; E-mail: info@unitedchem.com; Internet: www.unitedchem.com)

United Coconut Chemicals, Inc./Cocochem

UCPB Bldg., 17th Fl., Makati Ave., Makati City, Metro Manila, 1200, Philippines (Tel: 63 2 816 0371; FAX 63 2 817 2251; E-mail: cocchem@ibm.net)

United-Guardian

United-Guardian, Inc., 230 Marcus Blvd, PO Box 18050, Hauppauge, NY, 11788, USA (Tel: 631-273-0900; 800-645-5566; FAX 631-273-0858; E-mail: jvernice@u-g.com; Internet: www.u-g.com)

S. Black (Import & Export) Ltd., Distrib. for United-Guardian, Foxholes Business Park, John Tate Road, Hertford, Hertfordshire, SG13 7YH, UK (Tel: 44 1992 825555; FAX 44 1992 825566; E-mail: enquires@sblack.com; Internet: www.sblack.com)

United Mineral & Chemical Corp., UMC

Div. of ICD Group, 1100 Valley Brook Ave., Lyndhurst, NJ, 07071-3608, USA (Tel: 201-507-3300; 800-777-0505; FAX 201-507-1506; E-mail: inquiry@umccorp.com; Internet: www.umccorp.com)

United Pharma Ltd.

Laettichstr. 6, PO Box 2806, 6342 Baar, Switzerland (Tel: 41 41 768 40 10; FAX 41 41 768 40 11; E-mail: info@upltd.com; Internet: www.upltd.com)

U.S. Borax

U.S. Borax Inc., Member of The RTZ Corp. plc, 26877 Tourney Rd., Valencia, CA, 91355-1847, USA (Tel: 661-287-5400; 800-US BORAX; FAX 800-626-4872; E-mail: kkshen@compuserve.com; Internet: www.borax.com)

Borax Argentina S.A., Huaytiquina 227, 4407 Campo Quijano Salta, Argentina (Tel: 54 387 490 4030; FAX 54 387 490 4031)

Borax South America, Rua Cantagalo 74 cj. 1403, Sao Paulo SP, 03319-000, Brazil (Tel: 55 11 218 0497; FAX 55 11 6941 9891; E-mail: sergio.perrella@borax.com)

Borax Europe Ltd., 170 Priestley Rd., Guildford, Surrey, GU2 5RQ, UK (Tel: 44 1483 734000; FAX 44-1483 457676; E-mail: borax_europe@compuserve.com)

Borax Benelux SA NV, 43, Rue de Namur, 1000 Brussels, Belgium (Tel: 32 2 512 8858; FAX 32 2-514 0697)

Borax España S.A., C.N. 340, Km. 954, Apartado 197, 12520 Nules Castellon, Spain (Tel: 34 964 659 030; FAX 34 964 674 659)

Deutsche Borax GmbH, Postfach 1021, 65836 Sulzbach/Taunus, Germany (Tel: 49 6196 5000 50; FAX 49 6196 5000 60)

Borax Français S.A., 18-20 rue Jacques Dulud, 92521 Neuilly-sur-Seine Cedex, France (Tel: 33 1 41 43 00 60; FAX 33 1 41 43 00 79)

Borax Italia S.r.l., Palazzo Sirio 2, Centro Direzionale Colleoni, 20041 Agrate Brianza (MI), Italy (Tel: 39 39 605 6637; FAX 39 39 609 1855)

Borax Japan Ltd., 10th Fl., Mori Bldg., Shiroyama Hills, 3,1-Toranomon 4-chome, Minato-ku, Tokyo, 105-6010, Japan (Tel: 81 3 5401 2654; FAX 81 3 5473 7889)

Borax Singapore, #08-02 Wheelock Place, 501 Orchard Rd., Singapore, 238880, Singapore (Tel: 65 738 6068; FAX 65 738 6282)

U.S. Borax Beijing, Room 1105, 11/F, East Ocean Center, 24A Jian Guo Men Wai Da Jie, Beijing, 100004, P.R. China (Tel: 86 10 6515 5908; FAX 86 10 6515 5716)

U.S. Chemicals, Inc.

280 Elm St., New Canaan, CT, 06840, USA (Tel: 203-966-8777; FAX 203-966-3577; E-mail: sales@uschemicals.com; Internet: www.uschemicals.com)

U.S. Synthetics Corp.

PO Box 2236, Danbury, CT, 06813, USA (Tel: 860-567-2561; FAX 860-567-2561)

Unitex Chemical Corp.

PO Box 16344, 520 Broome Rd., Greensboro, NC, 27406, USA (Tel: 336-378-0965; FAX 336-272-4312; E-mail: sales@unitexchemical.com; Internet: www.unitexchemical.com)

Univar

Univar NV, Blaak 333, 3011 GB Rotterdam, The Netherlands

Univar E&E Ltd, Lakeside 5500, Cheadle Royal Business Park, Cheadle, Cheshire, SK8 3GR, UK (Tel: 44 161 741 7000; FAX 44 161 741 7047; Internet: www.univar.co.uk)

Univar USA Inc., PO Box 34325, Seattle, WA, 98124-1325, USA (Tel: 425-889-3400; Internet: www.univarusa.com)

Univar Canada Ltd., 9800 Van Horne Way, Richmond, BC, V6X 1W5, Canada (Tel: 604-273-1441; FAX 604-273-2046; Internet: www.vopakcanada.com)

Universal Preserv-A-Chem Inc./UPI

33 Truman Dr. South, Edison, NJ, 08817-2426, USA (Tel: 732-777-7338; FAX 732-777-7885; E-mail: patty@upichem.com (tech. support); Internet: www.upichem.com)

UOP

UOP Inc., 25 E. Algonquin Rd., PO Box 5017, Des Plaines, IL, 60017-5017, USA (Tel: 847-391-2000; 800-348-0832; FAX 847-391-2253; E-mail: unilink@uop.com; molsiv@uop.com; Internet: www.uop.com)

UOP Canada Inc., Suite 1860, 444 5th Ave. S.W., Calgary, Alberta, T2P 2T8, Canada (Tel: 403-777-3750; FAX 403-777-3755)

UOP Ltd., Liongate, Ladymeade, Guildford, Surrey, GU1 1AT, UK (Tel: 44 1483 304 848; FAX 44 1483 304 863)

UOP GmbH, Steinhof 39, 40699 Erkrath, Germany (Tel: 49 211 24903 22; FAX 40 211 249109)

UOP N.V., Noorderlaan 147, 2030 Antwerp, Belgium (Tel: 32 3 5409 911; FAX 32 3 5417 806)

UOP France s.a.r.l., 24, Rue Saarinen Silic 252, 94568 Rungis Cedex, France (Tel: 33 1 41 80 16 60; FAX 33 1 41 80 16 66)

UOP M.S., S.p.A., Viale Milanofiori, Strada 1, Palazzo E1, 20090 Assago MI, Italy (Tel: 39 2 892241; FAX 39 2 5750 0145)

UOP Services Ltd., 5th Fl. Marine House, Rakhamovsky per. 4, Bld. 1, 103051 Moscow, Russia (Tel: 7 95 258 2890; FAX 7 095 258 2989)

UOP Ltd., Al-Jamal St. 22, Mushrefah District 4, Jeddah, 21472, Saudi Arabia (Tel: 966 2 672 7033; FAX 966 2 671 7093)

UOP Middle East Company, Dubai World Trade Centre 25th Fl., PO Box 9248, Dubai, United Arab Emirates (Tel: 971 4 3313 841; FAX 971 4 3317 033)

UOP India Pvt. Ltd./UOP Asia Ltd., Industrial Park Sector-36, Pace City II, Gurgaon, Haryana, 122004, India (Tel: 91 124 6373288-91; FAX 91 124 6373283-84)

UOP Asia Pacific Pte. Ltd., 101 Thomson Rd. #16-03/04, United Square, Singapore, 307591, Singapore (Tel: 65 253 1652; FAX 65 253 0088)

UOP Processes International, Lido Commercial Bldg. A1-514, Jichang-Jiangtai Rd., Beijing, 100004, P.R. China (Tel: 86 10 6437 6766; FAX 86 10 6437 6616)

UOP K.K., NK Bldg. Ohsaki City, 6th Fl., 6-28, Kitashinagawa 5-chome, Shinagawa-ku, Tokyo, 141-001, Japan (Tel: 81 3 5421 2560; FAX 81 3 5421 2788)

Vamo-Fuji Specialities

Kuhlmannlaan 36, 9042 Gent, Belgium (Tel: 32 9 343 0202; FAX 32 9 343 0718; E-mail: sales@vamo-fuji.be)

Vam Organic Chemicals Ltd., R&D

Skyline House, 85-Nehru Place, New Delhi, 110 019, India (Tel: 6465772; FAX 6469088; E-mail: m/m@vamorganic.com; Internet: www.vamorganic.com)

Vanco Trading Inc.

53 Old King Hwy. N., Ste. 204, Darien, CT, 06820, USA (Tel: 203-656-2800; FAX 203-655-8307; E-mail: vanco.trading@prodigy.net)

Vandemoortele Bakery Products

Ottergemsesteenweg Zuid 806, 9000 Gent, Belgium (Tel: 32 9 240 1711; FAX 32 9 240 1876; E-mail: it.vamix@vandermoortele.com)

Van Den Burg Eiprodukten B.V.

Sluisweg 20, 5140 AE Waalwyk, The Netherlands (Tel: 31 416 671400; FAX 31 416 335285; E-mail: mail@cemave-eggproducts.nl)

R.T. Vanderbilt

R.T. Vanderbilt Co. Inc., 30 Winfield St, PO Box 5150, Norwalk, CT, 06856-5150, USA (Tel: 203-853-1400; 800-243-6064; FAX 203-853-1452; E-mail: sales@rtvanderbilt.com; Internet: www.rtvanderbilt.com)

R.T. Vanderbilt Co. (Canada), 1 Eva Rd., Suite 419, Etobicoke, Ontario, M9C 4Z5, Canada (Tel: 416-626-6027; FAX 416-626-6139; E-mail: rtvcan@rtvanderbilt.com)

Varsal Instruments, Inc.

363 Ivyland Rd., Warminster, PA, 18974, USA (Tel: 215-957-5880; FAX 215-957-9111; E-mail: info@varsal.com; Internet: www.varsal.com)

Veckridge Chemical Co. Inc.

60-70 Central Ave., South Kearny, NJ, 07032, USA (Tel: 973-344-1818; FAX 973-690-5936)

Veitsiluoto Oy/Forest Chemicals Industry

PO Box 196, 901010 Oulu, Finland (Tel: 358 81 316 3111; FAX 358 81 378 5755)

Velsicol Chemical

Velsicol Chemical Corp., 10400 W. Higgins Road, Suite 600, Rosemont, IL, 60018, USA (Tel: 847-298-9000; 800-843-7759; FAX 847-298-9014; E-mail: comms@velsicol.com; Internet: www.velsicol.com)

Velsicol Chemical Ltd., Viewpoint, Basing View, Basingstoke, Hampshire, RG21 4RG, UK (Tel: 44 1256 799766; FAX 44 1256 799767)

Veltek Associates Inc.

1039 W. Bridge St., Phoenixville, PA, 19460-4218, USA (Tel: 610-983-4949; FAX 610-983-9494)

Venchem Ltd.

Knotts Lane Works, Colne, Lancashire, BB8 8AA, UK (Tel: 44 1282 861198; FAX 44-1282-860020; E-mail: sales@venchem.co.uk; Internet: www.venchem.co.uk)

Venus Ethoxyethers Pvt. Ltd.

Hari Nivas, 1st Fl., 19 Mathew Rd., Mumbai, 400 004, India (Tel: 91 22 363 0466; FAX 91 22 361 1285; E-mail: info@venus-goa.com; Internet: www.venus-goa.com)

Vertex Chemical Corp.

9909 Clayton Rd., Suite 219, St. Louis, MO, 63124, USA (Tel: 314-991-4005)

Vevy Europe SpA

Via Semeria 18, 16131 Genova, Italy (Tel: 39 10 522 1515; FAX 39 10 522 1212; E-mail: info@vevy.com; Internet: www.vevy.com)

Vilax Corp.

33 Roberts St., Rockaway, NJ, 07866, USA (Tel: 973-625-2755; FAX 973-625-3543; E-mail: vilaxcorp@aol.com)

Vinchem Inc.

301 Main St., PO Box 639, Chatham, NJ, 07928, USA (Tel: 973-635-4841; FAX 973-635-1459)

Viobin

Viobin Corp., Subsid. of American Home Prods., 226 West Livingston, Monticello, IL, 61856, USA (Tel: 217-762-2561; FAX 217-762-2489; E-mail: viobin@monticello.net)

Viobin Corp., Subsid. of American Home Prods., 700 East Main Street, PO Box 158, Waunakee, WI, 53597, USA (Tel: 608-849-5944; FAX 608-849-4053)

Virginia Dare Extract Co., Inc.

882 Third Ave., Brooklyn, NY, 11232, USA (Tel: 718-788-1776; 800-847-4500; FAX 718-768-3978; Internet: www.virginiadare.com)

Vitamins, Inc.

200 E. Randolph Dr., Chicago, IL, 60601-7799, USA (Tel: 312-861-0700; FAX 312-861-0708)

Vivion Inc.

929 Bransten Rd., San Carlos, CA, 94070, USA (Tel: 650-595-3600; FAX 650-595-2094)

Vliegenthart B.V.

Postbus 6330, 4000 HH Tiel, The Netherlands (Tel: 31 344 633336; FAX 31 344 631616; E-mail: info@vliegenthart.com; Internet: www.vliegenthart.com)

Voigt Global Distribution LLC

PO Box 412762, Kansas City, MO, 64141-2762, USA (Tel: 816-471-9500; FAX 816-471-9502; E-mail: sales@voigtglobal.com; Internet: www.vgdllc.com)

Vopak USA. *See under* Royal Vopak

Vulcan Performance Chemicals

Vulcan Performance Chemicals, 1200 Urban Center Dr., Birmingham, AL, 35242, USA (Tel: 205-298-3000; 800-633-8280; FAX 205-298-2959; E-mail: contactus@vul.com; Internet: www.vul.com)

Vulcan Performance Chemicals Co., A Business Unit of Vulcan Materials Co., PO Box 9577, Columbus, GA, 31908-9577, USA (Tel: 706-576-2000; FAX 706-576-6455; Internet: www.vul.com)

VWR International

Company of the Merck Group, 1310 Goshen Pkwy., West Chester, PA, 19380, USA (Tel: 610-431-1700; 800-932-5000; FAX 610-431-9174; E-mail: solutions@vwrsp.com; Internet: www.vwrsp.com)

Wacker

Wacker-Chemie GmbH, Hanns-Seidel-Platz 4, 81737 München, Germany (Tel: 49 89 6279 01; FAX 49 89 6279 1770; E-mail: info@wacker.com; Internet: www.wacker.de)

Wacker Chemicals Ltd., Div. of Wacker-Chemie GmbH, 85 High St., Egham, Surrey, TW20 9HF, UK (Tel: 44 1784 487800; FAX 44 1784 487870; E-mail: info.uk@wacker.com; Internet: www.wacker.co.uk)

Wacker Chemicals Ltd./Dublin, Embassy House, Herbert Park Lane, Ballsbridge, Dublin, 4, Ireland (Tel: 353 1 637 5040; FAX 353 1 637 5050; E-mail: info.ireland@wacker.com; Internet: www.wacker.com/ir/)

Wacker-Chemie Italia SpA, Via L. Einaudi 6, 20068 Peschiera Borromeo (MI), Italy (Tel: 39 2 5 16 91 1; FAX 39 2 5 16 91 2 99)

Wacker-Chemie GmbH, Mariahilfer Strasse 54/Top 7+8, 1070 Wien, Austria (Tel: 43 1 5 36 52 0; FAX 43 1 5 36 52 33)

Wacker Chimie S.A., 99, Cours Gambetta, 69446 Lyon Cedex 03, France (Tel: 33 4 72 61 03 00; FAX 33 4 78 95 27 45; E-mail: info.france@wacker.com)

Wacker-Chemie Benelux B.V., Heiligeweg 166, 1561 DM Krommenie, The Netherlands (Tel: 31 75 647 6000; FAX 31 75 6 21 50 61; E-mail: info.netherlands@wacker.com)

Wacker Chemicals (USA) Inc., 3301 Sutton Rd., Adrian, MI, 49221, USA (Tel: 800-485-3686; FAX 517-264-8137)

Wacker Silicones Corp., Wholly owned subsid. of Wacker-Chemie GmbH, 3301 Sutton Rd., Adrian, MI, 49221-9397, USA (Tel: 517-264-8500; 800-248-0063; FAX 517-264-8246; E-mail: customercare@wackersilicones.com; Internet: www.wackersilicones.com)

Wacker Méxicana, S.A. de C.V., Calle Picacho-Ajusco No. 130-102, Col. Jardines en la Montana - Tlalpan, Mexico D.F., 14210, Mexico (Tel: 52 55 5446 4433; FAX 52 55 5630 6861)

Wacker Quimica do Brasil Ltda., Rua Municipal no. 100, Jd. Alvorada, 06612-060 Jandira - Sao Paulo, Brazil (Tel: 55 11 4789 8300; FAX 55 11 4789 8356; E-mail: info.brazil@wacker.com)

Wacker Chemicals Australia Pty. Ltd., Suite 3, 11 Leicester Ave., Glen Waverly, Victoria, 3150, Australia (Tel: 61 3 98 02 61 00; FAX 61 3 98 02 63 11; E-mail: info.australia@wacker.com)

Wacker Chemicals East Asia Ltd., TOHMA Nishi-Waseda Bldg., 2-14-1 Nishi-Waseda, Shinjuku-ku, Tokyo, 169-0051, Japan (Tel: 81 3 52 72 31 25; FAX 81 3 52 72 31 40)

Wacker Chemicals China Ltd., 6th Fl., HP Bldg., No. 112 Jian Guo Lu, Chaoyang District, Beijing, 100022, P.R. China (Tel: 86 10 65 66 83 40; FAX 86 10 65 66 83 63; Internet: www.wacker-stc.com.cn)

Wacker Chemicals Hong Kong Ltd., Suite 2408, 24/F, Nat West Tower, Times Sq., 1 Matheson St., Causeway Bay, Hong Kong (Tel: 852 2506 3228; FAX 852 2506 3280; Internet: www.wacker-stc.com.cn)

Wacker Chemicals (South Asia) Pte. Ltd., Jl. Jend. Sudiman 58, 7th Fl., Graha Niaga, Jakarta, 12190, Indonesia (Tel: 62 21 250 5528; FAX 62 21 250 5529)

Wacker Chemicals Korea Ltd., 15 Fl., Star Tower Bldg., 737 Yeoksam-dong, Kangnam-ku, Seoul, 135-080, Korea (Tel: 82 2 2112 1201; FAX 82 2 2112 1203; Internet: www.wacker.co.kr)

Wako

Wako Pure Chemical Industries Ltd., 3-1-2 Dosho-machi, Chuo-ku, Osaka, 540-8605, Japan (Tel: 81 6 6203 3741; FAX 81 6 6222 1203; Internet: www.wako-chem.co.jp/english/)

Wako Chemicals USA, Inc., 1600 Bellwood Rd., Richmond, VA, 23237, USA (Tel: 804-271-7677; FAX 804-271-7791)

Wako Chemicals GmbH, Div. of Wako Pure Chemical Industries Ltd., Nissanstrasse 2, 41468 Neuss, Germany (Tel: 49 2131 311 0; FAX 49 2131 311 100)

Wall Chemical Corp.

PO Box 583, Westfield, NJ, 07091, USA (Tel: 908-233-8188; FAX 908-233-8265; E-mail: sales@wallchemical.com; Internet: www.wallchemical.com)

Georges Walther AG

Grindelstrasse 6, 8304 Wallisellen, Switzerland (Tel: 41 839 20 20; FAX 41 839 20 10; E-mail: info@gwzag.ch; Internet: www.webyourworld.com/gwz/)

Wego Chemical & Mineral Corp.

265 Great Neck Rd., Great Neck, NY, 11021, USA (Tel: 516-487-3510; FAX 516-487-3794; E-mail: sales@wegochem.com; Internet: www.wegochem.com)

Welch, Holme & Clark Co. Inc.

7 Ave. L, Newark, NJ, 07105, USA (Tel: 973-465-1200; FAX 973-465-7332; E-mail: whc@welch-holme-clark.com; Internet: www.welch-holme-clark.com)

West Agro, Inc.

Part of DeLaval Organization, 11100 N. Congress Ave., Kansas City, MO, 64153-1296, USA (Tel: 816-891-1600; 800-421-1905; FAX 816-891-1595; E-mail: carol.moss@delaval.com; Internet: www.westagro.com)

Westco Chemicals, Inc.

12551-61 Saticoy St. South, North Hollywood, CA, 91605, USA (Tel: 818-255-3655; FAX 818-255-3650; E-mail: ddraney@westcochemicals.com; Internet: www.westcochemicals.com)

Westfalia Separator Inc.

100 Fairway Ct., Northvale, NJ, 07647, USA (Tel: 201-767-3900; 800-722-6622)

Westhove SA

39 rue Loucheur, BP 73, 62510 Arques, France (Tel: 33 3 21 88 38 38; FAX 33 3 21 88 38 39; E-mail: contact@westhove.com; Internet: www.westhove.com)

Westlake CA & O Corp.

2468 Industrial Pkwy., PO Box 527, Calvert City, KY, 42029, USA (Tel: 270-395-3260; 800-321-8550)

Westlake Plastics Co.

PO Box 127, W. Lenni Rd., Lenni, PA, 19052, USA (Tel: 610-459-1000; 800-999-1700; FAX 610-459-1084; E-mail: westlakeplastics@westlakeplastics.com; Internet: www.westlakeplastics.com)

Westpro

PO Box 927833, San Diego, CA, 92192-7833, USA (Tel: 866-412-9973; FAX 855-277-6726; E-mail: news@westprochem.com; Internet: www.westprochem.com)

Westtec Industries Inc.

PO Box 2595, 1161 Park Ave., Plainfield, NJ, 07060, USA (Tel: 908-769-9009; FAX 908-757-1010; E-mail: westteck@aol.com)

Weyerhaeuser Co.

33663 Weyerhaeuser Way South, Federal Way, WA, 98003, USA (Tel: 253-924-2345; FAX 253-924-3543)

White Cross Laboratories, Inc.

350 Theodore Fremd, Rye, NY, 10580, USA (Tel: 914-921-0600; FAX 914-921-0971)

Peter Whiting (Chemicals) Ltd.

1 Oil Mill Lane, Hammersmith, London, W6 9UA, UK (Tel: 44 20 8741 4025; FAX 44 20 8741 1737; E-mail: sales@whiting-chemicals.co.uk; Internet: www.whiting-chemicals.co.uk)

Whittaker, Clark & Daniels

Whittaker, Clark & Daniels, Member of Brenntag Group, 1000 Coolidge St., South Plainfield, NJ, 07080, USA (Tel: 908-561-6100; 800-732-0562; FAX 800-833-8139; E-mail: customerservice@wcdinc.com; Internet: www.wcdinc.com)

Whittaker, Clark & Daniels (Canada), 135 Bethridge Rd., Rexdale, Ontario, M9W 1N4, Canada (Tel: 416-744-6580; FAX 416-744-6582)

WCD International Sarl, 19, Rue Honore d'Estienne d'Orves, 93500 Pantin, Ontario, France (Tel: 33 1 41 83 60 00; FAX 33 1 41 83 60 01)

Whyte Chemicals Limited

Marlborough House, 298 Regents Park Rd., Finchley, London, N3 2UA, UK (Tel: 44 20 8346 5946; FAX 44 20 8349 4589; E-mail: sales@whytechemicals.co.uk; Internet: www.whytechemicals.co.uk)

Wilbur-Ellis Co.

345 California St., 27th Fl., San Francisco, CA, 94104, USA (Tel: 415-772-4000; E-mail: contactus@wecon.com; Internet: www.wilburellis.com)

Wilke International Inc.

14321 West 96th Terrace, Lanexa, KS, 66215-2052, USA (Tel: 913-438-5544; 800-779-5545; FAX 913-438-5544; E-mail: info@wilkeinternational.com; Internet: www.wilkeinternational.com)

Wilshire Chemical Co., Inc.

15324 S. Broadway, Gardena, CA, 90248, USA (Tel: 310-323-9232; FAX 310-770-4910; E-mail: WilshrChem@aol.com; Internet: http://users.aol.com/WilshrChem/Wilshire.htm)

J.C. Wilson Chemicals Ltd.

1900 Huron St., London, Ontario, N5V 4A3, Canada (Tel: 519-438-2433; FAX 519-438-6960; E-mail: wilsonchemicals@sympatico.ca; Internet: www.wilson chemicals.com)

Witco. *See* Crompton/Witco Refined Products

World Wide Chemnet, Inc. DEFUNCT

LAST KNOWN: 2100 S. Utica, Suite 200, Tulsa, OK, 74114, USA (Tel: 918-749-9056; 800-988-5782; FAX 918-747-1444)

Wright Corp.

PO Box 9009, 102 Orange St., Wilmington, NC, 28402, USA (Tel: 910-251-8952; FAX 910-762-9223)

WSI Industries L.L.C./WSI Chemical, Inc.

Div. of Washtech Systems, 300 John Q. Hammons Pkwy., Springfield, MO, 65806, USA (Tel: 417-865-2862; FAX 417-865-6978; E-mail: fwellman@solvent replacement.com; Internet: www.wsi-chem-sys.com)

Wutong Aroma Chemicals Co., Ltd.

Wenchang Rd. #151, Tengzhou, Shandong, 277500, China (Tel: 86 632 5583267; FAX 86 632 5571068; E-mail: tzwt@wu-tong.com; Internet: www.wu-tong.com)

Wuzhou International Co., Ltd.

6/F, West Block of Financial Bldg., 182 Qingchun Rd., Hangzhou, P.R. China (Tel: 86 571 8724 4967; FAX 86 571 8724 4707; E-mail: josephq@mail.hz.zj.cn)

Wychem Limited

Bury Rd., Stradishall, Near Newmarket, Suffolk, CB8 8YN, UK (Tel: 44 1440 820 338; FAX 44 1440 820 339; E-mail: terryr@wychem.freeserve.co.uk; Internet: www.wychem.co.uk)

Xinchem Company

204/5, 100 Fushan Rd., Pudong, Shanghai, 200120, P.R. China (Tel: 86 21 58203399; FAX 86 21 68671424; E-mail: info@finechemnet.com; Internet: www.finechemnet.com)

Yanagishima Pharmaceutical Co., Ltd.

2-5-5, Nihonbashi Muro-machi, Chuo-ku, Tokyo, 103, Japan (Tel: 81 3 3241 4584; FAX 81 3 3241 0309)

Yixing Fangqiao East Chemical

Yixing FangQiao Town, China (Tel: 86 510 7581290; FAX 86 510 7583694; E-mail: suming@public1.wx.js.cn; Internet: www.eastchemicals.com)

Yokkaichi Chemical Co., Ltd.

Joint venture of Mitsubishi Petrochem., Daiichi Kogyo Seiyaku, Lion Corp., 2-1, Miyahigashi-cho, Yokkaichi-shi, Mie, 510, Japan (Tel: 81 593 45 1161; FAX 81 593 45 1168)

Yorkshire

Yorkshire Chemicals plc, 27 Kirkstall Rd, Leeds, West Yorkshire, LS3 1LL, UK (Tel: 44 1132 443111; FAX 44 1132 421670; Internet: www.yorkshireamericas.com)

Yorkshire Americas, Wholly owned subsid. of Yorkshire Group plc, 3001 North Graham St., Charlotte, NC, 28206, USA (Tel: 704-372-5890; 800-438-4122; FAX 704-376-3091; E-mail: web@yorkshireamericas.com)

Yorkshire Americas Inc., 105 Wood St., Greenville, SC, 29611, USA (Tel: 864-295-1170; FAX 864-295-3851; Internet: www.yorkshireamericas.com)

Yorkshire Taiwan Ltd., 11F-3, No. 87, Sung Chiang Rd., Tapei, 104, Taiwan R.O.C. (Tel: 2 2506 0337; FAX 2 2506 7572; E-mail: yorkshire_tw@yorktw.com.tw)

Zaclon Inc.

2981 Independence Rd., Cleveland, OH, 44115, USA (Tel: 216-271-1715; 888-854-1605; FAX 216-271-1792; E-mail: zaclon@zaclon.com; Internet: www.zaclon.com)

Zeeland Chemicals, Inc.

A Cambrex Co., 215 N. Centennial St., Zeeland, MI, 49464, USA (Tel: 616-772-2193; 800-223-0453; FAX 616-772-7344; E-mail: jayne.perkins@cambrex.com; Internet: www.cambrex.com)

ZenPharm International Ltd.

9 Ammolite, Rancho Santa Margarita, CA, 92688, USA (Tel: 949-709-2709; FAX 949-589-6954; E-mail: junping@aol.com; Internet: www.zenpharm.com)

Zeochem

Owned by Chemie Uetikon AG, 1314 S. 12 St., PO Box 35940, Louisville, KY, 40232, USA (Tel: 502-634-7600; 800-626-5355; FAX 502-634-8133; E-mail: kdicks@zeochem.com; Internet: www.zeochem.com)

Zetapharm, Inc.

70 W. 36th St., Suite 600, New York, NY, 10088, USA (Tel: 212-643-2310; FAX 212-643-2316; E-mail: zeta@zetapharm.com; Internet: www.zetapharm.com)

Zhejiang Grand Chembay Co., Ltd.

P.R. China (Tel: 86 574 8773 1342; FAX 86 574 8770 0725; E-mail: webmaster@chembay.com; Internet: www.chembay.com)

Zhejiang Sanhuan Chemicals Co., Ltd.

No. 1007, Jiu Ling Xi Rd., Yongkang, Zhejiang, 321300, P.R. China (Tel: 86 579 7271588; FAX 86 579 7271589; E-mail: ykshc@mail.jhptt.zj.cn; Internet: www.sanhuanchemicals.com)

Zinkan Enterprises, Inc.

10574 Ravenna Rd., Twinsburg, OH, 44087, USA (Tel: 330-487-1500; 800-229-6801; FAX 330-425-8202; E-mail: sales@zinkan.com; Internet: www.zinkan.com)

Appendixes

CAS Number Index

75-69-4
Trichlorofluoromethane

75-71-8
Dichlorodifluoromethane

75-72-9
Chlorotrifluoromethane

75-73-0
Tetrafluoromethane

75-75-2
Methanesulfonic acid

75-76-3
Tetramethylsilane

75-83-2
Neohexane

75-84-3
2,2-Dimethylpropanol

75-85-4
t-Amyl alcohol

75-89-8
2,2,2-Trifluoroethanol

76-01-7
Pentachloroethane

76-03-9
Trichloroacetic acid

76-05-1
Trifluoroacetic acid

76-12-0
1,1,2,2-Tetrachloro-1,2-difluoroethane

76-13-1
Genesolv® D
Trichlorotrifluoroethane

76-14-2
1,2-Dichlorotetrafluoroethane

76-15-3
Chloropentafluoroethane

76-22-2
Camphor

76-37-9
2,2,3,3-Tetrafluoro-1-propanol

76-49-3
Bornyl acetate

77-71-4
DM hydantoin

77-75-8
Methyl pentynol

77-76-9
Dimethoxypropane

77-89-4
Acetyl triethyl citrate
Citroflex® A-2
Citrofol® AII

77-90-7
Acetyl tributyl citrate
Citroflex® A-4
Citrofol® BII

77-93-0
Citroflex® 2
Citrofol® AI
Triethyl citrate

77-94-1
Citroflex® 4
Citrofol® BI
Tributyl citrate

78-00-2
Tetraethyl lead

78-06-8
Dibutyltin β-mercaptopropionate

78-32-0
Tri-p-tolyl phosphate

78-40-0
Eastman® TEP
Eastman® Triethyl Phosphate
Triethyl phosphate

78-42-2
Reomol® TOP
Trioctyl phosphate

78-46-6
Dibutyl butyl phosphonate

78-50-2
Cyanex® 921
Trioctylphosphine oxide

78-51-3
Tributoxyethyl phosphate

78-59-1
Isophorone
Ucar® Isophorone

78-66-0
Dimethyl octynediol

78-78-4
Isopentane

78-83-1
Isobutyl alcohol
Ucar® Isobutanol

78-84-2
2-Methylpropanal

78-87-5
Propylene dichloride

78-92-2
2-Butanol

78-93-3
ExxonMobil MEK
Methyl ethyl ketone

78-97-7
Lactonitrile

79-00-5
1,1,2-Trichloroethane

79-01-6
Altene®
Hi-Tri
Neu-Tri E
PPG Trichlor
Tavoxene®
Trichloroethylene
Tristabil®

79-04-9
Chloracetyl chloride

79-16-3
N-Methylacetamide

79-20-9
Eastman® Methyl Acetate, High Purity
Methyl acetate

79-24-3
NE™
NE™ 101
Nitroethane

79-27-6
Tetrabromoethane

79-29-8
2,3-Dimethylbutane

79-31-2
Isobutyric acid

79-33-4
Lactic acid (L)

79-34-5
1,1,2,2-Tetrachloroethane

79-46-9
2-Nitropropane

80-56-8
α-Pinene

80-73-9
Dimethyl imidazolidinone

83-32-9
Acenaphthene

84-66-2
Diethyl phthalate
Eastman® DEP
Eastman® DEP, Fragrance Grade
Eastman® Diethyl Phthalate

84-69-5
Diisobutyl phthalate
Uniplex 155

84-74-2
Dibutyl phthalate
Eastman® DBP
Uniplex 150

84-78-6
Butyl octyl phthalate

85-01-8
Phenanthrene

85-61-0
Coenzyme A

85-68-7
Butyl benzyl phthalate

85-71-2
Methyl phthalyl ethyl glycolate

86-73-7
Fluorene

87-68-3
Hexachloro-1,3-butadiene

87-91-2
Diethyl tartrate

88-16-4
o-Chlorobenzotrifluoride

88-41-5
2-t-Butylcyclohexyl acetate

89-78-1
Menthol (DL)

89-83-8
Thymol

90-12-0
α-Methylnaphthalene
MN

90-13-1
1-Chloronaphthalene

91-17-8
Decahydronaphthalene
Decalin
Dekalin®

91-20-3
Naphthalene

91-22-5
Quinoline

91-58-7
2-Chloronaphthalene

92-50-2
Phenylethylethanolamine

92-52-4
Biphenyl

93-58-3
Methyl benzoate

93-89-0
Ethyl benzoate

94-51-9
Dipropylene glycol dibenzoate

94-96-2
Ethyl hexanediol

95-47-6
o-Xylene

95-48-7
o-Cresol

95-49-8
o-Chlorotoluene

95-50-1
o-Dichlorobenzene

95-51-2
o-Chloroaniline

95-63-6
1,2,4-Trimethylbenzene

95-65-8
3,4-Xylenol

95-73-8
2,4-Dichlorotoluene

95-87-4
2,5-Xylenol

107-15-3
Ethylenediamine (anhyd.)

107-18-6
Allyl alcohol

107-19-7
Propargyl alcohol

107-21-1
Ethylene glycol

107-31-3
Methyl formate

107-41-5
Hexylene glycol

107-46-0
Chemsil DM 0.65
Hexamethyldisiloxane

107-51-7
Octamethyltrisiloxane

107-54-0
Dimethyl hexynol

107-70-0
4-Methoxy-4-methylpentanone-2

107-83-5
Calumet Isohexane
Exxsol® Isohexane
Isohexane

107-84-6
Isoamyl chloride

107-87-9
Methyl propyl ketone

107-88-0
Butylene glycol

107-98-2
Arcosolv® PM
Dowanol® PM
Eastman® PM
Methoxyisopropanol
Methyl Proxitol
Poly-Solv® MPM
Solvenon® PM

108-01-0
Dimethylethanolamine

108-03-2
NiPar S-10™
1-Nitropropane

108-08-7
2,4-Dimethylpentane

108-10-1
Eastman® MIBK
Methyl isobutyl ketone
Ucar® Methyl Isobutyl Ketone

108-11-2
Methyl amyl alcohol

108-18-9
Diisopropylamine

108-20-3
Isopropyl ether

108-21-4
Eastman® Isopropyl Acetate
Isopropyl acetate
Ucar® Isopropyl Acetate

108-29-2
γ-Valerolactone

108-32-7
Arconate® PC-1000
Arconate® PC-HP
Jeffsol® PC
PCHP
Propylene carbonate
Solvenon® PC
UltraPure™ Propylene Carbonate

108-38-3
m-Xylene

108-48-5
2,6-Lutidine

108-65-6
Arcosolv® PMA
Dowanol® PMA
Eastman® PM Acetate
Methyl Proxitol Acetate
Propylene glycol methyl ether acetate

108-67-8
1,3,5-Trimethylbenzene

108-70-3
1,3,5-Trichlorobenzene

108-82-7
2,6-Dimethyl heptanol-4

108-83-8
Diisobutyl ketone
Ucar® Diisobutyl Ketone

108-84-9
Methyl amyl acetate

108-86-1
Bromobenzene

108-87-2
Methyl cyclohexane

108-88-3
Shell Toluene Solvent
Toluene

108-89-4
γ-Picoline

108-90-7
Chlorobenzene

108-91-8
Cyclohexylamine

108-93-0
Cyclohexanol
Hexalin®
Naxol®

108-94-1
Cyclohexanone
Nadone®

108-95-2
Phenol

108-99-6
β-Picoline

109-02-4
p-Methyl morpholine

109-06-8
α-Picoline

109-21-7
Butyl butyrate

109-38-6
Butoxyethyl stearate

109-59-1
Ethylene glycol isopropyl ether
i-PGE
2-Isopropoxyethanol
Isopropyl Oxitol

109-60-4
Eastman® n-Propyl Acetate
Propyl acetate
Ucar® n-Propyl Acetate

109-66-0
Exxsol® Isopentane S
Exxsol® Pentane 80
Hydrosol Pentane
n-Pentane

109-69-3
Butyl chloride

109-73-9
n-Butylamine

109-74-0
Butyronitrile

109-78-4
Ethylene cyanohydrin

109-79-5
n-Butyl mercaptan

109-83-1
N-Methylethanolamine

109-86-4
Hisolve MC
Methoxyethanol
Methyl Cellosolve®
MGE
Poly-Solv® EM

109-87-5
Methylal

109-89-7
Diethylamine

109-94-4
Ethyl formate

109-99-9
Lyondell Tetrahydrofuran (THF)
Tetrahydrofuran

110-00-9
Furan

110-01-0
Tetrahydrothiophene

110-02-1
Thiophene

110-12-3
Methyl isoamyl ketone

110-13-4
2,5-Hexanedione

110-19-0
Eastman® Isobutyl Acetate
Isobutyl acetate
Ucar® Isobutyl Acetate

110-27-0
Crodamol IPM
Estol IPM
Estol IPM 1509 (BIO-IPM)
HallStar™ IPM
Isopropyl myristate
Lexol® IPM
Lexol® IPM-NF
Polypax IPM
Promyr
STEPAN® IPM
STEPAN® IPM-NF
Super Refined® Crodamol IPM
Tegosoft® M
Thomil IS
Thomil IS-F
ZELEC® IPM

110-34-9
Isobutyl palmitate

110-38-3
Ethyl decanoate

110-42-9
Estasol™ MD10
Methyl caprate

110-43-0
Methyl n-amyl ketone

110-49-6
Methoxyethanol acetate
Methyl Cellosolve® Acetate

110-54-3
Calumet Hexane
Exxsol® Methylpentane Naphtha
Hexane

110-58-7
Pentylamine

110-63-4
1,4-Butanediol

110-65-6
But-2-yne-1,4-diol

110-67-8
3-Methoxypropionitrile

110-71-4
Ethylene glycol dimethyl ether
Monoglyme

110-80-5
Cellosolve®
EGE
Ethoxyethanol
Ethyl Icinol

110-82-7
Cyclohexane

110-86-1
Pyridine

110-88-3
Trioxane

110-89-4
Piperidine

110-91-8

N,N-Dimethylaniline

122-32-7
Arlypon SO
Triolein
Trioleina

122-62-3
Dioctyl sebacate
HallStar™ DOS

122-79-2
Phenyl acetate

122-99-6
Emery® 6705
Phenoxyethanol
Rewopal® MPG 10
Rewopal® MPG 40

123-17-1
2,6,8-Trimethylnonyl-4-alcohol

123-18-2
Ecosoft™ Solvent IK
Isobutyl heptyl ketone

123-19-3
4-Heptanone

123-25-1
Diethyl succinate

123-31-9
Hydroquinone

123-35-3
Myrcene

123-38-6
Propionaldehyde

123-39-7
N-Methylformamide

123-42-2
Diacetone alcohol
Ucar® Diacetone Alcohol

123-51-3
Isoamyl alcohol

123-54-6
Acetylacetone

123-63-7
Paraldehyde

123-66-0
Ethyl caproate

123-72-8
n-Butyraldehyde

123-86-4
n-Butyl acetate
Eastman® n-Butyl Acetate, Urethane Grade

123-91-1
1,4-Dioxane

123-92-2
Isoamyl acetate

123-94-4
Glyceryl stearate (pure grade)
Monestriol GP
Monestriol GP-35
Monestriol GP-40
Monestriol GP-40-DC
Monestriol GP-45-DC
Monestriol GP-60
Monestriol GP-AS
Protachem™ GMS 450
Protachem™ HMS

123-95-5
Butyl stearate
Exceparl BS
HallStar™ BST
Monestriol B
Norfox® B-54

123-96-6
2-Octanol

124-10-7
Emery® 2214
Methyl myristate

124-16-3
1-Butoxyethoxy-2-propanol

124-17-4
Butyl Carbitol® Acetate
Diethylene glycol butyl ether acetate
Eastman® DB Acetate

124-18-5
n-Decane

124-29-8
Cetyl alcohol

124-40-3
Dimethylamine

126-14-7
Sucrose octaacetate

126-33-0
Sulfolane
Sulfolane W

126-63-6
Dioctyl octylphosphonate

126-73-8
Albrite® TBP
Albrite® Tributyl Phosphate
Amgard® $TBPO_4$
Phosflex® 4
Reomol® TBP
TBP
Tributyl phosphate

126-92-1
Sodium 2-ethylhexyl sulfate

127-06-0
Acetone oxime

127-18-4
Dowper
Isoform
Perchloroethylene
Pertene®
PPG Perchlor

127-19-5
Dimethyl acetamide

127-25-3
Methyl abietate

127-63-9
Diphenyl sulfone

128-37-0
BHT

131-11-3
Dimethyl phthalate
Uniplex 110

132-34-9
Dibenzofuran

134-62-3
Diethyl toluamide

135-98-8
s-Butylbenzene

136-60-7
Butyl benzoate

137-03-1
2-Heptylcyclopentanone

137-32-6
2-Methyl-1-butanol

138-22-7
Butyl lactate

138-86-3
Dipentene No. 122
dl-Limonene

139-44-6
Trihydroxystearin

139-45-7
Glyceryl tripropionate

139-87-7
N-Ethyldiethanolamine

140-03-4
Methyl acetyl ricinoleate
Naturechem® MAR

140-11-4
Benzyl acetate

140-25-0
Benzyl laurate

140-31-8
Aminoethylpiperazine

141-04-8
Diisobutyl adipate
HallTress™ DIBA Special

141-08-2
Glyceryl ricinoleate

141-43-5
Ethanolamine

141-70-8
2-(2,2-Dimethylpropyl)-4,4-dimethylpent-1-ene

141-78-6
Eastman® Ethyl Acetate, Food Grade
Ethyl acetate

141-79-7
Mesityl oxide

141-94-6
Hexetidine

141-97-9
Eastman® EAA
Ethylacetoacetate

142-16-5
Bis (2-ethylhexyl) maleate
HallStar™ DOM

142-18-7
Glyceryl laurate
Protachem™ MLD

142-26-7
Acetamide MEA
Foamid AME-70
Foamid AME-75
Foamid AME-100
Incromectant AMEA-70
Upamide ACMEA

142-55-2
Imwitor® 412
Propylene glycol laurate

142-77-8
Butyl oleate
Tilol B

142-82-5
Exxsol® Heptane Anti-Static
Heptane
Hydrosol Heptane

142-84-7
Dipropylamine

142-91-6
Crodamol IPP
Isopropyl palmitate
Palmil IS
Polypax IPP
Propal NF
STEPAN® IPP
STEPAN® IPP-NF
Tegosoft® P

142-92-7
Hexyl acetate

142-96-1
Dibutyl ether

143-22-6
Butoxytriglycol

143-24-8
Dimethoxytetraglycol

143-28-2
Novol NF
Oleyl alcohol

144-19-4
2,2,4-Trimethyl-1,3-pentanediol

149-57-5
2-Ethylhexoic acid

150-76-5
Hydroquinone monomethyl ether

151-21-3
Sodium lauryl sulfate

156-59-2
cis-1,2-Dichloroethylene

156-60-5
trans-1,2-Dichloroethylene
VersaTRANS™ Solvent

1,3-Diisoamyl glycerol diether

627-83-8
Glycol distearate
Kemester® EGDS
Protachem™ EGDS

627-91-8
Methyl adipate

627-92-9
1-Isoamyl glycerol ether

627-93-0
DBE-6
Dimethyl adipate
Santosol® DMA

628-63-7
Amyl acetate
Ucar® Primary Amyl Acetate

629-11-8
Hexamethylene glycol

629-14-1
Ethylene glycol diethyl ether
Ethyl Glyme

629-38-9
Diethylene glycol methyl ether acetate

629-50-5
n-Tridecane

629-59-4
n-Tetradecane

629-62-9
Pentadecane

629-70-9
Cetyl acetate

629-82-3
Dioctyl ether
Texaprint® SDCE

629-96-9
Arachidyl alcohol
Cachalot® AR-20

630-20-6
1,1,1,2-Tetrachloroethane

631-41-4
Tetraethanol ammonium hydroxide

632-22-4
Tetramethylurea

635-46-1
1,2,3,4-Tetrahydroquinoline

636-53-3
Diethyl isophthalate
Morflex® DEIP

637-64-9
Tetrahydrofurfuryl acetate

638-04-0
cis-1,3-Dimethylcyclohexane

638-11-9
Isopropyl butyrate

638-49-3
Amyl formate

646-06-0
1,3-Dioxolane

657-84-1
Sodium toluenesulfonate

661-19-8
Behenyl alcohol
Cachalot® BE-22

678-26-2
Perfluoropentane
PF-5050

680-31-9
Hempa

687-47-8
Ethyl-(S)-lactate
Purasolv® EL
Purasolv® ELECT
Purasolv® ELS

693-65-2
Diamyl ether

696-29-7
Isopropylcyclohexane

717-74-8
1,3,5-Triisopropylbenzene

756-79-6
Amgard® DMMP
Antiblaze® DMMP
Dimethyl methylphosphonate

763-69-9
Eastman® EEP
Ethyl 3-ethoxypropionate
Ucar® Ester EEP

770-35-4
Phenoxyisopropanol

822-06-0
Hexamethylene diisocyanate

822-50-4
1,2-Dimethylcyclopentane (trans)

827-52-1
Phenylcyclohexane

872-50-4
Agsolex® 1
GAFsolver™ TMP
N-Methyl-2-pyrrolidone
M-Pyrol®
NMP
NMP-EL
Pharmasolve™

873-94-9
3,3,5-Trimethyl cyclohexanone

925-06-4
Diisobutyl succinate

928-96-1
cis-3-Hexenol

929-06-6
2-(2-Aminoethoxy) ethanol
Diglycolamine® Agent (DGA®)

929-77-1
Methyl behenate

994-05-8
t-Amyl methyl ether

999-97-3
Hexamethyldisilazane

1002-67-1
Diethylene glycol methyl ethyl ether

1118-92-9
Dimethyl octanamide

1119-40-0
DBE-5
Dimethyl glutarate
Santosol® DMG

1120-21-4
n-Undecane

1120-28-1
Kemester® 2050
Methyl eicosanoate

1120-36-1
Tetradecene-1

1190-63-2
Cetyl stearate
Monestriol C

1191-99-7
2,3-Dihydrofuran

1192-18-3
1,2-Dimethylcyclopentane (cis)

1300-71-6
Xylenol

1300-72-7
Pilot® SXS-40
Sodium xylenesulfonate

1310-58-3
Potassium hydroxide

1310-73-2
Sodium hydroxide

1317-38-0
Copper oxide (ic)

1319-77-3
Cresylic acid

1320-37-2
Dichlorotetrafluoroethane

1320-67-8
3-Methoxy-1-propanol (mixt.)

1321-67-1
Naphthol spirits

1321-87-5
Dimethyl octynediol

1321-94-4
Arosolve MN-LF
Methylnaphthalene

1323-38-2
Castor (Ricinus communis) oil

1323-39-3
Propylene glycol stearate
Protachem™ PGMS

1323-65-5
Dinonyl phenol

1330-20-7
Shell Xylene
Xylene

1330-78-5
Tricresyl phosphate

1331-22-2
Methylcyclohexanone

1335-12-2
Phenylpropanol

1336-21-6
Ammonium hydroxide

1338-24-5
Naphthenic acid

1338-39-2
Crill 1
Sorbitan laurate

1344-00-9
EZA®
Sodium silicoaluminate

1467-79-4
Dimethyl cyanamide

1490-04-6
Menthol

1501-27-5
Methyl glutarate

1559-35-9
Ethylene glycol 2-ethylhexyl ether

1559-36-0
Diethylene glycol 2-ethylhexyl ether

1569-01-3
Arcosolv® PNP
Dowanol® PnP
Propylene glycol n-propyl ether

1569-02-4
Ethyl Proxitol
Propylene glycol ethyl ether (α-isomer)

1587-20-8
Trimethyl citrate

1589-49-7
3-Methoxy-1-propanol (β-isomer)

1632-73-1
Fenchyl alcohol

1634-04-4
Arcopure® MTBE
Methyl t-butyl ether

1638-26-2
1,1-Dimethylcyclopentane

1640-89-7
Ethylcyclopentane

1678-91-7
Ethylcyclohexane

1678-92-8
Propylcyclohexane

Myristyl propionate

6283-86-9
2-Ethylhexyl lactate

6382-06-5
Amyl lactate

6440-58-0
DMDM hydantoin

6780-13-8
Ethylenediamine (monohydrate)

6789-88-4
Hexyl benzoate

6790-09-6
PEG-12

6837-24-7
CHP®
N-Cyclohexyl pyrrolidone

6876-23-9
trans-1,2-Dimethylcyclohexane

6881-94-3
Diethylene glycol propyl ether
Eastman® DP

6938-94-9
Crodamol DA
Diisopropyl adipate
Estalan DIA
HallStar™ DIPA
Schercemol DIA
Trivent DIA

7058-01-7
s-Butylcyclohexane

7311-27-5
Nonoxynol-4

7328-17-8
2-(2-Ethoxyethoxy) ethyl acrylate
SR 256

7360-38-5
Trioctanoin

7384-98-7
Nikkol Sefsol 228
Propylene glycol dicaprylate

7397-62-8
Butyl glycolate
Polysolvan® O

7425-87-8
N-Octyl pyrrolidone

7446-09-5
Sulfur dioxide

7491-02-3
Diisopropyl sebacate
Estalan-DIPS
HallStar™ DIPS
Schercemol DIS

7553-56-2
Iodine

7646-85-7
Zinc chloride

7664-38-2
Phosphoric acid

7664-41-7
Ammonia

7681-11-0
Potassium iodide

7681-82-5
Sodium iodide

7704-34-9
Sulfur

7705-14-8
dl-Limonene (±)

7722-84-1
Hydrogen peroxide
Oxypure® 50%

7726-95-6
Bromine

7732-18-5
Water

7774-34-7
Calcium chloride (hexahydrate)

7778-85-0
Propylene glycol dimethyl ether

7785-26-4
α-Pinene (L, -)

7785-70-8
α-Pinene (D, +)

8000-34-8
Clove (Eugenia caryophyllus) oil

8000-41-7
α-Terpineol (mixed isomers)

8000-42-8
Caraway (Carum carvi) oil

8000-48-4
Eucalyptus globulus oil

8001-21-6
EmCon™ SUN
Sunflower (Helianthus annuus) seed oil

8001-22-7
Cropure® Soybean
Oleocal® IVO-114
Soybean (Glycine soja) oil
Super Refined® Soybean USP

8001-23-8
NEOBEE® 18
Safflower (Carthamus tinctorius) oil
Super Refined® Safflower USP

8001-25-0
Cropure® Olive
EmCon™ Olive
Fanoliv™ Oil
Olive Oil, Refined
Olive (Olea europaea) oil
Super Refined® Olive NF

8001-29-4
Cottonseed (Gossypium) oil
Super Refined® Cottonseed NF

8001-30-7
Corn Oil, Refined
Corn (Zea mays) oil
Super Refined® Corn NF

8001-31-8
Coconut (Cocos nucifera) oil

8001-58-9
Creosote oil

8001-78-3
Hydrogenated castor oil

8001-79-4
Castor (Ricinus communis) oil
Crystal® O
Crystal® Crown
EmCon™ CO
Krystal Kleer Castor Oil
USP Castor Oil

8002-03-7
Peanut (Arachis hypogaea) oil
Super Refined® Peanut NF

8002-05-9
Petroleum distillates

8002-09-3
Herco® Pine Oil
Pine (Pinus palustris) oil

8002-13-9
Canola oil
Rapeseed (Brassica campestris) oil

8002-26-4
Tall oil

8002-31-1
Cocoa (Theobroma cacao) butter
Fancol™ CB
Fancol™ CB Extra

8002-50-4
Cropure® Menhaden
Menhaden oil

8002-64-0
Neatsfoot oil

8002-74-2
Paraffin

8005-44-5
Cetearyl alcohol

8006-54-0
Corona Lanolin
Coronet Lanolin
Lanolin (anhyd.)

8006-61-9
Gasoline

8006-64-2
Turpentine (steam distilled)

8006-81-3
Ylang ylang (Cananga odorata) oil

8006-90-4
Peppermint (Mentha piperita) oil

8006-95-9
EmCon™ W
Wheat (Triticum vulgare) germ oil

8007-24-7
Cashew nut shell oil

8007-69-0
Cropure® Almond
Super Refined® Almond NF
Sweet almond (Prunus amygdalus dulcis) oil

8008-20-6
Deodorized kerosene
Kerosene
PD-23

8008-51-3
Camphor (Cinnamomum camphora) oil

8008-74-0
Cropure® Sesame
EmCon™ SES
Lipovol SES
Sesame Oil USP/NF 16
Sesame (Sesamum indicum) oil
Super Refined® Sesame NF

8008-79-5
Spearmint (Mentha viridis) oil

8009-03-8
Petrolatum (NF)

8012-95-1
Mineral oil

8013-75-0
Fusel oil refined

8020-83-5
Mineral oil (wh.)

8020-84-6
Lanolin (hyd.)

8024-09-7
Walnut (Juglans regia) oil

8024-32-6
Avocado (Persea gratissima) oil
Phytol™ AVO

8027-32-5
Penreco Snow
Petrolatum (USP)

8027-33-6
Lanolin alcohol

8029-43-4
Corn syrup

8029-91-2
Acetylated hydrogenated lard glyceride
Axol® E 61

8029-92-3
Acetylated lard glyceride

8030-30-6
Hi-Sol® 70
Naphtha

8032-32-4
Calumet VM&P Naphtha < 1%
RotoSolv
Solvent 2218
Solvent 2268
Solvent 2391
Solvent 2409
Solvent 2423
Solvent 2472
Special Naphtholite 66/3

13380-94-4
TCD Ketone A
Tricyclo [5.2.1.0~2,6] decane-8-one

13429-07-7
PPG-2 methyl ether

13466-78-9
3-Carene

13475-82-6
Permethyl® 99A

13552-80-2
Captex® 8227
Triundecanoin

13952-84-6
s-Butylamine

14371-10-9
Cinnamal

14433-76-2
Dimethyl caproamide

15356-60-2
Menthol (+)

15356-70-4
Menthol (D, ±)

15764-24-6
Dipropylene glycol ethyl ether

16339-54-1
Methoxypropyl acetamide

16544-70-0
Trihexyl citrate

16853-85-3
Lithium aluminum hydride

17671-27-1
Behenyl behenate
Exceparl BB

17673-56-2
Dynacerin® 660
Oleyl erucate

18912-80-6
Diethylene glycol isobutyl ether

19224-26-1
Benzoflex® 284
Propylene glycol dibenzoate

20178-34-1
Tripropylene glycol ethyl ether

20298-69-5
2-t-Butylcyclohexyl acetate (cis)

20298-70-8
2-t-Butylcyclohexyl acetate (trans)

20324-32-7
PPG-2 methyl ether (α-isomer)

20324-33-8
1-(2-(2-Methoxy-1-methylethoxy)-1-methylethoxyl)-2-propanol (α-isomer]

20602-86-2
Diamyl oxalate

20636-48-0
Nonoxynol-5

21368-68-3
Camphor (DL)

22047-49-0
Exceparl EH-S
Octyl stearate

23336-36-9
PEG-5 oleate

23436-19-3
Propylene glycol isobutyl ether

23783-42-8
Tetraethylene glycol methyl ether

24133-73-1
Cumylphenyl acetate

24157-79-7
1,4-Diisopropyl naphthalene

24157-81-1
2,6-Diisopropyl naphthalene

24800-44-0
Tripropylene glycol

24938-91-8
Trideceth

24991-55-7
PEG dimethyl ether
PEG-3 dimethyl ether
PEG-4 dimethyl ether
Polyglyme
Tetraglyme

25013-15-4
Vinyltoluene monomer

25013-16-5
BHA

25103-12-2
Triisooctyl phosphite

25154-52-3
Nonylphenol

25168-05-2
Halso® 99
Monochlorotoluene

25190-05-0
Oleth-16

25231-21-4
Arlamol® E
Lipocol P-15
PPG-11 stearyl ether
PPG-15 stearyl ether
Prostearyl 15
Varonic® APS
Witconol™ APS

25265-77-4
Texanol® Ester-Alcohol
2,2,4-Trimethyl-1,3-pentanediol isobutyrate
Ucar® Filmer IBT

25322-68-3
Carbowax® E200
Carbowax® E300
Carbowax® E300 NF
Carbowax® E400
Carbowax® E400 NF
Carbowax® E600
Carbowax® E600 NF
Carbowax® E900
Carbowax® E900 NF
Carbowax® E1000
Carbowax® E1000 NF
Carbowax® E1450
Carbowax® E1450 NF
Carbowax® E3350
Carbowax® E3350 NF
Carbowax® E4500 NF
Carbowax® E8000
Carbowax® E8000 NF
Carbowax® PEG 300
Carbowax® Sentry® PEG 300
Carbowax® Sentry® PEG 400
Carbowax® Sentry® PEG 600
Carbowax® Sentry® PEG 900
Carbowax® Sentry® PEG 1000
Carbowax® Sentry® PEG 1450
Carbowax® Sentry® PEG 3350
Carbowax® Sentry® PEG 4600
Carbowax® Sentry® PEG 8000
DePEG 200
DePEG 300
DePEG 400
DePEG 600
Lutrol® E 300
Lutrol® E 400
Lutrol® E 600
Lutrol® E 1500
Lutrol® E 3350
Lutrol® E 4000
Lutrol® E 6000
PEG 200
PEG 300
PEG 400
PEG-4
PEG-6
PEG-8
PEG-9
PEG-10
PEG-12
PEG-14
PEG-16
PEG-18
PEG-20
PEG-32
PEG-40
PEG-75
PEG-90
PEG-100
PEG-125
PEG-135
PEG-150
PEG-200
PEG-350
Pluracol® E200
Pluracol® E300
Pluracol® E400
Pluracol® E400 NF
Pluracol® E600
Pluracol® E600 NF
Pluracol® E1450 NF
Pluracol® E4000
Pluracol® E6000
Pogol® 400
Polyethylene glycol
Renex® PEG 300
Renex® PEG 400
Renex® PEG 600
Rhodasurf® PEG 600

25322-69-4
Jeffox® PPG-400
Jeffox® PPG-2000
Polyglycol P425
Polyglycol P1000TB
Polyglycol P1200
Polyglycol P2000
Polyglycol P4000
Polypropylene glycol
PPG-9
PPG-17
PPG-20
PPG-26
PPG-30

25339-09-7
Isocetyl stearate
Monestriol IC

25339-17-7
Isodecyl alcohol

25340-17-4
Diethylbenzene
Sure Sol®-150ND

25395-31-7
Glyceryl diacetate

25496-72-4
Glyceryl oleate
Lumulse® GMO

25498-49-1
Arcosolv® TPM
Dowanol® TPM
Poly-Solv® TPM
PPG-3 methyl ether

25550-14-5
Ethyl toluene

25551-13-7
Trimethylbenzene

25637-84-7
Glyceryl dioleate

25639-42-3
Methylcyclohexanol
Methylhexalin®

25640-78-2
1,1´-Biphenyl (1-methylethyl)

25791-96-2
Polyglycol PT250
Polyglycol PT700
Polyglycol PT3000
PPG-10 glyceryl ether
PPG-55 glyceryl ether

26027-38-3
Nonoxynol-4
Nonoxynol-5
Nonoxynol-6

26183-52-8
Deceth-6

26264-02-8
Nonoxynol-5

26264-05-1
Isopropylamine dodecylbenzene-sulfonate
Rhodacal® 330

26266-57-9
Crill 2
Sorbitan palmitate

26266-58-0
Nikkol SO-30
Nikkol SO-30R
Sorbitan trioleate

26401-47-8
Dodoxynol-7

26402-22-2
Capmul® MCMC10
Glyceryl caprate

26402-26-6
Capmul® MCMC8
Glyceryl caprylate

26402-31-3
Propylene glycol ricinoleate

26403-17-8
Isodecanoic acid

26446-35-5
Glyceryl acetate

26446-38-8
Sucrose palmitate

26447-10-9
Ammonium xylenesulfonate
STEPANATE® AXS

26471-62-5
Toluene diisocyanate

26855-43-6
Polyglyceryl-3 stearate

26896-20-8
Neodecanoic acid

26942-95-0
Triisostearin

26952-21-6
Isooctyl alcohol

26962-29-8
Diethoxyethyl succinate

27138-31-4
Dipropylene glycol dibenzoate

27176-87-0
Dodecylbenzenesulfonic acid

27176-97-2
Nonoxynol-4

27177-01-1
Nonoxynol-6

27177-05-5
Nonoxynol-6

27194-74-7
Propylene glycol laurate
Schercemol PGML

27214-38-6
Glyceryl myristate

27233-00-7
Glyceryl triacetyl hydroxystearate

27321-72-8
Polyglyceryl-3 stearate

27321-96-6
Choleth-24

27342-88-7
Lauryl alcohol

27458-93-1
Diadol 18G-C
Isostearyl alcohol
Jarcol™ I-18EX

27638-00-2
Glicopol 2-G
Glyceryl dilaurate
Protachem™ GDL

27841-06-1
Estemol N-01
Neopentyl glycol dicaprate
Schercemol NGDC

27871-49-4
Methyl-(S)-lactate
Purasolv® ML

28510-23-8
Cosmol 525
Neopentyl glycol dioctanoate
Schercemol NGDO

29383-26-4
Octyl hydroxystearate
Wickenol® 171

29710-25-6
Octyl hydroxystearate

29806-73-3
Exceparl EH-P
Octyl palmitate
Palmil O
Trivent OP

29911-27-1
Arcosolv® DPNP
Dipropylene glycol n-propyl ether
Dowanol® DPnP

29911-28-2
Arcosolv® DPNB
Dipropylene glycol butyl ether (α-isomer)
Dowanol® DPnB
Solvenon® DPnB

30025-38-8
(2-Ethoxymethylethoxy) propanol

30136-13-1
Propoxypropanol

31160-64-2
Dihexyloctylphosphine oxide

31160-66-4
Dioctylhexylphosphine oxide

31335-74-7
Neopentyl glycol dioctanoate

31565-12-5
Imwitor® 408
Nikkol Sefsol 218
Propylene glycol caprylate

31565-37-4
Exceparl TD-S
Isotridecyl stearate

31566-31-1
Glyceryl stearate
Glyceryl stearate SE
Protachem™ GMS-D
Protachem™ GMS-T

31692-85-0
Glycofurol

31694-55-0
Glycereth-7
Glycereth-26
Volpo G26

31807-55-3
Isododecane

34316-64-8
Hexyl laurate

34451-19-9
Butyl-(S)-lactate
Purasolv® BL

34590-94-8
Arcosolv® DPM
Dowanol® DPM
Methyl Diproxitol
Poly-Solv® DPM
PPG-2 methyl ether (mixt.)

35274-05-6
Ceraphyl® 28
Cetyl lactate

35884-42-5
Dipropylene glycol butyl ether (mixt.)

36078-10-1
Exceparl L-OL
Lauryl oleate

36284-86-3
Stearamide AMP

36311-34-9
Ceraphyl® ICA
Isocetyl alcohol

36653-82-4
Cachalot® C-50
Cetyl alcohol

36936-60-4
Emery® 5794
PEG-3 triethanolamine

37200-49-0
Polysorbate 80

37205-87-1
Nonoxynol
Nonoxynol-4
Nonoxynol-5
Nonoxynol-6

37220-82-9
Glyceryl oleate

37311-01-6
PPG-5-ceteth-20

37311-04-9
PPG-3-myreth-3

37349-34-1
Polyglyceryl-3 stearate
Protachem™ G-5566

38178-38-0
Dibenzofuran

38640-62-9
Diisopropyl naphthalene

39355-35-6
Mineral oil

40482-44-8
3-Isopropyl-2-oxazolidinone

41395-83-9
Estalan-PGDP
Propylene glycol dipelargonate
Schercemol PGDP

41395-89-5
Isodecyl isononanoate
Wickenol® 152

41669-30-1
Estalan-ISI
Isostearyl isostearate
Schercemol 1818

42131-25-9
Isononyl isononanoate
Salacos® 99
Wickenol® 151

42131-27-1
Isotridecyl isononanoate

42131-28-2
Isostearyl lactate

50468-22-9
Isopentyldiol

51258-15-2
PPG-66-glycereth-12
PPG-24-glycereth-24

51609-41-7
Nonoxynol-9 phosphate

51730-94-0
Dipropylene glycol phenyl ether

51811-79-1
Nonoxynol-9 phosphate
Surphos 620
Surphos 645

52006-45-8
Isocetyl isostearate
Isostriol IC

52125-53-8
Arcosolv® PE
Propylene glycol ethyl ether (mixt.)

52581-71-2
PPG-23 oleyl ether
PPG-37 oleyl ether

52688-97-0
PEG-4 dioleate
PEG-8 dioleate
PEG-12 dioleate
PEG-20 dioleate

52845-07-5
Isoeicosane

52993-54-1
s-Butylcyclohexane

53637-25-5
Dow EP530
Poloxamer 181

53694-15-8
Sorbeth-20

54182-57-9
Carbomer

54839-24-6
1-Ethoxy-2-propanol acetate
Propylene glycol ethyl ether acetate (α-isomer)

55257-88-0
Lauryl pyrrolidone

55406-53-6
Iodopropynyl butylcarbamate

55934-93-5
Arcosolv® TPNB
Solvenon® TPnB
Tripropylene glycol n-butyl ether

56002-14-3
Olepal ISO
PEG-6 isostearate

56519-71-2
Captex® 355EP
Lexol® PG-800
Propylene glycol dioctanoate

56539-66-3
Methoxymethylbutanol

56554-53-1
Isodragol 2/050300
Triisononanoin

57018-52-7
Arcosolv® PTB
Propylene glycol t-butyl ether

57307-99-0
PEG-8 caprylic/capric glycerides

57533-90-1
Diisocetyl adipate

57569-76-3
Glycereth-7 triacetate
Pelemol® G7A

58262-41-2
Diisocetyl adipate

58670-89-6
2-Decyl-1-tetradecanol
Jarcol™ I-24

58748-27-9
Propylene glycol dicaprylate/dicaprate

59130-69-7
Cetyl octanoate
Schercemol CO

59130-70-0
Stearyl octanoate

59130-70-7
Cetearyl octanoate

59219-71-5
Isononyl isononanoate

59231-34-4
Ceraphyl® 140-A
Estalan-IDO
Isodecyl oleate

59231-35-5
Isodecyl isononanoate

59231-37-7
Isotridecyl isononanoate

Wickenol® 153

60164-51-4
Fomblin® Perfluorosolv PFS-1
Perfluoropolyether

60344-26-5
PEG-6 oleate

60908-77-2
Isohexadecane

61788-49-6
Acetylated lanolin alcohol

61788-59-8
Emery® 2253
Emery® 2254
Exceparl MC
Methyl cocoate

61788-60-1
Methyl cottonseed
Norfox® MCS

61788-61-2
Exceparl MT
Methyl tallowate
Norfox® MTL

61789-12-6
Glyceryl tallate
Lumulse® GMT
Lumulse® GMT-40

61790-81-6
PEG-75 lanolin
Solulan® 75
Solulan® L-575

61790-86-1
Polysorbate 80

61790-91-8
Laneth-5

61790-95-2
Polyglyceryl-3 stearate

61791-12-6
Cremophor® EL
Cremophor® ELP
Etocas 35 NF
Incrocas 30
Incrocas 40
PEG-30 castor oil
PEG-33 castor oil
PEG-35 castor oil
PEG-40 castor oil

61791-20-6
Aqualose W15
Laneth-5
Laneth-10
Laneth-15
Laneth-16
Nikkol BWA-5
Nikkol BWA-10
Polychol 5
Polychol 15

61791-31-9
Cocamide DEA

61849-72-7
Glucam® P-10
Glucam® P-20
PPG-10 methyl glucose ether
PPG-20 methyl glucose ether

63393-93-1
Isopropyl lanolate
Lanesta L

63449-39-8
Paraffin, chlorinated

63697-00-7
Isopropyl-S-lactate
Purasolv® IPL

63705-03-3
Polyglyceryl-3 diisostearate

63793-60-2
PPG-3 myristyl ether
Promyristyl PM-3
Witconol™ APM

64365-06-6
C9-13 isoparaffin
C20-40 isoparaffin

64475-85-0
Calumet Mineral Spirits < 1%
Mineral spirits
Shell OMS
Shell Mineral Spirits 7 EC
Shell Mineral Spirits 135

64741-65-7
Naphtha, heavy alkylate
Shellsol® OMS
Shellsol® T
Shellsol® TC
Shellsol® TD
Shellsol® TK

64741-66-8
Isopar® C
Isopar® E Fluid
Naphtha, light alkylate

64741-73-7
Parabase
Petroleum distillates, alkylate
Shellsol® TM

64741-77-1
Amsol 100
Petroleum distillates, light hydrocracked

64741-81-7
Petroleum distillates, heavy thermal cracked

64741-84-0
Naphtha, solvent-refined light
Textile Spirits

64741-85-1
Penreco 150-B Solvent
Petroleum raffinates, sorption process
Sasol MR Solvent

64741-86-2
EXK-385-MS0
Magie 470 Oil
Magie 500 Oil
Petroleum distillates, sweetened middle

64742-04-7
Paraffin distillate

64742-13-8
Hydroseal G3H
Hydroseal G232H
Hydroseal G240H
Hydroseal G250H
Hydroseal G400H
Petroleum distillates, acid-treated middle

64742-14-9
Drakesol® 165AT
Petroleum distillates

64742-46-7
Amsol 400
Amsol 500
Amsol 600
Conosol 260
Conosol 340
Drakesol® 205
Exx-Print® 588 D
Exxsol® D 130
Hydroseal G2
Isopar® V Fluid
Magie 5200-HTX
MagieSol 55LX
MagieSol 65LX
PCPL Solvent
PCP Solvent
Petroleum distillates, hydrotreated middle

64742-47-8
Actrel® 1111L
Actrel® 1160L
AM 321 Fluid
Amsol 650
Amsol 3800
Bayol 35
Conosol 215
Conosol C-145
Conosol C-170
Conosol C-200
Drakesol® 165
Escaid® 90
Escaid® 95
Escaid® 110
Exx-Print® 176 D
Exx-Print® 283 D
Exxsol® D 40
Exxsol® D 40 Anti-Static Fluid
Exxsol® D 60
Exxsol® D 60 Anti-Static Fluid
Exxsol® D 60 Naphtha Anti-Static
Exxsol® D 70
Exxsol® D 80
Exxsol® D 90/120
Exxsol® D 90/120 Anti-Static Solvent
Exxsol® D 95
Exxsol® D 110
Exxsol® D 3135
Exxsol® D 3135 Anti-Static Solvent
Exxsol® D 3135 Naphtha Anti-Static
Exxsol® DSP 90/120 Naphtha
Exxsol® S-K Premium Solvent
Isopar® M Fluid
Kerdane D65
Kerdane D75
Ketrul D70
Ketrul D80
Ketrul D85
Ketrul D211
Ketrul D220
MagieSol 44
MagieSol 47LX
Orfom® SX 7
Orfom® SX 12
Penreco 170 Exempt Solvent
Petroleum distillates, hydrotreated light
Sasol LPA Solvent
Sasol LPA-142 Solvent

Petroleum distillates, catalytic reformer fractionator residue, low boiling

68478-36-4
C10 alkyl acetate

68514-15-8
Gasoline

68515-32-2
C12-14 alkylbenzene
V-159L

68526-79-4
Hexyl alcohol

68526-85-2
Decyl alcohol

68551-16-6
C9-11 isoparaffin
HC-DCF Low Flash™ Dry Cleaning Solvent
Orfom® SX 18
Soltrol® 100

68551-17-7
C10-13 isoparaffin
EcoSolv™ Dry Cleaning Fluid
Soltrol® 130

68551-19-9
C12-14 isoparaffin
Orfom® SX 10
Soltrol® 170

68551-20-2
C13-16 isoparaffin
C13-17 isoparaffin
Orfom® SX 11
Soltrol® 220
Soltrol® 300

68553-81-1
EmCon™ Rice Bran
Rice (Oryza sativa) bran oil

68583-51-7
Captex® 200
Crodamol PC-PN
Lexol® PG-865
Liponate PC
Miglyol® 840
Myritol® PC
NEOBEE® M-20
Propylene glycol dicaprylate/dicaprate

68584-22-5
Dodecylbenzenesulfonic acid

68584-24-7
Isopropylamine dodecylbenzenesulfonate

68585-47-7
Sodium lauryl sulfate

68603-42-9
Cocamide DEA

68606-83-7
Ylang ylang (Cananga odorata) oil

68608-88-8
Dodecylbenzenesulfonic acid

68647-72-3
d-Limonene
Orange Terpenes

68648-87-3
Alkylate 215
Alkylate 229
C8-16 alkylbenzene

68855-18-5
LexFeel® 7
Neopentyl glycol diheptanoate

68917-73-7
Wheat (Triticum vulgare) germ oil

68920-06-9
Naphtha

68937-83-7
Methyl caprylate/caprate

68955-19-1
Sodium lauryl sulfate

68956-56-9
Dipentene Extra
Terpinolene

68988-72-7
Propylene glycol dicaprylate/dicaprate

68990-53-4
Mono- and diglycerides of fatty acids

68990-55-6
Acetylated mono- and diglycerides of fatty acids
Axol® E 66

68991-68-4
Caprylic/capric/lauric triglyceride
Captex® 350

69009-90-1
Bis (methylethyl)-1,1´-biphenyl

69430-24-6
Cyclomethicone

70321-63-0
Lanolin oil
Lipolan R

70657-70-4
2-Methoxy-1-propanol acetate

70914-20-4
Isoheptyl alcohol

71012-10-7
Polyglyceryl-4 oleate

71172-42-4
Methoxypropyl formamide

71195-64-7
Diisobutyl glutarate

71566-49-9
Octyl isononanoate

71808-49-6
Eastman® C-11 Ketone

73398-61-5
Aldo® MCT KFG
Caprylic/capric triglyceride
Myritol® 318
NEOBEE® M-5

73579-09-6
Methoxyisopropyl formamide

73891-99-3
Emery® 2232
Methyl canolate
Methyl rapeseedate

74623-31-7
PPG-5-buteth-7
PPG-12-buteth-16
PPG-20-buteth-30
PPG-33-buteth-45

76050-42-5
Carbomer

79956-36-8
Mineral oil

81846-81-3
1,1´-Biphenyl, butenylated

83046-05-3
Mineral oil

83411-71-6
Bis (2,4,4,-trimethylpentyl) phosphinic acid
Cyanex® 272

83708-66-1
Isocetyl myristate

84604-00-2
Walnut (Juglans regia) oil

84606-45-1
1-Isopropyl-2-methyl imidazole

84650-19-1
Laneth-15

84696-37-7
Rice (Oryza sativa) bran oil

84878-33-1
Cetearyl isononanoate
Tegosoft® CI

84961-70-6
Benzene, mono-C10-13-alkyl derivs., distillation residues
V-154L

84988-79-4
Excepari IB-OL
Isobutyl oleate
Vinycizer 30

85049-36-1
Ethyl oleate

85049-53-2
Neatsfoot oil

85116-87-6
Isopropyl oleate

85116-97-8
PEG-2 stearate

85117-49-3
Dodecylbenzenesulfonic acid

85186-88-5
Sorbitan trioleate

85251-77-0
Glyceryl stearate

85404-84-8
Polyglyceryl-3 diisostearate

85409-09-2
Caprylic/capric glycerides
Imwitor® 742

85536-04-5
Stearyl stearate

85536-07-8
Labrasol®
PEG-8 caprylic/capric glycerides

85536-14-7
Dodecylbenzenesulfonic acid

85536-46-5
Isopropanolamine lanolate

85586-21-6
Methyl stearate

85665-33-4
C10-18 triglycerides

85666-92-8
Glyceryl stearate
Glyceryl stearate SE

85736-49-8
PEG-12 dioleate

85940-31-4
Caraway (Carum carvi) oil

86398-53-0
Denatonium benzoate (monohydrate)

88230-35-7
Exxate® 600
Oxo-hexyl acetate

88917-22-0
Arcosolv® DPMA
Dowanol® DPMA
PPG-2 methyl ether acetate

90438-79-2
Exxate® 700
Oxo-heptyl acetate

90604-31-2
C13-15 alcohols

90622-46-1
C14-16 paraffins
Linpar® 15-V Normal Paraffin
Linpar® 1416-V Normal Paraffin

90622-47-2
C14-17 alkane

90622-56-3
C7-10 isoparaffin
Isane IP 130

90622-57-4
C9-12 isoparaffin
Isane IP 155
Isane IP 165
Isane IP 175
Isane IP 185

90622-58-5
C11-16 isoparaffin

91031-31-1
Glycol distearate

EINECS/ELINCS Number Index

202-627-7
Furfural
QO® Furfural

202-675-9
p-t-Butyl toluene

202-680-6
α-Terpineol

202-681-1
p-Chlorobenzotrifluoride
Oxsol® 100

202-707-1
α-Methylbenzyl alcohol

202-708-7
Acetophenone

202-773-1
1,3-Diisopropyl benzene

202-794-6
γ-Terpinene

202-796-7
p-Cymene

202-826-9
1,4-Diisopropyl benzene

202-845-2
Diethylaminoethanol

202-849-4
Ethylbenzene

202-851-5
Styrene

202-855-7
Benzonitrile

202-859-9
Benzyl alcohol

202-860-4
Benzaldehyde

202-869-3
N-Methylcyclohexylamine

202-870-9
N-Methylaniline

202-876-1
Anisole

202-882-4
Tetrahydropyran-2-methanol

202-885-0
Ethyl morpholine

202-888-7
2,2-Dimethyl-1,3-dioxolane-4-methanol

202-935-1
Glyceryl triacetyl ricinoleate
Naturechem® GTR

202-978-6
Diphenylmethane

202-980-7
Dicyclohexylamine

202-981-2
Diphenyl oxide

202-993-8
Ethyl phenylacetate

203-005-8
Diphenyl carbonate

203-041-4
Tetrahydroxypropyl ethylenediamine

203-049-8
Triethanolamine

203-051-9
Captex® 500
Captex® 500P
Eastman® Triacetin, Food Grade
Triacetin
Triacetina

203-063-4
Trilaurylamine

203-079-1
2-Ethylhexyl acetate

203-090-1
Dioctyl adipate
HallStar™ DOA
Wickenol® 158

203-104-6
Ethyl cinnamate

203-118-2
Benzyl ether

203-132-9
Propylbenzene

203-213-9
Cinnamal

203-214-4
Benzyl formate

203-219-1
γ-Nonalactone

203-225-4
γ-Undecalactone

203-234-3
2-Ethylhexanol
Ucar® 2-Ethylhexanol

203-291-4
Ethyl propionate

203-299-8
Eastman® MAA
Methyl acetoacetate

203-300-1
s-Butyl acetate

203-306-4
Ethyl butyrate

203-310-6
Acetal

203-311-1
Diethyl carbonate

203-312-7
Jefftreat® M Solvent
Jefftreat® MS-100
Jefftreat® MS-300
Methyldiethanolamine

203-320-0
Propyl butyrate

203-350-4
Cetiol® B
Dibutyl adipate

203-363-5
PEG-2 stearate
Protachem™ DGS

203-380-8
Isoamyl butyrate

203-385-5
Ethyl octanoate

203-388-1
3-Heptanone

203-389-7
Propyl propionate
Ucar® n-Propyl Propionate

203-396-5
p-Xylene

203-419-9
DBE-4
Dimethyl succinate
Santosol® DMS

203-423-0
3-Octanone

203-424-6
1,2,6-Hexanetriol

203-425-1
Methyl caproate

203-431-4
Dimethyl sebacate

203-439-8
Epichlorohydrin

203-445-0
AG 101
Leksol™
Leksol™ AL
n-Propyl bromide

203-448-7
Butane

203-453-4
Acrolein

203-458-1
Ethylene dichloride

203-459-7
Ethylene chlorohydrin

203-466-5
Acrylonitrile

203-468-6
Ethylenediamine

203-470-7
Allyl alcohol

203-471-2
Propargyl alcohol

203-473-3
Ethylene glycol
Polyethylene glycol

203-489-0
Hexylene glycol

203-492-7
Chemsil DM 0.65
Hexamethyldisiloxane

203-497-4
Octamethyltrisiloxane

203-500-9
Dimethyl hexynol

203-512-4
4-Methoxy-4-methylpentanone-2

203-523-4
Calumet Isohexane
Exxsol® Isohexane
Isohexane

203-528-1
Methyl propyl ketone

203-529-7
Butylene glycol

203-539-1
Arcosolv® PM
Dowanol® PM
Eastman® PM
Methoxyisopropanol
Methyl Proxitol
Poly-Solv® MPM
Solvenon® PM

203-542-8
Dimethylethanolamine

203-544-9
NiPar S-10™
1-Nitropropane

203-550-1
Eastman® MIBK
Methyl isobutyl ketone
Ucar® Methyl Isobutyl Ketone

203-551-7
Methyl amyl alcohol

203-558-5
Diisopropylamine

203-560-6
Isopropyl ether

203-561-1
Eastman® Isopropyl Acetate
Isopropyl acetate
Ucar® Isopropyl Acetate

203-569-5
γ-Valerolactone

203-572-1
Arconate® PC-1000
Arconate® PC-HP
Jeffsol® PC
PCHP
Propylene carbonate
Solvenon® PC
UltraPure™ Propylene Carbonate

203-587-3
2,6-Lutidine

203-919-7
Carbitol®
Carbitol® Low Gravity
Eastman® DE
EDGE
Ethoxydiglycol
Transcutol® CG
Transcutol® HP
Transcutol® P

203-924-4
Diethylene glycol dimethyl ether
Hisolve MDM

203-926-5
Thiodipropionitrile

203-931-2
Nonanoic acid

203-933-3
Butoxyethanol acetate
Butyl Cellosolve® Acetate
Eastman® EB Acetate

203-934-9
Isopropyl stearate
Monestriol IS-C
Wickenol® 127

203-935-4
Isopropyl oleate
Tilol IS

203-939-6
Octyl acetate

203-940-1
Carbitol® Acetate
Eastman® DE Acetate
Ethoxydiglycol acetate

203-951-1
Ethylene glycol hexyl ether
Hexyl Cellosolve®

203-953-2
Triethylene glycol

203-956-9
Decyl alcohol

203-961-6
BDGE
Butoxydiglycol
Butyl Carbitol®
Butyl Di-Icinol
Butyl Dioxitol®
Eastman® DB
Eastman® DB Stabilized

203-963-7
Diethylene glycol diethyl ether
Ethyl Diglyme

203-966-3
Emery® 2216
Methyl palmitate

203-967-9
n-Dodecane

203-975-2
1,10-Decanediol
Speziol C 10/2

203-978-9
Ethoxytriglycol

203-982-0
Cachalot® L-90
Lauryl alcohol

203-986-2
Tetraethylenepentamine

203-987-8
n-Hexyl ether

203-988-3
Diethylene glycol n-hexyl ether
Hexyl Carbitol®

203-989-9
Carbowax® E200
Carbowax® E300
Carbowax® E900
Carbowax® E1000
Carbowax® E1000 NF
Carbowax® E1450
Carbowax® E1450 NF
Carbowax® E3350
Carbowax® E3350 NF
Carbowax® E4500 NF
Carbowax® E8000
Carbowax® E8000 NF
Carbowax® Sentry® PEG 900
Carbowax® Sentry® PEG 1000
Carbowax® Sentry® PEG 1450
Carbowax® Sentry® PEG 3350
Carbowax® Sentry® PEG 4600
Carbowax® Sentry® PEG 8000
DePEG 200
Lutrol® E 1500
Lutrol® E 3350
Lutrol® E 4000
Lutrol® E 6000
PEG 200
PEG-4
PEG-6
PEG-20
PEG-32
PEG-75
PEG-90
PEG-100
PEG-135
PEG-150
Pluracol® E200
Pluracol® E1450 NF
Pluracol® E4000
Pluracol® E6000

203-990-4
Edenor® ME C18-90
Exceparl MS
Methyl stearate

203-992-5
Exceparl M-OL
Kemester® 205
Methyl oleate
Oleocal® ME-112

203-998-8
Tridecyl alcohol

204-000-3
Alfol® 14
Cachalot® M-43
Myristyl alcohol

204-001-9
Butyl Diglyme
Diethylene glycol dibutyl ether

204-007-1
Oleic acid

204-009-2
Erucamide

204-017-6
Cachalot® S-56
Stearyl alcohol

204-018-1
Eicosane

204-065-8
Dimethyl ether
Dymel® A

204-070-5
Methyl butynol

204-112-2
Triphenyl phosphate

204-122-7
Trimethylcyclohexanol

204-130-0
Triisopropyl phosphite

204-132-1
MDM hydantoin

204-211-0
n-Dioctyl phthalate

204-262-9
Benzylsalicylate

204-263-4
2-Ethylhexyl salicylate
HallBrite® OS

204-317-7
Methyl salicylate

204-340-2
Tetrahydronaphthalene
Tetralin®
Tetralin

204-341-8
Isoquinoline

204-371-1
Anthracene

204-402-9
Benzyl benzoate
Unichem BZBN

204-405-8
Benzyl laurate

204-407-6
Benzoflex® 2-45
Diethylene glycol dibenzoate

204-427-5
Pyrocatechol

204-428-0
1,2,4-Trichlorobenzene

204-468-9
Trimethyl borate

204-469-4
Triethylamine

204-493-5
N,N-Dimethylaniline

204-534-7
Arlypon SO
Triolein
Trioleina

204-558-8
Dioctyl sebacate
HallStar™ DOS

204-575-0
Phenyl acetate

204-589-7
Emery® 6705
Phenoxyethanol
Rewopal® MPG 10
Rewopal® MPG 40

204-607-3
Ecosoft™ Solvent IK
Isobutyl heptyl ketone

204-608-9
4-Heptanone

204-612-0
Diethyl succinate

204-617-8
Hydroquinone

204-622-5
Myrcene

204-623-0
Propionaldehyde

204-624-6
N-Methylformamide

204-626-7
Diacetone alcohol
Ucar® Diacetone Alcohol

204-633-5
Isoamyl alcohol

204-634-0
Acetylacetone

204-639-8
Paraldehyde

204-640-3
Ethyl caproate

204-646-6
n-Butyraldehyde

204-647-1
Crotonaldehyde

204-648-7
2-Pyrol®
2-Pyrrolidone
Soluphor® P

204-658-1
n-Butyl acetate
Eastman® n-Butyl Acetate, Urethane Grade

204-661-8
1,4-Dioxane

204-662-3
Isoamyl acetate

204-664-4
Glyceryl stearate

204-666-5
Butyl stearate
Exceparl BS
HallStar™ BST

208-878-9
n-Hexadecane

208-930-0
Methyl lactate

209-060-4
Methyl propionate

209-097-6
Tristearin

209-098-1
Tripalmitin

209-099-7
Trimyristin

209-136-7
Cyclomethicone
Octamethylcyclotetrasiloxane

209-189-6
Carbon tetrabromide

209-230-8
3,3-Dimethylpentane

209-264-3
Methyl isopropyl ketone

209-294-7
Diisopropyl ketone

209-400-1
2,6-Xylenol

209-526-7
3-Pentanol

209-529-3
Potassium carbonate

209-544-5
Toluene-2,4-diisocyanate

209-578-0
Dipentene Extra
Terpinolene

209-645-4
3-Hexanone

209-661-1
3-Heptanol

209-669-5
n-Butyl propionate
Ucar® n-Butyl Propionate

209-731-1
2-Hexanone

209-772-5
Butyl formate

209-790-3
N-Octadecane

209-854-0
1,1-Dichloro-1-nitroethane

209-954-4
Lactic acid (DL)

209-959-1
Methyl pivalate

210-254-6
Methyl 2-furoate

210-470-0
2,3-Dichloro-1-propanol

210-478-4
Methyl carbonate

210-702-0
Tricaprin

210-719-3
Ethylene glycol benzyl ether

210-736-6
Cyclohexyl acetate

210-791-6
3-Methoxy-1,2-propanediol

210-792-1
Methyl butyrate

210-808-7
Ethyl crotonate

210-817-6
Dowanol® PGDA
Propylene glycol diacetate

210-840-1
cis-1,4-Dimethylcyclohexane

210-852-7
Amyl propionate
Ucar® n-Pentyl Propionate

210-871-0
Methyl disulfide

211-014-3
Glycol distearate
Kemester® EGDS
Protachem™ EGDS

211-019-0
Methyl adipate

211-020-6
DBE-6
Dimethyl adipate
Santosol® DMA

211-047-3
Amyl acetate

211-074-0
Hexamethylene glycol

211-076-1
Ethylene glycol diethyl ether
Ethyl Glyme

211-093-4
n-Tridecane

211-096-0
n-Tetradecane

211-098-1
Pentadecane

211-103-7
Cetyl acetate

211-112-6
Dioctyl ether
Texaprint® SDCE

211-119-4
Arachidyl alcohol
Cachalot® AR-20

211-135-1
1,1,1,2-Tetrachloroethane
1,1,2,2-Tetrachloroethane

211-173-9
Tetramethylurea

211-237-6
1,2,3,4-Tetrahydroquinoline

211-296-8
Tetrahydrofurfuryl acetate

211-320-7
Isopropyl butyrate

211-340-6
Amyl formate

211-407-3
Ucar® Primary Amyl Acetate

211-546-6
Behenyl alcohol
Cachalot® BE-22

211-653-8
Hempa

211-694-1
Ethyl-(S)-lactate
Purasolv® EL
Purasolv® ELECT
Purasolv® ELS

211-756-8
Diamyl ether

211-792-4
Isopropylcyclohexane

211-941-3
1,3,5-Triisopropylbenzene

212-052-3
Amgard® DMMP
Antiblaze® DMMP
Dimethyl methylphosphonate

212-222-7
Phenoxyisopropanol

212-485-8
Hexamethylene diisocyanate

212-572-0
Phenylcyclohexane

212-828-1
Agsolex® 1
GAFsolver™ TMP
N-Methyl-2-pyrrolidone
M-Pyrol®
NMP
NMP-EL
Pharmasolve™

213-113-7
Diisobutyl succinate

213-192-8
cis-3-Hexenol

213-195-4
2-(2-Aminoethoxy) ethanol
Diglycolamine® Agent (DGA®)

213-207-8
Methyl behenate

213-611-4
t-Amyl methyl ether

213-668-5
Hexamethyldisilazane

214-272-5
Dimethyl octanamide

214-277-2
DBE-5
Dimethyl glutarate
Santosol® DMG

214-304-8
Kemester® 2050
Methyl eicosanoate

214-724-1
Cetyl stearate
Monestriol C

214-747-7
2,3-Dihydrofuran

215-090-9
Pilot® SXS-40
Sodium xylenesulfonate

215-181-3
Potassium hydroxide

215-185-5
Sodium hydroxide

215-269-1
Copper oxide (ic)

215-293-2
Cresylic acid

215-354-3
Propylene glycol stearate
Protachem™ PGMS

215-356-4
Dinonyl phenol

215-535-7
Shell Xylene
Xylene

215-548-8
Tricresyl phosphate

215-556-1
Methylcyclohexanone

215-621-4
Phenylpropanol

215-647-6
Ammonium hydroxide

215-662-8
Naphthenic acid

215-663-3
Crill 1
Sorbitan laurate

215-684-8
EZA®
Sodium silicoaluminate

215-991-7
Dimethyl cyanamide

216-074-4
Menthol

α-Terpineol (mixed isomers)

232-273-9
EmCon™ SUN
Sunflower (Helianthus annuus) seed oil

232-274-4
Cropure® Soybean
Oleocal® IVO-114
Soybean (Glycine soja) oil
Super Refined® Soybean USP

232-276-5
NEOBEE® 18
Safflower (Carthamus tinctorius) oil
Super Refined® Safflower USP

232-277-0
Cropure® Olive
EmCon™ Olive
Fanoliv™ Oil
Olive Oil, Refined
Olive (Olea europaea) oil
Super Refined® Olive NF

232-280-7
Cottonseed (Gossypium) oil
Super Refined® Cottonseed NF

232-281-2
Corn Oil, Refined
Corn (Zea mays) oil
Super Refined® Corn NF

232-282-8
Coconut (Cocos nucifera) oil

232-292-2
Hydrogenated castor oil

232-293-8
Castor (Ricinus communis) oil
Crystal® O
Crystal® Crown
EmCon™ CO
Krystal Kleer Castor Oil
USP Castor Oil

232-296-4
Peanut (Arachis hypogaea) oil
Super Refined® Peanut NF

232-298-5
Drakesol® 165AT
Petroleum distillates

232-299-0
Oleocal® C-102
Oleocal® C-102 A
Rapeseed (Brassica campestris) oil

232-304-6
Tall oil

232-311-4
Cropure® Menhaden
Menhaden oil

232-314-0
Neatsfoot oil

232-315-6
Paraffin

232-348-6
Corona Lanolin
Coronet Lanolin
Lanolin

232-349-1
Gasoline

232-350-7
Turpentine

232-355-4
Cashew nut shell oil

232-366-4
Deodorized kerosene
Kerosene
PD-23

232-370-6
Cropure® Sesame
EmCon™ SES
Lipovol SES
Sesame Oil USP/NF 16
Sesame (Sesamum indicum) oil
Super Refined® Sesame NF

232-373-2
Penreco Snow
Petrolatum

232-384-2
Drakeol® 7
Drakesol® 220
Drakesol® 260
Drakesol® 260 AT
Drakesol® 305
Mineral oil

232-395-2
Fusel oil refined

232-428-0
Avocado (Persea gratissima) oil
Phytol™ AVO

232-430-1
Lanolin alcohol

232-436-4
Corn syrup

232-443-2
Calumet 142 Flash Naphtha Rule 66
Calumet 180/210 Naphtha
Exxsol® D 40 Naphtha
Exxsol® D 60 Naphtha
Exxsol® D 3135 Naphtha
Exxsol® DSP 115/145 Naphtha
Exxsol® DSP 90/120 Naphtha Anti-Static
Exxsol® Hexane Naphtha
Hi-Sol® 15
Hi-Sol® 70
Kwik Dri
Lacolene™
Naphtha
1520 Naphtha

232-453-7
Calumet Mineral Spirits < 1%
Calumet VM&P Naphtha < 1%
Mineral spirits
RotoSolv
Shell OMS
Shell Mineral Spirits 7 EC
Shell Mineral Spirits 135
Solvent 2218
Solvent 2268
Solvent 2391
Solvent 2409
Solvent 2423
Solvent 2472
Special Naphtholite 66/3
VM&P naphtha

232-455-8
Mineral oil (wh.)

232-468-9
Pancreatin

232-489-3
0-Aromatic 142 Solvent
0-Aromatic Mineral Spirits
Mineral Spirits 066/3
Mineral Spirits 075
Mineral Spirits 150
Mineral Spirits 150 66/3
Naphthol Spirits 66/3
Solvent 2215
Solvent 2222
Solvent 2231
Solvent 2256
Solvent 2257
Solvent 2284
Solvent 2292
Solvent 2302
Solvent 2350
Solvent 2351
Solvent 2352
Solvent 2359
Solvent 2361
Solvent 2363
Solvent 2364
Solvent 2379
Solvent 2380
Solvent 2383
Solvent 2388
Solvent 2417
Stoddard solvent
Varsol® 1
Varsol® 1 Naphtha Anti-Static
Varsol® 3135 Naphtha
Varsol® 3139
Varsol® 3139 Anti-Static Solvent

232-688-5
Turpentine

233-095-4
2-Ethylbutyl acetate

233-113-0
Hydrobromic acid

233-126-1
Propylheptanol

233-139-2
Boric acid

233-140-8
Calcium chloride

233-293-0
PEG-4 oleate

233-411-0
Diethylene glycol ethylvinyl ether

233-560-1
Isopropyl laurate
Lexol® IPL

233-713-2
Lactic acid (D)

234-125-9
Ethyl crotonate (trans)

234-325-6
Glyceryl stearate
Glyceryl stearate SE

235-088-1
Sodium toluenesulfonate

235-940-2
2-Aminobutanol

236-458-5
TCD Ketone A
Tricyclo [5.2.1.0~2,6] decane-8-one

236-547-9
PPG-2 methyl ether

236-719-3
3-Carene

236-935-8
Captex® 8227
Triundecanoin

238-405-1
Dimethyl caproamide

239-388-3
Menthol (D)

240-877-9
Lithium aluminum hydride

241-646-5
Behenyl behenate
Exceparl BB

241-654-9
Dynacerin® 660
Oleyl erucate

243-718-1
2-t-Butylcyclohexyl acetate (cis)

243-734-9
Arcosolv® TPM
Dowanol® TPM
Poly-Solv® TPM
PPG-3 methyl ether

244-350-4
Camphor (DL)

244-754-0
Exceparl EH-S
Octyl stearate

245-663-9
Propylene glycol isobutyl ether

246-466-0
Tripropylene glycol

246-562-2
Vinyltoluene monomer

246-563-8
BHA

246-614-4
Triisooctyl phosphite

246-771-9
Texanol® Ester-Alcohol
2,2,4-Trimethyl-1,3-pentanediol isobutyrate
Ucar® Filmer IBT

246-868-6
Isocetyl stearate

EXK-385-MS0
Magie 470 Oil
Magie 500 Oil
Petroleum distillates, sweetened middle

265-113-1
Hydroseal G3H
Hydroseal G232H
Hydroseal G240H
Hydroseal G250H
Hydroseal G400H
Petroleum distillates, acid-treated middle

265-148-2
Amsol 400
Amsol 500
Amsol 600
Conosol 260
Conosol 340
Drakesol® 205
Exx-Print® 588 D
Exxsol® D 130
Hydroseal G2
Isopar® V Fluid
Magie 5200-HTX
MagieSol 55LX
MagieSol 65LX
PCPL Solvent
PCP Solvent
Petroleum distillates, hydrotreated middle

265-149-8
Actrel® 1111L
Actrel® 1160L
AM 321 Fluid
Amsol 650
Amsol 3800
Bayol 35
Conosol 215
Conosol C-145
Conosol C-170
Conosol C-200
Drakesol® 165
Escaid® 90
Escaid® 95
Escaid® 110
Exx-Print® 176 D
Exx-Print® 283 D
Exxsol® D 40
Exxsol® D 40 Anti-Static Fluid
Exxsol® D 60
Exxsol® D 60 Anti-Static Fluid
Exxsol® D 60 Naphtha Anti-Static
Exxsol® D 70
Exxsol® D 80
Exxsol® D 90/120
Exxsol® D 90/120 Anti-Static Solvent
Exxsol® D 95
Exxsol® D 110
Exxsol® D 3135
Exxsol® D 3135 Anti-Static Solvent
Exxsol® D 3135 Naphtha Anti-Static
Exxsol® DSP 90/120 Naphtha
Exxsol® S-K Premium Solvent
Isopar® M Fluid
Kerdane D65
Kerdane D75
Ketrul D70
Ketrul D80
Ketrul D85
Ketrul D211
Ketrul D220
MagieSol 44
MagieSol 47LX
Orfom® SX 7
Orfom® SX 12
Penreco 170 Exempt Solvent
Petroleum distillates, hydrotreated light
Sasol LPA Solvent
Sasol LPA-142 Solvent
Sasol LPA-150 Solvent
Sasol LPA-170 Solvent
Sasol LPA-210 Solvent
Shellsol® D70
Shellsol® D80
Shellsol® D90
Shellsol® D100
Shellsol® D100S
Solvent 170
Varsol® DX 3641

265-150-3
Actrel® 3338L
Actrel® 3360L
C8-9 isoparaffin
C9-11 isoparaffin
C9-12 isoparaffin
Isopar® G Fluid
Isopar® H Fluid
Isopar® K Fluid
Isopar® K Naphtha
Isopar® L Fluid
Naphtha, hydrotreated heavy
Shellsol® D40
Shellsol® D60 SC
Solvent 2217
Solvent 2324
Spirdane BT
Spirdane D30
Spirdane D30L
Spirdane D40
Spirdane D40L
Spirdane D60
Spirdane D60L
Spirdane K1
Spirdane K2
Spirdane L1

265-151-9
Exxsol® D 115/145
Exxsol® D 115/145 Anti-Static Solvent
Exxsol® D 115/145 Naphtha
Exxsol® Heptane
Exxsol® Hexane
Exxsol® Hexane Anti-Static
Exxsol® Hexane Naphtha Anti-Static
Exxsol® Isohexane Naphtha
Hydrosol 75/95 N
Hydrosol 100/130 N
Hydrosol Essence 60/95
Hydrosol Essence 80/110
Hydrosol Essence 100/140
Hydrosol Essence A
Hydrosol Essence C
Hydrosol Essence E<5
Hydrosol Essence F<5
Hydrosol Isohexane
Iosol RS Solvent
Isohexane LNH
Naphtha, hydrotreated light
SBP 100/140
SBP 140/165
Shellsol® D25
X-95 Solvent

265-155-0
Petroleum distillates, heavy hydrotreated naphthenic

265-156-6
Naphthenic oil

265-158-7
Petroleum distillates, hydrotreated light paraffinic
Shellsol® D120

265-184-9
Kerosene, hydrodesulfurized
Ketrul 212
Ketrul HT

265-185-4
Kerosol
Kristalloel 60
Naphtha, hydrodesulfurized heavy
Shellsol® 1495
Shellsol® 1626
Shellsol® 2046
Shellsol® H
Spirdane HT

265-191-7
Envirolite® Solvent
Naphtha, medium aliphatic
Shellsol® 9
Shellsol® 16
Shellsol® D38
Shellsol® D43
Shellsol® D60

265-192-2
Naphtha, light aliphatic
Shellsol® W HT
Special Lactolite
VM&P Naphtha HT

265-198-5
Aromatic 150
Aromatic 200
HAN® 906 Solvent
Naphtha, heavy aromatic
Naphthalene Depleted Aromatic 150 Fluid
Naphthalene Depleted Aromatic 200 Fluid
Petrinex® ASB
Shellsol® A200 Solvent
Solv G
Solvarex 10
Solvarex 90/180
Solvaro AFD
142 Solvent 66/3
Solvesso™ 150 Anti-Static Solvent
Solvesso™ 150 ND
Solvesso™ 200
Solvesso™ 200 ND
Ultra Low Naphthalene Aromatic 200
Ultra Low Naphthalene Solvesso 200 Solvent

265-199-0
Aromasol® H
Aromatic 100
Hi-Sol® 10
Naphtha, light aromatic
Shellsol® A100 Solvent
Solvarex 9
Solvarex 90/160
Solvesso™ 100
Solvesso™ 100 Anti-Static Solvent
Super Hi-Flash Naphtha

265-200-4
Naphtha, heavy aliphatic
Orfom® SX-1

265-724-3
Aldo® MCT KFG
Caprylic/capric triglyceride
Captex® 300
Crodamol GTCC
Crodamol GTCC-PN
Delios® S
Estalan M-5
Estasan™ GT 8-40 3578
Estasan™ GT 8-60 3575
Estasan™ GT 8-60 3580
Estasan™ GT 8-65 3577
Estasan™ GT 8-65 3581
Estasan™ GT 8-70 3579
Labrafac® Lipophile
Lexol® GT-855
Lexol® GT-865
Liponate GC-K
Miglyol® 810
Miglyol® 812
Miglyol® 812 N
Myritol® 318
NEOBEE® M-5
Tegosoft® CT

266-065-4
Crodamol W
Stearyl heptanoate

266-231-6
Nonoxynol-9 phosphate

266-948-4
Triolein

267-008-6
Cetearyl alcohol

267-009-1
Cetearyl alcohol

267-015-4
Methyl oleate

270-093-2
Amsol 10
Petroleum distillates, hydrotreated light, low-boiling

270-112-4
C12-15 alkyl benzoate
Crodamol AB
Finsolv® TN
Finsolv® TN-O
Tegosoft® TN

270-302-7
Lanolin acid

270-722-0
Petroleum distillates, catalytic reformer fractionator residue, low boiling

270-817-7
C10 alkyl acetate
Exxate® 1000

271-073-6
C12-14 alkylbenzene
V-159L

271-365-3
C9-11 isoparaffin
HC-DCF Low Flash™ Dry Cleaning Solvent
Orfom® SX 18
Soltrol® 100

271-366-9
C10-13 isoparaffin
EcoSolv™ Dry Cleaning Fluid
Soltrol® 130

FDA Number Index

21CFR §184.1666 Propylene glycol Substance affirmed as GRAS; anticaking agent, antioxidant, dough strengthener, emulsifier, flavor agent, formulation aid, humectant, processing aid, solvent, vehicle, stabilizer, thickener, surface-active agent, texturizer

21CFR §184.1901 Triacetin (Glyceryl triacetate) Substance affirmed as GRAS; flavoring agent/adjuvant, formulation aid, humectant, solvent, vehicle

KB Value Index

Glossary

Abel Closed Cup. One of several types of apparatuses used for determining flash points. The Abel closed tester is for products and liquid petroleum mixtures which have flash points between -30° and +70°C.

absorption. Assimilation of molecules or other substances into the physical structure of a liquid or solid without chemical reaction.

acetates. A group of organic solvents used in making lacquers and other coatings. Acetates are derived from the reaction of alcohols with acetic acid. The acetate usually takes its name from the alcohol, such as ethyl acetate from ethyl alcohol.

active solvent. A solvent that will readily dissolve or liquify a resin.

acute toxicity. The short-term toxicity of a product in a single dose; can be divided into oral, cutaneous, and respiratory toxicities.

adsorption. The adhesion of a substance to the surface of a solid or liquid. Adsorption is often used to extract pollutants by causing them to be attached to such adsorbents as activated carbon or silica gel.

aerosol. A suspension of a liquid or a solid in a gas or air. The suspended particles are so fine that they will never gravitationally settle.

aerosol spray. A means of atomizing liquids by propelling them from a pressurized container using a soluble gas.

alcohols. A series of hydrocarbon derivatives with at least one hydrogen atom replaced by an -OH group. The simplest alcohols (methanol, ethanol, n-propanol, and isopropanol) are good solvents for some organic soils, notably rosin, but are flammable and can form explosive mixtures with air.

aliphatic hydrocarbon. A hydrocarbon in which the carbon atoms are joined in open chains, rather than rings; used in coatings and insecticides and acts as an excellent degreaser and solvent for acrylics and epoxies; common aliphatics include mineral spirits, petroleum distillates, VM&P Naphtha, kerosene, gasoline, and heptane.

American Conference of Governmental and Industrial Hygienists (ACGIH). A U.S. nongovernmental body empowered to develop the Threshold Limit Values (TLVs) that represent permissible workplace exposure levels for many chemicals.

aniline point. Lowest temperature at which equal volumes of fresh aniline and an oil are completely miscible. A product of high aniline point will be low in aromatics and naphthenes and, therefore, high in paraffins. Aniline point is often specified for spray oils, cleaning solvents, and thinners, where effectiveness depends upon aromatic content.

anion. A negatively charged ion.

Antarctic "ozone hole." Refers to the seasonal depletion of stratospheric ozone in a large area over Antarctica.

anthropogenic. In the context of greenhouse gases, emissions that are produced as the result of human activities.

aprotic solvent. A solvent that has no –OH groups and therefore cannot donate a hydrogen bond.

aqueous cleaning. Cleaning parts with water to which may be added suitable detergents, saponifiers, or other additives.

aromatic hydrocarbon. Hydrocarbons having alternating single and double bonds between adjacent carbon atoms in a benzene-ring configuration. The name is derived from the 'pleasant' odor attributed to many of this group of substances; common aromatics include toluene, xylene, coal-tar naphtha, styrene, and benzene.

aromatic solvent. An unsaturated hydrocarbon solvent, such as toluene and xylene, originally derived from coal tar but now made from petroleum possessing excellent solvent properties.

autoignition temperature. The lowest temperature at which a material will ignite without an external source of ignition.

azeotrope. A mixture of chemicals is azeotropic if the vapor composition is identical to that of the liquid phase. This means that the distillate of an azeotrope is theoretically identical to the solvents from which it is distilled. In practice, the presence of contaminants in the solvent slightly upsets the azeotropy.

benzene. An aromatic hydrocarbon having the structure C_6H_6; it is present in small proportion in some crude oils and made commercially from petroleum by the catalytic reforming of naphthenes in petroleum naphtha; it is also made from coal in the manufacture of coke. It is used as a solvent in manufacturing detergents, synthetic fibers, and petrochemicals and as a component of high-octane gasoline.

biochemical oxygen demand (BOD). The amount of oxygen consumed by bacteria and other microorganisms.

biocide. An additive that destroys biological contaminants.

biodegradable. Capable of being broken down especially into innocuous products by the action of living things, e.g., microorganisms.

blistering. Defect caused by pressure under a paint film from vapor or moisture; the vapor may come from trapped solvents or from chemical reaction under the film.

brominated solvents. These are unregulated ozone-depleting solvents, represented mainly by chlorobromomethane and n-propyl bromide. They have ODPs similar to that of HCFCs.

buffer. A solution that minimizes changes in hydrogen ion concentration that would otherwise occur as a result of a chemical reaction.

buffering agent. Drives an acidic or alkaline solution to neutral.

California Proposition 65. Formally known as the California Safe Drinking Water and Toxic Enforcement Act, that establishes two lists, one of which contains compounds known to the state to cause cancer, while the other list contains compounds known to be fetotoxic.

carcinogen. A substance which causes cancer in animals or man when they are exposed to it. The carcinogenicity of a substance between species or even between strains of the same species may differ.

cation. A positively charged ion.

Cellosolves. A collective name for a family of synthetic solvents (manufactured by Dow) that are based on ethylene glycol ethers. They are manufactured from petroleum or natural gas and have a wide range of properties; they are fairly viscous and are used to make other solvents more viscous, or to make them more miscible with water; they are used in paints, lacquers, resins, printing inks, cleaning fluids; brake and clutch fluids; and in de-icing additives for aviation fuel.

chemical oxygen demand (COD). A test that measures the potential capacity for a microorganism to react with oxygen.

chlorinated hydrocarbons. A group of hydrocarbons which include: chloroform, carbon tetrachloride, ethylene dichloride, methylene chloride, tetrachloroethane, trichloroethylene. They are employed as solvents, degreasers, plasticizers, paint strippers, and monomers for plastics manufacture. Many are prohibited because of their toxicity. When exposed to heat or ultraviolet light many chlorinated hydrocarbons will form phosgene gas, which is extremely hazardous.

chlorinated solvent. A synthetic solvent with excellent solvent properties for fatty materials and for synthetic resins; although many of them are volatile they do not burn, and so are safer to use than hydrocarbon solvents. The main applications for chlorinated solvents are degreasing and dry cleaning.

chlorocarbon. An organic chemical composed of chlorine and carbon, e.g., carbon tetrachloride.

chlorofluorocarbons (CFC). Organic compounds composed of atoms of carbon, chlorine, and fluorine. They had widespread uses in many applications, including refrigeration and air conditioning, or the blowing of insulation foams; their chemical stability made them unreactive, however, the side effect of this chemical stability allows them, when released, to eventually reach the stratosphere and to be degraded there by strong UV radiation giving rise to ozone-depleting by-products.

chronic toxicity. The long-term toxicity of a product in small, repeated doses; it can often take many years to determine.

Clean Air Act. Enacted in 1970 and substantially amended in 1977 and 1990, the Clean Air Act is the longest and most complex of all U.S. environmental statutes. It regulates everything from industrial plants and refineries (so-called "stationary sources") to consumer products (ranging from paints to underarm deodorants) to engines and fuels. It also mandates programs designed to reduce acid rain, to protect the stratospheric ozone layer, and to control emissions of 188 compounds listed as hazardous air pollutants (HAPs).

cleaning. The removal of soil or unwanted matter from a surface to which it clings.

Cleveland Open Cup (COC). A test for determining the flash point and fire point of all petroleum products except fuel oil and products with flash points below 79°C (175°F). The oil sample is heated in a precisely specified brass cup containing a thermometer. At specified intervals a small flame is passed across the cup. The lowest temperature at which the vapors above the cup briefly ignite is the flash point; the temperature at which the vapors sustain combustion for at least five seconds is the fire point.

Closed Cup (CC). A method used to determine the flash point of fuels, solvents, and cutback asphalts, utilizing a covered container in which the test sample is heated and periodically exposed to a small flame introduced through a shuttered opening. The lowest temperature at which the vapors above the sample briefly ignite is the flash point.

coalescing aid. A solvent that causes the individual polymer molecules in the resin to unite into one unifed body, thereby giving a continuous film or coating.

coalescing solvent. A solvent which, when added in limited quantities to certain water-based coatings, facilitates the formation of a continuous film of the binder.

coal tar. A black liquid consisting of a complex mixture of hydrocarbons, obtained from coal during its conversion to coke. Coal tar is the basic raw material for many of the solvents, dyes, chemicals and resins used in the coating industry.

coating. The application of a protective polymeric layer to a surface; paint is a common form of coating. Some coatings rely on the evaporation of a solvent to form a film; others are solventless and rely on a chemical reaction.

COD. Chemical oxygen demand, a test that measures the potential capacity for a microorganism

Code of Federal Regulations (CFR). A multivolume compilation of federal regulations published by the U.S. Government Printing Office. Changes and additions to the CFR are made through notices published in the *Federal Register*, which is published daily. The CFR volumes that contain regulations established by EPA and Occupational Safety and Health Administration (OSHA) are normally updated annually.

colloid. A substance that remains suspended in a solution or fails to settle out of solution.

combustible. A combustible liquid is any liquid having a flash point at or above 37.8°C (100°F), but below 93.3°C (200°F), except any mixture having components with flash points of 93.3°C (200°F), or higher, the total volume of which make up 99 percent or more of the total volume of the mixture. Combustible solids are those capable of igniting and burning.

completely denatured alcohol (CDA). Alcohol that contains chemicals specified by the U.S. government (and other governments) making it completely unsuitable for human consumption with the intention of making it impossible to recover the alcohol from the mixture.

concrete. Extraction of fresh natural plant materials, usually with nonpolar organic solvents (e.g. hexane) which yield, after removal of the solvent by vacuum distillation, a solid or semisolid wax.

conformal coating. A protective material applied in a thin, uniform layer to all surfaces of a printed wiring assembly.

cosolvent process. A cleaning process in which a miscible mixture of a low-boiling point halogenated solvent and a high-boiling organic solvent is used in the first stage of a vapor-phase cleaning machine, followed by rinsing and drying with the pure halogenated solvent in subsequent steps.

cut. To dilute an ink, lacquer or varnish with solvents or with a clear base; to thin.

defluxing. The removal of flux residues after a soldering operation. Defluxing is a part of most high-reliability electronics production.

degreaser. Combination of solvents for the purpose of removing grease and oil from a surface.

denaturant. A material that is added to another substance to alter its effectiveness for certain purposes. For example, denatured alcohol is ethyl alcohol containing small quantities of other materials, which render it unfit for beverage purposes. Generally, most alcohol denaturants are poisonous.

detergency. Lifting soil from a surface by displacing it with surface active materials that have a greater affinity for the surface than they do for the soil.

detergent. A product designed to render dirt, oils, and greases soluble in water; usually made from synthetic surfactants.

diluent. A nonsolvent liquid that is blended with an active solvent to obtain certain properties, such as reduced viscosity, in a finishing material. A diluent may also be called a reducer or a thinner.

dipentene. A high boiling solvent for resins derived from the distillation of the sap of pine trees.

dissociation. The process by which a chemical compound breaks up into simpler constituents as a result of either added energy (dissociated by heat), or the effect of a solvent on a dissolved polar compound (electrolytic dissociation). It may occur in the gaseous, solid, or liquid state, or in solution.

distillation. A means for purifying solvents or water by boiling it into a vapor and collecting the vapor condensate.

dry cleaning. A common term for cleaning garments in organic solvents as opposed to water.

electrolyte. A molecule that separates into a cation and an anion when its dissolved in a solvent, usually water, e.g., NaCl separates into Na^+ and Cl^- in water.

emulsifier. A substance that prevents the separation of immiscible substances in an emulsion; helps to distribute evenly one substance in another.

emulsion. A suspension of very fine droplets of one liquid in another, where the two liquids are normally immiscible. A true emulsion is stable and will never separate, even if the density of the two liquids is very different.

Environmental Protection Agency (EPA). The U.S. federal agency that has been given primary responsibility for implementing and enforcing a broad range of pollution control laws enacted by Congress, including the Clean Air Act, the Clean Water Act, TSCA (Toxic Substances Control Act), FIFRA (Federal Insecticide, Fungicide, and Rodenticide Act), and EPCRA (Emergency Planning and Community Right-to-know Act).

ester. A class of organic compounds, sometimes used as solvents, produced by reacting an acid with an alcohol, e.g., ethyl acetate, isopropyl acetate.

evaporation. To convert into a gaseous state or vapor; the action of drying by removing moisture; lacquers, vinyl, and most latex finishes dry by the evaporation of solvents.

explosive limits. The minimum and maximum proportions of vapor of a flammable or combustible solvent in air, between which an explosion will occur if ignited; expressed as a percentage.

exposure limits. The limits to which an operator may be safely exposed to a gas or solvent vapor without incurring any risk to his health. Each country has an authority which sets the exposure limits with much disparity from country to country.

extract. A product derived from vegetable matter by the application of solvents, heat or pressure to separate it from the other constituent parts.

extraction. The process of treating a natural raw material with an organic solvent. The solvent portion containing the extracted material is filtered and the solvent removed. The extract will contain nonvolatile as well as volatile components. Oleoresins, resinoids, concretes, and absolutes are all produced by extraction.

exudate. Any material, usually of a resinous or balsamic nature, that filters through the walls of living vegetable cells and accumulates on the exterior tissues.

fixative. The material which slows down the rate of evaporation of the more volatile components in a perfume composition.

flammable. Anything that will ignite easily and burn rapidly. The National Fire Protection Association (NFPA) and the Department of Transportation (DOT) of the U.S. define flammable liquids as those having a flash point less than 37.8°C (100°F) and a vapor pressure of not over 40 psia at 37.8°C (100°F).

flash point. The temperature at which a liquid or volatile solid gives off vapor sufficient to form an ignitable mixture near the surface of the liquid or within the test vessel.

global warming. The progressive gradual rise of the earth's surface temperature believed to be caused by the enhancement of the greenhouse effect and responsible for changes in global climate patterns; an increase in the near surface temperature of the Earth. Global warming has occurred in the distant past as the result of natural influences, but the term is now most often used to refer to the warming predicted to occur as a result of increased emissions of greenhouse gases.

Global Warming Potential (GWP). This index expresses the climatic warming potential of a greenhouse gas relative to that of carbon dioxide, which by convention is set at 1. The GWP is calculated in terms of the 100 year warming potential of a kilogram (kg) of a gas relative to that of a kilogram of CO_2.

glycol. A loose term for a diol or a dihydric alcohol, i.e., one containing two -OH groups per molecule. Glycols and related compounds, esp. diglycol ethers, are often used as components in proprietary solvents and cleaning additives.

glycol ethers. Solvents that are used in coatings, resins and dyes as well as the semiconductor industry. Glycol ethers include 2-butoxyethanol, 2-ethoxyethanol, 2-methoxyethanol, and 2-ethoxyethyl acetate; most common glycol ethers are combustible.

greenhouse effect. A thermodynamic effect whereby energy absorbed at the earth's surface, which is normally able to radiate back into space in the form of long-wave infrared radiation, is retained by gases in the atmosphere, causing a rise in temperature. The gases in question are partially natural, but man-made pollution is thought to increasingly contribute to the effect. The same CFCs that cause ozone depletion are known to be "greenhouse gases," with a single CFC molecule having the same estimated effect as 10,000 carbon dioxide molecules.

greenhouse gas. A gas that absorbs radiation at specific wavelengths within the spectrum of radiation (infrared radiation) emitted by the Earth's surface and by clouds. The gas in turn emits infrared radiation from a level where the temperature is colder than the surface. The net effect is a local trapping of part of the absorbed energy and a tendency to warm the planetary surface. Water vapor (H_2O), carbon dioxide (CO_2), methane (CH_4), and nitrous oxide (N_2O) are the primary greenhouse gases in the earth's atmosphere; chlorofluorocarbon (CFC) halogenated fluorocarbons (HCFCs), ozone (O_3), perfluorinated carbons (PFCs), and hydrofluorocarbons (HFCs) are some other greenhouse gases.

halocarbon. Any organic substance in which at least one hydrogen atom in the hydrocarbon molecule has been replaced by a halogen atom (fluorine, chlorine, bromine, iodine, or astatine).

halogenated solvents. Solvents that contain halogen; they have improved solvency compared with the hydrocarbons from which they are derived and, in addition, flammability is reduced; some of these solvents are highly toxic, and inhalation of their vapors should be avoided.

halogens. A class of reactive, nonmetallic, chemical elements used in the manufacture of a group of solvents; the most usual halogen elements are bromine, chlorine, fluorine and iodine

halon. Fire-extinguishing agent with a high ODP.

Hansen Solubility Parameters. An indicator of the preferences for hydrogen bonding of solvents used to predict the solubility of polymers in solvents.

Hazardous Air Pollutants (HAPs). A list of 188 compounds and classes of compounds contained in Section 112 of the U.S. Clean Air Act. The U.S. Environmental Protection Agency (EPA) is required to develop MACT (maximum achievable control technology) standards for any facility that is classified as a "major source" of HAPs. The EPA has authority to add or remove compounds from the HAP list.

hydrocarbon. An organic chemical composed only of hydrogen and carbon; gaseous or volatile hydrocarbons are flammable.

hydrocarbon solvent. The most common group of solvents used in coatings; they are divided into three categories; aliphatic, naphthenic and aromatic.

hydrocarbon/surfactant solvents. A mixture of low-volatility hydrocarbon solvents with surfactants, allowing the use of a two-phase cleaning process. The first phase is solvent cleaning and the second phase is water washing and rinsing to remove the residues of the blend and any other water-soluble soils. The surfactant ensures the water-solubility of the otherwise insoluble hydrocarbon. They may be subdivided into three categories: miscible, separable and self-surfacting, depending on how they mix with water.

hydrochlorocarbon. An organic chemical composed of hydrogen, chlorine and carbon, e.g., trichloroethylene.

hydrochlorofluorocarbon. An organic chemical composed of hydrogen, chlorine, fluorine, and carbon atoms. These chemicals are less stable than pure CFCs, and therefore have generally lower ODPs.

hydroforming. A dehydrogenation process in which naphthas are passed over a solid catalyst at elevated temperatures and moderate pressures in the presence of hydrogen to form high-octane motor gasoline, high-grade aviation gasoline, or aromatic solvents.

hydrolysis. A chemical reaction of a substance with water, leading to either decomposition or some other change in the substance.

hydrophilic. Water-loving or having a strong affinity for water.

hydrophobic. Water-hating or having a strong aversion to water.

immiscible. Incapable of mixing or obtaining homogeneity.

ion. An atom or group of atoms that is not electrically neutral but instead carries a positive or negative electric charge.

Kauri Butanol (KB) Value. A measure of the solvency of a hydrocarbon solvent; values range from 20 (poor) to over 1000 (excellent).

ketone. An organic substance containing a C-O-C group. Alkane ketones, such as acetone (propanone) and methyl ethyl ketone (MEK) are flammable solvents; ketones are excellent solvents for fats, resins and lacquers.

lacquer. A fast-drying usually clear coating that is highly flammable and dries by solvent evaporation only; it can be reconstituted after drying by adding a solvent.

latent solvent. An organic liquid that dissolves a resin, such as nitrocellulose, only when mixed with a true solvent.

lipid. A loosely defined term for a substance of biological origin that is soluble in nonpolar solvents; this group is composed of saponifiable lipids, such as glycerides (fats and oils) and phospholipids, as well as nonsaponifiable lipids, principally steroids.

liquid drier. Solution of a soluble drier in an organic solvent.

lower explosive limit (LEL). The lowest concentration of a combustible solvent vapor or gas in air that will produce a flash fire if ignited.

lower flammability limit (LFL). The minimum concentrations at which ignition will occur .

low vapor pressure (LVP). A vapor pressure no greater than 0.1 mm Hg at 20°C. Any compound with a vapor pressure below this number is not counted against the applicable volatile organic compound (VOC)-content limit; the LVP cut-off is different for certain types of products.

lyophilic (lypophilic). Oil loving; marked by a strong affinity between a dispersed phase and the liquid in which it is dispersed.

miscible. Capable of being mixed in any ratio without separation of the two phases, e.g., water and ethyl alcohol are miscible.

molality (molal concentration). A concentration in which the amount of solute is stated in moles and the amount of solvent in kilograms.

molarity. The concentration of a solution in moles per liter.

mutagen. A substance which causes genetic mutation in animals or man when they are exposed to it. The mutagenicity of a substance between species or even between strains of the same species may differ

naphtha. A generic term applied to a petroleum fraction with an approximate boiling range between 50 and 204°C.

nonvolatile. The portion of a coating left after the solvent evaporates; sometimes called the solids content.

Occupational Health and Safety Administration (OSHA). The U.S. regulatory and enforcement agency set up to ensure the safety and health of workers in businesses that employ more than 10 persons.

oleoresin. Extraction, usually of natural spice or flavoring material, using selected solvents to remove the vital components; it contains the essential oil plus other important nonvolatile components which characterize the flavor, color and other aspects of the starting raw material.

organic. Describes a substance containing carbon covalently bonded to hydrogen or derived from such a substance.

organic solvent. A solvent composed of the elements carbon, hydrogen, oxygen and sometimes one or more of nitrogen, fluorine and chlorine.

organosol. A plastisol with an added solvent, which swells the resin particles, thereby increasing viscosity.

oxygenated solvent. A high purity synthetic compound that offers much stronger solvency than an organic solvent and is widely used as an active solvent for synthetic binders; the principle oxygenated solvents are ketones, esters, glycol esters and alcohols.

ozone. A gas formed when oxygen is ionized by the action of ultraviolet light or a strong electric field. It has the property of blocking the passage of dangerous wavelengths of ultraviolet light. Although it is a desirable gas in the stratosphere, it is toxic to living organisms at ground level.

ozone depleting substance (ODS). A family of man-made compounds that have been shown to deplete stratospheric ozone; they include, but are not limited to, chlorofluorocarbons (CFCs), bromofluorocarbons (halons), methyl chloroform, carbon tetrachloride, methyl bromide, and hydrochlorofluorocarbons (HCFCs).

ozone depletion. Accelerated chemical destruction of the stratospheric ozone layer by the presence of substances produced mostly by human activities. The most depleting species for the ozone layer are the chlorine and bromine free radicals generated from relatively stable chlorinated, fluorinated, and brominated products by ultraviolet radiation.

ozone depletion potential (ODP). A relative index indicating the extent to which a chemical product may cause ozone depletion. The reference level of 1 is the potential of CFC-11 and CFC-12 to cause ozone depletion. If a product has an ozone depletion potential of 0.5, a given weight of the product in the atmosphere would, in time, deplete half the ozone that the same weight of CFC-11 would deplete. The ozone depletion potentials are calculated with mathematical models that take into account factors such as the stability of the product, the rate of diffusion, the quantity of depleting atoms per molecule, and the effect of ultraviolet light and other radiation on the molecules.

ozone layer. The layer of gaseous ozone (O_3) in the stratosphere that protects life on earth by filtering out harmful ultraviolet radiation from the sun.

ozone precursors. Chemical compounds, such as carbon monoxide, methane, volatile hydrocarbons, and nitrogen oxides that, in the presence of solar radiation, react with other chemical compounds to form ozone, mainly in the troposphere.

paint. A coating including a resin, a solvent, additives, pigments and, in some products, a diluent. Paints are generally opaque, and commonly represent the portion of the industry known as architectural coatings.

paraffin. A hydrocarbon identified by saturated straight [normal or branched (iso)] carbon chains. A paraffin is relatively nonreactive and has excellent oxidation stability. A paraffinic solvent is generally lower in solvency than a naphthenic or aromatic solvent.

penetration. The ability of a liquid (ink, varnish or solvent) to be absorbed into a substrate.

Pensky-Martens Closed Cup (PMCC). One of several types of apparatuses used for determining flash points. The Pensky-Martens closed tester (ASTM D93-79) is used for liquids that have a viscosity of 45 SUS (Saybolt Universal Seconds) or more at 37.8°C (100°F), have flash points of 93.6°C (200°F) or higher, contain suspended solids, or form surface films.

perfluorocarbon. A perhalogenated substance whose molecule is composed of only fluorine and carbon. A perfluorocarbon is characterized by extremely long life, up to several thousand years, very high GWP, up to thousands, high chemical stability and inertness, low toxicity and a physical affinity to absorb oxygen. A perfluorocarbon solvent generally has poor solvency but may be used for vapor-phase reflow soldering, leak detection, as a blanket for flammable solvents, or as a secondary solvent in the cosolvent process. Their use is generally discouraged because of their high GWP and cost.

perhalogenation. An organic molecule in which the parent hydrogen atoms in a hydrocarbon are replaced with halogen atoms (astatine, bromine, chlorine, fluorine or iodine). For example, carbon tetrachloride (CCl_4) is perchlorinated methane (CH_4).

Permissible Exposure Limit (PEL). PELs are established by the U.S. Department of Labor acting through OSHA (Occupational Safety and Health Administration). A PEL represents the maximum worker exposure allowed for a substance over an eight hour day. While both threshold limit values (TLVs) and PEL values have been established for many substances, they are not necessarily the same. The TLVs are examined and modified more frequently than the PEL values. Consequently, in most cases, the TLV represents a more recent evaluation of the hazards associated with worker exposure to a substance.

petrochemical. Any chemical derived from crude oil, its products, or natural gas; it is basically a compound of carbon and hydrogen, but may incorporate many other elements; it is used in the manufacture of numerous products such as synthetic rubber, synthetic fibers (such as nylon and polyester), plastics, fertilizers, paints, detergents, and pesticides.

petroleum. An oily flammable bituminous liquid that is a complex mixture of hydrocarbons with small amounts of other substances and is prepared for use as gasoline, naphtha, or other products by various refining processes.

petroleum distillates. A broad range of compounds that are extracted by distillation during the refining of crude oil.

petroleum ether. Petroleum ether is the petroleum fraction consisting of C5 and C6 hydrocarbons and boiling in the range 35 to 60°C; commonly used as a laboratory solvent.

pH. A measure of acidity or alkalinity of an aqueous solution. The pH scale ranges from 0 (very acidic) to 14 (very alkaline), with a pH of 7 indicating a neutral solution equivalent to the pH of distilled water.

photoresist. A photomechanical product, in the form of a liquid or a laminated dry film, used in the manufacture of printed circuits. Certain types require the use of large quantities of ozone-depleting hydrochlorocarbon solvents, mostly 1,1,1-trichloroethane and methylene chloride, for their processing.

pickling. The removal of oxides (rust) with acidic materials.

plant-derived solvent. A natural oil, such as safflower oil, rapeseed oil, linseed oil, cottonseed oil, or turpentine; it is used either directly, or after chemical modification, in the food, paint and surface finish industry. A plant-derived solvent has the property of setting into a solid film on long exposure to air.

plasticizer. An organic compound added to a high polymer both to facilitate processing and to increase the flexibility and toughness of the final product by internal modification of the polymer molecule.

plastisol. A dispersion of finely divided resins in a plasticizer.

polar solvent. A solvent with oxygen in its molecule, e.g., water, alcohols, esters and ketones.

precision cleaning. Cleaning of high-precision mechanical parts and electronic sensory devices, as opposed to general metal cleaning. This is usually done under controlled atmospheres, notably in clean rooms with low particle contamination.

proprietary solvent alcohol. A completely denatured ethyl alcohol, mixed to government specifications and sold by vendors under a trade name.

protic solvent. A solvent that has an –OH group and therefore can donate a hydrogen bond.

raffinate. In solvent extraction, that portion of the oil which remains undissolved and is not removed by the selective solvent.

rectification. A second distillation of an essential oil to remove color, water, and resinous matter.

reduce. To lower the viscosity of a material or to thin a material by adding a solvent, thinner, varnish or oil.

resin. Any of a class of solid or semisolid organic products of natural or synthetic origin, generally of high molecular weight with no definite melting point.

resinoid. A perfumery material prepared from natural, resinous substances by extraction with a hydrocarbon solvent.

retarding agent. A slowly evaporating solvent that decreases the evaporation rate or slows the drying of lacquers and similar materials.

rosin. A solid resin obtained from pine trees which, in a pure form and usually with additives, is frequently used as a flux.

safety solvent. A solvent which is free from fire or toxicity hazards and is nondamaging to surfaces or materials being cleaned.

saponification. The breaking down of oils into very fine droplets called colloids; to hydrolyze a fat with alkali to form a soap and glycerol.

saponifier. A chemical designed to react with organic fatty acids, such as rosin, oils, and greases, to form a water-soluble soap.

saturated hydrocarbon. A hydrocarbon having the basic formula C_nH_{2n+2}; it is saturated with respect to hydrogen and cannot combine with the atoms of other elements without giving up hydrogen; a saturate is more chemically stable than an unsaturated hydrocarbon.

sesquiterpeneless. An essential oil which has had the sesquiterpenic hydrocarbons partially or completely removed to improve solubility in diluted alcohol or food grade solvents, improve odor and flavor of the essential oil, or lift the overall fragrance and flavor, since sesquiterpenes have a fixative effect.

Significant New Alternatives Policy (SNAP). Section 612 of the U.S. Clean Air Act Amendments of 1990 requiring the U.S. EPA to publish lists of acceptable and unacceptable alternatives for ozone-depleting substances (ODSs).

sol. A dispersion of small solid particles in a liquid. The particles may be macromolecules or clusters of small molecules. Lyophobic sols are those in which there is no affinity between the dispersed phase and the liquid (e.g., silver chloride dispersed in water). Lyophobic sols are inherently unstable; in time the particles aggregate and form a precipitate. However, a lyophilic sol is more like a true solution in which the solute molecules are large and have an affinity for the solvent (e.g., starch in water). An association colloid is a system in which the dispersed phase consists of clusters of molecules that have lyophobic and lyophilic parts (e.g., soap in water).

solubility. The maximum amount of solute that dissolves in a given quantity of solvent at a specific temperature. Generally, for a solid in a liquid, solubility increases with temperature; for a gas, the solubility decreases. Common measures of solubility include the mass of solute per unit mass of solution (mass fraction), mole fraction of solute, molality, and molarity.

solute. The substance that is dissolved in another substance (solvent) to form a solution; usually present in a smaller amount than the solvent.

solution. A mixture in which the components lose their identity and are uniformly dispersed. All solutions are composed of a solvent (water or other fluid) and the substance dissolved in it called the solute. A true solution is homogeneous, e.g., salt in water.

solvent. A compound that has the ability to dissolve, suspend, or extract other materials without causing a chemical change to the material or solvent . The most common petroleum solvents are mineral spirits, xylene, toluene, hexane, heptane, and naphthas. Aromatic-type solvents have the highest solvency for organic chemical materials, followed by naphthenes and paraffins. In most applications the solvent disappears, usually by evaporation, after it has served its purpose. Solvents are used in refining processes and the manufacture of paints, inks, cleaning products, adhesives, and petrochemicals.

solvent cementing. Joining pipe by the use of a solvent which dissolves the surface of the pipe and forms a continuous bond upon evaporation.

solvent containment. A means of preventing or reducing the emission of CFC or other solvents into the environment. This technique usually involves improving the design of the equipment in which the solvent is used and preventing solvent losses.

solvent extraction. A selective separation procedure for isolating and concentrating a substance from an aqueous solution with the aid of an organic solution. In this procedure the aqueous solution containing the substance to be isolated is mixed (extraction) with an organic solvent containing a reagent. The substance to be separated is then reacted with the reagent to form a chemical compound, which is more soluble in the organic than in the aqueous solution. As a consequence, the substance is transferred to the organic solution.

solvent neutral. High-quality paraffin-base oil refined by solvent extraction.

solvent release. The ability of a binder to influence the rate of evaporation of a solvent.

specially denatured alcohol (SDA). Ethyl alcohol that has been made unfit for drinking by the addition of other substances. SD alcohol is often followed by a number or a number-letter combination that indicates what substances were used to denature the alcohol.

surfactant. A product designed to reduce the surface tension of water. Also referred to as tensio-active agents/tensides. Detergents are made up principally from surfactants.

Tag Open Cup (TOC). Apparatus for determining the flash point of hydrocarbon liquids, usually solvents, having flash points between -17.8 and 168°C (0 to 325°F). The test sample is heated in an open cup at a slow, constant rate. A small flame is passed over the cup at specified intervals. The lowest temperature at which the vapors above the sample briefly ignite is the flash point.

teratogen. A substance which causes fetal malformation in animals or man when they are exposed to it. The teratogenicity of a substance between species or even between strains of the same species may differ.

terpene. Fraction of an essential oil consisting mainly of hydrocarbons, obtained as a by-product from either concentration or distillation of the oil.

terpeneless. Distillation of an essential oil to remove monoterpenes or hydrocarbons. This process improves solubility in diluted alcohol or food grade solvents as well as increasing the oils stability.

terpenes. Any of many homocyclic hydrocarbons with the empirical formula $C_{10}H_{16}$ and a characteristic odor. Turpentine is mainly a mixture of terpenes; other common terpenes are pinene and limonene.

thermoplastic. Plastic which is solid when cold, but which may flow and be re-formed with the application of heat.

thermoset. A polymer that solidifies when heated and cannot thereafter be changed. This occurs in some polymers because the heating process causes the chains of the polymer to bind to each other, via crosslinks, and these cannot then be broken.

Threshold Limit Value (TLV). The TLV, established by American Conference of Governmental and Industrial Hygienists, is the limit to which nearly all workers may be repeatedly exposed day after day without adverse effect. The airborne concentration is expressed as TLV/TWA (time weighted average) or TLV/STEL (Short-Term Exposure Limit).

tincture. A prepared perfumery material, flavor material, or pharmaceutical product; it is an alcoholic extract of natural raw materials; the solvent is left in the extract as a diluent.

TSCA (Toxic Substances Control Act). Enacted in 1976, TSCA gives U.S. Environmental Protection Agency (EPA) the ability to track the 75,000 industrial chemicals currently produced or imported into the United States. The EPA is supposed to screen these chemicals and can require reporting or testing of those that may pose an environmental or health hazard. Under TSCA, EPA can ban the manufacture and import of chemicals that pose an unreasonable risk.

ultrasonic cleaning. Immersion cleaning where mechanical energy formed by cavitational implosions close to the surfaces being cleaned significantly aids the cleaning operation. Ultrasonic cleaning is less effective if the solvent contains dissolved gases; it can damage some types of delicate parts.

unsaturated hydrocarbon. A hydrocarbon lacking a full complement of hydrogen atoms, and thus characterized by one or more double or triple bonds between carbon atoms; unsaturated hydrocarbons readily attract additional hydrogen, oxygen, or other atoms, and are therefore highly reactive.

unsaturated solution. A solution that contains less than the maximum possible equilibrium concentration of a solute.

upper explosive limits (UEL). The maximum concentration of a flammable gas or vapor (percent by volume in air) at which an explosion can occur if an ignition source is present.

upper flammability limit (UFL). The maximum concentrations at which ignition will occur.

vapor-phase cleaning. A cleaning process, usually with CFC-113 solvent or hydrochlorocarbon solvents, in which the final rinse of the parts being cleaned is achieved by condensing solvent vapors on the parts.

vapor pressure. The pressure exerted by a vapor in equilibrium with the solid or liquid phase of the same substance. Also, the partial pressure of the substance in the atmosphere above the solid or the liquid.

vehicle. An inert substance, such as a liquid, through which an active agent is administered or by which other ingredients are held together; used in paint technology to indicate the liquid portion of a paint, composed of a drying oil or resin, a solvent and a thinner in which the solid portions are dissolved or dispersed.

VOC-exempt compounds. A small number of organic compounds that are considered in the U.S. to not contribute to ozone formation.

volatile organic compound (VOC). Constituents that will evaporate at their temperature of use and which, by a photochemical reaction, will cause atmospheric oxygen to be converted into potential smog-promoting tropospheric ozone under favorable climatic conditions. Some areas classify a substance to be a VOC based on its vapor pressure, i.e., any hydrocarbon, except methane and ethane, with a vapor pressure equal to or greater than 0.1 mm Hg.

water miscible/dispersible. Can be uniformly mixed with water.

water soluble. Can be dissolved in water.

wetting. The action by which a cleaner penetrates and loosens the substrate-soil bond by lowering surface and interfacial tension.

zeolite. A natural or synthetic hydrated aluminosilicate with an open three-dimensional crystal structure in which water molecules are held; often called molecular sieves.

zeotrope. A solvent mixture whose vapor composition is different from its liquid composition; regeneration by distillation is therefore difficult and its use for vapor degreasing may be limited.